Biochemistry

SIXTH EDITION

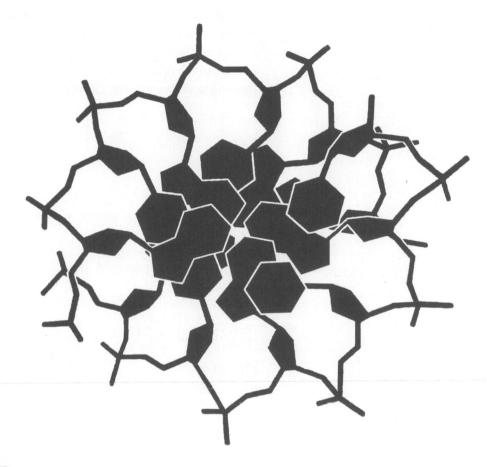

Jeremy M. Berg

John L. Tymoczko

Lubert Stryer

■■ W. H. Freeman and Company • New York

Publisher: Sara Tenney
Senior Acquisitions Editor: Kate Ahr
Marketing Managers: Sarah Martin, John Britch
Senior Developmental Editor: Susan Moran
Media Editor: Alysia Baker
Supplements Editors: Nick Tymoczko, Deena Goldman
Photo Editor: Bianca Moscatelli
Design Manager: Diana Blume
Text Designer: Patrice Sheridan
Senior Project Editor: Georgia Lee Hadler
Manuscript Editor: Patricia Zimmerman
Illustrations: Jeremy Berg with Network Graphics
Senior Illustration Coordinator: Bill Page
Production Coordinator: Susan Wein
Composition: Aptara®, Inc.
Printing and Binding: RR Donnelley

Library of Congress Cataloging-in-Publication Data

Berg, Jeremy Mark.
   Biochemistry / Jeremy M. Berg, John L. Tymoczko, Lubert Stryer.—6th ed.
     p. cm.
   Includes bibliographical references and index.
   ISBN 0-7167-8724-5 hardcover
    1. Biochemistry. I. Tymoczko, John L., II. Stryer, Lubert III. Title.

QP514.2.S66 2006
572—dc22

2005052751

ISBN-13: 978-0-7167-8724-2
ISBN-10: 0-7167-8724-5

Printed in the United States of America

Sixth printing

W. H. Freeman and Company
41 Madison Avenue
New York, NY 10010
Houndmills, Basingstoke RG21 6XS, England
www.whfreeman.com

To our teachers and our students

About the Authors

JEREMY M. BERG received his B.S. and M.S. degrees in Chemistry from Stanford (where he did research with Keith Hodgson and Lubert Stryer) and his Ph.D. in chemistry from Harvard with Richard Holm. He then completed a postdoctoral fellowship with Carl Pabo in Biophysics at Johns Hopkins University School of Medicine. He was an Assistant Professor in the Department of Chemistry at Johns Hopkins from 1986 to 1990. He then moved to Johns Hopkins University School of Medicine as Professor and Director of the Department of Biophysics and Biophysical Chemistry, where he remained until 2003. In 2003, he became the Director of the National Institute of General Medical Sciences at the National Institutes of Health. He is recipient of the American Chemical Society Award in Pure Chemistry (1994), the Eli Lilly Award for Fundamental Research in Biological Chemistry (1995), the Maryland Outstanding Young Scientist of the Year (1995), and the Harrison Howe Award (1997). While at Johns Hopkins, he received the W. Barry Wood Teaching Award (selected by medical students as award recipient), the Graduate Student Teaching Award, and the Professor's Teaching Award for the Preclinical Sciences. He is coauthor, with Stephen Lippard, of the textbook *Principles of Bioinorganic Chemistry.*

JOHN L. TYMOCZKO is Towsley Professor of Biology at Carleton College, where he has taught since 1976. He currently teaches Biochemistry, Biochemistry Laboratory, Oncogenes and the Molecular Biology of Cancer, and Exercise Biochemistry and coteaches an introductory course, Energy Flow in Biological Systems. Professor Tymoczko received his B.A. from the University of Chicago in 1970 and his Ph.D. in Biochemistry from the University of Chicago with Shutsung Liao at the Ben May Institute for Cancer Research. He then had a postdoctoral position with Hewson Swift of the Department of Biology at the University of Chicago. The focus of his research has been on steroid receptors, ribonucleoprotein particles, and proteolytic processing enzymes.

LUBERT STRYER is Winzer Professor of Cell Biology, Emeritus, in the School of Medicine and Professor of Neurobiology, Emeritus, at Stanford University, where he has been on the faculty since 1976. He received his M.D. from Harvard Medical School. Professor Stryer has received many awards for his research on the interplay of light and life, including the Eli Lilly Award for Fundamental Research in Biological Chemistry and the Distinguished Inventors Award of the Intellectual Property Owners' Association. He was elected to the National Academy of Sciences in 1984. He currently chairs the Scientific Advisory Boards of two biotechnology companies—Affymax, Inc., and Senomyx, Inc.—and serves on the Board of the McKnight Endowment Fund for Neuroscience. The publication of his first edition of *Biochemistry* in 1975 transformed the teaching of biochemistry.

PREFACE

The more we learn, the more we discover connections threading through our biochemical world. In writing the sixth edition, we have made every effort to present these connections in a way that will help first-time students of biochemistry understand the subject and how very relevant it is to their lives.

Emphasis on Physiological Relevance

Biochemistry is returning to its roots to renew the study of its role in physiology, with the tools of molecular biology and the information gained from gene sequencing in hand. In the sixth edition, we emphasize that an understanding of biochemical pathways is the underpinning for an understanding of physiological systems. Biochemical pathways make more sense to students when they understand how these pathways relate to the physiology of familiar activities such as digestion, respiration, and exercise. In this edition, particularly in the chapters on metabolism, we have taken several steps to ensure that students have a view of the bigger picture:

- Discussions of metabolic regulation emphasize the **everyday conditions** that determine regulation: exercise versus rest; fed versus fasting.

- New **pathway-integration figures** show how multiple pathways work together under a specific condition, such as during a fast.

- More **physiologically relevant examples** have been added throughout the book.

This physiological perspective is also evident in the new chapter on drug development. The use of a foreign compound to inhibit a specific enzyme sometimes has surprising physiological consequences that reveal new physiological principles.

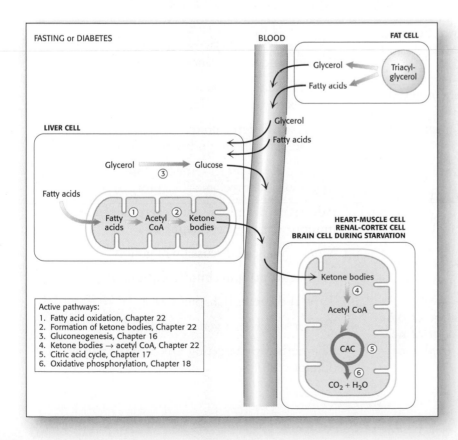

Active pathways:
1. Fatty acid oxidation, Chapter 22
2. Formation of ketone bodies, Chapter 22
3. Gluconeogenesis, Chapter 16
4. Ketone bodies → acetyl CoA, Chapter 22
5. Citric acid cycle, Chapter 17
6. Oxidative phosphorylation, Chapter 18

Figure 22.21 Pathway Integration: Liver supplies ketone bodies to the peripheral tissues. During fasting or in untreated diabetics, the liver converts fatty acids into ketone bodies, which are a fuel source for a number of tissues. Ketone bodies are the predominant fuel during starvation.

A Molecular Evolutionary Perspective

Evolutionary perspectives greatly enable and enhance the study of biochemistry. As Theodosius Dobzhansky noted, "nothing in biology makes sense except in the light of evolution." In the course of evolution, mutations altered many proteins and biochemical motifs so that they perform different functions while maintaining their core biochemical elements. By examining related proteins, we highlight essential chemical features as well as the specialization necessary for particular functions. The tracks of evolution are clear from the analysis of gene and protein sequences.

As sequence analysis becomes more important, the field of biochemistry is shifting from a science performed almost entirely in the laboratory to one that may also be explored through computers, by using information gathered from genomics and proteomics. This shift is manifest in the current edition and can be seen most clearly in Chapter 6, "Exploring Evolution and Bioinformatics," which develops the conceptual basis for comparing protein and nucleic acid sequences. Protein comparisons are a frequent source of insight throughout the book, especially for illuminating relations between structure and function.

New Chapters: Hemoglobin and Drug Development

Two new chapters illustrate the relation between structure and function by using a classic example and a contemporary one.

Chapter 7: Hemoglobin: Portrait of a Protein in Action. This classic example, used to convey the relation between structure and function, returns in an expanded treatment. New insights include:

- Oxygen transport during rest and during exercise
- The physiology of oxygen and CO_2 transport
- The molecular basis of sickle-cell anemia and thalassemia
- Balancing the production of α and β chains
- Newly discovered globins

Chapter 35: Drug Development. Knowledge of biochemical pathways is key to the development of new drugs such as Lipitor, Viagra, and Vioxx. In this new chapter, plentiful case studies illustrate:

- How drugs relate to other topics in the book— kinetics, enzyme inhibitors, membrane receptors, metabolic regulation, lipid synthesis, and signal transduction
- How the body's defenses respond to foreign compounds, especially the defenses provided by the biochemical pathways of xenobiotic metabolism
- The importance of administration, distribution, metabolism, excretion (ADME), and toxicology in drug development
- How the drug-development process works from target identification through clinical trials
- How the concepts and tools of genomics are used in the development of drugs

New Clinical Applications

 We have added a number of new examples from medical science to the already abundant selection of such examples (indicated by the icon above) (For a full list see p. x.) New topics include:

- Diseases of protein misfolding (Chapter 2)
- Human gene therapy (Chapter 5)
- Aggrecan and osteoarthritis (Chapter 11)
- The use of erythropoietin (EPO) to treat anemia and its abuse by athletes (Chapter 11)
- The use of monoclonal antibodies to target epidermal-growth-factor receptors in the treatment of colon and breast cancers (Chapter 14)
- Role of exercise in building defenses against superoxide radicals (Chapter 18)
- Diseases of altered ubiquination (Parkinson's disease, Angelman syndrome) (Chapter 23)
- The use of the proteasome inhibitor bortezomib to treat multiple myeloma (Chapter 23)
- Adenosine deaminase and severe combined immune deficiency (Chapter 25)
- Much enhanced discussion of gout (Chapter 25)
- Folic acid and spina bifida (Chapter 25)
- Type II diabetes (Chapter 27)
- Tumor suppressor genes and p53 (Chapter 28)
- Chemotherapy targeting DNA repair pathways (Chapter 28)

- Diseases of defective RNA splicing, including thalassemias and retinitis pigmentosa (Chapter 29)

- Innate immunity (Chapter 33)

Recent Advances

The sixth edition has been thoroughly updated throughout, including new discussions of the following recent advances:

- The **nucleation condensation model** of protein folding (Chapter 2)
- Using **MALDI-TOF mass spectrometry** to identify components of large protein complexes (Chapter 3)
- Update on the **human genome project** (Chapter 5)
- **Comparative genomics** (Chapter 5)
- Gene disruption by **RNA interference** (Chapter 5)
- Using **BLAST** searches (Chapter 6)
- **Lipid rafts** (Chapter 12)
- Mechanisms of action of several types of membrane channels and pumps such as the **acetylcholine receptor** (Chapter 13)
- **Aquaporin** (Chapter 13)
- The **insulin receptor pathway** (Chapter 14)

- The structure and function of the **EGF receptor** (Chapter 14)
- The structure of the **ATP-ADP translocase** (Chapter 18)
- Role of **glycogen synthase kinase** in glycogen regulation (Chapter 21)
- The role of **perlipin A** in fatty acid mobilization (Chapter 22)
- The newly revised structure of **fatty acid synthase** (Chapter 22)
- Prokaryotic and eukaryotic **replication initiation** (Chapter 28)
- DNA polymerase components (Chapter 28)
- The **trombone model** of DNA elongation (Chapter 28)
- **Promoter structure** in eukaryotes (Chapter 29)
- **Transcription initiation** in eukaryotes (Chapter 29)
- The **carboxy-terminal domain** (CTD) of RNA polymerase (Chapter 29)
- The role of **SNARE proteins** in protein targeting (Chapter 30)
- The structure of **taste receptors for detecting sweetness** (Chapter 32)

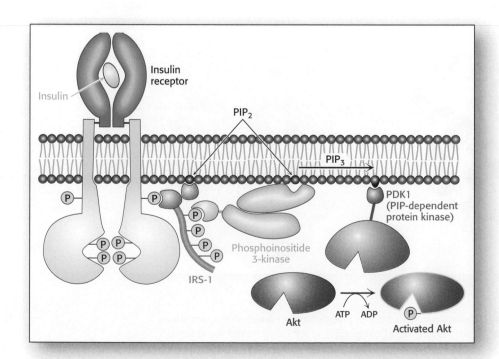

Figure 14.20 Insulin signaling. The binding of insulin to its receptor leads to a series of phosphorylations, resulting in the activation of the kinase Akt1. Activated Akt1 diffuses throughout the cell to continue the signal-transduction pathway.

Visualizing Molecular Structure

As in the fifth edition, all molecular structures have been selected and rendered by one of us, Jeremy Berg. The sixth edition includes new tools to help students read and understand molecular structures:

- A molecular model "primer" explains the different types of protein models and examines their strengths and weaknesses (appendices to Chapters 1 and 2).

- Figure legends direct students explicitly to the key features of a model.

- A greater variety of types of molecular structures are represented, including clearer renderings of membrane proteins.

- For most molecular models, the name of the file from the Protein Data Bank is given at the end of the figure legend. This file name (also known as a PDB ID) allows the reader easy access to the file used in generating the structure from the Protein Data Bank Web site (http://www.rcsb.org/pdb/). At this site, a variety of tools for visualizing and analyzing the structure are available.

- Living Figures for most molecular structures now appear on the Web site in Jmol to allow students to rotate 3-D molecules and view alternative renderings online.

End-of-Chapter Problems

In addition to general problems, the end-of-chapter problems include three categories to foster the development of specific skills.

- **Mechanism problems** ask students to suggest or elaborate a chemical mechanism.

- **Data interpretation problems** ask questions about a set of data provided in tabulated or graphic form. These problems give students a sense of how scientific conclusions are reached.

- **Chapter integration problems** require students to use information from several chapters to reach a solution. These problems reinforce a student's awareness of the interconnectedness of the different aspects of biochemistry.

Brief solutions to these problems are presented at the end of the book; expanded solutions are available in the accompanying *Student Companion*.

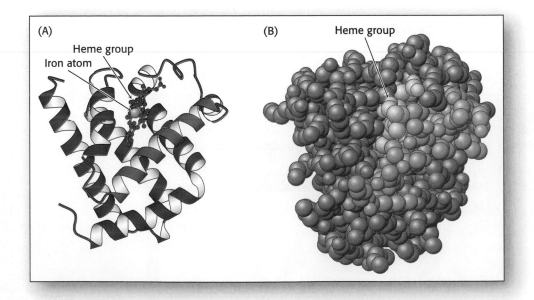

Figure 2.48 Three-dimensional structure of myoglobin. (A) A ribbon diagram shows that the protein consists largely of α helices. (B) A space-filling model in the same orientation shows how tightly packed the folded protein is. *Notice* that the heme group is nestled into a crevice in the compact protein with only an edge exposed. One helix is blue to allow comparison of the two structural depictions. [Drawn from 1A6N.pdb.]

Molecular Evolution

 This icon signals the start of many discussions that highlight protein commonalities or other molecular evolutionary insights that provide a framework to help students organize information.

Clinical Applications

 This icon signals the start of a clinical application in the text. Additional, briefer clinical correlations appear in the text as appropriate.

Tools and Techniques

The sixth edition of *Biochemistry* offers three chapters that present the tools and techniques of biochemistry: "Exploring Proteins and Proteomes" (Chapter 3), "Exploring Genes and Genomes" (Chapter 5), and "Exploring Evolution and Bioinformatics" (Chapter 6). Additional experimental techniques are presented throughout the book, as appropriate.

 Animated Techniques: Animated explanations of experimental techniques used for exploring genes and proteins are available at www.whfreeman.com/stryer

Living Figures

 This icon identifies molecular structures that are available in rotatable Jmol format on the companion Web site: www. whfreeman.com/stryer.

Media and Supplements

Companion Web site at
www.whfreeman.com/stryer

For students

- Living Figures. Every textbook illustration of a protein structure can also be viewed online in interactive 3-D using Jmol. Students can zoom and rotate the "live" structures to get a better understanding of their three-dimensional nature and can experiment with different display styles (space-filling, ball-and-stick, ribbon, backbone) by means of a user-friendly interface.

- Interactive structure-based tutorials in Jmol show how structure helps explain experimental data (such as the effect of mutations, sequence variation among homologs, the effects of chemical modification, and the results of spectroscopic experiments). *The tutorials were written by Neil D. Clarke, Johns Hopkins University School of Medicine.*

- Concept-based tutorials help students build intuitive understanding of some of the more difficult concepts covered in the text. *The tutorials were written by Neil D. Clarke, Johns Hopkins University School of Medicine.*

- Animated techniques help students grasp experimental techniques used for exploring genes and proteins.

- Self-assessment tool. Students can test their understanding by taking an online multiple-choice quiz provided for each chapter, as well as a general chemistry review.

- Glossary of key terms.

- Web links. This resource connects students with the world of biochemistry beyond the classroom.

For Instructors

All of the above plus:

- All illustrations and tables from the textbook, including structures from the *Glossary of Compounds*, in jpeg and PowerPoint formats, optimized for classroom projection.

- Lecture-ready Personal Response System ("clicker") questions. More than 100 questions for classroom use that will work seamlessly with any personal response system, including i-clicker, the new radio-frequency classroom response system being offered by W. H. Freeman and Company (www.iclicker.com).

- Assessment Bank, by Harvey Nikkel of Grand Valley State University and Susan Knock of Texas A&M University at Galveston, offers more than 1500 questions in editable Word format.

Instructor's Resource CD-ROM
[0-7167-4590-9]

The CD includes all the instructor's resources from the Web site.

Overhead Transparencies
[0-7167-6049-5]

200 full-color illustrations from the textbook, optimized for classroom projection

Student Companion
[0-7167-7067-9]

Richard I. Gumport, College of Medicine at Urbana-Champaign, University of Illinois

Frank H. Deis, Rutgers University

Nancy Counts Gerber, San Francisco State University

Expanded solutions to text problems provided by Roger E. Koeppe II, University of Arkansas

For each chapter of the textbook, the *Student Companion* includes:

- Chapter Learning Objectives and Summary

- Self-Assessment Problems, including multiple-choice, short-answer, matching questions, and challenge problems, and their answers

- Expanded Solutions to text end-of-chapter problems

Lecture Notebook
[0-7167-7157-8]

For students who find that they are too busy copying figures, equations, and diagrams to follow the lecture, the *Notebook* is an indispensable classroom companion, with:

- Illustrations and tables in the order in which they appear in the textbook, with plenty of room to take notes

- Three-hole punched, perforated pages so that students can reorganize the *Notebook* in any order necessary to follow lectures, and can insert instructor handouts

Acknowledgments

Thanks go first and foremost to our students. Not a word was written or an illustration constructed without the knowledge that bright, engaged students would immediately detect vagueness and ambiguity. We also thank our colleagues who supported, advised, instructed, and simply bore with us during this arduous task. We are also grateful to our colleagues throughout the world who patiently answered our questions and shared their insights into recent developments. We thank Susan J. Baserga and Erica A. Champion of the Yale University School of Medicine for their outstanding contributions in the revision of Chapter 29. Alan Mellors of the University of Guelph, Emeritus, deserves our thanks for reading every chapter of page proof to check for accuracy. We also especially thank those who served as reviewers for this new edition. Their thoughtful comments, suggestions, and encouragement have been of immense help to us in maintaining the excellence of the preceding editions. These reviewers are:

Steven Ackerman
University of Massachusetts

John S. Anderson
University of Minnesota

Kenneth Balazovich
University of Michigan

Susan J. Baserga
Yale University School of Medicine

Gail S. Begley
Northeastern University

Donald C. Beitz
Iowa State University

Peggy R. Borum
University of Florida

Amanda J. Bradley
University of British Columbia

Randy Brewton
University of Tennessee

Scott D. Briggs
Purdue University

Martin L. Brock
Eastern Kentucky University

Roger Brownsey
University of British Columbia

Michael F. Bruist
University of the Sciences in Philadelphia

John D. Brunstein
University of British Columbia

Mauricio Bustos
University of Maryland, Baltimore County

W. Malcolm Byrnes
Howard University College of Medicine

Larry D. Crouch
University of Nebraska Medical Center

David L. Daleke
Indiana University

Bansidhar Datta
Kent State University

Frank H. Deis
Rutgers University

Zachariah Dhanarajan
Florida A&M University

Preeti Dhar
State University of New York, New Paltz

Huangen Ding
Louisiana State University

Joseph Eichberg
University of Houston

Thomas Ellenberger
Harvard Medical School

Susan C. Evans
Ohio University

Ray Fall
University of Colorado

Wilson A. Francisco
Arizona State University

Terrence G. Frey
San Diego State University

K. Christopher Garcia
Stanford University School of Medicine

Ronald K. Gary
University of Nevada, Las Vegas

Alexandros Georgakilas
East Carolina University

Burt Goldberg
New York University

Michael R. Green
University of Massachusetts Medical School

E. M. Gregory
Virginia Tech

Charles B. Grissom
University of Utah

Anne Grove
Louisiana State University

Stuart Haring
University of Iowa Carver College of Medicine

Edward D. Harris
Texas A&M University

Newton Hilliard, Jr.
Eastern New Mexico University

Gerwald Jogl
Brown University

Konstantin V. Kandror
Boston University School of Medicine

A. Wali Karzai
Stony Brook University

Phillip E. Klebba
University of Oklahoma

Aileen F. Knowles
San Diego State University

John Koontz
University of Tennessee

Robert Kranz
Washington University

Min-Hao Kuo
Michigan State University

Patrick D. Larkin
Texas A&M University, Corpus Christi

Sylvia Lee-Huang
New York University School of Medicine

Glen B. Legge
University of Houston

Vince LiCata
Louisiana State University

Robley J. Light
Florida State University

Xuedong Liu
University of Colorado, Boulder

Andy LiWang
Texas A&M University

Timothy M. Logan
Florida State University

Michael A. Massiah
Oklahoma State University

Douglas D. McAbee
California State University, Long Beach

James McAfee
Pittsburg State University

Megan M. McEvoy
University of Arizona

Bryant W. Miles
Texas A&M University

Patricia M. Moroney
Louisiana State University

Mike Mossing
University of Mississippi

Michael P. Myers
California State University, Long Beach

Harry Noller
University of California, Santa Cruz

Mary Kay Orgill
University of Missouri

Oliver E. Owen
Retired clinical investigator, administrator, and academician

David C. Pendergrass
Kansas University

Cynthia B. Peterson
University of Tennessee

Philip A. Rea
University of Pennsylvania

Douglas D. Root
University of North Texas

Robert Rosenberg
Howard University

Richard L. Sabina
Medical College of Wisconsin

Robert Sanders
University of Kansas

Jamie L. Schlessman
United States Naval Academy

Todd P. Silverstein
Willamette University

Melanie A. Simpson
University of Nebraska, Lincoln

Kerry S. Smith
Clemson University

Deborah A. Spikes
Stony Brook University

Takita Felder Sumter
Winthrop University

David C. Teller
University of Washington

Marc E. Tischler
University of Arizona

Liang Tong
Columbia University

Michael Uhler
University of Michigan

Ronald Vale
University of California, San Francisco

Katherine Wall
University of Toledo

Malcom Watford
Rutgers University

Joachim Weber
Texas Tech University

Ian A. Wilson
The Scripps Research Institute

Marc S. Wold
University of Iowa Carver College of Medicine

Charles Yocum
University of Michigan

Robert Zand
University of Michigan

Brent M. Znosko
Saint Louis University

Working with our colleagues at W. H. Freeman and Company has been a wonderful experience. We would especially like to acknowledge the efforts of the following people. Our developmental editor, Susan Moran, has contributed immensely to the success of this project. Our project editor, Georgia Lee Hadler, managed the flow of the project—from final manuscript to final product—with admirable efficiency. The careful manuscript editor, Patricia Zimmerman, enhanced the text's literary consistency and clarity. Design manager Diana Blume produced a design and layout that are organizationally clear and esthetically pleasing. Our photo editor, Bianca Moscatelli, tenaciously tracked down new images. Bill Page, the illustration coordinator, ably oversaw the rendering of new illustrations, and Susan Wein, the production manager, astutely handled all the difficulties of scheduling, composition, and manufacturing. Media editor Alysia Baker and assistant editors Nick Tymoczko and Deena Goldman were invaluable in their management of the media and supplements program. We would also like to thank Timothy Driscoll for his work in converting our living figures into Jmol.

Our acquisitions editor, Kate Ahr, was an outstanding director of the project. Her enthusiasm, encouragement, patience, and good humor kept us going when we were tired, frustrated, and discouraged. Marketing mavens John Britch and Sarah Martin oversaw the introduction of this edition to the academic world. We also thank the sales people at W. H. Freeman and Company for their excellent suggestions and view of the market. We thank Elizabeth Widdicombe, President of W. H. Freeman and Company, for never losing faith in us.

Finally, the project would not have been possible without the unfailing support of our families—especially our wives, Wendie Berg and Alison Unger. Their patience, encouragement, and enthusiasm have made this endeavor possible. We also thank our children, Alex, Corey, and Monica Berg and Janina and Nicholas Tymoczko, for their forbearance and good humor and for constantly providing us a perspective on what is truly important in life.

Contents

Preface

Part I THE MOLECULAR DESIGN OF LIFE

Chapter 1 Biochemistry: An Evolving Science 1

Chapter 2 Protein Composition and Structure 25

Chapter 7 Hemoglobin: Portrait of a Protein in Action 183

Chapter 8 Enzymes: Basic Concepts and Kinetics 205

Open

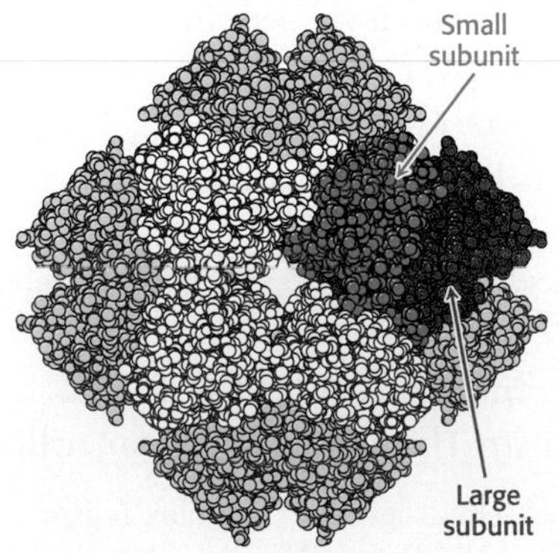

Small subunit

Large subunit

Chapter 23 Protein Turnover and Amino Acid Catabolism   649

Biochemistry: An Evolving Science

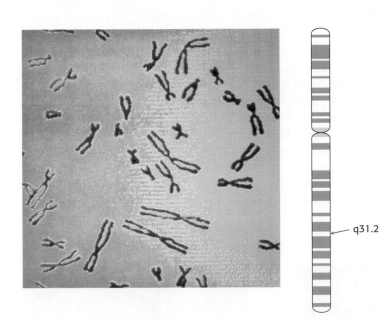

q31.2

Disease and the genome. Studies of the human genome are revealing disease origins and other biochemical mysteries. Human chromosomes, left, contain the DNA molecules that constitute the human genome. The staining pattern serves to identify specific regions of a chromosome. On the right is a diagram of human chromosome 7, with band q31.2 indicated by an arrow. A gene in this region encodes a protein that, when malfunctioning, causes cystic fibrosis. [(Left) Alfred Pasieka/Peter Arnold.]

Biochemistry is the study of the chemistry of life processes. Since the discovery that biological molecules such as urea could be synthesized from nonliving components in 1828, scientists have explored the chemistry of life with great intensity. Through these investigations, many of the most fundamental mysteries of how living things function at a biochemical level have now been solved. However, much remains to be investigated. As is often the case, each discovery raises at least as many new questions as it answers. Furthermore, we are now in an age of unprecedented opportunity for the application of our tremendous knowledge of biochemistry to problems in medicine, dentistry, agriculture, forensics, anthropology, environmental sciences, and many other fields. We begin our journey into biochemistry with one of the most startling discoveries of the past century: namely, the great unity of all living things at the biochemical level.

1.1 Biochemical Unity Underlies Biological Diversity

The biological world is magnificently diverse. The animal kingdom is rich with species ranging from nearly microscopic insects to elephants and whales. The plant kingdom includes species as small and relatively simple as algae and as large and complex as giant sequoias. Unlike animals that must eat to survive, plants have the remarkable ability to use sunlight to

1

convert carbon dioxide in the air into living tissues. This diversity extends further when we descend into the microscopic world. Single-celled organisms such as protozoa, yeast, and bacteria are present with great diversity in water, in soil, and on or within larger organisms. Some organisms can survive and even thrive in seemingly hostile environments such as hot springs and glaciers.

The microscope revealed a key unifying feature that underlies this diversity. Large organisms are built up of *cells,* resembling, to some extent, single-celled microscopic organisms. The construction of animals, plants, and microorganisms from cells suggested that these diverse organisms might have more in common than is apparent from their outward appearance. With the development of biochemistry, this suggestion has been tremendously supported and expanded. At the biochemical level, all organisms have many common features (Figure 1.1).

As mentioned earlier, biochemistry is the study of the chemistry of life processes. These processes entail the interplay of two different classes of molecules: large molecules such as proteins and nucleic acids, referred to as *biological macromolecules,* and low-molecular-weight molecules such as glucose and glycerol, referred to as *metabolites,* that are chemically transformed in biological processes. Members of both these classes of molecules are common, with minor variations, to all living things. For example, *deoxyribonucleic acid* (DNA) stores genetic information in all cellular organisms. *Proteins,* the macromolecules that are key participants in most biological processes, are built from a set of 20 building blocks that are the same in all

Glucose

Glycerol

Sulfolobus acidicaldarius

Arabidopsis thaliana

Homo sapiens

Figure 1.1 Biological diversity and similarity. The shape of a key molecule in gene regulation (the TATA-box-binding protein) is similar in three very different organisms that are separated from one another by billions of years of evolution. [(Left) Dr. T. J. Beveridge/Visuals Unlimited; (middle) Holt Studios/Photo Researchers; (right) Time Life Pictures/Getty Images.]

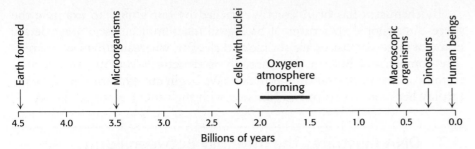

Figure 1.2 **A possible time line for biochemical evolution.** Selected key events are indicated. Note that life on Earth began approximately 3.5 billion years ago, whereas human beings emerged quite recently.

organisms. Furthermore, proteins that play similar roles in different organisms often have very similar three-dimensional structures (see Figure 1.1).

Key metabolic processes also are common to many organisms. For example, the set of chemical transformations that converts glucose and oxygen into carbon dioxide and water is essentially identical in simple bacteria such as *Escherichia coli (E. coli)* and human beings. Even processes that appear to be quite distinct often have common features at the biochemical level. Remarkably, the biochemical processes by which plants capture light energy and convert it into more-useful forms are strikingly similar to steps used in animals to capture energy released from the breakdown of glucose.

These observations overwhelmingly suggest that all living things on Earth have a common ancestor and that modern organisms have evolved from this ancestor into their present forms. Geological and biochemical findings support a time line for this evolutionary path (Figure 1.2). On the basis of their biochemical characteristics, the diverse organisms of the modern world can be divided into three fundamental groups called *domains: Eukarya* (eukaryotes), *Bacteria,* and *Archaea.* Eukarya comprise all multicellular organisms, including human beings as well as many microscopic, unicellular organisms such as yeast. The defining characteristic of *eukaryotes* is the presence of a well-defined nucleus within each cell. Unicellular organisms such as bacteria, which lack a nucleus, are referred to as *prokaryotes.* The prokaryotes were reclassified as two separate domains in response to Carl Woese's discovery in 1977 that certain bacteria-like organisms are biochemically quite distinct from other previously characterized bacterial species. These organisms, now recognized as having diverged from bacteria early in evolution, are the *archaea.* Evolutionary paths from a common ancestor to modern organisms can be deduced on the basis of biochemical information. One such path is shown in Figure 1.3.

Much of this book will explore the chemical reactions and the associated biological macromolecules and metabolites that are found in biological processes common to all organisms. The unity of life at the biochemical level makes this approach possible. At the same time, different organisms have specific needs, depending on the particular biological niche in which they evolved and live. By comparing and contrasting details of particular biochemical pathways in different organisms, we can learn how biological challenges are solved at the biochemical level. In most cases, these challenges are addressed by the adaptation of existing macromolecules to new roles rather than by the evolution of entirely new ones.

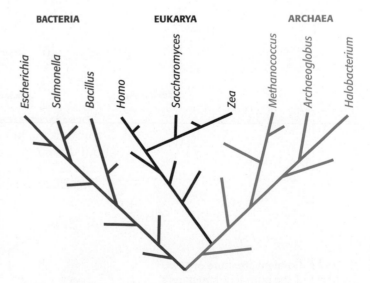

Figure 1.3 **The tree of life.** A possible evolutionary path from a common ancestor approximately 3.5 billion years ago at the bottom of the tree to organisms found in the modern world at the top.

Biochemistry has been greatly enriched by our ability to examine the three-dimensional structures of biological macromolecules in great detail. Some of these structures are simple and elegant, whereas others are incredibly complicated but, in any case, these structures provide an essential framework for understanding function. We begin our exploration of the interplay between structure and function with the genetic material, DNA.

1.2 DNA Illustrates the Interplay Between Form and Function

A fundamental biochemical feature common to all cellular organisms is the use of DNA for the storage of genetic information. The discovery that DNA plays this central role was first made in studies of bacteria in the 1940s. This discovery was followed by the elucidation of the three-dimensional structure of DNA in 1953, an event that set the stage for many of the advances in biochemistry and many other fields, extending to the present.

The structure of DNA powerfully illustrates a basic principle common to all biological macromolecules: the intimate relation between structure and function. The remarkable properties of this chemical substance allow it to function as a very efficient and robust vehicle for storing information. We start with an examination of the covalent structure of DNA and its extension into three dimensions.

DNA Is Constructed from Four Building Blocks

DNA is a *linear polymer* made up of four different types of monomers. It has a fixed backbone from which protrude variable substituents (Figure 1.4). The backbone is built of repeating sugar–phosphate units. The sugars are molecules of *deoxyribose* from which DNA receives its name. Each sugar is connected to two phosphate groups through different linkages. Moreover, each sugar is oriented in the same way, and so each DNA strand is polar, with one end distinguishable from the other. Joined to each deoxyribose is one of four possible bases: adenine (A), cytosine (C), guanine (G), and thymine (T).

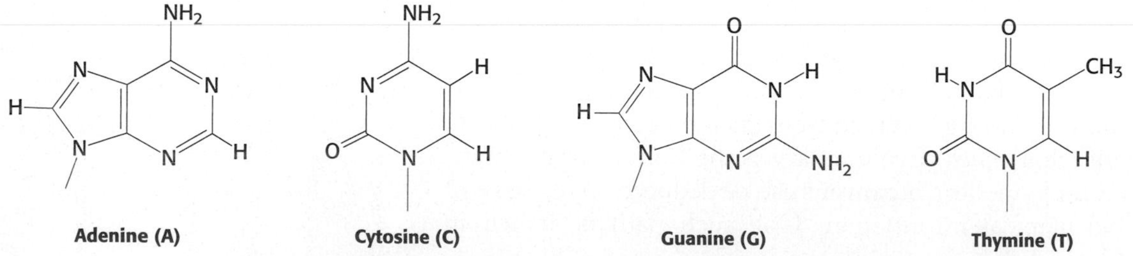

Adenine (A)   Cytosine (C)   Guanine (G)   Thymine (T)

These bases are connected to the sugar components in the DNA backbone through the bonds shown in black in Figure 1.4. All four bases are planar but differ significantly in other respects. Thus, each monomer of DNA consists of a sugar–phosphate unit and one of four bases attached to the sugar. These bases can be arranged in any order along a strand of DNA.

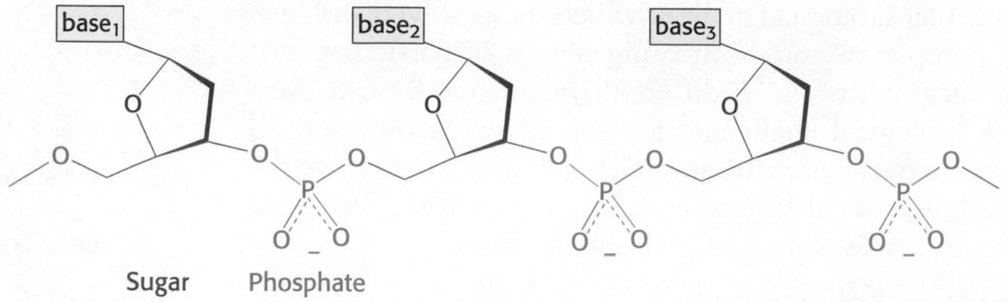

Sugar   Phosphate

Figure 1.4 Covalent structure of DNA. Each unit of the polymeric structure is composed of a sugar (deoxyribose), a phosphate, and a variable base that protrudes from the sugar–phosphate backbone.

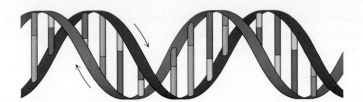

Figure 1.5 **The double helix.** The double-helical structure of DNA proposed by Watson and Crick. The sugar–phosphate backbones of the two chains are shown in red and blue, and the bases are shown in green, purple, orange, and yellow. The two strands are antiparallel, running in opposite directions with respect to the axis of the double helix, as indicated by the arrows.

Two Single Strands of DNA Combine to Form a Double Helix

Most DNA molecules consist of not one but two strands (Figure 1.5). In 1953, James Watson and Francis Crick deduced the arrangement of these strands and proposed a three-dimensional structure for DNA molecules. This structure is a *double helix* composed of two intertwined strands arranged such that the sugar–phosphate backbone lies on the outside and the bases on the inside. The key to this structure is that the bases form *specific base pairs* (bp) held together by *hydrogen bonds* (Section 1.3): adenine pairs with thymine (A–T) and guanine pairs with cytosine (G–C), as shown in Figure 1.6. Hydrogen bonds are much weaker than *covalent bonds* such as the carbon–carbon or carbon–nitrogen bonds that define the structures of the bases themselves. Such weak bonds are crucial to biochemical systems; they are weak enough to be reversibly broken in biochemical processes, yet they are strong enough, when many form simultaneously, to help stabilize specific structures such as the double helix.

DNA Structure Explains Heredity and the Storage of Information

The structure proposed by Watson and Crick has two properties of central importance to the role of DNA as the hereditary material. First, the structure is compatible with any sequence of bases. The base pairs have essentially the same shape (see Figure 1.6) and thus fit equally well into the center of the double-helical structure of any sequence. Without any constraints, the sequence of bases along a DNA strand can act as an efficient means of storing information. Indeed, the sequence of bases along DNA strands is how genetic information is stored. The DNA sequence determines the sequences of the ribonucleic acid (RNA) and protein molecules that carry out most of the activities within cells.

Second, because of base-pairing, the sequence of bases along one strand completely determines the sequence along the other strand. As Watson and Crick so coyly wrote: "It has not escaped our notice that the specific pairing we have postulated immediately suggests a possible copying mechanism for the genetic material." Thus, if the DNA double helix is separated into two single strands, each strand can act as a template for the generation of its

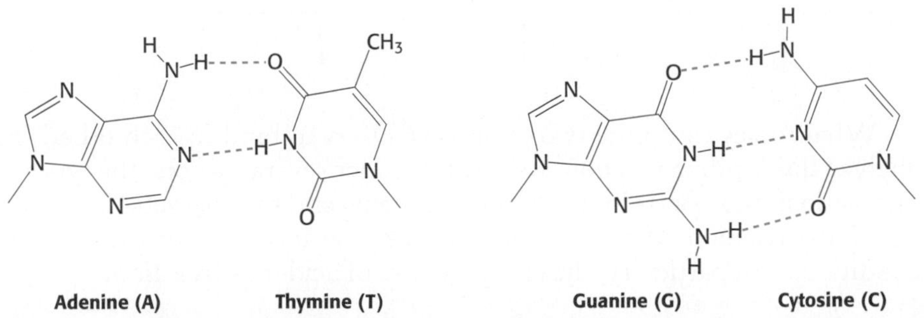

Adenine (A)   **Thymine (T)**   **Guanine (G)**   **Cytosine (C)**

Figure 1.6 **Watson–Crick base pairs.** Adenine pairs with thymine (A–T), and guanine with cytosine (G–C). The dashed green lines represent hydrogen bonds.

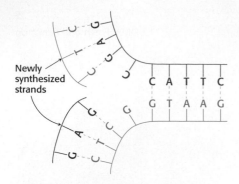

Figure 1.7 DNA replication. If a DNA molecule is separated into two strands, each strand can act as the template for the generation of its partner strand.

partner strand through specific base-pair formation (Figure 1.7). The three-dimensional structure of DNA beautifully illustrates the close connection between molecular form and function.

1.3 Concepts from Chemistry Explain the Properties of Biological Molecules

We have seen how a chemical insight, into the hydrogen-bonding capabilities of the bases of DNA, led to a deep understanding of a fundamental biological process. To lay the groundwork for the rest of the book, we begin our study of biochemistry by examining selected concepts from chemistry and showing how these concepts apply to biological systems. The concepts include the types of chemical bonds; the structure of water, the solvent in which most biochemical processes take place; the First and Second Laws of Thermodynamics; and the principles of acid–base chemistry. We will use these concepts to examine a archetypical biochemical process—namely, the formation of a DNA double helix from its two component strands. The process is but one of many examples that could have been chosen to illustrate these topics. Keep in mind that, although the specific discussion is about DNA and double-helix formation, the concepts considered are quite general and will apply to many other classes of molecules and processes that we will discuss in the remainder of the book.

The Double Helix Can Form from Its Component Strands

The discovery that DNA from natural sources exists in a double-helical form with Watson–Crick base pairs suggested, but did not prove, that such double helices would form spontaneously outside biological systems. Suppose that two short strands of DNA were chemically synthesized to have complementary sequences so that they could, in principle, form a double helix with Watson–Crick base pairs. Two such sequences are CGAT-TAAT and ATTAATCG. The structures of these molecules in solution can be examined by a variety of techniques. In isolation, each sequence exists almost exclusively as a single-stranded molecule. However, when the two sequences are mixed, a double helix with Watson–Crick base pairs does form (Figure 1.8). This reaction proceeds nearly to completion. If each of the strands are initially present at equal concentrations of 1 mM, then more than 99.99% of the strands are in the double helix at 25°C and in the presence of 1 M NaCl.

Figure 1.8 Formation of a double helix. When two DNA strands with appropriate, complementary sequences are mixed, they spontaneously assemble to form a double helix.

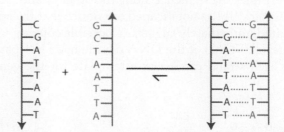

What forces cause the two strands of DNA to bind to each other? To analyze this binding reaction, we must consider several factors: the types of interactions and bonds in biochemical systems and the energetic favorability of the reaction. We also must consider the influence of the solution conditions—in particular, the consequences of acid–base reactions.

Covalent and Noncovalent Bonds Are Important for the Structure and Stability of Biological Molecules

Atoms interact with one another through chemical bonds. These bonds include the covalent bonds that define the structure of molecules as well as a variety of noncovalent bonds that are of great importance to biochemistry.

Covalent Bonds. The strongest bonds are covalent bonds, such as the bonds that hold the atoms together within the individual bases shown on page 4. A covalent bond is formed by the sharing of a pair of electrons between adjacent atoms. A typical carbon–carbon (C–C) covalent bond has a bond length of 1.54 Å and bond energy of 356 kJ mol^{-1} (85 kcal mol^{-1}). Because covalent bonds are so strong, considerable energy must be expended to break them. More than one electron pair can be shared between two atoms to form a multiple covalent bond. For example, three of the bases in Figure 1.6 include carbon–oxygen (C$=$O) double bonds. These bonds are even stronger than C$-$C single bonds, with energies near 730 kJ mol^{-1} (175 kcal mol^{-1}) and are somewhat shorter.

For some molecules, more than one pattern of covalent bonding can be written. For example, adenine can be written in two equivalent ways called *resonance structures.*

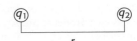

These adenine structures depict alternative arrangements of single and double bonds that are possible within the same structural framework. Resonance structures are shown connected by a double-headed arrow. Adenine's true structure is a composite of its two resonance structures. The composite structure is manifested in the bond lengths such as that for the bond joining carbon atoms C-4 and C-5. The observed bond length of 1.40 Å is between that expected for a C$-$C single bond (1.54 Å) and a C$=$C double bond (1.34 Å). A molecule that can be written as several resonance structures of approximately equal energies has greater stability than does a molecule without multiple resonance structures.

Noncovalent Bonds. Noncovalent bonds are weaker than covalent bonds but are crucial for biochemical processes such as the formation of a double helix. Four fundamental noncovalent bond types are *electrostatic interactions, hydrogen bonds, van der Waals interactions,* and *hydrophobic interactions.* They differ in geometry, strength, and specificity. Furthermore, these bonds are affected in vastly different ways by the presence of water. Let us consider the characteristics of each:

1. *Electrostatic Interactions.* A charged group on one molecule can attract an oppositely charged group on another molecule. The energy of an electrostatic interaction is given by *Coulomb's law:*

$$E = kq_1q_2/Dr$$

where E is the energy, q_1 and q_2 are the charges on the two atoms (in units of the electronic charge), r is the distance between the two atoms (in angstroms), D is the dielectric constant (which accounts for the effects of the

> **Distance and energy units**
>
> Interatomic distances and bond lengths are usually measured in angstrom (Å) units:
> $1\ \text{Å} = 10^{-10}\ \text{m} = 10^{-8}\ \text{cm} = 0.1\ \text{nm}.$
>
> Several energy units are in common use. One joule (J) is the amount of energy required to move 1 meter against a force of 1 newton. A kilojoule (kJ) is 1000 joules. One calorie is the amount of energy required to raise the temperature of 1 gram of water 1 degree Celsius. A kilocalorie (kcal) is 1000 calories. One joule is equal to 0.239 cal.

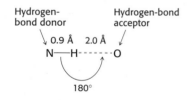

Figure 1.9 Hydrogen bonds. Hydrogen bonds are depicted by dashed green lines. The positions of the partial charges (δ^+ and δ^-) are shown.

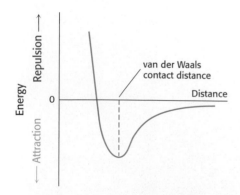

Figure 1.10 Energy of a van der Waals interaction as two atoms approach each other. The energy is most favorable at the van der Waals contact distance. Due to electron–electron repulsion, the energy rises rapidly as the atoms approach closer than this distance.

intervening medium), and k is a proportionality constant ($k = 1389$, for energies in units of kilojoules mol^{-1}, or 332 for energies in kcal mol^{-1}).

By convention, an attractive interaction has a negative energy. The electrostatic interaction between two ions bearing single opposite charges separated by 3 Å in water (which has a dielectric constant of 80) has an energy of -5.8 kJ mol^{-1} (-1.4 kcal mol^{-1}). Note how important the dielectric constant of the medium is. For the same ions separated by 3 Å in a nonpolar solvent such as hexane (which has a dielectric constant of 2), the energy of this interaction is -231 kJ mol^{-1} (-55 kcal mol^{-1}).

2. *Hydrogen Bonds.* These interactions are fundamentally electrostatic interactions. Hydrogen bonds are responsible for specific base-pair formation in the DNA double helix. The hydrogen atom in a hydrogen bond is shared by two electronegative atoms such as nitrogen or oxygen. The *hydrogen-bond donor* is the group that includes both the atom to which the hydrogen is more tightly linked and the hydrogen atom itself, whereas the *hydrogen-bond acceptor* is the atom less tightly linked to the hydrogen atom (Figure 1.9). The electronegative atom to which the hydrogen atom is covalently bonded pulls electron density away from the hydrogen atom, which thus develops a partial positive charge (δ^+). Thus, the hydrogen atom can interact with an atom having a partial negative charge (δ^-) through an electrostatic interaction.

Hydrogen bonds are much weaker than covalent bonds. They have energies ranging from 4 to 20 kJ mol^{-1} (1–5 kcal mol^{-1}). Hydrogen bonds are also somewhat longer than covalent bonds; their bond lengths (measured from the hydrogen atom) range from 1.5 Å to 2.6 Å; hence, a distance ranging from 2.4 Å to 3.5 Å separates the two nonhydrogen atoms in a hydrogen bond. The strongest hydrogen bonds have a tendency to be approximately straight, such that the hydrogen-bond donor, the hydrogen atom, and the hydrogen-bond acceptor lie along a straight line. Hydrogen-bonding interactions are responsible for many of the properties of water that make it such a special solvent, as will be described shortly.

3. *van der Waals Interactions.* The basis of a van der Waals interaction is that the distribution of electronic charge around an atom fluctuates with time. At any instant, the charge distribution is not perfectly symmetric. This transient asymmetry in the electronic charge about an atom acts through electrostatic interactions to induce a complementary asymmetry in the electron distribution within its neighboring atoms. The atom and its neighbors then attract one another. This attraction increases as two atoms come closer to each other, until they are separated by the van der Waals *contact distance* (Figure 1.10). At distances shorter than the van der Waals contact distance, very strong repulsive forces become dominant because the outer electron clouds of the two atoms overlap.

Energies associated with van der Waals interactions are quite small; typical interactions contribute from 2 to 4 kJ mol^{-1} (0.5 to 1 kcal mol^{-1}) per atom pair. When the surfaces of two large molecules come together, however, a large number of atoms are in van der Waals contact, and the net effect, summed over many atom pairs, can be substantial.

Properties of Water. Water is the solvent in which most biochemical reactions take place, and its properties are essential to the formation of macromolecular structures and the progress of chemical reactions. Two properties of water are especially relevant:

1. *Water is a polar molecule.* The water molecule is bent, not linear, and so the distribution of charge is asymmetric. The oxygen nucleus draws electrons away from the two hydrogen nuclei, which leaves the region around

each hydrogen nucleus with a net positive charge. The water molecule is thus an electrically polar structure.

2. *Water is highly cohesive.* Water molecules interact strongly with one another through hydrogen bonds. These interactions are apparent in the structure of ice (Figure 1.11). Networks of hydrogen bonds hold the structure together; similar interactions link molecules in liquid water and account for the cohesion of liquid water, although, in the liquid state, approximately one-fourth of the hydrogen bonds present in ice are broken. The polar nature of water is responsible for its high dielectric constant of 80. Molecules in aqueous solution interact with water molecules through the formation of hydrogen bonds and through ionic interactions. These interactions make water a versatile solvent, able to readily dissolve many species, especially polar and charged compounds that can participate in these interactions.

Electric dipole

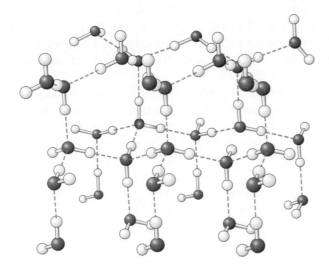

Figure 1.11 Structure of ice. Hydrogen bonds (shown as dashed green lines) are formed between water molecules to produce a highly ordered and open structure.

The Hydrophobic Effect. A final fundamental interaction called the *hydrophobic effect* is a manifestation of the properties of water. Some molecules (termed *nonpolar molecules*) cannot participate in hydrogen bonding or ionic interactions. The interactions of nonpolar molecules with water molecules are not as favorable as are interactions between the water molecules themselves. The water molecules in contact with these nonpolar molecules form "cages" around them, becoming more well ordered than water molecules free in solution. However, when two such nonpolar molecules come together, some of the water molecules are released, allowing them to interact freely with bulk water (Figure 1.12). The release of water from such cages is favorable for reasons to be considered shortly. The result is that

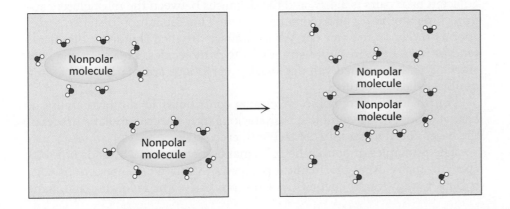

Figure 1.12 The hydrophobic effect. The aggregation of nonpolar groups in water leads to the release of water molecules, initially interacting with the nonpolar surface, into bulk water. The release of water molecules into solution makes the aggregation of nonpolar groups favorable.

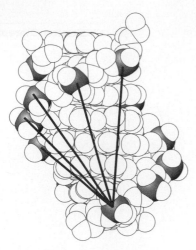

Figure 1.13 Electrostatic interactions in DNA. Each unit within the double helix includes a phosphate group (the phosphorus atom being shown in purple) that bears a negative charge. The unfavorable interactions of one phosphate (also known as a phosphoryl group) with several others are shown by red lines. These repulsive interactions oppose the formation of a double helix.

nonpolar molecules show an increased tendency to associate with one another in water compared with other, less polar and less self-associating, solvents. This tendency is called the hydrophobic effect and the associated interactions are called *hydrophobic interactions*.

The Double Helix Is an Expression of the Rules of Chemistry

Let us now see how these four noncovalent interactions work together in driving the association of two strands of DNA to form a double helix. First, each phosphate group in a DNA strand carries a negative charge. These negatively charged groups interact unfavorably with one another over distances. Thus, unfavorable electrostatic interactions take place when two strands of DNA come together. These phosphate groups are far apart in the double helix with distances greater than 10 Å, but many such interactions take place (Figure 1.13). Thus, electrostatic interactions oppose the formation of the double helix. The strength of these repulsive electrostatic interactions is diminished by the high dielectric constant of water and the presence of ionic species such as Na^+ or Mg^{2+} ions in solution. These positively charged species interact with the phosphate groups and partly neutralize their negative charges.

Second, we noted the importance of hydrogen bonds in determining the formation of specific base pairs in the double helix. However, in single-stranded DNA, the hydrogen-bond donors and acceptors are exposed to solution and can form hydrogen bonds with water molecules.

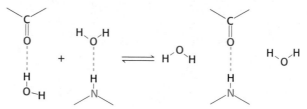

When two single strands come together, these hydrogen bonds with water are broken and new hydrogen bonds between the bases are formed. Because the number of hydrogen bonds broken is the same as the number formed, these hydrogen bonds do not contribute substantially to driving the overall process of double-helix formation. However, they contribute greatly to the specificity of binding. Suppose two bases that cannot form Watson–Crick base pairs are brought together. Hydrogen bonds with water must be broken as the bases come into contact. Because the bases are not complementary in structure, not all of these bonds can be simultaneously replaced by hydrogen bonds between the bases. Thus, the formation of a double helix between noncomplementary sequences is disfavored.

Third, within a double helix, the base pairs are parallel and stacked nearly on top of one another. The typical separation between the planes of adjacent base pairs is 3.4 Å, and the distances between the most closely approaching atoms are approximately 3.6 Å. This separation distance corresponds nicely to the van der Waals contact distance (Figure 1.14). Bases tend to stack even in single-stranded DNA molecules. However, the base stacking and associated van der Waals interactions are nearly optimal in a double-helical structure.

Fourth, the hydrophobic effect also contributes to the favorability of base stacking. More-complete base stacking moves the nonpolar surfaces of the bases out of water into contact with each other.

The principles of double-helix formation between two strands of DNA apply to many other biochemical processes. Many weak interactions contribute to the overall energetics of the process, some favorably and some

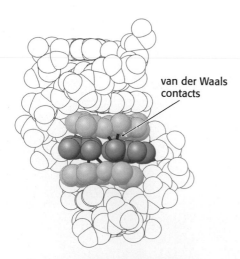

van der Waals contacts

Figure 1.14 Base stacking. In the DNA double helix, adjacent base pairs are stacked nearly on top of one another, and so many atoms in each base pair are separated by their van der Waals contact distance. The central base pair is shown in dark blue and the two adjacent base pairs in light blue. Several van der Waals contacts are shown in red.

unfavorably. Furthermore, surface complementarity is a key feature: when complementary surfaces meet, hydrogen-bond donors align with hydrogen-bond acceptors and nonpolar surfaces come together to maximize van der Waals interactions and minimize nonpolar surface area exposed to the aqueous environment. The properties of water play a major role in determining the importance of these interactions.

The Laws of Thermodynamics Govern the Behavior of Biochemical Systems

We can look at the formation of the double helix from a different perspective by examining the laws of thermodynamics. These laws are general principles that apply to all physical (and biological) processes. They are of great importance because they determine the conditions under which specific processes can or cannot take place. We will consider these laws from a general perspective first and then apply the principles that we have developed to the formation of the double helix.

The laws of thermodynamics distinguish between a system and its surroundings. A *system* refers to the matter within a defined region of space. The matter in the rest of the universe is called the *surroundings. The First Law of Thermodynamics states that the total energy of a system and its surroundings is constant.* In other words, the energy content of the universe is constant; energy can be neither created nor destroyed. Energy can take different forms, however. Heat, for example, is one form of energy. Heat is a manifestation of the *kinetic energy* associated with the random motion of molecules. Alternatively, energy can be present as *potential energy*—energy that will be released on the occurrence of some process. Consider, for example, a ball held at the top of a tower. The ball has considerable potential energy because, when it is released, the ball will develop kinetic energy associated with its motion as it falls. Within chemical systems, potential energy is related to the likelihood that atoms can react with one another. For instance, a mixture of gasoline and oxygen has a large potential energy because these molecules may react to form carbon dioxide and water and release energy as heat. The First Law requires that any energy released in the formation of chemical bonds must be used to break other bonds, released as heat, or stored in some other form.

Another important thermodynamic concept is that of *entropy*, a measure of the degree of randomness or disorder in a system. *The Second Law of Thermodynamics states that the total entropy of a system plus that of its surroundings always increases.* For example, the release of water from nonpolar surfaces responsible for the hydrophobic effect is favorable because water molecules free in solution are more disordered than they are when they are associated with nonpolar surfaces. At first glance, the Second Law appears to contradict much common experience, particularly about biological systems. Many biological processes, such as the generation of a well-defined structure such as a leaf from carbon dioxide gas and other nutrients, clearly increase the level of order and hence decrease entropy. Entropy may be decreased locally in the formation of such ordered structures only if the entropy of other parts of the universe is increased by an equal or greater amount. The local decrease in entropy is often accomplished by a release of heat, which increases the entropy of the environment.

We can analyze this process in quantitative terms. First, consider the system. The entropy (S) of the system may change in the course of a chemical reaction by an amount ΔS_{system}. If heat flows from the system to its surroundings, then the heat content, often referred to as the *enthalpy (H)*, of the system will be reduced by an amount ΔH_{system}. To apply the Second Law, we must determine the change in entropy of the surroundings. If heat flows from the system to the surroundings, then the entropy of the surroundings will increase. The precise change in the entropy of the surroundings depends on the temperature; the change in entropy is greater when heat is added to relatively cold surroundings than when heat is added to surroundings at high temperatures that are already in a high degree of disorder. To be even more specific, the change in the entropy of the surroundings will be proportional to the amount of heat transferred from the system and inversely proportional to the temperature (T) of the surroundings. In biological systems, T [in kelvins (K), absolute temperature] is usually assumed to be constant. Thus, a change in the entropy of the surroundings is given by

$$\Delta S_{surroundings} = -\Delta H_{system}/T \tag{1}$$

The total entropy change is given by the expression

$$\Delta S_{total} = \Delta S_{system} + \Delta S_{surroundings} \tag{2}$$

Substituting equation 1 into equation 2 yields

$$\Delta S_{total} = \Delta S_{system} - \Delta H_{system}/T \tag{3}$$

Multiplying by $-T$ gives

$$-T\Delta S_{total} = \Delta H_{system} - T\Delta S_{system} \tag{4}$$

The function $-T\Delta S$ has units of energy and is referred to as *free energy* or *Gibbs free energy*, after Josiah Willard Gibbs, who developed this function in 1878:

$$\Delta G = \Delta H_{system} - T\Delta S_{system} \tag{5}$$

The free-energy change, ΔG, will be used throughout this book to describe the energetics of biochemical reactions.

Recall that the Second Law of Thermodynamics states that, for a process to take place, the entropy of the universe must increase. Examination of equation 3 shows that the total entropy will increase if and only if

$$\Delta S_{system} > \Delta H_{system}/T \tag{6}$$

Rearranging gives $T\Delta S_{system} > \Delta H$, or, in other words, entropy will increase if and only if

$$\Delta G = \Delta H_{system} - T\Delta S_{system} < 0 \tag{7}$$

Thus, the free-energy change must be negative for a process to occur spontaneously. *There is negative free-energy change when and only when the overall entropy of the universe is increased.* The free energy represents a single term that takes into account both the entropy of the system and the entropy of the surroundings.

Heat Is Released in the Formation of the Double Helix

Let us see how the principles of thermodynamics apply to the formation of the double helix (Figure 1.15). Suppose solutions containing each of the two single strands are mixed. Before the double helix forms, each of the single strands is free to translate and rotate in solution, whereas each matched pair

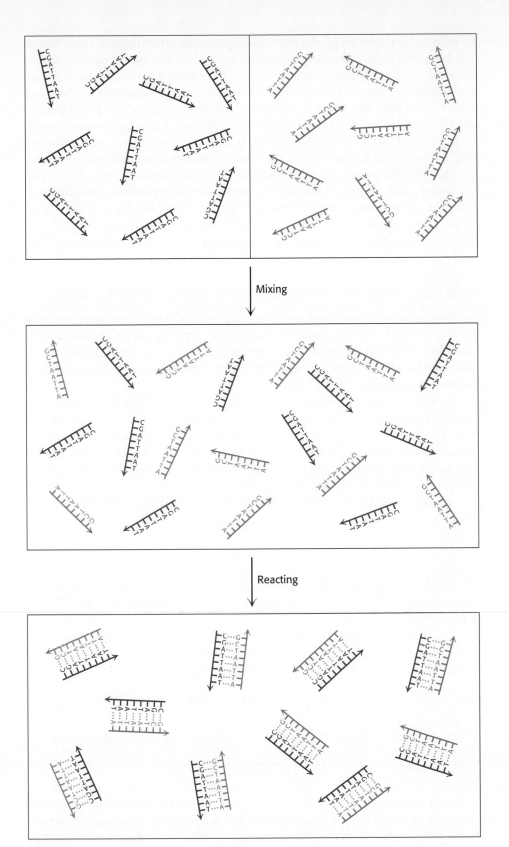

Figure 1.15 Double-helix formation and entropy. When solutions containing DNA strands with complementary sequences are mixed, the strands react to form double helices. This process results in a loss of entropy from the system, indicating that heat must be released to the surroundings to avoid violating the Second Law of Thermodynamics.

of strands in the double helix must move together. Furthermore, the free single strands exist in more conformations than possible when bound together in a double helix. Thus, the formation of a double helix from two single strands appears to result in an increase in order for the system.

On the basis of this analysis, we expect that the double helix cannot form without violating the Second Law of Thermodynamics unless heat is released to increase the entropy of the surroundings. Experimentally, we can measure the heat released by allowing the solutions containing the two single strands to come together within a water bath, which here corresponds to the surroundings. We then determine how much heat must be absorbed by the water bath or released from it to maintain it at a constant temperature. For these single DNA strands at $25°C$ and at pH 7.0 in 1 M NaCl, this experiment reveals that a substantial amount of heat is released—namely, approximately 250 kJ mol^{-1} (60 kcal mol^{-1}). This experimental result reveals that the change in enthalpy for the process is quite large, -250 kJ mol^{-1}, consistent with our expectation that significant heat would have to be released to the surroundings for the process not to violate the Second Law. We see in quantitative terms how order within a system can be increased by releasing sufficient heat to the surroundings to ensure that the entropy of the universe increases. We will encounter this general theme again and again throughout this book.

Acid–Base Reactions Are Central in Many Biochemical Processes

Throughout our consideration of the formation of the double helix, we have dealt only with the noncovalent bonds that are formed or broken in this process. Many biochemical processes entail the formation and cleavage of covalent bonds. Of these, a particularly important class of reactions prominent in biochemistry is *acid–base reactions.*

In acid and base reactions, hydrogen ions are added to molecules or removed from them. A hydrogen ion, often written as H^+, corresponds to a bare proton. In fact, hydrogen ions exist in solution bound to water molecules, thus forming what are known as *hydronium ions,* H_3O^+. For simplicity, we will continue to write H^+, but we should keep in mind that H^+ is shorthand for the actual species present.

The concentration of hydrogen ions in solution is expressed as the pH. Specifically, the *pH* of a solution is defined as

$$pH = -\log[H^+]$$

where $[H^+]$ is in units of molarity. Thus, pH 7.0 refers to a solution for which $-\log[H^+] = 7.0$, and so $\log[H^+] = -7.0$ and $[H^+] = 10^{\log[H^+]} = 10^{-7.0} = 1.0 \times 10^{-7} \text{ M}$.

The pH also indirectly expresses the concentration of hydroxide ions, $[OH^-]$, in solution. To see how, we must realize that water molecules dissociate to form H^+ and OH^- ions in an equilibrium process.

$$H_2O \rightleftharpoons H^+ + OH^-$$

The equilibrium constant (K) for the dissociation of water is defined as

$$K = [H^+][OH^-]/[H_2O]$$

and has a value of $K = 1.8 \times 10^{-16}$. Note that an equilibrium constant does not formally have units. Nonetheless, the value of the equilibrium constant given assumes that particular units are used for concentration; in this case and in most others, units of molarity (M) are assumed.

The concentration of water, $[H_2O]$, in pure water is 55.5 M, and this concentration is constant under most conditions. Thus, we can define a new constant, K_W:

$$K_W = K[H_2O] = [H^+][OH^-]$$

$$K[H_2O] = 1.8 \times 10^{-16} \times 55.5$$
$$= 1.0 \times 10^{-14}$$

Because $K_W = [H^+][OH^-] = 1.0 \times 10^{-14}$, we can calculate

$$[OH^-] = 10^{-14}/[H^+] \quad \text{and} \quad [H^+] = 10^{-14}/[OH^-].$$

With these relations in hand, we can easily calculate the concentration of hydroxide ions in an aqueous solution given the pH. For example, at pH = 7.0, we know that $[H^+] = 10^{-7}$ M and so $[OH^-] = 10^{-14}/10^{-7} = 10^{-7}$ M. In acidic solutions, the concentration of hydrogen ions is higher than 10^{-7} and, hence, the pH is below 7. For example, in 0.1 M HCl, $[H^+] = 10^{-1}$ M and so pH = 1.0 and $[OH^-] = 10^{-14}/10^{-1} = 10^{-13}$ M.

Acid–Base Reactions Can Disrupt the Double Helix

The reaction that we have been considering between two strands of DNA to form a double helix takes place readily at pH 7.0. Suppose that we take the solution containing the double-helical DNA and treat it with a solution of concentrated base (with a high concentration of OH^-). As the base is added, we monitor the pH and the fraction of DNA in double-helical form (Figure 1.16). When the first additions of base are made, the pH rises, but the concentration of the double-helical DNA does not change significantly. However, as the pH approaches 9, the DNA double helix begins to dissociate into its component single strands. As the pH continues to rise from 9 to 10, this dissociation becomes essentially complete. Why do the two strands dissociate? The hydroxide ions can react with bases in DNA base pairs to remove certain protons. The most susceptible proton is the one bound to the N-1 nitrogen atom in a guanine base.

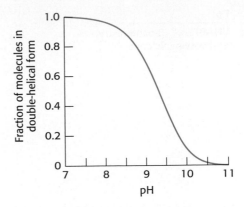

Figure 1.16 **DNA denaturation by the addition of a base.** The addition of base to a solution of double-helical DNA initially at pH 7 causes the double helix to separate into single strands. The process is half complete at slightly above pH 9.

Guanine (G)

Proton dissociation for a substance HA has an equilibrium constant defined by the expression

$$K_a = [H^+][A^-]/[HA]$$

The susceptibility of a proton to removal by reaction with base is described by its pK_a value:

$$pK_a = -\log(K_a)$$

When the pH is equal to the pK_a, we have

$$pH = pK_a$$

and so

$$-\log[H^+] = -\log([H^+][A^-]/[HA])$$

and

$$[H^+] = [H^+][A^-]/[HA]$$

Dividing by $[H^+]$ reveals that

$$1 = [A^-]/[HA]$$

15

and so

$$[A^-] = [HA]$$

Thus, when the pH equals the pK_a, the concentration of the deprotonated form of the group or molecule is equal to the concentration of the protonated form; the deprotonation process is halfway to completion.

The pK_a for the proton on N-1 of guanine is typically 9.7. When the pH approaches this value, the proton on N-1 is lost (see Figure 1.16). Because this proton participates in an important hydrogen bond, its loss substantially destabilizes the DNA double helix. The DNA double helix is also destabilized by *low* pH. Below pH 5, some of the hydrogen bond *acceptors* that participate in base-pairing become protonated. In their protonated forms, these bases can no longer form hydrogen bonds and the double helix separates. Thus, acid–base reactions that remove or donate protons at specific positions on the DNA bases can disrupt the double helix.

Buffers Regulate pH in Organisms and in the Laboratory

A significant change in pH can disrupt molecular structure and initiate harmful reactions. Thus, systems have evolved to mitigate changes in pH in biological systems. Solutions that resist such changes are called *buffers*. Specifically, when acid is added to an unbuffered aqueous solution, the pH drops in proportion to the amount of acid added. In contrast, when acid is added to a buffered solution, the pH drops more gradually. Buffers also mitigate the pH increase caused by the addition of base.

Compare the result of adding a 1 M solution of the strong acid HCl drop by drop to pure water with adding it to a solution containing 100 mM of the buffer sodium acetate ($Na^+CH_3COO^-$; Figure 1.17). The process of gradually adding known amounts of reagent to a solution with which the reagent reacts while monitoring the results is called a *titration*. For pure water, the pH drops from 7 to close to 2 on the addition of the first few drops of acid. However, for the sodium acetate solution, the pH first falls rapidly from its initial value near 10, then changes more gradually until the pH reaches 3.5, and then falls more rapidly again. Why does the pH decrease gradually in the middle of the titration? The answer is that, when hydrogen ions are added to this solution, they react with acetate ions to form acetic acid. This reaction consumes some of the added hydrogen ions so that the pH does not drop. Hydrogen ions continue reacting with acetate ions until essentially all the acetate ion is converted into acetic acid. After this point, added protons remain free in solution and the pH begins to fall sharply again.

We can analyze the effect of the buffer in quantitative terms. The equilibrium constant for the deprotonation of an acid is

$$K_a = [H^+][A^-]/[HA]$$

Taking logarithms of both sides yields

$$\log(K_a) = \log([H^+]) + \log([A^-]/[HA]).$$

Recalling the definitions of pK_a and pH and rearranging gives

$$pH = pK_a + \log([A^-]/[HA]).$$

This expression is referred to as the *Henderson–Hasselbalch equation*.

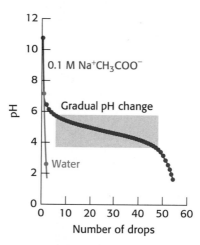

Figure 1.17 Buffer action. The addition of a strong acid, 1 M HCl, to pure water results in an immediate drop in pH to near 2. In contrast, the addition of the acid to a 0.1 M sodium acetate ($Na^+CH_3COO^-$) solution results in a much more gradual change in pH until the pH drops below 3.5.

We can apply the equation to our titration of sodium acetate. The pK_a of acetic acid is 4.75. We can calculate the ratio of the concentration of acetate ion to the concentration of acetic acid as a function of pH by using the Henderson–Hasselbalch equation, slightly rearranged.

$$[\text{Acetate ion}]/[\text{acetic acid}] = [A^-]/[HA] = 10^{pH - pK_a}$$

At pH 9, this ratio is approximately 18,000; very little acetic acid has been formed. At pH 4.75 (when the pH equals the pK_a), the ratio is 1. At pH 3, the ratio is approximately 0.02; almost all of the acetate ion has been converted into acetic acid. We can follow the conversion of acetate ion into acetic acid over the entire titration (Figure 1.18). The graph shows that the region of relatively constant pH corresponds precisely to the region in which acetate ion is being protonated to form acetic acid.

From this discussion, we see that buffers function best close to the pK_a values of their acid component. Physiological pH is typically about 7.4. An important buffer in biological systems is based on phosphoric acid (H_3PO_4). The acid can be deprotonated in three steps to form a phosphate ion.

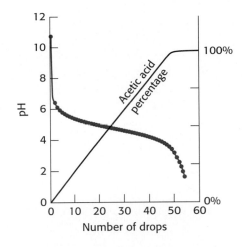

Figure 1.18 Buffer protonation. When acid is added to sodium acetate, the added hydrogen ions are used to convert acetate ion into acetic acid. Because the proton concentration does not increase significantly, the pH remains relatively constant until all of the acetate has been converted into acetic acid.

$$H_3PO_4 \underset{pK_a = 2.12}{\overset{H^+}{\rightleftharpoons}} H_2PO_4^- \underset{pK_a = 7.21}{\overset{H^+}{\rightleftharpoons}} HPO_4^{2-} \underset{pK_a = 12.67}{\overset{H^+}{\rightleftharpoons}} PO_4^{3-}$$

At about pH 7.4, inorganic phosphate exists primarily as a nearly equal mixture of $H_2PO_4^-$ and HPO_4^{2-}. Thus, phosphate solutions function as effective buffers near pH 7.4. The concentration of inorganic phosphate in blood is typically approximately 1 mM, providing a useful buffer against processes that produce either acid or base.

1.4 The Genomic Revolution Is Transforming Biochemistry and Medicine

Watson and Crick's discovery of the structure of DNA suggested the hypothesis that hereditary information is stored as a sequence of bases along long strands of DNA. This remarkable insight provided an entirely new way of thinking about biology. However, at the time that it was made, their discovery was full of potential but the practical consequences were unclear. Tremendously fundamental questions remained to be addressed. Is the hypothesis correct? How is the sequence information read and translated into action? What are the sequences of naturally occurring DNA molecules and how can such sequences be experimentally determined? Through advances in biochemistry and related sciences, we now have essentially complete answers to these questions. Indeed, in the past decade, scientists have determined the complete genome sequences of hundreds of different organisms, including simple microorganisms, plants, animals of varying degrees of complexity, and human beings. Comparisons of these genome sequences with the use of methods introduced in Chapter 6 have been sources of insight into many aspects of biochemistry. Because of these achievements, biochemistry has been transformed. In addition to its experimental and clinical aspects, biochemistry has now become an *information science*.

The Sequencing of the Human Genome Is a Landmark in Human History

The sequencing of the human genome was a daunting task because it contains approximately 3 billion (3×10^9) base pairs. For example, the sequence

ACATTTGCTTCTGACACAACTGTGTTCACTAGCAACCTC
AAACAGACACCATGGTGCATCTGACTCCTG**A**GGAGAAGT
CTGCCGTTACTGCCCTGTGGGGCAAGGTGAACGTGGA...

is a part of one of the genes that encodes hemoglobin, the oxygen carrier in our blood. This gene is found on the end of chromosome 9 among our 24 distinct chromosomes. If we were to include the complete sequence of our entire genome, this chapter would run for more than 500,000 pages. The sequencing of our genome is truly a landmark in human history. This sequence contains a vast amount of information, some of which we can now extract and interpret, but much of which we are only beginning to understand. For example, some human diseases have been linked to particular variations in genomic sequence. Sickle-cell anemia, discussed in detail in Chapter 7, is caused by a single base change of an A (noted in boldface type in the preceding sequence) to a T. We will encounter many other examples of diseases that have been linked to specific DNA sequence changes.

In addition to the implications for understanding human health and disease, the genome sequence is a source of deep insight into other aspects of human biology and culture. For example, by comparing the sequences of different individual persons and populations, we can learn a great deal about human history. On the basis of such analysis, a compelling case can be made that the human species originated in Africa, and the occurrence and even the timing of important migrations of groups of human beings can be demonstrated. Finally, comparisons of the human genome with the genomes of other organisms are confirming the tremendous unity that exists at the level of biochemistry and are revealing key steps that have been taken in the course of evolution from relatively simple, single-celled organisms to complex, multicellular organisms such as human beings. For example, many genes key to the function of the human brain and nervous system have evolutionary and functional relatives that can be recognized in the genomes of bacteria. Because many studies that are possible in model organisms are difficult or unethical to conduct in human beings, these discoveries have many practical implications. *Comparative genomics* has become a powerful science, linking evolution and biochemistry.

Genome Sequences Encode Proteins and Patterns of Expression

The structure of DNA revealed how information is stored in the base sequence along a DNA strand. But what information is stored and how is this information expressed? The most fundamental role of DNA is to encode the sequences of proteins. Like DNA, proteins are linear polymers. However, proteins differ from DNA in two important ways. First, proteins are built from 20 building blocks, called *amino acids*, rather than just four, as are present in DNA. The chemical complexity provided by this variety of building blocks enables proteins to perform a wide range of functions. Second, proteins spontaneously fold up into elaborate three-dimensional structures, determined only by their amino acid sequences (Figure 1.19). We have explored in depth how solutions containing two appropriate

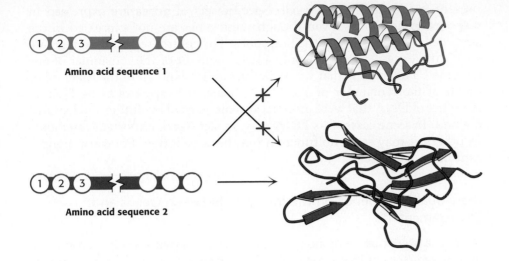

Amino acid sequence 1

Amino acid sequence 2

Figure 1.19 Protein folding. Proteins are linear polymers of amino acids that fold into elaborate structures. The sequence of amino acids determines the three-dimensional structure. Thus amino acid sequence 1 gives rise only to a protein with the shape depicted in blue, *not* the shape depicted in red.

strands of DNA come together to form a solution of double-helical molecules. A similar spontaneous folding process gives proteins their three-dimensional structure. A balance of hydrogen bonding, van der Waals interactions, and hydrophobic interactions overcome the entropy lost in going from an unfolded ensemble of proteins to a homogenous set of well-folded molecules. Proteins and protein folding will be discussed extensively in Chapter 2.

The fundamental unit of hereditary information, the *gene,* is becoming increasingly difficult to precisely define as our knowledge of the complexities of genetics and genomics increases. The genes that are simplest to define encode the sequences of proteins. For these protein-encoding genes, a block of DNA bases encodes the amino acid sequence of a specific protein molecule. A set of three bases along the DNA strand, called a *codon,* determines the identity of one amino acid within the protein sequence. The relation that links the DNA sequence to the encoded protein sequence is called the *genetic code.* One of the biggest surprises from the sequencing of the human genome is the small number of protein-encoding genes. Before the genome-sequencing project began, the consensus view was that the human genome would include approximately 100,000 protein-encoding genes. The current analysis suggests that the actual number is between 20,000 and 25,000. We shall use an estimate of 25,000 throughout this book. However, additional mechanisms allow many genes to encode more than one protein. For example, the genetic information in some genes is translated in more than one way to produce a set of proteins that differ from one another in parts of their amino acid sequences. In other cases, proteins are modified after they are synthesized through the addition of accessory chemical groups. Through these indirect mechanisms, much more complexity is encoded in our genomes than would be expected from the number of protein-encoding genes alone.

On the basis of current knowledge, the protein-encoding regions account for only about 3% of the human genome. What is the function of the rest of the DNA? Some of it contains information that regulates the expression of specific genes (i.e., the production of specific proteins) in particular cell types and physiological conditions. Essentially every cell contains the same DNA genome, yet cell types differ considerably in the proteins that they produce. For example, hemoglobin is expressed only in precursors of red blood cells, even though the genes for hemoglobin are present in essentially every cell. Specific sets of genes are expressed in response to hormones, even though these genes are not expressed in the same cell in the absence of the hormones. The control regions that regulate such differences account for only a small amount of the remainder of our genomes. The truth is that we do not yet understand all of the function of much of the remainder of the DNA. Some of it appears to be "junk," stretches of DNA that were inserted at some stage of evolution and have remained. In some cases, this DNA may, in fact, serve important functions. In others, it may serve no function but, because it does not cause significant harm, it has remained.

Individuality Depends on the Interplay Between Genes and Environment

With the exception of monozygotic ("identical") twins, each person has a unique sequence of DNA base pairs. How different are we from one another at the genomic level? An examination of variation across the genome reveals that, on average, each pair of individual people has a different base in one position per thousand; that is, the difference is approximately 0.1%. This person-to-person variation is quite substantial compared with differences in populations. Thus, the average difference between two people within one ethnic group is greater than the difference between the averages of two different ethnic groups.

The significance of much of this genetic variation is not understood. As noted earlier, variation in a single base within the genome can lead to a disease such as sickle-cell anemia. Scientists have now identified the genetic variations associated with more than 100 diseases for which the cause can be traced to a single gene. For other diseases and traits, we know that variation in many different genes contributes in significant and often complex ways. Many of the most prevalent human ailments such as heart disease are linked to variations in many genes. Furthermore, in most cases, the presence of a particular variation or set of variations does not inevitably result in the onset of a disease but, instead, leads to a *predisposition* to the development of the disease.

In addition to these genetic differences, *epigenetic factors* are important. They are factors associated with the genome but not simply represented in the sequence of DNA. For example, the consequences of some of this genetic variation depend, often dramatically, on whether the unusual gene sequence is inherited from the mother or from the father. This phenomenon, known as *genetic imprinting*, depends on the covalent modification of DNA, particularly the addition of methyl groups to particular bases. Epigenetics is a very active field of study and many novel discoveries can be expected.

Although our genetic makeup and associated epigenetic characteristics are important factors that contribute to disease susceptibility and to other traits, factors in a person's environment also are significant. What are these environmental factors? Perhaps the most obvious are chemicals that we eat or are exposed to in some other way. The adage "you are what you eat" has considerable validity; it applies both to substances that we ingest in significant quantities and to those that we ingest in only trace amounts. Throughout our study of biochemistry, we will encounter *vitamins* and *trace elements* and their derivatives that play crucial roles in many processes. In many cases, the roles of these chemicals were first revealed through investigation of *deficiency disorders* observed in people who do not take in a sufficient quantity of a particular vitamin or trace element. Despite the fact that the most important vitamins and trace elements have

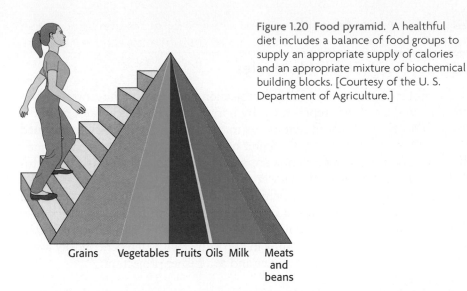

Figure 1.20 Food pyramid. A healthful diet includes a balance of food groups to supply an appropriate supply of calories and an appropriate mixture of biochemical building blocks. [Courtesy of the U. S. Department of Agriculture.]

Grains Vegetables Fruits Oils Milk Meats and beans

been known for some time, new roles for these essential dietary factors continue to be discovered.

A healthful diet requires a balance of major food groups (Figure 1.20). In addition to providing vitamins and trace elements, food provides calories in the form of substances that can be broken down to release energy to drive other biochemical processes. Proteins, fats, and carbohydrates provide the building blocks used to construct the molecules of life. Finally, it is possible to get too much of a good thing. Human beings evolved under circumstances in which food, particularly rich foods such as meat, was scarce. With the development of agriculture and modern economies, rich foods are now plentiful in parts of the world. Some of the most prevalent diseases in the so-called developed world, such as heart disease and diabetes, can be attributed to the large quantities of fats and carbohydrates that are present in modern diets. We are now developing a deeper understanding of the biochemical consequences of these diets and the interplay between diet and genetic factors.

Chemicals are only one important class of environmental factors. The behaviors in which we engage also have biochemical consequences. Through physical activity, we consume the calories that we take in, ensuring an appropriate balance between food intake and energy expenditure. Activities ranging from exercise to emotional responses such as fear and love may activate specific biochemical pathways, leading to changes in levels of gene expression, the release of hormones, and other consequences. For example, recent discoveries reveal that high stress levels are associated with the shortening of telomeres, structures at the ends of chromosomes. Furthermore, the interplay between biochemistry and behavior is bidirectional. Just as our biochemistry is affected by our behavior, so, too, our behavior is affected, although certainly not completely determined, by our genetic makeup and other aspects of our biochemistry. Genetic factors associated with a range of behavioral characteristics have been at least tentatively identified.

Just as vitamin deficiencies and genetic diseases revealed fundamental principles of biochemistry and biology, investigations of variations in behavior and their linkage to genetic and biochemical factors are potential sources of great insight into mechanisms within the brain. For example, studies of drug addiction have revealed neural circuits and biochemical pathways that greatly influence aspects of behavior. Unraveling the interplay between biology and behavior is one of the great challenges in modern science, and biochemistry is providing some of the most important concepts and tools for this endeavor.

APPENDIX: Visualizing Molecular Structures I: Small Molecules

The authors of a biochemistry textbook face the problem of trying to present three-dimensional molecules in the two dimensions available on the printed page. The interplay between the three-dimensional structures of biomolecules and their biological functions will be discussed extensively throughout this book. Toward this end, we will frequently use representations that, although of necessity are rendered in two dimensions, emphasize the three-dimensional structures of molecules.

Stereochemical Renderings

Most of the chemical formulas in this book are drawn to depict the geometric arrangement of atoms, crucial to chemical bonding and reactivity, as accurately as possible. For example, the carbon atom of methane is sp^3 hybridized and tetrahedral, with H–C–H angles of 109.5 degrees, whereas the carbon atom in formaldehyde is sp^2 hybridized with bond angles of 120 degrees.

Methane   **Formaldehyde**

To illustrate the correct *stereochemistry* about tetrahedral carbon atoms, wedges will be used to depict the direction of a bond into or out of the plane of the page. A solid wedge with the broad end away from the carbon atom denotes a bond coming toward the viewer out of the plane. A dashed wedge, with its broad end at the carbon atom, represents a bond going away from the viewer behind the plane of the page. The remaining two bonds are depicted as straight lines.

Fischer Projections

Although representative of the actual structure of a compound, stereochemical structures are often difficult to draw quickly. An alternative, less-representative method of depicting structures with tetrahedral carbon centers relies on the use of *Fischer projections*.

Fischer projection   **Stereochemical rendering**

In a Fischer projection, the bonds to the central carbon are represented by horizontal and vertical lines from the substituent atoms to the carbon atom, which is assumed to be at the center of the cross. By convention, the horizontal bonds are assumed to project out of the page toward the viewer, whereas the vertical bonds are assumed to project behind the page away from the viewer. The Glossary of Compounds found at the back of the book is a structural glossary of the key molecules in biochemistry, each presented in two forms: with stereochemically accurate bond angles and as a Fisher projection.

Molecular Models for Small Molecules

For depicting the molecular architecture of small molecules in more detail, two types of models will often be used: space filling and ball and stick. These models show structures at the atomic level.

1. *Space-Filling Models.* The space-filling models are the most realistic. The size and position of an atom in a space-filling model are determined by its bonding properties and van der Waals radius, or contact distance. A van der Waals radius describes how closely two atoms can approach each other when they are not linked by a covalent bond. The colors of the model are set by convention.

Carbon, black   Hydrogen, white   Nitrogen, blue
Oxygen, red   Sulfur, yellow   Phosphorus, purple

Space-filling models of several simple molecules are shown in Figure 1.21.

2. *Ball-and-Stick Models.* Ball-and-stick models are not as realistic as space-filling models, because the atoms are depicted as spheres of radii smaller than their van der Waals radii. However, the bonding arrangement is easier to see because the bonds are explicitly represented as sticks. In an illustration, the taper of a stick, representing parallax, tells which of a pair of bonded atoms is closer to the reader. A ball-and-stick model reveals a complex structure more clearly than a space-filling model does. Ball-and-stick models of several simple molecules are shown in Figure 1.21.

Molecular models for depicting large molecules will be discussed in the appendix to Chapter 2.

Key Terms

biological macromolecule (p. 2)

metabolite (p. 2)

deoxyribonucleic acid (DNA) (p. 2)

protein (p. 2)

Eukarya (p. 3)

Bacteria (p. 3)

Archaea (p. 3)

eukaryote (p. 3)

prokaryote (p. 3)

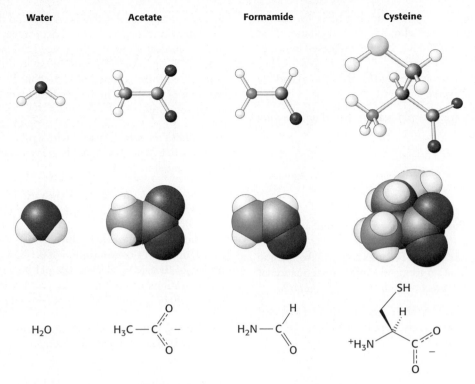

Figure 1.21 Molecular representations. Structural formulas (bottom), ball-and-stick models (middle), and space-filling representations (top) of selected molecules are shown. Black = carbon, red = oxygen, white = hydrogen, yellow = sulfur, blue = nitrogen.

Problems

1. *Donors and acceptors.* Identify the hydrogen-bond donors and acceptors in each of the four bases on page 4.

2. *Resonance structures.* The structure of an amino acid, tyrosine, is shown here. Draw an alternative resonance structure.

3. *It takes all types.* What types of noncovalent bonds hold together the following solids?

(a) Table salt (NaCl), which contains Na^+ and Cl^- ions.

(b) Graphite (C), which consists of sheets of covalently bonded carbon atoms.

4. *Don't break the law.* Given the following values for the changes in enthalpy (ΔH) and entropy (ΔS), which of the following processes can occur at 298 K without violating the Second Law of Thermodynamics?

(a) $\Delta H = -84 \text{ kJ mol}^{-1} (-20 \text{ kcal mol}^{-1})$,
$\Delta S = +125 \text{ J mol}^{-1} \text{ K}^{-1} (+30 \text{ cal mol}^{-1} \text{ K}^{-1})$

(b) $\Delta H = -84 \text{ kJ mol}^{-1} (-20 \text{ kcal mol}^{-1})$,
$\Delta S = -125 \text{ J mol}^{-1} \text{ K}^{-1} (-30 \text{ cal mol}^{-1} \text{ K}^{-1})$

(c) $\Delta H = +84 \text{ kJ mol}^{-1} (+20 \text{ kcal mol}^{-1})$,
$\Delta S = +125 \text{ J mol}^{-1} \text{ K}^{-1} (+30 \text{ cal mol}^{-1} \text{ K}^{-1})$

(d) $\Delta H = +84 \text{ kJ mol}^{-1} (+20 \text{ kcal mol}^{-1})$,
$\Delta S = -125 \text{ J mol}^{-1} \text{ K}^{-1} (-30 \text{ cal mol}^{-1} \text{ K}^{-1})$

5. *Double-helix-formation entropy.* For double-helix formation, ΔG can be measured to be -54 kJ mol^{-1} (-13 kcal mol^{-1}) at pH 7.0 in 1 M NaCl at 25°C (298 K). The heat released indicates an enthalpy change of -251 kJ mol^{-1} (-60 kcal mol^{-1}). For this process, calculate the entropy change for the system and the entropy change for the universe.

6. *Find the pH.* What are the pH values for the following solutions?

(a) 0.1 M HCl
(b) 0.1 M NaOH
(c) 0.05 M HCl
(d) 0.05 M NaOH

7. *A weak acid.* What is the pH of a 0.1 M solution of acetic acid ($pK_a = 4.75$)?
(Hint: Let x be the concentration of H^+ ions released from acetic acid when it dissociates. The solutions to a quadratic equation of the form $ax^2 + bx + c = 0$ are $x = (-b \pm \sqrt{b^2 - 4ac})/2a$.

8. *Find the pK_a.* For an acid HA, the concentrations of HA and A^- are 0.075 and 0.025, respectively, at pH 6.0. What is the pK_a value for HA?

9. *pH indicator.* A dye that is an acid and that appears as different colors in its protonated and deprotonated forms can be used as a pH indicator. Suppose that you have an 0.001 M solution of a dye with a pK_a of 7.2. From the color, the concentration of the protonated form is found to be 0.0002 M. Assume that the remainder of the dye is in the deprotonated form. What is the pH of the solution?

10. *What's the ratio?* An acid with a pK_a of 8.0 is present in a solution with a pH of 6.0. What is the ratio of the protonated to the deprotonated form of the acid?

11. *Phosphate buffer.* What is the ratio of the concentrations of $H_2PO_4^-$ and HPO_4^{2-} at (a) pH 7.0; (b) pH 7.5; (c) pH 8.0?

12. *Buffer capacity.* Two solutions of sodium acetate are prepared, one with a concentration of 0.1 M and one with a concentration of 0.01 M. Calculate the pH values when the following concentrations of HCl have been added to each of these solutions: 0.0025 M, 0.005 M, 0.01 M, and 0.05 M.

13. *Viva le difference.* On average, how many base differences are there between two human beings?

Protein Composition and Structure

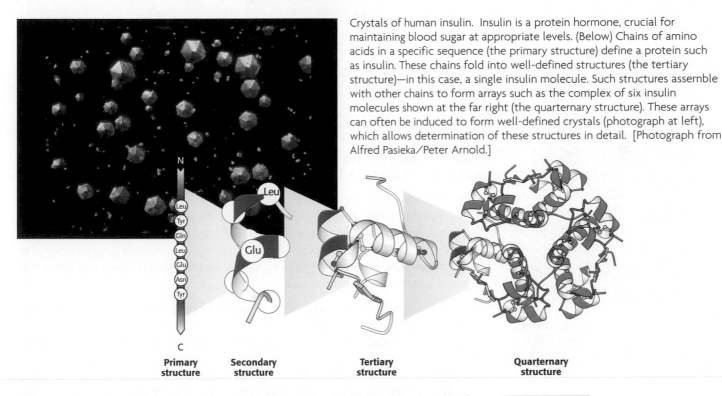

Crystals of human insulin. Insulin is a protein hormone, crucial for maintaining blood sugar at appropriate levels. (Below) Chains of amino acids in a specific sequence (the primary structure) define a protein such as insulin. These chains fold into well-defined structures (the tertiary structure)—in this case, a single insulin molecule. Such structures assemble with other chains to form arrays such as the complex of six insulin molecules shown at the far right (the quarternary structure). These arrays can often be induced to form well-defined crystals (photograph at left), which allows determination of these structures in detail. [Photograph from Alfred Pasieka/Peter Arnold.]

| Primary structure | Secondary structure | Tertiary structure | Quarternary structure |

Proteins are the most versatile macromolecules in living systems and serve crucial functions in essentially all biological processes. They function as catalysts, transport and store other molecules such as oxygen, provide mechanical support and immune protection, generate movement, transmit nerve impulses, and control growth and differentiation. Indeed, much of this book will focus on understanding what proteins do and how they perform these functions.

Several key properties enable proteins to participate in a wide range of functions.

1. *Proteins are linear polymers built of monomer units called amino acids*, which are linked end to end. Remarkably, proteins spontaneously fold up into three-dimensional structures that are determined by the sequence of amino acids in the protein polymer. Protein function is directly dependent on this three-dimensional structure (Figure 2.1). Thus, *proteins are the embodiment of the transition from the one-dimensional world of sequences to the three-dimensional world of molecules capable of diverse activities.*

2. *Proteins contain a wide range of functional groups.* These functional groups include alcohols, thiols, thioethers, carboxylic acids, carboxamides, and a variety of basic groups. Most of these groups are chemically reactive. When

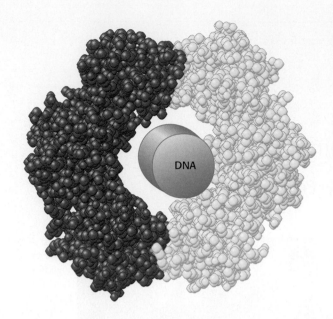

Figure 2.1 **Structure dictates function.** A protein component of the DNA replication machinery surrounds a section of DNA double helix depicted as a cylinder. The protein, which consists of two identical subunits (shown in red and yellow), acts as a clamp that allows large segments of DNA to be copied without the replication machinery dissociating from the DNA. [Drawn from 2POL.pdb.]

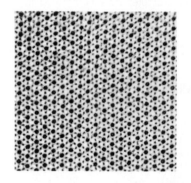

Figure 2.2 **A complex protein assembly.** An electron micrograph of insect flight tissue in cross section shows a hexagonal array of two kinds of protein filaments. [Courtesy of Dr. Michael Reedy.]

combined in various sequences, this array of functional groups accounts for the broad spectrum of protein function. For instance, their reactive properties are essential to the function of *enzymes,* the proteins that catalyze specific chemical reactions in biological systems (see Chapters 8 through 10).

3.   *Proteins can interact with one another and with other biological macromolecules to form complex assemblies.* The proteins within these assemblies can act synergistically to generate capabilities that individual proteins may lack (Figure 2.2). Examples of these assemblies include macromolecular machines that replicate DNA, transmit signals within cells, and carry out many other essential processes.

4.   *Some proteins are quite rigid, whereas others display a considerable flexibility.* Rigid units can function as structural elements in the cytoskeleton (the internal scaffolding within cells) or in connective tissue. Proteins with some flexibility may act as hinges, springs, or levers that are crucial to protein function, to the assembly of proteins with one another and with other molecules into complex units, and to the transmission of information within and between cells (Figure 2.3).

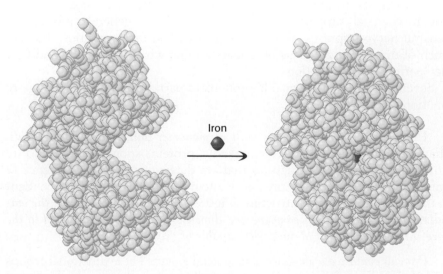

Figure 2.3 **Flexibility and function.** On binding iron, the protein lactoferrin undergoes a substantial change in conformation that allows other molecules to distinguish between the iron-free and the iron-bound forms. [Drawn from 1LFH.pdb and 1LFG.pdb.]

2.1 Proteins Are Built from a Repertoire of 20 Amino Acids

Amino acids are the building blocks of proteins. An *α-amino acid* consists of a central carbon atom, called the *α carbon,* linked to an amino group, a carboxylic acid group, a hydrogen atom, and a distinctive R group. The R group is often referred to as the *side chain.* With four different groups connected to the tetrahedral α-carbon atom, α-amino acids are *chiral:* they may exist in one or the other of two mirror-image forms, called the L isomer and the D isomer (Figure 2.4).

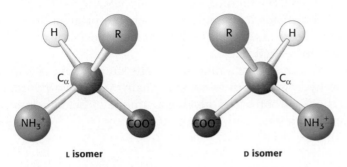

Figure 2.4 The L and D isomers of amino acids. The letter R refers to the side chain. The L and D isomers are mirror images of each other.

Only L amino acids are constituents of proteins. For almost all amino acids, the L isomer has *S* (rather than *R*) absolute configuration (Figure 2.5). Although considerable effort has gone into understanding why amino acids in proteins have the L absolute configuration, no satisfactory explanation has been reached. It seems plausible that the selection of L over D was arbitrary but, once made, the selection was fixed early in evolutionary history.

Amino acids in solution at neutral pH exist predominantly as *dipolar ions* (also called *zwitterions*). In the dipolar form, the amino group is protonated ($-NH_3^+$) and the carboxyl group is deprotonated ($-COO^-$). The ionization state of an amino acid varies with pH (Figure 2.6). In acid solution (e.g., pH 1), the amino group is protonated ($-NH_3^+$) and the carboxyl

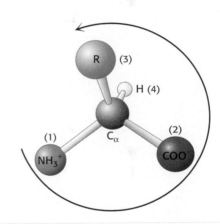

Figure 2.5 Only L amino acids are found in proteins. Almost all L amino acids have an *S* absolute configuration. The counterclockwise direction of the arrow from highest- to lowest-priority substituents indicates that the chiral center is of the *S* configuration.

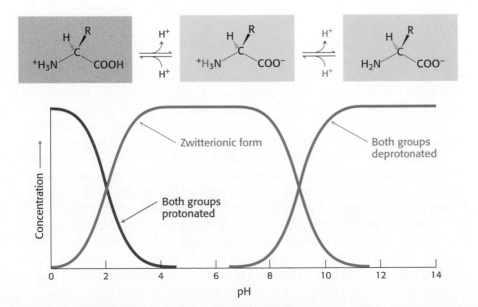

Figure 2.6 Ionization state as a function of pH. The ionization state of amino acids is altered by a change in pH. The zwitterionic form predominates near physiological pH.

group is not dissociated (—COOH). As the pH is raised, the carboxylic acid
is the first group to give up a proton, inasmuch as its pK_a is near 2. The
dipolar form persists until the pH approaches 9, when the protonated amino
group loses a proton.

Twenty kinds of side chains varying in *size, shape, charge, hydrogen-
bonding capacity, hydrophobic character,* and *chemical reactivity* are commonly found in proteins. Indeed, all proteins in all species—bacterial, archaeal, and eukaryotic—are constructed from the same set of 20 amino acids
with only a few exceptions. This fundamental alphabet for the construction
of proteins is several billion years old. The remarkable range of functions
mediated by proteins results from the diversity and versatility of these 20
building blocks. Understanding how this alphabet is used to create the intricate three-dimensional structures that enable proteins to carry out so
many biological processes is an exciting area of biochemistry and one that
we will return to in Section 2.6.

Let us look at this set of amino acids. The simplest one is *glycine,* which
has a single hydrogen atom as its side chain. With two hydrogen atoms
bonded to the α-carbon atom, glycine is unique in being *achiral. Alanine,*
the next simplest amino acid, has a methyl group (—CH$_3$) as its side chain
(Figure 2.7).

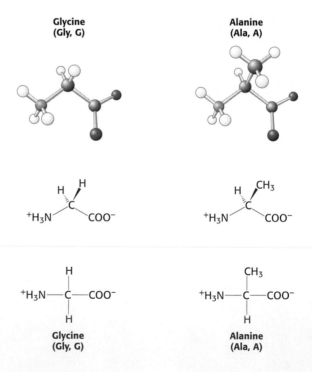

**Figure 2.7 Structures of glycine and
alanine.** (Top) Ball-and-stick models show
the arrangement of atoms and bonds in
space. (Middle) Stereochemically realistic
formulas show the geometric arrangement
of bonds around atoms (see p. 21).
(Bottom) Fischer projections show all
bonds as being perpendicular for a
simplified representation (see p. 22).

Larger hydrocarbon side chains are found in *valine, leucine,* and
isoleucine (Figure 2.8). *Methionine* contains a largely aliphatic side chain that
includes a *thioether* (—S—) group. The side chain of isoleucine includes an
additional chiral center; only the isomer shown in Figure 2.8 is found in proteins. The larger aliphatic side chains are *hydrophobic*—that is, they tend to
cluster together rather than contact water. The three-dimensional structures of water-soluble proteins are stabilized by this tendency of hydrophobic groups to come together, which is called *the hydrophobic effect* (p. 9). The
different sizes and shapes of these hydrocarbon side chains enable them to
pack together to form compact structures with little empty space. *Proline*
also has an aliphatic side chain, but it differs from other members of the set
of 20 in that its side chain is bonded to both the nitrogen and the α-carbon

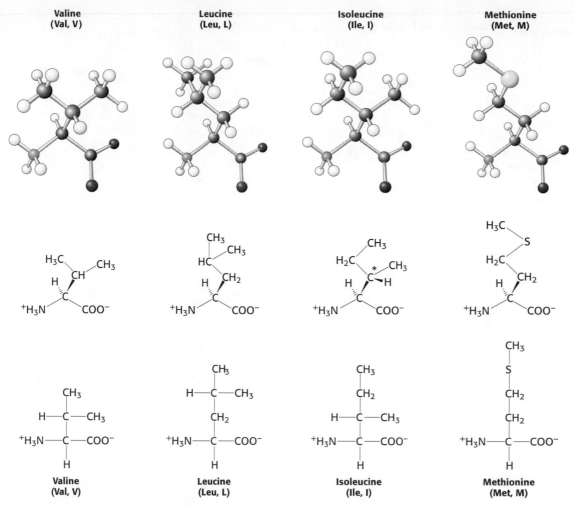

Valine
(Val, V)

Leucine
(Leu, L)

Isoleucine
(Ile, I)

Methionine
(Met, M)

Valine
(Val, V)

Leucine
(Leu, L)

Isoleucine
(Ile, I)

Methionine
(Met, M)

Figure 2.8 Amino acids with aliphatic side chains. The additional chiral center of isoleucine is indicated by an asterisk.

atoms (Figure 2.9). Proline markedly influences protein architecture because its ring structure makes it more conformationally restricted than the other amino acids.

Three amino acids with relatively simple *aromatic side chains* are part of the fundamental repertoire (Figure 2.10). *Phenylalanine,* as its name indicates, contains a phenyl ring attached in place of one of the hydrogens of alanine. The aromatic ring of *tyrosine* contains a hydroxyl group. This hydroxyl group is reactive, in contrast with the rather inert side chains of the other amino acids discussed thus far. *Tryptophan* has an indole group joined to a methylene ($-CH_2-$) group; the indole group comprises two

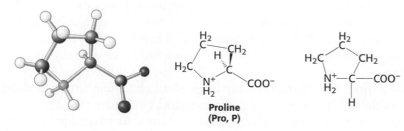

Proline
(Pro, P)

Figure 2.9 Cyclic structure of proline. The side chain is joined to both the α-carbon atom and the amino group.

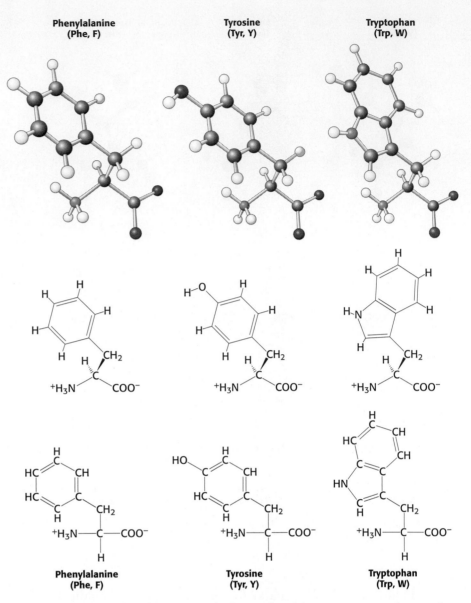

Figure 2.10 **Amino acids with aromatic side chains.** Phenylalanine, tyrosine, and tryptophan have hydrophobic character. Tyrosine and tryptophan also have hydrophilic properties because of their —OH and —NH— groups, respectively.

fused rings containing an NH group. Phenylalanine is purely hydrophobic, whereas tyrosine and tryptophan are less so because of their hydroxyl and NH groups.

Five amino acids are polar but uncharged. Two amino acids, *serine* and *threonine,* contain *hydroxyl groups* (—OH) attached to an aliphatic side chain (Figure 2.11). Serine can be thought of as a version of alanine with a hydroxyl group attached, whereas threonine resembles valine with a hydroxyl group in place of one of the valine methyl groups. The hydroxyl groups on serine and threonine make them much more *hydrophilic* (water loving) and *reactive* than alanine and valine. Threonine, like isoleucine, contains an additional asymmetric center; again only one isomer is present in proteins.

In addition, the set includes *asparagine* and *glutamine,* uncharged derivatives of the acidic amino acids aspartate and glutamate (see Figure 2.15). Each of these two amino acids contains a terminal *carboxamide* in place of a carboxylic acid (Figure 2.12). The side chain of glutamine is one methylene group longer than that of asparagine.

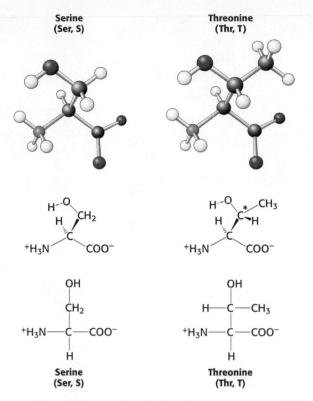

Serine
(Ser, S)

Threonine
(Thr, T)

Serine
(Ser, S)

Threonine
(Thr, T)

Figure 2.11 Amino acids containing aliphatic hydroxyl groups. Serine and threonine contain hydroxyl groups that render them hydrophilic. The additional chiral center in threonine is indicated by an asterisk.

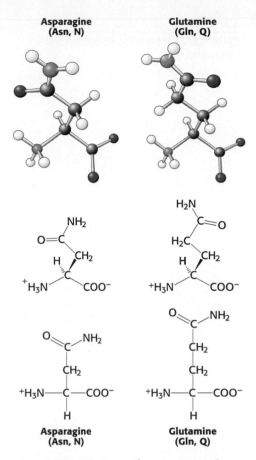

Asparagine
(Asn, N)

Glutamine
(Gln, Q)

Asparagine
(Asn, N)

Glutamine
(Gln, Q)

Figure 2.12 Structure of asparagine and glutamine, carboxamide-containing polar amino acids.

Cysteine is structurally similar to serine but contains a *sulfhydryl,* or *thiol* (—SH), group in place of the hydroxyl (—OH) group (Figure 2.13). The sulfhydryl group is much more reactive. Pairs of sulfhydryl groups may come together to form disulfide bonds, which are particularly important in stabilizing some proteins, as will be discussed shortly.

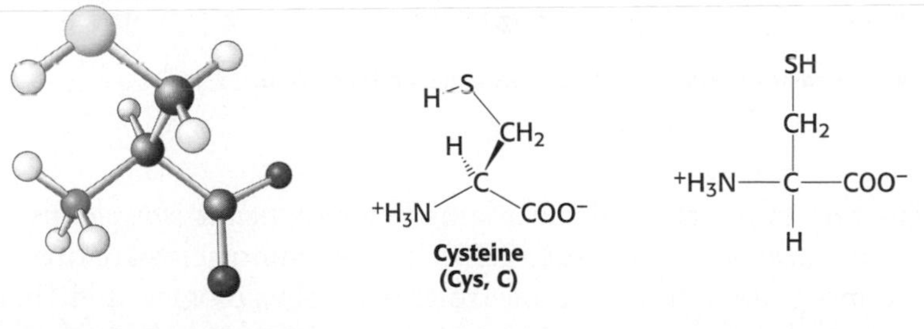

Cysteine
(Cys, C)

Figure 2.13 Structure of cysteine.

We turn now to amino acids with complete charges that render them highly hydrophilic. *Lysine* and *arginine* have relatively long side chains that terminate with groups that are *positively charged* at neutral pH. Lysine is capped by a primary amino group and arginine by a guanidinium group. *Histidine* contains an imidazole group, an aromatic ring that also can be positively charged (Figure 2.14).

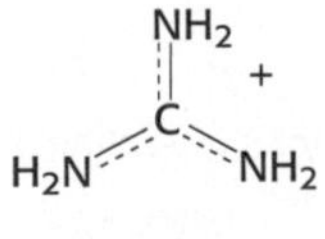

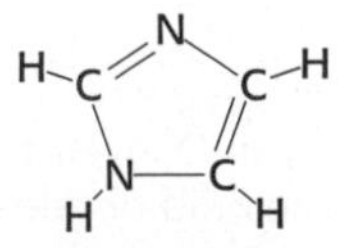

With a pK_a value near 6, the imidazole group can be uncharged or positively charged near neutral pH, depending on its local environment (Figure 2.15). Histidine is often found in the active sites of enzymes, where the imidazole ring can bind and release protons in the course of enzymatic reactions.

Guanidinium

Imidazole

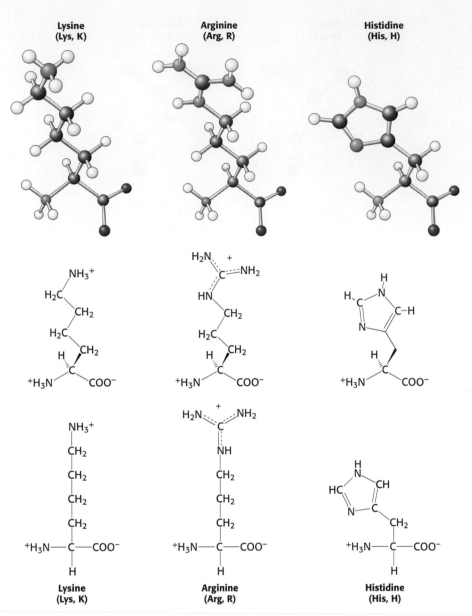

Figure 2.14 The basic amino acids lysine, arginine, and histidine.

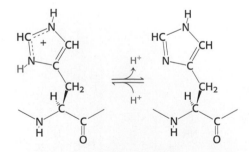

Figure 2.15 Histidine ionization. Histidine
can bind or release protons near
physiological pH.

This set of amino acids also contains two with *acidic side chains: aspartic acid* and *glutamic acid* (Figure 2.16). These amino acids are often called *aspartate* and *glutamate* to emphasize that at physiological pH their side chains usually lack a proton present in the acid form and hence are negatively charged. Nonetheless, in some proteins these side chains do accept protons, and this ability is often functionally important.

Seven of the 20 amino acids have readily ionizable side chains. These 7 amino acids are able to donate or accept protons to facilitate reactions as well as to form ionic bonds. Table 2.1 gives equilibria and typical pK_a values for ionization of the side chains of tyrosine, cysteine, arginine, lysine, histidine, and aspartic and glutamic acids in proteins. Two other groups in proteins—the terminal α-amino group and the terminal α-carboxyl group—can be ionized, and typical pK_a values also are included in Table 2.1.

Amino acids are often designated by either a three-letter abbreviation or a one-letter symbol (Table 2.2). The abbreviations for amino acids are the first three letters of their names, except for asparagine (Asn), glutamine (Gln), isoleucine (Ile), and tryptophan (Trp). The symbols for many amino

TABLE 2.1 Typical pK_a values of ionizable groups in proteins

Group	Acid $\rightleftharpoons$ Base	Typical pK_a*
Terminal α-carboxyl group		3.1
Aspartic acid Glutamic acid		4.1
Histidine		6.0
Terminal α-amino group		8.0
Cysteine		8.3
Tyrosine		10.9
Lysine		10.8
Arginine		12.5

*pK_a values depend on temperature, ionic strength, and the microenvironment of the ionizable group.

acids are the first letters of their names (e.g., G for glycine and L for leucine); the other symbols have been agreed on by convention. These abbreviations and symbols are an integral part of the vocabulary of biochemists.

How did this particular set of amino acids become the building blocks of proteins? First, as a set, they are diverse; their structural and chemical properties span a wide range, endowing proteins with the versatility to assume many functional roles. Second, many of these amino acids were probably available from prebiotic reactions, that is, from reactions that occurred before the origin of life. Finally, other possible amino acids may

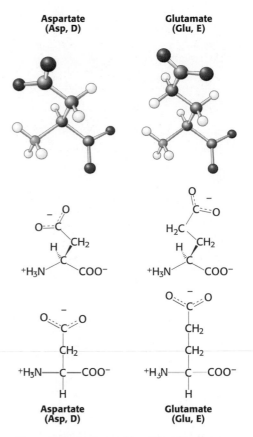

Aspartate (Asp, D) **Glutamate (Glu, E)**

Figure 2.16 Amino acids with side-chain carboxylates.

TABLE 2.2 Abbreviations for amino acids

Amino acid	Three-letter abbreviation	One-letter abbreviation	Amino acid	Three-letter abbreviation	One-letter abbreviation
Alanine	Ala	A	Methionine	Met	M
Arginine	Arg	R	Phenylalanine	Phe	F
Asparagine	Asn	N	Proline	Pro	P
Aspartic acid	Asp	D	Serine	Ser	S
Cysteine	Cys	C	Threonine	Thr	T
Glutamine	Gln	Q	Tryptophan	Trp	W
Glutamic acid	Glu	E	Tyrosine	Tyr	Y
Glycine	Gly	G	Valine	Val	V
Histidine	His	H	Asparagine or aspartic acid	Asx	B
Isoleucine	Ile	I			
Leucine	Leu	L	Glutamine or glutamic acid	Glx	Z
Lysine	Lys	K			

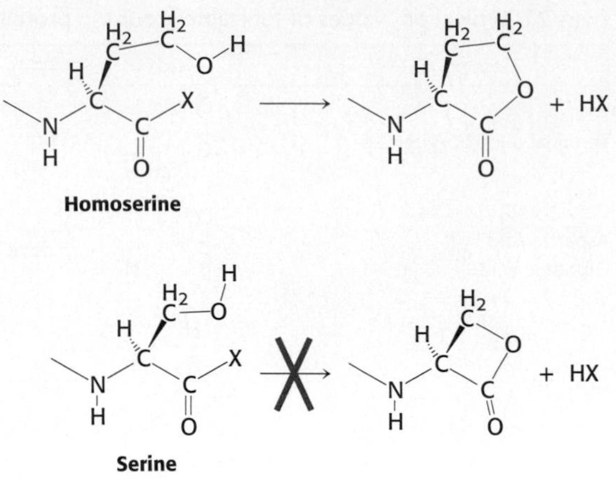

Figure 2.17 Undesirable
reactivity in amino acids.
Some amino acids are
unsuitable for proteins
because of undesirable
cyclization. Homoserine can
cyclize to form a stable, five-
membered ring, potentially
resulting in peptide-bond
cleavage. Cyclization of serine
would form a strained, four-
membered ring and thus is
disfavored. X can be an amino
group from a neighboring
amino acid or another
potential leaving group.

have simply been too reactive. For example, amino acids such as homoserine and homocysteine tend to form five-membered cyclic forms that limit their use in proteins; the alternative amino acids that are found in proteins—serine and cysteine—do not readily cyclize, because the rings in their cyclic forms are too small (Figure 2.17).

2.2 Primary Structure: Amino Acids Are Linked by Peptide Bonds to Form Polypeptide Chains

Proteins are *linear polymers* formed by linking the α-carboxyl group of one amino acid to the α-amino group of another amino acid. This type of linkage is called a *peptide bond* (or an *amide bond*). The formation of a dipeptide from two amino acids is accompanied by the loss of a water molecule (Figure 2.18). The equilibrium of this reaction lies on the side of hydrolysis rather than synthesis under most conditions. Hence, the biosynthesis of peptide bonds requires an input of free energy. Nonetheless, peptide bonds are quite *stable kinetically* because the rate of hydrolysis is extremely slow; the lifetime of a peptide bond in aqueous solution in the absence of a catalyst approaches 1000 years.

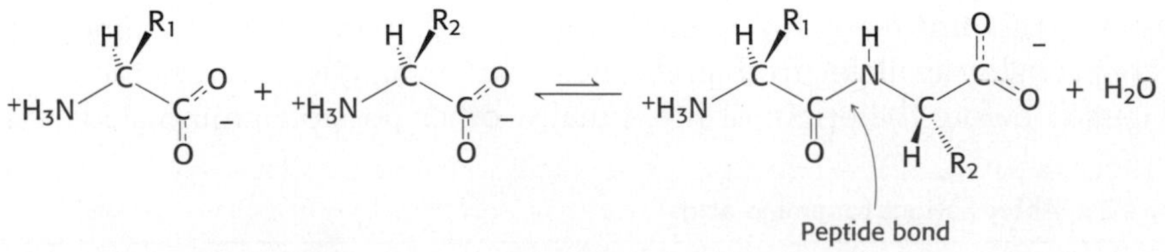

Figure 2.18 Peptide-bond formation. The linking of two amino acids is accompanied by the loss of a molecule of water.

A series of amino acids joined by peptide bonds form a *polypeptide chain*, and each amino acid unit in a polypeptide is called a *residue*. A *polypeptide chain has polarity* because its ends are different: an α-amino group is present at one end and an α-carboxyl group at the other. By convention, *the amino end is taken to be the beginning of a polypeptide chain*, and so the sequence of amino acids in a polypeptide chain is written starting with the amino-terminal residue. Thus, in the pentapeptide Tyr-Gly-Gly-Phe-Leu (YGGFL), tyrosine is the amino-terminal (N-terminal) residue

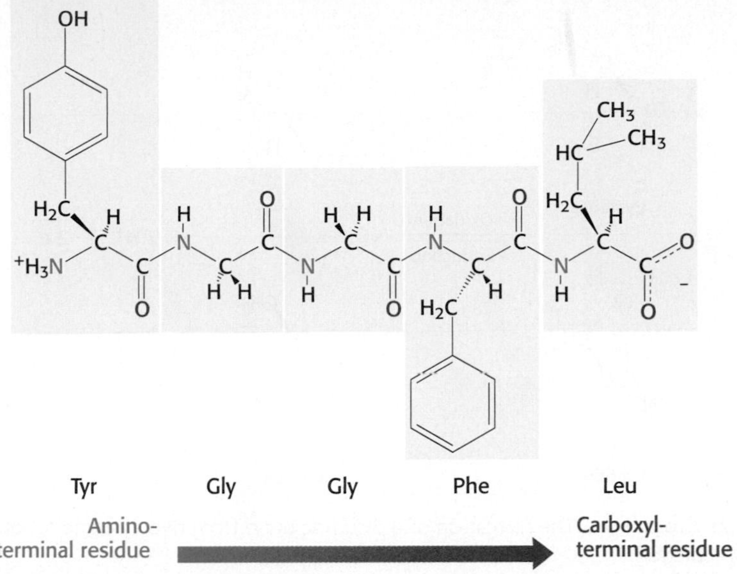

Tyr   Gly   Gly   Phe   Leu

Amino-
terminal residue   Carboxyl-
terminal residue

Figure 2.19 Amino acid sequences have direction. This illustration of the pentapeptide Try-Gly-Gly-Phe-Leu (YGGFL) shows the sequence from the amino terminus to the carboxyl terminus. This pentapeptide, Leu-enkephalin, is an opioid peptide that modulates the perception of pain. The reverse pentapeptide, Leu-Phe-Gly-Gly-Tyr (LFGGY), is a different molecule and shows no such effects.

and leucine is the carboxyl-terminal (C-terminal) residue (Figure 2.19). Leu-Phe-Gly-Gly-Tyr (LFGGY) is a different pentapeptide, with different chemical properties.

A polypeptide chain consists of a regularly repeating part, called the *main chain* or *backbone,* and a variable part, comprising the distinctive *side chains* (Figure 2.20). The polypeptide backbone is rich in hydrogen-bonding potential. Each residue contains a carbonyl group (C=O), which is a good hydrogen-bond acceptor and, with the exception of proline, an NH group, which is a good hydrogen-bond donor. These groups interact with each other and with functional groups from side chains to stabilize particular structures, as will be discussed in Section 2.3.

Most natural polypeptide chains contain between 50 and 2000 amino acid residues and are commonly referred to as *proteins.* The largest protein known is the muscle protein *titin,* which consists of more than 27,000 amino acids. Peptides made of small numbers of amino acids are called *oligopeptides* or simply *peptides.* The mean molecular weight of an amino acid residue is about 110 gm mol^{-1}, and so the molecular weights of most proteins are between 5500 and 220,000 gm mol^{-1}. We can also refer to the mass of a protein, which is expressed in units of daltons; one *dalton* is equal to one atomic mass unit. A protein with a molecular weight of 50,000 gm mol^{-1} has a mass of 50,000 daltons, or 50 kd (kilodaltons).

In some proteins, the linear polypeptide chain is cross-linked. The most common cross-links are *disulfide bonds,* formed by the oxidation of a pair of

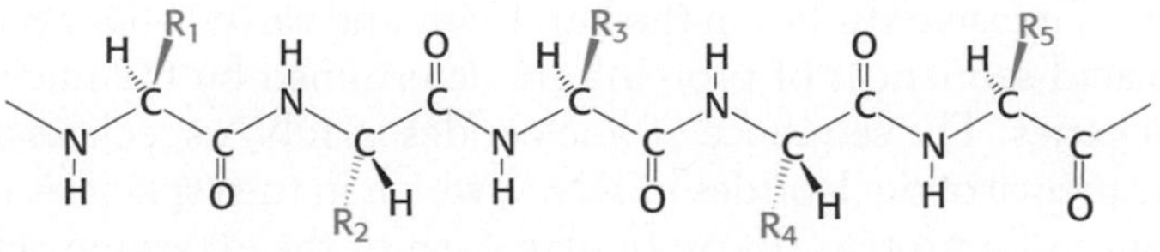

Dalton

A unit of mass very nearly equal to that of a hydrogen atom. Named after John Dalton (1766–1844), who developed the atomic theory of matter.

Kilodalton (kd)

A unit of mass equal to 1000 daltons.

Figure 2.20 Components of a polypeptide chain. A polypeptide chain consists of a constant backbone (shown in black) and variable side chains (shown in green).

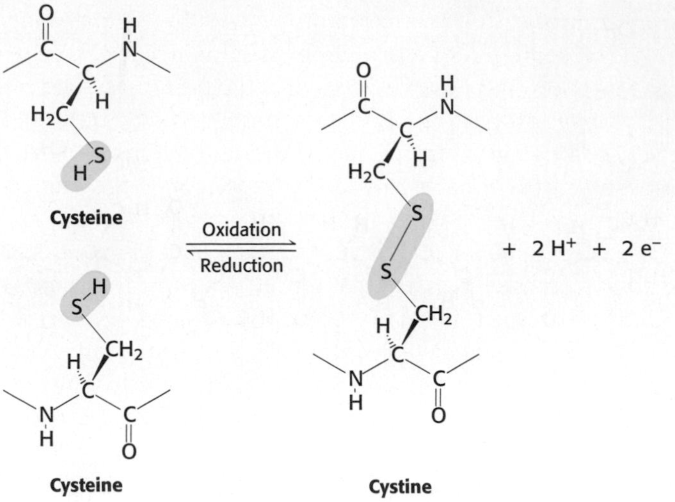

Cysteine

Oxidation ⇌ Reduction

$+ 2 H^+ + 2 e^-$

Cysteine

Cystine

Figure 2.21 Cross-links. The formation of a disulfide bond from two cysteine residues is an oxidation reaction.

cysteine residues (Figure 2.21). The resulting unit of two linked cysteines is called *cystine*. Extracellular proteins often have several disulfide bonds, whereas intracellular proteins usually lack them. Rarely, nondisulfide cross-links derived from other side chains are present in proteins. For example, collagen fibers in connective tissue are strengthened in this way, as are fibrin blood clots.

Proteins Have Unique Amino Acid Sequences Specified by Genes

In 1953, Frederick Sanger determined the amino acid sequence of insulin, a protein hormone (Figure 2.22). *This work is a landmark in biochemistry because it showed for the first time that a protein has a precisely defined amino acid sequence* consisting only of L amino acids linked by peptide bonds. This accomplishment stimulated other scientists to carry out sequence studies of a wide variety of proteins. Currently, the complete amino acid sequences of more than 2,000,000 proteins are known. *The striking fact is that each protein has a unique, precisely defined amino acid sequence.* The amino acid sequence of a protein is referred to as its *primary structure*.

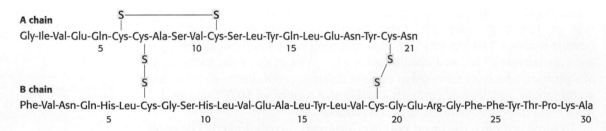

A chain
Gly-Ile-Val-Glu-Gln-Cys-Cys-Ala-Ser-Val-Cys-Ser-Leu-Tyr-Gln-Leu-Glu-Asn-Tyr-Cys-Asn
5 10 15 21

B chain
Phe-Val-Asn-Gln-His-Leu-Cys-Gly-Ser-His-Leu-Val-Glu-Ala-Leu-Tyr-Leu-Val-Cys-Gly-Glu-Arg-Gly-Phe-Phe-Tyr-Thr-Pro-Lys-Ala
5 10 15 20 25 30

Figure 2.22 Amino acid sequence of bovine insulin.

A series of incisive studies in the late 1950s and early 1960s revealed that the amino acid sequences of proteins are determined by the nucleotide sequences of genes. The sequence of nucleotides in DNA specifies a complementary sequence of nucleotides in RNA, which in turn specifies the amino acid sequence of a protein. In particular, each of the 20 amino acids of the repertoire is encoded by one or more specific sequences of three nucleotides (Section 5.5).

Knowing amino acid sequences is important for several reasons. First, knowledge of the sequence of a protein is usually essential to elucidating its mechanism of action (e.g., the catalytic mechanism of an enzyme). In fact, proteins with novel properties can be generated by varying the sequence of known proteins. Second, amino acid sequences determine the three-dimensional structures of proteins. Amino acid sequence is the link between the genetic message in DNA and the three-dimensional structure that performs a protein's biological function. Analyses of relations between amino acid sequences and three-dimensional structures of proteins are uncovering the rules that govern the folding of polypeptide chains. Third, sequence determination is a component of molecular pathology, a rapidly growing area of medicine. Alterations in amino acid sequence can produce abnormal function and disease. Severe and sometimes fatal diseases, such as sickle-cell anemia (195) and cystic fibrosis, can result from a change in a single amino acid within a protein. Fourth, the sequence of a protein reveals much about its evolutionary history (Chapter 6). Proteins resemble one another in amino acid sequence only if they have a common ancestor. Consequently, molecular events in evolution can be traced from amino acid sequences; molecular paleontology is a flourishing area of research.

Polypeptide Chains Are Flexible Yet Conformationally Restricted

Examination of the geometry of the protein backbone reveals several important features. First, *the peptide bond is essentially planar* (Figure 2.23). Thus, for a pair of amino acids linked by a peptide bond, six atoms lie in the same plane: the α-carbon atom and CO group of the first amino acid and the NH group and α-carbon atom of the second amino acid. The nature of the chemical bonding within a peptide accounts for the bond's planarity. The peptide bond has considerable *double-bond character*, which prevents rotation about this bond and thus constrains the conformation of the peptide backbone.

Peptide-bond resonance structures

The double-bond character is also expressed in the length of the bond between the CO and the NH groups. The C—N distance in a peptide bond is typically 1.32 Å, which is between the values expected for a C—N single

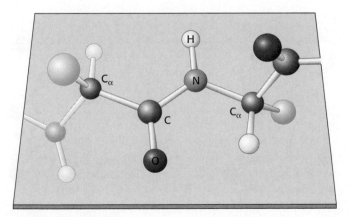

Figure 2.23 Peptide bonds are planar. In a pair of linked amino acids, six atoms (C_α, C, O, N, H, and C_α) lie in a plane. Side chains are shown as green balls.

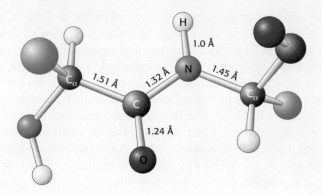

Figure 2.24 Typical bond lengths within a peptide unit. The peptide unit is shown in the trans configuration.

bond (1.49 Å) and a C=N double bond (1.27 Å), as shown in Figure 2.24. Finally, the peptide bond is uncharged, allowing polymers of amino acids linked by peptide bonds to form tightly packed globular structures.

Two configurations are possible for a planar peptide bond. In the *trans* configuration, the two α-carbon atoms are on opposite sides of the peptide bond. In the *cis* configuration, these groups are on the same side of the peptide bond. *Almost all peptide bonds in proteins are trans.* This preference for trans over cis can be explained by the fact that steric clashes between groups attached to the α-carbon atoms hinder formation of the cis form but do not occur in the trans configuration (Figure 2.25). By far the most common cis peptide bonds are X—Pro linkages. Such bonds show less preference for the trans configuration because the nitrogen of proline is bonded to two tetrahedral carbon atoms, limiting the steric differences between the trans and cis forms (Figure 2.26).

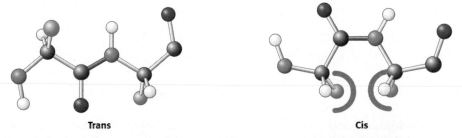

Trans   **Cis**

Figure 2.25 Trans and cis peptide bonds. The trans form is strongly favored because of steric clashes that occur in the cis form.

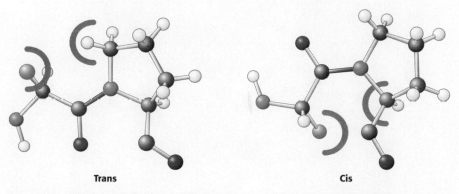

Trans   **Cis**

Figure 2.26 Trans and cis X–Pro bonds. The energies of these forms are similar to one another because steric clashes occur in both forms.

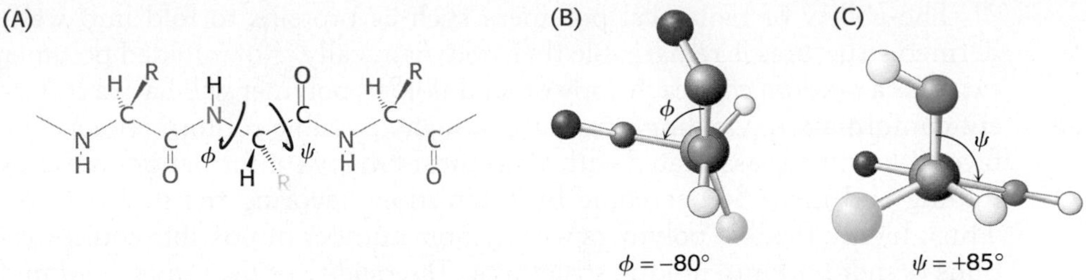

(A) (B) (C)

$\phi = -80°$ $\psi = +85°$

Figure 2.27 Rotation about bonds in a polypeptide. The structure of each amino acid in a polypeptide can be adjusted by rotation about two single bonds. (A) Phi (ϕ) is the angle of rotation about the bond between the nitrogen and the α-carbon atoms, whereas psi (ψ) is the angle of rotation about the bond between the α-carbon and the carbonyl carbon atoms. (B) A view down the bond between the nitrogen and the α-carbon atoms, showing how ϕ is measured. (C) A view down the bond between the α-carbon and the carbonyl carbon atoms, showing how ψ is measured.

In contrast with the peptide bond, the bonds between the amino group and the α-carbon atom and between the α-carbon atom and the carbonyl group are pure single bonds. The two adjacent rigid peptide units may rotate about these bonds, taking on various orientations. *This freedom of rotation about two bonds of each amino acid allows proteins to fold in many different ways.* The rotations about these bonds can be specified by *torsion angles* (Figure 2.27). The angle of rotation about the bond between the nitrogen and the α-carbon atoms is called *phi* (ϕ). The angle of rotation about the bond between the α-carbon and the carbonyl carbon atoms is called *psi* (ψ). A clockwise rotation about either bond as viewed from the nitrogen atom toward the α-carbon atom or from the carbonyl group toward the α-carbon atom corresponds to a positive value. The ϕ and ψ angles determine the path of the polypeptide chain.

Are all combinations of ϕ and ψ possible? Gopalasamudram Ramachandran recognized that many combinations are forbidden because of steric collisions between atoms. The allowed values can be visualized on a two-dimensional plot called a *Ramachandran diagram* (Figure 2.28). Three-quarters of the possible (ϕ, ψ) combinations are excluded simply by local steric clashes. *Steric exclusion, the fact that two atoms cannot be in the same place at the same time, can be a powerful organizing principle.*

> **Torsion angle**
> A measure of the rotation about a bond, usually taken to lie between -180 and $+180$ degrees. Torsion angles are sometimes called dihedral angles.

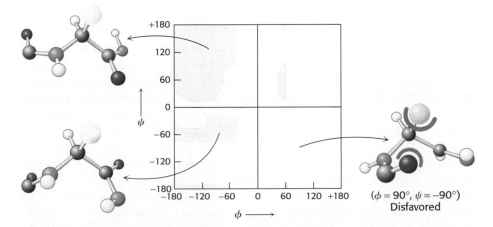

$(\phi = 90°, \psi = -90°)$
Disfavored

Figure 2.28 A Ramachandran diagram showing the values of ϕ and ψ. Not all ϕ and ψ values are possible without collisions between atoms. The most favorable regions are shown in dark green; borderline regions are shown in light green. The structure on the right is disfavored because of steric clashes.

The ability of biological polymers such as proteins to fold into well-defined structures is remarkable thermodynamically. An unfolded polymer exists as a random coil: each copy of an unfolded polymer will have a different conformation, yielding a mixture of many possible conformations. The favorable entropy associated with a mixture of many conformations opposes folding and must be overcome by interactions favoring the folded form. Thus, highly flexible polymers with a large number of possible conformations do not fold into unique structures. *The rigidity of the peptide unit and the restricted set of allowed ϕ and ψ angles limits the number of structures accessible to the unfolded form sufficiently to allow protein folding to occur.*

2.3 Secondary Structure: Polypeptide Chains Can Fold into Regular Structures Such As the Alpha Helix, the Beta Sheet, and Turns and Loops

Can a polypeptide chain fold into a regularly repeating structure? In 1951, Linus Pauling and Robert Corey proposed two periodic structures called the *α helix* (alpha helix) and the *β pleated sheet* (beta pleated sheet). Subsequently, other structures such as the *β turn* and *omega (Ω) loop* were identified. Although not periodic, these common turn or loop structures are well defined and contribute with α helices and β sheets to form the final protein structure. Alpha helices, β strands, and turns are formed by a regular pattern of hydrogen bonds between the peptide N—H and C=O groups of amino acids that are *near one another in the linear sequence*. Such folded segments are called *secondary structure*.

The Alpha Helix Is a Coiled Structure Stabilized by Intrachain Hydrogen Bonds

In evaluating potential structures, Pauling and Corey considered which conformations of peptides were sterically allowed and which most fully exploited the hydrogen-bonding capacity of the backbone NH and CO groups. The first of their proposed structures, the *α helix*, is a rodlike structure (Figure 2.29). A tightly coiled backbone forms the inner part of the rod and the side chains extend outward in a helical array. The α helix is stabilized by hydrogen bonds between the NH and CO groups of the main chain. In particular, the CO group of each amino acid forms a hydrogen bond with the NH group of the amino acid that is situated four residues ahead in the sequence (Figure 2.30). Thus, except for amino acids near the ends of an α helix, all *the main-chain CO and NH groups are hydrogen bonded*. Each residue is related to the next one by a *rise*, also called *translation*, of 1.5 Å along the helix axis and a rotation of 100 degrees, which gives 3.6 amino acid residues per turn of helix. Thus, amino acids spaced three and four apart in the sequence are spatially quite close to one another in an α helix. In contrast, amino acids spaced two apart in the sequence are situated on opposite sides of the helix and so are unlikely to make contact. The *pitch* of the α helix, which is equal to the product of the translation (1.5 Å) and the number of residues per turn (3.6), is 5.4 Å. The *screw sense* of a helix can be right-handed (clockwise) or left-handed (counterclockwise). The Ramachandran diagram reveals that both the right-handed and the left-handed helices are among allowed conformations (Figure 2.31). However,

Screw sense

Describes the direction in which a helical structure rotates with respect to its axis. If, viewed down the axis of a helix, the chain turns in a clockwise direction, it has a right-handed screw sense. If the turning is counterclockwise, the screw sense is left-handed.

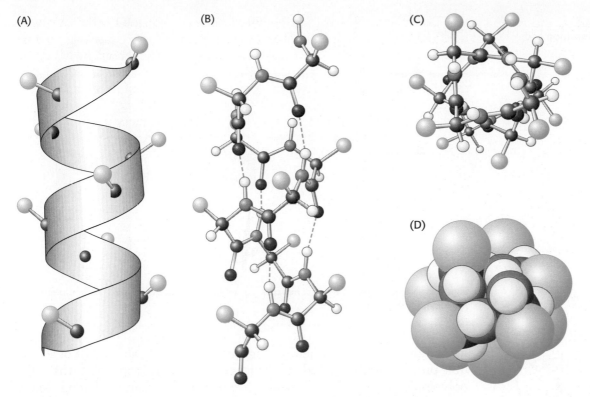

(A) **(B)** **(C)**

(D)

Figure 2.29 Structure of the α helix. (A) A ribbon depiction with the α-carbon atoms and side chains (green) shown. (B) A side view of a ball-and-stick version depicts the hydrogen bonds (dashed lines) between NH and CO groups. (C) An end view shows the coiled backbone as the inside of the helix and the side chains (green) projecting outward. (D) A space-filling view of part C shows the tightly packed interior core of the helix.

right-handed helices are energetically more favorable because there is less steric clash between the side chains and the backbone. *Essentially all α helices found in proteins are right-handed.* In schematic representations of proteins, α helices are depicted as twisted ribbons or rods (Figure 2.32).

Pauling and Corey predicted the structure of the α helix 6 years before it was actually seen in the x-ray reconstruction of the structure of myoglobin. *The elucidation of the structure of the α helix is a landmark in biochemistry because it demonstrated that the conformation of a polypeptide chain could be predicted if the properties of its components are rigorously and precisely known.*

The α-helical content of proteins ranges widely, from none to almost 100%. For example, about 75% of the residues in ferritin, a protein that

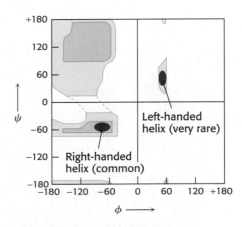

Figure 2.31 Ramachandran diagram for helices. Both right- and left-handed helices lie in regions of allowed conformations in the Ramachandran diagram. However, essentially all α helices in proteins are right-handed.

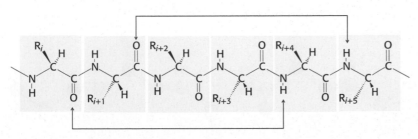

Figure 2.30 Hydrogen-bonding scheme for an α helix. In the α helix, the CO group of residue i forms a hydrogen bond with the NH group of residue $i + 4$.

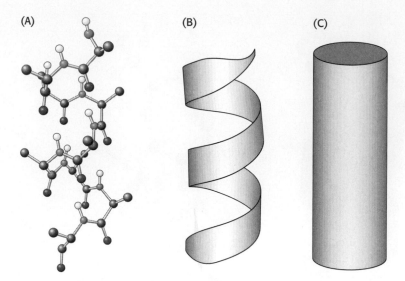

Figure 2.32 Schematic views of α helices. (A) A ball-and-stick model. (B) A ribbon depiction. (C) A cylindrical depiction.

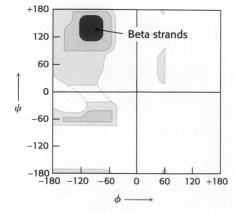

Figure 2.33 A largely α-helical protein. Ferritin, an iron-storage protein, is built from a bundle of α helices. [Drawn from 1AEW.pdb.]

helps store iron, are in α helices (Figure 2.33). Indeed, about 25% of all soluble proteins are composed of α helices connected by loops and turns of the polypeptide chain. Single α helices are usually less than 45 Å long. Many proteins that span biological membranes also contain α helices.

Beta Sheets Are Stabilized by Hydrogen Bonding Between Polypeptide Strands

Pauling and Corey proposed another periodic structural motif, which they named the *β pleated sheet* (β because it was the second structure that they elucidated, the α helix having been the first). The β pleated sheet (or, more simply, the β sheet) differs markedly from the rodlike α helix. It is composed of two or more polypeptide chains called *β strands*. A β strand is almost fully extended rather than being tightly coiled as in the α helix. A range of extended structures are sterically allowed (Figure 2.34).

The distance between adjacent amino acids along a β strand is approximately 3.5 Å, in contrast to a distance of 1.5 Å along an α helix. The side chains of adjacent amino acids point in opposite directions (Figure 2.35). A β sheet is formed by linking two or more β strands lying next to one another through hydrogen bonds. Adjacent chains in a β sheet can run in opposite directions (antiparallel β sheet) or in the same direction (parallel β sheet). In the antiparallel arrangement, the NH group and the CO group of each amino acid are respectively hydrogen bonded to the CO group and the NH group of a partner on the adjacent chain (Figure 2.36). In the parallel arrangement, the hydrogen-bonding scheme is slightly more complicated.

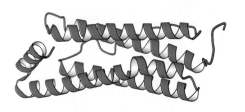

Figure 2.34 Ramachandran diagram for β strands. The red area shows the sterically allowed conformations of extended, β-strand-like structures.

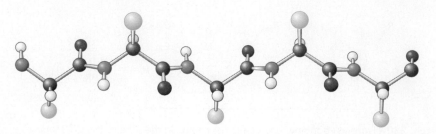

Figure 2.35 Structure of a β strand. The side chains (green) are alternately above and below the plane of the strand.

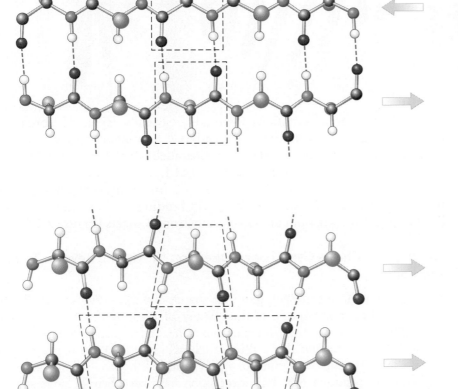

Figure 2.36 An antiparallel β sheet.
Adjacent β strands run in opposite
directions. Hydrogen bonds between NH
and CO groups connect each amino acid
to a single amino acid on an adjacent
strand, stabilizing the structure.

Figure 2.37 A parallel β sheet. Adjacent
β strands run in the same direction.
Hydrogen bonds connect each amino acid
on one strand with two different amino
acids on the adjacent strand.

For each amino acid, the NH group is hydrogen bonded to the CO group of
one amino acid on the adjacent strand, whereas the CO group is hydrogen
bonded to the NH group on the amino acid two residues farther along the
chain (Figure 2.37). Many strands, typically 4 or 5 but as many as 10 or
more, can come together in β sheets. Such β sheets can be purely antiparallel, purely parallel, or mixed (Figure 2.38).

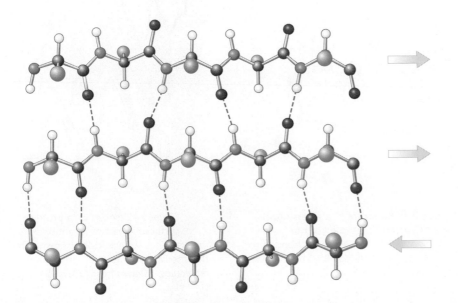

Figure 2.38 Structure of a mixed β sheet.

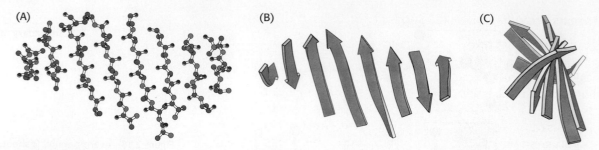

(A)   (B)   (C)

Figure 2.39 A twisted β sheet. (A) A ball-and-stick model. (B) A schematic model. (C) The schematic view rotated by 90 degrees to illustrate the twist more clearly.

In schematic representations, β strands are usually depicted by broad arrows pointing in the direction of the carboxyl-terminal end to indicate the type of β sheet formed—parallel or antiparallel. More structurally diverse than α helices, β sheets can be almost flat but most adopt a somewhat twisted shape (Figure 2.39). The β sheet is an important structural element in many proteins. For example, fatty acid-binding proteins, important for lipid metabolism, are built almost entirely from β sheets (Figure 2.40).

Polypeptide Chains Can Change Direction by Making Reverse Turns and Loops

Most proteins have compact, globular shapes owing to reversals in the direction of their polypeptide chains. Many of these reversals are accomplished by a common structural element called the *reverse turn* (also known as the *β turn* or *hairpin turn*), illustrated in Figure 2.41. In many reverse turns, the CO group of residue i of a polypeptide is hydrogen bonded to the NH group of residue $i + 3$. This interaction stabilizes abrupt changes in direction of the polypeptide chain. In other cases, more elaborate structures are responsible for chain reversals. These structures are called *loops* or sometimes *Ω loops* (omega loops) to suggest their overall shape. Unlike α helices and β strands, loops do not have regular, periodic structures. Nonetheless, loop structures are often rigid and well defined (Figure 2.42). Turns and loops invariably lie on the surfaces of proteins and thus often participate in interactions between proteins and other molecules.

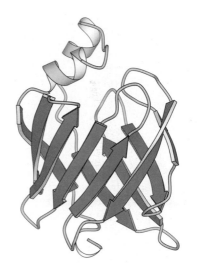

Figure 2.40 A protein rich in β sheets. The structure of a fatty acid-binding protein. [Drawn from 1FTP.pdb.]

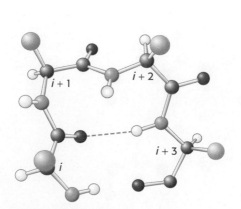

Figure 2.41 Structure of a reverse turn. The CO group of residue i of the polypeptide chain is hydrogen bonded to the NH group of residue $i +3$ to stabilize the turn.

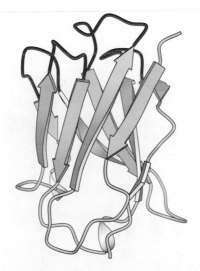

Figure 2.42 Loops on a protein surface. A part of an antibody molecule has surface loops (shown in red) that mediate interactions with other molecules. [Drawn from 7FTP.pdb.]

Fibrous Proteins Provide Structural Support for Cells and Tissues

Special types of helices are present in the two proteins α-keratin and collagen. These proteins form long fibers that serve a structural role.

α-Keratin, which is the primary component of wool and hair, consists of two right-handed α helices intertwined to form a type of left-handed superhelix called an *α coiled coil*. α-Keratin is a member of a superfamily of proteins referred to as *coiled-coil proteins* (Figure 2.43). In these proteins, two or more α helices can entwine to form a very stable structure, which can have a length of 1000 Å (100 nm, or 0.1 μm) or more. There are approximately 60 members of this family in humans, including intermediate filaments, proteins that contribute to the cell cytoskeleton (internal scaffolding in a cell), and the muscle proteins myosin and tropomyosin (Section 34.2). Members of this family are characterized by a central region of 300 amino acids that contains imperfect repeats of a sequence of seven amino acids called a *heptad repeat*.

(A)

(B)

Figure 2.43 An α-helical coiled coil. (A) Space-filling model. (B) Ribbon diagram. The two helices wind around one another to form a superhelix. Such structures are found in many proteins, including keratin in hair, quills, claws, and horns. [Drawn from 1CIG.pdb.]

The two helices in α-keratin are cross-linked by weak interactions such as van der Waals forces and ionic interactions. These interactions are facilitated by the fact that the left-handed supercoil alters the two right-handed α helices such that there are 3.5 residues per turn instead of 3.6. Thus, the pattern of side-chain interactions can be repeated every seven residues, forming the heptad repeats. Two helices with such repeats are able to interact with one another if the repeats are complementary (Figure 2.44). For example, the repeating residues may be hydrophobic, allowing van der Waals interactions, or have opposite charge, allowing ionic interactions. In addition, the two helices may be linked by disulfide bonds formed by neighboring cysteine residues. The bonding of the helices accounts for the physical properties of wool, an example of an α-keratin. Wool is extensible and can be stretched to nearly twice its length because the α helices stretch, breaking the weak interactions between neighboring helices. However, the covalent disulfide bonds resist breakage and return the fiber to its original state once the stretching force is released. The number of disulfide bond cross-links further defines the fiber's properties. Hair and wool, having fewer cross-links, are flexible. Horns, claws, and hooves, having more cross-links, are much harder.

A different type of helix is present in collagen, the most abundant protein of mammals. Collagen is the main fibrous component of skin, bone, tendon, cartilage, and teeth. This extracellular protein is a rod-shaped molecule, about 3000 Å long and only 15 Å in diameter. It contains three helical polypeptide chains, each nearly 1000 residues long. Glycine appears at every third residue in the amino acid sequence, and the sequence glycine-

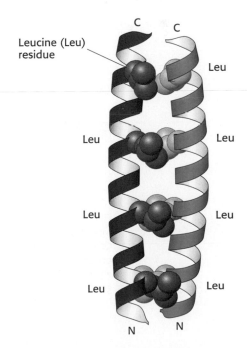

Figure 2.44 Heptad repeats in a coiled-coil protein. Every seventh residue in each helix is leucine. The two helices are held together by van der Waals interactions primarily between the leucine residues. [Drawn from 2ZTA.pdb.]

13
-Gly-Pro-Met-Gly-Pro-Ser-Gly-Pro-Arg-
22
-Gly-Leu-Hyp-Gly-Pro-Hyp-Gly-Ala-Hyp-
31
-Gly-Pro-Gln-Gly-Phe-Gln-Gly-Pro-Hyp-
40
-Gly-Glu-Hyp-Gly-Glu-Hyp-Gly-Ala-Ser-
49
-Gly-Pro-Met-Gly-Pro-Arg-Gly-Pro-Hyp-
58
-Gly-Pro-Hyp-Gly-Lys-Asn-Gly-Asp-Asp-

Figure 2.45 Amino acid sequence of a part of a collagen chain. Every third residue is a glycine. Proline and hydroxyproline also are abundant.

Figure 2.46 Conformation of a single strand of a collagen triple helix.

proline-hydroxyproline recurs frequently (Figure 2.45). Hydroxyproline is a derivative of proline that has a hydroxyl group in place of one of the hydrogens on the pyrrolidine rings.

The collagen helix has properties different from those of the α helix. Hydrogen bonds within a strand are absent. Instead, *the helix is stabilized by steric repulsion of the pyrrolidine rings of the proline and hydroxyproline residues* (Figure 2.46). The pyrrolidine rings keep out of each other's way when the polypeptide chain assumes its helical form, which has about three residues per turn. Three strands wind around each other to form a *superhelical cable* that is stabilized by hydrogen bonds between strands. The hydrogen bonds form between the peptide NH groups of glycine residues and the CO groups of residues on the other chains. The hydroxyl groups of hydroxyproline residues also participate in hydrogen bonding, and the absence of the hydroxyl groups results in the disease scurvy (Section 27.5).

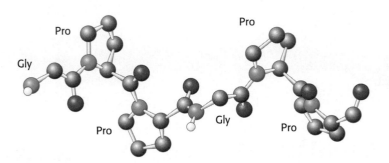

The inside of the triple-stranded helical cable is very crowded and accounts for the requirement that glycine be present at every third position on each strand (Figure 2.47A). *The only residue that can fit in an interior position is glycine.* The amino acid residue on either side of glycine is located on the outside of the cable, where there is room for the bulky rings of proline and hydroxyproline residues (Figure 2.47B).

(A)

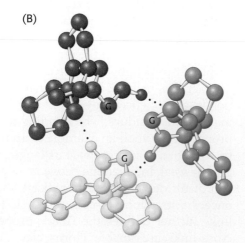

(B)

Figure 2.47 Structure of the protein collagen. (A) Space-filling model of collagen. Each strand is shown in a different color. (B) Cross section of a model of collagen. Each strand is hydrogen bonded to the other two strands. The α carbon atom of a glycine residue is labeled G. Every third residue must be glycine because there is no space in the center of the helix. *Notice* that the pyrrolidone rings are on the outside.

2.4 Tertiary Structure: Water-Soluble Proteins Fold into Compact Structures with Nonpolar Cores

Let us now examine how amino acids are grouped together in a complete protein. X-ray crystallographic and nuclear magnetic resonance studies (Section 3.6) have revealed the detailed three-dimensional structures of

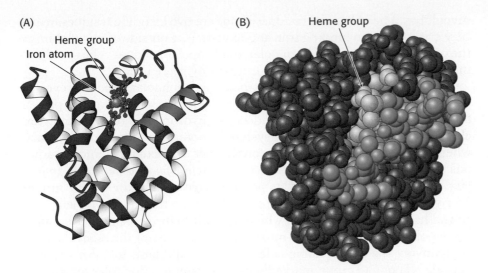

(A)

Heme group
Iron atom

(B)

Heme group

Figure 2.48 Three-dimensional structure of myoglobin. (A) A ribbon diagram shows that the protein consists largely of α helices. (B) A space-filling model in the same orientation shows how tightly packed the folded protein is. *Notice* that the heme group is nestled into a crevice in the compact protein with only an edge exposed. One helix is blue to allow comparison of the two structural depictions. [Drawn from 1A6N.pdb.]

thousands of proteins. We begin here with an examination of *myoglobin,* the first protein to be seen in atomic detail.

Myoglobin, the oxygen carrier in muscle, is a single polypeptide chain of 153 amino acids (see Chapter 7). The capacity of myoglobin to bind oxygen depends on the presence of *heme,* a nonpolypeptide *prosthetic (helper) group* consisting of protoporphyrin IX and a central iron atom. *Myoglobin is an extremely compact molecule.* Its overall dimensions are 45 × 35 × 25 Å, an order of magnitude less than if it were fully stretched out (Figure 2.48). About 70% of the main chain is folded into eight α helices, and much of the rest of the chain forms turns and loops between helices.

The folding of the main chain of myoglobin, like that of most other proteins, is complex and devoid of symmetry. The overall course of the polypeptide chain of a protein is referred to as its *tertiary structure.* A unifying principle emerges from the distribution of side chains. The striking fact is that *the interior consists almost entirely of nonpolar residues* such as leucine, valine, methionine, and phenylalanine (Figure 2.49). Charged residues such as aspartate, glutamate, lysine, and arginine are absent from the inside of

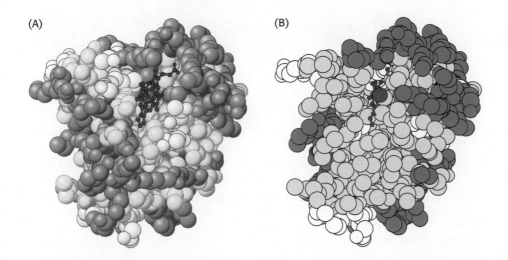

(A)

(B)

Figure 2.49 Distribution of amino acids in myoglobin. (A) A space-filling model of myoglobin with hydrophobic amino acids shown in yellow, charged amino acids shown in blue, and others shown in white. *Notice* that the surface of the molecule has many charged amino acids, as well as some hydrophobic amino acids. (B) In this cross-sectional view *notice* that mostly hydrophobic amino acids are found on the inside of the structure, whereas the charged amino acids are found on the protein surface. [Drawn from 1MBD.pdb.]

myoglobin. The only polar residues inside are two histidine residues, which play critical roles in binding iron and oxygen. The outside of myoglobin, on the other hand, consists of both polar and nonpolar residues. The space-filling model shows that there is very little empty space inside.

This contrasting distribution of polar and nonpolar residues reveals a key facet of protein architecture. In an aqueous environment, protein folding is driven by the strong tendency of hydrophobic residues to be excluded from water. Recall that a system is more thermodynamically stable when hydrophobic groups are clustered rather than extended into the aqueous surroundings. *The polypeptide chain therefore folds so that its hydrophobic side chains are buried and its polar, charged chains are on the surface.* Many α helices and β strands are amphipathic; that is, the α helix or β strand has a hydrophobic face, which points into the protein interior, and a more polar face, which points into solution. The fate of the main chain accompanying the hydrophobic side chains is important, too. An unpaired peptide NH or CO group markedly prefers water to a nonpolar milieu. The secret of burying a segment of main chain in a hydrophobic environment is to pair all the NH and CO groups by hydrogen bonding. This pairing is neatly accomplished in an α helix or β sheet. Van der Waals interactions between tightly packed hydrocarbon side chains also contribute to the stability of proteins. We can now understand why the set of 20 amino acids contains several that differ subtly in size and shape. They provide a palette from which to choose to fill the interior of a protein neatly and thereby maximize van der Waals interactions, which require intimate contact.

Some proteins that span biological membranes are "the exceptions that prove the rule" because they have the reverse distribution of hydrophobic and hydrophilic amino acids. For example, consider porins, proteins found in the outer membranes of many bacteria (Figure 2.50). Membranes are built largely of hydrophobic alkane chains (Section 12.2). Thus, porins are covered on the outside largely with hydrophobic residues that interact with

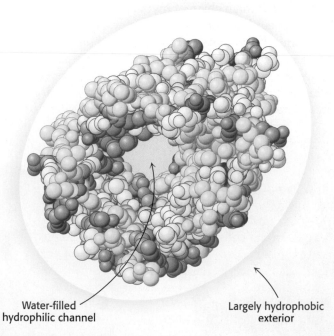

Water-filled
hydrophilic channel

Largely hydrophobic
exterior

Figure 2.50 "Inside out" amino acid distribution in porin. The outside of porin (which contacts hydrophobic groups in membranes) is covered largely with hydrophobic residues, whereas the center includes a water-filled channel lined with charged and polar amino acids. [Drawn from 1PRN.pdb.]

the neighboring alkane chains. In contrast, the center of the protein contains many charged and polar amino acids that surround a water-filled channel running through the middle of the protein. Thus, because porins function in hydrophobic environments, they are "inside out" relative to proteins that function in aqueous solution.

Certain combinations of secondary structure are present in many proteins and frequently exhibit similar functions. These combinations are called *motifs* or *supersecondary structure*. For example, an α helix separated from another α helix by a turn, called a *helix-turn-helix* unit, is found in many proteins that bind DNA (Figure 2.51).

Some polypeptide chains fold into two or more compact regions that may be connected by a flexible segment of polypeptide chain, rather like pearls on a string. These compact globular units, called *domains,* range in size from about 30 to 400 amino acid residues. For example, the extracellular part of CD4, the cell-surface protein on certain cells of the immune system to which the human immunodeficiency virus (HIV) attaches itself, comprises four similar domains of approximately 100 amino acids each (Figure 2.52). Proteins may have domains in common even if their overall tertiary structures are different.

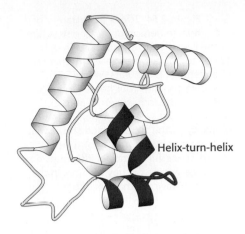

Figure 2.51 The helix-turn-helix motif, a supersecondary structural element. Helix turn-helix motifs are found in many DNA-binding proteins. [Drawn from 1LMB.pdb.]

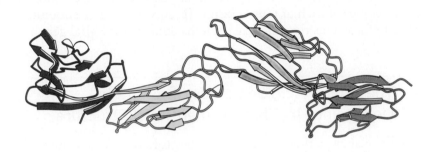

Figure 2.52 Protein domains. The cell-surface protein CD4 consists of four similar domains. [Drawn from 1WIO.pdb.]

2.5 Quaternary Structure: Polypeptide Chains Can Assemble into Multisubunit Structures

Four levels of structure are frequently cited in discussions of protein architecture. So far, we have considered three of them. *Primary structure* is the amino acid sequence. *Secondary structure* refers to the spatial arrangement of amino acid residues that are nearby in the sequence. Some of these arrangements are of a regular kind, giving rise to a periodic structure. The α helix and β strand are elements of secondary structure. *Tertiary structure* refers to the spatial arrangement of amino acid residues that are far apart in the sequence and to the pattern of disulfide bonds. We now turn to proteins containing more than one polypeptide chain. Such proteins exhibit a fourth level of structural organization. Each polypeptide chain in such a protein is called a *subunit*. *Quaternary structure* refers to the spatial arrangement of subunits and the nature of their interactions. The simplest sort of quaternary structure is a *dimer,* consisting of two identical subunits. This organization is present in the DNA-binding protein Cro found in a bacterial virus called λ (Figure 2.53). More complicated quaternary structures also are common. More than one type of subunit can be present, often in variable numbers. For example, human hemoglobin, the oxygen-carrying protein in blood, consists of two subunits of one type (designated α) and two subunits of another type (designated β), as illustrated in

Figure 2.53 Quaternary structure. The Cro protein of bacteriophage λ is a dimer of identical subunits. [Drawn from 5CRO.pdb.]

Figure 2.54 The α₂β₂ tetramer of human hemoglobin.

Figure 2.54 The $\alpha_2\beta_2$ tetramer of human hemoglobin. The structure of the two identical α subunits (red) is similar to but not identical with that of the two identical β subunits (yellow). The molecule contains four heme groups (gray with the iron atom shown in purple). (A) The ribbon diagram highlights the similarity of the subunits and shows that they are composed mainly of α helices. (B) The space-filling model illustrates how the heme groups occupy crevices in the protein. [Drawn from 1A3N.pdb.]

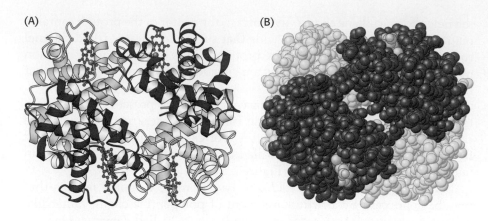

Figure 2.55 Complex quaternary structure. The coat of human rhinovirus, the cause of the common cold, comprises 60 copies of each of four subunits (shown in different colors).

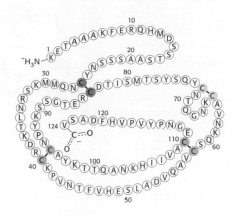

β-Mercaptoethanol

Figure 2.54. Thus, the hemoglobin molecule exists as an $\alpha_2\beta_2$ tetramer. Subtle changes in the arrangement of subunits within the hemoglobin molecule allow it to carry oxygen from the lungs to tissues with great efficiency (Chapter 7).

Viruses make the most of a limited amount of genetic information by forming coats that use the same kind of subunit repetitively in a symmetric array. The coat of rhinovirus, the virus that causes the common cold, includes 60 copies of each of four subunits (Figure 2.55). The subunits come together to form a nearly spherical shell that encloses the viral genome.

2.6 The Amino Acid Sequence of a Protein Determines Its Three-Dimensional Structure

How is the elaborate three-dimensional structure of proteins attained? The classic work of Christian Anfinsen in the 1950s on the enzyme ribonuclease revealed the relation between the amino acid sequence of a protein and its conformation. Ribonuclease is a single polypeptide chain consisting of 124 amino acid residues cross-linked by four disulfide bonds (Figure 2.56). Anfinsen's plan was to destroy the three-dimensional structure of the enzyme and to then determine what conditions were required to restore the structure.

Agents such as urea or guanidinium chloride effectively disrupt a protein's noncovalent bonds, although the mechanism of action of these agents is not fully understood. The disulfide bonds can be cleaved reversibly by reducing them with a reagent such as *β-mercaptoethanol* (Figure 2.57). In the presence of a large excess of β-mercaptoethanol, the disulfides (cystines) are fully converted into sulfhydryls (cysteines).

Most polypeptide chains devoid of cross-links assume a *random-coil conformation* in 8 M urea or 6 M guanidinium chloride. When ribonuclease was

Figure 2.56 Amino acid sequence of bovine ribonuclease. The four disulfide bonds are shown in color. [After C. H. W. Hirs, S. Moore, and W. H. Stein, *J. Biol. Chem.* 235(1960):633–647.]

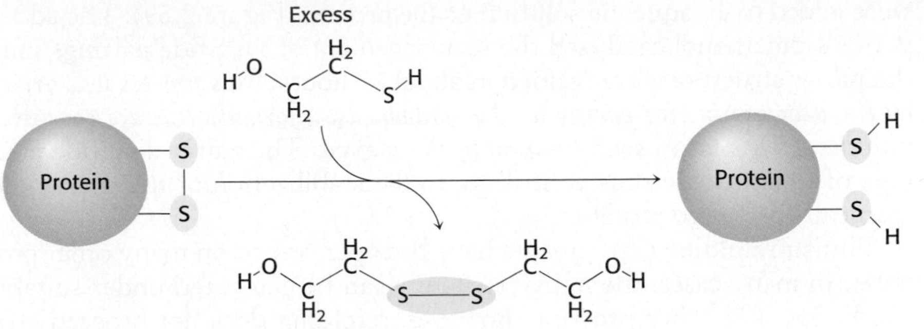

Excess

Figure 2.57 Role of β-mercaptoethanol in reducing disulfide bonds. Note that, as the disulfides are reduced, the β-mercaptoethanol is oxidized and forms dimers.

treated with β-mercaptoethanol in 8 M urea, the product was a fully reduced, randomly coiled polypeptide chain *devoid of enzymatic activity*. When a protein is converted into a randomly coiled peptide without its normal activity, it is said to be *denatured* (Figure 2.58).

Anfinsen then made the critical observation that the denatured ribonuclease, freed of urea and β-mercaptoethanol by dialysis, slowly regained enzymatic activity. He immediately perceived the significance of this chance finding: the sulfhydryl groups of the denatured enzyme became oxidized by air, and the enzyme spontaneously refolded into a catalytically active form. Detailed studies then showed that nearly all the original enzymatic activity was regained if the sulfhydryl groups were oxidized under suitable conditions. All the measured physical and chemical properties of the refolded enzyme were virtually identical with those of the native enzyme. These experiments showed that *the information needed to specify the catalytically active structure of ribonuclease is contained in its amino acid sequence*. Subsequent studies have established the generality of this central principle of biochemistry: *sequence specifies conformation*. The dependence of conformation on sequence is especially significant because of the intimate connection between conformation and function.

A quite different result was obtained when reduced ribonuclease was reoxidized while it was still in 8 M urea and the preparation was then dialyzed to remove the urea. Ribonuclease reoxidized in this way had only 1% of the enzymatic activity of the native protein. Why were the outcomes so different when reduced ribonuclease was reoxidized in the presence and absence of urea? The reason is that the wrong disulfides formed pairs in urea. There are 105 different ways of pairing eight cysteine molecules to form four disulfides; only one of these combinations is enzymatically active. The 104 wrong pairings have been picturesquely termed "scrambled" ribonuclease. Anfinsen found that scrambled ribonuclease spontaneously converted into fully active, native ribonuclease when trace amounts of β-mercaptoethanol

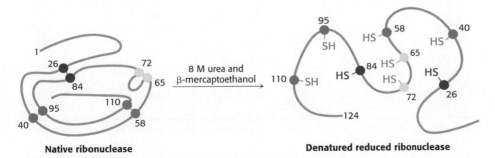

Figure 2.58 Reduction and denaturation of ribonuclease.

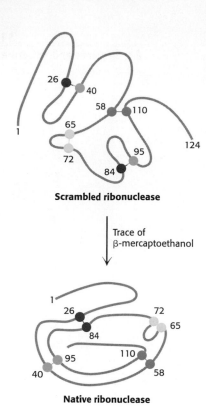

Scrambled ribonuclease

Trace of
β-mercaptoethanol

Native ribonuclease

Figure 2.59 Reestablishing correct disulfide pairing. Native ribonuclease can be re-formed from scrambled ribonuclease in the presence of a trace of β-mercaptoethanol.

were added to an aqueous solution of the protein (Figure 2.59). The added β-mercaptoethanol catalyzed the rearrangement of disulfide pairings until the native structure was regained in about 10 hours. *This process was driven by the decrease in free energy as the scrambled conformations were converted into the stable, native conformation of the enzyme.* The native disulfide pairings of ribonuclease thus contribute to the stabilization of the thermodynamically preferred structure.

Similar refolding experiments have been performed on many other proteins. In many cases, the native structure can be generated under suitable conditions. For other proteins, however, refolding does not proceed efficiently. In these cases, the unfolding protein molecules usually become tangled up with one another to form aggregates. Inside cells, proteins called *chaperones* block such illicit interactions.

Amino Acids Have Different Propensities for Forming Alpha Helices, Beta Sheets, and Beta Turns

How does the amino acid sequence of a protein specify its three-dimensional structure? How does an unfolded polypeptide chain acquire the form of the native protein? These fundamental questions in biochemistry can be approached by first asking a simpler one: What determines whether a particular sequence in a protein forms an α helix, a β strand, or a turn? One source of insight is to examine the frequency of occurrence of particular amino acid residues in these secondary structures (Table 2.3). Residues such as alanine, glutamate, and leucine tend to be present in α helices, whereas valine and isoleucine tend to be present in β strands. Glycine, asparagine, and proline have a propensity for being present in turns.

Studies of proteins and synthetic peptides have revealed some reasons for these preferences. The α helix can be regarded as the default conformation. Branching at the β-carbon atom, as in valine, threonine, and isoleucine, tends to destabilize α helices because of steric clashes. These residues are readily accommodated in β strands, in which their side chains project out of

TABLE 2.3 Relative frequencies of amino acid residues in secondary structures

Amino acid	α helix	β sheet	Reverse turn
Glu	1.59	0.52	1.01
Ala	1.41	0.72	0.82
Leu	1.34	1.22	0.57
Met	1.30	1.14	0.52
Gln	1.27	0.98	0.84
Lys	1.23	0.69	1.07
Arg	1.21	0.84	0.90
His	1.05	0.80	0.81
Val	0.90	1.87	0.41
Ile	1.09	1.67	0.47
Tyr	0.74	1.45	0.76
Cys	0.66	1.40	0.54
Trp	1.02	1.35	0.65
Phe	1.16	1.33	0.59
Thr	0.76	1.17	0.96
Gly	0.43	0.58	1.77
Asn	0.76	0.48	1.34
Pro	0.34	0.31	1.32
Ser	0.57	0.96	1.22
Asp	0.99	0.39	1.24

NOTE: The amino acids are grouped according to their preference for α helices (top group), β sheets (second group), or turns (third group).

SOURCE: T. E. Creighton, *Proteins: Structures and Molecular Properties*, 2d ed. (W. H. Freeman and Company, 1992), p. 256.

the plane containing the main chain. Serine, aspartate, and asparagine tend to disrupt α helices because their side chains contain hydrogen-bond donors or acceptors in close proximity to the main chain, where they compete for main-chain NH and CO groups. Proline tends to disrupt both α helices and β strands because it lacks an NH group and because its ring structure restricts its ϕ value to near 60 degrees. Glycine readily fits into all structures and for that reason does not favor helix formation in particular.

Can one predict the secondary structure of proteins by using this knowledge of the conformational preferences of amino acid residues? Predictions of secondary structure adopted by a stretch of six or fewer residues have proved to be from about 60% to 70% accurate. What stands in the way of more accurate prediction? Note that the conformational preferences of amino acid residues are not tipped all the way to one structure (see Table 2.3). For example, glutamate, one of the strongest helix formers, prefers α helix to β strand by only a factor of two. The preference ratios of most other residues are smaller. Indeed, some penta- and hexapeptide sequences have been found to adopt one structure in one protein and an entirely different structure in another (Figure 2.60). Hence, some amino acid sequences do not uniquely determine secondary structure. Tertiary interactions—interactions between residues that are far apart in the sequence—may be decisive in specifying the secondary structure of some segments. The context is often crucial in determining the conformational outcome. The conformation of a protein evolved to work in a particular environment or context. Substantial improvements in secondary structure prediction can be achieved by using families of related sequences, each of which adopts the same structure.

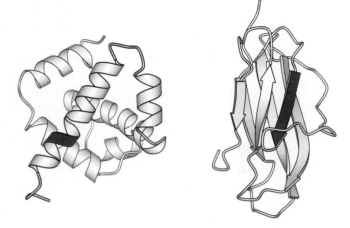

Figure 2.60 Alternative conformations of a peptide sequence. Many sequences can adopt alternative conformations in different proteins. Here the sequence VDLLKN shown in red assumes an α helix in one protein context (left) and a β strand in another (right). [Drawn from (left) 3WRP.pdb and (right) 2HLA.pdb.]

Protein Misfolding and Aggregation Are Associated with Some Neurological Diseases

Until quite recently, all infectious diseases were believed to be transmitted by either viruses or bacteria. In one of the great surprises in modern medicine, certain infectious neurological diseases were found to be transmitted by agents that were similar in size to viruses but consisted only of protein. These diseases include *bovine spongiform encephalopathy* (commonly referred to as *mad cow disease*) and the analogous diseases in other organisms, including *Creutzfeldt-Jakob disease* (CJD) in human beings and *scrapie* in sheep. The agents causing these diseases are termed *prions*. The leading proponent of the hypothesis that diseases can be transmitted purely by proteins, Stanley Prusiner, was awarded the Nobel Prize in physiology or medicine in 1997.

Examination of these infectious agents revealed the following characteristics:

1.   The transmissible agent consists of aggregated forms of a specific protein. The aggregates have a range of molecular weights.

2.   The protein aggregates are resistant to treatment with agents that degrade most proteins.

3.   The protein is largely or completely derived from a cellular protein, called PrP, that is normally present in the brain.

How does the structure of the protein in the aggregated form differ from that of the protein in its normal state in the brain? The structure of the normal cellular protein PrP contains extensive regions of α helix and relatively little β-strand structure. The structure of the form of the protein present in infected brains, termed PrP^{SC}, has not yet been determined because of challenges posed by its insoluble and heterogeneous nature. However, a variety of evidence indicates that some parts of the protein that had been in α-helical or turn conformations have been converted into β-strand conformations. The β strands of one protein link with those of another to form β sheets joining the two proteins and leading to the formation of aggregates. These fibrous protein aggregates are often referred to as *amyloid* forms.

With the realization that the infectious agent in prion diseases is an aggregated form of a protein that is already present in the brain, a model for disease transmission emerges (Figure 2.61). Protein aggregates built of abnormal forms of PrP act as nuclei to which other PrP molecules attach. Prion diseases can thus be transferred from one individual organism to another through the transfer of an aggregated nucleus, as likely happened in the mad cow disease outbreak in the United Kingdom in the 1990s. Cattle fed on animal feed containing material from diseased cows developed the disease in turn.

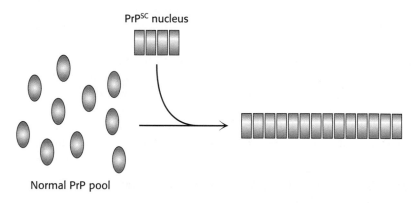

PrP^{SC} nucleus

Normal PrP pool

Figure 2.61 The protein-only model for prion-disease transmission. A nucleus consisting of proteins in an abnormal conformation grows by the addition of proteins from the normal pool.

Amyloid fibers are also seen in the brains of patients with certain noninfectious neurodegenerative diseases such as Alzheimer and Parkinson diseases. For example, the brains of patients with Alzheimer disease contain protein aggregates called *amyloid plaques* that consist primarily of a single polypeptide termed Aβ. This polypeptide is derived from a cellular protein *amyloid precursor protein,* or APP, through the action of specific proteases. Polypeptide Aβ is prone to form insoluble aggregates. Despite the difficulties posed by the protein's insolubility, a detailed structural model for Aβ has been derived through the use of NMR (nuclear magnetic resonance) techniques that can be applied to solids rather than materials in solution. As expected, the structure is rich in β strands, which come together to form extended parallel β-sheet structures (Figure 2.62).

Figure 2.62 A structure of amyloid fibers. A detailed model for Aβ fibrils deduced from solid-state NMR studies shows that protein aggregation is due to the formation of large parallel β sheets. [From A. T. Petkova, Y. Ishii, J. J. Balbach, O. N. Antzukin, R. D. Leapman, F. Delagio, and R. Tycko, *Proc, Natl. Acad. Sci. U.S.A.* 99(2002): 16742–16747.]

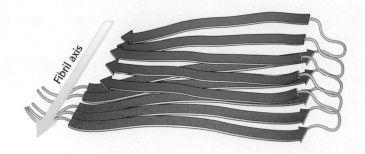

Fibril axis

How do such aggregates lead to the death of the cells that harbor them? The answer is still controversial. One hypothesis is that the large aggregates themselves are not toxic, but instead smaller aggregates of the same proteins may be the culprits, perhaps damaging cell membranes.

Protein Folding Is a Highly Cooperative Process

As stated earlier, proteins can be denatured by heat or by chemical denaturants such as urea or guanidinium chloride. For many proteins, a comparison of the degree of unfolding as the concentration of denaturant increases reveals a relatively sharp transition from the folded, or native, form to the unfolded, or denatured form, suggesting that only these two conformational states are present to any significant extent (Figure 2.63). A similar sharp transition is observed if denaturants are removed from unfolded proteins, allowing the proteins to fold.

The sharp transition seen in Figure 2.63 suggests that protein folding and unfolding is an *"all or none" process* that results from a *cooperative transition*. For example, suppose that a protein is placed in conditions under which some part of the protein structure is thermodynamically unstable. As this part of the folded structure is disrupted, the interactions between it and the remainder of the protein will be lost. The loss of these interactions, in turn, will destabilize the remainder of the structure. Thus, conditions that lead to the disruption of any part of a protein structure are likely to unravel the protein completely. The structural properties of proteins provide a clear rationale for the cooperative transition.

The consequences of cooperative folding can be illustrated by considering the contents of a protein solution under conditions corresponding to the middle of the transition between the folded and the unfolded forms. Under these conditions, the protein is "half folded." Yet the solution will contain no half-folded molecules but, instead, will be a 50/50 mixture of fully folded and fully unfolded molecules (Figure 2.64). Although the protein may appear to behave as if it exists in only two states, at an atomic level, this simple two-state existence is an impossibility. Unstable, transient intermediate structures must exist between the native and denatured state (p. 56). Determining the nature of these intermediate structures is an intense area of biochemical research.

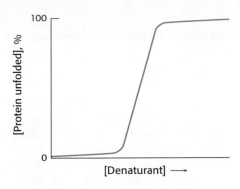

Figure 2.63 Transition from folded to unfolded state. Most proteins show a sharp transition from the folded to the unfolded form on treatment with increasing concentrations of denaturants.

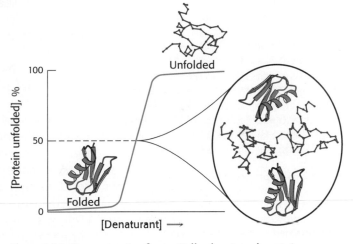

Figure 2.64 Components of a partially denatured protein solution. In a half-unfolded protein solution, half the molecules are fully folded and half are fully unfolded.

Proteins Fold by Progressive Stabilization of Intermediates Rather Than by Random Search

How does a protein make the transition from an unfolded structure to a unique conformation in the native form? One possibility a priori would be that all possible conformations are tried out to find the energetically most favorable one. How long would such a random search take? Consider a small protein with 100 residues. Cyrus Levinthal calculated that, if each residue can assume three different conformations, the total number of structures would be 3^{100}, which is equal to 5×10^{47}. If it takes 10^{-13} s to convert one structure into another, the total search time would be $5 \times 10^{47} \times 10^{-13}$ s, which is equal to 5×10^{34} s, or 1.6×10^{27} years. Clearly, it would take much too long for even a small protein to fold properly by randomly trying out all possible conformations. The enormous difference between calculated and actual folding times is called *Levinthal's paradox*. This paradox clearly reveals that proteins do not fold by trying every possible

```
 200  ?T(\G{+s x[A.N5~,#ATxSGpn`e□@
 400  oDr'Jh7s DFR:W4l'u+^v6zpJseOi
 600  e2ih'8zs n527x8l8d_ih=Hldseb.
 800  S#dh>}/s ]tZqC%lP%DK<|!^aseZ.
1000  VOth>nLs ut/isjl_kwojjwMasef.
1200  juth+nvs it is[lukh?SCw=ase5.
1400  Iithdn4s it isOl/ks/IxwLase~.
1600  M?thinrs it is lXk?T"_woasel.
1800  MSthinWs it is lwkN7□Kw(asel.
2000  Mhthin`s it is likv,aww_asel.
2200  MMthinns it is lik+5avwlasel.
2400  MethinXs it is likydaqw)asel.
2600  Methin4s it is lik2dasweasel.
2800  MethinHs it is like□aTweasel.
2883  Methinks it is like a weasel.
```

```
 200  )z~hg)W4{{cu!kO{d6jS!NlEyUx}p
 400  "W hi\kR.<&CfA%4-Y1G!iT$6({|6
 600  .L=hinkm4(uMGP^lAWoE6klwW=yiS
 800   AthinkaPa_vYH liR\Hb,Uo4\-"(
1000  OFthinksP)@fZO li8v] /+Eln26B
1200  6ithinksMVt -V likm+g1#K~}BFk
1400  vxthinksaEt □w like.SlGeutks.
1600  :Othinks<it MC likesN2[eaVe4.
1800  uxthinksqit Or likeQh)weaoeW.
2000  Y/thinks it id like7alwea)e&.
2200  Methinks it iW like a[weaWel.
2400  Methinks it is like a;weasel.
2431  Methinks it is like a weasel.
```

Figure 2.65 Typing-monkey analogy. A
monkey randomly poking a typewriter
could write a line from Shakespeare's
Hamlet, provided that correct keystrokes
were retained. In the two computer
simulations shown, the cumulative number
of keystrokes is given at the left of each
line.

conformation; instead, they must follow at least a partly defined folding
pathway consisting of intermediates between the fully denatured protein
and its native structure.

The way out of this paradox is to recognize the power of *cumulative selection*. Richard Dawkins, in *The Blind Watchmaker*, asked how long it
would take a monkey poking randomly at a typewriter to reproduce
Hamlet's remark to Polonius, "Methinks it is like a weasel" (Figure 2.65).
An astronomically large number of keystrokes, of the order of 10^{40}, would
be required. However, suppose that we preserved each correct character and
allowed the monkey to retype only the wrong ones. In this case, only a few
thousand keystrokes, on average, would be needed. The crucial difference
between these cases is that the first employs a completely random search,
whereas, in the second, *partly correct intermediates are retained*.

*The essence of protein folding is the tendency to retain partly correct inter-
mediates.* However, the protein-folding problem is much more difficult than
the one presented to our simian Shakespeare. First, the criterion of correct-
ness is not a residue-by-residue scrutiny of conformation by an omniscient
observer but rather the total free energy of the transient species. Second,
proteins are only marginally stable. The free-energy difference between the
folded and the unfolded states of a typical 100-residue protein is 42 kJ mol^{-1}
(10 kcal mol^{-1}), and thus each residue contributes on average only 0.42 kJ
mol^{-1} (0.1 kcal mol^{-1}) of energy to maintain the folded state. This amount
is less than the amount of thermal energy, which is 2.5 kJ mol^{-1} (0.6 kcal
mol^{-1}) at room temperature. This meager stabilization energy means that
correct intermediates, especially those formed early in folding, can be lost.
The analogy is that the monkey would be somewhat free to undo its correct
keystrokes. Nonetheless, the interactions that lead to cooperative folding
can stabilize intermediates as structure builds up. Thus, local regions that
have significant structural preference, though not necessarily stable on their
own, will tend to adopt their favored structures and, as they form, can interact with one other, leading to increasing stabilization. This conceptual
framework is often referred to as the *nucleation-condensation model*.

A simulation of the folding of a protein, based on the nucleation-
condensation model, is shown in Figure 2.66. This model suggests that
certain pathways may be preferred. Although Figure 2.66 suggests a discrete pathway, each of the intermediates shown represents an ensemble of
similar structures, and thus a protein follows a general rather than a precise
pathway in its transition from the unfolded to the native state. The energy
surface for the overall process of protein folding can be visualized as a funnel. The wide rim of the funnel represents the wide range of structures accessible to the ensemble of denatured protein molecules. As the free energy

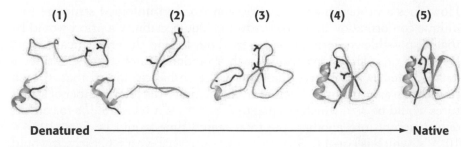

Figure 2.66 Proposed folding pathway of chymotrypsin inhibitor. Local regions with
sufficient structural preference tend to adopt their favored structures initially (1). These
structures come together to form a nucleus with a nativelike, but still mobile, structure (4).
This structure then fully condenses to form the native, more rigid structure (5). [From A. R.
Fersht and V. Daggett. *Cell* 108 (2002):573–582; with permission from Elsevier.]

of the population of protein molecules decreases, the proteins move down into narrower parts of the funnel and fewer conformations are accessible. At the bottom of the funnel is the folded state with its well-defined conformation. Many paths can lead to this same energy minimum.

Prediction of Three-Dimensional Structure from Sequence Remains a Great Challenge

The prediction of three-dimensional structure from sequence has proved to be extremely difficult. As we have seen, the local sequence appears to determine only between 60 and 70% of the secondary structure; long-range interactions are required to fix the full secondary structure and the tertiary structure.

Investigators are exploring two fundamentally different approaches to predicting three-dimensional structure from amino acid sequence. The first is *ab initio* (Latin, "from the beginning") *prediction,* which attempts to predict the folding of an amino acid sequence without prior knowledge about similar sequences in known protein structures. Computer-based calculations are employed that attempt to minimize the free energy of a structure with a given amino acid sequence or to simulate the folding process. The utility of these methods is limited by the vast number of possible conformations, the marginal stability of proteins, and the subtle energetics of weak interactions in aqueous solution. The second approach takes advantage of our growing knowledge of the three-dimensional structures of many proteins. In these *knowledge-based methods,* an amino acid sequence of unknown structure is examined for compatibility with known protein structures or fragments therefrom. If a significant match is detected, the known structure can be used as an initial model. Knowledge-based methods have been a source of many insights into the three-dimensional conformation of proteins of known sequence but unknown structure.

Protein Modification and Cleavage Confer New Capabilities

Proteins are able to perform numerous functions relying solely on the versatility of their 20 amino acids. In addition, many proteins are covalently modified, through the attachment of groups other than amino acids, to augment their functions (Figure 2.67). For example, *acetyl groups* are attached to the amino termini of many proteins, a modification that makes these proteins more resistant to degradation. As discussed earlier (p. 46), the addition of *hydroxyl groups* to many proline residues stabilizes fibers of newly synthesized collagen. The biological significance of this modification is evident in the disease scurvy: a deficiency of vitamin C

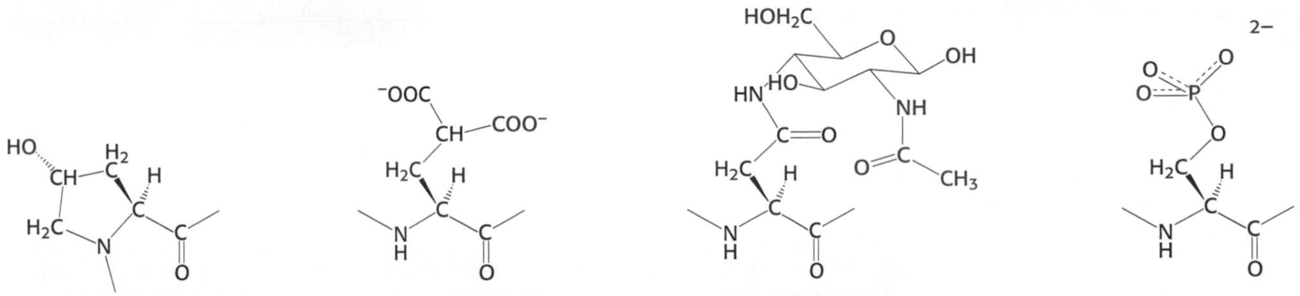

Hydroxyproline   **γ-Carboxyglutamate**   **Carbohydrate–asparagine adduct**   **Phosphoserine**

Figure 2.67 Finishing touches. Some common and important covalent modifications of amino acid side chains are shown.

results in insufficient hydroxylation of collagen, and the abnormal collagen fibers that result are unable to maintain normal tissue strength (Section 27.5). Another specialized amino acid produced by a finishing touch is *γ-carboxyglutamate*. In vitamin K deficiency, insufficient carboxylation of glutamate in prothrombin, a clotting protein, can lead to hemorrhage (p. 295). Many proteins, especially those that are present on the surfaces of cells or are secreted, acquire *carbohydrate units* on specific asparagine residues (see Chapter 11). The addition of sugars makes the proteins more hydrophilic and able to participate in interactions with other proteins. Conversely, the addition of a *fatty acid* to an α-amino group or a cysteine sulfhydryl group produces a more hydrophobic protein.

Many hormones, such as epinephrine (adrenaline), alter the activities of enzymes by stimulating the phosphorylation of the hydroxyl amino acids serine and threonine; *phosphoserine* and *phosphothreonine* are the most ubiquitous modified amino acids in proteins. Growth factors such as insulin act by triggering the phosphorylation of the hydroxyl group of tyrosine residues to form *phosphotyrosine*. The phosphoryl groups on these three modified amino acids are readily removed; thus they are able to act as reversible switches in regulating cellular processes. The roles of phosphorylation in signal transduction will be discussed extensively in Chapter 14.

The preceding modifications consist of the addition of special groups to amino acids. Other special groups are generated by chemical rearrangements of side chains and, sometimes, the peptide backbone. For example, certain jellyfish produce a green fluorescent protein (Figure 2.68). The

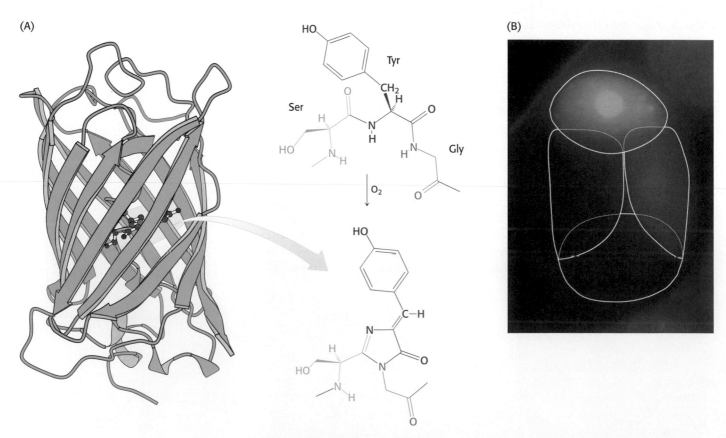

Figure 2.68 Chemical rearrangement in GFP. (A) The structure of green fluorescent protein (GFP). The rearrangement and oxidation of the sequence Ser-Tyr-Gly is the source of fluorescence. (B) Fluorescence micrograph of a four-cell embryo (cells are outlined) from the roundworm *Caenorhabditis elegans* containing a protein, PIE-1, labeled with GFP. The protein is expressed only in the cell (top) that will give rise to the germ line. [(A) Drawn from 1GFL.pdb; (B) courtesy of Dr. Geraldine Seydoux.]

source of the fluorescence is a group formed by the spontaneous rearrangement and oxidation of the sequence Ser-Tyr-Gly within the center of the protein. This protein is of great utility to researchers as a marker within cells (p. 89).

Finally, many proteins are cleaved and trimmed after synthesis. For example, digestive enzymes are synthesized as inactive precursors that can be stored safely in the pancreas. After release into the intestine, these precursors become activated by peptide-bond cleavage (p. 289). In blood clotting, peptide-bond cleavage converts soluble fibrinogen into insoluble fibrin (p. 293). A number of polypeptide hormones, such as adrenocorticotropic hormone, arise from the splitting of a single large precursor protein. Likewise, many viral proteins are produced by the cleavage of large polyprotein precursors. We shall encounter many more examples of modification and cleavage as essential features of protein formation and function. Indeed, these finishing touches account for much of the versatility, precision, and elegance of protein action and regulation.

Summary

Protein structure can be described at four levels. The primary structure refers to the amino acid sequence. The secondary structure refers to the conformation adopted by local regions of the polypeptide chain. Tertiary structure describes the overall folding of the polypeptide chain. Finally, quaternary structure refers to the specific association of multiple polypeptide chains to form multisubunit complexes.

2.1 Proteins Are Built from a Repertoire of 20 Amino Acids

Proteins are linear polymers of amino acids. Each amino acid consists of a central tetrahedral carbon atom linked to an amino group, a carboxylic acid group, a distinctive side chain, and a hydrogen atom. These tetrahedral centers, with the exception of that of glycine, are chiral; only the L isomer exists in natural proteins. All natural proteins are constructed from the same set of 20 amino acids. The side chains of these 20 building blocks vary tremendously in size, shape, and the presence of functional groups. They can be grouped as follows: (1) aliphatic side chains—glycine, alanine, valine, leucine, isoleucine, methionine, and proline; (2) aromatic side chains—phenylalanine, tyrosine, and tryptophan; (3) hydroxyl-containing aliphatic side chains—serine and threonine; (4) sulfhydryl-containing cysteine; (5) carboxamide-containing side chains—asparagine and glutamine; (6) basic side chains—lysine, arginine, and histidine; and (7) acidic side chains—aspartic acid and glutamic acid. These groupings are somewhat arbitrary and many other sensible groupings are possible.

2.2 Primary Structure: Amino Acids Are Linked by Peptide Bonds to Form Polypeptide Chains

The amino acids in a polypeptide are linked by amide bonds formed between the carboxyl group of one amino acid and the amino group of the next. This linkage, called a peptide bond, has several important properties. First, it is resistant to hydrolysis, and so proteins are remarkably stable kinetically. Second, the peptide group is planar because the C—N bond has considerable double-bond character. Third, each peptide bond has both a hydrogen-bond donor (the NH group) and a hydrogen-bond acceptor (the CO group). Hydrogen bonding between these backbone groups is a distinctive feature of protein structure. Finally, the peptide bond is uncharged, which allows proteins to form tightly packed globular structures having significant amounts of the backbone buried within the protein interior. Because they are linear polymers, proteins can be described as sequences of amino acids. Such sequences are written from the amino to the carboxyl terminus.

2.3 Secondary Structure: Polypeptide Chains Can Fold into Regular Structures Such As the Alpha Helix, the Beta Sheet, and Turns and Loops

Two major elements of secondary structure are the α helix and the β strand. In the α helix, the polypeptide chain twists into a tightly packed rod. Within the helix, the CO group of each amino acid is hydrogen bonded to the NH group of the amino acid four residues farther along the polypeptide chain. In the β strand, the polypeptide chain is nearly fully extended. Two or more β strands connected by NH-to-CO hydrogen bonds come together to form β sheets. The strands in β sheets can be antiparallel, parallel, or mixed.

2.4 Tertiary Structure: Water-Soluble Proteins Fold into Compact Structures with Nonpolar Cores

The compact, asymmetric structure that individual polypeptides attain is called tertiary structure. The tertiary structures of water-soluble proteins have features in common: (1) an interior formed of amino acids with hydrophobic side chains and (2) a surface formed largely of hydrophilic amino acids that interact with the aqueous environment. The driving force for the formation of the tertiary structure of water-soluble proteins is the hydrophobic interactions between the interior residues. Some proteins that exist in a hydrophobic environment, in membranes, display the inverse distribution of hydrophobic and hydrophilic amino acids. In these proteins, the hydrophobic amino acids are on the surface to interact with the environment, whereas the hydrophilic groups are shielded from the environment in the interior of the protein.

2.5 Quaternary Structure: Polypeptide Chains Can Assemble into Multisubunit Structures

Proteins consisting of more than one polypeptide chain display quaternary structure; each individual polypeptide chain is called a subunit. Quaternary structure can be as simple as two identical subunits or as complex as dozens of different subunits. In most cases, the subunits are held together by noncovalent bonds.

2.6 The Amino Acid Sequence of a Protein Determines Its Three-Dimensional Structure

The amino acid sequence completely determines the three-dimensional structure and, hence, all other properties of a protein. Some proteins can be unfolded completely yet refold efficiently when placed under conditions in which the folded form of the protein is stable. The amino acid sequence of a protein is determined by the sequences of bases in a DNA molecule. This one-dimensional sequence information is extended into the three-dimensional world by the ability of proteins to fold spontaneously. Protein folding is a highly cooperative process; structural intermediates between the unfolded and folded forms do not accumulate.

The versatility of proteins is further enhanced by covalent modifications. Such modifications can incorporate functional groups not present in the 20 amino acids. Other modifications are important to the regulation of protein activity. Through their structural stability, diversity, and chemical reactivity, proteins make possible most of the key processes associated with life.

APPENDIX: Visualizing Molecular Structures II: Proteins

Scientists have developed powerful techniques for the determination of protein structures, as will be considered in Chapter 3. In most cases, these techniques allow the positions of the thousands of atoms within a protein structure to be determined. The final results from such an experiment include the x, y, and z coordinates for each atom in the structure. These coordinate files are compiled in the Protein Data Bank (http://www. rcsb.org/pdb/) from which they can be readily downloaded. These structures comprise thousands or even tens of thousands of atoms. The complexity of proteins with thousands of atoms presents a challenge for the depiction of their structure. Several different types of representations are used to portray proteins, each with its own strengths and weaknesses. The types that you will see most often in this book are space-filling models, ball-and-stick models, backbone models, and ribbon diagrams. Where appropriate, we note structural features of particular importance or relevance in an illustration's legend.

Space-Filling Models

Space-filling models are the most realistic type of representation. Each atom is shown as a sphere with a size corresponding to the van der Waals radius of the atom (p. 8). Bonds are not shown explicitly but are represented by the intersection of the spheres shown when atoms are closer together than the sum of their van der Waals radii. All atoms are shown, including those that make up the backbone and those in the side chains. A space-filling model of lysozyme is depicted in Figure 2.69.

Space-filling models convey a sense of how little open space there is in a protein's structure, which always has many atoms in van der Waals contact with one another. These models are particularly useful in showing conformational changes in a protein from one set of circumstances to another. A disadvantage of space-filling models is that the secondary and tertiary structures of the protein are difficult to see. Thus, these models are not very effective in distinguishing one protein from another—many space-filling models of proteins look very much alike.

Ball-and-Stick Models

Ball-and-stick models are not as realistic as space-filling models. Realistically portrayed atoms occupy more space, determined by their van der Waals radii, than do the atoms depicted in ball-and-stick models. However, the bonding arrangement is easier to see because the bonds are explicitly represented as sticks (Figure 2.70). A ball-and-stick model reveals a complex structure more clearly than a space-filling model does.

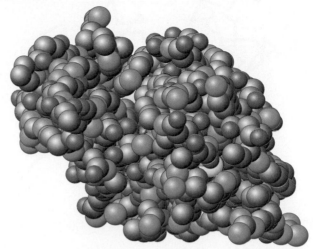

Figure 2.69 Space-filling model of lysozyme. *Notice* how tightly packed the atoms are, with little unfilled space. All atoms are shown with the exception of hydrogen atoms. Hydrogen atoms are often omitted because their positions are not readily determined by x-ray crystallographic methods and because their omission somewhat improves the clarity of the structure's depiction.

However, the depiction is so complicated that structural features such as α helices or potential binding sites are difficult to discern.

Because space-filling and ball-and-stick models depict protein structures at the atomic level, the large number of atoms in a complex structure makes it difficult to discern the relevant structural features. Thus, representations that are more schematic—such as backbone models and ribbon diagrams—have been developed for the depiction of macromolecular structures. In these representations, most or all atoms are not shown explicitly.

Figure 2.70 Ball-and-stick model of lysozyme. Again, hydrogen atoms are omitted.

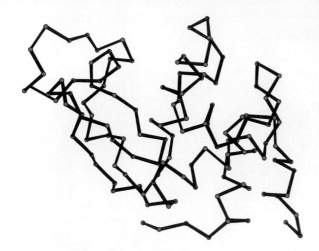

Figure 2.71 Backbone model of lysozyme.

Backbone Models

Backbone models show only the backbone atoms of a molecule's polypeptide or even only the α-carbon atom of each amino acid. Atoms are linked by lines representing bonds; if only α-carbon atoms are depicted, lines connect α-carbon atoms of amino acids that are adjacent in the amino acid sequence (Figure 2.71). In this book, backbone models show only the lines connecting the α-carbon atoms; other carbon atoms are not depicted.

A backbone model shows the overall course of the polypeptide chain much better than a space-filling or ball-and-stick model does. However, secondary structural elements are still difficult to see.

Ribbon Diagrams

Ribbon diagrams are highly schematic and most commonly used to accent a few dramatic aspects of protein structure, such as the α helix (depicted as a coiled ribbon or a cylinder), the β strand (a broad arrow), and loops (simple lines), to provide clear views of the folding patterns of proteins (Figure 2.72). The ribbon diagram allows the course of a polypeptide chain to be traced and readily shows the secondary structural elements. Thus, ribbon diagrams of proteins that are related to one another by evolutionary divergence appear similar (see Figure 6.14), whereas unrelated proteins are clearly distinct.

In this book, coiled ribbons will be generally used to depict α helices. However, for membrane proteins, which are often quite complex, cylinders will be used rather than coiled ribbons. This convention will also make membrane proteins with their membrane-spanning α helices easy to recognize (see Figure 12.18).

Bear in mind that the open appearance of ribbon diagrams is deceptive. As noted earlier, protein structures are tightly packed and have little open space. The openness of ribbon diagrams makes them particularly useful as frameworks in which to highlight additional aspects of protein structure. Active sites, substrates, bonds, and other structural fragments can be included in ball-and-stick or space-filling form within a ribbon diagram (Figure 2.73).

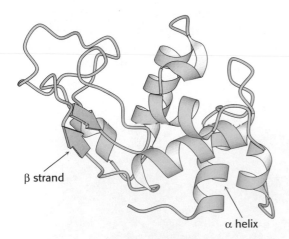

Figure 2.72 Ribbon diagram of lysozyme. The α helices are shown as coiled ribbons; β strands are depicted as arrows. More irregular structures are shown as thin tubes.

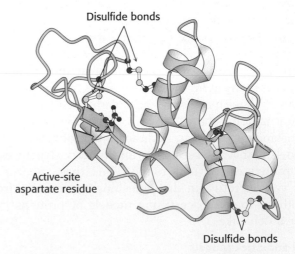

Figure 2.73 Ribbon diagram of lysozyme with highlights. Four disulfide bonds and a functionally important aspartate residue are shown in ball-and-stick form.

Key Terms

side chain (R group) (p. 27)

L amino acid (p. 27)

dipolar ion (zwitterion) (p. 27)

peptide bond (amide bond) (p. 34)

disulfide bond (p. 35)

primary structure (p. 36)

torsion angle (p. 39)

phi (ϕ) angle (p. 39)

psi (ψ) angle (p. 39)

Ramachandran diagram (p. 39)

secondary structure (p. 40)

α helix (p. 40)

rise (p. 40)

translation (p. 40)

β pleated sheet (p. 42)

β strand (p. 42)

reverse turn (β turn; hairpin turn) (p. 44)

coiled coil (p. 45)

heptad repeats (p. 45)

tertiary structure (p. 47)

motif (p. 49)

supersecondary structure (p. 49)

domain (p. 49)

subunit (p. 49)

quaternary structure (p. 49)

prion (p. 53)

cooperative transition (p. 55)

Selected Readings

Where to Start

Richardson, J. S. 1981. The anatomy and taxonomy of protein structure. *Adv. Protein Chem.* 34:167–339.

Doolittle, R. F. 1985. Proteins. *Sci. Am.* 253(4):88–99.

Richards, F. M. 1991. The protein folding problem. *Sci. Am.* 264(1): 54–57.

Weber, A. L., and Miller, S. L. 1981. Reasons for the occurrence of the twenty coded protein amino acids. *J. Mol. Evol.* 17:273–284.

Books

Petsko, G.A., and Ringe, D. 2004. *Protein Structure and Function.* New Science Press.

Branden, C., and Tooze, J. 1999. *Introduction to Protein Structure* (2d ed.). Garland.

Perutz, M. F. 1992. *Protein Structure: New Approaches to Disease and Therapy.* W. H. Freeman and Company.

Creighton, T. E. 1992. *Proteins: Structures and Molecular Principles* (2d ed.). W. H. Freeman and Company.

Conformation of Proteins

Dobson, C.M. 2003. Protein folding and misfolding. *Nature* 426: 884–890.

Richardson, J. S., Richardson, D. C., Tweedy, N. B., Gernert, K. M., Quinn, T. P., Hecht, M. H., Erickson, B. W., Yan, Y., McClain, R. D., Donlan, M. E., and Suries, M. C. 1992. Looking at proteins: Representations, folding, packing, and design. *Biophys. J.* 63:1186–1220.

Chothia, C., and Finkelstein, A. V. 1990. The classification and origin of protein folding patterns. *Annu. Rev. Biochem.* 59:1007–1039.

Alpha Helices, Beta Sheets, and Loops

O'Neil, K. T., and DeGrado, W. F. 1990. A thermodynamic scale for the helix-forming tendencies of the commonly occurring amino acids. *Science* 250:646–651.

Zhang, C., and Kim, S. H. 2000. The anatomy of protein beta-sheet topology. *J. Mol. Biol.* 299:1075–1089.

Regan, L. 1994. Protein structure: Born to be beta. *Curr. Biol.* 4:656–658.

Leszczynski, J. F., and Rose, G. D. 1986. Loops in globular proteins: A novel category of secondary structure. *Science* 234:849–855.

Srinivasan, R., and Rose, G. D. 1999. A physical basis for protein secondary structure. *Proc. Natl. Acad. Sci. U.S.A.* 96:14258–14263.

Domains

Bennett, M. J., Choe, S., and Eisenberg, D. 1994. Domain swapping: Entangling alliances between proteins. *Proc. Natl. Acad. Sci. U.S.A.* 91:3127–3131.

Bergdoll, M., Eltis, L. D., Cameron, A. D., Dumas, P., and Bolin, J. T. 1998. All in the family: Structural and evolutionary relationships among three modular proteins with diverse functions and variable assembly. *Protein Sci.* 7:1661–1670.

Hopfner, K. P., Kopetzki, E., Kresse, G. B., Bode, W., Huber, R., and Engh, R. A. 1998. New enzyme lineages by subdomain shuffling. *Proc. Natl. Acad. Sci. U.S.A* 95:9813–9818.

Ponting, C. P., Schultz, J., Copley, R. R., Andrade, M. A., and Bork, P. 2000. Evolution of domain families. *Adv. Protein Chem.* 54:185–244.

Protein Folding

Caughey, B., and Lansbury, P. T. 2003. Protofibrils, pores, fibrits, and neurodegeneration: Separating the responsible protein aggregates from innocent bystanders. *Annu. Rev. Neurosci.* 26: 267–298.

Daggett, V., and Fersht, A. R. 2003. Is there a unifying mechanism for protein folding? *Trends Biochem. Sci.* 28:18–25.

Selkoe, D. J. 2003. Folding proteins in fatal ways. *Nature* 426:900–904.

Anfinsen, C. B. 1973. Principles that govern the folding of protein chains. *Science* 181:223–230.

Baldwin, R. L., and Rose, G. D. 1999. Is protein folding hierarchic? I. Local structure and peptide folding. *Trends Biochem. Sci.* 24:26–33.

Baldwin, R. L., and Rose, G. D. 1999. Is protein folding hierarchic? II. Folding intermediates and transition states. *Trends Biochem. Sci.* 24:77–83.

Kuhlman, B., Dantas, G., Ireton, G. C., Varani, G., Stoddard, B. L., and Baker, D. 2003. Design of a novel globular protein with atomic-level accuracy. *Science* 302: 1364–1368.

Staley, J. P., and Kim, P. S. 1990. Role of a subdomain in the folding of bovine pancreatic trypsin inhibitor. *Nature* 344:685–688.

Covalent Modification of Proteins

Krishna, R. G., and Wold, F. 1993. Post-translational modification of proteins. *Adv. Enzymol. Relat. Areas. Mol. Biol.* 67:265–298.

Aletta, J. M., Cimato, T. R., and Ettinger, M. J. 1998. Protein methylation: A signal event in post-translational modification. *Trends Biochem. Sci.* 23:89–91.

Glazer, A. N., DeLange, R. J., and Sigman, D. S. 1975. *Chemical Modification of Proteins.* North-Holland.

Tsien, R. Y. 1998. The green fluorescent protein. *Annu. Rev. Biochem.* 67:509–544.

Molecular Graphics

Kraulis, P. 1991. MOLSCRIPT: A program to produce both detailed and schematic plots of protein structures. *J. Appl. Cryst.* 24:946–950.

Ferrin, T., Huang, C., Jarvis, L., and Langridge, R. 1988. The MIDAS display system. *J. Mol. Graphics* 6:13–27.

Richardson, D. C., and Richardson, J. S. 1994. Kinemages: Simple macromolecular graphics for interactive teaching and publication. *Trends Biochem. Sci.* 19:135–138.

Problems

1. *Shape and dimension.* (a) Tropomyosin, a 70-kd muscle protein, is a two-stranded α-helical coiled coil. Estimate the length of the molecule. (b) Suppose that a 40-residue segment of a protein folds into a two-stranded antiparallel β structure with a 4-residue hairpin turn. What is the longest dimension of this motif?

2. *Contrasting isomers*. Poly-L-leucine in an organic solvent such as dioxane is α helical, whereas poly-L-isoleucine is not. Why do these amino acids with the same number and kinds of atoms have different helix-forming tendencies?

3. *Active again*. A mutation that changes an alanine residue in the interior of a protein to valine is found to lead to a loss of activity. However, activity is regained when a second mutation at a different position changes an isoleucine residue to glycine. How might this second mutation lead to a restoration of activity?

4. *Shuffle test*. An enzyme that catalyzes disulfide–sulfhydryl exchange reactions, called protein disulfide isomerase (PDI), has been isolated. PDI rapidly converts inactive scrambled ribonuclease into enzymatically active ribonuclease. In contrast, insulin is rapidly inactivated by PDI. What does this important observation imply about the relation between the amino acid sequence of insulin and its three-dimensional structure?

5. *Stretching a target*. A protease is an enzyme that catalyzes the hydrolysis of the peptide bonds of target proteins. How might a protease bind a target protein so that its main chain becomes fully extended in the vicinity of the vulnerable peptide bond?

6. *Often irreplaceable*. Glycine is a highly conserved amino acid residue in the evolution of proteins. Why?

7. *Potential partners*. Identify the groups in a protein that can form hydrogen bonds or electrostatic bonds with an arginine side chain at pH 7.

8. *Permanent waves*. The shape of hair is determined in part by the pattern of disulfide bonds in keratin, its major protein. How can curls be induced?

9. *Location is everything*. Proteins that span biological membranes often contain α helices. Given that the insides of membranes are highly hydrophobic (Section 12.2), predict what type of amino acids would be in such a helix. Why is an α helix particularly suited to exist in the hydrophobic environment of the interior of a membrane?

10. *Issues of stability*. Proteins are quite stable. The lifetime of a peptide bond in aqueous solution is nearly 1000 years. However, the free energy of hydrolysis of proteins is negative and quite large. How can you account for the stability of the peptide bond in light of the fact that hydrolysis releases much energy?

11. *Minor species*. For an amino acid such as alanine, the major species in solution at pH 7 is the zwitterionic form. Assume a pK_a value of 8 for the amino group and a pK_a value of 3 for the carboxylic acid. Estimate the ratio of the concentration of the neutral amino acid species (with the carboxylic acid protonated and the amino group neutral) to that of the zwitterionic species at pH 7 (see p. 16).

12. *A matter of convention*. All L amino acids have an *S* absolute configuration except L-cysteine, which has the *R* configuration. Explain why L-cysteine is designated as having the *R* absolute configuration.

13. *Hidden message*. Translate the following amino acid sequence into one-letter code: Glu-Leu-Val-Ile-Ser-Ile-Ser-Leu-Ile-Val-Ile-Asn-Gly-Ile-Asn-Leu-Ala-Ser-Val-Glu-Gly-Ala-Ser.

14. *Who goes first?* Would you expect Pro–X peptide bonds to tend to have cis conformations like those of X–Pro bonds? Why or why not?

15. *Matching*. For each of the amino acid derivatives shown here (A–E), find the matching set of ϕ and ψ values (a–e).

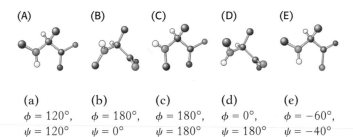

(A)	(B)	(C)	(D)	(E)

(a)	(b)	(c)	(d)	(e)
$\phi = 120°$,	$\phi = 180°$,	$\phi = 180°$,	$\phi = 0°$,	$\phi = -60°$,
$\psi = 120°$	$\psi = 0°$	$\psi = 180°$	$\psi = 180°$	$\psi = -40°$

Chapter 3

Exploring Proteins and Proteomes

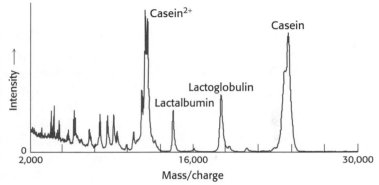

Milk, a source of nourishment for all mammals, is composed, in part, of a variety of proteins. The protein components of milk are revealed by the technique of MALDI–TOF mass spectrometry, which separates molecules on the basis of their mass to charge ratio. [(Left) Jean Paul Iris/FPG. (Right) Courtesy of Dr. Brian Chait.]

Proteins play crucial roles in nearly all biological processes—in catalysis, signal transmission, and structural support. This remarkable range of functions arises from the existence of thousands of proteins, each folded into a distinctive three-dimensional structure that enables it to interact with one or more of a highly diverse array of molecules. A major goal of biochemistry is to determine how amino acid sequences specify the conformations, and hence functions, of proteins. Other goals are to learn how individual proteins bind specific substrates and other molecules, mediate catalysis, and transduce energy and information.

Often a first step in these studies is the purification of the protein of interest. Proteins can be separated from one another on the basis of solubility, size, charge, and binding ability. After a protein has been purified, its amino acid sequence can be determined. Automated peptide sequencing and the application of recombinant DNA methods are providing a wealth of amino acid sequence data that are opening new vistas. Many protein sequences, often deduced from genome sequences, are now available in vast sequence databases. If the sequence of a purified protein is already in a database, then the job of the investigator becomes much easier. The investigator need determine only a small stretch of amino acid sequence of the protein to find its match in the database. Alternatively, such a protein might be identified by matching its mass to those deduced for proteins in the database. Mass spectrometry provides a powerful method for determining the mass of a protein.

To understand the physiological context of a protein, antibodies are choice probes for locating proteins in vivo and measuring their quantities. Monoclonal antibodies, able to probe for specific proteins, can be obtained in large amounts. These antibodies can be used to detect the protein in isolation and in cells and to quantify it. Peptides and proteins can be chemically synthesized, providing tools for research and, in some cases, highly pure proteins for use as drugs. Finally, x-ray crystallography and nuclear magnetic resonance (NMR) spectroscopy are the principal techniques for elucidating three-dimensional structure, the key determinant of function.

The exploration of proteins by this array of physical and chemical techniques has greatly enriched our understanding of the molecular basis of life. These techniques make it possible to tackle some of the most challenging questions of biology in molecular terms.

The Proteome Is the Functional Representation of the Genome

Many organisms are yielding the complete DNA base sequences of their genomes. For example, the roundworm *Caenorhabditis elegans* has a genome of 97 million bases and about 19,000 protein-encoding genes, whereas that of the fruit fly *Drosophilia melanogaster* contains 180 million bases and about 14,000 genes. The completely sequenced human genome contains 3 billion bases and about 25,000 genes. But this genomic knowledge is analogous to a list of parts for a car—it does not explain which parts are present in different components or how the parts work together. A new word has been coined, the *proteome,* to signify a more complex level of information content, the level of *functional* information, which encompasses the type, functions, and interactions of proteins that yield a functional unit.

The term *proteome* is derived from *prote*ins expressed by the gen*ome.* The genome provides a list of gene products that *could* be present, but only a subset of these gene products will actually be expressed in a given biological context. The proteome tells us what is functionally present—for example, which proteins interact to form a signal-transduction pathway or an ion channel in a membrane. The proteome is not a fixed characteristic of the cell. Rather, because it represents the functional expression of information, it varies with cell type, developmental stage, and environmental conditions, such as the presence of hormones. The proteome is much larger than the genome because almost all gene products are proteins that can be chemically modified in a variety of ways. Furthermore, these proteins do not exist in isolation; they often interact with one another to form complexes with specific functional properties. Unlike the genome, the proteome is not static.

An understanding of the proteome is acquired by investigating, characterizing, and cataloging proteins. In some, but not all, cases, this process begins by separating a particular protein from all other biomolecules in the cell.

3.1 The Purification of Proteins Is an Essential First Step in Understanding Their Function

An adage of biochemistry is, Never waste pure thoughts on an impure protein. Starting from pure proteins, we can determine amino acid sequences and investigate a protein's biochemical function. From the amino acid sequences, we can map evolutionary relationships between proteins in diverse organisms (Chapter 6). Using crystals grown from pure protein, we can obtain x-ray data that will provide us with a picture of the protein's tertiary structure—the shape that determines function.

The Assay: How Do We Recognize the Protein
That We Are Looking For?

67

3.1 The Purification of Proteins

Purification should yield a sample containing only one type of molecule—
the protein in which the biochemist is interested. This protein sample may
be only a fraction of 1% of the starting material, whether that starting material consists of cells in culture or a particular organ from a plant or animal.
How is the biochemist able to isolate a particular protein from a complex
mixture of proteins?

The biochemist needs a test, called an *assay,* for some unique identifying
property of the protein. A positive result on the assay indicates that the
protein is present. Determining an effective assay is often difficult; but, the
more specific the assay, the more effective the purification. For enzymes,
which are protein catalysts (Chapter 8), the assay usually measures *enzyme
activity*—that is, the ability of the enzyme to promote a particular chemical
reaction. This activity is often measured indirectly. Consider the enzyme
lactate dehydrogenase, which catalyzes the following reaction in the synthesis of glucose:

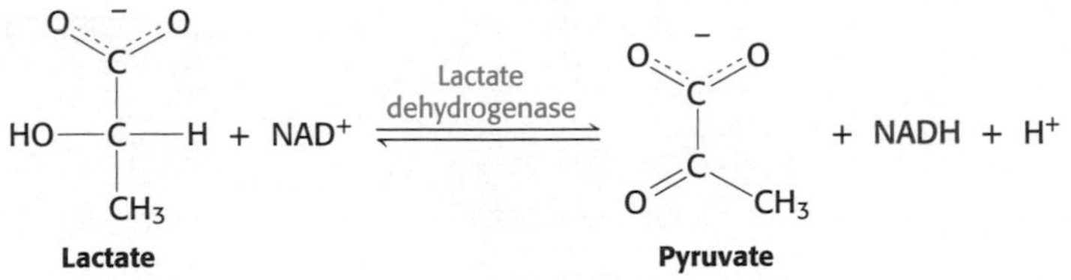

Lactate   **Pyruvate**

Reduced nicotinamide adenine dinucleotide (NADH, p. 420) absorbs light
at 340 nm, whereas oxidized nicotinamide adenine dinucleotide (NAD^+)
does not. Consequently, we can follow the progress of the reaction by examining how much light-absorbing ability is developed by a sample in a
given period of time—for instance, within 1 minute after the addition of the
enzyme. Our assay for enzyme activity during the purification of lactate
dehydrogenase is thus the increase in the absorbance of light at 340 nm
observed in 1 minute.

To analyze how our purification scheme is working, we need one additional piece of information—the amount of protein present in the mixture
being assayed. There are various rapid and reasonably accurate means of
determining protein concentration. With these two experimentally determined numbers—enzyme activity and protein concentration—we then
calculate the *specific activity,* the ratio of enzyme activity to the amount of
protein in the mixture. Ideally, the specific activity will rise as the purification proceeds and the protein mixture being assayed consists to a greater
and greater extent of lactate dehydrogenase. In essence, the overall goal of
the purification is to maximize the specific activity. For a pure enzyme, the
specific activity will have a constant value.

Proteins Must Be Released from the Cell to Be Purified

Having found an assay and chosen a source of protein, we now fractionate
the cell into components and determine which component is enriched in
the protein of interest. Such fractionation schemes are developed on the
basis of expected properties of the protein or by trial and error, guided by
previous experience. In the first step, a *homogenate* is formed by disrupting
the cell membrane, and the mixture is fractionated by centrifugation,
yielding a dense pellet of heavy material at the bottom of the centrifuge
tube and a lighter supernatant above (Figure 3.1). The supernatant is again

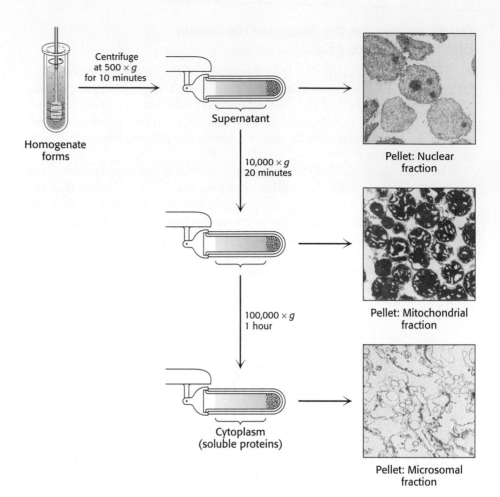

Figure 3.1 Differential centrifugation.
Cells are disrupted in a homogenizer
and the resulting mixture, called the
homogenate, is centrifuged in a step-
by-step fashion of increasing centrifugal
force. The denser material will form a
pellet at lower centrifugal force than will
the less-dense material. The isolated
fractions can be used for further
purification. [Photographs courtesy of
Dr. S. Fleischer and Dr. B. Fleischer.]

Homogenate
forms

Centrifuge
at 500 × *g*
for 10 minutes

Supernatant

Pellet: Nuclear
fraction

10,000 × *g*
20 minutes

Pellet: Mitochondrial
fraction

100,000 × *g*
1 hour

Cytoplasm
(soluble proteins)

Pellet: Microsomal
fraction

centrifuged at a greater force to yield yet another pellet and supernatant.
The procedure, called *differential centrifugation,* yields several fractions of
decreasing density, each still containing hundreds of different proteins.
The fractions are each separately assayed for the desired activity. Usually,
one fraction will be enriched for such activity, and it then serves as the
source of material to which more-discriminating purification techniques
are applied.

Proteins Can Be Purified According to Solubility, Size, Charge, and Binding Affinity

Several thousand proteins have been purified in active form on the basis of
such characteristics as *solubility, size, charge,* and *specific binding affinity.*
Usually, protein mixtures are subjected to a series of separations, each based
on a different property. At each step in the purification, the preparation is
assayed and the protein concentration is determined. A variety of purification techniques are available.

Salting Out. Most proteins are less soluble at high salt concentrations, an
effect called *salting out.* The salt concentration at which a protein precipitates differs from one protein to another. Hence, salting out can be used to
fractionate proteins. For example, 0.8 M ammonium sulfate precipitates
fibrinogen, a blood-clotting protein, whereas a concentration of 2.4 M is
needed to precipitate serum albumin. Salting out is also useful for concentrating dilute solutions of proteins, including active fractions obtained from
other purification steps. Dialysis can be used to remove the salt if necessary.

Dialysis. Proteins can be separated from small molecules by *dialysis* through a semipermeable membrane, such as a cellulose membrane with pores (Figure 3.2). Molecules having dimensions significantly greater than the pore diameter are retained inside the dialysis bag, whereas smaller molecules and ions traverse the pores of such a membrane and emerge in the dialysate outside the bag. This technique is useful for removing a salt or other small molecule, but it will not distinguish between proteins effectively.

Gel-Filtration Chromatography. More-discriminating separations on the basis of size can be achieved by the technique of *gel-filtration chromatography,* also known as molecular exclusion chromatography (Figure 3.3). The sample is applied to the top of a column consisting of porous beads made of an insoluble but highly hydrated polymer such as dextran or agarose (which are carbohydrates) or polyacrylamide. Sephadex, Sepharose, and Biogel are commonly used commercial preparations of these beads, which are typically 100 μm (0.1 mm) in diameter. Small molecules can enter these beads, but large ones cannot. The result is that small molecules are distributed in the aqueous solution both inside the beads and between them, whereas large molecules are located only in the solution between the beads. *Large molecules flow more rapidly through this column and emerge first because a smaller volume is accessible to them.* Molecules that are of a size to occasionally enter a bead will flow from the column at an intermediate position, and small molecules, which take a longer, tortuous path, will exit last.

Ion-Exchange Chromatography. Proteins can be separated on the basis of their net charge by *ion-exchange chromatography.* If a protein has a net positive charge at pH 7, it will usually bind to a column of beads containing

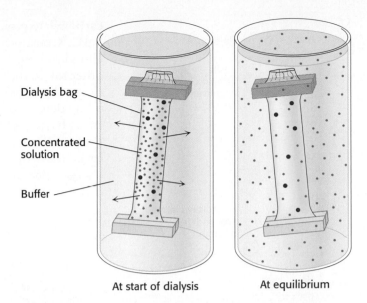

Figure 3.2 Dialysis. Protein molecules (red) are retained within the dialysis bag, whereas small molecules (blue) diffuse into the surrounding medium.

At start of dialysis   At equilibrium

Dialysis bag

Concentrated solution

Buffer

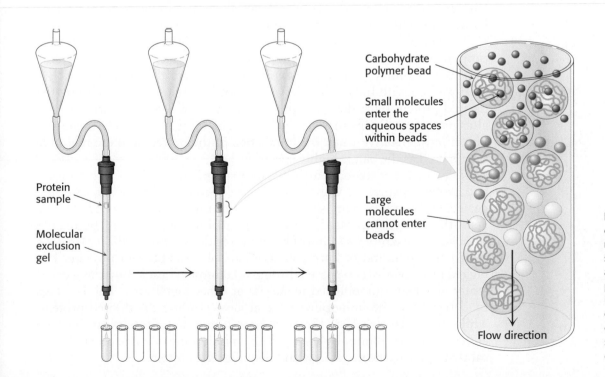

Carbohydrate polymer bead

Small molecules enter the aqueous spaces within beads

Large molecules cannot enter beads

Protein sample

Molecular exclusion gel

Flow direction

Figure 3.3 Gel-filtration chromatography. A mixture of proteins in a small volume is applied to a column filled with porous beads. Because large proteins cannot enter the internal volume of the beads, they emerge sooner than do small ones.

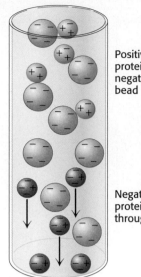

Figure 3.4 Ion-exchange chromatography. This technique separates proteins mainly according to their net charge.

Positively charged protein binds to negatively charged bead

Negatively charged protein flows through

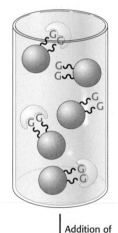

Glucose-binding protein attaches to glucose residues (G) on beads

Addition of glucose (G)

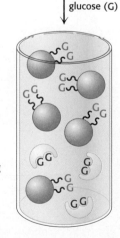

Glucose-binding proteins are released on addition of glucose

Figure 3.5 Affinity chromatography. Affinity chromatography of concanavalin A (shown in yellow) on a solid support containing covalently attached glucose residues (G).

carboxylate groups, whereas a negatively charged protein will not (Figure 3.4). A positively charged protein bound to such a column can then be eluted (released) by increasing the concentration of sodium chloride or another salt in the eluting buffer because sodium ions compete with positively charged groups on the protein for binding to the column. Proteins that have a low density of net positive charge will tend to emerge first, followed by those having a higher charge density. Positively charged proteins (cationic proteins) can be separated by chromatography on negatively charged carboxymethylcellulose (CM-cellulose) columns. Conversely, negatively charged proteins (anionic proteins) can be separated on positively charged diethylaminoethylcellulose (DEAE-cellulose) columns.

Carboxymethyl (CM) group (ionized form)

Diethylaminoethyl (DEAE) group (protonated form)

Affinity Chromatography. *Affinity chromatography* is another powerful and generally applicable means of purifying proteins. This technique takes advantage of the high affinity of many proteins for specific chemical groups. For example, the plant protein concanavalin A can be purified by passing a crude extract through a column of beads containing covalently attached glucose residues. Concanavalin A binds to such a column because it has affinity for glucose, whereas most other proteins do not. The bound concanavalin A can then be released from the column by adding a concentrated solution of glucose. The glucose in solution displaces the column-attached glucose residues from binding sites on concanavalin A (Figure 3.5). Affinity chromatography is a powerful means of isolating transcription factors—proteins that regulate gene expression by binding to specific DNA sequences (Chapters 29 and 31). A protein mixture is percolated through a column containing specific DNA sequences attached to a matrix; proteins with a high affinity for the sequence will bind and be retained. In this instance, the transcription factor is released by washing with a solution containing a high concentration of salt.

In general, affinity chromatography can be effectively used to isolate a protein that recognizes group X by (1) covalently attaching X or a derivative of it to a column; (2) adding a mixture of proteins to this column, which is then washed with buffer to remove unbound proteins; and (3) eluting the desired protein by adding a high concentration of a soluble form of X or altering the conditions to decrease binding affinity. Affinity chromatography is most effective when the interaction of the protein and the molecule that is used as the bait is highly specific.

The process of standard affinity chromatography can be reversed to isolate proteins expressed from cloned genes (Section 5.2). In this case, the gene encoding the protein is modified. Extra amino acids are encoded in the gene that, when expressed, serve as an affinity tag that can be readily trapped. For example, a string of histidine residues (called a *His tag*) may be added to the amino or carboxyl terminus of an expressed protein. The tagged proteins are then passed through a column of beads containing covalently attached, immobilized nickel(II) or other metal ions. The His tags bind tightly to the immobilized metal ions, binding the desired protein, while other proteins flow through the column. The protein can then be eluted from the column by the addition of imidazole or some other chemical that binds to the metal ions and displaces the protein.

High-Pressure Liquid Chromatography. A technique called *high-pressure liquid chromatography* (HPLC) is an enhanced version of the column techniques already discussed. The column materials are much more finely divided and, as a consequence, there are more interaction sites and thus greater resolving power. Because the column is made of finer material, pressure must be applied to the column to obtain adequate flow rates. The net result is both high resolution and rapid separation (Figure 3.6).

Proteins Can Be Separated by Gel Electrophoresis and Displayed

How can we tell that a purification scheme is effective? One way is to ascertain that the specific activity rises with each purification step. Another is to ascertain that the number of different proteins present declines at each step. The technique of electrophoresis makes the latter method possible.

Gel Electrophoresis. A molecule with a net charge will move in an electric field. This phenomenon, termed *electrophoresis,* offers a powerful means of separating proteins and other macromolecules, such as DNA and RNA. The velocity of migration (v) of a protein (or any molecule) in an electric field depends on the electric field strength (E), the net charge on the protein (z), and the frictional coefficient (f).

$$v = Ez/f \qquad (1)$$

The electric force Ez driving the charged molecule toward the oppositely charged electrode is opposed by the viscous drag fv arising from friction between the moving molecule and the medium. The frictional coefficient f depends on both the mass and shape of the migrating molecule and the viscosity (η) of the medium. For a sphere of radius r,

$$f = 6\pi\eta r \qquad (2)$$

Electrophoretic separations are nearly always carried out in gels (or on solid supports such as paper) because the gel serves as a molecular sieve that enhances separation (Figure 3.7). Molecules that are small compared with the pores in the gel readily move through the gel, whereas molecules much

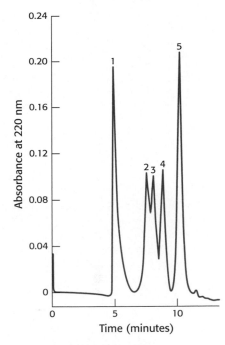

Figure 3.6 High-pressure liquid chromatography (HPLC). Gel filtration by HPLC clearly defines the individual proteins because of its greater resolving power: (1) thyroglobulin (669 kd), (2) catalase (232 kd), (3) bovine serum albumin (67 kd), (4) ovalbumin (43 kd), and (5) ribonuclease (13.4 kd). [After K. J. Wilson and T. D. Schlabach. In *Current Protocols in Molecular Biology*, vol. 2, suppl. 41, F. M. Ausubel, R. Brent, R. E. Kingston, D. D. Moore, J. G. Seidman, J. A. Smith, and K. Struhl, Eds. (Wiley, 1998), p. 10.14.1.]

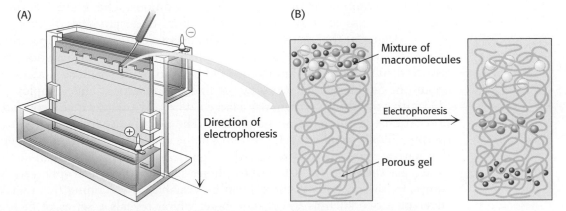

Figure 3.7 Polyacrylamide gel electrophoresis. (A) Gel-electrophoresis apparatus. Typically, several samples undergo electrophoresis on one flat polyacrylamide gel. A microliter pipette is used to place solutions of proteins in the wells of the slab. A cover is then placed over the gel chamber and voltage is applied. The negatively charged SDS (sodium dodecyl sulfate)–protein complexes migrate in the direction of the anode, at the bottom of the gel. (B) The sieving action of a porous polyacrylamide gel separates proteins according to size, with the smallest moving most rapidly.

larger than the pores are almost immobile. Intermediate-size molecules move through the gel with various degrees of facility. The direction of flow is from top to bottom. Electrophoresis is performed in a thin, vertical slab of polyacrylamide. Polyacrylamide gels are choice supporting media for electrophoresis because they are chemically inert and are readily formed by the polymerization of acrylamide with a small amount of the cross-linking agent methylenebisacrylamide included to make a three-dimensional mesh (Figure 3.8). Electrophoresis is distinct from gel filtration in that all of the molecules, regardless of size, are forced to move through the same matrix.

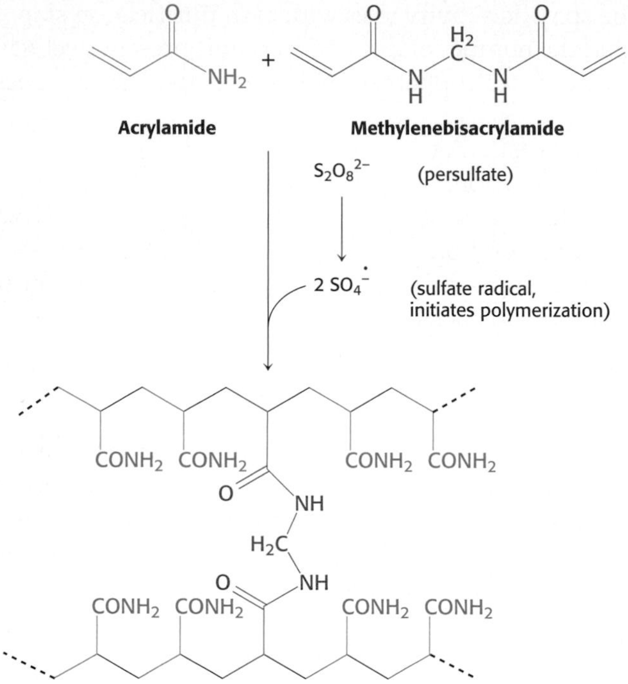

Figure 3.8 Formation of a polyacrylamide gel. A three-dimensional mesh is formed by copolymerizing activated monomer (blue) and cross-linker (red).

Sodium dodecyl sulfate (SDS)

Proteins can be separated largely on the basis of mass by electrophoresis in a polyacrylamide gel under denaturing conditions. The mixture of proteins is first dissolved in a solution of sodium dodecyl sulfate (SDS), an anionic detergent that disrupts nearly all noncovalent interactions in native proteins. Mercaptoethanol (2-thioethanol) or dithiothreitol also is added to reduce disulfide bonds. Anions of SDS bind to main chains at a ratio of about one SDS anion for every two amino acid residues. This complex of SDS with a denatured protein has a large net negative charge that is roughly proportional to the mass of the protein. The negative charge acquired on binding SDS is usually much greater than the charge on the native protein; this native charge is thus rendered insignificant. The SDS–protein complexes are then subjected to electrophoresis. When the electrophoresis is complete, the proteins in the gel can be visualized by staining them with silver or a dye such as Coomassie blue, which reveals a series of bands (Figure 3.9). Radioactive labels, if they have been incorporated into proteins, can be detected by placing a sheet of x-ray film over the gel, a procedure called *autoradiography*.

Small proteins move rapidly through the gel, whereas large proteins stay at the top, near the point of application of the mixture. The mobility of most polypeptide chains under these conditions is linearly proportional to the

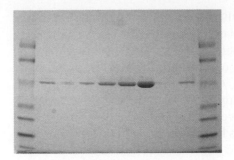

Figure 3.9 **Staining of proteins after electrophoresis.** Proteins subjected to electrophoresis on an SDS–polyacrylamide gel can be visualized by staining with Coomassie blue. [Courtesy of Kodak Scientific Imaging Systems.]

logarithm of their mass (Figure 3.10). Some carbohydrate-rich proteins and membrane proteins do not obey this empirical relation, however. SDS–polyacrylamide gel electrophoresis (often referred to as SDS-PAGE) is rapid, sensitive, and capable of a high degree of resolution. As little as 0.1 µg (~2 pmol) of a protein gives a distinct band when stained with Coomassie blue, and even less (~0.02 µg) can be detected with a silver stain. Proteins that differ in mass by about 2% (e.g., 50 and 51 kd, arising from a difference of about 10 amino acids) can usually be distinguished.

We can examine the efficacy of our purification scheme by analyzing a part of each fraction by electrophoresis. The initial fractions will display dozens to hundreds of proteins. As the purification progresses, the number of bands will diminish, and the prominence of one of the bands should increase. This band should correspond to the protein of interest.

Isoelectric Focusing. Proteins can also be separated electrophoretically on the basis of their relative contents of acidic and basic residues. The *isoelectric point* (pI) of a protein is the pH at which its net charge is zero. At this pH, its electrophoretic mobility is zero because z in equation 1 is equal to zero. For example, the pI of cytochrome c, a highly basic electron-transport protein, is 10.6, whereas that of serum albumin, an acidic protein in blood, is 4.8. Suppose that a mixture of proteins undergoes electrophoresis in a pH gradient in a gel in the absence of SDS. Each protein will move until it reaches a position in the gel at which the pH is equal to the pI of the protein. This method of separating proteins according to their isoelectric point is called *isoelectric focusing*. The pH gradient in the gel is formed first by subjecting a mixture of *polyampholytes* (small multi-charged polymers) having many different pI values to electrophoresis. Isoelectric focusing can readily resolve proteins that differ in pI by as little as 0.01, which means that proteins differing by one net charge can be separated (Figure 3.11).

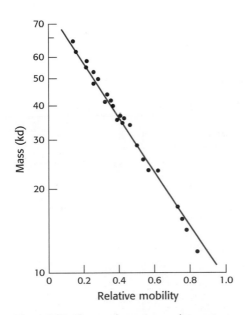

Figure 3.10 **Electrophoresis can determine mass.** The electrophoretic mobility of many proteins in SDS–polyacrylamide gels is inversely proportional to the logarithm of their mass. [After K. Weber and M. Osborn, *The Proteins*, vol. 1, 3d ed. (Academic Press, 1975), p. 179.]

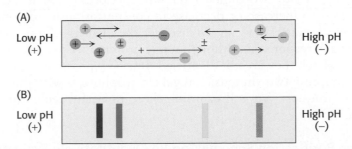

Figure 3.11 **The principle of isoelectric focusing.** A pH gradient is established in a gel before loading the sample. (A) The sample is loaded and voltage is applied. The proteins will migrate to their isoelectric pH, the location at which they have no net charge. (B) The proteins form bands that can be excised and used for further experimentation.

Two-Dimensional Electrophoresis. Isoelectric focusing can be combined with SDS-PAGE to obtain very high resolution separations. A single sample is first subjected to isoelectric focusing. This single-lane gel is then placed horizontally on top of an SDS–polyacrylamide slab. The proteins are thus spread across the top of the polyacrylamide gel according to how far they migrated during isoelectric focusing. They then undergo electrophoresis again in a perpendicular direction (vertically) to yield a two-dimensional pattern of spots. In such a gel, proteins have been separated in the horizontal direction on the basis of isoelectric point and in the vertical direction on the basis of mass. Remarkably, more than a thousand different proteins in the bacterium *Escherichia coli* can be resolved in a single experiment by two-dimensional electrophoresis (Figure 3.12).

(A)

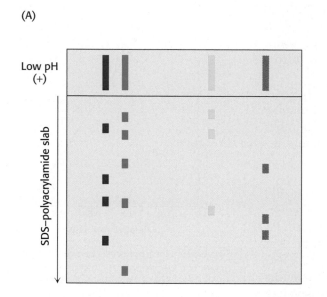

(B) Isoelectric focusing

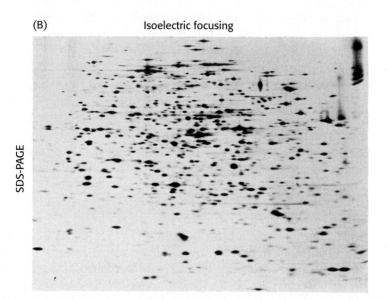

Figure 3.12 Two-dimensional gel electrophoresis. (A) A protein sample is initially fractionated in one dimension by isoelectric focusing as described in Figure 3.11. The isoelectric focusing gel is then attached to an SDS–polyacrylamide gel, and electrophoresis is performed in the second dimension, perpendicular to the original separation. Proteins with the same pI are now separated on the basis of mass. (B) Proteins from *E. coli* were separated by two-dimensional gel electrophoresis, resolving more than a thousand different proteins. The proteins were first separated according to their isoelectric pH in the horizontal direction and then by their apparent mass in the vertical direction. [(B) Courtesy of Dr. Patrick H. O'Farrell.]

Proteins isolated from cells under different physiological conditions can be subjected to two-dimensional electrophoresis. Particular proteins may be seen to increase or decrease in concentration in response to the physiological state. How can we discover the identity of a protein that is showing such responses? Although many proteins are displayed on a two-dimensional gel, they are not identified. It is now possible to identify proteins by coupling two-dimensional gel electrophoresis with mass spectrometric techniques. We will examine these powerful techniques shortly (Section 3.5).

A Protein Purification Scheme Can Be Quantitatively Evaluated

To determine the success of a protein purification scheme, we monitor the procedure at each step by determining specific activity and by performing an SDS-PAGE analysis. Consider the results for the purification of a ficti-

TABLE 3.1 Quantification of a purification protocol for a fictitious protein

Step	Total protein (mg)	Total activity (units)	Specific activity, (units mg^{-1})	Yield (%)	Purification level
Homogenization	15,000	150,000	10	100	1
Salt fractionation	4,600	138,000	30	92	3
Ion-exchange chromatography	1,278	115,500	90	77	9
Gel-filtration chromatography	68.8	75,000	1,100	50	110
Affinity chromatography	1.75	52,500	30,000	35	3,000

tious protein, summarized in Table 3.1 and Figure 3.13. At each step, the following parameters are measured:

Total Protein. The quantity of protein present in a fraction is obtained by determining the protein concentration of a part of each fraction and multiplying by the fraction's total volume.

Total Activity. The enzyme activity for the fraction is obtained by measuring the enzyme activity in the volume of fraction used in the assay and multiplying by the fraction's total volume.

Specific Activity. This parameter is obtained by dividing total activity by total protein.

Yield. This parameter is a measure of the activity retained after each purification step as a percentage of the activity in the crude extract. The amount of activity in the initial extract is taken to be 100%.

Purification Level. This parameter is a measure of the increase in purity and is obtained by dividing the specific activity, calculated after each purification step, by the specific activity of the initial extract.

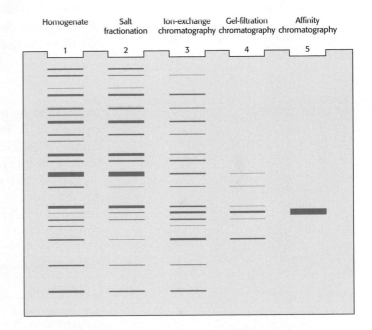

Figure 3.13 Electrophoretic analysis of a protein purification. The purification scheme in Table 3.1 was analyzed by SDS-PAGE. Each lane contained 50 μg of sample. The effectiveness of the purification can be seen as the band for the protein of interest becomes more prominent relative to other bands.

As we see in Table 3.1, the first purification step, salt fractionation, leads to an increase in purity of only 3-fold, but we recover nearly all the target protein in the original extract, given that the yield is 92%. After dialysis to lower the high concentration of salt remaining from the salt fractionation, the fraction is passed through an ion-exchange column. The purification now increases to 9-fold compared with the original extract, whereas the yield falls to 77%. Gel-filtration chromatography brings the level of purification to 110-fold, but the yield is now at 50%. The final step is affinity chromatography with the use of a ligand specific for the target enzyme. This step, the most powerful of these purification procedures, results in a purification level of 3000-fold but lowers the yield to 35%. The SDS-PAGE analysis in Figure 3.13 shows that, if we load a constant amount of protein onto each lane after each step, the number of bands decreases in proportion to the level of purification, and the amount of protein of interest increases as a proportion of the total protein present.

A good purification scheme takes into account both purification levels and yield. A high degree of purification and a poor yield leave little protein with which to experiment. A high yield with low purification leaves many contaminants (proteins other than the one of interest) in the fraction and complicates the interpretation of experiments.

Ultracentrifugation Is Valuable for Separating Biomolecules and Determining Their Masses

We have already seen that centrifugation is a powerful and generally applicable method for separating a crude mixture of cell components, but it is also useful for separating and analyzing biomolecules themselves. With this technique, we can determine such parameters as mass and density, learn something about the shape of a molecule, and investigate the interactions between molecules. To deduce these properties from the centrifugation data, we need a mathematical description of how a particle behaves in a centrifugal force.

A particle will move through a liquid medium when subjected to a centrifugal force. A convenient means of quantifying the rate of movement is to calculate the sedimentation coefficient, s, of a particle by using the following equation:

$$s = m(1 - \bar{v}\rho)/f$$

where m is the mass of the particle, $\bar{v}$ is the partial specific volume (the reciprocal of the particle density), ρ is the density of the medium, and f is the frictional coefficient (a measure of the shape of the particle). The $(1 - \bar{v}\rho)$ term is the buoyant force exerted by liquid medium.

Sedimentation coefficients are usually expressed in *Svedberg units* (S), equal to 10^{-13} s. The smaller the S value, the slower a molecule moves in a centrifugal field. The S values for a number of biomolecules and cellular components are listed in Table 3.2 and Figure 3.14.

TABLE 3.2 S values and molecular weights of sample proteins

Protein	S value (Svedberg units)	Molecular weight
Pancreatic trypsin inhibitor	1	6,520
Cytochrome *c*	1.83	12,310
Ribonuclease A	1.78	13,690
Myoglobin	1.97	17,800
Trypsin	2.5	23,200
Carbonic anhydrase	3.23	28,800
Concanavalin A	3.8	51,260
Malate dehydrogenase	5.76	74,900
Lactate dehydrogenase	7.54	146,200

Source: T. Creighton, *Proteins*, 2d ed. (W. H. Freeman and Company, 1993), Table 7.1.

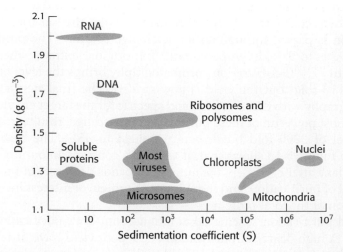

Figure 3.14 Density and sedimentation coefficients of cellular components. [After L. J. Kleinsmith and V. M. Kish, *Principles of Cell and Molecular Biology*, 2d ed. (HarperCollins, 1995), p. 138.]

Several important conclusions can be drawn from the preceding equation:

1. The sedimentation velocity of a particle depends in part on its mass. A more massive particle sediments more rapidly than does a less massive particle of the same shape and density.

2. Shape, too, influences the sedimentation velocity because it affects the viscous drag. The frictional coefficient f of a compact particle is smaller than that of an extended particle of the same mass. Hence, elongated particles sediment more slowly than do spherical ones of the same mass.

3. A dense particle moves more rapidly than does a less dense one because the opposing buoyant force $(1 - \bar{v}\rho)$ is smaller for the denser particle.

4. The sedimentation velocity also depends on the density of the solution (ρ). Particles sink when $\bar{v}\rho < 1$, float when $\bar{v}\rho > 1$, and do not move when $\bar{v}\rho = 1$.

A technique called *zonal, band,* or most commonly *gradient* centrifugation can be used to separate proteins with different sedimentation coefficients. The first step is to form a density gradient in a centrifuge tube. Differing proportions of a low-density solution (such as 5% sucrose) and a high-density solution (such as 20% sucrose) are mixed to create a linear gradient of sucrose concentration ranging from 20% at the bottom of the tube to 5% at the top (Figure 3.15). The role of the gradient is to prevent convective flow. A small volume of a solution containing the mixture of proteins to be separated is placed on top of the density gradient. When the rotor is spun, proteins move through the gradient and separate according to their sedimentation coefficients. The time and speed of the centrifugation is determined empirically. The separated bands, or zones, of protein can be harvested by making a hole in the bottom of the tube and collecting drops. The drops can be measured for protein content and catalytic activity or another functional property. This sedimentation-velocity technique readily separates proteins differing in sedimentation coefficient by a factor of two or more.

The mass of a protein can be directly determined by *sedimentation equilibrium,* in which a sample is centrifuged at relatively low speed so that sedimentation is counterbalanced by diffusion. *The sedimentation-equilibrium technique for determining mass is very accurate and can be applied without denaturing the protein. Thus the native quaternary structure of multimeric proteins is preserved.* In contrast, SDS–polyacrylamide gel electrophoresis

Figure 3.15 Zonal centrifugation. The steps are as follows: (A) form a density gradient, (B) layer the sample on top of the gradient, (C) place the tube in a swinging-bucket rotor and centrifuge it, and (D) collect the samples. [After D. Freifelder, *Physical Biochemistry,* 2d ed. (W. H. Freeman and Company, 1982), p. 397.]

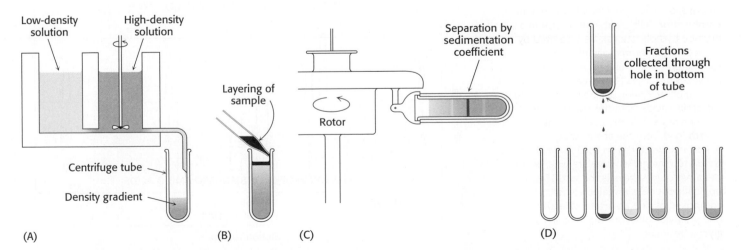

Low-density solution   High-density solution

Layering of sample

Separation by sedimentation coefficient

Rotor

Fractions collected through hole in bottom of tube

Centrifuge tube

Density gradient

(A)   (B)   (C)   (D)

provides an *estimate* of the mass of dissociated polypeptide chains under *denaturing* conditions. Note that, if we know the mass of the dissociated components of a multimeric protein as determined by SDS–polyacrylamide analysis and the mass of the intact multimeric protein as determined by sedimentation-equilibrium analysis, we can determine the number of copies of each polypeptide chain present in the multimeric protein.

3.2 Amino Acid Sequences Can Be Determined by Automated Edman Degradation

After a protein has been purified to homogeneity, a determination of the protein's amino acid sequence, or primary structure, is often desirable. Let us examine first how we can sequence a simple peptide, such as

$$Ala\text{-}Gly\text{-}Asp\text{-}Phe\text{-}Arg\text{-}Gly$$

The first step is to determine the *amino acid composition* of the peptide. The peptide is hydrolyzed into its constituent amino acids by heating it in 6 M HCl at 110°C for 24 hours. Amino acids in hydrolysates can be separated by ion-exchange chromatography. The identity of the amino acid is revealed by its elution volume, which is the volume of buffer used to remove the amino acid from the column (Figure 3.16), and its quantity is revealed by reaction with *ninhydrin*. Amino acids treated with ninhydrin give an intense blue color, except for proline, which gives a yellow color because it contains a secondary amino group. The concentration of an amino acid in a solution, after heating with ninhydrin, is proportional to the optical absorbance of the solution. This technique can detect a microgram (10 nmol) of an amino acid, which is about the amount present in a thumbprint. As little as a nanogram (10 pmol) of an amino acid can be detected by replacing ninhydrin with *fluorescamine,* which reacts with the α-amino group to form a highly fluorescent product (Figure 3.17). A comparison of the chromatographic patterns of our sample hydrolysate with that of a standard mixture of amino acids would show that the amino acid composition of the peptide is

$$(Ala, Arg, Asp, Gly_2, Phe)$$

The parentheses denote that this is the amino acid composition of the peptide, not its sequence.

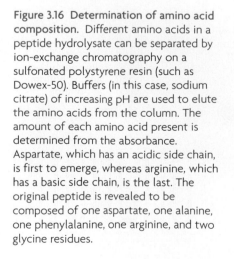

Figure 3.16 Determination of amino acid composition. Different amino acids in a peptide hydrolysate can be separated by ion-exchange chromatography on a sulfonated polystyrene resin (such as Dowex-50). Buffers (in this case, sodium citrate) of increasing pH are used to elute the amino acids from the column. The amount of each amino acid present is determined from the absorbance. Aspartate, which has an acidic side chain, is first to emerge, whereas arginine, which has a basic side chain, is the last. The original peptide is revealed to be composed of one aspartate, one alanine, one phenylalanine, one arginine, and two glycine residues.

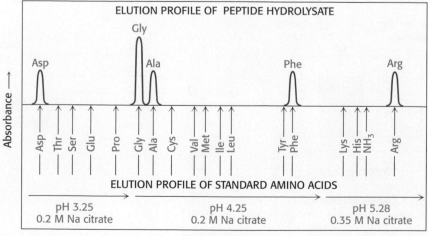

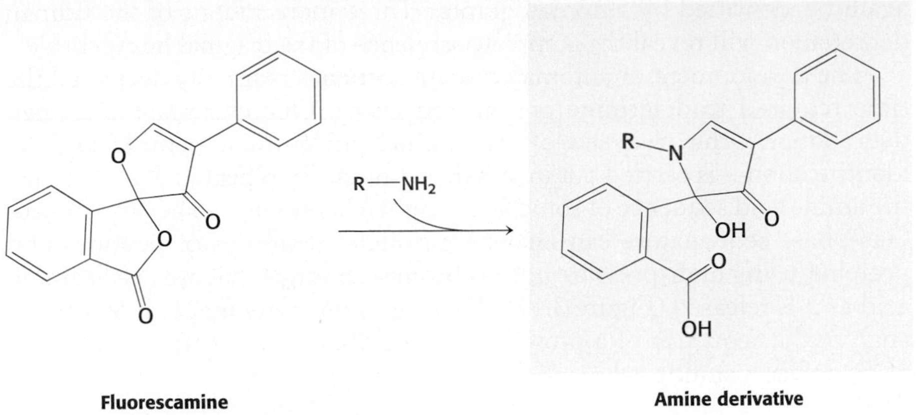

Fluorescamine

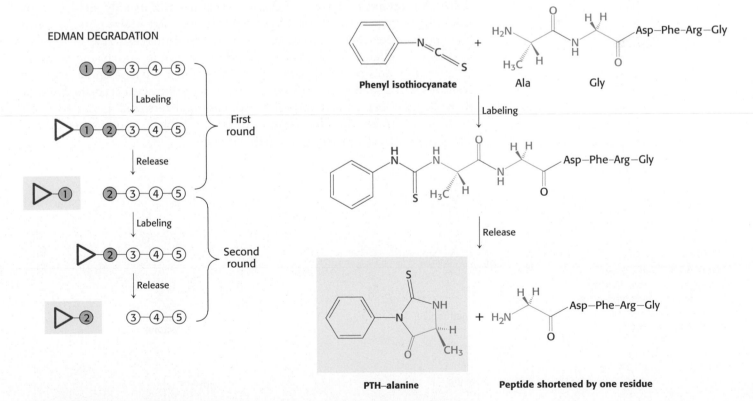

Amine derivative

Figure 3.17 Fluorescent derivatives of amino acids. Fluorescamine reacts with the α-amino group of an amino acid to form a fluorescent derivative.

The next step is to identify the N-terminal amino acid. Pehr Edman devised a method for labeling the amino-terminal residue and cleaving it from the peptide without disrupting the peptide bonds between the other amino acid residues. The *Edman degradation* sequentially removes one residue at a time from the amino end of a peptide (Figure 3.18). *Phenyl isothiocyanate* reacts with the uncharged terminal amino group of the peptide to form a phenylthiocarbamoyl derivative. Then, under mildly acidic conditions, a cyclic derivative of the terminal amino acid is liberated, which leaves an intact peptide shortened by one amino acid. The cyclic compound is a phenylthiohydantoin (PTH)–amino acid, which can be identified by chromatographic procedures. The Edman procedure can then be repeated on the shortened peptide, yielding another PTH–amino acid, which can

Ninhydrin

EDMAN DEGRADATION

First round

Second round

Phenyl isothiocyanate Ala Gly

PTH–alanine **Peptide shortened by one residue**

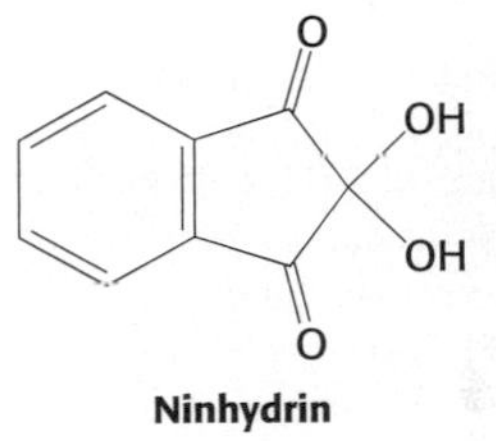

Figure 3.18 The Edman degradation. The labeled amino-terminal residue (PTH–alanine in the first round) can be released without hydrolyzing the rest of the peptide. Hence, the amino-terminal residue of the shortened peptide (Gly-Asp-Phe-Arg-Gly) can be determined in the second round. Three more rounds of the Edman degradation reveal the complete sequence of the original peptide.

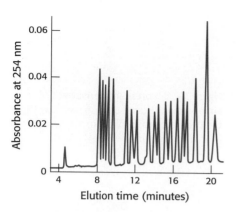

Figure 3.19 Separation of PTH–amino acids. PTH–amino acids can be rapidly separated by high-pressure liquid chromatography (HPLC). In this HPLC profile, a mixture of PTH–amino acids is clearly resolved into its components. An unknown amino acid can be identified by its elution position relative to the known ones.

again be identified by chromatography. Three more rounds of the Edman degradation will reveal the complete sequence of the original hexapeptide.

The development of automated sequencers has markedly decreased the time required to determine protein sequences. One cycle of the Edman degradation—the cleavage of an amino acid from a peptide and its identification—is carried out in less than 1 hour. By repeated degradations, the amino acid sequence of some 50 residues in a protein can be determined. Gas-phase sequenators can analyze picomole quantities of peptides and proteins, using high-pressure liquid chromatography to identify each amino acid as it is released (Figure 3.19). This high sensitivity makes it feasible to analyze the sequence of a protein sample eluted from a single band of an SDS–polyacrylamide gel.

Proteins Can Be Specifically Cleaved into Small Peptides to Facilitate Analysis

In principle, it should be possible to sequence an entire protein by using the Edman method. In practice, the peptides cannot be much longer than about 50 residues, because not all peptides in the reaction mixture release the amino acid derivative at each step. For instance, if the efficiency of release for each round were 98%, the proportion of "correct" amino acid released after 60 rounds would be (0.98^{60}), or 0.3—a hopelessly impure mix. This obstacle can be circumvented by cleaving a protein into smaller peptides that can be sequenced. In essence, the strategy is to *divide and conquer*.

The key is to cleave the protein into a small number of pure fragments. Specific cleavage at particular amino acid types can be achieved by chemical or enzymatic methods. For example, *cyanogen bromide* (CNBr) splits polypeptide chains only on the carboxyl side of methionine residues (Figure 3.20). A protein that has 10 methionine residues will usually yield 11 peptides on cleavage with CNBr. Highly specific cleavage is also obtained with *trypsin*, a proteolytic enzyme secreted by the pancreas. Trypsin cleaves polypeptide chains on the carboxyl side of arginine and lysine residues (Figure 3.21 and p. 246). A protein that contains 9 lysine and 7 arginine residues will usually yield 17 peptides on digestion with trypsin. Each of these tryptic peptides, except for the carboxyl-terminal peptide of the protein, will end with either arginine or lysine. Table 3.3 gives several other ways of specifically cleaving polypeptide chains.

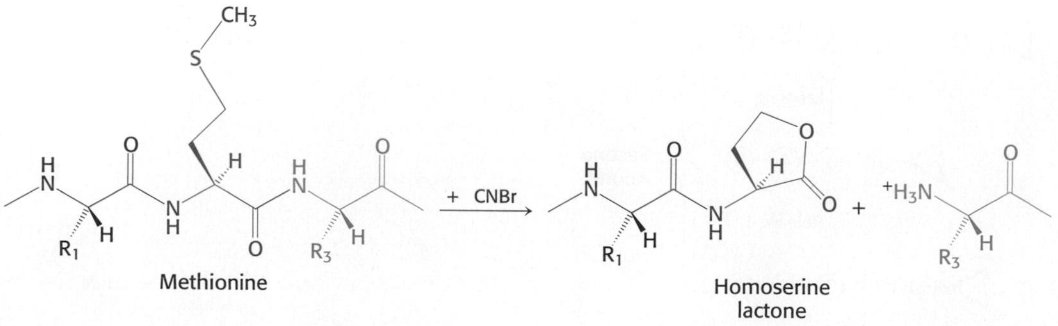

Figure 3.20 Cleavage by cyanogen bromide. Cyanogen bromide cleaves polypeptides on the carboxyl side of methionine residues.

Methionine   + CNBr   Homoserine lactone

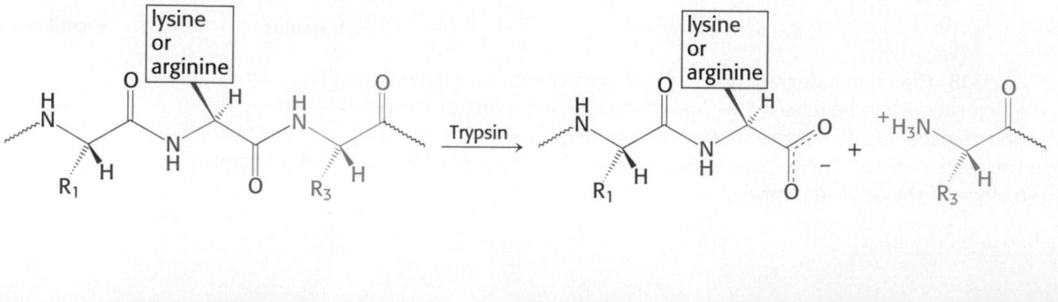

Figure 3.21 Cleavage by trypsin. Trypsin hydrolyzes polypeptides on the carboxyl side of arginine and lysine residues.

lysine or arginine   Trypsin   lysine or arginine

TABLE 3.3 Specific cleavage of polypeptides

Reagent	Cleavage site
Chemical cleavage	
Cyanogen bromide	Carboxyl side of methionine residues
O-Iodosobenzoate	Carboxyl side of tryptophan residues
Hydroxylamine	Asparagine–glycine bonds
2-Nitro-5-thiocyanobenzoate	Amino side of cysteine residues
Enzymatic cleavage	
Trypsin	Carboxyl side of lysine and arginine residues
Clostripain	Carboxyl side of arginine residues
Staphylococcal protease	Carboxyl side of aspartate and glutamate residues (glutamate only under certain conditions)
Thrombin	Carboxyl side of arginine
Chymotrypsin	Carboxyl side of tyrosine, tryptophan, phenylalanine, leucine, and methionine
Carboxypeptidase A	Amino side of C-terminal amino acid (not arginine, lysine, or proline)

The peptides obtained by specific chemical or enzymatic cleavage are separated by some type of chromatography. The sequence of each purified peptide is then determined by the Edman method. At this point, the amino acid sequences of segments of the protein are known, but the order of these segments is not yet defined. How can we order the peptides to obtain the primary structure of the original protein? The necessary additional information is obtained from *overlap peptides* (Figure 3.22). A second enzyme is used to split the polypeptide chain at different linkages. For example, chymotrypsin cleaves preferentially on the carboxyl side of aromatic and some other bulky nonpolar residues (p. 247). Because these chymotryptic peptides overlap two or more tryptic peptides, they can be used to establish the order of the peptides. The entire amino acid sequence of the polypeptide chain is then known.

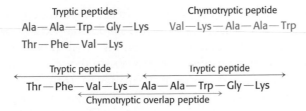

Figure 3.22 Overlap peptides. The peptide obtained by chymotryptic digestion overlaps two tryptic peptides, establishing their order.

Additional steps are necessary if the initial protein sample is actually several polypeptide chains. SDS–gel electrophoresis under reducing conditions should display the number of chains. Alternatively, the number of distinct N-terminal amino acids could be determined. After a protein has been identified as being made up of two or more polypeptide chains, denaturing agents, such as urea or guanidine hydrochloride, are used to dissociate chains held together by noncovalent bonds. The dissociated chains must be separated from one another before sequence determination can begin. Polypeptide chains linked by disulfide bonds are separated by reduction with thiols such as β-mercaptoethanol or dithiothreitol. To prevent the cysteine residues from recombining, they are then alkylated with iodoacetate to form stable *S*-carboxymethyl derivatives (Figure 3.23). Sequencing can then be performed as already described.

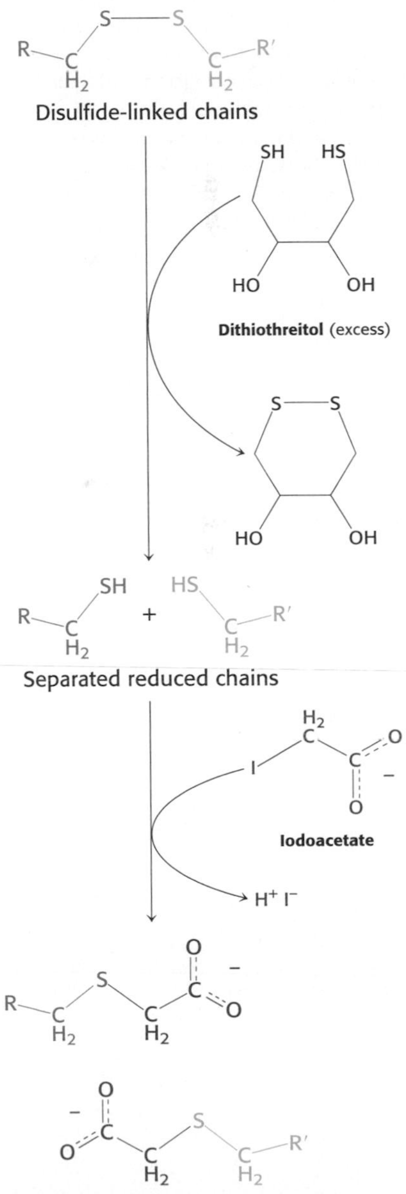

Disulfide-linked chains

Dithiothreitol (excess)

Separated reduced chains

Iodoacetate

Separated carboxymethylated chains

Figure 3.23 Disulfide-bond reduction. Polypeptides linked by disulfide bonds can be separated by reduction with dithiothreitol followed by alkylation to prevent reformation.

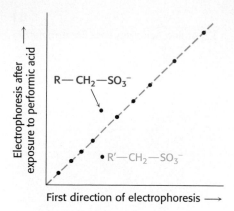

Figure 3.24 Diagonal electrophoresis. Peptides joined together by disulfide bonds can be detected by diagonal electrophoresis. The mixture of peptides is subjected to electrophoresis in a single lane in one direction (horizontal) and then treated with performic acid, which cleaves and oxidizes the disulfide bonds. The sample is then subjected to electrophoresis in the perpendicular direction (vertical).

To complete our understanding of the protein's structure, we need to determine the positions of the original disulfide bonds. This information can be obtained by using a *diagonal electrophoresis* technique to isolate the peptide sequences containing such bonds (Figure 3.24). First, the protein is specifically cleaved into peptides under conditions in which the disulfides remain intact. The mixture of peptides is applied to a corner of a sheet of paper and subjected to electrophoresis in a single lane along one side. The resulting sheet is exposed to vapors of performic acid, which cleaves disulfides and converts them into cysteic acid residues. Peptides originally linked by disulfides are now independent as well as more acidic because of the formation of an SO_3^- group.

Cystine → **Performic acid** → **Cysteic acid**

This mixture is subjected to electrophoresis in the perpendicular direction under the same conditions as those of the first electrophoresis. Peptides that were devoid of disulfides will have the same mobility as before, and consequently all will be located on a single diagonal line. In contrast, the newly formed peptides containing cysteic acid will usually migrate differently from their parent disulfide-linked peptides and hence will lie off the diagonal. These peptides can then be isolated and sequenced, and the location of the disulfide bond can be established.

Amino Acid Sequences Are Sources of Many Kinds of Insight

A protein's amino acid sequence is a valuable source of insight into the protein's function, structure, and history.

1. *The sequence of a protein of interest can be compared with all other known sequences to ascertain whether significant similarities exist. Does this protein belong to an established family?* A search for kinship between a newly sequenced protein and the millions of previously sequenced ones takes only a few seconds on a personal computer (Chapter 6). If the newly isolated protein is a member of an established class of protein, we can begin to infer information about the protein's structure and function. For instance, chymotrypsin and trypsin are members of the serine protease family, a clan of proteolytic enzymes that have a common catalytic mechanism based on a reactive serine residue (p. 245). If the sequence of the newly isolated protein shows sequence similarity with trypsin or chymotrypsin, the result suggests that it may be a serine protease.

2. *Comparison of sequences of the same protein in different species yields a wealth of information about evolutionary pathways.* Genealogical relations between species can be inferred from sequence differences between their proteins. We can even estimate the time at which two evolutionary lines diverged, thanks to the clocklike nature of random mutations. For example, a comparison of serum albumins found in primates indicates that human beings and African apes diverged 5 million years ago, not 30 million years ago as was once thought. Sequence analyses have opened a new perspective on the fossil record and the pathway of human evolution.

3. *Amino acid sequences can be searched for the presence of internal repeats.* Such internal repeats can reveal the history of an individual protein itself.

Many proteins apparently have arisen by duplication of primordial genes followed by their diversification. For example, calmodulin, a ubiquitous calcium sensor in eukaryotes, contains four similar calcium-binding modules that arose by gene duplication (Figure 3.25).

4. *Many proteins contain amino acid sequences that serve as signals designating their destinations or controlling their processing.* For example, a protein destined for export from a cell or for location in a membrane contains a *signal sequence*, a stretch of about 20 hydrophobic residues near the amino terminus that directs the protein to the appropriate membrane. Another protein may contain a stretch of amino acids that functions as a *nuclear localization signal*, directing the protein to the nucleus.

5. *Sequence data provide a basis for preparing antibodies specific for a protein of interest.* One or more parts of the amino acid sequence of a protein will elicit an antibody when injected into a mouse or rabbit. These specific antibodies can be very useful in determining the amount of a protein present in solution or in the blood, ascertaining its distribution within a cell, or cloning its gene (p. 85).

6. *Amino acid sequences are valuable for making DNA probes that are specific for the genes encoding the corresponding proteins* (p. 139). Knowledge of a protein's primary structure permits the use of reverse genetics. DNA sequences that correspond to a part of the amino acid sequence can be constructed on the basis of the genetic code. These DNA sequences can be used as probes to isolate the gene encoding the protein so that the entire sequence of the protein can be determined. The gene in turn can provide valuable information about the physiological regulation of the protein. Protein sequencing is an integral part of molecular genetics, just as DNA cloning is central to the analysis of protein structure and function. We will revisit some of these topics in more detail in Chapter 5.

Figure 3.25 **Repeating motifs in a protein chain.** Calmodulin, a calcium sensor, contains four similar units (shown in red, yellow, blue, and orange) in a single polypeptide chain. *Notice* that each unit binds a calcium ion (shown in green). [Drawn from 1CLL.pdb.]

Recombinant DNA Technology Has Revolutionized Protein Sequencing

Thousands of proteins have been sequenced by the Edman degradation of peptides derived from specific cleavages. Nevertheless, heroic effort is required to elucidate the sequence of large proteins, those with more than 1000 residues. For sequencing such proteins, a complementary experimental approach based on recombinant DNA technology is often more efficient. As will be discussed in Chapter 5, long stretches of DNA can be cloned and sequenced, and the nucleotide sequence can be translated to reveal the amino acid sequence of the protein encoded by the gene (Figure 3.26). Recombinant DNA technology is producing a wealth of amino acid sequence information at a remarkable rate.

Even with the use of the DNA base sequence to determine primary structure, there is still a need to work with isolated proteins. The amino acid sequence deduced by reading the DNA sequence is that of the *nascent* protein, the direct product of the translational machinery. Many proteins are

DNA sequence	GGG	TTC	TTG	GGA	GCA	GCA	GGA	AGC	ACT	ATG	GGC	GCA
Amino acid sequence	Gly	Phe	Leu	Gly	Ala	Ala	Gly	Ser	Thr	Met	Gly	Ala

Figure 3.26 **DNA sequence yields the amino acid sequence.** The complete nucleotide sequence of HIV-1 (human immunodeficiency virus), the cause of AIDS (acquired immune deficiency syndrome), was determined within a year after the isolation of the virus. A part of the DNA sequence specified by the RNA genome of the virus is shown here with the corresponding amino acid sequence (deduced from a knowledge of the genetic code).

modified after synthesis. Some have their ends trimmed, and others arise by cleavage of a larger initial polypeptide chain. Cysteine residues in some proteins are oxidized to form disulfide links, connecting either parts within a chain or separate polypeptide chains. Specific side chains of some proteins are altered. Amino acid sequences derived from DNA sequences are rich in information, but they do not disclose such posttranslational modifications. Chemical analyses of proteins in their mature form are needed to delineate the nature of these changes, which are critical for the biological activities of most proteins. *Thus, genomic and proteomic analyses are complementary approaches to elucidating the structural basis of protein function.*

3.3 Immunology Provides Important Techniques with Which to Investigate Proteins

Immunological techniques for studying proteins take advantage of the exquisite specificity of antibodies for their target proteins. Labeled antibodies provide a means to tag a specific protein so that it can be isolated, quantified, or visualized.

Antibodies to Specific Proteins Can Be Generated

Immunological techniques begin with the generation of antibodies to a particular protein. An *antibody* (also called an *immunoglobulin,* Ig) is itself a protein; it is synthesized by an animal in response to the presence of a foreign substance, called an *antigen*. Antibodies have specific and high affinity for the antigens that elicited their synthesis. The binding of antibody and antigen is a step in the immune response that protects the animal from infection (Chapter 33). Proteins, polysaccharides, and nucleic acids can be effective antigens. Small foreign molecules, such as synthetic peptides, also can elicit antibodies, provided that the small molecule is attached to a macromolecular carrier. An antibody recognizes a specific group or cluster of amino acids on the target molecule called an *antigenic determinant* or *epitope* (Figures 3.27 and 3.28). Animals have a very large repertoire of

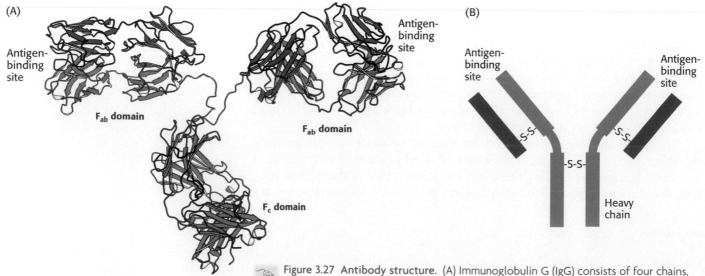

Figure 3.27 Antibody structure. (A) Immunoglobulin G (IgG) consists of four chains, two heavy chains (blue) and two light chains (red), linked by disulfide bonds. The heavy and light chains come together to form F_{ab} domains, which have the antigen-binding sites at the ends. The two heavy chains form the F_c domain. *Notice* that the F_{ab} domains are linked to the F_c domain by flexible linkers. (B) A more schematic representation of an IgG molecule. [Drawn from 1IGT.pdb.]

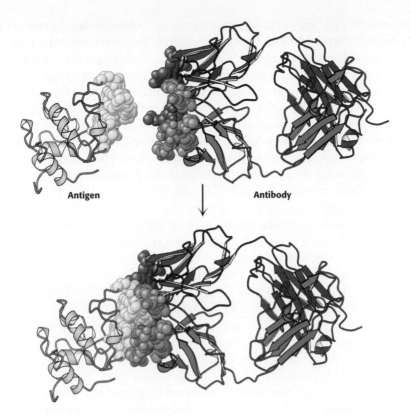

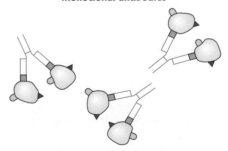

Figure 3.28 Antigen–antibody interactions. A protein antigen, in this case lysozyme, binds to the end of an F_{ab} domain of an antibody. *Notice* that the end of the antibody and the antigen have complementary shapes, allowing a large amount of surface to be buried on binding. [Drawn from 3HFL.pdb.]

antibody-producing cells, each producing an antibody of a single specificity. An antigen acts by stimulating the proliferation of the small number of cells that were already forming an antibody capable of recognizing the antigen.

Immunological techniques depend on the ability to generate antibodies to a specific antigen. To obtain antibodies that recognize a particular protein, a biochemist injects the protein into a rabbit twice, 3 weeks apart. The injected protein stimulates the reproduction of cells producing antibodies that recognize it. Blood is drawn from the immunized rabbit several weeks later and centrifuged to separate blood cells from the supernatant, or serum. The serum, called an *antiserum,* contains antibodies to all antigens to which the rabbit has been exposed. Only some of them will be antibodies to the injected protein. Moreover, antibodies of a given specificity are not a single molecular species. For instance, 2,4-dinitrophenol (DNP) was used as an antigen to generate antibodies. Analyses of anti-DNP antibodies revealed a wide range of binding affinities—the dissociation constants ranged from about 0.1 nM to 1 μM. Correspondingly, a large number of bands were evident when anti-DNP antibody was subjected to isoelectric focusing. These results indicate that cells are producing many different antibodies, each recognizing a different surface feature of the same antigen. The antibodies are heterogeneous, or *polyclonal* (Figure 3.29). This heterogeneity is a barrier that can complicate the use of these antibodies.

Monoclonal Antibodies with Virtually Any Desired Specificity Can Be Readily Prepared

The discovery of a means of producing *monoclonal antibodies* of virtually any desired specificity was a major breakthrough that intensified the power of immunological approaches. Just like working with impure proteins, working with an impure mixture of antibodies makes it difficult to interpret data. The ideal would be to isolate a clone of cells producing a single, identical antibody. The problem is that antibody-producing cells isolated from an organism die in a short time.

Polyclonal antibodies

Antigen

Monoclonal antibodies

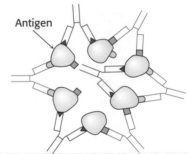

Figure 3.29 Polyclonal and monoclonal antibodies. Most antigens have several epitopes. Polyclonal antibodies are heterogeneous mixtures of antibodies, each specific for one of the various epitopes on an antigen. Monoclonal antibodies are all identical, produced by clones of a single antibody-producing cell. They recognize one specific epitope. [After R. A. Goldsby, T. J. Kindt, and B. A. Osborne, *Kuby Immunology,* 4th ed. (W. H. Freeman and Company, 2000), p. 154.]

Immortal cell lines that produce monoclonal antibodies do exist. These cell lines are derived from a type of cancer, *multiple myeloma*, which is a malignant disorder of antibody-producing cells. In this cancer, a single transformed plasma cell divides uncontrollably, generating a very large number of *cells of a single kind*. Such a group of cells is a *clone* because the cells are descended from the same cell and have identical properties. The identical cells of the myeloma secrete large amounts of normal *immunoglobulin of a single kind* generation after generation. These antibodies were useful for elucidating antibody structure, but nothing is known about their specificity and so they are useless for the immunological methods described in the next pages.

César Milstein and Georges Köhler discovered that *large amounts of homogeneous antibody of nearly any desired specificity can be obtained by fusing a short-lived antibody-producing cell with an immortal myeloma cell.* An antigen is injected into a mouse, and its spleen is removed several weeks later (Figure 3.30). A mixture of plasma cells from this spleen is fused in vitro with myeloma cells. Each of the resulting hybrid cells, called *hybridoma cells*, indefinitely produces the identical antibody specified by the parent cell from the spleen. Hybridoma cells can then be screened, by using some sort of specific assay for the antigen–antibody interaction, to determine which ones produce antibodies having the desired specificity. Collections of cells shown to produce the desired antibody are subdivided and reassayed. This process is repeated until a pure cell line, a clone producing a single antibody, is isolated. These positive cells can be grown in culture medium or injected into mice to induce myelomas. Alternatively, the cells can be frozen and stored for long periods.

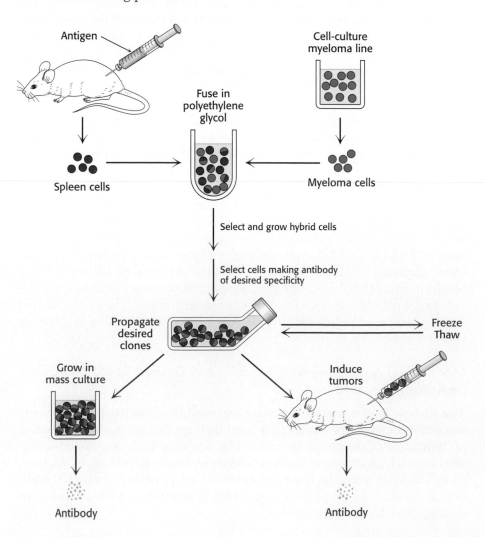

Figure 3.30 Preparation of monoclonal antibodies. Hybridoma cells are formed by the fusion of antibody-producing cells and myeloma cells. The hybrid cells are allowed to proliferate by growing them in selective medium. They are then screened to determine which ones produce antibody of the desired specificity. [After C. Milstein. Monoclonal antibodies. Copyright © 1980 by Scientific American, Inc. All rights reserved.]

The hybridoma method of producing monoclonal antibodies has opened new vistas in biology and medicine. *Large amounts of identical antibodies with tailor-made specificities can be readily prepared. They are sources of insight into relations between antibody structure and specificity. Moreover, monoclonal antibodies can serve as precise analytical and preparative reagents.* For example, a pure antibody can be obtained against an antigen that has not yet been isolated (p. 000). Proteins that guide development have been identified with the use of monoclonal antibodies as tags (Figure 3.31). Monoclonal antibodies attached to solid supports can be used as affinity columns to purify scarce proteins. This method has been used to purify interferon (an antiviral protein) 5000-fold from a crude mixture. *Clinical laboratories are using monoclonal antibodies in many assays.* For example, the detection in blood of isozymes that are normally localized in the heart points to a myocardial infarction (heart attack). Blood transfusions have been made safer by antibody screening of donor blood for viruses that cause AIDS (acquired immune deficiency syndrome), hepatitis, and other infectious diseases. Monoclonal antibodies also find uses as therapeutic agents, as in the treatment of cancer. For example, trastuzumab (Herceptin®) is a monoclonal antibody useful for treating some forms of breast cancer.

Figure 3.31 Fluorescence micrograph of a developing *Drosophila* embryo. The embryo was stained with a fluorescence-labeled monoclonal antibody for the DNA-binding protein encoded by *engrailed*, an essential gene in specifying the body plan. [Courtesy of Dr. Nipam Patel and Dr. Corey Goodman.]

Proteins Can Be Detected and Quantified by Using an Enzyme-Linked Immunosorbent Assay

Antibodies can be used as exquisitely specific analytic reagents to quantify the amount of a protein or other antigen. The technique is the *enzyme-linked immunosorbent assay* (ELISA). This method makes use of an enzyme that reacts with a colorless substrate to produce a colored product. The enzyme is covalently linked to a specific antibody that recognizes a target antigen. If the antigen is present, the antibody–enzyme complex will bind to it, and on the addition of the substrate the enzyme will catalyze the reaction generating the colored product. Thus, the presence of the colored product indicates the presence of the antigen. Such an enzyme-linked immunosorbent assay, which is rapid and convenient, can detect less than a nanogram (10^{-9} g) of a protein. ELISA can be performed with either polyclonal or monoclonal antibodies, but the use of monoclonal antibodies yields more-reliable results.

We will consider two among the several types of ELISA. *The indirect ELISA is used to detect the presence of antibody* and is the basis of the test for HIV infection. The HIV test detects the presence of antibodies that recognize viral core proteins, the antigen. Viral core proteins are adsorbed to the bottom of a well. Antibodies from the person being tested are then added to the coated well. Only someone infected with HIV will have antibodies that bind to the antigen. Finally, enzyme-linked antibodies to human antibodies (e.g., goat antibodies that recognize human antibodies) are allowed to react in the well, and unbound antibodies are removed by washing. Substrate is then applied. An enzyme reaction suggests that the enzyme-linked antibodies were bound to human antibodies, which in turn implies that the patient has antibodies to the viral antigen (Figure 3.32).

The sandwich ELISA is used to detect antigen rather than antibody. Antibody to a particular antigen is first adsorbed to the bottom of a well. Next, blood or urine containing the antigen is added to the well and binds to the antibody. Finally, a second, different antibody to the antigen is added. This antibody is enzyme linked and is processed as described for indirect ELISA. In this case, the extent of reaction is directly proportional to the amount of antigen present. Consequently, it permits the measurement of small quantities of antigen (see Figure 3.32).

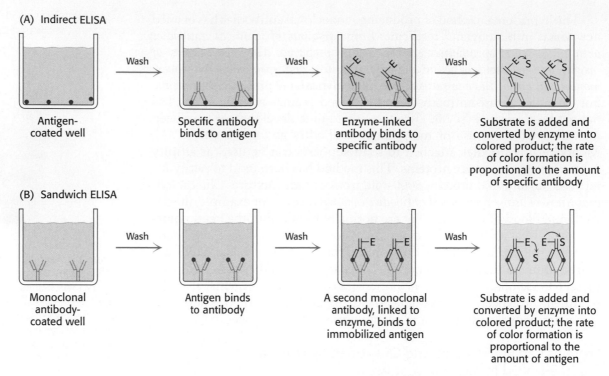

(A) Indirect ELISA

| Antigen-coated well | Wash → | Specific antibody binds to antigen | Wash → | Enzyme-linked antibody binds to specific antibody | Wash → | Substrate is added and converted by enzyme into colored product; the rate of color formation is proportional to the amount of specific antibody |

(B) Sandwich ELISA

| Monoclonal antibody-coated well | Wash → | Antigen binds to antibody | Wash → | A second monoclonal antibody, linked to enzyme, binds to immobilized antigen | Wash → | Substrate is added and converted by enzyme into colored product; the rate of color formation is proportional to the amount of antigen |

Figure 3.32 Indirect ELISA and sandwich ELISA. (A) In indirect ELISA, the production of color indicates the amount of an antibody to a specific antigen. (B) In sandwich ELISA, the production of color indicates the quantity of antigen. [After R. A. Goldsby, T. J. Kindt, and B. A. Osborne, *Kuby Immunology*, 4th ed. (W. H. Freeman and Company, 2000), p. 162.]

Western Blotting Permits the Detection of Proteins Separated by Gel Electrophoresis

Very small quantities of a protein of interest in a cell or in body fluid can be detected by an immunoassay technique called *Western blotting* (Figure 3.33). A sample is subjected to electrophoresis on an SDS–polyacrylamide gel. The resolved proteins on the gel are transferred to the surface of a polymer sheet by blotting to make them more accessible for reaction. An anti-

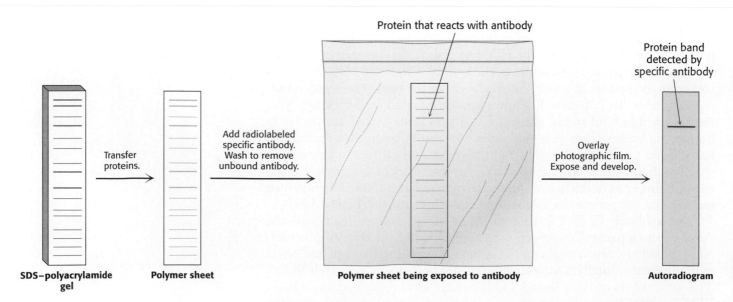

Figure 3.33 Western blotting. Proteins on an SDS–polyacrylamide gel are transferred to a polymer sheet and stained with radioactive antibody. A band corresponding to the protein to which the antibody binds appears in the autoradiogram.

body that is specific for the protein of interest is added to the sheet and reacts with the antigen. The antibody–antigen complex on the sheet can then be detected by rinsing the sheet with a second antibody specific for the first (e.g., goat antibody that recognizes mouse antibody). A radioactive label on the second antibody produces a dark band on x-ray film (an *autoradiogram*). Alternatively, an enzyme on the second antibody generates a colored product, as in the ELISA method. Western blotting makes it possible to find a protein in a complex mixture, the proverbial needle in a haystack. It is the basis for the test for infection by hepatitis C, where it is used to detect a core protein of the virus. This technique is also very useful in monitoring protein purification and in the cloning of genes.

Fluorescent Markers Make Possible the Visualization of Proteins in the Cell

Biochemistry is often performed in test tubes or polyacrylamide gels. However, most proteins function in the context of a cell. Fluorescent markers provide a powerful means of examining proteins in their biological context. Cells can be stained with fluorescence-labeled antibodies and examined by *fluorescence microscopy* to reveal the location of a protein of interest. For example, arrays of parallel bundles are evident in cells stained with antibody specific for actin, a protein that polymerizes into filaments (Figure 3.34). Actin filaments are constituents of the cytoskeleton, the internal scaffolding of cells that controls their shape and movement. By tracking protein location, fluorescent markers also provide clues to protein function. For instance, the glucocorticoid receptor protein binds to the steroid hormone cortisone. The receptor was linked to *green fluorescent protein* (GFP), a naturally fluorescent protein isolated from the jellyfish *Aequorea victoria* (p. 58). Fluorescence microscopy revealed that, in the absence of the hormone, the receptor is located in the cytoplasm (Figure 3.35A). On addition of the steroid, the receptor is translocated to the nucleus, where it binds to DNA (Figure 3.35B). These results suggested that glucocorticoid receptor protein is a transcription factor that controls gene expression.

The highest resolution of fluorescence microscopy is about 0.2 μm (200 nm, or 2000 Å), the wavelength of visible light. Finer spatial resolution can be achieved by electron microscopy if the antibodies are tagged with electron-dense markers. For example, antibodies conjugated to clusters of gold or to ferritin (which has an electron-dense core rich in iron) are highly visible under

Figure 3.34 Actin filaments. Fluorescence micrograph of actin filaments in a cell stained with an antibody specific to actin. [Courtesy of Dr. Elias Lazarides.]

(A)

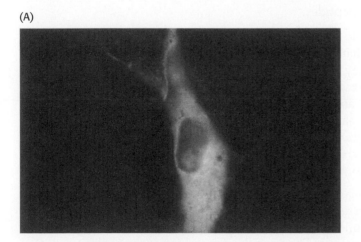

(B)

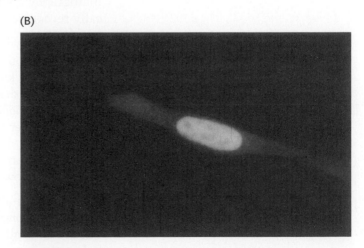

Figure 3.35 Nuclear localization of a steroid receptor. (A) The receptor, made visible by attachment of the green fluorescent protein, is located predominantly in the cytoplasm of the cultured cell. (B) Subsequent to the addition of corticosterone (a glucocorticoid steroid), the receptor moves into the nucleus. [Courtesy of Dr. William B. Pratt.]

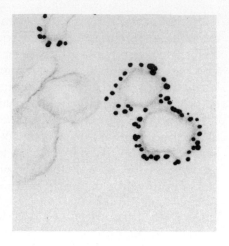

Figure 3.36 Immunoelectron microscopy. The opaque particles (150-Å, or 15-nm, diameter) in this electron micrograph are clusters of gold atoms bound to antibody molecules. These membrane vesicles from the synapses of neurons contain a channel protein that is recognized by the specific antibody. [Courtesy of Dr. Peter Sargent.]

the electron microscope. *Immunoelectron microscopy* can define the position of antigens to a resolution of 10 nm (100 Å) or finer (Figure 3.36).

3.4   Peptides Can Be Synthesized by Automated Solid-Phase Methods

Peptides of defined sequence can be synthesized to assist in biochemical analysis. These peptides are valuable tools for several purposes.

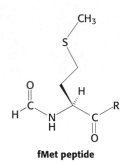

fMet peptide

1.   *Synthetic peptides can serve as antigens to stimulate the formation of specific antibodies.* Suppose we want to isolate the protein expressed by a specific gene. Peptides can be synthesized that match the translation of part of the gene's nucleic acid sequence, and antibodies can be generated that target these peptides. These antibodies can then be used to isolate the intact protein or localize it within the cell.

2.   *Synthetic peptides can be used to isolate receptors for many hormones and other signal molecules.* For example, white blood cells are attracted to bacteria by formylmethionyl (fMet) peptides released in the breakdown of bacterial proteins. Synthetic formylmethionyl peptides have been useful in identifying the cell-surface receptor for this class of peptide. Moreover, synthetic peptides can be attached to agarose beads to prepare affinity chromatography columns for the purification of receptor proteins that specifically recognize the peptides.

3.   *Synthetic peptides can serve as drugs.* Vasopressin is a peptide hormone that stimulates the reabsorption of water in the distal tubules of the kidney, leading to the formation of more-concentrated urine. Patients with diabetes insipidus are deficient in vasopressin (also called *antidiuretic hormone*), and so they excrete large volumes of dilute urine (more than 5 liters per day) and are continually thirsty. This defect can be treated by administering 1-desamino-8-D-arginine vasopressin, a synthetic analog of the missing hormone (Figure 3.37). This synthetic peptide is degraded in vivo much more slowly than vasopressin and does not increase blood pressure.

4.   Finally, *studying synthetic peptides can help define the rules governing the three-dimensional structure of proteins.* We can ask whether a particular sequence by itself tends to fold into an α helix, a β strand, a hairpin turn, or behaves as a random coil. The peptides created for such studies can incorporate amino acids not normally found in proteins, allowing more variation in chemical structure than is possible with the use of only 20 possible amino acids.

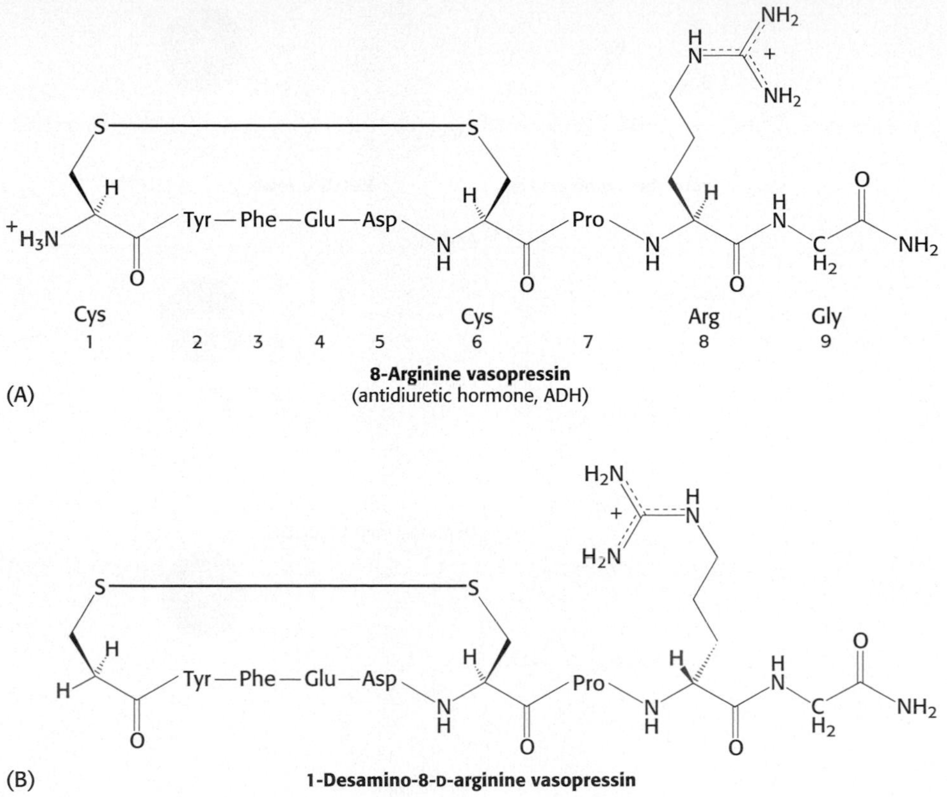

(A)

8-Arginine vasopressin
(antidiuretic hormone, ADH)

(B) **1-Desamino-8-D-arginine vasopressin**

Figure 3.37 Vasopressin and a synthetic vasopressin analog. Structural formulas of (A) vasopressin, a peptide hormone that stimulates water resorption, and (B) 1-desamino-8-D-arginine vasopressin, a more stable synthetic analog of this antidiuretic hormone.

How are these peptides constructed? The amino group of one amino acid is linked to the carboxyl group of another. However, a unique product is formed only if a single amino group and a single carboxyl group are available for reaction. Therefore, it is necessary to block some groups and to activate others to prevent unwanted reactions. The α-amino group of the first amino acid of the desired peptide is blocked with a protecting group such as a *tert*-butyloxycarbonyl (*t*-Boc) group, yielding a *t*-Boc amino acid. The carboxyl group of this same amino acid is activated by reaction with a reagent such as *dicyclohexylcarbodiimide* (DCC), as illustrated in Figure 3.38. The free amino group of the next amino acid to be linked attacks the activated carboxyl group, leading to the formation of a peptide bond and the release of dicyclohexylurea. The carboxyl group of the resulting dipeptide is activated with DCC and undergoes reaction with the free amino group of the amino acid that will be the third residue in the peptide. This process is repeated until the desired peptide is synthesized. Exposing the peptide to dilute acid removes the *t*-Boc protecting group from the first amino acid but leaves the peptide bonds intact.

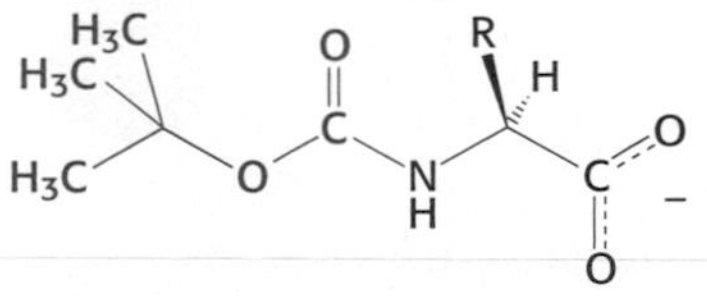

Peptides containing more than 50 amino acids can be synthesized by sequential repetition of the preceding reactions. Linking the growing peptide chain to an insoluble matrix, such as polystyrene beads, greatly enhances efficiency. A major advantage of this *solid-phase method,* first developed by R. Bruce Merrifield, is that the desired product at each stage is bound to beads that can be rapidly filtered and washed, and so there is no need to purify intermediates. All reactions are carried out in a single vessel, eliminating losses caused by repeated transfers of products. The carboxyl-terminal amino acid of the desired peptide sequence is first anchored to the polystyrene beads (Figure 3.39). The *t*-Boc protecting group of this amino acid is then removed. The next amino acid (in the protected *t*-Boc form) and DCC, the coupling agent, are added together. After the peptide bond forms, excess reagents and dicyclohexylurea are washed away, leaving the desired dipeptide product attached to the beads. Additional amino acids

t-Butyloxycarbonyl amino acid
(*t*-Boc amino acid)

Dicyclohexylcarbodiimide
(DCC)

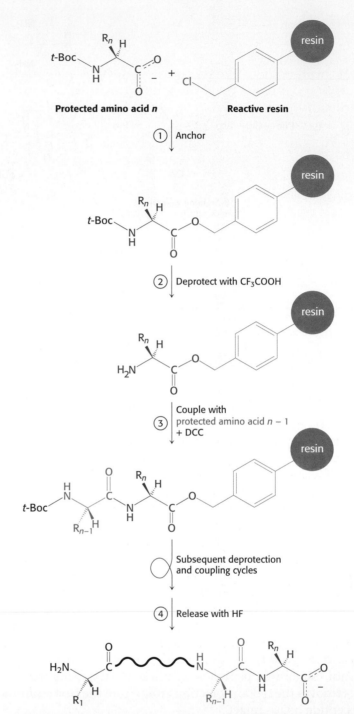

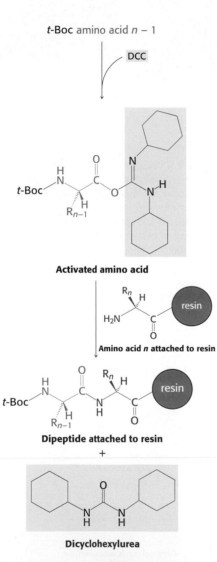

Figure 3.38 Amino acid activation.
Dicyclohexylcarbodiimide is used to
activate carboxyl groups for the formation
of peptide bonds.

Figure 3.39 Solid-phase peptide synthesis. The sequence of steps in solid-phase synthesis
is: (1) anchoring of the C-terminal amino acid, (2) deprotection of the amino terminus, and
(3) coupling of the next residue. Steps 2 and 3 are repeated for each added amino acid.
Finally, in step 4, the completed peptide is released from the resin.

are linked by the same sequence of reactions. At the end of the synthesis, the
peptide is released from the beads by adding hydrofluoric acid (HF), which
cleaves the carboxyl ester anchor without disrupting peptide bonds.
Protecting groups on potentially reactive side chains, such as that of lysine,
also are removed at this time. This cycle of reactions can be readily automated, which makes it feasible to routinely synthesize peptides containing
about 50 residues in good yield and purity. In fact, the solid-phase method
has been used to synthesize interferons (155 residues) that have antiviral

activity and ribonuclease (124 residues) that is catalytically active. The protecting groups and cleavage agents may be varied for increased flexibility or convenience.

Synthetic peptides may be linked to create even longer molecules. Using specially developed *peptide ligation* methods, proteins of 100 amino acids or more can by synthesized in very pure form. These methods enable the construction of even sharper tools for examining protein structure and function.

3.5 Mass Spectrometry Provides Powerful Tools for Protein Characterization and Identification

Modifications to the well-established technique of mass spectrometry now make it possible to determine protein masses with an accuracy of one mass unit or less in favorable cases. This ability has opened up many new horizons in the study of proteins. Most powerfully, the mass of a peptide or protein can be used as a name tag for picking out a specific molecule in a vast database of amino acid sequences.

The Mass of a Protein Can Be Precisely Determined by Mass Spectrometry

Mass spectrometry is a technique for analyzing ionized forms of molecules in the gas phase. It is most readily applied to gases or to volatile liquids that easily release gas-phase ions. Mass measurements are obtained by determining how readily an ion is accelerated in an applied electric field. Consider two ions with the same overall charge but with different masses. In a given electric field, the same force will act on each ion. However, the acceleration of the more massive ion due to this force will be less according to Newton's third law, $F = ma$, where F is the force, m is the mass, and a is the acceleration. Thus, a measurement of the acceleration in a known applied force provides the mass.

Because proteins and peptides are not volatile, generating a sufficiently high concentration of ionized but intact protein molecules in the gas phase is a great challenge. Two widely used methods, *matrix-assisted laser desorption–ionization* (MALDI) and *electrospray ionization* (ESI), have been developed to solve this problem. In MALDI, the protein or peptide under study is coprecipitated with an organic compound that absorbs laser light of an appropriate wavelength (the "matrix"). The flash of a laser on the preparation expels molecules from the surface. These molecules capture electrons as they exit the matrix and hence leave as negatively charged ions. In ESI, solutions containing the protein or peptide flow through a fine metallic tip held at a nonzero electrical potential. Fine electrically charged droplets are released containing both the protein and the solvent. The solvent evaporates from the droplet, concentrating the charge. The mutual repulsion of the like-charged molecules increases, and individual molecular ions fly free. These ions are usually multiply charged.

After gas-phase ions have been generated, several approaches may be used to determine their mass. In *time of flight* (TOF) analysis, the ions are accelerated in an electric field toward a detector (Figure 3.40). The lighter ions are accelerated more, travel faster, and arrive at the detector first. Tiny amounts of biomolecules, as small as a few picomoles (pmol) to femtomoles (fmol), can be analyzed in this manner. A MALDI-TOF mass spectrum for a mixture of the proteins insulin and β-lactoglobulin is shown in Figure 3.41. The masses determined by MALDI-TOF are 5733.9 and 18,364,

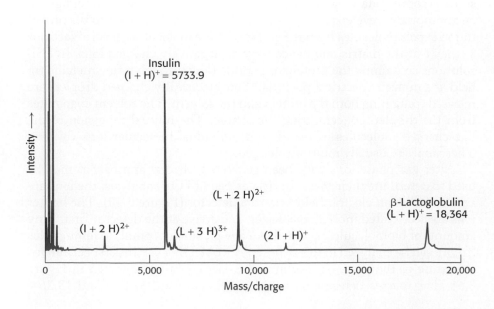

Figure 3.40 MALDI-TOF mass
spectrometry. (1) The protein sample,
embedded in an appropriate matrix, is
ionized by the application of a laser beam.
(2) An electrical field accelerates the ions
through the flight tube toward the
detector. (3) The lightest ions arrive first.
(4) The ionizing laser pulse also triggers a
clock that measures the time of flight
(TOF) for the ions. [After J. T. Watson,
Introduction to Mass Spectrometry, 3d ed.
(Lippincott-Raven, 1997), p. 279.]

respectively, compared with calculated values of 5733.5 and 18,388.
MALDI-TOF is indeed an accurate means of determining protein mass.

Individual Components of Large Protein Complexes Can Be Identified by MALDI-TOF Mass Spectrometry

Although protein masses serve as convenient name tags for distinguishing
proteins, the mass of a given protein is usually not enough to uniquely identify it among all possible proteins within a cell. However, the mass of the

Figure 3.41 MALDI-TOF mass spectrum of
insulin and β-lactoglobulin. A mixture of
5 pmol each of insulin (I) and β-
lactoglobulin (L) was ionized by MALDI,
which produces predominately singly
charged molecular ions from peptides and
proteins—the insulin ion $(I + H)^+$ and the
lactoglobulin ion $(L + H)^+$. However,
molecules with multiple charges as well as
small quantities of a singly charged dimer
of insulin $(2 I + H)^+$ also are produced.
[After J. T. Watson, *Introduction to Mass
Spectrometry*, 3d ed. (Lippincott-Raven,
1997), p. 282.]

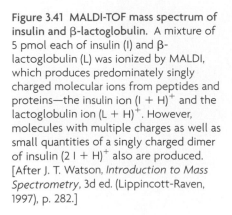

parent protein along with the masses of several protein fragments produced by a specific cleavage method can provide unique identification. Suppose we wish to identify proteins within a two-dimensional gel such as that described on page 74. After gel electrophoresis, the molecules in individual spots can be cleaved, often in the gel matrix itself, by using a protease such as trypsin. The mixture of fragments produced can then be analyzed by MALDI-TOF mass spectrometry. These peptide masses are matched against proteins in a database that have been "electronically cleaved" by a computer simulating the same fragmentation technique used for the experimental sample. In this way, the proteome within a given cell type or other sample can be analyzed in considerable detail.

As an example of the power of this proteomic approach, consider the analysis of the nuclear-pore complex from yeast, which facilitates transport of large molecules into and out of the nucleus. This huge macromolecular complex was purified intact from yeast cells by careful procedures. The purified complex was fractionated by HPLC followed by gel electrophoresis. Individual bands from the gel were isolated, cleaved with trypsin, and analyzed by MALDI-TOF mass spectrometry. The fragments produced were compared with amino acid sequences deduced from the DNA sequence of the yeast genome as shown in Figure 3.42. A total of 174 proteins were identified in this manner. Of these proteins, 40 were confirmed to be components of the nuclear-pore complex by other methods. Many of these proteins had not previously been identified as being associated with the nuclear pore despite years of study. Furthermore, mass spectrometric methods are sensitive enough to detect essentially all components of the pore if they are present in the samples used. Thus, a complete list of the components constituting this macromolecular complex could be obtained in a straightforward manner. Proteomic analysis of this type is growing in power as mass spectrometric and biochemical fractionation methods are refined.

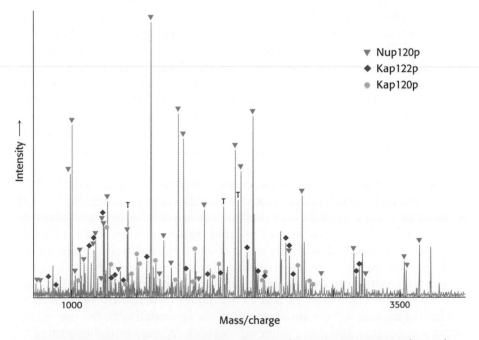

Figure 3.42 Proteomic analysis by mass spectrometry. This mass spectrum was obtained by analyzing a trypsin-treated band in a gel derived from a yeast nuclear-pore sample. Many of the peaks were found to match the masses predicted for peptide fragments from three proteins (Nup120p, Kap122p, and Kap120p) within the yeast genome. The band corresponded to an apparent molecular mass of 100 kd. [From M. P. Rout, J. D. Aitchison, A. Suprapto, K. Hjertaas, Y. Zhao, and B. T. Chait. *J. Cell Biol.* 148(2000):635–651.]

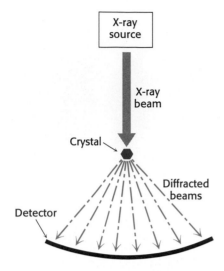

Figure 3.43 An x-ray crystallographic experiment. An x-ray source generates a beam, which is diffracted by a crystal. The resulting diffraction pattern is collected on a detector.

3.6 Three-Dimensional Protein Structure Can Be Determined by X-ray Crystallography and NMR Spectroscopy

A crucial question is, What is the three-dimensional structure of a specific protein? Protein structure is critical for determining protein function because the specificity of active sites and binding sites depends on the precise three-dimensional conformation. X-ray crystallography and nuclear magnetic resonance spectroscopy are the two most important techniques for elucidating the conformation of proteins.

X-ray Crystallography Reveals Three-Dimensional Structure in Atomic Detail

X-ray crystallography was the first method developed to determine protein structure in atomic detail and still provides the clearest visualization of protein structure currently available. This technique can reveal the precise three-dimensional positions of most atoms in a protein molecule. Of all forms of radiation, x-rays provide the best resolution because the wavelength of x-rays is about the same length as that of a covalent bond. The three components in an x-ray crystallographic analysis are a *protein crystal,* a *source of x-rays,* and a *detector* (Figure 3.43).

The technique requires a sample with all protein molecules oriented in a fixed way with respect to one another; crystals have this property. Thus, the first step is to obtain crystals of the protein of interest. Slowly adding ammonium sulfate or another salt to a concentrated solution of protein to reduce its solubility favors the formation of highly ordered crystals—the process of salting out discussed on page 68. For example, myoglobin crystallizes in 3 M ammonium sulfate. Some proteins crystallize readily, whereas others do so only after much effort has been expended in identifying the right conditions. Methods for screening many different crystallization conditions using a relatively small amount of protein sample have been developed. Typically, hundreds of conditions must be tested to obtain crystals fully suitable for crystallographic studies. Increasingly large and complex proteins are being crystallized. For example, poliovirus, an 8500-kd assembly of 240 protein subunits surrounding an RNA core, has been crystallized and its structure solved by x-ray methods. Crucially, proteins frequently crystallize in their biologically active configuration. Indeed, enzyme crystals may display catalytic activity if the crystals are suffused with substrate.

Next, a source of x-rays is required. A beam of x-rays of wavelength 1.54 Å is produced by accelerating electrons against a copper target. Equipment suitable for generating x-rays in this manner is available in many laboratories. Alternatively, x-rays can be produced by accelerating electrons in circular orbits at speeds close to the speed of light. Several facilities generating such *synchrotron radiation* are available throughout the world such as the Advanced Light Source at Argonne National Laboratory outside Chicago and the Photon Factory in Tsukuba City, Japan. Synchrotron-generated x-ray beams are much more intense than those generated with the use of laboratory apparatus and have other advantages. Whatever the source of x-rays, a narrow beam of x-rays strikes the protein crystal. Part of the beam goes straight through the crystal; the rest is *scattered* in various directions. Finally, these scattered, or *diffracted,* x-rays can be detected by x-ray film, the blackening of the emulsion being proportional to the intensity of the scattered x-ray beam, or by a solid-state electronic detector. The scattering pattern provides abundant information about protein structure. The basic physical principles underlying the technique are:

1.   *Electrons scatter x-rays.* The amplitude of the wave scattered by an atom is proportional to its number of electrons. Thus, a carbon atom scatters six times as strongly as a hydrogen atom does.

2.   *The scattered waves recombine.* Each atom contributes to each scattered beam. The scattered waves reinforce one another at the film or detector if they are in phase (in step) there, and they cancel one another if they are out of phase.

3.   *The way in which the scattered waves recombine depends only on the atomic arrangement.*

The protein crystal is mounted and positioned in a precise orientation with respect to the x-ray beam and the film. The crystal is rotated so that the beam can strike the crystal from many directions. This rotational motion results in an x-ray photograph consisting of a regular array of spots called *reflections*. The x-ray photograph shown in Figure 3.44 is a two-dimensional section through a three-dimensional array of 25,000 spots. The intensity of each spot is measured. These *intensities and their positions* are the basic experimental data of an x-ray crystallographic analysis. The next step is to reconstruct an image of the protein from the observed intensities. In light microscopy or electron microscopy, the diffracted beams are focused by lenses to directly form an image. However, appropriate lenses for focusing x-rays do not exist. Instead, the image is formed by applying a mathematical relation called a *Fourier transform*. For each spot, this operation yields a wave of electron density whose amplitude is proportional to the square root of the observed intensity of the spot. Each wave also has a *phase*—that is, the timing of its crests and troughs relative to those of other waves. The phase of each wave determines whether the wave reinforces or cancels the waves contributed by the other spots. Additional experiments or calculations must be performed to determine the phases corresponding to each diffraction spot.

Figure 3.44 An x-ray diffraction pattern. X-ray precession photograph from a crystal of myoglobin.

The stage is then set for the calculation of an electron-density map, which gives the density of electrons at a large number of regularly spaced points in the crystal. This three-dimensional electron-density distribution is represented by a series of parallel sections stacked on top of one another. Each section is a transparent plastic sheet (or, more recently, a layer in a computer image) on which the electron-density distribution is represented by contour lines (Figure 3.45), like the contour lines used in geological survey maps to depict altitude (Figure 3.46). The next step is to interpret the electron-density map. A critical factor is the *resolution* of the x-ray analysis, which is determined by the number of scattered intensities used in the Fourier synthesis. The fidelity of the image

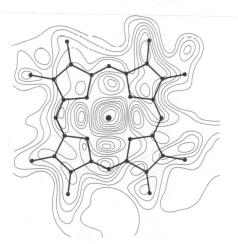

Figure 3.45 Section of the electron-density map of myoglobin. This section of the electron-density map shows the heme group. The peak of the center of this section corresponds to the position of the iron atom. [From J. C. Kendrew. The three-dimensional structure of a protein molecule. Copyright © 1961 by Scientific American, Inc. All rights reserved.]

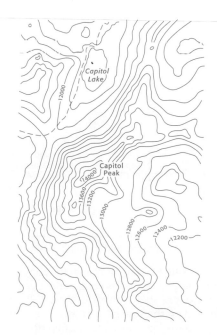

Figure 3.46 Section of a U.S. Geological Survey map. Capitol Peak Quadrangle, Colorado.

Figure 3.47 Resolution affects the quality of
an image. The effect of resolution on the
quality of a reconstructed image is shown
by an optical analog of x-ray diffraction:
(A) a photograph of the Parthenon;
(B) an optical diffraction pattern of the
Parthenon; (C and D) images reconstructed
from the pattern in part B. More data were
used to obtain image D than image C,
which accounts for the higher quality of
image D. [Courtesy of Dr. Thomas Steitz
(part A) and Dr. David DeRosier (part B).]

TABLE 3.4 Biologically important nuclei
giving NMR signals

Nucleus	Natural abundance (% by weight of the element)
^{1}H	99.984
^{2}H	0.016
^{13}C	1.108
^{14}N	99.635
^{15}N	0.365
^{17}O	0.037
^{23}Na	100.0
^{25}Mg	10.05
^{31}P	100.0
^{35}Cl	75.4
^{39}K	93.1

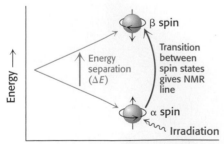

Figure 3.48 Basis of NMR spectroscopy.
The energies of the two orientations of
a nucleus of spin ½ (such as ^{31}P and ^{1}H)
depend on the strength of the
applied magnetic field. Absorption of
electromagnetic radiation of appropriate
frequency induces a transition from the
lower to the upper level.

depends on the resolution of the Fourier synthesis, as shown by the optical
analogy in Figure 3.47. A resolution of 6 Å reveals the course of the polypeptide chain but few other structural details. The reason is that polypeptide
chains pack together so that their centers are between 5 Å and 10 Å apart.
Maps at higher resolution are needed to delineate groups of atoms, which lie
between 2.8 Å and 4.0 Å apart, and individual atoms, which are between 1.0
Å and 1.5 Å apart. The ultimate resolution of an x-ray analysis is determined
by the degree of perfection of the crystal. For proteins, this limiting resolution is often about 2 Å.

Nuclear Magnetic Resonance Spectroscopy Can Reveal the Structures of Proteins in Solution

X-ray crystallography is the most powerful method for determining protein
structures. However, some proteins do not readily crystallize. Furthermore,
although structures present in crystallized proteins have been amply
demonstrated to very closely represent those of proteins free of the constraints imposed by the crystalline environment, structures in solution can
be sources of additional insights. *Nuclear magnetic resonance* (NMR) *spectroscopy* is unique in being able to reveal the atomic structure of macromolecules *in solution*, provided that highly concentrated solutions (~1 mM, or
15 mg ml^{-1} for a 15-kd protein) can be obtained. This technique depends
on the fact that certain atomic nuclei are intrinsically magnetic. Only a limited number of isotopes display this property, called *spin*, and the ones most
important to biochemistry are listed in Table 3.4. The simplest example is
the hydrogen nucleus (^{1}H), which is a proton. The spinning of a proton generates a magnetic moment. This moment can take either of two orientations,
or spin states (called α and β), when an external magnetic field is applied
(Figure 3.48). The energy difference between these states is proportional to
the strength of the imposed magnetic field. The α state has a slightly lower
energy and hence is slightly more populated (by a factor of the order of
1.00001 in a typical experiment) because it is aligned with the field. A spinning proton in an α state can be raised to an excited state (β state) by applying a pulse of electromagnetic radiation (a radio-frequency, or RF, pulse),
provided that the frequency corresponds to the energy difference between
the α and the β states. In these circumstances, the spin will change from α
to β; in other words, *resonance* will be obtained. A resonance spectrum for a
molecule is obtained by keeping the magnetic field constant and varying the
frequency of the electromagnetic radiation.

These properties can be used to examine the chemical surroundings of the hydrogen nucleus. The flow of electrons around a magnetic nucleus generates a small local magnetic field that opposes the applied field. The degree of such shielding depends on the surrounding electron density. Consequently, nuclei in different environments will change states, or resonate, at slightly different field strengths or radiation frequencies. The nuclei of the perturbed sample absorb electromagnetic radiation at a frequency that can be measured. The different frequencies, termed *chemical shifts*, are expressed in fractional units δ (parts per million, or ppm) relative to the shifts of a standard compound, such as a water-soluble derivative of tetramethysilane, that is added with the sample. For example, a $-CH_3$ proton typically exhibits a chemical shift (δ) of 1 ppm, compared with a chemical shift of 7 ppm for an aromatic proton. The chemical shifts of most protons in protein molecules fall between 0 and 9 ppm (Figure 3.49). Most protons in many proteins can be resolved by using this technique of *one-dimensional NMR*. With this information, we can then deduce changes to a particular chemical group under different conditions, such as the conformational change of a protein from a disordered structure to an α helix in response to a change in pH.

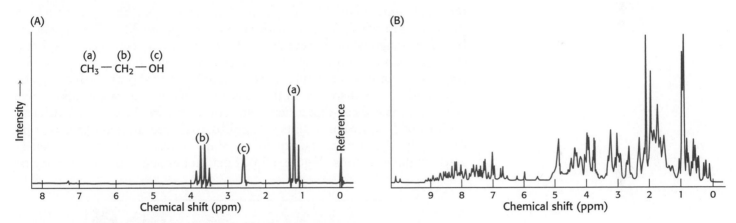

Figure 3.49 One-dimensional NMR spectra. (A) ^{1}H-NMR spectrum of ethanol (CH_3CH_2OH) shows that the chemical shifts for the hydrogen are clearly resolved. (B) ^{1}H-NMR spectrum from a 55 amino acid fragment of a protein with a role in RNA splicing shows a greater degree of complexity. A large number of peaks are present and many overlap. [(A) After C. Branden and J. Tooze, *Introduction to Protein Structure* (Garland, 1991), p. 280; (B) courtesy of Dr. Barbara Amann and Dr. Wesley McDermott.]

We can garner even more information by examining how the spins on different protons affect their neighbors. By inducing a transient magnetization in a sample through the application of a radio-frequency pulse, we can alter the spin on one nucleus and examine the effect on the spin of a neighboring nucleus. Especially revealing is a *two-dimensional spectrum obtained by nuclear Overhauser enhancement spectroscopy* (NOESY), *which graphically displays pairs of protons that are in close proximity*, even if they are not close together in the primary structure. The basis for this technique is the *nuclear Overhauser effect* (NOE), an interaction between nuclei that is proportional to the inverse sixth power of the distance between them. Magnetization is transferred from an excited nucleus to an unexcited one if the two nuclei are less than about 5 Å apart (Figure 3.50A). In other words, the effect provides a means of detecting the location of atoms relative to one another in the three-dimensional structure of the protein. The diagonal of a NOESY spectrum corresponds to a one-dimensional spectrum. The off-diagonal peaks provide crucial new information: *they identify pairs of protons that are less*

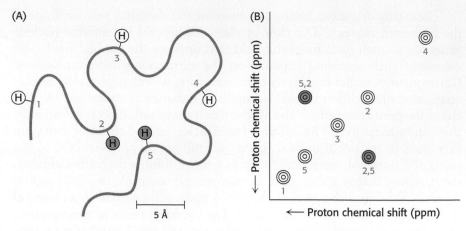

Figure 3.50 The nuclear Overhauser effect. The nuclear Overhauser effect (NOE) identifies pairs of protons that are in close proximity. (A) Schematic representation of a polypeptide chain highlighting five particular protons. Protons 2 and 5 are in close proximity (~4 Å apart), whereas other pairs are farther apart. (B) A highly simplified NOESY spectrum. The diagonal shows five peaks corresponding to the five protons in part A. The peak above the diagonal and the symmetrically related one below reveal that proton 2 is close to proton 5.

than 5 Å apart (Figure 3.50B). A two-dimensional NOESY spectrum for a protein comprising 55 amino acids is shown in Figure 3.51. The large number of off-diagonal peaks reveals short proton–proton distances. The three-dimensional structure of a protein can be reconstructed with the use of such proximity relations. Structures are calculated such that protons that must be separated by less than 5 Å on the basis of NOESY spectra are close to one another in the three-dimensional structure (Figure 3.52). If a sufficient number of distance constraints are applied, the three-dimensional structure can nearly be determined uniquely. A family of related structures is generated for three reasons (Figure 3.53). First, not enough constraints may be

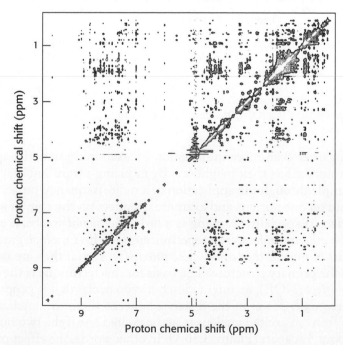

Figure 3.51 Detecting short proton–proton distances. A NOESY spectrum for a 55 amino acid domain from a protein having a role in RNA splicing. Each off-diagonal peak corresponds to a short proton–proton separation. This spectrum reveals hundreds of such short proton–proton distances, which can be used to determine the three-dimensional structure of this domain. [Courtesy of Dr. Barbara Amann and Dr. Wesley McDermott.]

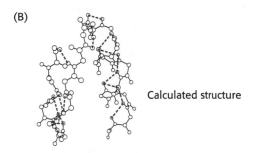

(B)

Calculated structure

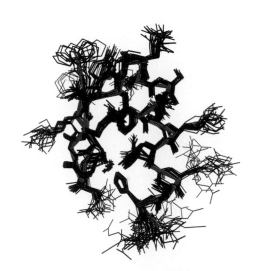

Figure 3.52 **Structures calculated on the basis of NMR constraints.** (A) NOESY observations show that protons (connected by dotted red lines) are close to one another in space. (B) A three-dimensional structure calculated with these proton pairs constrained to be close together.

experimentally accessible to fully specify the structure. Second, the distances obtained from analysis of the NOESY spectrum are only approximate. Finally, the experimental observations are made not on single molecules but on a large number of molecules in solution that may have slightly different structures at any given moment. Thus, the family of structures generated from NMR structure analysis indicates the range of conformations for the protein in solution. At present, NMR spectroscopy can determine the structures of only relatively small proteins (<40 kd), but its resolving power is certain to increase. The power of NMR has been greatly enhanced by the ability of recombinant DNA technology to produce proteins labeled uniformly or at specific sites with ^{13}C, ^{15}N, and ^{2}H (Chapter 5).

The structures of more than 30,000 proteins had been elucidated by x-ray crystallography and NMR spectroscopy by mid 2005, and several new structures are now determined each day. The coordinates are collected at the Protein Data Bank (http://www.rcsb.org/pdb), and the structures can be accessed for visualization and analysis. Knowledge of the detailed molecular architecture of proteins has been a source of insight into how proteins recognize and bind other molecules, how they function as enzymes, how they fold, and how they evolved. This extraordinarily rich harvest is continuing at a rapid pace and is greatly influencing the entire field of biochemistry as well as other biological and physical sciences.

Figure 3.53 **A family of structures.** A set of 25 structures for a 28 amino acid domain from a zinc-finger-DNA-binding protein. The red line traces the average course of the protein backbone. Each of these structures is consistent with hundreds of constraints derived from NMR experiments. The differences between the individual structures are due to a combination of imperfections in the experimental data and the dynamic nature of proteins in solution. [Courtesy of Dr. Barbara Amann.]

Summary

The rapid progress in gene sequencing has advanced another goal of biochemistry—elucidation of the proteome. The proteome is the complete set of proteins expressed and includes information about how they are modified, how they function, and how they interact with other molecules.

3.1 The Purification of Proteins Is an Essential First Step in Understanding Their Function

Proteins can be separated from one another and from other molecules on the basis of such characteristics as solubility, size, charge, and binding affinity. SDS–polyacrylamide gel electrophoresis separates the polypeptide chains of proteins under denaturing conditions largely according to mass. Proteins can also be separated electrophoretically on the basis of net charge by isoelectric focusing in a pH gradient. Ultracentrifugation and gel-filtration chromatography resolve proteins according to size, whereas ion-exchange chromatography separates them mainly on the basis of net charge. The high affinity of many proteins for specific chemical groups is exploited in affinity chromatography, in which proteins bind to columns containing beads bearing covalently linked substrates, inhibitors, or other specifically recognized groups. The mass of a protein can be determined by sedimentation-equilibrium measurements.

3.2 Amino Acid Sequences Can Be Determined by Automated Edman Degradation

The amino acid composition of a protein can be ascertained by hydrolyzing the protein into its constituent amino acids in 6 M HCl at 110°C. The amino acids can be separated by ion-exchange chromatography and quantitated by their reaction with ninhydrin or fluorescamine. Amino acid sequences can be determined by Edman degradation, which removes one amino acid at a time from the amino end of a peptide. Phenyl isothiocyanate reacts with the terminal amino group to form a phenylthiocarbamoyl derivative, which cyclizes under mildly acidic conditions to give a phenylthiohydantoin–amino acid and a peptide shortened by one residue. Automated repeated Edman degradations by a sequenator can analyze sequences of about 50 residues. Longer polypeptide chains are broken into shorter ones for analysis by specifically cleaving them with a reagent such as cyanogen bromide, which splits peptide bonds on the carboxyl side of methionine residues. Enzymes such as trypsin, which cleaves on the carboxyl side of lysine and arginine residues, also are very useful in splitting proteins. Amino acid sequences are rich in information concerning the kinship of proteins, their evolutionary relations, and diseases produced by mutations. Knowledge of a sequence provides valuable clues to conformation and function.

3.3 Immunology Provides Important Techniques with Which to Investigate Proteins

Proteins can be detected and quantitated by highly specific antibodies; monoclonal antibodies are especially useful because they are homogeneous. Enzyme-linked immunosorbent assays and Western blots of SDS–polyacrylamide gels are used extensively. Proteins can also be localized within cells by immunofluorescence microscopy and immunoelectron microscopy.

3.4 Peptides Can Be Synthesized by Automated Solid-Phase Methods

Polypeptide chains can be synthesized by automated solid-phase methods in which the carboxyl end of the growing chain is linked to an insoluble support. The carboxyl group of the incoming amino acid is activated by dicyclohexylcarbodiimide and joined to the amino group of the growing chain. Synthetic peptides can serve as drugs and as antigens to stimulate the formation of specific antibodies. They can also be sources of insight into the relation between amino acid sequence and conformation.

3.5 Mass Spectrometry Provides Powerful Tools for Protein Characterization and Identification

Techniques such as matrix-assisted laser desorption and ionization (MALDI) and electrospray ionization (ESI) allow the generation of ions of proteins and peptides in the gas phase. The mass of such protein ions can be determined with great accuracy and precision. Masses determined by these techniques act as protein name tags because the mass of a protein or peptide is precisely determined by its amino acid composition and, hence, by its sequence. Mass spectrometric techniques are central to proteomics because they make it possible to analyze the constituents of large macromolecular assemblies or other collections of proteins.

3.6 Three-Dimensional Protein Structure Can Be Determined by X-ray Crystallography and NMR Spectroscopy

X-ray crystallography and nuclear magnetic resonance spectroscopy have greatly enriched our understanding of how proteins fold, recognize other molecules, and catalyze chemical reactions. X-ray crystallography is possible because electrons scatter x-rays. The diffraction pattern produced can be analyzed to reveal the arrangement of atoms in a protein. The three-dimensional structures of tens of thousands of proteins are now known in atomic detail. Nuclear magnetic resonance spectroscopy reveals the structure and dynamics of proteins in solution. The chemical shift of nuclei depends on their local environment. Furthermore, the spins of neighboring nuclei interact with each other in ways that provide definitive structural information. This information can be used to determine complete three-dimensional structures of proteins.

Key Terms

proteome (p. 66)

assay (p. 67)

homogenate (p. 67)

salting out (p. 68)

dialysis (p. 69)

gel-filtration chromatography (p. 69)

ion-exchange chromatography (p. 70)

affinity chromatography (p. 70)

high-pressure liquid chromatography (HPLC) (p. 71)

gel electrophoresis (p. 71)

isoelectric point (p. 73)

isoelectric focusing (p. 73)

two-dimensional electrophoresis (p. 74)

sedimentation coefficient (Svedberg units, S) (p. 76)

Edman degradation (p. 79)

phenyl isothiocyanate (p. 79)

cyanogen bromide (CNBr) (p. 80)

overlap peptide (p. 81)

diagonal electrophoresis (p. 82)

antibody (p. 84)

antigen (p. 84)

antigenic determinant (epitope) (p. 84)

monoclonal antibody (p. 85)

enzyme-linked immunosorbent assay (ELISA) (p. 87)

Western blotting (p. 88)

fluorescence microscopy (p. 89)

green fluorescent protein (GFP) (p. 89)

solid-phase method (p. 90)

matrix-assisted laser desorption–ionization (MALDI) (p. 93)

electrospray ionization (ESI) (p. 93)

time of flight (TOF) (p. 93)

x-ray crystallography (p. 96)

Fourier transform (p. 97)

nuclear magnetic resonance (NMR) spectroscopy (p. 98)

Selected Readings

Where to Start

Hunkapiller, M. W., and Hood, L. E. 1983. Protein sequence analysis: Automated microsequencing. *Science* 219:650–659.

Merrifield, B. 1986. Solid phase synthesis. *Science* 232:341–347.

Sanger, F. 1988. Sequences, sequences, sequences. *Annu. Rev. Biochem.* 57:1–28.

Milstein, C. 1980. Monoclonal antibodies. *Sci. Am.* 243(4):66–74.

Moore, S., and Stein, W. H. 1973. Chemical structures of pancreatic ribonuclease and deoxyribonuclease. *Science* 180:458–464.

Books

Creighton, T. E. 1993. *Proteins: Structure and Molecular Properties* (2d ed.). W. H. Freeman and Company.

Kyte, J. 1994. *Structure in Protein Chemistry*. Garland.

Van Holde, K. E., Johnson, W. C., and Ho, P.-S. 1998. *Principles of Physical Biochemistry*. Prentice Hall.

Methods in Enzymology. Academic Press.

Wilson, K., and Walker, J. (Eds.). 2000. *Principles and Techniques of Practical Biochemistry* (5th ed.). Cambridge University Press.

Cantor, C. R., and Schimmel, P. R. 1980. *Biophysical Chemistry*. W. H. Freeman and Company.

Johnstone, R. A. W. 1996. *Mass Spectroscopy for Chemists and Biochemists* (2d ed.). Cambridge University Press.

Wilkins, M. R., Williams, K. L., Appel, R. D., and Hochstrasser, D. F. 1997. *Proteome Research: New Frontiers in Functional Genomics (Principles and Practice)*. Springer Verlag

Protein Purification and Analysis

Deutscher, M. (Ed.). 1997. *Guide to Protein Purification*. Academic Press.

Scopes, R. K., and Cantor, C. 1994. *Protein Purification: Principles and Practice* (3d ed.). Springer Verlag.

Dunn, M. J. 1997. Quantitative two-dimensional gel electrophoresis: From proteins to proteomes. *Biochem. Soc. Trans.* 25:248–254.

Aebersold, R., Pipes, G. D., Wettenhall, R. E., Nika, H., and Hood, L. E. 1990. Covalent attachment of peptides for high sensitivity solid-phase sequence analysis. *Anal. Biochem.* 187:56–65.

Blackstock, W. P., and Weir, M. P. 1999. Proteomics: Quantitative and physical mapping of cellular proteins. *Trends Biotechnol.* 17:121–127.

Ultracentrifugation and Mass Spectrometry

Schuster, T. M., and Laue, T. M. 1994. *Modern Analytical Ultracentrifugation.* Springer Verlag.

Arnott, D., Shabanowitz, J., and Hunt, D. F. 1993. Mass spectrometry of proteins and peptides: Sensitive and accurate mass measurement and sequence analysis. *Clin. Chem.* 39:2005–2010.

Chait, B. T., and Kent, S. B. H. 1992. Weighing naked proteins: Practical, high-accuracy mass measurement of peptides and proteins. *Science* 257:1885–1894.

Jardine, I. 1990. Molecular weight analysis of proteins. *Methods Enzymol.* 193:441–455.

Edmonds, C. G., Loo, J. A., Loo, R. R., Udseth, H. R., Barinaga, C. J., and Smith, R. D. 1991. Application of electrospray ionization mass spectrometry and tandem mass spectrometry in combination with capillary electrophoresis for biochemical investigations. *Biochem. Soc. Trans.* 19:943–947.

Li, L., Garden, R. W., and Sweedler, J. V. 2000. Single-cell MALDI: A new tool for direct peptide profiling. *Trends Biotechnol.* 18: 151–160.

Pappin, D. J. 1997. Peptide mass fingerprinting using MALDI-TOF mass spectrometry. *Methods Mol. Biol.* 64:165–173.

Yates, J. R., 3rd. 1998. Mass spectrometry and the age of the proteome. *J. Mass Spectrom.* 33:1–19.

Proteomics

Yates, J. R., 3rd. 2004. Mass spectral analysis in proteomics. *Annu. Rev. Biophys. Biomol. Struct.* 33:297–316.

Weston, A. D., and Hood, L. 2004. Systems biology, proteomics, and the future of health care: Toward predictive, preventative, and personalized medicine. *J. Proteome Res.* 3:179–196.

Pandey, A., and Mann, M. 2000. Proteomics to study genes and genomes. *Nature* 405:837–846.

Dutt, M. J., and Lee, K. H. 2000. Proteomic analysis. *Curr. Opin. Biotechnol.* 11:176–179.

Rout, M. P., Aitchison, J. D., Suprapto, A., Hjertaas, K., Zhao, Y., and Chait, B. T. 2000. The yeast nuclear pore complex: Composition, architecture, and transport mechanism. *J. Cell Biol.* 148:635–651.

X-ray Crystallography and NMR Spectroscopy

Glusker, J. P. 1994. X-ray crystallography of proteins. *Methods Biochem. Anal.* 37:1–72.

Moffat, K. 2003. The frontiers of time-resolved macromolecular crystallography: Movies and chirped X-ray pulses. *Faraday Discuss.* 122:65–88.

Bax, A. 2003. Weak alignment offers new NMR opportunities to study protein structure and dynamics. *Protein Sci.* 12:1–16.

Wery, J. P., and Schevitz, R. W. 1997. New trends in macromolecular x-ray crystallography. *Curr. Opin. Chem. Biol.* 1:365–369.

Wüthrich, K. 1989. Protein structure determination in solution by nuclear magnetic resonance spectroscopy. *Science* 243:45–50.

Clore, G. M., and Gronenborn, A. M. 1991. Structures of larger proteins in solution: Three- and four-dimensional heteronuclear NMR spectroscopy. *Science* 252:1390–1399.

Wüthrich, K. 1986. *NMR of Proteins and Nucleic Acids.* Wiley-Interscience.

Monoclonal Antibodies and Fluorescent Molecules

Köhler, G., and Milstein, C. 1975. Continuous cultures of fused cells secreting antibody of predefined specificity. *Nature* 256:495–497.

Goding, J. W. 1996. *Monoclonal Antibodies: Principles and Practice.* Academic Press.

Immunology Today. 2000. Volume 21, issue 8.

Tsien, R. Y. 1998. The green fluorescent protein. *Annu. Rev. Biochem.* 67:509–544.

Kendall, J. M., and Badminton, M. N. 1998. *Aequorea victoria* bioluminescence moves into an exciting era. *Trends Biotechnol.* 16:216–234.

Chemical Synthesis of Proteins

Bang, D., Chopra, N., and Kent, S. B. 2004. Total chemical synthesis of crambin. *J. Am. Chem. Soc.* 126:1377–1383.

Dawson, P. E., and Kent, S. B. 2000. Synthesis of native proteins by chemical ligation. *Annu. Rev. Biochem.* 69:923–960.

Mayo, K. H. 2000. Recent advances in the design and construction of synthetic peptides: For the love of basics or just for the technology of it. *Trends Biotechnol.* 18:212–217.

Problems

1. *Valuable reagents.* The following reagents are often used in protein chemistry:

CNBr	Trypsin	Ninhydrin
Urea	Performic acid	Phenyl isothiocyanate
Mercaptoethanol	6 N HCl	Chymotrypsin

Which one is the best suited for accomplishing each of the following tasks?

(a) Determination of the amino acid sequence of a small peptide.

(b) Reversible denaturation of a protein devoid of disulfide bonds. Which additional reagent would you need if disulfide bonds were present?

(c) Hydrolysis of peptide bonds on the carboxyl side of aromatic residues.

(d) Cleavage of peptide bonds on the carboxyl side of methionines.

(e) Hydrolysis of peptide bonds on the carboxyl side of lysine and arginine residues.

2. *Finding an end.* Anhydrous hydrazine ($H_2N—NH_2$) has been used to cleave peptide bonds in proteins. What are the reaction products? How might this technique be used to identify the carboxyl-terminal amino acid?

3. *Crafting a new breakpoint.* Ethyleneimine reacts with cysteine side chains in proteins to form *S*-aminoethyl derivatives. The peptide bonds on the carboxyl side of these modified cysteine residues are susceptible to hydrolysis by trypsin. Why?

4. *Spectrometry.* The absorbance A of a solution is defined as

$$A = \log_{10}(I_0/I)$$

in which I_0 is the incident light intensity and I is the transmitted light intensity. The absorbance is related to the molar absorption coefficient (extinction coefficient) ε (in M^{-1} cm^{-1}), concentration c (in M), and path length l (in cm) by

$$A = \varepsilon l c$$

The absorption coefficient of myoglobin at 580 nm is 15,000 M^{-1} cm^{-1}. What is the absorbance of a 1 mg ml^{-1} solution across a 1-cm path? What percentage of the incident light is transmitted by this solution?

5. *A slow mover.* Tropomyosin, a 70-kd muscle protein, sediments more slowly than does hemoglobin (65 kd). Their sedimentation coefficients are 2.6S and 4.31S, respectively. Which structural feature of tropomyosin accounts for its slow sedimentation?

6. *Sedimenting spheres.* What is the dependence of the sedimentation coefficient S of a spherical protein on its mass? How much more rapidly does an 80-kd protein sediment than does a 40-kd protein?

7. *Size estimate.* The relative electrophoretic mobilities of a 30-kd protein and a 92-kd protein used as standards on an SDS–polyacrylamide gel are 0.80 and 0.41, respectively. What is the apparent mass of a protein having a mobility of 0.62 on this gel?

8. *A new partnership?* The gene encoding a protein with a single disulfide bond undergoes a mutation that changes a serine residue into a cysteine residue. You want to find out whether the disulfide pairing in this mutant is the same as in the original protein. Propose an experiment to directly answer this question.

9. *Sorting cells.* Fluorescence-activated cell sorting (FACS) is a powerful technique for separating cells according to their content of particular molecules. For example, a fluorescence-labeled antibody specific for a cell-surface protein can be used to detect cells containing such a molecule. Suppose that you want to isolate cells that possess a receptor enabling them to detect bacterial degradation products. However, you do not yet have an antibody directed against this receptor. Which fluorescence-labeled molecule would you prepare to identify such cells?

10. *Column choice.* (a) The octapeptide AVGWRVKS was digested with the enzyme trypsin. Would ion-exchange or gel-filtration chromatography be most appropriate for separating the products? Explain. (b) Suppose that the peptide was digested with chymotrypsin. What would be the optimal separation technique? Explain.

11. *Making more enzyme?* In the course of purifying an enzyme, a researcher performs a purification step that results in an *increase* in the total activity to a value greater than that present in the original crude extract. Explain how the amount of total activity might increase.

12. *Divide and conquer.* The determination of the mass of a protein by mass spectrometry often does not allow its unique identification among possible proteins within a complete proteome, but determination of the masses of all fragments produced by digestion with trypsin almost always allows unique identification. Explain.

13. *Protein purification problem.* Complete the following table.

Purification procedure	Total protein (mg)	Total activity (units)	Specific activity (units mg^{-1})	Purification level	Yield (%)
Crude extract	20,000	4,000,000		1	100
(NH$_4$)$_2$SO$_4$ precipitation	5,000	3,000,000			
DEAE-cellulose chromatography	1,500	1,000,000			
Gel-filtration chromatography	500	750,000			
Affinity chromatography	45	675,000			

Chapter Integration Problems

14. *Quaternary structure.* A protein was purified to homogeneity. Determination of the mass by gel-filtration chromatography yields 60 kd. Chromatography in the presence of 6 M urea yields a 30-kd species. When the chromatography is repeated in the presence of 6 M urea and 10 mM β-mercaptoethanol, a single molecular species of 15 kd results. Describe the structure of the molecule.

15. *Helix–coil transitions.* (a) NMR measurements have shown that poly-L-lysine is a random coil at pH 7 but becomes α helical as the pH is raised above 10. Account for this pH-dependent conformational transition. (b) Predict the pH dependence of the helix–coil transition of poly-L-glutamate.

16. *Peptides on a chip.* Large numbers of different peptides can be synthesized in a small area on a solid support. This high-density array can then be probed with a fluorescence-labeled

Fluorescence scan of an array of 1024 peptides in a 1.6-cm^2 area. Each synthesis site is a 400-µm square. A fluorescently labeled monoclonal antibody was added to the array to identify peptides that are recognized. The height and color of each square denote the fluorescence intensity. [After S. P. A. Fodor, J. O. Read, M. C. Pirrung, L. Stryer, A. T. Lu, and D. Solas. *Science* 251(1991):767.]

protein to find out which peptides are recognized. The binding of an antibody to an array of 1024 different peptides occupying a total area the size of a thumbnail is shown in the figure on page 105. How would you synthesize such a peptide array? (Hint: Use light instead of acid to deprotect the terminal amino group in each round of synthesis.)

Data Interpretation Problems

17. *Protein sequencing 1.* Determine the sequence of hexapeptide on the basis of the following data. Note: When the sequence is not known, a comma separates the amino acids. (See Table 3.3)

Amino acid composition: (2R,A,S,V,Y)
N-terminal analysis of the hexapeptide: A
Trypsin digestion: (R,A,V) and (R,S,Y)
Carboxypeptidase digestion: No digestion.
Chymotrypsin digestion: (A,R,V,Y) and (R,S)

18. *Protein sequencing 2.* Determine the sequence of a peptide consisting of 14 amino acids on the basis of the following data.

Amino acid composition: (4S,2L,F,G,I,K,M,T,W,Y)
N-terminal analysis: S
Carboxypeptidase digestion: L
Trypsin digestion: (3S,2L,F,I,M,T,W) (G,K,S,Y)
Chymotrypsin digestion: (F,I,S) (G,K,L) (L,S) (M,T) (S,W) (S,Y)
N-terminal analysis of (F,I,S) peptide: S
Cyanogen bromide treatment: (2S,F,G,I,K,L,M*,T,Y) (2S,L,W)
M*, methionine detected as homoserine

19. *Edman degradation.* Alanine amide was treated with phenyl isothiocyanate to form PTH–alanine. Write a mechanism for this reaction.

DNA, RNA, and the Flow of Genetic Information

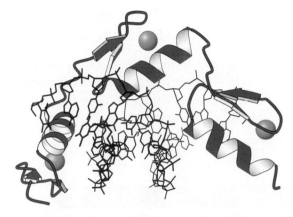

Having genes in common accounts for the resemblance of a mother to her daughters. Genes must be expressed to exert an effect, and proteins regulate such expression. One such regulatory protein, a zinc-finger protein (zinc ion is blue, protein is red), is shown bound to a control region of DNA (black). [(Left) Barnaby Hall/Photonica. (Right) Drawn from 1AAY.pdb.]

DNA and RNA are long linear polymers, called nucleic acids, that carry information in a form that can be passed from one generation to the next. These macromolecules consist of a large number of linked nucleotides, each composed of a sugar, a phosphate, and a base. Sugars linked by phosphates form a common backbone that plays a structural role, whereas *the sequence of bases along a nucleic acid chain carries genetic information.* The DNA molecule has the form of a *double helix,* a helical structure consisting of two complementary nucleic acid strands. *Each strand serves as the template for the other in DNA replication.* The genes of all cells and many viruses are made of DNA. Some viruses, however, use RNA as their genetic material.

Genes specify the kinds of proteins that are made by cells, but DNA is not the direct template for protein synthesis. Rather, a DNA strand is copied into a class of RNA molecules called *messenger RNA* (mRNA), which are the information-carrying intermediates in protein synthesis. This process of *transcription* is followed by *translation,* the synthesis of proteins according to instructions given by mRNA templates. Thus, the flow of genetic information, or *gene expression,* in normal cells is

$$\text{DNA} \xrightarrow{\text{Transcription}} \text{RNA} \xrightarrow{\text{Translation}} \text{Protein}$$

107

This flow of information depends on the genetic code, which defines the relation between the sequence of bases in DNA (or its mRNA transcript) and the sequence of amino acids in a protein. The code is nearly the same in all organisms: a sequence of three bases, called a *codon*, specifies an amino acid. There is another step, between transcription and translation, in the expression of most eukaryotic genes, which are mosaics of nucleic acid sequences called *introns* and *exons*. Both are transcribed, but introns are cut out of newly synthesized RNA molecules, leaving mature RNA molecules with continuous exons. The existence of introns and exons has crucial implications for the evolution of proteins.

4.1 A Nucleic Acid Consists of Four Kinds of Bases Linked to a Sugar–Phosphate Backbone

The nucleic acids DNA and RNA are well suited to function as the carriers of genetic information by virtue of their covalent structures. These macromolecules are *linear polymers* built up from similar units connected end to end (Figure 4.1). Each monomer unit within the polymer is a *nucleotide*. A single nucleotide unit consists of three components: a sugar, a phosphate, and one of four bases. The sequence of bases in the polymer uniquely characterizes a nucleic acid and constitutes a form of linear information.

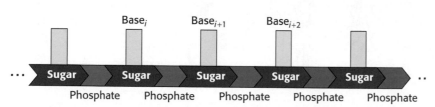

Figure 4.1 Polymeric structure of nucleic acids.

RNA and DNA Differ in the Sugar Component and One of the Bases

The sugar in *deoxyribonucleic acid* (DNA) is *deoxyribose*. The deoxy prefix indicates that the 2′-carbon atom of the sugar lacks the oxygen atom that is linked to the 2′-carbon atom of *ribose,* as shown in Figure 4.2. Note that sugar carbons are numbered with primes to differentiate them from atoms in the bases. The sugars in nucleic acids are linked to one another by phosphodiester bridges. Specifically, the 3′-hydroxyl (3′-OH) group of the sugar moiety of one nucleotide is esterified to a phosphate group, which is,

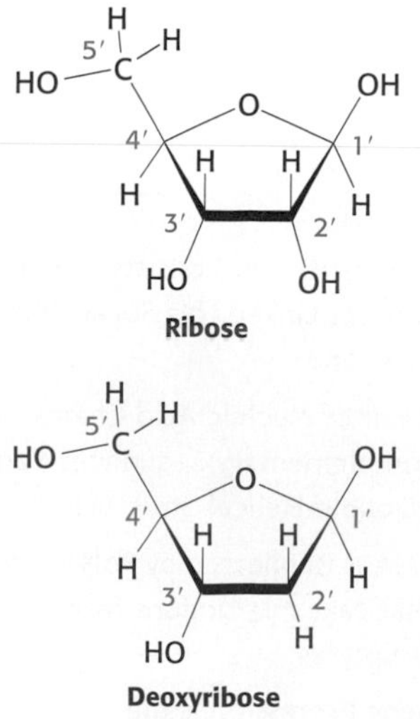

Figure 4.2 Ribose and deoxyribose. Atoms in sugar units are numbered with primes to distinguish them from atoms in bases (see Figure 4.4).

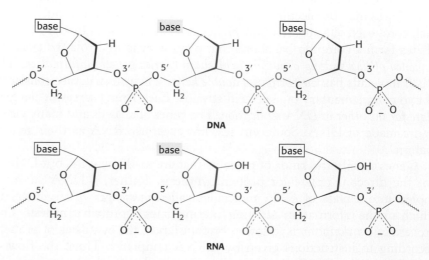

Figure 4.3 Backbones of DNA and RNA. The backbones of these nucleic acids are formed by 3′-to-5′ phosphodiester linkages. A sugar unit is highlighted in red and a phosphate group in blue.

in turn, joined to the 5′-hydroxyl group of the adjacent sugar. The chain of sugars linked by phosphodiester bridges is referred to as the *backbone* of the nucleic acid (Figure 4.3). Whereas the backbone is constant in a nucleic acid, the bases vary from one monomer to the next. Two of the bases of DNA are derivatives of *purine*—adenine (A) and guanine (G)—and two of *pyrimidine*—cytosine (C) and thymine (T), as shown in Figure 4.4.

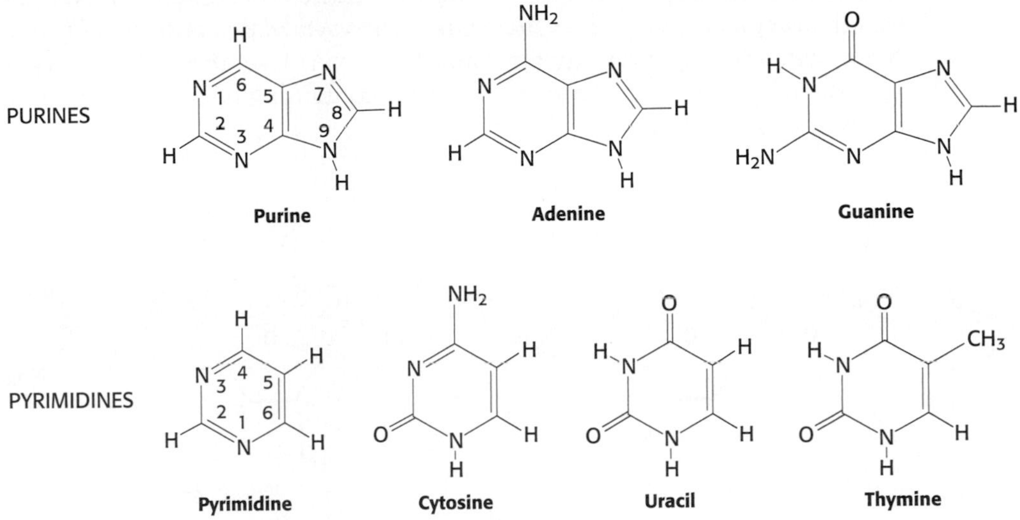

PURINES

Purine   Adenine   Guanine

PYRIMIDINES

Pyrimidine   Cytosine   Uracil   Thymine

Figure 4.4 Purines and pyrimidines. Atoms within bases are numbered without primes. Uracil is present in RNA instead of thymine.

Ribonucleic acid (RNA), like DNA, is a long unbranched polymer consisting of nucleotides joined by 3′-to-5′ phosphodiester bonds (see Figure 4.3). The covalent structure of RNA differs from that of DNA in two respects. First, the sugar units in RNA are riboses rather than deoxyriboses. Ribose contains a 2′-hydroxyl group not present in deoxyribose. Second, one of the four major bases in RNA is uracil (U) instead of thymine (T).

Note that each phosphodiester bridge has a negative charge. This negative charge repels nucleophilic species such as hydroxide ion; consequently, phosphodiester linkages are much less susceptible to hydrolytic attack than are other esters such as carboxylic acid esters. This resistance is crucial for maintaining the integrity of information stored in nucleic acids. The absence of the 2′-hydroxyl group in DNA further increases its resistance to hydrolysis. The greater stability of DNA probably accounts for its use rather than RNA as the hereditary material in all modern cells and in many viruses.

Nucleotides Are the Monomeric Units of Nucleic Acids

A unit consisting of a base bonded to a sugar is referred to as a *nucleoside*. The four nucleoside units in RNA are called *adenosine, guanosine, cytidine,* and *uridine,* whereas those in DNA are called *deoxyadenosine, deoxyguanosine, deoxycytidine,* and *thymidine.* In each case, N-9 of a purine or N-1 of a pyrimidine is attached to C-1′ of the sugar (Figure 4.5). The base lies above the plane of sugar when the structure is written in the standard orientation; that is, the configuration of the N-glycosidic linkage is β (p. 309). A *nucleotide* is a nucleoside joined to one or more phosphate groups by ester linkages. The most common site of attachment in naturally occurring nucleotides is the hydroxyl group attached to C-5′ of the sugar. A compound formed by the attachment of a phosphate group to the C-5′ of a nucleoside sugar is called a *nucleoside 5′-phosphate* or a *5′-nucleotide.* For example, ATP is *adenosine 5′-triphosphate.* This nucleotide is tremendously important since it is the most commonly used energy currency. The energy

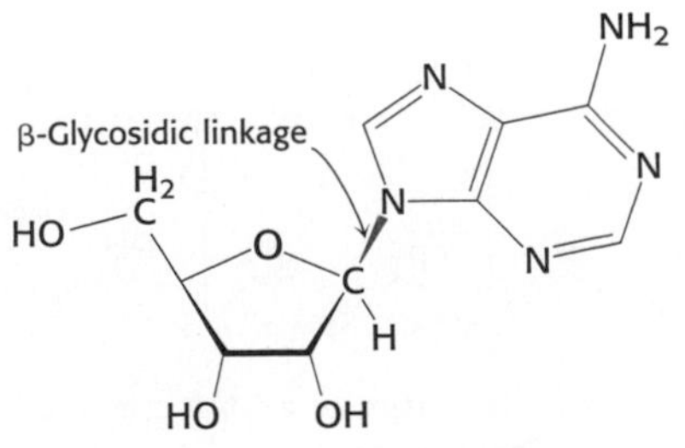

Figure 4.5 β-Glycosidic linkage in a nucleoside.

released from cleavage of the triphosphate group is used to power many
cellular processes (Chapter 15). Another nucleotide is deoxyguanosine
3'-monophosphate (3'-dGMP; Figure 4.6). This nucleotide differs from
ATP in that it contains guanine rather than adenine, contains deoxyribose
rather than ribose (indicated by the prefix "d"), contains one rather than
three phosphates, and has the phosphate esterified to the hydroxyl group in
the 3' rather than the 5' position. Nucleotides are the monomers that are
linked to form RNA and DNA. The four nucleotide units in DNA are
called *deoxyadenylate, deoxyguanylate, deoxycytidylate,* and *thymidylate.*
Note that, although thymidylate contains deoxyribose, the prefix deoxy is
not added by convention, because thymine-containing nucleotides are only
rarely found in RNA.

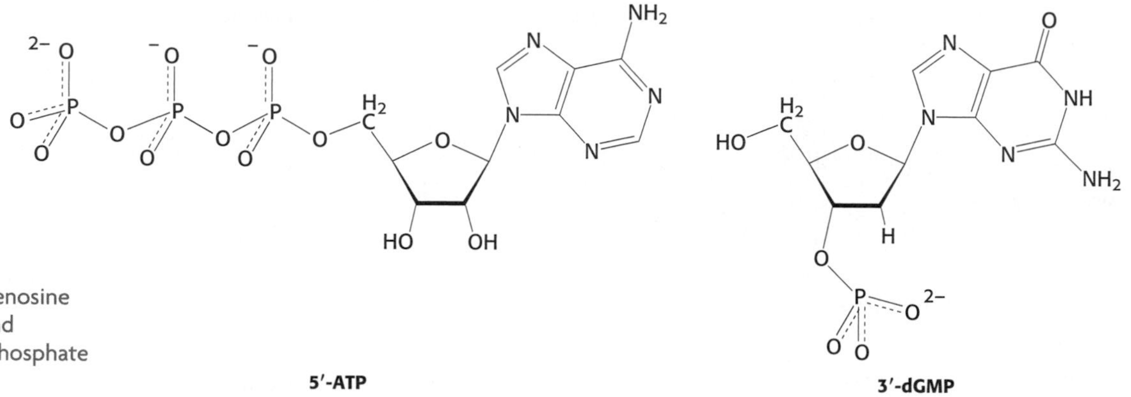

5'-ATP

3'-dGMP

Figure 4.6 Nucleotides adenosine
5'-triphosphate (5'-ATP) and
deoxyguanosine 3'-monophosphate
(3'-dGMP).

The abbreviated notations pApCpG or ACG denote a trinucleotide of
DNA consisting of the building blocks deoxyadenylate monophosphate,
deoxycytidylate monophosphate, and deoxyguanylate monophosphate linked
by a phosphodiester bridge, where "p" denotes a phosphate group (Figure
4.7). The 5' end will often have a phosphate group attached to the 5'-OH
group. Note that, like a polypeptide (Section 2.2), *a DNA chain has polarity.*
One end of the chain has a free 5'-OH group (or a 5'-OH group attached to a
phosphate), whereas the other end has a free 3'-OH group, neither of which is
linked to another nucleotide. By convention, *the base sequence is written in the
5'-to-3' direction.* Thus, the symbol
ACG indicates that the unlinked 5'-
OH group is on deoxyadenylate,
whereas the unlinked 3'-OH group is
on deoxyguanylate. Because of this
polarity, ACG and GCA correspond
to different compounds.

A striking characteristic of naturally occurring DNA molecules is
their length. A DNA molecule must
comprise many nucleotides to carry
the genetic information necessary for
even the simplest organisms. For
example, the DNA of a virus such as
polyoma, which can cause cancer in
certain organisms, is 5100 nucleotides
in length. The *E. coli* genome is a single DNA molecule consisting of two
chains of 4.6 million nucleotides each
(Figure 4.8).

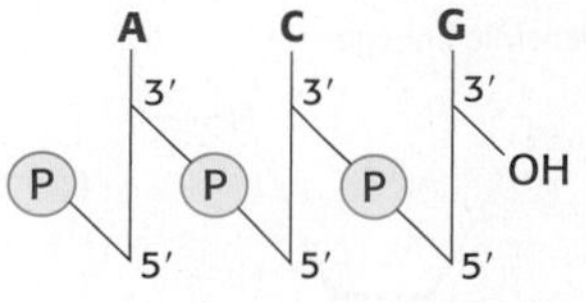

Figure 4.7 Structure of a DNA chain.
The chain has a 5' end, which is usually
attached to a phosphate group, and a 3'
end, which is usually a free hydroxyl group.

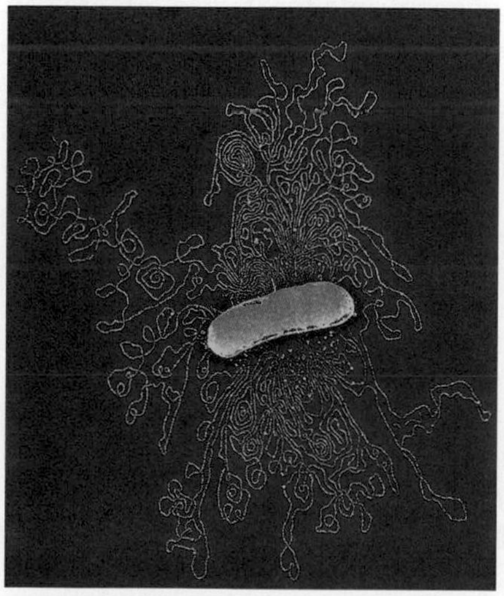

Figure 4.8 Electron micrograph of part
of the *E. coli* genome. [Dr. Gopal Murti/
Science Photo Library/Photo
Researchers.]

The DNA molecules of higher organisms can be much larger. The human genome comprises approximately 3 billion nucleotides in each chain of DNA, divided among 24 distinct chromosomes (22 autosomal chromosomes plus the X and Y sex chromosomes) of different sizes. One of the largest known DNA molecules is found in the Indian muntjak, an Asiatic deer; its genome is nearly as large as the human genome but is distributed on only 3 chromosomes (Figure 4.9). The largest of these chromosomes has two chains of more than 1 billion nucleotides each. If such a DNA molecule could be fully extended, it would stretch more than 1 foot in length. Some plants contain even larger DNA molecules.

 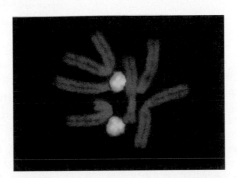

Figure 4.9 The Indian muntjak and its chromosomes. Cells from a female Indian muntjak (right) contain three pairs of very large chromosomes (stained orange). The cell shown is a hybrid containing a pair of human chromosomes (stained green) for comparison. [(Left) M. Birkhead, OSF/Animals Animals. (Right) J.–Y. Lee, M. Koi, E. J. Stanbridge, M. Oshimura, A. T. Kumamoto, and A. P. Feinberg. *Nature Genetics* 7(1994):30.]

4.2 A Pair of Nucleic Acid Chains with Complementary Sequences Can Form a Double-Helical Structure

As discussed in Chapter 1, the covalent structure of nucleic acids accounts for their ability to carry information in the form of a sequence of bases along a nucleic acid chain. The bases on the two separate nucleic acid strands form *specific base pairs* in such a way that a helical structure is formed. The double-helical structure of DNA facilitates the *replication* of the genetic material, that is, the generation of two copies of a nucleic acid from one.

The Double Helix Is Stabilized by Hydrogen Bonds and Hydrophobic Interactions

The formation of specific base pairs was discovered in the course of studies directed at determining the three-dimensional structure of DNA. Maurice Wilkins and Rosalind Franklin obtained x-ray diffraction photographs of fibers of DNA (Figure 4.10). The characteristics of these diffraction patterns indicated that DNA is formed of two chains that wind in a regular helical structure. From these data and others, James Watson and Francis Crick deduced a structural model for DNA that accounted for the diffraction pattern and was the source of some remarkable insights into the functional properties of nucleic acids (Figure 4.11).

The features of the Watson–Crick model of DNA deduced from the diffraction patterns are:

1. Two helical polynucleotide chains are coiled around a common axis. The chains run in opposite directions.

3.4-Å spacing

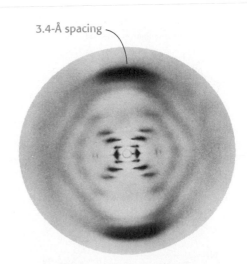

Figure 4.10 X-ray diffraction photograph of a hydrated DNA fiber. The central cross is diagnostic of a helical structure. The strong arcs on the meridian arise from the stack of nucleotide bases, which are 3.4 Å apart. [Courtesy of Dr. Maurice Wilkins.]

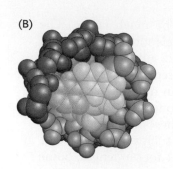
(A)

(B)

34Å

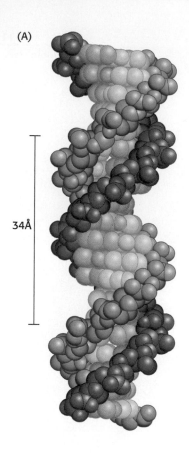

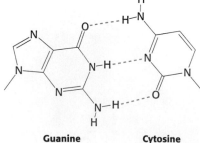

Figure 4.11 Watson–Crick model of double-helical DNA. One polynucleotide chain is shown in blue and the other in red. The purine and pyrimidine bases are shown in lighter colors than those of the sugar–phosphate backbone. (A) Axial view. The structure repeats along the helical axis (vertical) at intervals of 34 Å, which corresponds to 10 nucleotides on each chain. (B) Radial view, looking down the helix axis.

Guanine   Cytosine

Adenine   Thymine

Figure 4.12 Structures of the base pairs proposed by Watson and Crick.

2.   The sugar–phosphate backbones are on the outside and the purine and pyrimidine bases lie on the inside of the helix.

3.   The bases are nearly perpendicular to the helix axis, and adjacent bases are separated by 3.4 Å. This spacing is readily apparent in the DNA diffraction pattern (see Figure 4.10). The helical structure repeats every 34 Å, and so there are 10 bases (= 34 Å per repeat/3.4 Å per base) per turn of helix. There is a rotation of 36 degrees per base (360 degrees per full turn/10 bases per turn).

4.   The diameter of the helix is 20 Å.

How is such a regular structure able to accommodate an arbitrary sequence of bases, given the different sizes and shapes of the purines and pyrimidines? In attempting to answer this question, Watson and Crick discovered that guanine can be paired with cytosine and adenine with thymine to form base pairs that have essentially the same shape (Figure 4.12). These base pairs are held together by specific hydrogen bonds, which, although weak (4–21 kJ mol^{-1} or 1–5 kcal mol^{-1}), stabilize the helix because of their large numbers in a DNA molecule. These *base-pairing rules* account for the observation, originally made by Erwin Chargaff in 1950, that the ratios of adenine to thymine and of guanine to cytosine are nearly the same in all species studied, whereas the adenine-to-guanine ratio varies considerably (Table 4.1).

Inside the helix, the bases are essentially stacked one on top of another (Figure 4.13). The stacking of base pairs contributes to the stability of the double helix in two ways. First, the double helix is stabilized by the hydrophobic effect (p. 9). The hydrophobic bases cluster in the interior of the helix away from the surrounding water, whereas the more polar surfaces are exposed to water. This arrangement is reminiscent of protein folding, where hydrophobic amino acids are in the protein's interior and the hydrophilic amino acids are on the exterior (Section 2.4). The hydrophobic effect stacks the bases on top of one another. The stacked base pairs attract one another through van der Waals forces (p. 8). The energy associated with a single van der Waals interaction is quite small, typically from 2 to 4 kJ mol^{-1} (0.5–1.0 kcal mol^{-1}). In the double helix, however, a large number of atoms are in van der Waals contact, and the net effect, summed over these atom pairs,

TABLE 4.1 Base compositions experimentally determined for a variety of organisms

Species	A : T	G : C	A : G
Human being	1.00	1.00	1.56
Salmon	1.02	1.02	1.43
Wheat	1.00	0.97	1.22
Yeast	1.03	1.02	1.67
Escherichia coli	1.09	0.99	1.05
Serratia marcescens	0.95	0.86	0.70

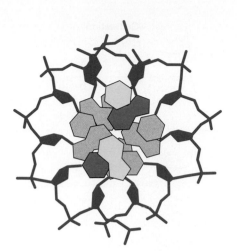

Figure 4.13 Axial view of DNA. Base pairs are stacked nearly one on top of another in the double helix.

is substantial. In addition, base stacking in DNA is favored by the conformations of the somewhat rigid five-membered rings of the backbone sugars.

The Double Helix Facilitates the Accurate Transmission of Hereditary Information

The double-helical model of DNA and the presence of specific base pairs immediately suggested how the genetic material might replicate. The sequence of bases of one strand of the double helix precisely determines the sequence of the other strand; a guanine base on one strand is always paired with a cytosine base on the other strand, and so on. Thus, separation of a double helix into its two component chains would yield two single-stranded templates onto which new double helices could be constructed, each of which would have the same sequence of bases as the parent double helix. Consequently, as DNA is replicated, one of the chains of each daughter DNA molecule is newly synthesized, whereas the other is passed unchanged from the parent DNA molecule. This distribution of parental atoms is achieved by *semiconservative replication*.

Matthew Meselson and Franklin Stahl carried out a critical test of this hypothesis in 1958. They labeled the parent DNA with ^{15}N, a heavy isotope of nitrogen, to make it denser than ordinary DNA. The labeled DNA was generated by growing *E. coli* for many generations in a medium that contained $^{15}NH_4Cl$ as the sole nitrogen source. After the incorporation of heavy nitrogen was complete, the bacteria were abruptly transferred to a medium that contained ^{14}N, the ordinary isotope of nitrogen. The question asked was: What is the distribution of ^{14}N and ^{15}N in the DNA molecules after successive rounds of replication?

The distribution of ^{14}N and ^{15}N was revealed by the technique of *density-gradient equilibrium sedimentation*. A small amount of DNA was dissolved in a concentrated solution of cesium chloride having a density close to that of the DNA (1.7 g cm^{-3}). This solution was centrifuged until it was nearly at equilibrium. At that point, the opposing processes of sedimentation and diffusion created a gradient in the concentration of cesium chloride across the centrifuge cell. The result was a stable density gradient ranging from 1.66 to 1.76 g cm^{-3}. The DNA molecules in this density gradient were driven by centrifugal force into the region where the solution's density was equal to their own. The genomic DNA yielded a narrow band that was detected by its absorption of ultraviolet light. A mixture of ^{14}N DNA and ^{15}N DNA molecules gave clearly separate bands because they differ in density by about 1% (Figure 4.14).

(A)

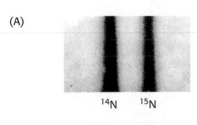

^{14}N ^{15}N

(B)

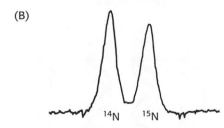

^{14}N ^{15}N

Figure 4.14 Resolution of ^{14}N DNA and ^{15}N DNA by density-gradient centrifugation. (A) Ultraviolet absorption photograph of a centrifuge cell showing the two distinct bands of DNA. (B) Densitometric tracing of the absorption photograph. [From M. Meselson and F. W. Stahl. *Proc. Natl. Acad. Sci. U. S. A.* 44(1958):671–682.]

Figure 4.15 Detection of semiconservative replication of *E. coli* DNA by density-gradient centrifugation. The position of a band of DNA depends on its content of ^{14}N and ^{15}N. After 1.0 generation, all of the DNA molecules were hybrids containing equal amounts of ^{14}N and ^{15}N. [From M. Meselson and F. W. Stahl. *Proc. Natl. Acad. Sci. U. S. A.* 44(1958):671–682.]

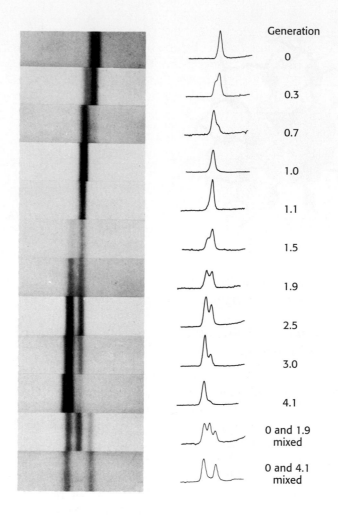

	Generation
	0
	0.3
	0.7
	1.0
	1.1
	1.5
	1.9
	2.5
	3.0
	4.1
	0 and 1.9 mixed
	0 and 4.1 mixed

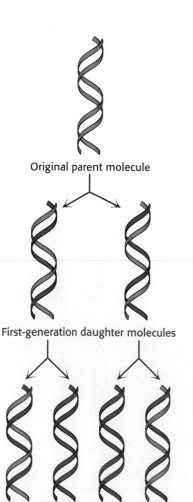

Original parent molecule

First-generation daughter molecules

Second-generation daughter molecules

Figure 4.16 Diagram of semiconservative replication. Parental DNA is shown in blue and newly synthesized DNA in red. [After M. Meselson and F. W. Stahl. *Proc. Natl. Acad. Sci. U. S. A.* 44(1958):671–682.]

DNA was extracted from the bacteria at various times after they were transferred from a ^{15}N to a ^{14}N medium. Analysis of these samples by the density-gradient technique showed that there was a single band of DNA after one generation. The density of this band was precisely halfway between the densities of the ^{14}N DNA and ^{15}N DNA bands (Figure 4.15). *The absence of ^{15}N DNA indicated that parental DNA was not preserved as an intact unit after replication.* The absence of ^{14}N DNA indicated that all the daughter DNA derived some of their atoms from the parent DNA. This proportion had to be half because the density of the hybrid DNA band was halfway between the densities of the ^{14}N DNA and ^{15}N DNA bands.

After two generations, there were equal amounts of two bands of DNA. One was hybrid DNA, and the other was ^{14}N DNA. Meselson and Stahl concluded from these incisive experiments that replication was semiconservative, and so each new double helix contains a parent strand and a newly synthesized strand. Their results agreed perfectly with the Watson–Crick model for DNA replication (Figure 4.16).

The Double Helix Can Be Reversibly Melted

In DNA replication and other processes, the two strands of the double helix must be separated from each other, at least in a local region. The two strands of a DNA helix readily come apart when the hydrogen bonds between base pairs are disrupted. In the laboratory, the double helix can be disrupted by heating a solution of DNA or by adding acid or alkali to ionize its bases. The dissociation of the double helix is called *melting* because it occurs abruptly at a certain temperature. The *melting temperature* (T_m) is defined as the temperature at which half the helical structure is lost. Inside cells, however, the double helix is not melted by the addition of heat. Instead, proteins called *helicases* use chemical energy (from ATP) to disrupt the helix.

Stacked bases in nucleic acids absorb less ultraviolet light than do unstacked bases, an effect called *hypochromism*. Thus, the melting of nucleic acids is readily monitored by measuring their absorption of light, which is maximal at a wavelength of 260 nm (Figure 4.17).

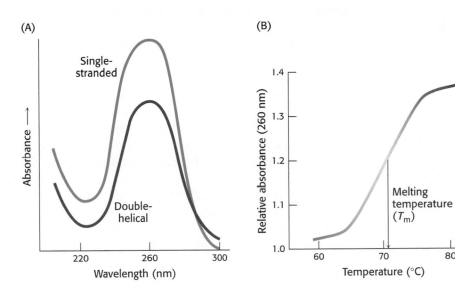

Figure 4.17 Hypochromism. (A) Single-stranded DNA absorbs light more effectively than does double-helical DNA. (B) The absorbance of a DNA solution at a wavelength of 260 nm increases when the double helix is melted into single strands.

Separated complementary strands of nucleic acids spontaneously reassociate to form a double helix when the temperature is lowered below T_m. This renaturation process is sometimes called *annealing*. The facility with which double helices can be melted and then reassociated is crucial for the biological functions of nucleic acids.

The ability to melt and reanneal DNA reversibly in the laboratory provides a powerful tool for investigating sequence similarity. For instance, DNA molecules from two different organisms can be melted and allowed to reanneal or *hybridize* in the presence of each other. If the sequences are similar, hybrid DNA duplexes, with DNA from each organism contributing a strand of the double helix, can form. The degree of hybridization is an indication of the relatedness of the genomes and hence the organisms. Similar hybridization experiments with RNA and DNA can locate genes in a cell's DNA that correspond to a particular RNA. We will return to this important technique in Chapter 5.

Some DNA Molecules Are Circular and Supercoiled

The DNA molecules in human chromosomes are linear. However, electron microscopic and other studies have shown that intact DNA molecules from some other organisms are circular (Figure 4.18A). The term *circular* refers to the continuity of the DNA chains, not to their geometric form. DNA molecules inside cells necessarily have a very compact shape. Note that the *E. coli* chromosome, fully extended, would be about 1000 times as long as the greatest diameter of the bacterium.

A closed DNA molecule has a property unique to circular DNA. The axis of the double helix can itself be twisted or supercoiled into a *superhelix* (Figure 4.18B). Supercoiling is biologically important for two reasons. First, *a supercoiled DNA molecule has a more compact shape than does its relaxed counterpart*. Second, *supercoiling may hinder or favor the capacity of the double helix to unwind and thereby affect the interactions between DNA and other*

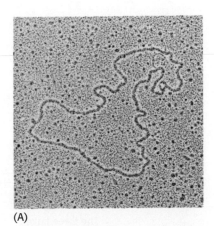

(A)

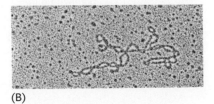

(B)

Figure 4.18 Electron micrographs of circular DNA from mitochondria. (A) Relaxed form. (B) Supercoiled form. [Courtesy of Dr. David Clayton.]

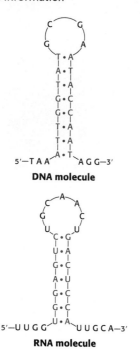

DNA molecule

RNA molecule

Figure 4.19 Stem-loop structures. Stem-
loop structures may be formed from
single-stranded DNA and RNA molecules.

molecules. These topological features of DNA will be considered further in
Chapter 28. A circular DNA molecule without any superhelical turns is
known as a *relaxed molecule.*

Single-Stranded Nucleic Acids Can Adopt Elaborate Structures

Single-stranded nucleic acids often fold back on themselves to form well-
defined structures. Such structures are important in entities such as the
ribosome—a large complex of RNAs and proteins on which proteins are
synthesized.

The simplest and most common structural motif formed is a *stem-loop*,
created when two complementary sequences within a single strand come
together to form double-helical structures (Figure 4.19). In many cases,
these double helices are made up entirely of Watson–Crick base pairs. In
other cases, however, the structures include mismatched base pairs or
unmatched bases that bulge out from the helix. Such mismatches destabilize the local structure but introduce deviations from the standard double-
helical structure that can be important for higher-order folding and for
function (Figure 4.20).

Single-stranded nucleic acids can adopt structures more complex than
simple stem-loops through the interaction of more widely separated bases.
Often, three or more bases may interact to stabilize these structures. In
such cases, hydrogen-bond donors and acceptors that do not participate in
Watson–Crick base pairs may participate in hydrogen bonds to form nonstandard pairings. Metal ions such as magnesium ion (Mg^{2+}) often assist
in the stabilization of these more elaborate structures. These complex
structures allow RNA to perform a host of functions that the double-

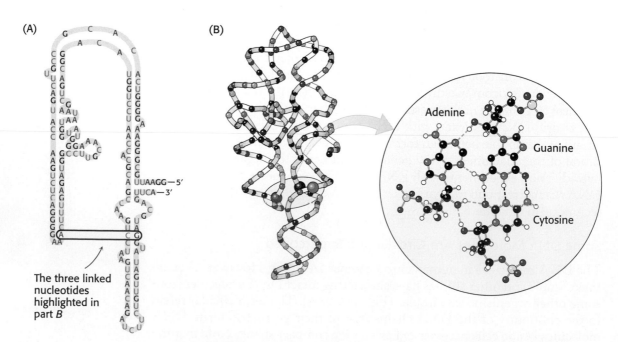

Figure 4.20 Complex structure of an RNA molecule. A single-stranded RNA molecule
may fold back on itself to form a complex structure. (A) The nucleotide sequence showing
Watson–Crick base pairs and other nonstandard base pairings in stem-loop structures.
(B) The three-dimensional structure and one important long-range interaction between
three bases. In the three-dimensional structure to the left, cytidine nucleotides are shown in
blue, adenosine in red, guanosine in black, and uridine in green. Hydrogen bonds within the
Watson–Crick base pair are shown as dashed black lines; additional hydrogen bonds are
shown as dashed green lines.

stranded DNA molecule cannot. Indeed, the complexity of some RNA molecules rivals that of proteins, and these RNA molecules perform a number of functions that had formerly been thought the private domain of proteins.

4.3 DNA Is Replicated by Polymerases That Take Instructions from Templates

We now turn to the molecular mechanism of DNA replication. The full replication machinery in a cell comprises more than 20 proteins engaged in intricate and coordinated interplay. In 1958, Arthur Kornberg and his colleagues isolated from *E. coli* the first known of the enzymes, called *DNA polymerases*, that promote the formation of the bonds joining units of the DNA backbone. *E. coli* has a number of DNA polymerases, designated by roman numerals, that participate in DNA replication and repair (Chapter 28).

DNA Polymerase Catalyzes Phosphodiester-Bond Formation

DNA polymerases catalyze the step-by-step addition of deoxyribonucleotide units to a DNA chain (Figure 4.21). The reaction catalyzed, in its simplest form, is

$$(\text{DNA})_n + \text{dNTP} \rightleftharpoons (\text{DNA})_{n+1} + \text{PP}_i$$

where dNTP stands for any deoxyribonucleotide and PP_i is a pyrophosphate ion.

DNA synthesis has the following characteristics:

1. The reaction requires all four activated precursors—that is, *the deoxynucleoside 5′-triphosphates dATP, dGTP, dCTP, and TTP*—as well as Mg^{2+} ion.

2. *The new DNA chain is assembled directly on a preexisting DNA template.* DNA polymerases catalyze the formation of a phosphodiester bond efficiently only if the base on the incoming nucleoside triphosphate is complementary to the base on the template strand. Thus, DNA polymerase is a *template-directed enzyme* that synthesizes a product with a base sequence complementary to that of the template.

3. *DNA polymerases require a primer to begin synthesis.* A *primer* strand having a free 3′-OH group must be already bound to the template strand. The chain-elongation reaction catalyzed by DNA polymerases is a nucleophilic attack by the 3′-OH terminus of the growing chain on the innermost

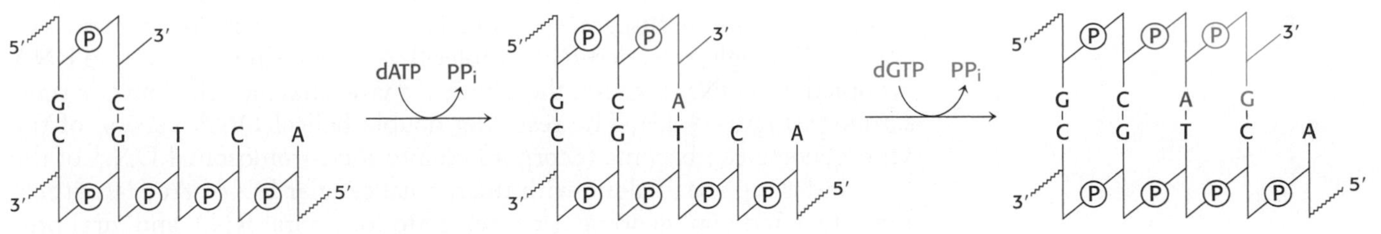

Figure 4.21 Polymerization reaction catalyzed by DNA polymerases.

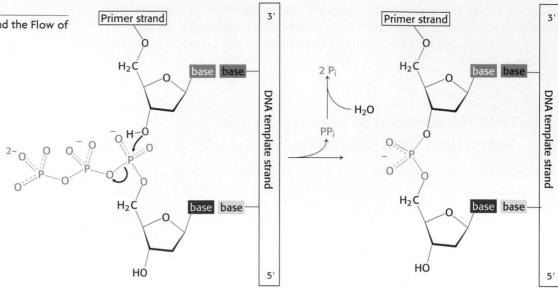

Figure 4.22 Chain elongation reaction. DNA polymerases catalyze the formation of a phosphodiester bridge.

phosphorus atom of the deoxynucleoside triphosphate (Figure 4.22). A phosphodiester bridge is formed and pyrophosphate is released. The subsequent hydrolysis of pyrophosphate to yield two ions of orthophosphate (P_i) by pyrophosphatase, a ubiquitous enzyme, helps drive the polymerization forward. *Elongation of the DNA chain proceeds in the 5'-to-3' direction.*

4. *Many DNA polymerases are able to correct mistakes in DNA by removing mismatched nucleotides.* These polymerases have a distinct nuclease activity that allows them to excise incorrect bases by a separate reaction. This nuclease activity contributes to the remarkably high fidelity of DNA replication, which has an error rate of less than 10^{-8} per base pair.

The Genes of Some Viruses Are Made of RNA

Genes in all cellular organisms are made of DNA. The same is true for some viruses but, for others, the genetic material is RNA. Viruses are genetic elements enclosed in protein coats that can move from one cell to another but are not capable of independent growth. One well-studied example of an RNA virus is the tobacco mosaic virus, which infects the leaves of tobacco plants. This virus consists of a single strand of RNA (6390 nucleotides) surrounded by a protein coat of 2130 identical subunits. An RNA polymerase that takes direction from an RNA template, called an *RNA-directed RNA polymerase*, copies the viral RNA.

Another important class of RNA virus comprises the *retroviruses*, so called because the genetic information flows from RNA to DNA rather than from DNA to RNA. This class includes human immunodeficiency virus 1 (HIV-1), the cause of AIDS, as well as a number of RNA viruses that produce tumors in susceptible animals. Retrovirus particles contain two copies of a single-stranded RNA molecule. On entering the cell, the RNA is copied into DNA through the action of a viral enzyme called *reverse transcriptase* (Figure 4.23). The resulting double-helical DNA version of the viral genome can become incorporated into the chromosomal DNA of the host and is replicated along with the normal cellular DNA. At a later time, the integrated viral genome is expressed to form viral RNA and viral proteins, which assemble into new virus particles.

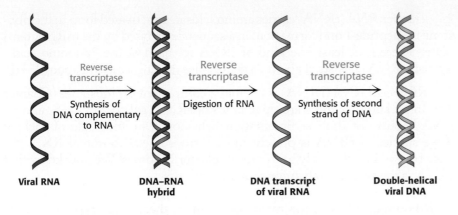

Figure 4.23 **Flow of information from RNA to DNA in retroviruses.** The RNA genome of a retrovirus is converted into DNA by reverse transcriptase, an enzyme brought into the cell by the infecting virus particle. Reverse transcriptase possesses several activities and catalyzes the synthesis of a complementary DNA strand, the digestion of the RNA, and the subsequent synthesis of the DNA strand.

4.4 Gene Expression Is the Transformation of DNA Information into Functional Molecules

The information stored as DNA becomes useful when it is expressed in the production of RNA and proteins. This rich and complex topic is the subject of several chapters later in this book, but here we introduce the basics of gene expression. DNA can be thought of as archival information, stored and manipulated judiciously to minimize damage (mutations). It is expressed in two steps. First, an RNA copy is made that encodes directions for protein synthesis. This messenger RNA can be thought of as a photocopy of the original information—it can be made in multiple copies, used, and then disposed of. Second, the information in messenger RNA is translated to synthesize functional proteins. Other types of RNA molecules exist to facilitate this translation.

Several Kinds of RNA Play Key Roles in Gene Expression

Scientists used to believe that RNA played a passive role in gene expression, as mere conveyors of information like messenger RNA. However, recent investigations have shown that RNA plays a variety of roles, from catalysis to regulation. Cells contain several kinds of RNA (Table 4.2):

1. *Messenger RNA* (mRNA) is the template for protein synthesis, or *translation*. An mRNA molecule may be produced for each gene or group of genes that is to be expressed in *E. coli*, whereas a distinct mRNA is produced for each gene in eukaryotes. Consequently, mRNA is a heterogeneous class of molecules. In prokaryotes, the average length of an mRNA molecule is about 1.2 kilobases (kb). In eukaryotes, mRNA has structural features, such as stem-loop structures, that regulate the efficiency of translation and lifetime of the mRNA.

TABLE 4.2 RNA molecules in *E. coli*

Type	Relative amount (%)	Sedimentation coefficient (S)	Mass (kd)	Number of nucleotides
Ribosomal RNA (rRNA)	80	23	1.2×10^3	3700
		16	0.55×10^3	1700
		5	3.6×10^1	120
Transfer RNA (tRNA)	15	4	2.5×10^1	75
Messenger RNA (mRNA)	5	Heterogeneous		

2. *Transfer RNA* (tRNA) carries amino acids in an activated form to the ribosome for peptide-bond formation, in a sequence dictated by the mRNA template. There is at least one kind of tRNA for each of the 20 amino acids. Transfer RNA consists of about 75 nucleotides (having a mass of about 25 kd).

3. *Ribosomal RNA* (rRNA) is the major component of ribosomes (Chapter 30). In prokaryotes there are three kinds of rRNA, called *23S, 16S,* and *5S RNA* because of their sedimentation behavior. One molecule of each of these species of rRNA is present in each ribosome. Ribosomal RNA was once believed to play only a structural role in ribosomes. We now know that rRNA is the actual catalyst for protein synthesis.

Ribosomal RNA is the most abundant of these three types of RNA. Transfer RNA comes next, followed by messenger RNA, which constitutes only 5% of the total RNA. Eukaryotic cells contain additional small RNA molecules.

4. *Small nuclear RNA* (snRNA) molecules participate in the splicing of RNA exons.

5. A small RNA molecule is an essential component of the *signal-recognition particle*, an RNA–protein complex in the cytoplasm that helps target newly synthesized proteins to intracellular compartments and extracellular destinations.

6. *Micro RNA* (miRNA) is a class of small (about 21 nucleotides) noncoding RNAs that bind to complementary mRNA molecules and inhibit their translation.

7. *Small interfering RNA* (siRNA) is a class of small RNA molecules that bind to mRNA and facilitate its degradation. Micro RNA and small interfering RNA also provide scientists with powerful experimental tools for inhibiting the expression of specific genes in the cell.

8. RNA is a component of *telomerase*, an enzyme that maintains the telomeres (ends) of chromosomes during DNA replication.

In this chapter, we will consider rRNA, mRNA, and tRNA.

All Cellular RNA Is Synthesized by RNA Polymerases

The synthesis of RNA from a DNA template is called *transcription* and is catalyzed by the enzyme *RNA polymerase* (Figure 4.24). RNA polymerase

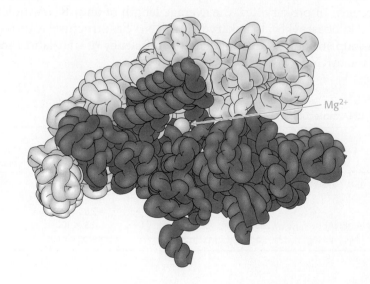

Figure 4.24 RNA polymerase. This large enzyme comprises many subunits, including β (red) and β′ (yellow), which form a "claw" that holds the DNA to be transcribed. *Notice* that the active site includes a Mg^{2+} ion (green) at the center of the structure. The curved tubes making up the protein in the image represent the backbone of the polypeptide chain. [Drawn from IL9Z.pdb.]

catalyzes the initiation and elongation of RNA chains. The reaction catalyzed by this enzyme is

$$(RNA)_n + \text{ribonucleoside triphosphate} \rightleftharpoons (RNA)_{n+1} + PP_i$$

RNA polymerase requires the following components:

1. *A template.* The preferred template is *double-stranded DNA.* Single-stranded DNA also can serve as a template. RNA, whether single or double stranded, is not an effective template; nor are RNA–DNA hybrids.

2. *Activated precursors.* All four *ribonucleoside triphosphates*—ATP, GTP, UTP, and CTP—are required.

3. *A divalent metal ion.* Either Mg^{2+} or Mn^{2+} is effective.

The synthesis of RNA is like that of DNA in several respects (Figure 4.25). First, the direction of synthesis is $5' \rightarrow 3'$. Second, the mechanism of elongation is similar: the 3'-OH group at the terminus of the growing chain makes a nucleophilic attack on the innermost phosphate of the incoming nucleoside triphosphate. Third, the synthesis is driven forward by the hydrolysis of pyrophosphate. In contrast with DNA polymerase, however, RNA polymerase does not require a primer. In addition, RNA polymerase lacks the ability of DNA polymerase to excise mismatched nucleotides.

All three types of cellular RNA—mRNA, tRNA, and rRNA—are synthesized in *E. coli* by the same RNA polymerase according to instructions given by a DNA template. In mammalian cells, there is a division of labor among several different kinds of RNA polymerases. We shall return to these RNA polymerases in Chapter 29.

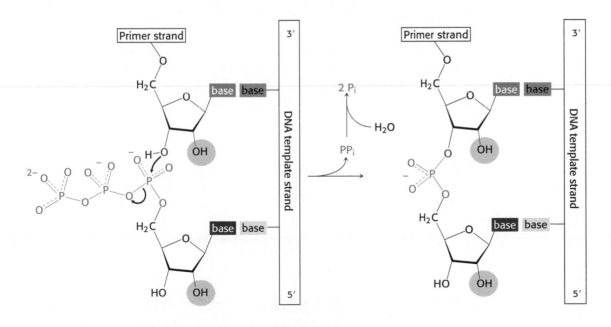

Figure 4.25
Transcription mechanism of the chain-elongation reaction catalyzed by RNA polymerase.

RNA Polymerases Take Instructions from DNA Templates

RNA polymerase, like the DNA polymerases described earlier, takes instructions from a DNA template. The earliest evidence was the finding that the *base composition* of newly synthesized RNA is the complement of that of the DNA template strand, as exemplified by the RNA synthesized from a template of single-stranded DNA from the φX174 virus (Table 4.3).

TABLE 4.3 Base composition (percentage) of RNA synthesized from a viral DNA template

DNA template (plus, or coding, strand of φX174)		RNA product	
A	25	25	U
T	33	32	A
G	24	23	C
C	18	20	G

Hybridization experiments also revealed that RNA synthesized by RNA polymerase is complementary to its DNA template. In these experiments, DNA is melted and allowed to reassociate in the presence of mRNA. RNA–DNA hybrids will form if the RNA and DNA have complementary sequences. The strongest evidence for the fidelity of transcription came from base-sequence studies showing that the RNA sequence is the precise complement of the DNA template sequence (Figure 4.26).

5′—GCGGCGACGCGCAGUUAAUCCCACAGCCGCCAGUUCCGCUGGCGGCAU—3′ **mRNA**
3′—CGCCGCTGCGCGTCAATTAGGGTGTCGGCGGTCAAGGCGACCGCCGTA—5′ **Template strand of DNA**
5′—GCGGCGACGCGCAGTTAATCCCACAGCCGCCAGTTCCGCTGGCGGCAT—3′ **Coding strand of DNA**

Figure 4.26 Complementarity between mRNA and DNA. The base sequence of mRNA (red) is the complement of that of the DNA template strand (blue). The sequence shown here is from the tryptophan operon, a segment of DNA containing the genes for five enzymes that catalyze the synthesis of tryptophan. The other strand of DNA (black) is called the coding strand because it has the same sequence as the RNA transcript except for thymine (T) in place of uracil (U).

Transcription Begins Near Promoter Sites and Ends at Terminator Sites

RNA polymerase must detect and transcribe discrete genes from within large stretches of DNA. What marks the beginning of the unit to be transcribed? DNA templates contain regions called *promoter sites* that specifically bind RNA polymerase and determine where transcription begins. In bacteria, two sequences on the 5′ (upstream) side of the first nucleotide to be transcribed function as promoter sites (Figure 4.27A). One of them, called the *Pribnow box*, has the consensus sequence TATAAT and is centered at −10 (10 nucleotides on the 5′ side of the first nucleotide transcribed, which is denoted by +1). The other, called the −35 *region*, has the consensus sequence TTGACA. The first nucleotide transcribed is usually a purine.

Eukaryotic genes encoding proteins have promoter sites with a TATAAA consensus sequence, called a *TATA box* or a *Hogness box,* centered at about −25 (Figure 4.27B). Many eukaryotic promoters also have a

---

Consensus sequence

Not all base sequences of promoter sites are identical. However, they do possess common features, which can be represented by an idealized consensus sequence. Each base in the consensus sequence TATAAT is found in most prokaryotic promoters. Nearly all promoter sequences differ from this consensus sequence at only one or two bases.

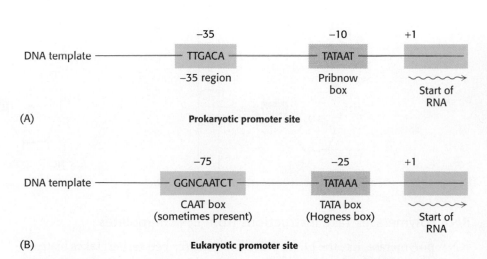

Figure 4.27 Promoter sites for transcription in (A) prokaryotes and (B) eukaryotes. Consensus sequences are shown. The first nucleotide to be transcribed is numbered +1. The adjacent nucleotide on the 5′ side is numbered −1. The sequences shown are those of the coding strand of DNA.

CAAT box with a G GN*CAAT*CT consensus sequence centered at about −75. The transcription of eukaryotic genes is further stimulated by *enhancer sequences,* which can be quite distant (as many as several kilobases) from the start site, on either its 5′ or its 3′ side.

RNA polymerase proceeds along the DNA template, transcribing one of its strands until it synthesizes a terminator sequence. This sequence encodes a termination signal, which in *E. coli* is a *base-paired hairpin* on the newly synthesized RNA molecule (Figure 4.28). This hairpin is formed by base-pairing of self-complementary sequences that are rich in G and C. Nascent RNA spontaneously dissociates from RNA polymerase when this hairpin is followed by a string of U residues. Alternatively, RNA synthesis can be terminated by the action of *rho*, a protein. Less is known about the termination of transcription in eukaryotes. A more detailed discussion of the initiation and termination of transcription will be given in Chapter 29. The important point now is that *discrete start and stop signals for transcription are encoded in the DNA template.*

In eukaryotes, the RNA transcript is modified (Figure 4.29). A "cap" structure is attached to the 5′ end, and a sequence of adenylates, the poly(A) tail, is added to the 3′ end. These modifications will be presented in detail in Chapter 29.

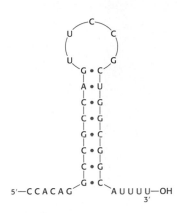

Figure 4.28 Base sequence of the 3′ end of an mRNA transcript in *E. coli*. A stable hairpin structure is followed by a sequence of uridine (U) residues.

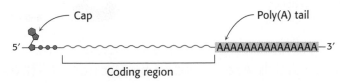

Figure 4.29 Modification of mRNA. Messenger RNA in eukaryotes is modified after transcription. A nucleotide "cap" structure is added to the 5′ end, and a poly(A) tail is added at the 3′ end.

Transfer RNAs Are the Adaptor Molecules in Protein Synthesis

We have seen that mRNA is the template for protein synthesis. How then does it direct amino acids to become joined in the correct sequence to form a protein? In 1958, Francis Crick wrote:

> RNA presents mainly a sequence of sites where hydrogen bonding could occur. One would expect, therefore, that whatever went onto the template in a *specific* way did so by forming hydrogen bonds. It is therefore a natural hypothesis that the amino acid is carried to the template by an adaptor molecule, and that the adaptor is the part that actually fits onto the RNA. In its simplest form, one would require twenty adaptors, one for each amino acid.

This highly innovative hypothesis soon became established as fact. *The adaptors in protein synthesis are transfer RNAs*. The structure and reactions of these remarkable molecules will be considered in detail in Chapter 30. For the moment, it suffices to note that tRNAs contain an *amino acid-attachment site* and a *template-recognition site*. A tRNA molecule carries a specific amino acid in an activated form to the site of protein synthesis. The carboxyl group of this amino acid is esterified to the 3′- or 2′-hydroxyl group of the ribose unit at the 3′ end of the tRNA chain (Figure 4.30). The joining of an amino acid to a tRNA molecule to form an *aminoacyl-tRNA* is catalyzed by a specific enzyme called an *aminoacyl-tRNA synthetase*. This esterification reaction is driven by ATP cleavage. There is at least one specific synthetase for each of the 20 amino acids. The template-recognition site on tRNA is a sequence of three bases called an *anticodon* (Figure 4.31). The anticodon on tRNA recognizes a complementary sequence of three bases, called a *codon*, on mRNA.

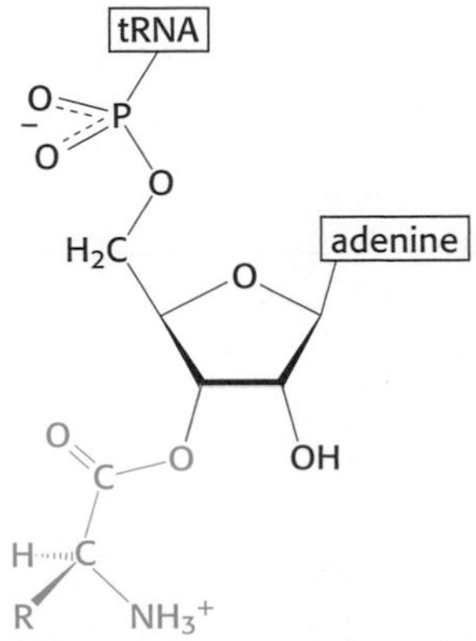

Figure 4.30 Attachment of an amino acid to a tRNA molecule. The amino acid (shown in blue) is esterified to the 3′-hydroxyl group of the terminal adenylate of tRNA.

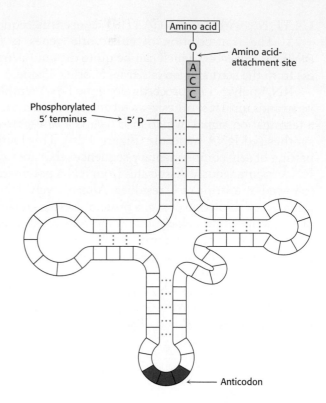

Figure 4.31 General
structure of an aminoacyl-
tRNA. The amino acid is
attached at the 3′ end of
the RNA. The anticodon is
the template-recognition
site. Notice that the tRNA
has a cloverleaf structure
with many hydrogen bonds
(green dots) between bases.

4.5   Amino Acids Are Encoded by Groups of Three Bases Starting from a Fixed Point

The *genetic code* is the relation between the sequence of bases in DNA (or its RNA transcripts) and the sequence of amino acids in proteins. Experiments by Marshall Nirenberg, Har Gobind Khorana, Francis Crick, Sydney Brenner, and others established the following features of the genetic code by 1961:

1.   *Three nucleotides encode an amino acid.* Proteins are built from a basic set of 20 amino acids, but there are only four bases. Simple calculations show that a minimum of three bases is required to encode at least 20 amino acids. Genetic experiments showed that *an amino acid is in fact encoded by a group of three bases, or codon.*

2.   *The code is nonoverlapping.* Consider a base sequence ABCDEF. In an overlapping code, ABC specifies the first amino acid, BCD the next, CDE the next, and so on. In a nonoverlapping code, ABC designates the first amino acid, DEF the second, and so forth. Genetic experiments again established the code to be nonoverlapping.

3.   *The code has no punctuation.* In principle, one base (denoted as Q) might serve as a "comma" between groups of three bases.

$$\ldots \text{QABCQDEFQGHIQJKLQ} \ldots$$

This is not the case. Rather, *the sequence of bases is read sequentially from a fixed starting point,* without punctuation.

4.   *The genetic code is degenerate.* Most amino acids are encoded by more than one codon. There are 64 possible base triplets and only 20 amino acids, and in fact 61 of the 64 possible triplets specify particular amino acids. Three triplets (called *stop codons*) designate the termination of translation. Thus, *for most amino acids, there is more than one code word.*

Major Features of the Genetic Code

All 64 codons have been deciphered (Table 4.4). Because the code is highly degenerate, only tryptophan and methionine are encoded by just one triplet each. Each of the other 18 amino acids is encoded by two or more. Indeed, leucine, arginine, and serine are specified by six codons each. The number of codons for a particular amino acid correlates with its frequency of occurrence in proteins.

TABLE 4.4 The genetic code

First position (5′ end)	Second Position				Third position (3′ end)
	U	C	A	G	
U	Phe	Ser	Tyr	Cys	U
	Phe	Ser	Tyr	Cys	C
	Leu	Ser	Stop	Stop	A
	Leu	Ser	Stop	Trp	G
C	Leu	Pro	His	Arg	U
	Leu	Pro	His	Arg	C
	Leu	Pro	Gln	Arg	A
	Leu	Pro	Gln	Arg	G
A	Ile	Thr	Asn	Ser	U
	Ile	Thr	Asn	Ser	C
	Ile	Thr	Lys	Arg	A
	Met	Thr	Lys	Arg	G
G	Val	Ala	Asp	Gly	U
	Val	Ala	Asp	Gly	C
	Val	Ala	Glu	Gly	A
	Val	Ala	Glu	Gly	G

Note: This table identifies the amino acid encoded by each triplet. For example, the codon 5′ AUG 3′ on mRNA specifies methionine, whereas CAU specifies histidine. UAA, UAG, and UGA are termination signals. AUG is part of the initiation signal, in addition to coding for internal methionine residues.

Codons that specify the same amino acid are called *synonyms*. For example, CAU and CAC are synonyms for histidine. Note that synonyms are not distributed haphazardly throughout the genetic code (depicted in Table 4.4). In the table, an amino acid specified by two or more synonyms occupies a single box (unless it is specified by more than four synonyms). The amino acids in a box are specified by codons that have the same first two bases but differ in the third base, as exemplified by GUU, GUC, GUA, and GUG. Thus, *most synonyms differ only in the last base of the triplet*. Inspection of the code shows that XYC and XYU always encode the same amino acid, whereas XYG and XYA usually encode the same amino acid. The structural basis for these equivalences of codons will become evident when we consider the nature of the anticodons of tRNA molecules (Section 30.3).

What is the biological significance of the extensive degeneracy of the genetic code? If the code were not degenerate, 20 codons would designate amino acids and 44 would lead to chain termination. The probability of mutating to chain termination would therefore be much higher with a nondegenerate code. Chain-termination mutations usually lead to inactive proteins, whereas substitutions of one amino acid for another are usually rather harmless. Moreover, the code is constructed such that a change to any single nucleotide base of a codon results in a synonym or an amino acid with similar chemical properties. Thus, *degeneracy minimizes the deleterious effects of mutations*.

Messenger RNA Contains Start and Stop Signals for Protein Synthesis

Messenger RNA is translated into proteins on *ribosomes*, large molecular complexes assembled from proteins and ribosomal RNA. How is mRNA interpreted by the translation apparatus? The start signal for protein synthesis is complex in bacteria. Polypeptide chains in bacteria start with a modified amino acid—namely, formylmethionine (fMet). A specific tRNA, the initiator tRNA, carries fMet. This fMet-tRNA recognizes the codon AUG or, less frequently, GUG. However, AUG is also the codon for an internal methionine residue, and GUG is the codon for an internal valine residue. Hence, the signal for the first amino acid in a prokaryotic polypeptide chain must be more complex than that for all subsequent ones. *AUG (or GUG) is only part of the initiation signal* (Figure 4.32). In bacteria, the initiating AUG (or GUG) codon is preceded several nucleotides away by a purine-rich sequence, called the *Shine–Dalgarno sequence*, that base-pairs with a complementary sequence in a ribosomal RNA molecule (Section 30.3). In eukaryotes, the AUG closest to the 5′ end of an mRNA molecule is usually the start signal for protein synthesis. This particular AUG is read by an initiator tRNA conjugated to methionine. After the initiator AUG has been located, the *reading frame* is established—groups of three nonoverlapping nucleotides are defined, beginning with the initiator AUG codon.

As already mentioned, *UAA, UAG, and UGA designate chain termination*. These codons are read not by tRNA molecules but rather by specific proteins called *release factors* (Section 30.3). Binding of the release factors to the ribosomes releases the newly synthesized protein.

fMet

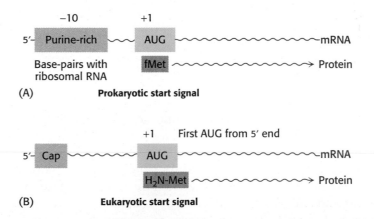

Figure 4.32 Initiation of protein synthesis. Start signals are required for the initiation of protein synthesis in (A) prokaryotes and (B) eukaryotes.

The Genetic Code Is Nearly Universal

Is the genetic code the same in all organisms? The base sequences of many wild-type and mutant genes are known, as are the amino acid sequences of their encoded proteins. For each mutant, the nucleotide change in the gene and the amino acid change in the protein are as predicted by the genetic code. Furthermore, mRNAs can be correctly translated by the protein-synthesizing machinery of very different species. For example, human hemoglobin mRNA is correctly translated by a wheat-germ extract, and bacteria efficiently express recombinant DNA molecules encoding human proteins such as insulin. These experimental findings strongly suggested that the genetic code is universal.

A surprise was encountered when the sequence of human mitochondrial DNA became known. Human mitochondria read UGA as a codon for tryptophan rather than as a stop signal (Table 4.5). Furthermore, AGA and AGG are read as stop signals rather than as codons for arginine, and AUA

TABLE 4.5 Distinctive codons of human mitochondria

Codon	Standard code	Mitochondrial code
UGA	Stop	Trp
UGG	Trp	Trp
AUA	Ile	Met
AUG	Met	Met
AGA	Arg	Stop
AGG	Arg	Stop

is read as a codon for methionine instead of isoleucine. Mitochondria of other species, such as those of yeast, also have genetic codes that differ slightly from the standard one. The genetic code of mitochondria can differ from that of the rest of the cell because mitochondrial DNA encodes a distinct set of tRNAs. Do any cellular protein-synthesizing systems deviate from the standard genetic code? At least 16 organisms deviate from the standard genetic code. Ciliated protozoa differ from most organisms in reading UAA and UAG as codons for amino acids rather than as stop signals; UGA is their sole termination signal. Thus, *the genetic code is nearly but not absolutely universal*. Variations clearly exist in mitochondria and in species, such as ciliates, that branched off very early in eukaryotic evolution. It is interesting to note that two of the codon reassignments in human mitochondria diminish the information content of the third base of the triplet (e.g., both AUA and AUG specify methionine). Most variations from the standard genetic code are in the direction of a simpler code.

Why has the code remained nearly invariant through billions of years of evolution, from bacteria to human beings? A mutation that altered the reading of mRNA would change the amino acid sequence of most, if not all, proteins synthesized by that particular organism. Many of these changes would undoubtedly be deleterious, and so there would be strong selection against a mutation with such pervasive consequences.

4.6 Most Eukaryotic Genes Are Mosaics of Introns and Exons

In bacteria, polypeptide chains are encoded by a continuous array of triplet codons in DNA. For many years, genes in higher organisms also were assumed to be continuous. This view was unexpectedly shattered in 1977, when investigators, including Philip Sharp and Richard Roberts, discovered that several genes are *discontinuous*. The mosaic nature of eukaryotic genes was revealed by electron microscopic studies of hybrids formed between mRNA and a segment of DNA containing the corresponding gene (Figure 4.33). For example, the gene for the β chain of hemoglobin is interrupted within its amino acid-coding sequence by a long *intron* of 550 base pairs and a short one of 120 base pairs. Thus, the *β-globin gene is split into three coding sequences*. The average human gene has 8 introns, and some have more than 100. The size ranges from 50 to 10,000 nucleotides.

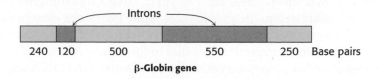

β-Globin gene

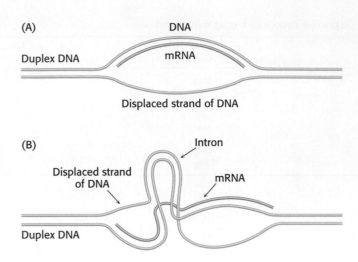

Figure 4.33 Detection of introns by electron microscopy. An mRNA molecule (shown in red) is hybridized to genomic DNA containing the corresponding gene. (A) A single loop of single-stranded DNA (shown in blue) is seen if the gene is continuous. (B) Two loops of single-stranded DNA (blue) and a loop of double-stranded DNA (blue and green) are seen if the gene contains an intron. Additional loops are evident if more than one intron is present.

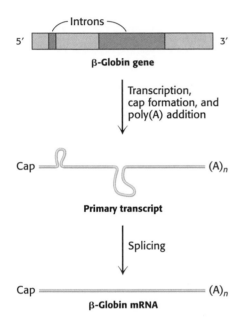

Figure 4.34 Transcription and processing of the β-globin gene. The gene is transcribed to yield the primary transcript, which is modified by cap and poly(A) addition. The introns in the primary RNA transcript are removed to form the mRNA.

RNA Processing Generates Mature RNA

At what stage in gene expression are introns removed? Newly synthesized RNA chains (pre-mRNA or primary transcript) isolated from nuclei are much larger than the mRNA molecules derived from them; in regard to β-globin RNA, the former sediment at 15S in zonal centrifugation experiments (p. 76) and the latter at 9S. In fact, the primary transcript of the β-globin gene contains two regions that are not present in the mRNA. *These regions in the 15S primary transcript are excised, and the coding sequences are simultaneously linked by a precise splicing enzyme to form the mature 9S mRNA* (Figure 4.34). Regions that are removed from the primary transcript are called *introns* (for *intervening* sequences), whereas those that are retained in the mature RNA are called *exons* (for *expressed* sequences). A common feature in the expression of split genes is that their exons are ordered in the same sequence in mRNA as in DNA. Thus, split genes, like continuous genes, are colinear with their polypeptide products.

Splicing is a complex operation that is carried out by *spliceosomes*, which are assemblies of proteins and small RNA molecules. RNA plays the catalytic role (Section 29.3). This enzymatic machinery recognizes signals in the nascent RNA that specify the splice sites. *Introns nearly always begin with GU and end with an AG that is preceded by a pyrimidine-rich tract* (Figure 4.35). *This consensus sequence is part of the signal for splicing.*

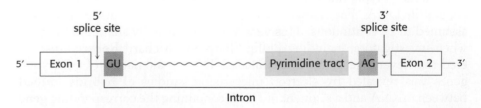

Figure 4.35 Consensus sequence for the splicing of mRNA precursors.

Many Exons Encode Protein Domains

Most genes of higher eukaryotes, such as birds and mammals, are split. Lower eukaryotes, such as yeast, have a much higher proportion of continuous genes. In prokaryotes, split genes are extremely rare. Have introns been inserted into genes in the evolution of higher organisms? Or have introns been removed from genes to form the streamlined genomes of prokaryotes and simple eukaryotes? Comparisons of the DNA sequences of genes encoding proteins that are highly conserved in evolution suggest that *introns were present in ancestral genes and were lost in the evolution of organisms that have become optimized for very rapid growth, such as prokaryotes.* The positions of introns in some genes are at least 1 billion years old. Furthermore, a common mechanism of splicing developed before the divergence of fungi, plants, and vertebrates, as shown by the finding that mammalian cell extracts can splice yeast RNA.

What advantages might split genes confer? *Many exons encode discrete structural and functional units of proteins.* An attractive hypothesis is that *new proteins arose in evolution by the rearrangement of exons encoding discrete structural elements, binding sites, and catalytic sites,* a process called *exon shuffling.* Because it preserves functional units but allows them to interact in new ways, exon shuffling is a rapid and efficient means of generating novel genes (Figure 4.36). Introns are extensive regions in which DNA can break and recombine with no deleterious effect on encoded proteins. In contrast, the exchange of sequences between different exons usually leads to loss of function.

Another advantage conferred by split genes is the potential for generating a series of related proteins by splicing a nascent RNA transcript in different ways. For example, a precursor of an antibody-producing cell forms an antibody that is anchored in the cell's plasma membrane (Figure 4.37). The attached antibody recognizes a specific foreign antigen, an event that leads to cell differentiation and proliferation. The activated antibody-producing cells then splice their nascent RNA transcript in an alternative manner to form soluble antibody molecules that are secreted rather than retained on the cell surface. We see here a clear-cut example of a benefit conferred by the complex arrangement of introns and exons in higher organisms. *Alternative splicing is a facile means of forming a set of proteins that are variations of a basic motif according to a developmental program without requiring a gene for each protein.*

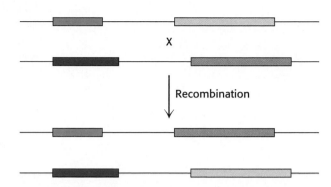

Figure 4.36 Exon shuffling. Exons can be readily shuffled by recombination of DNA to expand the genetic repertoire.

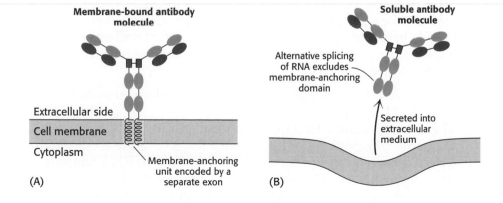

Figure 4.37 Alternative splicing. Alternative splicing generates mRNAs that are templates for different forms of a protein: (A) a membrane-bound antibody on the surface of a lymphocyte, and (B) its soluble counterpart, exported from the cell. The membrane-bound antibody is anchored to the plasma membrane by a helical segment (highlighted in yellow) that is encoded by its own exon.

Summary

4.1 A Nucleic Acid Consists of Four Kinds of Bases Linked to a Sugar–Phosphate Backbone

DNA and RNA are linear polymers of a limited number of monomers. In DNA, the repeating units are nucleotides, with the sugar being a deoxyribose and the bases being adenine (A), thymine (T), guanine (G), and cytosine (C). In RNA, the sugar is a ribose and the base uracil (U) is used in place of thymine. DNA is the molecule of heredity in all prokaryotic and eukaryotic organisms. In viruses, the genetic material is either DNA or RNA.

4.2 A Pair of Nucleic Acid Chains with Complementary Sequences Can Form a Double-Helical Structure

All cellular DNA consists of two very long, helical polynucleotide chains coiled around a common axis. The sugar–phosphate backbone of each strand is on the outside of the double helix, whereas the purine and pyrimidine bases are on the inside. The two chains are held together by hydrogen bonds between pairs of bases: adenine is always paired with thymine, and guanine is always paired with cytosine. Hence, one strand of a double helix is the complement of the other. The two strands of the double helix run in opposite directions. Genetic information is encoded in the precise sequence of bases along a strand. Most RNA molecules are single stranded, but many contain extensive double-helical regions that arise from the folding of the chain into hairpins.

4.3 DNA Is Replicated by Polymerases That Take Instructions from Templates

In the replication of DNA, the two strands of a double helix unwind and separate as new chains are synthesized. Each parent strand acts as a template for the formation of a new complementary strand. Thus, the replication of DNA is semiconservative—each daughter molecule receives one strand from the parent DNA molecule. The replication of DNA is a complex process carried out by many proteins, including several DNA polymerases. The activated precursors in the synthesis of DNA are the four deoxyribonucleoside 5′-triphosphates. The new strand is synthesized in the $5′ \rightarrow 3′$ direction by a nucleophilic attack by the 3′-hydroxyl terminus of the primer strand on the innermost phosphorus atom of the incoming deoxyribonucleoside triphosphate. Most important, DNA polymerases catalyze the formation of a phosphodiester bond only if the base on the incoming nucleotide is complementary to the base on the template strand. In other words, DNA polymerases are template-directed enzymes. The genes of some viruses, such as tobacco mosaic virus, are made of single-stranded RNA. An RNA-directed RNA polymerase mediates the replication of this viral RNA. Retroviruses, exemplified by HIV-1, have a single-stranded RNA genome that undergoes reverse transcription into double-stranded DNA by reverse transcriptase, an RNA-directed DNA polymerase.

4.4 Gene Expression Is the Transformation of DNA Information into Functional Molecules

The flow of genetic information in normal cells is from DNA to RNA to protein. The synthesis of RNA from a DNA template is called transcription, whereas the synthesis of a protein from an RNA template is termed translation. Cells contain several kinds of RNA, among which are messenger RNA (mRNA), transfer RNA (tRNA), and ribosomal RNA (rRNA), which vary in size from 75 to more than 5000 nucleotides. All cellular RNA is synthesized by RNA polymerases according to instructions given by DNA templates. The activated intermediates are ribonucleoside triphosphates and the direction of synthesis, like that of DNA, is $5′ \rightarrow 3′$. RNA polymerase differs from DNA polymerase in not requiring a primer.

4.5 Amino Acids Are Encoded by Groups of Three Bases Starting from a Fixed Point

The genetic code is the relation between the sequence of bases in DNA (or its RNA transcript) and the sequence of amino acids in proteins.

Amino acids are encoded by groups of three bases (called codons) starting from a fixed point. Sixty-one of the 64 codons specify particular amino acids, whereas the other 3 codons (UAA, UAG, and UGA) are signals for chain termination. Thus, for most amino acids, there is more than one code word. In other words, the code is degenerate. The genetic code is nearly the same in all organisms. Natural mRNAs contain start and stop signals for translation, just as genes do for directing where transcription begins and ends.

4.6 Most Eukaryotic Genes Are Mosaics of Introns and Exons

Most genes in higher eukaryotes are discontinuous. Coding sequences in these split genes, called exons, are separated by non-coding sequences, called introns, which are removed in the conversion of the primary transcript into mRNA and other functional mature RNA molecules. Split genes, like continuous genes, are colinear with their polypeptide products. A striking feature of many exons is that they encode functional domains in proteins. New proteins probably arose in the course of evolution by the shuffling of exons. Introns may have been present in primordial genes but were lost in the evolution of such fast-growing organisms such as bacteria and yeast.

Key Terms

double helix (p. 107)

deoxyribonucleic acid (DNA) (p. 108)

deoxyribose (p. 108)

ribose (p. 108)

purine (p. 109)

pyrimidine (p. 109)

ribonucleic acid (RNA) (p. 109)

nucleoside (p. 109)

nucleotide (p. 109)

semiconservative replication (p. 113)

DNA polymerase (p. 117)

template (p. 117)

primer (p. 117)

reverse transcriptase (p. 118)

messenger RNA (mRNA) (p. 119)

translation (p. 119)

transfer RNA (tRNA) (p. 120)

ribosomal RNA (rRNA) (p. 120)

small nuclear RNA (snRNA) (p. 120)

micro RNA (miRNA) (p. 120)

small interfering RNA (siRNA) (p. 120)

transcription (p. 120)

RNA polymerase (p. 120)

promoter site (p. 122)

codon (p. 123)

genetic code (p. 124)

ribosome (p. 126)

Shine–Dalgarno sequence (p. 126)

intron (p. 127)

exon (p. 127)

splicing (p. 128)

spliceosomes (p. 128)

exon shuffling (p. 129)

alternative splicing (p. 129)

Selected Readings

Where to Start

Felsenfeld, G. 1985. DNA. *Sci. Am.* 253(4):58–67.

Darnell, J. E., Jr. 1985. RNA. *Sci. Am.* 253(4):68–78.

Dickerson, R. E. 1983. The DNA helix and how it is read. *Sci. Am.* 249(6):94–111.

Crick, F. H. C. 1954. The structure of the hereditary material. *Sci. Am.*191(4): 54–61.

Chambon, P. 1981. Split genes. *Sci. Am.* 244(5):60–71.

Watson, J. D., and Crick, F. H. C. 1953. Molecular structure of nucleic acids: A structure for deoxyribose nucleic acid. *Nature* 171:737–738.

Watson, J. D., and Crick, F. H. C. 1953. Genetic implications of the structure of deoxyribonucleic acid. *Nature* 171:964–967.

Meselson, M., and Stahl, F. W. 1958. The replication of DNA in *Escherichia coli*. *Proc. Natl. Acad. Sci. U. S. A.* 44:671–682.

Books

Bloomfield, V. A., Crothers, D. M., Tinoco, I., and Hearst, J. 2000. *Nucleic Acids: Structures, Properties, and Functions*. University Science Books.

Singer, M., and Berg, P. 1991. *Genes and Genomes: A Changing Perspective*. University Science Books.

Lodish, H., Berk, A., Matsudaira, P., Kaiser, C. A., Krieger, M., Scott, M. P., Zipursky, L., and Darnell, J. 2004. *Molecular Cell Biology* (5th ed.). W. H. Freeman and Company.

Lewin, B. 2004. *Genes VIII*. Oxford University Press.

Watson, J. D., Baker, T. A., Bell, S. P., Gann, A., Levine, M., and Losick, R. 2004 *Molecular Biology of the Gene* (5th ed.). Benjamin Cummings.

DNA Structure

Saenger, W. 1984. *Principles of Nucleic Acid Structure*. Springer Verlag.

Dickerson, R. E., Drew, H. R., Conner, B. N., Wing, R. M., Fratini, A. V., and Kopka, M. L. 1982. The anatomy of A-, B-, and Z-DNA. *Science* 216:475–485.

Sinden, R. R. 1994. *DNA Structure and Function*. Academic Press.

DNA Replication

Lehman, I. R. 2003. Discovery of DNA polymerase. *J. Biol. Chem.* 278:34733–34738.

Hübscher, U., Maga, G., and Spardari, S. 2002. Eukaryotic DNA polymerases. *Annu. Rev. Biochem.* 71:133–163.

Hübscher, U., Nasheuer, H.-P., and Syväoja, J. E. 2000. Eukaryotic DNA polymerases: A growing family. *Trends Biochem. Sci.* 25:143–147.

Brautigam, C. A., and Steitz, T. A. 1998. Structural and functional insights provided by crystal structures of DNA polymerases and their substrate complexes. *Curr. Opin. Struct. Biol.* 8:54–63.

Kornberg, A., and Baker, T. A. 1992. *DNA Replication* (2d ed.). W. H. Freeman and Company.

Discovery of Messenger RNA

Jacob, F., and Monod, J. 1961. Genetic regulatory mechanisms in the synthesis of proteins. *J. Mol. Biol.* 3:318–356.

Brenner, S., Jacob, F., and Meselson, M. 1961. An unstable intermediate carrying information from genes to ribosomes for protein synthesis. *Nature* 190:576–581.

Hall, B. D., and Spiegelman, S. 1961. Sequence complementarity of T2-DNA and T2-specific RNA. *Proc. Natl. Acad. Sci. U. S. A.* 47:137–146.

Genetic Code

Freeland, S. J., and Hurst, L. D. 2004. Evolution encoded. *Sci. Am.* 290(4):84–91.

Crick, F. H. C., Barnett, L., Brenner, S., and Watts-Tobin, R. J. 1961. General nature of the genetic code for proteins. *Nature* 192:1227–1232.

Nirenberg, M. 1968. The genetic code. In *Nobel Lectures: Physiology or Medicine* (1963–1970), pp. 372–395. American Elsevier (1973).

Crick, F. H. C. 1958. On protein synthesis. *Symp. Soc. Exp. Biol.* 12:138–163.

Woese, C. R. 1967. *The Genetic Code.* Harper & Row.

Knight, R. D., Freeland, S. J., and Landweber L. F. 1999. Selection, history and chemistry: The three faces of the genetic code. *Trends Biochem. Sci.* 24(6):241–247.

Introns, Exons, and Split Genes

Sharp, P. A. 1988. RNA splicing and genes. *J. Am. Med. Assoc.* 260:3035–3041.

Dorit, R. L., Schoenbach, L., and Gilbert, W. 1990. How big is the universe of exons? *Science* 250:1377–1382.

Cochet, M., Gannon, F., Hen, R., Maroteaux, L., Perrin, F., and Chambon, P. 1979. Organization and sequence studies of the 17-piece chicken conalbumin gene. *Nature* 282:567–574.

Tilghman, S. M., Tiemeier, D. C., Seidman, J. G., Peterlin, B. M., Sullivan, M., Maizel, J. V., and Leder, P. 1978. Intervening sequence of DNA identified in the structural portion of a mouse β-globin gene. *Proc. Natl. Acad. Sci. U. S. A.* 75:725–729.

Reminiscences and Historical Accounts

Clayton, J., and Dennis, C. (Eds.). 2003. *50 Years of DNA.* Palgrave Macmillan.

Watson, J. D. 1968. *The Double Helix.* Atheneum.

McCarty, M. 1985. *The Transforming Principle: Discovering That Genes Are Made of DNA.* Norton.

Cairns, J., Stent, G. S., and Watson, J. D. 2000. *Phage and the Origins of Molecular Biology.* Cold Spring Harbor Laboratory.

Olby, R. 1974. *The Path to the Double Helix.* University of Washington Press.

Portugal, F. H., and Cohen, J. S. 1977. *A Century of DNA: A History of the Discovery of the Structure and Function of the Genetic Substance.* MIT Press.

Judson, H. F. 1996. *The Eighth Day of Creation.* Cold Spring Harbor Laboratory.

Sayre, A. 2000. *Rosalind Franklin and DNA.* Norton.

Problems

1. *Complements.* Write the complementary sequence (in the standard 5′ → 3′ notation) for (a) GATCAA, (b) TCGAAC, (c) ACGCGT, and (d) TACCAT.

2. *Compositional constraint.* The composition (in mole-fraction units) of one of the strands of a double-helical DNA molecule is [A] = 0.30 and [G] = 0.24. (a) What can you say about [T] and [C] for the same strand? (b) What can you say about [A], [G], [T], and [C] of the complementary strand?

3. *Lost DNA.* The DNA of a deletion mutant of λ bacteriophage has a length of 15 μm instead of 17 μm. How many base pairs are missing from this mutant?

4. *An unseen pattern.* What result would Meselson and Stahl have obtained if the replication of DNA were conservative (i.e., the parental double helix stayed together)? Give the expected distribution of DNA molecules after 1.0 and 2.0 generations for conservative replication.

5. *Tagging DNA.* (a) Suppose that you want to radioactively label DNA but not RNA in dividing and growing bacterial cells. Which radioactive molecule would you add to the culture medium? (b) Suppose that you want to prepare DNA in which the backbone phosphorus atoms are uniformly labeled with ^{32}P. Which precursors should be added to a solution containing DNA polymerase and primed template DNA? Specify the position of radioactive atoms in these precursors.

6. *Finding a template.* A solution contains DNA polymerase and the Mg^{2+} salts of dATP, dGTP, dCTP, and TTP. The follow-ing DNA molecules are added to aliquots of this solution. Which of them would lead to DNA synthesis? (a) A single-stranded closed circle containing 1000 nucleotide units. (b) A double-stranded closed circle containing 1000 nucleotide pairs. (c) A single-stranded closed circle of 1000 nucleotides base-paired to a linear strand of 500 nucleotides with a free 3′-OH terminus. (d) A double-stranded linear molecule of 1000 nucleotide pairs with a free 3′-OH group at each end.

7. *The right start.* Suppose that you want to assay reverse transcriptase activity. If polyriboadenylate is the template in the assay, what should you use as the primer? Which radioactive nucleotide should you use to follow chain elongation?

8. *Essential degradation.* Reverse transcriptase has ribonuclease activity as well as polymerase activity. What is the role of its ribonuclease activity?

9. *Virus hunting.* You have purified a virus that infects turnip leaves. Treatment of a sample with phenol removes viral proteins. Application of the residual material to scraped leaves results in the formation of progeny virus particles. You infer that the infectious substance is a nucleic acid. Propose a simple and highly sensitive means of determining whether the infectious nucleic acid is DNA or RNA.

10. *Mutagenic consequences.* Spontaneous deamination of cytosine bases in DNA occurs at low but measurable frequency. Cytosine is converted into uracil by loss of its amino group. After this conversion, which base pair occupies this position in

each of the daughter strands resulting from one round of replication? Two rounds of replication?

11. *Information content.* (a) How many different 8-mer sequences of DNA are there? (Hint: There are 16 possible dinucleotides and 64 possible trinucleotides.) We can quantify the information-carrying capacity of nucleic acids in the following way. Each position can be one of four bases, corresponding to two bits of information ($2^2 = 4$). Thus, a chain of 5100 nucleotides corresponds to $2 \times 5100 = 10,200$ bits, or 1275 bytes (1 byte = 8 bits). (b) How many bits of information are stored in an 8-mer DNA sequence? In the *E. coli* genome? In the human genome? (c) Compare each of these values with the amount of information that can be stored on a computer compact disc (CD).

12. *Key polymerases.* Compare DNA polymerase and RNA polymerase from *E. coli* in regard to each of the following features: (a) activated precursors, (b) direction of chain elongation, (c) conservation of the template, and (d) need for a primer.

13. *Encoded sequences.* (a) Write the sequence of the mRNA molecule synthesized from a DNA template strand having the sequence.

5′-ATCGTACCGTTA-3′

(b) What amino acid sequence is encoded by the following base sequence of an mRNA molecule? Assume that the reading frame starts at the 5′ end.

5′-UUGCCUAGUGAUUGGAUG-3′

(c) What is the sequence of the polypeptide formed on addition of poly(UUAC) to a cell-free protein-synthesizing system?

14. *A tougher chain.* RNA is readily hydrolyzed by alkali, whereas DNA is not. Why?

15. *A potent blocker.* How does cordycepin (3′-deoxyadenosine) block the synthesis of RNA?

16. *Silent RNA.* The code word GGG cannot be deciphered in the same way as can UUU, CCC, and AAA, because poly(G) does not act as a template. Poly(G) forms a triple-stranded helical structure. Why is it an ineffective template?

17. *Two from one.* Synthetic RNA molecules of defined sequence were instrumental in deciphering the genetic code. Their synthesis first required the synthesis of DNA molecules to serve as a template. H. Gobind Khorana synthesized, by organic-chemical methods, two complementary deoxyribonucleotides, each with nine residues: d(TAC)$_3$ and d(GTA)$_3$. Partly overlapping duplexes that formed on mixing these

oligonucleotides then served as templates for the synthesis by DNA polymerase of long, repeating double-helical DNA chains. The next step was to obtain long polyribonucleotide chains with a sequence complementary to only one of the two DNA strands. How did he obtain only poly(UAC)? Only poly(GUA)?

18. *Overlapping or not.* In a nonoverlapping triplet code, each group of three bases in a sequence ABCDEF . . . specifies only one amino acid—ABC specifies the first, DEF the second, and so forth—whereas, in a completely overlapping triplet code, ABC specifies the first amino acid, BCD the second, CDE the third, and so forth. Assume that you can mutate an individual nucleotide of a codon and detect the mutation in the amino acid sequence. Design an experiment that would establish whether the genetic code is overlapping or nonoverlapping.

19. *Triple entendre.* The RNA transcript of a region of T4 phage DNA contains the sequence 5′-AAAUGAGGA-3′. This sequence encodes three different polypeptides. What are they?

20. *Valuable synonyms.* Proteins generally have low contents of Met and Trp, intermediate ones of His and Cys, and high ones of Leu and Ser. What is the relation between the number of codons of an amino acid and its frequency of occurrence in proteins? What might be the selective advantage of this relation?

21. *A new translation.* A transfer RNA with a UGU anticodon is enzymatically conjugated to ^{14}C-labeled cysteine. The cysteine unit is then chemically modified to alanine (with the use of Raney nickel, which removes the sulfur atom of cysteine). The altered aminoacyl-tRNA is added to a protein-synthesizing system containing normal components except for this tRNA. The mRNA added to this mixture contains the following sequence:

5′-UUUUGCCAUGUUUGUGCU-3′

What is the sequence of the corresponding radiolabeled peptide?

Chapter Integration Problems

22. *Back to the bench.* A protein chemist told a molecular geneticist that he had found a new mutant hemoglobin in which aspartate replaced lysine. The molecular geneticist expressed surprise and sent his friend scurrying back to the laboratory. (a) Why did the molecular geneticist doubt the reported amino acid substitution? (b) Which amino acid substitutions would have been more palatable to the molecular geneticist?

23. *Eons apart.* The amino acid sequences of a yeast protein and a human protein having the same function are found to be 60% identical. However, the corresponding DNA sequences are only 45% identical. Account for this differing degree of identity.

Exploring Genes and Genomes

Processes such as the development from a caterpillar into a butterfly entail dramatic changes in patterns of gene expression. The expression levels of thousands of genes can be monitored through the use of DNA arrays. At right, a GeneChip reveals the expression levels of more than 12,000 human genes; the brightness of each spot indicates the expression level of the corresponding gene. [(Left) Roger Hart/Rainbow. (Right) GeneChip courtesy of Affymetrix.]

Since its emergence in the 1970s, recombinant DNA technology has revolutionized biochemistry. The genetic endowment of organisms can now be precisely changed in designed ways. Recombinant DNA technology is the fruit of several decades of basic research on DNA, RNA, and viruses. It depends, first, on having enzymes that can cut, join, and replicate DNA and reverse transcribe RNA. Restriction enzymes cut very long DNA molecules into specific fragments that can be manipulated; DNA ligases join the fragments together. Many kinds of restriction enzymes are available. By applying this assortment cleverly, researchers can treat DNA sequences as modules that can be moved at will from one DNA molecule to another. Thus, recombinant DNA technology is based on nucleic acid enzymology.

A second foundation is the base-pairing language that allows complementary sequences to recognize and bind to each other. Hybridization with complementary DNA (cDNA) or RNA probes is a sensitive means of detecting specific nucleotide sequences. In recombinant DNA technology, base-pairing is used to construct new combinations of DNA as well as to detect and amplify particular sequences. This revolutionary technology is also critically dependent on our understanding of viruses, the ultimate parasites. Viruses efficiently deliver their own DNA (or RNA) into hosts, subverting them either to replicate the viral genome and produce viral proteins or to incorporate viral DNA into the host genome. Likewise, plasmids, which are accessory chromosomes found in bacteria, have been indispensable in recombinant DNA technology.

Third, powerful methods have been developed for DNA sequencing. These methods have been harnessed to sequence complete genomes, first small genomes from viruses, then larger genomes from bacteria, and, finally eukaryotic genomes including the 3-billion-base-pair human genome. Scientists are just beginning to exploit the enormous information content of these genome sequences.

These new methods have wide-ranging benefits. The roles of particular genes and gene products can be explored in their natural context or in other ways. Novel proteins can be created by altering genes in specific ways to provide detailed views into protein function. Clinically useful proteins, such as hormones, are now synthesized by recombinant DNA techniques. Crops are being generated to resist pests and harsh conditions. The new opportunities opened by recombinant DNA technology promise to have even broader effects.

5.1 The Exploration of Genes Relies on Key Tools

The rapid progress in biotechnology—indeed its very existence—is a result of a few key techniques.

1. *Restriction-enzyme analysis.* Restriction enzymes are precise, molecular scalpels that allow an investigator to manipulate DNA segments.

2. *Blotting techniques.* The Southern and Northern blots are used to separate and characterize DNA and RNA, respectively. The Western blot, which uses antibodies to characterize proteins, was described on page 88.

3. *DNA sequencing.* The precise nucleotide sequence of a molecule of DNA can be determined. Sequencing has yielded a wealth of information concerning gene architecture, the control of gene expression, and protein structure.

4. *Solid-phase synthesis of nucleic acids.* Precise sequences of nucleic acids can be synthesized de novo and used to identify or amplify other nucleic acids.

5. *The polymerase chain reaction (PCR).* The polymerase chain reaction leads to a billionfold amplification of a segment of DNA. One molecule of DNA can be amplified to quantities that permit characterization and manipulation. This powerful technique is being used to detect pathogens and genetic diseases, to determine the source of a hair left at the scene of a crime, and to resurrect genes from fossils.

A final set of techniques, whose use we will highlight in Chapter 6, relies on the computer. Without the computer, it would be impossible to catalog, access, and characterize the abundant information, especially DNA sequence information, that the techniques just outlined are rapidly generating.

Restriction Enzymes Split DNA into Specific Fragments

Restriction enzymes, also called *restriction endonucleases,* recognize specific base sequences in double-helical DNA and cleave, at specific places, both strands of that duplex. To biochemists, these exquisitely precise scalpels are marvelous gifts of nature. They are indispensable for analyzing chromosome structure, sequencing very long DNA molecules, isolating genes, and

Palindrome
A word, sentence, or verse that reads the same from right to left as it does from left to right. 　　Radar 　　Senile felines 　　Do geese see God? 　　Roma tibi subito motibus ibit amor Derived from the Greek *palindromos*, "running back again."

creating new DNA molecules that can be cloned. Werner Arber and Hamilton Smith discovered restriction enzymes, and Daniel Nathans pioneered their use in the late 1960s.

Restriction enzymes are found in a wide variety of prokaryotes. Their biological role is to cleave foreign DNA molecules. The cell's own DNA is not degraded, because the sites recognized by its own restriction enzymes are methylated. Many restriction enzymes recognize specific sequences of four to eight base pairs and hydrolyze a phosphodiester bond in each strand in this region. A striking characteristic of these cleavage sites is that they almost always possess *twofold rotational symmetry*. In other words, the recognized sequence is *palindromic*, or an inverted repeat, and the cleavage sites are symmetrically positioned. For example, the sequence recognized by a restriction enzyme from *Streptomyces achromogenes* is

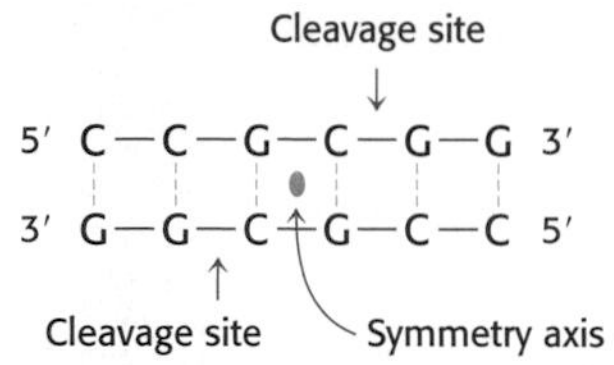

Cleavage site
↓
5′ C—C—G—C—G—G 3′
3′ G—G—C—G—C—C 5′
　　　　↑
Cleavage site　　Symmetry axis

In each strand, the enzyme cleaves the C–G phosphodiester bond on the 3′ side of the symmetry axis. As we shall see in Chapter 9, this symmetry corresponds to that of the structures of the restriction enzymes themselves.

Several hundred restriction enzymes have been purified and characterized. Their names consist of a three-letter abbreviation for the host organism (e.g., *Eco* for *Escherichia coli, Hin* for *Haemophilus influenzae, Hae* for *Haemophilus aegyptius*) followed by a strain designation (if needed) and a roman numeral (if more than one restriction enzyme from the same strain has been identified). The specificities of several of these enzymes are shown in Figure 5.1.

Restriction enzymes are used to cleave DNA molecules into specific fragments that are more readily analyzed and manipulated than the entire parent molecule. For example, the 5.1-kb circular duplex DNA of the tumor-producing SV40 virus is cleaved at 1 site by *Eco*RI, at 4 sites by *Hpa*I, and at 11 sites by *Hind*III. A piece of DNA produced by the action of one restriction enzyme can be specifically cleaved into smaller fragments by another restriction enzyme. The pattern of such fragments can serve as a *fingerprint* of a DNA molecule, as will be considered shortly. Indeed, complex chromosomes containing hundreds of millions of base pairs can be mapped by using a series of restriction enzymes.

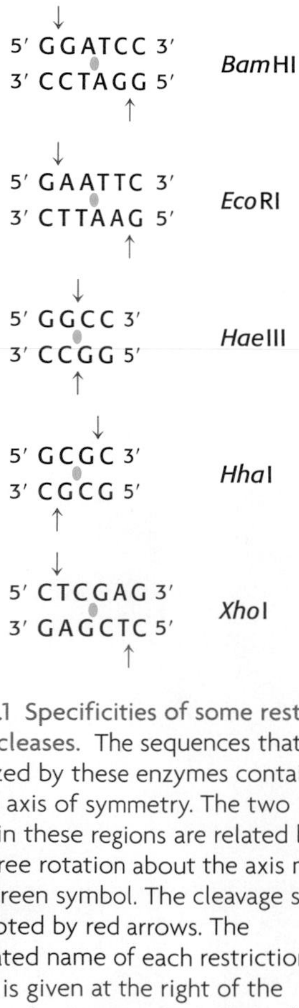

5′ GGATCC 3′　*Bam*HI
3′ CCTAGG 5′

5′ GAATTC 3′　*Eco*RI
3′ CTTAAG 5′

5′ GGCC 3′　*Hae*III
3′ CCGG 5′

5′ GCGC 3′　*Hha*I
3′ CGCG 5′

5′ CTCGAG 3′　*Xho*I
3′ GAGCTC 5′

Figure 5.1 Specificities of some restriction endonucleases. The sequences that are recognized by these enzymes contain a twofold axis of symmetry. The two strands in these regions are related by a 180-degree rotation about the axis marked by the green symbol. The cleavage sites are denoted by red arrows. The abbreviated name of each restriction enzyme is given at the right of the sequence that it recognizes. Note that the cuts may be staggered or even.

Restriction Fragments Can Be Separated by Gel Electrophoresis and Visualized

Small differences between related DNA molecules can be readily detected because their restriction fragments can be separated and displayed by gel electrophoresis. In many types of gels, the electrophoretic mobility of a DNA fragment is inversely proportional to the logarithm of the number of base pairs, up to a certain limit. Polyacrylamide gels are used to separate fragments containing up to 1000 base pairs, whereas more-porous agarose gels are used to resolve mixtures of larger fragments (up to 20 kb). An important feature of these gels is their high resolving power. In certain kinds of gels, fragments differing in length by just one nucleotide of several hundred can be distinguished. Moreover, entire chromosomes containing millions of nucleotides can be separated on agarose gels by applying pulsed electric fields (pulsed-field gel electrophoresis, PFGE) in different directions. The chromosomes are separated from one another by the differential stretching and relaxing of large DNA molecules as an electric field is turned off and on at short intervals. Bands or spots of radioactive DNA in gels can be visualized by autoradiography. Alternatively, a gel can be stained with ethidium bromide, which fluoresces an intense orange when bound to a double-helical DNA molecule (Figure 5.2). A band containing only 50 ng of DNA can be readily seen.

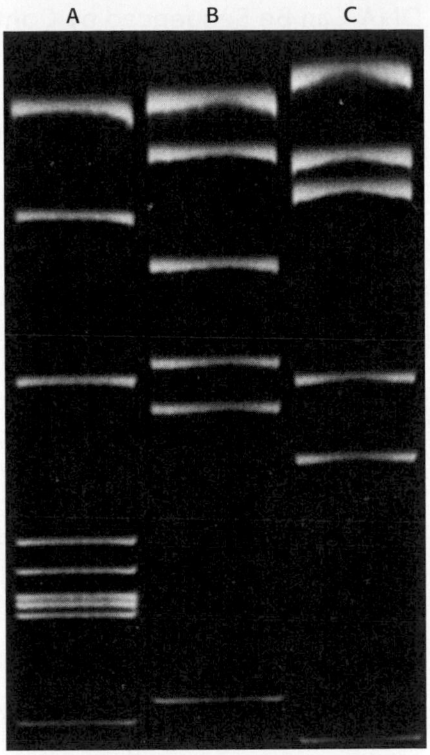

A restriction fragment containing a specific base sequence can be identified by hybridizing it with a labeled complementary DNA strand (Figure 5.3). A mixture of restriction fragments is separated by electrophoresis through an agarose gel, denatured to form single-stranded DNA, and transferred to a nitrocellulose sheet. The positions of the DNA fragments in the gel are preserved on the nitrocellulose sheet, where they are exposed to a ^{32}P-labeled single-stranded *DNA probe*. The probe hybridizes with a restriction fragment having a complementary sequence, and autoradiography then reveals the position of the restriction-fragment–probe duplex. A particular fragment amid a million others can be readily identified in this way, like finding a needle in a haystack. This powerful technique is known as *Southern blotting* because it was devised by Edwin Southern.

Similarly, RNA molecules can be separated by gel electrophoresis, and specific sequences can be identified by hybridization subsequent to their transfer to nitrocellulose. This analogous technique for the analysis of RNA has been whimsically termed *Northern blotting*. A further play on words accounts for the term *Western blotting*, which refers to a technique for detecting a particular protein by staining with specific antibody (p. 88). Southern, Northern, and Western blots are also known respectively as *DNA, RNA,* and *protein blots.*

Figure 5.2 Gel-electrophoresis pattern of a restriction digest. This gel shows the fragments produced by cleaving SV40 DNA with each of three restriction enzymes. These fragments were made fluorescent by staining the gel with ethidium bromide. [Courtesy of Dr. Jeffrey Sklar.]

Restriction-fragment-length polymorphism (RFLP)

Southern blotting can be used to follow the inheritance of selected genes. Mutations within restriction sites change the sizes of restriction fragments and hence the positions of bands in Southern-blot analyses. The existence of genetic diversity in a population is termed polymorphism. The detected mutation may itself cause disease or it may be closely linked to one that does. Genetic diseases such as sickle-cell anemia, cystic fibrosis, and Huntington chorea can be detected by RFLP analyses.

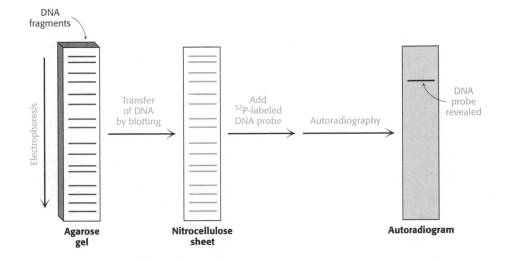

Figure 5.3 Southern blotting. A DNA fragment containing a specific sequence can be identified by separating a mixture of fragments by electrophoresis, transferring them to nitrocellulose, and hybridizing with a ^{32}P-labeled probe complementary to the sequence. The fragment containing the sequence is then visualized by autoradiography.

137

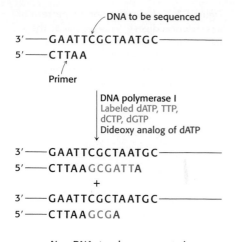

DNA to be sequenced

3'———GAATTCGCTAATGC———
5'———CTTAA

Primer

DNA polymerase I
Labeled dATP, TTP,
dCTP, dGTP
Dideoxy analog of dATP

3'———GAATTCGCTAATGC———
5'———CTTAAGCGATTA
+
3'———GAATTCGCTAATGC———
5'———CTTAAGCGA

**New DNA strands are separated
and subjected to electrophoresis**

Figure 5.4 Strategy of the chain-termination method for sequencing DNA. Fragments are produced by adding the 2',3'-dideoxy analog of a dNTP to each of four polymerization mixtures. For example, the addition of the dideoxy analog of dATP (shown in red) results in fragments ending in A. The strand cannot be extended past the dideoxy analog.

DNA Can Be Sequenced by Controlled Termination of Replication

The analysis of DNA structure and its role in gene expression also have been markedly facilitated by the development of powerful techniques for the *sequencing* of DNA molecules. The key to DNA sequencing is the generation of DNA fragments whose length depends on the last base in the sequence. Collections of such fragments can be generated through the *controlled termination of replication* (Sanger dideoxy method), a method developed by Frederick Sanger and coworkers. This technique has superseded alternative methods because of its simplicity. The same procedure is performed on four reaction mixtures at the same time. In all these mixtures, a DNA polymerase is used to make the complement of a particular sequence within a single-stranded DNA molecule. The synthesis is primed by a chemically synthesized fragment (p. 139) that is complementary to a part of the sequence known from other studies. In addition to the four deoxyribonucleoside triphosphates (radioactively labeled), each reaction mixture contains a small amount of the 2', 3'-*dideoxy analog* of one of the nucleotides, a different nucleotide for each reaction mixture.

2', 3'-Dideoxy analog

The incorporation of this analog blocks further growth of the new chain because it lacks the 3'-hydroxyl terminus needed to form the next phosphodiester bond. The concentration of the dideoxy analog is low enough that chain termination will take place only occasionally. The polymerase will insert the correct nucleotide sometimes and the dideoxy analog other times, stopping the reaction. For instance, if the dideoxy analog of dATP is present, fragments of various lengths are produced, but all will be terminated by the dideoxy analog (Figure 5.4). Importantly, this dideoxy analog of dATP will be inserted only where a T was located in the DNA being sequenced. Thus, the fragments of different length will correspond to the positions of T. Four such sets of *chain-terminated fragments* (one for each dideoxy analog) then undergo electrophoresis, and the base sequence of the new DNA is read from the autoradiogram of the four lanes.

Fluorescence detection is a highly effective alternative to autoradiography. A fluorescent tag is incorporated into each dideoxy analog—a differently colored one for each of the four chain terminators (e.g., a blue emitter for termination at A and a red one for termination at C). With the use of a mixture of terminators, a single reaction can be performed and the resulting fragments are then subjected to electrophoresis. The separated bands of DNA are detected by their fluorescence as they emerge following electrophoresis; the sequence of their colors directly gives the base sequence (Figure 5.5). Sequences of as many as 500 bases can be determined in this way. Fluorescence detection is attractive because it eliminates the use of radioactive reagents and can be readily automated. Indeed, modern

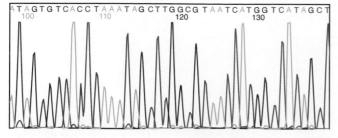

ATAGTGTCACCTAAAATAGCTTGGCGTAATCATGGTCATAGCT
100   110   120   130

Figure 5.5 Fluorescence detection of oligonucleotide fragments produced by the dideoxy method. A sequencing reaction is performed with four chain-terminating dideoxy nucleotides, each labeled with a tag that fluoresces at a different wavelength (e.g., red for T). Each of the four colors represents a different base in a chromatographic trace produced by fluorescence measurements at four wavelengths. [After A. J. F. Griffiths et al., *An Introduction to Genetic Analysis*, 8th ed. (W. H. Freeman and Company, 2005).]

DNA sequencing instruments can sequence more than 1 million bases per day with the use of this method.

DNA Probes and Genes Can Be Synthesized by Automated Solid-Phase Methods

DNA strands, like polypeptides (Section 3.4), can be synthesized by the sequential addition of activated monomers to a growing chain that is linked to an insoluble support. The activated monomers are protonated *deoxyribonucleoside 3′-phosphoramidites*. In step 1, the 3′-phosphorus atom of this incoming unit becomes joined to the 5′-oxygen atom of the growing chain to form a *phosphite triester* (Figure 5.6). The 5′-OH group of the activated monomer is unreactive because it is blocked by a dimethoxytrityl (DMT) protecting group, and the 3′-phosphoryl group is rendered unreactive by attachment of the β-cyanoethyl (βCE) group. Likewise, amino groups on the purine and pyrimidine bases are blocked.

Coupling is carried out under anhydrous conditions because water reacts with phosphoramidites. In step 2, the phosphite triester (in which P is trivalent) is oxidized by iodine to form a *phosphotriester* (in which P is pentavalent). In step 3, the DMT protecting group on the 5′-OH group of the growing chain is removed by the addition of dichloroacetic acid, which leaves other protecting groups intact. The DNA chain is now elongated by one unit and ready for another cycle of addition. Each cycle takes only about 10 minutes and usually elongates more than 99% of the chains.

A deoxyribonucleoside 3′-phosphoramidite with DMT and βCE attached

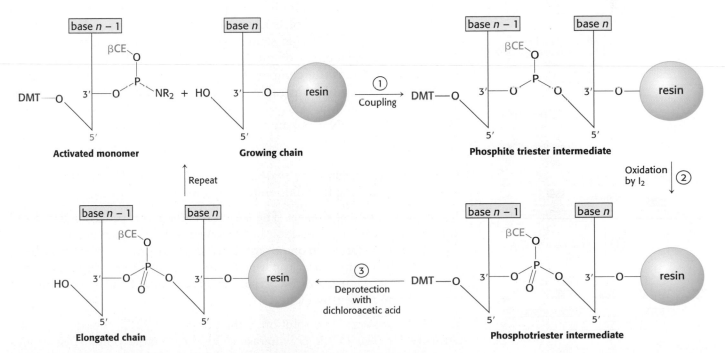

Figure 5.6 Solid-phase synthesis of a DNA chain by the phosphite triester method. The activated monomer added to the growing chain is a deoxyribonucleoside 3′-phosphoramidite containing a DMT protecting group on its 5′-oxygen atom, a β-cyanoethyl (βCE) protecting group on its 3′-phosphoryl oxygen atom, and a protecting group on the base.

This solid-phase approach is ideal for the synthesis of DNA, as it is for polypeptides, because the desired product stays on the insoluble support until the final release step. All the reactions take place in a single vessel, and excess soluble reagents can be added to drive reactions to completion. At the end of each step, soluble reagents and by-products are washed away from the resin that bears the growing chains. At the end of the synthesis, NH_3 is added to remove all protecting groups and release the oligonucleotide from the solid support. Because elongation is never 100% complete, the new DNA chains are of diverse lengths—the desired chain is the longest one. The sample can be purified by high-pressure liquid chromatography or by electrophoresis on polyacrylamide gels. DNA chains of as many as 100 nucleotides can be readily synthesized by this automated method.

The ability to rapidly synthesize DNA chains of any selected sequence opens many experimental avenues. For example, a synthesized oligonucleotide labeled at one end with ^{32}P or a fluorescent tag can be used to search for a complementary sequence in a very long DNA molecule or even in a genome consisting of many chromosomes. The use of labeled oligonucleotides as DNA probes is powerful and general. For example, a DNA probe that can base-pair to a known complementary sequence in a chromosome can serve as the starting point of an exploration of adjacent uncharted DNA. Such a probe can be used as a *primer* to initiate the replication of neighboring DNA by DNA polymerase. An exciting application of the solid-phase approach is the synthesis of new tailor-made genes. New proteins with novel properties can now be produced in abundance by the expression of synthetic genes.

Selected DNA Sequences Can Be Greatly Amplified by the Polymerase Chain Reaction

In 1984, Kary Mullis devised an ingenious method called the *polymerase chain reaction (PCR)* for amplifying specific DNA sequences. Consider a DNA duplex consisting of a target sequence surrounded by nontarget DNA. Millions of the target sequences can be readily obtained by PCR if the flanking sequences of the target are known. PCR is carried out by adding the following components to a solution containing the target sequence: (1) a pair of primers that hybridize with the flanking sequences of the target, (2) all four deoxyribonucleoside triphosphates (dNTPs), and (3) a heat-stable DNA polymerase. A PCR cycle consists of three steps (Figure 5.7).

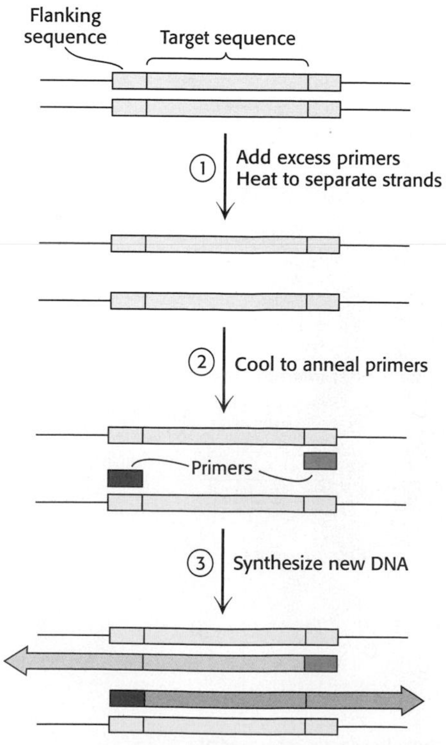

1. *Strand separation.* The two strands of the parent DNA molecule are separated by heating the solution to 95°C for 15 s.

2. *Hybridization of primers.* The solution is then abruptly cooled to 54°C to allow each primer to hybridize to a DNA strand. One primer hybridizes to the 3′ end of the target on one strand, and the other primer hybridizes to the 3′ end on the complementary target strand. Parent DNA duplexes do not form, because the primers are present in large excess. Primers are typically from 20 to 30 nucleotides long.

3. *DNA synthesis.* The solution is then heated to 72°C, the optimal temperature for *Taq* DNA polymerase. This heat-stable polymerase comes from *Thermus aquaticus*, a thermophilic bacterium that lives in hot springs. The polymerase elongates both primers in the direction of the target sequence because DNA synthesis is in the 5′-to-3′ direction. DNA synthesis takes place on both strands but extends beyond the target sequence.

Figure 5.7 The first cycle in the polymerase chain reaction (PCR). A cycle consists of three steps: strand separation, the hybridization of primers, and the extension of primers by DNA synthesis.

These three steps—strand separation, hybridization of primers, and DNA synthesis—constitute one cycle of the PCR amplification and can be carried out repetitively just by changing the temperature of the reaction mixture. The thermostability of the polymerase makes it feasible to carry out PCR in a closed container; no reagents are added after the first cycle. The duplexes are heated to begin the second cycle, which produces four duplexes, and then the third cycle is initiated (Figure 5.8). At the end of the third cycle, two short strands appear that constitute only the target sequence—the sequence including and bounded by the primers. Subsequent cycles will amplify the target sequence exponentially. The larger strands increase in number arithmetically and serve as a source for the synthesis of more short strands. Ideally, after n cycles, the desired sequence is amplified 2^n-fold. The amplification is a millionfold after 20 cycles and a billionfold after 30 cycles, which can be carried out in less than an hour.

Several features of this remarkable method for amplifying DNA are noteworthy. First, the sequence of the target need not be known. All that is required is knowledge of the flanking sequences. Second, the target can be much larger than the primers. Targets larger than 10 kb have been amplified by PCR. Third, primers do not have to be perfectly matched to flanking sequences to amplify targets. With the use of primers derived from a gene of known sequence, it is possible to search for variations on the theme. In this way, families of genes are being discovered by PCR. Fourth, PCR is highly specific because of the stringency of hybridization at relatively high temperature. *Stringency* is the required closeness of the match between primer and target, which can be controlled by temperature and salt. At high temperatures, the only DNA that is amplified is that situated between primers that have hybridized. A gene constituting less than a millionth of the total DNA of a higher organism is accessible by PCR. Fifth, PCR is exquisitely sensitive. A single DNA molecule can be amplified and detected.

PCR Is a Powerful Technique in Medical Diagnostics, Forensics, and Studies of Molecular Evolution

PCR can provide valuable diagnostic information in medicine. Bacteria and viruses can be readily detected with the use of specific primers. For example, PCR can reveal the presence of human immunodeficiency virus in people who have not mounted an immune response to this pathogen and would therefore be missed with an antibody assay. Finding *Mycobacterium tuberculosis* bacilli in tissue specimens is slow

Figure 5.8 Multiple cycles of the polymerase chain reaction. The two short strands produced at the end of the third cycle (along with longer stands not shown) represent the target sequence. Subsequent cycles will amplify the target sequence exponentially and the parent sequence arithmetically.

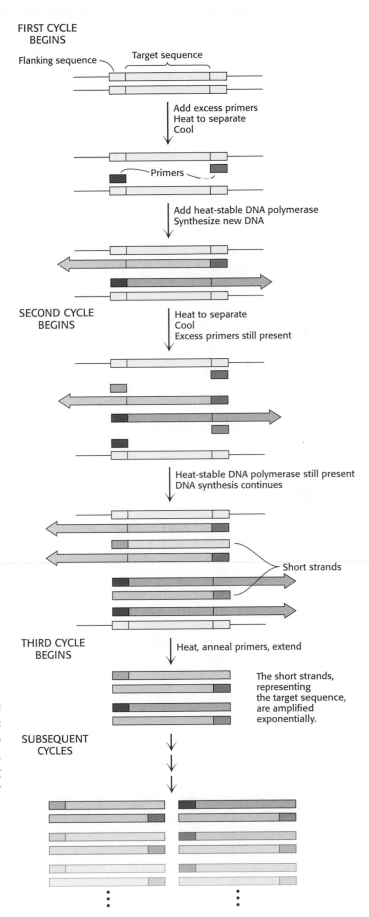

FIRST CYCLE BEGINS

Flanking sequence — Target sequence

Add excess primers
Heat to separate
Cool

Primers

Add heat-stable DNA polymerase
Synthesize new DNA

SECOND CYCLE BEGINS

Heat to separate
Cool
Excess primers still present

Heat-stable DNA polymerase still present
DNA synthesis continues

Short strands

THIRD CYCLE BEGINS

Heat, anneal primers, extend

The short strands, representing the target sequence, are amplified exponentially.

SUBSEQUENT CYCLES

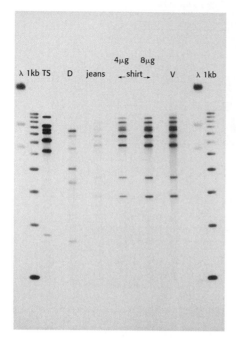

Figure 5.9 DNA and forensics. DNA isolated from bloodstains on the pants and shirt of a defendant was amplified by PCR, then compared with DNA from the victim as well as the defendant by using gel electrophoresis and autoradiography. DNA from the bloodstains on the defendant's clothing matched the pattern of the victim but not that of the defendant. The frequency of a coincidental match of the DNA pattern on the clothing and the victim is approximately 1 in 33 billion. Lanes λ, 1kb, and TS refer to control DNA samples; lane D, DNA from the defendant; jeans and shirt, DNA isolated from bloodstains on defendant's pants and shirt (two different amounts analyzed); V, DNA sample from victim's blood. [Courtesy of Cellmark Diagnostics, Germantown, Maryland.]

and laborious. With PCR, as few as 10 tubercle bacilli per million human cells can be readily detected. PCR is a promising method for the early detection of certain cancers. This technique can identify mutations of certain growth-control genes, such as the *ras* genes (p. 397). The capacity to greatly amplify selected regions of DNA can also be highly informative in monitoring cancer chemotherapy. Tests using PCR can detect when cancerous cells have been eliminated and treatment can be stopped; they can also detect a relapse and the need to immediately resume treatment. PCR is ideal for detecting leukemias caused by chromosomal rearrangements.

PCR is also having an effect in forensics and legal medicine. An individual DNA profile is highly distinctive because many genetic loci are highly variable within a population. For example, variations at a specific one of these locations determines a person's HLA type (human leukocyte antigen type; Section 33.5); organ transplants are rejected when the HLA types of the donor and recipient are not sufficiently matched. PCR amplification of multiple genes is being used to establish biological parentage in disputed paternity and immigration cases. Analyses of blood stains and semen samples by PCR have implicated guilt or innocence in numerous assault and rape cases. The root of a single shed hair found at a crime scene contains enough DNA for typing by PCR (Figure 5.9).

DNA is a remarkably stable molecule, particularly when relatively shielded from air, light, and water. Under such circumstances, large fragments of DNA can remain intact for thousands of years or longer. PCR provides an ideal method for amplifying such ancient DNA molecules so that they can be detected and characterized (p. 178). PCR can also be used to amplify DNA from microorganisms that have not yet been isolated and cultured. As will be discussed in the next chapter, sequences from these PCR products can be sources of considerable insight into evolutionary relationships between organisms.

5.2 Recombinant DNA Technology Has Revolutionized All Aspects of Biology

The pioneering work of Paul Berg, Herbert Boyer, and Stanley Cohen in the early 1970s led to the development of recombinant DNA technology, which has taken biology from an exclusively analytical science to a synthetic one. New combinations of unrelated genes can be constructed in the laboratory by applying recombinant DNA techniques. These novel combinations can be cloned—amplified many-fold—by introducing them into suitable cells, where they are replicated by the DNA-synthesizing machinery of the host. The inserted genes are often transcribed and translated in their new setting. What is most striking is that the genetic endowment of the host can be permanently altered in a designed way.

Restriction Enzymes and DNA Ligase Are Key Tools in Forming Recombinant DNA Molecules

Let us begin by seeing how novel DNA molecules can be constructed in the laboratory. A DNA fragment of interest is covalently joined to a DNA *vector*. The essential feature of a vector is that it can replicate autonomously in an appropriate host. *Plasmids* (naturally occurring circles of DNA that act as accessory chromosomes in bacteria) and bacteriophage lambda (λ phage), a virus, are choice vectors for cloning in *E. coli*. The vector can be prepared for accepting a new DNA fragment by cleaving it at a single specific site with a restriction enzyme. For example, the plasmid pSC101, a 9.9-kb double-helical circular DNA molecule, is split at a unique site by the *Eco*RI restriction enzyme. The staggered cuts made by this enzyme produce *complementary single-stranded ends*, which have specific affinity for each other and hence are known as *cohesive* or *sticky ends*. Any DNA fragment can be inserted into this plasmid if it has the same cohesive ends. Such a fragment can be prepared from a larger piece of DNA by using the same restriction enzyme as was used to open the plasmid DNA (Figure 5.10).

The single-stranded ends of the fragment are then complementary to those of the cut plasmid. The DNA fragment and the cut plasmid can be annealed and then joined by *DNA ligase*, which catalyzes the formation of a phosphodiester bond at a break in a DNA chain. DNA ligase requires a free 3'-hydroxyl group and a 5'-phosphoryl group. Furthermore, the chains joined by ligase must be in a double helix. An energy source such as ATP or NAD^+ is required for the joining reaction, as will be discussed in Chapter 28.

This cohesive-end method for joining DNA molecules can be made general by using a *short, chemically synthesized DNA linker* that can be cleaved by restriction enzymes. First, the linker is covalently joined to the ends of a DNA fragment or vector. For example, the 5' ends of a decameric linker and a DNA molecule are phosphorylated by polynucleotide kinase and then joined by the ligase from T4 phage (Figure 5.11). This ligase can form a covalent bond between blunt-ended (flush-ended) double-helical DNA molecules. Cohesive ends are produced when these terminal extensions are cut by an appropriate restriction enzyme. Thus, *cohesive ends corresponding to a particular restriction enzyme can be added to virtually any DNA molecule*. We see here the fruits of combining enzymatic and synthetic chemical approaches in crafting new DNA molecules.

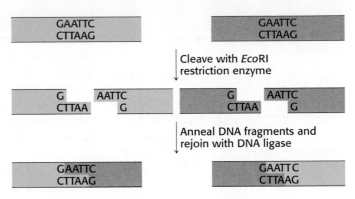

Figure 5.10 Joining of DNA molecules by the cohesive-end method. Two DNA molecules, cleaved with a common restriction enzyme such as *Eco*RI, can be ligated to form recombinant molecules.

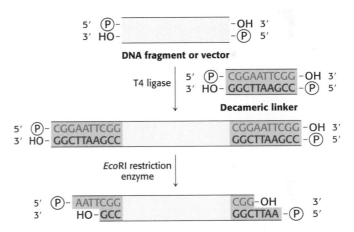

Figure 5.11 Formation of cohesive ends. Cohesive ends can be formed by the addition and cleavage of a chemically synthesized linker.

Plasmids and Lambda Phage Are Choice Vectors for DNA Cloning in Bacteria

Many plasmids and bacteriophages have been ingeniously modified to enhance the delivery of recombinant DNA molecules into bacteria and to facilitate the selection of bacteria harboring these vectors. As already mentioned, plasmids are circular double-stranded DNA molecules that occur naturally in some bacteria. They range in size from two to several hundred kilobases. Plasmids carry genes for the inactivation of antibiotics, the production of toxins, and the breakdown of natural products. These accessory chromosomes can replicate independently of the host chromosome. In contrast with the host genome, they are dispensable under certain conditions. A bacterial cell may have no plasmids at all or it may house as many as 20 copies of a plasmid.

pBR322 Plasmid. One of the first useful plasmids for cloning was *pBR322*, which contains genes for resistance to tetracycline and ampicillin (an antibiotic like penicillin). Different endonucleases can cleave this plasmid at a variety of unique sites, at which DNA fragments can be inserted. Insertion

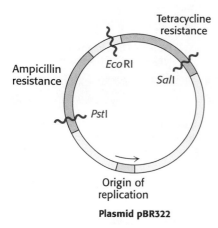

Figure 5.12 Genetic map of the plasmid pBR322. This plasmid carries two genes for antibiotic resistance. Like all other plasmids, it is a circular duplex DNA.

of DNA at the *Eco*RI restriction site does not alter either of the genes for antibiotic resistance (Figure 5.12). However, insertion at the *Sal*I or *Bam*HI restriction site inactivates the gene for tetracycline resistance, an effect called *insertional inactivation*. Cells containing pBR322 with a DNA insert at one of these restriction sites are resistant to ampicillin but sensitive to tetracycline, and so they can be readily selected. Cells that fail to take up the vector are sensitive to both antibiotics, whereas cells containing pBR322 without a DNA insert are resistant to both.

pUC18 Plasmid. Many newer plasmid vectors have additional features that increase their versatility compared with pBR322. pUC18 is a representative of one family of such plasmids (Figure 5.13). Like pBR322, this plasmid has an origin of replication and a selectable marker based on ampicillin resistance. However, this plasmid also contains a gene for β-galactosidase, an enzyme that degrades certain sugars (p. 311). In the presence of a specific substrate analog, this enzyme produces a blue pigment that can be easily seen. The gene for this enzyme has been engineered so that it contains a *polylinker* region that includes many unique restriction sites within its sequence. This polylinker can be cleaved with many different restriction enzymes or combinations of enzymes, providing great versatility in the DNA fragments that can be cloned. Insertion of a DNA fragment inactivates the β-galactosidase. The blue pigment is not generated, allowing identification of cells with recombinant DNA.

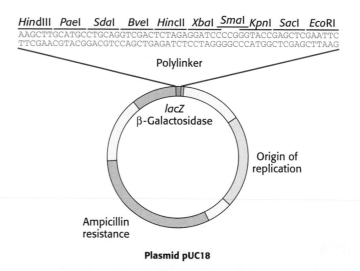

Figure 5.13 Genetic map of the plasmid pUC18. This plasmid includes a polylinker within a gene for β-galactosidase (often called the *lacZ* gene). Insertion of a DNA fragment into one of the many restriction sites within this polylinker can be detected by the absence of β-galactosidase activity.

Lambda Phage. Another widely used vector, λ *phage*, enjoys a choice of life styles: this bacteriophage can destroy its host or it can become part of its host (Figure 5.14). In the *lytic pathway*, viral functions are fully expressed: viral DNA and proteins are quickly produced and packaged into virus particles, leading to the lysis (destruction) of the host cell and the sudden appearance of about 100 progeny virus particles, or *virions*. In the *lysogenic pathway*, the phage DNA becomes inserted into the host-cell genome and can be replicated together with host-cell DNA for many generations, remaining inactive. Certain environmental changes can trigger the expression of this dormant viral DNA, which leads to the formation of progeny viruses and lysis of the host. Large segments of the 48-kb DNA of λ phage are not essential for productive infection and can be replaced by foreign DNA, thus making λ phage an ideal vector.

Mutant λ phages designed for cloning have been constructed. An especially useful one called λgt-λβ contains only two *Eco*RI cleavage sites

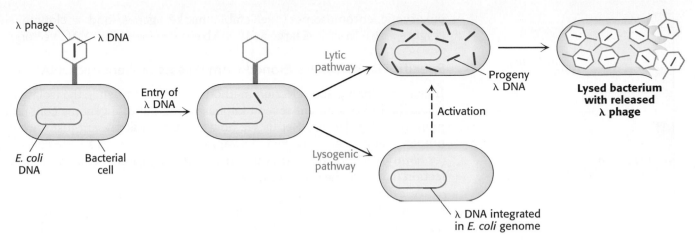

instead of the five normally present (Figure 5.15). After cleavage, the middle segment of this λ DNA molecule can be removed. The two remaining pieces of DNA (called arms) have a combined length equal to 72% of a normal genome length. This amount of DNA is too little to be packaged into a λ particle, because only DNA measuring from 75% to 105% of a normal genome in length can be readily packaged. However, a suitably long DNA insert (such as 10 kb) between the two ends of λ DNA enables such a recombinant DNA molecule (93% of normal length) to be packaged. Nearly all infectious λ particles formed in this way will contain an inserted piece of foreign DNA. Another advantage of using these modified viruses as vectors is that they enter bacteria much more easily than do plasmids. Among the variety of λ mutants that have been constructed for use as cloning vectors, one of them, called a *cosmid*, is essentially a hybrid of λ phage and a plasmid that can serve as a vector for large DNA inserts (as large as 45 kb).

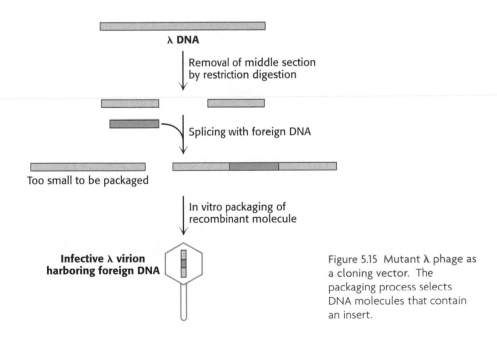

Figure 5.15 Mutant λ phage as a cloning vector. The packaging process selects DNA molecules that contain an insert.

Bacterial and Yeast Artificial Chromosomes

Much larger pieces of DNA can now be propagated in *bacterial artificial chromosomes* (BACs) or *yeast artificial chromosomes* (YACs). BACs are highly engineered versions of the *E. coli* fertility (F factor) that can include inserts as large as 300 kb. YACs contain a centromere, an *autonomously replicating sequence* (ARS, where replication begins), a pair of telomeres (normal ends of

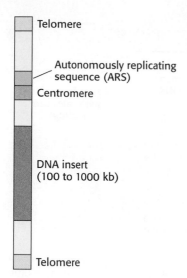

Figure 5.16 Diagram of a yeast artificial chromosome (YAC). These vectors include features necessary for replication and stability in yeast cells.

Telomere

Autonomously replicating sequence (ARS)

Centromere

DNA insert (100 to 1000 kb)

Telomere

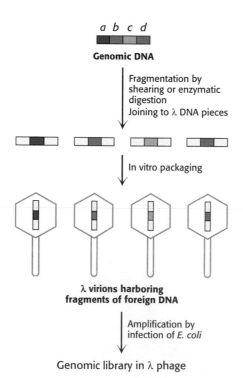

Figure 5.17 Creation of a genomic library. A genomic library can be created from a digest of a whole complex genome.

eukaryotic chromosomes), selectable marker genes, and a cloning site (Figure 5.16). Inserts as large as 1000 kb can be cloned into YAC vectors.

Specific Genes Can Be Cloned from Digests of Genomic DNA

Ingenious cloning and selection methods have made feasible the isolation of a specific DNA segment several kilobases long out of a genome containing more than 3×10^6 kb. Let us see how a gene that is present just once in a human genome can be cloned. The approach is to prepare a large collection (or *library*) of DNA fragments and then to identify those members of the collection that have the gene of interest.

A sample containing many copies of total genomic DNA is first mechanically sheared or partly digested by restriction enzymes into large fragments (Figure 5.17). This process yields a nearly random population of overlapping DNA fragments. These fragments are then separated by gel electrophoresis to isolate the set of all fragments that are about 15 kb long. Synthetic linkers are attached to the ends of these fragments, cohesive ends are formed, and the fragments are then inserted into a vector, such as λ phage DNA, prepared with the same cohesive ends. *E. coli* bacteria are then infected with these recombinant phages. These phages replicate themselves and then lyse their bacterial hosts. The resulting lysate contains fragments of human DNA housed in a sufficiently large number of virus particles to ensure that nearly the entire genome is represented. These phages constitute a *genomic library*. Phages can be propagated indefinitely, and so the library can be used repeatedly over long periods.

This genomic library is then screened to find the very small number of phages harboring the gene of interest. For the human genome, a calculation shows that a 99% probability of success requires screening about 500,000 clones; hence, a very rapid and efficient screening process is essential. Rapid screening can be accomplished by DNA hybridization.

A dilute suspension of the recombinant phages is first plated on a lawn of bacteria (Figure 5.18). Where each phage particle has landed and infected a bacterium, a *plaque* containing identical phages develops on the plate. A replica of this master plate is then made by applying a sheet of nitrocellulose. Infected bacteria and phage DNA released from lysed cells adhere to the sheet in a pattern of spots corresponding to the plaques. Intact bacteria on this sheet are lysed with NaOH, which also serves to denature the DNA so that it becomes accessible for hybridization with a ^{32}P-labeled probe. The presence of a specific DNA sequence in a single spot on the replica can be detected by using a radioactive complementary DNA or RNA molecule as a probe. Autoradiography then reveals the positions of spots harboring recombinant DNA. The corresponding plaques are picked out of the intact master plate and grown. A single investigator can readily screen a million clones in a day.

This method makes it possible to isolate virtually any gene, provided that a probe is available. How is a specific probe obtained? One approach is to start with the corresponding mRNA from cells in which it is abundant. For example, precursors of red blood cells contain large amounts of mRNA for hemoglobin, and plasma cells are rich in mRNAs for antibody molecules. The mRNAs from these cells can be fractionated by size to enrich for the mRNA

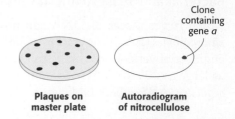

Figure 5.18 Screening a genomic library for a specific gene. Here, a plate is tested for plaques containing gene *a* of Figure 5.17.

Clone containing gene *a*

Plaques on master plate

Autoradiogram of nitrocellulose

of interest. As will be described shortly, a DNA complementary to this mRNA can be synthesized in vitro and cloned to produce a highly specific probe.

Alternatively, a probe for a gene can be prepared if part of the amino acid sequence of the protein encoded by the gene is known. A problem arises because a given peptide sequence can be encoded by a number of oligonucleotides (Figure 5.19). Thus, for this purpose, peptide sequences containing tryptophan and methionine are preferred, because these amino acids are specified by a single codon, whereas other amino acid residues have between two and six codons (see Table 4.4).

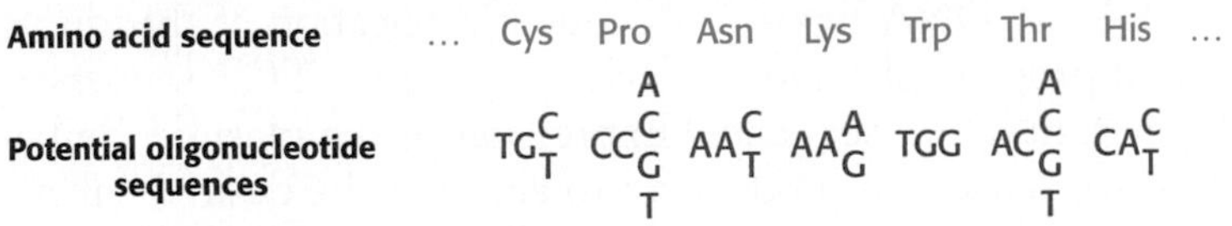

Amino acid sequence ... Cys Pro Asn Lys Trp Thr His ...

Potential oligonucleotide sequences

$$\text{TG}^{C}_{T} \quad \text{CC}^{A\,C\,G\,T} \quad \text{AA}^{C}_{T} \quad \text{AA}^{A}_{G} \quad \text{TGG} \quad \text{AC}^{A\,C\,G\,T} \quad \text{CA}^{C}_{T}$$

Figure 5.19 **Probes generated from a protein sequence.** A probe can be generated by synthesizing all possible oligonucleotides encoding a particular sequence of amino acids. Because of the degeneracy of the genetic code, 256 distinct oligonucleotides must be synthesized to ensure that the probe matching the sequence of seven amino acids in this example is present.

All the DNA sequences (or their complements) that encode the selected peptide sequence are synthesized by the solid-phase method and made radioactive by phosphorylating their 5′ ends with ^{32}P. The nitrocellulose membrane is exposed to a mixture containing all these probes and autoradiographed to identify clones with a complementary DNA sequence. Positive clones are then sequenced to determine which ones have a sequence matching that of the protein of interest. Some of them may contain the desired gene or a significant segment of it.

Proteins with New Functions Can Be Created Through Directed Changes in DNA

Much has been learned about genes and proteins by analyzing mutated genes selected from the repertoire offered by nature. In the classic genetic approach, mutations are generated randomly throughout the genome, and those exhibiting a particular phenotype are selected. Analysis of these mutants then reveals which genes are altered, and DNA sequencing identifies the precise nature of the changes. *Recombinant DNA technology now makes the creation of specific mutations feasible in vitro.* We can construct new genes with designed properties by making three kinds of directed changes: *deletions, insertions,* and *substitutions.*

Deletions. A specific deletion can be produced by cleaving a plasmid at two sites with a restriction enzyme and ligating to form a smaller circle. This simple approach usually removes a large block of DNA. A smaller deletion can be made by cutting a plasmid at a single site. The ends of the linear DNA are then digested by an exonuclease that removes nucleotides from both strands. The shortened piece of DNA is then ligated to form a circle that is missing a short length of DNA about the restriction site.

Substitutions: Oligonucleotide-Directed Mutagenesis. Mutant proteins with single amino acid substitutions can be readily produced by *oligonucleotide-directed mutagenesis* (Figure 5.20). Suppose that we want to replace a particular serine residue with cysteine. This mutation can be made if (1) we have a plasmid containing the gene or cDNA for the protein and (2) we know

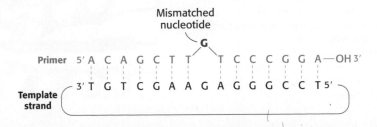

Figure 5.20 **Oligonucleotide-directed mutagenesis.** A primer containing a mismatched nucleotide is used to produce a desired change in the DNA sequence.

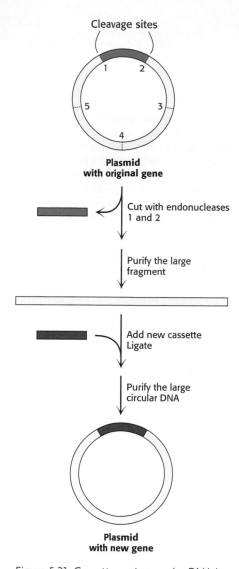

Cleavage sites

**Plasmid
with original gene**

Cut with endonucleases
1 and 2

Purify the large
fragment

Add new cassette
Ligate

Purify the large
circular DNA

**Plasmid
with new gene**

Figure 5.21 Cassette mutagenesis. DNA is cleaved at a pair of unique restriction sites by two different restriction endonucleases. A synthetic oligonucleotide with ends that are complementary to these sites (the cassette) is then ligated to the cleaved DNA. The method is highly versatile because the inserted DNA can have any desired sequence.

the base sequence around the site to be altered. If the serine of interest is encoded by TCT, we need to change the C to a G to get cysteine because cysteine is encoded by TGT. This type of mutation is called a *point mutation* because only one base is altered. The key to this mutation is to prepare an oligonucleotide primer that is complementary to this region of the gene except that it contains TGT instead of TCT. The two strands of the plasmid are separated, and the primer is then annealed to the complementary strand. The mismatch of 1 of 15 base pairs is tolerable if the annealing is carried out at an appropriate temperature. After annealing to the complementary strand, the primer is elongated by DNA polymerase, and the double-stranded circle is closed by adding DNA ligase. Subsequent replication of this duplex yields two kinds of progeny plasmid, half with the original TCT sequence and half with the mutant TGT sequence. Expression of the plasmid containing the new TGT sequence will produce a protein with the desired substitution of cysteine for serine at a unique site. We will encounter many examples of the use of oligonucleotide-directed mutagenesis to precisely alter regulatory regions of genes and to produce proteins with tailor-made features.

Insertions: Cassette Mutagenesis. In another valuable approach, *cassette mutagenesis,* plasmid DNA is cut with a pair of restriction enzymes to remove a short segment (Figure 5.21). A synthetic double-stranded oligonucleotide (the *cassette*) with cohesive ends that are complementary to the ends of the cut plasmid is then added and ligated. Each plasmid now contains the desired mutation.

Designer Genes. Novel proteins can also be created by splicing together gene segments that encode domains that are not associated in nature. For example, a gene for an antibody can be joined to a gene for a toxin to produce a chimeric protein that kills cells that are recognized by the antibody. These *immunotoxins* are being evaluated as anticancer agents. Furthermore, noninfectious coat proteins of viruses can be produced in large amounts by recombinant DNA methods. They can serve as *synthetic vaccines* that are safer than conventional vaccines prepared by inactivating pathogenic viruses. A subunit of the hepatitis B virus produced in yeast is proving to be an effective vaccine against this debilitating viral disease. Finally, entirely new genes can be synthesized de novo by the solid-phase method. These genes can encode proteins with no known counterparts in nature (Figure 5.22).

Design amino acid sequence

Design and synthesize gene

Produce and characterize protein

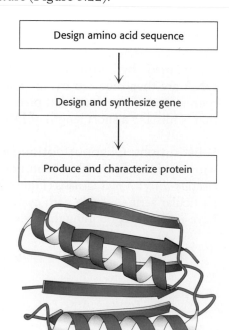

Figure 5.22 Protein design. New genes can be synthesized in their entirety. In some cases, these genes encode proteins that have been designed from scratch with the use of computer programs based on emerging principles of protein construction. The structure shown has no known natural counterpart. [Drawn from 1QYS.pdb.]

5.3 Complete Genomes Have Been Sequenced and Analyzed

The methods just described are extremely effective for the isolation and characterization of fragments of DNA. However, the genomes of organisms ranging from viruses to human beings comprise longer sequences of DNA, arranged in very specific ways crucial for their integrated functions. Is it possible to sequence complete genomes and analyze them? For small genomes, this sequencing was accomplished soon after DNA sequencing methods were developed. Sanger and coworkers determined the complete sequence of the 5386 bases in the DNA of the ϕ174 DNA virus in 1977, just a quarter century after Sanger's pioneering elucidation of the amino acid sequence of a protein. This tour de force was followed several years later by the determination of the sequence of human mitochondrial DNA, a double-stranded circular DNA molecule containing 16,569 base pairs. It encodes 2 ribosomal RNAs, 22 transfer RNAs, and 13 proteins. Many other viral genomes were sequenced in subsequent years. However, the genomes of free-living organisms presented a great challenge because even the simplest comprises more than 1 million base pairs. Thus, sequencing projects require both rapid sequencing techniques and efficient methods for assembling many short stretches of 300 to 500 base pairs into a complete sequence.

The Genomes of Organisms Ranging from Bacteria to Multicellular Eukaryotes Have Been Sequenced

With the development of automatic DNA sequencers based on fluorescent dideoxynucleotide chain terminators, high-volume, rapid DNA sequencing became a reality. The genome sequence of the bacterium *Haemophilus influenzae* was determined in 1995 by using a "shotgun" approach. The genomic DNA was sheared randomly into fragments that were then sequenced. Computer programs assembled the complete sequence by matching up overlapping regions between fragments. The *H. influenzae* genome comprises 1,830,137 base pairs and encodes approximately 1740 proteins (Figure 5.23). Using similar approaches, investigators have determined the sequences of more than 100 bacterial and archaeal species including key model organisms such as *E. coli*, *Salmonella typhimurium*, and *Archaeoglobus fulgidus*, as well as pathogenic organisms such as *Yersina pestis* (causing bubonic plague) and *Bacillus anthracis* (anthrax).

The first eukaryotic genome to be completely sequenced was that of baker's yeast, *Saccharomyces cerevisiae*, in 1996. The yeast genome comprises approximately 12 million base pairs, distributed on 16 chromosomes, and encodes more than 6000 proteins. This achievement was followed in 1998 by the first complete sequencing of the genome of a multicellular organism, the nematode *Caenorhabditis elegans*, which contains 97 million base pairs. This genome includes more than 19,000 genes. The genomes of many additional organisms widely used in biological and biomedical research have now been sequenced, including those of the fruit fly *Drosophila melanogaster*, the model plant *Arabidopsis thaliana*, the mouse, the rat, and the dog. Note that the sequencing of a complex genome proceeds in various stages from "draft"

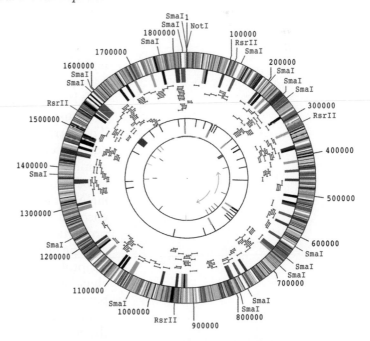

Figure 5.23 A complete genome. The diagram depicts the genome of *Haemophilus influenzae*, the first complete genome of a free-living organism to be sequenced. The genome encodes more than 1700 proteins and 70 RNA molecules. The likely function of approximately one-half of the proteins was determined by comparisons with sequences of proteins previously characterized in other species. [From R. D. Fleischmann et al., *Science* 269(1995):496–512; scan courtesy of The Institute for Genomic Research.]

through "completed" to "finished." Even after a sequence has been declared "finished," some sections, such as the repetitive sequences that make up heterochromatin, may be missing because these DNA sequences are very difficult to manipulate with the use of standard techniques.

The Sequencing of the Human Genome Has Been Finished

The ultimate goal of much of genomics research has been the sequencing and analysis of the human genome. Given that the human genome comprises approximately 3 billion base pairs of DNA distributed among 24 chromosomes, the challenge of producing a complete sequence was daunting. However, through an organized international effort of academic laboratories and private companies, the human genome has now progressed from a draft sequence first reported in 2001 to a finished sequence reported in late 2004 (Figure 5.24).

The human genome is a rich source of information about many aspects of humanity including biochemistry and evolution. Analysis of the genome will continue for many years to come. Developing an inventory of protein-encoding genes is one of the first tasks. At the beginning of the genome-sequencing project, the number of such genes was estimated to be approximately 100,000. With the availability of the completed (but not finished) genome, this estimate was reduced to 30,000 to 35,000. With the finished sequence, the estimate fell to 20,000 to 25,000. We will use the estimate of 25,000 throughout this book. The reduction in this estimate is due, in part, to the realization that there are a relatively large number of *pseudogenes,* many of which are formerly functional genes that have picked up mutations and are no longer expressed. For example, more than half of the genomic regions that correspond to olfactory receptors—key molecules responsible for our sense of smell—are pseudogenes (Section 32.1). The corresponding regions in the genomes of other primates and rodents encode functional olfactory receptors. Nonetheless, the surprisingly small number of genes belies the complexity of the human proteome. *Many genes encode more than one protein through mechanisms such as alternative splicing of mRNA and posttranslational modifications of proteins.* The different proteins encoded by a single gene often display important variations in functional properties.

The human genome contains a large amount of DNA that does not code for proteins. A great challenge in modern biochemistry and genetics is to elucidate the roles of this noncoding DNA. Much of this DNA is present because of the existence of *mobile genetic elements.* These elements, related to retroviruses (p. 118), have inserted themselves throughout the genome in the course of time. Most of these elements have accumulated mutations and are no longer functional. More than 1 million *Alu sequences,* each approximately 300 bases in length, are present in the human genome. *Alu* sequences are examples of *SINES, short interspersed elements.* The human genome also includes nearly 1 million *LINES, long interspersed elements,* DNA sequences that can be as long as 10 kilobase pairs (kbp). The roles of these elements as neutral genetic parasites or instruments of genome evolution are under current investigation.

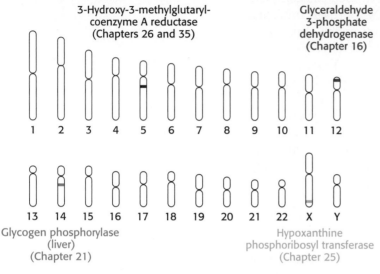

3-Hydroxy-3-methylglutaryl-
coenzyme A reductase
(Chapters 26 and 35)

Glyceraldehyde
3-phosphate
dehydrogenase
(Chapter 16)

1 2 3 4 5 6 7 8 9 10 11 12

13 14 15 16 17 18 19 20 21 22 X Y

Glycogen phosphorylase
(liver)
(Chapter 21)

Hypoxanthine
phosphoribosyl transferase
(Chapter 25)

Figure 5.24 The human genome. The human genome is arrayed on 46 chromosomes—22 pairs of autosomes and the X and Y sex chromosomes. The locations of several genes associated with important pathways in biochemistry are highlighted.

Comparative Genomics Has Become a Powerful Research Tool

Comparisons with genomes from other organisms are a source of insight into the human genome. The sequencing of the genome of the chimpanzee, our closest living relative, is nearing completion. The genomes of other mammals that are widely used in biological research, such as the mouse and the rat, have been completed. Comparisons reveal that an astonishing 99% of human genes have counterparts in these rodent genomes. However, these genes have been substantially reassorted among chromosomes in the estimated 75 million years of evolution since humans and rodents had a common ancestor (Figure 5.25).

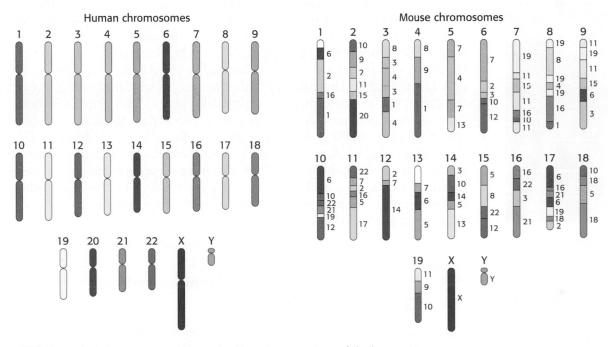

Figure 5.25 Genome comparison. A schematic comparison of the human genome and the mouse genome shows reassortment of large chromosomal fragments.

The genomes of other organisms also have been determined specifically for use in comparative genomics. For example, the genomes of two species of puffer fish, *Takifugu rubripes* and *Tetraodon nigroviridis,* have been determined. These genomes were selected because they are very small and lack much of the intergenic DNA present in such abundance in the human genome. The puffer fish genomes include fewer than 400 megabase pairs (Mbp), one eighth of the number in the human genome, yet it contains essentially the same number of genes. Comparison of the genomes of these species with that of humans revealed more than 1000 formerly unrecognized human genes. Furthermore, comparison of the two species of puffer fish, which had a common ancestor approximately 25 million years ago, is a source of insight into more recent events in evolution. Comparative genomics is a powerful tool, both for interpreting the human genome and for understanding major events in the origin of genera and species.

A puffer fish [Fred Bavendam/Peter Arnold.]

Gene-Expression Levels Can Be Comprehensively Examined

Most genes are present in the same quantity in every cell—namely, one copy per haploid cell or two copies per diploid cell. However, the level at which a gene is expressed, as indicated by mRNA quantities, can vary widely, ranging from no expression to hundreds of mRNA copies per cell. Gene-expression

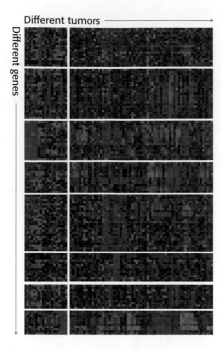

Figure 5.26 Gene-expression analysis using microarrays. The expression levels of thousands of genes can be simultaneously analyzed by using DNA microarrays (gene chips). Here, the analysis of 1733 genes in 84 breast-tumor samples reveals that the tumors can be divided into distinct classes on the basis of their gene-expression patterns. Red corresponds to gene induction and green corresponds to gene repression. [After C. M. Perou et al., *Nature* 406(2000):747–752.]

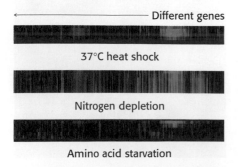

Figure 5.27 Monitoring changes in yeast gene expression. This microarray analysis shows levels of gene expression for yeast genes under different conditions. [After V. R. Iyer et al., *Nature* 409(2001):533–538.]

patterns vary from cell type to cell type, distinguishing, for example, a muscle cell from a nerve cell. Even within the same cell, gene-expression levels may vary as the cell responds to changes in physiological circumstances. Note that mRNA levels sometimes correlate with the levels of proteins expressed, but this correlation does not always hold. Thus, care must be exercised when interpreting the results of mRNA levels alone.

Using our knowledge of complete genome sequences, we can now analyze the pattern and level of expression of all genes in a particular cell or tissue. One of the most powerful methods developed to date for this purpose is based on hybridization. High-density arrays of oligonucleotides, called *DNA microarrays* or *gene chips,* can be constructed either through light-directed chemical synthesis carried out with photolithographic microfabrication techniques used in the semiconductor industry or by placing very small dots of oligonucleotides or cDNAs on a solid support such as a microscope slide. Fluorescently labeled cDNA is hybridized to the chip to reveal the expression level for each gene, identifiable by its known location on the chip (Figure 5.26). The intensity of the fluorescent spot on the chip reveals the extent of the transcription of a particular gene. DNA chips have been prepared that contain oligonucleotides complementary to all known open reading frames, 6200 in number, within the yeast genome (Figure 5.27). An analysis of mRNA pools with the use of these chips revealed, for example, that approximately 50% of all yeast genes are expressed at steady-state levels of between 0.1 and 1.0 mRNA copy per cell. This method readily detected variations in expression levels displayed by specific genes under different growth conditions.

5.4 Eukaryotic Genes Can Be Manipulated with Considerable Precision

Eukaryotic genes can be introduced into bacteria, and the bacteria can be used as factories to produce a desired protein product. It is also possible to introduce DNA into the cells of higher organisms. Genes introduced into animals are valuable tools for examining gene action, and they are the basis of gene therapy. Genes introduced into plants may make the plants resistant to pests, able to grow in harsh conditions, or carry greater quantities of essential nutrients. The manipulation of eukaryotic genes holds much promise as a source of medical and agricultural benefits, but it is also a source of controversy.

Complementary DNA Prepared from mRNA Can Be Expressed in Host Cells

How can mammalian DNA be cloned and expressed by *E. coli*? Recall that most mammalian genes are mosaics of introns and exons (p. 127). These interrupted genes cannot be expressed by bacteria, which lack the machinery to splice introns out of the primary transcript. However, this difficulty can be circumvented by causing bacteria to take up recombinant DNA that is complementary to mRNA. For example, proinsulin, a precursor of insulin, is synthesized by bacteria harboring plasmids that contain DNA complementary to mRNA for proinsulin (Figure 5.28). Indeed, bacteria produce much of the insulin used today by millions of diabetics.

The key to forming *complementary DNA* is the enzyme *reverse transcriptase.* As discussed on page 112, a retrovirus uses this enzyme to form a DNA–RNA hybrid in replicating its genomic RNA. Reverse transcriptase

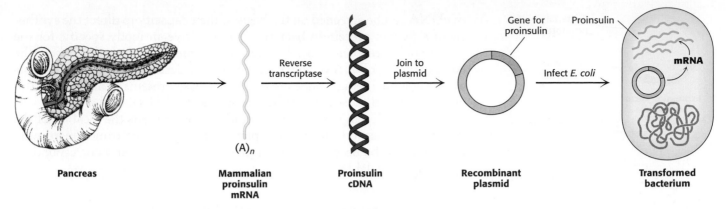

Pancreas | **Mammalian proinsulin mRNA** | **Proinsulin cDNA** | **Recombinant plasmid** | **Transformed bacterium**

Figure 5.28 Synthesis of proinsulin by bacteria. Proinsulin, a precursor of insulin, can be synthesized by transformed (genetically altered) clones of *E. coli*. The clones contain the mammalian proinsulin gene.

synthesizes a DNA strand complementary to an RNA template if the transcriptase is provided with a DNA primer that is base-paired to the RNA and contains a free 3′-OH group. We can use a simple sequence of linked thymidine [oligo(T)] residues as the primer. This oligo(T) sequence pairs with the poly(A) sequence at the 3′ end of most eukaryotic mRNA molecules (p. 123), as shown in Figure 5.29. The reverse transcriptase then synthesizes the rest of the cDNA strand in the presence of the four deoxyribonucleoside triphosphates. The RNA strand of this RNA–DNA hybrid is subsequently hydrolyzed by raising the pH. Unlike RNA, DNA is resistant to alkaline hydrolysis. The single-stranded DNA is converted into double-stranded DNA by creating another primer site. The enzyme *terminal transferase* adds nucleotides—for instance, several residues of dG—to the 3′ end of DNA. Oligo(dC) can bind to dG residues and prime the synthesis of the second DNA strand. Synthetic linkers can be added to this double-helical DNA for ligation to a suitable vector. Complementary DNA for all mRNA that a cell contains can be made, inserted into vectors, and then inserted into bacteria. Such a collection is called a *cDNA library*.

Complementary DNA molecules can be inserted into vectors that favor their efficient expression in hosts such as *E. coli*. Such plasmids or phages are called *expression vectors*. To maximize transcription, the cDNA is inserted into the vector in the correct reading frame near a strong bacterial promoter. In addition, these vectors ensure efficient translation by encoding a ribosome-binding site on the mRNA near the initiation codon. Clones

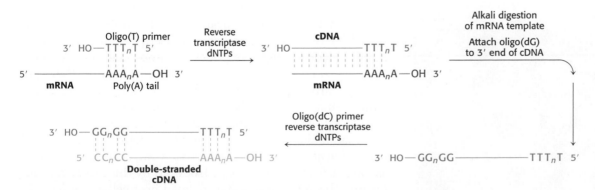

Figure 5.29 Formation of a cDNA duplex. A complementary DNA (cDNA) duplex is created from mRNA by using reverse transcriptase to synthesize a cDNA strand, first along the mRNA template and then, after digestion of the mRNA, along that same newly synthesized cDNA strand.

153

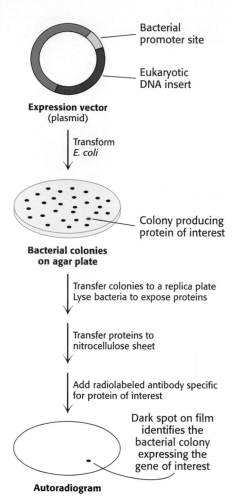

Expression vector
(plasmid)

Bacterial promoter site

Eukaryotic DNA insert

Transform
E. coli

Bacterial colonies
on agar plate

Colony producing
protein of interest

Transfer colonies to a replica plate
Lyse bacteria to expose proteins

Transfer proteins to
nitrocellulose sheet

Add radiolabeled antibody specific
for protein of interest

Dark spot on film
identifies the
bacterial colony
expressing the
gene of interest

Autoradiogram

Figure 5.30 Screening of cDNA clones. A method of screening for cDNA clones is to identify expressed products by staining with specific antibody.

of cDNA can be screened on the basis of their capacity to direct the synthesis of a foreign protein in bacteria. A radioactive antibody specific for the protein of interest can be used to identify colonies of bacteria that harbor the corresponding cDNA vector (Figure 5.30). As described on page 146, spots of bacteria on a replica plate are lysed to release proteins, which bind to an applied nitrocellulose filter. With the addition of ^{125}I-labeled antibody specific for the protein of interest, autoradiography reveals the location of the desired colonies on the master plate. This immunochemical screening approach can be used whenever a protein is expressed and corresponding antibody is available.

New Genes Inserted into Eukaryotic Cells Can Be Efficiently Expressed

Bacteria are ideal hosts for the amplification of DNA molecules. They can also serve as factories for the production of a wide range of prokaryotic and eukaryotic proteins. However, bacteria lack the necessary enzymes to carry out posttranslational modifications such as the specific cleavage of polypeptides and the attachment of carbohydrate units. Thus, many eukaryotic genes can be correctly expressed only in eukaryotic host cells. The introduction of recombinant DNA molecules into cells of higher organisms can also be a source of insight into how their genes are organized and expressed. How are genes turned on and off in embryological development? How does a fertilized egg give rise to an organism with highly differentiated cells that are organized in space and time? These central questions of biology can now be fruitfully approached by expressing foreign genes in mammalian cells.

Recombinant DNA molecules can be introduced into animal cells in several ways. In one method, foreign DNA molecules precipitated by calcium phosphate are taken up by animal cells. A small fraction of the imported DNA becomes stably integrated into the chromosomal DNA. The efficiency of incorporation is low, but the method is useful because it is easy to apply. In another method, DNA is *microinjected* into cells. A fine-tipped (0.1-mm-diameter) glass micropipet containing a solution of foreign DNA is inserted into a nucleus (Figure 5.31). A skilled investigator can inject hundreds of cells per hour. About 2% of injected mouse cells are viable and contain the new gene. In a third method, *viruses* are used to introduce new genes into animal cells. The most effective vectors are *retroviruses*. As described on page 118, retroviruses replicate through DNA intermediates, the reverse of the normal flow of information. A striking feature of the life cycle of a retrovirus is that the double-helical DNA form of its genome, produced by the action of reverse transcriptase, becomes randomly incorporated into host chromosomal DNA. This DNA version of the viral genome, called *proviral DNA*, can be efficiently expressed by the host cell and replicated along with normal cellular DNA. Retroviruses do not usually kill their hosts. Foreign genes have been efficiently introduced into mammalian cells by infecting them with vectors derived from *Moloney murine leukemia virus*, which can accept inserts as long as 6 kb. Some genes introduced by this retroviral vector into the genome of a transformed host cell are efficiently expressed.

Two other viral vectors are extensively used. *Vaccinia virus*, a large DNA-containing virus, replicates in the cyto-

Fertilized
mouse egg

Holding
pipette

Micropipette
with DNA
solution

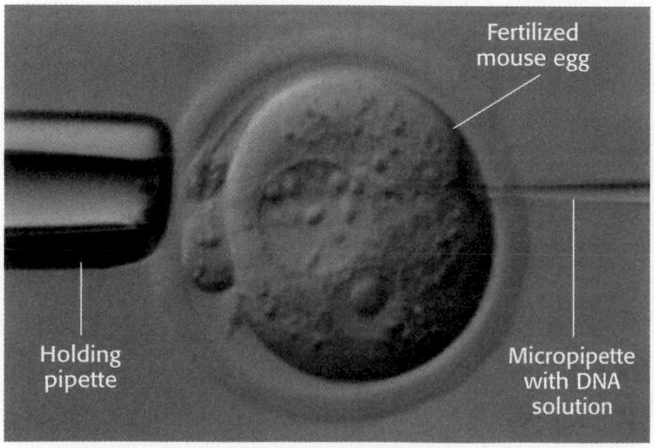

Figure 5.31 Microinjection of DNA. Cloned plasmid DNA is being microinjected into the male pronucleus of a fertilized mouse egg.

plasm of mammalian cells, where it shuts down host-cell protein synthesis. *Baculovirus* infects insect cells, which can be conveniently cultured. Insect larvae infected with this virus can serve as efficient protein factories. Vectors based on these large-genome viruses have been engineered to express DNA inserts efficiently.

Transgenic Animals Harbor and Express Genes That Were Introduced into Their Germ Lines

Genetically engineered giant mice illustrate the expression of foreign genes in mammalian cells (Figure 5.32). Giant mice were produced by introducing the gene for rat growth hormone into a fertilized mouse egg. *Growth hormone (somatotropin)*, a 21-kd protein, is normally synthesized by the pituitary gland. A deficiency of this hormone produces dwarfism, and an excess leads to gigantism. The gene for rat growth hormone was placed on a plasmid next to the mouse metallothionein *promoter* (Figure 5.33). This promoter site is normally located on a chromosome, where it controls the transcription of *metallothionein,* a cysteine-rich protein that has high affinity for heavy metals. Metallothionein binds to heavy metals, many of which are toxic for metabolic processes (p. 494), and sequesters them. The synthesis of this protective protein by the liver is induced by heavy-metal ions such as cadmium. Hence, if mice contain the new gene, its expression can be initiated by the addition of cadmium to the drinking water.

Figure 5.32 Transgenic mice. Injection of the gene for growth hormone into a fertilized mouse egg gave rise to a giant mouse (left), about twice the weight of his sibling (right). [Courtesy of Dr. Ralph Brinster.]

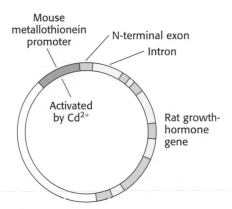

Figure 5.33 Rat growth hormone–metallothionein gene construct. The gene for rat growth hormone (shown in yellow) was inserted into a plasmid next to the metallothionein promoter, which is activated by the addition of heavy metals, such as cadmium ion.

Several hundred copies of the plasmid containing the promoter and growth-hormone gene were microinjected into the male pronucleus of a fertilized mouse egg, which was then inserted into the uterus of a foster-mother mouse. A number of mice that developed from such microinjected eggs contained the gene for rat growth hormone, as shown by Southern blots of their DNA. These *transgenic mice,* containing multiple copies (~30 per cell) of the rat growth-hormone gene, grew much more rapidly than did control mice. In the presence of cadmium, the level of growth hormone in these mice was 500 times as high as in normal mice, and their body weight at maturity was twice normal. The foreign DNA had been transcribed and its five introns correctly spliced out to form functional mRNA. These experiments strikingly demonstrate that a foreign gene under the control of a new promoter site can be integrated and efficiently expressed in mammalian cells.

Gene Disruption Provides Clues to Gene Function

A gene's function can also be probed by inactivating the gene and looking for resulting abnormalities. Powerful methods have been developed for accomplishing *gene disruption* (also called *gene knockout*) in organisms such as yeast

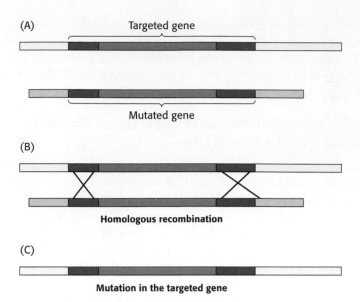

Figure 5.34 Gene disruption by homologous recombination. (A) A mutated version of the gene to be disrupted is constructed, maintaining some regions of homology with the normal gene (red). When the foreign mutated gene is introduced into an embryonic stem cell, (B) recombination takes place at regions of homology and (C) the normal (targeted) gene is replaced, or "knocked out," by the foreign gene. The cell is inserted into embryos, and mice lacking the gene (knockout mice) are produced.

and mice. These methods rely on the process of *homologous recombination.* Through this process, regions of strong sequence similarity exchange segments of DNA. Foreign DNA inserted into a cell can thus disrupt any gene that is at least in part homologous by exchanging segments (Figure 5.34). Specific genes can be targeted if their nucleotide sequences are known.

For example, the gene-knockout approach has been applied to the genes encoding gene-regulatory proteins (also called *transcription factors*) that control the differentiation of muscle cells. When both copies of the gene for the regulatory protein *myogenin* are disrupted, an animal dies at birth because it lacks functional skeletal muscle. Microscopic inspection reveals that the tissues from which muscle normally forms contain precursor cells that have failed to differentiate fully (Figure 5.35). Heterozygous mice containing one normal myogenin gene and one disrupted gene appear normal, suggesting that the level of gene expression is not essential for its function. Analogous studies have probed the function of many other genes to generate animal models for known human genetic diseases.

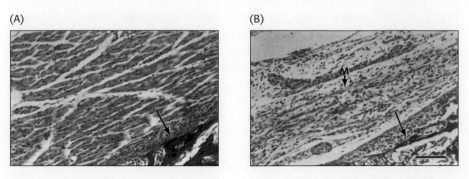

Figure 5.35 Consequences of gene disruption. Sections of muscle from normal (A) and gene-disrupted (B) mice, as viewed under the light microscope. Muscles do not develop properly in mice having both myogenin genes disrupted. [From P. Hasty, A. Bradley, J. H. Morris, D. G. Edmondson, J. M. Venuti, E. N. Olson, and W. H. Klein, *Nature* 364(1993):501–506.]

RNA Interference Provides an Additional Tool for Disrupting Gene Expression

An extremely powerful tool for disrupting gene expression was serendipitously discovered in the course of studies that required the introduction of RNA into a cell. The introduction of a specific double-stranded RNA molecule into a cell was found to suppress the transcription of genes that contained sequences present in the double-stranded RNA molecule. Thus, the introduction of a specific RNA molecule can interfere with the expression of a specific gene.

The mechanism of *RNA interference* has been largely established (Figure 5.36). When a double-stranded RNA molecule is introduced into an appropriate cell, the RNA is cleaved by an enzyme referred to as *Dicer* into fragments approximately 21 nucleotides in length. Each fragment consists of 19 bp of double-stranded RNA and 2 bases of unpaired RNA on each 5′ end. After separation, the two single strands of the RNA molecule, termed small interfering RNAs (siRNAs), are each incorporated into a different enzyme referred to as *RNA-induced silencing complex* (RISC). The single-stranded RNA segment incorporated into the enzyme acts as a guide that allows RISC to cleave mRNA molecules that include segments that are exact complements of the sequence. Thus, levels of such mRNA molecules are dramatically reduced.

The machinery necessary for RNA interference is found in many cells. In some organisms such as *C. elegans*, RNA interference is quite efficient. Indeed, RNA interference can be induced simply by feeding *C. elegans* strains of *E. coli* that have been engineered to produce appropriate double-stranded RNA molecules. Although not as efficient in mammalian cells, RNA interference has emerged as a powerful research tool for reducing the expression of specific genes. Moreover, initial clinical trials of therapies based on RNA interference are underway.

Tumor-Inducing Plasmids Can Be Used to Introduce New Genes into Plant Cells

The common soil bacterium *Agrobacterium tumefaciens* infects plants and introduces foreign genes into plants cells (Figure 5.37). A lump of tumor tissue called a *crown gall* grows at the site of infection. Crown galls synthesize opines, a group of amino acid derivatives that are metabolized by the infecting bacteria. In essence, the metabolism of the plant cell is diverted to satisfy the highly distinctive appetite of the intruder. *Tumor-inducing plasmids* (Ti plasmids) that are carried by *A. tumefaciens* carry instructions for the switch to the tumor state and the synthesis of opines. A small part of the

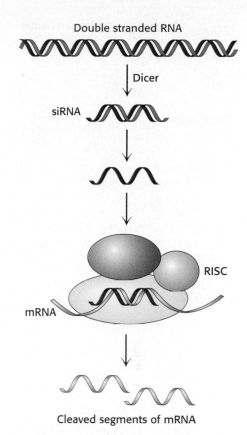

Figure 5.36 RNA interference mechanism. A double-stranded RNA molecule is cleaved into 21-bp fragments by the enzyme Dicer to produce siRNAs. These siRNAs are incorporated into the RNA-induced silencing complex (RISC), where the single-stranded RNAs guide the cleavage of mRNAs that contain complementary sequences.

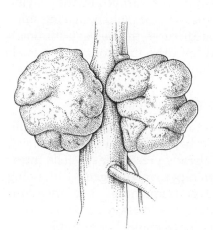

Figure 5.37 Tumors in plants. Crown gall, a plant tumor, is caused by a bacterium (*Agrobacterium tumefaciens*) that carries a tumor-inducing plasmid (Ti plasmid).

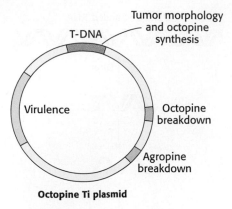

Octopine Ti plasmid

Figure 5.38 Ti plasmids. Agrobacteria containing Ti plasmids can deliver foreign genes into some plant cells. [After M. Chilton. A vector for introducing new genes into plants. Copyright © 1983 by Scientific American, Inc. All rights reserved.]

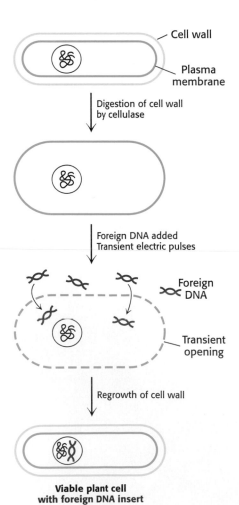

Figure 5.39 Electroporation. Foreign DNA can be introduced into plant cells by electroporation, the application of intense electric fields to make their plasma membranes transiently permeable.

Ti plasmid becomes integrated into the genome of infected plant cells; this 20-kb segment is called *T-DNA* (transferred DNA; Figure 5.38).

Ti plasmid derivatives can be used as vectors to deliver foreign genes into plant cells. First, a segment of foreign DNA is inserted into the T-DNA region of a small plasmid through the use of restriction enzymes and ligases. This synthetic plasmid is added to *A. tumefaciens* colonies harboring naturally occurring Ti plasmids. By recombination, Ti plasmids containing the foreign gene are formed. These Ti vectors hold great promise as tools for exploring the genomes of plant cells and modifying plants to improve their agricultural value and crop yield. However, they are not suitable for transforming all types of plants. Ti-plasmid transfer is effective with dicots (broad-leaved plants such as grapes) and a few kinds of monocots but not as effective with economically important cereal monocots.

Foreign DNA can be introduced into cereal monocots as well as dicots by applying intense electric fields, a technique called *electroporation* (Figure 5.39). First, the cellulose wall surrounding plant cells is removed by adding cellulase; this treatment produces *protoplasts,* plant cells with exposed plasma membranes. Electric pulses are then applied to a suspension of protoplasts and plasmid DNA. Because high electric fields make membranes transiently permeable to large molecules, plasmid DNA molecules enter the cells. The cell wall is then allowed to reform, and the plant cells are again viable. Maize cells and carrot cells have been stably transformed in this way with the use of plasmid DNA that includes genes for resistance to antibiotics. Moreover, the transformed cells efficiently express the plasmid DNA. Electroporation is also an effective means of delivering foreign DNA into animal and bacterial cells.

The most effective means of transforming plant cells is through the use of *"gene guns,"* or *bombardment-mediated transformation.* DNA is coated onto 1-mm-diameter tungsten pellets, and these microprojectiles are fired at the target cells with a velocity greater than 400 m s^{-1}. Despite its apparent crudeness, this technique is proving to be the most effective way of transforming plants, especially important crop species such as soybean, corn, wheat, and rice. The gene-gun technique affords an opportunity to develop genetically modified organisms (GMOs) with beneficial characteristics. Such characteristics could include the ability to grow in poor soils, resistance to natural climatic variation, resistance to pests, and nutritional fortification. These crops might be most useful in developing countries. The use of genetically modified organisms is highly controversial at this point because of fears of unexpected side effects.

The first GMO to come to market was a tomato characterized by delayed ripening, rendering it ideal for shipment. Pectin is a polysaccharide that gives tomatoes their firmness and is naturally destroyed by the enzyme *polygalacturonase.* As pectin is destroyed, the tomatoes soften, making shipment difficult. DNA was introduced that disrupts the polygalacturonase gene. Less of the enzyme was produced, and the tomatoes stayed fresh longer. However, the tomato's poor taste hindered its commercial success.

Human Gene Therapy Holds Great Promise for Medicine

The field of *gene therapy* attempts to express specific genes within the human body in such a way that beneficial results are obtained. The gene targeted for expression may be already present or specially introduced. Alternatively, gene therapy may attempt to modify genes containing sequence variations that have harmful consequences. A tremendous amount of research remains to be done before gene therapy becomes practical. Nonetheless, considerable progress has been made. For example, some people lack functional genes for *adenosine deaminase* and succumb to infections if exposed to a normal environment, a condition called *severe combined immunodeficiency* (SCID). Functional genes for this enzyme have been introduced by using gene-therapy vectors based on retroviruses. Although these vectors have produced functional enzyme and reduced the clinical symptoms, challenges remain. These challenges include increasing the longevity of the effects and eliminating unwanted side effects. Future research promises to transform gene therapy into an important tool for clinical medicine.

Summary

5.1 The Exploration of Genes Relies on Key Tools

The recombinant DNA revolution in biology is rooted in the repertoire of enzymes that act on nucleic acids. Restriction enzymes are a key group among them. These endonucleases recognize specific base sequences in double-helical DNA and cleave both strands of the duplex, forming specific fragments of DNA. These restriction fragments can be separated and displayed by gel electrophoresis. The pattern of these fragments on the gel is a fingerprint of a DNA molecule. A DNA fragment containing a particular sequence can be identified by hybridizing it with a labeled single-stranded DNA probe (Southern blotting).

Rapid sequencing techniques have been developed to further the analysis of DNA molecules. DNA can be sequenced by controlled interruption of replication. The fragments produced are separated by gel electrophoresis and visualized by autoradiography of a ^{32}P label at the 5′ end or by fluorescent tags.

DNA probes for hybridization reactions, as well as new genes, can be synthesized by the automated solid-phase method. The technique is to add deoxyribonucleoside 3′-phosphoramidites to one another to form a growing chain that is linked to an insoluble support. DNA chains a hundred nucleotides long can be readily synthesized. The polymerase chain reaction makes it possible to greatly amplify specific segments of DNA in vitro. The region amplified is determined by the placement of a pair of primers that are added to the target DNA along with a thermostable DNA polymerase and deoxyribonucleoside triphosphates. The exquisite sensitivity of PCR makes it a choice technique in detecting pathogens and cancer markers, in genotyping, and in reading DNA from fossils that are many thousands of years old.

5.2 Recombinant DNA Technology Has Revolutionized All Aspects of Biology

New genes can be constructed in the laboratory, introduced into host cells, and expressed. Novel DNA molecules are made by joining fragments that have complementary cohesive ends produced by the action of a restriction enzyme. DNA ligase seals breaks in DNA chains. Vectors for propagating the DNA include plasmids, λ phage, and bacterial and yeast artificial chromosomes. Specific genes can be cloned from a genomic library using a DNA or RNA probe. Foreign DNA can be expressed after insertion into prokaryotic and eukaryotic cells by the appropriate vector. Specific mutations can be generated in vitro to engineer novel proteins. A mutant protein with a single amino acid substitution can be produced by priming DNA replication with an oligonucleotide encoding the new amino acid. Plasmids can be engineered to permit the facile insertion of a DNA cassette containing any desired mutation. The techniques of protein and nucleic acid chemistry are highly synergistic. Investigators now move back and forth between gene and protein with great facility.

5.3 Complete Genomes Have Been Sequenced and Analyzed

The sequences of many important genomes are known in their entirety. More than 100 bacterial and archaeal genomes have been sequenced, including those from key model organisms and important pathogens. The sequence of the human genome has now been completed with nearly full coverage and high precision. Only from 20,000 to 25,000 protein-encoding genes appear to be present in the human genome, a substantially smaller number than earlier estimates. Comparative genomics has become a powerful tool for analyzing individual genomes and for exploring evolution. Genomewide gene-expression patterns can be examined through the use of DNA microarrays.

5.4 Eukaryotic Genes Can Be Manipulated with Considerable Precision

The production of giant mice by injecting the gene for rat growth hormone into fertilized mouse eggs vividly shows that mammalian cells can be genetically altered in an intended way. The functions of particular genes can be investigated by disruption. One method of disrupting the expression of a particular gene is through RNA interference, which depends on the introduction of specific double-stranded RNA molecules into eukaryotic cells. New DNA can be brought into plant cells by the soil bacterium *Agrobacterium tumefaciens,* which harbors Ti plasmids. DNA can also be introduced into plant cells by applying intense electric fields, which render them transiently permeable to very large molecules, or by bombarding them with DNA-coated microparticles. Gene therapy holds great promise for clinical medicine, but many challenges remain.

Key Terms

restriction enzyme (p. 135)

palindrome (p. 136)

DNA probe (p. 137)

Southern blotting (p. 137)

Northern blotting (p. 137)

controlled termination of replication (Sanger dideoxy method) (p. 138)

polymerase chain reaction (PCR) (p. 140)

vector (p. 142)

plasmid (p. 142)

sticky ends (p. 143)

DNA ligase (p. 143)

lambda (λ) phage (p. 144)

bacterial artificial chromosome (BAC) (p. 145)

yeast artificial chromosome (YAC) (p. 145)

genomic library (p. 146)

oligonucleotide-directed mutagenesis (p. 147)

cassette mutagenesis (p. 148)

pseudogene (p. 150)

mobile genetic element (p. 150)

short interspersed elements (SINES) (p. 150)

long interspersed elements (LINES) (p. 150)

DNA microarray (gene chip) (p. 152)

complementary DNA (cDNA) (p. 152)

reverse transcriptase (p. 152)

cDNA library (p. 153)

expression vector (p. 153)

transgenic mouse (p. 155)

gene disruption (gene knockout) (p. 155)

RNA interference (p. 157)

RNA-induced silencing complex (RISC) (p. 157)

tumor-inducing plasmid (Ti plasmid) (p. 157)

gene guns (bombardment-mediated transformation (p. 158)

Selected Readings

Where to Start

Berg, P. 1981. Dissections and reconstructions of genes and chromosomes. *Science* 213:296–303.

Gilbert, W. 1981. DNA sequencing and gene structure. *Science* 214:1305–1312.

Sanger, F. 1981. Determination of nucleotide sequences in DNA. *Science* 214:1205–1210.

Mullis, K. B. 1990. The unusual origin of the polymerase chain reaction. *Sci. Am.* 262(4):56–65.

Books on Recombinant DNA Technology

Watson, J. D., Gilman, M., Witkowski, J., and Zoller, M. 1992. *Recombinant DNA* (2d ed.). Scientific American Books.

Grierson, D. (Ed.). 1991. *Plant Genetic Engineering.* Chapman and Hall.

Mullis, K. B., Ferré, F., and Gibbs, R. A. (Eds.). 1994. *The Polymerase Chain Reaction.* Birkhaüser.

Russel, D., Sambrook, J., and Russel, D. 2000. *Molecular Cloning: A Laboratory Manual* (3d ed.). Cold Spring Harbor Laboratory Press.

Ausubel, F. M., Brent, R., Kingston, R. E., and Moore, D. D. (Eds.). 1999. *Short Protocols in Molecular Biology: A Compendium of Methods from Current Protocols in Molecular Biology.* Wiley.

Birren, B., Green, E. D., Klapholz, S., Myers, R. M., Roskams, J., Riethamn, H., and Hieter, P. (Eds.). 1999. *Genome Analysis* (vols. 1–4). Cold Spring Harbor Laboratory Press.

Methods in Enzymology. Academic Press. [Many volumes in this series deal with recombinant DNA technology.]

DNA Sequencing and Synthesis

Hunkapiller, T., Kaiser, R. J., Koop, B. F., and Hood, L. 1991. Large-scale and automated DNA sequence determination. *Science* 254:59–67.

Sanger, F., Nicklen, S., and Coulson, A. R. 1977. DNA sequencing with chain-terminating inhibitors. *Proc. Natl. Acad. Sci. U. S. A.* 74:5463–5467.

Maxam, A. M., and Gilbert, W. 1977. A new method for sequencing DNA. *Proc. Natl. Acad. Sci. U. S. A.* 74:560–564.

Smith, L. M., Sanders, J. Z., Kaiser, R. J., Hughes, P., Dodd, C., Connell, C. R., Heiner, C., Kent, S. B. H., and Hood, L. E. 1986. Fluorescence detection in automated DNA sequence analysis. *Nature* 321:674–679.

Pease, A. C., Solas, D., Sullivan, E. J., Cronin, M. T., Holmes, C. P., and Fodor, S. P. A. 1994. Light-generated oligonucleotide arrays for rapid DNA sequence analysis. *Proc. Natl. Acad. Sci. U. S. A.* 91:5022–5026.

Venter, J. C., Adams, M. D., Sutton, G. G., Kerlavage, A. R., Smith, H. O., and Hunkapiller, M. 1998. Shotgun sequencing of the human genome. *Science* 280:1540–1542.

Polymerase Chain Reaction

Arnheim, N., and Erlich, H. 1992. Polymerase chain reaction strategy. *Annu. Rev. Biochem.* 61:131–156.

Kirby, L. T. (Ed.). 1997. *DNA Fingerprinting: An Introduction.* Stockton Press.

Eisenstein, B. I. 1990. The polymerase chain reaction: A new method for using molecular genetics for medical diagnosis. *N. Engl. J. Med.* 322:178–183.

Foley, K. P., Leonard, M. W., and Engel, J. D. 1993. Quantitation of RNA using the polymerase chain reaction. *Trends Genet.* 9:380–386.

Pääbo, S. 1993. Ancient DNA. *Sci. Am.* 269(5):86–92.

Hagelberg, E., Gray, I. C., and Jeffreys, A. J. 1991. Identification of the skeletal remains of a murder victim by DNA analysis. *Nature* 352:427–429.

Lawlor, D. A., Dickel, C. D., Hauswirth, W. W., and Parham, P. 1991. Ancient HLA genes from 7500-year-old archaeological remains. *Nature* 349:785–788.

Krings, M., Geisert, H., Schmitz, R. W., Krainitzki, H., and Pääbo, S. 1999. DNA sequence of the mitochondrial hypervariable region II for the Neanderthal type specimen. *Proc. Natl. Acad. Sci. U. S. A.* 96:5581–5585.

Ovchinnikov, I. V., Götherström, A., Romanova, G. P., Kharitonov, V. M., Lidén, K., and Goodwin, W. 2000. Molecular analysis of Neanderthal DNA from the northern Caucasus. *Nature* 404:490–493.

Genome Sequencing

International Human Genome Sequencing Consortium. 2004. Finishing the euchromatic sequence of the human genome. *Nature* 431:931–945.

Lander, E. S., Linton, L. M., Birren, B., Nusbaum, C., Zody, M. C., Baldwin, J., Devon, K., Dewar, K., Doyle, M., FitzHugh, W., et al. 2001. Initial sequencing and analysis of the human genome. *Nature* 409:860–921.

Venter, J. C., Adams, M. D., Myers, E. W., Li, P. W., Mural, R. J., Sutton, G. G., Smith, H. O., Yandell, M., Evans, C. A., Holt, R. A., et al. 2001. The sequence of the human genome. *Science* 291:1304–1351.

Waterston, R. H., Lindblad-Toh, K., Birney, E., Rogers, J., Abril, J. F., Agarwal, P., Agarwala, R., Ainscough, R., Alexandersson, M., An, P., et al. 2002. Initial sequencing and comparative analysis of the mouse genome. *Nature* 420:520–562.

Koonin, E. V. 2003. Comparative genomics, minimal gene-sets and the last universal common ancestor. *Nat. Rev. Microbiol.* 1:127–236.

Gilligan, P., Brenner, S., and Venkatesh, B. 2002. Fugu and human sequence comparison identifies novel human genes and conserved non-coding sequences. *Gene* 294:35–44.

Enard, W., and Pääbo, S. 2004. Comparative primate genomics. *Annu. Rev. Genomics Hum. Genet.* 5:351–378.

DNA Arrays

Duggan, D. J., Bittner, J. M., Chen, Y., Meltzer, P., and Trent, J. M. 1999. Expression profiling using cDNA microarrays. *Nat. Genet.* 21:10–14.

Golub, T. R., Slonim, D. K., Tamayo, P., Huard, C., Gaasenbeek, M., Mesirov, J. P., Coller, H., Loh, M. L., Downing, J. R., Caligiuri, M. A., Bloomfield, C. D., and Lander, E. S. 1999. Molecular classification of cancer: Class discovery and class prediction by gene expression monitoring. *Science* 286:531–537.

Perou, C. M., Sørlie, T., Eisen, M. B., van de Rijn, M., Jeffery, S. S., Rees, C. A., Pollack, J. R., Ross, D. T., Johnsen, H., Akslen, L. A., Fluge, Ø., Pergamenschikov, A., Williams, C., Zhu, S. X., Lønning, P. E., Børresen-Dale, A.-L., Brown, P. O., and Botstein, D. 2000. Molecular portraits of human breast tumours. *Nature* 406:747–752.

Introduction of Genes into Animal Cells

Anderson, W. F. 1992. Human gene therapy. *Science* 256:808–813.

Friedmann, T. 1997. Overcoming the obstacles to gene therapy. *Sci. Am.* 277(6):96–101.

Blaese, R. M. 1997. Gene therapy for cancer. *Sci. Am.* 277 (6):111–115.

Brinster, R. L., and Palmiter, R. D. 1986. Introduction of genes into the germ lines of animals. *Harvey Lect.* 80:1–38.

Capecchi, M. R. 1989. Altering the genome by homologous recombination. *Science* 244:1288–1292.

Hasty, P., Bradley, A., Morris, J. H., Edmondson, D. G., Venuti, J. M., Olson, E. N., and Klein, W. H. 1993. Muscle deficiency and neonatal death in mice with a targeted mutation in the myogenin gene. *Nature* 364:501–506

Parkmann, R., Weinberg, K., Crooks, G., Nolta, J., Kapoor, N., and Kohn, D. 2000. Gene therapy for adenosine deaminase deficiency. *Annu. Rev. Med.* 51:33–47.

RNA Interference

Novina, C. D., and Sharp, P. A. 2004. The RNAi revolution. *Nature* 430:161–164.

Hannon, G. J., and Rossi, J. J. 2004. Unlocking the potential of the human genome with RNA interference. *Nature* 431:371–378.

Meister, G., and Tuschl, T. 2004. Mechanisms of gene silencing by double-stranded RNA. *Nature* 431:343–349.

Elbashir, S. M., Harborth, J., Lendeckel, W., Yalcin, A., Weber, K. and Tuschl, T. 2001. Duplexes of 21-nucleotide RNAs mediate RNA interference in cultured mammalian cells. *Nature* 411:494–498.

Fire, A., Xu, S., Montgomery, M. K., Kostas, S. A., Driver, S. E., and Mello, C. C. 1998. Potent and specific genetic interference by double-stranded RNA in *Caenorhabditis elegans*. *Nature* 391:806–811.

Genetic Engineering of Plants

Gasser, C. S., and Fraley, R. T. 1992. Transgenic crops. *Sci. Am.* 266(6):62–69.

Gasser, C. S., and Fraley, R. T. 1989. Genetically engineering plants for crop improvement. *Science* 244:1293–1299.

Shimamoto, K., Terada, R., Izawa, T., and Fujimoto, H. 1989. Fertile transgenic rice plants regenerated from transformed protoplasts. *Nature* 338:274–276.

Chilton, M.-D. 1983. A vector for introducing new genes into plants. *Sci. Am.* 248(6):50–59.

Hansen, G., and Wright, M. S. 1999. Recent advances in the transformation of plants. *Trends Plant Sci.* 4:226–231.

Hammond, J. 1999. Overview: The many uses of transgenic plants. *Curr. Top. Microbiol. Immunol.* 240:1–20.

Finer, J. J., Finer, K. R., and Ponappa, T. 1999. Particle bombardment mediated transformation. *Curr. Top. Microbiol. Immunol.* 240:60–80.

Problems

1. *Reading sequences.* An autoradiogram of a sequencing gel containing four lanes of DNA fragments is shown in the adjoining illustration. (a) What is the sequence of the DNA fragment? (b) Suppose that the Sanger dideoxy method shows that the template strand sequence is 5'-TGCAATGGC-3'. Sketch the gel pattern that would lead to this conclusion.

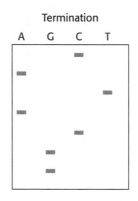

2. *The right template.* Ovalbumin is the major protein of egg white. The chicken ovalbumin gene contains eight exons separated by seven introns. Should one use ovalbumin cDNA or ovalbumin genomic DNA to form the protein in *E. coli*? Why?

3. *Cleavage frequency.* The restriction enzyme *Alu*I cleaves at the sequence 5'-AGCT-3', and *Not*I cleaves at 5'-GCGGCCGC-3'. What would be the average distance between cleavage sites for each enzyme on digestion of double-stranded DNA? Assume that the DNA contains equal proportions of A, G, C, and T.

4. *The right cuts.* Suppose that a human genomic library is prepared by exhaustive digestion of human DNA with the *Eco*RI restriction enzyme. Fragments averaging about 4 kb in length would be generated. Is this procedure suitable for cloning large genes? Why or why not?

5. *A revealing cleavage.* Sickle-cell anemia arises from a mutation in the gene for the β chain of human hemoglobin. The change from GAG to GTG in the mutant eliminates a cleavage site for the restriction enzyme *Mst*II, which recognizes the target sequence CCTGAGG. These findings form the basis of a diagnostic test for the sickle-cell gene. Propose a rapid procedure for distinguishing between the normal and the mutant gene. Would a positive result prove that the mutant contains GTG in place of GAG?

6. *Many melodies from one cassette.* Suppose that you have isolated an enzyme that digests paper pulp and have obtained its cDNA. The goal is to produce a mutant that is effective at high temperature. You have engineered a pair of unique restriction sites in the cDNA that flank a 30-bp coding region. Propose a rapid technique for generating many different mutations in this region.

7. *A blessing and a curse.* The power of PCR can also create problems. Suppose someone claims to have isolated dinosaur DNA by using PCR. What questions might you ask to determine if it is indeed dinosaur DNA?

8. *Questions of accuracy.* The stringency (p. 141) of PCR amplification can be controlled by altering the temperature at which the primers and the target DNA undergo hybridization. How would altering the temperature of hybridization affect the amplification? Suppose that you have a particular yeast gene *A* and that you wish to see if it has a counterpart in humans. How would controlling the stringency of the hybridization help you?

9. *Terra incognita.* PCR is typically used to amplify DNA that lies between two known sequences. Suppose that you want to explore DNA on both sides of a single known sequence. Devise a variation of the usual PCR protocol that would enable you to amplify entirely new genomic terrain.

10. *A puzzling ladder.* A gel pattern displaying PCR products shows four strong bands. The four pieces of DNA have lengths that are approximately in the ratio of 1:2:3:4. The largest band is cut out of the gel, and PCR is repeated with the same primers. Again, a ladder of four bands is evident in the gel. What does this result reveal about the structure of the encoded protein?

11. *Chromosome walking.* Propose a method for isolating a DNA fragment that is adjacent in the genome to a previously isolated DNA fragment. Assume that you have access to a complete library of DNA fragments in a BAC vector but that the sequence of the genome under study has not yet been determined.

12. *Man's best friend.* Why might the genomic analysis of dogs be particularly useful for investigating the genes responsible for body size and other physical characteristics?

Chapter Integration Problem

13. *Designing primers.* A successful PCR experiment often depends on designing the correct primers. In particular, the T_m for each primer should be approximately the same. What is the basis of this requirement?

Chapter Integration and Data Interpretation Problem

14. *Any direction but east.* A series of people are found to have difficulty eliminating certain types of drugs from their bloodstreams. The problem has been linked to a gene *X*, which

encodes an enzyme Y. Six people were tested with the use of various techniques of molecular biology. Person A is a normal control, person B is asymptomatic but some of his children have the metabolic problem, and persons C through F display the trait. Tissue samples from each person were obtained. Southern analysis was performed on the DNA after digestion with the restriction enzyme *Hind*III. Northern analysis of mRNA also was done. In both types of analysis, the gels were probed with labeled *X* cDNA. Finally, a Western blot with an enzyme-linked monoclonal antibody was used to test for the presence of protein Y. The results are shown here. Why is person B without symptoms? Suggest possible defects in the other people.

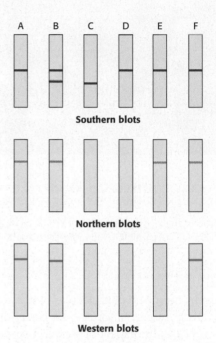

Data Interpretation Problem

15. *DNA diagnostics.* Representations of sequencing gels for variants of the α chain of human hemoglobin are shown here. What is the nature of the amino acid change in each of the variants? The first triplet encodes valine.

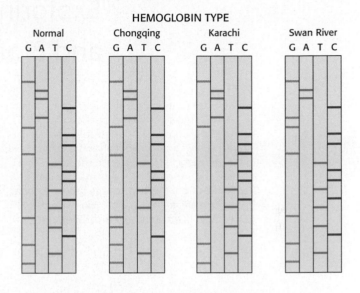

Animated Techniques

Visit www.whfreeman.com/Stryer to see animations of Dideoxy Sequencing of DNA, Polymerase Chain Reaction, Synthesizing an Oligonucleotide Array, Screening an Oligonucleotide Array for Patterns of Gene Expression, Plasmid Cloning, In Vitro Mutagenesis of Cloned Genes, Creating a Transgenic Mouse. [Courtesy of H. Lodish et al., *Molecular Cell Biology*, 5th ed. (W. H. Freeman and Company, 2004).]

Exploring Evolution and Bioinformatics

Evolutionary relationships are manifest in protein sequences. The close kinship between human beings and chimpanzees, hinted at by the mutual interest shown by Jane Goodall and a chimpanzee in the photograph, is revealed in the amino acid sequences of myoglobin. The human sequence (red) differs from the chimpanzee sequence (blue) in only one amino acid in a protein chain of 153 residues. [(Left) Kennan Ward/Corbis.]

GLSDGEWQLVLNVWGKVEADIPGHGQEVLIRLFKGHPETLEKFDKFKHLKSEDEMKASEDLKKHGATVLTALGGIL–
GLSDGEWQLVLNVWGKVEADIPGHGQEVLIRLFKGHPETLEKFDKFKHLKSEDEMKASEDLKKHGATVLTALGGIL–

KKKGHHEAEIKPLAQSHATKHKIPVKYLEFISECIIQVLHSKHPGDFGADAQGAMNKALELFRKDMASNYKELGFQG
KKKGHHEAEIKPLAQSHATKHKIPVKYLEFISECIIQVLQSKHPGDFGADAQGAMNKALELFRKDMASNYKELGFQG

Like members of a human family, members of molecular families often have features in common. Such family resemblance is most easily detected by comparing three-dimensional structure, the aspect of a molecule most closely linked to function. Consider as an example ribonuclease from cows, which was introduced in our consideration of protein folding (Section 2.6). Comparing structures reveals that the three-dimensional structure of this protein and that of a human ribonuclease are quite similar (Figure 6.1). Although this similarity is not unexpected, given the similarity in biological function, similarities revealed by comparisons are sometimes surprising. For example, angiogenin, a protein identified on the basis of its ability to stimulate the growth of new blood vessels, also turns out to be structurally similar to ribonuclease—so similar that it is clear that both angiogenin and ribonuclease are members of the same protein family (Figure 6.2). Angiogenin and ribonuclease must have had a common ancestor at some earlier stage of evolution.

Unfortunately, three-dimensional structures have been determined for only a small proportion of the total number of proteins. In contrast, gene sequences and the corresponding amino acid sequences are available for a great number of proteins, largely owing to the tremendous power of DNA cloning and sequencing techniques including applications to complete genome sequencing. Evolutionary relationships also are manifest in amino acid sequences. For example, 35% of the amino acids in corresponding positions are identical in the sequences of bovine ribonuclease and angiogenin. Is this level sufficiently high to ensure an evolutionary relationship? If not, what level is required? In this chapter, we shall examine the methods that are used to compare amino acid sequences and to deduce such evolutionary relationships.

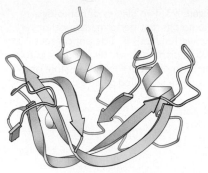

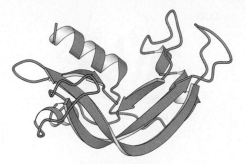

Bovine ribonuclease

Human ribonuclease

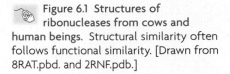

Figure 6.1 Structures of ribonucleases from cows and human beings. Structural similarity often follows functional similarity. [Drawn from 8RAT.pbd. and 2RNF.pdb.]

Sequence-comparison methods have become one of the most powerful tools in modern biochemistry. Sequence databases can be probed for matches to a newly elucidated sequence to identify related molecules. This information can often be a source of considerable insight into the function and mechanism of the newly sequenced molecule. When three-dimensional structures are available, they may be compared to confirm relationships suggested by sequence comparisons and to reveal others that are not readily detected at the level of sequence alone.

By examining the footprints present in modern protein sequences, the biochemist can become a molecular archeologist able to learn about events in the evolutionary past. Sequence comparisons can often reveal pathways of evolutionary descent and estimated dates of specific evolutionary landmarks. This information can be used to construct evolutionary trees that trace the evolution of a particular protein or nucleic acid in many cases from Archaea and Bacteria through Eukarya, including human beings. Molecular evolution can also be studied experimentally. In some cases, DNA from fossils can be amplified by PCR methods (p. 178) and sequenced, giving a direct view into the past. In addition, investigators can observe molecular evolution taking place in the laboratory, through experiments based on nucleic acid replication. The results of such studies are revealing more about how evolution proceeds.

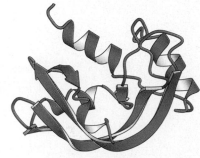

Angiogenin

Figure 6.2 Structure of angiogenin. The protein angiogenin, identified on the basis of its ability to stimulate blood-vessel growth, is highly similar in three-dimensional structure to ribonuclease. [Drawn from 2ANG.pdb.]

6.1 Homologs Are Descended from a Common Ancestor

The exploration of biochemical evolution consists largely of an attempt to determine how proteins, other molecules, and biochemical pathways have been transformed through time. The most fundamental relationship between two entities is *homology;* two molecules are said to be *homologous* if they have been derived from a common ancestor. Homologous molecules, or *homologs,* can be divided into two classes (Figure 6.3). *Paralogs* are homologs that are present within one species. Paralogs often differ in their detailed biochemical functions. *Orthologs* are homologs that are present within different species and have very similar or identical functions. Understanding the homology between molecules can reveal the evolutionary history of the molecules as well as information about their function; if a newly sequenced protein is homologous to an already characterized protein, we have a strong indication of the new protein's biochemical function.

How can we tell whether two human proteins are paralogs or whether a yeast protein is the ortholog of a human protein? As will be discussed in Section 6.2, *homology is often manifested by significant similarity in nucleotide or amino acid sequence and almost always manifested in three-dimensional structure.*

COW

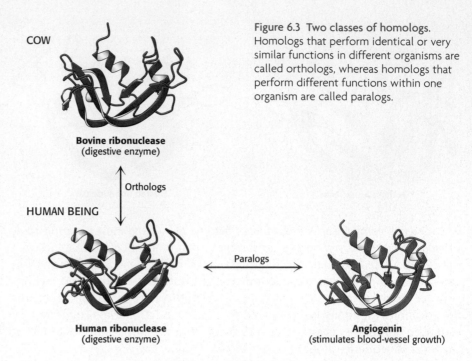

Bovine ribonuclease
(digestive enzyme)

Orthologs

HUMAN BEING

Human ribonuclease
(digestive enzyme)

Paralogs

Angiogenin
(stimulates blood-vessel growth)

Figure 6.3 Two classes of homologs.
Homologs that perform identical or very
similar functions in different organisms are
called orthologs, whereas homologs that
perform different functions within one
organism are called paralogs.

6.2   Statistical Analysis of Sequence Alignments Can Detect Homology

A significant sequence similarity between two molecules implies that they are likely to have the same evolutionary origin and, therefore, the same three-dimensional structure, function, and mechanism. Although both nucleic acid and protein sequences can be compared to detect homology, a comparison of protein sequences is much more effective for several reasons. Most notably, proteins are built from 20 different building blocks, whereas RNA and DNA are synthesized from only 4 building blocks.

To illustrate sequence-comparison methods, let us consider a class of proteins called the *globins*. Myoglobin is a protein that binds oxygen in muscle, whereas hemoglobin is the oxygen-carrying protein in blood (Chapter 7). Both proteins cradle a heme group, an iron-containing organic molecule that binds the oxygen. Each human hemoglobin molecule is composed of four heme-containing polypeptide chains, two identical α chains and two identical β chains. Here, we consider only the α chain. We wish to examine the similarity between the amino acid sequence of the human α chain and that of human myoglobin (Figure 6.4). To detect such similarity, methods have been developed for *sequence alignment*.

**Figure 6.4 Amino acid sequences of
human hemoglobin (α chain) and human
myoglobin.** α-Hemoglobin is composed
of 141 amino acids; myoglobin consists of
153 amino acids. (One-letter abbreviations
designating amino acids are used;
see Table 2.2.)

Human hemoglobin (α chain)

VLSPADKTNVKAAWGKVGAHAGEYGAEALERMFLSFPTTKTYFPHFDLSHG
SAQVKGHGKKVADALTNAVAHVDDMPNALSALSDLHAHKLRVDPVNFKLLS
HCLLVTLAAHLPAEFTPAVHASLDKFLASVSTVLTSKYR

Human myoglobin

GLSDGEWQLVLNVWGKVEADIPGHGQEVLIRLFKGHPETLEKFDKFKHLKS
EDEMKASEDLKKHGATVLTALGGILKKKGHHEAEIKPLAQSHATKHKIPVK
YLEFISECIIQVLQSKHPGDFGADAQGAMNKALELFRKDMASNYKELGFQG

(A)

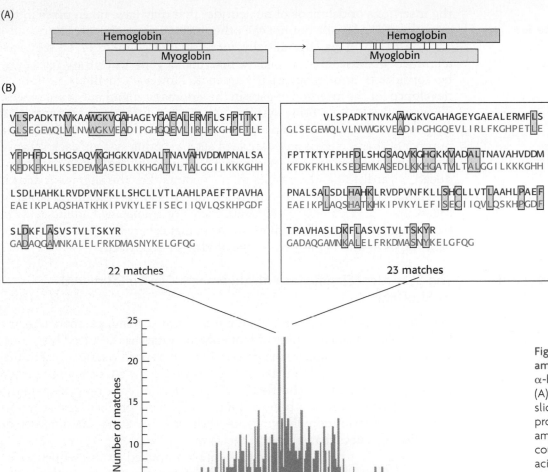

(B)

VLSPADKTNVKAAWGKVGAHAGEYGAEALERMFLSFPTTKT
GLSEGEWQLVLNWGKVEADIPGHGQEVLIRLFKGHPETLE

YFPHFDLSHGSAQVKGHGKKVADALTNAVAHVDDMPNALSA
KFDKFKHLKSEDEMKASEDLKKHGATVLTALGGILKKKGHH

LSDLHAHKLRVDPVNFKLLSHCLLVTLAAHLPAEFTPAVHA
EAEIKPLAQSHATKHKIPVKYLEFISECIIQVLQSKHPGDF

SLDKFLASVSTVLTSKYR
GADAQGAMNKALELFRKDMASNYKELGFQG

22 matches

VLSPADKTNVKAAWGKVGAHAGEYGAEALERMFLS
GLSEGEWQLVLNWGKVEADIPGHGQEVLIRLFKGHPETLE

FPTTKTYFPHFDLSHGSAQVKGHGKKVADALTNAVAHVDDM
KFDKFKHLKSEDEMKASEDLKKHGATVLTALGGILKKKGHH

PNALSALSDLHAHKLRVDPVNFKLLSHCLLVTLAAHLPAEF
EAEIKPLAQSHATKHKIPVKYLEFISECIIQVLQSKHPGDF

TPAVHASLDKFLASVSTVLTSKYR
GADAQGAMNKALELFRKDMASNYKELGFQG

23 matches

Figure 6.5 Comparing the amino acid sequences of α-hemoglobin and myoglobin. (A) A comparison is made by sliding the sequences of the two proteins past each other, one amino acid at a time, and counting the number of amino acid identities between the proteins. (B) The two alignments with the largest number of matches are shown above the graph, which plots the matches as a function of alignment.

How can we tell where to align the two sequences? The simplest approach is to compare all possible juxtapositions of one protein sequence with another, in each case recording the number of identical residues that are aligned with one another. This comparison can be accomplished by simply sliding one sequence past the other, one amino acid at a time, and counting the number of matched residues (Figure 6.5).

For α-hemoglobin and myoglobin, the best alignment reveals 23 sequence identities, spread throughout the central parts of the sequences. However, a nearby alignment showing 22 identities is nearly as good. In this alignment, the identities are concentrated toward the amino-terminal end of the sequences. The sequences can be aligned to capture most of the identities in *both* alignments by introducing a *gap* into one of the sequences (Figure 6.6). Such gaps must often be inserted to compensate for

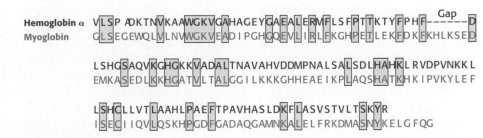

Hemoglobin α VLSPADKTNVKAAWGKVGAHAGEYGAEALERMFLSFPTTKTYFPHF------D
Myoglobin   GLSEGEWQLVLNWGKVEADIPGHGQEVLIRLFKGHPETLEKFDKFKHLKSED

LSHGSAQVKGHGKKVADALTNAVAHVDDMPNALSALSDLHAHKLRVDPVNKKL
EMKASEDLKKHGATVLTALGGILKKKGHHEAEIKPLAQSHATKHKIPVKYLEF

LSHCLLVTLAAHLPAEFTPAVHASLDKFLASVSTVLTSKYR
ISECIIQVLQSKHPGDFGADAQGAMNKALELFRKDMASNYKELGFQG

Figure 6.6 Alignment with gap insertion. The alignment of α-hemoglobin and myoglobin after a gap has been inserted into the hemoglobin α sequence.

the insertions or deletions of nucleotides that may have taken place in the gene for one molecule but not the other in the course of evolution.

The use of gaps substantially increases the complexity of sequence alignment because, in principle, the insertion of gaps of arbitrary sizes must be considered throughout each sequence. However, methods have been developed for the insertion of gaps in the automatic alignment of sequences. These methods use scoring systems to compare different alignments, and they include penalties for gaps to prevent the insertion of an unreasonable number of them. Here is an example of such a scoring system: each identity between aligned sequences results in +10 points, whereas each gap introduced, regardless of size, results in −25 points. For the alignment shown in Figure 6.6, there are 38 identities and 1 gap, producing a score of ($38 \times 10 - 1 \times 25 = 355$). Overall, there are 38 matched amino acids in an average length of 147 residues; so the sequences are 25.9% identical. The next step is to ask, Is this percentage of identity significant?

THISISTHEAUTHENTICSEQUENCE

Shuffling

SNUCSNSEATEEITUHEQIHHTTCEI

Figure 6.7 The generation of a shuffled sequence.

The Statistical Significance of Alignments Can Be Estimated by Shuffling

The similarities in sequence in Figure 6.5 appear striking, yet there remains the possibility that a grouping of sequence identities has occurred by chance alone. How can we estimate the probability that a specific series of identities is a chance occurrence? To make such an estimate, the amino acid sequence in one of the proteins is "shuffled"—that is, randomly rearranged—and the alignment procedure is repeated (Figure 6.7). This process is repeated to build up a distribution showing, for each possible score, the number of shuffled sequences that received that score.

When this procedure is applied to the sequences of myoglobin and α-hemoglobin the authentic alignment clearly stands out (Figure 6.8). Its score is far above the mean for the alignment scores based on shuffled sequences. The odds of the occurrence of such a deviation due to chance alone are approximately 1 in 10^{20}. Thus, we can comfortably conclude that the two sequences are genuinely similar; the simplest explanation for this similarity is that these sequences are homologous—that is, that the two molecules have descended by divergence from a common ancestor.

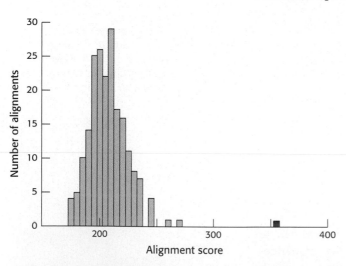

Figure 6.8 Statistical comparison of alignment scores. Alignment scores are calculated for many shuffled sequences, and the number of sequences generating a particular score is plotted against the score. The resulting plot is a distribution of alignment scores occurring by chance. The alignment score for α-hemoglobin and myoglobin (shown in red) is substantially greater than any of these scores, strongly suggesting that the sequence similarity is significant.

Distant Evolutionary Relationships Can Be Detected Through the Use of Substitution Matrices

The scoring scheme described above assigns points only to positions occupied by identical amino acids in the two sequences being compared. No credit is given for any pairing that is not an identity. However, not all substitutions are equivalent. Some are structurally *conservative substitutions,* replacing one amino acid with another that is similar in size and chemical properties. Such conservative amino acid substitutions may have minor effects on protein structure and can thus be tolerated without compromising function. In other substitutions, an amino acid replaces one that is dissimilar. Furthermore, some amino acid substitutions result from the replacement of only a single nucleotide in the gene sequence; whereas others require two or three replacements. Conservative and single-nucleotide substitutions are likely to be more common than are substitutions with more radical effects. How can we account for the type of substitution when comparing sequences? We can approach this problem by first examining the substitutions that have actually taken place in evolutionarily related proteins.

From an examination of appropriately aligned sequences, *substitution matrices* can be deduced. In these matrices, a large positive score corresponds to a substitution that occurs relatively frequently, whereas a large negative score corresponds to a substitution that occurs only rarely. The Blosum-62 substitution matrix illustrated in Figure 6.9 is an example. The highest scores in this substitution matrix indicate that amino acids such as

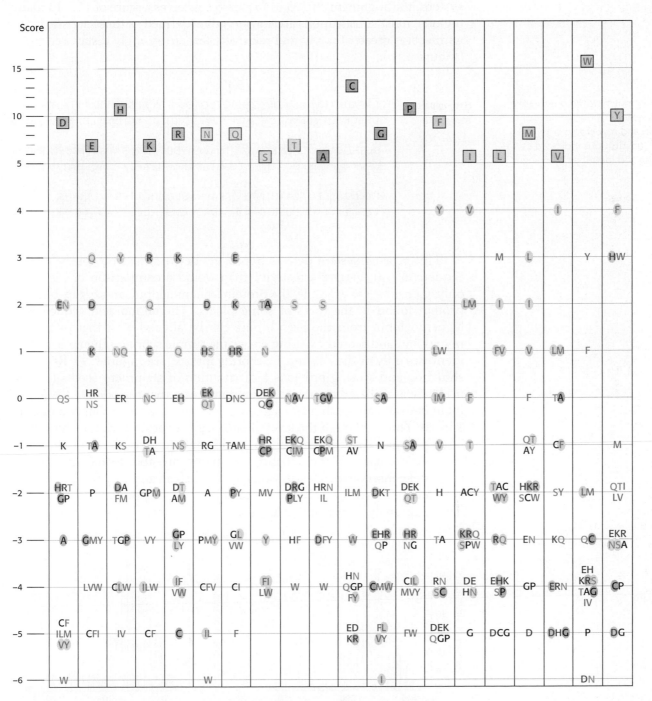

Figure 6.9 A graphic view of the Blosum-62 substitution matrix. This scoring scheme was derived by examining substitutions that occur within aligned sequence blocks in related proteins. Amino acids are classified into four groups (charged, red; polar, green; large and hydrophobic, blue; other, black). Substitutions that require the change of only a single nucleotide are shaded. To find the score for a substitution of, for instance, a Y for an H, you find the Y in the column having H (boxed) at the top and check the number at the left. In this case, the resulting score is 3.

cysteine (C) and tryptophan (W) tend to be conserved more than those such as serine (S) and alanine (A). Furthermore, structurally conservative substitutions such as lysine (K) for arginine (R) and isoleucine (I) for valine (V) have relatively high scores. When two sequences are compared, each substitution is assigned a score based on the matrix. In addition, a gap penalty is often assigned according to the size of the gap. For example, the introduction of a gap lowers the alignment score by 12 points and the extension of an existing gap costs 2 points per residue. With the use of this scoring system, the alignment shown in Figure 6.6 receives a score of 115. In many regions, most substitutions are conservative (defined as those substitutions with scores greater than 0) and relatively few are strongly disfavored types (Figure 6.10).

Figure 6.10 Alignment with conservative substitutions noted. The alignment of α-hemoglobin and myoglobin with conservative substitutions indicated by yellow shading and identities by orange.

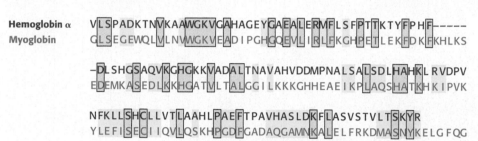

This scoring system detects homology between less obviously related sequences with greater sensitivity than would a comparison of identities only. Consider, for example, the protein leghemoglobin, an oxygen-binding protein found in the roots of some plants. The amino acid sequence of leghemoglobin from the herb lupine can be aligned with that of human myoglobin and scored by using either the simple scoring scheme based on identities only or the Blosum-62 scoring matrix (see Figure 6.9). Repeated shuffling and scoring provides a distribution of alignment scores (Figure 6.11). Scoring based on identities only indicates that the odds of the alignment between myoglobin and leghemoglobin occurring by chance alone are 1 in 20. Thus, although the level of similarity suggests a relationship, there is a 5% chance that the similarity is accidental on the basis of this analysis. In contrast, users of the substitution matrix are able to incorporate the effects of conservative substitutions. From such an analysis, the odds of the alignment occurring by chance are calculated to be approximately 1 in 300. Thus, an analysis performed by using the substitution matrix reaches a

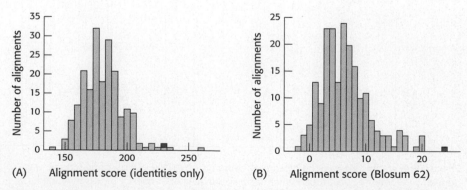

Figure 6.11 Alignment of identities only versus the Blosum 62 matrix. Repeated shuffling and scoring reveal the significance of sequence alignment for human myoglobin versus lupine leghemoglobin with the use of either (A) the simple, identity-based scoring system or (B) the Blosum-62 matrix. The scores for the alignment of the authentic sequences are shown in red. The Blosum matrix provides greater statistical power.

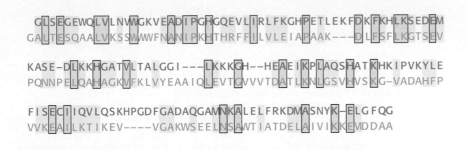

| Myoglobin | GLSEGEWQLVLNVWGKVEADIPGHGQEVLIRLFKGHPETLEKFDKFKHLKSEDEM |
| Leghemoglobin | GALTESQAALVKSSWWFNANIPKHTHRFFILVLEIAPAAK---DLFSFLKGTSEV |

KASE-DLKKHGATVLTALGGI---LKKKGH--HEAEIKPLAQSHATKHKIPVKYLE
PQNNPELQAHAGKVFKLVYEAAIQLEVTGVVVTDATLKNLGSVHVSKG-VADAHFP

FISECIIQVLQSKHPGDFGADAQGAMNKALELFRKDMASNYK-ELGFQG
VVKEAILKTIKEV----VGAKWSEELNSAWTIATDELAIVIKKEMDDAA

Figure 6.12 Alignment of human myoglobin and lupine leghemoglobin. The use of the Blosum-62 substitution matrix yields the alignment shown between human myoglobin and lupine leghemoglobin, illustrating identities (orange boxes) and conservative substitutions (yellow). These sequences are 23% identical.

much firmer conclusion about the evolutionary relationship between these proteins (Figure 6.12).

Experience with sequence analysis has led to the development of simpler rules of thumb. For sequences longer than 100 amino acids, sequence identities greater than 25% are almost certainly not the result of chance alone; such sequences are probably homologous. In contrast, if two sequences are less than 15% identical, pairwise comparison alone is unlikely to indicate statistically significant similarity. For sequences that are between 15 and 25% identical, further analysis is necessary to determine the statistical significance of the alignment. It must be emphasized that *the lack of a statistically significant degree of sequence similarity does not rule out homology*. The sequences of many proteins that have descended from common ancestors have diverged to such an extent that the relationship between the proteins can no longer be detected from their sequences alone. As we will see, such homologous proteins can often be detected by examining three-dimensional structures.

Databases Can Be Searched to Identify Homologous Sequences

When the sequence of a protein is first determined, comparing it with all previously characterized sequences can be a source of tremendous insight into its evolutionary relatives and, hence, its structure and function. Indeed, an extensive sequence comparison is almost always the first analysis performed on a newly elucidated sequence. The sequence-alignment methods just described are used to compare an individual sequence with all members of a database of known sequences.

Database searches for homologous sequences are most often accomplished by using resources available on the Internet at the National Center for Biotechnology Information (www.ncbi.nih.gov). The procedure used is referred to as a *BLAST* (Basic Local Alignment Search Tool) *search*. An amino acid sequence is typed or pasted into the Web browser, and a search is performed, most often against a nonredundant database of all known sequences. At the end of 2004, this database included more than 3 million sequences. A BLAST search yields a list of sequence alignments, each accompanied by an estimate giving the likelihood that the alignment occurred by chance (Figure 6.13).

In 1995, investigators reported the first complete sequence of the genome of a free-living organism, the bacterium *Haemophilus influenzae*. With the sequences available, they performed a BLAST search with each deduced protein sequence. Of 1743 identified open reading frames (p. 126), 1007 (58%) could be linked to some protein of known function that had been previously characterized in another organism. An additional 347 open reading frames could be linked to sequences in the database for which no function had yet been assigned ("hypothetical proteins"). The remaining 389 sequences did not match any sequence present in the database at that time. Thus, investigators were able to identify likely functions for more than half the proteins within this organism solely by sequence comparisons.

```
BLASTP 2.2.10 [Oct-19-2004]
```

Identifier of query sequence

```
Query= gi|12517444|gb|AAG58041.1|AE005521_9 ribosephosphate
isomerase, constitutive [Escherichia coli O157:H7]
        (219 letters)

Database: All non-redundant GenBank CDS
translations+PDB+SwissProt+PIR+PRF excluding environmental samples
        2,205,431 sequences; 747,548,157 total letters
```

Distribution of 268 Blast Hits on the Query Sequence

	Score (bits)	E Value
Sequences producing significant alignments:		

Identifier of homologous sequence bond in search

```
gi|42943|emb|CAA47309.1|  unnamed protein product [E. coli]    427    e-118
gi|12517444|gb|AAG58041.1| ribosephosphate isomerase, const... 427    e-118
```

Name [species] of homologous protein

```
gi|33126317|gb|AAK95569.1| ribose 5-phosphate isomerase [H. sapiens] 117  2e-25

gi|29897136|gb|AAP10413.1| Phosphoglycerate mutase [B. cereus]    37    0.35
```

```
gi|33126317|gb|AAK95569.1|   ribose 5-phosphate isomerase [Homo sapiens]
            Length = 237
```

Amino acid Sequence being queried

```
 Score = 117 bits (294), Expect = 2e-25
 Identities = 82/224 (36%), Positives = 118/224 (52%), Gaps = 15/224 (6%)

Query: 4   DELKKAVGWAALQ-YVQPGTIVGVGTGSTAAHFIDALGTMKGQIE---GAVSSSDASTEK 59
           +E KK  G AA++ +V+  ++G+G+GST H +  +      Q        + +S + +
Sbjct: 5   EEAKKLAGRAAVENHVRNNQVLGIGSGSTIVHAVQRIAERVKQENLNLVCIPTSFQARQL 64
```

Sequence of homologous protein from *Homo sapiens*

```
Query: 60  LKSLGIHVFDLNEVDSLGIYVDGADEINGHMQMIKGGGAALTREKIIASVAEKFICIADA 119
           +   G+  +DL+     + +DGADE++   + +IKGGG  LT+EKI+A  A +FI IAD
Sbjct: 65  ILQYGLTLSDLDRHPEIDLAIDGADEVDADLNLIKGGGGCLTQEKIVAGYASRFIVIADF 124
```

Plus sign = "positive," a frequent substitution

```
Query: 120 SKQVDILG---KFPLPVEVIPMARSAVARQL-VKLGGRPEYRQG------VVTDNGNVIL 169
           K    LG    +P+EVIPMA   V+R +  K GG  E R         VVTDNGN IL
Sbjct: 125 RKDSKNLGDQWHKGIPIEVIPMAYVPVSRAVSQKFGGVVELRMAVNKAGPVVTDNGNFIL 184
```

Letter = identity i.e the two sequences

```
Query: 170 DVHGMEILDPIAMENAINAIPGVVTVGLFANRGADVALIGTPDG 213
           D  +      + AI  IPGVV  GLF N  A+     G  DG
Sbjct: 185 DWKFDRVHKWSEVNTAIKMIPGVVDTGLFINM-AERVYFGMQDG 227
```

Figure 6.13 BLAST search results. Part of the results from a BLAST search of the nonredundant (nr) protein sequence database with the use of the sequence of ribose-5-phosphate isomerase (also called phosphopentose isomerase, pp. 572 and 581) from *E. coli* as a query. Among the 268 sequences found is the orthologous sequence from human beings, and the alignment between these sequences is shown (highlighted in yellow). The number of sequences with this level of similarity expected to be in the database by chance is 2×10^{-25} as shown by the E value (highlighted in red). Because this value is much less than 1, the observed sequence alignment is highly significant.

6.3 Examination of Three-Dimensional Structure Enhances Our Understanding of Evolutionary Relationships

Sequence comparison is a powerful tool for extending our knowledge of protein function and kinship. However, biomolecules generally function as intricate three-dimensional structures rather than as linear polymers. Mutations occur at the level of sequence, but the effects of the mutations are at the level of function, and function is directly related to tertiary structure. Consequently, to gain a deeper understanding of evolutionary relationships between proteins, we must examine three-dimensional structures, especially in conjunction with sequence information. The techniques of structural determination are presented in Chapter 3.

Tertiary Structure Is More Conserved Than Primary Structure

Because three-dimensional structure is much more closely associated with function than is sequence, tertiary structure is more evolutionarily conserved than is primary structure. This conservation is apparent in the tertiary structures of the globins (Figure 6.14), which are extremely similar even though the similarity between human myoglobin and lupine leghemoglobin is just barely detectable at the sequence level and that between human α-hemoglobin and lupine leghemoglobin is not statistically significant (15.6% identity). This structural similarity firmly establishes that the framework that binds the heme group and facilitates the reversible binding of oxygen has been conserved over a long evolutionary period.

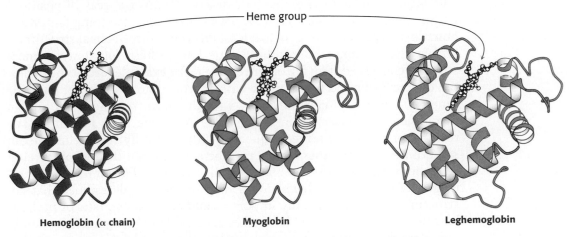

Hemoglobin (α chain)   **Myoglobin**   **Leghemoglobin**

Figure 6.14 Conservation of three-dimensional structure. The tertiary structures of human hemoglobin (α chain), human myoglobin, and lupine leghemoglobin are conserved. Each heme group contains an iron atom to which oxygen binds. [Drawn from 1HBB.pdb, 1MBD.pdb, and 1GDJ.pdb.]

Anyone aware of the similar biochemical functions of hemoglobin, myoglobin, and leghemoglobin could expect the structural similarities. In a growing number of other cases, however, a comparison of three-dimensional structures has revealed striking similarities between proteins that were *not* expected to be related. A case in point is the protein actin, a major component of the cytoskeleton (Section 34.2), and heat shock protein 70 (Hsp-70), which assists protein folding inside cells. These two proteins were found to be noticeably similar in structure despite only 15.6% sequence identity (Figure 6.15). On the basis of their three-dimensional structures,

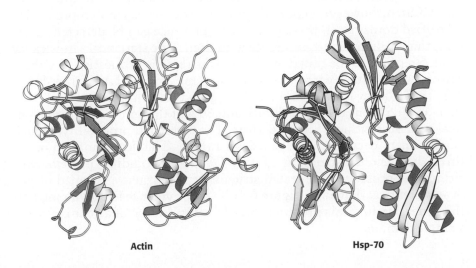

Actin   **Hsp-70**

Figure 6.15 Structures of actin and the large fragment of heat shock protein 70 (Hsp-70). A comparison of the identically colored elements of secondary structure reveals the overall similarity in structure despite the difference in biochemical activities. [Drawn from 1ATN.pdb and 1ATR.pdb.]

actin and Hsp-70 are paralogs. The level of structural similarity strongly suggests that, despite their different biological roles in modern organisms, these proteins descended from a common ancestor. As the three-dimensional structures of more proteins are determined, such unexpected kinships are being discovered with increasing frequency. The search for such kinships relies ever more frequently on computer-based searches that are able to compare the three-dimensional structure of any protein with all other known structures.

Knowledge of Three-Dimensional Structures Can Aid in the Evaluation of Sequence Alignments

The sequence-comparison methods described thus far treat all positions within a sequence equally. However, we know from examining families of homologous proteins for which at least one three-dimensional structure is known that regions and residues critical to protein function are more strongly conserved than are other residues. For example, each type of globin contains a bound heme group with an iron atom at its center. A histidine residue that interacts directly with this iron (residue 64 in human myoglobin) is conserved in all globins. After we have identified key residues or highly conserved sequences within a family of proteins, we can sometimes identify other family members even when the overall level of sequence similarity is below statistical significance. Thus it may be useful to generate a *sequence template*—a map of conserved residues that are structurally and functionally important and are characteristic of particular families of proteins, which makes it possible to recognize new family members that might be undetectable by other means. A variety of other methods for sequence classification that take advantage of known three-dimensional structures also are being developed. Still other methods are able to identify conserved residues within a family of homologous proteins, even without a known three-dimensional structure. These methods often use substitution matrices that differ at each position within a family of aligned sequences. Such methods can often detect quite distant evolutionary relationships.

Repeated Motifs Can Be Detected by Aligning Sequences with Themselves

More than 10% of all proteins contain sets of two or more domains that are similar to one another. Sequence search methods can often detect internally repeated sequences that have been characterized in other proteins. Often, however, repeated units do not correspond to previously identified domains. In these cases, their presence can be detected by attempting to align a given sequence with itself. The statistical significance of such repeats can be tested by aligning the regions in question as if these regions were sequences from separate proteins. For the TATA-box-binding protein, a key protein in controlling gene transcription (Section 29.2), such an alignment is highly significant: 30% of the amino acids are identical over 90 residues (Figure 6.16A). The estimated probability of such an alignment occurring by chance is 1 in 10^{13}. The determination of the three-dimensional structure of the TATA-box-binding protein confirmed the presence of repeated structures; the protein is formed of two nearly identical domains (Figure 6.16B). The evidence is convincing that the gene encoding this protein evolved by duplication of a gene encoding a single domain.

(A)

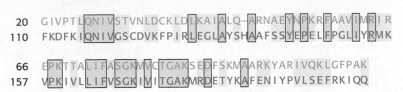

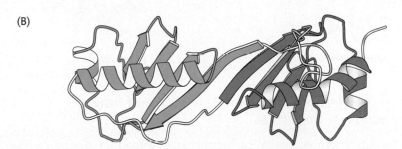

(B)

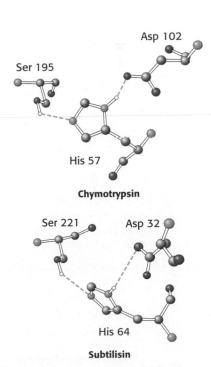

Figure 6.16 Sequence alignment of internal repeats. (A) An alignment of the sequences of the two repeats of the TATA-box-binding protein. The amino-terminal repeat is shown in green and the carboxyl-terminal repeat in blue. (B) Structure of the TATA-box-binding protein. The amino-terminal domain is shown in green and the carboxyl-terminal domain in blue. [Drawn from 1VOK.pdb.]

Convergent Evolution Illustrates Common Solutions to Biochemical Challenges

Thus far, we have been exploring proteins derived from common ancestors—that is, through *divergent evolution*. Other cases have been found of proteins that are structurally similar in important ways but are not descended from a common ancestor. How might two unrelated proteins come to resemble each other structurally? Two proteins evolving independently may have converged on a similar structure to perform a similar biochemical activity. Perhaps that structure was an especially effective solution to a biochemical problem that organisms face. The process by which very different evolutionary pathways lead to the same solution is called *convergent evolution*.

One example of convergent evolution is found among the serine proteases. These enzymes, to be considered in more detail in Chapter 9, cleave peptide bonds by hydrolysis. Figure 6.17 shows for two such enzymes the structure of the active sites—that is, the sites on the proteins at which the hydrolysis reaction takes place. These active-site structures are remarkably similar. In each case, a serine residue, a histidine residue, and an aspartic acid residue are positioned in space in nearly identical arrangements. As we will see, this is the case because chymotrypsin and subtilisin use the same mechanistic solution to the problem of peptide hydrolysis. At first glance, this similarity might suggest that these proteins are homologous. However, striking differences in the overall structures of these proteins make an evolutionary relationship extremely unlikely (Figure 6.18). Whereas chymotrypsin consists almost entirely of β sheets, subtilisin contains extensive α-helical structure. Moreover, the key serine, histidine, and aspartic acid residues do not occupy similar positions or even appear in the same order within the two sequences. It is extremely unlikely that two proteins evolving from a common ancestor could have retained similar active-site structures while other aspects of the structure changed so dramatically.

Figure 6.17 Convergent evolution of protease active sites. The relative positions of the three key residues shown are nearly identical in the active sites of the serine proteases chymotrypsin and subtilisin.

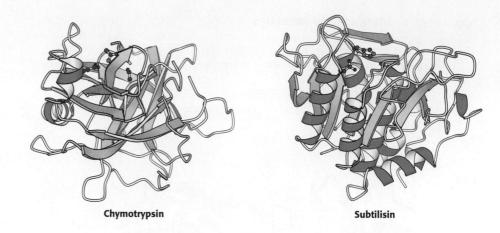

Chymotrypsin **Subtilisin**

Figure 6.18 Structures of mammalian chymotrypsin and bacterial subtilisin. The overall structures are quite dissimilar, in stark contrast with the active sites, shown at the top of each structure. The β strands are shown in yellow and α helices in blue. [Drawn from 1GCT.pdb. and 1SUP.pdb.]

Comparison of RNA Sequences Can Be a Source of Insight into RNA Secondary Structures

Homologous RNA sequences can be compared in a manner similar to that already described. Such comparisons can be a source of important insights into evolutionary relationships; in addition, they provide clues to the three-dimensional structure of the RNA itself. As noted in Chapter 4, single-stranded nucleic acid molecules fold back on themselves to form elaborate structures held together by Watson–Crick base-pairing and other interactions. In a family of sequences that form similar base-paired structures, base sequences may vary, but base-pairing ability is conserved. Consider, for example, a region from a large RNA molecule present in the ribosomes of all organisms (Figures 6.19). In the region shown, the *E. coli* sequence has a guanine (G) residue in position 9 and a cytosine (C) residue in position 22, whereas the human sequence has uracil (U) in position 9 and adenine (A) in position 22. Examination of the six sequences shown in Figure 6.19 (and many others) reveals that the bases in positions 9 and 22 retain the ability to form a Watson–Crick base pair even though the identities of the bases in these positions vary. Base-pairing ability is also conserved in neighboring positions; we can deduce that two segments with such compensating mutations are likely to form a double helix. Where sequences are known for several homologous RNA molecules, this type of sequence analysis can often suggest complete secondary structures as well as some additional interac-

Figure 6.19 Comparison of RNA sequences. (A) A comparison of sequences in a part of ribosomal RNA taken from a variety of species. (B) The implied secondary structure. Green bars indicate positions at which Watson–Crick base-pairing is completely conserved in the sequences shown, whereas dots indicate positions at which Watson–Crick base-pairing is conserved in most cases.

(A)

		9	22
BACTERIA	*Escherichia coli*	C A C A C G G C G G G U G C U A A C G U C C G U C G U G A A
	Pseudomonas aeruginosa	A C C A C G G C G G G U G C U A A C G U C C G U C G U G A A
ARCHAEA	*Halobacterium halobium*	C C G G U G U G C G G G G – U A A G C C U G U G C A C C G U
	Methanococcus vannielli	G A G G G C A U A C G G G – U A A G C U G U A U G U C C G A
EUKARYA	*Homo sapiens*	G G G C C A C U U U U G G – U A A G C A G A A C U G G C G C
	Saccharomyces cerevisiae	G G G C C A U U U U U G G – U A A G C A G A A C U G G C G A

(B)

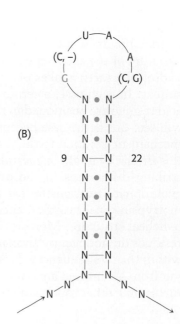

tions. For this ribosomal RNA, the subsequent determination of its three-dimensional structure (Section 30.3) confirmed the predicted secondary structure.

6.4 Evolutionary Trees Can Be Constructed on the Basis of Sequence Information

The observation that homology is often manifested as sequence similarity suggests that the evolutionary pathway relating the members of a family of proteins may be deduced by examination of sequence similarity. This approach is based on the notion that sequences that are more similar to one another have had less evolutionary time to diverge than have sequences that are less similar. This method can be illustrated by using the three globin sequences in Figures 6.10 and 6.12, as well as the sequence for the human hemoglobin β chain. These sequences can be aligned with the additional constraint that gaps, if present, should be at the same positions in all of the proteins. These aligned sequences can be used to construct an *evolutionary tree* in which the length of the branch connecting each pair of proteins is proportional to the number of amino acid differences between the sequences (Figure 6.20).

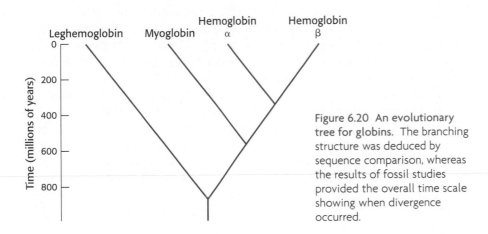

Figure 6.20 **An evolutionary tree for globins.** The branching structure was deduced by sequence comparison, whereas the results of fossil studies provided the overall time scale showing when divergence occurred.

Such comparisons reveal only the relative divergence times—for example, that myoglobin diverged from hemoglobin twice as long ago as the α chain diverged from the β chain. How can we estimate the approximate dates of gene duplications and other evolutionary events? Evolutionary trees can be calibrated by comparing the deduced branch points with divergence times determined from the fossil record. For example, the duplication leading to the two chains of hemoglobin appears to have occurred 350 million years ago. This estimate is supported by the observation that jawless fish such as the lamprey, which diverged from bony fish approximately 400 million years ago, contain hemoglobin built from a single type of subunit (Figure 6.21).

These methods can be applied to both relatively modern and very ancient molecules, such as the ribosomal RNAs that are found in all organisms. Indeed, such an RNA sequence analysis led to the realization that Archaea are a distinct group of organisms that diverged from Bacteria very early in evolutionary history.

Figure 6.21 **The lamprey.** A jawless fish whose ancestors diverged from bony fish approximately 400 million years ago, the lamprey contains hemoglobin molecules that contain only a single type of polypeptide chain. [Brent P. Kent.]

6.5 Modern Techniques Make the Experimental Exploration of Evolution Possible

Two techniques of biochemistry have made it possible to examine the course of evolution more directly and not simply by inference. The polymerase chain reaction (p. 140) allows the direct examination of ancient DNA sequences, releasing us, at least in some cases, from the constraints of being able to examine existing genomes from living organisms only. Molecular evolution may be investigated through the use of *combinatorial chemistry,* the process of producing large populations of molecules en masse and selecting for a biochemical property. This exciting process provides a glimpse into the types of molecules that may have existed very early in evolution.

Ancient DNA Can Sometimes Be Amplified and Sequenced

The tremendous chemical stability of DNA (p. 10) makes the molecule well suited to its role as the storage site of genetic information. So stable is the molecule that samples of DNA have survived for many thousands of years under appropriate conditions. With the development of PCR methods, such ancient DNA can sometimes be amplified and sequenced. This approach has been applied to mitochondrial DNA from a Neanderthal fossil estimated at between 30,000 and 100,000 years of age found near Düsseldorf, Germany, in 1856. Investigators managed to identify a total of 379 bases of sequence. Comparison with a number of the corresponding sequences from *Homo sapiens* revealed between 22 and 36 substitutions, considerably fewer than the average of 55 differences between human beings and chimpanzees over the common bases in this region. Further analysis suggested that the common ancestor of modern human beings and Neanderthals lived approximately 600,000 years ago. An evolutionary tree constructed by using these data and others revealed that the Neanderthal was not an intermediate between chimpanzees and human beings but, instead, was an evolutionary "dead end" that became extinct (Figure 6.22).

A few earlier studies claimed to determine the sequences of far more ancient DNA such as that found in insects trapped in amber, but these studies appear to have been flawed. The source of these sequences turned out to be contaminating modern DNA. Successful sequencing of ancient DNA requires sufficient DNA for reliable amplification and the rigorous exclusion of all sources of contamination.

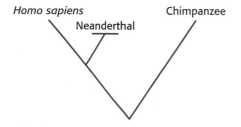

Figure 6.22 Placing Neanderthal on an evolutionary tree. Comparison of DNA sequences revealed that Neanderthal is not on the line of direct descent leading to *Homo sapiens* but, instead, branched off earlier and then became extinct.

Molecular Evolution Can Be Examined Experimentally

Evolution requires three processes: (1) the generation of a diverse population, (2) the selection of members based on some criterion of fitness, and (3) reproduction to enrich the population in these more-fit members. Nucleic acid molecules are capable of undergoing all three processes in vitro under appropriate conditions. The results of such studies enable us to glimpse how evolutionary processes might have generated catalytic activities and specific binding abilities—important biochemical functions in all living systems.

A diverse population of nucleic acid molecules can be synthesized in the laboratory by the process of combinatorial chemistry, which rapidly produces large populations of a particular type of molecule such as a nucleic acid. A population of molecules of a given size can be generated randomly so that many or all possible sequences are present in the mixture. When an initial population has been generated, it is subjected to a selection process that isolates specific molecules with desired binding or reactivity properties.

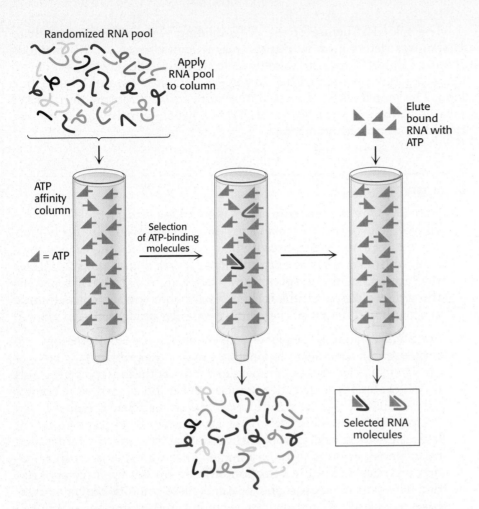

Randomized RNA pool

Apply RNA pool to column

ATP affinity column

◢ = ATP

Selection of ATP-binding molecules

Elute bound RNA with ATP

Selected RNA molecules

Figure 6.23 Evolution in the laboratory. A collection of RNA molecules of random sequences is synthesized by combinatorial chemistry. This collection is selected for the ability to bind ATP by passing the RNA through an ATP affinity column (Section 3.1). The ATP-binding RNA molecules are released from the column by washing with excess ATP and then replicated. The process of selection and replication is then repeated several times. The final RNA products with significant ATP-binding ability are isolated and characterized.

Finally, molecules that have survived the selection process are replicated through the use of PCR; primers are directed toward specific sequences included at the ends of each member of the population. Errors that occur naturally in the course of the replication process introduce additional variation into the population in each "generation."

As an example of this approach, consider an experiment that set a goal of creating an RNA molecule capable of binding adenosine triphosphate and related nucleotides. Such ATP-binding molecules are of interest because they might have been present very early in evolution, before the emergence of proteins, when RNA molecules may have played all major roles. An initial population of RNA molecules 169 nucleotides long was created; 120 of the positions differed randomly, with equimolar mixtures of adenine, cytosine, guanine, and uracil. The initial synthetic pool that was used contained approximately 10^{14} RNA molecules. Note that this number is a very small fraction of the total possible pool of random 120-base sequences. From this pool, those molecules that bound to ATP, which had been immobilized on a column, were selected (Figure 6.23).

The collection of molecules that were bound well by the ATP affinity column were allowed to replicate by reverse transcription into DNA, amplification by PCR, and transcription back into RNA. The somewhat error-prone replication processes introduced additional mutations into the population in each cycle. The new population was subjected to additional rounds of selection for ATP-binding activity. After eight generations, members of the selected population were characterized by sequencing. Seventeen different sequences were obtained, 16 of which could form the structure shown in Figure 6.24. Each of these molecules bound ATP with high affinity, as indicated by dissociation constants less than 50 μM.

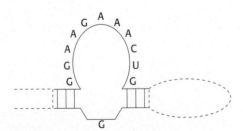

Figure 6.24 A conserved secondary structure. The secondary structure shown is common to RNA molecules selected for ATP binding.

179

(A)

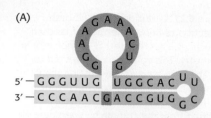

(B)

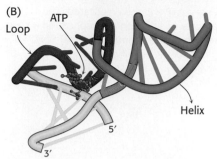

(C)

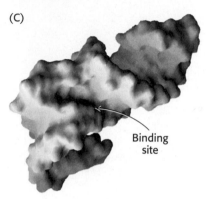

Figure 6.25 An evolved ATP-binding RNA molecule. (A) The Watson–Crick base-pairing pattern, (B) the folding pattern, and (C) a surface representation of an RNA molecule selected to bind adenosine nucleotides. The bound ATP is shown in part B, and the binding site is revealed as a deep pocket in part C.

The folded structure of the ATP-binding region from one of these RNAs was determined by nuclear magnetic resonance (p. 98) methods (Figure 6.25). As expected, this 40-nucleotide molecule is composed of two Watson–Crick base-paired helical regions separated by an 11-nucleotide loop. This loop folds back on itself in an intricate way to form a deep pocket into which the adenine ring can fit. Thus, a structure had evolved that was capable of a specific interaction.

Summary

6.1 Homologs Are Descended from a Common Ancestor

Exploring evolution biochemically often means searching for homology between molecules, because homologous molecules, or homologs, evolved from a common ancestor. Paralogs are homologous molecules that are found in one species and have acquired different functions through evolutionary time. Orthologs are homologous molecules that are found in different species and have similar or identical functions.

6.2 Statistical Analysis of Sequence Alignments Can Detect Homology

Protein and nucleic acid sequences are two of the primary languages of biochemistry. Sequence-alignment methods are the most powerful tools of the evolutionary detective. Sequences can be aligned to maximize their similarity, and the significance of these alignments can be judged by statistical tests. The detection of a statistically significant alignment between two sequences strongly suggests that two sequences are related by divergent evolution from a common ancestor. The use of substitution matrices makes the detection of more-distant evolutionary relationships possible. Any sequence can be used to probe sequence databases to identify related sequences present in the same organism or in other organisms.

6.3 Examination of Three-Dimensional Structure Enhances Our Understanding of Evolutionary Relationships

The evolutionary kinship between proteins may be even more strikingly evident in the conserved three-dimensional structures. The analysis of three-dimensional structure in combination with the analysis of especially conserved sequences has made it possible to determine evolutionary relationships that cannot be detected by other means. Sequence-comparison methods can also be used to detect imperfectly repeated sequences within a protein, indicative of linked similar domains.

6.4 Evolutionary Trees Can Be Constructed on the Basis of Sequence Information

Evolutionary trees can be constructed with the assumption that the number of sequence differences reflects the time since the two sequences diverged. Construction of an evolutionary tree based on sequence comparisons revealed approximate times for the gene-duplication events separating myoglobin and hemoglobin as well as the α and β subunits of hemoglobin. Evolutionary trees based on sequences can be compared with those based on fossil records.

6.5 Modern Techniques Make the Experimental Exploration of Evolution Possible

The exploration of evolution can also be a laboratory science. In favorable cases, PCR amplification of well-preserved samples allows the determination of nucleotide sequences from extinct organisms. Sequences so determined can help authenticate parts of an evolutionary tree constructed by other means. Molecular evolutionary experiments performed in the test tube can examine how molecules such as ligand-binding RNA molecules might have been generated.

Key Terms

homolog (p. 165)

paralog (p. 165)

ortholog (p. 165)

sequence alignment (p. 166)

conservative substitution (p. 168)

substitution matrix (p. 169)

BLAST search (p. 171)

sequence template (p. 174)

divergent evolution (p. 175)

convergent evolution (p. 175)

evolutionary tree (p. 177)

combinatorial chemistry (p. 178)

Selected Readings

Book

Claverie, J.-M., and Notredame, C. 2003. *Bioinformatics for Dummies.* Wiley.

Pevsner, J. 2003. *Bioinformatics and Functional Genomics.* Wiley-Liss.

Doolittle, R. F. 1987. *Of UFS and ORFS.* University Science Books.

Sequence Alignment

Schaffer, A. A., Aravind, L., Madden, T. L., Shavirin, S., Spouge, J. L., Wolf, Y. I., Koonin, E. V., and Altschul, S. F. 2001. Improving the accuracy of PSI-BLAST protein database searches with composition-based statistics and other refinements. *Nucleic Acids Res.* 29:2994–3005.

Henikoff, S., and Henikoff, J. G. 1992. Amino acid substitution matrices from protein blocks. *Proc. Natl. Acad. Sci. U. S. A.* 89:10915–10919.

Johnson, M. S., and Overington, J. P. 1993. A structural basis for sequence comparisons: An evaluation of scoring methodologies. *J. Mol. Biol.* 233:716–738.

Aravind, L., and Koonin, E. V. 1999. Gleaning non-trivial structural, functional and evolutionary information about proteins by iterative database searches. *J. Mol. Biol.* 287:1023–1040.

Altschul, S. F., Madden. T. L., Schaffer, A. A., Zhang, J., Zhang, Z., Miller, W., and Lipman, D. J. 1997. Gapped BLAST and PSI-BLAST: A new generation of protein database search programs. *Nucleic Acids Res.* 25:3389–3402.

Structure Comparison

Orengo, C. A., Bray, J. E., Buchan, D. W., Harrison, A., Lee, D., Pearl, F. M., Sillitoe, I., Todd, A. E., and Thornton, J. M. 2002. The CATH protein family database: A resource for structural and functional annotation of genomes. *Proteomics* 2:11–21.

Bashford, D., Chothia, C., and Lesk, A. M. 1987. Determinants of a protein fold: Unique features of the globin amino acid sequences. *J. Mol. Biol.* 196:199–216.

Harutyunyan, E. H., Safonova, T. N., Kuranova, I. P., Popov, A. N., Teplyakov, A. V., Obmolova, G. V., Rusakov, A. A., Vainshtein, B. K., Dodson, G. G., Wilson, J. C., et al. 1995. The structure of deoxy- and oxy-leghaemoglobin from lupin. *J. Mol. Biol.* 251:104–115.

Flaherty, K. M., McKay, D. B., Kabsch, W., and Holmes, K. C. 1991. Similarity of the three-dimensional structures of actin and the ATPase fragment of a 70-kDa heat shock cognate protein. *Proc. Natl. Acad. Sci. U. S. A.* 88:5041–5045.

Murzin, A. G., Brenner, S. E., Hubbard, T., and Chothia, C. 1995. SCOP: A structural classification of proteins database for the investigation of sequences and structures. *J. Mol. Biol.* 247:536–540.

Hadley, C., and Jones, D. T. 1999. A systematic comparison of protein structure classification: SCOP, CATH and FSSP. *Struct. Fold. Des.* 7:1099–1112.

Domain Detection

Marchler-Bauer, A., Anderson, J. B., DeWeese-Scott, C., Fedorova, N. D., Geer, L. Y., He, S., Hurwitz, D. I., Jackson, J. D., Jacobs, A. R., Lanczycki, C. J., Liebert, C. A., Liu, C., Madej, T., Marchler, G. H., Mazumder, R., Nikolskaya, A. N., Panchenko, A. R., Rao, B. S., Shoemaker, B. A., Simonyan, V., Song, J. S., Thiessen, P. A., Vasudevan, S., Wang, Y., Yamashita, R. A., Yin, J. J., and Bryant, S. H. 2003. CDD: A curated Entrez database of conserved domain alignments. *Nucleic Acids Res.* 31:383–387.

Ploegman, J. H., Drent, G., Kalk, K. H., and Hol, W. G. 1978. Structure of bovine liver rhodanese I: Structure determination at 2.5 Å resolution and a comparison of the conformation and sequence of its two domains. *J. Mol. Biol.* 123:557–594.

Nikolov, D. B., Hu, S. H., Lin, J., Gasch, A., Hoffmann, A., Horikoshi, M., Chua, N. H., Roeder, R. G., and Burley, S. K. 1992. Crystal structure of TFIID TATA-box binding protein. *Nature* 360:40–46.

Doolittle, R. F. 1995. The multiplicity of domains in proteins. *Annu. Rev. Biochem.* 64:287–314.

Heger, A., and Holm, L. 2000. Rapid automatic detection and alignment of repeats in protein sequences. *Proteins* 41:224–237.

Evolutionary Trees

Wolf, Y. I., Rogozin, I. B., Grishin, N. V., and Koonin, E. V. 2002. Genome trees and the tree of life. *Trends Genet.* 18:472–479.

Doolittle, R. F. 1992. Stein and Moore Award address. Reconstructing history with amino acid sequences. *Protein Sci.* 1:191–200.

Zuckerkandl, E., and Pauling, L. 1965. Molecules as documents of evolutionary history. *J. Theor. Biol.* 8:357–366.

Ancient DNA

Paabo, S., Poinar, H., Serre, D., Jaenicke-Despres, V., Hebler, J., Rohland, N., Kuch, M., Krause, J., Vigilant, L., and Hofreiter, M. 2004. Genetic analyses from ancient DNA. *Annu. Rev. Genet.* 38:645–679.

Krings, M., Stone, A., Schmitz, R. W., Krainitzki, H., Stoneking, M., and Pääbo, S. 1997. Neandertal DNA sequences and the origin of modern humans. *Cell* 90:19–30.

Krings, M., Geisert, H., Schmitz, R. W., Krainitzki, H., and Pääbo, S. 1999. DNA sequence of the mitochondrial hypervariable region II from the Neanderthal type specimen. *Proc. Natl. Acad. Sci. U. S. A.* 96:5581–5585.

Evolution in the Laboratory

Gold, L., Polisky, B., Uhlenbeck, O., and Yarus, M. 1995. Diversity of oligonucleotide functions. *Annu. Rev. Biochem.* 64:763–797.

Wilson, D. S., and Szostak, J. W. 1999. In vitro selection of functional nucleic acids. *Annu. Rev. Biochem.* 68:611–647.

Hermann, T., and Patel, D. J. 2000. Adaptive recognition by nucleic acid aptamers. *Science* 287:820–825.

Web Sites

The Protein Data Bank (PDB) site is the repository for three-dimensional macromolecular structures. It currently contains more than 30,000 structures. (http://www.rcsb.org/pdb/)

National Center for Biotechnology Information (NCBI) contains molecular biological databases and software for analysis. (http://www.ncbi.nlm.nih.gov/)

Problems

1. *What's the score?* Using the identity-based scoring system (Section 6.2), calculate the score for the following alignment. Do you think the score is statistically significant?

```
(1) WYLGKITRMDAEVLLKKPTVRDGHFLVTQCESSPGEF-
(2) WYFGKITRRESERLLLNPENPRGTFLVRESETTKGAY-

    SISVRFGDSVQ-----HFKVLRDQNGKYYLWAVK-FN-
    CLSVSDFDNAKGLNVKHYKIRKLDSGGFYITSRTQFS-

    SLNELVAYHRTASVSRTHTILLSDMNV
    SSLQQLVAYYSKHADGLCHRLTNV
```

2. *Sequence and structure.* A comparison of the aligned amino acid sequences of two proteins each consisting of 150 amino acids reveals them to be only 8% identical. However, their three-dimensional structures are very similar. Are these two proteins related evolutionarily? Explain.

3. *It depends on how you count.* Consider the following two sequence alignments:

```
(a) A-SNLFDIRLIG        (b) ASNLFDIRLI-G
    GSNDFYEVKIMD            GSNDFYEVKIMD
```

Which alignment has a higher score if the identity-based scoring system (Section 6.2) is used? Which alignment has a higher score if the Blosum-62 substitution matrix (Figure 6.9) is used?

4. *Discovering a new base pair.* Examine the ribosomal RNA sequences in Figure 6.19. In sequences that do not contain Watson–Crick base pairs, what base tends to be paired with G? Propose a structure for your new base pair.

5. *Overwhelmed by numbers.* Suppose that you wish to synthesize a pool of RNA molecules that contain all four bases at each of 40 positions. How much RNA must you have in grams if the pool is to have at least a single molecule of each sequence? The average molecular weight of a nucleotide is 330 g mol^{-1}.

6. *Form follows function.* The three-dimensional structure of biomolecules is more conserved evolutionarily than is sequence. Why?

7. *Shuffling.* Using the identity-based scoring system (Section 6.2), calculate the alignment score for the alignment of the following two short sequences:

```
(1) ASNFLDKAGK
(2) ATDYLEKAGK
```

Generate a shuffled version of sequence 2 by randomly reordering these 10 amino acids. Align your shuffled sequence with sequence 1 without allowing gaps, and calculate the alignment score between sequence 1 and your shuffled sequence.

8. *Interpreting the score.* Suppose that the sequences of two proteins each consisting of 200 amino acids are aligned and that the percentage of identical residues has been calculated. How would you interpret each of the following results in regard to the possible divergence of the two proteins from a common ancestor? (a) 80%, (b) 50%, (c) 20%, (d) 10%.

9. *A set of three.* The sequences of three proteins (A, B, and C) are compared with one another, yielding the following levels of identity:

	A	B	C
A	100%	65%	15%
B	65%	100%	55%
C	15%	55%	100%

Assume that the sequence matches are distributed uniformly along each aligned sequence pair. Would you expect protein A and protein C to have similar three-dimensional structures? Explain.

10. *RNA alignment.* Sequences of an RNA fragment from five species have been determined and aligned. Propose a likely secondary structure for these fragments.

```
(1) UUGGAGAUUCGGUAGAAUCUCCC
(2) GCCGGGAAUCGACAGAUUCCCCG
(3) CCCAAGUCCCGGCAGGGACUUAC
(4) CUCACCUGCCGAUAGGCAGGUCA
(5) AAUACCACCCGGUAGGGUGGUUC
```

11. *BLAST away.* Using the National Center for Biotechnology Information Web site (www.ncbi.nlm.nih.gov), find the sequence of the enzyme triose phosphate isomerase from *E. coli*. Use this sequence as the query for a protein–protein BLAST search. In the output, find the alignment with the sequence of triose phosphate isomerase from human beings (*Homo sapiens*). How many identities are observed in the alignment?

Chapter 7

Hemoglobin: Portrait of a Protein in Action

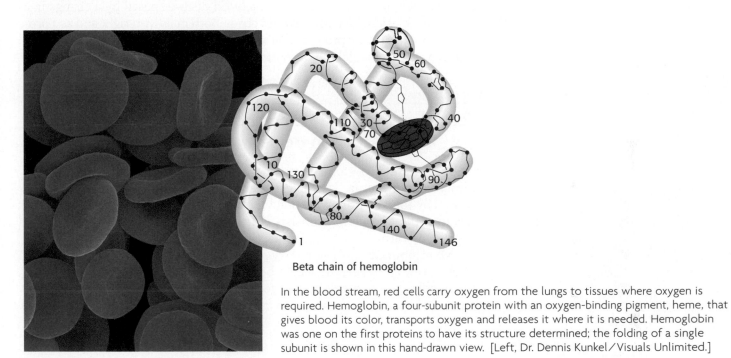

Beta chain of hemoglobin

In the blood stream, red cells carry oxygen from the lungs to tissues where oxygen is required. Hemoglobin, a four-subunit protein with an oxygen-binding pigment, heme, that gives blood its color, transports oxygen and releases it where it is needed. Hemoglobin was one on the first proteins to have its structure determined; the folding of a single subunit is shown in this hand-drawn view. [Left, Dr. Dennis Kunkel / Visuals Unlimited.]

The transition from anaerobic to aerobic life was a major step in evolution because it uncovered a rich reservoir of energy. Fifteen times as much energy is extracted from glucose in the presence of oxygen than in its absence. For single-celled and other small organisms, oxygen can be absorbed into actively metabolizing cells directly from the air or surrounding water. Vertebrates evolved two principal mechanisms for supplying their cells with an adequate supply of oxygen. The first is a circulatory system that actively delivers oxygen to cells throughout the body. The second is the use of the oxygen-transport and oxygen-storage proteins, hemoglobin and myoglobin. Hemoglobin, which is contained in red blood cells, is a fascinating protein, efficiently carrying oxygen from the lungs to the tissues while also contributing to the transport of carbon dioxide and hydrogen ions back to the lungs. Myoglobin, located in muscle, provides a reserve supply of oxygen available in time of need.

A comparison of myoglobin and hemoglobin illuminates some key aspects of protein structure and function. These two evolutionarily related proteins employ nearly identical structures for oxygen binding. However, hemoglobin is a remarkably efficient oxygen carrier, able to use as much as 90% of its potential oxygen-carrying capacity effectively. Under similar conditions, myoglobin would be able to use only 7% of its potential capacity. What accounts for this dramatic difference? Myoglobin exists as a single polypeptide, whereas hemoglobin comprises four polypeptide chains. The four chains in hemoglobin bind oxygen *cooperatively*, meaning that the

binding of oxygen to a site in one chain increases the likelihood that the remaining chains will bind oxygen. Furthermore, the oxygen-binding properties of hemoglobin are modulated by the binding of hydrogen ions and
carbon dioxide in a manner that enhances oxygen-carrying capacity. Both
cooperativity and the response to modulators are made possible by variations in the quaternary structure of hemoglobin when different combinations of molecules are bound.

Hemoglobin and myoglobin have played important roles in the history of
biochemistry. They were the first proteins for which three-dimensional structures were determined by x-ray crystallography. Furthermore, the possibility
that variations in protein sequence could lead to disease was first proposed and
demonstrated for sickle-cell anemia, a blood disease caused by a change in a
single amino acid in one hemoglobin chain. Both in its own right and as a prototype for many other proteins that we will encounter throughout our study of
biochemistry, hemoglobin is a source of wealth of knowledge and insight.

7.1 Myoglobin and Hemoglobin Bind Oxygen at Iron Atoms in Heme

Sperm whale myoglobin was the first protein for which the three-
dimensional structure was determined. X-ray crystallographic studies pioneered by John Kendrew revealed the structure of this protein in the 1950s
(Figure 7.1). Myoglobin consists largely of α helices that are linked to one
another by turns to form a globular structure.

Myoglobin can exist in an oxygen-free form called *deoxymyoglobin* or in
a form with an oxygen molecule bound called *oxymyoglobin*. The ability of
myoglobin, and hemoglobin as well, to bind oxygen depends on the presence of a bound prosthetic group called *heme*.

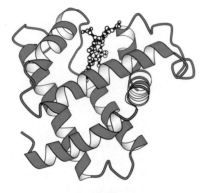

Myoglobin

Figure 7.1 Structure of myoglobin.
Notice that myoglobin consists of
a single polypeptide chain, formed of
α helices connected by turns, with one
oxygen-binding site. [Drawn from
1MBD.pdb.]

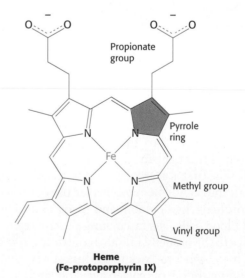

**Heme
(Fe-protoporphyrin IX)**

The heme group gives muscle and blood their distinctive red color. It consists of an organic component and a central iron atom. The organic component, called *protoporphyrin*, is made up of four pyrrole rings linked by
methine bridges to form a tetrapyrrole ring. Four methyl groups, two vinyl
groups, and two propionate side chains are attached.

The iron atom lies in the center of the protoporphyrin, bonded to the
four pyrrole nitrogen atoms. Under normal conditions, the iron is in the
ferrous (Fe^{2+}) oxidation state. The iron ion can form two additional bonds,

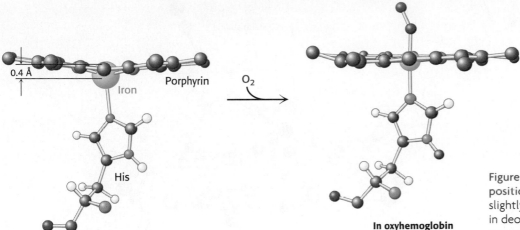

In deoxyhemoglobin

In oxyhemoglobin

Figure 7.2 **Oxygen binding changes the position of the iron ion.** The iron ion lies slightly outside the plane of the porphyrin in deoxyhemoglobin heme (left), but moves into the plane of the heme on oxygenation (right).

one on each side of the heme plane. These binding sites are called the fifth and sixth coordination sites. In myoglobin, the fifth coordination site is occupied by the imidazole ring of a histidine residue from the protein. This histidine is referred to as the *proximal histidine*. In deoxymyoglobin, the sixth coordination site remains unoccupied; this position is available for binding oxygen. The iron ion lies approximately 0.4 Å outside the porphyrin plane because an iron ion, in this form, is slightly too large to fit into the well-defined hole within the porphyrin ring (Figure 7.2, left).

The binding of the oxygen molecule at the sixth coordination site of the iron ion substantially rearranges the electrons within the iron so that the ion becomes effectively smaller, allowing it to move into the plane of the porphyrin (Figure 7.2, right). This change in electronic structure is paralleled by changes in the magnetic properties that are the basis for *functional magnetic resonance imaging* (fMRI), one of the most powerful methods for examining brain function (Section 32.1). Remarkably, the structural changes that take place on oxygen binding were predicted by Linus Pauling, on the basis of magnetic measurements in 1936, nearly 25 years before the three-dimensional structures of myoglobin and hemoglobin were elucidated.

The Structure of Myoglobin Prevents the Release of Reactive Oxygen Species

Oxygen binding to iron in heme is accompanied by the partial transfer of an electron from the ferrous ion to oxygen. In many ways, the structure is best described as a complex between ferric ion (Fe^{3+}) and *superoxide anion* (O_2^-), as illustrated in Figure 7.3. It is crucial that oxygen, when it is first released, leaves as dioxygen rather than superoxide. First, superoxide itself and other species that can be generated from it are reactive oxygen species (p. 517) that can be damaging to many biological materials. Second, the release of superoxide leaves the iron ion in the ferric state. This species, termed *metmyoglobin*, does not bind oxygen. Thus, potential oxygen-storage capacity is lost. Features of myoglobin stabilize the oxygen complex such that superoxide is less likely to be released. In particular, the binding

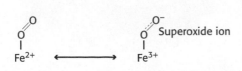

Figure 7.3 **Iron-oxygen bonding.** The interaction between iron and oxygen in hemoglobin can be described as a combination of resonance structures, one with Fe^{2+} and dioxygen and another with Fe^{3+} and superoxide ion.

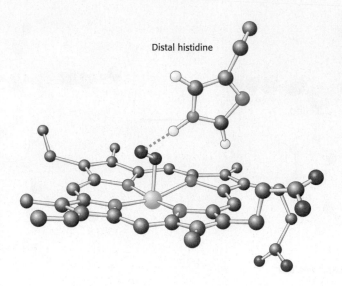

Distal histidine

Figure 7.4 Stabilizing bound oxygen.
A hydrogen bond (dotted green line)
donated by the distal histidine residue
to the bound oxygen molecule helps
stabilize oxymyoglobin.

pocket of myoglobin includes an additional histidine residue (termed the
distal histidine) that donates a hydrogen bond to the bound oxygen molecule (Figure 7.4). The superoxide character of the bound oxygen species
strengthens this interaction. Thus, *the protein component of myoglobin controls the intrinsic reactivity of heme,* making it more suitable for reversible
oxygen binding.

Human Hemoglobin Is an Assembly of Four Myoglobin-like Subunits

The three-dimensional structure of hemoglobin from horse heart was
solved by Max Perutz shortly after the determination of the myoglobin
structure. Since then, the structures of hemoglobins from other sources including human beings have been determined. These hemoglobin molecules
consist of four polypeptide chains, two (termed the α *chains*) with one, identical, amino acid sequence and two (termed the β *chains*) with another
(Figure 7.5). Each of the subunits comprises a set of α helices in the same
arrangement as the α helices in myoglobin (see Figure 6.13 for a comparison
of the structures). The recurring structure is called a *globin fold*. Consistent
with this structural similarity, the amino acid sequences of the α and β
chains of human hemoglobin are readily aligned with the amino acid sequence of sperm whale myoglobin with 25% and 24% identity, respectively,
and good conservation of key residues such as the proximal and distal histidine residues. Thus, the α and β chains are related to each other and to myoglobin by divergent evolution (see p. 175).

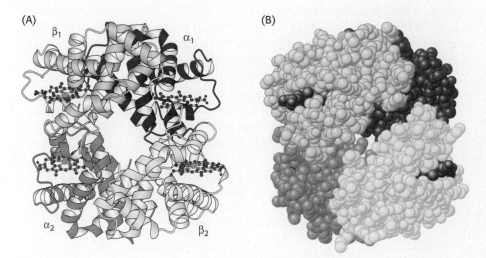

(A) β_1 α_1

(B)

α_2 β_2

**Figure 7.5 Quaternary structure of
deoxyhemoglobin.** Hemoglobin,
which is composed of two α chains and
two β chains, functions as a pair of $\alpha\beta$
dimers. (A) A ribbon diagram. (B) A space-
filling model. [Drawn from 1A3N.pdb.]

The hemoglobin tetramer, referred to as *hemoglobin A* (HbA), is best described as a pair of identical $\alpha\beta$ dimers ($\alpha_1\beta_1$ and $\alpha_2\beta_2$) that associate to form the tetramer. In deoxyhemoglobin, these $\alpha\beta$ dimers are linked by an extensive interface, which includes, among other regions, the carboxyl terminus of each chain. The heme groups are well separated in the tetramer by iron–iron distances ranging from 24 to 40 Å.

7.2 Hemoglobin Binds Oxygen Cooperatively

Let us compare the oxygen-binding properties of these proteins. The oxygen-binding characteristics are described by an *oxygen-binding curve*, a plot of the *fractional saturation* versus the concentration of oxygen. The fractional saturation, Y, is defined as the fraction of possible binding sites that contain bound oxygen. The value of Y can range from 0 (all sites empty) to 1 (all sites filled). The concentration of oxygen is most conveniently measured by its *partial pressure, pO_2*. For myoglobin, a binding curve indicating a simple chemical equilibrium is observed (Figure 7.6). Notice that the fraction of myoglobin molecules with bound oxygen rises sharply as pO_2 increases and then levels off. Half-saturation (P_{50} for 50% saturated) is at the relatively low value of 2 torr (mm Hg).

In contrast, the oxygen-binding curve for hemoglobin in red blood cells shows some remarkable features (Figure 7.7). It does not look like a simple binding curve such as that for myoglobin; instead, it resembles an "S." Such curves are referred to as *sigmoid* because of their S-like shape. Notice that oxygen binding is significantly weaker than that for myoglobin. Half-saturation is at the higher value of $P_{50} = 26$ torr. Note that this binding curve is derived from hemoglobin in red blood cells. Inside red cells, hemoglobin interacts with 2,3-bisphosphoglycerate, a molecule that significantly lowers hemoglobin's oxygen affinity, as will be considered in detail shortly.

A sigmoid binding curve indicates that a protein shows a special binding behavior. The shape of the curve reveals that the binding of oxygen at one site within the hemoglobin molecule increases the likelihood that oxygen binds at the remaining unoccupied sites. Conversely, the unloading of oxygen at one heme facilitates the unloading of oxygen at the others. This sort of binding behavior is *cooperative* (p. 189), because the binding reactions at individual sites in each hemoglobin molecule are not independent of one another. We will return to the mechanism of this cooperativity shortly.

What is the physiological significance of the cooperative binding of oxygen by hemoglobin? Oxygen must be transported in the blood from the lungs, where the partial pressure of oxygen is relatively high (approximately 100 torr), to the actively metabolizing tissues, where the partial pressure of

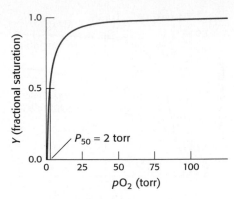

Figure 7.6 Oxygen binding by myoglobin. Half the myoglobin molecules have bound oxygen when the oxygen partial pressure is 2 torr.

Torr

A unit of pressure equal to that exerted by a column of mercury 1 mm high at 0°C and standard gravity (1 mm Hg). Named after Evangelista Torricelli (1608–1647), inventor of the mercury barometer.

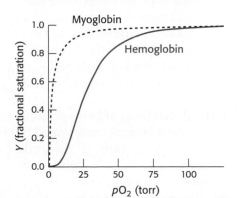

Figure 7.7 Oxygen binding by hemoglobin. This curve, obtained for hemoglobin in red blood cells, is shaped somewhat like an "S," indicating that distinct, but interacting, oxygen-binding sites are present in each hemoglobin molecule. For comparison, the binding curve for myoglobin is shown as a dashed black curve.

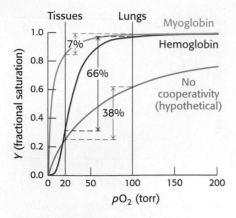

Figure 7.8 Cooperativity enhances oxygen delivery by hemoglobin. Because of cooperativity between O_2 binding sites, hemoglobin delivers more O_2 to tissues than would myoglobin or any noncooperative protein, even one with optimal O_2 affinity.

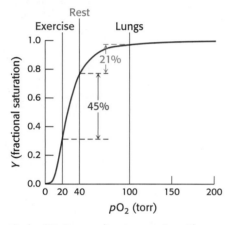

Figure 7.9 Responding to exercise. The drop in oxygen concentration from 40 torr in resting tissues to 20 torr in exercising tissues corresponds to the steepest part of its oxygen-binding curve. Consequently, hemoglobin is very effective in providing oxygen to exercising tissues.

oxygen is much lower (typically, 20 torr). Let us consider how the cooperative behavior indicated by the sigmoid curve leads to efficient oxygen transport (Figure 7.8). In the lungs, hemoglobin becomes nearly saturated with oxygen such that 98% of the oxygen-binding sites are occupied. When hemoglobin moves to the tissues and releases O_2, the saturation level drops to 32%. Thus, a total of $98 - 32 = 66\%$ of the potential oxygen-binding sites contribute to oxygen transport. The cooperative release of oxygen favors a more-complete unloading of oxygen in the tissues. If myoglobin were employed for oxygen transport, it would be 98% saturated in the lungs, but would remain 91% saturated in the tissues, and so only $98 - 91 = 7\%$ of the sites would contribute to oxygen transport; myoglobin binds oxygen too tightly to be useful in oxygen transport. The situation might have been improved without cooperativity by the evolution of a noncooperative oxygen carrier with an optimized affinity for oxygen. For such a protein, the most oxygen that could be transported from a region in which pO_2 is 100 torr to one in which it is 20 torr is $63 - 25 = 38\%$. Thus, the cooperative binding and release of oxygen by hemoglobin enables it to deliver nearly 10 times as much oxygen as could be delivered by myoglobin and more than 1.7 times as much as could be delivered by any noncooperative protein.

Closer examination of oxygen concentrations in tissues at rest and during exercise reveals how effective an oxygen provider hemoglobin can be (Figure 7.9). Under resting conditions, the oxygen concentration in muscle is approximately 40 torr but, during exercise, the concentration is reduced to 20 torr. In the decrease from 100 torr in the lungs to 40 torr in resting muscle, the oxygen saturation of hemoglobin is reduced from 98% to 77%, and so $98 - 77 = 21\%$ of the oxygen is released over a drop of 60 torr. In a decrease from 40 torr to 20 torr, the oxygen saturation is reduced from 77% to 32%, corresponding to an oxygen release of 45% over a drop of 20 torr. Thus, because the change in oxygen concentration from rest to exercise corresponds to the steepest part of the oxygen-binding curve, oxygen is effectively delivered to tissues where it is most needed. In Section 7.3, we shall examine other properties of hemoglobin that enhance its physiological responsiveness.

Oxygen Binding Markedly Changes the Quaternary Structure of Hemoglobin

The cooperative binding of oxygen by hemoglobin requires that the binding of oxygen at one site in the hemoglobin tetramer influence the oxygen-binding properties at the other sites. Given the large separation between the iron sites, direct interactions are not possible. Thus, indirect mechanisms for coupling the sites must be at work. These mechanisms are intimately related to the quaternary structure of hemoglobin.

Hemoglobin undergoes substantial changes in quaternary structure on oxygen binding: the $\alpha_1\beta_1$ and $\alpha_2\beta_2$ dimers rotate approximately 15 degrees with respect to one another (Figure 7.10). The dimers themselves are relatively unchanged, although there are localized conformational shifts. Thus, the interface between the $\alpha_1\beta_1$ and $\alpha_2\beta_2$ dimers is most affected by this structural transition. Recent studies have revealed that the $\alpha_1\beta_1$ and $\alpha_2\beta_2$ dimers are freer to move with respect to one another in the oxygenated state than they are in the deoxygenated state.

The quaternary structure observed in the deoxy form of hemoglobin is often referred to as the T (for tense) *state* because it is quite constrained by subunit–subunit interactions. The quaternary structure of the fully oxygenated form is referred to as the R (for relaxed) *state*. In light of the observation that the R form of hemoglobin is less constrained, the tense and relaxed designations seem particularly apt. Importantly, in the R state, the

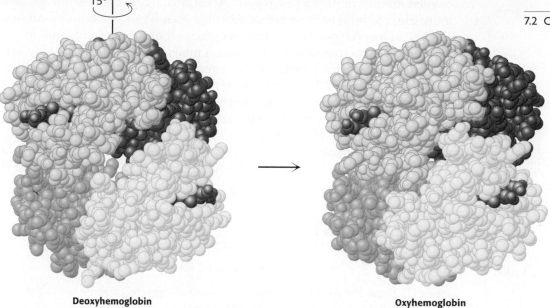

Deoxyhemoglobin   **Oxyhemoglobin**

Figure 7.10 **Quaternary structural changes on oxygen binding by hemoglobin.** *Notice that, on oxygenation, one αβ dimer shifts with respect to the other by a rotation of 15 degrees.* [Drawn from 1A3N.pdb and 1LFQ.pdb.]

oxygen-binding sites are free of strain and are capable of binding oxygen with higher affinity than are the sites in the T state. *By triggering the shift of the hemoglobin tetramer from the T state to the R state, the binding of oxygen to one site increases the binding affinity of other sites.*

Hemoglobin Cooperativity Can Be Potentially Explained by Several Models

Two limiting models have been developed to explain the cooperative binding of ligands to a multisubunit assembly such as hemoglobin. In the *concerted model,* also known as the *MWC model* after Jacques Monod, Jeffries Wyman, and Jean-Pierre Changeux who first proposed it, the overall assembly can exist only in two forms: the T state and the R state. The binding of ligands simply shifts the equilibrium between these two states (Figure 7.11). Thus, as a hemoglobin tetramer binds each oxygen molecule, the probability that the tetramer is in the R state increases. In the deoxy form, hemoglobin tetramers are almost exclusively in the T state. However, the binding of oxygen to one site in the molecule shifts the equilibrium toward the R state. If a molecule assumes the R quaternary structure, the

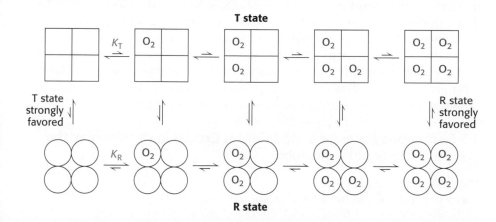

Figure 7.11 **Concerted model.** All molecules exist either in the T state or in the R state. At each level of oxygen loading, an equilibrium exists between the T and R states. The equilibrium shifts from strongly favoring the T state with no oxygen bound to strongly favoring the R state when the molecule is fully loaded with oxygen. The R state has a greater affinity for oxygen than does the T state.

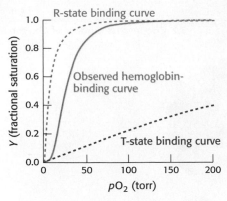

Figure 7.12 T-to-R transition. The observed binding curve for hemoglobin can be seen as a combination of the binding curves that would be observed if all molecules remained in the T state or if all of the molecules were in the R state. The sigmoidal curve is observed because molecules convert from the T state into the R state as oxygen molecules bind.

oxygen affinity of its sites increases. Additional oxygen molecules are now more likely to bind to the three unoccupied sites. Thus, the binding curve is shallow at low oxygen concentrations when all of the molecules are in the T state, becomes steeper as the fraction of molecules in the R state increases, and flattens out again when all of the sites within the R-state molecules become filled (Figure 7.12). These events produce the sigmoid binding curve so important for efficient oxygen transport.

In the concerted model, each tetramer can exist in only two states, the T state and the R state. In an alternative model, the *sequential model,* the binding of a ligand to one site in an assembly increases the binding affinity of neighboring sites without inducing a full conversion from the T into the R state (Figure 7.13).

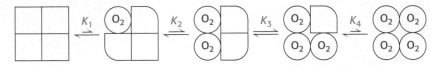

Figure 7.13 Sequential model. The binding of a ligand changes the conformation of the subunit to which it binds. This conformational change induces changes in neighboring subunits that increase their affinity for the ligand.

Is the cooperative binding of oxygen by hemoglobin best described by the concerted or the sequential model? Neither model in its pure form fully accounts for the behavior of hemoglobin. Instead, a combined model is required. Hemoglobin behavior is concerted in that hemoglobin with three sites occupied by oxygen is almost always in the quaternary structure associated with the R state. The remaining open binding site has an affinity for oxygen more than 20-fold greater than that of fully deoxygenated hemoglobin binding its first oxygen. However, the behavior is not fully concerted, because hemoglobin with oxygen bound to only one of four sites remains primarily in the T-state quaternary structure. Yet, this molecule binds oxygen three times as strongly as does fully deoxygenated hemoglobin, an observation consistent only with a sequential model. These results highlight the fact that the concerted and sequential models represent idealized limiting cases, which real systems may approach but rarely attain.

Structural Changes at the Heme Groups Are Transmitted to the $\alpha_1\beta_1$–$\alpha_2\beta_2$ Interface

We now examine how oxygen binding at one site is able to shift the equilibrium between the T and R states of the entire hemoglobin tetramer. As in myoglobin, oxygen binding causes each iron atom in hemoglobin to move from outside the plane of the porphyrin into the plane. When the iron atom moves, the histidine residue bound in the fifth coordination site moves with it. This histidine residue is part of an α helix, which also moves (Figure 7.14). The carboxyl terminal end of this α helix lies in the interface between the two $\alpha\beta$ dimers. The change in position of the carboxyl terminal end of the helix favors the T-to-R transition. Consequently, *the structural transition at the iron ion in one subunit is directly transmitted to the other subunits.* The rearrangement of the dimer interface provides a pathway for communication between subunits, enabling the cooperative binding of oxygen.

2,3-Bisphosphoglycerate in Red Cells Is Crucial in Determining the Oxygen Affinity of Hemoglobin

For hemoglobin to function efficiently, a requirement is that the T state remain stable until the binding of sufficient oxygen has converted it into the

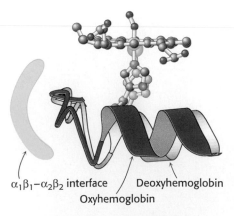

$\alpha_1\beta_1$–$\alpha_2\beta_2$ interface   Deoxyhemoglobin

Oxyhemoglobin

Figure 7.14 Conformational changes in hemoglobin. The movement of the iron ion on oxygenation brings the iron-associated histidine residue toward the porphyrin ring. The associated movement of the histidine-containing α helix alters the interface between the αβ dimers, instigating other structural changes. For comparison, the deoxyhemoglobin structure is shown in gray behind the oxyhemoglobin structure in color.

190

R state. The T state of hemoglobin is highly unstable, however, pushing the equilibrium so far toward the R state that little oxygen would be released in physiological conditions. Thus, an additional mechanism is needed to properly stabilize the T state. This mechanism was discovered by comparing the oxygen-binding properties of hemoglobin in red blood cells with fully purified hemoglobin (Figure 7.15). Pure hemoglobin binds oxygen much more tightly than does hemoglobin in red blood cells. This dramatic difference is due to the presence within these cells of *2,3-bisphosphoglycerate* (2,3-BPG; also known as 2,3-diphosphoglycerate or 2,3-DPG).

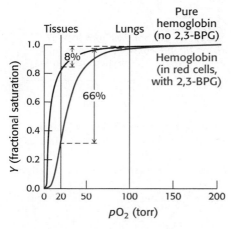

2,3-Bisphosphoglycerate (2,3-BPG)

This highly anionic compound is present in red blood cells at approximately the same concentration as that of hemoglobin (~2 mM). Without 2,3-BPG, hemoglobin would be an extremely inefficient oxygen transporter, releasing only 8% of its cargo in the tissues.

How does 2,3-BPG lower the oxygen affinity of hemoglobin so significantly? Examination of the crystal structure of deoxyhemoglobin in the presence of 2,3-BPG reveals that a single molecule of 2,3-BPG binds in the center of the tetramer, in a pocket present only in the T form (Figure 7.16). On T-to-R transition, this pocket collapses and 2,3-BPG is released. Thus, in order for the structural transition from T to R to take place, the bonds between hemoglobin and 2,3-BPG must be broken. In the presence of 2,3-BPG, more oxygen-binding sites within the hemoglobin tetramer must be occupied in order to induce the T-to-R transition, and so hemoglobin remains in the lower-affinity T state until higher oxygen concentrations are reached.

The regulation of hemoglobin by 2,3-BPG is remarkable because 2,3-BPG does not in any way resemble oxygen, the molecule on which hemoglobin

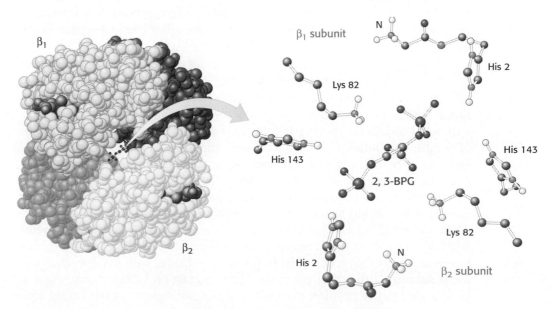

Figure 7.16 Mode of binding of 2,3-BPG to human deoxyhemoglobin. 2,3-Bisphosphoglycerate binds to the central cavity of deoxyhemoglobin (left). There, it interacts with three positively charged groups on each β chain (right). [Drawn from 1B86.pdb.]

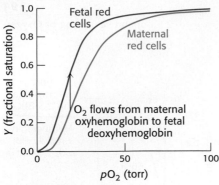

Figure 7.17 Oxygen affinity of fetal red blood cells. Fetal red blood cells have a higher oxygen affinity than maternal red blood cells because fetal hemoglobin does not bind 2,3-BPG as well as maternal hemoglobin does.

carries out its primary function. 2,3-BPG is referred to as an *allosteric effector* (from *allos*, "other," and *stereos*, "structure"). Regulation by a molecule structurally unrelated to oxygen is possible because the allosteric effector binds to a site that is completely distinct from that for oxygen. We will encounter allosteric effects again when we consider enzyme regulation in Chapter 10.

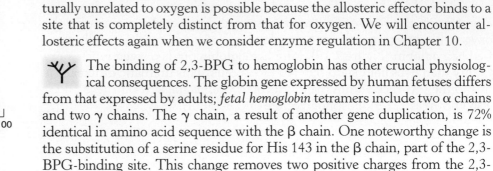 The binding of 2,3-BPG to hemoglobin has other crucial physiological consequences. The globin gene expressed by human fetuses differs from that expressed by adults; *fetal hemoglobin* tetramers include two α chains and two γ chains. The γ chain, a result of another gene duplication, is 72% identical in amino acid sequence with the β chain. One noteworthy change is the substitution of a serine residue for His 143 in the β chain, part of the 2,3-BPG-binding site. This change removes two positive charges from the 2,3-BPG-binding site (one from each chain) and reduces the affinity of 2,3-BPG for fetal hemoglobin. Consequently, the oxygen-binding affinity of fetal hemoglobin is higher than that of maternal (adult) hemoglobin (Figure 7.17). This difference in oxygen affinity allows oxygen to be effectively transferred from maternal to fetal red blood cells. We have here an example in which gene duplication and specialization produced a ready solution to a biological challenge—in this case, the transport of oxygen from mother to fetus.

7.3 Hydrogen Ions and Carbon Dioxide Promote the Release of Oxygen: The Bohr Effect

We have seen how hemoglobin's cooperative release of oxygen helps deliver oxygen to tissues where it is most needed, as revealed by their low oxygen partial pressure. This ability is enhanced by the ability of hemoglobin to respond to other cues in its physiological environment signaling the need for oxygen. Rapidly metabolizing tissues, such as contracting muscle, generate large amounts of hydrogen ions and carbon dioxide (pp. 447 and 448). So that oxygen is released where the need is greatest, hemoglobin has evolved to release oxygen more readily in response to higher levels of these substances. Like 2,3-BPG, hydrogen ions and carbon dioxide are *allosteric effectors* of hemoglobin that bind to sites on the molecule that are distinct from the oxygen-binding sites. The regulation of oxygen binding by hydrogen ions and carbon dioxide is called the *Bohr effect* after Christian Bohr, who described this phenomenon in 1904.

The oxygen affinity of hemoglobin decreases as pH decreases from a value of 7.4 (Figure 7.18). Consequently, as hemoglobin moves into a region of lower pH, its tendency to release oxygen increases. For example, transport from the lungs, with pH 7.4 and an oxygen partial pressure of 100 torr, to active muscle, with a pH of 7.2 and an oxygen partial pressure of 20 torr, results in a release of oxygen amounting to 77% of total carrying capacity. Only 66% of the oxygen would be released in the absence of any change in pH. Structural and chemical studies have revealed much about the chemical basis of the pH effect. At least two sets of chemical groups are responsible for the effect: the α-amino groups at the amino termini of the α chain and the side chains of histidines β146 and α122, all of which have pK_a values near pH 7. Consider histidine β146, the residue at the C terminus of the β chain. In deoxyhemoglobin, the terminal carboxylate group of β146 forms a salt bridge with a lysine residue in the α subunit of the other αβ dimer. This interaction locks the side chain of histidine β146 in a position from which it can participate in a salt bridge with negatively charged aspartate 94 in the same chain, provided that the imidazole group of the histidine residue is protonated (Figure 7.19).

The other groups also participate in salt bridges in the T state. *The formation of these salt bridges stabilizes the T state, leading to a greater tendency*

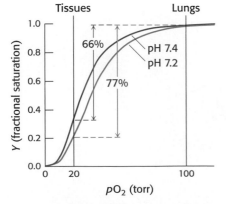

Figure 7.18 Effect of pH on the oxygen affinity of hemoglobin. Lowering the pH from 7.4 (red curve) to 7.2 (blue curve) results in the release of O_2 from oxyhemoglobin.

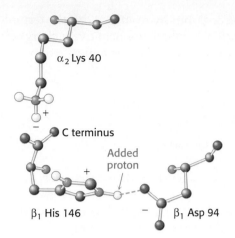

Figure 7.19 **Chemical basis of the Bohr effect.** In deoxyhemoglobin, three amino acid residues form two salt bridges that stabilize the T quaternary structure. The formation of one of the salt bridges depends on the presence of an added proton on histidine β146. The proximity of the negative charge on aspartate β94 in deoxyhemoglobin favors protonation of this histidine. *Notice* that the salt bridge between histidine β146 and aspartate β94 is stabilized by a hydrogen bond (green dashed line).

for oxygen to be released. For example, at high pH, the side chain of histidine β146 is not protonated and the salt bridge does not form. As the pH drops, however, the side chain of histidine β146 becomes protonated, the salt bridge with aspartate β94 forms, and the T state is stabilized.

Carbon dioxide, a neutral species, passes through the red-blood-cell membrane into the cell. This transport is also facilitated by membrane transporters including proteins associated with Rh blood types. Carbon dioxide stimulates oxygen release by two mechanisms. First, the presence of high concentrations of carbon dioxide leads to a drop in pH within the red blood cell (Figure 7.20).

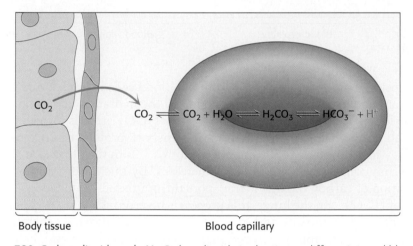

Figure 7.20 **Carbon dioxide and pH.** Carbon dioxide in the tissues diffuses into red blood cells. Inside a red blood cell, carbon dioxide reacts with water to form carbonic acid, in a reaction catalyzed by the enzyme carbonic anhydrase. Carbonic acid dissociates to form HCO_3^- and H^+, resulting in a drop in pH inside the red cell.

Carbon dioxide reacts with water to form carbonic acid, H_2CO_3. This reaction is accelerated by *carbonic anhydrase,* an enzyme abundant in red blood cells that will be considered extensively in Chapter 9. H_2CO_3 is a moderately strong acid with a pK_a of 3.5. Thus, once formed, carbonic acid dissociates to form bicarbonate ion, HCO_3^-, and H^+, resulting in a drop in pH. This drop in pH stabilizes the T state by the mechanism discussed previously.

In the second mechanism, a direct chemical interaction between carbon dioxide and hemoglobin stimulates oxygen release. The effect of carbon dioxide on oxygen affinity can be seen by comparing oxygen-binding curves in the absence and presence of carbon dioxide at a constant pH (Figure 7.21). In the presence of carbon dioxide at a partial pressure of 40 torr at pH 7.2, the amount of oxygen released approaches 90% of the maximum carrying capacity.

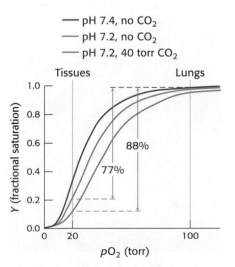

Figure 7.21 **Carbon dioxide effects.** The presence of carbon dioxide decreases the affinity of hemoglobin for oxygen even beyond the effect due to a decrease in pH, resulting in even more efficient oxygen transport from the tissues to the lungs.

Carbon dioxide stabilizes deoxyhemoglobin by reacting with the terminal amino groups to form *carbamate* groups, which are negatively charged, in contrast with the neutral or positive charges on the free amino groups.

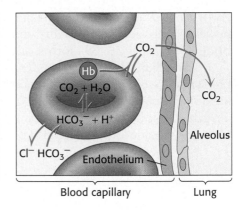

Carbamate

The amino termini lie at the interface between the αβ dimers, and these negatively charged carbamate groups participate in salt-bridge interactions that stabilize the T state, favoring the release of oxygen. This process also provides a mechanism for carbon dioxide transport, but it accounts for only about 14% of the total carbon dioxide transport.

Figure 7.22 Transport of CO_2 from tissues to lungs. Most carbon dioxide is transported to the lungs in the form of HCO_3^- produced in red blood cells and then released into the blood plasma. A lesser amount is transported by hemoglobin in the form of an attached carbamate.

Most carbon dioxide released from red blood cells is transported to the lungs in the form of HCO_3^- produced from the hydration of carbon dioxide inside the cell (Figure 7.22). Much of the HCO_3^- that is formed leaves the cell through a specific membrane-transport protein that exchanges HCO_3^- from one side of the membrane for Cl^- from the other side. Thus, the serum concentration of HCO_3^- increases. By this means, a large concentration of carbon dioxide is transported from tissues to the lungs in the form of HCO_3^-. In the lungs, this process is reversed: HCO_3^- is converted back into carbon dioxide and exhaled. Thus, carbon dioxide generated by active tissues contributes to a decrease in red-blood-cell pH and, hence, to oxygen release and is converted into a form that can be transported in the serum and released in the lungs.

7.4 Mutations in Genes Encoding Hemoglobin Subunits Can Result in Disease

In modern times, particularly after the sequencing of the human genome, to think of genetically encoded variations in protein sequence as a factor in specific diseases is routine. The notion that diseases might be caused by molecular defects was proposed by Linus Pauling in 1949 (4 years before Watson and Crick's proposal of the DNA double helix) to explain the blood disease *sickle-cell anemia*. The name of the disorder comes from the abnormal sickle shape of red blood cells deprived of oxygen that is seen in people suffering from this disease (Figure 7.23). Pauling proposed that sickle-cell disease might be caused by a specific variation in the amino acid sequence

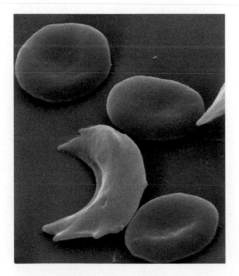

Figure 7.23 Sickled red blood cells. A micrograph showing a sickled red blood cell adjacent to normally shaped red blood cells. [Eye of Science/Photo Researchers.]

of one hemoglobin chain. Today, we know that this bold hypothesis is correct. In fact, approximately 7% of the world's population are carriers of some disorder of hemoglobin caused by a variation in the amino acid sequence. In concluding this chapter, we will focus on the two most important of these disorders, sickle-cell anemia and thalassemia.

Sickle-Cell Anemia Results from the Aggregation of Mutated Deoxyhemoglobin Molecules

People having sickled red blood cells experience a number of dangerous symptoms. Examination of the contents of these red cells reveals that the hemoglobin molecules have formed large fibrous aggregates (Figure 7.24). These fibers extend across the red blood cells, distorting them so that they clog small capillaries and impair blood flow. The results may be painful swelling of the extremities and a higher risk of stroke or bacterial infection (due to poor circulation). The sickled red cells also do not remain in circulation as long as normal cells do, leading to anemia.

What is the molecular defect associated with sickle-cell anemia? Using newly developed chromatographic techniques, Vernon Ingram demonstrated in 1956 that a single amino acid substitution in the β chain of hemoglobin is responsible—namely, the substitution of a valine residue for a glutamate residue in position 6. The mutated form is referred to as *hemoglobin S* (HbS). In people with sickle-cell anemia, both alleles of the hemoglobin β-chain gene (HbB) are mutated. The HbS substitution substantially decreases the solubility of deoxyhemoglobin, although it does not markedly alter the properties of oxyhemoglobin.

Examination of the structure of hemoglobin S reveals that the new valine residue lies on the surface of the T-state molecule (Figure 7.25). This new hydrophobic patch interacts with another hydrophobic patch formed by Phe 85 and Val 88 of the β chain of a neighboring molecule to initiate the aggregation process. More-detailed analysis reveals that a single hemoglobin S fiber is formed from 14 chains of multiple interlinked hemoglobin molecules. Why do these aggregates not form when hemoglobin S is oxygenated? Oxygenated hemoglobin S is in the R state, and residues Phe 85 and Val 88 on the β chain are largely buried inside the hemoglobin assembly. Without a partner with which to interact, the surface Val residue in position 6 is benign.

Approximately 1 in 100 West Africans suffer from sickle-cell anemia. Given the often devastating consequences of the disease, why is the HbS mutation so prevalent in Africa and in some other regions? Recall that both copies of the HbB gene are mutated in people with sickle-cell anemia. People with one copy of the HbB gene and one copy of the HbS are relatively unaffected. They are said to have *sickle-cell trait* because they can pass

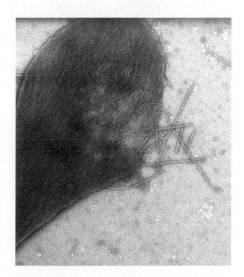

Figure 7.24 Sickle-cell hemoglobin fibers. An electron micrograph depicting a ruptured sickled red blood cell with fibers of sickle-cell hemoglobin emerging. [Courtesy of Robert Josephs and Thomas E. Wellems, University of Chicago.]

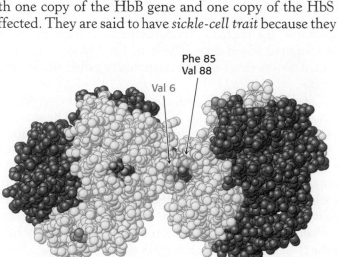

Figure 7.25 Deoxygenated hemoglobin S. The interaction between Val 6 (blue) on a β chain of one hemoglobin molecule and a hydrophobic patch formed by Phe 85 and Val 88 (gray) on a β chain of another deoxygenated hemoglobin molecule leads to hemoglobin aggregation. The exposed Val 6 residues of other β chains participate in other such interactions in hemoglobin S fibers. [Drawn from 2HBS.pdb.]

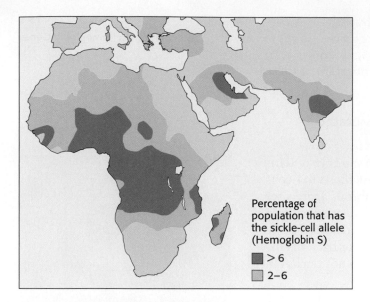 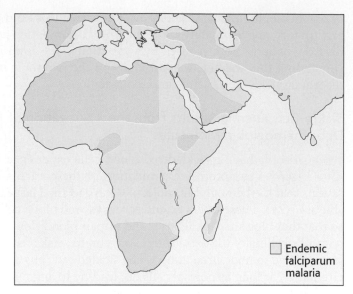

Percentage of population that has the sickle-cell allele (Hemoglobin S)

■ > 6
■ 2–6

□ Endemic falciparum malaria

Figure 7.26 Sickle-cell trait and malaria. A significant correlation is observed between regions with a high frequency of the HbS allele and regions with a high prevalence of malaria.

the HbS gene to their offspring. However, people with sickle-cell trait are resistant to *malaria,* a disease carried by a parasite, *Plasmodium falciparum,* that lives within red blood cells at one stage in its life cycle. The dire effect of malaria on health and reproductive likelihood in regions where malaria has been historically endemic has favored people with sickle-cell trait, increasing the prevalence of the HbS allele (Figure 7.26).

Thalassemia Is Caused by an Imbalanced Production of Hemoglobin Chains

Sickle-cell anemia is caused by the substitution of a single specific amino acid in one hemoglobin chain. *Thalassemia,* the other prevalent inherited disorder of hemoglobin, is caused by the loss or substantial reduction of a single hemoglobin *chain.* The result is low levels of functional hemoglobin and a decreased production of red blood cells, which may lead to anemia, fatigue, pale skin, and spleen and liver malfunction. Thalassemia is a set of related diseases. In α-thalassemia, the α chain of hemoglobin is not produced in sufficient quantity. Consequently, hemoglobin tetramers form that contain only the β chain. These tetramers, referred to as *hemoglobin H* (HbH), bind oxygen but with high affinity and no cooperativity. Thus, oxygen release in the tissues is poor. In β-thalassemia, the β chain of hemoglobin is not produced in sufficient quantity. In the absence of β chains, the α chains form insoluble aggregates that precipitate inside immature red blood cells. The loss of red blood cells results in anemia. The most severe form of β-thalassemia is called *thalassemia major* or *Cooley anemia.*

Both α- and β-thalassemia are associated with many different genetic variations and display a wide range of clinical severity. The most severe forms of α-thalassemia are usually fatal shortly before or just after birth. However, these forms are relatively rare. An examination of the repertoire of hemoglobin genes in the human genome provides one explanation. Normally, human beings have not two but four alleles for the α chain, arranged such that a pair of genes are located adjacent to each other on one end of each chromosome 16. Thus, the complete loss of α-chain expression requires the disruption of four alleles. β-Thalassemia is more common because we normally have only two alleles for the β chain, one on each copy of chromosome 11.

The Accumulation of Free Alpha-Hemoglobin Chains Is Prevented

The presence of four genes expressing the α chain, compared with two for the β chain, suggests that the α chain would be produced in excess (making the overly simple assumption that production from each gene is comparable). If this is correct, why doesn't the excess α chain precipitate? A recent discovery reveals one mechanism for keeping α chains in solution. Red blood cells produce an 11-kd protein, called α-*hemoglobin stabilizing protein* (AHSP), that binds specifically to α-chain monomers. This complex is soluble. The crystal structure of a complex between AHSP and α-hemoglobin reveals that AHSP binds to the same face of α-hemoglobin as does β-hemoglobin (Figure 7.27). AHSP binds the α chain in both the deoxygenated and oxygenated forms. In the complex with oxygen bound, the distal histidine, rather than the proximal histidine, binds the iron atom.

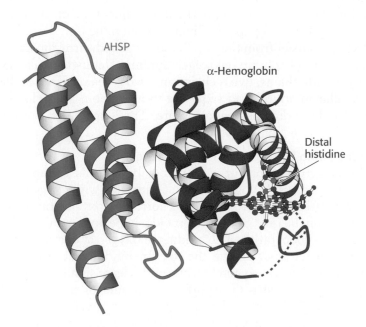

Figure 7.27 Stabilizing free α-hemoglobin. The structure of a complex between AHSP and α-hemoglobin is shown. In this complex, the iron atom is bound to oxygen and to the distal histidine. *Notice* that AHSP binds to the same surface of α-hemoglobin as does β-hemoglobin. [Drawn from 1Y01.pdb.]

AHSP serves to bind the α-hemoglobin as it is produced. As β-hemoglobin is produced, it displaces AHSP because the α-hemoglobin–β-hemoglobin dimer is more stable than the α-hemoglobin–AHSP complex. Thus, AHSP prevents the accumulation and precipitation of free α-hemoglobin. Studies are underway to determine if mutations in the gene encoding AHSP play a role in modulating the severity of β-thalassemia.

Additional Globins Are Encoded in the Human Genome

Does the haploid human genome contain globin genes in addition to the gene for myoglobin, the two genes for α-hemoglobin, and the one for β-hemoglobin? We have already encountered fetal hemoglobin, which contains the γ chain in place of the β chain. Several other genes encode other hemoglobin subunits that are expressed during development including the δ chain, the ε chain, and the ζ chain.

Recent examination of the human genome sequence has revealed two additional globins. Both of these proteins are monomeric proteins, more similar to myoglobin than to hemoglobin. The first, *neuroglobin*, is expressed primarily in the brain and at especially high levels in the retina. Neuroglobin may play a role in protecting neural tissues from hypoxia (insufficient oxygen). The second, *cytoglobin*, is expressed more widely throughout the body. Structural and spectroscopic studies reveal that, in both neuroglobin and cytoglobin, both the proximal and the distal histidines are coordinated to the iron atom in the deoxy form; the distal histidine is displaced on oxygen binding. Future studies should more completely elucidate the functions of these members of the globin family.

Summary

7.1 Myoglobin and Hemoglobin Bind Oxygen at Iron Atoms in Heme

Myoglobin is a largely α-helical protein that binds the prosthetic group heme. Heme consists of protoporphyrin, an organic component with four linked pyrrole rings, and a central iron ion in the Fe^{2+} state. The iron ion is coordinated to the side chain of a histidine residue in myoglobin, referred to as the proximal histidine. One of the oxygen atoms in O_2 binds to an open coordination site on the iron. Because of partial electron transfer from the iron to the oxygen, the iron ion moves into the plane of the porphyrin on oxygen binding. Hemoglobin consists of four polypeptide chains, two α chains and two β chains. Each of these chains is similar in amino acid sequence to myoglobin and folds into a very similar three-dimensional structure. The hemoglobin tetramer is best described as a pair of $\alpha\beta$ dimers.

7.2 Hemoglobin Binds Oxygen Cooperatively

The oxygen-binding curve for myoglobin reveals a simple equilibrium binding process. Myoglobin is half-saturated with oxygen at an oxygen concentration of approximately 2 torr. The oxygen-binding curve for hemoglobin has an "S"-like (sigmoid) shape, indicating that the oxygen binding is cooperative. The binding of oxygen at one site within the hemoglobin tetramer affects the oxygen-binding affinities of the other sites. Cooperative oxygen binding and release significantly increase the efficiency of oxygen transport. The amount of the potential oxygen-carrying capacity utilized in transporting oxygen from the lungs (with a partial pressure of oxygen of 100 torr) to tissues (with a partial pressure of oxygen of 20 torr) is 66% compared with 7% if myoglobin had been used as the oxygen carrier.

The quaternary structure of hemoglobin changes on oxygen binding. The structure of deoxyhemoglobin is referred to as the T state. The structure of oxyhemoglobin is referred to as the R state. The two $\alpha\beta$ dimers rotate by approximately 15 degrees with respect to one another in the transition from the T to the R state. Cooperative binding can be potentially explained by concerted and sequential models. In the concerted model, each hemoglobin adopts either the T state or the R state; the equilibrium between these two states is determined by the number of occupied oxygen-binding sites. Sequential models allow intermediate structures. Structural changes at the iron sites in response to oxygen binding are transmitted to the interface between $\alpha\beta$ dimers, influencing the T-to-R equilibrium.

Red blood cells contain 2,3-bisphosphoglycerate in concentrations approximately equal to that for hemoglobin. 2,3-BPG binds tightly to the T state but not to the R state, stabilizing the T state and lowering the oxygen affinity of hemoglobin. Fetal hemoglobin binds oxygen more tightly than does adult hemoglobin owing to weaker 2,3-BPG binding. This difference allows oxygen transfer from maternal to fetal blood.

7.3 Hydrogen Ions and Carbon Dioxide Promote the Release of Oxygen

The oxygen-binding properties of hemoglobin are markedly affected by pH and by the presence of carbon dioxide, a phenomenon known as the

Bohr effect. Increasing the concentration of hydrogen ions—that is, decreasing pH—decreases the oxygen affinity of hemoglobin, owing to the protonation of the amino termini and certain histidine residues. The protonated residues help stabilize the T state. Increasing concentrations of carbon dioxide decrease the oxygen affinity of hemoglobin by two mechanisms. First, carbon dioxide is converted into carbonic acid, which lowers the oxygen affinity of hemoglobin by decreasing the pH inside the red blood cell. Second, carbon dioxide adds to the amino termini of hemoglobin to form carbamates. These negatively charged groups stabilize deoxyhemoglobin through ionic interactions. Because hydrogen ions and carbon dioxide are produced in rapidly metabolizing tissues, the Bohr effect helps deliver oxygen to sites where it is most needed.

7.4 Mutations in Genes Encoding Hemoglobin Subunits Can Result in Disease

Sickle-cell disease is caused by a mutation in the β chain of hemoglobin that substitutes a valine residue for a glutamate residue. As a result, a hydrophobic patch forms on the surface of deoxy (T-state) hemoglobin that leads to the formation of fibrous polymers. These fibers distort red blood cells into sickle shapes. Sickle-cell disease was the first disease to be associated with a change in the amino acid sequence of a protein. Thalassemias are diseases caused by the reduced production of either the α or the β chains of hemoglobin. Hemoglobin tetramers are produced that contain only one type of hemoglobin chain. Such hemoglobin molecules are characterized by poor oxygen release and low solubility, leading to the destruction of red blood cells in the course of their development. Red-blood-cell precursors normally produce a slight excess of hemoglobin α chains compared with β chains. To prevent the aggregation of the excess α chains, they produce α-hemoglobin stabilizing protein, which binds specifically to α-chain monomers to form a soluble complex.

APPENDIX: Binding Models Can Be Formulated in Quantitative Terms: The Hill Plot and the Concerted Model

The Hill Plot

A useful way of quantitatively describing cooperative binding processes such as that for hemoglobin was developed by Archibald Hill in 1913. Consider the *hypothetical* equilibrium for a protein X binding a ligand S:

$$X + nS \rightleftharpoons X(S)_n \qquad (1)$$

where n is a variable that can take on both integral and fractional values. The parameter n is a measure of the degree of cooperativity in ligand binding, although it does not have deeper significance, because equation 1 does not represent an actual physical process. For X = hemoglobin and S = O_2, the maximum value of n is 4. The value of $n = 4$ would apply if oxygen binding by hemoglobin were completely cooperative. If oxygen binding were completely noncooperative, then n would be 1.

Analysis of the equilibrium in equation 1 yields the following expression for the fractional saturation, Y:

$$Y = \frac{[S]^n}{[S]^n + [S_{50}]^n}$$

where $[S_{50}]$ is the concentration at which X is half-saturated. For hemoglobin, this expression becomes

$$Y = \frac{pO_2{}^n}{pO_2{}^n + P_{50}{}^n}$$

where P_{50} is the partial pressure of oxygen at which hemoglobin is half-saturated. This expression can be rearranged to:

$$\frac{Y}{1 - Y} = \frac{pO_2{}^n}{P_{50}{}^n}$$

and so

$$\log\left(\frac{Y}{1 - Y}\right) = \log\left(\frac{pO_2{}^n}{P_{50}{}^n}\right) = n\log(pO_2) - n\log(P_{50})$$

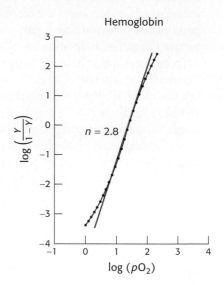

Figure 7.28 Hill plots for myoglobin and hemoglobin.

This equation predicts that a plot of $\log(Y/1 - Y)$ versus $\log(P_{50})$, called a *Hill plot,* should be linear with a slope of n.

Hill plots for myoglobin and hemoglobin are shown in Figure 7.28. For myoglobin, the Hill plot is linear with a slope of 1. For hemoglobin, the Hill plot is not completely linear, because the equilibrium on which the Hill plot is based is not entirely correct. However, the plot is approximately linear in the center with a slope of 2.8. The slope, often referred to as the *Hill coefficient,* is a measure of the cooperativity of oxygen binding. The utility of the Hill plot is that it provides a simply derived quantitative assessment of the degree of cooperativity in binding. With the use of the Hill equation and the derived Hill coefficient, a binding curve that closely resembles that for hemoglobin is produced (Figure 7.29).

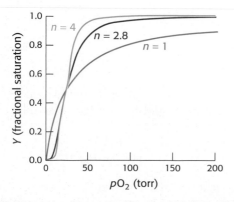

Figure 7.29 Oxygen-binding curves for several Hill coefficients. The curve labeled $n = 2.8$ closely resembles the curve for hemoglobin.

The Concerted Model

The concerted model can be formulated in quantitative terms. Only four parameters are required: (1) the number of binding sites (assumed to be equivalent) in the protein, (2) the ratio of the concentrations of the T and R states in the absence of bound ligands, (3) the affinity of sites in proteins in the R state for ligand binding, and (4) a measure of how much more tightly subunits in proteins in the R state bind ligands compared with subunits in the T state. The number of binding sites, n, is usually known from other information. For hemoglobin, $n = 4$. The ratio of the concentrations of the T and R states with no ligands bound is a constant:

$$L = [T_0]/[R_0]$$

where the subscript refers to the number of ligands bound (in this case, zero). The affinity of subunits in the R state is defined by the dissociation constant for a ligand binding to a single site in the R state, K_R. Similarly, the dissociation constant for a ligand binding to a single site in the T state is K_T. We can define the ratio of these two dissociation constant as

$$c = K_R/K_T$$

This is the measure of how much more tightly a subunit for a protein in the R state binds a ligand compared with a subunit for a protein in the T state. Note that $c < 1$ because K_R and K_T are dissociation constants and tight binding corresponds to a small dissociation constant.

What is the ratio of the concentration of T-state proteins with one ligand bound to the concentration of R-state proteins with one ligand bound? The dissociation constant for a single site in the R state is K_R. For a protein with n sites, there are n possible sites for the first ligand to bind. This statistical factor favors ligand binding compared with a single-site protein. Thus, $[R_1] = n[R_0][S]/K_R$. Similarly, $[T_1] = n[T_0][S]/K_T$. Thus,

$$[T_1]/[R_1] = \frac{n[T_0][S]/K_T}{n[R_0][S]/K_R} = \frac{[T_0]}{[R_0](K_R/K_T)} = cL$$

Similar analysis reveals that, for states with i ligands bound, $[T_i]/[R_i] = c^i L$. In other words, the ratio of the concentrations of the T state to the R state is reduced by a factor of c for each ligand that binds.

Let us define a convenient scale for the concentration of S:

$$\alpha = [S]/K_R$$

This definition is useful because it is the ratio of the concentration of S to the dissociation constant that determines the extent of binding. Using this definition, we see that

$$[R_1] = \frac{n[R_0][S]}{K_R} = n[R_0]\alpha$$

Similarly,

$$[T_1] = \frac{n[T_0][S]}{K_T} = ncL[R_0]\alpha$$

What is the concentration of R-state molecules with two ligands bound? Again, we must consider the statistical factor—that is, the number of ways in which a second ligand can bind to a molecule with one site occupied. The number of ways is $n - 1$. However, because which ligand is the "first" and which is the "second" does not matter, we must divide by a factor of 2. Thus,

$$[R_2] = \frac{\left(\dfrac{n-1}{2}\right)[R_1][S]}{K_R}$$

$$= \left(\frac{n-1}{2}\right)[R_1]\alpha$$

$$= \left(\frac{n-1}{2}\right)(n[R_0]\alpha)\alpha$$

$$= n\left(\frac{n-1}{2}\right)[R_0]\alpha^2$$

We can derive similar equations for the case with i ligands bound and for T states.

We can now calculate the fractional saturation, Y. This is the total concentration of sites with ligands bound divided by the total concentration of potential binding sites. Thus,

$$Y = \frac{([R_1] + [T_1]) + 2([R_2] + [T_2]) + \cdots + n([R_n] + [T_n])}{n([R_0] + [T_0] + [R_1] + [T_1] + \cdots + [R_n] + [T_n])}$$

Substituting into this equation, we find

$$Y = \frac{\begin{aligned}&n[R_0]\alpha + nc[T_0]\alpha + 2(n(n-1)/2)[R_0]\alpha^2 \\ &+ 2(n(n-1)/2)c^2[T_0]\alpha^2 + \cdots + n[R_0]\alpha^n + nc^n[T_0])\alpha^n\end{aligned}}{\begin{aligned}&n([R_0] + [T_0] + n[R_0]\alpha + nc[T_0]\alpha + \cdots \\ &+ [R_0]\alpha^n + c^n[T_0]\alpha^n)\end{aligned}}$$

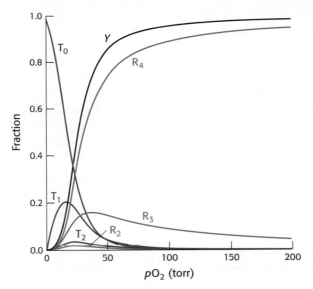

Figure 7.30 Modeling oxygen binding with the concerted model. The fractional saturation (Y) as a function pO_2: $L = 9000$, $c = 0.014$, and $K_R = 2.5$ torr. The fraction of molecules in the T state with zero, one, and two oxygen molecules bound (T_0, T_1, and T_2) and the fraction of molecules in the R state with two, three, and four oxygen molecules bound (R_2, R_3, and R_4) are shown. The fractions of molecules in other forms are too low to be shown.

Substituting $[T_0] = L[R_0]$ and summing these series yields

$$Y = \frac{\alpha(1 + \alpha)^{n-1} + Lc\alpha(1 + c\alpha)^{n-1}}{(1 + \alpha)^n + L(1 + c\alpha)^n}$$

We can now use this equation to fit the observed data for hemoglobin by varying the parameters L, c, and K_R (with $n = 4$). An excellent fit is obtained with $L = 9000$, $c = 0.014$, and $K_R = 2.5$ torr (Figure 7.30).

In addition to the fractional saturation, the concentrations of the species T_0, T_1, T_2, R_2, R_3, and R_4 are shown. The concentrations of all other species are very low. The addition of concentrations is a major difference between the analysis using the Hill equation and this analysis of the concerted model. The Hill equation gives only the fractional saturation, whereas the analysis of the concerted model yields concentrations for all species. In the present case, this analysis yields the expected ratio of T-state proteins to R-state proteins at each stage of binding. This ratio changes from 9000 to 126 to 1.76 to 0.025 to 0.00035 with zero, one, two, three, and four oxygen molecules bound. This ratio provides a quantitative measure of the switching of the population of hemoglobin molecules from the T state to the R state.

The sequential model can also be formulated in quantitative terms. However, the formulation entails many more parameters, and many different sets of parameters often yield similar fits to the experimental data.

Key Terms

heme (p. 184)

protoporphyrin (p. 184)

proximal histidine (p. 185)

functional magnetic resonance imaging (fMRI) (p. 185)

superoxide anion (p. 185)

metmyoglobin (p. 185)

distal histidine (p. 186)

α chain (p. 186)

β chain (p. 186)

globin fold (p. 186)

αβ dimer (p. 187)

oxygen-binding curve (p. 187)

fractional saturation (p. 187)

partial pressure (p. 187)

sigmoid (p. 187)

cooperative (p. 187)

T state (p. 188)

R state (p. 188)

concerted model (MWC model) (p. 189)

sequential model (p. 190)

2,3-bisphosphoglycerate (p. 191)

fetal hemoglobin (p. 192)

Bohr effect (p. 192)

carbonic anhydrase (p. 193)

carbamate (p. 194)

sickle-cell anemia (p. 194)

hemoglobin S (p. 195)

malaria (p. 196)

thalassemia (p. 196)

hemoglobin H (p. 196)

thalassemia major (Cooley anemia) (p. 196)

α-hemoglobin stabilizing protein (AHSP) (p. 197)

neuroglobin (p. 197)

cytoglobin (p. 197)

Hill plot (p. 200)

Hill coefficient (p. 200)

Selected Readings

Where to Start

Perutz, M. F. 1978. Hemoglobin structure and respiratory transport. *Sci. Am.* 239(6):92–125.

Perutz, M. F. 1980. Stereochemical mechanism of oxygen transport by haemoglobin. *Proc. R. Soc. Lond. Biol. Sci.* 208:135–162.

Kilmartin, J. V. 1976. Interaction of haemoglobin with protons, CO_2, and 2,3-diphosphoglycerate. *Brit. Med. Bull.* 32:209–222.

Structure

Kendrew, J. C., Bodo, G., Dintzis, H. M., Parrish, R. G., Wyckoff, H., and Phillips, D. C. 1958. A three-dimensional model of the myoglobin molecule obtained by x-ray analysis. *Nature* 181:662–666.

Shaanan, B. 1983. Structure of human oxyhaemoglobin at 2.1 Å resolution. *J. Mol. Biol.* 171:31–59.

Frier, J. A., and Perutz, M. F. 1977. Structure of human foetal deoxyhaemoglobin. *J. Mol. Biol.* 112:97–112.

Perutz, M. F. 1969. Structure and function of hemoglobin. *Harvey Lect.* 63:213–261.

Perutz, M. F. 1962. Relation between structure and sequence of haemoglobin. *Nature* 194:914–917.

Interaction of Hemoglobin with Allosteric Effectors

Benesch, R., and Beesch, R. E. 1969. Intracellular organic phosphates as regulators of oxygen release by haemoglobin. *Nature* 221:618–622.

Fang, T. Y., Zou, M., Simplaceanu, V., Ho, N. T., and Ho, C. 1999. Assessment of roles of surface histidyl residues in the molecular basis of the Bohr effect and of β 143 histidine in the binding of 2,3-bisphosphoglycerate in human normal adult hemoglobin. *Biochemistry* 38:13423–13432.

Arnone, A. 1992. X-ray diffraction study of binding of 2,3-diphosphoglycerate to human deoxyhaemoglobin. *Nature* 237:146–149.

Models for Cooperativity

Monod, J., Wyman, J., and Changeux, J.-P. 1965. On the nature of allosteric interactions: A plausible model. *J. Mol. Biol.* 12:88–118.

Koshland, D. L., Jr., Nemethy, G., and Filmer, D. 1966. Comparison of experimental binding data and theoretical models in proteins containing subunits. *Biochemistry* 5:365–385.

Ackers, G. K., Doyle, M. L., Myers, D., and Daugherty, M. A. 1992. Molecular code for cooperativity in hemoglobin. *Science* 255:54–63.

Sickle-Cell Anemia and Thalasssemia

Herrick, J. B. 1910. Peculiar elongated and sickle-shaped red blood corpuscles in a case of severe anemia. *Arch. Intern. Med.* 6:517–521.

Pauling, L., Itano, H. A., Singer, S. J., and Wells, L. C. 1949. Sickle cell anemia: A molecular disease. *Science* 110:543–548.

Ingram, V. M. 1957. Gene mutation in human hemoglobin: The chemical difference between normal and sickle cell haemoglobin. *Nature* 180:326–328.

Eaton, W. A., and Hofrichter, J. 1990. Sickle cell hemoglobin polymerization. *Adv. Prot. Chem.* 40:63–279.

Weatherall, D. J. 2001. Phenotype genotype relationships in monogenic disease: Lessons from the thalassemias. *Nat. Rev. Genet.* 2:245–255.

Globin-Binding Proteins and Other Globins

Kihm, A. J., Kong, Y., Hong, W., Russell, J. E., Rouda, S., Adachi, K., Simon, M. C., Blobel, G. A., and Weiss, M. J. 2002. An abundant erythroid protein that stabilizes free α-haemoglobin. *Nature* 417:758–763.

Feng, L., Zhou, S., Gu, L., Gell, D. A., Mackay, J. P., Weiss, M. J., Gow, A. J., and Shi, Y. 2005. Structure of oxidized α-haemoglobin bound to AHSP reveals a protective mechanism for haem. *Nature* 435:697–701.

Burmester, T., Haberkamp, M., Mitz, S., Roesner, A., Schmidt, M., Ebner, B., Gerlach, F., Fuchs, C., and Hankeln, T. 2004. Neuroglobin and cytoglobin: Genes, proteins and evolution. *IUBMB Life* 56:703–707.

Hankeln, T., Ebner, B., Fuchs, C., Gerlach, F., Haberkamp, M., Laufs, T. L., Roesner, A., Schmidt, M., Weich, B., Wystub, S., Saaler-Reinhardt, S., Reuss, S., Bolognesi, M., De Sanctis, D., Marden, M. C., Kiger, L., Moens, L., Dewilde, S., Nevo, E., Avivi, A., Weber, R. E., Fago, A., and Burmester, T. 2005. Neuroglobin and cytoglobin in search of their role in the vertebrate globin family. *J. Inorg. Biochem.* 99:110–119.

Burmester, T., Ebner, B., Weich, B., and Hankeln, T. 2002. Cytoglobin: A novel globin type ubiquitously expressed in vertebrate tissues. *Mol. Biol. Evol.* 19:416–421.

Zhang, C., Wang, C., Deng, M., Li, L., Wang, H., Fan, M., Xu, W., Meng, F., Qian, L., and He, F. 2002. Full-length cDNA cloning of human neuroglobin and tissue expression of rat neuroglobin. *Biochem. Biophys. Res. Commun.* 290:1411–9.

Problems

1. *Screening the biosphere.* The first protein structure to have its structure determined was myoglobin from sperm whale. Propose an explanation for the observation that sperm whale muscle is a rich source of this protein.

2. *Carrying a load.* Suppose that you are climbing a high mountain and the oxygen partial pressure in the air is reduced to 75 torr. Estimate the percentage of the oxygen-carrying capacity that will be utilized, assuming that the pH of both tissues and lungs is 7.4 and that the oxygen concentration in the tissues is 20 torr.

3. *High-altitude adaptation.* After spending a day or more at high altitude (with an oxygen partial pressure of 75 torr), the concentration of 2,3-bisphosphoglycerate (2,3-BPG) in red blood cells increases. What effect would an increased concentration of 2,3-BPG have on the oxygen-binding curve for hemoglobin Propose an explanation for why this adaptation would be beneficial for functioning well at high altitude.

4. *I'll take the lobster.* Arthropods such as lobsters have oxygen carriers quite different from hemoglobin. The oxygen-binding sites do not contain heme but, instead, are based on two copper(I) ions. The structural changes that accompany oxygen binding are shown here. How might these changes be used to facilitate cooperative oxygen binding?

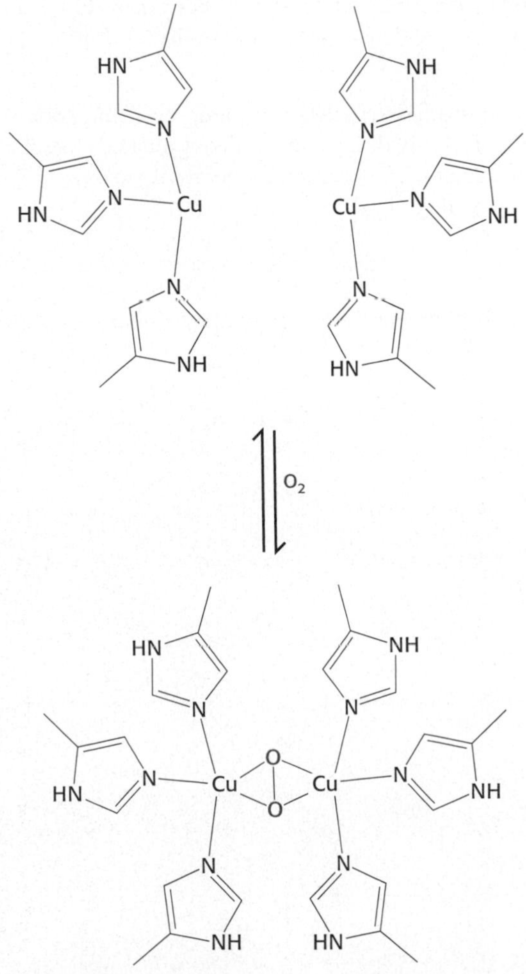

5. *A disconnect.* With the use of site-directed mutagenesis, hemoglobin has been prepared in which the proximal histidine residues in both the α and the β subunits have been replaced by glycine. The imidazole ring from the histidine residue can be replaced by adding free imidazole in solution. Would you expect this modified hemoglobin to show cooperativity in oxygen binding? Why or why not?

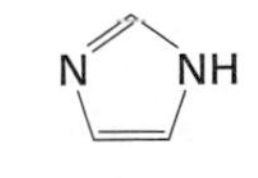

Imidazole

6. *Successful substitution.* Blood cells from some birds do not contain 2,3-bisphosphoglycerate but, instead, contain one of the compounds in parts *a* through *d*, which plays an analogous functional role. Which compound do you think is most likely to play this role? Explain briefly.

(a) 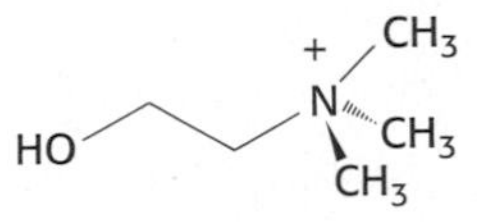 **Choline**

(b) 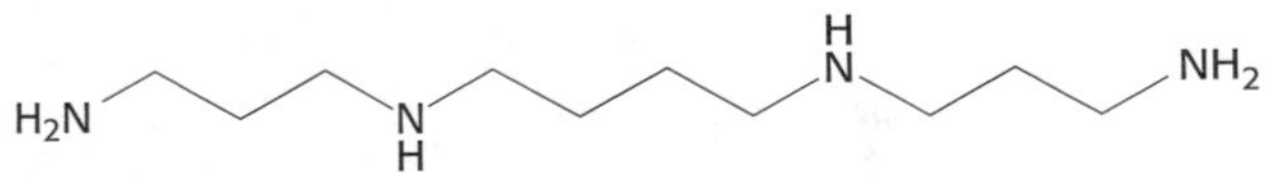 **Spermine**

(c) 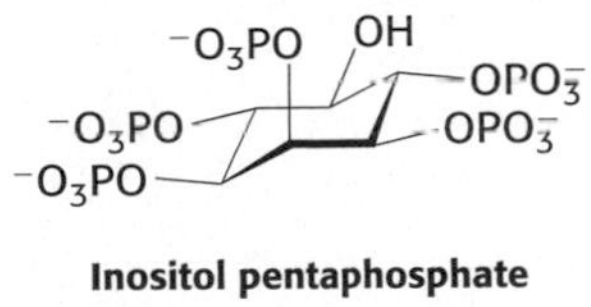 **Inositol pentaphosphate**

(d) 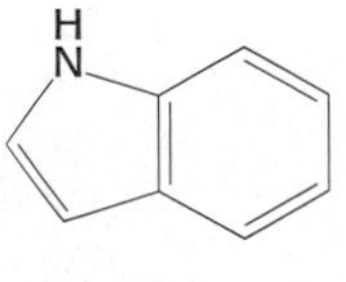 **Indole**

7. *Theoretical curves.* (a) Using the Hill equation, plot an oxygen-binding curve for a hypothetical two-subunit hemoglobin with $n = 1.8$ and $P_{50} = 10$ torr. (b) Repeat by using the concerted model with $n = 2$, $L = 1000$, $c = 0.01$, and $K_R = 1$ torr.

8. *Parasitic effect.* When *P. falciparum* lives inside red blood cells, the metabolism of the parasite tends to release acid. What effect is the presence of acid likely to have on the oxygen-carrying capacity of the red blood cells? On the likelihood that these cells sickle?

9. *Picket-fence porphyrin.* When free heme is exposed to oxygen, an oxo-bridged dimeric species forms readily and irreversibly.

$$2 \; \boxed{Fe^{2+}} \; + \; 1/_2O_2 \longrightarrow \boxed{Fe^{3+}} - O - \boxed{Fe^{3+}}$$

Heme   **Dimer**

Chemists have designed and synthesized a special heme derivative based on a "picket fence" porphyrin that binds oxygen reversibly without forming dimeric species.

Picket-fence porphyrin complex

Propose a basis for the design and an explanation for the reversible oxygen binding.

Data Interpretation Problem

10. *Primitive oxygen binding.* Lampreys are primitive organisms whose ancestors diverged from the ancestors of fish and mammals approximately 400 million years ago. Lamprey blood contains a hemoglobin related to mammalian hemoglobin. However, lamprey hemoglobin is monomeric in the oxygenated state. Oxygen-binding data for lamprey hemoglobin are as follows:

pO_2	Y	pO_2	Y	pO_2	Y
0.1	.0060	2.0	.112	50.0	.889
0.2	.0124	3.0	.170	60.0	.905
0.3	.0190	4.0	.227	70.0	.917
0.4	.0245	5.0	.283	80.0	.927
0.5	.0307	7.5	.420	90.0	.935
0.6	.0380	10.0	.500	100	.941
0.7	.0430	15.0	.640	150	.960
0.8	.0481	20.0	.721	200	.970
0.9	.0530	30.0	.812		
1.0	.0591	40.0	.865		

(a) Plot these data to produce an oxygen-binding curve. At what oxygen partial pressure is this hemoglobin half-saturated? On the basis of the appearance of this curve, does oxygen binding seem to be cooperative?

(b) Construct a Hill plot by using these data. Does the Hill plot show any evidence for cooperativity? What is the Hill coefficient?

(c) Further studies revealed that lamprey hemoglobin forms oligomers, primarily dimers, in the deoxygenated state. Propose a model to explain any observed cooperativity in oxygen binding by lamprey hemoglobin.

Enzymes: Basic Concepts and Kinetics

$$\text{O}_2, \text{Ca}^{2+} \xrightarrow{\text{Aequorin}}$$

$+ \text{CO}_2 + \text{light} \,(466\ \text{nm})$

The activity of an enzyme is responsible for the glow of the luminescent jellyfish at left. The enzyme aequorin catalyzes the oxidation of a compound by oxygen in the presence of calcium to release CO_2 and light. [(Left) Fred Bavendam/Peter Arnold.]

Enzymes, the catalysts of biological systems, are remarkable molecular devices that determine the patterns of chemical transformations. They also mediate the transformation of one form of energy into another. The most striking characteristics of enzymes are their *catalytic power* and *specificity*. Catalysis takes place at a particular site on the enzyme called the *active site. Nearly all known enzymes are proteins.* However, proteins do not have an absolute monopoly on catalysis; the discovery of catalytically active RNA molecules provides compelling evidence that RNA was a biocatalyst early in evolution.

Proteins as a class of macromolecules are highly effective catalysts for an enormous diversity of chemical reactions because of their capacity *to specifically bind a very wide range of molecules.* By utilizing the full repertoire of intermolecular forces, enzymes bring substrates together in an optimal orientation, the prelude to making and breaking chemical bonds. They catalyze reactions *by stabilizing transition states,* the highest-energy species in reaction pathways. By selectively stabilizing a transition state, an enzyme determines which one of several potential chemical reactions actually takes place.

8.1 Enzymes Are Powerful and Highly Specific Catalysts

Enzymes accelerate reactions by factors of as much as a million or more (Table 8.1). Indeed, most reactions in biological systems do not take place at perceptible rates in the absence of enzymes. Even a reaction as simple as the hydration of carbon dioxide is catalyzed by an enzyme—namely, carbonic anhydrase (Section 9.2). The transfer of CO_2 from the tissues into the blood and then to the alveolar air would be less complete in the absence of this enzyme. In fact, carbonic anhydrase is one of the fastest enzymes known. Each enzyme molecule can hydrate 10^6 molecules of CO_2 *per second*. This catalyzed reaction is 10^7 times as fast as the uncatalyzed one. We will consider the mechanism of carbonic anhydrase catalysis in Chapter 9.

Enzymes are highly specific both in the reactions that they catalyze and in their choice of reactants, which are called *substrates*. An enzyme usually catalyzes a single chemical reaction or a set of closely related reactions. Side reactions leading to the wasteful formation of by-products are rare in enzyme-catalyzed reactions, in contrast with uncatalyzed ones.

Let us consider *proteolytic enzymes* as an example. In vivo, these enzymes catalyze *proteolysis,* the hydrolysis of a peptide bond.

Peptide   **Carboxyl component**   **Amino component**

Most proteolytic enzymes also catalyze a different but related reaction in vitro—namely, the hydrolysis of an ester bond. Such reactions are more easily monitored than is proteolysis and are useful in experimental investigations of these enzymes (p. 244).

Ester   **Acid**   **Alcohol**

Proteolytic enzymes differ markedly in their degree of substrate specificity. Papain, which is found in papaya plants, is quite undiscriminating: it will cleave any peptide bond with little regard to the identity of the adjacent

TABLE 8.1 Rate enhancement by selected enzymes

Enzyme	Nonenzymatic half-life		Uncatalyzed rate $(k_{un} s^{-1})$	Catalyzed rate $(k_{cat} s^{-1})$	Rate enhancement $(k_{cat} s^{-1}/k_{un} s^{-1})$
OMP decarboxylase	78,000,000	years	2.8×10^{-16}	39	1.4×10^{17}
Staphylococcal nuclease	130,000	years	1.7×10^{-13}	95	5.6×10^{14}
AMP nucleosidase	69,000	years	1.0×10^{-11}	60	6.0×10^{12}
Carboxypeptidase A	7.3	years	3.0×10^{-9}	578	1.9×10^{11}
Ketosteroid isomerase	7	weeks	1.7×10^{-7}	66,000	3.9×10^{11}
Triose phosphate isomerase	1.9	days	4.3×10^{-6}	4,300	1.0×10^{9}
Chorismate mutase	7.4	hours	2.6×10^{-5}	50	1.9×10^{6}
Carbonic anhydrase	5	seconds	1.3×10^{-1}	1×10^{6}	7.7×10^{6}

Abbreviations: OMP, orotidine monophosphate; AMP, adenosine monophosphate.
Source: After A. Radzicka and R. Wolfenden. *Science* 267 (1995):90–93.

side chains. This lack of specificity accounts for its use in meat-tenderizing sauces. Trypsin, a digestive enzyme, is quite specific and catalyzes the splitting of peptide bonds only on the carboxyl side of lysine and arginine residues (Figure 8.1A). Thrombin, an enzyme that participates in blood clotting, is even more specific than trypsin (p. 293). It catalyzes the hydrolysis of Arg–Gly bonds in particular peptide sequences only (Figure 8.1B).

DNA polymerase I, a template-directed enzyme (Section 28.3), is another highly specific catalyst. To a DNA strand that is being synthesized, it adds nucleotides in a sequence determined by the sequence of nucleotides in another DNA strand that serves as a template. DNA polymerase I is remarkably precise in carrying out the instructions given by the template. It inserts the wrong nucleotide into a new DNA strand less than one in a thousand times. *The specificity of an enzyme is due to the precise interaction of the substrate with the enzyme. This precision is a result of the intricate three-dimensional structure of the enzyme protein.*

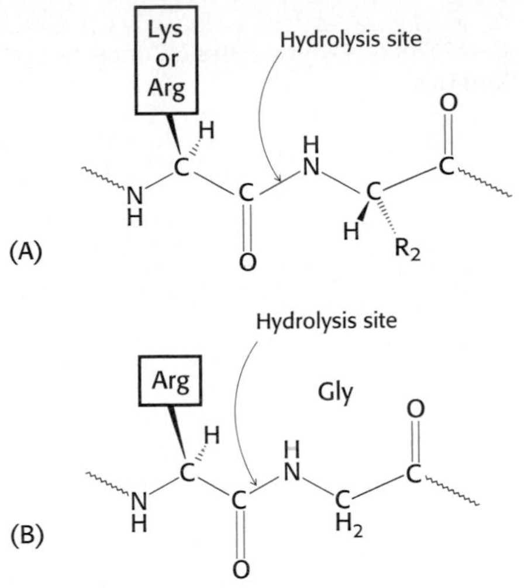

Many Enzymes Require Cofactors for Activity

The catalytic activity of many enzymes depends on the presence of small molecules termed *cofactors*, although the precise role varies with the cofactor and the enzyme. Generally, these cofactors are able to execute chemical reactions that cannot be performed by the standard set of twenty amino acids. An enzyme without its cofactor is referred to as an *apoenzyme;* the complete, catalytically active enzyme is called a *holoenzyme.*

$$\text{Apoenzyme} + \text{cofactor} = \text{holoenzyme}$$

Cofactors can be subdivided into two groups: (1) metals and (2) small organic molecules called *coenzymes* (Table 8.2). Often derived from vitamins, coenzymes can be either tightly or loosely bound to the enzyme. Tightly bound coenzymes are called *prosthetic groups.* Loosely associated coenzymes are more like cosubstrates because, like substrates and products, they bind to the enzyme and are released from it. The use of the same coenzyme by a variety of enzymes sets coenzymes apart from normal substrates, however, as does their source in vitamins. Enzymes that use the same coenzyme usually perform catalysis by similar mechanisms. In Chapter 9, we will examine the importance of metals to enzyme activity and, throughout the book, we will see how coenzymes and their enzyme partners operate in their biochemical context. A general discussion of vitamins can be found on page 423.

Enzymes May Transform Energy from One Form into Another

In many biochemical reactions, *the energy of the reactants is converted with high efficiency into a different form.* For example, in photosynthesis, light energy is converted into chemical-bond energy. In mitochondria, the free energy contained in small molecules derived from food is converted first into the free energy of an ion gradient and then into a different currency, the free energy of adenosine triphosphate. Enzymes may then use the chemical-bond energy of ATP in many ways, including to further transform energy. For instance, the enzyme myosin converts the energy of ATP into the mechanical energy of contracting muscles (Chapter 34). Pumps in the membranes of cells and organelles, which can be thought of as enzymes that

Figure 8.1 Enzyme specificity. (A) Trypsin cleaves on the carboxyl side of arginine and lysine residues, whereas (B) thrombin cleaves Arg–Gly bonds in particular sequences only.

TABLE 8.2 Enzyme cofactors

Cofactor	Enzyme
Coenzyme	
Thiamine pyrophosphate	Pyruvate dehydrogenase
Flavin adenine nucleotide	Monoamine oxidase
Nicotinamide adenine dinucleotide	Lactate dehydrogenase
Pyridoxal phosphate	Glycogen phosphorylase
Coenzyme A (CoA)	Acetyl CoA carboxylase
Biotin	Pyruvate carboxylase
5′-Deoxyadenosyl cobalamin	Methylmalonyl mutase
Tetrahydrofolate	Thymidylate synthase
Metal	
Zn^{2+}	Carbonic anhydrase
Zn^{2+}	Carboxypeptidase
Mg^{2+}	*Eco*RV
Mg^{2+}	Hexokinase
Ni^{2+}	Urease
Mo	Nitrate reductase
Se	Glutathione peroxidase
Mn	Superoxide dismutase
K^+	Propionyl CoA carboxylase

move substrates rather than chemically alter them, use the energy of ATP to transport molecules and ions across the membrane (Chapter 13). The chemical and electrical gradients resulting from the unequal distribution of these molecules and ions are themselves forms of energy that can be used for a variety of purposes, such as sending nerve impulses.

The molecular mechanisms of these energy-transducing enzymes are being unraveled. We will see in subsequent chapters how unidirectional cycles of discrete steps—binding, chemical transformation, and release—lead to the conversion of one form of energy into another.

8.2 Free Energy Is a Useful Thermodynamic Function for Understanding Enzymes

Some of the principles of thermodynamics were introduced in Chapter 1— notably the idea of *free energy (G)*. To understand how enzymes operate, we need to consider only two thermodynamic properties of the reaction: (1) the free-energy difference (ΔG) between the products and reactants and (2) the energy required to initiate the conversion of reactants into products. The former determines whether the reaction will occur spontaneously, whereas the latter determines the rate of the reaction. Enzymes affect only the latter. Let us review some of the principles of thermodynamics as they apply to enzymes.

The Free-Energy Change Provides Information About the Spontaneity but Not the Rate of a Reaction

As discussed on page 12, the free-energy change of a reaction (ΔG) tells us if the reaction can occur spontaneously:

1.   *A reaction can occur spontaneously only if ΔG is negative.* Such reactions are said to be *exergonic*.

2.   A system is at equilibrium and no *net* change can take place if ΔG is zero.

3.   A reaction cannot occur spontaneously if ΔG is positive. An input of free energy is required to drive such a reaction. These reactions are termed *endergonic*.

4.   The ΔG of a reaction depends only on the free energy of the products (the final state) minus the free energy of the reactants (the initial state). *The ΔG of a reaction is independent of the path (or molecular mechanism) of the transformation.* The mechanism of a reaction has no effect on ΔG. For example, the ΔG for the oxidation of glucose to CO_2 and H_2O is the same whether it occurs by combustion or by a series of enzyme-catalyzed steps in a cell.

5.   *The ΔG provides no information about the rate of a reaction.* A negative ΔG indicates that a reaction *can* occur spontaneously, but it does not signify whether it will proceed at a perceptible rate. As will be discussed shortly (Section 8.3), the rate of a reaction depends on the *free energy of activation* ($\Delta G^{\ddagger}$), which is largely unrelated to the ΔG of the reaction.

The Standard Free-Energy Change of a Reaction Is Related to the Equilibrium Constant

As for any reaction, we need to be able to determine ΔG for an enzyme-catalyzed reaction to know whether the reaction is spontaneous or an input of energy is required. To determine this important thermodynamic parameter, we need to take into account the nature of both the reactants and the products as well as their concentrations.

Consider the reaction

$$A + B \rightleftharpoons C + D$$

The ΔG of this reaction is given by

$$\Delta G = \Delta G^\circ + RT \ln \frac{[C][D]}{[A][B]} \tag{1}$$

in which ΔG° is the *standard free-energy change*, R is the gas constant, T is the absolute temperature, and [A], [B], [C], and [D] are the molar concentrations (more precisely, the activities) of the reactants. ΔG° is the free-energy change for this reaction under standard conditions—that is, when each of the reactants A, B, C, and D is present at a concentration of 1.0 M (for a gas, the standard state is usually chosen to be 1 atmosphere). Thus, the ΔG of a reaction depends on the *nature* of the reactants (expressed in the ΔG° term of equation 1) and on their *concentrations* (expressed in the logarithmic term of equation 1).

A convention has been adopted to simplify free-energy calculations for biochemical reactions. The standard state is defined as having a pH of 7. Consequently, when H^+ is a reactant, its activity has the value 1 (corresponding to a pH of 7) in equations 1 and 4 (below). The activity of water also is taken to be 1 in these equations. The *standard free-energy change at pH 7,* denoted by the symbol $\Delta G^{\circ\prime}$ will be used throughout this book. The *kilojoule* (abbreviated *kJ*) and the *kilocalorie* (*kcal*) will be used as the units of energy. One kilojoule is equivalent to 0.239 kilocalorie.

The relation between the standard free energy and the equilibrium constant of a reaction can be readily derived. This equation is important because it displays the energetic relation between products and reactants in terms of their concentrations. At equilibrium, $\Delta G = 0$. Equation 1 then becomes

$$0 = \Delta G^{\circ\prime} + RT \ln \frac{[C][D]}{[A][B]} \tag{2}$$

and so

$$\Delta G^{\circ\prime} = -RT \ln \frac{[C][D]}{[A][B]} \tag{3}$$

The equilibrium constant under standard conditions, K'_{eq}, is defined as

$$K'_{eq} = \frac{[C][D]}{[A][B]} \tag{4}$$

Substituting equation 4 into equation 3 gives

$$\Delta G^{\circ\prime} = -RT \ln K'_{eq} \tag{5}$$

$$\Delta G^{\circ\prime} = -2.303 \, RT \log_{10} K'_{eq} \tag{6}$$

which can be rearranged to give

$$K'_{eq} = 10^{-\Delta G^{\circ\prime}/2.303RT} \tag{7}$$

Units of energy

A *kilojoule* (kJ) is equal to 1000 J.

A *joule* (J) is the amount of energy needed to apply a 1-newton force over a distance of 1 meter.

A *kilocalorie* (kcal) is equal to 1000 cal.

A *calorie* (cal) is equivalent to the amount of heat required to raise the temperature of 1 gram of water from 14.5°C to 15.5°C.

1 kJ = 0.239 kcal

Dihydroxyacetone phosphate (DHAP)

Glyceraldehyde 3-phosphate (GAP)

TABLE 8.3 Relation between $\Delta G^{\circ\prime}$ and K'_{eq} (at 25°C)

K'_{eq}	$\Delta G^{\circ\prime}$	
	kJ mol^{-1}	kcal mol^{-1}
10^{-5}	28.53	6.82
10^{-4}	22.84	5.46
10^{-3}	17.11	4.09
10^{-2}	11.42	2.73
10^{-1}	5.69	1.36
1	0	0
10	−5.69	−1.36
10^{2}	−11.42	−2.73
10^{3}	−17.11	−4.09
10^{4}	−22.84	−5.46
10^{5}	−28.53	−6.82

Substituting $R = 8.315 \times 10^{-3}$ kJ mol^{-1} deg^{-1} and $T = 298$ K (corresponding to 25°C) gives

$$K'_{eq} = 10^{-\Delta G^{\circ\prime}/5.69} \qquad (8)$$

where $\Delta G^{\circ\prime}$ is here expressed in kilojoules per mole because of the choice of the units for R in equation 7. Thus, the standard free energy and the equilibrium constant of a reaction are related by a simple expression. For example, an equilibrium constant of 10 gives a standard free-energy change of -5.69 kJ mol^{-1} (-1.36 kcal mol^{-1}) at 25°C (Table 8.3). Note that, for each 10-fold change in the equilibrium constant, the $\Delta G^{\circ\prime}$ changes by 5.69 kJ mol^{-1} (1.36 kcal mol^{-1}).

As an example, let us calculate $\Delta G^{\circ\prime}$ and ΔG for the isomerization of dihydroxyacetone phosphate (DHAP) to glyceraldehyde 3-phosphate (GAP). This reaction takes place in glycolysis (p. 438). At equilibrium, the ratio of GAP to DHAP is 0.0475 at 25°C (298 K) and pH 7. Hence, $K'_{eq} = 0.0475$. The standard free-energy change for this reaction is then calculated from equation 6:

$$\begin{aligned} \Delta G^{\circ\prime} &= -2.303\, RT \log_{10} K'_{eq} \\ &= -2.303 \times 8.315 \times 10^{-3} \times 298 \times \log_{10}(0.0475) \\ &= +7.53 \text{ kJ mol}^{-1}\ (+1.80 \text{ kcal mol}^{-1}) \end{aligned}$$

Under these conditions, the reaction is endergonic. DHAP will not spontaneously convert into GAP.

Now let us calculate ΔG for this reaction when the initial concentration of DHAP is 2×10^{-4} M and the initial concentration of GAP is 3×10^{-6} M. Substituting these values into equation 1 gives

$$\begin{aligned} \Delta G &= 7.53 \text{ kJ mol}^{-1} + 2.303 RT \log_{10} \frac{3 \times 10^{-6}\text{ M}}{2 \times 10^{-4}\text{ M}} \\ &= 7.53 \text{ kJ mol}^{-1} - 10.42 \text{ kJ mol}^{-1} \\ &= -2.89 \text{ kJ mol}^{-1}\ (-0.69 \text{ kcal mol}^{-1}) \end{aligned}$$

This negative value for the ΔG indicates that the isomerization of DHAP to GAP is exergonic and can occur spontaneously when these species are present at the above concentrations. Note that ΔG for this reaction is negative, although $\Delta G^{\circ\prime}$ is positive. *It is important to stress that whether the ΔG for a reaction is larger, smaller, or the same as $\Delta G^{\circ\prime}$ depends on the concentrations of the reactants and products.* The criterion of spontaneity for a reaction is ΔG, not $\Delta G^{\circ\prime}$. This point is important because reactions that are not spontaneous based on $\Delta G^{\circ\prime}$ can be made spontaneous by adjusting the concentrations of reactants and products. This principle is the basis of the coupling of reactions to form metabolic pathways (Chapter 15).

Enzymes Alter Only the Reaction Rate and Not the Reaction Equilibrium

Because enzymes are such superb catalysts, it is tempting to ascribe to them powers that they do not have. An enzyme cannot alter the laws of thermodynamics and *consequently cannot alter the equilibrium of a chemical reaction.* Consider an enzyme-catalyzed reaction, the conversion of substrate, S, into product, P. Figure 8.2 shows the rate of product formation with time in the presence and absence of enzyme. Note that the amount of product formed

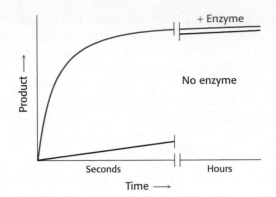

Figure 8.2 Enzymes accelerate the reaction rate. The same equilibrium point is reached but much more quickly in the presence of an enzyme.

is the same whether or not the enzyme is present but, in the present example, the amount of product formed in seconds when the enzyme is present might take hours (or centuries, see Table 8.1) to form if the enzyme were absent.

Why does the rate of product formation level off with time? The reaction has reached equilibrium. Substrate S is still being converted into product P, but P is being converted into S at a rate such that the amount of P present stays the same.

Let us examine the equilibrium in a more quantitative way. Suppose that, in the absence of enzyme, the forward rate constant (k_F) for the conversion of S into P is $10^{-4}\,\text{s}^{-1}$ and the reverse rate constant (k_R) for the conversion of P into S is $10^{-6}\,\text{s}^{-1}$. The equilibrium constant K is given by the ratio of these rate constants:

$$S \underset{10^{-6}\,\text{s}^{-1}}{\overset{10^{-4}\,\text{s}^{-1}}{\rightleftharpoons}} P$$

$$K = \frac{[P]}{[S]} = \frac{k_F}{k_R} = \frac{10^{-4}}{10^{-6}} = 100$$

The equilibrium concentration of P is 100 times that of S, whether or not enzyme is present. However, it might take a very long time to approach this equilibrium without enzyme, whereas equilibrium would be attained rapidly in the presence of a suitable enzyme (see Table 8.1). *Enzymes accelerate the attainment of equilibria but do not shift their positions. The equilibrium position is a function only of the free-energy difference between reactants and products.*

8.3 Enzymes Accelerate Reactions by Facilitating the Formation of the Transition State

The free-energy difference between reactants and products accounts for the equilibrium of the reaction, but enzymes accelerate how quickly this equilibrium is attained. How can we explain the rate enhancement in terms of thermodynamics? To do so, we have to consider not the end points of the reaction but the chemical pathway between the end points.

A chemical reaction of substrate S to form product P goes through a *transition state* $X^{\ddagger}$ that has a higher free energy than does either S or P.

$$S \longrightarrow X^{\ddagger} \longrightarrow P$$

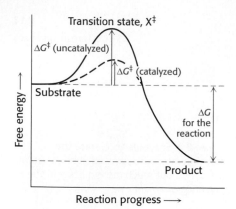

Figure 8.3 Enzymes decrease the activation energy. Enzymes accelerate reactions by decreasing $\Delta G^{\ddagger}$, the free energy of activation.

The double dagger denotes the transition state. The transition state is a transitory molecular structure that is no longer the substrate but is not yet the product. The transition state is the least-stable and most-seldom-occupied species along the reaction pathway because it is the one with the highest free energy. The difference in free energy between the transition state and the substrate is called the *Gibbs free energy of activation* or simply the *activation energy*, symbolized by $\Delta G^{\ddagger}$ (Figure 8.3).

$$\Delta G^{\ddagger} = G_{X^{\ddagger}} - G_S$$

Note that the energy of activation, or $\Delta G^{\ddagger}$, does not enter into the final ΔG calculation for the reaction, because the energy that had to be input to reach the transition state is released when the transition state forms the product. The activation-energy barrier immediately suggests how enzymes enhance reaction rate without altering ΔG of the reaction: enzymes function to lower the activation energy, or, in other words, *enzymes facilitate the formation of the transition state*.

One approach to understanding the increase in reaction rates achieved by enzymes is to assume that the transition state ($X^{\ddagger}$) and the substrate (S) are in equilibrium.

$$S \overset{K^{\ddagger}}{\rightleftharpoons} X^{\ddagger} \overset{v}{\longrightarrow} P$$

in which $K^{\ddagger}$ is the equilibrium constant for the formation of $X^{\ddagger}$, and v is the rate of formation of product from $X^{\ddagger}$. The rate of the reaction v is proportional to the concentration of $X^{\ddagger}$,

$$v \propto [X^{\ddagger}],$$

because only $X^{\ddagger}$ can be converted into product. The concentration of $X^{\ddagger}$ at equilibrium is in turn related to the free-energy difference $\Delta G^{\ddagger}$ between $X^{\ddagger}$ and S; the greater the difference between these two states, the smaller the amount of $X^{\ddagger}$. Thus, the overall rate of reaction V depends on $\Delta G^{\ddagger}$. Specifically,

$$V = v[X^{\ddagger}] = \frac{kT}{h}[S]e^{-\Delta G^{\ddagger}/RT}$$

In this equation, k is Boltzmann's constant, and h is Planck's constant. The value of kT/h at 25°C is $6.6 \times 10^{12}\,\mathrm{s}^{-1}$. Suppose that the free energy of activation is $28.53\,\mathrm{kJ\,mol}^{-1}$ ($6.82\,\mathrm{kcal\,mol}^{-1}$). As shown in Table 8.3, this free-energy difference will result when the ratio $[S^{\ddagger}]/[S]$ is 10^{-5}. If we assume for simplicity's sake that $[S] = 1$ M, then the reaction rate V is $6.2 \times 10^{7}\,\mathrm{s}^{-1}$. If $\Delta G^{\ddagger}$ were lowered by $5.69\,\mathrm{kJ\,mol}^{-1}$ ($1.36\,\mathrm{kcal\,mol}^{-1}$), the ratio $[X^{\ddagger}]/[S]$ would then be 10^{-4}, and the reaction rate would be $6.2 \times 10^{8}\,\mathrm{s}^{-1}$. A decrease of $5.69\,\mathrm{kJ\,mol}^{-1}$ in $\Delta G^{\ddagger}$ yields a 10-fold larger V. A relatively small decrease in $\Delta G^{\ddagger}$ (20% in this particular reaction) results in a much greater increase in V.

Thus, we see the key to how enzymes operate: *Enzymes accelerate reactions by decreasing $\Delta G^{\ddagger}$, the activation energy*. The combination of substrate and enzyme creates a reaction pathway whose transition-state energy is lower than that of the reaction in the absence of enzyme (see Figure 8.3). Because the activation energy is lower, more molecules have the energy required to reach the transition state. Decreasing the activation barrier is analogous to lowering the height of a high-jump bar; more athletes will be able to clear the bar. *The essence of catalysis is specific stabilization of the transition state*.

The Formation of an Enzyme–Substrate Complex Is the First Step in Enzymatic Catalysis

Much of the catalytic power of enzymes comes from their bringing substrates together in favorable orientations to promote the formation of the transition states. Enzymes bring together substrates in *enzyme–substrate* (ES) complexes. The substrates are bound to a specific region of the enzyme called the *active site*. Most enzymes are highly selective in the substrates that they bind. Indeed, the catalytic specificity of enzymes depends in part on the specificity of binding.

What is the evidence for the existence of an enzyme–substrate complex?

1. The first clue was the observation that, at a constant concentration of enzyme, the reaction rate increases with increasing substrate concentration until a maximal velocity is reached (Figure 8.4). In contrast, uncatalyzed reactions do not show this saturation effect. *The fact that an enzyme-catalyzed reaction has a maximal velocity suggests the formation of a discrete ES complex.* At a sufficiently high substrate concentration, all the catalytic sites are filled and so the reaction rate cannot increase. Although indirect, this evidence is the most general for the existence of ES complexes.

2. *X-ray crystallography* has provided high-resolution images of substrates and substrate analogs bound to the active sites of many enzymes (Figure 8.5). In Chapter 9, we will take a close look at several of these complexes. X-ray studies carried out at low temperatures (to slow reactions down) are providing revealing views of enzyme–substrate complexes and their subsequent reactions. A newer technique, *time-resolved crystallography*, depends on crystallizing a light-responsive (photolabile) substrate analog with the enzyme. Exposure to a pulse of light converts the substrate analog into substrate, and images of the enzyme–substrate complex are obtained in a fraction of a second by scanning the crystal with intense polychromatic x-rays from a synchrotron.

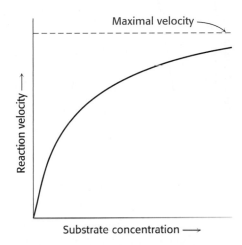

Figure 8.4 Reaction velocity versus substrate concentration in an enzyme-catalyzed reaction. An enzyme-catalyzed reaction approaches a maximal velocity.

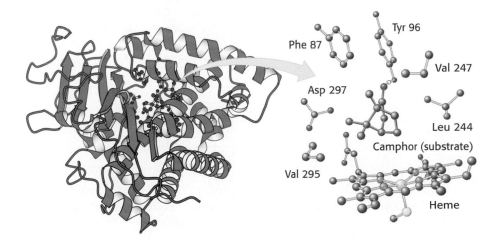

Figure 8.5 Structure of an enzyme–substrate complex. (Left) The enzyme cytochrome P450 is illustrated bound to its substrate camphor. (Right) *Notice* that, in the active site, the substrate is surrounded by residues from the enzyme. Note also the presence of a heme cofactor. [Drawn from 2CPP.pdb.]

3. The *spectroscopic characteristics* of many enzymes and substrates change on formation of an ES complex. These changes are particularly striking if the enzyme contains a colored prosthetic group. Tryptophan synthetase, a bacterial enzyme that contains a pyridoxal phosphate (PLP) prosthetic group, provides a nice illustration. This enzyme catalyzes the synthesis of L-tryptophan from L-serine and an indole derivative. The addition of L-serine to the enzyme produces a marked increase in the fluorescence

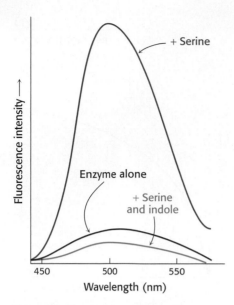

Figure 8.6 Change in spectroscopic characteristics with the formation of an enzyme–substrate complex. Fluorescence intensity of the pyridoxal phosphate group at the active site of tryptophan synthetase changes on addition of serine and indole, the substrates.

of the PLP group (Figure 8.6). The subsequent addition of indole, the second substrate, reduces this fluorescence to a level even lower than that produced by the enzyme alone. Thus, fluorescence spectroscopy reveals the existence of an enzyme–serine complex and of an enzyme–serine–indole complex. Other spectroscopic techniques, such as nuclear magnetic resonance and electron spin resonance, also are highly informative about ES interactions.

The Active Sites of Enzymes Have Some Common Features

The active site of an enzyme is the region that binds the substrates (and the cofactor, if any). It also contains the residues that directly participate in the making and breaking of bonds. These residues are called the *catalytic groups*. In essence, *the interaction of the enzyme and substrate at the active site promotes the formation of the transition state*. The active site is the region of the enzyme that most directly lowers the $\Delta G^{\ddagger}$ of the reaction, thus providing the rate-enhancement characteristic of enzyme action. Although enzymes differ widely in structure, specificity, and mode of catalysis, a number of generalizations concerning their active sites can be stated:

1. *The active site is a three-dimensional cleft or crevice* formed by groups that come from different parts of the amino acid sequence—indeed, residues far apart in the amino acid sequence may interact more strongly than adjacent residues in the sequence. In lysozyme, an enzyme that degrades the cell walls of some bacteria, the important groups in the active site are contributed by residues numbered 35, 52, 62, 63, 101, and 108 in the sequence of 129 amino acids (Figure 8.7).

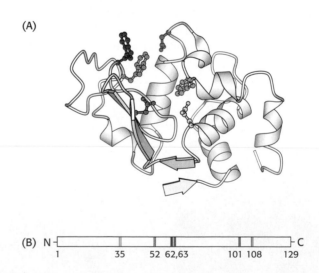

Figure 8.7 Active sites may include distant residues. (A) Ribbon diagram of the enzyme lysozyme with several components of the active site shown in color. (B) A schematic representation of the primary structure of lysozyme shows that the active site is composed of residues that come from different parts of the polypeptide chain. [Drawn from 6LYZ.pdb.]

2. *The active site takes up a relatively small part of the total volume of an enzyme.* Most of the amino acid residues in an enzyme are not in contact with the substrate, which raises the intriguing question of why enzymes are so big. Nearly all enzymes are made up of more than 100 amino acid residues, which gives them a mass greater than 10 kd and a diameter of more than 25 Å. The "extra" amino acids serve as a scaffold to create the three-dimensional active site from amino acids that are far apart in the primary structure. Amino acids near to one another in the primary structure are often sterically constrained from adopting the structural relations necessary to form the active site. In many proteins, the remaining amino acids also constitute regulatory sites, sites of interaction with other proteins, or channels to bring the substrates to the active sites.

3. *Active sites are unique microenvironments.* In all enzymes of known structure, substrate molecules are bound to a cleft or crevice. Water is usually excluded unless it is a reactant. The nonpolar microenvironment of the cleft enhances the binding of substrates as well as catalysis. Nevertheless, the cleft may also contain polar residues. In the nonpolar microenvironment of the active site, certain of these polar residues acquire special properties essential for substrate binding or catalysis. The internal positions of these polar residues are biologically crucial exceptions to the general rule that polar residues are exposed to water.

4. *Substrates are bound to enzymes by multiple weak attractions.* The noncovalent interactions in ES complexes are much weaker than covalent bonds, which have energies between -210 and -460 kJ mol^{-1} (between -50 and $-110 \text{ kcal mol}^{-1}$). In contrast, ES complexes usually have equilibrium constants that range from 10^{-2} to 10^{-8} M, corresponding to free energies of interaction ranging from about -13 to -50 kJ mol^{-1} (from -3 to $-12 \text{ kcal mol}^{-1}$). As discussed in Section 1.3, these weak reversible interactions are mediated by electrostatic interactions, hydrogen bonds, van der Waals forces, and hydrophobic interactions. Van der Waals forces become significant in binding only when numerous substrate atoms simultaneously come close to many enzyme atoms. Hence, the enzyme and substrate should have complementary shapes. The directional character of hydrogen bonds between enzyme and substrate often enforces a high degree of specificity, as seen in the RNA-degrading enzyme ribonuclease (Figure 8.8).

5. *The specificity of binding depends on the precisely defined arrangement of atoms in an active site.* Because the enzyme and the substrate interact by means of short-range forces that require close contact, a substrate must have a matching shape to fit into the site. Emil Fischer's analogy of the lock and key (Figure 8.9), expressed in 1890, has proved to be highly stimulating and fruitful. However, we now know that enzymes are flexible and that the shapes of the active sites can be markedly modified by the binding of substrate, as was postulated by Daniel E. Koshland, Jr., in 1958. The active site of some enzymes assume a shape that is complementary to that of the substrate only *after* the substrate is bound. This process of dynamic recognition is called *induced fit* (Figure 8.10).

The Binding Energy Between Enzyme and Substrate Is Important for Catalysis

Free energy is released by the formation of a large number of weak interactions between a complementary enzyme and substrate. The free energy released on binding is called the *binding energy*. Only the correct substrate can participate in most or all of the interactions with the enzyme and thus maximize binding energy, accounting for the exquisite substrate specificity exhibited by many enzymes. Furthermore, *the full complement of such interactions is formed only when the substrate is in the transition state.* Thus, the maximal binding energy is released when the enzyme facilitates the formation of the transition state. The energy released by the interactions between the

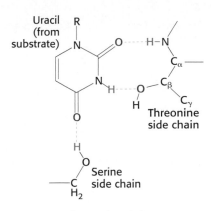

Figure 8.8 Hydrogen bonds between an enzyme and substrate. The enzyme ribonuclease forms hydrogen bonds with the uridine component of the substrate. [After F. M. Richards, H. W. Wyckoff, and N. Allewell. In *The Neurosciences: Second Study Program*, F. O. Schmidt, Ed. (Rockefeller University Press, 1970), p. 970.]

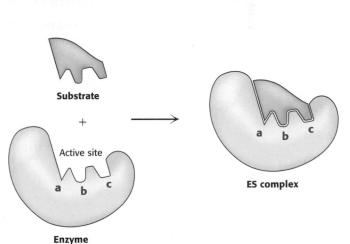

Figure 8.9 Lock-and-key model of enzyme–substrate binding. In this model, the active site of the unbound enzyme is complementary in shape to the substrate.

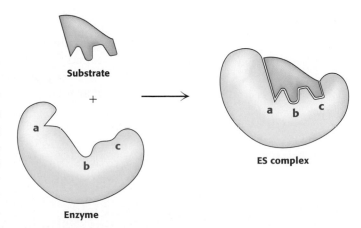

Figure 8.10 Induced-fit model of enzyme–substrate binding. In this model, the enzyme changes shape on substrate binding. The active site forms a shape complementary to the substrate only after the substrate has been bound.

enzyme and substrate can be thought of as lowering the activation energy. Paradoxically, the most stable interaction (maximum binding energy) takes place between the enzyme and the transition state, the least stable reaction intermediate. However, the transition state is too unstable to exist for long. It collapses to either substrate or product, but which of the two accumulates is determined only by the energy difference between the substrate and product—that is, by the ΔG of the reaction.

8.4 The Michaelis–Menten Equation Describes the Kinetic Properties of Many Enzymes

The study of the rates of chemical reactions is called *kinetics,* and the study of the rates of enzyme-catalyzed reactions is called *enzyme kinetics.* A kinetic description of enzyme activity will help us understand how enzymes function. We begin by briefly examining some of the basic principles of reaction kinetics.

Kinetics Is the Study of Reaction Rates

What do we mean when we say the "rate" of a chemical reaction? Consider a simple reaction:

$$A \rightarrow P$$

The rate V is the quantity of A that disappears in a specified unit of time. It is equal to the rate of the appearance of P, or the quantity of P that appears in a specified unit of time.

$$V = -\Delta A/\Delta T = \Delta P/\Delta T \tag{9}$$

If A is yellow and P is colorless, we can follow the decrease in the concentration of A by measuring the decrease in the intensity of yellow color with time. Consider only the change in the concentration of A for now. The rate of the reaction is directly related to the concentration of A by a proportionality constant, k, called the *rate constant.*

$$V = k[A] \tag{10}$$

Reactions that are directly proportional to the reactant concentration are called *first-order reactions.* First-order rate constants have the units of s^{-1}.

Many important biochemical reactions include two reactants. For example,

$$2\,A \rightarrow P$$

or

$$A + B \rightarrow P$$

They are called *bimolecular reactions* and the corresponding rate equations often take the form

$$V = k[A]^2 \tag{11}$$

and

$$V = k[A][B] \qquad (12)$$

The rate constants, called second-order rate constants, have the units $M^{-1}s^{-1}$.

Sometimes, second-order reactions can appear to be first-order reactions. For instance, in reaction 12, if B is present in excess and A is present at low concentrations, the reaction rate will be first order with respect to A and will not appear to depend on the concentration of B. These reactions are called *pseudo-first-order reactions,* and we will see them a number of times in our study of biochemistry.

Interestingly enough, under some conditions, a reaction can be zero order. In these cases, the rate is independent of reactant concentrations. Enzyme-catalyzed reactions can approximate zero-order reactions under some circumstances (p. 219).

The Steady-State Assumption Facilitates a Description of Enzyme Kinetics

The simplest way to investigate the reaction rate is to follow the increase in reaction product as a function of time. The extent of product formation is determined as a function of time for a series of substrate concentrations (Figure 8.11). As expected, in each case, the amount of product formed increases with time, although eventually a time is reached when there is *no net change* in the concentration of S or P. The enzyme is still actively converting substrate into product and visa versa, but the reaction equilibrium has been attained. However, enzyme kinetics is more readily comprehended if we consider only the forward reaction. We can define the rate of catalysis V_0 as the number of moles of product formed per second when the reaction is just beginning—that is, when $t \approx 0$ (see Figure 8.11). For many enzymes, V_0 varies with the substrate concentration [S] in a manner shown in Figure 8.12. The rate of catalysis rises linearly as substrate concentration increases and then begins to level off and approach a maximum at higher substrate concentrations.

In 1913, Leonor Michaelis and Maud Menten proposed a simple model to account for these kinetic characteristics. The critical feature in their treatment is that a specific ES complex is a necessary intermediate in catalysis. The model proposed, which is the simplest one that accounts for the kinetic properties of many enzymes, is

$$E + S \underset{k_{-1}}{\overset{k_1}{\rightleftharpoons}} ES \underset{k_{-2}}{\overset{k_2}{\rightleftharpoons}} E + P$$

An enzyme E combines with substrate S to form an ES complex, with a rate constant k_1. The ES complex has two possible fates. It can dissociate to E and S, with a rate constant k_{-1}, or it can proceed to form product P, with a rate constant k_2. The ES complex can also be re-formed from E and P by the reverse reaction with a rate constant k_{-2}. However, as before, we can simplify these reactions by considering the rate of reaction at times close to zero (hence, V_0) when there is negligible product formation and thus no back reaction ($k_{-2}[P] \approx 0$).

$$E + S \underset{k_{-1}}{\overset{k_1}{\rightleftharpoons}} ES \overset{k_2}{\longrightarrow} E + P \qquad (13)$$

Thus, for the graph in Figure 8.12, V_0 is determined for each substrate concentration by measuring the rate of product formation at early times before P accumulates (see Figure 8.11).

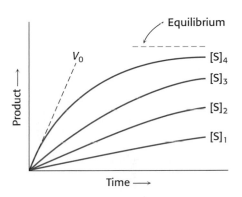

Figure 8.11 Determining initial velocity. The amount of product formed at different substrate concentrations is plotted as a function of time. The initial velocity (V_0) for each substrate concentration is determined from the slope of the curve at the beginning of a reaction, when the reverse reaction is insignificant.

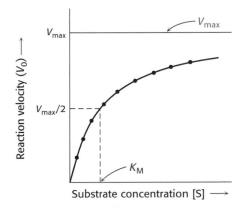

Figure 8.12 Michaelis–Menten kinetics. A plot of the reaction velocity (V_0) as a function of the substrate concentration [S] for an enzyme that obeys Michaelis–Menten kinetics shows that the maximal velocity (V_{max}) is approached asymptotically. The Michaelis constant (K_M) is the substrate concentration yielding a velocity of $V_{max}/2$.

We want an expression that relates the rate of catalysis to the concentrations of substrate and enzyme and the rates of the individual steps. Our starting point is that the catalytic rate is equal to the product of the concentration of the ES complex and k_2.

$$V_0 = k_2[\text{ES}] \tag{14}$$

Now we need to express [ES] in terms of known quantities. The rates of formation and breakdown of ES are given by

$$\text{Rate of formation of ES} = k_1[\text{E}][\text{S}] \tag{15}$$

$$\text{Rate of breakdown of ES} = (k_{-1} + k_2)[\text{ES}] \tag{16}$$

To simplify matters, George Briggs and John Haldane suggested the *steady-state assumption* in 1924. In a steady state, the concentrations of intermediates—in this case, [ES]—stay the same even if the concentrations of starting materials and products are changing. This steady state occurs when the rates of formation and breakdown of the ES complex are equal. Setting the right-hand sides of equations 15 and 16 equal gives

$$k_1[\text{E}][\text{S}] = (k_{-1} + k_2)[\text{ES}] \tag{17}$$

By rearranging equation 17, we obtain

$$[\text{E}][\text{S}]/[\text{ES}] = (k_{-1} + k_2)/k_1 \tag{18}$$

Equation 18 can be simplified by defining a new constant, K_M, called the *Michaelis constant:*

$$K_\text{M} = \frac{k_{-1} + k_2}{k_1} \tag{19}$$

Note that K_M has the units of concentration and is independent of enzyme and substrate concentrations. As we will explain (p. 220), K_M is an important characteristic of enzyme-substrate interactions.

Inserting equation 19 into equation 18 and solving for [ES] yields

$$[\text{ES}] = \frac{[\text{E}][\text{S}]}{K_\text{M}} \tag{20}$$

Now let us examine the numerator of equation 20. Because the substrate is usually present at much higher concentration than the enzyme, the concentration of uncombined substrate [S] is very nearly equal to the total substrate concentration. The concentration of uncombined enzyme [E] is equal to the total enzyme concentration $[\text{E}]_\text{T}$ minus the concentration of the ES complex:

$$[\text{E}] = [\text{E}]_\text{T} - [\text{ES}] \tag{21}$$

Substituting this expression for [E] in equation 20 gives

$$[\text{ES}] = \frac{([\text{E}]_\text{T} - [\text{ES}])[\text{S}]}{K_\text{M}} \tag{22}$$

Solving equation 22 for [ES] gives

$$[ES] = \frac{[E]_T[S]/K_M}{1 + [S]/K_M} \tag{23}$$

or

$$[ES] = [E]_T \frac{[S]}{[S] + K_M} \tag{24}$$

By substituting this expression for [ES] into equation 14, we obtain

$$V_0 = k_2[E]_T \frac{[S]}{[S] + K_M} \tag{25}$$

The maximal rate, V_{max}, is attained when the catalytic sites on the enzyme are saturated with substrate—that is, when $[ES] = [E]_T$. Thus,

$$V_{max} = k_2[E]_T \tag{26}$$

Substituting equation 26 into equation 25 yields the *Michaelis–Menten equation*:

$$V_0 = V_{max} \frac{[S]}{[S] + K_M} \tag{27}$$

This equation accounts for the kinetic data given in Figure 8.12. At very low substrate concentration, when [S] is much less than K_M, $V_0 = (V_{max}/K_M)[S]$; that is, the reaction is first order with the rate directly proportional to the substrate concentration. At high substrate concentration, when [S] is much greater than K_M, $V_0 = V_{max}$; that is, the rate is maximal. The reaction is zero order, independent of substrate concentration.

The meaning of K_M is evident from equation 27. When $[S] = K_M$, then $V_0 = V_{max}/2$. Thus, K_M *is equal to the substrate concentration at which the reaction rate is half its maximal value.* K_M is an important characteristic of an enzyme catalyzed reaction and is significant for its biological function.

The physiological consequence of K_M is illustrated by the sensitivity of some persons to ethanol. Such persons exhibit facial flushing and rapid heart rate (tachycardia) after ingesting even small amounts of alcohol. In the liver, alcohol dehydrogenase converts ethanol into acetaldehyde.

$$CH_3CH_2OH + NAD^+ \xrightleftharpoons[]{\substack{\text{Alcohol} \\ \text{dehydrogenase}}} CH_3CHO + H^+ + NADH$$

Normally, the acetaldehyde, which is the cause of the symptoms when present at high concentrations, is processed to acetate by aldehyde dehydrogenase.

$$CH_3CHO + NAD^+ + H_2O \xrightleftharpoons[]{\substack{\text{Aldehyde} \\ \text{dehydrogenase}}} CH_3COO^- + NADH + 2\,H^+$$

Most people have two forms of the aldehyde dehydrogenase, a low K_M mitochondrial form and a high K_M cytoplasmic form. In susceptible persons, the mitochondrial enzyme is less active owing to the substitution of a single amino acid, and acetaldehyde is processed only by the cytoplasmic enzyme.

Because this enzyme has a high K_M, it achieves a high rate of catalysis only at very high concentrations of acetaldehyde. Consequently, less acetaldehyde is converted into acetate; excess acetaldehyde escapes into the blood and accounts for the physiological effects.

K_M and V_{max} Values Can Be Determined by Several Means

The Michaelis constant, K_M, and the maximal rate, V_{max}, can be readily derived from rates of catalysis measured at a variety of substrate concentrations if an enzyme operates according to the simple scheme given in equation 27. The derivation of K_M and V_{max} is most commonly achieved with the use of curve-fitting programs on a computer. However, an older method, although rarely used because the data points at high and low concentrations are weighted differently and thus sensitive to errors, is a source of further insight into the meaning of K_M and V_{max}.

Before the availability of computers, the determination of K_M and V_{max} values required algebraic manipulation of the basic Michaelis–Menten equation. Because V_{max} is approached asymptotically (see Figure 8.12), it is impossible to obtain a definitive value from a Michaelis–Menten curve. Because K_M is the concentration of substrate at $V_{max}/2$, it is likewise impossible to determine an accurate value of K_M. However, V_{max} can be accurately determined if the Michaelis–Menten equation is transformed into one that gives a straight-line plot. Taking the reciprocal of both sides of equation 27 gives

$$\frac{1}{V_0} = \frac{K_M}{V_{max}} \cdot \frac{1}{S} + \frac{1}{V_{max}} \tag{28}$$

A plot of $1/V_0$ versus $1/[S]$, called a *Lineweaver–Burk* or *double-reciprocal plot*, yields a straight line with a *y*-intercept of $1/V_{max}$ and a slope of K_M/V_{max} (Figure 8.13). The intercept on the *x*-axis is $-1/K_M$.

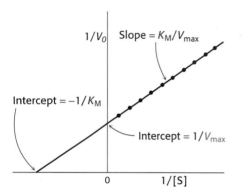

Figure 8.13 A double-reciprocal or Lineweaver–Burk plot. A double-reciprocal plot of enzyme kinetics is generated by plotting $1/V_0$ as a function $1/[S]$. The slope is K_M/V_{max}, the intercept on the vertical axis is $1/V_{max}$, and the intercept on the horizontal axis is $1/K_M$.

K_M and V_{max} Values Are Important Enzyme Characteristics

The K_M values of enzymes range widely (Table 8.4). For most enzymes, K_M lies between 10^{-1} and 10^{-7} M. The K_M value for an enzyme depends on the particular substrate and on environmental conditions such as pH, temperature, and ionic strength. The Michaelis constant, K_M, has two meanings. First, K_M is the concentration of substrate at which half the active sites are filled. Thus, K_M provides a measure of the substrate concentration required for significant catalysis to take place. For many enzymes, experimental evidence suggests that K_M provides an approximation of substrate concentration in vivo.

Second, K_M is related to the rate constants of the individual steps in the catalytic scheme given in equation 13. In equation 19, K_M is defined as $(k_{-1} + k_2)/k_1$. Consider a limiting case in which k_{-1} is much greater than k_2. Under such circumstances, the ES complex dissociates to E and S much more rapidly than product is formed. Under these conditions ($k_{-1} \gg k_2$),

TABLE 8.4 K_M values of some enzymes

Enzyme	Substrate	K_M (μM)
Chymotrypsin	Acetyl-L-tryptophanamide	5000
Lysozyme	Hexa-N-acetylglucosamine	6
β-Galactosidase	Lactose	4000
Threonine deaminase	Threonine	5000
Carbonic anhydrase	CO_2	8000
Penicillinase	Benzylpenicillin	50
Pyruvate carboxylase	Pyruvate	400
	HCO_3^-	1000
	ATP	60
Arginine-tRNA synthetase	Arginine	3
	tRNA	0.4
	ATP	300

$$K_M \approx \frac{k_{-1}}{k_1} \tag{29}$$

Equation 29 describes the dissociation constant of the ES complex.

$$K_{ES} = \frac{[E][S]}{[ES]} = \frac{k_{-1}}{k_1} \tag{30}$$

In other words, K_M *is equal to the dissociation constant of the ES complex if* k_2 *is much smaller than* k_{-1}. When this condition is met, K_M is a measure of the strength of the ES complex: a high K_M indicates weak binding; a low K_M indicates strong binding. It must be stressed that K_M indicates the affinity of the ES complex only when k_{-1} is much greater than k_2.

The maximal rate, V_{max}, reveals the *turnover number* of an enzyme, which is *the number of substrate molecules converted into product by an enzyme molecule in a unit time when the enzyme is fully saturated with substrate*. It is equal to the rate constant k_2, which is also called k_{cat}. The maximal rate, V_{max}, reveals the turnover number of an enzyme if the concentration of active sites $[E]_T$ is known, because

$$V_{max} = k_2[E]_T \tag{31}$$

and thus

$$k_2 = V_{max}/[E]_T \tag{32}$$

For example, a 10^{-6} M solution of carbonic anhydrase catalyzes the formation of 0.6 M H_2CO_3 per second when the enzyme is fully saturated with substrate. Hence, k_2 is $6 \times 10^5 \, s^{-1}$. This turnover number is one of the largest known. Each catalyzed reaction takes place in a time equal to, on average, $1/k_2$, which is 1.7 μs for carbonic anhydrase. The turnover numbers of most enzymes with their physiological substrates fall in the range from 1 to 10^4 per second (Table 8.5).

K_M and V_{max} also permit the determination of f_{ES}, the fraction of active sites filled. This relation of f_{ES} to K_M and V_{max} is given by the following equation:

$$f_{ES} = \frac{V}{V_{max}} = \frac{[S]}{[S] + K_M} \tag{33}$$

TABLE 8.5 Turnover numbers of some enzymes

Enzyme	Turnover number (per second)
Carbonic anhydrase	600,000
3-Ketosteroid isomerase	280,000
Acetylcholinesterase	25,000
Penicillinase	2,000
Lactate dehydrogenase	1,000
Chymotrypsin	100
DNA polymerase I	15
Tryptophan synthetase	2
Lysozyme	0.5

k_{cat}/K_M Is a Measure of Catalytic Efficiency

When the substrate concentration is much greater than K_M, the rate of catalysis is equal to V_{max}, which is a function of k_{cat}, the turnover number, as described above. However, most enzymes are not normally saturated with substrate. Under physiological conditions, the $[S]/K_M$ ratio is typically between 0.01 and 1.0. When $[S] \ll K_M$, the enzymatic rate is much less than k_{cat} because most of the active sites are unoccupied. Is there a number that characterizes the kinetics of an enzyme under these more typical cellular conditions? Indeed there is, as can be shown by combining equations 14 and 20 to give

$$V_0 = \frac{k_{cat}}{K_M}[E][S] \tag{34}$$

TABLE 8.6 Substrate preferences of chymotrypsin

Amino acid in ester	Amino acid side chain	k_{cat}/K_M ($s^{-1}M^{-1}$)
Glycine	$-H$	1.3×10^{-1}
Valine	$-\underset{\underset{CH_3}{\vert}}{\overset{\overset{CH_3}{\vert}}{CH}}$	2.0
Norvaline	$-CH_2CH_2CH_3$	3.6×10^2
Norleucine	$-CH_2CH_2CH_2CH_3$	3.0×10^3
Phenylalanine	$-CH_2-\hexagon$	1.0×10^5

Source: After A. Fersht, *Structure and Mechanism in Protein Science: A Guide to Enzyme Catalysis and Protein Folding* (W. H. Freeman and Company, 1999), Table 7.3.

When $[S] << K_M$, the concentration of free enzyme, $[E]$, is nearly equal to the total concentration of enzyme $[E]_T$; so

$$V_0 = \frac{k_{cat}}{K_M}[S][E]_T \qquad (35)$$

Thus, when $[S] << K_M$, the enzymatic velocity depends on the values of k_{cat}/K_M, $[S]$, and $[E]_T$. Under these conditions, k_{cat}/K_M is the rate constant for the interaction of S and E. The rate constant k_{cat}/K_M is a measure of catalytic efficiency because it takes into account both the rate of catalysis with a particular substrate (k_{cat}) and the strength of the enzyme–substrate interaction (K_M). For instance, by using k_{cat}/K_M values, one can compare an enzyme's preference for different substrates. Table 8.6 shows the k_{cat}/K_M values for several different substrates of chymotrypsin. Chymotrypsin clearly has a preference for cleaving next to bulky, hydrophobic side chains.

How efficient can an enzyme be? We can approach this question by determining whether there are any physical limits on the value of k_{cat}/K_M. Note that this ratio depends on k_1, k_{-1}, and k_{cat}, as can be shown by substituting for K_M.

$$k_{cat}/K_M = \frac{k_{cat}k_1}{k_{-1} + k_{cat}} = \left(\frac{k_{cat}}{k_{-1} + k_{cat}}\right)k_1 < k_1 \qquad (36)$$

TABLE 8.7 Enzymes for which k_{cat}/K_M is close to the diffusion-controlled rate of encounter

Enzyme	k_{cat}/K_M ($s^{-1}M^{-1}$)
Acetylcholinesterase	1.6×10^8
Carbonic anhydrase	8.3×10^7
Catalase	4×10^7
Crotonase	2.8×10^8
Fumarase	1.6×10^8
Triose phosphate isomerase	2.4×10^8
β-Lactamase	1×10^8
Superoxide dismutase	7×10^9

Source: After A. Fersht, *Structure and Mechanism in Protein Science: A Guide to Enzyme Catalysis and Protein Folding* (W. H. Freeman and Company, 1999), Table 4.5.

Suppose that the rate of formation of product (k_{cat}) is much faster than the rate of dissociation of the ES complex (k_{-1}). The value of k_{cat}/K_M then approaches k_1. Thus, the ultimate limit on the value of k_{cat}/K_M is set by k_1, the rate of formation of the ES complex. *This rate cannot be faster than the diffusion-controlled encounter of an enzyme and its substrate.* Diffusion limits the value of k_1 and so it cannot be higher than between 10^8 and 10^9 $s^{-1}M^{-1}$. Hence, the upper limit on k_{cat}/K_M is between 10^8 and 10^9 $s^{-1}M^{-1}$.

The k_{cat}/K_M ratios of the enzymes superoxide dismutase, acetylcholinesterase, and triosephosphate isomerase are between 10^8 and 10^9 $s^{-1}M^{-1}$. Enzymes that have k_{cat}/K_M ratios at the upper limits have attained *kinetic perfection. Their catalytic velocity is restricted only by the rate at which they encounter substrate in the solution* (Table 8.7). Any further gain in catalytic rate can come only by decreasing the time for diffusion. Remember that the active site is only a small part of the total enzyme structure. Yet, for catalytically perfect enzymes, every encounter between enzyme and substrate is productive. In these cases, there may be attractive electrostatic forces on the enzyme that entice the substrate to the active site. These forces are sometimes referred to poetically as *Circe effects*.

Diffusion in solution can also be partly overcome by confining substrates and products in the limited volume of a multienzyme complex. Indeed, some series of enzymes are associated into organized assemblies so that the product of one enzyme is very rapidly found by the next enzyme. In effect, products are channeled from one enzyme to the next, much as in an assembly line.

Most Biochemical Reactions Include Multiple Substrates

Most reactions in biological systems start with two substrates and yield two products. They can be represented by the bisubstrate reaction:

$$A + B \rightleftharpoons P + Q$$

Many such reactions transfer a functional group, such as a phosphoryl or an ammonium group, from one substrate to the other. Those that are oxidation–reduction reactions transfer electrons between substrates. Multiple substrate reactions can be divided into two classes: *sequential* reactions and *double-displacement* reactions.

Sequential Reactions. In sequential reactions, all substrates must bind to the enzyme before any product is released. Consequently, in a bisubstrate reaction, a *ternary complex* of the enzyme and both substrates forms. Sequential mechanisms are of two types: ordered, in which the substrates bind the enzyme in a defined sequence, and random.

Many enzymes that have NAD^+ or NADH as a substrate exhibit the ordered sequential mechanism. Consider lactate dehydrogenase, an important enzyme in glucose metabolism (p. 447). This enzyme reduces pyruvate to lactate while oxidizing NADH to NAD^+.

Pyruvate + NADH + H⁺ ⇌ HO—C—H + NAD⁺ **Lactate**

In the ordered sequential mechanism, the coenzyme always binds first and the lactate is always released first. This sequence can be represented by using a notation developed by W. Wallace Cleland:

E (NADH) (pyruvate) ⇌ E (lactate) (NAD⁺)

The enzyme exists as a ternary complex consisting of, first, the enzyme and substrates and, after catalysis, the enzyme and products.

In the random sequential mechanism, the order of the addition of substrates and the release of products is random. An example of a random sequential reaction is the formation of phosphocreatine and ADP from ATP and creatine, which is catalyzed by creatine kinase (p. 416).

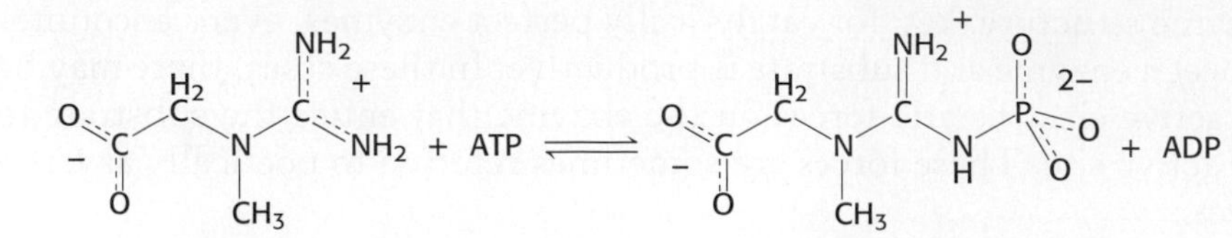

Creatine ⇌ **Phosphocreatine**

Either creatine or ATP may bind first, and either phosphocreatine or ADP may be released first. Phosphocreatine is an important energy source in muscle. Sequential random reactions also can be depicted in the Cleland notation.

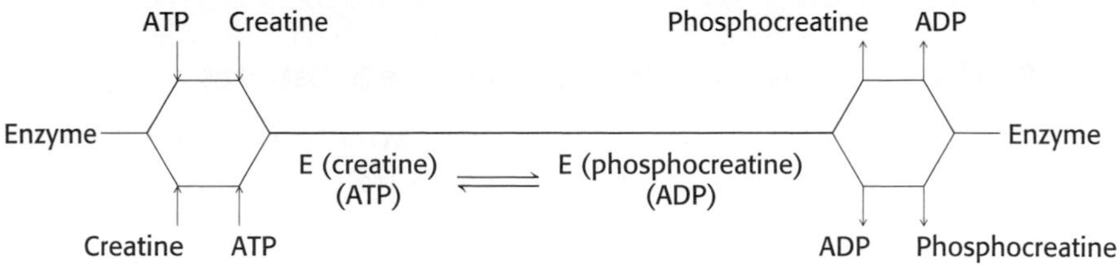

Although the order of certain events is random, the reaction still passes through the ternary complexes including, first, substrates and, then, products.

Double-Displacement (Ping-Pong) Reactions. In double-displacement, or ping-pong, reactions, one or more products are released before all substrates bind the enzyme. The defining feature of double-displacement reactions is the existence of a *substituted enzyme intermediate*, in which the enzyme is temporarily modified. Reactions that shuttle amino groups between amino acids and α-ketoacids are classic examples of double-displacement mechanisms. The enzyme aspartate aminotransferase (p. 659) catalyzes the transfer of an amino group from aspartate to α-ketoglutarate.

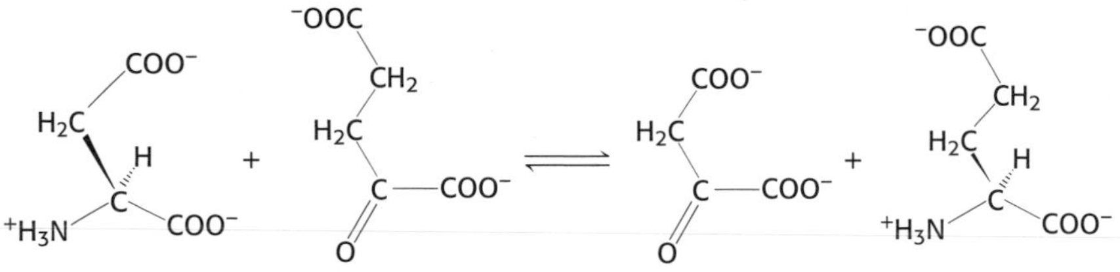

Aspartate + **α-Ketoglutarate** ⇌ **Oxaloacetate** + **Glutamate**

The sequence of events can be portrayed as the following diagram:

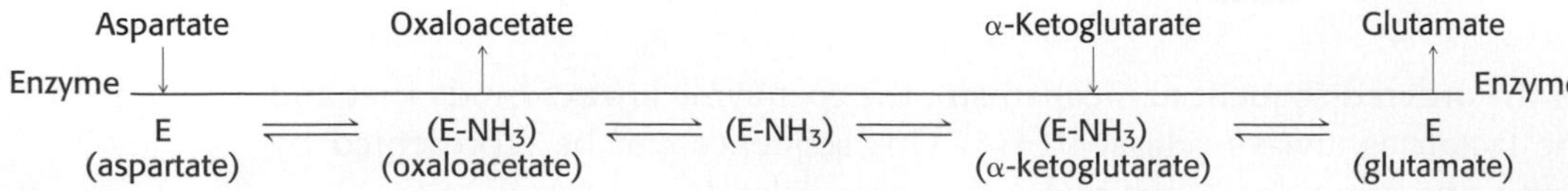

After aspartate binds to the enzyme, the enzyme accepts aspartate's amino group to form the substituted enzyme intermediate. The first product, oxaloacetate, subsequently departs. The second substrate, α-ketoglutarate, binds to the enzyme, accepts the amino group from the modified enzyme, and is then released as the final product, glutamate. In the Cleland notation, the substrates appear to bounce on and off the enzyme analogously to a Ping-Pong ball bouncing on a table.

Allosteric Enzymes Do Not Obey Michaelis–Menten Kinetics

The Michaelis–Menten model has greatly assisted the development of enzymology. Its virtues are simplicity and broad applicability. However, the

Michaelis–Menten model cannot account for the kinetic properties of many enzymes. An important group of enzymes that do not obey Michaelis–Menten kinetics are the *allosteric enzymes*. These enzymes consist of multiple subunits and multiple active sites.

Allosteric enzymes often display sigmoidal plots (Figure 8.14) of the reaction velocity V_0 versus substrate concentration [S], rather than the hyperbolic plots predicted by the Michaelis–Menten equation (equation 27). In allosteric enzymes, the binding of substrate to one active site can alter the properties of other active sites in the same enzyme molecule. A possible outcome of this interaction between subunits is that the binding of substrate becomes *cooperative;* that is, the binding of substrate to one active site facilitates the binding of substrate to the other active sites. As it does for hemoglobin (Chapter 7), such cooperativity results in a sigmoidal plot of V_0 versus [S]. In addition, the activity of an allosteric enzyme may be altered by regulatory molecules that are reversibly bound to specific sites other than the catalytic sites. The catalytic properties of allosteric enzymes can thus be adjusted to meet the immediate needs of a cell (Chapter 10). For this reason, allosteric enzymes are key regulators of metabolic pathways.

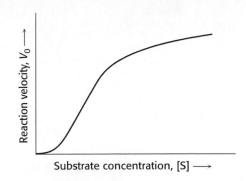

Figure 8.14 Kinetics for an allosteric enzyme. Allosteric enzymes display a sigmoidal dependence of reaction velocity on substrate concentration.

8.5 Enzymes Can Be Inhibited by Specific Molecules

The activity of many enzymes can be inhibited by the binding of specific small molecules and ions. This means of inhibiting enzyme activity serves as a major control mechanism in biological systems, typified by the regulation of allosteric enzymes. In addition, many drugs and toxic agents act by inhibiting enzymes (Chapter 35). Inhibition can be a source of insight into the mechanism of enzyme action: specific inhibitors can often be used to identify residues critical for catalysis. Transition-state analogs are especially potent inhibitors.

Enzyme inhibition can be either irreversible or reversible. An *irreversible inhibitor* dissociates very slowly from its target enzyme because it has become tightly bound to the enzyme, either covalently or noncovalently. Some irreversible inhibitors are important drugs. Penicillin acts by covalently modifying the enzyme transpeptidase, thereby preventing the synthesis of bacterial cell walls and thus killing the bacteria (p. 232). Aspirin acts by covalently modifying the enzyme cyclooxygenase, reducing the synthesis of signaling molecules involved in inflamation (p. 339).

Reversible inhibition, in contrast with irreversible inhibition, is characterized by a rapid dissociation of the enzyme–inhibitor complex. In the type of reversible inhibition called *competitive inhibition,* an enzyme can bind substrate (forming an ES complex) or inhibitor (EI) but not both (ESI). The competitive inhibitor often resembles the substrate and binds to the active site of the enzyme (Figure 8.15). The substrate is thereby prevented from binding to the same active site. *A competitive inhibitor diminishes the rate of catalysis by reducing the proportion of enzyme molecules bound to a substrate.* At any given inhibitor concentration, competitive inhibition can be relieved by increasing

(A) Substrate

(B) Competitive inhibitor
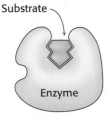

(C) Substrate   Uncompetitive inhibitor
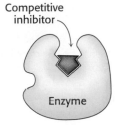

(D) Substrate
Noncompetitive inhibitor
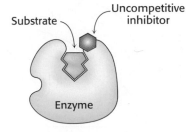

Figure 8.15 Distinction between reversible inhibitors. (A) Enzyme–substrate complex; (B) a competitive inhibitor binds at the active site and thus prevents the substrate from binding; (C) an uncompetitive inhibitor binds only to the enzyme–substrate complex; (D) a noncompetitive inhibitor does not prevent the substrate from binding.

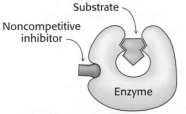

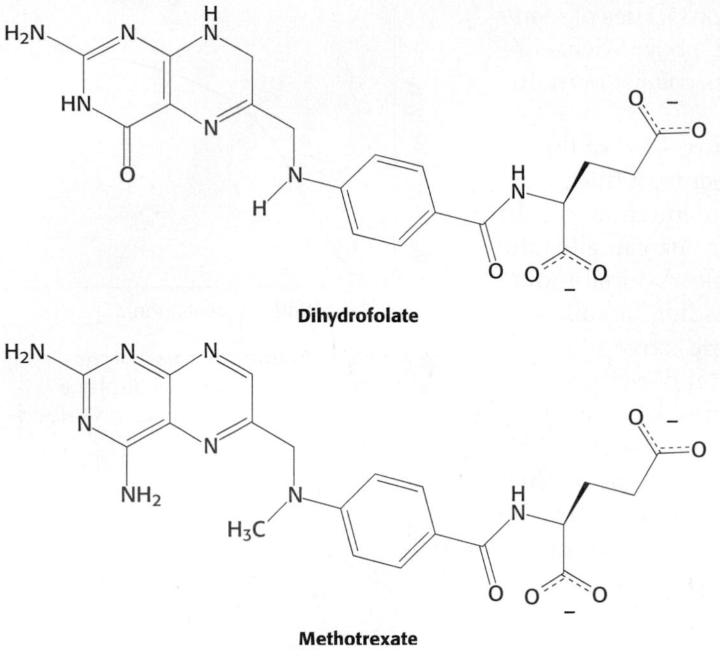

Figure 8.16 Enzyme inhibitors. The substrate dihydrofolate and its structural analog methotrexate. Regions with structural differences are shown in red.

the substrate concentration. Under these conditions, the substrate successfully competes with the inhibitor for the active site. Methotrexate is an especially potent competitive inhibitor of the enzyme dihydrofolate reductase, which plays a role in the biosynthesis of purines and pyrimidines. This compound is a structural analog of dihydrofolate, a substrate for dihydrofolate reductase (Figure 8.16). What makes it such a potent competitive inhibitor is that it binds to the enzyme 1000-fold more tightly than the natural substrate and inhibits nucleotide base synthesis. It is used to treat cancer.

Uncompetitive inhibition is distinguished by the fact that the inhibitor binds only to the enzyme–substrate complex. The uncompetitive inhibitor's binding site is created only on interaction of the enzyme and substrate (see Figure 8.15C). Uncompetitive inhibition cannot be overcome by the addition of more substrate.

In *noncompetitive inhibition,* the inhibitor and substrate can bind simultaneously to an enzyme molecule at different binding sites (see Figure 8.15D). A noncompetitive inhibitor acts by decreasing the turnover number rather than by diminishing the proportion of enzyme molecules that are bound to substrate. Noncompetitive inhibition, like uncompetitive inhibition, cannot be overcome by increasing the substrate concentration. A more complex pattern, called *mixed inhibition,* is produced when a single inhibitor both hinders the binding of substrate and decreases the turnover number of the enzyme.

Reversible Inhibitors Are Kinetically Distinguishable

How can we determine whether a reversible inhibitor acts by competitive or noncompetitive inhibition? Let us consider only enzymes that exhibit Michaelis–Menten kinetics. Measurements of the rates of catalysis at different concentrations of substrate and inhibitor serve to distinguish the three types of inhibition. In *competitive inhibition,* the inhibitor competes with the substrate for the active site. The dissociation constant for the inhibitor is given by

$$K_i = [\text{E}][\text{I}]/[\text{EI}]$$

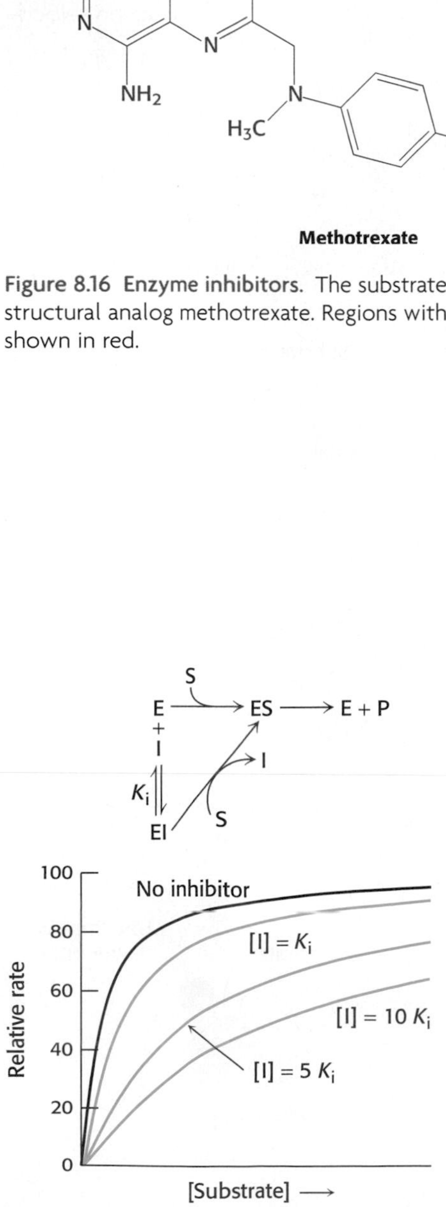

The smaller the K_i, the more potent the inhibition. *The hallmark of competitive inhibition is that it can be overcome by a sufficiently high concentration of substrate* (Figure 8.17). The effect of a competitive inhibitor is to increase the apparent value of K_M, meaning that more substrate is needed to obtain the same reaction rate. This new value of K_M, called K_M^{app}, is numerically equal to

$$K_M^{app} = K_M(1 + [\text{I}]/K_i)$$

Figure 8.17 Kinetics of a competitive inhibitor. As the concentration of a competitive inhibitor increases, higher concentrations of substrate are required to attain a particular reaction velocity. The reaction pathway suggests how sufficiently high concentrations of substrate can completely relieve competitive inhibition.

where [I] is the concentration of inhibitor and K_i is the dissociation constant for the enzyme–inhibitor complex. In the presence of a competitive inhibitor, an enzyme will have the same V_{max} as in the absence of an inhibitor. At a sufficiently high concentration, virtually all the active sites are filled by substrate, and the enzyme is fully operative.

Competitive inhibitors are commonly used as drugs. Drugs such as ibuprofen are competitive inhibitors of enzymes that participate in signaling pathways in the inflammatory response. Statins are drugs that reduce high cholesterol levels by competitively inhibiting a key enzyme in cholesterol biosynthesis (p. 339).

In *uncompetitive inhibition*, the inhibitor binds only to the ES, complex. This enzyme–substrate–inhibitor complex, ESI, does not go on to form any product. Because some unproductive ESI complex will always be present, V_{max} will be lower in the presence of inhibitor than in its absence (Figure 8.18). The uncompetitive inhibitor lowers that apparent value of K_M. This occurs since the inhibitor binds to ES to form ESI, depleting ES. To maintain the equilibrium between E and ES, more S binds to E. Thus, a lower concentration of S is required to form half of the maximal concentration of ES and the apparent value of K_M is reduced. The herbicide glyphosate, also known as Roundup, is an uncompetitive inhibitor of an enzyme in the biosynthetic pathway for aromatic amino acids.

In *noncompetitive inhibition* (Figure 8.19), substrate can still bind to the enzyme–inhibitor complex. However, the enzyme–inhibitor–substrate complex *does not* proceed to form product. The value of V_{max} is decreased to a new value called V_{max}^{app}, whereas the value of K_M is unchanged. The maximal velocity in the presence of a pure noncompetitive inhibitor, V_{max}^{app}, is given by

$$V_{max}^{app} = \frac{V_{max}}{1 + [I]/K_i} \qquad (37)$$

Why is V_{max} lowered though K_M remains unchanged? In essence, the inhibitor simply lowers the concentration of functional enzyme. The resulting solution behaves like a more dilute solution of enzyme. *Noncompetitive inhibition cannot be overcome by increasing the substrate concentration.* Deoxycycline, an antibiotic, functions at low concentrations as a noncompetitive inhibitor of a proteolytic enzyme (collagenase). It is used to treat periodontal disease. Some of the toxic effects of lead poisoning may be due to lead's ability to act as noncompetitive inhibitor of a host of enzymes. Lead reacts with crucial sulfhydryl groups in these enzymes.

Double-reciprocal plots are especially useful for distinguishing between competitive, uncompetitive, and noncompetitive inhibitors. In competitive inhibition, the intercept on the y-axis of the plot of $1/V_0$ versus $1/[S]$ is the

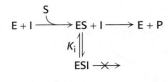

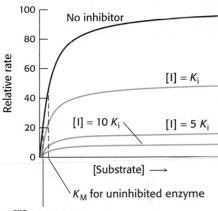

Figure 8.18 Kinetics of an uncompetitive inhibitor. The reaction pathway shows that the inhibitor binds only to the enzyme–substrate complex. Consequently, V_{max} cannot be attained, even at high substrate concentrations. The apparent value for K_M is lowered, becoming smaller as more inhibitor is added.

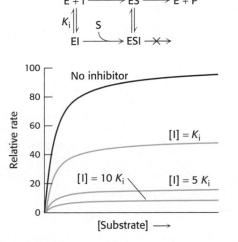

Figure 8.19 Kinetics of a noncompetitive inhibitor. The reaction pathway shows that the inhibitor binds both to free enzyme and to enzyme complex. Consequently, as with uncompetitive competition, V_{max} cannot be attained. K_M remains unchanged, and so the reaction rate increases more slowly at low substrate concentrations than is the case for uncompetitive competition.

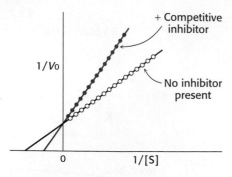

Figure 8.20 Competitive inhibition illustrated on a double-reciprocal plot. A double-reciprocal plot of enzyme kinetics in the presence and absence of a competitive inhibitor illustrates that the inhibitor has no effect on V_{max} but increases K_M.

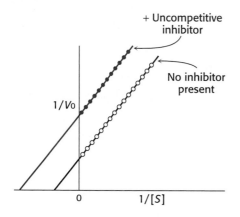

Figure 8.21 Uncompetitive inhibition illustrated by a double-reciprocal plot. An uncompetitive inhibitor does not effect the slope of the double-reciprocal plot. V_{max} and K_M are reduced by equivalent amounts.

same in the presence and in the absence of inhibitor, although the slope is increased (Figure 8.20). The intercept is unchanged because a competitive inhibitor does not alter V_{max}. The increase in the slope of the $1/V_0$ versus $1/[S]$ plot indicates the strength of binding of competitive inhibitor. In the presence of a competitive inhibitor, equation 28 is replaced by

$$\frac{1}{V_0} = \frac{1}{V_{max}} + \frac{K_M}{V_{max}}\left(1 + \frac{[I]}{K_i}\right)\left(\frac{1}{[S]}\right) \tag{38}$$

In other words, the slope of the plot is increased by the factor $(1 + [I]/K_i)$ in the presence of a competitive inhibitor. Consider an enzyme with a K_M of 10^{-4} M. In the absence of inhibitor, $V_0 = V_{max}/2$ when $[S] = 10^{-4}$ M. In the presence of 2×10^{-3} M competitive inhibitor that is bound to the enzyme with a K_i of 10^{-3} M, the apparent K_M (K_M^{app}) will be equal to $K_M(1 + [I]/K_i)$, or 3×10^{-4} M. Substitution of these values into equation 37 gives $V_0 = V_{max}/4$, when $[S] = 10^{-4}$ M. The presence of the competitive inhibitor thus cuts the reaction rate in half at this substrate concentration.

In uncompetitive inhibition (Figure 8.21), the inhibitor combines only with the enzyme–substrate complex. The equation that describes the double–reciprocal plot for an uncompetitive inhibitor is

$$\frac{1}{V_0} = \frac{K_M}{V_{max}}\frac{1}{[S]} + \frac{1}{V_{max}}\left(1 + \frac{[I]}{K_i}\right) \tag{39}$$

The slope of the line, K_M/V_{max}, is the same as that for the uninhibited enzyme, but the intercept on the y-axis will be increased by $1 + [I]/K_i$. Consequently, the lines in double-reciprocal plots will be parallel.

In noncompetitive inhibition (Figure 8.22), the inhibitor can combine with either the enzyme or the enzyme–substrate complex. In pure noncompetitive inhibition, the values of the dissociation constants of the inhibitor and enzyme and of the inhibitor and enzyme–substrate complex are equal. The value of V_{max} is decreased to the new value V_{max}^{app}, and so the intercept on the vertical axis is increased. The new slope, which is equal to K_M/V_{max}^{app}, is larger by the same factor. In contrast with V_{max}, K_M is not affected by pure noncompetitive inhibition.

Irreversible Inhibitors Can Be Used to Map the Active Site

In Chapter 9, we will examine the chemical details of how enzymes function. The first step in obtaining the chemical mechanism of an enzyme is to determine what functional groups are required for enzyme activity. How can we ascertain these functional groups? X-ray crystallography of the

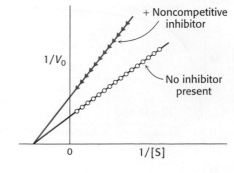

Figure 8.22 Noncompetitive inhibition illustrated on a double-reciprocal plot. A double-reciprocal plot of enzyme kinetics in the presence and absence of a noncompetitive inhibitor shows that K_M is unaltered and V_{max} is decreased.

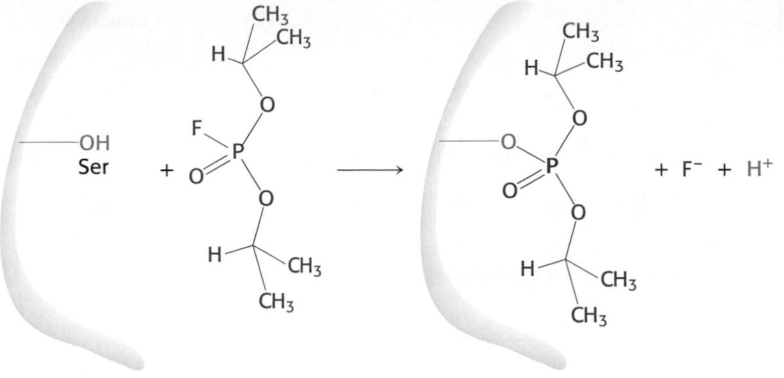

Figure 8.23 Enzyme inhibition by diisopropylphosphofluoridate (DIPF), a group-specific reagent. DIPF can inhibit an enzyme by covalently modifying a crucial serine residue.

enzyme bound to its substrate or substrate analogy provides one approach. Irreversible inhibitors that covalently bond to the enzyme provide an alternative and often complementary approach: the inhibitors modify the functional groups, which can then be identified. Irreversible inhibitors can be divided into three categories: group-specific reagents, reactive substrate analogs (also called affinity labels), and suicide inhibitors.

Group-specific reagents react with specific side chains of amino acids. Two examples of group-specific reagents are diisopropylphosphofluoridate (DIPF; Figure 8.23) and iodoacetamide (Figure 8.24). DIPF modifies only 1 of the 28 serine residues in the proteolytic enzyme chymotrypsin, implying that this serine residue is especially reactive. As we will see in Chapter 9, it is indeed the case that this serine residue is located at the active site. DIPF also revealed a reactive serine residue in acetylcholinesterase, an enzyme important in the transmission of nerve impulses (see Figure 8.23). Thus, DIPF and similar compounds that bind and inactivate acetylcholinesterase are potent nerve gases.

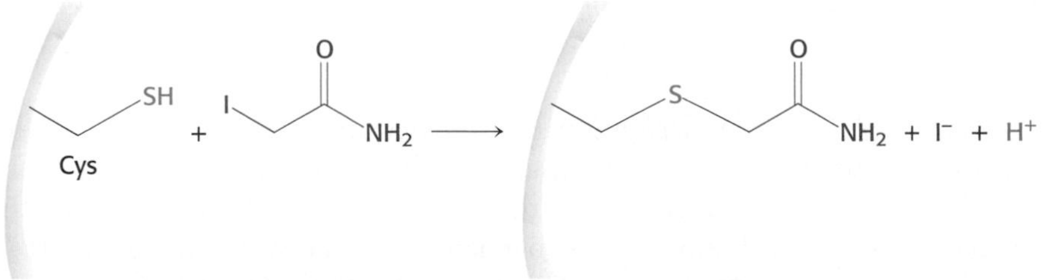

Figure 8.24 Enzyme inactivation by iodoacetamide, a group-specific reagent. Iodoacetamide can inactivate an enzyme by reacting with a critical cysteine residue.

Affinity labels, or *reactive substrate analogs,* are molecules that are structurally similar to the substrate for the enzyme and that covalently bind to active-site residues. They are thus more specific for the enzyme active site than are group-specific reagents. Tosyl-L-phenylalanine chloromethyl ketone (TPCK) is a substrate analog for chymotrypsin (Figure 8.25). TPCK binds at the active site and then reacts irreversibly with a histidine residue at that site, inhibiting the enzyme. The compound 3-bromoacetol phosphate is an affinity label for the enzyme triose phosphate isomerase (TPI). It mimics

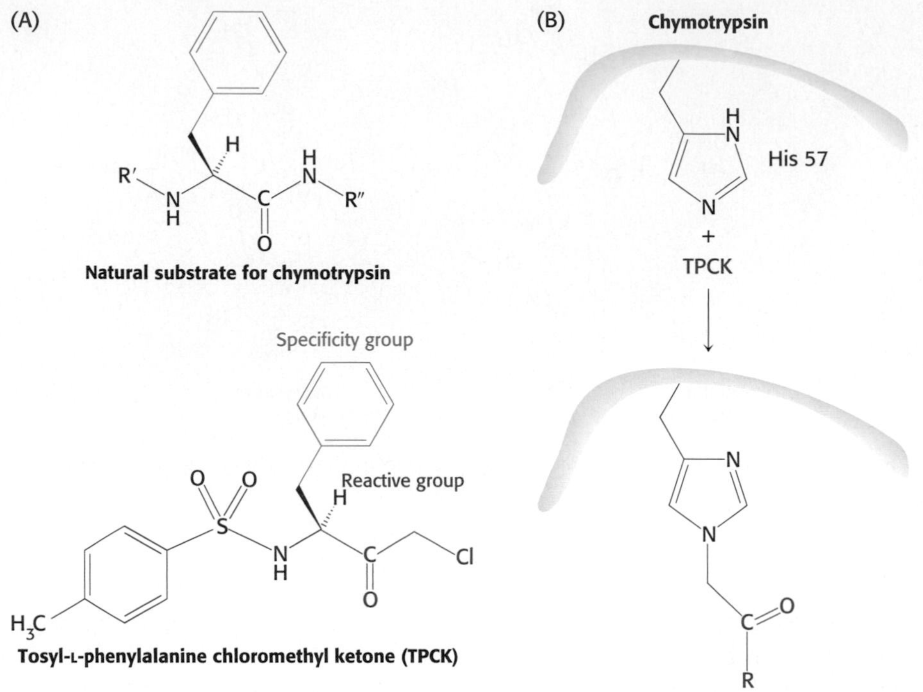

(A)

Natural substrate for chymotrypsin

(B) **Chymotrypsin**

His 57

TPCK

Figure 8.25 **Affinity labeling.** (A) Tosyl-L-phenylalanine chloromethyl ketone (TPCK) is a reactive analog of the normal substrate for the enzyme chymotrypsin. (B) TPCK binds at the active site of chymotrypsin and modifies an essential histidine residue.

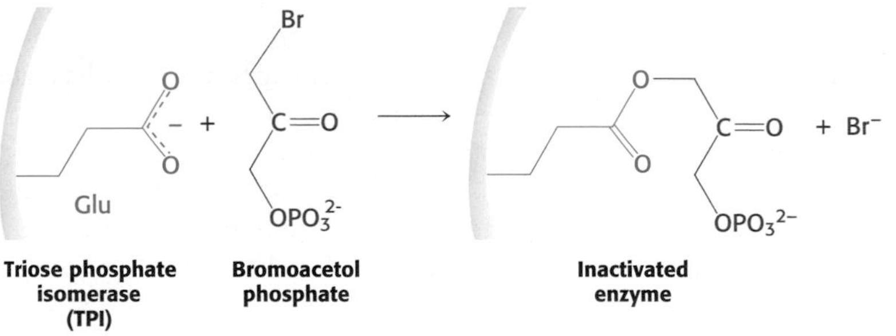

Specificity group

Reactive group

Tosyl-L-phenylalanine chloromethyl ketone (TPCK)

Figure 8.26 **Bromoacetol phosphate, an affinity label for triose phosphate isomerase (TPI).** Bromoacetol phosphate, an analog of dihydroxyacetone phosphate, binds at the active site of the enzyme and covalently modifies a glutamic acid residue required for enzyme activity.

Triose phosphate isomerase (TPI)

Bromoacetol phosphate

Inactivated enzyme

the normal substrate, dihydroxyacetone phosphate, by binding at the active site; then it covalently modifies the enzyme such that the enzyme is irreversibly inhibited (Figure 8.26).

Suicide inhibitors, or *mechanism-based inhibitors,* are modified substrates that provide the most specific means to modify an enzyme active site. The inhibitor binds to the enzyme as a substrate and is initially processed by the normal catalytic mechanism. The mechanism of catalysis then generates a chemically reactive intermediate that inactivates the enzyme through covalent modification. The fact that the enzyme participates in its own irreversible inhibition strongly suggests that the covalently modified group on the enzyme is vital for catalysis. One example of such an inhibitor is N,N-dimethylpropargylamine, an inhibitor of the enzyme monoamine oxidase (MAO). A flavin prosthetic group of monoamine oxidase oxidizes the N,N-dimethylpropargylamine, which in turn inactivates the enzyme by binding to N-5 of the flavin prosthetic group (Figure 8.27). Monoamine oxidase deaminates neurotransmitters such as dopamine and serotonin, lowering their levels in the brain. Parkinson disease is associated with low levels of dopamine, and depression is associated with low levels of serotonin. The drug (−)deprenyl, which is used to treat Parkinson disease and depression, is a suicide inhibitor of monoamine oxidase.

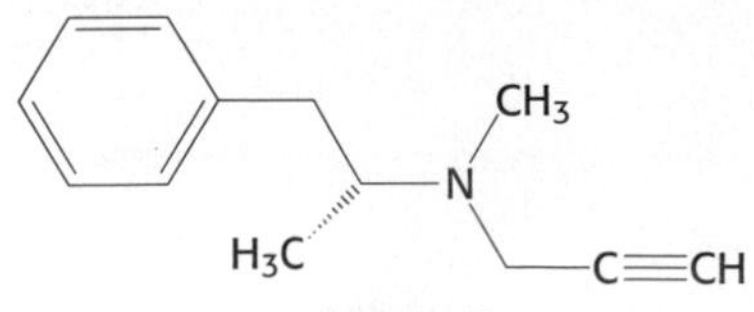

(−)Deprenyl

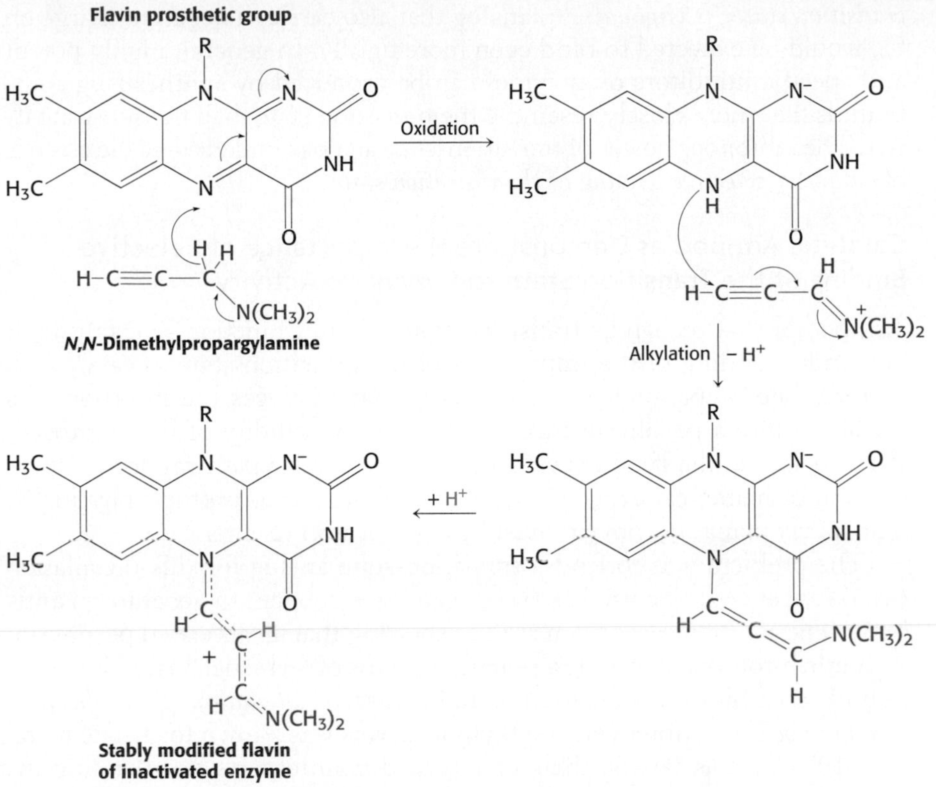

Flavin prosthetic group

N,N-Dimethylpropargylamine

**Stably modified flavin
of inactivated enzyme**

Figure 8.27 Mechanism-based (suicide) inhibition. Monoamine oxidase, an enzyme important for neurotransmitter synthesis, requires the cofactor FAD (flavin adenine dinucleotide). *N,N*-Dimethylpropargylamine inhibits monoamine oxidase by covalently modifying the flavin prosthetic group only after the inhibitor has been oxidized. The N-5 flavin adduct is stabilized by the addition of a proton.

Transition-State Analogs Are Potent Inhibitors of Enzymes

We turn now to compounds that provide the most intimate views of the catalytic process itself. Linus Pauling proposed in 1948 that compounds resembling the transition state of a catalyzed reaction should be very effective inhibitors of enzymes. These mimics are called *transition-state analogs*. The inhibition of proline racemase is an instructive example. *The racemization of proline proceeds through a transition state in which the tetrahedral α-carbon atom has become trigonal* (Figure 8.28). In the trigonal form, all three bonds are in the same plane; C_α also carries a net negative charge. This symmetric carbanion can be reprotonated on one side to give the L isomer or on the other side to give the D isomer. This picture is supported by the finding that the inhibitor pyrrole 2-carboxylate binds to the racemase 160 times more tightly than does proline. *The α-carbon atom of this inhibitor, like that of the*

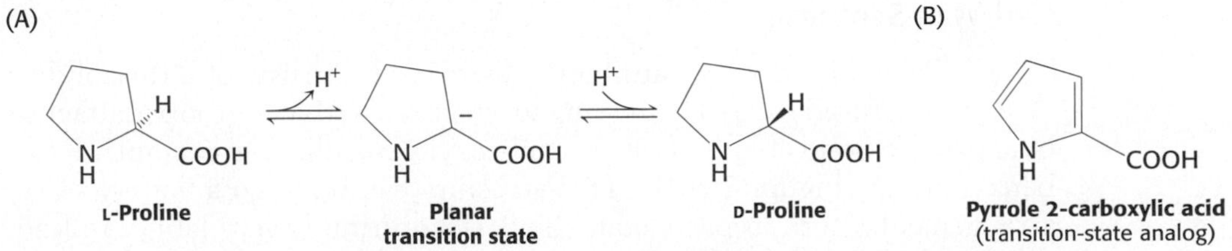

(A)

L-Proline

**Planar
transition state**

D-Proline

(B)

Pyrrole 2-carboxylic acid
(transition-state analog)

Figure 8.28 Inhibition by transition-state analogs. (A) The isomerization of L-proline to D-proline by proline racemase, a bacterial enzyme, proceeds through a planar transition state in which the α-carbon atom is trigonal rather than tetrahedral. (B) Pyrrole 2-carboxylate, a transition-state analog because of its trigonal geometry, is a potent inhibitor of proline racemase.

transition state, is trigonal. An analog that also carries a negative charge on C_α would be expected to bind even more tightly. In general, highly potent and specific inhibitors of enzymes can be produced by synthesizing compounds that more closely resemble the transition state than the substrate itself. The inhibitory power of transition-state analogs underscores the essence of catalysis: *selective binding of the transition state.*

Catalytic Antibodies Demonstrate the Importance of Selective Binding of the Transition State to Enzymatic Activity

Antibodies that recognize transition states should function as catalysts, if our understanding of the importance of the transition state to catalysis is correct. The preparation of an antibody that catalyzes the insertion of a metal ion into a porphyrin nicely illustrates the validity of this approach. Ferrochelatase, the final enzyme in the biosynthetic pathway for the production of heme, catalyzes the insertion of Fe^{2+} into protoporphyrin IX. The nearly planar porphyrin must be bent for iron to enter.

The problem was to find a transition-state analog for this metallation reaction that could be used as an antigen (immunogen) to generate an antibody. The solution came from studies showing that an alkylated porphyrin, N-methylprotoporphyrin, is a potent inhibitor of ferrochelatase. This compound resembles the transition state because N-*alkylation forces the porphyrin to be bent.* Moreover, N-alkylporphyrins was known to chelate metal ions 10^4 times as fast as their unalkylated counterparts do. Bending increases the exposure of the pyrrole nitrogen lone pairs of electrons to solvent, which enables the binding of the iron ion.

An antibody catalyst was produced with the use of an N-alkylporphyrin as the immunogen. The resulting antibody presumably distorts a planar porphyrin to facilitate the entry of a metal ion (Figure 8.29). On average, an antibody molecule metallated 80 porphyrin molecules per hour, a rate only 10-fold less than that of ferrochelatase, and 2500-fold faster than the uncatalyzed reaction. *Catalytic antibodies (abzymes) can indeed be produced by using transition-state analogs as antigens.* Antibodies catalyzing many other kinds of chemical reactions—exemplified by ester and amide hydrolysis, amide-bond formation, transesterification, photoinduced cleavage, photoinduced dimerization, decarboxylation, and oxidization—have been produced with the use of similar strategies. Studies with transition-state analogs provide strong evidence that enzymes can function by assuming a conformation in the active site that is complementary in structure to the transition state. *The power of transition-state analogs is now evident: (1) they are sources of insight into catalytic mechanisms, (2) they can serve as potent and specific inhibitors of enzymes, and (3) they can be used as immunogens to generate a wide range of novel catalysts.*

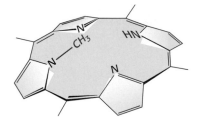

Figure 8.29 Use of transition-state analogs to generate catalytic antibodies. The insertion of a metal ion into a porphyrin by ferrochelatase proceeds through a transition state in which the porphyrin is bent. *N*-Methylmesoporphyrin, a bent porphyrin that resembles the transition state of the ferrochelatase-catalyzed reaction, was used to generate an antibody that also catalyzes the insertion of a metal ion into a porphyrin ring.

Penicillin Irreversibly Inactivates a Key Enzyme in Bacterial Cell-Wall Synthesis

Penicillin, the first antibiotic discovered, consists of a thiazolidine ring fused to a *β-lactam* ring to which a variable R group is attached by a peptide bond (Figure 8.30A). In benzyl penicillin, for example, R is a benzyl group (Figure 8.30B). This structure can undergo a variety of rearrangements, and, in particular, the β-lactam ring is very labile. Indeed, this instability is closely tied to the antibiotic action of penicillin, as will be evident shortly.

How does penicillin inhibit bacterial growth? In 1957, Joshua Lederberg showed that bacteria ordinarily susceptible to penicillin could be grown in its presence if a hypertonic medium were used. The organisms obtained in

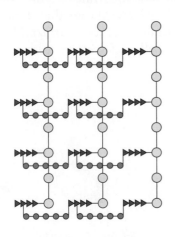

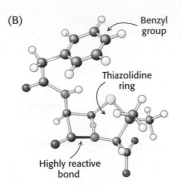

Figure 8.30 Structure of penicillin. The reactive site of penicillin is the peptide bond of its β-lactam ring. (A) Structural formula of penicillin. (B) Representation of benzyl penicillin.

this way, called *protoplasts,* are devoid of a cell wall and consequently lyse when transferred to a normal medium. Hence, penicillin was inferred to interfere with the synthesis of the bacterial cell wall. The cell-wall macromolecule, called a *peptidoglycan,* consists of linear polysaccharide chains that are cross-linked by short peptides (Figure 8.31). The enormous bag-shaped peptidoglycan confers mechanical support and prevents bacteria from bursting in response to their high internal osmotic pressure.

Figure 8.31 Schematic representation of the peptidoglycan in *Staphylococcus aureus*. The sugars are shown in yellow, the tetrapeptides in red, and the pentaglycine bridges in blue. The cell wall is a single, enormous, bag-shaped macromolecule because of extensive cross-linking.

In 1965, James Park and Jack Strominger independently deduced that penicillin blocks the last step in cell-wall synthesis—namely, the cross-linking of different peptidoglycan strands. In the formation of the cell wall of *Staphylococcus aureus,* the amino group at one end of a pentaglycine chain attacks the peptide bond between two D-alanine residues in another peptide unit (Figure 8.32). A peptide bond is formed between glycine and one of the D-alanine residues; the other D-alanine residue is released. This cross-linking reaction is catalyzed by *glycopeptide transpeptidase.* Bacterial cell walls are unique in containing D amino acids, which form cross-links by a mechanism different from that used to synthesize proteins.

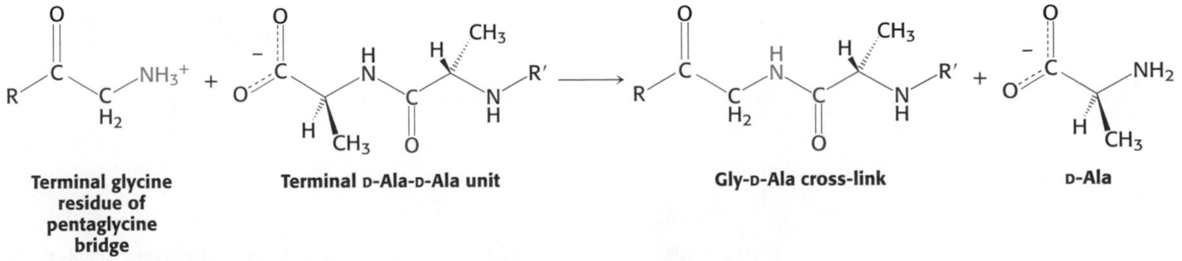

Terminal glycine residue of pentaglycine bridge   **Terminal D-Ala-D-Ala unit**   **Gly-D-Ala cross-link**   **D-Ala**

Figure 8.32 Formation of cross-links in *S. aureus* peptidoglycan. The terminal amino group of the pentaglycine bridge in the cell wall attacks the peptide bond between two D-alanine residues to form a cross-link.

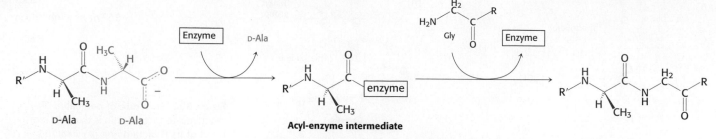

Figure 8.33 Transpeptidation reaction. An acyl-enzyme intermediate is formed in the transpeptidation reaction leading to cross-link formation.

Penicillin inhibits the cross-linking transpeptidase by the Trojan horse stratagem. The transpeptidase normally forms an *acyl intermediate* with the penultimate D-alanine residue of the D-Ala-D-Ala peptide (Figure 8.33). This covalent acyl-enzyme intermediate then reacts with the amino group of the terminal glycine in another peptide to form the cross-link. Penicillin is welcomed into the active site of the transpeptidase because it mimics the D-Ala-D-Ala moiety of the normal substrate (Figure 8.34). Bound penicillin then forms a covalent bond with a serine residue at the active site of the enzyme. *This penicilloyl-enzyme does not react further. Hence, the transpeptidase is irreversibly inhibited and cell-wall synthesis cannot take place.*

Figure 8.34 Conformations of penicillin and a normal substrate. The conformation of penicillin in the vicinity of its reactive peptide bond (A) resembles the postulated conformation of the transition state of R-D-Ala-D-Ala (B) in the transpeptidation reaction. [After B. Lee. *J. Mol. Biol.* 61 (1971):463–469.]

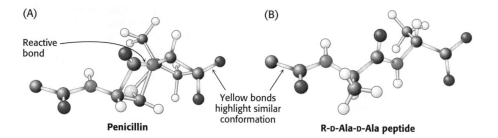

Why is penicillin such an effective inhibitor of the transpeptidase? The highly strained, four-membered β-lactam ring of penicillin makes it especially reactive. On binding to the transpeptidase, the serine residue at the active site attacks the carbonyl carbon atom of the lactam ring to form the penicilloyl-serine derivative (Figure 8.35). Because the peptidase participates in its own inactivation, penicillin acts as a suicide inhibitor.

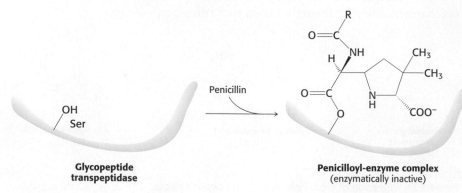

Figure 8.35 Formation of a penicilloyl-enzyme complex. Penicillin reacts with the transpeptidase to form an inactive complex, which is indefinitely stable.

Summary

8.1 Enzymes Are Powerful and Highly Specific Catalysts

The catalysts in biological systems are enzymes, and nearly all enzymes are proteins. Enzymes are highly specific and have great catalytic power. They can enhance reaction rates by factors of 10^6 or more. Many enzymes require cofactors for activity. Such cofactors can be metal ions or small, vitamin-derived organic molecules called coenzymes.

8.2 Free Energy Is a Useful Thermodynamic Function for Understanding Enzymes

Free energy (G) is the most valuable thermodynamic function for determining whether a reaction can take place and for understanding the energetics of catalysis. A reaction can occur spontaneously only if the change in free energy (ΔG) is negative. The free-energy change of a reaction that takes place when reactants and products are at unit activity is called the standard free-energy change (ΔG°). Biochemists usually use $\Delta G^{\circ\prime}$, the standard free-energy change at pH 7. Enzymes do not alter reaction equilibria; rather, they increase reaction rates.

8.3 Enzymes Accelerate Reactions by Facilitating the Formation of the Transition State

Enzymes serve as catalysts by decreasing the free energy of activation of chemical reactions. Enzymes accelerate reactions by providing a reaction pathway in which the transition state (the highest-energy species) has a lower free energy and hence is more rapidly formed than in the uncatalyzed reaction.

The first step in catalysis is the formation of an enzyme–substrate complex. Substrates are bound to enzymes at active-site clefts from which water is largely excluded when the substrate is bound. The specificity of enzyme–substrate interactions arises mainly from hydrogen bonding, which is directional, and from the shape of the active site, which rejects molecules that do not have a sufficiently complementary shape. The recognition of substrates by enzymes is often accompanied by conformational changes at active sites, and such changes facilitate the formation of the transition state.

8.4 The Michaelis–Menten Model Accounts for the Kinetic Properties of Many Enzymes

The kinetic properties of many enzymes are described by the Michaelis–Menten model. In this model, an enzyme (E) combines with a substrate (S) to form an enzyme–substrate (ES) complex, which can proceed to form a product (P) or to dissociate into E and S.

$$E + S \underset{k_{-1}}{\overset{k_1}{\rightleftharpoons}} ES \xrightarrow{k_2} E + P$$

The rate V_0 of formation of product is given by the Michaelis–Menten equation:

$$V_0 = V_{max}\frac{[S]}{[S] + K_M}$$

in which V_{max} is the reaction rate when the enzyme is fully saturated with substrate and K_M, the Michaelis constant, is the substrate concentration at which the reaction rate is half maximal. The maximal rate, V_{max}, is equal to the product of k_2 or k_{cat} and the total concentration of enzyme. The kinetic constant k_{cat}, called the turnover number, is the number of substrate molecules converted into product per unit time at a single catalytic site when the enzyme is fully saturated with substrate. Turnover numbers for most enzymes are between 1 and 10^4 per second. The ratio of k_{cat}/K_M provides a penetrating probe into enzyme efficiency.

Allosteric enzymes constitute an important class of enzymes whose catalytic activity can be regulated. These enzymes, which do not conform to Michaelis–Menten kinetics, have multiple active sites. These active sites display cooperativity, as evidenced by a sigmoidal dependence of reaction velocity on substrate concentration.

8.5 Enzymes Can Be Inhibited by Specific Molecules

Specific small molecules or ions can inhibit even nonallosteric enzymes. In irreversible inhibition, the inhibitor is covalently linked to the enzyme or bound so tightly that its dissociation from the enzyme is very slow. Covalent inhibitors provide a means of mapping the enzyme's active site. In contrast, reversible inhibition is characterized by a more rapid equilibrium between enzyme and inhibitor. A competitive inhibitor prevents the substrate from binding to the active site. It reduces the reaction velocity by diminishing the proportion of enzyme molecules that are bound to substrate. Competitive inhibition can be overcome by raising the substrate concentration. In uncompetitive inhibition, the inhibitor combines only with the enzyme–substrate complex. In noncompetitive inhibition, the inhibitor decreases the turnover number. Uncompetitive and noncompetitive inhibition cannot be overcome by raising the substrate concentration.

The essence of catalysis is selective stabilization of the transition state. Hence, an enzyme binds the transition state more tightly than the substrate. Transition-state analogs are stable compounds that mimic key features of this highest-energy species. They are potent and specific inhibitors of enzymes. Proof that transition-state stabilization is a key aspect of enzyme activity comes from the generation of catalytic antibodies. Transition-state analogs are used as antigens, or immunogens, in generating catalytic antibodies.

APPENDIX: Enzymes Are Classified on the Basis of the Types of Reactions That They Catalyze

Many enzymes have common names that provide little information about the reactions that they catalyze. For example, a proteolytic enzyme secreted by the pancreas is called trypsin. Most other enzymes are named for their substrates and for the reactions that they catalyze, with the suffix "ase" added. Thus, a peptide hydrolase is an enzyme that hyrolyzes peptide bonds, whereas ATP synthase is an enzyme that synthesizes ATP.

To bring some consistency to the classification of enzymes, in 1964 the International Union of Biochemistry established an Enzyme Commission to develop a nomenclature for enzymes. Reactions were divided into six major groups numbered 1 through 6 (Table 8.8). These groups were subdivided and further subdivided so that a four-digit number preceded by the letters EC for Enzyme Commission could precisely identify all enzymes.

Consider as an example nucleoside monophosphate (NMP) kinase, an enzyme that we will examine in detail in Section 9.4. It catalyzes the following reaction:

$$\text{ATP} + \text{NMP} \rightleftharpoons \text{ADP} + \text{NDP}$$

NMP kinase transfers a phosphoryl group from ATP to NMP to form a nucleoside diphosphate (NDP) and ADP. Consequently, it is a transferase, or member of group 2. Many groups other than phosphoryl groups, such as sugars and single-carbon units, can be transferred. Transferases that shift a phosphoryl group are designated 2.7. Various functional groups can accept the phosphoryl group. If a phosphate is the acceptor, the transferase is designated 2.7.4. The final number designates the acceptor more precisely. In regard to NMP kinase, a nucleoside monophosphate is the acceptor, and the enzyme's designation is EC 2.7.4.4. Although the common names are used routinely, the classification number is used when the precise identity of the enzyme might be ambiguous.

TABLE 8.8 Six major classes of enzymes

Class	Type of reaction	Example	Chapter
1. Oxidoreductases	Oxidation–reduction	Lactate dehydrogenase	16
2. Transferases	Group transfer	Nucleoside monophosphate kinase (NMP kinase)	9
3. Hydrolases	Hydrolysis reactions (transfer of functional groups to water)	Chymotrypsin	9
4. Lyases	Addition or removal of groups to form double bonds	Fumarase	17
5. Isomerases	Isomerization (intramolecular group transfer)	Triose phosphate isomerase	16
6. Ligases	Ligation of two substrates at the expense of ATP hydrolysis	Aminoacyl-tRNA synthetase	30

Key Terms

enzyme (p. 206)

substrate (p. 206)

cofactor (p. 207)

apoenzyme (p. 207)

holoenzyme (p. 207)

coenzyme (p. 207)

prosthetic group (p. 207)

transition state (p. 211)

free energy of activation (p. 211)

free energy (p. 212)

active site (p. 214)

induced fit (p. 215)

K_M (the Michaelis constant) (p. 218)

V_{max} (p. 219)

Michaelis-Menten equation (p. 219)

Lineweaver-Burk equation (p. 220)

turnover number (p. 221)

k_{cat}/K_M (p. 221)

sequential reaction (p. 223)

double-displacement (ping-pong) reaction (p. 224)

allosteric enzyme (p. 224)

competitive inhibition (p. 225)

uncompetitive inhibition (p. 226)

noncompetitive inhibition (p. 226)

group-specific reagent (p. 229)

affinity label (reactive substrate analog) (p. 229)

mechanism-based (suicide) inhibition (p. 230)

transition-state analog (p. 231)

catalytic antibody (abzyme) (p. 232)

Selected Readings

Where to Start

Koshland, D. E., Jr. 1987. Evolution of catalytic function. *Cold Spring Harbor Symp. Quant. Biol.* 52:1–7.

Jencks, W. P. 1987. Economics of enzyme catalysis. *Cold Spring Harbor Symp. Quant. Biol.* 52:65–73.

Lerner, R. A., and Tramontano, A. 1988. Catalytic antibodies. *Sci. Am.* 258(3):58–70.

Books

Fersht, A. 1999. *Structure and Mechanism in Protein Science: A Guide to Enzyme Catalysis and Protein Folding.* W. H. Freeman and Company.

Walsh, C. 1979. *Enzymatic Reaction Mechanisms.* W. H. Freeman and Company.

Page, M. I., and Williams, A. (Eds.). 1987. *Enzyme Mechanisms.* Royal Society of Chemistry.

Bender, M. L., Bergeron, R. J., and Komiyama, M. 1984. *The Bioorganic Chemistry of Enzymatic Catalysis.* Wiley-Interscience.

Abelson, J. N., and Simon, M. I. (Eds.). 1992. *Methods in Enzymology.* Academic Press.

Boyer, P. D. (Ed.). 1970. *The Enzymes* (3d ed.). Academic Press.

Friedmann, H. C. (Ed.). 1981. *Benchmark Papers in Biochemistry.* Vol. 1, *Enzymes.* Hutchinson Ross.

Transition-State Stabilization, Analogs, and Other Enzyme Inhibitors

Schramm, V. L. 1998. Enzymatic transition states and transition state analog design. *Annu. Rev. Biochem.* 67:693–720.

Pauling, L. 1948. Nature of forces between large molecules of biological interest. *Nature* 161:707–709.

Leinhard, G. E. 1973. Enzymatic catalysis and transition-state theory. *Science* 180:149–154.

Kraut, J. 1988. How do enzymes work? *Science* 242:533–540.

Waxman, D. J., and Strominger, J. L. 1983. Penicillin-binding proteins and the mechanism of action of β-lactam antibiotics. *Annu. Rev. Biochem.* 52:825–869.

Abraham, E. P. 1981. The β-lactam antibiotics. *Sci. Am.* 244:76–86.

Walsh, C. T. 1984. Suicide substrates, mechanism-based enzyme inactivators: Recent developments. *Annu. Rev. Biochem.* 53:493–535.

Catalytic Antibodies

Hilvert, D. 2000. Critical analysis of antibody catalysis. *Annu. Rev. Biochem.* 69:751–794.

Wade, H., and Scanlan, T. S. 1997. The structural and functional basis of antibody catalysis. *Annu. Rev. Biophys. Biomol. Struct.* 26:461–493.

Lerner, R. A., Benkovic, S. J., and Schultz, P. G. 1991. At the crossroads of chemistry and immunology: Catalytic antibodies. *Science* 252:659–667.

Cochran, A. G., and Schultz, P. G. 1990. Antibody-catalyzed porphyrin metallation. *Science* 249:781–783.

Enzyme Kinetics and Mechanisms

Benkovic, S. J., and Hammes-Schiller, S. 2003. A perspective on enzyme catalysis. *Science* 301:1196–1202.

Hur, S., and Bruice, T. C. 2003. The near attack conformation approach to the study of the chorismate to prephenate reaction. *Proc. Natl. Acad. Sci. U. S. A.* 100:12015–12020.

Xie, X. S., and Lu, H. P. 1999. Single-molecule enzymology. *J. Biol. Chem.* 274:15967–15970.

Miles, E. W., Rhee, S., and Davies, D. R. 1999. The molecular basis of substrate channeling. *J. Biol. Chem.* 274:12193–12196.

Warshel, A. 1998. Electrostatic origin of the catalytic power of enzymes and the role of preorganized active sites. *J. Biol. Chem.* 273:27035–27038.

Cannon, W. R., and Benkovic, S. J. 1999. Solvation, reorganization energy, and biological catalysis. *J. Biol. Chem.* 273:26257–26260.

Cleland, W. W., Frey, P. A., and Gerlt, J. A. 1998. The low barrier hydrogen bond in enzymatic catalysis. *J. Biol. Chem.* 273:25529–25532.

Romesberg, F. E., Santarsiero, B. D., Spiller, B., Yin, J., Barnes, D., Schultz, P. G., and Stevens, R. C. 1998. Structural and kinetic evidence for strain in biological catalysis. *Biochemistry* 37:14404–14409.

Lu, H. P., Xun, L., and Xie, X. S. 1998. Single-molecule enzymatic dynamics. *Science* 282:1877–1882.

Fersht, A. R., Leatherbarrow, R. J., and Wells, T. N. C. 1986. Binding energy and catalysis: A lesson from protein engineering of the tyrosyl-tRNA synthetase. *Trends Biochem. Sci.* 11:321–325.

Jencks, W. P. 1975. Binding energy, specificity, and enzymic catalysis: The Circe effect. *Adv. Enzymol.* 43:219–410.

Knowles, J. R., and Albery, W. J. 1976. Evolution of enzyme function and the development of catalytic efficiency. *Biochemistry* 15:5631–5640.

Problems

1. *Hydrolytic driving force.* The hydrolysis of pyrophosphate to orthophosphate is important in driving forward biosynthetic reactions such as the synthesis of DNA. This hydrolytic reaction is catalyzed in *Escherichia coli* by a pyrophosphatase that has a mass of 120 kd and consists of six identical subunits. For this enzyme, a unit of activity is defined as the amount of enzyme that hydrolyzes 10 μmol of pyrophosphate in 15 minutes at 37°C under standard assay conditions. The purified enzyme has a V_{max} of 2800 units per milligram of enzyme.

(a) How many moles of substrate are hydrolyzed per second per milligram of enzyme when the substrate concentration is much greater than K_M?

(b) How many moles of active sites are there in 1 mg of enzyme? Assume that each subunit has one active site.

(c) What is the turnover number of the enzyme? Compare this value with others mentioned in this chapter.

2. *Destroying the Trojan horse.* Penicillin is hydrolyzed and thereby rendered inactive by penicillinase (also known as β-lactamase), an enzyme present in some resistant bacteria. The mass of this enzyme in *Staphylococcus aureus* is 29.6 kd. The amount of penicillin hydrolyzed in 1 minute in a 10-ml solution containing 10^{-9} g of purified penicillinase was measured as a function of the concentration of penicillin. Assume that the concentration of penicillin does not change appreciably during the assay.

[Penicillin] μM	Amount hydrolyzed (nanomoles)
1	0.11
3	0.25
5	0.34
10	0.45
30	0.58
50	0.61

(a) Plot V_0 versus [S] and $1/V_0$ versus $1/$[S] for these data. Does penicillinase appear to obey Michaelis–Menten kinetics? If so, what is the value of K_M?

(b) What is the value of V_{max}?

(c) What is the turnover number of penicillinase under these experimental conditions? Assume one active site per enzyme molecule.

3. *Counterpoint.* Penicillinase (β-lactamase) hydrolyzes penicillin. Compare penicillinase with glycopeptide transpeptidase.

4. *Mode of inhibition.* The kinetics of an enzyme are measured as a function of substrate concentration in the presence and in the absence of 2 mM inhibitor (I).

[S] (μM)	Velocity (μmol/minute)	
	No inhibitor	Inhibitor
3	10.4	4.1
5	14.5	6.4
10	22.5	11.3
30	33.8	22.6
90	40.5	33.8

(a) What are the values of V_{max} and K_M in the absence of inhibitor? In its presence?

(b) What type of inhibition is it?

(c) What is the binding constant of this inhibitor?

(d) If [S] = 10 μM and [I] = 2 mM, what fraction of the enzyme molecules have a bound substrate? A bound inhibitor?

(e) If [S] = 30 μM, what fraction of the enzyme molecules have a bound substrate in the presence and in the absence of 2 mM inhibitor? Compare this ratio with the ratio of the reaction velocities under the same conditions.

5. *A different mode.* The kinetics of the enzyme considered in problem 4 are measured in the presence of a different inhibitor. The concentration of this inhibitor is 100 μM.

(a) What are the values of V_{max} and K_M in the presence of this inhibitor? Compare them with those obtained in problem 4.

(b) What type of inhibition is it?

(c) What is the dissociation constant of this inhibitor?

[S] (μM)	Velocity (μmol/minute)	
	No inhibitor	Inhibitor
3	10.4	2.1
5	14.5	2.9
10	22.5	4.5
30	33.8	6.8
90	40.5	8.1

(d) If [S] = 30 μM, what fraction of the enzyme molecules have a bound substrate in the presence and in the absence of 100 μM inhibitor?

6. *A fresh view.* The plot of $1/V_0$ versus $1/[S]$ is sometimes called a Lineweaver-Burk plot. Another way of expressing the kinetic data is to plot V_0 versus $V_0/[S]$, which is known as an Eadie–Hofstee plot.

(a) Rearrange the Michaelis–Menten equation to give V_0 as a function of $V_0/[S]$.

(b) What is the significance of the slope, the vertical intercept, and the horizontal intercept in a plot of V_0 versus $V_0/[S]$?

(c) Sketch a plot of V_0 versus $V_0/[S]$ in the absence of an inhibitor, in the presence of a competitive inhibitor, and in the presence of a noncompetitive inhibitor.

7. *Potential donors and acceptors.* The hormone progesterone contains two ketone groups. At pH 7, which side chains of the receptor might form hydrogen bonds with progesterone?

8. *Competing substrates.* Suppose that two substrates, A and B, compete for an enzyme. Derive an expression relating the ratio of the rates of utilization of A and B, V_A/V_B, to the concentrations of these substrates and their values of k_{cat} and K_M. (Hint: Express V_A as a function of k_{cat}/K_M for substrate A, and do the same for V_B.) Is specificity determined by K_M alone?

9. *A tenacious mutant.* Suppose that a mutant enzyme binds a substrate 100-fold more tightly than does the native enzyme. What is the effect of this mutation on catalytic rate if the binding of the transition state is unaffected?

10. *More Michaelis–Menten.* For an enzyme that follows simple Michaelis–Menten kinetics, what is the value of V_{max} if V_0 is equal to 1 μmol minute^{-1} at $1/10 \, K_M$?

Data Interpretation Problems

11. *Varying the enzyme.* For a one-substrate, enzyme-catalyzed reaction, double-reciprocal plots were determined for three different enzyme concentrations. Which of the following three families of curve would you expect to be obtained? Explain.

12. *Too much of a good thing.* A simple Michaelis–Menten enzyme, in the absence of any inhibitor, displayed the following

kinetic behavior. The expected value of V_{max} is shown on the y-axis.

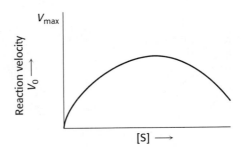

(a) Draw a double-reciprocal plot that corresponds to the velocity-versus-substrate curve.

(b) Provide an explanation for the kinetic results.

13. *Rate-limiting step.* In the conversion of A into D in the following biochemical pathway, enzymes E_A, E_B, and E_C have the K_M values indicated under each enzyme. If all of the substrates and products are present at a concentration of 10^{-4} M and the enzymes have approximately the same V_{max}, which step will be rate limiting and why?

$$A \underset{E_A}{\rightleftharpoons} B \underset{E_B}{\rightleftharpoons} C \underset{E_C}{\rightleftharpoons} D$$

$$K_M = \quad 10^{-2}\,M \quad\quad 10^{-4}\,M \quad\quad 10^{-4}\,M$$

Chapter Integration Problems

14. *Titration experiment.* The effect of pH on the activity of an enzyme was examined. At its active site, the enzyme has an ionizable group that must be negatively charged for substrate binding and catalysis to take place. The ionizable group has a pK_a of 6.0. The substrate is positively charged throughout the pH range of the experiment.

$$E^- + S^+ \rightleftharpoons E^- - S^+ \longrightarrow E^- + P^+$$
$$+$$
$$H^+$$
$$\Updownarrow$$
$$EH$$

(a) Draw the V_0-versus-pH curve when the substrate concentration is much greater than the enzyme K_M.

(b) Draw the V_0-versus-pH curve when the substrate concentration is much less than the enzyme K_M.

(c) At which pH will the velocity equal one-half of the maximal velocity attainable under these condition?

15. *A question of stability.* Pyridoxal phosphate (PLP) is a coenzyme for the enzyme ornithine aminotransferase. The enzyme was purified from cells grown in PLP-deficient media as well as from cells grown in media that contained pyridoxal phosphate. The stability of the enzyme was then measured by incubating

the enzyme at 37°C and assaying for the amount of enzyme activity remaining. The following results were obtained.

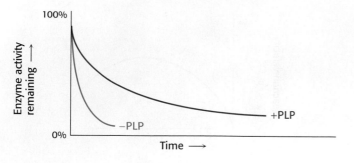

(a) Why does the amount of active enzyme decrease with the time of incubation?

(b) Why does the amount of enzyme from the PLP-deficient cells decline more rapidly?

Catalytic Strategies

Chess and enzymes have in common the use of strategy, consciously thought out in the game of chess and selected by evolution for the action of an enzyme. The three amino acid residues at the right, denoted by the white bonds, constitute a catalytic triad found in the active site of a class of enzymes that cleave peptide bonds. The substrate, represented by the molecule with the black bonds, is as hopelessly trapped as the king in the photograph of a chess match at the left and is sure to be cleaved. [(Left) Courtesy of Wendie Berg.]

What are the sources of the catalytic power and specificity of enzymes? This chapter presents the catalytic strategies used by four classes of enzymes: the serine proteases, carbonic anhydrases, restriction endonucleases, and nucleoside monophosphate (NMP) kinases. The first three classes of enzymes catalyze reactions that require the addition of water to a substrate, whereas the fourth class catalyzes reactions that require the *prevention* of the addition of water. The mechanisms of these enzymes have been revealed through the use of incisive experimental probes, including the techniques of protein structure determination (Chapter 3) and site-directed mutagenesis (Chapter 5). The mechanisms illustrate many important principles of catalysis. We shall see how these enzymes facilitate the formation of the transition state through the use of binding energy and induced fit as well as several specific catalytic strategies.

Each of the four enzymes in this chapter illustrates the use of such strategies to solve a different problem. For the serine proteases, exemplified by chymotrypsin, the challenge is to promote a reaction that is almost immeasurably slow at neutral pH in the absence of a catalyst. For carbonic anhydrases, the challenge is to achieve a high absolute rate of reaction, suitable for integration with other rapid physiological processes. For restriction endonucleases such as *Eco*RV, the challenge is to attain a high degree of specificity. Finally, for NMP kinases, the challenge is to transfer a phosphoryl group from ATP to a nucleotide and not to water. Each of the examples selected is a member of a large protein class. For each of these classes, comparison between class members reveal how enzyme active sites have evolved and been refined. Structural and mechanistic comparisons of enzyme action are thus the sources of insight into the evolutionary history of enzymes. In addition, our knowledge of catalytic strategies has been used to develop practical applications, including drugs that are potent and specific enzyme inhibitors. Finally, although we shall not consider catalytic RNA molecules explicitly in this chapter, the principles apply to these catalysts in addition to protein catalysts.

A Few Basic Catalytic Principles Are Used by Many Enzymes

In Chapter 8, we learned that enzymatic catalysis begins with substrate binding. The *binding energy* is the free energy released in the formation of a large number of weak interactions between the enzyme and the substrate. We can envision this binding energy as serving two purposes: it establishes substrate specificity and increases catalytic efficiency. Only the correct substrate can participate in most or all of the interactions with the enzyme and thus maximize binding energy, accounting for the exquisite substrate specificity exhibited by many enzymes. Furthermore, the full complement of such interactions is formed only when the substrate is in the transition state. Thus, interactions between the enzyme and the substrate stabilize the transition state, thereby lowering the activation energy. The binding energy can also promote structural changes in both the enzyme and the substrate that facilitate catalysis, a process referred to as *induced fit*.

Enzymes commonly employ one or more of the following strategies to catalyze specific reactions:

1. *Covalent Catalysis.* In covalent catalysis, the active site contains a reactive group, usually a powerful nucleophile, that becomes temporarily covalently attached to a part of the substrate in the course of catalysis. The proteolytic enzyme chymotrypsin provides an excellent example of this strategy (Section 9.1).

2. *General Acid–Base Catalysis.* In general acid–base catalysis, a molecule other than water plays the role of a proton donor or acceptor. Chymotrypsin uses a histidine residue as a base catalyst to enhance the nucleophilic power of serine, whereas a histidine residue in carbonic anhydrase facilitates the removal of a hydrogen ion from a zinc-bound water molecule to generate hydroxide ion (Section 9.2).

3. *Catalysis by Approximation.* Many reactions include two distinct substrates. In such cases, the reaction rate may be considerably enhanced by bringing the two substrates together along a single binding surface on an enzyme. NMP kinases bring two nucleotides together to facilitate the transfer of a phosphoryl group from one nucleotide to another (Section 9.4).

4. *Metal Ion Catalysis.* Metal ions can function catalytically in several ways. For instance, a metal ion may facilitate the formation of nucleophiles such as hydroxide ion by direct coordination. A zinc(II) ion serves this purpose in catalysis by carbonic anhydrase (Section 9.2). Alternatively, a metal ion may serve as an electrophile, stabilizing a negative charge on a reaction intermediate. A magnesium(II) ion plays this role in *Eco*RV (Section 9.3). Finally, a metal ion may serve as a bridge between enzyme and substrate, increasing the binding energy and holding the substrate in a conformation appropriate for catalysis. This strategy is used by NMP kinases (Section 9.4) and, indeed, by almost all enzymes that utilize ATP as a substrate.

9.1 Proteases Facilitate a Fundamentally Difficult Reaction

Protein turnover is an important process in living systems (Chapter 23). Proteins that have served their purpose must be degraded so that their constituent amino acids can be recycled for the synthesis of new proteins. Proteins ingested in the diet must be broken down into small peptides and amino acids for absorption in the gut. Furthermore, as described in detail in Chapter 10, proteolytic reactions are important in regulating the activity of certain enzymes and other proteins.

Proteases cleave proteins by a hydrolysis reaction—the addition of a molecule of water to a peptide bond:

$$R_1 \underset{\underset{H}{|}}{\overset{\overset{O}{\|}}{C}} N R_2 + H_2O \rightleftharpoons R_1 - \overset{\overset{O}{\|}}{C} - \underset{O}{} + R_2 - NH_3^+$$

Although the hydrolysis of peptide bonds is thermodynamically favored, such hydrolysis reactions are extremely slow. In the absence of a catalyst, the half-life for the hydrolysis of a typical peptide at neutral pH is estimated to be between 10 and 1000 years. Yet, peptide bonds must be hydrolyzed within milliseconds in some biochemical processes.

The chemical bonding in peptide bonds is responsible for their kinetic stability. Specifically, the resonance structure that accounts for the planarity of a peptide bond (p. 37) also makes such bonds resistant to hydrolysis. This resonance structure endows the peptide bond with partial double-bond character:

$$R_1 \overset{\overset{O}{\|}}{C} \underset{\underset{H}{|}}{N} R_2 \longleftrightarrow R_1 \overset{\overset{O^-}{|}}{C} \overset{+}{\underset{\underset{H}{|}}{N}} R_2$$

The carbon nitrogen bond is strengthened by its double-bond character, and the carbonyl carbon atom is less electrophilic and less susceptible to nucleophilic attack than are the carbonyl carbon atoms in compounds such as carboxylate esters. Consequently, to promote peptide-bond cleavage, an enzyme must facilitate nucleophilic attack at a normally unreactive carbonyl group.

Chymotrypsin Possesses a Highly Reactive Serine Residue

A number of proteolytic enzymes participate in the breakdown of proteins in the digestive systems of mammals and other organisms. One such enzyme, chymotrypsin, cleaves peptide bonds selectively on the carboxyl-terminal side of the large hydrophobic amino acids such as tryptophan, tyrosine, phenylalanine, and methionine (Figure 9.1). Chymotrypsin is a good example of the use of *covalent catalysis*. The enzyme employs a powerful nucleophile to attack the unreactive carbonyl carbon atom of the substrate. This nucleophile becomes covalently attached to the substrate briefly in the course of catalysis.

What is the nucleophile that chymotrypsin employs to attack the substrate carbonyl carbon atom? A clue came from the fact that chymotrypsin contains an extraordinarily reactive serine residue. Chymotrypsin molecules

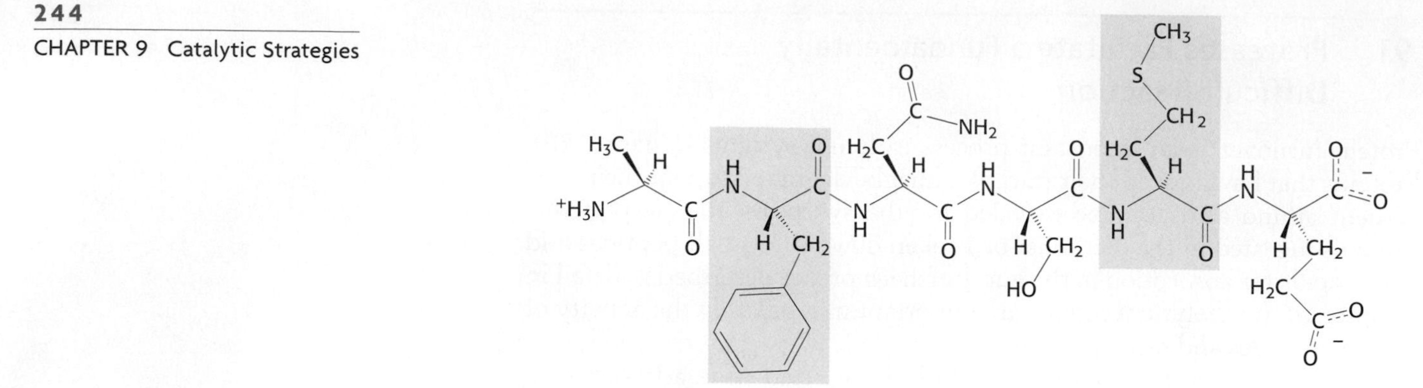

Figure 9.1 **Specificity of chymotrypsin.** Chymotrypsin cleaves proteins on the carboxyl side of aromatic or large hydrophobic amino acids (shaded orange). The likely bonds cleaved by chymotrypsin are indicated in red.

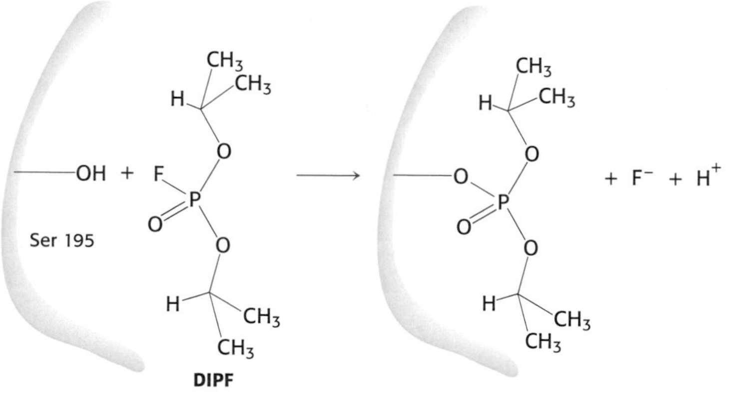

Figure 9.2 **An unusually reactive serine in chymotrypsin.** Chymotrypsin is inactivated by treatment with diisopropylphosphofluoridate (DIPF), which reacts only with serine 195 among 28 possible serine residues.

treated with organofluorophosphates such as diisopropylphosphofluoridate (DIPF; p. 229) lost all activity irreversibly (Figure 9.2). Only a single residue, serine 195, was modified. This *chemical modification reaction* suggested that this unusually reactive serine residue plays a central role in the catalytic mechanism of chymotrypsin.

Chymotrypsin Action Proceeds in Two Steps Linked by a Covalently Bound Intermediate

A study of the enzyme's kinetics provided a second clue to chymotrypsin's catalytic mechanism. The kinetics of enzyme action are often easily monitored by having the enzyme act on a substrate analog that forms a colored product. For chymotrypsin, such a *chromogenic substrate* is N-acetyl-L-phenylalanine p-nitrophenyl ester. This substrate is an ester rather than an amide, but many proteases will also hydrolyze esters. One of the products formed by chymotrypsin's cleavage of this substrate is p-nitrophenolate, which has a yellow color (Figure 9.3). Measurements of the absorbance of light revealed the amount of p-nitrophenolate being produced.

Under steady-state conditions, the cleavage of this substrate obeys Michaelis–Menten kinetics with a K_M of 20 μM and a k_{cat} of 77 s^{-1}. The

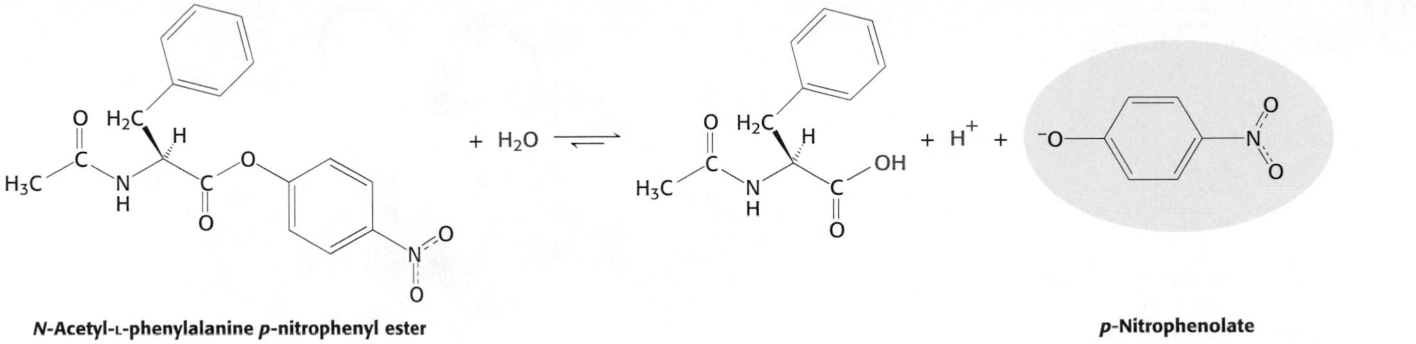

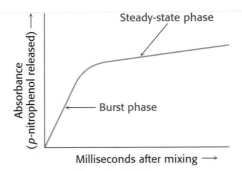

N-Acetyl-L-phenylalanine p-nitrophenyl ester

p-Nitrophenolate

Figure 9.3 Chromogenic substrate. N-Acetyl-L-phenylalanine p-nitrophenyl ester yields a yellow product, p-nitrophenolate, on cleavage by chymotrypsin. p-Nitrophenolate forms by deprotonation of p-nitrophenol at pH 7.

initial phase of the reaction was examined by using the stopped-flow method, which makes it possible to mix enzyme and substrate and monitor the results within a millisecond. This method revealed an initial rapid burst of colored product, followed by its slower formation as the reaction reached the steady state (Figure 9.4). These results suggest that hydrolysis proceeds in two steps. The burst is observed because the first step is more rapid than the second step.

The two steps are explained by the formation of a covalently bound enzyme–substrate intermediate (Figure 9.5). First, the acyl group of the substrate becomes covalently attached to the enzyme as p-nitrophenolate (or an amine if the substrate is an amide rather than an ester) is released. The enzyme–acyl group complex is called the *acyl-enzyme intermediate*. Second, the acyl-enzyme intermediate is hydrolyzed to release the carboxylic acid component of the substrate and regenerate the free enzyme. Thus, one molecule of p-nitrophenolate is produced rapidly from each enzyme molecule as the acyl-enzyme intermediate is formed. However, it takes longer for the enzyme to be "reset" by the hydrolysis of the acyl-enzyme intermediate, and both steps are required for enzyme turnover.

Figure 9.4 Kinetics of chymotrypsin catalysis. Two stages are evident in the cleaving of N-acetyl-L-phenylalanine p-nitrophenyl ester by chymotrypsin: a rapid burst phase (pre-steady state) and a steady-state phase.

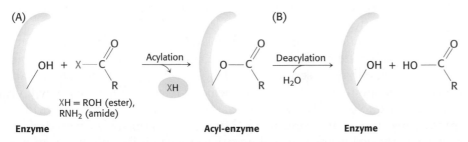

Figure 9.5 Covalent catalysis. Hydrolysis by chymotrypsin takes place in two stages: (A) acylation to form the acyl-enzyme intermediate followed by (B) deacylation to regenerate the free enzyme.

Serine Is Part of a Catalytic Triad That Also Includes Histidine and Aspartate

The three-dimensional structure of chymotrypsin was solved by David Blow in 1967. Overall, chymotrypsin is roughly spherical and comprises three polypeptide chains, linked by disulfide bonds. It is synthesized as a single polypeptide, termed *chymotrypsinogen*, which is activated by the

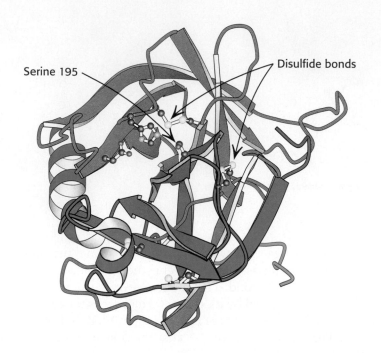

Figure 9.6 Location of the active site in chymotrypsin. Chymotrypsin consists of three chains, shown in ribbon form in orange, blue, and green. The side chains of the catalytic triad residues are shown as ball-and-stick representations. *Notice* these side chains, including serine 195, lining the active site in the upper half of the structure. *Also notice* two intrastrand and two interstrand disulfide bonds in various locations throughout the molecule. [Drawn from 1GCT.pdb.]

proteolytic cleavage of the polypeptide to yield the three chains (p. 289). The active site of chymotrypsin, marked by serine 195, lies in a cleft on the surface of the enzyme (Figure 9.6). The structure of the active site explained the special reactivity of serine 195 (Figure 9.7). The side chain of serine 195 is hydrogen bonded to the imidazole ring of histidine 57. The —NH group of this imidazole ring is, in turn, hydrogen bonded to the carboxylate group of aspartate 102. This constellation of residues is referred to as the *catalytic triad*. How does this arrangement of residues lead to the high reactivity of serine 195? The histidine residue serves to position the serine side chain and to polarize its hydroxyl group so that it is poised for deprotonation. In the presence of the substrate, it accepts the proton from the serine 195 hydroxyl group. In doing so, the residue acts as a general base catalyst. The withdrawal of the proton from the hydroxyl group generates an alkoxide ion, which is a much more powerful nucleophile than an alcohol is. The aspartate residue helps orient the histidine residue and make it a better proton acceptor through hydrogen bonding and electrostatic effects.

These observations suggest a mechanism for peptide hydrolysis (Figure 9.8). After substrate binding (step 1), the reaction begins with the oxygen atom of the side chain of serine 195 making a nucleophilic attack on the carbonyl carbon atom of the target peptide bond (step 2). There are now four atoms bonded to the carbonyl carbon, arranged as a tetrahedron, instead of three atoms in a planar arrangement. The inherently unstable *tetrahedral intermediate* formed bears a formal negative charge on the oxygen atom derived from the carbonyl group. This charge is stabilized by interactions with

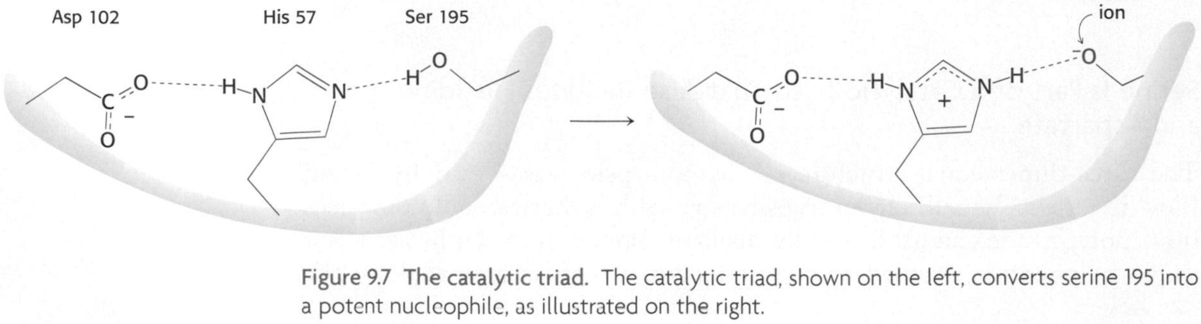

Figure 9.7 The catalytic triad. The catalytic triad, shown on the left, converts serine 195 into a potent nucleophile, as illustrated on the right.

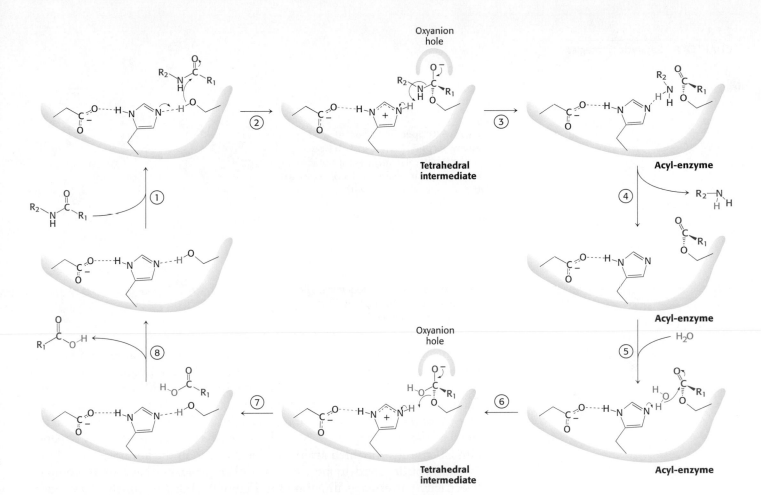

Figure 9.8 Peptide hydrolysis by chymotrypsin. The mechanism of peptide hydrolysis illustrates the principles of covalent and acid–base catalysis. The reaction proceeds in eight steps: (1) substrate binding, (2) nucleophilic attack of serine on the peptide carbonyl group, (3) collapse of the tetrahedral intermediate, (4) release of the amine component, (5) water binding, (6) nucleophilic attack of water on the acyl-enzyme intermediate, (7) collapse of the tetrahedral intermediate; and (8) release of the carboxylic acid component. The dashed green lines represent hydrogen bonds.

NH groups from the protein in a site termed the *oxyanion hole* (Figure 9.9). These interactions also help stabilize the transition state that precedes the formation of the tetrahedral intermediate. This tetrahedral intermediate collapses to generate the acyl-enzyme (step 3). This step is facilitated by the transfer of the proton being held by the positively charged histidine residue to the amino group formed by cleavage of the peptide bond. The amine component is now free to depart from the enzyme (step 4), completing the first stage of the hydrolytic reaction—acylation of the enzyme.

The next stage—deacylation—begins when a water molecule takes the place occupied earlier by the amine component of the substrate (step 5). The ester group of the acyl-enzyme is now hydrolyzed by a process that essentially repeats steps 2 through 4. Now acting as a general acid catalyst, histidine 57 draws a proton away from the water molecule. The resulting OH^- ion attacks the carbonyl carbon atom of the acyl group, forming a tetrahedral intermediate (step 6). This structure breaks down to form the carboxylic acid product (step 7). Finally, the release of the carboxylic acid product (step 8) readies the enzyme for another round of catalysis.

This mechanism accounts for all characteristics of chymotrypsin action except the observed preference for cleaving the peptide bonds just past residues with large, hydrophobic side chains. Examination of the three-

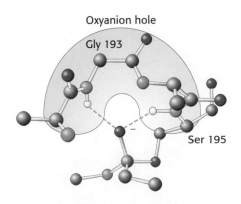

Figure 9.9 The oxyanion hole. The structure stabilizes the tetrahedral intermediate of the chymotrypsin reaction. *Notice* that hydrogen bonds (shown in green) link peptide NH groups and the negatively charged oxygen atom of the intermediate.

247

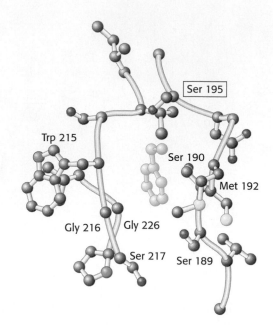

Figure 9.10 Specificity pocket of chymotrypsin. *Notice* that this pocket is lined with relatively hydrophobic residues and is relatively deep, favoring the binding of residues with long hydrophobic side chains such as phenylalanine (shown in green). *Also notice* that the active-site serine residue (serine 195) is positioned to cleave the peptide backbone between the residue bound in the pocket and the next residue in the sequence. The key amino acids that constitute the binding site are labeled.

dimensional structure of chymotrypsin with substrate analogs and enzyme inhibitors revealed the presence of a deep, relatively hydrophobic pocket, called the S1 pocket, into which the long, uncharged side chains of residues such as phenylalanine and tryptophan can fit. *The binding of an appropriate side chain into this pocket positions the adjacent peptide bond into the active site for cleavage* (Figure 9.10). The specificity of chymotrypsin depends almost entirely on which amino acid is directly on the amino-terminal side of the peptide bond to be cleaved. Other proteases have more complex specificity patterns, as illustrated in Figure 9.11. Such enzymes have additional pockets on their surfaces for the recognition of other residues in the substrate. Residues on the amino-terminal side of the scissile bond (the bond to be cleaved) are labeled P1, P2, P3, and so forth, heading away from the scissile bond. Likewise, residues on the carboxyl side of the scissile bond are labeled P1′, P2′, P3′, and so forth. The corresponding sites on the enzyme are referred to as S1, S2 or S1′, S2′, and so forth.

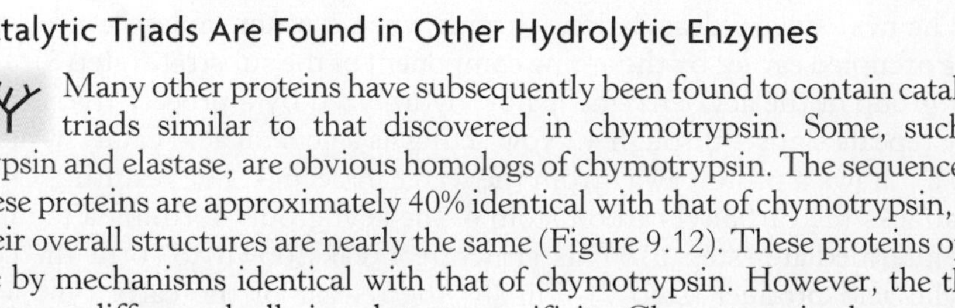

Figure 9.11 Specificity nomenclature for protease–substrate interactions. The potential sites of interaction of the substrate with the enzyme are designated P (shown in red), and corresponding binding sites on the enzyme are designated S. The scissile bond (also shown in red) is the reference point.

Catalytic Triads Are Found in Other Hydrolytic Enzymes

Many other proteins have subsequently been found to contain catalytic triads similar to that discovered in chymotrypsin. Some, such as trypsin and elastase, are obvious homologs of chymotrypsin. The sequences of these proteins are approximately 40% identical with that of chymotrypsin, and their overall structures are nearly the same (Figure 9.12). These proteins operate by mechanisms identical with that of chymotrypsin. However, the three enzymes differ markedly in substrate specificity. Chymotrypsin cleaves at the peptide bond after residues with an aromatic or long nonpolar side chain. Trypsin cleaves at the peptide bond after residues with long, positively charged side chains—namely, arginine and lysine. Elastase cleaves at the peptide bond after amino acids with small side chains—such as alanine and serine.

Comparison of the S1 pockets of these enzymes reveals that *these different specificities are due to small structural differences*. In trypsin, an aspartate residue (Asp 189) is present at the bottom of the S1 pocket in place of a serine residue in chymotrypsin. The aspartate residue attracts and stabilizes a positively charged arginine or lysine residue in the substrate. In elastase, two residues at the top of the pocket in chymotrypsin and trypsin are replaced with much bulkier valine residues (Val 190 and Val 216). These residues close off the mouth of the pocket so that only small side chains can enter (Figure 9.13).

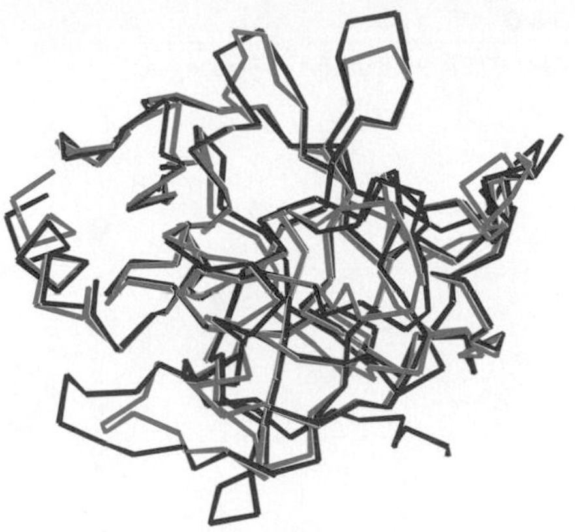

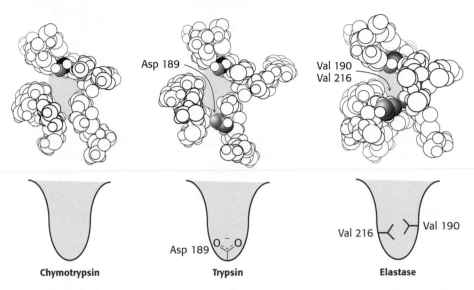

Figure 9.13 The S1 pockets of chymotrypsin, trypsin, and elastase. Certain residues play key roles in determining the specificity of these enzymes. The side chains of these residues, as well as those of the active-site serine residues, are shown in color.

Other members of the chymotrypsin family include a collection of proteins that take part in blood clotting, to be discussed in Chapter 10, as well as the tumor marker protein prostate specific antigen (PSA). In addition, a wide range of proteases found in bacteria, viruses, and plants belong to this clan.

Other enzymes that are not homologs of chymotrypsin have been found to contain very similar active sites. As noted in Chapter 6, the presence of very similar active sites in these different protein families is a consequence of convergent evolution. Subtilisin, a protease in bacteria such as *Bacillus amyloliquefaciens*, is a particularly well characterized example. The active site of this enzyme includes both the catalytic triad and the oxyanion hole. However, one of the NH groups that forms the oxyanion hole comes from the side chain of an asparagine residue rather than from the peptide backbone (Figure 9.14). Subtilisin is the founding member of another large family of proteases that includes representatives from Archaea, Bacteria, and Eukarya.

Figure 9.12 Structural similarity of trypsin and chymotrypsin. An overlay of the structure of chymotrypsin (red) on that of trypsin (blue) shows the high degree of similarity. Only α-carbon-atom positions are shown. The mean deviation in position between corresponding α-carbon atoms is 1.7 Å. [Drawn from 5PTP.pdb and 1GCT.pdb.]

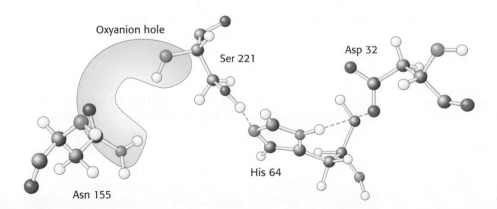

Figure 9.14 The catalytic triad and oxyanion hole of subtilisin. *Notice* the two enzyme NH groups (both in the backbone and in the side chain of Asn 155) located in the oxyanion hole. The NH groups will stabilize a negative charge that develops on the peptide bond attacked by nucleophilic serine 221 of the catalytic triad.

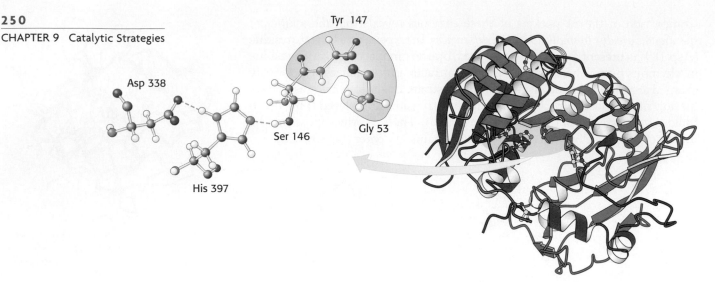

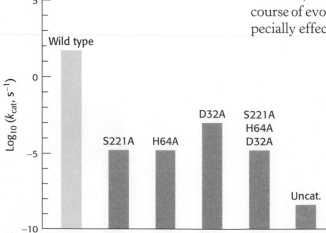

Figure 9.15 Carboxypeptidase II. The structure of carboxypeptidase II from wheat (right) is illustrated with its two chains (blue and red). *Notice* that the catalytic triad of carboxypeptidase II (left) is composed of the same amino acids as those in chymotrypsin, despite the fact that the enzymes display no structural similarity. The residues that form the oxyanion hole are highlighted in yellow. This protein is a member of an intriguing family of homologous proteins that includes esterases such as acetylcholine esterase and certain lipases. All these enzymes make use of histidine-activated nucleophiles, but the nucleophiles may be cysteine rather than serine. [Drawn from 1WHS.pdb.]

Yet another example of the catalytic triad has been found in carboxypeptidase II from wheat. The structure of this enzyme is not significantly similar to either chymotrypsin or subtilisin (Figure 9.15).

Finally, other proteases have been discovered that contain an active-site serine or threonine residue that is activated not by a histidine–aspartate pair but by a primary amino group from the side chain of lysine or by the N-terminal amino group of the polypeptide chain.

Thus, the catalytic triad in proteases has emerged at least three times in the course of evolution. We can conclude that this catalytic strategy must be an especially effective approach to the hydrolysis of peptides and related bonds.

The Catalytic Triad Has Been Dissected by Site-Directed Mutagenesis

How can we be sure that the mechanism proposed for the catalytic triad is correct? One way is to test the contribution of individual amino acid residues to the catalytic power of a protease by using site-directed mutagenesis (Section 5.2). Subtilisin has been extensively studied by this method. Each of the residues within the catalytic triad, consisting of aspartic acid 32, histidine 64, and serine 221, has been individually converted into alanine, and the ability of each mutant enzyme to cleave a model substrate has been examined (Figure 9.16).

As expected, the conversion of active-site serine 221 into alanine dramatically reduced catalytic power; the value of k_{cat} fell to less than *one-millionth* of its value for the wild-type enzyme. The value of K_M was essentially unchanged; its increase by no more than a factor of two indicated that substrate continued to bind normally. The mutation of histidine 64 to alanine reduced catalytic power to a similar degree. The conversion of aspartate 32 into alanine reduced

Figure 9.16 Site-directed mutagenesis of subtilisin. Residues of the catalytic triad were mutated to alanine, and the activity of the mutated enzyme was measured. Mutations in any component of the catalytic triad cause a dramatic loss of enzyme activity. Note that the activity is displayed on a logarithmic scale. The mutations are identified as follows: the first letter is the one-letter abbreviation for the amino acid being altered; the number identifies the position of the residue in the primary structure; and the second letter is the one-letter abbreviation for the amino acid replacing the original one. Uncat. refers to the estimated rate for the uncatalyzed reaction.

catalytic power by less, although the value of k_{cat} still fell to less than 0.005% of its wild-type value. The simultaneous conversion of all three residues into alanine was no more deleterious than conversion of serine or histidine alone. These observations support the notion that the catalytic triad and, particularly, the serine–histidine pair act together to generate a nucleophile of sufficient power to attack the carbonyl carbon atom of a peptide bond. Despite the reduction in their catalytic power, the mutated enzymes still hydrolyze peptides a thousand times as fast as buffer at pH 8.6.

Site-directed mutagenesis also offered a way to probe the importance of the oxyanion hole for catalysis by site-directed mutagenesis. The mutation of asparagine 155 to glycine eliminated the side-chain NH group from the oxyanion hole of subtilisin. The elimination of the NH group reduced the value of k_{cat} to 0.2% of its wild-type value but increased the value of K_M by only a factor of two. These observations demonstrate that the NH group of the asparagine residue plays a significant role in stabilizing the tetrahedral intermediate and the transition state leading to it.

Cysteine, Aspartyl, and Metalloproteases Are Other Major Classes of Peptide-Cleaving Enzymes

Not all proteases utilize strategies based on activated serine residues. Classes of proteins have been discovered that employ three alternative approaches to peptide-bond hydrolysis (Figure 9.17). These classes are

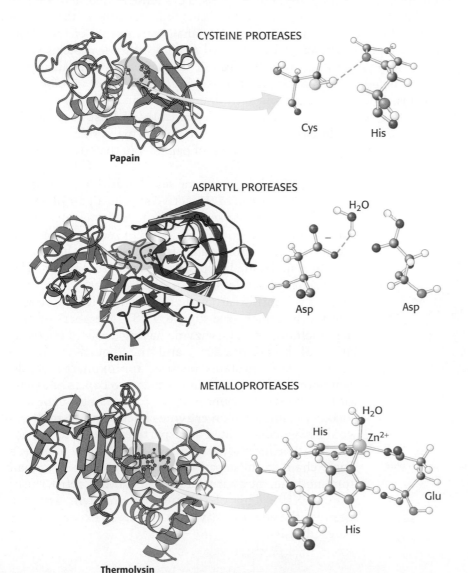

CYSTEINE PROTEASES

Cys   His

Papain

ASPARTYL PROTEASES

H_2O

Asp   Asp

Renin

METALLOPROTEASES

H_2O

His Zn^{2+}

Glu

His

Thermolysin

Figure 9.17 Three classes of proteases and their active sites. These examples of a cysteine protease, an aspartyl protease, and a metalloprotease use a histidine-activated cysteine residue, an aspartate-activated water molecule, and a metal-activated water molecule, respectively, as the nucleophile. The two halves of renin are in blue and red to highlight the approximate twofold symmetry of aspartyl proteases. [Drawn from 1PPN.pdb.; 1HRN.pdb; 1LND.pdb.]

(A) CYSTEINE PROTEASES   (B) ASPARTYL PROTEASES   (C) METALLOPROTEASES

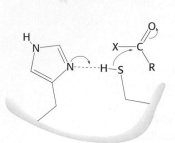
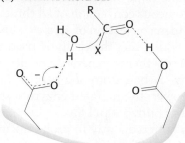
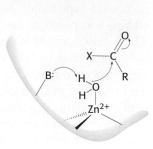

Figure 9.18 The activation strategies for three classes of proteases. The peptide carbonyl group is attacked by (A) a histidine-activated cysteine in the cysteine proteases, (B) an aspartate-activated water molecule in the aspartyl proteases, and (C) a metal-activated water molecule in the metalloproteases. For the metalloproteases, the letter B represents a base (often glutamate) that helps deprotonate the metal-bound water.

the (1) cysteine proteases, (2) aspartyl proteases, and (3) metalloproteases. In each case, the strategy is to generate a nucleophile that attacks the peptide carbonyl group (Figure 9.18).

The strategy used by the *cysteine proteases* is most similar to that used by the chymotrypsin family. In these enzymes, a cysteine residue, activated by a histidine residue, plays the role of the nucleophile that attacks the peptide bond (see Figure 9.18) in a manner quite analogous to that of the serine residue in serine proteases. A well-studied example of these proteins is papain, an enzyme purified from the fruit of the papaya. Mammalian proteases homologous to papain have been discovered, most notably the cathepsins, proteins having a role in the immune system and other systems. The cysteine-based active site arose independently at least twice in the course of evolution; the caspases, enzymes that play a major role in apoptosis (p. 535), have active sites similar to that of papain, but their overall structures are unrelated.

The second class comprises the *aspartyl proteases*. The central feature of the active sites is a pair of aspartic acid residues that act together to allow a water molecule to attack the peptide bond. One aspartic acid residue (in its deprotonated form) activates the attacking water molecule by poising it for deprotonation. The other aspartic acid residue (in its protonated form) polarizes the peptide carbonyl group so that it is more susceptible to attack (see Figure 9.18). Members of this class include renin, an enzyme having a role in the regulation of blood pressure, and the digestive enzyme pepsin. These proteins possess approximate twofold symmetry. A likely scenario is that two copies of a gene for the ancestral enzyme fused to form a single gene that encoded a single-chain enzyme. Each copy of the gene would have contributed an aspartate residue to the active site. The individual chains are now joined to make a single chain in the aspartyl proteases present in human immunodeficiency virus (HIV) and other retroviruses (Figure 9.19). This observation is consistent with the idea that the enzyme may have originally existed as separate subunits.

Flaps

Binding pocket

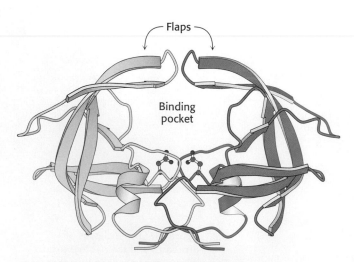

Figure 9.19 HIV protease, a dimeric aspartyl protease. The protease is a dimer of identical subunits, shown in blue and yellow, consisting of 99 amino acids each. *Notice* the placement of active-site aspartic acid residues, one from each chain, which are shown as ball-and-stick structures. The flaps will close down on the binding pocket after substrate has been bound. [Drawn from 3PHV.pdb.]

The *metalloproteases* constitute the final major class of peptide-cleaving enzymes. The active site of such a protein contains a bound metal ion, almost always zinc, that activates a water molecule to act as a nucleophile to attack the peptide carbonyl group. The bacterial enzyme thermolysin and the digestive enzyme carboxypeptidase A are classic examples of the zinc proteases. Thermolysin, but not carboxypeptidase A, is a member of a large and diverse family of homologous zinc proteases that includes the matrix metalloproteases, enzymes that catalyze the reactions in tissue remodeling and degradation.

In each of these three classes of enzymes, the active site includes features that act to (1) activate a water molecule or another nucleophile, (2) polarize the peptide carbonyl group, and (3) stabilize a tetrahedral intermediate (see Figure 9.18).

Protease Inhibitors Are Important Drugs

Several important drugs are protease inhibitors. For example, captopril, used to regulate blood pressure, is an inhibitor of the angiotensin-converting enzyme (ACE), a metalloprotease. Indinavir (Crixivan), retrovir, and more than 20 other compounds used in the treatment of AIDS are inhibitors of HIV protease, which is an aspartyl protease. This protease cleaves multidomain viral proteins into their active forms; blocking this process completely prevents the virus from being infectious (see Figure 9.19). To prevent unwanted side effects, protease inhibitors used as drugs must be specific for one enzyme without inhibiting other proteins within the body.

Indinavir resembles the peptide substrate of the HIV protease. Indinavir is constructed around an alcohol that mimics the tetrahedral intermediate; other groups are present to bind into the S2, S1, S1′, and S2′ recognition sites on the enzyme (Figure 9.20). X-ray crystallographic studies revealed that, in the active site, indinavir adopts a conformation that approximates the twofold symmetry of the enzyme (Figure 9.21). The active site of HIV protease is covered by two flexible flaps that fold down on top of the bound inhibitor. The OH group of the central alcohol interacts with the two aspartate residues of the active site. In addition, two carbonyl groups of the inhibitor are hydrogen bonded to a water molecule (not

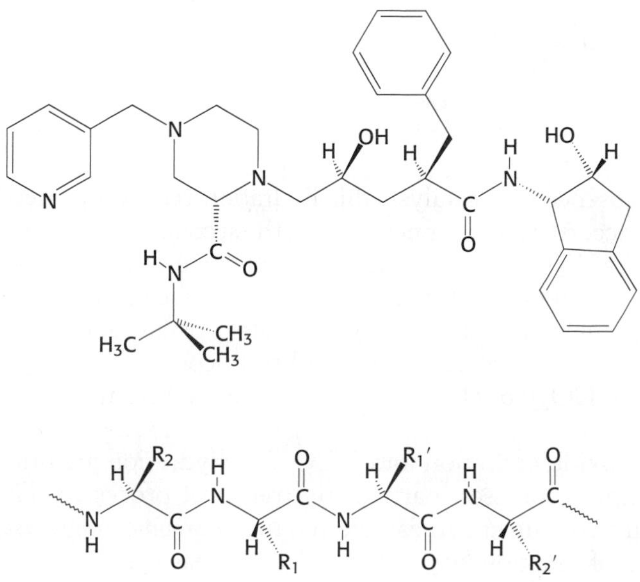

Figure 9.20 Indinavir, an HIV protease inhibitor. The structure of indinavir (Crixivan) is shown in comparison with that of a peptide substrate of HIV protease. The scissile bond in the substrate is highlighted in red.

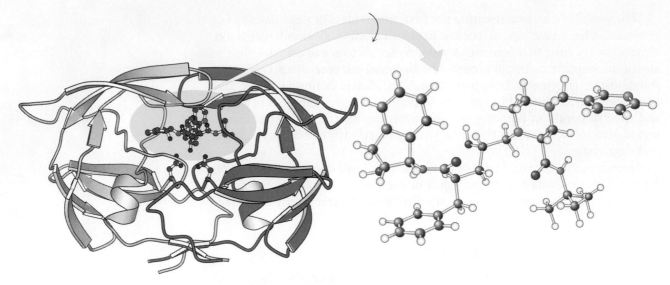

Figure 9.21 HIV protease–indinavir complex. (Left) The HIV protease is shown with the inhibitor indinavir bound at the active site. (Right) The drug has been rotated to reveal its approximately twofold symmetric conformation. [Drawn from 1HSH.pdb.]

shown in Figure 9.21), which, in turn, is hydrogen bonded to a peptide NH group in each of the flaps. This interaction of the inhibitor with water and the enzyme is not possible within cellular aspartyl proteases such as renin. Thus the interaction may contribute to the specificity of indinavir for HIV protease.

9.2 Carbonic Anhydrases Make a Fast Reaction Faster

Carbon dioxide is a major end product of aerobic metabolism. In mammals, this carbon dioxide is released into the blood and transported to the lungs for exhalation. While in the red blood cells, carbon dioxide reacts with water (Section 7.3). The product of this reaction is a moderately strong acid, carbonic acid ($pK_a = 3.5$), which is converted into bicarbonate ion (HCO_3^-) on the loss of a proton.

$$CO_2 + H_2O \underset{k_{-1}}{\overset{k_1}{\rightleftharpoons}} \underset{\substack{\text{Carbonic} \\ \text{acid}}}{H_2CO_3} \rightleftharpoons \underset{\substack{\text{Bicarbonate} \\ \text{ion}}}{HCO_3^-} + H^+$$

Even in the absence of a catalyst, this hydration reaction proceeds at a moderately fast pace. At 37°C near neutral pH, the second-order rate constant k_1 is 0.0027 M^{-1} s^{-1}. This corresponds to an effective first-order rate constant of 0.15 s^{-1} in water ($[H_2O] = 55.5$ M). The reverse reaction, the dehydration of HCO_3^-, is even more rapid, with a rate constant of $k_{-1} = 50$ s^{-1}. These rate constants correspond to an equilibrium constant of $K_1 = 5.4 \times 10^{-5}$ and a ratio of $[CO_2]$ to $[H_2CO_3]$ of 340 : 1 at equilibrium.

Carbon dioxide hydration and HCO_3^- dehydration are often coupled to rapid processes, particularly transport processes. Thus, almost all organisms contain enzymes, referred to as *carbonic anhydrases*, that increase the rate of reaction beyond the already reasonable spontaneous rate. For example, carbonic anhydrases dehydrate HCO_3^- in the blood to form CO_2 for exhalation as the blood passes through the lungs. Conversely, they convert CO_2 into HCO_3^- to generate the aqueous humor of the eye and other secretions. Furthermore, both CO_2 and HCO_3^- are substrates and products for a variety of enzymes, and the rapid interconversion of these species may be necessary to ensure appropriate substrate levels. So important are these enzymes in human beings that mutations in some carbonic anhydrases have been found to be associated with osteopetrosis (excessive formation of dense bones accompanied by anemia) and mental retardation.

Carbonic anhydrases accelerate CO_2 hydration dramatically. The most active enzymes hydrate CO_2 at rates as high as $k_{cat} = 10^6 \text{ s}^{-1}$, or a million times a second per enzyme molecule. Fundamental physical processes such as diffusion and proton transfer ordinarily limit the rate of hydration, and so the enzymes employ special strategies to attain such prodigious rates.

Carbonic Anhydrase Contains a Bound Zinc Ion Essential for Catalytic Activity

Less than 10 years after the discovery of carbonic anhydrase in 1932, this enzyme was found to contain a bound zinc ion. Moreover, the zinc ion appeared to be necessary for catalytic activity. This discovery, remarkable at the time, made carbonic anhydrase the first known zinc-containing enzyme. At present, hundreds of enzymes are known to contain zinc. In fact, more than one-third of all enzymes either contain bound metal ions or require the addition of such ions for activity. Metal ions have several properties that increase chemical reactivity: their positive charges, their ability to form relatively strong yet kinetically labile bonds, and, in some cases, their capacity to be stable in more than one oxidation state. The chemical reactivity of metal ions explains why catalytic strategies that employ metal ions have been adopted throughout evolution.

X-ray crystallographic studies have supplied the most detailed and direct information about the zinc site in carbonic anhydrase. At least seven carbonic anhydrases, each with its own gene, are present in human beings. They are all clearly homologous, as revealed by substantial sequence identity. Carbonic anhydrase II, a major protein component of red blood cells, has been the most extensively studied (Figure 9.22). It is also one of the most active carbonic anhydrases.

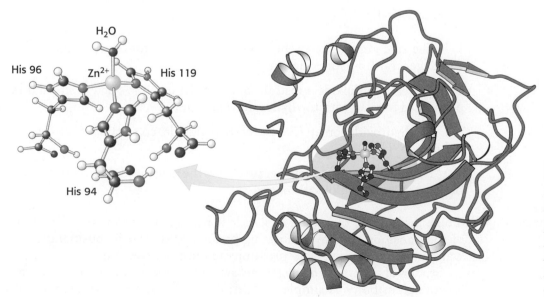

Figure 9.22 The structure of human carbonic anhydrase II and its zinc site. (Left) *Notice* that the zinc ion is bound to the imidazole rings of three histidine residues as well as to a water molecule. (Right) *Notice* the location of the zinc site in a cleft near the center of the enzyme. [Drawn from 1CA2.pdb.]

Zinc is found only in the +2 state in biological systems. A zinc atom is essentially always bound to four or more ligands; in carbonic anhydrase, three coordination sites are occupied by the imidazole rings of three histidine residues and an additional coordination site is occupied by a water molecule (or hydroxide ion, depending on pH). Because the molecules occupying the coordination sites are neutral, the overall charge on the $Zn(His)_3$ unit remains +2.

Catalysis Entails Zinc Activation of a Water Molecule

How does this zinc complex facilitate carbon dioxide hydration? A major clue comes from the pH profile of enzymatically catalyzed carbon dioxide hydration (Figure 9.23).

At pH 8, the reaction proceeds near its maximal rate. As the pH decreases, the rate of the reaction drops. The midpoint of this transition is near pH 7, suggesting that a group that loses a proton at pH 7 ($pK_a = 7$) plays an important role in the activity of carbonic anhydrase. Moreover, the curve suggests that the deprotonated (high pH) form of this group participates more effectively in catalysis. Although some amino acids, notably histidine, have pK_a values near 7, *a variety of evidence suggests that the group responsible for this transition is not an amino acid but is the zinc-bound water molecule.*

The binding of a water molecule to the positively charged zinc center reduces the pK_a of the water molecule from 15.7 to 7 (Figure 9.24).

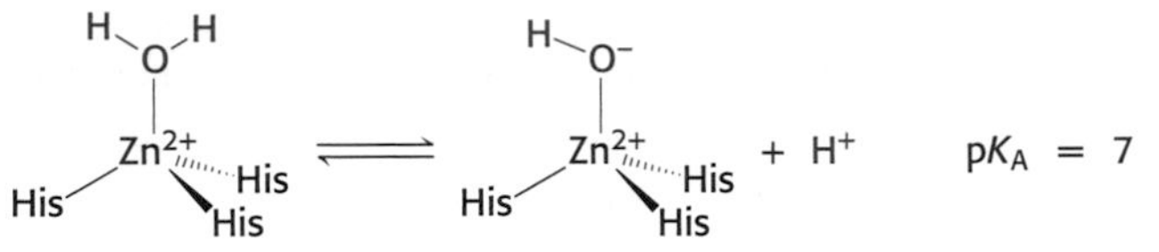

Figure 9.23 Effect of pH on carbonic anhydrase activity. Changes in pH alter the rate of carbon dioxide hydration catalyzed by carbonic anhydrase II. The enzyme is maximally active at high pH.

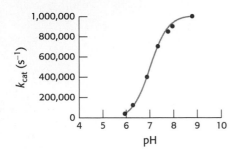

Figure 9.24 The pK_a of zinc-bound water. Binding to zinc lowers the pK_a of water from 15.7 to 7.

With the pK_a lowered, many water molecules lose a proton at neutral pH, generating a substantial concentration of hydroxide ion (bound to the zinc atom). A zinc-bound hydroxide ion (OH^-) is a potent nucleophile able to attack carbon dioxide much more readily than water does. The importance of the zinc-bound hydroxide ion suggests a simple mechanism for carbon dioxide hydration (Figure 9.25):

1. The zinc ion facilitates the release of a proton from a water molecule, which generates a hydroxide ion.

2. The carbon dioxide substrate binds to the enzyme's active site and is positioned to react with the hydroxide ion.

3. The hydroxide ion attacks the carbon dioxide, converting it into bicarbonate ion, HCO_3^-.

4. The catalytic site is regenerated with the release of HCO_3^- and the binding of another molecule of water.

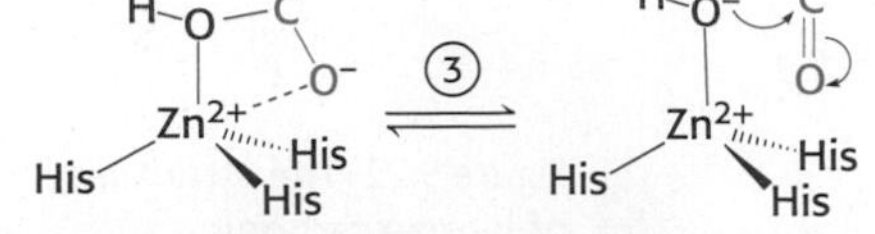

Figure 9.25 Mechanism of carbonic anhydrase. The zinc-bound hydroxide mechanism for the hydration of carbon dioxide reveals one aspect of metal ion catalysis. The reaction proceeds in four steps: (1) water deprotonation; (2) carbon dioxide binding; (3) nucleophilic attack of hydroxide on carbon dioxide; and (4) displacement of bicarbonate ion by water.

Thus, the binding of a water molecule to the zinc ion favors the formation of the transition state by facilitating proton release and by positioning the water molecule to be in close proximity to the other reactant.

Studies of a *synthetic analog model system* provide evidence for the mechanism's plausibility. A simple synthetic ligand binds zinc through four nitrogen atoms (compared with three histidine nitrogen atoms in the enzyme),

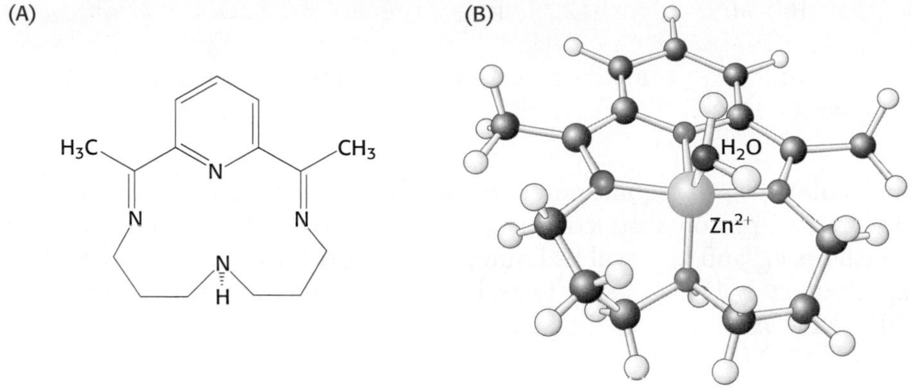

(A)

H₃C ... CH₃ (pyridine-based macrocyclic ligand structure)

(B)

Zn²⁺ ... H₂O complex structure

Figure 9.26 A synthetic analog model system for carbonic anhydrase. (A) An organic compound, capable of binding zinc, was synthesized as a model for carbonic anhydrase. The zinc complex of this ligand accelerates the hydration of carbon dioxide more than 100-fold under appropriate conditions. (B) The structure of the presumed active complex showing zinc bound to the ligand and to one water molecule.

as shown in Figure 9.26. One water molecule remains bound to the zinc ion in the complex. Direct measurements reveal that this water molecule has a pK_a value of 8.7, not as low as the value for the water molecule in carbonic anhydrase but substantially lower than the value for free water. At pH 9.2, this complex accelerates the hydration of carbon dioxide more than 100-fold. Although its rate of catalysis is much less efficient than catalysis by carbonic anhydrase, the model system strongly suggests that the zinc-bound hydroxide mechanism is likely to be correct. Carbonic anhydrases have evolved to employ the reactivity intrinsic to a zinc-bound hydroxide ion as a potent catalyst.

A Proton Shuttle Facilitates Rapid Regeneration of the Active Form of the Enzyme

As noted earlier, some carbonic anhydrases can hydrate carbon dioxide at rates as high as a million times a second (10^6 s^{-1}). The magnitude of this rate can be understood from the following observations. In the first step of a carbon dioxide hydration reaction, the zinc-bound water molecule must lose a proton to regenerate the active form of the enzyme (Figure 9.27). The rate of the reverse reaction, the protonation of the zinc-bound hydroxide ion, is limited by the rate of proton diffusion. Protons diffuse very rapidly with second-order rate constants near 10^{-11} M^{-1} s^{-1}. Thus, the backward rate constant k_{-1} must be less than 10^{11} M^{-1} s^{-1}. Because the equilibrium constant K is equal to k_1/k_{-1}, the forward rate constant is given by $k_1 = K \cdot k_{-1}$. Thus, if $k_{-1} \leq 10^{11}$ M^{-1} s^{-1} and $K = 10^{-7}$ M (because $pK_a = 7$), then k_1 must be less than or equal to 10^4 s^{-1}. In other words, the rate of proton diffusion limits the rate of proton release to less than 10^4 s^{-1} for a group with $pK_a = 7$. However, if carbon dioxide is hydrated at a rate of 10^6 s^{-1}, then every step in the mechanism (see Figure 9.25) must take place at least this fast. How is this apparent paradox resolved?

The answer became clear with the realization that *the highest rates of carbon dioxide hydration require the presence of buffer, suggesting that the buffer components participate in the reaction.* The buffer can bind or release protons. The advantage is that, whereas the concentrations of protons and hydroxide ions are limited to 10^{-7} M at neutral pH, the concentration of buffer components can be much higher, of the order of several millimolar. If the buffer component BH^+ has a pK_a of 7 (matching that for the zinc-bound water

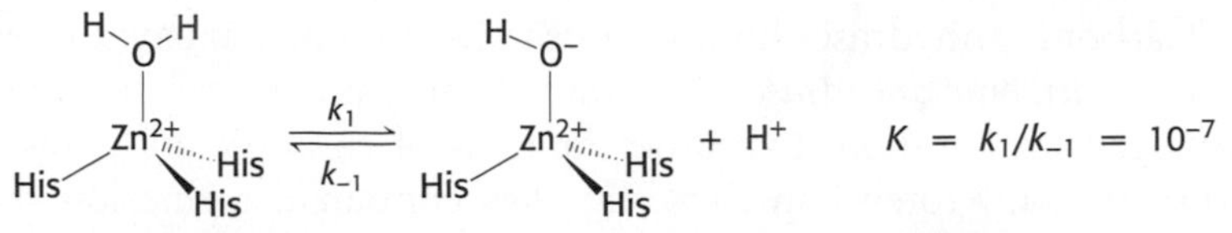

H H ... H O⁻
O
Zn²⁺ ··· His $\xrightleftharpoons[k_{-1}]{k_1}$ Zn²⁺ ··· His + H⁺ $K = k_1/k_{-1} = 10^{-7}$
His His   His His

Figure 9.27 Kinetics of water deprotonation. The kinetics of deprotonation and protonation of the zinc-bound water molecule in carbonic anhydrase.

Figure 9.28 The effect of buffer on deprotonation. The deprotonation of the zinc-bound water molecule in carbonic anhydrase is aided by buffer component B

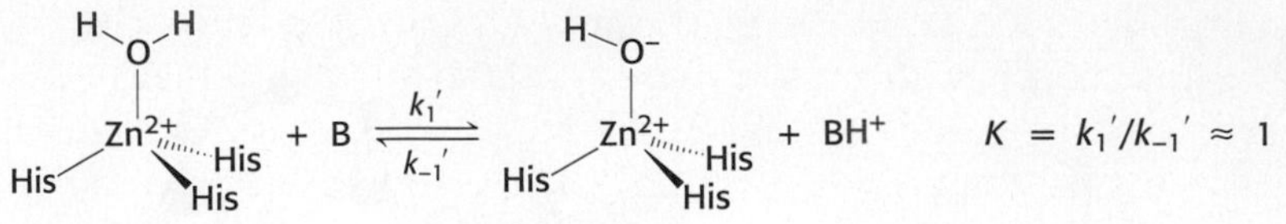

$$\text{His}-\underset{\text{His}}{\overset{\text{His}}{Zn^{2+}}}\cdots\overset{H\diagdown O\diagup H}{} + B \underset{k_{-1}'}{\overset{k_1'}{\rightleftharpoons}} \text{His}-\underset{\text{His}}{\overset{\text{His}}{Zn^{2+}}}\cdots\overset{H\diagdown O^-}{} + BH^+ \qquad K = k_1'/k_{-1}' \approx 1$$

molecule), then the equilibrium constant for the reaction in Figure 9.28 is 1. The rate of proton abstraction is given by $k_1' \cdot [B]$. The second-order rate constants k_1' and k_{-1}' will be limited by buffer diffusion to values less than approximately $10^9 \, M^{-1} \, s^{-1}$. Thus, buffer concentrations greater than $[B] = 10^{-3} \, M$ (1 mM) may be high enough to support carbon dioxide hydration rates of $10^6 \, M^{-1} \, s^{-1}$ because $k_1' \cdot [B] = (10^9 \, M^{-1} \, s^{-1}) \cdot (10^{-3} \, M) = 10^6 \, s^{-1}$. This prediction is confirmed experimentally (Figure 9.29).

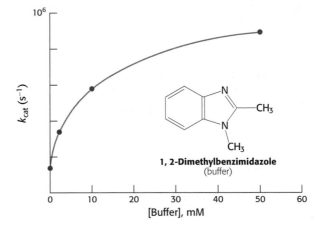

Figure 9.29 The effect of buffer concentration on the rate of carbon dioxide hydration. The rate of carbon dioxide hydration increases with the concentration of the buffer 1, 2-dimethylbenzimidazole. The buffer enables the enzyme to achieve its high catalytic rates.

The molecular components of many buffers are too large to reach the active site of carbonic anhydrase. Carbonic anhydrase II has evolved a *proton shuttle* to allow buffer components to participate in the reaction from solution. The primary component of this shuttle is histidine 64. This residue transfers protons from the zinc-bound water molecule to the protein surface and then to the buffer (Figure 9.30). Thus, catalytic function has been enhanced through the evolution of an apparatus for controlling proton transfer from and to the active site. Because protons participate in many biochemical reactions, the manipulation of the proton inventory within active sites is crucial to the function of many enzymes and explains the prominence of acid–base catalysis.

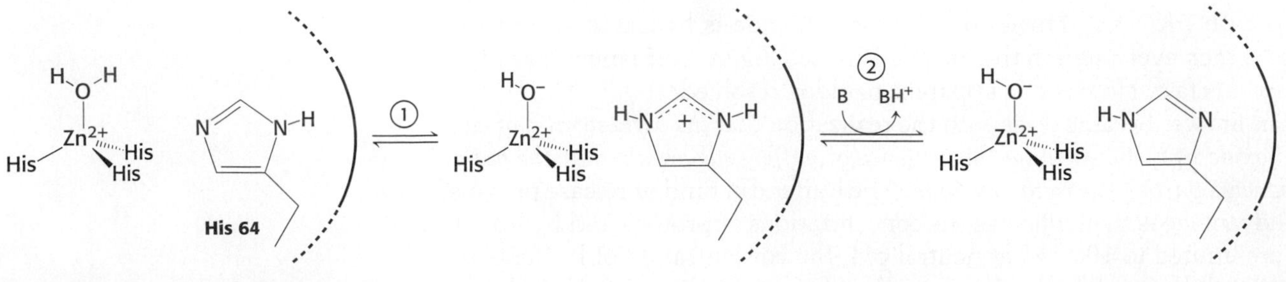

Figure 9.30 Histidine proton shuttle. (1) Histidine 64 abstracts a proton from the zinc-bound water molecule, generating a nucleophilic hydroxide ion and a protonated histidine. (2) The buffer (B) removes a proton from the histidine, regenerating the unprotonated form.

Convergent Evolution Has Generated Zinc-Based Active Sites in Different Carbonic Anhydrases

Carbonic anhydrases homologous to the human enzymes, referred to as *α-carbonic anhydrases*, are common in animals and in some bacteria and algae. In addition, two other families of carbonic anhydrases have been discovered. Proteins in these families contain the zinc ion required for catalytic activity but are not significantly similar in sequence to the α-carbonic anhydrases. The *β-carbonic anhydrases* are found in higher plants and in many bacterial species, including *E. coli.* Spectroscopic and structural studies reveal that the zinc ion is bound by one histidine residue and two cysteine residues. Moreover, the overall enzyme structures are unrelated to those of the α-carbonic anhydrases. In plants, these enzymes facilitate the accumulation of carbon dioxide, crucial for the Calvin cycle in photosynthesis. A third family, the *γ-carbonic anhydrases,* was initially identified in the archaeon *Methanosarcina thermophila.* The crystal structure of this enzyme reveals three zinc sites extremely similar to the zinc site in the α-carbonic anhydrases. In this case, however, the three zinc sites lie at the interfaces between the three subunits of a trimeric enzyme (Figure 9.31). The very striking left-handed β-helical structure (a β strand twisted into a left-handed helix) present in this enzyme is again different from any structure present in the α- and β-carbonic anhydrases. Thus, convergent evolution has generated carbonic anhydrases that rely on coordinated zinc ions at least three times.

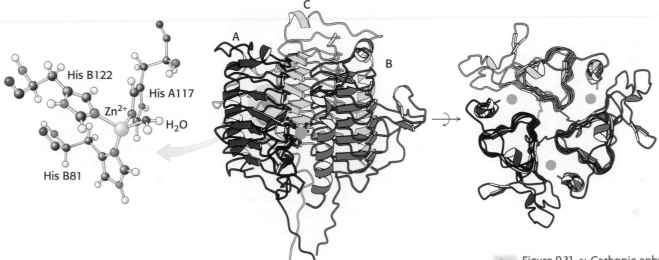

Figure 9.31 γ-Carbonic anhydrase. (Left) The zinc site of γ-carbonic anhydrase. *Notice* that the water-binding zinc ion is bound to three histidine residues. (Middle) The trimeric structure of the protein (individual chains are labeled A, B, and C). Each chain consists primarily of a left-handed β helix. (Right) The protein is rotated to show a top-down view that highlights its threefold symmetry. *Notice* the position of the zinc sites (green) at the interfaces between subunits. [Drawn from 1THJ.pdb.]

9.3 Restriction Enzymes Perform Highly Specific DNA-Cleavage Reactions

We next consider a hydrolytic reaction that results in the cleavage of DNA. Bacteria and archaea have evolved mechanisms to protect themselves from viral infections. Many viruses inject their DNA genomes into cells; once inside, the viral DNA hijacks the cell's machinery to drive the production of viral proteins and, eventually, of progeny virus. Often, a viral infection results in the death of the host cell. A major protective strategy for the host is to use *restriction endonucleases* (restriction enzymes) to degrade the viral DNA on its introduction into a cell. These enzymes recognize particular base sequences, called *recognition sequences* or *recognition sites,* in their target DNA and cleave that DNA at defined positions. We have already discussed the utility of these important enzymes for dissecting genes and genomes. The most well studied class of restriction enzymes comprises the type II restriction enzymes, which cleave DNA *within* their recognition sequences. Other types of restriction enzymes cleave DNA at positions somewhat distant from their recognition sites.

Restriction endonucleases must show tremendous specificity at two levels. First, they must not degrade host DNA containing the recognition sequences.

Second, they must cleave only DNA molecules that contain recognition sites (hereafter referred to as *cognate DNA*) without cleaving DNA molecules that lack these sites. How do these enzymes manage to degrade viral DNA while sparing their own? In *E. coli*, the restriction endonuclease *EcoRV* cleaves double-stranded viral DNA molecules that contain the sequence 5'-GATATC-3' but leaves intact host DNA containing hundreds of such sequences. The host DNA is protected by other enzymes called *methylases*, which methylate adenine bases within host recognition sequences (Figure 9.32). An endonuclease will not cleave DNA if its recognition sequence is methylated. For each restriction endonuclease, the host cell produces a corresponding methylase that marks the host DNA at the appropriate methylation site. These pairs of enzymes are referred to as *restriction-modification systems*.

Figure 9.32 Protection by methylation. The recognition sequence for *EcoRV* endonuclease (left) and the sites of methylation (right) in DNA protected from the catalytic action of the enzyme.

Restriction enzymes must cleave DNA only at unmodified recognition sites, without cleaving at other sites. Suppose that a recognition sequence is six base pairs long. Because there are 4^6, or 4096, sequences having six base pairs, the concentration of sites that must not be cleaved will be approximately 4000-fold higher than the concentration of sites that should be cleaved. Thus, to keep from damaging host-cell DNA, restriction enzymes must cleave cognate DNA molecules much more than 4000 times as efficiently as they cleave nonspecific sites. We shall return to the mechanism used to achieve the necessary high specificity after considering the chemistry of the cleavage process.

Cleavage Is by In-Line Displacement of 3'-Oxygen from Phosphorus by Magnesium-Activated Water

A restriction endonuclease catalyzes the hydrolysis of the phosphodiester backbone of DNA. Specifically, the bond between the 3'-oxygen atom and the phosphorus atom is broken. The products of this reaction are DNA strands with a free 3'-hydroxyl group and a 5'-phosphoryl group at the cleavage site (Figure 9.33). This reaction proceeds by nucleophilic attack at the phosphorus atom. We will consider two alternative mechanisms, suggested

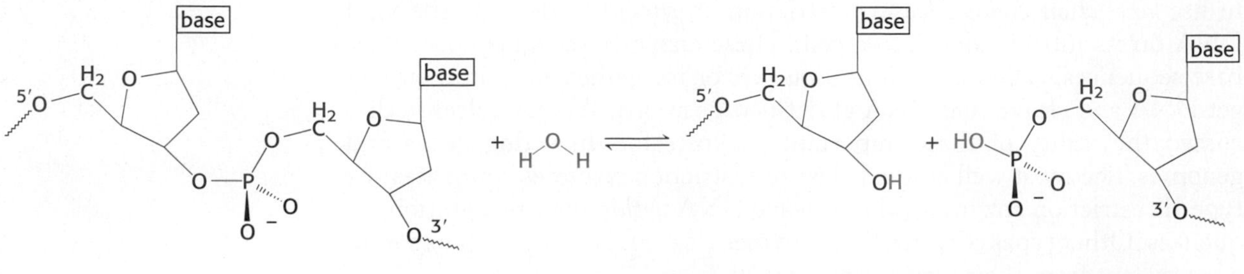

Figure 9.33 Hydrolysis of a phosphodiester bond. All restriction enzymes catalyze the hydrolysis of DNA phosphodiester bonds, leaving a phosphoryl group attached to the 5' end. The bond that is cleaved is shown in red.

by analogy with the proteases. The restriction endonuclease might cleave DNA by mechanism 1 through a covalent intermediate, employing a potent nucleophile (Nu), or by mechanism 2 through direct hydrolysis:

Mechanism 1 (covalent intermediate)

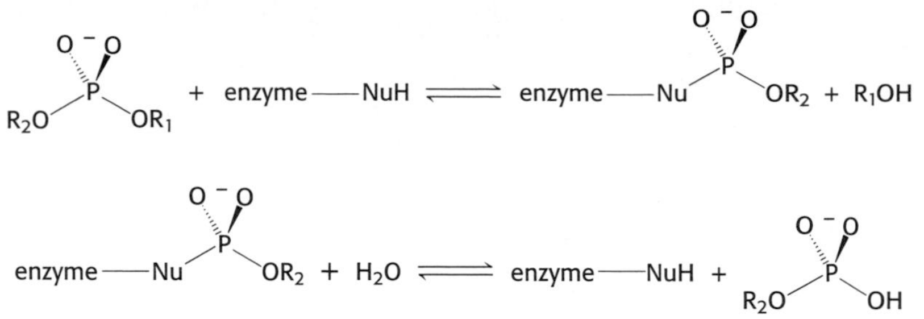

Mechanism 2 (direct hydrolysis)

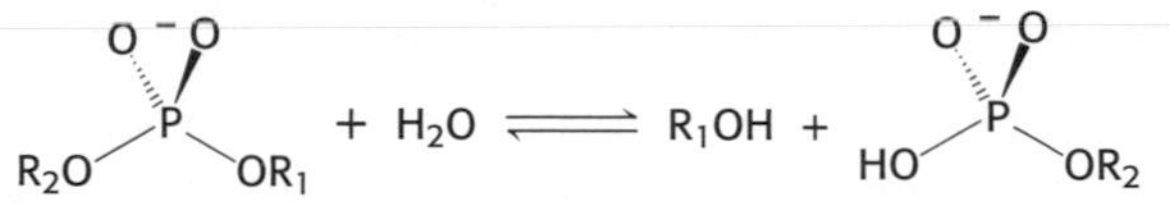

Each mechanism postulates a different nucleophile to attack the phosphorus. In either case, each reaction takes place by *in-line displacement*:

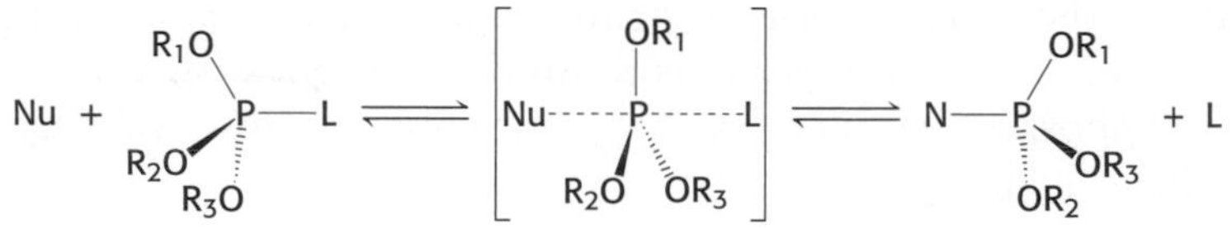

The incoming nucleophile attacks the phosphorus atom, and a pentacoordinate transition state is formed. This species has a trigonal bipyramidal geometry centered at the phosphorus atom, with the incoming nucleophile at one apex of the two pyramids and the group that is displaced (the leaving group, L) at the other apex. Note that the displacement inverts the stereochemical conformation at the tetrahedral phosphorous atom, analogous to the interconversion of the R and S configurations around a tetrahedral carbon center (Section 2.1).

The two mechanisms differ in the number of times that the displacement takes place in the course of the reaction. In the first type of mechanism, a nucleophile in the enzyme (analogous to serine 195 in chymotrypsin) attacks the phosphoryl group to form a covalent intermediate. In a second step, this intermediate is hydrolyzed to produce the final products. In this case, two displacement reactions take place at the phosphorus atom. Consequently, the stereochemical configuration at the phosphorus atom would be inverted and then inverted again, and the overall configuration would be *retained*. In the second type of mechanism, analogous to that used by the aspartyl- and metalloproteases, an activated water molecule attacks the phosphorus atom directly. In this mechanism, a single displacement reaction takes place at the phosphorus atom. Hence, the stereochemical configuration at the phosphorus atom is *inverted* after cleavage. To determine which mechanism is correct, we examine the stereochemistry at the phosphorus atom after cleavage.

A difficulty is that the stereochemistry is not easily observed, because two of the groups bound to the phosphorus atom are simple oxygen atoms, identical with each other. This difficulty can be circumvented by replacing

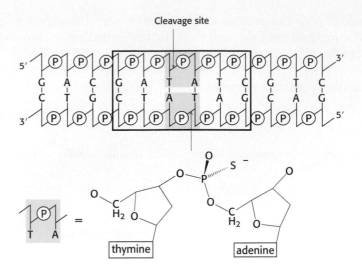

Figure 9.34 Labeling with phosphorothioates. Phosphorothioate groups, in which one of the nonbridging oxygen atoms is replaced by a sulfur atom, can be used to label specific sites in the DNA backbone to determine the overall stereochemical course of a displacement reaction. Here, a phosphorothioate is placed at sites that can be cleaved by *Eco*RV endonuclease.

one oxygen atom with sulfur (producing a species called a phosphorothioate). Let us consider *Eco*RV endonuclease. This enzyme cleaves the phosphodiester bond between the T and the A at the center of the recognition sequence 5′-GATATC-3′. The first step is to synthesize an appropriate substrate for *Eco*RV containing phosphorothioates at the sites of cleavage (Figure 9.34). The reaction is then performed in water that has been greatly enriched in ^{18}O to allow the incoming oxygen atom to be marked. The location of the ^{18}O label with respect to the sulfur atom indicates whether the reaction proceeds with inversion or retention of stereochemistry. *The analysis revealed that the stereochemical configuration at the phosphorus atom was inverted only once with cleavage.* This result is consistent with a direct attack by water at phosphorus and rules out the formation of any covalently bound intermediate (Figure 9.35).

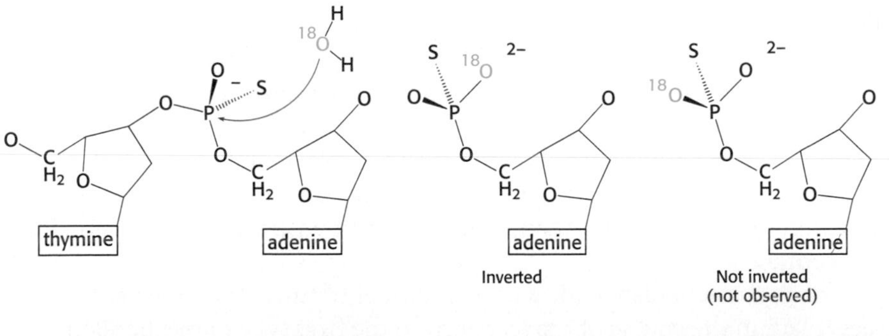

Figure 9.35 Stereochemistry of cleaved DNA. Cleavage of DNA by *Eco*RV endonuclease results in overall inversion of the stereochemical configuration at the phosphorus atom, as indicated by the stereochemistry of the phosphorus atom bound to one bridging oxygen atom, one ^{16}O, one ^{18}O, and one sulfur atom. This configuration strongly suggests that the hydrolysis takes place by water's direct attack at the phosphorus atom.

Restriction Enzymes Require Magnesium for Catalytic Activity

Many enzymes that act on phosphate-containing substrates require Mg^{2+} or some other similar divalent cation for activity. One or more Mg^{2+} (or similar) cations are essential to the function of restriction endonucleases. What are the functions of these metal ions?

It has not been possible to directly visualize the complex between *Eco*RV endonuclease and cognate DNA molecules in the presence of Mg^{2+} by crystallization because the enzyme cleaves the substrate under these circumstances. Nonetheless, it has been possible to visualize metal ion complexes through several approaches. In one approach, crystals of *Eco*RV endonuclease are prepared bound to oligonucleotides that contain the enzyme's recognition sequence. These crystals are grown in the absence of magnesium to prevent cleavage; after their preparation, the crystals are soaked in solutions containing the metal. Alternatively, crystals have been grown with the use of a mutated form of the enzyme that is less active. Finally, Mg^{2+} may be replaced by metal ions such as Ca^{2+} that bind but do not result in much catalytic activity. In all cases, no cleavage takes place, and so the locations of the metal ion-binding sites are readily determined.

As many as three metal ions have been found to be present per active site. The roles of these multiple metal ions is still under investigation. One ion-binding site is occupied in essentially all structures. This metal ion is coordinated to the protein through two aspartate residues and to one of the phosphoryl-group oxygen atoms near the site of cleavage. This metal ion binds the water molecule that attacks the phosphorus atom, helping to position and activate it in a manner similar to that for the Zn^{2+} ion of carbonic anhydrase (Figure 9.36).

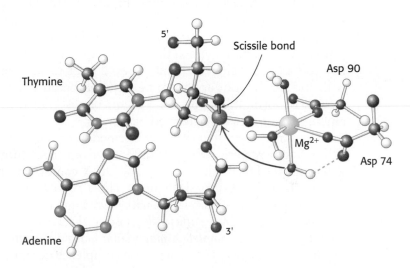

Figure 9.36 A magnesium ion-binding site in *Eco*RV endonuclease. The magnesium ion helps to activate a water molecule and positions it so that it can attack the phosphorus atom.

The Complete Catalytic Apparatus Is Assembled Only Within Complexes of Cognate DNA Molecules, Ensuring Specificity

We now return to the question of specificity, the defining feature of restriction enzymes. The recognition sequences for most restriction endonucleases are *inverted repeats*. This arrangement gives the three-dimensional structure of the recognition site a *twofold rotational symmetry* (Figure 9.37).

The restriction enzymes display a corresponding symmetry: they are dimers whose two subunits are related by twofold rotational symmetry. The matching symmetry of the recognition sequence and the enzyme facilitates the recognition of cognate DNA by the enzyme. This similarity in structure has been confirmed by the determination of the structure of the complex between *Eco*RV endonuclease and DNA fragments containing its

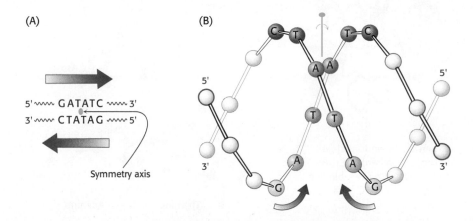

Figure 9.37 Structure of the recognition site of *Eco*RV endonuclease. (A) The sequence of the recognition site, which is symmetric around the axis of rotation designated in green. (B) The inverted repeat within the recognition sequence of *Eco*RV (and most other restriction endonucleases) endows the DNA site with twofold rotational symmetry.

recognition sequence (Figure 9.38). The enzyme surrounds the DNA in a tight embrace.

An enzyme's binding affinity for substrates often determines specificity. Surprisingly, however, binding studies performed in the absence of magnesium have demonstrated that the *Eco*RV endonuclease binds to all sequences, both cognate and noncognate, with approximately equal affinity. Why, then, does the enzyme cleave only cognate sequences? The answer lies in a unique set of interactions between the enzyme and a cognate DNA sequence.

Within the 5'-GATATC-3' sequence, the G and A bases at the 5' end of each strand and their Watson–Crick partners directly contact the enzyme by hydrogen bonding with residues that are located in two loops, one projecting from the surface of each enzyme subunit (Figure 9.39). The most striking feature of this complex is the *distortion of the DNA*, which is substantially kinked in the center (Figure 9.40). The central two TA base pairs in the recognition sequence play a key role in producing the kink. They do not make contact with the enzyme but appear to be required because of their ease of distortion. The 5'-TA-3' sequence is known to be among the most easily deformed base pairs.

The structures of complexes formed with noncognate DNA fragments are strikingly different from those formed with cognate DNA: the noncognate DNA conformation is not substantially distorted (Figure 9.41). *This lack of distortion has important consequences with regard to catalysis. No phosphate is positioned sufficiently close to the active-site aspartate residues to complete a magnesium ion-binding site* (see Figure 9.36). Hence, the nonspecific complexes do not bind the magnesium ions and the complete catalytic apparatus is never assembled. The distortion of the substrate and the subsequent binding of the magnesium ion account for the

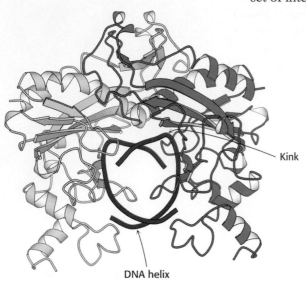

Figure 9.38 *Eco*RV embracing a cognate DNA molecule. This view of the structure of *Eco*RV endonuclease bound to a cognate DNA fragment is down the helical axis of the DNA. The two protein subunits are in yellow and blue, and the DNA backbone is in red. *Notice* that the twofold axes of the enzyme dimer and the DNA are aligned. [Drawn from 1RVB.pdb.]

(A)

(B)

(C)

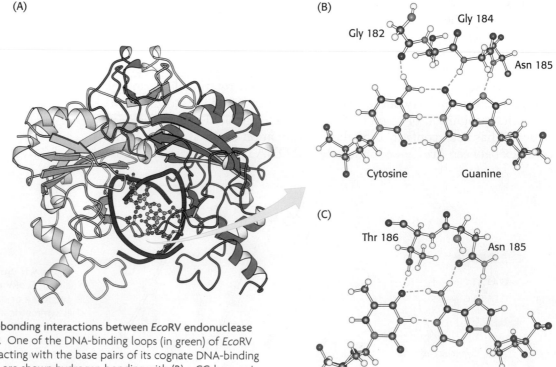

Figure 9.39 Hydrogen-bonding interactions between *Eco*RV endonuclease and its DNA substrate. One of the DNA-binding loops (in green) of *Eco*RV endonuclease is shown interacting with the base pairs of its cognate DNA-binding site. Key amino acid residues are shown hydrogen bonding with (B) a CG base pair and (C) an AT base pair. [Drawn from 1RVB.pdb.]

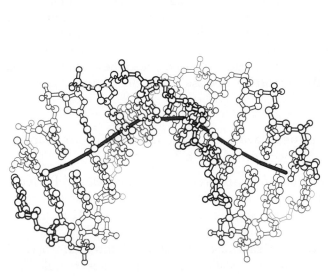

Figure 9.40 Distortion of the recognition site. The DNA is represented as a ball-and-stick model. The path of the DNA helical axis, shown in red, is substantially distorted on binding to the enzyme. For the B form of DNA, the axis is straight (not shown).

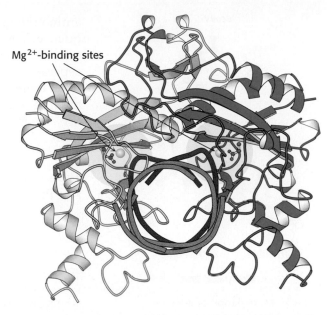

Mg²⁺-binding sites

Figure 9.41 Nonspecific and cognate DNA within *Eco*RV endonuclease. A comparison of the positions of the nonspecific (orange) and the cognate DNA (red) within *Eco*RV. *Notice* that, in the nonspecific complex, the DNA backbone is too far from the enzyme to complete the magnesium ion-binding sites. [Drawn from 1RVB.pdb.]

catalytic specificity of more than 1,000,000-fold that is observed for *Eco*RV endonculease. *Thus, enzyme specificity may be determined by the specificity of enzyme action rather than the specificity of substrate binding.*

We can now see the role of binding energy in this strategy for attaining catalytic specificity. The distorted DNA makes additional contacts with the enzyme, increasing the binding energy. However, the increase in binding energy is canceled by the energetic cost of distorting the DNA from its relaxed conformation (Figure 9.42). Thus, for *Eco*RV endonuclease, there is little difference in binding affinity for cognate and nonspecific DNA fragments. However, the distortion in the cognate complex dramatically affects catalysis by completing the magnesium ion-binding site. This example illustrates how enzymes can utilize available binding energy to deform substrates and poise them for chemical transformation. Interactions that take place within the distorted substrate complex stabilize the transition state leading to DNA hydrolysis.

The distortion in the DNA explains how methylation blocks catalysis and protects host-cell DNA. The host *E. coli* adds a methyl group to the

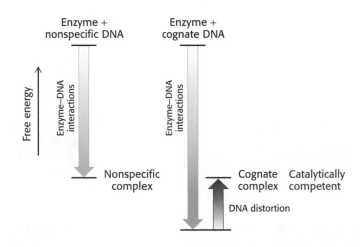

Figure 9.42 Greater binding energy of *Eco*RV endonuclease bound to cognate versus noncognate DNA. The additional interactions between *Eco*RV endonuclease and cognate DNA increase the binding energy, which can be used to drive DNA distortions necessary for forming a catalytically competent complex.

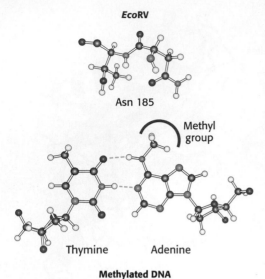

EcoRV

Asn 185

Methyl group

Thymine   Adenine

Methylated DNA

Figure 9.43 Methylation of adenine.
The methylation of adenine blocks the formation of hydrogen bonds between *Eco*RV endonuclease and cognate DNA molecules and prevents their hydrolysis.

Figure 9.44 A conserved structural core in type II restriction enzymes.
Four conserved structural elements, including the active-site region (in blue), are highlighted in color in these models of a single monomer from each dimeric enzyme. The positions of the amino acid sequences that form these elements within each overall sequence are represented schematically below each structure. [Drawn from 1RVB.pdb; 1ERI.pdb; 1BHM.pdb.]

amino group of the adenine nucleotide at the 5′ end of the recognition sequence. The presence of the methyl group blocks the formation of a hydrogen bond between the amino group and the side-chain carbonyl group of asparagine 185 (Figure 9.43). This asparagine residue is closely linked to the other amino acids that form specific contacts with the DNA. The absence of the hydrogen bond disrupts other interactions between the enzyme and the DNA substrate, and the distortion necessary for cleavage will not take place.

Type II Restriction Enzymes Have a Catalytic Core in Common and Are Probably Related by Horizontal Gene Transfer

Type II restriction enzymes are prevalent in Archaea and Bacteria. What can we tell of the evolutionary history of these enzymes? Comparison of the amino acid sequences of a variety of type II restriction endonucleases did not reveal significant sequence similarity between most pairs of enzymes. However, a careful examination of three-dimensional structures, taking into account the location of the active sites, revealed the presence of a core structure conserved in the different enzymes. This structure includes β strands that contain the aspartate (or, in some cases, glutamate) residues forming the magnesium ion-binding sites (Figure 9.44).

These observations indicate that many type II restriction enzymes are indeed evolutionarily related. Analyses of the sequences in greater detail suggest that bacteria may have obtained genes encoding these enzymes from other species by *horizontal gene transfer,* the passing between species of pieces of DNA (such as plasmids) that provide a selective advantage in a particular environment. For example, *Eco*RI (from *E. coli*) and *Rsr*I (from *Rhodobacter sphaeroides*) are 50% identical in sequence over 266 amino acids, clearly indicative of a close evolutionary relationship. However, these species of bacteria are not closely related. Thus, *these species appear to have obtained the gene for these restriction endonucleases from a common source more recently than the time of their evolutionary divergence.* Moreover, the codons used by the gene encoding *Eco*RI endonuclease to specify given amino acids are strikingly different from the codons used by most *E. coli* genes, which suggests that the gene did not originate in *E. coli*.

Horizontal gene transfer may be a common event. For example, genes that inactivate antibiotics are often transferred, leading to the transmission of antibiotic resistance from one species to another. For restriction-modification systems, protection against viral infections may have favored horizontal gene transfer.

EcoRV

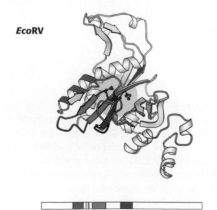

EcoRI

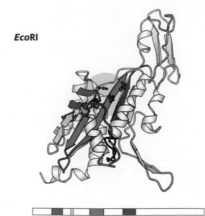

BamHI

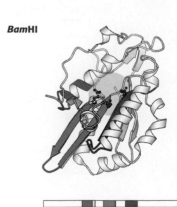

9.4 Nucleoside Monophosphate Kinases Catalyze Phosphoryl-Group Transfer Without Promoting Hydrolysis

The final enzymes that we will consider are the nucleoside monophosphate kinases (NMP kinases), typified by adenylate kinase. These enzymes catalyze the transfer of the terminal phosphoryl group from a nucleoside triphosphate (NTP), usually ATP, to the phosphoryl group on a nucleoside monophosphate (NMP; Figure 9.45). The challenge for NMP kinases is to promote the transfer of the phosphoryl group from NTP to NMP without promoting the competing reaction—the transfer of a phosphoryl group from NTP to water; that is, NTP hydrolysis. We shall see how these enzymes use induced fit to solve this problem. Moreover, a metal ion plays a crucial role in catalysis; but, in this case, the metal forms a complex with the substrate rather than the enzyme.

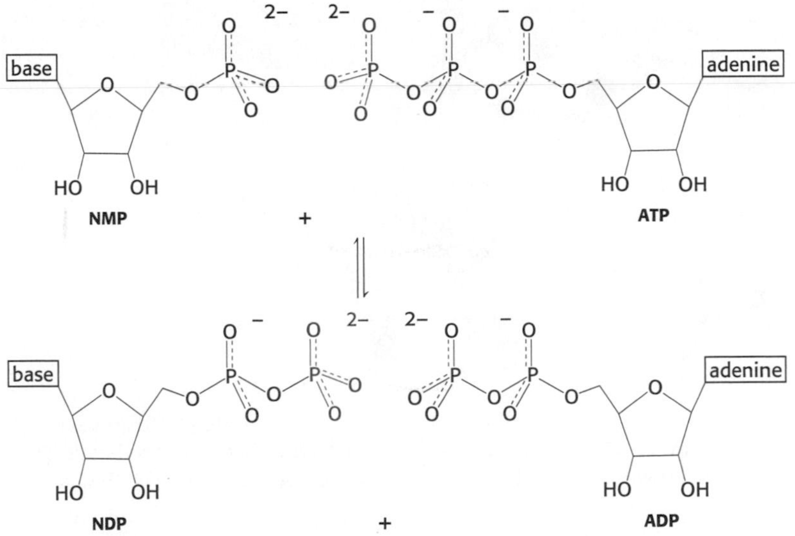

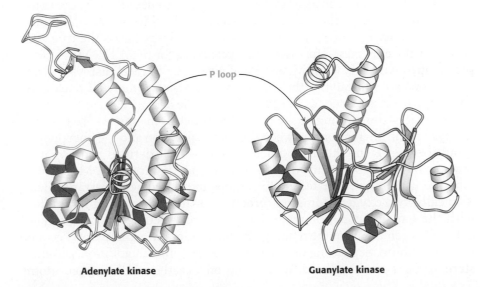

Figure 9.45 Phosphoryl-group transfer by nucleoside monophosphate kinases. These enzymes catalyze the interconversion of a nucleoside triphosphate (here, ATP) and a nucleoside monophosphate (NMP) into two nucleoside diphosphates by the transfer of a phosphoryl group (shown in red).

NMP Kinases Are a Family of Enzymes Containing P-Loop Structures

X-ray crystallography has yielded the three-dimensional structures of a number of different NMP kinases, both free and bound to substrates or substrate analogs. Comparison of these structures reveals that these enzymes form a family of homologous proteins (Figure 9.46). In particular, a

Figure 9.46 The similar structures of adenylate kinase and guanylate kinase. The nucleoside triphosphate-binding domain is a common feature in these homologous nucleotide kinases and others. *Notice* that the domain consists of a central β pleated sheet surrounded on both sides by α helices (highlighted in purple) as well as a key loop (shown in green). [Drawn from 4AKE.pdb and 1GKY.pdb.]

conserved NTP-binding domain is present. This domain consists of a central β sheet, surrounded on both sides by α helices (Figure 9.47). A characteristic feature of this domain is a loop between the first β strand and the first helix. This loop typically has an amino acid sequence of the form Gly-X-X-X-X-Gly-Lys. The loop is often referred to as the *P-loop* because it interacts with phosphoryl groups on the bound nucleotide (Figure 9.48). As described later, similar domains containing P-loops are present in a wide variety of important nucleotide-binding proteins.

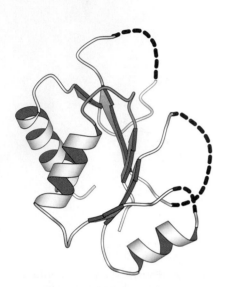

Figure 9.47 The core domain of NMP kinases. The P-loop is shown in green. The dashed lines represent the remainder of the protein structure. [Drawn from 1GKY.pdb.]

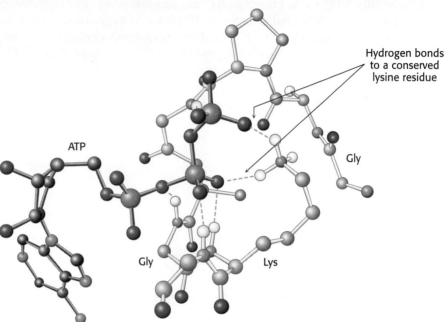

Figure 9.48 P-loop interaction with ATP. The P-loop of adenylate kinase interacts with the phosphoryl groups of ATP (shown with dark bonds). *Notice* that hydrogen bonds (green) link ATP to peptide NH groups as well as a lysine residue conserved among NMP kinases.

Magnesium or Manganese Complexes of Nucleoside Triphosphates Are the True Substrates for Essentially All NTP-Dependent Enzymes

Kinetic studies of NMP kinases, as well as many other enzymes having ATP or other nucleoside triphosphates as a substrate, reveal that these enzymes are essentially inactive in the absence of divalent metal ions such as magnesium (Mg^{2+}) or manganese (Mn^{2+}) but acquire activity on the addition of these ions. In contrast with the enzymes discussed so far, the metal is not a component of the active site. Rather, nucleotides such as ATP bind these ions, and it is the metal ion–nucleotide complex that is the true substrate for the enzymes. The dissociation constant for the ATP–Mg^{2+} complex is approximately 0.1 mM, and thus, given that intracellular Mg^{2+} concentrations are typically in the millimolar range, essentially all nucleoside triphosphates are present as NTP–Mg^{2+} complexes.

How does the binding of the magnesium ion to the nucleotide enhance catalysis? The interactions between the magnesium ion and the phosphoryl-group oxygen atoms hold the nucleotide in a well-defined conformation that can be bound by an enzyme in a specific way (Figure 9.49). Magnesium ions are usually coordinated to six groups in an octahedral arrangement. Typically, two oxygen atoms are directly coordinated to a magnesium ion, with the remaining coordination positions often occupied by water molecules. Oxygen atoms of the α and β, β and γ, or α and γ phosphoryl groups may contribute, depending on the particular enzyme. In addition, different stereoisomers are produced, depending on exactly which oxygen atoms

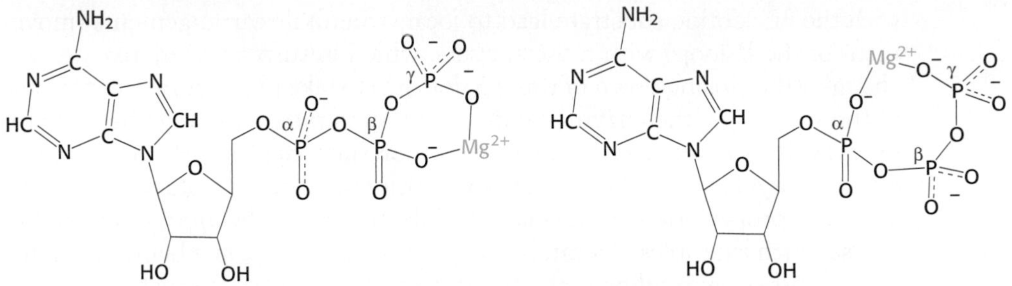

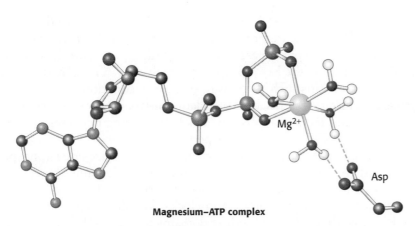

Figure 9.49 Structures of two isomeric forms of the ATP–Mg^{2+} complex. Other groups coordinated to the magnesium ion have been omitted for clarity. Each ATP-dependent enzyme will typically bind only one isomeric form.

Magnesium–ATP complex

Figure 9.50 ATP–Mg^{2+} complex bound to adenylate kinase. *Notice* that the magnesium ion is bound to the β and γ phosphoryl groups and to four water molecules at the remaining coordination positions. These water molecules interact with groups on the enzyme, including a conserved aspartate residue. Other interactions have been omitted for clarity.

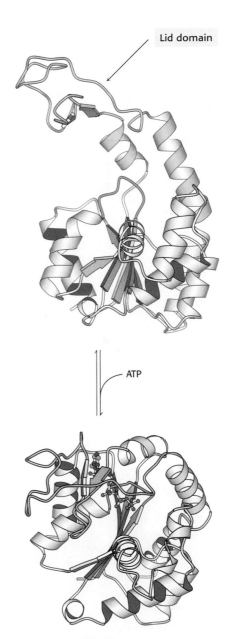

Figure 9.51 Conformational changes in adenylate kinase. Large conformational changes are associated with the binding of ATP by adenylate kinase. The P-loop is shown in green in each structure. *Notice* that the lid domain, highlighted in yellow, closes down on top of the substrate. [Drawn from 4AKE.pdb and 1AKE.pdb.]

bind to the metal ion. *The magnesium ion provides additional points of interaction between the ATP–Mg^{2+} complex and the enzyme, thus increasing the binding energy.* In some cases, such as the DNA polymerases (Section 28.3), side chains (often aspartate and glutamate residues) of the enzyme can bind directly to the magnesium ion. In other cases, the enzyme interacts indirectly with the magnesium ion through hydrogen bonds to the coordinated water molecules (Figure 9.50). Such interactions have been observed in adenylate kinases bound to ATP analogs.

ATP Binding Induces Large Conformational Changes

The action of adenylate kinase provides a classic example of induced fit. Comparison of the structure of adenylate kinase in the presence and absence of an ATP analog reveals that substrate binding induces large structural changes in the kinase (Figure 9.51). The P-loop closes down on top of the polyphosphate chain, interacting most extensively with the β phosphoryl group. The movement of the P-loop brings down the top domain of the enzyme to form a lid over the bound nucleotide. The ATP is held in position by the lid with the γ phosphoryl group positioned next to the binding site for the second substrate, NMP. The binding of NMP induces additional conformational changes. Both sets of changes ensure that a catalytically competent conformation is formed only when both the donor and the acceptor are bound, preventing wasteful transfer of the phosphoryl group to water.

The downward motion of the P loop and enzyme lid is favored by interactions between basic residues (conserved among the NMP kinases), the peptide backbone NH groups, and the nucleotide. In sum, the direct interactions with the nucleotide substrate lead to local structural rearrangements (movement of the P-loop) within the enzyme, which in turn allow more extensive changes (the closing down of the top domain) to take place. As a consequence of these changes, the enzyme holds its two substrates close together and appropriately oriented to stabilize the transition state that leads to the transfer of a phosphoryl group from ATP to NMP. Bringing the substrates into proximity at the proper orientation is an example of *catalysis by approximation*. We will see such examples of a catalytically competent active site being generated only on substrate binding many times in our study of biochemistry.

P-Loop NTPase Domains Are Present in a Range of Important Proteins

Domains similar (and almost certainly homologous) to those found in NMP kinases are present in a remarkably wide array of proteins, many of which participate in essential biochemical processes. Examples include ATP synthase, the key enzyme responsible for ATP generation; molecular motor proteins such as myosin; signal-transduction proteins such as G proteins; proteins essential for translating mRNA into proteins, such as elongation factor Tu; and DNA and RNA unwinding helicases. The wide utility of P-loop NTPase domains is perhaps best explained by their ability to undergo substantial conformational changes on nucleoside triphosphate binding and hydrolysis. We shall encounter these domains (hereafter referred to as P-loop NTPases) throughout the book and shall observe how they function as springs, motors, and clocks. To allow easy recognition of these domains, they, like the binding domains of the NMP kinases, will be depicted with the inner surfaces of the ribbons in a ribbon diagram shown in purple and the P-loop shown in green (Figure 9.52).

Figure 9.52 Three proteins containing P-loop NTPase domains. For the conserved domain, the inner surfaces of the ribbons are purple and the P-loops are green. [Drawn from 4AKE.pdb; 1TND.pdb; 1BMF.pdb.]

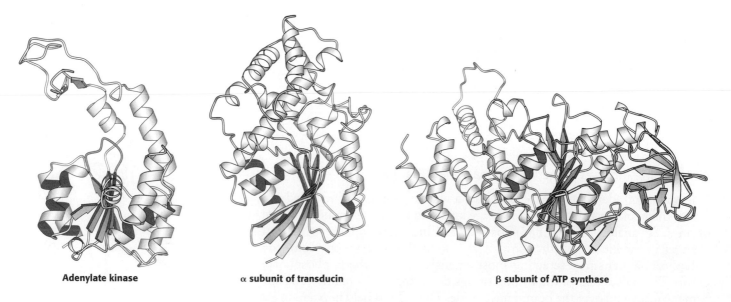

Adenylate kinase　　　**α subunit of transducin**　　　**β subunit of ATP synthase**

Summary

Enzymes adopt conformations that are structurally and chemically complementary to the transition states of the reactions that they catalyze. Sets of interacting amino acid residues make up sites with the special structural and chemical properties necessary to stabilize the transition state. Enzymes use five basic strategies to form and stabilize the transition state: (1) the use of binding energy, (2) covalent catalysis, (3) general acid–base catalysis, (4) metal ion catalysis, and (5) catalysis by approximation. Three of the classes of enzymes examined in this chapter catalyze the addition of water to their substrates but have different requirements for catalytic speed, specificity, and coupling to other processes. The remaining class makes use of induced fit to prevent the addition of water to substrate.

9.1 Proteases Facilitate a Fundamentally Difficult Reaction

The cleavage of peptide bonds by chymotrypsin is initiated by the attack by a serine residue on the peptide carbonyl group. The attacking hydroxyl group is activated by interaction with the imidazole group of a histidine residue, which is, in turn, linked to an aspartate residue. This Ser-His-Asp catalytic triad generates a powerful nucleophile. The product of this initial reaction is a covalent intermediate formed by the enzyme and an acyl group derived from the bound substrate. The hydrolysis of this acyl-enzyme intermediate completes the cleavage process. The tetrahedral intermediates for these reactions have a negative charge on the peptide carbonyl oxygen atom. This negative charge is stabilized by interactions with peptide NH groups in a region on the enzyme termed the oxyanion hole.

Other proteases employ the same catalytic strategy. Some of these proteases, such as trypsin and elastase, are homologs of chymotrypsin. Other proteases, such as subtilisin, contain a very similar catalytic triad that has arisen by convergent evolution. Active-site structures that differ from the catalytic triad are present in a number of other classes of proteases. These classes employ a range of catalytic strategies but, in each case, a nucleophile is generated that is sufficiently powerful to attack the peptide carbonyl group. In some enzymes, the nucleophile is derived from a side chain; whereas, in others, an activated water molecule attacks the peptide carbonyl directly.

9.2 Carbonic Anhydrases Make a Fast Reaction Faster

Carbonic anhydrases catalyze the reaction of water with carbon dioxide to generate carbonic acid. The catalysis can be extremely fast: some carbonic anhydrases hydrate carbon dioxide at rates as high as 1 million times per second. A tightly bound zinc ion is a crucial component of the active sites of these enzymes. Each zinc ion binds a water molecule and promotes its deprotonation to generate a hydroxide ion at neutral pH. This hydroxide ion attacks carbon dioxide to form bicarbonate ion, HCO_3^-. Because of the physiological roles of carbon dioxide and bicarbonate ions, speed is of the essence for this enzyme. To overcome limitations imposed by the rate of proton transfer from the zinc-bound water molecule, the most active carbonic anhydrases have evolved a proton shuttle to transfer protons to a buffer.

9.3 Restriction Enzymes Perform Highly Specific DNA-Cleavage Reactions

A high level of substrate specificity is often the key to biological function. Restriction endonucleases that cleave DNA at specific recognition sequences discriminate between molecules that contain these recognition sequences and those that do not. Within the enzyme–substrate complex, the DNA substrate is distorted in a manner that generates a magnesium ion-binding site between the enzyme and DNA. The magnesium ion binds and activates a water molecule, which attacks the phosphodiester backbone.

Some enzymes discriminate between potential substrates by binding them with different affinities. Others may bind many potential substrates but promote chemical reactions efficiently only on specific molecules. Restriction endonucleases such as *Eco*RV endonuclease employ the latter mechanism. Only molecules containing the proper recognition sequence are distorted in a manner that allows magnesium ion binding and, hence, catalysis. Restriction enzymes are prevented from acting on the DNA of a host cell by the methylation of key sites within its recognition sequences. The added methyl groups block specific interactions between the enzymes and the DNA such that the distortion necessary for cleavage does not take place.

9.4 Nucleoside Monophosphate Kinases Catalyze Phosphoryl-Group Transfer Without Promoting Hydrolysis

Finally, NMP kinases illustrate that induced fit—the alteration of enzyme structure on substrate binding—may be used to ensure phosphoryl transfer between nucleotides rather than to a water molecule. This class of enzyme displays a structural motif called the P-loop NTPase domain that is present in a wide array of nucleotide-binding proteins. The P-loop closes over a bound nucleoside triphosphate substrate such that the top domain of the enzyme forms a lid over the bound nucleotide. The structural shift positions the triphosphate near the monophosphate with which it will react—an example of catalysis by approximation. These enzymes depend on metal ions, but the ions bind to substrate instead of directly to the enzyme. The binding of the metal ion to the nucleoside triphosphate enhances the specificity of the enzyme–substrate interactions by holding the nucleotide in a well-defined conformation and providing additional points of interaction, thus increasing binding energy.

Key Terms

binding energy (p. 242)

induced fit (p. 242)

covalent catalysis (p. 242)

general acid–base catalysis (p. 242)

catalysis by approximation (p. 242)

metal ion catalysis (p. 242)

chemical modification reaction (p. 244)

catalytic triad (p. 245)

oxyanion hole (p. 247)

protease inhibitor (p. 253)

proton shuttle (p. 258)

recognition sequence (p. 259)

methylases (p. 260)

restriction-modification system (p. 260)

in-line displacement (p. 261)

horizontal gene transfer (p. 266)

P-loop (p. 268)

Selected Readings

Where to Start

Stroud, R. M. 1974. A family of protein-cutting proteins. *Sci. Am.* 231(1):74–88.

Kraut, J. 1977. Serine proteases: Structure and mechanism of catalysis. *Annu. Rev. Biochem.* 46:331–358.

Lindskog, S. 1997. Structure and mechanism of carbonic anhydrase. *Pharmacol. Ther.* 74:1–20.

Jeltsch, A., Alves, J., Maass, G., and Pingoud, A. 1992. On the catalytic mechanism of *Eco*RI and *Eco*RV: A detailed proposal based on biochemical results, structural data and molecular modelling. *FEBS Lett.* 304:4–8.

Yan, H., and Tsai, M.-D. 1999. Nucleoside monophosphate kinases: Structure, mechanism, and substrate specificity. *Adv. Enzymol. Relat. Areas Mol. Biol.* 73:103–134.

Lolis, E., and Petsko, G. A. 1990. Transition-state analogues in protein crystallography: Probes of the structural source of enzyme catalysis. *Annu. Rev. Biochem.* 59:597–630.

Books

Fersht, A. 1999. *Structure and Mechanism in Protein Science: A Guide to Enzyme Catalysis and Protein Folding.* W. H. Freeman and Company.

Silverman, R. B. 2000. *The Organic Chemistry of Enzyme-Catalyzed Reactions.* Academic Press.

Page, M., and Williams, A. 1997. *Organic and Bio-organic Mechanisms.* Addison Wesley Longman.

Chymotrypsin and Other Serine Proteases

Fastrez, J., and Fersht, A. R. 1973. Demonstration of the acyl-enzyme mechanism for the hydrolysis of peptides and anilides by chymotrypsin. *Biochemistry* 12:2025–2034.

Sigler, P. B., Blow, D. M., Matthews, B. W., and Henderson, R. 1968. Structure of crystalline-chymotrypsin II: A preliminary report including a hypothesis for the activation mechanism. *J. Mol. Biol.* 35:143–164.

Kossiakoff, A. A., and Spencer, S. A. 1981. Direct determination of the protonation states of aspartic acid-102 and histidine-57 in the tetrahedral intermediate of the serine proteases: Neutron structure of trypsin. *Biochemistry* 20:6462–6474.

Carter, P., and Wells, J. A. 1988. Dissecting the catalytic triad of a serine protease. *Nature* 332:564–568.

Carter, P., and Wells, J. A. 1990. Functional interaction among catalytic residues in subtilisin BPN'. *Proteins* 7:335–342.

Koepke, J., Ermler, U., Warkentin, E., Wenzl, G., and Flecker, P. 2000. Crystal structure of cancer chemopreventive Bowman-Birk inhibitor in ternary complex with bovine trypsin at 2.3 Å resolution: Structural basis of Janus-faced serine protease inhibitor specificity. *J. Mol. Biol.* 298:477–491.

Gaboriaud, C., Rossi, V., Bally, I., Arlaud, G. J., and Fontecilla-Camps, J. C. 2000. Crystal structure of the catalytic domain of human complement C1s: A serine protease with a handle. *EMBO J.* 19:1755–1765.

Other Proteases

Vega, S., Kang, L. W., Velazquez-Campoy, A., Kiso, Y., Amzel, L. M., and Freire, E. 2004. A structural and thermodynamic escape mechanism from a drug resistant mutation of the HIV-1 protease. *Proteins* 55:594–602.

Kamphuis, I. G., Kalk, K. H., Swarte, M. B., and Drenth, J. 1984. Structure of papain refined at 1.65 Å resolution. *J. Mol. Biol.* 179:233–256.

Kamphuis, I. G., Drenth, J., and Baker, E. N. 1985. Thiol proteases: Comparative studies based on the high-resolution structures of papain and actinidin, and on amino acid sequence information for cathepsins B and H, and stem bromelain. *J. Mol. Biol.* 182:317–329.

Sivaraman, J., Nagler, D. K., Zhang, R., Menard, R., and Cygler, M. 2000. Crystal structure of human procathepsin X: A cysteine protease with the proregion covalently linked to the active site cysteine. *J. Mol. Biol.* 295:939–951.

Davies, D. R. 1990. The structure and function of the aspartic proteinases. *Annu. Rev. Biophys. Biophys. Chem.* 19:189–215.

Dorsey, B. D., Levin, R. B., McDaniel, S. L., Vacca, J. P., Guare, J. P., Darke, P. L., Zugay, J. A., Emini, E. A., Schleif, W. A., Quintero, J. C., et al. 1994. L-735,524: The design of a potent and orally bioavailable HIV protease inhibitor. *J. Med. Chem.* 37:3443–3451.

Chen, Z., Li, Y., Chen, E., Hall, D. L., Darke, P. L., Culberson, C., Shafer, J. A., and Kuo, L. C. 1994. Crystal structure at 1.9-Å resolution of human immunodeficiency virus (HIV) II protease complexed with L-735,524, an orally bioavailable inhibitor of the HIV proteases. *J. Biol. Chem.* 269:26344–26348.

Ollis, D. L., Cheah, E., Cygler, M., Dijkstra, B., Frolow, F., Franken, S. M., Harel, M., Remington, S. J., Silman, I., Schrag, J., et al. 1992. The α/β hydrolase fold. *Protein Eng.* 5:197–211.

Carbonic Anhydrase

Strop, P., Smith, K. S., Iverson, T. M., Ferry, J. G., and Rees, D. C. 2001. Crystal structure of the "cab"-type beta class carbonic anhydrase from the archaeon *Methanobacterium thermoautotrophicum*. *J. Biol. Chem.* 276:10299–10305.

Lindskog, S., and Coleman, J. E. 1973. The catalytic mechanism of carbonic anhydrase. *Proc. Natl. Acad. Sci. U. S. A.* 70:2505–2508.

Kannan, K. K., Notstrand, B., Fridborg, K., Lovgren, S., Ohlsson, A., and Petef, M. 1975. Crystal structure of human erythrocyte carbonic anhydrase B: Three-dimensional structure at a nominal 2.2-Å resolution. *Proc. Natl. Acad. Sci. U. S. A.* 72:51–55.

Boriack-Sjodin, P. A., Zeitlin, S., Chen, H. H., Crenshaw, L., Gross, S., Dantanarayana, A., Delgado, P., May, J. A., Dean, T., and Christianson, D. W. 1998. Structural analysis of inhibitor binding to human carbonic anhydrase II. *Protein Sci.* 7:2483–2489.

Wooley, P. 1975. Models for metal ion function in carbonic anhydrase. *Nature* 258:677–682.

Jonsson, B. H., Steiner, H., and Lindskog, S. 1976. Participation of buffer in the catalytic mechanism of carbonic anhydrase. *FEBS Lett.* 64:310–314.

Sly, W. S., and Hu, P. Y. 1995. Human carbonic anhydrases and carbonic anhydrase deficiencies. *Annu. Rev. Biochem.* 64:375–401.

Maren, T. H. 1988. The kinetics of HCO_3^- synthesis related to fluid secretion, pH control, and CO_2 elimination. *Annu. Rev. Physiol.* 50:695–717.

Kisker, C., Schindelin, H., Alber, B. E., Ferry, J. G., and Rees, D. C. 1996. A left-hand beta-helix revealed by the crystal structure of a carbonic anhydrase from the archaeon *Methanosarcina thermophila*. *EMBO J.* 15:2323–2330.

Restriction Enzymes

Selvaraj, S., Kono, H., and Sarai, A. 2002. Specificity of protein-DNA recognition revealed by structure-based potentials: Symmetric/asymmetric and cognate/non-cognate binding. *J. Mol. Biol.* 322:907–915.

Winkler, F. K., Banner, D. W., Oefner, C., Tsernoglou, D., Brown, R. S., Heathman, S. P., Bryan, R. K., Martin, P. D., Petratos, K., and Wilson, K. S. 1993. The crystal structure of *Eco*RV endonuclease and of its complexes with cognate and non-cognate DNA fragments. *EMBO J.* 12:1781–1795.

Kostrewa, D., and Winkler, F. K. 1995. Mg^{2+} binding to the active site of *Eco*RV endonuclease: A crystallographic study of complexes with substrate and product DNA at 2 Å resolution. *Biochemistry* 34:683–696.

Athanasiadis, A., Vlassi, M., Kotsifaki, D., Tucker, P. A., Wilson, K. S., and Kokkinidis, M. 1994. Crystal structure of *Pvu*II endonuclease reveals extensive structural homologies to *Eco*RV. *Nat. Struct. Biol.* 1:469–475.

Sam, M. D., and Perona, J. J. 1999. Catalytic roles of divalent metal ions in phosphoryl transfer by *Eco*RV endonuclease. *Biochemistry* 38:6576–6586.

Jeltsch, A., and Pingoud, A. 1996. Horizontal gene transfer contributes to the wide distribution and evolution of type II restriction-modification systems. *J. Mol. Evol.* 42:91–96.

NMP Kinases

Krishnamurthy, H., Lou, H., Kimple, A., Vieille, C., and Cukier, R. I. 2005. Associative mechanism for phosphoryl transfer: A molecular dynamics simulation of *Escherichia coli* adenylate kinase complexed with its substrates. *Proteins* 58:88–100.

Byeon, L., Shi, Z., and Tsai, M. D. 1995. Mechanism of adenylate kinase: The "essential lysine" helps to orient the phosphates and the active site residues to proper conformations. *Biochemistry* 34:3172–3182.

Dreusicke, D., and Schulz, G. E. 1986. The glycine-rich loop of adenylate kinase forms a giant anion hole. *FEBS Lett.* 208:301–304.

Pai, E. F., Sachsenheimer, W., Schirmer, R. H., and Schulz, G. E. 1977. Substrate positions and induced-fit in crystalline adenylate kinase. *J. Mol. Biol.* 114:37–45.

Schlauderer, G. J., Proba, K., and Schulz, G. E. 1996. Structure of a mutant adenylate kinase ligated with an ATP-analogue showing domain closure over ATP. *J. Mol. Biol.* 256:223–227.

Vonrhein, C., Schlauderer, G. J., and Schulz, G. E. 1995. Movie of the structural changes during a catalytic cycle of nucleoside monophosphate kinases. *Structure* 3:483–490.

Muller-Dieckmann, H. J., and Schulz, G. E. 1994. The structure of uridylate kinase with its substrates, showing the transition state geometry. *J. Mol. Biol.* 236:361–367.

Problems

1. *No burst.* Examination of the cleavage of the *amide* substrate, A, by chymotrypsin with the use of stopped-flow kinetic methods reveals no burst. The reaction is monitored by noting the color produced by the release of the amino part of the substrate (highlighted in orange). Why is no burst observed?

A

2. *Contributing to your own demise.* Consider the subtilisin substrates A and B.

Phe-Ala-Gln-Phe-X   Phe-Ala-His-Phe-X

   A   **B**

These substrates are cleaved (between Phe and X) by native subtilisin at essentially the same rate. However, the His 64-to-Ala mutant of subtilisin cleaves substrate B more than 1000-fold more rapidly than it cleaves substrate A. Propose an explanation.

3. *1 + 1 ≠ 2.* Consider the following argument. In subtilisin, mutation of Ser 221 to Ala results in a 10^6-fold decrease in activity. Mutation of His 64 to Ala results in a similar 10^6-fold decrease. Therefore, simultaneous mutation of Ser 221 to Ala and His 64 to Ala should result in a $10^6 \times 10^6 = 10^{12}$-fold reduction in activity. Is this reduction correct? Why or why not?

4. *Adding a charge.* In chymotrypsin, a mutant was constructed with Ser 189, which is in the bottom of the substrate-specificity pocket, changed to Asp. What effect would you predict for this Ser 189→Asp 189 mutation?

5. *Conditional results.* In carbonic anhydrase II, mutation of the proton-shuttle residue His 64 to Ala was expected to result in a decrease in the maximal catalytic rate. However, in buffers such as imidazole with relatively small molecular components, no rate reduction was observed. In buffers with larger molecular components, significant rate reductions were observed. Propose an explanation.

6. *How many sites?* A researcher has isolated a restriction endonuclease that cleaves at only one particular 10-base-pair site. Would this enzyme be useful in protecting cells from viral infections, given that a typical viral genome is 50,000 base pairs long? Explain.

7. *Is faster better?* Restriction endonucleases are, in general, quite slow enzymes with typical turnover numbers of 1 s^{-1}. Suppose that endonucleases were faster with turnover numbers similar to those for carbonic anhydrase (10^6 s^{-1}). Would this increased rate be beneficial to host cells, assuming that the fast enzymes have similar levels of specificity?

8. *Adopting a new gene.* Suppose that one species of bacteria obtained one gene encoding a restriction endonuclease by horizontal gene transfer. Would you expect this acquisition to be beneficial?

9. *Chelation therapy.* Treatment of carbonic anhydrase with high concentrations of the metal chelator EDTA (ethylenediaminetetraacetic acid) results in the loss of enzyme activity. Propose an explanation.

10. *An aldehyde inhibitor.* Elastase is specifically inhibited by an aldehyde derivative of one of its substrates:

(a) Which residue in the active site of elastase is most likely to form a covalent bond with this aldehyde?
(b) What type of covalent link would be formed?

11. *Predict the product.* Adenylate kinase is treated with adenosine diphosphate (ADP).

(a) What products will be generated?
(b) If the initial concentration of ADP is 1 mM, estimate the concentrations of ADP and the products from part *a* after incubation with adenylate kinase for a long time.

12. *Identify the enzyme.* Consider the structure of molecule A. Which enzyme discussed in this chapter do you think molecule A will most effectively inhibit?

Molecule A

Mechanism Problem

13. *Complete the mechanism.* On the basis of the information provided in Figure 9.18, complete the mechanisms for peptide-bond cleavage by (a) a cysteine protease, (b) an aspartyl protease, and (c) a metalloprotease.

Regulatory Strategies

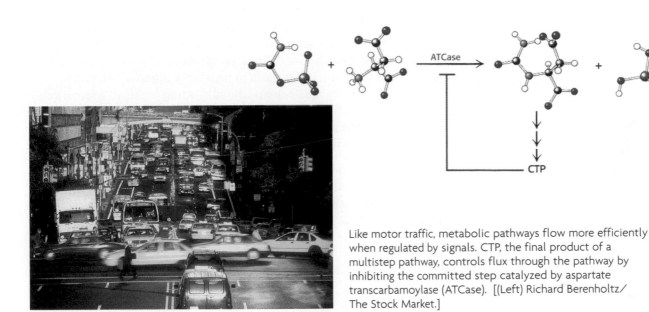

Like motor traffic, metabolic pathways flow more efficiently when regulated by signals. CTP, the final product of a multistep pathway, controls flux through the pathway by inhibiting the committed step catalyzed by aspartate transcarbamoylase (ATCase). [(Left) Richard Berenholtz/ The Stock Market.]

The activity of enzymes often must be regulated so that they function at the proper time and place. This regulation is essential for coordination of the vast array of biochemical processes taking place at any instant in an organism. Enzymatic activity is regulated in five principal ways:

1. *Allosteric Control.* Allosteric proteins contain distinct regulatory sites and multiple functional sites. The binding of small signal molecules at regulatory sites is a significant means of controlling the activity of these proteins. Moreover, allosteric proteins show the property of *cooperativity:* activity at one functional site affects the activity at others. Proteins displaying allosteric control are thus information transducers: their activity can be modified in response to signal molecules or to information shared among active sites. This chapter examines one of the best-understood allosteric proteins: the enzyme *aspartate transcarbamoylase* (ATCase). Catalysis by aspartate transcarbamoylase of the first step in pyrimidine biosynthesis is inhibited by cytidine triphosphate, the final product of that biosynthesis, in an example of *feedback inhibition.* We have already examined an allosteric protein—hemoglobin, the oxygen transport protein in the blood (Chapter 7).

2. *Multiple Forms of Enzymes.* Isozymes, or isoenzymes, provide an avenue for varying regulation of the same reaction at distinct locations or times. Isozymes are homologous enzymes within a single organism that catalyze the same reaction but differ slightly in structure and more obviously

in K_M and V_{max} values, as well as regulatory properties. Often, isozymes are expressed in a distinct tissue or organelle or at a distinct stage of development.

3. *Reversible Covalent Modification.* The catalytic properties of many enzymes are markedly altered by the covalent attachment of a modifying group, most commonly a phosphoryl group. ATP serves as the phosphoryl donor in these reactions, which are catalyzed by *protein kinases*. The removal of phosphoryl groups by hydrolysis is catalyzed by *protein phosphatases*. This chapter considers the structure, specificity, and control of *protein kinase A* (PKA), a ubiquitous eukaryotic enzyme that regulates diverse target proteins.

4. *Proteolytic Activation.* The enzymes controlled by some of these regulatory mechanisms cycle between active and inactive states. A different regulatory strategy is used to *irreversibly* convert an inactive enzyme into an active one. Many enzymes are activated by the hydrolysis of a few peptide bonds or even one such bond in inactive precursors called *zymogens* or *proenzymes*. This regulatory mechanism generates digestive enzymes such as chymotrypsin, trypsin, and pepsin. Blood clotting is due to a remarkable cascade of zymogen activations. Active digestive and clotting enzymes are switched off by the irreversible binding of specific inhibitory proteins that are irresistible lures to their molecular prey.

5. *Controlling the Amount of Enzyme Present.* Enzyme activity can also be regulated by adjusting the amount of enzyme present. This important form of regulation usually takes place at the level of transcription. We will consider the control of gene transcription in Chapter 31.

To begin here, we will consider the principles of allostery by examining the enzyme aspartate transcarbamoylase.

10.1 Aspartate Transcarbamoylase Is Allosterically Inhibited by the End Product of Its Pathway

Aspartate transcarbamoylase catalyzes the first step in the biosynthesis of pyrimidines: the condensation of aspartate and carbamoyl phosphate to form N-carbamoylaspartate and orthophosphate (Figure 10.1). This reaction is the committed step in the pathway that will ultimately yield pyrimidine

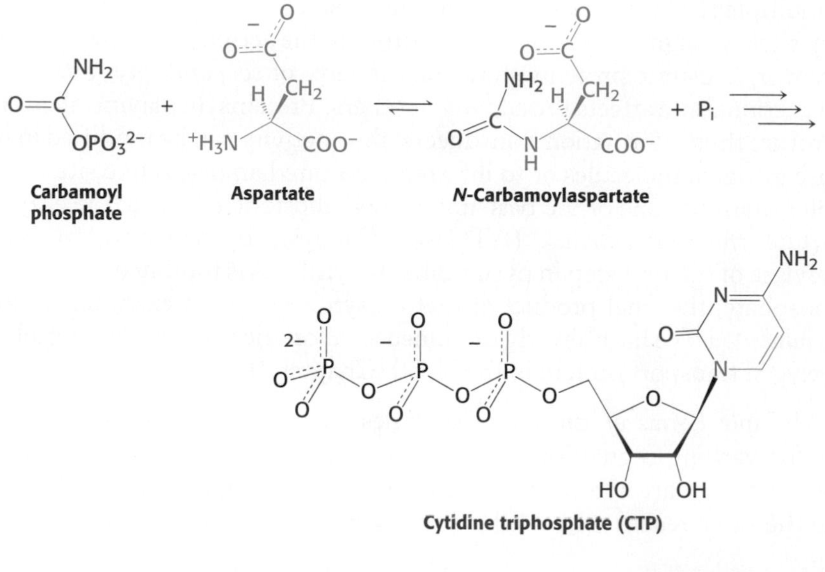

Figure 10.1 ATCase reaction. Aspartate transcarbamoylase catalyzes the committed step, the condensation of aspartate and carbamoyl phosphate to form N-carbamoylaspartate, in pyrimidine synthesis.

nucleotides such as cytidine triphosphate (CTP). How is this enzyme regulated to generate precisely the amount of CTP needed by the cell?

John Gerhart and Arthur Pardee found that ATCase is inhibited by CTP, the final product of the ATCase-initiated pathway. The rate of the reaction catalyzed by ATCase is fast at low concentrations of CTP but slows as CTP concentration increases (Figure 10.2). Thus, the pathway continues to make new pyrimidines until sufficient quantities of CTP have accumulated. The inhibition of ATCase by CTP is an example of *feedback inhibition,* the inhibition of an enzyme by the end product of the pathway. Feedback inhibition by CTP ensures that *N*-carbamoylaspartate and subsequent intermediates in the pathway are not needlessly formed when pyrimidines are abundant.

The inhibitory ability of CTP is remarkable because *CTP is structurally quite different from the substrates of the reaction* (see Figure 10.1). Thus CTP must bind to a site distinct from the active site where substrate binds. Such sites are called *allosteric* or *regulatory sites.* CTP is an example of an *allosteric inhibitor.* In ATCase (but not all allosterically regulated enzymes), the catalytic sites and the regulatory sites are on separate polypeptide chains.

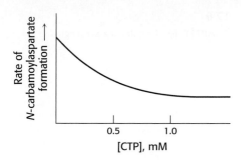

Figure 10.2 CTP inhibits ATCase. Cytidine triphosphate, an end product of the pyrimidine synthesis pathway, inhibits aspartate transcarbamoylase despite having little structural similarity to reactants or products.

Allosterically Regulated Enzymes Do Not Follow Michaelis–Menten Kinetics

Allosteric enzymes are distinguished by their response to changes in substrate concentration in addition to their susceptibility to regulation by other molecules. Let us examine the rate of product formation as a function of substrate concentration for ATCase (Figure 10.3). The curve differs from that expected for an enzyme that follows Michaelis–Menten kinetics. The observed curve is referred to as sigmoidal because it resembles an "S." The vast majority of allosteric enzymes display sigmoidal kinetics. Recall from the discussion of hemoglobin that sigmoidal curves result from cooperation between subunits: the binding of substrate to one active site in a molecule increases the likelihood that substrate will bind to other active sites. To understand the basis of sigmoidal enzyme kinetics and inhibition by CTP, we need to examine the structure of ATCase.

ATCase Consists of Separable Catalytic and Regulatory Subunits

What is the evidence that ATCase has distinct regulatory and catalytic sites? ATCase can be literally separated into regulatory (r) and catalytic (c) subunits by treatment with a mercurial compound such as *p*-hydroxymercuribenzoate, which reacts with sulfhydryl groups (Figure 10.4). Ultracentrifugation

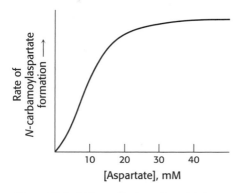

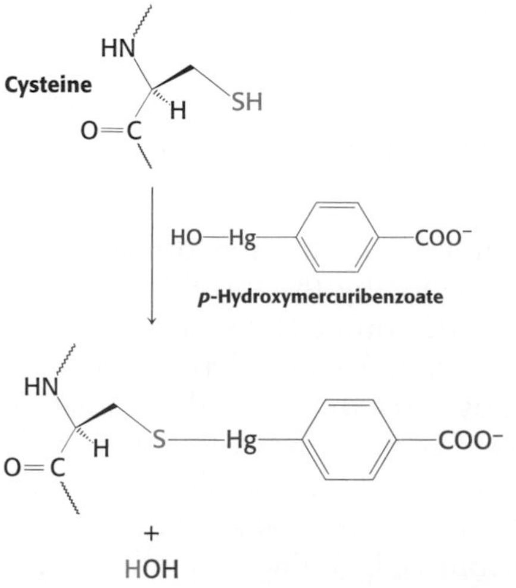

Figure 10.3 ATCase displays sigmoidal kinetics. A plot of product formation as a function of substrate concentration produces a sigmoidal curve because the binding of substrate to one active site increases the activity at the other active sites. Thus, the enzyme shows cooperativity.

Figure 10.4 Modification of cysteine residues. *p*-Hydroxymercuribenzoate reacts with crucial cysteine residues in aspartate transcarbamoylase.

studies carried out by John Gerhart and Howard Schachman showed that mercurials dissociate ATCase into two kinds of subunits (Figure 10.5). The subunits can be readily separated by ion-exchange chromatography because they differ markedly in charge or by centrifugation in a sucrose density gradient because they differ in size. These size differences are manifested in the sedimentation coefficients: that of the native enzyme is 11.6S, whereas those of the dissociated subunits are 2.8S and 5.8S. The attached p-mercuribenzoate groups can be removed from the separated subunits by adding an excess of mercaptoethanol, providing isolated subunits for study.

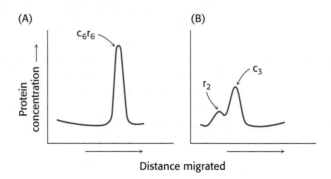

Figure 10.5 Ultracentrifugation studies of ATCase. Sedimentation velocity patterns of (A) native ATCase and (B) the enzyme after treatment with p-hydroxymercuribenzoate show that the enzyme can be dissociated into regulatory (r) and catalytic (c) subunits. [After J. C. Gerhart and H. K. Schachman. *Biochemistry* 4(1965):1054–1062.]

The larger subunit is called the *catalytic subunit*. This subunit displays catalytic activity but is unresponsive to CTP and does not display sigmoidal kinetics. The isolated smaller subunit can bind CTP, but has no catalytic activity. Hence, that subunit is called the *regulatory subunit*. The catalytic subunit (c_3) consists of three chains (34 kd each), and the regulatory subunit (r_2) consists of two chains (17 kd each). The catalytic and regulatory subunits combine rapidly when they are mixed. The resulting complex has the same structure, c_6r_6, as the native enzyme: two catalytic trimers and three regulatory dimers.

$$2\,c_3 + 3\,r_2 \rightarrow c_6r_6$$

Most strikingly, the reconstituted enzyme has the same allosteric and kinetic properties as those of the native enzyme. Thus, ATCase is composed of discrete catalytic and regulatory subunits, and *the interaction of the subunits in the native enzyme produces its regulatory and catalytic properties.*

Allosteric Interactions in ATCase Are Mediated by Large Changes in Quaternary Structure

What are the subunit interactions that account for the properties of ATCase? Significant clues have been provided by the three-dimensional structure of ATCase in various forms, first determined by x-ray crystallography in the laboratory of William Lipscomb. Two catalytic trimers are stacked one on top of the other, linked by three dimers of the regulatory chains (Figure 10.6). There are significant contacts between the catalytic and the regulatory subunits: each r chain within a regulatory dimer interacts with a c chain within a catalytic trimer. The c chain makes contact with a structural domain in the r chain that is stabilized by a zinc ion bound to four

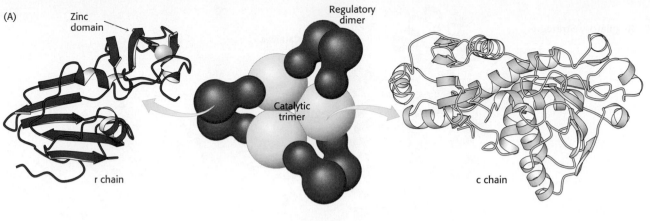

(A)

Zinc domain

Regulatory dimer

Catalytic trimer

r chain

c chain

(B)

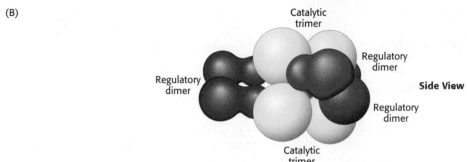

Catalytic trimer

Regulatory dimer

Regulatory dimer

Side View

Regulatory dimer

Catalytic trimer

Figure 10.6 Structure of ATCase.
(A) The quaternary structure of aspartate transcarbamoylase as viewed from the top. The drawing in the center is a simplified representation of the relations between subunits. A single catalytic trimer [catalytic (c) chains, shown in yellow] is visible; in this view, the second trimer is hidden behind the one visible. *Notice* that each r chain interacts with a c chain through the zinc domain. (B) A side view of the complex. [Drawn from 1 RAI.pdb.]

cysteine residues. The mercurial compound *p*-hydroxymercuribenzoate is able to dissociate the catalytic and regulatory subunits because mercury binds strongly to the cysteine residues, displacing the zinc and destabilizing this r-subunit domain.

To locate the active sites, the enzyme was crystallized in the presence of *N*-(phosphonacetyl)-L-aspartate (PALA), a bisubstrate analog (an analog of the two substrates) that resembles an intermediate along the pathway of catalysis (Figure 10.7). PALA is a potent competitive inhibitor of ATCase; it binds to the active sites and blocks them. The structure of the ATCase–PALA complex reveals that PALA binds at sites lying at the boundaries between pairs of c chains within a catalytic trimer (Figure 10.8). Each catalytic trimer contributes three active sites to the complete enzyme. Further examination of the ATCase–PALA complex reveals a remarkable change in quaternary structure on binding of PALA. The two catalytic

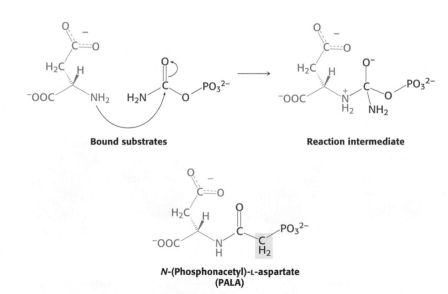

Bound substrates

Reaction intermediate

N-(Phosphonacetyl)-L-aspartate
(PALA)

Figure 10.7 PALA, a bisubstrate analog. (Top) Nucleophilic attack by the amino group of aspartate on the carbonyl carbon atom of carbamoyl phosphate generates an intermediate on the pathway to the formation of *N*-carbamoylaspartate. (Bottom) *N*-(Phosphonacetyl)-L-aspartate (PALA) is an analog of the reaction intermediate and a potent competitive inhibitor of aspartate transcarbamoylase.

279

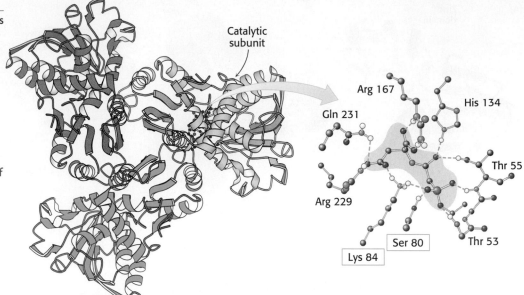

Figure 10.8 The active site of ATCase. Some of the crucial active-site residues are shown binding to the inhibitor PALA. *Notice* that the active site is composed mainly of residues from one subunit, but an adjacent subunit also contributes important residues (boxed in green). [Drawn from 8ATC.pdb.]

trimers move 12 Å farther apart and rotate approximately 10 degrees about their common threefold axis of symmetry. Moreover, the regulatory dimers rotate approximately 15 degrees to accommodate this motion (Figure 10.9). The enzyme literally expands on PALA binding. In essence, ATCase has two distinct quaternary forms: one that predominates in the absence of substrate or substrate analogs and another that predominates when substrates or analogs are bound. We call these forms the T (for tense) state and the R (for relaxed) state, respectively, as we did for the two quaternary states of hemoglobin.

How can we explain the enzyme's sigmoidal kinetics in light of the structural observations? Like hemoglobin (p. 188), the enzyme exists in an equilibrium between the T state and the R state. In the absence of substrate, almost all the enzyme molecules are in the T state. The T state has a low affinity for substrate and hence shows a low catalytic activity. The occasional binding of a substrate molecule to one active site in an enzyme increases the likelihood that the entire enzyme shifts to the R state with its higher binding affinity. The addition of more substrate has two effects. First, it increases the probability that each enzyme molecule will bind at least one substrate molecule. Second, it increases the average number of substrate molecules bound to each enzyme. The presence of additional substrate will increase the fraction of enzyme molecules in the more active R state because *the position of the equilibrium depends on the number of active sites that are occupied by substrate*. We considered this property, called

Figure 10.9 The T-to-R state transition in ATCase. Aspartate transcarbamoylase exists in two conformations: a compact, relatively inactive form called the tense (T) state and an expanded form called the relaxed (R) state. *Notice* that the structure of ATCase changes dramatically in the transition from the T state to the R State. PALA binding stabilizes the R state.

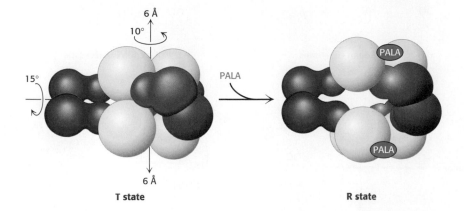

cooperativity because the subunits cooperate with one another, when we discussed the sigmoidal oxygen-binding curve of hemoglobin.

This mechanism for allosteric regulation is referred to as the *concerted mechanism* because the change in the enzyme is "all or none"; the entire enzyme is converted from T into R, affecting all of the catalytic sites equally (p. 189). In contrast, the *sequential model* assumes that the binding of ligand to one site on the complex can affect neighboring sites without causing all subunits to undergo the T-to-R transition. Although the concerted mechanism explains the behavior of ATCase well, most other allosteric enzymes have features of both models.

The sigmoidal curve for ATCase can be pictured as a composite of two Michaelis–Menten curves, one corresponding to the T state and the other to the R state. An increase in substrate concentration favors a transition from the T-state curve to the R-state curve (Figure 10.10). Note that such sigmoidal behavior has an additional consequence: in the concentration range at which the T-to-R transition is taking place, the curve depends quite steeply on the substrate concentration. The enzyme is switched from a less active state to a more active state within a narrow range of substrate concentration. This behavior is beneficial when a response to small changes in substrate concentration is physiologically important. The effects of substrates on allosteric enzymes are referred to as *homotropic effects* (from the Greek *homós*, "same").

In studies of the isolated catalytic trimer, the catalytic subunit shows the hyperbolic curve characteristic of Michaelis–Menten kinetics, which is indistinguishable from the curve deduced for the R state (see Figure 10.10). Thus, the term *tense* is apt: in the T state, the regulatory dimers hold the two catalytic trimers sufficiently close to each other that key loops on their surfaces collide and interfere with conformational adjustments necessary for high-affinity substrate binding and catalysis.

Allosteric Regulators Modulate the T-to-R Equilibrium

We now turn our attention to the effects of CTP. As noted earlier, CTP inhibits the action of ATCase. X-ray studies of ATCase in the presence of CTP revealed (1) that the enzyme is in the T state when bound to CTP and (2) that a binding site for this nucleotide exists in each regulatory chain in a domain that does not interact with the catalytic subunit (Figure 10.11). Each active site is more than 50 Å from the nearest CTP-binding site. The question naturally arises, How can CTP inhibit the catalytic activity of the enzyme when it does not interact with the catalytic chain?

The quaternary structural changes observed on substrate-analog binding suggest a mechanism for inhibition by CTP (Figure 10.12). *The binding of the inhibitor CTP shifts the equilibrium toward the T state, decreasing net enzyme activity.* The binding of CTP makes it more difficult for substrate

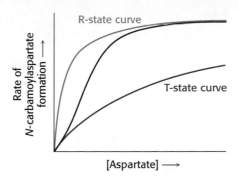

Figure 10.10 Basis for the sigmoidal curve. The generation of the sigmoidal curve by the property of cooperativity can be understood by imagining an allosteric enzyme as a mixture of two Michaelis–Menten enzymes, one with a high value of K_M that corresponds to the T state and another with a low value of K_M that corresponds to the R state. As the concentration of substrate is increased, the equilibrium shifts from the T state to the R state, which results in a steep rise in activity with respect to substrate concentration.

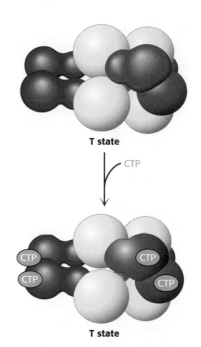

Figure 10.11 CTP stabilizes the T state. The binding of CTP to the regulatory subunit of aspartate transcarbamoylase stabilizes the T state.

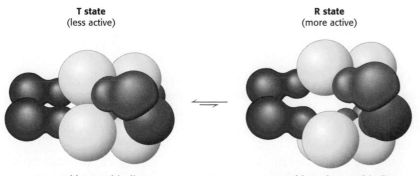

T state
(less active)

R state
(more active)

Favored by CTP binding

Favored by substrate binding

Figure 10.12 The R state and the T state are in equilibrium. Even in the absence of any substrate or regulators, aspartate transcarbamoylase exists in equilibrium between the R and the T states. Under these conditions, the T state is favored by a factor of approximately 200.

binding to convert the enzyme into the R state. Consequently, CTP increases the initial phase of the sigmoidal curve (Figure 10.13). More substrate is required to attain a given reaction rate.

Interestingly, ATP, too, is an allosteric effector of ATCase. However, the effect of ATP is to *increase* the reaction rate at a given aspartate concentration (Figure 10.14). At high concentrations of ATP, the kinetic profile shows a less-pronounced sigmoidal behavior. ATP competes with CTP for binding to regulatory sites. Consequently, high levels of ATP prevent CTP from inhibiting the enzyme. The effects of nonsubstrate molecules on allosteric enzymes (such as those of CTP and ATP on ATCase) are referred to as *heterotropic effects* (from the Greek *héteros*, "different").

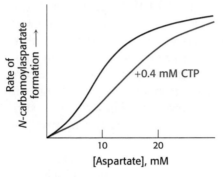

Figure 10.13 Effect of CTP on ATCase kinetics. Cytidine triphosphate (CTP) stabilizes the T state of aspartate transcarbamoylase, making it more difficult for substrate binding to convert the enzyme into the R state. As a result, the curve is shifted to the right, as shown in red.

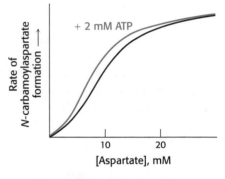

Figure 10.14 Effect of ATP on ATCase kinetics. ATP is an allosteric activator of aspartate transcarbamoylase because it stabilizes the R state, making it easier for substrate to bind. As a result, the curve is shifted to the left, as shown in blue.

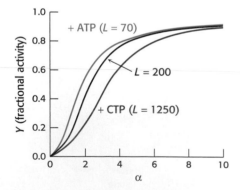

Figure 10.15 Quantitative description of the MWC model. In this description of the MWC (Monod, Wyman, and Changeaux) model, fractional activity, Y, is the fraction of active sites bound to substrate and is directly proportional to reaction velocity; α is the ratio of [S] to the dissociation constant of S with the enzyme in the R state; and L is the ratio of the concentration of enzyme in the T state to that in the R state. The binding of the regulators ATP and CTP to ATCase changes the value of L and thus the response to substrate concentration. To construct these curves, the formula on page 200 was used, with $c = 0.1$ and $n = 6$.

The increase in ATCase activity in response to increased ATP concentration has two potential physiological explanations. First, high ATP concentration signals a high concentration of purine nucleotides in the cell; the increase in ATCase activity will tend to balance the purine and pyrimidine pools. Second, a high concentration of ATP indicates that energy is available for mRNA synthesis and DNA replication and leads to the synthesis of pyrimidines needed for these processes.

In the Appendix to Chapter 7, we developed a quantitative description of the concerted model. Although developed to describe a binding process, the model also applies to enzyme activity because the fraction of enzyme active sites with substrate bound is proportional to enzyme activity. A key aspect of this model is the equilibrium between the T and the R states (p. 200). We defined L as the equilibrium constant between the R and the T forms.

$$R \rightleftharpoons T \qquad\qquad L = \frac{[T]}{[R]}$$

The effects of CTP and ATP can be modeled simply by changing the value of L. For the CTP-saturated form, the value of L increases from 250 to 1250. Thus, it takes more substrate to shift the equilibrium appreciably to the R form. For the ATP saturated form, the value of L decreases to 70 (Figure 10.15).

Thus, the concerted model provides us with a good description of the kinetic behavior of ATCase in the presence of its key regulators.

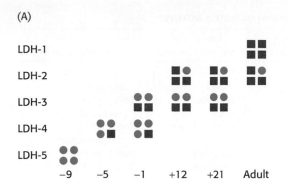

(A)

LDH-1						▪▪ / ▪▪
LDH-2				▪▪ / ▪▪	▪▪ / ▪▪	▪▪ / ▪▪
LDH-3			●● / ▪▪	●● / ▪▪	●● / ▪▪	
LDH-4		●● / ●▪	●● / ●▪			
LDH-5	●● / ●●					
	−9	−5	−1	+12	+21	Adult

(B)

	Heart	Kidney	Red blood cell	Brain	Leukocyte	Muscle	Liver
H_4	▬	▬	▬	▬	▬	—	—
H_3M	▬	▬	▬	▬	▬	—	—
H_2M_2	—	▬	▬	▬	▬	▬	—
HM_3	—	—	—	—	—	▬	—
M_4	—	—	—	—	—	▬	▬

Figure 10.16 Isozymes of lactate dehydrogenase. (A) The rat heart LDH isozyme profile changes in the course of development. The H isozyme is represented by squares and the M isozyme by circles. The negative and positive numbers denote the days before and after birth, respectively. (B) LDH isozyme content varies by tissue. [(A) After W.-H. Li, *Molecular Evolution* (Sinauer, 1997), p. 283; (B) after K. Urich, *Comparative Animal Biochemistry* (Springer Verlag, 1990), p. 542.]

10.2 Isozymes Provide a Means of Regulation Specific to Distinct Tissues and Developmental Stages

Isozymes, or *isoenzymes,* are enzymes that differ in amino acid sequence yet catalyze the same reaction. Usually, these enzymes display different kinetic parameters, such as K_M, or respond to different regulatory molecules. They are encoded by different genes, which usually arise through gene duplication and divergence. Isozymes can often be distinguished from one another by biochemical properties such as electrophoretic mobility.

The existence of isozymes permits the fine-tuning of metabolism to meet the needs of a given tissue or developmental stage. Consider the example of lactate dehydrogenase (LDH), an enzyme that catalyzes a step in anaerobic glucose metabolism and glucose synthesis. Human beings have two isozymic polypeptide chains for this enzyme: the H isozyme is highly expressed in heart muscle and the M isozyme is expressed in skeletal muscle. The amino acid sequences are 75% identical. Each functional enzyme is tetrameric, and many different combinations of the two isozymic polypeptide chains are possible. The H_4 isozyme, found in the heart, has a higher affinity for substrates than does the M_4 isozyme. The two isozymes also differ in that high levels of pyruvate allosterically inhibit the H_4 but not the M_4 isozyme. The other combinations, such as H_3M, have intermediate properties. We will consider these isozymes in their biological context in Chapter 16.

The M_4 isozyme functions optimally in the anaerobic environment of hard-working skeletal muscle, whereas the H_4 isozyme does so in the aerobic environment of heart muscle. Indeed, the proportions of these isozymes are altered in the development of the rat heart as the tissue switches from an anaerobic environment to an aerobic one (Figure 10.16A). Figure 10.16B shows the tissue-specific forms of lactate dehydrogenase in adult rat tissues.

The appearance of some isozymes in the blood is a sign of tissue damage, useful for clinical diagnosis. For instance, an increase in serum levels of H_4 relative to H_3M is an indication that a myocardial infarction, or heart attack, has damaged heart muscle cells, leading to the release of cellular material.

10.3 Covalent Modification Is a Means of Regulating Enzyme Activity

The covalent attachment of another molecule can modify the activity of enzymes and many other proteins. In these instances, a donor molecule

TABLE 10.1 Common covalent modifications of protein activity

Modification	Donor molecule	Example of modified protein	Protein function
Phosphorylation	ATP	Glycogen phosphorylase	Glucose homeostasis; energy transduction
Acetylation	Acetyl CoA	Histones	DNA packing; transcription
Myristoylation	Myristoyl CoA	Src	Signal transduction
ADP ribosylation	NAD^+	RNA polymerase	Transcription
Farnesylation	Farnesyl pyrophosphate	Ras	Signal transduction
γ-Carboxylation	HCO_3^-	Thrombin	Blood clotting
Sulfation	3'-Phosphoadenosine-5'-phosphosulfate	Fibrinogen	Blood-clot formation
Ubiquitination	Ubiquitin	Cyclin	Control of cell cycle

Acetylated lysine

provides the functional moiety being attached. Most modifications are reversible. Phosphorylation and dephosphorylation are the most common means of covalent modification. The attachment of acetyl groups and their removal are another. Histones—proteins that are packaged with DNA into chromosomes—are extensively acetylated and deacetylated in vivo (Section 31.3). More heavily acetylated histones are associated with genes that are being actively transcribed. The acetyltransferase and deacetylase enzymes are themselves regulated by phosphorylation, showing that the covalent modification of a protein may be controlled by the covalent modification of the modifying enzymes.

Modification is not readily reversible in some cases. The irreversible attachment of a lipid group causes some proteins in signal-transduction pathways, such as Ras (a GTPase) and Src (a protein tyrosine kinase), to become affixed to the cytoplasmic face of the plasma membrane. Fixed in this location, the proteins are better able to receive and transmit information that is being passed along their signaling pathways (Chapter 14). Mutations in both Ras and Src are seen in a wide array of cancers. The attachment of the small protein ubiquitin is a signal that a protein is to be destroyed, the ultimate means of regulation (Chapter 23). The protein cyclin must be ubiquitinated and destroyed before a cell can enter anaphase and proceed through the cell cycle.

Virtually all the metabolic processes that we will examine are regulated in part by covalent modification. Indeed, the allosteric properties of many enzymes are modified by covalent modification. Table 10.1 lists some of the common covalent modifications.

Phosphorylation Is a Highly Effective Means of Regulating the Activities of Target Proteins

We will see phosphorylation used as a regulatory mechanism in virtually every metabolic process in eukaryotic cells. Indeed, as much as 30% of eukaryotic proteins are phosphorylated. The enzymes catalyzing phosphorylation reactions are called *protein kinases*. These enzymes constitute one of the largest protein families known: there are more than 100 homologous protein kinases in yeast and more than 500 in human beings. This multiplicity of enzymes allows regulation to be fine-tuned according to a specific tissue, time, or substrate.

ATP is the most common donor of phosphoryl groups. The terminal (γ) phosphoryl group of ATP is transferred to a specific amino acid. In eukaryotes, the acceptor is always one of the three containing a hydroxyl group in its side chain. Transfers to *serine* and *threonine* residues are handled by one class of protein kinases and to *tyrosine* residues by another. Tyrosine kinases, which are unique to multicellular organisms, play pivotal roles in growth regulation, and mutations in these enzymes are commonly observed in cancer cells.

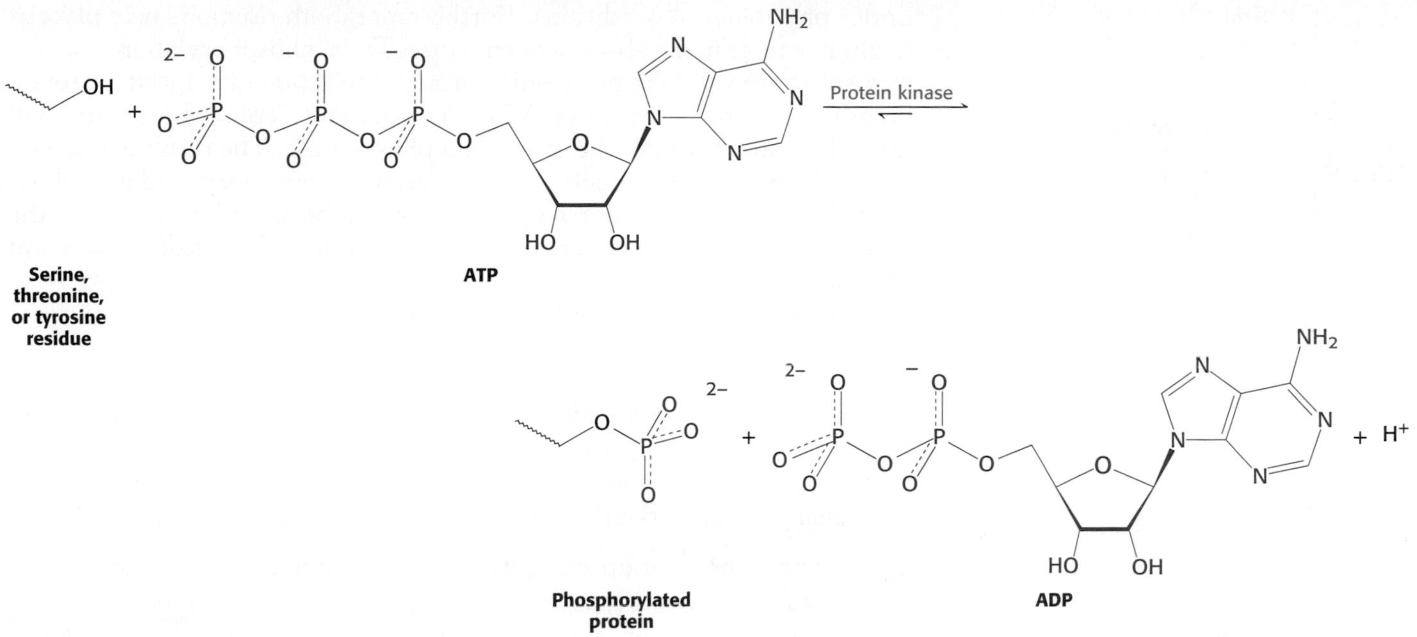

Serine, threonine, or tyrosine residue

ATP

Phosphorylated protein

ADP

Table 10.2 lists a few of the known serine and threonine protein kinases. The acceptors in protein-phosphorylation reactions are located inside cells, where the phosphoryl-group donor ATP is abundant. Proteins that are entirely extracellular are not regulated by reversible phosphorylation.

Protein phosphatases reverse the effects of kinases by catalyzing the removal of phosphoryl groups attached to proteins. The enzyme hydrolyzes the bond attaching the phosphoryl group.

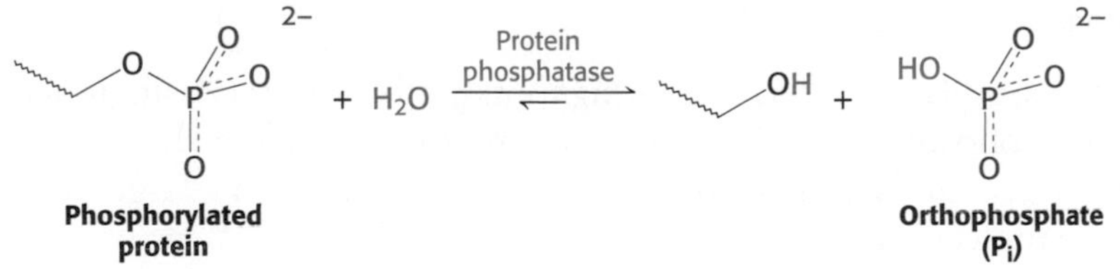

Phosphorylated protein

Orthophosphate (P_i)

The unmodified hydroxyl-containing side chain is regenerated and orthophosphate (P_i) is produced. These enzymes play a vital role in cells because they turn off the signaling pathways that are activated by kinases. One class of highly conserved phosphatase called PP2A suppresses the cancer-promoting activity of certain kinases.

It is important to note that the phosphorylation and dephosphorylation reactions are not the reverse of one another; each is essentially irreversible

TABLE 10.2 Examples of serine and threonine kinases and their activating signals

Signal	Enzyme
Cyclic nucleotides	Cyclic AMP-dependent protein kinase
	Cyclic GMP-dependent protein kinase
Ca^{2+} and calmodulin	Ca^{2+}–calmodulin protein kinase
	Phosphorylase kinase or glycogen synthase kinase 2
AMP	AMP-activated kinase
Diacylglycerol	Protein kinase C
Metabolic Intermediates and other "local" effectors	Many target-specific enzymes, such as pyruvate dehydrogenase kinase and branched-chain ketoacid dehydrogenase kinase

Source: After D. Fell, *Understanding the Control of Metabolism* (Portland Press, 1997), Table 7.2.

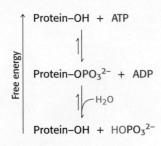

Free energy

Protein–OH + ATP

Protein–OPO$_3^{2-}$ + ADP

H$_2$O

Protein–OH + HOPO$_3^{2-}$

under physiological conditions. Furthermore, both reactions take place at negligible rates in the absence of enzymes. Thus, phosphorylation of a protein substrate will take place only through the action of a specific protein kinase and at the expense of ATP cleavage, and dephosphorylation will take place only through the action of a phosphatase. The result is that target proteins cycle unidirectionally between unphosphorylated and phosphorylated forms. The rate of cycling between the phosphorylated and the dephosphorylated states depends on the relative activities of kinases and phosphatases.

Phosphorylation is a highly effective means of controlling the activity of proteins for several reasons:

1. A phosphoryl group adds two negative charges to a modified protein. These new charges may disrupt electrostatic interactions in the unmodified protein and allow new electrostatic interactions to be formed. Such structural changes can markedly alter substrate binding and catalytic activity.

2. A phosphoryl group can form three or more hydrogen bonds. The tetrahedral geometry of a phosphoryl group makes these bonds highly directional, allowing for specific interactions with hydrogen-bond donors.

3. The free energy of phosphorylation is large. Of the $-50\,\text{kJ mol}^{-1}$ ($-12\,\text{kcal mol}^{-1}$) provided by ATP, about half is consumed in making phosphorylation irreversible; the other half is conserved in the phosphorylated protein. A free-energy change of $5.69\,\text{kJ mol}^{-1}$ ($1.36\,\text{kcal mol}^{-1}$) corresponds to a factor of 10 in an equilibrium constant (p. 210). Hence, phosphorylation can change the conformational equilibrium between different functional states by a large factor, of the order of 10^4.

4. Phosphorylation and dephosphorylation can take place in less than a second or over a span of hours. The kinetics can be adjusted to meet the timing needs of a physiological process.

5. Phosphorylation often evokes *highly amplified* effects. A single activated kinase can phosphorylate hundreds of target proteins in a short interval. If the target protein is an enzyme, it may in turn transform a large number of substrate molecules.

6. ATP is the cellular energy currency (Chapter 15). The use of this compound as a phosphoryl-group donor links the energy status of the cell to the regulation of metabolism.

Protein kinases vary in their degree of specificity. *Dedicated protein kinases* phosphorylate a single protein or several closely related ones. *Multifunctional protein kinases* modify many different targets; they have a wide reach and can coordinate diverse processes. Comparisons of amino acid sequences of many phosphorylation sites show that a multifunctional kinase recognizes related sequences. For example, the *consensus sequence* recognized by protein kinase A is Arg-Arg-X-*Ser*-Z or Arg-Arg-X-*Thr*-Z, in which X is a small residue, Z is a large hydrophobic one, and *Ser* or *Thr* is the site of phosphorylation. It should be noted that this sequence is not absolutely required. Lysine, for example, can substitute for one of the arginine residues but with some loss of affinity. Short synthetic peptides containing a consensus motif are nearly always phosphorylated by serine-threonine protein kinases. Thus, *the primary determinant of specificity is the amino acid sequence surrounding the serine or threonine phosphorylation site.* However, distant residues can contribute to specificity. For instance, a

Cyclic AMP Activates Protein Kinase A by Altering the Quaternary Structure

The "flight or fight" response is common to many animals presented with a dangerous or exciting situation. Muscle becomes primed for action. This priming is the result of the activity of a particular protein kinase. In this case, the hormone epinephrine (adrenaline) triggers the formation of cyclic AMP (cAMP), an intracellular messenger formed by the cyclization of ATP. Cyclic AMP subsequently activates a *key enzyme: protein kinase A* (PKA). The kinase alters the activities of target proteins by phosphorylating specific serine or threonine residues. The striking finding is that *most effects of cAMP in eukaryotic cells are achieved through the activation by cAMP of PKA.*

PKA provides a clear example of the integration of allosteric regulation and phosphorylation. PKA is activated by cAMP concentrations near 10 nM. The activation mechanism is reminiscent of that of aspartate transcarbamoylase. Like that enzyme, PKA in muscle consists of two kinds of subunits: a 49-kd regulatory (R) subunit and a 38-kd catalytic (C) subunit. In the absence of cAMP, the regulatory and catalytic subunits form an R_2C_2 complex that is enzymatically inactive (Figure 10.17). The binding of two molecules of cAMP to each of the regulatory subunits leads to the dissociation of R_2C_2 into an R_2 subunit and two C subunits. These free catalytic subunits are then enzymatically active. Thus, *the binding of cAMP to the regulatory subunit relieves its inhibition of the catalytic subunit.* PKA and most other kinases exist in isozymic forms for fine-tuning regulation to meet the needs of a specific cell or developmental stage.

How does the binding of cAMP activate the kinase? Each R chain contains the sequence Arg-Arg-Gly-*Ala*-Ile, which matches the consensus sequence for phosphorylation except for the presence of alanine in place of serine. In the R_2C_2 complex, this *pseudosubstrate sequence* of R occupies the catalytic site of C, thereby preventing the entry of protein substrates (see Figure 10.17). The binding of cAMP to the R chains allosterically moves the pseudosubstrate sequences out of the catalytic sites. The released C chains are then free to bind and phosphorylate substrate proteins.

Cyclic adenosine monophosphate (cAMP)

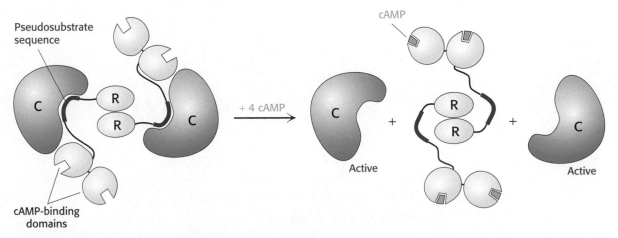

Figure 10.17 Regulation of protein kinase A. The binding of four molecules of cAMP activates protein kinase A by dissociating the inhibited holoenzyme (R_2C_2) into a regulatory subunit (R_2) and two catalytically active subunits (C). Each R chain includes cAMP-binding domains and a pseudosubstrate sequence.

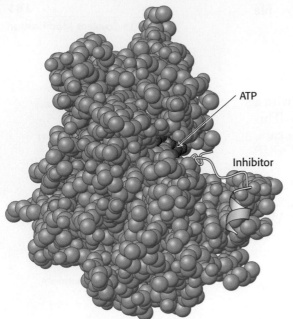

Figure 10.18 Protein kinase A bound to an inhibitor. This space-filling model shows a complex of the catalytic subunit of protein kinase A with an inhibitor bearing a pseudosubstrate sequence. *Notice* that the inhibitor (yellow) binds in a cleft between the domains of the enzyme. The bound ATP, shown in red, is in the active site adjacent to the inhibitor. [Drawn from 1ATP.pdb.]

ATP and the Target Protein Bind to a Deep Cleft in the Catalytic Subunit of Protein Kinase A

X-ray crystallography revealed the three-dimensional structure of the catalytic subunit of PKA containing a bound 20-residue peptide inhibitor. The 350-residue catalytic subunit has two lobes (Figure 10.18). ATP and part of the inhibitor fill a deep cleft between the lobes. The smaller lobe makes many contacts with ATP–Mg^{2+}, whereas the larger lobe binds the peptide and contributes the key catalytic residues. As with other kinases (p. 269), the two lobes move closer to one another on substrate binding; mechanisms that restrict this domain closure provide a means of regulating protein kinase activity. *The PKA structure has broad significance because residues 40 to 280 constitute a conserved catalytic core that is common to essentially all known protein kinases.* We see here an example of a successful biochemical solution to a problem (in this case, protein phosphorylation) being employed many times in the course of evolution.

The bound peptide in this crystal occupies the active site because it contains the pseudosubstrate sequence Arg-Arg-Asn-*Ala*-Ile (Figure 10.19). The structure of the complex reveals the interactions by which the enzyme recognizes the consensus sequence. The guanidinium group of the first arginine residue forms an ion pair with the carboxylate side chain of a glutamate residue (Glu 127) of the enzyme. The second arginine likewise interacts with two other carboxylate groups. The nonpolar side chain of isoleucine, which matches Z in the consensus sequence (see p. 285), fits snugly in a hydrophobic groove formed by two leucine residues of the enzyme.

Figure 10.19 Binding of pseudosubstrate to protein kinase A. *Notice* that the inhibitor makes multiple contacts with the enzyme. The two arginine side chains of the pseudosubstrate form salt bridges with three glutamate carboxylate groups. Hydrophobic interactions are also important in the recognition of substrate. The isoleucine residue of the pseudosubstrate is in contact with a pair of leucine residues of the enzyme.

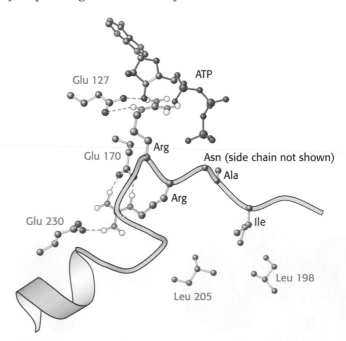

10.4 Many Enzymes Are Activated by Specific Proteolytic Cleavage

We turn now to a different mechanism of enzyme regulation. Many enzymes acquire full enzymatic activity as they spontaneously fold into their characteristic three-dimensional forms. In contrast, the folded forms of other enzymes are inactive until activated by cleavage of one or a few specific peptide bonds. The inactive precursor is called a *zymogen* or a *proenzyme*. An energy

TABLE 10.3 Gastric and pancreatic zymogens

Site of synthesis	Zymogen	Active enzyme
Stomach	Pepsinogen	Pepsin
Pancreas	Chymotrypsinogen	Chymotrypsin
Pancreas	Trypsinogen	Trypsin
Pancreas	Procarboxypeptidase	Carboxypeptidase
Pancreas	Proelastase	Elastase

source such as ATP is not needed for cleavage. Therefore, in contrast with reversible regulation by phosphorylation, even proteins located outside cells can be activated by this means. Another noteworthy difference is that proteolytic activation, in contrast with allosteric control and reversible covalent modification, occurs just once in the life of an enzyme molecule.

Specific proteolysis is a common means of activating enzymes and other proteins in biological systems. For example:

1. The *digestive enzymes* that hydrolyze proteins are synthesized as zymogens in the stomach and pancreas (Table 10.3).

2. *Blood clotting* is mediated by a cascade of proteolytic activations that ensures a rapid and amplified response to trauma.

3. Some protein hormones are synthesized as inactive precursors. For example, *insulin* is derived from *proinsulin* by proteolytic removal of a peptide.

4. The fibrous protein *collagen*, the major constituent of skin and bone, is derived from *procollagen*, a soluble precursor.

5. Many *developmental processes* are controlled by the activation of zymogens. For example, in the metamorphosis of a tadpole into a frog, large amounts of collagen are resorbed from the tail in the course of a few days. Likewise, much collagen is broken down in a mammalian uterus after delivery. The conversion of *procollagenase* into *collagenase*, the active protease, is precisely timed in these remodeling processes.

6. *Programmed cell death*, or *apoptosis*, is mediated by proteolytic enzymes called *caspases*, which are synthesized in precursor form as *procaspases*. When activated by various signals, caspases function to cause cell death in most organisms, ranging from *C. elegans* to human beings. Apoptosis provides a means of sculpting the shapes of body parts in the course of development and a means of eliminating damaged or infected cells.

We next examine the activation and control of zymogens, using as examples several digestive enzymes as well as blood-clot formation.

Chymotrypsinogen Is Activated by Specific Cleavage of a Single Peptide Bond

Chymotrypsin is a digestive enzyme that hydrolyzes proteins in the small intestine. Its mechanism of action was described in detail in Chapter 9. Its inactive precursor, *chymotrypsinogen*, is synthesized in the pancreas, as are several other zymogens and digestive enzymes. Indeed, the pancreas is one of the most active organs in synthesizing and secreting proteins. The enzymes and zymogens are synthesized in the acinar cells of the pancreas and stored inside membrane-bounded granules (Figure 10.20). The zymogen granules accumulate at the apex of the acinar cell; when the cell is stimulated by a hormonal signal or a nerve impulse, the contents of the granules are released into a duct leading into the duodenum.

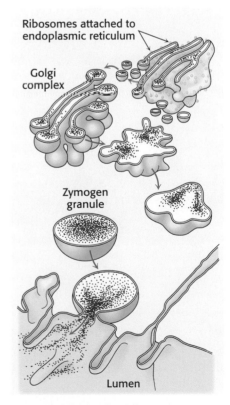

Figure 10.20 Secretion of zymogens by an acinar cell of the pancreas.

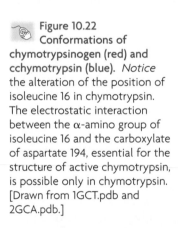

Chymotrypsinogen
(inactive)

| 1 | | | 245 |

↓ Trypsin

π-Chymotrypsin
(active)

| 1 | 15 | 16 | | 245 |

↓ π-Chymotrypsin

↘ Two dipeptides

α-Chymotrypsin
(active)

| 1 | 13 | 16 | 146 | 149 | 245 |

A chain   B chain   C chain

Figure 10.21 Proteolytic activation of chymotrypsinogen. The three chains of α-chymotrypsin are linked by two interchain disulfide bonds (A to B, and B to C).

Chymotrypsinogen, a single polypeptide chain consisting of 245 amino acid residues, is virtually devoid of enzymatic activity. It is converted into a fully active enzyme when the peptide bond joining arginine 15 and isoleucine 16 is cleaved by trypsin (Figure 10.21). The resulting active enzyme, called π-chymotrypsin, then acts on other π-chymotrypsin molecules. Two dipeptides are removed to yield α-chymotrypsin, the stable form of the enzyme. The three resulting chains in α-chymotrypsin remain linked to one another by two interchain disulfide bonds. The striking feature of this activation process is that *cleavage of a single specific peptide bond transforms the protein from a catalytically inactive form into one that is fully active.*

Proteolytic Activation of Chymotrypsinogen Leads to the Formation of a Substrate-Binding Site

How does cleavage of a single peptide bond activate the zymogen? The cleavage of the peptide bond between amino acids 15 and 16 triggers key conformational changes, which were revealed by the elucidation of the three-dimensional structure of chymotrypsinogen.

1. The newly formed *amino-terminal group of isoleucine 16 turns inward and forms an ionic bond with aspartate 194* in the interior of the chymotrypsin molecule (Figure 10.22).

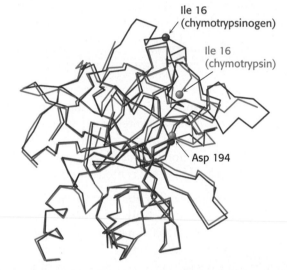

Ile 16
(chymotrypsinogen)

Ile 16
(chymotrypsin)

Asp 194

Figure 10.22 Conformations of chymotrypsinogen (red) and cchymotrypsin (blue). *Notice* the alteration of the position of isoleucine 16 in chymotrypsin. The electrostatic interaction between the α-amino group of isoleucine 16 and the carboxylate of aspartate 194, essential for the structure of active chymotrypsin, is possible only in chymotrypsin. [Drawn from 1GCT.pdb and 2GCA.pdb.]

2. This electrostatic interaction triggers a number of conformational changes. Methionine 192 moves from a deeply buried position in the zymogen to the surface of the active enzyme, and residues 187 and 193 become more extended. These changes result in the formation of the *substrate-specificity site* for aromatic and bulky nonpolar groups. One side of this site is made up of residues 189 through 192. *This cavity for binding part of the substrate is not fully formed in the zymogen.*

3. The tetrahedral transition state in catalysis by chymotrypsin is stabilized by hydrogen bonds between the negatively charged carbonyl oxygen atom of the substrate and two NH groups of the main chain of the enzyme (p. 247). One of these NH groups is not appropriately located in chymotrypsinogen, and so *the oxyanion hole is incomplete in the zymogen.*

4. The conformational changes elsewhere in the molecule are very small. Thus, *the switching on of enzymatic activity in a protein can be accomplished by discrete, highly localized conformational changes that are triggered by the hydrolysis of a single peptide bond.*

The Generation of Trypsin from Trypsinogen Leads to the Activation of Other Zymogens

The structural changes accompanying the activation of *trypsinogen,* the precursor of the proteolytic enzyme *trypsin,* are somewhat different from those in the activation of chymotrypsinogen. X-ray analyses have shown that the conformation of four stretches of polypeptide, constituting about 15% of the molecule, changes markedly on activation. *These regions are very flexible in the zymogen, whereas they have a well-defined conformation in trypsin.* Furthermore, the oxyanion hole (p. 247) in trypsinogen is too far from histidine 57 to promote the formation of the tetrahedral transition state.

The digestion of proteins in the duodenum requires the concurrent action of several proteolytic enzymes, because each is specific for a limited number of side chains. Thus, the zymogens must be switched on at the same time. Coordinated control is achieved by the action of *trypsin as the common activator of all the pancreatic zymogens*—trypsinogen, chymotrypsinogen, proelastase, procarboxypeptidase, and prolipase, a lipid degrading enzyme. To produce active trypsin, the cells that line the duodenum secrete an enzyme, *enteropeptidase,* which hydrolyzes a unique lysine–isoleucine peptide bond in trypsinogen as the zymogen enters the duodenum from the pancreas. The small amount of trypsin produced in this way activates more trypsinogen and the other zymogens (Figure 10.23). Thus, *the formation of trypsin by enteropeptidase is the master activation step.*

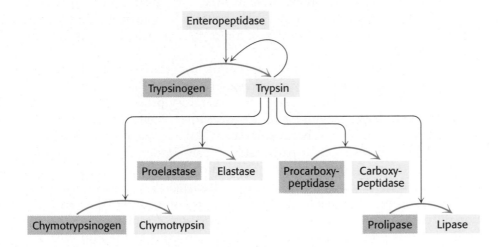

Figure 10.23 Zymogen activation by proteolytic cleavage. Enteropeptidase initiates the activation of the pancreatic zymogens by activating trypsin, which then activates other zymogens. Active enzymes are shown in yellow; zymogens are shown in orange.

Some Proteolytic Enzymes Have Specific Inhibitors

The conversion of a zymogen into a protease by cleavage of a single peptide bond is a precise means of switching on enzymatic activity. However, this activation step is irreversible, and so a different mechanism is needed to stop proteolysis. Specific protease inhibitors accomplish this task. For example, *pancreatic trypsin inhibitor,* a 6-kd protein, inhibits trypsin by binding very tightly to its active site. The dissociation constant of the complex is 0.1 pM, which corresponds to a standard free energy of binding of about -75 kJ mol^{-1} ($-18 \text{ kcal mol}^{-1}$). In contrast with nearly all known protein assemblies, this complex is not dissociated into its constituent chains by treatment with denaturing agents such as 8 M urea or 6 M guanidine hydrochloride.

The reason for the exceptional stability of the complex is that pancreatic trypsin inhibitor is a very effective substrate analog. X-ray analyses showed that the inhibitor lies in the active site of the enzyme, positioned such that

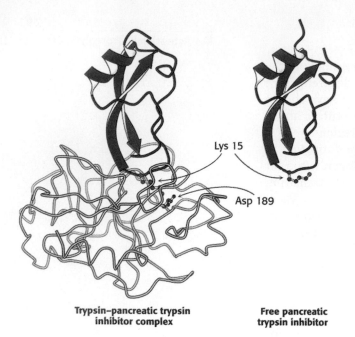

Trypsin–pancreatic trypsin inhibitor complex

Free pancreatic trypsin inhibitor

Lys 15

Asp 189

Figure 10.24 Interaction of trypsin with its inhibitor. Structure of a complex of trypsin (yellow) and pancreatic trypsin inhibitor (red). *Notice* that lysine 15 of the inhibitor penetrates into the active site of the enzyme. There it forms a salt bridge with aspartate 189 in the active site. Notice also that bound inhibitor and the free inhibitor are almost identical in structure. [Drawn from 18PI.pdb.]

the side chain of lysine 15 of this inhibitor interacts with the aspartate side chain in the specificity pocket of trypsin. In addition, there are many hydrogen bonds between the main chain of trypsin and that of its inhibitor. Furthermore, the carbonyl group of lysine 15 and the surrounding atoms of the inhibitor fit snugly in the active site of the enzyme. Comparison of the structure of the inhibitor bound to the enzyme with that of the free inhibitor reveals that *the structure is essentially unchanged on binding to the enzyme* (Figure 10.24). Thus, the inhibitor is preorganized into a structure that is highly complementary to the enzyme's active site. Indeed, the peptide bond between lysine 15 and alanine 16 in pancreatic trypsin inhibitor is cleaved but at a very slow rate: the half-life of the trypsin–inhibitor complex is several months. In essence, the inhibitor is a substrate, but its intrinsic structure is so nicely complementary to the enzyme's active site that it binds very tightly and is turned over slowly.

The amount of trypsin is much greater than the amount of inhibitor. Why does trypsin inhibitor exist? Recall that trypsin activates other zymogens. Consequently, the prevention of even small amounts of trypsin from initiating the inappropriately activated cascade prematurely is vital. Trypsin inhibitor binds to trypsin molecules in the pancreas or pancreatic ducts. This inhibition prevents severe damage to those tissues, which could lead to acute pancreatitis.

Pancreatic trypsin inhibitor is not the only important protease inhibitor. α_1-*Antitrypsin* (also called α_1-*antiproteinase*), a 53-kd plasma protein, protects tissues from digestion by elastase, a secretory product of neutrophils (white blood cells that engulf bacteria). *Antielastase* would be a more accurate name for this inhibitor, because it blocks *elastase* much more effectively than it blocks trypsin. Like pancreatic trypsin inhibitor, α_1-antitrypsin blocks the action of target enzymes by binding nearly irreversibly to their active sites. Genetic disorders leading to a deficiency of α_1-antitrypsin show that this inhibitor is physiologically important. For example, the substitution of lysine for glutamate at residue 53 in the type Z mutant slows the secretion of this inhibitor from liver cells. Serum levels of the inhibitor are about 15% of normal in people homozygous for this defect. The consequence is that excess elastase destroys alveolar walls in the lungs by digesting elastic fibers and other connective-tissue proteins.

The resulting clinical condition is called *emphysema* (also known as *destructive lung disease*). People with emphysema must breathe much harder than normal people to exchange the same volume of air because their alveoli are much less resilient than normal. Cigarette smoking markedly increases the likelihood that even a type Z heterozygote will develop emphysema. The reason is that smoke oxidizes methionine 358 of the inhibitor (Figure 10.25), a residue essential for binding elastase. Indeed, this methionine side chain is the bait that selectively traps elastase. The *methionine sulfoxide* oxidation product, in contrast, does not lure elastase, a striking consequence of the insertion of just one oxygen atom into a protein and a striking example of the effect of behavior on biochemistry. We will consider another protease inhibitor, antithrombin III, when we examine the control of blood clotting.

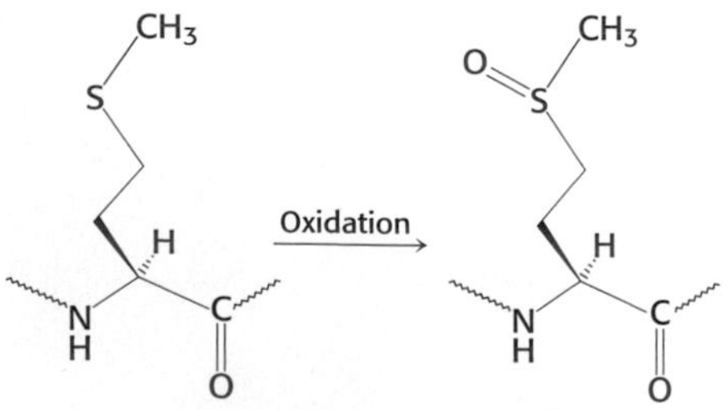

Figure 10.25 Oxidation of methionine to methionine sulfoxide.

Blood Clotting Is Accomplished by a Cascade of Zymogen Activations

Enzymatic cascades are often employed in biochemical systems to achieve a rapid response. In a cascade, an initial signal institutes a series of steps, each of which is catalyzed by an enzyme. At each step, the signal is amplified. For instance, if a signal molecule activates an enzyme that in turn activates 10 enzymes and each of the 10 enzymes in turn activates 10 additional enzymes, after four steps the original signal will have been amplified 10,000-fold. Blood clots are formed by a *cascade of zymogen activations:* the activated form of one clotting factor catalyzes the activation of the next (Figure 10.26). Thus, very small amounts of the initial factors suffice to trigger the cascade, ensuring a rapid response to trauma.

Two means of initiating blood clotting have been described, the *intrinsic pathway* and the *extrinsic pathway*. The intrinsic clotting pathway is activated by exposure of anionic surfaces on rupture of the endothelial lining of the blood vessels. The extrinsic pathway, which appears to be most crucial in blood clotting, is initiated when trauma exposes *tissue factor* (TF), an integral membrane glycoprotein. Shortly after the tissue factor is exposed, small amounts of *thrombin*, the key protease in clotting, are generated. Thrombin then activates enzymes and factors that lead to the generation of yet more thrombin, an example of positive feedback. The extrinsic and intrinsic pathways converge on a common sequence of final steps to form a clot composed of the protein fibrin (Figure 10.26). Note that the active forms of the clotting factors are designated with a subscript "a."

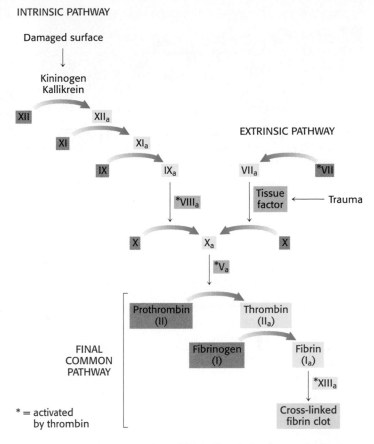

Figure 10.26 **Blood-clotting cascade.** A fibrin clot is formed by the interplay of the intrinsic, extrinsic, and final common pathways. The intrinsic pathway begins with the activation of factor XII (Hageman factor) by contact with abnormal surfaces produced by injury. The extrinsic pathway is triggered by trauma, which releases tissue factor (TF). TF forms a complex with VII, which initiates a cascade-activating thrombin. Inactive forms of clotting factors are shown in red; their activated counterparts (indicated by the subscript "a") are in yellow. Stimulatory proteins that are not themselves enzymes are shown in blue boxes. A striking feature of this process is that the activated form of one clotting factor catalyzes the activation of the next factor.

Fibrinogen Is Converted by Thrombin into a Fibrin Clot

The best-characterized part of the clotting process is the final step in the cascade: the conversion of *fibrinogen* into fibrin by thrombin, a proteolytic enzyme. Fibrinogen is made up of three globular units connected by two rods (Figure 10.27). This 340-kd protein consists of six chains: two each of Aα, Bβ, and γ. The rod regions are triple-stranded α-helical coiled coils, a recurring motif in proteins (p. 45). Thrombin cleaves four *arginine–glycine peptide bonds* in the central globular region of fibrinogen. On cleavage, an A peptide of 18 residues is released from each of the two Aα chains, as is

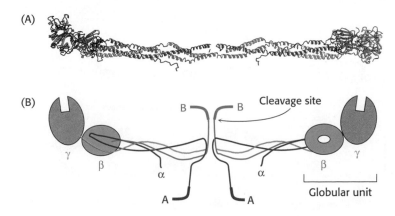

Figure 10.27 **Structure of a fibrinogen molecule.** (A) A ribbon diagram. The two rod regions are α-helical coiled coils, connected to a globular region at each end. The structure of the central globular region has not been determined. (B) A schematic representation showing the positions of the fibrinopeptides A and B. [Part A drawn from 1DEQ.pdb.]

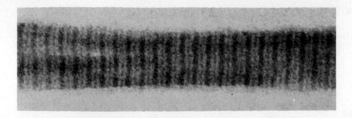

Figure 10.28 Electron micrograph of fibrin. The 23-nm period along the fiber axis is half the length of a fibrinogen molecule. [Courtesy of Dr. Henry Slayter.]

a B peptide of 20 residues from each of the two Bβ chains. These A and B peptides are called *fibrinopeptides*. A fibrinogen molecule devoid of these fibrinopeptides is called a *fibrin monomer* and has the subunit structure $(\alpha\beta\gamma)_2$.

Fibrin monomers spontaneously assemble into ordered fibrous arrays called *fibrin*. Electron micrographs and low-angle x-ray patterns show that fibrin has a periodic structure that repeats every 23 nm (Figure 10.28). Higher-resolution images reveal how the removal of the fibrinopeptides permits the fibrin monomers to come together to form fibrin. The homologous β and γ chains have globular domains at the carboxyl-terminal ends (Figure 10.29). These domains have binding "holes" that interact with peptides. The β domain is specific for sequences of the form H_3N^+-Gly-His-Arg-, whereas the γ domain binds H_3N^+-Gly-Pro-Arg-. Exactly these sequences (sometimes called "knobs") are exposed at the amino-terminal ends of the β and α chains, respectively, on thrombin cleavage. The knobs of the α subunits fit into the holes on the γ subunits of another monomer to form a protofibril. This protofibril is extended when the knobs of the β subunits fit into the holes of β subunits of other protofibrils. Thus, analogous to the activation of chymotrypsinogen, peptide-bond cleavage exposes new amino termini that can participate in specific interactions. The newly formed "soft clot" is stabilized by the formation of amide bonds between the side chains of lysine and glutamine residues in different monomers.

This cross-linking reaction is catalyzed by *transglutaminase (factor XIII$_a$)*, which itself is activated from the protransglutaminase form by thrombin.

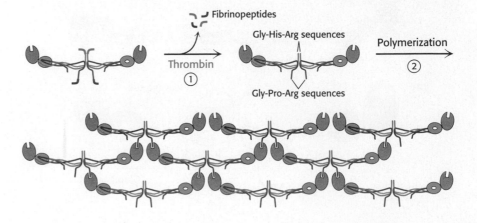

Figure 10.29 Formation of a fibrin clot. (1) Thrombin cleaves fibrinopeptides A and B from the central globule of fibrinogen. (2) Globular domains at the carboxyl-terminal ends of the β and γ chains interact with "knobs" exposed at the amino-terminal ends of the β and γ chains to form clots.

Prothrombin Is Readied for Activation by a Vitamin K-Dependent Modification

Thrombin is synthesized as a zymogen called *prothrombin*. The inactive molecule comprises four major domains, with the serine protease domain at its carboxyl terminus. The first domain is called a *gla domain* (a γ-carboxyglutamate-rich domain), and the second and third domains are called *kringle domains* (named after a Danish pastry that they resemble; Figure 10.30). These domains work in concert to keep prothrombin in an inactive form and to target it to appropriate sites for its activation by factor X_a (a serine protease) and factor V_a (a stimulatory protein). Activation is begun by proteolytic cleavage of the bond between arginine 274 and threonine 275 to release a fragment containing the first three domains. Cleavage of the bond between arginine 323 and isoleucine 324 (analogous to the key bond in chymotrypsinogen) yields active thrombin.

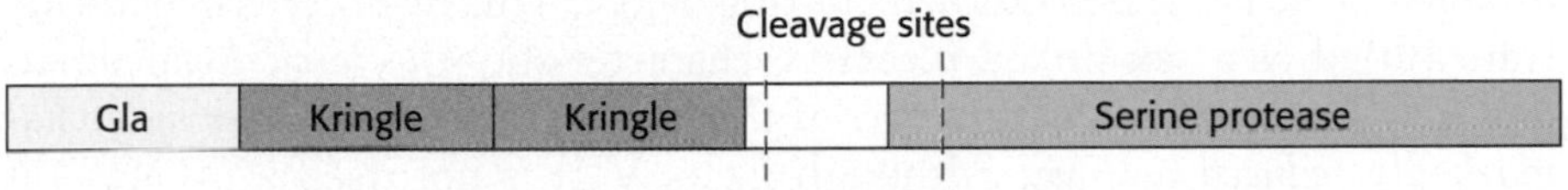

Figure 10.30 Modular structure of prothrombin. Cleavage of two peptide bonds yields thrombin. All the γ-carboxyglutamate residues are in the gla domain.

Vitamin K (p. 295 and Figure 10.31) has been known for many years to be essential for the synthesis of prothrombin and several other clotting factors. Indeed, it is called vitamin K because a deficiency in this vitamin results in defective blood *koagulation* (Scandinavian spelling). The results of studies of the abnormal prothrombin synthesized in the absence of vitamin K or in the presence of vitamin K antagonists, such as dicoumarol, revealed the vitamin's mode of action. *Dicoumarol* is found in spoiled sweet clover and causes a fatal hemorrhagic disease in cattle fed on this hay. This coumarin derivative is used clinically as an *anticoagulant* to prevent thromboses in patients prone to clot formation. Dicoumarol and such related vitamin K antagonists as *warfarin* also serve as effective rat poisons. Cows fed dicoumarol synthesize an abnormal prothrombin that does not bind Ca^{2+}, in contrast with normal prothrombin. This difference was puzzling for some time because abnormal prothrombin has the same number of amino acid residues as that of normal prothrombin and gives the same amino acid analysis after acid hydrolysis.

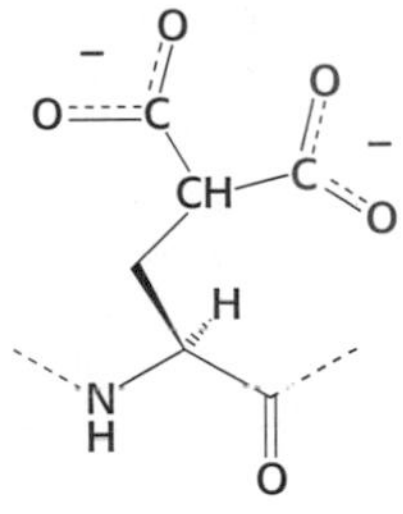

Nuclear magnetic resonance studies revealed that normal prothrombin contains *γ-carboxyglutamate,* a formerly unknown residue that evaded detection because its second carboxyl group is lost on acid hydrolysis during amino acid analysis. The abnormal prothrombin formed subsequent to the administration of anticoagulants lacks this modified amino acid. In fact, the first 10 glutamate residues in the amino-terminal region of prothrombin

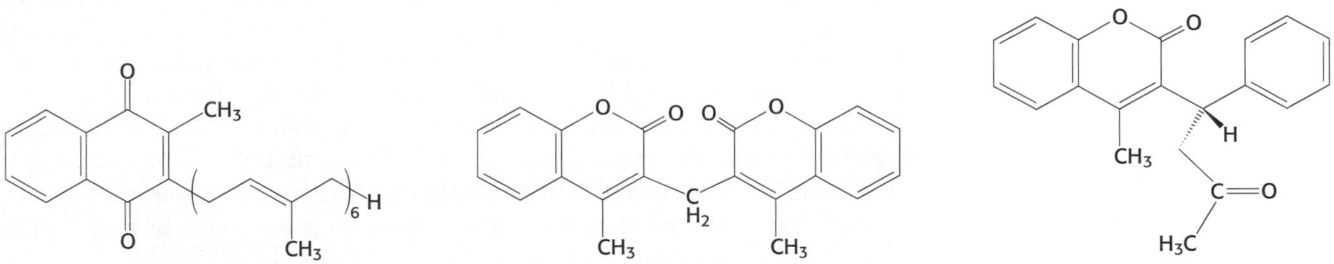

γ-Carboxyglutamate residue

Figure 10.31 Structures of vitamin K and two antagonists, dicoumarol and warfarin.

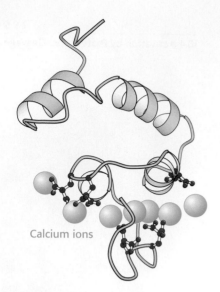

Figure 10.32 The calcium-binding region of prothrombin. Prothrombin binds calcium ions with the modified amino acid γ-carboxyglutamate (red). [Drawn from 2PF2.pdb.]

Calcium ions

are carboxylated to γ-carboxyglutamate by a vitamin K-dependent enzyme system (Figure 10.32). *The vitamin K-dependent carboxylation reaction converts glutamate, a weak chelator of Ca²⁺, into γ-carboxyglutamate, a much stronger chelator.* Prothrombin is thus able to bind Ca^{2+}, but what is the effect of this binding? The binding of Ca^{2+} by prothrombin anchors the zymogen to phospholipid membranes derived from blood platelets after injury. The binding of prothrombin to phospholipid surfaces is crucial because it brings prothrombin into close proximity to two clotting proteins that catalyze its conversion into thrombin. The calcium-binding domain is removed during activation, freeing the thrombin from the membrane so that it can cleave fibrinogen and other targets.

Hemophilia Revealed an Early Step in Clotting

Some important breakthroughs in the elucidation of clotting pathways have come from studies of patients with bleeding disorders. *Classic hemophilia*, or *hemophilia A*, is the best-known clotting defect. This disorder is genetically transmitted as a sex-linked recessive characteristic. *In classic hemophilia, factor VIII (antihemophilic factor) of the intrinsic pathway is missing or has markedly reduced activity.* Although factor VIII is not itself a protease, it markedly stimulates the activation of factor X, the final protease of the intrinsic pathway, by factor IX$_a$, a serine protease (Figure 10.33). Thus, activation of the intrinsic pathway is severely impaired in hemophilia.

In the past, hemophiliacs were treated with transfusions of a concentrated plasma fraction containing factor VIII. This therapy carried the risk of infection. Indeed, many hemophiliacs contracted hepatitis and, more recently, AIDS. A safer preparation of factor VIII was urgently needed. With the use of biochemical purification and recombinant DNA techniques, the gene for factor VIII was isolated and expressed in cells grown in culture. Recombinant factor VIII purified from these cells has largely replaced plasma concentrates in treating hemophilia.

Figure 10.33 Action of antihemophilic factor. Antihemophilic factor (Factor VIII) stimulates the activation of factor X by factor IX$_a$. It is interesting to note that the activity of factor VIII is markedly increased by limited proteolysis by thrombin. This positive feedback amplifies the clotting signal and accelerates clot formation after a threshold has been reached.

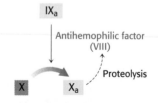

The Clotting Process Must Be Precisely Regulated

There is a fine line between hemorrhage and thrombosis. Clots must form rapidly yet remain confined to the area of injury. What are the mechanisms that normally limit clot formation to the site of injury? The lability of clotting factors contributes significantly to the control of clotting. Activated factors are short-lived because they are diluted by blood flow, removed by the liver, and degraded by proteases. For example, the stimulatory protein factors V$_a$ and VIII$_a$ are digested by protein C, a protease that is switched on by the action of thrombin. *Thus, thrombin has a dual function: it catalyzes the formation of fibrin and it initiates the deactivation of the clotting cascade.*

Specific inhibitors of clotting factors are also critical in the termination of clotting. For instance, *tissue factor pathway inhibitor* (TFPI) inhibits the complex of TF–VII$_a$–X$_a$. Separate domains in TFPI inhibit VII$_a$ and X$_a$. Another key inhibitor is *antithrombin III*, a plasma protein that inactivates thrombin by forming an irreversible complex with it. Antithrombin III resembles

α_1-antitrypsin except that it inhibits thrombin much more strongly than it inhibits elastase (see Figure 10.26). Antithrombin III also blocks other serine proteases in the clotting cascade—namely, factors XII_a, XI_a, IX_a, and X_a. The inhibitory action of antithrombin III is enhanced by *heparin,* a negatively charged polysaccharide found in mast cells near the walls of blood vessels and on the surfaces of endothelial cells (Figure 10.34). Heparin acts as an *anticoagulant* by increasing the rate of formation of irreversible complexes between antithrombin III and the serine protease clotting factors. Antitrypsin and antithrombin are *serpins,* a family of *serine protease inhibitors.*

The importance of the ratio of thrombin to antithrombin is illustrated in the case of a 14-year-old boy who died of a bleeding disorder because of a mutation in his α_1-antitrypsin, which normally inhibits elastase. Methionine 358 in α_1-antitrypsin's binding pocket for elastase was replaced by arginine, resulting in a change in specificity from an elastase inhibitor to a thrombin inhibitor. α_1-Antitrypsin activity normally increases markedly after injury to counteract excess elastase arising from stimulated neutrophils. The mutant α_1-antitrypsin caused the patient's thrombin activity to drop to such a low level that hemorrhage ensued. *We see here a striking example of how a change of a single residue in a protein can dramatically alter specificity and an example of the critical importance of having the right amount of a protease inhibitor.*

Antithrombin limits the extent of clot formation, but what happens to the clots themselves? Clots are not permanent structures but are designed to desolve when the structural integrity of damaged areas is restored. Fibrin is split by *plasmin,* a serine protease that hydrolyzes peptide bonds in the coiled-coil regions. Plasmin molecules can diffuse through aqueous channels in the porous fibrin clot to cut the accessible connector rods. Plasmin is formed by the proteolytic activation of *plasminogen,* an inactive precursor that has a high affinity for the fibrin clots. This conversion is carried out by *tissue-type plasminogen activator* (TPA), a 72-kd protein that has a domain structure closely related to that of prothrombin (Figure 10.35).

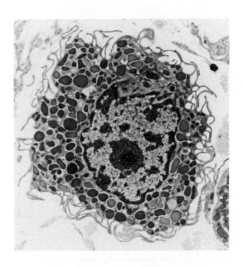

Figure 10.34 Electron micrograph of a mast cell. Heparin and other molecules in the dense granules are released into the extracellular space when the cell is triggered to secrete. [Courtesy of Lynne Mercer.]

Fibrin binding	Kringle	Kringle		Serine protease

Figure 10.35 Modular structure of tissue type plasminogen activator (TPA).

However, a domain that targets TPA to fibrin clots replaces the membrane-targeting gla domain of prothrombin. The TPA bound to fibrin clots swiftly activates adhering plasminogen. In contrast, TPA activates free plasminogen very slowly. The gene for TPA has been cloned and expressed in cultured mammalian cells. Clinical studies have shown that TPA administered intravenously within an hour of the formation of a blood clot in a coronary artery markedly increases the likelihood of surviving a heart attack (Figure 10.36).

(A)

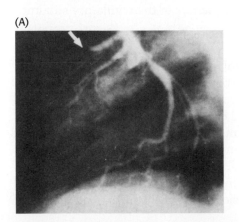

(B)

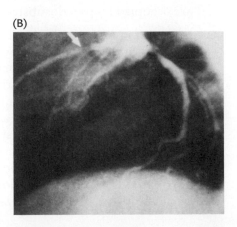

Figure 10.36 The effect of tissue-type plasminogen factor. TPA leads to the dissolution of blood clots, as shown by x-ray images of blood vessels in the heart (A) before and (B) 3 hours after the administration of TPA. The position of the clot is marked by the arrow in part A. [After F. Van de Werf, P. A. Ludbrook, S. R. Bergmann, A. J. Tiefenbrunn, K. A. A. Fox, H. de Geest, M. Verstraete, D. Collen, and B. E. Sobel. *New Engl. J. Med.* 310(1984):609–613.]

Summary

10.1 Aspartate Transcarbamoylase Is Allosterically Inhibited by the End Product of Its Pathway

Allosteric proteins constitute an important class of proteins whose biological activity can be regulated. Specific regulatory molecules can modulate the activity of allosteric proteins by binding to distinct regulatory sites, separate from the functional sites. These proteins have multiple functional sites, which display cooperation as evidenced by a sigmoidal dependence of function on substrate concentration. Aspartate transcarbamoylase (ATCase), one of the best-understood allosteric enzymes, catalyzes the synthesis of N-carbamoylaspartate, the first intermediate in the synthesis of pyrimidines. ATCase is feedback inhibited by cytidine triphosphate, the final product of the pathway. ATP reverses this inhibition. ATCase consists of separable catalytic (c_3) subunits (which bind the substrates) and regulatory (r_2) subunits (which bind CTP and ATP). The inhibitory effect of CTP, the stimulatory action of ATP, and the cooperative binding of substrates are mediated by large changes in quaternary structure. On binding substrates, the c_3 subunits of the c_6r_6 enzyme move apart and reorient themselves. This allosteric transition is highly concerted. All subunits of an ATCase molecule simultaneously interconvert from the T (low-affinity) to the R (high-affinity) state.

10.2 Isozymes Provide a Means of Regulation Specific to Distinct Tissues and Developmental Stages

Isozymes differ in structural characteristics but catalyze the same reaction. They provide a means of fine-tuning metabolism to meet the needs of a given tissue or developmental stage. The results of gene-duplication events provide the means for subtle regulation of enzyme function.

10.3 Covalent Modification Is a Means of Regulating Enzyme Activity

The covalent modification of proteins is a potent means of controlling the activity of enzymes and other proteins. Phosphorylation is the most common type of reversible covalent modification. Signals can be highly amplified by phosphorylation because a single kinase can act on many target molecules. The regulatory actions of protein kinases are reversed by protein phosphatases, which catalyze the hydrolysis of attached phosphoryl groups.

Cyclic AMP serves as an intracellular messenger in the transduction of many hormonal and sensory stimuli. Cyclic AMP switches on protein kinase A, a major multifunctional kinase, by binding to the regulatory subunit of the enzyme, thereby releasing the active catalytic subunits of PKA. In the absence of cAMP, the catalytic sites of PKA are occupied by pseudosubstrate sequences of the regulatory subunit.

10.4 Many Enzymes Are Activated by Specific Proteolytic Cleavage

The activation of an enzyme by the proteolytic cleavage of one or a few peptide bonds is a recurring control mechanism seen in processes as diverse as the activation of digestive enzymes and blood clotting. The inactive precursor is a zymogen (proenzyme). Trypsinogen is activated by enteropeptidase or trypsin, and trypsin then activates a host of other zymogens, leading to the digestion of foodstuffs. For instance, trypsin converts chymotrypsinogen, a zymogen, into active chymotrypsin by hydrolyzing a single peptide bond.

A striking feature of the clotting process is that it is accomplished by a cascade of zymogen conversions, in which the activated form of one clotting factor catalyzes the activation of the next precursor. Many of the activated clotting factors are serine proteases. In the final step of clot formation, fibrinogen, a highly soluble molecule in the plasma, is converted by thrombin into fibrin by the hydrolysis of four arginine–glycine bonds. The resulting fibrin monomer spontaneously forms long, insoluble fibers called fibrin. Zymogen activation is also essential in the lysis of clots. Plasminogen is converted into plasmin, a serine protease that cleaves fibrin, by tissue-type plasminogen activator. Although zymogen activation is irreversible, specific inhibitors of some proteases exert control. The irreversible protein inhibitor antithrombin III holds blood clotting in check in the clotting cascade.

Key Terms

cooperativity (p. 275)

feedback (end-product) inhibition (p. 277)

allosteric (regulatory) site (p. 277)

concerted mechanism (p. 281)

sequential model (p. 281)

homotropic effect (p. 281)

heterotropic effect (p. 282)

isozyme (isoenzyme) (p. 283)

covalent modification (p. 283)

protein kinase (p. 284)

protein phosphatase (p. 285)

consensus sequence (p. 286)

protein kinase A (PKA) (p. 287)

pseudosubstrate sequence (p. 287)

zymogen (proenzyme) (p. 288)

enzymatic cascade (p. 293)

intrinsic pathway (p. 293)

extrinsic pathway (p. 293)

Selected Readings

Where to Start

Kantrowitz, E. R., and Lipscomb, W. N. 1990. *Escherichia coli* aspartate transcarbamoylase: The molecular basis for a concerted allosteric transition. *Trends Biochem. Sci.* 15:53–59.

Schachman, H. K. 1988. Can a simple model account for the allosteric transition of aspartate transcarbamoylase? *J. Biol. Chem.* 263: 18583–18586.

Neurath, H. 1989. Proteolytic processing and physiological regulation. *Trends Biochem. Sci.* 14:268–271.

Bode, W., and Huber, R. 1992. Natural protein proteinase inhibitors and their interaction with proteinases. *Eur. J. Biochem.* 204: 433 451.

Aspartate Transcarbamoylase and Allosteric Interactions

West, J. M., Tsuruta, H., and Kantsrowitz, E. R. 2004. A fluorescent probe-labeled *Escherichia coli* aspartate transcarbamoylase that monitors the allosteric conformation state. *J. Biol. Chem.* 279: 945–951.

Endrizzi, J. A., Beernink, P. T., Alber, T., and Schachman, H. K. 2000. Binding of bisubstrate analog promotes large structural changes in the unregulated catalytic trimer of aspartate transcarbamoylase: Implications for allosteric regulation. *Proc. Natl. Acad. Sci. U. S. A.* 97:5077–5082.

Beernink, P. T., Endrizzi, J. A., Alber, T., and Schachman, H. K. 1999. Assessment of the allosteric mechanism of aspartate transcarbamoylase based on the crystalline structure of the unregulated catalytic subunit. *Proc. Natl. Acad. Sci. U. S. A.* 96:5388–5393.

Wales, M. E., Madison, L. L., Glaser, S. S., and Wild, J. R. 1999. Divergent allosteric patterns verify the regulatory paradigm for aspartate transcarbamoylase. *J. Mol. Biol.* 294:1387–1400.

LiCata, V. J., Burz, D. S., Moerke, N. J., and Allewell, N. M. 1998. The magnitude of the allosteric conformational transition of aspartate transcarbamoylase is altered by mutations. *Biochemistry* 37: 17381–17385.

Eisenstein, E., Markby, D. W., and Schachman, H. K. 1990. Heterotropic effectors promote a global conformational change in aspartate transcarbamoylase. *Biochemistry* 29:3724–3731.

Werner, W. E., and Schachman, H. K. 1989. Analysis of the ligand-promoted global conformational change in aspartate transcar-bamoylase: Evidence for a two-state transition from boundary spreading in sedimentation velocity experiments. *J. Mol. Biol.* 206:221–230.

Newell, J. O., Markby, D. W., and Schachman, H. K. 1989. Cooperative binding of the bisubstrate analog *N*-(phosphonacetyl)-L-aspartate to aspartate transcarbamoylase and the heterotropic effects of ATP and CTP. *J. Biol. Chem.* 264:2476–2481.

Stevens, R. C., Reinisch, K. M., and Lipscomb, W. N. 1991. Molecular structure of *Bacillus subtilis* aspartate transcarbamoylase at 3.0 Å resolution. *Proc. Natl. Acad. Sci. U. S. A.* 88:6087–6091.

Stevens, R. C., Gouaux, J. E., and Lipscomb, W. N. 1990. Structural consequences of effector binding to the T state of aspartate car-bamoyltransferase: Crystal structures of the unligated and ATP- and CTP-complexed enzymes at 2.6-Å resolution. *Biochemistry* 29:7691–7701.

Gouaux, J. E., and Lipscomb, W. N. 1990. Crystal structures of phos-phonoacetamide ligated T and phosphonoacetamide and malonate ligated R states of aspartate carbamoyltransferase at 2.8-Å resolution and neutral pH. *Biochemistry* 29:389–402.

Labedan, B., Boyen, A., Baetens, M., Charlier, D., Chen, P., Cunin, R., Durbeco, V., Glansdorff, N., Herve, G., Legrain, C., Liang, Z., Purcarea, C., Roovers, M., Sanchez, R., Toong, T. L., Van de Casteele, M., van Vliet, F., Xu, Y., and Zhang, Y. F. 1999. The evolutionary history of carbamoyltransferases: A complex set of paralogous genes was already present in the last universal common ancestor. *J. Mol. Evol.* 49:461–473.

Covalent Modification

Johnson, L. N., and Barford, D. 1993. The effects of phosphorylation on the structure and function of proteins. *Annu. Rev. Biophys. Biomol. Struct.* 22:199–232.

Ziegler, M. 2000. New functions of a long-known molecule: Emerging roles of NAD in cellular signaling. *Eur. J. Biochem.* 267:1550–1564.

Ng, H. H., and Bird, A. 2000. Histone deacetylases: Silencers for hire. *Trends Biochem. Sci.* 25:121–126.

Raju, R. V., Kakkar, R., Radhi, J. M., and Sharma, R. K. 1997. Biological significance of phosphorylation and myristoylation in the regulation of cardiac muscle proteins. *Mol. Cell. Biochem.* 176:135–143.

Jacobson, M. K., and Jacobson, E. L. 1999. Discovering new ADP-ribose polymer cycles: Protecting the genome and more. *Trends Biochem. Sci.* 24:415–417.

Barford, D., Das, A. K., and Egloff, M. P. 1998. The structure and mechanism of protein phosphatases: Insights into catalysis and regulation. *Annu. Rev. Biophys. Biomol. Struct.* 27:133–164.

Protein Kinase A

Taylor, S. S., Knighton, D. R., Zheng, J., Sowadski, J. M. Gibbs, C. S., and Zoller, M. J. 1993. A template for the protein kinase family. *Trends Biochem. Sci.* 18:84–89.

Gibbs, C. S., Knighton, D. R., Sowadski, J. M., Taylor, S. S., and Zoller, M. J. 1992. Systematic mutational analysis of cAMP-dependent protein kinase identifies unregulated catalytic subunits and defines regions important for the recognition of the regulatory subunit. *J. Biol. Chem.* 267:4806–4814.

Knighton, D. R., Zheng, J. H., TenEyck, L., Ashford, V. A., Xuong, N. H., Taylor, S. S., and Sowadski, J. M. 1991. Crystal structure of the catalytic subunit of cyclic adenosine monophosphate-dependent protein kinase. *Science* 253:407–414.

Knighton, D. R., Zheng, J. H., TenEyck, L., Xuong, N. H., Taylor, S. S., and Sowadski, J. M. 1991. Structure of a peptide inhibitor bound to the catalytic subunit of cyclic adenosine monophosphate-dependent protein kinase. *Science* 253:414–420.

Adams, S. R., Harootunian, A. T., Buechler, Y. J., Taylor, S. S., and Tsien, R. Y. 1991. Fluorescence ratio imaging of cyclic AMP in single cells. *Nature* 349:694–697.

Zymogen Activation

Neurath, H. 1986. The versatility of proteolytic enzymes. *J. Cell. Biochem.* 32:35–49.

Bode, W., and Huber, R. 1986. Crystal structure of pancreatic serine endopeptidases. In *Molecular and Cellular Basis of Digestion* (pp. 213–234), edited by P. Desnuelle, H. Sjostrom, and O. Noren. Elsevier.

Huber, R., and Bode, W. 1978. Structural basis of the activation and action of trypsin. *Acc. Chem. Res.* 11:114–122.

Stroud, R. M., Kossiakoff, A. A., and Chambers, J. L. 1977. Mechanism of zymogen activation. *Annu. Rev. Biophys. Bioeng.* 6:177–193.

Sielecki, A. R., Fujinaga, M., Read, R. J., and James, M. N. 1991. Refined structure of porcine pepsinogen at 1.8 Å resolution. *J. Mol. Biol.* 219:671–692.

Protease Inhibitors

Carrell, R., and Travis, J. 1985. α_1-Antitrypsin and the serpins: Variation and countervariation. *Trends Biochem. Sci.* 10:20–24.

Carp, H., Miller, F., Hoidal, J. R., and Janoff, A. 1982. Potential mechanism of emphysema: α_1-Proteinase inhibitor recovered from lungs of cigarette smokers contains oxidized methionine and has decreased elastase inhibitory capacity. *Proc. Natl. Acad. Sci. U. S. A.* 79:2041–2045.

Owen, M. C., Brennan, S. O., Lewis, J. H., and Carrell, R. W. 1983. Mutation of antitrypsin to antithrombin. *New Engl. J. Med.* 309:694–698.

Travis, J., and Salvesen, G. S. 1983. Human plasma proteinase inhibitors. *Annu. Rev. Biochem.* 52:655–709.

Clotting Cascade

Orfeo, T., Brufatto, N., Nesheim, M. E., Xu, H., Butenas, S., and Mann, K. G. 2004. The factor V activation paradox. *J. Biol. Chem.* 279:19580–19591.

Mann, K. G. 2003. Thrombin formation. *Chest* 124:4S–10S.

Rose, T., and Di Cera, E. 2002. Three-dimensional modeling of thrombin-fibrinogen interaction. *J. Biol. Chem.* 277:18875–18880.

Krem, M. M., and Di Cera, E. 2002. Evolution of cascades from embryonic development to blood coagulation. *Trends Biochem. Sci.* 27:67–74.

Fuentes-Prior, P., Iwanaga, Y., Huber, R., Pagila, R., Rumennik, G., Seto, M., Morser, J., Light, D. R., and Bode, W. 2000. Structural basis for the anticoagulant activity of the thrombin-thrombomodulin complex. *Nature* 404:518–525.

Herzog, R. W., and High, K. A. 1998. Problems and prospects in gene therapy for hemophilia. *Curr. Opin. Hematol.* 5:321–326.

Lawn, R. M., and Vehar, G. A. 1986. The molecular genetics of hemophilia. *Sci. Am.* 254(3):48–65.

Brown, J. H., Volkmann, N., Jun, G., Henschen-Edman, A. H., and Cohen, C. 2000. The crystal structure of modified bovine fibrinogen. *Proc. Natl. Acad. Sci. U. S. A.* 97:85–90.

Stubbs, M. T., Oschkinat, H., Mayr, I., Huber, R., Angliker, H., Stone, S. R., and Bode, W. 1992. The interaction of thrombin with fibrinogen: A structural basis for its specificity. *Eur. J. Biochem.* 206:187–195.

Problems

1. *Activity profile.* A histidine residue in the active site of aspartate transcarbamoylase is thought to be important in stabilizing the transition state of the bound substrates. Predict the pH dependence of the catalytic rate, assuming that this interaction is essential and dominates the pH-activity profile of the enzyme. (See equations on p. 16.)

2. *Allosteric switching.* A substrate binds 100 times as tightly to the R state of an allosteric enzyme as to its T state. Assume that the concerted (MWC) model applies to this enzyme. (See equations on p. 200.)

(a) By what factor does the binding of one substrate molecule per enzyme molecule alter the ratio of the concentrations of enzyme molecules in the R and T states?

(b) Suppose that L, the ratio of [T] to [R] in the absence of substrate, is 10^7 and that the enzyme contains four binding sites for substrate. What is the ratio of enzyme molecules in the R state to those in the T state in the presence of saturating amounts of substrate, assuming that the concerted model is obeyed?

3. *Allosteric transition.* Consider an allosteric protein that obeys the concerted model. Suppose that the ratio of T to R formed in the absence of ligand is 10^5, $K_T = 2$ mM, and $K_R = 5$ μM. The protein contains four binding sites for ligand. What is the fraction of molecules in the R form when 0, 1, 2, 3, and 4 ligands are bound? (See equations on p. 200.)

4. *Negative cooperativity.* You have isolated a dimeric enzyme that contains two identical active sites. The binding of substrate to one active site decreases the substrate affinity of the other active site. Can the concerted model account for this negative cooperativity?

5. *Paradoxical at first glance.* Recall that phosphonacetyl-L-aspartate (PALA) is a potent inhibitor of ATCase because it mimics the two physiological substrates. However, low concentrations of this unreactive bisubstrate analog *increase* the reaction velocity. On the addition of PALA, the reaction rate increases until an average of three molecules of PALA are bound per molecule of enzyme. This maximal velocity is 17-fold greater

than it is in the absence of PALA. The reaction rate then decreases to nearly zero on the addition of three more molecules of PALA per molecule of enzyme. Why do low concentrations of PALA activate ATCase?

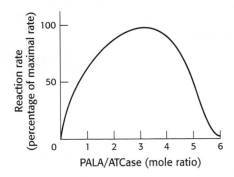

Effect of PALA on ATCase.

6. *R versus T.* An allosteric enzyme that follows the concerted mechanism has a T/R ratio of 300 in the absence of substrate. Suppose that a mutation reversed the ratio. How would this mutation affect the relation between the rate of the reaction and substrate concentration?

7. *Regulation energetics.* The phosphorylation and dephosphorylation of proteins is a vital means of regulation. Protein kinases attach phosphoryl groups, whereas only a phosphatase will remove the phosphoryl group from the target protein. What is the energy cost of this means of covalent regulation?

8. *Zymogen activation.* When very low concentrations of pepsinogen are added to acidic media, how does the half-time for activation depend on zymogen concentration?

9. *A revealing assay.* Suppose that you have just examined a young boy with a bleeding disorder highly suggestive of classic hemophilia (factor VIII deficiency). Because of the late hour, the laboratory that carries out specialized coagulation assays is closed. However, you happen to have a sample of blood from a classic hemophiliac whom you admitted to the hospital an hour earlier. What is the simplest and most rapid test that you can perform to determine whether your present patient also is deficient in factor VIII activity?

10. *Counterpoint.* The synthesis of factor X, like that of prothrombin, requires vitamin K. Factor X also contains γ-carboxyglutamate residues in its amino-terminal region. However, activated factor X, in contrast with thrombin, retains this region of the molecule. What is a likely functional consequence of this difference between the two activated species?

11. *A discerning inhibitor.* Antithrombin III forms an irreversible complex with thrombin but not with prothrombin. What is the most likely reason for this difference in reactivity?

12. *Repeating heptads.* Each of the three types of fibrin chains contains repeating heptapeptide units *(abcdefg)* in which residues *a* and *d* are hydrophobic. Propose a reason for this regularity.

13. *Drug design.* A drug company has decided to use recombinant DNA methods to prepare a modified α₁-antitrypsin that will be more resistant to oxidation than is the naturally occurring

inhibitor. Which single amino acid substitution would you recommend?

Data Interpretation Problems

14. *Distinguishing between models.* The following graph shows the fraction of an allosteric enzyme in the R state (f_R) and the fraction of active sites bound to substrate (Y) as a function of substrate concentration. Which model, the concerted or sequential, best explains these results?

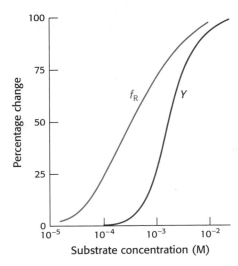

[From M. W. Kirschner and H. K. Schachman, *Biochemistry* 12(1966):2997–3004.]

15. *Reporting live from ATCase 1.* ATCase underwent reaction with tetranitromethane to form a colored nitrotyrosine group ($\lambda_{max} = 430$ nm) in each of its catalytic chains. The absorption by this reporter group depends on its immediate environment. An essential lysine residue at each catalytic site also was modified to block the binding of substrate. Catalytic trimers from this doubly modified enzyme were then combined with native trimers to form a hybrid enzyme. The absorption by the nitrotyrosine group was measured on addition of the substrate analog succinate. What is the significance of the alteration in the absorbance at 430 nm?

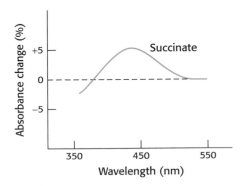

[After H. K. Schachman, *J. Biol. Chem.* 263(1988): 18583–18586.]

16. *Reporting live from ATCase 2.* A different ATCase hybrid was constructed to test the effects of allosteric activators and inhibitors. Normal regulatory subunits were combined with

nitrotyrosine-containing catalytic subunits. The addition of ATP in the absence of substrate increased the absorbance at 430 nm, the same change elicited by the addition of succinate (see the graph in Problem 15). Conversely, CTP in the absence of substrate decreased the absorbance at 430 nm. What is the significance of the changes in absorption of the reporter groups?

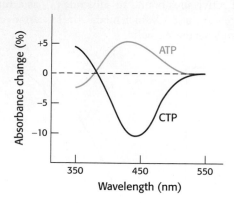

[After H. K. Schachman, *J. Biol. Chem.* 263 (1988): 18583–18586.]

Chapter Integration Problem

17. *Density matters.* The sedimenation value of aspartate transcarbamoylase decreases when the enzyme switches to the R state. On the basis of the allosteric properties of the enzyme, explain why the sedimentation value decreases.

Mechanism Problems

18. *Aspartate transcarbamoylase.* Write the mechanism (in detail) for the conversion of aspartate and carbamoyl phosphate into *N*-carbamoylaspartate. Include a role for the histidine residue present in the active site.

19. *Protein kinases.* Write a mechanism (in detail) for the phosphorylation of a serine residue by ATP catalyzed by a protein kinase. What groups might you expect to find in the enzyme's active site?

Carbohydrates

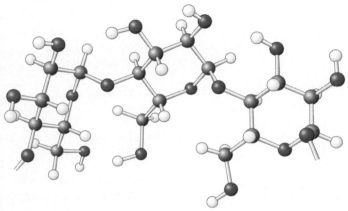

Carbohydrates in food are important sources of energy. Starch, found in plant-derived food such as pasta, consists of chains of linked glucose molecules. These chains are broken down into individual glucose molecules for eventual use in the generation of ATP and as building blocks for other molecules. [(Left) Superstock.]

L et us take an overview of carbohydrates, one of the four major classes of biomolecules along with proteins, nucleic acids, and lipids. Carbohydrates are aldehydes or ketones with multiple hydroxyl groups. They make up most of the organic matter on Earth because of their extensive roles in all forms of life. First, carbohydrates serve as *energy stores, fuels, and metabolic intermediates*. Second, ribose and deoxyribose sugars form part of the structural *framework of RNA and DNA*. Third, polysaccharides are structural *elements in the cell walls of bacteria and plants*. In fact, cellulose, the main constituent of plant cell walls, is one of the most abundant organic compounds in the biosphere. Fourth, carbohydrates are *linked to many proteins and lipids*. Such linked carbohydrates play key roles in cell–cell communication and in interactions between cells and other elements in the cellular environment.

A key property of carbohydrates in their role as mediators of cellular interactions is the tremendous structural *diversity* possible within this class of molecules. Carbohydrates are built from monosaccharides, small molecules, typically containing from three to nine carbon atoms, that vary in size and in the stereochemical configuration at one or more carbon centers. These monosaccharides may be linked together to form a large variety of oligosaccharide structures. The sheer number of possible oligosaccharides makes this class of molecules information rich. This information, when attached to proteins, can augment the already immense diversity of proteins.

Unraveling oligosaccharide structures and elucidating the effects of their attachment to proteins constitute a tremendous challenge in the field of proteomics. Indeed, this subfield has been given its own name, glycomics.

11.1 Monosaccharides Are Aldehydes or Ketones with Multiple Hydroxyl Groups

Monosaccharides, the simplest carbohydrates, are aldehydes or ketones that have two or more hydroxyl groups; the empirical formula of many is $(C—H_2O)n$, literally a "carbon hydrate." Monosaccharides are important fuel molecules as well as building blocks for nucleic acids. The smallest monosaccharides, for which $n = 3$, are dihydroxyacetone and D- and L-glyceraldehyde.

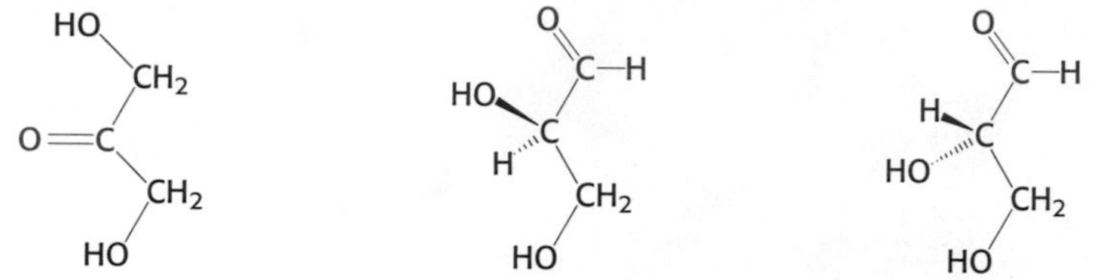

Dihydroxyacetone (a ketose)   **D-Glyceraldehyde** (an aldose)   **L-Glyceraldehyde** (an aldose)

They are referred to as *trioses* (tri- for 3). Dihydroxyacetone is a *ketose* because it contains a keto group, whereas glyceraldehyde is an *aldose* because it contains an aldehyde group. Glyceraldehyde has a single asymmetric carbon atom and, thus, there are two stereoisomers of this sugar: D-glyceraldehyde and L-glyceraldehyde. These two forms are *enantiomers,* or mirror images of each other. As mentioned in Chapter 2, the prefixes D and L designate the absolute configuration.

Monosaccharides and other sugars will often be represented in this book by *Fischer projections* (Figure 11.1). Recall that, in a Fischer projection of a molecule, atoms joined to an asymmetric carbon atom by horizontal bonds are in front of the plane of the page, and those joined by vertical bonds are behind the page (see the Appendix in Chapter 1). Fischer projections provide clear and simple views of the stereochemistry at each carbon center.

Simple monosaccharides with four, five, six, and seven carbon atoms are called *tetroses, pentoses, hexoses,* and *heptoses,* respectively. These molecules have multiple asymmetric carbon atoms and, for these monosaccharides, *the symbols* D *and* L *designate the absolute configuration of the asymmetric carbon*

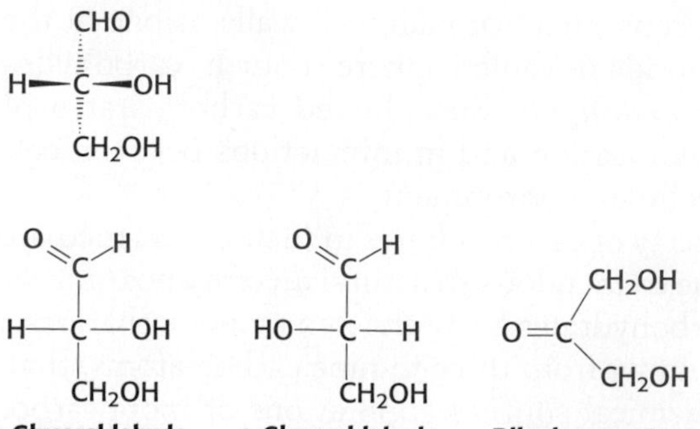

D-Glyceraldehyde   **L-Glyceraldehyde**   **Dihydroxyacetone**

Figure 11.1 Fischer projections of trioses. The top structure reveals the stereochemical relations assumed for Fischer projections.

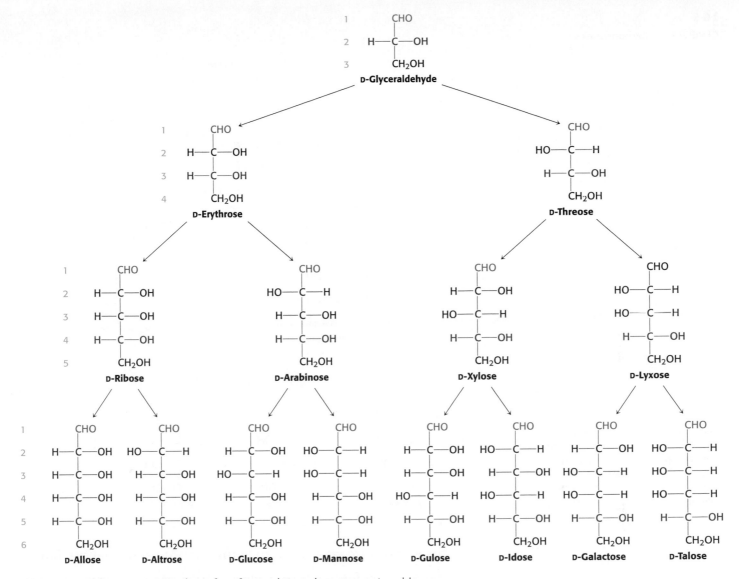

Figure 11.2 D-Aldoses containing three, four, five, and six carbon atoms. A D-aldose contains an aldehyde group (shown in blue) and has the absolute configuration of D-glyceraldehyde at the asymmetric center (shown in red) farthest from the aldehyde group. The numbers indicate the standard designations for each carbon atom.

atom farthest from the aldehyde or keto group. In Figure 11.2, for example, the four-carbon aldoses D-erythrose and D-threose have the same configuration at C-3 (because they are D sugars) but opposite configurations at C-2. They are *diastereoisomers,* not enantiomers, because they are not mirror images of each other. Figure 11.2 shows the common D-aldose sugars. D-Ribose, the carbohydrate component of RNA, is a five-carbon aldose. D-Glucose, D-mannose, and D-galactose are abundant six-carbon aldoses. Note that D-glucose and D-mannose differ in configuration only at C-2. Sugars differing in configuration at a single asymmetric center are called *epimers.* Thus, D-glucose and D-mannose are epimeric at C-2; D-glucose and D-galactose are epimeric at C-4.

Dihydroxyacetone is the simplest ketose. The stereochemical relations between D-ketoses containing as many as six carbon atoms are shown in Figure 11.3. Note that ketoses have one fewer asymmetric center than do aldoses with the same number of carbons. D-Fructose is the most abundant ketohexose.

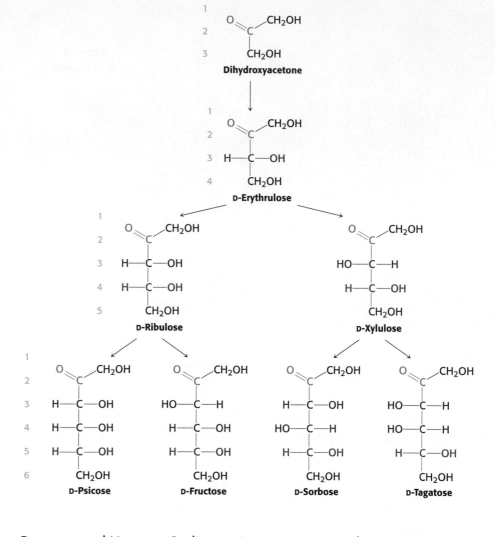

Figure 11.3 D-Ketoses containing three- four, five, and six carbon atoms. The keto group is shown in blue. The asymmetric center farthest from the keto group, which determines the D designation, is shown in red.

Pyran

Furan

Pentoses and Hexoses Cyclize to Form Furanose and Pyranose Rings

The predominant forms of ribose, glucose, fructose, and many other sugars in solution are not open chains. Rather, the open-chain forms of these sugars cyclize into rings because the ring forms are energetically more stable. The basis for ring formation is the fact that an aldehyde can react with an alcohol to form a *hemiacetal*.

For an aldohexose such as glucose, the C-1 aldehyde in the open-chain form of glucose reacts with the C-5 hydroxyl group to form an *intramolecular hemiacetal*. The resulting six-membered ring is called *pyranose* because of its similarity to pyran (Figure 11.4). Similarly, a ketone can react with an alcohol to form a *hemiketal*.

For a ketohexose such as fructose, the C-2 keto group in the open-chain form of fructose reacts with a hydroxyl group within the same molecule to

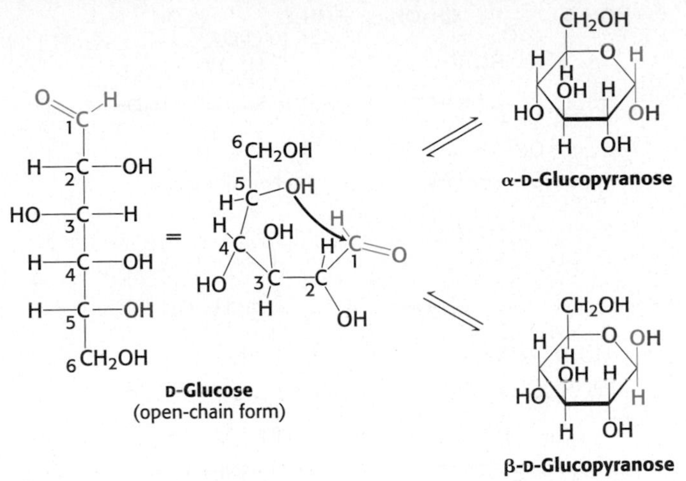

Figure 11.4 Pyranose formation. The open-chain form of glucose cyclizes when the C-5 hydroxyl group attacks the oxygen atom of the C-1 aldehyde group to form an intramolecular hemiacetal. Two anomeric forms, designated α and β, can result.

form an *intramolecular hemiketal*. The C-2 keto group can react with either the C-6 hydroxyl group to form a six-membered ring or the C-5 hydroxyl group to form a five-membered ring (Figure 11.5). The five-membered ring is called a *furanose* because of its similarity to furan.

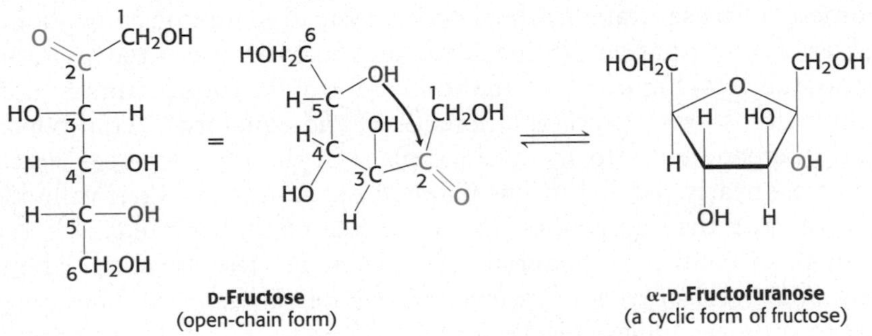

Figure 11.5 Furanose formation. The open-chain form of fructose cyclizes to a five-membered ring when the C-5 hydroxyl group attacks the C-2 ketone to form an intramolecular hemiketal. Two anomers are possible, but only the α anomer is shown.

The depictions of glucopyranose and fructofuranose shown in Figures 11.4 and 11.5 are *Haworth projections*. In such projections, the carbon atoms in the ring are not explicitly shown. The approximate plane of the ring is perpendicular to the plane of the paper, with the heavy line on the ring projecting toward the reader. Like Fischer projections, Haworth projections allow easy depiction of the stereochemistry of sugars.

An additional asymmetric center is created when a cyclic hemiacetal is formed. In glucose, C-1, the carbonyl carbon atom in the open-chain form, becomes an asymmetric center in the ring form. Thus, two ring structures can be formed: α-D-glucopyranose and β-D-glucopyranose (see Figure 11.4). For D sugars drawn as Haworth projections, the *designation α means that the hydroxyl group attached to C-1 is on the opposite side of the ring from the CH$_2$OH at the carbon atom that determines whether the sugar is designated D or L (the chiral center); β means that the hydroxyl group is on the same side as the CH$_2$OH at the chiral center.* The C-1 carbon atom is called the *anomeric carbon atom*, and the α and β forms are called *anomers*. An equilibrium mixture of glucose is approximately one-third α anomer, two-thirds β anomer, and < 1% open-chain form.

The same nomenclature applies to the furanose ring form of fructose, except that α and β refer to the hydroxyl groups attached to C-2, the anomeric carbon atom (see Figure 11.5). Fructose forms both pyranose and furanose rings. The pyranose form predominates in fructose free in solution, and the furanose form predominates in many fructose derivatives (Figure 11.6).

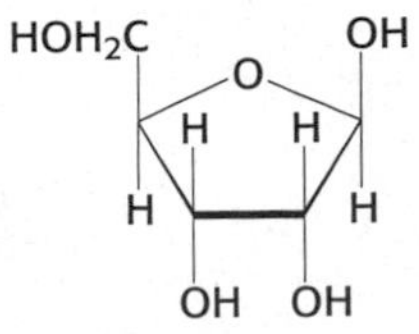

β-D-Ribose

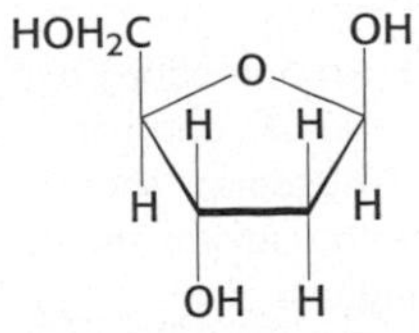

β-2-Deoxy-D-ribose

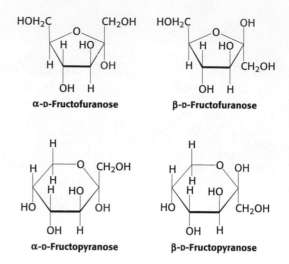

α-D-Fructofuranose

β-D-Fructofuranose

α-D-Fructopyranose

β-D-Fructopyranose

Figure 11.6 Ring structures of fructose. Fructose can form both five-membered furanose and six-membered pyranose rings. In each case, both α and β anomers are possible.

Pentoses such as D-ribose and 2-deoxy- D-ribose form furanose rings, as we have seen in the structure of these units in RNA and DNA.

Pyranose and Furanose Rings Can Assume Different Conformations

The six-membered pyranose ring is not planar, because of the tetrahedral geometry of its saturated carbon atoms. Instead, pyranose rings adopt two classes of conformations, termed chair and boat because of the resemblance to these objects (Figure 11.7). In the chair form, the substituents on the ring carbon atoms have two orientations: axial and equatorial. *Axial* bonds are nearly perpendicular to the average plane of the ring, whereas *equatorial* bonds are nearly parallel to this plane. Axial substituents sterically hinder each other if they emerge on the same side of the ring (e.g., 1,3-diaxial groups). In contrast, equatorial substituents are less crowded. The chair form of *β-D-glucopyranose predominates because all axial positions are occupied by hydrogen atoms.* The bulkier —OH and —CH$_2$OH groups emerge at the less-hindered periphery. The boat form of glucose is disfavored because it is quite sterically hindered.

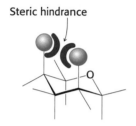

Steric hindrance

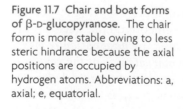

Figure 11.7 Chair and boat forms of β-D-glucopyranose. The chair form is more stable owing to less steric hindrance because the axial positions are occupied by hydrogen atoms. Abbreviations: a, axial; e, equatorial.

Chair form

Boat form

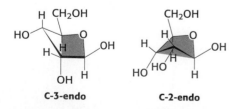

C-3-endo

C-2-endo

Figure 11.8 Envelope conformations of β-D-ribose. The C-3-endo and C-2-endo forms of β-D-ribose are shown. The color indicates the four atoms that lie approximately in a plane.

Furanose rings, like pyranose rings, are not planar. They can be puckered so that four atoms are nearly coplanar and the fifth is about 0.5 Å away from this plane (Figure 11.8). This conformation is called an *envelope form* because the structure resembles an opened envelope with the back flap raised. In the ribose moiety of most biomolecules, either C-2 or C-3 is out of the plane on the same side as C-5. These conformations are called C-2-endo and C-3-endo, respectively.

Monosaccharides Are Joined to Alcohols and Amines Through Glycosidic Bonds

Monosaccharides may react with alcohols and amines to form modified monosaccharides. For example, D-glucose will react with methanol in an acid-catalyzed process: the anomeric carbon atom C-1, which is part of a hemiacetal (p. 306), reacts with the hydroxyl group of methanol to form a *sugar acetal*, also called a *glycoside*. The reaction forms two glycosides: methyl α-D-glucopyranoside and methyl β-D-glucopyranoside. These two glucopyranosides differ in the configuration at the anomeric carbon atom. The bond formed between the anomeric carbon atom of a sugar and the hydroxyl oxygen atom of an alcohol is called a *glycosidic bond*—specifically, an O-*glycosidic bond*. Alternatively, the anomeric carbon atom of a sugar can be linked to the nitrogen atom of an amine to form an N-*glycosidic bond*. We have already encountered such reaction products; nucleosides are adducts between sugars, such as ribose, and amines, such as adenine (p. 109). Some other important modified sugars are shown in Figure 11.9.

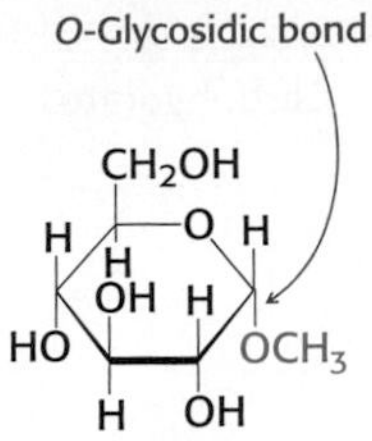

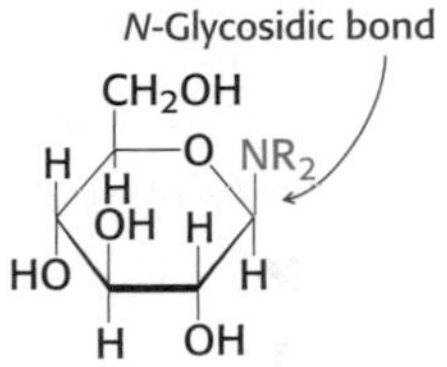

Compounds such as methyl glucopyranoside differ in reactivity from the parent monosaccharide. For example, unmodified glucose reacts with oxidizing agents such as cupric ion (Cu^{2+}) because the open-chain form has a free aldehyde group that is readily oxidized.

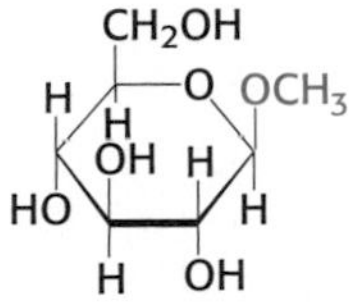

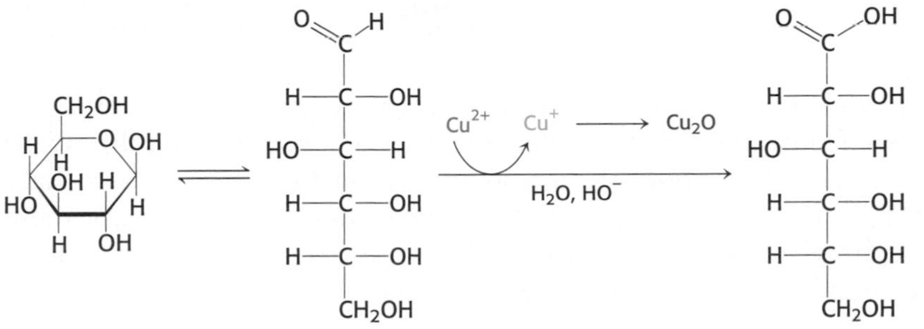

O-Glycosidic bond

Methyl α-D-glucopyranoside

Methyl β-D-glucopyranoside

N-Glycosidic bond

Glycosides such as methyl glucopyranoside do not react, because they are not readily interconverted with a form that includes a free aldehyde group. Solutions of cupric ion (known as Fehling's solution) provide a simple test for sugars, such as glucose, that can exist as a free aldehyde or ketone. Sugars that react are called *reducing sugars*; those that do not are called *nonreducing sugars*. Reducing sugars can often nonspecifically bind to other molecules. For instance, as a reducing sugar, glucose can react with hemoglobin to form

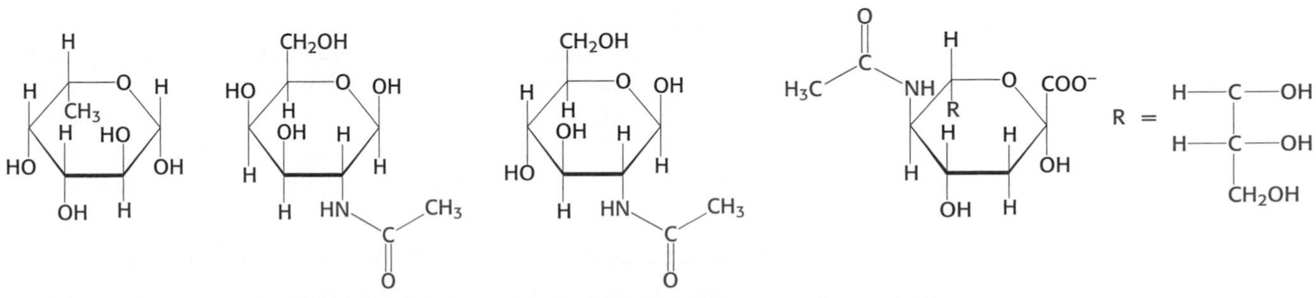

β-L-Fucose (Fuc)

β-D-Acetylgalactosamine (GalNAc)

β-D-Acetylglucosamine (GlcNAc)

Sialic acid (Sia) (N-Acetylneuraminate)

Figure 11.9 Modified monosaccharides. Carbohydrates can be modified by the addition of substituents (shown in red) other than hydroxyl groups. Such modified carbohydrates are often expressed on cell surfaces.

glycosylated hemoglobin. Changes in the amount of glycosylated hemoglobin can be used to monitor the effectiveness of treatments for diabetes mellitus, a condition characterized by high levels of blood glucose (p. 773). Reaction with glucose has no effect on the oxygen-binding ability of hemoglobin.

Phosphorylated Sugars Are Key Intermediates in Energy Generation and Biosyntheses

One sugar modification deserves special note because of its prominence in metabolism. The addition of phosphoryl groups is a common modification of sugars. For instance, the first step in the breakdown of glucose to obtain energy is its conversion into glucose 6-phosphate. Several subsequent intermediates in this metabolic pathway, such as dihydroxyacetone phosphate and glyceraldehyde 3-phosphate, are phosphorylated sugars.

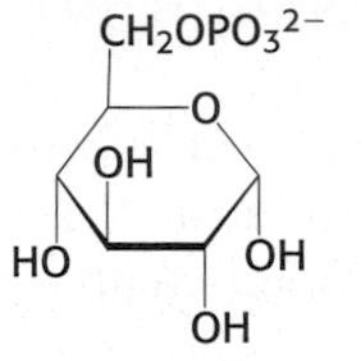

Glucose 6-phosphate (G-6P) 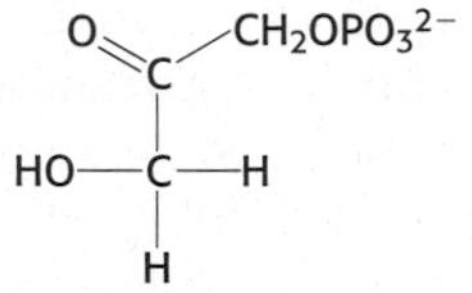   **Dihydroxyacetone phosphate (DHAP)** 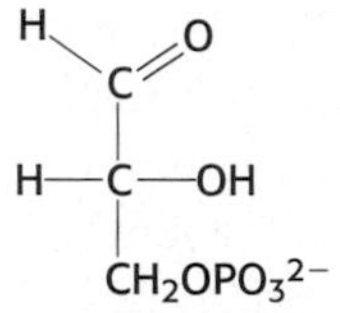   **Glyceraldehyde 3-phosphate (GAP)**

Phosphorylation makes sugars anionic; the negative charge prevents these sugars from spontaneously leaving the cell by crossing lipid-bilayer membranes. Phosphorylation also *creates reactive intermediates* that will more readily form linkages to other molecules. For example, a multiply phosphorylated derivative of ribose plays key roles in the biosyntheses of purine and pyrimidine nucleotides (p. 712).

11.2 Complex Carbohydrates Are Formed by the Linkage of Monosaccharides

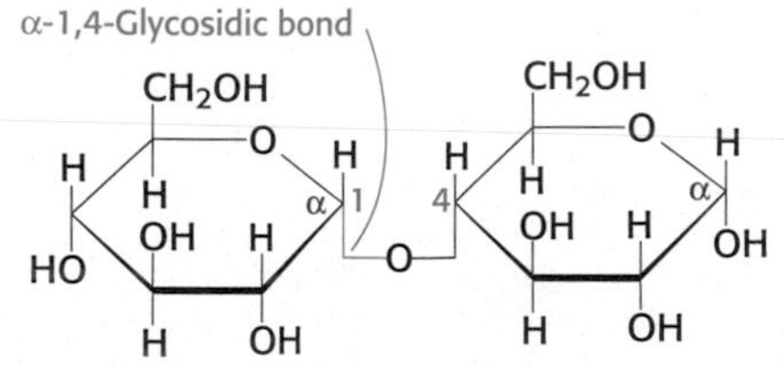

Glycosidic bonds can join one monosaccharide to another. *Oligosaccharides* are carbohydrates built by the linkage of two or more monosaccharides by *O*-glycosidic bonds (Figure 11.10). In maltose, for example, two D-glucose residues are joined by a glycosidic linkage between the C-1 carbon atom on one sugar and the hydroxyl oxygen atom on C-4 of the adjacent sugar. The sugar on the C-1 side of the link is in the α configuration. In other words, the bond emerging from C-1 lies below the plane of the ring when viewed in the standard orientation. Hence, the maltose linkage is called an α-1,4-glycosidic bond. Because monosaccharides have multiple hydroxyl groups, various glycosidic linkages are possible. Indeed, the wide array of these linkages in concert with the wide variety of monosaccharides and their many isomeric forms makes complex carbohydrates structurally diverse molecules.

Sucrose, Lactose, and Maltose Are the Common Disaccharides

A *disaccharide* consists of two sugars joined by an *O*-glycosidic bond. Three abundant disaccharides are sucrose, lactose, and maltose (Figure 11.11). *Sucrose* (common table sugar), a transport form of carbohydrates in plants, is obtained commercially from cane or beet. The anomeric carbon atoms of a glucose unit and a fructose unit are joined in this disaccharide; the configuration of this glycosidic linkage is α for glucose and β for fructose.

Figure 11.10 Maltose, a disaccharide. Two molecules of glucose are linked by an α-1,4-glycosidic bond to form the disaccharide maltose. The angles in the bonds to the central oxygen do not denote carbon atoms. The angles are added only for ease of illustration.

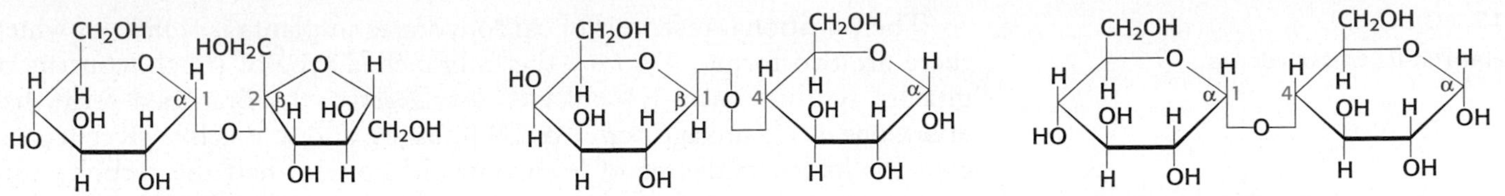

Sucrose
(α-D-Glucopyranosyl-(1→2)-β-D-fructofuranose)

Lactose
(β-D-Galactopyranosyl-(1→4)-α-D-glucopyranose)

Maltose
(α-D-Glucopyranosyl-(1→4)-α-D-glucopyranose)

Figure 11.11 Common disaccharides. Sucrose, lactose, and maltose are common dietary components. The angles in the bonds to the central oxygens do not denote carbon atoms.

Consequently, sucrose is not a reducing sugar, because neither component monosaccharide is readily converted into an aldehyde or ketone, in contrast with most other sugars. Sucrose can be cleaved into its component monosaccharides by the enzyme *sucrase*.

Lactose, the disaccharide of milk, consists of galactose joined to glucose by a β-1,4-glycosidic linkage. Lactose is hydrolyzed to these monosaccharides by *lactase* in human beings (p. 451) and by *β-galactosidase* in bacteria. In maltose, two glucose units are joined by an α-1,4-glycosidic linkage, as stated earlier. Maltose is produced by the hydrolysis of starch and is in turn hydrolyzed to glucose by *maltase.* Sucrase, lactase, and maltase are located on the outer surfaces of epithelial cells lining the small intestine (Figure 11.12).

Figure 11.12 Electron micrograph of a microvillus. Lactase and other enzymes that hydrolyze carbohydrates are present on microvilli that project from the outer face of the plasma membrane of intestinal epithelial cells. [From M. S. Mooseker and L. G. Tilney, *J. Cell. Biol.* 67(1975):725–743.]

Glycogen and Starch Are Mobilizable Stores of Glucose

Large polymeric oligosaccharides, formed by the linkage of multiple monosaccharides, are called *polysaccharides*. Polysaccharides play vital roles in energy storage and in maintaining the structural integrity of an organism. If all of the monosaccharides are the same, these polymers are called *homopolymers*. The most common homopolymer in animal cells is *glycogen*, the storage form of glucose. As will be considered in detail in Chapter 21, glycogen is a very large, branched polymer of glucose residues. Most of the glucose units in glycogen are linked by α-1,4-glycosidic bonds. Branches are formed by α-1,6-glycosidic bonds, present about once in 10 units (Figure 11.13).

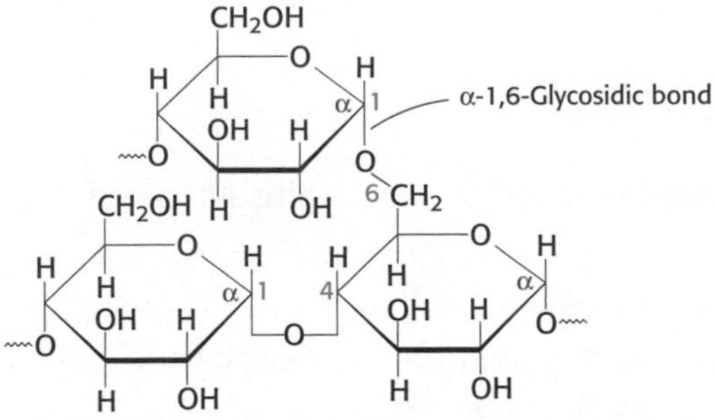

α-1,6-Glycosidic bond

Figure 11.13 Branch point in glycogen. Two chains of glucose molecules joined by α-1, 4-glycosidic bonds are linked by an α-1,6-glycosidic bond to create a branch point. Such an α-1,6-glycosidic bond forms at approximately every 10 glucose units, making glycogen a highly branched molecule.

The nutritional reservoir of carbohydrates in plants is *starch,* of which there are two forms. *Amylose,* the unbranched type of starch, consists of glucose residues in α-1,4 linkage. *Amylopectin,* the branched form, has about one α-1,6 linkage per 30 α-1,4 linkages, and so it is like glycogen except for its lower degree of branching. More than half the carbohydrate ingested by human beings is starch. Both amylopectin and amylose are rapidly hydrolyzed by *α-amylase,* an enzyme secreted by the salivary glands and the pancreas.

Cellulose, the Major Structural Polymer of Plants, Consists of Linear Chains of Glucose Units

Cellulose, the other major polysaccharide of glucose found in plants, serves a structural rather than a nutritional role. *Cellulose is one of the most abundant organic compounds in the biosphere.* Some 10^{15} kg of cellulose is synthesized and degraded on Earth each year. It is an unbranched polymer of glucose residues joined by β-1,4 linkages. The β configuration allows cellulose to form very long, straight chains. Fibrils are formed by parallel chains that interact with one another through hydrogen bonds. The α-1,4 linkages in glycogen and starch produce a very different molecular architecture from that of cellulose. A hollow helix is formed instead of a straight chain (Figure 11.14). These differing consequences of the α and β linkages are biologically important. The straight chain formed by β linkages is optimal for the construction of fibers having a high tensile strength. In contrast, the open helix formed by α linkages is well suited to forming an accessible store of sugar.

Although mammals lack cellulases and therefore cannot digest wood and vegetable fibers, cellulose and other plant fibers are still an important constituent of our diet as a component of dietary fiber. Dietary fiber produces a feeling of satiety. Soluble fiber such as *pectin* (polygalacturonic acid) slows the movement of food through the gastrointestinal tract, allowing better digestion and absorption of nutrients. Insoluble fibers, such as cellulose, increase the rate at which digestion products pass through the large intestine. This increase in rate may minimize exposure to toxins in our diet.

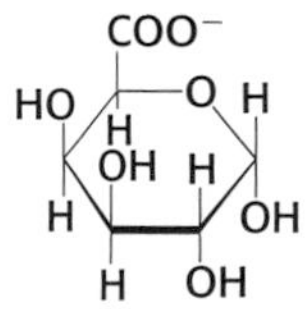

Galacturonic acid

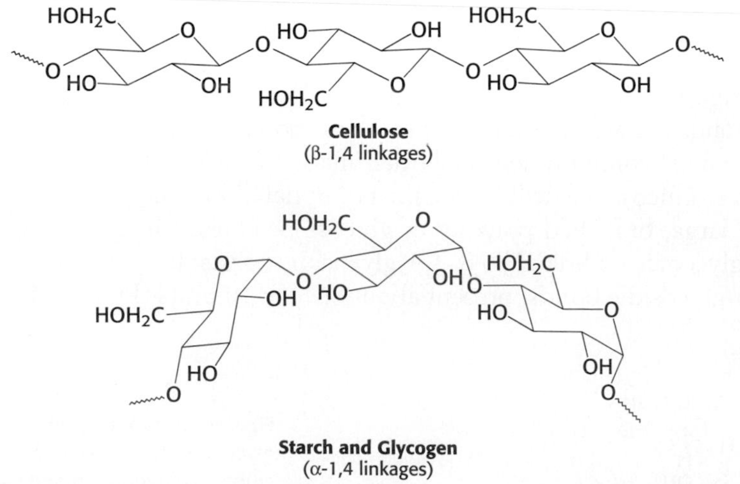

Cellulose
(β-1,4 linkages)

Starch and Glycogen
(α-1,4 linkages)

Figure 11.14 Glycosidic bonds determine polysaccharide structure. The β-1,4 linkages favor straight chains, which are optimal for structural purposes. The α-1,4 linkages favor bent structures, which are more suitable for storage.

Glycosaminoglycans Are Anionic Polysaccharide Chains Made of Repeating Disaccharide Units

Proteoglycans are proteins attached to a particular type of polysaccharide called *glycosaminoglycans.* Proteoglycans resemble polysaccharides more than proteins inasmuch as the glycosaminoglycan makes up as much as 95% of the biomolecule by weight. Proteoglycans function as joint lubricants and structural components in connective tissue. In other tissues, they mediate the adhesion of cells to the extracellular matrix and bind factors that stimulate cell proliferation.

The properties of proteoglycans are determined primarily by the glycosaminoglycan component. Many glycosaminoglycans are made of disaccharide repeating units containing a derivative of an *amino sugar,* either glucosamine or galactosamine (Figure 11.15). At least one of the sugars in the repeating unit has a *negatively charged carboxylate* or *sulfate group.* The major glycosaminoglycans in animals are chondroitin sulfate, keratan sulfate, heparin, dermatan sulfate, and hyaluronan. Proteoglycans were formerly called mucopolysaccharides, and so *mucopolysaccharidoses* is the name given to a collection of diseases that result from the inability to degrade glycosaminoglycans. Although precise clinical features vary with the disease, all mucopolysaccharidoses result in skeletal deformities and reduced life expectancies.

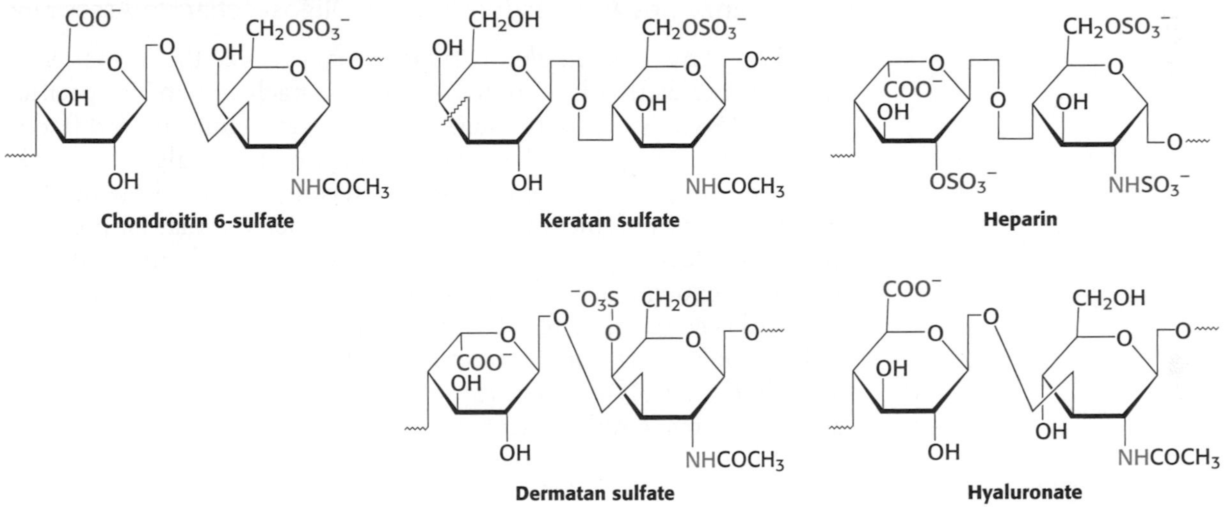

Figure 11.15 Repeating units in glycosaminoglycans. Structural formulas for five repeating units of important glycosaminoglycans illustrate the variety of modifications and linkages that are possible. Amino groups are shown in blue and negatively charged groups in red. Hydrogen atoms have been omitted for clarity. The right-hand structure is a glucosamine derivative in each case.

Among the best-characterized members of this diverse class is the proteoglycan in the extracellular matrix of cartilage. The proteoglycan *aggrecan* and the protein *collagen* are key components of cartilage. The triple helix of collagen (p. 45) provides structure and tensile strength, whereas aggrecan serves as a shock absorber. The protein component of aggrecan is a large molecule composed of 2397 amino acids. The protein has three globular domains, and the site of glycosaminoglycan attachment is the extended region between globular domains 2 and 3. This linear region contains highly repetitive amino acid sequences, which are sites for the attachment of keratan sulfate and chondroitin sulfate. Many molecules of aggrecan are in turn noncovalently bound through the first globular domain to a very long filament formed by linking together molecules of the glycosaminoglycan hyaluronan (Figure 11.16). Water is absorbed on the glycosaminoglycans, attracted by the many negative charges. Aggrecan can cushion compressive forces because the absorbed water enables it to spring back after having been deformed. When pressure is exerted, as when the foot hits the ground while walking, water is squeezed from the glycosaminoglycan, cushioning the impact. When the pressure is released, the water rebinds. *Osteoarthritis* can result from the proteolytic degradation of aggrecan and collagen in the cartilage.

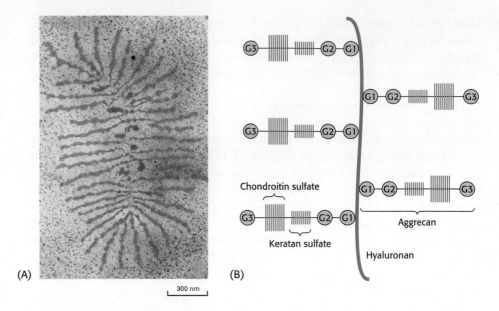

Figure 11.16 Structure of proteoglycan from cartilage. (A) Electron micrograph of a proteoglycan from cartilage (with false color added). Proteoglycan monomers emerge laterally at regular intervals from opposite sides of a central filament of hyaluronan. (B) Schematic representation. G = globular domain. [(A) Courtesy of Dr. Lawrence Rosenberg. From J. A. Buckwalter and L. Rosenberg. *Collagen Relat. Res.* 3(1983):489–504.]

(A)

(B)

Chondroitin sulfate

Keratan sulfate

Aggrecan

Hyaluronan

300 nm

Specific Enzymes Are Responsible for Oligosaccharide Assembly

Specific enzymes, called *glycosyltransferases,* catalyze the formation of the glycosidic bonds that link monosaccharides. Each enzyme is specific, to a greater or lesser extent, to the sugars being linked. Thus, many different enzymes are required to produce the diversity of known glycosidic linkages. Note that this mode of assembly stands in contrast with those used for the other biological polymers that we have considered—that is, polypeptides and oligonucleotides. As these polymers are assembled, information about monomer sequence is transferred from a template, and a single catalytic apparatus is responsible for all bond formation.

The general form of the reaction catalyzed by a glycosyltransferase is shown in Figure 11.17. The sugar to be added comes in the form of an activated sugar nucleotide. *Sugar nucleotides are important intermediates in*

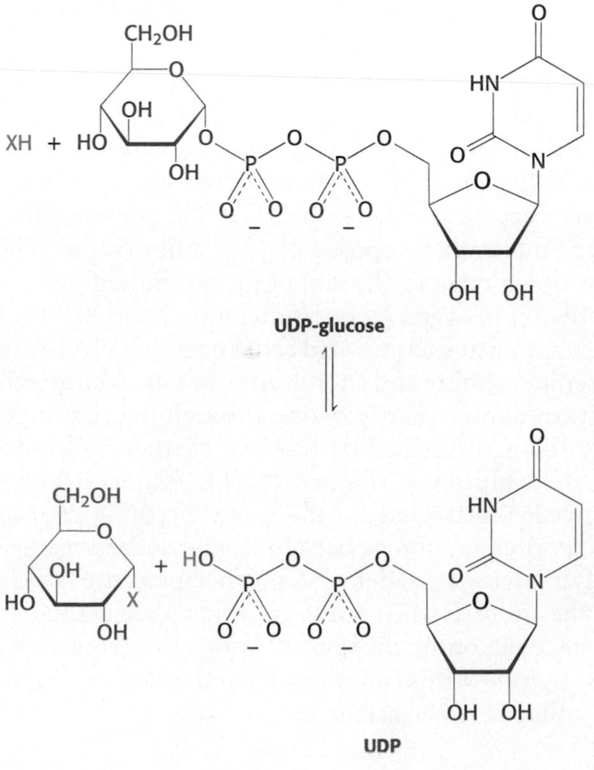

UDP-glucose

UDP

Figure 11.17 General form of a glycosyltransferase reaction. The sugar to be added comes from a sugar nucleotide— in this case, UDP-glucose. The acceptor, designated X in the figure, can be a simple monosaccharide, a complex polysaccharide, or a serine or threonine residue belonging to a protein.

314

many processes, and we will encounter these intermediates again in Chapters 16 and 21. Such reactions can proceed with either retention or inversion of configuration at the glycosidic carbon atom at which the new bond is formed; a given enzyme proceeds by one stereochemical path or the other.

The human ABO blood groups illustrate the effects of glycosyl-transferases. Carbohydrates are attached to proteins and lipids on the surfaces of red blood cells. For one type of blood group, one of the three different carbohydrate structures, termed A, B, and O, may be present (Figure 11.18). These structures have in common an oligosaccharide foundation called the O (or sometimes H) antigen. The A and B antigens are formed by the addition of one extra monosaccharide, either *N*-acetylgalac-tosamine (for A) or galactose (for B), through an α-1,3 linkage to a galactose moiety of the O antigen.

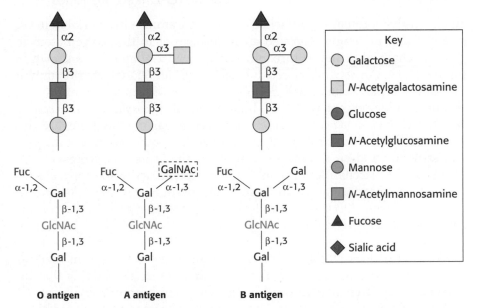

Figure 11.18 Structures of A, B, and O oligosaccharide antigens. The carbohydrate structures shown in the upper part of the figure are depicted symbolically by employing a scheme (shown in the key) that is becoming widely used. Abbreviations: Fuc, fucose; Gal, galactose; GalNAc, *N*-acetylgalactosamine; GlcNAc, *N*-acetylglucosamine.

Specific glycosyltransferases add the extra monosaccharide to the O antigen. Each person inherits the gene for one of these glycosyltransferases from each parent. The type A transferase adds *N*-acetylgalactosamine, whereas the type B transferase adds galactose. These enzymes are identical in all but 4 of 354 positions. The O phenotype is the result of mutations that prevent the synthesis of a glycosyltransferase required to add additional carbohydrates.

Use of the proper ABO blood group in blood transfusions and other transplantation procedures is crucial. Otherwise, such procedures can introduce an antigen not normally present in a person. The person's immune system recognizes the antigen as foreign and destroys the incompatible red blood cells, initiating adverse reactions.

Why are different blood types present in the human population? Suppose that a parasite or other pathogenic microorganism expresses on its cell surface a carbohydrate antigen similar to one of the blood-group antigens. This antigen may not be readily detected as foreign in a person with the blood type that matches the parasite antigen, and the parasite will flourish. However, other people with different blood types will be protected. Hence, there will be selective pressure on human beings to vary blood type to prevent parasitic mimicry and a corresponding selective pressure on parasites to enhance mimicry. This "arms race" between pathogenic microorganisms and human beings drives the evolution of diversity of surface antigens within the human population.

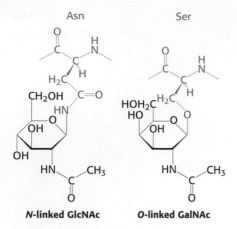

Figure 11.19 Glycosidic bonds between proteins and carbohydrates. A glycosidic bond links a carbohydrate to the side chain of asparagine (*N*-linked) or to the side chain of serine or threonine (*O*-linked). The glycosidic bonds are shown in red.

Abbreviations for sugars	
Fuc	Fucose
Gal	Galactose
GalNAc	*N*-Acetylgalactosamine
Glc	Glucose
GlcNAc	*N*-Acetylglucosamine
Man	Mannose
Sia	Sialic acid

11.3 Carbohydrates Can Be Attached to Proteins to Form Glycoproteins

A carbohydrate group can be covalently attached to a protein to form a *glycoprotein*. Carbohydrates are a much smaller percentage of the weight of glycoproteins than of proteoglycans. Many glycoproteins are components of cell membranes, where they play a variety of roles in processes such as cell adhesion and the binding of sperm to eggs. Other glycoproteins are formed by linking carbohydrates to soluble proteins. In particular, many of the proteins secreted from cells are glycosylated, including most proteins present in the serum component of blood.

Carbohydrates Can Be Linked to Proteins Through Asparagine (*N*-Linked) or Through Serine or Threonine (*O*-Linked) Residues

Sugars in glycoproteins are attached either to the amide nitrogen atom in the side chain of asparagine (termed an N-*linkage*) or to the oxygen atom in the side chain of serine or threonine (termed an O-*linkage*), as shown in Figure 11.19. An asparagine residue can accept an oligosaccharide only if the residue is part of an Asn-X-Ser or Asn-X-Thr sequence, in which X can be any residue, except proline. Thus, potential *glycosylation sites can be detected within amino acid sequences.* However, not all potential sites are glycosylated. Which sites are glycosylated depends on other aspects of the protein structure and on the cell type in which the protein is expressed. All N-linked oligosaccharides have in common a pentasaccharide core consisting of three mannose and two N-acetylglucosamine residues. Additional sugars are attached to this core to form the great variety of oligosaccharide patterns found in glycoproteins (Figure 11.20).

Let us look at a glycoprotein present in the blood serum that has dramatically improved treatment for anemia, particularly that induced by cancer chemotherapy. The glycoprotein hormone *erythropoietin* (EPO) is secreted by the kidneys and stimulates the production of red blood cells. EPO is composed of 165 amino acids and is N-glycosylated at three asparagine residues and O-glycosylated on a serine residue (Figure 11.21). The mature EPO is 40% carbohydrate by weight, and glycosylation en-

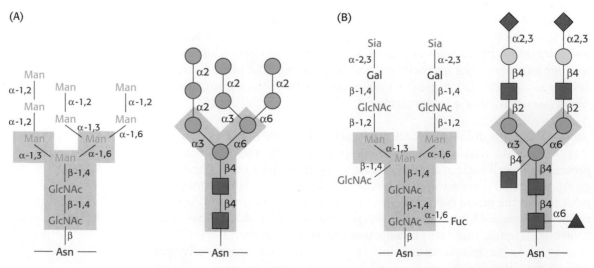

Figure 11.20 *N*-linked oligosaccharides. A pentasaccharide core (shaded gray) is common to all *N*-linked oligosaccharides and serves as the foundation for a wide variety of *N*-linked oligosaccharides, two of which are illustrated: (A) high-mannose type; (B) complex type. Detailed chemical formulas and schematic structures are shown for each type.

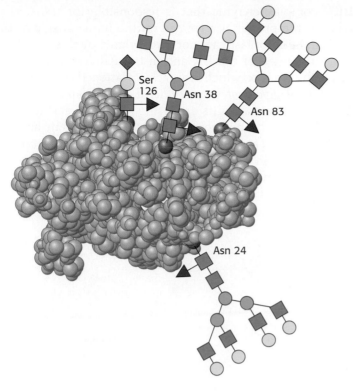

Figure 11.21 Oligosaccharides attached to erythropoietin. Erythropoietin has oligosaccharides linked to three asparagine residues and one serine residue. The structures shown are approximately to scale. See page 315 for the carbohydrate key. [Draw from 1BUY.pdf.]

hances the stability of the protein in the blood. Unglycosylated protein has only about 10% of the bioactivity of the glycosylated form because the protein is rapidly removed from the blood by the kidney. The availability of recombinant human EPO has greatly aided the treatment of anemias. However, some endurance athletes have used recombinant human EPO to increase the red-blood-cell count and hence their oxygen-carrying capacity. Drug-testing laboratories are able to distinguish some forms of prohibited human recombinant EPO from natural EPO in athletes by detecting differences in their glycosylation patterns.

Protein Glycosylation Takes Place in the Lumen of the Endoplasmic Reticulum and in the Golgi Complex

Protein glycosylation takes place inside the lumen of the *endoplasmic reticulum* (ER) and in the *Golgi complex*, organelles that play central roles in protein trafficking (Figure 11.22). The protein is synthesized by ribosomes attached to the cytoplasmic face of the ER membrane, and the peptide chain is inserted into the lumen of the ER (Section 30.6). The *N*-linked glycosylation begins in the ER and continues in the Golgi complex, whereas the *O*-linked glycosylation takes place exclusively in the Golgi complex.

A large oligosaccharide destined for attachment to the asparagine residue of a protein is assembled attached to *dolichol phosphate*, a specialized lipid molecule located in the ER membrane and containing about 20 isoprene (C$_5$) units (p. 740).

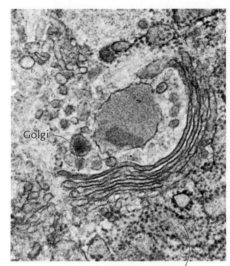

Endoplasmic reticulum

Figure 11.22 Golgi complex and endoplasmic reticulum. The electron micrograph shows the Golgi complex and adjacent endoplasmic reticulum. The black dots on the cytoplasmic surface of the ER membrane are ribosomes. [Micrograph courtesy of Lynne Mercer.]

Dolichol phosphate

$n = 15–19$

Isoprene

The terminal phosphate group is the site of attachment of the activated oligosaccharide, which is subsequently transferred en bloc to a specific asparagine residue of the growing polypeptide chain. Both the activated sugars and the complex enzyme that is responsible for transferring the oligosaccharide to the protein are located on the lumenal side of the ER. Thus, proteins in the cytoplasm are not glycosylated by this pathway.

Proteins in the lumen of the ER and in the ER membrane are transported to the Golgi complex, which is a stack of flattened membranous sacs. *Carbohydrate units of glycoproteins are altered and elaborated in the Golgi complex.* The *O*-linked sugar units are fashioned there, and the *N*-linked sugars, arriving from the ER as a component of a glycoprotein, are modified in many different ways. *The Golgi complex is the major sorting center of the cell.* Proteins proceed from the Golgi complex to lysosomes, secretory granules, or the plasma membrane, according to signals encoded within their amino acid sequences and three-dimensional structures (Figure 11.23).

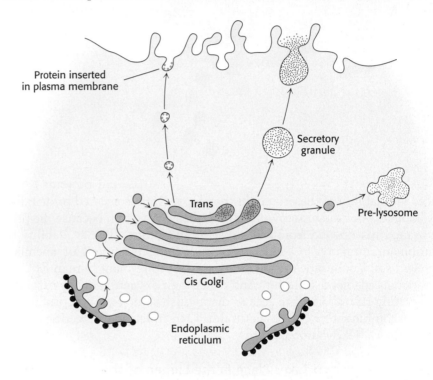

Figure 11.23 Golgi complex as sorting center. The Golgi complex is the sorting center in the targeting of proteins to lysosomes, secretory vesicles, and the plasma membrane. The cis face of the Golgi complex receives vesicles from the endoplasmic reticulum, and the trans face sends a different set of vesicles to target sites. Vesicles also transfer proteins from one compartment of the Golgi complex to another. [Courtesy of Dr. Marilyn Farquhar.]

Errors in Glycosylation Can Result in Pathological Conditions

Although the role of carbohydrate attachment to proteins is not known in detail in most cases, data indicate that carbohydrate attachment is important for the processing, stability, and targeting of these proteins. For instance, in humans, the attachment of a carbohydrate stabilizes a particular potassium channel (a protein that allows the regulated flow of potassium ions across the cell membrane; p. 364), preventing degradation. Certain types of muscular dystrophy can be traced to the improper glycosylation of membrane proteins. Indeed, an entire family of severe inherited human diseases called *congenital disorders of glycosylation* has been identified. These pathological conditions reveal the importance of the proper modification of proteins by carbohydrates and their derivatives.

An especially clear example of the role of glycosylation is provided by *I-cell disease* (also called *mucolipidosis II*), a lysosomal storage disease. *Lysosomes* are organelles that degrade and recycle damaged cellular components or material brought into the cell by endocytosis. A carbohydrate marker directs certain degradative enzymes from the Golgi complex to lysosomes. Patients having I-cell disease suffer severe psychomotor retardation and skeletal deformities. Their lysosomes contain large *inclusions* of undigested glycosaminoglycans and glycolipids (p. 331)—hence the "I" in the name of the disease. These inclusions are present because the enzymes responsible for the degradation of glycosaminoglycans are missing from affected lysosomes. Remarkably, the enzymes are present at very high levels in the blood and urine. Thus, active enzymes are synthesized, but they are exported instead of being sequestered in lysosomes. In other words, *a whole series of enzymes is incorrectly addressed and delivered to the wrong location in I-cell disease.* Normally, these enzymes contain a mannose 6-phosphate residue, but, in I-cell disease, the attached mannose is unmodified (Figure 11.24). *Mannose 6-phosphate is in fact the marker that normally directs many hydrolytic enzymes from the Golgi complex to lysosomes. I-cell patients are deficient in the phosphotransferase catalyzing the first step in the addition of the phosphoryl group; the consequence is the mistargeting of eight essential enzymes.* Paradoxically, the inability to correctly glycosylate a set of enzymes results in the accumulation of complex carbohydrates, the glycosaminoglycans, with the result being a pathological condition.

Oligosaccharides Can Be "Sequenced"

How is it possible to determine the structure of a glycoprotein—the oligosaccharide structures and their points of attachment? Most approaches make use of enzymes that cleave oligosaccharides at specific types of linkages.

The first step is to detach the oligosaccharide from the protein. For example, *N*-linked oligosaccharides can be released from proteins by an enzyme such as peptide *N*-glycosidase F, which cleaves the *N*-glycosidic bonds linking the oligosaccharide to the protein. The oligosaccharides can then be isolated and analyzed. MALDI-TOF or other mass spectrometric techniques (Section 3.5) provide the mass of an oligosaccharide fragment. However, many possible oligosaccharide structures are consistent with a given mass. More-complete information can be obtained by cleaving the oligosaccharide with enzymes of varying specificities. For example, β-1,4-galactosidase cleaves β-glycosidic bonds exclusively at galactose residues. The products can again be analyzed by mass spectrometry (Figure 11.25). The repetition of this process with the use of an array of enzymes of different specificity will eventually reveal the structure of the oligosaccharide.

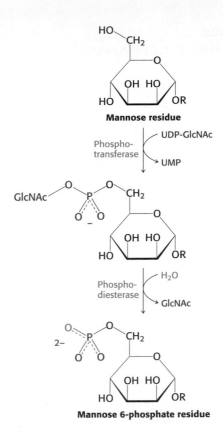

Figure 11.24 Formation of a mannose 6-phosphate marker. A glycoprotein destined for delivery to lysosomes acquires a phosphate marker in the Golgi compartment in a two-step process. First, a phosphotransferase adds a phospho-*N*-acetylglucosamine unit to the 6-OH group of a mannose, and then a phosphodiesterase removes the added sugar to generate a mannose 6-phosphate residue in the core oligosaccharide.

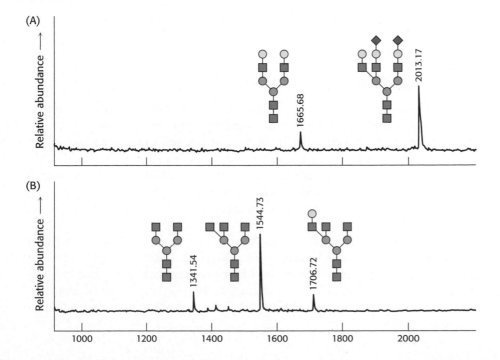

Figure 11.25 Mass spectrometric "sequencing" of oligosaccharides. Carbohydrate-cleaving enzymes were used to release and specifically cleave the oligosaccharide component of the glycoprotein fetuin from bovine serum. Parts A and B show the masses obtained with MALDI-TOF spectrometry as well as the corresponding structures of the oligosaccharide-digestion products (using the same scheme as that in Figure 11.18): (A) digestion with peptide *N*-glycosidase F (to release the oligosaccharide from the protein) and neuraminidase; (B) digestion with peptide *N*-glycosidase F, neuraminidase, and β-1,4-galactosidase. Knowledge of the enzyme specificities and the masses of the products permits the characterization of the oligosaccharide. See page 315 for the carbohydrate key. [After A. Varki, R. Cummings, J. Esko, H. Freeze, G. Hart, and J. Marth (Eds.), *Essentials of Glycobiology* (Cold Spring Harbor Laboratory Press, 1999), p. 596.]

Proteases applied to glycoproteins can reveal the points of oligosaccharide attachment. Cleavage by a specific protease yields a characteristic pattern of peptide fragments that can be analyzed chromatographically. Fragments attached to oligosaccharides can be picked out because their chromatographic properties will change on glycosidase treatment. Mass spectrometric analysis or direct peptide sequencing can reveal the identity of the peptide in question and, with additional effort, the exact site of oligosaccharide attachment.

Glycosylation greatly increases the complexity of the proteome. A given protein with several potential glycosylation sites can have many different glycosylated forms (sometimes called *glycoforms*), each of which may be generated only in a specific cell type or developmental stage. Now that the sequencing of the human genome is complete, the characterization of the much more complex proteome, including the biological roles of specifically modified proteins, can begin in earnest.

11.4 Lectins Are Specific Carbohydrate-Binding Proteins

The diversity and complexity of the carbohydrate units of glycoproteins suggest that they are functionally important. Nature does not construct complex patterns when simple ones suffice. Cellulose and starch, for example, are built solely from glucose units. In contrast, glycoproteins contain multiple types of residues joined by different kinds of glycosidic linkages. *An enormous number of patterns in the composition and structure of surface sugars are possible* because (1) different monosaccharides can be joined to one another through any of several OH groups, (2) the C-1 linkage can have either an α or a β configuration, and (3) extensive branching is possible. Indeed, *many more different oligosaccharides can be formed from four sugars than can oligopeptides from four amino acids.*

Why all this intricacy and diversity? It is becoming evident that carbohydrates are information-rich molecules that guide many biological processes. The diverse carbohydrate structures displayed on cell surfaces are well suited to serve as sites of interaction between cells and their environments. Proteins termed *lectins* (from the Latin *legere*, "to select") are the partners that bind specific carbohydrate structures on opposing cell surfaces. Lectins are ubiquitous: they are found in animals, plants, and microorganisms.

Lectins Promote Interactions Between Cells

The chief function of lectins in animals is to facilitate cell–cell contact. A lectin usually contains two or more binding sites for carbohydrate units. The binding sites of lectins on the surface of one cell interact with arrays of carbohydrates displayed on the surface of another cell. Lectins and carbohydrates are linked by a number of weak interactions that ensure specificity yet permit unlinking as needed. The interactions between one cell surface and another resemble the action of Velcro; each interaction is weak but the composite is strong.

Lectins can be divided into classes on the basis of their amino acid sequences and biochemical properties. One large class is the C type (for *calcium*-requiring) found in animals. These proteins each have a homologous domain of 120 amino acids that is responsible for carbohydrate binding. The structure of one such domain bound to a carbohydrate target is shown in Figure 11.26.

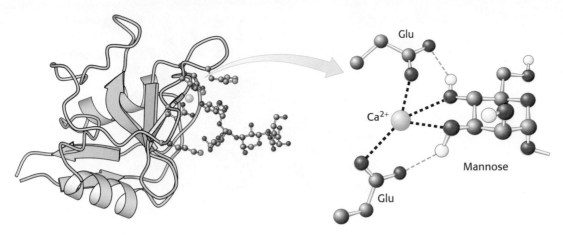

Figure 11.26 Structure of a C-type carbohydrate-binding domain of an animal lectin. *Notice* that a calcium ion links a mannose residue to the lectin. Selected interactions are shown, with some hydrogen atoms omitted for clarity.

A calcium ion on the protein acts as a bridge between the protein and the sugar through direct interactions with sugar OH groups. In addition, two glutamate residues in the protein bind to both the calcium ion and the sugar, and other protein side chains form hydrogen bonds with other OH groups on the carbohydrate. The carbohydrate-binding specificity of a particular lectin is determined by the amino acid residues that bind the carbohydrate.

Proteins termed *selectins* are members of the C-type family. Selectins bind immune-system cells to sites of injury in the inflammatory response (Figure 11.27). The L, E, and P forms of selectins bind specifically to carbohydrates on *lymph-node* vessels, *endothelium*, or activated blood *platelets*, respectively. New therapeutic agents that control inflammation may emerge from a deeper understanding of how selectins bind and distinguish different carbohydrates. L-Selectin, originally thought to participate only in the immune response, is produced by embryos when they are ready to attach to the endometrium of the mother's uterus. For a short period of time, the endometrial cells present an oligosaccharide on the cell surface. When the embryo attaches through lectins, the attachment activates signal pathways in the endometrium to make implantation of the embryo possible.

Plants also are rich in lectins. Although the exact role of lectins in plants is unclear, they can serve as potent insecticides. The binding specificities of lectins from plants have been well characterized (Figure 11.28). Bacteria, too, contain lectins. *Escherichia coli* bacteria are able to adhere to the epithelial cells of the gastrointestinal tract because lectins on the *E. coli* surface recognize oligosaccharide units on the surfaces of target cells. These lectins are located on slender hairlike appendages called fimbriae (*pili*).

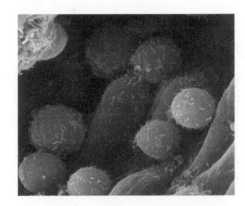

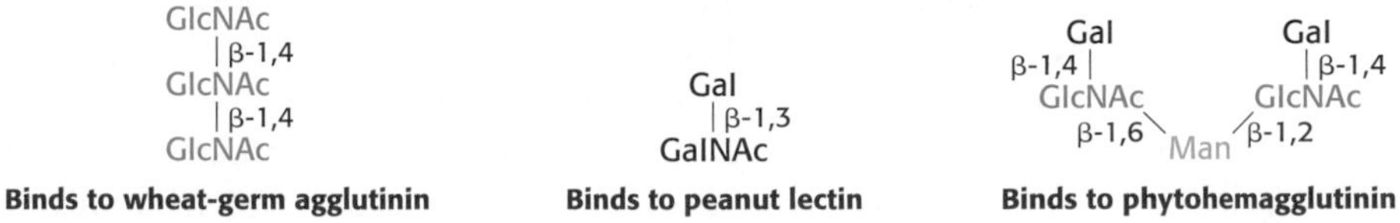

Figure 11.27 Selectins mediate cell–cell interactions. The scanning electron micrograph shows lymphocytes adhering to the endothelial lining of a lymph node. The L selectins on the lymphocyte surface bind specifically to carbohydrates on the lining of the lymph-node vessels. [Courtesy of Dr. Eugene Butcher.]

GlcNAc
| β-1,4
GlcNAc
| β-1,4
GlcNAc

Binds to wheat-germ agglutinin

Gal
| β-1,3
GalNAc

Binds to peanut lectin

Gal   Gal
β-1,4 | | β-1,4
GlcNAc   GlcNAc
β-1,6 \ Man / β-1,2

Binds to phytohemagglutinin

Figure 11.28 Binding selectivities of plant lectins. The plant lectins wheat-germ agglutinin, peanut lectin, and phytohemagglutinin recognize different oligosaccharides.

Influenza Virus Binds to Sialic Acid Residues

Some viruses gain entry into specific host cells by adhering to cell-surface carbohydrates. For example, influenza virus recognizes sialic acid residues present on cell-surface glycoproteins. The viral protein that binds to these sugars is called *hemagglutinin* (Figure 11.29).

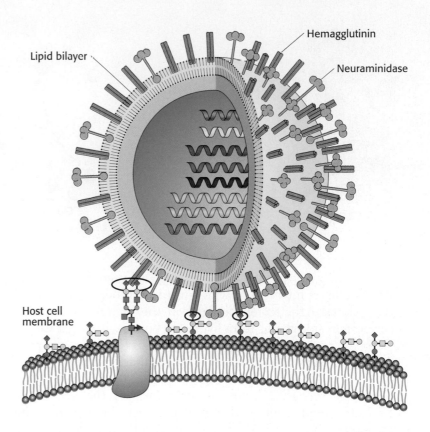

Figure 11.29 Viral receptors. Influenza virus targets cells by binding to sialic acid residues (purple diamonds) located at the termini of oligosaccharides present on cell-surface glycoproteins and glycolipids. These carbohydrates are bound by hemagglutinin (interaction circles), one of the major proteins expressed on the surface of the virus. The other major viral surface protein, neuraminidase, is an enzyme that cleaves oligosaccharide chains to release the viral particle at a later stage of the viral life cycle.

After the virus penetrates the cell membrane, another viral protein, neuraminidase (sialidase), cleaves the glycosidic bonds to the sialic acid residues, freeing the virus to infect the cell. Inhibitors of this enzyme such as oseltamivir (Tamiflu) and zanamivir (Relenza) are important anti-influenza agents.

Summary

11.1 Monosaccharides Are Aldehydes or Ketones with Multiple Hydroxyl Groups

An aldose is a carbohydrate with an aldehyde group (as in glyceraldehyde and glucose), whereas a ketose contains a keto group (as in dihydroxyacetone and fructose). A sugar belongs to the D series if the absolute configuration of its asymmetric carbon atom farthest from the aldehyde or keto group is the same as that of D-glyceraldehyde. Most naturally occurring sugars belong to the D series. The C-1 aldehyde in the open-chain form of glucose reacts with the C-5 hydroxyl group to form a six-membered pyranose ring. The C-2 keto group in the open-chain form of fructose reacts with the C-5 hydroxyl group to form a five-membered furanose ring. Pentoses such as ribose and deoxyribose also form furanose rings. An additional asymmetric center is formed at the anomeric carbon atom (C-1 in aldoses and C-2 in ketoses) in these cyclizations. The hydroxyl group attached to the anomeric carbon atom is on the opposite side of the ring from the CH_2OH group attached to the chiral center in the α anomer, whereas it is on the same side of the ring as the CH_2OH group in the β anomer. Not all the atoms in the rings lie in the same plane. Rather, pyranose rings usually adopt the chair conformation, and furanose rings usually adopt the envelope conformation. Sugars are joined to alcohols and amines by glycosidic bonds from the anomeric carbon atom. For example, N-glycosidic bonds link sugars to purines and pyrimidines in nucleotides, RNA, and DNA.

11.2 Complex Carbohydrates Are Formed by the Linkage of Monosaccharides

Sugars are linked to one another in disaccharides and polysaccharides by *O*-glycosidic bonds. Sucrose, lactose, and maltose are the common disaccharides. Sucrose (common table sugar) consists of α-glucose and β-fructose joined by a glycosidic linkage between their anomeric carbon atoms. Lactose (in milk) consists of galactose joined to glucose by a β-1,4 linkage. Maltose (in starch) consists of two glucoses joined by an α-1,4 linkage. Starch is a polymeric form of glucose in plants, and glycogen serves a similar role in animals. Most of the glucose units in starch and glycogen are in α-1,4 linkage. Glycogen has more branch points formed by α-1,6 linkages than does starch, and so glycogen is more soluble. Cellulose, the major structural polymer of plant cell walls, consists of glucose units joined by β-1,4 linkages. These β linkages give rise to long straight chains that form fibrils with high tensile strength. In contrast, the α linkages in starch and glycogen lead to open helices, in keeping with their roles as mobilizable energy stores. Cell surfaces and the extracellular matrices of animals contain polymers of repeating disaccharides called glycosaminoglycans. One of the units in each repeat is a derivative of glucosamine or galactosamine. These highly anionic carbohydrates have a high density of carboxylate or sulfate groups. Proteins bearing covalently linked glycosaminoglycans are proteoglycans.

11.3 Carbohydrates Can Attach to Proteins to Form Glycoproteins

Specific enzymes link the oligosaccharide units on proteins either to the side-chain oxygen atom of a serine or threonine residue or to the side-chain amide nitrogen atom of an asparagine residue. Protein glycosylation takes place in the lumen of the endoplasmic reticulum. The *N*-linked oligosaccharides are synthesized on dolichol phosphate and subsequently transferred to the protein acceptor. Additional sugars are attached in the Golgi complex to form diverse patterns.

11.4 Lectins Are Specific Carbohydrate-Binding Proteins

Carbohydrates on cell surfaces are recognized by proteins called lectins. In animals, the interplay of lectins and their sugar targets guides cell–cell contact. The viral protein hemagglutinin on the surface of the influenza virus recognizes sialic acid residues on the surfaces of cells invaded by the virus. A small number of carbohydrate residues can be joined in many different ways to form highly diverse patterns that can be distinguished by the lectin domains of protein receptors.

Key Terms

monosaccharide (p. 304)
triose (p. 304)
ketose (p. 304)
aldose (p. 304)
enantiomer (p. 304)
tetrose (p. 304)
pentose (p. 304)
hexose (p. 304)
heptose (p. 304)
diastereoisomer (p. 305)
epimer (p. 305)
hemiacetal (p. 306)

pyranose (p. 306)
hemiketal (p. 306)
furanose (p. 307)
anomer (p. 307)
glycosidic bond (p. 309)
reducing sugar (p. 309)
nonreducing sugar (p. 309)
oligosaccharide (p. 310)
disaccharide (p. 310)
polysaccharide (p. 311)
glycogen (p. 311)
starch (p. 311)

cellulose (p. 312)
proteoglycan (p. 312)
glycosaminoglycan (p. 312)
glycosyltransferase (p. 314)
glycoprotein (p. 316)
endoplasmic reticulum (p. 317)
Golgi complex (p. 317)
dolichol phosphate (p. 317)
lysosome (p. 318)
glycoform (p. 320)
lectin (p. 320)
selectin (p. 321)

Selected Readings

Where to Start

Sharon, N., and Lis, H. 1993. Carbohydrates in cell recognition. *Sci. Am.* 268(1):82–89.

Lasky, L. A. 1992. Selectins: Interpreters of cell-specific carbohydrate information during inflammation. *Science* 258:964–969.

Weiss, P., and Ashwell, G. 1989. The asialoglycoprotein receptor: Properties and modulation by ligand. *Prog. Clin. Biol. Res.* 300: 169–184.

Paulson, J. C. 1989. Glycoproteins: What are the sugar side chains for? *Trends Biochem. Sci.* 14:272–276.

Woods, R. J. 1995. Three-dimensional structures of oligosaccharides. *Curr. Opin. Struct. Biol.* 5:591–598.

Books

Varki, A., Cummings, R., Esko, J., Freeze, H., Hart, G., and Marth, J. 2002. *Essentials of Glycobiology.* Cold Spring Harbor Laboratory Press.

Fukuda, M., and Hindsgaul, O. 2000. *Molecular Glycobiology.* IRL Press at Oxford University Press.

El Khadem, H. S. 1988. *Carbohydrate Chemistry.* Academic Press.

Ginsburg, V., and Robbins, P. W. (Eds.). 1981. *Biology of Carbohydrates* (vols. 1–3). Wiley.

Fukuda, M. (Ed.). 1992. *Cell Surface Carbohydrates and Cell Development.* CRC Press.

Preiss, J. (Ed.). 1988. *The Biochemistry of Plants: A Comprehensive Treatise: Carbohydrates.* Academic Press.

Carbohydrate-Binding Proteins and Glycoproteins

Yan, A., and Lennarz, W. J. 2005. Unraveling the mechanism of protein *N*-glycosylation. *J. Biol. Chem.* 280:3121–3124.

Qasba, P. K., Ramakrishnan, B., and Boeggeman, E. 2005. Substrate-induced conformatioinal changes in glycosyltransferases. *Trends Biochem. Sci.* 30:53–62.

Pratta, M. A., Yao, W., Decicco, C., Tortorella, M., Liu, R.-Q., Copeland, R. A., Magolda, R., Newton, R. C., Trzaskos, J. M., and Arner, E. C. 2003. Aggrecan protects cartilage collagen from proteolytic cleavage. *J. Biol. Chem.* 278:45539–45545.

Fisher, J. W. 2003. Erythropoietin: Physiology and pharmacology update. *Exp. Biol. Med.* 228:1–14.

Ünligil, U., and Rini, J. M. 2000. Glycosyltransferase structure and mechanism. *Curr. Opin. Struct. Biol.* 10:510–517.

Cheetham, J. C., Smith, D. M., Aoki, K. H., Stevenson, J. L., Hoeffel, T. J., Syed, R. S., Egrie, J., and Harvey, T. S. 1998. NMR structure of human erythropoietin and a comparison with its receptor bound conformation. *Nat. Struct. Biol.* 5:861–866.

Bouckaert, J., Hamelryck, T., Wyns, L., and Loris, R. 1999. Novel structures of plant lectins and their complexes with carbohydrates. *Curr. Opin. Struct. Biol.* 9:572–577.

Weis, W. I., and Drickamer, K. 1996. Structural basis of lectin-carbohydrate recognition. *Annu. Rev. Biochem.* 65:441–473.

Vyas, N. K. 1991. Atomic features of protein-carbohydrate interactions. *Curr. Opin. Struct. Biol.* 1:732–740.

Weis, W. I., Drickamer, K., and Hendrickson, W. A. 1992. Structure of a C-type mannose-binding protein complexed with an oligosaccharide. *Nature* 360:127–134.

Wright, C. S. 1992. Crystal structure of a wheat germ agglutinin/glycophorin-sialoglycopeptide receptor complex: Structural basis for cooperative lectin-cell binding. *J. Biol. Chem.* 267:14345–14352.

Shaanan, B., Lis, H., and Sharon, N. 1991. Structure of a legume lectin with an ordered *N*-linked carbohydrate in complex with lactose. *Science* 254:862–866.

Glycoproteins

Bernfield, M., Götte, M., Park, P. W., Reizes, O., Fitzgerald, M. L., Lincecum, J., and Zako, M. 1999. Functions of cell surface heparan sulfate proteoglycans. *Annu. Rev. Biochem.* 68:729–777.

Iozzo, R. V. 1998. Matrix proteoglycans: From molecular design to cellular function. *Annu. Rev. Biochem.* 67:609–652.

Yanagishita, M., and Hascall, V. C. 1992. Cell surface heparan sulfate proteoglycans. *J. Biol. Chem.* 267:9451–9454.

Iozzo, R. V. 1999. The biology of small leucine-rich proteoglycans: Functional network of interactive proteins. *J. Biol. Chem.* 274: 18843–18846.

Carbohydrates in Recognition Processes

Weis, W. I. 1997. Cell-surface carbohydrate recognition by animal and viral lectins. *Curr. Opin. Struct. Biol.* 7:624–630.

Sharon, N., and Lis, H. 1989. Lectins as cell recognition molecules. *Science* 246:227–234.

Turner, M. L. 1992. Cell adhesion molecules: A unifying approach to topographic biology. *Biol. Rev. Camb. Philos. Soc.* 67:359–377.

Feizi, T. 1992. Blood group–related oligosaccharides are ligands in cell-adhesion events. *Biochem. Soc. Trans.* 20:274–278.

Jessell, T. M., Hynes, M. A., and Dodd, J. 1990. Carbohydrates and carbohydrate-binding proteins in the nervous system. *Annu. Rev. Neurosci.* 13:227–255.

Clothia, C., and Jones, E. V. 1997. The molecular structure of cell adhesion molecules. *Annu. Rev. Biochem.* 66:823–862.

Carbohydrate Sequencing

Venkataraman, G., Shriver, Z., Raman, R., and Sasisekharan, R. 1999. Sequencing complex polysaccharides. *Science* 286:537–542.

Zhao, Y., Kent, S. B. H., and Chait, B. T. 1997. Rapid, sensitive structure analysis of oligosaccharides. *Proc. Natl. Acad. Sci. U.S.A.* 94:1629–1633.

Rudd, P. M., Guile, G. R., Küster, B., Harvey, D. J., Opdenakker, G., and Dwek, R. A. 1997. Oligosaccharide sequencing technology. *Nature* 388:205–207.

Problems

1. *Word origin.* Account for the origin of the term *carbohydrate*.

2. *Diversity.* How many different oligosaccharides can be made by linking one glucose, one mannose, and one galactose? Assume that each sugar is in its pyranose form. Compare this number with the number of tripeptides that can be made from three different amino acids.

3. *Couples.* Indicate whether each of the following pairs of sugars consists of anomers, epimers, or an aldose–ketose pair:

(a) D-glyceraldehyde and dihydroxyacetone

(b) D-glucose and D-mannose

(c) D-glucose and D-fructose

(d) α-D-glucose and β-D-glucose

(e) D-ribose and D-ribulose

(f) D-galactose and D-glucose

4. *Mutarotation.* The specific rotations of the α and β anomers of D-glucose are +112 degrees and +18.7 degrees, respectively.

Specific rotation, $[\alpha]_D$, is defined as the observed rotation of light of wavelength 589 nm (the D line of a sodium lamp) passing through 10 cm of a 1 g ml^{-1} solution of a sample. When a crystalline sample of α-D-glucopyranose is dissolved in water, the specific rotation decreases from 112 degrees to an equilibrium value of 52.7 degrees. On the basis of this result, what are the proportions of the α and β anomers at equilibrium? Assume that the concentration of the open-chain form is negligible.

5. *Telltale marker.* Glucose reacts slowly with hemoglobin and other proteins to form covalent compounds. Why is glucose reactive? What is the nature of the adduct formed?

6. *Periodate cleavage.* Compounds containing hydroxyl groups on adjacent carbon atoms undergo carbon–carbon bond cleavage when treated with periodate ion (IO_4^-). How can this reaction be used to distinguish between pyranosides and furanosides?

7. *Oxygen source.* Does the oxygen atom attached to C-1 in methyl α-D-glucopyranoside come from glucose or methanol?

8. *Sugar lineup.* Identify the following four sugars.

9. *Cellular glue.* A trisaccharide unit of a cell-surface glycoprotein is postulated to play a critical role in mediating cell–cell adhesion in a particular tissue. Design a simple experiment to test this hypothesis.

10. *Mapping the molecule.* Each of the hydroxyl groups of glucose can be methylated with reagents such as dimethylsulfate under basic conditions. Explain how exhaustive methylation followed by the complete digestion of a known amount of glycogen would enable you to determine the number of branch points and reducing ends.

11. *Component parts.* Raffinose is a trisaccharide and a minor constituent in sugar beets.

(a) Is raffinose a reducing sugar? Explain.
(b) What are the monosaccharides that compose raffinose?

(c) β-Galactosidase is an enzyme that will remove galactose residues from an oligosaccharide. What are the products of β-galactosidase treatment of raffinose?

Raffinose

12. *Anomeric differences.* α-D-Mannose is a sweet-tasting sugar. β-D-Mannose, on the other hand, tastes bitter. A pure solution of α-D-mannose loses its sweet taste with time as it is converted into the β anomer. Draw the β anomer and explain how it is formed from the α anomer.

α-D-Mannose

13. *A taste of honey.* Fructose in its β-D-pyranose form accounts for the powerful sweetness of honey. The β-D-furanose form, although sweet, is not as sweet as the pyranose form. The furanose form is the more stable form. Draw the two forms and explain why it may not always be wise to cook with honey.

14. *Making ends meet.* (a) Compare the number of reducing ends to nonreducing ends in a molecule of glycogen. (b) As we will see in Chapter 21, glycogen is an important fuel-storage form that is rapidly mobilized. At which end—the reducing or nonreducing—would you expect most metabolism to take place?

15. *Carbohydrates and proteomics.* Suppose that a protein contains six potential N-linked glycosylation sites. How many possible proteins can be generated, depending on which of these sites is actually glycosylated? Do not include the effects of diversity within the carbohydrate added.

Chapter Integration Problem

16. *Stereospecificity.* Sucrose, a major product of photosynthesis in green leaves, is synthesized by a battery of enzymes. The substrates for sucrose synthesis, D-glucose and D-fructose, are a mixture of α and β anomers as well as acyclic compounds in solution. Nonetheless, sucrose consists of α-D-glucose linked by its carbon-1 atom to the carbon-2 atom of β-D-fructose. How can the specificity of sucrose be explained in light of the potential substrates?

Chapter 12

Lipids and Cell Membranes

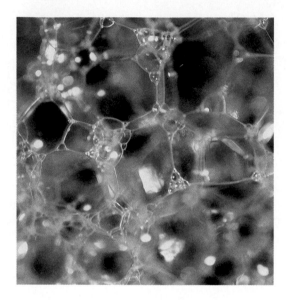

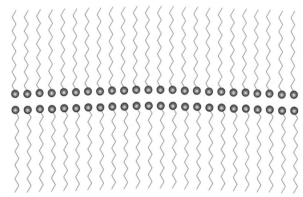

The surface of a soap bubble is a bilayer formed by detergent molecules. The polar heads (red) pack together, leaving the hydrophobic groups (green) in contact with air on the inside and outside of the bubble. Other bilayer structures define the boundary of a cell. [(Left) Photonica.]

The boundaries of cells are formed by *biological membranes*, the barriers that *define the inside and the outside of a cell* (Figure 12.1). These barriers prevent molecules generated inside the cell from leaking out and unwanted molecules from diffusing in; yet they also contain transport systems that allow the cell to take up specific molecules and remove unwanted ones. Such transport systems confer on membranes the important property of *selective permeability*.

Membranes are dynamic structures in which proteins float in a sea of lipids. The lipid components of the membrane form the barrier to permeability, and protein components act as a transport system of pumps and channels that allow selected molecules into and out of the cell. This transport system will be considered in the next chapter.

In addition to an external cell membrane (called the *plasma membrane*), eukaryotic cells also contain internal membranes that form the boundaries of organelles such as mitochondria, chloroplasts, peroxisomes, and lysosomes. Functional specialization in the course of evolution has been closely linked to the formation of such compartments. Specific systems have evolved to allow the targeting of selected proteins into or through particular internal membranes and, hence, into specific organelles. External and internal membranes have essential features in common, and these essential features are the subject of this chapter.

Biological membranes serve several additional functions indispensable for life, such as energy storage and information transduction, that are dictated by the proteins associated with them. In this chapter, we will examine the general properties of membrane proteins—how they can exist in the hydrophobic environment of the membrane while connecting two hydrophilic environments—and defer a discussion of the functions of these proteins until the next and later chapters.

Many Common Features Underlie the Diversity of Biological Membranes

Membranes are as diverse in structure as they are in function. However, they do have in common a number of important attributes:

1. Membranes are *sheetlike structures,* only two molecules thick, that form *closed boundaries* between different compartments. The thickness of most membranes is between 60 Å (6 nm) and 100 Å (10 nm).

2. Membranes consist mainly of *lipids* and *proteins.* The mass ratio of lipids to proteins ranges from 1:4 to 4:1. Membranes also contain *carbohydrates* that are linked to lipids and proteins.

3. Membrane lipids are small molecules that have both *hydrophilic* and *hydrophobic* moieties. These lipids spontaneously form *closed bimolecular sheets* in aqueous media. These *lipid bilayers* are barriers to the flow of polar molecules.

4. *Specific proteins mediate distinctive functions of membranes.* Proteins serve as pumps, channels, receptors, energy transducers, and enzymes. Membrane proteins are embedded in lipid bilayers, which create suitable environments for their action.

5. Membranes are *noncovalent assemblies.* The constituent protein and lipid molecules are held together by many noncovalent interactions, which act cooperatively.

6. Membranes are *asymmetric.* The two faces of biological membranes always differ from each other.

7. Membranes are *fluid structures.* Lipid molecules diffuse rapidly in the plane of the membrane, as do proteins, unless they are anchored by specific interactions. In contrast, lipid molecules and proteins do not readily rotate across the membrane. Membranes can be regarded as *two-dimensional solutions of oriented proteins and lipids.*

8. Most cell membranes are *electrically polarized,* such that the inside is negative [typically −60 millivolts (mV)]. Membrane potential plays a key role in transport, energy conversion, and excitability (Chapter 13).

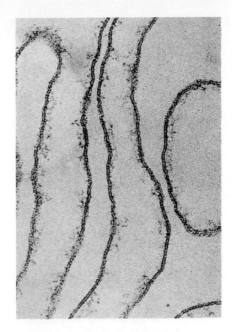

Figure 12.1 Red-blood-cell plasma membrane. An electron micrograph of a preparation of plasma membranes from red blood cells showing the membranes as seen "on edge," in cross section. [Courtesy of Dr. Vincent Marchesi.]

12.1 Fatty Acids Are Key Constituents of Lipids

The hydrophobic properties of lipids are essential to their ability to form membranes. Most lipids owe their hydrophobic properties to one component, their fatty acids.

Fatty Acid Names Are Based on Their Parent Hydrocarbons

Fatty acids are long hydrocarbon chains of various lengths and degrees of unsaturation terminated with carboxylic acid groups. The systematic name for a fatty acid is derived from the name of its parent hydrocarbon by the substitution of *oic* for the final *e.* For example, the C_{18} saturated fatty acid is called *octadecanoic acid* because the parent hydrocarbon is octadecane.

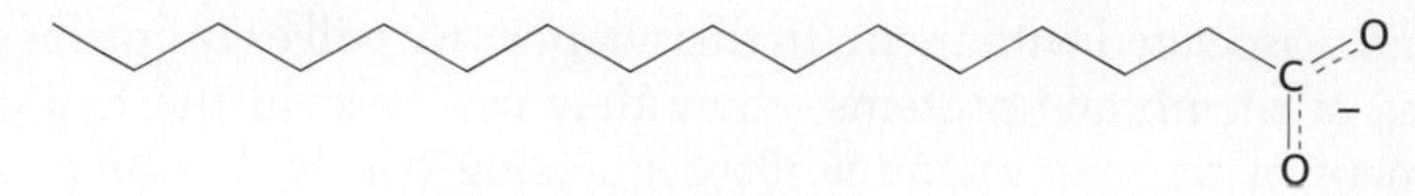

Palmitate
(ionized form of palmitic acid)

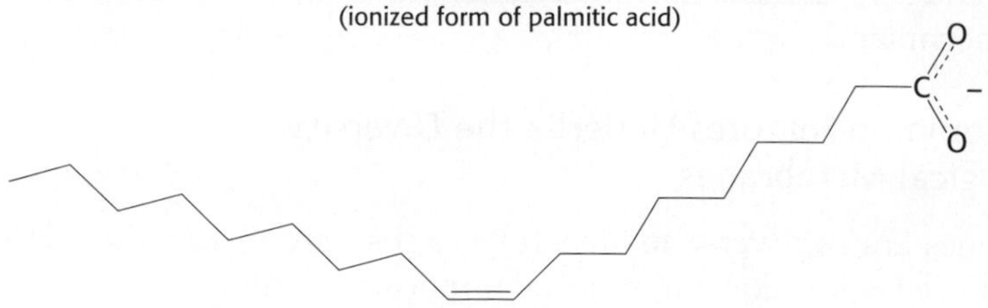

Oleate
(ionized form of oleic acid)

Figure 12.2 Structures of two fatty acids. Palmitate is a 16-carbon, saturated fatty acid, and oleate is an 18-carbon fatty acid with a single cis double bond.

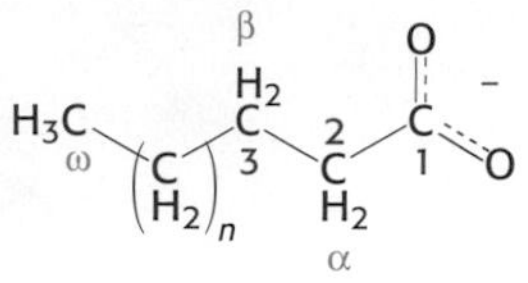

An ω-3 fatty acid

A C_{18} fatty acid with one double bond is called octadec*enoic* acid; with two double bonds, octadeca*dienoic* acid; and with three double bonds, octadeca-*trienoic* acid. The notation 18:0 denotes a C_{18} fatty acid with no double bonds, whereas 18:2 signifies that there are two double bonds. The structures of the ionized forms of two common fatty acids—palmitic acid (16:0) and oleic acid (18:1)—are shown in Figure 12.2.

Fatty acid carbon atoms are numbered starting at the carboxyl terminus, as shown in the margin. Carbon atoms 2 and 3 are often referred to as α and β, respectively. The methyl carbon atom at the distal end of the chain is called the *ω-carbon atom*. The position of a double bond is represented by the symbol Δ followed by a superscript number. For example, *cis*-Δ^9 means that there is a cis double bond between carbon atoms 9 and 10; *trans*-Δ^2 means that there is a trans double bond between carbon atoms 2 and 3. Alternatively, the position of a double bond can be denoted by counting from the distal end, with the ω-carbon atom (the methyl carbon) as number 1. An ω-3 fatty acid, for example, has the structure shown in the margin. Fatty acids are ionized at physiological pH, and so it is appropriate to refer to them according to their carboxylate form: for example, palmitate or hexadecanoate.

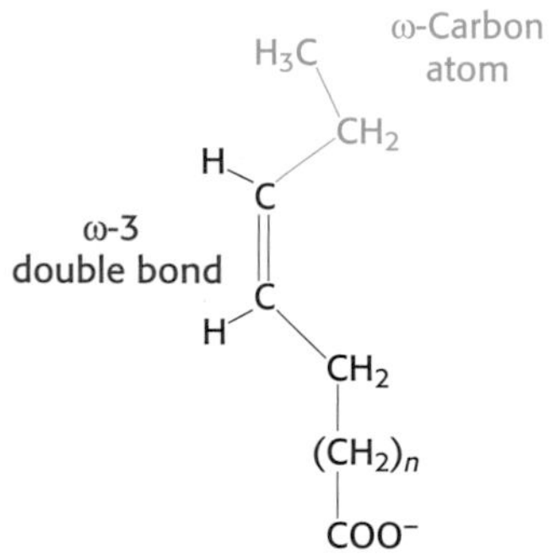

Fatty Acids Vary in Chain Length and Degree of Unsaturation

Fatty acids in biological systems usually contain an even number of carbon atoms, typically between 14 and 24 (Table 12.1). The 16- and 18-carbon fatty acids are most common. The dominance of fatty acid chains containing an even number of carbon atoms is in accord with the way in which

TABLE 12.1 Some naturally occurring fatty acids in animals

Number of carbons	Number of double bonds	Common name	Systematic name	Formula
12	0	Laurate	*n*-Dodecanoate	$CH_3(CH_2)_{10}COO^-$
14	0	Myristate	*n*-Tetradecanoate	$CH_3(CH_2)_{12}COO^-$
16	0	Palmitate	*n*-Hexadecanoate	$CH_3(CH_2)_{14}COO^-$
18	0	Stearate	*n*-Octadecanoate	$CH_3(CH_2)_{16}COO^-$
20	0	Arachidate	*n*-Eicosanoate	$CH_3(CH_2)_{18}COO^-$
22	0	Behenate	*n*-Docosanoate	$CH_3(CH_2)_{20}COO^-$
24	0	Lignocerate	*n*-Tetracosanoate	$CH_3(CH_2)_{22}COO^-$
16	1	Palmitoleate	*cis*-Δ^9-Hexadecenoate	$CH_3 (CH_2)_5CH{=}CH(CH_2)_7COO^-$
18	1	Oleate	*cis*-Δ^9-Octadecenoate	$CH_3 (CH_2)_7CH{=}CH(CH_2)_7COO^-$
18	2	Linoleate	*cis,cis*-Δ^9, Δ^{12}-Octadecadienoate	$CH_3 (CH_2)_4(CH{=}CHCH_2)_2(CH)_6COO^-$
18	3	Linolenate	all-*cis*-Δ^9, Δ^{12}, Δ^{15}-Octadecatrienoate	$CH_3CH_2(CH{=}CHCH_2)_3(CH_2)_6COO^-$
20	4	Arachidonate	all-*cis* Δ^5, Δ^8, Δ^{11}, -Δ^{14} Eicosatetraenoate	$CH_3(CH_2)_4(CH{=}CHCH_2)_4(CH_2)_2COO^-$

fatty acids are biosynthesized (Chapter 26). The hydrocarbon chain is almost invariably unbranched in animal fatty acids. The alkyl chain may be saturated or it may contain one or more double bonds. The configuration of the double bonds in most unsaturated fatty acids is cis. The double bonds in polyunsaturated fatty acids are separated by at least one methylene group.

The properties of fatty acids and of lipids derived from them are markedly dependent on chain length and degree of saturation. Unsaturated fatty acids have lower melting points than saturated fatty acids of the same length. For example, the melting point of stearic acid is 69.6°C, whereas that of oleic acid (which contains one cis double bond) is 13.4°C. The melting points of polyunsaturated fatty acids of the C_{18} series are even lower. Chain length also affects the melting point, as illustrated by the fact that the melting temperature of palmitic acid (C_{16}) is 6.5 degrees lower than that of stearic acid (C_{18}). Thus, *short chain length and unsaturation enhance the fluidity of fatty acids and of their derivatives.*

12.2 There Are Three Common Types of Membrane Lipids

By definition, *lipids are water-insoluble biomolecules that are highly soluble in organic solvents such as chloroform.* Lipids have a variety of biological roles: they serve as fuel molecules, highly concentrated energy stores, signal molecules and messengers in signal-transduction pathways, and components of membranes. The first three roles of lipids will be considered in later chapters. Here, our focus is on lipids as membrane constituents. The three major kinds of membrane lipids are *phospholipids, glycolipids,* and *cholesterol.* We begin with lipids found in eukaryotes and bacteria. The lipids in archaea are distinct, although they have many features related to membrane formation in common with lipids of other organisms.

Phospholipids Are the Major Class of Membrane Lipids

Phospholipids are abundant in all biological membranes. A phospholipid molecule is constructed from four components: one or more fatty acids, a platform to which the fatty acids are attached, a phosphate, and an alcohol attached to the phosphate (Figure 12.3). The fatty acid components provide a hydrophobic barrier, whereas the remainder of the molecule has hydrophilic properties to enable interaction with the aqueous environment.

The platform on which phospholipids are built may be *glycerol*, a three-carbon alcohol, or *sphingosine*, a more complex alcohol. Phospholipids derived from glycerol are called *phosphoglycerides*. A phosphoglyceride consists of a glycerol backbone to which are attached two fatty acid chains and a phosphorylated alcohol.

In phosphoglycerides, the hydroxyl groups at C-1 and C-2 of glycerol are esterified to the carboxyl groups of the two fatty acid chains. The C-3 hydroxyl group of the glycerol backbone is esterified to phosphoric acid. When no further additions are made, the resulting compound is *phosphatidate (diacylglycerol 3-phosphate)*, the simplest phosphoglyceride. Only small amounts of phosphatidate are present in membranes. However, the molecule is a key intermediate in the biosynthesis of the other phosphoglycerides (Section 26.1). The absolute configuration of the glycerol 3-phosphate moiety of membrane lipids is shown in Figure 12.4.

The major phosphoglycerides are derived from phosphatidate by the formation of an ester bond between the phosphate group of

Figure 12.3 Schematic structure of a phospholipid.

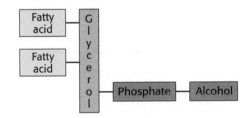

Phosphatidate (Diacylglycerol 3-phosphate)

Figure 12.4 Structure of phosphatidate (diacylglycerol 3-phosphate). The absolute configuration of the center carbon (C-2) is shown.

phosphatidate and the hydroxyl group of one of several alcohols. The common alcohol moieties of phosphoglycerides are the amino acid serine, ethanolamine, choline, glycerol, and inositol.

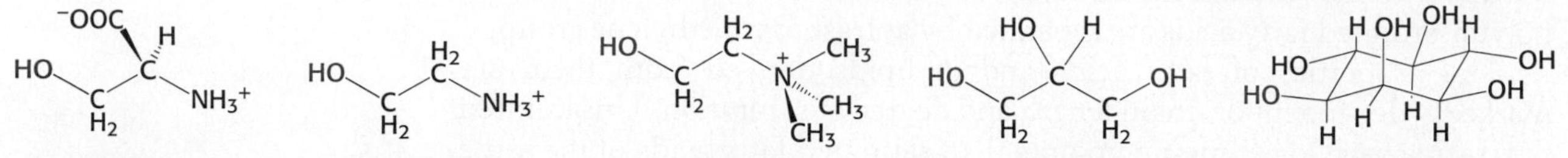

Serine   **Ethanolamine**   **Choline**   **Glycerol**   **Inositol**

The structural formulas of phosphatidylcholine and the other principal phosphoglycerides—namely, phosphatidylethanolamine, phosphatidylserine, phosphatidylinositol, and diphosphatidylglycerol—are given in Figure 12.5.

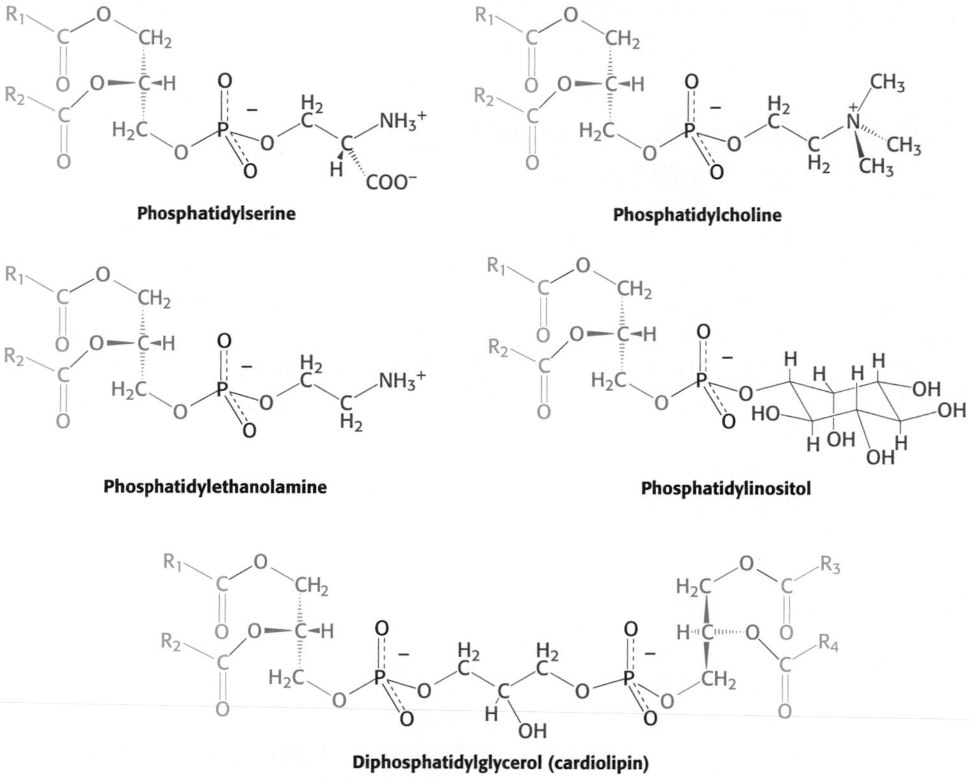

Phosphatidylserine   **Phosphatidylcholine**

Phosphatidylethanolamine   **Phosphatidylinositol**

Diphosphatidylglycerol (cardiolipin)

Figure 12.5 Some common phosphoglycerides found in membranes.

Sphingomyelin is a phospholipid found in membranes that is not derived from glycerol. Instead, the backbone in sphingomyelin is *sphingosine*, an amino alcohol that contains a long, unsaturated hydrocarbon chain (Figure 12.6). In sphingomyelin, the amino group of the sphingosine backbone is

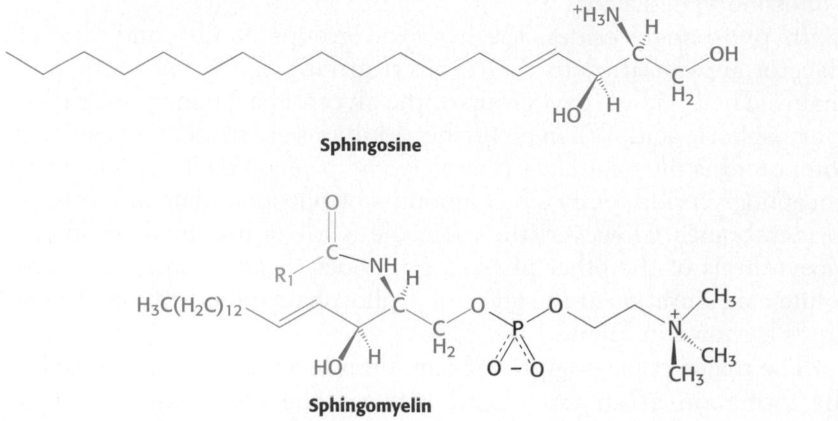

Sphingosine

Sphingomyelin

Figure 12.6 Structures of sphingosine and sphingomyelin. The sphingosine moiety of sphingomyelin is highlighted in blue.

linked to a fatty acid by an amide bond. In addition, the primary hydroxyl group of sphingosine is esterified to phosphorylcholine.

Membrane Lipids Can Include Carbohydrate Moieties

Glycolipids, as their name implies, are *sugar-containing lipids.* Like sphingomyelin, the glycolipids in animal cells are derived from sphingosine. The amino group of the sphingosine backbone is acylated by a fatty acid, as in sphingomyelin. Glycolipids differ from sphingomyelin in the identity of the unit that is linked to the primary hydroxyl group of the sphingosine backbone. In glycolipids, one or more sugars (rather than phosphorylcholine) are attached to this group. The simplest glycolipid, called a *cerebroside,* contains a single sugar residue, either glucose or galactose.

Cerebroside
(a glycolipid)

More-complex glycolipids, such as *gangliosides,* contain a branched chain of as many as seven sugar residues. Glycolipids are oriented in a completely asymmetric fashion with the *sugar residues always on the extracellular side of the membrane.*

Cholesterol Is a Lipid Based on a Steroid Nucleus

Cholesterol is a lipid with a structure quite different from that of phospholipids. It is a steroid, built from four linked hydrocarbon rings.

Cholesterol

A hydrocarbon tail is linked to the steroid at one end, and a hydroxyl group is attached at the other end. In membranes, the orientation of the molecule is parallel to the fatty acid chains of the phospholipids, and the hydroxyl group interacts with the nearby phospholipid head groups. Cholesterol is absent from prokaryotes but is found to varying degrees in virtually all animal membranes. It constitutes almost 25% of the membrane lipids in certain nerve cells but is essentially absent from some intracellular membranes.

Archaeal Membranes Are Built from Ether Lipids with Branched Chains

The membranes of archaea differ in composition from those of eukaryotes or bacteria in three important ways. Two of these differences clearly relate to the hostile living conditions of many archaea (Figure 12.7). First, the nonpolar chains are joined to a glycerol backbone by ether rather

Figure 12.7 An archaeon and its environment. Archaea can thrive in habitats as harsh as a volcanic vent. Here, the archaea form an orange mat surrounded by yellow sulfurous deposits. [Krafft-Explorer/Photo Researchers.]

than ester linkages. The ether linkage is more resistant to hydrolysis. Second, the alkyl chains are branched rather than linear. They are built up from repeats of a fully saturated five-carbon fragment. These branched, saturated hydrocarbons are more resistant to oxidation. The ability of archaeal lipids to resist hydrolysis and oxidation may help these organisms to withstand the extreme conditions, such as high temperature, low pH, or high salt concentration, under which some of these archaea grow. Finally, the stereochemistry of the central glycerol is inverted compared with that shown in Figure 12.4.

Membrane lipid from the archaeon *Methanococcus jannaschii*

A Membrane Lipid Is an Amphipathic Molecule Containing a Hydrophilic and a Hydrophobic Moiety

The repertoire of membrane lipids is extensive. However, these lipids possess a critical common structural theme: *membrane lipids are amphipathic molecules* (amphiphilic molecules). A membrane lipid contains both a *hydrophilic* and a *hydrophobic* moiety.

Let us look at a model of a phosphoglyceride, such as phosphatidylcholine. Its overall shape is roughly rectangular (Figure 12.8A). The two hydrophobic fatty acid chains are approximately parallel to each other, whereas the hydrophilic phosphorylcholine moiety points in the opposite direction. Sphingomyelin has a similar conformation, as does the archaeal lipid depicted. Therefore, the following shorthand has been adopted to represent these membrane lipids: the hydrophilic unit, also called the *polar head group,* is represented by a circle, whereas the hydrocarbon tails are depicted by straight or wavy lines (Figure 12.8B).

(A)

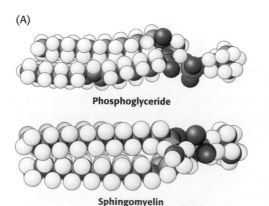

Phosphoglyceride

Sphingomyelin

Archaeal lipid

(B)

Shorthand depiction

Figure 12.8 Representations of membrane lipids. (A) Space-filling models of a phosphoglyceride, sphingomyelin, and an archaeal lipid show their shapes and distribution of hydrophilic and hydrophobic moieties. (B) A shorthand depiction of a membrane lipid.

12.3 Phospholipids and Glycolipids Readily Form Bimolecular Sheets in Aqueous Media

What properties enable phospholipids to form membranes? *Membrane formation is a consequence of the amphipathic nature of the molecules.* Their polar head groups favor contact with water, whereas their hydrocarbon tails interact with one another in preference to water. How can molecules with these preferences arrange themselves in aqueous solutions? One way is to form a globular structure called a *micelle.* The polar head groups form the outside surface of the micelle, which is surrounded by water, and the hydrocarbon tails are sequestered inside, interacting with one another (Figure 12.9).

Alternatively, the strongly opposed preferences of the hydrophilic and hydrophobic moieties of membrane lipids can be satisfied by forming a *lipid bilayer,* composed of two lipid sheets (Figure 12.10). A lipid bilayer is also called a *bimolecular sheet.* The hydrophobic tails of each individual sheet interact with one another, forming a hydrophobic interior that acts as a permeability barrier. The hydrophilic head groups interact with the aqueous medium on each side of the bilayer. The two opposing sheets are called leaflets.

The favored structure for most phospholipids and glycolipids in aqueous media is a bimolecular sheet rather than a micelle. The reason is that the two fatty acid chains of a phospholipid or a glycolipid are too bulky to fit into the interior of a micelle. In contrast, salts of fatty acids (such as sodium palmitate, a constituent of soap) readily form micelles because they contain only one chain. *The formation of bilayers instead of micelles by phospholipids is of critical biological importance.* A micelle is a limited structure, usually less than 200 Å (20 nm) in diameter. In contrast, a bimolecular sheet can extend to macroscopic dimensions, as much as a millimeter (10^7 Å, or 10^6 nm) or more. Phospholipids and related molecules are important membrane constituents because they readily form extensive bimolecular sheets (Figure 12.11).

Lipid bilayers form spontaneously by a *self-assembly process.* In other words, the structure of a bimolecular sheet is inherent in the structure of the constituent lipid molecules. The growth of lipid bilayers from phospholipids is rapid and spontaneous in water. *Hydrophobic interactions are the major driving force for the formation of lipid bilayers.* Recall that hydrophobic interactions also play a dominant role in the stacking of bases in nucleic

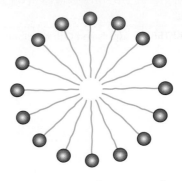

Figure 12.9 Diagram of a section of a micelle. Ionized fatty acids readily form such structures, but most phospholipids do not.

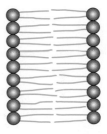

Figure 12.10 Diagram of a section of a bilayer membrane.

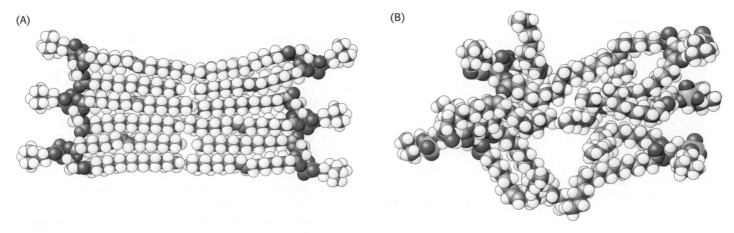

Figure 12.11 Space-filling model of a section of phospholipid bilayer membrane. (A) An idealized view showing regular structures. (B) A more realistic view of a fluid bilayer showing more irregular structures of the fatty acid chains.

acids and in the folding of proteins (pp. 10 and 48). Water molecules are released from the hydrocarbon tails of membrane lipids as these tails become sequestered in the nonpolar interior of the bilayer. Furthermore, *van der Waals attractive forces between the hydrocarbon tails favor close packing of the tails.* Finally, there are *electrostatic and hydrogen-bonding attractions between the polar head groups and water molecules.* Thus, lipid bilayers are stabilized by the full array of forces that mediate molecular interactions in biological systems. Because lipid bilayers are held together by many *reinforcing, noncovalent interactions (predominantly hydrophobic)*, they are *cooperative structures.* These hydrophobic interactions have three significant biological consequences: (1) lipid bilayers have an inherent tendency to be *extensive*; (2) lipid bilayers will tend to *close on themselves* so that there are no edges with exposed hydrocarbon chains, and so they form compartments; and (3) lipid bilayers are *self-sealing* because a hole in a bilayer is energetically unfavorable.

Lipid Vesicles Can Be Formed from Phospholipids

The propensity of phospholipids to form membranes has been used to create an important experimental and clinical tool. *Lipid vesicles,* or *liposomes,* are aqueous compartments enclosed by a lipid bilayer (Figure 12.12). These structures can be used to study membrane permeability or to deliver chemicals to cells. Liposomes are formed by suspending a suitable lipid, such as phosphatidylcholine, in an aqueous medium, and then *sonicating* (i.e., agitating by high-frequency sound waves) to give a dispersion of closed vesicles that are quite uniform in size. Vesicles formed by this method are nearly spherical in shape and have a diameter of about 500 Å (50 nm). Larger vesicles (of the order of 1 μm. or 10^4 Å, in diameter) can be prepared by slowly evaporating the organic solvent from a suspension of phospholipid in a mixed-solvent system.

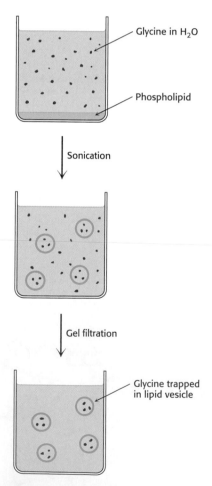

Figure 12.13 **Preparation of glycine-containing liposomes.** Liposomes containing glycine are formed by the sonication of phospholipids in the presence of glycine. Free glycine is removed by gel filtration.

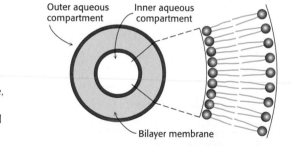

Figure 12.12 **Liposome.** A liposome, or lipid vesicle, is a small aqueous compartment surrounded by a lipid bilayer.

Ions or molecules can be trapped in the aqueous compartments of lipid vesicles by forming the vesicles in the presence of these substances (Figure 12.13). For example, 500-Å-diameter vesicles formed in a 0.1 M glycine solution will trap about 2000 molecules of glycine in each inner aqueous compartment. These glycine-containing vesicles can be separated from the surrounding solution of glycine by dialysis or by gel-filtration chromatography. The permeability of the bilayer membrane to glycine can then be determined by measuring the rate of efflux of glycine from the inner compartment of the vesicle to the ambient solution. Liposomes can be formed with specific membrane proteins embedded in them by solubilizing the proteins in the presence of detergents and then adding them to the phospholipids from which liposomes will be formed. Protein–liposome complexes provide valuable experimental tools for examining a range of membrane-protein functions.

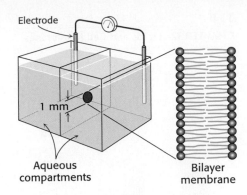

Figure 12.14 Experimental arrangement for the study of a planar bilayer membrane. A bilayer membrane is formed across a 1-mm hole in a septum that separates two aqueous compartments. This arrangement permits measurements of the permeability and electrical conductance of lipid bilayers.

Experiments are underway to develop clinical uses for liposomes. For example, liposomes containing drugs or DNA for gene-therapy experiments can be injected into patients. These liposomes fuse with the plasma membrane of many kinds of cells, introducing into the cells the molecules that they contain. Drug delivery with liposomes often lessens its toxicity. Less of the drug is distributed to normal tissues because long-circulating liposomes concentrate in regions of increased blood circulation, such as solid tumors and sites of inflammation. Moreover, the selective fusion of lipid vesicles with particular kinds of cells is a promising means of controlling the delivery of drugs to target cells.

Another well-defined synthetic membrane is a *planar bilayer membrane*. This structure can be formed across a 1-mm hole in a partition between two aqueous compartments by dipping a fine paintbrush into a membrane-forming solution, such as phosphatidylcholine in decane, and stroking the tip of the brush across the hole. The lipid film across the hole thins spontaneously into a lipid bilayer. The electrical conduction properties of this macroscopic bilayer membrane are readily studied by inserting electrodes into each aqueous compartment (Figure 12.14). For example, its permeability to ions is determined by measuring the current across the membrane as a function of the applied voltage.

Lipid Bilayers Are Highly Impermeable to Ions and Most Polar Molecules

Permeability studies of lipid vesicles and electrical-conductance measurements of planar bilayers have shown that *lipid bilayer membranes have a very low permeability for ions and most polar molecules*. Water is a conspicuous exception to this generalization; it traverses such membranes relatively easily because of its small size, high concentration, and lack of a complete charge. The range of measured permeability coefficients is very wide (Figure 12.15). For example, Na^+ and K^+ traverse these membranes 10^9 times as slowly as does H_2O. Tryptophan, a zwitterion at pH 7, crosses the membrane 10^3 times as slowly as does indole, a structurally related molecule that lacks ionic groups. In fact, *the permeability of small molecules is correlated with their solubility in a nonpolar solvent relative to their solubility in water*. This relation suggests that a small molecule might traverse a lipid bilayer membrane in the following way: first, it sheds its solvation shell of water; then, it becomes dissolved in the hydrocarbon core of the membrane; and, finally, it diffuses through this core to the other side of the membrane, where it becomes resolvated by water. An ion such as Na^+ traverses membranes very slowly because the replacement of its coordination shell of polar water molecules by nonpolar interactions with the membrane interior is highly unfavorable energetically.

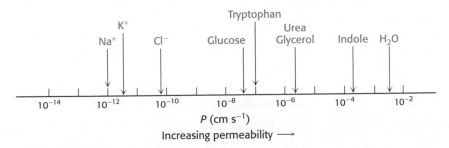

Figure 12.15 Permeability coefficients (P) of ions and molecules in a lipid bilayer. The ability of molecules to cross a lipid bilayer spans a wide range of values.

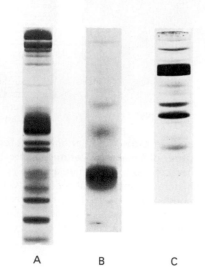

Figure 12.16 SDS–acrylamide gel patterns of membrane proteins. (A) The plasma membrane of erythrocytes. (B) The photoreceptor membranes of retinal rod cells. (C) The sarcoplasmic reticulum membrane of muscle cells. [Courtesy of Dr. Theodore Steck (part A) and Dr. David MacLennan (part C).]

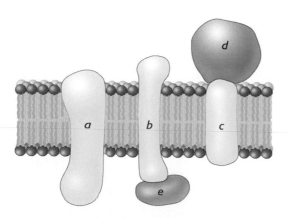

Figure 12.17 Integral and peripheral membrane proteins. Integral membrane proteins (*a*, *b*, and *c*) interact extensively with the hydrocarbon region of the bilayer. Most known integral membrane proteins traverse the lipid bilayer. Peripheral membrane proteins (*d* and *e*) bind to the surfaces of integral proteins. Some peripheral membrane proteins interact with the polar head groups of the lipids (not shown).

12.4 Proteins Carry Out Most Membrane Processes

We now turn to membrane proteins, which are responsible for most of the dynamic processes carried out by membranes. Membrane lipids form a permeability barrier and thereby establish compartments, whereas *specific proteins mediate nearly all other membrane functions*. In particular, proteins transport chemicals and information across a membrane. Membrane lipids create the appropriate environment for the action of such proteins.

Membranes differ in their protein content. Myelin, a membrane that serves as an electrical insulator around certain nerve fibers, has a low content of protein (18%). Relatively pure lipids are well suited for insulation. In contrast, the plasma membranes, or exterior membranes, of most other cells are much more metabolically active. They contain many pumps, channels, receptors, and enzymes. The protein content of these plasma membranes is typically 50%. Energy-transduction membranes, such as the internal membranes of mitochondria and chloroplasts, have the highest content of protein, typically 75%.

The protein components of a membrane can be readily visualized by *SDS–polyacrylamide gel electrophoresis*. As stated earlier (p. 71), the electrophoretic mobility of many proteins in SDS-containing gels depends on the mass rather than on the net charge of the protein. The gel-electrophoresis patterns of three membranes—the plasma membrane of erythrocytes, the photoreceptor membrane of retinal rod cells, and the sarcoplasmic reticulum membrane of muscle—are shown in Figure 12.16. It is evident that each of these three membranes contains many proteins but has a distinct protein composition. In general, *membranes performing different functions contain different repertoires of proteins*.

Proteins Associate with the Lipid Bilayer in a Variety of Ways

The ease with which a protein can be dissociated from a membrane indicates how intimately it is associated with the membrane. Some membrane proteins can be solubilized by relatively mild means, such as extraction by a solution of high ionic strength (e.g., 1 M NaCl). Other membrane proteins are bound much more tenaciously; they can be solubilized only by using a detergent or an organic solvent. Membrane proteins can be classified as being either *peripheral* or *integral* on the basis of this difference in dissociability (Figure 12.17). *Integral membrane proteins* interact extensively with the hydrocarbon chains of membrane lipids, and they can be released only by agents that compete for these nonpolar interactions. In fact, most integral membrane proteins span the lipid bilayer. In contrast, *peripheral membrane proteins* are bound to membranes primarily by electrostatic and hydrogen-bond interactions with the head groups of lipids. These polar interactions can be disrupted by adding salts or by changing the pH. Many peripheral membrane proteins are bound to the surfaces of integral proteins, on either the cytoplasmic or the extracellular side of the membrane. Others are anchored to the lipid bilayer by a covalently attached hydrophobic chain, such as a fatty acid.

Proteins Interact with Membranes in a Variety of Ways

Membrane proteins are more difficult to purify and crystallize than are water-soluble proteins. Nonetheless, researchers using x-ray crystallographic or electron microscopic methods have determined the three-dimensional structures of more than 100 such proteins at sufficiently high resolution to discern

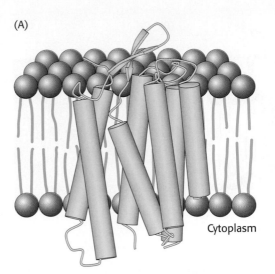

(A)

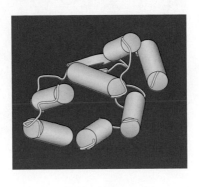

(B)

Cytoplasm

Figure 12.18 **Structure of bacteriorhodopsin.** *Notice* that bacteriorhodopsin consists largely of membrane-spanning α helices (represented by yellow cylinders). (A) View through the membrane bilayer. The interior of the membrane is green and the head groups are red. (B) View from the cytoplasmic side of the membrane. [Drawn from 1 BRX.pdb.]

the molecular details. As noted in Chapter 2, membrane proteins differ from soluble proteins in the distribution of hydrophobic and hydrophilic groups. We will consider the structures of three membrane proteins in some detail.

Proteins Can Span the Membrane with Alpha Helices.   The first membrane protein that we consider is the archaeal protein *bacteriorhodopsin*, shown in Figure 12.18. This protein uses light energy to transport protons from inside the cell to outside, generating a proton gradient used to form ATP. Bacteriorhodopsin is built almost entirely of α helices; seven closely packed α helices, arranged almost perpendicularly to the plane of the cell membrane, span its 45-Å width. Examination of the primary structure of bacteriorhodopsin reveals that most of the amino acids in these membrane-spanning α helices are nonpolar and only a very few are charged (Figure 12.19). This distribution of nonpolar amino acids is sensible because these residues are either in contact with the hydrocarbon core of the membrane or with one another. *Membrane-spanning α helices are the most common structural motif in membrane proteins.* As will be considered in Section 12.5, such regions can often be detected by examining amino acid sequence alone.

Figure 12.19 **Amino acid sequence of bacteriorhodopsin.** The seven helical regions are highlighted in yellow and the charged residues in red.

A Channel Protein Can Be Formed from Beta Strands.   Porin, a protein from the outer membranes of bacteria such as *E. coli* and *Rhodobacter capsulatus*, represents a class of membrane proteins with a completely different type of structure. Structures of this type are built from β strands and contain essentially no α helices (Figure 12.20).

   The arrangement of β strands is quite simple: each strand is hydrogen bonded to its neighbor in an antiparallel arrangement, forming a single β sheet. The β sheet curls up to form a hollow cylinder that, as its name suggests, forms a pore, or channel, in the membrane. The outside surface of porin is appropriately nonpolar, given that it interacts with the hydrocarbon core of the membrane. In contrast, the inside of the channel is quite hydrophilic and

Figure 12.20 **Structure of bacterial porin (from *Rhodopseudomonas blastica*).** *Notice* that this membrane protein is built entirely of beta strands. (A) Side view. (B) View from the periplasmic space. Only one monomer of the trimeric protein is shown. [Drawn from 1 PRN.pdb.]

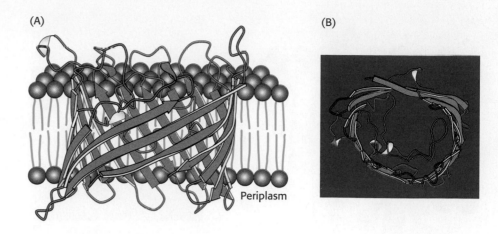

(A)

(B)

Periplasm

is filled with water. This arrangement of nonpolar and polar surfaces is accomplished by the alternation of hydrophobic and hydrophilic amino acids along each β strand (Figure 12.21).

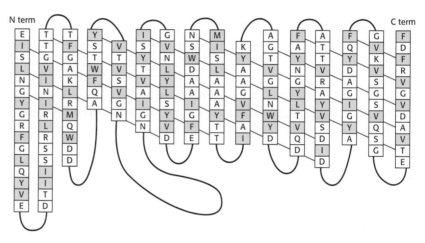

Figure 12.21 **Amino acid sequence of a porin.** Some membrane proteins such as porins are built from β strands that tend to have hydrophobic and hydrophilic amino acids in adjacent positions. The secondary structure of porin from *Rhodopseudomonas blastica* is shown, with the diagonal lines indicating the direction of hydrogen bonding along the β sheet. Hydrophobic residues (F, I, L, M, V, W, and Y) are shown in yellow. These residues tend to lie on the outside of the structure, in contact with the hydrophobic core of the membrane.

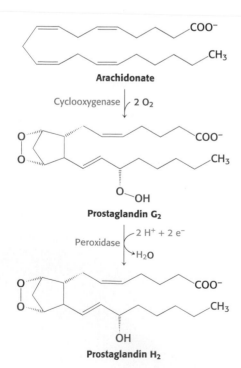

Arachidonate

Cyclooxygenase / 2 O$_2$

Prostaglandin G$_2$

Peroxidase / 2 H$^+$ + 2 e$^-$ → H$_2$O

Prostaglandin H$_2$

Embedding Part of a Protein in a Membrane Can Link the Protein to the Membrane Surface.

The structure of the endoplasmic reticulum membrane-bound enzyme prostaglandin H$_2$ synthase-1 reveals a rather different role for α helices in protein–membrane associations. This enzyme catalyzes the conversion of arachidonic acid into prostaglandin H$_2$ in two steps: (1) a cyclooxygenase reaction and (2) a peroxidase reaction (Figure 12.22). Prostaglandin H$_2$ promotes inflammation and modulates gastric acid secretion. The enzyme that produces prostaglandin H$_2$ is a homodimer with a rather complicated structure consisting primarily of α helices. Unlike bacteriorhodopsin, this protein is not largely embedded in the membrane. Instead, it lies along the outer surface of the membrane firmly bound by a set of α helices with hydrophobic surfaces that extend from the bottom of

Figure 12.22 **Formation of prostaglandin H$_2$.** Prostaglandin H$_2$ synthase-1 catalyzes the formation of prostaglandin H$_2$ from arachidonic acid in two steps.

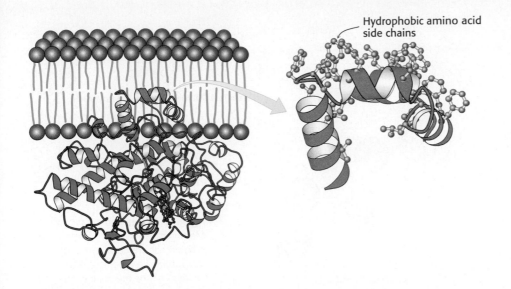

Hydrophobic amino acid side chains

Figure 12.23 Attachment of prostaglandin H$_2$ synthase-1 to the membrane. *Notice* that prostaglandin H$_2$ synthase-1 is held in the membrane by a set of α helices (orange) coated with hydrophobic side chains. One monomer of the dimeric enzyme is shown. [Drawn from 1 PTH.pdb.]

the protein into the membrane (Figure 12.23). This linkage is sufficiently strong that only the action of detergents can release the protein from the membrane. Thus, this enzyme is classified as an integral membrane protein, although it is not a membrane-spanning protein.

The localization of prostaglandin H$_2$ synthase-1 in the membrane is crucial to its function. The substrate for this enzyme, arachidonic acid, is a hydrophobic molecule generated by the hydrolysis of membrane lipids. Arachidonic acid reaches the active site of the enzyme from the membrane without entering an aqueous environment by traveling through a hydrophobic channel in the protein (Figure 12.24). Indeed, nearly all of us have experienced the importance of this channel: drugs such as aspirin and ibuprofen block the channel and prevent prostaglandin synthesis by inhibiting the cyclooxygenase activity of the synthase. In reference to aspirin, the drug acts through the transfer of an acetyl group from the aspirin to a serine residue (Ser 530) that lies along the path to the active site (Figure 12.25).

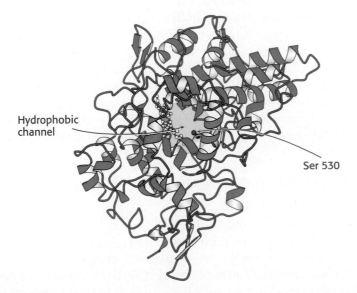

Hydrophobic channel

Ser 530

Figure 12.24 Hydrophobic channel of prostaglandin H$_2$ synthase-1. A view of prostaglandin H$_2$ synthase-1 from the membrane shows the hydrophobic channel that leads to the active site. The membrane-anchoring helices are shown in orange. [Drawn from 1PTH.pdb.]

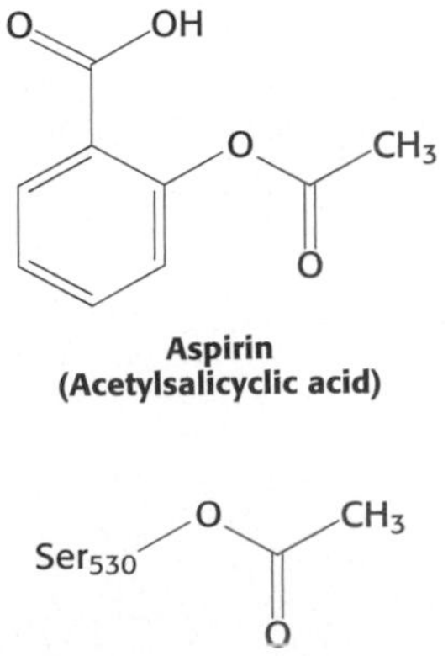

**Aspirin
(Acetylsalicyclic acid)**

Figure 12.25 Aspirin's effects on prostaglandin H$_2$ synthase-1. Aspirin acts by transferring an acetyl group to a serine residue in prostaglandin H$_2$ synthase-1.

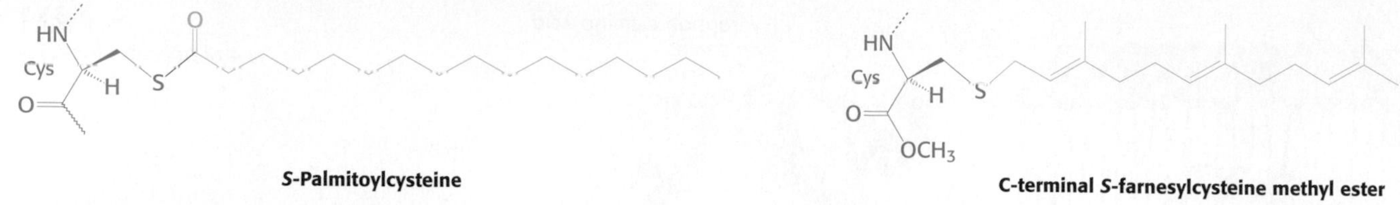

S-Palmitoylcysteine

C-terminal S-farnesylcysteine methyl ester

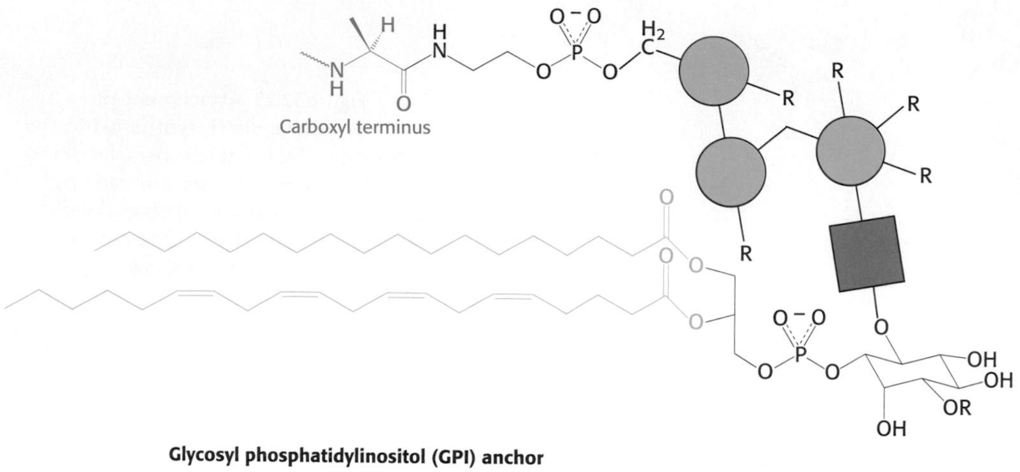

Carboxyl terminus

Figure 12.26 Membrane anchors.
Membrane anchors are hydrophobic groups that are covalently attached to proteins (in blue) and tether the proteins to the membrane. The green circles and blue square correspond to mannose and GlcNAc, respectively. R groups represent points of additional modification.

Glycosyl phosphatidylinositol (GPI) anchor

Two important features emerge from our examination of these three examples of membrane-protein structure. First, the parts of the protein that interact with the hydrophobic parts of the membrane are coated with nonpolar amino acid side chains, whereas those parts that interact with the aqueous environment are much more hydrophilic. Second, the structures positioned within the membrane are quite regular and, in particular, all backbone hydrogen-bond donors and acceptors participate in hydrogen bonds. *Breaking a hydrogen bond within a membrane is quite unfavorable, because little or no water is present to compete for the polar groups.*

Some Proteins Associate with Membranes Through Covalently Attached Hydrophobic Groups

The membrane proteins considered thus far associate with the membrane through surfaces generated by hydrophobic amino acid side chains. However, even otherwise soluble proteins can associate with membranes if hydrophobic groups are attached to the proteins. Three such groups are shown in Figure 12.26: (1) a palmitoyl group attached to a specific cysteine residue by a thioester bond, (2) a farnesyl group attached to a cysteine residue at the carboxyl terminus, and (3) a glycolipid structure termed a glycosylphosphatidylinositol (GPI) anchor attached to the carboxyl terminus. These modifications are attached by enzyme systems that recognize specific signal sequences near the site of attachment.

Transmembrane Helices Can Be Accurately Predicted from Amino Acid Sequences

Many membrane proteins, like bacteriorhodopsin, employ α helices to span the hydrophobic part of a membrane. As noted earlier, typically most of the residues in these α helices are nonpolar and almost none of them are charged. Can we use this information to identify likely membrane-spanning regions from sequence data alone? One approach to identifying transmembrane helices is to ask whether a postulated helical segment is likely to be most stable in a hydrocarbon environment or in water. Specifically, we want to estimate the free-energy change when a helical segment is transferred from the interior of a membrane to water. Free-energy changes for the transfer of individual amino acid residues from a hydrophobic to an aqueous environment are given in Table 12.2. For example, the transfer of a helix formed entirely of L-arginine residues, a positively charged amino acid, from the interior of a membrane to water would be highly favorable [-51.5 kJ mol^{-1} (-12.3 kcal mol^{-1}) per arginine residue in the helix]. In contrast, the transfer of a helix formed entirely of L-phenylalanine, a hydrophobic amino acid, would be unfavorable [$+15.5$ kJ mol^{-1} ($+3.7$ kcal mol^{-1}) per phenylalanine residue in the helix].

The hydrocarbon core of a membrane is typically 30 Å wide, a length that can be traversed by an α helix consisting of 20 residues. We can take the amino acid sequence of a protein and estimate the free-energy change that takes place when a hypothetical α helix formed of residues 1 through 20 is transferred from the membrane interior to water. The same calculation can be made for residues 2 through 21, 3 through 22, and so forth, until we reach the end of the sequence. The span of 20 residues chosen for this calculation is called a *window*. The free-energy change for each window is plotted against the first amino acid at the window to create a *hydropathy plot*. Empirically, a peak of $+84$ kJ mol^{-1} ($+20$ kcal mol^{-1}) or more in a hydropathy plot based on a window of 20 residues indicates that a polypeptide segment could be a membrane-spanning α helix. For example, glycophorin, a protein found in the membranes of red blood cells, is predicted by this criterion to have one membrane-spanning helix, in agreement with experimental findings (Figure 12.27). Note, however, that a peak in the hydropathy plot does not prove that a segment is a transmembrane helix. Even soluble proteins may have highly nonpolar regions. Conversely, some membrane proteins contain membrane-spanning features (such as a set of cylinder-forming β strands) that escape detection by these plots (Figure 12.28).

TABLE 12.2 Polarity scale for identifying transmembrane helices

Amino acid residue	Transfer free energy in kJ mol^{-1} (kcal mol^{-1})
Phe	15.5 (3.7)
Met	14.3 (3.4)
Ile	13.0 (3.1)
Leu	11.8 (2.8)
Val	10.9 (2.6)
Cys	8.4 (2.0)
Trp	8.0 (1.9)
Ala	6.7 (1.6)
Thr	5.0 (1.2)
Gly	4.2 (1.0)
Ser	2.5 (0.6)
Pro	−0.8 (−0.2)
Tyr	−2.9 (−0.7)
His	−12.6 (−3.0)
Gln	−17.2 (−4.1)
Asn	−20.2 (−4.8)
Glu	−34.4 (−8.2)
Lys	−37.0 (−8.8)
Asp	−38.6 (−9.2)
Arg	−51.7 (−12.3)

Source: After D. M. Engelman, T. A. Steitz, and A. Goldman. *Annu. Rev. Biophys. Biophys. Chem.* 15(1986):321–353.

Note: The free energies are for the transfer of an amino acid residue in an α helix from the membrane interior (assumed to have a dielectric constant of 2) to water.

(A)

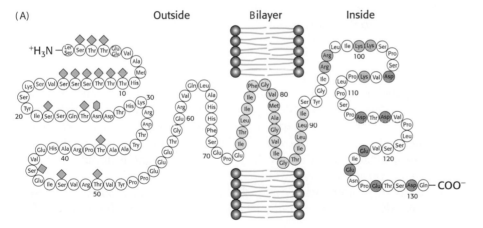

(B)

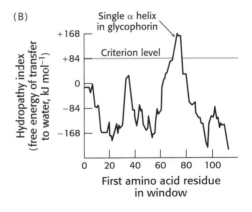

Figure 12.27 Locating the membrane-spanning helix of glycophorin. (A) Amino acid sequence and transmembrane disposition of glycophorin A from the red-blood-cell membrane. Fifteen O-linked carbohydrate units are shown as diamond shapes, and an N-linked unit is shown as a lozenge shape. The hydrophobic residues (yellow) buried in the bilayer form a transmembrane α helix. The carboxyl-terminal part of the molecule, located on the cytoplasmic side of the membrane, is rich in negatively charged (red) and positively charged (blue) residues. (B) Hydropathy plot for glycophorin. The free energy for transferring a helix of 20 residues from the membrane to water is plotted as a function of the position of the first residue of the helix in the sequence of the protein. Peaks of greater than $+84$ kJ mol^{-1} ($+20$ kcal mol^{-1}) in hydropathy plots are indicative of potential transmembrane helices. [(A) Courtesy of Dr. Vincent Marchesi; (B) after D. M. Engelman, T. A. Steitz, and A. Goldman. Identifying nonpolar transbilayer helices in amino acid sequences of membrane proteins. *Annu. Rev. Biophys. Biophys. Chem.* 15(1986):321–353. Copyright © 1986 by Annual Reviews, Inc. All rights reserved.]

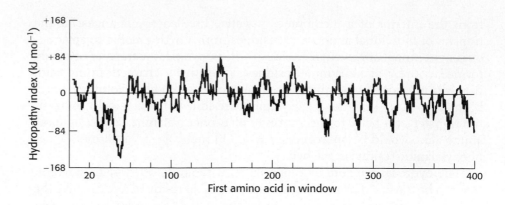

Figure 12.28 Hydropathy plot for porin. No strong peaks are observed for this intrinsic membrane protein, because it is constructed from membrane-spanning β strands rather than α helices.

12.5 Lipids and Many Membrane Proteins Diffuse Rapidly in the Plane of the Membrane

Biological membranes are not rigid, static structures. On the contrary, lipids and many membrane proteins are constantly in lateral motion, a process called *lateral diffusion*. The rapid lateral movement of membrane proteins has been visualized by means of fluorescence microscopy using the technique of *fluorescence recovery after photobleaching* (FRAP; Figure 12.29). First, a cell-surface component is specifically labeled with a fluorescent chromophore. A small region of the cell surface (~3 μm^2) is viewed through a fluorescence microscope. The fluorescent molecules in this region are then destroyed (bleached) by a very intense light pulse from a laser. The fluorescence of this region is subsequently monitored as a function of time by using a light level sufficiently low to prevent further bleaching. If the labeled component is mobile, bleached molecules leave and unbleached molecules enter the illuminated region, which results in an increase in the fluorescence intensity. The rate of recovery of fluorescence depends on the lateral mobility of the fluorescence-labeled component, which can be expressed in terms of a diffusion coefficient, D. The average distance S traversed in time t depends on D according to the expression

$$S = (4Dt)^{1/2}$$

The diffusion coefficient of lipids in a variety of membranes is about 1 μm^2 s^{-1}. Thus, a phospholipid molecule diffuses an average distance of 2 μm in 1 s. This rate means that *a lipid molecule can travel from one end of a bacterium to the other in a second.* The magnitude of the observed diffusion coefficient indicates that the viscosity of the membrane is about 100 times that of water, rather like that of olive oil.

In contrast, proteins vary markedly in their lateral mobility. *Some proteins are nearly as mobile as lipids, whereas others are virtually immobile.* For

Figure 12.29 Fluorescence recovery after photobleaching (FRAP) technique. (A) The cell surface fluoresces because of a labeled surface component. (B) The fluorescent molecules of a small part of the surface are bleached by an intense light pulse. (C) The fluorescence intensity recovers as bleached molecules diffuse out of the region and unbleached molecules diffuse into it. (D) The rate of recovery depends on the diffusion coefficient.

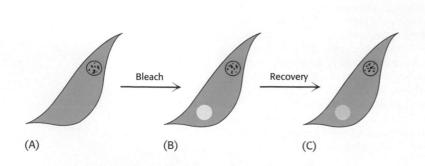

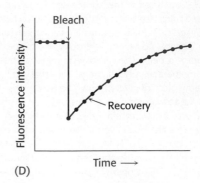

example, the photoreceptor protein rhodopsin (Section 32.3), a very mobile protein, has a diffusion coefficient of 0.4 $\mu m^2 s^{-1}$. The rapid movement of rhodopsin is essential for fast signaling. At the other extreme is fibronectin, a peripheral glycoprotein that interacts with the extracellular matrix. For fibronectin, D is less than $10^{-4} \mu m^2 s^{-1}$. Fibronectin has a very low mobility because it is anchored to actin filaments on the inside of the plasma membrane through *integrin*, a transmembrane protein that links the extracellular matrix to the cytoskeleton.

The Fluid Mosaic Model Allows Lateral Movement but Not Rotation Through the Membrane

On the basis of the mobility of proteins in membranes, S. Jonathan Singer and Garth Nicolson proposed a *fluid mosaic model* to describe the overall organization of biological membranes in 1972 (Figure 12.30). The essence of their model is that *membranes are two-dimensional solutions of oriented lipids and globular proteins*. The lipid bilayer has a dual role: it is both a *solvent* for integral membrane proteins and a *permeability barrier*. Membrane proteins are free to diffuse laterally in the lipid matrix unless restricted by special interactions.

Although the lateral diffusion of membrane components can be rapid, the spontaneous rotation of lipids from one face of a membrane to the other is a very slow process. The transition of a molecule from one membrane surface to the other is called *transverse diffusion* or *flip-flop* (Figure 12.31) The flip-flop of phospholipid molecules in phosphatidylcholine vesicles has been directly measured by electron spin resonance techniques, which show that *a phospholipid molecule flip-flops once in several hours*. Thus, a phospholipid molecule takes about 10^9 times as long to flip-flop across a membrane as it takes to diffuse a distance of 50 Å in the lateral direction. The free-energy barriers to flip-flopping are even larger for protein molecules than for lipids because proteins have more-extensive polar regions. In fact, the flip-flop of a protein molecule has not been observed. Hence, *membrane asymmetry can be preserved for long periods*.

Membrane Fluidity Is Controlled by Fatty Acid Composition and Cholesterol Content

Many membrane processes, such as transport or signal transduction, depend on the fluidity of the membrane lipids, which in turn depends on the properties of fatty acid chains. Fatty acid chains in membrane bilayers can exist in an ordered, rigid state or in a relatively disordered, fluid state. The transition from the rigid to the fluid state takes place rather abruptly as the temperature is raised above T_m, the melting temperature (Figure 12.32).

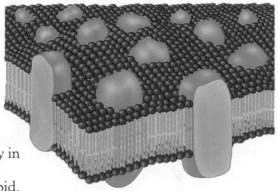

Figure 12.30 Fluid mosaic model. [After S. J. Singer and G. L. Nicolson. *Science* 175(1972):720–731.]

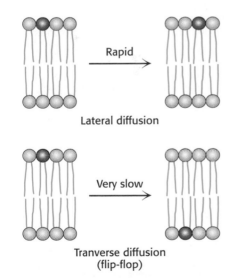

Lateral diffusion

Rapid

Tranverse diffusion
(flip-flop)

Very slow

Figure 12.31 Lipid movement in membranes. Lateral diffusion of lipids is much more rapid than transverse diffusion (flip-flop).

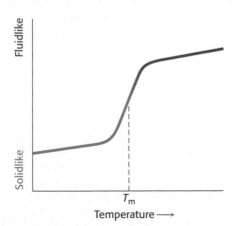

Figure 12.32 The phase-transition, or melting, temperature (T_m) for a phospholipid membrane. As the temperature is raised, the phospholipid membrane changes from a packed, ordered state to a more random one.

TABLE 12.3 The melting temperature of phosphatidylcholine containing different pairs of identical fatty acid chains

Number of carbons	Number of double bonds	FATTY ACID		
		Common name	Systematic name	T_m (°C)
22	0	Behenate	n-Docosanote	75
18	0	Stearate	n-Octadecanoate	58
16	0	Palmitate	n-Hexadecanoate	41
14	0	Myristate	n-Tetradecanoate	24
18	1	Oleate	cis-Δ^9-Octadecenoate	−22

This transition temperature depends on the length of the fatty acid chains and on their degree of unsaturation (Table 12.3). The presence of saturated fatty acid residues favors the rigid state because their straight hydrocarbon chains interact very favorably with one another. On the other hand, *a cis double bond produces a bend in the hydrocarbon chain. This bend interferes with a highly ordered packing of fatty acid chains, and so* T_m *is lowered* (Figure 12.33). The length of the fatty acid chain also affects the transition temperature. Long hydrocarbon chains interact more strongly than do short ones. Specifically, each additional —CH$_2$— group makes a favorable contribution of about −2 kJ mol^{-1} (−0.5 kcal mol^{-1}) to the free energy of interaction of two adjacent hydrocarbon chains.

Figure 12.33 Packing of fatty acid chains in a membrane. The highly ordered packing of fatty acid chains is disrupted by the presence of cis double bonds. The space-filling models show the packing of (A) three molecules of stearate (C$_{18}$, saturated) and (B) a molecule of oleate (C$_{18}$, unsaturated) between two molecules of stearate.

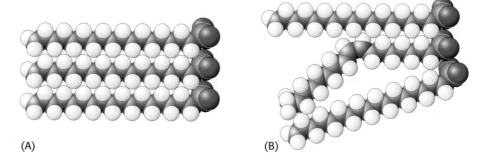

(A)

(B)

Bacteria regulate the fluidity of their membranes by varying the number of double bonds and the length of their fatty acid chains. For example, the ratio of saturated to unsaturated fatty acid chains in the *E. coli* membrane decreases from 1.6 to 1.0 as the growth temperature is lowered from 42°C to 27°C. This decrease in the proportion of saturated residues prevents the membrane from becoming too rigid at the lower temperature.

In animals, cholesterol is the key regulator of membrane fluidity. Cholesterol contains a bulky steroid nucleus with a hydroxyl group at one end and a flexible hydrocarbon tail at the other end. Cholesterol inserts into bilayers with its long axis perpendicular to the plane of the membrane. The hydroxyl group of cholesterol forms a hydrogen bond with a carbonyl oxygen atom of a phospholipid head group, whereas the hydrocarbon tail of cholesterol is located in the nonpolar core of the bilayer. The different shape of cholesterol compared with that of phospholipids disrupts the regular interactions between fatty acid chains. In addition, cholesterol appears to form specific complexes with some phospholipids. Such complexes may concentrate in specific regions within membranes. The resulting structures are often referred to as *lipid rafts.* One result of these interactions is the *moderation of membrane fluidity,* making membranes less fluid but at the same time less subject to phase transitions. Self-organizing structures such as lipid rafts may also play a role in concentrating proteins that participate in signal-transduction pathways, although this possibility is still controversial.

All Biological Membranes Are Asymmetric

Membranes are structurally and functionally asymmetric. The outer and inner surfaces of *all known biological membranes have different components and different enzymatic activities*. A clear-cut example is the pump that regulates the concentration of Na^+ and K^+ ions in cells (Figure 12.34). This transport protein is located in the plasma membrane of nearly all cells in higher organisms. The $Na^+–K^+$ pump is oriented so that it pumps Na^+ out of the cell and K^+ into it. Furthermore, ATP must be on the inside of the cell to drive the pump. Ouabain, a specific inhibitor of the pump, is effective only if it is located outside. We shall consider the mechanism of this important and fascinating pump and its family members in Chapter 13.

Membrane proteins have a unique orientation because, after synthesis, they are inserted into the membrane in an asymmetric manner. This absolute asymmetry is preserved because membrane proteins do not rotate from one side of the membrane to the other and because *membranes are always synthesized by the growth of preexisting membranes*. Lipids, too, are asymmetrically distributed as a consequence of their mode of biosynthesis, but this asymmetry is usually not absolute, except for glycolipids. In the red-blood-cell membrane, sphingomyelin and phosphatidylcholine are preferentially located in the outer leaflet of the bilayer, whereas phosphatidylethanolamine and phosphatidylserine are located mainly in the inner leaflet. Large amounts of cholesterol are present in both leaflets.

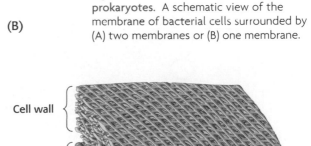

Figure 12.34 Asymmetry of the $Na^+–K^+$ transport system in plasma membranes. The $Na^+–K^+$ transport system pumps Na^+ out of the cell and K^+ into the cell.

12.6 Eukaryotic Cells Contain Compartments Bounded by Internal Membranes

Thus far, we have considered only the plasma membrane of cells. Some bacteria and archaea have only this single membrane, surrounded by a cell wall. Other bacteria such as *E. coli* have two membranes separated by a cell wall (made of proteins, peptides, and carbohydrates) lying between them (Figure 12.35). The inner membrane acts as the permeability barrier, and the outer

Figure 12.35 Cell membranes of prokaryotes. A schematic view of the membrane of bacterial cells surrounded by (A) two membranes or (B) one membrane.

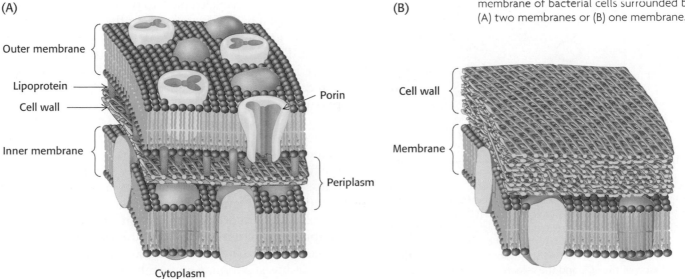

(A)

Outer membrane

Lipoprotein

Cell wall

Inner membrane

Porin

Periplasm

Cytoplasm

(B)

Cell wall

Membrane

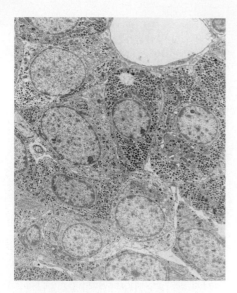

Figure 12.36 Internal membranes of eukaryotes. Electron micrograph of a thin section of a hormone-secreting cell for the rat pituitary, showing the presence of internal structures bounded by membranes. [Biophoto Associates/Photo Researchers.]

membrane and the cell wall provide additional protection. The outer membrane is quite permeable to small molecules, owing to the presence of porins. The region between the two membranes containing the cell wall is called the *periplasm*.

Eukaryotic cells, with the exception of plant cells, do not have cell walls, and their cell membranes consist of a single lipid bilayer. In plant cells, the cell wall is on the outside of the plasma membrane. Eukaryotic cells are distinguished from prokaryotic cells by the presence of membranes inside the cell that form internal compartments (Figure 12.36). For example, peroxisomes, organelles that play a major role in the oxidation of fatty acids for energy conversion, are defined by a single membrane. Mitochondria, the organelles in which ATP is synthesized, are surrounded by two membranes. Much like the case for a bacterium, the outer membrane is quite permeable to small molecules, whereas the inner membrane is not. Indeed, considerable evidence now indicates that mitochondria evolved from bacteria by *endosymbiosis* (p. 504). A double membrane also surrounds the nucleus. However, the *nuclear envelope* is not continuous but, instead, consists of a set of closed membranes that come together at structures called *nuclear pores*. These pores regulate transport into and out of the nucleus. The nuclear membranes are linked to another membrane-defined structure, the *endoplasmic reticulum*, which plays a host of cellular roles, including drug detoxification and the modification of proteins for secretion. Thus, a eukaryotic cell contains interacting compartments, and transport into and out of these compartments is essential to many biochemical processes.

Membranes must be able to separate or join together so that cells and compartments may take up, transport, and release molecules. Many cells take up molecules through the process of *receptor-mediated endocytosis* (Figure 12.37). Here, a protein or larger complex initially binds to a receptor on the cell surface. After the protein is bound, specialized proteins act to cause the membrane in the vicinity of the bound protein to invaginate. The invaginated membrane eventually breaks off and fuses to form a *vesicle*.

Receptor-mediated endocytosis plays a key role in cholesterol metabolism (p. 745). Some cholesterol in the blood is in the form of a lipid–protein complex called *low-density lipoprotein* (LDL). Low-density lipoprotein binds to an LDL receptor, an integral membrane protein. The segment of the plasma membrane containing the LDL–LDL-receptor complex then invaginates and buds off from the membrane. The LDL separates from the receptor, which is recycled back to the membrane in a separate vesicle. The vesicle containing the LDL fuses with a lysosome, an organelle containing an array of digestive enzymes. The cholesterol is released into the cell for storage or use in membrane biosynthesis, and the remaining protein components are degraded. Various hormones, transport proteins, and antibodies employ receptor-mediated endocytosis to gain entry into a cell. A less-advantageous consequence is that this pathway is available to viruses and toxins as a means of entry into cells. The reverse process—the fusion of a vesicle to a membrane—is a key step in the release of neurotransmitters from a neuron into the synaptic cleft (Figure 12.38).

Although budding and fusion appear deceptively simple, the structures of the intermediates in these processes and the detailed mechanisms remain on-going areas of investigation. Key membrane components called

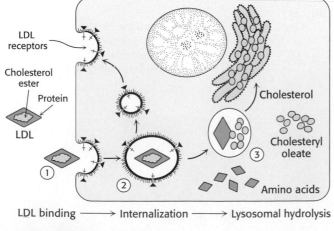

Figure 12.37 Receptor-mediated endocytosis. The process of receptor-mediated endocytosis is illustrated for the cholesterol-carrying complex, low-density lipoprotein (LDL): (1) LDL binds to a specific receptor, the LDL receptor; (2) this complex invaginates to form an internal vesicle; (3) after separation from its receptor, the LDL-containing vesicle fuses with a lysosome, leading to the degradation of the LDL and the release of the cholesterol.

SNARE (soluble *N*-ethylmaleimide-sensitive-factor attachment protein receptor) *proteins* help draw appropriate membranes together to initiate the fusion process. These proteins, encoded by gene families in all eukaryotic cells, largely determine the compartment with which a vesicle will fuse. The specificity of membrane fusion ensures the orderly trafficking of membrane vesicles and their cargos through eukaryotic cells.

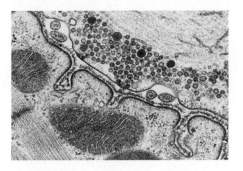

Figure 12.38 Neurotransmitter release. Neurotransmitter-containing synaptic vesicles are arrayed near the plasma membrane of a nerve cell. Synaptic vesicles fuse with the plasma membrane, releasing the neurotransmitter into the synaptic cleft. [T. Reese/Don Fawcett/ Photo Researchers.]

Summary

Biological membranes are sheetlike structures, typically from 60 to 100 Å thick, that are composed of protein and lipid molecules held together by noncovalent interactions. Membranes are highly selective permeability barriers. They create closed compartments, which may be entire cells or organelles within a cell. Proteins in membranes regulate the molecular and ionic compositions of these compartments. Membranes also control the flow of information between cells.

12.1 Fatty Acids Are Key Constituents of Lipids

Fatty acids are hydrocarbon chains of various lengths and degrees of unsaturation that terminate with a carboxylic acid group. The fatty acid chains in membranes usually contain between 14 and 24 carbon atoms; they may be saturated or unsaturated. Short chain length and unsaturation enhance the fluidity of fatty acids and their derivatives by lowering the melting temperature.

12.2 There Are Three Common Types of Membrane Lipids

The major classes of membrane lipids are phospholipids, glycolipids, and cholesterol. Phosphoglycerides, a type of phospholipid, consist of a glycerol backbone, two fatty acid chains, and a phosphorylated alcohol. Phosphatidylcholine, phosphatidylserine, and phosphatidylethanolamine are major phosphoglycerides. Sphingomyelin, a different type of phospholipid, contains a sphingosine backbone instead of glycerol. Glycolipids are sugar-containing lipids derived from sphingosine. Cholesterol, which modulates membrane fluidity, is constructed from a steroid nucleus. A common feature of these membrane lipids is that they are amphipathic molecules, having one hydrophobic and one hydrophilic end.

12.3 Phospholipids and Glycolipids Readily Form Bimolecular Sheets in Aqueous Media

Membrane lipids spontaneously form extensive bimolecular sheets in aqueous solutions. The driving force for membrane formation is the hydrophobic interactions among the fatty acid tails of membrane lipids. The hydrophilic head groups interact with the aqueous medium. Lipid bilayers are cooperative structures, held together by many weak bonds. These lipid bilayers are highly impermeable to ions and most polar molecules, yet they are quite fluid, which enables them to act as a solvent for membrane proteins.

12.4 Proteins Carry Out Most Membrane Processes

Specific proteins mediate distinctive membrane functions such as transport, communication, and energy transduction. Many integral membrane proteins span the lipid bilayer, whereas others are only partly embedded in the membrane. Peripheral membrane proteins are bound to membrane surfaces by electrostatic and hydrogen-bond interactions. Membrane-spanning proteins have regular structures, including β strands, although the α helix is the most common membrane-spanning structure. Sequences of 20 consecutive nonpolar amino acids can be diagnostic of a membrane-spanning α-helical region of a protein.

12.5 Lipids and Many Membrane Proteins Diffuse Rapidly in the Plane of the Membrane

Membranes are structurally and functionally asymmetric, as exemplified by the restriction of sugar residues to the external surface of mammalian plasma membranes. Membranes are dynamic structures in which proteins and lipids diffuse rapidly in the plane of the membrane (lateral diffusion), unless restricted by special interactions. In contrast, the rotation of lipids from one face of a membrane to the other (transverse diffusion, or flip-flop) is usually very slow. Proteins do not rotate across bilayers; hence, membrane asymmetry can be preserved. The degree of fluidity of a membrane partly depends on the chain length of its lipids and the extent to which their constituent fatty acids are unsaturated. In animals, cholesterol content also regulates membrane fluidity.

12.6 Eukaryotic Cells Contain Compartments Bounded by Internal Membranes

An extensive array of internal membranes in eukaryotes creates compartments within a cell for distinct biochemical functions. For instance, a double membrane surrounds the nucleus, the location of most of the cell's genetic material, and the mitochondria, the location of most ATP synthesis. A single membrane defines the other internal compartments, such as the endoplasmic reticulum. Some compartments can exchange material by the process of membrane budding and fusion.

Key Terms

fatty acid (p. 327)

phospholipid (p. 329)

sphingosine (p. 329)

phosphoglyceride (p. 329)

sphingomyelin (p. 330)

glycolipid (p. 331)

cerebroside (p. 331)

ganglioside (p. 331)

cholesterol (p. 331)

amphipathic molecule (p. 332)

lipid bilayer (p. 333)

liposome (p. 334)

integral membrane protein (p. 336)

peripheral membrane protein (p. 336)

hydropathy plot (p. 341)

lateral diffusion (p. 342)

fluid mosaic model (p. 343)

lipid raft (p. 344)

receptor-mediated endocytosis (p. 346)

SNARE (soluble *N*-ethylmaleimide-sensitive-factor attachment protein receptor) proteins (p. 347)

Selected Readings

Where to Start

De Weer, P. 2000. A century of thinking about cell membranes. *Annu. Rev. Physiol.* 62:919–926.

Bretscher, M. S. 1985. The molecules of the cell membrane. *Sci. Am.* 253(4):100–108.

Unwin, N., and Henderson, R. 1984. The structure of proteins in biological membranes. *Sci. Am.* 250(2):78–94.

Deisenhofer, J., and Michel, H. 1989. The photosynthetic reaction centre from the purple bacterium *Rhodopseudomonas viridis*. *EMBO J.* 8:2149–2170.

Singer, S. J., and Nicolson, G. L. 1972. The fluid mosaic model of the structure of cell membranes. *Science* 175:720–731.

Jacobson, K., Sheets, E. D., and Simson, R., 1995. Revisiting the fluid mosaic model of membranes. *Science* 268:1441–1442.

Books

Gennis, R. B. 1989. *Biomembranes: Molecular Structure and Function.* Springer Verlag.

Vance, D. E., and Vance, J. E. (Eds.). 1996. *Biochemistry of Lipids, Lipoproteins, and Membranes.* Elsevier.

Lipowsky, R., and Sackmann, E. 1995. *The Structure and Dynamics of Membranes.* Elsevier.

Racker, E. 1985. *Reconstitutions of Transporters, Receptors, and Pathological States.* Academic Press.

Tanford, C. 1980. *The Hydrophobic Effect: Formation of Micelles and Biological Membranes* (2d ed.). Wiley-Interscience.

Membrane Lipids and Dynamics

Simons, K., and Vaz, W. L. 2004. Model systems, lipid rafts, and cell membranes. *Annu. Rev. Biophys. Biomol. Struct.* 33:269–295.

Anderson, T. G., and McConnell, H. M. 2002. A thermodynamic model for extended complexes of cholesterol and phospholipid. *Biophys. J.* 83:2039–2052.

Saxton, M. J., and Jacobson, K. 1997. Single-particle tracking: Applications to membrane dynamics. *Annu. Rev. Biophys. Biomol. Struct.* 26:373–399.

Bloom, M., Evans, E., and Mouritsen, O. G. 1991. Physical properties of the fluid lipid-bilayer component of cell membranes: A perspective. *Q. Rev. Biophys.* 24:293–397.

Elson, E. L. 1986. Membrane dynamics studied by fluorescence correlation spectroscopy and photobleaching recovery. *Soc. Gen. Physiol. Ser.* 40:367–383.

Zachowski, A., and Devaux, P. F. 1990. Transmembrane movements of lipids. *Experientia* 46:644–656.

Devaux, P. F. 1992. Protein involvement in transmembrane lipid asymmetry. *Annu. Rev. Biophys. Biomol. Struct.* 21:417–439.

Silvius, J. R. 1992. Solubilization and functional reconstitution of biomembrane components. *Annu. Rev. Biophys. Biomol. Struct.* 21:323–348.

Yeagle, P. L., Albert, A. D., Boesze-Battaglia, K., Young, J., and Frye, J. 1990. Cholesterol dynamics in membranes. *Biophys. J.* 57:413–424.

Nagle, J. F., and Tristram-Nagle, S. 2000. Lipid bilayer structure. *Curr. Opin. Struct. Biol.* 10:474–480.

Dowhan, W. 1997. Molecular basis for membrane phospholipid diversity: Why are there so many lipids? *Annu. Rev. Biochem.* 66:199–232.

Huijbregts, R. P. H., de Kroon, A. I. P. M., and de Kruijff, B. 1998. Rapid transmembrane movement of newly synthesized phosphatidylethanolamine across the inner membrane of *Escherichia coli*. *J. Biol. Chem.* 273:18936–18942.

Structure of Membrane Proteins

Walian, P., Cross, T. A., and Jap, B. K. 2004. Structural genomics of membrane proteins. *Genome Biol.* 5:215.

Werten, P. J., Remigy, H. W., de Groot, B. L., Fotiadis, D., Philippsen, A., Stahlberg, H., Grubmuller, H., and Engel, A. 2002. Progress in the analysis of membrane protein structure and function. *FEBS Lett.* 529:65–72.

Popot, J.-L., and Engleman, D. M. 2000. Helical membrane protein folding, stability and evolution. *Annu. Rev. Biochem.* 69:881–922.

White, S. H., and Wimley, W. C. 1999. Membrane protein folding and stability: Physical principles. *Annu. Rev. Biophys. Biomol. Struct.* 28:319–365.

Marassi, F. M., and Opella, S. J. 1998. NMR structural studies of membrane proteins. *Curr. Opin. Struct. Biol.* 8:640–648.

Lipowsky, R. 1991. The conformation of membranes. *Nature* 349:475–481.

Altenbach, C., Marti, T., Khorana, H. G., and Hubbell, W. L. 1990. Transmembrane protein structure: Spin labeling of bacteriorhodopsin mutants. *Science* 248:1088–1092.

Fasman, G. D., and Gilbert, W. A. 1990. The prediction of transmembrane protein sequences and their conformation: An evaluation. *Trends Biochem. Sci.* 15:89–92.

Jennings, M. L. 1989. Topography of membrane proteins. *Annu. Rev. Biochem.* 58:999–1027.

Engelman, D. M., Steitz, T. A., and Goldman, A. 1986. Identifying non-polar transbilayer helices in amino acid sequences of membrane proteins. *Annu. Rev. Biophys. Biophys. Chem.* 15:321–353.

Udenfriend, S., and Kodukola, K. 1995. How glycosyl-phosphatidylinositol-anchored membrane proteins are made. *Annu. Rev. Biochem.* 64:563–591.

Intracellular Membranes

Skehel, J. J., and Wiley, D. C. 2000. Receptor binding and membrane fusion in virus entry: The influenza hemagglutinin. *Annu. Rev. Biochem.* 69:531–569.

Roth, M. G. 1999. Lipid regulators of membrane traffic through the Golgi complex. *Trends Cell Biol.* 9:174–179.

Jahn, R., and Sudhof, T. C. 1999. Membrane fusion and exocytosis. *Annu. Rev. Biochem.* 68:863–911.

Stroud, R. M., and Walter, P. 1999. Signal sequence recognition and protein targeting. *Curr. Opin. Struct. Biol.* 9:754–759.

Teter, S. A., and Klionsky, D. J. 1999. How to get a folded protein across a membrane. *Trends Cell Biol.* 9:428–431.

Hettema, E. H., Distel, B., and Tabak, H. F. 1999. Import of proteins into peroxisomes. *Biochim. Biophys. Acta* 1451:17–34.

Membrane Fusion

Sollner, T. H., and Rothman, J. E. 1996. Molecular machinery mediating vesicle budding, docking and fusion. *Experientia* 52:1021–1025.

Ungar, D., and Hughson, F. M. 2003. SNARE protein structure and function. *Annu. Rev. Cell Dev. Biol.* 19:493–517.

Problems

1. *Population density.* How many phospholipid molecules are there in a 1-μm^2 region of a phospholipid bilayer membrane? Assume that a phospholipid molecule occupies 70 Å^2 of the surface area.

2. *Lipid diffusion.* What is the average distance traversed by a membrane lipid in 1 μs, 1 ms, and 1 s? Assume a diffusion coefficient of 10^{-8} $cm^2 s^{-1}$.

3. *Protein diffusion.* The diffusion coefficient, D, of a rigid spherical molecule is given by

$$D = kT/6\pi\eta r$$

in which η is the viscosity of the solvent, r is the radius of the sphere, k is the Boltzman constant (1.38×10^{-16} erg degree^{-1}), and T is the absolute temperature. What is the diffusion coefficient at 37°C of a 100-kd protein in a membrane that has an ef-

fective viscosity of 1 poise (1 poise = 1 erg s^{-1} cm^{-3})? What is the average distance traversed by this protein in 1 μs, 1 ms, and 1 s? Assume that this protein is an unhydrated, rigid sphere of density 1.35 g cm^{-3}.

4. *Cold sensitivity.* Some antibiotics act as carriers that bind an ion on one side of a membrane, diffuse through the membrane, and release the ion on the other side. The conductance of a lipid-bilayer membrane containing a carrier antibiotic decreased abruptly when the temperature was lowered from 40°C to 36°C. In contrast, there was little change in conductance of the same bilayer membrane when it contained a channel-forming antibiotic. Why?

5. *Flip-flop 1.* The transverse diffusion of phospholipids in a bilayer membrane was investigated by using a paramagnetic

analog of phosphatidylcholine, called *spin-labeled phosphatidylcholine*.

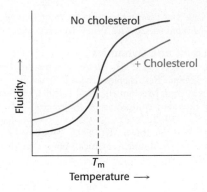

Spin-labeled phosphatidylcholine

The nitroxide (NO) group in spin-labeled phosphatidylcholine gives a distinctive paramagnetic resonance spectrum. This spectrum disappears when nitroxides are converted into amines by reducing agents such as ascorbate.

Lipid vesicles containing phosphatidylcholine (95%) and the spin-labeled analog (5%) were prepared by sonication and purified by gel-filtration chromatography. The outside diameter of these liposomes was about 250 Å (25 nm). The amplitude of the paramagnetic resonance spectrum decreased to 35% of its initial value within a few minutes of the addition of ascorbate. There was no detectable change in the spectrum within a few minutes after the addition of a second aliquot of ascorbate. However, the amplitude of the residual spectrum decayed exponentially with a half-time of 6.5 hours. How would you interpret these changes in the amplitude of the paramagnetic spectrum?

6. *Flip-flop 2*. Although proteins rarely if ever flip-flop across a membrane, the distribution of membrane lipids between the membrane leaflets is not absolute except for glycolipids. Why are glycosylated lipids less likely to flip-flop?

7. *Cis versus trans*. Why might most unsaturated fatty acids in phospholipids be in the cis rather than the trans conformation? Draw the structure of a 16-carbon fatty acid as (a) saturated, (b) trans monounsaturated, and (c) cis monounsaturated.

8. *A question of competition*. Would a homopolymer of alanine be more likely to form an α helix in water or in a hydrophobic medium? Explain.

9. *Maintaining fluidity*. A culture of bacteria growing at 37°C was shifted to 25°C. How would you expect this shift to alter the fatty acid composition of the membrane phospholipids? Explain.

10. *Let me count the ways*. Each intracellular fusion of a vesicle with a membrane requires a SNARE protein on the vesicle (called the v-SNARE) and a SNARE protein on the target membrane (called the t-SNARE). Assume that a genome encodes 21 members of the v-SNARE family and 7 members of the t-SNARE family. With the assumption of no specificity, how many potential v-SNARE–t-SNARE interactions could take place?

Data Interpretation Problems

11. *Cholesterol effects*. The red curve on the following graph shows the fluidity of the fatty acids of a phospholipid bilayer as a function of temperature. The blue curve shows the fluidity in the presence of cholesterol.

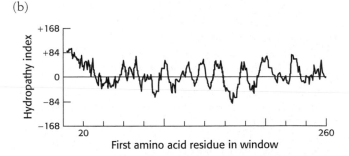

(a) What is the effect of cholesterol?

(b) Why might this effect be biologically important?

12. *Hydropathy plots*. On the basis of the following hydropathy plots for three proteins, predict which would be membrane proteins. What are the ambiguities with respect to using such plots to determine if a protein is a membrane protein?

(a)

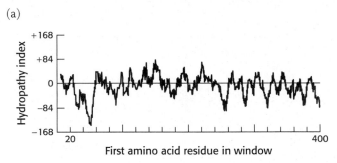

(b)

(c)

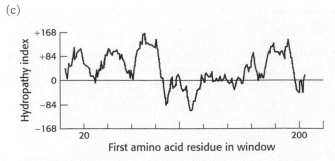

Chapter Integration Problem

13. *The proper environment*. An understanding of the structure and function of membrane proteins has lagged behind that of other proteins. The primary reason is that membrane proteins are more difficult to purify and crystallize. Why might this be the case?

Membrane Channels and Pumps

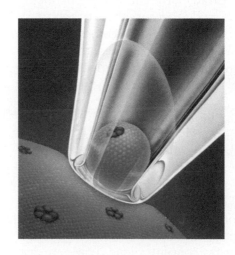

The flow of ions through a single membrane channel (channels are shown in red in the illustration at the left) can be detected by the patch-clamp technique, which records current changes as the channel transits between open and closed states. [(Left) After E. Neher and B. Sakmann. The patch clamp technique. Copyright © 1992 by Scientific American, Inc. All rights reserved. (Right) Courtesy of Dr. Mauricio Montal.]

The lipid bilayer of biological membranes is intrinsically impermeable to ions and polar molecules, yet certain such species must be able to cross these membranes for normal cell function. Permeability is conferred by two classes of membrane proteins, *pumps* and *channels*. Pumps use a source of free energy such as ATP hydrolysis or light absorption to drive the thermodynamically uphill transport of ions or molecules. Pump action is an example of *active transport*. Channels, in contrast, enable ions to flow rapidly through membranes in a thermodynamically downhill direction. Channel action illustrates *passive transport*, or *facilitated diffusion*.

Pumps are energy transducers in that they convert one form of free energy into another. Two types of *ATP-driven pumps*, P-type ATPases and the ATP-binding cassette (ABC) transporters, undergo conformational changes on ATP binding and hydrolysis that cause a bound ion to be transported across the membrane. A different mechanism of active transport utilizes the gradient of one ion to drive the active transport of another. An example of such a *secondary transporter* is the *E. coli* lactose transporter, a well-studied protein responsible for the uptake of a specific sugar from the environment of a bacterium. Many transporters of this class are present in the membranes of our cells. The expression of these transporters determines which metabolites a cell can import from the environment. Hence, adjusting the level of transporter expression is a primary means of controlling metabolism.

Pumps can establish persistent gradients of particular ions across membranes. Specific *ion channels* can allow these ions to flow rapidly across membranes down these gradients. These channels are among the most fascinating molecules in biochemistry in their ability to allow some ions to flow freely through a membrane while blocking the flow of even closely related species. These gated ion channels are central to the functioning of our nervous systems, acting as elaborately switched wires that allow the rapid flow of current.

We conclude with a discussion of a different class of channel: the cell-to-cell channel, or *gap junction*, allows the flow of metabolites or ions *between cells*. For example, gap junctions are responsible for synchronizing muscle-cell contraction in the beating heart.

The Expression of Transporters Largely Defines the Metabolic Activities of a Given Cell Type

Each cell type expresses a specific set of transporters in its plasma membrane. The set of transporters expressed is crucial because these transporters largely determine the ionic composition inside cells and the compounds that can be taken up from the cell's environment. In some senses, the array of transporters expressed by a cell determines the cell's characteristics because a cell can execute only those biochemical reactions for which it has taken up the substrates.

An example from glucose metabolism illustrates this point. As we will see in the discussion of glucose metabolism in Chapter 16, tissues differ in their ability to employ different molecules as energy sources. Which tissues can make use of glucose is largely governed by the expression of different members of a family of homologous glucose transporters called GLUT1, GLUT2, GLUT3, GLUT4, and GLUT5 in different cell types. GLUT3, for example, is expressed only on neurons and a few other cell types. This transporter binds glucose relatively tightly so that these cells have first call on glucose when it is present at relatively low concentrations. These are just the first of many examples that we will encounter demonstrating the critical role that transporter expression plays in the control and integration of metabolism.

13.1 The Transport of Molecules Across a Membrane May Be Active or Passive

We first consider some general principles of membrane transport. Two factors determine whether a molecule will cross a membrane: (1) the permeability of the molecule in a lipid bilayer and (2) the availability of an energy source.

Many Molecules Require Protein Transporters to Cross Membranes

As stated in Chapter 12, some molecules can pass through cell membranes because they dissolve in the lipid bilayer. Such molecules are called *lipophilic molecules*. The steroid hormones provide a physiological example. These cholesterol relatives can pass through a membrane in their path of movement, but what determines the direction in which they will move? Such molecules will pass through a membrane down their concentration gradient in a process called *simple diffusion*. In accord with the Second Law of Thermodynamics, molecules spontaneously move from a region of higher concentration to one of lower concentration.

Matters become more complicated when the molecule is highly polar. For example, sodium ions are present at 143 mM outside a typical cell and at 14 mM inside the cell, yet sodium does not freely enter the cell, because the charged ion cannot pass through the hydrophobic membrane interior.

In some circumstances, as during a nerve impulse, sodium ions must enter the cell. How are they able to do so? Sodium ions pass through specific channels in the hydrophobic barrier formed by membrane proteins. This means of crossing the membrane is called *facilitated diffusion*, because the diffusion across the membrane is facilitated by the channel. It is also called *passive transport*, because the energy driving the ion movement originates from the ion gradient itself, without any contribution by the transport system. Channels, like enzymes, display substrate specificity in that they facilitate the transport of some ions, but not other, even closely related, ions.

How is the sodium gradient established in the first place? In this case, sodium must move, or be pumped, *against* a concentration gradient. Because moving the ion from a low concentration to a higher concentration results in a decrease in entropy, it requires an input of free energy. Protein transporters embedded in the membrane are capable of using an energy source to move the molecule up a concentration gradient. Because an input of energy from another source is required, this means of crossing the membrane is called *active transport*.

Free Energy Stored in Concentration Gradients Can Be Quantified

An unequal distribution of molecules is an energy-rich condition because free energy is minimized when all concentrations are equal. Consequently, to attain such an unequal distribution of molecules, or *concentration gradient*, requires an input of free energy. Can we quantify the amount of energy required to generate a concentration gradient (Figure 13.1)? Consider an uncharged solute molecule. The free-energy change in transporting this species from side 1, where it is present at a concentration of c_1, to side 2, where it is present at concentration c_2, is

$$\Delta G = RT \ln (c_2/c_1) = 2.303RT \log_{10}(c_2/c_1)$$

where R is the gas constant (8.315×10^{-3} kJ mol^{-1}, or 1.987×10^{-3} kcal mol^{-1}) and T is the temperature in kelvins. For a charged species, the unequal distribution across the membrane generates an electrical potential that also must be considered because the ions will be repelled by the like charges. The sum of the concentration and electrical terms is called the *electrochemical potential* or *membrane potential*. The free-energy change is then given by

$$\Delta G = RT \ln (c_2/c_1) + ZF\Delta V = 2.303RT \log_{10}(c_2/c_1) + ZF\Delta V$$

in which Z is the electrical charge of the transported species, ΔV is the potential in volts across the membrane, and F is the Faraday constant (96.5 kJ V^{-1} mol^{-1} (or 23.1 kcal V^{-1} mol^{-1}).

A transport process must be active when ΔG is positive, whereas it can be passive when ΔG is negative. For example, consider the transport of an uncharged molecule from $c_1 = 10^{-3}$ M to $c_2 = 10^{-1}$ M.

$$\Delta G = 2.303RT \log_{10}(10^{-1}/10^{-3})$$
$$= 2.303 \times 8.315 \times 298 \times 2$$
$$= +11.4 \text{ kJ mol}^{-1} (+2.7 \text{ kcal mol}^{-1})$$

At 25 °C (298 K), ΔG is $+11.4$ kJ mol^{-1} ($+2.7$ kcal mol^{-1}), indicating that this transport process requires an input of free energy.

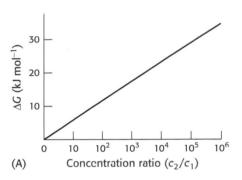

(A)

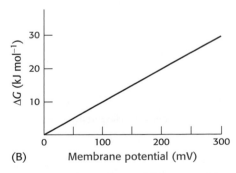

(B)

Figure 13.1 Free energy and transport. The free-energy change in transporting (A) an uncharged solute from a compartment at concentration c_1 to one at c_2 and (B) a singly charged species across a membrane to the side having the same charge as that of the transported ion. Note that the free-energy change imposed by a membrane potential of 59 mV is equivalent to that imposed by a concentration ratio of 10 for a singly charged ion at 25°C.

13.2   Two Families of Membrane Proteins Use ATP Hydrolysis to Pump Ions and Molecules Across Membranes

The extracellular fluid of animal cells has a salt concentration similar to that of seawater. However, cells must control their intracellular salt concentrations to prevent unfavorable interactions with high concentrations of ions such as Ca^{2+} and to facilitate specific processes. For instance, most animal cells contain a high concentration of K^+ and a low concentration of Na^+ relative to the external medium. These ionic gradients are generated by a specific transport system, an enzyme that is called the *Na^+–K^+ pump* or the *Na^+–K^+ ATPase*. The hydrolysis of ATP by the pump provides the energy needed for the active transport of Na^+ out of the cell and K^+ into the cell, generating the gradients. The pump is called the Na^+–K^+ ATPase because the hydrolysis of ATP takes place only when Na^+ and K^+ are present. This ATPase, like all such enzymes, requires Mg^{2+}.

The change in free energy accompanying the transport of Na^+ and K^+ can be calculated. Suppose that the concentrations of Na^+ outside and inside the cell are 143 and 14 mM, respectively, and the corresponding values for K^+ are 4 and 157 mM. At a membrane potential of -50 mV and a temperature of 37°C, we can use the equation on page 353 to determine that the free-energy change in transporting 3 mol of Na^+ out of the cell and 2 mol of K^+ into it is $3(5.99) + 2(9.46) = +36.9$ kJ mol^{-1} (+8.8 kcal mol^{-1}). The hydrolysis of a single ATP molecule per transport cycle provides sufficient free energy, about -50 kJ mol^{-1} (-12 kcal mol^{-1}) under typical cellular conditions, to drive the uphill transport of these ions. The active transport of Na^+ and K^+ is of great physiological significance. Indeed, more than a third of the ATP consumed by a resting animal is used to pump these ions. The Na^+–K^+ gradient in animal cells controls cell volume, renders neurons and muscle cells electrically excitable, and drives the active transport of sugars and amino acids.

The subsequent purification of other ion pumps has revealed a large family of evolutionarily related ion pumps including proteins from bacteria, archaea, and all eukaryotes. These pumps are specific for an array of ions. Two are of particular interest: the *Ca^{2+} ATPase* transports Ca^{2+} out of the cytoplasm and into the sarcoplasmic reticulum of muscle cells, and the *gastric H^+–K^+ ATPase* is the enzyme responsible for pumping sufficient protons into the stomach to lower the pH below 1.0. These enzymes and the hundreds of known homologs, including the Na^+–K^+ ATPase, are referred to as *P-type ATPases* because they form a key *phosphorylated intermediate*. In the formation of this intermediate, a phosphoryl group obtained from ATP is linked to the side chain of a specific conserved aspartate residue in the ATPase to form phosphorylaspartate.

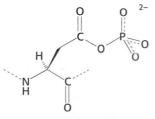

Phosphorylaspartate

P-Type ATPases Couple Phosphorylation and Conformational Changes to Pump Calcium Ions Across Membranes

Membrane pumps function by mechanisms that are simple in principle but often complex in detail. Fundamentally, each pump protein can exist in two principal conformational states, one with ion-binding sites open to one side of the membrane and the other with ion-binding sites open to the other side (Figure 13.2). To pump ions in a single direction across a membrane, free energy must be provided in a manner that can be coupled to the interconversion between these conformational states.

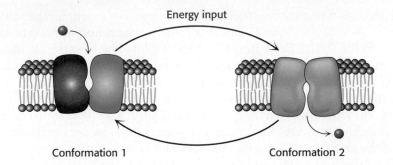

Energy input

Conformation 1   Conformation 2

Figure 13.2 Pump action. A simple scheme for the pumping of a molecule across a membrane. The pump interconverts to two conformational states, each with a binding site accessible to a different side of the membrane.

We will consider the structural and mechanistic features of P-type ATPases by examining the Ca^{2+} ATPase found in the sarcoplasmic reticulum (*SR Ca^{2+} ATPase,* or *SERCA*) of muscle cells. The properties of this family member have been established in great detail, by relying on crystal structures of the pump in five different states. This enzyme, which constitutes 80% of the protein in the sarcoplasmic reticulum membrane, plays an important role in muscle contraction, a process triggered by an abrupt rise in the cytoplasmic calcium ion level. Muscle relaxation depends on the rapid removal of Ca^{2+} from the cytoplasm into the sarcoplasmic reticulum, a specialized compartment for Ca^{2+} storage, by SERCA. This pump maintains a Ca^{2+} concentration of approximately 0.1 μM in the cytoplasm compared with 1.5 mM in the sarcoplasmic reticulum.

The first structure of SERCA to be determined had Ca^{2+} bound, but no nucleotides present (Figure 13.3). SERCA is a single 110-kd polypeptide with a transmembrane domain consisting of 10 α helices. The transmembrane domain includes sites for binding two calcium ions. Each calcium ion is coordinated to seven oxygen atoms coming from a combination of side-chain glutamate, aspartate, threonine, and asparagine residues, backbone carbonyl groups, and water molecules. A large cytoplasmic headpiece constitutes nearly half the molecular weight of the protein and consists of three distinct domains. The three cytoplasmic domains of the SERCA have distinct functions. One domain (N) binds the ATP *n*ucleotide, another (P) accepts the *p*hosphoryl group on a conserved aspartate residue, and the third (A) serves as an *a*ctuator, linking changes in the N and P domains to the transmembrane part of the enzyme.

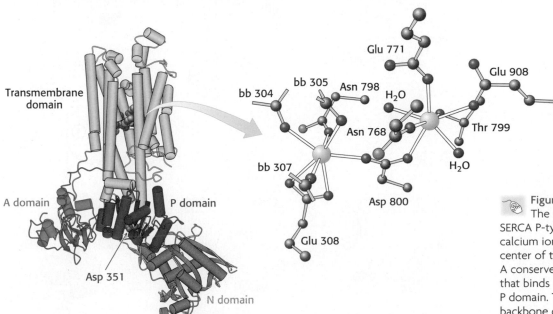

Transmembrane domain

A domain   P domain

Asp 351

N domain

Glu 771
bb 305   Asn 798
bb 304 H_2O Glu 908
                  Asn 768   Thr 799
bb 307 H_2O
          Asp 800
Glu 308

Figure 13.3 Calcium-pump structure. The overall structure of the SERCA P-type ATPase. *Notice* the two calcium ions (green) that lie in the center of the transmembrane domain. A conserved aspartate residue (Asp 351) that binds a phosphoryl group lies in the P domain. The designation bb refers to backbone carbonyl groups. [Drawn from 1SU4.pdb.]

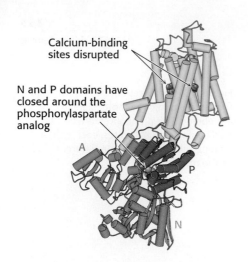

Calcium-binding
sites disrupted

N and P domains have
closed around the
phosphorylaspartate
analog

A

P

N

Figure 13.4 Conformational changes
associated with calcium pumping.
This structure was determined in the
absence of bound calcium and with a
phosphorylaspartate analog present in
the P domain. *Notice* how different this
structure is from the calcium-bound
form shown in Figure 13.3: both the
transmembrane part (yellow) and the
A, P, and N domains have substantially
rearranged. [Drawn from 1WPG.pdb.]

SERCA is remarkably structurally dynamic. For example, the structure of SERCA without bound Ca^{2+} and with a phosphorylaspartate analog present in the P domain is shown in Figure 13.4. The N and P domains are now closed around the phosphorylaspartate analog, and the A domain has rotated substantially relative to its position in SERCA with Ca^{2+} bound and without the phosphoryl analog. Furthermore, the transmembrane part of the enzyme has rearranged substantially and the well-organized Ca^{2+}-binding sites are disrupted. These sites are now accessible from the side of the membrane opposite the N, P, and A domains.

The structural results can be combined with other studies to construct a detailed mechanism for Ca^{2+} pumping by SERCA (Figure 13.5).

1.   The catalytic cycle begins with the enzyme in its unphosphorylated state with two calcium ions bound. We will refer to the overall enzyme conformation in this state as E_1; with Ca^{2+} bound, it is E_1-$(Ca^{2+})_2$. In this conformation, SERCA can exchange calcium ions but only with calcium ions from the cytoplasmic side of the membrane. This conformation is shown in Figure 13.3.

2.   In the E_1 conformation, the enzyme can bind ATP. The N, P, and A domains undergo substantial rearrangement as they close around the bound ATP, but there is no substantial conformational change in the transmembrane domain. The calcium ions are now trapped inside the enzyme.

3.   The phosphoryl group is then transferred from ATP to Asp 351.

4.   Upon ADP release, the enzyme again changes its overall conformation, including the membrane domain this time. This new conformation is referred to as E_2 or E_2-P in its phosphorylated form. The process of interconverting the E_1 and E_2 conformations is sometimes referred to as *eversion*.

In the E_2-P conformation, the Ca^{2+}-binding sites become disrupted and the calcium ions are released to the side of the membrane opposite that at

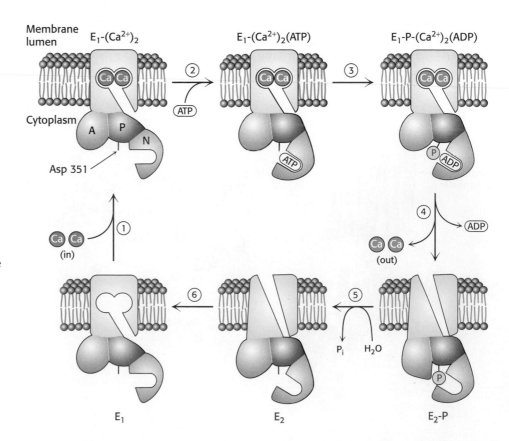

Figure 13.5 Pumping calcium. Ca^{2+} ATPase transports Ca^{2+} through the membrane by a mechanism that includes (1) Ca^{2+} binding from the cytoplasm, (2) ATP binding, (3) ATP cleavage with the transfer of a phosphoryl group to Asp 351 on the enzyme, (4) ADP release and eversion of the enzyme to release Ca^{2+} on the opposite side of the membrane, (5) hydrolysis of the phosphorylaspartate residue, and (6) eversion to prepare for the binding of Ca^{2+} from cytoplasm.

which they entered; ion transport has been achieved. This conformation is shown in Figure 13.4.

5. The phosphorylaspartate residue is hydrolyzed to release inorganic phosphate.

6. With the release of phosphate, the interactions stabilizing the E_2 conformation are lost, and the enzyme everts back to the E_1 conformation.

The binding of two calcium ions from the cytoplasmic side of the membrane completes the cycle.

This mechanism likely applies to other P-type ATPases. For example, $Na^+–K^+$ ATPase is an $\alpha_2\beta_2$ tetramer. Its α subunit is homologous to SERCA and includes a key aspartate residue analogous to Asp 351. The β subunit does not directly take part in ion transport. A mechanism analogous to that shown in Figure 13.5 applies, with three Na^+ ions binding from the inside of the cell to the E_1 conformation and two K^+ ions binding from outside the cell to the E_2 conformation.

Digitalis Specifically Inhibits the $Na^+–K^+$ Pump by Blocking Its Dephosphorylation

Certain steroids derived from plants are potent inhibitors ($K_i \approx 10$ nM) of the $Na^+–K^+$ pump. Digitoxigenin and ouabain are members of this class of inhibitors, which are known as *cardiotonic steroids* because of their strong effects on the heart (Figure 13.6). These compounds inhibit the dephosphorylation of the E_2-P form of the ATPase when applied on the *extracellular* face of the membrane.

Digitalis is a mixture of cardiotonic steroids derived from the dried leaf of the foxglove plant (*Digitalis purpurea*). The compound increases the force of contraction of heart muscle and is consequently a choice drug in the treatment of congestive heart failure. Inhibition of the $Na^+–K^+$ pump by digitalis leads to a higher level of Na^+ inside the cell. The diminished Na^+ gradient results in slower extrusion of Ca^{2+} by the sodium–calcium exchanger. The subsequent increase in the intracellular level of Ca^{2+} enhances the ability of cardiac muscle to contract. It is interesting to note that digitalis was used effectively long before the discovery of the $Na^+–K^+$ ATPase. In 1785, William Withering, a British physician, heard tales of an elderly woman, known as "the old woman of Shropshire," who cured people of "dropsy" (which today would be recognized as congestive heart failure) with an extract of foxglove. Withering conducted the first scientific study of the effects of foxglove on congestive heart failure and documented its effectiveness.

Foxglove (*Digitalis purpurea*) is the source of digitalis, one of the most widely used drugs. [Inga Spence/Visuals Unlimited.]

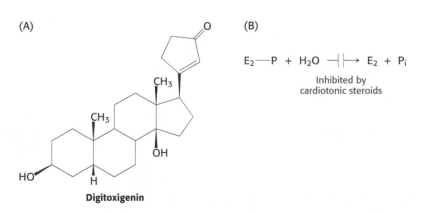

(A)

Digitoxigenin

(B)

$$E_2—P + H_2O \longrightarrow E_2 + P_i$$

Inhibited by cardiotonic steroids

Figure 13.6 Digitoxigenin. Cardiotonic steroids such as digitoxigenin inhibit the $Na^+–K^+$ pump by blocking the dephosphorylation of E_2-P.

P-Type ATPases Are Evolutionarily Conserved and Play a Wide Range of Roles

Analysis of the complete yeast genome revealed the presence of 16 proteins that clearly belong to the P-type ATPase family. More-detailed sequence analysis suggests that 2 of these proteins transport H^+ ions, 2 transport Ca^{2+}, 3 transport Na^+, and 2 transport metals such as Cu^{2+}. In addition, 5 members of this family appear to participate in the transport of phospholipids with amino acid head groups. These 5 proteins help maintain membrane asymmetry by transporting lipids such as phosphatidylserine from the inner to the outer leaflet of the bilayer membrane. Such enzymes have been termed "flippases." More impressively, the human genome encodes 70 P-type ATPases. All members of this protein family employ the same fundamental mechanism. The free energy of ATP hydrolysis drives membrane transport by means of conformational changes, which are induced by the addition and removal of a phosphoryl group at an analogous aspartate site in each protein.

Multidrug Resistance Highlights a Family of Membrane Pumps with ATP-Binding Cassette Domains

Studies of human disease revealed another large and important family of active-transport proteins, with structures and mechanisms quite different from those of the P-type ATPase family. Tumor cells in culture often become resistant to drugs that were initially quite toxic to the cells. Remarkably, the development of resistance to one drug also makes the cells less sensitive to a range of other compounds. This phenomenon is known as *multidrug resistance*. In a significant discovery, the onset of multidrug resistance was found to correlate with the expression and activity of a membrane protein with an apparent molecular mass of 170 kd. This protein acts as an ATP-dependent pump that extrudes a wide range of small molecules from cells that express it. The protein is called the *multidrug-resistance* (MDR) *protein* or *P-glycoprotein* ("glyco" because it includes a carbohydrate moiety). Thus, when cells are exposed to a drug, the MDR pumps the drug out of the cell before the drug can exert its effects.

Analysis of the amino acid sequences of MDR and homologous proteins revealed a common architecture (Figure 13.7). Each protein comprises four domains: two membrane-spanning domains and two ATP-binding domains. The ATP-binding domains of these proteins are called *ATP-binding cassettes* (ABCs) and are homologous to domains in a large family of transport proteins of bacteria and archaea. Transporters that include these domains are called *ABC transporters*. With 79 members, the ABC transporters are the largest single family identified in the *E. coli* genome. The human genome includes more than 150 ABC transporter genes.

The ABC proteins are members of the P-loop NTPase superfamily (p. 267). The three-dimensional structures of several members of the ABC transporter family have now been determined. One such structure, that of a lipid transporter from *Vibrio cholerae* is shown in Figure 13.8. This protein is a dimer of 62-kd chains. The amino-terminal half of each protein is the membrane-spanning domain, and the carboxyl-terminal half is the ATP-binding cassette. In contrast with the eukaryotic MDR protein, some ABC proteins, particularly those of prokaryotes, are multisubunit proteins, either dimers as in this case or heterotetramers with two membrane-spanning-domain subunits and two ATP-binding-cassette subunits. The consolidation of the enzymatic activities of several polypeptide chains in prokaryotes to a single chain in eukaryotes is a theme that we will see again.

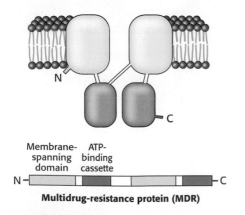

Figure 13.7 ABC transporters. The multidrug-resistance protein (MDR) is a representative of a large family of homologous proteins composed of two transmembrane domains and two ATP-binding domains, called ATP-binding cassettes (ABCs).

The two ATP-binding cassettes are in contact, but they do not interact strongly in the absence of bound ATP. On the basis of this structure and others, as well as on other experiments, a mechanism for active transport by these proteins has been developed (Figure 13.9).

1. The catalytic cycle begins with the transporter free of both ATP and substrate. The transporter can interconvert between closed and open forms.

2. Substrate enters the central cavity of the open form of the transporter from inside the cell. Substrate binding induces conformational changes in the ATP-binding cassettes that increase their affinity for ATP.

3. ATP binds to the ATP-binding cassettes, changing their conformations so that the two domains interact strongly with one another.

4. The strong interaction between the ATP-binding cassettes induces a change in the relation between the two membrane-spanning domains, releasing the substrate to the outside of the cell.

5. The hydrolysis of ATP and the release of ADP and inorganic phosphate reset the transporter for another cycle.

Whereas eukaryotic ABC transporters generally act to export molecules from inside the cell, prokaryotic ABC transporters often act to import specific molecules from *outside* the cell. A specific binding protein acts in concert with the bacterial ABC transporter, delivering the substrate to the transporter and stimulating ATP hydrolysis inside the cell. These binding proteins are present in the periplasm of bacterial cells, the compartment between the two membranes that surround some bacterial cells.

Thus, ABC transporters use a substantially different mechanism from the P-type ATPases to couple the ATP hydrolysis reaction to conformational changes. Nonetheless, the net result is the same: the transporters are

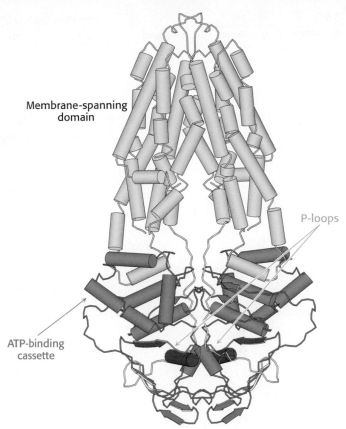

Figure 13.8 ABC transporter structure. The structure of a lipid transporter from *Vibrio cholerae*, a representative ABC transporter. The two ATP-binding casettes (blue) are related to the P-loop NTPases and, like them, contain P-loops (green). The surrounding β strand and α helix are shown in purple. [Drawn from 1PF4.pdb.]

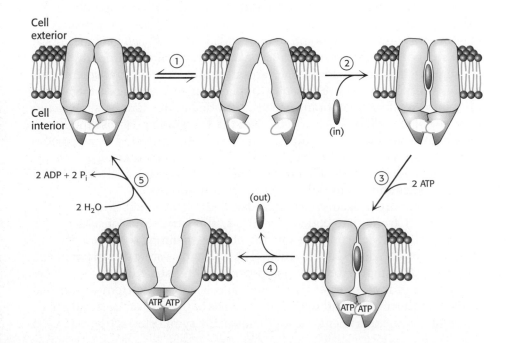

Figure 13.9 ABC transporter mechanism. The mechanism includes the following steps: (1) opening of the channel toward the inside of the cell; (2) substrate binding and conformational changes in the ATP-binding cassettes; (3) ATP binding and further conformational changes; (4) separation of the membrane-binding domains and release of the substrate to the other side of the membrane; and (5) ATP hydrolysis to reset the transporter to its initial state.

converted from one conformation capable of binding substrate from one side of the membrane to another that releases the substrate on the other side.

13.3 Lactose Permease Is an Archetype of Secondary Transporters That Use One Concentration Gradient to Power the Formation of Another

Many active-transport processes are not directly driven by the hydrolysis of ATP. Instead, the thermodynamically unfavorable flow of one species of ion or molecule *up* a concentration gradient is coupled to the favorable flow of a different species *down* a concentration gradient. Membrane proteins that pump ions or molecules "uphill" by this means are termed *secondary transporters* or *cotransporters*. These proteins can be classified as either *antiporters* or *symporters*. Antiporters couple the downhill flow of one species to the uphill flow of another in the *opposite direction* across the membrane; symporters use the flow of one species to drive the flow of a different species in the *same direction* across the membrane. Other related proteins are *uniporters*, which, like ion channels, are able to transport a specific species in either direction governed only by concentrations of that species on either side of the membrane (Figure 13.10).

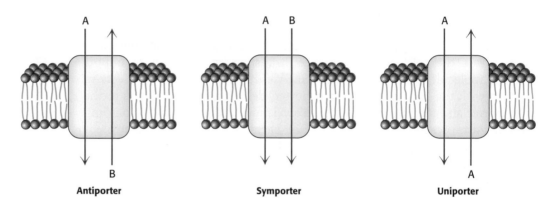

Figure 13.10 Antiporters, symporters, and uniporters. Secondary transporters can transport two substrates in opposite directions (antiporters), two substrates in the same direction (symporters), or one substrate in either direction (uniporter).

Antiporter **Symporter** **Uniporter**

Secondary transporters are ancient molecular machines, common today in bacteria and archaea as well as in eukaryotes. For example, approximately 160 (of approximately 4000) proteins encoded by the *E. coli* genome appear to be secondary transporters. Sequence comparison and hydropathy analysis suggest that members of the largest family have 12 transmembrane helices that appear to have arisen by duplication and fusion of a membrane protein with 6 transmembrane helices. Included in this family is the *lactose permease* of *E. coli*. This symporter uses the H^+ gradient across the *E. coli* membrane (outside H^+ has higher concentration) generated by the oxidation of fuel molecules to drive the uptake of lactose and other sugars against a concentration gradient. This transporter has been extensively studied for many decades and is a useful archetype for this family.

The structure of lactose permease has been determined (Figure 13.11). As expected from the sequence analysis, this structure consists of two halves, each of which comprises six membrane-spanning α helices. Some of these helices are somewhat irregular. The two halves are well separated and are joined by a single stretch of polypeptide. In this structure, the sugar lies in a pocket in the center of the protein and is accessible from a path that leads from the interior of the cell. On the basis of these structures and a wide range of other experiments, a mechanism for symporter action has been

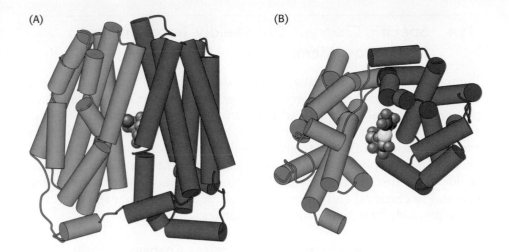

Figure 13.11 **Structure of lactose permease with a bound lactose analog.** The amino-terminal half of the protein is shown in blue and the carboxyl-terminal half in red. (A) Side view. (B) Bottom view (from inside the cell). *Notice* that the structure consists of two halves that surround the sugar and are linked to one another by only a single stretch of polypeptide. [Drawn from 1PV7.pdb.]

developed. This mechanism (Figure 13.12) has many features similar to those for P-type ATPases and ABC transporters.

1.   The cycle begins with the two halves oriented so that the opening to the binding pocket faces outside the cell, in a conformation different from that observed in the structures solved to date. A proton from outside the cell binds to a residue in the permease, quite possibly Glu 269.

2.   In the protonated form, the permease binds lactose from outside the cell.

3.   The structure everts to the form observed in the crystal structure (see Figure 13.11).

4.   The permease releases lactose to the inside of the cell.

5.   The permease releases a proton to the inside of the cell.

6.   The permease everts to complete the cycle.

The site of protonation likely changes in the course of this cycle.

   The same eversion mechanism very likely applies to all classes of secondary transporters, which appear to resemble the lactose permease in overall architecture.

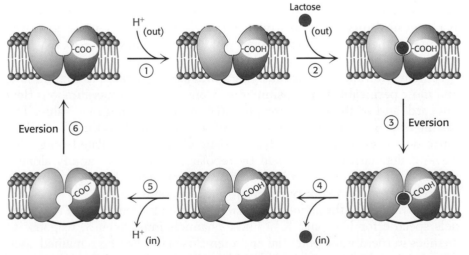

Figure 13.12 **Lactose permease mechanism.** The mechanism begins with the permease open to the outside of the cell (upper left). The permease binds a proton from the outside of the cell (1) and then binds its substrate (2). The permease everts (3) and then releases its substrate (4) and a proton (5) to the inside of the cell. It then everts (6) to complete the cycle.

13.4 Specific Channels Can Rapidly Transport Ions Across Membranes

Pumps can transport ions at rates approaching several thousand ions per second. Other membrane proteins, the passive-transport systems called *ion channels*, are capable of ion-transport rates that are more than 1000 times as fast. These rates of transport through ion channels are close to rates expected for ions diffusing freely through aqueous solution. Yet, ion channels are not simply tubes that span membranes through which ions can rapidly flow. Instead, they are highly sophisticated molecular machines that respond to chemical and physical changes in their environments and undergo precisely timed conformational changes.

Action Potentials Are Mediated by Transient Changes in Na^+ and K^+ Permeability

One of the most important manifestations of ion-channel action is the nerve impulse, which is the fundamental means of communication in the nervous system. A *nerve impulse* is an electrical signal produced by the flow of ions across the plasma membrane of a neuron. The interior of a neuron, like that of most other cells, contains a high concentration of K^+ and a low concentration of Na^+. These ionic gradients are generated by the Na^+–K^+ ATPase. The cell membrane has an electrical potential determined by the ratio of the internal to the external concentration of ions. In the resting state, the membrane potential is typically -60 mV. A nerve impulse, or *action potential*, is generated when the membrane potential is depolarized beyond a critical threshold value (e.g., from -60 to -40 mV). The membrane potential becomes positive within about a millisecond and attains a value of about $+30$ mV before turning negative again. This amplified depolarization is propagated along the nerve terminal (Figure 13.13).

Ingenious experiments carried out by Alan Hodgkin and Andrew Huxley revealed that action potentials arise from large, transient changes in the permeability of the axon membrane to Na^+ and K^+ ions. The conductance of the membrane to Na^+ changes first. Depolarization of the membrane beyond the threshold level leads to an increase in permeability to sodium ions. Sodium ions begin to flow into the cell because of the large electrochemical gradient across the plasma membrane. The entry of Na^+ further depolarizes the membrane, leading to a further increase in Na^+ permeability. This positive feedback leads to a very rapid and large change in membrane potential, from about -60 mV to $+30$ mV in a millisecond.

The membrane spontaneously becomes less permeable to sodium ions and more permeable to potassium ions. Consequently, potassium ions flow outward, and so the membrane potential returns to a negative value. The resting level of -60 mV is restored in a few milliseconds as the K^+ conductance decreases to the value characteristic of the unstimulated state. The wave of depolarization followed by repolarization moves rapidly along a nerve cell. The propagation of these waves allows a touch at the tip of your toe to be detected in your brain in a few milliseconds.

This model for the action potential postulated the existence of ion channels specific for Na^+ and K^+. These channels must open in response to changes in membrane potential and then close after having remained open for a brief period of time. This bold hypothesis predicted the existence of molecules with a well-defined set of properties long before tools existed for their direct detection and characterization.

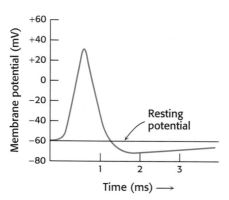

Figure 13.13 Action potential. Signals are sent along neurons by the transient depolarization and repolarization of the membrane.

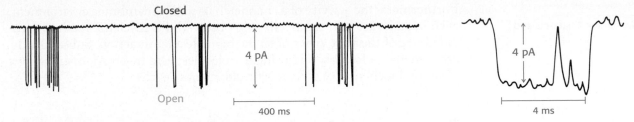

Closed

4 pA

Open

400 ms

4 pA

4 ms

Figure 13.14 Observing single channels. The results of a patch-clamp experiment revealing a single ion channel undergoing transitions between closed and open states.

Patch-Clamp Conductance Measurements Reveal the Activities of Single Channels

Direct evidence for the existence of these channels was provided by the *patch-clamp technique*, which was introduced by Erwin Neher and Bert Sakmann in 1976 (Figure 13.14). This powerful technique enables the measurement of the ion conductance through a small patch of cell membrane. Remarkably, stepwise changes in membrane conductance are observed, corresponding to the opening and closing of individual ion channels. In this technique, a clean glass pipette with a tip diameter of about 1 μm is pressed against an intact cell to form a seal (Figure 13.15). Slight suction leads to the formation of a very tight seal so that the resistance between the inside of the pipette and the bathing solution is many gigaohms (1 gigaohm is equal to 10^9 ohms). Thus, a gigaohm seal (called a *gigaseal*) ensures that an electric current flowing through the pipette is identical with the current flowing through the membrane covered by the pipette. The gigaseal makes possible high-resolution current measurements while a known voltage is applied across the membrane. The flow of ions through a single channel and transitions between the open and the closed states of a channel can be monitored with a time resolution of microseconds.

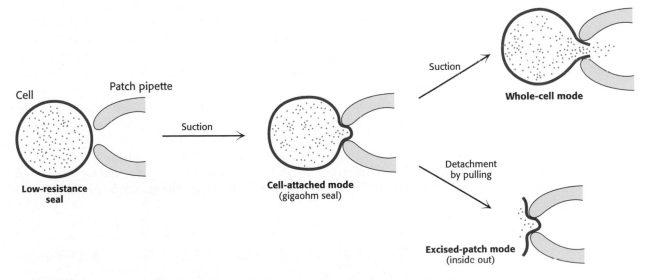

Cell

Patch pipette

Low-resistance seal

Suction

Cell-attached mode (gigaohm seal)

Suction

Whole-cell mode

Detachment by pulling

Excised-patch mode (inside out)

Figure 13.15 Patch-clamp modes. The patch-clamp technique for monitoring channel activity is highly versatile. A high-resistance seal (gigaseal) is formed between the pipette and a small patch of plasma membrane. This configuration is called *cell attached mode*. The breaking of the membrane patch by increased suction produces a low-resistance pathway between the pipette and the interior of the cell. The activity of the channels in the entire plasma membrane can be monitored in this *whole-cell mode*. To prepare a membrane in the *excised-patch mode*, the pipette is pulled away from the cell. A piece of plasma membrane with its cytoplasmic side now facing the medium is monitored by the patch pipette.

Furthermore, the activity of a channel in its native membrane environment, even in an intact cell, can be directly observed. Patch-clamp methods provided one of the first views of single biomolecules in action. Subsequently, other methods for observing single molecules were invented, opening new vistas on biochemistry at its most fundamental level.

The Structure of a Potassium Ion Channel Is an Archetype for Many Ion-Channel Structures

With the existence of ion channels firmly established by patch-clamp methods, scientists sought to identify the molecules that form ion channels. The Na^+ channel was first purified from the electric organ of electric eel, which is a rich source of the protein forming this channel. That protein was purified on the basis of its ability to bind a specific neurotoxin. Tetrodotoxin, an organic compound isolated from the puffer fish, binds to Na^+ channels with great avidity ($K_i \approx 1$ nM). The lethal dose of this poison for an adult human being is about 10 ng.

The isolated protein is a single 260-kd chain. Subsequently, cDNAs encoding Na^+ channels were cloned and sequenced. Interestingly, the channel contains four internal repeats, each having a similar amino acid sequence, suggesting that gene duplication and divergence have produced the gene for this channel. Hydrophobicity profiles indicate that each repeat contains five hydrophobic segments (S1, S2, S3, S5, and S6). Each repeat also contains a highly positively charged S4 segment; positively charged arginine or lysine residues are present at nearly every third residue. It was proposed that segments S1 through S6 are membrane-spanning α helices. The positively charged residues in S4 were proposed to act as the voltage sensors of the channel.

The purification of K^+ channels proved to be much more difficult because of their low abundance and the lack of known high-affinity ligands comparable to tetrodotoxin. The breakthrough came in studies of mutant fruit flies that shake violently when anesthetized with ether. The mapping and cloning of the gene, termed *shaker*, responsible for this defect revealed the amino acid sequence encoded by a K^+-channel gene. *Shaker* cDNA encodes a 70-kd protein that has four subunits. Remarkably, each polypeptide contains sequences corresponding to segments S1 through S6 in one of the repeated units of the Na^+ channel. Thus, a K^+-channel subunit is homologous to one of the repeated units of Na^+ channels. Consistent with this hypothesis, four K^+-channel subunits come together to form a functional channel. More recently, bacterial K^+ channels were discovered that contain only the two membrane-spanning regions corresponding to segments S5 and S6. This and other information suggested that S5 and S6, including the region between them, form the actual pore

Tetrodotoxin

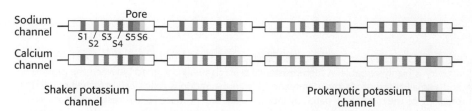

Figure 13.16 Sequence relations of ion channels. Like colors indicate structurally similar regions of the sodium, calcium, and potassium channels. Each of these channels exhibits approximate fourfold symmetry, either within one chain (sodium, calcium channels) or by forming tetramers (potassium channels).

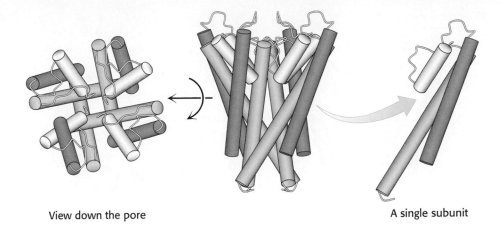

View down the pore   A single subunit

Figure 13.17 Structure of the potassium ion channel. The K$^+$ channel, composed of four identical subunits, is cone shaped, with the larger opening facing the inside of the cell (center). A view down the pore, looking toward the outside of the cell, shows the relations of the individual subunits (left). One of the four identical subunits of the pore is illustrated at the right, with the pore-forming region shown in gray. [Drawn from 1K4C.pdb.]

in the K$^+$ channel. Segments S1 through S4 contain the apparatus that opens the pore. The sequence relations between these ion channels are summarized in Figure 13.16.

In 1998, Roderick MacKinnon and coworkers determined the structure of a bacterial K$^+$ channel (from *Streptomyces lividans*) by x-ray crystallography. This channel contains only the pore-forming segments S5 and S6. As expected, the K$^+$ channel is a tetramer of identical subunits, each of which includes two membrane-spanning α helices (Figure 13.17). The four subunits come together to form a pore in the shape of a cone that runs through the center of the structure.

The Structure of the Potassium Ion Channel Reveals the Basis of Ion Specificity

The structure presented in Figure 13.17 probably represents the K$^+$ channel in a closed form. Nonetheless, it suggests how the channel is able to exclude all but K$^+$ ions. Beginning from the inside of the cell, the pore starts with a diameter of approximately 10 Å and then constricts to a smaller cavity with a diameter of 8 Å. Both the opening to the outside and the central cavity of the pore are filled with water, and a K$^+$ ion can fit in the pore without losing its shell of bound water molecules. Approximately two-thirds of the way through the membrane, the pore becomes more constricted (3-Å diameter). At that point, any K$^+$ ions must give up their water molecules and interact directly with groups from the protein. The channel structure effectively reduces the thickness of the membrane from 34 Å to 12 Å by allowing the solvated ions to penetrate into the membrane before the ions must directly interact with the channel (Figure 13.18).

For potassium ions to relinquish their water molecules, other polar interactions must replace those with water. The restricted part of the pore is built from residues contributed by the two transmembrane α helices. In particular, a five-amino-acid stretch within this region functions as the *selectivity filter* that determines the preference for K$^+$ over other ions (Figure 13.19). The stretch has the sequence Thr-Val-Gly-Tyr-Gly (TVGYG). This sequence is nearly completely conserved in all K$^+$ channels and had already been identified as a signature sequence useful for identifying potential K$^+$ channels. The region of the strand containing the conserved sequence lies in an

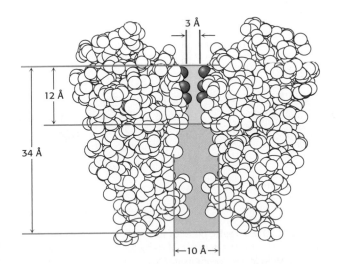

Figure 13.18 Path through a channel. A potassium ion entering the K$^+$ channel can pass a distance of 22 Å into the membrane while remaining solvated with water (blue). At this point, the pore diameter narrows to 3 Å (yellow), and potassium ions must shed their water and interact with carbonyl groups (red) of the pore amino acids.

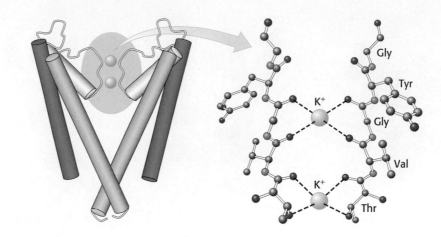

Figure 13.19 Selectivity filter of the potassium ion channel. Potassium ions interact with the carbonyl groups of the TVGYG sequence of the selectivity filter, located at the 3-Å-diameter pore of the K⁺ channel. Only two of the four channel subunits are shown.

TABLE 13.1 Properties of alkali cations

Ion	Ionic radius (Å)	Hydration free energy in kJ mol⁻¹ (kcal mol⁻¹)
Li^+	0.60	−410 (−98)
Na^+	0.95	−301 (−72)
K^+	1.33	−230 (−55)
Rb^+	1.48	−213 (−51)
Cs^+	1.69	−197 (−47)

extended conformation and is oriented such that the peptide carbonyl groups are directed into the channel, in good position to interact with the potassium ions.

Potassium ion channels are 100-fold more permeable to K^+ than to Na^+. How is this high degree of selectivity achieved? Ions having a radius larger than 1.5 Å cannot pass into the narrow diameter (3 Å) of the selectivity filter of the K^+ channel. However, a bare Na^+ is small enough (Table 13.1) to pass through the pore. Indeed, the ionic radius of Na^+ is substantially smaller than that of K^+. How then is Na^+ rejected?

The key point is that the free-energy costs of dehydrating these ions are considerable [Na^+, 301 kJ mol⁻¹ (72 kcal mol⁻¹), and K^+, 230 kJ mol⁻¹ (55 kcal mol⁻¹)]. *The channel pays the cost of dehydrating K^+ by providing compensating interactions with the carbonyl oxygen atoms lining the selectivity filter.* However, these oxygen atoms are positioned such that they do not interact favorably with Na^+, because the ion is too small (Figure 13.20). Sodium ions are rejected because the higher cost of dehydrating them would be unrecovered. The potassium ion channel avoids closely embracing sodium ions, which must stay hydrated and hence cannot pass through the channel.

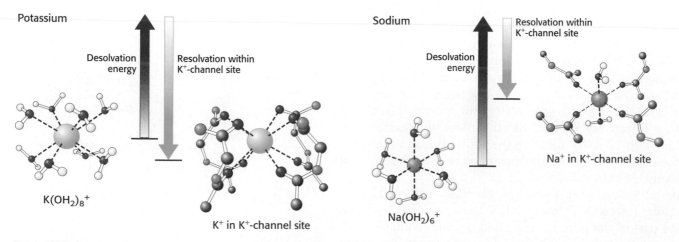

Potassium

Desolvation energy   Resolvation within K⁺-channel site

$K(OH_2)_8^+$

K^+ in K⁺-channel site

Sodium

Desolvation energy   Resolvation within K⁺-channel site

$Na(OH_2)_6^+$

Na^+ in K⁺-channel site

Figure 13.20 Energetic basis of ion selectivity. The energy cost of dehydrating a potassium ion is compensated by favorable interactions with the selectivity filter. Because a sodium ion is too small to interact favorably with the selectivity filter, the free energy of desolvation cannot be compensated and the sodium ion does not pass through the channel.

The structure determined for K^+ channels is a good start for considering the structure and function of Na^+ and Ca^{2+} channels because of their homology to K^+ channels. Sequence comparisons and the results of mutagenesis experiments have implicated the region between segments S5 and S6 in ion selectivity in the Ca^{2+} channel. In Ca^{2+} channels, one glutamate residue of this region in each of the four repeated units plays a major role in determining ion selectivity. Residues in the positions corresponding to the glutamate residues in Ca^{2+} channels are major components of the selectivity filter of the Na^+ channel. These residues are aspartate, glutamate, lysine, and alanine in units 1, 2, 3, and 4, respectively (the DEKA locus). Thus, the potential fourfold symmetry of the channel is clearly broken in this region, which explains why Na^+ channels consist of a single large polypeptide chain rather than a noncovalent assembly of four identical subunits. The Na^+ channel's selection of Na^+ over K^+ depends on ionic radius; the diameter of the pore determined by these residues and others is sufficiently restricted that small ions such as Na^+ and Li^+ can pass through the channel, but larger ions such as K^+ are significantly hindered.

The Structure of the Potassium Ion Channel Explains Its Rapid Rate of Transport

The tight binding sites required for ion selectivity should slow the progress of ions through a channel, yet ion channels achieve rapid rates of ion transport. How is this apparent paradox resolved? A structural analysis of the K^+ channel at high resolution provides an appealing explanation. Four K^+-binding sites crucial for rapid ion flow are present in the constricted region of the K^+ channel. Consider the process of ion conductance starting from inside the cell (Figure 13.21). A hydrated potassium ion proceeds into the channel and through the relatively unrestricted part of the channel. The ion then gives up its coordinated water molecules and binds to a site within the selectivity-filter region. The ion can move between the four sites within the selectivity filter because they have similar ion affinities. As each subsequent potassium ion moves into the selectivity filter, its positive charge will repel the potassium ion at the nearest site, causing it to shift to a site farther up the channel and in turn push upward any potassium ion already bound to a site farther up. Thus, each ion that binds anew favors the release of an ion from the other side of the channel. This multiple-binding-site mechanism solves the apparent paradox of high ion selectivity and rapid flow.

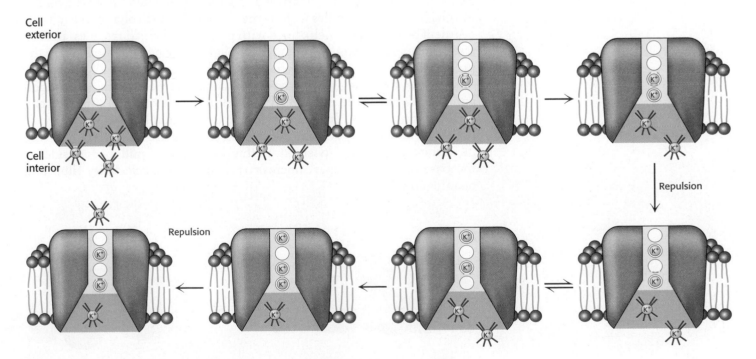

Figure 13.21 Model for K^+-channel ion transport. The selectivity filter has four binding sites. Hydrated potassium ions can enter these sites, one at a time, losing their hydration shells. When two ions occupy adjacent sites, electrostatic repulsion forces them apart. Thus, as ions enter the channel from one side, other ions are pushed out the other side.

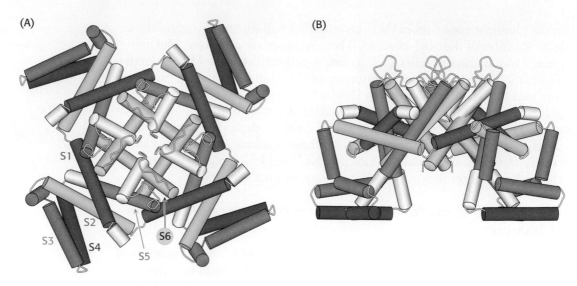

(A)

S1

S2

S3   S4

S5   S6

(B)

Figure 13.22 Structure of a voltage-gated potassium channel. (A) A view looking down through the pore. (B) A side view. *Notice* that the positively charged S4 region (red) lies on the outside of the structure at the bottom of the pore. [Drawn from 1ORQ.pdb.]

Voltage Gating Requires Substantial Conformational Changes in Specific Ion-Channel Domains

Some Na^+ and K^+ channels are gated by membrane potential; that is, they change conformation to a highly conducting form in response to changes in voltage across the membrane. As already noted, these *voltage-gated channels* include segments S1 through S4 in addition to the pore itself formed by S5 and S6. The structure of a voltage-gated K^+ channel from *Aeropyrum pernix* has been determined by x-ray crystallography (Figure 13.22). The segments S1 through S4 form domains termed "paddles" that extend from the core of the channel. These paddles include the segment S4, the voltage sensor itself. Segment S4 forms an α helix lined with positively charged residues. In contrast with expectations, segments S1 through S4 are not enclosed within the protein but, instead, are positioned to lie in the membrane itself.

A model for voltage gating has been proposed by Roderick MacKinnon and coworkers on the basis of this structure and a range of other experiments (Figure 13.23). In the closed state, the paddles lie in a "down" position. On membrane depolarization, the paddles are pulled through the membrane into an "up" position. In this position, they pull the four sides of the base on the pore apart, increasing access to the selectivity filter and opening the channel.

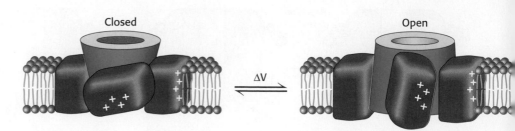

Closed   ΔV   Open

Figure 13.23 A model for voltage gating of ion channels. The voltage sensing paddles lie in the "down" position below the closed channel (left). Membrane depolarization pulls these paddles through the membrane. The motion pulls the base of the channel apart, opening the channel (right).

A Channel Can Be Inactivated by Occlusion of the Pore: The Ball-and-Chain Model

The K^+ channel and the Na^+ channel undergo inactivation within milliseconds of opening (Figure 13.24). A first clue to the mechanism of inactivation came from exposing the cytoplasmic side of either channel to trypsin; cleavage by trypsin produced trimmed channels that stayed persistently open after depolarization. Furthermore, a mutant Shaker channel lacking 42 amino acids near the amino terminus opened in response to depolarization but did not inactivate. Remarkably, inactivation was restored by adding a synthetic peptide corresponding to the first 20 residues of the native channel.

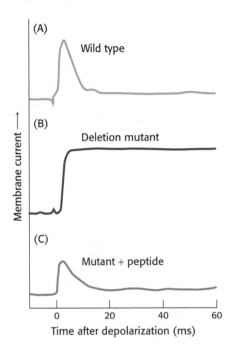

(A) Wild type

(B) Deletion mutant

(C) Mutant + peptide

Membrane current →

Time after depolarization (ms)

0   20   40   60

Figure 13.24 Inactivation of the potassium ion channel. The amino-terminal region of the K^+ chain is critical for inactivation. (A) The wild-type Shaker K^+ channel displays rapid inactivation after opening. (B) A mutant channel lacking residues 6 through 46 does not inactivate. (C) Inactivation can be restored by adding a peptide consisting of residues 1 through 20 at a concentration of 100 μM. [After W. N. Zagotta, T. Hoshi, and R. W. Aldrich. *Science* 250(1990):568–571.]

These experiments strongly support the *ball-and-chain model* for channel inactivation that had been proposed years earlier (Figure 13.25). According to this model, the first 20 residues of the K^+ channel form a cytoplasmic unit (the *ball*) that is attached to a flexible segment of the polypeptide (the *chain*). When the channel is closed, the ball rotates freely in the aqueous solution. When the channel opens, the ball quickly finds a complementary site in the open pore and occludes it. Hence, the channel opens for only a brief interval before it undergoes inactivation by occlusion.

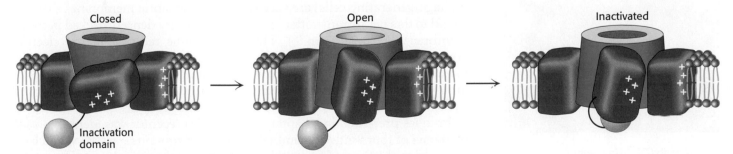

Closed

Inactivation domain

Open

Inactivated

Figure 13.25 Ball-and-chain model for channel inactivation. The inactivation domain, or "ball" (gray), is tethered to the channel by a flexible "chain." In the closed state, the ball is located in the cytoplasm. Depolarization opens the channel and creates a binding site for the positively charged ball in the mouth of the pore. Movement of the ball into this site inactivates the channel by occluding it.

Shortening the chain speeds inactivation because the ball finds its target more quickly. Conversely, lengthening the chain slows inactivation. Thus, the duration of the open state can be controlled by the length and flexibility of the tether. In some senses, the "ball" domains, which include substantial regions of positive charge, can be thought of as large, tethered cations that are pulled into the open channel but get stuck and block further ion conductance.

The Acetylcholine Receptor Is an Archetype for Ligand-Gated Ion Channels

Nerve impulses are communicated across synapses by small, diffusible molecules called *neurotransmitters*. One neurotransmitter is *acetylcholine*. The presynaptic membrane of a synapse is separated from the postsynaptic membrane by a gap of about 50 nm, called the *synaptic cleft*. The arrival of a nerve impulse at the end of an axon leads to the synchronous export of the contents of some 300 vesicles of acetylcholine into the cleft (Figure 13.26). The binding of acetylcholine to the postsynaptic membrane markedly changes its ionic permeability, triggering an action potential. Acetylcholine opens a single kind of cation channel, called the *acetylcholine receptor,* which is almost equally permeable to Na^+ and K^+.

Acetylcholine

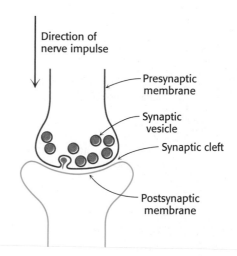

Figure 13.26 Schematic representation of a synapse.

The acetylcholine receptor is the best-understood *ligand-gated channel.* This type of channel is gated not by voltage but by the presence of specific ligands. The binding of acetylcholine to the channel is followed by its transient opening. The electric organ of *Torpedo marmorata,* an electric ray, is a choice source of acetylcholine receptors for study because its electroplaxes (voltage-generating cells) are very rich in postsynaptic membranes that respond to this neurotransmitter. The receptor is very densely packed in these membranes ($\sim$20,000 μm^{-2}). The acetylcholine receptor of the electric organ has been solubilized by adding a nonionic detergent to a postsynaptic membrane preparation and purified by affinity chromatography on a column bearing covalently attached cobratoxin, a small protein toxin from snakes that has a high affinity for acetylcholine receptors. With the use of techniques presented in Chapter 3, the 268-kd receptor was identified as a pentamer of four kinds of membrane-spanning subunits—α_2, β, γ, and δ— arranged in the form of a ring that creates a pore through the membrane.

The cloning and sequencing of the cDNAs for the four kinds of subunits (50–58 kd) showed that they have clearly similar sequences; the genes for the α, β, γ, and δ subunits arose by duplication and divergence of a common ancestral gene. Each subunit has a large extracellular domain, followed at

The torpedo (*Torpedo marmorata,* also known as the electric ray) has an electric organ, rich in acetylcholine receptors, that can deliver a shock of as much as 200 V for approximately 1 s. [Yves Gladu/Jacana/ Photo Researchers.]

(A)

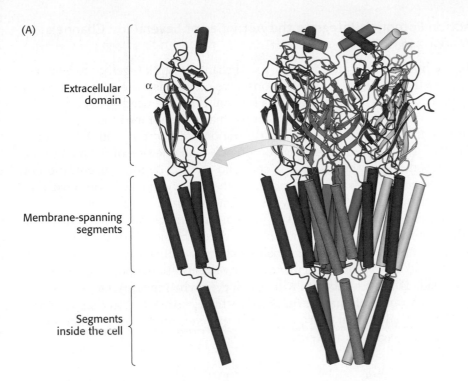

Extracellular domain

α

Membrane-spanning segments

Segments inside the cell

(B)

β

α

γ

α

δ

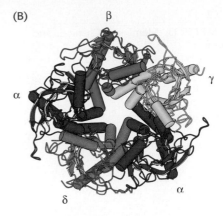

Figure 13.27 Structure of the acetylcholine receptor. A model for the structure of the acetylcholine receptor deduced from high-resolution electron microscopic studies reveals that each subunit consists of a large extracellular domain consisting primarily of β strands, four membrane-spanning α helices, and a final α helix inside the cell. (A) A side view shows the pentameric receptor with each subunit type in a different color. One copy of the α subunit is shown in isolation. (B) A view down the channel from outside the cell. [Drawn from 2BG9.pdb.]

the carboxyl end by four predominantly hydrophobic segments that span the bilayer membrane. Acetylcholine binds at the α–γ and α–δ interfaces. Electron microscopic studies of purified acetylcholine receptors demonstrated that the structure has approximate fivefold symmetry, in harmony with the similarity of its five constituent subunits (Figure 13.27).

What is the basis of channel opening? A comparison of the structures of the closed and open forms of the channel would be highly revealing, but such comparisons have been difficult to obtain. Cryoelectron micrographs indicate that the binding of acetylcholine to the extracellular domain causes a structural alteration, which initiates rotations of the α-helical rods lining the membrane-spanning pore. The amino acid sequences of these helices point to the presence of alternating ridges of small polar or neutral residues (serine, threonine, glycine) and large nonpolar ones (isoleucine, leucine, phenylalanine). In the closed state, the large residues may occlude the channel by forming a tight hydrophobic ring (Figure 13.28). Indeed, each subunit has a bulky leucine residue at a critical position. The binding of acetylcholine could allosterically rotate the membrane-spanning helices so that the pore would be lined by small polar residues rather than by large hydrophobic ones. The wider, more polar pore would then be open to the passage of Na^+ and K^+ ions.

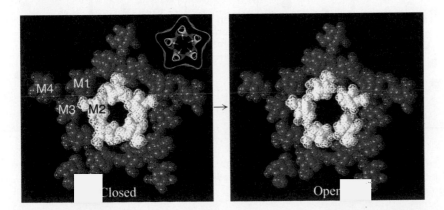

M4 M1

M3 M2

Closed

Open

Figure 13.28 Opening the acetylcholine receptor. Cross sections from electron microscopic reconstructions of the acetylcholine receptor in (left) its closed form and (right) its open form. (The open form corresponds to the structure shown in Figure 13.27). The areas labeled M1, M2, M3, and M4 correspond to the four membrane-spanning α helices of one subunit. The cross section of the open channel was generated by treating the receptor with acetylcholine and freezing the sample within 20 ms. *Notice* that the hole in the center of the channel is substantially larger in the open structure. The enlargement of the hole is due to the rotation of the M2 helices by approximately 15 degrees along their long axes. [Courtesy of Nigel Unwin.]

Action Potentials Integrate the Activities of Several Ion Channels Working in Concert

To see how ligand-gated and ion-gated channels work together to generate a sophisticated physiological response, we now revisit the action potential introduced at the beginning of this section. First, we need to introduce the concept of *equilibrium potential*. Suppose that a membrane separates two solutions that contain different concentrations of some cation X^+ (Figure 13.29). Let $[X^+]_{in}$ be the concentration of X^+ on one side of the membrane (corresponding to the inside of a cell) and $[X^+]_{out}$ be the concentration of X^+ on the other side (corresponding to the outside of a cell). Suppose that an ion channel opens that allows X^+ to move across the membrane. What will happen? It seems clear that X^+ will move through the channel from the side with the higher concentration (side A) to the side with the lower concentration (side B). However, positive charges will start to accumulate on side B, making it more difficult to move each additional positively charged ion to side B. An equilibrium will be achieved when the driving force due to the concentration gradient is balanced by the electrostatic force resisting the motion of a additional charge. Under these circumstances, the membrane potential is given by the *Nernst equation*:

$$V_{eq} = -(RT/zF)\ln([X]_{in}/[X]_{out})$$
$$= -(2.303)(RT/zF)\log_{10}([X]_{in}/[X]_{out})$$

where R is the gas constant and F is the Faraday constant (p. 353) and z is the charge on the ion X (e.g., $+1$ for X^+).

The membrane potential at equilibrium is called the equilibrium potential for a given ion at a given concentration ratio across a membrane. For sodium with $[Na^+]_{in} = 14$ mM and $[Na^+]_{out} = 143$ mM, the equilibrium potential is $+62$ mV at 37°C. Similarly, for potassium with $[K^+]_{in} = 157$ mM and $[K^+]_{out} = 4$ mM, the equilibrium potential is -98 mV. In the absence of stimulation, the resting potential for a typical neuron is -60 mV. This value is close to the equilibrium potential for K^+ owing to the fact that a small number of K^+ channels are open.

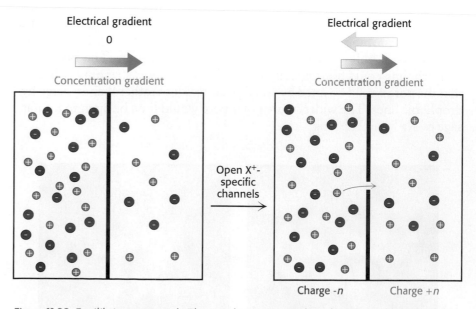

Figure 13.29 Equilibrium potential. The membrane potential reaches an equilibrium when the driving force due to the concentration gradient is exactly balanced by the opposing force due to the repulsion of like charges.

We are now prepared to consider what happens in the generation of an action potential (Figure 13.30). Initially, a neurotransmitter such as acetylcholine is released from an adjacent cell. The released acetylcholine binds to the acetylcholine receptor, causing it to open within less than a millisecond. The acetylcholine receptor is a nonspecific cation channel. Sodium ions flow into the cell and potassium ions flow out of the cell. Without any further events, the membrane potential would move to a value corresponding to the average of the equilibrium potentials for Na^+ and K^+, approximately -20 mV. However, as the membrane potential approaches -40 mV, the voltage-sensing paddles of Na^+ channels are pulled into the membrane, opening the Na^+ channels. With these channels open, sodium ions flow rapidly into the cell and the membrane potential rises rapidly toward the Na^+ equilibrium potential. The voltage-sensing paddles of K^+ channels also are pulled into the membrane by the changed membrane potential, but more slowly than Na^+ channel paddles. Nonetheless, after approximately 1 ms, many K^+ channels start to open. At the same time, inactivation "ball" domains plug the open Na^+ channels, decreasing the Na^+ current. The acetylcholine receptors that initiated these events are also inactivated on this time scale. With the Na^+ channels inactivated and only the K^+ channels open, the membrane potential drops rapidly toward the K^+ equilibrium potential. The open K^+ channels are susceptible to inactivation by their "ball" domains, and these K^+ currents, too, are blocked. With the membrane potential returned to close to its initial value, the inactivation domains are released and the channels return to their original closed states. These events propagate along the neuron as the depolarization of the membrane opens channels in nearby patches of membrane.

How much current is there in these processes? This question can be addressed from two complementary directions. First, a typical nerve cell contains 100 Na^+ channels per square micrometer. At a membrane potential of $+20$ mV, each channel conducts 10^7 ions per second. Thus, in a period of 1 millisecond, approximately 10^5 ions flow through each square micrometer of membrane surface. Assuming a cell volume of 10^4 μm^3 and a surface area of 10^4 μm^2, this rate of ion flow corresponds to an increase in the Na^+ concentration of less than 1%. How can this be? A robust action potential is generated because the membrane potential is very sensitive to even a slight change in the distribution of charge. This sensitivity makes the action potential a very efficient means of signaling over long distances and with rapid repetition rates.

13.5 Gap Junctions Allow Ions and Small Molecules to Flow Between Communicating Cells

The ion channels that we have considered thus far have narrow pores and are moderately to highly selective in the ions that they allow to pass through them. They are closed in the resting state and have short lifetimes in the open state, typically a millisecond, that enable them to transmit highly frequent neural signals. We turn now to a channel with a very different role. *Gap junctions*, also known as *cell-to-cell channels*, serve as passageways between the interiors of contiguous cells. Gap junctions are clustered in discrete regions of the plasma membranes of apposed cells. Electron micrographs of sheets of gap junctions show them tightly packed in a regular hexagonal array (Figure 13.31). A 20-Å central hole, the lumen of the channel, is prominent in each gap junction. These channels span the intervening space, or gap, between apposed cells (hence, the name gap junction). The width of the gap between the cytoplasms of the two cells is about 35 Å.

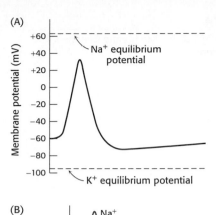

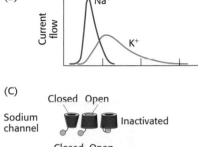

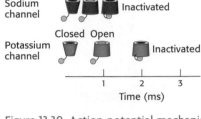

Figure 13.30 Action-potential mechanism. (A) On the initation of an action potential, the membrane potential moves from the resting potential upward toward the Na^+ equilibrium potential and then downward toward the K^+ equilibrium potential. (B) The currents through the Na^+ and K^+ channels underlying the action potential. (C) The states of the Na^+ and K^+ channels during the action potential.

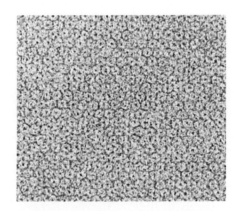

Figure 13.31 Gap junctions. This electron micrograph shows a sheet of isolated gap junctions. The cylindrical connexons form a hexagonal lattice having a unit-cell length of 85 Å. The densely stained central hole has a diameter of about 20 Å. [Courtesy of Dr. Nigel Unwin and Dr. Guido Zampighi.]

Small hydrophilic molecules as well as ions can pass through gap junctions. The pore size of the junctions was determined by microinjecting a series of fluorescent molecules into cells and observing their passage into adjoining cells. All polar molecules with a mass of less than about 1 kd can readily pass through these cell-to-cell channels. Thus, *inorganic ions and most metabolites (e.g., sugars, amino acids, and nucleotides) can flow between the interiors of cells joined by gap junctions.* In contrast, proteins, nucleic acids, and polysaccharides are too large to traverse these channels. *Gap junctions are important for intercellular communication.* Cells in some excitable tissues, such as heart muscle, are coupled by the rapid flow of ions through these junctions, which ensures a rapid and synchronous response to stimuli. Gap junctions are also essential for the nourishment of cells that are distant from blood vessels, as in lens and bone. Moreover, communicating channels are important in development and differentiation. For example, a pregnant uterus is transformed from a quiescent protector of the fetus to a forceful ejector at the onset of labor; the formation of functional gap junctions at that time creates a syncytium of muscle cells that contract in synchrony.

A cell-to-cell channel is made of 12 molecules of *connexin,* one of a family of transmembrane proteins with molecular masses ranging from 30 to 42 kd. Each connexin molecule appears to have four membrane-spanning helices. Six connexin molecules are hexagonally arrayed to form a half channel, called a *connexon* or *hemichannel* (Figure 13.32). Two connexons join end to end in the intercellular space to form a functional channel between the communicating cells. Cell-to-cell channels differ from other membrane channels in three respects: (1) they traverse *two* membranes rather than one; (2) they connect cytoplasm to cytoplasm, rather than to the extracellular space or the lumen of an organelle; and (3) the connexons forming a channel are synthesized by different cells. Gap junctions form readily when cells are brought together. A cell-to-cell channel, once formed, tends to stay open for seconds to minutes. They are closed by high concentrations of calcium ion and by low pH. *The closing of gap junctions by Ca^{2+} and H^+ serves to seal normal cells from traumatized or dying neighbors.* Gap junctions are also controlled by membrane potential and by hormone-induced phosphorylation.

The human genome encodes 21 distinct connexins. Different members of this family are expressed in different tissues. For example, connexin 26 is expressed in key tissues in the ear. Mutations in this connexin are associated with hereditary deafness. The mechanistic basis for this deafness appears to be insufficient transport of ions or second-messenger molecules, such as inositol trisphosphate, between sensory cells.

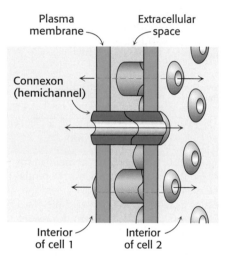

Plasma membrane

Extracellular space

Connexon (hemichannel)

Interior of cell 1

Interior of cell 2

Figure 13.32 Schematic representation of a gap junction. [Courtesy of Dr. Werner Loewenstein.]

13.6 Specific Channels Increase the Permeability of Some Membranes to Water

One more important class of channels does not take part in ion transport at all. Instead, these channels increase the rate at which water flows through membranes. As noted in Chapter 12, membranes are reasonably permeable to water. Why, then, are water-specific channels required? In certain tissues under some circumstances, rapid water transport through membranes is necessary. In the kidney, for example, water must be rapidly reabsorbed into the bloodstream after filtration. Similarly, in the secretion of saliva and tears, water must flow quickly through membranes. These observations suggested the existence of specific water channels, but initially the channels could not be identified.

The channels (now called *aquaporins*) were discovered serendipitously. Peter Agre noticed a protein present at high levels in red-blood-cell membranes that had been missed because the protein does not stain well with Coomassie brilliant blue. This protein was found to be present in large

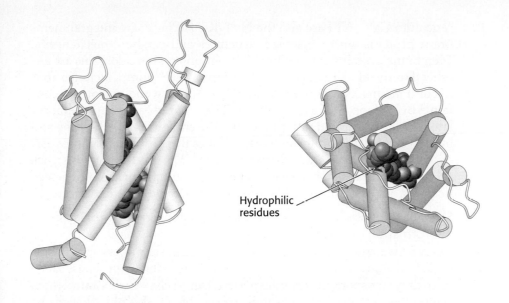

Hydrophilic
residues

Figure 13.33 Structure of aquaporin. The structure of aquaporin viewed from the side (left) and from the top (right). *Notice* the hydrophilic residues (shown as space-filling models) that line the water channel. [Drawn from 1J4N.pdb.]

quantities in red blood cells as well as in tissues such as kidneys and corneas, precisely the tissues thought to contain water channels. On the basis of this observation, further studies were designed that revealed that this 24-kd membrane protein is, indeed, a water channel.

The structure of aquaporin has been determined (Figure 13.33). The protein consists of six membrane-spanning α helices. Two loops containing hydrophilic residues line the actual channel. Water molecules pass through in single file at a rate of 10^6 molecules per second. Importantly, specific positively charged residues toward the center of the channel prevent the transport of protons through aquaporin. Thus, aquaporin channels will not disrupt proton gradients, which play fundamental roles in energy transduction, as we will see in Chapter 18. The aquaporins reveal that channels can evolve that specifically do not conduct ions, as can those that do.

Summary

13.1 The Transport of Molecules Across a Membrane May Be Active or Passive

For a net movement of molecules across a membrane, two features are required: (1) the molecule must be able to cross a hydrophobic barrier and (2) an energy source must power the movement. Lipophilic molecules can pass through a membrane's hydrophobic interior by simple diffusion. These molecules will move down their concentration gradients. Polar or charged molecules require proteins to form passages through the hydrophobic barrier. Passive transport or facilitated diffusion takes place when an ion or polar molecule moves down its concentration gradient. If a molecule moves against a concentration gradient, an external energy source is required; this movement is referred to as active transport and results in the generation of concentration gradients. The electrochemical potential measures the combined ability of a concentration gradient and an uneven distribution of charge to drive species across a membrane.

13.2 Two Families of Membrane Proteins Use ATP Hydrolysis to Pump Ions Across Membranes

Active transport is often carried out at the expense of ATP hydrolysis. P-type ATPases pump ions against a concentration gradient and become transiently phosphorylated on an aspartic acid residue in the process of transport. P-type ATPases, which include the sarcoplasmic reticulum Ca^{2+} ATPase and the Na^+–K^+ ATPase, are integral membrane proteins with conserved structures and catalytic mechanisms. Membrane proteins containing ATP-binding cassette domains are another family of ATP-dependent pumps. Each pump includes four major domains: two domains span the membrane and two others contain ABC P-loop ATPase structures. These pumps are not phosphorylated during pumping; rather, they use the energy of ATP binding and hydrolysis to drive conformational changes that result in the transport of specific substrates across membranes. The multidrug-resistance proteins confer resistance on cancer cells by pumping chemotherapeutic drugs out of a cancer cell before the drugs can exert their effects.

13.3 Secondary Transporters Use One Concentration Gradient to Power the Formation of Another

Many active-transport systems couple the uphill flow of one ion or molecule to the downhill flow of another. These membrane proteins, called secondary transporters or cotransporters, can be classified as antiporters, symporters, and uniporters. Antiporters couple the downhill flow of one type of ion in one direction to the uphill flow of another in the opposite direction. Symporters move both ions in the same direction. Uniporters transport a substrate in either direction, determined by the concentration differences. Studies of the lactose permease from *E. coli* have been a source of insight into both the structures and the mechanisms of secondary transporters.

13.4 Specific Channels Can Rapidly Transport Ions Across Membranes

Ion channels allow the rapid movement of ions across the hydrophobic barrier of the membrane. The activity of individual ion-channel molecules can be observed by using patch-clamp techniques. Many ion channels have a common structural framework. In regard to K^+ channels, hydrated potassium ions must transiently lose their coordinated water molecules as they move to the narrowest part of the channel, termed the selectivity filter. In the selectivity filter, peptide carbonyl groups coordinate the ions. Rapid ion flow through the selectivity filter is facilitated by ion–ion repulsion, with one ion pushing the next ion through the channel. Some ion channels are voltage gated: changes in membrane potential induce conformational changes that open these channels. Many channels spontaneously inactivate after having been open for a short period of time. In some cases, inactivation is due to the binding of a domain of the channel termed the "ball" in the mouth of the channel to block it. Other channels, typified by the acetylcholine receptor, are opened or closed by the binding of ligands. Ligand-gated and voltage-gated channels work in concert to generate action potentials.

13.5 Gap Junctions Allow Ions and Small Molecules to Flow Between Communicating Cells

In contrast with many channels, which connect the cell interior with the environment, gap junctions, or cell-to-cell channels, serve to connect the interiors of contiguous groups of cells. A cell-to-cell channel is composed of 12 molecules of connexin, which associate to form two 6-membered connexons.

13.6 Specific Channels Increase the Permeability of Some Membranes to Water

Some tissues contain proteins that increase the permeability of membranes to water. Each water-channel-forming protein, termed an aquaporin, consists of six membrane-spanning α helices and a central channel lined with hydrophilic residues that allow water molecules to pass in single file. Aquaporins do not transport protons.

Key Terms

pump (p. 351)

channel (p. 351)

active transport (p. 351)

facilitated diffusion (passive transport) (p. 351)

ATP-driven pump (p. 351)

secondary transporter (p. 351)

ion channel (p. 351)

simple diffusion (p. 352)

electrochemical potential (membrane potential) (p. 353)

Na^+–K^+ pump (Na^+–K^+ATPase) (p. 354)

gastric H^+–K^+ ATPase (p. 354)

P-type ATPase (p. 354)

Ca^{2+} ATPase (SR Ca^{2+} ATPase, or SERCA) (p. 355)

eversion (p. 356)

cardiotonic steroid (p. 357)

digitalis (p. 357)

multidrug resistance (p. 358)

multidrug-resistance (MDR) protein (P-glycoprotein) (p. 358)

ATP-binding cassette (ABC) domain (p. 358)

ABC transporter (p. 358)

cotransporter (p. 360)

antiporter (p. 360)

symporter (p. 360)

uniporter (p. 360)

lactose permease (p. 360)

ion channel (p. 362)

nerve impulse (p. 362)

action potential (p. 362)

patch-clamp technique (p. 363)

gigaseal (p. 363)

selectivity filter (p. 365)

voltage-gated channel (p. 368)

ball-and-chain model (p. 369)

neurotransmitter (p. 369)

acetylcholine (p. 369)

synaptic cleft (p. 369)

acetylcholine receptor (p. 369)

ligand-gated channel (p. 369)

equilibrium potential (p. 372)

Nernst equation (p. 372)

gap junction (p. 373)

connexin (p. 374)

connexon (hemichannel) (p. 374)

aquaporin (p. 374)

Selected Readings

Where to Start

Lancaster, C. R. 2004. Structural biology: Ion pump in the movies. *Nature* 432:286–287.

Unwin, N. 2003. Structure and action of the nicotinic acetylcholine receptor explored by electron microscopy. *FEBS Lett.* 555:91–95.

Abramson, J., Smirnova, I., Kasho, V., Verner, G., Iwata, S., and Kaback, H. R. 2003. The lactose permease of *Escherichia coli*: Overall structure, the sugar-binding site and the alternating access model for transport. *FEBS Lett.* 555:96–101.

Lienhard, G. E., Slot, J. W., James, D. E., and Mueckler, M. M. 1992. How cells absorb glucose. *Sci. Am.* 266(1):86–91.

King, L. S., Kozono, D., and Agre, P. 2004. From structure to disease: The evolving tale of aquaporin biology. *Nat. Rev. Mol. Cell Biol.* 5:687–698.

Neher, E., and Sakmann, B. 1992. The patch clamp technique. *Sci. Am.* 266(3):28–35.

Sakmann, B. 1992. Elementary steps in synaptic transmission revealed by currents through single ion channels. *Science* 256:503–512.

Books

Ashcroft, F. M. 2000. *Ion Channels and Disease.* Academic Press.

Conn, P. M. (Ed.). 1998. *Ion Channels*, vol. 293, *Methods in Enzymology.* Academic Press.

Aidley, D. J., and Stanfield, P. R. 1996. *Ion Channels: Molecules in Action.* Cambridge University Press.

Hille, B. 2001. *Ionic Channels of Excitable Membranes* (3d ed.). Sinauer.

Läuger, P. 1991. *Electrogenic Ion Pumps.* Sinauer.

Stein, W. D. 1990. *Channels, Carriers, and Pumps: An Introduction to Membrane Transport.* Academic Press.

Hodgkin, A. 1992. *Chance and Design: Reminiscences of Science in Peace and War.* Cambridge University Press.

P-Type ATPases

Sorensen, T. L., Moller, J. V., and Nissen, P. 2004. Phosphoryl transfer and calcium ion occlusion in the calcium pump. *Science* 304:1672–1675.

Sweadner, K. J., and Donnet, C. 2001. Structural similarities of Na,K-ATPase and SERCA, the Ca^{2+}-ATPase of the sarcoplasmic reticulum. *Biochem. J.* 356:685–704.

Toyoshima, C., and Mizutani, T. 2004. Crystal structure of the calcium pump with a bound ATP analogue. *Nature* 430:529–535.

Toyoshima, C., Nakasako, M., Nomura, H., and Ogawa, H. 2000. Crystal structure of the calcium pump of sarcoplasmic reticulum at 2.6 Å resolution. *Nature* 405:647–655.

Auer, M., Scarborough, G. A., and Kuhlbrandt, W. 1998. Three-dimensional map of the plasma membrane H^+-ATPase in the open conformation. *Nature* 392:840–843.

Axelsen, K. B., and Palmgren, M. G. 1998. Evolution of substrate specificities in the P-type ATPase superfamily. *J. Mol. Evol.* 46:84–101.

Pedersen, P. A., Jorgensen, J. R., and Jorgensen, P. L. 2000. Importance of conserved α-subunit ^{709}GDGVND for Mg^{2+} binding, phosphorylation, energy transduction in Na, K-ATPase. *J. Biol. Chem.* 275:37588–37595.

Blanco, G., and Mercer, R. W. 1998. Isozymes of the Na-K-ATPase: Heterogeneity in structure, diversity in function. *Am. J. Physiol.* 275:F633–F650.

Estes, J. W., and White, P. D. 1965. William Withering and the purple foxglove. *Sci. Am.* 212(6):110–117.

ATP-Binding Cassette Proteins

Locher, K. P. 2004. Structure and mechanism of ABC transporters. *Curr. Opin. Struct. Biol.* 14:426–431.

Locher, K. P., Lee, A. T., and Rees, D. C. 2002. The *E. coli* BtuCD structure: A framework for ABC transporter architecture and mechanism. *Science* 296:1091–1098.

Reyes, C. L., and Chang, G. 2005. Structure of the ABC transporter MsbA in complex with ADP.vanadate and lipopolysaccharide. *Science* 308:1028–31.

Borths, E. L., Locher, K. P., Lee, A. T., and Rees, D. C. 2002. The structure of *Escherichia coli* BtuF and binding to its cognate ATP binding cassette transporter. *Proc. Natl. Acad. Sci. U.S.A.* 99:16642–16647.

Chang, G. 2003. Structure of MsbA from *Vibrio cholera*: A multidrug resistance ABC transporter homolog in a closed conformation. *J. Mol. Biol.* 330:419–430.

Dong, J., Yang, G., and McHaourab, H. S. 2005. Structural basis of energy transduction in the transport cycle of MsbA. *Science* 308:1023–1028.

Akabas, M. H. 2000. Cystic fibrosis transmembrane conductance regulator: Structure and function of an epithelial chloride channel. *J. Biol. Chem.* 275:3729–3732.

Chen, J., Sharma, S., Quiocho, F. A., and Davidson, A. L. 2001. Trapping the transition state of an ATP-binding cassette transporter: Evidence for a concerted mechanism of maltose transport. *Proc. Natl. Acad. Sci. U. S. A.* 98:1525–1530.

Sheppard, D. N., and Welsh, M. J. 1999. Structure and function of the CFTR chloride channel. *Physiol. Rev.* 79:S23–S45.

Jones, P. M., and George, A. M. 2000. Symmetry and structure in P-glycoprotein and ABC transporters: What goes around comes around. *Eur. J. Biochem.* 287:5298–5305.

Chen, Y., and Simon, S. M. 2000. In situ biochemical demonstration that P-glycoprotein is a drug efflux pump with broad specificity. *J. Cell Biol.* 148:863–870.

Saier, M. H., Jr., Paulsen, I. T., Sliwinski, M. K., Pao, S. S., Skurray, R. A., and Nikaido, H. 1998. Evolutionary origins of multidrug and drug-specific efflux pumps in bacteria. *FASEB J.* 12:265–274.

Symporters and Antiporters

Abramson, J., Smirnova, I., Kasho, V., Verner, G., Kaback, H. R., and Iwata, S. 2003. Structure and mechanism of the lactose permease of *Escherichia coli. Science* 301:610–615.

Philipson, K. D., and Nicoll, D. A. 2000. Sodium-calcium exchange: A molecular perspective. *Annu. Rev. Physiol.* 62:111–133.

Pao, S. S., Paulsen, I. T., and Saier, M. H., Jr. 1998. Major facilitator superfamily. *Microbiol. Mol. Biol. Rev.* 62:1–34.

Wright, E. M., Hirsch, J. R., Loo, D. D., and Zampighi, G. A. 1997. Regulation of Na^+/glucose cotransporters. *J. Exp. Biol.* 200:287–293.

Kaback, H. R., Bibi, E., and Roepe, P. D. 1990. β-Galactoside transport in *E. coli:* A functional dissection of lac permease. *Trends Biochem. Sci.* 8:309–314.

Hilgemann, D. W., Nicoll, D. A., and Philipson, K. D. 1991. Charge movement during Na^+ translocation by native and cloned cardiac Na^+/Ca^{2+} exchanger. *Nature* 352:715–718.

Hediger, M. A., Turk, E., and Wright, E. M. 1989. Homology of the human intestinal Na^+/glucose and *Escherichia coli* Na^+/proline cotransporters. *Proc. Natl. Acad. Sci. U. S. A.* 86:5748–5752.

Ion Channels

Zhou, Y., and MacKinnon, R. 2003. The occupancy of ions in the K^+ selectivity filter: Charge balance and coupling of ion binding to a protein conformational change underlie high conduction rates. *J. Mol. Biol.* 333:965–975.

Zhou, Y., Morais-Cabral, J. H., Kaufman, A., and MacKinnon, R. 2001. Chemistry of ion coordination and hydration revealed by a K^+ channel-Fab complex at 2.0 Å resolution. *Nature* 414:43–48.

Jiang, Y., Lee, A., Chen, J., Cadene, M., Chait, B. T., and MacKinnon, R. 2002. The open pore conformation of potassium channels. *Nature* 417:523–526.

Jiang, Y., Lee, A., Chen, J., Ruta, V., Cadene, M., Chait, B. T., and MacKinnon, R. 2003. X-ray structure of a voltage-dependent K^+ channel. *Nature* 423:33–41.

Jiang, Y., Ruta, V., Chen, J., Lee, A., and MacKinnon, R. 2003. The principle of gating charge movement in a voltage-dependent K^+ channel. *Nature* 423:42–48.

Mackinnon, R. 2004. Structural biology: Voltage sensor meets lipid membrane. *Science* 306:1304–1305.

Bezanilla, F. 2000. The voltage sensor in voltage-dependent ion channels. *Physiol. Rev.* 80:555–592.

Shieh, C.-C., Coghlan, M., Sullivan, J. P., and Gopalakrishnan, M. 2000. Potassium channels: Molecular defects, diseases, and therapeutic opportunities. *Pharmacol. Rev.* 52:557–594.

Horn, R. 2000. Conversation between voltage sensors and gates of ion channels. *Biochemistry* 39:15653–15658.

Perozo, E., Cortes, D. M., and Cuello, L. G. 1999. Structural rearrangements underlying K^+-channel activation gating. *Science* 285:73–78.

Doyle, D. A., Morais Cabral, J., Pfuetzner, R. A., Kuo, A., Gulbis, J. M., Cohen, S. L., Chait, B. T., and MacKinnon R. 1998. The structure of the potassium channel: Molecular basis of K^+ conduction and selectivity. *Science* 280:69–77.

Marban, E., Yamagishi, T., and Tomaselli, G. F. 1998. Structure and function of the voltage-gated Na^+ channel. *J. Physiol.* 508:647–657.

Miller, R. J. 1992. Voltage-sensitive Ca^{2+} channels. *J. Biol. Chem.* 267:1403–1406.

Catterall, W. A. 1991. Excitation-contraction coupling in vertebrate skeletal muscle: A tale of two calcium channels. *Cell* 64:871–874.

Ligand-Gated Ion Channels

Unwin, N. 2005. Refined structure of the nicotinic acetylcholine receptor at 4 Å resolution. *J. Mol. Biol.* 346:967–989.

Miyazawa, A., Fujiyoshi, Y., Stowell, M., and Unwin, N. 1999. Nicotinic acetylcholine receptor at 4.6 Å resolution: Transverse tunnels in the channel wall. *J. Mol. Biol.* 288:765–786.

Jiang, Y., Lee, A., Chen, J., Cadene, M., Chait, B. T., and MacKinnon, R. 2002. Crystal structure and mechanism of a calcium-gated potassium channel. *Nature* 417:515–522.

Barrantes, F. J., Antollini, S. S., Blanton, M. P., and Prieto, M. 2000. Topography of the nicotinic acetylcholine receptor membrane-embedded domains. *J. Biol. Chem.* 275:37333–37339.

Cordero-Erausquin, M., Marubio, L. M., Klink, R., and Changeux, J. P. 2000. Nicotinic receptor function: New perspectives from knockout mice. *Trends Pharmacol. Sci.* 21:211–217.

Le Novère, N., and Changeux, J. P. 1995. Molecular evolution of the nicotinic acetylcholine receptor: An example of multigene family in excitable cells. *J. Mol. Evol.* 40:155–172.

Kunishima, N., Shimada, Y., Tsuji, Y., Sato, T., Yamamoto, M., Kumasaka, T., Nakanishi, S., Jingami, H., and Morikawa, K. 2000. Structural basis of glutamate recognition by dimeric metabotropic glutamate receptor. *Nature* 407:971–978.

Betz, H., Kuhse, J., Schmieden, V., Laube, B., Kirsch, J., and Harvey, R. J. 1999. Structure and functions of inhibitory and excitatory glycine receptors. *Ann. N. Y. Acad. Sci.* 868:667–676.

Unwin, N. 1995. Acetylcholine receptor channel imaged in the open state. *Nature* 373:37–43.

Colquhoun, D., and Sakmann, B. 1981. Fluctuations in the microsecond time range of the current through single acetylcholine receptor ion channels. *Nature* 294:464–466.

Gap Junctions

Saez, J. C., Berthoud, V. M., Branes, M. C., Martinez, A. D., and Beyer, E. C. 2003. Plasma membrane channels formed by connexins: Their regulation and functions. *Physiol. Rev.* 83:1359–1400.

Revilla, A., Bennett, M. V. L., and Barrio, L. C. 2000. Molecular determinants of membrane potential dependence in vertebrate gap junction channels. *Proc. Natl. Acad. Sci. U. S. A.* 97:14760–14765.

Unger, V. M., Kumar, N. M., Gilula, N. B., and Yeager, M. 1999. Three-dimensional structure of a recombinant gap junction membrane channel. *Science* 283:1176–1180.

Simon, A. M. 1999. Gap junctions: More roles and new structural data. *Trends Cell Biol.* 9:169–170.

Beltramello, M., Piazza, V., Bukauskas, F. F., Pozzan, T., and Mammano, F. 2005. Impaired permeability to Ins(1,4,5)P_3 in a mutant connexin underlies recessive hereditary deafness. *Nat. Cell Biol.* 7:63–69.

White, T. W., and Paul, D. L. 1999. Genetic diseases and gene knockouts reveal diverse connexin functions. *Annu. Rev. Physiol.* 61:283–310.

Water Channels

Agre, P., King, L. S., Yasui, M., Guggino, W. B., Ottersen, O. P., Fujiyoshi, Y., Engel, A., and Nielsen, S. 2002. Aquaporin water channels: From atomic structure to clinical medicine. *J. Physiol.* 542:3–16.

Agre, P., and Kozono, D. 2003. Aquaporin water channels: Molecular mechanisms for human diseases. *FEBS Lett.* 555:72–78.

de Groot, B. L., Engel, A., and Grubmuller, H. 2003. The structure of the aquaporin-1 water channel: A comparison between cryo-electron microscopy and X-ray crystallography. *J. Mol. Biol.* 325:485–493.

Problems

1. *The price of extrusion.* What is the free-energy cost of pumping Ca^{2+} out of a cell when the cytoplasmic concentration is 0.4 μM, the extracellular concentration is 1.5 mM, and the membrane potential is -60 mV?

2. *How sweet it is.* Glucose is pumped into some animal cells by a symporter powered by the simultaneous entry of Na^+. The entry of Na^+ provides a free-energy input of 10.8 kJ mol^{-1} (2.6 kcal mol^{-1}) under typical cellular conditions (external $[Na^+]$ = 143 mM, internal $[Na^+]$ = 14 mM, and membrane potential = -50 mV). How large a concentration of glucose can be generated by this free-energy input?

3. *Variations on a theme.* Write a detailed mechanism for transport by the Na^+–K^+ ATPase based on analogy with the mechanism of the Ca^{2+} ATPase shown in Figure 13.5.

4. *Pumping protons.* Design an experiment to show that the action of lactose permease can be reversed in vitro to pump protons.

5. *Different directions.* The K^+ channel and the Na^+ channel have similar structures and are arranged in the same orientation in the cell membrane. Yet, the Na^+ channel allows sodium ions to flow into the cell and the K^+ channel allows potassium ions to flow out of the cell. Explain.

6. *Structure–activity relations.* On the basis of the structure of tetrodotoxin, propose a mechanism by which the toxin inhibits Na^+ flow through the Na^+ channel.

7. *A dangerous snail.* Cone snails are carnivores that inject a powerful set of toxins into their prey, leading to rapid paralysis. Many of these toxins are found to bind to specific ion-channel proteins. Why are such molecules so toxic? How might such toxins be useful for biochemical studies?

8. *Only a few.* Why do only a small number of sodium ions need to flow through the Na^+ channel to significantly change the membrane potential?

9. *Mechanosensitive channels.* Many species contain ion channels that respond to mechanical stimuli. On the basis of the properties of other ion channels, would you expect the flow of ions through a single open mechanosensitive channel to increase in response to an appropriate stimulus? Why or why not?

10. *Concerted opening.* Suppose that a channel obeys the concerted allosteric model (MWC model, p. 200). The binding of ligand to the R state (the open form) is 20 times as tight as that to the T state (the closed form). In the absence of ligand, the ratio of closed to open channels is 10^5. If the channel is a tetramer, what is the fraction of open channels when 1, 2, 3, and 4 ligands are bound?

11. *Respiratory paralysis.* The neurotransmitter acetylcholine is degraded by a specific enzyme that is inactivated by Tabun, sarin, and parathion. Based on the structures below, propose a possible basis for their lethal actions.

Tabun

Sarin

Parathion

12. *Ligand-induced channel opening.* The ratio of open to closed forms of the acetylcholine receptor channel containing zero, one, and two bound acetylcholine molecules is 5×10^{-6}, 1.2×10^{-3}, and 14, respectively.

(a) By what factor is the open-to-closed ratio increased by the binding of the first acetylcholine molecule? The second acetylcholine molecule?
(b) What are the corresponding free-energy contributions to channel opening at 25°C?
(c) Can the allosteric transition be accounted for by the MWC concerted model (p. 200)?

13. *Frog poison.* Batrachotoxin (BTX) is a steroidal alkaloid from the skin of *Phyllobates terribilis*, a poisonous Colombian frog (source of the poison used on blowgun darts). In the presence of BTX, Na^+ channels in an excised patch stay persistently open when the membrane is depolarized. They close when the membrane is repolarized. Which transition is blocked by BTX?

14. *Valium target.* γ-Aminobutyric acid (GABA) opens channels that are specific for chloride ions. The $GABA_A$ receptor channel is pharmacologically important because it is the target of Valium, which is used to diminish anxiety.

(a) The extracellular concentration of Cl^- is 123 mM and the intracellular concentration is 4 mM. In which direction does Cl^- flow through an open channel when the membrane potential is in the -60 mV to $+30$ mV range?
(b) What is the effect of Cl^--channel opening on the excitability of a neuron?

(c) The hydropathy profile of the GABA$_A$ receptor resembles that of the acetylcholine receptor. Predict the number of subunits in this Cl$^-$ channel.

Chapter Integration Problem

15. *Speed and efficiency matter.* Acetylcholine is rapidly destroyed by the enzyme acetylcholinesterase. This enzyme, which has a turnover number of 25,000 per second, has attained catalytic perfection with a k_{cat}/K_M of 2×10^8 M^{-1}s^{-1}. Why is the efficiency of this enzyme physiologically crucial?

Mechanism Problem

16. *Remembrance of mechanisms past.* Acetylcholinesterase converts acetylcholine into acetate and choline. Like serine proteases, acetylcholinesterase is inhibited by DIPF. Propose a catalytic mechanism for acetylcholine digestion by acetylcholinesterase. Show the reaction as chemical structures.

Data Interpretation Problems

17. *Tarantula toxin.* Acid sensing is associated with pain, tasting, and other biological activities (Chapter 32). Acid sensing is carried out by a ligand-gated channel that permits Na$^+$ influx in response to H$^+$. This family of acid-sensitive ion channels (ASICs) comprises a number of members. Psalmotoxin 1 (PcTX1), a venom from the tarantula, inhibits some members of this family. The following electrophysiological recordings of cells containing several members of the ASIC family were made in the presence of the toxin at a concentration of 10 nM. The channels were opened by changing the pH from 7.4 to the indicated values. The PcTX1 was present for a short time (indicated by the black bar above the recordings) after which time it was rapidly washed from the system.

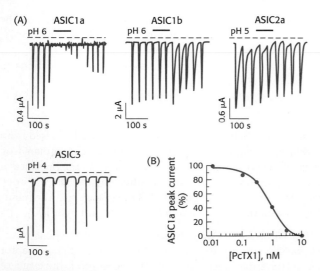

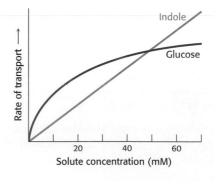

(A) Electrophysiological recordings of cells exposed to tarantula toxin. (B) Plot of peak current of a cell containing the ASIC1a protein versus the toxin concentration. [From P. Escoubas et al. *J. Biol. Chem.* 275(2000):25116–25121.]

(a) Which of the ASIC family members—ASIC1a, ASIC1b, ASIC2a, or ASIC3—is most sensitive to the toxin?

(b) Is the effect of the toxin reversible? Explain.

(c) What concentration of PcTX1 yields 50% inhibition of the sensitive channel?

18. *Channel problems 1.* A number of pathological conditions result from mutations in the acetylcholine receptor channel. One such mutation in the β subunit, βV266M, causes muscle weakness and rapid fatigue. An investigation of the acetylcholine-generated currents through the acetylcholine receptor channel for both a control and a patient yielded the following results. What is the effect of the mutation on channel function? Suggest some possible biochemical explanations for the effect.

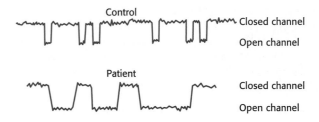

19. *Channel problems 2.* The acetylcholine receptor channel can also undergo mutation leading to fast-channel syndrome (FCS), with clinical manifestations similar to those of slow-channel syndrome (SCS). What would the recordings of ion movement look like in this syndrome? Suggest a biochemical explanation.

20. *Transport differences.* The rate of transport of two molecules, indole and glucose, across a cell membrane is shown below. What are the differences between the transport mechanisms of the two molecules? Suppose that ouabain inhibited the transport of glucose. What would this inhibition suggest about the mechanism of transport?

Signal-Transduction Pathways

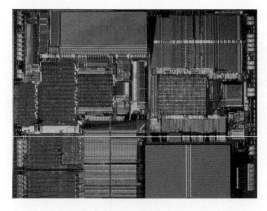

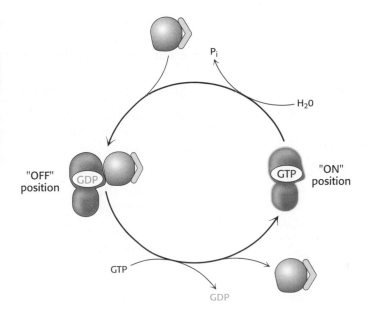

Signal-transduction circuits in biological systems have molecular on–off switches that, like those in a computer chip (above), transmit information when "on." Common among these circuits are those including G proteins (right), which transmit a signal when bound to GTP and are silent when bound to GDP. [(Left) Courtesy of Intel.]

A cell is highly responsive to specific chemicals in its environment: it may adjust its metabolism or alter gene-expression patterns on sensing their presence. In multicellular organisms, these chemical signals are crucial to coordinating physiological responses (Figure 14.1). Three examples of molecular signals that stimulate a physiological response are epinephrine (sometimes called adrenaline), insulin, and epidermal growth factor (EGF). When an individual organism is threatened, the adrenal glands, located over the kidneys, release the hormone epinephrine, which stimulates the mobilization of energy stores and leads to improved cardiac function. After a full meal, the β cells in the pancreas release insulin, which stimulates the uptake of glucose from the bloodstream and leads to other physiological changes. The release of epidermal growth factor in response to a wound stimulates specific cells to grow and divide. In all these cases, the cell receives information that a certain molecule within its environment is present above some threshold concentration. The chain of events that converts the message "this molecule is present" into the ultimate physiological response is called *signal transduction*.

Signal-transduction pathways are often characterized by many components and branches. They can thus be immensely complicated and confusing. Nonetheless, strategies and classes of molecules recur in many signal-transduction pathways. Examining the principles that underlie these common features can reveal the logic of signal-transduction pathways. We

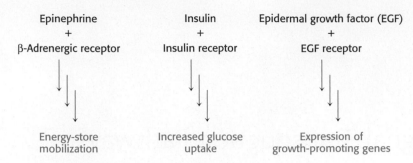

Figure 14.1 Three signal-transduction pathways. The binding of signaling molecules to their receptors initiates pathways that lead to important physiological responses.

introduce these principles of signal transduction here because signal-transduction pathways affect essentially all of the metabolic pathways that we will be exploring throughout the rest of the book.

Signal Transduction Depends on Molecular Circuits

Signal-transduction pathways follow a broadly similar course that can be viewed as a molecular circuit (Figure 14.2). All such circuits contain certain key steps:

1. *Release of the Primary Messenger*. A stimulus such as a wound or digested meal triggers the release of the signal molecule, also called the *primary messenger*.

2. *Reception of the Primary Messenger*. Most signal molecules do not enter cells. Instead, proteins in the cell membrane act as *receptors* that bind the signal molecules and transfer the information that the molecule has bound from the environment to the cell's interior. Receptors span the cell membrane and, thus, have both extracellular and intracellular components. A binding site on the extracellular side specifically recognizes the signal molecule (often referred to as the *ligand*). Such binding sites are analogous to enzyme active sites except that no catalysis takes place within them. The interaction of the ligand and the receptor alters the tertiary or quaternary structure of the receptor, including the intracellular parts.

3. *Delivery of the Message Inside the Cell by the Second Messenger*. Other small molecules, called *second messengers*, are used to relay information from receptor–ligand complexes. Second messengers are intracellular molecules that change in concentration in response to environmental signals. These changes in concentration constitute the next step in the molecular information circuit. Some particularly important second messengers are cyclic AMP and cyclic GMP, calcium ion, inositol 1,4,5-trisphosphate (IP_3), and diacylglycerol (DAG; Figure 14.3).

The use of second messengers has several consequences. First, the signal may be amplified significantly in the generation of second messengers. Only a small number of receptor molecules may be activated by the direct binding of signal molecules, but each activated receptor molecule can lead to the generation of many second messengers. Thus, *a low concentration of signal in the environment, even as little as a single molecule, can yield a large intracellular signal and response*. Second, second messengers are often free to diffuse to other cellular compartments where they can influence processes throughout the cell. Third, the use of common second messengers in multiple signaling pathways creates both opportunities and potential problems. Input from several signaling pathways, often called *cross talk*, may alter the concentration of a common second messenger. Cross talk permits more

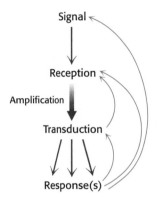

Figure 14.2 Principles of signal transduction. An environmental signal is first received by interaction with a cellular component, most often a cell-surface receptor. The information that the signal has arrived is then converted into other chemical forms, or *transduced*. The transduction process often comprises many steps. The signal is often amplified before evoking a response. Feedback pathways regulate the entire signaling process.

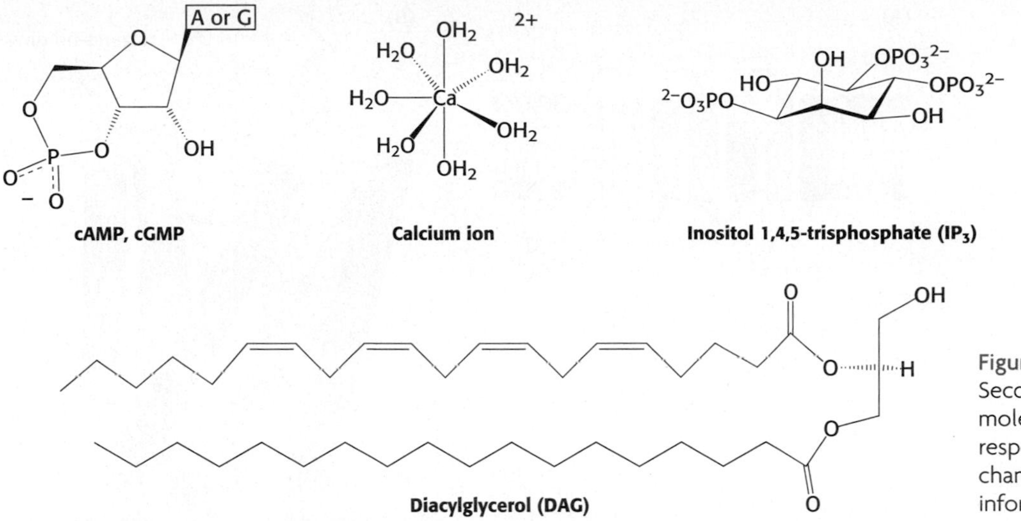

cAMP, cGMP

Calcium ion

Inositol 1,4,5-trisphosphate (IP$_3$)

Diacylglycerol (DAG)

Figure 14.3 Common second messengers. Second messengers are intracellular molecules that change in concentration in response to environmental signals. That change in concentration conveys information inside the cell.

finely tuned regulation of cell activity than would the action of individual independent pathways. However, inappropriate cross talk can cause changes in second-messenger concentration to be misinterpreted.

4. *Activation of Effectors That Directly Alter the Physiological Response.* The ultimate effect of the signal pathway is to activate (or inhibit) the pumps, enzymes, and gene-transcription factors that directly control metabolic pathways, gene activation, and processes such as nerve transmission.

5. *Termination of the Signal.* After a cell has completed its response to a signal, the signaling process must be terminated or the cell loses its responsiveness to new signals. Moreover, signaling processes that fail to terminate properly can have highly undesirable consequences. As we will see, many cancers are associated with signal-transduction processes that are not properly terminated, especially processes that control cell growth.

In this chapter, we will examine components of the three signal-transduction pathways shown in Figure 14.1. In doing so, we will see several classes of adaptor domains present in signal-transduction proteins. These domains usually recognize specific classes of molecules and help transfer information from one protein to another. The components described in the context of these three pathways recur in many other signal-transduction pathways; so bear in mind that the specific examples are representative of many such pathways.

14.1 Heterotrimeric G Proteins Transmit Signals and Reset Themselves

Epinephrine signaling begins with ligand binding to a protein called *the β-adrenergic receptor* (β-AR). The β-adrenergic receptor is a member of the largest class of cell-surface receptors, called the *seven-transmembrane-helix* (7TM) *receptors*. Members of this family are responsible for transmitting information initiated by signals as diverse as hormones, neurotransmitters, odorants, tastants, and even photons (Table 14.1). Several thousand such

TABLE 14.1 Biological functions mediated by 7TM receptors

- Hormone action
- Hormone secretion
- Neurotransmission
- Chemotaxis
- Exocytosis
- Control of blood pressure
- Embryogenesis
- Cell growth and differentiation
- Development
- Smell
- Taste
- Vision
- Viral infection

Source: After J. S. Gutkind, *J. Biol. Chem.* 273(1998): 1839–1842.

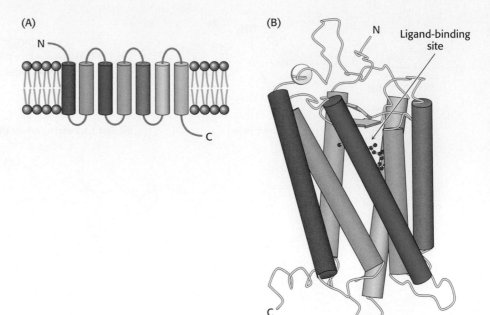

Figure 14.4 Structure of 7TM
receptors. (A) Schematic
representation of a 7TM receptor showing
how it passes through the membrane
seven times. (B) Three-dimensional
structure of rhodopsin, a 7TM receptor
taking part in visual signal transduction.
Notice the ligand-binding site near the
extracellular surface. As the first 7TM
receptor whose structure was determined,
rhodopsin provides a framework for
understanding other 7TM receptors.
[Drawn from 1F88.pdb.]

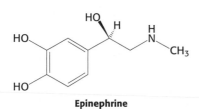

Epinephrine

receptors are now known. Furthermore, approximately 50% of the therapeutic drugs that we use target receptors of this class. As the name indicates,
these receptors contain seven helices that span the membrane bilayer. The
receptors are sometimes referred to as *serpentine receptors* because the single
polypeptide chain "snakes" through the membrane seven times (Figure
14.4A). The first member of this family to have its three-dimensional structure determined was *rhodopsin* (Figure 14.4B and Section 32.3), which plays
an essential role in vision. Whereas many 7TM receptors are quite similar
in structure to rhodopsin, others have larger extracellular domains.

The binding of epinephrine to β-AR triggers conformational changes in
the cytoplasmic loops and the carboxyl terminus, although the details of
these conformational changes remain to be established. Thus, the binding
of a ligand from outside the cell induces a conformational change in the part
of the 7TM receptor that is positioned inside the cell.

Ligand Binding to 7TM Receptors Leads to the Activation of Heterotrimeric G Proteins

What is the next step in the pathway? The conformational change in the receptor's cytoplasmic domain activates a protein called a *G protein* because it
binds guanyl nucleotides. The activated G protein stimulates the activity of
adenylate cyclase, an enzyme that increases the concentration of cAMP by
forming it from ATP. The G protein and adenylate cyclase remain attached
to the membrane, whereas cAMP can travel throughout the cell carrying the
signal originally brought by the binding of epinephrine. Figure 14.5 provides a broad overview of these steps.

In the G protein's unactivated state, the guanyl nucleotide bound to the
G protein is GDP. In this form, the G protein exists as a heterotrimer consisting of α, β, and γ subunits; the α subunit (referred to as G_α) binds the
nucleotide (Figure 14.6). The α subunit is a member of the P-loop NTPase
family (Section 9.4), and the P-loop participates in nucleotide binding. The
α and γ subunits are usually anchored to the membrane by covalently
attached fatty acids. *The role of the hormone-bound receptor is to catalyze the
exchange of GTP for bound GDP.* The hormone–receptor complex interacts
with the heterotrimeric G protein and opens the nucleotide-binding site

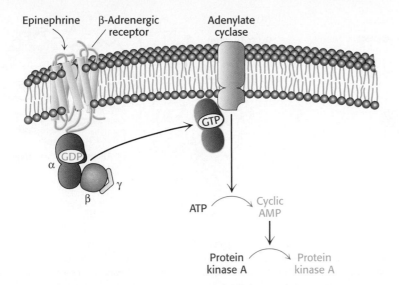

Figure 14.5 Activation of protein kinase A by a G-protein pathway. Hormone binding to a 7TM receptor initiates a signal-transduction pathway that acts through a G protein and cAMP to activate protein kinase A.

so that GDP can depart and GTP from solution can bind. The α subunit simultaneously dissociates from the $\beta\gamma$ dimer ($G_{\beta\gamma}$).

The dissociation of the G-protein heterotrimer into G_α and $G_{\beta\gamma}$ units transmits the signal that the receptor has bound its ligand. *A single hormone–receptor complex can stimulate nucleotide exchange in many G-protein heterotrimers.* Thus, hundreds of G_α molecules are converted from their GDP into their GTP forms for each bound molecule of hormone, giving an amplified response. Because they signal through G proteins, 7TM receptors are often called *G-protein-coupled receptors* or GPCRs.

Activated G Proteins Transmit Signals by Binding to Other Proteins

In the GTP form, the surface of the G protein that had been bound to the $\beta\gamma$ dimer has changed its conformation from the GDP form so that it no longer has a high affinity for the $\beta\gamma$ dimer. However, this surface is now exposed for binding to other proteins. In the β-AR pathway, the new binding partner is *adenylate cyclase,* the enzyme that converts ATP into cAMP. This enzyme is a membrane protein that contains 12 membrane spanning helices; two large cytoplasmic domains form the catalytic part of the enzyme. The interaction of $G_{\alpha s}$ with adenylate cyclase favors a more catalytically

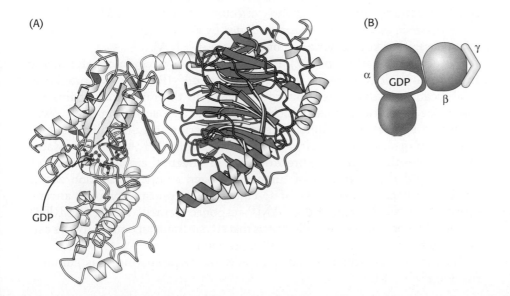

(A)

(B)

Figure 14.6 A heterotrimeric G protein. (A) A ribbon diagram shows the relation between the three subunits. In this complex, the α subunit (gray and purple) is bound to GDP. *Notice* that GDP is bound in a pocket close to the surface at which the α subunit interacts with the $\beta\gamma$ dimer. (B) A schematic representation of the heterotrimeric G protein. [Drawn from 1GOT.pdb.]

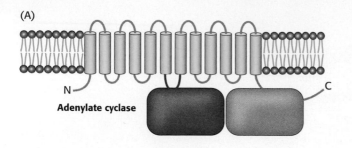

(A)

Adenylate cyclase

N   C

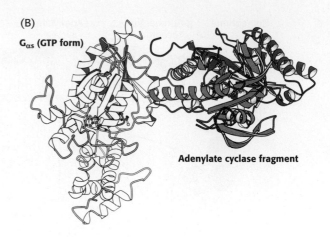

(B)

$G_{\alpha s}$ (GTP form)

Adenylate cyclase fragment

Figure 14.7 Adenylate cyclase activation. (A) Adenylate cyclase is a membrane protein with two large intracellular domains that contain the catalytic apparatus. (B) The structure of a complex between G_α in its GTP form bound to a catalytic fragment from adenylate cyclase. *Notice* that the surface of G_α that had been bound to the βγ dimer now binds adenylate cyclase. [Drawn from 1AZS.pdb.]

active conformation of the enzyme, thus stimulating cAMP production (Figure 14.7). Indeed, the G protein subunit that participates in the β-AR pathway is called $G_{\alpha s}$ where "s" stands for stimulatory. *The net result is that the binding of epinephrine to the receptor on the cell surface increases the rate of cAMP production inside the cell.* The generation of cAMP by adenylate cyclase provides a second level of amplification because each activated adenylate cyclase can convert many molecules of ATP into cAMP.

Cyclic AMP Stimulates the Phosphorylation of Many Target Proteins by Activating Protein Kinase A

Let us continue to follow the information flow down this signal-transduction pathway. The increased concentration of cAMP can affect a wide range of cellular processes. In regard to epinephrine in the muscle, cAMP stimulates the production of ATP for muscle contraction. In other cells, cAMP enhances the degradation of storage fuels, increases the secretion of acid by the gastric mucosa, leads to the dispersion of melanin pigment granules, diminishes the aggregation of blood platelets, and induces the opening of chloride channels. How does cAMP influence so many cellular processes? Is there a common denominator for its diverse effects? Indeed there is. *Most effects of cyclic AMP in eukaryotic cells are mediated by the activation of a single protein kinase.* This key enzyme is *protein kinase A* (PKA).

As described earlier (p. 287), PKA consists of two regulatory (R) chains and two catalytic (C) chains. In the absence of cAMP, the R_2C_2 complex is catalytically inactive. The binding of cAMP to the regulatory chains releases the catalytic chains, which are catalytically active on their own. Activated PKA then phosphorylates specific serine and threonine residues in many targets to alter their activity. For instance, PKA phosphorylates two enzymes that lead to the breakdown of glycogen, the polymeric store of glucose, and the inhibition of further glycogen synthesis (Section 21.3). PKA stimulates the *expression of specific genes* by phosphorylating a transcriptional activator called the cAMP-response element binding (CREB) protein. This activity of PKA illustrates that signal-transduction pathways can extend into the nucleus to alter gene expression.

The signal-transduction pathway initiated by epinephrine is summarized in Figure 14.8.

Epinephrine
+
β-Adrenergic receptor

Binding

Activated
receptor

GTP for GDP
exchange | Amplification

Activated
G protein

Protein–protein
interaction

Activated
adenylate cyclase

Enzymatic
reaction | Amplification

Increased
[cAMP]

Activated
protein kinase A
and other effectors

Figure 14.8 Epinephrine signaling pathway. The binding of epinephrine to the β-adrenergic receptor initiates the signal-transduction pathway. The process in each step is indicated (in black) to the left of each arrow. Steps that have the potential for signal amplification are indicated to the right in green.

G Proteins Spontaneously Reset Themselves Through GTP Hydrolysis

How is the signal initiated by epinephrine switched off? *G_α subunits have intrinsic GTPase activity,* which is used to hydrolyze bound GTP to GDP and P_i. This hydrolysis reaction is slow, however, requiring from seconds to minutes. Thus, the GTP form of G_α is able to activate downstream components of the signal-transduction pathway before GTP hydrolysis deactivates the subunit. In essence, *the bound GTP acts as a built-in clock that spontaneously resets the G_α subunit after a short time period.* After GTP hydrolysis and the release of P_i, the GDP-bound form of G_α then reassociates with $G_{\beta\gamma}$ to re-form the inactive heterotrimeric protein (Figure 14.9).

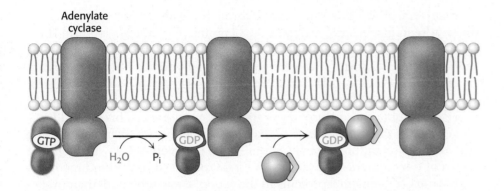

Adenylate cyclase

GTP H_2O P_i GDP   GDP

Figure 14.9 Resetting G_α. On hydrolysis of the bound GTP by the intrinsic GTPase activity of G_α, G_α reassociates with the $\beta\gamma$ dimer to form the heterotrimeric G protein, thereby terminating the activation of adenylate cyclase.

The hormone-bound activated receptor must be reset as well to prevent the continuous activation of G proteins. This resetting is accomplished by two processes (Figure 14.10). First, the hormone dissociates, returning the receptor to its initial, unactivated state. The likelihood that the receptor remains in its unbound state depends on the concentration of hormone. Second, the hormone–receptor complex is deactivated by the phosphorylation of serine and threonine residues in the carboxyl-terminal tail. In the example under consideration, *β-adrenergic-receptor kinase* (also called G-protein receptor kinase 2, GRK2) phosphorylates the carboxyl-terminal tail of the hormone–receptor complex but not the unoccupied receptor. Finally, the

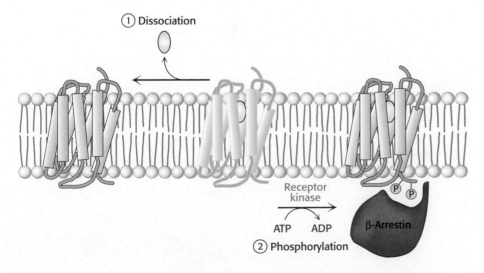

① Dissociation

Receptor kinase

ATP   ADP   β-Arrestin

② Phosphorylation

Figure 14.10 Signal termination. Signal transduction by the 7TM receptor is halted (1) by dissociation of the signal molecule from the receptor and (2) by phosphorylation of the cytoplasmic C-terminal tail of the receptor and the subsequent binding of β-arrestin.

molecule *β-arrestin* binds to the phosphorylated receptor and further diminishes its ability to activate G proteins.

Some 7TM Receptors Activate the Phosphoinositide Cascade

We turn now to another common second-messenger cascade, also employing a 7TM receptor, that is used by many hormones to evoke a variety of responses. The *phosphoinositide cascade*, like the cAMP cascade, converts extracellular signals into intracellular ones. The intracellular messengers formed by activation of this pathway arise from the cleavage of *phosphatidylinositol 4,5-bisphosphate* (PIP_2), a phospholipid present in cell membranes. An example of a signaling pathway based on the phosphoinositide cascade is the one triggered by the angiotensin II receptor, which binds a peptide hormone taking part in the control of blood pressure.

Each type of 7TM receptor signals through a distinct G protein. Whereas the β-adrenergic receptor activates the G protein $G_{\alpha s}$, the angiotensin II receptor activates a G protein called $G_{\alpha q}$. In its GTP-form, $G_{\alpha q}$ binds to and activates the β isoform of the enzyme *phospholipase C*. This enzyme catalyzes the cleavage of PIP_2 to form the two second messengers inositol 1,4,5-trisphosphate (IP_3) and diacylglycerol (DAG; Figure 14.11).

IP_3 is soluble and diffuses away from the membrane. This second messenger causes the rapid release of Ca^{2+} from the intracellular stores in the endoplasmic reticulum, which accumulates a reservoir of Ca^{2+} through the action of transporters such as Ca^{2+} ATPase (p. 354). On the binding of IP_3 to specific Ca^{2+}-channel proteins in the endoplasmic reticulum membrane, these IP_3 receptors open to allow calcium ions to flow from the endoplasmic reticulum into the cytoplasm. Calcium ion is itself a signaling molecule: it can bind proteins, including a ubiquitous signaling protein called calmodulin and enzymes such as protein kinase C. By such means, the elevated level of Ca^{2+} triggers processes such as smooth-muscle contraction, glycogen breakdown, and vesicle release.

DAG glycerol remains in the plasma membrane. There, it activates *protein kinase C*, a protein kinase that phosphorylates serine and threonine residues in many target proteins. The specialized DAG-binding domains of

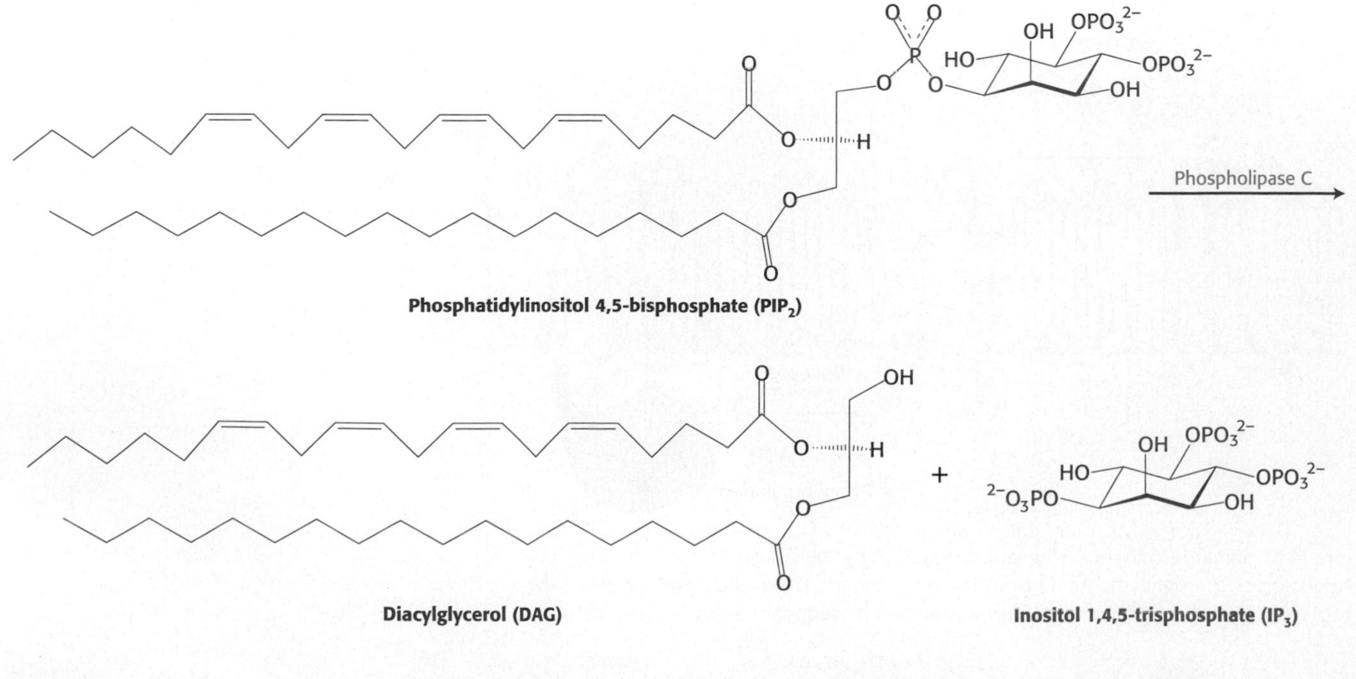

Figure 14.11 Phospholipase C reaction. Phospholipase C cleaves the membrane lipid phosphatidylinositol 4,5-bisphosphate into two second messengers: diacylglycerol, which remains in the membrane, and inositol 1,4,5-trisphosphate, which diffuses away from the membrane.

Phospholipase C

Phosphatidylinositol 4,5-bisphosphate (PIP₂)

Diacylglycerol (DAG)

Inositol 1,4,5-trisphosphate (IP₃)

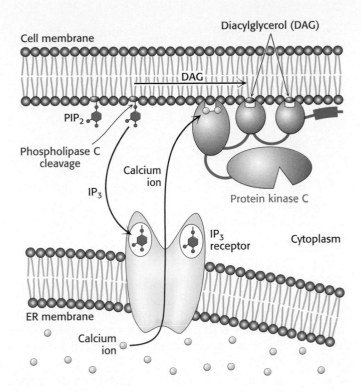

Figure 14.12 Phosphoinositide cascade. The cleavage of phosphatidylinositol 4,5-bisphosphate (PIP_2) into diacylglycerol (DAG) and inositol 1,4,5-trisphosphate (IP_3) results in the release of calcium ions (due to the opening of the IP_3 receptor ion channels) and the activation of protein kinase C (due to the binding of protein kinase C to free DAG in the membrane). Calcium ions bind to protein kinase C and help facilitate its activation.

this kinase require bound calcium to bind to DAG. Note that diacylglycerol and IP_3 work in tandem: IP_3 increases the Ca^{2+} concentration, and Ca^{2+} facilitates the activation of protein kinase C. The phosphoinositide cascade is summarized in Figure 14.12. Both IP_3 and DAG act transiently because they are converted into other species by phosphorylation or other processes.

Calcium Ion Is a Widely Used Second Messenger

Calcium ion participates in many signaling processes in addition to the phosphoinositide cascade. Several properties of this ion account for its widespread use as an intracellular messenger. First, fleeting changes in Ca^{2+} concentration are readily detected. At steady state, intracellular levels of Ca^{2+} must be kept low to prevent the precipitation of carboxylated and phosphorylated compounds, which form poorly soluble salts with Ca^{2+}. Calcium ion levels are kept low by transport systems that extrude Ca^{2+} from the cell. Because of their action, the cytoplasmic level of Ca^{2+} is approximately 100 nM, several orders of magnitude lower than the concentration in the extracellular medium. Given this low steady-state level, transient increases in Ca^{2+} content produced by signaling events can be readily sensed.

A second property of Ca^{2+} that makes it a highly suitable intracellular messenger is that it can bind tightly to proteins and induce conformational changes (Figure 14.13). Calcium ions bind well to negatively charged oxygen atoms (from the side chains of glutamate and aspartate) and uncharged oxygen atoms (main-chain carbonyl groups and side-chain oxygen atoms from glutamine and asparagine). *The capacity of Ca^{2+} to be coordinated to multiple ligands—from six to eight oxygen atoms—enables it to cross-link different segments of a protein and induce significant conformational changes.*

Our understanding of the role of Ca^{2+} in cellular processes has been greatly enhanced by our ability to detect changes in Ca^{2+} concentrations inside cells and even monitor the changing concentration in real time. This ability depends on the use of specialized designed dyes such as Fura-2 that bind Ca^{2+} and change their fluorescent properties on Ca^{2+} binding.

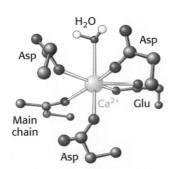

Figure 14.13 Calcium-binding site. In one common mode of binding, calcium is coordinated to six oxygen atoms from a protein and one (top) of water.

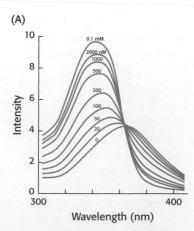

Fluorescent
component
(yellow)

Fura-2

Fura-2 binds Ca^{2+} through appropriately positioned oxygen atoms (shown in red) within its structure. When such a dye is introduced into cells, changes in available Ca^{2+} concentration can be monitored with microscopes that detect changes in fluorescence (Figure 14.14). Probes for sensing other second messengers such as cAMP also have been developed. These *molecular-imaging agents* are greatly enhancing our understanding of signal-transduction processes.

Calcium Ion Often Activates the Regulatory Protein Calmodulin

Calmodulin (CaM), a 17-kd protein with four Ca^{2+}-binding sites, serves as a Ca^{2+} sensor in nearly all eukaryotic cells. *Calmodulin is activated by the binding of Ca^{2+}, which takes place when the cytoplasmic Ca^{2+} level is raised above about 500 nM.* Calmodulin is a member of the *EF-hand protein family.* The *EF hand* is a Ca^{2+}-binding motif that consists of a helix, a loop, and a second helix. This motif, originally discovered in the protein parvalbumin,

Figure 14.14 Calcium imaging. (A) The fluorescence spectra of the calcium-binding dye Fura-2 can be used to measure available calcium ion concentrations in solution and in cells. (B) A series of images show Ca^{2+} spreading across a cell. These images were obtained through the use of a fluorescent calcium-binding dye. The images are false colored: red represents high Ca^{2+} concentrations, and blue represents low Ca^{2+} concentrations. [(A) After S. J. Lippard and J. M. Berg, *Principles of Bioinorganic Chemistry* (University Science Books, 1994), p. 193; (B) courtesy of Dr. Masashi Isshiki, Department of Nephrology, University of Tokyo, and Dr. G. W. Anderson, Department of Cell Biology, University of Texas Southwestern Medical School.]

(A)

(B)

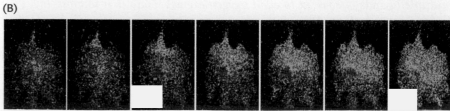

was named the EF hand because the two key helices designated E and F in parvalbumin are positioned like the forefinger and thumb of the right hand (Figure 14.15). These two helices and the intervening loop form the Ca^{2+}-binding motif. Seven oxygen atoms are coordinated to each Ca^{2+}, six from the protein and one from a bound water molecule.

The binding of Ca^{2+} to calmodulin induces substantial conformational changes in the EF hands. These conformational changes expose hydrophobic surfaces that can be used to bind other proteins. Using its two sets of two EF hands, calmodulin clamps down around specific regions of target proteins, usually exposed α helices with appropriately positioned hydrophobic and charged groups (Figure 14.16). By inducing conformational changes, the Ca^{2+}–calmodulin complex stimulates a wide variety of enzymes, pumps, and other target proteins. One especially noteworthy set of targets is several *calmodulin-dependent protein kinases* (CaM kinases) that phosphorylate many different proteins. These enzymes regulate fuel metabolism, ionic permeability, neurotransmitter synthesis, and neurotransmitter release. We see here a recurring theme in signal-transduction pathways: the concentration of a second messenger is increased (in this case, Ca^{2+}); the signal is sensed by a second-messenger-binding protein (in this case, calmodulin); and the second-messenger-binding protein acts to generate changes in enzymes (in this case, calmodulin-dependent kinases) that control effectors.

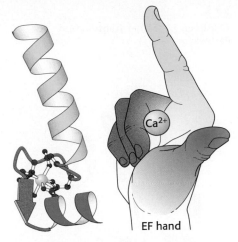

Figure 14.15 EF hand. Formed by a helix-loop-helix unit, an EF hand is a binding site for Ca^{2+} in many calcium-sensing proteins. Here, the E helix is yellow, the F helix is blue, and calcium is represented by the green sphere. *Notice* that the calcium ion is bound in a loop connecting two nearly perpendicular helices. [Drawn from 1CLL.pdb.]

14.2 Insulin Signaling: Phosphorylation Cascades Are Central to Many Signal-Transduction Processes

The signaling pathways that we have examined so far have activated a protein kinase as a downstream component of the pathway. We now turn to a class of signal-transduction pathways that are *initiated by receptors that include protein kinases as part of their structures*. The activation of these protein kinases sets in motion other processes that ultimately modify the effectors of these pathways.

An example is the signal-transduction pathway initiated by *insulin*, the hormone released after eating a full meal. In all of its detail, this many-branched pathway is quite complex; so we will focus solely on the major

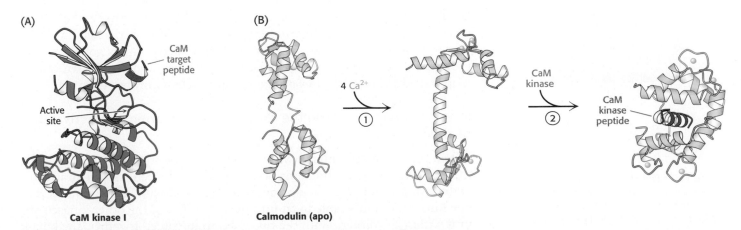

Figure 14.16 Calmodulin binds to α helices. (A) An α helix (purple) in CaM kinase I is a target for calmodulin. (B) After Ca^{2+} binding (1), the two halves of calmodulin clamp down around the target helix (2), binding it through hydrophobic and ionic interactions. In CaM kinase I, this interaction allows the enzyme to adopt an active conformation. [Drawn from 1A06, 1CFD, 1CLL, and 1CM1.pdb.]

branch. This branch leads to the mobilization of glucose transporters to the cell surface. As already mentioned, these transporters allow the cell to take up the glucose that is plentiful in the bloodstream after a meal.

The Insulin Receptor Is a Dimer That Closes Around a Bound Insulin Molecule

Insulin is a peptide hormone that consists of two chains, linked by three disulfide bonds (p. 36 and Figure 14.17). Its receptor has a quite different structure from that of the epinephrine receptor β-AR (Figure 14.18). The *insulin receptor* is a dimer of two identical units. Each unit consists of one α chain and one β chain linked to one another by a single disulfide bond. Each α subunit lies completely outside the cell, whereas each β subunit lies primarily inside the cell, spanning the membrane with a single transmembrane segment. The two α subunits move together to form a binding site for a single insulin molecule, a surprising occurrence because two different surfaces on the insulin molecule must interact with the two identical insulin receptor chains. The moving together of the dimeric units in the presence of an insulin molecule sets the signaling pathway in motion. *The closing up of an oligomeric receptor or the oligomerization of monomeric receptors around a bound ligand is a strategy used by many receptors to initiate a signal, particularly receptors containing a protein kinase.*

Each β subunit consists primarily of a protein kinase domain, homologous to protein kinase A. This kinase differs from protein kinase A in two important ways. First, the insulin receptor kinase is a *tyrosine kinase;* that is, it catalyzes the transfer of a phosphoryl group from ATP to the hydroxyl group of tyrosine, rather than serine or threonine, as is the case for protein kinase A.

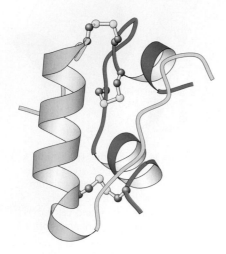

Figure 14.17 Insulin structure. *Notice* that insulin consists of two chains (shown in blue and yellow) linked by two interchain disulfide bonds. The α chain (blue) also has an intrachain disulfide bond. [Drawn from IB2F.pdb.]

Because this tyrosine kinase is a component of the receptor itself, the insulin receptor is referred to as a *receptor tyrosine kinase.* Second, the insulin receptor kinase is in an inactive conformation when the domain is not covalently modified. The kinase is rendered inactive by the position of an unstructured loop (called the *activation loop*) that lies in the center of the structure.

Insulin Binding Results in the Cross-Phosphorylation and Activation of the Insulin Receptor

When the two α subunits move together to surround an insulin molecule, the two protein kinase domains on the inside of the cell also are drawn together. Importantly, as they come together, the flexible activation loop of one kinase subunit is able to fit into the active site of the other kinase subunit within the dimer. With the two β subunits forced together, the kinase domains catalyze the addition of phosphoryl groups from ATP to tyrosine residues in the activation loops. When these tyrosine residues are phosphorylated, a striking conformational change takes place (Figure 14.19). The activation loop changes conformation dramatically and the kinase converts

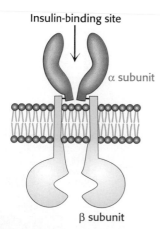

Insulin-binding site

α subunit

β subunit

Figure 14.18 The insulin receptor. The receptor consists of two units, each of which consists of an α subunit and a β subunit linked by a disulfide bond. The α subunit lies outside the cell and two α subunits come together to form a binding site for insulin. Each β subunit lies primarily inside the cell and includes a protein kinase domain.

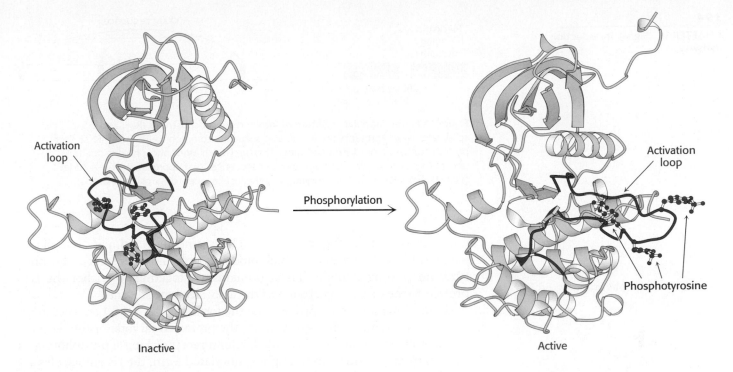

Activation loop

Phosphorylation →

Phosphotyrosine

Activation loop

Inactive

Active

into an active conformation. Thus, *insulin binding on the outside of the cell results in the activation of a membrane-associated kinase within the cell.*

The Activated Insulin Receptor Kinase Initiates a Kinase Cascade

On phosphorylation, the insulin receptor tyrosine kinase is activated. Because the two units of the receptor are held in close proximity to one another, additional sites within the receptor also are phosphorylated. These phosphorylated sites act as docking sites for other substrates, including a class of molecules referred to as *insulin-receptor substrates* (IRS). From the IRS protein, the signal is conveyed through a series of membrane-anchored molecules to a protein kinase that finally leaves the membrane (Figure 14.20). IRS-1 and IRS-2 are two homologous proteins with a common

Figure 14.19 Activation of the insulin receptor by phosphorylation. The activation loop is shown in red in this model of the protein kinase domain of the β subunit of the insulin receptor. The unphosphorylated structure on the left is not catalytically active. *Notice* that, when three tyrosine residues in the activation loop are phosphorylated, the activation loop swings across the structure and the kinase structure adopts a more compact conformation. This conformation is catalytically active. [Drawn from 1IRK.pdb and 1IR3.pdb.]

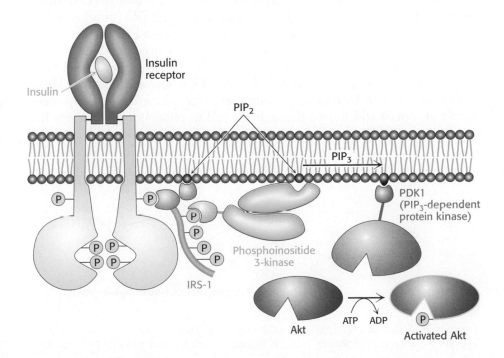

Insulin

Insulin receptor

PIP₂

PIP₃

PDK1 (PIP₃-dependent protein kinase)

Phosphoinositide 3-kinase

IRS-1

Akt

ATP ADP

Activated Akt

Figure 14.20 Insulin signaling. The binding of insulin results in the cross-phosphorylation and activation of the insulin receptor. Phosphorylated sites on the receptor act as binding sites for insulin receptor substrates such as IRS-1. The lipid kinase phosphoinositide 3-kinase binds to phosphorylated sites on IRS-1 through its regulatory domain, then converts PIP₂ into PIP₃. Binding to PIP₃ activates PIP₃-dependent protein kinase, which phosphorylates and activates kinases such as Akt1. Activated Akt1 can then diffuse throughout the cell to continue the signal-transduction pathway.

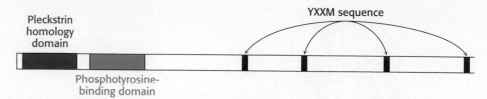

Figure 14.21 The modular structure of insulin receptor substrates IRS-1 and IRS-2. This schematic view represents the amino acid sequence common to IRS-1 and IRS-2. Each protein contains a pleckstrin homology domain (which binds phosphoinositide lipids), a phosphotyrosine-binding domain, and four sequences that approximate Tyr-X-X-Met (YXXM). The latter are phosphorylated by the insulin receptor tyrosine kinase.

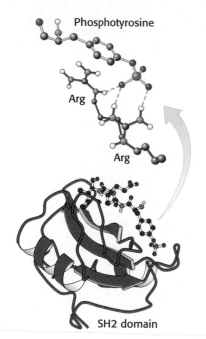

Figure 14.22 Structure of the SH2 domain. The domain is shown bound to a phosphotyrosine-containing peptide. *Notice* at the top that the negatively charged phosphotyrosine residue interacts with two Arg residues that are conserved in essentially all SH2 domains. [Drawn from 1SPS.pdb.]

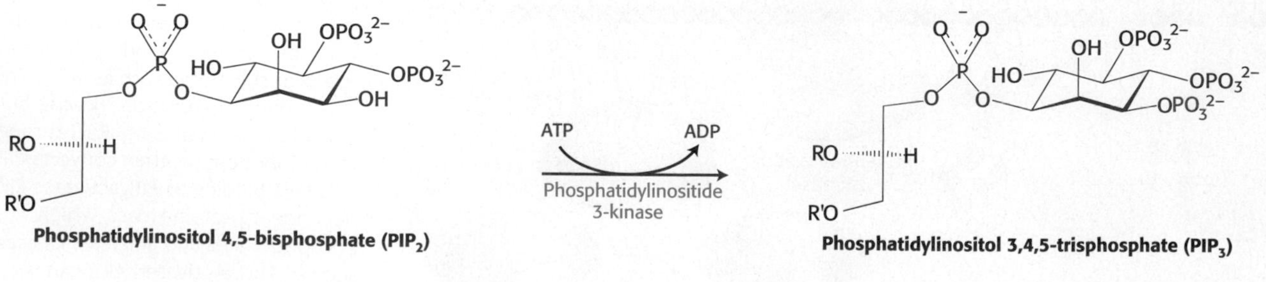

modular structure (Figure 14.21). The amino-terminal part includes a *pleckstrin homology domain*, which binds phosphoinositide, and a phospho-tyrosine-binding domain. These domains act together to anchor the IRS protein to the insulin receptor and the associated membrane.

Four sequences that approximate the form Tyr-X-X-Met are present in each IRS protein. Such sequences are the preferred targets for the receptor tyrosine kinase. Thus, the activated insulin receptor kinase phosphorylates these tyrosine residues. In their phosphorylated form, the IRS molecules act as *adaptor proteins:* they do not activate the next component of the pathway, a lipid kinase; rather, they bind to the lipid kinase and bring it to the membrane so that it can act on its substrate, a membrane lipid.

The phosphotyrosine residues in the IRS proteins are recognized by other proteins that contain a class of domain called a *Src homology 2* (SH2) domain (Figure 14.22). These domains, present in many signal-transduction proteins, bind to stretches of polypeptide that contain phosphotyrosine residues. Each specific SH2 domain shows a binding preference for phos-photyrosine in a particular sequence context. Which proteins contain SH2 domains that would bind to sequences in the IRS proteins? The most important of them are in a class of lipid kinases that add a phosphoryl group to the 3-position of inositol in phosphatidylinositol 4,5-bisphosphate (Figure 14.23). These enzymes are heterooligomers that consist of 110-kd catalytic subunits and 85-kd regulatory subunits. Through SH2 domains in the regulatory subunits, these enzymes bind to the IRS proteins and are drawn to the membrane where they can phosphorylate PIP_2 to form phos-phatidyl-inositol 3,4,5-trisphosphate (PIP_3). This lipid product, in turn, activates a protein kinase, PDK1, by virtue of a PIP_3-specific pleckstrin homology domain present in this kinase. This activated protein kinase phosphorylates and activates Akt, another protein kinase. Akt is not membrane anchored and moves through the cell to phosphorylate targets that

Phosphatidylinositol 4,5-bisphosphate (PIP₂)

ATP → ADP

Phosphatidylinositide 3-kinase

Phosphatidylinositol 3,4,5-trisphosphate (PIP₃)

Figure 14.23 Action of a lipid kinase in insulin signaling. Phosphorylated IRS-1 and IRS-2 activate the enzyme phosphatidylinositide 3-kinase, an enzyme that converts PIP₂ into PIP₃.

include components that control the trafficking of the glucose transporter GLUT4 to the cell surface as well as enzymes that stimulate glycogen synthesis (Section 21.4).

The cascade initiated by the binding of insulin to the insulin receptor is summarized in Figure 14.24. The signal is amplified at several stages along this pathway. Because the activated insulin receptor itself is a protein kinase, each activated receptor can phosphorylate multiple IRS molecules. Activated enzymes further amplify the signal in at least two of the subsequent steps. Thus, a small increase in the concentration of circulating insulin can produce a robust intracellular response. Note that, as complicated as the pathway described here is, it is substantially less elaborate than the full network of pathways initiated by insulin.

Insulin Signaling Is Terminated by the Action of Phosphatases

We have seen that the activated G protein promotes its own inactivation by the release of a phosphoryl group from GTP. In contrast, proteins phosphorylated on serine, threonine, or tyrosine residues do not hydrolyze spontaneously; they are extremely stable kinetically. Specific enzymes are required to hydrolyze these phosphorylated proteins and convert them back into the states that they were in before the initiation of signaling. Similarly, lipid phosphatases are required to remove phosphoryl groups from inositol lipids that had been phosphorylated as part of a signaling cascade. In insulin signaling, three classes of enzymes are of particular importance: protein tyrosine phosphatases that remove phosphoryl groups from tyrosine residues on the insulin receptor, lipid phosphatases that hydrolyze phosphatidylinositol 3,4,5-trisphosphate to phosphatidylinositol 3,4-bisphosphate, and protein serine phosphatases that remove phosphoryl groups from activated protein kinases such as Akt. Many of these phosphatases are activated or recruited as part of the response to insulin. Thus, the binding of the initial signal sets the stage for the eventual termination of the response.

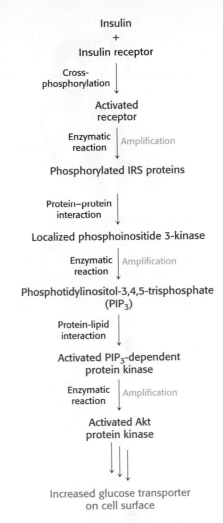

Figure 14.24 Insulin signaling pathway. Key steps in the signal-transduction pathway initiated by insulin binding to the insulin receptor.

14.3 EGF Signaling: Signal-Transduction Pathways Are Poised to Respond

Our consideration of the signal-transduction cascades initiated by epinephrine and insulin included examples of how components of signal-transduction pathways are poised for action, ready to be activated by minor modifications. For example, G-protein α subunits require only the binding of GTP in exchange for GDP to transmit a signal. This exchange reaction is thermodynamically favorable, but it is quite slow in the absence of an appropriate activated 7TM receptor. Similarly, the tyrosine kinase domains of the dimeric insulin receptor are ready for phosphorylation and activation but require the presence of insulin bound between two α subunits to draw the activation loop of one tyrosine kinase into the active site of a partner tyrosine kinase to initiate this process.

We now examine a signal-transduction pathway that reveals another clear example of how many signal-transduction pathways are poised to respond. This pathway is activated by the signal molecule *epidermal growth factor*. Like that of the insulin receptor, the initiator of this pathway is a receptor tyrosine kinase. Both the extracellular and the intracellular domains of this receptor are ready for action, held in check only by

Epidermal growth factor (EGF)

Figure 14.25 Structure of epidermal growth factor. *Notice* that three intrachain disulfide bonds stabilize the compact three-dimensional structure of the growth factor. [Drawn from 1EGF.pdb.]

a specific structure that prevents receptors from coming together. We will also encounter several additional signaling components that participate in many pathways.

EGF Binding Results in the Dimerization of the EGF Receptor

Epidermal growth factor is a 6-kd polypeptide that stimulates the growth of epidermal and epithelial cells (Figure 14.25). The *EGF receptor*, like the insulin receptor, is a protein tyrosine kinase that participates in cross-phosphorylation reactions (Figure 14.26). Like the insulin receptor, the EGF receptor is a dimer of two identical units. Unlike those of the insulin receptor, these units exist as monomers until EGF ligands bind to them. Moreover, each EGF receptor monomer binds a molecule of EGF in its extracellular domain (Figure 14.27). Thus the dimer binds two ligand molecules, in contrast to the one bound by the insulin receptor dimer. Note that each EGF molecule lies far away from the dimer interface. This interface includes a so-called *dimerization arm* from each monomer that reaches out and inserts into a binding pocket on the other monomer.

Although this structure nicely reveals the interactions that support the formation of a receptor dimer favoring cross-phosphorylation, it raises another question: Why doesn't the receptor dimerize and signal in

| EGF-binding domain | Transmembrane helix | Kinase domain | C-terminal tail (tyrosine-rich) |

Figure 14.26 Modular structure of the EGF receptor. This schematic view of the amino acid sequence of the EGF receptor shows the EGF-binding domain that lies outside the cell, a single transmembrane helix-forming region, the intracellular tyrosine kinase domain, and the tyrosine-rich domain at the carboxyl terminus.

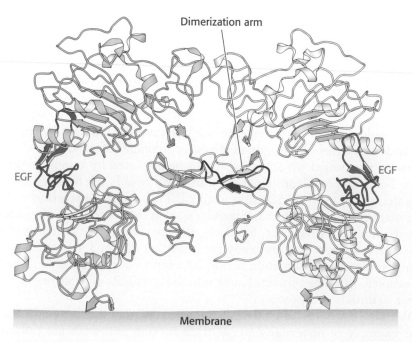

Figure 14.27 EGF receptor dimerization. The structure of the extracellular region of the EGF receptor is shown bound to EGF. *Notice* that the structure is dimeric with one EGF molecule bound to each receptor molecule and that the dimerization is mediated by a dimerization arm that extends from each receptor molecule.

the absence of EGF? This question has been addressed by examining the structure of the EGF receptor in the absence of bound ligand (Figure 14.28). This structure is, indeed, monomeric and each monomer is in a conformation that is quite different from that observed in the ligand-bound dimer. In particular, the dimerization arm binds to a domain *within the same monomer* that holds the receptor in a cyclic configuration. In essence, the receptor is poised in a spring-loaded conformation held in position by the contact between the interaction loop and another part of the structure, ready to bind ligand and change into a conformation active for dimerization and signaling.

This observation suggests that a receptor that exists in the extended conformation even in the absence of bound ligand would be constitutively active. Remarkably, such a receptor exists. This receptor, Her2, is approximately 50% identical in amino acid sequence with the EGF receptor and has the same domain structure. Her2 does not bind any known ligand, yet crystallographic studies reveal that it adopts an extended structure very similar to that observed for the EGF receptor bound to EGF. Under normal conditions, Her2 forms heterodimers with the EGF receptor and other members of the EGF receptor family and participates in cross-phosphorylation reactions with these receptors. Her2 is overexpressed in some cancers, and this overexpression contributes to tumor growth, presumably by forming homodimers that signal even in the absence of ligand. We will return to Her2 when we consider approaches to cancer treatment based on knowledge of signaling pathways (Section 14.5).

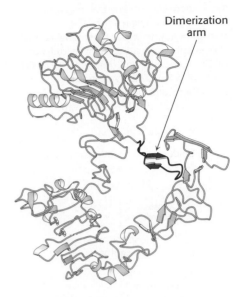

Figure 14.28 Structure of the unactivated EGF receptor. The extracellular domain of the EGF receptor is shown in the absence of bound EGF. *Notice* that the dimerization arm is bound to a part of the receptor that makes it unavailable for interaction with the other receptor.

The EGF Receptor Undergoes Phosphorylation of Its Carboxyl-Terminal Tail

Like the insulin receptor, the EGF receptor undergoes cross-phosphorylation of one unit by another unit within a dimer. However, unlike that of the insulin receptor, the site of this phosphorylation is not the activation loop of the kinase, but rather a region that lies on the C-terminal site of the kinase domain. As many as five tyrosines residues in this region are phosphorylated. The dimerization of the EGF receptor brings the C-terminal region on one receptor into the active site of its partner's kinase. The kinase itself is in an active conformation without phosphorylation, revealing again how this signaling system is poised to respond.

EGF Signaling Leads to the Activation of Ras, a Small G Protein

What happens after the carboxyl terminus of the receptor is phosphorylated? The phosphotyrosines on the receptors act as docking sites for SH2 domains on other proteins. The chain begins with the binding of a key adaptor protein, called *Grb-2* (Figure 14.29). On phosphorylation of the receptor, the SH2 domain of Grb-2 binds to the phosphotyrosine residues of the receptor tyrosine kinase. Grb-2 then recruits a protein called *Sos* through two *Src homology 3 (SH3) domains* that bind proline-rich stretches of polypeptide. Sos, in turn, binds to *Ras*, a very prominent signal-transduction component, and activates it. Ras is a member of a class of proteins called the *small G proteins*. Like the G proteins described in Section 14.1, the small G proteins contain bound GDP in their unactivated forms. Sos opens up the nucleotide-binding pocket of Ras, allowing GDP to escape and GTP to enter in its place. Because of its role, Sos is referred to as a *guanine-nucleotide-exchange factor* (GEF). Thus, the binding of EGF to the EGF receptor leads to the conversion of Ras into its GTP form through the intermediacy of Grb-2 and Sos (Figure 14.30).

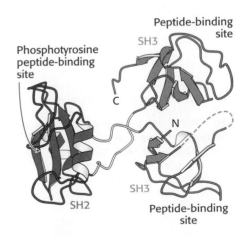

Figure 14.29 Structure of Grb-2, an adaptor protein. Grb-2 consists of a central SH2 domain surrounded by two SH3 domains. The SH2 domain binds phosphotyrosine residues in the activated receptor, and the SH3 domains bind proline-rich regions on other proteins. [Drawn from 1GRI.pdb.]

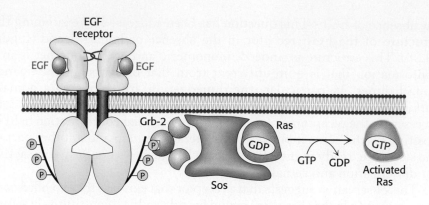

EGF
receptor

EGF   EGF

Grb-2   Ras

                                              GDP

                                        GTP   GDP   GTP

Sos   Activated
                                                        Ras

Figure 14.30 Ras activation mechanism. The dimerization of the EGF receptor due to EGF binding leads to the phosphorylation of the C-terminal tails of the receptor, the subsequent recruitment of Grb-2 and Sos, and the exchange of GTP for GDP in Ras. This signal-transduction pathway results in the conversion of Ras into its activated GTP-bound form.

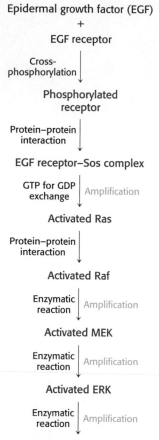

Epidermal growth factor (EGF)
+
EGF receptor

Cross-
phosphorylation

Phosphorylated
receptor

Protein–protein
interaction

EGF receptor–Sos complex

GTP for GDP
exchange   Amplification

Activated Ras

Protein–protein
interaction

Activated Raf

Enzymatic
reaction   Amplification

Activated MEK

Enzymatic
reaction   Amplification

Activated ERK

Enzymatic
reaction   Amplification

Phosphorylated transcription factors
Changes in gene expression

Figure 14.31 EGF signaling pathway. The key steps in the pathway initiated by EGF binding to the EGF receptor. A kinase cascade leads to the phosphorylation of transcription factors and concomitant changes in gene expression.

Activated Ras Initiates a Protein Kinase Cascade

Ras changes conformation when it is transformed from its GDP into its GTP form. In the GTP form, Ras binds other proteins, including a protein kinase termed *Raf*. When Raf binds to Ras, Raf undergoes a conformational change that activates the Raf protein kinase domain. Like Ras, Raf is anchored to the membrane through a covalently bound isoprene lipid. Activated Raf then phosphorylates other proteins, including protein kinases termed MEKs. In turn, MEKs activate kinases called *extracellular signal-regulated kinases* (ERKs). ERKs then phosphorylate numerous substrates, including transcription factors in the nucleus as well as other protein kinases. The complete flow of information from the arrival of EGF at the cell surface to changes in gene expression is summarized in Figure 14.31.

As stated earlier, the signal-transduction protein Ras belongs to a superfamily of signal proteins referred to as small G proteins or small GTPases. This large superfamily of proteins—grouped into subfamilies called Ras, Rho, Arf, Rab, and Ran—plays a major role in a host of cell functions including growth, differentiation, cell motility, cytokinesis, and the transport of materials throughout the cell (Table 14.2). Like their relatives the heterotrimeric G proteins, the small G proteins cycle between an active GTP-bound form and an inactive GDP-bound form. They differ from the heterotrimeric G proteins in being smaller (20–25 kd versus 30–35 kd) and monomeric. Nonetheless, the two families are related by divergent evolution, and small G proteins have many key mechanistic and structural motifs in common with the G_{α} subunit of the heterotrimeric G proteins.

EGF Signaling Is Terminated by Protein Phosphatases and the Intrinsic GTPase Activity of Ras

Because so many components of the EGF signal-transduction pathway are activated by phosphorylation, we can expect protein phosphatases to play key roles in the termination of EGF signaling. Indeed, crucial phosphatases

TABLE 14.2 Ras superfamily of GTPases

Subfamily	Function
Ras	Regulates cell growth through serine-threonine protein kinases
Rho	Reorganizes cytoskeleton through serine-threonine protein kinases
Arf	Activates the ADP-ribosyltransferase of the cholera toxin A subunit; regulates vesicular trafficking pathways; activates phospholipase D
Rab	Plays a key role in secretory and endocytotic pathways
Ran	Functions in the transport of RNA and protein into and out of the nucleus

remove phosphoryl groups from tyrosine residues on the EGF receptor and from serine, threonine, and tyrosine residues in the protein kinases that participate in the signaling cascade. The signaling process itself sets in motion the events that activate many of these phosphatases. Consequently, signal activation also initiates signal termination.

Like the G proteins activated by 7TM receptors, Ras possesses intrinsic GTPase activity. Thus, the activated GTP form of Ras spontaneously converts into the inactive GDP form. The rate of conversion accelerates in the presence of *GTPase-activating proteins* (GAPs), proteins that interact with Ras in the GTP form and facilitate GTP hydrolysis. Thus, the lifetime of activated Ras is regulated by accessory proteins in the cell. The GTPase activity of Ras is crucial for shutting off signals leading to cell growth, and so it is not surprising that mutations in Ras are found in many types of cancer, as we will see in Section 14.5.

14.4 Many Elements Recur with Variation in Different Signal-Transduction Pathways

We can begin to make sense of the complexity of signal-transduction pathways by taking note of several recurring elements. These elements have appeared consistently in the pathways described in this chapter and underlie many additional signaling pathways not considered herein.

1. *Protein kinases are central to many signal-transduction pathways.* Protein kinases are central to all three signal-transduction pathways described in this chapter. In the epinephrine-initiated pathway, cAMP-dependent protein kinase (PKA) lies at the end of the pathway, transducing information represented by an increase in cAMP concentration into covalent modifications that alter the activity of key metabolic enzymes. In the insulin- and EGF-initiated pathways, the receptors themselves are protein kinases and several additional protein kinases participate downstream in the pathways. Signal amplification due to protein kinase cascades are common features of each of these pathways and many others. Furthermore, protein kinases often phosphorylate multiple substrates, including many not considered herein, and by this means are able to generate a diversity of responses.

2. *Second messengers participate in many signal-transduction pathways.* We have encountered several second messengers, including cAMP, Ca^{2+}, IP_3, and the lipid DAG. Because second messengers are activated by enzymes or by the opening of ion channels, their concentrations can be tremendously amplified compared with the signals that lead to their generation. Specialized proteins sense the concentrations of these second messengers and continue the flow of information along signal-transduction pathways.

The second messengers that we have seen recur in many additional signal-transduction pathways. For example, in a consideration of the sensory systems in Chapter 32, we will see how Ca^{2+}-based signaling and cyclic nucleotide-based signaling play key roles in vision and olfaction.

3. *Specialized domains that mediate specific interactions are present in many signaling proteins.* The "wiring" of many signal-transduction pathways is based on particular protein domains, present in signal-transduction proteins, that bring components of signal-transduction pathways closer together. We have encountered several of them, including pleckstrin homology domains (which facilitate protein interactions with the lipid PIP_3) and SH_2 domains (which facilitate interactions with polypeptides containing phosphorylated tyrosine residues). *Signal-transduction pathways have evolved in large part by*

*the incorporation of DNA fragments encoding these domains into genes encoding
pathway components.*

The presence of these domains is tremendously helpful to scientists trying to unravel signal-transduction pathways. When a protein in a signal-
transduction pathway is identified, its amino acid sequence can be analyzed
for the presence of these specialized domains by the methods described in
Chapter 6. If one or more domains of known function is found, it is often
possible to develop clear hypotheses about potential binding partners and
signal-transduction mechanisms.

14.5 Defects in Signal-Transduction Pathways Can Lead to Cancer and Other Diseases

In light of their complexity, it comes as no surprise that signal-
transduction pathways occasionally fail, leading to pathological or
disease states. Cancer, a set of diseases characterized by uncontrolled or
inappropriate cell growth, is strongly associated with defects in signal-
transduction proteins. Indeed, the study of cancer, particularly cancers
caused by certain viruses, has contributed greatly to our understanding of
signal-transduction proteins and pathways.

For example, Rous sarcoma virus is a retrovirus that causes sarcoma
(a cancer of tissues of mesodermal origin such as muscle or connective tissue) in chickens. In addition to the genes necessary for viral replication,
this virus carries a gene termed v-*src*. The v-*src* gene is an *oncogene;* it leads
to the transformation of susceptible cell types—that is, the generation of
cancerlike characteristics in the cell types. The protein encoded
by v-*src* is a protein tyrosine kinase that includes SH2 and SH3
domains (Figure 14.32). The v-Src protein is similar in amino
acid sequence to a protein normally found in chicken muscle cells
referred to as c-Src (for cellular Src). The c-*src* gene does not induce cell transformation and is termed a *proto-oncogene.* The
protein that it encodes is a signal-transduction protein that regulates cell growth.

Why is the biological activity of the v-Src protein so different
from that of c-Src? In v-Src, the C-terminal 19 amino acids of c-Src
are replaced by a completely different stretch of 11 amino acids that
lacks the key tyrosine residue that is phosphorylated to inactivate
c-Src. Thus, v-Src is always active. Since the discovery of Src, many
other mutated protein kinases have been identified as oncogenes.

The gene encoding Ras, a component of the EGF-initiated
pathway, is one of the genes most commonly mutated in human
tumors. Mammalian cells contain three 21-kd Ras proteins (H-,
K-, and N-Ras), each of which cycles between inactive GDP and
active GTP forms. The most common mutations in tumors lead to
a loss of the ability to hydrolyze GTP. Thus, the Ras protein is
trapped in the "on" position and continues to stimulate cell
growth, even in the absence of a continuing signal.

Other genes can contribute to cancer development only when
both copies of the gene normally present in a cell are deleted or
otherwise damaged. Such genes are called *tumor-suppressor genes.*
For example, genes for some of the phosphatases that participate in
the termination of EGF signaling are tumor suppressors. Without
any functional phosphatase present, EGF signaling persists once
initiated, stimulating inappropriate cell growth.

(A)

(B)

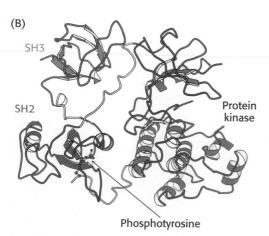

Figure 14.32 Src structure. (A) Cellular Src
includes an SH3 domain, an SH2 domain, a
protein kinase domain, and a carboxyl-terminal tail that
includes a key tyrosine residue. (B) Structure of c-Src
in an inactivated form with the key tyrosine residue
phosphorylated. *Notice* how the three domains
work together to keep the enzyme in an inactive
conformation: phosphotyrosine residue is bound in the
SH2 domain and the linker between the SH2 domain
and the protein kinase domain is bound by the SH3
domain. [Drawn from 2PTK.pdb.]

Monoclonal Antibodies Can Be Used to Inhibit Signal-Transduction Pathways Activated in Tumors

Mutated or overexpressed receptor tyrosine kinases are frequently observed in tumors. For instance, the epidermal-growth-factor receptor (EGFR) is overexpressed in some human epithelial cancers, including breast, ovarian, and colorectal cancer. Because some small amount of the receptor can dimerize and activate the signaling pathway even without binding to EGF, overexpression of the receptor increases the likelihood that a "grow and divide" signal will be inappropriately sent to the cell. This understanding of cancer-related signal-transduction pathways has led to a therapeutic approach that targets the EGFR. The strategy is to produce monoclonal antibodies to the extracellular domains of the offending receptors. One such antibody, cetuximab (Erbitux), has effectively targeted the EGFR in colorectal cancers. Cetuximab inhibits the EGFR by competing with EGF for the binding site on the receptor. Because the antibody sterically blocks the change in conformation that exposes the dimerization arm, the antibody itself cannot induce dimerization. The result is that the EGFR-controlled pathway is not initiated.

Cetuximab is not the only monoclonal antibody that has been developed to target a receptor tyrosine kinase. Trastuzumab (Herceptin) inhibits another EGFR family member, Her2, that is overexpressed in approximately 30% of breast cancers. Recall that this protein can signal even in the absence of ligand; so it is especially likely that overexpression will stimulate cell proliferation. Breast-cancer patients are now being screened for Her2 overexpression and treated with Herceptin as appropriate. Thus, this cancer treatment is tailored to the genetic characteristics of the tumor.

Protein Kinase Inhibitors Can Be Effective Anticancer Drugs

The widespread occurrence of overactive protein kinases in cancer cells suggests that molecules that inhibit these enzymes might act as antitumor agents. Recent results have dramatically supported this concept. For example, more that 90% of patients with chronic myelogenous leukemia (CML) show a specific chromosomal defect in cancer cells (Figure 14.33). The translocation of genetic material between chromosomes 9 and 22 causes the c-*abl* gene, which encodes a tyrosine kinase of the Src family, to be inserted into the *bcr* gene on chromosome 22. The result is the production of a fusion protein called Bcr-Abl that consists primarily of sequences for the c-Abl kinase. However, the *bcr-abl* gene is not regulated appropriately; it is expressed at higher levels than the gene encoding the normal c-Abl kinase, stimulating a growth-promoting pathway. Because of this overexpression, leukemia cells express a unique target for chemotherapy. A specific inhibitor of the Bcr-Abl kinase, Gleevec (STI-571, imatinib mesylate), has proved to be a highly effective treatment for patients suffering from CML. This approach to cancer chemotherapy is fundamentally distinct from most approaches, which target all rapidly growing cells, including normal ones. *Thus, our understanding of signal-transduction pathways is leading to conceptually new disease treatments.*

Cholera and Whooping Cough Are Due to Altered G-Protein Activity

Although defects in signal-transduction pathways have been most extensively studied in the context of cancer, such defects are important in many other diseases. Cholera and whooping cough are two pathologies of the G-protein-dependent signal pathways. Let us first consider the mechanism of action of the cholera toxin, secreted by the intestinal bacterium *Vibrio cholerae*. Cholera is a potentially life-threatening, acute diarrheal disease transmitted through contaminated water and food. It causes the voluminous

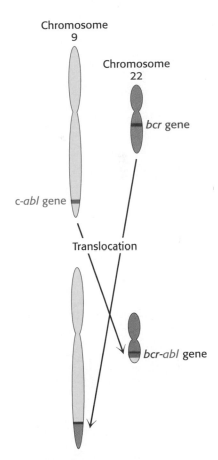

Figure 14.33 Formation of the *bcr-abl* gene by translocation. In chronic myelogenous leukemia, parts of chromosomes 9 and 22 are reciprocally exchanged, causing the *bcr* and *abl* genes to fuse. The protein kinase encoded by the *bcr-abl* gene is expressed at higher levels in cells having this translocation than is the c-*abl* gene in normal cells.

secretion of electrolytes and fluids from the intestines of infected persons. The cholera toxin, *choleragen*, is a protein composed of two functional units—a B subunit that binds to G_{M1} gangliosides (p. 738) of the intestinal epithelium and a catalytic A subunit that enters the cell. The A subunit catalyzes the covalent modification of a $G_{\alpha s}$ protein: the α subunit is modified by the attachment of an ADP-ribose to an arginine residue. This modification stabilizes the GTP-bound form of $G_{\alpha s}$, trapping the molecule in the active conformation. The active G protein, in turn, continuously activates protein kinase A. PKA opens a chloride channel and inhibits sodium absorption by the Na^+–H^+ exchanger by phosphorylating both the channel and the exchanger. The net result of the phosphorylation is an excessive loss of NaCl and the loss of large amounts of water into the intestine. Patients suffering from cholera for 4 to 6 days may pass as much as twice their body weight in fluid. Treatment consists of rehydration with a glucose–electrolyte solution.

Whereas cholera is a result of a G protein trapped in the active conformation, causing the signal-transduction pathway to be perpetually stimulated, pertussis, or whooping cough, is a result of the opposite situation. Pertussis toxin also adds an ADP-ribose moiety—in this case, to a $G_{\alpha i}$ protein, a G_α protein that inhibits adenylate cyclase, closes Ca^{2+} channels, and opens K^+ channels. The effect of this modification, however, is to lower the G protein's affinity for GTP, effectively trapping it in the "off" conformation. The pulmonary symptoms have not yet been traced to a particular target of the $G_{\alpha i}$ protein. Pertussis toxin is secreted by *Bordetella pertussis*, the bacterium responsible for whooping cough.

Summary

In human beings and other multicellular organisms, specific signal molecules are released from cells in one organ and are sensed by cells in other organs throughout the body. The message "a signal molecule is present" is converted into specific changes in metabolism or gene expression by means of often complex networks referred to as signal-transduction pathways. These pathways amplify the initial signal and lead to changes in the properties of specific effector molecules.

14.1   Heterotrimeric G Proteins Transmit Signals and Reset Themselves

Epinephrine binds to a cell-surface protein called the β-adrenergic receptor. This receptor is a member of the seven-transmembrane-helix receptor family, so named because each receptor has seven α helices that span the cell membrane. When epinephrine binds to the β-adrenergic receptor on the outside of the cell, the receptor undergoes a conformational change that is sensed inside the cell by a signaling protein termed a heterotrimeric G protein. The α subunit of the G protein exchanges a bound GDP molecule for GTP and concomitantly releases the heterodimer consisting of the β and γ subunits. The α subunit in the GTP form then binds to adenylate cyclase and activates it, leading to an increase in the concentration of the second messenger cyclic AMP. This increase in cyclic AMP concentration, in turn, activates protein kinase A. Other 7TM receptors also signal through heterotrimeric G proteins, although these pathways often include enzymes other than adenylate cyclase. One prominent pathway, the phosphoinositide pathway, leads to the activation of phospholipase C, which cleaves a membrane lipid to produce two secondary messengers, diacylglycerol and inositol 1,4,5-trisphosphate. An increased IP_3 concentration leads to the release of calcium ion, another important second messenger, into the cell. G-protein signaling is terminated by the hydrolysis of the bound GTP to GDP.

14.2 Insulin Signaling: Phosphorylation Cascades Are Central to Many Signal-Transduction Processes

Protein kinases are key components in many signal-transduction pathways, including some for which the protein kinase is an integral component of the initial receptor. An example of such a receptor is the membrane tyrosine kinase bound by insulin. Insulin binding causes one subunit within the dimeric receptor to phosphorylate specific tyrosine residues in the other subunit. The resulting conformational changes dramatically increase the kinase activity of the receptor. The activated receptor kinase initiates a kinase cascade that includes both lipid kinases and protein kinases. This cascade eventually leads to the mobilization of glucose transporters to the cell surface, increasing glucose uptake. Insulin signaling is terminated through the action of phosphatases.

14.3 EGF Signaling: Signal-Transduction Systems Are Poised to Respond

Only minor modifications are necessary to transform many signal-transduction proteins from their inactive into their active forms. Epidermal growth factor also signals through a receptor tyrosine kinase. EGF binding induces a conformational change that allows receptor dimerization and cross-phosphorylation. The phosphorylated receptor binds adaptor proteins that mediate the activation of Ras, a small G protein. Activated Ras initiates a protein kinase cascade that eventually leads to the phosphorylation of transcription factors and changes in gene expression. EGF signaling is terminated by the action of phosphatases and the hydrolysis of GTP by Ras.

14.4 Many Elements Recur with Variation in Different Signal-Transduction Pathways

Protein kinases are components of many signal-transduction pathways, both as components of receptors and in other roles. Second messengers, including cyclic nucleotides, calcium, and lipid derivatives, are common in many signaling pathways. The changes in the concentrations of second messengers are often much larger than the changes associated with the initial signal owing to amplification processes. Small domains that recognize phosphotyrosine residues or specific lipids are present in many signaling proteins and are essential to determining the specificity of interactions.

14.5 Defects in Signal-Transduction Pathways Can Lead to Cancer and Other Diseases

Genes encoding components of signal-transduction pathways that control cell growth are often mutated in cancer. Some genes can be mutated to forms called oncogenes that are active regardless of appropriate signals. Monoclonal antibodies directed against cell-surface receptors that participate in signaling have been developed for use in cancer treatment. Our understanding of the molecular basis of cancer is leading to the development of anticancer drugs directed against specific targets, such as the specific kinase inhibitor Gleevec.

Key Terms

primary messenger (p. 382)

ligand (p. 382)

second messenger (p. 382)

cross talk (p. 382)

β-adrenergic receptor (β-AR) (p. 383)

seven-transmembrane-helix (7TM) receptor (p. 383)

rhodopsin (p. 384)

G protein (p. 385)

G-protein-coupled receptor (GPCR) (p. 385)

adenylate cyclase (p. 385)

Selected Readings

Where to Start

Scott, J. D., and Pawson, T. 2000. Cell communication: The inside story. *Sci. Am.* 282(6):7279.

Pawson, T. 1995. Protein modules and signalling networks. *Nature* 373:573–580.

Okada, T., Ernst, O. P., Palczewski, K., and Hofmann, K. P. 2001. Activation of rhodopsin: New insights from structural and biochemical studies. *Trends Biochem. Sci.* 26:318–324.

Tsien, R. Y. 1992. Intracellular signal transduction in four dimensions: From molecular design to physiology. *Am. J. Physiol.* 263:C723–C728.

Loewenstein, W. R. 1999. *Touchstone of Life: Molecular Information, Cell Communication, and the Foundations of Life.* Oxford University Press.

G Proteins and 7TM Receptors

Palczewski, K., Kumasaka, T., Hori, T., Behnke, C. A., Motoshima, H., Fox, B. A., Le Trong, I., Teller, D. C., Okada, T., Stenkamp, R. E., Yamamoto, M., and Miyano, M. 2000. Crystal structure of rhodopsin: A G protein-coupled receptor. *Science* 289:739–745.

Lefkowitz, R. J. 2000. The superfamily of heptahelical receptors. *Nat. Cell Biol.* 2:E133–E136.

Bourne, H. R., Sanders, D. A., and McCormick, F. 1991. The GTPase superfamily: Conserved structure and molecular mechanism. *Nature* 349:117–127.

Lambright, D. G., Noel, J. P., Hamm, H. E., and Sigler, P. B. 1994. Structural determinants for activation of the α-subunit of a heterotrimeric G protein. *Nature* 369:621–628.

Noel, J. P., Hamm, H. E., and Sigler, P. B. 1993. The 2.2 Å crystal structure of transducin-α complexed with GTPγS. *Nature* 366:654–663.

Sondek, J., Lambright, D. G., Noel, J. P., Hamm, H. E., and Sigler, P. B. 1994. GTPase mechanism of G proteins from the 1.7-Å crystal structure of transducin α-GDP-AIF$_4^-$. *Nature* 372:276–279.

Sondek, J., Bohm, A., Lambright, D. G., Hamm, H. E., and Sigler, P. B. 1996. Crystal structure of a G-protein βγ dimer at 2.1 Å resolution. *Nature* 379:369–374.

Wedegaertner, P. B., Wilson, P. T., and Bourne, H. R. 1995. Lipid modifications of trimeric G proteins. *J. Biol. Chem.* 270:503–506.

Farfel, Z., Bourne, H. R., and Iiri, T. 1999. The expanding spectrum of G protein diseases. *N. Engl. J. Med.* 340:1012–1020.

Bockaert, J., and Pin, J. P. 1999. Molecular tinkering of G protein-coupled receptors: An evolutionary success. *EMBO J.* 18:1723–1729.

cAMP Cascade

Hurley, J. H. 1999. Structure, mechanism, and regulation of mammalian adenylyl cyclase. *J. Biol. Chem.* 274:7599–7602.

Kim, C., Xuong, N. H., and Taylor, S. S. 2005. Crystal structure of a complex between the catalytic and regulatory (RIα) subunits of PKA. *Science* 307:690–696.

Tesmer, J. J., Sunahara, R. K., Gilman, A. G., and Sprang, S. R. 1997. Crystal structure of the catalytic domains of adenylyl cyclase in a complex with G$_{s\alpha}$-GTPγS. *Science* 278:1907–1916.

Smith, C. M., Radzio-Andzelm, E., Madhusudan, Akamine, P., and Taylor, S. S. 1999. The catalytic subunit of cAMP-dependent protein kinase: Prototype for an extended network of communication. *Prog. Biophys. Mol. Biol.* 71:313–341.

Taylor, S. S., Buechler, J. A., and Yonemoto, W. 1990. cAMP-dependent protein kinase: Framework for a diverse family of regulatory enzymes. *Annu. Rev. Biochem.* 59:971–1005.

Phosphoinositide Cascade

Berridge, M. J., and Irvine, R. F. 1989. Inositol phosphates and cell signalling. *Nature* 341:197–205.

Berridge, M. J. 1993. Inositol trisphosphate and calcium signalling. *Nature* 361:315–325.

Essen, L. O., Perisic, O., Cheung, R., Katan, M., and Williams, R. L. 1996. Crystal structure of a mammalian phosphoinositide-specific phospholipase C δ. *Nature* 380:595–602.

Ferguson, K. M., Lemmon, M. A., Schlessinger, J., and Sigler, P. B. 1995. Structure of the high affinity complex of inositol trisphosphate with a phospholipase C pleckstrin homology domain. *Cell* 83:1037–1046.

Baraldi, E., Carugo, K. D., Hyvonen, M., Surdo, P. L., Riley, A. M., Potter, B. V., O'Brien, R., Ladbury, J. E., and Saraste, M. 1999. Structure of the PH domain from Bruton's tyrosine kinase in complex with inositol 1,3,4,5-tetrakisphosphate. *Structure Fold Des.* 7:449–460.

Calcium

Ikura, M., Clore, G. M., Gronenborn, A. M., Zhu, G., Klee, C. B., and Bax, A. 1992. Solution structure of a calmodulin-target peptide complex by multidimensional NMR. *Science* 256:632–638.

Kuboniwa, H., Tjandra, N., Grzesiek, S., Ren, H., Klee, C. B., and Bax, A. 1995. Solution structure of calcium-free calmodulin. *Nat. Struct. Biol.* 2:768–776.

Grynkiewicz, G., Poenie, M., and Tsien, R. Y. 1985. A new generation of Ca^{2+} indicators with greatly improved fluorescence properties. *J. Biol. Chem.* 260:3440–3450.

Kerr, R., Lev-Ram, V., Baird, G., Vincent, P., Tsien, R. Y., and Schafer, W. R. 2000. Optical imaging of calcium transients in neurons and pharyngeal muscle of *C. elegans. Neuron* 26:583–594.

Chin, D., and Means, A. R. 2000. Calmodulin: A prototypical calcium sensor. *Trends Cell Biol.* 10:322–328.

Dawson, A. P. 1997. Calcium signalling: How do IP$_3$ receptors work? *Curr. Biol.* 7:R544–R547.

Protein Kinases, Including Receptor Tyrosine Kinases

Riedel, H., Dull, T. J., Honegger, A. M., Schlessinger, J., and Ullrich, A. 1989. Cytoplasmic domains determine signal specificity, cellular routing characteristics and influence ligand binding of epidermal growth factor and insulin receptors. *EMBO J.* 8:2943–2954.

Taylor, S. S., Knighton, D. R., Zheng, J., Sowadski, J. M., Gibbs, C. S., and Zoller, M. J. 1993. A template for the protein kinase family. *Trends Biochem. Sci.* 18:84–89.

Sicheri, F., Moarefi, I., and Kuriyan, J. 1997. Crystal structure of the Src family tyrosine kinase Hck. *Nature* 385:602–609.

Waksman, G., Shoelson, S. E., Pant, N., Cowburn, D., and Kuriyan, J. 1993. Binding of a high affinity phosphotyrosyl peptide to the Src SH2 domain: Crystal structures of the complexed and peptide-free forms. *Cell* 72:779–790.

Schlessinger, J. 2000. Cell signaling by receptor tyrosine kinases. *Cell* 103:211–225.

Simon, M. A. 2000. Receptor tyrosine kinases: Specific outcomes from general signals. *Cell* 103:13–15.

Robinson, D. R., Wu, Y. M., and Lin, S. F. 2000. The protein tyrosine kinase family of the human genome. *Oncogene* 19:5548–5557.

Hubbard, S. R. 1999. Structural analysis of receptor tyrosine kinases. *Prog. Biophys. Mol. Biol.* 71:343–358.

Carter-Su, C., and Smit, L. S. 1998. Signaling via JAK tyrosine kinases: Growth hormone receptor as a model system. *Recent Prog. Horm. Res.* 53:61–82.

Insulin Signaling Pathway

Khan, A. H., and Pessin, J. E. 2002. Insulin regulation of glucose uptake: A complex interplay of intracellular signalling pathways. *Diabetologia* 45:1475–1483.

Bevan, P. 2001. Insulin signalling. *J. Cell Sci.* 114:1429–1430.

De Meyts, P., and Whittaker, J. 2002. Structural biology of insulin and IGF1 receptors: Implications for drug design. *Nat. Rev. Drug Discov.* 1:769–783.

Dhe-Paganon, S., Ottinger, E. A., Nolte, R. T., Eck, M. J., and Shoelson, S. E. 1999. Crystal structure of the pleckstrin homology-phosphotyrosine binding (PH-PTB) targeting region of insulin receptor substrate 1. *Proc. Natl. Acad. Sci. U. S. A.* 96:8378–8383.

Domin, J., and Waterfield, M. D. 1997. Using structure to define the function of phosphoinositide 3-kinase family members. *FEBS Lett.* 410:91–95.

Hubbard, S. R. 1997. Crystal structure of the activated insulin receptor tyrosine kinase in complex with peptide substrate and ATP analog. *EMBO J.* 16:5572–5581.

Hubbard, S. R., Wei, L., Ellis, L., and Hendrickson, W. A. 1994. Crystal structure of the tyrosine kinase domain of the human insulin receptor. *Nature* 372:746–754.

EGF Signaling Pathway

Burgess, A. W., Cho, H. S., Eigenbrot, C., Ferguson, K. M., Garrett, T. P., Leahy, D. J., Lemmon, M. A., Sliwkowski, M. X., Ward, C. W., and Yokoyama, S. 2003. An open-and-shut case? Recent insights into the activation of EGF/ErbB receptors. *Mol. Cell* 12:541–552.

Cho, H. S., Mason, K., Ramyar, K. X., Stanley, A. M., Gabelli, S. B., Denney, D. W., Jr., and Leahy, D. J. 2003. Structure of the extracellular region of HER2 alone and in complex with the Herceptin Fab. *Nature* 421:756–760.

Chong, H., Vikis, H. G., and Guan, K. L. 2003. Mechanisms of regulating the Raf kinase family. *Cell. Signal.* 15:463–469.

Stamos, J., Sliwkowski, M. X., and Eigenbrot, C. 2002. Structure of the epidermal growth factor receptor kinase domain alone and in complex with a 4-anilinoquinazoline inhibitor. *J. Biol. Chem.* 277:46265–46272.

Ras

Milburn, M. V., Tong, L., deVos, A. M., Brunger, A., Yamaizumi, Z., Nishimura, S., and Kim, S. H. 1990. Molecular switch for signal transduction: Structural differences between active and inactive forms of protooncogenic Ras proteins. *Science* 247:939–945.

Boriack-Sjodin, P. A., Margarit, S. M., Bar-Sagi, D., and Kuriyan, J. 1998. The structural basis of the activation of Ras by Sos. *Nature* 394:337–343.

Maignan, S., Guilloteau, J. P., Fromage, N., Arnoux, B., Becquart, J., and Ducruix, A. 1995. Crystal structure of the mammalian Grb2 adaptor. *Science* 268:291–293.

Takai, Y., Sasaki, T., and Matozaki, T. 2001. Small GTP-binding proteins. *Physiol. Rev.* 81:153–208.

Cancer

Druker, B. J., Sawyers, C. L., Kantarjian, H., Resta, D. J., Reese, S. F., Ford, J. M., Capdeville, R., and Talpaz, M. 2001. Activity of a specific inhibitor of the BCR-ABL tyrosine kinase in the blast crisis of chronic myeloid leukemia and acute lymphoblastic leukemia with the Philadelphia chromosome. *N. Engl. J. Med.* 344:1038–1042.

Vogelstein, B., and Kinzler, K. W. 1993. The multistep nature of cancer. *Trends Genet.* 9:138–141.

Ellis, C. A., and Clark, G. 2000. The importance of being K-Ras. *Cell. Signal.* 12:425–434.

Hanahan, D., and Weinberg, R. A. 2000. The hallmarks of cancer. *Cell* 100:57–70.

McCormick, F. 1999. Signalling networks that cause cancer. *Trends Cell Biol.* 9:M53–M56.

Problems

1. *Active mutants.* Some protein kinases are inactive unless they are phosphorylated on key serine or threonine residues. In some cases, active enzymes can be generated by mutating these serine or threonine residues to glutamate. Propose an explanation.

2. *In the pocket.* SH2 domains bind phosphotyrosine residues in deep pockets on their surfaces. Would you expect SH2 domains to bind phosphoserine or phosphothreonine with high affinity? Why or why not?

3. *Antibodies mimicking hormones.* Antibodies have two identical antigen-binding sites. Remarkably, antibodies to the extracellular parts of growth-factor receptors often lead to the same cellular effects as does exposure to growth factors. Explain this observation.

4. *Facile exchange.* A mutated form of the α subunit of the heterotrimeric G protein has been identified; this form readily exchanges nucleotides even in the absence of an activated receptor. What would be the effect on a signaling pathway containing the mutated α subunit?

5. *Diffusion rates.* Normally, rates of diffusion vary inversely with molecular weights; so smaller molecules diffuse faster than do larger ones. In cells, however, calcium ion diffuses more slowly than does cAMP. Propose a possible explanation.

6. *Awash with glucose.* Glucose is mobilized for ATP generation in muscle in response to epinephrine, which activates $G_{\alpha s}$. Cyclic AMP phosphodiesterase is an enzyme that converts

cAMP into AMP. How would inhibitors of cAMP phosphodiesterase affect glucose mobilization in muscle?

7. *Many defects.* Considerable effort has been directed toward determining the genes in which sequence variation contributes to the development of type 2 diabetes. Approximately 800 genes have been implicated. Propose an explanation for this observation.

8. *Growth-factor signaling.* Human growth hormone binds to a cell-surface membrane protein that is not a receptor tyrosine kinase. The intracellular domain of the receptor can bind other proteins inside the cell. Furthermore, studies indicate that the receptor is monomeric in the absence of hormone but dimerizes on hormone binding. Propose a possible mechanism for growth-hormone signaling.

9. *Hybrid.* Suppose that, through genetic manipulations, a chimeric receptor is produced that consists of the extracellular domain of the insulin receptor and the transmembrane and intracellular domains of the EGF receptor. Cells expressing this receptor are exposed to insulin and the level of phosphorylation of the chimeric receptor is examined. What would you expect to observe and why? What would you expect to observe if these cells were exposed to EGF?

10. *Total amplification.* Suppose that each β-adrenergic receptor bound to epinephrine converts 100 molecules of $G_{\alpha s}$ into their GTP forms and that each molecule of activated adenylate cyclase produces 1000 molecules of cAMP per second. With the assumption of a full response, how many molecules of cAMP will be produced in 1 s after the formation of a single complex between epinephrine and the β-adrenergic receptor?

Chapter Integration Problem

11. *Nerve-growth-factor pathway.* Nerve-growth factor (NGF) binds to a protein tyrosine kinase receptor. The amount of diacylglycerol in the plasma membrane increases in cells expressing this receptor when treated with NGF. Propose a simple signaling pathway and identify the isoform of any participating enzymes. Would you expect the concentrations of any other common second messengers to increase on NGF treatment?

Mechanism Problem

12. *Distant relatives.* The structure of adenylate cyclase is similar to the structures of some types of DNA polymerases, suggesting that these enzymes derived from a common ancestor. Compare the reactions catalyzed by these two enzymes. In what ways are they similar?

Data Interpretation Problems

13. *Establishing specificity.* You wish to determine the hormone-binding specificity of a newly identified membrane receptor. Three different hormones, X, Y, and Z, were mixed with the receptor in separate experiments, and the percentage of binding capacity of the receptor was determined as a function of hormone concentration, as shown in graph A.

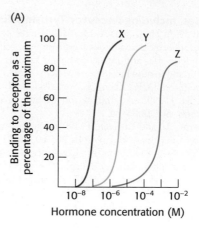

(A)

(a) What concentrations of each hormone yield 50% maximal binding?
(b) Which hormone shows the highest binding affinity for the receptor?

You next wish to determine whether the hormone–receptor complex stimulates the adenylate cyclase cascade. To do so, you measure adenylate cyclase activity as a function of hormone concentration, as shown in graph B.

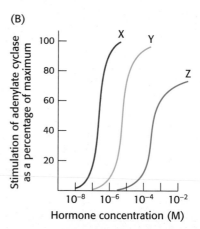

(B)

(c) What is the relation between the binding affinity of the hormone–receptor complex and the ability of the hormone to enhance adenylate cyclase activity? What can you conclude about the mechanism of action of the hormone–receptor complex?
(d) Suggest experiments that would determine whether a $G_{\alpha s}$ protein is a component of the signal-transduction pathway.

14. *Binding issues.* A scientist wishes to determine the number of receptors specific for a ligand X, which he has in both radioactive and nonradioactive form. In one experiment, he adds increasing amounts of the radioactive X and measures how much of it is bound to the cells. The result is shown as total activity in the graph at the top of page 407. Next, he performs the same experiment, except that he includes a several hundredfold excess of nonradioactive X. This result is shown as nonspecific binding. The difference between the two curves is the specific binding.

(a) Why is the total binding not an accurate representation of the number of receptors on the cell surface?

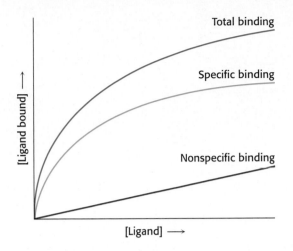

(b) What is the purpose of performing the experiment in the presence of excess nonradioactive ligand?

(c) What is the significance of the fact that specific binding attains a plateau?

15. *Counting receptors.* With the use of experiments such as those described in problems 13 and 14, it is possible to calculate the number of receptors in the cell membrane. Suppose that the specific activity of the ligand is 10^{12} cpm per millimole and that the maximal specific binding is 10^4 cpm per milligram of membrane protein. There are 10^{10} cells per milligram of membrane protein. Assume that one ligand binds per receptor. Calculate the number of receptor molecules present per cell.

Metabolism: Basic Concepts and Design

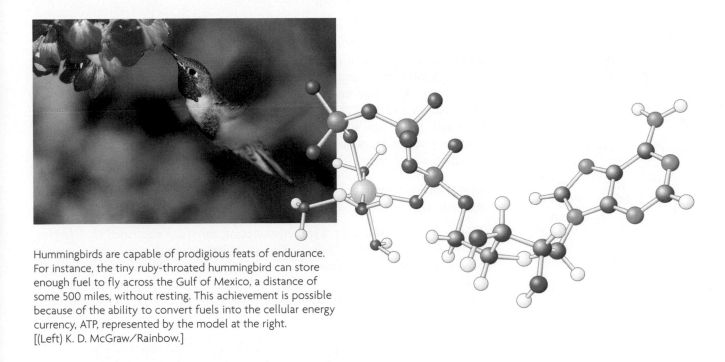

Hummingbirds are capable of prodigious feats of endurance. For instance, the tiny ruby-throated hummingbird can store enough fuel to fly across the Gulf of Mexico, a distance of some 500 miles, without resting. This achievement is possible because of the ability to convert fuels into the cellular energy currency, ATP, represented by the model at the right. [(Left) K. D. McGraw/Rainbow.]

The concepts of conformation and dynamics developed in Part I—especially those dealing with the specificity and catalytic power of enzymes, the regulation of their catalytic activity, and the transport of molecules and ions across membranes—enable us to now ask questions fundamental to biochemistry:

1. *How does a cell extract energy and reducing power from its environment?*

2. *How does a cell synthesize the building blocks of its macromolecules and then the macromolecules themselves?*

These processes are carried out by a highly integrated network of chemical reactions that are collectively known as *metabolism* or *intermediary metabolism*.

   More than a thousand chemical reactions take place in even as simple an organism as *Escherichia coli*. The array of reactions may seem overwhelming at first glance. However, closer scrutiny reveals that metabolism has a *coherent design containing many common motifs*. These motifs include the use of an energy currency and the repeated appearance of a limited number of activated intermediates. In fact, a group of about 100 molecules play central roles in all forms of life. Furthermore, although the number of reactions in metabolism is large, the number of *kinds* of reactions is small and the

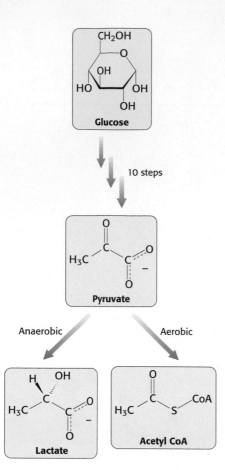

Glucose

10 steps

Pyruvate

Anaerobic   Aerobic

Lactate   Acetyl CoA

Figure 15.1 Glucose metabolism. Glucose is metabolized to pyruvate in 10 linked reactions. Under anaerobic conditions, pyruvate is metabolized to lactate and, under aerobic conditions, to acetyl CoA. The glucose-derived carbons of acetyl CoA are subsequently oxidized to CO_2.

mechanisms of these reactions are usually quite simple. Metabolic pathways are also regulated in common ways. The purpose of this chapter is to introduce some general principles and motifs of metabolism to provide a foundation for the more detailed studies to follow.

15.1 Metabolism Is Composed of Many Coupled, Interconnecting Reactions

Living organisms require a continual input of free energy for three major purposes: (1) the performance of mechanical work in muscle contraction and cellular movements, (2) the active transport of molecules and ions, and (3) the synthesis of macromolecules and other biomolecules from simple precursors. The free energy used in these processes, which maintain an organism in a state that is far from equilibrium, is derived from the environment. Photosynthetic organisms, or *phototrophs,* obtain this energy by trapping sunlight, whereas *chemotrophs,* which include animals, obtain energy through the oxidation of foodstuffs generated by phototrophs.

Metabolism Consists of Energy-Yielding and Energy-Requiring Reactions

Metabolism is essentially a linked series of chemical reactions that begins with a particular molecule and converts it into some other molecule or molecules in a carefully defined fashion (Figure 15.1). There are many such defined pathways in the cell (Figure 15.2), and we will examine a few of them in some detail later. These pathways are interdependent, and their

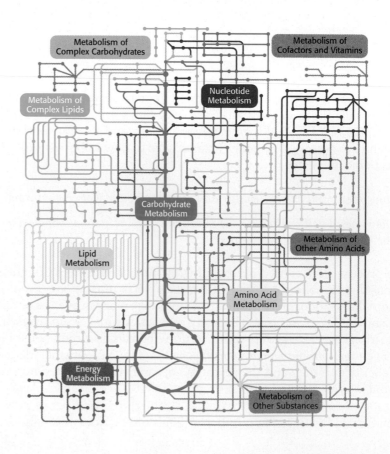

Figure 15.2 Metabolic pathways. [From the Kyoto Encyclopedia of Genes and Genomes (www.genome.ad.jp/kegg).]

activity is coordinated by exquisitely sensitive means of communication in which allosteric enzymes are predominant (Section 10.1). We considered the principles of this communication in Chapter 14.

We can divide metabolic pathways into two broad classes: (1) those that convert energy from fuels into biologically useful forms and (2) those that require inputs of energy to proceed. Although this division is often imprecise, it is nonetheless a useful distinction in an examination of metabolism. Those reactions that transform fuels into cellular energy are called *catabolic reactions* or, more generally, *catabolism*.

$$\text{Fuel (carbohydrates, fats)} \xrightarrow{\text{Catabolism}} CO_2 + H_2O + \text{useful energy}$$

Those reactions that require energy—such as the synthesis of glucose, fats, or DNA—are called *anabolic reactions* or *anabolism*. The useful forms of energy that are produced in catabolism are employed in anabolism to generate complex structures from simple ones, or energy-rich states from energy-poor ones.

$$\text{Useful energy} + \text{simple precursors} \xrightarrow{\text{Anabolism}} \text{complex molecules}$$

Some pathways can be either anabolic or catabolic, depending on the energy conditions in the cell. They are referred to as *amphibolic pathways*.

An important general principle of metabolism is that *biosynthetic and degradative pathways are almost always distinct*. This separation is necessary for energetic reasons, as will be evident in subsequent chapters. It also facilitates the control of metabolism.

A Thermodynamically Unfavorable Reaction Can Be Driven by a Favorable Reaction

How are specific pathways constructed from individual reactions? A pathway must satisfy minimally two criteria: (1) the individual reactions must be *specific* and (2) the entire set of reactions that constitute the pathway must be *thermodynamically favored*. A reaction that is specific will yield only one particular product or set of products from its reactants. As discussed in Chapter 8, a function of enzymes is to provide this specificity. The thermodynamics of metabolism is most readily approached in relation to free energy, which was discussed on pages 11, 208, and 211. A reaction can occur spontaneously only if ΔG, the change in free energy, is negative. Recall that ΔG for the formation of products C and D from substrates A and B is given by

$$\Delta G = \Delta G^{\circ\prime} + RT \ln \frac{[C][D]}{[A][B]}$$

Thus, the ΔG of a reaction depends on the *nature* of the reactants and products (expressed by the $\Delta G^{\circ\prime}$ term, the standard free-energy change) and on their *concentrations* (expressed by the second term).

An important thermodynamic fact is that *the overall free-energy change for a chemically coupled series of reactions is equal to the sum of the free-energy changes of the individual steps*. Consider the following reactions:

$$
\begin{array}{lll}
A \rightleftharpoons B + C & \Delta G^{\circ\prime} = +21 \text{ kJ mol}^{-1} \, (+5 \text{ kcal mol}^{-1}) \\
B \rightleftharpoons D & \Delta G^{\circ\prime} = -34 \text{ kJ mol}^{-1} \, (-8 \text{ kcal mol}^{-1}) \\
\hline
A \rightleftharpoons C + D & \Delta G^{\circ\prime} = -13 \text{ kJ mol}^{-1} \, (-3 \text{ kcal mol}^{-1})
\end{array}
$$

Under standard conditions, A cannot be spontaneously converted into B and C, because $\Delta G°'$ is positive. However, the conversion of B into D under standard conditions is thermodynamically feasible. Because free-energy changes are additive, the conversion of A into C and D has a $\Delta G°'$ of -13 kJ mol^{-1} (-3 kcal mol^{-1}), which means that it can occur spontaneously under standard conditions. Thus, *a thermodynamically unfavorable reaction can be driven by a thermodynamically favorable reaction to which it is coupled.* In this example, the reactions are coupled by the shared chemical intermediate B. Thus, metabolic pathways are formed by the coupling of enzyme-catalyzed reactions such that the overall free energy of the pathway is negative.

15.2 ATP Is the Universal Currency of Free Energy in Biological Systems

Just as commerce is facilitated by the use of a common currency, the commerce of the cell—metabolism—is facilitated by the use of a common energy currency, *adenosine triphosphate* (ATP). Part of the free energy derived from the oxidation of foodstuffs and from light is transformed into this highly accessible molecule, which acts as the free-energy donor in most energy-requiring processes such as motion, active transport, or biosynthesis. Indeed, most of catabolism consists of reactions that extract energy from fuels such as carbohydrates and fats and convert it into ATP.

ATP Hydrolysis Is Exergonic

ATP is a nucleotide consisting of adenine, a ribose, and a triphosphate unit (Figure 15.3). The active form of ATP is usually a complex of ATP with Mg^{2+} or Mn^{2+} (p. 268). In considering the role of ATP as an energy carrier, we can focus on its triphosphate moiety. *ATP is an energy-rich molecule because its triphosphate unit contains two phosphoanhydride bonds.* A large amount of free energy is liberated when ATP is hydrolyzed to adenosine diphosphate (ADP) and orthophosphate (P_i) or when ATP is hydrolyzed to adenosine monophosphate (AMP) and pyrophosphate (PP_i).

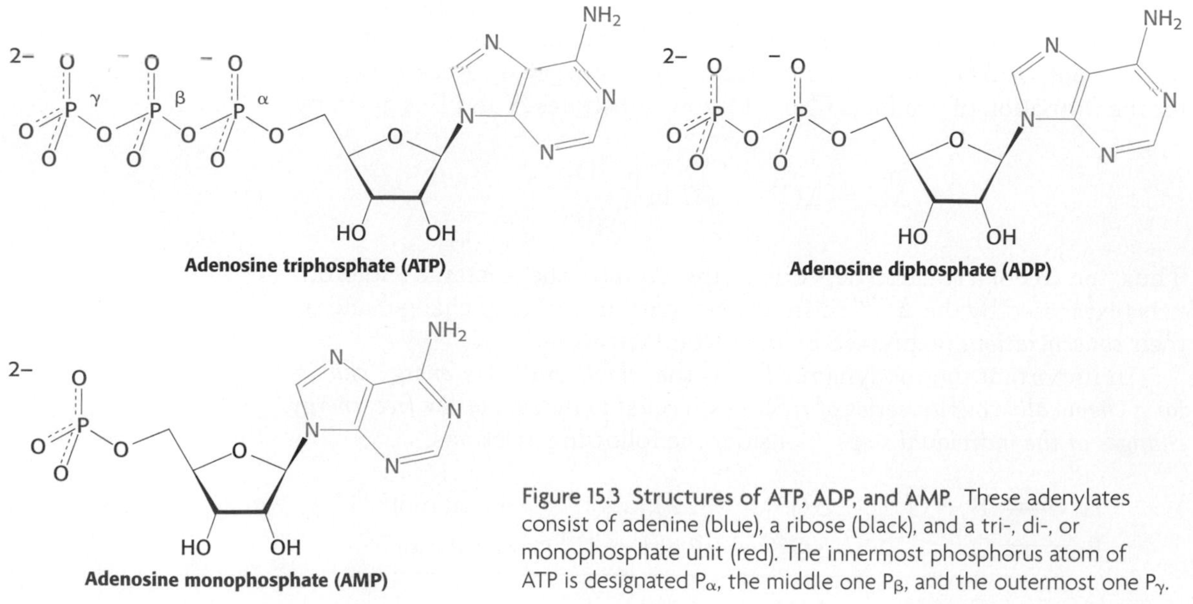

Adenosine triphosphate (ATP)

Adenosine diphosphate (ADP)

Adenosine monophosphate (AMP)

Figure 15.3 Structures of ATP, ADP, and AMP. These adenylates consist of adenine (blue), a ribose (black), and a tri-, di-, or monophosphate unit (red). The innermost phosphorus atom of ATP is designated P_α, the middle one P_β, and the outermost one P_γ.

$$ATP + H_2O \rightleftharpoons ADP + P_i$$
$$\Delta G^{\circ\prime} = -30.5 \text{ kJ mol}^{-1} (-7.3 \text{ kcal mol}^{-1})$$

$$ATP + H_2O \rightleftharpoons AMP + PP_i$$
$$\Delta G^{\circ\prime} = -45.6 \text{ kJ mol}^{-1} (-10.9 \text{ kcal mol}^{-1})$$

The precise $\Delta G^{\circ\prime}$ for these reactions depends on the ionic strength of the medium and on the concentrations of Mg^{2+} and other metal ions. Under typical cellular concentrations, the actual ΔG for these hydrolyses is approximately $-50 \text{ kJ mol}^{-1} (-12 \text{ kcal mol}^{-1})$.

The free energy liberated in the hydrolysis of ATP is harnessed to drive reactions that require an input of free energy, such as muscle contraction. In turn, ATP is formed from ADP and P_i when fuel molecules are oxidized in chemotrophs or when light is trapped by phototrophs. *This ATP–ADP cycle is the fundamental mode of energy exchange in biological systems.*

Some biosynthetic reactions are driven by the hydrolysis of nucleoside triphosphates that are analogous to ATP—namely, guanosine triphosphate (GTP), uridine triphosphate (UTP), and cytidine triphosphate (CTP). The diphosphate forms of these nucleotides are denoted by GDP, UDP, and CDP, and the monophosphate forms are denoted by GMP, UMP, and CMP. Enzymes catalyze the transfer of the terminal phosphoryl group from one nucleotide to another. The phosphorylation of nucleoside monophosphates is catalyzed by a family of *nucleoside monophosphate kinases*, as discussed in Section 9.4. The phosphorylation of nucleoside diphosphates is catalyzed by *nucleoside diphosphate kinase*, an enzyme with broad specificity.

$$\underset{\substack{\text{Nucleoside}\\\text{monophosphate}}}{NMP} + ATP \underset{}{\overset{\substack{\text{Nucleoside monophosphate}\\\text{kinase}}}{\rightleftharpoons}} NDP + ADP$$

$$\underset{\substack{\text{Nucleoside}\\\text{diphosphate}}}{NDP} + ATP \underset{}{\overset{\substack{\text{Nucleoside diphosphate}\\\text{kinase}}}{\rightleftharpoons}} NTP + ADP$$

It is intriguing to note that, although all of the nucleotide triphosphates are energetically equivalent, ATP is nonetheless the primary cellular energy carrier. In addition, two important electron carriers, NAD^+ and FAD, are derivatives of ATP. *The role of ATP in energy metabolism is paramount.*

ATP Hydrolysis Drives Metabolism by Shifting the Equilibrium of Coupled Reactions

An otherwise unfavorable reaction can be made possible by coupling to ATP hydrolysis. Consider a chemical reaction that is thermodynamically unfavorable without an input of free energy, a situation common to many biosynthetic reactions. Suppose that the standard free energy of the conversion of compound A into compound B is $+16.7 \text{ kJ mol}^{-1} (+4.0 \text{ kcal mol}^{-1})$:

$$A \rightleftharpoons B \qquad \Delta G^{\circ\prime} = +16.7 \text{ kJ mol}^{-1} (+4.0 \text{ kcal mol}^{-1})$$

The equilibrium constant K'_{eq} of this reaction at 25°C is related to $\Delta G^{\circ\prime}$ (in units of kilojoules per mole) by

$$K'_{eq} = [B]_{eq}/[A]_{eq} = 10^{-\Delta G^{\circ\prime}/5.69} = 1.15 \times 10^{-3}$$

Thus, net conversion of A into B cannot take place when the molar ratio of B to A is equal to or greater than 1.15×10^{-3}. However, A can be converted into B under these conditions if the reaction is coupled to the hydrolysis of ATP. Under standard conditions, the $\Delta G^{\circ\prime}$ of hydrolysis is approximately $-30.5 \text{ kJ mol}^{-1} (-7.3 \text{ kcal mol}^{-1})$. The new overall reaction is

$$A + ATP + H_2O \rightleftharpoons B + ADP + P_i$$
$$\Delta G^{\circ\prime} = -13.8 \text{ kJ mol}^{-1} (-3.3 \text{ kcal mol}^{-1})$$

Its free-energy change of $-13.8 \text{ kJ mol}^{-1} (-3.3 \text{ kcal mol}^{-1})$ is the sum of the value of $\Delta G^{\circ\prime}$ for the conversion of A into B $\left[+16.7 \text{ kJ mol}^{-1} (+4.0 \text{ kcal mol}^{-1})\right]$ and the value of $\Delta G^{\circ\prime}$ for the hydrolysis of ATP $\left[-30.5 \text{ kJ mol}^{-1} (-7.3 \text{ kcal mol}^{-1})\right]$. At pH 7, the equilibrium constant of this coupled reaction is

$$K'_{eq} = \frac{[B]_{eq}}{[A]_{eq}} \times \frac{[ADP]_{eq}[P_i]_{eq}}{[ATP]_{eq}} = 10^{13.8/5.69} = 2.67 \times 10^2$$

At equilibrium, the ratio of [B] to [A] is given by

$$\frac{[B]_{eq}}{[A]_{eq}} = K'_{eq} \frac{[ATP]_{eq}}{[ADP]_{eq}[P_i]_{eq}}$$

which means that the hydrolysis of ATP enables A to be converted into B until the [B]/[A] ratio reaches a value of 2.67×10^2. This equilibrium ratio is strikingly different from the value of 1.15×10^{-3} for the reaction $A \rightarrow B$ in the absence of ATP hydrolysis. In other words, coupling the hydrolysis of ATP with the conversion of A into B under standard conditions has changed the equilibrium ratio of B to A by a factor of about 10^5. If we were to use the ΔG of hydrolysis of ATP under cellular conditions $\left[-50.2 \text{ kJ mol}^{-1} (-12 \text{ kcal mol}^{-1})\right]$ in our calculations instead of $\Delta G^{\circ\prime}$, the change in the equilibrium ratio would be even more dramatic, on the order of 10^8.

We see here the thermodynamic essence of ATP's action as an *energy-coupling agent*. Cells maintain a high level of ATP by using oxidizable substrates or light as sources of free energy for synthesizing the molecule. In the cell, the hydrolysis of an ATP molecule in a coupled reaction then changes the equilibrium ratio of products to reactants by a very large factor, of the order of 10^8. More generally, the hydrolysis of n ATP molecules changes the equilibrium ratio of a coupled reaction (or sequence of reactions) by a factor of 10^{8n}. For example, the hydrolysis of three ATP molecules in a coupled reaction changes the equilibrium ratio by a factor of 10^{24}. Thus, *a thermodynamically unfavorable reaction sequence can be converted into a favorable one by coupling it to the hydrolysis of a sufficient number of ATP molecules in a new reaction*. It should also be emphasized that A and B in the preceding coupled reaction may be interpreted very generally, not only as different chemical species. For example, A and B may represent activated and unactivated conformations of a protein that is activated by phosphorylation with ATP. Through such changes in protein conformation, molecular motors such as myosin, kinesin, and dynein convert the chemical energy of ATP into mechanical energy (Chapter 34). Indeed, this conversion is the basis of muscle contraction.

Alternatively, A and B may refer to the concentrations of an ion or molecule on the outside and inside of a cell, as in the active transport of a nutrient. The active transport of Na^+ and K^+ across membranes is driven by the phosphorylation of the sodium–potassium pump by ATP and its subsequent dephosphorylation (p. 357).

The High Phosphoryl Potential of ATP Results from Structural Differences Between ATP and Its Hydrolysis Products

What makes ATP a particularly efficient phosphoryl-group donor? Let us compare the standard free energy of hydrolysis of ATP with that of a phosphate ester, such as glycerol 3-phosphate:

$$ATP + H_2O \rightleftharpoons ADP + P_i$$
$$\Delta G^{\circ\prime} = -30.5 \text{ kJ mol}^{-1} (-7.3 \text{ kcal mol}^{-1})$$
$$\text{Glycerol 3-phosphate} + H_2O \rightleftharpoons \text{glycerol} + P_i$$
$$\Delta G^{\circ\prime} = -9.2 \text{ kJ mol}^{-1} (-2.2 \text{ kcal mol}^{-1})$$

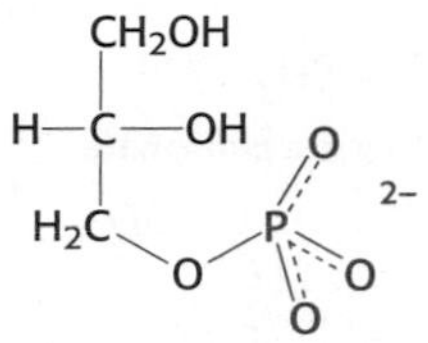

The magnitude of $\Delta G^{\circ\prime}$ for the hydrolysis of glycerol 3-phosphate is much smaller than that of ATP, which means that ATP has a stronger tendency to transfer its terminal phosphoryl group to water than does glycerol 3-phosphate. In other words, ATP has a higher *phosphoryl-transfer potential (phosphoryl-group-transfer potential)* than does glycerol 3-phosphate.

The high phosphoryl-transfer potential of ATP can be explained by features of the ATP structure. Because $\Delta G^{\circ\prime}$ depends on the *difference* in free energies of the products and reactants, we need to examine the structures of both ATP and its hydrolysis products, ADP and P_i, to answer this question. Three factors are important: *resonance stabilization, electrostatic repulsion, and stabilization due to hydration.*

1. *Resonance Stabilization.* ADP and, particularly, P_i, have greater resonance stabilization than does ATP. Orthophosphate has a number of resonance forms of similar energy (Figure 15.4), whereas the γ phosphoryl group of ATP has a smaller number. Forms like that shown in Figure 15.5 are unfavorable because a positively charged oxygen atom is adjacent to a positively charged phosphorus atom, an electrostatically unfavorable juxtaposition.

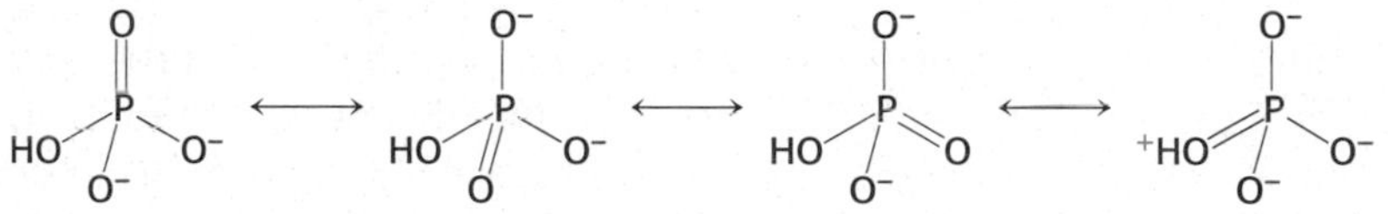

Glycerol 3-phosphate

Figure 15.4 Resonance structures of orthophosphate.

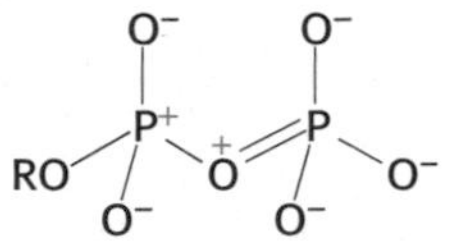

Figure 15.5 Improbable resonance structure. The structure contributes little to the terminal part of ATP, because two positive charges are placed adjacent to each other.

2. *Electrostatic Repulsion.* At pH 7, the triphosphate unit of ATP carries about four negative charges. These charges repel one another because they are in close proximity. The repulsion between them is reduced when ATP is hydrolyzed.

3. *Stabilization Due to Hydration.* More water can bind more effectively to ADP and P_i than can bind to the phosphoanhydride part of ATP, stabilizing the ADP and P_i by hydration.

ATP is often called a high-energy phosphate compound, and its phosphoanhydride bonds are referred to as high-energy bonds. Indeed, a "squiggle" (~P) is often used to indicate such a bond. Nonetheless, there is nothing special about the bonds themselves. *They are high-energy bonds in the sense that much free energy is released when they are hydrolyzed*, for the reasons listed in factors 1 through 3.

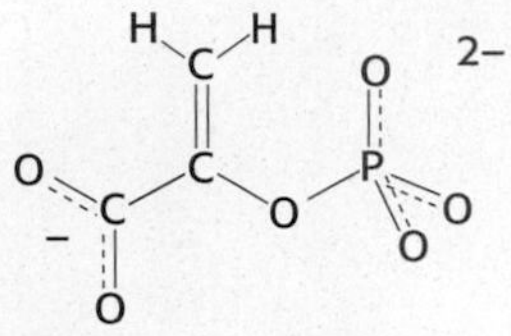

Phosphoenolpyruvate (PEP)

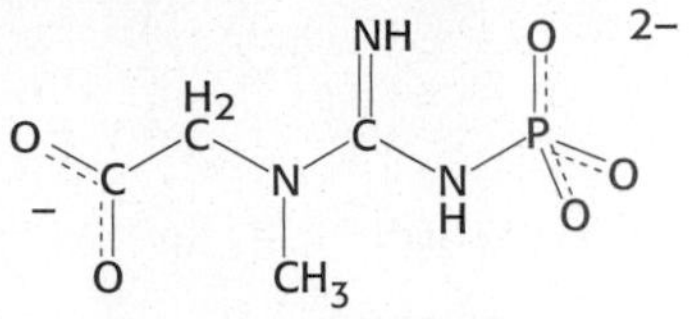

Creatine phosphate

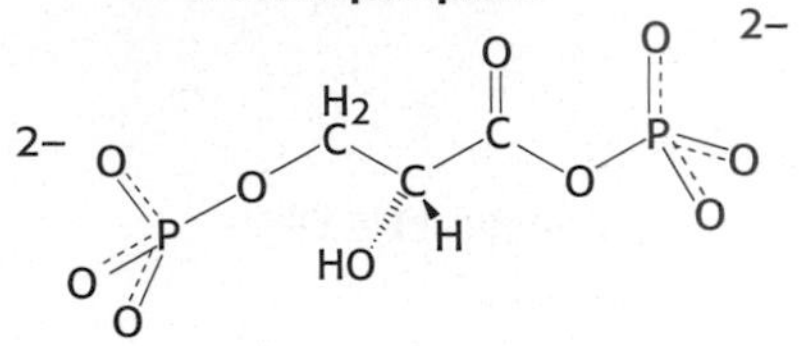

1,3-Bisphosphoglycerate (1,3-BPG)

Figure 15.6 Compounds with high phosphoryl-transfer potential. These compounds have a higher phosphoryl-transfer potential than that of ATP and can be used to phosphorylate ADP to form ATP.

Phosphoryl-Transfer Potential Is an Important Form of Cellular Energy Transformation

The standard free energies of hydrolysis provide a convenient means of comparing the phosphoryl-transfer potential of phosphorylated compounds. Such comparisons reveal that ATP is not the only compound with a high phosphoryl-transfer potential. In fact, some compounds in biological systems have a higher phosphoryl-transfer potential than that of ATP. These compounds include phosphoenolpyruvate (PEP), 1,3-bisphosphoglycerate (1,3-BPG), and creatine phosphate (Figure 15.6). Thus, PEP can transfer its phosphoryl group to ADP to form ATP. Indeed, this transfer is one of the ways in which ATP is generated in the breakdown of sugars (pp. 436 and 444). It is significant that ATP has a phosphoryl-transfer potential that is intermediate among the biologically important phosphorylated molecules (Table 15.1). *This intermediate position enables ATP to function efficiently as a carrier of phosphoryl groups.*

The amount of ATP in muscle suffices to sustain contractile activity for less than a second. Creatine phosphate in vertebrate muscle serves as a reservoir of high-potential phosphoryl groups that can be readily transferred to ADP. Indeed, we use creatine phosphate to regenerate ATP from ADP every time that we exercise strenuously. This reaction is catalyzed by *creatine kinase.*

$$\text{Creatine phosphate} + \text{ADP} \underset{}{\overset{\text{Creatine kinase}}{\rightleftharpoons}} \text{ATP} + \text{creatine}$$

At pH 7, the standard free energy of hydrolysis of creatine phosphate is -43.1 kJ mol^{-1} (-10.3 kcal mol^{-1}), compared with -30.5 kJ mol^{-1} (-7.3 kcal mol^{-1}) for ATP. Hence, the standard free-energy change in forming ATP from creatine phosphate is -12.6 kJ mol^{-1} (-3.0 kcal mol^{-1}), which corresponds to an equilibrium constant of 162.

$$K_{\text{eq}} = \frac{[\text{ATP}][\text{creatine}]}{[\text{ADP}][\text{creatine phosphate}]} = 10^{-\Delta G^{\circ\prime}/2.303RT} = 10^{12.6/5.69} = 162$$

In resting muscle, typical concentrations of these metabolites are $[\text{ATP}] = 4$ mM, $[\text{ADP}] = 0.013$ mM, $[\text{creatine phosphate}] = 25$ mM, and $[\text{creatine}] = 13$ mM. Because of its abundance and high phosphoryl-transfer potential relative to that of ATP, creatine phosphate is a highly effective phosphoryl buffer. Indeed, creatine phosphate is the major source of phosphoryl groups for ATP regeneration for a runner during the first 4 seconds of a 100-meter sprint. The fact that creatine phosphate can replenish ATP pools is the basis of creatine's use as a dietary supplement by athletes in sports requiring short bursts of intense activity. After the creatine phosphate pool is depleted, ATP must be generated through metabolism (Figure 15.7).

TABLE 15.1 Standard free energies of hydrolysis of some phosphorylated compounds

Compound	kJ mol^{-1}	kcal mol^{-1}
Phosphoenolpyruvate	−61.9	−14.8
1,3-Bisphosphoglycerate	−49.4	−11.8
Creatine phosphate	−43.1	−10.3
ATP (to ADP)	−30.5	− 7.3
Glucose 1-phosphate	−20.9	− 5.0
Pyrophosphate	−19.3	− 4.6
Glucose 6-phosphate	−13.8	− 3.3
Glycerol 3-phosphate	−9.2	− 2.2

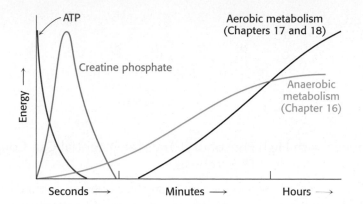

Figure 15.7 **Sources of ATP during exercise.** In the initial seconds, exercise is powered by existing high-phosphoryl-transfer compounds (ATP and creatine phosphate). Subsequently, the ATP must be regenerated by metabolic pathways.

15.3 The Oxidation of Carbon Fuels Is an Important Source of Cellular Energy

ATP serves as the principal *immediate donor of free energy* in biological systems rather than as a long-term storage form of free energy. In a typical cell, an ATP molecule is consumed within a minute of its formation. Although the total quantity of ATP in the body is limited to approximately 100 g, *the turnover of this small quantity of ATP is very high*. For example, a resting human being consumes about 40 kg of ATP in 24 hours. During strenuous exertion, the rate of utilization of ATP may be as high as 0.5 kg/minute. For a 2-hour run, 60 kg (132 pounds) of ATP is utilized. Clearly, having mechanisms for regenerating ATP is vital. Motion, active transport, signal amplification, and biosynthesis can take place only if ATP is continually regenerated from ADP (Figure 15.8). The generation of ATP is one of the primary roles of catabolism. The carbon in fuel molecules—such as glucose and fats—is oxidized to CO_2, and the energy released is used to regenerate ATP from ADP and P_i.

In aerobic organisms, the ultimate electron acceptor in the oxidation of carbon is O_2 and the oxidation product is CO_2. Consequently, the more reduced a carbon is to begin with, the more free energy is released by its oxidation. Figure 15.9 shows the $\Delta G^{\circ\prime}$ of oxidation for one-carbon compounds.

Although fuel molecules are more complex (Figure 15.10) than the single-carbon compounds depicted in Figure 15.9, when a fuel is oxidized, the oxidation takes place one carbon at a time. The carbon-oxidation energy is used in some cases to create a compound with high phosphoryl-transfer potential and in other cases to create an ion gradient. In either case, the end point is the formation of ATP.

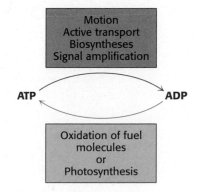

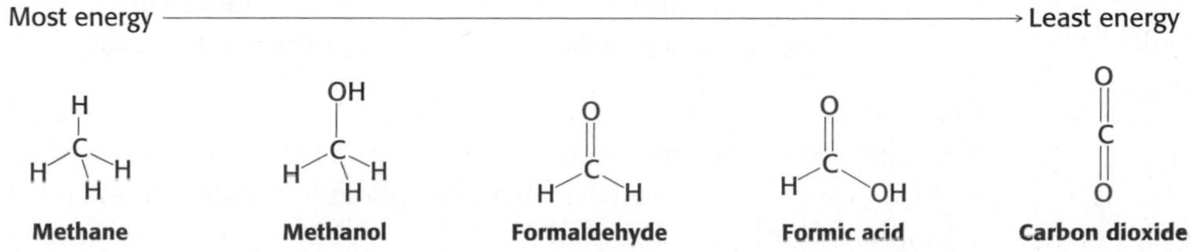

Figure 15.8 **ATP–ADP cycle.** This cycle is the fundamental mode of energy exchange in biological systems.

Most energy $\longrightarrow$				Least energy
Methane	**Methanol**	**Formaldehyde**	**Formic acid**	**Carbon dioxide**
$\Delta G^{\circ\prime}_{oxidation}$ (kJ mol^{-1}) −820	−703	−523	−285	0
$\Delta G^{\circ\prime}_{oxidation}$ (kcal mol^{-1}) −196	−168	−125	−68	0

Figure 15.9 **Free energy of oxidation of single-carbon compounds.**

Figure 15.10 **Prominent fuels.** Fats are a more efficient fuel source than carbohydrates such as glucose because the carbon in fats is more reduced.

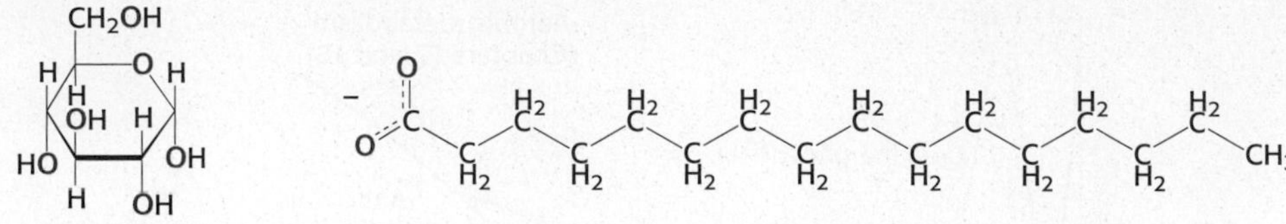

Glucose

Fatty acid

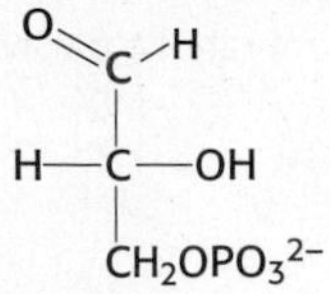

Glyceraldehyde 3-phosphate (GAP)

Compounds with High Phosphoryl-Transfer Potential Can Couple Carbon Oxidation to ATP Synthesis

How is the energy released in the oxidation of a carbon compound converted into ATP? As an example, consider glyceraldehyde 3-phosphate (shown in the margin), which is a metabolite of glucose formed in the oxidation of that sugar. The C-1 carbon (shown in red) is at the aldehyde-oxidation level and is not in its most oxidized state. Oxidation of the aldehyde to an acid will release energy.

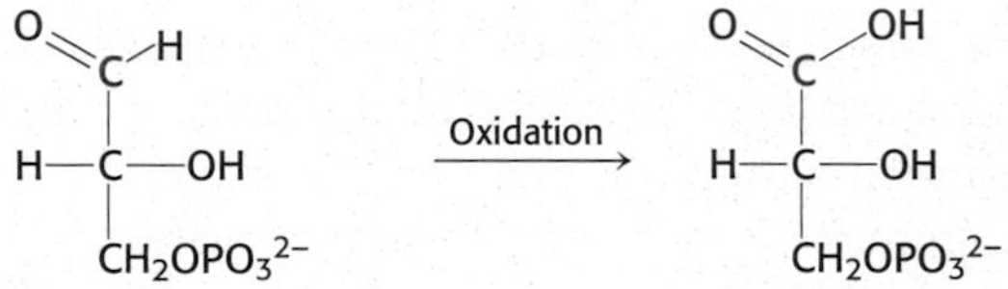

Glyceraldehyde 3-phosphate $\xrightarrow{\text{Oxidation}}$ **3-Phosphoglyceric acid**

However, the oxidation does not take place directly. Instead, the carbon oxidation generates an acyl phosphate, 1,3-bisphosphoglycerate. The electrons released are captured by NAD^+, which we will consider shortly.

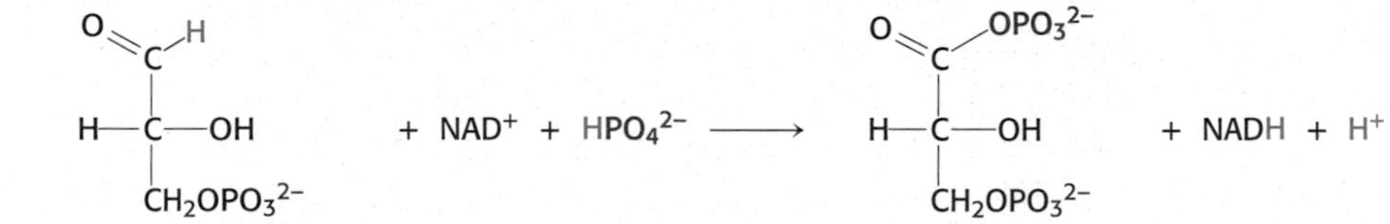

Glyceraldehyde 3-phosphate (GAP) $+ NAD^+ + HPO_4^{2-} \longrightarrow$ **1,3-Bisphosphoglycerate (1,3-BPG)** $+ NADH + H^+$

For reasons similar to those discussed for ATP, 1,3-bisphosphoglycerate has a high phosphoryl-transfer potential. Thus, the cleavage of 1,3-BPG can be coupled to the synthesis of ATP.

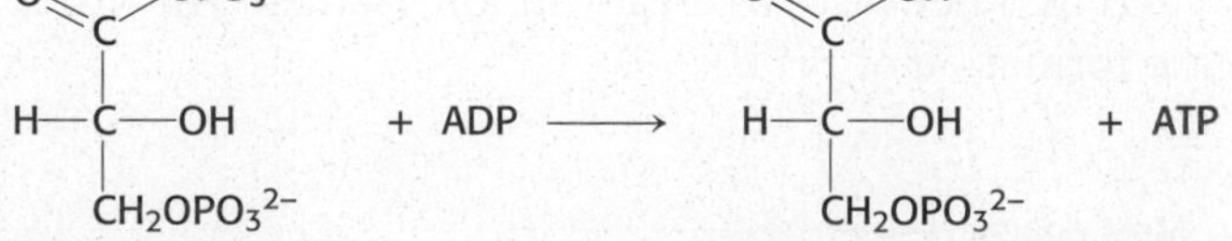

1,3-Bisphosphoglycerate $+ ADP \longrightarrow$ **3-Phosphoglyceric acid** $+ ATP$

The energy of oxidation is initially trapped as a high-phosphoryl-transfer-potential compound and then used to form ATP. The oxidation energy of a carbon atom is transformed into phosphoryl-transfer potential, first as 1, 3-bisphosphoglycerate and ultimately as ATP. We will consider these reactions in mechanistic detail on p. 440.

Ion Gradients Across Membranes Provide an Important Form of Cellular Energy That Can Be Coupled to ATP Synthesis

As described in Chapter 13, electrochemical potential is an effective means of storing free energy. Indeed, the electrochemical potential of *ion gradients*

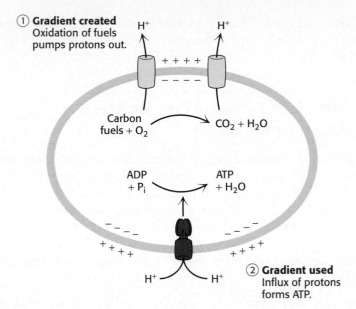

① **Gradient created**
Oxidation of fuels pumps protons out.

Carbon fuels + O_2 → CO_2 + H_2O

ADP + P_i → ATP + H_2O

② **Gradient used**
Influx of protons forms ATP.

Figure 15.11 Proton gradients. The oxidation of fuels can power the formation of proton gradients by the action of specific proton pumps. These proton gradients can in turn drive the synthesis of ATP when the protons flow through an ATP synthesizing enzyme.

across membranes, produced by the oxidation of fuel molecules or by photosynthesis, ultimately powers the synthesis of most of the ATP in cells. In general, ion gradients are versatile means of coupling thermodynamically unfavorable reactions to favorable ones. Indeed, in animals, *proton gradients* generated by the oxidation of carbon fuels account for more than 90% of ATP generation (Figure 15.11). This process is called *oxidative phosphorylation* (Chapter 18). ATP hydrolysis can then be used to form ion gradients of different types and functions. The electrochemical potential of a Na^+ gradient, for example, can be tapped to pump Ca^{2+} out of cells or to transport nutrients such as sugars and amino acids into cells.

Energy from Foodstuffs Is Extracted in Three Stages

Let us take an overall view of the processes of energy conversion in higher organisms before considering them in detail in subsequent chapters. Hans Krebs described three stages in the generation of energy from the oxidation of foodstuffs (Figure 15.12).

In the first stage, large molecules in food are broken down into smaller units. This process is *digestion.* Proteins are hydrolyzed to their 20 different amino acids, polysaccharides are hydrolyzed to simple sugars such as glucose, and fats are hydrolyzed to glycerol and fatty acids. This stage is strictly a preparation stage; no useful energy is captured in this phase.

In the second stage, these numerous small molecules are degraded to a few simple units that play a central role in metabolism. In fact, most of them—sugars, fatty acids, glycerol, and several amino acids—are converted into the acetyl unit of acetyl CoA (p. 422). Some ATP is generated in this stage, but the amount is small compared with that obtained in the third stage.

In the third stage, ATP is produced from the complete oxidation of the acetyl unit of acetyl CoA. The third stage consists of the citric acid cycle and oxidative phosphorylation, which are *the final common pathways in the oxidation of fuel molecules.* Acetyl CoA brings acetyl units into the citric acid cycle [also called the tricarboxylic acid (TCA) cycle or Krebs cycle], where they are completely oxidized to CO_2. Four pairs of electrons are transferred (three to NAD^+ and one to FAD) for each acetyl

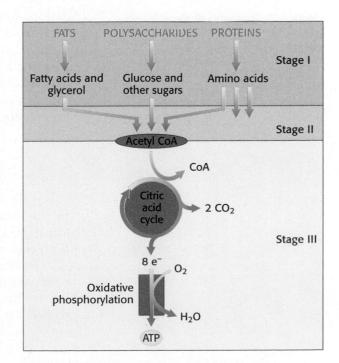

FATS   POLYSACCHARIDES   PROTEINS

Stage I

Fatty acids and glycerol   Glucose and other sugars   Amino acids

Stage II

Acetyl CoA

CoA

Citric acid cycle → 2 CO_2

Stage III

8 e^- O_2

Oxidative phosphorylation

H_2O

ATP

Figure 15.12 Stages of catabolism. The extraction of energy from fuels can be divided into three stages.

group that is oxidized. Then, a proton gradient is generated as electrons flow from the reduced forms of these carriers to O_2, and this gradient is used to synthesize ATP.

15.4 Metabolic Pathways Contain Many Recurring Motifs

At first glance, metabolism appears intimidating because of the sheer number of reactants and reactions. Nevertheless, there are unifying themes that make the comprehension of this complexity more manageable. These unifying themes include common metabolites, reactions, and regulatory schemes that stem from a common evolutionary heritage.

Activated Carriers Exemplify the Modular Design and Economy of Metabolism

We have seen that phosphoryl transfer can be used to drive otherwise endergonic reactions, alter the energy of conformation of a protein, or serve as a signal to alter the activity of a protein. The phosphoryl-group donor in all of these reactions is ATP. In other words, *ATP is an activated carrier of phosphoryl groups because phosphoryl transfer from ATP is an exergonic process.* The use of activated carriers is a recurring motif in biochemistry, and we will consider several such carriers here. Many such activated carriers function as coenzymes:

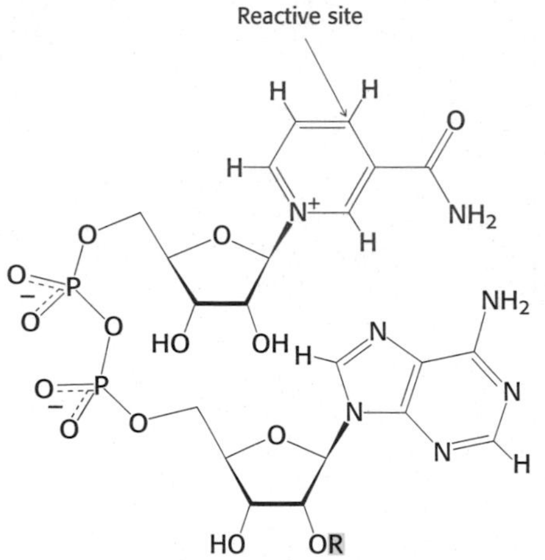

1. *Activated Carriers of Electrons for Fuel Oxidation.* In aerobic organisms, the ultimate electron acceptor in the oxidation of fuel molecules is O_2. However, electrons are not transferred directly to O_2. Instead, fuel molecules transfer electrons to special carriers, which are either *pyridine nucleotides* or *flavins*. The reduced forms of these carriers then transfer their high-potential electrons to O_2.

Nicotinamide adenine dinucleotide is a major electron carrier in the oxidation of fuel molecules (Figure 15.13). The reactive part of NAD^+ is its nicotinamide ring, a pyridine derivative synthesized from the vitamin niacin. *In the oxidation of a substrate, the nicotinamide ring of NAD^+ accepts a hydrogen ion and two electrons, which are equivalent to a hydride ion* ($H:^-$). The reduced form of this carrier is called *NADH*. In the oxidized form, the nitrogen atom carries a positive charge, as indicated by NAD^+. NAD^+ is the electron acceptor in many reactions of the type

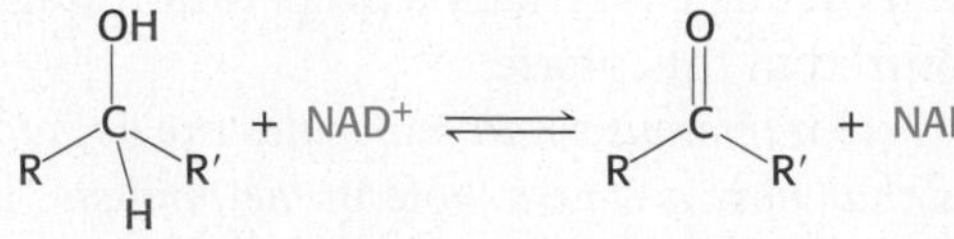

Figure 15.13 **Structures of the oxidized forms of nicotinamide-derived electron carriers.** Nicotinamide adenine dinucleotide (NAD^+) and nicotinamide adenine dinucleotide phosphate ($NADP^+$) are prominent carriers of high-energy electrons. In NAD^+, R = H; in $NADP^+$, R = PO_3^{2-}.

In this dehydrogenation, one hydrogen atom of the substrate is directly transferred to NAD^+, whereas the other appears in the solvent as a proton. Both electrons lost by the substrate are transferred to the nicotinamide ring.

The other major electron carrier in the oxidation of fuel molecules is the coenzyme *flavin adenine dinucleotide* (Figure 15.14). The abbreviations for the oxidized and reduced forms of this carrier are FAD and $FADH_2$, respectively. FAD is the electron acceptor in reactions of the type

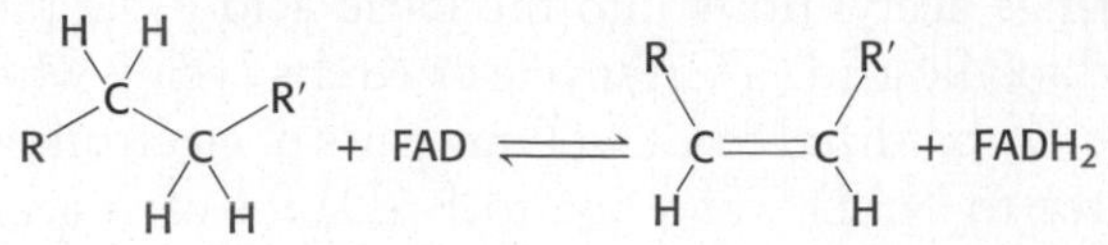

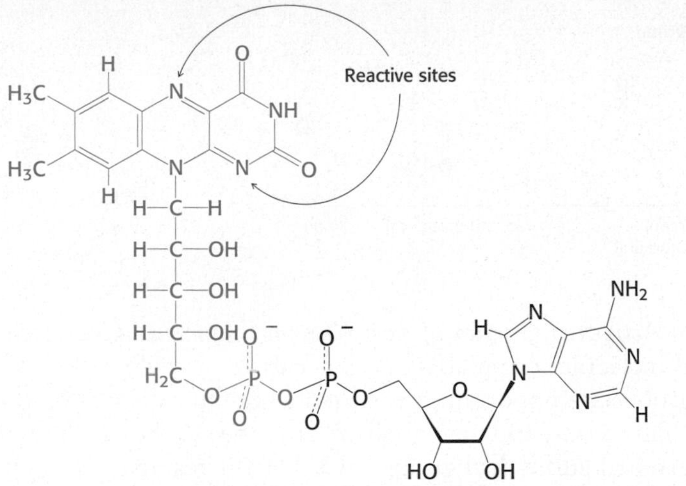

Figure 15.14 **Structure of the oxidized form of flavin adenine dinucleotide (FAD).** This electron carrier consists of a flavin mononucleotide (FMN) unit (shown in blue) and an AMP unit (shown in black).

The reactive part of FAD is its isoalloxazine ring, a derivative of the vitamin riboflavin (Figure 15.15). FAD, like NAD^+, can accept two electrons. In doing so, FAD, unlike NAD^+, takes up two protons. These carriers of high-potential electrons as well as flavin mononucleotide (FMN), an electron carrier related to FAD, will be considered further in Chapter 18.

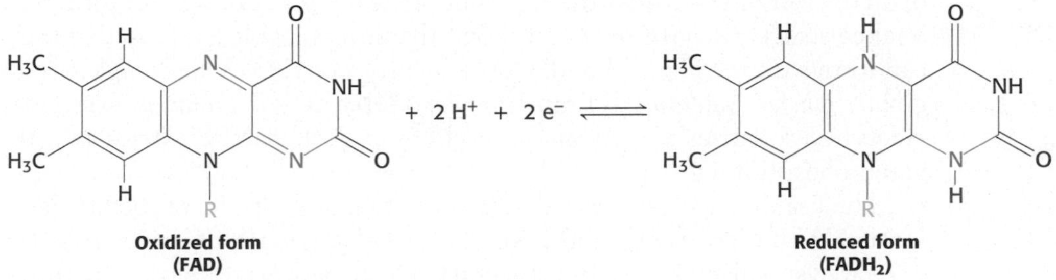

Figure 15.15 **Structures of the reactive parts of FAD and FADH$_2$.** The electrons and protons are carried by the isoalloxazine ring component of FAD and FADH$_2$.

2. *An Activated Carrier of Electrons for Reductive Biosynthesis.* High-potential electrons are required in most biosyntheses because the precursors are more oxidized than the products. Hence, reducing power is needed in addition to ATP. For example, in the biosynthesis of fatty acids, the keto group of an added two-carbon unit is reduced to a methylene group in several steps. This sequence of reactions requires an input of four electrons.

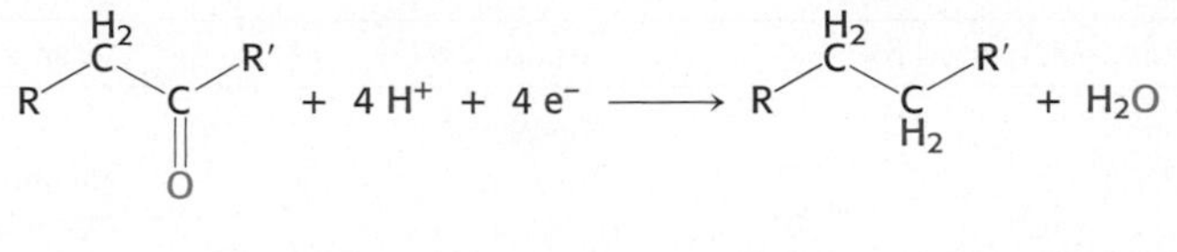

The electron donor in most reductive biosyntheses is NADPH, the reduced form of nicotinamide adenine dinucleotide phosphate ($NADP^+$; see Figure 15.13). NADPH differs from NADH in that the 2′-hydroxyl group of its adenosine moiety is esterified with phosphate. NADPH carries electrons in the same way as NADH. However, *NADPH is used almost exclusively for reductive biosyntheses, whereas NADH is used primarily for the generation of ATP.* The extra phosphoryl group on NADPH is a tag that enables enzymes to distinguish between high-potential electrons to be used in anabolism and those to be used in catabolism.

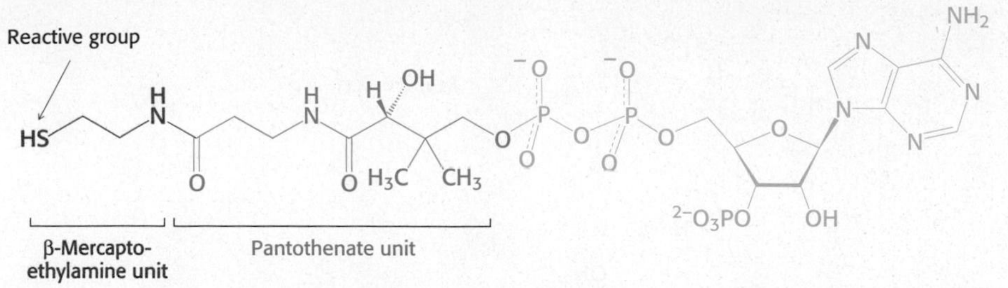

Reactive group

β-Mercapto-ethylamine unit | Pantothenate unit

Figure 15.16 Structure of coenzyme A
(CoA-SH).

Acyl CoA   **Acetyl CoA**

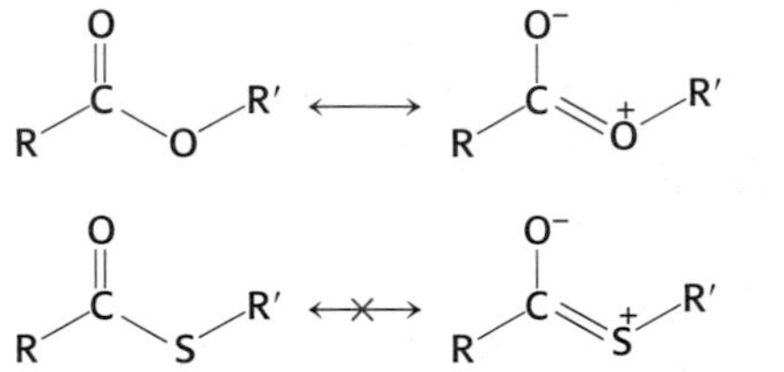

**Oxygen esters are stabilized by resonance
structures not available to thioesters.**

3.   *An Activated Carrier of Two-Carbon Fragments*. Coenzyme A, another central molecule in metabolism, is a carrier of acyl groups (Figure 15.16). Acyl groups are important constituents both in catabolism, as in the oxidation of fatty acids, and in anabolism, as in the synthesis of membrane lipids. The terminal sulfhydryl group in CoA is the reactive site. Acyl groups are linked to CoA by thioester bonds. The resulting derivative is called an *acyl CoA*. An acyl group often linked to CoA is the acetyl unit; this derivative is called *acetyl CoA*. The $\Delta G^{\circ\prime}$ for the hydrolysis of acetyl CoA has a large negative value:

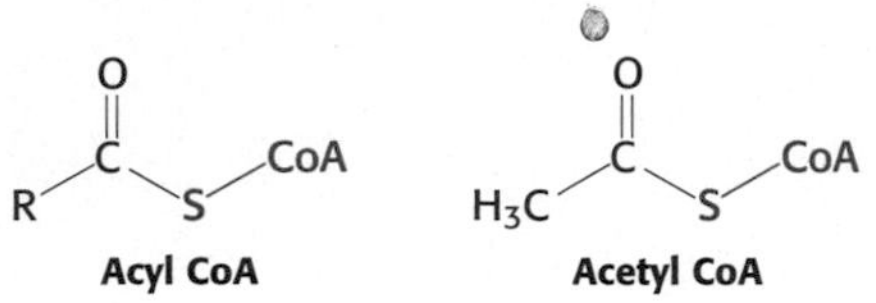

$$\text{Acetyl CoA} + H_2O \rightleftharpoons \text{acetate} + \text{CoA} + H^+$$
$$\Delta G^{\circ\prime} = -31.4 \text{ kJ mol}^{-1} \, (-7.5 \text{ kcal mol}^{-1})$$

The hydrolysis of a thioester is thermodynamically more favorable than that of an oxygen ester because the electrons of the C=O bond cannot form resonance structures with the C—S bond that are as stable as those that they can form with the C—O bond. Consequently, *acetyl CoA has a high acetyl-group-transfer potential because transfer of the acetyl group is exergonic.* Acetyl CoA carries an activated acetyl group, just as ATP carries an activated phosphoryl group.

The use of activated carriers illustrates two key aspects of metabolism. First, NADH, NADPH, and FADH$_2$ react slowly with O$_2$ in the absence of a catalyst. Likewise, ATP and acetyl CoA are hydrolyzed slowly (in times of many hours or even days) in the absence of a catalyst. These molecules are kinetically quite stable in the face of a large thermodynamic driving force for reaction with O$_2$ (in regard to the electron carriers) and H$_2$O (for ATP and acetyl CoA). *The kinetic stability of these molecules in the absence of specific catalysts is essential for their biological function because it enables enzymes to control the flow of free energy and reducing power.*

Second, *most interchanges of activated groups in metabolism are accomplished by a rather small set of carriers* (Table 15.2). The existence of a recurring

TABLE 15.2   Some activated carriers in metabolism

Carrier molecule in activated form	Group carried	Vitamin precursor
ATP	Phosphoryl	
NADH and NADPH	Electrons	Nicotinate (niacin)
FADH$_2$	Electrons	Riboflavin (vitamin B$_2$)
FMNH$_2$	Electrons	Riboflavin (vitamin B$_2$)
Coenzyme A	Acyl	Pantothenate
Lipoamide	Acyl	
Thiamine pyrophosphate	Aldehyde	Thiamine (vitamin B$_1$)
Biotin	CO$_2$	Biotin
Tetrahydrofolate	One-carbon units	Folate
S-Adenosylmethionine	Methyl	
Uridine diphosphate glucose	Glucose	
Cytidine diphosphate diacylglycerol	Phosphatidate	
Nucleoside triphosphates	Nucleotides	

NOTE: Many of the activated carriers are coenzymes that are derived from water-soluble vitamins.

TABLE 15.3 The B vitamins

Vitamin	Coenzyme	Typical reaction type	Consequences of deficiency
Thiamine (B$_1$)	Thiamine pyrophosphate	Aldehyde transfer	Beriberi (weight loss, heart problems, neurological dysfunction)
Riboflavin (B$_2$)	Flavin adenine dinucleotide (FAD)	Oxidation–reduction	Cheliosis and angular stomatitis (lesions of the mouth), dermatitis
Pyridoxine (B$_6$)	Pyridoxal phosphate	Group transfer to or from amino acids	Depression, confusion, convulsions
Nicotinic acid (niacin)	Nicotinamide adenine dinucleotide (NAD$^+$)	Oxidation–reduction	Pellagra (dermatitis, depression, diarrhea)
Pantothenic acid	Coenzyme A	Acyl-group transfer	Hypertension
Biotin	Biotin–lysine adducts (biocytin)	ATP-dependent carboxylation and carboxyl-group transfer	Rash about the eyebrows, muscle pain, fatigue (rare)
Folic acid	Tetrahydrofolate	Transfer of one-carbon components; thymine synthesis	Anemia, neural-tube defects in development
B$_{12}$	5′-Deoxyadenosyl cobalamin	Transfer of methyl groups; intramolecular rearrangements	Anemia, pernicious anemia, methylmalonic acidosis

set of activated carriers in all organisms is one of the unifying motifs of biochemistry. Furthermore, it illustrates the modular design of metabolism. A small set of molecules carries out a very wide range of tasks. Metabolism is readily comprehended because of the economy and elegance of its underlying design.

Many Activated Carriers Are Derived from Vitamins

Almost all the activated carriers that act as coenzymes are derived from *vitamins*. Vitamins are organic molecules that are needed in small amounts in the diets of some higher animals. Table 15.3 lists the vitamins that act as coenzymes (Figure 15.17). This series of vitamins is known as the vitamin B group. Note that, in all cases, the vitamin must be modified before it can serve its function. We have already touched on the roles of niacin, riboflavin, and pantothenate. We will see these three and the other B vitamins many times in our study of biochemistry.

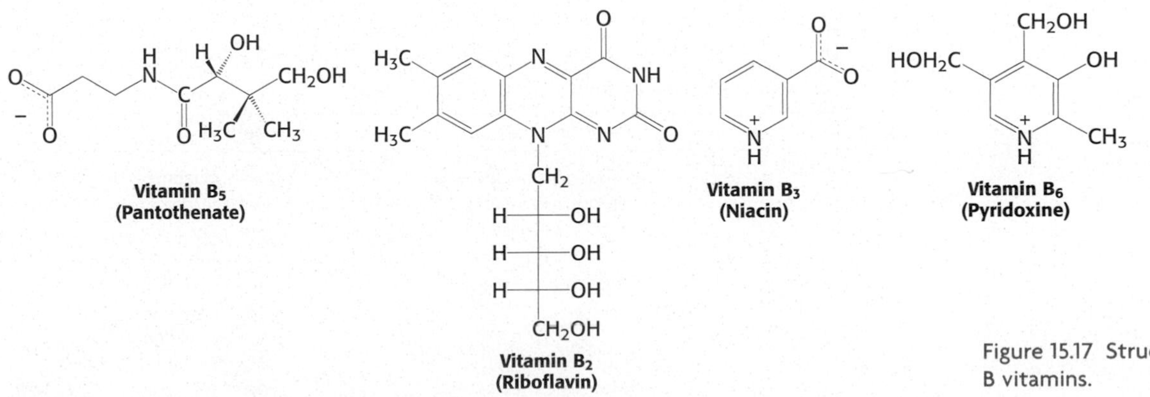

Figure 15.17 Structures of some of the B vitamins.

Vitamins serve the same roles in nearly all forms of life, but higher animals lost the capacity to synthesize them in the course of evolution. For instance, whereas *E. coli* can thrive on glucose and organic salts, human beings require at least 12 vitamins in their diet. The biosynthetic pathways for vitamins can be complex; thus, it is biologically more efficient

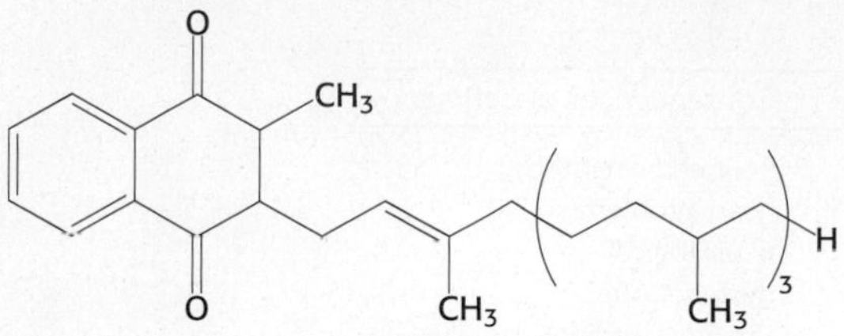

Vitamin K$_1$

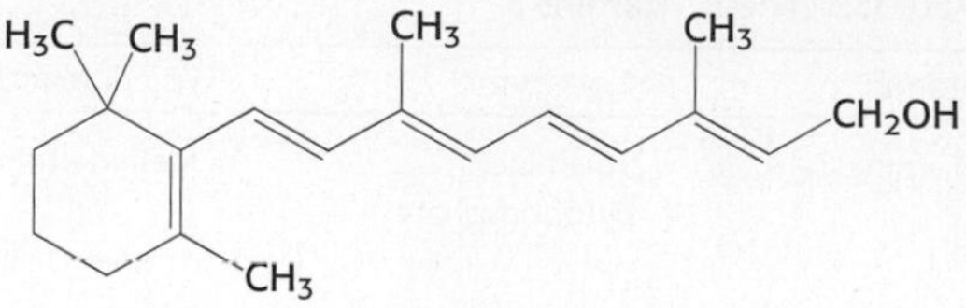

Vitamin A (Retinol)

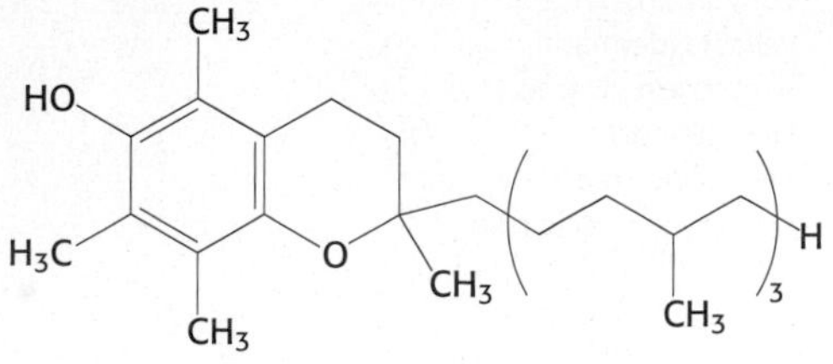

Vitamin E (α-Tocopherol)

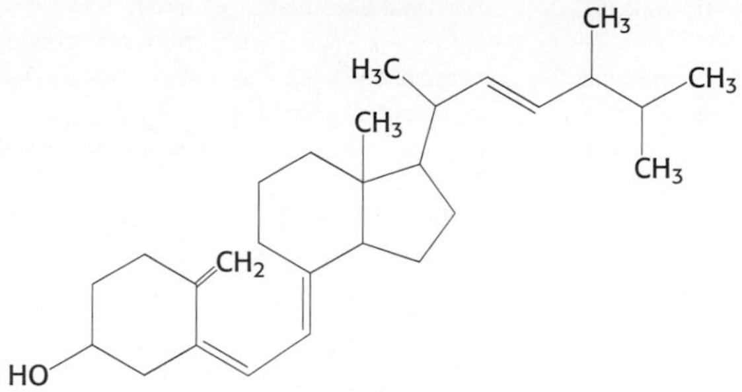

Vitamin D$_2$ (Calciferol)

Figure 15.18 Structures of some vitamins that do not function as coenzymes.

to ingest vitamins than to synthesize the enzymes required to construct them from simple molecules. This efficiency comes at the cost of dependence on other organisms for chemicals essential for life. Indeed, vitamin deficiency can generate diseases in all organisms requiring these molecules (see Tables 15.3 and 15.4).

Not all vitamins function as coenzymes. Vitamins designated by the letters A, C, D, E, and K (Figure 15.18 and Table 15.4) have a diverse array of functions. Vitamin A (retinol) is the precursor of retinal, the light-sensitive group in rhodopsin and other visual pigments (Section 32.3), and retinoic acid, an important signalling molecule. A deficiency of this vitamin leads to night blindness. In addition, young animals require vitamin A for growth. Vitamin C, or ascorbate, acts as an antioxidant. A deficiency in vitamin C can lead to scurvy, a disease due to malformed collagen and characterized by skin lesions and blood-vessel fragility (p. 778). A metabolite of vitamin D is a hormone that regulates the metabolism of calcium and phosphorus. A

TABLE 15.4 Noncoenzyme vitamins

Vitamin	Function	Deficiency
A	Roles in vision, growth, reproduction	Night blindness, cornea damage, damage to respiratory and gastrointestinal tract
C (ascorbic acid)	Antioxidant	Scurvy (swollen and bleeding gums, subdermal hemorrhaging)
D	Regulation of calcium and phosphate metabolism	Rickets (children): skeletal deformaties, impaired growth Osteomalacia (adults): soft, bending bones
E	Antioxidant	Inhibition of sperm production; lesions in muscles and nerves (rare)
K	Blood coagulation	Subdermal hemorrhaging

deficiency in vitamin D impairs bone formation in growing animals. Infertility in rats is a consequence of vitamin E (α-tocopherol) deficiency. This vitamin reacts with reactive oxygen species such as hydroxyl radicals and inactivates them before they can oxidize unsaturated membrane lipids, damaging cell structures. Vitamin K is required for normal blood clotting (p. 295).

Key Reactions Are Reiterated Throughout Metabolism

Just as there is an economy of design in the use of activated carriers, so is there an economy of design in biochemical reactions. The thousands of metabolic reactions, bewildering at first in their variety, can be subdivided into just six types (Table 15.5). Specific reactions of each type appear repeatedly, reducing the number of reactions that a student needs to learn.

1. *Oxidation–reduction reactions* are essential components of many pathways. Useful energy is often derived from the oxidation of carbon compounds. Consider the following two reactions:

Succinate   Fumarate   (1)

Malate   Oxaloacetate   (2)

These two oxidation–reduction reactions are components of the citric acid cycle (Chapter 17), which completely oxidizes the activated two-carbon fragment of acetyl CoA to two molecules of CO_2. In reaction 1, $FADH_2$ carries the electrons, whereas, in reaction 2, electrons are carried by NADH.

2. *Ligation reactions* form bonds by using free energy from ATP cleavage. Reaction 3 illustrates the ATP-dependent formation of a carbon–carbon bond, necessary to combine smaller molecules to form larger ones. Oxaloacetate is formed from pyruvate and CO_2.

TABLE 15.5   Types of chemical reactions in metabolism

Type of reaction	Description
Oxidation–reduction	Electron transfer
Ligation requiring ATP cleavage	Formation of covalent bonds (i.e., carbon–carbon bonds)
Isomerization	Rearrangement of atoms to form isomers
Group transfer	Transfer of a functional group from one molecule to another
Hydrolytic	Cleavage of bonds by the addition of water
Addition or removal of functional groups	Addition of functional groups to double bonds or their removal to form double bonds

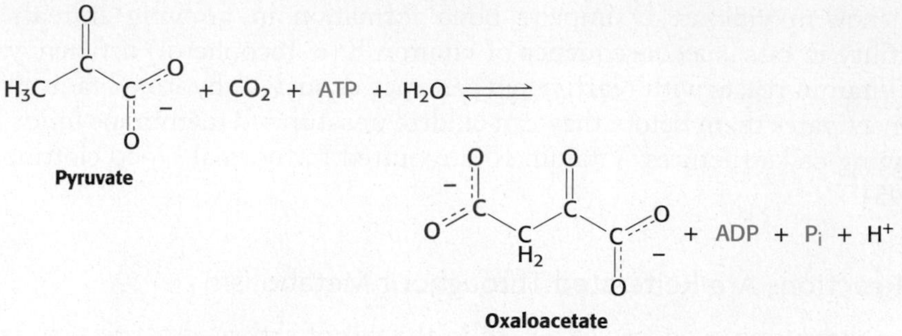

Pyruvate + CO_2 + ATP + H_2O $\rightleftharpoons$

+ ADP + P_i + H^+ (3)

Oxaloacetate

The oxaloacetate can be used in the citric acid cycle or converted into amino acids such as aspartic acid.

3. *Isomerization reactions* rearrange particular atoms within a molecule. Their role is often to prepare the molecule for subsequent reactions such as the oxidation–reduction reactions described in point 1.

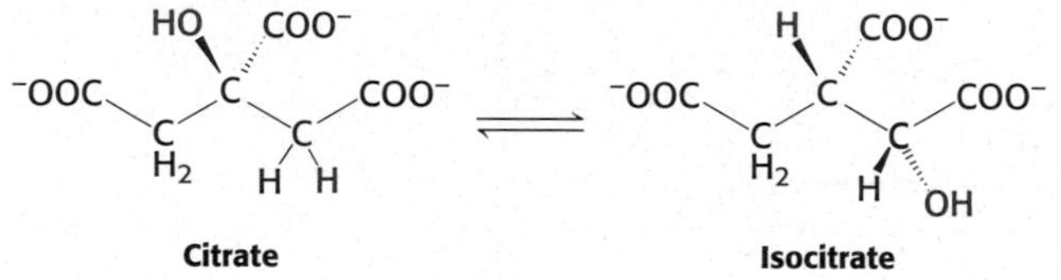

(4)

Citrate   **Isocitrate**

Reaction 4 is, again, a component of the citric acid cycle. This isomerization prepares the molecule for subsequent oxidation and decarboxylation by moving the hydroxyl group of citrate from a tertiary to a secondary position.

4. *Group-transfer reactions* play a variety of roles. Reaction 5 is representative of such a reaction. A phosphoryl group is transferred from the activated phosphoryl-group carrier, ATP, to glucose, the initial step in glycolysis, a key pathway for extracting energy from glucose (Chapter 16). This reaction traps glucose in the cell so that further catabolism can take place.

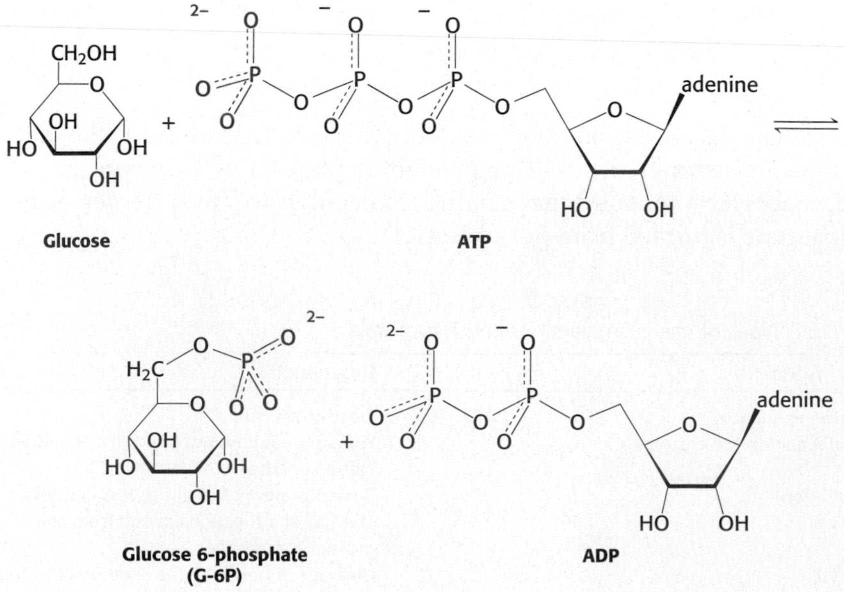

Glucose   **ATP**

(5)

Glucose 6-phosphate   **ADP**
(G-6P)

As stated earlier, group-transfer reactions are used to synthesize ATP (p. 418). We also saw examples of their use in signaling pathways (Chapter 14).

5. *Hydrolytic reactions* cleave bonds by the addition of water. Hydrolysis is a common means employed to break down large molecules, either to facilitate further metabolism or to reuse some of the components for biosynthetic purposes. Proteins are digested by hydrolytic cleavage (Chapters 9 and 10). Reaction 6 illustrates the hydrolysis of a peptide to yield two smaller peptides.

$$\text{(6)}$$

6. *Functional groups may be added to double bonds to form single bonds or removed from single bonds to form double bonds.* The enzymes that catalyze these types of reaction are classified as *lyases*. An important example, illustrated in reaction 7, is the conversion of the six-carbon molecule fructose 1,6-bisphosphate into two three-carbon fragments: dihydroxyacetone phosphate and glyceraldehyde 3-phosphate.

$$\text{(7)}$$

Fructose 1,6-bisphosphate   Dihydroxyacetone phosphate   Glyceraldehyde 3-phosphate
(F-1,6-BP)   (DHAP)   (GAP)

This reaction is a critical step in glycolysis (Chapter 16). Dehydrations to form double bonds, such as the formation of phosphoenolpyruvate (see Table 15.1) from 2-phosphoglycerate (reaction 8), are important reactions of this type.

$$\text{(8)}$$

2-Phosphoglycerate   Phosphoenolpyruvate
(PEP)

The dehydration sets up the next step in the pathway, a group-transfer reaction that uses the high phosphoryl-transfer potential of the product PEP to form ATP from ADP.

These six fundamental reaction types are the basis of metabolism. Remember that all six types can proceed in either direction, depending on the standard free energy for the specific reaction and the concentrations of the reactants and products inside the cell. An effective way to learn is to look for commonalities in the diverse metabolic pathways that we will be examining. There is a chemical logic that, when exposed, renders the complexity of the chemistry of living systems more manageable and reveals its elegance.

Metabolic Processes Are Regulated in Three Principal Ways

It is evident that the complex network of metabolic reactions must be rigorously regulated. At the same time, metabolic control must be flexible, to adjust metabolic activity to the constantly changing external environments of cells. Metabolism is regulated through control of (1) *the amounts of enzymes,* (2) *their catalytic activities,* and (3) *the accessibility of substrates.*

Controlling the Amounts of Enzymes.

The amount of a particular enzyme depends on both its rate of synthesis and its rate of degradation. The level of many enzymes is adjusted primarily by a change in the *rate of transcription* of the genes encoding them (Chapters 29 and 31). In *E. coli,* for example, the presence of lactose induces within minutes a more than 50-fold increase in the rate of synthesis of β-galactosidase, an enzyme required for the breakdown of this disaccharide.

Controlling Catalytic Activity.

The catalytic activity of enzymes is controlled in several ways. *Reversible allosteric control* is especially important. For example, the first reaction in many biosynthetic pathways is allosterically inhibited by the ultimate product of the pathway. The inhibition of aspartate transcarbamoylase by cytidine triphosphate (Section 10.1) is a well-understood example of *feedback inhibition.* This type of control can be almost instantaneous. Another recurring mechanism is *reversible covalent modification.* For example, glycogen phosphorylase, the enzyme catalyzing the breakdown of glycogen, a storage form of sugar, is activated by the phosphorylation of a particular serine residue when glucose is scarce (p. 559).

Hormones coordinate metabolic relations between different tissues, often by regulating the reversible modification of key enzymes. For instance, the hormone epinephrine triggers a signal-transduction cascade in muscle, resulting in the phosphorylation and activation of key enzymes and leading to the rapid degradation of glycogen to glucose, which is then used to supply ATP for muscle contraction (p. 601). As described in Chapter 14, many hormones act through intracellular messengers, such as cyclic AMP and calcium ion, that coordinate the activities of many target proteins.

Many reactions in metabolism are controlled by the *energy status* of the cell. One index of the energy status is the *energy charge,* which is proportional to the mole fraction of ATP plus half the mole fraction of ADP, given that ATP contains two anhydride bonds, whereas ADP contains one. Hence, the energy charge is defined as

$$\text{Energy charge} = \frac{[\text{ATP}] + \frac{1}{2}[\text{ADP}]}{[\text{ATP}] + [\text{ADP}] + [\text{AMP}]}$$

The energy charge can have a value ranging from 0 (all AMP) to 1 (all ATP). Daniel Atkinson showed that *ATP-generating (catabolic) pathways are inhibited by a high energy charge, whereas ATP-utilizing (anabolic) pathways are stimulated by a high energy charge.* In plots of the reaction rates of such pathways versus the energy charge, the curves are steep near an energy charge of 0.9, where they usually intersect (Figure 15.19). It is evident that the control of these pathways has evolved to maintain the energy charge within rather narrow limits. In other words, *the energy charge, like the pH of a cell, is buffered.* The energy charge of most cells ranges from 0.80 to 0.95. An alternative index of the energy status is the *phosphorylation potential,* which is defined as

$$\text{Phosphorylation potential} = \frac{[\text{ATP}]}{[\text{ADP}] + [\text{P}_i]}$$

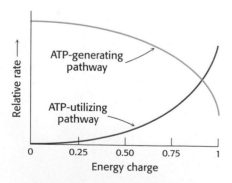

Figure 15.19 Energy charge regulates metabolism. High concentrations of ATP inhibit the relative rates of a typical ATP-generating (catabolic) pathway and stimulate the typical ATP-utilizing (anabolic) pathway.

The phosphorylation potential, in contrast with the energy charge, depends on the concentration of P_i and is directly related to the free-energy storage available from ATP.

Controlling the Accessibility of Substrates.

In eukaryotes, metabolic regulation and flexibility are enhanced by compartmentalization. For example, fatty acid oxidation takes place in mitochondria, whereas fatty acid synthesis takes place in the cytoplasm. *Compartmentalization segregates opposed reactions.*

Controlling the *flux of substrates* is another means of regulating metabolism. Glucose breakdown can take place in many cells only if insulin is present to promote the entry of glucose into the cell. The transfer of substrates from one compartment of a cell to another (e.g., from the cytoplasm to mitochondria) can serve as a control point.

Aspects of Metabolism May Have Evolved from an RNA World

How did the complex pathways that constitute metabolism evolve? The current thinking is that RNA was an early biomolecule and that, in an early RNA world, RNA served as catalysts and information-storage molecules.

Why do activated carriers such as ATP, NADH, FADH$_2$, and coenzyme A contain adenosine diphosphate units (Figure 15.20)? A possible explanation is that these molecules evolved from the early RNA catalysts. Non-RNA units such as the isoalloxazine ring may have been recruited to serve as efficient carriers of activated electrons and chemical units, a function not readily performed by RNA itself. We can picture the adenine ring of FADH$_2$ binding to a uracil unit in a niche of an RNA enzyme (ribozyme) by base-pairing, whereas the isoalloxazine ring protrudes and functions as an electron carrier. When the more versatile proteins replaced RNA as the major catalysts, the ribonucleotide coenzymes stayed essentially unchanged because they were already well suited to their metabolic roles. The nicotinamide unit of NADH, for example, can readily transfer electrons irrespective of whether the adenine unit interacts with a base in an RNA enzyme or with amino acid residues in a protein enzyme. With the advent of protein enzymes, these important cofactors evolved as free molecules without losing the adenosine diphosphate vestige of their RNA-world ancestry. That molecules and motifs of metabolism are common to all forms of life testifies to their common origin and to the retention of functioning modules through billions of years of evolution. Our understanding of metabolism, like that of other biological processes, is enriched by inquiry into how these beautifully integrated patterns of reactions came into being.

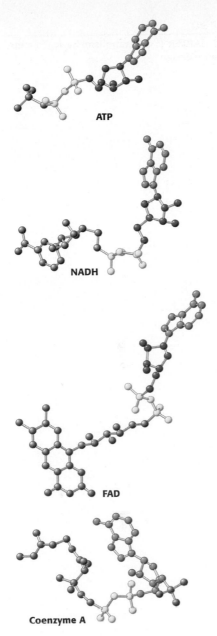

Figure 15.20 Adenosine diphosphate (ADP) is an ancient module in metabolism. This fundamental building block is present in key molecules such as ATP, NADH, FAD, and coenzyme A. The adenine unit is shown in blue, the ribose unit in red, and the diphosphate unit in yellow.

ATP

NADH

FAD

Coenzyme A

Summary

All cells transform energy. They extract energy from their environment and use this energy to convert simple molecules into cellular components.

15.1 Metabolism Is Composed of Many Coupled, Interconnecting Reactions

The process of energy transduction takes place through metabolism, a highly integrated network of chemical reactions. Metabolism can be subdivided into catabolism (reactions employed to extract energy from fuels) and anabolism (reactions that use this energy for biosynthesis). The most valuable thermodynamic concept for understanding bioenergetics is free energy. A reaction can occur spontaneously only if the change in free energy (ΔG) is negative. A thermodynamically unfavorable reaction can be driven by a thermodynamically favorable one, which is the hydrolysis of ATP in many cases.

15.2 ATP Is the Universal Currency of Free Energy in Biological Systems

The energy derived from catabolism is transformed into adenosine triphosphate. ATP hydrolysis is exergonic and the energy released can be used to power cellular processes, including motion, active transport, and biosynthesis. Under cellular conditions, the hydrolysis of ATP shifts the equilibrium of a coupled reaction by a factor of 10^8. ATP, the universal currency of energy in biological systems, is an energy-rich molecule because it contains two phosphoanhydride bonds.

15.3 The Oxidation of Carbon Fuels Is an Important Source of Cellular Energy

ATP formation is coupled to the oxidation of carbon fuels, either directly or through the formation of ion gradients. Photosynthetic organisms can use light to generate such gradients. ATP is consumed in muscle contraction and other motions of cells, in active transport, in signal-transduction processes, and in biosyntheses. The extraction of energy from foodstuffs by aerobic organisms comprises three stages. In the first stage, large molecules are broken down into smaller ones, such as amino acids, sugars, and fatty acids. In the second stage, these small molecules are degraded to a few simple units that have pervasive roles in metabolism. One of them is the acetyl unit of acetyl CoA, a carrier of activated acyl groups. The third stage of metabolism is the citric acid cycle and oxidative phosphorylation, in which ATP is generated as electrons flow to O_2, the ultimate electron acceptor, and fuels are completely oxidized to CO_2.

15.4 Metabolic Pathways Contain Many Recurring Motifs

Metabolism is characterized by common motifs. A small number of recurring activated carriers, such as ATP, NADH, and acetyl CoA, transfer activated groups in many metabolic pathways. NADPH, which carries two electrons at a high potential, provides reducing power in the biosynthesis of cell components from more-oxidized precursors. Many activated carriers are derived from vitamins, small organic molecules required in the diets of many higher organisms. Moreover, key reaction types are used repeatedly in metabolic pathways.

Metabolism is regulated in a variety of ways. The amounts of some critical enzymes are controlled by regulation of the rate of synthesis and degradation. In addition, the catalytic activities of many enzymes are regulated by allosteric interactions (as in feedback inhibition) and by covalent modification. The movement of many substrates into cells and subcellular compartments also is controlled. The energy charge, which depends on the relative amounts of ATP, ADP, and AMP, plays a role in metabolic regulation. A high energy charge inhibits ATP-generating (catabolic) pathways, whereas it stimulates ATP-utilizing (anabolic) pathways.

Key Terms

metabolism or intermediary metabolism (p. 409)

phototroph (p. 410)

chemotroph (p. 410)

catabolism (p. 411)

anabolism (p. 411)

amphibolic pathway (p. 411)

adenosine triphosphate (ATP) (p. 412)

phosphoryl-transfer potential (p. 415)

oxidative phosphorylation (p. 419)

activated carrier (p. 420)

vitamin (p. 423)

oxidation–reduction reaction (p. 425)

ligation reaction (p. 425)

isomerization reaction (p. 426)

group-transfer reaction (p. 426)

hydrolytic reaction (p. 427)

addition to or formation of double-bond reaction (p. 427)

lyase (p. 427)

energy charge (p. 428)

phosphorylation potential (p. 428)

Selected Readings

Where to Start

McGrane, M. M., Yun, J. S., Patel, Y. M., and Hanson, R. W. 1992. Metabolic control of gene expression: In vivo studies with transgenic mice. *Trends Biochem. Sci.* 17:40–44.

Westheimer, F. H. 1987. Why nature chose phosphates. *Science* 235:1173–1178.

Books

Harold, F. M. 1986. *The Vital Force: A Study of Bioenergetics.* W. H. Freeman and Company.

Krebs, H. A., and Kornberg, H. L. 1957. *Energy Transformations in Living Matter.* Springer Verlag.

Linder, M. C. (Ed.). 1991. *Nutritional Biochemistry and Metabolism* (2d ed.). Elsevier.

Gottschalk, G. 1986. *Bacterial Metabolism* (2d ed.). Springer Verlag.

Nicholls, D. G., and Ferguson, S. J. 1997. *Bioenergetics 2* (2d ed.) Academic Press.

Martin, B. R. 1987. *Metabolic Regulation: A Molecular Approach.* Blackwell Scientific.

Frayn, K. N. 1996. *Metabolic Regulation: A Human Perspective.* Portland Press.

Fell, D. 1997. *Understanding the Control of Metabolism.* Portland Press.

Harris, D. A. 1995. *Bioenergetics at a Glance.* Blackwell Scientific.

Von Baeyer, H. C. 1999. *Warmth Disperses and Time Passes: A History of Heat.* Modern Library.

Edsall, J. T., and Gutfreund, H. 1983. *Biothermodynamics: The Study of Biochemical Processes at Equilibrium.* Wiley.

Klotz, I. M. 1967. *Energy Changes in Biochemical Reactions.* Academic Press.

Hill, T. L. 1977. *Free Energy Transduction in Biology.* Academic Press.

Atkinson, D. E. 1977. *Cellular Energy Metabolism and Its Regulation.* Academic Press.

Thermodynamics

Alberty, R. A. 1993. Levels of thermodynamic treatment of biochemical reaction systems. *Biophys. J.* 65:1243–1254.

Alberty, R. A., and Goldberg, R. N. 1992. Standard thermodynamic formation properties for the adenosine 5′-triphosphate series. *Biochemistry* 31:10610–10615.

Alberty, R. A. 1968. Effect of pH and metal ion concentration on the equilibrium hydrolysis of adenosine triphosphate to adenosine diphosphate. *J. Biol. Chem.* 243:1337–1343.

Goldberg, R. N. 1984. *Compiled Thermodynamic Data Sources for Aqueous and Biochemical Systems: An Annotated Bibliography (1930–1983).* National Bureau of Standards Special Publication 685, U.S. Government Printing Office.

Frey, P. A., and Arabshahi, A. 1995. Standard free energy change for the hydrolysis of the α,β-phosphoanhydride bridge in ATP. *Biochemistry* 34:11307–11310.

Bioenergetics and Metabolism

Schilling, C. H., Letscher, D., and Palsson, B. O. 2000. Theory for the systemic definition of metabolic pathways and their use in interpreting metabolic function from a pathway-oriented perspective. *J. Theor. Biol.* 203:229–248.

DeCoursey, T. E., and Cherny, V. V. 2000. Common themes and problems of bioenergetics and voltage-gated proton channels. *Biochim. Biophys. Acta* 1458:104–119.

Giersch, C. 2000. Mathematical modelling of metabolism. *Curr. Opin. Plant Biol.* 3:249–253.

Rees, D. C., and Howard, J. B. 1999. Structural bioenergetics and energy transduction mechanisms. *J. Mol. Biol.* 293:343–350.

Regulation of Metabolism

Kemp, G. J. 2000. Studying metabolic regulation in human muscle. *Biochem. Soc. Trans.* 28:100–103.

Towle, H. C., Kaytor, E. N., and Shih, H. M. 1996. Metabolic regulation of hepatic gene expression. *Biochem. Soc. Trans.* 24:364–368.

Hofmeyr, J. H. 1995. Metabolic regulation: A control analytic perspective. *J. Bioenerg. Biomembr.* 27:479–490.

Erecińska, M., and Wilson, D. F. 1978. Homeostatic regulation of cellular energy metabolism. *Trends Biochem. Sci.* 3:219–223.

Historical Aspects

Kalckar, H. M. 1991. 50 years of biological research: From oxidative phosphorylation to energy requiring transport regulation. *Annu. Rev. Biochem.* 60:1–37.

Kalckar, H. M. (Ed.). 1969. *Biological Phosphorylations.* Prentice Hall.

Fruton, J. S. 1972. *Molecules and Life.* Wiley-Interscience.

Lipmann, F. 1971. *Wanderings of a Biochemist.* Wiley-Interscience.

Problems

1. *Energy flow.* What is the direction of each of the following reactions when the reactants are initially present in equimolar amounts? Use the data given in Table 15.1.

(a) ATP + creatine $\rightleftharpoons$ creatine phosphate + ADP
(b) ATP + glycerol $\rightleftharpoons$ glycerol 3-phosphate + ADP
(c) ATP + pyruvate $\rightleftharpoons$ phosphoenolpyruvate + ADP
(d) ATP + glucose $\rightleftharpoons$ glucose 6-phosphate + ADP

2. *A proper inference.* What information do the $\Delta G^{\circ\prime}$ data given in Table 15.1 provide about the relative rates of hydrolysis of pyrophosphate and acetyl phosphate?

3. *A potent donor.* Consider the following reaction:

ATP + pyruvate $\rightleftharpoons$ phosphoenolpyruvate + ADP

(a) Calculate $\Delta G^{\circ\prime}$ and K'_{eq} at 25°C for this reaction by using the data given in Table 15.1.

(b) What is the equilibrium ratio of pyruvate to phosphoenolpyruvate if the ratio of ATP to ADP is 10?

4. *Isomeric equilibrium.* Calculate $\Delta G^{\circ\prime}$ for the isomerization of glucose 6-phosphate to glucose 1-phosphate. What is the equilibrium ratio of glucose 6-phosphate to glucose 1-phosphate at 25°C?

5. *Activated acetate.* The formation of acetyl CoA from acetate is an ATP-driven reaction:

Acetate + ATP + CoA $\rightleftharpoons$ acetyl CoA + AMP + PP$_i$

(a) Calculate $\Delta G^{\circ\prime}$ for this reaction by using data given in this chapter.

(b) The PP$_i$ formed in the preceding reaction is rapidly hydrolyzed in vivo because of the ubiquity of inorganic pyrophosphatase. The $\Delta G^{\circ\prime}$ for the hydrolysis of PP$_i$ is

-19.2 kJ mol^{-1} (-4.6 kcal mol^{-1}). Calculate the $\Delta G^{\circ\prime}$ for the overall reaction, including pyrophosphate hydrolysis. What effect does the hydrolysis of PP$_i$ have on the formation of acetyl CoA?

6. *Acid strength.* The pK of an acid is a measure of its proton-group-transfer potential.

(a) Derive a relation between $\Delta G^{\circ\prime}$ and pK.
(b) What is the $\Delta G^{\circ\prime}$ for the ionization of acetic acid, which has a pK of 4.8?

7. *Raison d'être.* The muscles of some invertebrates are rich in *arginine phosphate* (phosphoarginine). Propose a function for this amino acid derivative.

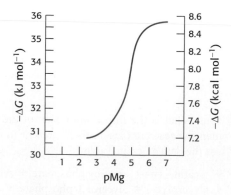

Arginine phosphate

8. *Recurring motif.* What is the structural feature common to ATP, FAD, NAD$^+$, and CoA?

9. *Ergogenic help or hindrance?* Creatine is a popular, but untested, dietary supplement.

(a) What is the biochemical rationale for the use of creatine?
(b) What type of exercise would most benefit from creatine supplementation?

10. *Standard conditions versus real life 1.* The enzyme aldolase catalyzes the following reaction in the glycolytic pathway:

Fructose 1,6-bisphosphate $\xrightleftharpoons{\text{Aldolase}}$
    dihydroxyacetone phosphate + glyceraldehyde 3-phosphate

The $\Delta G^{\circ\prime}$ for the reaction is $+23.8$ kJ mol^{-1} ($+5.7$ kcal mol^{-1}) whereas the ΔG in the cell is -1.3 kJ mol^{-1} (-0.3 kcal mol^{-1}). Calculate the ratio of reactants to products under equilibrium and intracellular conditions. Using your results, explain how the reaction can be endergonic under standard conditions and exergonic under intracellular conditions.

11. *Standard conditions versus real life 2.* On page 413, we showed that a reaction, A $\rightleftharpoons$ B, with a $\Delta G' = +13$ kJ mol^{-1} ($+4.0$ kcal mol^{-1}) has an K_{eq} of 1.15×10^{-3}. The K_{eq} is increased to 2.67×10^2 if the reaction is coupled to ATP hydrolysis under standard conditions. The ATP-generating system of cells maintains the [ATP]/[ADP][P$_i$] ratio at a high level, typi-

cally of the order of 500 M^{-1}. Calculate the ratio of B/A under cellular conditions.

12. *Not all alike.* The concentrations of ATP, ADP, and P$_i$ differ with cell type. Consequently, the release of free energy with the hydrolysis of ATP will vary with cell type. Using the following table, calculate the ΔG for the hydrolysis of ATP in liver, muscle, and brain cells. In which cell type is the free energy of ATP hydrolysis most negative?

	ATP (mM)	ADP (mM)	P$_i$ (mM)
Liver	3.5	1.8	5.0
Muscle	8.0	0.9	8.0
Brain	2.6	0.7	2.7

13. *Running downhill.* Glycolysis is a series of 10 linked reactions that convert one molecule of glucose into two molecules of pyruvate with the concomitant synthesis of two molecules of ATP (Chapter 16). The $\Delta G^{\circ\prime}$ for this set of reactions is -35.6 kJ mol^{-1} (-8.5 kcal mol^{-1}), whereas the ΔG is -76.6 kJ mol^{-1} (-18.3 kcal mol^{-1}). Explain why the free-energy release is so much greater under intracellular conditions than under standard conditions.

Chapter Integration Problem

14. *Activated sulfate.* Fibrinogen contains tyrosine-*O*-sulfate. Propose an activated form of sulfate that could react in vivo with the aromatic hydroxyl group of a tyrosine residue in a protein to form tyrosine-*O*-sulfate.

Data Interpretation Problem

15. *Opposites attract.* The following graph shows how the ΔG for the hydrolysis of ATP varies as a function of the Mg^{2+} concentration (pMg $= -\log[\text{Mg}^{2+}]$).

(a) How does decreasing [Mg^{2+}] affect the ΔG of hydrolysis for ATP?
(b) Explain this effect.

Chapter 16

Glycolysis and Gluconeogenesis

Michael Johnson sprints to another victory in the 200-meter semifinals of the Olympics. Glucose metabolism can generate the ATP to power muscle contraction. During a sprint, when the ATP needs outpace oxygen delivery, glucose is metabolized to lactate (A). When oxygen delivery is adequate, glucose is metabolized more efficiently to carbon dioxide and water (B). [(Left) Simon Bruty/Allsport.]

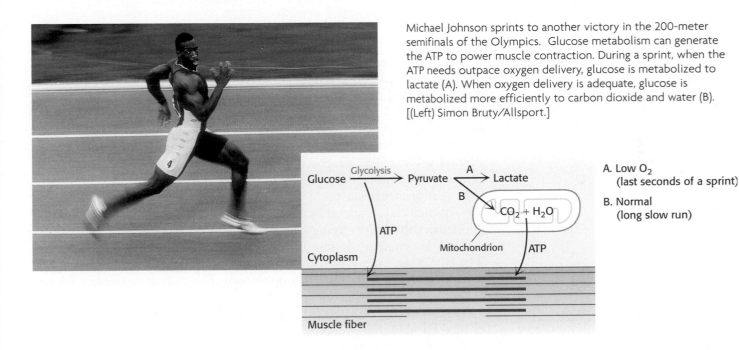

A. Low O_2
(last seconds of a sprint)

B. Normal
(long slow run)

The first metabolic pathway that we encounter is *glycolysis,* an ancient pathway employed by a host of organisms. *Glycolysis is the sequence of reactions that metabolizes one molecule of glucose to two molecules of pyruvate with the concomitant net production of two molecules of ATP.* This process is anaerobic (i.e., it does not require O_2) inasmuch as it evolved before the accumulation of substantial amounts of oxygen in the atmosphere. Pyruvate can be further processed anaerobically to lactate (*lactic acid fermentation*) or ethanol (*alcoholic fermentation*). Under aerobic conditions, pyruvate can be completely oxidized to CO_2, generating much more ATP, as will be described in Chapters 17 and 18. Figure 16.1 shows some possible fates of pyruvate produced by glycolysis.

Because glucose is such a precious fuel, metabolic products, such as pyruvate and lactate, are salvaged to synthesize glucose in the process of *gluconeogenesis.* Although glycolysis and gluconeogenesis have some enzymes in common, the two pathways are not simply the reverse of each other. In particular, the highly exergonic, irreversible steps of glycolysis are bypassed in gluconeogenesis. The two pathways are reciprocally regulated so that glycolysis and gluconeogenesis do not take place simultaneously in the same cell to a significant extent.

Our understanding of glucose metabolism, especially glycolysis, has a rich history. Indeed, the development of biochemistry and the delineation

Glycolysis

Derived from the Greek stem *glyk-,* "sweet," and the word *lysis,* "dissolution."

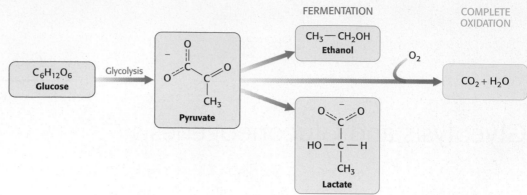

Figure 16.1 Some fates of glucose.

of glycolysis went hand in hand. A key discovery was made by Hans Buchner and Eduard Buchner in 1897, quite by accident. The Buchners were interested in manufacturing cell-free extracts of yeast for possible therapeutic use. These extracts had to be preserved without the use of antiseptics such as phenol, and so they decided to try sucrose, a commonly used preservative in kitchen chemistry. They obtained a startling result: sucrose was rapidly fermented into alcohol by the yeast juice. The significance of this finding was immense. *The Buchners demonstrated for the first time that fermentation could take place outside living cells.* The accepted view of their day, asserted by Louis Pasteur in 1860, was that fermentation is inextricably tied to living cells. The chance discovery by the Buchners refuted this vitalistic dogma and opened the door to modern biochemistry. The Buchners' discovery inspired the search for the biochemicals that catalyze the conversion of sucrose into alcohol. *The study of metabolism became the study of chemistry.*

Studies of muscle extracts then showed that many of the reactions of lactic acid fermentation were the same as those of alcoholic fermentation. *This exciting discovery revealed an underlying unity in biochemistry.* The complete glycolytic pathway was elucidated by 1940, largely through the pioneering contributions of Gustav Embden, Otto Meyerhof, Carl Neuberg, Jacob Parnas, Otto Warburg, Gerty Cori, and Carl Cori. Glycolysis is also known as the *Embden–Meyerhof pathway.*

Enzyme
A term coined by Friedrich Wilhelm Kühne in 1878 to designate catalytically active substances that had formerly been called ferments. Derived from the Greek words *en*, "in," and *zyme*, "leaven."

Glucose Is Generated from Dietary Carbohydrates

We typically consume in our diets a generous amount of starch and a smaller amount of glycogen. These complex carbohydrates must be converted into simpler carbohydrates for absorption by the intestine and transport in the blood. Starch and glycogen are digested primarily by the pancreatic enzyme *α-amylase* and to a lesser extent by salivary α-amylase. Amylase cleaves the α-1,4 bonds of starch and glycogen, but not the α-1,6 bonds. The products are the di- and trisaccharides maltose and maltotriose. The material not digestible because of the α-1,6 bonds is called the *limit dextrin*.

Maltase cleaves maltose into two glucose molecules, whereas *α-glucosidase* digests maltotriose and any other oligosaccharides that may have escaped digestion by the amylase. *α-Dextrinase* further digests the limit dextrin. Maltase and α-glucosidase are located on the surface of the intestinal cells, as is *sucrase*, an enzyme that degrades the sucrose contributed by vegetables to fructose and glucose. The enzyme *lactase* is responsible for degrading the milk sugar lactose into glucose and galactose. The monosaccharides are transported into the cells lining the intestine and then into the bloodstream.

Glucose Is an Important Fuel for Most Organisms

Glucose is a common and important fuel. In mammals, glucose is the only fuel that the brain uses under nonstarvation conditions and the only fuel that red blood cells can use at all. Indeed, almost all organisms use glucose, and most that do process it in a similar fashion. Recall from Chapter 11 that there are many carbohydrates. Why is glucose instead of some other monosaccharide such a prominent fuel? We can speculate on the reasons. First, glucose is one of several monosaccharides formed from formaldehyde under prebiotic conditions, and so it may have been available as a fuel source for primitive biochemical systems. Second, glucose has a low tendency, relative to other monosaccharides, to nonenzymatically glycosylate proteins. In their open-chain forms, monosaccharides contain carbonyl groups that can react with the amino groups of proteins to form Schiff bases, which rearrange to form a more stable amino–ketone linkage. Such nonspecifically modified proteins often do not function effectively. Glucose has a strong tendency to exist in the ring formation and, consequently, relatively little tendency to modify proteins. Recall that all the hydroxyl groups in the ring conformation of β-glucose are equatorial, contributing to the sugar's high relative stability (p. 308).

16.1 Glycolysis Is an Energy-Conversion Pathway in Many Organisms

We now begin our consideration of the glycolytic pathway. This pathway is common to virtually all cells, both prokaryotic and eukaryotic. In eukaryotic cells, glycolysis takes place in the cytoplasm. This pathway can be thought of as comprising three stages (Figure 16.2). Stage 1, which is the conversion of glucose into fructose 1,6-bisphosphate, consists of three steps: a phosphorylation, an isomerization, and a second phosphorylation reaction. *The strategy of these initial steps in glycolysis is to trap the glucose in the cell and form a compound that can be readily cleaved into phosphorylated three-carbon units.* Stage 2 is the cleavage of the fructose 1,6-bisphosphate into two three-carbon fragments. These resulting three-carbon units are readily interconvertible. In stage 3, ATP is harvested when the three-carbon fragments are oxidized to pyruvate.

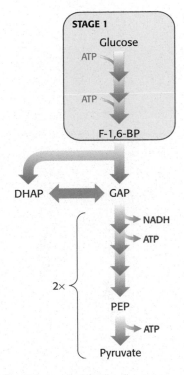

Stage 1 of glycolysis. The three steps of stage 1 begin with the phosphorylation of glucose by hexokinase.

Hexokinase Traps Glucose in the Cell and Begins Glycolysis

Glucose enters cells through specific transport proteins (p. 456) and has one principal fate: *it is phosphorylated by ATP to form glucose 6-phosphate.* This step is notable for two reasons: (1) glucose 6-phosphate cannot pass through the membrane because it is not a substrate for the glucose transporters, and (2) the addition of the phosphoryl group acts to destabilize glucose, thus facilitating its further metabolism. The transfer of the phosphoryl group from ATP to the hydroxyl group on carbon 6 of glucose is catalyzed by *hexokinase*.

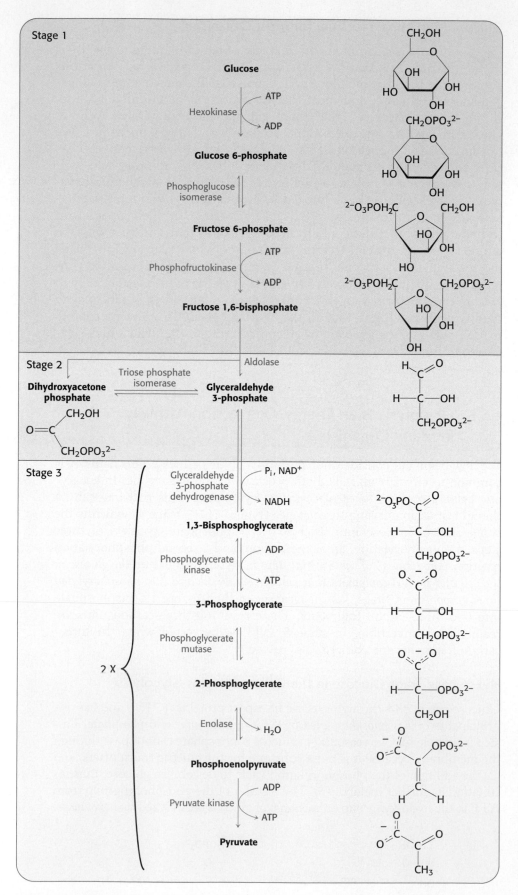

Figure 16.2 Stages of glycolysis. The glycolytic pathway can be divided into three stages: (1) glucose is trapped and destabilized; (2) two interconvertible three-carbon molecules are generated by cleavage of six-carbon fructose; and (3) ATP is generated.

Phosphoryl transfer is a fundamental reaction in biochemistry. *Kinases are enzymes that catalyze the transfer of a phosphoryl group from ATP to an acceptor.* Hexokinase, then, catalyzes the transfer of a phosphoryl group from ATP to a variety of six-carbon sugars (*hexoses*), such as glucose and mannose. *Hexokinase, like adenylate kinase* (p. 267) *and all other kinases, requires* Mg^{2+} *(or another divalent metal ion such as* Mn^{2+}*) for activity.* The divalent metal ion forms a complex with ATP.

X-ray crystallographic studies of yeast hexokinase revealed that the binding of glucose induces a large conformational change in the enzyme. Hexokinase consists of two lobes, which move toward each other when glucose is bound (Figure 16.3). On glucose binding, one lobe rotates 12 degrees with respect to the other, resulting in movements of the polypeptide backbone of as much as 8 Å. The cleft between the lobes closes, and the bound glucose becomes surrounded by protein, except for the hydroxyl group of carbon 6, which will accept the phosphoryl group from ATP. The closing of the cleft in hexokinase is a striking example of the role of *induced fit* in enzyme action (p. 215).

The glucose-induced structural changes are significant in two respects. First, the environment around the glucose becomes more nonpolar, which favors the donation of the terminal phosphoryl group of ATP. Second, as noted on page 269, the conformational changes enable the kinase to discriminate against H_2O as a substrate. The closing of the cleft keeps water molecules away from the active site. If hexokinase were rigid, a molecule of H_2O occupying the binding site for the —CH_2OH of glucose could attack the γ phosphoryl group of ATP, forming ADP and P_i. In other words, a rigid kinase would likely also be an ATPase. It is interesting to note that other kinases taking part in glycolysis—phosphofructokinase, phosphoglycerate kinase, and pyruvate kinase—also contain clefts between lobes that close when substrate is bound, although the structures of these enzymes are different in other regards. *Substrate-induced cleft closing is a general feature of kinases.*

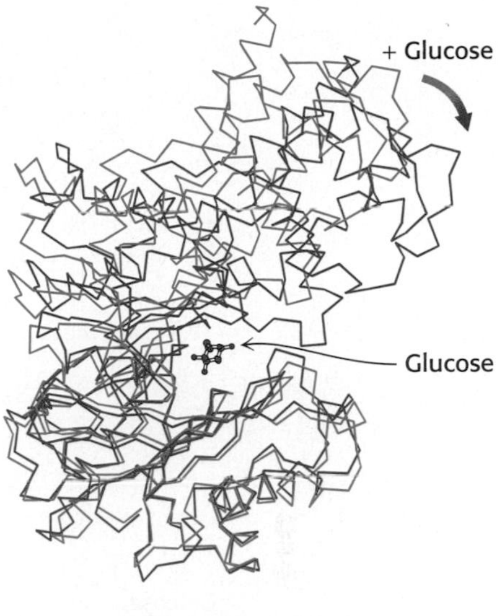

Figure 16.3 **Induced fit in hexokinase.** As shown in blue, the two lobes of hexokinase are separated in the absence of glucose. The conformation of hexokinase changes markedly on binding glucose, as shown in red. *Notice that two lobes of the enzyme come together and surround the substrate, creating the necessary environment for catalysis.*[Courtesy of Dr. Thomas Steitz.]

Fructose 1,6-bisphosphate Is Generated from Glucose 6-phosphate

The next step in glycolysis is the *isomerization of glucose 6-phosphate to fructose 6-phosphate.* Recall that the open-chain form of glucose has an aldehyde group at carbon 1, whereas the open-chain form of fructose has a keto group at carbon 2. Thus, the isomerization of glucose 6-phosphate to fructose 6-phosphate is a *conversion of an aldose into a ketose.* The reaction catalyzed by *phosphoglucose isomerase* includes additional steps because both glucose 6-phosphate and fructose 6-phosphate are present primarily in the cyclic forms. The enzyme must first open the six-membered ring of glucose 6-phosphate, catalyze the isomerization, and then promote the formation of the five-membered ring of fructose 6-phosphate.

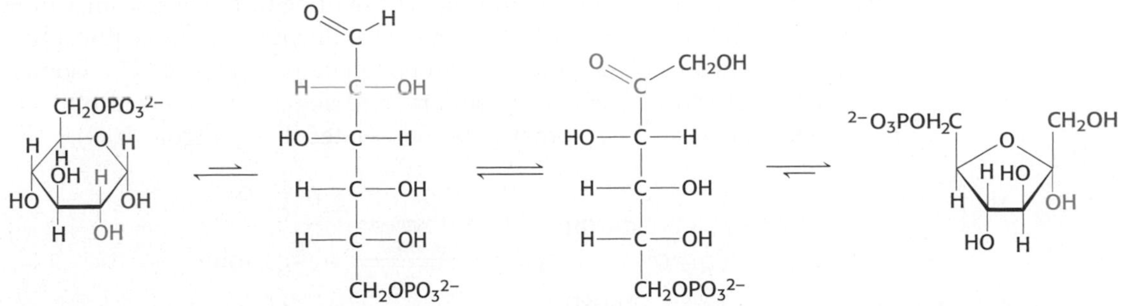

Glucose 6-phosphate (G-6P) ⇌ **Glucose 6-phosphate (open-chain form)** ⇌ **Fructose 6-phosphate (open-chain form)** ⇌ **Fructose 6-phosphate (F-6P)**

A second phosphorylation reaction follows the isomerization step. *Fructose 6-phosphate is phosphorylated by ATP to fructose 1,6-bisphosphate* (F-1,6-BP). The prefix *bis-* in bisphosphate means that two separate monophosphoryl groups are present, whereas the prefix *di-* in diphosphate (as in adenosine diphosphate) means that two phosphoryl groups are present and are connected by an anhydride bond.

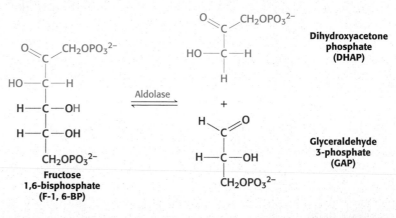

This reaction is catalyzed by *phosphofructokinase* (PFK), an allosteric enzyme that sets the pace of glycolysis (p. 454). As we will learn, this enzyme plays a central role in the integration of much of metabolism.

The Six-Carbon Sugar Is Cleaved into Two Three-Carbon Fragments

The second stage of glycolysis begins with the splitting of fructose 1,6-bisphosphate into *glyceraldehyde 3-phosphate* (GAP) and *dihydroxyacetone phosphate* (DHAP). The products of the remaining steps in glycolysis consist of three-carbon units rather than six-carbon units.

This reaction, which is readily reversible, is catalyzed by *aldolase*. This enzyme derives its name from the nature of the reverse reaction, an aldol condensation.

Glyceraldehyde 3-phosphate is on the direct pathway of glycolysis, whereas dihydroxyacetone phosphate is not. Unless a means exists to convert dihydroxyacetone phosphate into glyceraldehyde 3-phosphate, a three-carbon fragment useful for generating ATP will be lost. These compounds are isomers that can be readily interconverted: dihydroxyacetone phosphate is a ketose, whereas glyceraldehyde 3-phosphate is an aldose. The isomerization of these three-carbon phosphorylated sugars is catalyzed by *triose phosphate isomerase* (TPI, sometimes abbreviated TIM; Figure 16.4).

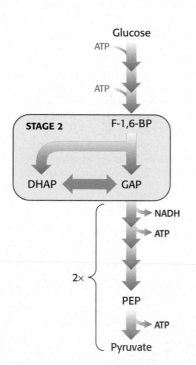

Stage 2 of glycolysis. Two three-carbon fragments are produced from one six-carbon sugar.

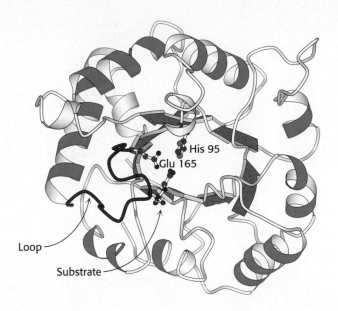

Figure 16.4 **Structure of triose phosphate isomerase.** This enzyme consists of a central core of eight parallel β strands (orange) surrounded by eight α helices (blue). This structural motif, called an αβ barrel, is also found in the glycolytic enzymes aldolase, enolase, and pyruvate kinase. *Notice* that histidine 95 and glutamate 165, essential components of the active site of triose phosphate isomerase, are located in the barrel. A loop (red) closes off the active site on substrate binding. [Drawn from 2YPI.pdb.]

This reaction is rapid and reversible. At equilibrium, 96% of the triose phosphate is dihydroxyacetone phosphate. However, the reaction proceeds readily from dihydroxyacetone phosphate to glyceraldehyde 3-phosphate because the subsequent reactions of glycolysis remove this product.

We now see the significance of the isomerization of glucose 6-phosphate to fructose 6-phosphate and its subsequent phosphorylation to form fructose 1,6-bisphosphate. Had the aldol cleavage occurred in the aldose glucose, a two-carbon and a four-carbon fragment would have resulted. Two different metabolic pathways, one to process the two-carbon fragment and one for the four-carbon fragment, would have been required to extract energy. Isomerization to the ketose fructose followed by aldol cleavage yields two phosphorylated interconvertible three-carbon fragments that will be oxidized in the later steps of glycolysis to capture energy in the form of ATP.

Mechanism: Triose Phosphate Isomerase Salvages a Three-Carbon Fragment

Much is known about the catalytic mechanism of triose phosphate isomerase. TPI catalyzes the transfer of a hydrogen atom from carbon 1 to carbon 2, an intramolecular oxidation–reduction. This isomerization of a ketose into an aldose proceeds through an *enediol intermediate* (Figure 6.5).

X-ray crystallographic and other studies showed that glutamate 165 plays the role of a general acid–base catalyst: it abstracts a proton (H^+) from carbon 1, and then donates it to carbon 2. However, the carboxylate group of glutamate 165 by itself is not basic enough to pull a proton away from a carbon atom adjacent to a carbonyl group. Histidine 95 assists catalysis by donating a proton to stabilize the negative charge that develops on the C-2 carbonyl group.

Two features of this enzyme are noteworthy. First, TPI displays great catalytic prowess. It accelerates isomerization by a factor of 10^{10} compared with the rate obtained with a simple base catalyst such as acetate ion. Indeed, the k_{cat}/K_M ratio for the isomerization of glyceraldehyde 3-phosphate is $2 \times 10^8 \, M^{-1} \, s^{-1}$, which is close to the diffusion-controlled limit. In other words, catalysis takes place every time that enzyme and substrate meet. The diffusion-controlled encounter of substrate and enzyme is thus the rate-limiting step in catalysis. TPI is an example of a

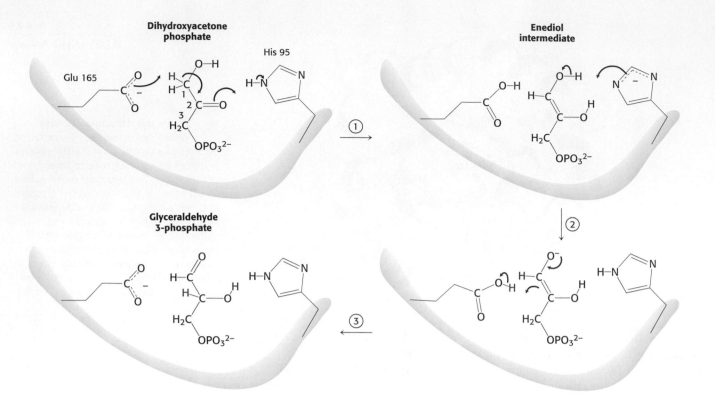

Figure 16.5 Catalytic mechanism of triose phosphate isomerase. (1) Glutamate 165 acts as a general base by abstracting a proton (H⁺) from carbon 1. Histidine 95, acting as a general acid, donates a proton to the oxygen atom bonded to carbon 2, forming the enediol intermediate. (2) Glutamic acid, now acting as a general acid, donates a proton to C-2 while histidine removes a proton from the OH of C-1. (3) The product is formed, and glutamate and histidine are returned to their ionized and neutral forms, respectively.

kinetically perfect enzyme (p. 221). Second, TPI suppresses an undesired side reaction, the decomposition of the enediol intermediate into methyl glyoxal and orthophosphate.

Enediol intermediate

Methyl glyoxal

In solution, this physiologically useless reaction is 100 times faster than isomerization. Hence, TPI must prevent the enediol from leaving the enzyme. This labile intermediate is trapped in the active site by the movement of a loop of 10 residues (see Figure 16.4). This loop serves as a lid on the active site, shutting it when the enediol is present and reopening it when isomerization is completed. *We see here a striking example of one means of preventing an undesirable alternative reaction: the active site is kept closed until the desirable reaction takes place.*

Thus, two molecules of glyceraldehyde 3-phosphate are formed from one molecule of fructose 1,6-bisphosphate by the sequential action of aldolase and triose phosphate isomerase. The economy of metabolism is evident in this reaction sequence. The isomerase funnels dihydroxyacetone phosphate into the main glycolytic pathway; a separate set of reactions is not needed.

The Oxidation of an Aldehyde to an Acid Powers the Formation of a Compound with High Phosphoryl-Transfer Potential

The preceding steps in glycolysis have transformed one molecule of glucose into two molecules of glyceraldehyde 3-phosphate, but no energy has yet been extracted. On the contrary, thus far, two molecules of ATP have been

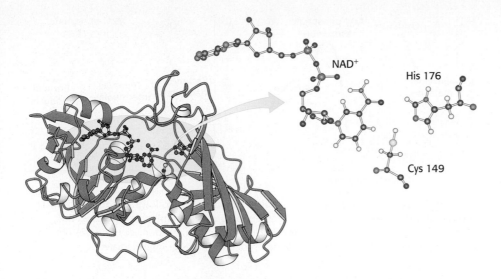

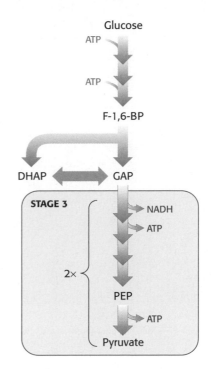

Figure 16.6 Structure of glyceraldehyde 3-phosphate dehydrogenase. *Notice* that the active site includes a cysteine residue and a histidine residue adjacent to a bound NAD$^+$ molecule. The sulfur atom of cysteine will link with the substrate to form a transitory thioester intermediate. [Drawn from 1GAD.pdb.]

invested. We come now to the final stage of glycolysis, a series of steps that harvest some of the energy contained in glyceraldehyde 3-phosphate as ATP. The initial reaction in this sequence is the *conversion of glyceraldehyde 3-phosphate into 1,3-bisphosphoglycerate* (1,3-BPG), a reaction catalyzed by *glyceraldehyde 3-phosphate dehydrogenase* (Figure 16.6).

Glyceraldehyde 3-phosphate (GAP) + NAD$^+$ + P$_i$ $\rightleftharpoons$ **1,3-Bisphosphoglycerate (1,3-BPG)** + NADH + H$^+$

1,3-Bisphosphoglycerate is an acyl phosphate, which is a mixed anhydride of phosphoric acid and a carboxylic acid. Such compounds have a high phosphoryl-transfer potential; one of its phosphoryl groups is transferred to ADP in the next step in glycolysis.

The reaction catalyzed by glyceraldehyde 3-phosphate dehydrogenase can be viewed as the sum of two processes: the *oxidation* of the aldehyde to a carboxylic acid by NAD$^+$ and the *joining* of the carboxylic acid and orthophosphate to form the acyl-phosphate product.

(Oxidation) + NAD$^+$ + H$_2$O $\rightleftharpoons$ + NADH + H$^+$

(Acyl-phosphate formation (dehydration)) + P$_i$ $\rightleftharpoons$ + H$_2$O

The first reaction is thermodynamically quite favorable, with a standard free-energy change, $\Delta G°'$, of approximately -50 kJ mol^{-1} (-12 kcal mol^{-1}), whereas the second reaction is quite unfavorable, with a standard free-

Stage 3 of glycolysis. The oxidation of three-carbon fragments yields ATP.

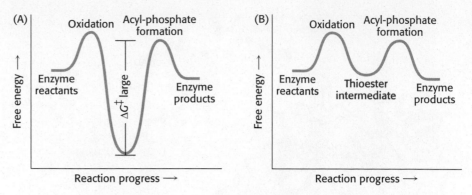

Figure 16.7 Free-energy profiles for glyceraldehyde oxidation followed by acyl-phosphate formation. (A) A hypothetical case with no coupling between the two processes. The second step must have a large activation barrier, making the reaction very slow. (B) The actual case with the two reactions coupled through a thioester intermediate.

energy change of the same magnitude but the opposite sign. If these two re-actions simply took place in succession, the second reaction would have a very large activation energy and thus not take place at a biologically signifi-cant rate. These two processes *must be coupled* so that the favorable aldehyde oxidation can be used to drive the formation of the acyl phosphate. How are these reactions coupled? *The key is an intermediate, formed as a result of the aldehyde oxidation, that is linked to the enzyme by a thioester bond.* Thioesters are high-energy compounds found in many biochemical pathways (p. 422). This intermediate reacts with orthophosphate to form the high-energy compound 1,3-bisphosphoglycerate.

The thioester intermediate is higher in free energy than the free carboxylic acid is. The favorable oxidation and unfavorable phosphorylation reactions are coupled by the thioester intermediate, which preserves much of the free energy released in the oxidation reaction. We see here the *use of a covalent enzyme-bound intermediate as a mechanism of energy coupling.* A free-energy profile of the glyceraldehyde 3-phosphate dehydrogenase reaction, com-pared with a hypothetical process in which the reaction proceeds without this intermediate, reveals how this intermediate allows a favorable process to drive an unfavorable one (Figure 16.7).

Mechanism: Phosphorylation Is Coupled to the Oxidation of Glyceraldehyde 3-phosphate by a Thioester Intermediate

Let us consider the mechanism of glyceraldehyde 3-phosphate dehydro-genase in detail (Figure 16.8). In step 1, the aldehyde substrate reacts with the sulfhydryl group of cysteine 149 on the enzyme to form a hemithioac-etal. Step 2 is the *transfer of a hydride ion to a molecule of NAD^+ that is tightly bound to the enzyme and is adjacent to the cysteine residue.* This re-action is favored by the deprotonation of the hemithioacetal by histidine 176. The products of this reaction are the reduced coenzyme NADH and a thioester intermediate. *This thioester intermediate has a free energy close to that of the reactants* (see Figure 16.7). In step 3, the NADH formed from the aldehyde oxidation leaves the enzyme and is replaced by a sec-ond molecule of NAD^+. This step is important because the positive charge on NAD^+ polarizes the thioester intermediate to facilitate the at-tack by orthophosphate. In step 4, orthophosphate attacks the thioester to form 1,3-BPG and free the cysteine residue. This example illustrates the essence of energy transformations and of metabolism itself: energy released by carbon oxidation is converted into high phosphoryl-transfer potential.

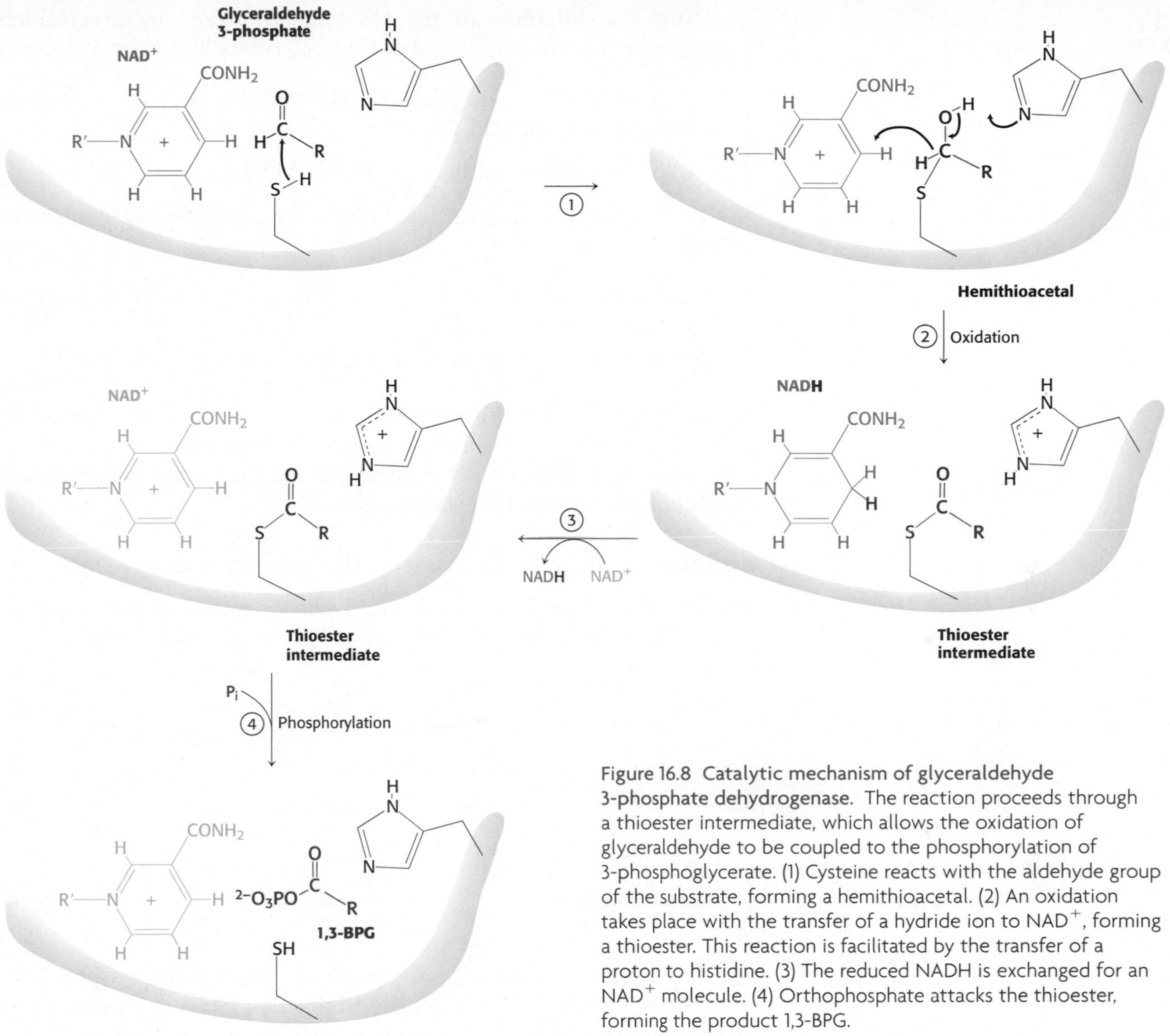

Figure 16.8 Catalytic mechanism of glyceraldehyde 3-phosphate dehydrogenase. The reaction proceeds through a thioester intermediate, which allows the oxidation of glyceraldehyde to be coupled to the phosphorylation of 3-phosphoglycerate. (1) Cysteine reacts with the aldehyde group of the substrate, forming a hemithioacetal. (2) An oxidation takes place with the transfer of a hydride ion to NAD^+, forming a thioester. This reaction is facilitated by the transfer of a proton to histidine. (3) The reduced NADH is exchanged for an NAD^+ molecule. (4) Orthophosphate attacks the thioester, forming the product 1,3-BPG.

ATP Is Formed by Phosphoryl Transfer from 1,3-Bisphosphoglycerate

1,3-Bisphosphoglycerate is an energy-rich molecule with a greater phosphoryl-transfer potential than that of ATP (p. 416). Thus, 1,3-BPG can be used to power the synthesis of ATP from ADP. *Phosphoglycerate kinase* catalyzes the transfer of the phosphoryl group from the acyl phosphate of 1,3-bisphosphoglycerate to ADP. ATP and 3-phosphoglycerate are the products.

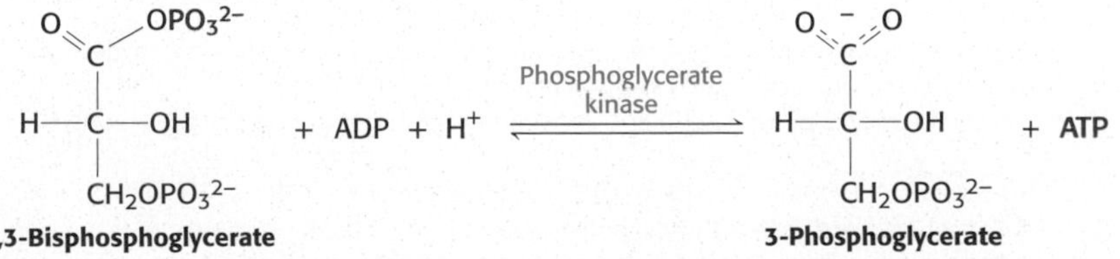

The formation of ATP in this manner is referred to as *substrate-level phosphorylation* because the phosphate donor, 1,3-BPG, is a substrate with high phosphoryl-transfer potential. We will contrast this manner of ATP formation with the formation of ATP from ionic gradients in Chapters 18 and 19.

Thus, the outcomes of the reactions catalyzed by glyceraldehyde 3-phosphate dehydrogenase and phosphoglycerate kinase are as follows:

1. Glyceraldehyde 3-phosphate, an aldehyde, is oxidized to 3-phosphoglycerate, a carboxylic acid.

2. NAD^+ is concomitantly reduced to NADH.

3. ATP is formed from P_i and ADP at the expense of carbon-oxidation energy.

In essence, the energy released during the oxidation of glyceraldehyde 3-phosphate to 3-phosphoglycerate is temporarily trapped as 1,3-bisphosphoglycerate. This energy powers the transfer of a phosphoryl group from 1,3-bisphosphoglycerate to ADP to yield ATP. Keep in mind that, because of the actions of aldolase and triose phosphate isomerase, two molecules of glyceraldehyde 3-phosphate were formed and hence two molecules of ATP were generated. These ATP molecules make up for the two molecules of ATP consumed in the first stage of glycolysis.

Additional ATP Is Generated with the Formation of Pyruvate

In the remaining steps of glycolysis, 3-phosphoglycerate is converted into pyruvate, and a second molecule of ATP is formed from ADP.

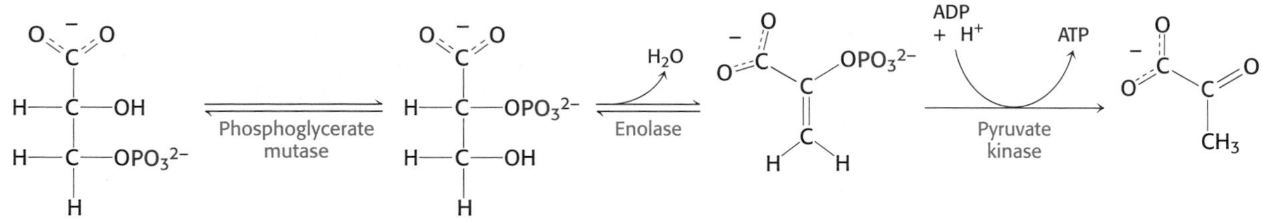

3-Phosphoglycerate   **2-Phosphoglycerate**   **Phosphenolpyruvate**   **Pyruvate**

The first reaction is a rearrangement. The position of the phosphoryl group shifts in the *conversion of 3-phosphoglycerate into 2-phosphoglycerate*, a reaction catalyzed by *phosphoglycerate mutase*. In general, a *mutase* is an enzyme that catalyzes the intramolecular shift of a chemical group, such as a phosphoryl group. The phosphoglycerate mutase reaction has an interesting mechanism: the phosphoryl group is not simply moved from one carbon to

TABLE 16.1 Reactions of glycolysis

Step	Reaction
1	Glucose + ATP → glucose 6-phosphate + ADP + H^+
2	Glucose 6-phosphate ⇌ fructose 6-phosphate
3	Fructose 6-phosphate + ATP → fructose 1,6-bisphosphate + ADP + H^+
4	Fructose 1,6-bisphosphate ⇌ dihydroxyacetone phosphate + glyceraldehyde 3-phosphate
5	Dihydroxyacetone phosphate ⇌ glyceraldehyde 3-phosphate
6	Glyceraldehyde 3-phosphate + P_i + NAD^+ ⇌ 1,3-bisphosphoglycerate + NADH + H^+
7	1,3-Bisphosphoglycerate + ADP ⇌ 3-phosphoglycerate + ATP
8	3-Phosphoglycerate ⇌ 2-phosphoglycerate
9	2-Phosphoglycerate ⇌ phosphoenolpyruvate + H_2O
10	Phosphoenolpyruvate + ADP + H^+ → pyruvate + ATP

Note: ΔG, the actual free-energy change, has been calculated from ΔG°′ and known concentrations of reactants under typical physiological conditions. Glycolysis can proceed only if the ΔG values of all reactions are negative. The small positive ΔG values of three of the above reactions indicate that the concentrations of metabolites in vivo in cells undergoing glycolysis are not precisely known.

another. This enzyme requires catalytic amounts of 2,3-bisphosphoglycerate (2,3-BPG) to maintain an active-site histidine residue in a phosphorylated form. This phosphoryl group is transferred to 3-phosphoglycerate to re-form 2,3-bisphosphoglycerate.

Enz-His-phosphate + 3-phosphoglycerate $\rightleftharpoons$
$$\text{Enz-His} + \text{2,3-bisphosphoglycerate}$$

The mutase then functions as a phosphatase: it converts 2,3-bisphosphoglycerate into 2-phosphoglycerate. The mutase retains the phosphoryl group to regenerate the modified histidine.

Enz-His + 2,3-bisphosphoglycerate $\rightleftharpoons$
$$\text{Enz-His-phosphate} + \text{2-phosphoglycerate}$$

The sum of these reactions yields the mutase reaction:

$$\text{3-Phosphoglycerate} \rightleftharpoons \text{2-phosphoglycerate}$$

In the next reaction, the dehydration of 2-phosphoglycerate introduces a double bond, creating an *enol*. *Enolase* catalyzes this formation of the enol phosphate *phosphoenolpyruvate* (PEP). This dehydration markedly elevates the transfer potential of the phosphoryl group. An *enol phosphate* has a high phosphoryl-transfer potential, whereas the phosphate ester of an ordinary alcohol, such as 2-phosphoglycerate, has a low one. The $\Delta G^{\circ\prime}$ of the hydrolysis of a phosphate ester of an ordinary alcohol is -13 kJ mol^{-1} (-3 kcal mol^{-1}), whereas that of phosphoenolpyruvate is -62 kJ mol^{-1} ($-15 \text{ kcal mol}^{-1}$).

Why does phosphoenolpyruvate have such a high phosphoryl-transfer potential? The phosphoryl group traps the molecule in its unstable enol form. When the phosphoryl group has been donated to ATP, the enol undergoes a conversion into the more stable ketone—namely, pyruvate.

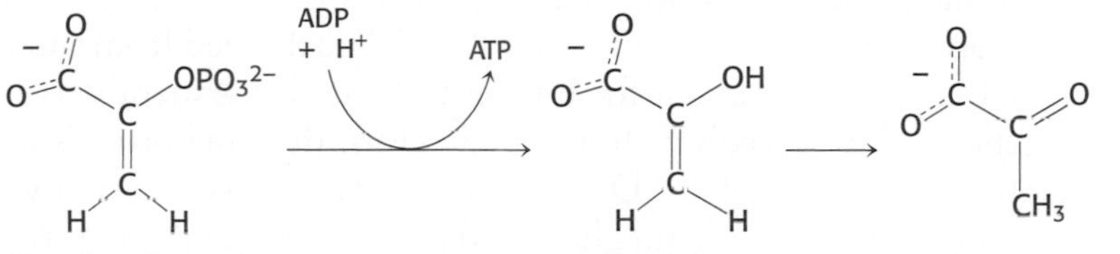

Phosphenolpyruvate   **Pyruvate**
(enol form)   **Pyruvate**

Enzyme	Reaction type	$\Delta G^{\circ\prime}$ in kJ mol^{-1} (kcal mol^{-1})	ΔG in kJ mol^{-1} (kcal mol^{-1})
Hexokinase	Phosphoryl transfer	−16.7 (−4.0)	−33.5 (−8.0)
Phosphoglucose isomerase	Isomerization	+1.7 (+0.4)	−2.5 (−0.6)
Phosphofructokinase	Phosphoryl transfer	−14.2 (−3.4)	−22.2 (−5.3)
Aldolase	Aldol cleavage	+23.8 (+5.7)	−1.3 (−0.3)
Triose phosphate isomerase	Isomerization	+7.5 (+1.8)	+2.5 (+0.6)
Glyceraldehyde 3-phosphate dehydrogenase	Phosphorylation coupled to oxidation	+6.3 (+1.5)	−1.7 (−0.4)
Phosphoglycerate kinase	Phosphoryl transfer	−18.8 (−4.5)	+1.3 (+0.3)
Phosphoglycerate mutase	Phosphoryl shift	+4.6 (+1.1)	+0.8 (+0.2)
Enolase	Dehydration	+1.7 (+0.4)	−3.3 (−0.8)
Pyruvate kinase	Phosphoryl transfer	−31.4 (−7.5)	−16.7 (−4.0)

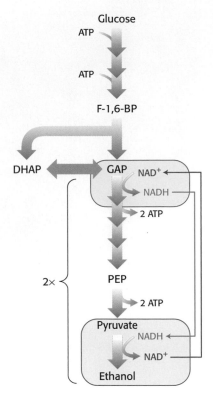

Location of redox-balance steps. The
generation and consumption of NADH,
located within the glycolytic pathway.

Thus, *the high phosphoryl-transfer potential of phosphoenolpyruvate arises primarily from the large driving force of the subsequent enol–ketone conversion.* Hence, pyruvate is formed, and ATP is generated concomitantly. The virtually irreversible transfer of a phosphoryl group from phosphoenolpyruvate to ADP is catalyzed by *pyruvate kinase*. Because the molecules of ATP used in forming fructose 1,6-bisphosphate have already been regenerated, the two molecules of ATP generated from phosphoenolpyruvate are "profit."

Two ATP Molecules Are Formed in the Conversion of Glucose into Pyruvate

The net reaction in the transformation of glucose into pyruvate is

$$\text{Glucose} + 2\,P_i + 2\,ADP + 2\,NAD^+ \longrightarrow$$
$$2\,\text{pyruvate} + 2\,ATP + 2\,NADH + 2\,H^+ + 2\,H_2O$$

Thus, *two molecules of ATP are generated in the conversion of glucose into two molecules of pyruvate.* The reactions of glycolysis are summarized in Table 16.1.

Note that the energy released in the anaerobic conversion of glucose into two molecules of pyruvate is about -96 kJ mol^{-1} (-23 kcal mol^{-1}). We shall see in Chapters 17 and 18 that much more energy can be released from glucose in the presence of oxygen.

NAD$^+$ Is Regenerated from the Metabolism of Pyruvate

The conversion of glucose into two molecules of pyruvate has resulted in the net synthesis of ATP. However, an energy-converting pathway that stops at pyruvate will not proceed for long, because redox balance has not been maintained. As we have seen, the activity of glyceraldehyde 3-phosphate dehydrogenase, in addition to generating a compound with high phosphoryl-transfer potential, of necessity leads to the reduction of NAD$^+$ to NADH. In the cell, there are limited amounts of NAD$^+$, which is derived from the vitamin niacin, a dietary requirement for human beings. Consequently, NAD$^+$ must be regenerated for glycolysis to proceed. Thus, the final process in the pathway is the regeneration of NAD$^+$ through the metabolism of pyruvate.

The sequence of reactions from glucose to pyruvate is similar in most organisms and most types of cells. In contrast, the fate of pyruvate is variable. Three reactions of pyruvate are of primary importance: conversion into ethanol, lactate, or carbon dioxide (Figure 16.9). The first two reactions are fermentations that take place in the absence of oxygen. In the presence of oxygen, the most common situation in multicellular organisms and in many unicellular ones, pyruvate is metabolized to carbon dioxide and water through the citric acid cycle and the electron-transport chain. We now take a closer look at these three possible fates of pyruvate.

1. *Ethanol* is formed from pyruvate in yeast and several other microorganisms. The first step is the decarboxylation of pyruvate. This reaction is catalyzed by *pyruvate decarboxylase*, which requires the coenzyme thiamine pyrophosphate. This coenzyme, derived from the vitamin thiamine (B$_1$), also participates in reactions catalyzed by other enzymes (p. 478). The second step is the reduction of acetaldehyde to ethanol by NADH, in a reaction catalyzed by *alcohol dehydrogenase*. This process regenerates NAD$^+$.

Figure 16.9 Diverse fates of pyruvate. Ethanol and lactate can be formed by reactions that include NADH. Alternatively, a two-carbon unit from pyruvate can be coupled to coenzyme A (see p. 420) to form acetyl CoA.

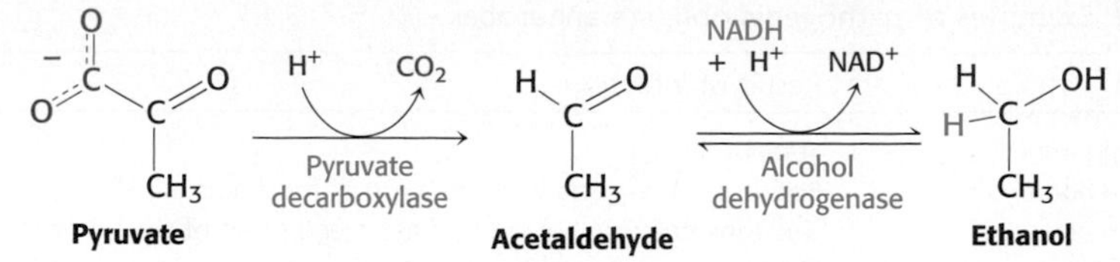

The top of the page shows reaction schemes for Pyruvate → Acetaldehyde → Ethanol.

Pyruvate — (via Pyruvate decarboxylase, H⁺, CO₂) → Acetaldehyde — (via Alcohol dehydrogenase, NADH + H⁺, NAD⁺) → Ethanol

Pyruvate **Acetaldehyde** **Ethanol**

The active site of alcohol dehydrogenase contains a zinc ion that is coordinated to the sulfur atoms of two cysteine residues and a nitrogen atom of histidine (Figure 16.10). This zinc ion polarizes the carbonyl group of the substrate to favor the transfer of a hydride from NADH.

The conversion of glucose into ethanol is an example of *alcoholic fermentation*. The net result of this anaerobic process is

$$\text{Glucose} + 2\,P_i + 2\,\text{ADP} + 2\,H^+ \longrightarrow$$
$$2\text{ ethanol} + 2\,CO_2 + 2\,\text{ATP} + 2\,H_2O$$

Note that NAD⁺ and NADH do not appear in this equation, even though they are crucial for the overall process. NADH generated by the oxidation of glyceraldehyde 3-phosphate is consumed in the reduction of acetaldehyde to ethanol. Thus, *there is no net oxidation–reduction in the conversion of glucose into ethanol* (Figure 16.11). The ethanol formed in alcoholic fermentation provides a key ingredient for brewing and winemaking.

2. *Lactate* is formed from pyruvate in a variety of microorganisms in a process called *lactic acid fermentation*. The reaction also takes place in the cells of higher organisms when the amount of oxygen is limiting, as in muscle cells during intense activity. The reduction of pyruvate by NADH to form lactate is catalyzed by *lactate dehydrogenase*.

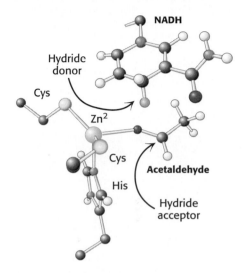

Figure 16.10 Active site of alcohol dehydrogenase. The active site contains a zinc ion bound to two cysteine residues and one histidine residue. *Notice* that the zinc ion binds the acetaldehyde substrate through its oxygen atom, polarizing the substrate so that it more easily accepts a hydride from NADH. Only the nicotinamide ring of NADH is shown.

Pyruvate — (via Lactate dehydrogenase, NADH + H⁺ ⇌ NAD⁺) → Lactate

Pyruvate **Lactate**

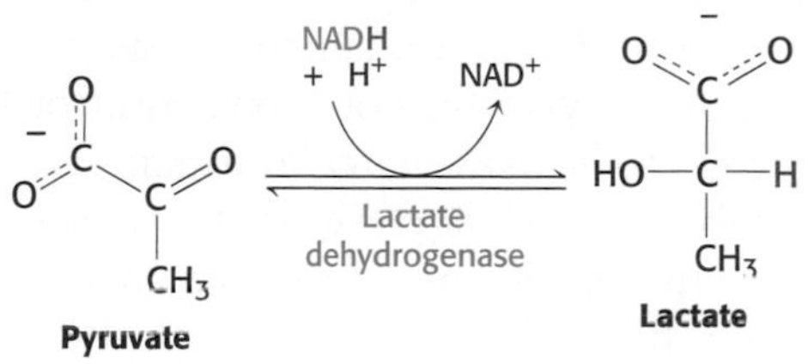

The overall reaction in the conversion of glucose into lactate is

$$\text{Glucose} + 2\,P_i + 2\,\text{ADP} \longrightarrow 2\text{ lactate} + 2\,\text{ATP} + 2\,H_2O$$

As in alcoholic fermentation, there is no net oxidation–reduction. The NADH formed in the oxidation of glyceraldehyde 3-phosphate is consumed in the reduction of pyruvate. *The regeneration of NAD⁺ in the reduction of*

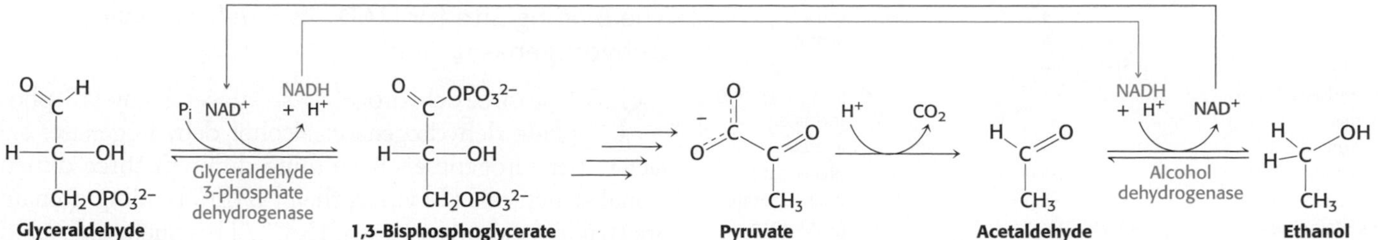

Figure 16.11 Maintaining redox balance. The NADH produced by the glyceraldehyde 3-phosphate dehydrogenase reaction must be reoxidized to NAD⁺ for the glycolytic pathway to continue. In alcoholic fermentation, alcohol dehydrogenase oxidizes NADH and generates ethanol. In lactic acid fermentation (not shown), lactate dehydrogenase oxidizes NADH while generating lactic acid.

Glyceraldehyde 3-phosphate — (Glyceraldehyde 3-phosphate dehydrogenase, P_i, NAD⁺ ⇌ NADH + H⁺) → 1,3-Bisphosphoglycerate (1,3-BPG) → ⟶ Pyruvate — (H⁺, CO₂) → Acetaldehyde — (Alcohol dehydrogenase, NADH + H⁺ ⇌ NAD⁺) → Ethanol

Glyceraldehyde 3-phosphate **1,3-Bisphosphoglycerate (1,3-BPG)** **Pyruvate** **Acetaldehyde** **Ethanol**

TABLE 16.2 Examples of pathogenic obligate anaerobes

Bacterium	Result of infection
Clostridium tetani	Tetanus (lockjaw)
Clostridium botulinum	Botulism (an especially severe type of food poisoning)
Clostridium perfringens	Gas gangrene (gas is produced as an end point of the fermentation, distorting and destroying the tissue)
Bartonella hensela	Cat scratch fever (flu-like symptoms)
Bacteroides fragilis	Abdominal, pelvic, pulmonary, and blood infections

pyruvate to lactate or ethanol sustains the continued process of glycolysis under anaerobic conditions.

3.   Only a fraction of the energy of glucose is released in its anaerobic conversion into ethanol or lactate. Much more energy can be extracted aerobically by means of the citric acid cycle and the electron-transport chain. The entry point to this oxidative pathway is *acetyl coenzyme A* (acetyl CoA), which is formed inside mitochondria by the oxidative decarboxylation of pyruvate.

$$\text{Pyruvate} + \text{NAD}^+ + \text{CoA} \longrightarrow \text{acetyl CoA} + \text{CO}_2 + \text{NADH}$$

This reaction, which is catalyzed by the pyruvate dehydrogenase complex, will be considered in detail in Chapter 17. The NAD^+ required for this reaction and for the oxidation of glyceraldehyde 3-phosphate is regenerated when NADH ultimately transfers its electrons to O_2 through the electron-transport chain in mitochondria.

Fermentations Provide Usable Energy in the Absence of Oxygen

Fermentations yield only a fraction of the energy available from the complete combustion of glucose. Why is a relatively inefficient metabolic pathway so extensively used? The fundamental reason is that oxygen is not required. The ability to survive without oxygen affords a host of living accommodations such as soils, deep water, and skin pores. Some organisms, called *obligate anaerobes*, cannot survive in the presence of O_2, a highly reactive compound. The bacterium *Clostridium perfringens*, the cause of gangrene, is an example of an obligate anaerobe. Other pathogenic obligate anaerobes are listed in Table 16.2. Skeletal muscles in most animals can function anaerobically for short periods. For example, when animals perform bursts of intense exercise, their ATP needs rise faster than the ability of the body to provide oxygen to the muscle. The muscle functions anaerobically until fatigue sets in, which is caused, in part, by lactate buildup.

Although we have considered only lactic acid and alcoholic fermentation, microorganisms are capable of generating a wide array of molecules as end points to fermentation (Table 16.3). Indeed, many food products, including sour cream, yogurt, various cheeses, beer, wine, and sauerkraut, result from fermentation.

TABLE 16.3 Starting and ending points of various fermentations

Glucose	→	lactate
Lactate	→	acetate
Glucose	→	ethanol
Ethanol	→	acetate
Arginine	→	carbon dioxide
Pyrimidines	→	carbon dioxide
Purines	→	formate
Ethylene glycol	→	acetate
Threonine	→	propionate
Leucine	→	2-alkylacetate
Phenylalanine	→	propionate

Note: The products of some fermentations are the substrates for others.

The Binding Site for NAD⁺ Is Similar in Many Dehydrogenases

The three dehydrogenases—glyceraldehyde 3-phosphate dehydrogenase, alcohol dehydrogenase, and lactate dehydrogenase—have quite different three-dimensional structures. However, their NAD^+ binding domains are strikingly similar (Figure 16.12). This nucleotide-binding region is made up of four α helices and a sheet of six

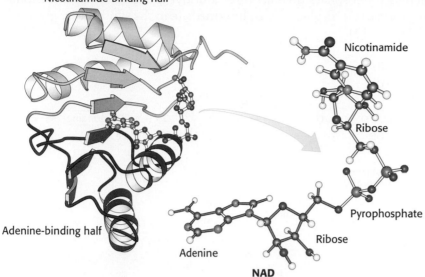

Nicotinamide-binding half

Nicotinamide

Ribose

Pyrophosphate

Ribose

Adenine-binding half

Adenine

NAD

Figure 16.12 NAD⁺-binding region
in dehydrogenases. *Notice* that the
nicotinamide-binding half (yellow) is
structurally similar to the adenine-binding
half (red). The two halves together form a
structural motif called a Rossmann fold.
The NAD⁺ molecule binds in an extended
conformation. [Drawn from 3LDH.pdb.]

parallel β strands. Moreover, in all cases, the bound NAD^+ displays nearly the same conformation. This common structural domain was one of the first recurring structural domains to be discovered. It is often called a *Rossmann fold* after Michael Rossmann, who first recognized it. This fold likely represents a primordial dinucleotide-binding domain that recurs in the dehydrogenases of glycolysis and other enzymes because of their descent from a common ancestor.

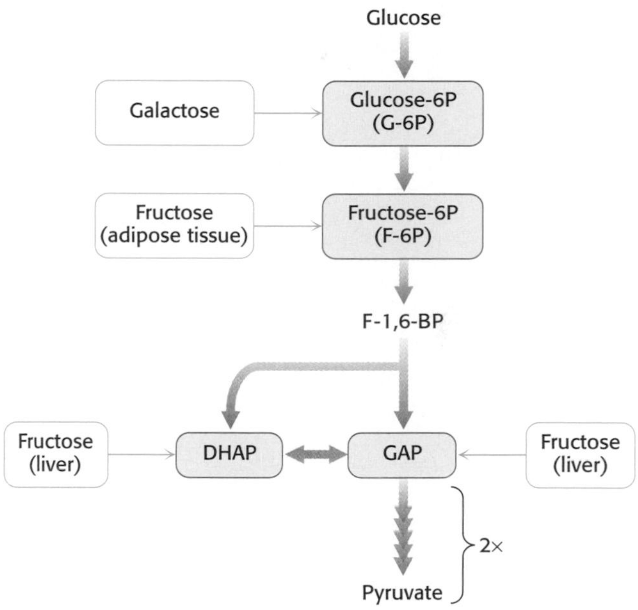

Fructose and Galactose Are Converted into Glycolytic Intermediates

Although glucose is the most widely used monosaccharide, others also are important fuels. Let us consider how two abundant sugars—fructose and galactose—can be funneled into the glycolytic pathway (Figure 16.13). There are no catabolic pathways for metabolizing fructose or galactose, and so the strategy is to convert these sugars into a metabolite of glucose.

Fructose can take one of two pathways to enter the glycolytic pathway. Much of the ingested fructose is metabolized by the liver, using the *fructose 1-phosphate pathway* (Figure 16.14). The first step is the phosphorylation of *fructose* to *fructose 1-phosphate* by *fructokinase*. Fructose 1-phosphate is then split into *glyceraldehyde* and *dihydroxyacetone phosphate,* an intermediate in glycolysis. This aldol cleavage is catalyzed by a specific *fructose 1-phosphate aldolase*. Glyceraldehyde is then phosphorylated to *glyceraldehyde 3-phosphate*, a glycolytic intermediate, by *triose kinase*. In other tissues, *fructose can be phosphorylated to fructose 6-phosphate by hexokinase.*

Galactose is converted into *glucose 6-phosphate* in four steps. The first reaction in the *galactose–glucose interconversion pathway* is the phosphorylation of galactose to galactose 1-phosphate by *galactokinase*.

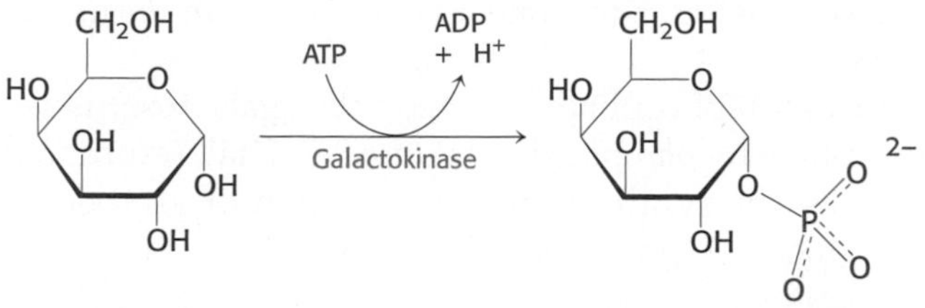

Glucose

Galactose → Glucose-6P
(G-6P)

Fructose
(adipose tissue) → Fructose-6P
(F-6P)

F-1,6-BP

Fructose
(liver) → DHAP ⇌ GAP ← Fructose
(liver)

2×

Pyruvate

Figure 16.13 Entry points in glycolysis for
galactose and fructose.

CH₂OH — O — OH — HO — OH — OH

Galactose

ATP → ADP + H⁺

Galactokinase

CH₂OH — O — OH — HO — OH — O — P — O — O — 2−

Galactose 1-phosphate

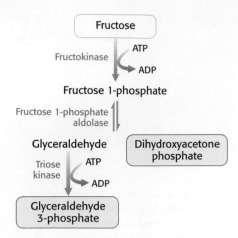

Figure 16.14 Fructose metabolism. Fructose enters the glycolytic pathway in the liver through the fructose 1-phosphate pathway.

Galactose 1-phosphate then acquires a uridyl group from uridine diphosphate glucose (UDP-glucose), an intermediate in the synthesis of glycosidic linkages (p. 314).

Galactose 1-phosphate

UDP-glucose

Galactose 1-phosphate uridyl transferase

UDP-galactose

Glucose 1-phosphate

UDP-galactose 4-epimerase

UDP-glucose

The products of this reaction, which is catalyzed by *galactose 1-phosphate uridyl transferase,* are UDP-galactose and glucose 1-phosphate. The galactose moiety of UDP-galactose is then epimerized to glucose. The configuration of the hydroxyl group at carbon 4 is inverted by *UDP-galactose 4-epimerase.*

The sum of the reactions catalyzed by galactokinase, the transferase, and the epimerase is

$$\text{Galactose} + \text{ATP} \longrightarrow \text{glucose 1-phosphate} + \text{ADP} + \text{H}^+$$

Note that UDP-glucose is not consumed in the conversion of galactose into glucose, because it is regenerated from UDP-galactose by the epimerase. This reaction is reversible, and the product of the reverse direction also is important. *The conversion of UDP-glucose into UDP-galactose is essential for the synthesis of galactosyl residues in complex polysaccharides and glycoproteins if the amount of galactose in the diet is inadequate to meet these needs.*

Finally, glucose 1-phosphate, formed from galactose, is isomerized to glucose 6-phosphate by *phosphoglucomutase.* We shall return to this reaction when we consider the synthesis and degradation of glycogen, which proceeds through glucose 1-phosphate, in Chapter 21.

Many Adults Are Intolerant of Milk Because They Are Deficient in Lactase

Many adults are unable to metabolize the milk sugar lactose and experience gastrointestinal disturbances if they drink milk. *Lactose intolerance,* or hypolactasia, is most commonly caused by a deficiency of the enzyme lactase, which cleaves lactose into glucose and galactose.

Lactose + H_2O ⇌ (Lactase) **Galactose** + **Glucose**

"Deficiency" is not quite the appropriate term, because a decrease in lactase is normal in the course of development in all mammals. As children are weaned and milk becomes less prominent in their diets, lactase activity normally declines to about 5 to 10% of the level at birth. This decrease is not as pronounced with some groups of people, most notably Northern Europeans, and people from these groups can continue to ingest milk without gastrointestinal difficulties. With the appearance of milk-producing domesticated animals, an adult with active lactase would hypothetically have a selective advantage in being able to consume calories from the readily available milk.

What happens to the lactose in the intestine of a lactase-deficient person? The lactose is a good energy source for microorganisms in the colon, and they ferment it to lactic acid while also generating methane (CH_4) and hydrogen gas (H_2). The gas produced creates the uncomfortable feeling of gut distension and the annoying problem of flatulence. The lactate produced by the microorganisms is osmotically active and draws water into the intestine, as does any undigested lactose, resulting in diarrhea. If severe enough, the gas and diarrhea hinder the absorption of other nutrients such as fats and proteins. The simplest treatment is to avoid the consumption of products containing much lactose. Alternatively, the enzyme lactase can be ingested with milk products.

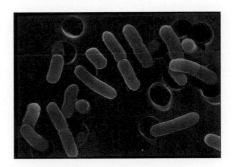

Scanning electron micrograph of *Lactobacillus.* The anaerobic bacterium *Lactobacillus* is shown here (artificially colored) at a magnification of 22,245×. As suggested by its name, this genus of bacteria ferments glucose into lactic acid and is widely used in the food industry. *Lactobacillus* is also a component of the normal human bacterial flora of the urogenital tract where, because of its ability to generate an acidic environment, it prevents the growth of harmful bacteria. [Dr. Dennis Kunkel/PhotoTake.]

Galactose Is Highly Toxic If the Transferase Is Missing

Less common than lactose intolerance are disorders that interfere with the metabolism of galactose. The disruption of galactose metabolism is referred to as *galactosemia.* The most common form, called classic galactosemia, is an inherited deficiency in galactose 1-phosphate uridyl transferase activity. Afflicted infants fail to thrive. They vomit or have diarrhea after consuming milk, and enlargement of the liver and jaundice are common, sometimes progressing to cirrhosis. Cataracts will form, and lethargy and retarded mental development also are common. The blood-galactose level is markedly elevated, and galactose is found in the urine. The absence of the transferase in red blood cells is a definitive diagnostic criterion.

The most common treatment is to remove galactose (and lactose) from the diet. An enigma of galactosemia is that, although elimination of galactose from the diet prevents liver disease and cataract development, the majority of patients still suffer from central nervous system malfunction, most commonly a delayed acquisition of language skills. Female patients also display ovarian failure.

Cataract formation is better understood. A cataract is the clouding of the normally clear lens of the eye. If the transferase is not active in the lens of the eye, the presence of aldose reductase causes the accumulating galactose to be reduced to galactitol.

Galactitol is osmotically active, and water will diffuse into the lens, instigating the formation of cataracts. In fact, there is a high incidence of cataract formation with age in populations that consume substantial amounts of milk into adulthood.

16.2 The Glycolytic Pathway Is Tightly Controlled

The glycolytic pathway has a dual role: it degrades glucose to generate ATP and it provides building blocks for synthetic reactions, such as the formation of fatty acids. The rate of conversion of glucose into pyruvate is regulated to meet these two major cellular needs. *In metabolic pathways, enzymes catalyzing essentially irreversible reactions are potential sites of control.* In glycolysis, the reactions catalyzed by hexokinase, phosphofructokinase, and pyruvate kinase are virtually irreversible; hence, these enzymes would be expected to have regulatory as well as catalytic roles. In fact, each of them serves as a control site. These enzymes become more active or less so in response to the reversible binding of allosteric effectors or covalent modification. In addition, the amounts of these important enzymes are varied by the regulation of transcription to meet changing metabolic needs. The time required for reversible allosteric control, regulation by phosphorylation, and transcriptional control is measured typically in milliseconds, seconds, and hours, respectively. We will consider the control of glycolysis in two different tissues—skeletal muscle and liver.

Glycolysis in Muscle Is Regulated to Meet the Need for ATP

Glycolysis in skeletal muscle provides ATP primarily to power contraction. Consequently, the primary control of muscle glycolysis is the energy charge of the cell—the ratio of ATP to AMP. Let us examine how each of the key regulatory enzymes responds to changes in the amounts of ATP and AMP present in the cell.

Phosphofructokinase. *Phosphofructokinase is the most important control site in the mammalian glycolytic pathway* (Figure 16.15). High levels of ATP allosterically inhibit the enzyme (a 340-kd tetramer). ATP binds to a specific regulatory site that is distinct from the catalytic site. The binding of ATP lowers the enzyme's affinity for fructose 6-phosphate. Thus, a high concentration of ATP converts the hyperbolic binding curve of fructose

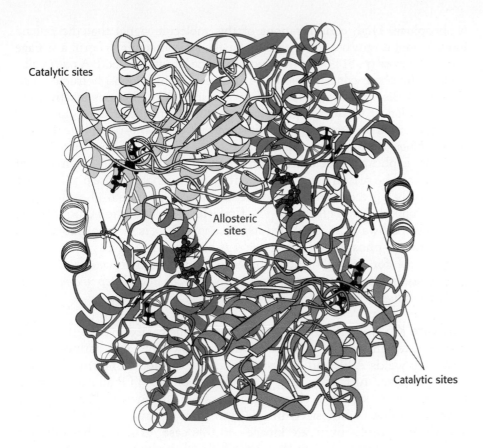

Catalytic sites

Allosteric sites

Catalytic sites

Figure 16.15 Structure of phosphofructokinase. The structure of phosphofructokinase from *E. coli* comprises a tetramer of four identical subunits. *Notice* the separation of the catalytic and allosteric sites. Each subunit of the human liver enzyme consists of two domains that are similar to the *E. coli* enzyme. [Drawn from 1PFK.pdb.]

6-phosphate into a sigmoidal one (Figure 16.16). AMP reverses the inhibitory action of ATP, and so *the activity of the enzyme increases when the ATP/AMP ratio is lowered*. In other words, *glycolysis is stimulated as the energy charge falls*. A decrease in pH also inhibits phosphofructokinase activity by augmenting the inhibitory effect of ATP. The pH might fall when muscle is functioning anaerobically, producing excessive quantities of lactic acid. The inhibitory effect protects the muscle from damage that would result from the accumulation of too much acid.

Why is AMP and not ADP the positive regulator of phosphofructokinase? When ATP is being utilized rapidly, the enzyme *adenylate kinase* (Section 9.4) can form ATP from ADP by the following reaction:

$$\text{ADP} + \text{ADP} \rightleftharpoons \text{ATP} + \text{AMP}$$

Thus, some ATP is salvaged from ADP, and AMP becomes the signal for the low-energy state. Moreover, the use of AMP as an allosteric regulator provides an especially sensitive control. We can understand why by considering, first, that the total adenylate pool ([ATP], [ADP], [AMP]) in a cell is constant over the short term and, second, that the concentration of ATP is greater than that of ADP and the concentration of ADP is, in turn, greater than that of AMP. Consequently, small-percentage changes in [ATP] result in larger-percentage changes in the concentrations of the other adenylate nucleotides. This magnification of small changes in [ATP] to larger changes in [AMP] leads to tighter control by increasing the range of sensitivity of phosphofructokinase.

Hexokinase. Phosphofructokinase is the most prominent regulatory enzyme in glycolysis, but it is not the only one. Hexokinase, the enzyme catalyzing the first step of glycolysis, is inhibited by its product, glucose

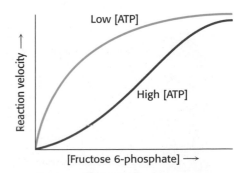

Low [ATP]

High [ATP]

Reaction velocity →

[Fructose 6-phosphate] →

Figure 16.16 Allosteric regulation of phosphofructokinase. A high level of ATP inhibits the enzyme by decreasing its affinity for fructose 6-phosphate. AMP diminishes and citrate enhances the inhibitory effect of ATP.

6-phosphate. High concentrations of this molecule signal that the cell no longer requires glucose for energy or for the synthesis of glycogen, a storage form of glucose (p. 311), and the glucose will be left in the blood. A rise in glucose 6-phosphate concentration is a means by which phosphofructokinase communicates with hexokinase. When phosphofructokinase is inactive, the concentration of fructose 6-phosphate rises. In turn, the level of glucose 6-phosphate rises because it is in equilibrium with fructose 6-phosphate. Hence, *the inhibition of phosphofructokinase leads to the inhibition of hexokinase.*

Why is phosphofructokinase rather than hexokinase the pacemaker of glycolysis? The reason becomes evident on noting that glucose 6-phosphate is not solely a glycolytic intermediate. In muscle, glucose 6-phosphate can also be converted into glycogen. The first irreversible reaction unique to the glycolytic pathway, the *committed step* (Section 10.1), is the phosphorylation of fructose 6-phosphate to fructose 1,6-bisphosphate. Thus, it is highly appropriate for phosphofructokinase to be the primary control site in glycolysis. In general, *the enzyme catalyzing the committed step in a metabolic sequence is the most important control element in the pathway.*

Pyruvate Kinase. Pyruvate kinase, the enzyme catalyzing the third irreversible step in glycolysis, controls the outflow from this pathway. This final step yields ATP and pyruvate, a central metabolic intermediate that can be oxidized further or used as a building block. ATP allosterically inhibits pyruvate kinase to slow glycolysis when the energy charge is high. Finally, alanine (synthesized in one step from pyruvate, p. 686) also allosterically inhibits pyruvate kinase—in this case, to signal that building blocks are abundant. When the pace of glycolysis increases, fructose 1,6-bisphosphate, the product of the preceding irreversible step in glycolysis, activates the kinase to enable it to keep pace with the oncoming high flux of intermediates. A summary of the regulation of glycolysis in resting and active muscle is shown in Figure 16.17.

Figure 16.17 Regulation of glycolysis in muscle. At rest (left), glycolysis is not very active (thin arrows). The high concentration of ATP inhibits phosphofructokinase (PFK), pyruvate kinase, and hexokinase. Glucose 6-phosphate is converted into glycogen (Chapter 21). During exercise (right), the decrease in the ATP/AMP ratio resulting from muscle contraction activates phosphofructokinase and hence glycolysis. The flux down the pathway is increased, as represented by the thick arrows.

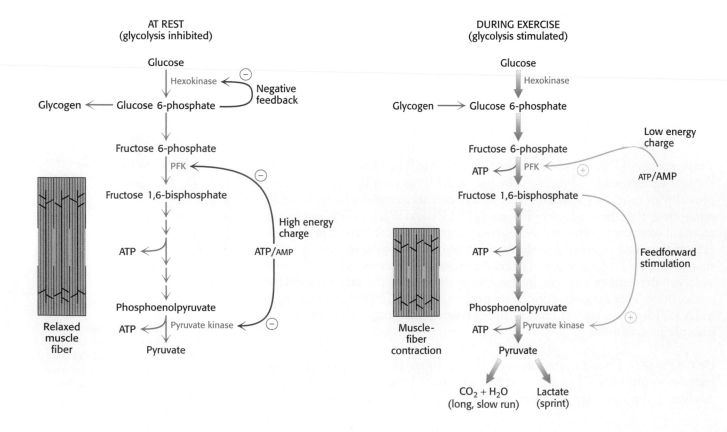

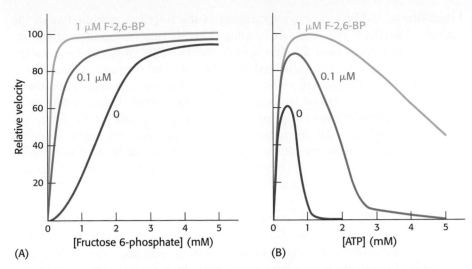

(A)

(B)

Figure 16.18 Activation of phosphofructokinase by fructose 2, 6-bisphosphate. (A) The sigmoidal dependence of velocity on substrate concentration becomes hyperbolic in the presence of 1 μM fructose 2,6-bisphosphate. (B) ATP, acting as a substrate, initially stimulates the reaction. As the concentration of ATP increases, it acts as an allosteric inhibitor. The inhibitory effect of ATP is reversed by fructose 2,6-bisphosphate. [After E. Van Schaftingen, M. F. Jett, L. Hue, and H. G. Hers. *Proc. Natl. Acad. Sci.* U.S.A. 78(1981):3483–3486.]

The Regulation of Glycolysis in the Liver Reflects the Biochemical Versatility of the Liver

The liver has more-diverse biochemical functions than muscle. Significantly, the liver maintains blood-glucose levels: it stores glucose as glycogen when glucose is plentiful, and it releases glucose when supplies are low. It also uses glucose to generate reducing power for biosynthesis (p. 577) as well as to synthesize a host of biochemicals. So, although the liver has many of the regulatory features of muscle glycolysis, the regulation of glycolysis in the liver is more complex.

Phosphofructokinase.   Regulation with respect to ATP is the same in the liver as in muscle. Low pH is not a metabolic signal for the liver enzyme, because lactate is not normally produced in the liver. Indeed, as we will see, lactate is converted into glucose in the liver.

Glycolysis also furnishes carbon skeletons for biosyntheses, and so a signal indicating whether building blocks are abundant or scarce should also regulate phosphofructokinase. In the liver, *phosphofructokinase is inhibited by citrate*, an early intermediate in the citric acid cycle (p. 482). A high level of citrate in the cytoplasm means that biosynthetic precursors are abundant, and so there is no need to degrade additional glucose for this purpose. Citrate inhibits phosphofructokinase by enhancing the inhibitory effect of ATP.

One means by which glycolysis in the liver responds to changes in blood glucose is through the signal molecule *fructose 2,6-bisphosphate* (F-2,6-BP), a potent activator of phosphofructokinase (Figure 16.18). In the liver, the concentration of fructose 6-phosphate rises when blood-glucose concentration is high, and the abundance of fructose 6-phosphate accelerates the synthesis of F-2,6-BP (Figure 16.19). Hence, *an abundance of fructose 6-phosphate leads to a higher concentration of F-2,6-BP*. The binding of fructose 2,6-bisphosphate increases the affinity of phosphofructokinase for fructose 6-phosphate and diminishes the inhibitory effect of ATP. Glycolysis is thus accelerated when glucose is abundant. Such a process is called *feedforward stimulation*. We will turn to the synthesis and degradation of this important regulatory molecule after we have considered gluconeogenesis.

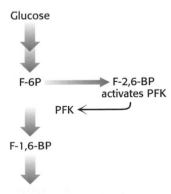

Figure 16.19 Regulation of phosphofructokinase by fructose 2,6-bisphosphate. In high concentrations, fructose 6-phosphate (F-6P) activates the enzyme phosphofructokinase (PFK) through an intermediary, fructose 2,6-bisphosphate (F-2,6-BP).

$^{2-}O_3POH_2C$ $OPO_3{}^{2-}$

**Fructose 2,6-bisphosphate
(F-2,6-BP)**

Hexokinase. The hexokinase reaction in the liver is controlled as in the muscle. However, the liver, in keeping with its role as monitor of blood-glucose levels, possesses another specialized isozyme of hexokinase, called *glucokinase,* that is not inhibited by glucose 6-phosphate. Glucokinase phosphorylates glucose only when glucose is abundant because the affinity of glucokinase for glucose is about 50-fold lower than that of hexokinase. The role of glucokinase is to provide glucose 6-phosphate for the synthesis of glycogen and for the formation of fatty acids (Section 22.1). The low affinity of glucokinase for glucose in the liver gives the brain and muscles first call on glucose when its supply is limited, and it ensures that glucose will not be wasted when it is abundant.

Pyruvate Kinase. Several isozymic forms of pyruvate kinase (a tetramer of 57-kd subunits) encoded by different genes are present in mammals: the L type predominates in liver, and the M type in muscle and brain. The L and M forms of pyruvate kinase have many properties in common. Indeed, the liver enzyme behaves much like the muscle enzyme with regard to allosteric regulation. However, the isozymic forms differ in their susceptibility to covalent modification. The catalytic properties of the L form—but not of the M form—are also controlled by reversible phosphorylation (Figure 16.20). When the blood-glucose level is low, the glucagon-triggered cyclic AMP cascade (p. 466) leads to the phosphorylation of pyruvate kinase, which diminishes its activity. This hormone-triggered phosphorylation prevents the liver from consuming glucose when it is more urgently needed by brain and muscle (Section 27.3). We see here a clear-cut example of how isoenzymes contribute to the metabolic diversity of different organs. We will return to the control of glycolysis after considering gluconeogenesis.

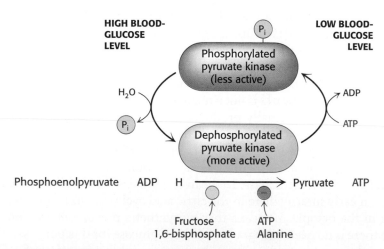

Figure 16.20 Control of the catalytic activity of pyruvate kinase. Pyruvate kinase is regulated by allosteric effectors and covalent modification.

A Family of Transporters Enables Glucose to Enter and Leave Animal Cells

Several glucose transporters mediate the thermodynamically downhill movement of glucose across the plasma membranes of animal cells. Each member of this protein family, named GLUT1 to GLUT5, consists of a single polypeptide chain about 500 residues long (Table 16.4). Each glucose transporter has a 12-transmembrane-helix structure similar to that of lactose permease (Section 13.3).

The members of this family have distinctive roles:

TABLE 16.4 Family of glucose transporters

Name	Tissue location	K_M	Comments
GLUT1	All mammalian tissues	1 mM	Basal glucose uptake
GLUT2	Liver and pancreatic β cells	15–20 mM	In the pancreas, plays a role in the regulation of insulin In the liver, removes excess glucose from the blood
GLUT3	All mammalian tissues	1 mM	Basal glucose uptake
GLUT4	Muscle and fat cells	5 mM	Amount in muscle plasma membrane increases with endurance training
GLUT5	Small intestine	—	Primarily a fructose transporter

1.   GLUT1 and GLUT3, present in nearly all mammalian cells, are responsible for basal glucose uptake. Their K_M value for glucose is about 1 mM, significantly less than the normal serum-glucose level, which typically ranges from 4 mM to 8 mM. Hence, GLUT1 and GLUT3 continually transport glucose into cells at an essentially constant rate.

2.   GLUT2, present in liver and pancreatic β cells, is distinctive in having a very high K_M value for glucose (15–20 mM). Hence, glucose enters these tissues at a biologically significant rate only when there is much glucose in the blood. The pancreas can thereby sense the glucose level and accordingly adjust the rate of insulin secretion. Insulin signals the need to remove glucose from the blood for storage as glycogen or conversion into fat (Section 27.3). The high K_M value of GLUT2 also ensures that glucose rapidly enters liver cells only in times of plenty.

3.   GLUT4, which has a K_M value of 5 mM, transports glucose into muscle and fat cells. The number of GLUT4 transporters in the plasma membrane increases rapidly in the presence of insulin, which signals the fed state. Hence, insulin promotes the uptake of glucose by muscle and fat. Endurance exercise training increases the amount of this transporter present in muscle membranes.

4.   GLUT5, present in the small intestine, functions primarily as a fructose transporter.

This family of transporters vividly illustrates how isoforms of a single protein can significantly shape the metabolic character of cells and contribute to their diversity and functional specialization. The transporters are members of a superfamily of transporters called the major facilitator (MF) superfamily. Members of this family transport sugars in organisms as diverse as *E. coli, Trypanosoma brucei* (a parasitic protozoan that causes sleeping sickness), and human beings. An elegant solution to the problem of fuel transport evolved early and has been tailored to meet the needs of different organisms and even different tissues.

Cancer and Exercise Training Affect Glycolysis in a Similar Fashion

Tumors have been known for decades to display enhanced rates of glucose uptake and glycolysis. We now know that these enhanced rates of glucose processing are not fundamental to the development of cancer, but we can ask what selective advantage they might confer on cancer cells.

Cancer cells grow more rapidly than the blood vessels to nourish them; thus, as solid tumors grow, they are unable to obtain oxygen efficiently. In other words, they begin to experience *hypoxia*, a deficiency of oxygen. Under this condition, glycolysis leading to lactic acid fermentation becomes

TABLE 16.5 Proteins in glucose metabolism encoded by genes regulated by
hypoxia-inducible factor

GLUT1
GLUT3
Hexokinase
Phosphofructokinase
Aldolase
Glyceraldehyde 3-phosphate dehydrogenase
Phosphoglycerate kinase
Enolase
Pyruvate kinase

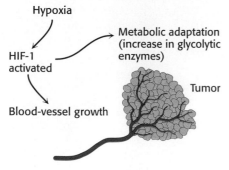

Figure 16.21 Alteration of gene expression in tumors owing to hypoxia. The hypoxic conditions inside a tumor mass lead to the activation of the hypoxia-inducible transcription factor (HIF-1), which induces metabolic adaptation (an increase in glycolytic enzymes) and activates angiogenic factors that stimulate the growth of new blood vessels. [After C. V. Dang and G. L. Semenza. *Trends Biochem. Sci.* 24(1999):68–72.]

the primary source of ATP. Glycolysis is made more efficient in hypoxic tumors by the action of a transcription factor, *hypoxia-inducible transcription factor* (HIF-1). In the absence of oxygen, HIF-1 increases the expression of most glycolytic enzymes and the glucose transporters GLUT1 and GLUT3 (Table 16.5). In fact, tumors with a high glucose uptake are particularly aggressive, and the cancer is likely to have a poor prognosis. These adaptations by the cancer cells enable a tumor to survive until blood vessels can grow. HIF-1 also increases the expression of signal molecules, such as vascular endothelial growth factor (VEGF), that facilitate the growth of blood vessels (Figure 16.21). Without new blood vessels, a tumor would cease to grow and either die or remain harmlessly small. Efforts are underway to develop drugs that inhibit the growth of blood vessels in tumors.

Interestingly, anaerobic exercise training also activates HIF-1 with the same effects seen in the tumor—enhanced ability to generate ATP anaerobically and a stimulation of blood-vessel growth. These biochemical effects account for the improved athletic performance that results from training and demonstrate how behavior can affect biochemistry.

16.3   Glucose Can Be Synthesized from Noncarbohydrate Precursors

We now turn to the *synthesis of glucose from noncarbohydrate precursors*, a process called *gluconeogenesis*. Maintaining levels of glucose is important because the brain depends on glucose as its primary fuel and red blood cells use glucose as their only fuel. The daily glucose requirement of the brain in a typical adult human being is about 120 g, which accounts for most of the 160 g of glucose needed daily by the whole body. The amount of glucose present in body fluids is about 20 g, and that readily available from glycogen is approximately 190 g. Thus, the direct glucose reserves are sufficient to meet glucose needs for about a day. Gluconeogenesis is especially important during a longer period of fasting or starvation (p. 771).

The gluconeogenic pathway converts pyruvate into glucose. Noncarbohydrate precursors of glucose are first converted into pyruvate or enter the pathway at later intermediates such as oxaloacetate and dihydroxyacetone phosphate (Figure 16.22). The major noncarbohydrate precursors are *lactate, amino acids,* and *glycerol.* Lactate is formed by active skeletal muscle when the rate of glycolysis exceeds the rate of oxidative metabolism. Lactate is readily converted into pyruvate by the action of lactate dehydrogenase (p. 447). Amino acids are derived from proteins in the diet and, during starvation, from the breakdown of proteins in skeletal muscle (p. 650). The hydrolysis of triacylglycerols (p. 621) in fat cells yields glycerol and fatty acids.

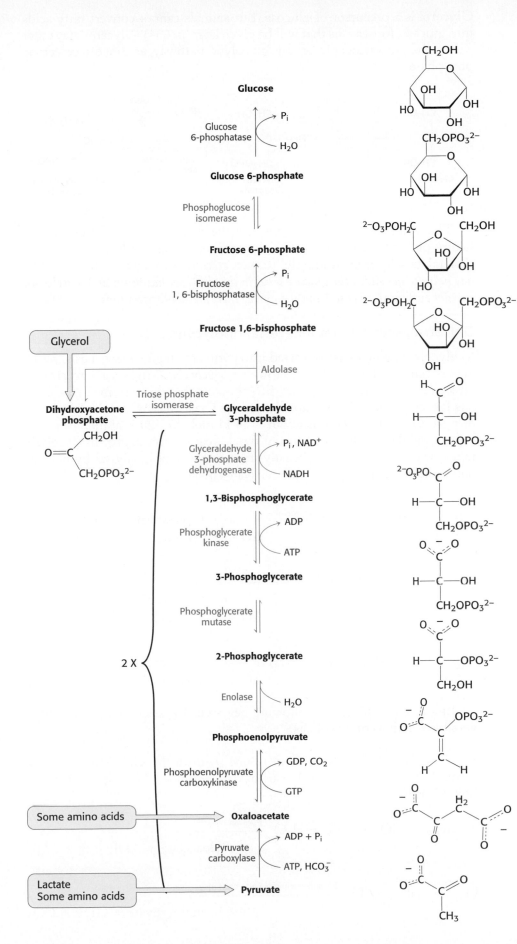

Figure 16.22 **Pathway of gluconeogenesis.** The distinctive reactions and enzymes of this pathway are shown in red. The other reactions are common to glycolysis. The enzymes for gluconeogenesis are located in the cytoplasm, except for pyruvate carboxylase (in the mitochondria) and glucose 6-phosphatase (membrane bound in the endoplasmic reticulum). The entry points for lactate, glycerol, and amino acids are shown.

Glycerol is a precursor of glucose, but animals cannot convert fatty acids into glucose, for reasons that will be given later (p. 634). Glycerol may enter either the gluconeogenic or the glycolytic pathway at dihydroxyacetone phosphate.

The major site of gluconeogenesis is the *liver*, with a small amount also taking place in the *kidney*. Little gluconeogenesis takes place in the brain, skeletal muscle, or heart muscle. Rather, *gluconeogenesis in the liver and kidney helps to maintain the glucose level in the blood so that brain and muscle can extract sufficient glucose from it to meet their metabolic demands.*

Gluconeogenesis Is Not a Reversal of Glycolysis

In glycolysis, glucose is converted into pyruvate; in gluconeogenesis, pyruvate is converted into glucose. However, *gluconeogenesis is not a reversal of glycolysis*. Several reactions must differ because the equilibrium of glycolysis lies far on the side of pyruvate formation. The actual ΔG for the formation of pyruvate from glucose is about -84 kJ mol^{-1} (-20 kcal mol^{-1}) under typical cellular conditions. Most of the decrease in free energy in glycolysis takes place in the three essentially irreversible steps catalyzed by hexokinase, phosphofructokinase, and pyruvate kinase.

$$\text{Glucose} + \text{ATP} \xrightarrow{\text{Hexokinase}} \text{glucose 6-phosphate} + \text{ADP}$$
$$\Delta G = -33 \text{ kJ mol}^{-1} (-8.0 \text{ kcal mol}^{-1})$$

$$\text{Fructose 6-phosphate} + \text{ATP} \xrightarrow{\text{Phosphofructokinase}}$$
$$\text{fructose 1,6-bisphosphate} + \text{ADP}$$
$$\Delta G = -22 \text{ kJ mol}^{-1} (-5.3 \text{ kcal mol}^{-1})$$

$$\text{Phosphoenolpyruvate} + \text{ADP} \xrightarrow{\text{Pyruvate kinase}} \text{pyruvate} + \text{ATP}$$
$$\Delta G = -17 \text{ kJ mol}^{-1} (-4.0 \text{ kcal mol}^{-1})$$

In gluconeogenesis, the following new steps bypass these virtually irreversible reactions of glycolysis:

1. *Phosphoenolpyruvate is formed from pyruvate by way of oxaloacetate* through the action of pyruvate carboxylase and phosphoenolpyruvate carboxykinase.

$$\text{Pyruvate} + \text{CO}_2 + \text{ATP} + \text{H}_2\text{O} \xrightarrow{\text{Pyruvate carboxylase}}$$
$$\text{oxaloacetate} + \text{ADP} + \text{P}_i + 2 \text{ H}^+$$

$$\text{Oxaloacetate} + \text{GTP} \xrightarrow{\text{Phosphoenolpyruvate carboxykinase}}$$
$$\text{phosphoenolpyruvate} + \text{GDP} + \text{CO}_2$$

2. *Fructose 6-phosphate is formed from fructose 1,6-bisphosphate by hydrolysis of the phosphate ester at carbon 1.* Fructose 1,6-bisphosphatase catalyzes this exergonic hydrolysis.

$$\text{Fructose 1,6-bisphosphate} + H_2O \longrightarrow \text{fructose 6-phosphate} + P_i$$

3. *Glucose is formed by the hydrolysis of glucose 6-phosphate* in a reaction catalyzed by glucose 6-phosphatase.

$$\text{Glucose 6-phosphate} + H_2O \longrightarrow \text{glucose} + P_i$$

We will examine each of these steps in turn.

The Conversion of Pyruvate into Phosphoenolpyruvate Begins with the Formation of Oxaloacetate

The first step in gluconeogenesis is the carboxylation of pyruvate to form oxaloacetate at the expense of a molecule of ATP.

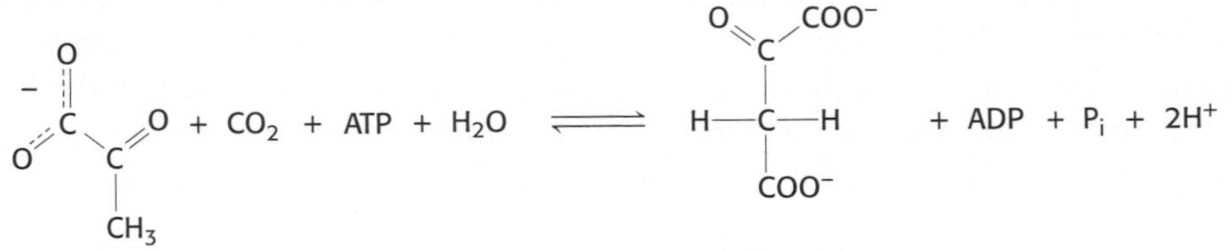

$$\text{Pyruvate} + CO_2 + ATP + H_2O \rightleftharpoons \text{Oxaloacetate} + ADP + P_i + 2H^+$$

Then, oxaloacetate is decarboxylated and phosphorylated to yield phosphoenolpyruvate, at the expense of the high phosphoryl-transfer potential of GTP.

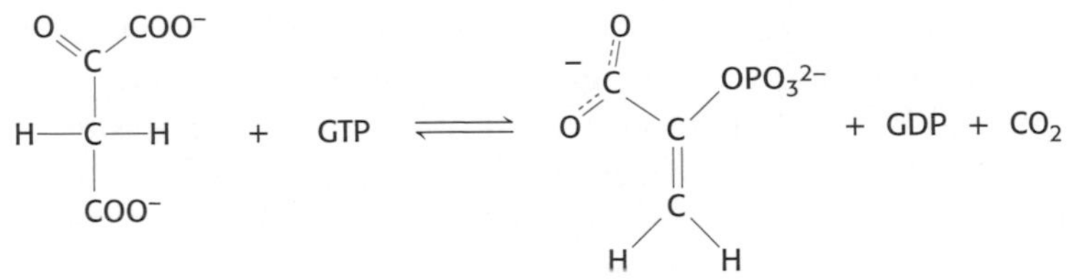

$$\text{Oxaloacetate} + GTP \rightleftharpoons \text{Phosphoenolpyruvate} + GDP + CO_2$$

The first of these reactions takes place inside the mitochondria.

The first reaction is catalyzed by *pyruvate carboxylase* and the second by *phosphoenolpyruvate carboxykinase*. The sum of these reactions is

$$\text{Pyruvate} + ATP + GTP + H_2O \rightleftharpoons$$
$$\text{phosphoenolpyruvate} + ADP + GDP + P_i + 2\,H^+$$

Pyruvate carboxylase is of special interest because of its structural, catalytic, and allosteric properties. The N-terminal 300 to 350 amino acids form an *ATP-grasp domain* (Figure 16.23), which is a widely used ATP-activating

ATP-grasp domain   Biotin-binding domain

1   350   1100   1180

Figure 16.23 Domain structure of pyruvate carboxylase. The ATP-grasp domain activates HCO_3^- and transfers CO_2 to the biotin-binding domain. From there, the CO_2 is transferred to pyruvate generated in the central domain.

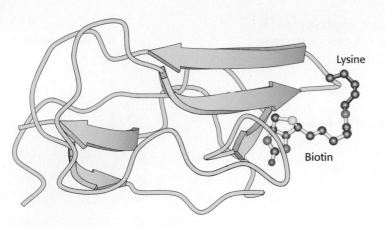

Lysine

Biotin

Figure 16.24 Biotin-binding domain of pyruvate carboxylase. This likely structure is based on the structure of the homologous domain of the enzyme acetyl CoA carboxylase (p. 635). *Notice* that the biotin is on a flexible tether, allowing it to move between the ATP-bicarbonate site and the pyruvate site. [Drawn from 1BDO.pdb.]

domain to be considered in more detail when we discuss nucleotide biosynthesis (p. 711). The C-terminal 80 amino acids constitute a biotin-binding domain (Figure 16.24) that we will see again in fatty acid synthesis (p. 635). *Biotin* is a covalently attached prosthetic group, which serves as a *carrier of activated* CO_2. The carboxylate group of biotin is linked to the ε-amino group of a specific lysine residue by an amide bond (Figure 16.25). Note that biotin is attached to pyruvate carboxylase by a *long, flexible chain*.

The carboxylation of pyruvate takes place in three stages:

$$HCO_3^- + ATP \rightleftharpoons HOCO_2\text{-}PO_3^{2-} + ADP$$

$$\text{Biotin–enzyme} + HOCO_2\text{-}PO_3^{2-} \rightleftharpoons CO_2\text{–biotin–enzyme} + P_i$$

$$CO_2\text{–biotin–enzyme} + \text{pyruvate} \rightleftharpoons \text{biotin–enzyme} + \text{oxaloacetate}$$

Recall that, in aqueous solutions, CO_2 exists primarily as HCO_3^- with the aid of carbonic anhydrase (Section 9.2). HCO_3^- is activated to carboxyphosphate. This activated CO_2 is subsequently bonded to the N-1 atom of the biotin ring to form the carboxybiotin–enzyme intermediate (see Figure 16.25). The CO_2 attached to biotin is quite activated. The $\Delta G°'$ for its cleavage

$$CO_2\text{–biotin–enzyme} + H^+ \longrightarrow CO_2 + \text{biotin–enzyme}$$

is -20 kJ mol^{-1} ($-4.7 \text{ kcal mol}^{-1}$). This negative $\Delta G°'$ indicates that carboxybiotin is able to transfer CO_2 to acceptors without the input of additional free energy.

The activated carboxyl group is then transferred from carboxybiotin to pyruvate to form oxaloacetate. The long, flexible link between biotin and the enzyme enables this prosthetic group to rotate from one active site of the enzyme (the ATP-bicarbonate site) to the other (the pyruvate site).

The first partial reaction of pyruvate carboxylase, the formation of carboxybiotin, depends on the presence of acetyl CoA. *Biotin is not carboxylated unless acetyl CoA is bound to the enzyme.* Acetyl CoA has no effect on the second partial reaction. The allosteric activation of pyruvate carboxylase by acetyl CoA is an important physiological control mechanism that will be discussed on page 493.

Oxaloacetate Is Shuttled into the Cytoplasm and Converted into Phosphoenolpyruvate

Pyruvate carboxylase is a mitochondrial enzyme, whereas the other enzymes of gluconeogenesis are present primarily in the cytoplasm. Oxaloacetate, the product of the pyruvate carboxylase reaction, must thus be transported to the cytoplasm to complete the pathway. Oxaloacetate is transported from a mitochondrion in the form of malate: oxaloacetate is reduced to malate inside the mitochondrion by an NADH-linked malate dehydrogenase. After malate has been transported across the mitochondrial membrane, it is reoxidized to oxaloacetate by an NAD^+-linked malate dehydrogenase in the cytoplasm (Figure 16.26). The formation of oxaloacetate from malate also provides NADH for use in subsequent steps in gluconeogenesis. Finally, oxaloacetate is simultaneously *decarboxylated* and *phosphorylated* by phosphoenolpyruvate carboxykinase to generate phosphoenol pyruvate. The phosphoryl donor is GTP. The CO_2 that was added to pyruvate by pyruvate carboxylase comes off in this step.

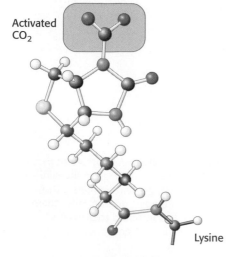

Activated CO_2

Lysine

Figure 16.25 Structure of carboxybiotin.

Why is a carboxylation and a decarboxylation required to form phosphoenolpyruvate from pyruvate? Recall that, in glycolysis, the presence of a phosphoryl group traps the unstable enol isomer of pyruvate as phosphoenolpyruvate (p. 445). However, the addition of a phosphoryl group to pyruvate is a highly unfavorable reaction: the $\Delta G^{\circ\prime}$ of the reverse of the glycolytic reaction catalyzed by pyruvate kinase is $+31$ kJ mol^{-1} ($+7.5$ kcal mol^{-1}). In gluconeogenesis, the use of the carboxylation and decarboxylation steps results in a much more favorable $\Delta G^{\circ\prime}$. The formation of phosphoenolpyruvate from pyruvate in the gluconeogenic pathway has a $\Delta G^{\circ\prime}$ of $+0.8$ kJ mol^{-1} ($+0.2$ kcal mol^{-1}). A molecule of ATP is used to power the addition of a molecule of CO_2 to pyruvate in the carboxylation step. That CO_2 is then removed to power the formation of phosphoenolpyruvate in the decarboxylation step. *Decarboxylations often drive reactions that are otherwise highly endergonic.* This metabolic motif is used in the citric acid cycle (p. 485), the pentose phosphate pathway (p. 577), and fatty acid synthesis (p. 636).

The Conversion of Fructose 1,6-bisphosphate into Fructose 6-phosphate and Orthophosphate Is an Irreversible Step

On formation, phosphoenolpyruvate is metabolized by the enzymes of glycolysis but in the reverse direction. These reactions are near equilibrium under intracellular conditions; so, when conditions favor gluconeogenesis, the reverse reactions will take place until the next irreversible step is reached. This step is the hydrolysis of fructose 1,6-bisphosphate to fructose 6-phosphate and P_i.

$$\text{Fructose 1,6-bisphosphate} + H_2O \xrightarrow{\text{Fructose 1,6-bisphosphatase}} \text{fructose 6-phosphate} + P_i$$

The enzyme responsible for this step is fructose 1,6-bisphosphatase. Like its glycolytic counterpart, it is an allosteric enzyme that participates in the regulation of gluconeogenesis. We will return to its regulatory properties later in the chapter.

The Generation of Free Glucose Is an Important Control Point

The fructose 6-phosphate generated by fructose 1,6-bisphosphatase is readily converted into glucose 6-phosphate. In most tissues, gluconeogenesis ends here. Free glucose is not generated; rather, the glucose 6-phosphate is processed in some other fashion, notably to form glycogen. One advantage to ending gluconeogenesis at glucose 6-phosphate is that, unlike free glucose, the molecule is not transported out of the cell. To keep glucose inside the cell, the generation of free glucose is controlled in two ways. First, the enzyme responsible for the conversion of glucose 6-phosphate into glucose, *glucose 6-phosphatase,* is regulated. Second, the enzyme is present only in tissues whose metabolic duty is to maintain blood-glucose homeostasis— tissues that release glucose into the blood. These tissues are the liver and to a lesser extent the kidney.

This final step in the generation of glucose does not take place in the cytoplasm. Rather, glucose 6-phosphate is transported into the lumen of the endoplasmic reticulum, where it is hydrolyzed to glucose by glucose 6-phosphatase, which is bound to the membrane (Figure 16.27). An associated Ca^{2+}-binding stabilizing protein is essential for phosphatase activity. Glucose and P_i are then shuttled back to the cytoplasm by a pair of transporters. The glucose transporter in the endoplasmic reticulum membrane is like those found in the plasma membrane. It is striking that five proteins are needed to transform cytoplasmic glucose 6-phosphate into glucose.

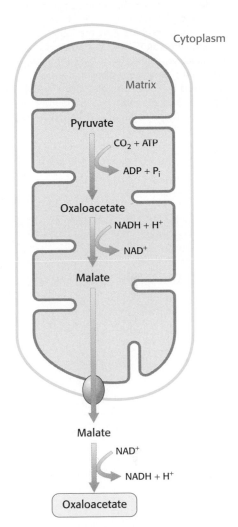

Figure 16.26 Compartmental cooperation. Oxaloacetate used in the cytoplasm for gluconeogenesis is formed in the mitochondrial matrix by the carboxylation of pyruvate. Oxaloacetate leaves the mitochondrion by a specific transport system (not shown) in the form of malate, which is reoxidized to oxaloacetate in the cytoplasm.

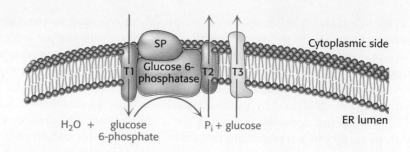

Figure 16.27 Generation of glucose from glucose 6-phosphate. Several endoplasmic reticulum (ER) proteins play a role in the generation of glucose from glucose 6-phosphate. T1 transports glucose 6-phosphate into the lumen of the ER, whereas T2 and T3 transport P_i and glucose, respectively, back into the cytoplasm. Glucose 6-phosphatase is stabilized by a Ca^{2+}-binding protein (SP). [After A. Buchell and I. D. Waddel. *Biochem. Biophys. Acta* 1092(1991):129–137.]

Six High-Transfer-Potential Phosphoryl Groups Are Spent in Synthesizing Glucose from Pyruvate

The formation of glucose from pyruvate is energetically unfavorable unless it is coupled to reactions that are favorable. Compare the stoichiometry of gluconeogenesis with that of the reverse of glycolysis.

The stoichiometry of gluconeogenesis is

$$2 \text{ Pyruvate} + 4 \text{ ATP} + 2 \text{ GTP} + 2 \text{ NADH} + 6 \text{ H}_2\text{O} \longrightarrow$$
$$\text{glucose} + 4 \text{ ADP} + 2 \text{ GDP} + 6 \text{ P}_i + 2 \text{ NAD}^+ + 2 \text{ H}^+$$
$$\Delta G^{\circ\prime} = -38 \text{ kJ mol}^{-1}(-9 \text{ kcal mol}^{-1})$$

In contrast, the stoichiometry for the reversal of glycolysis is

$$2 \text{ Pyruvate} + 2 \text{ ATP} + \text{NADH} + 2 \text{ H}_2\text{O} \longrightarrow$$
$$\text{glucose} + 2 \text{ ADP} + 2 \text{ P}_i + 2 \text{NAD}^+ + 2 \text{ H}^+$$
$$\Delta G^{\circ\prime} = +84 \text{ kJ mol}^{-1}(+20 \text{ kcal mol}^{-1})$$

Note that *six* nucleoside triphosphate molecules are hydrolyzed to synthesize glucose from pyruvate in gluconeogenesis, whereas only *two* molecules of ATP are generated in glycolysis in the conversion of glucose into pyruvate. Thus, the extra cost of gluconeogenesis is four high phosphoryl-transfer potential molecules for each molecule of glucose synthesized from pyruvate. The four additional molecules having high phosphoryl-transfer potential are needed to turn an energetically unfavorable process (the reversal of glycolysis) into a favorable one (gluconeogenesis). Here we have a clear example of the coupling of reactions: NTP hydrolysis is used to power an energetically unfavorable reaction.

16.4 Gluconeogenesis and Glycolysis Are Reciprocally Regulated

Gluconeogenesis and glycolysis are coordinated so that, within a cell, one pathway is relatively inactive while the other is highly active. If both sets of reactions were highly active at the same time, the net result would be the hydrolysis of four nucleoside triphosphates (two ATP molecules plus two GTP molecules) per reaction cycle. Both glycolysis and gluconeogenesis are highly exergonic under cellular conditions, and so there is no thermodynamic barrier to such simultaneous activity. However, the *amounts* and *activities* of the distinctive enzymes of each pathway are controlled so that both pathways are not highly active at the same time. The rate of glycolysis is also determined by the concentration of glucose, and the rate of gluconeogenesis by the concentrations of lactate and other precursors of glucose. The basic premise of the reciprocal regulation is that, when energy is needed, glycolysis will predominate. When there is a surplus of energy, gluconeogenesis will take over.

Energy Charge Determines Whether Glycolysis or Gluconeogenesis Will Be Most Active

The first important regulation site is the interconversion of fructose 6-phosphate and fructose 1,6-bisphosphate (Figure 16.28). Consider first a situation in which energy is needed. In this case, the concentration of AMP is high. Under this condition, AMP stimulates phosphofructokinase but inhibits fructose 1,6-bisphosphatase. Thus, glycolysis is turned on and gluconeogenesis is inhibited. Conversely, high levels of ATP and citrate indicate that the energy charge is high and that biosynthetic intermediates are abundant. ATP and citrate inhibit phosphofructokinase, whereas citrate activates fructose 1,6 bisphosphatase. Under these conditions, glycolysis is nearly switched off and gluconeogenesis is promoted. Why does citrate take part in this regulatory scheme? As we will see in Chapter 17, citrate reports on the status of the citric acid cycle, the primary pathway for oxidizing fuels in the presence of oxygen. High levels of citrate indicate an energy-rich situation and the presence of precursors for biosynthesis.

Glycolysis and gluconeogenesis are also reciprocally regulated at the interconversion of phosphoenolpyruvate and pyruvate in the liver. The glycolytic enzyme pyruvate kinase is inhibited by allosteric effectors ATP and alanine, which signal that the energy charge is high and that building blocks

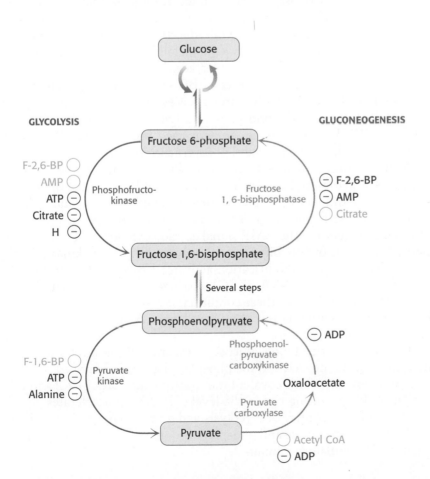

Figure 16.28 Reciprocal regulation of gluconeogenesis and glycolysis in the liver. The level of fructose 2, 6-bisphosphate is high in the fed state and low in starvation. Another important control is the inhibition of pyruvate kinase by phosphorylation during starvation.

are abundant. Conversely, pyruvate carboxylase, which catalyzes the first step in gluconeogenesis from pyruvate, is inhibited by ADP. Likewise, ADP inhibits phosphoenolpyruvate carboxykinase. Pyruvate carboxylase is activated by acetyl CoA, which, like citrate, indicates that the citric acid cycle is producing energy and biosynthetic intermediates (Chapter 17). Hence, gluconeogenesis is favored when the cell is rich in biosynthetic precursors and ATP.

The Balance Between Glycolysis and Gluconeogenesis in the Liver Is Sensitive to Blood-Glucose Concentration

In the liver, rates of glycolysis and gluconeogenesis are adjusted to maintain blood-glucose levels. *Fructose 2,6-bisphosphate strongly stimulates phosphofructokinase and inhibits fructose 1,6-bisphosphatase* (p. 455). When blood glucose is low, fructose 2,6-bisphosphate loses a phosphoryl group to form fructose 6-phosphate, which no longer binds to PFK. How is the concentration of fructose 2,6-bisphosphate controlled to rise and fall with blood-glucose levels? Two enzymes regulate the concentration of this molecule: one phosphorylates fructose 6-phosphate and the other dephosphorylates fructose 2,6-bisphosphate. Fructose 2,6-bisphosphate is formed in a reaction catalyzed by *phosphofructokinase 2* (PFK2), a different enzyme from phosphofructokinase. Fructose 6-phosphate is formed through hydrolysis of fructose 2,6-bisphosphate by a specific phosphatase, *fructose bisphosphatase 2* (FBPase2). The striking finding is that *both PFK2 and FBPase2 are present in a single 55-kd polypeptide chain* (Figure 16.29). This *bifunctional enzyme* contains an N-terminal *regulatory domain*, followed by a *kinase domain* and a *phosphatase domain*. PFK2 resembles adenylate kinase in having a P-loop NTPase domain (p. 267), whereas FBPase2 resembles phosphoglycerate mutase (p. 444). Recall that the mutase is essentially a phosphatase. In the bifunctional enzyme, the phosphatase activity evolved to become specific for F-2,6-BP. The bifunctional enzyme itself probably arose by the fusion of genes encoding the kinase and phosphatase domains.

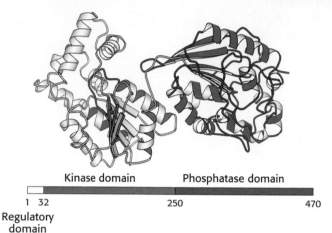

Kinase domain   Phosphatase domain

1 32   250   470
Regulatory
domain

Figure 16.29 Domain structure of the bifunctional enzyme phosphofructokinase 2. The kinase domain (purple) is fused to the phosphatase domain (red). The kinase domain is a P-loop NTP hydrolase domain, as indicated by the purple shading (p. 267). The bar represents the amino acid sequence of the enzyme. [Drawn from 1BIF.pdb.]

What controls whether PFK2 or FBPase2 dominates the bifunctional enzyme's activities in the liver? The activities of PFK2 and FBPase2 are reciprocally controlled by *phosphorylation of a single serine residue*. When glucose is scarce, as during a night's fast, a rise in the blood level of the hormone glucagon triggers a cyclic AMP signal cascade (Section 14.1), leading to the phosphorylation of this bifunctional enzyme by protein kinase A (Figure 16.30). This covalent modification activates FBPase2 and inhibits PFK2, lowering the level of F-2,6-BP. Gluconeogenesis predominates. Glucose formed by the liver under these conditions is essential for the viability of the brain. Glucagon stimulation of protein kinase A also inactivates pyruvate kinase in the liver (p. 456).

Conversely, when blood-glucose levels are high, as after a meal, gluconeogenesis is not needed and the phosphoryl group is removed from the bifunctional enzyme. This covalent modification activates PFK2 and inhibits FBPase2. The resulting rise in the level of F-2,6-BP accelerates glycolysis. The coordinated control of glycolysis and gluconeogenesis is facilitated by the location of the kinase and phosphatase domains on the same polypeptide chain as the regulatory domain.

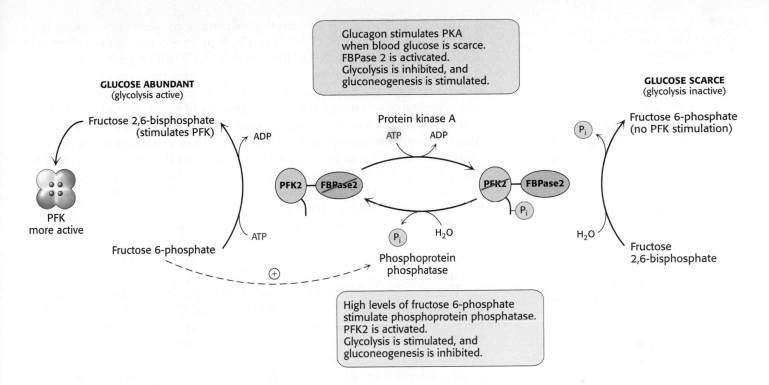

GLUCOSE ABUNDANT
(glycolysis active)

Fructose 2,6-bisphosphate
(stimulates PFK)

PFK
more active

Fructose 6-phosphate

Glucagon stimulates PKA
when blood glucose is scarce.
FBPase 2 is activcated.
Glycolysis is inhibited, and
gluconeogenesis is stimulated.

Protein kinase A

ATP   ADP

PFK2 — FBPase2   PFK2 — FBPase2

P_i H_2O H_2O

Phosphoprotein
phosphatase

ADP

ATP

P_i

GLUCOSE SCARCE
(glycolysis inactive)

Fructose 6-phosphate
(no PFK stimulation)

Fructose
2,6-bisphosphate

High levels of fructose 6-phosphate
stimulate phosphoprotein phosphatase.
PFK2 is activated.
Glycolysis is stimulated, and
gluconeogenesis is inhibited.

The hormones insulin and glucagon also regulate the amounts of essential enzymes. These hormones alter gene expression primarily by changing the rate of transcription. Insulin levels rise subsequent to eating, when there is plenty of glucose for glycolysis. To encourage glycolysis, insulin stimulates the expression of phosphofructokinase, pyruvate kinase, and the bifunctional enzyme that makes and degrades F-2,6-BP. Glucagon rises during fasting, when gluconeogenesis is needed to replace scarce glucose. To encourage gluconeogenesis, glucagon inhibits the expression of the three regulated glycolytic enzymes and stimulates instead the production of two key gluconeogenic enzymes, phosphoenolpyruvate carboxykinase and fructose 1,6-bisphosphatase. Transcriptional control in eukaryotes is much slower than allosteric control, taking hours or days instead of seconds to minutes. The richness and complexity of hormonal control are graphically displayed by the promoter of the phosphoenolpyruvate carboxykinase gene, which contains regulatory sequences that respond to insulin, glucagon (through the cAMP response elements), glucocorticoids, and thyroid hormone (Figure 16.31).

Figure 16.30 Control of the synthesis and degradation of fructose 2,6-bisphosphate. A low blood-glucose level as signaled by glucagon leads to the phosphorylation of the bifunctional enzyme and hence to a lower level of fructose 2,6-bisphosphate, slowing glycolysis. High levels of fructose 6-phosphate accelerate the formation of fructose 2,6-bisphosphate by facilitating the dephosphorylation of the bifunctional enzyme.

Substrate Cycles Amplify Metabolic Signals and Produce Heat

A pair of reactions such as the phosphorylation of fructose 6-phosphate to fructose 1,6-bisphosphate and its hydrolysis back to fructose 6-phosphate is called a *substrate cycle*. As already mentioned, both reactions are not simultaneously fully active in most cells, because of reciprocal allosteric controls. However, isotope-labeling studies have shown that some fructose 6-phosphate is phosphorylated to fructose 1,6-bisphosphate even during gluconeogenesis. There also is a limited degree of cycling in other pairs of opposed irreversible reactions. This cycling was regarded as an imperfection in metabolic control, and so substrate cycles have sometimes been called

Figure 16.31 The promoter of the phosphoenolpyruvate carboxykinase gene. This promoter is approximately 500 bp in length and contains regulatory sequences (response elements) that mediate the action of several hormones. Abbreviations: IRE, insulin response element; GRE, glucocorticoid response element; TRE, thyroid hormone response element; CREI and CREII, cAMP response elements. [After M. M. McGrane, J. S. Jun, Y. M. Patel, and R. W. Hanson. *Trends Biochem. Sci.* 17(1992):40–44.]

−500   1

IRE   GRE   TRE   CREI   CREII

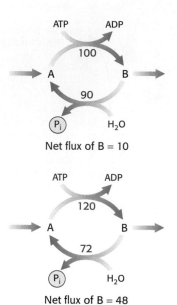

Net flux of B = 10

Net flux of B = 48

Figure 16.32 Substrate cycle. This ATP-driven cycle operates at two different rates. A small change in the rates of the two opposing reactions results in a large change in the *net* flux of product B

futile cycles. Indeed, there are pathological conditions, such as malignant hyperthermia, in which control is lost and both pathways proceed rapidly. One result is the rapid, uncontrolled hydrolysis of ATP, which generates heat.

Despite such extraordinary circumstances, substrate cycles now seem likely to be biologically important. One possibility is that *substrate cycles amplify metabolic signals.* Suppose that the rate of conversion of A into B is 100 and of B into A is 90, giving an initial net flux of 10. Assume that an allosteric effector increases the A → B rate by 20% to 120 and reciprocally decreases the B → A rate by 20% to 72. The new net flux is 48, and so a 20% change in the rates of the opposing reactions has led to a 380% increase in the net flux. In the example shown in Figure 16.32, this amplification is made possible by the rapid hydrolysis of ATP. The flux down the glycolytic pathway has been suggested to increase as much as 1000-fold at the initiation of intense exercise. Because the allosteric activation of enzymes alone seem unlikely to explain this increased flux, the existence of substrate cycles may partly account for the rapid rise in the rate of glycolysis.

The other potential biological role of substrate cycles is the *generation of heat produced by the hydrolysis of ATP.* A striking example is provided by bumblebees, which must maintain a thoracic temperature of about 30°C to fly. A bumblebee is able to maintain this high thoracic temperature and forage for food even when the ambient temperature is only 10°C, in part, because phosphofructokinase and fructose 1,6-bisphosphatase in its flight muscle are simultaneously highly active; the continuous hydrolysis of ATP generates heat. This bisphosphatase is not inhibited by AMP, which suggests that the enzyme specifically evolved for the generation of heat.

Lactate and Alanine Formed by Contracting Muscle Are Used by Other Organs

Lactate produced by active skeletal muscle and erythrocytes is a source of energy for other organs. Erythrocytes lack mitochondria and can never oxidize glucose completely. In contracting skeletal muscle during vigorous exercise, the rate at which glycolysis produces pyruvate exceeds the rate at which the citric acid cycle oxidizes it. In these cells, lactate dehydrogenase reduces excess pyruvate to lactate to restore redox balance (p. 447). However, lactate is a dead end in metabolism. It must be converted back into pyruvate before it can be metabolized. Both pyruvate and lactate diffuse out of these cells through carriers into the blood. *In contracting skeletal muscle, the formation and release of lactate lets the muscle generate ATP in the absence of oxygen and shifts the burden of metabolizing lactate from muscle to other organs.* The pyruvate and lactate in the bloodstream have two fates. In one fate, the plasma membranes of some cells, particularly cells in cardiac muscle, contain carriers that make the cells highly permeable to lactate and pyruvate. These molecules diffuse from the blood into such permeable cells. Once inside these well-oxygenated cells, lactate can be reverted back to pyruvate and metabolized through the citric acid cycle and oxidative phosphorylation to generate ATP. The use of lactate in place of glucose by these cells makes more circulating glucose available to the active muscle cells. In the other fate, excess lactate enters the liver and is converted first into pyruvate and then into glucose by the gluconeogenic pathway. *Contracting skeletal muscle supplies lactate to the liver, which uses it to synthesize and release glucose. Thus, the liver restores the level of glucose necessary for active muscle cells, which derive ATP from the glycolytic conversion of glucose into lactate. These reactions constitute the Cori cycle* (Figure 16.33).

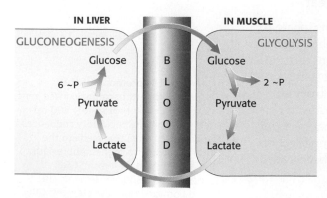

Figure 16.33 The Cori cycle. Lactate formed by active muscle is converted into glucose by the liver. This cycle shifts part of the metabolic burden of active muscle to the liver.

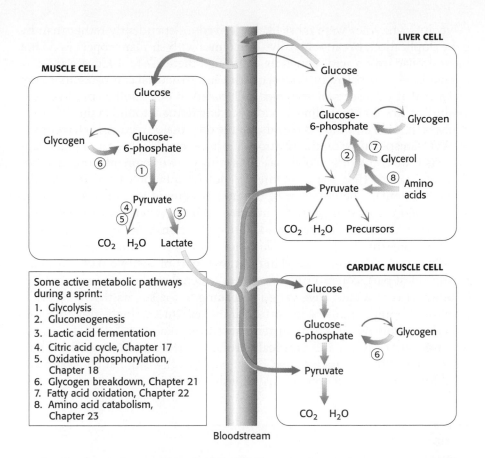

Figure 16.34 PATHWAY INTEGRATION: Cooperation between glycolysis and gluconeogenesis during a sprint. Glycolysis and gluconeogenesis are coordinated, in a tissue-specific fashion, to ensure that the energy needs of all cells are met. Consider a sprinter. In skeletal leg muscle, glucose will be metabolized aerobically to CO_2 and H_2O or, more likely (thick arrows) during a sprint, anaerobically to lactate. In cardiac muscle, the lactate can be converted into pyruvate and used as a fuel, along with glucose, to power the heartbeats to keep the sprinter's blood flowing. Gluconeogenesis, a primary function of the liver, will be taking place rapidly (thick arrows) to ensure that enough glucose is present in the blood for skeletal and cardiac muscle, as well as for other tissues. Glycogen, glycerol, and amino acids are other sources of energy that we will learn about in later chapters.

Studies have shown that alanine, like lactate, is a major precursor of glucose in the liver. The alanine is generated in muscle when the carbon skeletons of some amino acids are used as fuels. The nitrogens from these amino acids are transferred to pyruvate to form alanine. (p. 660); the reverse reaction takes place in the liver. This process also helps maintain nitrogen balance. The interplay between glycolysis and gluconeogenesis is summarized in Figure 16.34, which shows how these pathways help meet the energy needs of different cell types.

Isozymic forms of lactate dehydrogenase in different tissues catalyze the interconversions of pyruvate and lactate. Lactate dehydrogenase is a tetramer of two kinds of 35-kd subunits encoded by similar genes: the H type predominates in the heart, and the homologous M type in skeletal muscle and the liver. These subunits associate to form five types of tetramers: H_4, H_3M_1, H_2M_2, H_1M_3, and M_4. The H_4 isozyme (type 1) has higher affinity for substrates than that of the M_4 isozyme (type 5) and, unlike M_4, is allosterically inhibited by high levels of pyruvate. The other isozymes have intermediate properties, depending on the ratio of the two kinds of chains. The H_4 isozyme oxidizes lactate to pyruvate, which is then used as a fuel by the heart through aerobic metabolism. Indeed, heart muscle never functions anaerobically. In contrast, M_4 is optimized to operate in the reverse direction, to convert pyruvate into lactate to allow glycolysis to proceed under anaerobic conditions. We see here an example of how gene duplication and divergence generate a series of homologous enzymes that foster metabolic cooperation between organs.

Glycolysis and Gluconeogenesis Are Evolutionarily Intertwined

The metabolism of glucose has ancient origins. Organisms living in the early biosphere depended on the anaerobic generation of energy until significant amounts of oxygen began to accumulate 2 billion years ago.

Glycolytic enzymes were most likely derived independently rather than by gene duplication, because glycolytic enzymes with similar properties do not have similar amino acid sequences. Although there are four kinases and two isomerases in the pathway, both sequence and structural comparisons do not suggest that these sets of enzymes are related to one another by divergent evolution. The common dinucleotide-binding domain found in the dehydrogenases (p. 449) and the αβ barrels are the only major recurring elements.

We can speculate on the relationship between glycolysis and gluconeogenesis if we think of glycolysis as consisting of two segments: the metabolism of hexoses (the upper segment) and the metabolism of trioses (the lower segment). The enzymes of the upper segment are different in some species and are missing entirely in some archaea, whereas enzymes of the lower segment are quite conserved. In fact, four enzymes of the lower segment are present in all species. *This lower part of the pathway is common to glycolysis and gluconeogenesis.* This common part of the two pathways may be the oldest part, constituting the core to which the other steps were added. The upper part would have varied according to the sugars that were available to evolving organisms in particular niches. Interestingly, this core part of carbohydrate metabolism can generate triose precursors for ribose sugars, a component of RNA and a critical requirement for the RNA world. Thus, we are left with the unanswered question, Was the original core pathway used for energy conversion or biosynthesis?

Summary

16.1 Glycolysis Is an Energy-Conversion Pathway in Many Organisms

Glycolysis is the set of reactions that converts glucose into pyruvate. The 10 reactions of glycolysis take place in the cytoplasm. In the first stage, glucose is converted into fructose 1,6-bisphosphate by a phosphorylation, an isomerization, and a second phosphorylation reaction. Two molecules of ATP are consumed per molecule of glucose in these reactions, which are the prelude to the net synthesis of ATP. In the second stage, fructose 1,6-bisphosphate is cleaved by aldolase into dihydroxyacetone phosphate and glyceraldehyde 3-phosphate, which are readily interconvertible. In the third stage, ATP is generated. Glyceraldehyde 3-phosphate is oxidized and phosphorylated to form 1,3-bisphosphoglycerate, an acyl phosphate with a high phosphoryl-transfer potential. This molecule transfers a phosphoryl group to ADP to form ATP and 3-phosphoglycerate. A phosphoryl shift and a dehydration form phosphoenolpyruvate, a second intermediate with a high phosphoryl-transfer potential. Another molecule of ATP is generated as phosphoenolpyruvate is converted into pyruvate. There is a net gain of two molecules of ATP in the formation of two molecules of pyruvate from one molecule of glucose.

The electron acceptor in the oxidation of glyceraldehyde 3-phosphate is NAD^+, which must be regenerated for glycolysis to continue. In aerobic organisms, the NADH formed in glycolysis transfers its electrons to O_2 through the electron-transport chain, which thereby regenerates NAD^+. Under anaerobic conditions and in some microorganisms, NAD^+ is regenerated by the reduction of pyruvate to lactate. In other microorganisms, NAD^+ is regenerated by the reduction of pyruvate to ethanol. These two processes are examples of fermentations.

16.2 The Glycolytic Pathway Is Tightly Controlled

The glycolytic pathway has a dual role: it degrades glucose to generate ATP, and it provides building blocks for the synthesis of cellular components. The rate of conversion of glucose into pyruvate is regulated to meet these two major cellular needs. Under physiological conditions, the reactions of glycolysis are readily reversible except for those catalyzed by hexokinase, phosphofructokinase, and pyruvate kinase. Phosphofructokinase, the most important control element in glycolysis, is inhibited by high levels of ATP and citrate, and it is activated by AMP and fructose 2,6-bisphosphate. In the liver, this bisphosphate signals that glucose is abundant. Hence, phosphofructokinase is active when either energy or building blocks are needed. Hexokinase is inhibited by glucose 6-phosphate, which accumulates when phosphofructokinase is inactive. ATP and alanine allosterically inhibit pyruvate kinase, the other control site, and fructose 1,6-bisphosphate activates the enzyme. Consequently, pyruvate kinase is maximally active when the energy charge is low and glycolytic intermediates accumulate.

16.3 Glucose Can Be Synthesized from Noncarbohydrate Precursors

Gluconeogenesis is the synthesis of glucose from noncarbohydrate sources, such as lactate, amino acids, and glycerol. Several of the reactions that convert pyruvate into glucose are common to glycolysis. Gluconeogenesis, however, requires four new reactions to bypass the essential irreversibility of three reactions in glycolysis. In two of the new reactions, pyruvate is carboxylated in mitochondria to oxaloacetate, which in turn is decarboxylated and phosphorylated in the cytoplasm to phosphoenolpyruvate. Two molecules having high phosphoryl-transfer potential are consumed in these reactions, which are catalyzed by pyruvate carboxylase and phosphoenolpyruvate carboxykinase. Pyruvate carboxylase contains a biotin prosthetic group. The other distinctive reactions of gluconeogenesis are the hydrolyses of fructose 1,6-bisphosphate and glucose 6-phosphate, which are catalyzed by specific phosphatases. The major raw materials for gluconeogenesis by the liver are lactate and alanine produced from pyruvate by active skeletal muscle. The formation of lactate during intense muscular activity buys time and shifts part of the metabolic burden from muscle to the liver.

16.4 Gluconeogenesis and Glycolysis Are Reciprocally Regulated

Gluconeogenesis and glycolysis are reciprocally regulated so that one pathway is relatively inactive while the other is highly active. Phosphofructokinase and fructose 1,6-bisphosphatase are key control points. Fructose 2,6-bisphosphate, an intracellular signal molecule present at higher levels when glucose is abundant, activates glycolysis and inhibits gluconeogenesis by regulating these enzymes. Pyruvate kinase and pyruvate carboxylase are regulated by other effectors so that both are not maximally active at the same time. Allosteric regulation and reversible phosphorylation, which are rapid, are complemented by transcriptional control, which takes place in hours or days.

Key Terms

glycolysis (p. 433)

lactic acid fermentation (p. 433)

alcoholic fermentation (p. 433)

gluconeogenesis (p. 433)

α-amylase (p. 434)

hexokinase (p. 435)

kinase (p. 437)

phosphofructokinase (PFK) (p. 438)

thioester intermediate (p. 442)

substrate-level phosphorylation (p. 443)

mutase (p. 444)

enol phosphate (p. 445)

pyruvate kinase (p. 446)

obligate anaerobe (p. 448)

Rossmann fold (p. 449)

committed step (p. 454)

feedforward stimulation (p. 455)

pyruvate carboxylase (p. 461)

biotin (p. 462)

glucose 6-phosphatase (p. 463)

bifunctional enzyme (p. 466)

substrate cycle (p. 467)

Cori cycle (p. 468)

Selected Readings

Where to Start

Knowles, J. R. 1991. Enzyme catalysis: Not different, just better. *Nature* 350:121–124.

Granner, D., and Pilkis, S. 1990. The genes of hepatic glucose metabolism. *J. Biol. Chem.* 265:10173–10176.

McGrane, M. M., Yun, J. S., Patel, Y. M., and Hanson, R. W. 1992. Metabolic control of gene expression: In vivo studies with transgenic mice. *Trends Biochem. Sci.* 17:40–44.

Pilkis, S. J., and Granner, D. K. 1992. Molecular physiology of the regulation of hepatic gluconeogenesis and glycolysis. *Annu. Rev. Physiol.* 54:885–909.

Books

Fell, D. 1997. *Understanding the Control of Metabolism.* Portland.

Fersht, A. 1999. *Structure and Mechanism in Protein Science: A Guide to Enzyme Catalysis and Protein Folding.* W. H. Freeman and Company.

Frayn, K. N. 1996. *Metabolic Regulation: A Human Perspective.* Portland.

Hargreaves, M., and Thompson, M. (Eds.). 1999. *Biochemistry of Exercise X.* Human Kinetics.

Poortmans, J. R. (Ed.). 2004. *Principles of Exercise Biochemistry.* Krager.

Structure of Glycolytic and Gluconeogenic Enzymes

Aleshin, A. E., Kirby, C., Liu, X., Bourenkov, G. P., Bartunik, H. D., Fromm, H. J., and Honzatko, R. B. 2000. Crystal structures of mutant monomeric hexokinase I reveal multiple ADP binding sites and conformational changes relevant to allosteric regulation. *J. Mol. Biol.* 296:1001–1015.

Jeffery, C. J., Bahnson, B. J., Chien, W., Ringe, D., and Petsko, G. A. 2000. Crystal structure of rabbit phosphoglucose isomerase, a glycolytic enzyme that moonlights as neuroleukin, autocrine motility factor, and differentiation mediator. *Biochemistry* 39:955–964.

Schirmer, T., and Evans, P. R. 1990. Structural basis of the allosteric behaviour of phosphofructokinase. *Nature* 343:140–145.

Cooper, S. J., Leonard, G. A., McSweeney, S. M., Thompson, A. W., Naismith, J. H., Qamar, S., Plater, A., Berry, A., and Hunter, W. N. 1996. The crystal structure of a class II fructose-1,6-bisphosphate aldolase shows a novel binuclear metal-binding active site embedded in a familiar fold. *Structure* 4:1303–1315.

Davenport, R. C., Bash, P. A., Seaton, B. A., Karplus, M., Petsko, G. A., and Ringe, D. 1991. Structure of the triosephosphate isomerase–phosphoglycolohydroxamate complex: An analogue of the intermediate on the reaction pathway. *Biochemistry* 30:5821–5826.

Skarzynski, T., Moody, P. C., and Wonacott, A. J. 1987. Structure of holo-glyceraldehyde-3-phosphate dehydrogenase from *Bacillus stearothermophilus* at 1.8 Å resolution. *J. Mol. Biol.* 193:171–187.

Bernstein, B. E., and Hol, W. G. 1998. Crystal structures of substrates and products bound to the phosphoglycerate kinase active site reveal the catalytic mechanism. *Biochemistry* 37:4429–4436.

Rigden, D. J., Alexeev, D., Phillips, S. E. V., and Fothergill-Gilmore, L. A. 1998. The 2.3 Å X-ray crystal structure of *S. cerevisiae* phosphoglycerate mutase. *J. Mol. Biol.* 276:449–459.

Zhang, E., Brewer, J. M., Minor, W., Carreira, L. A., and Lebioda, L. 1997. Mechanism of enolase: The crystal structure of asymmetric dimer enolase-2-phospho-D-glycerate/enolase-phosphoenolpyruvate at 2.0 Å resolution. *Biochemistry* 36:12526–12534.

Mattevi, A., Valentini, G., Rizzi, M., Speranza, M. L., Bolognesi, M., and Coda, A. 1995. Crystal structure of *Escherichia coli* pyruvate kinase type I: Molecular basis of the allosteric transition. *Structure* 3:729–741.

Hasemann, C. A., Istvan E. S., Uyeda, K., and Deisenhofer, J. 1996. The crystal structure of the bifunctional enzyme 6-phosphofructo-2-kinase/fructose-2,6-biphosphatase reveals distinct domain homologies. *Structure* 4:1017–1029.

Tari, L. W., Matte, A., Pugazhenthi, U., Goldie, H., and Delbaere, L. T. J. 1996. Snapshot of an enzyme reaction intermediate in the structure of the ATP-Mg^{2+}-oxalate ternary complex of *Escherichia coli* PEP carboxykinase. *Nat. Struct. Biol.* 3:355–363.

Catalytic Mechanisms

Soukri, A., Mougin, A., Corbier, C., Wonacott, A., Branlant, C., and Branlant, G. 1989. Role of the histidine 176 residue in glyceraldehyde-3-phosphate dehydrogenase as probed by site-directed mutagenesis. *Biochemistry* 28:2586–2592.

Bash, P. A., Field, M. J., Davenport, R. C., Petsko, G. A., Ringe, D., and Karplus, M. 1991. Computer simulation and analysis of the reaction pathway of triosephosphate isomerase. *Biochemistry* 30:5826–5832.

Knowles, J. R., and Albery, W. J. 1977. Perfection in enzyme catalysis: The energetics of triosephosphate isomerase. *Acc. Chem. Res.* 10:105–111.

Rose, I. A. 1981. Chemistry of proton abstraction by glycolytic enzymes (aldolase, isomerases, and pyruvate kinase). *Philos. Trans. R. Soc. Lond. B Biol. Sci.* 293:131–144.

Regulation

Iancu, C. V., Mukund, S., Fromm, H. J., and Honzatko, R. B. 2005. R-state AMP complex reveals initial steps of the quaternary transition of fructose-1,6-bisphosphatase. *J. Biol. Chem.* 280:19737–19745.

Wilson, J. E. 2003. Isozymes of mammalian hexokinase: Structure, function and subcellular location. *J. Exp. Biol.* 206:2049–2057.

Lee, Y. H., Li, Y., Uyeda, K., and Hasemann, C. A. 2003. Tissue-specific structure/function differentiation of the five isoforms of 6-phosphofructo-2-kinase/fructose-2,6-bisphosphatase. *J. Biol. Chem.* 278:523–530.

Dang, C. V., and Semenza, G. L. 1999. Oncogenic alterations of metabolism. *Trends Biochem. Sci.* 24:68–72.

Depre, C., Rider, M. H., and Hue, L. 1998. Mechanisms of control of heart glycolysis. *Eur. J. Biochem.* 258:277–290.

Harrington, G. N., and Bush, D. R. 2003. The bifunctional role of hexokinase in metabolism and glucose signaling. *Plant Cell* 15:2493–2496.

Gleeson, T. T. 1996. Post-exercise lactate metabolism: A comparative review of sites, pathways, and regulation. *Annu. Rev. Physiol.* 58:556–581.

Hers, H.-G., and Van Schaftingen, E. 1982. Fructose 2,6-bisphosphate two years after its discovery. *Biochem. J.* 206:1–12.

Middleton, R. J. 1990. Hexokinases and glucokinases. *Biochem. Soc. Trans.* 18:180–183.

Nordlie, R. C., Foster, J. D., and Lange, A. J. 1999. Regulation of glucose production by the liver. *Annu. Rev. Nutr.* 19:379–406.

Jitrapakdee, S., and Wallace, J. C. 1999. Structure, function and regulation of pyruvate carboxylase. *Biochem. J.* 340:1–16.

Pilkis, S. J., and Claus, T. H. 1991. Hepatic gluconeogenesis/glycolysis: Regulation and structure/function relationships of substrate cycle enzymes. *Annu. Rev. Nutr.* 11:465–515.

Plaxton, W. C. 1996. The organization and regulation of plant glycolysis. *Annu. Rev. Plant Physiol. Plant Mol. Biol.* 47:185–214.

van de Werve, G., Lange, A., Newgard, C., Mechin, M. C., Li, Y., and Berteloot, A. 2000. New lessons in the regulation of glucose metabolism taught by the glucose 6-phosphatase system. *Eur. J. Biochem.* 267:1533–1549.

Sugar Transporters

Czech, M. P., and Corvera, S. 1999. Signaling mechanisms that regulate glucose transport. *J. Biol. Chem.* 274:1865–1868.

Silverman, M. 1991. Structure and function of hexose transporters. *Annu. Rev. Biochem.* 60:757–794.

Thorens, B., Charron, M. J., and Lodish, H. F. 1990. Molecular physiology of glucose transporters. *Diabetes Care* 13:209–218.

Genetic Diseases

Nakajima, H., Raben, N., Hamaguchi, T., and Yamasaki, T. 2002. Phosphofructokinase deficiency, past, present, and future. *Curr. Mol. Med.* 2:197–212.

Scriver, C. R., Beaudet, A. L., Valle, D., Sly, W. S., Childs, B., Kinzler, K., and Vogelstein, B. (Eds.). 2001. *The Metabolic and Molecular Basis of Inherited Disease* (8th ed.). McGraw-Hill.

Evolution

Dandekar, T., Schuster, S., Snel, B., Huynen, M., and Bork, P. 1999. Pathway alignment: Application to the comparative analysis of glycolytic enzymes. *Biochem. J.* 343:115–124.

Heinrich, R., Melendez-Hevia, E., Montero, F., Nuno, J. C., Stephani, A., and Waddell, T. G. 1999. The structural design of glycolysis: An evolutionary approach. *Biochem. Soc. Trans.* 27:294–298.

Walmsley, A. R., Barrett, M. P., Bringaud, F., and Gould, G. W. 1998. Sugar transporters from bacteria, parasites and mammals: Structure-activity relationships. *Trends Biochem. Sci.* 23:476–480.

Maes, D., Zeelen, J. P., Thanki, N., Beaucamp, N., Alvarez, M., Thi, M. H., Backmann, J., Martial, J. A., Wyns, L., Jaenicke, R., and Wierenga, R. K. 1999. The crystal structure of triosephosphate isomerase (TIM) from *Thermotoga maritima*: A comparative thermostability structural analysis of ten different TIM structures. *Proteins* 37:441–453.

Historical Aspects

Friedmann, H. C. 2004. From *Butyribacterium* to *E. coli:* An Essay on Unity in Biochemistry. *Perspect. Biol. Med.* 47:47–66.

Fruton, J. S. 1999. *Proteins, Enzymes, Genes: The Interplay of Chemistry and Biology.* Yale University Press.

Kalckar, H. M. (Ed.). 1969. *Biological Phosphorylations: Development of Concepts.* Prentice Hall.

Problems

1. *Kitchen chemistry*. Sucrose is commonly used to preserve fruits. Why is glucose not suitable for preserving foods?

2. *Tracing carbon atoms 1*. Glucose labeled with ^{14}C at C-1 is incubated with the glycolytic enzymes and necessary cofactors.

(a) What is the distribution of ^{14}C in the pyruvate that is formed? (Assume that the interconversion of glyceraldehyde 3-phosphate and dihydroxyacetone phosphate is very rapid compared with the subsequent step.)
(b) If the specific activity of the glucose substrate is 10 mCi $mmol^{-1}$, what is the specific activity of the pyruvate that is formed?

3. *Lactic acid fermentation*. (a) Write a balanced equation for the conversion of glucose into lactate. (b) Calculate the standard free-energy change of this reaction by using the data given in Table 16.3 and the fact that $\Delta G^{\circ\prime}$ is -25 kJ mol^{-1} (-6 kcal mol^{-1}) for the following reaction:

$$\text{Pyruvate} + \text{NADH} + \text{H}^+ \rightleftharpoons \text{lactate} + \text{NAD}^+$$

What is the free-energy change (ΔG, not $\Delta G^{\circ\prime}$) of this reaction when the concentrations of reactants are: glucose, 5 mM; lactate, 0.05 mM; ATP, 2 mM; ADP, 0.2 mM; and P_i, 1 mM?

4. *High potential*. What is the equilibrium ratio of phosphoenolpyruvate to pyruvate under standard conditions when [ATP]/[ADP] = 10?

5. *Hexose–triose equilibrium*. What are the equilibrium concentrations of fructose 1,6-bisphosphate, dihydroxyacetone phosphate, and glyceraldehyde 3-phosphate when 1 mM fructose 1,6-bisphosphate is incubated with aldolase under standard conditions?

6. *Double labeling*. 3-Phosphoglycerate labeled uniformly with ^{14}C is incubated with 1,3-BPG labeled with ^{32}P at C-1. What is the radioisotope distribution of the 2,3-BPG that is formed on addition of BPG mutase?

7. *An informative analog*. Xylose has the same structure as that of glucose except that it has a hydrogen atom at C-5 in place of a hydroxymethyl group. The rate of ATP hydrolysis by hexokinase is markedly enhanced by the addition of xylose. Why?

8. *Distinctive sugars*. The intravenous infusion of fructose into healthy volunteers leads to a two- to fivefold increase in the level of lactate in the blood, a far greater increase than that observed after the infusion of the same amount of glucose.

(a) Why is glycolysis more rapid after the infusion of fructose?
(b) Fructose has been used in place of glucose for intravenous feeding. Why is this use of fructose unwise?

9. *Metabolic mutants*. Predict the effect of each of the following mutations on the pace of glycolysis in liver cells:

(a) Loss of the allosteric site for ATP in phosphofructokinase.
(b) Loss of the binding site for citrate in phosphofructokinase.
(c) Loss of the phosphatase domain of the bifunctional enzyme that controls the level of fructose 2,6-bisphosphate.
(d) Loss of the binding site for fructose 1,6-bisphosphate in pyruvate kinase.

10. *Metabolic mutant*. What are the likely consequences of a genetic disorder rendering fructose 1,6-bisphosphatase in the liver less sensitive to regulation by fructose 2,6-bisphosphate?

11. *Biotin snatcher*. Avidin, a 70-kd protein in egg white, has very high affinity for biotin. In fact, it is a highly specific inhibitor of biotin enzymes. Which of the following conversions would be blocked by the addition of avidin to a cell homogenate?

(a) Glucose → pyruvate
(b) Pyruvate → glucose
(c) Oxaloacetate → glucose
(d) Malate → oxaloacetate
(e) Pyruvate → oxaloacetate
(f) Glyceraldehyde 3-phosphate → fructose 1,6-bisphosphate

12. *Tracing carbon atoms 2*. If cells synthesizing glucose from lactate are exposed to CO_2 labeled with ^{14}C, what will be the distribution of label in the newly synthesized glucose?

13. *Arsenate poisoning*. Arsenate (AsO_4^{3-}) closely resembles P_i in structure and reactivity. In the reaction catalyzed by glyceraldehyde 3-phosphate dehydrogenase, arsenate can replace phosphate in attacking the energy-rich thioester intermediate. The product of this reaction, 1-arseno-3-phosphoglycerate, is unstable. It and other acyl arsenates are rapidly and sponta-

neously hydrolyzed. What is the effect of arsenate on energy generation in a cell?

14. *Reduce, reuse, recycle.* In the conversion of glucose into two molecules of lactate, the NADH generated earlier in the pathway is oxidized to NAD^+. Why is it not to the cell's advantage to simply make more NAD^+ so that the regeneration would not be necessary? After all, the cell would save much energy because it would no longer need to synthesize lactic acid dehydrogenase.

15. *Adenylate kinase again.* Adenylate kinase, an enzyme considered in great detail in Chapter 9, is responsible for interconverting the adenylate nucleotide pool:

$$ADP + ADP \xrightarrow{\text{Adenylate kinase}} ATP + AMP$$

The equilibrium constant for this reaction is close to 1, inasmuch as the number of phosphoanhydride bonds is the same on each side of the equation. Using the equation for the equilibrium constant for this reaction, show why changes in [AMP] are a more effective indicator of the adenylate pool than [ATP].

16. *Working at cross-purposes?* Gluconeogenesis takes place during intense exercise, which seems counterintuitive. Why would an organism synthesize glucose and at the same time use glucose to generate energy?

17. *Powering pathways.* Compare the stoichiometries of glycolysis and gluconeogenesis. Recall that the input of one ATP equivalent changes the equilibrium constant of a reaction by a factor of about 10^8 (p. 414). By what factor do the additional high-phosphoryl-transfer compounds alter the equilibrium constant of gluconeogenesis?

Mechanism Problem

18. *Argument by analogy.* Propose a mechanism for the conversion of glucose 6-phosphate into fructose 6-phosphate by phosphoglucose isomerase based on the mechanism of triose phosphate isomerase.

Chapter Integration Problem

19. *Not just for energy.* People with galactosemia display central nervous system abnormalities even if galactose is eliminated from the diet. The precise reason for it is not known. Suggest a plausible explanation.

Data Interpretation Problem

20. *Now, that's unusual.* Phosphofructokinase has recently been isolated from the hyperthermophilic archaeon *Pyrococcus furiosus*. It was subjected to standard biochemical analysis to determine basic catalytic parameters. The processes under study were of the form

$$\text{Fructose 6-phosphate} + (x - P_i) \longrightarrow$$
$$\text{fructose 1,6-bisphosphate} + (x)$$

The assay measured the increase in fructose 1,6-bisphosphate. Selected results are shown in the adjoining graph.

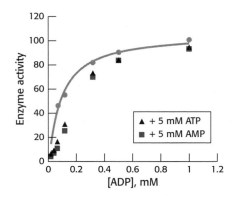

[Data from J. E. Tuininga et al. *J. Biol. Chem.* 274(1999):21023–21028.]

(a) How does the *P. furiosus* phosphofructokinase differ from the phosphofructokinase considered in this chapter?
(b) What effects do AMP and ATP have on the reaction with ADP?

The Citric Acid Cycle

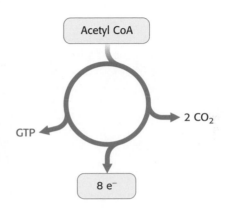

Roundabouts, or traffic circles, function as hubs to facilitate traffic flow. The citric acid cycle is the biochemical hub of the cell, oxidizing carbon fuels, usually in the form of acetyl CoA, as well as serving as a source of precursors for biosynthesis. [(Above) Chris Warren/International Stock.]

The metabolism of glucose to pyruvate in glycolysis, an anaerobic process, harvests but a fraction of the ATP available from glucose. Most of the ATP generated in metabolism is provided by the *aerobic* processing of glucose. This process starts with the complete oxidation of glucose derivatives to carbon dioxide. This oxidation takes place in a series of reactions called the *citric acid cycle,* also known as the *tricarboxylic acid* (TCA) *cycle* or the *Krebs cycle.* The citric acid cycle is the *final common pathway for the oxidation of fuel molecules*—carbohydrates, fatty acids, and amino acids. Most fuel molecules enter the cycle as *acetyl coenzyme A.*

Acetyl coenzyme A (Acetyl CoA)

Figure 17.1 Mitochondrion. The double membrane of the mitochondrion is evident in this electron micrograph. The numerous invaginations of the inner mitochondrial membrane are called cristae. The oxidative decarboxylation of pyruvate and the sequence of reactions in the citric acid cycle take place within the matrix. [(Left) Omikron/Photo Researchers.]

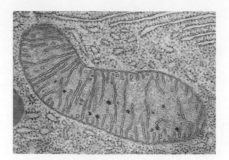
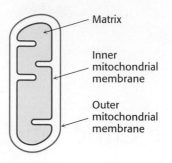

Matrix

Inner mitochondrial membrane

Outer mitochondrial membrane

Under aerobic conditions, the pyruvate generated from glucose is oxidatively decarboxylated to form acetyl CoA. In eukaryotes, the reactions of the citric acid cycle take place inside mitochondria (Figure 17.1), in contrast with those of glycolysis, which take place in the cytoplasm.

The Citric Acid Cycle Harvests High-Energy Electrons

The citric acid cycle is the central metabolic hub of the cell. It is the gateway to the aerobic metabolism of any molecule that can be transformed into an acetyl group or a component of the citric acid cycle. The cycle is also an important source of precursors for the building blocks of many other molecules such as amino acids, nucleotide bases, and porphyrin (the organic component of heme). The citric acid cycle component, oxaloacetate, is also an important precursor to glucose (p. 460).

What is the function of the citric acid cycle in transforming fuel molecules into ATP? Recall that fuel molecules are carbon compounds that are capable of being oxidized—of losing electrons (Chapter 15). The citric acid cycle includes a series of oxidation–reduction reactions that result in the oxidation of an acetyl group to two molecules of carbon dioxide. This oxidation generates high-energy electrons that will be used to power the synthesis of ATP. *The function of the citric acid cycle is the harvesting of high-energy electrons from carbon fuels.*

The overall pattern of the citric acid cycle is shown in Figure 17.2. A four-carbon compound (oxaloacetate) condenses with a two-carbon acetyl unit to yield a six-carbon tricarboxylic acid. The six-carbon compound releases CO_2 twice in two successive oxidative decarboxylations that yield high-energy electrons. A four-carbon compound remains. This four-carbon compound is further processed to regenerate oxaloacetate, which can initiate another round of the cycle. Two carbon atoms enter the cycle as an acetyl unit and two carbon atoms leave the cycle in the form of two molecules of CO_2.

Note that the citric acid cycle itself neither generates a large amount of ATP nor includes oxygen as a reactant (Figure 17.3). Instead, the citric acid cycle removes electrons from acetyl CoA and uses these electrons to form NADH and $FADH_2$. Three hydride ions (hence, six electrons) are transferred to three molecules of nicotinamide adenine dinucleotide (NAD^+), and one pair of hydrogen atoms (hence, two electrons) is transferred to one molecule of flavin adenine dinucleotide (FAD). These electron carriers yield nine molecules of ATP when they are oxidized by O_2 in *oxidative phosphorylation* (Chapter 18). Electrons released in the reoxidation of NADH and $FADH_2$ flow through a series of membrane proteins (referred to as the *electron-transport chain*) to generate a proton gradient across the membrane. These protons then flow through ATP synthase to generate ATP from ADP and inorganic phosphate.

The citric acid cycle, in conjunction with oxidative phosphorylation, provides the vast preponderance of energy used by aerobic cells—in human beings, greater than 90%. It is highly efficient because the oxidation of a

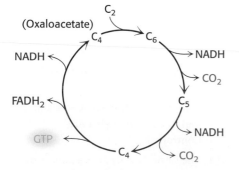

Figure 17.2 Overview of the citric acid cycle. The citric acid cycle oxidizes two-carbon units, producing two molecules of CO_2, one molecule of GTP, and high-energy electrons in the form of NADH and $FADH_2$.

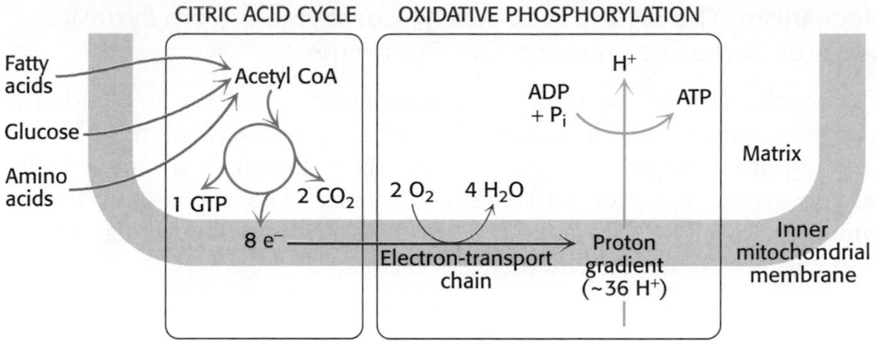

CITRIC ACID CYCLE   OXIDATIVE PHOSPHORYLATION

Figure 17.3 **Cellular respiration.** The citric acid cycle constitutes the first stage in cellular respiration, the removal of high-energy electrons from carbon fuels (left). These electrons reduce O_2 to generate a proton gradient (red pathway), which is used to synthesize ATP (green pathway). The reduction of O_2 and the synthesis of ATP constitute oxidative phosphorylation.

limited number of citric acid cycle molecules can generate large amounts of NADH and $FADH_2$. Note in Figure 17.2 that the four-carbon molecule, oxaloacetate, that initiates the first step in the citric acid cycle is regenerated at the end of one passage through the cycle. Thus, one molecule of oxaloacetate is capable of participating in the oxidation of many acetyl molecules.

17.1 Pyruvate Dehydrogenase Links Glycolysis to the Citric Acid Cycle

Carbohydrates, most notably glucose, are processed by glycolysis into pyruvate (Chapter 16). Under anaerobic conditions, the pyruvate is converted into lactate or ethanol, depending on the organism. Under aerobic conditions, the pyruvate is transported into mitochondria by a specific carrier protein embedded in the mitochondrial membrane. In the mitochondrial matrix, pyruvate is oxidatively decarboxylated by the *pyruvate dehydrogenase complex* to form acetyl CoA.

$$\text{Pyruvate} + \text{CoA} + \text{NAD}^+ \rightarrow \text{acetyl CoA} + CO_2 + \text{NADH} + H^+$$

This irreversible reaction is the link between glycolysis and the citric acid cycle (Figure 17.4). Note that the pyruvate dehydrogenase complex produces CO_2 and captures high-transfer-potential electrons in the form of NADH. Thus, the pyruvate dehydrogenase reaction has many of the key features of the reactions of the citric acid cycle itself.

The pyruvate dehydrogenase complex is a large, highly integrated complex of three distinct enzymes (Table 17.1). Pyruvate dehydrogenase complex is a member of a family of homologous complexes that include the citric acid cycle enzyme α-ketoglutarate dehydrogenase complex (p. 485). These complexes are giant, with molecular masses ranging from 4 million to 10 million daltons (Figure 17.5). As we will see, their elaborate structures allow groups to travel from one active site to another, connected by tethers to the core of the structure.

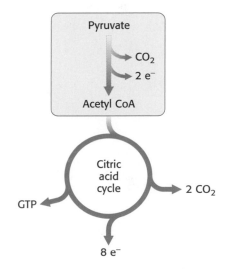

Figure 17.4 **The link between glycolysis and the citric acid cycle.** Pyruvate produced by glycolysis is converted into acetyl CoA, the fuel of the citric acid cycle.

TABLE 17.1 **Pyruvate dehydrogenase complex of *E. coli***

Enzyme	Abbreviation	Number of chains	Prosthetic group	Reaction catalyzed
Pyruvate dehydrogenase component	E_1	24	TPP	Oxidative decarboxylation of pyruvate
Dihydrolipoyl transacetylase	E_2	24	Lipoamide	Transfer of acetyl group to CoA
Dihydrolipoyl dehydrogenase	E_3	12	FAD	Regeneration of the oxidized form of lipoamide

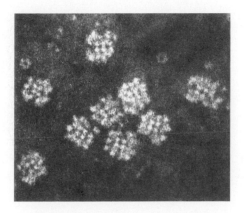

Figure 17.5 **Electron micrograph of the pyruvate dehydrogenase complex from *E. coli*.** [Courtesy of Dr. Lester Reed.]

Mechanism: The Synthesis of Acetyl Coenzyme A from Pyruvate Requires Three Enzymes and Five Coenzymes

The mechanism of the pyruvate dehydrogenase reaction is wonderfully complex, more so than is suggested by its simple stoichiometry. The reaction requires the participation of the three enzymes of the pyruvate dehydrogenase complex and five coenzymes. The coenzymes *thiamine pyrophosphate* (TPP), *lipoic acid,* and *FAD* serve as catalytic cofactors, and CoA and NAD^+ are stoichiometric cofactors.

Thiamine pyrophosphate (TPP)   Lipoic acid

The conversion of pyruvate into acetyl CoA consists of three steps: decarboxylation, oxidation, and transfer of the resultant acetyl group to CoA.

Pyruvate   Acetyl CoA

These steps must be coupled to preserve the free energy derived from the decarboxylation step to drive the formation of NADH and acetyl CoA.

1. *Decarboxylation.* Pyruvate combines with TPP and is then decarboxylated to yield hydroxyethyl-TPP (Figure 17.6).

Carbanion   Pyruvate   Hydroxyethyl-TPP
of TPP

This reaction is catalyzed by the *pyruvate dehydrogenase component* (E_1) of the multienzyme complex. TPP is the prosthetic group of the pyruvate dehydrogenase component.

2. *Oxidation.* The hydroxyethyl group attached to TPP is *oxidized* to form an acetyl group while being simultaneously transferred to lipoamide, a derivative of lipoic acid that is linked to the side chain of a lysine residue by an amide linkage. Note that this transfer results in the formation of an energy-rich thioester bond.

Hydroxyethyl-TPP   Lipoamide   Carbanion of TPP   Acetyllipoamide
(ionized form)

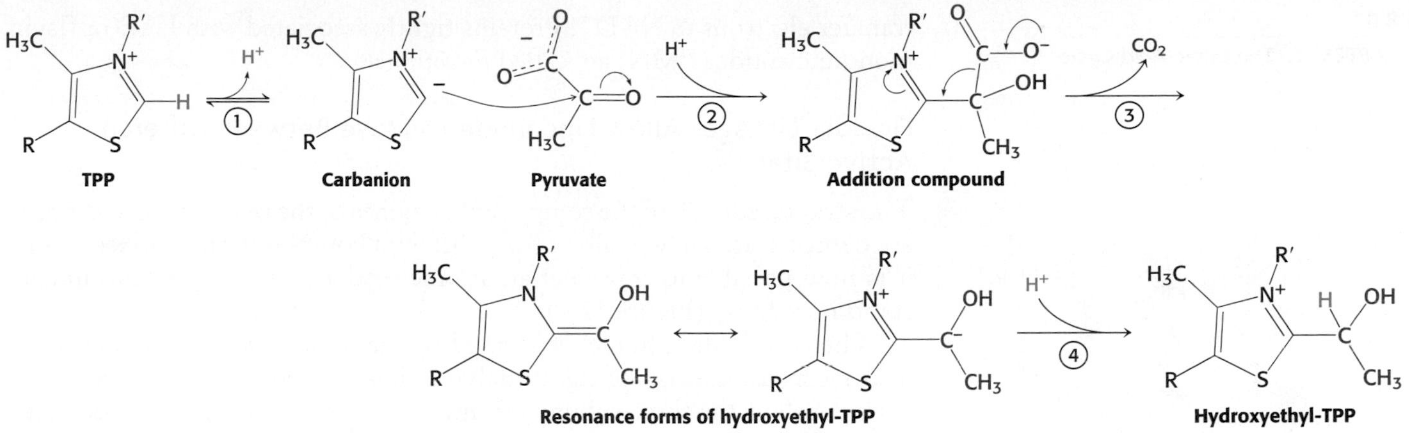

Figure 17.6 Mechanism of the E_1 decarboxylation reaction. E_1 is the pyruvate dehydrogenase component of the pyruvate dehydrogenese complex. A key feature of the prosthetic group, TPP, is that the carbon atom between the nitrogen and sulfur atoms in the thiazole ring is much more acidic than most =CH— groups, with a pK_a value near 10. (1) This carbon center ionizes to form a *carbanion*. (2) The carbanion readily adds to the carbonyl group of pyruvate. (3) This addition is followed by the decarboxylation of pyruvate. The positively charged ring of TPP acts as an electron sink that stabilizes the negative charge that is transferred to the ring as part of the decarboxylation. (4) Protonation yields hydroxyethyl-TPP.

The oxidant in this reaction is the disulfide group of lipoamide, which is re-duced to its disulfhydryl form. This reaction, also catalyzed by the pyruvate dehydrogenase component E_1, yields *acetyllipoamide*.

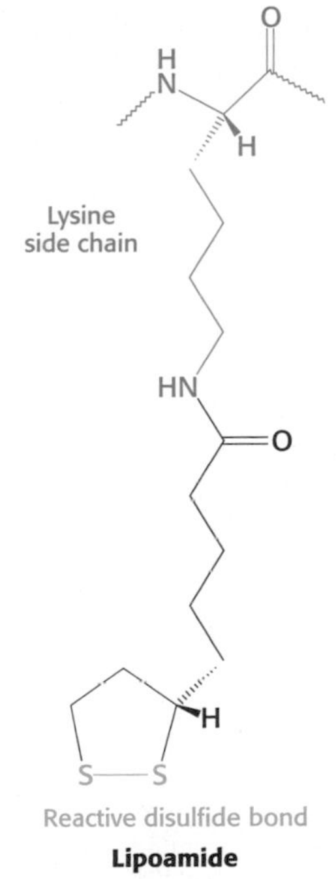

3. *Formation of Acetyl CoA. The acetyl group is transferred from acetyl-lipoamide to CoA to form acetyl CoA.*

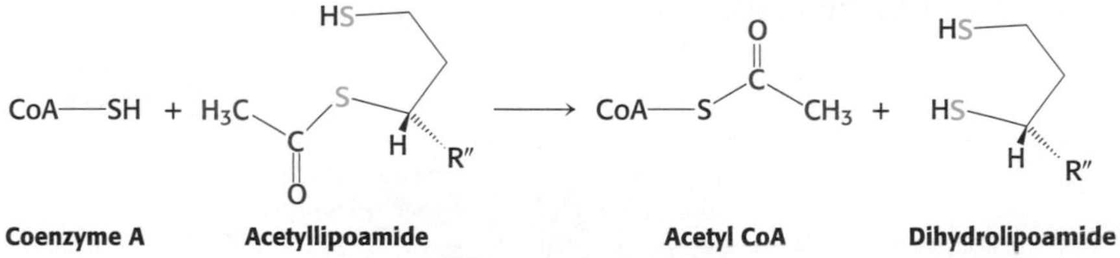

Dihydrolipoyl transacetylase (E_2) catalyzes this reaction. The energy-rich thioester bond is preserved as the acetyl group is transferred to CoA. Recall that CoA serves as a carrier of many activated acyl groups, of which acetyl is the simplest (p. 422). Acetyl CoA, the fuel for the citric acid cycle, has now been generated from pyruvate.

The pyruvate dehydrogenase complex cannot complete another catalytic cycle until the dihydrolipoamide is oxidized to lipoamide. In a fourth step, *the oxidized form of lipoamide is regenerated by dihydrolipoyl dehydrogenase* (E_3). Two electrons are transferred to an FAD prosthetic group of the en-zyme and then to NAD^+.

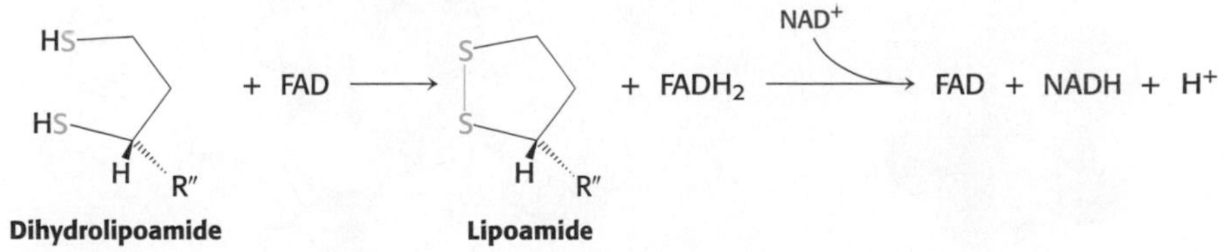

This electron transfer from FAD to NAD^+ is unusual, because the common role for FAD is to receive electrons from NADH. The electron-transfer po-tential of FAD is increased by its association with the enzyme, enabling it to

Lysine
side chain

Reactive disulfide bond
Lipoamide

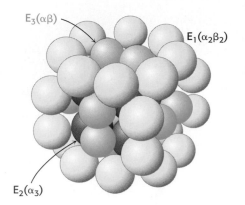

Figure 17.7 Schematic representation of the pyruvate dehydrogenase complex. The transacetylase core (E₂) is shown in red, the pyruvate dehydrogenase component (E₁) in yellow, and the dihydrolipoyl dehydrogenase (E₃) in green.

transfer electrons to NAD^+. Proteins tightly associated with FAD or flavin mononucleotide (FMN) are called *flavoproteins*.

Flexible Linkages Allow Lipoamide to Move Between Different Active Sites

The structures of all of the component enzymes of the pyruvate dehydrogenase complex are known, albeit from different complexes and species. Thus, it is now possible to construct an atomic model of the complex to understand its activity (Figure 17.7).

The core of the complex is formed by the transacetylase component E_2. Transacetylase consists of eight catalytic trimers assembled to form a hollow cube. Each of the three subunits forming a trimer has three major domains (Figure 17.8). At the amino terminus is a small domain that contains a bound flexible lipoamide cofactor attached to a lysine residue. This domain is homologous to biotin-binding domains such as that of pyruvate carboxylase (see Figure 16.26). The lipoamide domain is followed by a small domain that interacts with E_3 within the complex. A larger transacetylase domain completes an E_2 subunit. E_1 is an $\alpha_2\beta_2$ tetramer, and E_3 is a $\alpha\beta$ dimer. Twenty-four copies of E_1 and 12 copies of E_3 surround the E_2 core. How do the three distinct active sites work in concert (Figure 17.9)? The key is the long, flexible lipoamide arm of the E_2 subunit, which carries substrate from active site to active site.

1. Pyruvate is decarboxylated at the active site of E_1, forming the hydroxyethyl-TPP intermediate, and CO_2 leaves as the first product. This active site lies deep within the E_1 complex, connected to the enzyme surface by a 20-Å-long hydrophobic channel.

2. E_2 inserts the lipoamide arm of the lipoamide domain into the deep channel in E_1 leading to the active site.

Figure 17.8 Structure of the transacetylase (E₂) core. Each red ball represents a trimer of three E₂ subunits. *Notice* that each subunit consists of three domains: a lipoamide-binding domain, a small domain for interaction with E₃, and a large transacetylase catalytic domain. The transacetylase domain has three subunits, with one depicted in red and the others in white in the ribbon representation.

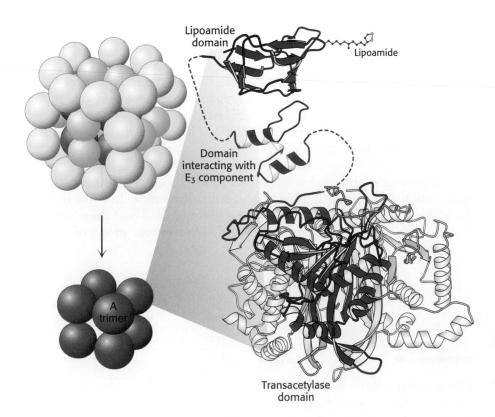

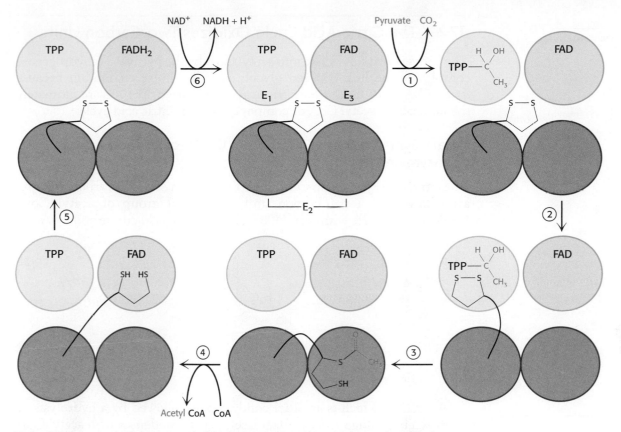

Figure 17.9 Reactions of the pyruvate dehydrogenase complex. At the top (center), the enzyme (represented by a yellow, a green, and two red spheres) is unmodified and ready for a catalytic cycle. (1) Pyruvate is decarboxylated to form hydroxyethyl-TPP. (2) The lipoamide arm of E_2 moves into the active site of E_1. (3) E_1 catalyzes the transfer of the two-carbon group to the lipoamide group to form the acetyl–lipoamide complex. (4) E_2 catalyzes the transfer of the acetyl moiety to CoA to form the product acetyl CoA. The dihydrolipoamide arm then swings to the active site of E_3. E_3 catalyzes (5) the oxidation of the dihydrolipoamide acid and (6) the transfer of the protons and electrons to NAD^+ to complete the reaction cycle.

3. E_1 catalyzes the transfer of the acetyl group to the lipoamide. The acetylated arm then leaves E_1 and enters the E_2 cube to visit the active site of E_2, located deep in the cube at the subunit interface.

4. The acetyl moiety is then transferred to CoA, and the second product, acetyl CoA, leaves the cube. The reduced lipoamide arm then swings to the active site of the E_3 flavoprotein.

5. At the E_3 active site, the lipoamide is oxidized by coenzyme FAD. The reactivated lipoamide is ready to begin another reaction cycle.

6. The final product, NADH, is produced with the reoxidation of $FADH_2$ to FAD.

   The structural integration of three kinds of enzymes and the long flexible lipoamide arm makes the coordinated catalysis of a complex reaction possible. The proximity of one enzyme to another increases the overall reaction rate and minimizes side reactions. All the intermediates in the oxidative decarboxylation of pyruvate remain bound to the complex throughout the reaction sequence and are readily transferred as the flexible arm of E_2 calls on each active site in turn.

17.2 The Citric Acid Cycle Oxidizes Two-Carbon Units

The conversion of pyruvate into acetyl CoA by the pyruvate dehydrogenase complex is the link between glycolysis and cellular respiration because *acetyl CoA is the fuel for the citric acid cycle*. Indeed, all fuels are ultimately metabolized to acetyl CoA or components of the citric acid cycle.

Citrate Synthase Forms Citrate from Oxaloacetate and Acetyl Coenzyme A

The citric acid cycle begins with the condensation of a four-carbon unit, oxaloacetate, and a two-carbon unit, the acetyl group of acetyl CoA. Oxaloacetate reacts with acetyl CoA and H_2O to yield citrate and CoA.

Oxaloacetate   **Acetyl CoA**   **Citryl CoA**   **Citrate**

Synthase

An enzyme catalyzing a synthetic reaction in which two units are joined usually without the direct participation of ATP (or another nucleoside triphosphate).

This reaction, which is an aldol condensation followed by a hydrolysis, is catalyzed by *citrate synthase*. Oxaloacetate first condenses with acetyl CoA to form *citryl CoA,* an energy-rich molecule because it contains the thioester bond that originated in acetyl CoA. The hydrolysis of citryl CoA thioester to citrate and CoA drives the overall reaction far in the direction of the synthesis of citrate. In essence, the hydrolysis of the thioester powers the synthesis of a new molecule from two precursors.

Mechanism: The Mechanism of Citrate Synthase Prevents Undesirable Reactions

Because the condensation of acetyl CoA and oxaloacetate initiates the citric acid cycle, it is very important that side reactions be minimized. Let us briefly consider how the citrate synthase prevents wasteful processes such as the hydrolysis of acetyl CoA.

Mammalian citrate synthase is a dimer of identical 49-kd subunits. Each active site is located in a cleft between the large and the small domains of a subunit, adjacent to the subunit interface. X-ray crystallographic studies of citrate synthase and its complexes with several substrates and inhibitors revealed that the enzyme undergoes large conformational changes in the course of catalysis. Citrate synthase exhibits sequential, ordered kinetics: oxaloacetate binds first, followed by acetyl CoA. The reason for the ordered binding is that *oxaloacetate induces a major structural rearrangement leading to the creation of a binding site for acetyl CoA*. The binding of oxaloacetate converts the open form of the enzyme into a closed form (Figure 17.10). In each subunit, the small domain rotates 19 degrees relative to the large domain. *Movements as large as 15 Å are produced by the rotation of α helices elicited by quite small shifts of side chains around bound oxaloacetate.* These structural changes create a binding site for acetyl CoA. This conformational transition is reminiscent of the cleft closure in hexokinase induced by the binding of glucose (p. 437).

Citrate synthase catalyzes the condensation reaction by bringing the substrates into close proximity, orienting them, and polarizing certain

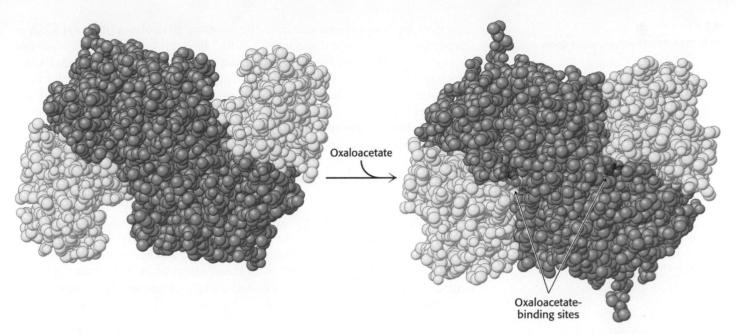

Oxaloacetate

Oxaloacetate-
binding sites

Figure 17.10 Conformational changes in citrate synthase on binding oxaloacetate. The small domain of each subunit of the homodimer is shown in yellow; the large domains are shown in blue. (Left) Open form of enzyme alone. (Right) Closed form of the liganded enzyme. [Drawn from 5CSC.pdb and 4CTS.pdb.]

bonds (Figure 17.11). The donation and removal of protons transforms acetyl CoA into an *enol intermediate*. The enol attacks oxaloacetate to form a carbon–carbon double bond linking acetyl CoA and oxaloacetate. The newly formed citryl CoA induces additional structural changes in the enzyme, causing the active site to become completely enclosed. The enzyme cleaves the citryl CoA thioester by hydrolysis. CoA leaves the enzyme, followed by citrate, and the enzyme returns to the initial open conformation.

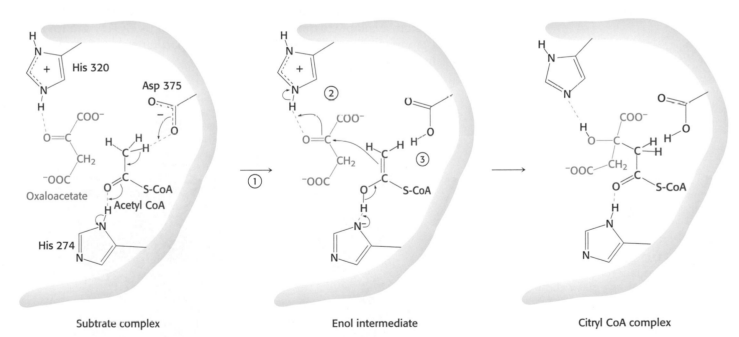

Subtrate complex

Enol intermediate

Citryl CoA complex

Figure 17.11 Mechanism of synthesis of citryl CoA by citrate synthase. (1) In the substrate complex (left), His 274 donates a proton to the carbonyl oxygen of acetyl CoA to promote the removal of a methyl proton by Asp 375 to form the enol intermediate (center). (2) Oxaloacetate is activated by the transfer of a proton from His 320 to its carbonyl carbon atom. (3) Simultaneously, the enol of acetyl CoA attacks the carbonyl carbon of oxaloacetate to form a carbon–carbon bond linking acetyl CoA and oxaloacetate. His 274 is reprotonated. Citryl CoA is formed. His 274 participates again as a proton donor to hydrolyze the thioester (not shown), yielding citrate and CoA.

We can now understand how the wasteful hydrolysis of acetyl CoA is prevented. Citrate synthase is well suited to hydrolyze *citryl* CoA but not *acetyl* CoA. How is this discrimination accomplished? First, acetyl CoA does not bind to the enzyme until oxaloacetate is bound and ready for condensation. Second, the catalytic residues crucial for the hydrolysis of the thioester linkage are not appropriately positioned *until citryl CoA is formed.* As with hexokinase (p. 437) and triose phosphate isomerase (p. 440), *induced fit prevents an undesirable side reaction.*

Citrate Is Isomerized into Isocitrate

The hydroxyl group is not properly located in the citrate molecule for the oxidative decarboxylations that follow. Thus, citrate is isomerized into isocitrate to enable the six-carbon unit to undergo oxidative decarboxylation. The isomerization of citrate is accomplished by a *dehydration* step followed by a *hydration* step. The result is an interchange of an H and an OH. The enzyme catalyzing both steps is called *aconitase* because cis-*aconitate* is an intermediate.

$$
\begin{array}{ccccc}
\text{Citrate} & \overset{H_2O}{\rightleftharpoons} & \textit{cis-}\text{Aconitate} & \overset{H_2O}{\rightleftharpoons} & \text{Isocitrate}
\end{array}
$$

Aconitase is an *iron–sulfur protein,* or *nonheme-iron protein,* in that it contains iron that is not bonded to heme. Rather, its four iron atoms are complexed to four inorganic sulfides and three cysteine sulfur atoms, leaving one iron atom available to bind citrate through one of its COO^- groups and an OH group (Figure 17.12). This Fe-S cluster participates in dehydrating and rehydrating the bound substrate.

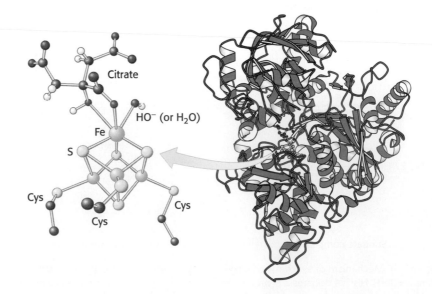

Figure 17.12 Binding of citrate to the iron–sulfur complex of aconitase. A 4Fe-4S iron–sulfur cluster is a component of the active site of aconitase. *Notice* that one of the iron atoms of the cluster binds to a COO^- group and an OH group of citrate. [Drawn from 1C96.pdb.]

Isocitrate Is Oxidized and Decarboxylated to α-Ketoglutarate

We come now to the first of four oxidation–reduction reactions in the citric acid cycle. The oxidative decarboxylation of isocitrate is catalyzed by *isocitrate dehydrogenase.*

$$\text{Isocitrate} + \text{NAD}^+ \longrightarrow \alpha\text{-ketoglutarate} + CO_2 + \text{NADH}$$

The intermediate in this reaction is oxalosuccinate, an unstable β-ketoacid. While bound to the enzyme, it loses CO_2 to form α-ketoglutarate.

The rate of formation of α-ketoglutarate is important in determining the overall rate of the cycle, as will be discussed on page 492. This oxidation generates the first high-transfer-potential electron carrier, NADH, in the cycle.

Succinyl Coenzyme A Is Formed by the Oxidative Decarboxylation of α-Ketoglutarate

The conversion of isocitrate into α-ketoglutarate is followed by a second oxidative decarboxylation reaction, the formation of succinyl CoA from α-ketoglutarate.

This reaction is catalyzed by the *α-ketoglutarate dehydrogenase complex*, an organized assembly of three kinds of enzymes that is homologous to the pyruvate dehydrogenase complex. In fact, the oxidative decarboxylation of α-ketoglutarate closely resembles that of pyruvate, also an α-ketoacid.

$$\text{Pyruvate} + \text{CoA} + \text{NAD}^+ \xrightarrow{\text{Pyruvate dehydrogenase complex}} \text{acetyl CoA} + CO_2 + \text{NADH}$$

$$\alpha\text{-Ketoglutarate} + \text{CoA} + \text{NAD}^+ \xrightarrow{\alpha\text{-Ketoglutarate dehydrogenase complex}} \text{succinyl CoA} + CO_2 + \text{NADH}$$

Both reactions include the decarboxylation of an α-ketoacid and the subsequent formation of a thioester linkage with CoA that has a high transfer potential. The reaction mechanisms are entirely analogous (p. 478).

A Compound with High Phosphoryl-Transfer Potential Is Generated from Succinyl Coenzyme A

Succinyl CoA is an energy-rich thioester compound. The $\Delta G^{\circ\prime}$ for the hydrolysis of succinyl CoA is about $-33.5\ \text{kJ mol}^{-1}$ ($-8.0\ \text{kcal mol}^{-1}$), which is comparable to that of ATP ($-30.5\ \text{kJ mol}^{-1}$, or $-7.3\ \text{kcal mol}^{-1}$). In the citrate synthase reaction, the cleavage of the thioester bond powers the synthesis of the six-carbon citrate from the four-carbon oxaloacetate and the two-carbon fragment. *The cleavage of the thioester bond of succinyl CoA is coupled to the phosphorylation of a purine nucleoside diphosphate, usually GDP.* This reaction is catalyzed by *succinyl CoA synthetase* (succinate thiokinase).

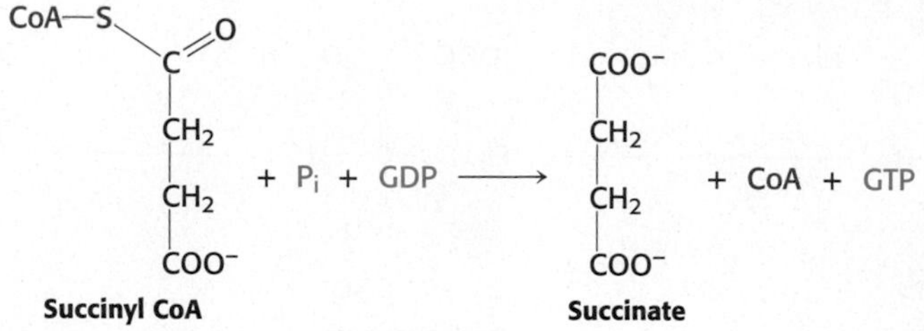

This reaction is the only step in the citric acid cycle that directly yields a compound with high phosphoryl-transfer potential. Some mammalian succinyl CoA synthetases are specific for GDP and others for ADP. The *E. coli* enzyme uses either GDP or ADP as the phosphoryl-group acceptor. We have already seen that GTP is an important component of signal-transduction systems (Chapter 14). Alternatively, its γ phosphoryl group can be readily transferred to ADP to form ATP, in a reaction catalyzed by *nucleoside diphosphokinase*.

$$GTP + ADP \rightleftharpoons GDP + ATP$$

Mechanism: Succinyl Coenzyme A Synthetase Transforms Types of Biochemical Energy

The mechanism of this reaction is a clear example of an energy transformation: energy inherent in the thioester molecule is transformed into phosphoryl-group-transfer potential (Figure 17.13). The first step is the displacement of coenzyme A by orthophosphate, which generates another energy-rich compound, succinyl phosphate. A histidine residue plays a key role as a moving arm that detaches the phosphoryl group, then swings over to a bound nucleoside diphosphate and transfers the group to form the nucleoside triphosphate. The participation of high-energy compounds in all the steps is attested to by the fact that the reaction is readily reversible: $\Delta G^{\circ\prime} = -3.4$ kJ mol^{-1} (-0.8 kcal mol^{-1}). The formation of GTP at the expense of succinyl CoA is an example of substrate-level phosphorylation.

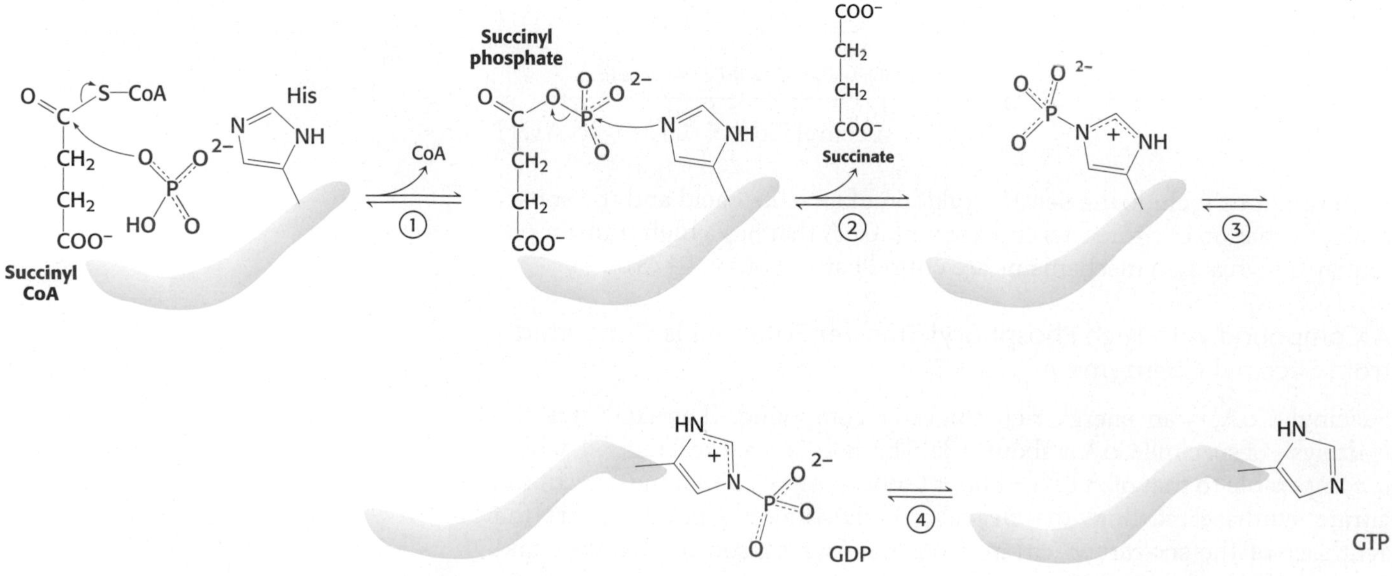

Figure 17.13 Reaction mechanism of succinyl CoA synthetase. The reaction proceeds through a phosphorylated enzyme intermediate. (1) Orthophosphate displaces coenzyme A, which generates another energy-rich compound, succinyl phosphate. (2) A histidine residue removes the phosphoryl group with the concomitant generation of succinate and phosphohistidine. (3) The phosphohistidine residue then swings over to a bound nucleoside diphosphate, and (4) the phosphoryl group is transferred to form the nucleoside triphosphate.

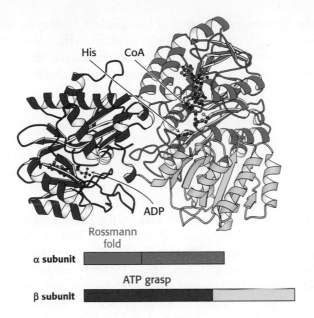

α subunit

Rossmann fold

β subunit

ATP grasp

Succinyl CoA synthetase is an $\alpha_2\beta_2$ heterodimer; the functional unit is one $\alpha\beta$ pair. The enzyme mechanism shows that a phosphoryl group is transferred first to succinyl CoA bound in the α subunit and then to a nucleoside diphosphate bound in the β subunit. Examination of the three-dimensional structure of succinyl CoA synthetase reveals that each subunit comprises two domains (Figure 17.14). The amino-terminal domains of the two subunits have different structures, each characteristic of its role in the mechanism. The amino-terminal domain of the α subunit forms a Rossmann fold (p. 449), which binds the ADP component of succinyl CoA. The amino-terminal domain of the β subunit is an ATP-grasp domain, found in many enzymes, which here binds and activates GDP. Succinyl CoA synthetase has evolved by adopting these domains and harnessing them to capture the energy associated with succinyl CoA cleavage, which is used to drive the generation of a nucleoside triphosphate.

Oxaloacetate Is Regenerated by the Oxidation of Succinate

Reactions of four-carbon compounds constitute the final stage of the citric acid cycle: the regeneration of oxaloacetate.

Succinate → (FAD → FADH$_2$) → **Fumarate** → (H$_2$O) → **Malate** → (NAD$^+$ → NADH + H$^+$) → **Oxaloacetate**

The reactions constitute a metabolic motif that we will see again in fatty acid synthesis and degradation as well as in the degradation of some amino acids. A methylene group (CH$_2$) is converted into a carbonyl group (C=O) in three steps: an oxidation, a hydration, and a second oxidation reaction. Oxaloacetate is thereby regenerated for another round of the cycle, and more energy is extracted in the form of FADH$_2$ and NADH.

Succinate is oxidized to fumarate by *succinate dehydrogenase*. The hydrogen acceptor is FAD rather than NAD$^+$, which is used in the other three oxidation reactions in the cycle. FAD is the hydrogen acceptor in this reaction because the free-energy change is insufficient to reduce NAD^+. FAD is nearly always the electron acceptor in oxidations that remove two hydrogen *atoms* from a substrate. In succinate dehydrogenase, the isoalloxazine ring of FAD is covalently attached to a histidine side chain of the enzyme (denoted E-FAD).

$$E\text{-}FAD + \text{succinate} \rightleftharpoons E\text{-}FADH_2 + \text{fumarate}$$

Succinate dehydrogenase, like aconitase, is an iron–sulfur protein. Indeed, succinate dehydrogenase contains three different kinds of iron–sulfur clusters, 2Fe-2S (two iron atoms bonded to two inorganic sulfides), 3Fe-4S, and 4Fe-4S. Succinate dehydrogenase—which consists of a 70-kd and a 27-kd subunit—differs from other enzymes in the citric acid cycle in being embedded in the inner mitochondrial membrane. In fact, *succinate dehydrogenase is directly associated with the electron-transport chain, the link between the citric acid cycle and ATP formation.* $FADH_2$ produced by the oxidation of succinate does not dissociate from the enzyme, in contrast with NADH produced in other oxidation–reduction reactions. Rather, two electrons are transferred from $FADH_2$ directly to iron–sulfur clusters of the enzyme, which in turn passes the electrons to coenzyme Q (CoQ). Coenzyme Q, an important member of the electron-transport chain, passes electrons to the ultimate acceptor, molecular oxygen, as we shall see in Chapter 18.

The next step is the hydration of fumarate to form L-malate. *Fumarase* catalyzes a stereospecific trans addition of H^+ and OH^-. The OH^- group adds to only one side of the double bond of fumarate; hence, only the L isomer of malate is formed.

Finally, malate is oxidized to form oxaloacetate. This reaction is catalyzed by *malate dehydrogenase*, and NAD^+ is again the hydrogen acceptor.

$$\text{Malate} + NAD^+ \rightleftharpoons \text{oxaloacetate} + NADH + H^+$$

The standard free energy for this reaction, unlike that for the other steps in the citric acid cycle, is significantly positive ($\Delta G^{\circ\prime} = +29.7$ kJ mol^{-1}, or $+7.1$ kcal mol^{-1}). The oxidation of malate is driven by the use of the products—oxaloacetate by citrate synthase and NADH by the electron-transport chain.

The Citric Acid Cycle Produces High-Transfer-Potential Electrons, GTP, and CO_2

The net reaction of the citric acid cycle is

$$\text{Acetyl CoA} + 3\,NAD^+ + FAD + GDP + P_i + 2\,H_2O \rightarrow$$
$$2\,CO_2 + 3\,NADH + FADH_2 + GTP + 2\,H^+ + CoA$$

Let us recapitulate the reactions that give this stoichiometry (Figure 17.15 and Table 17.2):

1. Two carbon atoms enter the cycle in the condensation of an acetyl unit (from acetyl CoA) with oxaloacetate. Two carbon atoms leave the cycle in the form of CO_2 in the successive decarboxylations catalyzed by isocitrate dehydrogenase and α-ketoglutarate dehydrogenase.

2. Four pairs of hydrogen atoms leave the cycle in four oxidation reactions. Two NAD^+ molecules are reduced in the oxidative decarboxylations of isocitrate and α-ketoglutarate, one FAD molecule is reduced in the oxidation of succinate, and one NAD^+ molecule is reduced in the oxidation of

Fumarate

L-Malate

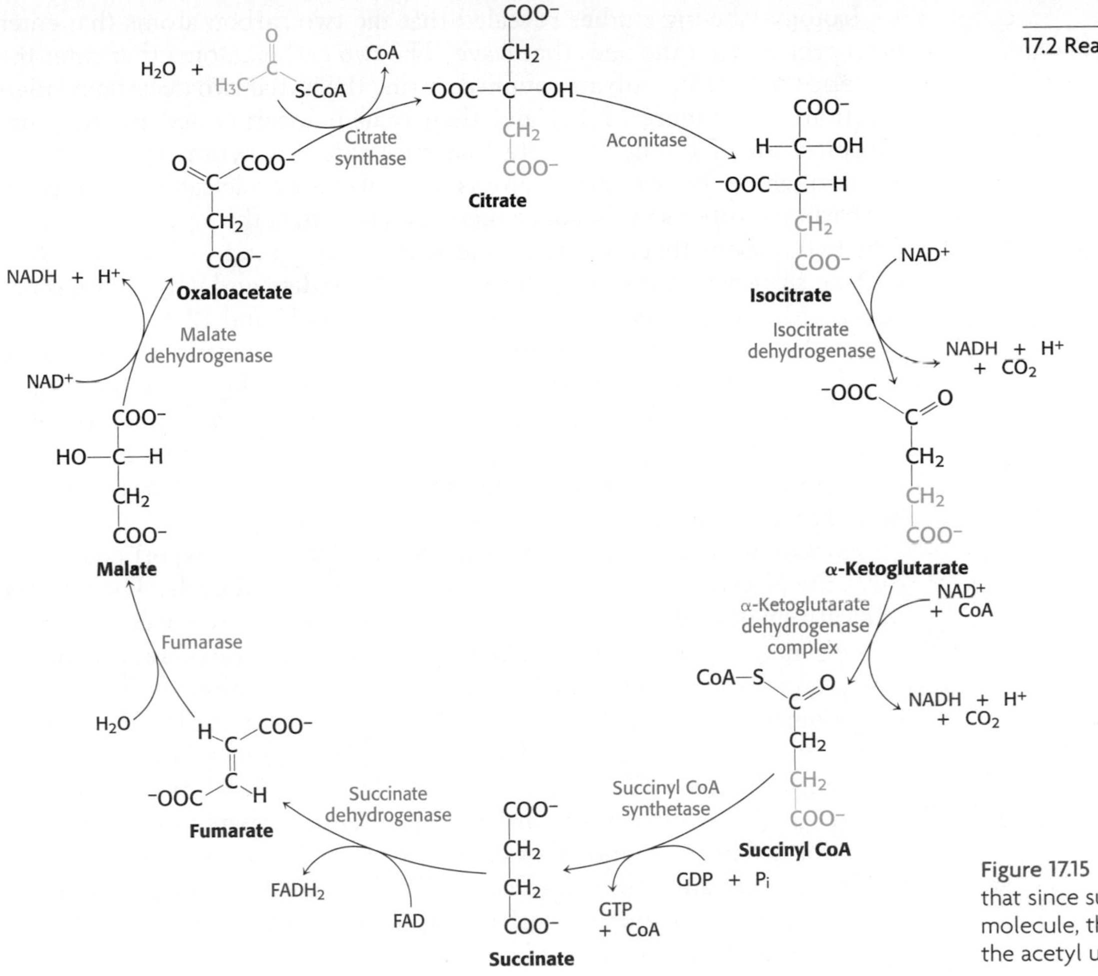

Figure 17.15 The citric acid cycle. *Notice that since succinate is a symmetric molecule, the identity of the carbons from the acetyl unit is lost.*

malate. Recall also that one NAD^+ molecule is reduced in the oxidative decarboxylation of pyruvate to form acetyl CoA.

3. One compound with high phosphoryl-transfer potential, usually GTP, is generated from the cleavage of the thioester linkage in succinyl CoA.

4. Two water molecules are consumed: one in the synthesis of citrate by the hydrolysis of citryl CoA and the other in the hydration of fumarate.

TABLE 17.2 Citric acid cycle

Step	Reaction	Enzyme	Prosthetic group	Type*	$\Delta G^{\circ\prime}$ kJ mol^{-1}	kcal mol^{-1}
1	Acetyl CoA + oxaloacetate + $H_2O \rightarrow$ citrate + CoA + H^+	Citrate synthase		a	−31.4	−7.5
2a	Citrate $\rightleftharpoons$ cis-aconitate + H_2O	Aconitase	Fe-S	b	+8.4	+2.0
2b	cis-Aconitate + $H_2O \rightleftharpoons$ isocitrate	Aconitase	Fe-S	c	−2.1	−0.5
3	Isocitrate + $NAD^+ \rightleftharpoons$ α-ketoglutarate + CO_2 + NADH	Isocitrate dehydrogenase		d + e	−8.4	−2.0
4	α-Ketoglutarate + NAD^+ + CoA $\rightleftharpoons$ succinyl CoA + CO_2 + NADH	α-Ketoglutarate dehydrogenase complex	Lipoic acid, FAD, TPP	d + e	−30.1	−7.2
5	Succinyl CoA + P_i + GDP $\rightleftharpoons$ succinate + GTP + CoA	Succinyl CoA synthetase		f	−3.3	−0.8
6	Succinate + FAD (enzyme-bound) $\rightleftharpoons$ fumarate + $FADH_2$ (enzyme-bound)	Succinate dehydrogenase	FAD, Fe-S	e	0	0
7	Fumarate + $H_2O \rightleftharpoons$ L-malate	Fumarase		c	−3.8	−0.9
8	L-Malate + $NAD^+ \rightleftharpoons$ oxaloacetate + NADH + H^+	Malate dehydrogenase		e	+29.7	+7.1

*Reaction type: (a) condensation; (b) dehydration; (c) hydration; (d) decarboxylation; (e) oxidation; (f) substrate-level phosphorylation.

Isotope-labeling studies revealed that the two carbon atoms that enter each cycle are not the ones that leave. The two carbon atoms that enter the cycle as the acetyl group are retained during the initial two decarboxylation reactions (see Figure 17.15) and then remain incorporated in the four-carbon acids of the cycle. Note that succinate is a symmetric molecule. Consequently, the two carbon atoms that enter the cycle can occupy any of the carbon positions in the subsequent metabolism of the four-carbon acids. The two carbons that enter the cycle as the acetyl group will be released as CO_2 in *subsequent* trips through the cycle. To understand why citrate is not processed as a symmetric molecule, see problems 11 and 12.

Evidence is accumulating that the enzymes of the citric acid cycle are physically associated with one another. The close arrangement of enzymes enhances the efficiency of the citric acid cycle because a reaction product can pass directly from one active site to the next through connecting channels, a process called *substrate channeling*. The word *metabolon* has been suggested as the name for such multienzyme complexes.

As will be considered in Chapter 18, the electron-transport chain oxidizes the NADH and $FADH_2$ formed in the citric acid cycle. The transfer of electrons from these carriers to O_2, the ultimate electron acceptor, leads to the generation of a proton gradient across the inner mitochondrial membrane. This proton-motive force then powers the generation of ATP; the net stoichiometry is about 2.5 ATP per NADH, and 1.5 ATP per $FADH_2$. Consequently, nine high-transfer-potential phosphoryl groups are generated when the electron-transport chain oxidizes 3 NADH molecules and 1 $FADH_2$ molecule, and one high-transfer-potential phosphoryl group is directly formed in one round of the citric acid cycle. Thus, one acetyl unit generates approximately 10 molecules of ATP. In dramatic contrast, the anaerobic glycolysis of 1 glucose molecule generates only 2 molecules of ATP (and 2 molecules of lactate).

Recall that molecular oxygen does not participate directly in the citric acid cycle. However, the cycle operates only under aerobic conditions because NAD^+ and FAD can be regenerated in the mitochondrion only by the transfer of electrons to molecular oxygen. *Glycolysis has both an aerobic and an anaerobic mode, whereas the citric acid cycle is strictly aerobic.* Glycolysis can proceed under anaerobic conditions because NAD^+ is regenerated in the conversion of pyruvate into lactate or ethanol.

17.3 Entry to the Citric Acid Cycle and Metabolism Through It Are Controlled

The citric acid cycle is the final common pathway for the aerobic oxidation of fuel molecules. Moreover, as we will see shortly (Section 17.4) and repeatedly elsewhere in our study of biochemistry, the cycle is an important source of building blocks for a host of important biomolecules. As befits its role as the metabolic hub of the cell, entry into the cycle and the rate of the cycle itself are controlled at several stages.

The Pyruvate Dehydrogenase Complex Is Regulated Allosterically and by Reversible Phosphorylation

As stated earlier, glucose can be formed from pyruvate (Section 16.3). *However, the formation of acetyl CoA from pyruvate is an irreversible step in animals and thus they are unable to convert acetyl CoA back into glucose.* The oxidative decarboxylation of pyruvate to acetyl CoA commits the carbon atoms of glucose to two principal fates: oxidation to CO_2 by the citric acid cycle, with the concomitant generation of energy, or incorporation into lipid (Figure 17.16). As expected of an enzyme at a critical branch point in metabolism, the activity of the pyruvate dehydrogenase complex is stringently controlled. High concentrations of reaction products inhibit the reaction: acetyl CoA inhibits the transacetylase component (E_2) by direct binding, whereas NADH inhibits the dihydrolipoyl dehydrogenase (E_3). High concentrations of NADH and acetyl CoA inform the enzyme that the energy needs of the cell have been met or that fatty acids are being degraded to produce acetyl CoA and NADH (p. 624). In either case, there is no need to metabolize pyruvate to acetyl CoA. This inhibition has the effect of sparing glucose, because most pyruvate is derived from glucose by glycolysis (Section 16.1).

The key means of regulation of the complex in eukaryotes is covalent modification (Figure 17.17). *Phosphorylation of the pyruvate dehydrogenase component (E_1) by a specific kinase switches off the activity of the complex. Deactivation is reversed by the action of a specific phosphatase.* The kinase is associated with the transacetylase component (E_2), again highlighting the structural and mechanistic importance of this core. Both the kinase and the phosphatase are regulated. To see how this regulation works in biological conditions, consider muscle that is becoming active after a period of rest (Figure 17.18). At rest, the muscle will not have significant energy demands. Consequently, the NADH/NAD$^+$, acetyl CoA/CoA, and ATP/ADP ratios will be high. These high ratios promote phosphorylation and, hence, deactivation of the pyruvate dehydrogenase complex. In other words, high concentrations of immediate (acetyl CoA and NADH) and ultimate (ATP) products inhibit the activity. Thus, *pyruvate dehydrogenase is switched off when the energy charge is high.*

As exercise begins, the concentrations of ADP and pyruvate will increase as muscle contraction consumes ATP and glucose is converted into

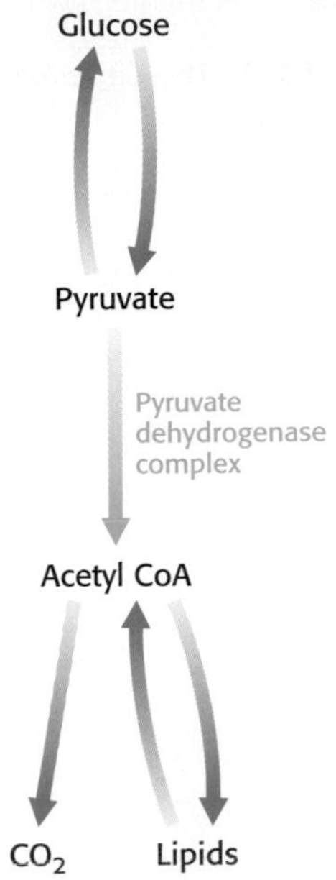

Figure 17.16 From glucose to acetyl CoA. The synthesis of acetyl CoA by the pyruvate dehydrogenase complex is a key irreversible step in the metabolism of glucose.

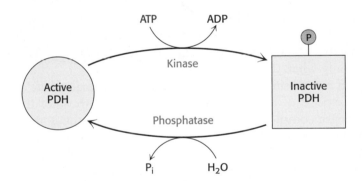

Figure 17.17 Regulation of the pyruvate dehydrogenase complex. A specific kinase phosphorylates and inactivates pyruvate dehydrogenase (PDH), and a phosphatase activates the dehydrogenase by removing the phosphoryl group. The kinase and the phosphatase also are highly regulated enzymes.

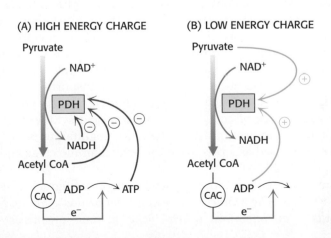

Figure 17.18 Response of the pyruvate dehydrogenase complex to the energy charge. The pyruvate dehydrogenase complex is regulated to respond to the energy charge of the cell. (A) The complex is inhibited by its immediate products, NADH and acetyl CoA, as well as by the ultimate product of cellular respiration, ATP. (B) The complex is activated by pyruvate and ADP, which inhibit the kinase that phosphorylates PDH.

491

pyruvate to meet the energy demands. Both ADP and pyruvate activate the dehydrogenase by inhibiting the kinase. Moreover, the phosphatase is stimulated by Ca^{2+}, the same signal that initiates muscle contraction. A rise in the cytoplasmic Ca^{2+} level (p. 355) elevates the mitochondrial Ca^{2+} level. The rise in mitochondrial Ca^{2+} activates the phosphatase, enhancing pyruvate dehydrogenase activity.

In some tissues, the phosphatase is regulated by hormones. In liver, epinephrine binds to the α-adrenergic receptor to initiate the phosphatidyl inositol pathway (p. 388), causing an increase in Ca^{2+} concentration that activates the phosphatase. In tissues capable of fatty acid synthesis, such as the liver and adipose tissue, insulin, the hormone that signifies the fed state, stimulates the phosphatase, increasing the conversion of pyruvate into acetyl CoA. Acetyl CoA is the precursor for fatty acid synthesis (p. 635). In these tissues, the pyruvate dehydrogenase complex is activated to funnel glucose to pyruvate and then to acetyl CoA and ultimately to fatty acids.

In people with a phosphatase deficiency, pyruvate dehydrogenase is always phosphorylated and thus inactive. Consequently, glucose is processed to lactate rather than acetyl CoA. This condition results in unremitting lactic acidosis—high blood levels of lactic acid. In such an acidic environment, many tissues malfunction, most notably the central nervous system.

The Citric Acid Cycle Is Controlled at Several Points

The rate of the citric acid cycle is precisely adjusted to meet an animal cell's needs for ATP (Figure 17.19). The primary control points are the allosteric enzymes isocitrate dehydrogenase and α-ketoglutarate dehydrogenase, the first two enzymes to generate high-energy electrons in the cycle.

The first control site is isocitrate dehydrogenase. The enzyme is allosterically stimulated by ADP, which enhances the enzyme's affinity for substrates. The binding of isocitrate, NAD^+, Mg^{2+}, and ADP is mutually cooperative. In contrast, ATP is inhibitory. The reaction product NADH inhibits isocitrate dehydrogenase by directly displacing NAD^+. It is important to note that several steps in the cycle require NAD^+ or FAD, which are abundant only when the energy charge is low.

A second control site in the citric acid cycle is α-ketoglutarate dehydrogenase. Some aspects of this enzyme's control are like those of the pyruvate dehydrogenase complex, as might be expected from the homology of the two enzymes. α-Ketoglutarate dehydrogenase is inhibited by succinyl CoA and NADH, the products of the reaction that it catalyzes. In addition, α-ketoglutarate dehydrogenase is inhibited by a high energy charge. Thus, *the rate of the cycle is reduced when the cell has a high level of ATP.*

The use of isocitrate dehydrogenase and α-ketoglutarate dehydrogenase as control points integrates the citric acid cycle with other pathways and highlights the central role of the citric acid cycle in metabolism. For instance, the inhibition of isocitrate dehydrogenase leads to a buildup of citrate, because the interconversion of isocitrate and citrate is readily reversible under intracellular conditions. Citrate can be transported to the cytoplasm where it signals phosphofructokinase to halt glycolysis (p. 455) and where it can serve as a source of acetyl CoA for fatty acid synthesis (p. 638). The α-ketoglutarate that accumulates when α-ketoglutarate dehydrogenase is inhibited can be used as a precursor for several amino acids and the purine bases (pp. 683 and 714).

In many bacteria, the funneling of two-carbon fragments into the cycle also is controlled. *The synthesis of citrate from oxaloacetate and acetyl CoA carbon units is an important control point in these organisms.* ATP is an

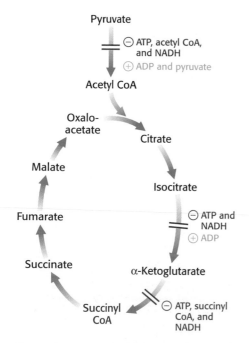

Figure 17.19 Control of the citric acid cycle. The citric acid cycle is regulated primarily by the concentration of ATP and NADH. The key control points are the enzymes isocitrate dehydrogenase and α-ketoglutarate dehydrogenase.

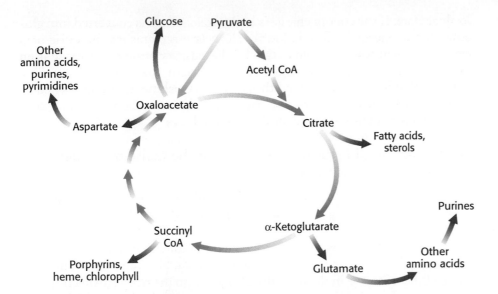

Figure 17.20 Biosynthetic roles of the citric acid cycle. Intermediates are drawn off for biosyntheses (shown by red arrows) when the energy needs of the cell are met. Intermediates are replenished by the formation of oxaloacetate from pyruvate.

allosteric inhibitor of citrate synthase. The effect of ATP is to increase the value of K_M for acetyl CoA. Thus, as the level of ATP increases, less of this enzyme is saturated with acetyl CoA and so less citrate is formed.

17.4 The Citric Acid Cycle Is a Source of Biosynthetic Precursors

Thus far, discussion has focused on the citric acid cycle as the *major degradative pathway for the generation of ATP*. As a major metabolic hub of the cell, the citric acid cycle also *provides intermediates for biosyntheses* (Figure 17.20). For example, most of the carbon atoms in porphyrins come from *succinyl CoA*. Many of the amino acids are derived from *α-ketoglutarate* and *oxaloacetate*. These biosynthetic processes will be considered in subsequent chapters.

The Citric Acid Cycle Must Be Capable of Being Rapidly Replenished

The important point now is that *citric acid cycle intermediates must be replenished if any are drawn off for biosyntheses*. Suppose that much oxaloacetate is converted into amino acids for protein synthesis and, subsequently, the energy needs of the cell rise. The citric acid cycle will operate to a reduced extent unless new oxaloacetate is formed, because acetyl CoA cannot enter the cycle unless it condenses with oxaloacetate. Even though oxaloacetate is recycled, a minimal level must be maintained to allow the cycle to function.

How is oxaloacetate replenished? Mammals lack the enzymes for the net conversion of acetyl CoA into oxaloacetate or any other citric acid cycle intermediate. Rather, oxaloacetate is formed by the carboxylation of pyruvate, in a reaction catalyzed by the biotin-dependent enzyme *pyruvate carboxylase* (Figure 17.21).

$$\text{Pyruvate} + CO_2 + \text{ATP} + H_2O \rightarrow \text{oxaloacetate} + \text{ADP} + P_i + 2\,H^+$$

Recall that this enzyme plays a crucial role in gluconeogenesis (p. 460). It is active only in the presence of acetyl CoA, which signifies the need for more

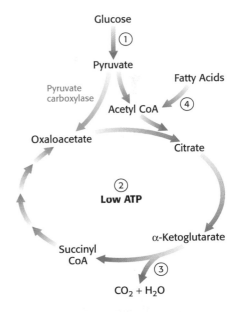

Active pathways

1. Glycolysis, Ch. 16
2. Citric acid cycle, Ch. 17
3. Oxidative phosphorylation, Ch. 18
4. Fatty acid oxidation, Ch. 22

Figure 17.21 PATHWAY INTEGRATION: Pathways active during exercise after a night's rest. The rate of the citric acid cycle increases during exercise, requiring the replenishment of oxaloacetate and acetyl CoA. Oxaloacetate is replenished by its formation from pyruvate. Acetyl CoA may be produced from the metabolism of both pyruvate and fatty acids.

oxaloacetate. If the energy charge is high, oxaloacetate is converted into glucose. If the energy charge is low, oxaloacetate replenishes the citric acid cycle. The synthesis of oxaloacetate by the carboxylation of pyruvate is an example of an *anaplerotic reaction* (of Greek origin, meaning to "fill up"), a reaction that leads to the net synthesis, or replenishment, of pathway components. Note that, because the citric acid cycle is a cycle, it can be replenished by the generation of any of the intermediates.

The Disruption of Pyruvate Metabolism Is the Cause of Beriberi and Poisoning by Mercury and Arsenic

Beriberi, a neurologic and cardiovascular disorder, is caused by a dietary deficiency of thiamine (also called *vitamin B₁*). The disease has been and continues to be a serious health problem in the Far East because rice, the major food, has a rather low content of thiamine. This deficiency is partly ameliorated if the whole rice grain is soaked in water before milling; some of the thiamine in the husk then leaches into the rice kernel. The problem is exacerbated if the rice is polished, because only the outer layer contains significant amounts of thiamine. Beriberi is also occasionally seen in alcoholics who are severely malnourished and thus thiamine deficient. The disease is characterized by neurologic and cardiac symptoms. Damage to the peripheral nervous system is expressed as pain in the limbs, weakness of the musculature, and distorted skin sensation. The heart may be enlarged and the cardiac output inadequate.

Which biochemical processes might be affected by a deficiency of thiamine? Thiamine is the precursor of the cofactor thiamine pyrophosphate. *This cofactor is the prosthetic group of three important enzymes: pyruvate dehydrogenase, α-ketoglutarate dehydrogenase, and transketolase.* Transketolase functions in the pentose phosphate pathway, which will be considered in Chapter 20. *The common feature of enzymatic reactions utilizing TPP is the transfer of an activated aldehyde unit.* In beriberi, the levels of pyruvate and α-ketoglutarate in the blood are higher than normal. The increase in the level of pyruvate in the blood is especially pronounced after the ingestion of glucose. A related finding is that the activities of the pyruvate and α-ketoglutarate dehydrogenase complexes in vivo are abnormally low. The low transketolase activity of red blood cells in beriberi is an easily measured and reliable diagnostic indicator of the disease.

Why does TPP deficiency lead primarily to neurological disorders? The nervous system relies essentially on glucose as its only fuel. The product of aerobic glycolysis, pyruvate, can enter the citric acid cycle only through the pyruvate dehydrogenase complex. With that enzyme deactivated, the nervous system has no source of fuel. In contrast, most other tissues can use fats as a source of fuel for the citric acid cycle.

Symptoms similar to those of beriberi appear in organisms exposed to mercury or arsenite (AsO_3^{3-}). Both materials have a high affinity for neighboring sulfhydryls, such as those in the reduced dihydrolipoyl groups of the E_3 component of the pyruvate dehydrogenase complex (Figure 17.22). The binding of mercury or arsenite to the dihydrolipoyl groups inhibits the complex and leads to central nervous system pathologies. The proverbial phrase "mad as a hatter" refers to the strange behavior of poisoned hat makers who used mercury nitrate to soften and shape animal furs. This form of mercury is absorbed through the skin. Similar symptoms afflicted the early photographers, who used vaporized mercury to create daguerreotypes.

Treatment for these poisons is the administration of sulfhydryl reagents with adjacent sulfhydryl groups to compete with the dihydrolipoyl residues for binding with the metal ion. The reagent–metal complex is then excreted

Beriberi

A vitamin-deficiency disease first described in 1630 by Jacob Bonitus, a Dutch physician working in Java:

A certain very troublesome affliction, which attacks men, is called by the inhabitants Beriberi (which means sheep). I believe those, whom this same disease attacks, with their knees shaking and the legs raised up, walk like sheep. It is a kind of paralysis, or rather Tremor: for it penetrates the motion and sensation of the hands and feet indeed sometimes of the whole body.

[The Granger Collection.]

Figure 17.22 Arsenite poisoning. Arsenite inhibits the pyruvate dehydrogenase complex by inactivating the dihydrolipoamide component of the transacetylase. Some sulfhydryl reagents, such as 2,3-dimercaptoethanol, relieve the inhibition by forming a complex with the arsenite that can be excreted.

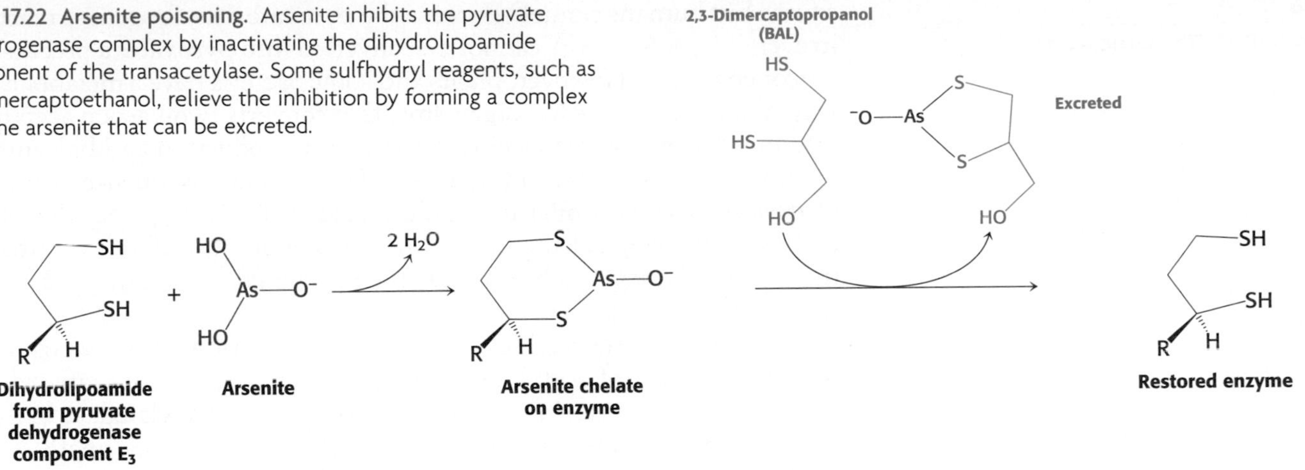

Dihydrolipoamide from pyruvate dehydrogenase component E₃

Arsenite

Arsenite chelate on enzyme

2,3-Dimercaptopropanol (BAL)

Excreted

Restored enzyme

in the urine. Indeed, 2,3-dimercaptopropanol (see Figure 17.22) was developed after World War I as an antidote to lewisite, an arsenic-based chemical weapon. This compound was initially called BAL, for British anti-lewisite.

The Citric Acid Cycle May Have Evolved From Preexisting Pathways

How did the citric acid cycle come into being? Although definitive answers are elusive, informed speculation is possible. We can perhaps begin to comprehend how evolution might work at the level of biochemical pathways.

The citric acid cycle was most likely assembled from preexisting reaction pathways. As noted earlier, many of the intermediates formed in the citric acid cycle are used in metabolic pathways for amino acids and porphyrins. Thus, compounds such as pyruvate, α-ketoglutarate, and oxaloacetate were likely present early in evolution for biosynthetic purposes. The oxidative decarboxylation of these α-ketoacids is quite favorable thermodynamically and can be used to drive the synthesis of both acyl CoA derivatives and NADH. These reactions almost certainly formed the core of processes that preceded the citric acid cycle evolutionarily. Interestingly, α-ketoglutarate can be directly converted into oxaloacetate by transamination of the respective amino acids by aspartate aminotransferase, another key biosynthetic enzyme. Thus, cycles comprising smaller numbers of intermediates used for a variety of biochemical purposes could have existed before the present form evolved.

The manuscript proposing the citric acid cycle was submitted for publication to *Nature* but was rejected. It was subsequently published in *Enzymologia.* Dr. Krebs proudly displayed the rejection letter throughout his career as encouragement for young scientists.

"June 1937

The editor of NATURE presents his compliments to Dr. H. A. Krebs and regrets that as he has already sufficient letters to fill the correspondence columns of NATURE for seven or eight weeks, it is undesirable to accept further letters at the present time on account of the time delay which must occur In their publication.

If Dr. Krebs does not mind much delay the editor is prepared to keep the letter until the congestion is relieved in the hope of making use of it.

He returns it now, in case Dr. Krebs prefers to submit it for early publication to another periodical."

17.5 The Glyoxylate Cycle Enables Plants and Bacteria to Grow on Acetate

Many plants and bacteria are able to subsist on acetate or other compounds that yield acetyl CoA. Acetyl CoA can be synthesized from acetate and CoA by an ATP-driven reaction that is catalyzed by *acetyl CoA synthetase.*

$$\text{Acetate} + \text{CoASH} + \text{ATP} \rightleftharpoons \text{acetyl CoA} + \text{AMP} + \text{PP}_i$$

Pyrophosphate is then hydrolyzed to orthophosphate, and so the equivalents of two compounds having high phosphoryl-transfer potential are consumed in the activation of acetate.

Acetyl CoA synthetase exists in many organisms, including humans. Why can plants and some bacteria exist with acetate as the sole energy source but humans cannot? Because the pyruvate dehydrogenase reaction is irreversible, acetyl CoA cannot be converted into pyruvate, a precursor for gluconeogenesis. However, plants and some bacteria have a metabolic pathway, absent in most other organisms, that converts two-carbon acetyl units into four-carbon units (succinate) for energy production and biosyntheses, including the synthesis of glucose. This reaction sequence, called the *glyoxylate cycle*, is similar to the citric acid cycle but bypasses the two decarboxylation steps of the cycle. Another important difference is that two molecules of acetyl CoA enter per turn of the glyoxylate cycle, compared with one in the citric acid cycle.

The glyoxylate cycle (Figure 17.23), like the citric acid cycle, begins with the condensation of acetyl CoA and oxaloacetate to form citrate, which is then isomerized to isocitrate. Instead of being decarboxylated, as in the citric acid cycle, isocitrate is cleaved by *isocitrate lyase* into succinate and glyoxylate. The ensuing steps regenerate oxaloacetate from glyoxylate. First, acetyl CoA condenses with glyoxylate to form malate in a reaction catalyzed by *malate synthase*, and then malate is oxidized to oxaloacetate, as in the citric acid cycle. The sum of these reactions is

$$2 \text{ Acetyl CoA} + \text{NAD}^+ + 2 \text{ H}_2\text{O} \longrightarrow$$
$$\text{succinate} + 2 \text{ CoASH} + \text{NADH} + 2 \text{ H}^+$$

In plants, these reactions take place in organelles called *glyoxysomes*. This cycle is especially prominent in oil-rich seeds, such as those from sunflowers, cucumbers, and castor beans. Succinate, released midcycle, can be converted into carbohydrates by a combination of the citric acid cycle and

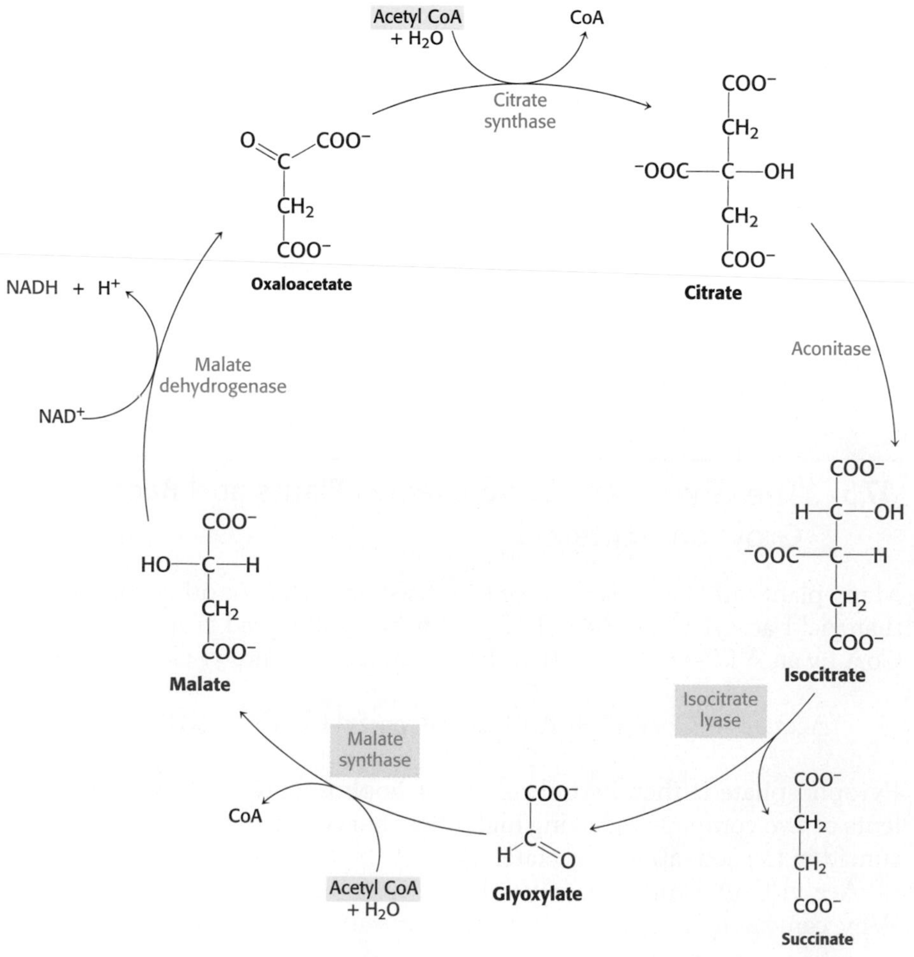

Figure 17.23 The glyoxylate pathway. The glyoxylate cycle allows plants and some microorganisms to grow on acetate because the cycle bypasses the decarboxylation steps of the citric acid cycle. The reactions of this cycle are the same as those of the citric acid cycle except for the ones catalyzed by isocitrate lyase and malate synthase, which are boxed in blue.

gluconeogenesis. The carbohydrates power seedling growth until the cell can begin photosynthesis. Thus, organisms with the glyoxylate cycle gain a metabolic versatility because they can use acetyl CoA as a precursor of glucose and other biomolecules.

Summary

The citric acid cycle is the final common pathway for the oxidation of fuel molecules. It also serves as a source of building blocks for biosyntheses.

17.1 Pyruvate Dehydrogenase Links Glycolysis to the Citric Acid Cycle

Most fuel molecules enter the cycle as acetyl CoA. The link between glycolysis and the citric acid cycle is the oxidative decarboxylation of pyruvate to form acetyl CoA. In eukaryotes, this reaction and those of the cycle take place inside mitochondria, in contrast with glycolysis, which takes place in the cytoplasm.

17.2 The Citric Acid Cycle Oxidizes Two-Carbon Units

The cycle starts with the condensation of oxaloacetate (C_4) and acetyl CoA (C_2) to give citrate (C_6), which is isomerized to isocitrate (C_6). Oxidative decarboxylation of this intermediate gives α-ketoglutarate (C_5). The second molecule of carbon dioxide comes off in the next reaction, in which α-ketoglutarate is oxidatively decarboxylated to succinyl CoA (C_4). The thioester bond of succinyl CoA is cleaved by orthophosphate to yield succinate, and a high-phosphoryl-transfer-potential compound in the form of GTP is concomitantly generated. Succinate is oxidized to fumarate (C_4), which is then hydrated to form malate (C_4). Finally, malate is oxidized to regenerate oxaloacetate (C_4). Thus, two carbon atoms from acetyl CoA enter the cycle, and two carbon atoms leave the cycle as CO_2 in the successive decarboxylations catalyzed by isocitrate dehydrogenase and α-ketoglutarate dehydrogenase. In the four oxidation–reduction reactions in the cycle, three pairs of electrons are transferred to NAD^+ and one pair to FAD. These reduced electron carriers are subsequently oxidized by the electron-transport chain to generate approximately 9 molecules of ATP. In addition, 1 molecule of a compound having a high phosphoryl-transfer potential is directly formed in the citric acid cycle. Hence, a total of 10 molecules of compounds having high phosphoryl-transfer potential are generated for each two-carbon fragment that is completely oxidized to H_2O and CO_2.

17.3 Entry to the Citric Acid Cycle and Metabolism Through It Are Controlled

The citric acid cycle operates only under aerobic conditions because it requires a supply of NAD^+ and FAD. The irreversible formation of acetyl CoA from pyruvate is an important regulatory point for the entry of glucose-derived pyruvate into the citric acid cycle. The activity of the pyruvate dehydrogenase complex is stringently controlled by reversible phosphorylation. The electron acceptors are regenerated when NADH and $FADH_2$ transfer their electrons to O_2 through the electron-transport chain, with the concomitant production of ATP. Consequently, the rate of the citric acid cycle depends on the need for ATP. In eukaryotes, the regulation of two enzymes in the cycle also is important for control. A high energy charge diminishes the activities of isocitrate dehydrogenase and α-ketoglutarate dehydrogenase. These mechanisms complement each other in reducing the rate of formation of acetyl CoA when the energy charge of the cell is high and when biosynthetic intermediates are abundant.

17.4 The Citric Acid Cycle Is a Source of Biosynthetic Precursors

When the cell has adequate energy available, the citric acid cycle can also provide a source of building blocks for a host of important biomolecules, such as nucleotide bases, proteins, and heme groups. This use depletes the cycle of intermediates. When the cycle again needs to metabolize fuel, anaplerotic reactions replenish the cycle intermediates.

17.5 The Glyoxylate Cycle Enables Plants and Bacteria to Grow on Acetate

The glyoxylate cycle enhances the metabolic versatility of many plants and bacteria. This cycle, which uses some of the reactions of the citric acid cycle, enables these organisms to subsist on acetate because it bypasses the two decarboxylation steps of the citric acid cycle.

Key Terms

citric acid (tricarboxylic acid, TCA; Krebs) cycle (p. 475)

acetyl CoA (p. 475)

oxidative phosphorylation (p. 476)

pyruvate dehydrogenase complex (p. 477)

flavoprotein (p. 480)

citrate synthase (p. 482)

iron–sulfur (nonheme iron) protein (p. 484)

isocitrate dehydrogenase (p. 484)

α-ketoglutarate dehydrogenase (p. 485)

metabolon (p. 490)

anaplerotic reaction (p. 494)

beriberi (p. 494)

glyoxylate cycle (p. 495)

isocitrate lyase (p. 496)

malate synthase (p. 496)

glyoxysome (p. 496)

Selected Readings

Where to Start

Sugden, M. C., and Holness, M. J. 2003. Recent advances in mechanisms regulating glucose oxidation at the level of the pyruvate dehydrogenase complex by PDKs. *Am. J. Physiol. Endocrinol. Metab.* 284:E855–E862.

Owen, O. E., Kalhan, S. C., and Hanson, R. W. 2002. The key role of anaplerosis and cataplerosis for citric acid function. *J. Biol. Chem.* 277:30409–30412.

Pyruvate Dehydrogenase Complex

Hiromasa, Y., Fujisawa, T., Aso, Y., and Roche, T. E. 2004. Organization of the cores of the mammalian pyruvate dehydrogenase complex formed by E2 and E2 plus the E3-binding proteins and their capacities to bind the E1 and E3 components. *J. Biol Chem.* 279:6921–6933.

Izard, T., Ævarsson, A., Allen, M. D., Westphal, A. H., Perham, R. N., De Kok, A., and Hol, W. G. 1999. Principles of quasi-equivalence and Euclidean geometry govern the assembly of cubic and dodecahedral cores of pyruvate dehydrogenase complexes. *Proc. Natl. Acad. Sci. U. S. A.* 96:1240–1245.

Ævarsson, A., Seger, K., Turley, S., Sokatch, J. R., and Hol, W. G. 1999. Crystal structure of 2-oxoisovalerate dehydrogenase and the architecture of 2-oxo acid dehydrogenase multiple enzyme complexes. *Nat. Struct. Biol.* 6:785–792.

Domingo, G. J., Chauhan, H. J., Lessard, I. A., Fuller, C., and Perham, R. N. 1999. Self-assembly and catalytic activity of the pyruvate dehydrogenase multienzyme complex from *Bacillus stearothermophilus*. *Eur. J. Biochem.* 266:1136–1146.

Jones, D. D., Horne, H. J., Reche, P. A., and Perham, R. N. 2000. Structural determinants of post-translational modification and catalytic specificity for the lipoyl domains of the pyruvate dehydrogenase multienzyme complex of *Escherichia coli*. *J. Mol. Biol.* 295:289–306.

McCartney, R. G., Rice, J. E., Sanderson, S. J., Bunik, V., Lindsay, H., and Lindsay, J. G. 1998. Subunit interactions in the mammalian α-ketoglutarate dehydrogenase complex: Evidence for direct association of the α-ketoglutarate dehydrogenase and dihydrolipoamide dehydrogenase components. *J. Biol. Chem.* 273: 24158–24164.

Structure of Citric Acid Cycle Enzymes

Yankovskaya, V., Horsefield, R., Törnroth, S., Luna-Chavez, C., Miyoshi, H., Léger, C., Byrne, B., Cecchini, G., and Iowata, S. 2003. Architecture of succinate dehydrogenase and reactive oxygen species generation. *Science* 299:700–704.

Chapman, A. D., Cortes, A., Dafforn, T. R., Clarke, A. R., and Brady, R. L. 1999. Structural basis of substrate specificity in malate dehydrogenases: Crystal structure of a ternary complex of porcine cytoplasmic malate dehydrogenase, α-ketomalonate and tetrahydoNAD. *J. Mol. Biol.* 285:703–712.

Fraser, M. E., James, M. N., Bridger, W. A., and Wolodko, W. T. 1999. A detailed structural description of *Escherichia coli* succinyl-CoA synthetase. *J. Mol. Biol.* 285:1633–1653. [Published erratum appears in May 7, 1999, issue of *J. Mol. Biol.* 288(3):501.]

Lloyd, S. J., Lauble, H., Prasad, G. S., and Stout, C. D. 1999. The mechanism of aconitase: 1.8 Å resolution crystal structure of the S642a:citrate complex. *Protein Sci.* 8:2655–2662.

Remington, S. J. 1992. Structure and mechanism of citrate synthase. *Curr. Top. Cell. Regul.* 33:209–229.

Rose, I. A. 1998. How fumarase recycles after the malate → fumarate reaction: Insights into the reaction mechanism. *Biochemistry* 37:17651–17658.

Johnson, J. D., Muhonen, W. W., and Lambeth, D. O. 1998. Characterization of the ATP- and GTP-specific succinyl-CoA synthetases in pigeon: The enzymes incorporate the same subunit. *J. Biol. Chem.* 273:27573 27579.

Karpusas, M., Branchaud, B., and Remington, S. J. 1990. Proposed mechanism for the condensation reaction of citrate synthase: 1.9-Å structure of the ternary complex with oxaloacetate and carboxymethyl coenzyme A. *Biochemistry* 29:2213–2219.

Lauble, H., Kennedy, M. C., Beinert, H., and Stout, C. D. 1992. Crystal structures of aconitase with isocitrate and nitroisocitrate bound. *Biochemistry* 31:2735–2748.

Organization of the Citric Acid Cycle

Lambeth, D. O., Tews, K. N., Adkins, S., Frohlich, D., and Milavetz, B. I. 2004. Expression of two succinyl-CoA specificities in mammalian tissues. *J. Biol. Chem.* 279:36621–36624.

Velot, C., Mixon, M. B., Teige, M., and Srere, P. A. 1997. Model of a quinary structure between Krebs TCA cycle enzymes: A model for the metabolon. *Biochemistry* 36:14271–14276.

Barnes, S. J., and Weitzman, P. D. 1986. Organization of citric acid cycle enzymes into a multienzyme cluster. *FEBS Lett.* 201:267–270.

Haggie, P. M., and Brindle, K. M. 1999. Mitochondrial citrate synthase is immobilized in vivo. *J. Biol. Chem.* 274:3941–3945.

Morgunov, I., and Srere, P. A. 1998. Interaction between citrate synthase and malate dehydrogenase: Substrate channeling of oxaloacetate. *J. Biol. Chem.* 273:29540–29544.

Regulation

Hiromasa, Y., and Roche, T. E. 2003. Facilitated interaction between the pyruvate dehydrogenase kinase isoform 2 and the dihydrolipoyl acetyltransferases. *J. Biol. Chem.* 278:33681–33693.

Huang, B., Gudi, R., Wu, P., Harris, R. A., Hamilton, J., and Popov, K. M. 1998. Isoenzymes of pyruvate dehydrogenase phosphatase: DNA-derived amino acid sequences, expression, and regulation. *J. Biol. Chem.* 273:17680–17688.

Bowker-Kinley, M., and Popov, K. M. 1999. Evidence that pyruvate dehydrogenase kinase belongs to the ATPase/kinase superfamily. *Biochem. J.* 1:47–53.

Jitrapakdee, S., and Wallace, J. C. 1999. Structure, function and regulation of pyruvate carboxylase. *Biochem. J.* 340:1–16.

Evolutionary Aspects

Meléndez-Hevia, E., Waddell, T. G., and Cascante, M. 1996. The puzzle of the Krebs citric acid cycle: Assembling the pieces of chemically feasible reactions, and opportunism in the design of metabolic pathways in evolution. *J. Mol. Evol.* 43:293–303.

Baldwin, J. E., and Krebs, H. 1981. The evolution of metabolic cycles. *Nature* 291:381–382.

Gest, H. 1987. Evolutionary roots of the citric acid cycle in prokaryotes. *Biochem. Soc. Symp.* 54:3–16.

Weitzman, P. D. J. 1981. Unity and diversity in some bacterial citric acid cycle enzymes. *Adv. Microbiol. Physiol.* 22:185–244.

Discovery of the Citric Acid Cycle

Kornberg, H. 2000. Krebs and his trinity of cycles. *Nat. Rev. Mol. Cell. Biol.* 1:225–228.

Krebs, H. A., and Johnson, W. A. 1937. The role of citric acid in intermediate metabolism in animal tissues. *Enzymologia* 4:148–156.

Krebs, H. A. 1970. The history of the tricarboxylic acid cycle. *Perspect. Biol. Med.* 14:154–170.

Krebs, H. A., and Martin, A. 1981. *Reminiscences and Reflections.* Clarendon Press.

Problems

1. *Flow of carbon atoms.* What is the fate of the radioactive label when each of the following compounds is added to a cell extract containing the enzymes and cofactors of the glycolytic pathway, the citric acid cycle, and the pyruvate dehydrogenase complex? (The ^{14}C label is printed in red.)

(a)

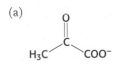

(b)

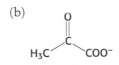

(c)

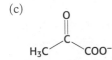

(d)

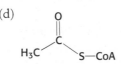

(e) Glucose 6-phosphate labeled at C-1.

2. $C_2 + C_2 \rightarrow C_4$.

(a) Which enzymes are required to get *net synthesis* of oxaloacetate from acetyl CoA?

(b) Write a balanced equation for the net synthesis.

(c) Do mammalian cells contain the requisite enzymes?

3. *Driving force.* What is the $\Delta G^{\circ\prime}$ for the complete oxidation of the acetyl unit of acetyl CoA by the citric acid cycle?

4. *Acting catalytically.* The citric acid cycle itself, which is composed of enzymatically catalyzed steps, can be thought of essentially as a supramolecular enzyme. Explain.

5. *A potent inhibitor.* Thiamine thiazolone pyrophosphate binds to pyruvate dehydrogenase about 20,000 times as strongly as does thiamine pyrophosphate, and it competitively inhibits the enzyme. Why?

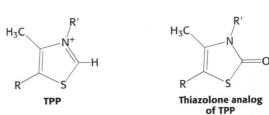

TPP | **Thiazolone analog of TPP**

6. *Lactic acidosis.* Patients in shock often suffer from lactic acidosis owing to a deficiency of O_2. Why does a lack of O_2 lead to lactic acid accumulation? One treatment for shock is to administer dichloroacetate, which inhibits the kinase associated with the pyruvate dehydrogenase complex. What is the biochemical rationale for this treatment?

7. *Coupling reactions.* The oxidation of malate by NAD^+ to form oxaloacetate is a highly endergonic reaction under standard conditions [$\Delta G^{\circ\prime} = 29$ kJ mol^{-1} (7 kcal mol^{-1})]. The reaction proceeds readily under physiological conditions.

(a) Why?

(b) Assuming an $[NAD^+]/[NADH]$ ratio of 8 and a pH of 7, what is the lowest [malate]/[oxaloacetate] ratio at which oxaloacetate can be formed from malate?

8. *Synthesizing α-ketoglutarate.* It is possible, with the use of the reactions and enzymes considered in this chapter, to convert pyruvate into α-ketoglutarate without depleting any of the citric acid cycle components. Write a balanced reaction scheme for this conversion, showing cofactors and identifying the required enzymes.

Chapter Integration Problem

9. *Fats into glucose?* Fats are usually metabolized into acetyl CoA and then further processed through the citric acid cycle. In Chapter 16, we learned that glucose could be synthesized from oxaloacetate, a citric acid cycle intermediate. Why, then, after a long bout of exercise depletes our carbohydrate stores, do we need to replenish those stores by eating carbohydrates? Why do we not simply replace them by converting fats into carbohydrates?

Mechanism Problems

10. *Theme and variation.* Propose a reaction mechanism for the condensation of acetyl CoA and glyoxylate in the glyoxylate cycle of plants and bacteria.

11. *Symmetry problems.* In experiments carried out in 1941 to investigate the citric acid cycle, oxaloacetate labeled with ^{14}C in the carboxyl carbon atom farthest from the keto group was introduced to an active preparation of mitochondria.

Oxaloacetate

Analysis of the α-ketoglutarate formed showed that none of the radioactive label had been lost. Decarboxylation of α-ketoglutarate then yielded succinate devoid of radioactivity. All the label was in the released CO_2. Why were the early investigators of the citric acid cycle surprised that *all* the label emerged in the CO_2?

12. *Symmetric molecules reacting asymmetrically.* The interpretation of the experiments described in problem 11 was that citrate (or any other symmetric compound) cannot be an intermediate in the formation of α-ketoglutarate, because of the asymmetric fate of the label. This view seemed compelling until Alexander Ogston incisively pointed out in 1948 that "it is possible that *an asymmetric enzyme which attacks a symmetrical compound can distinguish between its identical groups.*" For simplicity, consider a molecule in which two hydrogen atoms, a group X, and a different group Y are bonded to a tetrahedral carbon atom as a model for citrate. Explain how a symmetric molecule can react with an enzyme in an asymmetric way.

Data Interpretation Problems

13. *A little goes a long way.* As will become clearer in Chapter 18, the activity of the citric acid cycle can be monitored by measuring the amount of O_2 consumed. The greater the rate of O_2 consumption, the faster the rate of the cycle. Hans Krebs used this assay to investigate the cycle in 1937. He used as his exper-

imental system minced pigeon-breast muscle, which is rich in mitochondria. In one set of experiments, Krebs measured the O_2 consumption in the presence of carbohydrate only and in the presence of carbohydrate and citrate. The results are shown in the following table.

Effect of citrate on oxygen consumption by minced pigeon-breast muscle

Time (min)	Micromoles of oxygen consumed	
	Carbohydrate only	Carbohydrate plus 3 μmol of citrate
10	26	28
60	43	62
90	46	77
150	49	85

(a) How much O_2 would be absorbed if the added citrate were completely oxidized to H_2O and CO_2?

(b) On the basis of your answer to part *a*, what do the results given in the table suggest?

14. *Arsenite poisoning.* The effect of arsenite on the experimental system of problem 13 was then examined. Experimental data (not presented here) showed that the amount of citrate present did not change in the course of the experiment in the absence of arsenite. However, if arsenite was added to the system, different results were obtained, as shown in the following table.

Disappearance of citric acid in pigeon-breast muscle in the presence of arsenite

Micromoles of citrate added	Micromoles of citrate found after 40 minutes	Micromoles of citrate used
22	00.6	21
44	20.0	24
90	56.0	34

(a) What is the effect of arsenite on the disappearance of citrate?

(b) How is the arsenite's action altered by the addition of more citrate?

(c) What do these data suggest about the site of action of arsenite?

15. *Isocitrate lyase and tuberculosis.* The bacterium *Mycobacterium tuberculosis*, the cause of tuberculosis, can invade the lungs and persist in a latent state for years. During this time, the bacteria reside in granulomas—nodular scars containing bacteria and host-cell debris in the center and surrounded by immune cells. The granulomas are lipid-rich, oxygen-poor environments. How these bacteria manage to persist is something of a mystery. The results of recent research suggest that the glyoxylate cycle is required for the persistence. The following data show the amount of bacteria [presented as colony-forming units (cfu)] in mice lungs in the weeks after an infection.

In graph A, the black circles represent the results for wild-type bacteria and the red circles represent the results for bacteria from which the gene for isocitrate lyase was deleted.

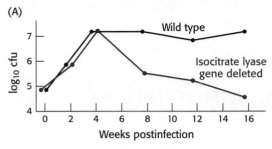

(A)

(a) What is the effect of the absence of isocitrate lyase?

The techniques described in Chapter 6 were used to reinsert the gene encoding isocitrate lyase into bacteria from which it had previously been deleted.

In graph B, black circles represent bacteria into which the gene was reinserted and red circles represent bacteria in which the gene was still missing.

(b) Do these results support those obtained in part *a*?
(c) What is the purpose of the experiment in part *b*?

(d) Why do these bacteria perish in the absence of the glyoxylate cycle?

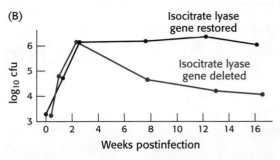

(B)

[Data after McKinney et al., *Nature* 406(2000):735–738.]

 NEED EXTRA HELP? Purchase chapters of the Student Companion with complete solutions online at www.whfreeman.com/stryer.

Chapter 18

Oxidative Phosphorylation

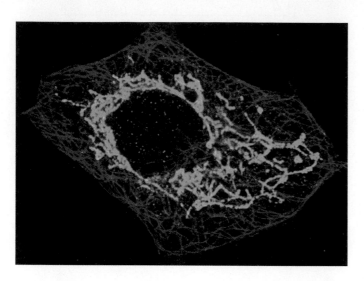

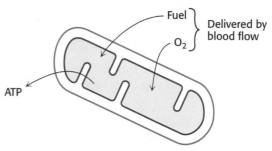

Mitochondria, stained green, form a network inside a fibroblast cell (left). Mitochondria oxidize carbon fuels to form cellular energy in the form of ATP. [(Left) Courtesy of Michael P. Yaffee, Department of Biology, University of California at San Diego.]

The amount of ATP that human beings require to go about their lives is staggering. A sedentary male of 70 kg (154 lbs) requires about 8400 kJ (2000 kcal) for a day's worth of activity. To provide this much energy requires 83 kg of ATP. However, human beings possess only about 250 g of ATP. The disparity between the amount of ATP that we have and the amount that we require is solved by recycling ADP back to ATP. Each ATP molecule is recycled approximately 300 times per day. This recycling takes place primarily through *oxidative phosphorylation. Oxidative phosphorylation is the process in which ATP is formed as a result of the transfer of electrons from NADH or FADH$_2$ to O$_2$ by a series of electron carriers.* This process, which takes place in mitochondria, is the major source of ATP in aerobic organisms. For example, oxidative phosphorylation generates 26 of the 30 molecules of ATP that are formed when glucose is completely oxidized to CO$_2$ and H$_2$O.

Oxidative Phosphorylation Couples the Oxidation of Carbon Fuels to ATP Synthesis with a Proton Gradient

The NADH and FADH$_2$ formed in glycolysis, fatty acid oxidation, and the citric acid cycle are energy-rich molecules because each contains a pair of electrons having a high transfer potential. When these electrons are used to reduce molecular oxygen to water in oxidative phosphorylation, a large amount of free energy is liberated, which can be used to generate ATP.

Oxidative phosphorylation is conceptually simple and mechanistically complex. Indeed, the unraveling of the mechanism of oxidative phosphorylation has been one of the most challenging problems of biochemistry. The flow of electrons from NADH or FADH$_2$ to O$_2$ through protein complexes located in the mitochondrial inner membrane leads to the pumping of

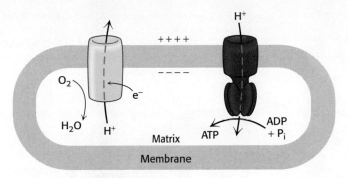

Figure 18.1 Essence of oxidative phosphorylation. Oxidation and ATP synthesis are coupled by transmembrane proton fluxes.

protons out of the mitochondrial matrix. The resulting unequal distribution of protons generates a pH gradient and a transmembrane electrical potential that creates a *proton-motive force*. ATP is synthesized when protons flow back to the mitochondrial matrix through an enzyme complex. Thus, *the oxidation of fuels and the phosphorylation of ADP are coupled by a proton gradient across the inner mitochondrial membrane* (Figure 18.1).

　　Oxidative phosphorylation is the culmination of a series of energy transformations that are called *cellular respiration* or simply *respiration* in their entirety. Carbon fuels are first oxidized in the citric acid cycle to yield electrons with high transfer potential. In oxidative phosphorylation, this electron-motive force is converted into proton-motive force and then into phosphoryl-transfer potential. The conversion of electron-motive force into proton-motive force is carried out by a *respiratory chain* consisting of three electron-driven proton pumps—NADH-Q oxidoreductase, Q-cytochrome *c* oxidoreductase, and cytochrome *c* oxidase. These large transmembrane complexes contain multiple oxidation–reduction centers, including quinones, flavins, iron–sulfur clusters, hemes, and copper ions. The final phase of oxidative phosphorylation is carried out by *ATP synthase*, an ATP-synthesizing assembly that is driven by the flow of protons back into the mitochondrial matrix. Components of this remarkable enzyme rotate as part of its catalytic mechanism. Oxidative phosphorylation vividly shows that *proton gradients are an interconvertible currency of free energy in biological systems.*

Respiration

An ATP-generating process in which an inorganic compound (such as molecular oxygen) serves as the ultimate electron acceptor. The electron donor can be either an organic compound or an inorganic one.

18.1 Oxidative Phosphorylation in Eukaryotes Takes Place in Mitochondria

Mitochondria are oval-shaped organelles, typically about 2 μm in length and 0.5 μm in diameter, about the size of a bacterium. Eugene Kennedy and Albert Lehninger discovered more than a half-century ago that *mitochondria contain the respiratory assembly, the enzymes of the citric acid cycle, and the enzymes of fatty acid oxidation.*

Mitochondria Are Bounded by a Double Membrane

Electron microscopic studies by George Palade and Fritjof Sjöstrand revealed that mitochondria have two membrane systems: an *outer membrane* and an extensive, highly folded *inner membrane.* The inner membrane is folded into a series of internal ridges called *cristae.* Hence, there are two compartments in mitochondria: (1) the *intermembrane space* between the outer and the inner membranes and (2) the *matrix,* which is bounded by the

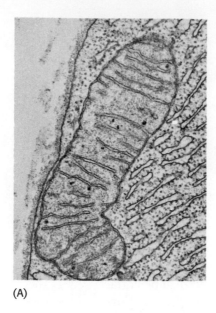

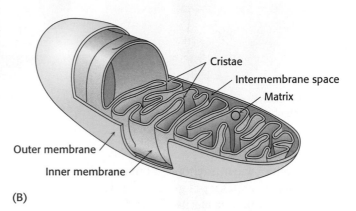

Figure 18.2 Electron micrograph (A) and diagram (B) of a mitochondrion. [(A) Courtesy of George Palade. (B) After *Biology of the Cell* by Stephen L. Wolfe. © 1972 by Wadsworth Publishing Company, Inc., Belmont, California 94002. Adapted by permission of the publisher.]

Cristae

Intermembrane space

Matrix

Outer membrane

Inner membrane

(A)

(B)

Rickettsia
(a bacterium)

Arabidopsis
(a plant)

Plasmodium
(a protozoan)

Homo sapiens

Figure 18.3 Sizes of mitochondrial genomes. The sizes of three mitochondrial genomes compared with the genome of *Rickettsia*, a relative of the the presumed ancestor of all mitochondria. For genomes of more than 60 kbp, the DNA coding region for genes with known function is shown in red.

inner membrane (Figure 18.2). The mitochondrial matrix is the site of most of the reactions of the citric acid cycle and fatty acid oxidation. In contrast, oxidative phosphorylation takes place in the inner mitochondrial membrane. The increase in surface area of the inner mitochondrial membrane provided by the cristae creates more sites for oxidative phosphorylation than would be the case with a simple, unfolded membrane. Humans contain an estimated 14,000 m^2 of inner mitochondrial membrane, which is the approximate equivalent of three football fields in the United States.

The outer membrane is quite permeable to most small molecules and ions because it contains many copies of *mitochondrial porin,* a 30 to 35 kd pore-forming protein also known as VDAC, for *v*oltage-*d*ependent *a*nion *c*hannel. VDAC plays a role in the regulated flux of metabolites—usually anionic species such as phosphate, chloride, organic anions, and the adenine nucleotides—across the outer membrane. In contrast, the inner membrane is intrinsically impermeable to nearly all ions and polar molecules. A large family of transporters shuttles metabolites such as ATP, pyruvate, and citrate across the inner mitochondrial membrane. The two faces of this membrane will be referred to as the *matrix side* and the *cytoplasmic side* (the latter because it is freely accessible to most small molecules in the cytoplasm). They are also called the *N* and *P* sides, respectively, because the membrane potential is negative on the matrix side and positive on the cytoplasmic side.

In prokaryotes, the electron-driven proton pumps and ATP-synthesizing complex are located in the cytoplasmic membrane, the inner of two membranes. The outer membrane of bacteria, like that of mitochondria, is permeable to most small metabolites because of the presence of porins.

Mitochondria Are the Result of an Endosymbiotic Event

Mitochondria are semiautonomous organelles that live in an endosymbiotic relation with the host cell. These organelles contain their own DNA, which encodes a variety of different proteins and RNAs. Mitochondrial DNA is usually portrayed as being circular, but recent research suggests that the mitochondrial DNA of many organisms may be linear. The genomes of mitochondria range broadly in size across species. The mitochondrial genome of the protist *Plasmodium falciparum* consists of fewer than 6000 base pairs, whereas those of some land plants comprise more than 200,000 bp (Figure 18.3). Human mitochondrial DNA comprises 16,569 bp and encodes 13 respiratory-chain proteins as well as the small and large ribosomal RNAs and enough tRNAs to translate all codons. However, mitochondria also contain many proteins encoded by nuclear DNA. Cells that contain mitochondria depend on these organelles for oxidative phosphorylation, and the mitochondria in turn depend on the cell for their very existence. How did this intimate symbiotic relation come to exist?

An *endosymbiotic event* is thought to have occurred whereby a free-living organism capable of oxidative phosphorylation was engulfed by another cell. The double-membrane, circular DNA (with exceptions), and mitochondrial-specific transcription and translation machinery all point to this conclusion. Thanks to the rapid accumulation of sequence data for mitochondrial and bacterial genomes, speculation on the origin of the "original" mitochondrion with some authority is now possible. The most mitochondrial-like bacterial genome is that of *Rickettsia prowazekii*, the cause of louse-borne typhus. The genome for this organism is more than 1 million base pairs in size and contains 834 protein-encoding genes. Sequence data suggest that all extant mitochondria are derived from an ancestor of *R. prowazekii* as the result of a single endosymbiotic event.

The evidence that modern mitochondria result from a single event comes from examination of the most bacteria-like mitochondrial genome, that of the protozoan *Reclinomonas americana*. Its genome contains 97 genes, of which 62 specify proteins. The genes encoding these proteins include all of the protein-coding genes found in all of the sequenced mitochondrial genomes (Figure 18.4). Yet, this genome encodes less than 2% of the protein-coding genes in the bacterium *E. coli*. It seems unlikely that mitochondrial genomes resulting from several endosymbiotic events could have been independently reduced to the same set of genes found in *R. americana*.

Note that transient engulfment of prokaryotic cells by larger cells is not uncommon in the microbial world. In the case of mitochondria, such a transient relation became permanent as the bacterial cell lost DNA, making it incapable of independent living, and the host cell became dependent on the ATP generated by its tenant.

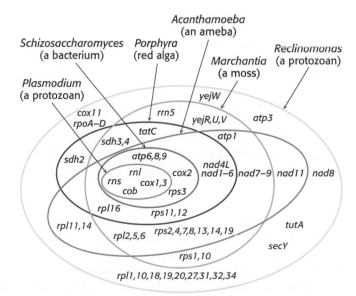

Figure 18.4 Overlapping gene complements of mitochondria. The genes present within each oval are those present within the organism represented by the oval. Only rRNA- and protein-coding genes are shown. The genome of *Reclinomonas* contains all the protein-coding genes found in all the sequenced mitochondrial genomes. [After M. W. Gray, G. Burger, and B. F. Lang. *Science* 283(1999):1476–1481.]

18.2 Oxidative Phosphorylation Depends on Electron Transfer

In Chapter 17, the primary function of the citric acid cycle was identified as the generation of NADH and $FADH_2$ by the oxidation of acetyl CoA. In oxidative phosphorylation, electrons from NADH and $FADH_2$ are used to reduce molecular oxygen to water. The highly exergonic reduction of molecular oxygen by NADH and $FADH_2$ takes place in a number of electron-transfer reactions, which take place in a set of membrane proteins known as the *electron-transport chain*.

The Electron-Transfer Potential of an Electron Is Measured As Redox Potential

In oxidative phosphorylation, the *electron-transfer potential* of NADH or $FADH_2$ is converted into the *phosphoryl-transfer potential* of ATP. We need quantitative expressions for these forms of free energy. The measure of phosphoryl-transfer potential is already familiar to us: it is given by $\Delta G^{\circ\prime}$ for the hydrolysis of the activated phosphoryl compound. The corresponding expression for the electron-transfer potential is E_0', the *reduction potential* (also called the *redox potential* or *oxidation–reduction potential*).

The reduction potential is an electrochemical concept. Consider a substance that can exist in an oxidized form X and a reduced form X^-. Such a pair is called a *redox couple*. The reduction potential of this couple can be determined by measuring the electromotive force generated by a *sample half-cell* connected to a *standard reference half-cell* (Figure 18.5). The sample half-cell consists of an electrode immersed in a solution of 1 M oxidant (X) and 1 M reductant (X^-). The standard reference half-cell consists of an electrode immersed in a 1 M H^+ solution that is in equilibrium with H_2 gas at 1 atmosphere (1 atm) of pressure. The electrodes are connected to a voltmeter, and an agar bridge establishes electrical continuity between the half-cells. Electrons then flow from one half-cell to the other. If the reaction proceeds in the direction

$$X^- + H^+ \longrightarrow X + \tfrac{1}{2} H_2$$

the reactions in the half-cells (referred to as *half-reactions* or *couples*) must be

$$X^- \longrightarrow X + e^- \qquad H^+ + e^- \longrightarrow \tfrac{1}{2} H_2$$

Thus, electrons flow from the sample half-cell to the standard reference half-cell, and the sample-cell electrode is taken to be negative with respect to the standard-cell electrode. *The reduction potential of the $X:X^-$ couple is the observed voltage at the start of the experiment* (when X, X^-, and H^+ are 1 M with 1 atm of H_2). *The reduction potential of the $H^+:H_2$ couple is defined to be 0 volts.*

The meaning of the reduction potential is now evident. A negative reduction potential means that the oxidized form of a substance has lower affinity for electrons than does H_2, as in the preceding example. A positive reduction potential means that the oxidized form of a substance has higher affinity for electrons than does H_2. These comparisons refer to standard conditions—namely, 1 M oxidant, 1 M reductant, 1 M H^+, and 1 atm H_2. Thus, *a strong reducing agent (such as NADH) is poised to donate electrons and has a negative reduction potential, whereas a strong oxidizing agent (such as O_2) is ready to accept electrons and has a positive reduction potential.*

The reduction potentials of many biologically important redox couples are known (Table 18.1). Table 18.1 is like those presented in chemistry

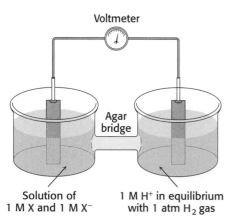

Voltmeter

Agar bridge

Solution of 1 M X and 1 M X^-

1 M H^+ in equilibrium with 1 atm H_2 gas

Figure 18.5 Measurement of redox potential. Apparatus for the measurement of the standard oxidation–reduction potential of a redox couple. Electrons flow through the wire connecting the cells, while ions flow through the agar bridge.

TABLE 18.1 Standard reduction potentials of some reactions

Oxidant	Reductant	n	E_0' (V)
Succinate + CO_2	α-Ketoglutarate	2	−0.67
Acetate	Acetaldehyde	2	−0.60
Ferredoxin (oxidized)	Ferredoxin (reduced)	1	−0.43
2 H^+	H_2	2	−0.42
NAD^+	NADH + H^+	2	−0.32
$NADP^+$	NADPH + H^+	2	−0.32
Lipoate (oxidized)	Lipoate (reduced)	2	−0.29
Glutathione (oxidized)	Glutathione (reduced)	2	−0.23
FAD	$FADH_2$	2	−0.22
Acetaldehyde	Ethanol	2	−0.20
Pyruvate	Lactate	2	−0.19
Fumarate	Succinate	2	+0.03
Cytochrome b (+3)	Cytochrome b (+2)	1	+0.07
Dehydroascorbate	Ascorbate	2	+0.08
Ubiquinone (oxidized)	Ubiquinone (reduced)	2	+0.10
Cytochrome c (+3)	Cytochrome c (+2)	1	+0.22
Fe (+3)	Fe (+2)	1	+0.77
½ O_2 + 2 H^+	H_2O	2	+0.82

Note: E_0' is the standard oxidation–reduction potential (pH 7, 25°C) and n is the number of electrons transferred. E_0' refers to the partial reaction written as

$$\text{Oxidant} + e^- \rightarrow \text{reductant}$$

textbooks, except that a hydrogen ion concentration of 10^{-7} M (pH 7) instead of 1 M (pH 0) is the standard state adopted by biochemists. This difference is denoted by the prime in E_0'. Recall that the prime in $\Delta G^{\circ\prime}$ denotes a standard free-energy change at pH 7.

The standard free-energy change $\Delta G^{\circ\prime}$ is related to the change in reduction potential $\Delta E_0'$ by

$$\Delta G^{\circ\prime} = -nF\Delta E_0'$$

in which n is the number of electrons transferred, F is a proportionality constant called the *faraday* [96.48 kJ mol^{-1} V^{-1} (23.06 kcal mol^{-1} V^{-1})], $\Delta E_0'$ is in volts, and $\Delta G^{\circ\prime}$ is in kilojoules or kilocalories per mole.

The free-energy change of an oxidation–reduction reaction can be readily calculated from the reduction potentials of the reactants. For example, consider the reduction of pyruvate by NADH, catalyzed by lactate dehydrogenase.

$$\text{Pyruvate} + \text{NADH} + H^+ \rightleftharpoons \text{lactate} + NAD^+ \qquad (A)$$

The reduction potential of the NAD^+:NADH couple, or half-reaction, is −0.32 V, whereas that of the pyruvate:lactate couple is −0.19 V. By convention, reduction potentials (as in Table 18.1) refer to partial reactions written as reductions: oxidant + $e^- \rightarrow$ reductant. Hence,

$$\text{Pyruvate} + 2\,H^+ + 2\,e^- \longrightarrow \text{lactate} \qquad E_0' = -0.19\,\text{V} \qquad (B)$$

$$NAD^+ + H^+ + 2\,e^- \longrightarrow \text{NADH} \qquad E_0' = -0.32\,\text{V} \qquad (C)$$

To obtain reaction A from reactions B and C, we need to reverse the direction of reaction C so that NADH appears on the left side of the arrow. In doing so, the sign of E_0' must be changed.

$$\text{Pyruvate} + 2\,H^+ + 2\,e^- \longrightarrow \text{lactate} \qquad E_0' = -0.19\,\text{V} \qquad (B)$$

$$\text{NADH} \longrightarrow NAD^+ + H^+ + 2\,e^- \qquad E_0' = +0.32\,\text{V} \qquad (D)$$

For reaction B, the free energy can be calculated with $n = 2$.

$$\Delta G^{\circ\prime} = -2 \times 96.48 \text{ kJ mol}^{-1}\text{V}^{-1} \times -0.19 \text{ V}$$
$$= +36.7 \text{ kJ mol}^{-1}\,(+8.8 \text{ kcal mol}^{-1})$$

Likewise, for reaction D,

$$\Delta G^{\circ\prime} = -2 \times 96.48 \text{ kJ mol}^{-1}\text{V}^{-1} \times +0.32 \text{ V}$$
$$= -61.8 \text{ kJ mol}^{-1}\,(-14.8 \text{ kcal mol}^{-1})$$

Thus, the free energy for reaction A is given by

$$\Delta G^{\circ\prime} = \Delta G^{\circ\prime}\,(\text{for reaction B}) + \Delta G^{\circ\prime}\,(\text{for reaction D})$$
$$= +36.7 \text{ kJ mol}^{-1} + (-61.8 \text{ kJ mol}^{-1})$$
$$= -25.1 \text{ kJ mol}^{-1}\,(-6.0 \text{ kcal mol}^{-1})$$

A 1.14-Volt Potential Difference Between NADH and Molecular Oxygen Drives Electron Transport Through the Chain and Favors the Formation of a Proton Gradient

The driving force of oxidative phosphorylation is the electron-transfer potential of NADH or $FADH_2$ relative to that of O_2. How much energy is released by the reduction of O_2 with NADH? Let us calculate $\Delta G^{\circ\prime}$ for this reaction. The pertinent half-reactions are

$$\tfrac{1}{2}\,O_2 + 2\,H^+ + 2\,e^- \longrightarrow H_2O \qquad E_0' = +0.82 \text{ V} \qquad \text{(A)}$$
$$NAD^+ + H^+ + 2\,e^- \longrightarrow NADH \qquad E_0' = -0.32 \text{ V} \qquad \text{(B)}$$

Subtracting reaction B from reaction A yields

$$\tfrac{1}{2}\,O_2 + NADH + H^+ \longrightarrow H_2O + NAD^+ \qquad \text{(C)}$$

The standard free energy for this reaction is then given by

$$\Delta G^{\circ\prime} = -2 \times 96.48 \text{ kJ mol}^{-1}\text{V}^{-1} \times +0.82 \text{ V} -$$
$$(-2 \times 96.48 \text{ kJ mol}^{-1}\text{V}^{-1} \times -0.32 \text{ V})$$
$$= -158.2 \text{ kJ mol}^{-1} - 61.9 \text{ kJ mol}^{-1}$$
$$= -220.1 \text{ kJ mol}^{-1}\,(-52.6 \text{ kcal mol}^{-1})$$

This release of free energy is substantial. Recall that $\Delta G^{\circ\prime}$ for the hydrolysis of ATP is $-30.5 \text{ kJ mol}^{-1}$ ($-7.3 \text{ kcal mol}^{-1}$). The released energy is initially used to generate a proton gradient that is then used for the synthesis of ATP and the transport of metabolites across the mitochondrial membrane.

How can the energy associated with a proton gradient be quantified? Recall that the free-energy change for a species moving from one side of a membrane where it is at concentration c_1 to the other side where it is at a concentration c_2 is given by

$$\Delta G = RT\ln(c_2/c_1) + ZF\Delta V = 2.303RT\log_{10}(c_2/c_1) + ZF\Delta V$$

in which Z is the electrical charge of the transported species and ΔV is the potential in volts across the membrane (p. 353). Under typical conditions for the inner mitochondrial membrane, the pH outside is 1.4 units lower than inside [corresponding to $\log_{10}(c_2/c_1)$ of 1.4] and the membrane potential is 0.14 V, the outside being positive. Because $Z = +1$ for protons, the free-energy change is $(2.303 \times 8.32 \times 10^{-3}\ \text{kJ mol}^{-1}\,\text{K}^{-1} \times 310\ \text{K} \times 1.4) + (+1 \times 96.48\ \text{kJ mol}^{-1}\,\text{V}^{-1} \times 0.14\ \text{V}) = 21.8\ \text{kJ mol}^{-1}$ (5.2 kcal mol^{-1}). Thus, each proton that is transported out of the matrix to the cytoplasmic side corresponds to $21.8\ \text{kJ mol}^{-1}$ of free energy.

18.3 The Respiratory Chain Consists of Four Complexes: Three Proton Pumps and a Physical Link to the Citric Acid Cycle

Electrons are transferred from NADH to O_2 through a chain of three large protein complexes called *NADH-Q oxidoreductase, Q-cytochrome c oxidoreductase,* and *cytochrome c oxidase* (Figure 18.6 and Table 18.2). *Electron flow within these transmembrane complexes leads to the transport of protons across the inner mitochondrial membrane.* A fourth large protein complex, called *succinate-Q reductase,* contains the succinate dehydrogenase that generates $FADH_2$ in the citric acid cycle. Electrons from this $FADH_2$ enter the electron-transport chain at Q-cytochrome oxidoreductase. Succinate-Q reductase, in contrast with the other complexes, does not pump protons. NADH-Q oxidoreductase, succinate-Q reductase, Q-cytochrome *c* oxidoreductase, and cytochrome *c* oxidase are also called *Complex I, II, III,* and *IV,* respectively. Complexes I, II, and IV appear to be associated in a supramolecular complex termed the *respirasome.* As we have seen before, such supramolecular complexes facilitate the rapid transfer of substrate and prevent the release of reaction intermediates.

Special electron carriers ferry the electrons from one complex to the next. Electrons are carried from NADH-Q oxidoreductase to Q-cytochrome *c* oxidoreductase, the second complex of the chain, by the reduced form of *coenzyme Q* (Q), also known as *ubiquinone* because it is a *ubiquitous quinone* in biological systems. Ubiquinone is a hydrophobic quinone that diffuses rapidly within the inner mitochondrial membrane. Cytochrome *c*, a small soluble protein, shuttles electrons from Q-cytochrome *c* oxidoreductase to cytochrome *c* oxidase, the final component in the chain and the one that catalyzes the reduction of O_2. Electrons from the $FADH_2$ generated by

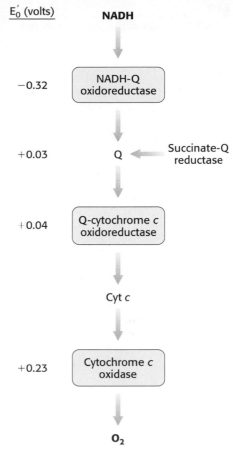

Figure 18.6 Sequence of electron carriers in the respiratory chain. *Notice* that the electron affinity of the components increases as electrons move down the chain.

TABLE 18.2 Components of the mitochondrial electron-transport chain

				OXIDANT OR REDUCTANT		
Enzyme complex	Mass (kd)	Subunits	Prosthetic group	Matrix side	Membrane core	Cytoplasmic side
NADH-Q oxidoreductase	>900	46	FMN Fe-S	NADH	Q	
Succinate-Q reductase	140	4	FAD Fe-S	Succinate	Q	
Q-cytochrome c oxidoreductase	250	11	Heme b_H Heme b_L Heme c_1 Fe-S		Q	Cytochrome c
Cytochrome c oxidase	160	13	Heme a Heme a_3 Cu_A and Cu_B			Cytochrome c

Sources: J. W. DePierre and L. Ernster. *Annu. Rev. Biochem.* 46(1977):215; Y. Hatefi. *Annu Rev. Biochem.* 54(1985):1015; and J. E. Walker. *Q. Rev. Biophys.* 25(1992):253.

succinate dehydrogenase of the citric acid cycle are transferred first to ubiquinone and then to the Q-cytochrome oxidoreductase complex.

Coenzyme Q is a quinone derivative with a long tail consisting of five-carbon isoprene units. The number of isoprene units in the tail depends on the species. The most common form in mammals contains 10 isoprene units (coenzyme Q_{10}). For simplicity, the subscript will be omitted from this abbreviation because all varieties function in an identical manner. Quinones can exist in three oxidation states. In the fully oxidized state (Q), coenzyme Q has two keto groups (Figure 18.7). The addition of one electron and one proton results in the semiquinone form (QH·). The semiquinone can lose a proton to form a semiquinone radical anion (Q·⁻). The addition of a second electron and proton to the semiquinone generates ubiquinol (QH₂), the fully reduced form of coenzyme Q, which holds its protons more tightly. Thus, *for quinones, electron-transfer reactions are coupled to proton binding and release,* a property that is key to transmembrane proton transport. Because ubiquinone is soluble in the membrane, a pool of Q and QH₂—the Q *pool*—is thought to exist in the inner mitochonrial membrane.

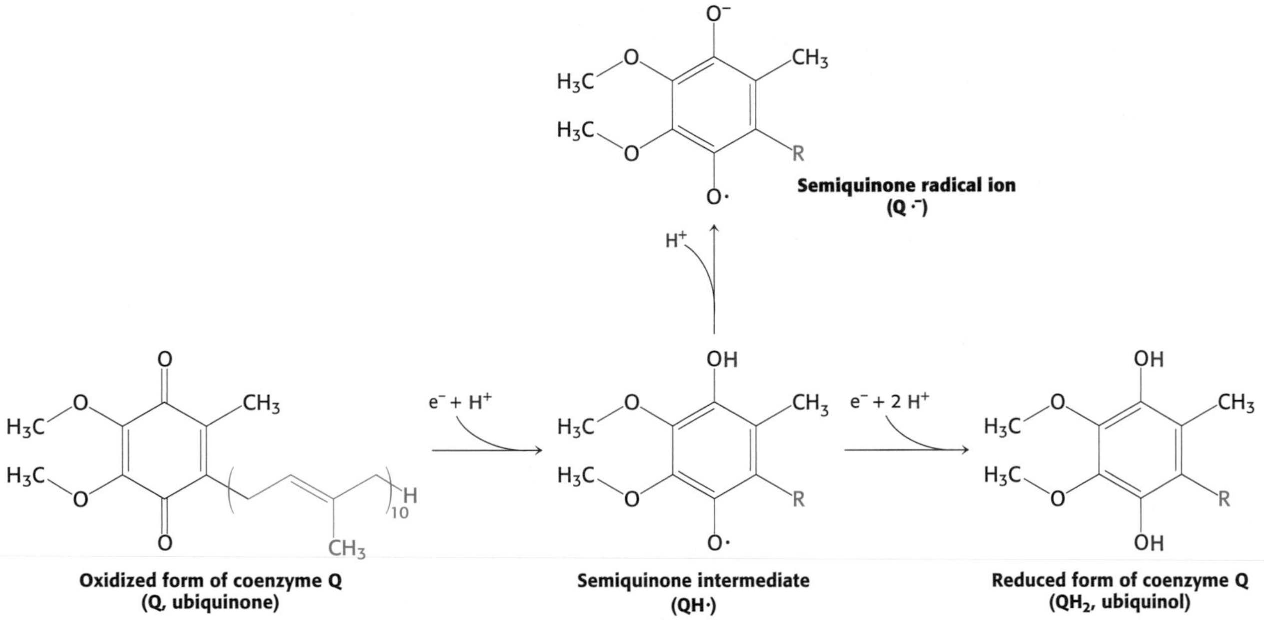

Figure 18.7 Oxidation states of quinones. The reduction of ubiquinone (Q) to ubiquinol (QH₂) proceeds through a semiquinone intermediate (QH·).

The High-Potential Electrons of NADH Enter the Respiratory Chain at NADH-Q Oxidoreductase

The electrons of NADH enter the chain at *NADH-Q oxidoreductase* (also called *Complex I* and *NADH dehydrogenase*), an enormous enzyme (>900 kd) consisting of approximately 46 polypeptide chains. This proton pump, like that of the other two in the respiratory chain, is encoded by genes residing in both the mitochondria and the nucleus. NADH-Q oxidoreductase is L-shaped, with a horizontal arm lying in the membrane and a vertical arm that projects into the matrix.

The reaction catalyzed by this enzyme appears to be

$$\text{NADH} + \text{Q} + 5\,\text{H}^+_{\text{matrix}} \longrightarrow \text{NAD}^+ + \text{QH}_2 + 4\,\text{H}^+_{\text{cytoplasm}}$$

The initial step is the binding of NADH and the transfer of its two high-potential electrons to the *flavin mononucleotide* (FMN) prosthetic group of

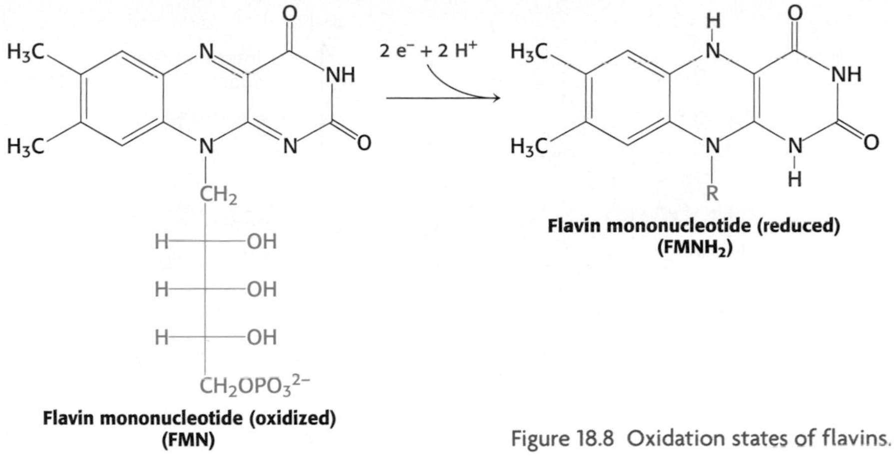

**Flavin mononucleotide (oxidized)
(FMN)**

$2 e^- + 2 H^+$

**Flavin mononucleotide (reduced)
(FMNH$_2$)**

Figure 18.8 Oxidation states of flavins.

this complex to give the reduced form, FMNH$_2$ (Figure 18.8). The electron acceptor of FMN, the isoalloxazine ring, is identical with that of FAD.

Electrons are then transferred from FMNH$_2$ to a series of *iron–sulfur clusters*, the second type of prosthetic group in NADH-Q oxidoreductase. Fe-S clusters in *iron–sulfur proteins* (also called *nonheme iron proteins*) play a critical role in a wide range of reduction reactions in biological systems. Several types of Fe-S clusters are known (Figure 18.9). In the simplest kind, a single iron ion is tetrahedrally coordinated to the sulfhydryl groups of four cysteine residues of the protein. A second kind, denoted by 2Fe-2S, contains two iron ions, two inorganic sulfides, and usually four cysteine residues. A third type, designated 4Fe-4S, contains four iron ions, four inorganic sulfides, and four cysteine residues. NADH-Q oxidoreductase contains both 2Fe-2S and 4Fe-4S clusters. Iron ions in these Fe-S complexes cycle between Fe^{2+} (reduced) and Fe^{3+} (oxidized) states. Unlike quinones and flavins, iron–sulfur clusters generally undergo oxidation–reduction reactions without releasing or binding protons.

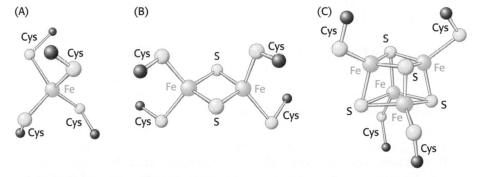

Figure 18.9 Iron–sulfur clusters. (A) A single iron ion bound by four cysteine residues. (B) 2Fe-2S cluster with iron ions bridged by sulfide ions. (C) 4Fe-4S cluster. Each of these clusters can undergo oxidation–reduction reactions.

All the redox reactions take place in the extramembranous portion of NADH-Q oxidoreductase. Although the details of electron transfer through this complex remain the subject of on-going investigation, NADH clearly binds to a site in the extramembranous domain. NADH transfers its two electrons to FMN. These electrons flow through a series of Fe-S centers and then to coenzyme Q. *The flow of two electrons from NADH to coenzyme Q through NADH-Q oxidoreductase leads to the pumping of four hydrogen ions out of the matrix of the mitochondrion. In accepting two electrons, Q takes up two protons*

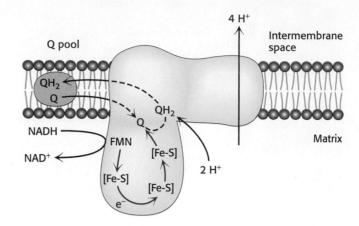

Figure 18.10 Coupled electron–proton transfer reactions through NADH-Q oxidoreductase. Electrons flow in Complex I from NADH through FMN and a series of iron–sulfur cluster to ubiquinone (Q). The electron flow results in the pumping of four protons and the uptake of two protons from the mitochondrial matrix. [Based on U. Brandt et al. *FEBS Letters* 545(2003):9–17, Figure 2.]

from the matrix as it is reduced to QH_2 (Figure 18.10). The QH_2 leaves the enzyme for the hydrophobic interior of the membrane.

Ubiquinol Is the Entry Point for Electrons from FADH$_2$ of Flavoproteins

Recall that $FADH_2$ is formed in the citric acid cycle, in the oxidation of succinate to fumarate by succinate dehydrogenase (p. 487). This enzyme is part of the *succinate-Q reductase complex (Complex II)*, an integral membrane protein of the inner mitochondrial membrane. $FADH_2$ does not leave the complex. Rather, its electrons are transferred to Fe-S centers and then to Q for entry into the electron-transport chain. The succinate-Q reductase complex, in contrast with NADH-Q oxidoreductase, does not transport protons. Consequently, less ATP is formed from the oxidation of $FADH_2$ than from NADH.

Two other enzymes that we will encounter later, *glycerol phosphate dehydrogenase* (p. 528) and *fatty acyl CoA dehydrogenase* (p. 624), likewise transfer their high-potential electrons from $FADH_2$ to Q to form ubiquinol (QH_2), the reduced state of ubiquinone. These enzymes oxidize glycerol and fats, respectively, providing electrons for oxidative phosphorylation. These enzymes also do not pump protons.

Electrons Flow from Ubiquinol to Cytochrome *c* Through Q-Cytochrome *c* Oxidoreductase

The second of the three proton pumps in the respiratory chain is *Q-cytochrome c oxidoreductase* (also known as *Complex III* and as cytochrome reductase). The function of Q-cytochrome *c* oxidoreductase is to catalyze the transfer of electrons from QH_2 to oxidized *cytochrome* c (Cyt *c*), a water-soluble protein, and concomitantly pump protons out of the mitochondrial matrix. The flow of a pair of electrons through this complex leads to the effective net transport of 2 H^+ to the cytoplasmic side, half the yield obtained with NADH-Q reductase because of a smaller thermodynamic driving force.

$$QH_2 + 2\,\text{Cyt}\,c_{ox} + 2\,H^+_{matrix} \longrightarrow Q + 2\,\text{Cyt}\,c_{red} + 4\,H^+_{cytoplasm}$$

Q-cytochrome *c* oxidoreductase itself contains two types of cytochromes, named *b* and c_1 (Figure 18.11). *A cytochrome is an electron-transferring protein that contains a heme prosthetic group.* The iron ion of a cytochrome alternates between a reduced ferrous ($+2$) state and an oxidized ferric ($+3$) state during electron transport. The two cytochrome subunits of Q-cytochrome *c* oxidoreductase contain a total of three hemes: two hemes,

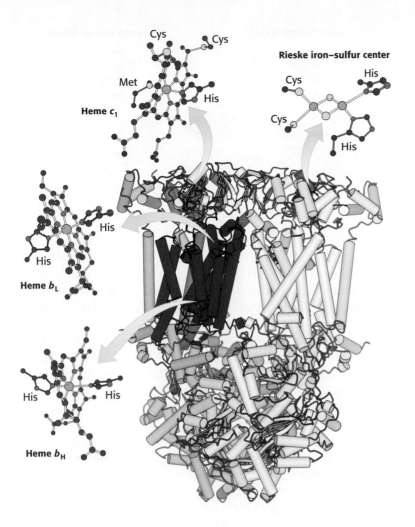

Figure 18.11 **Structure of Q-cytochrome *c* oxidoreductase (cytochrome *bc₁*).** This enzyme is a homodimer with 11 distinct polypeptide chains. *Notice* that the major prosthetic groups, three hemes and a 2Fe-2S cluster, are located either near the cytoplasmic edge of the complex bordering the intermembrane space (top) or in the region embedded in the membrane (α helices represented by vertical tubes). They are well positioned to mediate the electron-transfer reactions between quinones in the membrane and cytochrome *c* in the intermembrane space. [Drawn from 1BCC.pdb.]

termed heme b_L (L for low affinity) and heme b_H (H for high affinity), within cytochrome b, and one heme within cytochrome c_1. The heme prosthetic group in cytochromes b, c_1, and c is iron-protoporphyrin IX, the same heme present in myoglobin and hemoglobin (p. 184). These identical hemes have different electron affinities because they are in different polypeptide environments. For example, heme b_L, which is located near the cytoplasmic face of the membrane, has lower affinity for an electron than does heme b_H, which is near the matrix side. This enzyme is also known as cytochrome bc_1 after its cytochrome groups.

In addition to the hemes, the enzyme contains an iron–sulfur protein with an 2Fe-2S center. This center, termed the *Rieske center,* is unusual in that one of the iron ions is coordinated by two histidine residues rather than two cysteine residues. This coordination stabilizes the center in its reduced form, raising its reduction potential so that it can readily accept electrons from QH_2.

The Q Cycle Funnels Electrons from a Two-Electron Carrier to a One-Electron Carrier and Pumps Protons

The mechanism for the coupling of electron transfer from Q to cytochrome c to transmembrane proton transport is known as the Q *cycle* (Figure 18.12). Two QH_2 molecules bind to the complex consecutively, each giving up two electrons and two H^+. *These protons are released to the cytoplasmic side of the membrane.* The two electrons travel through the complex to different destinations. One electron flows first to the Rieske 2Fe-2S cluster, then to cytochrome c_1, and finally to a molecule of oxidized cytochrome c, converting

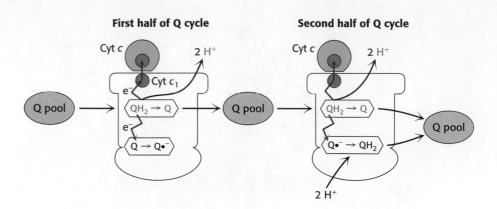

Figure 18.12 Q cycle. In the first half of the cycle, two electrons of a bound QH_2 are transferred, one to cytochrome c and the other to a bound Q in a second binding site to form the semiquinone radical anion $Q\cdot^-$. The newly formed Q dissociates and enters the Q pool. In the second half of the cycle, a second QH_2 also gives up its electrons, one to a second molecule of cytochrome c and the other to reduce $Q\cdot^-$ to QH_2. This second electron transfer results in the uptake of two protons from the matrix. The path of electron transfer is shown in red.

it into its reduced form. The reduced cytochrome c molecule is free to diffuse away from the enzyme to continue down the respiratory chain.

The second electron passes through the two heme groups of cytochrome b to an oxidized ubiquinone bound in a second binding site. This quinone (Q) molecule is reduced to a semiquinone radical anion $(Q\cdot^-)$ by the electron from the first QH_2 molecule.

On the addition of the electron from the second QH_2 molecule, this quinone radical anion takes up two protons from the matrix side to form QH_2. *The removal of these two protons from the matrix contributes to the formation of the proton gradient.* In sum, four protons are released on the cytoplasmic side, and two protons are removed from the mitochondrial matrix.

In one Q cycle, two QH_2 molecules are oxidized to form two Q molecules, and then one Q molecule is reduced to QH_2. Why this complexity? The formidable problem solved here is to efficiently funnel electrons from a two-electron carrier (QH_2) to a one-electron carrier (cytochrome c). The cytochrome b component of the reductase is in essence a recycling device that enables both electrons of QH_2 to be used effectively.

Cytochrome c Oxidase Catalyzes the Reduction of Molecular Oxygen to Water

The last of the three proton-pumping assemblies of the respiratory chain is *cytochrome* c *oxidase (Complex IV)*. Cytochrome oxidase catalyzes the transfer of electrons from the reduced form of cytochrome c to molecular oxygen, the final acceptor.

$$4\,\text{Cyt}\,c_{\text{red}} + 8\,H^+_{\text{matrix}} + O_2 \longrightarrow 4\,\text{Cyt}\,c_{\text{ox}} + 2\,H_2O + 4\,H^+_{\text{cytoplasm}}$$

The requirement of oxygen for this reaction is what makes "aerobic" organisms aerobic. To obtain oxygen for this reaction is the reason that human beings must breath. Four electrons are funneled to O_2 to completely reduce it to H_2O, and, concomitantly, protons are pumped from the matrix to the cytoplasmic side of the inner mitochondrial membrane. This reaction is quite thermodynamically favorable. From the reduction potentials in Table 18.1, the standard free-energy change for this reaction is calculated to be $\Delta G^{\circ\prime} = -231.8\text{ kJ mol}^{-1}\,(-55.4\text{ kcal mol}^{-1})$. As much of this free energy as possible must be captured in the form of a proton gradient for subsequent use in ATP synthesis.

Bovine cytochrome c oxidase is reasonably well understood at the structural level (Figure 18.13). It consists of 13 subunits, of which 3 are encoded by the mitochondrion's own genome. Cytochrome c oxidase contains two *heme A* groups and three *copper ions*, arranged as two copper centers, designated A and B. One center, Cu_A/Cu_A, contains two copper ions linked by two bridging cysteine residues. This center initially accepts electrons from

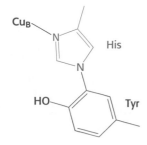

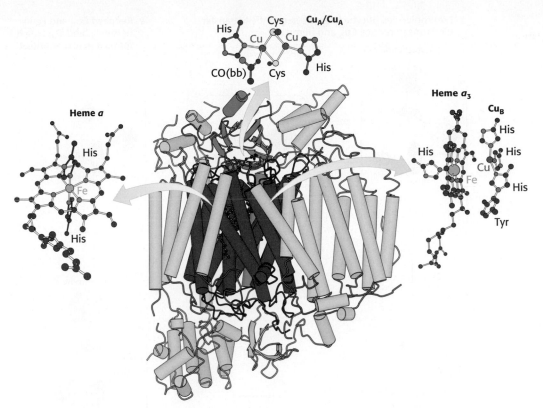

Figure 18.13 **Structure of cytochrome *c* oxidase.** This enzyme consists of 13 polypeptide chains. *Notice* that most of the complex, as well as two major prosthetic groups (heme *a* and heme a_3–Cu_B) are embedded in the membrane (α helices represented by vertical tubes). Heme a_3–Cu_B is the site of the reduction of oxygen to water. The Cu_A/Cu_A prosthetic group is positioned near the intermembrane space to better accept electrons from cytochrome *c*. CO(bb) is a carbonyl group of the peptide backbone. [Drawn from 2OCC.pdb.]

reduced cytochrome *c*. The remaining copper ion, Cu_B, is coordinated by three histidine residues, one of which is modified by covalent linkage to a tyrosine residue. The copper centers alternate between the reduced Cu^+ (cuprous) form and the oxidized Cu^{2+} (cupric) form as they accept and donate electrons.

There are two heme A molecules, called *heme* a and *heme* a_3, in cytochrome *c* oxidase. Heme A differs from the heme in cytochrome *c* and c_1 in three ways: (1) a formyl group replaces a methyl group, (2) a C_{17} hydrocarbon chain replaces one of the vinyl groups, and (3) the heme is not covalently attached to the protein.

Heme A

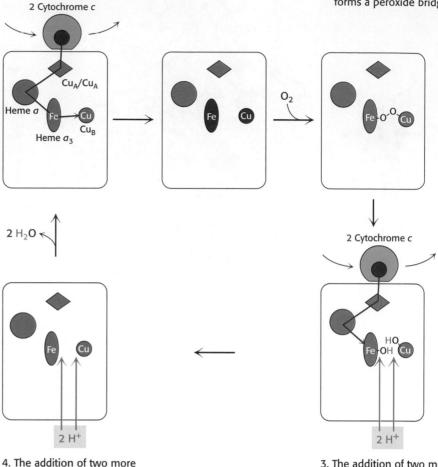

1. Two molecules of cytochrome *c* sequentially transfer electrons to reduce Cu_B and heme a_3.

2. Reduced Cu_B and Fe in heme a_3 bind O_2, which forms a peroxide bridge.

4. The addition of two more protons leads to the release of water.

3. The addition of two more electrons and two more protons cleaves the peroxide bridge.

Figure 18.14 Cytochrome oxidase mechanism. The cycle begins and ends with all prosthetic groups in their oxidized forms (shown in blue). Reduced forms are in red. Four cytochrome *c* molecules donate four electrons, which, in allowing the binding and cleavage of an O_2 molecule, also makes possible the import of four H^+ from the matrix to form two molecules of H_2O, which are released from the enzyme to regenerate the initial state.

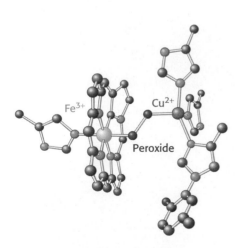

Figure 18.15 Peroxide bridge. The oxygen bound to heme a_3 is reduced to peroxide by the presence of Cu_B.

Heme *a* and heme a_3 have distinct redox potentials because they are located in different environments within cytochrome *c* oxidase. An electron flows from cytochrome *c* to Cu_A/Cu_A, to heme *a* to heme a_3 to Cu_B, and finally to O_2. Heme a_3 and Cu_B are directly adjacent. Together, *heme a_3 and Cu_B form the active center at which O_2 is reduced to H_2O.*

Four molecules of cytochrome *c* bind consecutively to the enzyme and transfer an electron to reduce one molecule of O_2 to H_2O (Figure 18.14).

1. Electrons from two molecules of reduced cytochrome *c* flow down the electron-transfer pathway, one stopping at Cu_B and the other at heme a_3. With both centers in the reduced state, they together can now bind an oxygen molecule.

2. As molecular oxygen binds, it abstracts an electron from each of the nearby ions in the active center to form a peroxide (O_2^{2-}) bridge between them (Figure 18.15).

3. Two more molecules of cytochrome *c* bind and release electrons that travel to the active center. The addition of an electron as well as H^+ to each oxygen atom reduces the two ion–oxygen groups to Cu_B^{2+}—OH and Fe^{3+}—OH.

4. Reaction with two more H^+ ions allows the release of two molecules of H_2O and resets the enzyme to its initial, fully oxidized form.

$$4 \, \text{Cyt} \, c_{\text{red}} + 4 \, \text{H}^+_{\text{matrix}} + \text{O}_2 \longrightarrow 4 \, \text{Cyt} \, c_{\text{ox}} + 2 \, \text{H}_2\text{O}$$

The four protons in this reaction come exclusively from the matrix. Thus, the consumption of these four protons contributes directly to the proton gradient. Recall that each proton contributes 21.8 kJ mol^{-1} (5.2 kcal mol^{-1}) to the free energy associated with the proton gradient; so these four protons contribute 87.2 kJ mol^{-1} (20.8 kcal mol^{-1}), an amount substantially less than the free energy available from the reduction of oxygen to water. What is the fate of this missing energy? Remarkably, *cytochrome c oxidase uses this energy to pump four additional protons from the matrix to the cytoplasmic side of the membrane in the course of each reaction cycle for a total of eight protons removed from the matrix* (Figure 18.16). The details of how these protons are transported through the protein is still under study. However, two effects contribute to the mechanism. First, charge neutrality tends to be maintained in the interior of proteins. Thus, the addition of an electron to a site inside a protein tends to favor the binding of H$^+$ to a nearby site. Second, conformational changes take place, particularly around the heme a_3–Cu$_B$ center, in the course of the reaction cycle. Presumably, in one conformation, protons may enter the protein exclusively from the matrix side, whereas, in another, they may exit exclusively to the cytoplasmic side. Thus, the overall process catalyzed by cytochrome *c* oxidase is

$$4 \, \text{Cyt} \, c_{\text{red}} + 8 \, \text{H}^+_{\text{matrix}} + \text{O}_2 \longrightarrow 4 \, \text{Cyt} \, c_{\text{ox}} + 2 \, \text{H}_2\text{O} + 4 \, \text{H}^+_{\text{cytoplasm}}$$

Figure 18.17 summarizes the flow of electrons from NADH and FADH$_2$ through the respiratory chain. This series of exergonic reactions is coupled to the pumping of protons from the matrix. As we will see shortly, the energy inherent in the proton gradient will be used to synthesize ATP.

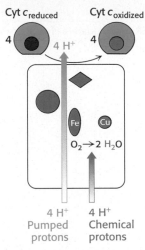

Figure 18.16 Proton transport by cytochrome *c* oxidase. Four protons are taken up from the matrix side to reduce one molecule of O$_2$ to two molecules of H$_2$O. These protons are called "chemical protons" because they participate in a clearly defined reaction with O$_2$. Four additional "pumped" protons are transported out of the matrix and released on the cytoplasmic side in the course of the reaction. The pumped protons double the efficiency of free-energy storage in the form of a proton gradient for this final step in the electron-transport chain.

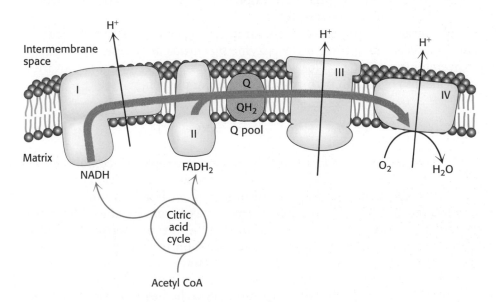

Figure 18.17 The electron-transport chain. High-energy electrons in the form of NADH and FADH$_2$ are generated by the citric acid cycle. These electrons flow through the respiratory chain, which powers proton pumping and results in the reduction of O$_2$.

Toxic Derivatives of Molecular Oxygen Such As Superoxide Radical Are Scavenged by Protective Enzymes

As discussed earlier, molecular oxygen is an ideal terminal electron acceptor, because its high affinity for electrons provides a large thermodynamic driving force. However, *danger lurks in the reduction of O$_2$.* The transfer of four electrons leads to safe products (two molecules of H$_2$O), but partial reduction generates hazardous compounds. In particular, *the transfer of a*

TABLE 18.3 Pathological and other conditions that may entail free-radical injury

Atherogenesis
Emphysema; bronchitis
Parkinson disease
Duchenne muscular dystrophy
Cervical cancer
Alcoholic liver disease
Diabetes
Acute renal failure
Down syndrome
Retrolental fibroplasia (conversion of the retina into a fibrous mass in premature infants)
Cerebrovascular disorders
Ischemia; reperfusion injury

Source: After D. B. Marks, A. D. Marks, and C. M. Smith, *Basic Medical Biochemistry: A Clinical Approach* (Williams & Wilkins, 1996), p. 331.

single electron to O_2 forms superoxide anion, whereas the transfer of two electrons yields peroxide.

$$O_2 \xrightarrow{e^-} \underset{\substack{\text{Superoxide} \\ \text{ion}}}{O_2^{\cdot-}} \xrightarrow{e^-} \underset{\text{Peroxide}}{O_2^{2-}}$$

Both compounds are potentially destructive. The strategy for the safe reduction of O_2 is clear: *the catalyst does not release partly reduced intermediates.* Cytochrome *c* oxidase meets this crucial criterion by holding O_2 tightly between Fe and Cu ions.

Although cytochrome *c* oxidase and other proteins that reduce O_2 are remarkably successful in not releasing intermediates, small amounts of superoxide anion and hydrogen peroxide are unavoidably formed. Superoxide, hydrogen peroxide, and species that can be generated from them such as OH· are collectively referred to as *reactive oxygen species* or *ROS*. Oxidative damage caused by ROS has been implicated in the aging process as well as in a growing list of diseases (Table 18.3).

What are the cellular defense strategies against oxidative damage by ROS? Chief among them is the enzyme *superoxide dismutase*. This enzyme scavenges superoxide radicals by catalyzing the conversion of two of these radicals into hydrogen peroxide and molecular oxygen.

> **Dismutation**
>
> A reaction in which a single reactant is converted into two different products.

$$2\,O_2^{\cdot-} + 2\,H^+ \underset{}{\overset{\text{Superoxide}\atop\text{dismutase}}{\rightleftharpoons}} O_2 + H_2O_2$$

Eukaryotes contain two forms of this enzyme, a manganese-containing version located in mitochondria and a copper- and zinc-dependent cytoplasmic form. These enzymes perform the dismutation reaction by a similar mechanism (Figure 18.18). The oxidized form of the enzyme is reduced by superoxide to form oxygen. The reduced form of the enzyme, formed in this reaction, then reacts with a second superoxide ion to form peroxide, which takes up two protons along the reaction path to yield hydrogen peroxide.

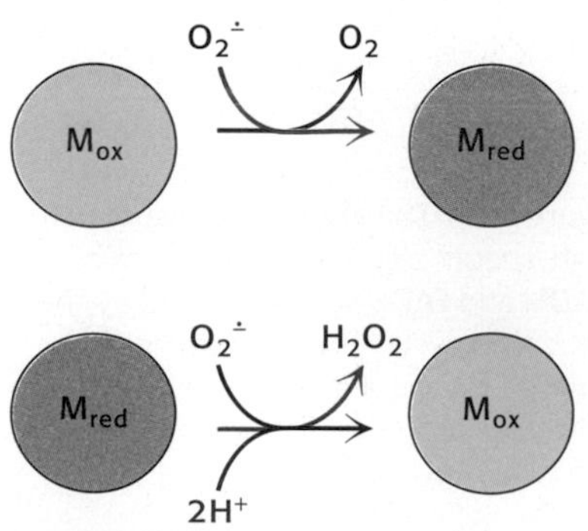

The hydrogen peroxide formed by superoxide dismutase and by other processes is scavenged by *catalase*, a ubiquitous heme protein that catalyzes the dismutation of hydrogen peroxide into water and molecular oxygen.

$$2\,H_2O_2 \xrightarrow{\text{Catalase}} O_2 + 2\,H_2O$$

Figure 18.18 Superoxide dismutase mechanism. The oxidized form of superoxide dismutase (M_{ox}) reacts with one superoxide ion to form O_2 and generate the reduced form of the enzyme (M_{red}). The reduced form then reacts with a second superoxide and two protons to form hydrogen peroxide and regenerate the oxidized form of the enzyme.

Superoxide dismutase and catalase are remarkably efficient, performing their reactions at or near the diffusion-limited rate (p. 221). Glutathione peroxidase also plays a role in scavenging H_2O_2 (p. 587). Other cellular defenses against oxidative damage include the antioxidant vitamins, vitamins

E and C. Because it is lipophilic, vitamin E is especially useful in protecting membranes from lipid peroxidation.

One of the long-term benefits of exercise may be to increase the amount of superoxide dismutase in the cell. The elevated aerobic metabolism during exercise causes more ROS to be generated. In response, the cell synthesizes more protective enzymes. The net effect is one of protection, because the increase in superoxide dismutase more effectively protects the cell during periods of rest.

Electrons Can Be Transferred Between Groups That Are Not in Contact

How are electrons transferred between electron-carrying groups of the respiratory chain? This question is intriguing because these groups are frequently buried in the interior of a protein in fixed positions and are therefore not directly in contact with one another. Electrons can move through space, even through a vacuum. However, the rate of electron transfer through space falls off rapidly as the electron donor and electron acceptor move apart from each other, decreasing by a factor of 10 for each increase in separation of 0.8 Å. The protein environment provides more-efficient pathways for electron conduction: typically, the rate of electron transfer decreases by a factor of 10 every 1.7 Å (Figure 18.19). For groups in contact, electron-transfer reactions can be quite fast, with rates of approximately 10^{13} s^{-1}. Within proteins in the electron-transport chain, electron-carrying groups are typically separated by 15 Å beyond their van der Waals contact distance. For such separations, we expect electron-transfer rates of approximately 10^4 s^{-1} (i.e., electron transfer in less than 1 ms), assuming that all other factors are optimal. Without the mediation of the protein, an electron transfer over this distance would take approximately 1 day.

The case is more complicated when electrons must be transferred between two distinct proteins, such as when cytochrome c accepts electrons from Complex III or passes them on to Complex IV. A series of hydrophobic interactions bring the heme groups of cytochrome c and c_1 to within 4.5 Å of each other, with the iron atoms separated by 17.4 Å. This distance could allow cytochrome c reduction at a rate of 8.3×10^6 s^{-1}.

Another important factor in determining the rate of electron transfer is the driving force, the free-energy change associated with the reaction (Figure 18.20). Like the rates of most reactions, those of electron-transfer reactions tend to increase as the free-energy change for the reaction becomes more favorable. Interestingly, however, each electron-transfer reaction has

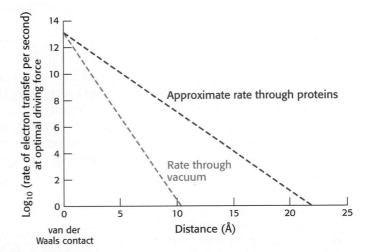

Figure 18.19 Distance dependence of electron-transfer rate. The rate of electron transfer decreases as the electron donor and the electron acceptor move apart. In a vacuum, the rate decreases by a factor of 10 for every increase of 0.8 Å. In proteins, the rate decreases more gradually, by a factor of 10 for every increase of 1.7 Å. This rate is only approximate because variations in the structure of the intervening protein medium can affect the rate.

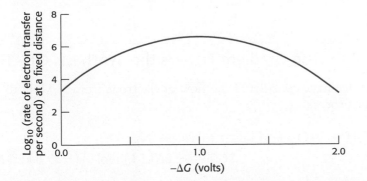

Figure 18.20 Free-energy dependence of electron-transfer rate. The rate of an electron-transfer reaction at first increases as the driving force for the reaction increases. The rate reaches a maximum and then decreases at very large driving forces.

an optimal driving force; making the reaction more favorable beyond this point decreases the rate of electron transfer. This so-called *inverted region* is important for the light reactions of photosynthesis, to be discussed in Chapter 19.

The Conformation of Cytochrome *c* Has Remained Essentially Constant for More Than a Billion Years

Cytochrome c is present in all organisms having mitochondrial respiratory chains: plants, animals, and eukaryotic microorganisms. This electron carrier evolved more than 1.5 billion years ago, before the divergence of plants and animals. Its function has been conserved throughout this period, as evidenced by the fact that *the cytochrome* c *of any eukaryotic species reacts in vitro with the cytochrome* c *oxidase of any other species tested thus far.* For example, wheat-germ cytochrome *c* reacts with human cytochrome oxidase. Additionally, some prokaryotic cytochromes, such as cytochrome c_2 from a photosynthetic bacterium and cytochrome c_{550} from a denitrifying bacterium, closely resemble cytochrome *c* from tuna-heart mitochondria (Figure 18.21). This evidence attests to an efficient evolutionary solution to electron transfer bestowed by the structural and functional characteristics of cytochrome *c*.

Figure 18.21 Conservation of the three-dimensional structure of cytochrome *c*. The side chains are shown for the 21 conserved amino acids and the heme. [Drawn from 3CYT.pdb, 3C2C.pdb, and 1SSC.pdb.]

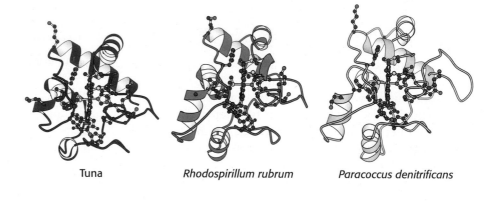

Tuna   *Rhodospirillum rubrum*   *Paracoccus denitrificans*

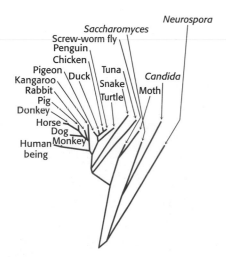

Figure 18.22 Evolutionary tree constructed from sequences of cytochrome *c*. Branch lengths are proportional to the number of amino acid changes that are believed to have occurred. This drawing is an adaptation of the work of Walter M. Fitch and Emanuel Margoliash.

The resemblance among cytochrome *c* molecules extends to the level of amino acid sequence. Because of the molecule's relatively small size and ubiquity, the amino acid sequences of cytochrome *c* from more than 80 widely ranging eukaryotic species have been determined by direct protein sequencing by Emil Smith, Emanuel Margoliash, and others. The striking finding is that *21 of 104 residues have been invariant for more than one and a half billion years of evolution.* A phylogenetic tree, constructed from the amino acid sequences of cytochrome *c*, reveals the evolutionary relationships between many animal species (Figure 18.22).

18.4 A Proton Gradient Powers the Synthesis Of ATP

Thus far, we have considered the flow of electrons from NADH to O_2, an exergonic process.

$$\text{NADH} + \tfrac{1}{2}O_2 + H^+ \rightleftharpoons H_2O + \text{NAD}^+$$
$$\Delta G^{\circ\prime} = -220.1 \text{ kJ mol}^{-1} \, (-52.6 \text{ kcal mol}^{-1})$$

Next, we consider how this process is coupled to the synthesis of ATP, an endergonic process.

$$ADP + P_i + H^+ \rightleftharpoons ATP + H_2O$$
$$\Delta G^{\circ\prime} = +30.5 \ \mathrm{kJ\,mol^{-1}} \ (+7.3 \ \mathrm{kcal\ mol^{-1}})$$

A molecular assembly in the inner mitochondrial membrane carries out the synthesis of ATP. This enzyme complex was originally called the *mitochondrial ATPase* or F_1F_0 *ATPase* because it was discovered through its catalysis of the reverse reaction, the hydrolysis of ATP. *ATP synthase*, its preferred name, emphasizes its actual role in the mitochondrion. It is also called *Complex V*.

How is the oxidation of NADH coupled to the phosphorylation of ADP? Electron transfer was first suggested to lead to the formation of a covalent high-energy intermediate that serves as a compound having a high phosphoryl-transfer potential, analogous to the generation of ATP by the formation of 1,3-bisphosphoglycerate in glycolysis. An alternative proposal was that electron transfer aids the formation of an activated protein conformation, which then drives ATP synthesis. The search for such intermediates for several decades proved fruitless.

In 1961, Peter Mitchell suggested a radically different mechanism, *the chemiosmotic hypothesis*. He proposed that electron transport and ATP synthesis are coupled by *a proton gradient across the inner mitochondrial membrane*. In his model, the transfer of electrons through the respiratory chain leads to the pumping of protons from the matrix to the cytoplasmic side of the inner mitochondrial membrane. The H^+ concentration becomes lower in the matrix, and an electric field with the matrix side negative is generated (Figure 18.23). Protons then flow back into the matrix to equalize the distribution. Mitchell's idea was that this flow of protons drives the synthesis of ATP by ATP synthase. The energy-rich unequal distribution of protons is called the *proton-motive force*. The proton-motive force can be thought of as being composed of two components: a chemical gradient and a charge gradient. The chemical gradient for protons can be represented as a pH gradient. The charge gradient is created by the positive charge on the unequally distributed protons forming the chemical gradient. Mitchell proposed that both components power the synthesis of ATP.

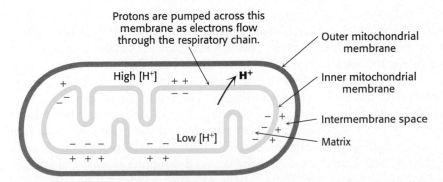

Figure 18.23 Chemiosmotic hypothesis. Electron transfer through the respiratory chain leads to the pumping of protons from the matrix to the cytoplasmic side of the inner mitochondrial membrane. The pH gradient and membrane potential constitute a proton-motive force that is used to drive ATP synthesis.

Figure 18.24 Testing the chemiosmotic hypothesis. ATP is synthesized when reconstituted membrane vesicles containing bacteriorhodopsin (a light-driven proton pump) and ATP synthase are illuminated. The orientation of ATP synthase in this reconstituted membrane is the reverse of that in the mitochondrion.

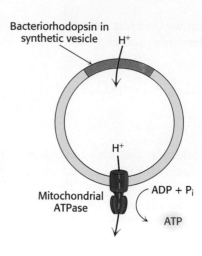

Figure 18.24 Testing the chemiosmotic hypothesis. ATP is synthesized when reconstituted membrane vesicles containing bacteriorhodopsin (a light-driven proton pump) and ATP synthase are illuminated. The orientation of ATP synthase in this reconstituted membrane is the reverse of that in the mitochondrion.

$$\text{Proton-motive force } (\Delta p) = \text{chemical gradient } (\Delta\text{pH}) + \text{charge gradient } (\Delta\psi)$$

Mitchell's highly innovative hypothesis that oxidation and phosphorylation are coupled by a proton gradient is now supported by a wealth of evidence. Indeed, electron transport does generate a proton gradient across the inner mitochondrial membrane. The pH outside is 1.4 units lower than inside, and the membrane potential is 0.14 V, the outside being positive. As calculated on page 508, this membrane potential corresponds to a free energy of 21.8 kJ (5.2 kcal) per mole of protons.

An artificial system was created to elegantly demonstrate the basic principle of the chemiosmotic hypothesis. The role of the respiratory chain was played by bacteriorhodopsin. This purple membrane protein from halobacteria pumps protons when illuminated. Synthetic vesicles containing bacteriorhodopsin and mitochondrial ATP synthase purified from beef heart were created (Figure 18.24). When the vesicles were exposed to light, ATP was formed. This key experiment clearly showed that *the respiratory chain and ATP synthase are biochemically separate systems, linked only by a proton-motive force.*

ATP Synthase Is Composed of a Proton-Conducting Unit and a Catalytic Unit

Biochemical, electron microscopic, and crystallographic studies of ATP synthase have revealed many details of its structure (Figure 18.25). It is a large, complex enzyme that looks like a ball on a stick. Much of the "stick" part, called the F_0 subunit, is embedded in the inner mitochondrial membrane. The 85-Å-diameter ball, called the F_1 subunit, protrudes into the mitochondrial matrix. The F_1 subunit contains the catalytic activity of the synthase. In fact, isolated F_1 subunits display ATPase activity.

The F_1 subunit consists of five types of polypeptide chains (α_3, β_3, γ, δ, and ϵ) with the indicated stoichiometry. The α and β subunits, which make up the bulk of the F_1, are arranged alternately in a hexameric ring; they are homologous to one another and are members of the P-loop NTPase family (p. 267). Both bind nucleotides but only the β subunits participate directly in catalysis. Beginning just below the α and β subunits is a central stalk consisting of the γ and ϵ proteins. The γ subunit includes a long helical coiled coil (p. 45) that extends into the center of the $\alpha_3\beta_3$ hexamer. *The γ subunit breaks the symmetry of the $\alpha_3\beta_3$ hexamer: each of the β subunits is distinct by virtue of its interaction with a different face of γ.* Distinguishing the three β subunits is crucial for understanding the mechanism of ATP synthesis.

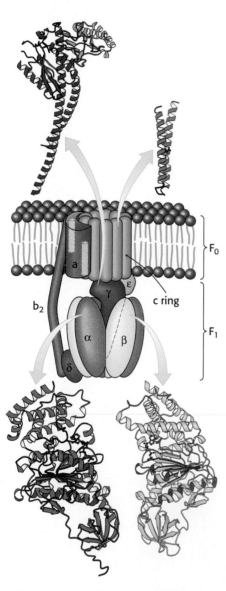

Figure 18.25 Structure of ATP synthase. A schematic structure is shown along with representations of the components for which structures have been determined to high resolution. The P-loop NTPase domains of the α and β subunits are indicated by purple shading. *Notice* that part of the enzyme complex is embedded in the inner mitochondrial membrane, whereas the remainder resides in the matrix. [Drawn from 1E79.pdb and 1COV.pdb.]

The F_0 subunit is a hydrophobic segment that spans the inner mitochondrial membrane. *F_0 contains the proton channel of the complex.* This channel consists of a ring comprising from 10 to 14 **c** subunits that are embedded in the membrane. A single **a** subunit binds to the outside of the ring. The F_0 and F_1 subunits are connected in two ways: by the central $\gamma\varepsilon$ stalk and by an exterior column. The exterior column consists of one **a** subunit, two **b** subunits, and the δ subunit. As will be discussed shortly, we can think of the enzyme as consisting of a moving part and a stationary part: (1) the moving unit, or *rotor*, consists of the **c** ring and the $\gamma\varepsilon$ stalk and (2) the stationary unit, or *stator*, is composed of the remainder of the molecule.

Proton Flow Through ATP Synthase Leads to the Release of Tightly Bound ATP: The Binding-Change Mechanism

ATP synthase catalyzes the formation of ATP from ADP and orthophosphate.

$$ADP^{3-} + HPO_4^{2-} + H^+ \rightleftharpoons ATP^{4-} + H_2O$$

The actual substrates are Mg^{2+} complexes of ADP and ATP, as in all known phosphoryl-transfer reactions with these nucleotides. A terminal oxygen atom of ADP attacks the phosphorus atom of P_i to form a pentacovalent intermediate, which then dissociates into ATP and H_2O (Figure 18.26). The attacking oxygen atom of ADP and the departing oxygen atom of P_i occupy the apices of a trigonal bipyramid.

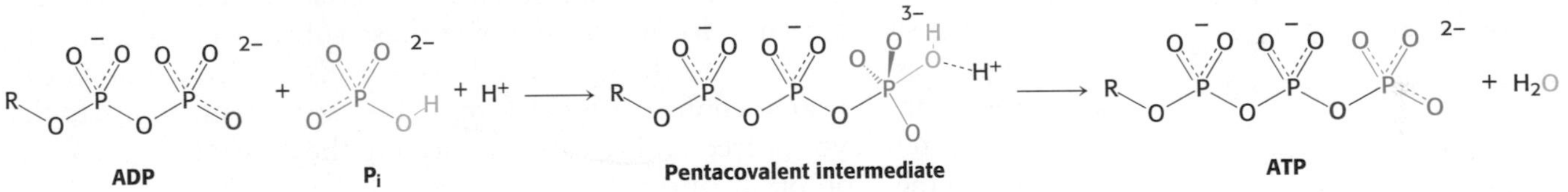

Figure 18.26 ATP-synthesis mechanism. One of the oxygen atoms of ADP attacks the phosphorus atom of P_i to form a pentacovalent intermediate, which then forms ATP and releases a molecule of H_2O.

How does the flow of protons drive the synthesis of ATP? Isotopic-exchange experiments unexpectedly revealed that *enzyme-bound ATP forms readily in the absence of a proton-motive force.* When ADP and P_i were added to ATP synthase in $H_2^{18}O$, ^{18}O became incorporated into P_i through the synthesis of ATP and its subsequent hydrolysis (Figure 18.27). The rate of incorporation of ^{18}O into P_i showed that about equal amounts of bound ATP and ADP are in equilibrium at the catalytic site, even in the absence of a proton gradient. However, ATP does not leave the catalytic site unless

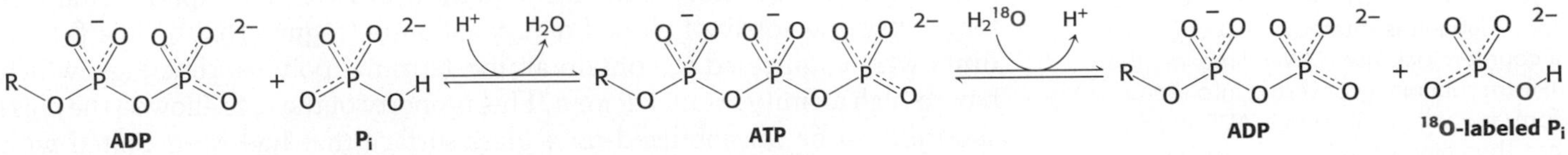

Figure 18.27 ATP forms without a proton-motive force but is not released. The results of isotopic-exchange experiments indicate that enzyme-bound ATP is formed from ADP and P_i in the absence of a proton-motive force.

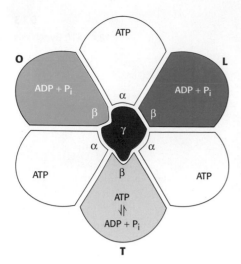

Figure 18.28 ATP synthase nucleotide-binding sites are not equivalent. The γ subunit passes through the center of the α₃β₃ hexamer and makes the nucleotide-binding sites in the β subunits distinct from one another. Note that each α subunit contains bound ATP, but these nucleotides do not participate in any reactions.

Progressive alteration of the forms of the three active sites of ATP synthase

Subunit 1	L → T → O → L → T → O.........
Subunit 2	O → L → T → O → L → T.........
Subunit 3	T → O → L → T → O → L.......

protons flow through the enzyme. Thus, *the role of the proton gradient is not to form ATP but to release it from the synthase.*

On the basis of these observations and others, Paul Boyer proposed a *binding-change mechanism* for proton-driven ATP synthesis. This proposal states that a β subunit can perform each of three sequential steps in the function of ATP synthesis by changing conformation. These steps are (1) ADP and P_i binding, (2) ATP synthesis, and (3) ATP release. The concepts of this initial proposal refined by more-recent crystallographic and other data yield a satisfying mechanism for ATP synthesis. As already noted, interactions with the γ subunit make the three β subunits unequivalent (Figure 18.28). One β subunit can be in the L, or loose, conformation. This conformation binds ADP and P_i. A second subunit can be in the T, or tight, conformation. This conformation binds ATP with great avidity, so much so that it will convert bound ADP and P_i into ATP. Both the T and L conformations are sufficiently constrained that they cannot release bound nucleotides. The final subunit will be in the O, or open, form. This form can exist with a bound nucleotide in a structure that is similar to those of the T and L forms, but it can also convert to form a more open conformation and release a bound nucleotide.

The rotation of the γ subunit drives the interconversion of these three forms (Figure 18.29). ADP and P_i bound in the subunit in the T form are transiently combining to form ATP. Suppose that the γ subunit is rotated by 120 degrees in a counterclockwise direction (as viewed from the top). This rotation converts the T-form site into an O-form site with the nucleotide bound as ATP. Concomitantly, the L-form site is converted into a T-form site, enabling the transformation of an additional ADP and P_i into ATP. The ATP in the O-form site can now depart from the enzyme to be replaced by ADP and P_i. An additional 120-degree rotation converts this O-form site into an L-form site, trapping these substrates. Each subunit progresses from the T to the O to the L form with no two subunits ever present in the same conformational form. This mechanism suggests that ATP can be synthesized and released by driving the rotation of the γ subunit in the appropriate direction.

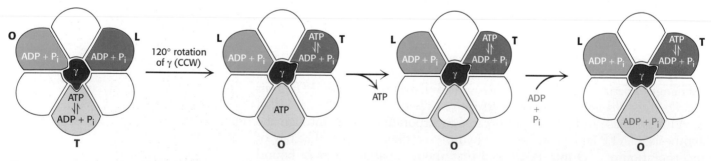

Figure 18.29 Binding-change mechanism for ATP synthase. The rotation of the γ subunit interconverts the three β subunits. The subunit in the T (tight) form interconverts ADP and P_i and ATP but does not allow ATP be released. When the γ subunit is rotated by 120 degrees in a counterclockwise (CCW) direction, the T-form subunit is converted into the O form, allowing ATP release. ADP and P_i can then bind to the O-form subunit. An additional 120-degree rotation (not shown) traps these substrates in an L-form subunit.

Rotational Catalysis Is the World's Smallest Molecular Motor

Is it possible to observe the proposed rotation directly? Elegant experiments have demonstrated the rotation through the use of a simple experimental system consisting solely of cloned α₃β₃γ subunits (Figure 18.30). The β subunits were engineered to contain amino-terminal polyhistidine tags, which have a high affinity for nickel ions. This property of the tags allowed the α₃β₃ assembly to be immobilized on a glass surface that had been coated with nickel ions. The γ subunit was linked to a fluorescently labeled actin filament to provide a long segment that could be observed under a fluorescence microscope. Remarkably, the addition of ATP caused the actin filament to

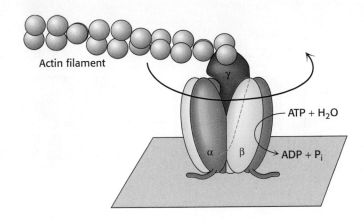

Actin filament

γ

ATP + H₂O

α β

ADP + Pᵢ

Figure 18.30 Direct observation of ATP-driven rotation in ATP synthase. The $\alpha_3\beta_3$ hexamer of ATP synthase is fixed to a surface, with the γ subunit projecting upward and linked to a fluorescently labeled actin filament. The addition and subsequent hydrolysis of ATP result in the counterclockwise rotation of the γ subunit, which can be directly seen under a fluorescence microscope.

rotate unidirectionally in a counterclockwise direction. *The γ subunit was rotating, driven by the hydrolysis of ATP.* Thus, the catalytic activity of an individual molecule could be observed. The counterclockwise rotation is consistent with the predicted mechanism for hydrolysis because the molecule was viewed from below relative to the view shown in Figure 18.30.

More-detailed analysis in the presence of lower concentrations of ATP revealed that the γ subunit rotates in 120-degree increments. Each increment corresponds to the hydrolysis of a single ATP molecule. In addition, from the results obtained by varying the length of the actin filament and measuring the rate of rotation, the enzyme appears to operate near 100% efficiency; that is, essentially all of the energy released by ATP hydrolysis is converted into rotational motion.

Proton Flow Around the c Ring Powers ATP Synthesis

The direct observation of rotary motion of the γ subunit is strong evidence for the rotational mechanism for ATP synthesis. The last remaining question is: How does proton flow through F_0 drive the rotation of the γ subunit? Howard Berg and George Oster proposed an elegant mechanism that provides a clear answer to this question. The mechanism depends on the structures of the **a** and **c** subunits of F_0 (Figure 18.31). The **a** subunit directly abuts the membrane-spanning ring formed by 10 to 14 **c** subunits. Although the structure of the **a** subunit has not yet been experimentally determined, a variety of evidence is consistent with a structure that includes two hydrophilic half-channels that do not span the membrane (see Figure 18.31). Thus, protons can pass into either of these channels, but they cannot move completely across the membrane. The **a** subunit is positioned such that each half-channel directly interacts with one **c** subunit.

The structure of the **c** subunit was determined both by NMR methods and by x-ray crystallography. Each polypeptide chain forms a pair of α helices that span the membrane. An aspartic acid residue (Asp 61) is found in the middle of one of the helices. When the Asp 61 residues of the two **c** subunits are in contact with the hydrophilic environment of a half-channel, they can give up their protons so that they are in the charged aspartate form (Figure 18.32). The key to proton movement across the membrane is that, in a proton-rich environment, such as the cytoplasmic side of the mitochondrial membrane, a proton will enter a channel and bind the aspartate residue. The subunit with the bound proton then rotates through the membrane until the aspartic acid is in a proton-poor environment of the other half channel, where the proton is released. *The movement of protons through the half-channels from the high proton concentration of the cytoplasm to the low proton concentration of the matrix powers the rotation of the c ring.* Its rotation is favored by the ability of the newly protonated (neutralized)

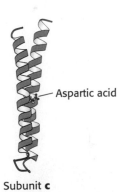

Aspartic acid

Subunit **c**

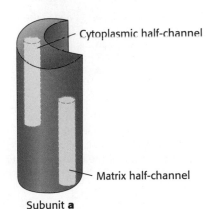

Cytoplasmic half-channel

Matrix half-channel

Subunit **a**

Figure 18.31 Components of the proton-conducting unit of ATP synthase. The **c** subunit consists of two α helices that span the membrane. An aspartic acid residue in one of the helices lies on the center of the membrane. The structure of the **a** subunit has not yet been directly observed, but it appears to include two half-channels that allow protons to enter and pass partway but not completely through the membrane.

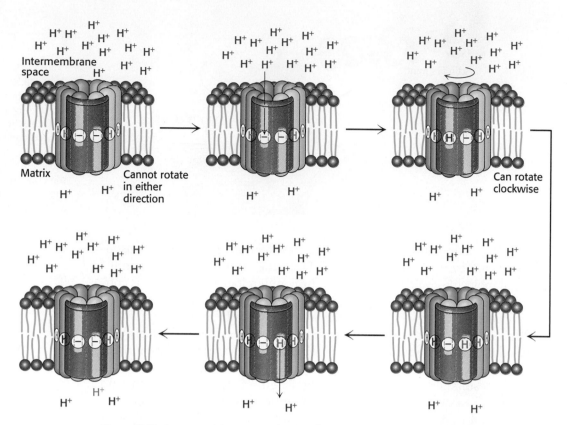

Figure 18.32 Proton motion across the membrane drives rotation of the **c** ring. A proton enters from the intermembrane space into the cytoplasmic half-channel to neutralize the charge on an aspartate residue in a **c** subunit. With this charge neutralized, the **c** ring can rotate clockwise by one **c** subunit, moving an aspartic acid residue out of the membrane into the matrix half-channel. This proton can move into the matrix, resetting the system to its initial state.

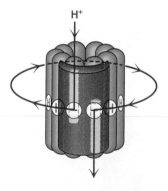

Figure 18.33 Proton path through the membrane. Each proton enters the cytoplasmic half-channel, follows a complete rotation of the **c** ring, and exits through the other half-channel into the matrix.

A little goes a long way

Despite the various molecular machinations and the vast numbers of ATPs synthesized and protons pumped, a resting human being requires surprisingly little power. Approximately 116 watts, the energy output of a typical light bulb, provides enough energy to sustain a resting person.

---

aspartic acid residue to occupy the hydrophobic environment of the membrane. Thus, the **c** subunit with the newly protonated aspartic acid moves from contact with the cytoplasmic half-channel into the membrane, and the other **c** subunits move in unison. Each proton that enters the cytoplasmic half-channel moves through the membrane by riding around on the rotating **c** ring to exit through the matrix half-channel into the proton-poor environment of the matrix (Figure 18.33).

How does the rotation of the **c** ring lead to the synthesis of ATP? The **c** ring is tightly linked to the γ and ε subunits. Thus, as the **c** ring turns, these subunits are turned inside the $\alpha_3\beta_3$ hexamer unit of F_1. The rotation of the γ subunit in turn promotes the synthesis of ATP through the binding-change mechanism. The exterior column formed by the two **b** chains and the δ subunit prevents the $\alpha_3\beta_3$ hexamer from rotating. Recall that the number of **c** subunits in the **c** ring appears to range between 10 and 14. This number is significant because it determines the number of protons that must be transported to generate a molecule of ATP. Each 360-degree rotation of the γ subunit leads to the synthesis and release of three molecules of ATP. Thus, if there are 10 **c** subunits in the ring (as was observed in a crystal structure of yeast mitochondrial ATP synthase), each ATP generated requires the transport of $10/3 = 3.33$ protons. For simplicity, we will assume that three protons must flow into the matrix for each ATP formed, but we must keep in mind that the true value may differ. As we will see, the electrons from NADH pump enough protons to generate 2.5 molecules of ATP, whereas those from $FADH_2$ yield 1.5 molecules of ATP.

Let us return for a moment to the example with which we began this chapter. If a resting human being requires 85 kg of ATP per day for bodily

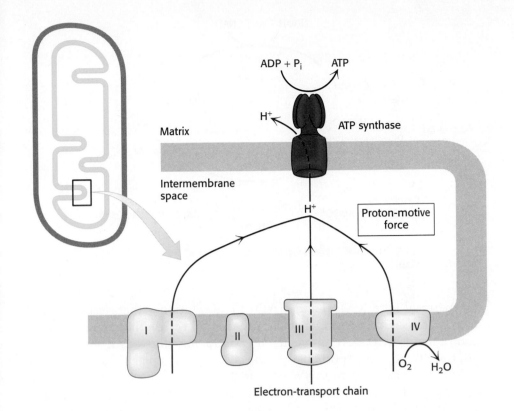

Figure 18.34 Overview of oxidative phosphorylation. The electron-transport chain generates a proton gradient, which is used to synthesize ATP.

functions, then 3.3×10^{25} protons must flow through the ATP synthase per day, or 3.3×10^{21} protons per second. Figure 18.34 summarizes the process of oxidative phosphorylation.

ATP Synthase and G Proteins Have Several Common Features

The α and β subunits of ATP synthase are members of the P-loop NTPase family of proteins. In Chapter 14, we learned that the signaling properties of other members of this family, the G proteins, depend on their ability to bind nucleoside triphosphates and diphosphates with great tenacity. They do not exchange nucleotides unless they are stimulated to do so by interaction with other proteins. The binding-change mechanism of ATP synthase is a variation on this theme. The P-loop regions of the β subunits will bind either ADP or ATP (or release ATP), depending on which of three different faces of the γ subunit they interact with. The conformational changes take place in an orderly way, driven by the rotation of the γ subunit.

18.5 Many Shuttles Allow Movement Across The Mitochondrial Membranes

The inner mitochondrial membrane must be impermeable to most molecules, yet much exchange has to take place between the cytoplasm and the mitochondria. This exchange is mediated by an array of membrane-spanning transporter proteins (Section 13.4).

Electrons from Cytoplasmic NADH Enter Mitochondria by Shuttles

One function of the respiratory chain is to regenerate NAD^+ for use in glycolysis. How is cytoplasmic NADH reoxidized to NAD^+ under aerobic conditions? NADH cannot simply pass into mitochondria for oxidation by the respiratory chain, because the inner mitochondrial membrane

NADH + H^+ +   E–FAD

Cytoplasmic   Mitochondrial

NAD$^+$ +   E–FADH$_2$

Cytoplasmic   Mitochondrial

Glycerol 3-phosphate shuttle

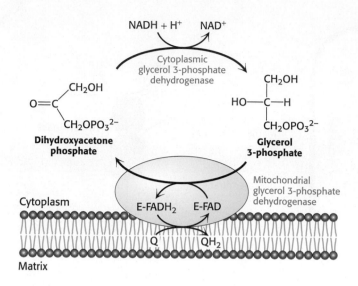

Figure 18.35 Glycerol 3-phosphate shuttle. Electrons from NADH can enter the mitochondrial electron-transport chain by being used to reduce dihydroxyacetone phosphate to glycerol 3-phosphate. Glycerol 3-phosphate is reoxidized by electron transfer to an FAD prosthetic group in a membrane-bound glycerol 3-phosphate dehydrogenase. Subsequent electron transfer to Q to form QH_2 allows these electrons to enter the electron-transport chain.

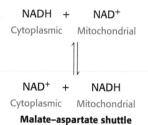

is impermeable to NADH and NAD^+. The solution is that *electrons from NADH,* rather than NADH itself, are carried across the mitochondrial membrane. One of several means of introducing electrons from NADH into the electron-transport chain is the *glycerol 3-phosphate shuttle* (Figure 18.35). The first step in this shuttle is the transfer of a pair of electrons from NADH to dihydroxyacetone phosphate, a glycolytic intermediate, to form glycerol 3-phosphate. This reaction is catalyzed by a glycerol 3-phosphate dehydrogenase in the cytoplasm. Glycerol 3-phosphate is reoxidized to dihydroxyacetone phosphate on the outer surface of the inner mitochondrial membrane by a membrane-bound isozyme of glycerol 3-phosphate dehydrogenase. An electron pair from glycerol 3-phosphate is transferred to an FAD prosthetic group in this enzyme to form $FADH_2$. This reaction also regenerates dihydroxyacetone phosphate.

The reduced flavin transfers its electrons to the electron carrier Q, which then enters the respiratory chain as QH_2. *When cytoplasmic NADH transported by the glycerol 3-phosphate shuttle is oxidized by the respiratory chain, 1.5 rather than 2.5 molecules of ATP are formed.* The yield is lower because FAD rather than NAD^+ is the electron acceptor in mitochondrial glycerol 3-phosphate dehydrogenase. The use of FAD enables electrons from cytoplasmic NADH to be transported into mitochondria against an NADH concentration gradient. The price of this transport is one molecule of ATP per two electrons. This glycerol 3-phosphate shuttle is especially prominent in muscle and enables it to sustain a very high rate of oxidative phosphorylation. Indeed, some insects lack lactate dehydrogenase and are completely dependent on the glycerol 3-phosphate shuttle for the regeneration of cytoplasmic NAD^+.

In the heart and liver, electrons from cytoplasmic NADH are brought into mitochondria by the *malate–aspartate shuttle,* which is mediated by two membrane carriers and four enzymes (Figure 18.36). Electrons are transferred from NADH in the cytoplasm to oxaloacetate, forming malate, which traverses the inner mitochondrial membrane in exchange for α-ketoglutarate and is then reoxidized by NAD^+ in the matrix to form NADH in a reaction catalyzed by the citric acid cycle enzyme malate dehydrogenase. The resulting oxaloacetate does not readily cross the inner mitochondrial membrane and so a transamination reaction (p. 656) is needed to form aspartate, which can be transported to the cytoplasmic side in exchange for glutamate. Glutamate donates an amino group to oxaloacetate, forming aspartate and α-ketoglutarate. In the cytoplasm, aspartate is then deaminated to form oxaloacetate and the cycle is restarted.

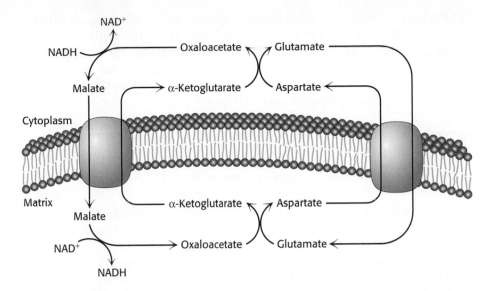

Figure 18.36 Malate–aspartate shuttle.

The Entry of ADP into Mitochondria Is Coupled to the Exit of ATP by ATP-ADP Translocase

The major function of oxidative phosphorylation is to generate ATP from ADP. ATP and ADP do not diffuse freely across the inner mitochondrial membrane. How are these highly charged molecules moved across the inner membrane into the cytoplasm? A specific transport protein, *ATP-ADP translocase,* enables these molecules to transverse this permeability barrier. Most important, the flows of ATP and ADP are coupled. *ADP enters the mitochondrial matrix only if ATP exits, and vice versa.* This process is carried out by the translocase, an antiporter:

$$ADP^{3-}_{cytoplasm} + ATP^{4-}_{matrix} \longrightarrow ADP^{3-}_{matrix} + ATP^{4-}_{cytoplasm}$$

ATP-ADP translocase is highly abundant, constituting about 15% of the protein in the inner mitochondrial membrane. The abundance is a manifestation of the fact that human beings exchange the equivalent of their weight in ATP each day. The 30-kd translocase contains a single nucleotide-binding site that alternately faces the matrix and the cytoplasmic sides of the membrane (Figure 18.37). ATP and ADP bind to the translocase without Mg^{2+}, and ATP has one more negative charge than that of ADP. Thus, in an actively respiring mitochondrion with a positive membrane potential,

Figure 18.37 Mechanism of mitochondrial ATP-ADP translocase. The translocase catalyzes the coupled entry of ADP into the matrix and the exit of ATP from it. The binding of ADP (1) from the cytoplasm favors eversion of the transporter (2) to release ADP into the matrix (3). Subsequent binding of ATP from the matrix to the everted form (4) favors eversion back to the original conformation (5), releasing ATP into the cytoplasm (6).

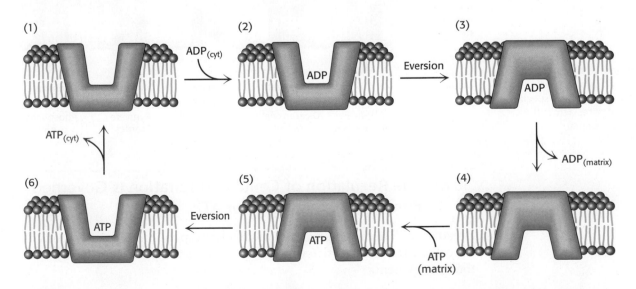

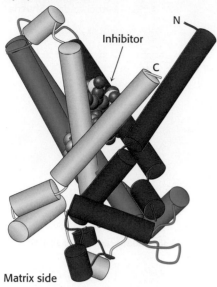

Cytoplasmic side

Inhibitor

N

C

Matrix side

Figure 18.38 Structure of mitochondrial transporters. The structure of the ATP-ADP translocase is shown. *Notice* that this structure comprises three similar units (shown in red, blue, and yellow) that come together to form a binding site, here occupied by an inhibitor of this transporter. Other members of the mitochondrial transporter family adopt similar tripartite structures. [Drawn from 1OKC.pdb.]

ATP transport out of the mitochondrial matrix and ADP transport into the matrix are favored. This ATP–ADP exchange is energetically expensive; about a quarter of the energy yield from electron transfer by the respiratory chain is consumed to regenerate the membrane potential that is tapped by this exchange process. The inhibition of this process leads to the subsequent inhibition of cellular respiration as well (p. 534).

Mitochondrial Transporters for Metabolites Have a Common Tripartite Structure

Examination of the amino acid sequence of the ATP-ADP translocase revealed that this protein consists of three tandem repeats of a 100-amino-acid module, each of which appears to have two transmembrane segments. This tripartite structure has recently been confirmed by the determination of the three-dimensional structure of this transporter (Figure 18.38). The transmembrane helices form a teepeelike structure with the nucleotide-binding site (marked by a bound inhibitor) lying in the center. Each of the three repeats adopts a similar structure.

ATP-ADP translocase is but one of many mitochondrial transporters for ions and charged metabolites (Figure 18.39). The *phosphate carrier,* which works in concert with ATP-ADP translocase mediates the electroneutral exchange of $H_2PO_4^-$ for OH^-. The combined action of these two transporters leads to the exchange of cytoplasmic ADP and P_i for matrix ATP at the cost of the influx of one H^+ (owing to the transport of one OH^- out of the matrix). These two transporters, which provide ATP synthase with its substrates, are associated with the synthase to form a large complex called the *ATP synthasome.*

Other homologous carriers also are present in the inner mitochondrial membrane. The dicarboxylate carrier enables malate, succinate, and fumarate to be exported from the mitochondrial matrix in exchange for P_i. The tricarboxylate carrier exchanges citrate and H^+ for malate. Pyruvate in the cytoplasm enters the mitochondrial membrane in exchange for OH^- by means of the pyruvate carrier. In all, more than 40 such carriers are encoded in the human genome.

Figure 18.39 Mitochondrial transporters. Transporters (also called carriers) are transmembrane proteins that carry specific ions and charged metabolites across the inner mitochondrial membrane.

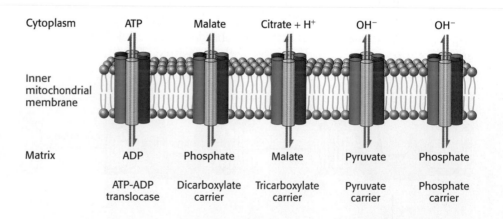

Cytoplasm

| ATP | Malate | Citrate + H^+ | OH^- | OH^- |

Inner mitochondrial membrane

Matrix

| ADP | Phosphate | Malate | Pyruvate | Phosphate |
| ATP-ADP translocase | Dicarboxylate carrier | Tricarboxylate carrier | Pyruvate carrier | Phosphate carrier |

18.6 The Regulation of Cellular Respiration Is Governed Primarily by the Need for ATP

Because ATP is the end product of cellular respiration, the ATP needs of the cell are the ultimate determinant of the rate of respiratory pathways and their components.

The Complete Oxidation of Glucose Yields
About 30 Molecules of ATP

We can now estimate how many molecules of ATP are formed when glucose is completely oxidized to CO_2. The number of ATP (or GTP) molecules formed in glycolysis and the citric acid cycle is unequivocally known because it is determined by the stoichiometries of chemical reactions. In contrast, the ATP yield of oxidative phosphorylation is less certain because the stoichiometries of proton pumping, ATP synthesis, and metabolite-transport processes need not be integer numbers or even have fixed values. As stated earlier, the best current estimates for the number of protons pumped out of the matrix by NADH-Q oxidoreductase, Q-cytochrome c oxidoreductase, and cytochrome c oxidase per electron pair are four, two, and four, respectively. The synthesis of a molecule of ATP is driven by the flow of about three protons through ATP synthase. An additional proton is consumed in transporting ATP from the matrix to the cytoplasm. Hence, about 2.5 molecules of cytoplasmic ATP are generated as a result of the flow of a pair of electrons from NADH to O_2. For electrons that enter at the level of Q-cytochrome c oxidoreductase, such as those from the oxidation of succinate or cytoplasmic NADH, the yield is about 1.5 molecules of ATP per electron pair. Hence, as tallied in Table 18.4, *about 30 molecules of ATP are formed*

TABLE 18.4 ATP yield from the complete oxidation of glucose

Reaction sequence	ATP yield per glucose molecule
Glycolysis: Conversion of glucose into pyruvate	
(in the cytoplasm)	
Phosphorylation of glucose	−1
Phosphorylation of fructose 6-phosphate	−1
Dephosphorylation of 2 molecules of 1,3-BPG	+2
Dephosphorylation of 2 molecules of phosphoenolpyruvate	+2
2 molecules of NADH are formed in the oxidation of 2 molecules of glyceraldehyde 3-phosphate	
Conversion of pyruvate into acetyl CoA	
(inside mitochondria)	
2 molecules of NADH are formed	
Citric acid cycle (inside mitochondria)	
2 molecules of guanosine triphosphate are formed from 2 molecules of succinyl CoA	+2
6 molecules of NADH are formed in the oxidation of 2 molecules each of isocitrate, α-ketoglutarate, and malate	
2 molecules of $FADH_2$ are formed in the oxidation of 2 molecules of succinate	
Oxidative phosphorylation (inside mitochondria)	
2 molecules of NADH formed in glycolysis; each yields 1.5 molecules of ATP (assuming transport of NADH by the glycerol 3-phosphate shuttle)	+3
2 molecules of NADH formed in the oxidative decarboxylation of pyruvate; each yields 2.5 molecules of ATP	+5
2 molecules of $FADH_2$ formed in the citric acid cycle; each yields 1.5 molecules of ATP	+3
6 molecules of NADH formed in the citric acid cycle; each yields 2.5 molecules of ATP	+15
NET YIELD PER MOLECULE OF GLUCOSE	+30

Source: The ATP yield of oxidative phosphorylation is based on values given in P. C. Hinkle, M. A. Kumar, A. Resetar, and D. L. Harris. *Biochemistry* 30(1991):3576.
Note: The current value of 30 molecules of ATP per molecule of glucose supersedes the earlier one of 36 molecules of ATP. The stoichiometries of proton pumping, ATP synthesis, and metabolite transport should be regarded as estimates. About two more molecules of ATP are formed per molecule of glucose oxidized when the malate–aspartate shuttle rather than the glycerol 3-phosphate shuttle is used.

when glucose is completely oxidized to CO_2; this value supersedes the traditional estimate of 36 molecules of ATP. Most of the ATP, 26 of 30 molecules formed, is generated by oxidative phosphorylation. Recall that the anaerobic metabolism of glucose yields only 2 molecules of ATP. The efficiency of cellular respiration is manifested in the fact that one of the effects of endurance exercise, a practice that calls for much ATP over an extended period of time, is to increase the number of mitochondria and blood vessels in muscle and so increase the extent of ATP generation by oxidative phosphorylation.

The Rate of Oxidative Phosphorylation Is Determined by the Need for ATP

How is the rate of the electron-transport chain controlled? Under most physiological conditions, electron transport is tightly coupled to phosphorylation. *Electrons do not usually flow through the electron-transport chain to O_2 unless ADP is simultaneously phosphorylated to ATP.* When ADP concentration rises, as would be the case in active muscle, the rate of oxidative phosphorylation increases to meet the ATP needs of the muscle. The regulation of the rate of oxidative phosphorylation by the ADP level is called *respiratory control* or *acceptor control*. Experiments on isolated mitochondria demonstrate the importance of ADP level (Figure 18.40). The rate of oxygen consumption by mitochondria increases markedly when ADP is added and then returns to its initial value when the added ADP has been converted into ATP.

The level of ADP likewise affects the rate of the citric acid cycle. At low concentrations of ADP, as in a resting muscle, NADH and $FADH_2$ are not consumed by the electron-transport chain. The citric acid cycle slows because there is less NAD^+ and FAD to feed the cycle. As the ADP level rises and oxidative phosphorylation speeds up, NADH and $FADH_2$ are oxidized, and the citric acid cycle becomes more active. *Electrons do not flow from fuel molecules to O_2 unless ATP needs to be synthesized.* We see here another example of the regulatory significance of the energy charge (Figure 18.41).

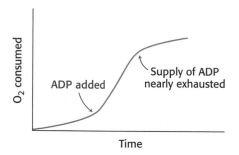

Figure 18.40 Respiratory control. Electrons are transferred to O_2 only if ADP is concomitantly phosphorylated to ATP.

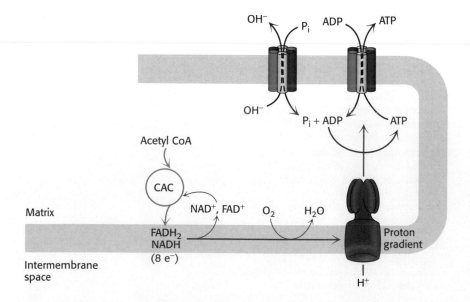

Figure 18.41 Energy charge regulates the use of fuels. The synthesis of ATP from ADP and P_i controls the flow of electrons from NADH and $FADH_2$ to oxygen. The availability of NAD^+ and FAD in turn control the rate of the citric acid cycle (CAC).

Regulated Uncoupling Leads to the Generation of Heat

Some organisms possess the ability to uncouple oxidative phosphorylation from ATP synthesis to generate heat. Such uncoupling is a means to maintain body temperature in hibernating animals, in some newborn animals (including

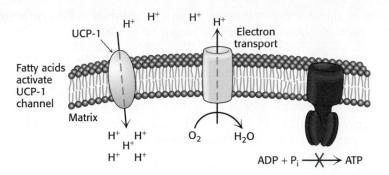

Figure 18.42 Action of an uncoupling
protein. Uncoupling protein (UCP-1)
generates heat by permitting the influx of
protons into the mitochondria without
the synthesis of ATP.

human beings), and in mammals adapted to cold. The skunk cabbage uses an analogous mechanism to heat its floral spikes in early spring, increasing the evaporation of odoriferous molecules that attract insects to fertilize its flowers. In animals, *brown fat* (brown adipose tissue) is specialized tissue for this process of *nonshivering thermogenesis*. Brown adipose tissue is very rich in mitochondria, often called *brown fat mitochondria*. The tissue appears brown from the combination of the greenish-colored cytochromes in the numerous mitochondria and the red hemoglobin present in the extensive blood supply, which helps to carry the heat through the body. The inner mitochondrial membrane of these mitochondria contains a large amount of *uncoupling protein* (UCP-1), or *thermogenin*, a dimer of 33-kd subunits that resembles ATP-ADP translocase. UCP-1 forms a pathway for the flow of protons from the cytoplasm to the matrix. In essence, *UCP-1 generates heat by short-circuiting the mitochondrial proton battery.* The energy of the proton gradient, normally captured as ATP, is released as heat as the protons flow through UCP-1 to the mitochondrial matrix. This dissipative proton pathway is activated when the core body temperature begins to fall. In response to a temperature drop, the release of hormones leads to the liberation of free fatty acids from triacylglycerols that in turn activate thermogenin (Figure 18.42).

In addition to UCP-1, two other uncoupling proteins have been identified. UCP-2, which is 56% identical in sequence with UCP-1, is found in a wide variety of tissues. UCP-3 (57% identical with UCP-1 and 73% identical with UCP-2) is localized to skeletal muscle and brown fat. This family of uncoupling proteins, especially UCP-2 and UCP-3, may play a role in energy homeostasis. In fact, the genes for UCP-2 and UCP-3 map to regions of the human and mouse chromosomes that have been linked to obesity, supporting the notion that they function as a means of regulating body weight.

Oxidative Phosphorylation Can Be Inhibited at Many Stages

Many potent and lethal poisons exert their effect by inhibiting oxidative phosphorylation at one of a number of different locations.

1. *Inhibition of the electron-transport chain. Rotenone,* which is used as a fish and insect poison, and *amytal,* a barbituate sedative, block electron transfer in NADH-Q oxidoreductase and thereby prevent the utilization of NADH as a substrate (Figure 18.43). In contrast, electron flow resulting from the oxidation of succinate is unimpaired, because these electrons enter through QH_2, beyond the block. *Antimycin A* interferes with electron flow from cytochrome b_H in Q-cytochrome c oxidoreductase. Furthermore, electron flow in cytochrome c oxidase can be blocked by *cyanide* (CN^-), *azide* (N_3^-), and *carbon monoxide* (CO). Cyanide and azide react with the ferric form of heme a_3, whereas carbon monoxide inhibits the ferrous form.

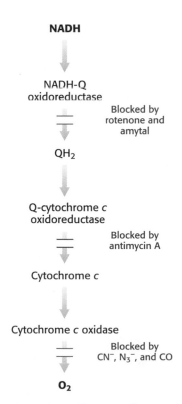

Figure 18.43 Sites of action of some
inhibitors of electron transport.

Inhibition of the electron-transport chain also inhibits ATP synthesis because the proton-motive force can no longer be generated.

2. *Inhibition of ATP synthase.* Oligomycin, an antibiotic used as an antifungal agent, and dicyclohexylcarbodiimide (DCCD; p. 91) prevent the influx of protons through ATP synthase. If actively respiring mitochondria are exposed to an inhibitor of ATP synthase, the electron-transport chain ceases to operate. This observation clearly illustrates that electron transport and ATP synthesis are normally tightly coupled.

3. *Uncoupling electron transport from ATP synthesis.* The tight coupling of electron transport and phosphorylation in mitochondria can be uncoupled by 2,4-dinitrophenol (DNP) and certain other acidic aromatic compounds. These substances carry protons across the inner mitochondrial membrane, down their concentration gradient. In the presence of these uncouplers, electron transport from NADH to O_2 proceeds in a normal fashion, but ATP is not formed by mitochondrial ATP synthase, because the proton-motive force across the inner mitochondrial membrane is continuously dissipated. This loss of respiratory control leads to increased oxygen consumption and oxidation of NADH. Indeed, in the accidental ingestion of uncouplers, large amounts of metabolic fuels are consumed, but no energy is captured as ATP. Rather, energy is released as heat. DNP is the active ingredient in some herbicides and fungicides. Remarkably, some people consume DNP as a weight-loss drug, despite the fact that the FDA banned its use in 1938. There are also reports that Soviet soldiers were given DNP to keep them warm during the long Russian winters. Chemical uncouplers are nonphysiological, unregulated counterparts of uncoupling proteins.

4. *Inhibition of ATP export.* ATP-ADP translocase is specifically inhibited by very low concentrations of *atractyloside* (a plant glycoside) or *bongkrekic acid* (an antibiotic from a mold). Atractyloside binds to the translocase when its nucleotide site faces the cytoplasm, whereas bongkrekic acid binds when this site faces the mitochondrial matrix. Oxidative phosphorylation stops soon after either inhibitor is added, showing that ATP-ADP translocase is essential for maintaining adequate amounts of ADP to accept the energy associated with the proton-motive force.

2,4-Dinitrophenol (DNP)

Mitochondrial Diseases Are Being Discovered

The number of diseases that can be attributed to mitochondrial mutations is steadily growing in step with our growing understanding of the biochemistry and genetics of mitochondria. The prevalence of mitochondrial diseases is estimated to be from 10 to 15 per 100,000 people, roughly equivalent to the prevalence of the muscular dystrophies. The first mitochondrial disease to be understood was *Leber hereditary optic neuropathy* (LHON), a form of blindness that strikes in midlife as a result of mutations in Complex I. Some of these mutations impair NADH utilization, whereas others block electron transfer to Q. Mutations in Complex I are the most frequent cause of mitochondrial diseases. The accumulation of mutations in mitochondrial genes in a span of several decades may contribute to aging, degenerative disorders, and cancer.

A human egg harbors several hundred thousand molecules of mitochondrial DNA, whereas a sperm contributes only a few hundred and thus has little effect on the mitochondrial genotype. Because the maternally inherited mitochondria are present in large numbers and not all of the mitochondria may be affected, the pathologies of mitochondrial mutants can be quite complex. Even within a single family carrying an identical mutation, chance fluctuations in the percentage of mitochondria with the mutation lead to large variations in the nature and severity of the symptoms of the pathological condition as well as the time of onset. As the percentage of defective mitochondria increases, energy-generating capacity diminishes until, at some threshold, the cell can no longer function properly. Defects in cellular respiration are doubly dangerous. Not only does energy transduction decrease, but the likelihood that reactive oxygen species will be generated increases. Organs that are highly dependent on oxidative phosphorylation, such as the nervous system and the heart, are most vulnerable to mutations in mitochondrial DNA.

Mitochondria Play a Key Role in Apoptosis

In the course of development or in cases of significant cell damage, individual cells within multicellular organisms undergo *programmed cell death*, or *apoptosis*. Mitochondria act as control centers regulating this process. Although the details have not yet been established, a pore called the mitochondrial permeability transition pore (mtPTP) forms in damaged mitochondria. This pore appears to consist of VDAC, the adenine nucleotide translocase, and several other mitochondrial proteins, including members of a family of proteins (Bcl family) that were initially discovered because of their role in cancer. One of the most potent activators of apoptosis, cytochrome *c*, exits the mitochondria through the mtPTP. Its presence in the cytoplasm activates a cascade of proteolytic enzymes called *caspases*, members of the cysteine protease family (p. 251). Cytochrome *c*, in conjunction with other proteins, initiates the cascade by activating procaspase 9 to form caspase 9, which then activates other caspases. Each caspase type destroys a particular target, such as the proteins that maintain cell structure. Another target is a protein that inhibits an enzyme that destroys DNA (caspase-activated DNAse, CAD), freeing CAD to cleave the genetic material. This cascade of proteolytic enzymes has been called "death by a thousand tiny cuts."

Power Transmission by Proton Gradients Is a Central Motif of Bioenergetics

The main concept presented in this chapter is that mitochondrial electron transfer and ATP synthesis are linked by a transmembrane proton gradient. ATP synthesis in bacteria and chloroplasts also is driven by proton gradients. In fact, proton gradients power a variety of energy-requiring processes such as the active transport of calcium ions by mitochondria, the entry of some amino acids and sugars into bacteria, the rotation of bacterial flagella, and the transfer of electrons from $NADP^+$ to NADPH. Proton gradients can also be used to generate heat, as in hibernation. It is evident that *proton gradients are a central interconvertible currency of free energy in biological systems* (Figure 18.44). Mitchell noted that the proton-motive force is a marvelously simple and effective store of free energy because it requires only a thin, closed lipid membrane between two aqueous phases.

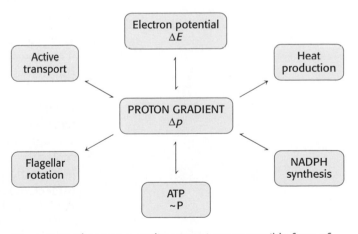

Figure 18.44 The proton gradient is an interconvertible form of free energy.

***

Summary

18.1 Oxidative Phosphorylation in Eukaryotes Takes Place in Mitochondria

Mitochondria generate most of the ATP required by aerobic cells through a joint endeavor of the reactions of the citric acid cycle, which take place in the mitochondrial matrix, and oxidative phosphorylation, which takes place in the inner mitochondrial membrane. Mitochondria are descendents of a free-living bacterium that established a symbiotic relation with another cell.

18.2   Oxidative Phosphorylation Depends on Electron Transfer

In oxidative phosphorylation, the synthesis of ATP is coupled to the flow of electrons from NADH or $FADH_2$ to O_2 by a proton gradient across the inner mitochondrial membrane. Electron flow through three asymmetrically oriented transmembrane complexes results in the pumping of protons out of the mitochondrial matrix and the generation of a membrane potential. ATP is synthesized when protons flow back to the matrix through a channel in an ATP-synthesizing complex, called ATP synthase (also known as F_0F_1-ATPase). Oxidative phosphorylation exemplifies a fundamental theme of bioenergetics: the transmission of free energy by proton gradients.

18.3   The Respiratory Chain Consists of Four Complexes: Three Proton Pumps and a Physical Link to the Citric Acid Cycle

The electron carriers in the respiratory assembly of the inner mitochondrial membrane are quinones, flavins, iron–sulfur complexes, heme groups of cytochromes, and copper ions. Electrons from NADH are transferred to the FMN prosthetic group of NADH-Q oxidoreductase (Complex I), the first of four complexes. This oxidoreductase also contains Fe-S centers. The electrons emerge in QH_2, the reduced form of ubiquinone (Q). The citric acid cycle enzyme succinate dehydrogenase is a component of the succinate-Q reductase complex (Complex II), which donates electrons from $FADH_2$ to Q to form QH_2. This highly mobile hydrophobic carrier transfers its electrons to Q-cytochrome c oxidoreductase (Complex III), a complex that contains cytochromes b and c_1 and an Fe-S center. This complex reduces cytochrome c, a water-soluble peripheral membrane protein. Cytochrome c, like Q, is a mobile carrier of electrons, which it then transfers to cytochrome c oxidase (Complex IV). This complex contains cytochromes a and a_3 and three copper ions. A heme iron ion and a copper ion in this oxidase transfer electrons to O_2, the ultimate acceptor, to form H_2O.

18.4   A Proton Gradient Powers the Synthesis of ATP

The flow of electrons through Complexes I, III, and IV leads to the transfer of protons from the matrix side to the cytoplasmic side of the inner mitochondrial membrane. A proton-motive force consisting of a pH gradient (matrix side basic) and a membrane potential (matrix side negative) is generated. The flow of protons back to the matrix side through ATP synthase drives ATP synthesis. The enzyme complex is a molecular motor made of two operational units: a rotating component and a stationary component. The rotation of the γ subunit induces structural changes in the β subunit that result in the synthesis and release of ATP from the enzyme. Proton influx provides the force for the rotation.

The flow of two electrons through NADH-Q oxidoreductase, Q-cytochrome c oxidoreductase, and cytochrome c oxidase generates a gradient sufficient to synthesize 1, 0.5, and 1 molecule of ATP, respectively. Hence, 2.5 molecules of ATP are formed per molecule of NADH oxidized in the mitochondrial matrix, whereas only 1.5 molecules of ATP are made per molecule of $FADH_2$ oxidized, because its electrons enter the chain at QH_2, after the first proton-pumping site.

18.5   Many Shuttles Allow Movement Across the Mitochondrial Membranes

Mitochondria employ a host of transporters, or carriers, to move molecules across the inner mitochondrial membrane. The electrons of cytoplasmic NADH are transferred into the mitochondria by the glycerol phosphate shuttle to form $FADH_2$ from FAD or by the malate–aspartate shuttle to form mitochondrial NADH. The entry of ADP into the mitochondrial matrix is coupled to the exit of ATP by ATP-ADP translocase, a transporter driven by membrane potential.

18.6 The Regulation of Oxidative Phosphorylation Is Governed Primarily by the Need for ATP

About 30 molecules of ATP are generated when a molecule of glucose is completely oxidized to CO_2 and H_2O. Electron transport is normally tightly coupled to phosphorylation. NADH and $FADH_2$ are oxidized only if ADP is simultaneously phosphorylated to ATP, a form of regulation called acceptor or respiratory control. Proteins have been identified that uncouple electron transport and ATP synthesis for the generation of heat. Uncouplers such as DNP also can disrupt this coupling; they dissipate the proton gradient by carrying protons across the inner mitochondrial membrane.

Key Terms

oxidative phosphorylation (p. 502)

proton-motive force (p. 503)

cellular respiration (p. 503)

electron-transport chain (p. 506)

reduction (redox, oxidation–reduction, E_0') potential (p. 506)

coenzyme Q (Q, ubiquinone) (p. 510)

Q pool (p. 510)

NADH-Q oxidoreductase (Complex I) (p. 510)

flavin mononucleotide (FMN) (p. 510)

iron–sulfur (nonheme iron) protein (p. 511)

succinate-Q reductase (Complex II) (p. 512)

Q-cytochrome c oxidoreductase (Complex III) (p. 512)

cytochrome c (Cyt c) (p. 512)

Rieske center (p. 513)

Q cycle (p. 513)

cytochrome c oxidase (Complex IV) (p. 514)

superoxide dismutase (p. 518)

catalase (p. 518)

inverted region (p. 519)

ATP synthase (Complex V, F_1F_0 ATPase) (p. 521)

glycerol 3-phosphate shuttle (p. 528)

malate–aspartate shuttle (p. 528)

ATP-ADP translocase (adenine nucleotide translocase, ANT) (p. 529)

respiratory (acceptor) control (p. 532)

uncoupling protein (UCP) (p. 533)

programmed cell death (apoptosis) (p. 535)

caspase (p. 535)

Selected Readings

Where to Start

Gray, M. W., Burger, G., and Lang, B. F. 1999. Mitochondrial evolution. *Science* 283:1476–1481.

Wallace, D. C. 1997. Mitochondrial DNA in aging and disease. *Sci. Am.* 277(2):40–47.

Saraste, M. 1999. Oxidative phosphorylation at the *fin de siècle*. *Science* 283:1488–1493.

Shultz, B. E., and Chan, S. I. 2001. Structures and proton-pumping strategies of mitochondrial respiratory enzymes. *Annu. Rev. Biophys. Biomol. Struct.* 30:23–65.

Moser, C. C., Keske, J. M., Warncke, K., Farid, R. S., and Dutton, P. L. 1992. Nature of biological electron transfer. *Nature* 355:796–802.

Books

Scheffler, I. E. 1999. *Mitochondria.* Wiley.

Nicholls, D. G., and Ferguson, S. J. 2002. *Bioenergetics 3.* Academic Press.

Electron-Transport Chain

Sun, F., Huo, X., Zhai, Y., Wang, A., Xu, J., Su, D., Bartlam, M., and Ral, Z. 2005. Crystal structure of mitochondrial respiratory membrane protein complex II. *Cell* 121:1043–1057.

Crofts, A. R. 2004. The cytochrome bc_1 complex: Function in the context of structure. *Annu. Rev. Physiol.* 66:689–733.

Bianchi, C., Genova, M. L., Castelli, G. P., and Lenaz, G. 2004. The mitochondrial respiratory chain is partially organized in a supramolecular complex. *J. Biol. Chem.* 279:36562–36569.

Ugalde, C., Vogel, R., Huijbens, R., van den Heuvel, B., Smeitink, J., and Nijtmans, L. 2004. Human mitochondrial complex I assembles through a combination of evolutionary conserved modules: A framework to interpret complex I deficiencies. *Human Molecular Genetics* 13:2461–2472.

Yagi, T., and Matsuno-Yagi, A. 2003. The proton-translocating NADH-quinone oxidoreductase in the respiratory chain: The secret unlocked. *Biochemistry* 42:2266–2274.

Cecchini, G. 2003. Function and structure of complex II of the respiratory chain. *Annu. Rev. Biochem.* 72:77–109.

Lange, C., and Hunte, C. 2002. Crystal structue of the yeast cytochrome bc_1 complex with its bond substrate cytochrome c. *Proc. Natl. Acad. Sci. U. S. A.* 99:2800–2805.

Zaslavsky, D., and Gennis, R. B. 2000. Proton pumping by cytochrome oxidase: Progress, problems and postulates. *Biochim. Biophys. Acta* 1458:164–179.

Grigorieff, N. 1999. Structure of the respiratory NADH:ubiquinone oxidoreductase (complex I). *Curr. Opin. Struct. Biol.* 9:476–483.

Ackrell, B. A. 2000. Progress in understanding structure–function relationships in respiratory chain complex II. *FEBS Lett.* 466:1–5.

Grigorieff, N. 1998. Three-dimensional structure of bovine NADH:ubiquinone oxidoreductase (complex I) at 22 Å in ice. *J. Mol. Biol.* 277:1033–1046.

Dutton, P. L., Moser, C. C., Sled, V. D., Daldal, F., and Ohnishi, T. 1998. A reductant-induced oxidation mechanism for complex I. *Biochim. Biophys. Acta* 1364:245–257.

Xia, D., Yu, C. A., Kim, H., Xia, J. Z., Kachurin, A. M., Zhang, L., Yu, L., and Deisenhofer, J. 1997. Crystal structure of the cytochrome bc_1 complex from bovine heart mitochondria. *Science* 277:60–66.

Michel, H., Behr, J., Harrenga, A., and Kannt, A. 1998. Cytochrome *c* oxidase: Structure and spectroscopy. *Annu. Rev. Biophys. Biomol. Struct.* 27:329–356.

Verkhovsky, M. I., Jasaitis, A., Verkhovskaya, M. L., Morgan, J. E., and Wikstrom, M. 1999. Proton translocation by cytochrome *c* oxidase. *Nature* 400:480–483.

ATP Synthase

Chen, C., Ko, Y., Delannoy, M., Ludtke, S. J., Chiu, W., and Pedersen, P. L. 2004. Mitochondrial ATP synthasome: Three-dimensional structure by electron microscopy of the ATP synthase in complex formation with the carriers for P_i and ADP/ATP. *J. Biol. Chem.* 279:31761–31768.

Noji, H., and Yoshida, M. 2001. The rotary machine in the cell: ATP synthase. *J. Biol. Chem.* 276:1665–1668.

Yasuda, R., Noji, H., Kinosita, K., Jr., and Yoshida, M. 1998. F1-ATPase is a highly efficient molecular motor that rotates with discrete 120 degree steps. *Cell* 93:1117–1124.

Kinosita, K., Jr., Yasuda, R., Noji, H., Ishiwata, S., and Yoshida, M. 1998. F1-ATPase: A rotary motor made of a single molecule. *Cell* 93:21–24.

Noji, H., Yasuda, R., Yoshida, M., and Kinosita, K., Jr., 1997. Direct observation of the rotation of F1-ATPase. *Nature* 386:299–302.

Tsunoda, S. P., Aggeler, R., Yoshida, M., and Capaldi, R. A. 2001. Rotation of the *c* subunit oligomer in fully functional $F_1 F_0$ ATP synthase. *Proc. Natl. Acad. Sci. U. S. A.* 987:898–902.

Gibbons, C., Montgomery, M. G., Leslie, A. G. W., and Walker, J. 2000. The structure of the central stalk in F_1-ATPase at 2.4 Å resolution. *Nat. Struct. Biol.* 7:1055–1061.

Boyer, P. D. 2000. Catalytic site forms and controls in ATP synthase catalysis. *Biochim. Biophys. Acta* 1458:252–262.

Stock, D., Leslie, A. G., and Walker, J. E. 1999. Molecular architecture of the rotary motor in ATP synthase. *Science* 286:1700–1705.

Sambongi, Y., Iko, Y., Tanabe, M., Omote, H., Iwamoto-Kihara, A., Ueda, I., Yanagida, T., Wada, Y., and Futai, M. 1999. Mechanical rotation of the *c* subunit oligomer in ATP synthase (F_0F_1): Direct observation. *Science* 286:1722–1724.

Abrahams, J. P., Leslie, A. G., Lutter, R., and Walker, J. E. 1994. Structure at 2.8 Å resolution of F1-ATPase from bovine heart mitochondria. *Nature* 370:621–628.

Bianchet, M. A., Hullihen, J., Pedersen, P. L., and Amzel, L. M. 1998. The 2.8-Å structure of rat liver F_1-ATPase: Configuration of a critical intermediate in ATP synthesis/hydrolysis. *Proc. Natl. Acad. Sci. U. S. A.* 95:11065–11070.

Translocators

Pebay-Peyroula, E., Dahout, C., Kahn, R., Trézéguet, V., Lauquin, G. J.-M., and Brandolin, G. 2003. Structure of mitochondrial ADP/ATP carrier in complex with carboxyatractyloside. *Nature* 246:39–44.

Klingenberg, M., and Huang, S. G. 1999. Structure and function of the uncoupling protein from brown adipose tissue. *Biochim. Biophys. Acta* 1415:271–296.

Nicholls, D. G., and Rial, E. 1999. A history of the first uncoupling protein, UCP1. *J. Bioenerg. Biomembr.* 31:399–406.

Ricquier, D., and Bouillaud, F. 2000. The uncoupling protein homologues: UCP1, UCP2, UCP3, StUCP and AtUCP. *Biochem. J.* 345:161–179.

Walker, J. E. 1992. The mitochondrial transporter family. *Curr. Opin. Struct. Biol.* 2:519–526.

Klingenberg, M. 1992. Structure-function of the ADP/ATP carrier. *Biochem. Soc. Trans.* 20:547–550.

Superoxide Dismutase and Catalase

Culotta, V. C. 2000. Superoxide dismutase, oxidative stress, and cell metabolism. *Curr. Top. Cell Regul.* 36:117–132.

Morrison, B. M., Morrison, J. H., and Gordon, J. W. 1998. Superoxide dismutase and neurofilament transgenic models of amyotrophic lateral sclerosis. *J. Exp. Zool.* 282:32–47.

Tainer, J. A., Getzoff, E. D., Richardson, J. S., and Richardson, D. C. 1983. Structure and mechanism of copper, zinc superoxide dismutase. *Nature* 306:284–287.

Reid, T. J., Murthy, M. R., Sicignano, A., Tanaka, N., Musick, W. D., and Rossmann, M. G. 1981. Structure and heme environment of beef liver catalase at 2.5 Å resolution. *Proc. Natl. Acad. Sci. U. S. A.* 78:4767–4771.

Stallings, W. C., Pattridge, K. A., Strong, R. K., and Ludwig, M. L. 1984. Manganese and iron superoxide dismutases are structural homologs. *J. Biol. Chem.* 259:10695–10699.

Hsieh, Y., Guan, Y., Tu, C., Bratt, P. J., Angerhofer, A., Lepock, J. R., Hickey, M. J., Tainer, J. A., Nick, H. S., and Silverman, D. N. 1998. Probing the active site of human manganese superoxide dismutase: The role of glutamine 143. *Biochemistry* 37:4731–4739.

Mitochondrial Diseases

DiMauro, S., and Schon, E. A. 2003. Mitochondrial respiratory-chain disease. *New Engl. J. Med.* 348:2656–2668.

Smeitink, J., van den Heuvel, L., and DiMauro, S. 2001. The genetics and pathology of oxidative phosphorylation. *Nat. Rev. Genet.* 2:342–352.

Wallace, D. C. 1999. Mitochondrial diseases in man and mouse. *Science* 283:1482–1488.

Wallace, D. C. 1992. Diseases of the mitochondrial DNA. *Annu. Rev. Biochem.* 61:1175–1212.

Benecke, R., Strumper, P., and Weiss, H. 1992. Electron transfer Complex I defect in idiopathic dystonia. *Ann. Neurol.* 32:683–686.

Apoptosis

Joza, N., Susin, S. A., Daugas, E., Stanford, W. L., Cho, S. K., Li, C. Y. J., Sasaki, T., Elia, A. J., Cheng, H.-Y. M., Ravagnan, L., Ferri, K. F., Zamzami, N., Wakeham, A., Hakem, R., Yoshida, H., Kong, Y.-Y., Mak, T. W., Zúñiga-Pflücker, J. C., Kroemer, G., and Penninger, J. M. 2001. Essential role of the mitochondrial *apoptosis*-inducing factor in programmed cell death. *Nature* 410:549–554.

Desagher, S., and Martinou, J. C. 2000. Mitochondria as the central control point of apoptosis. *Trends Cell Biol.* 10:369–377.

Hengartner, M. O. 2000. The biochemistry of apoptosis. *Nature* 407:770–776.

Earnshaw, W. C., Martins, L. M., and Kaufmann, S. H. 1999. Mammalian caspases: Structure, activation, substrates, and functions during apoptosis. *Annu. Rev. Biochem.* 68:383–424.

Wolf, B. B., and Green, D. R. 1999. Suicidal tendencies: Apoptotic cell death by caspase family proteinases. *J. Biol. Chem.* 274:20049–20052.

Historical Aspects

Mitchell, P. 1979. Keilin's respiratory chain concept and its chemiosmotic consequences. *Science* 206:1148–1159.

Mitchell, P. 1976. Vectorial chemistry and the molecular mechanics of chemiosmotic coupling: Power transmission by proticity. *Biochem. Soc. Trans.* 4:399–430.

Racker, E. 1980. From Pasteur to Mitchell: A hundred years of bioenergetics. *Fed. Proc.* 39:210–215.

Keilin, D. 1966. *The History of Cell Respiration and Cytochromes.* Cambridge University Press.

Kalckar, H. M. (Ed.). 1969. *Biological Phosphorylations: Development of Concepts.* Prentice Hall.

Kalckar, H. M. 1991. Fifty years of biological research: From oxidative phosphorylation to energy requiring transport and regulation. *Annu. Rev. Biochem.* 60:1–37.

Fruton, J. S. 1972. *Molecules and Life: Historical Essays on the Interplay of Chemistry and Biology.* Wiley-Interscience.

Problems

1. *Energy harvest.* What is the yield of ATP when each of the following substrates is completely oxidized to CO_2 by a mammalian cell homogenate? Assume that glycolysis, the citric acid cycle, and oxidative phosphorylation are fully active.

(a) Pyruvate   (e) Galactose
(b) Lactate   (f) Dihydroxyacetone
(c) Fructose 1,6-bisphosphate   phosphate
(d) Phosphoenolpyruvate

2. *Reference states.* The standard oxidation–reduction potential for the reduction of O_2 to H_2O is given as 0.82 V in Table 18.1. However, the value given in textbooks of chemistry is 1.23 V. Account for this difference.

3. *Potent poisons.* What is the effect of each of the following inhibitors on electron transport and ATP formation by the respiratory chain?

(a) Azide   (d) DNP
(b) Atractyloside   (e) Carbon monoxide
(c) Rotenone   (f) Antimycin A

4. *A question of coupling.* What is the mechanistic basis for the observation that the inhibitors of ATP synthase also lead to an inhibition of the electron-transport chain?

5. *O_2 consumption.* Using the axes in the adjoining illustration, draw an oxygen-uptake curve ([O_2] versus time) for a suspension of isolated mitochondria when the following compounds are added in the order from *a* to *h*. With the addition of each compound, all of the previously added compounds remain present. The experiment starts with the amount of oxygen indicated by the arrow on the *y*-axis. [O_2] can only decrease or be unaffected.

(a) Glucose
(b) ADP + P_i
(c) Citrate
(d) Oligomycin
(e) Succinate
(f) Dinitrophenol
(g) Rotenone
(h) Cyanide

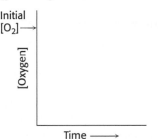

6. *P:O ratios.* The number of molecules of inorganic phosphate incorporated into organic form per atom of oxygen consumed, termed the *P:O ratio*, was frequently used as an index of oxidative phosphorylation.

(a) What is the relation of the P:O ratio to the ratio of the number of protons translocated per electron pair ($H^+/2\ e^-$) and the ratio of the number of protons needed to synthesize ATP and transport it to the cytoplasm (P/H^+)?

(b) What are the P:O ratios for electrons donated by matrix NADH and by succinate?

7. *Thermodynamic constraint.* Compare the $\Delta G^{\circ\prime}$ values for the oxidation of succinate by NAD^+ and by FAD. Use the data given in Table 18.1, and assume that E_0' for the FAD–$FADH_2$ redox couple is nearly 0 V. Why is FAD rather than NAD^+ the electron acceptor in the reaction catalyzed by succinate dehydrogenase?

8. *Cyanide antidote.* The immediate administration of nitrite is a highly effective treatment for cyanide poisoning. What is the basis for the action of this antidote? (Hint: Nitrite oxidizes ferrohemoglobin to ferrihemoglobin.)

9. *Currency exchange.* For a proton-motive force of 0.2 V (matrix negative), what is the maximum [ATP]/[ADP][P_i] ratio compatible with ATP synthesis? Calculate this ratio three times, assuming that the number of protons translocated per ATP formed is two, three, and four and that the temperature is 25°C.

10. *Runaway mitochondria 1.* Suppose that the mitochondria of a patient oxidize NADH irrespective of whether ADP is present. The P:O ratio for oxidative phosphorylation by these mitochondria is less than normal. Predict the likely symptoms of this disorder.

11. *An essential residue.* The conduction of protons by the F_0 unit of ATP synthase is blocked by the modification of a single side chain by dicyclohexylcarbodiimide. What are the most likely targets of action of this reagent? How might you use site-specific mutagenesis to determine whether this residue is essential for proton conduction?

12. *Recycling device.* The cytochrome *b* component of Q-cytochrome *c* oxidoreductase enables both electrons of QH_2 to be effectively utilized in generating a proton-motive force. Cite another recycling device in metabolism that brings a potentially dead end reaction product back into the mainstream.

13. *Crossover point.* The precise site of action of a respiratory-chain inhibitor can be revealed by the *crossover technique.* Britton Chance devised elegant spectroscopic methods for determining the proportions of the oxidized and reduced forms of each carrier. This determination is feasible because the forms have distinctive absorption spectra, as illustrated in the adjoining graph for cytochrome *c*. You are given a new inhibitor and find that its addition to respiring mitochondria causes the carriers between NADH and QH_2 to become more reduced and those between cytochrome *c* and O_2 to become more oxidized. Where does your inhibitor act?

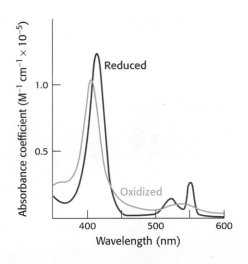

14. *Runaway mitochondria 2.* Years ago, it was suggested that uncouplers would make wonderful diet drugs. Explain why this idea was proposed and why it was rejected. Why might the producers of antiperspirants be supportive of the idea?

15. *Coupled processes.* If actively respiring mitochondria are exposed to an inhibitor of ATP synthase, the electron-transport chain ceases to operate. Why?

16. *Identifying the inhibition.* You are asked to determine whether a chemical is an electron-transport-chain inhibitor or an inhibitor of ATP synthase. Design an experiment to make this determination.

17. *Obeying the laws of thermodynamics.* Why will isolated F_1 subunits display ATPase activity and not ATP synthase activity?

18. *Opposites attract.* An arginine residue (Arg 210) in the **a** subunit of ATP synthase is near the aspartate residue (Asp 61) in the matrix-side proton channel. How might Arg 210 assist proton flow?

19. *Variable c subunits.* Recall that the number of **c** subunits in the **c** ring appears to range between 10 and 14. This number is significant because it determines the number of protons that must be transported to generate a molecule of ATP. Each 360-degree rotation of the γ subunit leads to the synthesis and release of three molecules of ATP. Thus, if there are 10 **c** subunits in the ring (as was observed in a crystal structure of yeast mitochondrial ATP synthase), each ATP generated requires the transport of $10/3 = 3.33$ protons. How many ATPs would form if the ring had 12 **c** subunits? 14?

20. *Exaggerating the difference.* Why must the ATP-ADP translocase (also called adenine nucleotide translocase or ANT) use Mg^{2+}-free forms of ATP and ADP?

Chapter Integration Problems

21. *The right location.* Some cytoplasmic kinases, enzymes that phosphorylate substrates at the expense of ATP, bind to VDAC. What advantage might this binding be?

22. *No exchange.* Mice that are completely lacking ATP-ADP translocase (ANT^-/ANT^-) can be made by the knockout technique. Remarkably, these mice are viable but have the following pathological conditions: (1) high serum levels of lactate, alanine, and succinate; (2) little electron transport; and (3) a six- to eight-fold increase in the level of mitochondrial H_2O_2 compared with that in normal mice. Provide a possible biochemical explanation for each of these conditions.

Data Interpretation Problem

23. *Mitochondrial disease.* A mutation in a mitochondrial gene encoding a component of ATP synthase has been identified. People who have this mutation suffer from muscle weakness, ataxia, and retinitis pigmentosa. A tissue biopsy was performed on each of three patients having this mutation, and submitochondrial particles were isolated that were capable of succinate-sustained ATP synthesis. First, the activity of the ATP synthase was measured on the addition of succinate and the following results were obtained.

	ATP synthase activity (nmol of ATP formed min^{-1} mg^{-1})
Controls	3.0
Patient 1	0.25
Patient 2	0.11
Patient 3	0.17

(a) What was the purpose of the addition of succinate?

(b) What is the effect of the mutation on succinate-coupled ATP synthesis?

Next, the ATPase activity of the enzyme was measured by incubating the submitochondrial particles with ATP in the absence of succinate.

	ATP hydrolysis (nmol of ATP hydrolyzed min^{-1} mg^{-1})
Controls	33
Patient 1	30
Patient 2	25
Patient 3	31

(c) Why was succinate omitted from the reaction?

(d) What is the effect of the mutation on ATP hydrolysis?

(e) What do these results, in conjunction with those obtained in the first experiment, tell you about the nature of the mutation?

Mechanism Problem

24. *Chiral clue.* ATPγS, a slowly hydrolyzed analog of ATP, can be used to probe the mechanism of phosphoryl-transfer reactions. Chiral ATPγS has been synthesized containing ^{18}O in a specific γ position and ordinary ^{16}O elsewhere in the molecule. The hydrolysis of this chiral molecule by ATP synthase in ^{17}O-enriched water yields inorganic $[^{16}O, ^{17}O, ^{18}O]$thiophosphate having the following absolute configuration. In contrast, the hydrolysis of this chiral ATPγS by a calcium-pumping ATPase from muscle gives thiophosphate of the opposite configuration. What is the simplest interpretation of these data?

ATPγS   **Thiophosphate**

The Light Reactions of Photosynthesis

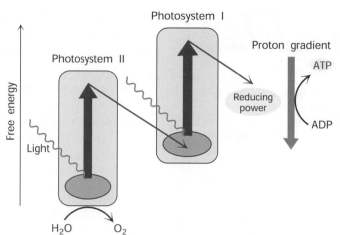

Chloroplasts (left) convert light energy into chemical energy. High-energy electrons in chloroplasts are transported through two photosystems (right). In this transit, which culminates in the generation of reducing power, ATP is synthesized in a manner analogous to mitochondrial ATP synthesis. In contrast with mitochondrial electron transport, however, electrons in chloroplasts are energized by light. [(Left) Herb Charles Ohlmeyer/Fran Heyl Associates.]

Essentially all free energy utilized by biological systems arises from solar energy that is trapped by the process of photosynthesis. The basic equation of photosynthesis is deceptively simple. Water and carbon dioxide combine to form carbohydrates and molecular oxygen.

$$CO_2 + H_2O \xrightarrow{\text{Light}} (CH_2O) + O_2$$

In this equation, (CH_2O) represents carbohydrate, primarily sucrose and starch. The mechanism of photosynthesis is complex and requires the interplay of many proteins and small molecules. Photosynthesis in green plants takes place in *chloroplasts* (Figure 19.1). *The energy of light is captured by pigment molecules, called chlorophylls, located in chloroplasts. This captured energy excites certain electrons to higher energies. In essence, light is used to create reducing potential.*

The excited electrons are used to produce NADPH as well as ATP in a series of reactions called the *light reactions* because they require light. NADPH and ATP formed by the action of light then reduce carbon dioxide and convert it into *3-phosphoglycerate* by a series of reactions called the *Calvin cycle* or the *dark reactions*. The Calvin cycle will be discussed in Chapter 20. The amount of energy stored by photosynthesis is enormous.

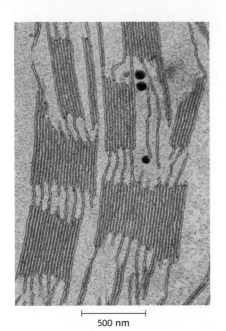

Figure 19.1 Electron micrograph of a chloroplast from a spinach leaf. The thylakoid membranes pack together to form grana. [Courtesy of Dr. Kenneth Miller.]

500 nm

More than 4.2×10^{17} kJ (10^{17} kcal) of free energy is stored annually by photosynthesis on Earth, which corresponds to the assimilation of more than 10^{10} tons of carbon into carbohydrate and other forms of organic matter.

As animals ourselves, we perhaps easily overlook the ultimate primacy of photosynthesis for our biosphere. Photosynthesis is the source of essentially all the carbon compounds and all the oxygen that makes aerobic metabolism possible.

Photosynthesis Converts Light Energy into Chemical Energy

The light reactions of photosynthesis closely resemble the events of oxidative phosphorylation. In both, the flow of high-energy electrons through an electron-transport chain generates a proton-motive force. This force drives ATP synthesis through the action of an ATP synthase. In photosynthesis, the electrons are also used directly to reduce $NADP^+$ to NADPH.

A principal difference between oxidative phosphorylation and photosynthesis is the source of the high-energy electrons. In oxidative phosphorylation, these electrons come from the oxidation of carbon fuels to CO_2. In photosynthesis, electrons are excited to a higher energy level by the energy from *photons* (Figure 19.2). The electronic excitation passes from one chlorophyll molecule to another in a light-harvesting complex until the excitation is trapped by a chlorophyll with special properties. At such a reaction center, the energy of the excited electron is converted into a separation of charge with useful reducing properties.

Photosynthetic yield

"If a year's yield of photosynthesis were amassed in the form of sugar cane, it would form a heap over two miles high and with a base 43 square miles."

—G. E. FOGGE

If all of this sugar cane were converted into sugar cubes (0.5 inch, or 1.27 cm, on a side) and stacked end to end, the sugar cubes would extend 1.6×10^{10} miles (2.6×10^{10} kilometers) or to the planet Pluto.

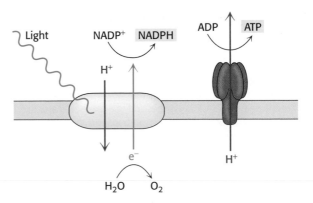

Figure 19.2 The light reactions of photosynthesis. Light is absorbed and the energy is used to drive electrons from water to generate NADPH and to drive protons across a membrane. These protons return through ATP synthase to make ATP.

Photosynthetic catastrophe

If photosynthesis were to cease, all higher forms of life would be extinct in about 25 years. A milder version of such a catastrophe ended the Cretaceous period 65.1 million years ago when a large asteroid struck the Yucatan Peninsula of Mexico. Enough dust was sent into the atmosphere that photosynthetic capacity was greatly diminished, which apparently led to the disappearance of the dinosaurs and allowed the mammals to rise to prominence.

Photosynthesis in green plants is mediated by two kinds of light reactions. Photosystem I generates reducing power in the form of NADPH. Photosystem II transfers the electrons of water to a quinone. A side product of these reactions is O_2. Electron flow within each photosystem and between the two photosystems generates the transmembrane proton gradient that drives the synthesis of ATP. In the dark reactions, the NADPH and ATP formed by the action of light drive the reduction of CO_2 to more-useful organic compounds.

19.1 Photosynthesis Takes Place in Chloroplasts

As described in Chapter 18, oxidative phosphorylation, the predominant means of generating ATP from fuel molecules, is compartmentalized into

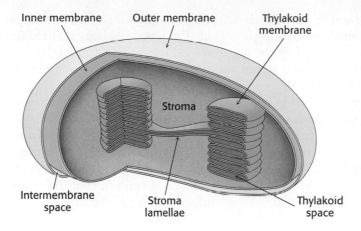

Inner membrane   Outer membrane   Thylakoid membrane

Stroma

Intermembrane space   Stroma lamellae   Thylakoid space

Figure 19.3 Diagram of a chloroplast. [After S. L. Wolfe, *Biology of the Cell,* p. 130. © 1972 by Wadsworth Publishing Company, Inc. Adapted by permission of the publisher.]

mitochondria. Likewise, photosynthesis, the means of converting light into chemical energy, is sequestered into organelles called *chloroplasts*, typically 5 μm long. Like a mitochondrion, a chloroplast has an outer membrane and an inner membrane, with an intervening intermembrane space (Figure 19.3). The inner membrane surrounds a space called the *stroma*, which is the site of the carbon chemistry of photosynthesis (Section 20.1). In the stroma are membranous structures called *thylakoids,* which are flattened sacs, or discs. The thylakoid sacs are stacked to form a *granum.* Different grana are linked by regions of thylakoid membrane called *stroma lamellae.* The thylakoid membranes separate the thylakoid space from the stroma space. Thus, chloroplasts have three different membranes (*outer, inner,* and *thylakoid membranes*) and three separate spaces (*intermembrane, stroma,* and *thylakoid spaces*). In developing chloroplasts, thylakoids are believed to arise from infoldings of the inner membrane, and so they are analogous to the mitochondrial cristae. Like the mitochondrial cristae, they are the site of coupled oxidation–reduction reactions that generate the proton-motive force.

The Primary Events of Photosynthesis Take Place in Thylakoid Membranes

The thylakoid membranes contain the energy-transforming machinery: light-harvesting proteins, reaction centers, electron-transport chains, and ATP synthase. These membranes contain nearly equal amounts of lipids and proteins. The lipid composition is highly distinctive: about 40% of the total lipids are *galactolipids* and 4% are *sulfolipids*, whereas only 10% are phospholipids. The thylakoid membrane and the inner membrane, like the inner mitochondrial membrane, are impermeable to most molecules and ions. The outer membrane of a chloroplast, like that of a mitochondrion, is highly permeable to small molecules and ions. The stroma contains the soluble enzymes that utilize the NADPH and ATP synthesized by the thylakoids to convert CO_2 into sugar. Plant-leaf cells contain between 1 and 100 chloroplasts, depending on the species, cell type, and growth conditions.

Chloroplasts Arose from an Endosymbiotic Event

Chloroplasts contain their own DNA and the machinery for replicating and expressing it. However, chloroplasts are not autonomous: they also contain many proteins encoded by nuclear DNA. How did the intriguing relation between the cell and its chloroplasts develop? We now believe that, in a manner analogous to the evolution of mitochondria (p. 504), chloroplasts are the result of endosymbiotic events in which a photosynthetic

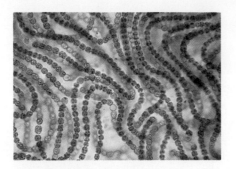

Figure 19.4 Cyanobacteria. A colony of the photosynthetic filamentous cyanobacterium *Anabaena* shown at 450× magnification. Ancestors of these bacteria are thought to have evolved into present-day chloroplasts. [James W. Richardson/Visuals Unlimited.]

microorganism, most likely an ancestor of a cyanobacterium (Figure 19.4), was engulfed by a eukaryotic host. Evidence suggests that chloroplasts in higher plants and green algae are derived from a single endosymbiotic event, whereas those in red and brown algae are derived from at least one additional event.

The chloroplast genome is smaller than that of a cyanobacterium, but the two genomes have key features in common. Both are circular and have a single start site for DNA replication, and the genes of both are arranged in operons—sequences of functionally related genes under common control (Chapter 31). In the course of evolution, many of the genes of the chloroplast ancestor were transferred to the plant cell's nucleus or, in some cases, lost entirely, thus establishing a fully dependent relation.

19.2 Light Absorption by Chlorophyll Induces Electron Transfer

The trapping of light energy is the key to photosynthesis. The first event is the absorption of light by a photoreceptor molecule. The principal photoreceptor in the chloroplasts of most green plants is *chlorophyll* a, a substituted tetrapyrrole (Figure 19.5). The four nitrogen atoms of the pyrroles are coordinated to a magnesium ion. Unlike a porphyrin such as heme, chlorophyll has a reduced pyrrole ring and an additional 5-carbon ring fused to one of the pyrrole rings. Another distinctive feature of chlorophyll is the presence of *phytol*, a highly hydrophobic 20-carbon alcohol, esterified to an acid side chain.

Figure 19.5 Chlorophyll. Like heme, chlorophyll *a* is a cyclic tetrapyrrole. One of the pyrrole rings (shown in red) is reduced, and an additional five-carbon ring (shown in blue) is fused to another pyrrole ring. A phytol chain (shown in green) is connected by an ester linkage. Magnesium ion binds at the center of the structure.

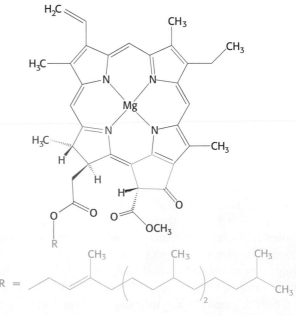

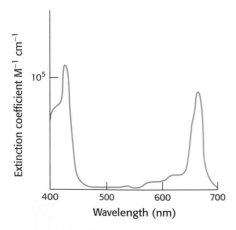

Figure 19.6 Light absorption by chlorophyll *a*. Chlorophyll *a* absorbs visible light efficiently as judged by the extinction coefficients near $10^5 \text{ M}^{-1} \text{ cm}^{-1}$.

Chlorophylls are very effective photoreceptors because they contain networks of alternating single and double bonds. Such compounds are called conjugated *polyenes*. Chlorophylls have very strong absorption bands in the visible region of the spectrum, where the solar output reaching Earth is maximal (Figure 19.6). Chlorophyll *a*'s peak molar extinction coefficient (ϵ), a measure of a compound's ability to absorb light, is higher than $10^5 \text{ M}^{-1} \text{ cm}^{-1}$, among the highest observed for organic compounds.

What happens when light is absorbed by a pigment molecule such as chlorophyll? The energy from the light excites an electron from its ground energy level to an excited energy level (Figure 19.7). This high-energy electron can have one of two fates. For most compounds that absorb light, the electron simply returns to the ground state and the absorbed energy is converted into heat. In regard to chlorophyll, however, a suitable electron acceptor is nearby, and so the excited electron can move from the initial molecule to the acceptor (Figure 19.8). A positive charge forms on the initial molecule, owing to the loss of an electron, and a negative charge forms on the acceptor, owing to the gain of an electron. Hence, this process is referred to as *photoinduced charge separation*. The site at which the charge separation takes place is called the *reaction center*. The photosynthetic apparatus is arranged to maximize photoinduced charge separation and minimize an unproductive return of the electron to its ground state. The electron, extracted from its initial site by the absorption of light, now has reducing power: it can reduce other molecules to store the energy originally obtained from light in chemical forms.

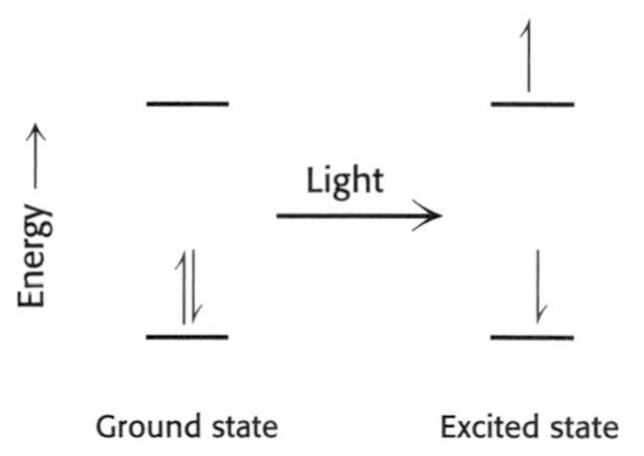

Figure 19.7 **Light absorption.** The absorption of light leads to the excitation of an electron from its ground state to a higher energy level.

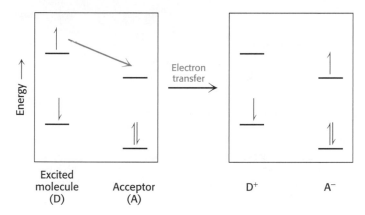

Figure 19.8 **Photoinduced charge separation.** If a suitable electron acceptor is nearby, an electron that has been moved to a high energy level by light absorption can move from the excited molecule to the acceptor.

A Special Pair of Chlorophylls Initiate Charge Separation

Photosynthetic bacteria such as *Rhodopseudomonas viridis* contain a photosynthetic reaction center that has been revealed at atomic resolution. The bacterial reaction center consists of four polypeptides: L (31 kd), M (36 kd), and H (28 kd) subunits and C, a *c*-type cytochrome with four *c*-type hemes (Figure 19.9). *Sequence comparisons and low-resolution structural studies have revealed that the bacterial reaction center is homologous to the more complex plant systems.* Thus, many of our observations of the bacterial system will apply to plant systems as well.

The L and M subunits form the structural and functional core of the bacterial photosynthetic reaction center (see Figure 19.9). Each of these homologous subunits contains five transmembrane helices, in contrast with the H subunit, which has just one. The H subunit lies on the cytoplasmic side of the cell membrane, and the cytochrome subunit lies on the exterior face of the cell membrane, called the periplasmic side because it faces the periplasm, the space between the cell membrane and the cell wall. Four bacteriochlorophyll *b* (BChl-*b*) molecules, two bacteriopheophytin *b* (BPh) molecules, two quinones (Q$_A$ and Q$_B$), and a ferrous ion are associated with the L and M subunits.

Bacteriochlorophylls are similar to chlorophylls, except for the reduction of an additional pyrrole ring and other minor differences that shift their absorption maxima to the near infrared, to wavelengths as long as 1000 nm.

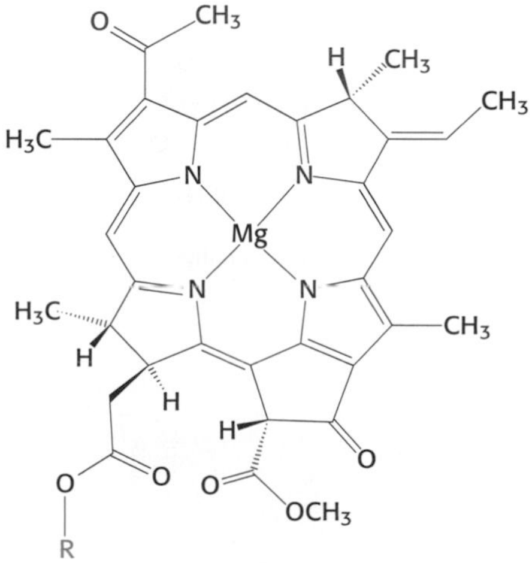

Bacteriochlorophyll *b*
(BChl-*b*)

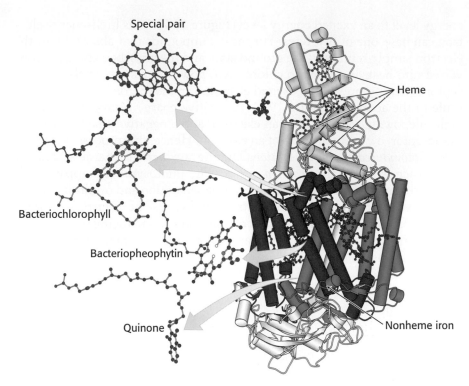

Figure 19.9 **Bacterial photosynthetic reaction center.** The core of the reaction center from *Rhodopseudomonas viridis* consists of two similar chains: L (red) and M (blue). An H chain (white) and a cytochrome subunit (yellow) complete the structure. *Notice* that the L and M subunits are composed largely of α helices that span the membrane. *Notice* also that a chain of electron-carrying prosthetic groups, beginning with a special pair of bacteriochlorophylls and ending at a bound quinone, runs through the structure from top to bottom in this view. [Drawn from 1PRC.pdb.]

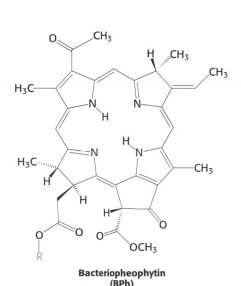

**Bacteriopheophytin
(BPh)**

Bacteriopheophytin is the term for a bacteriochlorophyll that has two protons instead of a magnesium ion at its center.

The reaction begins with light absorption by a pair of BChl-*b* molecules that lie near the periplasmic side of the membrane in the L-M dimer. The pair of BChl-*b* molecules is called the *special pair* because of its fundamental role in photosynthesis. The special pair absorbs light maximally at 960 nm, in the infrared near the edge of the visible region and, for this reason, is often called *P960* (P stands for pigment). After absorbing light, the excited special pair ejects an electron, which is transferred through another BChl-*b* to a bacteriopheophytin (Figure 19.10, steps 1 and 2). This initial charge separation yields a positive charge on the special pair (P960$^+$) and a negative charge on BPh. The electron ejection and transfer takes place in less than 10 picoseconds (10^{-11} seconds).

A nearby electron acceptor, a tightly bound quinone (Q_A), quickly grabs the electron away from BPh$^-$ before the electron has a chance to fall back to the P960 special pair. From Q_A, the electron moves to a more loosely associated quinone, Q_B. The absorption of a second photon and the movement of a second electron down the path from the special pair completes the two-electron reduction of Q_B from Q to QH$_2$. Because the Q_B-binding site lies near the cytoplasmic side of the membrane, *two protons are taken up from the cytoplasm, contributing to the development of a proton gradient across the cell membrane* (Figure 19.10, steps 5, 6, and 7).

In their high-energy states, P960$^+$ and BPh$^-$ could undergo charge recombination; that is, the electron on BPh$^-$ could move back to neutralize the positive charge on the special pair. Its return to the special pair would

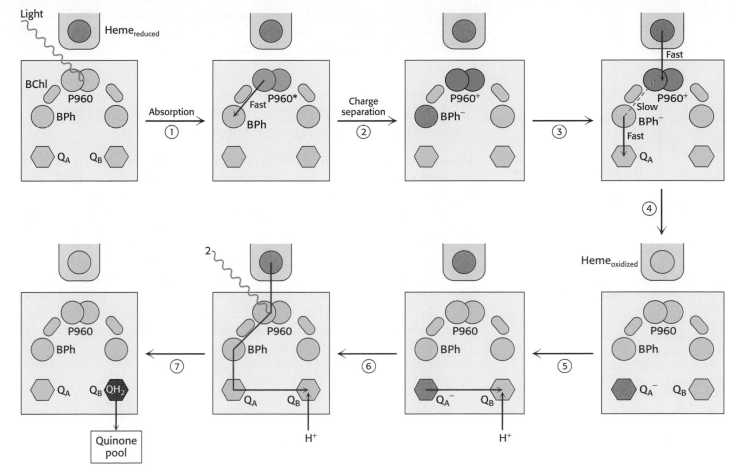

Figure 19.10 Electron chain in the photosynthetic bacterial reaction center. The absorption of light by the special pair (P960) results in the rapid transfer of an electron from this site to a bacteriopheophytin (BPh), creating a photoinduced charge separation (steps 1 and 2). (The asterisk on P960 stands for excited state.) The possible return of the electron from the pheophytin to the oxidized special pair is suppressed by the "hole" in the special pair being refilled with an electron from the cytochrome subunit and the electron from the pheophytin being transferred to a quinone (Q_A) that is farther away from the special pair (steps 3 and 4). Q_A passes the electron to Q_B. The reduction of a quinone (Q_B) on the cytoplasmic side of the membrane results in the uptake of two protons from the cytoplasm (steps 5 and 6). The reduced quinone can move into the quinone pool in the membrane (step 7).

waste a valuable high-energy electron and simply convert the absorbed light energy into heat. How is charge recombination prevented? Three factors in the structure of the reaction center work together to suppress charge recombination nearly completely (Figure 19.10, steps 3 and 4). First, the next electron acceptor (Q_A) is less than 10 Å away from BPh^-, and so the electron is rapidly transferred farther away from the special pair. Second, one of the hemes of the cytochrome subunit is less than 10 Å away from the special pair, and so the positive charge on P960 is neutralized by the transfer of an electron from the reduced cytochrome. Finally, the electron transfer from BPh^- to the positively charged special pair is especially slow: the transfer is so thermodynamically favorable that it takes place in the inverted region where electron-transfer rates become slower (p. 519). Thus, electron transfer proceeds efficiently from BPh^- to Q_A.

Cyclic Electron Flow Reduces the Cytochrome of the Reaction Center

The cytochrome subunit of the reaction center must regain an electron to complete the cycle. It does so by taking back two electrons from reduced quinone (QH_2). QH_2 first enters the Q pool in the membrane where it is reoxidized to Q by complex bc_1, which is homologous to complex III of the respiratory electron-transport chain (p. 512). Complex bc_1 transfers the electrons from QH_2 to cytochrome c_2, a water-soluble protein in the periplasm, and in the process pumps protons into the periplasmic space. The electrons now on cytochrome c_2 flow to the cytochrome subunit of the

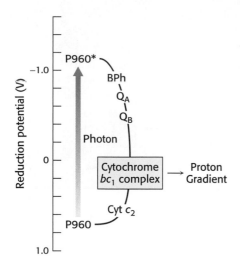

Figure 19.11 Cyclic electron flow in the bacterial reaction center. Excited electrons from the P960 reaction center flow through bacteriopheophytin (BPh), a pair of quinone molecules (Q_A and Q_B), cytochrome bc_1 complex, and finally through cytochrome c_2 to the reaction center. The cytochrome bc_1 complex pumps protons as a result of electron flow, which powers the formation of ATP.

reaction center. The flow of electrons is thus cyclic (Figure 19.11). The proton gradient generated in the course of this cycle drives the generation of ATP through the action of ATP synthase.

19.3 Two Photosystems Generate a Proton Gradient and NADPH in Oxygenic Photosynthesis

Photosynthesis by green plants depends on the interplay of two kinds of membrane-bound, light-sensitive complexes—*photosystem I* (PS I) and *photosystem II* (PS II), as shown in Figure 19.12. Photosystem I, which responds to light with wavelengths shorter than 700 nm, uses light-derived high-energy electrons to create biosynthetic reducing power in the form of NADPH, a versatile reagent for driving biosynthetic processes. The electrons to create one molecule of NADPH are derived from two molecules of water by photosystem II, which responds to wavelengths shorter than 680 nm. A molecule of O_2 is generated as a side product. The electrons travel from photosystem II to photosystem I through cytochrome *bf*, a membrane-bound complex homologous to Complex III in oxidative phosphorylation (p. 512). Cytochrome *bf* generates a proton gradient across the thylakoid membrane that drives the formation of ATP. Thus, the two photosystems cooperate to produce NADPH and ATP.

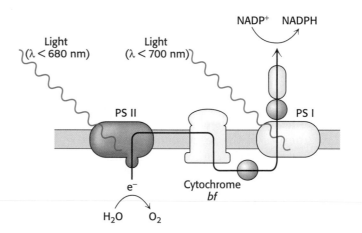

Figure 19.12 Two photosystems. The absorption of photons by two distinct photosystems (PS I and PS II) is required for complete electron flow from water to $NADP^+$.

Photosystem II Transfers Electrons from Water to Plastoquinone and Generates a Proton Gradient

Photosystem II, an enormous transmembrane assembly of more than 20 subunits, catalyzes the light-driven transfer of electrons from water to plastoquinone. This electron acceptor closely resembles ubiquinone, a component of the mitochondrial electron-transport chain. Plastoquinone cycles between an oxidized form (Q) and a reduced form (QH_2, plastoquinol). The overall reaction catalyzed by photosystem II is

$$2\,Q + 2\,H_2O \xrightarrow{\text{Light}} O_2 + 2\,QH_2$$

The electrons in QH_2 are at a higher redox potential than those in H_2O. Recall that, in oxidative phosphorylation, electrons flow from ubiquinol to an acceptor, O_2, that is at a *lower* potential. Photosystem II drives the reaction in a thermodynamically uphill direction by using the free energy of light.

This reaction is similar to one catalyzed by the bacterial system in that a quinone is converted from its oxidized into its reduced form. Photosystem II is reasonably similar to the bacterial reaction center (Figure 19.13). The

Plastoquinone
(oxidized form, Q)

$(n = 6 \text{ to } 10)$

Plastoquinol
(reduced form, QH_2)

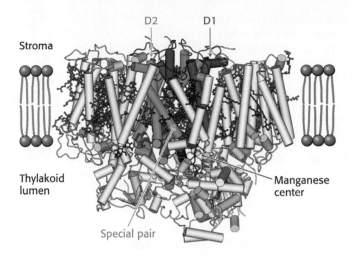

Stroma

Thylakoid lumen

D2

D1

Manganese center

Special pair

Figure 19.13 The structure of photosystem II. The D1 and D2 subunits are shown in red and blue, and the numerous bound chlorophyll molecules are shown in green. *Notice* the special pair and the manganese center (the site of oxygen evolution) lie toward the thylakoid-lumen side of the membrane. [Drawn from 1S5L.pdb.]

core of the photosystem is formed by D1 and D2, a pair of similar 32-kd subunits that span the thylakoid membrane. These subunits are homologous to the L and M chains of the bacterial reaction center. Unlike the bacterial system, photosystem II contains a large number of additional subunits that bind more than 30 chlorophyll molecules altogether and increase the efficiency with which light energy is absorbed and transferred to the reaction center (Section 19.5).

The photochemistry of photosystem II begins with excitation of a special pair of chlorophyll molecules that are bound by the D1 and D2 subunits (Figure 19.14). Because the chlorophyll *a* molecules of the special pair absorb light at 680 nm, the special pair is often called *P680*. On excitation, P680 rapidly transfers an electron to a nearby pheophytin. From there, the electron is transferred first to a tightly bound plastoquinone at site Q_A and then to a mobile plastoquinone at site Q_B. This electron flow is entirely analogous to that in the bacterial system. With the arrival of a second electron and the uptake of two protons, the mobile plastoquinone is reduced to QH_2. At this point, the energy of two photons has been safely and efficiently stored in the reducing potential of QH_2.

The major difference between the bacterial system and photosystem II is the source of the electrons that are used to neutralize the positive charge formed on the special pair. *$P680^+$, a very strong oxidant, extracts electrons from water molecules bound at the manganese center.* The structure of this center includes a calcium ion and four manganese ions. Manganese was apparently evolutionarily selected for this role because of its ability to exist in multiple oxidation states (Mn^{2+}, Mn^{3+}, Mn^{4+}, Mn^{5+}) and to form strong

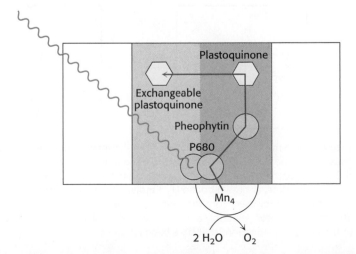

Plastoquinone

Exchangeable plastoquinone

Pheophytin

P680

Mn_4

$2 H_2O$ O_2

Figure 19.14 Electron flow through photosystem II. Light absorption induces electron transfer from P680 down an electron-transfer pathway to an exchangeable plastoquinone. The positive charge on P680 is neutralized by electron flow from water molecules bound at the manganese center.

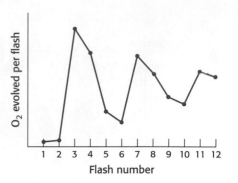

Figure 19.15 Four photons are required to generate one oxygen molecule. When dark-adapted chloroplasts are exposed to a brief flash of light, one electron passes through photosystem II. Monitoring the O_2 released after each flash reveals that four flashes are required to generate each O_2 molecule. The peaks in O_2 release occur after the 3rd, 7th, and 11th flashes because the dark-adapted chloroplasts start in the S_1 state—that is, the one-electron reduced state.

Evolution of oxygen is evident by the generation of bubbles in the aquatic plant *Elodea*. [Colin Milkins/Oxford Scientific Films.]

bonds with oxygen-containing species. The manganese center, in its reduced form, oxidizes two molecules of water to form a single molecule of oxygen. Each time the absorbance of a photon kicks an electron out of P680, the positively charged special pair extracts an electron from the manganese center (Figure 19.15). However, the electrons do not come directly from the manganese ions. A tyrosine residue (often designated Z) of subunit D1 in photosystem II is the immediate electron donor, forming a tyrosine radical. The tyrosine radical removes electrons from the manganese ions, which in turn removes electrons from H_2O to generate O_2 and H^+. Thus four photons must be absorbed to extract four electrons from a water molecule (Figure 19.16). The four electrons harvested from water are used to reduce two molecules of Q to QH_2.

Photosystem II spans the thylakoid membrane such that the site of quinone reduction is on the side of the stroma, whereas the manganese center, hence the site of water oxidation, lies in the thylakoid lumen. Thus, the

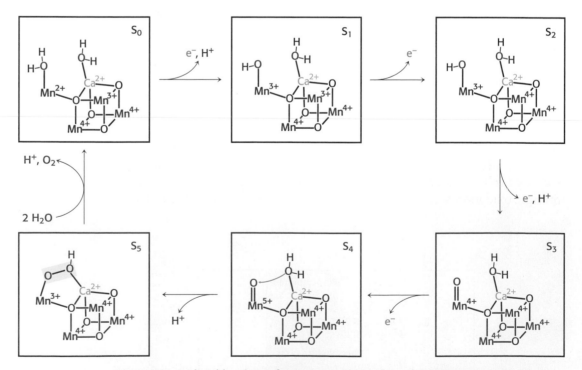

Figure 19.16 A plausible scheme for oxygen evolution from the manganese center. The deduced core structure of the manganese center including four manganese ions and one calcium ion is shown, although many additional ligands are omitted for clarity. The center is oxidized, one electron at a time, until two bound H_2O molecules are linked to form a molecule of O_2, which is then released from the center. A tyrosine residue (not shown) participates in the coupled proton–electron transfer steps. The structures are designated S_0 to S_4 to indicate the number of electrons that have been removed.

two protons that are taken up with the reduction of Q to QH_2 come from the stroma, and the four protons that are liberated in the course of water oxidation are released into the lumen. This distribution of protons generates a proton gradient across the thylakoid membrane characterized by an excess of protons in the thylakoid lumen compared with the stroma (Figure 19.17).

Cytochrome *bf* Links Photosystem II to Photosystem I

Electrons flow from photosystem I to photosystem II through the cytochrome *bf* complex. This complex catalyzes the transfer of electrons from plastoquinol (QH_2) to plastocyanin (Pc), a small, soluble copper protein in the thylakoid lumen.

$$QH_2 + 2\,Pc(Cu^{2+}) \longrightarrow Q + 2\,Pc(Cu^+) + 2\,H^+_{\text{thylakoid lumen}}$$

The two protons from plastoquinol are released into the thylakoid lumen. This reaction is reminiscent of that catalyzed by Complex III in oxidative phosphorylation, and most components of the *cytochrome* bf complex are homologous to those of Complex III. The cytochrome *bf* complex includes four subunits: a 23-kd cytochrome with two *b*-type hemes, a 20-kd Rieske-type Fe-S protein, a 33-kd cytochrome *f* with a *c*-type cytochrome, and a 17-kd chain.

This complex catalyzes the reaction by proceeding through the Q cycle (p. 513). In the first half of the Q cycle, plastoquinol is oxidized to plastoquinone, one electron at a time. The electrons from plastoquinol flow through the Fe-S protein to convert oxidized plastocyanin into its reduced form.

In the second half of the Q cycle, cytochrome *bf* reduces a molecule of plastoquinone from the Q pool to plastoquinol, taking up two protons from one side of the membrane, and then reoxidizes plastoquinol to release these protons on the other side. The enzyme is oriented so that protons are released into the thylakoid lumen and taken up from the stroma, contributing further to the proton gradient across the thylakoid membrane (Figure 19.18).

Photosystem I Uses Light Energy to Generate Reduced Ferredoxin, a Powerful Reductant

The final stage of the light reactions is catalyzed by photosystem I, a transmembrane complex consisting of about 14 polypeptide chains and multiple associated proteins and cofactors (Figure 19.19). The core of this system is a

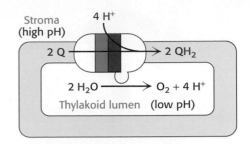

Figure 19.17 Proton-gradient direction. Photosystem II releases protons into the thylakoid lumen and takes them up from the stroma. The result is a pH gradient across the thylakoid membrane with an excess of protons (low pH) inside.

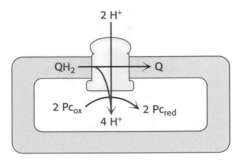

Figure 19.18 Cytochrome *bf* contribution to proton gradient. The cytochrome *bf* complex oxidizes QH_2 to Q through the Q cycle. Four protons are released into the thylakoid lumen in each cycle.

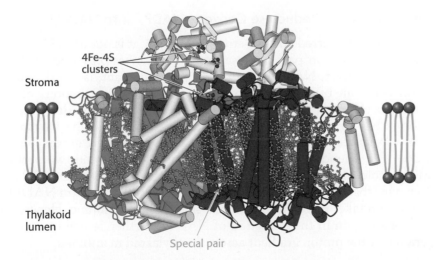

Figure 19.19 The structure of photosystem I. The psaA and psaB subunits are shown in red and blue. *Notice* the numerous bound chlorophyll molecules, shown in green, including the special pair, as well as the iron–sulfur clusters that facilitate electron transfer from the stroma. [Drawn from 1JB0.pdb.]

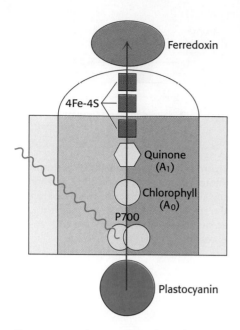

Figure 19.20 Electron flow through photosystem I to ferredoxin. Light absorption induces electron transfer from P700 down an electron-transfer pathway that includes a chlorophyll molecule, a quinone molecule, and three 4Fe-4S clusters to reach ferredoxin. The positive charge left on P700 is neutralized by electron transfer from reduced plastocyanin.

pair of similar subunits psaA (83 kd) and psaB (82 kd). These subunits are quite a bit larger than the core subunits of photosystem II and the bacterial reaction center. Nonetheless, they appear to be homologous; the terminal 40% of each subunit is similar to a corresponding subunit of photosystem II. A special pair of chlorophyll *a* molecules lie at the center of the structure and absorb light maximally at 700 nm. This center, called *P700*, initiates photoinduced charge separation (Figure 19.20). The electron travels from P700 down a pathway through chlorophyll at site A_0 and quinone at site A_1 to a set of 4Fe-4S clusters. The next step is the transfer of the electron to ferredoxin (Fd), a soluble protein containing a 2Fe-2S cluster coordinated to four cysteine residues (Figure 19.21). Ferredoxin transfers electrons to $NADP^+$. Meanwhile, $P700^+$ captures an electron from reduced plastocyanin to return to P700 so that P700 can be excited again. Thus, the overall reaction catalyzed by photosystem I is a simple one-electron oxidation–reduction reaction.

$$Pc(Cu^+) + Fd_{ox} \xrightarrow{\text{Light}} Pc(Cu^{2+}) + Fd_{red}$$

Given that the reduction potentials for plastocyanin and ferredoxin are $+0.37\,V$ and $-0.45\,V$, respectively, the standard free energy for this reaction is $+79.1\,kJ\,mol^{-1}$ ($+18.9\,kcal\,mol^{-1}$). This uphill reaction is driven by the absorption of a 700-nm photon, which has an energy of $171\,kJ\,mol^{-1}$ ($40.9\,kcal\,mol^{-1}$).

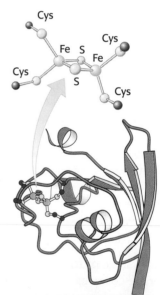

Figure 19.21 Structure of ferredoxin. In plants, ferredoxin contains a 2Fe-2S cluster. This protein accepts electrons from photosystem I and carries them to ferredoxin-NADP reductase. [Drawn from 1FXA.pdb.]

Ferredoxin-NADP$^+$ Reductase Converts NADP$^+$ into NADPH

Although reduced ferredoxin is a strong reductant, it is not useful for driving many reactions, in part because ferredoxin carries only one available electron. In contrast, NADPH, a two-electron reductant, is a widely used electron donor in biosynthetic processes, including the reactions of the Calvin cycle (Chapter 20). How is reduced ferredoxin used to drive the reduction of $NADP^+$ to NADPH? This reaction is catalyzed by *ferredoxin-NADP$^+$ reductase*, a flavoprotein with an FAD prosthetic group (Figure 19.22A). The bound FAD moiety accepts two electrons and two protons from two molecules of reduced ferredoxin to form $FADH_2$ (Figure 19.22B). The enzyme then transfers a hydride ion (H^-) to $NADP^+$ to form NADPH. This reaction takes place on the stromal side of the membrane. Hence, the uptake of a proton in the reduction of $NADP^+$ further contributes to the generation of the proton gradient across the thylakoid membrane.

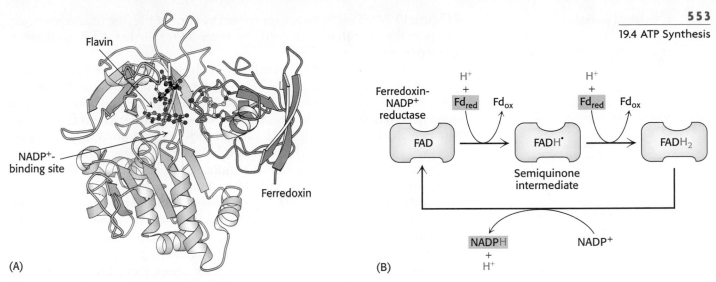

(A)

(B)

Figure 19.22 Structure and function of ferredoxin-NADP$^+$ reductase. (A) Structure of ferredoxin-NADP$^+$ reductase. This enzyme accepts electrons, one at a time, from ferredoxin (shown in orange). (B) Ferredoxin-NADP$^+$ reductase first accepts two electrons and two protons from two molecules of reduced ferridoxin (Fd) to form FADH$_2$, which then transfers two electrons and a proton to NADP$^+$ to form NADPH. [Drawn from 1EWY.pdb.]

The cooperation between photosystem I and photosystem II creates a flow of electrons from H$_2$O to NADP$^+$. The pathway of electron flow is called the *Z scheme of photosynthesis* because the redox diagram from P680 to P700* looks like the letter Z (Figure 19.23).

19.4 A Proton Gradient Across the Thylakoid Membrane Drives ATP Synthesis

In 1966, André Jagendorf showed that chloroplasts synthesize ATP in the dark when an artificial pH gradient is imposed across the thylakoid membrane. To create this transient pH gradient, he soaked chloroplasts in a pH 4 buffer for several hours and then rapidly mixed them with a pH 8 buffer containing ADP and P$_i$. The pH of the stroma suddenly increased to 8, whereas the pH of the thylakoid space remained at 4. *A burst of ATP synthesis then accompanied the disappearance of the pH gradient across the thylakoid membrane*

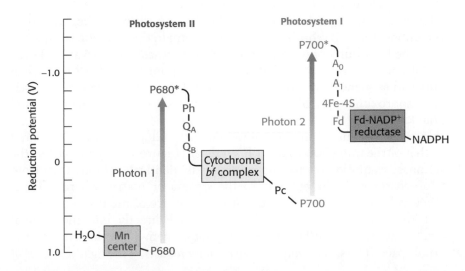

Figure 19.23 Pathway of electron flow from H$_2$O to NADP$^+$ in photosynthesis. This endergonic reaction is made possible by the absorption of light by photosystem II (P680) and photosystem I (P700). Abbreviations: Ph, pheophytin; Q$_A$ and Q$_B$, plastoquinone-binding proteins; Pc, plastocyanin; A$_0$ and A$_1$, acceptors of electrons from P700*; Fd, ferredoxin; Mn, manganese.

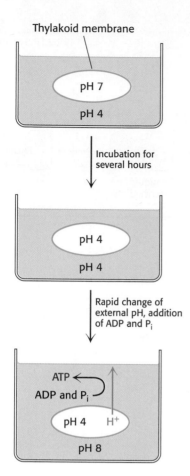

Figure 19.24 Jagendorf's demonstration. Chloroplasts synthesize ATP after the imposition of a pH gradient.

(Figure 19.24). This incisive experiment was one of the first to unequivocally support the hypothesis put forth by Peter Mitchell that ATP synthesis is driven by proton-motive force.

The principles of ATP synthesis in chloroplasts are nearly identical with those in mitochondria. *ATP formation is driven by a proton-motive force in both photophosphorylation and oxidative phosphorylation.* We have seen how light induces electron transfer through photosystems II and I and the cytochrome *bf* complex. At various stages in this process, protons are released into the thylakoid lumen or taken up from the stroma, generating a proton gradient. The gradient is maintained because the thylakoid membrane is essentially impermeable to protons. *The thylakoid space becomes markedly acidic, with the pH approaching 4. The light-induced transmembrane proton gradient is about 3.5 pH units.* As discussed in Section 18.4, energy inherent in the proton gradient, called the *proton-motive force (Δp)*, is described as the sum of two components: a charge gradient and a chemical gradient. In chloroplasts, nearly all of Δp arises from the pH gradient, whereas, in mitochondria, the contribution from the membrane potential is larger. The reason for this difference is that the thylakoid membrane is quite permeable to Cl^- and Mg^{2+}. The light-induced transfer of H^+ into the thylakoid space is accompanied by the transfer of either Cl^- in the same direction or Mg^{2+} (1 Mg^{2+} per 2 H^+) in the opposite direction. Consequently, electrical neutrality is maintained and no membrane potential is generated. A pH gradient of 3.5 units across the thylakoid membrane corresponds to a proton-motive force of 0.20 V or a ΔG of -20.0 kJ mol^{-1} (-4.8 kcal mol^{-1}).

The ATP Synthase of Chloroplasts Closely Resembles Those of Mitochondria and Prokaryotes

The proton-motive force generated by the light reactions is converted into ATP by the *ATP synthase* of chloroplasts, also called the *CF_1–CF_0 complex* (*C* stands for chloroplast and *F* for factor). CF_1–CF_0 ATP synthase closely resembles the F_1–F_0 complex of mitochondria (p. 522). CF_0 conducts protons across the thylakoid membrane, whereas CF_1 catalyzes the formation of ATP from ADP and P_i.

CF_0 is embedded in the thylakoid membrane. It consists of four different polypeptide chains known as I (17 kd), II (16.5 kd), III (8 kd), and IV (27 kd) having an estimated stoichiometry of 1:2:12:1. Subunits I and II have sequence similarity to subunit **b** of the mitochondrial F_0 subunit, III corresponds to subunit **c** of the mitochondrial complex, and subunit IV is similar in sequence to subunit **a**. CF_1, the site of ATP synthesis, has a subunit composition $\alpha_3\beta_3\gamma\delta\varepsilon$. The β subunits contain the catalytic sites, similarly to the F_1 subunit of mitochondrial ATP synthase. Remarkably, β subunits of corn chloroplast ATP synthase are more than 60% identical in amino acid sequence with those of human ATP synthase, despite the passage of approximately 1 billion years since the separation of the plant and animal kingdoms.

Note that the membrane orientation of CF_1–CF_0 is reversed compared with that of the mitochondrial ATP synthase (Figure 19.25). However, the functional orientation of the two synthases is identical: protons flow from the lumen through the enzyme to the stroma or matrix where ATP is synthesized. Because CF_1 is on the stromal surface of the thylakoid membrane, the newly synthesized ATP is released directly into the stromal space. Likewise, NADPH formed by photosystem I is released into the stromal space. Thus, *ATP and NADPH, the products of the light reactions of*

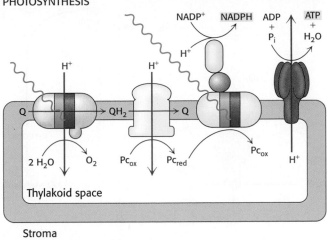

OXIDATIVE PHOSPHORYLATION

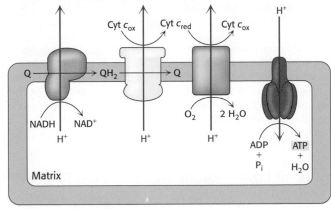

Figure 19.25 Comparison of photosynthesis and oxidative phosphorylation. The light-induced electron transfer in photosynthesis drives protons into the thylakoid lumen. The excess protons flow out of the lumen through ATP synthase to generate ATP in the stroma. In oxidative phosphorylation, electron flow down the electron-transport chain pumps protons out of the mitochondrial matrix. Excess protons from the intermembrane space flow into the matrix through ATP synthase to generate ATP in the matrix.

photosynthesis, are appropriately positioned for the subsequent dark reactions, in which CO_2 is converted into carbohydrate.

Cyclic Electron Flow Through Photosystem I Leads to the Production of ATP Instead of NADPH

On occasion, when the ratio of $NADPH$ to $NADP^+$ is very high, $NADP^+$ may be unavailable to accept electrons from reduced ferredoxin. In this case, electrons arising from P700, the reaction center of photosystem I, may take an alternative pathway that does not end at NADPH. The electron in reduced ferredoxin is transferred to the cytochrome *bf* complex rather than to $NADP^+$. This electron then flows back through the cytochrome *bf* complex to reduce plastocyanin, which can then be reoxidized by P700$^+$ to complete a cycle. The net outcome of this cyclic flow of electrons is the pumping of protons by the cytochrome *bf* complex. The resulting proton gradient then drives the synthesis of ATP. In this process, called *cyclic photophosphorylation, ATP is generated without the concomitant formation of NADPH* (Figure 19.26). Photosystem II does not participate in cyclic photophosphorylation, and so O_2 is not formed from H_2O.

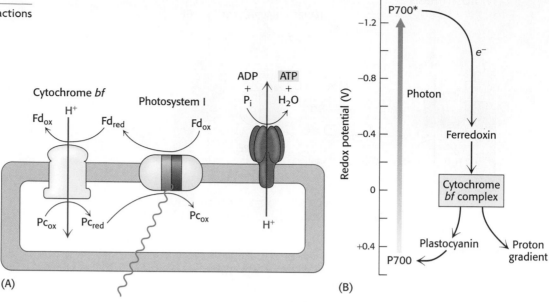

Figure 19.26 Cyclic photophosphorylation. (A) In this pathway, electrons from reduced ferredoxin are transferred to cytochrome *bf* rather than to ferredoxin-$NADP^+$ reductase. The flow of electrons through cytochrome *bf* pumps protons into the thylakoid lumen. These protons flow through ATP synthase to generate ATP. Neither NADPH nor O_2 is generated by this pathway. (B) A scheme showing the energetic basis for cyclic photophosphorylation. Abbreviations: Fd, ferredoxin; Pc, plastocyanin.

The Absorption of Eight Photons Yields One O_2, Two NADPH, and Three ATP Molecules

We can now estimate the overall stoichiometry for the light reactions. The absorption of four photons by photosystem II generates one molecule of O_2 and releases 4 protons into the thylakoid lumen. The two molecules of plastoquinol are oxidized by the Q cycle of the cytochrome *bf* complex to release 8 protons into the lumen. Finally, the electrons from four molecules of reduced plastocyanin are driven to ferredoxin by the absorption of four additional photons. The four molecules of reduced ferredoxin generate two molecules of NADPH. Thus, the overall reaction is

$$2\,H_2O + 2\,NADP^+ + 10\,H^+_{stroma} \longrightarrow O_2 + 2\,NADPH + 12\,H^+_{lumen}$$

The 12 protons released in the lumen can then flow through ATP synthase. Given that there are apparently 12 subunit III components in CF_0, we expect that 12 protons must pass through CF_0 to complete one full rotation of CF_1. A single rotation generates three molecules of ATP. Given the ratio of 3 ATP for 12 protons, the overall reaction is

$$2\,H_2O + 2\,NADP^+ + 10\,H^+_{stroma} \longrightarrow O_2 + 2\,NADPH + 12\,H^+_{lumen}$$
$$\underline{3\,ADP^{3-} + 3\,P_i^{2-} + 3\,H^+ + 12\,H^+_{lumen} \longrightarrow 3\,ATP^{4-} + 3\,H_2O + 12\,H^+_{stroma}}$$
$$2\,NADP^+ + 3\,ADP^{3-} + 3\,P_i^{2-} + H^+ \longrightarrow O_2 + 2\,NADPH + 3\,ATP^{4-} + H_2O$$

Thus, eight photons are required to yield three molecules of ATP (2.7 photons/ATP).

Cyclic photophosphorylation is a somewhat more productive way to synthesize ATP. The absorption of four photons by photosystem I leads to the release of 8 protons into the lumen by the cytochrome *bf* system. These protons flow through ATP synthase to yield two molecules of ATP. Thus, each two absorbed photons yield one molecule of ATP. No NADPH is produced.

19.5 Accessory Pigments Funnel Energy into Reaction Centers

A light-harvesting system that relied only on the chlorophyll *a* molecules of the special pair would be rather inefficient for two reasons. First, chlorophyll *a* molecules absorb light only at specific wavelengths (see Figure 19.6). A large gap is present in the middle of the visible region between approximately 450 and 650 nm. This gap falls right at the peak of the solar spectrum, and so failure to collect this light would constitute a considerable lost opportunity. Second, even many photons that do not fall in the gap pass through without being absorbed, because the density of chlorophyll *a* molecules in a reaction center is not very great. Accessory pigments, both additional chlorophylls as well as other classes of molecules, are closely associated with reaction centers. *These pigments absorb light and funnel the energy to the reaction center for conversion into chemical forms.*

Resonance Energy Transfer Allows Energy to Move from the Site of Initial Absorbance to the Reaction Center

How is energy funneled from an associated pigment to a reaction center? The absorption of a photon does not always lead to electron excitation and transfer. More commonly, excitation energy is transferred from one molecule to a nearby molecule through electromagnetic interactions through space (Figure 19.27). The rate of this process, called *resonance energy transfer,* depends strongly on the distance between the energy donor and the energy acceptor molecules; an increase in the distance between the donor and the acceptor by a factor of two typically results in a decrease in the energy-transfer rate by a factor of $2^6 = 64$. For reasons of conservation of energy, energy transfer must be from a donor in the excited state to an acceptor of equal or lower energy. *The excited state of the special pair of chlorophyll molecules is lower in energy than that of single chlorophyll molecules, allowing reaction centers to trap the energy transferred from other molecules (Figure 19.28).*

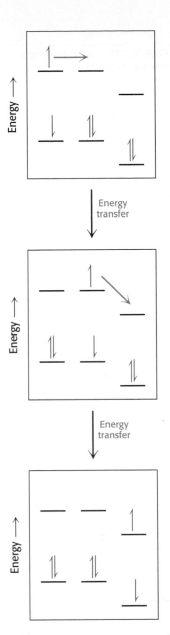

Figure 19.27 Resonance energy transfer. Energy absorbed by one molecule can be transferred to nearby molecules with excited states of equal or lower energy.

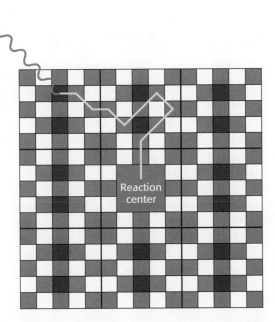

Figure 19.28 Energy transfer from accessory pigments to reaction centers. Light energy absorbed by accessory chlorophyll molecules or other pigments can be transferred to reaction centers, where it drives photoinduced charge separation. The green squares represent accessory chlorophyll molecules and the red squares represent carotenoid molecules; the white squares designate protein.

Light-Harvesting Complexes Contain Additional Chlorophylls and Carotenoids

Chlorophyll b and *carotenoids* are important light-harvesting molecules that funnel energy to the reaction center. Chlorophyll *b* differs from chlorophyll *a* in having a formyl group in place of a methyl group. This small difference shifts its two major absorption peaks toward the center of the visible region. In particular, chlorophyll *b* efficiently absorbs light with wavelengths between 450 and 500 nm (Figure 19.29).

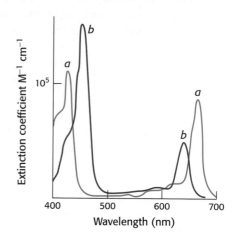

Figure 19.29 Absorption spectra of chlorophylls *a* and *b*.

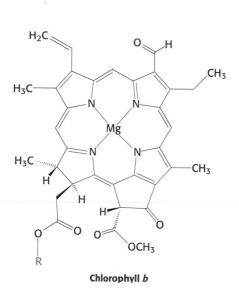

Chlorophyll *b*

Carotenoids are extended polyenes that absorb light between 400 and 500 nm. The carotenoids are responsible for most of the yellow and red colors of fruits and flowers, and they provide the brilliance of fall, when the chlorophyll molecules are degraded to reveal the carotenoids.

Lycopene

β-Carotene

In addition to their role in transferring energy to reaction centers, the carotenoids serve a safeguarding function. Carotenoids suppress damaging photochemical reactions, particularly those including oxygen that can be induced by bright sunlight. This protection may be especially important in the fall when the primary pigment chlorophyll is being degraded and thus not able to absorb light energy. Plants lacking carotenoids are quickly killed on exposure to light and oxygen.

The accessory pigments are arranged in numerous *light-harvesting complexes* that completely surround the reaction center. The 26-kd subunit of light-harvesting complex II (LHC-II) is the most abundant membrane protein in chloroplasts. This subunit binds seven chlorophyll *a* molecules, six chlorophyll *b* molecules, and two carotenoid molecules. Similar light-harvesting assemblies exist in photosynthetic bacteria (Figure 19.30).

Figure 19.30 **Structure of a bacterial light-harvesting complex.** Eight polypeptides, each of which binds three chlorophyll molecules (green) and a carotenoid molecule (red), surround a central cavity that contains the reaction center (not shown). *Notice* the high concentration of accessory pigments that surround the reaction center. [Drawn from 1LGH.pdb.]

The Components of Photosynthesis Are Highly Organized

The complexity of photosynthesis, seen already in the elaborate interplay of complex components, extends even to the placement of the components in the thylakoid membranes. *Thylakoid membranes of most plants are differentiated into stacked (appressed) and unstacked (nonappressed) regions* (see Figures 19.1 and 19.3). Stacking increases the amount of thylakoid membrane in a given chloroplast volume. Both regions surround a common internal thylakoid space, but only unstacked regions make direct contact with the chloroplast stroma. Stacked and unstacked regions differ in the nature of their photosynthetic assemblies (Figure 19.31). Photosystem I and ATP synthase are located almost exclusively in unstacked regions, whereas photosystem II is present mostly in stacked regions. The cytochrome *bf* complex is found in both regions. Indeed, this complex rapidly moves back and forth between the stacked and the unstacked regions. Plastoquinone and

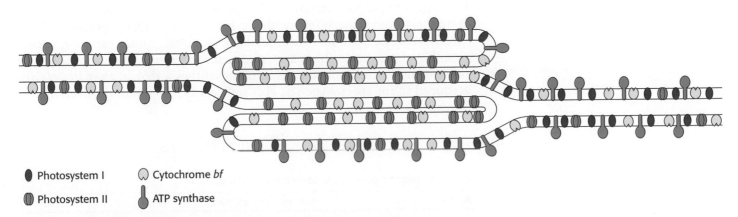

● Photosystem I   ◖ Cytochrome *bf*

▥ Photosystem II   ⬤ ATP synthase

Figure 19.31 **Location of photosynthesis components.** Photosynthetic assemblies are differentially distributed in the stacked (appressed) and unstacked (nonappressed) regions of thylakoid membranes. [After a drawing kindly provided by Dr. Jan M. Anderson and Dr. Bertil Andersson.]

Diuron

Atrazine

plastocyanin are the mobile carriers of electrons between assemblies located in different regions of the thylakoid membrane. A common internal thylakoid space enables protons liberated by photosystem II in stacked membranes to be utilized by ATP synthase molecules that are located far away in unstacked membranes.

What is the functional significance of this lateral differentiation of the thylakoid membrane system? The positioning of photosystem I in the unstacked membranes also gives it direct access to the stroma for the reduction of $NADP^+$. ATP synthase, too, is located in the unstacked region to provide space for its large CF_1 globule and to give access to ADP. In contrast, the tight quarters of the appressed region pose no problem for photosystem II, which interacts with a small polar electron donor (H_2O) and a highly lipid soluble electron carrier (plastoquinone).

Many Herbicides Inhibit the Light Reactions of Photosynthesis

Many commercial herbicides kill weeds by interfering with the action of photosystem II or photosystem I. Inhibitors of photosystem II block electron flow, whereas inhibitors of photosystem I divert electrons from the terminal part of this photosystem. Photosystem II inhibitors include urea derivatives such as *diuron* and triazine derivatives such as *atrazine*. These chemicals bind to the Q_B site of the D1 subunit of photosystem II and block the formation of plastoquinol (QH_2).

Paraquat (1,1'-dimethyl-4-4'-bipyridinium) is an inhibitor of photosystem I. Paraquat, a dication, can accept electrons from photosystem I to become a radical. This radical reacts with O_2 to produce reactive oxygen species such as superoxide ($O_2^{\cdot-}$) and hydroxyl radical ($OH\cdot$). Such reactive oxygen species react with double bonds in membrane lipids, damaging the membrane (p. 517).

19.6 The Ability to Convert Light into Chemical Energy Is Ancient

The ability to convert light energy into chemical energy is a tremendous evolutionary advantage. Geological evidence suggests that oxygenic photosynthesis became important approximately 2 billion years ago. Anoxygenic photosynthetic systems arose much earlier in the 3.5-billion-year history of life on Earth (Table 19.1). The photosynthetic system of the nonsulfur purple bacterium *Rhodopseudomonas viridis* has most features common to oxygenic photosynthetic systems and clearly predates them. Green sulfur bacteria such as *Chlorobium thiosulfatophilum* carry out a reaction that also seems to have appeared before oxygenic photosynthesis and is even more similar to oxygenic photosynthesis than that of *R. viridis*. Reduced sulfur species such as H_2S are electron donors in the overall photosynthetic reaction:

$$CO_2 + 2\,H_2S \xrightarrow{\text{Light}} (CH_2O) + 2\,S + H_2O$$

TABLE 19.1 Major groups of photosynthetic prokaryotes

Bacteria	Photosynthetic electron donor	O_2 use
Green sulfur	H_2, H_2S, S	Anoxygenic
Green nonsulfur	Variety of amino acids and organic acids	Anoxygenic
Purple sulfur	H_2, H_2S, S	Anoxygenic
Purple nonsulfur	Usually organic molecules	Anoxygenic
Cyanobacteria	H_2O	Oxygenic

Nonetheless, photosynthesis did not evolve immediately at the origin of life. No photosynthetic organisms have been discovered in the domain of Archaea, implying that photosynthesis evolved in the domain of Bacteria after Archaea and Bacteria diverged from a common ancestor. All domains of life do have electron-transport chains in common, however. As we have seen, components such as the ubiquinone–cytochrome c oxidoreductase and cytochrome bf family are present in both respiratory and photosynthetic electron-transport chains. These components were the foundations on which light-energy-capturing systems evolved.

Summary

19.1 Photosynthesis Takes Place in Chloroplasts

The proteins that participate in the light reactions of photosynthesis are located in the thylakoid membranes of chloroplasts. The light reactions result in (1) the creation of reducing power for the production of NADPH, (2) the generation of a transmembrane proton gradient for the formation of ATP, and (3) the production of O_2.

19.2 Light Absorption by Chlorophyll Induces Electron Transfer

Chlorophyll molecules—tetrapyrroles with a central magnesium ion—absorb light quite efficiently because they are polyenes. An electron excited to a high-energy state by the absorption of a photon can move to nearby electron acceptors. In photosynthesis, an excited electron leaves a pair of associated chlorophyll molecules known as the special pair. The functional core of photosynthesis, a reaction center, from a photosynthetic bacterium has been studied in great detail. In this system, the electron moves from the special pair (containing bacteriochlorophyll) to a bacteriopheophytin (a bacteriochlorophyll lacking the central magnesium ion) to quinones. The reduction of quinones leads to the generation of a proton gradient, which drives ATP synthesis in a manner analogous to that of oxidative phosphorylation.

19.3 Two Photosystems Generate a Proton Gradient and NADPH in Oxygenic Photosynthesis.

Photosynthesis in green plants is mediated by two linked photosystems. In photosystem II, the excitation of a special pair of chlorophyll molecules called P680 leads to electron transfer to plastoquinone in a manner analogous to that of the bacterial reaction center. The electrons are replenished by the extraction of electrons from a water molecule at a center containing four manganese ions. One molecule of O_2 is generated at this center for each four electrons transferred. The plastoquinol produced at photosystem II is reoxidized by the cytochrome bf complex, which transfers the electrons to plastocyanin, a soluble copper protein. From plastocyanin, the electrons enter photosystem I. In photosystem I, the excitation of special pair P700 releases electrons that flow to ferredoxin, a powerful reductant. Ferredoxin-NADP$^+$ reductase, a flavoprotein located on the stromal side of the membrane, then catalyzes the formation of NADPH. A proton gradient is generated as electrons pass through photosystem II, through the cytochrome bf complex, and through ferredoxin-NADP$^+$ reductase.

19.4 A Proton Gradient Across the Thylakoid Membrane Drives ATP Synthesis

The proton gradient across the thylakoid membrane creates a proton-motive force, used by ATP synthase to form ATP. The ATP synthase of chloroplasts (also called CF_0–CF_1) closely resembles the ATP-synthesizing assemblies of bacteria and mitochondria (F_0–F_1). If the $NADPH:NADP^+$ ratio is high, electrons transferred to ferredoxin by photosystem I can reenter the cytochrome *bf* complex. This process, called cyclic photophosphorylation, leads to the generation of a proton gradient by the cytochrome *bf* complex without the formation of NADPH or O_2.

19.5 Accessory Pigments Funnel Energy into Reaction Centers

Light-harvesting complexes that surround the reaction centers contain additional molecules of chlorophyll *a*, as well as carotenoids and chlorophyll *b* molecules, which absorb light in the center of the visible spectrum. These accessory pigments increase the efficiency of light capture by absorbing light and transferring the energy to reaction centers through resonance energy transfer.

19.6 The Ability to Convert Light into Chemical Energy Is Ancient

The photosystems have structural features in common that suggest a common evolutionary origin. Similarities in organization and molecular structure to those of oxidative phosphorylation suggest that the photosynthetic apparatus evolved from an early energy-transduction system.

Key Terms

light reactions (p. 541)

chloroplast (p. 542)

stroma (p. 543)

thylakoid (p. 543)

granum (p. 543)

chlorophyll *a* (p. 543)

photoinduced charge separation (p. 545)

reaction center (p. 545)

special pair (p. 546)

P960 (p. 546)

photosystem I (PS I) (p. 548)

photosystem II (PS II) (p. 548)

P680 (p. 549)

manganese center (p. 549)

cytochrome *bf* (p. 551)

P700 (p. 552)

Z scheme of photosynthesis (p. 553)

proton-motive force (Δp) (p. 554)

ATP synthase (CF_1–CF_0 complex) (p. 554)

cyclic photophosphorylation (p. 555)

carotenoid (p. 558)

light-harvesting complex (p. 558)

Selected Readings

Where to Start

Huber, R. 1989. A structural basis of light energy and electron transfer in biology. *EMBO J.* 8:2125–2147.

Deisenhofer, J., and Michel, H. 1989. The photosynthetic reaction centre from the purple bacterium *Rhodopseudomonas viridis*. *EMBO J.* 8:2149–2170.

Barber, J., and Andersson, B. 1994. Revealing the blueprint of photosynthesis. *Nature* 370:31–34.

Books and General Reviews

Raghavendra, A. S. 1998. *Photosynthesis: A Comprehensive Treatise.* Cambridge University Press.

Cramer, W. A., and Knaff, D. B. 1991. *Energy Transduction in Biological Membranes: A Textbook of Bioenergetics.* Springer Verlag.

Nicholls, D. G., and Ferguson, S. J. 2002. *Bioenergetics* (3rd ed.). Academic Press.

Harold, F. M. 1986. *The Vital Force: A Study of Bioenergetics.* W. H. Freeman and Company.

Electron-Transfer Mechanisms

Beratan, D., and Skourtis, S. 1998. Electron transfer mechanisms. *Curr. Opin. Chem. Biol.* 2:235–243.

Moser, C. C., Keske, J. M., Warncke, K., Farid, R. S., and Dutton, P. L. 1992. Nature of biological electron transfer. *Nature* 355:796–802.

Boxer, S. G. 1990. Mechanisms of long-distance electron transfer in proteins: Lessons from photosynthetic reaction centers. *Annu. Rev. Biophys. Biophys. Chem.* 19:267–299.

Photosystem II

Kirchhoff, H., Tremmel, I., Haase, W., and Kubitscheck, U. 2004. Supramolecular photosystem II organization in grana of thylakoid membranes: Evidence for a structured arrangement. *Biochemistry* 43:9204–9213.

Diner, B. A., and Rappaport, F. 2002. Structure, dynamics, and energetics of the primary photochemistry of photosystem II of oxygenic photosynthesis. *Annu. Rev. Plant Biol.* 54:551–580.

Zouni, A., Witt, H. T., Kern, J., Fromme, P., Krauss, N., Saenger, W., and Orth, P. 2001. Crystal structure of photosystem II from *Synechococcus elongatus* at 3.8 Å resolution. *Nature* 409:739–743.

Rhee, K. H. 2001. Photosystem II: The solid structural era. *Annu. Rev. Biophys. Biomolec. Struct.* 30:307–328.

Rhee, K. H., Morris, E. P., Barber, J., and Kuhlbrandt, W. 1998. Three-dimensional structure of the plant photosystem II reaction centre at 8 Å resolution. *Nature* 396:283–286.

Morris, E. P., Hankamer, B., Zheleva, D., Friso, G., and Barber, J. 1997. The three-dimensional structure of a photosystem II core complex determined by electron crystallography. *Structure* 5:837–849.

Deisenhofer, J., and Michel, H. 1991. High-resolution structures of photosynthetic reaction centers. *Annu. Rev. Biophys. Biophys. Chem.* 20:247–266.

Oxygen Evolution

Ferreira, K. N., Iverson, T. M., Maghlaoui, K., Barber, J., and Iwata, S. 2004. Architecture of the photosynthetic oxygen-evolving center. *Science* 303:1831–1838.

Hoganson, C. W., and Babcock, G. T. 1997. A metalloradical mechanism for the generation of oxygen from water in photosynthesis. *Science* 277:1953–1956.

Yamachandra, V. K., DeRose, V. J., Latimer, M. J., Mukerji, I., Sauer, K., and Klein, M. P. 1993. Where plants make oxygen: A structural model for the photosynthetic oxygen-evolving manganese complex. *Science* 260:675–679.

Brudvig, G. W., Beck, W. F., and de Paula, J. C. 1989. Mechanism of photosynthetic water oxidation. *Annu. Rev. Biophys. Biophys. Chem.* 18:25–46.

Peloquin, J. M., and Britt, R. D. 2001. EPR/ENDOR characterization of the physical and electronic structure of the OEC Mn cluster. *Biochim. Biophys. Acta* 1503:96–111.

Photosystem I and Cytochrome *bf*

Cramer, W. A., Zhang, H., Yan, J., Kurisu, G., and Smith, J. L. 2004. Evolution of photosynthesis: Time-independent structure of the cytochrome b_6f complex. *Biochemistry* 43:5921–5929.

Kargul, J., Nield, J., and Barber, J. 2003. Three-dimensional reconstruction of a light-harvesting complex I-photosystem I (LHCI-PSI) supercomplex from the green alga *Chlamydomonas reinhardtii*. *J. Biol. Chem.* 278:16135–16141.

Schubert, W. D., Klukas, O., Saenger, W., Witt, H. T., Fromme, P., and Krauss, N. 1998. A common ancestor for oxygenic and anoxygenic photosynthetic systems: A comparison based on the structural model of photosystem I. *J. Mol. Biol.* 280:297–314.

Fotiadis, D., Muller, D. J., Tsiotis, G., Hasler, L., Tittmann, P., Mini, T., Jeno, P., Gross, H., and Engel, A. 1998. Surface analysis of the photosystem I complex by electron and atomic force microscopy. *J. Mol. Biol.* 283:83–94.

Klukas, O., Schubert, W. D., Jordan, P., Krauss, N., Fromme, P., Witt, H. T., and Saenger, W. 1999. Photosystem I, an improved model of the stromal subunits PsaC, PsaD, and PsaE. *J. Biol. Chem.* 274:7351–7360.

Jensen, P. E., Gilpin, M., Knoetzel, J., and Scheller, H. V. 2000. The PSI-K subunit of photosystem I is involved in the interaction between light-harvesting complex I and the photosystem I reaction center core. *J. Biol. Chem.* 275:24701–24708.

Kitmitto, A., Mustafa, A. O., Holzenburg, A., and Ford, R. C. 1998. Three-dimensional structure of higher plant photosystem I determined by electron crystallography. *J. Biol. Chem.* 273:29592–29599.

Krauss, N., Hinrichs, W., Witt, I., Fromme, P., Pritzkow, W., Dauter, Z., Betzel, C., Wilson, K. S., Witt, H. T., and Saenger, W. 1993. Three-dimensional structure of system I photosynthesis at 6 Å resolution. *Nature* 361:326–331.

Malkin, R. 1992. Cytochrome bc_1 and b_6f complexes of photosynthetic membranes. *Photosynth. Res.* 33:121–136.

Karplus, P. A., Daniels, M. J., and Herriott, J. R. 1991. Atomic structure of ferredoxin-NADP$^+$ reductase: Prototype for a structurally novel flavoenzyme family. *Science* 251:60–66.

ATP Synthase

Richter, M. L., Hein, R., and Huchzermeyer, B. 2000. Important subunit interactions in the chloroplast ATP synthase. *Biochim. Biophys. Acta* 1458:326–329.

Oster, G., and Wang, H. 1999. ATP synthase: Two motors, two fuels. *Structure* 7:R67–R72.

Junge, W., Lill, H., and Engelbrecht, S. 1997. ATP synthase: An electrochemical transducer with rotatory mechanics. *Trends Biochem. Sci.* 22:420–423.

Weber, J., and Senior, A. E. 2000. ATP synthase: What we know about ATP hydrolysis and what we do not know about ATP synthesis. *Biochim. Biophys. Acta* 1458:300–309.

Light-Harvesting Assemblies

Conroy, M. J., Westerhuis, W. H., Parkes-Loach, P. S., Loach, P. A., Hunter, C. N., and Williamson, M. P. 2000. The solution structure of *Rhodobacter sphaeroides* LH1β reveals two helical domains separated by a more flexible region: Structural consequences for the LH1 complex. *J. Mol. Biol.* 298:83–94.

Koepke, J., Hu, X., Muenke, C., Schulten, K., and Michel, H. 1996. The crystal structure of the light-harvesting complex II (B800–850) from *Rhodospirillum molischianum*. *Structure* 4:581–597.

Grossman, A. R., Bhaya, D., Apt, K. E., and Kehoe, D. M. 1995. Light-harvesting complexes in oxygenic photosynthesis: Diversity, control, and evolution. *Annu. Rev. Genet.* 29:231–288.

Kühlbrandt, W., Wang, D.-N., and Fujiyoshi, Y. 1994. Atomic model of plant light-harvesting complex by electron crystallography. *Nature* 367:614–621.

Glazer, A. N. 1983. Comparative biochemistry of photosynthetic light-harvesting systems. *Annu. Rev. Biochem.* 52:125–157.

Evolution

Cavalier-Smith, T. 2002. Chloroplast evolution: Secondary symbiogenesis and multiple losses. *Curr. Biol.* 12:R62–64.

Green, B. R. 2001. Was "molecular opportunism" a factor in the evolution of different photosynthetic light-harvesting pigment systems? *Proc. Natl. Acad. Sci. U. S. A.* 98:2119–2121.

Dismukes, G. C., Klimov, V. V., Baranov, S. V., Nozlov, Y. N., DasGupta, J., and Tyryshkin, A. 2001. The origin of atmospheric oxygen on Earth: The innovation of oxygenic photosynthesis. *Proc. Natl. Acad. Sci. U. S. A.* 98:2170–2175.

Moreira, D., Le Guyader, H., and Phillippe, H. 2000. The origin of red algae and the evolution of chloroplasts. *Nature* 405:69–72.

Cavalier-Smith, T. 2000. Membrane heredity and early chloroplast evolution. *Trends Plant Sci.* 5:174–182.

Blankenship, R. E., and Hartman, H. 1998. The origin and evolution of oxygenic photosynthesis. *Trends Biochem. Sci.* 23:94–97.

Problems

1. *Electron transfer.* Calculate the $\Delta E_0'$ and $\Delta G^{\circ\prime}$ for the reduction of NADP$^+$ by ferredoxin. Use data given in Table 18.1.

2. *To boldly go.* (a) It can be argued that, if life were to exist elsewhere in the universe, it would require some process like photosynthesis. Why is this argument reasonable? (b) If the *Enterprise* were to land on a distant plant and find no measurable oxygen in the atmosphere, could the crew conclude that photosynthesis is not taking place?

3. *Weed killer 1.* Dichlorophenyldimethylurea (DCMU), a herbicide, interferes with photophosphorylation and O$_2$ evolution.

However, it does not block O_2 evolution in the presence of an artificial electron acceptor such as ferricyanide. Propose a site for the inhibitory action of DCMU.

4. *Weed killer 2.* Predict the effect of the herbicide dichlorophenyldimethylurea (DCMU) on a plant's ability to perform cyclic photophosphorylation.

5. *Infrared harvest.* Consider the relation between the energy of a photon and its wavelength.

(a) Some bacteria are able to harvest 1000-nm light. What is the energy (in kilojoules or kilocalories) of a mole (also called an einstein) of 1000-nm photons?
(b) What is the maximum increase in redox potential that can be induced by a 1000-nm photon?
(c) What is the minimum number of 1000-nm photons needed to form ATP from ADP and P_i? Assume a ΔG of 50 kJ mol^{-1} (12 kcal mol^{-1}) for the phosphorylation reaction.

6. *Missing acceptors.* Suppose that a bacterial reaction center containing only the special pair and the quinones has been prepared. Given the separation of 22 Å between the special pair and the closest quinone, estimate the rate of electron transfer between the excited special pair and this quinone.

7. *Close approach.* Suppose that energy transfer between two chlorophyll *a* molecules separated by 10 Å takes place in 10 picoseconds. Suppose that this distance is increased to 20 Å with all other factors remaining the same. How long would energy transfer take?

Mechanism Problem

8. *Hill reaction.* In 1939, Robert Hill discovered that chloroplasts evolve O_2 when they are illuminated in the presence of an artificial electron acceptor such as ferricyanide $[Fe^{3+}(CN)_6]^{3-}$. Ferricyanide is reduced to ferrocyanide $[Fe^{2+}(CN)_6]^{4-}$ in this process. No NADPH or reduced plastocyanin is produced. Propose a mechanism for the Hill reaction.

Data Interpretation and Chapter Integration Problem

9. *The same, but different.* The $\alpha_3\beta_3\gamma$ complex of mitochondrial or chloroplast ATP synthase will function as an ATPase in vitro. The chloroplast enzyme (both synthase and ATPase activity) is sensitive to redox control, whereas the mitochondrial enzyme is not. To determine where the enzymes differ, a segment of the mitochondrial γ subunit was removed and replaced with the equivalent segment from the chloroplast γ subunit. The ATPase activity of the modified enzyme was then measured as a function of redox conditions.

(a) What is the redox regulator of the ATP synthase in vivo? The adjoining graph shows the ATPase activity of modified and control enzymes under various redox conditions.

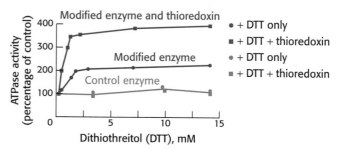

[Data from O. Bald et al. *J. Biol. Chem.* 275(2000):12757–12762.]

(b) What is the effect of increasing the reducing power of the reaction mixture for the control and the modified enzymes?
(c) What is the effect of the addition of thioredoxin? How do these results differ from those in the presence of DTT alone? Suggest a possible explanation for the difference.
(d) Did the researchers succeed in identifying the region of the γ subunit responsible for redox regulation?
(e) What is the biological rationale of regulation by high concentrations of reducing agents?
(f) What amino acids in the γ subunit are most likely affected by the reducing conditions?
(g) What experiments might confirm your answer to part *e*?

The Calvin Cycle and the Pentose Phosphate Pathway

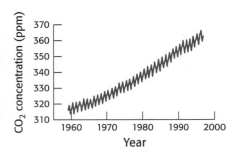

Atmospheric carbon dioxide measurements at Mauna Loa, Hawaii. These measurements show annual cycles resulting from seasonal variation in carbon dioxide fixation by the Calvin cycle in terrestrial plants. Much of this fixation takes place in rain forests, which account for approximately 50% of terrestrial fixation. [Dennis Potokar/Photo Researchers.]

Photosynthesis proceeds in two parts: the light reactions and the dark reactions. The light reactions, discussed in Chapter 19, transform light energy into ATP and biosynthetic reducing power, NADPH. The dark reactions use the ATP and NADPH produced by the light reactions to reduce carbon atoms from their fully oxidized state as carbon dioxide to a more reduced state as a hexose. Carbon dioxide is thereby trapped in a form that is useful for many processes but most especially as a fuel. Together, *the light reactions and dark reactions of photosynthesis cooperate to transform light energy into carbon fuel.* The dark reactions are also called the *Calvin cycle,* after Melvin Calvin, the biochemist who elucidated the pathway. The components of the Calvin cycle are called the dark reactions because, in contrast with the light reactions, these reactions do not directly depend on the presence of light.

The second half of this chapter examines a pathway common to all organisms, known variously as the pentose phosphate pathway, the hexose monophosphate pathway, the phosphogluconate pathway, or the pentose shunt. The pathway provides a means by which glucose can be oxidized to generate NADPH, *the currency of readily available reducing power in cells.* The phosphoryl group on the 2′-hydroxyl group of one of the ribose units of NADPH distinguishes NADPH from NADH. *There is a fundamental distinction between NADPH and NADH in biochemistry: NADH is oxidized by the respiratory chain to generate ATP, whereas NADPH serves as a reductant in biosynthetic processes.* The pentose phosphate pathway can also be used for the catabolism of pentose sugars from the diet, the synthesis of pentose sugars for nucleotide biosynthesis, and the catabolism and synthesis of less common four- and seven-carbon sugars. The pentose phosphate pathway and the Calvin cycle have in common several enzymes and intermediates that attest to an evolutionary kinship. Like glycolysis and gluconeogenesis, these pathways are mirror images of each other: the Calvin cycle uses NADPH to reduce carbon dioxide to generate hexoses, whereas the pentose phosphate pathway breaks down glucose into carbon dioxide to generate NADPH.

20.1   The Calvin Cycle Synthesizes Hexoses from Carbon Dioxide and Water

As stated in Chapter 16, glucose can be formed from noncarbohydrate precursors, such as lactate and amino acids, by gluconeogenesis. The energy powering gluconeogenesis ultimately comes from previous catabolism of carbon fuels. In contrast, photosynthetic organisms can use the Calvin cycle to synthesize glucose from carbon dioxide gas and water, by using sunlight as an energy source. The Calvin cycle introduces into life all of the carbon atoms that will be used as fuel and as the carbon backbones of biomolecules. Photosynthetic organisms are called *autotrophs* (literally "self-feeders") because they can convert sunlight into chemical energy, which they subsequently use to power their biosynthetic processes. Organisms that obtain energy from chemical fuels only are called *heterotrophs,* and such organisms ultimately depend on autotrophs for their fuel.

The Calvin cycle comprises three stages (Figure 20.1):

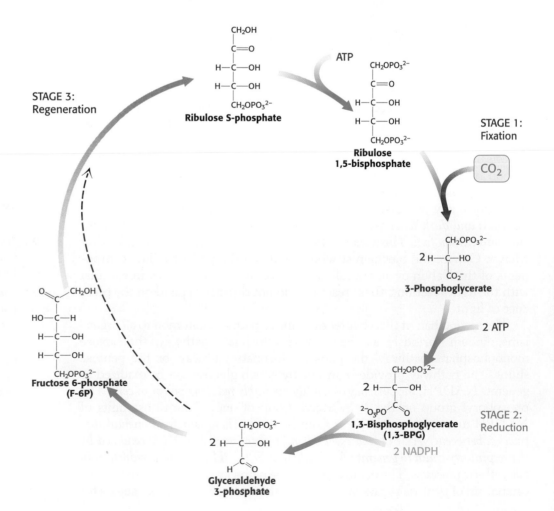

Figure 20.1 Calvin cycle. The Calvin cycle consists of three stages. Stage 1 is the fixation of carbon by the carboxylation of ribulose 1,5-bisphosphate. Stage 2 is the reduction of the fixed carbon to begin the synthesis of hexose. Stage 3 is the regeneration of the starting compound, ribulose 1,5- bisphosphate.

1. the fixation of CO_2 by ribulose 1,5-bisphosphate to form two molecules of 3-phosphoglycerate;

2. the reduction of 3-phosphoglycerate to form hexose sugars; and

3. the regeneration of ribulose 1,5-bisphosphate so that more CO_2 can be fixed.

This set of reactions takes place in the stroma of chloroplasts, the photosynthetic organelles.

Carbon Dioxide Reacts with Ribulose 1,5-bisphosphate to Form Two Molecules of 3-Phosphoglycerate

The first step in the Calvin cycle is the fixation of CO_2. The CO_2 molecule condenses with ribulose 1,5-bisphosphate to form an unstable six-carbon compound, which is rapidly hydrolyzed to two molecules of 3-phosphoglycerate.

The initial incorporation of CO_2 into 3-phosphoglycerate was revealed through the use of a carbon-14 radioactive tracer (Figure 20.2). This highly exergonic reaction $[\Delta G^{\circ\prime} = -51.9 \text{ kJ mol}^{-1} (-12.4 \text{ kcal mol}^{-1})]$ is catalyzed by *ribulose 1,5-bisphosphate carboxylase/oxygenase* (usually called *rubisco*), an enzyme located on the stromal surface of the thylakoid membranes of chloroplasts. This important reaction is the rate-limiting step in hexose synthesis. Rubisco in chloroplasts consists of eight large (L, 55-kd) subunits and eight small (S, 13-kd) ones (Figure 20.3). Each L chain contains a catalytic site and a regulatory site. The S chains enhance the catalytic activity of the L chains. This enzyme is very abundant in chloroplasts, accounting for approximately 30% of the total leaf protein in some plants. In fact, rubisco is the most abundant enzyme and probably the most abundant protein in the biosphere. Large amounts are present because rubisco is a slow enzyme; its maximal catalytic rate is only 3 s^{-1}.

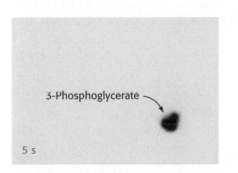

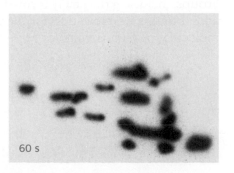

Figure 20.2 Tracing the fate of carbon dioxide. Radioactivity from $^{14}CO_2$ is incorporated into 3-phosphoglycerate within 5 s in irradiated cultures of algae. After 60 s, the radioactivity appears in many compounds, the intermediates within the Calvin cycle. [Courtesy of Dr. J. A. Bassham.]

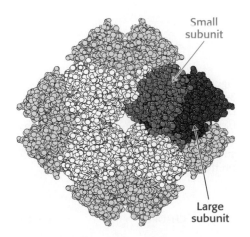

Figure 20.3 Structure of rubisco. The enzyme ribulose 1,5-bisphosphate carboxylase/oxygenase (rubisco) comprises eight large subunits (one shown in red and the others in yellow) and eight small subunits (one shown in blue and the others in white). The active sites lie in the large subunits. [Drawn from 1RXO.pdb.]

Lysine side chain

Carbamate

Rubisco Activity Depends on Magnesium and Carbamate

Rubisco requires a bound divalent metal ion for activity, usually magnesium ion. Like the zinc ion in the active site of carbonic anhydrase (p. 256), this metal ion serves to activate a bound substrate molecule by stabilizing a negative charge. Interestingly, a CO_2 molecule other than the substrate is required to complete the assembly of the Mg^{2+}-binding site in rubisco. This CO_2 molecule adds to the uncharged ε-amino group of lysine 201 to form a *carbamate*. This negatively charged adduct then binds the Mg^{2+} ion. The formation of the carbamate is facilitated by the enzyme *rubisco activase*, although it will also form spontaneously at a lower rate.

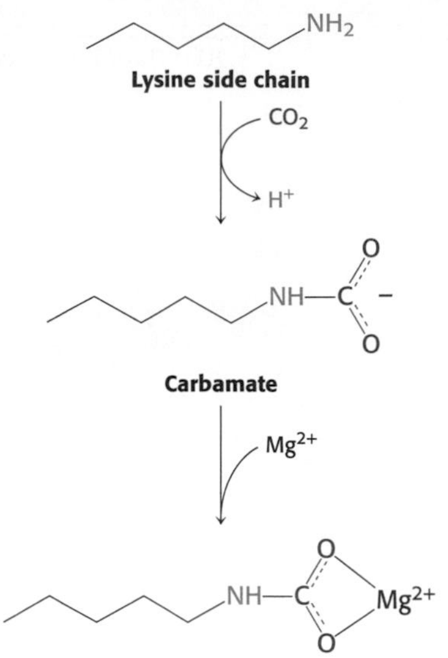

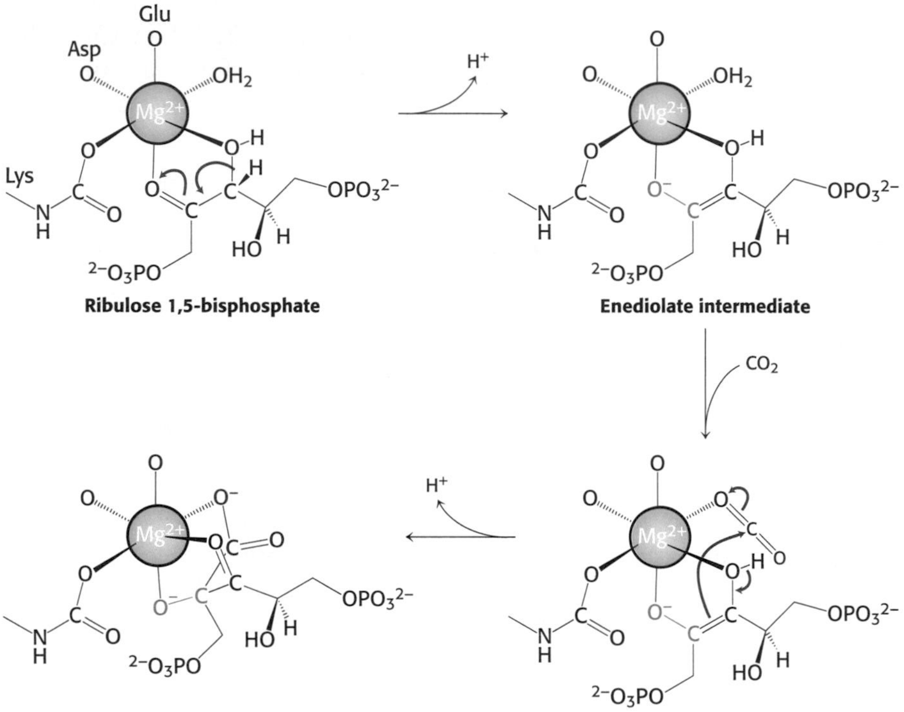

Ribulose 1,5-bisphosphate

Enediolate intermediate

Figure 20.4 Role of the magnesium ion in the rubisco mechanism. Ribulose 1,5-bisphosphate binds to a magnesium ion that is linked to rubisco through a glutamate residue, an aspartate residue, and the lysine carbamate. The coordinated ribulose 1,5-bisphosphate gives up a proton to form a reactive enediolate species that reacts with CO_2 to form a new carbon–carbon bond.

The metal center plays a key role in binding ribulose 1,5-bisphosphate and activating it so that it will react with CO_2 (Figure 20.4). Ribulose 1,5-bisphosphate binds to Mg^{2+} through its keto group and an adjacent hydroxyl group. This complex is readily deprotonated to form an enediolate intermediate. This reactive species, analogous to the zinc–hydroxide species in carbonic anhydrase, couples with CO_2, forming the new carbon–carbon bond. The resulting product is coordinated to the Mg^{2+} ion through three groups, including the newly formed carboxylate. A molecule of H_2O is then added to this β-ketoacid to form an intermediate that cleaves to form two molecules of 3-phosphoglycerate (Figure 20.5).

Figure 20.5 Formation of 3-phosphoglycerate. The overall pathway for the conversion of ribulose 1,5 bisphosphate and CO_2 into two molecules of 3-phosphoglycerate. Although the free species are shown, these steps take place on the magnesium ion.

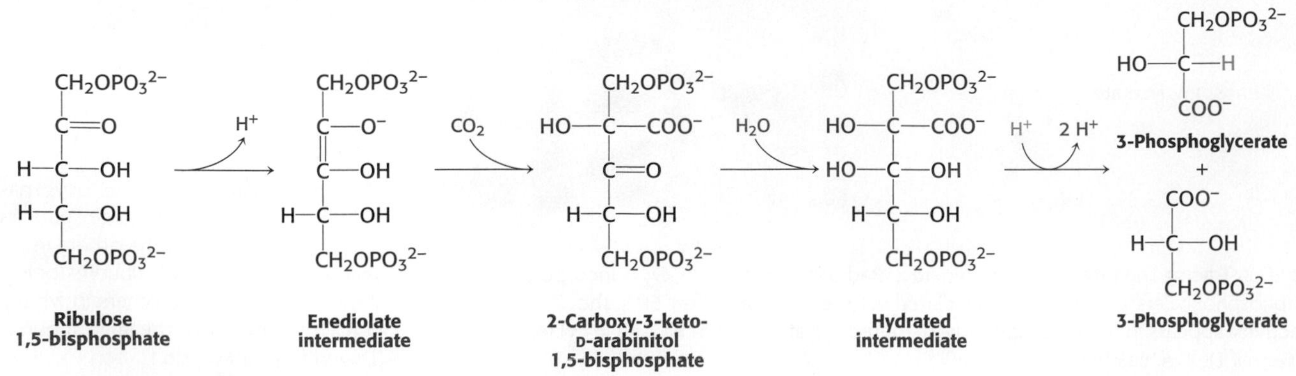

Ribulose 1,5-bisphosphate — Enediolate intermediate — 2-Carboxy-3-keto-D-arabinitol 1,5-bisphosphate — Hydrated intermediate — 3-Phosphoglycerate

Rubisco Also Catalyzes a Wasteful Oxygenase Reaction: Catalytic Imperfection

The reactive intermediate generated on the Mg^{2+} ion sometimes reacts with O_2 instead of CO_2. Thus, rubisco also catalyzes a deleterious oxygenase reaction. The products of this reaction are *phosphoglycolate* and *3-phosphoglycerate* (Figure 20.6). The rate of the carboxylase reaction is four times that of the oxygenase reaction under normal atmospheric conditions at 25°C; the stromal concentration of CO_2 is then 10 μM and that of O_2 is 250 μM. The oxygenase reaction, like the carboxylase reaction, requires that lysine 201 be in the carbamate form. Because this carbamate forms only in the presence of CO_2, rubisco is prevented from catalyzing the oxygenase reaction exclusively when CO_2 is absent.

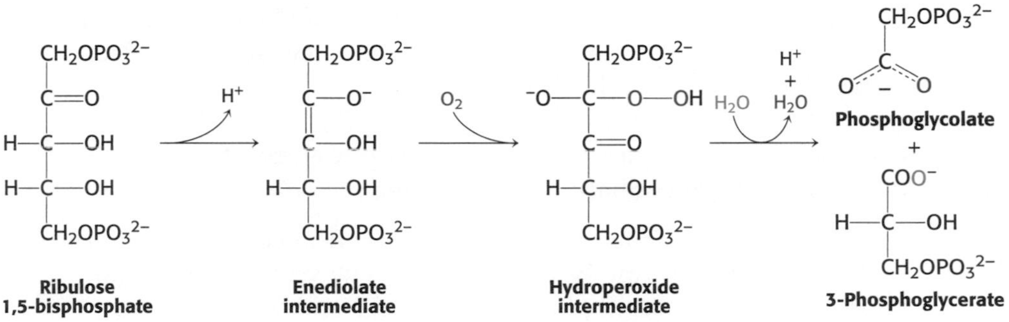

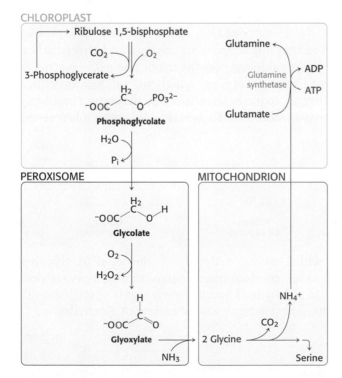

Figure 20.6 A wasteful side reaction. The reactive enediolate intermediate on rubisco also reacts with molecular oxygen to form a hydroperoxide intermediate, which then proceeds to form one molecule of 3-phosphoglycerate and one molecule of phosphoglycolate.

Phosphoglycolate is not a versatile metabolite. A salvage pathway recovers part of its carbon skeleton (Figure 20.7). A specific phosphatase converts phosphoglycolate into *glycolate*, which enters *peroxisomes* (also called *microbodies*; Figure 20.8). Glycolate is then oxidized to *glyoxylate* by

Figure 20.7 Photorespiratory reactions. Phosphoglycolate is formed as a product of the oxygenase reaction in chloroplasts. After dephosphorylation, glycolate is transported into peroxisomes where it is converted into glyoxylate and then glycine. In mitochondria, two glycines are converted into serine, after losing a carbon as CO_2 and ammonium ion. The ammonium ion is salvaged in chloroplasts.

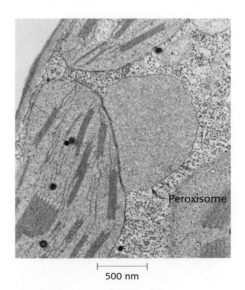

Figure 20.8 Electron micrograph of a peroxisome nestled between two chloroplasts. [Courtesy of Dr. Sue Ellen Frederick.]

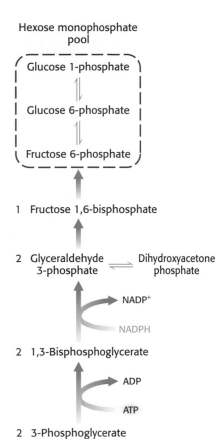

Figure 20.9 Hexose phosphate formation.
3-Phosphoglycerate is converted into
fructose 6-phosphate in a pathway parallel
to that of gluconeogenesis.

glycolate oxidase, an enzyme with a flavin mononucleotide prosthetic group. The H_2O_2 produced in this reaction is cleaved by catalase to H_2O and O_2. Transamination of glyoxylate then yields *glycine*. Two glycine molecules can unite to form serine, a potential precursor of glucose, with the release of CO_2 and ammonium ion (NH_4^+). The ammonium ion, used in the synthesis of nitrogen-containing compounds, is salvaged by a glutamine synthetase reaction (see Figure 20.7 and p. 661).

This salvage pathway serves to recycle three of the four carbon atoms of two molecules of glycolate. However, one carbon atom is lost as CO_2. This process is called *photorespiration* because O_2 is consumed and CO_2 is released. Photorespiration is wasteful because organic carbon is converted into CO_2 without the production of ATP, NADPH, or another energy-rich metabolite. Evolutionary processes have presumably enhanced the preference of rubisco for carboxylation. For instance, the rubisco of higher plants is eightfold as specific for carboxylation as that of photosynthetic bacteria. However, some oxygenase activity may be an inevitable side effect of the carboxylase reaction mechanism.

Hexose Phosphates Are Made from Phosphoglycerate, and Ribulose 1,5-bisphosphate Is Regenerated

The 3-phosphoglycerate product of rubisco is next converted into fructose 6-phosphate, which readily interconverts between its isomers glucose 1-phosphate and glucose 6-phosphate (pp. 450 and 595). The steps in this conversion (Figure 20.9) are like those of the gluconeogenic pathway (p. 458), except that glyceraldehyde 3-phosphate dehydrogenase in chloroplasts, which generates glyceraldehyde 3-phosphate (GAP), is specific for NADPH rather than NADH. These reactions and that catalyzed by rubisco bring CO_2 to the level of a hexose, converting CO_2 into a chemical fuel at the expense of NADPH and ATP generated from the light reactions.

The third phase of the Calvin cycle is the regeneration of ribulose 1,5-bisphosphate, the acceptor of CO_2 in the first step. The problem is to construct a five-carbon sugar from six-carbon and three-carbon sugars. A transketolase and an aldolase play the major role in the rearrangement of the carbon atoms. The *transketolase*, which we will see again in the pentose phosphate pathway, requires the coenzyme thiamine pyrophosphate (TPP) to transfer a two-carbon unit ($CO—CH_2OH$) from a ketose to an aldose.

Ketose (*n* carbons) + **Aldose** (*m* carbons) ⇌ (Transketolase) **Aldose** (*n* − 2 carbons) + **Ketose** (*m* + 2 carbons)

Aldolase, which we have already encountered in glycolysis (p. 438), catalyzes an aldol condensation between dihydroxyacetone phosphate (DHAP) and an aldehyde. This enzyme is highly specific for dihydroxyacetone phosphate, but it accepts a wide variety of aldehydes.

Aldose (*n* carbons) + **Dihydroxyacetone phosphate** ⇌ (Aldolase) **Ketose** (*n* + 3 carbons)

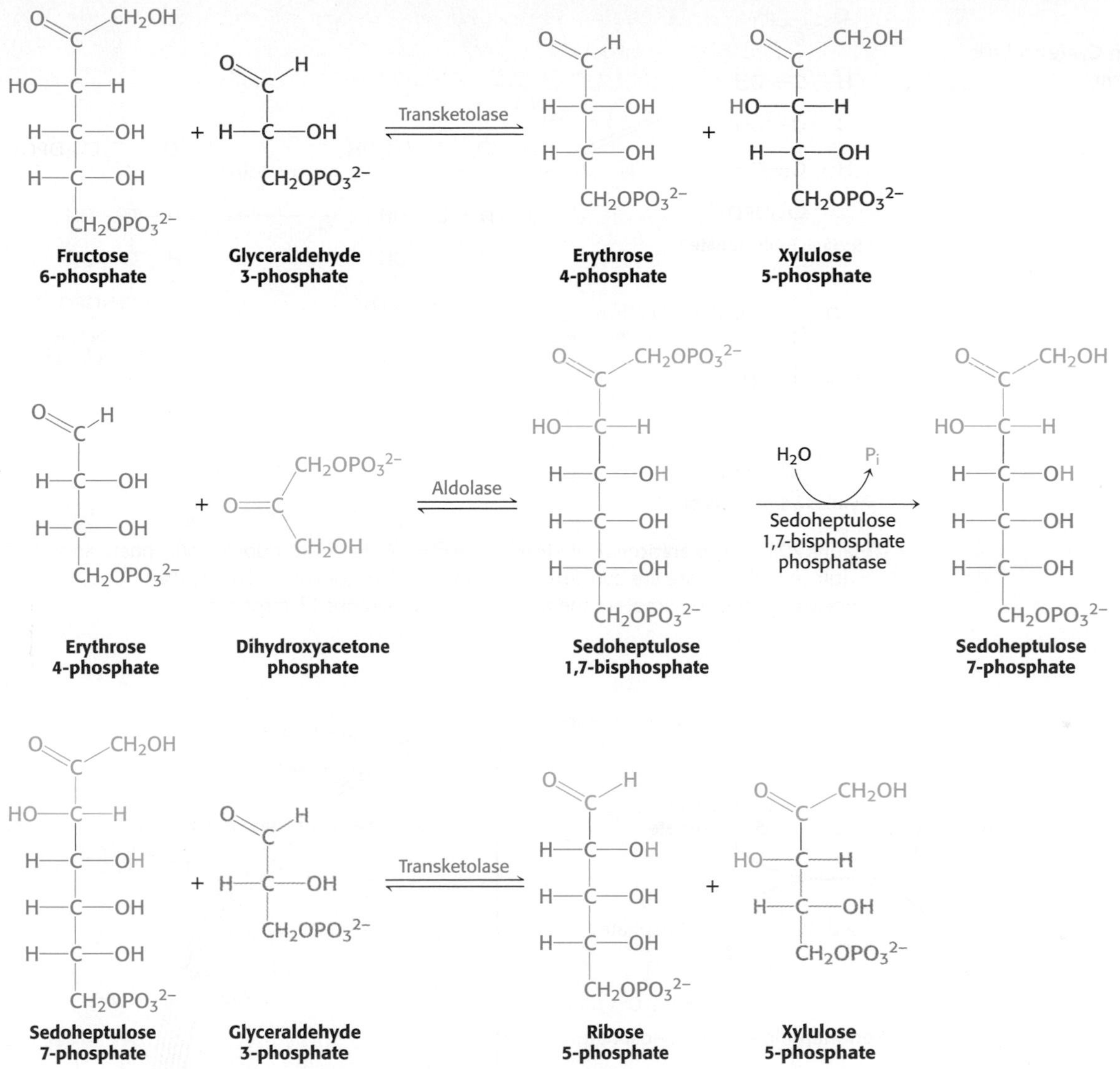

**Fructose
6-phosphate**   **Glyceraldehyde
3-phosphate**   Transketolase   **Erythrose
4-phosphate**   **Xylulose
5-phosphate**

**Erythrose
4-phosphate**   **Dihydroxyacetone
phosphate**   Aldolase   **Sedoheptulose
1,7-bisphosphate**   Sedoheptulose
1,7-bisphosphate
phosphatase   **Sedoheptulose
7-phosphate**

**Sedoheptulose
7-phosphate**   **Glyceraldehyde
3-phosphate**   Transketolase   **Ribose
5-phosphate**   **Xylulose
5-phosphate**

Figure 20.10 Formation of five-carbon sugars. First, transketolase converts a six-carbon sugar and a three-carbon sugar into a four-carbon sugar and a five-carbon sugar. Then, aldolase combines the four-carbon product and a three-carbon sugar to form a seven-carbon sugar. Finally, this seven-carbon sugar reacts with another three-carbon sugar to form two additional five-carbon sugars.

With these enzymes, the construction of the five-carbon sugar proceeds as shown in Figure 20.10.

Finally, ribose 5-phosphate is converted into ribulose 5-phosphate by *phosphopentose isomerase* while xylulose 5-phosphate is converted into ribulose 5-phosphate by *phosphopentose epimerase*. Ribulose 5-phosphate is converted into ribulose 1,5-bisphosphate through the action of *phosphoribulose kinase* (Figure 20.11). The sum of these reactions is

Fructose 6-phosphate + 2 glyceraldehyde 3-phosphate
    + dihydroxyacetone phosphate + 3 ATP $\longrightarrow$
        3 ribulose 1,5-bisphosphate + 3 ADP

This series of reactions completes the Calvin cycle (Figure 20.12). The sum of all the reactions results in the generation of a hexose and the regeneration of the starting compound, ribulose 1,5-bisphosphate. Thus, only a limited amount of ribulose 1,5-bisphosphate is needed to incorporate many molecules of CO_2 into hexoses.

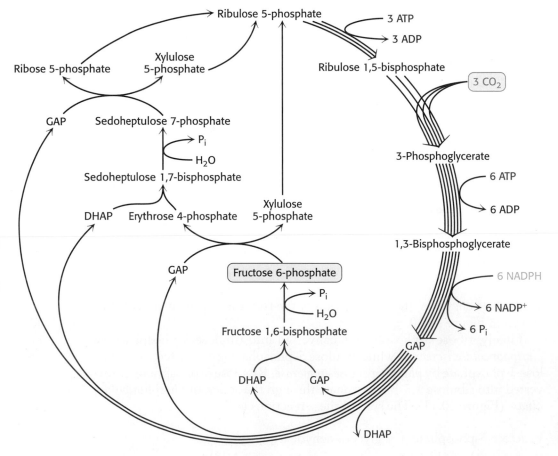

Figure 20.11 Regeneration of ribulose 1,5-bisphosphate. Both ribose 5-phosphate and xylulose 5-phosphate are converted into ribulose 5-phosphate, which is then phosphorylated to complete the regeneration of ribulose 1,5-bisphosphate.

Figure 20.12 Calvin cycle. The diagram shows the reactions necessary with the correct stoichiometry to convert three molecules of CO_2 into one molecule of dihydroxyacetone phosphate (DHAP). The cycle is not as simple as presented in Figure 20.1; rather, it entails many reactions that lead ultimately to the synthesis of glucose and the regeneration of ribulose 1,5-bisphosphate. [After J. R. Bowyer and R. C. Leegood. "Photosynthesis," in *Plant Biochemistry*, P. M. Dey and J. B. Harborne, Eds. (Academic Press, 1997), p. 85.]

Three ATP and Two NADPH Molecules Are Used to Bring Carbon Dioxide to the Level of a Hexose

What is the energy expenditure for synthesizing a hexose? Six rounds of the Calvin cycle are required, because one carbon atom is reduced in each round. Twelve molecules of ATP are expended in phosphorylating 12 molecules of 3-phosphoglycerate to 1,3-bisphosphoglycerate, and 12 molecules of NADPH are consumed in reducing 12 molecules of 1,3-bisphosphoglycerate to glyceraldehyde 3-phosphate. An additional six molecules of ATP are spent in regenerating ribulose 1,5-bisphosphate.

We can now write a balanced equation for the net reaction of the Calvin cycle:

$$6\,CO_2 + 18\,ATP + 12\,NADPH + 12\,H_2O \longrightarrow$$
$$C_6H_{12}O_6 + 18\,ADP + 18\,P_i + 12\,NADP^+ + 6\,H^+$$

Thus, three molecules of ATP and two molecules of NADPH are consumed in incorporating a single CO_2 molecule into a hexose such as glucose or fructose.

Starch and Sucrose Are the Major Carbohydrate Stores in Plants

Plants contain two major storage forms of sugar: *starch* and *sucrose*. Starch, like its animal counterpart glycogen, is a polymer of glucose residues, but it is less branched than glycogen because it contains a smaller proportion of α-1,6-glycosidic linkages (p. 311). Another difference is that ADP-glucose, not UDP-glucose, is the activated precursor. Starch is synthesized and stored in chloroplasts.

In contrast, sucrose (common table sugar), a disaccharide, is synthesized in the cytoplasm. Plants lack the ability to transport hexose phosphates across the chloroplast membrane, but they are able to transport *triose* phosphates from chloroplasts to the cytoplasm. Triose phosphate intermediates such as glyceraldehyde 3-phosphate cross into the cytoplasm in exchange for phosphate through the action of an abundant phosphate translocator. Fructose 6-phosphate formed from triose phosphates joins the glucose unit of UDP-glucose to form sucrose 6-phosphate (Figure 20.13). The hydrolysis of the phosphate ester yields sucrose, a readily transportable and mobilizable sugar that is stored in many plant cells, as in sugar beets and sugar cane.

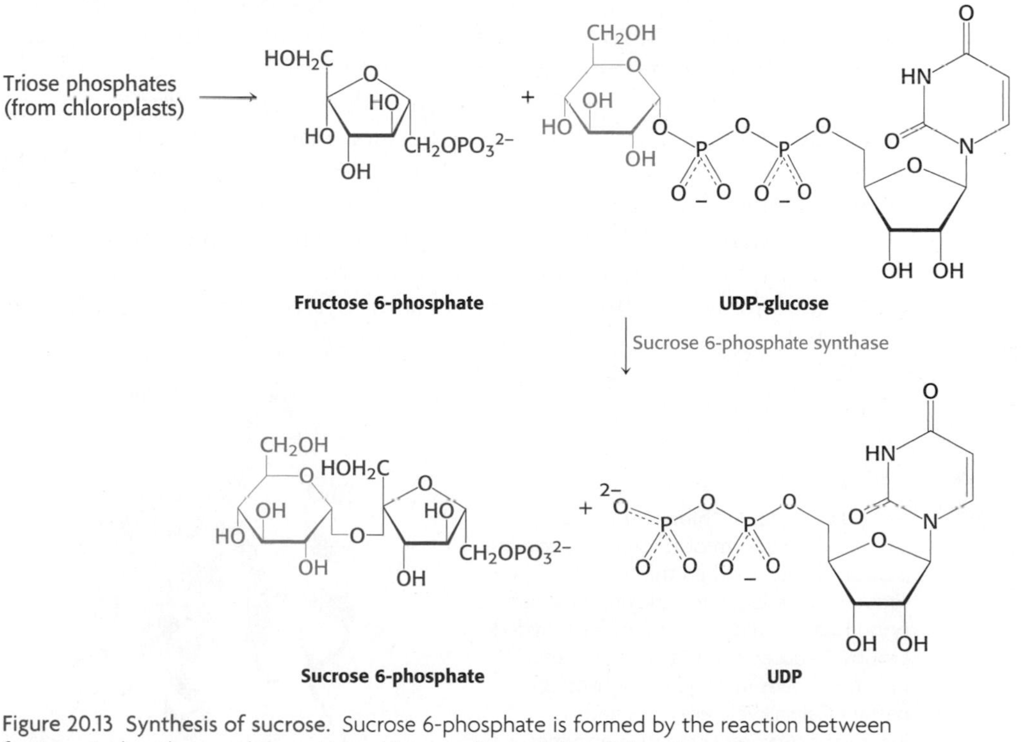

Figure 20.13 Synthesis of sucrose. Sucrose 6-phosphate is formed by the reaction between fructose 6-phosphate and the activated intermediate uridine diphosphate glucose (UDP-glucose).

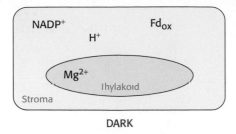

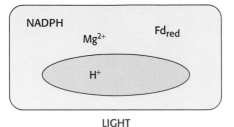

Figure 20.14 Light regulation of the Calvin cycle. The light reactions of photosynthesis transfer electrons out of the thylakoid lumen into the stroma and they transfer protons from the stroma into the thylakoid lumen. As a consequence of these processes, the concentrations of NADPH, reduced ferredoxin (Fd), and Mg^{2+} in the stroma are higher in the light than in the dark, whereas the concentration of H^+ is lower in the dark. Each of these concentration changes helps couple the Calvin cycle reactions to the light reactions.

20.2 The Activity of the Calvin Cycle Depends on Environmental Conditions

The Calvin cycle operates during the day, whereas carbohydrate degradation to yield energy takes place primarily at night. How are synthesis and degradation coordinately controlled? The light reactions lead to changes in the stroma—namely, an increase in pH and in Mg^{2+}, NADPH, and reduced ferredoxin concentration—all of which contribute to the activation of certain Calvin cycle enzymes (Figure 20.14).

Rubisco Is Activated by Light-Driven Changes in Proton and Magnesium Ion Concentrations

As stated earlier, the rate-limiting step in the Calvin cycle is the carboxylation of ribulose 1,5-bisphosphate to form two molecules of 3-phosphoglycerate. *The activity of rubisco increases markedly on illumination because light facilitates the carbamate formation necessary to enzyme activity.* In the stroma, the pH increases from 7 to 8, and the level of Mg^{2+} rises. Both effects are consequences of the light-driven pumping of protons into the thylakoid space. Mg^{2+} ions from the thylakoid space are released into the stroma to compensate for the influx of protons. Carbamate formation is favored at alkaline pH. CO_2 adds to a deprotonated from of lysine 201 of rubisco, and Mg^{2+} ion binds to the carbamate to generate the active form of the enzyme. Thus, light leads to the generation of regulatory signals as well as ATP and NADPH.

Thioredoxin Plays a Key Role in Regulating the Calvin Cycle

Light-driven reactions lead to electron transfer from water to ferredoxin and, eventually, to NADPH. The presence of reduced ferredoxin and NADPH are good signals that conditions are right for biosynthesis. One way in which this information is conveyed to biosynthetic enzymes is by *thioredoxin,* a 12-kd protein containing neighboring cysteine residues that cycle between a reduced sulfhydryl and an oxidized disulfide form (Figure 20.15). The reduced form of thioredoxin activates many biosynthetic enzymes by reducing disulfide bridges that control their activity and inhibits several degradative enzymes by the same means (Table 20.1). In chloroplasts, oxidized thioredoxin is reduced by ferredoxin in a reaction catalyzed by *ferredoxin–thioredoxin reductase.* This enzyme contains a 4Fe-4S cluster that couples two one-electron oxidations of reduced ferredoxin to the two-electron reduction of thioredoxin. Thus, *the activities of the light and dark reactions of photosynthesis are coordinated through electron transfer from reduced ferredoxin to thioredoxin and*

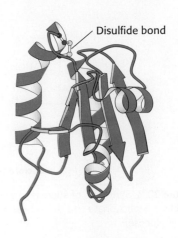

Disulfide bond

Figure 20.15 Thioredoxin. The oxidized form of thioredoxin contains a disulfide bond. When thioredoxin is reduced by reduced ferredoxin, the disulfide bond is converted into two free sulfhydryl groups. Reduced thioredoxin can cleave disulfide bonds in enzymes, activating certain Calvin cycle enzymes and inactivating some degradative enzymes. [Drawn from 1F9M.pdb.]

TABLE 20.1 Enzymes regulated by thioredoxin

Enzyme	Pathway
Rubisco	Carbon fixation in the Calvin cycle
Fructose 1,6-bisphosphatase	Gluconeogenesis
Glyceraldehyde 3-phosphate dehydrogenase	Calvin cycle, gluconeogenesis, glycolysis
Sedoheptulose 1,7-bisphosphatase	Calvin cycle
Glucose 6-phosphate dehydrogenase	Pentose phosphate pathway
Phenylalanine ammonia lyase	Lignin synthesis
Ribulose 5′-phosphate kinase	Calvin cycle
$NADP^+$-malate dehydrogenase	C_4 pathway

then to component enzymes containing regulatory disulfide bonds (Figure 20.16). We shall return to thioredoxin when we consider the reduction of ribonucleotides (Section 25.3).

NADPH is a signal molecule that activates two biosynthetic enzymes, phosphoribulose kinase and glyceraldehyde 3-phosphate dehydrogenase. In the dark, these enzymes are inhibited by association with a small protein called CP12. NADPH disrupts this association, leading to the release of the active enzymes.

The C_4 Pathway of Tropical Plants Accelerates Photosynthesis by Concentrating Carbon Dioxide

The oxygenase activity of rubisco presents a biochemical challenge to tropical plants because the oxygenase activity increases more rapidly with temperature than does the carboxylase activity. How, then, do plants, such as sugar cane, that grow in hot climates prevent very high rates of wasteful photorespiration? Their solution to this problem is to achieve a high local concentration of CO_2 at the site of the Calvin cycle in their photosynthetic cells. The essence of this process, which was elucidated by Marshall Davidson Hatch and C. Roger Slack, is that *four-carbon* (C_4) *compounds such as oxaloacetate and malate carry CO_2 from mesophyll cells, which are in contact with air, to bundle-sheath cells, which are the major sites of photosynthesis* (Figure 20.17). The decarboxylation of the four-carbon compound in a bundle-sheath cell maintains a high concentration of CO_2 at the site of the Calvin cycle. The three-carbon product returns to the mesophyll cell for another round of carboxylation.

The C_4 *pathway* for the transport of CO_2 starts in a mesophyll cell with the condensation of CO_2 and phosphoenolpyruvate to form *oxaloacetate*, in a reaction catalyzed by *phosphoenolpyruvate carboxylase*. In some species, oxaloacetate is converted into *malate* by an $NADP^+$-linked

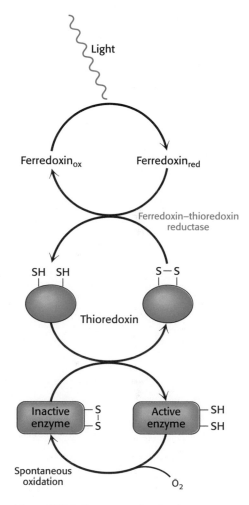

Figure 20.16 Enzyme activation by thioredoxin. Reduced thioredoxin activates certain Calvin cycle enzymes by cleaving regulatory disulfide bonds.

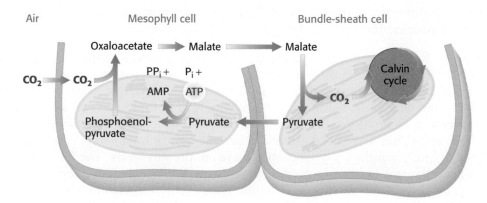

Figure 20.17 C_4 pathway. Carbon dioxide is concentrated in bundle-sheath cells by the expenditure of ATP in mesophyll cells.

malate dehydrogenase. Malate enters the bundle-sheath cell and is oxidatively decarboxylated within the chloroplasts by an $NADP^+$-linked malate dehydrogenase. The released CO_2 enters the Calvin cycle in the usual way by condensing with ribulose 1,5-bisphosphate. Pyruvate formed in this decarboxylation reaction returns to the mesophyll cell. Finally, phosphoenolpyruvate is formed from pyruvate by *pyruvate-P_i dikinase*.

The net reaction of this C_4 pathway is

$$CO_2 \text{ (in mesophyll cell)} + ATP + 2\,H_2O \longrightarrow$$
$$CO_2 \text{ (in bundle-sheath cell)} + AMP + 2\,P_i + 2\,H^+$$

Thus, *the energetic equivalent of two ATP molecules is consumed in transporting CO_2 to the chloroplasts of the bundle-sheath cells.* In essence, this process is active transport: the pumping of CO_2 into the bundle-sheath cell is driven by the hydrolysis of one molecule of ATP to one molecule of AMP and two molecules of orthophosphate. The CO_2 concentration can be 20-fold as great in the bundle-sheath cells as in the mesophyll cells.

When the C_4 pathway and the Calvin cycle operate together, the net reaction is

$$6\,CO_2 + 30\,ATP + 12\,NADPH + 24\,H_2O \longrightarrow$$
$$C_6H_{12}O_6 + 30\,ADP + 30\,P_i + 12\,NADP^+ + 18\,H^+$$

Note that 30 molecules of ATP are consumed per hexose molecule formed when the C_4 pathway delivers CO_2 to the Calvin cycle, in contrast with 18 molecules of ATP per hexose molecule in the absence of the C_4 pathway. The high concentration of CO_2 in the bundle-sheath cells of C_4 plants, which is due to the expenditure of the additional 12 molecules of ATP, is critical for their rapid photosynthetic rate, because CO_2 is limiting when light is abundant. A high CO_2 concentration also minimizes the energy loss caused by photorespiration.

Tropical plants with a C_4 pathway do little photorespiration because the high concentration of CO_2 in their bundle-sheath cells accelerates the carboxylase reaction relative to the oxygenase reaction. This effect is especially important at higher temperatures. The geographic distribution of plants having this pathway (C_4 plants) and those lacking it (C_3 plants) can now be understood in molecular terms. *C_4 plants have the advantage in a hot environment and under high illumination, which accounts for their prevalence in the tropics. C_3 plants,* which consume only 18 molecules of ATP per hexose molecule formed in the absence of photorespiration (compared with 30 molecules of ATP for C_4 plants), are more efficient at temperatures lower than about 28°C, and so they predominate in temperate environments.

Rubisco is present in bacteria, eukaryotes, and even archaea, though other photosynthetic components have not been found in archaea. Thus, rubisco emerged early in evolution, when the atmosphere was rich in CO_2 and almost devoid of O_2. The enzyme was not originally selected to operate in an environment like the present one, which is almost devoid of CO_2 and rich in O_2. Photorespiration became significant about 600 million years ago, when the CO_2 concentration fell to present levels. The C_4 pathway is thought to have evolved in response no more than 30 million years ago and possibly as recently as 7 million years ago. It is interesting that none of the enzymes are unique to C_4 plants, suggesting that this pathway made use of already existing enzymes.

Crassulacean Acid Metabolism Permits Growth in Arid Ecosystems

Many plants growing in hot, dry climates keep the stomata of their leaves closed in the heat of the day to prevent water loss (Figure 20.18). As a consequence, CO_2 cannot be absorbed during the daylight hours when it is needed for glucose synthesis. Rather, CO_2 enters the leaf when the stomata open at the cooler temperatures of night. To store the CO_2 until it can be used during the day, such plants make use of an adaptation called *crassulacean acid metabolism* (CAM), named after the genus *Crassulacea* (the succulents). Carbon dioxide is fixed by the C_4 pathway into malate, which is stored in vacuoles. During the day, malate is decarboxylated and the CO_2 becomes available to the Calvin cycle. In contrast with C_4 plants, CAM plants separate CO_2 accumulation from CO_2 utilization temporally rather than spatially.

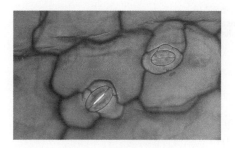

Figure 20.18 Electron micrograph of an open stoma and a closed stoma. [Herb Charles Ohlmeyer/Fran Heyl Associates.]

20.3 The Pentose Phosphate Pathway Generates NADPH and Synthesizes Five-Carbon Sugars

The pentose phosphate pathway meets the need of all organisms for a source of NADPH to use in reductive biosynthesis (Table 20.2). This pathway consists of two phases: (1) the oxidative generation of NADPH and (2) the nonoxidative interconversion of sugars (Figure 20.19). In the oxidative phase, NADPH is generated when glucose 6-phosphate is oxidized to ribose 5-phosphate. This five-carbon sugar and its derivatives are components of RNA and DNA, as well as ATP, NADH, FAD, and coenzyme A.

$$\text{Glucose 6-phosphate} + 2\,\text{NADP}^+ + H_2O \longrightarrow$$
$$\text{ribose 5-phosphate} + 2\,\text{NADPH} + 2\,H^+ + CO_2$$

In the nonoxidative phase, the pathway catalyzes the interconversion of three-, four-, five-, six-, and seven-carbon sugars in a series of nonoxidative reactions. Excess five-carbon sugars may be converted into intermediates of the glycolytic pathway. All these reactions take place in the cytoplasm. These interconversions rely on the same reactions that lead to the regeneration of ribulose 1,5-bisphosphate in the Calvin cycle.

Two Molecules of NADPH Are Generated in the Conversion of Glucose 6-phosphate into Ribulose 5-phosphate

The oxidative phase of the pentose phosphate pathway starts with the dehydrogenation of glucose 6-phosphate at carbon 1, a reaction catalyzed by *glucose 6-phosphate dehydrogenase* (Figure 20.20). This enzyme is highly specific for $NADP^+$; the K_M for NAD^+ is about a thousand times as great as that for $NADP^+$. The product is *6-phosphoglucono-δ-lactone*, which is an intramolecular ester between the C-1 carboxyl group and the C-5 hydroxyl

TABLE 20.2 Pathways requiring NADPH

Synthesis
Fatty acid biosynthesis
Cholesterol biosynthesis
Neurotransmitter biosynthesis
Nucleotide biosynthesis

Detoxification
Reduction of oxidized glutathione
Cytochrome P450 monooxygenases

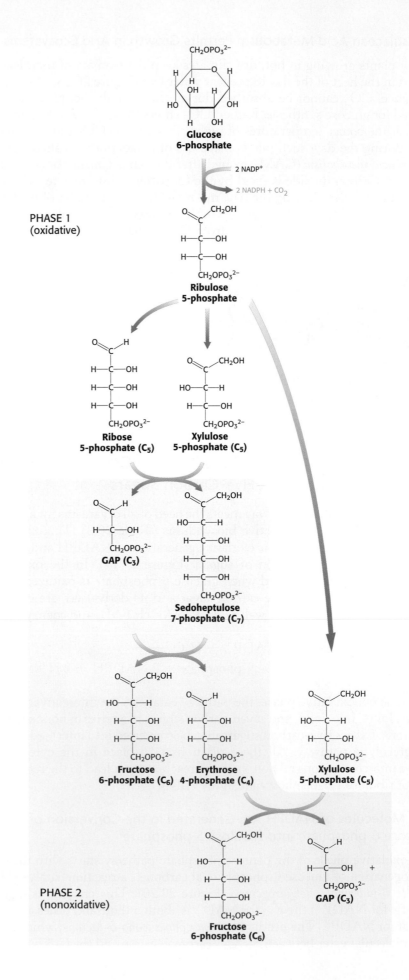

Figure 20.19 Pentose phosphate pathway.
The pathway consists of (1) an oxidative
phase that generates NADPH and (2) a
nonoxidative phase that interconverts
phosphorylated sugars.

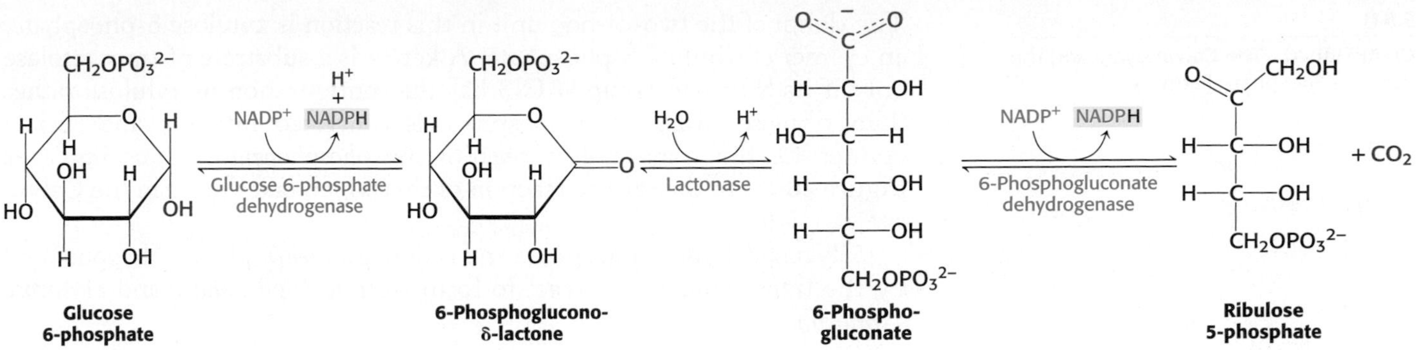

Glucose 6-phosphate → **6-Phosphoglucono-δ-lactone** → **6-Phospho-gluconate** → **Ribulose 5-phosphate**

Figure 20.20 Oxidative phase of the pentose phosphate pathway. Glucose 6-phosphate is oxidized to 6-phosphoglucono-δ-lactone to generate one molecule of NADPH. The lactone product is hydrolyzed to 6-phosphogluconate, which is oxidatively decarboxylated to ribulose 5-phosphate with the generation of a second molecule of NADPH.

group. The next step is the hydrolysis of 6-phosphoglucono-δ-lactone by a specific *lactonase* to give *6-phosphogluconate*. This six-carbon sugar is then oxidatively decarboxylated by *6-phosphogluconate dehydrogenase* to yield *ribulose 5-phosphate*. NADP$^+$ is again the electron acceptor. The final step in the synthesis of ribose 5-phosphate is the isomerization of ribulose 5-phosphate by phosphopentose isomerase (see Figure 20.11)

The Pentose Phosphate Pathway and Glycolysis Are Linked by Transketolase and Transaldolase

The preceding reactions yield two molecules of NADPH and one molecule of ribose 5-phosphate for each molecule of glucose 6-phosphate oxidized. However, many cells need NADPH for reductive biosyntheses much more than they need ribose 5-phosphate for incorporation into nucleotides and nucleic acids. In these cases, ribose 5-phosphate is converted into glyceraldehyde 3-phosphate and fructose 6-phosphate by *transketolase* and *transaldolase*. *These enzymes create a reversible link between the pentose phosphate pathway and glycolysis by catalyzing these three successive reactions.*

$$C_5 + C_5 \xrightleftharpoons{\text{Transketolase}} C_3 + C_7$$
$$C_3 + C_7 \xrightleftharpoons{\text{Transaldolase}} C_6 + C_4$$
$$C_4 + C_5 \xrightleftharpoons{\text{Transketolase}} C_6 + C_3$$

The net result of these reactions is the *formation of two hexoses and one triose from three pentoses:*

$$3\,C_5 \rightleftharpoons 2\,C_6 + C_3$$

The first of the three reactions linking the pentose phosphate pathway and glycolysis is the formation of *glyceraldehyde 3-phosphate* and *sedoheptulose 7-phosphate* from two pentoses.

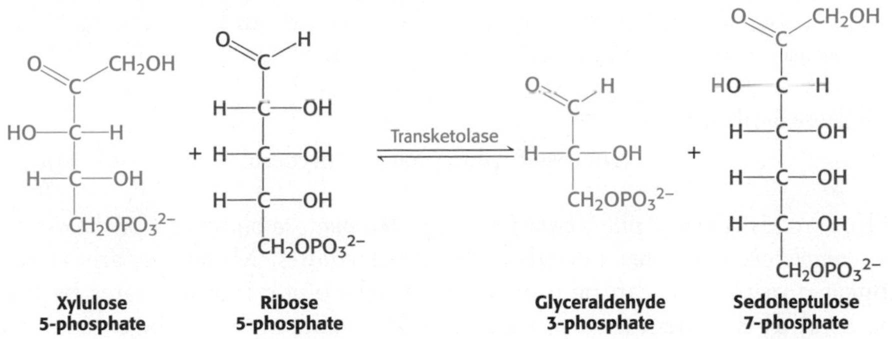

Xylulose 5-phosphate + **Ribose 5-phosphate** → (Transketolase) → **Glyceraldehyde 3-phosphate** + **Sedoheptulose 7-phosphate**

The donor of the two-carbon unit in this reaction is xylulose 5-phosphate, an epimer of ribulose 5-phosphate. A ketose is a substrate of transketolase only if its hydroxyl group at C-3 has the configuration of xylulose rather than ribulose. Ribulose 5-phosphate is converted into the appropriate epimer for the transketolase reaction by *phosphopentose epimerase* (see Figure 20.11) in the reverse reaction of that which takes place in the Calvin cycle.

Glyceraldehyde 3-phosphate and sedoheptulose 7-phosphate generated by the transketolase then react to form *fructose 6-phosphate* and *erythrose 4-phosphate*.

Glyceraldehyde 3-phosphate   **Sedoheptulose 7-phosphate**   **Fructose 6-phosphate**   **Erythrose 4-phosphate**

This synthesis of a four-carbon sugar and a six-carbon sugar is catalyzed by *transaldolase*.

In the third reaction, transketolase catalyzes the synthesis of *fructose 6-phosphate* and *glyceraldehyde 3-phosphate* from erythrose 4-phosphate and xylulose 5-phosphate.

Erythrose 4-phosphate   **Xylulose 5-phosphate**   **Fructose 6-phosphate**   **Glyceraldehyde 3-phosphate**

The sum of these reactions is

2 Xylulose 5-phosphate + ribose 5-phosphate $\rightleftharpoons$

2 fructose 6-phosphate + glyceraldehyde 3-phosphate

Xylulose 5-phosphate can be formed from ribose 5-phosphate by the sequential action of phosphopentose isomerase and phosphopentose epimerase, and so the net reaction starting from ribose 5-phosphate is

3 Ribose 5-phosphate $\rightleftharpoons$

2 fructose 6-phosphate + glyceraldehyde 3-phosphate

Thus, *excess ribose 5-phosphate formed by the pentose phosphate pathway can be completely converted into glycolytic intermediates.* Moreover, any ribose ingested in the diet can be processed into glycolytic intermediates by this pathway. It is evident that the carbon skeletons of sugars can be extensively rearranged to meet physiological needs (Table 20.3).

TABLE 20.3 Pentose phosphate pathway

Reaction	Enzyme
Oxidative phase	
Glucose 6-phosphate + NADP$^+$ $\longrightarrow$ 6-phosphoglucono-δ-lactone + NADPH + H$^+$	Glucose 6-phosphate dehydrogenase
6-Phosphoglucono-δ-lactone + H$_2$O $\longrightarrow$ 6-phosphogluconate + H$^+$	Lactonase
6-Phosphogluconate + NADP$^+$ $\longrightarrow$ ribulose 5-phosphate + CO$_2$ + NADPH + H$^+$	6-Phosphogluconate dehydrogenase
Nonoxidative Phase	
Ribulose 5-phosphate $\rightleftharpoons$ ribose 5-phosphate	Phosphopentose isomerase
Ribulose 5-phosphate $\rightleftharpoons$ xylulose 5-phosphate	Phosphopentose epimerase
Xylulose 5-phosphate + ribose 5-phosphate $\rightleftharpoons$	Transketolase
$\qquad$ sedoheptulose 7-phosphate + glyceraldehyde 3-phosphate	
Sedoheptulose 7-phosphate + glyceraldehyde 3-phosphate $\rightleftharpoons$	Transaldolase
$\qquad$ fructose 6-phosphate + erythrose 4-phosphate	
Xylulose 5-phosphate + erythrose 4-phosphate $\rightleftharpoons$	Transketolase
$\qquad$ fructose 6-phosphate + glyceraldehyde 3-phosphate	

Mechanism: Transketolase and Transaldolase Stabilize Carbanionic Intermediates by Different Mechanisms

The reactions catalyzed by transketolase and transaldolase are distinct yet similar in many ways. One difference is that transketolase transfers a two-carbon unit, whereas transaldolase transfers a three-carbon unit. Each of these units is transiently attached to the enzyme in the course of the reaction.

Transketolase Reaction. Transketolase contains a tightly bound thiamine pyrophosphate as its prosthetic group. The enzyme transfers a two-carbon glycoaldehyde from a ketose donor to an aldose acceptor. The site of the addition of the two-carbon unit is the thiazole ring of TPP. Transketolase is homologous to the E$_1$ subunit of the pyruvate dehydrogenase complex (p. 478) and the reaction mechanism is similar (Figure 20.21).

The C-2 carbon atom of bound TPP readily ionizes to give a *carbanion*. The negatively charged carbon atom of this reactive intermediate attacks the carbonyl group of the ketose substrate. The resulting addition compound releases the aldose product to yield an *activated glycoaldehyde unit*. The positively charged nitrogen atom in the thiazole ring acts as an *electron sink* in the development of this activated intermediate. The carbonyl group of a suitable aldose acceptor then condenses with the activated glycoaldehyde unit to form a new ketose, which is released from the enzyme.

Figure 20.21 Transketolase mechanism. (1) Thiamine pyrophosphate (TPP) ionizes to form a carbanion. (2) The carbanion of TPP attacks the ketose substrate. (3) Cleavage of a carbon–carbon bond frees the aldose product and leaves a two-carbon fragment joined to TPP. (4) This activated glycoaldehyde intermediate attacks the aldose substrate to form a new carbon–carbon bond. (5) The ketose product is released, freeing the TPP for the next reaction cycle.

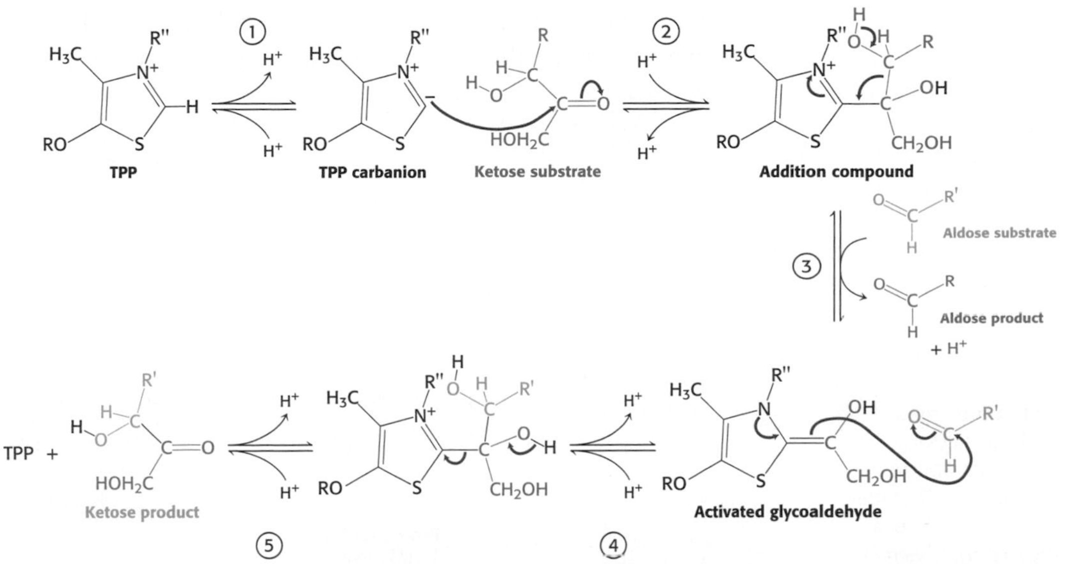

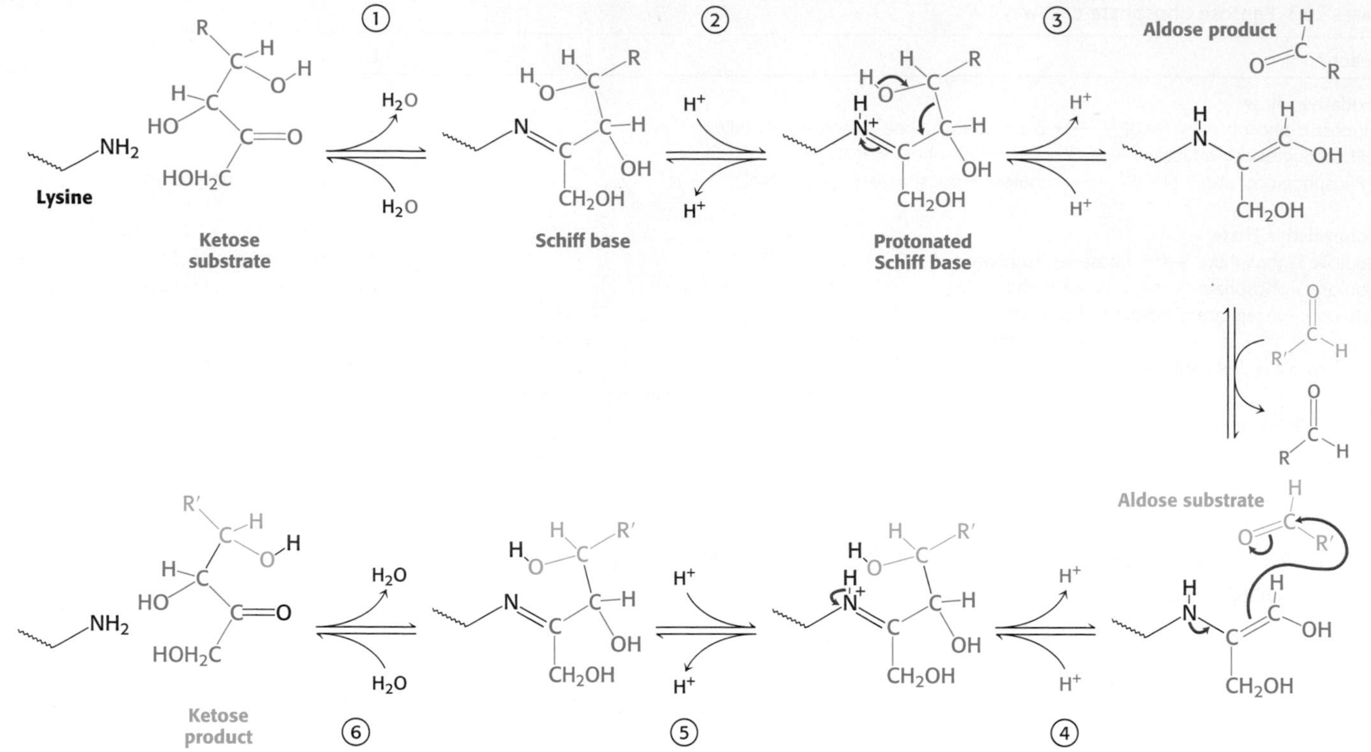

Figure 20.22 Transaldolase mechanism.
(1) The reaction begins with the formation of a Schiff base between a lysine residue in transaldolase and the ketose substrate. Protonation of the Schiff base (2) leads to the release of the aldose product (3), leaving a three-carbon fragment attached to the lysine residue. (4) This intermediate adds to the aldose substrate, with a concomitant protonation to form a new carbon–carbon bond. Subsequent deprotonation (5) and hydrolysis of the Schiff base (6) release the ketose product from the lysine side chain, completing the reaction cycle.

Transaldolase Reaction. Transaldolase transfers a three-carbon *dihydroxyacetone* unit from a ketose donor to an aldose acceptor. Transaldolase, in contrast with transketolase, does not contain a prosthetic group. Rather, *a Schiff base is formed between the carbonyl group of the ketose substrate and the ε-amino group of a lysine residue at the active site of the enzyme* (Figure 20.22). This kind of covalent enzyme–substrate intermediate is like that formed in fructose 1,6-bisphosphate aldolase in the glycolytic pathway (p. 438) and, indeed, the enzymes are homologous. The Schiff base becomes protonated, the bond between C-3 and C-4 is split, and an aldose is released. The negative charge on the Schiff-base carbanion moiety is stabilized by resonance. The positively charged nitrogen atom of the protonated Schiff base acts as an electron sink. The Schiff-base adduct is stable until a suitable aldose becomes bound. The dihydroxyacetone moiety then reacts with the carbonyl group of the aldose. The ketose product is released by hydrolysis of the Schiff base. The nitrogen atom of the protonated Schiff base plays the same role in transaldolase as the thiazole-ring nitrogen atom does in transketolase. In

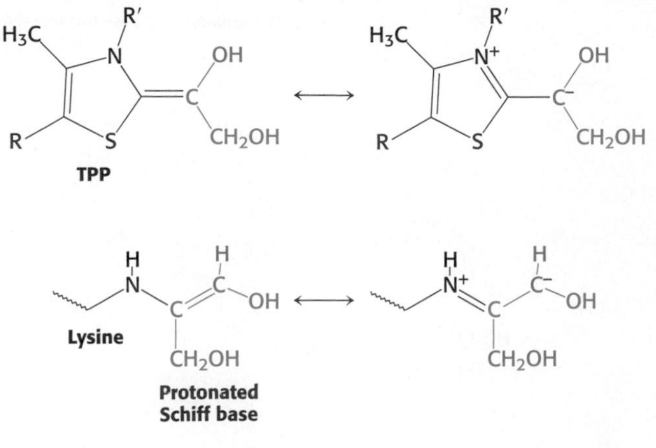

Figure 20.23 Carbanion intermediates.
For transketolase and transaldolase, a carbanion intermediate is stabilized by resonance. In transketolase, TPP stabilizes this intermediate; in transaldolase, a protonated Schiff base plays this role.

each enzyme, a group within an intermediate reacts like a carbanion in attacking a carbonyl group to form a new carbon–carbon bond. In each case, the charge on the carbanion is stabilized by resonance (Figure 20.23).

20.4 The Metabolism of Glucose 6-phosphate by the Pentose Phosphate Pathway Is Coordinated with Glycolysis

Glucose 6-phosphate is metabolized by both the glycolytic pathway (Chapter 16) and the pentose phosphate pathway. How is the processing of this important metabolite partitioned between these two metabolic routes? The cytoplasmic concentration of $NADP^+$ plays a key role in determining the fate of glucose 6-phosphate.

The Rate of the Pentose Phosphate Pathway Is Controlled by the Level of $NADP^+$

The first reaction in the oxidative branch of the pentose phosphate pathway, the dehydrogenation of glucose 6-phosphate, is essentially irreversible. In fact, this reaction is rate limiting under physiological conditions and serves as the control site. The most important regulatory factor is the level of $NADP^+$. Low levels of $NADP^+$ inhibit the dehydrogenation of glucose 6-phosphate because it is needed as the electron acceptor. The inhibitory effect of low levels of $NADP^+$ is intensified by the fact that NADPH competes with $NADP^+$ in binding to the enzyme. The ratio of $NADP^+$ to NADPH in the cytoplasm of a liver cell from a well-fed rat is about 0.014, several orders of magnitude lower than the ratio of NAD^+ to NADH, which is 700 under the same conditions. The marked effect of the $NADP^+$ level on the rate of the oxidative phase ensures that NADPH is not generated unless the supply needed for reductive biosyntheses is low. The nonoxidative phase of the pentose phosphate pathway is controlled primarily by the availability of substrates.

The Flow of Glucose 6-phosphate Depends on the Need for NADPH, Ribose 5-phosphate, and ATP

We can grasp the intricate interplay between glycolysis and the pentose phosphate pathway by examining the metabolism of glucose 6-phosphate in four different metabolic situations (Figure 20.24).

Mode 1. *Much more ribose 5-phosphate than NADPH is required.* For example, rapidly dividing cells need ribose 5-phosphate for the synthesis of nucleotide precursors of DNA. Most of the glucose 6-phosphate is converted into fructose 6-phosphate and glyceraldehyde 3-phosphate by the glycolytic pathway. Transaldolase and transketolase then convert two molecules of fructose 6-phosphate and one molecule of glyceraldehyde 3-phosphate into three molecules of ribose 5-phosphate by a reversal of the reactions described earlier. The stoichiometry of mode 1 is

$$5 \text{ Glucose 6-phosphate} + \text{ATP} \longrightarrow 6 \text{ ribose 5-phosphate} + \text{ADP} + 2 \text{ H}^+$$

Mode 2. *The needs for NADPH and ribose 5-phosphate are balanced.* The predominant reaction under these conditions is the formation of two molecules of NADPH and one molecule of ribose 5-phosphate from one molecule

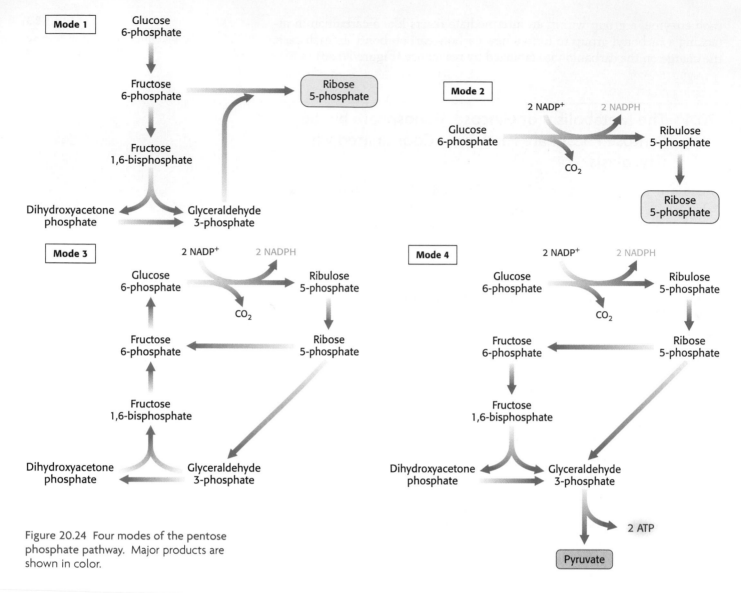

Figure 20.24 Four modes of the pentose phosphate pathway. Major products are shown in color.

of glucose 6-phosphate in the oxidative phase of the pentose phosphate pathway. The stoichiometry of mode 2 is

$$\text{Glucose 6-phosphate} + 2\,\text{NADP}^+ + \text{H}_2\text{O} \longrightarrow$$
$$\text{ribose 5-phosphate} + 2\,\text{NADPH} + 2\,\text{H}^+ + \text{CO}_2$$

Mode 3. *Much more NADPH than ribose 5-phosphate is required.* For example, adipose tissue requires a high level of NADPH for the synthesis of fatty acids (Table 20.4). In this case, glucose 6-phosphate is completely oxidized to CO_2. Three groups of reactions are active in this situation. First, the oxidative phase of the pentose phosphate pathway forms two molecules of NADPH and one molecule of ribose 5-phosphate. Then, ribose 5-phosphate

TABLE 20.4 Tissues with active pentose phosphate pathways

Tissue	Function
Adrenal gland	Steroid synthesis
Liver	Fatty acid and cholesterol synthesis
Testes	Steroid synthesis
Adipose tissue	Fatty acid synthesis
Ovary	Steroid synthesis
Mammary gland	Fatty acid synthesis
Red blood cells	Maintenance of reduced glutathione

is converted into fructose 6-phosphate and glyceraldehyde 3-phosphate by transketolase and transaldolase. Finally, glucose 6-phosphate is resynthesized from fructose 6-phosphate and glyceraldehyde 3-phosphate by the gluconeogenic pathway. The stoichiometries of these three sets of reactions are

$$6 \text{ Glucose 6-phosphate} + 12 \text{ NADP}^+ + 6 \text{ H}_2\text{O} \longrightarrow$$
$$6 \text{ ribose 5-phosphate} + 12 \text{ NADPH} + 12 \text{ H}^+ + 6 \text{ CO}_2$$

$$6 \text{ Ribose 5-phosphate} \longrightarrow$$
$$4 \text{ fructose 6-phosphate} + 2 \text{ glyceraldehyde 3-phosphate}$$

$$4 \text{ Fructose 6-phosphate} + 2 \text{ glyceraldehyde 3-phosphate} + \text{H}_2\text{O} \longrightarrow$$
$$5 \text{ glucose 6-phosphate} + \text{P}_i$$

The sum of the mode 3 reactions is

$$\text{Glucose 6-phosphate} + 12 \text{ NADP}^+ + 7 \text{ H}_2\text{O} \longrightarrow$$
$$6 \text{ CO}_2 + 12 \text{ NADPH} + 12 \text{ H}^+ + \text{P}_i$$

Thus, *the equivalent of glucose 6-phosphate can be completely oxidized to CO_2 with the concomitant generation of NADPH.* In essence, ribose 5-phosphate produced by the pentose phosphate pathway is recycled into glucose 6-phosphate by transketolase, transaldolase, and some of the enzymes of the gluconeogenic pathway.

Mode 4. *Both NADPH and ATP are required.* Alternatively, ribose 5-phosphate formed by the oxidative phase of the pentose phosphate pathway can be converted into pyruvate. Fructose 6-phosphate and glyceraldehyde 3-phosphate derived from ribose 5-phosphate enter the glycolytic pathway rather than reverting to glucose 6-phosphate. In this mode, *ATP and NADPH are concomitantly generated, and five of the six carbons of glucose 6-phosphate emerge in pyruvate.*

$$3 \text{ Glucose 6-phosphate} + 6 \text{ NADP}^+ + 5 \text{ NAD}^+ + 5 \text{ P}_i + 8 \text{ ADP} \longrightarrow$$
$$5 \text{ pyruvate} + 3 \text{ CO}_2 + 6 \text{ NADPH} + 5 \text{ NADH}$$
$$+ 8 \text{ ATP} + 2 \text{ H}_2\text{O} + 8 \text{ H}^+$$

Pyruvate formed by these reactions can be oxidized to generate more ATP or it can be used as a building block in a variety of biosyntheses.

Through the Looking Glass: The Calvin Cycle and the Pentose Phosphate Pathway Are Mirror Images

The complexities of the Calvin cycle and the pentose phosphate pathway are easier to comprehend if we consider them mirror images of each other. The Calvin cycle begins with the fixation of CO_2 and proceeds to use NADPH in the synthesis of glucose. The pentose phosphate pathway begins with the oxidation of a glucose-derived carbon atom to CO_2 and concomitantly generates NADPH. The regeneration phase of the Calvin cycle converts C_6 and C_3 molecules back into the starting material—the C_5 molecule ribulose 1,5-bisphosphate. The pentose phosphate pathway converts a C_5 molecule, ribose 5-phosphate, into C_6 and C_3 intermediates of the glycolytic pathway. Not surprisingly, in photosynthetic organisms, many enzymes are common to the two pathways. We see the economy of evolution: the use of identical enzymes for similar reactions with different ends.

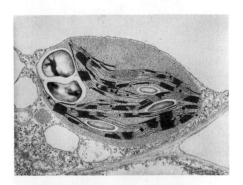

Electron micrograph of a chloroplast. The thylakoid membranes course throughout the stroma of a chloroplast from a cell of *Phleum pratense*, a grass. The dark areas of stacked thylakoid membrane are grana. Several large starch granules, which store the newly synthesized glucose, are also obvious. [Biophoto Associates/Photo Researchers.]

γ-Glutamate

Cysteine

Glycine

Glutathione (reduced)
(γ-Glutamylcysteinylglycine)

Vicia faba. The Mediterranean plant *Vicia faba* is a source of fava beans that contain the purine glycoside vicine. [Inga Spence/Visuals Unlimited.]

20.5 Glucose 6-phosphate Dehydrogenase Plays a Key Role in Protection Against Reactive Oxygen Species

Reactive oxygen species (ROS) generated in oxidative metabolism inflict damage on all classes of macromolecules. Ultimately, their oxidative power can lead to cell death. Indeed, reactive oxygen species are implicated in a number of human diseases (p. 517). Reduced *glutathione* (GSH), a tripeptide with a free sulfhydryl group, combats oxidative stress by reducing reactive oxygen species to harmless forms. Its task accomplished, the glutathione is now in the oxidized form (GSSG) and must itself be reduced to regenerate GSH. The reducing power is supplied by the NADPH generated by glucose 6-phosphate dehydrogenase in the pentose phosphate pathway. Consequently, the pentose phosphate pathway is important to reduce oxidative stress.

A Deficiency of Glucose 6-phosphate Dehydrogenase Causes a Drug-Induced Hemolytic Anemia

At its introduction in 1926, an antimalarial drug, pamaquine, was associated with the appearance of severe and mysterious ailments. Most patients tolerated the drug well, but a few developed severe symptoms within a few days after therapy was started. The urine turned black, jaundice developed, and the hemoglobin content of the blood dropped sharply. In some cases, massive destruction of red blood cells caused death.

This drug-induced hemolytic anemia was shown 30 years later to be caused by a *deficiency of glucose 6-phosphate dehydrogenase*, the enzyme catalyzing the first step in the oxidative branch of the pentose phosphate pathway. This defect, which is inherited on the X chromosome, is the most common enzymopathy, affecting hundreds of millions of people. Cells with reduced levels of glucose 6-phosphate dehydrogenase are especially sensitive to oxidative stress because they produce less NADPH, needed to restore the antioxidant reduced glutathione. This stress is most acute in red blood cells because, lacking mitochondria, they have no alternative means of generating reducing power. In fact, the major role of NADPH in red blood cells is to reduce the disulfide form of glutathione to the sulfhydryl form in a reaction catalyzed by the flavoprotein *glutathione reductase*. The reduced form of glutathione serves as a *sulfhydryl buffer* that maintains the cysteine residues of hemoglobin and other red-blood-cell proteins in the reduced state. Normally, the ratio of the reduced to oxidized forms of glutathione in red blood cells is 500.

How is reduced glutathione regenerated by glutathione reductase? The electrons from NADPH are not directly transferred to the disulfide bond in oxidized glutathione. Rather, they are transferred from NADPH to a tightly bound flavin adenine dinucleotide (FAD) on the reductase, then to a disulfide bridge between two cysteine residues in the enzyme subunit, and finally to oxidized glutathione.

Reduced glutathione is essential for maintaining the normal structure of red blood cells. Red blood cells with a lowered level of reduced glutathione are more susceptible to lysis. How can we explain this phenomenon biochemically? We can approach the question by asking another: Why does the antimalarial drug pamaquine destroy red blood cells? Pamaquine, a purine glycoside of fava beans, is an oxidative agent that leads to the generation of peroxides, reactive oxygen species that can damage membranes as well as other biomolecules. Peroxides are normally eliminated by glutathione peroxidase, which uses reduced glutathione as a reducing agent.

$$2\,\text{GSH} + \text{ROOH} \xrightarrow{\text{Glutathione peroxidase}} \text{GSSG} + \text{H}_2\text{O} + \text{ROH}$$

In the absence of glucose 6-phosphate dehydrogenase, peroxides continue to damage membranes because no NADPH is being produced to restore reduced glutathione. Moreover, the hemoglobin sulfhydryl groups can no longer be maintained in the reduced form. Hemoglobin molecules then cross-link with one another to form aggregates called *Heinz bodies* on cell membranes (Figure 20.25). Membranes damaged by Heinz bodies and reactive oxygen species become deformed, and the cell is likely to undergo lysis. Thus, the answer to our question is that glucose 6-phosphate dehydrogenase is required to maintain reduced glutathione levels to protect against oxidative stress. In the absence of oxidative stress, however, the deficiency is quite benign. The sensitivity to pamaquine of people having this dehydrogenase deficiency also clearly demonstrates that *atypical reactions to drugs may have a genetic basis.*

A Deficiency of Glucose 6-phosphate Dehydrogenase Confers an Evolutionary Advantage in Some Circumstances

The incidence of the most common form of glucose 6-phosphate dehydrogenase deficiency, characterized by a tenfold reduction in enzymatic activity in red blood cells, is 11% among Americans of African heritage. This high frequency suggests that the deficiency may be advantageous under certain environmental conditions. Indeed, *glucose 6-phosphate dehydrogenase deficiency protects against falciparum malaria.* The parasites causing this disease require reduced glutathione and the products of the pentose phosphate pathway for optimal growth. Thus, glucose 6-phosphate dehydrogenase deficiency is a mechanism of protection against malaria, which accounts for its high frequency in malaria-infested regions of the world. We see here once again the interplay of heredity and environment in the production of disease.

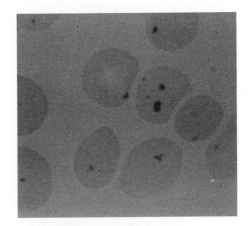

Figure 20.25 Red blood cells with Heinz bodies. The light micrograph shows red blood cells obtained from a person deficient in glucose 6-phosphate dehydrogenase. The dark particles, called Heinz bodies, inside the cells are clumps of denatured protein that adhere to the plasma membrane and stain with basic dyes. Red blood cells in such people are highly susceptible to oxidative damage. [Courtesy of Dr. Stanley Schrier.]

Summary

20.1 The Calvin Cycle Synthesizes Hexoses from Carbon Dioxide and Water

ATP and NADPH formed in the light reactions of photosynthesis are used to convert CO_2 into hexoses and other organic compounds. The dark phase of photosynthesis, called the Calvin cycle, starts with the reaction of CO_2 and ribulose 1,5-bisphosphate to form two molecules of 3-phosphoglycerate. The steps in the conversion of 3-phosphoglycerate into fructose 6-phosphate and glucose 6-phosphate are like those of gluconeogenesis, except that glyceraldehyde 3-phosphate dehydrogenase in chloroplasts is specific for NADPH rather than NADH. Ribulose 1,5-bisphosphate is regenerated from fructose 6-phosphate, glyceraldehyde 3-phosphate, and dihydroxyacetone phosphate by a complex series of reactions. Several of the steps in the regeneration of ribulose 1,5-bisphosphate are like those of the pentose phosphate pathway. Three molecules of ATP and two molecules of NADPH are consumed for each molecule of CO_2 converted into a hexose. Starch in chloroplasts and sucrose in the cytoplasm are the major carbohydrate stores in plants.

20.2 The Activity of the Calvin Cycle Depends on Environmental Conditions

Reduced thioredoxin formed by the light-driven transfer of electrons from ferredoxin activates enzymes of the Calvin cycle by reducing disulfide bridges. The light-induced increase in pH and Mg^{2+} level of the stroma is important in stimulating the carboxylation of ribulose 1,5-bisphosphate by rubisco. This enzyme also catalyzes a competing oxygenase reaction, which produces phosphoglycolate and 3-phosphoglycerate. The recycling of phosphoglycolate leads to the release of CO_2 and further consumption of O_2 in a process called photorespiration. This wasteful side reaction is minimized in tropical plants, which have an accessory pathway—the C_4 pathway—for concentrating CO_2 at the site of the Calvin cycle. This pathway enables tropical plants to take advantage of high levels of light and minimize the oxygenation of ribulose 1,5-bisphosphate. Plants in arid ecosystems employ crassulacean acid metabolism to prevent dehydration. In CAM plants, the C_4 pathway is active during the night when the plant exchanges gases with the air. During the day, gas exchange is eliminated and CO_2 is generated from malate stored in vacuoles.

20.3 The Pentose Phosphate Pathway Generates NADPH and Synthesizes Five-Carbon Sugars

Whereas the Calvin cycle is present only in photosynthetic organisms, the pentose phosphate pathway is present in all organisms. The pentose phosphate pathway generates NADPH and ribose 5-phosphate in the cytoplasm. NADPH is used in reductive biosyntheses, whereas ribose 5-phosphate is used in the synthesis of RNA, DNA, and nucleotide coenzymes. The pentose phosphate pathway starts with the dehydrogenation of glucose 6-phosphate to form a lactone, which is hydrolyzed to give 6-phosphogluconate and then oxidatively decarboxylated to yield ribulose 5-phosphate. $NADP^+$ is the electron acceptor in both of these oxidations. The last step is the isomerization of ribulose 5-phosphate (a ketose) to ribose 5-phosphate (an aldose). A different mode of the pathway is active when cells need much more NADPH than ribose 5-phosphate. Under these conditions, ribose 5-phosphate is converted into glyceraldehyde 3-phosphate and fructose 6-phosphate by *transketolase* and *transaldolase*. These two enzymes create a reversible link between the pentose phosphate pathway and gluconeogenesis. Xylulose 5-phosphate, sedoheptulose 7-phosphate, and erythrose 4-phosphate are intermediates in these interconversions. In this way, 12 molecules of NADPH can be generated for each molecule of glucose 6-phosphate that is completely oxidized to CO_2.

20.4 The Metabolism of Glucose 6-phosphate by the Pentose Phosphate Pathway Is Coordinated with Glycolysis

Only the nonoxidative branch of the pathway is significantly active when much more ribose 5-phosphate than NADPH needs to be synthesized. Under these conditions, fructose 6-phosphate and glyceraldehyde 3-phosphate (formed by the glycolytic pathway) are converted into ribose 5-phosphate without the formation of NADPH. Alternatively, ribose 5-phosphate formed by the oxidative branch can be converted into pyruvate through fructose 6-phosphate and glyceraldehyde 3-phosphate. In this mode, ATP and NADPH are generated, and five of the six carbons of glucose 6-phosphate emerge in pyruvate. The interplay of the glycolytic and pentose phosphate pathways enables the levels of NADPH, ATP, and building blocks such as ribose 5-phosphate and pyruvate to be continuously adjusted to meet cellular needs.

20.5 Glucose 6-phosphate Dehydrogenase Plays a Key Role in Protection Against Reactive Oxygen Species

NADPH generated by glucose 6-phosphate dehydrogenase maintains the appropriate levels of reduced glutathione required to combat oxidative stress and maintain the proper reducing environment in the cell. Cells with diminished glucose 6-phosphate dehydrogenase activity are especially sensitive to oxidative stress.

Key Terms

Calvin cycle (dark reactions) (p. 565)

autotroph (p. 566)

heterotroph (p. 566)

rubisco (ribulose 1,5-bisphosphate carboxylase/oxygenase) (p. 567)

peroxisome (microbody) (p. 569)

photorespiration (p. 570)

transketolase (p. 570)

aldolase (p. 570)

starch (p. 573)

sucrose (p. 573)

thioredoxin (p. 574)

C_4 pathway (p. 576)

C_4 plant (p. 576)

C_3 plant (p. 576)

crassulacean acid metabolism (CAM) (p. 577)

glucose 6-phosphate dehydrogenase (p. 577)

pentose phosphate pathway (p. 577)

glutathione (p. 586)

Selected Readings

Where to Start

Horecker, B. L. 1976. Unravelling the pentose phosphate pathway. In *Reflections on Biochemistry* (pp. 65–72), edited by A. Kornberg, L. Cornudella, B. L. Horecker, and J. Oro. Pergamon.

Levi, P. 1984. Carbon. In *The Periodic Table.* Random House.

Melendez-Hevia, E., and Isidoro, A. 1985. The game of the pentose phosphate cycle. *J. Theor. Biol.* 117:251–263.

Barber, J., and Andersson, B. 1994. Revealing the blueprint of photosynthesis. *Nature* 370:31–34.

Rawsthorne, S. 1992. Towards an understanding of C_3-C_4 photosynthesis. *Essays Biochem.* 27:135–146.

Books and General Reviews

Parry, M. A. J., Andralojc, P. J., Mitchell, R. A. C., Madgwick, P. J., and Keys, A. J. 2003. Manipulation of rubisco: The amount, activity, function and regulation. *J. Exp. Bot.* 54:1321–1333.

Spreitzer, R. J., and Salvucci, M. E. 2002. Rubisco: Structure, regulatory interactions, and possibilities for a better enzyme. *Annu. Rev. Plant Biol.* 53:449–475.

Wood, T. 1985. *The Pentose Phosphate Pathway.* Academic Press.

Buchanan, B. B., Gruissem, W., and Jones, R. L. 2000. *Biochemistry and Molecular Biology of Plants.* American Society of Plant Physiologists.

Enzymes and Reaction Mechanisms

Harrison, D. H., Runquist, J. A., Holub, A., and Miziorko, H. M. 1998. The crystal structure of phosphoribulokinase from *Rhodobacter sphaeroides* reveals a fold similar to that of adenylate kinase. *Biochemistry* 37:5074–5085.

Miziorko, H. M. 2000. Phosphoribulokinase: Current perspectives on the structure/function basis for regulation and catalysis. *Adv. Enzymol. Relat. Areas Mol. Biol.* 74:95–127.

Thorell, S., Gergely, P., Jr., Banki, K., Perl, A., and Schneider, G. 2000. The three-dimensional structure of human transaldolase. *FEBS Lett.* 475:205–208.

Lindqvist, Y., Schneider, G., Ermler, U., and Sundstrom, M. 1992. Three-dimensional structure of transketolase, a thiamine diphosphate dependent enzyme, at 2.5 Å resolution. *EMBO J.* 11:2373–2379.

Robinson, B. H., and Chun, K. 1993. The relationships between transketolase, yeast pyruvate decarboxylase and pyruvate dehydrogenase of the pyruvate dehydrogenase complex. *FEBS Lett.* 328:99–102.

Carbon Dioxide Fixation and Rubisco

Sugawara, H., Yamamoto, H., Shibata, N., Inoue, T., Okada, S., Miyake, C., Yokota, A., and Kai, Y. 1999. Crystal structure of carboxylase reaction-oriented ribulose 1,5-bisphosphate carboxylase/oxygenase from a thermophilic red alga, *Galdieria partita. J. Biol. Chem.* 274:15655–15661.

Hansen, S., Vollan, V. B., Hough, E., and Andersen, K. 1999. The crystal structure of rubisco from *Alcaligenes eutrophus* reveals a novel central eight-stranded β-barrel formed by β-strands from four subunits. *J. Mol. Biol.* 288:609–621.

Knight, S., Andersson, I., and Branden, C. I. 1990. Crystallographic analysis of ribulose 1,5-bisphosphate carboxylase from spinach at 2.4 Å resolution: Subunit interactions and active site. *J. Mol. Biol.* 215:113–160.

Taylor, T. C., and Andersson, I. 1997. The structure of the complex between rubisco and its natural substrate ribulose 1,5-bisphosphate. *J. Mol. Biol.* 265:432–444.

Cleland, W. W., Andrews, T. J., Gutteridge, S., Hartman, F. C., and Lorimer, G. H. 1998. Mechanism of rubisco: The carbamate as general base. *Chem. Rev.* 98:549–561.

Buchanan, B. B. 1992. Carbon dioxide assimilation in oxygenic and anoxygenic photosynthesis. *Photosynth. Res.* 33:147–162.

Hatch, M. D. 1987. C_4 photosynthesis: A unique blend of modified biochemistry, anatomy, and ultrastructure. *Biochim. Biophys. Acta* 895:81–106.

Regulation

Graciet, E., Lebreton, S., and Gontero, B. 2004. The emergence of new regulatory mechanisms in the Benson-Calvin pathway via protein-protein interactions: A glyceraldehyde-3-phosphate dehydrogenase/CP12/phosphoribulokinase complex. *J. Exp. Bot.* 55:1245–1254.

Balmer, Y., Koller, A., del Val, G., Manieri, W., Schürmann, P., and Buchanan, B. B. 2003. Proteomics gives insight into the regulatory function of chloroplast thioredoxins. *Proc. Natl. Acad. Sci. U. S. A.* 100:370–375.

Rokka, A., Zhang, L., and Aro, E.-M. 2001. Rubisco activase: An enzyme with a temperature-dependent dual function? *Plant J.* 25:463–472.

Zhang, N., and Portis, A. R., Jr. 1999. Mechanism of light regulation of rubisco: A specific role for the larger rubisco activase isoform involving reductive activation by thioredoxin-f. *Proc. Natl. Acad. Sci. U. S. A.* 96:9438–9443.

Wedel, N., Soll, J., and Paap, B. K. 1997. CP12 provides a new mode of light regulation of Calvin cycle activity in higher plants. *Proc. Natl. Acad. Sci. U. S. A.* 94:10479–10484.

Avilan, L., Lebreton, S., and Gontero, B. 2000. Thioredoxin activation of phosphoribulokinase in a bi-enzyme complex from *Chlamydomonas reinhardtii* chloroplasts. *J. Biol. Chem.* 275:9447–9451.

Irihimovitch, V., and Shapira, M. 2000. Glutathione redox potential modulated by reactive oxygen species regulates translation of rubisco large subunit in the chloroplast. *J. Biol. Chem.* 275:16289–16295.

Glucose 6-phosphate Dehydrogenase

Au, S. W., Gover, S., Lam, V. M., and Adams, M. J. 2000. Human glucose-6-phosphate dehydrogenase: The crystal structure reveals a structural NADP(+) molecule and provides insights into enzyme deficiency. *Struct. Fold. Des.* 8:293–303.

Salvemini, F., Franze, A., Iervolino, A., Filosa, S., Salzano, S., and Ursini, M. V. 1999. Enhanced glutathione levels and oxidoresistance mediated by increased glucose-6-phosphate dehydrogenase expression. *J. Biol. Chem.* 274:2750–2757.

Tian, W. N., Braunstein, L. D., Apse, K., Pang, J., Rose, M., Tian, X., and Stanton, R. C. 1999. Importance of glucose-6-phosphate dehydrogenase activity in cell death. *Am. J. Physiol.* 276:C1121–C1131.

Tian, W. N., Braunstein, L. D., Pang, J., Stuhlmeier, K. M., Xi, Q. C., Tian, X., and Stanton, R. C. 1998. Importance of glucose-6-phosphate dehydrogenase activity for cell growth. *J. Biol. Chem.* 273:10609–10617.

Ursini, M. V., Parrella, A., Rosa, G., Salzano, S., and Martini, G. 1997. Enhanced expression of glucose-6-phosphate dehydrogenase in human cells sustaining oxidative stress. *Biochem. J.* 323:801–806.

Evolution

Coy, J. F., Dubel, S., Kioschis, P., Thomas, K., Micklem, G., Delius, H., and Poustka, A. 1996. Molecular cloning of tissue-specific transcripts of a transketolase-related gene: Implications for the evolution of new vertebrate genes. *Genomics* 32:309–316.

Schenk, G., Layfield, R., Candy, J. M., Duggleby, R. G., and Nixon, P. F. 1997. Molecular evolutionary analysis of the thiamine-diphosphate-dependent enzyme, transketolase. *J. Mol. Evol.* 44:552–572.

Notaro, R., Afolayan, A., and Luzzatto L. 2000. Human mutations in glucose 6-phosphate dehydrogenase reflect evolutionary history. *FASEB J.* 14:485–494.

Wedel, N., and Soll, J. 1998. Evolutionary conserved light regulation of Calvin cycle activity by NADPH-mediated reversible phosphoribulokinase/CP12/glyceraldehyde-3-phosphate dehydrogenase complex dissociation. *Proc. Natl. Acad. Sci. U. S. A.* 95:9699–9704.

Martin, W., and Schnarrenberger, C. 1997. The evolution of the Calvin cycle from prokaryotic to eukaryotic chromosomes: A case study of functional redundancy in ancient pathways through endosymbiosis. *Curr. Genet.* 32:1–18.

Ku, M. S., Kano-Murakami, Y., and Matsuoka, M. 1996. Evolution and expression of C_4 photosynthesis genes. *Plant Physiol.* 111:949–957.

Pereto, J. G., Velasco, A. M., Becerra, A., and Lazcano, A. 1999. Comparative biochemistry of CO_2 fixation and the evolution of autotrophy. *Int. Microbiol.* 2:3–10.

Problems

1. *Variation on a theme.* Sedoheptulose 1,7-bisphosphate is an intermediate in the Calvin cycle but not in the pentose phosphate pathway. What is the enzymatic basis of this difference?

2. *Total eclipse.* An illuminated suspension of *Chlorella* is actively carrying out photosynthesis. Suppose that the light is suddenly switched off. How would the levels of 3-phosphoglycerate and ribulose 1,5-bisphosphate change in the next minute?

3. *CO_2 deprivation.* An illuminated suspension of *Chlorella* is actively carrying out photosynthesis in the presence of 1% CO_2. The concentration of CO_2 is abruptly reduced to 0.003%. What effect would this reduction have on the levels of 3-phosphoglycerate and ribulose 1,5-bisphosphate in the next minute?

4. *A potent analog.* 2-Carboxyarabinitol 1,5-bisphosphate (CABP) has been useful in studies of rubisco.

(a) Write the structural formula of CABP.
(b) Which catalytic intermediate does it resemble?
(c) Predict the effect of CABP on rubisco.

5. *Salvage operation.* Write a balanced equation for the transamination of glyoxylate to yield glycine.

6. *When one equals two.* In the C_4 pathway, one ATP molecule is used in combining the CO_2 with phosphoenolpyruvate to form oxaloacetate (Figure 20.17), but, in the computation of energetics bookkeeping, two ATP molecules are said to be consumed. Explain.

7. *Dog days of August.* Before the days of pampered lawns, most homeowners practiced horticultural Darwinism. A result was that the lush lawns of early summer would often convert into robust cultures of crabgrass in the dog days of August. Provide a possible biochemical explanation for this transition.

8. *Global warming.* C_3 plants are most common in higher latitudes and become less common at latitudes near the equator. The reverse is true of C_4 plants. How might global warming affect this distribution?

9. *Tracing glucose.* Glucose labeled with ^{14}C at C-6 is added to a solution containing the enzymes and cofactors of the oxidative phase of the pentose phosphate pathway. What is the fate of the radioactive label?

10. *Recurring decarboxylations.* Which reaction in the citric acid cycle is most analogous to the oxidative decarboxylation of 6-phosphogluconate to ribulose 5-phosphate? What kind of enzyme-bound intermediate is formed in both reactions?

11. *Carbon shuffling.* Ribose 5-phosphate labeled with ^{14}C at C-1 is added to a solution containing transketolase, transaldolase, phosphopentose epimerase, phosphopentose isomerase, and glyceraldehyde 3-phosphate. What is the distribution of the radioactive label in the erythrose 4-phosphate and fructose 6-phosphate that are formed in this reaction mixture?

12. *Synthetic stoichiometries.* What is the stoichiometry of the synthesis of (a) ribose 5-phosphate from glucose 6-phosphate without the concomitant generation of NADPH? (b) NADPH from glucose 6-phosphate without the concomitant formation of pentose sugars?

13. *Trapping a reactive lysine.* Design a chemical experiment to identify the lysine residue that forms a Schiff base at the active site of transaldolase.

14. *Reductive power.* What ratio of NADPH to $NADP^+$ is required to sustain [GSH] = 10 mM and [GSSG] = 1 mM? Use the redox potentials given in Table 18.1.

Mechanism Problems

15. *An alternative approach.* The mechanisms of some aldolases do not include Schiff-base intermediates. Instead, these enzymes require bound metal ions. Propose such a mechanism for the conversion of dihydroxyacetone phosphate and glyceraldehyde 3-phosphate into fructose 1,6-bisphosphate.

16. *A recurring intermediate.* Phosphopentose isomerase interconverts the aldose ribose 5-phosphate and the ketose ribulose 5-phosphate. Propose a mechanism.

Chapter Integration Problems

17. *Catching carbons.* Radioactive-labeling experiments can yield estimates of how much glucose 6-phosphate is metabolized by the pentose phosphate pathway and how much is metabolized by the combined action of glycolysis and the citric acid cycle. Suppose that you have samples of two different tissues as well as two radioactively labeled glucose samples, one with glucose labeled with ^{14}C at C-1 and the other with glucose labeled with ^{14}C at C-6. Design an experiment that would enable you to determine the relative activity of the aerobic metabolism of glucose compared with metabolism by the pentose phosphate pathway.

18. *Photosynthetic efficiency.* Use the following information to estimate the efficiency of photosynthesis.

The $\Delta G^{\circ\prime}$ for the reduction of CO_2 to the level of hexose is +477 kJ mol^{-1} (+114 kcal mol^{-1}).
A mole of 600-nm photons has an energy content of 199 kJ (47.6 kcal).
Assume that the proton gradient generated in producing the required NADPH is sufficient to drive the synthesis of the required ATP.

Data Interpretation Problem

19. Graph A shows the photosynthetic activity of two species of plant, one a C_4 plant and the other a C_3 plant, as a function of leaf temperature.

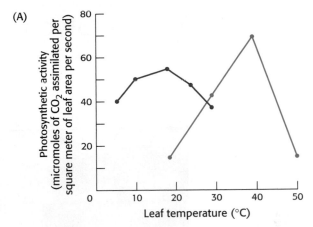

(a) Which data were most likely generated by the C_4 plant and which by the C_3 plant? Explain.
(b) Suggest some possible explanations for why the photosynthetic activity falls at higher temperatures.

Graph B illustrates how the photosynthetic activity of C_3 and C_4 plants varies with CO_2 concentration when temperature (30°C) and light intensity (high) are constant.

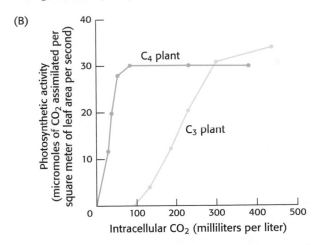

(c) Why can C_4 plants thrive at CO_2 concentrations that do not support the growth of C_3 plants?
(d) Suggest a plausible explanation for why C_3 plants continue to increase photosynthetic activity at higher CO_2 concentrations, whereas C_4 plants reach a plateau.

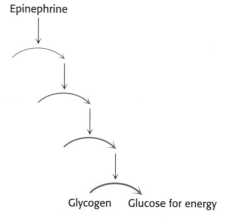

Epinephrine

Glycogen   Glucose for energy

Signaling cascades lead to the mobilization of glycogen to produce glucose, an energy source for runners. [(Left) Mike Powell/Allsport.]

Glycogen Metabolism

Glycogen is a *readily mobilized storage form of glucose.* It is a very large, branched polymer of glucose residues that can be broken down to yield glucose molecules when energy is needed (Figure 21.1). Most of the glucose residues in glycogen are linked by α-1,4-glycosidic bonds. Branches at about every tenth residue are created by α-1,6-glycosidic bonds. Recall that α-glycosidic linkages form open helical polymers, whereas β linkages produce nearly straight strands that form structural fibrils, as in cellulose (p. 312).

Glycogen is not as reduced as fatty acids are and consequently not as energy rich. Why isn't all excess fuel stored as fatty acids rather than as glycogen? The controlled release of glucose from glycogen maintains blood-glucose levels between meals. The circulating blood keeps the brain supplied

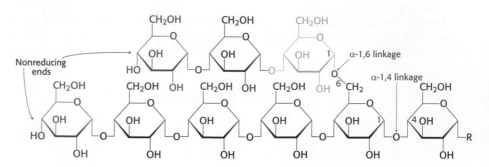

Figure 21.1 Glycogen structure. In this structure of two outer branches of a glycogen molecule, the residues at the nonreducing ends are shown in red and the residue that starts a branch is shown in green. The rest of the glycogen molecule is represented by R.

with glucose, which is virtually the only fuel used by the brain, except during prolonged starvation. Moreover, the readily mobilized glucose from glycogen is a good source of energy for sudden, strenuous activity. Unlike fatty acids, the released glucose can provide energy in the absence of oxygen and can thus supply energy for anaerobic activity.

The two major sites of glycogen storage are the liver and skeletal muscle. The concentration of glycogen is higher in the liver than in muscle (10% versus 2% by weight), but more glycogen is stored in skeletal muscle overall because of muscle's much greater mass. Glycogen is present in the cytoplasm in the form of granules ranging in diameter from 10 to 40 nm (Figure 21.2). In the liver, glycogen synthesis and degradation are regulated to maintain blood-glucose levels as required to meet the needs of the organism as a whole. In contrast, in muscle, these processes are regulated to meet the energy needs of the muscle itself.

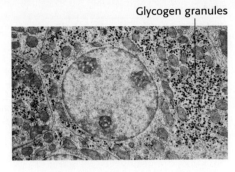

Figure 21.2 Electron micrograph of a liver cell. The dense particles in the cytoplasm are glycogen granules. [Courtesy of Dr. George Palade.]

Glycogen Metabolism Is the Regulated Release and Storage of Glucose

Glycogen degradation and synthesis are simple biochemical processes. Glycogen degradation consists of three steps: (1) the release of glucose 1-phosphate from glycogen, (2) the remodeling of the glycogen substrate to permit further degradation, and (3) the conversion of glucose 1-phosphate into glucose 6-phosphate for further metabolism. The glucose 6-phosphate derived from the breakdown of glycogen has three fates (Figure 21.3): (1) it is the initial substrate for glycolysis, (2) it can be converted into free glucose for release into the bloodstream, and (3) it can be processed by the pentose phosphate pathway to yield NADPH and ribose derivatives. The conversion into free glucose takes place mainly in the liver.

Glycogen synthesis requires an activated form of glucose, uridine diphosphate glucose (UDP-glucose), which is formed by the reaction of UTP and glucose 1-phosphate. UDP-glucose is added to the nonreducing ends of glycogen molecules. As is the case for glycogen degradation, the glycogen molecule must be remodeled for continued synthesis.

The regulation of glycogen degradation and synthesis is complex. Several enzymes taking part in glycogen metabolism allosterically respond to metabolites that signal the energy needs of the cell. *Through these allosteric responses, enzyme activity is adjusted to meet the needs of the cell.* In addition, hormones may initiate signal cascades that lead to the reversible phosphorylation of enzymes, which alters their catalytic rates. *Regulation by hormones adjusts glycogen metabolism to meet the needs of the entire organism.*

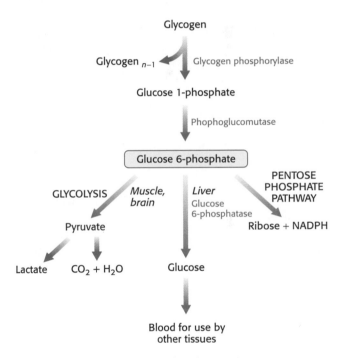

Figure 21.3 Fates of glucose 6-phosphate. Glucose 6-phosphate derived from glycogen can (1) be used as a fuel for anaerobic or aerobic metabolism as in, for instance, muscle; (2) be converted into free glucose in the liver and subsequently released into the blood; (3) be processed by the pentose phosphate pathway to generate NADPH or ribose in a variety of tissues.

21.1 Glycogen Breakdown Requires the Interplay of Several Enzymes

The efficient breakdown of glycogen to provide glucose 6-phosphate for further metabolism requires four enzyme activities: one to degrade glycogen, two to remodel glycogen so that it remains a substrate for degradation, and one to convert the product of glycogen breakdown into a form suitable for further metabolism. We will examine each of these activities in turn.

Phosphorylase Catalyzes the Phosphorolytic Cleavage of Glycogen to Release Glucose 1-phosphate

Glycogen phosphorylase, the key enzyme in glycogen breakdown, cleaves its substrate by the addition of orthophosphate (P_i) to yield *glucose 1-phosphate*. The cleavage of a bond by the addition of orthophosphate is referred to as *phosphorolysis*.

$$\text{Glycogen} + P_i \rightleftharpoons \text{glucose 1-phosphate} + \text{glycogen}$$
$$(n \text{ residues}) \qquad\qquad\qquad\qquad\qquad (n - 1 \text{ residues})$$

Phosphorylase catalyzes the sequential removal of glucosyl residues from the nonreducing ends of the glycogen molecule (the ends with a free OH group on carbon 4; p. 309). Orthophosphate splits the glycosidic linkage between C-1 of the terminal residue and C-4 of the adjacent one. Specifically, it cleaves the bond between the C-1 carbon atom and the glycosidic oxygen atom, and the α configuration at C-1 is retained.

Glycogen *(n residues)* → **Glucose 1-phosphate** + **Glycogen** *(n – 1 residues)*

Glucose 1-phosphate released from glycogen can be readily converted into glucose 6-phosphate (p. 595), an important metabolic intermediate, by the enzyme phosphoglucomutase.

The reaction catalyzed by phosphorylase is readily reversible in vitro. At pH 6.8, the equilibrium ratio of orthophosphate to glucose 1-phosphate is 3.6. The value of $\Delta G°'$ for this reaction is small because a glycosidic bond is replaced by a phosphoryl ester bond that has a nearly equal transfer potential. However, phosphorolysis proceeds far in the direction of glycogen breakdown in vivo because the $[P_i]/[\text{glucose 1-phosphate}]$ ratio is usually greater than 100, substantially favoring phosphorolysis. We see here an example of how the cell can alter the free-energy change to favor a reaction's occurrence by altering the ratio of substrate and product.

The phosphorolytic cleavage of glycogen is energetically advantageous because the released sugar is already phosphorylated. In contrast, a hydrolytic cleavage would yield glucose, which would then have to be phosphorylated at the expense of a molecule of ATP to enter the glycolytic pathway. An additional advantage of phosphorolytic cleavage for muscle cells is that no transporters exist for glucose 1-phosphate, negatively charged under physiological conditions, so it cannot be transported out of the cell.

A Debranching Enzyme Also Is Needed for the Breakdown of Glycogen

Glycogen phosphorylase acting alone degrades glycogen to a limited extent. However, the enzyme soon encounters an obstacle. The α-1,6-glycosidic bonds at the branch points are not susceptible to cleavage by phosphorylase. Indeed, phosphorylase stops cleaving α-1,4 linkages when it reaches a terminal residue four residues away from a branch point. Because about 1 in 10 residues is branched, cleavage by the phosphorylase alone would come to a halt after the release of six glucose molecules per branch.

How can the remainder of the glycogen molecule be mobilized for use as a fuel? Two additional enzymes, a *transferase* and *α-1,6-glucosidase*, remodel

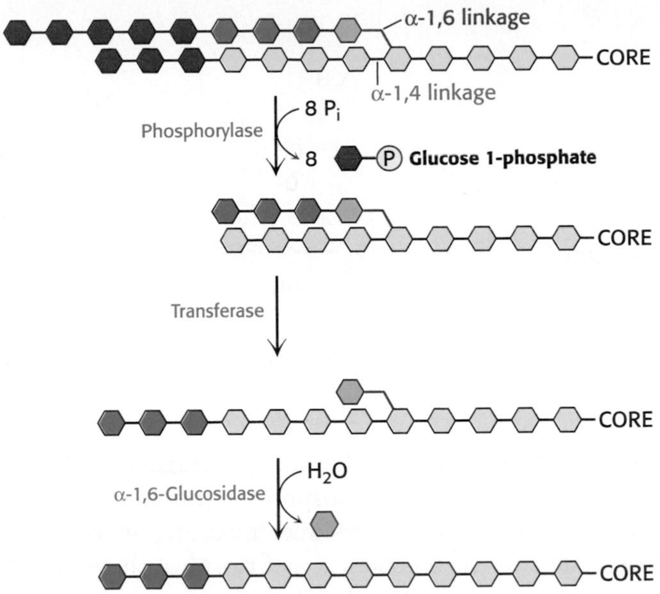

Figure 21.4 Glycogen remodeling. First, α-1,4-glycosidic bonds on each branch are cleaved by phosphorylase, leaving four residues along each branch. The transferase shifts a block of three glucosyl residues from one outer branch to the other. In this reaction, the α-1,4-glycosidic link between the blue and the green residues is broken and a new α-1,4 link between the blue and the yellow residues is formed. The green residue is then removed by α-1,6-glucosidase, leaving a linear chain with all α-1,4 linkages, suitable for further cleavage by phosphorylase.

the glycogen for continued degradation by the phosphorylase (Figure 21.4). *The transferase shifts a block of three glucosyl residues from one outer branch to the other.* This transfer exposes a single glucose residue joined by an α-1,6-glycosidic linkage. α-1,6-Glucosidase, also known as the debranching enzyme, hydrolyzes the α-1,6-glycosidic bond.

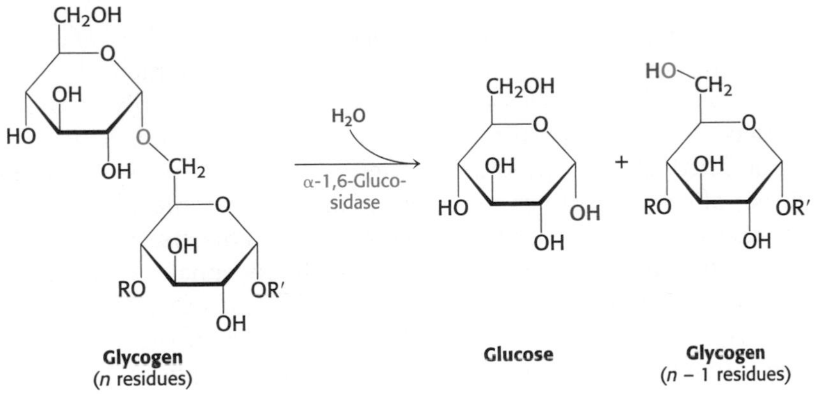

A free glucose molecule is released and then phosphorylated by the glycolytic enzyme hexokinase. Thus, the transferase and α-1,6-glucosidase convert the branched structure into a linear one, which paves the way for further cleavage by phosphorylase. It is noteworthy that, in eukaryotes, the transferase and the α-1,6-glucosidase activities are present in a single 160-kd polypeptide chain, providing yet another example of a bifunctional enzyme (p. 466). Furthermore, these enzymes may have additional features in common (p. 606).

Phosphoglucomutase Converts Glucose 1-phosphate into Glucose 6-phosphate

Glucose 1-phosphate formed in the phosphorolytic cleavage of glycogen must be converted into glucose 6-phosphate to enter the metabolic mainstream. This shift of a phosphoryl group is catalyzed by *phosphoglucomutase*. Recall that this enzyme is also used in galactose metabolism (p. 450). To effect this shift, the enzyme exchanges a phosphoryl group with the substrate (Figure 21.5). The catalytic site of an active mutase molecule contains a phosphorylated serine residue. The phosphoryl group is transferred from

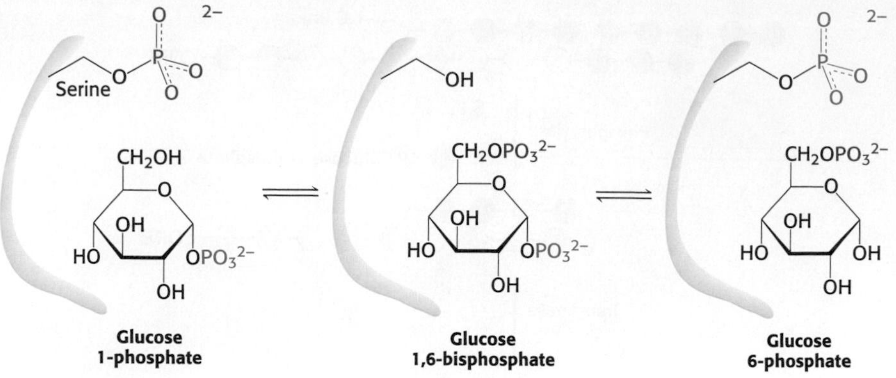

Figure 21.5 Reaction catalyzed by phosphoglucomutase. A phosphoryl group is transferred from the enzyme to the substrate, and a different phosphoryl group is transferred back to restore the enzyme to its initial state.

Glucose 1-phosphate

Glucose 1,6-bisphosphate

Glucose 6-phosphate

the serine residue to the C-6 hydroxyl group of glucose 1-phosphate to form glucose 1,6-bisphosphate. The C-1 phosphoryl group of this intermediate is then shuttled to the same serine residue, resulting in the formation of glucose 6-phosphate and the regeneration of the phosphoenzyme.

These reactions are like those of *phosphoglycerate mutase*, a glycolytic enzyme (p. 444). The role of glucose 1,6-bisphosphate in the interconversion of the phosphoglucoses is like that of 2,3-bisphosphoglycerate (2,3-BPG) in the interconversion of 2-phosphoglycerate and 3-phosphoglycerate in glycolysis. A phosphoenzyme intermediate participates in both reactions.

The Liver Contains Glucose 6-phosphatase, a Hydrolytic Enzyme Absent from Muscle

A major function of the liver is to maintain a nearly constant level of glucose in the blood. The liver releases glucose into the blood during muscular activity and between meals. The released glucose is taken up primarily by the brain and skeletal muscle. In contrast with unmodified glucose, however, the phosphorylated glucose produced by glycogen breakdown is not transported out of cells. The liver contains a hydrolytic enzyme, *glucose 6-phosphatase* that enables glucose to leave that organ. This enzyme cleaves the phosphoryl group to form free glucose and orthophosphate. This glucose 6-phosphatase is the same enzyme that releases free glucose at the conclusion of gluconeogenesis. It is located on the lumenal side of the smooth endoplasmic reticulum membrane. Recall that glucose 6-phosphate is transported into the endoplasmic reticulum; glucose and orthophosphate formed by hydrolysis are then shuttled back into the cytoplasm (p. 463).

$$\text{Glucose 6-phosphate} + H_2O \longrightarrow \text{glucose} + P_i$$

Glucose 6-phosphatase is absent from most other tissues. These tissues retain glucose 6-phosphate for the generation of ATP. In contrast, glucose is not a major fuel for the liver.

Mechanism: Pyridoxal Phosphate Participates in the Phosphorolytic Cleavage of Glycogen

We now examine the catalytic mechanism of glycogen phosphorylase. This enzyme is a dimer of two identical 97-kd subunits. Each subunit is compactly folded into an *amino-terminal domain* (480 residues) containing a *glycogen-binding site* and a *carboxyl-terminal domain* (360 residues; Figure 21.6). The catalytic site in each subunit is located in a deep crevice formed by residues from both domains. The special challenge faced by phosphorylase is to cleave glycogen phosphorolytically rather than hydrolytically to save the ATP required to phosphorylate free glucose. Thus, water must be excluded from the active site.

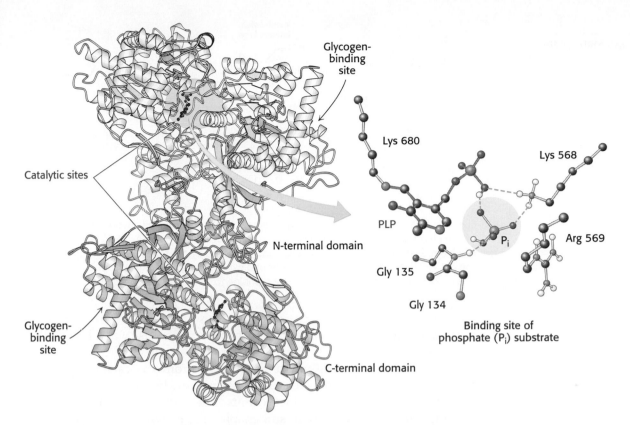

Figure 21.6 **Structure of glycogen phosphorylase.** This enzyme forms a homodimer: one subunit is shown in white and the other in yellow. Each catalytic site includes a pyridoxal phosphate (PLP) group, linked to lysine 680 of the enzyme. The binding site for the phosphate (P$_i$) substrate is shown. *Notice* that the catalytic site lies between the C-terminal domain and the glycogen-binding site. A narrow crevice, which binds four or five glucose units of glycogen, connects the two sites. The separation of the sites allows the catalytic site to phosphorolyze several glucose units before the enzyme must rebind the glycogen substrate. [Drawn from 1NOI.pdb.]

Several clues suggest a mechanism by which phosphorylase achieves the exclusion of water. First, both the glycogen substrate and the glucose 1-phosphate product have an α configuration at C-1. A direct attack of phosphate on C-1 of a sugar would invert the configuration at this carbon because the reaction would proceed through a pentacovalent transition state. The fact that the glucose 1-phosphate formed has an α rather than a β configuration suggests that an even number of steps (most simply, two) is required. The most likely explanation for these results is that a *carbonium ion intermediate* is formed.

A second clue to the catalytic mechanism of phosphorylase is its requirement for *pyridoxal phosphate* (PLP), a derivative of pyridoxine (vitamin B$_6$, p. 423). The aldehyde group of this coenzyme forms a Schiff base with a specific lysine side chain of the enzyme (Figure 21.7). Structural studies indicate that the reacting orthophosphate group takes a position between the 5′-phosphate group of PLP and the glycogen substrate (Figure 21.8). *The 5′-phosphate group of PLP acts in tandem with orthophosphate by serving as a proton donor and then as a proton acceptor (that is, as a general acid–base catalyst).* Orthophosphate (in the HPO$_4^{2-}$ form) donates a proton to the oxygen atom attached to carbon 4 of the departing glycogen chain and simultaneously acquires a proton from PLP. The carbocation (carbonium ion) intermediate formed in this step is then attacked by orthophosphate to form α-glucose 1-phosphate, with the concomitant return of a hydrogen atom to pyridoxal phosphate. The special role of

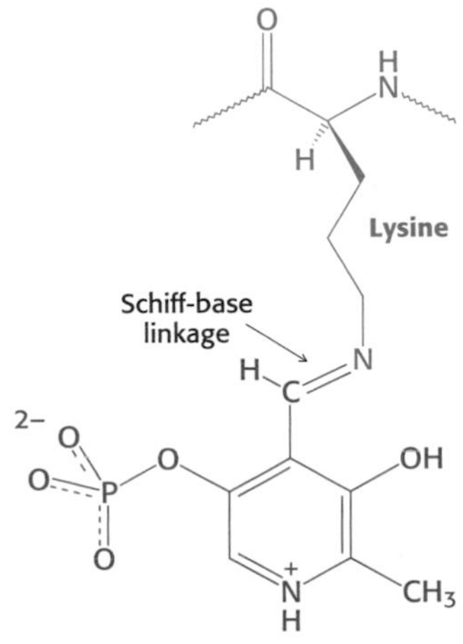

Figure 21.7 **PLP–Schiff-base linkage.** A pyridoxal phosphate (PLP) group (red) forms a Schiff base with a lysine residue (blue) at the active site of phosphorylase.

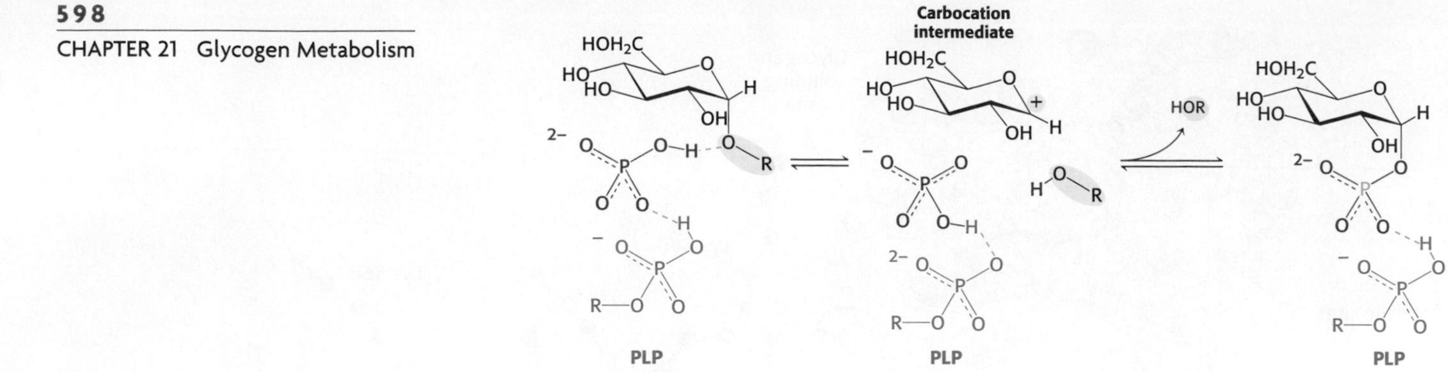

Figure 21.8 Phosphorylase mechanism. A bound HPO_4^{2-} group (red) favors the cleavage of the glycosidic bond by donating a proton to the departing glucose (black). This reaction results in the formation of a carbocation and is favored by the transfer of a proton from the protonated phosphate group of the bound pyridoxal phosphate (PLP) group (blue). The carbocation and the orthophosphate combine to form glucose 1-phosphate.

pyridoxal phosphate in the reaction is necessary because water is excluded from the active site.

The glycogen-binding site is 30 Å away from the catalytic site (see Figure 21.6), but it is connected to the catalytic site by a narrow crevice able to accommodate four or five glucose units. The large separation between the binding site and the catalytic site enables the enzyme to phosphorolyze many residues without having to dissociate and reassociate after each catalytic cycle. An enzyme that can catalyze many reactions without having to dissociate and reassociate after each catalytic step is said to be *processive*—a property of enzymes that synthesize and degrade large polymers. We will see such enzymes again when we consider DNA and RNA synthesis.

21.2 Phosphorylase Is Regulated by Allosteric Interactions and Reversible Phosphorylation

Glycogen metabolism is precisely controlled by multiple interlocking mechanisms. The focus of this control is the enzyme glycogen phosphorylase. *Phosphorylase is regulated by several allosteric effectors that signal the energy state of the cell as well as by reversible phosphorylation, which is responsive to hormones such as insulin, epinephrine, and glucagon.* We will examine the differences in the control of glycogen metabolism in two tissues: skeletal muscle and liver. These differences are due to the fact that *the muscle uses glucose to produce energy for itself, whereas the liver maintains glucose homeostasis of the organism as a whole.*

Muscle Phosphorylase Is Regulated by the Intracellular Energy Charge

The dimeric skeletal-muscle phosphorylase exists in two interconvertible forms: a *usually active* phosphorylase *a* and a *usually inactive* phosphorylase *b* (Figure 21.9). Each of these two forms exists in equilibrium between an active relaxed (R) state and a much less active tense (T) state, but the equilibrium for phosphorylase *a* favors the R state, whereas the equilibrium for phosphorylase *b* favors the T state (Figure 21.10). Muscle phosphorylase *b* is active only in the presence of high concentrations

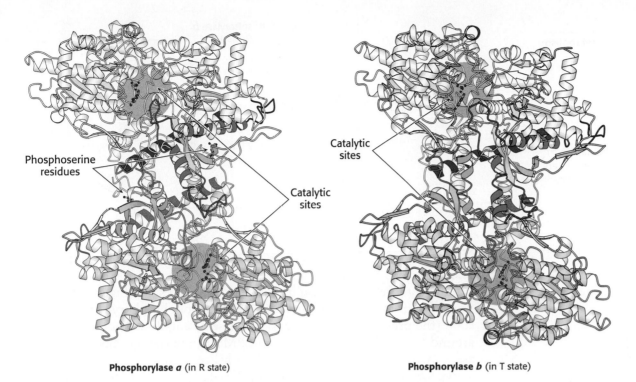

Phosphorylase *a* (in R state)

Phosphorylase *b* (in T state)

Figure 21.9 Structures of phosphorylase *a* and phosphorylase *b*. Phosphorylase *a* is phosphorylated on serine 14 of each subunit. This modification favors the structure of the more active R state. One subunit is shown in white, with helices and loops important for regulation shown in blue and red. The other subunit is shown in yellow, with the regulatory structures shown in orange and green. Phosphorylase *b* is not phosphorylated and exists predominantly in the T state. *Notice* that the catalytic sites are partly occluded in the T state. [Drawn from 1GPA.pdb and 1NOJ.pdb.]

of AMP, which binds to a nucleotide-binding site and stabilizes the conformation of phosphorylase *b* in the active state (Figure 21.11). ATP acts as a negative allosteric effector by competing with AMP. Thus, *the transition of phosphorylase* b *between the active R state and the less-active T state is controlled by the energy charge of the muscle cell.* Glucose 6-phosphate also favors the less-active state of phosphorylase *b,* an example of feedback inhibition.

Phosphorylase *b* is converted into phosphorylase *a* by the phosphorylation of a single serine residue (serine 14) in each subunit. This conversion is initiated by hormones. Fear or the excitement of exercise will cause levels of the hormone epinephrine to increase. The increase in hormone levels and the electrical stimulation of muscle result in phosphorylation of the enzyme to the phosphorylase *a* form. The regulatory enzyme *phosphorylase kinase* catalyzes this covalent modification.

Under most physiological conditions, *phosphorylase* b *is inactive because of the inhibitory effects of ATP and glucose 6-phosphate.* In contrast, *phosphorylase* a *is fully active*, regardless of the levels of AMP, ATP, and glucose 6-phosphate. In resting muscle, nearly all the enzyme is in the inactive *b* form. When exercise commences, the elevated level of AMP leads to the activation of phosphorylase *b*. Exercise will also result in hormone release that generates the phosphorylated *a* form of the enzyme. The absence of glucose 6-phosphatase in muscle ensures that glucose 6-phosphate derived from glycogen remains within the cell for energy transformation.

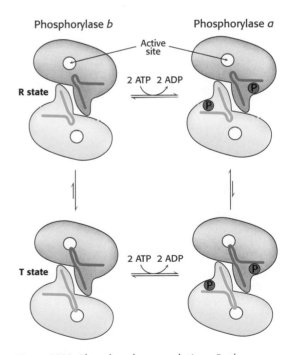

Figure 21.10 Phosphorylase regulation. Both phosphorylase *b* and phosphorylase *a* exist as equilibria between an active R state and a less-active T state. Phosphorylase *b* is usually inactive because the equilibrium favors the T state. Phosphorylase *a* is usually active because the equilibrium favors the R state. Regulatory structures are shown in blue and green.

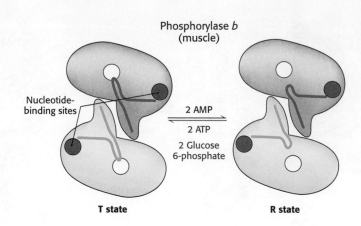

Figure 21.11 Allosteric regulation of muscle phosphorylase. A low energy charge, represented by high concentrations of AMP, favors the transition to the R state.

Comparison of the structures of phosphorylase *a* in the R state and phosphorylase *b* in the T state reveals that subtle structural changes at the subunit interfaces are transmitted to the active sites (see Figure 21.9). The transition from the T state (the prevalent state of phosphorylase *b*) to the R state (the prevalent state of phosphorylase *a*) entails a 10-degree rotation around the twofold axis of the dimer. Most importantly, this transition is associated with structural changes in α helices that move a loop out of the active site of each subunit. Thus, the T state is less active because the catalytic site is partly blocked. In the R state, the catalytic site is more accessible and a binding site for orthophosphate is well organized.

Liver Phosphorylase Produces Glucose for Use by Other Tissues

The role of glycogen degradation in the liver is to form glucose for *export to other tissues* when the blood-glucose level is low. Hence, if free glucose is present from some other source such as diet, there is no need to mobilize glycogen. Consequently, in the liver, the action of phosphorylase is sensitive to the presence of glucose: the binding of glucose deactivates the enzyme.

In human beings, liver phosphorylase and muscle phosphorylase are approximately 90% identical in amino acid sequence. The differences result in subtle but important shifts in the stability of various forms of the enzyme. In contrast with the muscle enzyme, liver phosphorylase *a* but not *b* exhibits the most responsive R-to-T transition. The binding of glucose shifts the allosteric equilibrium of the *a* form from the R to the T state, deactivating the enzyme (Figure 21.12). Unlike the enzyme in muscle, the liver phosphorylase is insensitive to regulation by AMP because the liver does not undergo the dramatic changes in energy charge seen in a contracting muscle. We see here a clear example of the use of isozymic forms of the same enzyme to establish the tissue-specific biochemical properties of muscle and the liver.

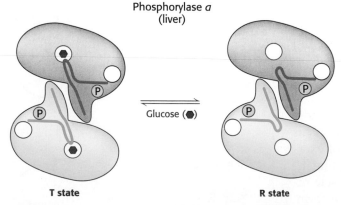

Figure 21.12 Allosteric regulation of liver phosphorylase. The binding of glucose to phosphorylase *a* shifts the equilibrium to the T state and inactivates the enzyme. Thus, glycogen is not mobilized when glucose is already abundant.

Phosphorylase Kinase Is Activated by Phosphorylation and Calcium Ions

Phosphorylase kinase is the enzyme that activates phosphorylase *b* by attaching a phosphoryl group. Its subunit composition in skeletal muscle is $(\alpha\beta\gamma\delta)_4$, and the mass of this very large protein is 1200 kd. The catalytic activity resides in the γ subunit, whereas the other subunits serve regulatory functions. This kinase is under dual control: it is activated both by phos-

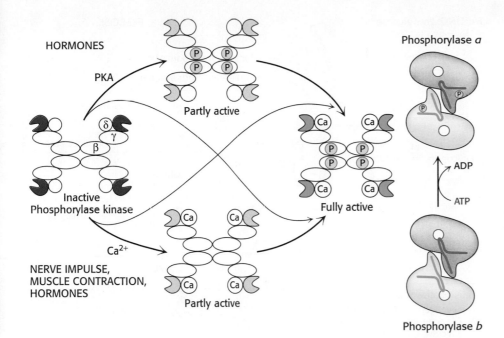

Figure 21.13 **Activation of phosphorylase kinase.** Phosphorylase kinase, an $(\alpha\beta\gamma\delta)_4$ assembly, is activated by hormones that lead to the phosphorylation of the β subunit and by Ca^{2+} binding of the δ subunit. Both types of stimulation are required for maximal enzyme activity. When active, the enzyme converts phosphorylase b into phosphorylase a.

phorylation and by increases in Ca^{2+} levels (Figure 21.13). Like its own substrate, phosphorylase kinase is activated by phosphorylation: the kinase is converted from *a low-activity form into a high-activity one by phosphorylation of its β subunit.* The activation of phosphorylase kinase is one step in a signal-transduction cascade initiated by hormones.

Phosphorylase kinase can also be partly activated by Ca^{2+} levels of the order of 1 μM. Its δ subunit is *calmodulin,* a calcium sensor that stimulates many enzymes in eukaryotes (p. 390). This mode of activation of the kinase is especially noteworthy in muscle, where contraction is triggered by the release of Ca^{2+} from the sarcoplasmic reticulum. Phosphorylase kinase attains maximal activity only after both phosphorylation of the β subunit and activation of the δ subunit by Ca^{2+} binding.

21.3 Epinephrine and Glucagon Signal the Need for Glycogen Breakdown

Protein kinase A activates phosphorylase kinase, which in turn activates glycogen phosphorylase. What activates protein kinase A? What is the signal that ultimately triggers an increase in glycogen breakdown?

G Proteins Transmit the Signal for the Initiation of Glycogen Breakdown

Several hormones greatly affect glycogen metabolism. Glucagon and epinephrine trigger the breakdown of glycogen. Muscular activity or its anticipation leads to the release of *epinephrine (adrenaline),* a catecholamine derived from tyrosine, from the adrenal medulla. Epinephrine markedly stimulates glycogen breakdown in muscle and, to a lesser extent, in the liver. The liver is more responsive to *glucagon,* a polypeptide hormone that is secreted by the α cells of the pancreas when the blood-sugar level is low. Physiologically, glucagon signifies the starved state (Figure 21.14).

How do hormones trigger the breakdown of glycogen? They initiate a cyclic AMP signal-transduction cascade, already discussed in Section 16.1 (Figure 21.15).

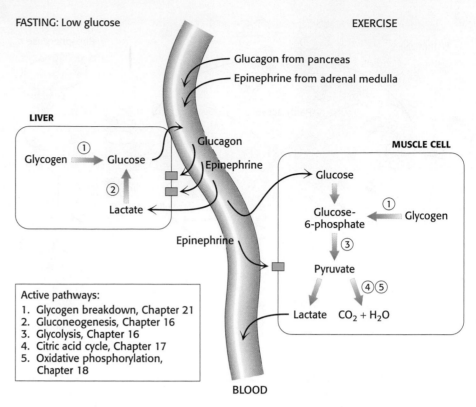

Figure 21.14 Pathway Integration: Hormonal control of glycogen breakdown. Glucagon stimulates liver glycogen breakdown when blood glucose is low. Epinephrine enhances glycogen breakdown in muscle and the liver to provide fuel for muscle contraction.

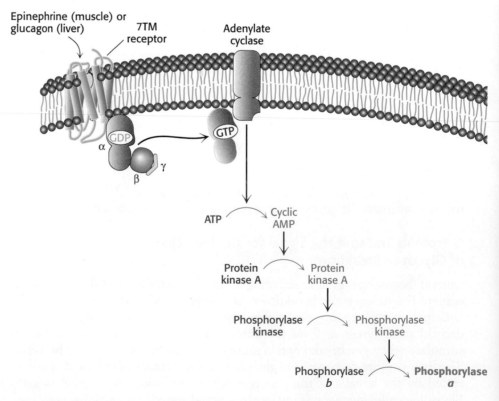

Figure 21.15 Regulatory cascade for glycogen breakdown. Glycogen degradation is stimulated by hormone binding to 7TM receptors. Hormone binding initiates a G-protein-dependent signal-transduction pathway that results in the phosphorylation and activation of glycogen phosphorylase.

1. The signal molecules epinephrine and glucagon bind to specific seven-transmembrane (7TM) receptors in the plasma membranes of target cells (p. 383). Epinephrine binds to the β-adrenergic receptor in muscle, whereas glucagon binds to the glucagon receptor in the liver. These binding events activate the G_S protein. *A specific external signal has been transmitted into the cell through structural changes*, first in the receptor and then in the G protein.

Epinephrine

5   10
^+H_3N–His–Ser–Glu–Gly–Thr–Phe–Thr–Ser–Asp–Tyr–

15   20
–Ser–Lys–Tyr–Leu–Asp–Ser–Arg–Arg–Ala–Gln–

25   29
–Asp–Phe–Val–Gln–Trp–Leu–Met–Asn–Thr–COO⁻

Glucagon

2. The GTP-bound subunit of G_S activates the transmembrane protein adenylate cyclase. This enzyme catalyzes the formation of the second messenger cyclic AMP from ATP.

3. The elevated cytoplasmic level of cyclic AMP activates *protein kinase A* (p. 287). The binding of cyclic AMP to inhibitory regulatory subunits triggers their dissociation from the catalytic subunits. The free catalytic subunits are now active.

4. Protein kinase A phosphorylates phosphorylase kinase, which subsequently activates glycogen phosphorylase.

The cyclic AMP cascade highly amplifies the effects of hormones. The binding of a small number of hormone molecules to cell-surface receptors leads to the release of a very large number of sugar units. Indeed, much of the stored glycogen would be mobilized within seconds were it not for a counterregulatory system.

The signal-transduction processes in the liver are more complex than those in muscle. Epinephrine can also elicit glycogen degradation in the liver. However, in addition to binding to the β-adrenergic receptor, it binds to the 7TM α-adrenergic receptor, which then initiates the *phosphoinositide cascade* (Section 16.2) that induces the release of Ca^{2+} from endoplasmic reticulum stores. Recall that the δ subunit of phosphorylase kinase is the Ca^{2+} sensor calmodulin. The binding of Ca^{2+} to calmodulin leads to a partial activation of phosphorylase kinase. Stimulation by both glucagon and epinephrine leads to maximal mobilization of liver glycogen.

Glycogen Breakdown Must Be Rapidly Turned Off When Necessary

There must be a way to shut down the high-gain system of glycogen breakdown quickly to prevent the wasteful depletion of glycogen after energy needs have been met. When glucose needs have been satisfied, phosphorylase kinase and glycogen phosphorylase are dephosphorylated and inactivated. Simultaneously, glycogen synthesis is activated.

The signal-transduction pathway leading to the activation of glycogen phosphorylase is shut down automatically when the initiating hormone is no longer present. The inherent GTPase activity of the G protein converts the bound GTP into inactive GDP, and phosphodiesterases always present in the cell convert cyclic AMP into AMP. Protein kinase A sets the stage for the shutdown of glycogen degradation by adding a phosphoryl group to the α subunit of phosphorylase kinase after first phosphorylating the β subunit. This addition of a phosphoryl group renders the enzyme a better substrate for dephosphorylation and consequent inactivation by the enzyme *protein phosphatase 1* (PP1). Protein phosphatase 1 also removes the phosphoryl group from glycogen phosphorylase, converting the enzyme into the usually inactive *b* form.

The Regulation of Glycogen Phosphorylase Became More Sophisticated As the Enzyme Evolved

Analyses of the primary structures of glycogen phosphorylase from human beings, rats, *Dictyostelium* (slime mold), yeast, potatoes, and *E. coli* have enabled inferences to be made about the evolution of this important enzyme. The 16 residues that come into contact with glucose at the active site are identical in nearly all the enzymes. There is more variation but still substantial conservation of the 15 residues at the pyridoxal phosphate-binding site. Likewise, the glycogen-binding site is well conserved in all the enzymes. The high degree of similarity among these three sites shows that the catalytic mechanism has been maintained throughout evolution.

Differences arise, however, when we compare the regulatory sites. The simplest type of regulation would be feedback inhibition by glucose 6-phosphate. Indeed, the glucose 6-phosphate regulatory site is highly conserved among most of the phosphorylases. The crucial amino acid residues that participate in regulation by phosphorylation and nucleotide binding are well conserved only in the mammalian enzymes. Thus, this level of regulation was a later evolutionary acquisition.

21.4 Glycogen Is Synthesized and Degraded by Different Pathways

As we have seen in glycolysis and gluconeogenesis, biosynthetic and degradative pathways rarely operate by precisely the same reactions in the forward and reverse directions. Glycogen metabolism provided the first known example of this important principle. *Separate pathways afford much greater flexibility, both in energetics and in control.*

In 1957, Luis Leloir and his coworkers showed that glycogen is synthesized by a pathway that utilizes *uridine diphosphate glucose* (UDP-glucose) rather than glucose 1-phosphate as the activated glucose donor.

$$\text{Synthesis: Glycogen}_n + \text{UDP-glucose} \longrightarrow \text{glycogen}_{n+1} + \text{UDP}$$

$$\text{Degradation: Glycogen}_{n+1} + \text{P}_i \longrightarrow \text{glycogen}_n + \text{glucose 1-phosphate}$$

UDP-Glucose Is an Activated Form of Glucose

UDP-glucose, the glucose donor in the biosynthesis of glycogen, is an *activated form of glucose,* just as ATP and acetyl CoA are activated forms of orthophosphate and acetate, respectively. The C-1 carbon atom of the glucosyl unit of UDP-glucose is activated because its hydroxyl group is esterified to the diphosphate moiety of UDP.

UDP-glucose is synthesized from glucose 1-phosphate and uridine triphosphate (UTP) in a reaction catalyzed by *UDP-glucose pyrophosphorylase.* This reaction liberates the outer two phosphoryl residues of UTP as pyrophosphate.

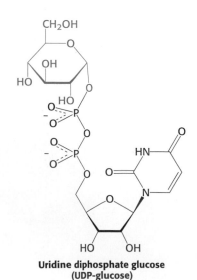

**Uridine diphosphate glucose
(UDP-glucose)**

Glucose 1-phosphate   **UTP**   **UDP-glucose**

This reaction is readily reversible. However, pyrophosphate is rapidly hydrolyzed in vivo to orthophosphate by an inorganic pyrophosphatase. The essentially irreversible hydrolysis of pyrophosphate drives the synthesis of UDP-glucose.

$$\text{Glucose 1-phosphate} + \text{UTP} \rightleftharpoons \text{UDP-glucose} + \text{PP}_i$$
$$\text{PP}_i + \text{H}_2\text{O} \longrightarrow 2\,\text{P}_i$$

$$\overline{\text{Glucose 1-phosphate} + \text{UTP} + \text{H}_2\text{O} \longrightarrow \text{UDP-glucose} + 2\,\text{P}_i}$$

The synthesis of UDP-glucose exemplifies another recurring theme in biochemistry: *many biosynthetic reactions are driven by the hydrolysis of pyrophosphate.*

Glycogen Synthase Catalyzes the Transfer of Glucose from UDP-Glucose to a Growing Chain

New glucosyl units are added to the nonreducing terminal residues of glycogen. The activated glucosyl unit of UDP-glucose is transferred to the hydroxyl group at C-4 of a terminal residue to form an α-1,4-glycosidic linkage. UDP is displaced by the terminal hydroxyl group of the growing glycogen molecule. This reaction is catalyzed by *glycogen synthase, the key regulatory enzyme in glycogen synthesis.*

UDP-glucose   **Glycogen**
(*n* residues)

UDP   **Glycogen**
(*n* + 1 residues)

Glycogen synthase can add glucosyl residues only to a polysaccharide chain already containing more than four residues. Thus, glycogen synthesis requires a *primer*. This priming function is carried out by *glycogenin*, a glycosyltransferase (p. 314) composed of two identical 37-kd subunits. Each subunit of glycogenin catalyzes the addition of eight glucosyl units to the other subunit. These glucosyl units form short α-1,4-glucose polymers, which are covalently attached to the phenolic hydroxyl group of a specific tyrosine residue in each glycogenin subunit. UDP-glucose is the donor in this autoglycosylation. At this point, glycogen synthase takes over to extend the glycogen molecule.

Despite no detectable sequence similarity, structural studies have revealed that glycogen synthase is homologous to glycogen phosphorylase. The binding site for UDP-glucose in glycogen synthase corresponds in position to the pyridoxal phosphate in glycogen phosphorylase.

A Branching Enzyme Forms α-1,6 Linkages

Glycogen synthase catalyzes only the synthesis of α-1,4 linkages. Another enzyme is required to form the α-1,6 linkages that make glycogen a branched polymer. Branching takes place after a number of glucosyl residues are joined in α-1,4 linkages by glycogen synthase. A branch is created by the breaking of an α-1,4 link and the formation of an α-1,6 link: this reaction is different from debranching. A block of residues, typically 7 in number, is transferred to a more interior site. The *branching enzyme* that catalyzes this reaction is quite exacting. The block of 7 or so residues must include the nonreducing terminus and come from a chain at least 11 residues long. In addition, the new branch point must be at least 4 residues away from a preexisting one.

Branching is important because it increases the solubility of glycogen. Furthermore, branching creates a large number of terminal residues, the sites of action of glycogen phosphorylase and synthase (Figure 21.16). Thus, *branching increases the rate of glycogen synthesis and degradation.*

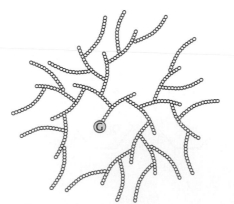

Figure 21.16 Cross section of a glycogen molecule. The component labeled G is glycogenin.

Glycogen branching requires a single transferase activity. Glycogen debranching requires two enzyme activities: a transferase and an α-1,6 glucosidase. Sequence analysis suggests that the two transferases and, perhaps, the α-1,6 glucosidase are members of the same enzyme family, termed the *α-amylase family*. An enzyme of this family catalyzes a reaction by forming a covalent intermediate attached to a conserved aspartate residue. Thus, the branching enzyme appears to transfer a chain of glucose molecules from an α-1,4 linkage to an aspartate residue on the enzyme and then from this site to a more interior location on the glycogen molecule to form an α-1,6 linkage.

Glycogen Synthase Is the Key Regulatory Enzyme in Glycogen Synthesis

The activity of glycogen synthase, like that of phosphorylase, is regulated by covalent modification. Glycogen synthase is phosphorylated at multiple sites by several protein kinases, notably protein kinase A and *glycogen synthase kinase* (GSK). The resulting alteration of the charges in the protein lead to its inactivation. *Phosphorylation has opposite effects on the enzymatic activities of glycogen synthase and phosphorylase.* Phosphorylation converts the active *a* form of the synthase into a usually inactive *b* form. The phosphorylated *b* form is active only if a high level of the allosteric activator glucose 6-phosphate is present, whereas the *a* form is active whether or not glucose 6-phosphate is present.

Glycogen Is an Efficient Storage Form of Glucose

What is the cost of converting glucose 6-phosphate into glycogen and back into glucose 6-phosphate? The pertinent reactions have already been described, except for reaction 5, which is the regeneration of UTP. ATP phosphorylates UDP in a reaction catalyzed by *nucleoside diphosphokinase*.

$$\text{Glucose 6-phosphate} \longrightarrow \text{glucose 1-phosphate} \tag{1}$$

$$\text{Glucose 1-phosphate} + \text{UTP} \longrightarrow \text{UDP-glucose} + \text{PP}_i \tag{2}$$

$$\text{PP}_i + \text{H}_2\text{O} \longrightarrow 2\,\text{P}_i \tag{3}$$

$$\text{UDP-glucose} + \text{glycogen}_n \longrightarrow \text{glycogen}_{n+1} + \text{UDP} \tag{4}$$

$$\text{UDP} + \text{ATP} \longrightarrow \text{UTP} + \text{ADP} \tag{5}$$

$$\text{Sum: Glucose 6-phosphate} + \text{ATP} + \text{glycogen}_n + \text{H}_2\text{O} \longrightarrow$$
$$\text{glycogen}_{n+1} + \text{ADP} + 2\,\text{P}_i$$

Thus, one ATP is hydrolyzed to incorporate glucose 6-phosphate into glycogen. The energy yield from the breakdown of glycogen is highly efficient. About 90% of the residues are phosphorolytically cleaved to glucose 1-phosphate, which is converted at no cost into glucose 6-phosphate. The other 10% are branch residues, which are hydrolytically cleaved. One molecule of ATP is then used to phosphorylate each of these glucose molecules to glucose 6-phosphate. The complete oxidation of glucose 6-phosphate yields about 31 molecules of ATP, and storage consumes slightly more than 1 molecule of ATP per molecule of glucose 6-phosphate; so *the overall efficiency of storage is nearly 97%.*

21.5 Glycogen Breakdown and Synthesis Are Reciprocally Regulated

An important control mechanism prevents glycogen from being synthesized at the same time that it is being broken down. *The same glucagon- and epinephrine-triggered cAMP cascades that initiate glycogen breakdown in the liver and muscle, respectively, also shut off glycogen synthesis. Glucagon and epinephrine control both glycogen breakdown and glycogen synthesis through protein kinase A* (Figure 21.17). Recall that protein kinase A adds a phosphoryl group to phosphorylase kinase, activating that enzyme and initiating

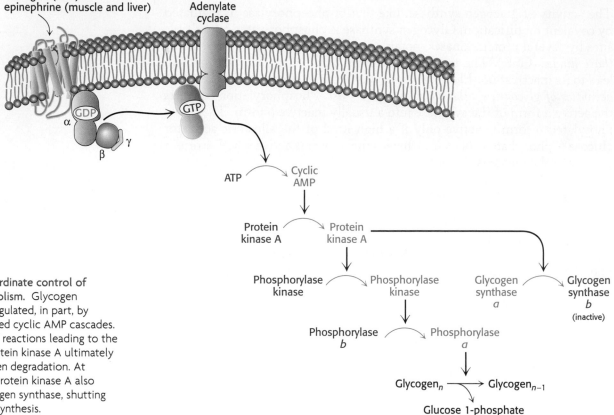

DURING EXERCISE OR FASTING

Glucagon (liver) or epinephrine (muscle and liver)

Adenylate cyclase

ATP → Cyclic AMP

Protein kinase A → Protein kinase A

Phosphorylase kinase → Phosphorylase kinase

Glycogen synthase *a* → Glycogen synthase *b* (inactive)

Phosphorylase *b* → Phosphorylase *a*

Glycogen$_n$ → Glycogen$_{n-1}$

Glucose 1-phosphate

Figure 21.17 Coordinate control of glycogen metabolism. Glycogen metabolism is regulated, in part, by hormone-triggered cyclic AMP cascades. The sequence of reactions leading to the activation of protein kinase A ultimately activates glycogen degradation. At the same time, protein kinase A also inactivates glycogen synthase, shutting down glycogen synthesis.

glycogen breakdown. Likewise, protein kinase A adds a phosphoryl group to glycogen synthase, but this phosphorylation leads to a *decrease* in enzymatic activity. Other kinases, such as glycogen synthase kinase, help to inactive the synthase. In this way, glycogen breakdown and synthesis are reciprocally regulated. How is the enzymatic activity reversed so that glycogen breakdown halts and glycogen synthesis begins?

Protein Phosphatase 1 Reverses the Regulatory Effects of Kinases on Glycogen Metabolism

After a bout of exercise, muscle must shift from a glycogen-degrading mode to one of glycogen replenishment. A first step in this metabolic task is to shut down the phosphorylated proteins that stimulate glycogen breakdown. This task is accomplished by *protein phosphatases* that catalyze the hydrolysis of phosphorylated serine and threonine residues in proteins. *Protein phosphatase 1 plays key roles in regulating glycogen metabolism* (Figure 21.18). PP1 inactivates phosphorylase *a* and phosphorylase kinase by dephosphorylating them. PP1 decreases the rate of glycogen breakdown; it reverses the effects of the phosphorylation cascade. Moreover, *PP1 also removes phosphoryl groups from glycogen synthase* b *to convert it into the much more active glycogen synthase* a *form.* Here, PP1 also accelerates glycogen synthesis. PP1 is yet another molecular device for coordinating carbohydrate storage.

The catalytic subunit of PP1 is a 37-kd single-domain protein. This subunit is usually bound to one of a family of regulatory subunits with masses of approximately 120 kd; in skeletal muscle and heart, the most prevalent regulatory subunit is called G_M, whereas, in the liver, the most prevalent subunit is G_L. These regulatory subunits have modular structures with domains that participate in interactions with glycogen, with the catalytic subunit, and

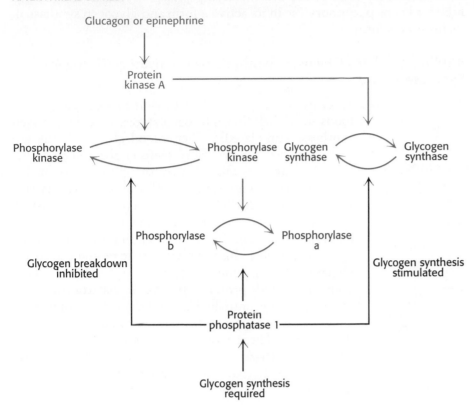

Figure 21.18 Regulation of glycogen synthesis by protein phosphatase 1. Protein phosphatase 1 stimulates glycogen synthesis while inhibiting glycogen breakdown.

with target enzymes. Thus, *these regulatory subunits act as scaffolds, bringing together the phosphatase and its substrates in the context of a glycogen particle.*

The phosphatase activity of PP1 must be reduced when glycogen degradation is called for (Figure 21.19). In such cases, epinephrine or glucagon has activated the cAMP cascade and protein kinase A is active. Protein kinase A reduces the activity of PP1 by two mechanisms. First, in muscle, G_M is phosphorylated in the domain responsible for binding the catalytic subunit. The catalytic subunit is released from glycogen and from its substrates and dephosphoryation is greatly reduced. Second, almost all tissues contain small proteins that, when phosphorylated, bind to the catalytic subunit of PP1 and inhibit it. Thus, when glycogen degradation is

DURING EXERCISE OR FASTING

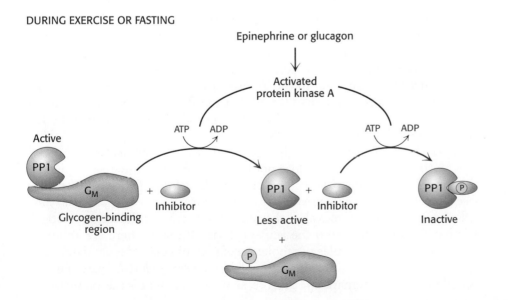

Figure 21.19 Regulation of protein phosphatase 1 (PP1) in muscle takes place in two steps. Phosphorylation of G_M by protein kinase A dissociates the catalytic subunit from its substrates in the glycogen particle. Phosphorylation of the inhibitor subunit by protein kinase A inactivates the catalytic unit of PP1.

switched on by cAMP, the accompanying phosphorylation of these inhibitors keeps phosphorylase in its active *a* form and glycogen synthase in its inactive *b* form.

Insulin Stimulates Glycogen Synthesis by Inactivating Glycogen Synthase Kinase

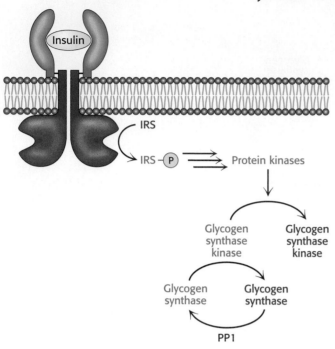

After exercise, people often consume carbohydrate-rich foods to restock their glycogen stores. How is glycogen synthesis stimulated? When blood-glucose levels are high, *insulin stimulates the synthesis of glycogen by inactivating glycogen synthase kinase*, the enzyme that maintains glycogen synthase in its phosphorylated, inactive state (Figure 21.20). The first step in the action of insulin is its binding to a receptor tyrosine kinase in the plasma membrane (Section 14.2). The binding of insulin activates the tyrosine kinase activity of the receptor so that it phosphorylates insulin-receptor substrates (IRSs). These phosphorylated proteins trigger signal-transduction pathways that eventually lead to the activation of protein kinases that phosphorylate and inactivate glycogen synthase kinase. The inactive kinase can no longer maintain glycogen synthase in its phosphorylated, inactive state. Protein phosphatase 1 dephosphorylates glycogen synthase, activating it, and restoring glycogen reserves.

Figure 21.20 Insulin inactivates glycogen synthase kinase. Insulin triggers a cascade that leads to the phosphorylation and inactivation of glycogen synthase kinase and prevents the phosphorylation of glycogen synthase. Protein phosphatase 1 (PP1) removes the phosphates from glycogen synthase, thereby activating the enzyme and allowing glycogen synthesis. IRS, insulin-receptor substrate.

Glycogen Metabolism in the Liver Regulates the Blood-Glucose Level

After a meal rich in carbohydrates, blood-glucose levels rise, and glycogen synthesis is stepped up in the liver. Although insulin is the primary signal for glycogen synthesis, another is the concentration of glucose in the blood, which normally ranges from about 80 to 120 mg per 100 ml (4.4–6.7 mM). The liver senses the concentration of glucose in the blood and takes up or releases glucose accordingly. The amount of liver phosphorylase *a* decreases rapidly when glucose is infused (Figure 21.21). After a lag period, the amount of glycogen synthase *a* increases, which results in glycogen synthesis. In fact, *phosphorylase a is the glucose sensor in liver cells.* The binding of glucose to phosphorylase *a* shifts its allosteric equilibrium from the active R form to the inactive T form. This conformational change *renders the phosphoryl group on serine 14 a substrate for protein phosphatase 1.* PP1 binds tightly to phosphorylase *a* only when the phosphorylase is in the R state but is inactive when bound. When glucose induces the transition to the T form, PP1 dissociates from the phosphorylase and becomes active. Recall that the R ⟷ T transition of muscle phosphorylase *a* is unaffected by glucose and is thus unaffected by the rise in blood-glucose levels (p. 598).

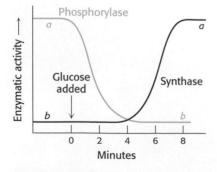

Figure 21.21 Blood glucose regulates liver-glycogen metabolism. The infusion of glucose into the bloodstream leads to the inactivation of phosphorylase, followed by the activation of glycogen synthase, in the liver. [After W. Stalmans, H. De Wulf, L. Hue, and H.-G. Hers. *Eur. J. Biochem.* 41(1974):117–134.]

How does glucose activate glycogen synthase? Phosphorylase *b*, in contrast with phosphorylase *a*, does not bind the phosphatase. Consequently, the conversion of *a* into *b* is accompanied by the *release of PP1, which is then free to activate glycogen synthase* and *dephosphorylate glycogen phosphorylase* (Figure 21.22). The removal of the phosphoryl group of inactive glycogen synthase *b* converts it into the active *a* form. Initially, there are about 10 phosphorylase *a* molecules per molecule of phosphatase. Hence, *the activity of glycogen synthase begins to increase only after most of phosphorylase* a *is converted into* b. This remarkable glucose-sensing system depends on three key

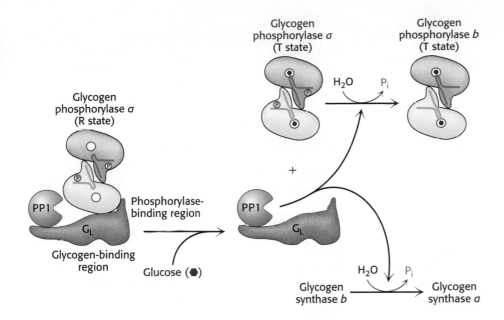

Glycogen
phosphorylase *a*
(R state)

Phosphorylase-
binding region

PP1

G_L

Glycogen-binding
region

Glucose (●)

Glycogen
phosphorylase *a*
(T state)

H_2O P_i

Glycogen
phosphorylase *b*
(T state)

+

PP1

G_L

H_2O P_i

Glycogen
synthase *b*

Glycogen
synthase *a*

Figure 21.22 Glucose regulation of liver-glycogen metabolism. Glucose binds to and inhibits glycogen phosphorylase *a* in the liver, facilitating the formation of the T state of phosphorylase *a*. The T state of phosphorylase *a* does not bind protein phosphate 1 (PP1), leading to the dissociation and activation of PP1 from glycogen phosphorylase *a*. The free PP1 dephosphorylates glycogen phosphorylase *a* and glycogen synthase *b*, leading to the inactivation of glycogen breakdown and the activation of glycogen synthesis.

elements: (1) communication between the allosteric site for glucose and the serine phosphate, (2) the use of PP1 to inactivate phosphorylase and activate glycogen synthase, and (3) the binding of the phosphatase to phosphorylase *a* to prevent the premature activation of glycogen synthase.

A Biochemical Understanding of Glycogen-Storage Diseases Is Possible

Edgar von Gierke described the first glycogen-storage disease in 1929. A patient with this disease has a huge abdomen caused by a *massive enlargement of the liver.* There is a pronounced *hypoglycemia* between meals. Furthermore, the blood-glucose level does not rise on administration of epinephrine and glucagon. An infant with this glycogen-storage disease may have convulsions because of the low blood-glucose level.

The enzymatic defect in von Gierke disease was elucidated in 1952 by Carl and Gerty Cori. They found that *glucose 6-phosphatase is missing from the liver of a patient with this disease.* This finding was the first demonstration of an inherited deficiency of a liver enzyme. The liver glycogen is normal in structure but present in abnormally large amounts. The absence of glucose 6-phosphatase in the liver causes hypoglycemia because glucose cannot be formed from glucose 6-phosphate. This phosphorylated sugar does not leave the liver, because it cannot cross the plasma membrane. The presence of excess glucose 6-phosphate triggers an increase in glycolysis in the liver, leading to a high level of lactate and pyruvate in the blood. Patients who have von Gierke disease also have an increased dependence on fat metabolism. This disease can also be produced by a mutation in the gene that encodes the *glucose 6-phosphate transporter.* Recall that glucose 6-phosphate must be transported into the lumen of the endoplasmic reticulum to be hydrolyzed by phosphatase (p. 463). Mutations in the other three essential proteins of this system can likewise lead to von Gierke disease.

Seven other glycogen-storage diseases have been characterized (Table 21.1). In Pompe disease (type II), lysosomes become engorged with glycogen because they lack α-1,4-glucosidase, a hydrolytic enzyme confined to these organelles (Figure 21.23). The Coris elucidated the biochemical defect in another glycogen-storage disease (type III), which cannot be distinguished from von Gierke disease (type I) by physical examination alone. In type III disease, the structure of liver and muscle glycogen is abnormal and

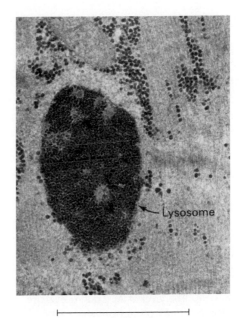

← Lysosome

|⟵ 1 μm ⟶|

Figure 21.23 Glycogen-engorged lysosome. This electron micrograph shows skeletal muscle from an infant with type II glycogen-storage disease (Pompe disease). The lysosomes are filled with glycogen because of a deficiency in α-1,4-glucosidase, a hydrolytic enzyme confined to lysosomes. The amount of glycogen in the cytoplasm is normal. [From H.-G. Hers and F. Van Hoof, Eds., *Lysosomes and Storage Diseases* (Academic Press, 1973), p. 205.]

TABLE 21.1 Glycogen-storage diseases

Type	Defective enzyme	Organ affected	Glycogen in the affected organ	Clinical features
I Von Gierke disease	Glucose 6-phosphatase or transport system	Liver and kidney	Increased amount; normal structure.	Massive enlargement of the liver. Failure to thrive. Severe hypoglycemia, ketosis, hyperuricemia, hyperlipemia.
II Pompe disease	α-1,4-Glucosidase (lysosomal)	All organs	Massive increase in amount; normal structure.	Cardiorespiratory failure causes death, usually before age 2.
III Cori disease	Amylo-1,6-glucosidase (debranching enzyme)	Muscle and liver	Increased amount; short outer branches.	Like type I, but milder course.
IV Andersen disease	Branching enzyme (α-1,4 ⟶ α-1,6)	Liver and spleen	Normal amount; very long outer branches.	Progressive cirrhosis of the liver. Liver failure causes death, usually before age 2.
V McArdle disease	Phosphorylase	Muscle	Moderately increased amount; normal structure.	Limited ability to perform strenuous exercise because of painful muscle cramps. Otherwise patient is normal and well developed.
VI Hers disease	Phosphorylase	Liver	Increased amount.	Like type I, but milder course.
VII	Phosphofructokinase	Muscle	Increased amount; normal structure.	Like type V.
VIII	Phosphorylase kinase	Liver	Increased amount; normal structure.	Mild liver enlargement. Mild hypoglycemia.

Note: Types I through VII are inherited as autosomal recessives. Type VIII is sex linked.

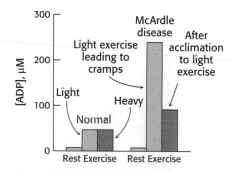

Figure 21.24 NMR study of human arm muscle. The level of ADP during exercise increases much more in a patient with McArdle glycogen-storage disease (type V) than in normal controls. [After G. K. Radda. *Biochem. Soc. Trans.* 14(1986):517–525.]

the amount is markedly increased. Most striking, the outer branches of the glycogen are very short. *Patients having this type lack the debranching enzyme (α-1,6-glucosidase),* and so only the outermost branches of glycogen can be effectively utilized. Thus, only a small fraction of this abnormal glycogen is functionally active as an accessible store of glucose.

A defect in glycogen metabolism confined to muscle is found in McArdle disease (type V). *Muscle phosphorylase activity is absent,* and a patient's capacity to perform strenuous exercise is limited because of painful muscle cramps. The patient is otherwise normal and well developed. Thus, effective utilization of muscle glycogen is not essential for life. Phosphorus-31 nuclear magnetic resonance studies of these patients have been very informative. The pH of skeletal-muscle cells of normal people drops during strenuous exercise because of the production of lactate. In contrast, the muscle cells of patients with McArdle disease become more alkaline during exercise because of the breakdown of creatine phosphate (p. 416). Lactate does not accumulate in these patients, because the glycolytic rate of their muscle is much lower than normal; their glycogen cannot be mobilized. NMR studies have also shown that the painful cramps in this disease are correlated with high levels of ADP (Figure 21.24). NMR spectroscopy is a valuable, noninvasive technique for assessing dietary and exercise therapy for this disease.

Summary

Glycogen, a readily mobilized fuel store, is a branched polymer of glucose residues. Most of the glucose units in glycogen are linked by α-1,4-glycosidic bonds. At about every tenth residue, a branch is created by an α-1,6-glycosidic bond. Glycogen is present in large amounts in muscle cells and in liver cells, where it is stored in the cytoplasm in the form of hydrated granules.

21.1 Glycogen Breakdown Requires the Interplay of Several Enzymes

Most of the glycogen molecule is degraded to glucose 1-phosphate by the action of glycogen phosphorylase, the key enzyme in glycogen breakdown. The glycosidic linkage between C-1 of a terminal residue and C-4 of the adjacent one is split by orthophosphate to give glucose 1-phosphate, which can be reversibly converted into glucose 6-phosphate. Branch points are degraded by the concerted action of an oligosaccharide transferase and an α-1,6-glucosidase.

21.2 Phosphorylase Is Regulated by Allosteric Interactions and Reversible Phosphorylation

Phosphorylase *b*, which is usually inactive, is converted into active phosphorylase *a* by the phosphorylation of a single serine residue in each subunit. This reaction is catalyzed by phosphorylase kinase. The *b* form in muscle can also be activated by the binding of AMP, an effect counteracted by ATP and glucose 6-phosphate. The *a* form in the liver is inhibited by glucose. The AMP-binding sites and phosphorylation sites are located at the subunit interface. In muscle, phosphorylase is activated to generate glucose for use inside the cell as a fuel for contractile activity. In contrast, liver phosphorylase is activated to liberate glucose for export to other organs, such as skeletal muscle and the brain.

21.3 Epinephrine and Glucagon Signal the Need for Glycogen Breakdown

Epinephrine and glucagon stimulate glycogen breakdown through specific 7TM receptors. Muscle is the primary target of epinephrine, whereas the liver is responsive to glucagon. Both signal molecules initiate a kinase cascade that leads to the activation of glycogen phosphorylase.

21.4 Glycogen Is Synthesized and Degraded by Different Pathways

Glycogen is synthesized by a different pathway from that of glycogen breakdown. UDP-glucose, the activated intermediate in glycogen synthesis, is formed from glucose 1-phosphate and UTP. Glycogen synthase catalyzes the transfer of glucose from UDP-glucose to the C-4 hydroxyl group of a terminal residue in the growing glycogen molecule. Synthesis is primed by glycogenin, an autoglycosylating protein that contains a covalently attached oligosaccharide unit on a specific tyrosine residue. A branching enzyme converts some of the α-1,4 linkages into α-1,6 linkages to increase the number of ends so that glycogen can be made and degraded more rapidly.

21.5 Glycogen Breakdown and Synthesis Are Reciprocally Regulated

Glycogen synthesis and degradation are coordinated by several amplifying reaction cascades. Epinephrine and glucagon stimulate glycogen breakdown and inhibit its synthesis by increasing the cytoplasmic level of cyclic AMP, which activates protein kinase A. Protein kinase A activates glycogen breakdown by attaching a phosphate to phosphorylase kinase and inhibits glycogen synthesis by phosphorylating glycogen synthase.

The glycogen-mobilizing actions of protein kinase A are reversed by protein phosphatase 1, which is regulated by several hormones. Epinephrine inhibits this phosphatase by blocking its attachment to glycogen molecules and by turning on an inhibitor. Insulin, in contrast, triggers a cascade that phosphorylates and inactivates glycogen synthase kinase, one of the enzymes that inhibits glycogen synthase.

Hence, glycogen synthesis is decreased by epinephrine and increased by insulin. Glycogen synthase and phosphorylase are also regulated by noncovalent allosteric interactions. In fact, phosphorylase is a key part of the glucose-sensing system of liver cells. Glycogen metabolism exemplifies the power and precision of reversible phosphorylation in regulating biological processes.

Key Terms

glycogen phosphorylase (p. 594)

phosphorolysis (p. 594)

pyridoxal phosphate (PLP) (p. 596)

phosphorylase kinase (p. 600)

calmodulin (p. 601)

epinephrine (adrenaline) (p. 601)

glucagon (p. 601)

protein kinase A (PKA) (p. 603)

uridine diphosphate glucose
(UDP-glucose) (p. 604)

glycogen synthase (p. 605)

glycogenin (p. 606)

protein phosphatase 1 (PP1)
(p. 608)

insulin (p. 610)

Selected Readings

Where to Start

Krebs, E. G. 1993. Protein phosphorylation and cellular regulation I. *Biosci. Rep.* 13:127–142.

Fischer, E. H. 1993. Protein phosphorylation and cellular regulation II. *Angew. Chem. Int. Ed.* 32:1130–1137.

Johnson, L. N. 1992. Glycogen phosphorylase: Control by phosphorylation and allosteric effectors. *FASEB J.* 6:2274–2282.

Browner, M. F., and Fletterick, R. J. 1992. Phosphorylase: A biological transducer. *Trends Biochem. Sci.* 17:66–71.

Books and General Reviews

Shulman, R. G., and Rothman, D. L. 1996. Enzymatic phosphorylation of muscle glycogen synthase: A mechanism for maintenance of metabolic homeostasis. *Proc. Natl. Acad. Sci. U. S. A.* 93:7491–7495.

Roach, P. J., Cao, Y., Corbett, C. A., DePaoli, R. A., Farkas, I., Fiol, C. J., Flotow, H., Graves, P. R., Hardy, T. A., and Hrubey, T. W. 1991. Glycogen metabolism and signal transduction in mammals and yeast. *Adv. Enzyme Regul.* 31:101–120.

Shulman, G. I., and Landau, B. R. 1992. Pathways of glycogen repletion. *Physiol. Rev.* 72:1019–1035.

X-ray Crystallographic Studies

Buschiazzo, A., Ugalde, J. E., Guerin, M. E., Shepard, W., Ugalde, R. A., and Alzari, P. M. 2004. Crystal structure of glycogen synthase: Homologous enzymes catalyze glcogen synthesis and degradation. *EMBO J.* 23:3196–3205.

Gibbons, B. J., Roach, P. J., and Hurley, T. D. 2002. Cyrstal structure of the autocatalytic initiator of glycogen biosynthesis, glycogenin. *J. Mol. Biol.* 319:463–477.

Lowe, E. D., Noble, M. E., Skamnaki, V. T., Oikonomakos, N. G., Owen, D. J., and Johnson, L. N. 1997. The crystal structure of a phosphorylase kinase peptide substrate complex: Kinase substrate recognition. *EMBO J.* 16:6646–6658.

Barford, D., Hu, S. H., and Johnson, L. N. 1991. Structural mechanism for glycogen phosphorylase control by phosphorylation and AMP. *J. Mol. Biol.* 218:233–260.

Sprang, S. R., Withers, S. G., Goldsmith, E. J., Fletterick, R. J., and Madsen, N. B. 1991. Structural basis for the activation of glycogen phosphorylase *b* by adenosine monophosphate. *Science* 254:1367–1371.

Johnson, L. N., and Barford, D. 1990. Glycogen phosphorylase: The structural basis of the allosteric response and comparison with other allosteric proteins. *J. Biol. Chem.* 265:2409–2412.

Browner, M. F., Fauman, E. B., and Fletterick, R. J. 1992. Tracking conformational states in allosteric transitions of phosphorylase. *Biochemistry* 31:11297–11304.

Martin, J. L., Johnson, L. N., and Withers, S. G. 1990. Comparison of the binding of glucose and glucose 1-phosphate derivatives to T-state glycogen phosphorylase *b*. *Biochemistry* 29:10745–10757.

Priming of Glycogen Synthesis

Lomako, J., Lomako, W. M., and Whelan, W. J. 2004. Glycogenin: The primer for mammalian and yeast glycogen synthesis. *Biochim. Biophys. Acta* 1673:45–55.

Lin, A., Mu, J., Yang, J., and Roach, P. J. 1999. Self-glucosylation of glycogenin, the initiator of glycogen biosynthesis, involves an inter-subunit reaction. *Arch. Biochem. Biophys.* 363:163–170.

Roach, P. J., and Skurat, A. V. 1997. Self-glucosylating initiator proteins and their role in glycogen biosynthesis. *Prog. Nucleic Acid Res. Mol. Biol.* 57:289–316.

Smythe, C., and Cohen, P. 1991. The discovery of glycogenin and the priming mechanism for glycogen biogenesis. *Eur. J. Biochem.* 200:625–631.

Catalytic Mechanisms

Skamnaki, V. T., Owen, D. J., Noble, M. E., Lowe, E. D., Lowe, G., Oikonomakos, N. G., and Johnson, L. N. 1999. Catalytic mechanism of phosphorylase kinase probed by mutational studies. *Biochemistry* 38:14718–14730.

Buchbinder, J. L., and Fletterick, R. J. 1996. Role of the active site gate of glycogen phosphorylase in allosteric inhibition and substrate binding. *J. Biol. Chem.* 271:22305–22309.

Palm, D., Klein, H. W., Schinzel, R., Buehner, M., and Helmreich, E. J. M. 1990. The role of pyridoxal 5′-phosphate in glycogen phosphorylase catalysis. *Biochemistry* 29:1099–1107.

Regulation of Glycogen Metabolism

Jope, R. S., and Johnson, G. V. W. 2004. The glamour and gloom of glycogen synthase kinase-3. *Trends Biochem. Sci.* 29:95–102.

Doble, B. W., and Woodgett, J. R. 2003. GSK-3: Tricks of the trade for a multi-tasking kinase. *J. Cell Sci.* 116:1175–1186.

Pederson, B. A., Cheng, C., Wilson, W. A., and Roach, P. J. 2000. Regulation of glycogen synthase: Identification of residues involved in regulation by the allosteric ligand glucose-6-P and by phosphorylation. *J. Biol. Chem.* 275:27753–27761.

Melendez, R., Melendez-Hevia, E., and Canela, E. I. 1999. The fractal structure of glycogen: A clever solution to optimize cell metabolism. *Biophys. J.* 77:1327–1332.

Franch, J., Aslesen, R., and Jensen, J. 1999. Regulation of glycogen synthesis in rat skeletal muscle after glycogen-depleting contractile activity: Effects of adrenaline on glycogen synthesis and activation of glycogen synthase and glycogen phosphorylase. *Biochem. J.* 344(pt.1):231–235.

Aggen, J. B., Nairn, A. C., and Chamberlin, R. 2000. Regulation of protein phosphatase-1. *Chem. Biol.* 7:R13–R23.

Egloff, M. P., Johnson, D. F., Moorhead, G., Cohen, P. T., Cohen, P., and Barford, D. 1997. Structural basis for the recognition of regulatory subunits by the catalytic subunit of protein phosphatase 1. *EMBO J.* 16:1876–1887.

Wu, J., Liu, J., Thompson, I., Oliver, C. J., Shenolikar, S., and Brautigan, D. L. 1998. A conserved domain for glycogen binding in protein phosphatase-1 targeting subunits. *FEBS Lett.* 439:185–191.

Genetic Diseases

Chen, Y.-T., and Burchell, A. 1995. Glycogen storage diseases. In *The Metabolic Basis of Inherited Diseases* (7th ed., pp. 935–965), edited by C. R. Scriver., A. L. Beaudet, W. S. Sly, D. Valle, J. B. Stanbury, J. B. Wyngaarden, and D. S. Fredrickson. McGraw-Hill.

Burchell, A., and Waddell, I. D. 1991. The molecular basis of the hepatic microsomal glucose-6-phosphatase system. *Biochim. Biophys. Acta* 1092:129–137.

Lei, K. J., Shelley, L. L., Pan, C. J., Sidbury, J. B., and Chou, J. Y. 1993. Mutations in the glucose-6-phosphatase gene that cause glycogen storage disease type Ia. *Science* 262:580–583.

Ross, B. D., Radda, G. K., Gadian, D. G., Rocker, G., Esiri, M., and Falconer-Smith, J. 1981. Examination of a case of suspected McArdle's syndrome by ^{31}P NMR. *N. Engl. J. Med.* 304:1338–1342.

Evolution

Holm, L., and Sander, C. 1995. Evolutionary link between glycogen phosphorylase and a DNA modifying enzyme. *EMBO J.* 14:1287–1293.

Hudson, J. W., Golding, G. B., and Crerar, M. M. 1993. Evolution of allosteric control in glycogen phosphorylase. *J. Mol. Biol.* 234:700–721.

Rath, V. L., and Fletterick, R. J. 1994. Parallel evolution in two homologues of phosphorylase. *Nat. Struct. Biol.* 1:681–690.

Melendez, R., Melendez-Hevia, E., and Cascante, M. 1997. How did glycogen structure evolve to satisfy the requirement for rapid mobilization of glucose? A problem of physical constraints in structure building. *J. Mol. Evol.* 45:446–455.

Rath, V. L., Lin, K., Hwang, P. K., and Fletterick, R. J. 1996. The evolution of an allosteric site in phosphorylase. *Structure* 4:463 473.

Problems

1. *Carbohydrate conversion.* Write a balanced equation for the formation of glycogen from galactose.

2. *If a little is good, a lot is better.* α-Amylose is an unbranched glucose polymer. Why would this polymer not be as effective a storage form of glucose as glycogen?

3. *Telltale products.* A sample of glycogen from a patient with liver disease is incubated with orthophosphate, phosphorylase, the transferase, and the debranching enzyme (α-1,6-glucosidase). The ratio of glucose 1-phosphate to glucose formed in this mixture is 100. What is the most likely enzymatic deficiency in this patient?

4. *Excessive storage.* Suggest an explanation for the fact that the amount of glycogen in type I glycogen-storage disease (von Gierke disease) is increased.

5. *A shattering experience.* Crystals of phosphorylase *a* grown in the presence of glucose shatter when a substrate such as glucose 1-phosphate is added. Why?

6. *Recouping an essential phosphoryl.* The phosphoryl group on phosphoglucomutase is slowly lost by hydrolysis. Propose a mechanism that utilizes a known catalytic intermediate for restoring this essential phosphoryl group. How might this phosphoryl donor be formed?

7. *Hydrophobia.* Why is water excluded from the active site of phosphorylase? Predict the effect of a mutation that allows water molecules to enter.

8. *Removing all traces.* In human liver extracts, the catalytic activity of glycogenin was detectable only after treatment with α-amylase (p. 606). Why was α-amylase necessary to reveal the glycogenin activity?

9. *Two in one.* A single polypeptide chain houses the transferase and debranching enzyme. Cite a potential advantage of this arrangement.

10. *How did they do that?* A strain of mice has been developed that lack the enzyme phosphorylase kinase. Yet, after strenuous exercise, the glycogen stores of a mouse of this strain are depleted. Explain how this depletion is possible.

11. *Metabolic mutants.* Predict the major consequence of each of the following mutations:

(a) Loss of the AMP-binding site in muscle phosphorylase.
(b) Mutation of Ser 14 to Ala 14 in liver phosphorylase.
(c) Overexpression of phosphorylase kinase in the liver.
(d) Loss of the gene that encodes inhibitor 1 of protein phosphatase 1.
(e) Loss of the gene that encodes the glycogen-targeting subunit of protein phosphatase 1.
(f) Loss of the gene that encodes glycogenin.

12. *More metabolic mutants.* Briefly, predict the major consequences of each of the following mutations affecting glycogen utilization.

(a) Loss of GTPase activity of the G-protein α subunit.
(b) Loss of the gene that encodes inhibitor 1 of protein phosphatase 1.
(c) Loss of phosphodiesterase activity.

13. *Multiple phosphorylations.* Protein kinase A activates muscle phosphorylase kinase by rapidly phosphorylating its β subunits. The α subunits of phosphorylase kinase are then slowly phosphorylated, which makes the α and β subunits susceptible to the action of protein phosphatase 1. What is the functional significance of the slow phosphorylation of α?

14. *The wrong switch.* What would be the consequences for glycogen mobilization of a mutation in phosphorylase kinase that leads to the phosphorylation of the α subunit before that of the β subunit?

Mechanism Problem

15. *Family resemblance.* Propose mechanisms for the two enzymes catalyzing steps in glycogen debranching on the basis of their potential membership in the α-amylase family.

Chapter Integration and Data Interpretation Problems

16. *Glycogen isolation 1.* The liver is a major storage site for glycogen. Purified from two samples of human liver, glycogen was either treated or not treated with α-amylase and subsequently analyzed by SDS-PAGE and Western blotting with the use of antibodies to glycogenin. The results are presented in the adjoining illustration

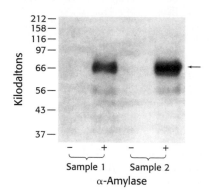

Glycogen isolation 1. [Courtesy of Dr. Peter J. Roach, Indiana University School of Medicine.]

(a) Why are no proteins visible in the lanes without amylase treatment?
(b) What is the effect of treating the samples with α-amylase? Explain the results.
(c) List other proteins that you might expect to be associated with glycogen. Why are other proteins not visible?

17. *Glycogen isolation 2.* The gene for glycogenin was transfected into a cell line that normally stores only small amounts of glycogen. The cells were then manipulated according to the following protocol, and glycogen was isolated and analyzed by SDS-PAGE and Western blotting by using an antibody to glycogenin with and without α-amylase treatment. The results are presented in the adjoining illustration.

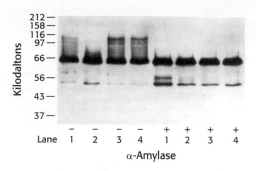

Glycogen isolation 2. [Courtesy of Dr. Peter J. Roach, Indiana University School of Medicine.]

The protocol: Cells cultured in growth medium and 25 mM glucose (lane 1) were switched to medium containing no glucose for 24 hours (lane 2). Glucose-starved cells were refed with medium containing 25 mM glucose for 1 hour (lane 3) or 3 hours (lane 4). Samples (12 μg of protein) were either treated or not treated with α-amylase, as indicated, before being loaded on the gel.

(a) Why did the Western analysis produce a "smear"—that is, the high-molecular-weight staining in lane 1(−)?
(b) What is the significance of the decrease in high-molecular-weight staining in lane 2(−)?
(c) What is the significance of the difference between lanes 2(−) and 3(−)?
(d) Suggest a plausible reason why there is essentially no difference between lanes 3(−) and 4(−)?
(e) Why are the bands at 66 kd the same in the lanes treated with amylase, despite the fact that the cells were treated differently?

Chapter 22

Fatty Acid Metabolism

Fats provide efficient means for storing energy for later use. (Right) The processes of fatty acid synthesis (preparation for energy storage) and fatty acid degradation (preparation for energy use) are, in many ways, the reverse of each other. (Above) Studies of mice are revealing the interplay between these pathways and the biochemical bases of appetite and weight control. [(Above) © Jackson/Visuals Unlimited.]

We turn now from the metabolism of carbohydrates to that of fatty acids. A fatty acid contains a long hydrocarbon chain and a terminal carboxylate group. Fatty acids have four major physiological roles. First, *fatty acids are fuel molecules.* They are stored as *triacylglycerols* (also called *neutral fats* or *triglycerides*), which are uncharged esters of fatty acids with glycerol (Figure 22.1). Fatty acids mobilized from triacylglycerols are oxidized to meet the energy needs of a cell or organism. During rest or moderate exercise, such as walking, fatty acids are our primary source of energy. Second, *fatty acids are building blocks of phospholipids and glycolipids.* These amphipathic molecules are important components of biological membranes, as discussed in Chapter 12. Third, many proteins are modified by the *covalent attachment of fatty acids, which targets them to membrane locations* (p. 340). Fourth, *fatty acid derivatives serve as hormones and intracellular messengers.* In this chapter, we focus on the degradation and synthesis of fatty acids.

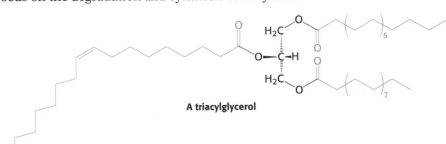

A triacylglycerol

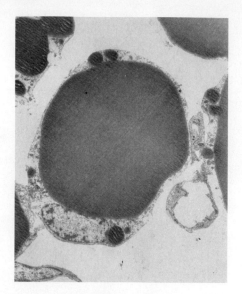

Figure 22.1 Electron micrograph of an adipocyte. A small band of cytoplasm surrounds the large deposit of triacylglycerols. [Biophoto Associates/ Photo Researchers.]

Fatty Acid Degradation and Synthesis Mirror Each Other in Their Chemical Reactions

Fatty acid degradation and synthesis consist of four steps that are the reverse of each other in their basic chemistry. Degradation is an oxidative process that converts a fatty acid into a set of activated acetyl units (acetyl CoA) that can be processed by the citric acid cycle (Figure 22.2). An activated fatty acid is oxidized to introduce a double bond; the double bond is hydrated to introduce a hydroxyl group; the alcohol is oxidized to a ketone; and, finally, the fatty acid is cleaved by coenzyme A to yield acetyl CoA and a fatty acid chain two carbons shorter. If the fatty acid has an even number of carbon atoms and is saturated, the process is simply repeated until the fatty acid is completely converted into acetyl CoA units.

Fatty acid synthesis is essentially the reverse of this process. The process starts with the individual units to be assembled—in this case with an activated acyl group (most simply, an acetyl unit) and a malonyl unit (see Figure 22.2). The malonyl unit condenses with the acetyl unit to form a four-carbon fragment. To produce the required hydrocarbon chain, the carbonyl group is reduced to a methylene group in three steps: a reduction,

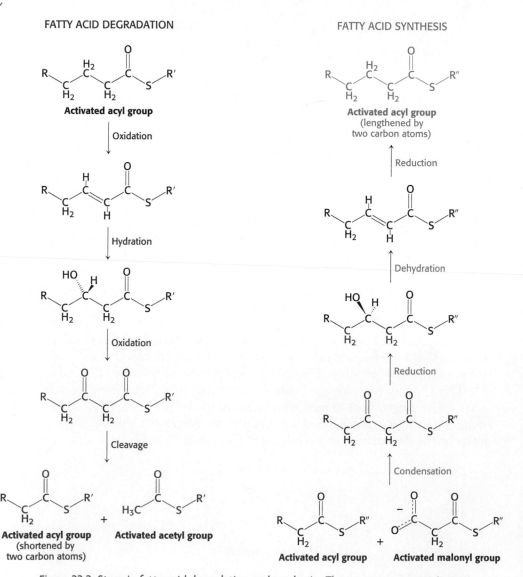

Figure 22.2 Steps in fatty acid degradation and synthesis. The two processes are in many ways mirror images of each other.

a dehydration, and another reduction, exactly the opposite of degradation. The product of the reduction is butyryl CoA. Another activated malonyl group condenses with the butyryl unit, and the process is repeated until a C_{16} fatty acid is synthesized.

22.1 Triacylglycerols Are Highly Concentrated Energy Stores

Triacylglycerols are highly concentrated stores of metabolic energy because they are *reduced* and *anhydrous*. The yield from the complete oxidation of fatty acids is about 38 kJ g^{-1} (9 kcal g^{-1}), in contrast with about 17 kJ g^{-1} (4 kcal g^{-1}) for carbohydrates and proteins. The basis of this large difference in caloric yield is that fatty acids are much more reduced. Furthermore, triacylglycerols are nonpolar, and so they are stored in a nearly anhydrous form, whereas much more polar carbohydrates are more highly hydrated. In fact, 1 g of dry glycogen binds about 2 g of water. Consequently, *a gram of nearly anhydrous fat stores 6.75 times as much energy as a gram of hydrated glycogen,* which is likely the reason that triacylglycerols rather than glycogen were selected in evolution as the major energy reservoir. Consider a typical 70-kg man, who has fuel reserves of 420,000 kJ (100,000 kcal) in triacylglycerols, 100,000 kJ (24,000 kcal) in protein (mostly in muscle), 2500 kJ (600 kcal) in glycogen, and 170 kJ (40 kcal) in glucose. Triacylglycerols constitute about 11 kg of his total body weight. If this amount of energy were stored in glycogen, his total body weight would be 64 kg greater. The glycogen and glucose stores provide enough energy to sustain physiological function for about 24 hours, whereas the triacylglycerol stores allow survival for several weeks.

In mammals, the major site of triacylglycerol accumulation is the cytoplasm of *adipose cells (fat cells)*. Droplets of triacylglycerol coalesce to form a large globule, which may occupy most of the cell volume (see Figure 22.1). Adipose cells are specialized for the synthesis and storage of triacylglycerols and for their mobilization into fuel molecules that are transported to other tissues by the blood. Muscle also stores triacylglycerols for its own energy needs. Indeed, triacylglycerols are evident as the "marbling" of expensive cuts of beef.

The utility of triacylglycerols as an energy source is dramatically illustrated by the abilities of migratory birds, which can fly great distances without eating after having stored energy as triacylglycerols. Examples are the American golden plover and the ruby-throated hummingbird. The golden plover flies from Alaska to the southern tip of South America; a large segment of the flight (3800 km, or 2400 miles) is over open ocean, where the birds cannot feed. The ruby-throated hummingbird can fly nonstop across the Gulf of Mexico. Fatty acids provide the energy source for both these prodigious feats.

Triacylglycerols fuel the long migration flights of the American golden plover *(Pluvialis dominica).* [Gerard Fuehrer/ Visuals Unlimited.]

Dietary Lipids Are Digested by Pancreatic Lipases

Most lipids are ingested in the form of triacylglycerols and must be degraded to fatty acids for absorption across the intestinal epithelium. Intestinal enzymes called *lipases,* secreted by the pancreas, degrade triacylglycerols to free fatty acids and monoacylglycerol (Figure 22.3). Lipids present a special problem because, unlike carbohydrates and proteins, these molecules are not soluble in water. How are they made accessible to the lipases, which are in aqueous solution? The solution is to wrap lipids in a

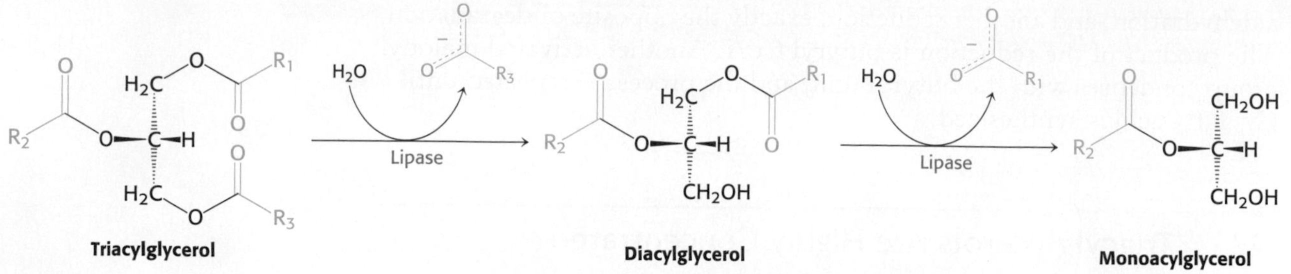

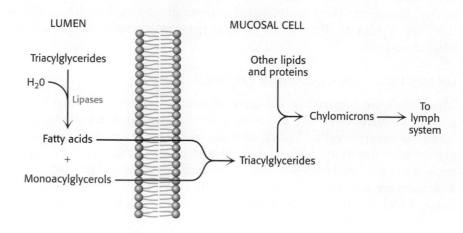

Triacylglycerol ... **Diacylglycerol** ... **Monoacylglycerol**

Figure 22.3 Action of pancreatic lipases. Lipases secreted by the pancreas convert triacylglycerols into fatty acids and monoacylglycerol for absorption into the intestine.

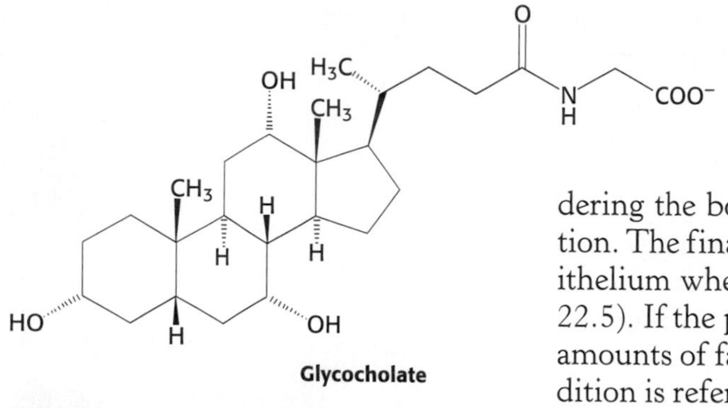

Glycocholate

Figure 22.4 Glycocholate. Bile salts, such as glycocholate, facilitate lipid digestion in the intestine.

soluble container. Triacylglycerols in the intestinal lumen are incorporated into micelles composed of *bile salts* (Figure 22.4), amphipathic molecules synthesized from cholesterol in the liver and secreted from the gall bladder. The ester bond of each lipid is oriented toward the surface of the micelle, rendering the bond more susceptible to digestion by lipases in aqueous solution. The final digestion products are carried in micelles to the intestinal epithelium where they are transported across the plasma membrane (Figure 22.5). If the production of bile salts is inadequate due to liver disease, large amounts of fats (as much as 30 g day^{-1}) are excreted in the feces. This condition is referred to as *steatorrhea*, after stearic acid, a common fatty acid.

Dietary Lipids Are Transported in Chylomicrons

In the intestinal mucosal cells, the triacylglycerols are resynthesized from fatty acids and monoacylglycerols and then packaged into lipoprotein transport particles called *chylomicrons*, stable particles approximately 2000 Å (200 nm) in diameter (see Figure 22.5). These particles are composed mainly of triacylglycerols, with apoliprotein B-48 (apo B-48) as the main protein component. Protein constituents of lipoprotein particles are called *apolipoproteins*. Chylomicrons also transport fat-soluble vitamins and cholesterol.

The chylomicrons are released into the lymph system and then into the blood. These particles bind to membrane-bound lipases, primarily at adipose tissue and muscle, where the triacylglycerols are once again degraded into free fatty acids and monoacylglycerol for transport into the tissue. The triacylglycerols are then resynthesized inside the cell and stored. In the muscle, they can be oxidized to provide energy.

Figure 22.5 Chylomicron formation. Free fatty acids and monoacylglycerols are absorbed by intestinal epithelial cells. Triacylglycerols are resynthesized and packaged with other lipids and apolipoprotein B-48 to form chylomicrons, which are then released into the lymph system.

LUMEN ... MUCOSAL CELL

Triacylglycerides

H_2O

Lipases

Fatty acids

+

Monoacylglycerols

Other lipids and proteins

Chylomicrons → To lymph system

Triacylglycerides

22.2 The Utilization of Fatty Acids as Fuel Requires Three Stages of Processing

Peripheral tissues gain access to the lipid energy reserves stored in adipose tissue through three stages of processing. First, the lipids must be mobilized. In this process, triacylglycerols are degraded to fatty acids and glycerol, which are released from the adipose tissue and transported to the energy-requiring tissues. Second, at these tissues, the fatty acids must be activated and transported into mitochondria for degradation. Third, the fatty acids are broken down in a step-by-step fashion into acetyl CoA, which is then processed in the citric acid cycle.

Triacylglycerols Are Hydrolyzed by Hormone-Stimulated Lipases

Consider someone who has just awakened from a night's sleep and begins a bout of exercise. Glycogen stores will be low, but lipids are readily available. How are these lipid stores mobilized?

Before fats can be used as fuels, the triacylglycerol storage form must be hydrolyzed to yield isolated fatty acids. This reaction is catalyzed by a hormonally controlled lipase. Under the physiological conditions facing an early-morning runner, glucagon and epinephrine will be present. In adipose tissue, these hormones trigger 7TM receptors that activate adenylate cyclase (p. 383). The increased level of cyclic AMP then stimulates protein kinase A, which phosphorylates two key proteins: *perilipin A*, a fat-droplet-associated protein, and hormone-sensitive lipase. The phosphorylation of perilipin A restructures the fat droplet so that the triacylglycerols are more accessible to the hormone-sensitive lipase. The phosphorylated lipase hydrolyzes triacylglycerols to free fatty acids. Thus, *epinephrine and glucagon induce lipolysis* (Figure 22.6). Although their role in muscle is not as firmly established, these hormones probably also regulate the use of triacylglycerol stores in that tissue.

The released fatty acids are not soluble in blood plasma, and so serum albumin in the bloodstream binds the fatty acids and serves as a carrier. By these means, free fatty acids are made accessible as a fuel in other tissues.

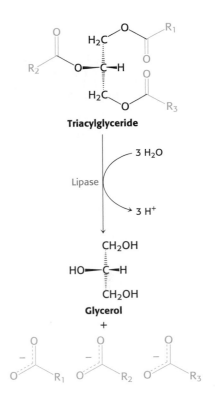

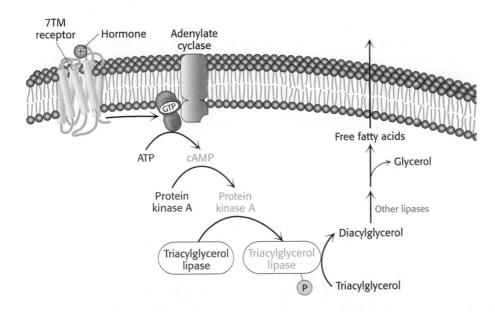

Figure 22.6 Mobilization of triacylglycerols. Triacylglycerols in adipose tissue are converted into free fatty acids and glycerol for release into the bloodstream in response to hormonal signals. A hormone-sensitive lipase initiates the process. Abbreviation: 7TM, seven-transmembrane helix.

Glycerol formed by lipolysis is absorbed by the liver and phosphory-lated. It is then oxidized to dihydroxyacetone phosphate, which is isomer-ized to glyceraldehyde 3-phosphate. This molecule is an intermediate in both the glycolytic and the gluconeogenic pathways.

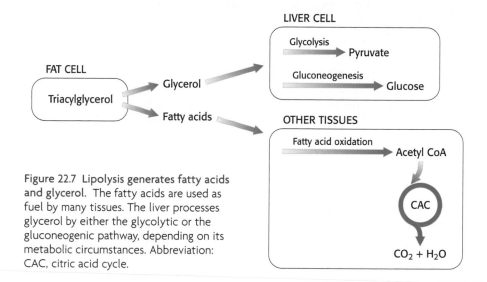

Glycerol → L-Glycerol 3-phosphate (Glycerol kinase, ATP → ADP) → Dihydroxyacetone phosphate (Glycerol phosphate dehydrogenase, NAD⁺ → NADH + H⁺) ⇌ D-Glyceraldehyde 3-phosphate

Hence, glycerol can be converted into pyruvate or glucose in the liver, which contains the appropriate enzymes (Figure 22.7). The reverse process can take place by the reduction of dihydroxyacetone phosphate to glycerol 3-phosphate. Hydrolysis by a phosphatase then gives glycerol. Thus, glycerol and glycolytic intermediates are readily interconvertible.

Figure 22.7 Lipolysis generates fatty acids and glycerol. The fatty acids are used as fuel by many tissues. The liver processes glycerol by either the glycolytic or the gluconeogenic pathway, depending on its metabolic circumstances. Abbreviation: CAC, citric acid cycle.

Fatty Acids Are Linked to Coenzyme A Before They Are Oxidized

Eugene Kennedy and Albert Lehninger showed in 1949 that fatty acids are oxidized in mitochondria. Subsequent work demonstrated that they are first activated through the formation of a thioester linkage to coenzyme A before they enter the mitochondrial matrix. Adenosine triphosphate drives the formation of the thioester linkage between the carboxyl group of a fatty acid and the sulfhydryl group of coenzyme A. This activation reaction takes place on the outer mitochondrial membrane, where it is catalyzed by *acyl CoA synthetase* (also called *fatty acid thiokinase*).

Acyl adenylate

Paul Berg showed that the activation of a fatty acid is accomplished in two steps. First, the fatty acid reacts with ATP to form an *acyl adenylate*. In this mixed anhydride, the carboxyl group of a fatty acid is bonded to the phosphoryl group of AMP. The other two phosphoryl groups of the ATP substate are released as pyrophosphate. In the second step, the sulfhydryl group of coenzyme A attacks the acyl adenylate, which is tightly bound to the enzyme, to form acyl CoA and AMP.

$$\text{Fatty acid} + \text{ATP} \rightleftharpoons \text{Acyl adenylate} + \text{PP}_i \quad (1)$$

$$\text{R-C(O)-AMP} + \text{HS-CoA} \rightleftharpoons \text{Acyl CoA} + \text{AMP} \quad (2)$$

These partial reactions are freely reversible. In fact, the equilibrium constant for the sum of these reactions is close to 1. One high-transfer-potential compound is cleaved (between PP$_i$ and AMP) and one high-transfer-potential compound is formed (the thioester acyl CoA). How is the overall reaction driven forward? The answer is that pyrophosphate is rapidly hydrolyzed by a pyrophosphatase, and so the complete reaction is

$$\text{RCOO}^- + \text{CoA} + \text{ATP} + \text{H}_2\text{O} \longrightarrow$$
$$\text{RCO-CoA} + \text{AMP} + 2\,\text{P}_i + 2\,\text{H}^+$$

This reaction is quite favorable because the equivalent of two molecules of ATP is hydrolyzed, whereas only one high-transfer-potential compound is formed. We see here another example of a recurring theme in biochemistry: *many biosynthetic reactions are made irreversible by the hydrolysis of inorganic pyrophosphate.*

Another motif recurs in this activation reaction. The enzyme-bound acyladenylate intermediate is not unique to the synthesis of acyl CoA. *Acyl adenylates are frequently formed when carboxyl groups are activated in biochemical reactions.* Amino acids are activated for protein synthesis by a similar mechanism (p. 862), although the enzymes that catalyze this process are not homologous to acyl CoA synthetase. Thus, *activation by adenylation recurs in part because of convergent evolution.*

Carnitine Carries Long-Chain Activated Fatty Acids into the Mitochondrial Matrix

Fatty acids are activated on the outer mitochondrial membrane, whereas they are oxidized in the mitochondrial matrix. A special transport mechanism is needed to carry activated long-chain fatty acids across the inner mitochondrial membrane. These fatty acids must be conjugated to *carnitine*, a zwitterionic alcohol. The acyl group is transferred from the sulfur atom of coenzyme A to the hydroxyl group of carnitine to form *acyl carnitine*. This reaction is catalyzed by *carnitine acyltransferase I*, also called *carnitine palmitoyl transferase I* (CPTI), which is bound to the outer mitochondrial membrane.

$$\text{Acyl CoA} + \text{Carnitine} \rightleftharpoons \text{Acyl carnitine} + \text{HS-CoA}$$

Acyl carnitine is then shuttled across the inner mitochondrial membrane by a translocase (Figure 22.8). The acyl group is transferred back to coenzyme

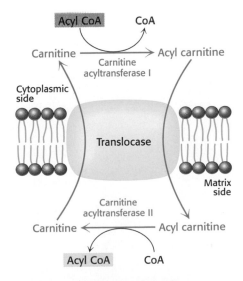

Figure 22.8 Acyl carnitine translocase. The entry of acyl carnitine into the mitochondrial matrix is mediated by a translocase. Carnitine returns to the cytoplasmic side of the inner mitochondrial membrane in exchange for acyl carnitine.

A on the matrix side of the membrane. This reaction, which is catalyzed by *carnitine acyltransferase II (carnitine palmitoyl transferase II)*, is simply the reverse of the reaction that takes place in the cytoplasm. The reaction is thermodynamically feasible because of the zwitterionic nature of carnitine. The O-acyl link in carnitine has a high group-transfer potential, apparently because, being zwitterions, carnitine and its esters are solvated differently from most other alcohols and their esters. Finally, the translocase returns carnitine to the cytoplasmic side in exchange for an incoming acyl carnitine.

A number of diseases have been traced to a deficiency of carnitine, the transferase, or the translocase. The symptoms of carnitine deficiency range from mild muscle cramping to severe weakness and even death. Muscle, kidney, and heart are the tissues primarily impaired. Muscle weakness during prolonged exercise is a symptom of a deficiency of carnitine acyltransferases because muscle relies on fatty acids as a long-term source of energy. Medium-chain (C_8–C_{10}) fatty acids are oxidized normally in these patients because these fatty acids do not require carnitine to enter the mitochondria. These diseases illustrate that *the impaired flow of a metabolite from one compartment of a cell to another can lead to a pathological condition.*

Acetyl CoA, NADH, and FADH₂ Are Generated in Each Round of Fatty Acid Oxidation

A saturated acyl CoA is degraded by a recurring sequence of four reactions: oxidation by flavin adenine dinucleotide (FAD), hydration, oxidation by NAD^+, and thiolysis by coenzyme A (Figure 22.9). The fatty acid chain is shortened by two carbon atoms as a result of these reactions, and $FADH_2$, NADH, and acetyl CoA are generated. Because oxidation takes place at the β carbon, this series of reactions is called the *β-oxidation pathway.*

The first reaction in each round of degradation is the *oxidation* of acyl CoA by an *acyl CoA dehydrogenase* to give an enoyl CoA with a trans double bond between C-2 and C-3.

$$\text{Acyl CoA} + \text{E-FAD} \longrightarrow \textit{trans}\text{-}\Delta^2\text{-enoyl CoA} + \text{E-FADH}_2$$

As in the dehydrogenation of succinate in the citric acid cycle, FAD rather than NAD^+ is the electron acceptor because the ΔG for this reaction is insufficient to drive the reduction of NAD^+. Electrons from the $FADH_2$ prosthetic group of the reduced acyl CoA dehydrogenase are transferred to a second flavoprotein called *electron-transferring flavoprotein* (ETF). In turn, ETF donates electrons to *ETF:ubiquinone reductase*, an iron–sulfur protein. Ubiquinone is thereby reduced to ubiquinol, which delivers its high-potential electrons to the second proton-pumping site of the respiratory chain (p. 512). Consequently, 1.5 molecules of ATP are generated per molecule of $FADH_2$ formed in this dehydrogenation step, as in the oxidation of succinate to fumarate.

The next step is the *hydration* of the double bond between C-2 and C-3 by *enoyl CoA hydratase.*

$$\textit{trans}\text{-}\Delta^2\text{-Enoyl CoA} + H_2O \longrightarrow \text{L-3-hydroxyacyl CoA}$$

The hydration of enoyl CoA is stereospecific. Only the L isomer of 3-hydroxyacyl CoA is formed when the trans-Δ^2 double bond is hydrated.

The enzyme also hydrates a cis-Δ^2 double bond, but the product then is the D isomer. We shall return to this point shortly in considering how unsaturated fatty acids are oxidized.

The hydration of enoyl CoA is a prelude to the second *oxidation* reaction, which converts the hydroxyl group at C-3 into a keto group and generates NADH. This oxidation is catalyzed by *L-3-hydroxyacyl CoA dehydrogenase*, which is specific for the L isomer of the hydroxyacyl substrate.

$$\text{L-3-Hydroxyacyl CoA} + \text{NAD}^+ \rightleftharpoons \text{3-ketoacyl CoA} + \text{NADH} + \text{H}^+$$

The preceding reactions have oxidized the methylene group at C-3 to a keto group. The final step is the *cleavage* of 3-ketoacyl CoA by the thiol group of a second molecule of coenzyme A, which yields acetyl CoA and an acyl CoA shortened by two carbon atoms. This thiolytic cleavage is catalyzed by *β-ketothiolase*.

$$\begin{array}{cc}\text{3-Ketoacyl CoA} + \text{HS-CoA} \rightleftharpoons & \text{acetyl CoA} + \text{acyl CoA} \\ (n \text{ carbons}) & (n-2 \text{ carbons})\end{array}$$

Table 22.1 summarizes the reactions in fatty acid oxidation.

The shortened acyl CoA then undergoes another cycle of oxidation, starting with the reaction catalyzed by acyl CoA dehydrogenase (Figure 22.10). Fatty acid chains containing from 12 to 18 carbon atoms are oxidized by the long-chain acyl CoA dehydrogenase. The medium-chain acyl CoA dehydrogenase oxidizes fatty acid chains having from 14 to 4 carbons, whereas the short-chain acyl CoA dehydrogenase acts only on 4- and 6-carbon fatty acid chains. In contrast, β-ketothiolase, hydroxyacyl dehydrogenase, and enoyl CoA hydratase act on fatty acid molecules of almost any length.

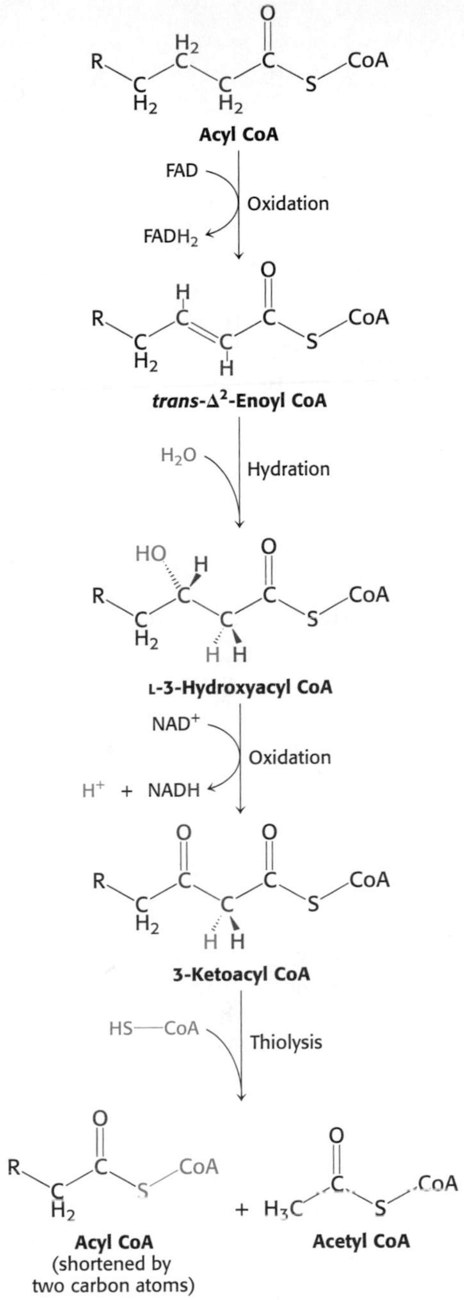

The Complete Oxidation of Palmitate Yields 106 Molecules of ATP

We can now calculate the energy yield derived from the oxidation of a fatty acid. In each reaction cycle, an acyl CoA is shortened by two carbon atoms, and one molecule each of $FADH_2$, NADH, and acetyl CoA are formed.

$$\begin{array}{l}\text{C}_n\text{-acyl CoA} + \text{FAD} + \text{NAD}^+ + \text{H}_2\text{O} + \text{CoA} \longrightarrow \\ \quad \text{C}_{n-2}\text{-acyl CoA} + \text{FADH}_2 + \text{NADH} + \text{acetyl CoA} + \text{H}^+\end{array}$$

The degradation of palmitoyl CoA (C_{16}-acyl CoA) requires seven reaction cycles. In the seventh cycle, the C_4-ketoacyl CoA is thiolyzed to two molecules of acetyl CoA. Hence, the stoichiometry of the oxidation of palmitoyl CoA is

$$\begin{array}{l}\text{Palmitoyl CoA} + 7\,\text{FAD} + 7\,\text{NAD}^+ + 7\,\text{CoA} + 7\,\text{H}_2\text{O} \longrightarrow \\ \quad 8\,\text{acetyl CoA} + 7\,\text{FADH}_2 + 7\,\text{NADH} + 7\,\text{H}^+\end{array}$$

Figure 22.9 Reaction sequence for the degradation of fatty acids. Fatty acids are degraded by the repetition of a four-reaction sequence consisting of oxidation, hydration, oxidation, and thiolysis.

TABLE 22.1 Principal reactions in fatty acid oxidation

Step	Reaction	Enzyme
1	Fatty acid + CoA + ATP $\rightleftharpoons$ acyl CoA + AMP + PP$_i$	Acyl CoA synthetase [also called fatty acid thiokinase and fatty acid:CoA ligase]*
2	Carnitine + acyl CoA $\rightleftharpoons$ acyl carnitine + CoA	Carnitine acyltransferase (also called carnitine palmitoyl transferase)
3	Acyl CoA + E-FAD $\longrightarrow$ trans-Δ^2-enoyl CoA + E-FADH$_2$	Acyl CoA dehydrogenases (several isozymes having different chain-length specificity)
4	trans-Δ^2-Enoyl CoA + H$_2$O $\rightleftharpoons$ L-3-hydroxyacyl CoA	Enoyl CoA hydratase (also called crotonase or 3-hydroxyacyl CoA hydrolyase)
5	L-3-Hydroxyacyl CoA + NAD$^+$ $\rightleftharpoons$ 3-ketoacyl CoA + NADH + H$^+$	L-3-Hydroxyacyl CoA dehydrogenase
6	3-Ketoacyl CoA + CoA $\rightleftharpoons$ acetyl CoA + acyl CoA (shortened by C$_2$)	β-Ketothiolase (also called thiolase)

* An AMP-forming ligase.

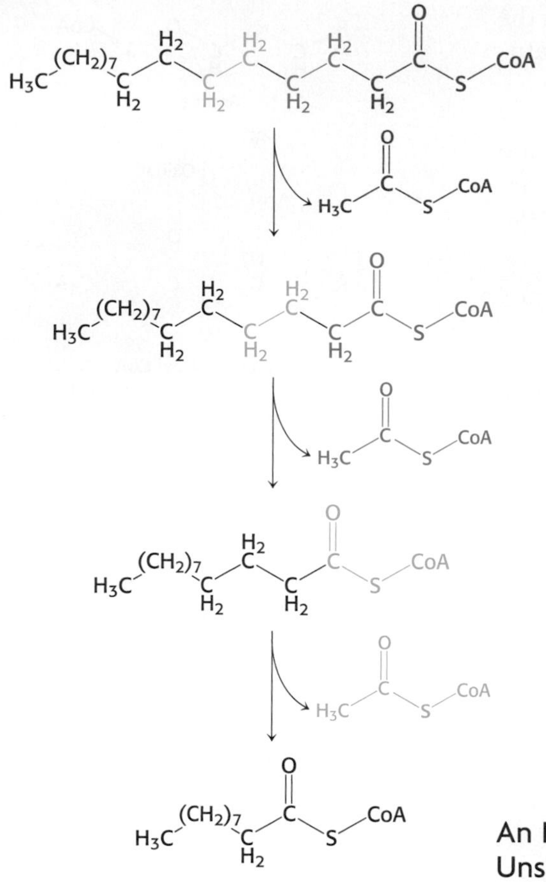

Figure 22.10 **First three rounds in the degradation of palmitate.** Two-carbon units are sequentially removed from the carboxyl end of the fatty acid.

Palmitoleoyl CoA

cis-Δ^3-**Enoyl CoA**

cis-Δ^3-Enoyl CoA isomerase

trans-Δ^2-**Enoyl CoA**

Approximately 2.5 molecules of ATP are generated when the respiratory chain oxidizes each of these NADH molecules, whereas 1.5 molecules of ATP are formed for each $FADH_2$ because their electrons enter the chain at the level of ubiquinol. Recall that the oxidation of acetyl CoA by the citric acid cycle yields 10 molecules of ATP. Hence, the number of ATP molecules formed in the oxidation of palmitoyl CoA is 10.5 from the seven $FADH_2$, 17.5 from the seven NADH, and 80 from the eight acetyl CoA molecules, which gives a total of 108. The equivalent of 2 molecules of ATP is consumed in the activation of palmitate, in which ATP is split into AMP and two molecules of orthophosphate. Thus, *the complete oxidation of a molecule of palmitate yields 106 molecules of ATP.*

22.3 Unsaturated and Odd-Chain Fatty Acids Require Additional Steps for Degradation

The β-oxidation pathway accomplishes the complete degradation of saturated fatty acids having an even number of carbon atoms. Most fatty acids have such structures because of their mode of synthesis (p. 636). However, not all fatty acids are so simple. The oxidation of fatty acids containing double bonds requires additional steps, as does the oxidation of fatty acids containing an odd number of carbon atoms.

An Isomerase and a Reductase Are Required for the Oxidation of Unsaturated Fatty Acids

The oxidation of unsaturated fatty acids presents some difficulties, yet many such fatty acids are available in the diet. Most of the reactions are the same as those for saturated fatty acids. In fact, only two additional enzymes—an isomerase and a reductase—are needed to degrade a wide range of unsaturated fatty acids.

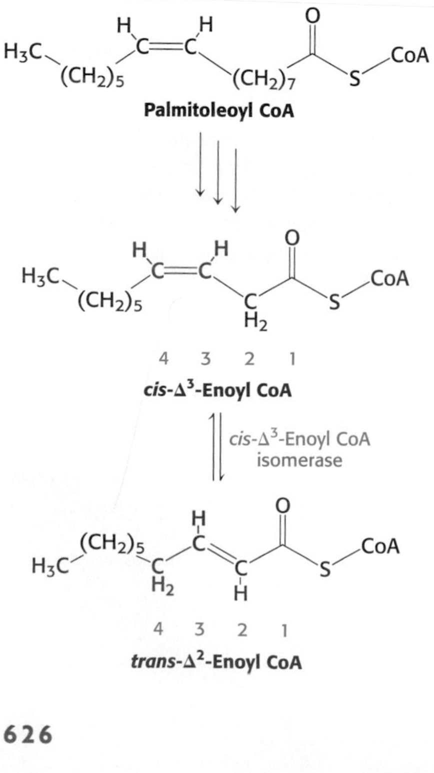

Consider the oxidation of palmitoleate. This C_{16} unsaturated fatty acid, which has one double bond between C-9 and C-10, is activated and transported across the inner mitochondrial membrane in the same way as saturated fatty acids. Palmitoleoyl CoA then undergoes three cycles of degradation, which are carried out by the same enzymes as those in the oxidation of saturated fatty acids. However, the *cis*-Δ^3-enoyl CoA formed in the third round is not a substrate for acyl CoA dehydrogenase. The presence of a double bond between C-3 and C-4 prevents the formation of another double bond between C-2 and C-3. This impasse is resolved by a new reaction that shifts the position and configuration of the *cis*-Δ^3 double bond. *cis*-Δ^3 *Enoyl CoA isomerase converts this double bond into a trans-*Δ^2 *double bond.* The subsequent reactions are those of the saturated fatty acid oxidation pathway, in which the *trans*-Δ^2-enoyl CoA is a regular substrate.

Another problem arises with the oxidation of polyunsaturated fatty acids. Consider linoleate, a C_{18} polyunsaturated fatty acid with *cis*-Δ^9 and *cis*-Δ^{12} double bonds (Figure 22.11). The *cis*-Δ^3 double bond formed after three rounds of β oxidation is converted into a *trans*-Δ^2 double bond by the aforementioned isomerase. The acyl CoA produced by another round of β oxidation contains a *cis*-Δ^4 double bond. Dehydrogenation of this species by acyl CoA dehydrogenase yields a *2,4-dienoyl intermediate*, which is not a substrate for the next enzyme in the β-oxidation pathway. This impasse is circumvented by *2,4-dienoyl CoA reductase*, an enzyme that uses NADPH to reduce the 2,4-dienoyl intermediate to *trans*-Δ^3-enoyl CoA. *cis*-Δ^3-Enoyl CoA isomerase then converts *trans*-Δ^3-enoyl CoA into the *trans*-Δ^2 form, a customary intermediate in the β-oxidation pathway. These catalytic

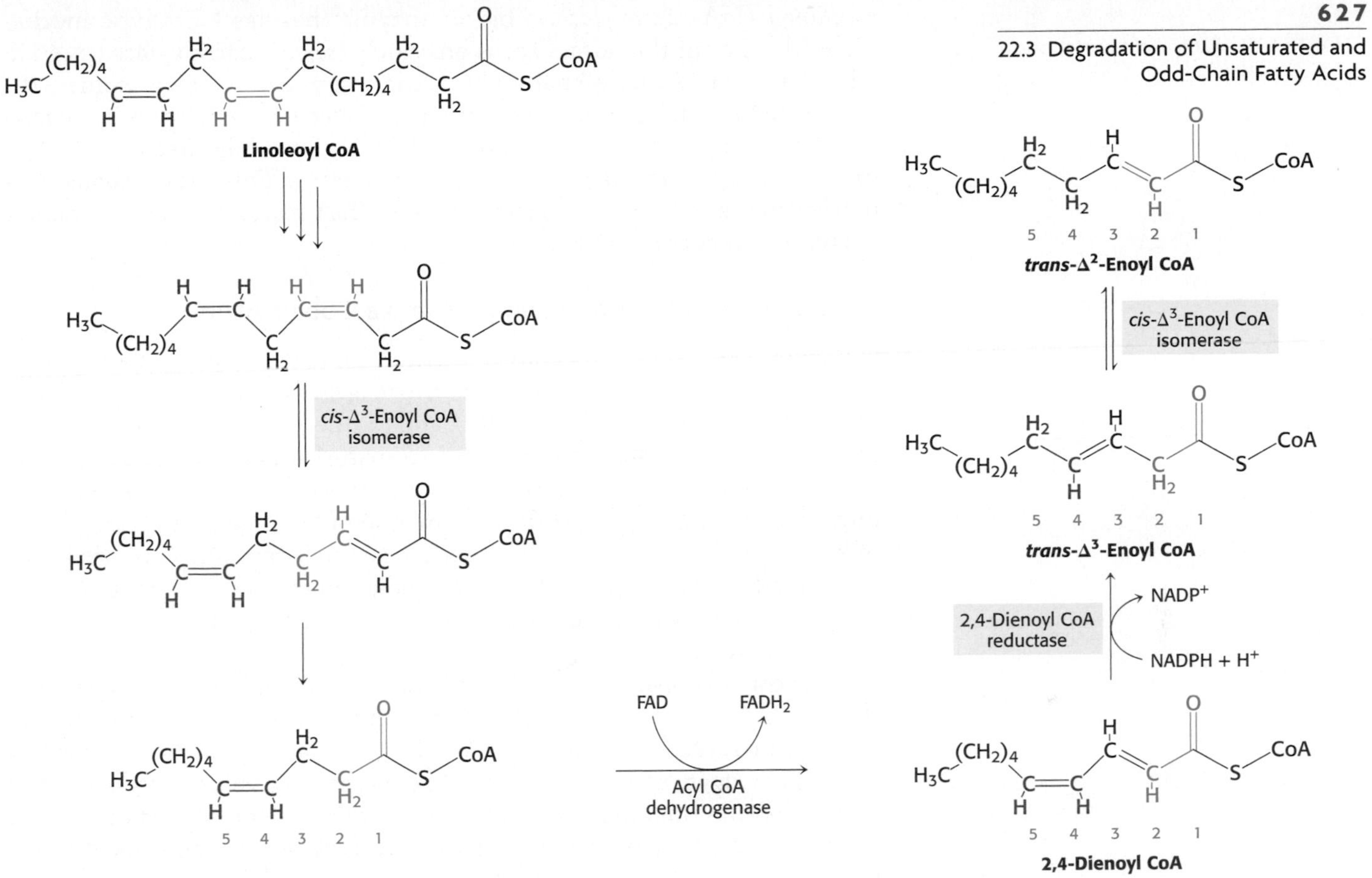

Linoleoyl CoA

cis-Δ³-Enoyl CoA
isomerase

Acyl CoA
dehydrogenase

FAD   FADH₂

trans-Δ²-Enoyl CoA

cis-Δ³-Enoyl CoA
isomerase

trans-Δ³-Enoyl CoA

2,4-Dienoyl CoA
reductase

NADP⁺

NADPH + H⁺

2,4-Dienoyl CoA

Figure 22.11 Oxidation of linoleoyl CoA.
The complete oxidation of the
diunsaturated fatty acid linoleate is
facilitated by the activity of enoyl CoA
isomerase and 2,4-dienoyl CoA reductase.

strategies are elegant and economical. Only two extra enzymes are needed
for the oxidation of *any* polyunsaturated fatty acid. *Odd-numbered double
bonds are handled by the isomerase, and even-numbered ones by the reductase
and the isomerase.*

Odd-Chain Fatty Acids Yield Propionyl CoA in the Final Thiolysis Step

Fatty acids having an odd number of carbon atoms are minor species. They
are oxidized in the same way as fatty acids having an even number, except
that propionyl CoA and acetyl CoA, rather than two molecules of acetyl
CoA, are produced in the final round of degradation. The activated three-
carbon unit in propionyl CoA enters the citric acid cycle after it has been
converted into succinyl CoA.

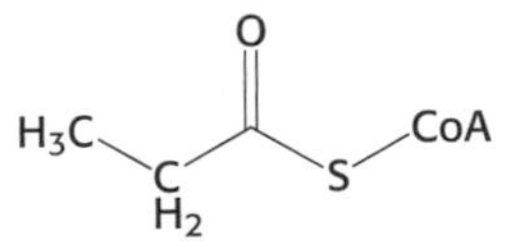

The pathway from propionyl CoA to succinyl CoA is especially interesting because it entails a rearrangement that requires *vitamin B₁₂* (also
known as *cobalamin*). Propionyl CoA is carboxylated at the expense of
the hydrolysis of a molecule of ATP to yield the D isomer of methylmalonyl CoA (Figure 22.12). This carboxylation reaction is catalyzed by

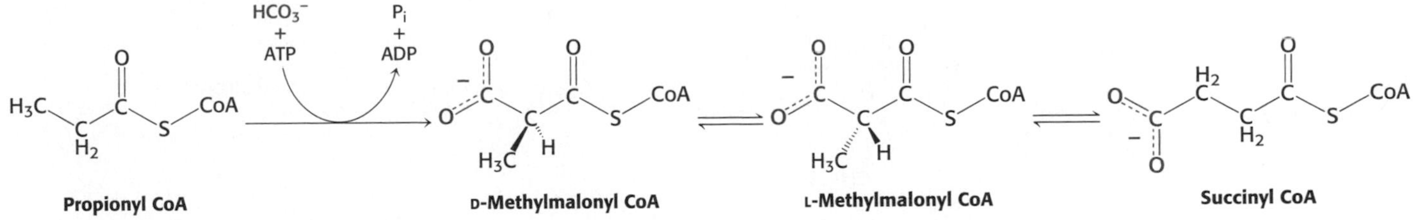

Propionyl CoA

HCO₃⁻
+
ATP

Pᵢ
+
ADP

Propionyl CoA   **D-Methylmalonyl CoA**   **L-Methylmalonyl CoA**   **Succinyl CoA**

Figure 22.12 Conversion of propionyl CoA into succinyl CoA. Propionyl CoA, generated
from fatty acids with an odd number of carbons as well as some amino acids, is converted
into the citric acid cycle intermediate succinyl CoA.

propionyl CoA carboxylase, a biotin enzyme that has a catalytic mechanism like that of the homologous enzyme pyruvate carboxylase (p. 462). The D isomer of methylmalonyl CoA is racemized to the L isomer, the substrate for a mutase that converts it into *succinyl CoA* by an *intramolecular rearrangement*. The —CO—S—CoA group migrates from C-2 to methyl group in exchange for a hydrogen atom. This very unusual isomerization is catalyzed by *methylmalonyl CoA mutase,* which contains a derivative of cobalamin as its coenzyme.

Vitamin B$_{12}$ Contains a Corrin Ring and a Cobalt Atom

Cobalamin enzymes, which are present in most organisms, catalyze three types of reactions: (1) *intramolecular rearrangements;* (2) *methylations,* as in the synthesis of methionine (p. 691); and (3) the *reduction of ribonucleotides to deoxyribonucleotides* (Section 25.3). In mammals, only two reactions are known to require coenzyme B$_{12}$. The conversion of L-methylmalonyl CoA into succinyl CoA is one, and the formation of methionine by methylation of homocysteine is the other. The latter reaction is especially important because methionine is required for the generation of coenzymes that participate in the synthesis of purines and thymine, which are needed for nucleic acid synthesis.

The core of cobalamin consists of a *corrin ring with a central cobalt atom* (Figure 22.13). The corrin ring, like a porphyrin, has *four pyrrole units.* Two of them are directly bonded to each other, whereas the others are joined by methine bridges, as in porphyrins. The corrin ring is more reduced than that of porphyrins and the substituents are different. A cobalt atom is bonded to the four pyrrole nitrogens. The fifth substituent linked to the cobalt atom is a derivative of *dimethylbenzimidazole* that contains ribose 3-phosphate and aminoisopropanol. One of the nitrogen atoms of dimethylbenzimidazole is linked to the cobalt atom. In coenzyme B$_{12}$, *the sixth substituent* linked to the cobalt atom is a *5′-deoxyadenosyl unit.* This position can also be occupied by a cyano group, a methyl group, or other ligands. In all of these compounds, the cobalt is in the +3 oxidation state.

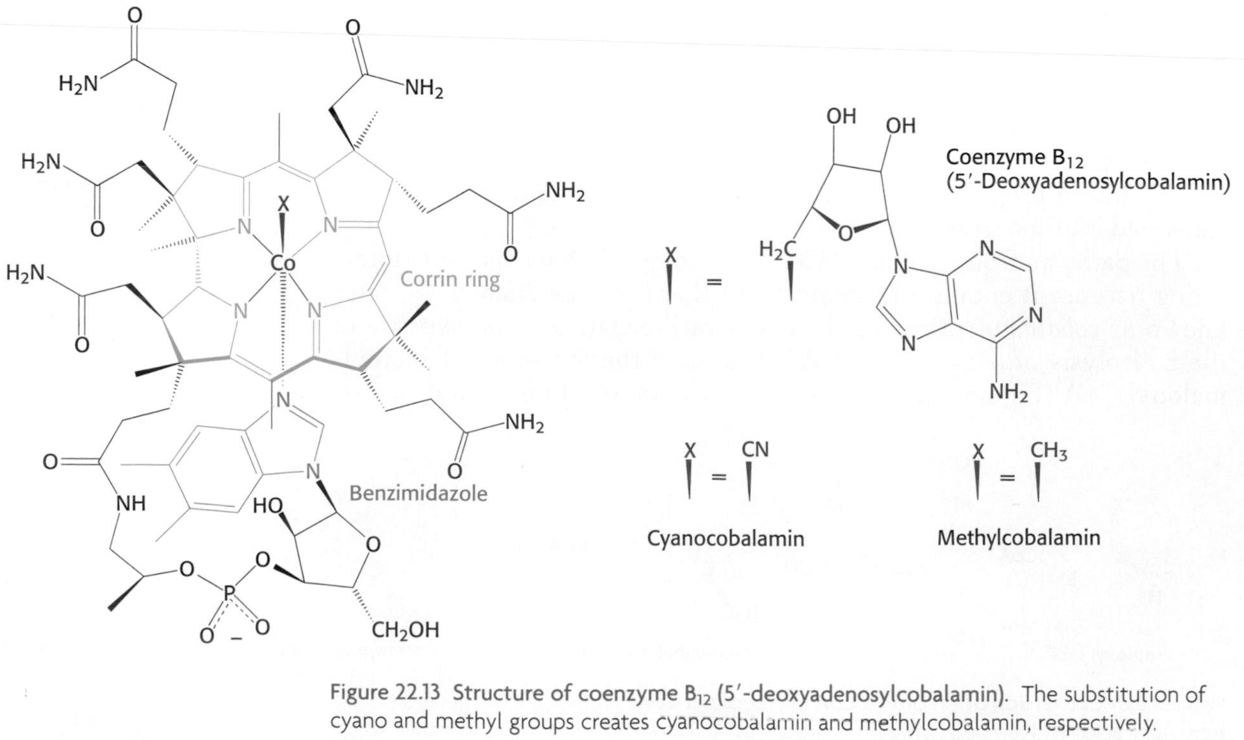

Figure 22.13 Structure of coenzyme B$_{12}$ (5′-deoxyadenosylcobalamin). The substitution of cyano and methyl groups creates cyanocobalamin and methylcobalamin, respectively.

Mechanism: Methylmalonyl CoA Mutase Catalyzes a Rearrangement to Form Succinyl CoA

The rearrangement reactions catalyzed by coenzyme B_{12} are exchanges of two groups attached to adjacent carbon atoms (Figure 22.14). A hydrogen atom migrates from one carbon atom to the next, and an R group (such as the —CO—S—CoA group of methylmalonyl CoA) concomitantly moves in the reverse direction. The first step in these intramolecular rearrangements is the cleavage of the carbon–cobalt bond of 5′-deoxyadenosylcobalamin to generate the Co^{2+} form of coenzyme B_{12} and a 5′-deoxyadenosyl radical, —CH_2· (Figure 22.15). In this *homolytic cleavage reaction,* one electron of the Co–C bond stays with Co (reducing it from the +3 to the +2 oxidation state) while the other stays with the carbon atom, generating a free radical. In contrast, nearly all other cleavage reactions in biological systems are *heterolytic*—an electron *pair* is transferred to one of the two atoms that were bonded together.

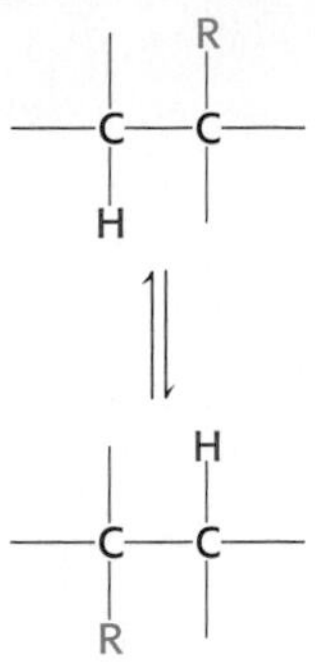

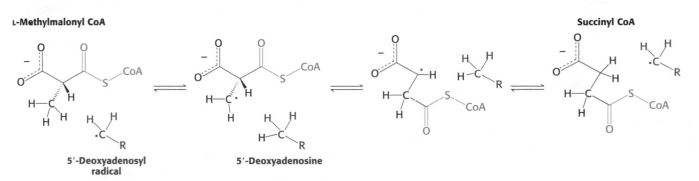

Figure 22.14 Rearrangement reaction catalyzed by cobalamin enzymes. The R group can be an amino group, a hydroxyl group, or a substituted carbon.

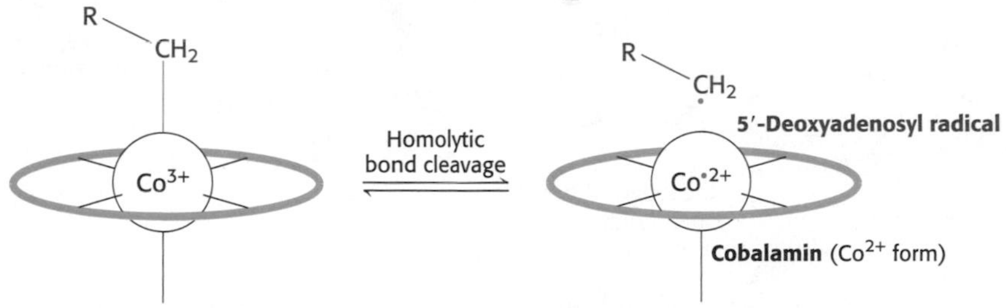

Figure 22.15 Formation of a 5′-deoxyadenosyl radical. The methylmalonyl CoA mutase reaction begins with the homolytic cleavage of the bond joining Co^{3+} to a carbon atom of the ribose of the adenosine moiety. The cleavage generates a 5′-deoxyadenosyl radical and leads to the reduction of Co^{3+} to Co^{2+}.

What is the role of this very unusual —CH_2· radical? This highly reactive species abstracts a *hydrogen atom* from the substrate to form 5′-deoxyadenosine and a substrate radical (Figure 22.16). This substrate radical spontaneously rearranges: the carbonyl CoA group migrates to the position formerly occupied by H on the neighboring carbon atom to produce a different radical. This product radical abstracts a hydrogen atom from the methyl group of 5′-deoxyadenosine to complete the rearrangement and return the deoxyadenosyl unit to the radical form. *The role of coenzyme B_{12} in such intramolecular migrations is to serve as a source of free radicals for the abstraction of hydrogen atoms.*

An essential property of coenzyme B_{12} is the weakness of its cobalt–carbon bond, which is readily cleaved to generate a radical. To facilitate the cleavage of this bond, enzymes such as methylmalonyl CoA mutase displace the benzimidazole group from the cobalamin and bind to the cobalt atom

Figure 22.16 Formation of succinyl CoA by a rearrangement reaction. A free radical abstracts a hydrogen atom in the rearrangement of methylmalonyl CoA to succinyl CoA.

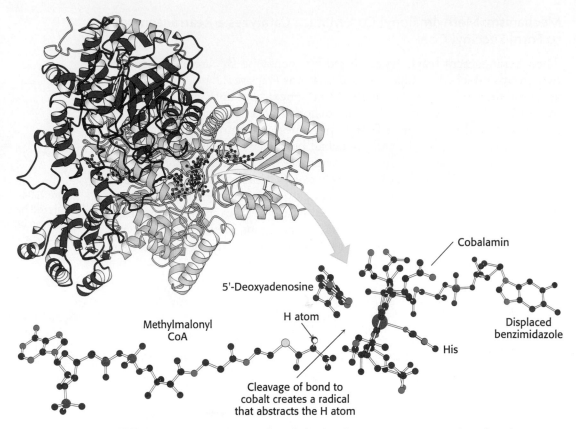

Figure 22.17 Active site of methylmalonyl CoA mutase. *Notice* that a histidine residue from the enzyme binds to cobalt in place of benzimidazole. This arrangement of substrate and coenzyme in the active site facilitates the cleavage of the cobalt–carbon bond and the subsequent abstraction of a hydrogen atom from the substrate. [Drawn from 4REQ.pdb.]

through a histidine residue (Figure 22.17). The steric crowding around the cobalt–carbon bond within the corrin ring system contributes to the bond weakness.

Fatty Acids Are Also Oxidized in Peroxisomes

Although most fatty acid oxidation takes place in mitochondria, some oxidation of fatty acids can take place in cellular organelles called *peroxisomes* (Figure 22.18). These organelles are small membrane-bounded compartments that are present in the cells of most eukaryotes. Fatty acid oxidation in these organelles, which halts at octanoyl CoA, may serve to shorten long chains to make them better substrates of β oxidation in mitochondria. Peroxisomal oxidation differs from β oxidation in the initial dehydrogenation reaction (Figure 22.19). In peroxisomes, a flavoprotein dehydrogenase transfers electrons to O_2 to yield H_2O_2 instead of capturing high-energy electrons as $FADH_2$, as in mitochondrial β oxidation. Peroxisomes contain high concentrations of the enzyme catalase to degrade H_2O_2 into water and O_2. Subsequent steps are identical with those of their mitochondrial counterparts, although they are carried out by different isoforms of the enzymes.

Peroxisomes do not function in patients with Zellweger syndrome. Liver, kidney, and muscle abnormalities usually lead to death by age six. The syndrome is caused by a defect in the import of enzymes into the peroxisomes. Here we see a pathological condition resulting from an inappropriate cellular distribution of enzymes.

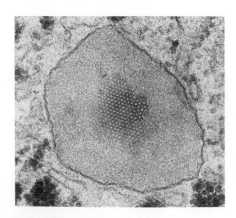

Figure 22.18 Electron micrograph of a peroxisome in a liver cell. A crystal of urate oxidase is present inside the organelle, which is bounded by a single bilayer membrane. The dark granular structures outside the peroxisome are glycogen particles. [Courtesy of Dr. George Palade.]

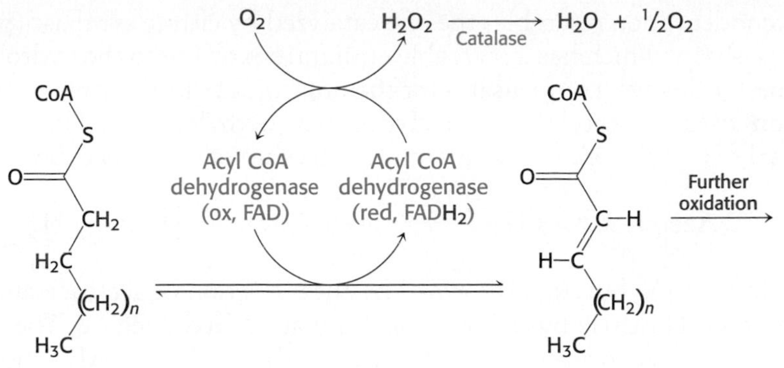

Figure 22.19 **Initiation of peroxisomal fatty acid degradation.** The first dehydration in the degradation of fatty acids in peroxisomes requires a flavoprotein dehydrogenase that transfers electrons to O_2 to yield H_2O_2.

Ketone Bodies Are Formed from Acetyl CoA When Fat Breakdown Predominates

The acetyl CoA formed in fatty acid oxidation enters the citric acid cycle only if fat and carbohydrate degradation are appropriately balanced. Acetyl CoA must combine with oxaloacetate to gain entry to the citric acid cycle. The availability of oxaloacetate, however, depends on an adequate supply of carbohydrate. Recall that oxaloacetate is normally formed from pyruvate, the product of glucose degradation in glycolysis. If carbohydrate is unavailable or improperly utilized, the concentration of oxaloacetate is lowered and acetyl CoA cannot enter the citric acid cycle. This dependency is the molecular basis of the adage that *fats burn in the flame of carbohydrates*.

In fasting or diabetes, oxaloacetate is consumed to form glucose by the gluconeogenic pathway (p. 461) and hence is unavailable for condensation with acetyl CoA. Under these conditions, acetyl CoA is diverted to the formation of acetoacetate and D-3-hydroxybutyrate. Acetoacetate, D-3-hydroxybutyrate, and acetone are often referred to as *ketone bodies*. Abnormally high levels of ketone bodies are present in the blood of untreated diabetics (p. 773).

Acetoacetate is formed from acetyl CoA in three steps (Figure 22.20). Two molecules of acetyl CoA condense to form acetoacetyl CoA. This reaction, which is catalyzed by thiolase, is the reverse of the thiolysis step in the oxidation of fatty acids. Acetoacetyl CoA then reacts with acetyl CoA and water to give 3-hydroxy-3-methylglutaryl CoA (HMG-CoA) and CoA.

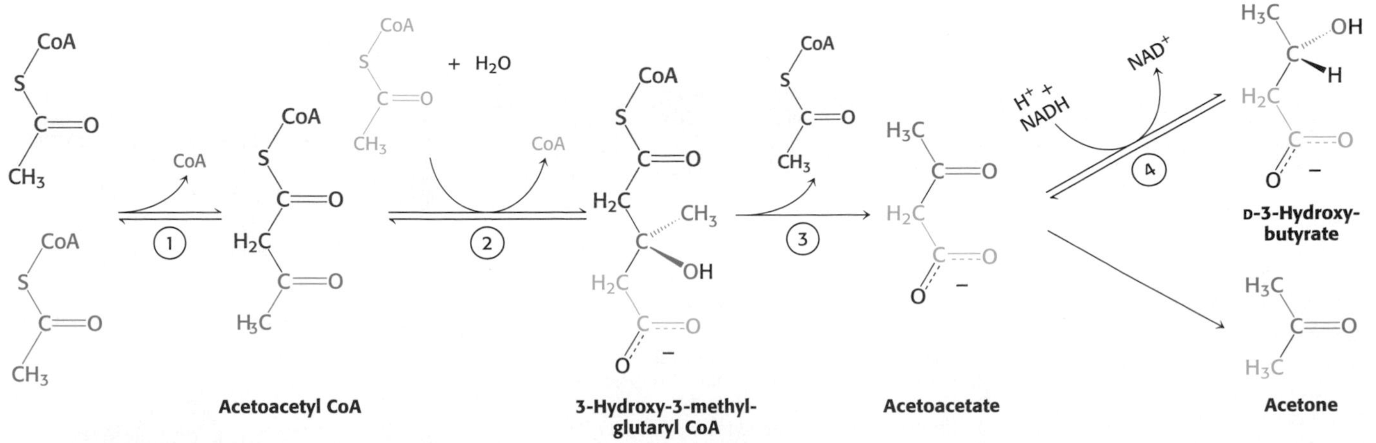

Figure 22.20 **Formation of ketone bodies.** The ketone bodies—acetoacetate, D-3-hydroxybutyrate, and acetone from acetyl CoA—are formed primarily in the liver. Enzymes catalyzing these reactions are (1) 3-ketothiolase, (2) hydroxymethylglutaryl CoA synthase, (3) hydroxymethylglutaryl CoA cleavage enzyme, and (4) D-3-hydroxybutyrate dehydrogenase. Acetoacetate spontaneously decarboxylates to form acetone.

This condensation resembles the one catalyzed by citrate synthase (p. 482). This reaction, which has a favorable equilibrium owing to the hydrolysis of a thioester linkage, compensates for the unfavorable equilibrium in the formation of acetoacetyl CoA. 3-Hydroxy-3-methylglutaryl CoA is then cleaved to acetyl CoA and acetoacetate. The sum of these reactions is

$$2 \text{ Acetyl CoA} + H_2O \longrightarrow \text{acetoacetate} + 2 \text{ CoA} + H^+$$

D-3-Hydroxybutyrate is formed by the reduction of acetoacetate in the mitochondrial matrix by D-3-hydroxybutyrate dehydrogenase. The ratio of hydroxybutyrate to acetoacetate depends on the NADH/NAD ratio inside mitochondria.

Because it is a β-ketoacid, acetoacetate also undergoes a slow, spontaneous decarboxylation to acetone. The odor of acetone may be detected in the breath of a person who has a high level of acetoacetate in the blood.

Ketone Bodies Are a Major Fuel in Some Tissues

The major site of the production of acetoacetate and 3-hydroxybutyrate is the liver. These substances diffuse from the liver mitochondria into the blood and are transported to peripheral tissues (Figure 22.21). Acetoacetate and 3-hydroxybutyrate are normal fuels of respiration and are quantitatively important as sources of energy. Indeed, heart muscle and the renal cortex use acetoacetate in preference to glucose. In contrast, glucose is the major fuel for the brain and red blood cells in well-nourished people on a

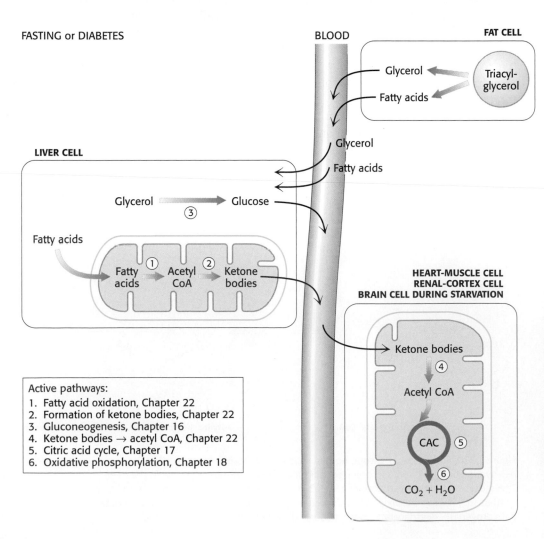

Figure 22.21 PATHWAY INTEGRATION: Liver supplies ketone bodies to the peripheral tissues. During fasting or in untreated diabetics, the liver converts fatty acids into ketone bodies, which are a fuel source for a number of tissues. Ketone-body production is especially important during starvation, when they are the predominant fuel.

Active pathways:
1. Fatty acid oxidation, Chapter 22
2. Formation of ketone bodies, Chapter 22
3. Gluconeogenesis, Chapter 16
4. Ketone bodies → acetyl CoA, Chapter 22
5. Citric acid cycle, Chapter 17
6. Oxidative phosphorylation, Chapter 18

balanced diet. However, the brain adapts to the utilization of acetoacetate during starvation and diabetes (pp. 772 and 773). In prolonged starvation, 75% of the fuel needs of the brain are met by ketone bodies.

Acetoacetate is converted into acetyl CoA in two steps. First, acetoacetate is activated by the transfer of CoA from succinyl CoA in a reaction catalyzed by a specific CoA transferase. Second, acetoacetyl CoA is cleaved by thiolase to yield two molecules of acetyl CoA, which can then enter the citric acid cycle (Figure 22.22). The liver has acetoacetate available to supply to other organs because it lacks this particular CoA transferase. 3-Hydroxybutyrate requires an additional step to yield acetyl CoA. It is first oxidized to produce acetoacetate, which is processed as heretofore described, as well as NADH for use in oxidative phosphorylation.

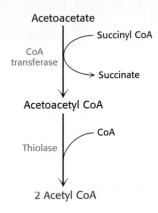

Figure 22.22 Utilization of acetoacetate as a fuel. Acetoacetate can be converted into two molecules of acetyl CoA, which then enter the citric acid cycle.

D-3-Hydroxybutyrate   Acetoacetate

Ketone bodies can be regarded as a water-soluble, transportable form of acetyl units. Fatty acids are released by adipose tissue and converted into acetyl units by the liver, which then exports them as acetoacetate. As might be expected, acetoacetate also has a regulatory role. *High levels of acetoacetate in the blood signify an abundance of acetyl units and lead to a decrease in the rate of lipolysis in adipose tissue.*

High blood levels of ketone bodies, the result of certain pathological conditions, can be life threatening. The most common of these conditions is diabetic ketosis in patients with insulin-dependent diabetes mellitus. These patients are unable to produce insulin. As stated earlier, this hormone, normally released after meals, signals tissues to take up glucose (p. 610). In addition, it curtails fatty acid mobilization by adipose tissue. The absence of insulin has two major biochemical consequences (Figure 22.23). First, the liver cannot absorb glucose and consequently cannot provide oxaloacetate to process fatty acid-derived acetyl CoA (p. 482). Second, adipose

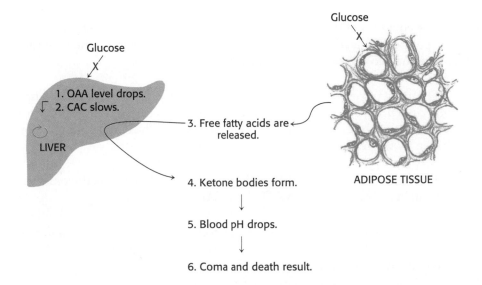

Figure 22.23 Diabetic ketosis results when insulin is absent. In the absence of insulin, fats are released from adipose tissue, and glucose cannot be absorbed by the liver or adipose tissue. The liver degrades the fatty acids by β oxidation but cannot process the acetyl CoA, because of a lack of glucose-derived oxaloacetate (OAA). Excess ketone bodies are formed and released into the blood.

633

cells continue to release fatty acids into the bloodstream, which are taken up by the liver and converted into ketone bodies. The liver thus produces large amounts of ketone bodies, which are moderately strong acids. The result is severe acidosis. The decrease in pH impairs tissue function, most importantly in the central nervous system.

Animals Cannot Convert Fatty Acids into Glucose

It is important to note that *animals are unable to effect the net synthesis of glucose from fatty acids.* Specifically, acetyl CoA cannot be converted into pyruvate or oxaloacetate in animals. Recall that the reaction that generates acetyl CoA from pyruvate is irreversible (p. 477). The two carbon atoms of the acetyl group of acetyl CoA enter the citric acid cycle, but two carbon atoms leave the cycle in the decarboxylations catalyzed by isocitrate dehydrogenase and α-ketoglutarate dehydrogenase. Consequently, oxaloacetate is regenerated, but it is not formed de novo when the acetyl unit of acetyl CoA is oxidized by the citric acid cycle. In contrast, plants have two additional enzymes enabling them to convert the carbon atoms of acetyl CoA into oxaloacetate (Section 18.4.).

22.4 Fatty Acids Are Synthesized and Degraded by Different Pathways

Although fatty acid synthesis is the reversal of the degradative pathway in regard to basic chemical reactions, the synthetic and degradative pathways are different mechanistically, again exemplifying the principle that *synthetic and degradative pathways are almost always distinct.* Some important differences between the pathways are as follows:

1.   Synthesis takes place in the *cytoplasm,* in contrast with degradation, which takes place primarily in the mitochondrial matrix.

2.   Intermediates in fatty acid synthesis are covalently linked to the sulfhydryl groups of an *acyl carrier protein* (ACP), whereas intermediates in fatty acid breakdown are covalently attached to the sulfhydryl group of coenzyme A.

3.   The enzymes of fatty acid synthesis in higher organisms are joined in a *single polypeptide chain* called *fatty acid synthase.* In contrast, the degradative enzymes do not seem to be associated.

4.   The growing fatty acid chain is elongated by the *sequential addition of two-carbon units* derived from acetyl CoA. The activated donor of two-carbon units in the elongation step is *malonyl ACP.* The elongation reaction is driven by the release of CO_2.

5.   The reductant in fatty acid synthesis is *NADPH,* whereas the oxidants in fatty acid degradation are NAD^+ and *FAD.*

6.   Elongation by the fatty acid synthase complex stops on the formation of *palmitate* (C_{16}). Further elongation and the insertion of double bonds are carried out by other enzyme systems.

The Formation of Malonyl CoA Is the Committed Step in Fatty Acid Synthesis

Fatty acid synthesis starts with the carboxylation of acetyl CoA to *malonyl CoA*. This irreversible reaction is the committed step in fatty acid synthesis.

Acetyl CoA + ATP + HCO₃⁻ ⟶ Malonyl CoA + ADP + Pᵢ + H⁺

The synthesis of malonyl CoA is catalyzed by *acetyl CoA carboxylase*, which contains a biotin prosthetic group. The carboxyl group of biotin is covalently attached to the ε amino group of a lysine residue, as in pyruvate carboxylase (p. 462) and propionyl CoA carboxylase (p. 627). As with these other enzymes, a carboxybiotin intermediate is formed at the expense of the hydrolysis of a molecule of ATP. The activated CO_2 group in this intermediate is then transferred to acetyl CoA to form malonyl CoA.

$$\text{Biotin-enzyme} + \text{ATP} + \text{HCO}_3^- \rightleftharpoons$$
$$\text{CO}_2\text{-biotin-enzyme} + \text{ADP} + \text{P}_i$$

$$\text{CO}_2\text{-biotin-enzyme} + \text{acetyl CoA} \longrightarrow \text{malonyl CoA} + \text{biotin-enzyme}$$

This enzyme is also the essential regulatory enzyme for fatty acid metabolism (Section 22.5).

Intermediates in Fatty Acid Synthesis Are Attached to an Acyl Carrier Protein

The intermediates in fatty acid synthesis are linked to an acyl carrier protein. Specifically, they are linked to the sulfhydryl terminus of a phosphopantetheine group. In the degradation of fatty acids, this unit is present as part of coenzyme A (p. 422), whereas, in their synthesis, it is attached to a serine residue of the acyl carrier protein (Figure 22.24). Thus, ACP, a single polypeptide chain of 77 residues, can be regarded as a giant prosthetic group, a "macro CoA."

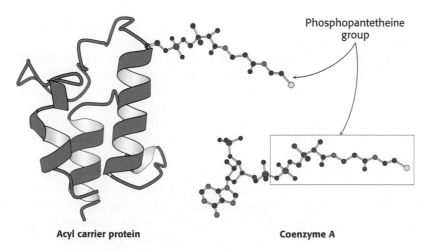

Phosphopantetheine group

Acyl carrier protein   **Coenzyme A**

Figure 22.24 Phosphopantetheine. Both acyl carrier protein and coenzyme A include phosphopantetheine as their reactive units.

TABLE 22.2 Principal reactions in fatty acid synthesis in bacteria

Step	Reaction	Enzyme
1	Acetyl CoA + HCO_3^- + ATP $\longrightarrow$ malonyl CoA + ADP + P_i + H^+	Acetyl CoA carboxylase
2	Acetyl CoA + ACP $\rightleftharpoons$ acetyl ACP + CoA	Acetyl transacylase
3	Malonyl CoA + ACP $\rightleftharpoons$ malonyl ACP + CoA	Malonyl transacylase
4	Acetyl ACP + malonyl ACP $\longrightarrow$ acetoacetyl ACP + ACP + CO_2	Acyl-malonyl ACP condensing enzyme
5	Acetoacetyl ACP + NADPH + H^+ $\rightleftharpoons$ D-3-hydroxybutyryl ACP + $NADP^+$	β-Ketoacyl ACP reductase
6	D-3-Hydroxybutyryl ACP $\rightleftharpoons$ crotonyl ACP + H_2O	3-Hydroxyacyl ACP dehydratase
7	Crotonyl ACP + NADPH + H^+ $\longrightarrow$ butyryl ACP + $NADP^+$	Enoyl ACP reductase

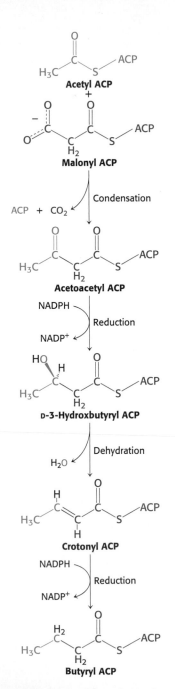

Figure 22.25 Fatty acid synthesis. Fatty acids are synthesized by the repetition of the following reaction sequence: condensation, reduction, dehydration, and reduction. The intermediates shown here are produced in the first round of synthesis.

Fatty Acid Synthesis Consists of a Series of Condensation, Reduction, Dehydration, and Reduction Reactions

The enzyme system that catalyzes the synthesis of saturated long-chain fatty acids from acetyl CoA, malonyl CoA, and NADPH is called the *fatty acid synthase*. The synthase is actually a complex of distinct enzymes. The fatty acid synthase complex in bacteria is readily dissociated into individual enzymes when the cells are broken apart. The availability of these isolated enzymes has helped biochemists elucidate the steps in fatty acid synthesis (Table 22.2). In fact, the reactions leading to fatty acid synthesis in higher organisms are very much like those of bacteria.

The elongation phase of fatty acid synthesis starts with the formation of acetyl ACP and malonyl ACP. *Acetyl transacylase* and *malonyl transacylase* catalyze these reactions.

$$\text{Acetyl CoA} + \text{ACP} \rightleftharpoons \text{acetyl ACP} + \text{CoA}$$

$$\text{Malonyl CoA} + \text{ACP} \rightleftharpoons \text{malonyl ACP} + \text{CoA}$$

Malonyl transacylase is highly specific, whereas acetyl transacylase can transfer acyl groups other than the acetyl unit, though at a much slower rate. The synthesis of fatty acids with an odd number of carbon atoms starts with propionyl ACP, which is formed from propionyl CoA by acetyl transacylase.

Acetyl ACP and malonyl ACP react to form acetoacetyl ACP (Figure 22.25). The *acyl-malonyl ACP condensing enzyme* catalyzes this condensation reaction.

$$\text{Acetyl ACP} + \text{malonyl ACP} \longrightarrow \text{acetoacetyl ACP} + \text{ACP} + CO_2$$

In the condensation reaction, a four-carbon unit is formed from a two-carbon unit and a three-carbon unit, and CO_2 is released. Why is the four-carbon unit not formed from two two-carbon units—say, two molecules of acetyl ACP? The answer is that the equilibrium for the synthesis of acetoacetyl ACP from two molecules of acetyl ACP is highly unfavorable. In contrast, *the equilibrium is favorable if malonyl ACP is a reactant because its decarboxylation contributes a substantial decrease in free energy.* In effect, ATP drives the condensation reaction, though ATP does not directly participate in the condensation reaction. Instead, ATP is used to carboxylate acetyl CoA to malonyl CoA. The free energy thus stored in malonyl CoA is released in the decarboxylation accompanying the formation of acetoacetyl ACP. Although HCO_3^- is required for fatty acid synthesis, its carbon atom does not appear in the product. Rather, *all the carbon atoms of fatty acids containing an even number of carbon atoms are derived from acetyl CoA.*

The next three steps in fatty acid synthesis reduce the keto group at C-3 to a methylene group (see Figure 22.25). First, acetoacetyl ACP is reduced to D-3-hydroxybutyryl ACP. This reaction differs from the corresponding one in fatty acid degradation in two respects: (1) the D rather than the L isomer is formed; and (2) NADPH is the reducing agent, whereas NAD^+ is the oxidizing agent in β oxidation. This difference exemplifies the general principle that *NADPH is consumed in biosynthetic reactions, whereas NADH is generated in energy-yielding reactions.* Then D-3-hydroxybutyryl ACP is *dehydrated* to form crotonyl ACP, which is a *trans*-Δ^2-enoyl ACP. The final step in the cycle *reduces* crotonyl ACP to butyryl ACP. NADPH is again the reductant, whereas FAD is the oxidant in the corresponding reaction in β oxidation. The enzyme that catalyzes this step, *enoyl ACP reductase*, is inhibited by *triclosan*, a broad-spectrum antibacterial agent that is added to a variety of products such as toothpaste, soaps, and skin creams. These last three reactions—a reduction, a dehydration, and a second reduction—convert acetoacetyl ACP into butyryl ACP, which completes the first elongation cycle.

In the second round of fatty acid synthesis, butyryl ACP condenses with malonyl ACP to form a C_6-β-ketoacyl ACP. This reaction is like the one in the first round, in which acetyl ACP condenses with malonyl ACP to form a C_4-β-ketoacyl ACP. Reduction, dehydration, and a second reduction convert the C_6-β-ketoacyl ACP into a C_6-acyl ACP, which is ready for a third round of elongation. The elongation cycles continue until C_{16}-acyl ACP is formed. This intermediate is a good substrate for a thioesterase that hydrolyzes C_{16}-acyl ACP to yield palmitate and ACP. *The thioesterase acts as a ruler to determine fatty acid chain length.* The synthesis of longer-chain fatty acids is discussed in Section 22.6.

Fatty Acids Are Synthesized by a Multifunctional Enzyme Complex in Animals

Although the basic biochemical reactions in fatty acid synthesis are very similar in *E. coli* and eukaryotes, the structure of the synthase varies considerably. The component enzymes of animal fatty acid synthases, in contrast with those of *E. coli* and plants, are linked in a large polypeptide chain.

Mammalian fatty acid synthase is a dimer of identical 272-kd subunits. Each chain is folded into three domains joined by flexible regions that allow domain movements that are required for cooperation between the enzyme's active sites (Figure 22.26). *Domain 1, the substrate-entry*

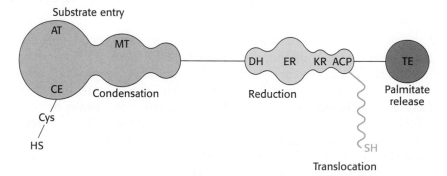

Figure 22.26 Schematic representation of a single chain of animal fatty acid synthase. The active enzyme consists of a dimer of two identical chains, each of which contains three domains. Domain 1 (blue) contains acetyl transferase (AT), malonyl transferase (MT), and condensing enzyme (CE). Domain 2 (yellow) contains acyl carrier protein (ACP), β-ketoacyl reductase (KR), dehydratase (DH), and enoyl reductase (ER). Domain 3 (red) contains thioesterase (TE). The flexible phosphopantetheinyl group (green) carries the fatty acid chain from one catalytic site on a chain to another. [After Y. Tsukamoto, H. Wong, J. S. Mattick, and S. J. Wakil. *J. Biol. Chem.* 258(1983):15312–15322.]

and -condensation unit, contains acetyl transferase, malonyl transferase, and β-ketoacyl synthase (condensing enzyme). *Domain 2, the reduction unit*, contains the acyl carrier protein, β-ketoacyl reductase, dehydratase, and enoyl reductase. *Domain 3, the palmitate release unit*, contains the thioesterase. Thus, *seven different catalytic sites are present on a single polypeptide chain*. Despite the fact that each chain possesses all of the enzymes required for fatty acid synthesis, the monomers are not active. A dimer is required.

Many eukaryotic multienzyme complexes are multifunctional proteins in which different enzymes are linked covalently. An advantage of this arrangement is that the synthetic activity of different enzymes is coordinated. In addition, intermediates can be efficiently handed from one active site to another without leaving the assembly. Furthermore, a complex of covalently joined enzymes is more stable than one formed by noncovalent attractions. Each of the component enzymes is recognizably homologous to its bacterial counterpart. It seems likely that multifunctional enzymes such as fatty acid synthase arose in eukaryotic evolution by fusion of the individual genes of evolutionary ancestors.

The Synthesis of Palmitate Requires 8 Molecules of Acetyl CoA, 14 Molecules of NADPH, and 7 Molecules of ATP

The stoichiometry of the synthesis of palmitate is

$$\text{Acetyl CoA} + 7\,\text{malonyl CoA} + 14\,\text{NADPH} + 20\,\text{H}^+ \longrightarrow$$
$$\text{palmitate} + 7\,\text{CO}_2 + 14\,\text{NADP}^+ + 8\,\text{CoA} + 6\,\text{H}_2\text{O}$$

The equation for the synthesis of the malonyl CoA used in the preceding reaction is

$$7\,\text{Acetyl CoA} + 7\,\text{CO}_2 + 7\,\text{ATP} \longrightarrow$$
$$7\,\text{malonyl CoA} + 7\,\text{ADP} + 7\,\text{P}_i + 14\,\text{H}^+$$

Hence, the overall stoichiometry for the synthesis of palmitate is

$$8\,\text{Acetyl CoA} + 7\,\text{ATP} + 14\,\text{NADPH} + 6\,\text{H}^+ \longrightarrow$$
$$\text{palmitate} + 14\,\text{NADP}^+ + 8\,\text{CoA} + 6\,\text{H}_2\text{O} + 7\,\text{ADP} + 7\,\text{P}_i$$

Citrate Carries Acetyl Groups from Mitochondria to the Cytoplasm for Fatty Acid Synthesis

Fatty acids are synthesized in the cytoplasm, whereas acetyl CoA is formed from pyruvate in mitochondria. Hence, acetyl CoA must be transferred from mitochondria to the cytoplasm. Mitochondria, however, are not readily permeable to acetyl CoA. Recall that carnitine carries only long-chain fatty acids. *The barrier to acetyl CoA is bypassed by citrate, which carries acetyl groups across the inner mitochondrial membrane.* Citrate is formed in the mitochondrial matrix by the condensation of acetyl CoA with oxaloacetate (Figure 22.27). When present at high levels, citrate is transported to the cytoplasm, where it is cleaved by *ATP-citrate lyase*.

$$\text{Citrate} + \text{ATP} + \text{CoA} + \text{H}_2\text{O} \longrightarrow$$
$$\text{acetyl CoA} + \text{ADP} + \text{P}_i + \text{oxaloacetate}$$

Lyases

Enzymes catalyzing the cleavage of C–C, C–O, or C–N bonds by elimination. A double bond is formed in these reactions.

Thus, acetyl CoA and oxaloacetate are transferred from mitochondria to the cytoplasm at the expense of the hydrolysis of a molecule of ATP.

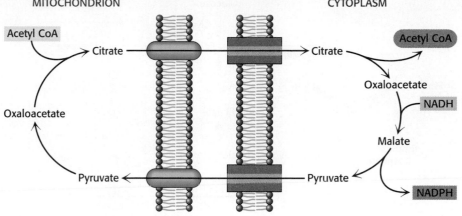

MITOCHONDRION   CYTOPLASM

Figure 22.27 Transfer of acetyl CoA to the cytoplasm. Acetyl CoA is transferred from mitochondria to the cytoplasm, and the reducing potential of NADH is concomitantly converted into that of NADPH by this series of reactions.

Several Sources Supply NADPH for Fatty Acid Synthesis

Oxaloacetate formed in the transfer of acetyl groups to the cytoplasm must now be returned to the mitochondria. The inner mitochondrial membrane is impermeable to oxaloacetate. Hence, a series of bypass reactions are needed. Most important, these reactions generate much of the NADPH needed for fatty acid synthesis. First, oxaloacetate is reduced to malate by NADH. This reaction is catalyzed by a *malate dehydrogenase* in the cytoplasm.

$$\text{Oxaloacetate} + \text{NADH} + \text{H}^+ \rightleftharpoons \text{malate} + \text{NAD}^+$$

Second, malate is oxidatively decarboxylated by an *NADP$^+$-linked malate enzyme* (also called *malic enzyme*).

$$\text{Malate} + \text{NADP}^+ \longrightarrow \text{pyruvate} + \text{CO}_2 + \text{NADPH}$$

The pyruvate formed in this reaction readily enters mitochondria, where it is carboxylated to oxaloacetate by pyruvate carboxylase.

$$\text{Pyruvate} + \text{CO}_2 + \text{ATP} + \text{H}_2\text{O} \longrightarrow$$
$$\text{oxaloacetate} + \text{ADP} + \text{P}_i + 2\,\text{H}^+$$

The sum of these three reactions is

$$\text{NADP}^+ + \text{NADH} + \text{ATP} + \text{H}_2\text{O} \longrightarrow$$
$$\text{NADPH} + \text{NAD}^+ + \text{ADP} + \text{P}_i + \text{H}^+$$

Thus, *one molecule of NADPH is generated for each molecule of acetyl CoA that is transferred from mitochondria to the cytoplasm.* Hence, eight molecules of NADPH are formed when eight molecules of acetyl CoA are transferred to the cytoplasm for the synthesis of palmitate. *The additional six molecules of NADPH required for this process come from the pentose phosphate pathway* (Section 20.3).

The accumulation of the precursors for fatty acid synthesis is a wonderful example of the coordinated use of multiple pathways. The citric acid cycle, transport of oxaloacetate from the mitochondria, and pentose phosphate pathway provide the carbon atoms and reducing power, whereas glycolysis and oxidative phosphorylation provide the ATP to meet the needs for fatty acid synthesis (Figure 22.28).

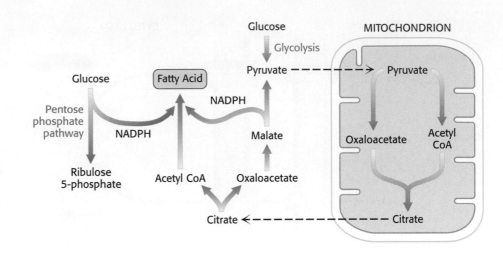

Figure 22.28 PATHWAY INTEGRATION: Fatty acid synthesis. Fatty acid synthesis requires the cooperation of various metabolic pathways located in different cellular compartments.

Fatty Acid Synthase Inhibitors May Be Useful Drugs

Fatty acid synthase is overexpressed in a number of cancers. Researchers intrigued by this observation have tested inhibitors of fatty acid synthase on mice to see if the inhibitors slow tumor growth. These inhibitors do indeed slow tumor growth, apparently by inducing apoptosis. However, another startling observation was made: *mice treated with inhibitors of the condensing enzyme showed remarkable weight loss* because they ate less. Thus, fatty acid synthase inhibitors are exciting candidates both as antitumor and as antiobesity drugs.

22.5   Acetyl CoA Carboxylase Plays a Key Role in Controlling Fatty Acid Metabolism

Fatty acid metabolism is stringently controlled so that synthesis and degradation are highly responsive to physiological needs. Fatty acid synthesis is maximal when carbohydrates and energy are plentiful and when fatty acids are scarce. *Acetyl CoA carboxylase plays an essential role in regulating fatty acid synthesis and degradation.* Recall that this enzyme catalyzes the committed step in fatty acid synthesis: the production of malonyl CoA (the activated two-carbon donor). This important enzyme is subject to both local and hormonal regulation. We will examine each of these levels of regulation in turn.

Acetyl CoA Carboxylase Is Regulated by Conditions in the Cell

Acetyl CoA carboxylase responds to changes in its immediate environment. *Acetyl CoA carboxylase is switched off by phosphorylation* and activated by dephosphorylation (Figure 22.29). *AMP-dependent protein kinase* (AMPK) converts the carboxylase into an inactive form by modifying a single serine residue. AMPK is essentially a fuel gauge; it is activated by AMP and in-

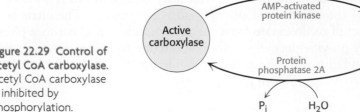

Figure 22.29 Control of acetyl CoA carboxylase. Acetyl CoA carboxylase is inhibited by phosphorylation.

hibited by ATP. Thus, the carboxylase is inactivated when the energy charge is low. Fats are not synthesized when energy is required.

The carboxylase is also allosterically stimulated by citrate. Citrate acts in an unusual manner on inactive acetyl CoA carboxylase, which exists as isolated dimers. Citrate facilitates the polymerization of the inactive dimers into active filaments (Figure 22.30). Citrate-induced polymerization can partly reverse the inhibition produced by phosphorylation (Figure 22.31). The level of citrate is high when both acetyl CoA and ATP are abundant, signifying that raw materials and energy are available for fatty acid synthesis. The stimulatory effect of citrate on the carboxylase is counteracted by *palmitoyl CoA*, which is abundant when there is an excess of fatty acids. Palmitoyl CoA causes the filaments to disassemble into the inactive subunits. Palmitoyl CoA also inhibits the translocase that transports citrate from mitochondria to the cytoplasm, as well as glucose 6-phosphate dehydrogenase, which generates NADPH in the pentose phosphate pathway.

Acetyl CoA carboxylase also plays a role in the regulation of fatty acid degradation. Malonyl CoA, the product of the carboxylase reaction, is present at a high level when fuel molecules are abundant. *Malonyl CoA inhibits carnitine acyltransferase I, preventing the entry of fatty acyl CoAs into the mitochondrial matrix in times of plenty.* Malonyl CoA is an especially effective inhibitor of carnitine acyltransferase I in heart and muscle, tissues that have little fatty acid synthesis capacity of their own. In these tissues, acetyl CoA carboxylase may be a purely regulatory enzyme.

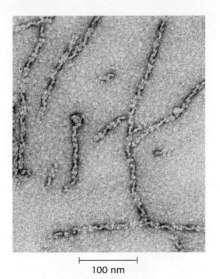

Figure 22.30 Filaments of acetyl CoA carboxylase. The electron micrograph shows the enzymatically active filamentous form of acetyl CoA carboxylase from chicken liver. The inactive form is a dimer of 265-kd subunits. [Courtesy of Dr. M. Daniel Lane.]

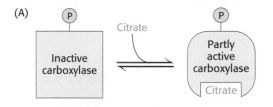

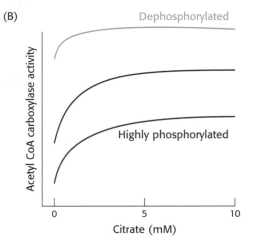

Figure 22.31 Dependence of the catalytic activity of acetyl CoA carboxylase on the concentration of citrate. (A) Citrate can partly activate the phosphorylated carboxylase. (B) The dephosphorylated form of the carboxylase is highly active even when citrate is absent. Citrate partly overcomes the inhibition produced by phosphorylation. [After G. M. Mabrouk, I. M. Helmy, K. G. Thampy, and S. J. Wakil. *J. Biol. Chem.* 265(1990):6330–6338.]

Acetyl CoA Carboxylase Is Regulated by a Variety of Hormones

Carboxylase is controlled by the hormones glucagon, epinephrine, and insulin, which reflect the overall energy status of the organism. *Insulin stimulates fatty acid synthesis by activating the carboxylase, whereas glucagon and epinephrine have the reverse effect.*

Regulation by Glucagon and Epinephrine. Consider, again, a person who has just awakened from a night's sleep and begins a bout of exercise. As mentioned, glycogen stores will be low, but lipids are readily available for mobilization.

As stated earlier, the hormones glucagon and epinephrine, present under conditions of fasting and exercise, will stimulate the release of fatty acids from triacylglycerols in fat cells, which will be released into the blood, and probably from muscle cells, where they will be used immediately as fuel. These same hormones will inhibit fatty acid synthesis by inhibiting acetyl CoA carboxylase. Although the exact mechanism by which these hormones exert their effects is not known, the net result is to augment the inhibition by the AMP-dependent kinase. This result makes sound physiological sense: when the energy level of the cell is low, as signified by high concentration of AMP, and the energy level of the organism is low, as signaled by glucagon, fats should not be synthesized. Epinephrine, which signals the need for immediate energy, enhances this effect. Hence, *these catabolic hormones switch off fatty acid synthesis by keeping the carboxylase in the inactive phosphorylated state.*

Regulation by Insulin. Now consider the situation after the exercise has ended and the runner has had a meal. In this case, the hormone insulin inhibits the mobilization of fatty acids and stimulates their accumulation as triacylglycerols by muscle and adipose tissue. Insulin also stimulates fatty acid synthesis by activating acetyl CoA carboxylase. Insulin stimulates the carboxylase by stimulating the activity of a protein phosphatase that dephosphorylates and activates acetyl CoA carboxylase. Thus, the signal molecules glucagon, epinephrine, and insulin act in concert on triacylglycerol metabolism and acetyl CoA carboxylase to carefully regulate the utilization and storage of fatty acids.

Response to Diet. *Long-term control is mediated by changes in the rates of synthesis and degradation of the enzymes participating in fatty acid synthesis.* Animals that have fasted and are then fed high-carbohydrate, low-fat diets show marked increases in their amounts of acetyl CoA carboxylase and fatty acid synthase within a few days. This type of regulation is known as *adaptive control.* This regulation, which is mediated both by insulin and glucose, is at the level of gene transcription.

22.6 The Elongation and Unsaturation of Fatty Acids Are Accomplished by Accessory Enzyme Systems

The major product of the fatty acid synthase is palmitate. In eukaryotes, longer fatty acids are formed by elongation reactions catalyzed by enzymes on the cytoplasmic face of the *endoplasmic reticulum membrane.* These reactions add two-carbon units sequentially to the carboxyl ends of both saturated and unsaturated fatty acyl CoA substrates. Malonyl CoA is the two-carbon donor in the elongation of fatty acyl CoAs. Again, condensation is driven by the decarboxylation of malonyl CoA.

Membrane-Bound Enzymes Generate Unsaturated Fatty Acids

Endoplasmic reticulum systems also introduce double bonds into long-chain acyl CoAs. For example, in the conversion of stearoyl CoA into oleoyl CoA, a cis-Δ^9 double bond is inserted by an oxidase that employs *molecular oxygen* and *NADH* (or *NADPH*).

$$\text{Stearoyl CoA} + \text{NADH} + \text{H}^+ + \text{O}_2 \longrightarrow$$
$$\text{oleoyl CoA} + \text{NAD}^+ + 2\,\text{H}_2\text{O}$$

This reaction is catalyzed by a complex of three membrane-bound proteins: *NADH-cytochrome* b_5 *reductase, cytochrome* b_5, and a *desaturase* (Figure 22.32). First, electrons are transferred from NADH to the FAD moiety of NADH-cytochrome b_5 reductase. The heme iron atom of cytochrome b_5 is then reduced to the Fe^{2+} state. The nonheme iron atom of the desaturase is subsequently converted into the Fe^{2+} state, which enables it to interact with O_2 and the saturated fatty acyl CoA substrate. A double bond is

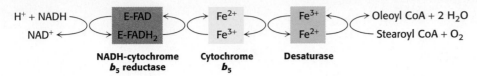

Figure 22.32 Electron-transport chain in the desaturation of fatty acids.

formed and two molecules of H_2O are released. Two electrons come from NADH and two from the single bond of the fatty acyl substrate.

A variety of unsaturated fatty acids can be formed from oleate by a combination of elongation and desaturation reactions. For example, oleate can be elongated to a 20:1 cis-Δ^{11} fatty acid. Alternatively, a second double bond can be inserted to yield an 18:2 cis-Δ^6,Δ^9 fatty acid. Similarly, palmitate (16:0) can be oxidized to palmitoleate (16:1 cis-Δ^9), which can then be elongated to *cis*-vaccenate (18:1 cis-Δ^{11}).

Unsaturated fatty acids in mammals are derived from either palmitoleate (16:1), oleate (18:1), linoleate (18:2), or linolenate (18:3). The number of carbon atoms from the ω end of a derived unsaturated fatty acid to the nearest double bond identifies its precursor.

Mammals lack the enzymes to introduce double bonds at carbon atoms beyond C-9 in the fatty acid chain. Hence, mammals cannot synthesize linoleate (18:2 cis-Δ^9,Δ^{12}) and linolenate (18:3 cis-$\Delta^9,\Delta^{12},\Delta^{15}$). *Linoleate and linolenate are the two essential fatty acids.* The term *essential* means that they must be supplied in the diet because they are required by an organism and cannot be synthesized by the organism itself. Linoleate and linolenate furnished by the diet are the starting points for the synthesis of a variety of other unsaturated fatty acids.

Precursor	Formula
Linolenate (ω-3)	$CH_3-(CH_2)_1-CH=CH-R$
Linoleate (ω-6)	$CH_3-(CH_2)_4-CH=CH-R$
Palmitoleate (ω-7)	$CH_3-(CH_2)_5-CH=CH-R$
Oleate (ω-9)	$CH_3-(CH_2)_7-CH=CH-R$

Eicosanoid Hormones Are Derived from Polyunsaturated Fatty Acids

Arachidonate, a 20:4 fatty acid derived from linoleate, is the major precursor of several classes of signal molecules: prostaglandins, prostacyclins, thromboxanes, and leukotrienes (Figure 22.33).

A prostaglandin is a 20-carbon fatty acid containing a 5-carbon ring (Figure 22.34). A series of prostaglandins are fashioned by reductases and isomerases. The major classes are designated PGA through PGI; a subscript denotes the number of carbon–carbon double bonds outside the ring. Prostaglandins with two double bonds, such as PGE_2, are derived from arachidonate; the other two double bonds of this precursor are lost in forming a five-membered ring. *Prostacyclin* and *thromboxanes* are related compounds that arise from a nascent prostaglandin. They are generated by *prostacyclin synthase* and *thromboxane synthase,* respectively. Alternatively,

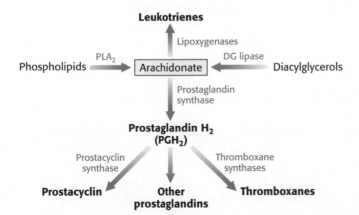

Figure 22.33 Arachidonate is the major precursor of eicosanoid hormones. Prostaglandin synthase catalyzes the first step in a pathway leading to prostaglandins, prostacyclins, and thromboxanes. Lipoxygenase catalyzes the initial step in a pathway leading to leukotrienes.

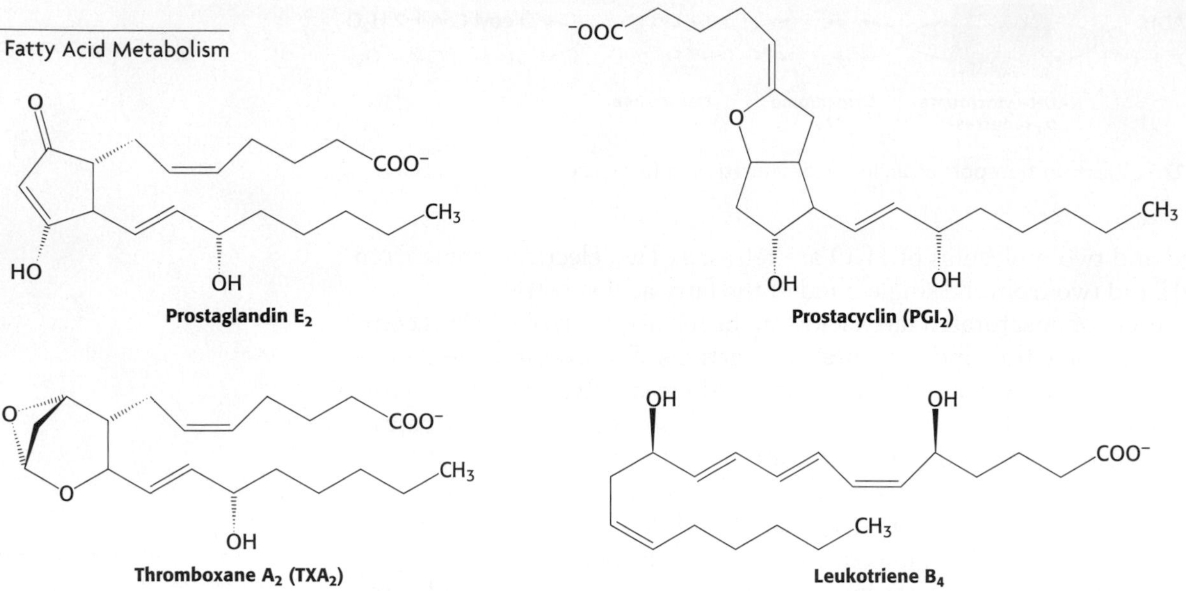

Figure 22.34 Structures of several eicosanoids.

arachidonate can be converted into *leukotrienes* by the action of *lipoxygenase*. Leukotrienes, first found in leukocytes, contain three conjugated double bonds—hence, the name. Prostaglandins, prostacyclin, thromboxanes, and leukotrienes are called *eicosanoids* (from the Greek *eikosi*, "twenty") because they contain 20 carbon atoms.

Prostaglandins and other eicosanoids are *local hormones* because they are short-lived. They alter the activities both of the cells in which they are synthesized and of adjoining cells by binding to 7TM receptors. Their effects may vary from one type of cell to another, in contrast with the more uniform actions of global hormones such as insulin and glucagon. Prostaglandins stimulate inflammation, regulate blood flow to particular organs, control ion transport across membranes, modulate synaptic transmission, and induce sleep.

Recall that aspirin blocks access to the active site of the enzyme that converts arachidonate into prostaglandin H_2 (p. 339). Because arachidonate is the precursor of other prostaglandins, prostacyclin, and thromboxanes, blocking this step interferes with many signaling pathways. Aspirin's ability to obstruct these pathways accounts for its wide-ranging effects on inflammation, fever, pain, and blood clotting.

Summary

22.1 Triacylglycerols Are Highly Concentrated Energy Stores

Fatty acids are physiologically important as (1) fuel molecules, (2) components of phospholipids and glycolipids, (3) hydrophobic modifiers of proteins, and (4) hormones and intracellular messengers. They are stored in adipose tissue as triacylglycerols (neutral fat).

22.2 The Utilization of Fatty Acids as Fuel Requires Three Stages of Processing

Triacylglycerols can be mobilized by the hydrolytic action of lipases that are under hormonal control. Glucagon and epinephrine stimulate triacylglycerol breakdown by activating the lipase. Insulin, in contrast, inhibits lipolysis. Fatty acids are activated to acyl CoAs, transported across the inner mitochondrial membrane by carnitine, and degraded in the mitochondrial matrix by a recurring sequence of four reactions: oxidation by FAD, hydration, oxidation by NAD^+, and thiolysis by coenzyme A. The $FADH_2$ and NADH formed in the oxidation steps transfer their electrons to O_2 by means of the respiratory chain, whereas the acetyl CoA formed in the thiolysis step normally enters the citric acid cycle by condensing with oxaloacetate. Mammals are unable to convert fatty acids into glucose, because they lack a pathway for the net production of oxaloacetate, pyruvate, or other gluconeogenic intermediates from acetyl CoA.

22.3 Unsaturated and Odd-Chain Fatty Acids Require Additional Steps for Degradation

Fatty acids that contain double bonds or odd numbers of carbon atoms require ancillary steps to be degraded. An isomerase and a reductase are required for the oxidation of unsaturated fatty acids, whereas propionyl CoA derived from chains with odd numbers of carbon atoms requires a vitamin B_{12}-dependent enzyme to be converted into succinyl CoA.

22.4 Fatty Acids Are Synthesized and Degraded by Different Pathways

Fatty acids are synthesized in the cytoplasm by a different pathway from that of β oxidation. Synthesis starts with the carboxylation of acetyl CoA to malonyl CoA, the committed step. This ATP-driven reaction is catalyzed by acetyl CoA carboxylase, a biotin enzyme. The intermediates in fatty acid synthesis are linked to an acyl carrier protein. Acetyl ACP is formed from acetyl CoA, and malonyl ACP is formed from malonyl CoA. Acetyl ACP and malonyl ACP condense to form acetoacetyl ACP, a reaction driven by the release of CO_2 from the activated malonyl unit. A reduction, a dehydration, and a second reduction follow. NADPH is the reductant in these steps. The butyryl ACP formed in this way is ready for a second round of elongation, starting with the addition of a two-carbon unit from malonyl ACP. Seven rounds of elongation yield palmitoyl ACP, which is hydrolyzed to palmitate. In higher organisms, the enzymes catalyzing fatty acid synthesis are covalently linked in a multifunctional enzyme complex. A reaction cycle based on the formation and cleavage of citrate carries acetyl groups from mitochondria to the cytoplasm. NADPH needed for synthesis is generated in the transfer of reducing equivalents from mitochondria by the malate–pyruvate shuttle and by the pentose phosphate pathway.

22.5 Acetyl CoA Carboxylase Plays a Key Role in Controlling Fatty Acid Metabolism

Fatty acid synthesis and degradation are reciprocally regulated so that both are not simultaneously active. Acetyl CoA carboxylase, the essential control site, is phosphorylated and inactivated by AMP-dependent kinase. The phosphorylation is reversed by a protein phosphatase. Citrate, which signals an abundance of building blocks and energy, partly reverses the inhibition by phosphorylation. Carboxylase activity is stimulated by insulin and inhibited by glucagon and epinephrine. In times of plenty, fatty acyl CoAs do not enter the mitochondrial matrix, because malonyl CoA inhibits carnitine acyltransferase I.

22.6 The Elongation and Unsaturation of Fatty Acids Are Accomplished by Accessory Enzyme Systems

Fatty acids are elongated and desaturated by enzyme systems in the endoplasmic reticulum membrane. Desaturation requires NADH and O_2 and is carried out by a complex consisting of a flavoprotein, a cytochrome, and a nonheme iron protein. Mammals lack the enzymes to introduce double bonds distal to C-9, and so they require linoleate and linolenate in their diets.

Arachidonate, an essential precursor of prostaglandins and other signal molecules, is derived from linoleate. This 20:4 polyunsaturated fatty acid is the precursor of several classes of signal molecules— prostaglandins, prostacyclins, thromboxanes, and leukotrienes—that act as messengers and local hormones because of their transience. They are called eicosanoids because they contain 20 carbon atoms. Aspirin (acetylsalicylate), an anti-inflammatory and antithrombotic drug, irreversibly blocks the synthesis of these eicosanoids.

Key Terms

triacylglycerol (neutral fat, triglyceride) (p. 617)

acyl adenylate (p. 622)

carnitine (p. 623)

β-oxidation pathway (p. 624)

vitamin B_{12} (cobalamin) (p. 627)

peroxisome (p. 630)

ketone body (p. 631)

acyl carrier protein (ACP) (p. 634)

fatty acid synthase (p. 634)

malonyl CoA (p. 635)

acetyl CoA carboxylase (p. 635)

AMP-dependent protein kinase (AMPK) (p. 640)

arachidonate (p. 643)

prostaglandin (p. 643)

eicosanoid (p. 644)

Selected Readings

Where to Start

Rinaldo, P., Matern, D., and Bennet, M. J. 2002. Fatty acid oxidation disorders. *Annu. Rev. Physiol.* 64:477–502.

Rasmussen, B. B., and Wolfe, R. R. 1999. Regulation of fatty acid oxidation in skeletal muscle. *Annu. Rev. Nutr.* 19: 463–484.

Semenkovich, C. F. 1997. Regulation of fatty acid synthase (FAS). *Prog. Lipid Res.* 36:43–53.

Sul, H. S., Smas, C. M., Wang, D., and Chen, L. 1998. Regulation of fat synthesis and adipose differentiation. *Prog. Nucleic Acid Res. Mol. Biol.* 60:317–345.

Wolf, G. 1996. Nutritional and hormonal regulation of fatty acid synthase. *Nutr. Rev.* 54:122–123.

Munday, M. R., and Hemingway, C. J. 1999. The regulation of acetyl-CoA carboxylase: A potential target for the action of hypolipidemic agents. *Adv. Enzyme Regul.* 39:205–234.

Books

Vance, D. E., and Vance, J. E. (Eds.). 1996. *Biochemistry of Lipids, Lipoproteins, and Membranes.* Elsevier.

Stipanuk, M. H. (Ed.). 2000. *Biochemical and Physiological Aspects of Human Nutrition.* Saunders.

Fatty Acid Oxidation

Saha, P. K., Kojima, H., Marinez-Botas, J., Sunehag, A. L., and Chan, L. 2004. Metabolic adaptations in absence of perilipin. *J. Biol. Chem.* 279:35150–35158.

Barycki, J. J., O'Brien, L. K., Strauss, A. W., and Banaszak, L. J. 2000. Sequestration of the active site by interdomain shifting: Crystallographic and spectroscopic evidence for distinct conformations of L-3-hydroxya-cyl-CoA dehydrogenase. *J. Biol. Chem.* 275:27186–27196.

Ramsay, R. R. 2000. The carnitine acyltransferases: Modulators of acyl-CoA-dependent reactions. *Biochem. Soc. Trans.* 28:182–186.

Eaton, S., Bartlett, K., and Pourfarzam, M. 1996. Mammalian mito-chondrial β-oxidation. *Biochem. J.* 320:345–357.

Thorpe, C., and Kim, J. J. 1995. Structure and mechanism of action of the acyl-CoA dehydrogenases. *FASEB J.* 9:718–725.

Fatty Acid Synthesis

Witkowski, A., Ghosal, A., Joshi, A. K., Witkowska, H. E., Asturias, F. J., and Smith, S. 2004. Head-to-head arrangement of the subunits of the animal fatty acid synthase. *Chem. Biol.* 11:1667–1676.

Ming, D., Kong, Y., Wakil, S. J., Brink, J., and Ma, J. 2002. Domain movements in human fatty acid synthase by quantized elastic de-formational model. *Proc. Natl. Acad. Sci. U. S A.* 99:7895–7899.

Zhang, Y.-M., Rao, M. S., Heath, R. J., Price, A. C., Olson, A. J., Rock, C. O., and White, S. W. 2001. Identification and analysis of the acyl carrier protein (ACP) docking site on β-ketoacyl-ACP synthase III. *J. Biol. Chem.* 276:8231–8238.

Davies, C., Heath, R. J., White, S. W., and Rock, C. O. 2000. The 1.8 Å crystal structure and active-site architecture of β-ketoacyl-acyl car-rier protein synthase III (FabH) from *Escherichia coli. Structure Fold Des.* 8:185–195.

Denton, R. M., Heesom, K. J., Moule, S. K., Edgell, N. J., and Burnett, P. 1997. Signalling pathways involved in the stimulation of fatty acid synthesis by insulin. *Biochem. Soc. Trans.* 25:1238–1242.

Stoops, J. K., Kolodziej, S. J., Schroeter, J. P., Bretaudiere, J. P., and Wakil, S. J. 1992. Structure-function relationships of the yeast fatty acid synthase: Negative-stain, cryo-electron microscopy, and image analysis studies of the end views of the structure. *Proc. Natl. Acad. Sci. U. S. A.* 89:6585–6589.

Loftus, T. M., Jaworsky, D. E., Frehywot, G. L., Townsend, C. A., Ronnett, G. V., Lane, M. D., and Kuhajda, F. P. 2000. Reduced food intake and body weight in mice treated with fatty acid synthase inhibitors. *Science* 288:2379–2381.

Acetyl CoA Carboxylase

Munday, M. R. 2002. Regulation of acetyl CoA carboxylase. *Biochem. Soc. Trans.* 30: 1059–1064.

Thoden, J. B., Blanchard, C. Z., Holden, H. M., and Waldrop, G. L. 2000. Movement of the biotin carboxylase B-domain as a result of ATP binding. *J. Biol. Chem.* 275:16183–16190.

Kim, K. H. 1997. Regulation of mammalian acetyl-coenzyme A car-boxylase. *Annu. Rev. Nutr.* 17:77–99.

Mabrouk, G. M., Helmy, I. M., Thampy, K. G., and Wakil, S. J. 1990. Acute hormonal control of acetyl-CoA carboxylase: The roles of in-sulin, glucagon, and epinephrine. *J. Biol. Chem.* 265:6330–6338.

Kim, K. H., Lopez, C. F., Bai, D. H., Luo, X., and Pape, M. E. 1989. Role of reversible phosphorylation of acetyl-CoA carboxylase in long-chain fatty acid synthesis. *FASEB J.* 3:2250–2256.

Witters, L. A., and Kemp, B. E. 1992. Insulin activation of acetyl-CoA carboxylase accompanied by inhibition of the 5'-AMP-activated protein kinase. *J. Biol. Chem.* 267:2864–2867.

Cohen, P., and Hardie, D. G. 1991. The actions of cyclic AMP on biosynthetic processes are mediated indirectly by cyclic AMP-dependent protein kinases. *Biochim. Biophys. Acta* 1094: 292–299.

Moore, F., Weekes, J., and Hardie, D. G. 1991. Evidence that AMP triggers phosphorylation as well as direct allosteric activation of rat liver AMP-activated protein kinase: A sensitive mechanism to protect the cell against ATP depletion. *Eur. J. Biochem.* 199:691–697.

Eicosanoids

Nakamura, M. T., and Nara, T. Y. 2004. Structure, function, and dietary regulation of $\Delta 6$, $\Delta 5$, and $\Delta 9$ desaturases. *Annu. Rev. Nutr.* 24:345–376.

Malkowski, M. G., Ginell, S. L., Smith, W. L., and Garavito, R. M. 2000. The productive conformation of arachidonic acid bound to prostaglandin synthase. *Science* 289:1933–1937.

Smith, T., McCracken, J., Shin, Y.-K., and DeWitt, D. 2000. Arachidonic acid and nonsteroidal anti-inflammatory drugs induce conformational changes in the human prostaglandin endoperoxide H$_2$ synthase-2 (cyclooxygenase-2). *J. Biol. Chem.* 275:40407–40415.

Kalgutkar, A. S., Crews, B. C., Rowlinson, S. W., Garner, C., Seibert, K., and Marnett L. J. 1998. Aspirin-like molecules that covalently inactivate cyclooxygenase-2. *Science* 280:1268–1270.

Lands, W. E. 1991. Biosynthesis of prostaglandins. *Annu. Rev. Nutr.* 11:41–60.

Sigal, E. 1991. The molecular biology of mammalian arachidonic acid metabolism. *Am. J. Physiol.* 260:L13–L28.

Weissmann, G. 1991. Aspirin. *Sci. Am.* 264(1):84–90.

Vane, J. R., Flower, R. J., and Botting, R. M. 1990. History of aspirin and its mechanism of action. *Stroke* (12 suppl.):IV12–IV23.

Genetic Diseases

Brivet, M., Boutron, A., Slama, A., Costa, C., Thuillier, L., Demaugre, F., Rabier, D., Saudubray, J. M., and Bonnefont, J. P. 1999. Defects in activation and transport of fatty acids. *J. Inherited Metab. Dis.* 22:428–441.

Wanders, R. J., van Grunsven, E. G., and Jansen, G. A. 2000. Lipid metabolism in peroxisomes: Enzymology, functions and dysfunctions of the fatty acid α- and β-oxidation systems in humans. *Biochem. Soc. Trans.* 28:141–149.

Wanders, R. J., Vreken, P., den Boer, M. E., Wijburg, F. A., van Gennip, A. H., and IJlst, L. 1999. Disorders of mitochondrial fatty acyl-CoA β-oxidation. *J. Inherited Metab. Dis.* 22:442–487.

Kerner, J., and Hoppel, C. 1998. Genetic disorders of carnitine metabolism and their nutritional management. *Annu. Rev. Nutr.* 18:179–206.

Bartlett, K., and Pourfarzam, M. 1998. Recent developments in the detection of inherited disorders of mitochondrial β-oxidation. *Biochem. Soc. Trans.* 26:145–152.

Pollitt, R. J. 1995. Disorders of mitochondrial long-chain fatty acid oxidation. *J. Inherited Metab. Dis.* 18:473–490.

Roe, C. R., and Coates, P. M. 1995. Mitochondrial fatty acid oxidation disorders. In *The Metabolic Basis of Inherited Diseases* (7th ed., pp. 1501–1534), edited by C. R. Scriver, A. L. Beaudet, W. S. Sly, D. Valle, J. B. Stanbury, J. B. Wyngaarden, and D. S. Fredrickson. McGraw-Hill.

Problems

1. *After lipolysis.* Write a balanced equation for the conversion of glycerol into pyruvate. Which enzymes are required in addition to those of the glycolytic pathway?

2. *From fatty acid to ketone body.* Write a balanced equation for the conversion of stearate into acetoacetate.

3. *Counterpoint.* Compare and contrast fatty acid oxidation and synthesis with respect to

(a) site of the process.
(b) acyl carrier.
(c) reductants and oxidants.
(d) stereochemistry of the intermediates.
(e) direction of synthesis or degradation.
(f) organization of the enzyme system.

4. *Sources.* For each of the following unsaturated fatty acids, indicate whether the biosynthetic precursor in animals is palmitoleate, oleate, linoleate, or linolenate.

(a) 18:1 cis-Δ^{11}
(b) 18:3 cis-Δ^6, Δ^9, Δ^{12}
(c) 20:2 cis-Δ^{11}, Δ^{14}
(d) 20:3 cis-Δ^5, Δ^8, Δ^{11}
(e) 22:1 cis-Δ^{13}
(f) 22:6 cis-Δ^4, Δ^7, Δ^{10}, Δ^{13}, Δ^{16}, Δ^{19}

5. *Tracing carbons.* Consider a cell extract that actively synthesizes palmitate. Suppose that a fatty acid synthase in this preparation forms one molecule of palmitate in about 5 minutes. A large amount of malonyl CoA labeled with ^{14}C in each carbon atom of its malonyl unit is suddenly added to this system, and fatty acid synthesis is stopped a minute later by altering the pH. The fatty acids in the supernatant are analyzed for radioactivity. Which carbon atom of the palmitate formed by this system is more radioactive—C-1 or C-14?

6. *Driven by decarboxylation.* What is the role of decarboxylation in fatty acid synthesis? Name another key reaction in a metabolic pathway that employs this mechanistic motif.

7. *Kinase surfeit.* Suppose that a promoter mutation leads to the overproduction of protein kinase A in adipose cells. How might fatty acid metabolism be altered by this mutation?

8. *An unaccepting mutant.* The serine residue in acetyl CoA carboxylase that is the target of the AMP-dependent protein kinase is mutated to alanine. What is a likely consequence of this mutation?

9. *Blocked assets.* The presence of a fuel molecule in the cytoplasm does not ensure that the fuel molecule can be effectively used. Give two examples of how impaired transport of metabolites between compartments leads to disease.

10. *Elegant inversion.* Peroxisomes have an alternative pathway for oxidizing polyunsaturated fatty acids. They contain a hydratase that converts D-3-hydroxyacyl CoA into *trans*-Δ^2-enoyl CoA. How can this enzyme be used to oxidize CoAs containing a cis double bond at an even-numbered carbon atom (e.g., the cis-Δ^{12} double bond of linoleate)?

11. *Covalent catastrophe.* What is a potential disadvantage of having many catalytic sites together on one very long polypeptide chain?

12. *Missing acyl CoA dehydrogenases.* A number of genetic deficiencies in acyl CoA dehydrogenases have been described. This deficiency presents early in life after a period of fasting. Symptoms include vomiting, lethargy, and sometimes coma. Not only are blood levels of glucose low (hypoglycemia), but starvation-induced ketosis is absent. Provide a biochemical explanation for these last two observations.

13. *Effects of clofibrate.* High blood levels of triacylglycerides are associated with heart attacks and strokes. Clofibrate, a drug that increases the activity of peroxisomes, is sometimes used to treat patients with such a condition. What is the biochemical basis for this treatment?

14. *A different kind of enzyme.* Figure 22.31 shows the response of acetyl CoA carboxylase to varying amounts of citrate. Explain this effect in light of the allosteric effects that citrate has on the enzyme. Predict the effects of increasing concentrations of palmitoyl CoA.

Mechanism Problems

15. *Variation on a theme.* Thiolase is homologous in structure to the condensing enzyme. On the basis of this observation, propose a mechanism for the cleavage of 3-ketoacyl CoA by CoA.

16. *Two plus three to make four.* Propose a reaction mechanism for the condensation of an acetyl unit with a malonyl unit to form an acetoacetyl unit in fatty acid synthesis.

Chapter Integration Problems

17. *Ill-advised diet.* Suppose that, for some bizarre reason, you decided to exist on a diet of whale and seal blubber, exclusively.

(a) How would lack of carbohydrates affect your ability to utilize fats?
(b) What would your breath smell like?
(c) One of your best friends, after trying unsuccessfully to convince you to abandon this diet, makes you promise to consume a healthy dose of odd-chain fatty acids. Does your friend have your best interests at heart? Explain.

18. *Fats to glycogen.* An animal is fed stearic acid that is radioactively labeled with [^{14}C]carbon. A liver biopsy reveals the presence of ^{14}C-labeled glycogen. How is this possible in light of the fact that animals cannot convert fats into carbohydrates?

Data Interpretation Problem

19. *Mutant enzyme.* Carnitine palmitoyl transferase I (CPTI) catalyzes the conversion of long-chain acyl CoA into acyl carnitine, a prerequisite for transport into mitochondria and subsequent degradation. A mutant enzyme was constructed with a single amino acid change at position 3 of glutamic acid for alanine. Figures A through C show data from studies performed to identify the effect of the mutation [data from J. Shi, H. Zhu, D. N. Arvidson, and G. J. Woldegiorgis. *J. Biol. Chem.* 274(1999):9421–9426].

(a) What is the effect of the mutation on enzyme activity when the concentration of carnitine is varied (Figure A)? What are the K_M and V_{max} values for the wild-type and mutant enzymes?

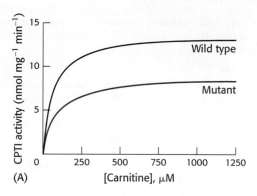

(A)

(b) What is the effect when the experiment is repeated with varying concentrations of palmitoyl CoA (Figure B)? What are the K_M and V_{max} values for the wild-type and mutant enzymes?

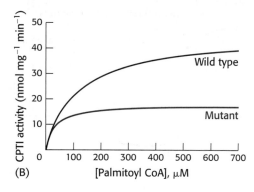

(B)

(c) Figure C shows the inhibitory effect of malonyl CoA on the wild-type and mutant enzymes. Which enzyme is more sensitive to malonyl CoA inhibition?

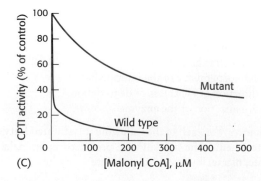

(C)

(d) Suppose that the concentration of palmitoyl CoA = 100 μM, that of carnitine = 100 μM, and that of malonyl CoA = 10 μM. Under these conditions, what is the most prominent effect of the mutation on the properties of the enzyme?
(e) What can you conclude about the role of glutamate 3 in carnitine acyltransferase I function?

Protein Turnover and Amino Acid Catabolism

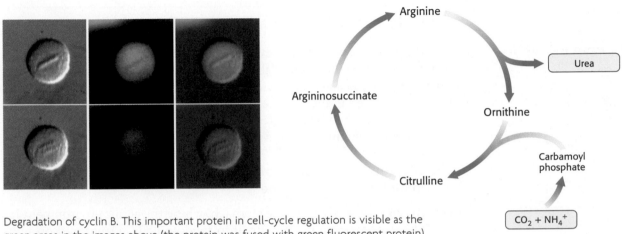

Degradation of cyclin B. This important protein in cell-cycle regulation is visible as the green areas in the images above (the protein was fused with green fluorescent protein). Cyclin B is prominent during metaphase but is degraded in anaphase to prevent the premature initiation of another cell cycle. A large protease complex called the proteasome digests the protein into peptides, which are then degraded into amino acids. These amino acids are either reused or further processed so that the carbon skeletons can be used as fuels or building blocks. The released amino group is converted into urea for excretion by the urea cycle. [(Left) Courtesy of Dr. Jonathan Pines, University of Cambridge, Wellcome/CRC Institute of Cancer and Developmental Biology.]

The digestion of dietary proteins in the intestine and the degradation of proteins within the cell provide a steady supply of amino acids to the cell. Many cellular proteins are constantly degraded and resynthesized in response to changing metabolic demands. Others are misfolded or become damaged and they, too, must be degraded. Unneeded or damaged proteins are marked for destruction by the covalent attachment of chains of a small protein called *ubiquitin* and then degraded by a large, ATP-dependent complex called the *proteasome. The primary use of amino acids provided through degradation or digestion is as building blocks for the synthesis of proteins and other nitrogenous compounds such as nucleotide bases.*

Amino acids in excess of those needed for biosynthesis can neither be stored, in contrast with fatty acids and glucose, nor excreted. Rather, surplus amino acids are used as metabolic fuel. *The α-amino group is removed, and the resulting carbon skeleton is converted into a major metabolic intermediate.* Most of the amino groups harvested from surplus amino acids are converted into urea through the *urea cycle,* whereas their carbon skeletons are transformed into acetyl CoA, acetoacetyl CoA, pyruvate, or one of the intermediates of the citric acid cycle. The principal fate of the carbon skeletons is conversion into glucose and glycogen.

Several coenzymes play key roles in amino acid degradation, foremost among them is *pyridoxal phosphate.* This coenzyme forms Schiff-base intermediates that allow α-amino groups to be shuttled between amino acids and ketoacids. We will consider several genetic errors of amino acid degradation that lead to brain damage and mental retardation unless remedial action is initiated soon after birth. *Phenylketonuria,* which is caused by a block in the conversion of phenylalanine into tyrosine, is readily diagnosed and can be treated by removing phenylalanine from the diet. The study of amino acid metabolism is especially rewarding because it is rich in connections between basic biochemistry and clinical medicine.

23.1 Proteins Are Degraded to Amino Acids

Dietary protein is a vital source of amino acids. Especially important dietary proteins are those containing the essential amino acids—amino acids that cannot be synthesized and must be acquired in the diet (Table 23.1). Proteins ingested in the diet are digested into amino acids or small peptides that can be absorbed by the intestine and transported in the blood. Another crucial source of amino acids is the degradation of cellular proteins.

The Digestion of Dietary Proteins Begins in the Stomach and Is Completed in the Intestine

Protein digestion begins in the stomach, where the acidic environment favors the denaturation of proteins into random coils. Denatured proteins are more accessible as substrates for proteolysis than are native proteins. The primary proteolytic enzyme of the stomach is *pepsin,* a nonspecific protease that, remarkably, is maximally active at pH 2. Thus, pepsin can function in the highly acidic environment of the stomach that disables other proteins.

Protein degradation continues in the lumen of the intestine. The pancreas secretes a variety of proteolytic enzymes into the intestinal lumen as inactive zymogens that are then converted into active enzymes (Sections 9.1 and 10.4). The battery of enzymes displays a wide array of specificity, and so the substrates are degraded into free amino acids as well as di- and tripeptides. Digestion is further enhanced by proteolytic enzymes, such as aminopeptidase N, that are located in the plasma membrane of the intestinal cells. Aminopeptidases digest proteins from the amino-terminal end. Single amino acids, as well as di- and tripeptides, are transported into the intestinal cells from the lumen and subsequently released into the blood for absorption by other tissues (Figure 23.1).

TABLE 23.1 Essential amino acids in human beings

Histidine
Isoleucine
Leucine
Lysine
Methionine
Phenylalanine
Threonine
Tryptophan
Valine

Figure 23.1 Digestion and absorption of proteins. Protein digestion is primarily a result of the activity of enzymes secreted by the pancreas. Aminopeptidases associated with the intestinal epithelium further digest proteins. The amino acids and di- and tripeptides are absorbed into the intestinal cells by specific transporters. Free amino acids are then released into the blood for use by other tissues.

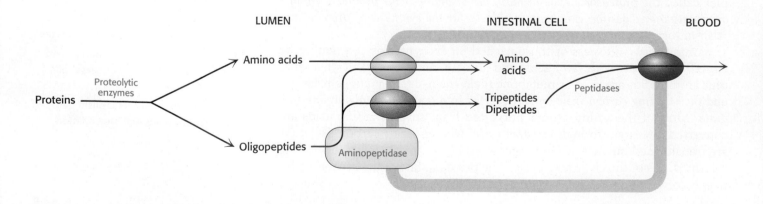

Cellular Proteins Are Degraded at Different Rates

Protein turnover—the degradation and resynthesis of proteins—takes place constantly in cells. Although some proteins are very stable, many proteins are short lived, particularly those that participate in metabolic regulation. These proteins can be quickly degraded to activate or shut down a signaling pathway. In addition, cells must eliminate damaged proteins. A significant proportion of newly synthesized protein molecules are defective because of errors in translation or misfolding. Even proteins that are normal when first synthesized may undergo oxidative damage or be altered in other ways with the passage of time. These proteins must be removed before they accumulate and aggregate. Indeed, a number of pathological conditions such as certain forms of Parkinson disease and Huntington disease are associated with protein aggregation.

The half-lives of proteins range over several orders of magnitude (see Table 23.2). Ornithine decarboxylase, at approximately 11 minutes, has one of the shortest half-lives of any mammalian protein. This enzyme participates in the synthesis of polyamines, which are cellular cations essential for growth and differentiation. The life of hemoglobin, on the other hand, is limited only by the life of the red blood cell, and the lens protein, crystallin, by the life of the organism.

TABLE 23.2 Dependence of the half-lives of cytoplasmic yeast proteins on the identity of their amino-terminal residues

Highly stabilizing residues ($t_{1/2} >$ 20 hours)			
Ala	Cys	Gly	Met
Pro	Ser	Thr	Val

Intrinsically destabilizing residues ($t_{1/2} =$ 2 to 30 minutes)			
Arg	His	Ile	Leu
Lys	Phe	Trp	Tyr

Destabilizing residues after chemical modification ($t_{1/2} =$ 3 to 30 minutes)			
Asn	Asp	Gln	Glu

Source: J. W. Tobias, T. E. Schrader, G. Rocap, and A. Varshavsky. *Science* 254(1991):1374–1377.

23.2 Protein Turnover Is Tightly Regulated

How can a cell distinguish proteins that should be degraded? *Ubiquitin* (Ub), a small (8.5-kd) protein present in all eukaryotic cells, is a tag that marks proteins for destruction (Figure 23.2). Ubiquitin is the cellular equivalent of the "black spot" of Robert Louis Stevenson's *Treasure Island:* the signal for death.

Ubiquitin Tags Proteins for Destruction

Ubiquitin is highly conserved in eukaryotes: yeast and human ubiquitin differ at only 3 of 76 residues. The carboxyl-terminal glycine residue of ubiquitin becomes covalently attached to the ε-amino groups of several lysine residues on a protein destined to be degraded. The energy for the formation of these *isopeptide bonds* (*iso* because ε- rather than α-amino groups are targeted) comes from ATP hydrolysis.

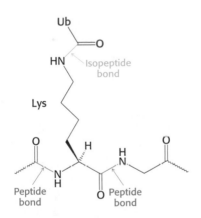

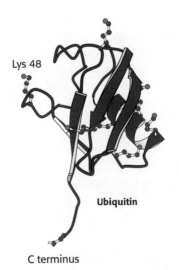

Figure 23.2 Structure of ubiquitin. *Notice* that ubiquitin has an extended carboxyl terminus, which is activated and linked to proteins targeted for destruction. Lysine residues are shown as ball-and-stick models, including lysine 48, the major site for linking additional ubiquitin molecules. [Drawn from 1UBI.pdb.]

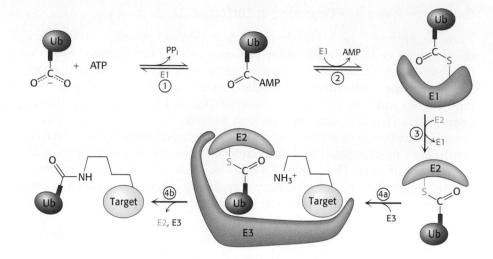

Figure 23.3 Ubiquitin conjugation. The ubiquitin-activating enzyme E1 adenylates ubiquitin (Ub) (1) and transfers the ubiquitin to one of its own cysteine residues (2). Ubiquitin is then transferred to a cysteine residue in the ubiquitin-conjugating enzyme E2 (3). Finally, the ubiquitin–protein ligase E3 transfers the ubiquitin to a lysine residue on the target protein (4a and 4b).

Three enzymes participate in the attachment of ubiquitin to a protein (Figure 23.3): ubiquitin-activating enzyme, or E1; ubiquitin-conjugating enzyme, or E2; and ubiquitin–protein ligase, or E3. First, the C-terminal carboxylate group of ubiquitin becomes linked to a sulfhydryl group of E1 by a thioester bond. This ATP-driven reaction is reminiscent of fatty acid activation (p. 622). In this reaction, ATP is linked to the C-terminal carboxylate of ubiquitin with the release of pyrophosphate, and the ubiquitin is transferred to a sulfhydryl group of a key cysteine residue in E1. The activated ubiquitin is then shuttled to a sulfhydryl group of E2. Finally, E3 catalyzes the transfer of ubiquitin from E2 to an ε-amino group on the target protein.

A chain of four or more ubiquitin molecules is especially effective in signaling the need for degradation (Figure 23.4). The ubiquitination reaction is processive: a chain of ubiquitin molecules can be generated by the linkage of the ε-amino group of lysine residue 48 of one ubiquitin molecule to the terminal carboxylate of another.

What determines whether a protein becomes ubiquitinated? One signal turns out to be unexpectedly simple. *The half-life of a cytoplasmic protein is determined to a large extent by its amino-terminal residue* (Table 23.2). This dependency is referred to as the *N-terminal rule.* A yeast protein with methionine at its N terminus typically has a half-life of more than 20 hours, whereas one with arginine at this position has a half-life of about 2 minutes. A highly destabilizing N-terminal residue such as arginine or leucine favors rapid ubiquitination, whereas a stabilizing residue such as methionine or proline does not. Other signals thought to identify proteins for degradation include *cyclin destruction boxes,* which are amino acid sequences that mark cell-cycle proteins for destruction, and PEST sequences, which contain the amino acid sequence proline (P, single-letter abbreviation), glutamic acid (E), serine (S), and threonine (T).

E3 enzymes are the readers of N-terminal residues. Although most eukaryotes have only one or a small number of distinct E1 enzymes, all eukaryotes have many distinct E2 and E3 enzymes. Moreover, there appears to be only a single family of evolutionarily related E2 proteins but three distinct families of E3 proteins, altogether consisting of

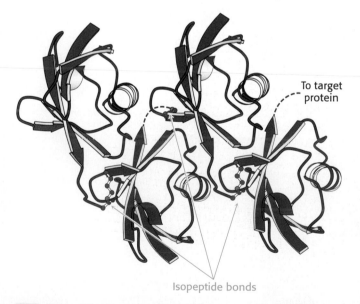

Figure 23.4 Structure of tetraubiquitin. Four ubiquitin molecules are linked by isopeptide bonds. *Notice* that each isopeptide bond is formed by the linkage of the carboxylate group at the end of the extended C terminus with the ε-amino group of a lysine residue. Dashed lines indicate the positions of the extended C-termini that were not observed in the crystal structure. This unit is the primary signal for degradation when linked to a target protein. [Drawn from 1TBE.pdb.]

hundreds of members. Indeed, the E3 family is one of the largest gene families in human beings. The diversity of target proteins that must be tagged for destruction requires a large number of E3 proteins as readers.

Three examples demonstrate the importance of E3 proteins to normal cell function. Proteins that are not broken down owing to a defective E3 may accumulate to create a disease of protein aggregation such as juvenile and early-onset Parkinson disease. A defect in another member of the E3 family causes Angelman syndrome, a severe neurological disorder characterized by mental retardation, absence of speech, uncoordinated movement, and hyperactivity. Conversely, uncontrolled protein turnover also can create dangerous pathological conditions. For example, human papilloma virus (HPV) encodes a protein that activates a specific E3 enzyme. The enzyme ubiquitinates the tumor suppressor p53 and other proteins that control DNA repair, which are then destroyed. The activation of this E3 enzyme is observed in more than 90% of cervical carcinomas. Thus, the inappropriate marking of key regulatory proteins for destruction can trigger further events, leading to tumor formation.

The Proteasome Digests the Ubiquitin-Tagged Proteins

If ubiquitin is the mark of death, what is the executioner? *A large protease complex called the proteasome or the 26S proteasome digests the ubiquitinated proteins.* This ATP-driven multisubunit protease spares ubiquitin, which is then recycled. The 26S proteasome is a complex of two components: a 20S catalytic unit and a 19S regulatory unit.

The 20S unit is constructed from two copies each of 14 homologous subunits and has a mass of 700 kd (Figure 23.5). The subunits are arranged in four rings of 7 subunits that stack to form a structure resembling a barrel. The outer two rings of the barrel are made up of α subunits and the inner two rings of β subunits. The 20S catalytic core is a sealed barrel. *Access to its interior is controlled by a 19S regulatory unit,* itself a 700-kd complex made up of 20 subunits. Two such 19S complexes bind to the 20S proteasome core, one at each end, to form the complete 26S proteasome (Figure 23.6). The 19S unit binds specifically to polyubiquitin chains, thereby ensuring that only ubiquitinated proteins are degraded. Key components of the 19S complex are six ATPases of a type called the AAA class (ATPase *a*ssociated with various cellular *a*ctivities). ATP hydrolysis likely assists the 19S complex to unfold the substrate and induce conformational changes in the 20S catalytic core so that the substrate can be passed into the center of the complex.

The proteolytic active sites are sequestered in the interior of the barrel to protect potential substrates until they are directed into the barrel. There are three types of active sites in the β subunits, each with a different specificity, but all employ an N-terminal threonine. The hydroxyl group of the threonine residue is converted into a nucleophile that attacks the carbonyl groups of peptide bonds to form acyl-enzyme intermediates (p. 244). Substrates are degraded in a processive manner without the release of degradation intermediates, until the substrate is reduced to peptides ranging in length from seven to nine residues. Finally, an isopeptidase in the 19S unit cleaves off intact ubiquitin molecules from these peptides. The ubiquitin is recycled and the peptide products are further degraded by other cellular proteases to yield individual amino acids. Thus, the ubiquitination pathway and the proteasome cooperate to degrade unwanted proteins. Figure 23.7 presents an overview of the fates of amino acids following proteasomal digestion.

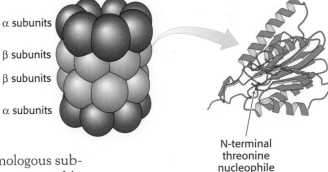

α subunits

β subunits

β subunits

α subunits

N-terminal threonine nucleophile

Figure 23.5 20S proteasome. The 20S proteasome comprises 28 homologous subunits (α, red; β, blue), arranged in four rings of 7 subunits each. Some of the β subunits include protease active sites at their amino termini. [Subunit drawn from 1RYP.pdb.]

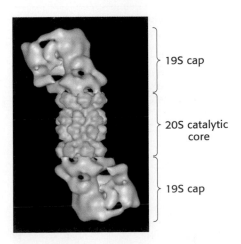

19S cap

20S catalytic core

19S cap

Figure 23.6 26S proteasome. A 19S cap is attached to each end of the 20S catalytic unit. [From W. Baumeister, J. Walz, F. Zuhl, and E. Seemuller. *Cell* 92(1998):367–380; courtesy of Dr. Wolfgang Baumeister.]

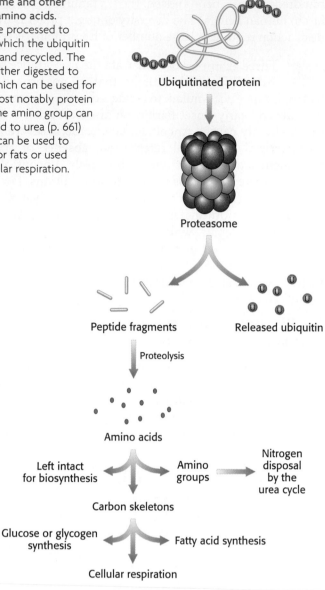

Figure 23.7 The proteasome and other proteases generate free amino acids. Ubiquitinated proteins are processed to peptide fragments from which the ubiquitin is subsequently removed and recycled. The peptide fragments are further digested to yield free amino acids, which can be used for biosynthetic reactions, most notably protein synthesis. Alternatively, the amino group can be removed and processed to urea (p. 661) and the carbon skeleton can be used to synthesize carbohydrate or fats or used directly as a fuel for cellular respiration.

TABLE 23.3   Processes regulated by protein degradation

Gene transcription
Cell-cycle progression
Organ formation
Circadian rhythms
Inflammatory response
Tumor suppression
Cholesterol metabolism
Antigen processing

Bortezomib
(a dipeptidyl boronic acid)

Protein Degradation Can Be Used to Regulate Biological Function

Table 23.3 lists a number of physiological processes that are controlled at least in part by protein degradation through the ubiquitin–proteasome pathway. In each case, the proteins being degraded are regulatory proteins. Consider, for example, control of the inflammatory response. A transcription factor called *NF-κB* (NF for nuclear factor) initiates the expression of a number of the genes that take part in this response. This factor is itself activated by the degradation of an attached inhibitory protein, *I-κB* (I for inhibitor). In response to inflammatory signals that bind to membrane-bound receptors, I-κB is phosphorylated at two serine residues, creating an E3 binding site. The binding of E3 leads to the ubiquitination and degradation of I-κB, unleashing NF-κB. The liberated transcription factor migrates to the nucleus to stimulate the transcription of the target genes. The NF-κB–I-κB system illustrates the interplay of several key regulatory motifs: receptor-mediated signal transduction, phosphorylation, compartmentalization, controlled and specific degradation, and selective gene expression. The importance of the ubiquitin–proteasome system for the regulation of gene expression is highlighted by the recent approval of bortezomib (Velcade), a potent inhibitor of the proteasome, as a therapy for multiple myeloma. Bortezomib is a dipeptidyl boronic acid inhibitor of the proteasome.

Archaeal proteasome   **Eukaryotic proteasome**

Figure 23.8 Proteasome evolution. The archaeal proteasome consists of 14 identical α subunits and 14 identical β subunits. In the eukaryotic proteasome, gene duplication and specialization has led to 7 distinct subunits of each type. The overall architecture of the proteasome is conserved.

The Ubiquitin Pathway and the Proteasome Have Prokaryotic Counterparts

Both the ubiquitin pathway and the proteasome appear to be present in all eukaryotes. Homologs of the proteasome are found in prokaryotes, although the physiological roles of these homologs have not been well established. The proteasomes of some archaea are quite similar in overall structure to their eukaryotic counterparts and similarly have 28 subunits (Figure 23.8). In the archaeal proteasome, however, all α outer-ring subunits and all β inner-ring subunits are identical; in eukaryotes, each α or β subunit is one of seven different isoforms. This specialization provides distinct substrate specificity.

Although ubiquitin has not been found in prokaryotes, ubiquitin's molecular ancestors were recently identified in prokaryotes. Remarkably, these proteins take part not in protein modification but in biosynthesis of the coenzyme thiamine (p. 423). A key enzyme in thiamine biosynthesis is ThiF, which activates the protein ThiS as an acyl adenylate and then adds a sulfide ion derived from cysteine (Figure 23.9). ThiF is homologous to human E1, which includes two tandem regions of 160 amino acids that are 28% identical in amino acid sequence with a region of ThiF from *E. coli.*

Figure 23.9 Biosynthesis of thiamine. The biosynthesis of thiamine begins with the addition of sulfide to the carboxyl terminus of the protein ThiS. This protein is activated by adenylation and conjugated in a manner analogous to the first steps in the ubiquitin pathway.

The evolutionary relationships between these two pathways were cemented by the determination of the three-dimensional structure of ThiS, which revealed a structure very similar to that of ubiquitin, despite being only 14% identical in amino acid sequence (Figure 23.10). Thus, a eukaryotic system for protein modification evolved from a preexisting prokaryotic pathway for coenzyme biosynthesis.

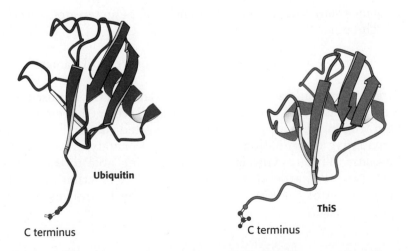

Ubiquitin

C terminus

ThiS

C terminus

Figure 23.10 Structures of ThiS and ubiquitin compared. *Notice* that ThiS is structurally similar to ubiquitin despite only 14% sequence identity. This observation suggests that a prokaryotic protein such as ThiS evolved into ubiquitin. [Drawn from 1UBI.pdb and 1FOZ.pdb.]

23.3 The First Step in Amino Acid Degradation Is the Removal of Nitrogen

What is the fate of amino acids released on protein digestion or turnover? The first call is for use as building blocks for biosynthetic reactions. However, any not needed as building blocks are degraded to compounds able to enter the metabolic mainstream. The amino group is first removed, and then the remaining carbon skeleton is metabolized to glucose, one of several citric acid cycle intermediates, or to acetyl CoA. The major site of amino acid degradation in mammals is the liver, although muscles readily degrade the branched-chain amino acids (Leu, Ile, and Val). The fate of the α-amino group will be considered first, followed by that of the carbon skeleton (Section 23.5).

Alpha-Amino Groups Are Converted into Ammonium Ions by the Oxidative Deamination of Glutamate

The α-amino group of many amino acids is transferred to α-ketoglutarate to form *glutamate*, which is then oxidatively deaminated to yield ammonium ion (NH_4^+).

Aminotransferases catalyze the transfer of an α-amino group from an α-amino acid to an α-ketoacid. These enzymes, also called *transaminases*, generally funnel α-amino groups from a variety of amino acids to α-ketoglutarate for conversion into NH_4^+.

Aspartate aminotransferase, one of the most important of these enzymes, catalyzes the transfer of the amino group of aspartate to α-ketoglutarate.

$$\text{Aspartate} + \text{α-ketoglutarate} \rightleftharpoons \text{oxaloacetate} + \text{glutamate}$$

Alanine aminotransferase catalyzes the transfer of the amino group of alanine to α-ketoglutarate.

$$\text{Alanine} + \text{α-ketoglutarate} \rightleftharpoons \text{pyruvate} + \text{glutamate}$$

These transamination reactions are reversible and can thus be used to synthesize amino acids from α-ketoacids, as we shall see in Chapter 24.

The nitrogen atom in glutamate is converted into free ammonium ion by oxidative deamination. This reaction is catalyzed by *glutamate dehydrogenase*. This enzyme is unusual in being able to utilize either NAD^+ or $NADP^+$, at least in some species. The reaction proceeds by dehydrogenation of the C–N bond, followed by hydrolysis of the resulting Schiff base.

This reaction is close to equilibrium in the liver, and the direction of the reaction is determined by the concentrations of reactants and products. Normally, the reaction is driven forward by the rapid removal of ammonium ion. Glutamate dehydrogenase is located in mitochondria, as are some of the other enzymes required for the production of urea. This compartmentalization sequesters free ammonium ion, which is toxic.

The sum of the reactions catalyzed by aminotransferases and glutamate dehydrogenase is

$$\text{α-Amino acid} + NAD^+ \rightleftharpoons \text{α-ketoacid} + NH_4^+ + NADH + H^+$$
$$\text{(or } NADP^+) \qquad\qquad\qquad \text{(or NADPH)}$$

In most terrestrial vertebrates, NH_4^+ is converted into urea, which is excreted.

Mechanism: Pyridoxal Phosphate Forms Schiff-Base Intermediates in Aminotransferases

All aminotransferases contain the prosthetic group *pyridoxal phosphate* (PLP), which is derived from *pyridoxine (vitamin B_6)*. Pyridoxal phosphate includes a pyridine ring that is slightly basic to which is attached an OH group that is slightly acidic. Thus, pyridoxal phosphate derivatives can form a stable tautomeric form in which the pyridine nitrogen atom is protonated and, hence, positively charged while the OH group loses a proton and hence is negatively charged, forming a phenolate.

Pyridoxine (Vitamin B_6)

Pyridoxal phosphate (PLP)

PLP (protonated)   **PLP (phenolate)**

The most important functional group on PLP is the aldehyde. This group forms covalent Schiff-base intermediates with amino acid substrates. Indeed, even in the absence of substrate, the aldehyde group of PLP usually forms a Schiff-base linkage with the ε-amino group of a specific lysine residue at the enzyme's active site. A new Schiff-base linkage is formed on addition of an amino acid substrate.

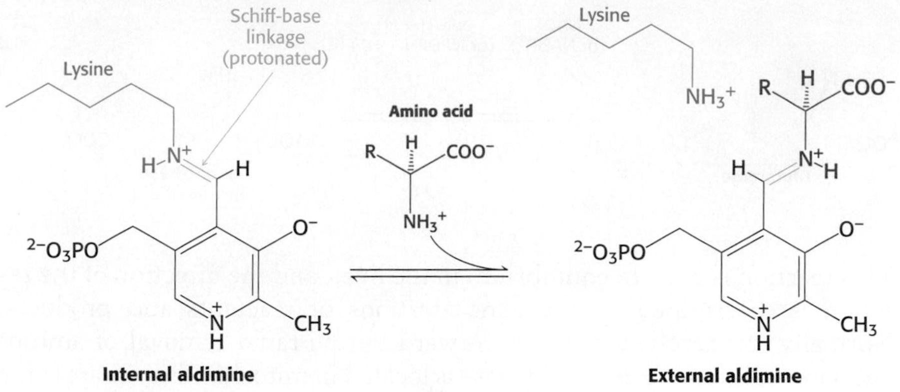

Internal aldimine

External aldimine

The α-amino group of the amino acid substrate displaces the ε-amino group of the active-site lysine residue. In other words, an *internal* aldimine becomes an *external* aldimine. The amino acid–PLP Schiff base that is formed remains tightly bound to the enzyme by multiple noncovalent interactions. The Schiff-base linkage often accepts a proton at the N, with the positive charge stabilized by interaction with the negatively charged phenolate group of PLP.

The Schiff base between the amino acid substrate and PLP, the external *aldimine*, loses a proton from the α-carbon atom of the amino acid to form a *quinonoid* intermediate (Figure 23.11). Reprotonation of this intermediate at the aldehyde carbon atom yields a *ketimine*. The ketimine is then hydrolyzed to an α-ketoacid and *pyridoxamine phosphate* (PMP). These steps constitute half of the transamination reaction.

$$\text{Amino acid}_1 + \text{E-PLP} \rightleftharpoons \alpha\text{-ketoacid}_1 + \text{E-PMP}$$

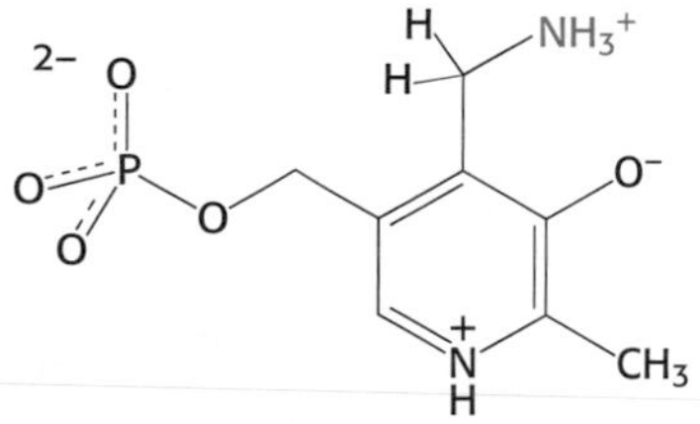

The second half takes place by the reverse of the preceding pathway. A second α-ketoacid reacts with the enzyme–pyridoxamine phosphate complex (E-PMP) to yield a second amino acid and regenerate the enzyme–pyridoxal phosphate complex (E-PLP).

$$\alpha\text{-Ketoacid}_2 + \text{E-PMP} \rightleftharpoons \alpha\text{-amino acid}_2 + \text{E-PLP}$$

The sum of these partial reactions is

$$\text{Amino acid}_1 + \alpha\text{-ketoacid}_2 \rightleftharpoons \text{amino acid}_2 + \alpha\text{-ketoacid}_1$$

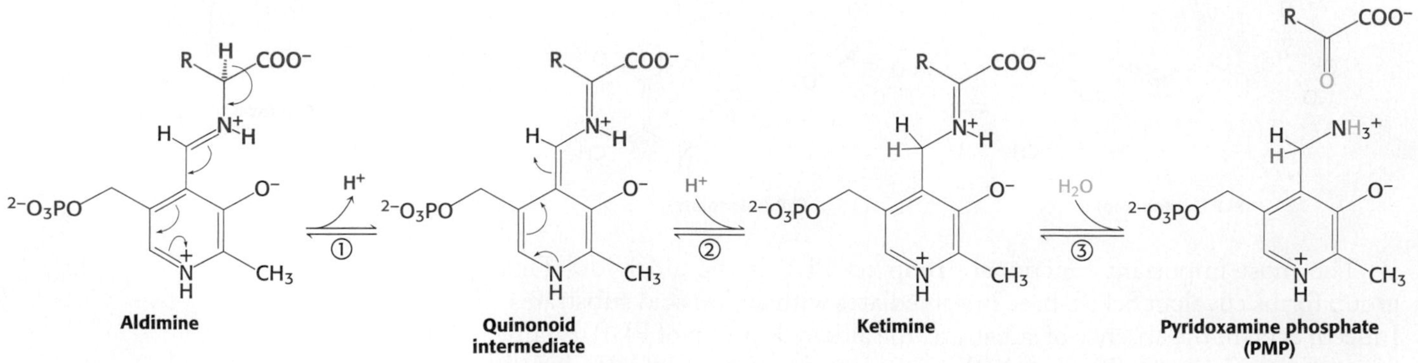

**Pyridoxamine phosphate
(PMP)**

Aldimine

**Quinonoid
intermediate**

Ketimine

**Pyridoxamine phosphate
(PMP)**

Figure 23.11 Transamination mechanism. (1) The external aldimine loses a proton to form a quinonoid intermediate. (2) Reprotonation of this intermediate at the aldehyde carbon atom yields a ketimine. (3) This intermediate is hydrolyzed to generate the α-ketoacid product and pyridoxamine phosphate.

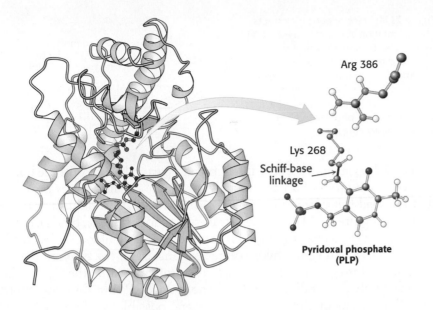

Arg 386

Lys 268

Schiff-base
linkage

**Pyridoxal phosphate
(PLP)**

Figure 23.12 Aspartate aminotransferase. The active site of this prototypical PLP-dependent enzyme includes pyridoxal phosphate attached to the enzyme by a Schiff-base linkage with lysine 258. An arginine residue in the active site helps orient substrates by binding to their α-carboxylate groups. Only one of the enzyme's two subunits is shown. [Drawn from 1AAW.pdb.]

Aspartate Aminotransferase Is an Archetypal Pyridoxal-Dependent Transaminase

The mitochondrial enzyme aspartate aminotransferase provides an especially well studied example of PLP as a coenzyme for transamination reactions (Figure 23.12). X-ray crystallographic studies provided detailed views of how PLP and substrates are bound and confirmed much of the proposed catalytic mechanism. Each of the identical 45-kd subunits of this dimer consists of a large domain and a small one. PLP is bound to the large domain, in a pocket near the subunit interface. In the absence of substrate, the aldehyde group of PLP is in a Schiff-base linkage with lysine 258, as expected. Adjacent to the coenzyme's binding site is a conserved arginine residue that interacts with the α-carboxylate group of the amino acid substrate, helping to orient the substrate appropriately in the active site. A base is necessary to remove a proton from the α-carbon group of the amino acid and to transfer it to the aldehyde carbon atom of PLP (see Figure 23.11, steps 1 and 2). The lysine amino group that was initially in Schiff-base linkage with PLP appears to serve as the proton donor and acceptor.

Pyridoxal Phosphate Enzymes Catalyze a Wide Array of Reactions

Transamination is just one of a wide range of amino acid transformations that are catalyzed by PLP enzymes. The other reactions catalyzed by PLP enzymes at the α-carbon atom of amino acids are decarboxylations, deaminations, racemizations, and aldol cleavages (Figure 23.13). In addition, PLP enzymes catalyze elimination and replacement reactions at the β-carbon atom (e.g., tryptophan synthetase; p. 696) and the γ-carbon atom (e.g., cystathionine β-synthase, p. 693) of amino acid substrates. Three common features of PLP catalysis underlie these diverse reactions.

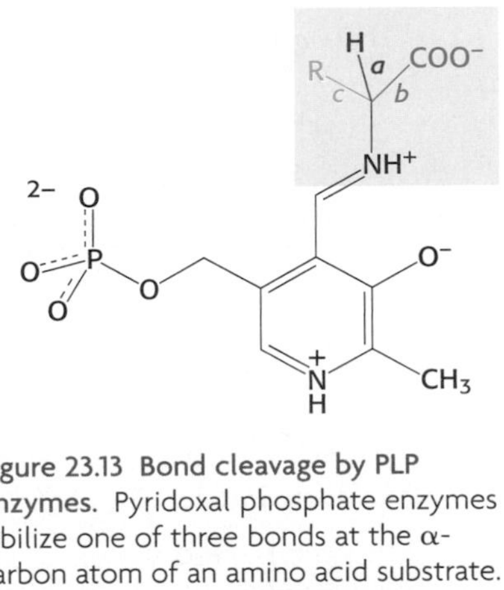

1. A Schiff base is formed by the amino acid substrate (the amine component) and PLP (the carbonyl component).

2. The protonated form of PLP acts as an *electron sink* to stabilize catalytic intermediates that are negatively charged. Electrons from these intermediates are attracted to the positive charge on the ring nitrogen atom. In other words, PLP is an *electrophilic catalyst*.

3. The product Schiff base is cleaved at the completion of the reaction.

Figure 23.13 Bond cleavage by PLP enzymes. Pyridoxal phosphate enzymes labilize one of three bonds at the α-carbon atom of an amino acid substrate. For example, bond *a* is labilized by aminotransferases, bond *b* by decarboxylases, and bond *c* by aldolases (such as threonine aldolases). PLP enzymes also catalyze reactions at the β- and γ-carbon atoms of amino acids.

Figure 23.14 Stereoelectronic effects.
The orientation about the NH—C_α bond
determines the most favored reaction
catalyzed by a pyridoxal phosphate
enzyme. The bond that is most nearly
perpendicular to the plane of delocalized
π orbitals (represented by dashed lines)
of the pyridoxal phosphate electron sink
is most easily cleaved.

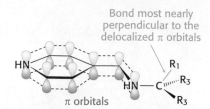

Bond most nearly
perpendicular to the
delocalized π orbitals

π orbitals

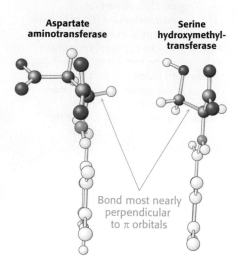

**Aspartate
aminotransferase**

**Serine
hydroxymethyl-
transferase**

Bond most nearly
perpendicular
to π orbitals

Figure 23.15 Reaction choice. In aspartate
aminotransferase, the C_α—H bond is most
nearly perpendicular to the π-orbital
system and is cleaved. In serine
hydroxymethyltransferase, a small rotation
about the N—C_α bond places the C_α—C_β
bond perpendicular to the π system,
favoring its cleavage.

How does an enzyme selectively break a particular one of three bonds at
the α-carbon atom of an amino acid substrate? An important principle is
that *the bond being broken must be perpendicular to the π orbitals of the electron sink* (Figure 23.14). An aminotransferase, for example, binds the amino
acid substrate so that the C_α—H bond is perpendicular to the PLP ring
(Figure 23.15). In serine hydroxymethyltransferase, the enzyme that con-
verts serine into glycine, the N—C_α bond is rotated so that the C_α—C_β bond
is most nearly perpendicular to the plane of the PLP ring, favoring its cleavage. This means of choosing one of several possible catalytic outcomes is
called *stereoelectronic control*.

Many of the PLP enzymes that catalyze amino acid transformations,
such as serine hydroxymethyltransferase, have a similar structure and
are clearly related by divergent evolution. Others, such as tryptophan synthetase, have quite different overall structures. Nonetheless, the active sites
of these enzymes are remarkably similar to that of aspartate aminotransferase, revealing the effects of convergent evolution.

Serine and Threonine Can Be Directly Deaminated

The α-amino groups of serine and threonine can be directly converted into
NH_4^+, without first being transferred to α-ketoglutarate. These direct
deaminations are catalyzed by *serine dehydratase* and *threonine dehydratase*,
in which PLP is the prosthetic group.

$$\text{Serine} \longrightarrow \text{pyruvate} + NH_4^+$$
$$\text{Threonine} \longrightarrow \alpha\text{-ketobutyrate} + NH_4^+$$

These enzymes are called *dehydratases* because *dehydration precedes
deamination*. Serine loses a hydrogen ion from its α-carbon atom and a hydroxide ion group from its β-carbon atom to yield aminoacrylate. This unstable compound reacts with H_2O to give pyruvate and NH_4^+. Thus, the
presence of a hydroxyl group attached to the β-carbon atom in each of these
amino acids permits the direct deamination.

Peripheral Tissues Transport Nitrogen to the Liver

Although most amino acid degradation takes place in the liver, other tissues
can degrade amino acids. For instance, muscle uses branched-chain amino
acids as a source of fuel during prolonged exercise and fasting. How is the
nitrogen processed in these other tissues? As in the liver, the first step is the
removal of the nitrogen from the amino acid. However, muscle lacks the enzymes of the urea cycle, and so the nitrogen must be released in a form that
can be absorbed by the liver and converted into urea.

Nitrogen is transported from muscle to the liver in two principal transport forms. Glutamate is formed by transamination reactions, but the nitrogen is then transferred to pyruvate to form alanine, which is released into the
blood (Figure 23.16). The liver takes up the alanine and converts it back into
pyruvate by transamination. The pyruvate can be used for gluconeogenesis

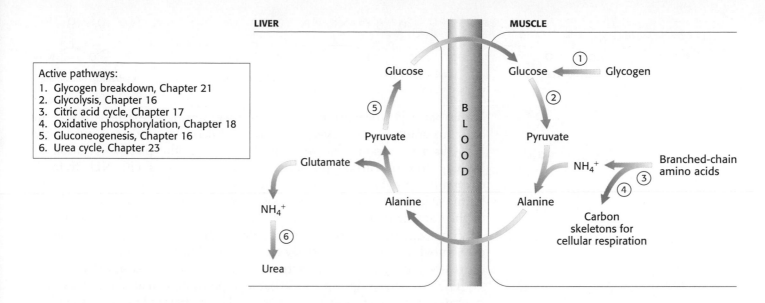

LIVER | MUSCLE

Active pathways:
1. Glycogen breakdown, Chapter 21
2. Glycolysis, Chapter 16
3. Citric acid cycle, Chapter 17
4. Oxidative phosphorylation, Chapter 18
5. Gluconeogenesis, Chapter 16
6. Urea cycle, Chapter 23

and the amino group eventually appears as urea. This transport is referred to as the *glucose–alanine cycle*. It is reminiscent of the Cori cycle discussed earlier (p. 468). However, in contrast with the Cori cycle, pyruvate is not reduced to lactate by NADH, and thus more high-energy electrons are available for oxidative phosphorylation.

Nitrogen can also be transported as glutamine. Glutamine synthetase (p. 683) catalyzes the synthesis of glutamine from glutamate and NH_4^+ in an ATP-dependent reaction:

$$NH_4^+ + \text{glutamate} + \text{ATP} \xrightarrow{\text{Glutamine synthetase}} \text{glutamine} + \text{ADP} + P_i$$

The nitrogens of glutamine can be converted into urea in the liver.

Figure 23.16 **PATHWAY INTEGRATION: The glucose–alanine cycle.** During prolonged exercise and fasting, muscle uses branched-chain amino acids as fuel. The nitrogen removed is transferred (through glutamate) to alanine, which is released into the bloodstream. In the liver, alanine is taken up and converted into pyruvate for the subsequent synthesis of glucose.

23.4 Ammonium Ion Is Converted into Urea in Most Terrestrial Vertebrates

Some of the NH_4^+ formed in the breakdown of amino acids is consumed in the biosynthesis of nitrogen compounds. In most terrestrial vertebrates, the excess NH_4^+ is converted into *urea* and then excreted. Such organisms are referred to as *ureotelic*.

In terrestrial vertebrates, urea is synthesized by the *urea cycle* (Figure 23.17). The urea cycle, proposed by Hans Krebs and Kurt Henseleit in 1932, was the first cyclic metabolic pathway to be discovered. One of the nitrogen atoms of urea is transferred from an amino acid, aspartate. The other nitrogen atom is derived directly from free NH_4^+, and the carbon atom comes from HCO_3^- (derived by the hydration of CO_2; see Section 9.2).

The Urea Cycle Begins with the Formation of Carbamoyl Phosphate

The urea cycle begins with the coupling of free NH_3 with HCO_3^- to form carbamoyl phosphate. Carbamoyl phosphate is a simple molecule, but its synthesis is complex. *Carbamoyl phosphate synthetase* catalyzes the required three steps.

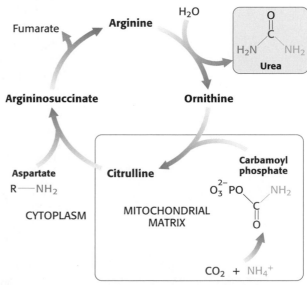

Figure 23.17 The urea cycle.

661

Bicarbonate → **Carboxyphosphate** → **Carbamic acid** → **Carbamoyl phosphate**

Note that NH_3, because it is a strong base, normally exists as NH_4^+ in aqueous solution. However, carbamoyl phosphate synthetase uses only NH_3 as a substrate. The reaction begins with the phosphorylation of HCO_3^- to form carboxyphosphate, which then reacts with NH_3 to form carbamic acid. Finally, a second molecule of ATP phosphorylates carbamic acid to form carbamoyl phosphate. The structure and mechanism of the enzyme that catalyzes these reactions will be presented in Chapter 25. The consumption of two molecules of ATP makes this synthesis of carbamoyl phosphate essentially irreversible. The mammalian enzyme requires N-*acetylglutamate* for activity, as will be described shortly.

The carbamoyl group of carbamoyl phosphate has a high transfer potential because of its anhydride bond. The carbamoyl group is transferred to *ornithine* to form *citrulline*, in a reaction catalyzed by *ornithine transcarbamoylase*.

Ornithine + **Carbamoyl phosphate** → (Ornithine transcarbamoylase) → **Citrulline**

Ornithine and citrulline are amino acids, but they are not used as building blocks of proteins. The formation of NH_4^+ by glutamate dehydrogenase, its incorporation into carbamoyl phosphate as NH_3, and the subsequent synthesis of citrulline take place in the mitochondrial matrix. In contrast, the next three reactions of the urea cycle, which lead to the formation of urea, take place in the cytoplasm.

Citrulline is transported to the cytoplasm where it condenses with aspartate, the donor of the second amino group of urea. This synthesis of *argininosuccinate*, catalyzed by *argininosuccinate synthetase*, is driven by the cleavage of ATP into AMP and pyrophosphate and by the subsequent hydrolysis of pyrophosphate.

Citrulline + **Aspartate** → (Argininosuccinate synthetase) → **Argininosuccinate**

Argininosuccinase cleaves argininosuccinate into *arginine* and *fumarate*. Thus, the carbon skeleton of aspartate is preserved in the form of fumarate.

Argininosuccinate → (Argininosuccinase) → **Arginine** + **Fumarate**

Finally, arginine is hydrolyzed to generate urea and ornithine in a reaction catalyzed by *arginase*. Ornithine is then transported back into the mitochondrion to begin another cycle. The urea is excreted. Indeed, human beings excrete about 10 kg (22 pounds) of urea per year.

Arginine + H_2O → (Arginase) → **Ornithine** + **Urea**

In ancient Rome, urine was a valuable commodity. Vessels were placed on street corners for passers-by to urinate into. Bacteria would degrade the urea, releasing ammonium ion, which would act as a bleach to brighten togas.

The Urea Cycle Is Linked to Gluconeogenesis

The stoichiometry of urea synthesis is

$$CO_2 + NH_4^+ + 3\,ATP + \text{aspartate} + 2H_2O \longrightarrow$$
$$\text{urea} + 2\,ADP + P_i + AMP + PP_i + \text{fumarate}$$

Pyrophosphate is rapidly hydrolyzed, and so the equivalent of four molecules of ATP are consumed in these reactions to synthesize one molecule of urea. The synthesis of fumarate by the urea cycle is important because it is a precursor for glucose synthesis (Figure 23.18). Fumarate is hydrated to malate, which is in turn oxidized to oxaloacetate. Oxaloacetate can be converted into glucose by gluconeogenesis or transaminated to aspartate.

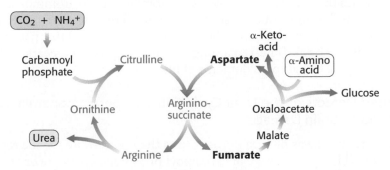

Figure 23.18 Metabolic integration of nitrogen metabolism. The urea cycle, gluconeogenesis, and the transamination of oxaloacetate are linked by fumarate and aspartate.

Figure 23.19 Homologous enzymes. The structure of the catalytic subunit of ornithine transcarbamoylase (blue) is quite similar to that of the catalytic subunit of aspartate transcarbamoylase (red), indicating that these two enzymes are homologs. [Drawn from 1RKM.pdb and 1RAI.pdb.]

N-Acetylglutamate

Urea-Cycle Enzymes Are Evolutionarily Related to Enzymes in Other Metabolic Pathways

Carbamoyl phosphate synthetase generates carbamoyl phosphate for both the urea cycle and the first step in pyrimidine biosynthesis (p. 711). In mammals, two distinct isozymes of the enzyme are present. The carbamoyl phosphate synthetase used in pyrimidine biosynthesis differs in two important ways from its urea-cycle counterpart. First, this enzyme utilizes glutamine as a nitrogen source rather than NH_3. The side-chain amide of glutamine is hydrolyzed within one domain of the enzyme, and the ammonia generated moves through a tunnel in the enzyme to a second active site, where it reacts with carboxyphosphate. Second, this enzyme is part of a large polypeptide called CAD that comprises three distinct enzymes: *car*bamoyl phosphate synthetase, *a*spartate transcarbamoylase, and *d*ihydroorotase. All three enzymes catalyze steps in pyrimidine biosynthesis (Section 25.1). Interestingly, the domain in which glutamine hydrolysis takes place is largely preserved in the urea-cycle enzyme, although that domain is catalytically inactive. This site binds N-acetylglutamate, an allosteric activator of the enzyme. N-Acetylglutamate is synthesized whenever the rate of amino acid catabolism increases and, consequently, signals that the ammonium ion generated in the catabolism of the free amino acids must be disposed of. *A catalytic site in one isozyme has been adapted to act as an allosteric site in another isozyme having a different physiological role.*

Can we find homologs for the other enzymes in the urea cycle? Ornithine transcarbamoylase is homologous to aspartate transcarbamoylase, which catalyzes the first step in pyrimidine biosynthesis, and the structures of their catalytic subunits are quite similar (Figure 23.19). Thus, two consecutive steps in the pyrimidine biosynthetic pathway were adapted for urea synthesis. The next step in the urea cycle is the addition of aspartate to citrulline to form argininosuccinate, and the subsequent step is the removal of fumarate. These two steps together accomplish the net addition of an amino group to citrulline to form arginine. Remarkably, these steps are analogous to two consecutive steps in the purine biosynthetic pathway (p. 715).

The enzymes that catalyze these steps are homologous to argininosuccinate synthetase and argininosuccinase, respectively. Thus, four of the five enzymes in the urea cycle were adapted from enzymes taking part in nucleotide biosynthesis. The remaining enzyme, arginase, appears to be an ancient enzyme found in all domains of life.

Inherited Defects of the Urea Cycle Cause Hyperammonemia and Can Lead to Brain Damage

The synthesis of urea in the liver is the major route for the removal of NH_4^+. A blockage of carbamoyl phosphate synthesis or of any of the four steps of the urea cycle has devastating consequences because there is no alternative pathway for the synthesis of urea. *All defects in the urea* cycle lead to an elevated level of NH_4^+ in the blood (hyperammonemia). Some of these genetic defects become evident a day or two after birth, when the afflicted infant becomes lethargic and vomits periodically. Coma and irreversible brain damage may soon follow. Why are high levels of NH_4^+ toxic? The answer to this question is not yet known. One possibility is that elevated levels of glutamine, formed from NH_4^+ and glutamate, produce osmotic effects that lead directly to brain swelling.

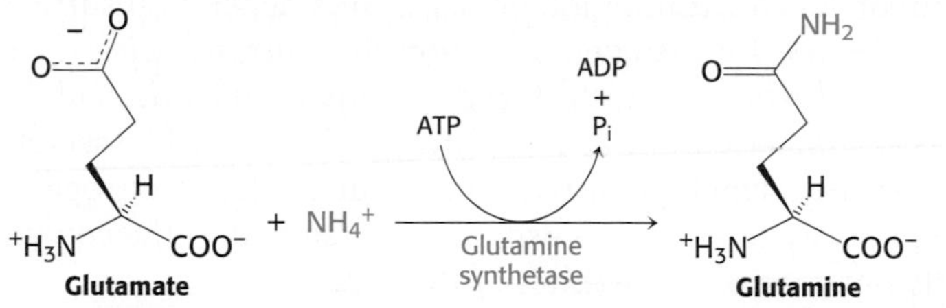

Ingenious strategies for coping with deficiencies in urea synthesis have been devised on the basis of a thorough understanding of the underlying biochemistry. Consider, for example, *argininosuccinase deficiency*. This defect can be partly bypassed by *providing a surplus of arginine in the diet and restricting the total protein intake*. In the liver, arginine is split into urea and ornithine, which then reacts with carbamoyl phosphate to form citrulline (Figure 23.20). This urea-cycle intermediate condenses with aspartate to yield argininosuccinate, which is then excreted. Note that two nitrogen atoms—one from carbamoyl phosphate and the other from aspartate—are eliminated from the body per molecule of arginine provided in the diet. In essence, *argininosuccinate substitutes for urea in carrying nitrogen out of the body*.

The treatment of *carbamoyl phosphate synthetase deficiency* or *ornithine transcarbamoylase deficiency* illustrates a different strategy for circumventing a metabolic block. Citrulline and argininosuccinate cannot be used to dispose of nitrogen atoms, because their formation is impaired. Under these conditions, excess nitrogen accumulates in glycine and glutamine. The challenge then is to rid the body of the nitrogen accumulating in these two amino acids. That goal is accomplished by supplementing a protein-restricted diet with *large amounts of benzoate and phenylacetate*. Benzoate is activated to benzoyl CoA, which reacts with glycine to form hippurate (Figure 23.21). Likewise, phenylacetate is activated to phenylacetyl CoA, which reacts with glutamine to form phenylacetylglutamine. These conjugates substitute for urea in the disposal of nitrogen. Thus, *latent biochemical pathways can be activated to partly bypass a genetic defect*.

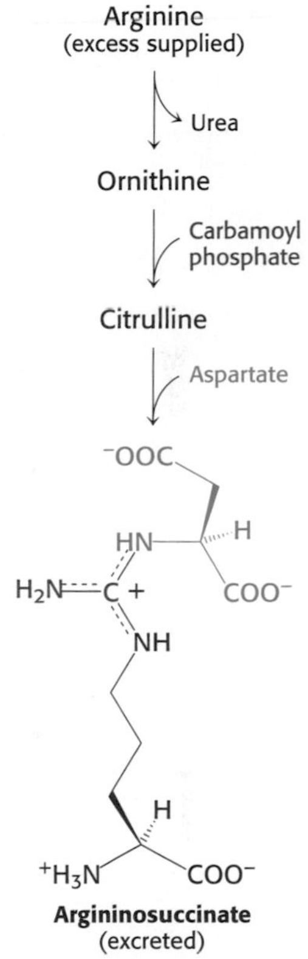

Figure 23.20 Treatment of argininosuccinase deficiency. Argininosuccinase deficiency can be managed by supplementing the diet with arginine. Nitrogen is excreted in the form of argininosuccinate.

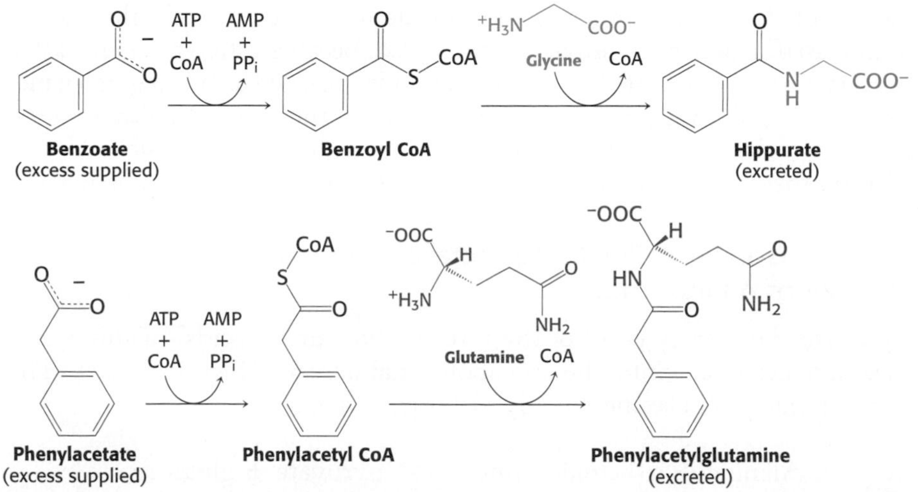

Figure 23.21 Treatment of carbamoyl phosphate synthetase and ornithine transcarbamoylase deficiencies. Both deficiencies can be treated by supplementing the diet with benzoate and phenylacetate. Nitrogen is excreted in the form of hippurate and phenylacetylglutamine.

Urea Is Not the Only Means of Disposing of Excess Nitrogen

As stated earlier, most terrestrial vertebrates are ureotelic; they excrete excess nitrogen as urea. However, urea is not the only excretable form of nitrogen. *Ammoniotelic organisms, such as aquatic vertebrates and invertebrates, release nitrogen as* NH_4^+ and rely on the aqueous environment to dilute this toxic substance. Interestingly, lungfish, which are normally ammoniotelic, become ureotelic in time of drought, when they live out of the water.

Both ureotelic and ammoniotelic organisms depend on sufficient water, to varying degrees, for nitrogen excretion. *In contrast, uricotelic organisms, such as birds and reptiles, secrete nitrogen as the purine uric acid.* Uric acid is secreted as an almost solid slurry requiring little water. The secretion of uric acid also has the advantage of removing four atoms of nitrogen per molecule. The pathway for nitrogen excretion developed in the course of evolution clearly depends on the habitat of the organism.

23.5   Carbon Atoms of Degraded Amino Acids Emerge As Major Metabolic Intermediates

We now turn to the fates of the carbon skeletons of amino acids after the removal of the α-amino group. *The strategy of amino acid degradation is to transform the carbon skeletons into major metabolic intermediates that can be converted into glucose or oxidized by the citric acid cycle.* The conversion pathways range from extremely simple to quite complex. The carbon skeletons of the diverse set of 20 fundamental amino acids are funneled into only seven molecules: *pyruvate, acetyl CoA, acetoacetyl CoA, α-ketoglutarate, succinyl CoA, fumarate,* and *oxaloacetate.* We see here an example of the remarkable economy of metabolic conversions.

Amino acids that are degraded to acetyl CoA or acetoacetyl CoA are termed *ketogenic* amino acids because they can give rise to ketone bodies or fatty acids. Amino acids that are degraded to pyruvate, α-ketoglutarate, succinyl CoA, fumarate, or oxaloacetate are termed *glucogenic* amino acids. The net synthesis of glucose from these amino acids is feasible because these citric acid cycle intermediates and pyruvate can be converted into phosphoenolpyruvate and then into glucose (p. 460). Recall that mammals lack a pathway for the net synthesis of glucose from acetyl CoA or acetoacetyl CoA.

Of the basic set of 20 amino acids, only leucine and lysine are solely ketogenic (Figure 23.22). Isoleucine, phenylalanine, tryptophan, and tyrosine are both ketogenic and glucogenic. Some of their carbon atoms emerge in acetyl CoA or acetoacetyl CoA, whereas others appear in potential precursors of glucose. The other 14 amino acids are classed as solely glucogenic. This classification is not universally accepted, because different quantitative criteria are applied. Whether an amino acid is regarded as being glucogenic, ketogenic, or both depends partly on the eye of the beholder, although the majority of amino acid carbons clearly end up in glucose or glycogen. We will identify the degradation pathways by the entry point into metabolism.

Pyruvate Is an Entry Point into Metabolism for a Number of Amino Acids

Pyruvate is the entry point of the three-carbon amino acids—alanine, serine, and cysteine—into the metabolic mainstream (Figure 23.23). The transamination of alanine directly yields pyruvate.

$$\text{Alanine} + \alpha\text{-ketoglutarate} \rightleftharpoons \text{pyruvate} + \text{glutamate}$$

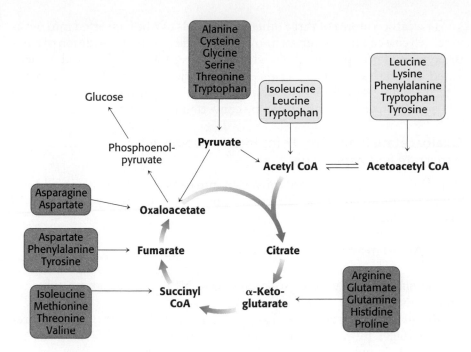

Figure 23.22 **Fates of the carbon skeletons of amino acids.** Glucogenic amino acids are shaded red, and ketogenic amino acids are shaded yellow. Several amino acids are both glucogenic and ketogenic.

As already mentioned (p. 656), glutamate is then oxidatively deaminated, yielding NH_4^+ and regenerating α-ketoglutarate. The sum of these reactions is

$$\text{Alanine} + NAD(P)^+ + H_2O \longrightarrow \text{pyruvate} + NH_4^+ + NAD(P)H + H^+$$

Another simple reaction in the degradation of amino acids is the *deamination of serine to pyruvate* by *serine dehydratase* (p. 660).

$$\text{Serine} \longrightarrow \text{pyruvate} + NH_4^+$$

Cysteine can be converted into pyruvate by several pathways, with its sulfur atom emerging in H_2S, SCN^-, or SO_3^{2-}.

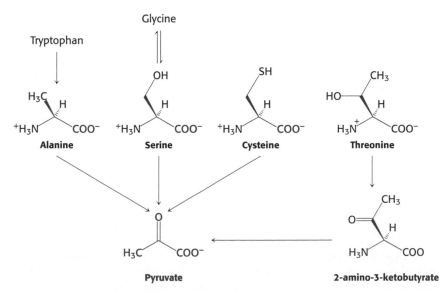

Figure 23.23 **Pyruvate formation from amino acids.** Pyruvate is the point of entry for alanine, serine, cysteine, glycine, threonine, and tryptophan.

The carbon atoms of three other amino acids can be converted into pyruvate. *Glycine* can be converted into serine by the enzymatic addition of a hydroxymethyl group or it can be cleaved to give CO_2, NH_4^+, and an activated one-carbon unit. *Threonine* can give rise to pyruvate through the intermediate 2-amino-3-ketobutyrate. Three carbon atoms of *tryptophan* can emerge in alanine, which can be converted into pyruvate.

Oxaloacetate Is an Entry Point into Metabolism for Aspartate and Asparagine

Aspartate and asparagine are converted into oxaloacetate, a citric acid cycle intermediate. *Aspartate,* a four-carbon amino acid, is directly *transaminated to oxaloacetate.*

$$\text{Aspartate} + \alpha\text{-ketoglutarate} \rightleftharpoons \text{oxaloacetate} + \text{glutamate}$$

Asparagine is hydrolyzed by *asparaginase* to NH_4^+ and aspartate, which is then transaminated.

Recall that aspartate can also be converted into *fumarate* by the urea cycle (p. 663). Fumarate is a point of entry for half the carbon atoms of tyrosine and phenylalanine, as will be discussed shortly.

Alpha-Ketoglutarate Is an Entry Point into Metabolism for Five-Carbon Amino Acids

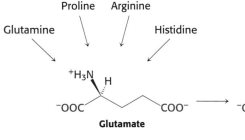

Glutamate → **α-Ketoglutarate**

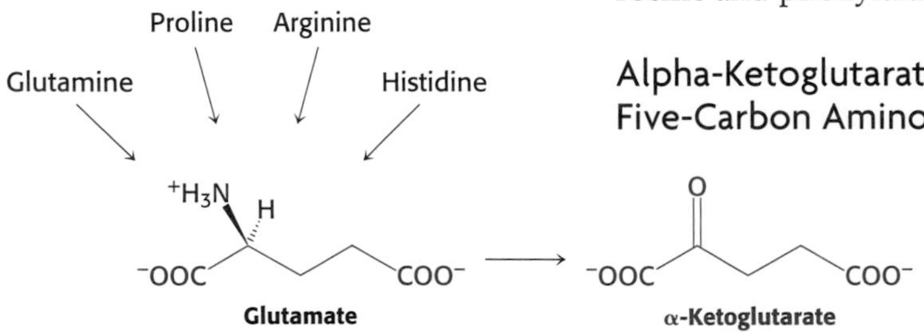

Figure 23.24 α-Ketoglutarate formation from amino acids. α-Ketoglutarate is the point of entry of several five-carbon amino acids that are first converted into glutamate.

The carbon skeletons of several five-carbon amino acids enter the citric acid cycle at α-*ketoglutarate.* These amino acids are first converted into *glutamate,* which is then oxidatively deaminated by glutamate dehydrogenase to yield α-ketoglutarate (Figure 23.24).

Histidine is converted into 4-imidazolone 5-propionate (Figure 23.25). The amide bond in the ring of this intermediate is hydrolyzed to the N-formimino derivative of glutamate, which is then converted into glutamate by the transfer of its formimino group to tetrahydrofolate, a carrier of activated one-carbon units (p. 689).

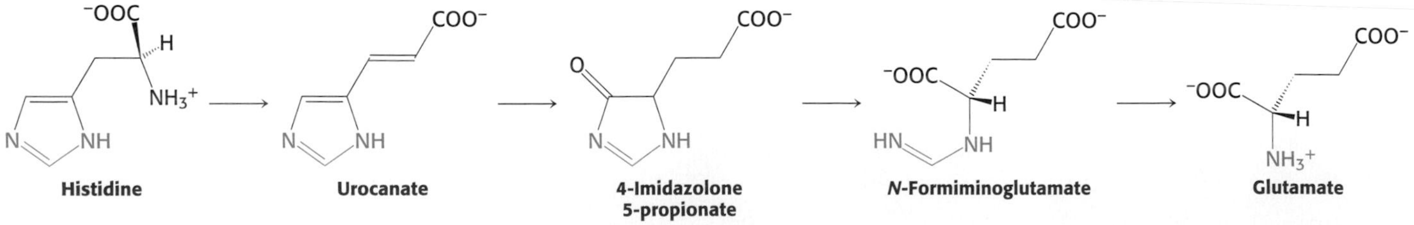

Histidine → **Urocanate** → **4-Imidazolone 5-propionate** → **N-Formiminoglutamate** → **Glutamate**

Figure 23.25 Histidine degradation. Conversion of histidine into glutamate.

Glutamine is hydrolyzed to glutamate and NH_4^+ by *glutaminase*. *Proline* and *arginine* are each converted into glutamate γ-semialdehyde, which is then oxidized to glutamate (Figure 23.26).

Succinyl Coenzyme A Is a Point of Entry for Several Nonpolar Amino Acids

Succinyl CoA is a point of entry for some of the carbon atoms of methionine, isoleucine, and valine. Propionyl CoA and then methylmalonyl CoA are intermediates in the breakdown of these three nonpolar amino acids (Figure 23.27). The mechanism for the interconversion of propionyl CoA and methylmalonyl CoA was presented on page 629. This pathway from

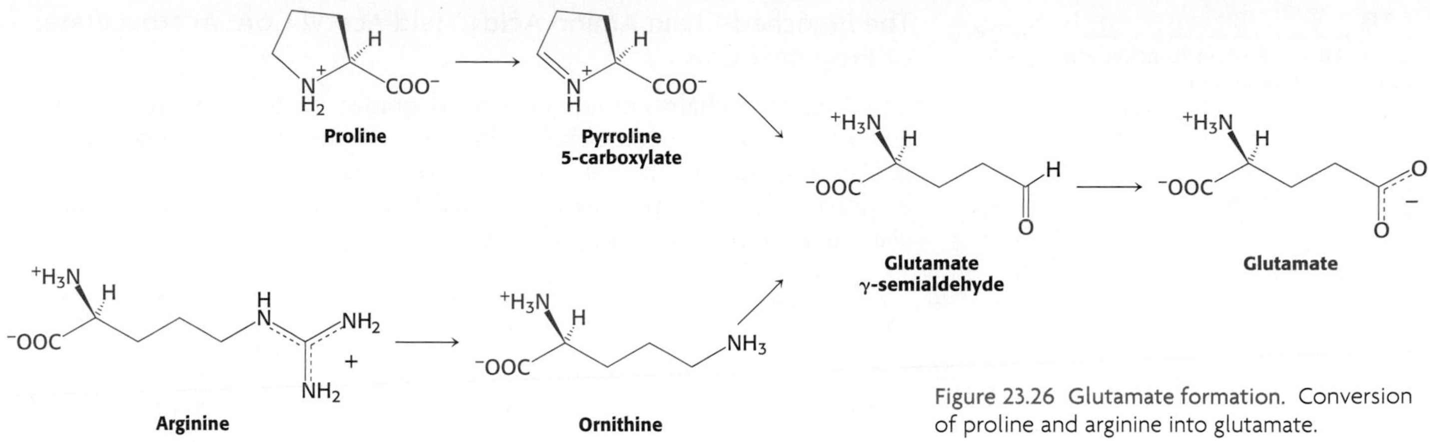

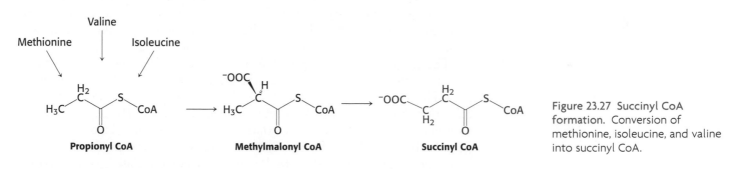

Figure 23.26 Glutamate formation. Conversion of proline and arginine into glutamate.

propionyl CoA to succinyl CoA is also used in the oxidation of fatty acids that have an odd number of carbon atoms (p. 627).

Figure 23.27 Succinyl CoA formation. Conversion of methionine, isoleucine, and valine into succinyl CoA.

Methionine Degradation Requires the Formation of a Key Methyl Donor, S-Adenosylmethionine

Methionine is converted into succinyl CoA in nine steps (Figure 23.28). The first step is the adenylation of methionine to form S-*adenosylmethionine* (SAM), a common methyl donor in the cell (p. 691). Loss of the methyl and adenosyl groups yields homocysteine, which is eventually processed to α-*ketobutyrate*. This α-ketoacid is oxidatively decarboxylated by the α-ketoacid dehydrogenase complex to *propionyl CoA*, which is processed to *succinyl CoA*, as described on page 627.

Figure 23.28 Methionine metabolism. The pathway for the conversion of methionine into succinyl CoA. S-Adenosylmethionine, formed along this pathway, is an important molecule for transferring methyl groups.

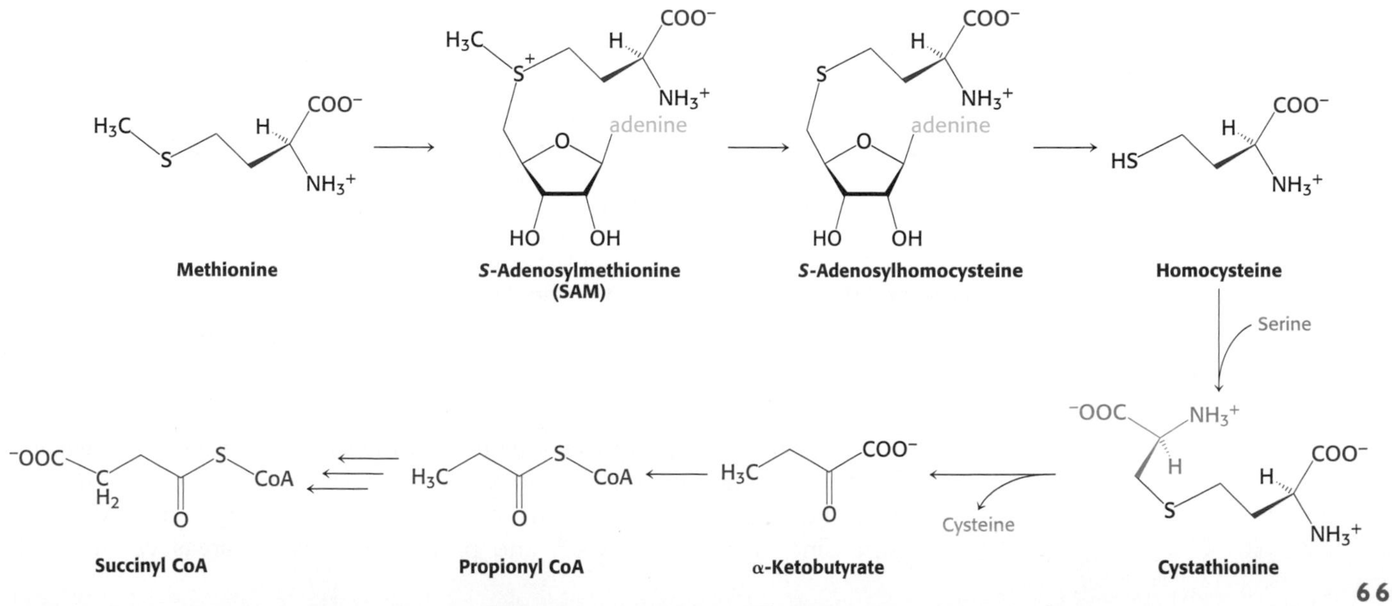

The Branched-Chain Amino Acids Yield Acetyl CoA, Acetoacetate, or Propionyl CoA

The branched-chain amino acids are degraded by reactions that we have already encountered in the citric acid cycle and fatty acid oxidation. Leucine is transaminated to the corresponding α-ketoacid, *α-ketoisocaproate*. This α-ketoacid is *oxidatively decarboxylated to isovaleryl CoA by the branched-chain α-ketoacid dehydrogenase complex.*

The α-ketoacids of valine and isoleucine, the other two branched-chain aliphatic amino acids, also are substrates (as is α-ketobutyrate derived from methionine). The oxidative decarboxylation of these α-ketoacids is analogous to that of pyruvate to acetyl CoA and of α-ketoglutarate to succinyl CoA. The branched-chain α-ketoacid dehydrogenase, a multienzyme complex, is a homolog of pyruvate dehydrogenase (p. 477) and α-ketoglutarate dehydrogenase (p. 485). Indeed, the E3 components of these enzymes, which regenerate the oxidized form of lipoamide, are identical.

The isovaleryl CoA derived from leucine is *dehydrogenated* to yield *β-methylcrotonyl CoA.* This oxidation is catalyzed by *isovaleryl CoA dehydrogenase.* The hydrogen acceptor is FAD, as in the analogous reaction in fatty acid oxidation that is catalyzed by acyl CoA dehydrogenase. *β-Methylglutaconyl CoA* is then formed by the *carboxylation* of β-methylcrotonyl CoA at the expense of the hydrolysis of a molecule of ATP. As might be expected, the carboxylation mechanism of β-methylcrotonyl CoA carboxylase is similar to that of pyruvate carboxylase and acetyl CoA carboxylase.

β-Methylglutaconyl CoA is then *hydrated* to form *3-hydroxy-3-methylglutaryl CoA,* which is cleaved into *acetyl CoA* and *acetoacetate.* This reaction has already been discussed in regard to the formation of ketone bodies from fatty acids (p. 632).

The degradative pathways of valine and isoleucine resemble that of leucine. After transamination and oxidative decarboxylation to yield a CoA derivative, the subsequent reactions are like those of fatty acid oxidation. Isoleucine yields acetyl CoA and propionyl CoA, whereas valine yields

CO_2 and propionyl CoA. The degradation of leucine, valine, and isoleucine validate a point made earlier (Chapter 15): the number of reactions in metabolism is large, but the number of *kinds* of reactions is relatively small. The degradation of leucine, valine, and isoleucine provides a striking illustration of the underlying simplicity and elegance of metabolism.

Oxygenases Are Required for the Degradation of Aromatic Amino Acids

The degradation of the aromatic amino acids yields the common intermediates acetoacetate, fumarate, and pyruvate. The degradation pathway is not as straightforward as that of the amino acids previously discussed. For the aromatic amino acids, *molecular oxygen is used to break an aromatic ring*.

The degradation of phenylalanine begins with its hydroxylation to tyrosine, a reaction catalyzed by *phenylalanine hydroxylase*. This enzyme is called a *monooxygenase* (or *mixed-function oxygenase*) because *one atom of O_2 appears in the product and the other in H_2O*.

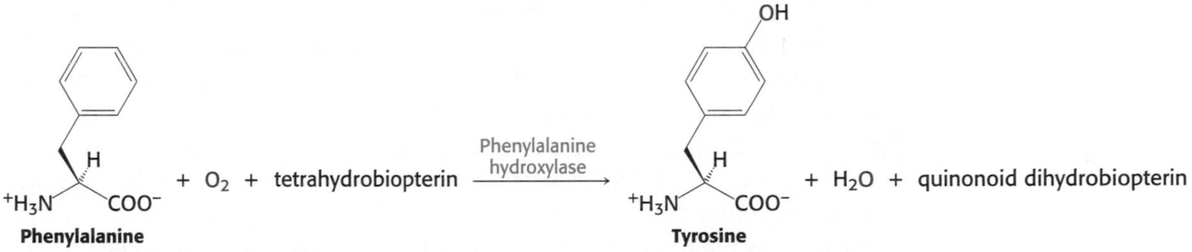

The reductant here is *tetrahydrobiopterin*, an electron carrier that has not been previously discussed and is derived from the cofactor *biopterin*. Because biopterin is synthesized in the body, it is not a vitamin. Tetrahydrobiopterin is initially formed by the reduction of dihydrobiopterin by NADPH in a reaction catalyzed by *dihydrofolate reductase* (Figure 23.29). The quinonoid form of dihydrobiopterin is produced in the hydroxylation of phenylalanine. It is reduced back to tetrahydrobiopterin by NADPH in a reaction catalyzed by *dihydropteridine reductase*. The sum of the reactions catalyzed by phenylalanine hydroxylase and dihydropteridine reductase is

$$\text{Phenylalanine} + O_2 + \text{NADPH} + H^+ \longrightarrow \text{tyrosine} + \text{NADP}^+ + H_2O$$

Note that these reactions can also be used to synthesize tyrosine from phenylalanine.

The next step in the degradation of phenylalanine and tyrosine is the transamination of tyrosine to p-*hydroxyphenylpyruvate* (Figure 23.30). This α-ketoacid then reacts with O_2 to form *homogentisate*. The enzyme catalyzing this complex reaction, p-*hydroxyphenylpyruvate hydroxylase*, is called a

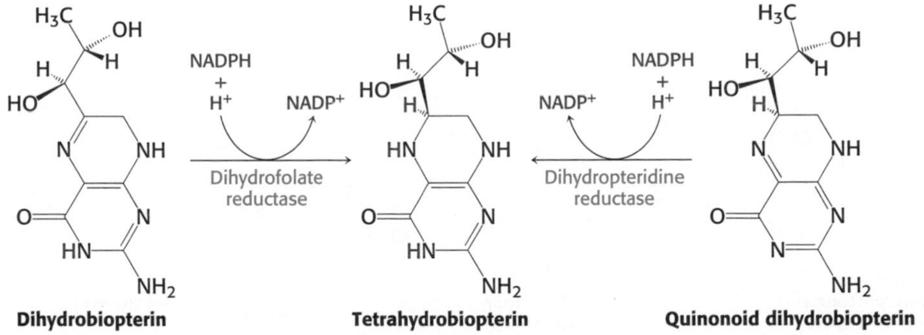

Figure 23.29 Formation of tetrahydrobiopterin, an important electron carrier. Tetrahydrobiopterin can be formed by the reduction of either of two forms of dihydrobiopterin.

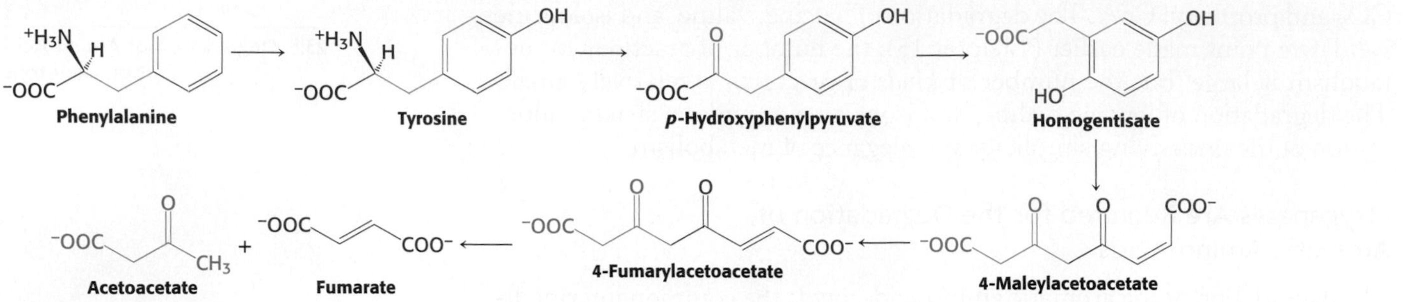

Figure 23.30 Phenylalanine and tyrosine degradation. The pathway for the conversion of phenylalanine into acetoacetate and fumarate.

dioxygenase because *both atoms of O$_2$ become incorporated into the product*, one on the ring and one in the carboxyl group. The aromatic ring of homogentisate is then cleaved by O$_2$, which yields 4-maleylacetoacetate. This reaction is catalyzed by *homogentisate oxidase,* another dioxygenase. 4-Maleylacetoacetate is then isomerized to *4-fumarylacetoacetate* by an enzyme that uses glutathione as a cofactor. Finally, 4-fumarylacetoacetate is hydrolyzed to *fumarate* and *acetoacetate.*

Tryptophan degradation requires several oxygenases (Figure 23.31). Tryptophan 2,3-dioxygenase cleaves the pyrrole ring, and kynureinine 3-monooxygenase hydroxylates the remaining benzene ring, a reaction similar to the hydroxylation of phenylalanine to form tyrosine. Alanine is removed and the 3-hydroxyanthranilic acid is cleaved by another dioxygenase and subsequently processed to acetoacetyl CoA. *Nearly all cleavages of aromatic rings in biological systems are catalyzed by dioxygenases,* a class of enzymes discovered by Osamu Hayaishi. The active sites of these enzymes contain iron that is not part of heme or an iron–sulfur cluster.

23.6 Inborn Errors of Metabolism Can Disrupt Amino Acid Degradation

Errors in amino acid metabolism provided some of the first examples of biochemical defects linked to pathological conditions. For instance, *alcaptonuria* is an inherited metabolic disorder caused by the

Figure 23.31 Tryptophan degradation. The pathway for the conversion of tryptophan into alanine and acetoacetate.

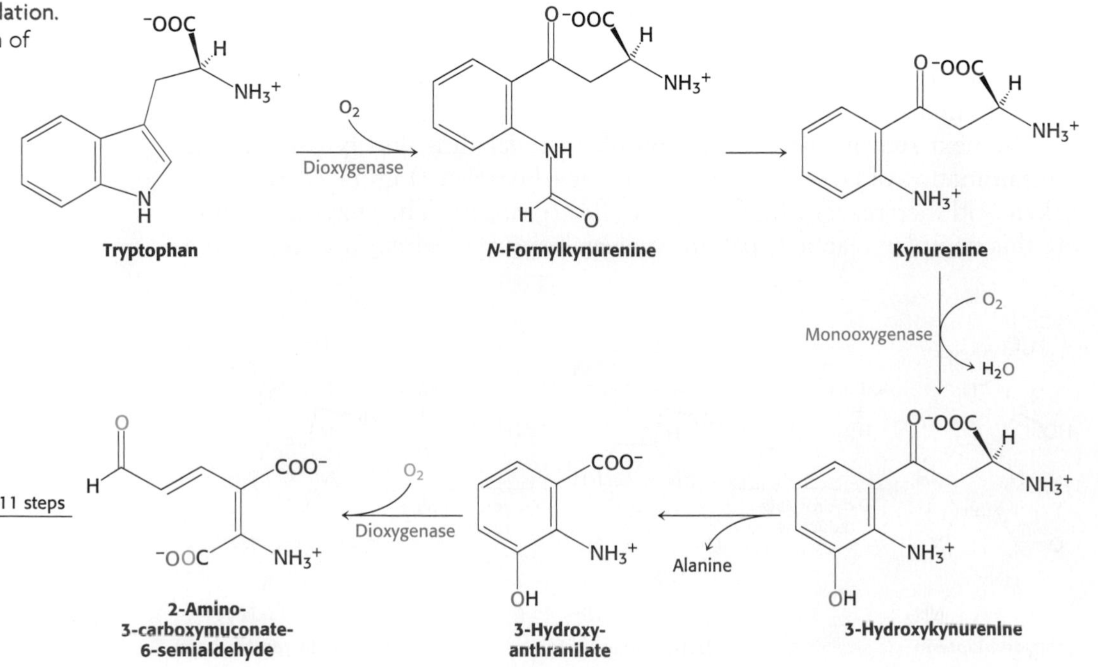

absence of homogentisate oxidase. In 1902, Archibald Garrod showed that alcaptonuria is transmitted as a single recessive Mendelian trait. Furthermore, he recognized that homogentisate is a normal intermediate in the degradation of phenylalanine and tyrosine (see Figure 23.30) and that it accumulates in alcaptonuria because its degradation is blocked. He concluded that "the splitting of the benzene ring in normal metabolism is the work of a special enzyme, that in congenital alcaptonuria this enzyme is wanting." Homogentisate accumulates and is excreted in the urine, which turns dark on standing as homogentisate is oxidized and polymerized to a melanin-like substance.

Although alcaptonuria is a relatively harmless condition, such is not the case with other errors in amino acid metabolism. In *maple syrup urine disease*, the oxidative decarboxylation of α-ketoacids derived from valine, isoleucine, and leucine is blocked because the branched-chain dehydrogenase is missing or defective. Hence, the levels of these α-ketoacids and the branched-chain amino acids that give rise to them are markedly elevated in both blood and urine. The urine of patients has the odor of maple syrup—hence the name of the disease (also called *branched-chain ketoaciduria*). Maple syrup urine disease usually leads to mental and physical retardation unless the patient is placed on a diet low in valine, isoleucine, and leucine early in life. The disease can be readily detected in newborns by screening urine samples with 2,4-dinitrophenylhydrazine, which reacts with α-ketoacids to form 2,4-dinitrophenylhydrazone derivatives. A definitive diagnosis can be made by mass spectrometry.

Phenylketonuria is perhaps the best known of the diseases of amino acid metabolism. Phenylketonuria is caused by an *absence or deficiency of phenylalanine hydroxylase* or, more rarely, of its tetrahydrobiopterin cofactor. *Phenylalanine accumulates in all body fluids because it cannot be converted into tyrosine.* Normally, three-quarters of phenylalanine molecules are converted into tyrosine, and the other quarter become incorporated into proteins. Because the major outflow pathway is blocked in phenylketonuria, the blood level of phenylalanine is typically at least 20-fold as high as in normal people. Minor fates of phenylalanine in normal people, such as the formation of phenylpyruvate, become major fates in phenylketonurics. Indeed, the initial description of phenylketonuria in 1934 was made by observing the reaction of phenylpyruvate in the urine of phenylketonurics with $FeCl_3$, which turns the urine olive green.

Almost all untreated phenylketonurics are severely mentally retarded. In fact, about 1% of patients in mental institutions have phenylketonuria. The brain weight of these people is below normal, myelination of their nerves is defective, and their reflexes are hyperactive. The life expectancy of untreated

TABLE 23.4 Inborn errors of amino acid metabolism

Disease	Enzyme deficiency	Symptoms
Citrullinema	Arginosuccinate lyase	Lethargy, seizures, reduced muscle tension
Tyrosinemia	Various enzymes of tyrosine degradation	Weakness, self-mutilation, liver damage, mental retardation
Albinism	Tyrosinase	Absence of pigmentation
Homocystinuria	Cystathionine β-synthase	Scoliosis, muscle weakness, mental retardation, thin blond hair
Hyperlysinemia	α-Aminoadipic semialdehyde dehydrogenase	Seizures, mental retardation, lack of muscle tone, ataxia

phenylketonurics is drastically shortened. Half die by age 20 and three-quarters by age 30. *The biochemical basis of their mental retardation is an enigma.*

Phenylketonurics appear normal at birth but are severely defective by age 1 if untreated. The therapy for phenylketonuria is a *low-phenylalanine diet.* The aim is to provide just enough phenylalanine to meet the needs for growth and replacement. Proteins that have a low content of phenylalanine, such as casein from milk, are hydrolyzed and phenylalanine is removed by adsorption. A low-phenylalanine diet must be started very soon after birth to prevent irreversible brain damage. In one study, the average IQ of phenylketonurics treated within a few weeks after birth was 93; a control group treated starting at age 1 had an average IQ of 53.

Early diagnosis of phenylketonuria is essential and has been accomplished by mass screening programs. The phenylalanine level in the blood is the preferred diagnostic criterion because it is more sensitive and reliable than the $FeCl_3$ test. Prenatal diagnosis of phenylketonuria with DNA probes has become feasible because the gene has been cloned and the exact locations of many mutations have been discovered in the protein. Interestingly, whereas some mutations lower the activity of the enzyme, others decrease the enzyme concentration instead. These latter mutations lead to degradation of the enzyme, at least in part by the ubiquitin–proteasome pathway (p. 653).

The incidence of phenylketonuria is about 1 in 20,000 newborns. The disease is inherited as an *autosomal recessive.* Heterozygotes, who make up about 1.5% of a typical population, appear normal. Carriers of the phenylketonuria gene have a reduced level of phenylalanine hydroxylase, as indicated by an increased level of phenylalanine in the blood. However, this criterion is not absolute, because the blood levels of phenylalanine in carriers and normal people overlap to some extent. The measurement of the kinetics of the disappearance of intravenously administered phenylalanine is a more definitive test for the carrier state. It should be noted that a high blood level of phenylalanine in a pregnant woman can result in abnormal development of the fetus. This is a striking example of maternal–fetal relationships at the molecular level. Table 23.4 lists some other diseases of amino acid metabolism.

Summary

23.1 Proteins Are Degraded to Amino Acids

Dietary protein is digested in the intestine, producing amino acids that are transported throughout the body. Cellular proteins are degraded at widely variable rates, ranging from minutes to the life of the organism.

23.2 Protein Turnover Is Tightly Regulated

The turnover of cellular proteins is a regulated process requiring complex enzyme systems. Proteins to be degraded are conjugated with ubiquitin, a small conserved protein, in a reaction driven by ATP hydrolysis. The ubiquitin-conjugating system is composed of three distinct enzymes. A large, barrel-shaped complex called the proteasome digests the ubiquitinated proteins. The proteasome also requires ATP hydrolysis to function. The resulting amino acids provide a source of precursors for protein, nucleotide bases, and other nitrogenous compounds.

23.3 The First Step in Amino Acid Degradation Is the Removal of Nitrogen

Surplus amino acids are used as metabolic fuel. The first step in their degradation is the removal of their α-amino groups by transamination to α-ketoacids. Pyridoxal phosphate is the coenzyme in all aminotransferases and in many other enzymes catalyzing amino acid transformations. The α-amino group funnels into α-ketoglutarate to form glutamate, which is then oxidatively deaminated by glutamate dehydrogenase to give NH_4^+ and α-ketoglutarate. NAD^+ or $NADP^+$ is the electron acceptor in this reaction.

23.4 Ammonium Ion Is Converted into Urea in Most Terrestrial Vertebrates

The first step in the synthesis of urea is the formation of carbamoyl phosphate, which is synthesized from HCO_3^-, NH_3, and two molecules of ATP by carbamoyl phosphate synthetase. Ornithine is then carbamoylated to citrulline by ornithine transcarbamoylase. These two reactions take place in mitochondria. Citrulline leaves the mitochondrion and condenses with aspartate to form argininosuccinate, which is cleaved into arginine and fumarate. The other nitrogen atom of urea comes from aspartate. Urea is formed by the hydrolysis of arginine, which also regenerates ornithine. Some enzymatic deficiencies of the urea cycle can be bypassed by supplementing the diet with arginine or compounds that form conjugates with glycine and glutamine.

23.5 Carbon Atoms of Degraded Amino Acids Emerge as Major Metabolic Intermediates

The carbon atoms of degraded amino acids are converted into pyruvate, acetyl CoA, acetoacetate, or an intermediate of the citric acid cycle. Most amino acids are solely glucogenic, two are solely ketogenic, and a few are both ketogenic and glucogenic. Alanine, serine, cysteine, glycine, threonine, and tryptophan are degraded to pyruvate. Asparagine and aspartate are converted into oxaloacetate. α-Ketoglutarate is the point of entry for glutamate and four amino acids (glutamine, histidine, proline, and arginine) that can be converted into glutamate. Succinyl CoA is the point of entry for some of the carbon atoms of three amino acids (methionine, isoleucine, and valine) that are degraded through the intermediate methylmalonyl CoA. Leucine is degraded to acetoacetate and acetyl CoA. The breakdown of valine and isoleucine is like that of leucine. Their α-ketoacid derivatives are oxidatively decarboxylated by the branched-chain α-ketoacid dehydrogenase.

The rings of aromatic amino acids are degraded by oxygenases. Phenylalanine hydroxylase, a monooxygenase, uses tetrahydrobiopterin as the reductant. One of the oxygen atoms of O_2 emerges in tyrosine and the other in water. Subsequent steps in the degradation of these aromatic amino acids are catalyzed by dioxygenases, which catalyze the insertion of both atoms of O_2 into organic products. Four of the carbon atoms of phenylalanine and tyrosine are converted into fumarate, and four emerge in acetoacetate.

23.6 Inborn Errors of Metabolism Can Disrupt Amino Acid Degradation

Errors in amino acid metabolism were sources of some of the first insights into the correlation between pathology and biochemistry. Phenylketonuria is the best known of the many hereditary errors of amino acid metabolism. This condition results from the accumulation of high levels of phenylalanine in the body fluids. By unknown mechanisms, this accumulation leads to mental retardation unless the afflicted are placed on low-phenylalanine diets immediately after birth.

Key Terms

ubiquitin (p. 651)

proteasome (p. 653)

aminotransferase (transaminase) (p. 656)

glutamate dehydrogenase (p. 656)

pyridoxal phosphate (PLP) (p. 657)

pyridoxamine phosphate (PMP) (p. 658)

glucose-alanine cycle (p. 661)

urea cycle (p. 661)

carbamoyl phosphate synthetase (p. 661)

N-acetylglutamate (p. 664)

ketogenic amino acid (p. 666)

glucogenic amino acid (p. 666)

biopterin (p. 671)

phenylketonuria (p. 673)

Selected Readings

Where to Start

Torchinsky, Y. M. 1989. Transamination: Its discovery, biological and chemical aspects. *Trends Biochem. Sci.* 12:115–117.

Eisensmith, R. C., and Woo, S. L. C. 1991. Phenylketonuria and the phenylalanine hydroxylase gene. *Mol. Biol. Med.* 8:3–18.

Levy, H. L. 1989. Nutritional therapy for selected inborn errors of metabolism. *J. Am. Coll. Nutr.* 8:54S–60S.

Schwartz, A. L., and Ciechanover, A. 1999. The ubiquitin-proteasome pathway and pathogenesis of human diseases. *Annu. Rev. Med.* 50:57–74.

Watford, M. 2003. The urea cycle. *Biochem. Mol. Biol. Ed.* 31:289–297.

Books

Bender, D. A. 1985. *Amino Acid Metabolism* (2d ed.). Wiley.

Lippard, S. J., and Berg, J. M. 1994. *Principles of Bioinorganic Chemistry.* University Science Books.

Schauder, P., Wahren, J., Paoletti, R., Bernardi, R., and Rinetti, M. (Eds.). 1992. *Branched-Chain Amino Acids: Biochemistry, Physiopathology, and Clinical Sciences.* Raven Press.

Walsh, C. 1979. *Enzymatic Reaction Mechanisms.* W. H. Freeman and Company.

Christen, P., and Metzler, D. E. 1985. *Transaminases.* Wiley.

Ubiquitin and the Proteasome

Cooper, E. M., Hudson, A. W., Amos, J., Wagstaff, J., and Howley, P. M. 2004. Biochemical analysis of Angelman syndrome-associated mutation in the E3 ubiquitin ligase E6-associated protein. *J. Biol. Chem.* 279:41208–41217.

Giasson, B. I. and Lee, V. M.-Y. 2003. Are ubiquitination pathways central to Parkinson's disease? *Cell* 114:1–8.

Pagano, M., and Benmaamar, R. 2003. When protein destruction runs amok, malignancy is on the loose. *Cancer Cell* 4:251–256.

Bochtler, M., Ditzel, L., Groll, M., Hartmann, C., and Huber, R. 1999. The proteasome. *Annu. Rev. Biophys. Biomol. Struct.* 28:295–317.

Thrower, J. S., Hoffman, L., Rechsteiner, M., and Pickart, C. M. 2000. Recognition of the polyubiquitin proteolytic signal. *EMBO J.* 19:94–102.

Hochstrasser, M. 2000. Evolution and function of ubiquitin-like protein-conjugation systems. *Nat. Cell Biol.* 2:E153–E157.

Jentsch, S., and Pyrowolakis, G. 2000. Ubiquitin and its kin: How close are the family ties? *Trends Cell Biol.* 10:335–342.

Laney, J. D., and Hochstrasser, M. 1999. Substrate targeting in the ubiquitin system. *Cell* 97:427–430.

Hartmann-Petersen, R., Tanaka, K., and Hendil, K. B. 2001. Quaternary structure of the ATPase complex of human 26S proteasomes determined by chemical cross-linking. *Arch. Biochem. Biophys.* 386:89–94.

Pyridoxal Phosphate-Dependent Enzymes

Eliot, A. C., and Kirsch, J. F. 2004. Pyridoxal phosphate enzymes: Mechanistic, structural, and evolutionary considerations. *Annu. Rev. Biochem.* 73:383–415.

Mehta, P. K., and Christen, P. 2000. The molecular evolution of pyridoxal-5′-phosphate-dependent enzymes. *Adv. Enzymol. Relat. Areas Mol. Biol.* 74:129–184.

Schneider, G., Kack, H., and Lindqvist, Y. 2000. The manifold of vitamin B_6 dependent enzymes. *Structure Fold Des.* 8:R1–R6.

Jager, J., Moser, M., Sauder, U., and Jansonius, J. N. 1994. Crystal structures of *Escherichia coli* aspartate aminotransferase in two conformations: Comparison of an unliganded open and two liganded closed forms. *J. Mol. Biol.* 239:285–305.

Malashkevich, V. N., Toney, M. D., and Jansonius, J. N. 1993. Crystal structures of true enzymatic reaction intermediates: Aspartate and glutamate ketimines in aspartate aminotransferase. *Biochemistry* 32:13451–13462.

McPhalen, C. A., Vincent, M. G., Picot, D., Jansonius, J. N., Lesk, A. M., and Chothia, C. 1992. Domain closure in mitochondrial aspartate aminotransferases. *J. Mol. Biol.* 227:197–213.

Urea Cycle Enzymes

Morris, S. M., Jr. 2002. Regulation of enzymes of the urea cycle and arginine metabolism. *Annu. Rev. Nutr.* 22:87–105.

Huang, X., and Raushel, F. M. 2000. Restricted passage of reaction intermediates through the ammonia tunnel of carbamoyl phosphate synthetase. *J. Biol. Chem.* 275:26233–26240.

Lawson, F. S., Charlebois, R. L., and Dillon, J. A. 1996. Phylogenetic analysis of carbamoylphosphate synthetase genes: Complex evolutionary history includes an internal duplication within a gene which can root the tree of life. *Mol. Biol. Evol.* 13:970–977.

McCudden, C. R., and Powers-Lee, S. G. 1996. Required allosteric effector site for N-acetylglutamate on carbamoyl-phosphate synthetase I. *J. Biol. Chem.* 271:18285–18294.

Turner, M. A., Simpson, A., McInnes, R. R., and Howell, P. L. 1997. Human argininosuccinate lyase: A structural basis for intragenic complementation. *Proc. Natl. Acad. Sci. U. S. A.* 94:9063–9068.

Amino Acid Degradation

Fusetti, F., Erlandsen, H., Flatmark, T., and Stevens, R. C. 1998. Structure of tetrameric human phenylalanine hydroxylase and its implications for phenylketonuria. *J. Biol. Chem.* 273:16962–16967.

Sugimoto, K., Senda, T., Aoshima, H., Masai, E., Fukuda, M., and Mitsui, Y. 1999. Crystal structure of an aromatic ring opening dioxygenase LigAB, a protocatechuate 4,5-dioxygenase, under aerobic conditions. *Structure Fold Des.* 7:953–965.

Titus, G. P., Mueller, H. A., Burgner, J., Rodriguez De Cordoba, S., Penalva, M. A., and Timm, D. E. 2000. Crystal structure of human homogentisate dioxygenase. *Nat. Struct. Biol.* 7:542–546.

Erlandsen, H., and Stevens, R. C. 1999. The structural basis of phenylketonuria. *Mol. Genet. Metab.* 68:103–125.

Genetic Diseases

Scriver, C. R., and Sly, W. S. (Eds.), Childs, B., Beaudet, A. L., Valle, D., Kinzler, K. W., and Vogelstein, B. 2000. *The Metabolic Basis of Inherited Disease* (8th ed.). McGraw-Hill.

Nyhan, W. L. (Ed.). 1984. *Abnormalities in Amino Acid Metabolism in Clinical Medicine.* Appleton-Century-Crofts.

Historical Aspects and the Process of Discovery

Cooper, A. J. L., and Meister, A. 1989. An appreciation of Professor Alexander E. Braunstein: The discovery and scope of enzymatic transamination. *Biochimie* 71:387–404.

Garrod, A. E. 1909. *Inborn Errors in Metabolism.* Oxford University Press (reprinted in 1963 with a supplement by H. Harris).

Childs, B. 1970. Sir Archibald Garrod's conception of chemical individuality: A modern appreciation. *N. Engl. J. Med.* 282:71–78.

Holmes, F. L. 1980. Hans Krebs and the discovery of the ornithine cycle. *Fed. Proc.* 39:216–225.

Problems

1. *Wasted energy?* Protein hydrolysis is an exergonic process, yet the 26S proteasome is dependent on ATP hydrolysis for activity.

(a) Although the exact function of the ATPase activity is not known, suggest some likely functions.

(b) Small peptides can be hydrolyzed without the expenditure of ATP. How does this information concur with your answer to part *a*?

2. *Keto counterparts.* Name the α-ketoacid that is formed by the transamination of each of the following amino acids:

(a) Alanine   (d) Leucine
(b) Aspartate   (e) Phenylalanine
(c) Glutamate   (f) Tyrosine

3. *A versatile building block.* (a) Write a balanced equation for the conversion of aspartate into glucose through the intermediate oxaloacetate. Which coenzymes participate in this transformation? (b) Write a balanced equation for the conversion of aspartate into oxaloacetate through the intermediate fumarate.

4. *The benefits of specialization.* The archaeal proteasome contains 14 identical active β subunits, whereas the eukaryotic proteasome has 7 distinct β subunits. What are the potential benefits of having several distinct active subunits?

5. *Propose a structure.* The 19S subunit of the proteasome contains 6 subunits that are members of the AAA ATPase family. Other members of this large family are associated into homohexamers with sixfold symmetry. Propose a structure for the AAA ATPases within the 19S proteasome. How might you test and refine your prediction?

6. *Effective electron sinks.* Pyridoxal phosphate stabilizes carbanionic intermediates by serving as an electron sink. Which other prosthetic group catalyzes reactions in this way?

7. *Helping hand.* Propose a role for the positively charged guanidinium nitrogen atom in the cleavage of argininosuccinate into arginine and fumarate.

8. *Completing the cycle.* Four high-transfer-potential phosphoryl groups are consumed in the synthesis of urea according to the stoichiometry given on page 663. In this reaction, aspartate is converted into fumarate. Suppose that fumarate is converted back into aspartate. What is the resulting stoichiometry of urea synthesis? How many high-transfer-potential phosphoryl groups are spent?

9. *Inhibitor design.* Compound A has been synthesized as a potential inhibitor of a urea-cycle enzyme. Which enzyme do you think compound A might inhibit?

Compound A

10. *Ammonia toxicity.* Glutamate is an important neurotransmitter whose levels must be carefully regulated in the brain. Explain how a high concentration of ammonia might disrupt this regulation. How might a high concentration of ammonia alter the citric acid cycle?

11. *A precise diagnosis.* The urine of an infant gives a positive reaction with 2,4-dinitrophenylhydrazine. Mass spectrometry shows abnormally high blood levels of pyruvate, α-ketoglutarate, and the α-ketoacids of valine, isoleucine, and leucine. Identify a likely molecular defect and propose a definitive test of your diagnosis.

12. *Therapeutic design.* How would you treat an infant who is deficient in argininosuccinate synthetase? Which molecules would carry nitrogen out of the body?

13. *Sweet hazard.* Why should phenylketonurics avoid using aspartame, an artificial sweetener? (Hint: Aspartame is L-aspartyl-L-phenylalanine methyl ester.)

14. *Déjà vu.* N-Acetylglutamate is required as a cofactor in the synthesis of carbamoyl phosphate. How might N-acetylglutamate be synthesized from glutamate?

Mechanism Problems

15. *Serine dehydratase.* Write out a complete mechanism for the conversion of serine into aminoacrylate catalyzed by serine dehydratase.

16. *Serine racemase.* The nervous system contains a substantial amount of D-serine, which is generated from L-serine by serine racemase, a PLP-dependent enzyme. Propose a mechanism for this reaction. What is the equilibrium constant for the reaction L-serine $\rightleftharpoons$ D-serine?

Chapter Integration Problems

17. *Double duty.* Degradation signals are commonly located in protein regions that also facilitate protein–protein interactions. Explain why this coexistence of two functions in the same domain might be useful.

18. *Fuel choice.* Within a few days after a fast begins, nitrogen excretion accelerates to a higher-than-normal level. After a few weeks, the rate of nitrogen excretion falls to a lower level and continues at this low rate. However, after the fat stores have been depleted, nitrogen excretion rises to a high level.

(a) What events trigger the initial surge of nitrogen excretion?
(b) Why does nitrogen excretion fall after several weeks of fasting?
(c) Explain the increase in nitrogen excretion when the lipid stores have been depleted.

19. *Isoleucine degradation.* Isoleucine is degraded to acetyl CoA and succinyl CoA. Suggest a plausible reaction sequence, based on reactions discussed in the text, for this degradation pathway.

Data Interpretation Problem

20. *Another helping hand.* In eukaryotes, the 20S proteasome component in conjunction with the 19S component degrades ubiquitinated proteins with the hydrolysis of a molecule of ATP. Archaea lack ubiquitin and the 26S proteasome but do contain a 20S proteasome. Some archaea also contain an ATPase that is homologous to the ATPases of the eukaryotic 19S component. This archaeal ATPase activity was isolated as a 650-kd complex (called PAN) from the archaeon *Thermoplasma*, and experiments were performed to determine if PAN could enhance the activity of the 20S proteasome from *Thermoplasma* as well as other 20S proteasomes.

Protein degradation was measured as a function of time and in the presence of various combinations of components. Graph A shows the results.

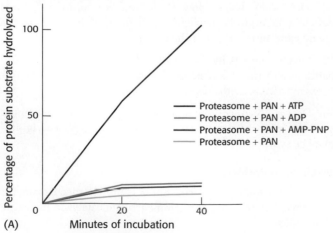

(A) Minutes of incubation

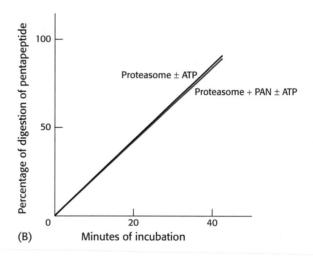

AMP-PNP

(a) What is the effect of PAN on archaeal proteasome activity in the absence of nucleotides?
(b) What is the nucleotide requirement for protein digestion?
(c) What evidence suggests that ATP hydrolysis, and not just the presence of ATP, is required for digestion?

A similar experiment was performed with a small peptide as a substrate for the proteasome instead of a protein. The results obtained are shown in graph B.

(B) Minutes of incubation

(d) How do the requirements for peptide digestion differ from those of protein digestion?
(e) Suggest some reasons for the difference.

The ability of PAN from the archaeon *Thermoplasma* to support protein degradation by the 20S proteasomes from the archaeon *Methanosarcina* and rabbit muscle was then examined.

PERCENTAGE OF DIGESTION OF PROTEIN SUBSTRATE (20S PROTEASOME SOURCE)			
Additions	Thermoplasma	Methanosarcina	Rabbit muscle
None	11	10	10
PAN	8	8	8
PAN + ATP	100	40	30
PAN + ADP	12	9	10

[Data from P. Zwickl, D. Ng, K. M. Woo, H.-P. Klenk, and A. L. Goldberg. An archaebacterial ATPase, homologous to ATPase in the eukaryotic 26S proteasome, activates protein breakdown by 20S proteasomes. *J. Biol. Chem.* 274(1999): 26008–26014.]

(f) Can the *Thermoplasma* PAN augment protein digestion by the proteasomes from other organisms?
(g) What is the significance of the stimulation of rabbit-muscle proteasome by *Thermoplasma* PAN?

The Biosynthesis of Amino Acids

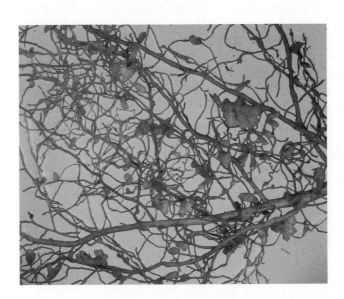

Glutamate

Nitrogen is a key component of amino acids. The atmosphere is rich in nitrogen gas (N_2), a very unreactive molecule. Certain organisms, such as bacteria that live in the root nodules of yellow clover, can convert nitrogen gas into ammonia, which can then be used to synthesize, first, glutamate and then other amino acids. [(Left) Runk/Schoenberger from Grant Heilman.]

The assembly of biological molecules, including proteins and nucleic acids, requires the generation of appropriate starting materials. We have already considered the assembly of carbohydrates in discussions of the Calvin cycle and the pentose phosphate pathway (Chapter 20). The present chapter and the next two examine the assembly of the other important building blocks—namely, amino acids, nucleotides, and lipids.

The pathways for the biosynthesis of these molecules are extremely ancient, going back to the last common ancestor of all living things. Indeed, these pathways probably predate many of the pathways of energy transduction discussed in Part II and may have provided key selective advantages in early evolution. Many of the intermediates in energy-transduction pathways play a role in biosynthesis as well. These common intermediates allow efficient interplay between energy-transduction (catabolic) and biosynthetic (anabolic) pathways. Thus, cells are able to balance the degradation of compounds for energy mobilization and the synthesis of starting materials for macromolecular construction.

We begin our consideration of biosynthesis with amino acids—the building blocks of proteins and the nitrogen source for many other important molecules, including nucleotides, neurotransmitters, and prosthetic groups such as porphyrins. Amino acid biosynthesis is intimately connected with nutrition because many higher organisms, including human beings, have lost the ability to synthesize some amino acids and must therefore

Anabolism
Biosynthetic processes.
Catabolism
Degradative processes.
Derived from the Greek *ana*, "up"; *kata*, "down"; *ballein*, "to throw."

obtain adequate quantities of these essential amino acids in their diets. Furthermore, because some amino acid biosynthetic enzymes are absent in mammals but present in plants and microorganisms, they are useful targets for herbicides and antibiotics.

Amino Acid Synthesis Requires Solutions to Three Key Biochemical Problems

Nitrogen is an essential component of amino acids. Earth has an abundant supply of nitrogen, but it is primarily in the form of atmospheric nitrogen gas, a remarkably inert molecule. Thus, a fundamental problem for biological systems is to obtain nitrogen in a more usable form. This problem is solved by certain microorganisms capable of reducing the inert $N\equiv N$ molecule of nitrogen gas to two molecules of ammonia in one of the most amazing reactions in biochemistry. Nitrogen in the form of ammonia is the source of nitrogen for all the amino acids. The carbon backbones come from the glycolytic pathway, the pentose phosphate pathway, or the citric acid cycle.

In amino acid production, we encounter an important problem in biosynthesis—namely, stereochemical control. Because all amino acids except glycine are chiral, biosynthetic pathways must generate the correct isomer with high fidelity. In each of the 19 pathways for the generation of chiral amino acids, the stereochemistry at the α-carbon atom is established by a transamination reaction that includes pyridoxal phosphate (PLP). Almost all the transaminases that catalyze these reactions descend from a common ancestor, illustrating once again that effective solutions to biochemical problems are retained throughout evolution.

Biosynthetic pathways are often highly regulated such that building blocks are synthesized only when supplies are low. Very often, a high concentration of the final product of a pathway inhibits the activity of enzymes that function early in the pathway. Often present are allosteric enzymes capable of sensing and responding to concentrations of regulatory species. These enzymes are similar in functional properties to aspartate transcarbamylase and its regulators (Section 10.1). Feedback and allosteric mechanisms ensure that all 20 amino acids are maintained in sufficient amounts for protein synthesis and other processes.

24.1 Nitrogen Fixation: Microorganisms Use ATP and a Powerful Reductant to Reduce Atmospheric Nitrogen to Ammonia

The nitrogen in amino acids, purines, pyrimidines, and other biomolecules ultimately comes from atmospheric nitrogen, N_2. The biosynthetic process starts with the reduction of N_2 to NH_3 (ammonia), a process called *nitrogen fixation*. The extremely strong $N\equiv N$ bond, which has a bond energy of $940\ kJ\ mol^{-1}$ ($225\ kcal\ mol^{-1}$), is highly resistant to chemical attack. Indeed, Antoine Lavoisier named nitrogen gas "azote," meaning "without life," because it is so unreactive. Nevertheless, the conversion of nitrogen and hydrogen to form ammonia is thermodynamically favorable; the reaction is difficult kinetically because intermediates along the reaction pathway are unstable.

Although higher organisms are unable to fix nitrogen, this conversion is carried out by some bacteria and archaea. Symbiotic *Rhizobium* bacteria invade the roots of leguminous plants and form root nodules in which they fix nitrogen, supplying both the bacteria and the plants. The amount of N_2

fixed by *diazotrophic (nitrogen-fixing) microorganisms* has been estimated to be 10^{11} kilograms per year, about 60% of Earth's newly fixed nitrogen. Lightning and ultraviolet radiation fix another 15%; the other 25% is fixed by industrial processes. The industrial process for nitrogen fixation devised by Fritz Haber in 1910 is still being used in fertilizer factories.

$$N_2 + 3\,H_2 \rightleftharpoons 2\,NH_3$$

The fixation of N_2 is typically carried out by mixing with H_2 gas over an iron catalyst at about 500°C and a pressure of 300 atmospheres.

To meet the kinetic challenge, the biological process of nitrogen fixation requires a complex enzyme with multiple redox centers. The *nitrogenase complex,* which carries out this fundamental transformation, consists of two proteins: a *reductase,* which provides electrons with high reducing power, and *nitrogenase,* which uses these electrons to reduce N_2 to NH_3. The transfer of electrons from the reductase to the nitrogenase component is coupled to the hydrolysis of ATP by the reductase (Figure 24.1). The nitrogenase complex is exquisitely sensitive to inactivation by O_2. Leguminous plants maintain a very low concentration of free O_2 in their root nodules by binding O_2 to *leghemoglobin,* a homolog of hemoglobin (p. 170).

In principle, the reduction of N_2 to NH_3 is a six-electron process.

$$N_2 + 6\,e^- + 6\,H^+ \longrightarrow 2\,NH_3$$

However, the biological reaction always generates at least 1 mol of H_2 in addition to 2 mol of NH_3 for each mol of $N{\equiv}N$. Hence, an input of two additional electrons is required.

$$N_2 + 8\,e^- + 8\,H^+ \longrightarrow 2\,NH_3 + H_2$$

In most nitrogen-fixing microorganisms, *the eight high-potential electrons come from reduced ferredoxin,* generated by photosynthesis or oxidative processes. Two molecules of ATP are hydrolyzed for each electron transferred. Thus, *at least 16 molecules of ATP are hydrolyzed for each molecule of N_2 reduced.*

$$N_2 + 8\,e^- + 8\,H^+ + 16\,ATP + 16\,H_2O \longrightarrow$$
$$2\,NH_3 + H_2 + 16\,ADP + 16\,P_i$$

Again, ATP hydrolysis is not required to make nitrogen reduction favorable thermodynamically. Rather, it is essential to reduce the heights of activation barriers along the reaction pathway, thus making the reaction kinetically feasible.

The Iron–Molybdenum Cofactor of Nitrogenase Binds and Reduces Atmospheric Nitrogen

Both the reductase and the nitrogenase components of the complex are *iron–sulfur proteins,* in which iron is bonded to the sulfur atom of a cysteine residue and to inorganic sulfide. Recall that iron–sulfur clusters act as electron carriers (p. 511). The *reductase* (also called the *iron protein* or the *Fe protein*) is a dimer of identical 30-kd subunits bridged by a 4Fe-4S cluster (Figure 24.2).

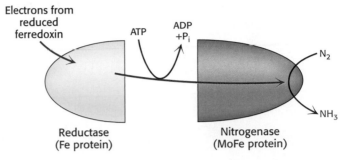

Figure 24.1 Nitrogen fixation. Electrons flow from ferredoxin to the reductase (iron protein, or Fe protein) to nitrogenase (molybdenum–iron protein, or MoFe protein) to reduce nitrogen to ammonia. ATP hydrolysis within the reductase drives conformational changes necessary for the efficient transfer of electrons.

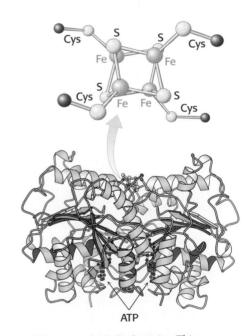

Figure 24.2 Fe Protein. This protein is a dimer composed of two polypeptide chains linked by a 4Fe-4S cluster. *Notice* that each monomer is a member of the P-loop NTPase family and contains an ATP-binding site. [Drawn from 1N2C.pdb.]

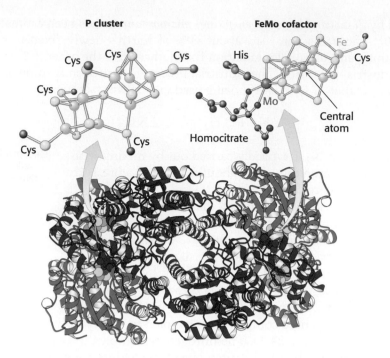

Figure 24.3 MoFe protein. This protein is a heterotetramer composed of two α subunits (red) and two β subunits (blue). *Notice* that the protein contains two copies each of two types of clusters: P clusters and FeMo cofactors. Each P cluster contains eight iron atoms (green) and seven sulfides linked to the protein by six cysteinate residues. Each FeMo cofactor contains one molybdenum atom, seven iron atoms, nine sulfides, a central atom, and a homocitrate, and is linked to the protein by one cysteinate residue and one histidine residue. [Drawn from 1M1N.pdb.]

The role of the reductase is to transfer electrons from a suitable donor, such as reduced ferredoxin, to the nitrogenase component. The 4Fe-4S cluster carries the electrons, one at a time, to nitrogenase. The binding and hydrolysis of ATP triggers a conformational change that moves the reductase closer to the nitrogenase component from whence it is able to transfer its electron to the center of nitrogen reduction. The structure of the ATP-binding region reveals it to be a member of the P-loop NTPase family (Section 9.4) that is clearly related to the nucleotide-binding regions fround in G proteins and related proteins. Thus, we see another example of how this domain has been recruited in evolution because of its ability to couple nucleoside triphosphate hydrolysis to conformational changes.

The nitrogenase component is an $\alpha_2\beta_2$ tetramer (240 kd), in which the α and β subunits are homologous to each other and structurally quite similar (Figure 24.3). Because molybdenum is present in this cluster, the nitrogenase component is also called the *molybdenum–iron protein* (MoFe protein). The FeMo cofactor consists of two M-3Fe-3S clusters, in which molybdenum occupies the M site in one cluster and iron occupies it in the other. The two clusters are joined by three sulfide ions and a central atom, the identity of which has not yet been conclusively established. The FeMo cofactor is also coordinated to a homocitrate moiety and to the α subunit through one histidine residue and one cysteinate residue. This cofactor is distinct from the molybdenum-containing cofactor found in sulfite oxidase and apparently all other molybdenum-containing enzymes.

Electrons from the reductase enter at the *P clusters*, which are located at the α–β interface. The role of the P clusters is to store electrons until they can be used productively to reduce nitrogen at the FeMo cofactor. *The FeMo cofactor is the site of nitrogen fixation.* One face of the FeMo cofactor is likely to be the site of nitrogen reduction. The electron-transfer reactions from the P cluster take place in concert with the binding of hydrogen ions to nitrogen as it is reduced. Further studies are underway to elucidate the mechanism of this remarkable reaction.

Ammonium Ion Is Assimilated into an Amino Acid Through Glutamate and Glutamine

The next step in the assimilation of nitrogen into biomolecules is the entry of NH_4^+ into amino acids. *Glutamate* and *glutamine* play pivotal roles in this regard. The α-amino group of most amino acids comes from the α-amino group of glutamate by transamination (p. 656). Glutamine, the other major nitrogen donor, contributes its side-chain nitrogen atom in the biosynthesis of a wide range of important compounds, including the amino acids tryptophan and histidine.

Glutamate is synthesized from NH_4^+ and α-ketoglutarate, a citric acid cycle intermediate, by the action of *glutamate dehydrogenase*. We have already encountered this enzyme in the degradation of amino acids (p. 656). Recall that NAD^+ is the oxidant in catabolism, whereas NADPH is the reductant in biosyntheses. Glutamate dehydrogenase is unusual in that it does not discriminate between NADH and NADPH, at least in some species.

$$NH_4^+ + \alpha\text{-ketoglutarate} + NADPH + H^+ \rightleftharpoons$$
$$\text{glutamate} + NADP^+ + H_2O$$

The reaction proceeds in two steps. First, a Schiff base forms between ammonia and α-ketoglutarate. The formation of a Schiff base between an amine and a carbonyl compound is a key reaction that takes place at many stages of amino acid biosynthesis and degradation.

Schiff bases are easily protonated. In the second step, the protonated Schiff base is reduced by the transfer of a hydride ion from NADPH to form glutamate.

This reaction is crucial because it establishes the stereochemistry of the α-carbon atom (S absolute configuration) in glutamate. The enzyme binds the α-ketoglutarate substrate in such a way that hydride transferred from NAD(P)H is added to form the L isomer of glutamate (Figure 24.4). As we shall see, this stereochemistry is established for other amino acids by transamination reactions that rely on pyridoxal phosphate.

A second ammonium ion is incorporated into glutamate to form glutamine by the action of *glutamine synthetase*. This amidation is driven by the hydrolysis of ATP. ATP participates directly in the reaction by

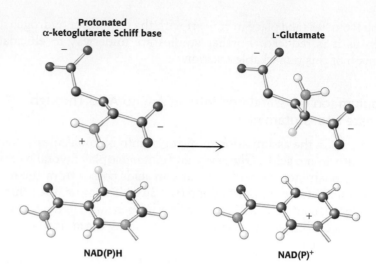

**Protonated
α-ketoglutarate Schiff base**

L-Glutamate

NAD(P)H

NAD(P)⁺

Figure 24.4 **Establishment of Chirality.** In
the active site of glutamate dehydrogenase,
hydride transfer (green) from NAD(P)H to a
specific face of the achiral protonated
Schiff base of α-ketoglutarate establishes
the L configuration of glutamate.

phosphorylating the side chain of glutamate to form an acyl-phosphate
intermediate, which then reacts with ammonia to form glutamine.

Glutamate → **Acyl-phosphate intermediate** → **Glutamine**

(ATP, ADP / NH₃, Pᵢ)

A high-affinity ammonia-binding site is formed in the enzyme only after
the formation of the acyl-phosphate intermediate. A specific site for ammonia binding is required to prevent attack by water from hydrolyzing
the intermediate and wasting a molecule of ATP. The regulation of glutamine synthetase plays a critical role in controlling nitrogen metabolism
(Section 24.3).

Glutamate dehydrogenase and glutamine synthetase are present in all
organisms. Most prokaryotes also contain an evolutionarily unrelated
enzyme, *glutamate synthase,* which catalyzes the reductive amination of
α-ketoglutarate to glutamate. Glutamine is the nitrogen donor.

$$\alpha\text{-Ketoglutarate} + \text{glutamine} + \text{NADPH} + \text{H}^+ \rightleftharpoons$$
$$2\text{ glutamate} + \text{NADP}^+$$

The side-chain amide of glutamine is hydrolyzed to generate ammonia
within the enzyme, a recurring theme throughout nitrogen metabolism.
*When NH₄⁺ is limiting, most of the glutamate is made by the sequential
action of glutamine synthetase and glutamate synthase.* The sum of these
reactions is

$$\text{NH}_4^+ + \alpha\text{-ketoglutarate} + \text{NADPH} + \text{ATP} \longrightarrow$$
$$\text{glutamate} + \text{NADP}^+ + \text{ADP} + \text{P}_i$$

Note that this stoichiometry differs from that of the glutamate dehydrogenase reaction in that ATP is hydrolyzed. Why do prokaryotes sometimes
use this more expensive pathway? The answer is that the value of K_M of glutamate dehydrogenase for NH₄⁺ is high (~ 1 mM), and so this enzyme is
not saturated when NH₄⁺ is limiting. In contrast, glutamine synthetase has
very high affinity for NH₄⁺. Thus, ATP hydrolysis is required to capture
ammonia when it is scarce.

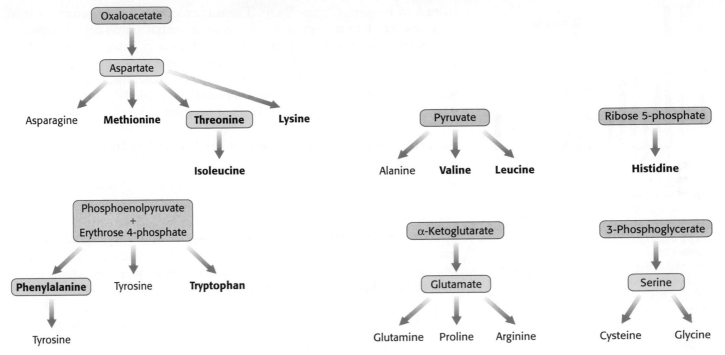

Figure 24.5 Biosynthetic families of amino acids in bacteria and plants. Major metabolic precursors are shaded blue. Amino acids that give rise to other amino acids are shaded yellow. Essential amino acids are in boldface type.

24.2 Amino Acids Are Made from Intermediates of the Citric Acid Cycle and Other Major Pathways

Thus far, we have considered the conversion of N_2 into NH_4^+ and the assimilation of NH_4^+ into glutamate and glutamine. We turn now to the biosynthesis of the other amino acids. The pathways for the biosynthesis of amino acids are diverse. However, they have an important common feature: *their carbon skeletons come from intermediates of glycolysis, the pentose phosphate pathway, or the citric acid cycle.* On the basis of these starting materials, amino acids can be grouped into six biosynthetic families (Figure 24.5).

Human Beings Can Synthesize Some Amino Acids but Must Obtain Others from the Diet

Most microorganisms such as *E. coli* can synthesize the entire basic set of 20 amino acids, whereas human beings cannot make 9 of them. The amino acids that must be supplied in the diet are called *essential amino acids*, whereas the others are termed *nonessential amino acids* (Table 24.1). These designations refer to the needs of an organism under a particular set of conditions. For example, enough arginine is synthesized by the urea cycle to meet the needs of an adult but perhaps not those of a growing child. A deficiency of even one amino acid results in a *negative nitrogen balance*. In this state, more protein is degraded than is synthesized, and so more nitrogen is excreted than is ingested.

The nonessential amino acids are synthesized by quite simple reactions, whereas the pathways for the formation of the essential amino acids are quite complex. For example, the nonessential amino acids *alanine* and *aspartate* are synthesized in a single step from pyruvate and oxaloacetate, respectively. In contrast, the pathways for the essential amino acids require

TABLE 24.1 Basic set of 20 amino acids

Nonessential	Essential
Alanine	Histidine
Arginine	Isoleucine
Asparagine	Leucine
Aspartate	Lysine
Cysteine	Methionine
Glutamate	Phenylalanine
Glutamine	Threonine
Glycine	Tryptophan
Proline	Valine
Serine	
Tyrosine	

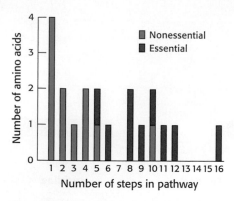

Figure 24.6 Essential and nonessential amino acids. Some amino acids are nonessential to human beings because they can be biosynthesized in a small number of steps. Those amino acids requiring a large number of steps for their synthesis are essential in the diet because some of the enzymes for these steps have been lost in the course of evolution.

Figure 24.7 Amino acid biosynthesis by transamination. (1) Within a transaminase, the internal aldimine is converted into pyridoxamine phosphate (PMP) by reaction with glutamate in a multistep process not shown. (2) PMP then reacts with an α-ketoacid to generate a ketimine. (3) This intermediate is converted into a quinonoid intermediate (4), which in turn yields an external aldimine. (5) The aldimine is cleaved to release the newly formed amino acid to complete the cycle.

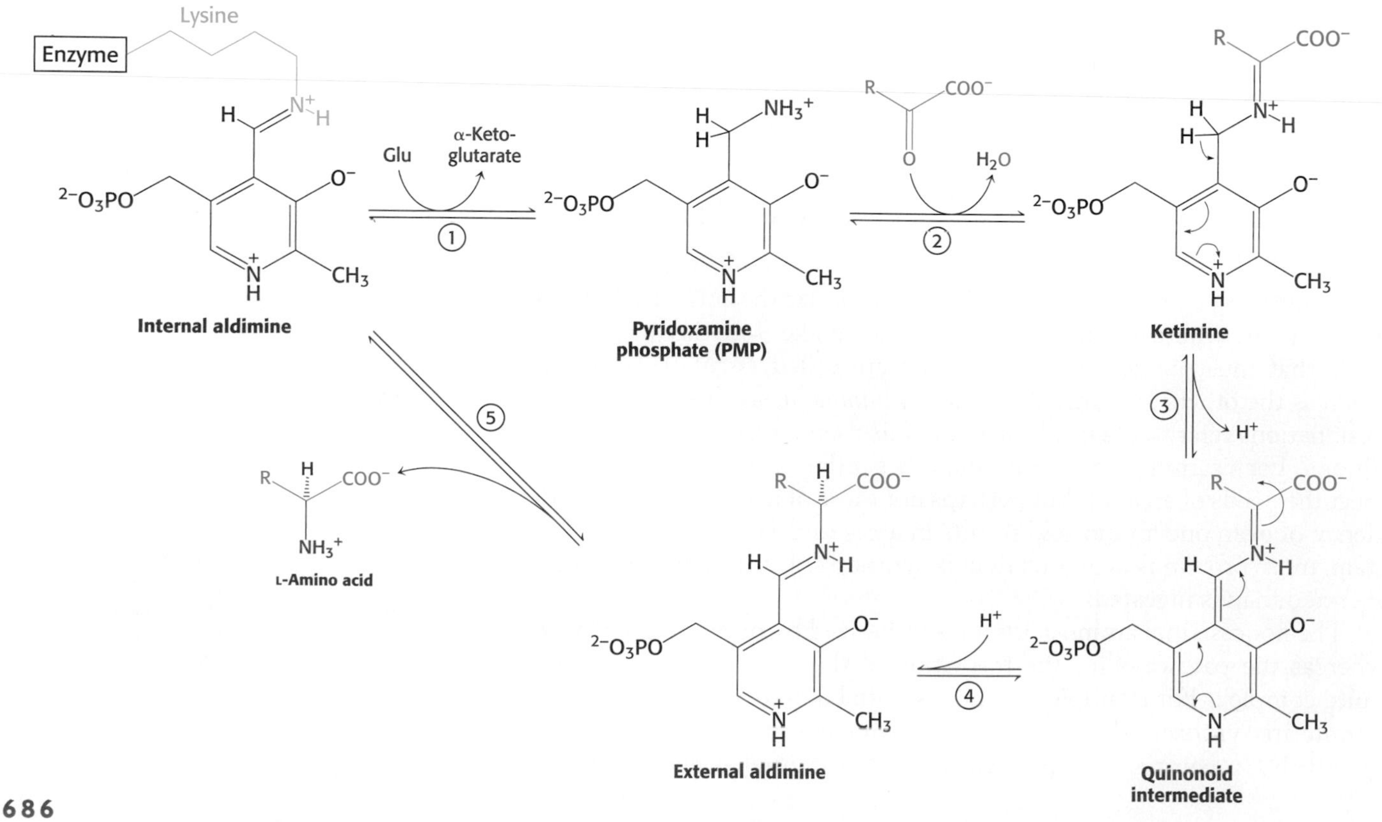

from 5 to 16 steps (Figure 24.6). The sole exception to this pattern is arginine, inasmuch as the synthesis of this nonessential amino acid de novo requires 10 steps. Typically, though, it is made in only 3 steps from ornithine as part of the urea cycle. Tyrosine, classified as a nonessential amino acid because it can be synthesized in 1 step from phenylalanine, requires 10 steps to be synthesized from scratch and is essential if phenylalanine is not abundant. We begin with the biosynthesis of nonessential amino acids.

Aspartate, Alanine, and Glutamate Are Formed by the Addition of an Amino Group to an α-Ketoacid

Three α-ketoacids—α-ketoglutarate, oxaloacetate, and pyruvate—can be converted into amino acids in one step through the addition of an amino group. We have seen that α-ketoglutarate can be converted into glutamate by reductive amination (p. 683). The amino group from glutamate can be transferred to other α-ketoacids by transamination reactions. Thus, aspartate and alanine can be made from the addition of an amino group to oxaloacetate and pyruvate, respectively.

$$\text{Oxaloacetate} + \text{glutamate} \rightleftharpoons \text{aspartate} + \alpha\text{-ketoglutarate}$$

$$\text{Pyruvate} + \text{glutamate} \rightleftharpoons \text{alanine} + \alpha\text{-ketoglutarate}$$

These reactions are carried out by *pyridoxal phosphate-dependent transaminases*. Transamination reactions are required for the synthesis of most amino acids.

A Common Step Determines the Chirality of All Amino Acids

We shall review the transaminase mechanism (p. 657) as it applies to amino acid biosynthesis (see Figure 23.11). The reaction pathway begins with *pyridoxal phosphate* in a Schiff-base linkage with lysine at the transaminase active site, forming an internal aldimine (Figure 24.7). An amino group is transferred from glutamate to form pyridoxamine phosphate (PMP), the actual amino donor, in a multistep process. Pyridoxamine phosphate then reacts with an incoming α-ketoacid to form a ketimine. Proton loss forms a quinonoid intermediate that then accepts a proton at a different site to form an external aldimine. The newly formed amino acid is released with the concomitant formation of the internal aldimine.

Aspartate aminotransferase is the prototype of a large family of PLP-dependent enzymes. Comparisons of amino acid sequences as well as several three-dimensional structures reveal that almost all transaminases having roles in amino acid biosynthesis are related to aspartate aminotransferase by divergent evolution. An examination of the aligned amino acid sequences reveals that two residues are completely conserved. These residues are the lysine residue that forms the Schiff base with the PLP cofactor (lysine 258 in aspartate aminotransferase) and an arginine residue that interacts with the α-carboxylate group of the ketoacid (see Figure 23.12).

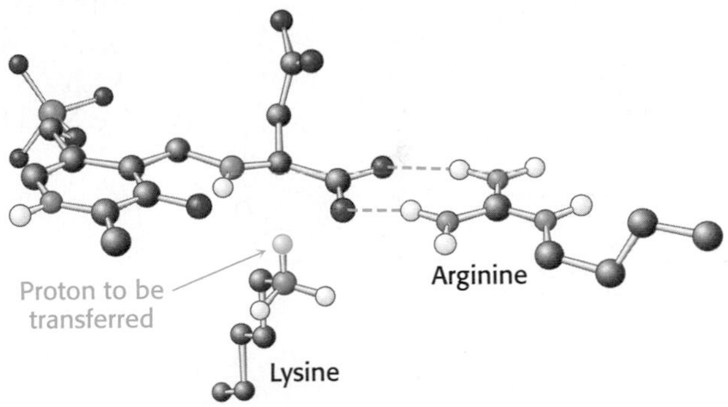

An essential step in the transamination reaction is the protonation of the quinonoid intermediate to form the external aldimine. *The chirality of the amino acid formed is determined by the direction from which this proton is added to the quinonoid form* (Figure 24.8). The interaction between the conserved arginine residue and the α-carboxylate group helps orient the substrate so that the lysine residue transfers a proton to the bottom face of the quinonoid intermediate, generating an aldimine with an L configuration at the C_α center.

Proton to be transferred

Arginine

Lysine

Figure 24.8 Stereochemistry of proton addition. In a transaminase active site, the addition of a proton from the lysine residue to the bottom face of the quinonoid intermediate determines the L configuration of the amino acid product. The conserved arginine residue interacts with the α-carboxylate group and helps establish the appropriate geometry of the quinonoid intermediate.

The Formation of Asparagine from Aspartate Requires an Adenylated Intermediate

The formation of asparagine from aspartate is chemically analogous to the formation of glutamine from glutamate. Both transformations are amidation reactions and both are driven by the hydrolysis of ATP. The actual reactions are different, however. In bacteria, the reaction for the asparagine synthesis is

$$\text{Asparate} + NH_4^+ + ATP \longrightarrow \text{asparagine} + AMP + PP_i + H^+$$

Thus, the products of ATP hydrolysis are AMP and PP_i rather than ADP and P_i. Aspartate is activated by adenylation rather than by phosphorylation.

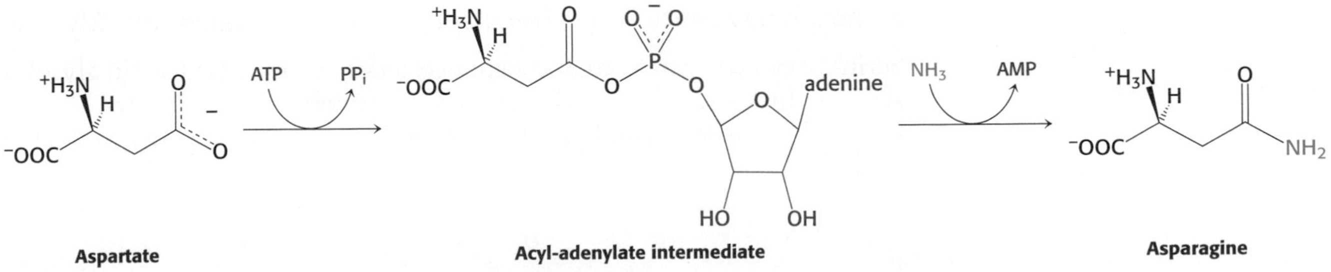

Aspartate   **Acyl-adenylate intermediate**   **Asparagine**

We have encountered this mode of activation in fatty acid degradation and will see it again in lipid and protein synthesis.

In mammals, the nitrogen donor for asparagine is glutamine rather than ammonia as in bacteria. Ammonia is generated by hydrolysis of the side chain of glutamine and directly transferred to activated aspartate, bound in the active site. An advantage is that the cell is not directly exposed to NH_4^+, which is toxic at high levels to human beings and other mammals. *The use of glutamine hydrolysis as a mechanism for generating ammonia for use within the same enzyme is a motif common throughout biosynthetic pathways.*

Glutamate Is the Precursor of Glutamine, Proline, and Arginine

The synthesis of glutamate by the reductive amination of α-ketoglutarate has already been discussed, as has the conversion of glutamate into glutamine (p. 684). Glutamate is the precursor of two other nonessential amino acids: *proline* and *arginine*. First, the γ-carboxyl group of glutamate reacts with ATP to form an acyl phosphate. This mixed anhydride is then reduced by NADPH to an aldehyde.

Glutamic γ-semialdehyde cyclizes with a loss of H_2O in a nonenzymatic process to give Δ^1-pyrroline 5-carboxylate, which is reduced by NADPH to proline. Alternatively, the semialdehyde can be transaminated to ornithine, which is converted in several steps into arginine (p. 661).

3-Phosphoglycerate Is the Precursor of Serine, Cysteine, and Glycine

Serine is synthesized from 3-phosphoglycerate, an intermediate in glycolysis. The first step is an oxidation to 3-phosphohydroxypyruvate. This α-ketoacid is transaminated to 3-phosphoserine, which is then hydrolyzed to serine.

Serine is the precursor of *glycine* and *cysteine*. In the formation of glycine, the side-chain methylene group of serine is transferred to *tetrahydrofolate,* a carrier of one-carbon units that will be discussed shortly.

$$\text{Serine} + \text{tetrahydrofolate} \longrightarrow$$
$$\text{glycine} + \text{methylenetetrahydrofolate} + H_2O$$

This interconversion is catalyzed by *serine hydroxymethylaseferase,* a PLP enzyme that is homologous to aspartate aminotransferase. The bond between the α- and β-carbon atoms of serine is labilized by the formation of a Schiff base between serine and PLP (p. 657). The side-chain methylene group of serine is then transferred to tetrahydrofolate. The conversion of serine into cysteine requires the substitution of a sulfur atom derived from methionine for the side-chain oxygen atom.

Tetrahydrofolate Carries Activated One-Carbon Units at Several Oxidation Levels

Tetrahydrofolate (also called *tetrahydropteroylglutamate*) is a highly versatile carrier of activated one-carbon units. This cofactor consists of three groups: a substituted pteridine, *p*-aminobenzoate, and a chain of one or more glutamate residues (Figure 24.9). Mammals can synthesize the pteridine ring, but they are unable to conjugate it to the other two units. They obtain tetrahydrofolate from their diets or from microorganisms in their intestinal tracts.

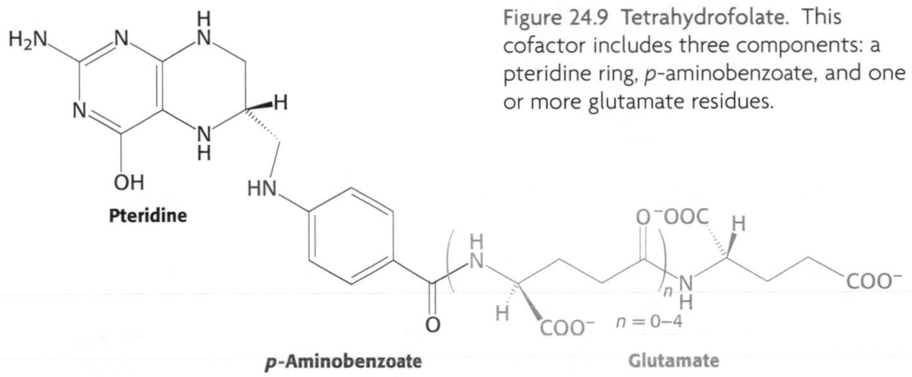

Figure 24.9 Tetrahydrofolate. This cofactor includes three components: a pteridine ring, *p*-aminobenzoate, and one or more glutamate residues.

The one-carbon group carried by tetrahydrofolate is bonded to its N-5 or N-10 nitrogen atom (denoted as N^5 and N^{10}) or to both. This unit can exist in three oxidation states (Table 24.2). The most-reduced form carries a *methyl* group, whereas the intermediate form carries a *methylene* group. More-oxidized forms carry a *formyl, formimino,* or *methenyl* group. The fully oxidized one-carbon unit, CO_2, is carried by biotin rather than by tetrahydrofolate.

The one-carbon units carried by tetrahydrofolate are interconvertible (Figure 24.10). N^5,N^{10}-*Methylenetetrahydrofolate* can be reduced to N^5-*methyltetrahydrofolate* or oxidized to N^5,N^{10}-*methenyltetrahydrofolate.* N^5,N^{10}-*Methenyltetrahydrofolate* can be converted into N^5-*formiminotetrahydrofolate* or N^{10}-*formyl-tetrahydrofolate,* both of which are at the same oxidation level. N^{10}-*Formyltetrahydrofolate* can also be synthesized from tetrahydrofolate, formate, and ATP.

$$\text{Formate} + \text{ATP} + \text{tetrahydrofolate} \longrightarrow$$
$$N^{10}\text{-formyltetrahydrofolate} + \text{ADP} + P_i$$

TABLE 24.2 One-carbon groups carried by tetrahydrofolate

Oxidation state	Group	
	Formula	Name
Most reduced (= methanol)	—CH_3	Methyl
Intermediate (= formaldehyde)	—CH_2—	Methylene
Most oxidized (= formic acid)	—CHO	Formyl
	—CHNH	Formimino
	—CH=	Methenyl

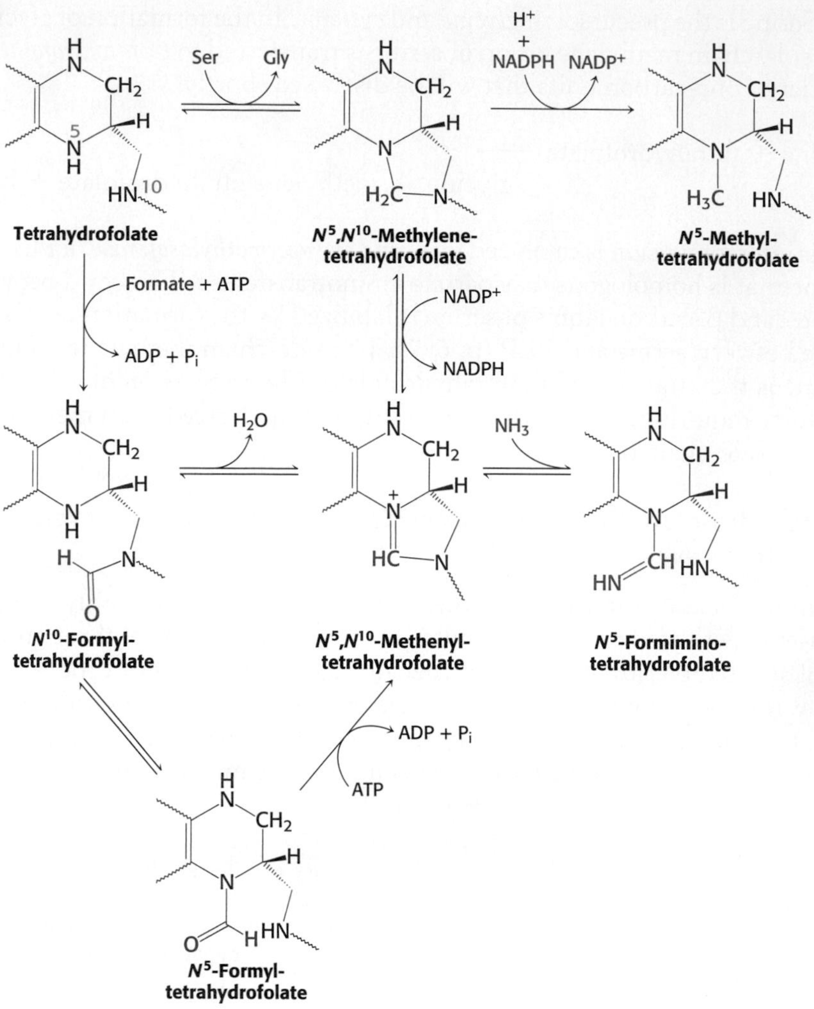

Figure 24.10 Conversions of one-carbon units attached to tetrahydrofolate.

N^5-*Formyl*tetrahydrofolate can be reversibly isomerized to N^{10}-*formyl*-tetrahydrofolate or it can be converted into N^5,N^{10}-*methenyl*tetrahydrofolate.

These tetrahydrofolate derivatives serve as donors of one-carbon units in a variety of biosyntheses. Methionine is regenerated from homocysteine by transfer of the methyl group of N^5-methyltetrahydrofolate, as will be discussed shortly. We shall see in Chapter 25 that some of the carbon atoms of *purines* are acquired from derivatives of N^{10}-formyltetrahydrofolate. The methyl group of *thymine*, a pyrimidine, comes from N^5,N^{10}-methylenetetrahydrofolate. This tetrahydrofolate derivative can also donate a one-carbon unit in an alternative synthesis of *glycine* that starts with CO_2 and NH_4^+, a reaction catalyzed by *glycine synthase* (called the *glycine cleavage enzyme* when it operates in the reverse direction).

$$CO_2 + NH_4^+ + N^5,N^{10}\text{-methylenetetrahydrofolate} + NADH \rightleftharpoons$$
$$\text{glycine} + \text{tetrahydrofolate} + NAD^+$$

Thus, one-carbon units at each of the three oxidation levels are utilized in biosyntheses. Furthermore, *tetrahydrofolate serves as an acceptor of one-carbon units in degradative reactions.* The major source of one-carbon units is the facile conversion of serine into glycine, which yields N^5,N^{10}-methylenetetrahydrofolate. Serine can be derived from 3-phosphoglycerate, and so *this pathway enables one-carbon units to be formed de novo from carbohydrates.*

S-Adenosylmethionine Is the Major Donor of Methyl Groups

Tetrahydrofolate can carry a methyl group on its N-5 atom, but its transfer potential is not sufficiently high for most biosynthetic methylations. Rather, the activated methyl donor is usually S-*adenosylmethionine* (SAM), which is synthesized by the transfer of an adenosyl group from ATP to the sulfur atom of methionine.

Methionine   **S-Adenosylmethionine (SAM)**

The methyl group of the methionine unit is activated by the positive charge on the adjacent sulfur atom, which makes the molecule much more reactive than N^5-methyltetrahydrofolate. The synthesis of S-adenosylmethionine is unusual in that the triphosphate group of ATP is split into pyrophosphate and orthophosphate; the pyrophosphate is subsequently hydrolyzed to two molecules of P_i. S-*Adenosylhomocysteine* is formed when the methyl group of S-adenosylmethionine is transferred to an acceptor. S-Adenosylhomocysteine is then hydrolyzed to *homocysteine* and adenosine.

S-Adenosylmethionine (SAM)   **S-Adenosylhomocysteine**   **Homocysteine**

Methionine can be regenerated by the transfer of a methyl group to homocysteine from N^5-methyltetrahydrofolate, a reaction catalyzed by *methionine synthase* (also known as *homocysteine methyltransferase*).

Homocysteine   **N^5-Methyl-tetrahydrofolate**   **Methionine**   **Tetrahydrofolate**

The coenzyme that mediates this transfer of a methyl group is *methylcobalamin*, derived from vitamin B_{12}. In fact, this reaction and the rearrangement of L-methylmalonyl CoA to succinyl CoA (p. 628), catalyzed by a

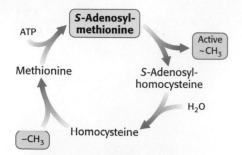

Figure 24.11 Activated methyl cycle. The methyl group of methionine is activated by the formation of *S*-adenosylmethionine.

homologous enzyme, are the only two B_{12}-dependent reactions known to take place in mammals. Another enzyme that converts homocysteine into methionine without vitamin B_{12} also is present in many organisms.

These reactions constitute the *activated methyl cycle* (Figure 24.11). Methyl groups enter the cycle in the conversion of homocysteine into methionine and are then made highly reactive by the addition of an adenosyl group, which makes the sulfur atoms positively charged and the methyl groups much more electrophilic. The high transfer potential of the *S*-methyl group enables it to be transferred to a wide variety of acceptors.

Among the acceptors modified by *S*-adenosylmethionine are specific bases in DNA. The methylation of DNA protects bacterial DNA from cleavage by restriction enzymes (Section 9.3). The base to be methylated is flipped out of the DNA double helix into the active site of a DNA methylase where it can accept a methyl group from *S*-adenosylmethionine (Figure 24.12). A recurring *S*-adenosylmethionine-binding domain is present in many SAM-dependent methylases.

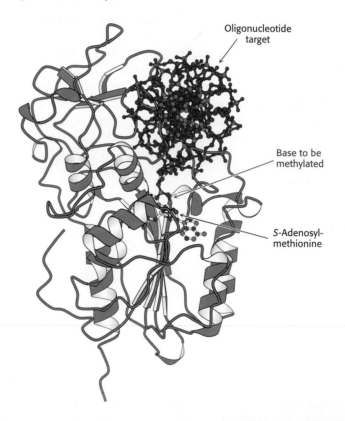

Figure 24.12 DNA methylation. The structure of a DNA methylase bound to an oligonucleotide target shows that the base to be methylated is flipped out of the DNA helix into the active site of a SAM-dependent methylase. [Drawn from 1OMH.pdb.]

S-Adenosylmethionine is also the precursor of *ethylene*, a gaseous plant hormone that induces the ripening of fruit. *S*-Adenosylmethionine is cyclized to a cyclopropane derivative that is then oxidized to form ethylene. The Greek philosopher Theophrastus recognized more than 2000 years ago that sycamore figs do not ripen unless they are scraped with an iron claw. The reason is now known: *wounding triggers ethylene production, which in turn induces ripening.* Much effort is being made to understand this biosynthetic pathway because ethylene is a culprit in the spoilage of fruit.

S-Adenosylmethionine $\xrightarrow{\text{ACC synthase}}$ 1-Aminocyclopropane-1-carboxylate (ACC) $\xrightarrow{\text{ACC oxidase}}$ $H_2C{=}CH_2$ **Ethylene**

Cysteine Is Synthesized from Serine and Homocysteine

In addition to being a precursor of methionine in the activated methyl cycle, homocysteine is an intermediate in the synthesis of cysteine. Serine and homocysteine condense to form *cystathionine*. This reaction is catalyzed by *cystathionine β-synthase*. Cystathionine is then deaminated and cleaved to cysteine and α-ketobutyrate by *cystathioninase*. Both of these enzymes utilize PLP and are homologous to aspartate aminotransferase. The net reaction is

$$\text{Homocysteine} + \text{serine} \rightleftharpoons \text{cysteine} + \alpha\text{-ketobutyrate} + NH_4^+$$

Note that the sulfur atom of cysteine is derived from homocysteine, whereas the carbon skeleton comes from serine.

Homocysteine · Serine · Cystathionine · α-Ketobutyrate · Cysteine

High Homocysteine Levels Correlate with Vascular Disease

People with elevated serum levels of homocysteine or the disulfide-linked dimer homocystine have an unusually high risk for coronary heart disease and arteriosclerosis. The most common genetic cause of high homocysteine levels is a mutation within the gene encoding cystathionine β-synthase. The molecular basis of homocysteine's action has not been clearly identified, although it appears to damage cells lining blood vessels and to increase the growth of vascular smooth muscle. The amino acid raises oxidative stress as well. Vitamin treatments are effective in reducing homocysteine levels in some people. Treatment with vitamins maximizes the activity of the two major metabolic pathways processing homocysteine. Pyridoxal phosphate, a vitamin B_6 derivative, is necessary for the activity of cystathionine β-synthase, which converts homocysteine into cystathionine; tetrahydrofolate, as well as vitamin B_{12}, supports the methylation of homocysteine to methionine.

Shikimate and Chorismate Are Intermediates in the Biosynthesis of Aromatic Amino Acids

We turn now to the biosynthesis of essential amino acids. These amino acids are synthesized by plants and microorganisms, and those in the human diet are ultimately derived primarily from plants. The essential amino acids are formed by much more complex routes than are the nonessential amino acids. The pathways for the synthesis of aromatic amino acids in bacteria have been selected for discussion here because they are well understood and exemplify recurring mechanistic motifs.

Phenylalanine, tyrosine, and tryptophan are synthesized by a common pathway in *E. coli* (Figure 24.13). The initial step is the condensation of phosphoenolpyruvate (a glycolytic intermediate) with erythrose 4-phosphate (a pentose phosphate pathway intermediate). The resulting seven-carbon open-chain sugar is oxidized, loses its phosphoryl group, and cyclizes to 3-dehydroquinate. Dehydration then yields 3-dehydroshikimate, which is reduced by NADPH to shikimate. The phosphorylation of shikimate by ATP gives shikimate 3-phosphate, which condenses with a second molecule of phosphoenolpyruvate. The resulting 5-enolpyruvyl intermediate loses its

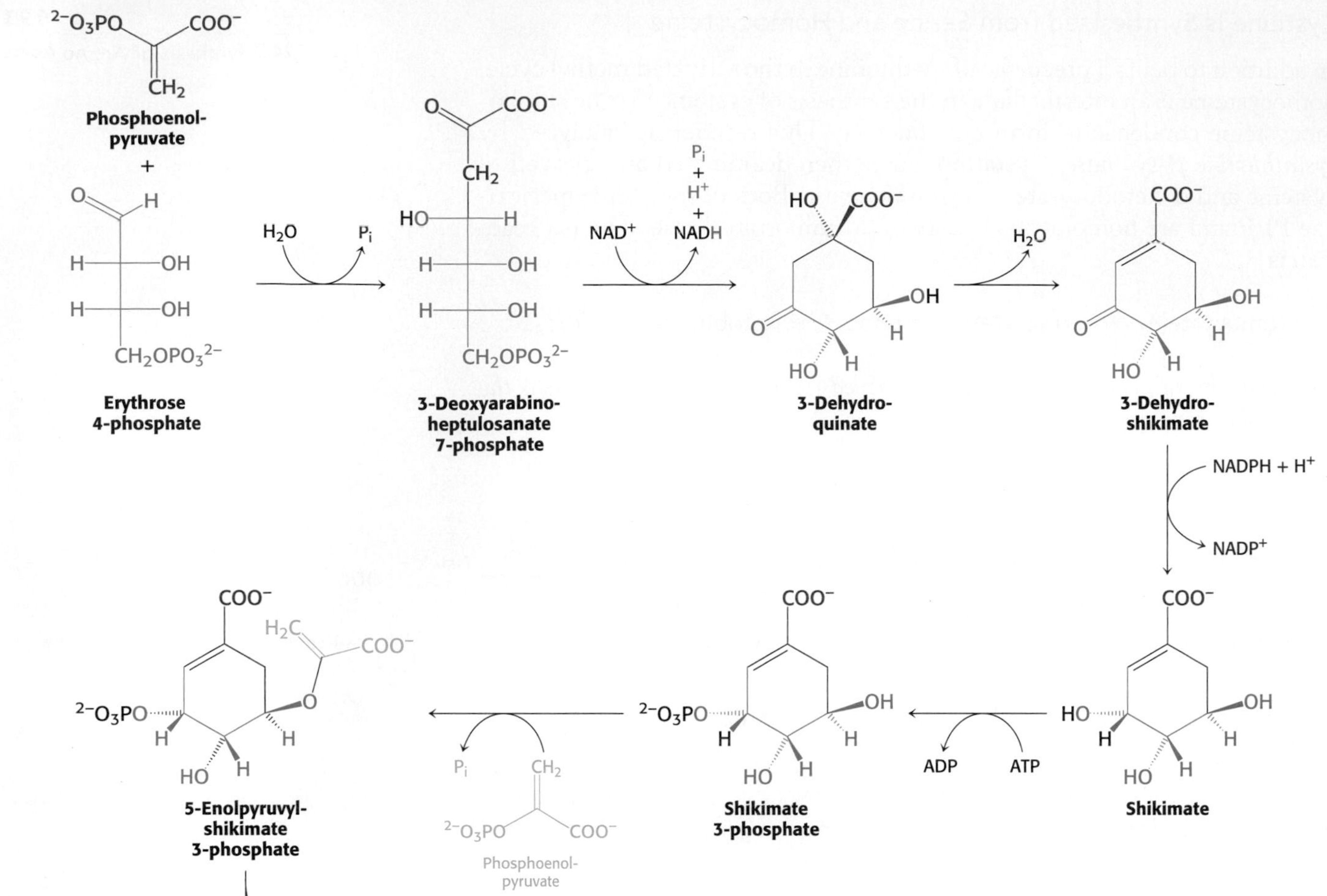

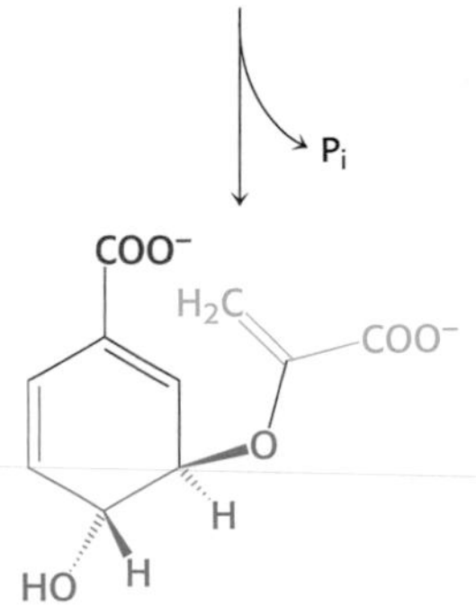

Figure 24.13 Pathway to chorismate. Chorismate is an intermediate in the biosynthesis of phenylalanine, tyrosine, and tryptophan.

phosphoryl group, yielding chorismate, the common precursor of all three aromatic amino acids. The importance of this pathway is revealed by the effectiveness of glyphosate (Roundup), a broad-spectrum herbicide. This compound is an uncompetitive inhibitor of the enzyme that produces 5-enolpyruvylshikimate 3-phosphate. It blocks aromatic amino acid biosynthesis in plants but is fairly nontoxic in animals because they lack the enzyme.

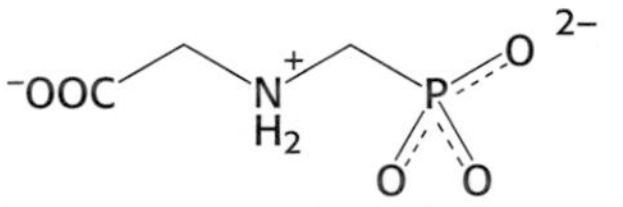

Glyphosate
(Roundup)

The pathway bifurcates at chorismate. Let us first follow the *prephenate branch* (Figure 24.14). A mutase converts chorismate into prephenate, the immediate precursor of the aromatic ring of phenylalanine and tyrosine. This fascinating conversion is a rare example of an electrocyclic reaction in biochemistry, mechanistically similar to the well-known Diels–Alder reaction from organic chemistry. Dehydration and decarboxylation yield *phenylpyruvate*. Alternatively, prephenate can be oxidatively decarboxylated to p-*hydroxyphenylpyruvate*. These α-ketoacids are then transaminated to form *phenylalanine* and *tyrosine*.

The branch starting with *anthranilate* leads to the synthesis of tryptophan (Figure 24.15). Chorismate acquires an amino group derived from the hydrolysis of the side chain of glutamine and releases pyruvate to form anthranilate. Then anthranilate condenses with *5-phosphoribosyl-1-pyrophosphate* (PRPP), *an activated form of ribose phosphate.* PRPP is also

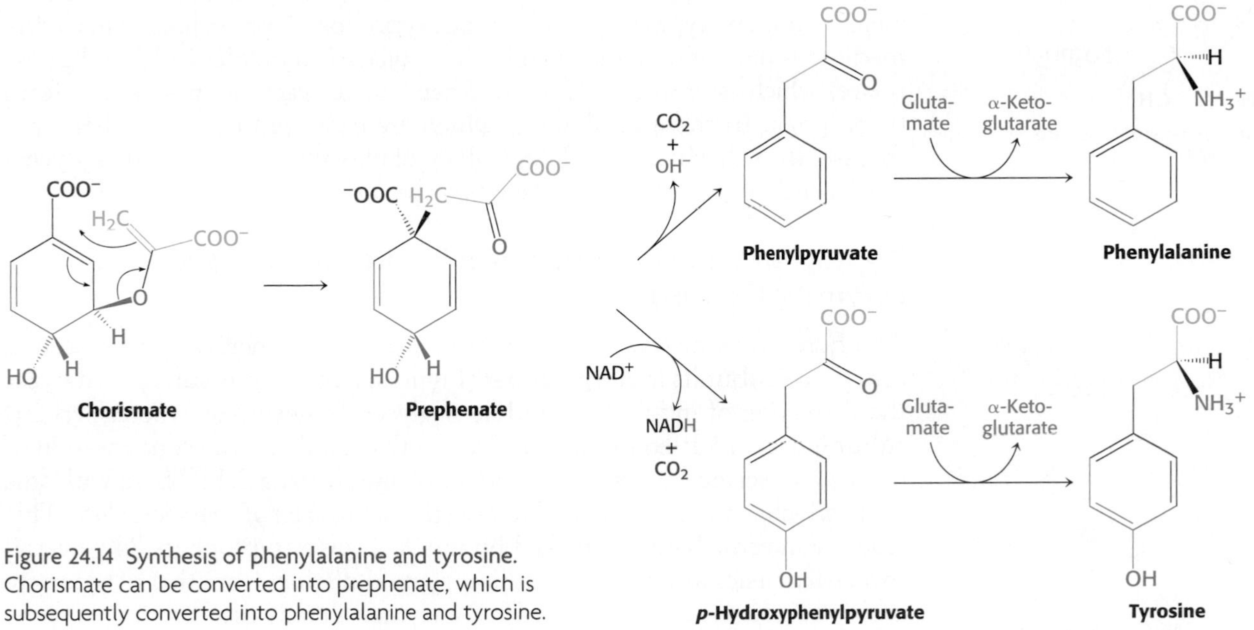

Figure 24.14 Synthesis of phenylalanine and tyrosine. Chorismate can be converted into prephenate, which is subsequently converted into phenylalanine and tyrosine.

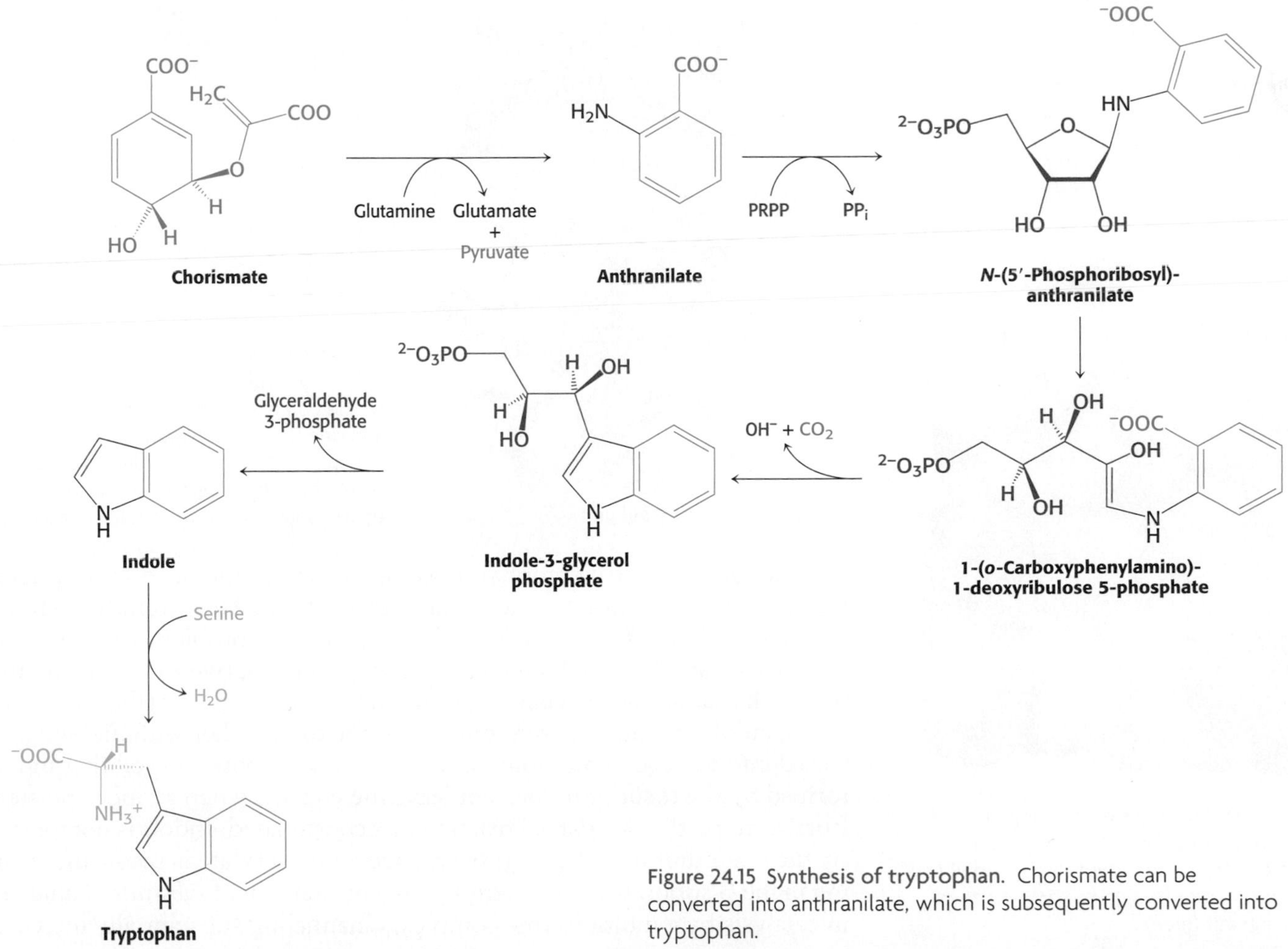

Figure 24.15 Synthesis of tryptophan. Chorismate can be converted into anthranilate, which is subsequently converted into tryptophan.

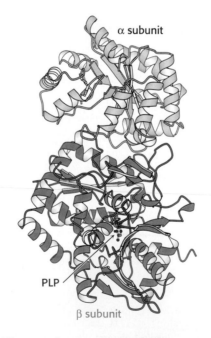

5-Phosphoribosyl-1-pyrophosphate
(PRPP)

Indole

Schiff base of aminoacrylate
(derived from serine)

an important intermediate in the synthesis of histidine, pyrimidine nucleotides, and purine nucleotides (pp. 712 and 714). The C-1 atom of ribose 5-phosphate becomes bonded to the nitrogen atom of anthranilate in a reaction that is driven by the release and hydrolysis of pyrophosphate. The ribose moiety of phosphoribosylanthranilate undergoes rearrangement to yield 1-(o-carboxyphenylamino)-1-deoxyribulose 5-phosphate. This intermediate is dehydrated and then decarboxylated to indole-3-glycerol phosphate, which is cleaved to indole. Then indole reacts with serine to form tryptophan. In these final steps, which are catalyzed by tryptophan synthetase, the side chain of indole-3-glycerol phosphate is removed as glyceraldehyde 3-phosphate and replaced by the carbon skeleton of serine.

Tryptophan Synthase Illustrates Substrate Channeling in Enzymatic Catalysis

The *E. coli* enzyme *tryptophan synthase*, an $\alpha_2\beta_2$ tetramer, can be dissociated into two α subunits and a β_2 dimer (Figure 24.16). The α subunit catalyzes the formation of indole from indole-3-glycerol phosphate, whereas each β subunit has a PLP-containing active site that catalyzes the condensation of indole and serine to form tryptophan. Serine forms a Schiff base with this PLP, which is then dehydrated to give the *Schiff base of aminoacrylate*. This reactive intermediate is attacked by indole to give tryptophan. The overall three-dimensional structure of this enzyme is distinct from that of aspartate aminotransferase and the other PLP enzymes already discussed.

α subunit

PLP

β subunit

Figure 24.16 Structure of tryptophan synthase. The structure of the complex formed by one α subunit and one β subunit. *Notice* that pyridoxal phosphate (PLP) is bound deeply inside the β subunit, a considerable distance from the α subunit. [Drawn from 1BKS.pdb.]

The synthesis of tryptophan poses a challenge. Indole, a hydrophobic molecule, readily traverses membranes and would be lost from the cell if it were allowed to diffuse away from the enzyme. This problem is solved in an ingenious way. A 25-Å-long channel connects the active site of the α subunit with that of the adjacent β subunit in the $\alpha_2\beta_2$ tetramer (Figure 24.17). Thus, indole can diffuse from one active site to the other without being released into bulk solvent. Isotopic-labeling experiments showed that indole formed by the α subunit does not leave the enzyme when serine is present. Furthermore, the two partial reactions are coordinated. Indole is not formed by the α subunit until the highly reactive aminoacrylate is ready and waiting in the β subunit. We see here a clear-cut example of *substrate channeling* in catalysis by a multienzyme complex. Channeling substantially increases

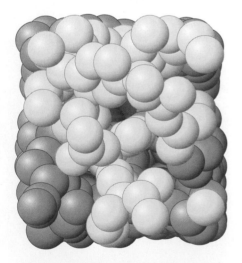

Figure 24.17 Substrate channeling. A 25-Å tunnel runs from the active site of the α subunit of tryptophan synthase (yellow) to the PLP cofactor (red) in the active site of the β subunit (blue).

the catalytic rate. Furthermore, a deleterious side reaction—in this case, the potential loss of an intermediate—is prevented. We shall encounter other examples of substrate channeling in Chapter 25.

24.3 Feedback Inhibition Regulates Amino Acid Biosynthesis

The rate of synthesis of amino acids depends mainly on the *amounts* of the biosynthetic enzymes and on their *activities*. We now consider the control of enzymatic activity. The regulation of enzyme synthesis will be discussed in Chapter 31.

In a biosynthetic pathway, the first irreversible reaction, called the *committed step*, is usually an important regulatory site. *The final product of the pathway (Z) often inhibits the enzyme that catalyzes the committed step (A → B).*

This kind of control is essential for the conservation of building blocks and metabolic energy. Consider the biosynthesis of serine (p. 688). The committed step in this pathway is the oxidation of 3-phosphoglycerate, catalyzed by the enzyme *3-phosphoglycerate dehydrogenase*. The *E. coli* enzyme is a tetramer of four identical subunits, each comprising a catalytic domain and a serine-binding regulatory domain (Figure 24.18). The binding of serine to a regulatory site reduces the value of V_{max} for the enzyme; an enzyme bound to four molecules of serine is essentially inactive. Thus, if serine is abundant in the cell, the enzyme activity is inhibited, and so 3-phosphoglycerate, a key building block that can be used for other processes, is not wasted.

Branched Pathways Require Sophisticated Regulation

The regulation of branched pathways is more complicated because the concentration of two products must be accounted for. In fact, several intricate feedback mechanisms have been found in branched biosynthetic pathways.

Feedback Inhibition and Activation. Two pathways with a common initial step may each be inhibited by its own product and activated by the product of the other pathway. Consider, for example, the biosynthesis of the amino acids valine, leucine, and isoleucine. A common intermediate, hydroxyethyl thiamine pyrophosphate (hydroxyethyl-TPP; p. 478), initiates the pathways leading to all three of these amino acids. Hydroxyethyl-TPP reacts with α-ketobutyrate in the initial step for the synthesis of isoleucine. Alternatively, hydroxyethyl-TPP reacts with pyruvate in the committed step for the pathways leading to valine and leucine. Thus, the relative concentrations of α-ketobutyrate and pyruvate determine how much isoleucine is produced compared with valine and leucine. *Threonine deaminase*, the PLP enzyme that catalyzes the formation of α-ketobutyrate, is allosterically inhibited by isoleucine (Figure 24.19). This enzyme is also

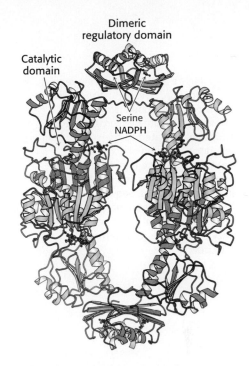

Figure 24.18 Structure of 3-phosphoglycerate dehydrogenase. This enzyme, which catalyzes the committed step in the serine biosynthetic pathway, is inhibited by serine. *Notice* the two serine-binding dimeric regulatory domains—one at the top and the other at the bottom of the structure. [Drawn from 1PSD.pdb.]

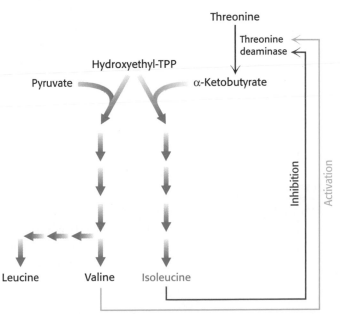

Figure 24.19 Regulation of threonine deaminase. Threonine is converted into α-ketobutyrate in the committed step, leading to the synthesis of isoleucine. The enzyme that catalyzes this step, threonine deaminase, is inhibited by isoleucine and activated by valine, the product of a parallel pathway.

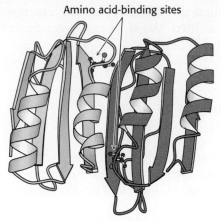

Amino acid-binding sites

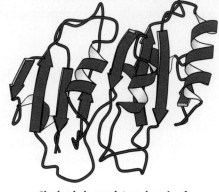

Dimeric regulatory domain of phosphoglycerate dehydrogenase

Single-chain regulatory domain of threonine deaminase

Figure 24.20 A recurring regulatory domain. The regulatory domain formed by two subunits of 3-phosphoglycerate dehydrogenase is structurally related to the single-chain regulatory domain of threonine deaminase. *Notice* that both structures have four α helices and eight β strands in similar locations. Sequence analyses have revealed this amino acid-binding regulatory domain to be present in other enzymes as well. [Drawn from 1PSD and 1TDJ.pdb.]

allosterically activated by valine. Thus, this enzyme is inhibited by the end product of the pathway that it initiates and is activated by the end product of a competitive pathway. This mechanism balances the amounts of different amino acids that are synthesized.

The regulatory domain in threonine deaminase is very similar in structure to the regulatory domain in 3-phosphoglycerate dehydrogenase (Figure 24.20). In the latter enzyme, regulatory domains of two subunits interact to form a dimeric serine-binding regulatory unit, and so the tetrameric enzyme contains two such regulatory units. Each unit is capable of binding two serine molecules. In threonine deaminase, the two regulatory domains are fused into a single unit with two differentiated amino acid-binding sites, one for isoleucine and the other for valine. Sequence analysis shows that similar regulatory domains are present in other amino acid biosynthetic enzymes. *The similarities suggest that feedback-inhibition processes may have evolved by the linkage of specific regulatory domains to the catalytic domains of biosynthetic enzymes.*

Enzyme Multiplicity. The committed step can be catalyzed by two or more enzymes with different regulatory properties. For example, the phosphorylation of aspartate is the committed step in the biosynthesis of threonine, methionine, and lysine. Three distinct aspartokinases catalyze this reaction in *E. coli* (Figure 24.21). The catalytic domains of these enzymes show approximately 30% sequence identity. Although the mechanisms of catalysis are essentially identical, their activities are regulated differently: one enzyme is not subject to feedback inhibition, another is inhibited by threonine, and the third is inhibited by lysine. Thus, sophisticated regulation can also evolve by duplication of the genes encoding the biosynthetic enzymes.

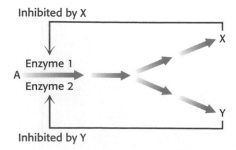

Figure 24.21 Domain structures of three aspartokinases. Each catalyzes the committed step in the biosynthesis of a different amino acid: (top) methionine, (middle) threonine, and (bottom) lysine. They have a catalytic domain in common but differ in their regulatory domains.

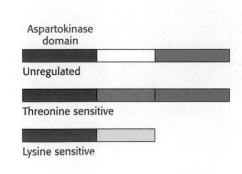

Aspartokinase domain

Unregulated

Threonine sensitive

Lysine sensitive

Cumulative Feedback Inhibition.

Cumulative Feedback Inhibition. A common step is partly inhibited by each of the final products, acting independently. The regulation of glutamine synthetase in *E. coli* is a striking example of cumulative feedback inhibition. Recall that glutamine is synthesized from glutamate, NH_4^+, and ATP. *Glutamine synthetase* consists of 12 identical 50-kd subunits arranged in two hexagonal rings that face each other. Earl Stadtman showed that this enzyme regulates the flow of nitrogen and hence plays a key role in controlling bacterial metabolism. The amide group of glutamine is a source of nitrogen in the biosyntheses of a variety of compounds, such as tryptophan, histidine, carbamoyl phosphate, glucosamine 6-phosphate, cytidine triphosphate, and adenosine monophosphate. Glutamine synthetase is cumulatively inhibited by each of these final products of glutamine metabolism, as well as by alanine and glycine. *In cumulative inhibition, each inhibitor can reduce the activity of the enzyme, even when other inhibitors are bound at saturating levels.* The enzymatic activity of glutamine synthetase is switched off almost completely when all final products are bound to the enzyme.

An Enzymatic Cascade Modulates the Activity of Glutamine Synthetase

The activity of glutamine synthetase is also controlled by *reversible covalent modification*—the attachment of an *AMP unit* by a phosphodiester bond to the hydroxyl group of a specific tyrosine residue in each subunit (Figure 24.22). *This adenylylated enzyme is less active and more susceptible to cumulative feedback inhibition than is the de-adenylylated form.* The covalently attached AMP unit is removed from the adenylylated enzyme by phosphorolysis. The attachment of an AMP unit is the final step in an enzymatic cascade that is initiated several steps back by reactants and immediate products in glutamine synthesis.

The adenylylation and phosphorolysis reactions are catalyzed by the same enzyme, *adenylyl transferase*. Sequence analysis indicates that this adenylyl transferase comprises two homologous halves, suggesting that one half catalyzes the adenylation reaction and the other half the phospholytic de-adenylylation reaction. What determines whether an AMP unit is added or removed? The specificity of adenylyl transferase is controlled by a *regulatory protein* (designated P or P_{II}), a trimeric protein that can exist in two forms, P_A and P_D. The complex of P_A and adenylyl transferase catalyzes the attachment of an AMP unit to glutamine synthetase, which reduces its activity. Conversely, the complex of P_D and adenylyl transferase removes AMP from the adenylylated enzyme.

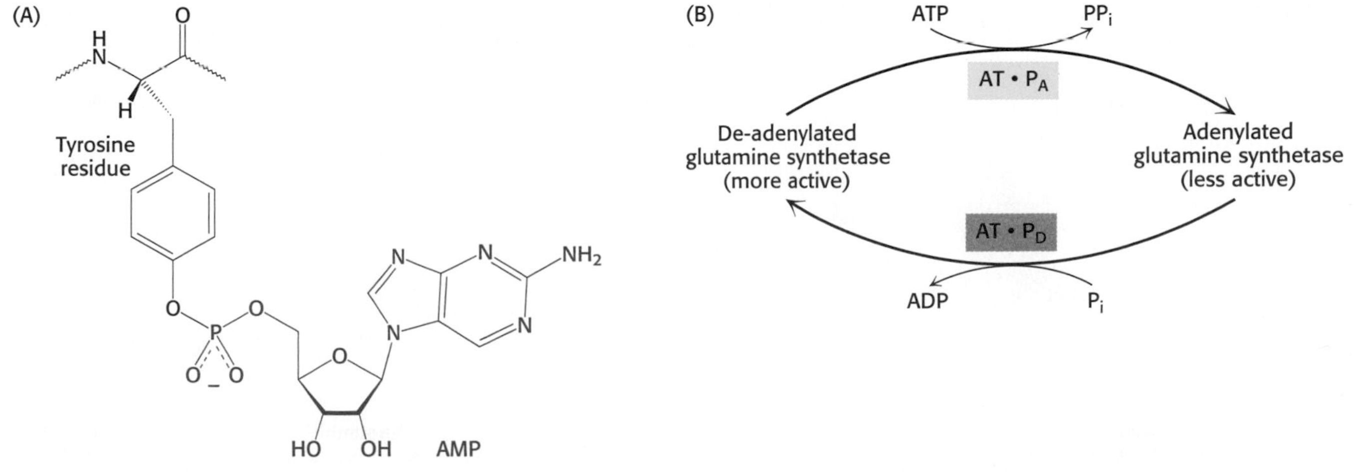

Figure 24.22 Regulation by adenylation. (A) A specific tyrosine residue in each subunit in glutamine synthetase is modified by adenylylation. (B) Adenylylation of tyrosine is catalyzed by a complex of adenylyl transferase (AT) and one form of a regulatory protein (P_A). The same enzyme catalyzes de-adenylylation when it is complexed with the other form (P_D) of the regulatory protein.

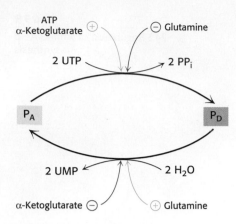

Figure 24.23 A higher level in the regulatory cascade of glutamine synthetase. P_A and P_D, the regulatory proteins that control the specificity of adenylyl transferase, are interconvertible. P_A is converted into P_D by uridylylation, which is reversed by hydrolysis. The enzymes catalyzing these reactions are regulated by the concentrations of metabolic intermediates.

This brings us to another level of reversible covalent modification. P_A is converted into P_D by the attachment of uridine monophosphate to a specific tyrosine residue (Figure 24.23). This reaction, which is catalyzed by *uridylyl transferase*, is stimulated by ATP and α-ketoglutarate, whereas it is inhibited by glutamine. In turn, the UMP units on P_D are removed by hydrolysis, a reaction promoted by glutamine and inhibited by α-ketoglutarate. These opposing catalytic activities are present on a single polypeptide chain, homologous to adenylyl transferase, and are controlled so that the enzyme does not simultaneously catalyze uridylylation and hydrolysis.

Why is an enzymatic cascade used to regulate glutamine synthetase? One advantage of a cascade is that it *amplifies signals*, as in blood clotting and the control of glycogen metabolism. Another advantage is that the *potential for allosteric control is markedly increased when each enzyme in the cascade is an independent target for regulation.* The integration of nitrogen metabolism in a cell requires that a large number of input signals be detected and processed. In addition, the regulatory protein P also participates in regulating the transcription of genes for glutamine synthetase and other enzymes taking part in nitrogen metabolism. The evolution of a cascade provided many more regulatory sites and made possible a finer tuning of the flow of nitrogen in the cell.

24.4 Amino Acids Are Precursors of Many Biomolecules

In addition to being the building blocks of proteins and peptides, amino acids serve as precursors of many kinds of small molecules that have important and diverse biological roles. Let us briefly survey some of the biomolecules that are derived from amino acids (Figure 24.24).

Purines and *pyrimidines* are derived largely from amino acids. The biosynthesis of these precursors of DNA, RNA, and numerous coenzymes will be discussed in detail in Chapter 25. The reactive terminus of *sphingosine,* an intermediate in the synthesis of sphingolipids, comes from serine. *Histamine,* a potent vasodilator, is derived from histidine by decarboxylation. Tyrosine is a precursor of the hormones *thyroxine* (tetraiodothyronine) and *epinephrine* and of *melanin,* a complex polymeric pigment. The

Figure 24.24 Selected biomolecules derived from amino acids. The atoms contributed by amino acids are shown in blue.

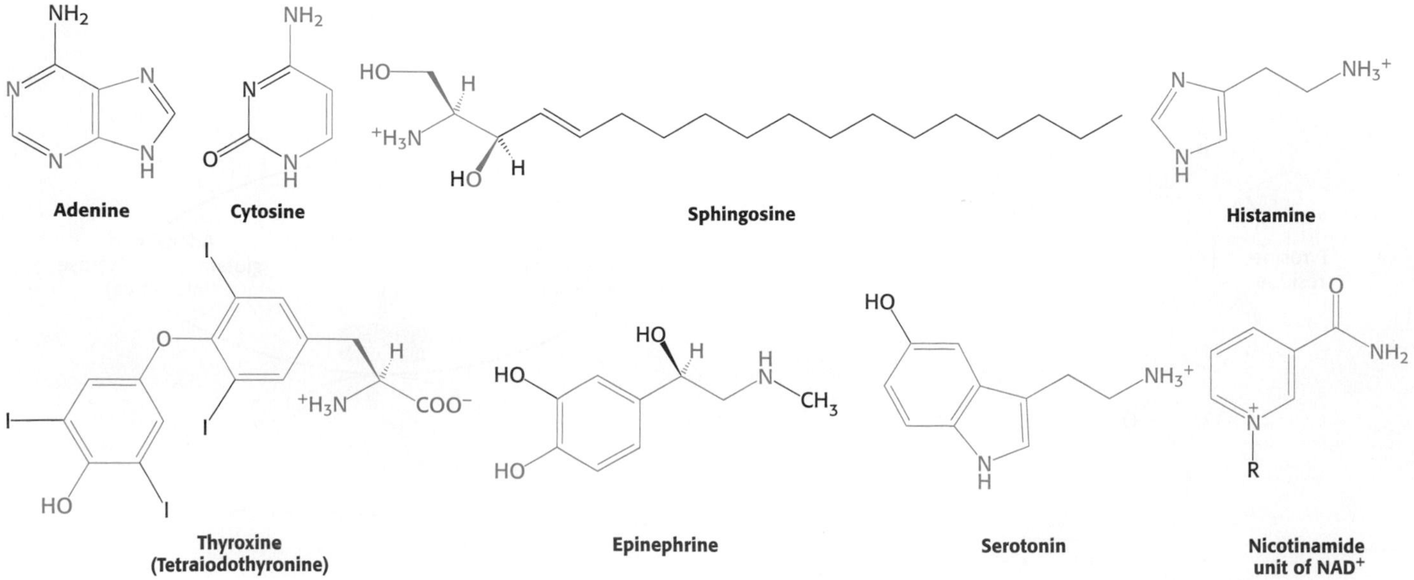

Adenine   **Cytosine**   **Sphingosine**   **Histamine**

Thyroxine (Tetraiodothyronine)   **Epinephrine**   **Serotonin**   **Nicotinamide unit of NAD⁺**

neurotransmitter *serotonin* (5-hydroxytryptamine) and the *nicotinamide ring* of NAD$^+$ are synthesized from tryptophan. Let us now consider in more detail three particularly important biochemicals derived from amino acids.

Glutathione, a Gamma-Glutamyl Peptide, Serves As a Sulfhydryl Buffer and an Antioxidant

Glutathione, a tripeptide containing a sulfhydryl group, is a highly distinctive amino acid derivative with several important roles (Figure 24.25).

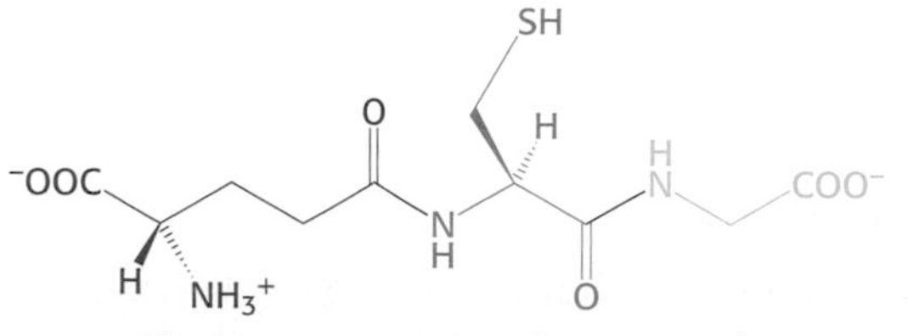

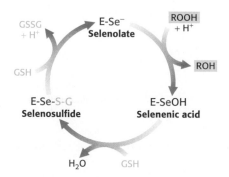

Figure 24.25 Glutathione. This tripeptide consists of a cysteine residue flanked by a glycine residue and a glutamate residue that is linked to cysteine by an isopeptide bond between glutamate's side-chain carboxylate group and cysteine's amino group.

For example, glutathione, present at high levels (~5 mM) in animal cells, protects red blood cells from oxidative damage by serving as a sulfhydryl buffer (p. 586). It cycles between a reduced thiol form (GSH) and an oxidized form (GSSG) in which two tripeptides are linked by a disulfide bond.

$$2\,GSH + RO{-}OH \rightleftharpoons GSSG + H_2O + ROH$$

GSSG is reduced to GSH by *glutathione reductase,* a flavoprotein that uses NADPH as the electron source. The ratio of GSH to GSSG in most cells is greater than 500. *Glutathione plays a key role in detoxification by reacting with hydrogen peroxide and organic peroxides, the harmful by-products of aerobic life.*

Glutathione peroxidase, the enzyme catalyzing this reaction, is remarkable in having a modified amino acid containing a *selenium* (Se) atom (Figure 24.26). Specifically, its active site contains the selenium analog of cysteine, in which selenium has replaced sulfur. The selenolate (E-Se$^-$) form of this residue reduces the peroxide substrate to an alcohol and is in turn oxidized to selenenic acid (E-SeOH). Glutathione then comes into action by forming a selenosulfide adduct (E-Se-S-G). A second molecule of glutathione then regenerates the active form of the enzyme by attacking the selenosulfide to form oxidized glutathione (Figure 24.27).

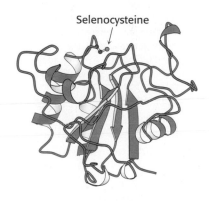

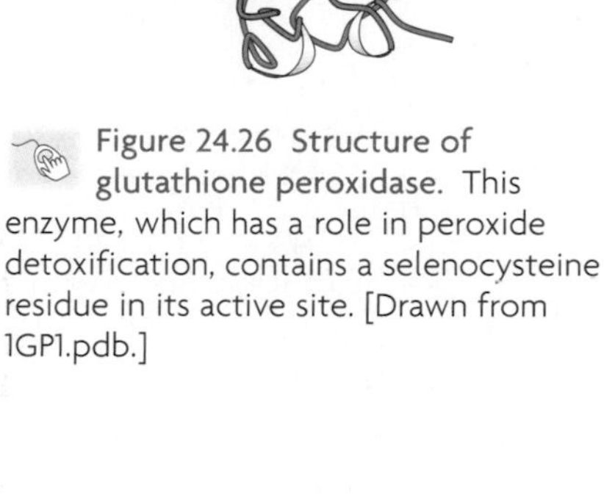

Selenocysteine

Figure 24.26 Structure of glutathione peroxidase. This enzyme, which has a role in peroxide detoxification, contains a selenocysteine residue in its active site. [Drawn from 1GP1.pdb.]

Figure 24.27 Catalytic cycle of glutathione peroxidase. [After O. Epp, R. Ladenstein, and A. Wendel. *Eur. J. Biochem.* 133(1983):51–69.]

Figure 24.28 **Formation of nitric oxide.** NO is generated by the oxidation of arginine.

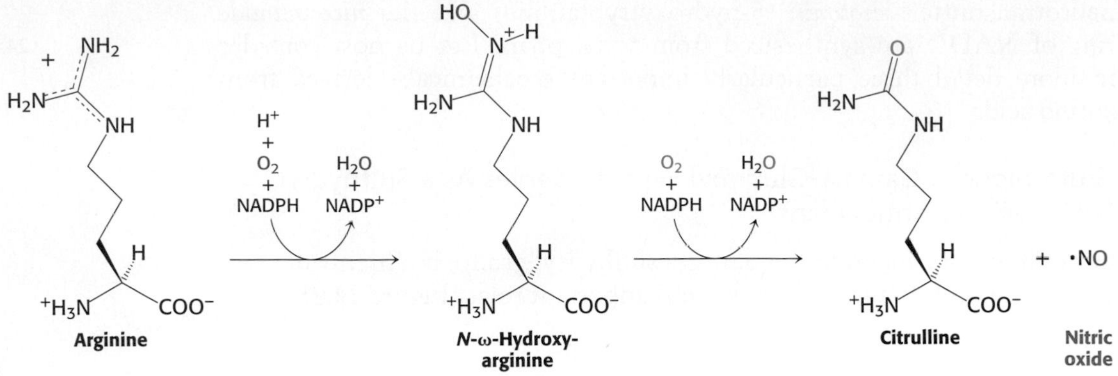

^{15}N labeling: A pioneer's account

"Myself as a Guinea Pig

. . . in 1944, I undertook, together with David Rittenberg, an investigation on the turnover of blood proteins of man. To this end I synthesized 66 g of glycine labeled with 35 percent ^{15}N at a cost of \$1000 for the ^{15}N. On 12 February 1945, I started the ingestion of the labeled glycine. Since we did not know the effect of relatively large doses of the stable isotope of nitrogen and since we believed that the maximum incorporation into the proteins could be achieved by the administration of glycine in some continual manner, I ingested 1 g samples of glycine at hourly intervals for the next 66 hours. . . . At stated intervals, blood was withdrawn and after proper preparation the ^{15}N concentrations of different blood proteins were determined."

—David Shemin
Bioessays 10(1989):30

Nitric Oxide, a Short-Lived Signal Molecule, Is Formed from Arginine

Nitric oxide (NO) is an important messenger in many vertebrate signal-transduction processes. For instance, NO stimulates mitochondrial biogenesis. This free-radical gas is produced endogenously from *arginine* in a complex reaction that is catalyzed by *nitric oxide synthase.* NADPH and O_2 are required for the synthesis of nitric oxide (Figure 24.28). Nitric oxide acts by binding to and activating soluble guanylate cyclase, an important enzyme in signal transduction (Section 32.3). This enzyme is homologous to adenylate cyclase but includes a heme-containing domain that binds NO.

Porphyrins Are Synthesized from Glycine and Succinyl Coenzyme A

The participation of an amino acid in the biosynthesis of the porphyrin rings of hemes and chlorophylls was first revealed by isotope-labeling experiments carried out by David Shemin and his colleagues. In 1945, they showed that the nitrogen atoms of heme were labeled after the feeding of [^{15}N]glycine to human subjects (of whom Shemin was the first), whereas the ingestion of [^{15}N]glutamate resulted in very little labeling.

Using ^{14}C, which had just become available, they discovered that 8 of the carbon atoms of heme in nucleated duck erythrocytes are derived from the α-carbon atom of glycine and none from the carboxyl carbon atom. Subsequent studies demonstrated that the other 26 carbon atoms of heme can arise from acetate. Moreover, the ^{14}C in methyl-labeled acetate emerged in 24 of these carbon atoms, whereas the ^{14}C in carboxyl-labeled acetate appeared only in the other 2 (Figure 24.29).

This highly distinctive labeling pattern led Shemin to propose that a heme precursor is formed by the condensation of glycine with an activated succinyl compound. In fact, *the first step in the biosynthesis of porphyrins in mammals is the condensation of glycine and succinyl CoA to form δ-aminolevulinate.*

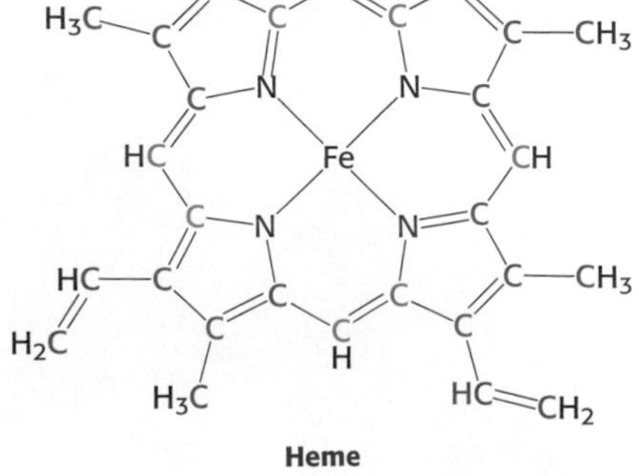

Figure 24.29 **Heme labeling.** The origins of atoms in heme revealed by the results of isotopic labeling studies.

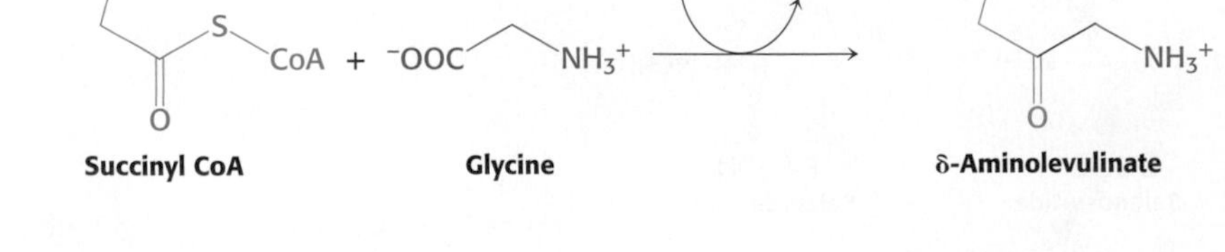

This reaction is catalyzed by *δ-aminolevulinate synthase,* a PLP enzyme present in mitochondria.

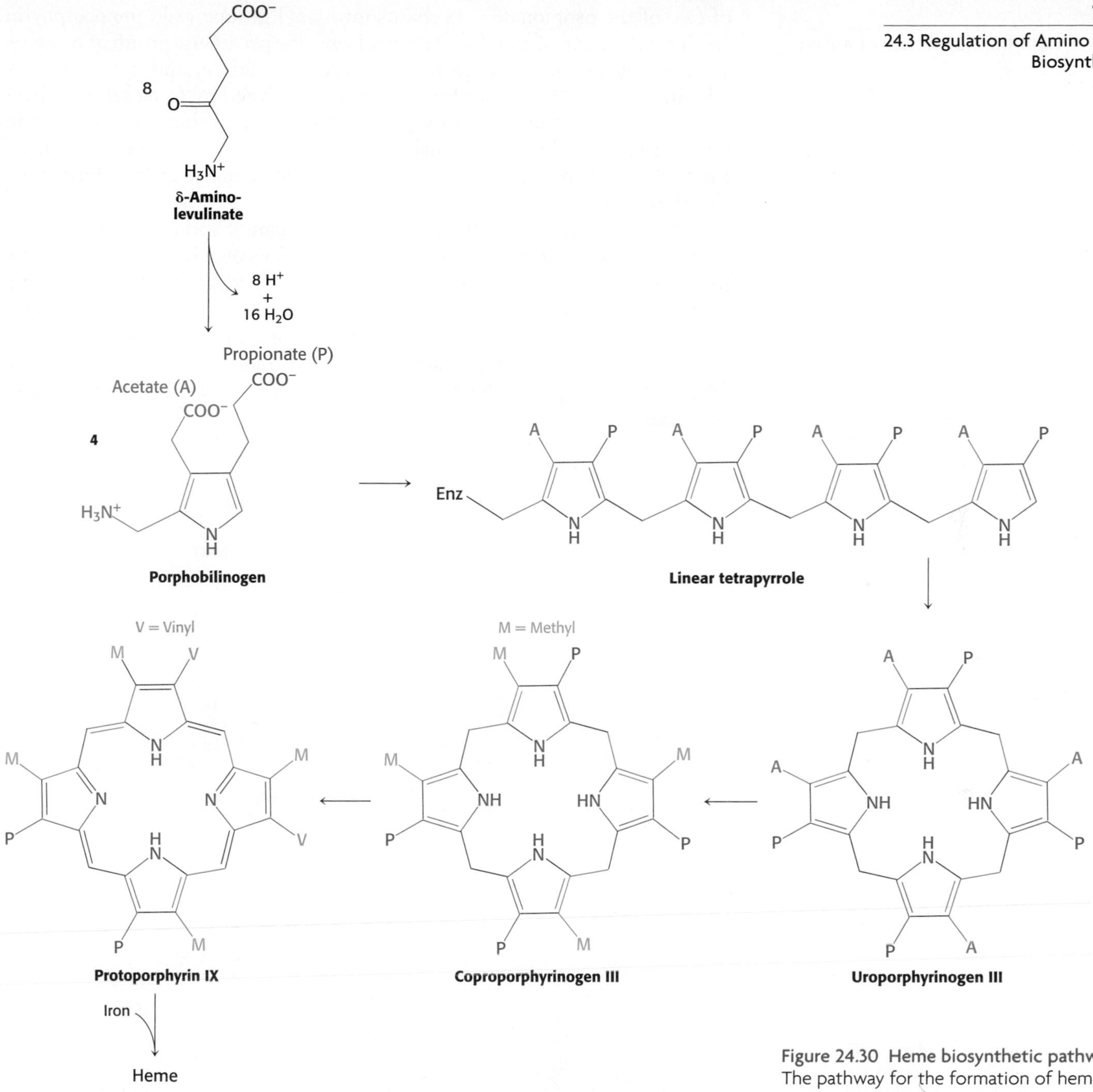

Figure 24.30 Heme biosynthetic pathway. The pathway for the formation of heme starts with eight molecules of δ-aminolevulinate.

Two molecules of δ-aminolevulinate condense to form *porphobilinogen*, the next intermediate. Four molecules of porphobilinogen then condense head to tail to form a linear *tetrapyrrole* in a reaction catalyzed by *porphobilinogen deaminase*. The enzyme-bound linear tetrapyrrole then cyclizes to form uroporphyrinogen III, which has an asymmetric arrangement of side chains. This reaction requires a *cosynthase*. In the presence of synthase alone, uroporphyrinogen I, the nonphysiological symmetric isomer, is produced. Uroporphyrinogen III is also a key intermediate in the synthesis of vitamin B_{12} by bacteria and that of chlorophyll by bacteria and plants (Figure 24.30).

The porphyrin skeleton is now formed. Subsequent reactions alter the side chains and the degree of saturation of the porphyrin ring (see Figure 24.29). *Coproporphyrinogen III* is formed by the decarboxylation of the acetate side chains. The desaturation of the porphyrin ring and the conversion of two of the propionate side chains into vinyl groups yield *protoporphyrin IX*. The chelation of iron finally gives *heme*, the prosthetic group of proteins such as myoglobin, hemoglobin, catalase, peroxidase, and cytochrome *c*. The insertion of the *ferrous* form of iron is catalyzed by *ferrochelatase*. Iron is transported in the plasma by *transferrin*, a protein that binds two ferric ions, and is stored in tissues inside molecules of *ferritin*. The large internal cavity (~80 Å in diameter) of ferritin can hold as many as 4500 ferric ions (Section 31.4).

The normal human erythrocyte has a life span of about 120 days, as was first shown by the time course of ^{15}N in Shemin's own hemoglobin after he ingested ^{15}N-labeled glycine. The first step in the degradation of the heme group is the cleavage of its α-methine bridge to form the green pigment *biliverdin*, a linear tetrapyrrole. The central methine bridge of biliverdin is then reduced by *biliverdin reductase* to form *bilirubin*, a red pigment (Figure 24.31). The changing color of a bruise is a highly graphic indicator of these degradative reactions.

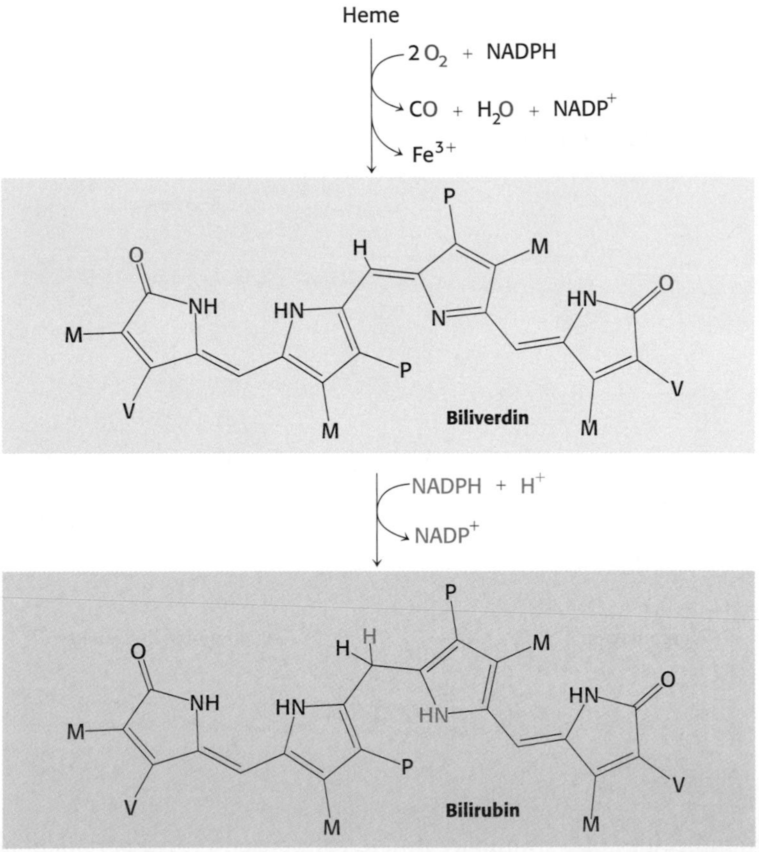

Figure 24.31 Heme degradation. The formation of the heme-degradation products biliverdin and bilirubin is responsible for the color of bruises. Abbreviations: M, methyl; V, vinyl.

Porphyrins Accumulate in Some Inherited Disorders of Porphyrin Metabolism

Porphyrias are inherited or acquired disorders caused by a deficiency of enzymes in the heme biosynthetic pathway. Porphyrin is synthesized in both the erythroblasts and the liver, and either one may be the site of a disorder. *Congenital erythropoietic porphyria*, for example, prematurely destroys eythrocytes. This disease results from insufficient cosynthase. In this porphyria, the synthesis of the required amount of uroporphyrinogen III is accompanied by the formation of very large quantities of uroporphyrinogen I, the useless symmetric isomer. Uroporphyrin I, coproporphyrin I, and other symmetric derivatives also accumulate. The urine of patients having this disease is red because of the excretion of large amounts of uroporphyrin I. Their teeth exhibit a strong red fluorescence under ultraviolet light because of the deposition of porphyrins. Furthermore, their *skin is usually very sensitive to light* because photoexcited porphyrins are quite reactive. *Acute intermittent porphyria* is the most prevalent of the porphyrias affecting the liver. This porphyria is characterized by the overproduction of porphobilinogen and δ-aminolevulinate, which results in severe abdominal pain and neurological dysfunction. The "madness" of George III, King of England during the American Revolution, is believed to have been due to this porphyria.

Summary

24.1 Nitrogen Fixation: Microorganisms Use ATP and a Powerful Reductant to Reduce Atmospheric Nitrogen to Ammonia

Microorganisms use ATP and reduced ferredoxin, a powerful reductant, to reduce N_2 to NH_3. An iron–molybdenum cluster in nitrogenase deftly catalyzes the fixation of N_2, a very inert molecule. Higher organisms consume the fixed nitrogen to synthesize amino acids, nucleotides, and other nitrogen-containing biomolecules. The major points of entry of NH_4^+ into metabolism are glutamine or glutamate.

24.2 Amino Acids Are Made from Intermediates of the Citric Acid Cycle and Other Major Pathways

Human beings can synthesize 11 of the basic set of 20 amino acids. These amino acids are called nonessential, in contrast with the essential amino acids, which must be supplied in the diet. The pathways for the synthesis of nonessential amino acids are quite simple. Glutamate dehydrogenase catalyzes the reductive amination of α-ketoglutarate to glutamate. A transamination reaction takes place in the synthesis of most amino acids. At this step, the chirality of the amino acid is established. Alanine and aspartate are synthesized by the transamination of pyruvate and oxaloacetate, respectively. Glutamine is synthesized from NH_4^+ and glutamate, and asparagine is synthesized similarly. Proline and arginine are derived from glutamate. Serine, formed from 3-phosphoglycerate, is the precursor of glycine and cysteine. Tyrosine is synthesized by the hydroxylation of phenylalanine, an essential amino acid. The pathways for the biosynthesis of essential amino acids are much more complex than those for the nonessential ones.

Tetrahydrofolate, a carrier of activated one-carbon units, plays an important role in the metabolism of amino acids and nucleotides. This coenzyme carries one-carbon units at three oxidation states, which are interconvertible: most reduced—methyl; intermediate—methylene; and most oxidized—formyl, formimino, and methenyl. The major donor of activated methyl groups is S-adenosylmethionine, which is synthesized by the transfer of an adenosyl group from ATP to the sulfur atom of methionine. S-Adenosylhomocysteine is formed when the activated methyl group is transferred to an acceptor. It is hydrolyzed to adenosine and homocysteine, and the latter is then methylated to methionine to complete the activated methyl cycle.

24.3 Feedback Inhibition Regulates Amino Acid Biosynthesis

Most of the pathways of amino acid biosynthesis are regulated by feedback inhibition, in which the committed step is allosterically inhibited by the final product. The regulation of branched pathways requires extensive interaction among the branches that includes both negative and positive regulation. The regulation of glutamine synthetase in *E. coli* is a striking demonstration of cumulative feedback inhibition and of control by a cascade of reversible covalent modifications.

24.4 Amino Acids Are Precursors of Many Biomolecules

Amino acids are precursors of a variety of biomolecules. Glutathione (γ-Glu-Cys-Gly) serves as a sulfhydryl buffer and detoxifying agent. Glutathione peroxidase, a selenoenzyme, catalyzes the reduction of hydrogen peroxide and organic peroxides by glutathione. Nitric oxide, a short-lived messenger, is formed from arginine. Porphyrins are synthesized from glycine and succinyl CoA, which condense to give δ-aminolevulinate. Two molecules of this intermediate become linked to form porphobilinogen. Four molecules of porphobilinogen combine to form a linear tetrapyrrole, which cyclizes to uroporphyrinogen III. Oxidation and side-chain modifications lead to the synthesis of protoporphyrin IX, which acquires an iron atom to form heme.

Key Terms

nitrogen fixation (p. 680)

nitrogenase complex (p. 681)

essential amino acids (p. 685)

nonessential amino acids (p. 685)

pyridoxal phosphate (p. 686)

tetrahydrofolate (p. 689)

S-adenosylmethionine (SAM) (p. 691)

activated methyl cycle (p. 692)

substrate channeling (p. 696)

committed step (p. 697)

enzyme multiplicity (p. 698)

cumulative feedback inhibition (p. 699)

Selected Readings

Where to Start

Kim, J., and Rees, D. C. 1989. Nitrogenase and biological nitrogen fixation. *Biochemistry* 33:389–397.

Christen, P., Jaussi, R., Juretic, N., Mehta, P. K., Hale, T. I., and Ziak, M. 1990. Evolutionary and biosynthetic aspects of aspartate aminotransferase isoenzymes and other aminotransferases. *Ann. N. Y. Acad. Sci.* 585:331–338.

Schneider, G., Kack, H., and Lindqvist, Y. 2000. The manifold of vitamin B6 dependent enzymes. *Structure Fold Des.* 8:R1–R6.

Rhee, S. G., Chock, P. B., and Stadtman, E. R. 1989. Regulation of *Escherichia coli* glutamine synthetase. *Adv. Enzymol. Mol. Biol.* 62:37–92.

Shemin, D. 1989. An illustration of the use of isotopes: The biosynthesis of porphyrins. *Bioessays* 10:30–35.

Books

Bender, D. A. 1985. *Amino Acid Metabolism* (2d ed.). Wiley.

Jordan, P. M. (Ed.). 1991. *Biosynthesis of Tetrapyrroles*. Elsevier.

Scriver, C. R. (Ed.), Sly, W. S. (Ed.), Childs, B., Beaudet, A. L., Valle, D., Kinzler, K. W., and Vogelstein, B. 2000. *The Metabolic Basis of Inherited Disease* (8th ed.). McGraw-Hill.

Meister, A. 1965. *Biochemistry of the Amino Acids* (vols. 1 and 2, 2d ed.). Academic Press.

Blakley, R. L., and Benkovic, S. J. 1989. *Folates and Pterins* (vol. 2). Wiley.

Walsh, C. 1979. *Enzymatic Reaction Mechanisms*. W. H. Freeman and Company.

Nitrogen Fixation

Halbleib, C. M., and Ludden, P. W. 2000. Regulation of biological nitrogen fixation. *J. Nutr.* 130:1081–1084.

Einsle, O., Tezcan, F. A., Andrade, S. L., Schmid, B., Yoshida, M., Howard, J. B., and Rees, D. C. 2002. Nitrogenase MoFe-protein at 1.16 Å resolution: A central ligand in the FeMo-cofactor. *Science* 297:1696–1700.

Benton, P. M., Laryukhin, M., Mayer, S. M., Hoffman, B. M., Dean, D. R., and Seefeldt, L. C. 2003. Localization of a substrate binding site on the FeMo-cofactor in nitrogenase: Trapping propargyl alcohol with an α-70-substituted MoFe protein. *Biochemistry* 42: 9102–9109.

Peters, J. W., Fisher, K., and Dean, D. R. 1995. Nitrogenase structure and function: A biochemical-genetic perspective. *Annu. Rev. Microbiol.* 49:335–366.

Leigh, G. J. 1995. The mechanism of dinitrogen reduction by molybdenum nitrogenases. *Eur. J. Biochem.* 229:14–20.

Georgiadis, M. M., Komiya, H., Chakrabarti, P., Woo, D., Kornuc, J. J., and Rees, D. C. 1992. Crystallographic structure of the nitrogenase iron protein from *Azotobacter vinelandii*. *Science* 257:1653–1659.

Regulation of Amino Acid Biosynthesis

Eisenberg, D., Gill, H. S., Pfluegl, G. M., and Rotstein, S. H. 2000. Structure-function relationships of glutamine synthetases. *Biochim. Biophys. Acta* 1477:122–145.

Purich, D. L. 1998. Advances in the enzymology of glutamine synthesis. *Adv. Enzymol. Relat. Areas Mol. Biol.* 72:9–42.

Yamashita, M. M., Almassy, R. J., Janson, C. A., Cascio, D., and Eisenberg, D. 1989. Refined atomic model of glutamine synthetase at 3.5 Å resolution. *J. Biol. Chem.* 264:17681–17690.

Schuller, D. J., Grant, G. A., and Banaszak, L. J. 1995. The allosteric ligand site in the V_{max}-type cooperative enzyme phosphoglycerate dehydrogenase. *Nat. Struct. Biol.* 2:69–76.

Rhee, S. G., Park, R., Chock, P. B., and Stadtman, E. R. 1978. Allosteric regulation of monocyclic interconvertible enzyme cascade systems: Use of *Escherichia coli* glutamine synthetase as an experimental model. *Proc. Natl. Acad. Sci. U. S. A.* 75:3138–3142.

Wessel, P. M., Graciet, E., Douce, R., and Dumas, R. 2000. Evidence for two distinct effector-binding sites in threonine deaminase by site-directed mutagenesis, kinetic, and binding experiments. *Biochemistry* 39:15136–15143.

James, C. L., and Viola, R. E. 2002. Production and characterization of bifunctional enzymes: Domain swapping to produce new bifunctional enzymes in the aspartate pathway. *Biochemistry* 41: 3720–3725.

Xu, Y., Carr, P. D., Huber, T., Vasudevan, S. G., and Ollis, D. L. 2001. The structure of the PII-ATP complex. *Eur. J. Biochem.* 268: 2028–2037.

Krappmann, S., Lipscomb, W. N., and Braus, G. H. 2000. Coevolution of transcriptional and allosteric regulation at the chorismate metabolic branch point of *Saccharomyces cerevisiae. Proc. Natl. Acad. Sci. U. S. A.* 97:13585–13590.

Aromatic Amino Acid Biosynthesis

Brown, K. A., Carpenter, E. P., Watson, K. A., Coggins, J. R., Hawkins, A. R., Koch, M. H., and Svergun, D. I. 2003. Twists and turns: A tale of two shikimate-pathway enzymes. *Biochem. Soc. Trans.* 31:543–547.

Pan, P., Woehl, E., and Dunn, M. F. 1997. Protein architecture, dynamics and allostery in tryptophan synthase channeling. *Trends Biochem. Sci.* 22:22–27.

Sachpatzidis, A., Dealwis, C., Lubetsky, J. B., Liang, P. H., Anderson, K. S., and Lolis, E. 1999. Crystallographic studies of phosphonate-based α-reaction transition-state analogues complexed to tryptophan synthase. *Biochemistry* 38:12665–12674.

Weyand, M., and Schlichting, I. 1999. Crystal structure of wild-type tryptophan synthase complexed with the natural substrate indole-3-glycerol phosphate. *Biochemistry* 38:16469–16480.

Crawford, I. P. 1989. Evolution of a biosynthetic pathway: The tryptophan paradigm. *Annu. Rev. Microbiol.* 43:567–600.

Carpenter, E. P., Hawkins, A. R., Frost, J. W., and Brown, K. A. 1998. Structure of dehydroquinate synthase reveals an active site capable of multistep catalysis. *Nature* 394:299–302.

Schlichting, I., Yang, X. J., Miles, E. W., Kim, A. Y., and Anderson, K. S. 1994. Structural and kinetic analysis of a channel-impaired mutant of tryptophan synthase. *J. Biol. Chem.* 269:26591–26593.

Glutathione

Edwards, R., Dixon, D. P., and Walbot, V. 2000. Plant glutathione S-transferases: Enzymes with multiple functions in sickness and in health. *Trends Plant Sci.* 5:193–198.

Lu, S. C. 2000. Regulation of glutathione synthesis. *Curr. Top. Cell Regul.* 36:95–116.

Schulz, J. B., Lindenau, J., Seyfried, J., and Dichgans, J. 2000. Glutathione, oxidative stress and neurodegeneration. *Eur. J. Biochem.* 267:4904–4911.

Lu, S. C. 1999. Regulation of hepatic glutathione synthesis: Current concepts and controversies. *FASEB J.* 13:1169–1183.

Salinas, A. E., and Wong, M. G. 1991. Glutathione S-transferases: A review. *Curr. Med. Chem.* 6:279–309.

Ethylene and Nitric Oxide

Nisoli, E., Falcone, S., Tonello, C., Cozzi, V., Palomba, L., Fiorani, M., Pisconti, A., Brunelli, S., Cardile, A., Francolini, M., Cantoni, O., Carruba, M. O., Moncada, S., and Clementi, E. 2004. Mitochondrial biogenesis by NO yields functionally active mitochondria in mammals. *Proc. Natl. Acad. Sci U. S. A.* 101:16507–16512.

Bretscher, L. E., Li, H., Poulos, T. L. and Griffith, O. W. 2003. Structural characterization and kinetics of nitric oxide synthase inhibition by novel N^5-(iminoalkyl)- and N^5-(iminoalkenyl)-ornithines. *J. Biol. Chem.* 278:46789–46797.

Haendeler, J., Zeiher, A. M., and Dimmeler, S. 1999. Nitric oxide and apoptosis. *Vitam. Horm.* 57:49–77.

Capitani, G., Hohenester, E., Feng, L., Storici, P., Kirsch, J. F., and Jansonius, J. N. 1999. Structure of 1-aminocyclopropane-1-carboxylate synthase, a key enzyme in the biosynthesis of the plant hormone ethylene. *J. Mol. Biol.* 294:745–756.

Hobbs, A. J., Higgs, A., and Moncada, S. 1999. Inhibition of nitric oxide synthase as a potential therapeutic target. *Annu. Rev. Pharmacol. Toxicol.* 39:191–220.

Stuehr, D. J. 1999. Mammalian nitric oxide synthases. *Biochim. Biophys. Acta* 1411:217–230.

Chang, C., and Shockey, J. A. 1999. The ethylene-response pathway: Signal perception to gene regulation. *Curr. Opin. Plant Biol.* 2:352–358.

Theologis, A. 1992. One rotten apple spoils the whole bushel: The role of ethylene in fruit ripening. *Cell* 70:181–184.

Biosynthesis of Porphyrins

Kaasik, K. and Lee, C. C. 2004. Reciprocal regulation of haem biosynthesis and the circadian clock in mammals. *Nature* 430:467–471.

Leeper, F. J. 1989. The biosynthesis of porphyrins, chlorophylls, and vitamin B$_{12}$. *Nat. Prod. Rep.* 6:171–199.

Porra, R. J., and Meisch, H.-U. 1984. The biosynthesis of chlorophyll. *Trends Biochem. Sci.* 9:99–104.

Problems

1. *From sugar to amino acid.* Write a balanced equation for the synthesis of alanine from glucose.

2. *From air to blood.* What are the intermediates in the flow of nitrogen from N_2 to heme?

3. *One-carbon transfers.* Which derivative of folate is a reactant in the conversion of (a) glycine into serine? (b) homocysteine into methionine?

4. *Telltale tag.* In the reaction catalyzed by glutamine synthetase, an oxygen atom is transferred from the side chain of glutamate to orthophosphate, as shown by the results of ^{18}O-labeling studies. Account for this finding.

5. *Therapeutic glycine.* Isovaleric acidemia is an inherited disorder of leucine metabolism caused by a deficiency of isovaleryl CoA dehydrogenase. Many infants having this disease die in the first month of life. The administration of large amounts of glycine sometimes leads to marked clinical improvement. Propose a mechanism for the therapeutic action of glycine.

6. *Deprived bacteria.* Blue-green algae (cyanobacteria) form *heterocysts* when deprived of ammonia and nitrate. In this form, the cyanobacteria lack nuclei and are attached to adjacent vegetative cells. Heterocysts have photosystem I activity but are entirely devoid of photosystem II activity. What is their role?

7. *Cysteine and cystine.* Most cytoplasmic proteins lack disulfide bonds, whereas extracellular proteins usually contain them. Why?

8. *Through the looking glass.* Suppose that aspartate aminotransferase were chemically synthesized with the use of D-amino acids only. What products would you expect if this mirror-image enzyme were treated with (a) L-asparate and α-ketoglutarate; (b) D-aspartate and α-ketoglutarate?

9. *To and fro.* The synthesis of δ-aminolevulinate takes place in the mitochondrial matrix, whereas the formation of porphobilinogen takes place in the cytoplasm. Propose a reason for the mitochondrial location of the first step in heme synthesis.

10. *Direct synthesis.* Which of the 20 amino acids can be synthesized directly from a common metabolic intermediate by a transamination reaction?

11. *Lines of communication.* For the following example of a branched pathway, propose a feedback inhibition scheme that would result in the production of equal amounts of Y and Z.

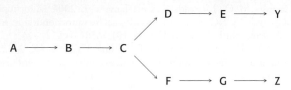

12. *Cumulative feedback inhibition.* Consider the branched pathway in problem 11. The first common step $(A \rightarrow B)$ is partly inhibited by both of the final products, each acting independently of the other. Suppose that a high level of Y alone decreased the rate of the $A \rightarrow B$ step from 100 to 60 s^{-1} and that a high level of Z alone decreased the rate from 100 to 40 s^{-1}. What would the rate be in the presence of high levels of both Y and Z?

Mechanism Problems

13. *Ethylene formation.* Propose a mechanism for the conversion of S-adenosylmethionine into 1-aminocyclopropane-1-carboxylate (ACC) by ACC synthase, a PLP enzyme. What is the other product?

14. *Mirror-image serine.* Brain tissue contains substantial amounts of D-serine which is generated from L-serine by serine racemase, a PLP enzyme. Propose a mechanism for the interconversion of L- and D-serine. What is the equilibrium constant for the reaction L-serine $\rightleftharpoons$ D-serine?

Chapter Integration Problems

15. *Connections.* How might increased synthesis of aspartate and glutamate affect energy production in a cell? How would the cell respond to such an effect?

16. *Protection required.* Suppose that a mutation in bacteria resulted in the diminished activity of methionine adenosyltransferase, the enzyme responsible for the synthesis of SAM from methionine and ATP. Predict how this diminished activity might affect the stability of the mutated bacteria's DNA.

Chapter Integration and Data Interpretation Problem

17. *Light effects.* The adjoining graph shows the concentration of several free amino acids in light- and dark-adapted plants.

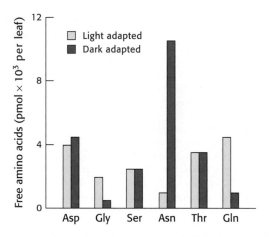

[After B. B. Buchanan, W. Gruissem, and R. L. Jones, *Biochemistry and Molecular Biology of Plants* (American Society of Plant Physiology, 2000), Fig. 8.3, p. 363.]

(a) Of the amino acids shown, which are most affected by light–dark adaptation?

(b) Suggest a plausible biochemical explanation for the difference observed.

(c) White asparagus, a culinary delicacy, is the result of growing asparagus plants in the dark. What chemical might you think enhances the taste of white asparagus?

Chapter 25

Nucleotide Biosynthesis

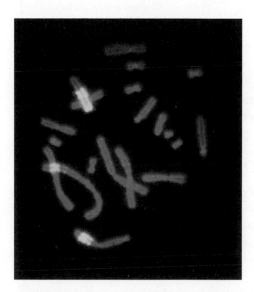

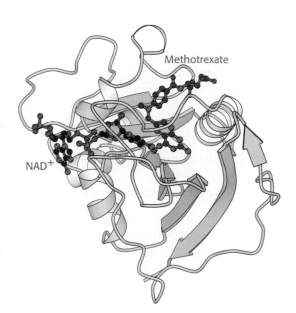

Methotrexate

NAD$^+$

Nucleotides are required for cell growth and replication. A key enzyme for the synthesis of one nucleotide is dihydrofolate reductase (right). Cells grown in the presence of methotrexate, a reductase inhibitor, respond by increasing the number of copies of the reductase gene. The bright yellow regions visible on three of the chromosomes in the fluorescence micrograph (left), which were grown in the presence of methotrexate, contain hundreds of copies of the reductase gene. [(Left) Courtesy of Dr. Barbara Trask and Dr. Joyce Hamlin.]

An ample supply of nucleotides is essential for many life processes. First, nucleotides are the *activated precursors of nucleic acids*. As such, they are necessary for the replication of the genome and the transcription of the genetic information into RNA. Second, an adenine nucleotide, ATP, is *the universal currency of energy*. A guanine nucleotide, GTP, also serves as an energy source for a more select group of biological processes. Third, nucleotide derivatives such as UDP-glucose *participate in biosynthetic processes* such as the formation of glycogen. Fourth, nucleotides are *essential components of signal-transduction pathways*. Cyclic nucleotides such as cyclic AMP and cyclic GMP are second messengers that transmit signals both within and between cells. ATP acts as the donor of phosphoryl groups transferred by protein kinases.

In this chapter, we continue along the path begun in Chapter 24, which described the incorporation of nitrogen into amino acids from inorganic sources such as nitrogen gas. The amino acids glycine and aspartate are the scaffolds on which the ring systems present in nucleotides are assembled. Furthermore, aspartate and the side chain of glutamine serve as sources of NH_2 groups in the formation of nucleotides.

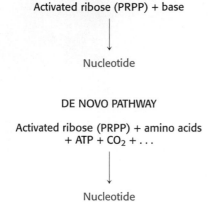

SALVAGE PATHWAY

Activated ribose (PRPP) + base

↓

Nucleotide

DE NOVO PATHWAY

**Activated ribose (PRPP) + amino acids
+ ATP + CO₂ + . . .**

↓

Nucleotide

Figure 25.1 Salvage and de novo pathways. In a salvage pathway, a base is reattached to a ribose, activated in the form of 5-phosphoribosyl-1-pyrophosphate (PRPP). In de novo synthesis, the base itself is synthesized from simpler starting materials, including amino acids. ATP hydrolysis is required for de novo synthesis.

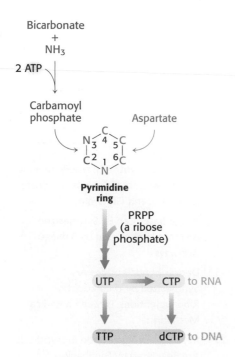

Figure 25.2 De novo pathway for pyrimidine nucleotide synthesis. The C-2 and N-3 atoms in the pyrimidine ring come from carbamoyl phosphate, whereas the other atoms of the ring come from aspartate.

TABLE 25.1 Nomenclature of bases, nucleosides, and nucleotides

RNA

Base	Ribonucleoside	Ribonucleotide (5'-monophosphate)
Adenine (A)	Adenosine	Adenylate (AMP)
Guanine (G)	Guanosine	Guanylate (GMP)
Uracil (U)	Uridine	Uridylate (UMP)
Cytosine (C)	Cytidine	Cytidylate (CMP)

DNA

Base	Deoxyribonucleoside	Deoxyribonucleotide (5'-monophosphate)
Adenine (A)	Deoxyadenosine	Deoxyadenylate (dAMP)
Guanine (G)	Deoxyguanosine	Deoxyguanylate (dGMP)
Thymine (T)	Thymidine	Thymidylate (TMP)
Cytosine (C)	Deoxycytidine	Deoxycytidylate (dCMP)

Nucleotide biosynthetic pathways are tremendously important as intervention points for therapeutic agents. Many of the most widely used drugs in the treatment of cancer block steps in nucleotide biosynthesis, particularly steps in the synthesis of DNA precursors.

Nucleotides Can Be Synthesized by de Novo or Salvage Pathways

The pathways for the biosynthesis of nucleotides fall into two classes: *de novo* pathways and *salvage* pathways (Figure 25.1). In de novo (from scratch) pathways, the nucleotide bases are assembled from simpler compounds. The framework for a *pyrimidine* base is assembled first and then attached to ribose. In contrast, the framework for a *purine* base is synthesized piece by piece directly onto a ribose-based structure. These pathways comprise a small number of elementary reactions that are repeated with variation to generate different nucleotides, as might be expected for pathways that appeared very early in evolution. In salvage pathways, preformed bases are recovered and reconnected to a ribose unit.

Both de novo and salvage pathways lead to the synthesis of *ribo*nucleotides. However, DNA is built from *deoxyribo*nucleotides. Consistent with the notion that RNA preceded DNA in the course of evolution, all deoxyribonucleotides are synthesized from the corresponding ribonucleotides. The deoxyribose sugar is generated by the reduction of ribose within a fully formed nucleotide. Furthermore, the methyl group that distinguishes the thymine of DNA from the uracil of RNA is added at the last step in the pathway.

The nomenclature of nucleotides and their constituent units was presented in Chapter 4. Recall that a *nucleoside* is a purine or pyrimidine base linked to a sugar and that a *nucleotide* is a phosphate ester of a nucleoside. The names of the major bases of RNA and DNA, and of their nucleoside and nucleotide derivatives, are given in Table 25.1.

25.1 In de Novo Synthesis, the Pyrimidine Ring Is Assembled from Bicarbonate, Aspartate, and Glutamine

In de novo synthesis of pyrimidines, the ring is synthesized first and then it is attached to a ribose phosphate to form a *pyrimidine nucleotide* (Figure 25.2). Pyrimidine rings are assembled from bicarbonate, aspartate, and ammonia. Although an ammonia molecule already present in solution can be used, the ammonia is usually produced from the hydrolysis of the side chain of glutamine.

Bicarbonate and Other Oxygenated Carbon Compounds Are Activated by Phosphorylation

The first step in de novo pyrimidine biosynthesis is the synthesis of *carbamoyl phosphate* from bicarbonate and ammonia in a multistep process, requiring the cleavage of two molecules of ATP. This reaction is catalyzed by *carbamoyl phosphate synthetase* (CPS; p. 661). Analysis of the structure of CPS reveals two homologous domains, each of which catalyzes an ATP-dependent step (Figure 25.3).

In the first step, bicarbonate is phosphorylated by ATP to form carboxyphosphate and ADP. Ammonia then reacts with carboxyphosphate to form carbamic acid and inorganic phosphate.

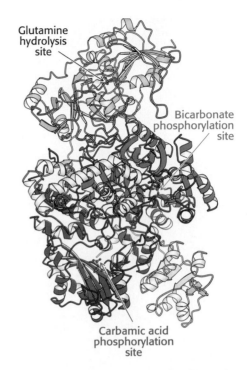

Figure 25.3 **Structure of carbamoyl phosphate synthetase.** *Notice* that the enzyme contains sites for three reactions. This enzyme consists of two chains. The smaller chain (yellow) contains a site for glutamine hydrolysis to generate ammonia. The larger chain includes two ATP-grasp domains (blue and red). In one ATP-grasp domain (blue), bicarbonate is phosphorylated to carboxyphosphate, which then reacts with ammonia to generate carbamic acid. In the other ATP-grasp domain, the carbamic acid is phosphorylated to produce carbamoyl phosphate. [Drawn from 1JDB.pdb.]

Bicarbonate   **Carboxyphosphate**   **Carbamic acid**

The active site for this reaction lies in a domain formed by the amino-terminal third of CPS. This domain forms a structure, called an *ATP-grasp fold,* that surrounds ATP and holds it in an orientation suitable for nucleophilic attack at the γ phosphoryl group. Proteins containing ATP-grasp folds catalyze the formation of carbon–nitrogen bonds through acyl-phosphate intermediates. Such ATP-grasp folds are widely used in nucleotide biosynthesis.

In the second step catalyzed by carbamoyl phosphate synthetase, carbamic acid is phosphorylated by another molecule of ATP to form carbamoyl phosphate.

Carbamic acid   **Carbamoyl phosphate**

This reaction takes place in a second ATP-grasp domain within the enzyme. The active sites leading to carbamic acid formation and carbamoyl phosphate formation are very similar, revealing that this enzyme evolved by a gene duplication event. Indeed, duplication of a gene encoding an ATP-grasp domain followed by specialization was central to the evolution of nucleotide biosynthetic processes (p. 715).

The Side Chain of Glutamine Can Be Hydrolyzed to Generate Ammonia

Glutamine is the primary source of ammonia for carbamoyl phosphate synthetase. In this case, a second polypeptide component of the enzyme hydrolyzes glutamine to form ammonia and glutamate. The active site of the glutamine-hydrolyzing component contains a catalytic dyad comprising a cysteine and a histidine residue. Such a catalytic dyad, reminiscent of the active site of cysteine proteases (p. 251), is conserved in a family of amidotransferases, including CTP synthetase and GMP synthetase.

Intermediates Can Move Between Active Sites by Channeling

Carbamoyl phosphate synthetase contains three different active sites (see Figure 25.3), separated from one another by a total of 80 Å (Figure 25.4).

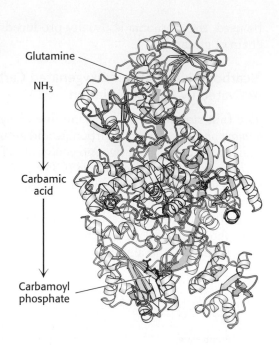

Figure 25.4 Substrate channeling. The three active sites of carbamoyl phosphate synthetase are linked by a channel (yellow) through which intermediates pass. Glutamine enters one active site, and carbamoyl phosphate, which includes the nitrogen atom from the glutamine side chain, leaves another 80 Å away. [Drawn from 1JDB.pdb.]

Intermediates generated at one site move to the next without leaving the enzyme. These intermediates move within the enzyme by means of substrate channeling, similar to the process described for tryptophan synthetase (p. 696). The ammonia generated in the glutamine-hydrolysis active site travels 45 Å through a channel within the enzyme to reach the site at which carboxyphosphate has been generated. The carbamic acid generated at this site diffuses an additional 35 Å through an extension of the channel to reach the site at which carbamoyl phosphate is generated. This channeling serves two roles: (1) intermediates generated at one active site are captured with no loss caused by diffusion and (2) labile intermediates, such as carboxyphosphate and carbamic acid (which decompose in less than 1 s at pH 7), are protected from hydrolysis. We will see additional examples of substrate channeling later in this chapter.

Orotate Acquires a Ribose Ring from PRPP to Form a Pyrimidine Nucleotide and Is Converted into Uridylate

Carbamoyl phosphate reacts with aspartate to form carbamoylaspartate in a reaction catalyzed by *aspartate transcarbamoylase* (Section 10.1). Carbamoylaspartate then cyclizes to form dihydroorotate, which is then oxidized by NAD^+ to form orotate.

Carbamoyl phosphate → (Aspartate, P_i) → **Carbamoylaspartate** → (H^+, H_2O) → **Dihydroorotate** → (NAD^+, NADH + H^+) → **Orotate**

At this stage, orotate couples to ribose, in the form of *5-phosphoribosyl-1-pyrophosphate* (PRPP), a form of ribose activated to accept nucleotide bases. PRPP is synthesized from ribose-5-phosphate, formed by the pentose phosphate pathway, by the addition of pyrophosphate from ATP. Orotate reacts with PRPP to form *orotidylate*, a pyrimidine nucleotide. This reaction is driven by the hydrolysis of pyrophosphate. The enzyme that catalyzes this addition, *pyrimidine phosphoribosyltransferase*, is homologous to a number of other phosphoribosyltransferases that add different groups to PRPP to form the other nucleotides. Orotidylate is then decarboxylated to form *uridylate* (UMP), a major pyrimidine nucleotide that is a precursor to RNA. This reaction is catalyzed by *orotidylate decarboxylase*.

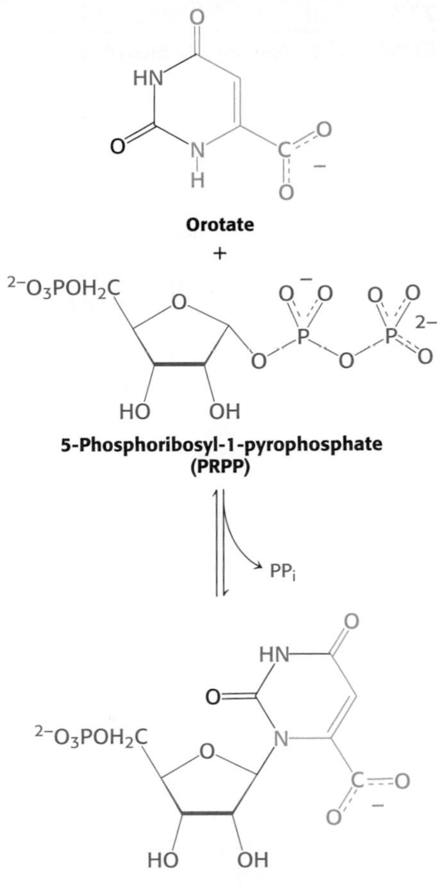

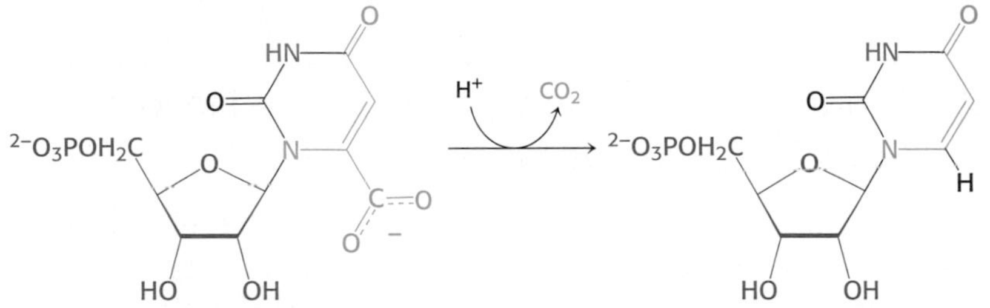

Orotidylate → **Uridylate**

Orotidylate decarboxylase is one of the most proficient enzymes known. In its absence, decarboxylation is extremely slow and is estimated to take place once every 78 million years; with the enzyme present, it takes place approximately once per second, a rate enhancement of 10^{17}-fold.

Nucleotide Mono-, Di-, and Triphosphates Are Interconvertible

How is the other major pyrimidine ribonucleotide, cytidine, formed? It is synthesized from the uracil base of UMP, but the synthesis can take place only after UMP has been converted into UTP. Recall that the diphosphates and triphosphates are the active forms of nucleotides in biosynthesis and energy conversions. Nucleoside monophosphates are converted into nucleoside triphosphates in stages. First, nucleoside monophosphates are converted into diphosphates by specific *nucleoside monophosphate kinases* that utilize ATP as the phosphoryl-group donor. For example, UMP is phosphorylated to UDP by *UMP kinase*.

$$\text{UMP} + \text{ATP} \rightleftharpoons \text{UDP} + \text{ADP}$$

Nucleoside diphosphates and triphosphates are interconverted by *nucleoside diphosphate kinase,* an enzyme that has broad specificity, in contrast with the monophosphate kinases. X and Y represent any of several ribonucleosides or even deoxyribonucleosides:

$$\text{XDP} + \text{YTP} \rightleftharpoons \text{XTP} + \text{YDP}$$

CTP Is Formed by Amination of UTP

After uridine triphosphate has been formed, it can be transformed into *cytidine triphosphate* by the replacement of a carbonyl group by an amino group.

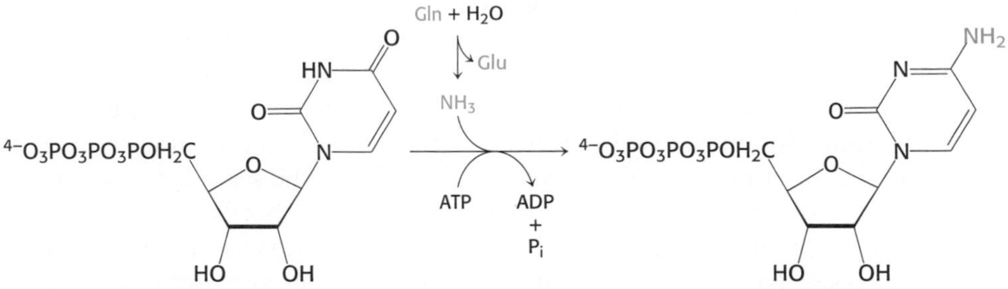

Orotate
+

5-Phosphoribosyl-1-pyrophosphate (PRPP)

PP$_i$

Orotidylate

UTP → **CTP**

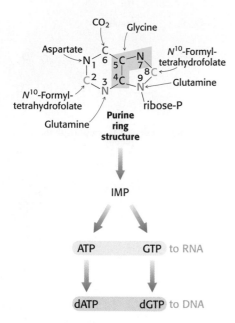

Figure 25.5 De novo pathway for purine nucleotide synthesis. The origins of the atoms in the purine ring are indicated.

Like the synthesis of carbamoyl phosphate, this reaction requires ATP and uses glutamine as the source of the amino group. The reaction proceeds through an analogous mechanism in which the O-4 atom is phosphorylated to form a reactive intermediate, and then the phosphate is displaced by ammonia, freed from glutamine by hydrolysis. CTP can then be used in many biochemical processes, including RNA synthesis.

25.2 Purine Bases Can Be Synthesized de Novo or Recycled by Salvage Pathways

Purine nucleotides can be synthesized in two distinct pathways. First, purines are synthesized de novo, beginning with simple starting materials such as amino acids and bicarbonate (Figure 25.5). Unlike the bases of pyrimidines, the purine bases are assembled already attached to the ribose ring. Alternatively, purine bases, released by the hydrolytic degradation of nucleic acids and nucleotides, can be salvaged and recycled. Purine salvage pathways are especially noted for the energy that they save and the remarkable effects of their absence (p. 725).

Salvage Pathways Economize Intracellular Energy Expenditure

Free purine bases, derived from the turnover of nucleotides or from the diet, can be attached to PRPP to form purine nucleoside monophosphates, in a reaction analogous to the formation of orotidylate. Two salvage enzymes with different specificities recover purine bases. *Adenine phosphoribosyltransferase* catalyzes the formation of adenylate (AMP):

$$\text{Adenine} + \text{PRPP} \longrightarrow \text{adenylate} + \text{PP}_i$$

whereas *hypoxanthine-guanine phosphoribosyltransferase* (HGPRT) catalyzes the formation of guanylate (GMP) as well as *inosinate* (inosine monophosphate, IMP), a precursor of guanylate and adenylate.

$$\text{Guanine} + \text{PRPP} \longrightarrow \text{guanylate} + \text{PP}_i$$
$$\text{Hypoxanthine} + \text{PRPP} \longrightarrow \text{inosinate} + \text{PP}_i$$

Similar salvage pathways exist for pyrimidines. Pyrimidine phosphoribosyltransferase will reconnect uracil, but not cytosine, to PRPP.

The Purine Ring System Is Assembled on Ribose Phosphate

De novo purine biosynthesis, like pyrimidine biosynthesis, requires PRPP but, for purines, PRPP provides the foundation on which the bases are constructed step by step. The initial committed step is the displacement of pyrophosphate by ammonia, rather than by a preassembled base, to produce *5-phosphoribosyl-1-amine*, with the amine in the β configuration.

Glutamine phosphoribosyl amidotransferase catalyzes this reaction. This enzyme comprises two domains: the first is homologous to the phosphoribosyltransferases in salvage pathways, whereas the second produces ammonia from glutamine by hydrolysis. However, this glutamine-hydrolysis domain is distinct from the domain that performs the same function in carbamoyl phosphate synthetase. In glutamine phosphoribosyl amidotransferase, a cysteine residue located at the amino terminus facilitates glutamine hydrolysis. To prevent wasteful hydrolysis of either substrate, the amidotransferase assumes the active configuration only on binding of both PRPP and glutamine. As is the case with carbamoyl phosphate synthetase, the ammonia generated at the glutamine-hydrolysis active site passes through a channel to reach PRPP without being released into solution.

The Purine Ring Is Assembled by Successive Steps of Activation by Phosphorylation Followed by Displacement

Nine additional steps are required to assemble the purine ring. Remarkably, the first six steps are analogous reactions. Most of these steps are catalyzed by enzymes with ATP-grasp domains that are homologous to those in carbamoyl phosphate synthetase. *Each step consists of the activation of a carbon-bound oxygen atom (typically a carbonyl oxygen atom) by phosphorylation, followed by the displacement of the phosphoryl group by ammonia or an amine group acting as a nucleophile* (Nu).

De novo purine biosynthesis proceeds as follows (Figure 25.6).

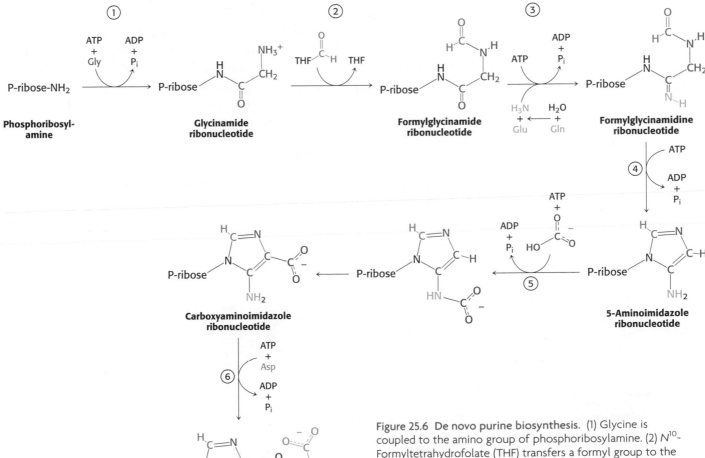

Figure 25.6 De novo purine biosynthesis. (1) Glycine is coupled to the amino group of phosphoribosylamine. (2) N^{10}-Formyltetrahydrofolate (THF) transfers a formyl group to the amino group of the glycine residue. (3) The inner amide group is phosphorylated and converted into an amidine by the addition of ammonia derived from glutamine. (4) An intramolecular coupling reaction forms the five-membered imidazole ring. (5) Bicarbonate adds first to the exocyclic amino group and then to a carbon atom of the imidazole ring. (6) The imidazole carboxylate is phosphorylated, and the phosphate is displaced by the amino group of aspartate.

1. The carboxylate group of a glycine residue is activated by phosphorylation and then coupled to the amino group of phosphoribosylamine. A new amide bond is formed, and the amino group of glycine is free to act as a nucleophile in the next step.

2. Formate is activated and then added to this amino group to form formylglycinamide ribonucleotide. In some organisms, two distinct enzymes can catalyze this step. One enzyme transfers the formyl group from N^{10}-formyltetrahydrofolate (p. 690). The other enzyme activates formate as formyl phosphate, which is added directly to the glycine amino group.

3. The inner amide group is activated by phosphorylation and then converted into an amidine by the addition of ammonia derived from glutamine.

4. The product of this reaction, formylglycinamidine ribonucleotide, cyclizes to form the five-membered imidazole ring found in purines. Although this cyclization is likely to be favorable thermodynamically, a molecule of ATP is consumed to ensure irreversibility. The familiar pattern is repeated: a phosphoryl group from the ATP molecule activates the carbonyl group and is displaced by the nitrogen atom attached to the ribose molecule. Cyclization is thus an intramolecular reaction in which the nucleophile and phosphate-activated carbon atom are present within the same molecule.

5. Bicarbonate is activated by phosphorylation and then attacked by the exocyclic amino group. The product of the reaction in step 5 rearranges to transfer the carboxylate group to the imidazole ring. Interestingly, mammals do not require ATP for this step; bicarbonate apparently attaches directly to the exocyclic amino group and is then transferred to the imidazole ring.

6. The imidazole carboxylate group is phosphorylated again and the phosphate group is displaced by the amino group of aspartate. Thus, a six-step process links glycine, formate, ammonia, bicarbonate, and aspartate to form an intermediate that contains all but two of the atoms necessary for the formation of the purine ring.

Three more steps complete ring construction (Figure 25.7). Fumarate, an intermediate in the citric acid cycle, is eliminated, leaving the nitrogen atom from aspartate joined to the imidazole ring. The use of aspartate as an

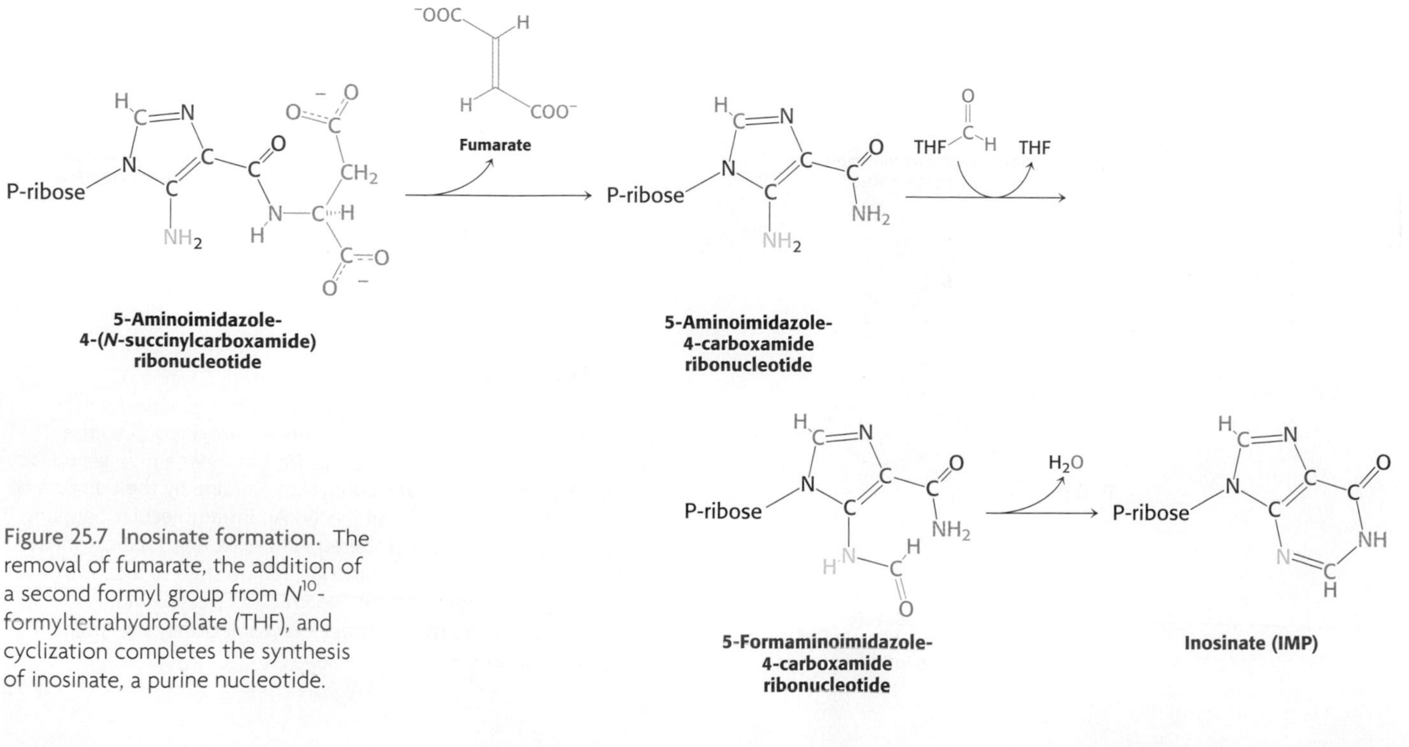

Figure 25.7 Inosinate formation. The removal of fumarate, the addition of a second formyl group from N^{10}-formyltetrahydrofolate (THF), and cyclization completes the synthesis of inosinate, a purine nucleotide.

5-Aminoimidazole-4-(N-succinylcarboxamide) ribonucleotide

5-Aminoimidazole-4-carboxamide ribonucleotide

5-Formaminoimidazole-4-carboxamide ribonucleotide

Inosinate (IMP)

amino-group donor and the concomitant release of fumarate are reminiscent of the conversion of citrulline into arginine in the urea cycle, and these steps are catalyzed by homologous enzymes in the two pathways (p. 663). A formyl group from N^{10}-formyltetrahydrofolate is added to this nitrogen atom to form a final intermediate that cyclizes with the loss of water to form inosinate. Many of the intermediates in the de novo purine biosynthesis pathway degrade rapidly in water. Their instability in water suggests that the product of one enzyme must be channeled directly to the next enzyme along the pathway. Yet these enzymes, at least in the absence of bound substrates and products, do not readily form complexes with one another. An appealing hypothesis currently under investigation is that complexes are formed, but only when the appropriate intermediate is bound. This behavior reveals further sophistication in substrate channeling and highlights the importance of substrate-induced conformational changes.

AMP and GMP Are Formed from IMP

A few steps convert inosinate into either AMP or GMP (Figure 25.8). *Adenylate* is synthesized from inosinate by the substitution of an amino group for the carbonyl oxygen atom at C-6. Again, the addition of aspartate followed by the elimination of fumarate contributes the amino group. GTP, rather than ATP, is the phosphoryl-group donor in the synthesis of the adenylosuccinate intermediate from inosinate and aspartate. In accord with the use of GTP, the enzyme that promotes this conversion, *adenylosuccinate synthase,* is structurally related to the G-protein family and does not contain an ATP-grasp domain. The same enzyme catalyzes the removal of fumarate from adenylosuccinate in the synthesis of adenylate and from 5-aminoimidazole-4-N-succinocarboxamide ribonucleotide in the synthesis of inosinate.

Guanylate is synthesized by the oxidation of inosinate to xanthylate (XMP), followed by the incorporation of an amino group at C-2. NAD^+ is the hydrogen acceptor in the oxidation of inosinate. Xanthylate is activated by the transfer of an AMP group (rather than a phosphoryl group) from ATP to the oxygen atom in the newly formed carbonyl group. Ammonia,

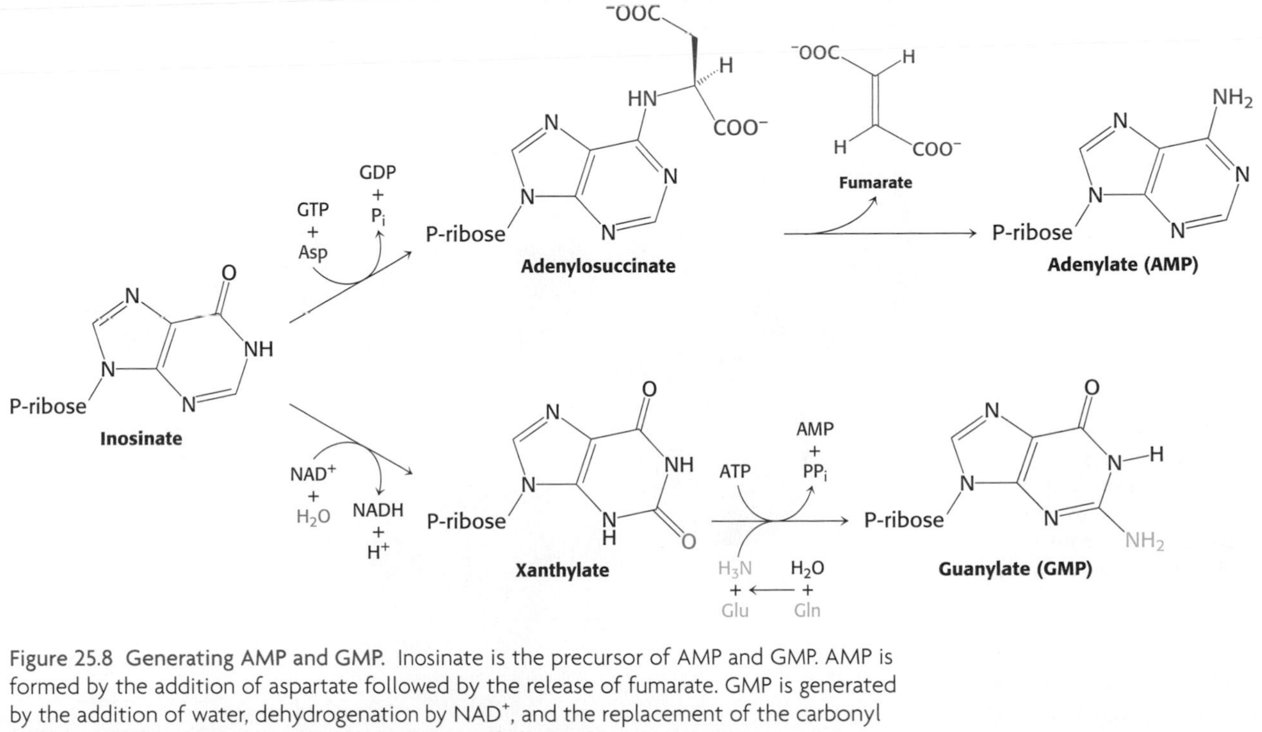

Figure 25.8 Generating AMP and GMP. Inosinate is the precursor of AMP and GMP. AMP is formed by the addition of aspartate followed by the release of fumarate. GMP is generated by the addition of water, dehydrogenation by NAD^+, and the replacement of the carbonyl oxygen atom by $-NH_2$ derived by the hydrolysis of glutamine.

generated by the hydrolysis of glutamine, then displaces the AMP group to form guanylate, in a reaction catalyzed by *GMP synthetase*. Note that the synthesis of adenylate requires GTP, whereas the synthesis of guanylate requires ATP. This reciprocal use of nucleotides by the pathways creates an important regulatory opportunity (Section 25.4).

25.3   Deoxyribonucleotides Are Synthesized by the Reduction of Ribonucleotides Through a Radical Mechanism

We turn now to the synthesis of deoxyribonucleotides. These precursors of DNA are formed by the reduction of ribonucleotides; specifically, the $2'$-hydroxyl group on the ribose moiety is replaced by a hydrogen atom. The substrates are ribonucleoside diphosphates, and the ultimate reductant is NADPH. The enzyme *ribonucleotide reductase* is responsible for the reduction reaction for all four ribonucleotides. The ribonucleotide reductases of different organisms are a remarkably diverse set of enzymes. Yet detailed studies have revealed that they have a common reaction mechanism, and their three-dimensional structural features indicate that these enzymes are homologous. We will focus on the best understood of these enzymes, that of *E. coli* living aerobically.

Mechanism: A Tyrosyl Radical Is Critical to the Action of Ribonucleotide Reductase

The ribonucleotide reductase of *E. coli* consists of two subunits: R1 (an 87-kd dimer) and R2 (a 43-kd dimer). The R1 subunit contains the active site as well as two allosteric control sites (Section 25.4). This subunit includes three conserved cysteine residues and a glutamate residue, all four of which participate in the reduction of ribose to deoxyribose (Figure 25.9). The R2 subunit's role in catalysis is to generate a remarkable free radical in each of its two chains. Each R2 chain contains a stable *tyrosyl radical* with an unpaired electron delocalized onto its aromatic ring (Figure 25.10). This very unusual free radical is generated by a nearby *iron center* consisting of two ferric (Fe^{3+}) ions bridged by an oxide (O^{2-}) ion.

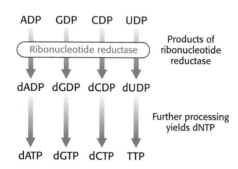

Products of ribonucleotide reductase

Further processing yields dNTP

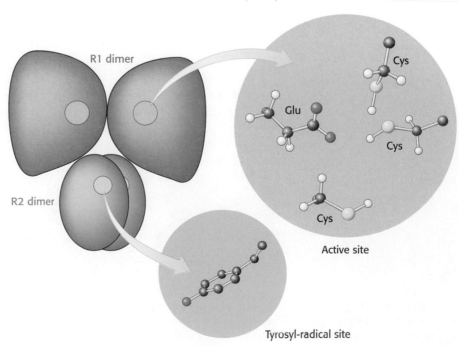

Figure 25.9 Ribonucleotide reductase. Ribonucleotide reductase reduces ribonucleotides to deoxyribonucleotides in its active site, which contains three key cysteine residues and one glutamate residue. Each R2 subunit contains a tyrosyl radical that accepts an electron from one of the cysteine residues in the active site to initiate the reduction reaction. Two R1 subunits come together to form a dimer as do two R2 subunits.

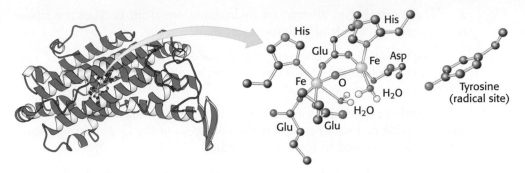

Figure 25.10 Ribonucleotide reductase R2 subunit. This subunit contains a stable free radical on a tyrosine residue. This radical is generated by the reaction of oxygen (not shown) at a nearby site containing two iron atoms. Two R2 subunits come together to form a dimer. [Drawn from 1RIB.pdb.]

In the synthesis of a deoxyribonucleotide, the OH bonded to C-2′ of the ribose ring is replaced by H, with retention of the configuration at the C-2′ carbon atom (Figure 25.11).

1. The reaction begins with the transfer of an electron from a cysteine residue on R1 to the tyrosyl radical on R2. The loss of an electron generates a highly reactive *cysteine thiyl radical* within the active site of R1.

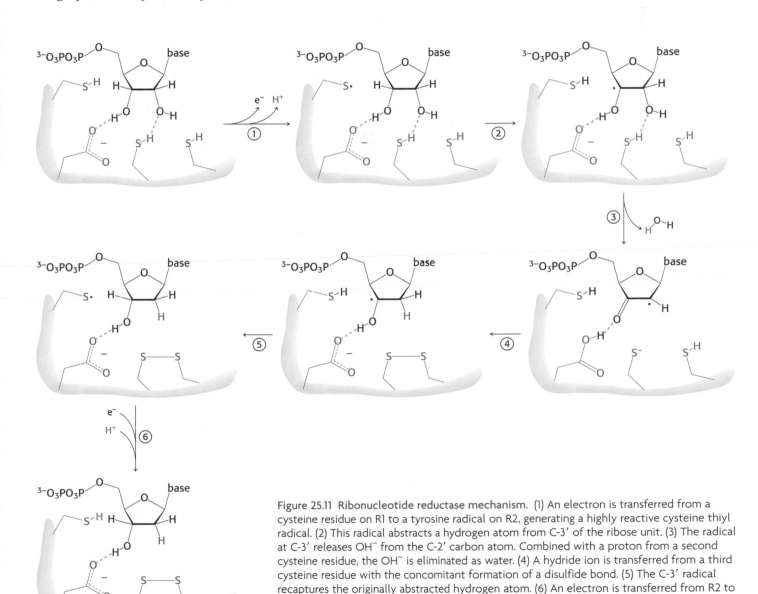

Figure 25.11 Ribonucleotide reductase mechanism. (1) An electron is transferred from a cysteine residue on R1 to a tyrosine radical on R2, generating a highly reactive cysteine thiyl radical. (2) This radical abstracts a hydrogen atom from C-3′ of the ribose unit. (3) The radical at C-3′ releases OH⁻ from the C-2′ carbon atom. Combined with a proton from a second cysteine residue, the OH⁻ is eliminated as water. (4) A hydride ion is transferred from a third cysteine residue with the concomitant formation of a disulfide bond. (5) The C-3′ radical recaptures the originally abstracted hydrogen atom. (6) An electron is transferred from R2 to reduce the thiyl radical, which also accepts a proton. The deoxyribonucleotide is free to leave R1. The disulfide formed in the active site must be reduced to begin another cycle.

2. This radical then abstracts a hydrogen atom from C-3' of the ribose unit, generating a radical at that carbon atom.

3. The radical at C-3' promotes the release of the OH^- from the C-2' carbon atom. Protonated by a second cysteine residue, the departing OH^- leaves as a water molecule.

4. A hydride ion (a proton with two electrons) is then transferred from a third cysteine residue to complete the reduction of the C-2' position, form a disulfide bond, and re-form a C-3' radical.

5. This C-3' radical recaptures the same hydrogen atom originally abstracted by the first cysteine residue, and the deoxyribonucleotide is free to leave the enzyme.

6. R2 provides an electron to reduce the thiyl radical. The disulfide bond generated in the enzyme's active site is then reduced by a specific disulfide-containing protein, such as thioredoxin, to regenerate the active enzyme.

To complete the overall reaction, the oxidized thioredoxin generated by this process is reduced by NADH in a reaction catalyzed by *thioredoxin reductase*.

Ribonucleotide reductases that do not contain tyrosyl radicals have been characterized in other organisms. Instead, these enzymes contain other stable radicals that are generated by other processes. For example, in one class of reductases, the coenzyme adenosylcobalamin is the radical source. Despite differences in the stable radical employed, the active sites of these enzymes are similar to that of the *E. coli* ribonucleotide reductase, and they appear to act by the same mechanism, based on the exceptional reactivity of cysteine radicals. Thus, these enzymes have a common ancestor but evolved a range of mechanisms for generating stable radical species that function well under different growth conditions. The primordial enzymes appear to have been inactivated by oxygen, whereas enzymes such as the *E. coli* enzyme make use of oxygen to generate the initial tyrosyl radical. Note that the reduction of ribonucleotides to deoxyribonucleotides is a difficult reaction chemically, likely to require a sophisticated catalyst. The existence of a common protein enzyme framework for this process strongly suggests that proteins joined the RNA world before the evolution of DNA as a stable storage form for genetic information.

Thymidylate Is Formed by the Methylation of Deoxyuridylate

Uracil, produced by the pyrimidine synthesis pathway, is not a component of DNA. Rather, DNA contains *thymine*, a methylated analog of uracil. Another step is required to generate thymidylate from uracil. *Thymidylate synthase* catalyzes this finishing touch: deoxyuridylate (dUMP) is methylated to thymidylate (TMP). As will be described in Chapter 28, the methylation of this nucleotide marks sites of DNA damage for repair and, hence, helps preserve the integrity of the genetic information stored in DNA. The methyl donor in this reaction is N^5,N^{10}-methylenetetrahydrofolate rather than *S*-adenosylmethionine.

The methyl group becomes attached to the C-5 atom within the aromatic ring of dUMP, but this carbon atom is not a good nucleophile and cannot itself attack the appropriate group on the methyl donor. Thymidylate synthase promotes methylation by adding a thiolate from a cysteine side chain to this ring to generate a nucleophilic species that can attack the methylene group of N^5,N^{10}-methylenetetrahydrofolate (Figure 25.12). This methylene group, in turn, is activated by distortions imposed by the enzyme that favor opening the five-membered ring. The activated

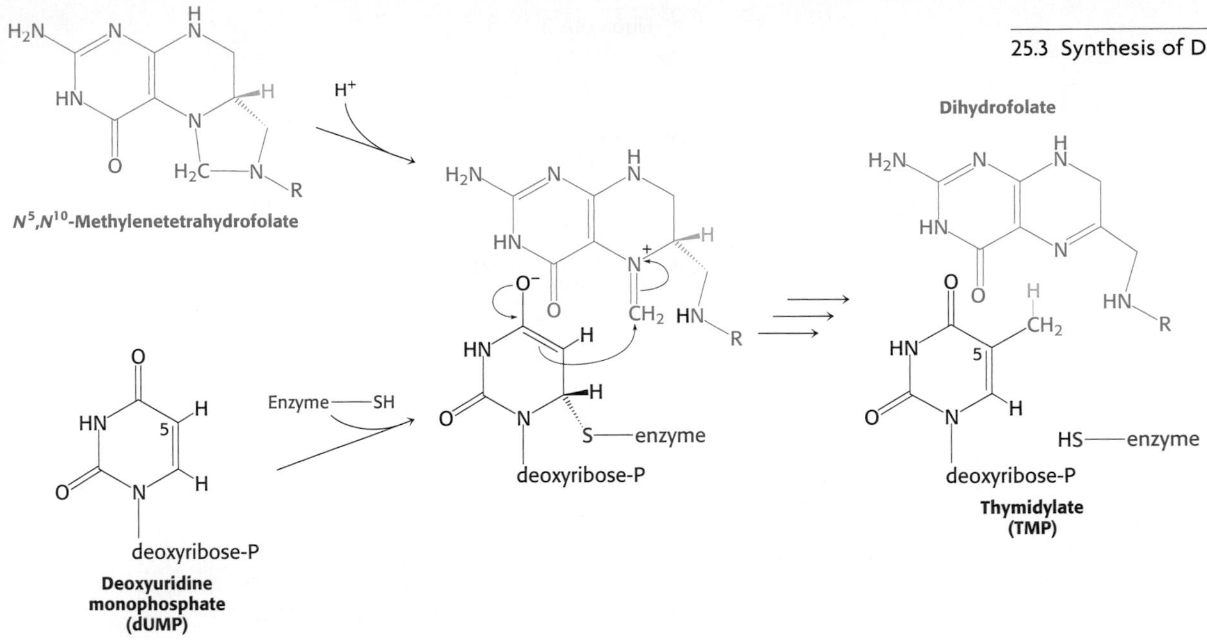

Figure 25.12 **Thymidylate synthesis.** Thymidylate synthase catalyzes the addition of a methyl group (derived from N^5,N^{10}-methylenetetrahydrofolate) to dUMP to form TMP. The addition of a thiolate from the enzyme activates dUMP. Opening the five-membered ring of the THF derivative prepares the methylene group for nucleophilic attack by the activated dUMP. The reaction is completed by the transfer of a hydride ion to form dihydrofolate.

dUMP's attack on the methylene group forms the new carbon–carbon bond. The intermediate formed is then converted into product: a hydride ion is transferred from the tetrahydrofolate ring to transform the methylene group into a methyl group, and a proton is abstracted from the carbon atom bearing the methyl group to eliminate the cysteine and regenerate the aromatic ring. The tetrahydrofolate derivative loses both its methylene group and a hydride ion and, hence, is oxidized to dihydrofolate. For the synthesis of more thymidylate, tetrahydrofolate must be regenerated.

Dihydrofolate Reductase Catalyzes the Regeneration of Tetrahydrofolate, a One-Carbon Carrier

Tetrahydrofolate is regenerated from the dihydrofolate that is produced in the synthesis of thymidylate. This regeneration is accomplished by *dihydrofolate reductase* with the use of NADPH as the reductant.

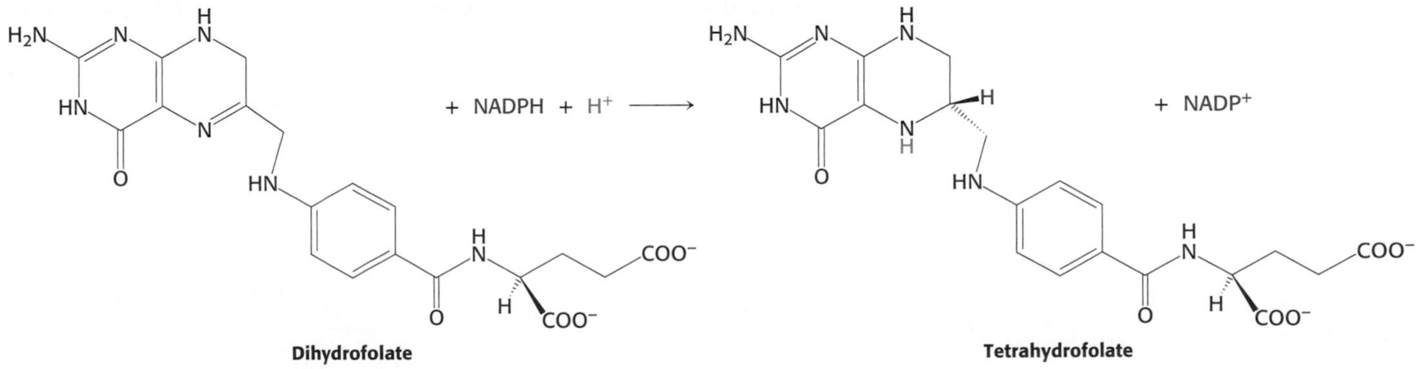

A hydride ion is directly transferred from the nicotinamide ring of NADPH to the pteridine ring of dihydrofolate. The bound dihydrofolate and NADPH are held in close proximity to facilitate the hydride transfer.

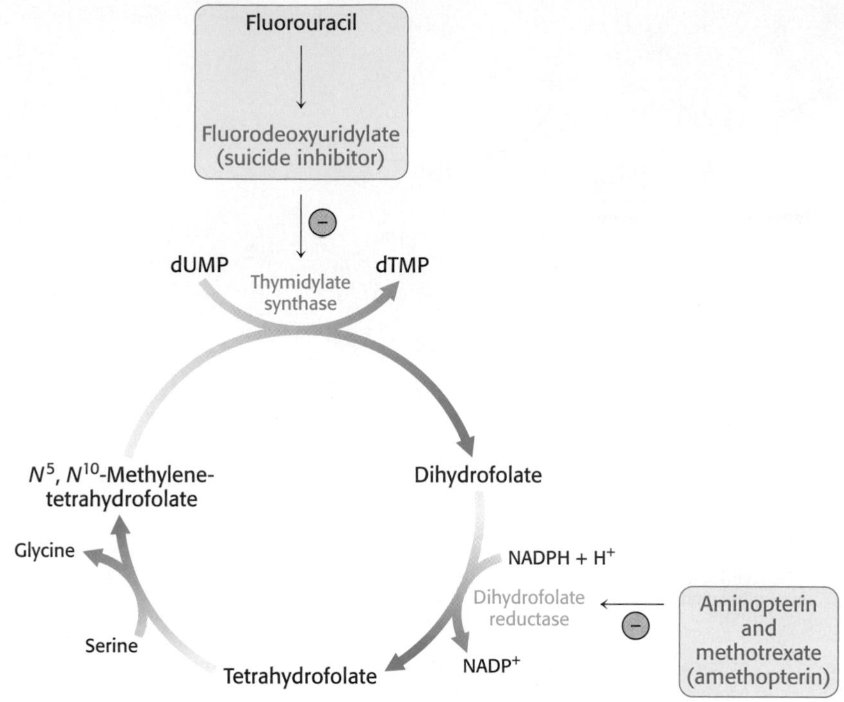

Figure 25.13 Anticancer drug targets. Thymidylate synthase and dihydrofolate reductase are choice targets in cancer chemotherapy because the generation of large quantities of precursors for DNA synthesis is required for rapidly dividing cancer cells.

Several Valuable Anticancer Drugs Block the Synthesis of Thymidylate

Rapidly dividing cells require an abundant supply of thymidylate for the synthesis of DNA. The vulnerability of these cells to the inhibition of TMP synthesis has been exploited in the treatment of cancer. Thymidylate synthase and dihydrofolate reductase are choice targets of chemotherapy (Figure 25.13).

Fluorouracil, a clinically useful anticancer drug, is converted in vivo into *fluorodeoxyuridylate* (F-dUMP). This analog of dUMP irreversibly inhibits thymidylate synthase after acting as a normal substrate through part of the catalytic cycle. Recall that the formation of TMP requires the removal of a proton (H^+) from C-5 of the bound nucleotide (see Figure 25.12). However, the enzyme cannot abstract F^+ from F-dUMP, and so catalysis is blocked at the stage of the covalent complex formed by F-dUMP, methylenetetrahydrofolate, and the sulfhydryl group of the enzyme (Figure 25.14). We see here an example of *suicide inhibition,* in which an enzyme converts a substrate into a reactive inhibitor that halts the enzyme's catalytic activity (p. 230).

The synthesis of TMP can also be blocked by inhibiting the regeneration of tetrahydrofolate. Analogs of dihydrofolate, such as *aminopterin* and

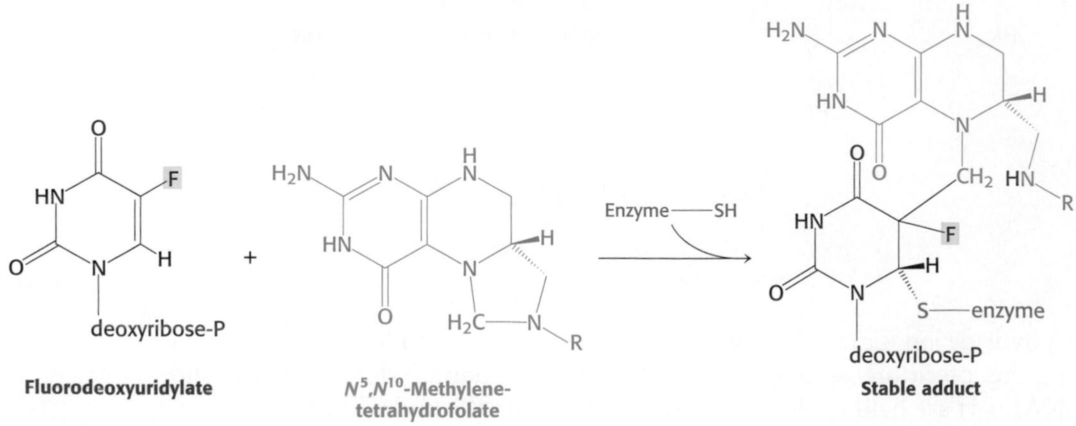

Figure 25.14 Suicide inhibition. Fluorodeoxyuridylate (generated from fluorouracil) traps thymidylate synthase in a form that cannot proceed down the reaction pathway.

Fluorodeoxyuridylate N^5,N^{10}-Methylene-tetrahydrofolate   **Stable adduct**

methotrexate (amethopterin), are potent competitive inhibitors ($K_i < 1$ nM) of dihydrofolate reductase.

Aminopterin (R = H) or methotrexate (R = CH₃)

Methotrexate is a valuable drug in the treatment of many rapidly growing tumors, such as those in acute leukemia and choriocarcinoma, a cancer derived from placental cells. However, methotrexate kills rapidly replicating cells whether they are malignant or not. Stem cells in bone marrow, epithelial cells of the intestinal tract, and hair follicles are vulnerable to the action of this folate antagonist, accounting for its toxic side effects, which include weakening of the immune system, nausea, and hair loss.

Folate analogs such as *trimethoprim* have potent antibacterial and antiprotozoal activity. Trimethoprim binds 10^5-fold less tightly to mammalian dihydrofolate reductase than it does to reductases of susceptible microorganisms. Small differences in the active-site clefts of these enzymes account for the highly selective antimicrobial action. The combination of trimethoprim and sulfamethoxazole (an inhibitor of folate synthesis) is widely used to treat infections.

Trimethoprim

25.4 Key Steps in Nucleotide Biosynthesis Are Regulated by Feedback Inhibition

Nucleotide biosynthesis is regulated by feedback inhibition in a manner similar to the regulation of amino acid biosynthesis (Section 24.3). These regulatory pathways ensure that the various nucleotides are produced in the required quantities.

Pyrimidine Biosynthesis Is Regulated by Aspartate Transcarbamoylase

Aspartate transcarbamoylase, one of the key enzymes for the regulation of pyrimidine biosynthesis in bacteria, was described in detail in Chapter 10. Recall that *ATCase is inhibited by CTP, the final product of pyrimidine biosynthesis,* and stimulated by ATP.

$$\text{Aspartate} + \text{carbamoyl phosphate} \xrightarrow{\text{ATCase}} \text{carbamoylaspartate} \rightarrow \rightarrow \rightarrow \text{UMP} \longrightarrow \text{UDP} \longrightarrow \text{UTP} \longrightarrow \text{CTP}$$

Carbamoyl phosphate synthetase is also a site of feedback inhibition in both prokaryotes and eukaryotes.

The Synthesis of Purine Nucleotides Is Controlled by Feedback Inhibition at Several Sites

The regulatory scheme for purine nucleotides is more complex than that for pyrimidine nucleotides (Figure 25.15).

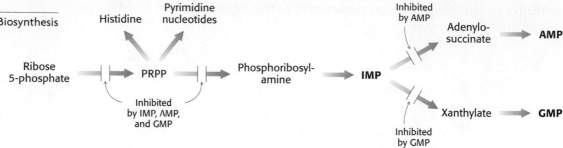

Figure 25.15 Control of purine biosynthesis. Feedback inhibition controls both the overall rate of purine biosynthesis and the balance between AMP and GMP production.

1.   The committed step in purine nucleotide biosynthesis is the conversion of PRPP into phosphoribosylamine by *glutamine phosphoribosyl amidotransferase*. This important enzyme is feedback-inhibited by many purine ribonucleotides. It is noteworthy that AMP and GMP, the final products of the pathway, are synergistic in inhibiting the amidotransferase.

2.   Inosinate is the branch point in the synthesis of AMP and GMP. *The reactions leading away from inosinate are sites of feedback inhibition.* AMP inhibits the conversion of inosinate into adenylosuccinate, its immediate precursor. Similarly, GMP inhibits the conversion of inosinate into xanthylate, its immediate precursor.

3.   As already noted, GTP is a substrate in the synthesis of AMP, whereas ATP is a substrate in the synthesis of GMP. This *reciprocal substrate relation* tends to balance the synthesis of adenine and guanine ribonucleotides.

The Synthesis of Deoxyribonucleotides Is Controlled by the Regulation of Ribonucleotide Reductase

The reduction of ribonucleotides to deoxyribonucleotides is precisely controlled by allosteric interactions. Each polypeptide of the R1 subunit of the aerobic *E. coli* ribonucleotide reductase contains two allosteric sites: one of them controls the *overall activity* of the enzyme, whereas the other regulates *substrate specificity* (Figure 25.16). The overall catalytic activity of ribonucleotide reductase is diminished by the binding of dATP, which signals an abundance of deoxyribonucleotides. The binding of ATP reverses this feedback inhibition. The binding of dATP or ATP to the substrate-specificity

Figure 25.16 Regulation of ribonucleotide reductase. (A) Each subunit in the R1 dimer contains two allosteric sites in addition to the active site. One site regulates the overall activity and the other site regulates substrate specificity. (B) The patterns of regulation with regard to different nucleoside diphosphates demonstrated by ribonucleotide reductase.

(A)

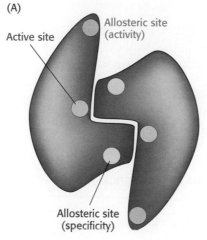

Active site

Allosteric site (activity)

Allosteric site (specificity)

(B) Regulation of overall activity

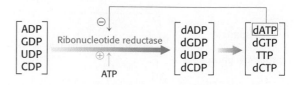

Regulation of substrate specificity

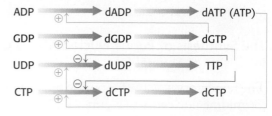

control site enhances the reduction of UDP and CDP, the pyrimidine nucleotides. The binding of thymidine triphosphate (TTP) promotes the reduction of GDP and inhibits the further reduction of pyrimidine ribonucleotides. The subsequent increase in the level of dGTP stimulates the reduction of ATP to dATP. This complex pattern of regulation supplies the appropriate balance of the four deoxyribonucleotides needed for the synthesis of DNA.

25.5 Disruptions in Nucleotide Metabolism Can Cause Pathological Conditions

Nucleotides are vital to a host of biochemical processes. It is not surprising, then, that disruption of nucleotide metabolism would have a variety of physiological effects. The nucleotides of a cell undergo continual turnover. Nucleotides are hydrolytically degraded to nucleosides by *nucleotidases*. The phosphorolytic cleavage of nucleosides to free bases and ribose 1-phosphate (or deoxyribose 1-phosphate) is catalyzed by *nucleoside phosphorylases*. Ribose 1-phosphate is isomerized by *phosphoribomutase* to ribose 5-phosphate, a substrate in the synthesis of PRPP. Some of the bases are reused to form nucleotides by salvage pathways. Others are degraded to products that are excreted (Figure 25.17). A deficiency of an enzyme can disrupt these pathways, leading to a pathological condition.

The Loss of Adenosine Deaminase Activity Results in Severe Combined Immunodeficiency

The pathway for the degradation of AMP includes an extra step. First, the phosphate is removed by a nucleotidase to yield the nucleoside adenosine (see Figure 25.17). In the extra step, adenosine is deaminated by adenosine deaminase to form inosine. Finally, the ribose is removed by nucleoside phophorylase, generating hypoxanthine and ribose 1-phosphate.

A deficiency in adenosine deaminase activity is associated with some forms of *severe combined immunodeficiency* (SCID), an immunological disorder. Persons with the disorder have severe recurring infections, often leading to death at an early age. SCID is characterized by a loss of T cells, which are crucial to the immune response (Section 33.5). Although the biochemical basis of the disorder is not clearly established, a lack of adenosine

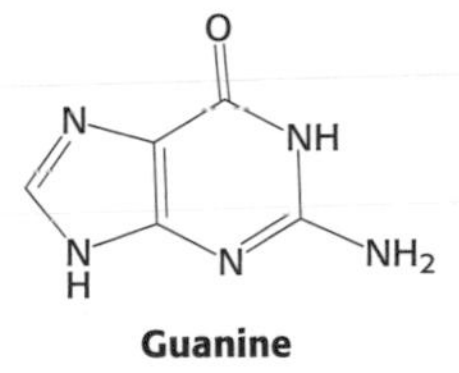

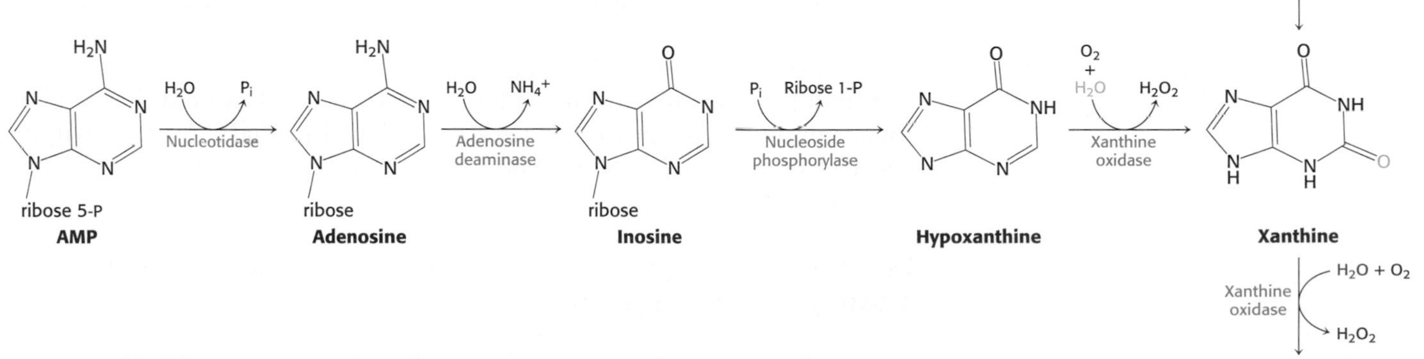

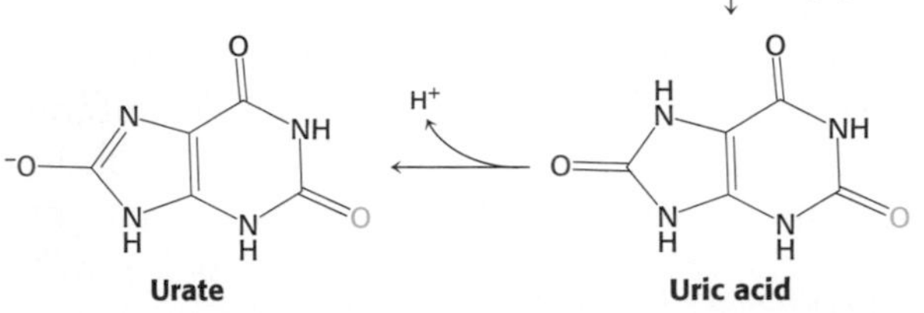

Figure 25.17 Purine catabolism. Purine bases are converted first into xanthine and then into urate for excretion. Xanthine oxidase catalyzes two steps in this process.

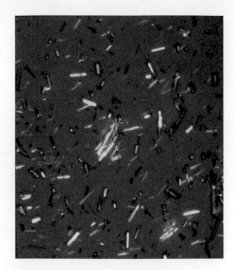

Figure 25.18 Micrograph of sodium urate crystals. The accumulation of these crystals damages joints and kidneys. [Courtesy of Dr. James McGuire.]

OH

Allopurinol

deaminase results in an increase of 50 to 100 times the normal level of dATP, which inhibits ribonucleotide reductase and, consequently, DNA synthesis. SCID is often called the "bubble boy disease" because its treatment may include complete isolation of the patient from the environment. Adenosine deaminase deficiency has been successfully treated by gene therapy.

Gout Is Induced by High Serum Levels of Urate

After the production of hypoxanthine, *xanthine oxidase*, a molybdenum- and iron-containing flavoprotein, oxidizes hypoxanthine to *xanthine* and then to *uric acid*. Molecular oxygen, the oxidant in both reactions, is reduced to H_2O_2, which is decomposed to H_2O and O_2 by catalase. Uric acid loses a proton at physiological pH to form *urate. In human beings, urate is the final product of purine degradation and is excreted in the urine.*

High serum levels of urate (hyperuricemia) induce the painful joint disease *gout*. In this disease, the sodium salt of urate crystallizes in the fluid and lining of the joints (Figure 25.18). The small joint at the base of the big toe is a common site for sodium urate buildup, although the salt accumulates at other joints also. Painful inflammation results when cells of the immune system engulf the sodium urate crystals. The kidneys, too, may be damaged by the deposition of urate crystals. Gout is a common medical problem, affecting 1% of the population of Western countries. It is nine times as common in men as in women.

Administration of *allopurinol,* an analog of hypoxanthine, is one treatment for gout. The mechanism of action of allopurinol is interesting: it acts *first as a substrate* and *then as an inhibitor* of xanthine oxidase. The oxidase hydroxylates allopurinol to *alloxanthine (oxipurinol),* which then remains tightly bound to the active site. The binding of alloxanthine keeps the molybdenum atom of xanthine oxidase in the +4 oxidation state instead of it returning to the +6 oxidation state as in a normal catalytic cycle. We see here another example of *suicide inhibition.*

The synthesis of urate from hypoxanthine and xanthine decreases soon after the administration of allopurinol. The serum concentrations of hypoxanthine and xanthine rise, and that of urate drops.

The average serum level of urate in human beings is close to the solubility limit. In contrast, prosimians (such as lemurs) have 10-fold lower levels. A striking increase in urate levels occurred in the evolution of primates. What is the selective advantage of a urate level so high that it teeters on the brink of gout in many people? It turns out that urate has a markedly beneficial action. Urate is a highly effective scavenger of reactive oxygen species. Indeed, urate is about as effective as ascorbate (vitamin C) as an antioxidant. The increased level of urate in human beings compared with prosimians and other lower primates may contribute significantly to the longer human life span and to lowering the incidence of human cancer.

Lesch–Nyhan Syndrome Is a Dramatic Consequence of Mutations in a Salvage-Pathway Enzyme

Mutations in genes that encode nucleotide biosynthetic enzymes can reduce levels of needed nucleotides and can lead to an accumulation of intermediates. A nearly total absence of hypoxanthine-guanine phosphoribosyltransferase has unexpected and devastating consequences. The most striking expression of this inborn error of metabolism, called the *Lesch–Nyhan syndrome,* is *compulsive self-destructive behavior.* At age 2 or 3, children with this disease begin to bite their fingers and lips and will chew them off if unrestrained. These children also behave aggressively toward others. *Mental deficiency* and *spasticity* are other characteristics of the Lesch–Nyhan syndrome. Elevated levels of urate in the serum lead to the formation of kidney stones early in life, followed by the symptoms of gout years later. The disease is inherited as a sex-linked recessive disorder.

The biochemical consequences of the virtual absence of hypoxanthine-guanine phosphoribosyl transferase are *an elevated concentration of PRPP, a marked increase in the rate of purine biosynthesis by the de novo pathway, and an overproduction of urate.* The relation between the absence of the transferase and the bizarre neurological signs is an enigma. Specific cells in the brain may depend on the salvage pathway for the synthesis of IMP and GMP. Indeed, transporters of the neurotransmitter dopamine are present at lower levels in affected persons. Alternatively, cells may be damaged by the accumulation of intermediates to abnormal levels. The Lesch–Nyhan syndrome demonstrates that the salvage pathway for the synthesis of IMP and GMP is not gratuitous. Moreover, the Lesch–Nyhan syndrome reveals that *abnormal behavior such as self-mutilation and extreme hostility can be caused by the absence of a single enzyme.* Psychiatry will no doubt benefit from the unraveling of the molecular basis of such mental disorders.

Folic Acid Deficiency Promotes Birth Defects Such As Spina Bifida

Spina bifida is one of a class of birth defects characterized by the incomplete or incorrect formation of the neural tube early in development. In the United States, the prevalence of *neural-tube defects* is approximately 1 case per 1000 births. A variety of studies have demonstrated that the prevalence of neural-tube defects is reduced by as much as 70% when women take folic acid as a dietary supplement before and during the first trimester of pregnancy. One hypothesis is that more folate derivatives are needed for the synthesis of DNA precursors when cell division is frequent and substantial amounts of DNA must be synthesized.

Summary

25.1 In de Novo Synthesis, the Pyrimidine Ring Is Assembled from Bicarbonate, Aspartate, and Glutamine

The pyrimidine ring is assembled first and then linked to ribose phosphate to form a pyrimidine nucleotide. 5-Phosphoribosyl-1-pyrophosphate is the donor of the ribose phosphate moiety. The synthesis of the pyrimidine ring starts with the formation of carbamoylaspartate from carbamoyl phosphate and aspartate, a reaction catalyzed by aspartate transcarbamoylase. Dehydration, cyclization, and oxidation yield orotate, which reacts with PRPP to give orotidylate. Decarboxylation of this pyrimidine nucleotide yields UMP. CTP is then formed by the amination of UTP.

25.2 Purine Bases Can Be Synthesized de Novo or Recycled by Salvage Pathways

The purine ring is assembled from a variety of precursors: glutamine, glycine, aspartate, N^{10}-formyltetrahydrofolate, and CO_2. The committed step in the de novo synthesis of purine nucleotides is the formation of 5-phosphoribosylamine from PRPP and glutamine. The purine ring is assembled on ribose phosphate, in contrast with the de novo synthesis of pyrimidine nucleotides. The addition of glycine, followed by formylation, amination, and ring closure, yields 5-aminoimidazole ribonucleotide. This intermediate contains the completed five-membered ring of the purine skeleton. The addition of CO_2, the nitrogen atom of aspartate, and a formyl group, followed by ring closure, yields inosinate, a purine ribonucleotide. AMP and GMP are formed from IMP. Purine ribonucleotides can also be synthesized by a salvage pathway in which a preformed base reacts directly with PRPP.

25.3 Deoxyribonucleotides Are Synthesized by the Reduction of Ribonucleotides Through a Radical Mechanism

Deoxyribonucleotides, the precursors of DNA, are formed in *E. coli* by the reduction of ribonucleoside diphosphates. These conversions are catalyzed by ribonucleotide reductase. Electrons are transferred from NADPH to sulfhydryl groups at the active sites of this enzyme by thioredoxin or glutaredoxin. A tyrosyl free radical generated by an iron center in the reductase initiates a radical reaction on the sugar, leading to the exchange of H for OH at C-2′. TMP is formed by the methylation of dUMP. The donor of a methylene group and a hydride in this reaction is N^5,N^{10}-methylenetetrahydrofolate, which is converted into dihydrofolate. Tetrahydrofolate is regenerated by the reduction of dihydrofolate by NADPH. Dihydrofolate reductase, which catalyzes this reaction, is inhibited by folate analogs such as aminopterin and methotrexate. These compounds and fluorouracil, an inhibitor of thymidylate synthase, are used as anticancer drugs.

25.4 Key Steps in Nucleotide Biosynthesis Are Regulated by Feedback Inhibition

Pyrimidine biosynthesis in *E. coli* is regulated by the feedback inhibition of aspartate transcarbamoylase, the enzyme that catalyzes the committed step. CTP inhibits and ATP stimulates this enzyme. The feedback inhibition of glutamine-PRPP amidotransferase by purine nucleotides is important in regulating their biosynthesis.

25.5 Disruptions in Nucleotide Metabolism Can Cause Pathological Conditions

Purines are degraded to urate in human beings. Gout, a disease that affects joints and leads to arthritis, is associated with an excessive accumulation of urate. The Lesch–Nyhan syndrome, a genetic disease characterized by self-mutilation, mental deficiency, and gout, is caused by the absence of hypoxanthine-guanine phosphoribosyltransferase. This enzyme is essential for the synthesis of purine nucleotides by the salvage pathway. Neural-tube defects are more frequent when a pregnant woman is deficient in folate derivatives early in pregnancy, possibly because of the important role of these derivatives in the synthesis of DNA precursors.

Key Terms

nucleotide (p. 709)

pyrimidine (p. 710)

carbamoyl phosphate synthetase (CPS) (p. 711)

ATP-grasp fold (p. 711)

5-phosphoribosyl-1-pyrophosphate (PRPP) (p. 712)

orotidylate (p. 712)

purine (p. 714)

salvage pathway (p. 714)

inosinate (p. 714)

hypoxanthine (p. 714)

glutamine phosphoribosyl amidotransferase (p. 714)

ribonucleotide reductase (p. 718)

thymidylate synthase (p. 720)

dihydrofolate reductase (p. 721)

severe combined immunodeficiency (SCID) (p. 725)

gout (p. 726)

Lesch–Nyhan syndrome (p. 726)

spina bifida (p. 727)

neural-tube defect (p. 727)

Selected Readings

Where to Start

Kappock, T. J., Ealick, S. E., and Stubbe, J. 2000. Modular evolution of the purine biosynthetic pathway. *Curr. Opin. Chem. Biol.* 4:567–572.

Galperin, M. Y., and Koonin, E. V. 1997. A diverse superfamily of enzymes with ATP-dependent carboxylate-amine/thiol ligase activity. *Protein Sci.* 6:2639–2643.

Jordan, A., and Reichard, P. 1998. Ribonucleotide reductases. *Annu. Rev. Biochem.* 67:71–98.

Seegmiller, J. E. 1989. Contributions of Lesch-Nyhan syndrome to the understanding of purine metabolism. *J. Inherited Metab. Dis.* 12:184–196.

Pyrimidine Biosynthesis

Raushel, F. M., Thoden, J. B., Reinhart, G. D., and Holden, H. M. 1998. Carbamoyl phosphate synthetase: A crooked path from substrates to products. *Curr. Opin. Chem. Biol.* 2:624–632.

Huang, X., Holden, H. M., and Raushel, F. M. 2001. Channeling of substrates and intermediates in enzyme-catalyzed reactions. *Annu. Rev. Biochem.* 70:149–180.

Begley, T. P., Appleby, T. C., and Ealick, S. E. 2000. The structural basis for the remarkable proficiency of orotidine 5′-monophosphate decarboxylase. *Curr. Opin. Struct. Biol.* 10:711–718.

Traut, T. W., and Temple, B. R. 2000. The chemistry of the reaction determines the invariant amino acids during the evolution and divergence of orotidine 5′-monophosphate decarboxylase. *J. Biol. Chem.* 275:28675–28681.

Lee, L., Kelly, R. E., Pastra-Landis, S. C., and Evans, D. R. 1985. Oligomeric structure of the multifunctional protein CAD that initiates pyrimidine biosynthesis in mammalian cells. *Proc. Natl. Acad. Sci. U. S. A.* 82:6802–6806.

Purine Biosynthesis

Thoden, J. B., Firestine, S., Nixon, A., Benkovic, S. J., and Holden, H. M. 2000. Molecular structure of *Escherichia coli* PurT-encoded glycinamide ribonucleotide transformylase. *Biochemistry* 39:8791–8802.

McMillan, F. M., Cahoon, M., White, A., Hedstrom, L., Petsko, G. A., and Ringe, D. 2000. Crystal structure at 2.4 Å resolution of *Borrelia burgdorferi* inosine 5′-monophosphate dehydrogenase: Evidence of a substrate-induced hinged-lid motion by loop 6. *Biochemistry* 39:4533–4542.

Levdikov, V. M., Barynin, V. V., Grebenko, A. I., Melik-Adamyan, W. R., Lamzin, V. S., and Wilson, K. S. 1998. The structure of SAICAR synthase: An enzyme in the de novo pathway of purine nucleotide biosynthesis. *Structure* 6:363–376.

Smith, J. L., Zaluzec, E. J., Wery, J. P., Niu, L., Switzer, R. L., Zalkin, H., and Satow, Y. 1994. Structure of the allosteric regulatory enzyme of purine biosynthesis. *Science* 264:1427–1433.

Weber, G., Nagai, M., Natsumeda, Y., Ichikawa, S., Nakamura, H., Eble, J. N., Jayaram, H. N., Zhen, W. N., Paulik, E., and Hoffman, R. 1991. Regulation of de novo and salvage pathways in chemotherapy. *Adv. Enzyme Regul.* 31:45–67.

Ribonucleotide Reductases

Eklund, H., Uhlin, U., Farnegardh, M., Logan, D. T. and Nordlund, P. 2001. Structure and function of the radical enzyme ribonucleotide reductase. *Prog. Biophys. Mol. Biol.* 77:177–268.

Reichard, P. 1997. The evolution of ribonucleotide reduction. *Trends Biochem. Sci.* 22:81–85.

Stubbe, J. 2000. Ribonucleotide reductases: The link between an RNA and a DNA world? *Curr. Opin. Struct. Biol.* 10:731–736.

Logan, D. T., Andersson, J., Sjoberg, B. M., and Nordlund, P. 1999. A glycyl radical site in the crystal structure of a class III ribonucleotide reductase. *Science* 283:1499–1504.

Tauer, A., and Benner, S. A. 1997. The B$_{12}$-dependent ribonucleotide reductase from the archaebacterium *Thermoplasma acidophila:* An evolutionary solution to the ribonucleotide reductase conundrum. *Proc. Natl. Acad. Sci. U. S. A.* 94:53–58.

Stubbe, J., Nocera, D. G., Yee, C. S. and Chang, M. C. 2003. Radical initiation in the class I ribonucleotide reductase: Long-range proton-coupled electron transfer? *Chem. Rev.* 103:2167–2201.

Stubbe, J., and Riggs-Gelasco, P. 1998. Harnessing free radicals: Formation and function of the tyrosyl radical in ribonucleotide reductase. *Trends Biochem. Sci.* 23:438–443.

Thymidylate Synthase and Dihydrofolate Reductase

Schnell, J. R., Dyson, H. J., and Wright, P. E. 2004. Structure, dynamics, and catalytic function of dihydrofolate reductase. *Annu. Rev. Biophys. Biomol. Struct.* 33:119–140.

Li, R., Sirawaraporn, R., Chitnumsub, P., Sirawaraporn, W., Wooden, J., Athappilly, F., Turley, S., and Hol, W. G. 2000. Three-dimensional structure of *M. tuberculosis* dihydrofolate reductase reveals opportunities for the design of novel tuberculosis drugs. *J. Mol. Biol.* 295:307–323.

Liang, P. H., and Anderson, K. S. 1998. Substrate channeling and domain-domain interactions in bifunctional thymidylate synthase-dihydrofolate reductase. *Biochemistry* 37:12195–12205.

Miller, G. P., and Benkovic, S. J. 1998. Stretching exercises: Flexibility in dihydrofolate reductase catalysis. *Chem. Biol.* 5:R105–R113.

Blakley, R. L. 1995. Eukaryotic dihydrofolate reductase. *Adv. Enzymol. Relat. Areas Mol. Biol.* 70:23–102.

Carreras, C. W., and Santi, D. V. 1995. The catalytic mechanism and structure of thymidylate synthase. *Annu. Rev. Biochem.* 64:721–762.

Schweitzer, B. I., Dicker, A. P., and Bertino, J. R. 1990. Dihydrofolate reductase as a therapeutic target. *FASEB J.* 4:2441–2452.

Bystroff, C., Oatley, S. J., and Kraut, J. 1990. Crystal structures of *Escherichia coli* dihydrofolate reductase: The NADP$^+$ holoenzyme and the folate NADP$^+$ ternary complex—Substrate binding and a model for the transition state. *Biochemistry* 29:3263–3277.

Defects in Nucleotide Biosynthesis

Scriver, C. R., Beaudet, A. L., Sly, W. S., Valle, D., Stanbury, J. B., Wyngaarden, J. B., and Fredrickson, D. S. (Eds.). 1995. *The Metabolic Basis of Inherited Diseases* (7th ed., pp. 1655–1840). McGraw-Hill.

Nyhan, W. L. 1997. The recognition of Lesch-Nyhan syndrome as an inborn error of purine metabolism. *J. Inherited Metab. Dis.* 20:171–178.

Wong, D. F., Harris, J. C., Naidu, S., Yokoi, F., Marenco, S., Dannals, R. F., Ravert, H. T., Yaster, M., Evans, A., Rousset, O., Bryan, R. N., Gjedde, A., Kuhar, M. J., and Breese, G. R. 1996. Dopamine transporters are markedly reduced in Lesch-Nyhan disease in vivo. *Proc. Natl. Acad. Sci. U. S. A.* 93:5539–5543.

Resta, R., and Thompson, L. F. 1997. SCID: The role of adenosine deaminase deficiency. *Immunol. Today* 18:371–374.

Davidson, B. L., Pashmforoush, M., Kelley, W. N., and Palella, T. D. 1989. Human hypoxanthine-guanine phosphoribosyltransferase deficiency: The molecular defect in a patient with gout (HPRTAshville). *J. Biol. Chem.* 264:520–525.

Sculley, D. G., Dawson, P. A., Emerson, B. T., and Gordon, R. B. 1992. A review of the molecular basis of hypoxanthine-guanine phosphoribosyltransferase (HPRT) deficiency. *Hum. Genet.* 90:195–207.

Neychev, V. K., and Mitev, V. I. 2004. The biochemical basis of the neurobehavioral abnormalities in the Lesch-Nyhan syndrome: A hypothesis. *Med. Hypotheses* 63:131–134.

Daly, L. E., Kirke, P. N., Molloy, A., Weir, D. G., and Scott, J. M. 1995. Folate levels and neural tube defects: Implications for prevention. *JAMA* 274:1698–1702.

Problems

1. *Activated ribose phosphate.* Write a balanced equation for the synthesis of PRPP from glucose through the oxidative branch of the pentose phosphate pathway.

2. *Making a pyrimidine.* Write a balanced equation for the synthesis of orotate from glutamine, CO_2, and aspartate.

3. *Identifying the donor.* What is the activated reactant in the biosynthesis of each of the following compounds?

 (a) Phosphoribosylamine
 (b) Carbamoylaspartate
 (c) Orotidylate (from orotate)
 (d) Phosphoribosylanthranilate

4. *Inhibiting purine biosynthesis.* Amidotransferases are inhibited by the antibiotic azaserine (*O*-diazoacetyl-L-serine), which is an analog of glutamine.

Azaserine

Which intermediates in purine biosynthesis would accumulate in cells treated with azaserine?

5. *The price of methylation.* Write a balanced equation for the synthesis of TMP from dUMP that is coupled to the conversion of serine into glycine.

6. *Sulfa action.* Bacterial growth is inhibited by sulfanilamide and related sulfa drugs, and there is a concomitant accumulation of 5-aminoimidazole-4-carboxamide ribonucleotide. This inhibition is reversed by the addition of *p*-aminobenzoate.

Sulfanilamide

Propose a mechanism for the inhibitory effect of sulfanilamide.

7. *A generous donor.* What major biosynthetic reactions utilize PRPP?

8. *HAT medium.* Mutant cells unable to synthesize nucleotides by salvage pathways are very useful tools in molecular and cell biology. Suppose that cell A lacks thymidine kinase, the enzyme catalyzing the phosphorylation of thymidine to thymidylate, and that cell B lacks hypoxanthine-guanine phosphoribosyl transferase.

(a) Cell A and cell B do not proliferate in a HAT medium containing *hypoxanthine*, *aminopterin* or *amethopterin* (methotrexate), and thymine. However, cell C formed by the fusion of cells A and B grows in this medium. Why?

(b) Suppose that you wanted to introduce foreign genes into cell A. Devise a simple means of distinguishing between cells that have taken up foreign DNA and those that have not.

9. *Find the label.* Suppose that cells are grown on amino acids that have all been labeled at the α carbons with ^{13}C. Identify the atoms in cytosine and guanine that will be labeled with ^{13}C.

10. *Different strokes.* Human beings contain two different carbamoyl phosphate synthetase enzymes. One uses glutamine as a substrate, whereas the other uses ammonia. What are the functions of the these two enzymes?

11. *Adjunct therapy.* Allopurinol is sometimes given to patients with acute leukemia who are being treated with anticancer drugs. Why is allopurinol used?

12. *A hobbled enzyme.* Both side-chain oxygen atoms of aspartate 27 at the active site of dihydrofolate reductase form hydrogen bonds with the pteridine ring of folates. The importance of this interaction was assessed by studying two mutants at this position, Asn 27 and Ser 27. The dissociation constant of methotrexate was 0.07 nM for the wild type, 1.9 nM for the Asn 27 mutant, and 210 nM for the Ser 27 mutant, at 25°C. Calculate the standard free energy of the binding of methotrexate by these three proteins. What is the decrease in binding energy resulting from each mutation?

13. *Correcting deficiencies.* Suppose that a person is found who is deficient in an enzyme required for IMP synthesis. How might this person be treated?

14. *Labeled nitrogen.* Purine biosynthesis is allowed to take place in the presence of $[^{15}N]$aspartate, and the newly synthesized GTP and ATP are isolated. What positions are labeled in the two nucleotides?

15. *Changed inhibitor.* Xanthine oxidase treated with allopurinol results in the formation of a new compound that is an extremely potent inhibitor of the enzyme. Propose a structure for this compound.

Mechanism Problems

16. *The same and not the same.* Write out mechanisms for the conversion of phosphoribosylamine into glycinamide ribonucleotide and of xanthylate into guanylate.

17. *Closing the ring.* Propose a mechanism for the conversion of 5-formamidoimidazole-4-carboxamide ribonucleotide into inosinate.

Chapter Integration Problems

18. *They're everywhere!* Nucleotides play a variety of roles in the cell. Give an example of a nucleotide that acts in each of the following roles or processes.

 (a) Second messenger
 (b) Phosphoryl-group transfer
 (c) Activation of carbohydrates
 (d) Activation of acetyl groups
 (e) Transfer of electrons
 (f) DNA sequencing
 (g) Chemotherapy
 (h) Allosteric effector

19. *Pernicious anemia.* Purine biosynthesis is impaired by vitamin B_{12} deficiency. Why? How might fatty acid and amino acid metabolism also be affected by a vitamin B_{12} deficiency?

20. *Hyperuricemia.* Many patients with glucose 6-phosphatase deficiency have high serum levels of urate. Hyperuricemia can be induced in normal people by the ingestion of alcohol or by strenuous exercise. Propose a common mechanism that accounts for these findings.

21. *Labeled carbon.* Succinate uniformly labeled with ^{14}C is added to cells actively engaged in pyrimidine biosynthesis. Propose a mechanism by which carbon atoms from succinate could be incorporated into a pyrimidine. At what positions is the pyrimidine labeled?

22. *Exercising muscle.* Some interesting reactions take place in muscle tissue to facilitate the generation of ATP for contraction.

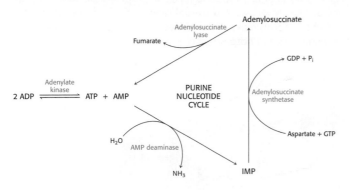

In muscle contraction, ATP is converted into ADP. Adenylate kinase converts two molecules of ADP into a molecule of ATP and AMP.

(a) Why is this reaction beneficial to contracting muscle?
(b) Why is the equilibrium for the adenylate kinase approximately equal to 1?

Muscle can metabolize AMP by using the purine nucleotide cycle. The initial step in this cycle, catalyzed by AMP deaminase, is the conversion of AMP into IMP.

(c) Why might the deamination of AMP facilitate ATP formation in muscle?
(d) How does the purine nucleotide cycle assist the aerobic generation of ATP?

The Biosynthesis of Membrane Lipids and Steroids

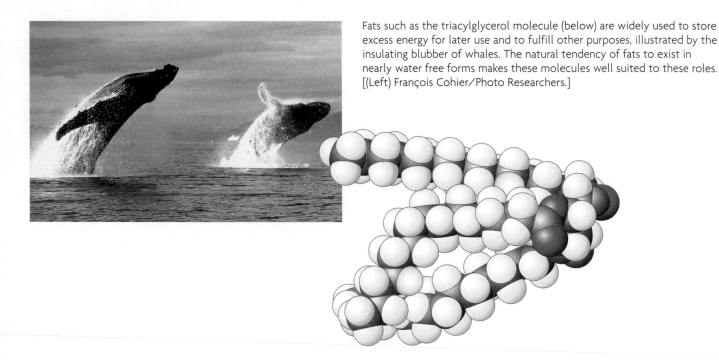

Fats such as the triacylglycerol molecule (below) are widely used to store excess energy for later use and to fulfill other purposes, illustrated by the insulating blubber of whales. The natural tendency of fats to exist in nearly water free forms makes these molecules well suited to these roles. [(Left) François Cohier/Photo Researchers.]

This chapter examines the biosynthesis of three important components of biological membranes—phospholipids, sphingolipids, and cholesterol (Chapter 12). Triacylglycerols also are considered here because the pathway for their synthesis overlaps that of phospholipids. Cholesterol is of interest both as a membrane component and as a precursor of many signal molecules, including the steroid hormones progesterone, testosterone, estrogen, and cortisol.

Cholesterol is transported in blood by the low-density lipoprotein (LDL) and taken up into cells by the LDL receptor on the cell surface. The transport and uptake of cholesterol vividly illustrate a recurring mechanism for the entry of metabolites and signal molecules into cells. The LDL receptor is absent in people with *familial hypercholesterolemia,* a genetic disease. People lacking the receptor have markedly elevated cholesterol levels in the blood and cholesterol deposits on blood vessels, and they are prone to childhood heart attacks. Indeed, cholesterol is implicated in the development of atherosclerosis in people who do not have genetic defects. Thus, the regulation of cholesterol synthesis and transport can be a source of especially clear insight into the role that our understanding of biochemistry plays in medicine.

26.1 Phosphatidate Is a Common Intermediate in the Synthesis of Phospholipids and Triacylglycerols

Figure 26.1 provides a broad overview of lipid synthesis. The first step in the synthesis of both phospholipids for membranes and triacylglycerols for energy storage is the synthesis of *phosphatidate* (diacylglycerol 3-phosphate). In mammalian cells, phosphatidate is synthesized in the endoplasmic reticulum and the outer mitochondrial membrane. The pathway begins with *glycerol 3-phosphate*, which is formed primarily by the reduction of dihydroxyacetone phosphate (DHAP), a glycolytic intermediate, and to a lesser extent by the phosphorylation of glycerol. The addition of two fatty acids to glycerol-3-phosphate yields phosphatidate. First, acyl coenzyme A contributes a fatty acid chain to form *lysophosphatidate* and, then, a second acyl CoA contributes a fatty acid chain to yield phosphatidate.

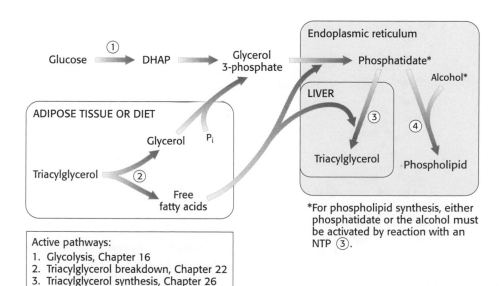

Glycerol 3-phosphate   **Lysophosphatidate**   **Phosphatidate**

These acylations are catalyzed by *glycerol phosphate acyltransferase.* In most phosphatidates, the fatty acid chain attached to the C-1 atom is saturated, whereas the one attached to the C-2 atom is unsaturated.

The pathways diverge at phosphatidate. In the synthesis of triacylglycerols, phosphatidate is hydrolyzed by a specific phosphatase to give a *diacylglycerol* (DAG). This intermediate is acylated to a *triacylglycerol* through the addition of a third fatty acid chain in a reaction that is catalyzed by *diglyceride acyltransferase.* Both enzymes are associated in a *triacylglycerol synthetase complex* that is bound to the endoplasmic reticulum membrane.

*For phospholipid synthesis, either phosphatidate or the alcohol must be activated by reaction with an NTP ③.

Active pathways:
1. Glycolysis, Chapter 16
2. Triacylglycerol breakdown, Chapter 22
3. Triacylglycerol synthesis, Chapter 26
4. Phospholipid synthesis, Chapter 26

Figure 26.1 PATHWAY INTEGRATION: Sources of intermediates in the synthesis of triacylglycerols and phospholipids. Phosphatidate, synthesized from dihydroxyacetone phosphate (DHAP) produced in glycolysis and fatty acids, can be further processed to produce triacylglycerol or phospholipids. Phospholipids and other membrane lipids are continuously produced in all cells.

Phosphatidate → **Diacylglycerol (DAG)** → **Triacylglycerol**

The liver is the primary site of triacylglycerol synthesis. From the liver, the triacylglycerols are transported to the muscles for energy conversion or to the adipose cells for storage.

The Synthesis of Phospholipids Requires an Activated Intermediate

Membrane-lipid synthesis continues in the endoplasmic reticulum. *Phospholipid* synthesis requires the combination of a diacylglycerol with an alcohol. As in most anabolic reactions, one of the components must be activated. In this case, either the diacylglerol or the alcohol may be activated, depending on the source of the reactants.

Synthesis from an Activated Diacylglycerol. The de novo pathway starts with the reaction of phosphatidate with cytidine triphosphate (CTP) to form the activated diacylglycerol, *cytidine diphosphodiacylglycerol* (CDP-diacylglycerol; Figure 26.2). This reaction, like those of many biosyntheses, is driven forward by the hydrolysis of pyrophosphate.

Phosphatidate → **CDP-diacylglycerol**

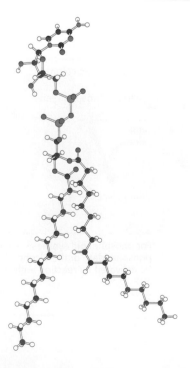

Figure 26.2 Structure of CDP-diacylglycerol. A key intermediate in the synthesis of phospholipids consists of phosphatidate and cytidine monophosphate joined by a pyrophosphate linkage.

The activated phosphatidyl unit then reacts with the hydroxyl group of an alcohol to form a phosphodiester linkage. If the alcohol is inositol, the products are *phosphatidylinositol* and cytidine monophosphate (CMP).

CDP-diacylglycerol   **Inositol**

Phosphatidylinositol   **CMP**

Subsequent phosphorylations catalyzed by specific kinases lead to the synthesis of *phosphatidylinositol 4,5-bisphosphate*, the precursor of two intracellular messengers—diacylglycerol and inositol 1,4,5-trisphosphate (Section 14.2). If the alcohol is phosphatidylglycerol, the products are diphosphatidylglycerol (cardiolipin) and CMP. In eukaryotes, cardiolipin is located exclusively in inner mitochondrial membranes and plays an important role in the organization of the protein components of oxidative phosphorylation.

Diphosphatidylglycerol (Cardiolipin)

   The fatty acid components of phospholipids may vary, and thus cardiolipin, as well as most other phospholipids, represents a class of molecules rather than a single species. As a result, a single mammalian cell may contain thousands of distinct phospholipids. Phosphatidylinositol is unusual in that it has a nearly fixed fatty acid composition. Stearic acid usually occupies the C-1 position and arachidonic acid the C-2 position.

Synthesis from an Activated Alcohol.

Phosphatidylethanolamine, a common phospholipid in mammals, is synthesized from the alcohol ethanolamine. To activate the alcohol, ethanolamine is phosphorylated by ATP to form the precursor, *phosphorylethanolamine*. This precursor then reacts with CTP to form the activated alcohol, *CDP-ethanolamine*. The

Ethanolamine

ATP
ADP

Phosphorylethanolamine

CTP
PP$_i$

CDP-ethanolamine

Diacylglycerol
CMP

Phosphatidylethanolamine

phosphorylethanolamine unit of CDP-ethanolamine is transferred to a diacylglycerol to form *phosphatidylethanolamine.*

The most common phospholipid in mammals is phosphatidylcholine. In this case, dietary choline is activated in a series of reactions analogous to those in the activation of ethanolamine. Interestingly, the liver possesses an enzyme, *phosphatidylethanolamine methyltransferase,* that synthesizes phosphatidylcholine from phosphatidylethanolamine when dietary choline is insufficient. The amino group of this phosphatidylethanolamine is methylated three times to form *phosphatidylcholine.* S-*Adenosylmethionine* is the methyl donor.

Phosphatidyl-ethanolamine → **Phosphatidyl-choline**

(3 S-Adenosyl methionine, 3 S-Adenosyl homocysteine)

Thus, phosphatidylcholine can be produced by two distinct pathways in mammals, ensuring that this phospholipid can be synthesized even if the components for one pathway are in limited supply.

Phosphatidylserine makes up 10% of phospholipids in mammals. This phospholipid is synthesized in a *base-exchange reaction* of serine with phosphatidylcholine or phosphatidylethanolamine. In the reaction, serine replaces choline or ethanolamine.

Phosphatidylcholine + serine ⟶ choline + phosphatidylserine

Phosphatidylethanolamine + serine ⟶
$\qquad\qquad\qquad$ ethanolamine + phosphatidylserine

Phosphatidylserine is normally located in the inner leaflet of the plasma membrane bilayer but is moved to the outer leaflet in apoptosis. There, it may serve to attracted phagocytes to consume the cell remnants after apoptosis is complete.

Note that a cytidine nucleotide plays the same role in the synthesis of these phosphoglycerides as a uridine nucleotide does in the formation of glycogen (p. 604). In all of these biosyntheses, an activated intermediate (UDP-glucose, CDP-diacylglycerol, or CDP-alcohol) is formed from a phosphorylated substrate (glucose 1-phosphate, phosphatidate, or a phosphorylalcohol) and a nucleoside triphosphate (UTP or CTP). The activated intermediate then reacts with a hydroxyl group (the terminus of glycogen, the side chain of serine, or a diacylglycerol).

Sphingolipids Are Synthesized from Ceramide

We turn now from glycerol-based phospholipids to another class of membrane lipid—the *sphingolipids.* These lipids are found in the plasma membranes of all eukaryotic cells, although the concentration is highest in the cells of the central nervous system. The backbone of a sphingolipid is *sphingosine,* rather than glycerol. Palmitoyl CoA and serine condense to form 3-ketosphinganine, which is reduced to dihydrosphingosine before conversion into *ceramide,* a lipid consisting of a fatty acid chain attached to the amino group of a sphingosine backbone. The enzyme catalyzing this reaction requires pyridoxal phosphate, revealing again the dominant role of this cofactor in transformations that include amino acids.

Sphingosine

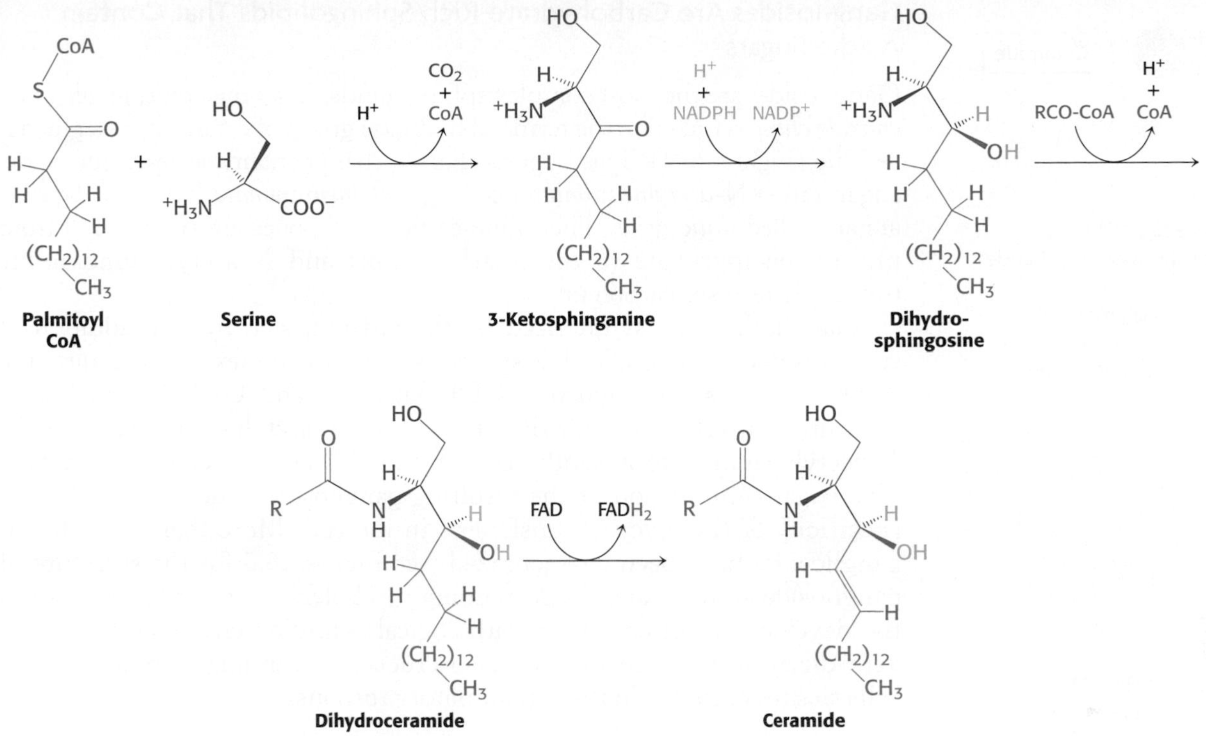

In all sphingolipids, the amino group of ceramide is acylated (Figure 26.3). The terminal hydroxyl group also is substituted. In *sphingomyelin,* a component of the myelin sheath covering many nerve fibers, the substituent is phosphorylcholine, which comes from phosphatidylcholine. In a *cerebroside,* the substituent is glucose or galactose. UDP-glucose or UDP-galactose is the sugar donor.

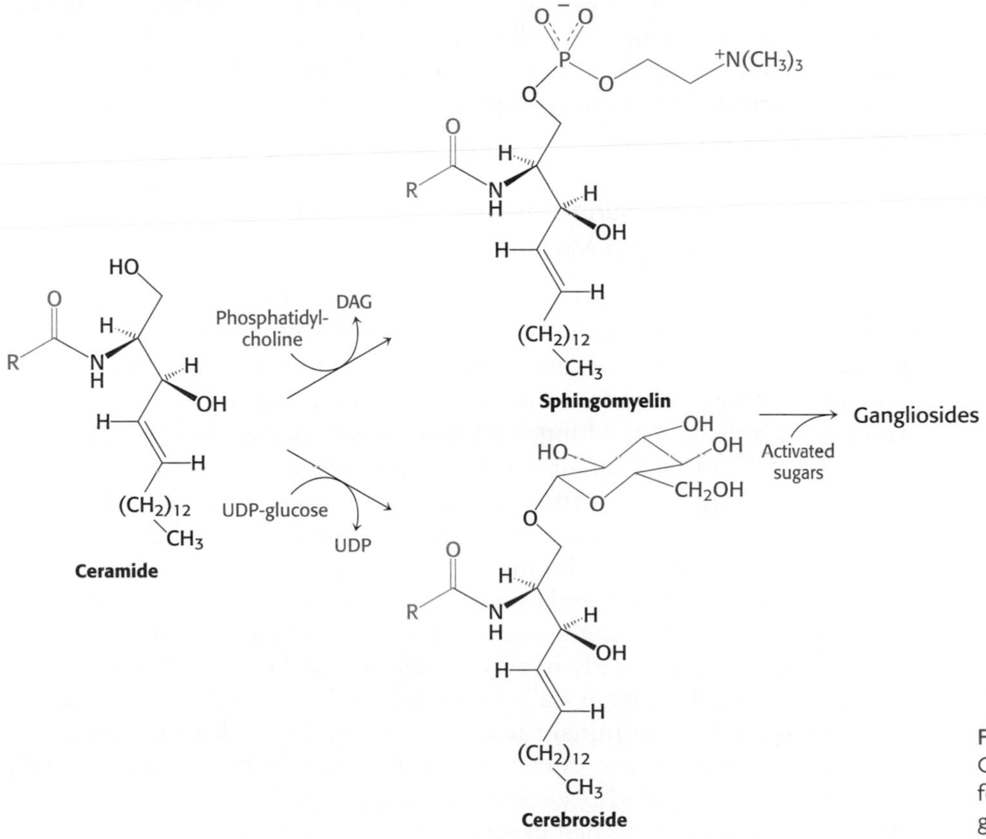

Figure 26.3 Synthesis of sphingolipids. Ceramide is the starting point for the formation of sphingomyelin and gangliosides.

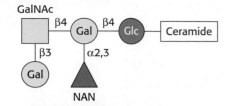

GalNAc

β4 — Gal — β4 — Glc — Ceramide

β3

Gal

α2,3

NAN

Figure 26.4 Ganglioside G$_{M1}$. This ganglioside consists of five monosaccharides linked to ceramide: one glucose (Glc) molecule, two galactose (Gal) molecules, one N-acetylgalactosamine (GalNAc) molecule, and one N-acetylneuraminate (NAN) molecule.

R$_2$ = H, N-acetylneuraminate
R$_2$ = OH, N-glycolylneuraminate

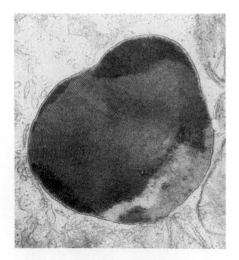

Figure 26.5 Lysosome with lipids. An electron micrograph of a lysosome containing an abnormal amount of lipid. [Courtesy of Dr. George Palade.]

Gangliosides Are Carbohydrate-Rich Sphingolipids That Contain Acidic Sugars

Gangliosides are the most complex sphingolipids. In a *ganglioside,* an *oligosaccharide chain* is linked to the terminal hydroxyl group of ceramide by a glucose residue (Figure 26.4). This oligosaccharide chain contains at least one acidic sugar, either N-*acetylneuraminate* or N-*glycolylneuraminate.* These acidic sugars are called *sialic acids.* Their nine-carbon backbones are synthesized from phosphoenolpyruvate (a three-carbon unit) and N-acetylmannosamine 6-phosphate (a six-carbon unit).

Gangliosides are synthesized by the ordered, step-by-step addition of sugar residues to ceramide. The synthesis of these complex lipids requires the activated sugars UDP-glucose, UDP-galactose, and UDP-N-acetylgalactosamine, as well as the CMP derivative of N-acetylneuraminate. CMP-N-acetylneuraminate is synthesized from CTP and N-acetylneuraminate. The sugar composition of the resulting ganglioside is determined by the specificity of the glycosyltransferases in the cell. More than 60 different gangliosides have been characterized (see Figure 26.3 for the structure of ganglioside G$_{M1}$). Ganglioside-binding by cholera toxin is the first step in the development of cholera, a pathological condition characterized by severe diarrhea. Gangliosides are also crucial for binding immune-system cells to sites of injury in the inflammatory response.

Sphingolipids Confer Diversity on Lipid Structure and Function

The structures of sphingolipids and the more abundant glycerophospholipids are very similar. Given the structural similarity of these two types of lipids, why are sphingolipids required at all? Indeed, the prefix "sphingo" was applied to capture the "sphinxlike" properties of this enigmatic class of lipids. Although the precise role of sphingolipids is not firmly established, progress toward solving the riddle of their function is being made. Most notably, sphingolipids may serve as a source of second messengers. For instance, ceramide derived from a sphingolipid may initiate programmed cell death in some cell types.

Respiratory Distress Syndrome and Tay-Sachs Disease Result from the Disruption of Lipid Metabolism

Respiratory distress syndrome is a pathological condition resulting from a failure in the biosynthetic pathway for dipalmitoyl phosphatidylcholine. This phospholipid, in conjunction with specific proteins and other phospholipids, is found in the extracellular fluid that surrounds the alveoli of the lung. Its function is to decrease the surface tension of the fluid to prevent lung collapse at the end of the expiration phase of breathing. Premature infants may suffer from respiratory distress syndrome because their immature lungs do not synthesize enough dipalmitoyl phosphatidylcholine.

Tay-Sachs disease is caused by a failure of lipid degradation: an inability to degrade gangliosides. Gangliosides are found in highest concentration in the nervous system, particularly in gray matter, where they constitute 6% of the lipids. Gangliosides are normally degraded inside lysosomes by the sequential removal of their terminal sugars but, in Tay-Sachs disease, this degradation does not take place. As a consequence, neurons become significantly swollen with lipid-filled lysosomes (Figure 26.5). An affected infant displays weakness and retarded psychomotor skills before 1 year of age. The child is demented and blind by age 2 and usually dies before age 3.

The ganglioside content of the brain of an infant with Tay-Sachs disease is greatly elevated. *The concentration of ganglioside G$_{M2}$ is many times higher than normal because its terminal N-acetylgalactosamine residue is removed very slowly or not at all.* The missing or deficient enzyme is a specific β-N-acetylhexosaminidase.

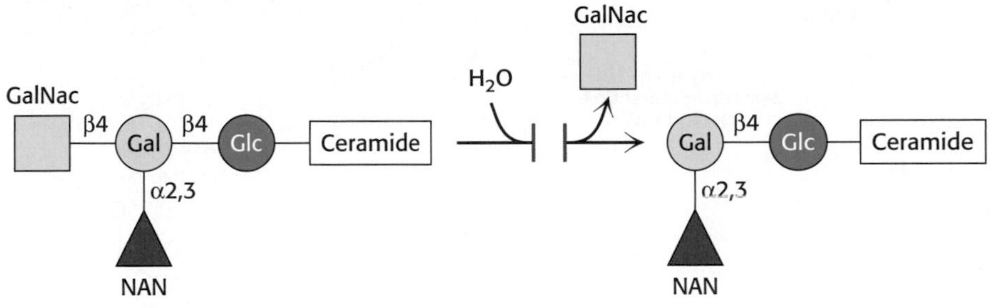

Ganglioside G$_{M2}$ **Ganglioside G$_{M3}$**

Tay-Sachs disease can be diagnosed in the course of fetal development. Amniotic fluid is obtained by amniocentesis and assayed for β-*N*-acetylhexosaminidase activity.

26.2 Cholesterol Is Synthesized from Acetyl Coenzyme A in Three Stages

We now turn our attention to the synthesis of the fundamental lipid *cholesterol*. This steroid modulates the fluidity of animal cell membranes (p. 343) and is the precursor of steroid hormones such as progesterone, testosterone, estradiol, and cortisol. *All 27 carbon atoms of cholesterol are derived from acetyl CoA in a three-stage synthetic process* (Figure 26.6).

1. Stage one is the synthesis of isopentenyl pyrophosphate, an activated isoprene unit that is the key building block of cholesterol.

2. Stage two is the condensation of six molecules of isopentenyl pyrophosphate to form squalene.

3. In stage three, squalene cyclizes and the tetracyclic product is subsequently converted into cholesterol.

The first stage takes place in the cytoplasm, and the second two in the endoplasmic reticulum.

The Synthesis of Mevalonate, Which Is Activated as Isopentenyl Pyrophosphate, Initiates the Synthesis of Cholesterol

The first stage in the synthesis of cholesterol is the formation of isopentenyl pyrophosphate from acetyl CoA. This set of reactions starts with the formation of 3-hydroxy-3-methylglutaryl CoA (HMG CoA) from acetyl CoA and acetoacetyl CoA. This intermediate is reduced to *mevalonate* for the synthesis of cholesterol (Figure 26.7). Recall that, alternatively, mitochondrial 3-hydroxy-3-methylglutaryl CoA may be processed to form ketone bodies (p. 631).

The synthesis of mevalonate is the committed step in cholesterol formation. The enzyme catalyzing this irreversible step, *3-hydroxy-3-methylglutaryl CoA reductase* (HMG-CoA reductase), is an important control site in cholesterol biosynthesis, as will be discussed shortly.

3-Hydroxy-3-methylglutaryl CoA + 2 NADPH + 2 H$^+$ $\longrightarrow$
mevalonate + 2 NADP$^+$ + CoA

> **Cholesterol**
>
> "Cholesterol is the most highly decorated small molecule in biology. Thirteen Nobel Prizes have been awarded to scientists who devoted major parts of their careers to cholesterol. Ever since it was isolated from gallstones in 1784, cholesterol has exerted an almost hypnotic fascination for scientists from the most diverse areas of science and medicine. . . . Cholesterol is a Janus-faced molecule. The very property that makes it useful in cell membranes, namely its absolute insolubility in water, also makes it lethal."
>
> —MICHAEL BROWN AND
> JOSEPH GOLDSTEIN
> *Nobel Lectures (1985)*
> © The Nobel Foundation, 1985

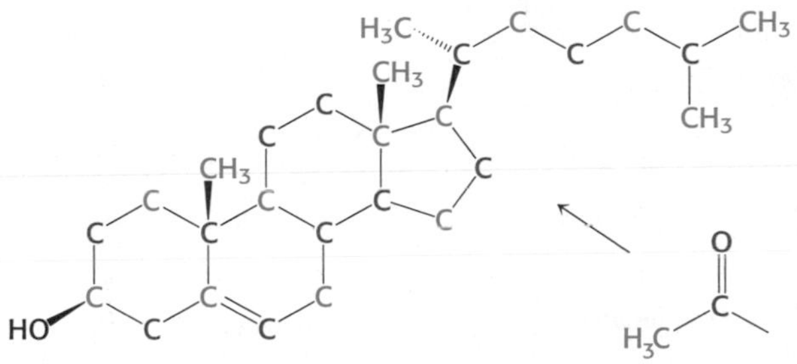

Figure 26.6 Labeling of cholesterol. Isotope-labeling experiments reveal the source of carbon atoms in cholesterol synthesized from acetate labeled in its methyl group (blue) or carboxylate atom (red).

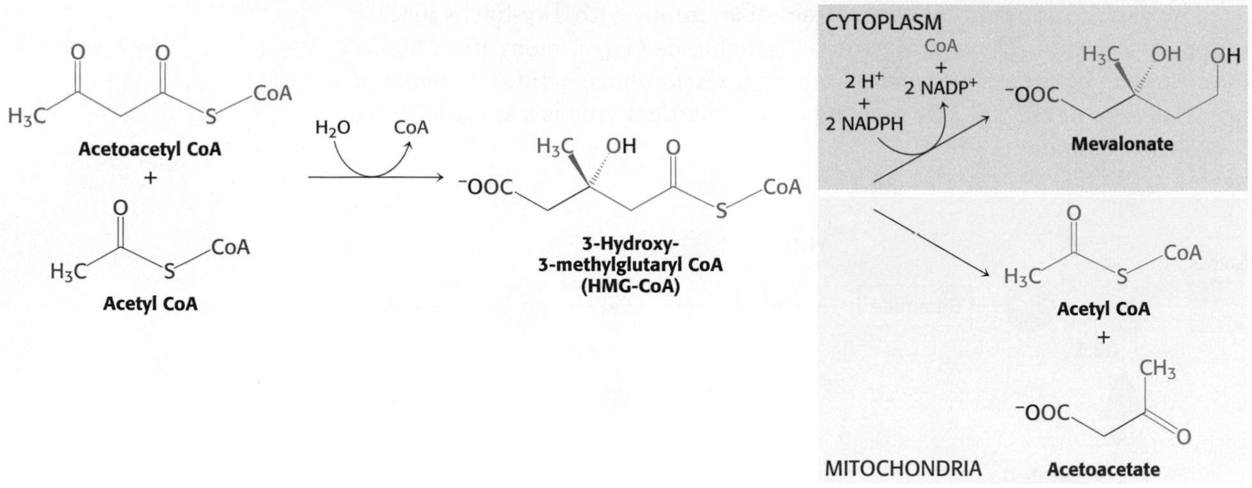

Figure 26.7 Fates of 3-hydroxy-3-methylglutaryl CoA. In the cytoplasm, HMG-CoA is converted into mevalonate. In mitochondria, it is converted into acetyl CoA and acetoacetate.

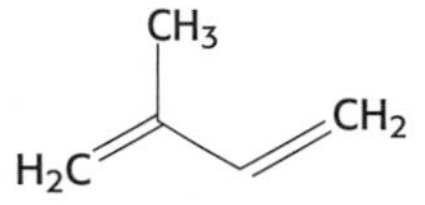

HMG-CoA reductase is an integral membrane protein in the endoplasmic reticulum.

Mevalonate is converted into *3-isopentenyl pyrophosphate* in three consecutive reactions requiring ATP (Figure 26.8). In the last step, the release of CO_2 yields isopentenyl pyrophosphate, an activated isoprene unit that is a key building block for many important biomolecules throughout the kingdoms of life.

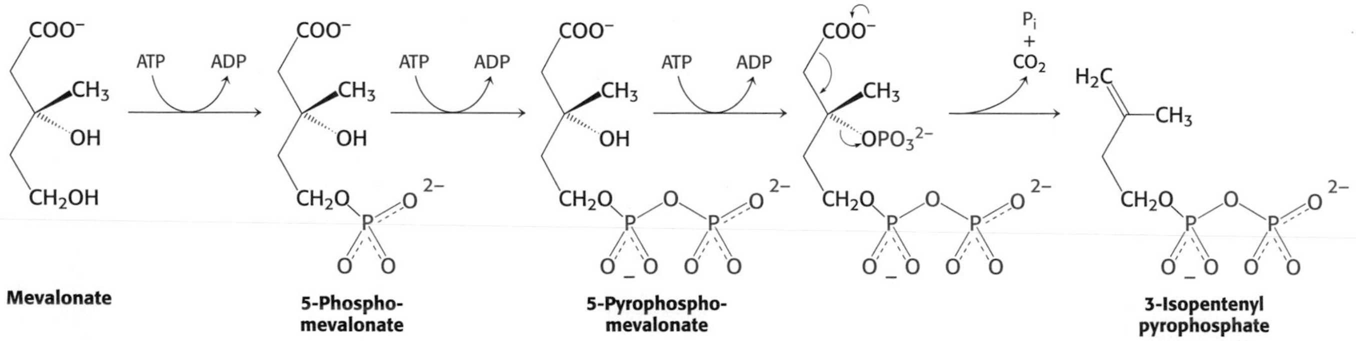

Figure 26.8 Synthesis of isopentenyl pyrophosphate. This activated intermediate is formed from mevalonate in three steps, the last of which includes a decarboxylation.

Squalene (C_{30}) Is Synthesized from Six Molecules of Isopentenyl Pyrophosphate (C_5)

Squalene is synthesized from isopentenyl pyrophosphate by the reaction sequence

$$C_5 \longrightarrow C_{10} \longrightarrow C_{15} \longrightarrow C_{30}$$

This stage in the synthesis of cholesterol starts with the isomerization of *isopentenyl pyrophosphate* to *dimethylallyl pyrophosphate*.

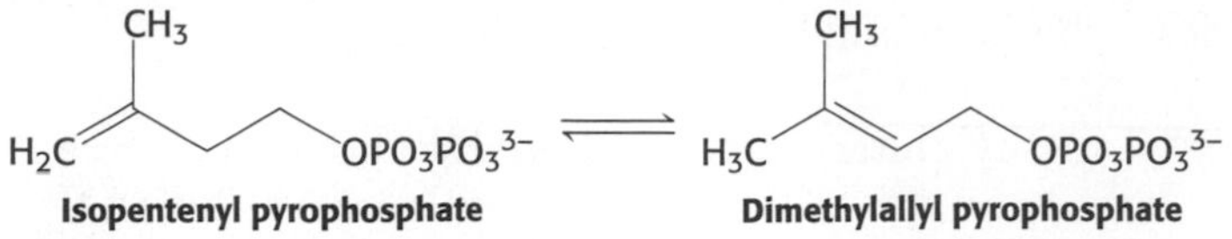

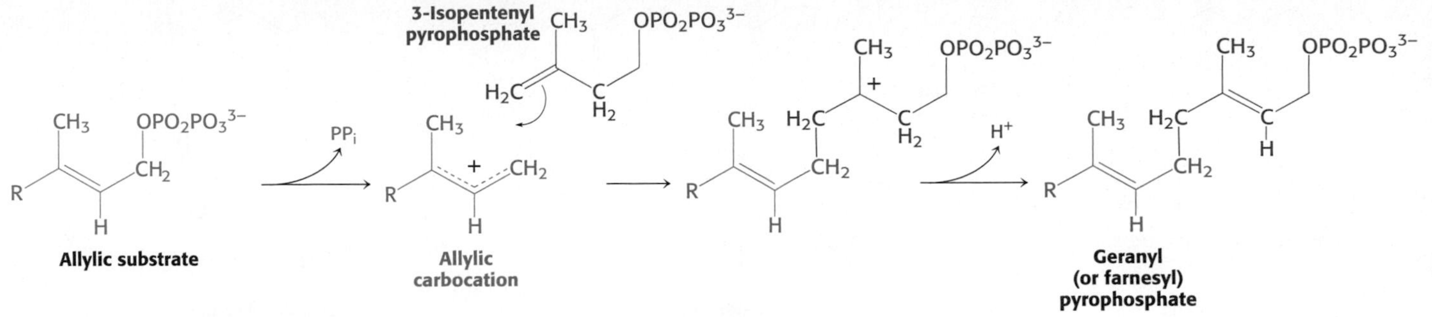

Figure 26.9 Condensation mechanism in cholesterol synthesis. The mechanism for joining dimethylallyl pyrophosphate and isopentenyl pyrophosphate to form geranyl pyrophosphate. The same mechanism is used to add an additional isopentenyl pyrophosphate to form farnesyl pyrophosphate.

These two isomeric C_5 units (one of each type) condense to form a C_{10} compound: isopentenyl pyrophosphate attacks an allylic carbocation formed from dimethylallyl pyrophosphate to yield *geranyl pyrophosphate* (Figure 26.9). The same kind of reaction takes place again: geranyl pyrophosphate is converted into an allylic carbonium ion, which is attacked by isopentenyl pyrophosphate. The resulting C_{15} compound is called *farnesyl pyrophosphate*. The same enzyme, *geranyl transferase*, catalyzes each of these condensations.

The last step in the synthesis of *squalene* is a reductive tail-to-tail condensation of two molecules of farnesyl pyrophosphate catalyzed by the endoplasmic reticulum enzyme *squalene synthase*.

$$2 \text{ Farnesyl pyrophosphate } (C_{15}) + \text{NADPH} \longrightarrow$$
$$\text{squalene } (C_{30}) + 2 \text{ PP}_i + \text{NADP}^+ + \text{H}^+$$

The reactions leading from C_5 units to squalene, a C_{30} isoprenoid, are summarized in Figure 26.10.

Squalene Cyclizes to Form Cholesterol

The final stage of cholesterol biosynthesis starts with the cyclization of squalene (Figure 26.11). Squalene is first activated by conversion into squalene epoxide (2,3-oxidosqualene) in a reaction that uses O_2 and NADPH. Squalene epoxide is then cyclized to *lanosterol* by *oxidosqualene cyclase*. This remarkable transformation proceeds in a concerted fashion. The enzyme holds squalene epoxide in an appropriate conformation and initiates the reaction by protonating the epoxide oxygen. The carbocation formed spontaneously rearranges to produce lanosterol. Lanosterol is converted into cholesterol in a

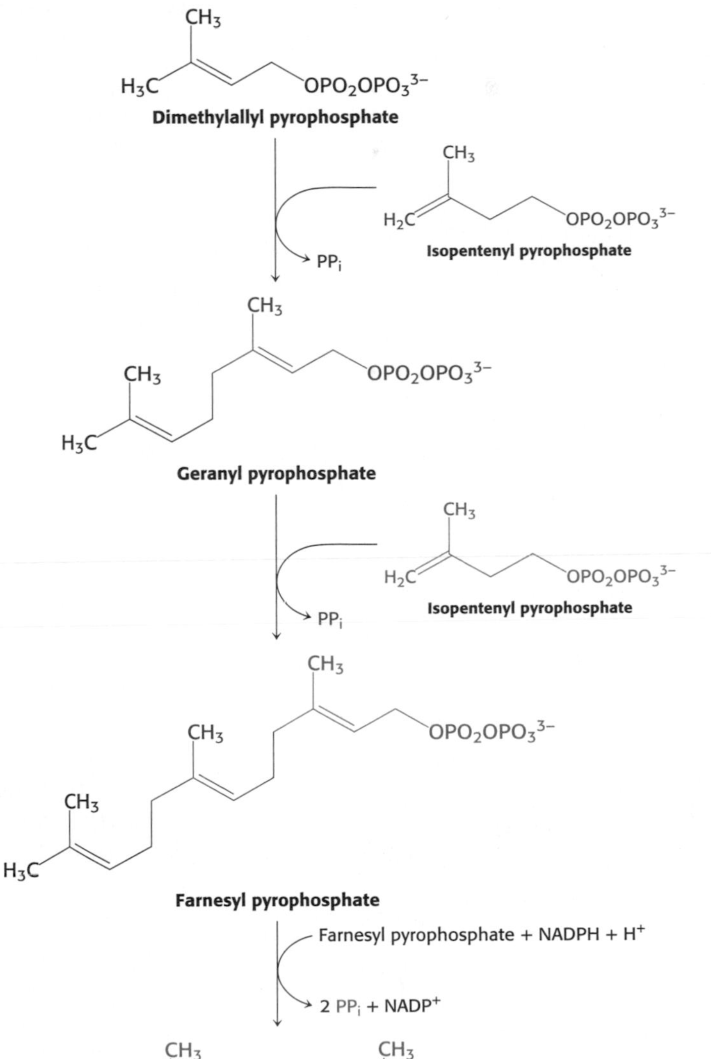

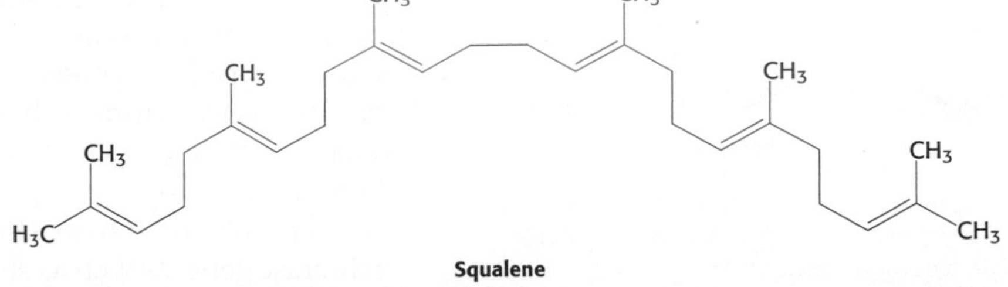

Figure 26.10 Squalene synthesis. One molecule of dimethyallyl pyrophosphate and two molecules of isopentenyl pyrophosphate condense to form farnesyl pyrophosphate. The tail-to-tail coupling of two molecules of farnesyl pyrophosphate yields squalene.

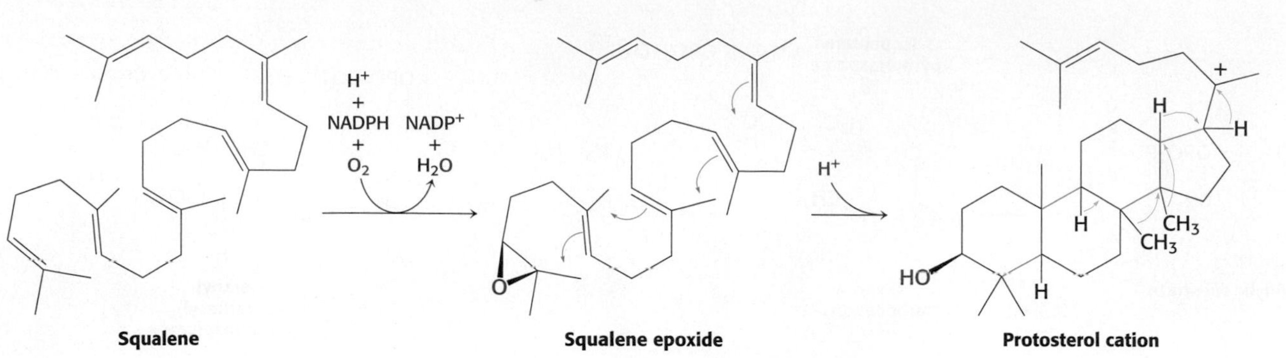

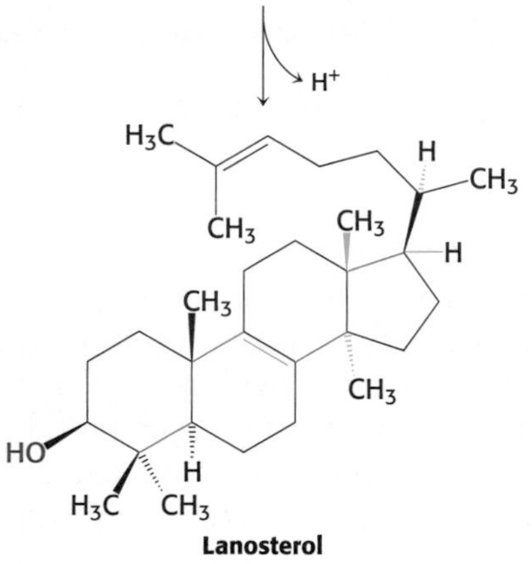

Figure 26.11 Squalene cyclization. The formation of the steroid nucleus from squalene begins with the formation of squalene epoxide. This intermediate is protonated to form a carbocation that cyclizes to form a tetracyclic structure, which rearranges to form lanosterol.

multistep process by the removal of three methyl groups, the reduction of one double bond by NADPH, and the migration of the other double bond (Figure 26.12).

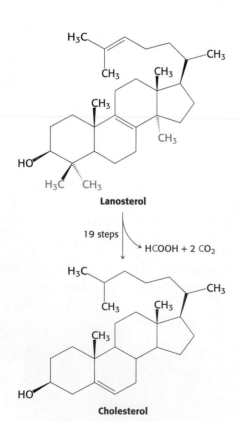

Figure 26.12 Cholesterol formation. Lanosterol is converted into cholesterol in a complex process.

26.3 The Complex Regulation of Cholesterol Biosynthesis Takes Place at Several Levels

Cholesterol can be obtained from the diet or it can be synthesized de novo. An adult on a low-cholesterol diet typically synthesizes about 800 mg of cholesterol per day. The liver is the major site of cholesterol synthesis in mammals, although the intestine also forms significant amounts. The rate of cholesterol formation by these organs is highly responsive to the cellular level of cholesterol. *This feedback regulation is mediated primarily by changes in the amount and activity of 3-hydroxy-3-methylglutaryl CoA reductase.* As described earlier (p. 739), this enzyme catalyzes the formation of mevalonate, the committed step in cholesterol biosynthesis. HMG CoA reductase is controlled in multiple ways:

1. The rate of *synthesis of reductase mRNA* is controlled by the *sterol regulatory element binding protein* (SREBP). This transcription factor binds to a short DNA sequence called the *sterol regulatory element* (SRE) on the 5′ side of the reductase gene. It binds to the SRE when cholesterol levels are low and enhances transcription. In its inactive state, the SREBP resides in the endoplasmic reticulum membrane, where it is associated with the SREBP cleavage activating protein (SCAP), an integral membrane protein. SCAP is the cholesterol sensor. When cholesterol levels fall, SCAP escorts SREBP in small membrane vesicles to the Golgi complex, where it is released from the membrane by two specific proteolytic cleavages (Figure 26.13). The released protein migrates to the nucleus and binds the SRE of the HMG-CoA reductase gene, as well as several other genes in the cholesterol biosynthetic

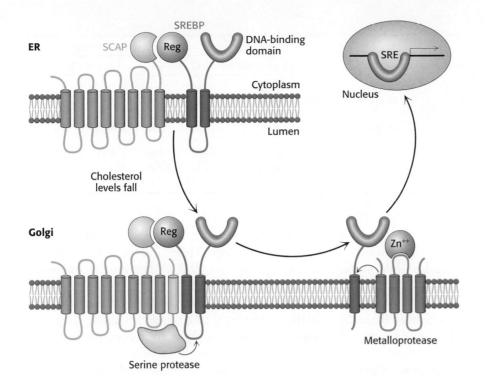

Figure 26.13 The SREBP pathway. SREBP resides in the endoplasmic reticulum, where it is bound to SCAP by its regulatory (Reg) domain. When cholesterol levels fall, SCAP and SREBP move to the Golgi complex, where SREBP undergoes successive proteolytic cleavages by a serine protease and a metalloprotease. The released DNA-binding domain moves to the nucleus to alter gene expression. [After an illustration provided by Dr. Michael Brown and Dr. Joseph Goldstein.]

pathway, to enhance transcription. When cholesterol levels rise, the proteolytic release of the SREBP is blocked, and the SREBP in the nucleus is rapidly degraded. These two events halt the transcription of genes of the cholesterol biosynthetic pathways.

2.   The rate of *translation of reductase mRNA* is inhibited by nonsterol metabolites derived from mevalonate as well as by dietary cholesterol.

3.   The *degradation of the reductase* is stringently controlled. The enzyme is bipartite: its cytoplasmic domain carries out catalysis and *its membrane domain senses signals that lead to its degradation.* The membrane domain may undergo structural changes in *response to increasing concentrations of sterols such as cholesterol* that make the enzyme more susceptible to proteolysis. The reductase may be further degraded by ubiquitination and targeting to the 26S proteasome under some conditions (Section 23.2). A combination of these three regulatory devices can alter the amount of enzyme in the cell more than 200-fold.

4.   *Phosphorylation decreases the activity of the reductase.* This enzyme, like acetyl CoA carboxylase (which catalyzes the committed step in fatty acid synthesis, Section 22.5), is switched off by an AMP-activated protein kinase. Thus, cholesterol synthesis ceases when the ATP level is low.

As we will see shortly, all four regulatory mechanisms are modulated by receptors that sense the presence of cholesterol in the blood.

Lipoproteins Transport Cholesterol and Triacylglycerols Throughout the Organism

Cholesterol and triacylglycerols are transported in body fluids in the form of *lipoprotein particles.* Each particle consists of a core of hydrophobic lipids surrounded by a shell of more-polar lipids and proteins. The protein components of these macromolecular aggregates, called apoproteins, have two roles: *they solubilize hydrophobic lipids and contain cell-targeting signals.* Apolipoproteins are synthesized and secreted by the liver and the intestine.

TABLE 26.1 Properties of plasma lipoproteins

Plasma lipoproteins	Density (g ml^{-1})	Diameter (nm)	Apolipoprotein	Physiological role	TAG	CE	C	PL	P
					COMPOSITION (%)				
Chylomicron	<0.95	75–1200	B48, C, E	Dietary fat transport	86	3	1	8	2
Very low density lipoprotein	0.95–1.006	30–80	B100, C, E	Endogenous fat transport	52	14	7	18	8
Intermediate-density lipoprotein	1.006–1.019	15–35	B100, E	LDL precursor	38	30	8	23	11
Low-density lipoprotein	1.019–1.063	18–25	B100	Cholesterol transport	10	38	8	22	21
High-density lipoprotein	1.063–1.21	7.5–20	A	Reverse cholesterol transport	5–10	14–21	3–7	19–29	33–57

Abbreviations: TAG, triacylglycerol; CE, cholesterol ester; C, free cholesterol; PL, phospholipid; P, protein.

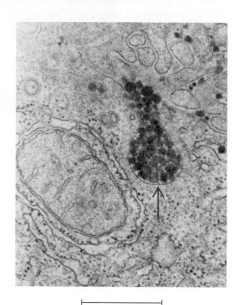

|—————— 500 nm ——————|

Figure 26.14 Site of cholesterol synthesis. Electron micrograph of a part of a liver cell actively engaged in the synthesis and secretion of very low density lipoprotein (VLDL). The arrow points to a vesicle that is releasing its content of VLDL particles. [Courtesy of Dr. George Palade.]

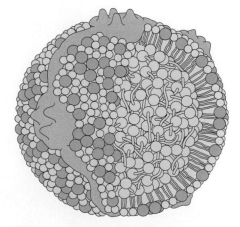

■ Unesterified cholesterol

■ Phospholipid

■ Cholesteryl ester

■ Apoprotein B-100

Lipoprotein particles are classified according to increasing density (Table 26.1): *chylomicrons, chylomicron remnants, very low density lipoproteins* (VLDLs), *intermediate-density lipoproteins* (IDLs), *low-density lipoproteins* (LDLs), and *high-density lipoproteins* (HDLs).

Triacylglycerols, cholesterol, and other lipids obtained from the diet are carried away from the intestine in the form of large *chylomicrons*. These particles have a very low density because triacylglycerols constitute ~90% of their content. Apolipoprotein B-48 (apo B-48), a large protein (240 kd), forms an amphipathic spherical shell around the fat globule; the external face of this shell is hydrophilic. The triacylglycerols in chylomicrons are released through hydrolysis by *lipoprotein lipases*. These enzymes are located on the lining of blood vessels in muscle and other tissues that use fatty acids as fuels and in the synthesis of lipids. The liver then takes up the cholesterol-rich residues, known as *chylomicron remnants*.

The liver is a major site of triacylglycerol and cholesterol synthesis (Figure 26.14). Triacylglycerols and cholesterol in excess of the liver's own needs are exported into the blood in the form of very low density lipoproteins. These particles are stabilized by two apolipoproteins—apo B-100 and apo E (34 kd). Apo B-100, one of the largest proteins known (513 kd), is a longer version of apo B-48. Both apo B proteins are encoded by the same gene and produced from the same initial RNA transcript. In the intestine, RNA editing (Section 29.3) modifies the transcript to generate the mRNA for apo B-48, the truncated form. Triacylglycerols in very low density lipoproteins, as in chylomicrons, are hydrolyzed by lipases on capillary surfaces. The resulting remnants, which are rich in cholesteryl esters, are called *intermediate-density lipoproteins*. These particles have two fates. Half of them are taken up by the liver for processing, and half are converted into low-density lipoprotein by the removal of more triacylglycerol.

Low-density lipoprotein is the major carrier of cholesterol in blood (Figure 26.15). It contains a core of some 1500 cholesterol molecules esterified to fatty acids; the most common fatty acid chain in these esters is linoleate, a polyunsaturated fatty acid. A shell of phospholipids and unesterified cholesterol molecules surrounds this highly hydrophobic core. The shell also contains a single copy of apo B-100, which is recognized by target cells. *The role of LDL is to transport cholesterol to peripheral tissues and regulate de novo cholesterol synthesis at these sites,* as described on page 745. A different purpose is served by *high-density lipoprotein,* which picks up cholesterol released into the plasma from dying cells and from membranes undergoing turnover, a process termed *reverse cholesterol transport.* An acyltransferase in HDL esterifies these cholesterols, which are then returned by HDL to the liver (Figure 26.16).

◀ Figure 26.15 Schematic model of low-density lipoprotein. The LDL particle is approximately 22 nm (220 Å) in diameter.

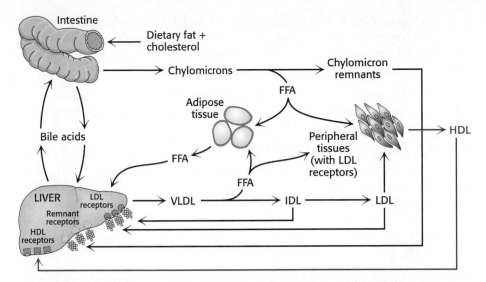

Figure 26.16 An overview of lipoprotein particle metabolism. Fatty acids are abbreviated FFA. [After J. G. Hardman (Ed.), L. L. Limbird (Ed.), and A. G. Gilman (Consult. Ed.), *Goodman and Gilman's The Pharmacological Basis of Therapeutics,* 10th ed. (McGraw-Hill, 2001), p. 975, Fig. 36.1.]

The Blood Levels of Certain Lipoproteins Can Serve Diagnostic Purposes

High serum levels of cholesterol cause disease and death by contributing to the formation of atherosclerotic plaques in arteries throughout the body. This excess cholesterol is present in the form of the low-density lipoprotein particle, so-called bad cholesterol.

High-density lipoprotein is sometimes referred to as "good cholesterol." HDL functions as a shuttle that moves cholesterol throughout the body. HDL binds and esterifies cholesterol released from the peripheral tissues and then transfers cholesteryl esters to the liver or to tissues that use cholesterol to synthesize steroid hormones. A specific receptor mediates the docking of the HDL to these tissues. The exact nature of the protective effect of HDL levels is not known; however, a possible mechanism will be examined on page 747.

The ratio of cholesterol in the form of LDL to that in the form of HDL can be used to evaluate susceptibility to the development of heart disease. For a healthy person, the HDL/LDL ratio is 3.5.

Low-Density Lipoproteins Play a Central Role in Cholesterol Metabolism

Cholesterol metabolism must be precisely regulated to prevent atherosclerosis. The mode of control in the liver, the primary site of cholesterol synthesis, has already been considered: dietary cholesterol reduces the activity and amount of 3-hydroxy-3-methylglutaryl CoA reductase, the enzyme catalyzing the committed step. Studies by Michael Brown and Joseph Goldstein are sources of insight into the control of cholesterol metabolism in nonhepatic cells. In general, cells outside the liver and intestine obtain cholesterol from the plasma rather than synthesizing it de novo. Specifically, *their primary source of cholesterol is the low-density lipoprotein.* The process of LDL uptake, called *receptor-mediated endocytosis,* serves as a paradigm for the uptake of many molecules.

The steps in the receptor-mediated endocytosis of LDL are as follows (see Figure 12.37).

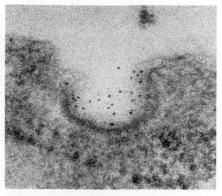

(A)

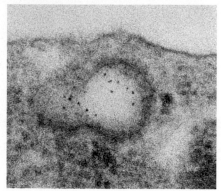

(B)

Figure 26.17 Endocytosis of LDL bound to its receptor. (A) Electron micrograph showing LDL (conjugated to ferritin for visualization, dark spots) bound to a coated-pit region on the surface of a cultured human fibroblast cell. (B) Micrograph showing this region invaginating and fusing to form an endocytic vesicle. [From R. G. W. Anderson, M. S. Brown, and J. L. Goldstein. *Cell* 10(1977):351–364.]

Figure 26.18 LDL receptor domains. A schematic representation of the amino acid sequence of the LDL receptor showing six types of domain.

1. Apolipoprotein B-100 on the surface of an LDL particle binds to a specific receptor protein on the plasma membrane of nonhepatic cells. The receptors for LDL are localized in specialized regions called *coated pits,* which contain a specialized protein called *clathrin.*

2. The receptor–LDL complex is internalized by *endocytosis;* that is, the plasma membrane in the vicinity of the complex invaginates and then fuses to form an endocytic vesicle (Figure 26.17).

3. These vesicles, containing LDL, subsequently fuse with *lysosomes,* acidic vesicles that carry a wide array of degradative enzymes. The protein component of LDL is hydrolyzed to free amino acids. The cholesteryl esters in LDL are hydrolyzed by a lysosomal acid lipase. The LDL receptor itself usually returns unscathed to the plasma membrane. The round-trip time for a receptor is about 10 minutes; in its lifetime of about a day, it may bring many LDL particles into the cell.

4. *The released unesterified cholesterol can then be used for membrane biosynthesis.* Alternatively, it can be *reesterified for storage inside the cell.* In fact, free cholesterol activates *acyl CoA:cholesterol acyltransferase* (ACAT), the enzyme catalyzing this reaction. Reesterified cholesterol contains mainly oleate and palmitoleate, which are monounsaturated fatty acids, in contrast with the cholesterol esters in LDL, which are rich in linoleate, a polyunsaturated fatty acid (see Table 12.1). It is imperative that the cholesterol be reesterified. High concentrations of unesterified cholesterol disrupt the integrity of cell membranes.

The synthesis of the LDL receptor is itself subject to feedback regulation. Studies of cultured fibroblasts show that, *when cholesterol is abundant inside the cell, new LDL receptors are not synthesized, and so the uptake of additional cholesterol from plasma LDL is blocked.* The gene for the LDL receptor, like that for the reductase, is regulated by SREBP, which binds to a sterol regulatory element that controls the rate of mRNA synthesis.

The LDL Receptor Is a Transmembrane Protein Having Six Different Functional Regions

The amino acid sequence of the human LDL receptor reveals the mosaic structure of this 115-kd protein, which is composed of six different types of domains (Figure 26.18). The amino-terminal region of the mature receptor is the site of LDL binding. It consists of a cysteine-rich sequence of about 40 residues that is repeated, with some variation, seven times.

A second type of domain in the LDL receptor is homologous to one found in the epidermal growth factor (EGF). This domain is repeated three times, and in between the second and third repeat are six repeats of a third domain that is similar to the blades of the transducin β subunit (p. 270). The six repeats form a propeller-like structure that packs against one of the EGF-like domains (Figure 26.19). An aspartate residue forms hydrogen bonds that hold each blade to the rest of the structure. Exposure to the low-pH environment of the lysosomes causes the propeller-like structures to interact with the LDL-binding domain. This interaction displaces the LDL, which is then digested by the lysosome.

Cysteine-rich   EGF-like   Blade   O-linked glycosylated
Hydrophobic   Cytoplasmic

The final three domains are represented by a single copy apiece. The fourth domain, which is very rich in serine and threonine residues, contains *O*-linked sugars. These oligosaccharides may function as struts to keep the receptor extended from the membrane so that the LDL-binding domain is accessible to LDL. The fifth type of domain consists of 22 hydrophobic residues that span the plasma membrane. The sixth and final domain consists of 50 residues and emerges on the cytoplasmic side of the membrane, where it controls the interaction of the receptor with coated pits and participates in endocytosis. The gene for the LDL receptor consists of 18 exons, which correspond closely to the structural units of the protein. *The LDL receptor is a striking example of a mosaic protein encoded by a gene that was assembled by exon shuffling.*

The Absence of the LDL Receptor Leads to Hypercholesterolemia and Atherosclerosis

Brown and Goldstein's pioneering studies of *familial hypercholesterolemia* revealed the physiological importance of the LDL receptor. The total concentration of cholesterol and LDL in the blood plasma is markedly elevated in this genetic disorder, which results from a mutation at a single autosomal locus. The cholesterol level in the plasma of homozygotes is typically 680 mg dl^{-1}, compared with 300 mg dl^{-1} in heterozygotes (clinical assay results are often expressed in milligrams per deciliter, which is equal to milligrams per 100 milliliters). A value of < 200 mg dl^{-1} is regarded as desirable, but many people have higher levels. *In familial hypercholesterolemia, cholesterol is deposited in various tissues because of the high concentration of LDL cholesterol in the plasma.* Nodules of cholesterol called *xanthomas* are prominent in skin and tendons. Of particular concern is the oxidation of the excess blood LDL to form oxidized LDL (oxLDL). The oxLDL is taken up by immune-system cells called macrophages, which become engorged to form foam cells. These foam cells become trapped in the walls of the blood vessels and contribute to the formation of atherosclerotic plaques that cause arterial narrowing and lead to heart attacks (Figure 26.20). In fact, *most homozygotes die of coronary artery disease in childhood.* The disease in heterozygotes (1 in 500 people) has a milder and more variable clinical course. A serum esterase that degrades oxidized lipids is found in association with HDL. Possibly, the HDL-associated

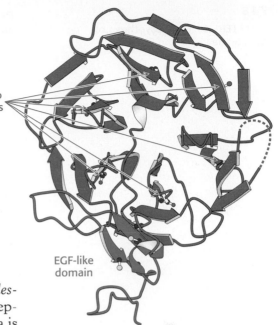

Figure 26.19 Structure of propeller domain. *Notice* the six-bladed propeller-like shape of this domain (red). The propeller domain is adjacent to an EGF-like domain (blue) of the LDL receptor. [Drawn from 1IJQ.pdb.]

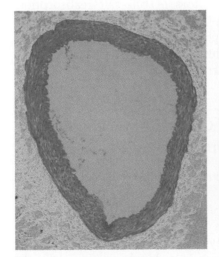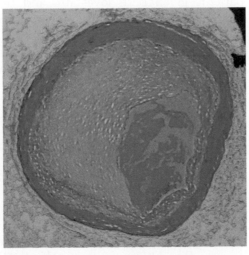

(A) (B)

Figure 26.20 The effects of excess cholesterol. Cross section of (A) a normal artery and (B) an artery blocked by a cholesterol-rich plaque. [SPL/Photo Researchers.]

protein destroys the oxLDL, accounting for HDL's ability to protect against coronary disease.

The molecular defect in most cases of familial hypercholesterolemia is an absence or deficiency of functional receptors for LDL. Receptor mutations that disrupt each of the stages in the endocytotic pathway have been identified. Homozygotes have almost no functional receptors for LDL, whereas heterozygotes have about half the normal number. Consequently, the entry of LDL into liver and other cells is impaired, leading to an increased level of LDL in the blood plasma. Furthermore, less IDL enters liver cells because IDL entry, too, is mediated by the LDL receptor. Consequently, IDL stays in the blood longer in familial hypercholesterolemia, and more of it is converted into LDL than in normal people. All deleterious consequences of an absence or deficiency of the LDL receptor can be attributed to the ensuing elevated level of LDL cholesterol in the blood.

The Clinical Management of Cholesterol Levels Can Be Understood at a Biochemical Level

Homozygous familial hypercholesterolemia can be treated only by a liver transplant. A more generally applicable therapy is available for heterozygotes and others with high levels of cholesterol. *The goal is to reduce the amount of cholesterol in the blood by stimulating the single normal gene to produce more than the customary number of LDL receptors.* We have already observed that the production of LDL receptors is controlled by the cell's need for cholesterol. Therefore, in essence, the strategy is to deprive the cell of ready sources of cholesterol. When cholesterol is required, the amount of mRNA for the LDL receptor rises and more receptor is found on the cell surface. This state can be induced by a two-pronged approach. First, the reabsorption of bile salts from the intestine is inhibited. Bile salts are cholesterol derivatives that promote the absorption of dietary cholesterol and dietary fats (p. 619). Second, de novo synthesis of cholesterol is blocked.

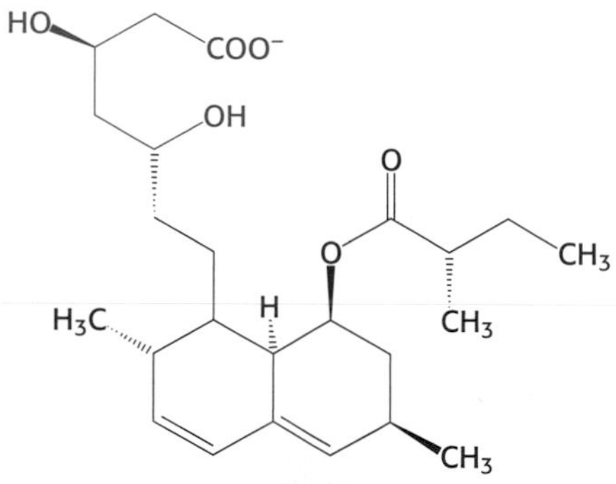

The reabsorption of bile is impeded by oral administration of positively charged polymers, such as cholestyramine, that bind negatively charged bile salts and are not themselves absorbed. Cholesterol synthesis can be effectively blocked by a class of compounds called *statins*. A well-known example of such a compound is lovastatin, which is also called mevacor (Figure 26.21). These compounds are potent competitive inhibitors ($K_i = 1\,\text{nM}$) of HMG-CoA reductase, the essential control point in the biosynthetic pathway. Plasma cholesterol levels decrease by 50% in many patients given both lovastatin and inhibitors of bile-salt reabsorption. Lovastatin and other inhibitors of HMG-CoA reductase are widely used to lower the plasma-cholesterol level in people who have atherosclerosis, which is the leading cause of death in industrialized societies. The development of statins as effective drugs is further described in Chapter 35.

Figure 26.21 Lovastatin, a competitive inhibitor of HMG-CoA reductase. The part of the structure that resembles the 3-hydroxy-3-methylglutaryl moiety is shown in red.

26.4 Important Derivatives of Cholesterol Include Bile Salts and Steroid Hormones

Cholesterol is a precursor for other important steroid molecules: the bile salts, steroid hormones, and vitamin D.

Bile Salts. *Bile salts* are polar derivatives of cholesterol. These compounds are highly effective *detergents* because they contain both polar and nonpolar regions. Bile salts are synthesized in the liver, stored and concentrated in the

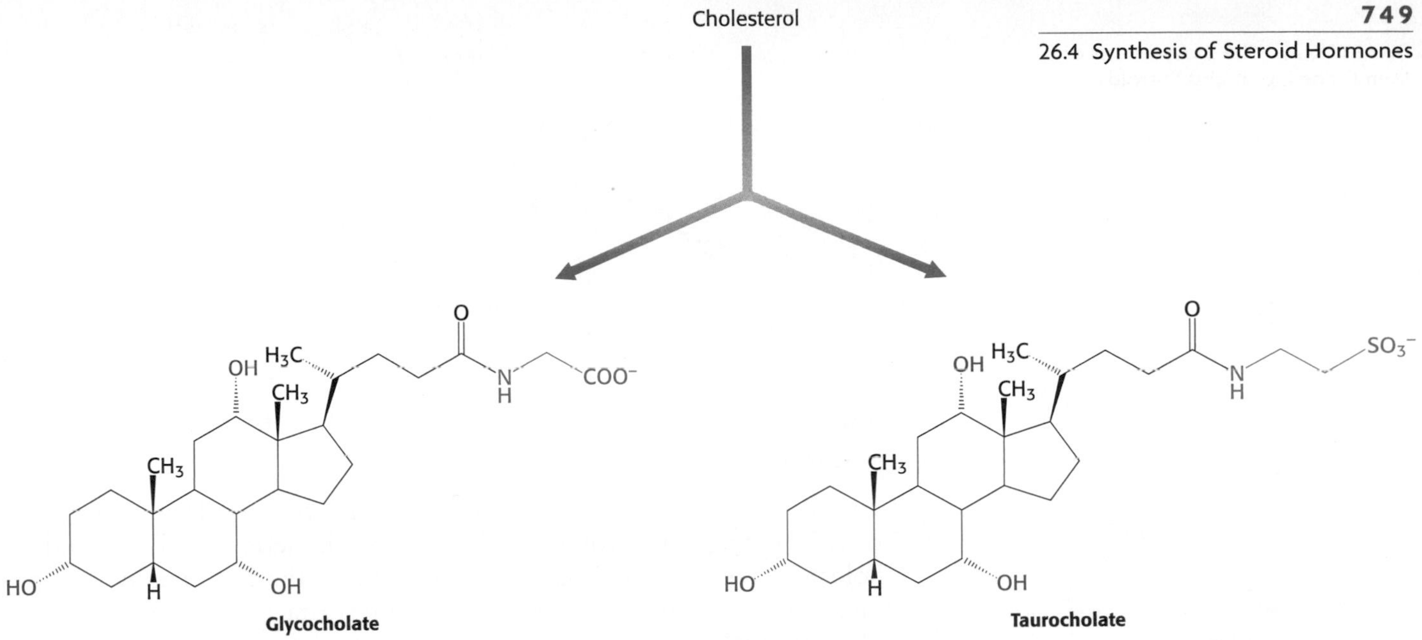

Cholesterol

Glycocholate

Taurocholate

Figure 26.22 **Synthesis of bile salts.** The OH groups in red are added to cholesterol, as are the groups shown blue.

gall bladder, and then released into the small intestine. Bile salts, the major constituent of bile, *solubilize dietary lipids* (p. 619). Solubilization increases the effective surface area of lipids with two consequences: (1) more surface area is exposed to the digestive action of lipases and (2) lipids are more readily absorbed by the intestine. Bile salts are also the major breakdown products of cholesterol. The bile salts glycocholate, the primary bile salt, and taurocholate are shown in Figure 26.22.

Steroid Hormones. Cholesterol is the precursor of the five major classes of *steroid hormones:* progestagens, glucocorticoids, mineralocorticoids, androgens, and estrogens (Figure 26.23). These hormones are powerful signal molecules that regulate a host of organismal functions. *Progesterone, a progestagen,* prepares the lining of the uterus for the implantation of an ovum. Progesterone is also essential for the maintenance of pregnancy. *Androgens* (such as *testosterone*) are responsible for the development of male secondary sex characteristics, whereas *estrogens* (such as *estrone*) are required for the development of female secondary sex characteristics. Estrogens, along with progesterone, also participate in the ovarian cycle. *Glucocorticoids* (such as *cortisol*) promote gluconeogenesis and the formation of glycogen, enhance the degradation of fat and protein, and inhibit the inflammatory response. They enable animals to respond to stress; indeed, the absence of glucocorticoids can be fatal. *Mineralocorticoids* (primarily *aldosterone*) act on the distal tubules of the kidney to increase the reabsorption of Na^+ and the excretion of K^+ and H^+, which leads to an increase in blood volume and blood pressure. The major sites of synthesis of these classes of hormones are the corpus luteum, for progestagens; the testes, for androgens; the ovaries, for estrogens; and the adrenal cortex, for glucocorticoids and mineralocorticoids.

Steroid hormones bind to and activate receptor molecules that serve as transcription factors to regulate gene expression (Section 31.3). These small similar molecules are able to have greatly differing effects because the slight structural differences among them allow interactions with specific receptor molecules.

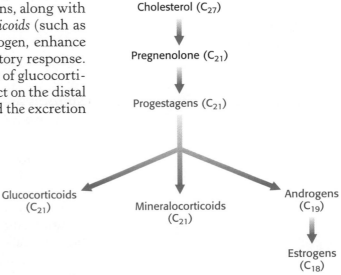

Cholesterol (C_{27})

Pregnenolone (C_{21})

Progestagens (C_{21})

Glucocorticoids (C_{21})

Mineralocorticoids (C_{21})

Androgens (C_{19})

Estrogens (C_{18})

Figure 26.23 **Biosynthetic relations of classes of steroid hormones and cholesterol.**

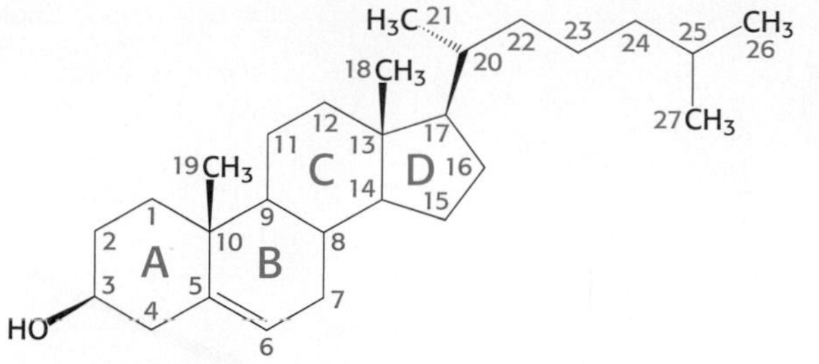

Figure 26.24 Cholesterol carbon numbering. The numbering scheme for the carbon atoms in cholesterol and other steroids.

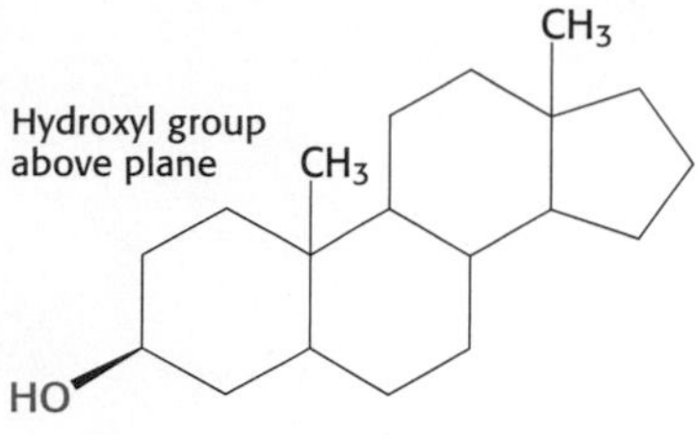

3β-Hydroxy

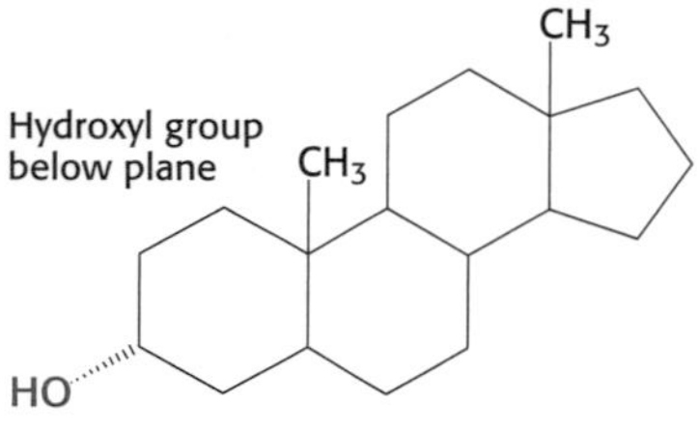

3α-Hydroxy

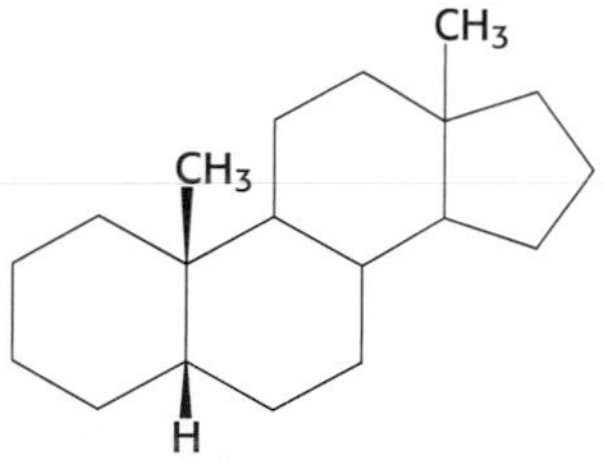

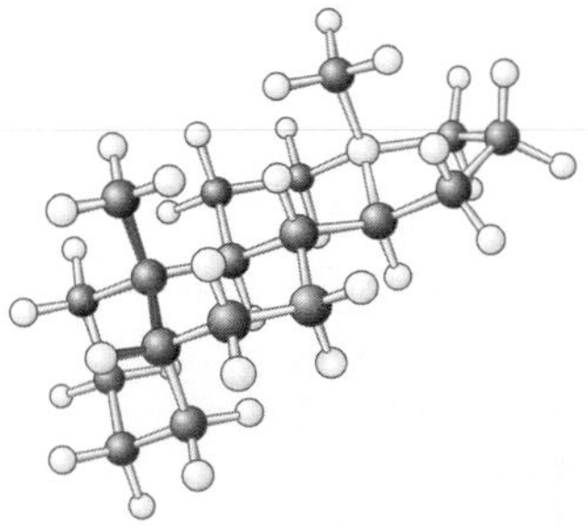

5β-Hydrogen
(cis fusion)

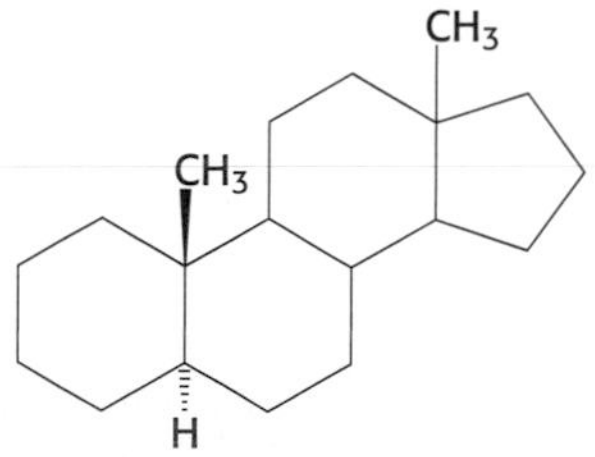

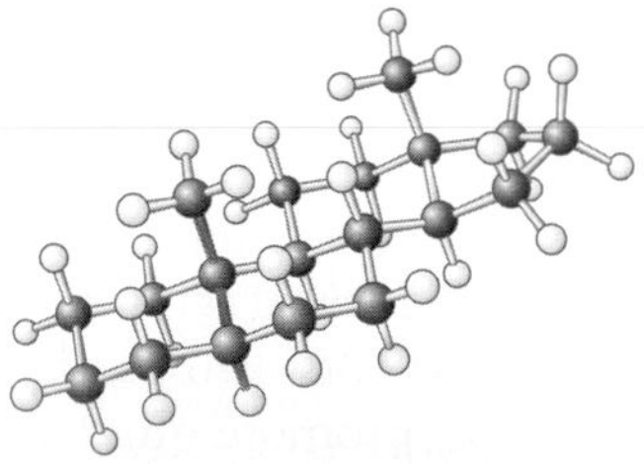

5α-Hydrogen
(trans fusion)

Letters Identify the Steroid Rings and Numbers Identify the Carbon Atoms

Carbon atoms in steroids are numbered, as shown for cholesterol in Figure 26.24. The rings in steroids are denoted by the letters A, B, C, and D. Cholesterol contains two angular methyl groups: the C-19 methyl group is attached to C-10, and the C-18 methyl group is attached to C-13. The C-18 and C-19 methyl groups of cholesterol lie *above* the plane containing the four rings. A substituent that is above the plane is termed β *oriented*, whereas a substituent that is below the plane is α *oriented*.

If a hydrogen atom is attached to C-5, it can be either α or β oriented. The A and B steroid rings are fused in a *trans* conformation if the C-5 hydrogen is α *oriented*, and *cis* if it is β *oriented*. The absence of a Greek letter for the C-5 hydrogen atom on the steroid nucleus implies a trans fusion. The C-5 hydrogen atom is α oriented in all steroid hormones that contain a hydrogen atom in that position. In contrast, bile salts have a β-oriented hydrogen atom at C-5. Thus, *a cis fusion is characteristic of the bile salts, whereas a trans fusion is characteristic of all steroid hormones that possess a hydrogen atom at C-5.* A trans fusion yields a nearly planar structure, whereas a cis fusion gives a buckled structure.

Steroids Are Hydroxylated by Cytochrome P450 Monooxygenases That Use NADPH and O$_2$

The addition of OH groups plays an important role in the synthesis of cholesterol from squalene and in the conversion of cholesterol into steroid hormones and bile salts. All these hydroxylations require *NADPH* and O_2. The oxygen atom of the incorporated hydroxyl group comes from O_2 rather than from H_2O. Whereas one oxygen atom of the O_2 molecule goes into the substrate, the other is reduced to water. The enzymes catalyzing these reactions are called *monooxygenases* (or *mixed-function oxygenases*). Recall that a monooxygenase also participates in the hydroxylation of aromatic amino acids (p. 671).

$$RH + O_2 + NADPH + H^+ \longrightarrow ROH + H_2O + NADP^+$$

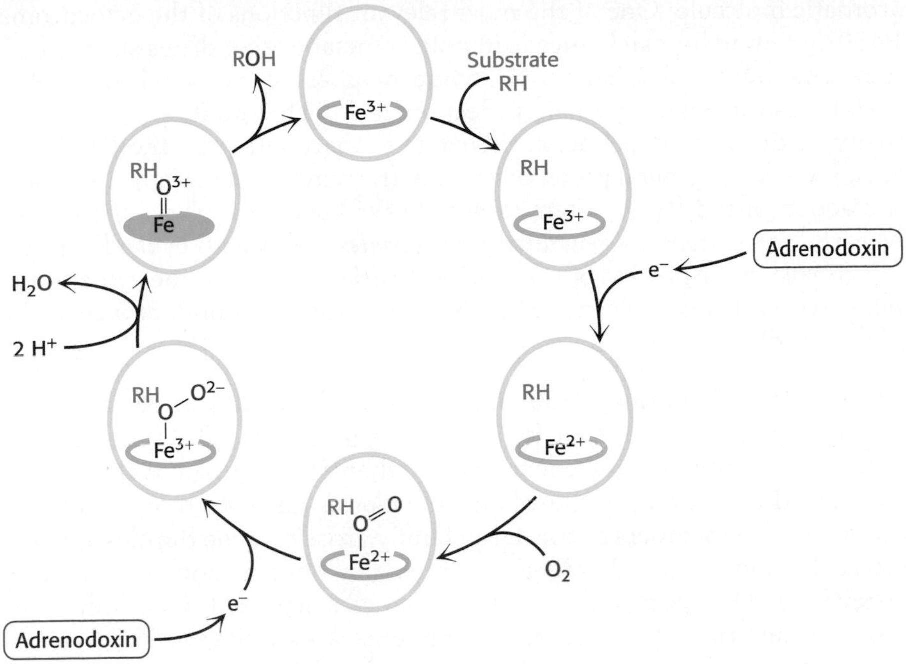

Figure 26.25 Cytochrome P450 mechanism. These enzymes bind O_2 and use one oxygen atom to hydroxylate their substrates.

Hydroxylation requires the activation of oxygen. In the synthesis of steroid hormones and bile salts, activation is accomplished by members of the cytochrome P450 family, a family of cytochromes that absorb light maximally at 450 nm when complexed in vitro with exogenous carbon monoxide. These membrane-anchored proteins ($\sim$50 kd) contain a heme prosthetic group. Oxygen is activated through its binding to the iron atom in the heme group.

Because the hydroxylation reactions promoted by P450 enzymes are oxidation reactions, it is at first glance surprising that they also consume the reductant NADPH. NADPH transfers its high-potential electrons to a flavoprotein, which transfers them, one at a time, to *adrenodoxin*, a nonheme iron protein. Adrenodoxin transfers one electron to reduce the ferric (Fe^{3+}) form of P450 to the ferrous (Fe^{2+}) form (Figure 26.25).

Without the addition of this electron, P450 will not bind oxygen. Recall that only the ferrous form of myoglobin binds oxygen (p. 185). The binding of O_2 to the heme is followed by the acceptance of a second electron from adrenodoxin. The acceptance of this second electron leads to cleavage of the O–O bond. One of the oxygen atoms is then protonated and released as water. The remaining oxygen atom forms a highly reactive ferryl Fe=O intermediate. This intermediate abstracts a hydrogen atom from the substrate RH to form R·. This transient free radical captures the OH group from the iron atom to form ROH, the hydroxylated product, returning the iron atom to the ferric state.

The Cytochrome P450 System Is Widespread and Performs a Protective Function

The cytochrome P450 system, which in mammals is located primarily in the endoplasmic reticulum of the liver and small intestine, is also important in the *detoxification of foreign substances* (xenobiotic compounds). For example, the hydroxylation of phenobarbital, a barbiturate, *increases its solubility* and *facilitates its excretion*. Likewise, polycyclic aromatic hydrocarbons that are ingested by drinking contaminated water are hydroxylated by P450, providing sites for conjugation with highly polar units (e.g., glucuronate or sulfate) that markedly increase the solubility of the modified aromatic molecule. One of the most relevant functions of the cytochrome P450 system to human beings is its role in metabolizing drugs such as caffeine and ibuprofen (Chapter 35). Some members of the cytochrome P450 system also metabolize ethanol (Section 27.5). The duration of action of many medications depends on their rate of inactivation by the P450 system. Despite its general protective role in the removal of foreign chemicals, the action of the P450 system is not always beneficial. *Some of the most powerful carcinogens are generated from harmless compounds by the P450 system in vivo* in the process of *metabolic activation*. In plants, the cytochrome P450 system plays a role in the synthesis of toxic compounds as well as the pigments of flowers.

The cytochrome P450 system is a ubiquitous superfamily of monooxygenases that is present in plants, animals, and prokaryotes. The human genome encodes more than 50 members of the family, whereas the genome of the plant *Arabidopsis* encodes more than 250 members. All members of this large family arose by gene duplication followed by subsequent divergence, which generated a range of substrate specificity. The specificity of these enzymes is encoded in delimited regions of the primary structure, and the substrate specificity of closely related members is often defined by a few critical residues or even a single amino acid.

Pregnenolone, a Precursor for Many Other Steroids, Is Formed from Cholesterol by Cleavage of Its Side Chain

Steroid hormones contain 21 or fewer carbon atoms, whereas cholesterol contains 27. Thus, the first stage in the synthesis of steroid hormones is the removal of a six-carbon unit from the side chain of cholesterol to form *pregnenolone*. The side chain of cholesterol is hydroxylated at C-20 and then at C-22, and the bond between these carbon atoms is subsequently cleaved by *desmolase*. Three molecules of NADPH and three molecules of O_2 are consumed in this remarkable six-electron oxidation.

Cholesterol   20α,22β-Dihydroxycholesterol   Pregnenolone

Progesterone and Corticosteroids Are Synthesized from Pregnenolone

Progesterone is synthesized from pregnenolone in two steps. The 3-hydroxyl group of pregnenolone is oxidized to a 3-keto group, and the Δ^5 double bond is isomerized to a Δ^4 double bond (Figure 26.26). *Cortisol*, the major glucocorticoid, is synthesized from progesterone by hydroxylations at C-17, C-21, and C-11; C-17 must be hydroxylated before C-21 is hydroxylated, whereas C-11 can be hydroxylated at any stage. The enzymes catalyzing these hydroxylations are highly specific, as shown by some inherited disorders. The initial step in the synthesis of *aldosterone*, the major mineralocorticoid, is the hydroxylation of progesterone at C-21. The resulting deoxycorticosterone is hydroxylated at C-11. The oxidation of the C-18 angular methyl group to an aldehyde then yields aldosterone.

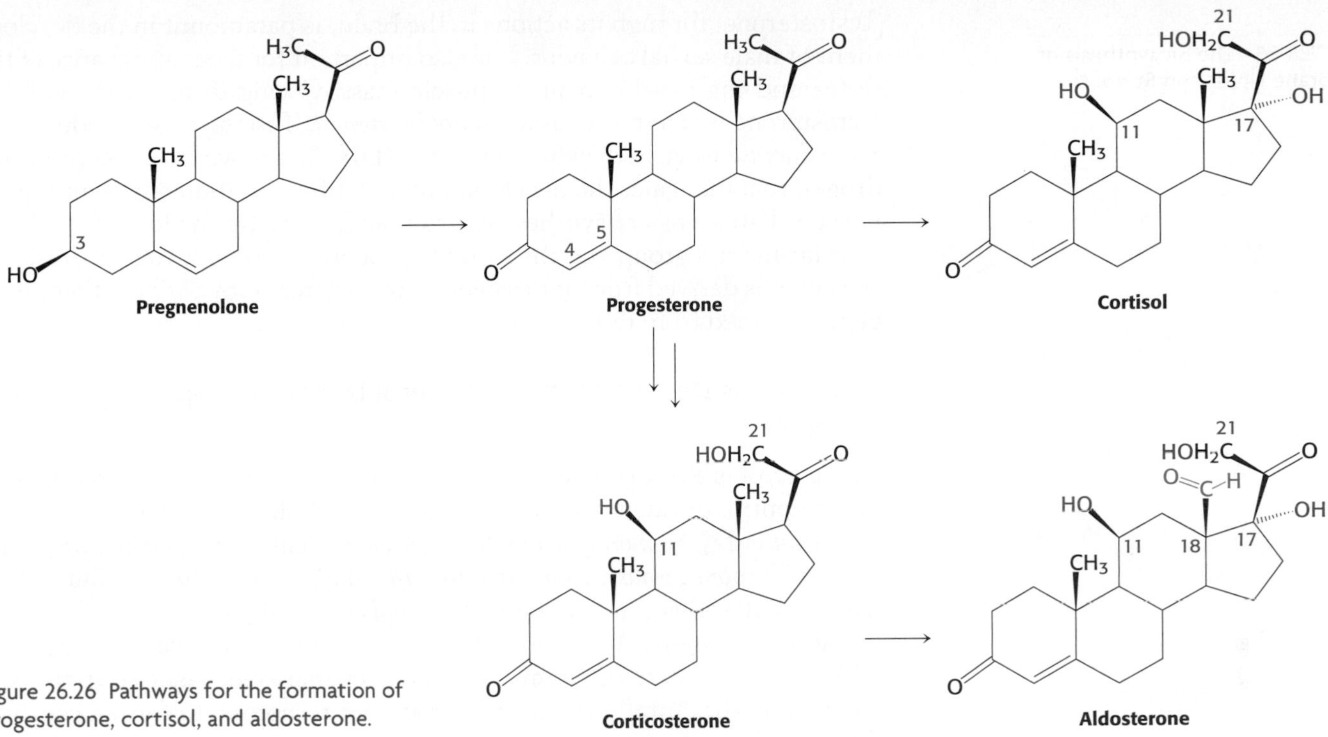

Figure 26.26 Pathways for the formation of progesterone, cortisol, and aldosterone.

Androgens and Estrogens Are Synthesized from Pregnenolone

Androgens and estrogens also are synthesized from pregnenolone through the intermediate progesterone. Androgens contain 19 carbon atoms. The synthesis of androgens starts with the hydroxylation of progesterone at C-17 (Figure 26.27). The side chain consisting of C-20 and C-21 is then cleaved to yield *androstenedione*, an androgen. *Testosterone*, another androgen, is formed by the reduction of the 17-keto group of androstenedione.

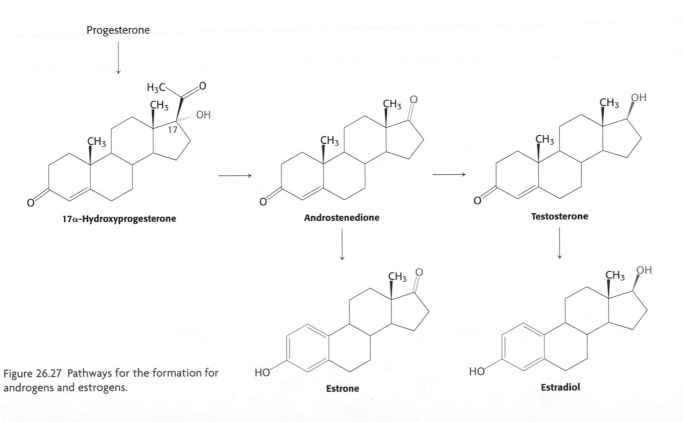

Figure 26.27 Pathways for the formation for androgens and estrogens.

753

Testosterone, through its actions in the brain, is paramount in the development of male sexual behavior. It is also important for the maintenance of the testes and the development of muscle mass. Owing to the latter activity, testosterone is referred to as an *anabolic steroid*. Testosterone is reduced by *5α-reductase* to yield *dihydrotestosterone* (DHT), a powerful embryonic androgen that instigates the development and differentiation of the male phenotype. Estrogens are synthesized from androgens by the loss of the C-19 angular methyl group and the formation of an aromatic A ring. *Estrone*, an estrogen, is derived from androstenedione, whereas *estradiol*, another estrogen, is formed from testosterone.

Vitamin D Is Derived from Cholesterol by the Ring-Splitting Activity of Light

Cholesterol is also the precursor of vitamin D, which plays an essential role in the control of calcium and phosphorus metabolism. *7-Dehydrocholesterol (provitamin D₃) is photolyzed by the ultraviolet light of sunlight to previtamin D₃, which spontaneously isomerizes to vitamin D₃* (Figure 26.28). Vitamin D₃ (cholecalciferol) is converted into *calcitriol* (1,25-dihydroxycholecalciferol), the active hormone, by hydroxylation reactions in the liver and kidneys. Although not a steroid, vitamin D acts in an analogous fashion. It binds to a receptor, structurally similar to the steroid receptors, to form a complex that functions as a transcription factor, regulating gene expression.

Vitamin D deficiency in childhood produces *rickets*, a disease characterized by inadequate calcification of cartilage and bone. Rickets was so common in seventeenth-century England that it was called the "chil-

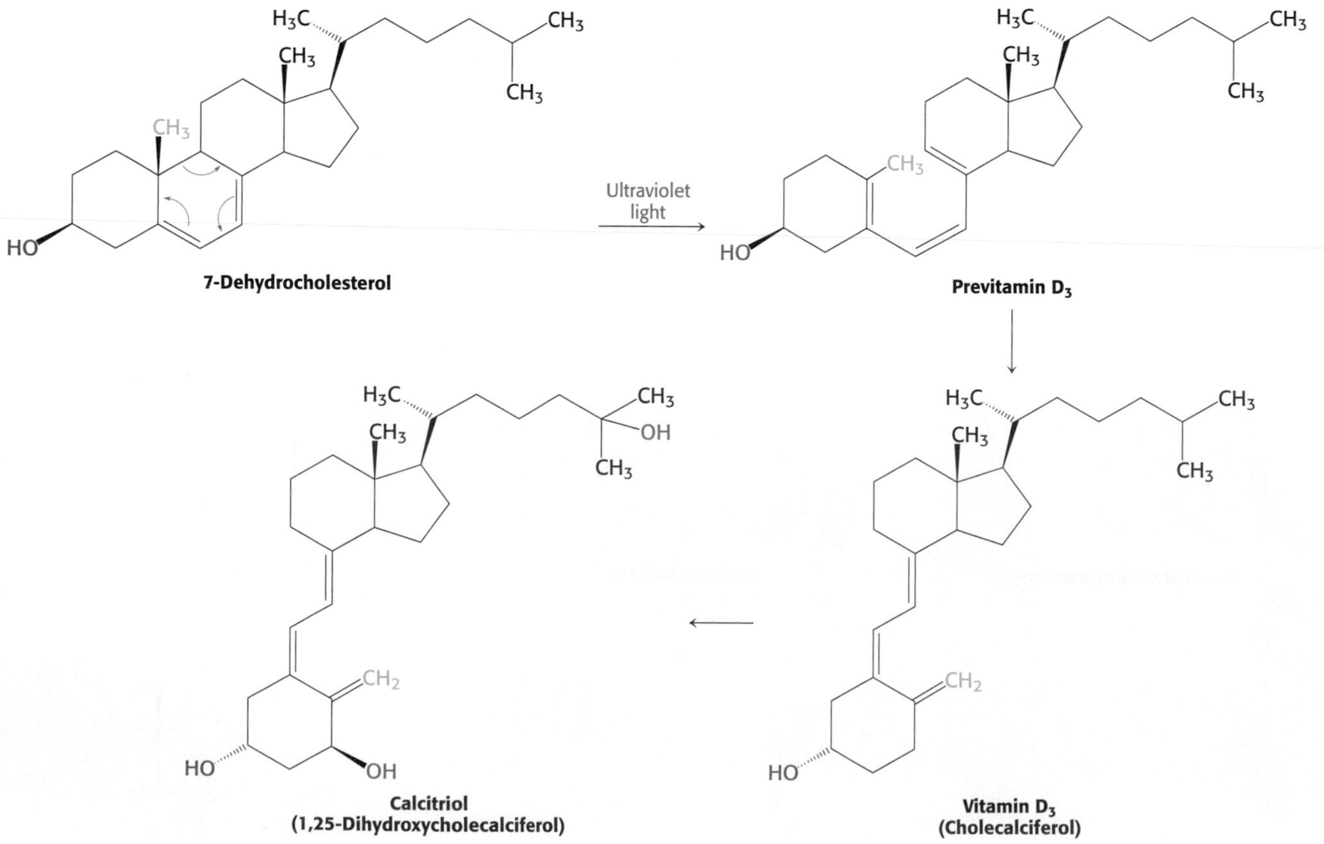

Figure 26.28 Vitamin D synthesis. The pathway for the conversion of 7-dehydrocholesterol into vitamin D₃ and then into calcitriol, the active hormone.

dren's disease of the English." The 7-dehydrocholesterol in the skin of these children was not photolyzed to previtamin D_3, because there was little sunlight for many months of the year. Furthermore, their diets provided little vitamin D, because most naturally occurring foods have a low content of this vitamin. Fish-liver oils are a notable exception. Cod-liver oil, abhorred by generations of children because of its unpleasant taste, was used in the past as a rich source of vitamin D. Today, the most reliable dietary sources of vitamin D are fortified foods. Milk, for example, is fortified to a level of 400 international units per quart (10 μg per quart). The recommended daily intake of vitamin D is 200 international units until age 50, after which it increases with age. In adults, vitamin D deficiency leads to softening and weakening of bones, a condition called *osteomalacia*. The occurrence of osteomalacia in Muslim women who are clothed so that only their eyes are exposed to sunlight is a striking reminder that vitamin D is needed by adults as well as by children.

Summary

26.1 Phosphatidate Is a Common Intermediate in the Synthesis of Phospholipids and Triacylglycerols

Phosphatidate is formed by successive acylations of glycerol 3-phosphate by acyl CoA. Hydrolysis of its phosphoryl group followed by acylation yields a triacylglycerol. CDP-diacylglycerol, the activated intermediate in the de novo synthesis of several phospholipids, is formed from phosphatidate and CTP. The activated phosphatidyl unit is then transferred to the hydroxyl group of a polar alcohol, such as inositol, to form a phospholipid such as phosphatidylinositol. In mammals, phosphatidylethanolamine is formed by CDP-ethanolamine and diacylglycerol. Phosphatidylethanolamine is methylated by S-adenosylmethionine to form phosphatidylcholine. In mammals, this phosphoglyceride can also be synthesized by a pathway that utilizes dietary choline. CDP-choline is the activated intermediate in this route.

Sphingolipids are synthesized from ceramide, which is formed by the acylation of sphingosine. Gangliosides are sphingolipids that contain an oligosaccharide unit having at least one residue of N-acetylneuraminate or a related sialic acid. They are synthesized by the step-by-step addition of activated sugars, such as UDP-glucose, to ceramide.

26.2 Cholesterol Is Synthesized from Acetyl Coenzyme A in Three Stages

Cholesterol is a steroid component of animal membranes and a precursor of steroid hormones. The committed step in its synthesis is the formation of mevalonate from 3-hydroxy-3-methylglutaryl CoA (derived from acetyl CoA and acetoacetyl CoA). Mevalonate is converted into isopentenyl pyrophosphate (C_5), which condenses with its isomer, dimethylallyl pyrophosphate (C_5), to form geranyl pyrophosphate (C_{10}). The addition of a second molecule of isopentenyl pyrophosphate yields farnesyl pyrophosphate (C_{15}), which condenses with itself to form squalene (C_{30}). This intermediate cyclizes to lanosterol (C_{30}), which is modified to yield cholesterol (C_{27}).

26.3 The Complex Regulation of Cholesterol Biosynthesis Takes Place at Several Levels

In the liver, cholesterol synthesis is regulated by changes in the amount and activity of 3-hydroxy-3-methylglutaryl CoA reductase.

Transcription of the gene, translation of the mRNA, and degradation of the enzyme are stringently controlled. In addition, the activity of the reductase is regulated by phosphorylation.

Triacylglycerols exported by the intestine are carried by chylomicrons and then hydrolyzed by lipases lining the capillaries of target tissues. Cholesterol and other lipids in excess of those needed by the liver are exported in the form of very low density lipoprotein. After delivering its content of triacylglycerols to adipose tissue and other peripheral tissue, VLDL is converted into intermediate-density lipoprotein and then into low-density lipoprotein. IDL and LDL carry cholesteryl esters, primarily cholesteryl linoleate. Liver and peripheral tissue cells take up LDL by receptor-mediated endocytosis. The LDL receptor, a protein spanning the plasma membrane of the target cell, binds LDL and mediates its entry into the cell. Absence of the LDL receptor in the homozygous form of familial hypercholesterolemia leads to a markedly elevated plasma level of LDL cholesterol and the deposition of cholesterol on blood-vessel walls, which in turn may result in childhood heart attacks. Apolipoprotein B, a very large protein, is a key structural component of chylomicrons, VLDL, and LDL. High-density lipoproteins transport cholesterol from the peripheral tissues to the liver.

26.4 Important Derivatives of Cholesterol Include Bile Salts and Steroid Hormones

In addition to bile salts, which facilitate the digestion of lipids, five major classes of steroid hormones are derived from cholesterol: progestagens, glucocorticoids, mineralocorticoids, androgens, and estrogens. Hydroxylations by P450 monooxygenases that use NADPH and O_2 play an important role in the synthesis of steroid hormones and bile salts from cholesterol. P450 enzymes, a large superfamily, also participate in the detoxification of drugs and other foreign substances.

Pregnenolone (C_{21}) is an essential intermediate in the synthesis of steroids. This steroid is formed by scission of the side chain of cholesterol. Progesterone (C_{21}), synthesized from pregnenolone, is the precursor of cortisol and aldosterone. Hydroxylation of progesterone and cleavage of its side chain yields androstenedione, an androgen (C_{19}). Estrogens (C_{18}) are synthesized from androgens by the loss of an angular methyl group and the formation of an aromatic A ring. Vitamin D, which is important in the control of calcium and phosphorus metabolism, is formed from a derivative of cholesterol by the action of light.

Key Terms

phosphatidate (p. 733)

triacylglycerol (p. 733)

phospholipid (p. 734)

cytidine diphosphodiacylglycerol (CDP-diacylglycerol) (p. 734)

sphingolipid (p. 736)

ceramide (p. 736)

cerebroside (p. 737)

ganglioside (p. 738)

cholesterol (p. 739)

mevalonate (p. 739)

3-hydroxy-3-methylglutaryl CoA reductase (HMG-CoA reductase) (p. 739)

3-isopentenyl pyrophosphate (p. 740)

sterol regulatory element binding protein (SREBP) (p. 742)

lipoprotein particles (p. 743)

low-density lipoprotein (LDL) (p. 744)

high-density lipoprotein (HDL) (p. 744)

reverse cholesterol transport (p.744)

receptor-mediated endocytosis (p. 745)

bile salt (p. 748)

steroid hormone (p. 749)

cytochrome P450 monooxygenase (p. 750)

pregnenolone (p. 752)

Selected Readings

Where to Start

Gimpl, G., Burger, K., and Fahrenholz, F. 2002. A closer look at the cholesterol sensor. *Trends Biochem. Sci.* 27:595–599.

Oram, J. F. 2002. Molecular basis of cholesterol homeostasis: Lessons from Tangier disease and ABCA1. *Trends Mol. Med.* 8:168–173.

Vance, D. E., and Van den Bosch, H. 2000. Cholesterol in the year 2000. *Biochim. Biophys. Acta* 1529:1–8.

Brown, M. S., and Goldstein, J. L. 1986. A receptor-mediated pathway for cholesterol homeostasis. *Science* 232:34–47.

Brown, M. S., and Goldstein, J. L. 1984. How LDL receptors influence cholesterol and atherosclerosis. *Sci. Am.* 251(5):58–66.

Endo, A. 1992. The discovery and development of HMG-CoA reductase inhibitors. *J. Lipid Res.* 33:1569–1582.

Books

Vance, D. E., and Vance, J. E. (Eds.). 1996. *Biochemistry of Lipids, Lipoproteins and Membranes.* Elsevier.

Scriver, C. R. (Ed.), Sly, W. S. (Ed.), Childs, B., Beaudet, A. L., Valle, D., Kinzler, K. W., and Vogelstein, B. 2000. *The Metabolic Basis of Inherited Diseases* (7th ed.). McGraw-Hill.

Phospholipids and Sphingolipids

Huwiler, A., Kolter, T., Pfeilschifter, J., and Sandhoff, K. 2000. Physiology and pathophysiology of sphingolipid metabolism and signaling. *Biochim. Biophys. Acta* 1485:63–99.

Lykidis, A., and Jackowski, S. 2000. Regulation of mammalian cell membrane biosynthesis. *Prog. Nucleic Acid Res. Mol. Biol.* 65:361–393.

Carman, G. M., and Zeimetz, G. M. 1996. Regulation of phospholipid biosynthesis in the yeast *Saccharomyces cerevisiae*. *J. Biol. Chem.* 271:13293–13296.

Henry, S. A., and Patton-Vogt, J. L. 1998. Genetic regulation of phospholipid metabolism: Yeast as a model eukaryote. *Prog. Nucleic Acid Res. Mol. Biol.* 61:133–179.

Kent, C. 1995. Eukaryotic phospholipid biosynthesis. *Annu. Rev. Biochem.* 64:315–343.

Prescott, S. M., Zimmerman, G. A., Stafforini, D. M., and McIntyre, T. M. 2000. Platelet-activating factor and related lipid mediators. *Annu. Rev. Biochem.* 69:419–445.

Biosynthesis of Cholesterol and Steroids

Hampton, R. Y. 2002. Proteolysis and sterol regulation. *Annu. Rev. Cell Dev. Biol.* 18:345–378.

Kelley, R. I., and Herman, G. E. 2001. Inborn errors of sterol biosynthesis. *Annu. Rev. Genom.* 2:299–341.

Goldstein, J. L., and Brown, M. S. 1990. Regulation of the mevalonate pathway. *Nature* 343:425–430.

Gardner R. G., Shan, H., Matsuda, S. P. T., and Hampton, R. Y. 2001. An oxysterol-derived positive signal for 3-hydroxy-3-methylglutaryl-CoA reductase degradation in yeast. *J. Biol. Chem.* 276:8681–8694.

Istvan, E. S., and Deisenhofer, J. 2001. Structural mechanism for statin inhibition of HMG-CoA reductase. *Science* 292:1160–1164.

Ness, G. C., and Chambers, C. M. 2000. Feedback and hormonal regulation of hepatic 3-hydroxy-3-methylglutaryl coenzyme A reductase: The concept of cholesterol buffering capacity. *Proc. Soc. Exp. Biol. Med.* 224:8–19.

Libby, P., Aikawa, M., and Schonbeck, U. 2000. Cholesterol and atherosclerosis. *Biochim. Biophys. Acta* 1529:299–309.

Yokoyama, S. 2000. Release of cellular cholesterol: Molecular mechanism for cholesterol homeostasis in cells and in the body. *Biochim. Biophys. Acta* 1529:231–244.

Cronin, S. R., Khoury, A., Ferry, D. K., and Hampton, R. Y. 2000. Regulation of HMG-CoA reductase degradation requires the P-type ATPase Cod1p/Spf1p. *J. Cell Biol.* 148:915–924.

Edwards, P. A., Tabor, D., Kast, H. R., and Venkateswaran, A. 2000. Regulation of gene expression by SREBP and SCAP. *Biochim. Biophys. Acta* 1529:103–113.

Istvan, E. S., Palnitkar, M., Buchanan, S. K., and Deisenhofer, J. 2000. Crystal structure of the catalytic portion of human HMG-CoA reductase: Insights into regulation of activity and catalysis. *EMBO J.* 19:819–830.

Tabernero, L., Bochar, D. A., Rodwell, V. W., and Stauffacher, C. V. 1999. Substrate-induced closure of the flap domain in the ternary complex structures provides insights into the mechanism of catalysis by 3-hydroxy-3-methylglutaryl-CoA reductase. *Proc. Natl. Acad. Sci. U. S. A.* 96:7167–7171.

Fass, D., Blacklow, S., Kim, P. S., and Berger, J. M. 1997. Molecular basis of familial hypercholesterolaemia from structure of LDL receptor module. *Nature* 388:691–693.

Jeon, H., Meng, W., Takagi, J., Eck, M. J., Springer, T. A., and Blacklow, S. C. 2001. Implications for familial hypercholesterolemia from the structure of the LDL receptor YWTD-EGF domain pair. *Nat. Struct. Biol.* 8:499–504.

Lipoproteins and Their Receptors

Jeon, H., and Blacklow, S. C. 2005. Structure and physiologic function of the low-density lipoprotein receptor. *Annu. Rev. Biochem.* 74:535–562.

Brouillette, C. G., Anantharamaiah, G. M., Engler, J. A., and Borhani, D. W. 2001. Structural models of human apolipoprotein A-I: A critical analysis and review. *Biochem. Biophys. Acta* 1531:4–46.

Hevonoja, T., Pentikainen, M. O., Hyvonen, M. T., Kovanen, P. T., and Ala-Korpela, M. 2000. Structure of low density lipoprotein (LDL) particles: Basis for understanding molecular changes in modified LDL. *Biochim. Biophys. Acta* 1488:189–210.

Silver, D. L., Jiang, X. C., Arai, T., Bruce, C., and Tall, A. R. 2000. Receptors and lipid transfer proteins in HDL metabolism. *Ann. N. Y. Acad. Sci.* 902:103–111.

Nimpf, J., and Schneider, W. J. 2000. From cholesterol transport to signal transduction: Low density lipoprotein receptor, very low density lipoprotein receptor, and apolipoprotein E receptor-2. *Biochim. Biophys. Acta* 1529:287–298.

Borhani, D. W., Rogers, D. P., Engler, J. A., and Brouillette, C. G. 1997. Crystal structure of truncated human apolipoprotein A-I suggests a lipid-bound conformation. *Proc. Natl. Acad. Sci. U. S. A.* 94:12291–12296.

Wilson, C., Wardell, M. R., Weisgraber, K. H., Mahley, R. W., and Agard, D. A. 1991. Three-dimensional structure of the LDL receptor-binding domain of human apolipoprotein E. *Science* 252:1817–1822.

Plump, A. S., Smith, J. D., Hayek, T., Aalto-Setälä, K., Walsh, A., Verstuyft, J. G., Rubin, E. M., and Breslow, J. L. 1992. Severe hypercholesterolemia and atherosclerosis in apolipoprotein E-deficient mice created by homologous recombination in ES cells. *Cell* 71:343–353.

Oxygen Activation and P450 Catalysis

Williams, P. A., Cosme, J., Vinkovic, D. M., Ward, A., Angove, H. C., Day, P. J., Vonrhein, C., Tickle, I. J., and Jhoti, H. 2004. Crystal structure of human cytochrome P450 3A4 bound to metyrapone and progesterone. *Science* 305:683–686.

Ingelman-Sundberg, M., Oscarson, M., and McLellan, R. A. 1999. Polymorphic human cytochrome P450 enzymes: An opportunity for individualized drug treatment. *Trends Pharmacol. Sci.* 20:342–349.

Nelson, D. R. 1999. Cytochrome P450 and the individuality of species. *Arch. Biochem. Biophys.* 369:1–10.

Wong, L. L. 1998. Cytochrome P450 monooxygenases. *Curr. Opin. Chem. Biol.* 2:263–268.

Denison, M. S., and Whitlock, J. P. 1995. Xenobiotic-inducible transcription of cytochrome P450 genes. *J. Biol. Chem.* 270: 18175–18178.

Poulos, T. L. 1995. Cytochrome P450. *Curr. Opin. Struct. Biol.* 5:767–774.

Vaz, A. D., and Coon, M. J. 1994. On the mechanism of action of cytochrome P450: Evaluation of hydrogen abstraction in oxygen-dependent alcohol oxidation. *Biochemistry* 33:6442–6449.

Gonzalez, F. J., and Nebert, D. W. 1990. Evolution of the P450 gene superfamily: Animal–plant "warfare," molecular drive and human genetic differences in drug oxidation. *Trends Genet.* 6:182–186.

Problems

1. *Making fat.* Write a balanced equation for the synthesis of a triacylglycerol, starting from glycerol and fatty acids.

2. *Making a phospholipid.* Write a balanced equation for the synthesis of phosphatidylethanolamine by the de novo pathway, starting from ethanolamine, glycerol, and fatty acids.

3. *Activated donors.* What is the activated reactant in each of the following biosyntheses?

(a) Phosphatidylinositol from inositol
(b) Phosphatidylethanolamine from ethanolamine
(c) Ceramide from sphingosine
(d) Sphingomyelin from ceramide
(e) Cerebroside from ceramide
(f) Ganglioside G_{M1} from ganglioside G_{M2}
(g) Farnesyl pyrophosphate from geranyl pyrophosphate

4. *Telltale labels.* What is the distribution of isotopic labeling in cholesterol synthesized from each of the following precursors?

(a) Mevalonate labeled with ^{14}C in its carboxyl carbon atom
(b) Malonyl CoA labeled with ^{14}C in its carboxyl carbon atom

5. *Familial hypercholesterolemia.* Several classes of LDL-receptor mutations have been identified as causes of this disease. Suppose that you have been given cells from patients with different mutations, an antibody specific for the LDL receptor that can be seen with an electron microscope, and access to an electron microscope. What differences in antibody distribution might you expect to find in the cells from different patients?

6. *RNA editing.* A shortened version (apo B-48) of apolipoprotein B is formed by the intestine, whereas the full-length protein (apo B-100) is synthesized by the liver. A glutamine codon (CAA) is changed into a stop codon. Propose a simple mechanism for this change.

7. *Inspiration for drug design.* Some actions of androgens are mediated by dihydrotestosterone, which is formed by the reduction of testosterone. This finishing touch is catalyzed by an NADPH-dependent 5α-reductase.

5α-Dihydrotestosterone

Chromosomal XY males with a genetic deficiency of this reductase are born with a male internal urogenital tract but predominantly female external genitalia. These people are usually reared as girls. At puberty, they masculinize because the testosterone level rises. The testes of these reductase-deficient men are normal, whereas their prostate glands remain small. How might this information be used to design a drug to treat *benign prostatic hypertrophy,* a common consequence of the normal aging process in men? A majority of men older than age 55 have some degree of prostatic enlargement, which often leads to urinary obstruction.

8. *Drug idiosyncrasies.* Debrisoquine, a β-adrenergic blocking agent, has been used to treat hypertension. The optimal dose varies greatly (20–400 mg daily) in a population of patients. The urine of most patients taking the drug contains a high level of 4-hydroxydebrisoquine. However, those most sensitive to the drug (about 8% of the group studied) excrete debrisoquine and very little of the 4-hydroxy derivative. Propose a molecular basis for this drug idiosyncrasy. Why should caution be exercised in giving other drugs to patients who are very sensitive to debrisoquine?

Debrisoquine

9. *Removal of odorants.* Many odorant molecules are highly hydrophobic and concentrate within the olfactory epithelium. They would give a persistent signal independent of their concentration in the environment if they were not rapidly modified. Propose a mechanism for converting hydrophobic odorants into water-soluble derivatives that can be rapidly eliminated.

10. *Development difficulties.* Propecia (finasteride) is a synthetic steroid that functions as a competitive and specific inhibitor of 5α-reductase, the enzyme responsible for the synthesis of dihydrotestosterone from testosterone.

Finasteride

It is now widely used to retard the development of male pattern hair loss. Pregnant women are advised to avoid handling this drug. Why is it vitally important that pregnant women avoid contact with Propecia?

11. *Life-style consequences.* Human beings and the plant *Arabidopsis* evolved from the same distant ancestor possessing a

small number of cytochrome P450 genes. Human beings have approximately 50 such genes, whereas *Arabidopsis* has more than 250 of them. Propose a role for the large number of P450 isozymes in plants.

12. *Personalized medicine.* The cytochrome P450 system metabolizes many medicinally useful drugs. Although all human beings have the same number of P450 genes, individual polymorphisms exist that alter the specificity and efficiency of the proteins encoded by the genes. How could knowledge of individual polymorphisms be useful clinically?

Mechanism Problems

13. *An interfering phosphate.* In the course of the overall reaction catalyzed by HMG-CoA reductase, a histidine residue protonates a coenzyme A thiolate, CoA–S$^-$, generated in a previous step.

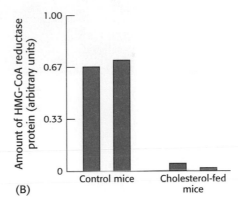

A nearby serine residue can be phosphorylated by AMP-dependent kinase, which results in a loss of activity. Propose an explanation for why phosphorylation of the serine residue inhibits enzyme activity.

14. *Demethylation.* Methyl amines are often demethylated by cytochrome P450 enzymes. Propose a mechanism for the formation of methylamine from dimethylamine catalyzed by cytochrome P450. What is the other product?

Data Interpretation and Chapter Integration Problem

15. *Cholesterol feeding.* Mice were divided into four groups, two of which were fed a normal diet and two of which were fed a cholesterol-rich diet. HMG-CoA reductase mRNA and protein from liver were then isolated and quantified. Graph A shows the results of the mRNA isolation.

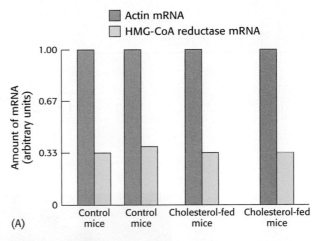

(a) What is the effect of cholesterol feeding on the amount of HMG-CoA reductase mRNA?

(b) What is the purpose of also isolating the mRNA for the protein actin, which is not under the control of the sterol regulatory element?

HMG-CoA reductase protein was isolated by precipitation with a monoclonal antibody to HMG-CoA reductase. The amount of HMG-CoA protein in each group is shown in graph B.

(c) What is the effect of the cholesterol diet on the amount of HMG-CoA reductase protein?

(d) Why is this result surprising in light of the results in graph A?

(e) Suggest possible explanations for the results in graph B.

The Integration of Metabolism

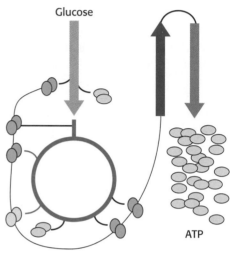

Glucose

ATP

The image at the left shows a detail of runners on a Greek amphora painted in the sixth century B.C. Athletic feats, as well as others as seemingly simple as the maintenance of blood-glucose levels, require elaborate metabolic integration. The schematic representation illustrates the oxidation of glucose to yield ATP in a process requiring interplay between glycolysis, the citric acid cycle, and oxidative phosphorylation. These are a few of the many metabolic pathways that must be coordinated to meet the demands of living. [(Left) Metropolitan Museum of Art, Rogers Fund, 1914 (14.130.12). Copyright © 1977 by the Metropolitan Museum of Art.]

We have been examining the biochemistry of metabolism one pathway at a time but, in living systems, many pathways are operating simultaneously. Each pathway must be able to sense the status of the others to function optimally to meet the needs of an organism. How is the intricate network of reactions in metabolism coordinated? This chapter presents some of the principles underlying the *integration of metabolism* in mammals. We begin with a recapitulation of the strategy of metabolism and of recurring motifs in its regulation. We then turn to the interplay of different pathways as we examine the flow of molecules at three key crossroads: glucose 6-phosphate, pyruvate, and acetyl CoA. We consider the differences in the metabolic patterns of the brain, muscle, adipose tissue, kidney, and liver. Finally, we examine how the interplay between these tissues is altered in a variety of metabolic perturbations. These considerations of metabolism will illustrate how biochemical knowledge illuminates the functioning of the organism.

27.1 Metabolism Consists of Highly Interconnected Pathways

The basic strategy of catabolic metabolism is to form ATP, reducing power, and building blocks for biosyntheses. Let us briefly review these central themes:

1. *ATP is the universal currency of energy.* The high phosphoryl-transfer potential of ATP enables it to serve as the energy source in muscle contraction, active transport, signal amplification, and biosyntheses. In the cell, the hydrolysis of an ATP molecule changes the equilibrium ratio of products to reactants in a coupled reaction by a factor of about 10^8. Hence, *a thermodynamically unfavorable reaction sequence can be made highly favorable by coupling it to the hydrolysis of a sufficient number of ATP molecules.*

2. *ATP is generated by the oxidation of fuel molecules such as glucose, fatty acids, and amino acids.* The common intermediate in most of these oxidations is acetyl CoA. The carbon atoms of the acetyl unit are completely oxidized to CO_2 by the citric acid cycle with the concomitant formation of NADH and $FADH_2$. These electron carriers then transfer their high-potential electrons to the respiratory chain. The subsequent flow of electrons to O_2 leads to the pumping of protons across the inner mitochondrial membrane (Figure 27.1). This proton gradient is then used to synthesize ATP. Glycolysis also generates ATP, but the amount formed is much smaller than that formed by oxidative phosphorylation. The oxidation of glucose to pyruvate yields only 2 molecules of ATP, whereas the complete oxidation of glucose to CO_2 yields 30 molecules of ATP.

3. *NADPH is the major electron donor in reductive biosyntheses.* In most biosyntheses, the products are more reduced than the precursors, and so reductive power is needed as well as ATP. The high-potential electrons required to drive these reactions are usually provided by NADPH. The pentose phosphate pathway supplies much of the required NADPH.

4. *Biomolecules are constructed from a small set of building blocks.* The highly diverse molecules of life are synthesized from a much smaller number of precursors. The metabolic pathways that generate ATP and NADPH also provide building blocks for the biosynthesis of more-complex molecules. For example, acetyl CoA, the common intermediate in the breakdown of most fuels, supplies a two-carbon unit in a wide variety of biosyntheses, such as those leading to fatty acids, prostaglandins, and cholesterol. Thus, *the central metabolic pathways have anabolic as well as catabolic roles.*

5. *Biosynthetic and degradative pathways are almost always distinct.* For example, the pathway for the synthesis of fatty acids is different from that for their degradation. This separation enables both biosynthetic and degradative pathways to be thermodynamically favorable at all times. A biosynthetic pathway is made exergonic by coupling it to the hydrolysis of a sufficient number of ATP molecules. The separation of biosynthetic and degradative pathways contributes greatly to the effectiveness of metabolic control.

Recurring Motifs are Common in Metabolic Regulation

Anabolism and catabolism must be precisely coordinated. Metabolic networks sense and respond to information on the status of their component pathways. The information is received and metabolism is controlled in several ways:

"To every thing there is a season, and a time to every purpose under the heaven:
    A time to be born, and a time to die; a time to plant, and a time to pluck up that which is planted;
    A time to kill, and a time to heal; a time to break down, and a time to build up."
                                    ECCLESIASTES 3:1–3

Pasteur effect

The inhibition of glycolysis by respiration, discovered by Louis Pasteur in studying fermentation by yeast. The consumption of carbohydrate is about sevenfold lower under aerobic conditions than under anaerobic ones. The inhibition of phosphofructokinase by citrate and ATP accounts for much of the Pasteur effect.

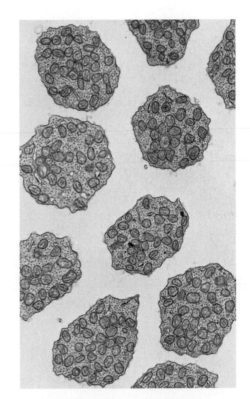

Figure 27.1 Electron micrograph of mitochondria. Numerous mitochondria occupy the inner segment of retinal rod cells. These photoreceptor cells generate large amounts of ATP and are highly dependent on a continuous supply of O_2. [Courtesy of Dr. Michael Hogan.]

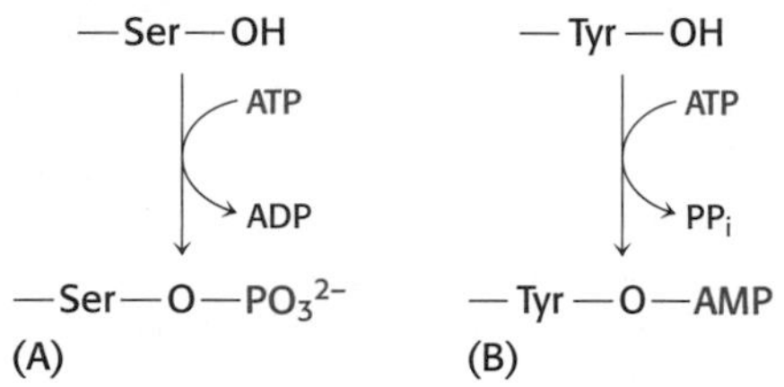

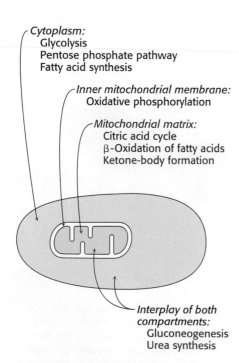

Figure 27.2 Covalent modifications.
Covalent modifications. Examples of
reversible covalent modifications
of proteins: (A) phosphorylation and
(B) adenylation.

Cytoplasm:
Glycolysis
Pentose phosphate pathway
Fatty acid synthesis

Inner mitochondrial membrane:
Oxidative phosphorylation

Mitochondrial matrix:
Citric acid cycle
β-Oxidation of fatty acids
Ketone-body formation

*Interplay of both
compartments:*
Gluconeogenesis
Urea synthesis

**Figure 27.3 Compartmentation of the
major pathways of metabolism.**

1. *Allosteric Interactions.* The flow of molecules in most metabolic pathways is determined primarily by the activities of certain enzymes rather than by the amount of substrate available. Enzymes that catalyze essentially irreversible reactions are likely control sites, and the first irreversible reaction in a pathway (the committed step) is nearly always tightly controlled. Enzymes catalyzing committed steps are allosterically regulated, as exemplified by phosphofructokinase (PFK) in glycolysis and acetyl CoA carboxylase in fatty acid synthesis. *Allosteric interactions enable such enzymes to rapidly detect diverse signals and to adjust their activity accordingly.*

2. *Covalent Modification.* Some regulatory enzymes are controlled by covalent modification in addition to allosteric interactions. For example, the catalytic activity of glycogen phosphorylase is enhanced by phosphorylation, whereas that of glycogen synthase is diminished. Specific enzymes catalyze the addition and removal of these modifying groups (Figure 27.2). Phosphorylation often takes place in response to hormonal signals. For instance, insulin, glucagon, and epinephrine stimulate protein kinases. Why is covalent modification used in addition to noncovalent allosteric control? The covalent modification of an essential enzyme in a pathway is often the final step in an amplifying cascade and allows metabolic pathways to be rapidly switched on or off by very low concentrations of triggering signals. In addition, covalent modifications usually last longer (from seconds to minutes) than do reversible allosteric interactions (from milliseconds to seconds).

3. *Adjustment of Enzyme Levels.* The amounts of enzymes, as well as their activities, are controlled. The rates of synthesis and degradation of many regulatory enzymes are altered by hormones. The basics of this control will be considered in Chapter 29; we will return to the topic in Chapter 31.

4. *Compartmentation.* The metabolic patterns of eukaryotic cells are markedly affected by the presence of compartments (Figure 27.3). The fates of certain molecules depend on whether they are in the cytoplasm or in mitochondria, and so their flow across the inner mitochondrial membrane is often regulated. For example, fatty acids are transported into mitochondria for degradation only when energy is required, whereas fatty acids in the cytoplasm are esterified or exported.

5. *Metabolic Specializations of Organs.* Regulation in higher eukaryotes is enhanced by the existence of organs with different metabolic roles. Metabolic specialization is the result of differential gene expression.

Major Metabolic Pathways Have Specific Control Sites

Let us now review the roles of the major pathways of metabolism and the principal sites for their control:

1. *Glycolysis.* This sequence of reactions in the cytoplasm converts one molecule of glucose into two molecules of pyruvate with the concomitant generation of two molecules each of ATP and NADH. The NAD^+ consumed in the reaction catalyzed by glyceraldehyde 3-phosphate dehydrogenase must be regenerated for glycolysis to proceed. Under anaerobic conditions, as in highly active skeletal muscle, this regeneration is accomplished by the reduction of pyruvate to lactate. Alternatively, under aerobic conditions, NAD^+ is regenerated by the transfer of electrons from NADH to O_2 through the electron-transport chain. Glycolysis serves two main purposes: it degrades glucose to generate ATP, and it provides carbon skeletons for biosyntheses.

Phosphofructokinase, which catalyzes the committed step in glycolysis, is the most important control site. ATP is both a substrate in the phosphoryl-group-transfer reaction and a regulatory molecule. A high level of ATP inhibits phosphofructokinase: the regulatory sites are distinct from the substrate-binding sites and have a lower affinity for the nucleotide. This inhibitory effect is enhanced by citrate and reversed by AMP (Figure 27.4). Thus, the rate of glycolysis depends on the need for ATP, as signaled by the ATP/AMP ratio, and on the availability of building blocks, as signaled by the level of citrate. *In the liver, the most important regulator of phosphofructokinase activity is fructose 2,6-bisphosphate* (F-2,6-BP). Recall that the level of F-2,6-BP is determined by the activity of the kinase that forms it from fructose 6-phosphate and of the phosphatase that hydrolyzes the 2-phosphoryl group (p. 466). When the blood-glucose level is low, a glucagon-triggered cascade leads to the activation of the phosphatase and the inhibition of the kinase in the liver. The level of F-2,6-BP declines and, consequently, so does phosphofructokinase activity. Hence, glycolysis is slowed, and the spared glucose is released into the blood for use by other tissues. Pyruvate kinase, which controls the outflow of glycolysis, also is an important regulatory site. It is stimulated by fructose 1,6-bisphosphate, a product of the PFK reaction, and inhibited by ATP. In the liver, pyruvate kinase is phosphorylated by the glucagon-stimulated cAMP cascade, diminishing the enzyme's activity.

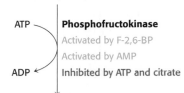

Figure 27.4 Regulation of glycolysis. Phosphofructokinase is the key enzyme in the regulation of glycolysis.

2. *Citric Acid Cycle and Oxidative Phosphorylation.* The reactions of this common pathway for the oxidation of fuel molecules—carbohydrates, fatty acids, and amino acids—take place inside mitochondria. Most fuels enter the cycle as acetyl CoA. The complete oxidation of an acetyl unit by the citric acid cycle generates one molecule of GTP and results in the reduction of three molecules of NAD^+ to NADH and one molecule of FAD to $FADH_2$. These electrons are transferred to O_2 through the electron-transport chain, which results in the formation of a proton gradient that drives the synthesis of nine molecules of ATP. The electron donors are oxidized and recycled back to the citric acid cycle only if ADP is simultaneously phosphorylated to ATP. *This tight coupling, called respiratory control, ensures that the rate of the citric acid cycle matches the need for ATP.* An abundance of ATP also diminishes the activities of two enzymes in the cycle—*isocitrate dehydrogenase* and *α-ketoglutarate dehydrogenase.* The citric acid cycle has an anabolic role as well. In concert with *pyruvate carboxylase,* the citric acid cycle provides intermediates for biosyntheses, such as succinyl CoA for the formation of porphyrins and citrate for the formation of fatty acids.

3. *Pyruvate Dehydrogenase Complex.* This enzyme complex is a key regulatory site because it is the irreversible link between glycolysis and the citric acid cycle. The pyruvate dehydrogenase complex catalyzes the conversion of pyruvate into acetyl CoA. *This reaction, which takes place inside mitochondria, is a decisive reaction in metabolism: it commits the carbon atoms of carbohydrates and amino acids to oxidation by the citric acid cycle or to the synthesis of lipids.* The pyruvate dehydrogenase complex is stringently regulated by multiple allosteric interactions and covalent modifications. Pyruvate is rapidly converted into acetyl CoA only if ATP is needed or if two-carbon fragments are required for the synthesis of lipids.

4. *Pentose Phosphate Pathway.* This series of reactions, which takes place in the cytoplasm, consists of two stages. The first stage is the oxidative decarboxylation of glucose 6-phosphate. Its purpose is the production of NADPH for reductive biosyntheses and the formation of ribose 5-phosphate for the synthesis of nucleotides. Two molecules of

Glucose 6-phosphate

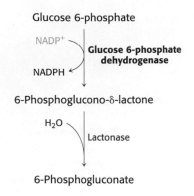

NADP$^+$

**Glucose 6-phosphate
dehydrogenase**

NADPH

6-Phosphoglucono-δ-lactone

H$_2$O

Lactonase

6-Phosphogluconate

**Figure 27.5 Regulation of the pentose
phosphate pathway.** The dehydrogenation
of glucose 6-phosphate is the committed
step in the pentose phosphate pathway.

NADPH are generated in the conversion of glucose 6-phosphate into ribose 5-phosphate. The dehydrogenation of glucose 6-phosphate is the committed step in this pathway. This reaction is controlled by the level of NADP$^+$, the electron acceptor (Figure 27.5).

The second stage of the pentose phosphate pathway is the nonoxidative, reversible metabolism of five-carbon phosphosugars into phosphorylated three-carbon and six-carbon glycolytic intermediates. Thus, the nonoxidative branch can either introduce riboses into glycolysis for catabolism or generate riboses from glycolytic intermediates for biosyntheses.

5.   *Gluconeogenesis.* Glucose can be synthesized primarily by the liver from noncarbohydrate precursors such as lactate, glycerol, and amino acids. The major entry point of this pathway is pyruvate, which is carboxylated to oxaloacetate in mitochondria. Oxaloacetate is then metabolized in the cytoplasm to form phosphoenolpyruvate. The other distinctive reactions used by gluconeogenesis are two hydrolytic steps that bypass the irreversible reactions of glycolysis. *Gluconeogenesis and glycolysis are usually reciprocally regulated so that one pathway is minimally active while the other is highly active.* For example, AMP inhibits and citrate activates fructose 1,6-bisphosphatase, an essential enzyme in gluconeogenesis, whereas these molecules have opposite effects on phosphofructokinase, the pacemaker of glycolysis (Figure 27.6). Fructose-2,6-bisphosphate also coordinates these processes by inhibiting fructose 1,6-bisphosphatase. Hence, when glucose is abundant, the high level of F-2,6-BP inhibits gluconeogenesis and activates glycolysis.

**Figure 27.6 Regulation of
gluconeogenesis.** Fructose 1,6-
bisphosphatase is the principal
enzyme controlling the rate of
gluconeogenesis.

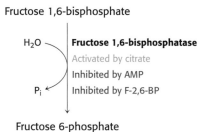

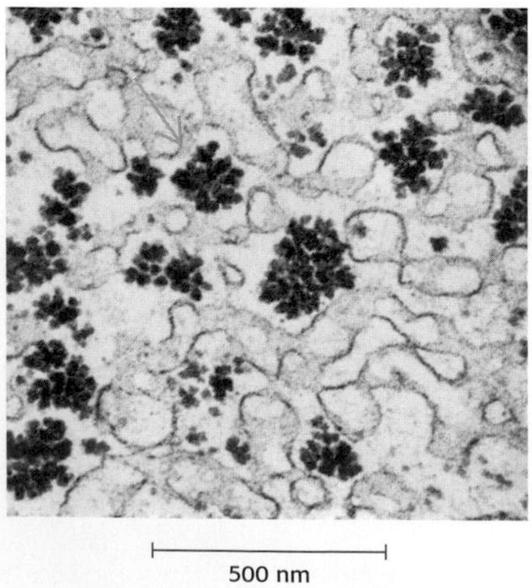

Fructose 1,6-bisphosphate

H$_2$O

Fructose 1,6-bisphosphatase

Activated by citrate

Inhibited by AMP

P$_i$

Inhibited by F-2,6-BP

Fructose 6-phosphate

6.   *Glycogen Synthesis and Degradation.* Glycogen, a readily mobilizable fuel store, is a branched polymer of glucose residues (Figure 27.7). In glycogen degradation, a phosphorylase catalyzes the cleavage of glycogen by orthophosphate to yield glucose 1-phosphate, which is rapidly converted into glucose 6-phosphate for further metabolism. In glycogen synthesis, the activated intermediate is UDP-glucose, which is formed from glucose 1-phosphate and UTP. Glycogen synthase catalyzes the transfer of glucose from UDP-glucose to the terminal glucose residue of a growing strand. *Glycogen degradation and synthesis are coordinately controlled by a hormone-triggered amplifying cascade so that the phosphorylase is active when synthase is inactive and vice versa.* Phosphorylation and noncovalent allosteric interactions regulate these enzymes.

7.   *Fatty Acid Synthesis and Degradation.* Fatty acids are synthesized in the cytoplasm by the addition of two-carbon units to a growing chain on an acyl carrier protein. Malonyl CoA, the activated intermediate, is formed by the carboxylation of acetyl CoA. Acetyl groups are carried from mitochondria to the cytoplasm as citrate by the citrate–malate shuttle. In the cytoplasm, citrate is cleaved to yield acetyl CoA. In addition to transporting acetyl CoA, *citrate in the cytoplasm stimulates acetyl CoA carboxylase, the enzyme*

500 nm

Figure 27.7 Glycogen granules. The
electron micrograph shows part of a
liver cell containing glycogen particles.
[Courtesy of Dr. George Palade.]

catalyzing the committed step. When ATP and acetyl CoA are abundant, the level of citrate increases, which accelerates the rate of fatty acid synthesis (Figure 27.8).

A different pathway in a different compartment degrades fatty acids. Carnitine transports fatty acids into mitochondria, where they are degraded to acetyl CoA in the mitochondrial matrix by β oxidation. The acetyl CoA then enters the citric acid cycle if the supply of oxaloacetate is sufficient. Alternatively, acetyl CoA can give rise to ketone bodies. The $FADH_2$ and NADH formed in the β-oxidation pathway transfer their electrons to O_2 through the electron-transport chain. Like the citric acid cycle, β oxidation can continue only if NAD^+ and FAD are regenerated. Hence, *the rate of fatty acid degradation also is coupled to the need for ATP.* Malonyl CoA, the precursor for fatty acid synthesis, inhibits fatty acid degradation by inhibiting the formation of acyl carnitine by carnitine acyltransferase I, thus preventing the translocation of fatty acids into mitochondria (Figure 27.9).

Glucose 6-phosphate, Pyruvate, and Acetyl CoA Are Key Junctions in Metabolism

The factors governing the flow of molecules in metabolism can be further understood by examining three important molecules: glucose 6-phosphate, pyruvate, and acetyl CoA. Each of these molecules has several contrasting fates:

1. *Glucose 6-phosphate.* Glucose entering a cell is rapidly phosphorylated to glucose 6-phosphate and is subsequently stored as glycogen, degraded to pyruvate, or converted into ribose 5-phosphate (Figure 27.10). Glycogen is formed when glucose 6-phosphate and ATP are abundant. In contrast, glucose 6-phosphate flows into the glycolytic pathway when ATP or carbon skeletons for biosyntheses are required. Thus, the conversion of glucose 6-phosphate into pyruvate can be anabolic as well as catabolic. The third major fate of glucose 6-phosphate, to flow through the pentose phosphate pathway, provides NADPH for reductive biosyntheses and ribose 5-phosphate for the synthesis of nucleotides. Glucose 6-phosphate can be formed by the mobilization of glycogen or it can be synthesized from pyruvate and glucogenic amino acids by the gluconeogenic pathway.

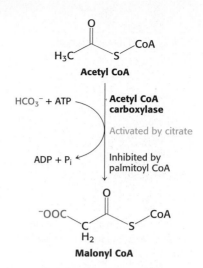

Figure 27.8 Regulation of fatty acid synthesis. Acetyl CoA carboxylase is the key control site in fatty acid synthesis.

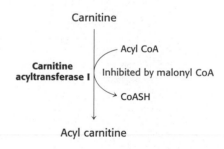

Figure 27.9 Control of fatty acid degradation. Malonyl CoA inhibits fatty acid degradation by inhibiting the formation of acyl carnitine.

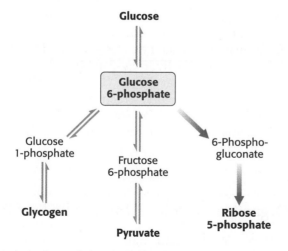

Figure 27.10 Metabolic fates of glucose 6-phosphate.

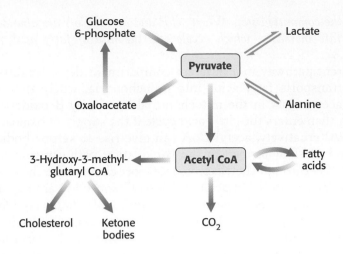

Figure 27.11 Major metabolic fates of
pyruvate and acetyl CoA in mammals.

2. *Pyruvate.* This three-carbon α-ketoacid is another major metabolic junction (Figure 27.11). Pyruvate is derived primarily from glucose 6-phosphate, alanine, and lactate. Pyruvate can be reduced to lactate by lactate dehydrogenase to regenerate NAD^+. This reaction enables glycolysis to proceed transiently under anaerobic conditions in active tissues such as contracting muscle. The lactate formed in active tissue is subsequently oxidized back to pyruvate in other tissues. The essence of this interconversion is that it buys time and shifts part of the metabolic burden of active muscle to other tissues. Another readily reversible reaction in the cytoplasm is the transamination of pyruvate, an α-ketoacid, to alanine, the corresponding amino acid. Conversely, several amino acids can be converted into pyruvate. Thus, *transamination is a major link between amino acid and carbohydrate metabolism.*

A third fate of pyruvate is its carboxylation to oxaloacetate inside mitochondria, the first step in gluconeogenesis. This reaction and the subsequent conversion of oxaloacetate into phosphoenolpyruvate bypass an irreversible step of glycolysis and hence enable glucose to be synthesized from pyruvate. The carboxylation of pyruvate is also important for replenishing intermediates of the citric acid cycle. Acetyl CoA activates pyruvate carboxylase, enhancing the synthesis of oxaloacetate, when the citric acid cycle is slowed by a paucity of this intermediate. A fourth fate of pyruvate is its oxidative decarboxylation to acetyl CoA, as described on page 763.

3. *Acetyl CoA.* The major sources of this activated two-carbon unit are the oxidative decarboxylation of pyruvate and the β oxidation of fatty acids (see Figure 27.11). Acetyl CoA is also derived from ketogenic amino acids. The fate of acetyl CoA, in contrast with that of many molecules in metabolism, is quite restricted. The acetyl unit can be completely oxidized to CO_2 by the citric acid cycle. Alternatively, 3-hydroxy-3-methylglutaryl CoA can be formed from three molecules of acetyl CoA. This six-carbon unit is a precursor of *cholesterol* and of *ketone bodies,* which are transport forms of acetyl units released from the liver for use by some peripheral tissues. A third major fate of acetyl CoA is its export to the cytoplasm in the form of citrate for the synthesis of fatty acids.

27.2 Each Organ Has a Unique Metabolic Profile

The metabolic patterns of the brain, muscle, adipose tissue, kidney, and liver are strikingly different. Let us consider how these organs differ in their use of fuels to meet their energy needs:

1. *Brain. Glucose is virtually the sole fuel for the human brain, except during prolonged starvation.* The brain lacks fuel stores and hence requires a continuous supply of glucose. It consumes about 120 g daily, which corresponds to an energy input of about 1760 kJ (420 kcal), accounting for some 60% of the utilization of glucose by the whole body in the resting state. Much of the energy, estimates suggest from 60% to 70%, is used to power transport mechanisms that maintain the Na^+–K^+ membrane potential required for the transmission of the nerve impulses. The brain must also synthesize neurotransmitters and their receptors to propagate nerve impulses. Overall, glucose metabolism remains unchanged during mental activity, although local increases are detected when a subject performs certain tasks.

Glucose is transported into brain cells by the glucose transporter GLUT3. This transporter has a low value of K_M for glucose (1.6 mM), which means that it is nearly saturated under most conditions, given that the plasma concentration of glucose during fasting is 4.7 mM (84.7 mg/dl). Under these conditions, the concentration of glucose in the brain is about 1 mM. Glycolysis slows down when the glucose level approaches the K_M value of hexokinase ($\sim$50 μM), the enzyme that traps glucose in the cell (p. 435). This danger point is reached when the plasma-glucose level drops below about 2.2 mM (39.6 mg/dl) and thus approaches the K_M value of GLUT3.

Fatty acids do not serve as fuel for the brain but rather are utilized for membrane synthesis. In starvation, *ketone bodies generated by the liver partly replace glucose as fuel for the brain.*

2. *Muscle. The major fuels for muscle are fatty acids, glucose, and ketone bodies.* In resting muscle, fatty acids are the major fuel, meeting 85% of the energy needs. Muscle differs from the brain in having a large store of glycogen (5000 kJ, or 1200 kcal). In fact, about three-fourths of all the glycogen in the body is stored in muscle (Table 27.1). This glycogen is readily converted into glucose 6-phosphate for use within muscle cells. Muscle, like the brain, lacks glucose 6-phosphatase, and so it does not export glucose. Rather, *muscle retains glucose, its preferred fuel for bursts of activity.*

In vigorously contracting skeletal muscle, the rate of glycolysis far exceeds that of the citric acid cycle, and much of the pyruvate formed is reduced to lactate, some of which flows to the liver, where it is converted into glucose (Figure 27.12). These interchanges, known as the Cori cycle (p. 468), shift part of the metabolic burden of muscle to the liver. In addition, a large amount of alanine is formed in active muscle by the transamination of pyruvate. Alanine, like lactate, can be converted into glucose by the liver. Why does the muscle release alanine? Muscle can absorb and transaminate branched-chain amino acids in order to use the carbon skeletons as fuel; however, it cannot form urea. Consequently, the nitrogen is released into the blood as alanine. The liver absorbs the alanine, removes the nitrogen for disposal as urea, and processes the pyruvate to glucose or fatty acids.

TABLE 27.1 Fuel reserves in a typical 70-kg man

Organ	Glucose or glycogen	Triacylglycerols	Mobilizable proteins
Blood	250 (60)	200 (45)	0 (0)
Liver	1700 (400)	2000 (450)	1700 (400)
Brain	30 (8)	0 (0)	0 (0)
Muscle	5000 (1200)	2000 (450)	100,000 (24,000)
Adipose tissue	330 (80)	560,000 (135,000)	170 (40)

Table header spanning: AVAILABLE ENERGY IN KILOJOULES (KCAL)

Source: After G. F. Cahill, Jr. *Clin. Endocrinol. Metab.* 5(1976):398.

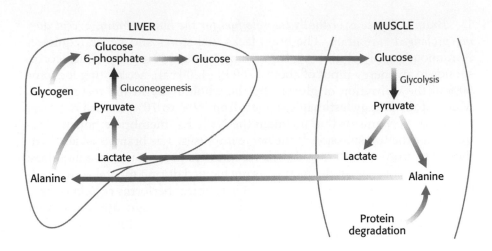

Figure 27.12 Metabolic interchanges
between muscle and the liver.

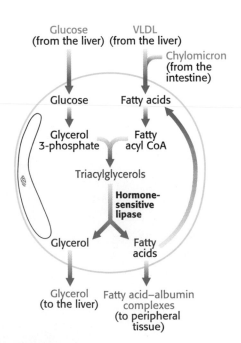

Figure 27.13 Synthesis and degradation
of triacylglycerols by adipose tissue.
Fatty acids from the liver are delivered
to adipose cells in the form of
triacylglycerols contained in very low
density lipoproteins (VLDLs). Fatty acids
from the diet are transported in
chylomicrons.

Unlike skeletal muscle, heart muscle functions almost exclusively aerobically, as evidenced by the density of mitochondria in heart muscle. Moreover, the heart has virtually no glycogen reserves. Fatty acids are the heart's main source of fuel, although ketone bodies as well as lactate can serve as fuel for heart muscle.

3. *Adipose Tissue. The triacylglycerols stored in adipose tissue are an enormous reservoir of metabolic fuel* (see Table 27.1). In a typical 70-kg man, the 15 kg of triacylglycerols have an energy content of 565,000 kJ (135,000 kcal). Adipose tissue is specialized for the esterification of fatty acids to form triacylglycerols and for their release from triacylglycerols. In human beings, the liver is the major site of fatty acid synthesis, although, in the developed world, most people obtain most of their fatty acids from their diets. Dietary fats are delivered to the adipose tissue from the intestines by chylomicrons. Fatty acids in the liver are esterified to glycerol phosphate to form triacylglycerol and are transported to the adipose tissue in lipoprotein particles, such as very low density lipoproteins (p. 743). Triacylglycerols are not taken up by adipocytes; rather, they are first hydrolyzed by an extracellular lipoprotein lipase for uptake. This lipase is stimulated by processes initiated by insulin. After the fatty acids enter the cell, the principal task of adipose tissue is to activate these fatty acids and transfer the resulting CoA derivatives to glycerol in the form of glycerol 3-phosphate. This essential intermediate in lipid biosynthesis comes from the reduction of the glycolytic intermediate dihydroxyacetone phosphate. Thus, *adipose cells need glucose for the synthesis of triacylglycerols* (Figure 27.13).

Triacylglycerols are hydrolyzed to fatty acids and glycerol by intracellular lipases. The release of the first fatty acid from a triacylglycerol, the rate-limiting step, is catalyzed by a hormone-sensitive lipase that is reversibly phosphorylated. The hormone epinephrine stimulates the formation of cyclic AMP, the intracellular messenger in the amplifying cascade, which activates a protein kinase—a recurring theme in hormone action. Triacylglycerols in adipose cells are continually being hydrolyzed and resynthesized. Glycerol derived from their hydrolysis is exported to the liver. Most of the fatty acids formed on hydrolysis are reesterified if glycerol 3-phosphate is abundant. In contrast, they are released into the plasma if glycerol 3-phosphate is scarce because of a paucity of glucose. Thus, *the availability of glucose inside adipose cells is a major factor in determining whether fatty acids are released into the blood.*

4. *The Kidneys. The major purpose of the kidneys is to produce urine,* which serves as a vehicle for excreting metabolic waste products and for maintaining the osmolarity of the body fluids. The blood plasma is filtered nearly 60

times each day in the renal tubules. Most of the material filtered out of the blood is reabsorbed; so only 1 to 2 liters of urine is produced. Water-soluble materials in the plasma, such as glucose, and water itself are reabsorbed to prevent wasteful loss. The kidneys require large amounts of energy to accomplish the reabsorption. Although constituting only 0.5% of body mass, the kidneys consume 10% of the oxygen used in cellular respiration. Much of the glucose that is reabsorbed is carried into the kidney cells by the sodium–glucose cotransporter. This transporter is powered by the $Na^+–K^+$ gradient, which is itself maintained by the $Na^+–K^+$ ATPase (Section 13.4). During starvation, the kidney becomes an important site of gluconeogenesis and may contribute as much as half of the blood glucose.

5. *The Liver. The metabolic activities of the liver are essential for providing fuel to the brain, muscle, and other peripheral organs.* Indeed, the liver, which can be from 2% to 4% of body weight, is an organism's metabolic hub (Figure 27.14). Most compounds absorbed by the intestine first pass through the liver, which is thus able to regulate the level of many metabolites in the blood.

Let us first consider how the liver metabolizes carbohydrates. The liver removes two-thirds of the glucose from the blood and all of the remaining monosaccharides after meals. Some glucose is left in the blood for use by other tissues. The absorbed glucose is converted into glucose 6-phosphate by hexokinase and the liver-specific glucokinase. Glucose 6-phosphate, as already stated, has a variety of fates, although the liver uses little of it to meet its own energy needs. Much of the glucose 6-phosphate is converted into glycogen. As much as 1700 kJ (400 kcal) can be stored in this form in the liver. Excess glucose 6-phosphate is metabolized to acetyl CoA, which is used to form fatty acids, cholesterol, and bile salts. The pentose phosphate pathway, another means of processing glucose 6-phosphate, supplies the NADPH for these reductive biosyntheses. The liver can produce glucose for release into the blood by breaking down its store of glycogen and by carrying out gluconeogenesis. The main precursors for gluconeogenesis are lactate and alanine from muscle, glycerol from adipose tissue, and glucogenic amino acids from the diet.

The liver also plays a central role in the regulation of lipid metabolism. When fuels are abundant, fatty acids derived from the diet or synthesized by the liver are esterified and secreted into the blood in the form of very low density lipoprotein (see Figure 27.13). However, in the fasting state, the liver converts fatty acids into ketone bodies. How is the fate of liver fatty acids determined? The selection is made according to whether the fatty acids enter the mitochondrial matrix. Recall that long-chain fatty acids traverse the inner mitochondrial membrane only if they are esterified to carnitine. Carnitine acyltransferase I (also known as carnitine palmitoyl transferase I), which catalyzes the formation of acyl carnitine, is inhibited by malonyl CoA, the committed intermediate in the synthesis of fatty acids (see Figure 27.9). Thus, *when malonyl CoA is abundant, long-chain fatty acids are prevented from entering the mitochondrial matrix, the compartment of β oxidation and ketone-body formation. Instead, fatty acids are exported to adipose tissue for incorporation into triacylglycerols.* In contrast, the level of malonyl CoA is low when fuels are scarce. Under these conditions, fatty acids liberated from adipose tissues enter the mitochondrial matrix for conversion into ketone bodies.

The liver also plays an essential role in dietary amino acid metabolism. The liver absorbs the majority of amino acids, leaving some in the blood for peripheral tissues. The priority use of amino acids is for protein synthesis rather than catabolism. By what means are amino acids directed to protein

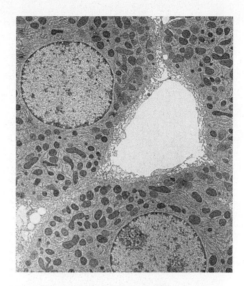

Figure 27.14 Electron micrograph of liver cells. The liver plays an essential role in the integration of metabolism. [Courtesy of Dr. Ann Hubbard.]

Fatty acyl carnitine

synthesis in preference to use as a fuel? The K_M value for the aminoacyl-tRNA synthetases is lower than that for the enzymes taking part in amino acid catabolism. Thus, amino acids are used to synthesize aminoacyl-tRNAs before they are catabolized. When catabolism does take place, the first step is the removal of nitrogen, which is subsequently processed to urea. The liver secretes from 20 to 30 g of urea a day. The α-ketoacids are then used for gluconeogenesis or fatty acid synthesis. Interestingly, the liver cannot remove nitrogen from the branched-chain amino acids (leucine, isoleucine, and valine). Transamination of these amino acids takes place in the muscle.

How does the liver meet its own energy needs? α-Ketoacids derived from the degradation of amino acids are the liver's own fuel. In fact, the main role of glycolysis in the liver is to form building blocks for biosyntheses. Furthermore, the liver cannot use acetoacetate as a fuel, because it has little of the transferase needed for acetoacetate's activation to acetyl CoA. Thus, the liver eschews the fuels that it exports to muscle and the brain.

27.3 Food Intake and Starvation Induce Metabolic Changes

We shall now consider the biochemical responses to a series of physiological conditions. Our first example is the *starved–fed cycle*, which we all experience in the hours after an evening meal and through the night's fast. This nightly starved–fed cycle has three stages: the postabsorptive state after a meal, the early fasting during the night, and the refed state after breakfast. A major goal of the many biochemical alterations in this period is to maintain *glucose homeostasis*—that is, a constant blood-glucose level.

1. *The Well-Fed, or Postprandial, State.* After we consume and digest an evening meal, glucose and amino acids are transported from the intestine to the blood. The dietary lipids are packaged into chylomicrons and transported to the blood by the lymphatic system. This fed condition leads to the secretion of *insulin*, which is one of the two most important regulators of fuel metabolism, the other regulator being *glucagon*. The secretion of the hormone insulin by the β cells of the pancreas is stimulated by glucose and the parasympathetic nervous system (Figure 27.15). *In essence, insulin signals the fed state; it stimulates the storage of fuels and the synthesis of proteins in a variety of ways.* For instance, insulin initiates protein kinase cascades. These cascades stimulate glycogen synthesis in both muscle and the liver and suppresses gluconeogenesis by the liver. Insulin also accelerates glycolysis in the liver, which in turn increases the synthesis of fatty acids.

The liver helps to limit the amount of glucose in the blood during times of plenty by storing it as glycogen so as to be able to release glucose in times of scarcity. How is the excess blood glucose present after a meal removed? The liver is able to trap large quantities of glucose because it possesses an isozyme of hexokinase called *glucokinase*. Recall that glucokinase has a high K_M value and is thus active only when blood-glucose levels are high. Furthermore, glucokinase is not inhibited by glucose 6-phosphate as is the hexokinase with the low K_M value. Consequently, *the liver forms glucose 6-phosphate more rapidly as the blood-glucose level rises. The increase in glucose 6-phosphate coupled with insulin action leads to a buildup of glycogen stores.* The hormonal effects on glycogen synthesis and storage are reinforced by a direct action of glucose itself. *Phosphorylase a is a glucose sensor in addition to being the enzyme that cleaves glycogen.* When the glucose level is high, the

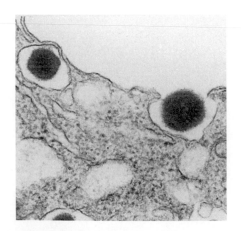

Figure 27.15 Insulin secretion. The electron micrograph shows the release of insulin from a pancreatic β cell. One secretory granule is on the verge of fusing with the plasma membrane and releasing insulin into the extracellular space, and the other has already released the hormone. [Courtesy of Dr. Lelio Orci. L. Orci, J.-D. Vassalli, and A. Perrelet. *Sci. Am.* 259 (September 1988):85–94.]

binding of glucose to phosphorylase *a* renders the enzyme susceptible to the action of a phosphatase that converts it into phosphorylase *b*, which does not readily degrade glycogen. Thus, *glucose allosterically shifts the glycogen system from a degradative to a synthetic mode.*

The high insulin level in the fed state also promotes *the entry of glucose into muscle and adipose tissue.* Insulin stimulates the synthesis of glycogen by muscle as well as by the liver. The entry of glucose into adipose tissue provides glycerol 3-phosphate for the synthesis of triacylglycerols. The action of insulin also extends to amino acid and protein metabolism. Insulin promotes the uptake of branched-chain amino acids (valine, leucine, and isoleucine) by muscle. Indeed, insulin has a general stimulating effect on protein synthesis, which favors a building up of muscle protein. In addition, it inhibits the intracellular degradation of proteins.

2.   *The Early Fasting, or Postabsorptive, State.* The blood-glucose level begins to drop several hours after a meal, leading to a decrease in insulin secretion and a rise in *glucagon* secretion; glucagon is secreted by the α cells of the pancreas in response to a *low blood-sugar level in the fasting state.* Just as insulin signals the fed state, glucagon signals the starved state. It serves to mobilize glycogen stores when there is no dietary intake of glucose. *The main target organ of glucagon is the liver.* Glucagon stimulates glycogen breakdown and inhibits glycogen synthesis by triggering the cyclic AMP cascade leading to the phosphorylation and activation of phosphorylase and the inhibition of glycogen synthase (Section 21.5). Glucagon also inhibits fatty acid synthesis by diminishing the production of pyruvate and by lowering the activity of acetyl CoA carboxylase by maintaining it in a phosphorylated state. In addition, glucagon stimulates gluconeogenesis in the liver and blocks glycolysis by lowering the level of F-2,6-BP.

All known actions of glucagon are mediated by protein kinases that are activated by cyclic AMP. The activation of the cyclic AMP cascade results in a higher level of phosphorylase *a* activity and a lower level of glycogen synthase *a* activity. Glucagon's effect on this cascade is reinforced by the diminished binding of glucose to phosphorylase *a*, which makes the enzyme less susceptible to the hydrolytic action of the phosphatase. Instead, the phosphatase remains bound to phosphorylase *a*, and so the synthase stays in the inactive phosphorylated form. Consequently, there is a rapid mobilization of glycogen.

The large amount of glucose formed by the hydrolysis of glucose 6-phosphate derived from glycogen is then released from the liver into the blood. The entry of glucose into muscle and adipose tissue decreases in response to a low insulin level. The diminished utilization of glucose by muscle and adipose tissue also contributes to the maintenance of the blood-glucose level. The net result of these actions of glucagon is to *markedly increase the release of glucose by the liver.* Both muscle and the liver use fatty acids as fuel when the blood-glucose level drops. Thus, *the blood-glucose level is kept at or above 4.4 M (80 mg/dl) by three major factors:* (1) *the mobilization of glycogen and the release of glucose by the liver,* (2) *the release of fatty acids by adipose tissue, and* (3) *the shift in the fuel used from glucose to fatty acids by muscle and the liver.*

What is the result of the depletion of the liver's glycogen stores? Gluconeogenesis from lactate and alanine continues, but this process merely replaces glucose that had already been converted into lactate and alanine by the peripheral tissues. Moreover, the brain oxidizes glucose completely to CO_2 and H_2O. Thus, for the net synthesis of glucose to take place, another source of carbon is required. Glycerol released from adipose tissue on lipolysis provides some of the carbon atoms, with the remaining carbon atoms coming from the hydrolysis of muscle proteins.

3. *The Refed State.* What are the biochemical responses to a hearty break-fast? Fat is processed exactly as it is processed in the normal fed state. However, this is not the case for glucose. The liver does not initially absorb glucose from the blood, but, instead, leaves it for the peripheral tissues. Moreover, the liver remains in a gluconeogenic mode. Now, however, the newly synthesized glucose is used to replenish the liver's glycogen stores. As the blood-glucose levels continue to rise, the liver completes the replenishment of its glycogen stores and begins to process the remaining excess glucose for fatty acid synthesis.

Metabolic Adaptations in Prolonged Starvation Minimize Protein Degradation

What are the adaptations if fasting is prolonged to the point of starvation? A typical well-nourished 70-kg man has fuel reserves totaling about 670,000 kJ (161,000 kcal; see Table 27.1). The energy need for a 24-hour period ranges from about 6700 kJ (1600 kcal) to 25,000 kJ (6000 kcal), depending on the extent of activity. Thus, stored fuels suffice to meet caloric needs in starvation for 1 to 3 months. However, the carbohydrate reserves are exhausted in only a day.

Even under starvation conditions, the blood-glucose level must be maintained above 2.2 mM (40 mg/dl). *The first priority of metabolism in starvation is to provide sufficient glucose to the brain and other tissues (such as red blood cells) that are absolutely dependent on this fuel.* However, precursors of glucose are not abundant. Most energy is stored in the fatty acyl moieties of triacylglycerols. Recall that fatty acids cannot be converted into glucose, because acetyl CoA cannot be transformed into pyruvate (p. 634). The glycerol moiety of triacylglycerol can be converted into glucose, but only a limited amount is available. The only other potential source of glucose is amino acids derived from the breakdown of proteins. However, proteins are not stored, and so any breakdown will necessitate a loss of function. Thus, *the second priority of metabolism in starvation is to preserve protein, which is accomplished by shifting the fuel being used from glucose to fatty acids and ketone bodies* (Figure 27.16).

The metabolic changes on the first day of starvation are like those after an overnight fast. The low blood-sugar level leads to decreased secretion of insulin and increased secretion of glucagon. *The dominant metabolic processes are the mobilization of triacylglycerols in adipose tissue and gluconeogenesis by the liver. The liver obtains energy for its own needs by oxidizing fatty acids released from adipose tissue.* The concentrations of acetyl CoA and citrate consequently increase, which switches off glycolysis. The uptake of glucose by muscle is markedly diminished because of the low insulin level, whereas fatty acids enter freely. Consequently, *muscle uses no glucose and relies exclusively on fatty acids for fuel.* The β oxidation of fatty acids by muscle halts the conversion of pyruvate into acetyl CoA, because acetyl CoA stimulates the phosphorylation of the pyruvate dehydrogenase complex, which renders it inactive (p. 490). Hence, pyruvate, lactate, and alanine are exported to the liver for conversion into glucose. Glycerol derived from the cleavage of triacylglycerols is another raw material for the synthesis of glucose by the liver.

Proteolysis also provides carbon skeletons for gluconeogenesis. During starvation, degraded proteins are not replenished and serve as carbon sources for glucose synthesis. Initial sources of protein are those that turn over rapidly, such as proteins of the intestinal epithelium and the secretions of the pancreas. Proteolysis of muscle protein provides some of the three-carbon precursors of glucose. However, survival for most animals depends on being able to move rapidly, which requires a large muscle mass, and so muscle loss must be minimized.

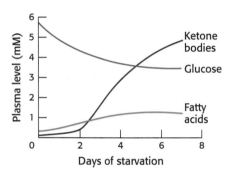

Figure 27.16 Fuel choice during starvation. The plasma levels of fatty acids and ketone bodies increase in starvation, whereas that of glucose decreases.

TABLE 27.2 Fuel metabolism in starvation

Fuel exchanges and consumption	AMOUNT FORMED OR CONSUMED IN 24 HOURS (GRAMS)	
	3d day	40th day
Fuel use by the brain		
Glucose	100	40
Ketone bodies	50	100
All other use of glucose	50	40
Fuel mobilization		
Adipose-tissue lipolysis	180	180
Muscle-protein degradation	75	20
Fuel output of the liver		
Glucose	150	80
Ketone bodies	150	150

How is the loss of muscle curtailed? After about 3 days of starvation, the liver forms large amounts of acetoacetate and D-3-hydroxybutyrate (ketone bodies; Figure 27.17). Their synthesis from acetyl CoA increases markedly because the citric acid cycle is unable to oxidize all the acetyl units generated by the degradation of fatty acids. Gluconeogenesis depletes the supply of oxaloacetate, which is essential for the entry of acetyl CoA into the citric acid cycle. Consequently, the liver produces large quantities of ketone bodies, which are released into the blood. At this time, *the brain begins to consume significant amounts of acetoacetate in place of glucose*. After 3 days of starvation, about a third of the energy needs of the brain are met by ketone bodies (Table 27.2). The heart also uses ketone bodies as fuel.

After several weeks of starvation, ketone bodies become the major fuel of the brain. Acetoacetate is activated by the transfer of CoA from succinyl CoA to give acetoacetyl CoA (Figure 27.18). Cleavage by thiolase then yields two molecules of acetyl CoA, which enter the citric acid cycle. In essence, *ketone bodies are equivalents of fatty acids that are an accessible fuel source for the brain.* Only 40 g of glucose is then needed per day for the brain, compared with about 120 g in the first day of starvation. *The effective conversion of fatty acids into ketone bodies by the liver and their use by the brain markedly diminishes the need for glucose. Hence, less muscle is degraded than in the first days of starvation.* The breakdown of 20 g of muscle daily compared with 75 g early in starvation is most important for survival. A person's survival time is mainly determined by the size of the triacylglycerol depot.

What happens after depletion of the triacylglycerol stores? The only source of fuel that remains is protein. Protein degradation accelerates, and death inevitably results from a loss of heart, liver, or kidney function.

Metabolic Derangements in Diabetes Result from Relative Insulin Insufficiency and Glucagon Excess

We now consider *diabetes mellitus,* a complex disease characterized by grossly abnormal fuel usage: *glucose is overproduced by the liver and underutilized by other organs.* The incidence of diabetes mellitus (usually referred to simply as *diabetes*) is about 5% of the population. Indeed, diabetes is the most common serious metabolic disease in the world; it affects hundreds of millions. *Type 1 diabetes,* or *insulin-dependent diabetes mellitus* (IDDM), is caused by the autoimmune destruction of the insulin-secreting β cells in the pancreas and usually begins before age 20. Insulin dependency means that the affected person requires the administration of insulin to live. Most diabetics, in contrast, have a normal or even higher level of insulin in their blood, but

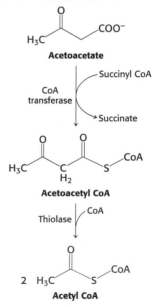

Figure 27.17 Synthesis of ketone bodies by the liver.

Figure 27.18 Entry of ketone bodies into the citric acid cycle.

they are quite unresponsive to the hormone. This form of the disease, known as *type 2,* or *non-insulin-dependent diabetes mellitus* (NIDDM), typically arises later in life than does the insulin-dependent form.

In type 1 diabetes, insulin production is insufficient and consequently glucagon is present at higher-than-normal levels. In essence, the diabetic person is in biochemical starvation mode despite a high concentration of blood glucose. Because insulin is deficient, *the entry of glucose into adipose and muscle cells is impaired.* The liver becomes stuck in a gluconeogenic and ketogenic state. The excessive level of glucagon relative to that of insulin leads to a decrease in the amount of F-2,6-BP in the liver. Hence, glycolysis is inhibited and gluconeogenesis is stimulated because of the opposite effects of F-2,6-BP on phosphofructokinase and fructose-1,6-bisphosphatase (Section 16.4; see also Figures 27.4 and 27.6). The high glucagon/insulin ratio in diabetes also promotes glycogen breakdown. Hence, *an excessive amount of glucose is produced by the liver and released into the blood.* Glucose is excreted in the urine (hence the name *mellitus*) when its concentration in the blood exceeds the reabsorptive capacity of the renal tubules. Water accompanies the excreted glucose, and so an untreated diabetic in the acute phase of the disease is hungry and thirsty.

Because carbohydrate utilization is impaired, a lack of insulin leads to the uncontrolled breakdown of lipids and proteins. Large amounts of acetyl CoA are then produced by β oxidation. However, much of the acetyl CoA cannot enter the citric acid cycle, because there is insufficient oxaloacetate for the condensation step. Recall that mammals can synthesize oxaloacetate from pyruvate, a product of glycolysis, but not from acetyl CoA; instead, they generate ketone bodies. *A striking feature of diabetes is the shift in fuel usage from carbohydrates to fats; glucose, more abundant than ever, is spurned.* In high concentrations, ketone bodies overwhelm the kidney's capacity to maintain acid–base balance. The untreated diabetic can go into a coma because of a lowered blood-pH level and dehydration.

Type 2, or non-insulin-dependent, diabetes accounts for more than 90% of the diabetes cases and is the most common metabolic disease in the world. In the United States, it is the leading cause of blindness, kidney failure, and amputation. The hallmark of type 2 diabetes is insulin resistance. The β cells of the pancreas secrete normal or even greater-than-normal amounts of insulin, but the tissues do not respond to the hormone despite the fact that the insulin receptor is functional. Sometimes the β cells fail, leading to type 1 diabetes. The exact cause of type 2 diabetes remains to be elucidated; obesity is a significant predisposing factor.

Caloric Homeostasis Is a Means of Regulating Body Weight

In the United States, obesity has become an epidemic, with nearly 30% of adults classified as obese. Obesity is identified as a risk factor in a host of pathological conditions including diabetes mellitus, hypertension, and cardiovascular disease. The cause of obesity is quite simple in the vast majority of cases: more food is consumed than is needed, and the excess calories are stored as fat.

Although the proximal cause of obesity is simple, the biochemical means by which caloric homeostasis and appetite control are usually maintained is enormously complex, but two important signal molecules are insulin and leptin. A protein consisting of 146 amino acids, *leptin* is a hormone secreted by adipocytes in direct proportion to fat mass. Leptin acts through a membrane receptor, related in structure and mechanism of action to the growth-hormone receptor, in the hypothalamus to generate satiation signals. During periods when more energy is expended than ingested (the starved state), adipose tissue loses mass. Under these conditions, the secretion of both leptin and insulin declines, fuel utilization is increased, and energy stores are used. The converse is true when calories are consumed in excess.

The importance of leptin to obesity is dramatically illustrated in mice. Mice lacking leptin are obese and will lose weight if given leptin. Mice that lack the leptin receptor are insensitive to leptin administration. Preliminary evidence indicates that leptin and its receptor play a role in human obesity, but the results are not as clear-cut as in the mouse.

As stated earlier, obesity is a predisposing factor for type 2 diabetes. What is the biochemical basis for this relation? Although much remains to be determined, recent research suggests that the adipocytes secrete a hormone called *resistin* (*resist*ance to *in*sulin) that renders tissues insensitive to insulin. Moreover, the amount of resistin secreted is directly proportional to fat mass. The precise physiological role of resistin remains to be determined.

Insulin and leptin can be thought of as long-term regulators of caloric homeostasis, but short-duration hormones also have a role. For instance, *cholecystokinin* is released by the gastrointestinal tract during eating and binds to specific receptors in the brain, promoting a sense of fullness. On the other hand, the appetite-stimulating gastric peptide *gherlin* is secreted when the stomach is empty. The complex interplay of the genes and their products of the neuroendocrine system that control energy balance will be an exciting area of research for some time to come.

27.4 Fuel Choice During Exercise Is Determined by the Intensity and Duration of Activity

The fuels used in anaerobic exercises—sprinting, for example—differ from those used in aerobic exercises—such as distance running. The selection of fuels during these different forms of exercise illustrates many important facets of energy transduction and metabolic integration. ATP directly powers myosin, the protein immediately responsible for converting chemical energy into movement (Chapter 34). However, the amount of ATP in muscle is small. Hence, the power output and, in turn, the velocity of running depend on the rate of ATP production from other fuels. As shown in Table 27.3, *creatine phosphate* (phosphocreatine) can swiftly transfer its high-potential phosphoryl group to ADP to generate ATP (p. 416). However, the amount of creatine phosphate, like that of ATP itself, is limited. Creatine phosphate and ATP can power intense muscle contraction for 5 to 6 s. Maximum speed in a sprint can thus be maintained for only 5 to 6 s (see Figure 15.7). Thus, the winner in a 100-meter sprint is the runner who both achieves the highest initial velocity and then slows down the least.

TABLE 27.3 Fuel sources for muscle contraction

Fuel source	Maximal rate of ATP production (mmol s^{-1})	Total ~P available (mmol)
Muscle ATP		223
Creatine phosphate	73.3	446
Conversion of muscle glycogen into lactate	39.1	6,700
Conversion of muscle glycogen into CO_2	16.7	84,000
Conversion of liver glycogen into CO_2	6.2	19,000
Conversion of adipose-tissue fatty acids into CO_2	6.7	4,000,000

Note: Fuels stored are estimated for a 70-kg person having a muscle mass of 28 kg.

Source: After E. Hultman and R. C. Harris. In *Principles of Exercise Biochemistry*, edited by J. R. Poortmans (Karger, 2004), pp. 78–119.

A 100-meter sprint is powered by stored ATP, creatine phosphate, and the anaerobic glycolysis of muscle glycogen. The conversion of muscle glycogen into lactate can generate a good deal more ATP, but the rate is slower than that of phosphoryl-group transfer from creatine phosphate. During a ~10-second sprint, the ATP level in muscle drops from 5.2 to 3.7 mM, and that of creatine phosphate decreases from 9.1 to 2.6 mM. The essential role of anaerobic glycolysis is manifested in the elevation of the blood-lactate level from 1.6 to 8.3 mM. The release of H^+ from the intensely active muscle concomitantly lowers the blood pH from 7.42 to 7.24. This pace cannot be sustained in a 1000-meter run (~132 s) for two reasons. First, creatine phosphate is consumed within a few seconds. Second, the lactate produced would cause acidosis. Thus, alternative fuel sources are needed.

The complete oxidation of muscle glycogen to CO_2 substantially increases the energy yield, but this aerobic process is a good deal slower than anaerobic glycolysis. However, as the distance of a run increases, aerobic respiration, or oxidative phosphorylation, becomes increasingly important. For instance, *part of the ATP consumed in a 1000-meter run must come from oxidative phosphorylation.* Because ATP is produced more slowly by oxidative phosphorylation than by glycolysis (see Table 27.3), the pace is necessarily slower than in a 100-meter sprint. The championship velocity for the 1000-meter run is about 7.6 m/s, compared with approximately 10.2 m/s for the 100-meter event (Figure 27.19).

The running of a marathon (26 miles 385 yards, or 42,200 meters) requires a different selection of fuels and is characterized by cooperation between muscle, liver, and adipose tissue. Liver glycogen complements muscle glycogen as an energy store that can be tapped. However, the total body glycogen stores (103 mol of ATP at best) are insufficient to provide the 150 mol of ATP needed for this grueling ~2-hour event. Much larger quantities of ATP can be obtained by the oxidation of fatty acids derived from the breakdown of *fat in adipose tissue,* but the maximal rate of ATP generation is slower yet than that of glycogen oxidation and is more than 10-fold slower than that with creatine phosphate. Thus, *ATP is generated much more slowly from high-capacity stores than from limited ones,* accounting for the different velocities of anaerobic and aerobic events.

ATP generation from fatty acids is essential for distance running. However, a marathon would take about 6 hours to run if all the ATP came from fatty acid oxidation, because it is much slower than glycogen oxidation. Elite runners consume about equal amounts of glycogen and fatty acids during a marathon to achieve a mean velocity of 5.5 m/s, about half that of a 100-meter sprint. How is an optimal mix of these fuels achieved? *A low blood-sugar level leads to a high glucagon/insulin ratio, which in turn mobilizes fatty*

Figure 27.19 Dependence of the velocity of running on the duration of the race. The values shown are world track records.

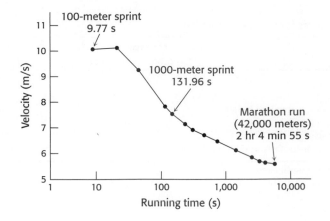

acids from adipose tissue. Fatty acids readily enter muscle, where they are degraded by β oxidation to acetyl CoA and then to CO_2. The elevated acetyl CoA level decreases the activity of the pyruvate dehydrogenase complex to block the conversion of pyruvate into acetyl CoA. Hence, fatty acid oxidation decreases the funneling of glucose into the citric acid cycle and oxidative phosphorylation. Glucose is spared so that just enough remains available at the end of the marathon. The simultaneous use of both fuels gives a higher mean velocity than would be attained if glycogen were totally consumed before the start of fatty acid oxidation.

27.5 Ethanol Alters Energy Metabolism in the Liver

Ethanol has been a part of the human diet for centuries. However, its consumption in excess can result in a number of health problems, most notably liver damage. What is the biochemical basis of these health problems?

Ethanol Metabolism Leads to an Excess of NADH

Ethanol cannot be excreted and must be metabolized, primarily by the liver. This metabolism is accomplished by two pathways. The first pathway comprises two steps. The first step, catalyzed by the enzyme *alcohol dehydrogenase,* takes place in the cytoplasm:

$$CH_3CH_2OH + NAD^+ \xrightarrow[\text{dehydrogenase}]{\text{Alcohol}} CH_3CHO + NADH + H^+$$

$$\text{\textbf{Ethanol}} \qquad\qquad\qquad\qquad \text{\textbf{Acetaldehyde}}$$

The second step, catalyzed by *aldehyde dehydrogenase,* takes place in mitochondria:

$$CH_3CHO + NAD^+ + H_2O \xrightarrow[\text{dehydrogenase}]{\text{Aldehyde}} CH_3COO^- + NADH + H^+$$

$$\text{\textbf{Acetaldehyde}} \qquad\qquad\qquad\qquad \text{\textbf{Acetate}}$$

Note that *ethanol consumption leads to an accumulation of NADH.* This high concentration of NADH inhibits gluconeogenesis by preventing the oxidation of lactate to pyruvate. In fact, the high concentration of NADH will cause the reverse reaction to predominate, and lactate will accumulate. The consequences may be hypoglycemia and lactic acidosis.

The overabundance of NADH also inhibits fatty acid oxidation. The metabolic purpose of fatty acid oxidation is to generate NADH for ATP generation by oxidative phosphorylation, but an alcohol consumer's NADH needs are met by ethanol metabolism. In fact, the excess NADH signals that conditions are right for fatty acid synthesis. Hence, triacylglycerols accumulate in the liver, leading to a condition known as "fatty liver."

The second pathway for ethanol metabolism is called the ethanol-inducible *microsomal ethanol-oxidizing system* (MEOS). This cytochrome P450-dependent pathway (p. 750) generates acetaldehyde and subsequently acetate while oxidizing biosynthetic reducing power, NADPH, to $NADP^+$. Because it uses oxygen, this pathway generates free radicals that damage tissues. Moreover, because the system consumes NADPH, the antioxidant glutathione cannot be regenerated (Section 20.5), exacerbating the oxidative stress.

What are the effects of the other metabolites of ethanol? Liver mitochondria can convert acetate into acetyl CoA in a reaction requiring ATP. The enzyme is the thiokinase that normally activates short-chain fatty acids.

$$\text{Acetate} + \text{coenzyme A} + \text{ATP} \longrightarrow \text{acetyl CoA} + \text{AMP} + \text{PP}_i$$

$$\text{PP}_i \longrightarrow 2\,\text{P}_i$$

However, further processing of the acetyl CoA by the citric acid cycle is blocked, because NADH inhibits two important regulatory enzymes—isocitrate dehydrogenase and α-ketoglutarate dehydrogenase. The accumulation of acetyl CoA has several consequences. First, ketone bodies will form and be released into the blood, aggravating the acidic condition already resulting from the high lactate concentration. The processing of the acetate in the liver becomes inefficient, leading to a buildup of acetaldehyde. This very reactive compound forms covalent bonds with many important functional groups in proteins, impairing protein function. If ethanol is consistently consumed at high levels, the acetaldehyde can significantly damage the liver, eventually leading to cell death.

Liver damage from excessive ethanol consumption occurs in three stages. The first stage is the aforementioned development of fatty liver. In the second stage—alcoholic hepatitis—groups of cells die and inflammation results. This stage can itself be fatal. In stage three—cirrhosis—fibrous structure and scar tissue are produced around the dead cells. Cirrhosis impairs many of the liver's biochemical functions. The cirrhotic liver is unable to convert ammonia into urea, and blood levels of ammonia rise. Ammonia is toxic to the nervous system and can cause coma and death. Cirrhosis of the liver arises in about 25% of alcoholics, and about 75% of all cases of liver cirrhosis are the result of alcoholism. Viral hepatitis is a nonalcoholic cause of liver cirrhosis.

Excess Ethanol Consumption Disrupts Vitamin Metabolism

The adverse effects of ethanol are not limited to the metabolism of ethanol itself. Vitamin A (retinol) is converted into retinoic acid, an important signal molecule for growth and development in vertebrates, by the same dehydrogenases that metabolize ethanol. Consequently, this activation does not take place in the presence of ethanol, which acts as a competitive inhibitor. Moreover, the MEOS system induced by ethanol inactivates retinoic acid. These disruptions in the retinoic acid signaling pathway are believed to be responsible, at least in part, for fetal alcohol syndrome as well as the development of a variety of cancers.

The disruption of vitamin A metabolism is a direct result of the biochemical changes induced by excess ethanol consumption. Other disruptions in metabolism result from another common characteristic of alcoholics—malnutrition. A dramatic neurological disorder, referred to as *Wernicke–Korsakoff syndrome*, results from insufficient intake of the vitamin thiamine. Symptoms include mental confusion, unsteady gait, and lack of fine motor skills. The symptoms of Wernicke–Korsakoff syndrome are similar to those of beriberi because both conditions result from a lack of thiamine. Thiamine is converted into the coenzyme thiamine pyrophosphate, a key constituent of the pyruvate dehydrogenase complex. Recall that this complex links glycolysis with the citric acid cycle. Disruptions in the pyruvate dehydrogenase complex are most evident as neurological disorders because the brain is normally dependent on glucose for energy generation.

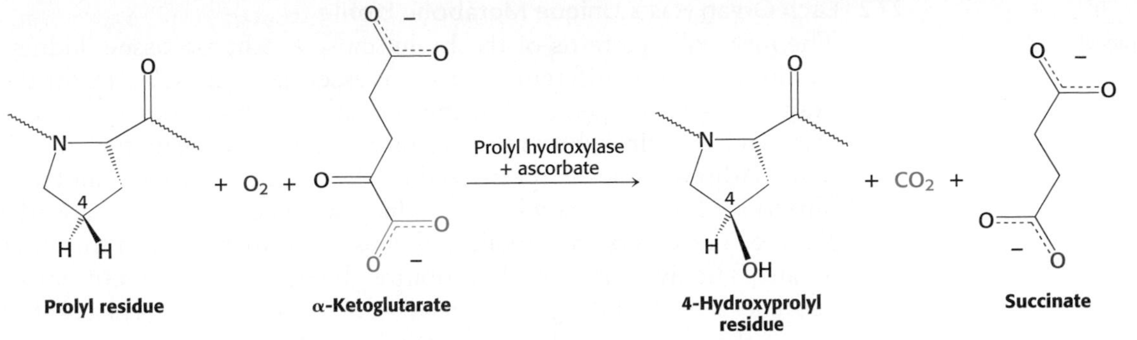

Figure 27.20 Formation of 4-hydroxyproline. Proline is hydroxylated at C-4 by the action of prolyl hydroxylase, an enzyme that activates molecular oxygen.

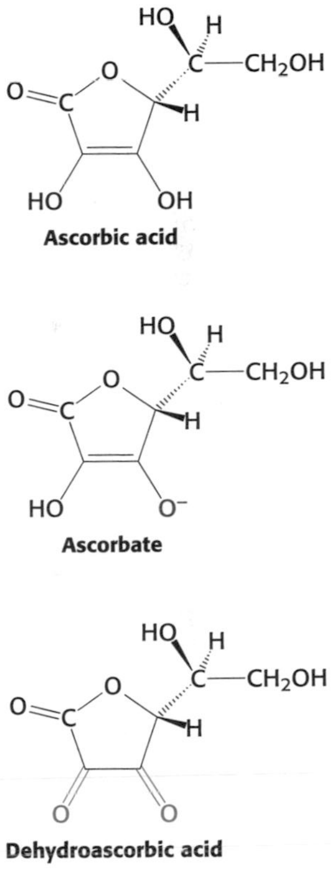

Alcoholic scurvy is occasionally observed because of an insufficient ingestion of vitamin C. Vitamin C is required for the formation of stable collagen fibers. The symptoms of scurvy include skin lesions and blood-vessel fragility. Most notable are bleeding gums, the loss of teeth, and periodontal infections. Gums are especially sensitive to a lack of vitamin C because the collagen in gums turns over rapidly. What is the biochemical basis for scurvy? Vitamin C is required for the continued activity of prolyl hydroxylase. This enzyme synthesizes 4-hydroxyproline, an amino acid that is required in collagen. To form this unusual amino acid, proline residues on the amino side of glycine residues in nascent collagen chains become hydroxylated. One oxygen atom from O_2 becomes attached to C-4 of proline while the other oxygen atom is taken up by α-ketoglutarate, which is converted into succinate (Figure 27.20). This reaction is catalyzed by *prolyl hydroxylase*, a *dioxygenase*, which requires an Fe^{2+} ion to activate O_2. The enzyme also converts α-ketoglutarate into succinate without hydroxylating proline. In this partial reaction, an oxidized iron complex is formed, which inactivates the enzyme. How is the active enzyme regenerated? *Ascorbate (vitamin C)* comes to the rescue by reducing the ferric ion of the inactivated enzyme. In the recovery process, ascorbate is oxidized to dehydroascorbic acid (Figure 27.21). Thus, ascorbate serves here as a specific *antioxidant*. Why does impaired hydroxylation have such devastating consequences? *Collagen synthesized in the absence of ascorbate is less stable than the normal protein.* Hydroxyproline stabilizes the collagen triple helix by forming interstrand hydrogen bonds. The abnormal fibers formed by insufficiently hydroxylated collagen account for the symptoms of scurvy.

Figure 27.21 Forms of ascorbic acid (vitamin C). Ascorbate is the ionized form of vitamin C, and dehydroascorbic acid is the oxidized form of ascorbate.

Summary

27.1 Metabolism Consists of Highly Interconnected Pathways

The basic strategy of metabolism is simple: the formation of ATP, reducing power, and building blocks for biosyntheses. This complex network of reactions is controlled by the allosteric interactions and reversible covalent modifications of enzymes and changes in their amounts, by compartmentation, and by interactions between metabolically distinct organs. The enzyme catalyzing the committed step in a pathway is usually the most important control site. Opposing pathways such as gluconeogenesis and glycolysis are reciprocally regulated so that one pathway is usually less active when the other is highly active.

27.2 Each Organ Has a Unique Metabolic Profile

The metabolic patterns of the brain, muscle, adipose tissue, kidney, and liver are very different. Glucose is essentially the sole fuel for the brain in a well-fed person. During starvation, ketone bodies (acetoacetate and 3-hydroxybutyrate) become the predominant fuel of the brain. Adipose tissue is specialized for the synthesis, storage, and mobilization of triacylglycerols. The kidney produces urine and reabsorbs glucose. The diverse metabolic activities of the liver support the other organs. The liver can rapidly mobilize glycogen and carry out gluconeogenesis to meet the glucose needs of other organs. It plays a central role in the regulation of lipid metabolism. When fuels are abundant, fatty acids are synthesized, esterified, and sent from the liver to adipose tissue. In the fasting state, however, fatty acids are converted into ketone bodies by the liver.

27.3 Food Intake and Starvation Induce Metabolic Changes

Insulin signals the fed state; it stimulates the formation of glycogen and triacylglycerols and the synthesis of proteins. In contrast, glucagon signals a low blood-glucose level; it stimulates glycogen breakdown and gluconeogenesis by the liver and triacylglycerol hydrolysis by adipose tissue. After a meal, the rise in the blood-glucose level leads to an increased secretion of insulin and decreased secretion of glucagon. Consequently, glycogen is synthesized in muscle and the liver. When the blood-glucose level drops several hours later, glucose is then formed by the degradation of glycogen and by the gluconeogenic pathway, and fatty acids are released by the hydrolysis of triacylglycerols. The liver and muscle then increasingly use fatty acids instead of glucose to meet their own energy needs so that glucose is conserved for use by the brain and the red blood cells.

The metabolic adaptations in starvation serve to minimize protein degradation. Large amounts of ketone bodies are formed by the liver from fatty acids and released into the blood within a few days after the onset of starvation. After several weeks of starvation, ketone bodies become the major fuel of the brain. The diminished need for glucose decreases the rate of muscle breakdown, and so the likelihood of survival is enhanced.

Diabetes mellitus, the most common serious metabolic disease, is due to metabolic derangements resulting in an insufficiency of insulin and an excess of glucagon relative to the needs of the person. The result is an elevated blood-glucose level, the mobilization of triacylglycerols, and excessive ketone-body formation. Accelerated ketone-body formation can lead to acidosis, coma, and death in untreated insulin-dependent diabetics.

27.4 Fuel Choice During Exercise Is Determined by the Intensity and Duration of Activity

Sprinting and marathon running are powered by different fuels to maximize power output. The 100-meter sprint is powered by stored ATP, creatine phosphate, and anaerobic glycolysis. In contrast, the oxidation of both muscle glycogen and fatty acids derived from adipose tissue is essential in the running of a marathon, a highly aerobic process.

27.5 Ethanol Alters Energy Metabolism in the Liver

The oxidation of ethanol results in an unregulated overproduction of NADH, which has several consequences. A rise in the blood levels of lactic acid and ketone bodies causes a fall in blood pH, or acidosis. The liver is damaged because the excess NADH causes excessive fat formation as well as the generation of acetaldehyde, a reactive molecule. Severe liver damage can result.

Key Terms

allosteric interaction (p. 762)

covalent modification (p. 762)

glycolysis (p. 762)

phosphofructokinase (p. 763)

citric acid cycle (p. 763)

oxidative phosphorylation (p. 763)

pyruvate dehydrogenase complex (p. 763)

pentose phosphate pathway (p. 763)

gluconeogenesis (p. 764)

glycogen synthesis and degradation (p. 764)

fatty acid synthesis and degradation (p. 764)

glucose 6-phosphate (p. 765)

pyruvate (p. 766)

acetyl CoA (p. 766)

ketone body (p. 767)

starved–fed cycle (p. 770)

glucose homeostasis (p. 770)

insulin (p. 770)

glucagon (p. 770)

caloric homeostasis (p. 774)

leptin (p. 774)

resistin (p. 775)

creatine phosphate (p. 775)

Selected Readings

Books

Fell, D. 1997. *Understanding the Control of Metabolism.* Portland Press.

Frayn, K. N. 1996. *Metabolic Regulation: A Human Perspective.* Portland Press.

Hargreaves, M., and Thompson, M. (Eds.). 1999. *Biochemistry of Exercise X.* Human Kinetics.

Poortmans, J. R. (Ed.). 2004. *Principles of Exercise Biochemistry.* Karger.

Harris, R. A., and Crabb, D. W. 2006. Metabolic interrelationships. In *Textbook of Biochemistry with Clinical Correlations* (pp. 849–890), edited by T. M. Devlin. Wiley-Liss.

Fuel Metabolism

Rolland, F., Winderickx, J., and Thevelein, J. M. 2001. Glucose-sensing mechanism in eukaryotic cells. *Trends Biochem. Sci.* 26:310–317.

Rasmussen, B. B., and Wolfe, R. R. 1999. Regulation of fatty acid oxidation in skeletal muscle. *Annu. Rev. Nutr.* 19:463–484.

Hochachka, P. W. 2000. Oxygen, homeostasis, and metabolic regulation. *Adv. Exp. Med. Biol.* 475:311–335.

Holm, E., Sedlaczek, O., and Grips, E. 1999. Amino acid metabolism in liver disease. *Curr. Opin. Clin. Nutr. Metab. Care* 2:47–53.

Wagenmakers, A. J. 1998. Protein and amino acid metabolism in human muscle. *Adv. Exp. Med. Biol.* 441:307–319.

Metabolic Adaptations in Starvation

Baverel, G., Ferrier, B., and Martin, M. 1995. Fuel selection by the kidney: Adaptation to starvation. *Proc. Nutr. Soc.* 54:197–212.

MacDonald, I. A., and Webber, J. 1995. Feeding, fasting and starvation: Factors affecting fuel utilization. *Proc. Nutr. Soc.* 54:267–274.

Cahill, G. F., Jr. 1976. Starvation in man. *Clin. Endocrinol. Metab.* 5:397–415.

Sugden, M. C., Holness, M. J., and Palmer, T. N. 1989. Fuel selection and carbon flux during the starved-to-fed transition. *Biochem. J.* 263:313–323.

Diabetes Mellitus

Lowel, B. B., and Shulman, G. 2005. Mitochondrial dysfunction and type 2 diabetes. *Science* 307:384–387.

Rutter, G. A. 2000. Diabetes: The importance of the liver. *Curr. Biol.* 10:R736–R738.

Saltiel, A. R. 2001. New perspectives into the molecular pathogenesis and treatment of type 2 diabetes. *Cell* 104:517–529.

Bell, G. I., Pilikis, S. J., Weber, I. T., and Polonsky, K. S. 1996. Glucokinase mutations, insulin secretion, and diabetes mellitus. *Annu. Rev. Physiol.* 58:171–186.

Withers, D. J., and White, M. 2000. Perspective: The insulin signaling system—a common link in the pathogenesis of type 2 diabetes. *Endocrinology* 141:1917–1921.

Taylor, S. I. 1995. Diabetes mellitus. In *The Metabolic Basis of Inherited Diseases* (7th ed., pp. 935–936), edited by C. R. Scriver, A. L. Beaudet, W. S. Sly, D. Valle, J. B. Stanbury, J. B. Wyngaarden, and D. S. Fredrickson. McGraw-Hill.

Exercise Metabolism

Shulman, R. G., and Rothman, D. L. 2001. The "glycogen shunt" in exercising muscle: A role for glycogen in muscle energetics and fatigue. *Proc. Natl. Acad. Sci. U. S. A.* 98:457–461.

Gleason, T. 1996. Post-exercise lactate metabolism: A comparative review of sites, pathways, and regulation. *Annu. Rev. Physiol.* 58:556–581.

Holloszy, J. O., and Kohrt, W. M. 1996. Regulation of carbohydrate and fat metabolism during and after exercise. *Annu. Rev. Nutr.* 16:121–138.

Hochachka, P. W., and McClelland, G. B. 1997. Cellular metabolic homeostasis during large-scale change in ATP turnover rates in muscles. *J. Exp. Biol.* 200:381–386.

Horowitz, J. F., and Klein, S. 2000. Lipid metabolism during endurance exercise. *Am. J. Clin. Nutr.* 72:558S–563S.

Wagenmakers, A. J. 1999. Muscle amino acid metabolism at rest and during exercise. *Diabetes Nutr. Metab.* 12:316–322.

Ethanol Metabolism

Molotkov, A., and Duester, G. 2002. Retinol/ethanol drug interaction during acute alcohol intoxication involves inhibition of retinol metabolism to retinoic acid by alcohol dehydrogenase. *J. Biol. Chem.* 277:22553–22557.

Stewart, S., Jones, D., and Day, C. P. 2001. Alcoholic liver disease: New insights into mechanisms and preventive strategies. *Trends Mol. Med.* 7:408–413.

Lieber, C. S. 2000. Alcohol: Its metabolism and interaction with nutrients. *Annu. Rev. Nutr.* 20:395–430.

Niemela, O. 1999. Aldehyde-protein adducts in the liver as a result of ethanol-induced oxidative stress. *Front. Biosci.* 1:D506–D513.

Riveros-Rosas, H., Julian-Sanchez, A., and Pina, E. 1997. Enzymology of ethanol and acetaldehyde metabolism in mammals. *Arch. Med. Res.* 28:453–471.

Problems

1. *Distinctive organs.* What are the key enzymatic differences between the liver, kidney, muscle, and brain that account for their differing utilization of metabolic fuels?

2. *Missing enzymes.* Predict the major consequence of each of the following enzymatic deficiencies:

(a) Hexokinase in adipose tissue
(b) Glucose 6-phosphatase in liver
(c) Carnitine acyltransferase I in skeletal muscle
(d) Glucokinase in liver
(e) Thiolase in brain
(f) Kinase in liver that synthesizes fructose 2,6-bisphosphate

3. *Contrasting milieux.* Cerebrospinal fluid has a low content of albumin and other proteins compared with plasma.

(a) What effect does this lower content have on the concentration of fatty acids in the extracellular medium of the brain?
(b) Propose a plausible reason for the selection by the brain of glucose rather than fatty acids as the prime fuel.
(c) How does the fuel preference of muscle complement that of the brain?

4. *Metabolic energy and power.* The rate of energy expenditure of a typical 70-kg person at rest is about 70 watts (W), like that of a light bulb.

(a) Express this rate in kilojoules per second and in kilocalories per second.
(b) How many electrons flow through the mitochondrial electron-transport chain per second under these conditions?
(c) Estimate the corresponding rate of ATP production.
(d) The total ATP content of the body is about 50 g. Estimate how often an ATP molecule turns over in a person at rest.

5. *Respiratory quotient (RQ).* This classic metabolic index is defined as the volume of CO_2 released divided by the volume of O_2 consumed.

(a) Calculate the RQ values for the complete oxidation of glucose and of tripalmitoylglycerol.
(b) What do RQ measurements reveal about the contributions of different energy sources during intense exercise? (Assume that protein degradation is negligible.)

6. *Camel's hump.* Compare the H_2O yield from the complete oxidation of 1 g of glucose with that of 1 g of tripalmitoylglycerol. Relate these values to the evolutionary selection of the contents of a camel's hump.

7. *The wages of sin.* How long does one have to jog to offset the calories obtained from eating 10 macadamia nuts (75 kJ, or 18 kcal, per nut)? (Assume an incremental power consumption of 400 W.)

8. *Sweet hazard.* Ingesting large amounts of glucose before a marathon might seem to be a good way of increasing the fuel stores. However, experienced runners do not ingest glucose before a race. What is the biochemical reason for their avoidance of this potential fuel? (Hint: Consider the effect of glucose ingestion on the level of insulin.)

9. *An effect of diabetes.* Insulin-dependent diabetes is often accompanied by hypertriglyceridemia, which is an excess blood level of triacylglycerols in the form of very low density lipoproteins. Suggest a biochemical explanation.

10. *Sharing the wealth.* The hormone glucagon signifies the starved state, yet it inhibits glycolysis in the liver. How does this inhibition of an energy-production pathway benefit the organism?

11. *Compartmentation.* Glycolysis takes place in the cytoplasm, whereas fatty acid degradation takes place in mitochondria. What metabolic pathways depend on the interplay of reactions that take place in both compartments?

12. *Kwashiorkor.* The most common form of malnutrition in children in the world, kwashiorkor, is caused by a diet having ample calories but little protein. The high levels of carbohydrate result in high levels of insulin. What is the effect of high levels of insulin on

(a) lipid utilization?
(b) protein metabolism?
(c) Children suffering from kwashiorkor often have large distended bellies caused by water from the blood leaking into extracellular spaces. Suggest a biochemical basis for this condition.

13. *Oxygen deficit.* After light exercise, the oxygen consumed in recovery is approximately equal to the oxygen deficit, which is the amount of additional oxygen that would have been consumed had oxygen consumption reached steady state immediately. How is the oxygen consumed in recovery used?

14. *Excess postexercise oxygen consumption.* The oxygen consumed after strenuous exercise stops is significantly greater than the oxygen deficit and is termed *excess postexercise oxygen consumption* (EPOC). Why is so much more oxygen required after intense exercise?

15. *Psychotropic effects.* Ethanol is unusual in that it is freely soluble in both water and lipids. Thus, it has access to all regions of the highly vascularized brain. Although the molecular basis of ethanol action in the brain is not clear, ethanol evidently influences a number of neurotransmitter receptors and ion channels. Suggest a biochemical explanation for the diverse effects of ethanol.

16. *Fiber type.* Skeletal muscle has several distinct fiber types. Type I is used primarily for aerobic activity, whereas type II is specialized for short, intense bursts of activity. How could you distinguish between these types of muscle fiber if you viewed them with an electron microscope?

17. *Tour de France.* Cyclists in the Tour de France (more than 2000 miles in 3 weeks) require about 836, 000 kJ (200,000 kcal) of energy, or 41,840 kJ (10,000 kcal) day^{-1} (a resting male requires $\approx$ 8368 kJ, or 2000 kcal, day^{-1}).

(a) With the assumptions that the energy yield of ATP is about 50.2 kJ (12 kcal) mol^{-1} and that ATP has a molecular weight of 503 g mol^{-1}, how much ATP would be expended by a Tour de France cyclist?
(b) Pure ATP can be purchased at the cost of approximately $150 per gram. How much would it cost to power a cyclist through the Tour de France if the ATP had to be purchased?

DNA Replication, Repair, and Recombination

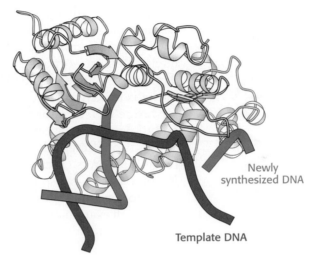

Newly synthesized DNA

Template DNA

Faithful copying is essential to the storage of genetic information. With the precision of a diligent monk copying an illuminated manuscript, a DNA polymerase (above) copies DNA strands, preserving the precise sequence of bases with very few errors. [(Left)The Pierpont Morgan Library/Art Resource.]

Perhaps the most exciting aspect of the structure of DNA deduced by Watson and Crick was, as expressed in their words, that the "specific pairing we have postulated immediately suggests a possible copying mechanism for the genetic material." A double helix separated into two single strands can be replicated because each strand serves as a template on which its complementary strand can be assembled (Figure 28.1). To preserve the information encoded in DNA through many cell divisions, copying of the genetic information must be extremely faithful. To replicate the human genome without mistakes, an error rate of less than 1 bp per 3×10^9 bp must be achieved. Such remarkable accuracy is achieved through a multilayered system of accurate DNA synthesis (which has an error rate of 1 per 10^3–10^4 bases inserted), proofreading during DNA synthesis (which reduces that error rate to approximately 1 per 10^6–10^7 bp), and postreplication mismatch repair (which reduces the error rate to approximately 1 per 10^9–10^{10} bp).

Even after DNA has been initially replicated, the genome is still not safe. Although DNA is remarkably robust, ultraviolet light as well as a range of chemical species can damage DNA, introducing changes in the DNA sequence (mutations) or lesions that can block further DNA replication

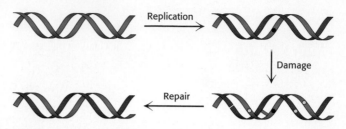

Figure 28.2 DNA Replication, damage, and repair. Some errors (shown as a black dot) may occur in the replication processes. Additional defects (shown in yellow) including modified bases, crosslinks, and single- and double-strand breaks are introduced into DNA by subsequent DNA-damaging reactions. Many of the errors are detected and subsequently repaired.

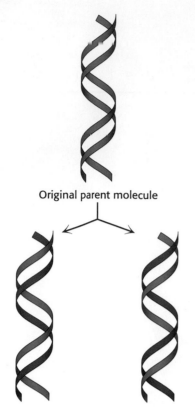

Original parent molecule

First-generation daughter molecules

Figure 28.1 DNA replication. Each strand of one double helix (shown in blue) acts as a template for the synthesis of a new complementary strand (shown in red).

(Figure 28.2). All organisms contain DNA repair systems that detect DNA damage and act to preserve the original sequence. Mutations in genes that encode components of DNA repair systems are key factors in the development of cancer. Among the most potentially devastating types of DNA damage are double-stranded breaks in DNA. With both strands of the double helix broken in a local region, neither strand is intact to act as a template for future DNA synthesis. A mechanism used to repair such lesions relies on DNA recombination—that is, the reassortment of DNA sequences present on two different double helices. In addition to its role in DNA repair, recombination is crucial for the generation of genetic diversity in meiosis. Recombination is also the key to generating a highly diverse repertoire of genes for key molecules in the immune system (Chapter 33). We begin with a thorough examination of the structural properties of DNA.

28.1 DNA Can Assume a Variety of Structural Forms

The double-helical structure of DNA deduced by Watson and Crick immediately suggested how genetic information is stored and replicated. As was discussed earlier (Section 4.2), the essential features of their model are:

1. Two polynucleotide chains running in opposite directions coil around a common axis to form a right-handed double helix.

2. Purine and pyrimidine bases are on the inside of the helix, whereas phosphate and deoxyribose units are on the outside.

3. Adenine (A) is paired with thymine (T), and guanine (G) with cytosine (C). An A–T base pair is held together by two hydrogen bonds, whereas a G–C base pair is held together by three such bonds.

The A-DNA Double Helix Is Shorter and Wider Than the More Common B-DNA Double Helix

Watson and Crick based their model (known as the *B-DNA helix*) on x-ray diffraction patterns of highly hydrated DNA fibers, which provided information about properties of the double helix that are averaged over its constituent residues. X-ray diffraction studies of less-hydrated DNA fibers revealed a different form called *A-DNA,* which appears when the relative humidity is reduced to less than about 75%. A-DNA, like B-DNA, is a right-handed double helix made up of antiparallel strands held together by Watson–Crick base-pairing. The A helix is wider and shorter than the B helix, and its base pairs are tilted rather than perpendicular to the helix axis (Figure 28.3).

Top
view

Side
view

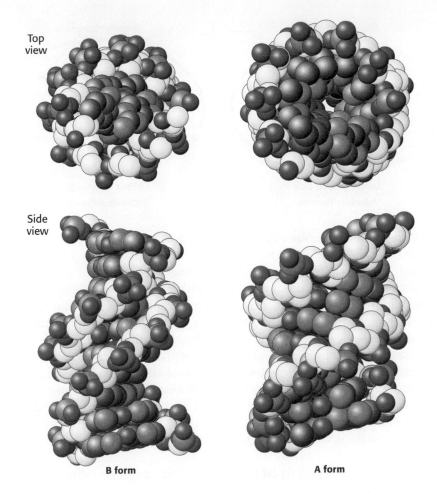

B form   **A form**

Figure 28.3 B-form and A-form DNA. Space-filling models of ten base pairs of B-form and A-form DNA depict their right-handed helical structures. *Notice* that the B-form helix is longer and narrower than the A-form helix. The carbon atoms of the backbone are shown in white. [Drawn from 1BNA.pdb and 1DNZ.pdb.]

Many of the structural differences between B-DNA and A-DNA arise from different puckerings of their ribose units (Figure 28.4). In A-DNA, C-3′ lies out of the plane (a conformation referred to as C-3′ endo) formed by the other four atoms of the furanose ring; in B-DNA, C-2′ lies out of the plane (a conformation called C-2′ endo). The C-3′-endo puckering in A-DNA leads to an 11-degree tilting of the base pairs away from the normal to the helix. The phosphates and other groups in the A helix bind fewer H_2O molecules than do those in B-DNA. Hence, dehydration favors the A form.

Cellular DNA is generally B form. However, the A helix is not confined to dehydrated DNA. Double-stranded regions of RNA and at least some RNA–DNA hybrids adopt a double-helical form very similar to that of A-DNA. The position of the 2′-hydroxyl group of ribose prevents RNA from forming a classic Watson–Crick B helix because of steric hindrance: the 2′-oxygen atom would come too close to three atoms of the adjoining phosphate group and one atom in the next base. In an A-type helix, in contrast, the 2′-oxygen projects outward, away from other atoms.

The Major and Minor Grooves Are Lined by Sequence-Specific Hydrogen-Bonding Groups

Double-helical nucleic acid molecules contain two grooves, called the *major groove* and the *minor groove*. These grooves arise because the glycosidic bonds of a base pair are not diametrically opposite each other (Figure 28.5). The minor groove contains the pyrimidine O-2 and the purine N-3 of the base pair, and the major groove is on the opposite side of the pair. The methyl group of thymine lies in the major groove. In B-DNA, the major

C-3′ endo (A form)

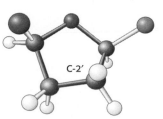

C-2′ endo (B form)

Figure 28.4 Sugar pucker. In A-form DNA, the C-3′ carbon atom lies above the approximate plane defined by the four other sugar nonhydrogen atoms (called C-3′ endo). In B-form DNA, each ribose is in a C-2′-endo conformation, in which C-2′ lies out of the plane.

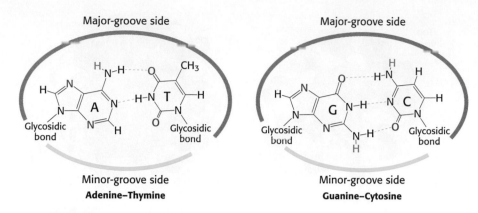

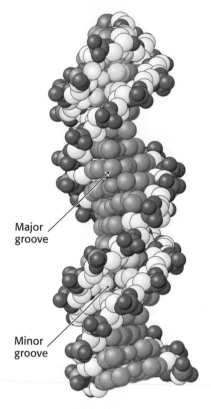

Figure 28.5 Major- and minor-groove sides. Because the two glycosidic bonds are not diametrically opposite each other, each base pair has a larger side that defines the major groove and a smaller side that defines the minor groove. The grooves are lined by potential hydrogen-bond donors (blue) and acceptors (red).

Figure 28.6 Major and minor grooves in B-form DNA. Notice the presence of the major groove (depicted in orange) and the narrower minor groove (depicted in yellow). The carbon atoms of the backbone are shown in white.

groove is wider (12 versus 6 Å) and deeper (8.5 versus 7.5 Å) than the minor groove (Figure 28.6).

Each groove is lined by potential hydrogen-bond donor and acceptor atoms that enable specific interactions with proteins (see Figure 28.5). In the minor groove, N-3 of adenine or guanine and O-2 of thymine or cytosine can serve as hydrogen-bond acceptors, and the amino group attached to C-2 of guanine can be a hydrogen-bond donor. In the major groove, N-7 of guanine or adenine is a potential acceptor, as are O-4 of thymine and O-6 of guanine. The amino groups attached to C-6 of adenine and C-4 of cytosine can serve as hydrogen-bond donors. Note that the major groove displays more features that distinguish one base pair from another than does the minor groove. The larger size of the major groove in B-DNA makes it more accessible for interactions with proteins that recognize specific DNA sequences.

Studies of Single Crystals of DNA Revealed Local Variations in DNA Structure

X-ray analyses of single crystals of DNA oligomers had to await the development of techniques for synthesizing large amounts of DNA fragments with defined base sequences. X-ray analyses of single crystals of DNA at atomic resolution revealed that *DNA exhibits much more structural variability and diversity than formerly envisaged.*

The x-ray analysis of a crystallized DNA dodecamer by Richard Dickerson and his coworkers revealed that its overall structure is very much like a B-form Watson–Crick double helix. However, the dodecamer differs from the Watson–Crick model in not being uniform; there are rather large local deviations from the average structure. The Watson–Crick model has 10 residues per complete turn, and so a residue is related to the next along a chain by a rotation of 36 degrees. In Dickerson's dodecamer, the rotation angles range from 28 degrees (less tightly wound) to 42 degrees (more tightly wound). Furthermore, the two bases of many base pairs are not perfectly coplanar (Figure 28.7). Rather, they are arranged like the blades of a propeller. This deviation from the idealized structure, called *propeller twist,* enhances the stacking of bases along a strand. These local variations of the

Figure 28.7 Propeller twist. The bases of a DNA base pair are often not precisely coplanar. They are twisted with respect to each other, like the blades of a propeller.

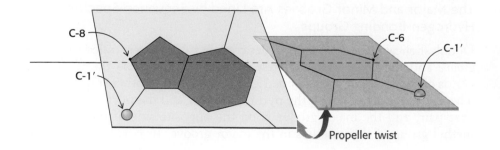

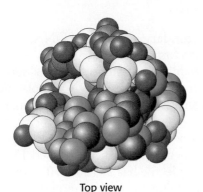

Top view

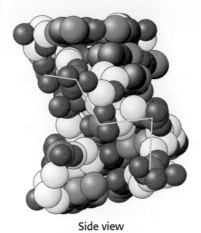

Side view

Figure 28.8 Z-DNA. DNA oligomers such as dCGCGCG adopt an alternative conformation under some conditions. This conformation is called Z-DNA because the phosphate groups zigzag along the backbone. [Drawn from 131D.pdb.]

double helix and others depend on base sequence. A protein searching for a specific target sequence in DNA may sense its presence partly through its effect on the precise shape of the double helix.

Z-DNA Is a Left-Handed Double Helix in Which Backbone Phosphates Zigzag

Alexander Rich and his associates discovered a third type of DNA helix when they solved the structure of dCGCGCG. They found that this hexanucleotide forms a duplex of antiparallel strands held together by Watson–Crick base-pairing, as expected. What was surprising, however, was that this double helix was *left-handed*, in contrast with the *right-handed* screw sense of the A and B helices. Furthermore, the phosphates in the backbone *zigzagged;* hence, they called this new form *Z-DNA* (Figure 28.8).

The Z-DNA form is adopted by short oligonucleotides that have *sequences of alternating pyrimidines and purines.* High salt concentrations are required to reduce electrostatic repulsion between the backbone phosphates, which are closer to one another than in A- and B-DNA. Under physiological conditions, *most DNA is in the B form*. Nonetheless, protein domains have been discovered that bind nucleic acids specifically in the Z-form. This observation strongly suggests that such structures are present in cells and perform specific functions. The properties of A-, B-, and Z-DNA are compared in Table 28.1.

TABLE 28.1 Comparison of A-, B-, and Z-DNA

	HELIX TYPE		
	A	B	Z
Shape	Broadest	Intermediate	Narrowest
Rise per base pair	2.3 Å	3.4 Å	3.8 Å
Helix diameter	25.5 Å	23.7 Å	18.4 Å
Screw sense	Right-handed	Right-handed	Left-handed
Glycosidic bond*	anti	anti	Alternating *anti* and *syn*
Base pairs per turn of helix	11	10.4	12
Pitch per turn of helix	25.3 Å	35.4 Å	45.6 Å
Tilt of base pairs from normal to helix axis	19°	1°	9°
Major groove	Narrow and very deep	Wide and quite deep	Flat
Minor groove	Very broad and shallow	Narrow and quite deep	Very narrow and deep

Syn and *anti* refer to the orientation of the *N*-glycosidic bond between the base and deoxyribose. In the *anti* orientation, the base extends away from the deoxyribose. In the *syn* orientation, the base is above the deoxyribose. Pyrimidine can be only in *anti* orientations, while purines can be *anti* or *syn*.

28.2 Double-Stranded DNA Can Wrap Around Itself to Form Supercoiled Structures

Thus far, we have been considering the secondary structure of DNA. DNA double helices can fold up on themselves to form tertiary structures created by *supercoiling*. Supercoiling is most readily understood by considering covalently closed DNA molecules, but it also applies to DNA molecules constrained to be in loops by other means. Most DNA molecules inside cells are subject to supercoiling.

Consider a linear 260-bp DNA duplex in the B-DNA form (Figure 28.9). Because the number of base pairs per turn in an unstressed DNA molecule averages 10.4, this linear DNA molecule has 25 (260/10.4) turns. The ends of this helix can be joined to produce a *relaxed* circular DNA (Figure 28.9B). A different circular DNA can be formed by unwinding the linear duplex by two turns before joining its ends (Figure 28.9C). What is the structural consequence of unwinding before ligation? Two limiting con-

Figure 28.9 Linking number. The relations between the linking number *(Lk)*, twisting number *(Tw)*, and writhing number *(Wr)* of a circular DNA molecule revealed schematically. [After W. Saenger, *Principles of Nucleic Acid Structure* (Springer Verlag, 1984), p. 452.]

(A)

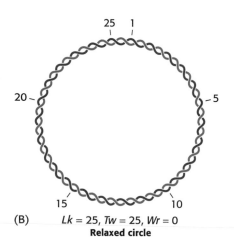

(B) *Lk* = 25, *Tw* = 25, *Wr* = 0
Relaxed circle

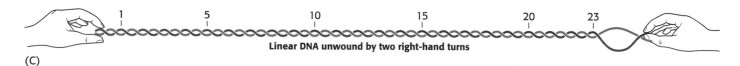

Linear DNA unwound by two right-hand turns

(C)

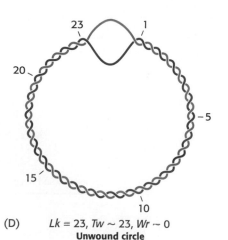

(D) *Lk* = 23, *Tw* ~ 23, *Wr* ~ 0
Unwound circle

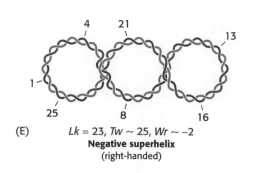

(E) *Lk* = 23, *Tw* ~ 25, *Wr* ~ −2
Negative superhelix
(right-handed)

formations are possible. The DNA can fold into a structure containing 23 turns of B helix and an unwound loop (Figure 28.9D). Alternatively, the double helix can fold up to cross itself. Such crossings are called *supercoils*. In particular, a supercoiled structure with 25 turns of B helix and 2 turns of *right-handed* (termed *negative*) superhelix can be formed (Figure 28.9E).

Supercoiling markedly alters the overall form of DNA. *A supercoiled DNA molecule is more compact than a relaxed DNA molecule of the same length.* Hence, supercoiled DNA moves faster than relaxed DNA when analyzed by centrifugation or electrophoresis. Unwinding will cause supercoiling in circular DNA molecules, whether covalently closed or constrained in closed configurations by other means.

The Linking Number of DNA, a Topological Property, Determines the Degree of Supercoiling

Our understanding of the conformation of DNA is enriched by concepts drawn from topology, a branch of mathematics dealing with structural properties that are unchanged by deformations such as stretching and bending. A key topological property of a circular DNA molecule is its *linking number (Lk)*, which is equal to the number of times that a strand of DNA winds in the right-handed direction around the helix axis when the axis is constrained to lie in a plane. For the relaxed DNA shown in Figure 28.9B, $Lk = 25$. For the partly unwound molecule shown in part D and the supercoiled one shown in part E, $Lk = 23$ because the linear duplex was unwound two complete turns before closure. Molecules differing only in linking number are *topological isomers*, or *topoisomers*, of one another. *Topoisomers of DNA can be interconverted only by cutting one or both DNA strands and then rejoining them.*

The unwound DNA and supercoiled DNA shown in Figure 28.9D and E are topologically identical but geometrically different. They have the same value of Lk but differ in *twist (Tw)* and *writhe (Wr)*. Although the rigorous definitions of twist and writhe are complex, twist is a measure of the helical winding of the DNA strands around each other, whereas writhe is a measure of the coiling of the axis of the double helix—that is, supercoiling. A right-handed coil is assigned a negative number (negative supercoiling) and a left-handed coil is assigned a positive number (positive supercoiling).

Is there a relation between Tw and Wr? Indeed, there is. Topology tells us that the sum of Tw and Wr is equal to Lk.

$$Lk = Tw + Wr$$

In Figure 28.9, the partly unwound circular DNA has $Tw \sim 23$ and $Wr \sim 0$, whereas the supercoiled DNA has $Tw \sim 25$ and $Wr \sim -2$. These forms can be interconverted without cleaving the DNA chain because they have the same value of Lk—namely, 23. The partitioning of Lk (which must be an integer) between Tw and Wr (which need not be integers) is determined by energetics. The free energy is minimized when about 70% of the change in Lk is expressed in Wr and 30% is expressed in Tw. Hence, the most stable form would be one with $Tw = 24.4$ and $Wr = -1.4$. Thus, *a lowering of* Lk *causes both right-handed (negative) supercoiling of the DNA axis and unwinding of the duplex.* Topoisomers differing by just 1 in Lk, and consequently by 0.7 in Wr, can be readily separated by agarose gel electrophoresis because their hydrodynamic volumes are quite different; *supercoiling condenses DNA* (Figure 28.10).

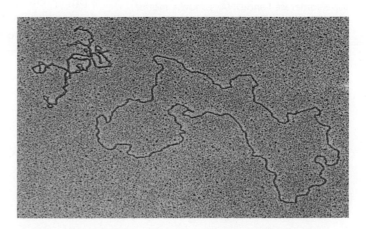

Figure 28.10 Topoisomers. An electron micrograph showing negatively supercoiled and relaxed DNA. [Courtesy of Dr. Jack Griffith.]

Topoisomerases Prepare the Double Helix for Unwinding

Most naturally occurring DNA molecules are negatively supercoiled. What is the basis for this prevalence? As already stated, negative supercoiling arises from the unwinding or underwinding of the DNA. In essence, negative supercoiling prepares DNA for processes requiring separation of the DNA strands, such as replication. Positive supercoiling condenses DNA as effectively, but it makes strand separation more difficult.

The presence of supercoils in the immediate area of unwinding would, however, make unwinding difficult. Therefore, negative supercoils must be continuously removed, and the DNA relaxed, as the double helix unwinds.

Specific enzymes called *topoisomerases* that introduce or eliminate supercoils were discovered by James Wang and Martin Gellert. *Type I topoisomerases* catalyze the relaxation of supercoiled DNA, a thermodynamically favorable process. *Type II topoisomerases* utilize free energy from ATP hydrolysis to add negative supercoils to DNA. Both type I and type II topoisomerases play important roles in DNA replication as well as in transcription and recombination.

These enzymes alter the linking number of DNA by catalyzing a three-step process: (1) the *cleavage* of one or both strands of DNA, (2) the *passage* of a segment of DNA through this break, and (3) the *resealing* of the DNA break. Type I topoisomerases cleave just one strand of DNA, whereas type II enzymes cleave both strands. The two types of enzymes have several common features, including the use of key tyrosine residues to form covalent links to the polynucleotide backbone that is transiently broken.

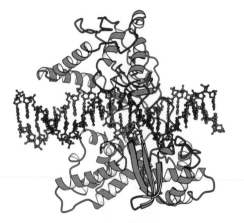

Figure 28.11 Structure of topoisomerase I. The structure of a complex between a fragment of human topoisomerase I and DNA is shown. *Notice that DNA lies in a central cavity within the enzyme.* [Drawn from 1EJ9.pdb.]

Type I Topoisomerases Relax Supercoiled Structures

The three-dimensional structures of several type I topoisomerases have been determined (Figure 28.11). These structures reveal many features of the reaction mechanism. Human type I topoisomerase comprises four domains, which are arranged around a central cavity having a diameter of 20 Å, just the correct size to accommodate a double-stranded DNA molecule. This cavity also includes a tyrosine residue (Tyr 723), which acts as a nucleophile to cleave the DNA backbone in the course of catalysis.

From analyses of these structures and the results of other studies, the relaxation of negatively supercoiled DNA molecules is known to proceed in the following manner (Figure 28.12). First, the DNA molecule binds inside the cavity of the topoisomerase. The hydroxyl group of tyrosine 723 attacks a phosphate group on one strand of the DNA backbone to form a phosphodiester linkage between the enzyme and the DNA, cleaving the DNA and releasing a free 5′-hydroxyl group.

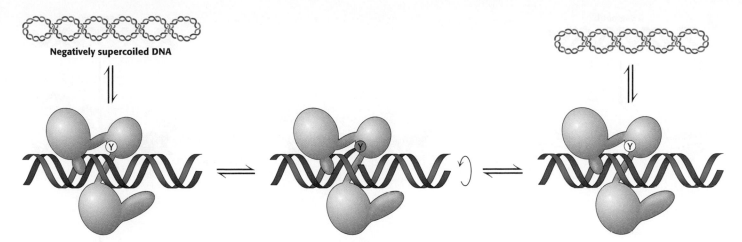

Negatively supercoiled DNA

Figure 28.12 Topoisomerase I mechanism. On binding to DNA, topoisomerase I cleaves one strand of the DNA by means of a tyrosine (Y) residue attacking a phosphate. When the strand has been cleaved, it rotates in a controlled manner around the other strand. The reaction is completed by religation of the cleaved strand. This process results in partial or complete relaxation of a supercoiled plasmid.

With the backbone of one strand cleaved, the DNA can now rotate around the remaining strand, its movement driven by the release of energy stored because of the supercoiling. The rotation of the DNA unwinds the supercoils. The enzyme controls the rotation so that the unwinding is not rapid. The free hydroxyl group of the DNA attacks the phosphotyrosine residue to reseal the backbone and release tyrosine. The DNA is then free to dissociate from the enzyme. Thus, reversible cleavage of one strand of supercoiled DNA allows controlled rotation to partly relax the supercoils.

Type II Topoisomerases Can Introduce Negative Supercoils Through Coupling to ATP Hydrolysis

Supercoiling requires an input of energy because a supercoiled molecule, in contrast with its relaxed counterpart, is torsionally stressed. The introduction of an additional supercoil into a 3000-bp plasmid typically requires about 30 kJ mol^{-1} (7 kcal mol^{-1}).

Supercoiling can be catalyzed by type II topoisomerases. These elegant molecular machines couple the binding and hydrolysis of ATP to the directed passage of one DNA double helix through another, temporarily cleaved DNA double helix. These enzymes have several mechanistic features in common with the type I topoisomerases.

The topoisomerase II from yeast is a dimer with two internal cavities (Figure 28.13). The larger cavity has gates at both the top and the bottom that are crucial to topoisomerase action. The reaction begins with the binding of one double helix (hereafter referred to as the G, for gate, segment) to the enzyme (Figure 28.14). Each strand is positioned next to a tyrosine residue, one from each monomer, capable of forming a covalent linkage with the DNA backbone. This complex then loosely binds a second DNA double helix (hereafter referred to as the T, for transported, segment). Each monomer of the enzyme has a domain that binds ATP; this ATP binding leads to a conformational change that strongly favors the coming together of the two domains. As these domains come closer together, they trap the bound T segment. This conformational change also forces the separation and cleavage of the two strands of the G segment. Each

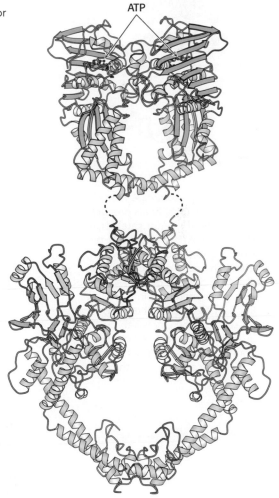

Figure 28.13 Structure of topoisomerase II. A composite structure of topoisomerase II formed from the amino-terminal ATP-binding domain of *E. coli* topoisomerase II (green) and the carboxyl-terminal fragment from yeast topoisomerase II (yellow) is shown. Both units form dimeric structures as shown. *Notice* the central cavity in each dimeric fragment. [Drawn from 1EI1.pdb and 1BJT.pdb.]

ATP

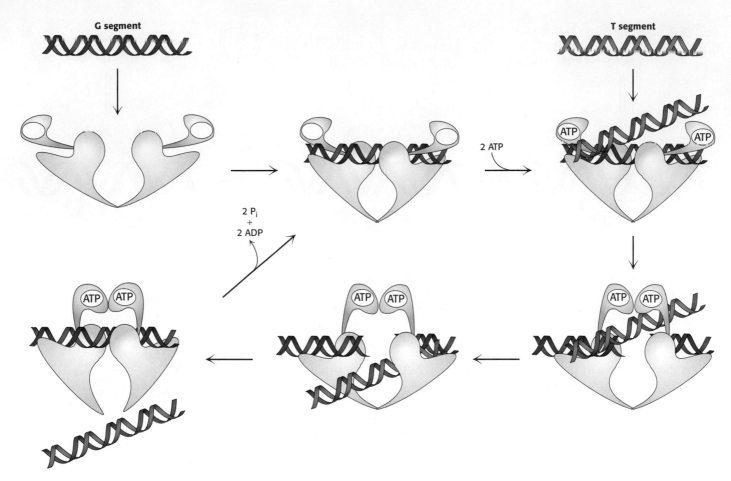

Figure 28.14 Topoisomerase II mechanism. Topoisomerase II first binds one DNA duplex termed the G (for gate) segment. The binding of ATP to the two N-terminal domains brings these two domains together. This conformational change leads to the cleavage of both strands of the G segment and the binding of an additional DNA duplex, the T segment. This T segment then moves through the break in the G segment and out the bottom of the enzyme. The hydrolysis of ATP resets the enzyme with the G segment still bound.

strand is joined to the enzyme by a tyrosine–phosphodiester linkage. Unlike the type I enzymes, the type II topoisomerases hold the DNA tightly so that it cannot rotate. The T segment then passes through the cleaved G segment and into the large central cavity. The ligation of the G segment leads to the release of the T segment through the gate at the bottom of the enzyme. The hydrolysis of ATP and the release of ADP and orthophosphate allow the ATP-binding domains to separate, preparing the enzyme to bind another T segment. The overall process leads to a decrease in the linking number by two.

The bacterial topoisomerase II (often called DNA gyrase) is the target of several antibiotics that inhibit the prokaryotic enzyme much more than the eukaryotic one. *Novobiocin* blocks the binding of ATP to gyrase. *Nalidixic acid* and *ciprofloxacin*, in contrast, interfere with the breakage and rejoining of DNA chains. These two gyrase inhibitors are widely used to treat urinary tract and other infections including those due to *Bacillus anthracis* (anthrax). *Camptothecin*, an antitumor agent, inhibits human topoisomerase I by stabilizing the form of the enzyme covalently linked to DNA.

Nalidixic acid

Ciprofloxacin

28.3 DNA Replication Proceeds by the Polymerization of Deoxyribonucleoside Triphosphates Along a Template

The base sequences of newly synthesized DNA must faithfully match the sequences of parent DNA. To achieve faithful replication, each strand within the parent double helix acts as a *template* for the synthesis of a new DNA strand with a complementary sequence. The building blocks for the synthesis of the new strands are deoxyribonucleoside triphosphates. They are added, one at a time, to the 3′ end of an existing strand of DNA.

Although this reaction is in principle quite simple, it is significantly complicated by specific features of the DNA double helix. First, the two strands of the double helix run in opposite directions. Because DNA strand synthesis always proceeds in the 5′-to-3′ direction, the DNA replication process must have special mechanisms to accommodate the oppositely directed strands. Second, the two strands of the double helix interact with one another in such a way that the edges of the bases on which the newly synthesized DNA is to be assembled are occupied. Thus, the two strands must be separated from each other so as to generate appropriate templates. Finally, the two strands of the double helix wrap around each other. Thus, strand separation also entails the unwinding of the double helix. This unwinding creates supercoils that must themselves be resolved as replication continues, as described in Section 28.2. We begin with a consideration of the chemistry that underlies the formation of the phosphodiester backbone of newly synthesized DNA.

DNA Polymerases Require a Template and a Primer

DNA polymerases catalyze the formation of polynucleotide chains. Each incoming nucleoside triphosphate first forms an appropriate base pair with a base in this template. Only then does the DNA polymerase link the incoming base with the predecessor in the chain. Thus, *DNA polymerases are template-directed enzymes.*

DNA polymerases add nucleotides to the 3′ end of a polynucleotide chain. The polymerase catalyzes the nucleophilic attack by the 3′-hydroxyl-group terminus of the polynucleotide chain on the α phosphoryl group of the nucleoside triphosphate to be added (see Figure 4.22). To initiate this reaction, DNA polymerases require a *primer* with a free 3′-hydroxyl group already base-paired to the template. They cannot start from scratch by adding nucleotides to a free single-stranded DNA template. RNA polymerase, in contrast, can initiate RNA synthesis without a primer (p. 827).

All DNA Polymerases Have Structural Features in Common

The three-dimensional structures of a number of DNA polymerase enzymes are known. The first such structure was elucidated by Tom Steitz and coworkers, who determined the structure of the so-called *Klenow fragment* of DNA polymerase I from *E. coli* (Figure 28.15). This fragment comprises two main parts of the full enzyme, including the polymerase unit. This unit approximates the shape of a right hand with domains that are referred to as the

| **Primer** |
| The initial segment of a polymer that is to be extended on which elongation depends. |

| **Template** |
| A sequence of DNA or RNA that directs the synthesis of a complementary sequence. |

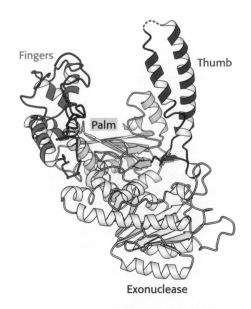

Figure 28.15 DNA polymerase structure. The first DNA polymerase structure determined was that of a fragment of *E. coli* DNA polymerase I called the Klenow fragment. *Notice* that, like other DNA polymerases, the polymerase unit resembles a right hand with fingers (blue), palm (yellow), and thumb (red). The Klenow fragment also includes an exonuclease domain that removes incorrect nucleotide bases. [Drawn from 1DPI.pdb.]

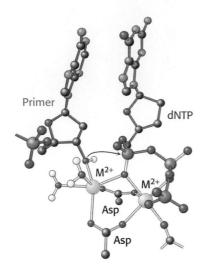

Figure 28.16 DNA polymerase mechanism.
Two metal ions (typically, Mg^{2+})
participate in the DNA polymerase
reaction. One metal ion coordinates the
3'-hydroxyl group of the primer, whereas
the other metal ion interacts only with the
dNTP. The phosphate group of the
nucleoside triphosphate bridges between
the two metal ions. The hydroxyl group of
the primer attacks the phosphate group to
form a new O–P bond.

fingers, the thumb, and the palm. In addition to the polymerase, the Klenow
fragment includes a domain with $3' \rightarrow 5'$ *exonuclease* activity that participates
in proofreading and correcting the polynucleotide product (Section 28.4).

DNA polymerases are remarkably similar in overall shape, although
they differ substantially in detail. At least five structural classes have
been identified; some of them are clearly homologous, whereas others are
probably the products of convergent evolution. In all cases, the finger and
thumb domains wrap around DNA and hold it across the enzyme's active site,
which comprises residues primarily from the palm domain. Furthermore, all
DNA polymerases use similar strategies to catalyze the polymerase reaction,
making use of a mechanism in which two metal ions take part.

Two Bound Metal Ions Participate in the Polymerase Reaction

Like all enzymes with nucleoside triphosphate substrates, DNA polymerases
require metal ions for activity. Examination of the structures of DNA polymerases with bound substrates and substrate analogs reveals the presence of
two metal ions in the active site. One metal ion binds both the deoxynucleoside triphosphate (dNTP) and the 3'-hydroxyl group of the primer, whereas
the other interacts only with the dNTP (Figure 28.16). The two metal ions are
bridged by the carboxylate groups of two aspartate residues in the palm
domain of the polymerase. These side chains hold the metal ions in the proper
positions and orientations. The metal ion bound to the primer activates the
3'-hydroxyl group of the primer, facilitating its attack on the α phosphoryl
group of the dNTP substrate in the active site. The two metal ions together
help stabilize the negative charge that accumulates on the pentacoordinate
transition state. The metal ion initially bound to dNTP stabilizes the negative
charge on the pyrophosphate product.

The Specificity of Replication Is Dictated by Complementarity of Shape Between Bases

DNA must be replicated with high fidelity. Each base added to the growing
chain should, with high probability, be the Watson–Crick complement of
the base in the corresponding position in the template strand. The binding
of the dNTP containing the proper base is favored by the formation of a
base pair with its partner on the template strand. Although hydrogen bonding contributes to the formation of this base pair, overall shape complementarity is crucial. Studies show that a nucleotide with a base that is very
similar in shape to adenine but lacks the ability to form base-pairing hydrogen bonds can still direct the incorporation of thymidine, both in vitro and
in vivo (Figure 28.17).

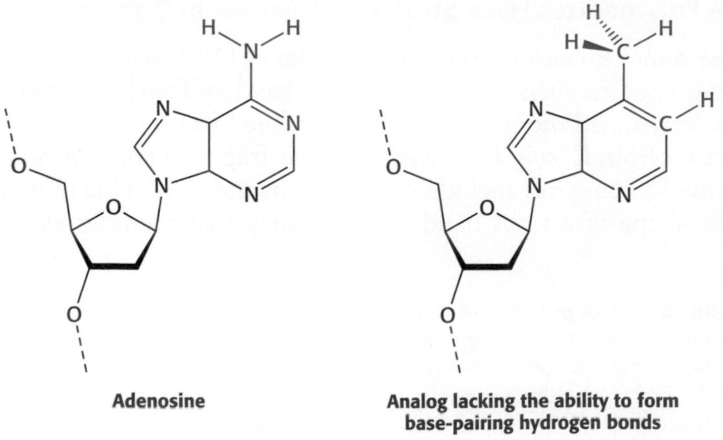

Figure 28.17 Shape complementarity. The
base analog on the right has the same
shape as adenosine, but groups that form
hydrogen bonds between base pairs have
been replaced by groups (shown in red)
not capable of hydrogen bonding.
Nonetheless, studies reveal that, when
incorporated into the template strand, this
analog directs the insertion of thymidine
in DNA replication.

Adenosine

**Analog lacking the ability to form
base-pairing hydrogen bonds**

An examination of the crystal structures of various DNA polymerases reveals why shape complementarity is so important. First, residues of the enzyme form hydrogen bonds with *the minor-groove side of the base pair in the active site* (Figure 28.18). In the minor groove, hydrogen-bond acceptors are present in the same positions for all Watson–Crick base pairs. These interactions act as a "ruler" that measures whether a properly spaced base pair has formed in the active site.

Second, DNA polymerases close down around the incoming dNTP (Figure 28.19). The binding of a deoxyribonucleoside triphosphate into the active site of a DNA polymerase triggers a conformational change: the finger domain rotates to form a tight pocket into which only a properly shaped base pair will readily fit. Many of the residues lining this pocket are important to ensure the efficiency and fidelity of DNA synthesis. For example, mutation of a conserved tyrosine residue that forms part of the pocket results in a polymerase that is approximately 40 times as error prone as the parent polymerase.

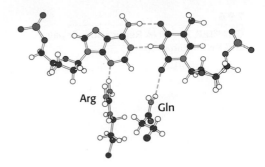

Figure 28.18 Minor-groove interactions. DNA polymerases donate two hydrogen bonds to base pairs in the minor groove. Hydrogen-bond acceptors are present in these two positions for all Watson–Crick base pairs including the A–T base pair shown.

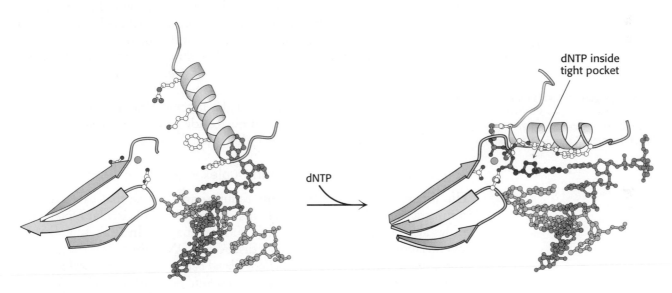

Figure 28.19 Shape selectivity. The binding of a deoxyribonucleoside triphosphate (dNTP) to DNA polymerase induces a conformational change, generating a tight pocket for the base pair consisting of the dNTP and its partner on the template strand. Such a conformational change is possible only when the dNTP corresponds to the Watson–Crick partner of the template base. [Drawn from 2BDP.pdb and 1T7P.pdb.]

An RNA Primer Synthesized by Primase Enables DNA Synthesis to Begin

DNA polymerases cannot initiate DNA synthesis without a primer, a section of nucleic acid having a free 3′ end that forms a double helix with the template. How is this primer formed? An important clue came from the observation that RNA synthesis is essential for the initiation of DNA synthesis. In fact, *RNA primes the synthesis of DNA.* An RNA polymerase called *primase* synthesizes a short stretch of RNA (about five nucleotides) that is complementary to one of the template DNA strands (Figure 28.20). Primase, like other RNA polymerases, can initiate synthesis without a primer. After DNA synthesis has been initiated, the short stretch of RNA is removed by hydrolysis and replaced by DNA.

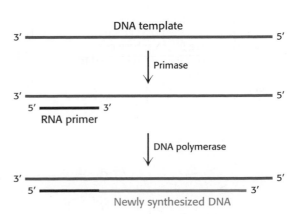

Figure 28.20 Priming. DNA replication is primed by a short stretch of RNA that is synthesized by primase, an RNA polymerase. The RNA primer is removed at a later stage of replication.

One Strand of DNA Is Made Continuously, Whereas the Other Strand Is Synthesized in Fragments

Both strands of parental DNA serve as templates for the synthesis of new DNA. The site of DNA synthesis is called the *replication fork* because the complex formed by the newly synthesized daughter helices arising from the parental duplex resembles a two-pronged fork. Recall that the two strands are antiparallel; that is, they run in opposite directions. During DNA replication, both daughter strands appear on cursory examination to grow in the same direction. However, all known DNA polymerases synthesize DNA in the $5' \rightarrow 3'$ direction but not in the $3' \rightarrow 5'$ direction. How then does one of the daughter DNA strands appear to grow in the $3' \rightarrow 5'$ direction?

This dilemma was resolved by Reiji Okazaki, who found that *a significant proportion of newly synthesized DNA exists as small fragments.* These units of about a thousand nucleotides (called *Okazaki fragments*) are present briefly in the vicinity of the replication fork (Figure 28.21).

As replication proceeds, these fragments become covalently joined through the action of the enzyme DNA ligase to form a continuous daughter strand. The other new strand is synthesized continuously. The strand formed from Okazaki fragments is termed the *lagging strand,* whereas the one synthesized without interruption is the *leading strand.* The discontinuous assembly of the lagging strand enables $5' \rightarrow 3'$ polymerization at the nucleotide level to give rise to overall growth in the $3' \rightarrow 5'$ direction.

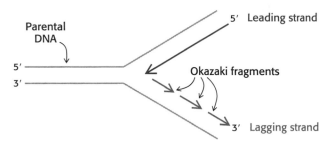

Figure 28.21 Okazaki fragments. At a replication fork, both strands are synthesized in the $5' \rightarrow 3'$ direction. The leading strand is synthesized continuously, whereas the lagging strand is synthesized in short pieces termed Okazaki fragments.

DNA Ligase Joins Ends of DNA in Duplex Regions

The joining of Okazaki fragments requires an enzyme that catalyzes the joining of the ends of two DNA chains. The existence of circular DNA molecules also points to the existence of such an enzyme. In 1967, scientists in several laboratories simultaneously discovered *DNA ligase. This enzyme catalyzes the formation of a phosphodiester bond between the 3' hydroxyl group at the end of one DNA chain and the 5'-phosphate group at the end of the other* (Figure 28.22). An energy source is required to drive this thermodynamically uphill reaction. In eukaryotes and archaea, ATP is the energy source. In bacteria, NAD^+ typically plays this role.

DNA ligase cannot link two molecules of single-stranded DNA or circularize single-stranded DNA. Rather, *ligase seals breaks in double-stranded DNA molecules.* The enzyme from *E. coli* ordinarily forms a phosphodiester bridge only if there are at least a few bases of single-stranded DNA on the end of a double-stranded fragment that can come together with those on another fragment to form base pairs. Ligase encoded by T4 bacteriophage can link two blunt-ended double-helical fragments, a capability that is exploited in recombinant DNA technology.

Figure 28.22 DNA ligase reaction. DNA ligase catalyzes the joining of one DNA strand with a free 3'-hydroxyl group to another with a free 5'-phosphate group. In eukaryotes and archaea, ATP is cleaved to AMP and PP_i to drive this reaction. In bacteria, NAD^+ is cleaved to AMP and nicotinamide mononucleotide (NMN).

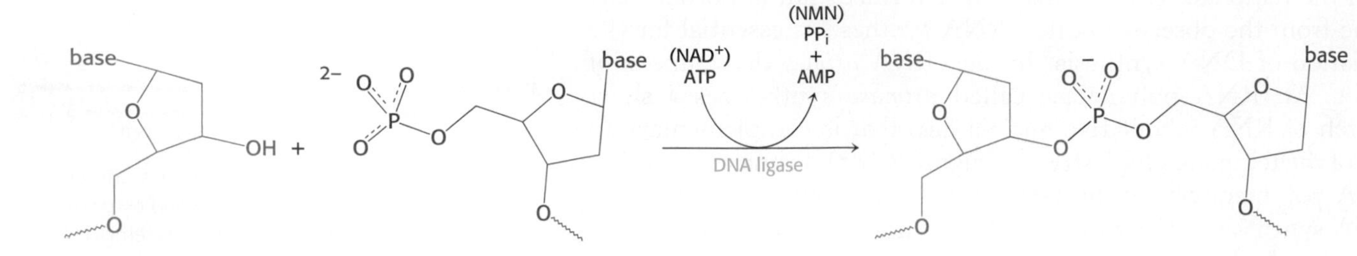

The Separation of DNA Strands Requires Specific Helicases and ATP Hydrolysis

For a double-stranded DNA molecule to replicate, the two strands of the double helix must be separated from each other, at least locally. This separation allows each strand to act as a template on which a new polynucleotide chain can be assembled. Specific enzymes, termed *helicases,* utilize the energy of ATP hydrolysis to power strand separation.

The detailed mechanisms of helicases are still under investigation. However, the determination of the three-dimensional structures of several helicases has been a source of insight. We will begin with a bacterial helicase called *PcrA* because it has been extensively studied, even though it differs from most helicases important to DNA replication in being a monomer. PcrA comprises four domains, hereafter referred to as domains A1, A2, B1, and B2 (Figure 28.23). Domain A1 contains a P-loop NTPase fold, as was expected from amino acid sequence analysis. This domain participates in ATP binding and hydrolysis. Domain B1 is homologous to domain A1 but lacks a P-loop. Domains A2 and B2 have unique structures.

From an analysis of a set of helicase crystal structures bound to nucleotide analogs and appropriate double- and single-stranded DNA molecules, a mechanism for the action of these enzymes was proposed (Figure 28.24). Domains A1 and B1 are capable of binding single-stranded DNA. In the absence of bound ATP, both domains are bound to DNA. The binding of ATP triggers conformational changes in the P-loop and adjacent regions that lead to the closure of the cleft between these two domains. To achieve this movement, domain A1 releases the DNA and slides along the DNA strand, moving closer to domain B1. The enzyme then catalyzes the hydrolysis of ATP to form ADP and orthophosphate. On product release, the cleft between domains A and B springs open. In this state, however, domain A1 has a tighter grip on the DNA than does domain B1, and so the DNA is pulled across domain B1 toward domain A1. The result is the translocation of the enzyme along the DNA strand in a manner similar to the way in which an inchworm moves. The PcrA enzyme translocates in the $3' \rightarrow 5'$ direction. When the helicase encounters a region of double-stranded DNA, it continues to move along one strand and displaces the opposite DNA strand as it progresses. Interactions with specific pockets on the helicase help destabilize the DNA duplex, aided by ATP-induced conformational changes.

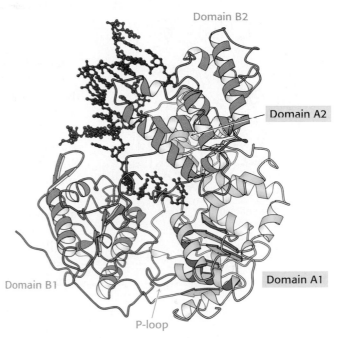

Figure 28.23 Helicase structure. The bacterial helicase PcrA comprises four domains: A1, A2, B1, and B2. *Notice* that the A1 domain includes a P-loop NTPase fold (indicated by the purple shading with the P loop shown in green), whereas the B1 domain has a similar overall structure but lacks a P-loop and does not bind nucleotides. Single-stranded DNA binds to the A1 and B1 domains near the interfaces with domains A2 and B2. [Drawn from 3PJR.pdb]

Figure 28.24 Helicase mechanism. Initially, both domains A1 and B1 of PcrA bind single-stranded DNA. On binding of ATP, the cleft between these domains closes and domain A1 slides along the DNA. On ATP hydrolysis, the cleft opens up, pulling the DNA from domain B1 toward domain A1. As this process is repeated, double-stranded DNA is unwound. *Notice* that the dots on the red strand, representing two locations on the strand, move as the double helix is unwound.

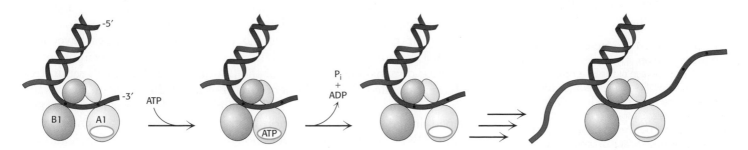

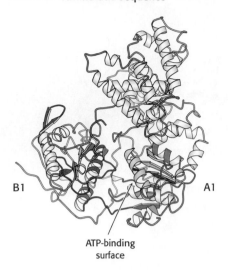

Amino acid sequence

ATP-binding
surface

B1   A1

Figure 28.25 Conserved residues among helicases. A comparison of the amino acid sequences of hundreds of helicases revealed seven regions of strong sequence conservation (shown in color). *Notice* that, when mapped onto the structure of PcrA, these conserved regions lie along the interface between the A1 and B1 domains and along the ATP-binding surface. [Drawn from 3PJR.pdb.]

Helicases constitute a large and diverse class of enzymes. Some of these enzymes move in a $5' \rightarrow 3'$ direction, whereas others unwind RNA rather than DNA and participate in processes such as RNA splicing and the initiation of mRNA translation. A comparison of the amino acid sequences of hundreds of these enzymes reveals seven regions of striking conservation (Figure 28.25). Mapping these regions onto the PcrA structure shows that they line the ATP-binding site and the cleft between the two domains, consistent with the notion that other helicases undergo conformational changes analogous to those found in PcrA. However, whereas PcrA appears to function as a monomer, other members of the helicase family function as oligomers. The hexameric structures of one important group are similar to that of the F_1 component of ATP synthase (Section 18.4), suggesting potential mechanistic similarities. In particular, using a mechanism similar to the binding-change mechanism, the subunits within the helicase may act in a concerted fashion to unwind double-stranded DNA as one strand is pulled through the center of the hexameric ring while the other remains outside. These hexameric helicases include P-loops and are members of a class of ATPases called the AAA family.

Processive enzyme

From the Latin *procedere,* "to go forward."
    An enzyme that catalyzes multiple rounds of the elongation or digestion of a polymer while the polymer stays bound. A *distributive enzyme,* in contrast, releases its polymeric substrate between successive catalytic steps.

28.4 DNA Replication Is Highly Coordinated

DNA replication must be very rapid, given the sizes of the genomes and the rates of cell division. The *E. coli* genome contains 4.6 million base pairs and is copied in less than 40 minutes. Thus, 2000 bases are incorporated per second. Enzyme activities must be highly coordinated to replicate entire genomes precisely and rapidly.

We begin our consideration of the coordination of DNA replication by looking at *E. coli,* which has been extensively studied. For this organism with a relatively small genome, replication begins at a single site and continues around the circular chromosome. The coordination of eukaryotic DNA replication is more complex, because there are many initiation sites throughout the genome and because an additional enzyme is needed to replicate the ends of linear chromosomes.

DNA Replication Requires Highly Processive Polymerases

Replicative polymerases are characterized by their *very high catalytic potency, fidelity, and processivity. Processivity* refers to the ability of an enzyme to catalyze many consecutive reactions without releasing its substrate. These polymerases are assemblies of many subunits that have evolved to grasp their templates and not let go until many nucleotides have been added. The source of the processivity was revealed by the determination of the three-dimensional structure of the β_2 subunit of the *E. coli* replicative polymerase called DNA polymerase III (Figure 28.26). This unit keeps the polymerase associated with the DNA double helix. It has the form of a star-shaped ring. A 35-Å-diameter hole in its center can readily accommodate a duplex DNA molecule, yet leaves enough space between the DNA and the protein to allow rapid sliding during replication. To achieve a catalytic rate

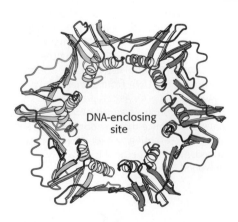

DNA-enclosing
site

Figure 28.26 Structure of a sliding DNA clamp. The dimeric β subunit of DNA polymerase III forms a ring that surrounds the DNA duplex. *Notice* the central cavity through which the DNA template slides. Clasping the DNA molecule in the ring, the polymerase enzyme is able to move without falling off the DNA substrate. [Drawn from 2POL.pdb.]

of 1000 nucleotides polymerized per second requires that 100 turns of duplex DNA (a length of 3400 Å, or 0.34 μm) slide through the central hole of β_2 per second. Thus, β_2 *plays a key role in replication by serving as a sliding DNA clamp.*

How does DNA become entrapped inside the sliding clamp? Replicative polymerases also include assemblies of subunits that function as *clamp loaders*. These enzymes grasp the sliding clamp and, utilizing the energy of ATP binding, pull apart one of the interfaces between the two subunits of the sliding clamp. DNA can move through the gap, inserting itself through the central hole. ATP hydrolysis then releases the clamp, which closes around the DNA.

The Leading and Lagging Strands Are Synthesized in a Coordinated Fashion

Replicative polymerases such as DNA polymerase III synthesize the leading and lagging strands simultaneously at the replication fork (Figure 28.27). DNA polymerase III begins the synthesis of the leading strand starting from the RNA primer formed by primase. The duplex DNA ahead of the polymerase is unwound by a hexameric helicase called DnaB. Copies of single-strand-binding protein (SSB) bind to the unwound strands, keeping the strands separated so that both strands can serve as templates. The leading strand is synthesized continuously by polymerase III. Topoisomerase II concurrently introduces right-handed (negative) supercoils to avert a topological crisis.

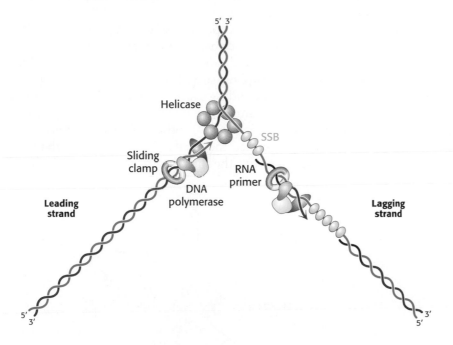

Figure 28.27 Replication fork. A schematic view of the arrangement of DNA polymerase III and associated enzymes and proteins present in replicating DNA. The helicase separates the two strands of the parent double helix, allowing DNA polymerases to use each strand as a template for DNA synthesis. Abbreviation: SSB, single-strand-binding protein.

The mode of synthesis of the lagging strand is necessarily more complex. As mentioned earlier, the lagging strand is synthesized in fragments so that $5' \rightarrow 3'$ polymerization leads to overall growth in the $3' \rightarrow 5'$ direction. Yet the synthesis of the lagging strand is coordinated with the synthesis of the leading strand. How is this coordination accomplished? Examination of the subunit composition of the DNA polymerase III holoenzyme reveals an elegant solution (Figure 28.28). The holoenzyme includes two copies of the polymerase core enzyme, which consists of the DNA polymerase itself (the α subunit); the ε subunit, a $3'$-to-$5'$ proofreading exonuclease (see p. 807); another subunit called θ, and two copies of the dimeric β-subunit sliding

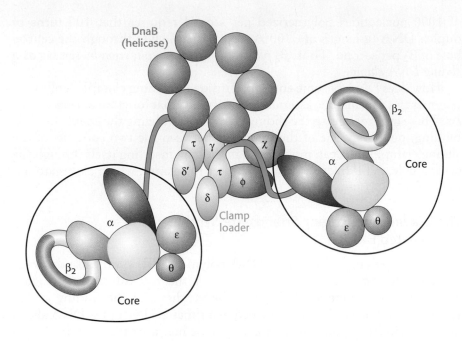

Figure 28.28 DNA polymerase holoenzyme. Each holoenzyme consists of two copies of the polymerase core enzyme, which comprises the α, ε, and θ subunits and two copies of the β subunit, linked to a central structure. The central structure includes the clamp-loader complex and the hexameric helicase DnaB.

clamp. The core enzymes are linked to a central structure having the subunit composition $\gamma\tau_2\delta\delta'\chi\phi$. The $\gamma\tau_2\delta\delta'$ complex is the clamp loader, and the χ and ϕ subunits interact with the single-strand-DNA–binding protein. The entire apparatus interacts with the hexameric helicase DnaB. Eukaryotic replicative polymerases have similar, albeit slightly more complicated, subunit compositions and structures.

The lagging-strand template is looped out so that it passes through the polymerase site in one subunit of a dimeric polymerase III in the same direction as that of the leading-strand template in the other subunit. DNA polymerase III lets go of the lagging-strand template after adding about 1000 nucleotides by releasing the sliding clamp. A new loop is then formed, a sliding clamp is added, and primase again synthesizes a short stretch of RNA primer to initiate the formation of another Okazaki fragment. This mode of replication has been termed the *trombone model* because the size of the loop lengthens and shortens like the slide on a trombone (Figure 28.29).

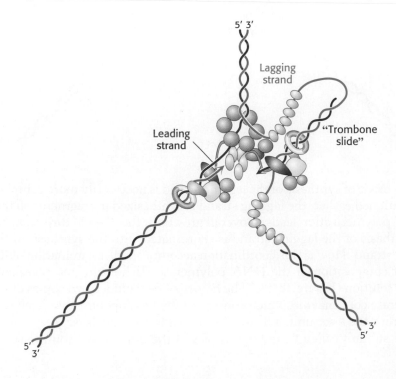

Figure 28.29 Trombone model. The replication of the leading and lagging strands is coordinated by the looping out of the lagging strand to form a structure that acts somewhat like a trombone slide, growing as the replication fork moves forward. When the polymerase on the lagging strand reaches a region that has been replicated, the sliding clamp is released and a new loop is formed.

The gaps between fragments of the nascent lagging strand are filled by DNA polymerase I. This essential enzyme also uses its $5' \rightarrow 3'$ exonuclease activity to remove the RNA primer lying ahead of the polymerase site. The primer cannot be erased by DNA polymerase III, because the enzyme lacks $5' \rightarrow 3'$ editing capability. Finally, DNA ligase connects the fragments.

DNA Replication in *Escherichia coli* Begins at a Unique Site

In *E. coli,* DNA replication starts at a unique site within the entire 4.6×10^6 bp genome. This *origin of replication,* called the oriC *locus,* is a 245-bp region that has several unusual features (Figure 28.30). The *oriC* locus contains five copies of a sequence that are preferred binding sites for the origin-recognition protein DnaA. In addition, the locus contains a tandem array of 13-bp sequences that are rich in AT base pairs. Several steps are required to prepare for the start of replication:

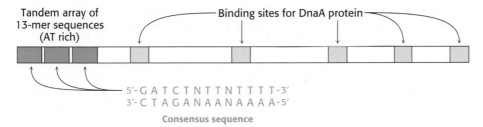

Figure 28.30 Origin of replication in *E. coli*. The *oriC* locus has a length of 245 bp. It contains a tandem array of three nearly identical 13-nucleotide sequences (green) and five binding sites (yellow) for the DnaA protein.

1. *The binding of DnaA proteins to DNA is the first step in the preparation for replication.* DnaA is a member of the P-loop NTPase family and, more specifically, an AAA ATPase (p. 653). Each DnaA monomer comprises an ATPase domain linked to a DNA-binding domain at its C-terminus. DnaA molecules are able to bind to each other through their ATPase domains; a group of bound DnaA molecules will break apart on the binding and hydrolysis of ATP. The binding of DnaA molecules to one another signals the start of the preparatory phase, and their breaking apart signals the end of that phase. The DnaA proteins bind to the five high-affinity sites in *oriC* and then come together with DnaA molecules bound to lower-affinity sites to form an oligomer, possibly a cyclic hexamer. The DNA is wrapped around the outside of the DnaA hexamer (Figure 28.31).

2. *Single DNA strands are exposed in the prepriming complex.* With DNA wrapped around a DnaA hexamer, additional proteins are brought into play. The hexameric helicase DnaB is loaded around the DNA with the help of the helicase loader protein DnaC. Local regions of *oriC*, including the AT regions, are unwound and trapped by single-strand-binding protein. The result of this process is the generation of a structure called the *prepriming complex,* which makes single-stranded DNA accessible to other proteins. Significantly, the primase, DnaG, is now able to insert the RNA primer.

3. *The polymerase holoenzyme assembles.* The DNA polymerase III holoenzyme assembles on the prepriming complex, initiated by interactions between DnaB and the sliding clamp subunit of DNA polymerase III. These interactions also trigger ATP hydrolysis within the DnaA subunits, signaling the initiation of DNA replication. The breakup of the DnaA assembly prevents additional rounds of replication from beginning at the replication origin.

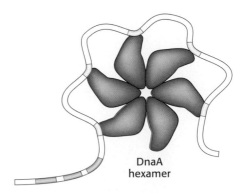

Figure 28.31 Assembly of dnaA. Monomers of DnaA bind to their binding sites (shown in yellow) in *oriC* and come together to form a complex structure, possibly the cyclic hexamer shown here. This structure marks the origin of replication and favors DNA strand separation in the AT-rich sites (green).

DNA Synthesis in Eukaryotes Is Initiated at Multiple Sites

Replication in eukaryotes is mechanistically similar to replication in prokaryotes but is more challenging for a number of reasons. One of them is sheer size: *E. coli* must replicate 4.6 million base pairs, whereas a human diploid cell must replicate more than 6 billion base pairs. Second, the genetic information for *E. coli* is contained on 1 chromosome, whereas, in human beings, 23 pairs of chromosomes must be replicated. Finally, whereas the *E. coli* chromosome is circular, human chromosomes are linear. Unless countermeasures are taken, linear chromosomes are subject to shortening with each round of replication.

The first two challenges are met by the use of multiple origins of replication. In human beings, replication requires about 30,000 origins of replication, with each chromosome containing several hundred. Each origin of replication is the starting site for a replication unit, or *replicon*. In contrast with *E. coli,* the origins of replication in human beings do not contain regions of sharply defined sequence. Instead, more broadly defined AT-rich sequences are the sites around which the *origin of replication complexes* (ORCs) are assembled.

1.   *The assembly of the ORC is the first step in the preparation for replication.* In human beings, the ORC is composed of six different proteins, each homologous to DnaA. These proteins likely come together to form a hexameric structure analogous to the assembly formed by DnaA.

2.   *Licensing factors recruit a helicase that exposes single strands of DNA.* After the ORC has been assembled, additional proteins are recruited, including Cdc6, a homolog of the ORC subunits, and Cdt1. These proteins, in turn, recruit a hexameric helicase with six distinct subunits called Mcm2-7. These proteins, including the helicase, are sometimes called *licensing factors* because they permit the formation of the initiation complex. After the initiation complex has formed, Mcm2-7 separates the parental DNA strands, and the single strands are stabilized by the binding of *replication protein A,* a single-stranded-DNA–binding protein.

3.   *Two distinct polymerases are needed to copy a eukaryotic replicon.* An initiator polymerase called *polymerase α* begins replication but is soon replaced by a more processive enzyme. This process is called *polymerase switching* because one polymerase has replaced another. This second enzyme, called *DNA polymerase δ,* is the principal replicative polymerase in eukaryotes (Table 28.2).

TABLE 28.2 Some types of DNA polymerases

Name	Function
Prokaryotic Polymerases	
DNA polymerase I	Erases primer and fills in gaps on lagging strand
DNA polymerase II (error-prone polymerase)	DNA repair
DNA polymerase III	Primary enzyme of DNA synthesis
Eukaryotic Polymerases	
DNA polymerase α	Initiator polymerase
Primase subunit	Synthesizes the RNA primer
DNA polymerase unit	Adds stretch of about 20 nucleotides to the primer
DNA polymerase β (error-prone polymerase)	DNA repair
DNA polymerase δ	Primary enzyme of DNA synthesis

Replication begins with the binding of DNA polymerase α. This enzyme includes a primase subunit, used to synthesize the RNA primer, as well as an active DNA polymerase. After this polymerase has added a stretch of about 20 deoxynucleotides to the primer, another replication protein, called *replication factor C* (RFC), displaces DNA polymerase α. RFC attracts a sliding clamp called *proliferating cell nuclear antigen* (PCNA), which is homologous to the β_2 subunit of *E. coli* polymerase III. The binding of PCNA to DNA polymerase δ renders the enzyme highly processive and suitable for long stretches of replication. Replication continues in both directions from the origin of replication until adjacent replicons meet and fuse. RNA primers are removed and the DNA fragments are ligated by DNA ligase.

The use of multiple origins of replication requires mechanisms for ensuring that each sequence is replicated once and only once. The events of eukaryotic DNA replication are linked to the eukaryotic *cell cycle* (Figure 28.32). The processes of DNA synthesis and cell division are coordinated in the cell cycle so that the replication of all DNA sequences is complete before the cell progresses into the next phase of the cycle. This coordination requires several *checkpoints* that control the progression along the cycle. A family of small proteins termed *cyclins* are synthesized and degraded by proteasomal digestion in the course of the cell cycle. Cyclins act by binding to specific *cyclic-dependent protein kinases* and activating them. One such kinase, cyclin-dependent kinase 2 (cdk2) binds to assemblies at origins of replication and regulates replication through a number of interlocking mechanisms.

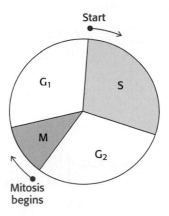

Figure 28.32 Eukaryotic cell cycle. DNA replication and cell division must take place in a highly coordinated fashion in eukaryotes. Mitosis (M) takes place only after DNA synthesis (S). Two gaps (G_1 and G_2) in time separate the two processes.

Telomeres Are Unique Structures at the Ends of Linear Chromosomes

Whereas the genomes of essentially all prokaryotes are circular, the chromosomes of human beings and other eukaryotes are linear. The free ends of linear DNA molecules introduce several complications that must be resolved by special enzymes. In particular, complete replication of DNA ends is difficult because polymerases act only in the $5' \rightarrow 3'$ direction. The lagging strand would have an incomplete 5' end after the removal of the RNA primer. Each round of replication would further shorten the chromosome.

The first clue to how this problem is resolved came from sequence analyses of the ends of chromosomes, which are called *telomeres* (from the Greek *telos*, "an end"). Telomeric DNA contains hundreds of tandem repeats of a six-nucleotide sequence. One of the strands is G rich at the 3' end, and it is slightly longer than the other strand. In human beings, the repeating G-rich sequence is AGGGTT.

The structure adopted by telomeres has been extensively investigated. Recent evidence suggests that they may form large duplex loops (Figure 28.33). It has been proposed that the single-stranded region at the very end of the structure loops back to form a DNA duplex with another part of the repeated sequence, displacing a part of the original telomeric duplex. This looplike structure is formed and stabilized by specific telomere-binding proteins. Such structures would nicely mask and protect the end of the chromosome.

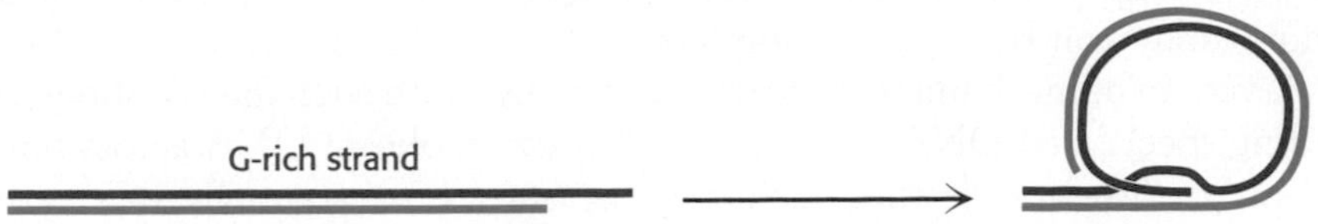

G-rich strand

Figure 28.33 Proposed model for telomeres. A single-stranded segment of the G-rich strand extends from the end of the telomere. In one model for telomeres, this single-stranded region invades the duplex to form a large duplex loop.

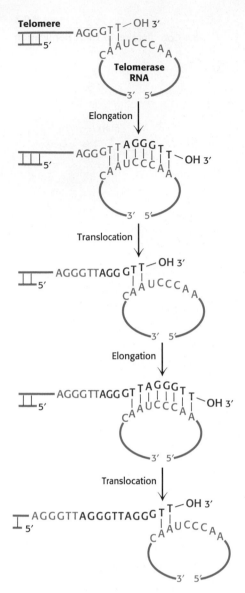

Figure 28.34 Telomere formation. Mechanism of synthesis of the G-rich strand of telomeric DNA. The RNA template of telomerase is shown in blue and the nucleotides added to the G-rich strand of the primer are shown in red. [After E. H. Blackburn. *Nature* 350(1991):569–573.]

Telomeres Are Replicated by Telomerase, a Specialized Polymerase That Carries Its Own RNA Template

How are the repeated sequences generated? An enzyme, termed *telomerase*, that executes this function has been purified and characterized. When a primer ending in GGTT is added to the human enzyme in the presence of deoxynucleoside triphosphates, the sequences GGTTAGGGTT and GGTTAGGGTTAGGGTT, as well as longer products, are generated. Elizabeth Blackburn and Carol Greider discovered that the enzyme adding the repeats contains an RNA molecule that serves as the template for elongation of the G-rich strand (Figure 28.34). Thus, the enzyme carries the information necessary to generate the telomere sequences. The exact number of repeated sequences is not crucial.

Subsequently, a protein component of telomerases also was identified. From its amino acid sequence, this component is clearly related to reverse transcriptases, enzymes first discovered in retroviruses that copy RNA into DNA. Thus, *telomerase is a specialized reverse transcriptase that carries its own template.* Telomerase is generally expressed at high levels only in rapidly growing cells. Thus, telomeres and telomerase can play important roles in cancer-cell biology and in cell aging.

28.5 Many Types of DNA Damage Can Be Repaired

We have examined how even very large and complex genomes can, in principle, be replicated with considerable fidelity. However, DNA does become damaged, both in the course of replication and through other processes. Damage to DNA can be as simple as the misincorporation of a single base or it can take more complex forms such as the chemical modification of bases, chemical cross-links between the two strands of the double helix, or breaks in one or both of the phosphodiester backbones. The results may be cell death or cell transformation, changes in the DNA sequence that can be inherited by future generations, or blockage of the DNA replication process itself. A variety of DNA-repair systems have evolved that can recognize these defects and, in many cases, restore the DNA molecule to its undamaged form. We begin with some of the sources of DNA damage.

Errors Can Occur in DNA Replication

Errors introduced in the replication process are the simplest source of damage in the double helix. With the addition of each base, there is the possibility that an incorrect base might be incorporated, forming a non-Watson–Crick base pair. These non-Watson–Crick base pairs can locally distort the DNA double helix. Furthermore, such mismatches can be *mutagenic;* that is, they can result in permanent changes in the DNA sequence. When a double helix containing a non-Watson–Crick base pair is replicated, the two daughter double helices will have different sequences because the mismatched base is very likely to pair with its Watson–Crick partner. Errors other than mismatches include insertions, deletions, and breaks in one or both strands. Furthermore, replicative polymerases can stall or even fall off a damaged template entirely. As a consequence, replication of the genome may halt before it is complete.

A variety of mechanisms have evolved to deal with such interruptions, including specialized DNA polymerases that can replicate DNA across many lesions. A drawback is that such polymerases are substantially more error prone than are normal replicative polymerases. Nonetheless, these *error-prone polymerases* allow the completion of a draft sequence of the genome that can be at least partly repaired by DNA-repair processes. DNA recombination (Section 28.6) provides an additional mechanism for salvaging interruptions in DNA replication.

Some Genetic Diseases Are Caused by the Expansion of Repeats of Three Nucleotides

Some genetic diseases are caused by the presence of DNA sequences that are inherently prone to errors in the course of replication. A particularly important class of such diseases is characterized by the presence of long tandem arrays of repeats of three nucleotides. An example is *Huntington disease,* an autosomal dominant neurological disorder with a variable age of onset. The mutated gene in this disease expresses a protein in the brain called *huntingtin,* which contains a stretch of consecutive glutamine residues. These glutamine residues are encoded by a tandem array of CAG sequences within the gene. In unaffected persons, this array is between 6 and 31 repeats, whereas, in those with the disease, the array is between 36 and 82 repeats or longer. Moreover, the array tends to become longer from one generation to the next. The consequence is a phenomenon called *anticipation:* the children of an affected parent tend to show symptoms of the disease at an earlier age than did the parent.

The tendency of these *trinucleotide repeats* to expand is explained by the formation of alternative structures in DNA replication (Figure 28.35). Part of the array within a template strand can loop out without disrupting base-pairing outside this region. In replication, DNA polymerase extends this strand through the remainder of the array by a poorly understood mechanism, leading to an increase in the number of copies of the trinucleotide sequence.

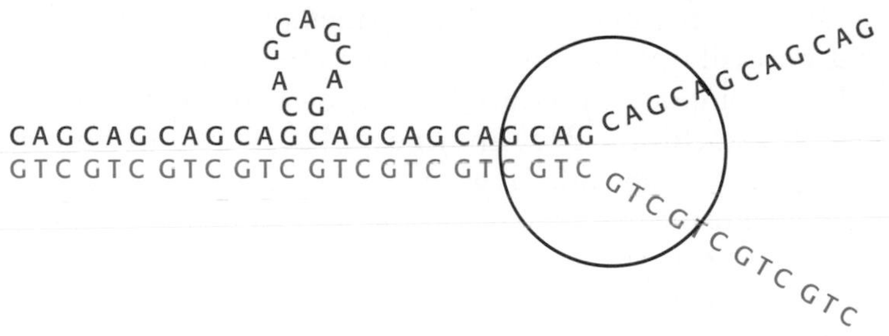

Figure 28.35 Triplet repeat expansion. Sequences containing tandem arrays of repeated triplet sequences can be expanded to include more repeats by the looping out of some of the repeats before replication. The double helix formed from the red template strand will contain additional sequences encompassing the looped-out region.

A number of other neurological diseases are characterized by expanding arrays of trinucleotide repeats. How do these long stretches of repeated amino acids cause disease? For huntingtin, it appears that the polyglutamine stretches become increasingly prone to aggregate as their length increases; the additional consequences of such aggregation are still under investigation.

Bases Can Be Damaged by Oxidizing Agents, Alkylating Agents, and Light

A variety of chemical agents can alter specific bases within DNA after replication is complete. Such *mutagens* include reactive oxygen species such as hydroxyl radical. For example, hydroxyl radical reacts with guanine to form 8-oxoguanine. 8-Oxoguanine is mutagenetic because it often pairs with adenine rather than cytosine in DNA replication. Its choice of pairing partner differs from that of guanine because it uses a different edge of the base

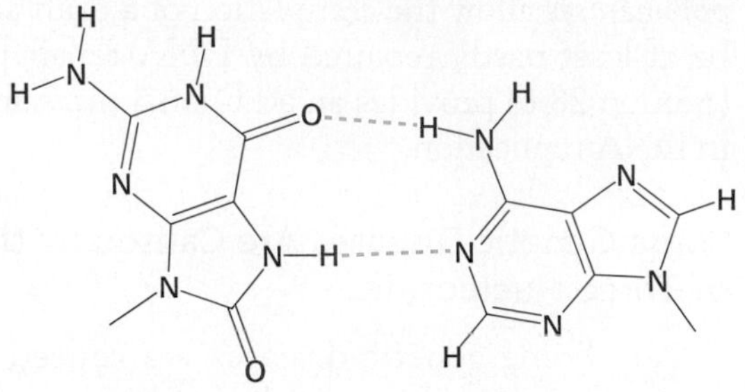

Figure 28.36 Oxoguanine–adenine base pair. When guanine is oxidized to 8-oxoguanine, the damaged base can form a base pair with adenine through an edge of the base that does not normally participate in base-pair formation.

8-Oxoguanine   **Adenine**

to form base pairs (Figure 28.36). Deamination is another potentially deleterious process. For example, adenine can be deaminated to form hypoxanthine (Figure 28.37). This process is mutagenic because hypoxanthine pairs with cytosine rather than thymine. Guanine and cytosine also can be deaminated to yield bases that pair differently from the parent base.

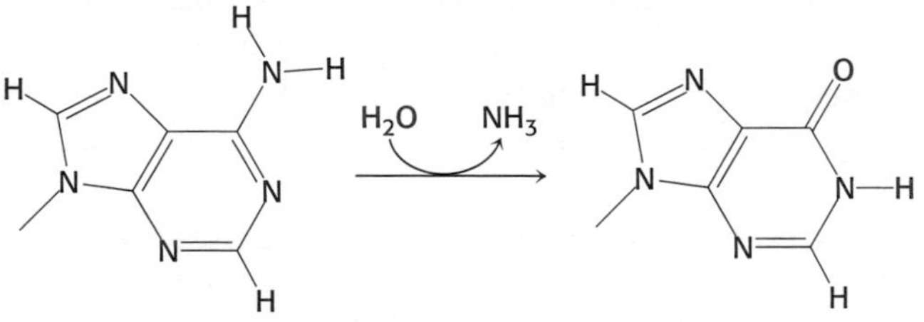

Figure 28.37 Adenine deamination. The base adenine can be deaminated to form hypoxanthine. Hypoxanthine forms base pairs with cytosine in a manner similar to that of guanine, and so the deamination reaction can result in mutation.

Adenine   **Hypoxanthine**

In addition to oxidation and deamination, nucleotide bases are subject to alkylation. Electrophilic centers can be attacked by nucleophiles such as N-7 of guanine and adenine to form alkylated adducts. Some compounds are converted into highly active electrophiles through the action of enzymes that normally play a role in detoxification. A striking example is aflatoxin B_1, a compound produced by molds that grow on peanuts and other foods. A cytochrome P450 enzyme (p. 750) converts this compound into a highly reactive epoxide (Figure 28.38). This agent reacts with the N-7 atom of guanosine to form a mutagenic adduct that frequently leads to a G–C-to-T–A transversion.

The ultraviolet component of sunlight is a ubiquitous DNA-damaging agent. Its major effect is to covalently link adjacent pyrimidine residues along a DNA strand (Figure 28.39). Such a pyrimidine dimer cannot fit into a double helix, and so replication and gene expression are blocked until the lesion is removed.

A thymine dimer is an example of an *intra*strand cross-link because both participating bases are in the same strand of the double helix. Cross-links

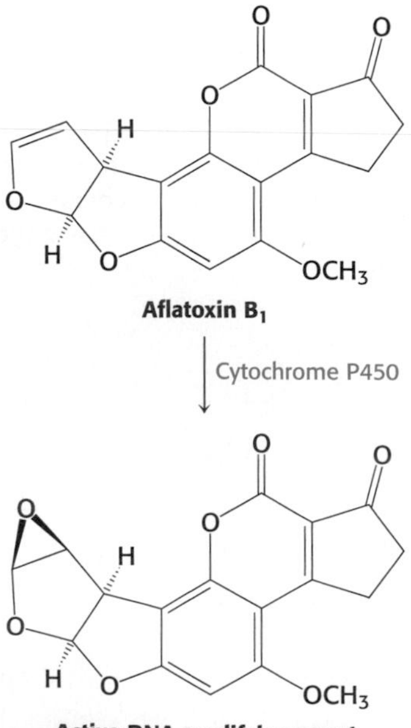

Aflatoxin B₁

Cytochrome P450

Active DNA-modifying agent

Figure 28.38 Aflatoxin activation. The compound, produced by molds that grow on peanuts, is activated by cytochrome P450 to form a highly reactive species that modifies bases such as guanine in DNA, leading to mutations.

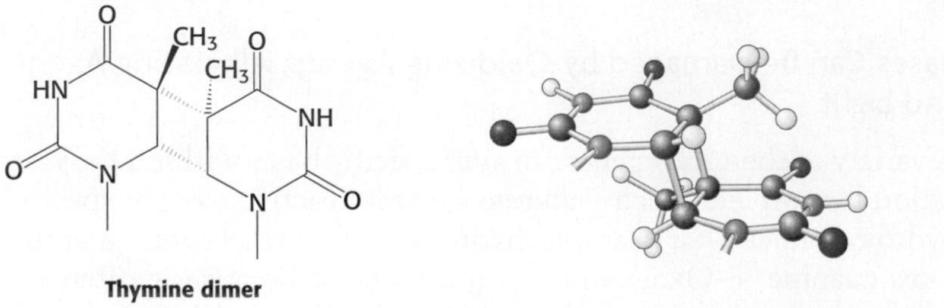

Thymine dimer

Figure 28.39 Cross-linked dimer of two thymine bases. Ultraviolet light induces cross-links between adjacent pyrimidines along one strand of DNA.

between bases on opposite strands also can be introduced by various agents. Psoralens are compounds produced by a Chinese herb that form such *interstrand* cross-links (Figure 28.40). Interstrand cross-links disrupt replication because they prevent strand separation.

High-energy electromagnetic radiation such as x-rays can damage DNA by producing high concentrations of reactive species in solution. X-ray exposure can induce several types of DNA damage including single- and double-stranded breaks in DNA. This ability to induce such DNA damage led Hermann Muller to discover the mutagenic effects of x-rays in *Drosophila* in 1927. This discovery contributed to the development of *Drosophila* as one of the premier organisms for genetic studies.

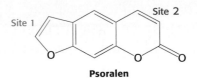

Figure 28.40 A cross-linking agent. The compound psoralen and its derivatives can form interstrand cross-links through two reactive sites that can form adducts with nucleotide bases.

DNA Damage Can Be Detected and Repaired by a Variety of Systems

To protect the genetic message, a wide range of DNA-repair systems are present in most organisms. Many systems repair DNA by using sequence information from the uncompromised strand. Such single-strand replication systems follow a similar mechanistic outline:

1. Recognize the offending base(s).

2. Remove the offending base(s).

3. Repair the resulting gap with a DNA polymerase and DNA ligase.

We will briefly consider examples of several repair pathways. Although many of these examples are taken from *E. coli*, corresponding repair systems are present in most other organisms including human beings.

The replicative DNA polymerases themselves are able to correct many DNA mismatches produced in the course of replication. For example, the ε subunit of *E. coli* DNA polymerase III functions as a 3′-to-5′ exonuclease. This domain removes mismatched nucleotides from the 3′ end of DNA by hydrolysis. How does the enzyme sense whether a newly added base is correct? As a new strand of DNA is synthesized, it is *proofread*. If an incorrect base is inserted, then DNA synthesis slows down owing to the difficulty of threading a non-Watson–Crick base pair into the polymerase. In addition, the mismatched base is weakly bound and therefore able to fluctuate in position. The delay from the slowdown allows time for these fluctuations to take the newly synthesized strand out of the polymerase active site and into the exonuclease active site (Figure 28.41). There, the DNA is degraded, one nucleotide at a time, until it moves back into the polymerase active site and synthesis continues.

Figure 28.41 Proofreading. The growing polynucleotide chain occasionally leaves the polymerase site and migrates to the active site of exonuclease. There, one or more nucleotides are excised from the newly synthesized chain, removing potentially incorrect bases.

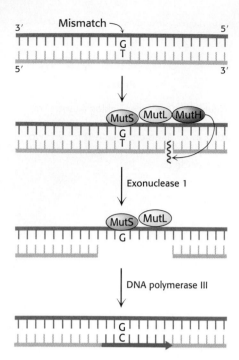

Figure 28.42 **Mismatch repair.** DNA mismatch repair in *E. coli* is initiated by the interplay of MutS, MutL, and MutH proteins. A G–T mismatch is recognized by MutS. MutH cleaves the backbone in the vicinity of the mismatch. A segment of the DNA strand containing the erroneous T is removed by exonuclease I and synthesized anew by DNA polymerase III. [After R. F. Service. *Science* 263(1994):1559–1560.]

A second mechanism is present in essentially all cells to correct errors made in replication that are not corrected by proofreading (Figure 28.42). *Mismatch-repair* systems consist of at least two proteins, one for detecting the mismatch and the other for recruiting an endonuclease that cleaves the newly synthesized DNA strand close to the lesion to facilitate repair. In *E. coli,* these proteins are called MutS and MutL and the endonuclease is called MutH.

An example of *direct repair* is the photochemical cleavage of pyrimidine dimers. Nearly all cells contain a *photoreactivating enzyme* called *DNA photolyase.* The *E. coli* enzyme, a 35-kd protein that contains bound N^5, N^{10}-methenyltetrahydrofolate and flavin adenine dinucleotide (FAD) cofactors, binds to the distorted region of DNA. The enzyme uses light energy—specifically, the absorption of a photon by the N^5,N^{10}-methenyltetrahydrofolate coenzyme—to form an excited state that cleaves the dimer into its component bases.

The excision of modified bases such as 3-methyladenine by the *E. coli* enzyme AlkA is an example of *base-excision repair.* The binding of this enzyme to damaged DNA flips the affected base out of the DNA double helix and into the active site of the enzyme (Figure 28.43). The enzyme then acts as *a glycosylase,* cleaving the glycosidic bond to release the damaged base. At this stage, the DNA backbone is intact, but a base is missing. This hole is called an *AP site* because it is apurinic (devoid of A or G) or apyrimidinic (devoid of C or T). An *AP endonuclease* recognizes this defect and nicks the backbone adjacent to the missing base. *Deoxyribose phosphodiesterase* excises the residual deoxyribose phosphate unit, and DNA polymerase I inserts an undamaged nucleotide, as dictated by the base on the undamaged complementary strand. Finally, the repaired strand is sealed by DNA ligase.

One of the best-understood examples of *nucleotide-excision repair* is utilized for the excision of a pyrimidine dimer. Three enzymatic activities are essential for this repair process in *E. coli* (Figure 28.44). First, an enzyme complex consisting of the proteins encoded by the *uvrABC* genes detects the distortion produced by the DNA damage. The uvrABC enzyme then cuts the damaged DNA strand at two sites, 8 nucleotides away from the damaged site on the 5′ side and 4 nucleotides away on the 3′ side. The 12-residue oligonucleotide excised by this highly specific

Figure 28.43 **Structure of DNA-repair enzyme.** A complex between the DNA-repair enzyme AlkA and an analog of a DNA molecule missing a purine base (an apurinic site) is shown. *Notice* that the backbone sugar in the apurinic site is flipped out of the double helix into the active site of the enzyme. [Drawn from 1BNK.pdb.]

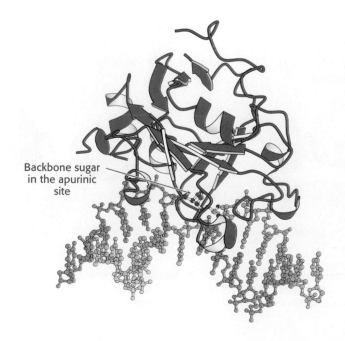

Backbone sugar in the apurinic site

Figure 28.44 **Nucleotide-excision repair.** Repair of a region of DNA containing a thymine dimer by the sequential action of a specific excinuclease, a DNA polymerase, and a DNA ligase. The thymine dimer is shown in blue and the new region of DNA is in red. [After P. C. Hanawalt. *Endevour* 31(1982):83.]

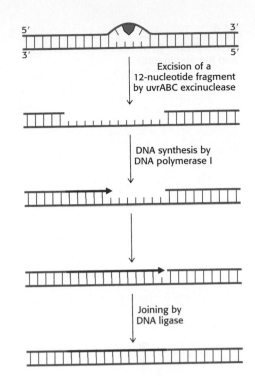

excinuclease (from the Latin *exci*, "to cut out") then diffuses away. DNA polymerase I enters the gap to carry out repair synthesis. The 3′ end of the nicked strand is the primer, and the intact complementary strand is the template. Finally, the 3′ end of the newly synthesized stretch of DNA and the original part of the DNA chain are joined by DNA ligase.

DNA ligase is able to seal simple breaks in one strand of the DNA backbone. However, alternative mechanisms are required to repair breaks on both strands that are close enough together to separate the DNA into two double helices. Several distinct mechanisms are able to repair such damage. One mechanism, *nonhomologous end joining* (NHEJ), does not depend on other DNA molecules in the cell. In NHEJ, the free double-stranded ends are bound by a heterodimer of two proteins, Ku70 and Ku80. These proteins stabilize the ends and mark them for subsequent manipulations. Through mechanisms that are not yet well understood, the Ku70/80 heterodimers act as handles used by other proteins to draw the two double-stranded ends close together so that enzymes can seal the break.

Alternative mechanisms of double-stranded-break repair can operate if an intact stretch of double-stranded DNA with an identical or very similar sequence is present in the cell. These repair processes use homologous recombination, presented in Section 28.6.

The Presence of Thymine Instead of Uracil in DNA Permits the Repair of Deaminated Cytosine

The presence in DNA of thymine rather than uracil was an enigma for many years. Both bases pair with adenine. The only difference between them is a methyl group in thymine in place of the C-5 hydrogen atom in uracil. Why is a methylated base employed in DNA and not in RNA? The existence of an active repair system to correct the deamination of cytosine provides a convincing solution to this puzzle.

Cytosine in DNA spontaneously deaminates at a perceptible rate to form uracil. The deamination of cytosine is potentially mutagenic because uracil pairs with adenine, and so one of the daughter strands will contain a U–A base pair rather than the original C–G base pair. This mutation is prevented by a repair system that recognizes uracil to be foreign to DNA (Figure 28.45). The repair enzyme, *uracil DNA glycosylase*, is homologous to AlkA. The enzyme hydrolyzes the glycosidic bond between the uracil and deoxyribose moieties but does not attack thymine-containing nucleotides. The AP site generated is repaired to reinsert cytosine. Thus, *the methyl group on thymine is a tag that distinguishes thymine from deaminated cytosine.* If thymine were not used in DNA, uracil correctly in place would be indistinguishable from uracil formed by deamination. The defect would persist unnoticed, and so a C–G base pair would necessarily be mutated to U–A in one of the daughter DNA molecules. This mutation is prevented by a repair system that searches for uracil and leaves thymine alone. *Thymine is used instead of uracil in DNA to enhance the fidelity of the genetic message.*

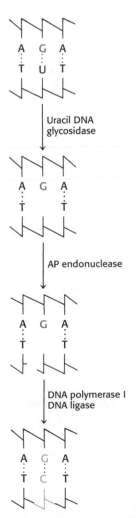

Figure 28.45 **Uracil repair.** Uridine bases in DNA, formed by the deamination of cytidine, are excised and replaced by cytidine.

Many Cancers Are Caused by the Defective Repair of DNA

As described in Chapter 14, cancers are caused by mutations in genes associated with growth control. Defects in DNA-repair systems increase the overall frequency of mutations and, hence, the likelihood of cancer-causing mutations. Indeed, the synergy between studies of mutations that predispose people to cancer and studies of DNA repair in model organisms has been tremendous in revealing the biochemistry of DNA-repair pathways. Genes for DNA-repair proteins are often *tumor-suppressor genes;* that is, they suppress tumor development when at least one copy of the gene is free of a deleterious mutation. When both copies of a gene are mutated, however, tumors develop at rates greater than those for the population at large. People who inherit defects in a single tumor-suppressor allele do not necessarily develop cancer but are susceptible to developing the disease because only the one remaining normal copy of the gene must develop a new defect to further the development of cancer.

Consider, for example, *xeroderma pigmentosum,* a rare human skin disease. The skin in an affected person is extremely sensitive to sunlight or ultraviolet light. In infancy, severe changes in the skin become evident and worsen with time. The skin becomes dry, and there is a marked atrophy of the dermis. Keratoses appear, the eyelids become scarred, and the cornea ulcerates. Skin cancer usually develops at several sites. Many patients die before age 30 from metastases of these malignant skin tumors. Studies of xeroderma pigmentosum patients have revealed that mutations occur in genes for a number of different proteins. These proteins are components of the human nucleotide-excision-repair pathway, including homologs of the UvrABC subunits.

Defects in other repair systems can increase the frequency of other tumors. For example, *hereditary nonpolyposis colorectal cancer* (HNPCC, or *Lynch syndrome*) results from defective DNA mismatch repair. HNPCC is not rare—as many as 1 in 200 people will develop this form of cancer. Mutations in two genes, called *hMSH2* and *hMLH1,* account for most cases of this hereditary predisposition to cancer. The striking finding is that these genes encode the human counterparts of MutS and MutL of *E. coli.* Mutations in *hMSH2* and *hMLH1* seem likely to allow mutations to accumulate throughout the genome. In time, genes important in controlling cell proliferation become altered, resulting in the onset of cancer.

Not all tumor-suppressor genes are specific to particular types of cancer. *The gene for a protein called p53 is mutated in more than half of all tumors.* The p53 protein helps control the fate of damaged cells. First, it plays a central role in sensing DNA damage, especially double-stranded breaks. Then, after sensing damage, the protein either promotes a DNA-repair pathway or activates the apoptosis pathway, leading to cell death. Most mutations in the p53 gene are sporatic; that is, they occur in somatic cells rather than being inherited. People who inherit a deleterious mutation in one copy of the p53 gene suffer from *Li-Fraumeni syndrome* and have a high probability of developing several types of cancer.

Cancer cells often have two characteristics that make them especially vulnerable to agents that damage DNA molecules. First, they divide frequently, and so their DNA replication pathways are more active than they are in most cells. Second, as already noted, cancer cells often have defects in DNA-repair pathways. Several agents widely used in cancer chemotherapy, including cyclophosphamide and cisplatin, act by damaging DNA.

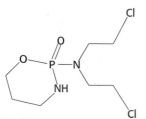

Cyclophosphamide

Cisplatin

Many Potential Carcinogens Can Be Detected by Their Mutagenic Action on Bacteria

Many human cancers are caused by exposure to chemicals that cause mutations. It is important to identify such compounds that can cause mutations and ascertain their potency so that human exposure to them can be minimized. Bruce Ames devised a simple and sensitive test for detecting chemical mutagens. In the *Ames test,* a thin layer of agar containing about 10^9 bacteria of a specially constructed tester strain of *Salmonella* is placed on a petri plate. These bacteria are unable to grow in the absence of histidine, because a mutation is present in one of the genes for the biosynthesis of this amino acid. The addition of a chemical mutagen to the center of the plate results in many new mutations. A small proportion of them reverse the original mutation, and histidine can be synthesized. These *revertants* multiply in the absence of an external source of histidine and appear as discrete colonies after the plate has been incubated at 37°C for 2 days (Figure 28.46). For example, 0.5 μg of 2-aminoanthracene gives 11,000 revertant colonies, compared with only 30 spontaneous revertants in its absence. A series of concentrations of a chemical can be readily tested to generate a dose–response curve. These curves are usually linear, which suggests that there is no threshold concentration for mutagenesis.

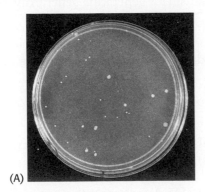 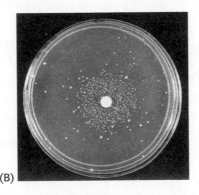

(A)　　　　(B)

Figure 28.46 Ames test. (A) A petri plate containing about 10^9 *Salmonella* bacteria that cannot synthesize histidine and (B) a petri plate containing a filter-paper disc with a mutagen, which produces a large number of revertants that can synthesize histidine. After 2 days, the revertants appear as rings of colonies around the disc. The small number of visible colonies in plate A are spontaneous revertants. [From B. N. Ames, J. McCann, and E. Yamasake. *Mutat. Res.* 31(1975):347–364.]

Some of the tester strains are responsive to *base-pair substitutions,* whereas others detect *deletions or additions of base pairs (frameshifts).* The sensitivity of these specially designed strains has been enhanced by the genetic deletion of their excision-repair systems. Potential mutagens enter the tester strains easily because the lipopolysaccharide barrier that normally coats the surface of *Salmonella* is incomplete in these strains. A key feature of this detection system is the inclusion of a *mammalian liver homogenate.* Recall that some potential carcinogens such as aflatoxin are converted into their active forms by enzyme systems in the liver or other mammalian tissues. Bacteria lack these enzymes, and so the test plate requires a few milligrams of a liver homogenate to activate this group of mutagens.

The *Salmonella* test is extensively used to help evaluate the mutagenic and carcinogenic risks of a large number of chemicals. This rapid and inexpensive bacterial assay for mutagenicity complements epidemiological surveys and animal tests that are necessarily slower, more laborious, and far more expensive. The *Salmonella* test for mutagenicity is an outgrowth of studies of gene–protein relations in bacteria. It is a striking example of how fundamental research in molecular biology can lead directly to important advances in public health.

28.6   DNA Recombination Plays Important Roles in Replication, Repair, and Other Processes

Most processes associated with DNA replication function to copy the genetic message as faithfully as possible. However, several biochemical processes require the *recombination* of genetic material between two DNA molecules. In genetic recombination, two daughter molecules are formed by the exchange of genetic material between two parent molecules (Figure 28.47). Recombination is essential in the following processes.

1.   When replication stalls, recombination processes can reset the replication machinery so that replication can continue.

2.   Some double-stranded breaks in DNA are repaired by recombination.

3.   In meiosis, the limited exchange of genetic material between paired chromosomes provides a simple mechanism for generating genetic diversity in a population.

4.   As we shall see in Chapter 33, recombination plays a crucial role in generating molecular diversity for antibodies and some other molecules in the immune system.

5.   Some viruses employ recombination pathways to integrate their genetic material into the DNA of a host cell.

6.   Recombination is used to manipulate genes in, for example, the generation of "gene knockout" mice (p. 155).

Recombination is most efficient between DNA sequences that are similar in sequence. In homologous recombination, parent DNA duplexes align at regions of sequence similarity, and new DNA molecules are formed by the breakage and joining of homologous segments.

Recombination

Figure 28.47 Recombination. Two DNA molecules can recombine with each other to form new DNA molecules that have segments from both parent molecules.

RecA Can Initiate Recombination by Promoting Strand Invasion

In many recombination pathways, a DNA molecule with a free end recombines with a DNA molecule having no free ends available for interaction. DNA molecules with free ends are the common result of double-stranded DNA breaks, but they may also be generated in DNA replication if the replication complex stalls. This type of recombination has been studied extensively in *E. coli*, but it also takes place in other organisms through the action of proteins homologous to those of *E. coli*. Often dozens of proteins participate in the complete recombination process. However, the key protein is *RecA*, another member of the AAA

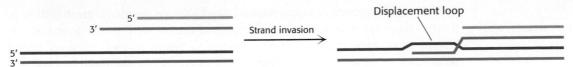

Figure 28.48 Strand invasion. This process, promoted by proteins such as RecA, can initiate recombination.

ATPase family. To accomplish the exchange, the single-stranded DNA displaces one of the strands of the double helix (Figure 28.48). The resulting three-stranded structure is called a *displacement loop* or *D-loop*. This process is often referred to as *strand invasion*. Because a free 3′ end is now bound to a contiguous strand of DNA, the 3′ end can act as a primer to initiate new DNA synthesis. Strand invasion can initiate many processes, including the repair of double-stranded breaks and the reinitiation of replication after the replication apparatus has come off its template. In the repair of a break, the recombination partner is an intact DNA molecule with an overlapping sequence.

Some Recombination Reactions Proceed Through Holliday-Junction Intermediates

In recombination pathways for meiosis and some other processes, intermediates form that are composed of four polynucleotide chains in a crosslike structure. Intermediates with these crosslike structures are often referred to as *Holliday junctions*, after Robin Holliday, who proposed their role in recombination in 1964. Such intermediates have been characterized by a wide range of techniques including x-ray crystallography.

Specific enzymes, termed *recombinases*, bind to these structures and resolve them into separated DNA duplexes. The Cre recombinase from bacteriophage P1 has been extensively studied. The mechanism begins with the recombinase binding to the DNA substrates (Figure 28.49).

Figure 28.49 Recombination mechanism. Recombination begins as two DNA molecules come together to form a recombination synapse. One strand from each duplex is cleaved by the recombinase enzyme; the 3′ end of each of the cleaved strands is linked to a tyrosine (Y) residue on the recombinase enzyme. New phosphodiester bonds are formed when a 5′ end of the other cleaved strand in the complex attacks these tyrosine–DNA adducts. After isomerization, these steps are repeated to form the recombined products.

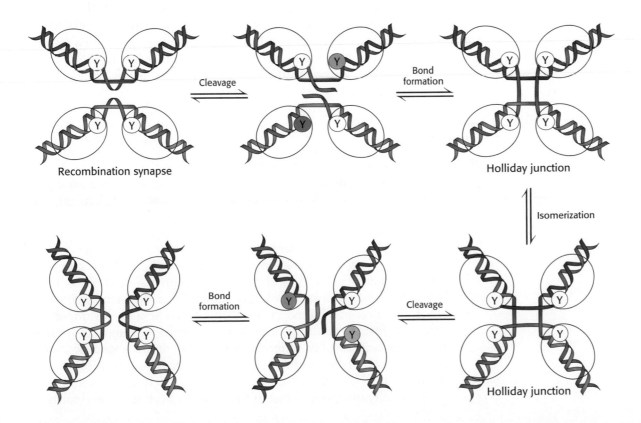

Recombination synapse

Holliday junction

Isomerization

Holliday junction

Cleavage

Bond formation

Bond formation

Cleavage

Four molecules of the enzyme and two DNA molecules come together to form a *recombination synapse*. The reaction begins with the cleavage of one strand from each duplex. The 5′-hydroxyl group of each cleaved strand remains free, whereas the 3′-phosphoryl group becomes linked to a specific tyrosine residue in the recombinase. The free 5′ ends invade the other duplex in the synapse and attack the DNA–tyrosine units to form new phosphodiester bonds and free the tyrosine residues. These reactions result in the formation of a Holliday junction. This junction can then isomerize to form a structure in which the polynucleotide chains in the center of the structure are reoriented. From this junction, the processes of strand cleavage and phosphodiester-bond formation repeat. The result is a synapse containing the two recombined duplexes. Dissociation of this complex generates the final recombined products.

Cre catalyzes the formation of Holliday junctions as well as their resolution. In contrast, other proteins bind to Holliday junctions that have already been formed by other processes and resolve them into separate duplexes. In many cases, these proteins also promote the process of branch migration whereby a Holliday junction is moved along the two component double helices. Branch migration can affect which segments of DNA are exchanged in a recombination process.

Some Recombinases Are Evolutionarily Related to Topoisomerases

The intermediates that form in the Cre recombination reaction, with their tyrosine adducts possessing 3′-phosphoryl groups, are reminiscent of the intermediates that form in the reactions catalyzed by topoisomerases. This mechanistic similarity uncovers deeper evolutionary relationships. Examination of the three-dimensional structures of Cre-like recombinases and type I topoisomerases reveals that these proteins are related by divergent evolution despite little amino acid sequence similarity (Figure 28.50). From this perspective, the action of a recombinase can be viewed as an intermolecular topoisomerase reaction. In each case, a tyrosine–DNA adduct is formed. In a topoisomerase reaction, this adduct is resolved when the 5′-hydroxyl group of the same duplex attacks to reform the same phosphodiester bond that was initially cleaved. In a recombinase reaction, the attacking 5′-hydroxyl group comes from a DNA chain that was not initially linked to the phosphoryl group participating in the phosphodiester bond.

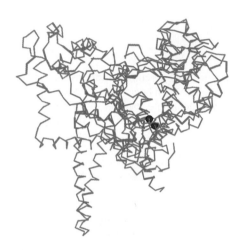

Figure 28.50 Recombinases and topoisomerase I. A superposition of Cre recombinase (blue) and topoisomerase I (orange) reveals that these two enzymes have a common structural core. The positions of the tyrosine residues that participate in DNA cleavage reactions are shown as red spheres for both enzymes. [Drawn from 2CRX.pdb and 1A31.pdb.]

Summary

28.1 DNA Can Assume a Variety of Structural Forms

DNA is a structurally dynamic molecule that can exist in a variety of helical forms: A-DNA, B-DNA (the classic Watson–Crick helix), and Z-DNA. DNA can be bent, kinked, and unwound. In A-, B-, and Z-DNA, two antiparallel chains are held together by Watson–Crick base pairs and stacking interactions between bases in the same strand. The sugar–phosphate backbone is on the outside, and the bases are inside the double helix. A- and B-DNA are right-handed helices. In B-DNA, the base pairs are nearly perpendicular to the helix axis. In A-DNA, the bases are tilted rather than perpendicular. An important structural feature of the B helix is the presence of major and minor grooves, which display different potential hydrogen-bond acceptors and donors according to the base sequence. X-ray analysis of a single crystal of B-DNA reveals that the structure is much more variable than was originally imagined. Most of the DNA in a cell is in the B form.

28.2 Double-Stranded DNA Can Wrap Around Itself to Form Supercoiled Structures

A key topological property of DNA is its linking number *(Lk)*, which is defined as the number of times one strand of DNA winds around the other in the right-hand direction when the DNA axis is constrained to lie in a plane. Molecules differing in linking number are topoisomers of one another and can be interconverted only by cutting one or both DNA strands; these reactions are catalyzed by topoisomerases. Changes in linking number generally lead to changes in both the number of turns of double helix and the number of turns of superhelix. Topoisomerase II catalyzes the ATP-driven introduction of negative supercoils, which leads to the compaction of DNA and renders it more susceptible to unwinding. Supercoiled DNA can be relaxed by topoisomerase I or topoisomerase II. Topoisomerase I acts by transiently cleaving one strand of DNA in a double helix, whereas topoisomerase II transiently cleaves both strands simultaneously.

28.3 DNA Replication Proceeds by the Polymerization of Deoxyribonucleoside Triphosphates Along a Template

DNA polymerases are template-directed enzymes that catalyze the formation of phosphodiester bonds by nucleophilic attack by the 3'-hydroxyl group on the innermost phosphorus atom of a deoxyribonucleoside 5'-triphosphate. The complementarity of shape between correctly matched nucleotide bases is crucial to ensuring the fidelity of base incorporation. DNA polymerases cannot start chains de novo; a primer with a free 3'-hydroxyl group is required. Thus DNA synthesis is initiated by the synthesis of an RNA primer, the task of a specialized primase enzyme. After serving as a primer, the RNA is degraded and replaced by DNA. DNA polymerases always synthesize a DNA strand in the 5'-to-3' direction. So that both strands of the double helix may be synthesized in the same direction simultaneously, one strand is synthesized continuously while the other is synthesized in fragments called Okazaki fragments. Gaps between the fragments are sealed by DNA ligases. ATP-driven helicases prepare the way for DNA replication by separating the strands of the double helix.

28.4 DNA Replication Is Highly Coordinated

Replicative DNA polymerases are processive; that is, they catalyze the addition of many nucleotides without dissociating from the template. A major contributor to processivity is the DNA sliding clamp, such as the dimeric β subunit of the *E. coli* replicative polymerase. The sliding clamp has a ring structure that encircles the DNA double helix and keeps the enzyme and DNA associated. The DNA polymerase holoenzyme is a large DNA-copying machine formed by two DNA polymerase enzymes, one to act on each template strand, associated with other subunits including a sliding clamp and clamp loader.

The synthesis of the leading and lagging strands of a double-stranded DNA template is coordinated. As a replicative polymerase moves along a DNA template, the leading strand is copied smoothly while the lagging strand forms loops that change length in the course of the synthesis of each Okazaki fragment. The mode of action is referred to as the trombone model.

DNA replication is initiated at a single site within the *E. coli* genome. A set of specific proteins recognize this origin of replication and assemble the enzymes needed for DNA synthesis, including a helicase that promotes strand separation. The initiation of replication in eukaryotes is more complex. DNA synthesis is initiated at thousands of sites throughout the genome. Assemblies homologous to those in *E. coli*, but more complicated, are assembled at each eukaryotic origin of replication. A special polymerase called telomerase that relies on an RNA template synthesizes specialized structures called telomeres at the ends of linear chromosomes.

28.5 Many Types of DNA Damage Can Be Repaired

A wide variety of DNA damage can occur. For example, mismatched bases may be incorporated in the course of DNA replication or individual bases may be damaged by oxidation or alkylation after DNA replication. Other forms of damage are the formation of cross-links and the introduction of single- or double-stranded breaks in the DNA backbone. Several different repair systems detect and repair DNA damage. Repair begins with the process of proofreading in DNA replication: mismatched bases that were incorporated in the course of synthesis are excised by exonuclease activity present in replicative polymerases. Some DNA lesions such as thymine dimers can be directly reversed through the action of specific enzymes. Other DNA-repair pathways act through the excision of single damaged bases (base-excision repair) or short segments of nucleotides (nucleotide-excision repair). Double-stranded breaks in DNA can be repaired by homologous or nonhomologous end-joining processes. Defects in DNA-repair components are associated with susceptibility to many different sorts of cancer. Such defects are a common target of cancer treatments. Many potential carcinogens can be detected by their mutagenic action on bacteria (the Ames test).

28.6 DNA Recombination Plays Important Roles in Replication, Repair, and Other Processes

Recombination is the exchange of segments between two DNA molecules. Recombination is important in some types of DNA repair as well as other processes such as meiosis, the generation of antibody diversity, and the life cycles of some viruses. Some recombination pathways are initiated by strand invasion, in which a single strand at the end of a DNA double helix forms base pairs with one strand of DNA in another double helix and displaces the other strand. A common intermediate formed in other recombination pathways is the Holliday junction, which consists of four strands of DNA that come together to form a crosslike structure. Recombinases promote recombination reactions through the introduction of specific DNA breaks and the formation and resolution of Holliday-junction intermediates.

Key Terms

B-DNA helix (p. 784)
A-DNA helix (p. 784)
major groove (p. 785)
minor groove (p. 785)
Z-DNA helix (p. 787)
supercoil (p. 789)
linking number (p. 789)
topoisomer (p. 789)
twist (p. 789)
writhe (p. 789)

topoisomerase (p. 790)
DNA polymerase (p. 793)
template (p. 793)
primer (p. 793)
exonuclease (p. 794)
primase (p. 795)
replication fork (p. 796)
Okazaki fragment (p. 796)
lagging strand (p. 796)
leading stand (p. 796)

DNA ligase (p. 796)
helicase (p. 797)
processivity (p. 798)
sliding clamp (p. 799)
trombone model (p. 800)
origin of replication (p. 801)
origin of replication complex (ORC) (p. 802)
cell cycle (p. 803)
telomere (p. 803)

Selected Readings

Where to Start

Johnson, A., and O'Donnell, M. 2005. Cellular DNA replicases: Components and dynamics at the replication fork. *Annu. Rev. Biochem.* 74:283–315.

Kornberg, A. 1988. DNA replication. *J. Biol. Chem.* 263:1–4.

Dickerson, R. E. 1983. The DNA helix and how it is read. *Sci. Am.* 249(6):94–111.

Wang, J. C. 1982. DNA topoisomerases. *Sci. Am.* 247(1):94–109.

Lindahl, T. 1993. Instability and decay of the primary structure of DNA. *Nature* 362:709–715.

Greider, C. W., and Blackburn, E. H. 1996. Telomeres, telomerase, and cancer. *Sci. Am.* 274(2):92–97.

Books

Kornberg, A., and Baker, T. A. 1992. *DNA Replication* (2d ed.). W. H. Freeman and Company.

Bloomfield, V. A., Crothers, D., Tinoco, I., and Hearst, J. 2000. *Nucleic Acids: Structures, Properties and Functions.* University Science Books.

Friedberg, E. C., Walker, G. C., and Siede, W. 1995. *DNA Repair and Mutagenesis.* American Society for Microbiology.

Cozzarelli, N. R., and Wang, J. C. (Eds.). 1990. *DNA Topology and Its Biological Effects.* Cold Spring Harbor Laboratory Press.

DNA Structure

Chiu, T. K., and Dickerson, R. E. 2000. 1 Å crystal structures of B-DNA reveal sequence-specific binding and groove-specific bending of DNA by magnesium and calcium. *J. Mol. Biol.* 301:915–945.

Herbert, A., and Rich, A. 1999. Left-handed Z-DNA: Structure and function. *Genetica* 106:37–47.

Dickerson, R. E. 1992. DNA structure from A to Z. *Methods Enzymol.* 211:67–111.

Quintana, J. R., Grzeskowiak, K., Yanagi, K., and Dickerson, R. E. 1992. Structure of a B-DNA decamer with a central T-A step: C-G-A-T-T-A-A-T-C-G. *J. Mol. Biol.* 225:379–395.

Verdaguer, N., Aymami, J., Fernandez, F. D., Fita, I., Coll, M., Huynh, D. T., Igolen, J., and Subirana, J. A. 1991. Molecular structure of a complete turn of A-DNA. *J. Mol. Biol.* 221:623–635.

DNA Topology and Topoisomerases

Charvin, G., Strick, T.R., Bensimon, D., and Croquette, V. 2005. Tracking topoisomerase activity at the single-molecule level. *Annu. Rev. Biophys. Biomol. Struct.* 34:201–219.

Sikder, D., Unniraman, S., Bhaduri, T., and Nagaraja, V. 2001. Functional cooperation between topoisomerase I and single strand DNA-binding protein. *J. Mol. Biol.* 306:669–679.

Yang, Z., and Champoux, J. J. 2001. The role of histidine 632 in catalysis by human topoisomerase I. *J. Biol. Chem.* 276:677–685.

Fortune, J. M., and Osheroff, N. 2000. Topoisomerase II as a target for anticancer drugs: When enzymes stop being nice. *Prog. Nucleic Acid Res. Mol. Biol.* 64:221–253.

Isaacs, R. J., Davies, S. L., Sandri, M. I., Redwood, C., Wells, N. J., and Hickson, I. D. 1998. Physiological regulation of eukaryotic topoisomerase II. *Biochim. Biophys. Acta* 1400:121–137.

Wang, J. C. 1996. DNA topoisomerases. *Annu. Rev. Biochem.* 65:635–692.

Wang, J. C. 1998. Moving one DNA double helix through another by a type II DNA topoisomerase: The story of a simple molecular machine. *Q. Rev. Biophys.* 31:107–144.

Baird, C. L., Harkins, T. T., Morris, S. K., and Lindsley, J. E. 1999. Topoisomerase II drives DNA transport by hydrolyzing one ATP. *Proc. Natl. Acad. Sci. U. S. A.* 96:13685–13690.

Vologodskii, A. V., Levene, S. D., Klenin, K. V., Frank, K. M., and Cozzarelli, N. R. 1992. Conformational and thermodynamic properties of supercoiled DNA. *J. Mol. Biol.* 227:1224–1243.

Fisher, L. M., Austin, C. A., Hopewell, R., Margerrison, M., Oram, M., Patel, S., Wigley, D. B., Davies, G. J., Dodson, E. J., Maxwell, A., and Dodson, G. 1991. Crystal structure of an N-terminal fragment of the DNA gyrase B protein. *Nature* 351:624–629.

Mechanism of Replication

Davey, M. J., and O'Donnell, M. 2000. Mechanisms of DNA replication. *Curr. Opin. Chem. Biol.* 4:581–586.

Keck, J. L., and Berger, J. M. 2000. DNA replication at high resolution. *Chem. Biol.* 7:R63–R71.

Kunkel, T. A., and Bebenek, K. 2000. DNA replication fidelity. *Annu. Rev. Biochem.* 69:497–529.

Waga, S., and Stillman, B. 1998. The DNA replication fork in eukaryotic cells. *Annu. Rev. Biochem.* 67:721–751.

Marians, K. J. 1992. Prokaryotic DNA replication. *Annu. Rev. Biochem.* 61:673–719.

DNA Polymerases and Other Enzymes of Replication

Toth, E. A., Li, Y., Sawaya, M. R., Cheng, Y., and Ellenberger, T. 2003. The crystal structure of the bifunctional primase-helicase of bacteriophage T7. *Mol. Cell* 12:1113–1123.

Hubscher, U., Maga, G., and Spadari, S. 2002. Eukaryotic DNA polymerases. *Annu. Rev. Biochem.* 71:133–163.

Doublié, S., Tabor, S., Long, A. M., Richardson, C. C., and Ellenberger, T. 1998. Crystal structure of a bacteriophage T7 DNA replication complex at 2.2 Å resolution. *Nature* 391:251–258.

Arezi, B., and Kuchta, R. D. 2000. Eukaryotic DNA primase. *Trends Biochem. Sci.* 25:572–576.

Jager, J., and Pata, J. D. 1999. Getting a grip: Polymerases and their substrate complexes. *Curr. Opin. Struct. Biol.* 9:21–28.

Steitz, T. A. 1999. DNA polymerases: Structural diversity and common mechanisms. *J. Biol. Chem.* 274:17395–17398.

Beese, L. S., Derbyshire, V., and Steitz, T. A. 1993. Structure of DNA polymerase I Klenow fragment bound to duplex DNA. *Science* 260:352–355.

McHenry, C. S. 1991. DNA polymerase III holoenzyme: Components, structure, and mechanism of a true replicative complex. *J. Biol. Chem.* 266:19127–19130.

Kong, X. P., Onrust, R., O'Donnell, M., and Kuriyan, J. 1992. Three-dimensional structure of the β subunit of *E. coli* DNA polymerase III holoenzyme: A sliding DNA clamp. *Cell* 69:425–437.

Polesky, A. H., Steitz, T. A., Grindley, N. D., and Joyce, C. M. 1990. Identification of residues critical for the polymerase activity of the Klenow fragment of DNA polymerase I from *Escherichia coli.* *J. Biol. Chem.* 265:14579–14591.

Lee, J. Y., Chang, C., Song, H. K., Moon, J., Yang, J. K., Kim, H. K., Kwon, S. T., and Suh, S. W. 2000. Crystal structure of NAD^+-

dependent DNA ligase: Modular architecture and functional implications. *EMBO J.* 19:1119–1129.

Timson, D. J., and Wigley, D. B. 1999. Functional domains of an NAD$^+$-dependent DNA ligase. *J. Mol. Biol.* 285:73–83.

Doherty, A. J., and Wigley, D. B. 1999. Functional domains of an ATP-dependent DNA ligase. *J. Mol. Biol.* 285:63–71.

von Hippel, P. H., and Delagoutte, E. 2001. A general model for nucleic acid helicases and their "coupling" within macromolecular machines. *Cell* 104:177–190.

Tye, B. K., and Sawyer, S. 2000. The hexameric eukaryotic MCM helicase: Building symmetry from nonidentical parts. *J. Biol. Chem.* 275:34833–34836.

Marians, K. J. 2000. Crawling and wiggling on DNA: Structural insights to the mechanism of DNA unwinding by helicases. *Structure Fold Des.* 5:R227–R235.

Soultanas, P., and Wigley, D. B. 2000. DNA helicases: "Inching forward." *Curr. Opin. Struct. Biol.* 10:124–128.

Bachand, F., and Autexier, C. 2001. Functional regions of human telomerase reverse transcriptase and human telomerase RNA required for telomerase activity and RNA-protein interactions. *Mol. Cell Biol.* 21:1888–1897.

Bryan, T. M., and Cech, T. R. 1999. Telomerase and the maintenance of chromosome ends. *Curr. Opin. Cell Biol.* 11:318–324.

Griffith, J. D., Comeau, L., Rosenfield, S., Stansel, R. M., Bianchi, A., Moss, H., and de Lange, T. 1999. Mammalian telomeres end in a large duplex loop. *Cell* 97:503–514.

McEachern, M. J., Krauskopf, A., and Blackburn, E. H. 2000. Telomeres and their control. *Annu. Rev. Genet.* 34:331–358.

Recombination and Recombinases

Singleton, M. R., Dillingham, M. S., Gaudier, M., Kowalczykowski, S. C., and Wigley, D. B. 2004. Crystal structure of RecBCD enzyme reveals a machine for processing DNA breaks. *Nature* 432:187–193.

Spies, M., Bianco, P. R., Dillingham, M. S., Handa, N., Baskin, R. J., and Kowalczykowski, S. C. 2003. A molecular throttle: The recombination hotspot chi controls DNA translocation by the RecBCD helicase. *Cell* 114:647–654.

Kowalczykowski, S. C. 2000. Initiation of genetic recombination and recombination-dependent replication. *Trends Biochem. Sci.* 25:156–165.

Prevost, C., and Takahashi, M. 2003. Geometry of the DNA strands within the RecA nucleofilament: Role in homologous recombination. *Q. Rev. Biophys.* 36:429–453.

Van Duyne, G. D. 2001. A structural view of cre-loxp site-specific recombination. *Annu. Rev. Biophys. Biomol. Struct.* 30:87–104.

Chen, Y., Narendra, U., Iype, L. E., Cox, M. M., and Rice, P. A. 2000. Crystal structure of a Flp recombinase-Holliday junction complex: Assembly of an active oligomer by helix swapping. *Mol. Cell* 6:885–897.

Craig, N. L. 1997. Target site selection in transposition. *Annu. Rev. Biochem.* 66:437–474.

Gopaul, D. N., Guo, F., and Van Duyne, G. D. 1998. Structure of the Holliday junction intermediate in Cre-loxP site-specific recombination. *EMBO J.* 17:4175–4187.

Gopaul, D. N., and Duyne, G. D. 1999. Structure and mechanism in site-specific recombination. *Curr. Opin. Struct. Biol.* 9:14–20.

Mutations and DNA Repair

Yang, W. 2003. Damage repair DNA polymerases Y. *Curr. Opin. Struct. Biol.* 13:23–30.

Wood, R. D., Mitchell, M., Sgouros, J., and Lindahl, T. 2001. Human DNA repair genes. *Science* 291:1284–1289.

Shin, D. S., Chahwan, C., Huffman, J. L., and Tainer, J. A. 2004. Structure and function of the double-strand break repair machinery. *DNA Repair (Amst.)* 3:863–873.

Michelson, R. J., and Weinert, T. 2000. Closing the gaps among a web of DNA repair disorders. *Bioessays* 22:966–969.

Aravind, L., Walker, D. R., and Koonin, E. V. 1999. Conserved domains in DNA repair proteins and evolution of repair systems. *Nucleic Acids Res.* 27:1223–1242.

Mol, C. D., Parikh, S. S., Putnam, C. D., Lo, T. P., and Tainer, J. A. 1999. DNA repair mechanisms for the recognition and removal of damaged DNA bases. *Annu. Rev. Biophys. Biomol. Struct.* 28:101–128.

Parikh, S. S., Mol, C. D., and Tainer, J. A. 1997. Base excision repair enzyme family portrait: Integrating the structure and chemistry of an entire DNA repair pathway. *Structure* 5:1543–1550.

Vassylyev, D. G., and Morikawa, K. 1997. DNA-repair enzymes. *Curr. Opin. Struct. Biol.* 7:103–109.

Verdine, G. L., and Bruner, S. D. 1997. How do DNA repair proteins locate damaged bases in the genome? *Chem. Biol.* 4:329–334.

Bowater, R. P., and Wells, R. D. 2000. The intrinsically unstable life of DNA triplet repeats associated with human hereditary disorders. *Prog. Nucleic Acid Res. Mol. Biol.* 66:159–202.

Cummings, C. J., and Zoghbi, H. Y. 2000. Fourteen and counting: Unraveling trinucleotide repeat diseases. *Hum. Mol. Genet.* 9:909–916.

Defective DNA Repair and Cancer

Berneburg, M., and Lehmann, A. R. 2001. Xeroderma pigmentosum and related disorders: Defects in DNA repair and transcription. *Adv. Genet.* 43:71–102.

Lambert, M. W., and Lambert, W. C. 1999. DNA repair and chromatin structure in genetic diseases. *Prog. Nucleic Acid Res. Mol. Biol.* 63:257–310.

Buys, C. H. 2000. Telomeres, telomerase, and cancer. *N. Engl. J. Med.* 342:1282–1283.

Urquidi, V., Tarin, D., and Goodison, S. 2000. Role of telomerase in cell senescence and oncogenesis. *Annu. Rev. Med.* 51:65–79.

Lynch, H. T., Smyrk, T. C., Watson, P., Lanspa, S. J., Lynch, J. F., Lynch, P. M., Cavalieri, R. J., and Boland, C. R. 1993. Genetics, natural history, tumor spectrum, and pathology of hereditary nonpolyposis colorectal cancer: An updated review. *Gastroenterology* 104:1535–1549.

Fishel, R., Lescoe, M. K., Rao, M. R. S., Copeland, N. G., Jenkins, N. A., Garber, J., Kane, M., and Kolodner, R. 1993. The human mutator gene homolog *MSH2* and its association with hereditary nonpolyposis colon cancer. *Cell* 75:1027–1038.

Ames, B. N., and Gold, L. S. 1991. Endogenous mutagens and the causes of aging and cancer. *Mutat. Res.* 250:3–16.

Ames, B. N. 1979. Identifying environmental chemicals causing mutations and cancer. *Science* 204:587–593.

Problems

1. *Activated intermediates.* DNA polymerase I, DNA ligase, and topoisomerase I catalyze the formation of phosphodiester bonds. What is the activated intermediate in the linkage reaction catalyzed by each of these enzymes? What is the leaving group?

2. *Life in a hot tub.* An archaeon (*Sulfolobus acidocaldarius*) found in acidic hot springs contains a topoisomerase that catalyzes the ATP-driven introduction of positive supercoils into DNA. How might this enzyme be advantageous to this unusual organism?

3. *A cooperative transition.* The transition from B-DNA to Z-DNA takes place over a small change in the superhelix density, which shows that the transition is highly cooperative.

(a) Consider a DNA molecule at the midpoint of this transition. Are B- and Z-DNA regions frequently intermingled or are there long stretches of each?
(b) What does this finding reveal about the energetics of forming a junction between the two kinds of helices?
(c) Would you expect the transition from B- to A-DNA to be more or less cooperative than the one from B- to Z-DNA? Why?

4. *Molecular motors in replication.* (a) How fast does template DNA spin (expressed in revolutions per second) at an *E. coli* replication fork? (b) What is the velocity of movement (in micrometers per second) of DNA polymerase III holoenzyme relative to the template?

5. *Wound tighter than a drum.* Why would replication come to a halt in the absence of topoisomerase II?

6. *Telomeres and cancer.* Telomerase is not active in most human cells. Some cancer biologists have suggested that activation of the telomerase gene would be a requirement for a cell to become cancerous. Explain why this might be the case.

7. *Nick translation.* Suppose that you wish to make a sample of DNA duplex highly radioactive to use as a DNA probe. You have a DNA endonuclease that cleaves the DNA internally to generate 3'-OH and 5'-phosphoryl groups, intact DNA polymerase I, and radioactive dNTPs. Suggest a means for making the DNA radioactive.

8. *Revealing tracks.* Suppose that replication is initiated in a medium containing *moderately* radioactive tritiated thymine. After a few minutes of incubation, the bacteria are transferred to a medium containing *highly* radioactive tritiated thymidine. Sketch the autoradiographic pattern that would be seen for (a) undirectional replication and (b) bidirectional replication, each from a single origin.

9. *Mutagenic trail.* Suppose that the single-stranded RNA from tobacco mosaic virus was treated with a chemical mutagen, that mutants were obtained having serine or leucine instead of proline at a specific position, and that further treatment of these mutants with the same mutagen yielded phenylalanine at this position.

(a) What are the plausible codon assignments for these four amino acids?
(b) Was the mutagen 5-bromouracil, nitrous acid, or an acridine dye?

10. *Induced spectrum.* DNA photolyases convert the energy of light in the near-ultraviolet or visible region (300–500 nm) into chemical energy to break the cyclobutane ring of pyrimidine dimers. In the absence of substrate, these photoreactivating en-

zymes do not absorb light of wavelengths longer than 300 nm. Why is the substrate-induced absorption band advantageous?

Mechanism Problem

11. *A revealing analog.* AMP-PNP, the β,γ-imido analog of ATP, is hydrolyzed very slowly by most ATPases.

AMP-PNP

The addition of AMP-PNP to topoisomerase II and circular DNA leads to the negative supercoiling of a single molecule of DNA per enzyme. DNA remains bound to the enzyme in the presence of this analog. What does this finding reveal about the catalytic mechanism?

Data Interpretation and Chapter Integration Problems

12. *Like a ladder.* Circular DNA from SV40 virus was isolated and subjected to gel electrophoresis. The results are shown in lane A (the control) of the adjoining gel patterns.

(a) Why does the DNA separate in agarose gel electrophoresis? How does the DNA in each band differ?
The DNA was then incubated with topoisomerase I for 5 minutes and again analyzed by gel electrophoresis with the results shown in lane B.
(b) What types of DNA do the various bands represent?
Another sample of DNA was incubated with topoisomerase I for 30 minutes and again analyzed as shown in lane C.
(c) What is the significance of the fact that more of the DNA is in slower-moving forms?

13. *Ames test.* The adjoining illustration shows four petri plates used for the Ames test. A piece of filter paper (white circle in the center of each plate) was soaked in one of four preparations and then placed on a petri plate. The four preparations contained

(A) purified water (control), (B) a known mutagen, (C) a chemical whose mutagenicity is under investigation, and (D) the same chemical after treatment with liver homogenate. The number of revertants, visible as colonies on the petri plates, was determined in each case.

(a)   What was the purpose of the control plate, which was exposed only to water?

(b)   Why was it wise to use a known mutagen in the experimental system?

(c)   How would you interpret the results obtained with the experimental compound?

(d)   What liver components would you think are responsible for the effects observed in preparation D?

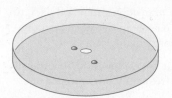
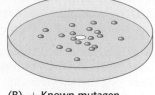

(A) Control: No mutagen   (B) + Known mutagen

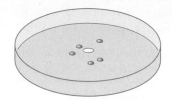
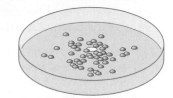

(C) + Experimental sample   (D) + Experimental sample
                                        after treatment with
                                        liver homogenate

RNA Synthesis and Processing

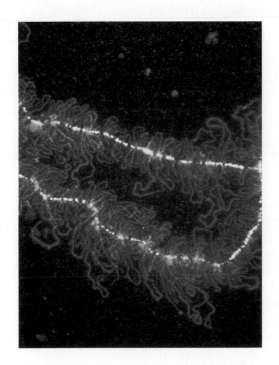

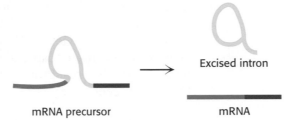

mRNA precursor → Excised intron

mRNA

RNA synthesis is a key step in the expression of genetic information. For eukaryotic cells, the initial RNA transcript (the mRNA precursor) is often spliced, removing introns that do not encode protein sequences. Often, the same pre-mRNA is spliced differently in different cell types or at different developmental stages. In the image at the left, proteins associated with RNA splicing (stained with a fluorescent antibody) highlight regions of the newt genome that are being actively transcribed. [(Left) courtesy of Dr. Mark B. Roth and Dr. Joseph G. Gall.]

DNA stores genetic information in a stable form that can be readily replicated. However, the expression of this genetic information requires its flow from DNA to RNA to protein, as was introduced in Chapter 4. The present chapter deals with how RNA is synthesized and then modified to prepare for its translation into protein. We begin with transcription in prokaryotes and focus on the three stages of transcription: promoter binding and initiation, elongation of the nascent RNA transcript, and termination at the end of the gene.

We then turn to transcription in eukaryotes, focusing on the distinctions between prokaryotic and eukaryotic transcription. Eukaryotes use three different polymerases to transcribe their coding and noncoding RNAs. Eukaryotic transcription is highly regulated by the binding of transcription-factor proteins that control promoter activity and by the presence of enhancer sequences that can stimulate transcriptional initiation more than a thousand base pairs away from the start site. Primary transcripts in eukaryotes are extensively modified, as exemplified by the capping of the 5′ end of an mRNA precursor and the addition of a long poly(A) tail to its 3′ end.

Chapter 29 revised by Susan J. Baserga and Erica A. Champion, Yale University.

Most striking is the splicing of mRNA precursors, which is catalyzed by spliceosomes consisting of small nuclear ribonucleoprotein particles (snRNPs). The small nuclear RNA (snRNA) molecules in these complexes play a key role in directing the alignment of splice sites and in mediating catalysis. Indeed, some RNA molecules can splice themselves in the absence of protein. This landmark discovery by Thomas Cech and Sidney Altman revealed that RNA molecules can serve as catalysts and greatly influenced our view of molecular evolution.

RNA splicing is not merely a curiosity. At least 15% of all genetic diseases are caused by mutations that affect RNA splicing. Moreover, the same pre-mRNA can be spliced differently in various cell types, at different stages of development, or in response to other biological signals. In addition, individual bases in some pre-mRNA molecules are changed, in a process called *RNA editing*. One of the biggest surprises of the sequencing of the human genome was that only about 25,000 genes were identified compared with previous estimates of 100,000 or more. The ability of one gene to encode more than one distinct mRNA by alternative splicing and, hence, more than one protein, may play a key role in expanding the repertoire of our genomes.

RNA Synthesis Comprises Three Stages: Initiation, Elongation, and Termination

RNA synthesis, or *transcription*, is the process of transcribing DNA nucleotide sequence information into RNA sequence information. RNA synthesis is catalyzed by a large enzyme called *RNA polymerase*. The basic biochemistry of RNA synthesis is common to prokaryotes and eukaryotes, although its regulation is more complex in eukaryotes. The close connection between prokaryotic and eukaryotic transcription has been beautifully illustrated by the three-dimensional structures of representative RNA polymerases from prokaryotes and eukaryotes (Figure 29.1). Despite substantial differences in size and number of polypeptide subunits, the overall structures of these enzymes are quite similar, revealing a common evolutionary origin.

RNA synthesis, like nearly all biological polymerization reactions, takes place in three stages: *initiation, elongation,* and *termination*. RNA polymerase performs multiple functions in this process:

Figure 29.1 RNA polymerase structures. The three-dimensional structures of RNA polymerases from a prokaryote *(Thermus aquaticus)* and a eukaryote *(Saccharomyces cerevisiae)*. The two largest subunits for each structure are shown in dark red and dark blue. *Notice* that both structures contain a central metal ion (green) in their active sites, near a large cleft on the right. The similarity of these structures reveals that these enzymes have the same evolutionary origin and have many mechanistic features in common. [Drawn from 1I6V.pdb and 1I6H.pdb.]

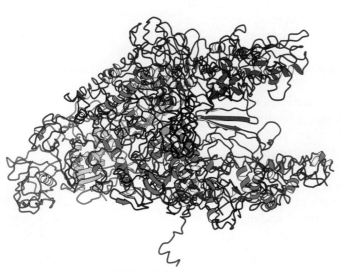

Prokaryotic RNA polymerase

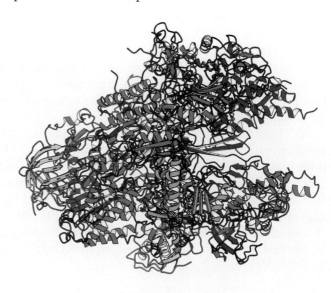

Eukaryotic RNA polymerase

1. It searches DNA for initiation sites, also called *promoter sites* or simply *promoters*. For instance, *E. coli* DNA has about 2000 promoter sites in its 4.8×10^6 bp genome. Because these sequences are on the *same* molecule of DNA as the genes being transcribed, they are called *cis-acting elements*.

2. It unwinds a short stretch of double-helical DNA to produce a single-stranded DNA template from which it takes instructions.

3. It selects the correct ribonucleoside triphosphate and catalyzes the formation of a phosphodiester bond. This process is repeated many times as the enzyme moves unidirectionally along the DNA template. RNA polymerase is completely processive—a transcript is synthesized from start to end by a single RNA polymerase molecule.

4. It detects termination signals that specify where a transcript ends.

5. It interacts with activator and repressor proteins that modulate the rate of transcription initiation over a wide dynamic range. These proteins, which play a more prominent role in eukaryotes than in prokaryotes, are called *transcription factors* or *trans-acting factors*. Gene expression is controlled mainly at the level of transcription, as will be discussed in detail in Chapter 31.

The fundamental reaction of RNA synthesis is the formation of a phosphodiester bond. The 3′-hydroxyl group of the last nucleotide in the chain nucleophilically attacks the α phosphoryl group of the incoming nucleoside triphosphate with the concomitant release of a pyrophosphate (see Figure 4.25). This reaction is thermodynamically favorable, and the subsequent degradation of the pyrophosphate to orthophosphate locks the reaction in the direction of RNA synthesis.

The chemistry of RNA synthesis is identical for all forms of RNA, including messenger RNA, transfer RNA, and ribosomal RNA. The basic steps just outlined also apply to all forms. Their synthetic processes differ mainly in regulation, posttranscriptional processing, and the specific polymerase that participates.

29.1 RNA Polymerase Catalyzes Transcription

We begin our consideration of transcription by examining the process in bacteria such as *E. coli*. The *E. coli* RNA polymerase is a very large (~400 kd) and complex enzyme consisting of four kinds of subunits (Table 29.1). The subunit composition of the entire enzyme, called the *holoenzyme*, is $\alpha_2\beta\beta'\sigma$. The σ subunit helps find a promoter site where transcription begins, participates in the initiation of RNA synthesis, and then dissociates from the rest of the enzyme. RNA polymerase without this subunit ($\alpha_2\beta\beta'$) is called the *core enzyme*, which contains the catalytic site.

This catalytic site resembles that of DNA polymerase (p. 794) in that it includes two metal ions in its active form (Figure 29.2). One metal ion remains bound to the enzyme, whereas the other appears to come in with the nucleoside triphosphate and leave with the pyrophosphate. Three conserved aspartate residues of the enzyme participate in binding these metal ions. Note that the overall structures of DNA polymerase and RNA polymerase are quite different; their similar active sites are the products of convergent evolution.

TABLE 29.1 Subunits of RNA polymerase from *E. coli*

Subunit	Gene	Number	Mass (kd)
α	*rpoA*	2	37
β	*rpoB*	1	151
β'	*rpoC*	1	155
σ^{70}	*rpoD*	1	70

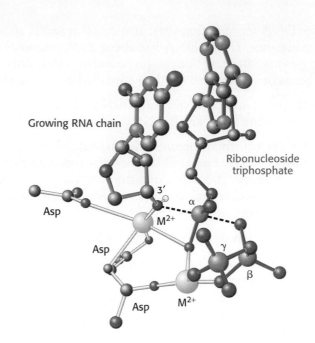

Figure 29.2 RNA polymerase active site.
A model of the transition state for
phosphodiester bond formation in the
active site of RNA polymerase. The 3′-
hydroxyl group of the growing RNA chain
attacks the α-phosphoryl group of the
incoming nucleoside triphosphate. This
transition state is structurally similar to
that in DNA polymerase (see Figure 28.12).

RNA Polymerase Binds to Promoter Sites on the DNA Template to Initiate Transcription

Transcription starts at *promoters* on the DNA template. *Promoters are sequences of DNA that direct the RNA polymerase to the proper initiation site for transcription.* Promoter sites can be identified and characterized by a combination of techniques. One powerful technique for characterizing these protein-binding sites and others on DNA is called *footprinting* (Figure 29.3). First, one of the strands of a DNA fragment under investigation is labeled on one end with ^{32}P. RNA polymerase is added to the labeled DNA, and *the complex is digested with DNase just long enough to make an average of one cut in each chain.* A part of the radioactive DNA is treated in the same way but without the addition of RNA polymerase to serve as a control. The resulting DNA fragments are separated according to size by electrophoresis. The gel pattern is highly revealing: a series of bands present in the control sample is absent from the sample containing RNA polymerase. These bands are missing because RNA polymerase bound to the promoter shields promoter DNA from cleavages that would give rise to the corresponding fragments.

Figure 29.3 Footprinting. One end of a
DNA chain is labeled with ^{32}P (shown as
a red circle). This labeled DNA is then
treated with DNase I such that each
fragment is cut only once. The same
cleavage is carried out after a protein that
binds to specific sites on the DNA has
been added. The bound protein protects
a segment on the DNA from the action of
DNase I. Hence, certain fragments present
in the reaction without protein will be
missing. These missing bands in the gel
pattern identify the binding site on DNA.

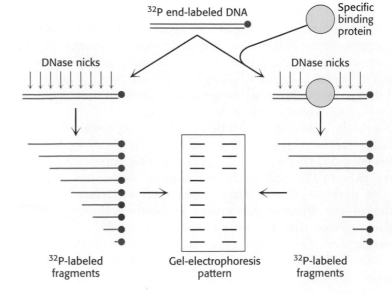

A striking pattern is evident when the sequences of many prokaryotic promoters are compared. *Two common motifs are present on the 5' (upstream) side of the transcription start site.* They are known as the *−10 sequence* and the *−35 sequence* because they are centered at about 10 and 35 nucleotides upstream of the start site. The region containing these sequences, which spans as many as 40 nucleotides upstream from the start site, is called the *core promoter*. The −10 and −35 sequences are each 6 bp long. Their *consensus (average) sequences*, deduced from analyses of many promoters (Figure 29.4), are

−35   −10   +1

5'~~~T T G A C A~~~~~~~~~~~T A T A A T~~~~Start site

The first nucleotide (the start site) of a transcribed DNA sequence is denoted as +1 and the second one as +2; the nucleotide preceding the start site is denoted as −1. These designations refer to the coding strand of DNA. Recall that the sequence of the *template strand of DNA* is the *complement* of that of the RNA transcript (see Figure 4.26). In contrast, the *coding strand of DNA* has the *same* sequence as that of the RNA transcript except for thymine (T) in place of uracil (U). The coding strand is also known as the *sense (+) strand*, and the template strand as the *antisense (−) strand*.

Promoters differ markedly in their efficacy. Genes with strong promoters are transcribed frequently—as often as every 2 seconds in *E. coli*. In contrast, genes with very weak promoters are transcribed about once in 10 minutes. The −10 and −35 regions of most strong promoters have sequences that correspond closely to the consensus sequences, whereas weak promoters tend to have multiple substitutions at these sites. Indeed, mutation of a single base in either the −10 sequence or the −35 sequence can diminish promoter activity. The distance between these conserved sequences also is important; a separation of 17 nucleotides is optimal. Thus, *the efficiency or strength of a promoter sequence serves to regulate transcription.* Regulatory proteins that bind to specific sequences near promoter sites and interact with RNA polymerase (Chapter 31) also markedly influence the frequency of transcription of many genes.

Outside the core promoter in a subset of highly expressed genes is the *upstream element* (also called the UP element for *upstream element*). This sequence is present from 40 to 60 nucleotides upstream of the transcription start site. The UP element is bound by the α subunit of RNA polymerase and serves to increase the efficiency of transcription by creating an additional binding site for the polymerase.

Sigma Subunits of RNA Polymerase Recognize Promoter Sites

To initiate transcription, the $\alpha_2\beta\beta'$ core of RNA polymerase must bind the promoter. However, it is the σ subunit that makes this binding possible by enabling *RNA polymerase to recognize promoter sites.* In the presence of the σ subunit, the RNA polymerase binds weakly to the DNA and slides along the double helix until it dissociates or encounters a promoter. The σ subunit recognizes the promoter through several interactions with the nucleotide bases of the promoter DNA. Although each interaction by itself is weak, the combined effect is a strong sequence-specific interaction overall. A recent crystal structure of the RNA polymerase holoenzyme bound to a promoter site shows the σ subunit interacting with DNA at the −10 and −35 regions essential to promoter recognition. (Figure 29.5). Therefore, *the σ subunit is responsible for the specific binding of the RNA polymerase to a promoter site on the template DNA.*

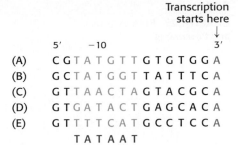

Transcription starts here
↓

	5'	−10		3'
(A)	C G T A T G T T	G T G T G G A
(B)	G C T A T G G T	T A T T T C A
(C)	G T T A A C T A	G T A C G C A
(D)	G T G A T A C T	G A G C A C A
(E)	G T T T T C A T	G C C T C C A
		T A T A A T

Figure 29.4 Prokaryotic promoter sequences. A comparison of five sequences from prokaryotic promoters reveals a recurring sequence of TATAAT centered on position −10. The −10 consensus sequence (in red) was deduced from a large number of promoter sequences. The sequences are from the (A) *lac*, (B) *gal*, and (C) *trp* operons of *E. coli*; from (D) λ phage; and from (E) φX174 phage.

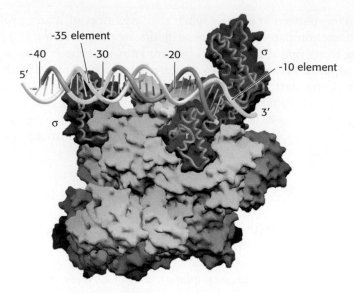

Figure 29.5 RNA polymerase holoenzyme complex. *Notice* that the σ subunit (orange) of the bacterial RNA polymerase holoenzyme makes sequence-specific contacts with the −10 and −35 promoter sequences (yellow). [From K. S. Murakami, S. Masuda, E. A. Campbell, O. Muzzin, and S. A. Darst. *Science* 296(2002):1285–1290.]

As the holoenzyme moves along the double helix in search of a promoter, it forms transient hydrogen bonds with exposed hydrogen-bond donor and acceptor groups on the base pairs. The search is rapid because RNA polymerase slides along DNA instead of repeatedly binding and dissociating from it. *In other words, the promoter site is encountered by a random walk in one dimension rather than in three dimensions.* The observed rate constant for the binding of the RNA polymerase holoenzyme to promoter sequences is $10^{10}\,\mathrm{M}^{-1}\mathrm{s}^{-1}$, more than 100 times larger than that expected for repeated encounters moving on and off the DNA. The σ subunit is released when the nascent RNA chain reaches 9 or 10 nucleotides in length. After its release, it can assist initiation by another core enzyme. Thus, the σ *subunit acts catalytically.*

E. coli contains multiple σ factors to recognize several types of promoter sequences contained in *E. coli* DNA. The type that recognizes the consensus sequences described earlier is called σ^{70} because it has a mass of 70 kd. A different σ factor comes into play when the temperature is raised abruptly. *E. coli* responds by synthesizing σ^{32}, which recognizes the promoters of *heat-shock genes*. These promoters exhibit −10 sequences that are somewhat different from the −10 sequence for standard promoters (Figure 29.6). The increased transcription of heat-shock genes leads to the coordinated synthesis of a series of protective proteins. Other σ factors respond to environmental conditions, such as nitrogen starvation. These findings demonstrate that *σ plays a key role in determining where RNA polymerase initiates transcription.*

Figure 29.6 Alternative promoter sequences. A comparison of the consensus sequences of standard, heat-shock, and nitrogen-starvation promoters of *E. coli*. These promoters are recognized by σ^{70}, σ^{32}, and σ^{54}, respectively.

	−35		−10		
5′〰〰TTGACA〰〰〰〰〰〰TATAAT〰〰〰3′					Standard promoter
5′〰〰TNNCNCNCTTGAA〰〰〰〰CCCATNT〰〰〰3′					Heat-shock promoter
5′〰〰CTGGGNA〰〰〰〰〰〰〰TTGCA〰〰〰3′					Nitrogen-starvation promoter

RNA Polymerase Must Unwind the Template Double Helix for Transcription to Take Place

Although RNA polymerase can search for promoter sites when bound to double-helical DNA, a segment of the helix must be unwound before synthesis can begin. A region of duplex DNA must be unpaired so that nucleotides on one of its strands become accessible for base-pairing with incoming ribonucleoside triphosphates. The DNA template strand selects the correct ribonucleoside triphosphate by forming a Watson–Crick base pair with it (p. 112), as in DNA synthesis.

How much of the template DNA is unwound by the polymerase? Because unwinding increases the negative supercoiling of the DNA (p. 789), this question was answered by analyzing the supercoiling of a circular duplex DNA exposed to varying amounts of RNA polymerase. Topoisomerase I, an enzyme catalyzing the concerted cleavage and resealing of duplex DNA (p. 790), was then added to relax the part of circular DNA not in contact with polymerase molecules. These DNA samples were analyzed by gel electrophoresis after the removal of bound protein. *The degree of negative supercoiling increased in proportion to the number of RNA polymerase molecules bound per template DNA, showing that the enzyme unwinds DNA. Each bound polymerase molecule unwinds a 17-bp segment of DNA, which corresponds to 1.6 turns of B-DNA helix* (Figure 29.7).

Negative supercoiling of circular DNA favors the transcription of genes because it facilitates unwinding. Thus, the introduction of negative supercoils into DNA by topoisomerase II can increase the efficiency of promoters located at distant sites. However, not all promoter sites are stimulated by negative supercoiling. The promoter site for topoisomerase II itself is a noteworthy exception. Negative supercoiling decreases the rate of transcription of this gene, an elegant feedback control ensuring that DNA does not become excessively supercoiled. Negative supercoiling could decrease the efficiency of this promoter by changing the structural relation of the -10 and -35 regions.

The transition from the *closed promoter complex* (in which DNA is double helical) to the *open promoter complex* (in which a DNA segment is unwound) is an essential event in transcription. The stage is now set for the formation of the first phosphodiester bond of the new RNA chain.

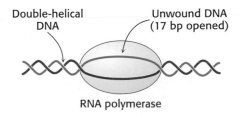

Figure 29.7 DNA unwinding. RNA polymerase unwinds about 17 base pairs of template DNA.

RNA Chains Are Formed de Novo and Grow in the 5′-to-3′ Direction

In contrast with DNA synthesis, *RNA synthesis can start de novo, without the requirement for a primer.* Most newly synthesized RNA chains carry a highly distinctive tag on the 5′ end: the first base at that end is either *pppG* or *pppA*.

The presence of the triphosphate moiety suggests that RNA synthesis starts at the 5′ end. The results of labeling experiments with γ-^{32}P substrates confirmed that RNA chains, like DNA chains, grow in the 5′ $\rightarrow$ 3′ direction.

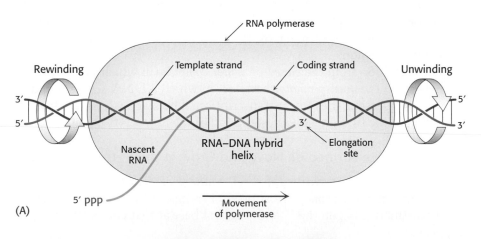

5′ → 3′ growth

Elongation Takes Place at Transcription Bubbles That Move Along the DNA Template

The elongation phase of RNA synthesis begins after the formation of the first phosphodiester bond. An important change is the loss of σ; without σ, the core enzyme binds more strongly to the DNA template. Indeed, RNA polymerase stays bound to its template until a termination signal is reached. The region containing RNA polymerase, DNA, and nascent RNA is called a *transcription bubble* because it contains a locally melted "bubble" of DNA (Figure 29.8). The newly synthesized RNA forms a hybrid helix with the template DNA strand. This RNA–DNA helix is about 8 bp long, which corresponds to nearly one turn of a double helix (p. 112). The 3′-hydroxyl group of the RNA in this hybrid helix is positioned so that it can attack the α-phosphorus atom of an incoming ribonucleoside triphosphate. The core enzyme also contains a binding site for the other DNA strand. About 17 bp of DNA are unwound throughout the elongation phase, as in the initiation phase. The transcription bubble moves a distance of 170 Å (17 nm) in a second, which corresponds to a rate of elongation of about 50 nucleotides per second. Although rapid, it is much slower than the rate of DNA synthesis, which is 800 nucleotides per second.

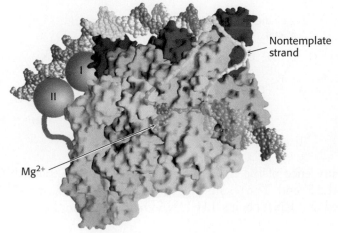

Figure 29.8 Transcription bubble. (A) A schematic representation of a transcription bubble in the elongation of an RNA transcript. Duplex DNA is unwound at the forward end of RNA polymerase and rewound at its rear end. The RNA–DNA hybrid rotates during elongation. (B) A surface model based on the crystal structure of the RNA polymerase holoenzyme shows the unwound DNA (yellow and green) forming the transcription bubble. *Notice* that the template strand (green) is in contact with the catalytic Mg^{2+} (pink). [(B) From K. S. Murakami, S. Masuda, E. A. Campbell, O. Muzzin, and S. A. Darst. *Science* 296(2002):1285–1290.]

The lengths of the RNA–DNA hybrid and of the unwound region of DNA stay rather constant as RNA polymerase moves along the DNA template. This finding indicates that DNA is rewound at about the same rate at the rear of RNA polymerase as it is unwound at the front of the enzyme. The RNA–DNA hybrid must also rotate each time a nucleotide is added so that the 3'-OH end of the RNA stays at the catalytic site. The length of the RNA–DNA hybrid is determined by a structure within the enzyme that forces the RNA–DNA hybrid to separate, allowing the RNA chain to exit from the enzyme and the DNA chain to rejoin its DNA partner (Figure 29.9).

For many years, RNA polymerase was thought not to proofread the RNA transcript. However, recent studies have indicated that RNA polymerases do show proofreading nuclease activity, particularly in the presence of accessory proteins. Studies of single molecules of RNA polymerase reveal that the enzymes hesitate and backtrack to correct errors. The error rate of the order of one mistake per 10^4 or 10^5 nucleotides is higher than that for DNA replication, including all error-correcting mechanisms. The lower fidelity of RNA synthesis can be tolerated because mistakes are not transmitted to progeny. For most genes, many RNA transcripts are synthesized; a few defective transcripts are unlikely to be harmful.

Sequences Within the Newly Transcribed RNA Signal Termination

The termination of transcription is as precisely controlled as its initiation. In the termination phase of transcription, the formation of phosphodiester bonds ceases, the RNA–DNA hybrid dissociates, the melted region of DNA rewinds, and RNA polymerase releases the DNA. What determines where transcription is terminated? *The transcribed regions of DNA templates contain stop signals.* The simplest one is a *palindromic GC-rich region followed by an AT-rich region.* The RNA transcript of this DNA palindrome is self-complementary (Figure 29.10). Hence, its bases can pair to form a hairpin structure with a stem and loop, a structure favored by its high content of G and C residues. Guanine–cytosine base pairs are more stable than adenine–thymine pairs because of the extra hydrogen bond in the base pair. This stable hairpin is followed by a sequence of four or more uracil residues, which also are crucial for termination. The RNA transcript ends within or just after them.

How does this combination hairpin–oligo(U) structure terminate transcription? First, it seems likely that RNA polymerase pauses immediately after it has synthesized a stretch of RNA that folds into a hairpin. Furthermore, the RNA–DNA hybrid helix produced after the hairpin is unstable because its rU–dA base pairs are the weakest of the four kinds.

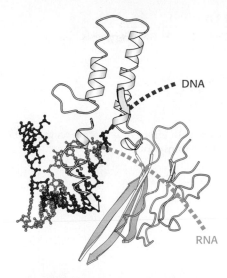

Figure 29.9 RNA–DNA hybrid separation. A structure within RNA polymerase forces the separation of the RNA–DNA hybrid. *Notice* that the DNA strand exits in one direction and the RNA product exits in another. [Drawn from 1I6H.pdb.]

Figure 29.10 Termination signal. A termination signal found at the 3' end of an mRNA transcript consists of a series of bases that form a stable stem-loop structure and a series of U residues.

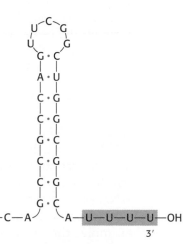

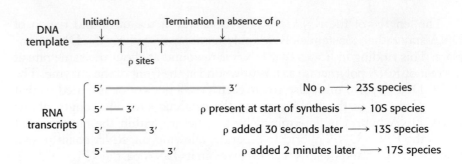

Figure 29.11 Effect of ρ protein on the size
of RNA transcripts.

Hence, the pause in transcription caused by the hairpin permits the weakly bound *nascent RNA to dissociate from the DNA template and then from the enzyme*. The solitary DNA template strand rejoins its partner to re-form the DNA duplex, and the transcription bubble closes.

The *rho* Protein Helps to Terminate the Transcription of Some Genes

RNA polymerase needs no help to terminate transcription at a hairpin followed by several U residues. At other sites, however, termination requires the participation of an additional factor. This discovery was prompted by the observation that some RNA molecules synthesized in vitro by RNA polymerase acting alone are *longer* than those made in vivo. The missing factor, a protein that caused the correct termination, was isolated and named *rho* (ρ). Additional information about the action of ρ was obtained by adding this termination factor to an incubation mixture at various times after the initiation of RNA synthesis (Figure 29.11). RNAs with sedimentation coefficients of 10S, 13S, and 17S were obtained when ρ was added at initiation, a few seconds after initiation, and 2 minutes after initiation, respectively. If no ρ was added, transcription yielded a 23S RNA product. It is evident that the template contains at least three termination sites that respond to ρ (yielding 10S, 13S, and 17S RNA) and one termination site that does not (yielding 23S RNA). Thus, specific termination at a site producing 23S RNA can occur in the absence of ρ. However, ρ detects additional termination signals that are not recognized by RNA polymerase alone.

How does ρ provoke the termination of RNA synthesis? *A key clue is the finding that ρ hydrolyzes ATP in the presence of single-stranded RNA but not in the presence of DNA or duplex RNA.* Hexameric ρ, which is structurally similar and homologous to ATP synthase (p. 522), specifically binds single-stranded RNA; a stretch of 72 nucleotides is bound in such a way that the RNA passes through the center of the structure (Figure 29.12). *The ρ protein is brought into action by sequences located in the nascent RNA that are rich in cytosine and poor in guanine.* The ATPase activity of ρ enables the protein to pull the nascent RNA while pursuing RNA polymerase. When ρ catches RNA polymerase at the transcription bubble, it breaks the RNA–DNA hybrid helix by functioning as an RNA–DNA helicase. Given the structural and evolutionary connection, the mechanism of action of ρ may be similar to that of ATP synthase, with the single-stranded RNA playing the role of the γ subunit.

Proteins in addition to ρ may provoke termination. For example, the *nusA protein* enables RNA polymerase in

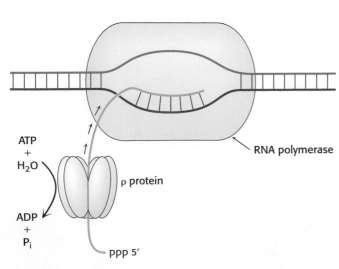

Figure 29.12 Mechanism for the termination of transcription by ρ protein. This protein is an ATP-dependent helicase that binds the nascent RNA chain and pulls it away from RNA polymerase and the DNA template.

E. coli to recognize a characteristic class of termination sites. In *E. coli*, specialized termination signals called *attenuators* are regulated to meet the nutritional needs of the cell (Section 31.4). *A common feature of protein-independent and protein-dependent termination is that the functioning signals lie in newly synthesized RNA rather than in the DNA template.*

Some Antibiotics Inhibit Transcription

Many antibiotics are highly specific inhibitors of biological processes. Rifampicin and actinomycin are two antibiotics that inhibit transcription, although in quite different ways. *Rifampicin* is a semisynthetic derivative of *rifamycins*, which are compounds derived from a strain of *Streptomyces*.

Rifampicin

This antibiotic *specifically inhibits the initiation of RNA synthesis.* Rifampicin does not block the binding of RNA polymerase to the DNA template; rather, it interferes with the formation of the first few phosphodiester bonds in the RNA chain. The structure of a complex between a prokaryotic RNA polymerase and rifampicin reveals that the antibiotic blocks the channel into which the RNA–DNA hybrid generated by the enzyme must pass (Figure 29.13). The binding site is 12 Å from the active site itself. Rifampicin does not hinder chain elongation once initiated, because the RNA–DNA hybrid present in the enzyme prevents the antibiotic from binding. The pocket in which rifampicin binds is conserved among bacterial RNA polymerases, but not eukaryotic polymerases, and so rifampicin can be used as an antibiotic in antituberculosis therapy.

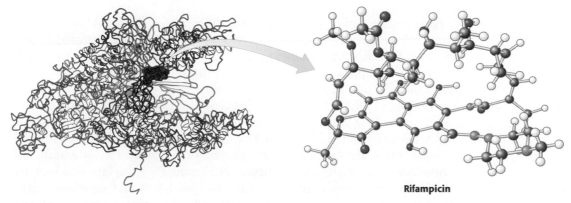

Rifampicin

Figure 29.13 Antibiotic action. Rifampicin binds to a pocket in the channel that is normally occupied by the newly formed RNA–DNA hybrid. Thus, the antibiotic blocks elongation after only two or three nucleotides have been added.

Actinomycin D, a polypeptide-containing antibiotic from a different strain of *Streptomyces,* inhibits transcription by an entirely different mechanism. *Actinomycin D binds tightly and specifically to double-helical DNA and thereby prevents it from being an effective template for RNA synthesis.* It does not bind to single-stranded DNA or RNA, double-stranded RNA, or RNA–DNA hybrids. The results of spectroscopic and hydrodynamic studies of complexes of actinomycin D and DNA suggested that the phenoxazone ring of actinomycin slips in between neighboring base pairs in DNA. This mode of binding is called *intercalation.* At low concentrations, actinomycin D inhibits transcription without significantly affecting DNA replication or protein synthesis. Hence, *actinomycin D is extensively used as a highly specific inhibitor of the formation of new RNA in both prokaryotic and eukaryotic cells.* Its ability to inhibit the growth of rapidly dividing cells makes it an effective therapeutic agent in the treatment of some cancers.

Precursors of Transfer and Ribosomal RNA Are Cleaved and Chemically Modified After Transcription in Prokaryotes

In prokaryotes, messenger RNA molecules undergo little or no modification after synthesis by RNA polymerase. Indeed, many mRNA molecules are translated while they are being transcribed. In contrast, *transfer RNA and ribosomal RNA molecules are generated by cleavage and other modifications of nascent RNA chains.* For example, in *E. coli,* the three rRNAs and a tRNA are excised from a single primary RNA transcript that also contains spacer regions (Figure 29.14). Other transcripts contain arrays of several kinds of tRNA or of several copies of the same tRNA. The nucleases that cleave and trim these precursors of rRNA and tRNA are highly precise. *Ribonuclease P* (RNase P), for example, generates the correct 5′ terminus of all tRNA molecules in *E. coli.* Sidney Altman and his coworkers showed that this interesting enzyme contains a catalytically active RNA molecule. *Ribonuclease III* (RNase III) excises 5S, 16S, and 23S rRNA precursors from the primary transcript by cleaving double-helical hairpin regions at specific sites.

H₃C
        N—CH₃

6-Dimethyladenine
(in prokaryotes)

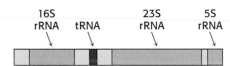

Figure 29.14 Primary transcript. Cleavage of this transcript produces 5S, 16S, and 23S rRNA molecules and a tRNA molecule. Spacer regions are shown in yellow.

A second type of processing is the *addition of nucleotides to the termini of some RNA chains.* For example, CCA, a terminal sequence required for the function of all tRNAs, is added to the 3′ ends of tRNA molecules for which this terminal sequence is not encoded in the DNA. The enzyme that catalyzes the addition of CCA is atypical for an RNA polymerase in that it does not use a DNA template. A third type of processing is the *modification of bases and ribose units* of ribosomal RNAs. In prokaryotes, some bases of rRNA are methylated. Unusual bases are found in all tRNA molecules (p. 860). They are formed by the enzymatic modification of a standard ribonucleotide in a tRNA precursor. For example, uridylate residues are modified after transcription to form *ribothymidylate* and *pseudouridylate.* These modifications generate diversity, allowing greater structural and functional versatility.

Ribothymidylate

Uridylate

Pseudouridylate

29.2 Transcription in Eukaryotes Is Highly Regulated

We turn now to transcription in eukaryotes, a much more complex process than in prokaryotes. Eukaryotic cells have a remarkable ability to precisely regulate the time at which each gene is transcribed and how much RNA is produced. This ability has allowed some eukaryotes to evolve into multi-cellular organisms, with distinct tissues. *That is, multicellular eukaryotes use differential transcriptional regulation to create different cell types.* Eukaryotic cells achieve their precision through more complex transcriptional regulation. In addition, gene expression is influenced by three important characteristics unique to eukaryotes: the nuclear membrane, more complex transcriptional regulation, and RNA processing.

1. *The nuclear membrane. In eukaryotes, transcription and translation take place in different cellular compartments*: transcription takes place in the membrane-bounded nucleus, whereas translation takes place outside the nucleus in the cytoplasm. In prokaryotes, the two processes are closely coupled (Figure 29.15). Indeed, the translation of bacterial mRNA begins while

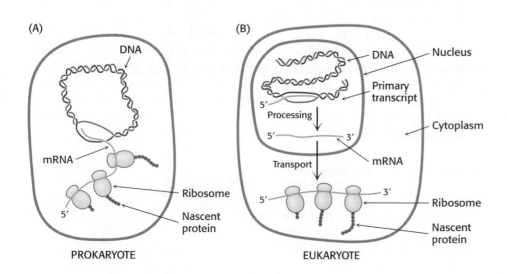

(A) PROKARYOTE

(B) EUKARYOTE

Figure 29.15 Transcription and translation. These two processes are closely coupled in prokaryotes, whereas they are spatially and temporally separate in eukaryotes. (A) In prokaryotes, the primary transcript serves as mRNA and is used immediately as the template for protein synthesis. (B) In eukaryotes, mRNA precursors are processed and spliced in the nucleus before being transported to the cytoplasm for translation into protein. [After J. Darnell, H. Lodish, and D. Baltimore. *Molecular Cell Biology,* 2d ed. (Scientific American Books, 1990), p. 230.]

the transcript is still being synthesized. *The spatial and temporal separation of transcription and translation enables eukaryotes to regulate gene expression in much more intricate ways, contributing to the richness of eukaryotic form and function.*

2. *More complex transcriptional regulation.* Like prokaryotes, eukaryotes rely on conserved sequences in DNA to regulate the initiation of transcription. But prokaryotes have only three promoter elements (the −10, −35, and UP elements), whereas eukaryotes use a variety of types of promoter elements, each identified by its own conserved sequence. Not all possible types will be present together in the same promoter. *In eukaryotes, elements that regulate transcription can be found at a variety of locations in DNA,* upstream or downstream of the start site and sometimes at distances much farther from the start site than in prokaryotes. For example, enhancer elements located on DNA far from the start site increase the promoter activity of specific genes.

3. *RNA processing.* Although both prokaryotes and eukaryotes modify RNA, *eukaryotes very extensively process nascent RNA destined to become mRNA.* This processing includes modifications to both ends and, most significantly, splicing out segments of the primary transcript. RNA processing is described in Section 29.3.

Three Types of RNA Polymerase Synthesize RNA in Eukaryotic Cells

In prokaryotes, RNA is synthesized by a single kind of polymerase. In contrast, the nucleus of a eukaryote contains three types of RNA polymerase differing in template specificity, location in the nucleus, and susceptibility to inhibitors (Table 29.2). All these polymerases are large proteins, containing from 8 to 14 subunits and having total molecular masses greater than 500 kd. *RNA polymerase I* is located in nucleoli, where it transcribes the tandem array of genes for 18S, 5.8S, and 28S ribosomal RNA (p. 839). The other ribosomal RNA molecule (5S rRNA, p. 839) and all the transfer RNA molecules (p. 840) are synthesized by *RNA polymerase III*, which is located in the nucleoplasm rather than in nucleoli. *RNA polymerase II*, which also is located in the nucleoplasm, synthesizes the precursors of messenger RNA as well as several small RNA molecules, such as those of the splicing apparatus (p. 844).

Although all eukaryotic RNA polymerases are homologous to one another and to prokaryotic RNA polymerase, RNA polymerase II contains a unique *carboxyl-terminal domain* on the 220-kd subunit called the CTD; this domain is unusual because it contains multiple repeats of a YSPTSPS consensus sequence. The activity of RNA polymerase II is regulated by phosphorylation mainly on the serine residues of the carboxyl-terminal domain. Another major distinction among the polymerases lies in their responses to the toxin α-*amanitin*, a cyclic octapeptide that contains several modified amino acids.

TABLE 29.2 Eukaryotic RNA polymerases

Type	Location	Cellular transcripts	Effects of α-amanitin
I	Nucleolus	18S, 5.8S, and 28S rRNA	Insensitive
II	Nucleoplasm	mRNA precursors and snRNA	Strongly inhibited
III	Nucleoplasm	tRNA and 5S rRNA	Inhibited by high concentrations

α-Amanitin

α-Amanitin is produced by the poisonous mushroom *Amanita phalloides*, which is also called the *death cup* or the *destroying angel* (Figure 29.16). More than a hundred deaths result worldwide each year from the ingestion of poisonous mushrooms. α-Amanitin binds very tightly (K_d = 10 nM) to RNA polymerase II and thereby blocks the elongation phase of RNA synthesis. Higher concentrations of α-amanitin (1 μM) inhibit polymerase III, whereas polymerase I is insensitive to this toxin. This pattern of sensitivity is highly conserved throughout the animal and plant kingdoms.

Eukaryotic genes, like prokaryotic genes, require promoters for transcription initiation. Like prokaryotic promoters, eukaryotic promoters consist of conserved sequences that serve to attract the polymerase to the start site. However, eukaryotic promoters differ distinctly in sequence and position, depending on the type of RNA polymerase to which they bind (Figure 29.17).

1. *RNA Polymerase I.* The ribosomal DNA (rDNA) transcribed by polymerase I is arranged in several hundred tandem repeats, each containing a copy of each of three rRNA genes. The promoter sequences are located in stretches of DNA separating the genes. At the transcriptional start site lies a TATA-like sequence called the *ribosomal initiator element* (rInr). Farther upstream, 150 to 200 bp from the start site, is the *upstream promoter element* (UPE). Both elements aid transcription by binding proteins that serve to recruit RNA polymerase I.

2. *RNA Polymerase II.* Promoters for RNA polymerase II, like prokaryotic promoters, include a set of conserved-sequence elements that define the start site and recruit the polymerase. However, the promoter can contain any combination of a number of possible elements. Unique to eukaryotes, they also include enhancer elements that can be very distant (more than 1 kb) from the start site (p. 838).

3. *RNA Polymerase III.* Promoters for RNA polymerase III are *within* the transcribed sequence, downstream of the start site. There are two types of intergenic promoters for

Figure 29.16 RNA polymerase poison. *Amanita phalloides,* a poisonous mushroom that produces α-amanitin. [After G. Lincoff and D. H. Mitchel, *Toxic and Hallucinogenic Mushroom Poisoning* (Van Nostrand Reinhold, 1977), p. 30.]

RNA polymerase I promoter

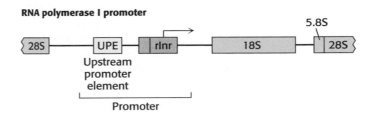

RNA polymerase II promoter

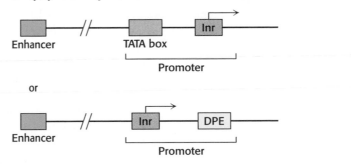

RNA polymerase III promoter

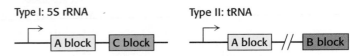

Figure 29.17 Common eukaryotic promoter elements. Each eukaryotic RNA polymerase recognizes a set of promoter elements—sequences in DNA that promote transcription. The RNA polymerase I promoter consists of a ribosomal initator (rInr) and an upstream promoter element (UPE). The RNA polymerase II promoter likewise includes an initator element (Inr) and may also include either a TATA box or a downstream promoter element (DPE). Separate from the promoter region, enhancer elements bind specific transcription factors. RNA polymerase III promoters consist of conserved sequences that lie within the transcribed genes.

$5'$ T_{82} A_{97} T_{93} A_{85} A_{63} A_{88} A_{50} $3'$
TATA box

Figure 29.18 TATA box. Comparisons of the sequences of more than 100 eukaryotic promoters led to the consensus sequence shown. The subscripts denote the frequency (%) of the base at that position.

$5'$ G G N C A A T C T $3'$
CAAT box

$5'$ G G G C G G $3'$
GC box

Figure 29.19 CAAT box and GC box. Consensus sequences for the CAAT and GC boxes of eukaryotic promoters for mRNA precursors.

RNA polymerase III. Type I promoters, found in the 5S rRNA gene, contain two short conserved sequences known as the A block and the C block. Type II promoters, found in tRNA genes, consist of two 11-bp sequences, the A block and the B block, situated about 15 bp from either end of the gene.

Three Common Elements Can Be Found in the RNA Polymerase II Promoter Region

Promoters for RNA polymerase II, like those for bacterial polymerases, are generally located on the $5'$ side of the start site for transcription. The results of mutagenesis experiments, footprinting studies, and comparisons of many higher eukaryotic genes have demonstrated the importance of several upstream regions. The most commonly recognized cis-acting element for genes transcribed by RNA polymerase II is called the *TATA box* on the basis of its consensus sequence (Figure 29.18). The TATA box is usually centered between positions -30 and -100. Note that the eukaryotic TATA box closely resembles the prokaryotic -10 sequence (TATAAT) but is farther from the start site. The mutation of a single base in the TATA box markedly impairs promoter activity. Thus, the precise sequence, not just a high content of AT pairs, is essential.

The TATA box is often paired with an *initiator element* (Inr), a sequence found at the transcriptional start site, between positions -3 and $+5$. This sequence defines the start site, because the other promoter elements are at variable distances from that site. Its presence increases transcriptional activity.

A third element, the *downstream core promoter element* (DPE), is commonly found in conjunction with the Inr in transcripts that lack the TATA box. In contrast with the TATA box, the DPE is found downstream of the start site, between positions $+28$ and $+32$.

Additional regulatory sequences are located between -40 and -150. Many promoters contain a *CAAT box*, and some contain a *GC box* (Figure 29.19). Constitutive genes (genes that are continuously expressed rather than regulated) tend to have GC boxes in their promoters. The positions of these upstream sequences vary from one promoter to another, in contrast with the quite constant location of the -35 region in prokaryotes. Another difference is that the CAAT box and the GC box can be effective when present on the template (antisense) strand, unlike the -35 region, which must be present on the coding (sense) strand. These differences between prokaryotes and eukaryotes correspond to fundamentally different mechanisms for the recognition of cis-acting elements. The -10 and -35 sequences in prokaryotic promoters are binding sites for RNA polymerase and its associated σ factor. In contrast, the TATA, CAAT, and GC boxes and other cis-acting elements in eukaryotic promoters are recognized by proteins other than RNA polymerase itself.

The TFIID Protein Complex Initiates the Assembly of the Active Transcription Complex

Cis-acting elements constitute only part of the puzzle of eukaryotic gene expression. Transcription factors that bind to these elements also are required. For example, RNA polymerase II is guided to the start site by a set of transcription factors known collectively as *TFII* (*TF* stands for transcription factor, and *II* refers to RNA polymerase II). Individual TFII factors are called TFIIA, TFIIB, and so on. Initiation begins with the binding of TFIID to the TATA box (Figure 29.20).

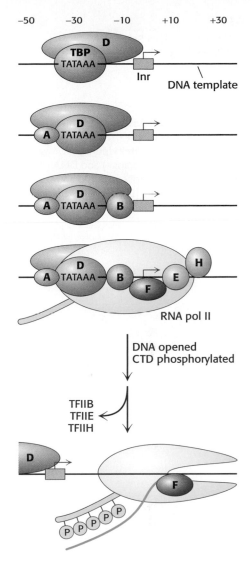

Figure 29.20 Transcription initation. Transcription factors TFIIA, B, D, E, F, and H are essential in initiating transcription by RNA polymerase II. The step-by-step assembly of these general transcription factors begins with the binding of TFIID (purple) to the TATA box. [The TATA-box-binding protein (TBP), a component of TFIID, recognizes the TATA box.] After assembly, TFIIH opens the DNA double helix and phosphorylates the carboxyl-terminal domain (CTD), allowing the polymerase to leave the promoter and begin transcription. The red arrow marks the transcription start site.

In TATA-box promoters, the key initial event is the recognition of the TATA box by the TATA-box-binding protein (TBP), a 30-kd component of the 700-kd TFIID complex. In TATA-less promoters, other proteins in the TFIID complex bind the core promoter elements but, because less is known about these interactions, we will consider only the TATA-box–TBP binding interaction. TBP binds 10^5 times as tightly to the TATA box as to noncognate sequences; the dissociation constant of the specific complex is approximately 1 nM. TBP is a saddle-shaped protein consisting of two similar domains (Figure 29.21). The TATA box of DNA binds to the concave surface of TBP. This binding induces large conformational changes in the bound DNA. The double helix is substantially unwound to widen its *minor groove*, enabling it to make extensive contact with the antiparallel β strands on the concave side of TBP. Hydrophobic interactions are prominent at this interface. Four phenylalanine residues, for example, are intercalated between base pairs of the TATA box. The flexibility of AT-rich sequences is generally exploited here in bending the DNA. Immediately outside the TATA box, classical B-DNA resumes. This complex is distinctly asymmetric. The asymmetry is crucial for specifying a unique start site and ensuring that transcription proceeds unidirectionally.

TBP bound to the TATA box is the heart of the initiation complex (see Figure 29.20). The surface of the TBP saddle provides docking sites for the binding of other components (Figure 29.22). Additional transcription factors assemble on this nucleus in a defined sequence. TFIIA is recruited, followed by TFIIB; then TFIIF, RNA polymerase II, TFIIE, and TFIIH join the other factors to form a complex called the *basal transcription apparatus*. During the formation of the basal transcription apparatus, the carboxyl-terminal domain (CTD) is unphosphorylated and plays a role in transcription regulation through its binding to an enhancer-associated complex called mediator (see Section 31.3). *Phosphorylation of the CTD by TFIIH marks the transition from initiation to elongation.* The phosphorylated CTD stabilizes transcription elongation by RNA polymerase II and recruits RNA-processing enzymes that act during elongation (p. 846). The importance of the carboxyl-terminal domain is highlighted by the finding that

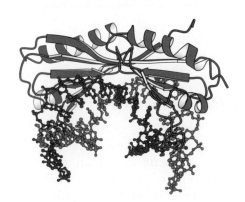

Figure 29.21 Complex formed by TATA-box-binding protein and DNA. The saddlelike structure of the protein sits atop a DNA fragment. *Notice* that the DNA is significantly unwound and bent. [Drawn from 1CDW.pdb.]

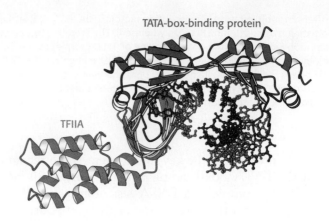

Figure 29.22 Assembly of the initiation complex. *Notice* that the TATA-box-binding protein (purple) forms a complex with TFIIA (orange) and DNA. TFIIA interacts primarily with the TATA-box-binding protein. [Drawn from 1YTF.pdb.]

yeast containing mutant polymerase II with fewer than 10 repeats in the CTD is not viable. Most of the factors are released before the polymerase leaves the promoter and can then participate in another round of initiation.

Although bacteria lack TBP, archaea utilize a TBP molecule that is structurally quite similar to the eukaryotic protein. In fact, transcriptional control processes in archaea are, in general, much more similar to those in eukaryotes than are the processes in bacteria. Many components of the eukaryotic transcriptional machinery evolved from an ancestor of archaea.

Multiple Transcription Factors Interact with Eukaryotic Promoters

The basal transcription complex described in the preceding section initiates transcription at a low frequency. Additional transcription factors that bind to other sites are required to achieve a high rate of mRNA synthesis. Their role is to selectively stimulate specific genes. Upstream stimulatory sites in eukaryotic genes are diverse in sequence and variable in position. Their variety suggests that they are recognized by many different specific proteins. Indeed, many transcription factors have been isolated, and their binding sites have been identified by footprinting experiments. For example, *heat-shock transcription factor* (HSTF) is expressed in *Drosophila* after an abrupt increase in temperature. This 93-kd DNA-binding protein binds to the consensus sequence

$$5'\text{-CNNGAANNTCCNNG-}3'$$

Several copies of this sequence, known as the *heat-shock response element,* are present starting at a site 15 bp upstream of the TATA box.

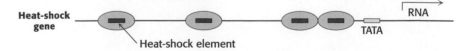

HSTF differs from σ^{32}, a heat-shock protein of *E. coli* (p. 826), in binding directly to response elements in heat-shock promoters rather than first becoming associated with RNA polymerase.

Enhancer Sequences Can Stimulate Transcription at Start Sites Thousands of Bases Away

The activities of many promoters in higher eukaryotes are greatly increased by another type of cis-acting element called an *enhancer.* Enhancer sequences have no promoter activity of their own *yet can exert their stimulatory actions over distances of several thousand base pairs. They can be upstream, downstream, or even in the midst of a transcribed gene.* Moreover, enhancers are effective when present on *either DNA strand* (equivalently, in either orientation). Enhancers in yeast are known as *upstream activator sequences* (UASs).

A particular enhancer is effective only in certain cells. For example, the immunoglobulin enhancer functions in B lymphocytes but not elsewhere. Cancer can result if the relation between genes and enhancers is disrupted. In Burkitt lymphoma and B-cell leukemia, a chromosomal translocation brings the proto-oncogene *myc* (a transcription factor itself) under the control of a powerful immunoglobin enhancer. The consequent dysregulation of the *myc* gene is believed to play a role in the progression of the cancer.

Transcription factors and other proteins that bind to regulatory sites on DNA can be regarded as passwords that cooperatively open multiple locks, giving RNA polymerase access to specific genes. The discovery of promoters and enhancers has opened the door to understanding how genes are selectively expressed in eukaryotic cells. The regulation of gene transcription, discussed in Chapter 31, is the fundamental means of controlling gene expression.

29.3 The Transcription Products of All Three Eukaryotic Polymerases Are Processed

Virtually all the initial products of transcription are further processed in eukaryotes. For example, primary transcripts (pre-mRNA molecules), the products of RNA polymerase II action, acquire a cap at their 5′ ends and a poly(A) tail at their 3′ ends. Most importantly, *nearly all mRNA precursors in higher eukaryotes are spliced* (p. 127). Introns are precisely excised from primary transcripts, and exons are joined to form mature mRNAs with continuous messages. Some mRNAs are only a tenth the size of their precursors, which can be as large as 30 kb or more. The pattern of splicing can be regulated in the course of development to generate variations on a theme, such as membrane-bound or secreted forms of antibody molecules. Alternative splicing enlarges the repertoire of proteins in eukaryotes and is a clear illustration of why the proteome is more complex than the genome. The particular processing steps and the factors taking part vary according to the type of RNA polymerase.

RNA Polymerase I Produces Three Ribosomal RNAs

RNA polymerase I transcription results in a single precursor (45S in mammals) that encodes three RNA components of the ribosome: the 18S rRNA, the 28S rRNA, and the 5.8S rRNA (Figure 29.23). The 18S rRNA is the RNA component of the small ribosomal subunit (40S), and the 28S and 5.8S rRNAs are two RNA components of the large ribosomal subunit (60S). The other RNA component of the large ribosomal subunit, the 5S rRNA, is transcribed by RNA polymerase III as a separate transcript.

The cleavage of the precursor into three separate rRNAs is actually the final step in its processing. First, the nucleotides of the pre-rRNA sequences destined for the ribosome undergo extensive modification, on both ribose and base components, directed by many *small nucleolar ribonucleoproteins* (snoRNPs), each of which consists of one snoRNA and several proteins. The pre-rRNA is assembled with ribosomal proteins, as guided by

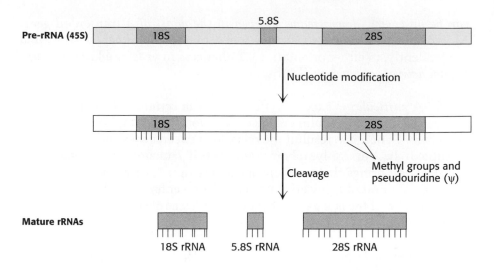

Pre-rRNA (45S) 18S 5.8S 28S

Nucleotide modification

18S 28S

Methyl groups and pseudouridine (ψ)

Cleavage

Mature rRNAs

18S rRNA   5.8S rRNA   28S rRNA

Figure 29.23 Processing of eukaryotic pre-rRNA. The mammalian pre-rRNA transcript contains the RNA sequences destined to become the 18S, 5.8S, and 28S rRNAs of the small and large ribosomal subunits. First, nucleotides are modified: small nucleolar ribonucleoproteins methylate specific ribose groups and convert selected uridines into pseudouridines (indicated by red lines). Next, the pre-rRNA is cleaved and packaged to form mature ribosomes, in a highly regulated process in which more than 200 proteins take part.

rDNA

Pre-rRNA

SSU processome

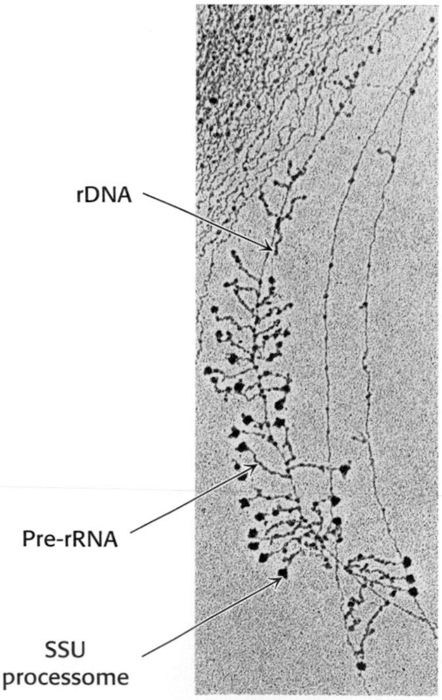

Figure 29.24 Visualization of rRNA transcription and processing in eukaryotes. Transcription of rRNA and its assembly into precursor-ribosomes can be visualized by electron microscopy. The structures resemble Christmas trees: the trunk is the rDNA and each branch is a pre-rRNA transcript. Transcription starts at the top of the tree, where the shortest transcripts can be seen, and progresses down the rDNA to the end of the gene. The terminal knobs visible at the end of some pre-rRNA transcripts likely correspond to the SSU processome, a large ribonucleoprotein required for processing the pre-rRNA. [From F. Dragon et. al. *Nature* 417(2002):967–970.]

processing factors, in a large ribonucleoprotein. For instance, the small-subunit (SSU) processome is required for 18S rRNA biogenesis and can be visualized in electronmicrographs as a terminal knob at the 5′ ends of the nascent rRNAs (Fig. 29.24). Finally, rRNA cleavage (sometimes coupled with additional processing steps) releases the mature rRNAs assembled with ribosomal proteins as ribosomes. Like those of RNA polymerase I transcription itself, most of these processing steps take place in the cell nucleolus, a nuclear subcompartment.

RNA Polymerase III Produces Transfer RNA

Nascently transcribed eukaryotic tRNAs are among the most processed of all RNA polymerase III transcripts. Like those of prokaryotic tRNAs, the 5′ leader is cleaved by RNase P, the 3′ trailer is removed, and CCA is added by the CCA-adding enzyme (Figure 29.25). Eukaryotic tRNAs are also heavily modified on base and ribose moieties; these modifications are important for function. In contrast with prokaryotic tRNAs, many eukaryotic pre-tRNAs are also spliced by an endonuclease and a ligase to remove an intron.

The Product of RNA Polymerase II, the Pre-mRNA Transcript, Acquires a 5′ Cap and a 3′ Poly(A) Tail

Perhaps the most extensively studied transcription product is the product of RNA polymerase II: most of this RNA will be processed to mRNA. The immediate product of RNA polymerase II is sometimes referred to as *pre-mRNA*. Most pre-mRNA molecules are spliced to remove the introns. Moreover, both the 5′ and the 3′ ends are modified, and both modifications are retained as the pre-mRNA is converted into mRNA.

As in prokaryotes, eukaryotic transcription usually begins with A or G. However, the 5′ triphosphate end of the nascent RNA chain is immediately modified. First, a phosphoryl group is released by hydrolysis. The diphosphate 5′ end then attacks the α-phosphorus atom of GTP to form a very unusual 5′–5′ triphosphate linkage. This distinctive terminus is called a *cap* (Figure 29.26). The N-7 nitrogen of the terminal guanine is then methylated by S-adenosylmethionine to form *cap 0*. The adjacent riboses may be methylated to form *cap 1* or *cap 2*. Transfer RNA and ribosomal RNA molecules, in contrast with messenger RNAs and with small RNAs that participate in splicing, do not have caps. Caps contribute to the stability of mRNAs by protecting their 5′ ends from phosphatases and nucleases. In addition, caps enhance the translation of mRNA by eukaryotic protein-synthesizing systems (p. 879).

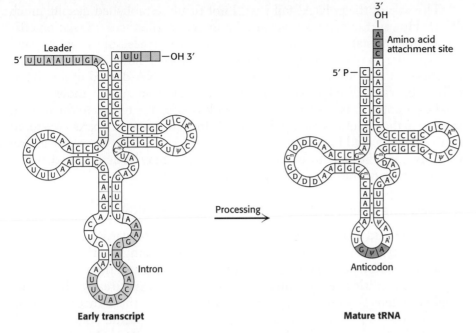

Early transcript　　　　　　　　**Mature tRNA**

Figure 29.25 **Transfer RNA precursor processing.** The conversion of a yeast tRNA precursor into a mature tRNA requires the removal of a 14-nucleotide intron (yellow), the cleavage of a 5′ leader (green), and the removal of UU and the attachment of CCA at the 3′ end (red). In addition, several bases are modified.

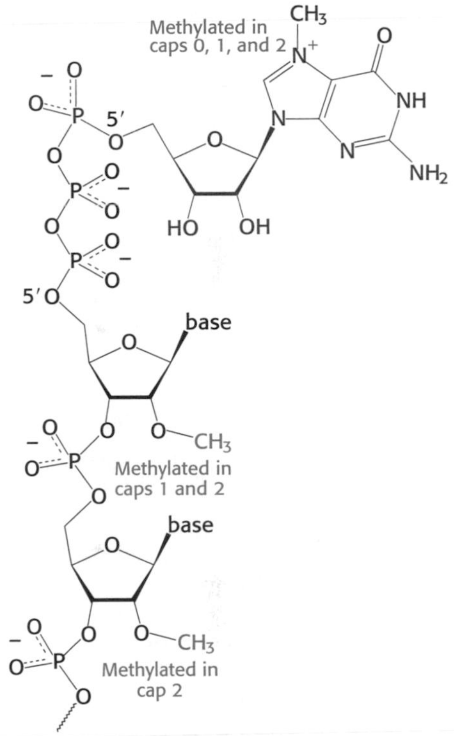

As mentioned earlier, pre-mRNA is also modified at the 3′ end. *Most eukaryotic mRNAs contain a polyadenylate, poly(A), tail at that end,* added after transcription has ended. Thus, DNA does not encode this poly(A) tail. Indeed, the nucleotide preceding poly(A) is not the last nucleotide to be transcribed. Some primary transcripts contain hundreds of nucleotides beyond the 3′ end of the mature mRNA.

How is the 3′ end of the pre-mRNA given its final form? *Eukaryotic primary transcripts are cleaved by a specific endonuclease that recognizes the sequence AAUAAA* (Figure 29.27). Cleavage does not take place if this sequence or a segment of some 20 nucleotides on its 3′ side is deleted. The presence of internal AAUAAA sequences in some mature mRNAs indicates that AAUAAA is only part of the cleavage signal; its context also is important. After cleavage of the pre-RNA by the endonuclease, a *poly(A) polymerase* adds about 250 adenylate residues to the 3′ end of the transcript; ATP is the donor in this reaction.

Figure 29.26 **Capping the 5′ end.** Caps at the 5′ end of eukaryotic mRNA include 7-methylguanylate (red) attached by a triphosphate linkage to the ribose at the 5′ end. None of the riboses are methylated in cap 0, one is methylated in cap 1, and both are methylated in cap 2.

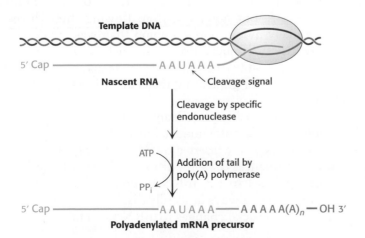

Figure 29.27 **Polyadenylation of a primary transcript.** A specific endonuclease cleaves the RNA downstream of AAUAAA. Poly(A) polymerase then adds about 250 adenylate residues.

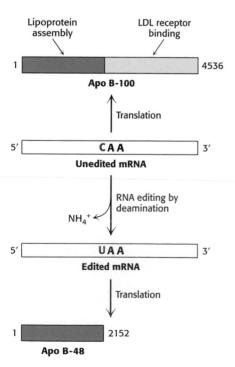

Figure 29.28 RNA editing. Enzyme-catalyzed deamination of a specific cytidine residue in the mRNA for apolipoprotein B-100 changes a codon for glutamine (CAA) to a stop codon (UAA). Apolipoprotein B-48, a truncated version of the protein lacking the LDL receptor-binding domain, is generated by this posttranscriptional change in the mRNA sequence. [After P. Hodges and J. Scott. *Trends Biochem. Sci.* 17(1992):77.]

The role of the poly(A) tail is still not firmly established despite much effort. However, evidence is accumulating that it enhances translation efficiency and the stability of mRNA. Blocking the synthesis of the poly(A) tail by exposure to *3'-deoxyadenosine (cordycepin)* does not interfere with the synthesis of the primary transcript. Messenger RNA devoid of a poly(A) tail can be transported out of the nucleus. However, an mRNA molecule devoid of a poly(A) tail is usually a much less effective template for protein synthesis than is one with a poly(A) tail. Indeed, some mRNAs are stored in an unadenylated form and receive the poly(A) tail only when translation is imminent. The half-life of an mRNA molecule may be determined in part by the rate of degradation of its poly(A) tail.

RNA Editing Changes the Proteins Encoded by mRNA

The amino acid sequence information encoded by some mRNAs is altered after transcription. *RNA editing* is the term for a change in the nucleotide sequence of RNA after transcription by processes other than RNA splicing. RNA editing is prominent in some systems already discussed. *Apolipoprotein B* (apo B) plays an important role in the transport of triacylglycerols and cholesterol by forming an amphipathic spherical shell around the lipids carried in lipoprotein particles (p. 743). Apo B exists in two forms, a 512-kd *apo B-100* and a 240-kd *apo B-48*. The larger form, synthesized by the liver, participates in the transport of lipids synthesized in the cell. The smaller form, synthesized by the small intestine, carries dietary fat in the form of chylomicrons. Apo B-48 contains the 2152 N-terminal residues of the 4536-residue apo B-100. This truncated molecule can form lipoprotein particles but cannot bind to the low-density-lipoprotein receptor on cell surfaces. What is the biosynthetic relation of these two forms of apo B? One possibility a priori is that apo B-48 is produced by proteolytic cleavage of apo B-100, and another is that the two forms arise from alternative splicing. Experiments show that neither event takes place. A totally unexpected and new mechanism for generating diversity is at work: *the changing of the nucleotide sequence of mRNA after its synthesis* (Figure 29.28). *A specific cytidine residue of mRNA is deaminated to uridine, which changes the codon at residue 2153 from CAA (Gln) to UAA (stop).* The deaminase that catalyzes this reaction is present in the small intestine, but not in the liver, and is expressed only at certain developmental stages.

RNA editing is not confined to apolipoprotein B. Glutamate opens cation-specific channels in the vertebrate central nervous system by binding to receptors in postsynaptic membranes. RNA editing changes a single glutamine codon (CAG) in the mRNA for the glutamate receptor to the codon for arginine (read as CGG). The substitution of Arg for Gln in the receptor prevents Ca^{2+}, but not Na^+, from flowing through the channel. RNA editing is likely much more common than was previously thought. The chemical reactivity of nucleotide bases, including the susceptibility to deamination that necessitates complex DNA-repair mechanisms, has been harnessed as an engine for generating molecular diversity at the RNA and, hence, protein levels.

In trypanosomes (parasitic protozoans), a different kind of RNA editing markedly changes several mitochondrial mRNAs. Nearly half the uridine residues in these mRNAs are *inserted* by RNA editing. A *guide RNA molecule* identifies the sequences to be modified, and a *poly(U) tail* on the guide donates uridine residues to the mRNAs undergoing editing. DNA sequences evidently do not always faithfully disclose the sequence of encoded proteins: functionally crucial changes to mRNA can take place.

Sequences at the Ends of Introns Specify Splice Sites in mRNA Precursors

Most genes in higher eukaryotes are composed of exons and introns. The introns must be excised and the exons must be linked to form the final mRNA in a process called *splicing*. This splicing must be exquisitely sensitive: a one-nucleotide shift would alter the reading frame on the 3′ side of the splice to give an entirely different amino acid sequence, likely including a premature stop codon. Thus, the correct splice site must be clearly marked. Does a particular sequence denote the splice site? The sequences of thousands of intron–exon junctions within RNA transcripts are known. In eukaryotes from yeast to mammals, these sequences have a common structural motif: *the intron begins with GU and ends with AG.* The consensus sequence at the 5′ splice in vertebrates is AG<u>GU</u>AAGU, where the GU is invariant (Figure 29.29). At the 3′ end of an intron, the consensus sequence is a stretch of *10 pyrimidines* (U or C; termed the *polypyrimidine tract*), followed by any base and then by C, and ending with the invariant AG. Introns also have an important internal site located between 20 and 50 nucleotides upstream of the 3′ splice site; it is called the *branch site* for reasons that will be evident shortly. In yeast, the branch-site sequence is nearly always UACUAAC, whereas in mammals a variety of sequences are found.

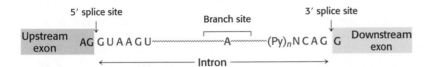

The 5′ and 3′ splice sites and the branch site are essential for determining where splicing takes place. Mutations in each of these three critical regions lead to aberrant splicing. Introns vary in length from 50 to 10,000 nucleotides, and so the splicing machinery may have to find the 3′ site several thousand nucleotides away. Specific sequences near the splice sites (in both the introns and the exons) play an important role in splicing regulation, particularly in designating splice sites when there are many alternatives (p. 847). These cis-acting sequences that contribute to splice-site selection are in the process of being characterized for individual mRNAs. Despite our knowledge of splice-site sequences, predicting pre-mRNAs and their protein products from genomic DNA sequence information remains a challenge.

Splicing Consists of Two Sequential Transesterification Reactions

The splicing of nascent mRNA molecules is a complicated process. It requires the cooperation of several small RNAs and proteins that form a large complex called a *spliceosome*. However, the chemistry of the splicing process is simple. Splicing begins with the cleavage of the phosphodiester bond between the upstream exon (exon 1) and the 5′ end of the intron (Figure 29.30). The attacking group in this reaction is the 2′-OH group of an adenylate residue in the branch site. A 2′–5′ phosphodiester bond is formed between this A residue and the 5′ terminal phosphate of the intron. This reaction is a transesterification.

Figure 29.29 Splice sites. Consensus sequences for the 5′ splice site and the 3′ splice site are shown. Py stands for pyrimidine.

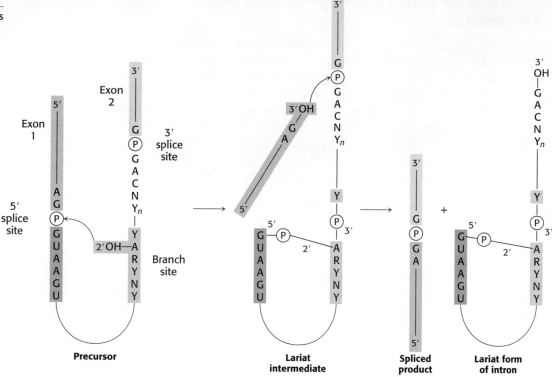

Precursor　　　　　　**Lariat**　　　**Spliced**　　　**Lariat form**
　　　　　　　　　　　　intermediate　　**product**　　**of intron**

Figure 29.30 Splicing mechanism used for mRNA precursors. The upstream (5′) exon is shown in blue, the downstream (3′) exon in green, and the branch site in yellow. Y stands for a pyrimidine nucleotide, R for a purine nucleotide, and N for any nucleotide. The 5′ splice site is attacked by the 2′-OH group of the branch-site adenosine residue. The 3′ splice site is attacked by the newly formed 3′-OH group of the upstream exon. The exons are joined, and the intron is released in the form of a lariat. [After P. A. Sharp. *Cell* 42(1985):397–408.]

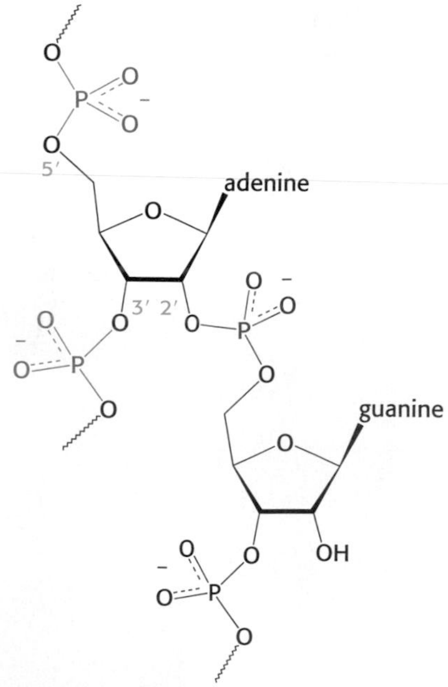

Note that this adenylate residue is also joined to two other nucleotides by normal 3′–5′ phosphodiester bonds (Figure 29.31). Hence a *branch* is generated at this site, and a *lariat intermediate* is formed.

The 3′-OH terminus of exon 1 then attacks the phosphodiester bond between the intron and exon 2. Exons 1 and 2 become joined, and the intron is released in lariat form. Again, this reaction is a transesterification. Splicing is thus accomplished by two *transesterification reactions* rather than by hydrolysis followed by ligation. The first reaction generates a free 3′-OH group at the 3′ end of exon 1, and the second reaction links this group to the 5′-phosphate of exon 2. *The number of phosphodiester bonds stays the same during these steps,* which is crucial because it allows the splicing reaction itself to proceed without an energy source such as ATP or GTP.

Small Nuclear RNAs in Spliceosomes Catalyze the Splicing of mRNA Precursors

The nucleus contains many types of small RNA molecules with fewer than 300 nucleotides, referred to as *snRNAs* (small nuclear RNAs). A few of them—designated U1, U2, U4, U5, and U6—are essential for splicing mRNA precursors. The secondary structures of these RNAs are highly conserved in organisms ranging from yeast to human beings. These RNA molecules are associated with specific proteins to form complexes termed *snRNPs* (small nuclear ribonucleoprotein particles); investigators often speak of them as "snurps." Spliceosomes are large (60S) dynamic assemblies composed of snRNPs, hundreds of other proteins called *splicing factors,* and the mRNA precursors being processed (Table 29.3).

Figure 29.31 Splicing branch point. The structure of the branch point in the lariat intermediate in which the adenylate residue is joined to three nucleotides by phosphodiester bonds. The new 2′-to-5′ linkage is shown in red, and the usual 3′-to-5′ linkages are shown in blue.

TABLE 29.3 Small nuclear ribonucleoprotein particles (snRNPs) in the splicing of mRNA precursors

snRNP	Size of snRNA (nucleotides)	Role
U1	165	Binds the 5′ splice site and then the 3′ splice site
U2	185	Binds the branch site and forms part of the catalytic center
U5	116	Binds the 5′ splice site
U4	145	Masks the catalytic activity of U6
U6	106	Catalyzes splicing

In mammalian cells, splicing begins with the recognition of the 5′ splice site by the U1 snRNP (Figure 29.32). In fact, U1 snRNA contains a highly conserved six-nucleotide sequence, not covered by protein in the snRNP, that base-pairs to the 5′ splice site of the pre-mRNA. This binding initiates spliceosome assembly on the pre-mRNA molecule.

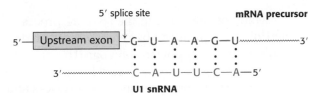

U1 snRNA

U2 snRNP then binds the branch site in the intron by base-pairing between a highly conserved sequence in U2 snRNA and the pre-mRNA. U2 snRNP binding requires ATP hydrolysis. A preassembled U4-U5-U6 tri-snRNP joins this complex of U1, U2, and the mRNA precursor to form the spliceosome. This association also requires ATP hydrolysis.

A revealing view of the interplay of RNA molecules in this assembly came from examining the pattern of cross-links formed by *psoralen*, a photoactivable reagent that joins neighboring pyrimidines in base-paired regions. These cross-links suggest that splicing takes place in the following way. First, U5 interacts with exon sequences in the 5′ splice site and subsequently with the 3′ exon. Next, U6 disengages from U4 and undergoes an intramolecular rearrangement that permits base-pairing with U2 as well as interaction with the 5′ end of the intron, displacing U1 from the spliceosome. The U2-U6 helix is indispensable for splicing, suggesting that *U2 and U6 snRNAs probably form the catalytic center of the spliceosome* (Figure 29.33). U4 serves as an inhibitor that masks U6 until the specific splice sites are aligned. These rearrangements result in the first transesterification reaction, cleaving the 5′ exon and generating the lariat intermediate.

Further rearrangements of RNA in the spliceosome facilitate the second transesterification. In these rearrangements, U5 aligns the free 5′ exon with the 3′ exon such that the 3′-hydroxyl group of the 5′ exon is positioned to nucleophilically attack the 3′ splice site to generate the spliced product. U2, U5, and U6 bound to the excised lariat intron are released to complete the splicing reaction.

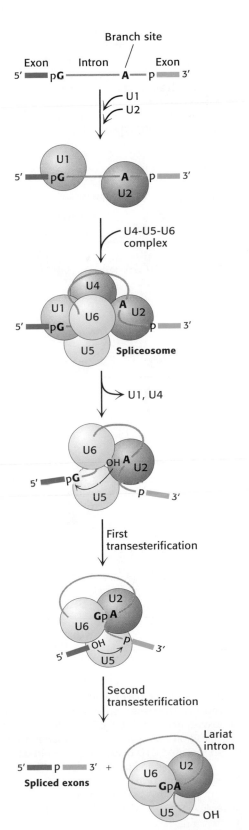

Figure 29.32 Spliceosome assembly and action. U1 binds the 5′ splice site and U2 binds to the branch point. A preformed U4-U5-U6 complex then joins the assembly to form the complete spliceosome. The U6 snRNA re-folds and binds the 5′ splice site, displacing U1. Extensive interactions between U6 and U2 displace U4. Then, in the first transesterification step, the branch-site adenosine attacks the 5′ splice site, making a lariat intermediate. U5 holds the two exons in close proximity, and the second transesterification takes place, with the 5′ splice-site hydroxyl group attacking the 3′ splice site. These reactions result in the mature spliced mRNA and a lariat form of the intron bound by U2, U5, and U6. [After T. Villa, J. A. Pleiss, and C. Guthrie, *Cell* 109(2002):149–152.]

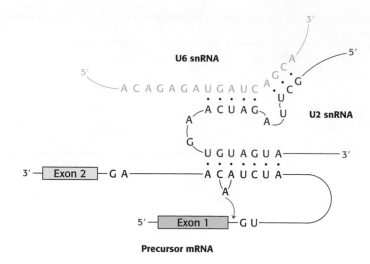

Figure 29.33 Splicing catalytic center.
The catalytic center of the spliceosome is
formed by U2 snRNA (red) and U6 snRNA
(green), which are base paired. U2 is also
base paired to the branch site of the
mRNA precursor. [After H. D. Madhani and
C. Guthrie. *Cell* 71(1992):803–817.]

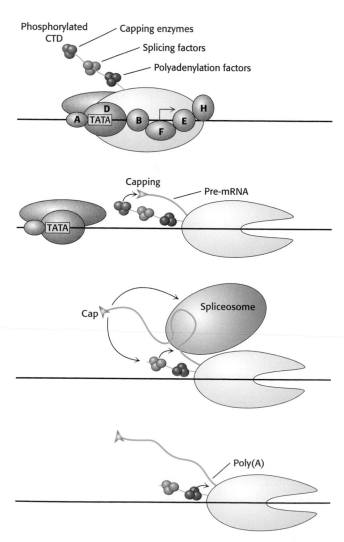

**Figure 29.34 The CTD: Coupling transcription to pre-mRNA
processing.** The transcription factor TFIIH phosphorylates the
carboxyl-terminal domain (CTD) of RNA polymerase II, signaling
the transition from transcription initiation to elongation. The
phosphorylated CTD binds factors required for pre-mRNA
capping, splicing, and polyadenylation. These proteins are
brought in close proximity to their sites of action on the
nascent pre-mRNA as it is transcribed during elongation.
[After P. A. Sharp. *TIBS* 30(2005):279–281.]

Many of the steps in the splicing process require ATP
hydrolysis. How is the free energy associated with ATP hydrolysis used to power splicing? To achieve the well-or-
dered rearrangements necessary for splicing, ATP-powered
RNA helicases must unwind RNA helices and allow alternative base-pairing arrangements to form. Thus, two features of the splicing process are noteworthy. First, *RNA
molecules play key roles in directing the alignment of splice
sites and in carrying out catalysis.* Second, *ATP-powered helicases unwind RNA duplex intermediates that facilitate
catalysis and induce the release of snRNPs from the mRNA.*

Transcription and Processing of mRNA Are Coupled

Although we have described the transcription and processing of mRNAs as separate events in gene expression, experimental evidence suggests that the two steps are coordinated by the carboxyl-terminal domain of RNA
polymerase II. We have seen that the CTD consists of a
unique repeated seven-amino-acid sequence, YSPTSPS.
Either S_2 or S_5 or both may be phosphorylated in the various repeats. The phosphorylation state of the CTD is controlled by a number of kinases and phosphatases and leads
the CTD to bind many of the proteins having roles in
RNA transcription and processing. The CTD contributes
to efficient transcription by recruiting these proteins to the
pre-mRNA (Figure 29.34), including:

1.   capping enzymes, which methylate the 5′ guanine on
the pre-mRNA immediately after transcription begins;

2.   components of the splicing machinery, which initiate
the excision of each intron as it is synthesized; and

3.   an endonuclease that cleaves the transcript at the
poly(A) addition site, creating a free 3′-OH group that is
the target for 3′ adenylation.

These events take place sequentially, directed by the phosphorylation state of the CTD.

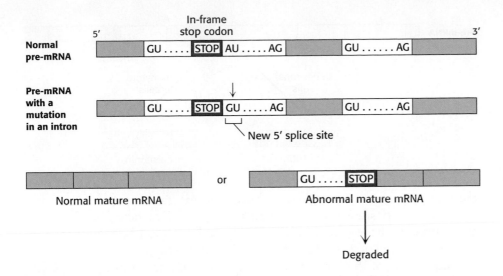

Figure 29.35 A splicing mutation that causes thalassemia. An A-to-G mutation within the first intron of the gene for the human hemoglobin β chain creates a new 5' splice site (GU). Both 5' splice sites are recognized by the U1 snRNP; so splicing may sometimes create a normal mature mRNA and an abnormal mature mRNA that contains intron sequences. The normal mature mRNA is translated into a hemoglobin β chain. Because it includes intron sequences, the abnormal mature mRNA now has a premature stop codon and is degraded.

Mutations That Affect Pre-mRNA Splicing Cause Disease

Mutations in either the pre-mRNA (cis-acting) or the splicing factors (trans-acting) can cause defective pre-mRNA splicing. Mutations in the pre-mRNA cause some forms of thalassemia, a group of hereditary anemias characterized by the defective synthesis of hemoglobin (p. 196). Cis-acting mutations that cause aberrant splicing can occur at the 5' or 3' splice sites in either of the two introns of the hemoglobin β chain or in its exons. The mutations usually result in an incorrectly spliced pre-mRNA that, because of a premature stop codon, cannot encode a full-length protein. The defective mRNA is normally degraded rather than translated. Mutations in the 5' splice site may alter that site so that the splicing machinery cannot recognize it, forcing the machinery to find another 5' splice site in the intron and introducing the potential for a premature stop codon. Mutations in the intron itself may create a new 5' splice site; in this case, either one of the two splice sites may be recognized (Figure 29.35). Consequently, some normal protein can be made, and so the disease is less severe. *Mutations affecting splicing have been estimated to cause at least 15% of all genetic diseases.*

Disease-causing mutations may also appear in splicing factors. Retinitis pigmentosa is a disease of acquired blindness, first described in 1857, with an incidence of 1/3500. About 5% of the autosomal dominant form of retinitis pigmentosa is likely due to mutations in the hPrp8 protein, a pre-mRNA splicing factor that is a component of the U4-U5-U6 tri-snRNP. How a mutation in a splicing factor that is present in all cells causes disease only in the retina is not clear; nevertheless, retinitis pigmentosa is a good example of how mutations that disrupt spliceosome function can cause disease.

Most Human Pre-mRNAs Can Be Spliced in Alternative Ways to Yield Different Proteins

Alternative splicing is a widespread mechanism for generating protein diversity. Different combinations of exons from the same gene may be spliced into a mature RNA, producing distinct forms of a protein for specific tissues, developmental stages, or signaling pathways. What controls which splicing sites are selected? The selection is determined by the binding of trans-acting splicing factors to cis-acting sequences in the pre-mRNA. Most alternative splicing leads to changes in the coding sequence, resulting in proteins with different functions. *Alternative splicing provides a powerful mechanism for expanding*

Figure 29.36 An example of alternative splicing. In human beings, two very different hormones are produced from a single calcitonin/CGRP pre-mRNA. Alternative splicing produces the mature mRNA for either calcitonin or CGRP (calcitonin-gene-related protein), depending on the cell type in which the gene is expressed. Each alternative transcript incorporates one of two alternative polyadenylation signals (A) present in the pre-mRNA.

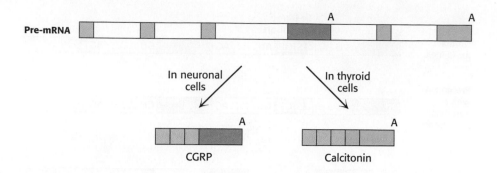

the versatility of genomic sequences through combinatorial control. Consider a gene with five positions at which alternative splicing can take place. With the assumption that these alternative splicing pathways can be regulated independently, a total of $2^5 = 32$ different mRNAs can be generated.

Sequencing of the human genome has revealed that most pre-mRNAs are alternatively spliced, leading to a much greater number of proteins than would be predicted from the number of genes. An example of alternative splicing leading to the expression of two different proteins, each in a different tissue, is provided by the gene encoding both calcitonin and calcitonin-gene-related peptide (CGRP; Figure 29.36). In the thyroid gland, the inclusion of exon 4 in one splicing pathway produces calcitonin, a peptide hormone that regulates calcium and phosphorus metabolism. In neuronal cells, the exclusion of exon 4 in another splicing pathway produces CGRP, a peptide hormone that acts as a vasodilator. A single pre-mRNA thus yields two very different peptide hormones, depending on cell type. In this case, only two proteins result from alternative splicing; however, in other cases, many more can be produced. An extreme example is the *Drosophila* pre-mRNA that encodes DSCAM, a neuronal protein affecting axon connectivity. Alternative splicing of this pre-mRNA has the potential to produce 38,016 different combinations of exons, a greater number than the total number of genes in the *Drosophila* genome. Several human diseases that can be attributed to defects in alternative splicing are listed in Table 29.4. Further understanding of alternative splicing and the mechanisms of splice-site selection will be crucial to understanding how the proteome is represented by the human genome.

29.4 The Discovery of Catalytic RNA Was Revealing in Regard to Both Mechanism and Evolution

RNAs form a surprisingly versatile class of molecules. As we have seen, splicing is catalyzed largely by RNA molecules, with proteins playing a secondary role. Another enzyme that contains a key RNA component is

TABLE 29.4 Selected human diseases attributed to defects in alternative splicing

Disorder	Gene or its product
Acute intermittent porphyria	Porphobilinogen deaminase
Breast and ovarian cancer	*BRCA1*
Cystic fibrosis	*CFTR*
Frontotemporal dementia	τ protein
Hemophilia A	Factor VIII
HGPRT deficiency (Lesch–Nyhan syndrome)	Hypoxanthine-guanine phosphoribosyltransferase
Leigh encephalomyelopathy	Pyruvate dehydrogenase E1α
Severe combined immunodeficiency	Adenosine deaminase
Spinal muscle atrophy	*SMN1* or *SMN2*

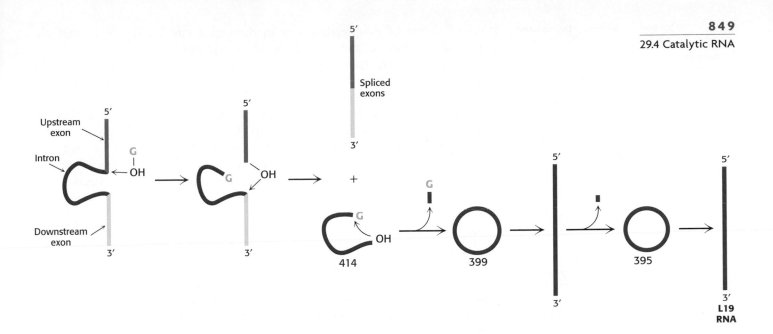

ribonuclease P, which catalyzes the maturation of tRNA by endonucleolytic cleavage of nucleotides from the 5′ end of the precursor molecule (p. 840). Finally, as we shall see in Chapter 30, the RNA component of ribosomes is the catalyst that carries out protein synthesis.

The versatility of RNA first became clear from observations of the processing of ribosomal RNA in a single-cell eukaryote. In *Tetrahymena* (a ciliated protozoan), a 414-nucleotide intron is removed from a 6.4-kb precursor to yield the mature 26S rRNA molecule (Figure 29.37). In an elegant series of studies of this splicing reaction, Thomas Cech and his coworkers established that the RNA spliced itself to precisely excise the intron. These remarkable experiments demonstrated that an RNA molecule can *splice itself* in the absence of protein. Indeed, the RNA alone is catalytic and, under certain conditions, is thus a *ribozyme*. More than 1500 similar introns have since been found in species as widely dispersed as bacteria and eukaryotes, though not in vertebrates. Collectively, they are referred to as *group I introns*.

The *self-splicing* reaction in the group I intron requires an added guanosine nucleotide. Nucleotides were originally included in the reaction mixture because it was thought that ATP or GTP might be needed as an energy source. Instead, the nucleotides were found to be necessary as cofactors. The required cofactor proved to be a guanosine unit, in the form of guanosine, GMP, GDP, or GTP. G (denoting any one of these species) serves not as an energy source but as an attacking group that becomes transiently incorporated into the RNA (see Figure 29.37). G binds to the RNA and then attacks the 5′ splice site to form a phosphodiester bond with the 5′ end of the intron. This transesterification reaction generates a 3′-OH group at the end of the upstream exon. This newly attached 3′-OH group then attacks the 3′ splice site. This second transesterification reaction joins the two exons and leads to the release of the 414-nucleotide intron.

Self-splicing depends on the structural integrity of the RNA precursor. Much of the group I intron is needed for self-splicing. This molecule, like many RNAs, has a folded structure formed by many double-helical stems and loops (Figure 29.38), with a well-defined pocket for binding the guanosine. Examination of the three-dimensional structure of a catalytically active group I intron determined by x-ray crystallography reveals the

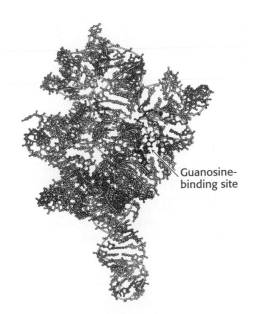

Guanosine-binding site

Figure 29.38 Structure of a self-splicing intron. The structure of a large fragment of the self-splicing intron from *Tetrahymena* reveals a complex folding pattern of helices and loops. Bases are shown in green, A; yellow, C; purple, G; and orange, U.

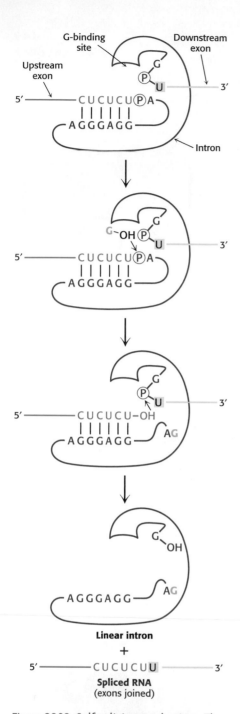

Figure 29.39 Self-splicing mechanism. The catalytic mechanism of the group I intron includes a series of transesterification reactions. [After T. Cech. RNA as an enzyme. Copyright © 1986 by Scientific American, Inc. All rights reserved.]

coordination of magnesium ions in the active site analogous to that observed in protein enzymes such as DNA polymerase.

Analysis of the base sequence of the rRNA precursor suggested that the splice sites are aligned with the catalytic residues by base-pairing between the *internal guide sequence* (IGS) in the intron and the 5′ and 3′ exons (Figure 29.39). The IGS first brings together the guanosine cofactor and the 5′ splice site so that the 3′-OH group of G can nucleophilically attack the phosphorus atom at this splice site. The IGS then holds the downstream exon in position for attack by the newly formed 3′-OH group of the upstream exon. A phosphodiester bond is formed between the two exons, and the intron is released as a linear molecule. Like catalysis by protein enzymes, self-catalysis of bond formation and breakage in this rRNA precursor is highly specific.

The finding of enzymatic activity in the self-splicing intron and in the RNA component of RNase P has opened new areas of inquiry and changed the way in which we think about molecular evolution. The discovery that RNA can be a catalyst as well as an information carrier suggests that an RNA world may have existed early in the evolution of life, before the appearance of DNA and protein.

Messenger RNA precursors in the mitochondria of yeast and fungi also undergo self-splicing, as do some RNA precursors in the chloroplasts of unicellular organisms such as *Chlamydomonas*. Self-splicing reactions can be classified according to the nature of the unit that attacks the upstream splice site. Group I self-splicing is mediated by a guanosine cofactor, as in *Tetrahymena*. The attacking moiety in group II splicing is the 2′-OH group of a specific adenylate of the intron (Figure 29.40).

Group I and group II self-splicing resembles spliceosome-catalyzed splicing in two respects. First, in the initial step, a ribose hydroxyl group attacks the 5′ splice site. The newly formed 3′-OH terminus of the upstream exon then attacks the 3′ splice site to form a phosphodiester bond with the downstream exon. Second, both reactions are transesterifications in which the phosphate moieties at each splice site are retained in the products. The number of phosphodiester bonds stays constant. Group II splicing is like the spliceosome-catalyzed splicing of mRNA precursors in several additional ways. The attack at the 5′ splice site is carried out by a part of the intron itself (the 2′-OH group of adenosine) rather than by an external cofactor (G). In both cases, the intron is released in the form of a lariat. Moreover, in some instances, the group II intron is transcribed in pieces that assemble through hydrogen bonding to the catalytic intron, in a manner analogous to the assembly of the snRNAs in the spliceosome.

These similarities have led to the suggestion that the spliceosome-catalyzed splicing of mRNA precursors evolved from RNA-catalyzed self-splicing. Group II splicing may well be an intermediate between group I splicing and the splicing in the nuclei of higher eukaryotes. *A major step in this transition was the transfer of catalytic power from the intron itself to other molecules.* The formation of spliceosomes gave genes a new freedom because introns were no longer constrained to provide the catalytic center for splicing. Another advantage of external catalysts for splicing is that they can be more readily regulated. However, it is important to note that similarities do not establish ancestry. The similarities between group II introns and mRNA splicing may be a result of convergent evolution. Perhaps there are only a limited number of ways to carry out efficient, specific intron excision. The determination of whether these similarities stem from ancestry or from chemistry will require expanding our understanding of RNA biochemistry.

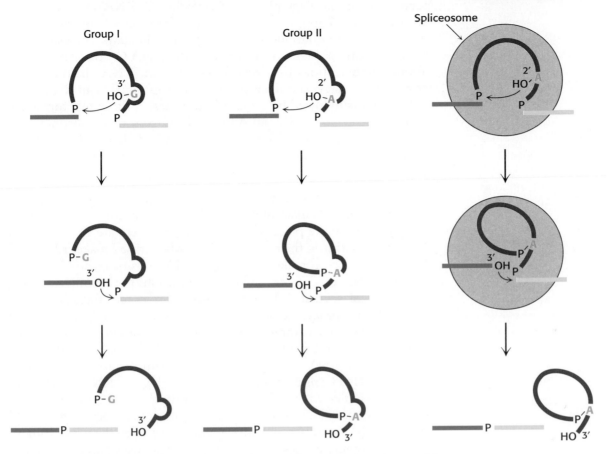

SELF-SPLICING INTRONS

Group I

Group II

SPLICEOSOME-CATALYZED SPLICING
OF NUCLEAR mNRA

Spliceosome

Figure 29.40 Comparison of splicing pathways. The exons being joined are shown in blue and yellow and the attacking unit is shown in green. The catalytic site is formed by the intron itself (red) in group I and group II splicing. In contrast, the splicing of nuclear mRNA precursors is catalyzed by snRNAs and their associated proteins in the spliceosome. [After P. A. Sharp. *Science* 235(1987):766–771.]

Summary

29.1 RNA Polymerase Catalyzes Transcription

All cellular RNA molecules are synthesized by RNA polymerases according to instructions given by DNA templates. The activated monomer substrates are ribonucleoside triphosphates. The direction of RNA synthesis is $5' \rightarrow 3'$, as in DNA synthesis. RNA polymerases, unlike DNA polymerases, do not need a primer.

RNA polymerase in *E. coli* is a multisubunit enzyme. The subunit composition of the ~500-kd holoenzyme is $\alpha_2\beta\beta'\sigma$ and that of the core enzyme is $\alpha_2\beta\beta'$. Transcription is initiated at promoter sites consisting of two sequences, one centered near -10 and the other near -35; that is, 10 and 35 nucleotides away from the start site in the 5' (upstream) direction. The consensus sequence of the -10 region is TATAAT. The σ subunit enables the holoenzyme to recognize promoter sites. When the growth temperature is raised, *E. coli* expresses a special σ subunit that selectively binds the distinctive promoter of heat-shock genes. RNA polymerase must unwind the template double helix for transcription to take place. Unwinding exposes some 17 bases on the template strand and sets the stage for the formation of the first phosphodiester bond. RNA chains usually start with pppG or pppA. The σ subunit dissociates from the holoenzyme after the initiation of the new chain. Elongation takes place at transcription bubbles that move along the DNA template at a rate of about 50 nucleotides per second. The nascent RNA chain contains stop signals that end transcription. One stop signal is an RNA hairpin, which is followed by several U residues. A different stop signal is read by the *rho* protein, an ATPase. In *E. coli,* precursors of transfer RNA and ribosomal RNA are cleaved and chemically modified after transcription, whereas messenger RNA is used unchanged as a template for protein synthesis.

29.2 Transcription in Eukaryotes Is Highly Regulated

RNA synthesis in eukaryotes takes place in the nucleus, whereas protein synthesis takes place in the cytoplasm. There are three types of RNA polymerase in the nucleus: RNA polymerase I makes ribosomal RNA precursors, II makes messenger RNA precursors, and III makes transfer RNA precursors. Eukaryotic promoters are complex, being composed of several different elements. Promoters for RNA polymerase II may be located on the 5′ side or the 3′ side of the start site for transcription. One common type of eukaryotic promoter consists of a TATA box centered between −30 and −100 and paired with an initiator element (Inr). Eukaryotic promoter elements are recognized by proteins called transcription factors rather than by RNA polymerase II. The saddle-shaped TATA-box-binding protein unwinds and sharply bends DNA at TATA-box sequences and serves as a focal point for the assembly of transcription complexes. The TATA-box-binding protein initiates the assembly of the active transcription complex. The activity of many promoters is greatly increased by enhancer sequences that have no promoter activity of their own. Enhancer sequences can act over distances of several kilobases, and they can be located either upstream or downstream of a gene.

29.3 The Transcription Products of All Three Eukaryotic Polymerases Are Processed

The 5′ ends of mRNA precursors become capped and methylated in the course of transcription. A 3′ poly(A) tail is added to most mRNA precursors after the nascent chain has been cleaved by an endonuclease. RNA editing processes alter the nucleotide sequence of some mRNAs, such as the one for apolipoprotein B.

The splicing of mRNA precursors is carried out by spliceosomes, which consist of small nuclear ribonucleoprotein particles (snRNPs). Splice sites in mRNA precursors are specified by sequences at ends of introns and by branch sites near their 3′ ends. The 2′-OH group of an adenosine residue in the branch site attacks the 5′ splice site to form a lariat intermediate. The newly generated 3′-OH terminus of the upstream exon then attacks the 3′ splice site to become joined to the downstream exon. Splicing thus consists of two transesterification reactions, with the number of phosphodiester bonds remaining constant during reactions. Small nuclear RNAs in spliceosomes catalyze the splicing of mRNA precursors. In particular, U2 and U6 snRNAs form the active centers of spliceosomes.

The events in posttranscriptional processing of mRNA are controlled by the phosphorylation state of the carboxy-terminal domain (CTD), part of RNA polymerase II.

Some RNA molecules, such as those containing the group I intron, undergo self-splicing in the absence of protein. A self-modified version of this rRNA intron displays true catalytic activity and is thus a ribozyme. Spliceosome-catalyzed splicing may have evolved from self-splicing. The discovery of catalytic RNA has opened new vistas in our exploration of early stages of molecular evolution and the origins of life.

Key Terms

transcription (p. 822)

RNA polymerase (p. 822)

promoter site (p. 823)

transcription factor (p. 823)

footprinting (p. 824)

consensus sequence (p. 825)

sigma (σ) subunit (p. 825)

transcription bubble (p. 828)

rho (ρ) protein (p. 830)

carboxy-terminal domain (CTD) (p. 834)

TATA box (p. 836)

enhancer (p. 838)

small nucleolar ribonucleoprotein
(snoRNP) (p. 839)

pre-mRNA (p. 840)

5′ cap (p. 840)

poly(A) tail (p. 841)

RNA editing (p. 842)

RNA splicing (p. 843)

spliceosome (p. 843)

small nuclear RNA (snRNA) (p. 844)

small nuclear ribonucleoprotein particles
(snRNP) (p.844)

alternative splicing (p. 847)

catalytic RNA (p. 848)

self-splicing (p. 849)

Selected Readings

Where to Start

Woychik, N. A. 1998. Fractions to functions: RNA polymerase II thirty years later. *Cold Spring Harbor Symp. Quant. Biol.* 63:311–317.

Losick, R. 1998. Summary: Three decades after sigma. *Cold Spring Harbor Symp. Quant. Biol.* 63:653–666.

Ast, G. 2005. The alternative genome. *Sci. Am.* 292(4):40–47.

Sharp, P. A. 1994. Split genes and RNA splicing (Nobel Lecture). *Angew. Chem. Int. Ed. Engl.* 33:1229–1240.

Cech, T. R. 1990. Nobel lecture: Self-splicing and enzymatic activity of an intervening sequence RNA from *Tetrahymena*. *Biosci. Rep.* 10: 239–261.

Villa, T., Pleiss, J. A., and Guthrie, C. 2002. Spliceosomal snRNAs: Mg^{2+} dependent chemistry at the catalytic core? *Cell* 109:149–152.

Books

Lewin, B. 2000. *Genes* (7th ed.). Oxford University Press.

Kornberg, A., and Baker, T. A. 1992. *DNA Replication* (2d ed.). W. H. Freeman and Company.

Lodish, H., Berk, A., Matsudaira, P., Kaiser, C. A., Krieger, M., Scott, M. P., Zipursky, S. L., and Darnell, J. 2004. *Molecular Cell Biology* (5th ed.). W. H. Freeman and Company.

Watson, J. D., Baker, T. A., Bell, S. P., Gann, A., Levine, M., and Losick, R. 2004. *Molecular Biology of the Gene* (5th ed.). Pearson/Benjamin Cummings.

Gesteland, R. F., Cech, T., and Atkins, J. F. 2006. *The RNA World: The nature of Modern RNA Suggests a Prebiotic RNA* (3d ed.) Cold Spring Harbor Laboratory Press.

RNA Polymerases

Darst, S. A. 2001. Bacterial RNA polymerase. *Curr. Opin. Struct. Biol.* 11:155–162.

Ross, W., Gosink, K. K., Salomon, J., Igarashi, K., Zou, C., Ishihama, A., Severinov, K., and Gourse, R. L. 1993. A third recognition element in bacterial promoters: DNA binding by the alpha subunit of RNA polymerase. *Science* 262:1407–1413.

Cramer, P., Bushnell, D. A., and Kornberg, R. D. 2001. Structural basis of transcription: RNA polymerase II at 2.8 Å resolution. *Science* 292:1863–1875.

Gnatt, A. L., Cramer, P., Fu, J., Bushnell, D. A., and Kornberg, R. D. 2001. Structural basis of transcription: An RNA polymerase II elongation complex at 3.3 Å resolution. *Science* 292:1876–1882.

Zhang, G., Campbell, E. A., Minakhin, L., Richter, C., Severinov, K., and Darst, S. A. 1999. Crystal structure of *Thermus aquaticus* core RNA polymerase at 3.3 Å resolution. *Cell* 98:811–824.

Campbell, E. A., Korzheva, N., Mustaev, A., Murakami, K., Nair, S., Goldfarb, A., and Darst, S. A. 2001. Structural mechanism for rifampicin inhibition of bacterial RNA polymerase. *Cell* 104:901–912.

Darst, S. A. 2004. New inhibitors targeting bacterial RNA polymerase. *Trends Biochem. Sci.* 29:159–160.

Cheetham, G. M., and Steitz, T. A. 1999. Structure of a transcribing T7 RNA polymerase initiation complex. *Science* 286:2305–2309.

Ebright, R. H. 2000. RNA polymerase: Structural similarities between bacterial RNA polymerase and eukaryotic RNA polymerase II. *J. Mol. Biol.* 304:687–698.

Paule, M. R., and White, R. J. 2000. Survey and summary: Transcription by RNA polymerases I and III. *Nucleic Acids Res.* 28:1283–1298.

Initiation and Elongation

Murakami, K. S., and Darst, S. A. 2003. Bacterial RNA polymerases: The whole story. *Curr. Opin. Struct. Biol.* 13:31–39.

Buratowski, S. 2000. Snapshots of RNA polymerase II transcription initiation. *Curr. Opin. Cell Biol.* 12:320–325.

Conaway, J. W., and Conaway, R. C. 1999. Transcription elongation and human disease. *Annu. Rev. Biochem.* 68:301–319.

Conaway, J. W., Shilatifard, A., Dvir, A., and Conaway, R. C. 2000. Control of elongation by RNA polymerase II. *Trends Biochem. Sci.* 25:375–380.

Korzheva, N., Mustaev, A., Kozlov, M., Malhotra, A., Nikiforov, V., Goldfarb, A., and Darst, S. A. 2000. A structural model of transcription elongation. *Science* 289:619–625.

Reines, D., Conaway, R. C., and Conaway, J. W. 1999. Mechanism and regulation of transcriptional elongation by RNA polymerase II. *Curr. Opin. Cell Biol.* 11:342–346.

Promoters, Enhancers, and Transcription Factors

Merika, M., and Thanos, D. 2001. Enhanceosomes. *Curr. Opin. Genet. Dev.* 11:205–208.

Park, J. M., Gim, B. S., Kim, J. M., Yoon, J. H., Kim, H. S., Kang, J. G., and Kim, Y. J. 2001. *Drosophila* mediator complex is broadly utilized by diverse gene-specific transcription factors at different types of core promoters. *Mol. Cell. Biol.* 21:2312–2323.

Smale, S. T., and Kadonaga, J. T. 2003. The RNA polymerase II core promoter. *Annu. Rev. Biochem.* 72:449–479.

Gourse, R. L., Ross, W., and Gaal, T. 2000. Ups and downs in bacterial transcription initiation: The role of the alpha subunit of RNA polymerase in promoter recognition. *Mol. Microbiol.* 37:687–695.

Fiering, S., Whitelaw, E., and Martin, D. I. 2000. To be or not to be active: The stochastic nature of enhancer action. *Bioessays* 22:381–387.

Hampsey, M., and Reinberg, D. 1999. RNA polymerase II as a control panel for multiple coactivator complexes. *Curr. Opin. Genet. Dev.* 9:132–139.

Chen, L. 1999. Combinatorial gene regulation by eukaryotic transcription factors. *Curr. Opin. Struct. Biol.* 9:48–55.

Muller, C. W. 2001. Transcription factors: Global and detailed views. *Curr. Opin. Struct. Biol.* 11:26–32.

Reese, J. C. 2003. Basal transcription factors. *Curr. Opin. Genet. Dev.* 13:114–118.

Kadonaga, J. T. 2004. Regulation of RNA polymerase II transcription by sequence-specific DNA binding factors. *Cell* 116:247–257.

Harrison, S. C. 1991. A structural taxonomy of DNA-binding domains. *Nature* 353:715–719.

Sakurai, H., and Fukasawa, T. 2000. Functional connections between mediator components and general transcription factors of *Saccharomyces cerevisiae*. *J. Biol. Chem.* 275:37251–37256.

Droge, P., and Muller-Hill, B. 2001. High local protein concentrations at promoters: Strategies in prokaryotic and eukaryotic cells. *Bioessays* 23:179–183.

Smale, S. T., Jain, A., Kaufmann, J., Emami, K. H., Lo, K., and Garraway, I. P. 1998. The initiator element: A paradigm for core promoter heterogeneity within metazoan protein-coding genes. *Cold Spring Harbor Symp. Quant. Biol.* 63:21–31.

Kim, Y., Geiger, J. H., Hahn, S., and Sigler, P. B., 1993. Crystal structure of a yeast TBP/TATA-box complex. *Nature* 365:512–520.

Kim, J. L., Nikolov, D. B., and Burley, S. K., 1993. Co-crystal structure of TBP recognizing the minor groove of a TATA element. *Nature* 365:520–527.

White, R. J., and Jackson, S. P., 1992. The TATA-binding protein: A central role in transcription by RNA polymerases I, II and III. *Trends Genet.* 8:284–288.

Martinez, E. 2002. Multi-protein complexes in eukaryotic gene transcription. *Plant Mol. Biol.* 50:925–947.

Meinhart, A., Kamenski, T., Hoeppner, S., Baumli, S., and Cramer, P. 2005. A structural perspective of CTD function. *Genes Dev.* 19:1401–1415.

Palancade, B. and Bensaude, O. 2003. Investigating RNA polymerase II carboxyl-terminal domain (CTD) phosphorylation. *Eur. J. Biochem.* 270:3859–3870.

Termination

Burgess, B. R., and Richardson, J. P. 2001. RNA passes through the hole of the protein hexamer in the complex with *Escherichia coli* Rho factor. *J. Biol. Chem.* 276:4182–4189.

Yu, X., Horiguchi, T., Shigesada, K., and Egelman, E. H. 2000. Three-dimensional reconstruction of transcription termination factor rho: Orientation of the N-terminal domain and visualization of an RNA-binding site. *J. Mol. Biol.* 299:1279–1287.

Stitt, B. L. 2001. *Escherichia coli* transcription termination factor Rho binds and hydrolyzes ATP using a single class of three sites. *Biochemistry* 40:2276–2281.

Henkin, T. M. 2000. Transcription termination control in bacteria. *Curr. Opin. Microbiol.* 3:149–153.

Gusarov, I., and Nudler, E. 1999. The mechanism of intrinsic transcription termination. *Mol. Cell* 3:495–504.

Noncoding RNA

Peculis, B. A. 2002. Ribosome biogenesis: Ribosomal RNA synthesis as a package deal. *Curr. Biol.* 12:R623–R624.

Decatur, W. A., and Fournier, M. J. 2002. rRNA modifications and ribosome function. *Trends Biochem. Sci.* 27:344–351.

Hopper, A. K., and Phizicky, E. M. 2003. tRNA transfers to the limelight. *Genes Dev.* 17:162–180.

Weiner, A. M. 2004. tRNA maturation: RNA polymerization without a nucleic acid template. *Curr. Biol.* 14:R883–R885.

5'-Cap Formation and Polyadenylation

Shatkin, A. J., and Manley, J. L. 2000. The ends of the affair: Capping and polyadenylation. *Nat. Struct. Biol.* 7:838–842.

Bentley, D. L. 2005. Rules of engagement: Co-transcriptional recruitment of pre-mRNA processing factors. *Curr. Opin. Cell Biol.* 17:251–256.

Aguilera, A. 2005. Cotranscriptional mRNP assembly: From the DNA to the nuclear pore. *Curr. Opin. Cell Biol.* 17:242–250.

Ro-Choi, T. S. 1999. Nuclear snRNA and nuclear function (discovery of 5' cap structures in RNA). *Crit. Rev. Eukaryotic Gene Expr.* 9:107–158.

Bard, J., Zhelkovsky, A. M., Helmling, S., Earnest, T. N., Moore, C. L., and Bohm, A. 2000. Structure of yeast poly(A) polymerase alone and in complex with 3'-dATP. *Science* 289:1346–1349.

Martin, G., Keller, W., and Doublie, S. 2000. Crystal structure of mammalian poly(A) polymerase in complex with an analog of ATP. *EMBO J.* 19:4193–4203.

Zhao, J., Hyman, L., and Moore, C. 1999. Formation of mRNA 3' ends in eukaryotes: Mechanism, regulation, and interrelationships with other steps in mRNA synthesis. *Microbiol. Mol. Biol. Rev.* 63:405–445.

Minvielle-Sebastia, L., and Keller, W. 1999. mRNA polyadenylation and its coupling to other RNA processing reactions and to transcription. *Curr. Opin. Cell Biol.* 11:352–357.

RNA Editing

Gott, J. M., and Emeson, R. B. 2000. Functions and mechanisms of RNA editing. *Annu. Rev. Genet.* 34:499–531.

Simpson, L., Thiemann, O. H., Savill, N. J., Alfonzo, J. D., and Maslov, D. A. 2000. Evolution of RNA editing in trypanosome mitochondria. *Proc. Natl. Acad. Sci. U. S. A.* 97:6986–6993.

Chester, A., Scott, J., Anant, S., and Navaratnam, N. 2000. RNA editing: Cytidine to uridine conversion in apolipoprotein B mRNA. *Biochim. Biophys. Acta* 1494:1–3.

Maas, S., and Rich, A. 2000. Changing genetic information through RNA editing. *Bioessays* 22:790–802.

Splicing of mRNA Precursors

Caceres, J. F., and Kornblihtt, A. R. 2002. Alternative splicing: Multiple control mechanisms and involvement in human disease. *Trends Genet.* 18:186–193.

Faustino, N. A., and Cooper, T. A. 2003. Pre-mRNA splicing and human disease. *Genes Dev.* 17:419–437.

Lou, H., and Gagel, R. F. 1998. Alternative RNA processing: Its role in regulating expression of calcitonin/calcitonin gene-related peptide. *J. Endocrinol.* 156:401–405.

Matlin, A. J., Clark, F., and Smith, C. W. 2005. Understanding alternative splicing: Towards a cellular code. *Nat. Rev. Mol. Cell Biol.* 6:386–398.

McKie, A. B., McHale, J. C., Keen, T. J., Tarttelin, E. E., Goliath, R., et al. 2001. Mutations in the pre-mRNA splicing factor gene PRPC8 in autosomal dominant retinitis pigmentosa (RP13). *Hum. Mol. Genet.* 10:1555–1562.

Nilsen, T. W. 2003. The spliceosome: The most complex macromolecular machine in the cell? *Bioessays* 25:1147–1149.

Rund, D., and Rachmilewitz, E. 2005. β-Thalassemia. *N. Engl. J. Med.* 353:1135–1146.

Patel, A. A., and Steitz, J. A. 2003. Splicing double: Insights from the second spliceosome. *Nat. Rev. Mol. Cell Biol.* 4:960–970.

Sharp, P. A. 2005. The discovery of split genes and RNA splicing. *Trends Biochem. Sci.* 30:279–281.

Valadkhan, S., and Manley, J. L. 2001. Splicing-related catalysis by protein-free snRNAs. *Nature* 413:701–707.

Zhou, Z., Licklider, L. J., Gygi, S. P., and Reed, R. 2002. Comprehensive proteomic analysis of the human spliceosome. *Nature* 419:182–185.

Stark, H., Dube, P., Luhrmann, R., and Kastner, B. 2001. Arrangement of RNA and proteins in the spliceosomal U1 small nuclear ribonucleoprotein particle. *Nature* 409:539–542.

Strehler, E. E., and Zacharias, D. A. 2001. Role of alternative splicing in generating isoform diversity among plasma membrane calcium pumps. *Physiol. Rev.* 81:21–50.

Graveley, B. R. 2001. Alternative splicing: Increasing diversity in the proteomic world. *Trends Genet.* 17:100–107.

Newman, A. 1998. RNA splicing. *Curr. Biol.* 8:R903–R905.

Reed, R. 2000. Mechanisms of fidelity in pre-mRNA splicing. *Curr. Opin. Cell Biol.* 12:340–345.

Sleeman, J. E., and Lamond, A. I. 1999. Nuclear organization of pre-mRNA splicing factors. *Curr. Opin. Cell Biol.* 11:372–377.

Black, D. L. 2000. Protein diversity from alternative splicing: A challenge for bioinformatics and post-genome biology. *Cell* 103:367–370.

Collins, C. A., and Guthrie, C. 2000. The question remains: Is the spliceosome a ribozyme? *Nat. Struct. Biol.* 7:850–854.

Self-Splicing and RNA Catalysis

Adams, P. L., Stanley, M. R., Kosek, A. B., Wang, J., and Strobel, S. A. 2004. Crystal structure of a self-splicing group I intron with both exons. *Nature* 430:45–50.

Adams, P. L., Stanley, M. R., Gill, M. L., Kosek, A. B., Wang, J., and Strobel, S. A. 2004. Crystal structure of a group I intron splicing intermediate. *RNA* 10(12):1867–1887.

Stahley, M. R., and Strobel, S. A. 2005. Structural evidence for a two-metal-ion mechanism of group I intron splicing. *Science* 309:1587–1590.

Carola, C., and Eckstein, F. 1999. Nucleic acid enzymes. *Curr. Opin. Chem. Biol.* 3:274–283.

Doherty, E. A., and Doudna, J. A. 2000. Ribozyme structures and mechanisms. *Annu. Rev. Biochem.* 69:597–615.

Fedor, M. J. 2000. Structure and function of the hairpin ribozyme. *J. Mol. Biol.* 297:269–291.

Hanna, R., and Doudna, J. A. 2000. Metal ions in ribozyme folding and catalysis. *Curr. Opin. Chem. Biol.* 4:166–170.

Scott, W. G. 1998. RNA catalysis. *Curr. Opin. Struct. Biol.* 8:720–726.

Problems

1. *Complements.* The sequence of part of an mRNA is

5′-AUGGGGAACAGCAAGAGUGGGGCCCUGUCCAAGGAG-3′

What is the sequence of the DNA coding strand? Of the DNA template strand?

2. *Checking for errors.* Why is RNA synthesis not as carefully monitored for errors as is DNA synthesis?

3. *Speed is not of the essence.* Why is it advantageous for DNA synthesis to be more rapid than RNA synthesis?

4. *Potent inhibitor.* Heparin inhibits transcription by binding to RNA polymerase. What properties of heparin allow it to bind so effectively to RNA polymerase?

5. *A loose cannon.* Sigma protein by itself does not bind to promoter sites. Predict the effect of a mutation enabling σ to bind to the -10 region in the absence of other subunits of RNA polymerase.

6. *Stuck sigma.* What would be the likely effect of a mutation that would prevent σ from dissociating from the RNA polymerase core?

7. *Transcription time.* What is the minimum length of time required for the synthesis by *E. coli* polymerase of an mRNA encoding a 100-kd protein?

8. *Between bubbles.* How far apart are transcription bubbles on *E. coli* genes that are being transcribed at a maximal rate?

9. *A revealing bubble.* Consider the synthetic RNA–DNA transcription bubble illustrated here. Let us refer to the coding DNA strand, the template strand, and the RNA strand as strands 1, 2, and 3, respectively.

(1) DNA Coding strand

5′-GGATACTTACAGCCAT GGA CACGGC GAA TACTCCATT...3

3′-CCTATGAATGTCGGTACCTGTGCCGCTTATGAGGTAA...5

(2) Template strand

5′-UUUUUUUU UGGACACGGCGAA

(3) RNA strand

(a) Suppose that strand 3 is labeled with ^{32}P at its 5′ end and that polyacrylamide gel electrophoresis is carried out under nondenaturing conditions. Predict the autoradiographic pattern for (i) strand 3 alone, (ii) strands 1 and 3, (iii) strands 2 and 3, (iv) strands 1, 2, and 3, and (v) strands 1, 2, and 3 and core RNA polymerase.

(b) What is the likely effect of rifampicin on RNA synthesis in this system?

(c) Heparin blocks elongation of the RNA primer if it is added to core RNA polymerase before the onset of transcription but not if added after transcription starts. Account for this difference.

(d) Suppose that synthesis is carried out in the presence of ATP, CTP, and UTP. Compare the length of the longest product obtained with that expected when all four ribonucleoside triphosphates are present.

10. *Abortive cycling.* Di- and trinucleotides are occasionally released from RNA polymerase at the very start of transcription, a process called abortive cycling. This process requires the restart of transcription. Suggest a plausible explanation for abortive cycling.

11. *Polymerase inhibition.* Cordycepin inhibits poly(A) synthesis at low concentrations and RNA synthesis at higher concentrations.

Cordycepin (3'-deoxyadenosine)

(a) What is the basis of inhibition by cordycepin?
(b) Why is poly(A) synthesis more sensitive to the presence of cordycepin?
(c) Does cordycepin need to be modified to exert its effect?

12. *An extra piece.* In one type of mutation leading to a form of thalassemia, the mutation of a single base (G to A) generates a new 3' splice site (blue in the illustration below) akin to the normal one (yellow) but farther upstream.

Normal 3' end of intron

5' CCTATT**G**GTCTATTTTCCACCC**TTAG**GCTGCTG 3'

5' CCTA**TTAG**TCTATTTTCCACCCTTAGGCTGCTG 3'

What is the amino acid sequence of the extra segment of protein synthesized in a thalassemic patient having a mutation leading to aberrant splicing? The reading frame after the splice site begins with TCT.

13. *A long-tailed messenger.* Another thalassemic patient had a mutation leading to the production of an mRNA for the β chain of hemoglobin that was 900 nucleotides longer than the normal one. The poly(A) tail of this mutant mRNA was located a few nucleotides after the only AAUAAA sequence in the additional sequence. Propose a mutation that would lead to the production of this altered mRNA.

Mechanism Problem

14. *RNA editing.* Many uridine molecules are inserted into some mitochondrial mRNAs in trypanosomes. The uridine residues come from the poly(U) tail of a donor strand. Nucleoside triphosphates do not participate in this reaction. Propose a reaction mechanism that accounts for these findings. (Hint: Relate RNA editing to RNA splicing.)

Chapter Integration Problems

15. *Proteome complexity.* What processes considered in this chapter make the proteome more complex than the genome? What processes might further enhance this complexity?

16. *Separation technique.* Suggest a means by which you could separate mRNA from the other types of RNA in a eukaryotic cell.

Data Interpretation Problems

17. *Run-off experiment.* Nuclei were isolated from brain, liver, and muscle. The nuclei were then incubated with α-[³²P]UTP under conditions that allow RNA synthesis, except that an inhibitor of RNA initiation was present. The radioactive RNA was isolated and annealed to various DNA sequences that had been attached to a gene chip. In the adjoining graphs, the intensity of the shading indicates roughly how much mRNA was attached to each DNA sequence.

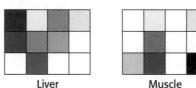
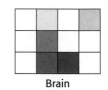

Liver   Muscle   Brain

(a) Why does the intensity of hybridization differ between genes?
(b) What is the significance of the fact that some of the RNA molecules display different hybridization patterns in different tissues?
(c) Some genes are expressed in all three tissues. What would you guess is the nature of these genes?
(d) Suggest a reason why an initiation inhibitor was included in the reaction mixture.

18. *Christmas trees.* The adjoining autoradiograph depicts several bacterial genes undergoing transcription. Identify the DNA. What are the strands of increasing length? Where is the beginning of transcription? The end of transcription? On the page, what is the direction of RNA synthesis? What can you conclude about the number of enzymes participating in RNA synthesis on a given gene?

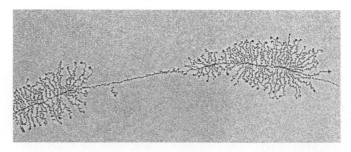

Protein Synthesis

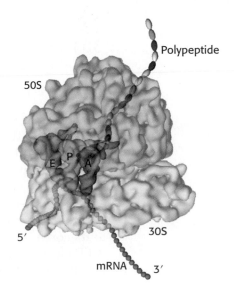

Polypeptide
50S
E P A
5'
30S
mRNA 3'

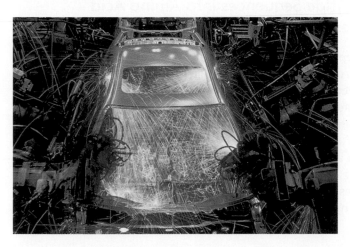

The ribosome, shown at the right, is a factory for the manufacture of polypeptides. Amino acids are carried into the ribosome, one at a time, connected to transfer RNA molecules. Each amino acid is joined to the growing polypeptide chain, which detaches from the ribosome only after the polypeptide has been completed. This assembly-line approach allows even very long polypeptide chains to be assembled rapidly and with impressive accuracy. [(Left) Doug Martin/Photo Researchers.]

Genetic information is most important because of the proteins that it encodes, in that proteins play most of the functional roles in cells. In Chapters 28 and 29, we examined how DNA is replicated and how DNA is transcribed into RNA. We now turn to the mechanism of protein synthesis, a process called *translation* because the four-letter alphabet of nucleic acids is translated into the entirely different twenty-letter alphabet of proteins. Translation is a conceptually more complex process than either replication or transcription, both of which take place within the framework of a common base-pairing language. As befits its position linking the nucleic acid and protein languages, the process of protein synthesis depends critically on both nucleic acid and protein factors. Protein synthesis takes place on *ribosomes*—enormous complexes containing three large RNA molecules and more than 50 proteins. Among the great triumphs in biochemistry in recent years has been the determination of the structure of the ribosome and its components so that its function can be examined in atomic detail. Perhaps the most significant conclusion from these studies is that *the ribosome is a ribozyme;* that is, the RNA components play the most fundamental roles. These observations strongly support the notion that life evolved through an RNA world, and the ribosome is a surviving inhabitant of that world.

Transfer RNA molecules (tRNAs), messenger RNA (mRNA) and many proteins participate in protein synthesis along with ribosomes. The link between amino acids and nucleic acids is first made by enzymes called aminoacyl-tRNA synthetases. By specifically linking a particular amino acid to each tRNA, these enzymes translate the genetic code. This chapter focuses primarily on protein synthesis in prokaryotes because it illustrates many general principles and is well understood. Some distinctive features of protein synthesis in eukaryotes also are presented.

30.1 Protein Synthesis Requires the Translation of Nucleotide Sequences into Amino Acid Sequences

The basics of protein synthesis are the same across all kingdoms of life—evidence that the protein-synthesis system arose very early in evolution. A protein is synthesized in the amino-to-carboxyl direction by the sequential addition of amino acids to the carboxyl end of the growing peptide chain (Figure 30.1). The amino acids arrive at the growing chain in activated form as aminoacyl-tRNAs, created by joining the carboxyl group of an amino acid to the 3′ end of a transfer RNA molecule. The linking of an amino acid to its corresponding tRNA is catalyzed by an *aminoacyl-tRNA synthetase*. ATP cleavage drives this activation reaction. For each amino acid, there is usually one activating enzyme and at least one kind of tRNA.

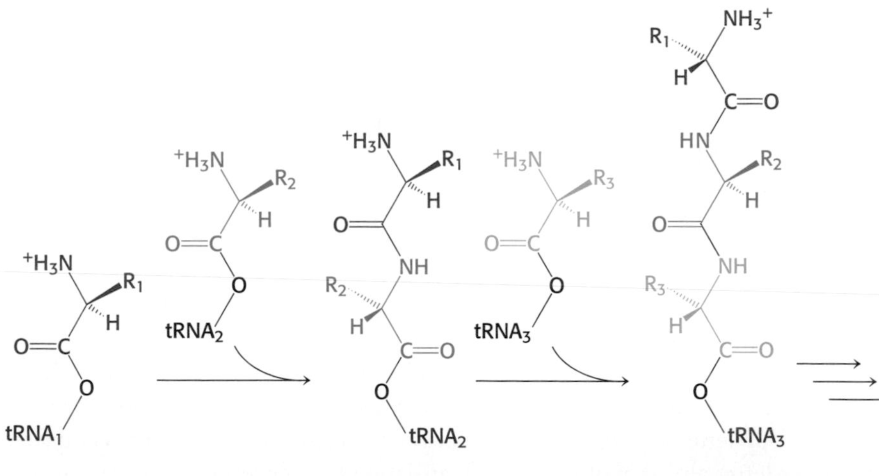

Figure 30.1 Polypeptide-chain growth. Proteins are synthesized by the successive addition of amino acids to the carboxyl terminus.

The Synthesis of Long Proteins Requires a Low Error Frequency

The process of transcription is analogous to copying, word for word, a page from a book. There is no change of alphabet or vocabulary; so the likelihood of a change in meaning is small. Translating the base sequence of an mRNA molecule into a sequence of amino acids is similar to translating the page of a book into another language. Translation is a complex process, entailing many steps and dozens of molecules. The potential for error exists at each step. The complexity of translation creates a conflict between two requirements: the process must be both accurate and fast enough to meet a cell's needs. In *E. coli,* translation takes place at a rate of approximately 20 amino acids per second, a truly impressive speed, considering the complexity of the process.

TABLE 30.1 Accuracy of protein synthesis

Frequency of inserting an incorrect amino acid	PROBABILITY OF SYNTHESIZING AN ERROR-FREE PROTEIN		
	NUMBER OF AMINO ACID RESIDUES		
	100	300	1000
10^{-2}	0.366	0.049	0.000
10^{-3}	0.905	0.741	0.368
10^{-4}	0.990	0.970	0.905
10^{-5}	0.999	0.997	0.990

Note: The probability p of forming a protein with no errors depends on n, the number of amino acids, and ε, the frequency of insertion of a wrong amino acid: $p = (1 - \varepsilon)^n$.

How accurate must protein synthesis be? Let us consider error rates. The probability of forming a protein with no errors depends on the number of amino acid residues and on the frequency (ε) of insertion of a wrong amino acid. As Table 30.1 shows, an error frequency of 10^{-2} would be intolerable, even for quite small proteins. An ε value of 10^{-3} would usually lead to the error-free synthesis of a 300-residue protein ($\sim$33 kd) but not of a 1000-residue protein ($\sim$110 kd). Thus, the error frequency must not exceed approximately 10^{-4} to produce the larger proteins effectively. Lower error frequencies are conceivable; however, except for the largest proteins, they will not dramatically increase the percentage of proteins with accurate sequences. In addition, such lower error rates are likely to be possible only by a reduction in the rate of protein synthesis because additional time for proofreading will be required. *In fact, the observed values of ε are close to 10^{-4}.* An error frequency of about 10^{-4} per amino acid residue was selected in the course of evolution to accurately produce proteins consisting of as many as 1000 amino acids while maintaining a remarkably rapid rate for protein synthesis.

Transfer RNA Molecules Have a Common Design

The fidelity of protein synthesis requires the accurate recognition of three-base *codons* on messenger RNA. Recall that the genetic code relates each amino acid to a three-letter codon (p. 125). An amino acid cannot itself recognize a codon. Consequently, an amino acid is attached to a specific tRNA molecule that can recognize the codon by Watson–Crick base-pairing. *Transfer RNA serves as the adapter molecule that binds to a specific codon and brings with it an amino acid for incorporation into the polypeptide chain.*

Anticodon

$$
\begin{array}{ccc}
3' & & 5' \\
-\text{C} & -\text{G} & -\text{I}- \\
\vdots & \vdots & \vdots \\
-\text{G} & -\text{C} & -\text{C}- \\
5' & & 3'
\end{array}
$$

Codon

Robert Holley first determined the base sequence of a tRNA molecule in 1965, as the culmination of 7 years of effort. Indeed, his study of yeast alanyl-tRNA provided the first complete sequence of any nucleic acid. This adapter molecule is a single chain of 76 ribonucleotides (Figure 30.2). The 5' terminus is phosphorylated (pG), whereas the 3' terminus has a free hydroxyl group. The *amino acid-attachment site* is the 3'-hydroxyl group of the adenosine residue at the 3' terminus of the molecule. The sequence 5'-IGC-3' in the middle of the molecule is the *anticodon*, where I is the purine base inosine. It is complementary to 5'-GCC-3', one of the codons for alanine.

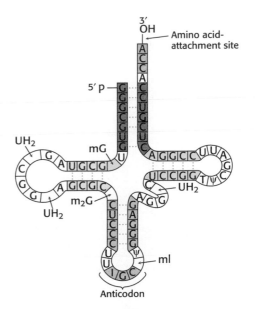

Figure 30.2 Alanyl-tRNA sequence. The base sequence of yeast alanyl-tRNA and the deduced cloverleaf secondary structure are shown. Modified nucleosides are abbreviated as follows: methylinosine (mI), dihydrouridine (UH$_2$), ribothymidine (T), pseudouridine (ψ), methylguanosine (mG), and dimethylguanosine (m$_2$G). Inosine (I), another modified nucleoside, is part of the anticodon.

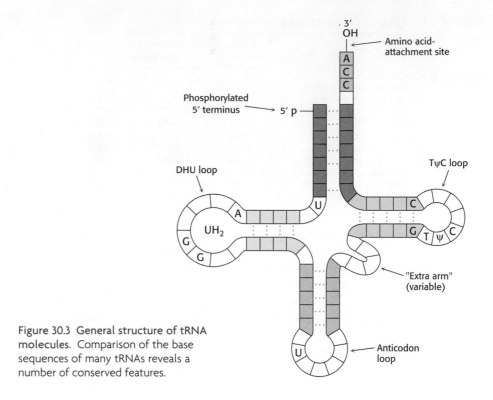

Figure 30.3 **General structure of tRNA molecules.** Comparison of the base sequences of many tRNAs reveals a number of conserved features.

The sequences of several other tRNA molecules were determined a short time later. Thousands of sequences are now known. The striking finding is that all of them can be arranged in a cloverleaf pattern in which about half the residues are base-paired (Figure 30.3). Hence, *tRNA molecules have many common structural features.* This finding is not unexpected, because all tRNA molecules must be able to interact in nearly the same way with the ribosomes, mRNAs, and protein factors that participate in translation.

All known transfer RNA molecules have the following features:

1.   Each is a single chain containing between *73 and 93 ribonucleotides* (~25 kd).

2.   They contain *many unusual bases,* typically between 7 and 15 per molecule. Some of these bases are methylated or dimethylated derivatives of A, U, C, and G formed by enzymatic modification of a precursor tRNA. Some methylations prevent the formation of certain base pairs, thereby rendering some of the bases accessible for other interactions. In addition, methylation imparts a hydrophobic character to some regions of tRNAs, which may be important for their interaction with synthetases and ribosomal proteins. Other modifications alter codon recognition, as will be described shortly.

3.   About half the nucleotides in tRNAs are base-paired to form double helices. Five groups of bases are not base-paired in this way: the 3′ *CCA terminal region,* which is part of a region called the *acceptor stem;* the *TψC loop,* which acquired its name from the sequence ribothymine-pseudouracil-cytosine; the *"extra arm,"* which contains a variable number of residues; the *DHU loop,* which contains several dihydrouracil residues; and the *anticodon loop.* The structural diversity generated by this combination of helices and loops containing modified bases ensures that the tRNAs can be uniquely distinguished, though structurally similar overall.

5-Methylcytidine (mC)

Dihydrouridine (UH₂)

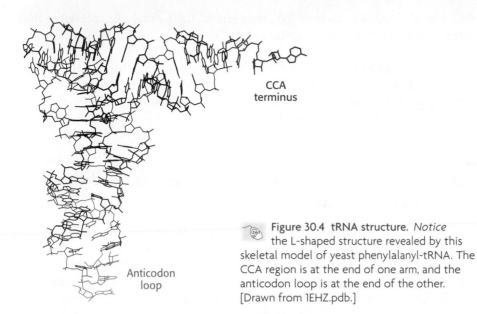

Figure 30.4 **tRNA structure.** *Notice* the L-shaped structure revealed by this skeletal model of yeast phenylalanyl-tRNA. The CCA region is at the end of one arm, and the anticodon loop is at the end of the other. [Drawn from 1EHZ.pdb.]

4.   The 5′ end of a tRNA is phosphorylated. The 5′ terminal residue is usually pG.

5.   The activated amino acid is attached to a hydroxyl group of the adenosine residue located at the end of the 3′ CCA component of the acceptor stem. This region is single stranded at the 3′ end of mature tRNAs.

6.   The anticodon is present in a loop near the center of the sequence.

The Activated Amino Acid and the Anticodon of tRNA Are at Opposite Ends of the L-Shaped Molecule

The three-dimensional structure of a tRNA molecule was first determined in 1974 through x-ray crystallographic studies carried out in the laboratories of Alexander Rich and Aaron Klug. The structure determined, that of yeast phenylalanyl-tRNA, is highly similar to all structures subsequently determined for other tRNA molecules. The most important properties of the tRNA structure are:

1.   The molecule is *L-shaped* (Figure 30.4).

2.   The four helical regions are arranged to form two apparently continuous segments of double helix. These segments are like A-form DNA, as expected for an RNA helix (p. 785). One helix, containing the 5′ and 3′ ends, runs horizontally in the model shown in Figure 30.5. The other helix, which contains the anticodon and runs vertically in Figure 30.5, forms the other arm of the L.

3.   Most of the bases in the nonhelical regions participate in hydrogen-bonding interactions, even if the interactions are not like those in Watson–Crick base pairs.

4.   The CCA terminus containing the *amino acid-attachment site* extends from one end of the L. This single-stranded region can change conformation in the course of amino acid activation and protein synthesis.

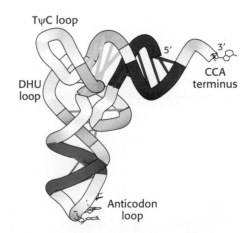

Figure 30.5 **Helix stacking in tRNA.** The four double-stranded regions of the tRNA (see Figure 30.3) stack to form an L-shaped structure. [Drawn from 1EHZ.pdb.]

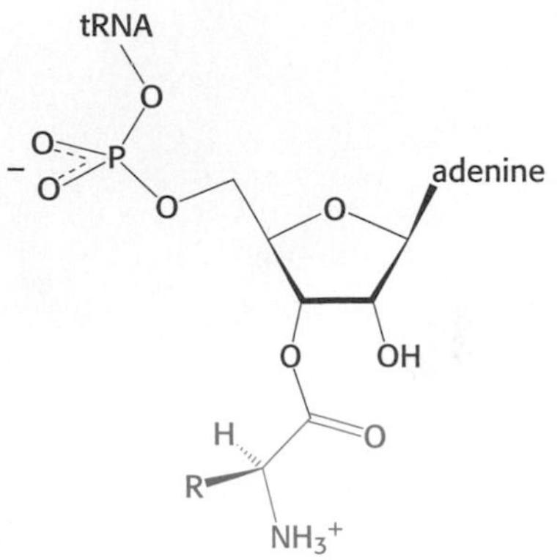

Figure 30.6 Aminoacyl-tRNA. Amino acids are coupled to tRNAs through ester linkages to either the 2′- or the 3′-hydroxyl group of the 3′-adenosine residue. A linkage to the 3′-hydroxyl group is shown.

5. The anticodon loop is at the other end of the L, making accessible the three bases that make up the anticodon.

Thus, the architecture of the tRNA molecule is well suited to its role as adaptor: the anticodon is available to interact with an appropriate codon on mRNA while the end that is linked to an activated amino acid is well positioned to participate in peptide-bond formation.

30.2 Aminoacyl-Transfer RNA Synthetases Read the Genetic Code

The linkage of an amino acid to a tRNA is crucial for two reasons. *First, the attachment of a given amino acid to a particular tRNA establishes the genetic code.* When an amino acid has been linked to a tRNA, it will be incorporated into a growing polypeptide chain at a position dictated by the anticodon of the tRNA. *Second, because the formation of a peptide bond between free amino acids is not thermodynamically favorable, the amino acid must first be activated for protein synthesis to proceed. The activated intermediates in protein synthesis are amino acid esters,* in which the carboxyl group of an amino acid is linked to either the 2′- or the 3′-hydroxyl group of the ribose unit at the 3′ end of tRNA. An amino acid ester of tRNA is called an *aminoacyl-tRNA* or sometimes a *charged tRNA* (Figure 30.6).

Amino Acids Are First Activated by Adenylation

The activation reaction is catalyzed by specific *aminoacyl-tRNA synthetases,* which are also called *activating enzymes.* The first step is the formation of an *aminoacyl adenylate* from an amino acid and ATP.

$$\text{Amino acid} + \text{ATP} \rightleftharpoons \text{aminoacyl-AMP} + \text{PP}_i$$

This activated species is a mixed anhydride in which the carboxyl group of the amino acid is linked to the phosphoryl group of AMP; hence, it is also known as *aminoacyl-AMP.*

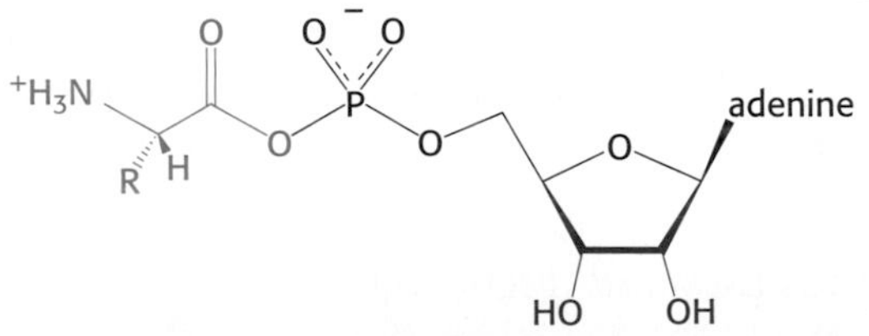

Aminoacyl adenylate

The next step is the transfer of the aminoacyl group of aminoacyl-AMP to a particular tRNA molecule to form *aminoacyl-tRNA.*

$$\text{Aminoacyl-AMP} + \text{tRNA} \rightleftharpoons \text{aminoacyl-tRNA} + \text{AMP}$$

The sum of these activation and transfer steps is

$$\text{Amino acid} + \text{ATP} + \text{tRNA} \rightleftharpoons \text{aminoacyl-tRNA} + \text{AMP} + \text{PP}_i$$

The $\Delta G^{\circ\prime}$ of this reaction is close to 0, because the free energy of hydrolysis of the ester bond of aminoacyl-tRNA is similar to that for the hydrolysis of ATP to AMP and PP$_i$. As we have seen many times, the reaction is driven by the hydrolysis of pyrophosphate. The sum of these three reactions is highly exergonic:

$$\text{Amino acid} + \text{ATP} + \text{tRNA} + H_2O \longrightarrow$$
$$\text{aminoacyl-tRNA} + \text{AMP} + 2P_i$$

Thus, *the equivalent of two molecules of ATP is consumed in the synthesis of each aminoacyl-tRNA.* One of them is consumed in forming the ester linkage of aminoacyl-tRNA, whereas the other is consumed in driving the reaction forward.

The activation and transfer steps for a particular amino acid are catalyzed by the same aminoacyl-tRNA synthetase. Indeed, *the aminoacyl-AMP intermediate does not dissociate from the synthetase.* Rather, it is tightly bound to the active site of the enzyme by noncovalent interactions. Aminoacyl-AMP is normally a transient intermediate in the synthesis of aminoacyl-tRNA, but it is relatively stable and readily isolated if tRNA is absent from the reaction mixture.

We have already encountered an acyl adenylate intermediate in fatty acid activation (p. 622). The major difference between these reactions is that the acceptor of the acyl group is CoA in fatty acid activation and tRNA in amino acid activation. The energetics of these biosyntheses are very similar: both are made irreversible by the hydrolysis of pyrophosphate.

Aminoacyl-tRNA Synthetases Have Highly Discriminating Amino Acid Activation Sites

Each aminoacyl-tRNA synthetase is highly specific for a given amino acid. Indeed, a synthetase will incorporate the incorrect amino acid only once in 10^4 or 10^5 catalytic reactions. How is this level of specificity achieved? Each aminoacyl-tRNA synthetase takes advantage of the properties of its amino acid substrate. Let us consider the challenge faced by threonyl-tRNA synthetase. Threonine is particularly similar to two other amino acids—namely, valine and serine. Valine has almost exactly the same shape as threonine, except that valine has a methyl group in place of a hydroxyl group. Serine has a hydroxyl group as does threonine but lacks the methyl group. How can the threonyl-tRNA synthetase avoid coupling these incorrect amino acids to threonyl-tRNA?

The structure of the amino acid-binding site of threonyl-tRNA synthetase reveals how valine is avoided (Figure 30.7). The enzyme contains a zinc ion, bound to the enzyme by two histidine residues and one cysteine

Acyl adenylate intermediate

Aminoacyl-tRNA

Fatty acyl CoA

Threonine

Valine

Serine

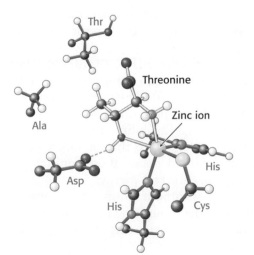

Figure 30.7 **Active site of threonyl-tRNA synthetase.** *Notice* that the amino acid-binding site includes a zinc ion (green ball) that coordinates threonine through its amino and hydroxyl groups.

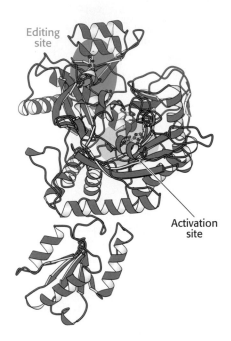

Editing
site

Activation
site

✋ **Figure 30.8 Editing site.**
Mutagenesis studies revealed the
position of the editing site (shown in
green) in threonyl-tRNA synthetase. Only
one subunit of the dimeric enzyme is
shown here and in subsequent illustrations.
[Drawn from 1QF6.pdb.]

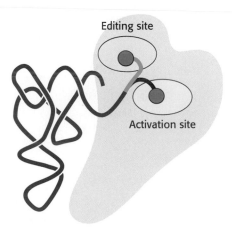

Editing site

Activation site

Figure 30.9 Editing of aminoacyl-tRNA.
The flexible CCA arm of an aminoacyl-
tRNA can move the amino acid between
the activation site and the editing site. If
the amino acid fits well into the editing
site, the amino acid is removed by
hydrolysis.

residue. Like carbonic anhydrase (p. 255), the remaining coordination sites are available for substrate binding. Threonine coordinates to the zinc ion through its amino group and its side-chain hydroxyl group. The side-chain hydroxyl group is further recognized by an aspartate residue that hydrogen bonds to it. The methyl group present in valine in place of this hydroxyl group cannot participate in these interactions; it is excluded from this active site and, hence, does not become adenylated and transferred to threonyl-tRNA (abbreviated tRNA^{Thr}). The use of a zinc ion appears to be unique to threonyl-tRNA synthetase; other aminoacyl-tRNA synthetases have different strategies for recognizing their cognate amino acids. The carboxylate group of the correctly positioned threonine is available to attack the α phosphoryl group of ATP to form the aminoacyl adenylate.

The zinc site is less well suited to discrimination against serine because this amino acid does have a hydroxyl group that can bind to the zinc ion. Indeed, with only this mechanism available, threonyl-tRNA synthetase does mistakenly couple serine to threonyl-tRNA at a rate 10^{-2} to 10^{-3} times that for threonine. As noted on page 858, this error rate is likely to lead to many translation errors. How is a higher level of specificity achieved?

Proofreading by Aminoacyl-tRNA Synthetases Increases the Fidelity of Protein Synthesis

Threonyl-tRNA synthetase can be incubated with tRNA^{Thr} that has been covalently linked with serine (Ser-tRNA^{Thr}); the tRNA has been "mischarged." The reaction is immediate: a rapid hydrolysis of the aminoacyl-tRNA forms serine and free tRNA. In contrast, incubation with correctly charged Thr-tRNA^{Thr} results in no reaction. Thus, threonyl-tRNA synthetase contains an additional functional site that hydrolyzes Ser-tRNA^{Thr} but not Thr-tRNA^{Thr}. This editing site provides an opportunity for the synthetase to correct its mistakes and improve its fidelity to less than one mistake in 10^4. The results of structural and mutagenesis studies revealed that the editing site is more than 20 Å from the activation site (Figure 30.8). This site readily accepts and cleaves Ser-tRNA^{Thr} but does not cleave Thr-tRNA^{Thr}. The discrimination of serine from threonine is easy because threonine contains an *extra* methyl group; a site that conforms to the structure of serine will sterically exclude threonine.

Most aminoacyl-tRNA synthetases contain editing sites in addition to activation sites. These complementary pairs of sites function as a *double sieve* to ensure very high fidelity. In general, the acylation site rejects amino acids that are *larger* than the correct one because there is insufficient room for them, whereas the hydrolytic site cleaves activated species that are *smaller* than the correct one.

The structure of the complex between threonyl-tRNA synthetase and its substrate reveals that the aminoacylated CCA can swing out of the activation site and into the editing site (Figure 30.9). Thus, the aminoacyl-tRNA can be edited without dissociating from the synthetase. This proofreading, which depends on the conformational flexibility of a short stretch of polynucleotide sequence, is entirely analogous to that of DNA polymerase (p. 807). In both cases, editing without dissociation significantly improves fidelity with only modest costs in time and energy.

A few synthetases achieve high accuracy without editing. For example, tyrosyl-tRNA synthetase has no difficulty discriminating between tyrosine and phenylalanine; the hydroxyl group on the tyrosine ring enables tyrosine to bind to the enzyme 10^4 times as strongly as phenylalanine. *Proof-reading has been selected in evolution only when fidelity must be enhanced beyond what can be obtained through an initial binding interaction.*

Synthetases Recognize Various Features of Transfer RNA Molecules

How do synthetases choose their tRNA partners? This enormously important step is the point at which "translation" takes place—at which the correlation between the amino acid and the nucleic acid worlds is made. In a sense, aminoacyl-tRNA synthetases are the only molecules in biology that "know" the genetic code. Their precise recognition of tRNAs is as important for high-fidelity protein synthesis as is the accurate selection of amino acids. In general, tRNA recognition by the synthetase is different for each synthetase and tRNA pair. Consequently, generalities are difficult to make. We will examine the interaction of two synthetases with their tRNA partners.

A priori, the anticodon of tRNA would seem to be a good identifier because each type of tRNA has a different one. Indeed, *some synthetases recognize their tRNA partners primarily on the basis of their anticodons,* although they may also recognize other aspects of tRNA structure. The most direct evidence comes from crystallographic studies of complexes formed between synthetases and their cognate tRNAs. Consider, for example, the structure of the complex between threonyl-tRNA synthetase and tRNA^{Thr} (Figure 30.10). As expected, the CCA arm extends into the zinc-containing activation site, where it is well positioned to accept threonine from threonyl adenylate. The enzyme interacts extensively not only with the acceptor stem of the tRNA, but also with the anticodon loop. The interactions with the anticodon loop are particularly revealing. Each base within the sequence 5'-CGU-3' of the anticodon participates in hydrogen bonds with the enzyme; those in which G and U take part appear to be more important because the C can be replaced by G or U with no loss of acylation efficiency.

Mutagenesis studies established that not all synthetases interact with the anticodons of their cognate tRNAs. For example, *E. coli* tRNA^{Cys} differs from tRNA^{Ala} at 40 positions and contains a C · G base pair at the 3:70 position. When this C · G base pair is changed to the non-Watson–Crick G · U base pair, tRNA^{Cys} is recognized by alanyl-tRNA synthetase as though it were tRNA^{Ala}. This finding raised the question whether a fragment of tRNA suffices for aminoacylation by alanyl-tRNA synthetase. Indeed, a "microhelix" containing just 24 of the 76 nucleotides of the native tRNA is specifically aminoacylated by the alanyl-tRNA synthetase. This microhelix contains only the acceptor stem and a hairpin loop (Figure 30.11). Thus, specific aminoacylation is possible for some synthetases even if the anticodon loop is completely lacking.

Aminoacyl-tRNA Synthetases Can Be Divided into Two Classes

At least one aminoacyl-tRNA synthetase exists for each amino acid. The diverse sizes, subunit composition, and sequences of these enzymes were bewildering for many years. Could it be that essentially all

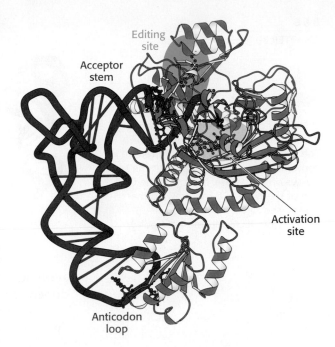

Figure 30.10 Threonyl-tRNA synthetase complex. The structure shows the complex between threonyl-tRNA synthetase and tRNA^{Thr}. *Notice* that the synthetase binds to both the acceptor stem and the anticodon loop. [Drawn from 1QF6.pdb.]

```
              A 76
              C
              C
              A
        1 G · C
          G · C
        3 G · U 70
          G · C
          C · G
          U · A
          A · U 66
        U       C 13
      A       U
        G   C
       10
```

Figure 30.11 Microhelix recognized by alanyl-tRNA synthetase. A stem-loop containing just 24 nucleotides corresponding to the acceptor stem is aminoacylated by alanyl-tRNA synthetase.

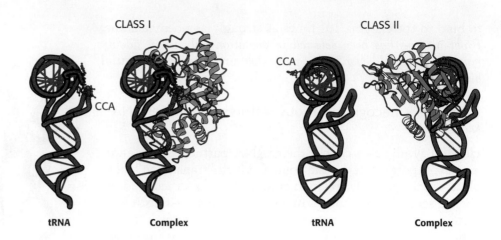

CLASS I **CLASS II**

tRNA Complex tRNA Complex

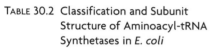
Figure 30.12 Classes of aminoacyl-tRNA synthetases. *Notice* that class I and class II synthetases recognize different faces of the tRNA molecule. The CCA arm of tRNA adopts different conformations in complexes with the two classes of synthetase. Note that the CCA arm of the tRNA is turned toward the viewer (see Figures 30.4 and 30.5). [Drawn from 1EUY.pdb and 1QF6.pdb.]

Table 30.2 Classification and Subunit Structure of Aminoacyl-tRNA Synthetases in *E. coli*

Class I	Class II
Arg (α)	Ala (α_4)
Cys (α)	Asn (α_2)
Gln (α)	Asp (α_2)
Glu (α)	Gly ($\alpha_2\beta_2$)
Ile (α)	His (α_2)
Leu (α)	Lys (α_2)
Met (α)	Phe ($\alpha_2\beta_2$)
Trp (α_2)	Ser (α_2)
Tyr (α_2)	Pro (α_2)
Val (α)	Thr (α_2)

synthetases evolved independently? The determination of the three-dimensional structures of several synthetases followed by more-refined sequence comparisons revealed that different synthetases are, in fact, related. Specifically, synthetases fall into two classes, termed *class I* and *class II*, each of which includes enzymes specific for 10 of the 20 amino acids (Table 30.2). Intriguingly, synthetases from the two classes bind to different faces of the tRNA molecule (Figure 30.12). The CCA arm of tRNA adopts different conformations to accommodate these interactions; the arm is in the helical conformation observed for free tRNA (see Figures 30.4 and 30.5) for class II enzymes and in a hairpin conformation for class I enzymes. These two classes also differ in other ways.

1. Class I enzymes acylate the 2′-hydroxyl group of the terminal adenosine of tRNA, whereas class II enzymes (except the enzyme for Phe-tRNA) acylate the 3′-hydroxyl group.

2. These two classes bind ATP in different conformations.

3. Most class I enzymes are monomeric, whereas most class II enzymes are dimeric.

Why did two distinct classes of aminoacyl-tRNA synthetases evolve? The observation that the two classes bind to distinct faces of tRNA suggests a possibility. Recognition sites on both faces of tRNA may have been required to allow the recognition of 20 different tRNAs.

30.3 A Ribosome Is a Ribonucleoprotein Particle (70S) Made of a Small (30S) and a Large (50S) Subunit

We turn now to ribosomes, the molecular machines that coordinate the interplay of charged tRNAs, mRNA, and proteins that leads to protein synthesis. An *E. coli* ribosome is a ribonucleoprotein assembly with a mass of about 2500 kd, a diameter of approximately 250 Å, and a sedimentation coefficient of 70S. The 20,000 ribosomes in a bacterial cell constitute nearly a fourth of its mass.

A ribosome can be dissociated into a *large subunit (50S)* and a *small subunit (30S)*. These subunits can be further split into their constituent proteins and RNAs. The 30S subunit contains 21 different proteins (referred to as S1 through S21) and a 16S RNA molecule. The 50S subunit contains 34 different proteins (L1 through L34) and two RNA molecules, a 23S and a 5S

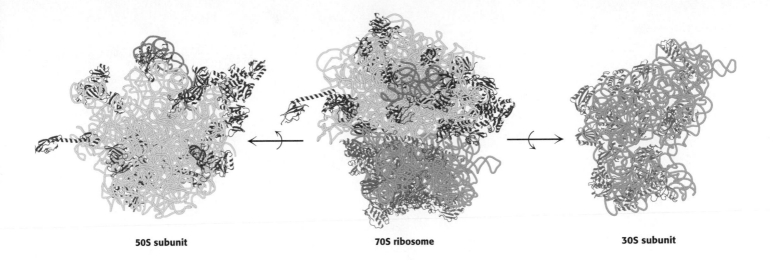

50S subunit　　　　　　**70S ribosome**　　　　　　**30S subunit**

species. A ribosome contains one copy of each RNA molecule, two copies each of the L7 and L12 proteins, and one copy of each of the other proteins. The L7 protein is identical with L12 except that its amino terminus is acetylated. Both the 30S and the 50S subunits can be reconstituted in vitro from their constituent proteins and RNA, as was first achieved by Masayasu Nomura in 1968. *This reconstitution is an outstanding example of the principle that supramolecular complexes can form spontaneously from their macromolecular constituents.*

Electron microscopic studies of the ribosome at increasingly high resolution provided views of the overall structure and revealed the positions of tRNA-binding sites. Astounding progress on the structure of the ribosome has been made by x-ray crystallographic methods, after the pioneering work by Ada Yonath. The structures of both the 30S and the 50S subunits have been determined at or close to atomic resolution, and the elucidation of the structure of intact 70S ribosomes at a similar resolution is following rapidly (Figure 30.13). The determination of this structure requires the positioning of more than 100,000 atoms. The features of these structures are in remarkable agreement with interpretations of less-direct experimental probes. These structures provide an invaluable framework for examining the mechanism of protein synthesis.

Figure 30.13 The ribosome at high resolution. Detailed models of the ribosome based on the results of x-ray crystallographic studies of the 70S ribosome and the 30S and 50S subunits. 23S RNA is shown in yellow, 5S RNA in orange, 16S RNA in green, proteins of the 50S subunit in red, and proteins of the 30S subunit in blue. *Notice* that the interface between the 50S and the 30S subunits consists entirely of RNA. [Drawn from 1GIX.pdb and 1GIY.pdb.]

Ribosomal RNAs (5S, 16S, and 23S rRNA) Play a Central Role in Protein Synthesis

The prefix *ribo* in the name *ribosome* is apt, because RNA constitutes nearly two-thirds of the mass of these large molecular assemblies. The three RNAs present—5S, 16S, and 23S—are critical for ribosomal architecture and function. They are formed by cleavage of primary 30S transcripts and further processing. These molecules fold to form structures that allow them to form internal base pairs. Their base-pairing patterns were deduced by comparing the nucleotide sequences of many species to detect conserved sequences as well as conserved base pairings. For instance, one species of RNA may have a G–C base pair, whereas another may have an A–U base pair, but the location of the base pair is the same in both molecules. Chemical modification and digestion experiments supported the structures deduced from sequence comparisons (Figure 30.14). The striking finding is that *ribosomal RNAs* (rRNAs) *are folded into defined structures with many short duplex regions.* This conclusion and essentially all features of the secondary structure have been confirmed by the x-ray crystallographically determined structures.

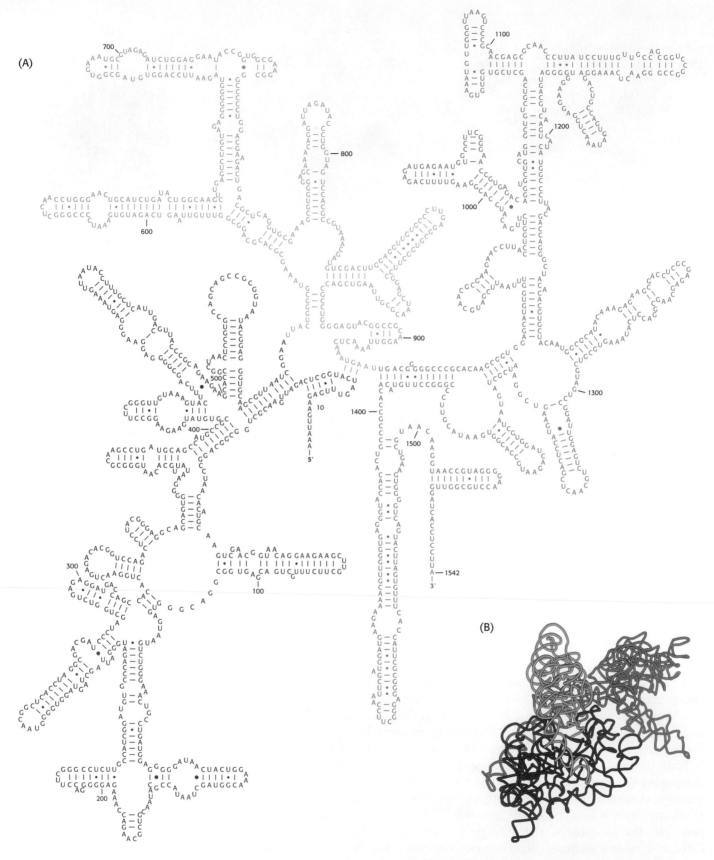

(A)

(B)

Figure 30.14 Ribosomal RNA folding pattern. (A) The secondary structure of 16S ribosomal RNA deduced from sequence comparison and the results of chemical studies. (B)The tertiary structure of 16S RNA determined by x-ray crystallography. [(A) Courtesy of Dr. Bryn Weiser and Dr. Harry Noller; (B) drawn from 1FJG.pdb.]

For many years, ribosomal proteins were presumed to orchestrate protein synthesis and ribosomal RNAs were presumed to serve primarily as structural scaffolding. The current view is almost the reverse. The discovery of catalytic RNA made biochemists receptive to the possibility that RNA plays a much more active role in ribosomal function. The detailed structures make it clear that the key sites in the ribosome are composed almost entirely of RNA. Contributions from the proteins are minor. Many of the proteins have elongated structures that "snake" their way into the RNA matrix. The almost inescapable conclusion is that the ribosome initially consisted only of RNA and that the proteins were added later to fine-tune its functional properties. This conclusion has the pleasing consequence of dodging a "chicken and egg" question: How can complex proteins be synthesized if complex proteins are required for protein synthesis?

Proteins Are Synthesized in the Amino-to-Carboxyl Direction

Before the mechanism of protein synthesis could be examined, several key facts had to be established. Pulse-labeling studies by Howard Dintzis established that protein synthesis proceeds sequentially from the amino terminus. Reticulocytes (young red blood cells) that were actively synthesizing hemoglobin were treated with [^{3}H]leucine. In a period of time shorter than that required to synthesize a complete chain, samples of hemoglobin were taken, separated into α and β chains, and analyzed for the distribution of ^{3}H within their sequences. In the earliest samples, only regions near the carboxyl ends contained radioactivity. In later samples, radioactivity was present closer to the amino terminus as well. This distribution is the one expected if the amino-terminal regions of some chains had already been partly synthesized before the addition of the radioactive amino acid. Thus, *protein synthesis begins at the amino terminus and extends toward the carboxyl terminus.*

Messenger RNA Is Translated in the 5′-to-3′ Direction

The sequence of amino acids in a protein is translated from the nucleotide sequence in mRNA. In which direction is the message read? The answer was established by using the synthetic polynucleotide

$$\overset{5'}{A}-A-A+(A-A-A)_n-A-A-\overset{3'}{C}$$

as the template in a cell-free protein-synthesizing system. AAA encodes lysine, whereas AAC encodes asparagine. The polypeptide product was

$$^+H_3N-Lys-(Lys)_n-Asn-\overset{O}{\underset{O}{C}}-$$

Because asparagine was the carboxyl-terminal residue, we can conclude that the codon AAC was the last to be read. Hence, *the direction of translation is 5′ → 3′.*

The direction of translation has important consequences. Recall that transcription also is in the 5′ → 3′ direction (p. 827). If the direction of translation were opposite that of transcription, only fully synthesized mRNA could be translated. In contrast, because the directions are the same, mRNA can be translated while it is being synthesized. In prokaryotes, almost no time is lost between transcription and translation. The 5′ end of mRNA interacts with ribosomes very soon after it is made, much before the

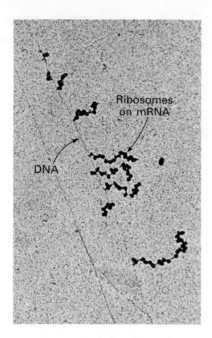

Figure 30.15 Polysomes. Transcription of a segment of DNA from *E. coli* generates mRNA molecules that are immediately translated by multiple ribosomes. [From O. L. Miller, Jr., B. A. Hamkalo, and C. A. Thomas, Jr. *Science* 169(1970):392–395.]

3′ end of the mRNA molecule is finished. *An important feature of prokaryotic gene expression is that translation and transcription are closely coupled in space and time.* Many ribosomes can be translating an mRNA molecule simultaneously. This parallel synthesis markedly increases the efficiency of mRNA translation. The group of ribosomes bound to an mRNA molecule is called a *polyribosome* or a *polysome* (Figure 30.15).

The Start Signal Is Usually AUG Preceded by Several Bases That Pair with 16S rRNA

How does protein synthesis start? The simplest possibility would be for the first 3 nucleotides of each mRNA to serve as the first codon; no special start signal would then be needed. However, experiments show that translation does not begin immediately at the 5′ terminus of mRNA. Indeed, the first translated codon is nearly always more than 25 nucleotides away from the 5′ end. Furthermore, in prokaryotes, many mRNA molecules are *polycistronic*, or polygenic—that is, they encode two or more polypeptide chains. For example, a single mRNA molecule about 7000 nucleotides long specifies five enzymes in the biosynthetic pathway for tryptophan in *E. coli*. Each of these five proteins has its own start and stop signals on the mRNA. In fact, *all known mRNA molecules contain signals that define the beginning and end of each encoded polypeptide chain.*

A clue to the mechanism of initiation was the finding that nearly half the amino-terminal residues of proteins in *E. coli* are methionine. In fact, the initiating codon in mRNA is AUG (methionine) or, less frequently, GUG (valine) or, rarely UUG (leucine). What additional signals are necessary to specify a translation start site? The first step toward answering this question was the isolation of initiator regions from a number of mRNAs. This isolation was accomplished by using pancreatic ribonuclease to digest mRNA–ribosome complexes (formed under conditions of chain initiation but not elongation). In each case, a sequence of about 30 nucleotides was protected from digestion. As expected, each initiator region displays an AUG (or GUG or UUG) codon (Figure 30.16). In addition, each initiator region contains a purine-rich sequence centered about 10 nucleotides on the 5′ side of the initiator codon.

The role of this purine-rich region, called the *Shine–Dalgarno sequence*, became evident when the sequence of 16S rRNA was elucidated. The 3′ end of this rRNA component of the 30S subunit contains a sequence of several bases that is complementary to the purine-rich region in the initiator sites of mRNA. Mutagenesis of the CCUCC sequence near the 3′ end of 16S rRNA to ACACA markedly interferes with the recognition of start sites in mRNA. This result and other evidence show that the initiator region of mRNA binds to the 16S rRNA very near its 3′ end. The number of base

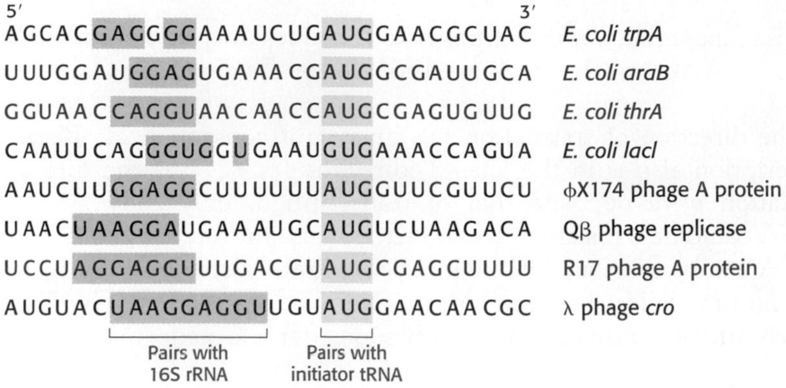

Figure 30.16 Initiation sites. Sequences of mRNA initiation sites for protein synthesis in some bacterial and viral mRNA molecules. Comparison of these sequences reveals some recurring features.

```
5′                                              3′
AGCACGAGGGGAAAUCUGAUGGAACGCUAC      E. coli trpA
UUUGGAUGGAGUGAAACGAUGGCGAUUGCA      E. coli araB
GGUAACCAGGUAACAACCAUGCGAGUGUUG      E. coli thrA
CAAUUCAGGGUGGUGAAUGUGAAACCAGUA      E. coli lacI
AAUCUUGGAGGCUUUUUUUAUGGUUCGUUCU     φX174 phage A protein
UAACUAAGGAUGAAAUGCAUGUCUAAGACA      Qβ phage replicase
UCCUAGGAGGUUUGACCUAUGCGAGCUUUU      R17 phage A protein
AUGUACUAAGGAGGUUGUAUGGAACAACGC      λ phage cro
```
      Pairs with   Pairs with
      16S rRNA   initiator tRNA

pairs linking mRNA and 16S rRNA ranges from three to nine. Thus, *two kinds of interactions determine where protein synthesis starts: (1) the pairing of mRNA bases with the 3' end of 16S rRNA and (2) the pairing of the initiator codon on mRNA with the anticodon of an initiator tRNA molecule.*

Bacterial Protein Synthesis Is Initiated by Formylmethionyl Transfer RNA

The methionine residue found at the amino-terminal end of *E. coli* proteins is usually modified. In fact, *protein synthesis in bacteria starts with N-formylmethionine* (fMet). A special tRNA brings formylmethionine to the ribosome to initiate protein synthesis. This *initiator tRNA* (abbreviated as tRNA$_f$) differs from the tRNA that inserts methionine in internal positions (abbreviated as tRNA$_m$). The subscript "f" indicates that methionine attached to the initiator tRNA can be formylated, whereas it cannot be formylated when attached to tRNA$_m$. Transfer RNA$_f$ can bind to all three possible initiation codons, but with decreasing affinity (AUG > GUG > UUG). In approximately one-half of *E. coli* proteins, *N*-formylmethionine is removed when the nascent chain exits the ribosome.

Methionine is linked to these two kinds of tRNAs by the same aminoacyl-tRNA synthetase. A specific enzyme then formylates the amino group of methionine attached to tRNA$_f$ (Figure 30.17). The activated formyl donor in this reaction is N^{10}-formyltetrahydrofolate (p. 689). It is significant that free methionine and methionyl-tRNA$_m$ are not substrates for this transformylase.

Ribosomes Have Three tRNA-Binding Sites That Bridge the 30S and 50S Subunits

A snapshot of a significant moment in protein synthesis was obtained by determining the structure of the 70S ribosome bound to three tRNA molecules and a fragment of mRNA (Figure 30.18). As expected, the mRNA fragment is bound within the 30S subunit. Each of the tRNA molecules bridges between the 30S and the 50S subunits. At the 30S end, two of the three tRNA molecules are bound to the mRNA fragment through

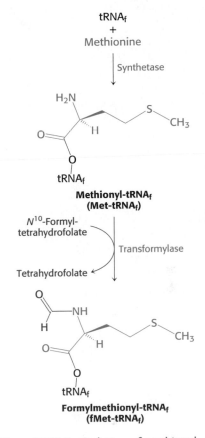

Figure 30.17 Formylation of methionyl-tRNA. Initiator tRNA (tRNA$_f$) is first charged with methionine, and then a formyl group is transferred to the methionyl-tRNA$_f$ from N^{10}-formyltetrahydrofolate.

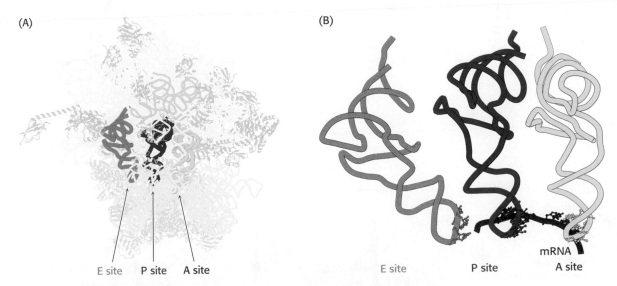

Figure 30.18 Transfer RNA-binding sites. (A) Three tRNA-binding sites are present on the 70S ribosome. They are called the A (for aminoacyl), P (for peptidyl), and E (for exit) sites. Each tRNA molecule contacts both the 30S and the 50S subunit. (B)The tRNA molecules in sites A and P are base-paired with mRNA. [(B) Drawn from 1JGP.pdb.]

anticodon–codon base pairs. These binding sites are called the A site (for *aminoacyl*) and the P site (for *peptidyl*). The third tRNA molecule is bound to an adjacent site called the E site (for *exit*).

The other end of each tRNA molecule interacts with the 50S subunit. The acceptor stems of the tRNA molecules occupying the A site and the P site converge at a site where a peptide bond is formed. Further examination of this site reveals that a tunnel connects this site to the back of the ribosome. *The growing polypeptide chain escapes the ribosome through this tunnel during synthesis.*

The Growing Polypeptide Chain Is Transferred Between tRNAs on Peptide-Bond Formation

Protein synthesis begins with the interaction of the 30S subunit and mRNA through the Shine–Dalgarno sequence. On formation of this complex, the initiator tRNA charged with formylmethionine binds to the initiator AUG codon, and the 50S subunit binds to the 30S subunit to form the complete 70S ribosome. How does the polypeptide chain increase in length (Figure 30.19)? The three sites in our snapshot of protein synthesis provide a clue. The initiator tRNA is bound in the P site on the ribosome. A charged tRNA with an anticodon complementary to the codon in the A site then binds. The stage is set for the formation of a peptide bond: the formylmethionine molecule linked to the initiator tRNA will be transferred to the amino group of the amino acid in the A site. The formation of the peptide bond, one of the most important reactions in life, is a thermodynamically spontaneous reaction that is catalyzed by a site on the 23S rRNA called the *peptidyl transferase center*.

The amino group of the aminoacyl-tRNA in the A site is well positioned to attack the ester linkage between the initiator tRNA and the formylmethionine molecule (Figure 30.20). The peptidyl transferase center includes bases that promote this reaction by helping to form an —NH_2 group on the A-site aminoacyl-tRNA and by helping to stabilize the tetrahedral intermediate

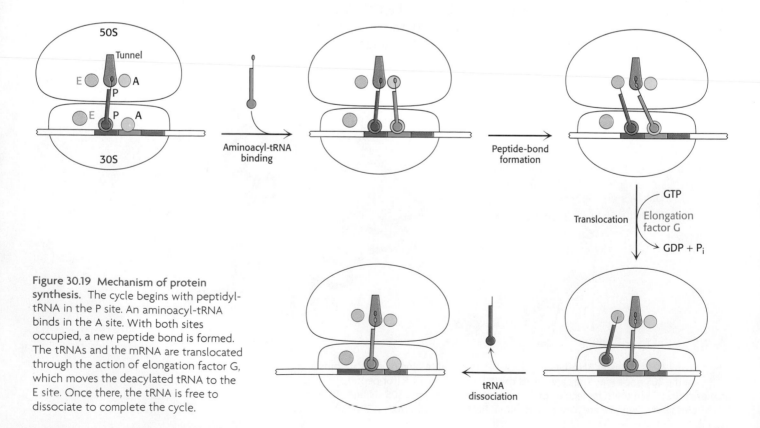

Figure 30.19 Mechanism of protein synthesis. The cycle begins with peptidyl-tRNA in the P site. An aminoacyl-tRNA binds in the A site. With both sites occupied, a new peptide bond is formed. The tRNAs and the mRNA are translocated through the action of elongation factor G, which moves the deacylated tRNA to the E site. Once there, the tRNA is free to dissociate to complete the cycle.

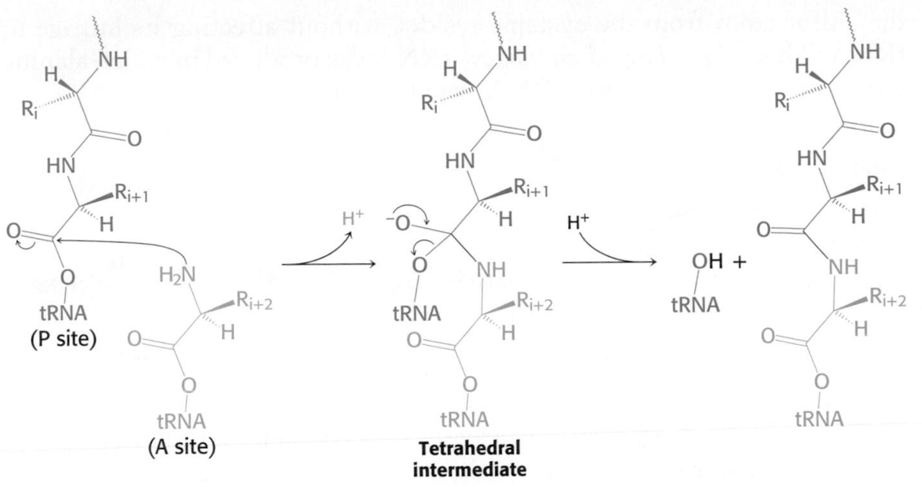

Figure 30.20 Peptide-bond formation. The amino group of the aminoacyl-tRNA attacks the carbonyl group of the ester linkage of the peptidyl-tRNA to form a tetrahedral intermediate. This intermediate collapses to form the peptide bond and release the deacylated tRNA.

that forms. This reaction is, in many ways, analogous to the reverse of the reaction catalyzed by serine proteases such as chymotrypsin (p. 247). The peptidyl-tRNA is analogous to the acyl-enzyme form of a serine protease. In a serine protease, the acyl-enzyme is generated with the use of the free energy associated with cleaving an amide bond. In the ribosome, the free energy necessary to form the analogous species, an aminoacyl-tRNA, comes from the ATP that is cleaved by the aminoacyl-tRNA synthetase before the arrival of the tRNA at the ribosome.

With the peptide bond formed, the peptide chain is now attached to the tRNA in the A site on the 30S subunit while a change in the interaction with the 50S subunit has placed that tRNA and its peptide in the P site of the large subunit. The tRNA in the P site of the 30S subunit is now uncharged. For translation to proceed, the mRNA must be moved (or *translocated*) so that the codon for the next amino acid to be added is in the A site. This translocation is assisted by a protein enzyme called *elongation factor G* (p. 876), driven by the hydrolysis of GTP. On completion of this step, the peptidyl-tRNA is now fully in the P site, and the uncharged initiator tRNA is in the E site and has been disengaged from the mRNA. On dissociation of the initiator tRNA, the ribosome has returned to its initial state except that the peptide chain is attached to a different tRNA, the one corresponding to the first codon past the initiating AUG. Note that *the peptide chain remains in the P site on the 50S subunit, at the entrance to the exit channel, throughout this cycle*, presumably growing into the tunnel. This cycle is repeated as new aminoacyl-tRNAs move into the A site, allowing the polypeptide to be elongated until the cycle is terminated.

Only the Codon–Anticodon Interactions Determine the Amino Acid That Is Incorporated

On the basis of the mechanism described on page 871, the base-pairing interaction between the anticodon on the incoming tRNA and the codon in the A site on mRNA determines which amino acid is added to the polypeptide chain. Does the amino acid attached to the tRNA play any role in this process? This question was answered in the following way. First, cysteine was attached to its cognate tRNA. The attached cysteine unit was then converted into alanine by removing the sulfor atom from the side chain in cysteine in a reaction catalyzed by Raney nickel; the reaction removed the sulfur atom from the cysteine residue without affecting its linkage to tRNA. Thus, a *mischarged aminoacyl-tRNA* was produced in which alanine was covalently attached to a tRNA specific for cysteine.

Cysteine + tRNA^{Cys} → Cys-tRNA^{Cys} → Ala-tRNA^{Cys}

H_2O + ATP → 2 P_i + AMP, Cysteinyl-tRNA synthetase; H_2 Raney nickel

Does this mischarged tRNA recognize the codon for cysteine or for alanine? The answer came when the tRNA was added to a cell-free protein-synthesizing system. The template was a random copolymer of U and G in the ratio of 5:1, which normally incorporates cysteine (encoded by UGU) but not alanine (encoded by GCN). However, alanine was incorporated into a polypeptide when Ala-tRNA^{Cys} was added to the incubation mixture. The same result was obtained when mRNA for hemoglobin served as the template and [^{14}C]alanyl-tRNA^{Cys} was used as the mischarged aminoacyl-tRNA. When the hemoglobin was digested with trypsin, the only radioactive peptide produced was one that normally contained cysteine but not alanine. Thus, *the amino acid in aminoacyl-tRNA does not play a role in selecting a codon.*

In recent years, the ability of mischarged tRNAs to transfer their amino acid cargo to a growing polypeptide chain has been used to synthesize peptides with amino acids not found in proteins incorporated into specific sites in a protein. Aminoacyl-tRNAs are first linked to these unnatural amino acids by chemical methods. These mischarged aminoacyl-tRNAs are added to a cell-free protein-synthesizing system along with specially engineered mRNA that contains codons corresponding to the anticodons of the mischarged aminoacyl-tRNAs in the desired positions. The proteins produced have unnatural amino acids in the expected positions. More than 100 different unnatural amino acids have been incorporated in this way. However, only L-amino acids can be used; apparently this stereochemistry is required for peptide-bond formation to take place.

Some Transfer RNA Molecules Recognize More Than One Codon Because of Wobble in Base-Pairing

What are the rules that govern the recognition of a codon by the anticodon of a tRNA? A simple hypothesis is that each of the bases of the codon forms a Watson–Crick type of base pair with a complementary base on the anticodon. The codon and anticodon would then be lined up in an antiparallel fashion. In the diagram in the margin, the prime denotes the complementary base. Thus X and X′ would be either A and U (or U and A) or G and C (or C and G). According to this model, a particular anticodon can recognize only one codon.

The facts are otherwise. As found experimentally, *some pure tRNA molecules can recognize more than one codon.* For example, the yeast alanyl-tRNA binds to *three* codons: GCU, GCC, and GCA. The first two bases of these codons are the same, whereas the third is different. Could it be that recognition of the third base of a codon is sometimes less discriminating than recognition of the other two? The pattern of degeneracy of the genetic code indicates that this might be so. XYU and XYC always encode the same amino acid; XYA and XYG usually do. Francis Crick surmised from these

Anticodon

$$-\overset{3'}{X'}-\overset{}{Y'}-\overset{5'}{Z'}-$$
$$-\underset{5'}{X}-\underset{}{Y}-\underset{3'}{Z}-$$

Codon

data that the steric criteria might be less stringent for pairing of the third base than for the other two. Models of various base pairs were built to determine which ones are similar to the standard A · U and G · C base pairs with regard to the distance and angle between the glycosidic bonds. Inosine was included in this study because it appeared in several anticodons. With the assumption of some steric freedom ("wobble") in the pairing of the third base of the codon, the combinations shown in Table 30.3 seemed plausible.

The *wobble hypothesis* is now firmly established. The anticodons of tRNAs of known sequence bind to the codons predicted by this hypothesis. For example, the anticodon of yeast alanyl-tRNA is IGC. This tRNA recognizes the codons GCU, GCC, and GCA. Recall that, by convention, nucleotide sequences are written in the 5′ → 3′ direction unless otherwise noted. Hence, I (the 5′ base of this anticodon) pairs with U, C, or A (the 3′ base of the codon), as predicted.

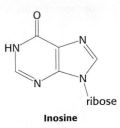

Inosine

TABLE 30.3 Allowed pairings at the third base of the codon according to the wobble hypothesis

First base of anticodon	Third base of codon
C	G
A	U
U	A or G
G	U or C
I	U, C, or A

Inosine–cytidine base pair

Inosine–uridine base pair

Inosine–adenosine base pair

Two generalizations concerning the codon–anticodon interaction can be made:

1.   The first two bases of a codon pair in the standard way. Recognition is precise. Hence, *codons that differ in either of their first two bases must be recognized by different tRNAs.* For example, both UUA and CUA encode leucine but are read by different tRNAs.

2.   The first base of an anticodon determines whether a particular tRNA molecule reads one, two, or three kinds of codons: C or A (one codon), U or G (two codons), or I (three codons). Thus, *part of the degeneracy of the genetic code arises from imprecision (wobble) in the pairing of the third base of the codon with the first base of the anticodon.* We see here a strong reason for the frequent appearance of inosine, one of the unusual nucleosides, in anticodons. *Inosine maximizes the number of codons that can be read by a particular tRNA molecule.* The inosines in tRNA are formed by the deamination of adenosine after the synthesis of the primary transcript.

Why is wobble tolerated in the third position of the codon but not in the first two? The 30S subunit has three universally conserved bases—adenine 1492, adenine 1493, and guanine 530—in the 16S RNA that form hydrogen bonds on the minor-groove side but only with correctly formed base pairs of the codon–anticodon duplex (Figure 30.21). These interactions serve to check whether Watson–Crick base pairs are present in the first two

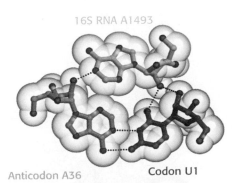

Figure 30.21 16S rRNA monitors base-pairing between the codon and the anticodon. Adenine 1493, one of three universally conserved bases in 16S rRNA, forms hydrogen bonds with the bases in both the codon and the anticodon only if the codon and anticodon are correctly paired. [From J. M. Ogle and V. Ramakrishnan. *Annu. Rev. Biochem.* 74 (2005):129–177, Fig. 2a.]

30S ribosomal subunit

↓ — Initiation factors

30S·IF1·IF3

↓ IF2 **(GTP)**·fMet-tRNA_f
+ mRNA

fMet

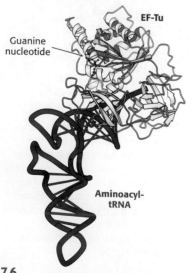

5′ —AUG— mRNA

30S initiation complex

↓ — IF1 + IF3
↓ — 50S subunit + H₂O
↓ — IF2, GDP + P_i

fMet

—AUG—

70S initiation complex

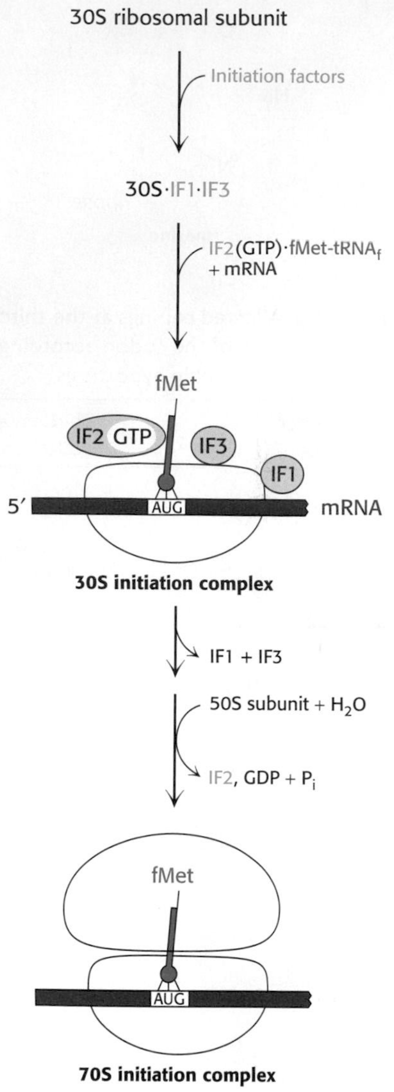

Figure 30.22 Translation initiation in prokaryotes. Initiation factors aid the assembly first of the 30S initiation complex and then of the 70S initiation complex.

EF-Tu

Guanine
nucleotide

Aminoacyl-
tRNA

Figure 30.23 Structure of elongation factor Tu. The structure of a complex between elongation factor Tu (EF-Tu) and an aminoacyl-tRNA. *Notice* the P-loop NTPase domain (purple shading) at the amino-terminal end of EF-Tu. This NTPase domain is similar to those in other G proteins. [Drawn from 1B23.pdb.]

positions of the codon–anticodon duplex. No such inspection device is present for the third position; so more-varied base pairs are tolerated. This mechanism for ensuring fidelity is analogous to the minor-groove interactions utilized by DNA polymerase for a similar purpose (p. 794). *Thus, the ribosome plays an active role in decoding the codon–anticodon interactions.*

30.4 Protein Factors Play Key Roles in Protein Synthesis

Although rRNA is paramount in the process of translation, protein factors also are required for the efficient synthesis of a protein. Protein factors participate in the initiation, elongation, and termination of protein synthesis. P-loop NTPases of the G-protein family play particularly important roles. Recall that these proteins serve as molecular switches as they cycle between a GTP-bound form and a GDP-bound form (p. 387).

Formylmethionyl-tRNA_f Is Placed in the P Site of the Ribosome in the Formation of the 70S Initiation Complex

Messenger RNA and formylmethionyl-tRNA_f must be brought to the ribosome for protein synthesis to begin. How is this accomplished? Three protein *initiation factors* (IF1, IF2, and IF3) are essential. The 30S ribosomal subunit first forms a complex with IF1 and IF3 (Figure 30.22). Binding of IF3 to the 30S subunit prevents it from prematurely joining the 50S subunit to form a dead-end 70S complex, devoid of mRNA and fMet-tRNA_f. IF1 binds near the A site and thereby directs the fMet-RNA_f to the P site. IF2, a member of the G-protein family, binds GTP, and the concomitant conformational change enables IF2 to associate with formylmethionyl-tRNA_f. The IF2–GTP–initiator-tRNA complex binds with mRNA (correctly positioned by the Shine–Dalgarno sequence interaction with the 16S rRNA) and the 30S subunit to form the *30S initiation complex*. Structural changes then lead to the ejection of IF1 and IF3. IF2 stimulates the association of the 50S subunit to the complex. The GTP bound to IF2 is hydrolyzed, leading to the release of IF2. The result is a *70S initiation complex*.

When the 70S initiation complex has been formed, the ribosome is ready for the elongation phase of protein synthesis. The fMet-tRNA_f molecule occupies the P site on the ribosome. The other two sites for tRNA molecules, the A site and the E site, are empty. Formylmethionyl-tRNA_f is positioned so that its anticodon pairs with the initiating AUG (or GUG or UUG) codon on mRNA. This interaction sets the reading frame for the translation of the entire mRNA.

Elongation Factors Deliver Aminoacyl-tRNA to the Ribosome

The second phase of protein synthesis is the elongation cycle. This phase begins with the insertion of an aminoacyl-tRNA into the empty A site on the ribosome. The particular species inserted depends on the mRNA codon in the A site. The cognate aminoacyl-tRNA does not simply leave the synthetase and diffuse to the A site. Rather, it is delivered to the A site in association with a 43-kd protein called *elongation factor Tu* (EF-Tu). Elongation factor Tu, another member of the G-protein family, requires GTP to bind aminoacyl-tRNA (Figure 30.23) and to bind the ribosome. The binding of

EF-Tu to aminoacyl-tRNA serves two functions. First, EF-Tu protects the delicate ester linkage in aminoacyl-tRNA from hydrolysis. Second, the GTP in EF-Tu is hydrolyzed to GDP when an appropriate complex between the EF-Tu–aminoacyl-tRNA complex and the ribosome has formed. If the anticodon is not properly paired with the codon, hydrolysis does not take place and the aminoacyl-tRNA is not transferred to the ribosome. This mechanism allows the free energy of GTP hydrolysis to contribute to the fidelity of protein synthesis. GTP hydrolysis also releases EF-Tu from the ribosome.

EF-Tu in the GDP form must be reset to the GTP form to bind another aminoacyl-tRNA. *Elongation factor Ts*, a second elongation factor, joins the EF-Tu complex and induces the dissociation of GDP. Finally, GTP binds to EF-Tu, and EF-Ts is concomitantly released. It is noteworthy that *EF-Tu does not interact with fMet-tRNA$_f$*. Hence, this initiator tRNA is not delivered to the A site. In contrast, Met-tRNA$_m$, like all other aminoacyl-tRNAs, does bind to EF-Tu. These findings account for the fact that *internal AUG codons are not read by the initiator tRNA*. Conversely, IF2 recognizes fMet-tRNA$_f$ but no other tRNA.

This GTP–GDP cycle of EF-Tu is reminiscent of those of the heterotrimeric G proteins in signal transduction (p. 387) and the Ras proteins in growth control (p. 398). This similarity is due to their shared evolutionary heritage, seen in the homology of the amino-terminal domain of EF-Tu to the P-loop NTPase domains in the other G proteins. The other two domains of the tripartite EF-Tu are distinctive; they mediate interactions between aminoacyl-tRNA and the ribosome. In all these related enzymes, the change in conformation between the GTP and the GDP forms leads to a change in interaction partners. A further similarity is the requirement that an additional protein catalyzes the exchange of GTP for GDP; ET-Ts catalyzes the exchange for ET-Tu, just as an activated receptor does for a heterotrimeric G protein.

The Formation of a Peptide Bond Is Followed by the GTP-Driven Translocation of tRNAs and mRNA

After the correct aminoacyl-tRNA has been placed in the A site, the transfer of the polypeptide chain from the tRNA in the P site is a thermodynamically spontaneous process, driven by the formation of the stronger peptide bond in place of the ester linkage. However, protein synthesis cannot continue without the translocation of the mRNA and the tRNAs within the ribosome. The mRNA must move by a distance of three nucleotides so that the next codon is positioned in the A site for interaction with the incoming aminoacyl-tRNA. At the same time, the deacylated tRNA moves out of the P site into the E site on the 30S subunit and the peptidyl-tRNA moves out of the A site into the P site on the 30S subunit. The movement of the peptidyl-tRNA into the P site shifts the mRNA by one codon, exposing the next codon to be translated in the A site.

The three-dimensional structure of the ribosome undergoes significant change during translocation, and evidence suggests that translocation may result from properties of the ribosome itself. However, protein factors accelerate the process. Translocation is enhanced by *elongation factor G* (EF-G, also called *translocase*). A possible mechanism for accelerating the translocation process in shown in Figure 30.24. First, EF-G in the GTP form binds to the ribosome near the A site, interacting with the 23S rRNA of the 50S subunit. The binding of EF-G to the ribosome stimulates the GTPase activity of EF-G. On GTP hydrolysis, EF-G undergoes a conformational change that displaces the peptidyl-tRNA in the A site to the P site, carrying

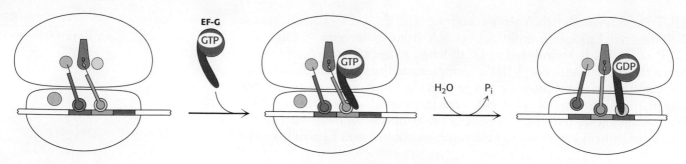

Figure 30.24 Translocation mechanism. In the GTP form, EF-G binds to the EF-Tu-binding site on the 50S subunit. This stimulates GTP hydrolysis, inducing a conformational change in EF-G that forces the tRNAs and mRNA to move through the ribosome by a distance corresponding to one codon.

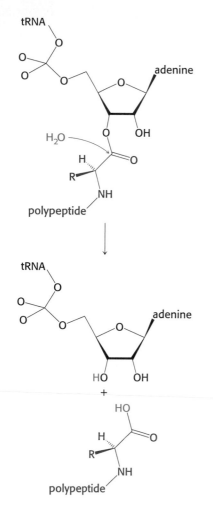

the mRNA and the deacylated tRNA with it. The dissociation of EF-G leaves the ribosome ready to accept the next aminoacyl-tRNA into the A site.

Protein Synthesis Is Terminated by Release Factors That Read Stop Codons

The final phase of translation is termination. How does the synthesis of a polypeptide chain come to an end when a stop codon is encountered? Aminoacyl-tRNA does not normally bind to the A site of a ribosome if the codon is UAA, UGA, or UAG, because normal cells do not contain tRNAs with anticodons complementary to these stop signals. Instead, these *stop codons are recognized by release factors* (RFs), *which are proteins that promote the release of the completed protein from the last tRNA.* One of these release factors, RF1, recognizes UAA or UAG. A second factor, RF2, recognizes UAA or UGA. A third factor, RF3, mediates interactions between RF1 or RF2 and the ribosome. RF3 is another G protein homologous to EF-Tu.

RF1 and RF2 are compact proteins that in eukaryotes resemble a tRNA molecule. When bound to the ribosome, the proteins unfold to bridge the gap between the stop codon on the mRNA and the peptidyl transferase center on the 50S subunit. Although the precise mechanism of release is not known, the release factor may promote, assisted by the peptidyl transferase, a water molecule's attack on the ester linkage, freeing the polypeptide chain. The detached polypeptide leaves the ribosome. Transfer RNA and messenger RNA remain briefly attached to the 70S ribosome until the entire complex is dissociated in a GTP-dependent fashion in response to the binding of EF-G and another factor, called the *ribosome release factor* (RRF) (Figure 30.25).

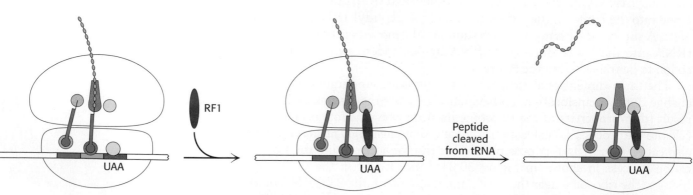

Figure 30.25 Termination of protein synthesis. A release factor recognizes a stop codon in the A site and stimulates the release of the completed protein from the tRNA in the P site.

30.5 Eukaryotic Protein Synthesis Differs from Prokaryotic Protein Synthesis Primarily in Translation Initiation

The basic plan of protein synthesis in eukaryotes and archaea is similar to that in bacteria. The major structural and mechanistic themes recur in all domains of life. However, eukaryotic protein synthesis entails more protein components than does prokaryotic protein synthesis, and some steps are more intricate. Some noteworthy similarities and differences are as follows:

1. *Ribosomes.* Eukaryotic ribosomes are larger. They consist of a 60S large subunit and a 40S small subunit, which come together to form an 80S particle having a mass of 4200 kd, compared with 2700 kd for the prokaryotic 70S ribosome. The 40S subunit contains an 18S RNA that is homologous to the prokaryotic 16S RNA. The 60S subunit contains three RNAs: the 5S RNA, which is homologous to the prokaryotic 5S rRNA; the 28S RNA, which is homologous to the prokaryotic 23S molecules; and the 5.8S RNA, which is homologous to the 5′ end of the 23 S RNA of prokaryotes.

2. *Initiator tRNA.* In eukaryotes, the initiating amino acid is methionine rather than N-formylmethionine. However, as in prokaryotes, a special tRNA participates in initiation. This aminoacyl-tRNA is called Met-tRNA$_i$ or Met-tRNA$_f$ (the subscript "i" stands for initiation, and "f" indicates that it can be formylated in vitro).

3. *Initiation.* The initiating codon in eukaryotes is always AUG. Eukaryotes, in contrast with prokaryotes, do not have a specific purine-rich sequence on the 5′ side to distinguish initiator AUGs from internal ones. Instead, the AUG nearest the 5′ end of mRNA is usually selected as the start site. A 40S ribosome, with a bound Met-tRNA$_i$, attaches to the cap at the 5′ end of eukaryotic mRNA (p. 846) and searches for an AUG codon by moving step-by-step in the 3′ direction (Figure 30.26). This scanning process is catalyzed by helicases that move along the mRNA powered by ATP hydrolysis. Pairing of the anticodon of Met-tRNA$_i$ with the AUG codon of mRNA signals that the target has been found. In almost all cases, eukaryotic mRNA has only one start site and hence is the template for a single protein. In contrast, a prokaryotic mRNA can have multiple Shine–Dalgarno sequences and, hence, start sites, and it can serve as a template for the synthesis of several proteins.

Eukaryotes utilize many more initiation factors than do prokaryotes, and their interplay is much more intricate. The prefix *eIF* denotes a eukaryotic initiation factor. For example, eIF-4E is a protein that binds directly to the 7-methylguanosine cap (p. 846), whereas eIF-2, in association with GTP, delivers the met-tRNA$_i$ to the ribosome. The difference in initiation mechanism between prokaryotes and eukaryotes is, in part, a consequence of the difference in RNA processing. The 5′ end of mRNA is readily available to ribosomes immediately after transcription in prokaryotes. In contrast, pre-mRNA must be processed and transported to the cytoplasm in eukaryotes before translation is initiated. The 5′ cap provides an easily recognizable starting point. In addition, the complexity of eukaryotic translation initiation provides another mechanism for regulation of gene expression that we shall explore further in Chapter 31.

4. *The Structure of mRNA.* Eukaryotic mRNA is circular. The eIF-4E protein that binds to the mRNA cap structure also binds to the poly(A) tail through two protein intermediaries. The protein binds first to the

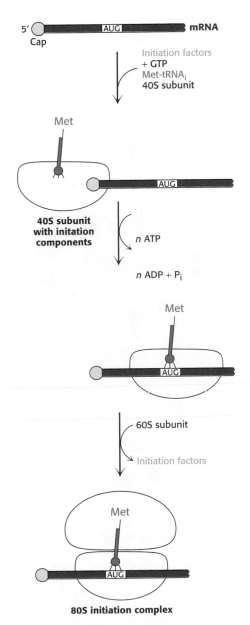

Figure 30.26 Eukaryotic translation initiation. In eukaryotes, translation initiation starts with the assembly of a complex on the 5′ cap that includes the 40S subunit and Met-tRNA$_i$. Driven by ATP hydrolysis, this complex scans the mRNA until the first AUG is reached. The 60S subunit is then added to form the 80S initiation complex.

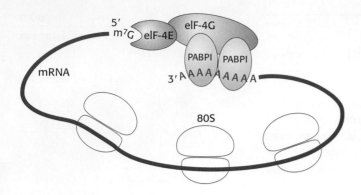

5'
m⁷G eIF-4E
eIF-4G
PABPI PABPI
mRNA
3' AAAAAAAAA
80S

Figure 30.27 Protein interactions circularize eukaryotic mRNA. [After H. Lodish et al., *Molecular Cell Biology*, 5th ed. (W. H. Freeman and Company, 2004), Fig. 4.31.]

eIF-4G protein, which in turn binds to a protein associated with the poly(A) tail, the poly(A)-binding protein (PABPI; Figure 30.27). Cap and tail are thus brought together to form a circle of mRNA. The circular structure may facilitate the rebinding of the ribosomes following protein-synthesis termination.

5. *Elongation and Termination.* Eukaryotic elongation factors EF1 α and EF1 βγ are the counterparts of prokaryotic EF-Tu and EF-Ts. The GTP form of EF1 α delivers aminoacyl-tRNA to the A site of the ribosome, and EF1 βγ catalyzes the exchange of GTP for bound GDP. Eukaryotic EF2 mediates GTP-driven translocation in much the same way as does prokaryotic EF-G. Termination in eukaryotes is carried out by a single release factor, eRF1, compared with two in prokaryotes. Finally, eIF-3, like its prokaryotic counterpart IF3, prevents the reassociation of ribosomal subunits in the absence of an initiation complex.

30.6 Ribosomes Bound to the Endoplasmic Reticulum Manufacture Secretory and Membrane Proteins

A newly synthesized protein in *E. coli* can stay in the cytoplasm or it can be sent to the plasma membrane, the outer membrane, the space between them, or the extracellular medium. Eukaryotic cells can direct proteins to internal sites such as lysosomes, mitochondria, chloroplasts, and the nucleus. How is sorting accomplished? In eukaryotes, a key choice is made soon after the synthesis of a protein begins. The ultimate destination of a protein depends broadly on the location of the ribosome on which it is being synthesized.

In eukaryotic cells, a ribosome remains free in the cytoplasm unless it is directed to the *endoplasmic reticulum* (ER), the extensive membrane system that comprises about half the total membrane of a cell. The region that binds ribosomes is called the *rough ER* because of its studded appearance, in contrast with the *smooth ER,* which is devoid of ribosomes (Figure 30.28). Free ribosomes synthesize proteins that remain within the cell, either within the cytoplasm or directed to organelles bounded by a double membrane, such as the nucleus, mitochondria and chloroplasts. Ribosomes bound to the ER usually synthesize proteins destined to leave the cell or to at least contact the cell exterior from a position in the cell membrane. These proteins fall into three major classes: *secretory proteins* (proteins exported by the cell), *lysosomal proteins,* and *proteins spanning the plasma membrane.* Virtually all integral membrane proteins of the cell, except those located in the membranes of mitochondria and chloroplasts, are formed by ribosomes bound to the ER.

A variety of strategies are used to send proteins synthesized by free ribosomes to the nucleus, peroxisomes, mitochondria, and chloroplasts of eukaryotic cells. However, in this section, we will focus on the targeting of proteins produced by ribosomes bound to the endoplasmic reticulum.

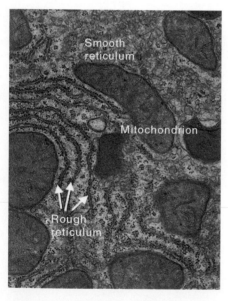

Figure 30.28 Ribosomes are bound to the endoplasmic reticulum. In this electron micrograph, ribosomes appear as small black dots binding to the cytoplasmic side of the endoplasmic reticulum to give a rough appearance. In contrast, the smooth endoplasmic reticulum is devoid of ribosomes. [From G. K. Voletz, M. M. Rolls, and T. A. Rapoport, *EMBO Rep.* 3(2002): 944–950.]

Signal Sequences Mark Proteins for Translocation Across the Endoplasmic Reticulum Membrane

The synthesis of proteins destined to leave the cell or become embedded in the plasma membrane begins on a free ribosome but, shortly after synthesis begins, it is halted until the ribosome is directed to the cytoplasmic side of the endoplasmic reticulum. When the ribosome docks with the membrane, protein synthesis begins again. As the newly forming peptide chain exits the ribosome, it is transported, cotranslationally, through the membrane into the lumen of the endoplasmic reticulum.

Free ribosomes that are synthesizing proteins for use in the cell are identical with those attached to the ER. What is the process that directs the ribosome synthesizing a protein destined to enter the ER to bind to the ER? The translocation consists of four components.

		Cleavage site
Human growth hormone	M A T G S R T S L L L A F G L L C L P W L Q E G S A	F P T
Human proinsulin	M A L W M R L L P L L A L L A L W G P D P A A A	F V N
Bovine proalbumin	M K W V T F I S L L L F S S A Y S	R G V
Mouse antibody H chain	M K V L S L L Y L L T A I P H I M S	D V Q
Chicken lysozyme	M R S L L I L V L C F L P K L A A L G	K V F
Bee promellitin	M K F L V N V A L V F M V V Y I S Y I Y A	A P E
Drosophila glue protein	M K L L V V A V I A C M L I G F A D P A S G	C K D
Zea maize protein 19	M A A K I F C L I M L L G L S A S A A T A	S I F
Yeast invertase	M L L Q A F L F L L A G F A A K I S A	S M T
Human influenza virus A	M K A K L L V L L Y A F V A G	D Q I

Figure 30.29 Amino-terminal signal sequences of some eukaryotic secretory and plasma-membrane proteins. The hydrophobic core (yellow) is preceded by basic residues (blue) and followed by a cleavage site (red) for signal peptidase.

1. *The Signal Sequence.* The signal sequence is *a sequence of 9 to 12 hydrophobic amino acid residues, sometimes containing positively charged amino acids* (Figure 30.29). This sequence is usually near the amino terminus of the nascent polypeptide chain. The presence of the signal sequence identifies the nascent peptide as one that must cross the ER membrane. Some signal sequences are maintained in the mature protein, whereas others are cleaved by a *signal peptidase* on the lumenal side of the ER membrane (see Figure 30.29).

2. *The Signal-Recognition Particle (SRP).* The signal-recognition particle recognizes the signal sequence and binds the sequence and the ribosome as soon as the signal sequence exits the ribosome. SRP then sheperds the ribosome and its nascent polypeptide chain to the ER membrane. SRP is a ribonucleoprotein consisting of a 7S RNA and six different proteins (Figure 30.30). One protein, SRP54, is a GTPase that is crucial for SRP function. SRP binds all ribosomes but binds tightly only to ribosomes that display the signal sequence. Thus, SRP samples ribosomes until it locates one exhibiting a signal sequence. After SRP is bound to the signal sequence, interactions between the ribosome and the SRP occlude the elongation-factor-binding site, thereby halting protein synthesis.

3. *The SRP Receptor (SR).* The SRP–ribosome complex diffuses to the endoplasmic reticulum, where SRP binds the SRP receptor, an integral membrane protein consisting of two subunits, SRα and SRβ. SRα is, like SRP54, a GTPase.

4. *The Translocon.* The SRP–SR complex delivers the ribosome to the ER membrane. There it docks with the translocation machinery, called the *translocon,* a multisubunit assembly of integral and peripheral membrane proteins. The translocon is a protein-conducting channel. This channel opens when the translocon and ribosome bind to each other. Protein synthesis resumes with the growing polypeptide chain passing through the translocon channel into the lumen of the ER.

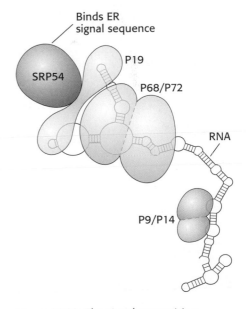

Figure 30.30 The signal-recognition particle. The signal-recognition particle (SRP) consists of six proteins (one of which is SRP54) and one 300-nucleotide RNA molecule. The RNA has a complex structure with many double-helical stretches punctuated by single-stranded regions, shown as circles. [After H. Lodish et al., *Molecular Cell Biology*, 5th ed. (W. H. Freeman and Company, 2004). See K. Strub et al., *Mol. Cell Biol.* 11(1991):3949–3959, and S. High and B. Dobberstein, *J. Cell Biol.* 113(1991): 229–233.]

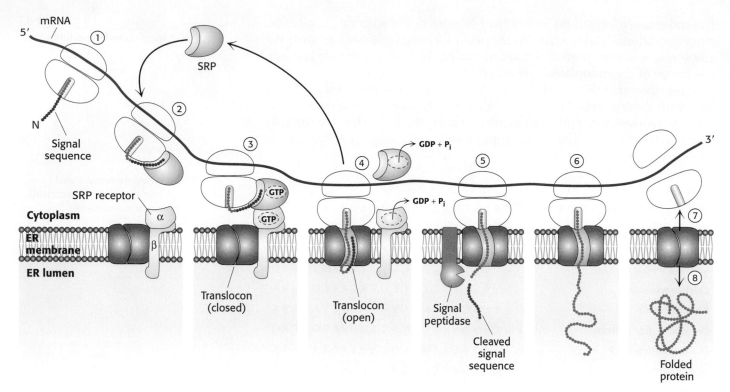

Figure 30.31 The SRP targeting cycle. (1) Protein synthesis begins on free ribosomes. (2) After the signal sequence has exited the ribosome, it is bound by the SRP, and protein synthesis halts. (3) The SRP–ribosome complex docks with the SRP receptor in the ER membrane. (4) The SRP and SRP receptor simultaneously hydrolyze bound GTPs. Protein synthesis resumes and the SRP is free to bind another signal sequence. (5) The signal peptidase may remove the signal sequence as it enters the lumen of the ER. (6) Protein synthesis continues as the protein is synthesized directly into the ER. (7) On completion of protein synthesis, the ribosome is released and the protein tunnel in the translocon closes. [After H. Lodish et al., *Molecular Cell Biology*, 5th ed. (W. H. Freeman and Company, 2004), Fig. 16.6.]

The interactions of the components of the translocation machinery are shown in Figure 30.31. For the SRP–SR complex to form, both the SRP54 and the SRα subunits of SR must bind GTP. For the SRP–SR complex to then deliver the ribosome to the translocon, the two GTP molecules—one in SRP and the other in SR—are aligned in what is essentially an active site shared by the two proteins. After the ribosome has been passed along to the translocon, the GTPs are hydrolyzed, SRP and SR dissociate, and SRP is free to search for another signal sequence to begin the cycle anew. Thus, SRP acts catalytically. The signal peptidase, which is associated with the translocon in the lumen of the ER, removes the signal sequence from most proteins.

Transport Vesicles Carry Cargo Proteins to Their Final Destination

As the proteins are synthesized, they fold to form their three-dimensional structures in the lumen of the ER. Some proteins are modified by the attachment of N-linked carbohydrates. Finally, the proteins must be sorted and transported to their final destinations. Regardless of the destination, the principles of transport are the same. Transport is mediated by *transport vesicles* that bud off the endoplasmic reticulum (Figure 30.32). Transport vesicles from the ER carry their cargo (the proteins) to the Golgi complex, where the vesicles fuse and deposit the cargo inside the complex. There the cargo proteins are modified—for instance, by the attachment of carbohydrates (p. 317). From the Golgi complex, transport vesicles carry the cargo proteins to their final destinations, as shown in Figure 30.32.

How does a protein end up at the correct destination? A newly synthesized protein will float inside the ER lumen until it binds to an integral membrane protein called a *cargo receptor*. This binding sequesters the cargo protein into a small region of the membrane that can subsequently form a membrane bud. The bud will carry the protein to a specific destination—plasma membrane, lysosome, or cell exterior. The key to assuring that the protein reaches the proper destination is that the protein must bind to a receptor in the ER region associated with the protein's destination. To ensure the proper match of protein with ER region, cargo receptors recognize various characteristics of the cargo protein, such as a particular amino acid sequence or an added carbohydrate.

The formation of buds is facilitated by the binding of *coat proteins* (COPs) to the cytoplasmic side of the bud. The coat proteins associate with one another to pinch off the vesicle. After the transport vesicle has formed and is released, the coat proteins are shed to reveal another integral protein called *v-SNARE* ("v" for *vesicle*). v-SNARE will bind to a particular *t-SNARE* ("t" for *target*) in the target membrane. This binding leads to the fusion of the transport vesicle to the target membrane, and the cargo is delivered. Thus, the assignment of identical v-SNARE proteins to the same region of the ER membrane causes an ER region to be associated with a particular destination.

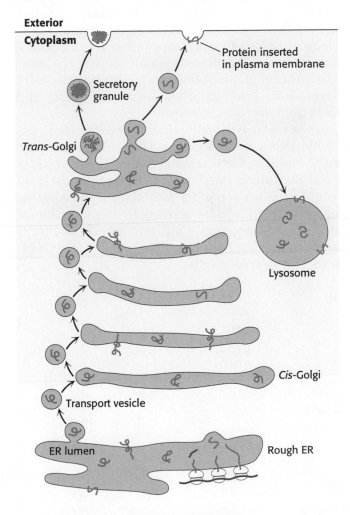

Figure 30.32 Protein-sorting pathways. Newly synthesized proteins in the lumen of the ER are collected into membrane buds. These buds pinch off to form transport vesicles. The transport vesicles carry the cargo proteins to the Golgi complex, where the cargo proteins are modified. Transport vesicles then carry the cargo to the final destination as directed by the v-SNARE and t-SNARE proteins.

TABLE 30.4 Antibiotic inhibitors of protein synthesis

Antibiotic	Action
Streptomycin and other aminoglycosides	Inhibit initiation and cause the misreading of mRNA (prokaryotes)
Tetracycline	Binds to the 30S subunit and inhibits the binding of aminoacyl-tRNAs (prokaryotes)
Chloramphenicol	Inhibits the peptidyl transferase activity of the 50S ribosomal subunit (prokaryotes)
Cycloheximide	Inhibits translocation (eukaryotes)
Erythromycin	Binds to the 50S subunit and inhibits translocation (prokaryotes)
Puromycin	Causes premature chain termination by acting as an analog of aminoacyl-tRNA (prokaryotes and eukaryotes)

30.7 A Variety of Antibiotics and Toxins Can Inhibit Protein Synthesis

The differences between eukaryotic and prokaryotic ribosomes can be exploited for the development of antibiotics (Table 30.4). For example, the antibiotic *puromycin* inhibits protein synthesis by causing nascent polypeptide chains to be released before their synthesis is completed. Puromycin is an analog of the terminal aminoacyl adenylate part of aminoacyl-tRNA (Figure 30.33). It binds to the A site on the ribosome and inhibits the entry of aminoacyl-tRNA. Furthermore, puromycin contains an α-amino group. This amino group, like the one on aminoacyl-tRNA, forms a peptide bond with the carboxyl group of the growing peptide chain. The product, a peptide having a covalently attached puromycin residue at its carboxyl end, dissociates from the ribosome.

Streptomycin, a highly basic trisaccharide, interferes with the binding of formylmethionyl-tRNA to ribosomes and thereby prevents the correct initiation of protein synthesis. Other *aminoglycoside antibiotics* such as neomycin, kanamycin, and gentamycin interfere with the *decoding site* located near nucleotide 1492 in 16S rRNA of the 30S subunit (p. 875). *Chloramphenicol* acts by inhibiting peptidyl transferase activity. *Erythromycin* binds to the 50S subunit and blocks translocation. Finally, *cycloheximide* blocks translocation in eukaryotic ribosomes, making a useful laboratory tool for blocking protein synthesis in eukaryotic cells.

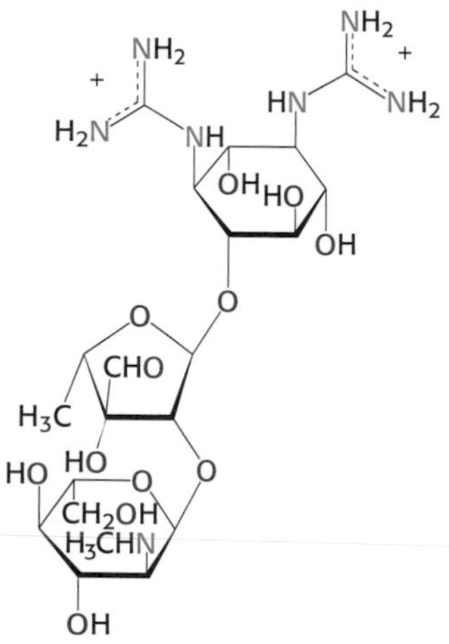

Streptomycin

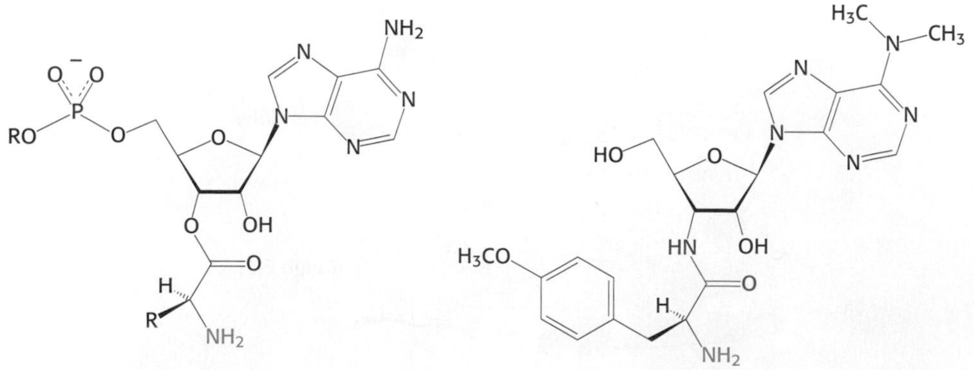

Aminoacyl-tRNA

Puromycin

Figure 30.33 **Antibiotic action of puromycin.** Puromycin resembles the aminoacyl terminus of an aminoacyl-tRNA. Its amino group joins the carbonyl group of the growing polypeptide chain to form an adduct that dissociates from the ribosome. This adduct is stable because puromycin has an amide (shown in red) rather than an ester linkage.

Diphtheria Toxin Blocks Protein Synthesis in Eukaryotes by Inhibiting Translocation

Diphtheria was a major cause of death in childhood before the advent of effective immunization. The lethal effects of this disease are due mainly to a protein toxin produced by *Corynebacterium diphtheriae*, a bacterium that grows in the upper respiratory tract of an infected person. The gene that encodes the toxin comes from a lysogenic phage that is harbored by some strains of *C. diphtheriae*. A few micrograms of diphtheria toxin is usually lethal in an unimmunized person because it inhibits protein synthesis. The toxin is cleaved shortly after entering a target cell into a 21-kd A fragment and a 40-kd B fragment. *The A fragment of the toxin catalyzes the covalent modification of an important component of the protein-synthesizing machinery, whereas the B fragment enables the A fragment to enter the cytoplasm of its target cell.*

A single A fragment of the toxin in the cytoplasm can kill a cell. Why is it so lethal? The target of the A fragment is EF2, the elongation factor catalyzing translocation in eukaryotic protein synthesis. EF2 contains *diphthamide*, an unusual amino acid residue of unknown function that is formed by the posttranslational modification of histidine. The A fragment catalyzes the transfer of the adenosine diphosphate ribose unit of NAD^+ to a nitrogen atom of the diphthamide ring (Figure 30.34). *This ADP ribosylation of a single side chain of EF2 blocks EF2's capacity to carry out the translocation of the growing polypeptide chain.* Protein synthesis ceases, accounting for the remarkable toxicity of diphtheria toxin.

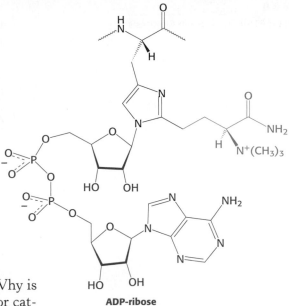

ADP-ribose

Figure 30.34 Blocking of translocation by diphtheria toxin. Diphtheria toxin blocks protein synthesis in eukaryotes by catalyzing the transfer of an ADP-ribose unit from NAD^+ to diphthalamide, a modified amino acid residue in elongation factor 2 (translocase). Diphthamide is formed by the posttranslational modification (blue) of a histidine residue.

Ricin Is an *N*-Glycosidase That Inhibits Protein Synthesis

Ricin is a biomolecule frequently in the news because of its potential use as a bioterrorism agent. Ricin is a small protein (65 kd) found in the seeds of the castor oil plant, *Ricinus communis*. It is indeed a deadly molecule, because as little as 500 μg is lethal to an adult human, and a single molecule can inhibit all protein synthesis in a cell, resulting in cell death.

Ricin is a heterodimeric protein composed of a catalytic A chain joined by a single disulfide bond to a B chain, which binds to galactose on the surface of a target cell. The B chain allows the toxin to bind to the cell, and this binding leads to an endocytotic uptake of the dimer and the eventual release of the A chain into the cytoplasm. The A chain is an *N*-glycosidase, whose substrate is the universally conserved adenosine nucleotide 4324 in the 28S rRNA. Removal of the adenine base completely inactivates the ribosome by preventing the binding of elongation factors. Thus, ricin and diphtheria toxin both act by inhibiting protein-synthesis elongation; ricin does so by covalently modifying rRNA, and diphtheria does so by covalently modifying the elongation factor.

Summary

30.1 Protein Synthesis Requires the Translation of Nucleotide Sequences into Amino Acid Sequences

Protein synthesis is called translation because information present as a nucleic acid sequence is translated into a different language, the sequence of amino acids in a protein. This complex process is mediated by the coordinated interplay of more than a hundred macromolecules, including mRNA, rRNAs, tRNAs, aminoacyl-tRNA synthetases, and protein factors. Given that proteins typically comprise from 100 to 1000 amino acids, the frequency at which an incorrect amino acid is incorporated in the course of protein synthesis must be less than 10^{-4}. Transfer RNAs are the adaptors that make the link between a nucleic acid and an amino acid. These molecules, single chains of about 80 nucleotides, have an L-shaped structure.

30.2 Aminoacyl-Transfer RNA Synthetases Read the Genetic Code

Each amino acid is activated and linked to a specific transfer RNA by an enzyme called an aminoacyl-tRNA synthetase. Such an enzyme links the carboxyl group of an amino acid to the 2'- or 3'-hydroxyl group of the adenosine unit of a CCA sequence at the 3' end of the tRNA by an ester linkage. There is at least one specific aminoacyl-tRNA synthetase and at least one specific tRNA for each amino acid. A synthetase utilizes both the functional groups and the shape of its cognate amino acid to prevent the attachment of an incorrect amino acid to a tRNA. Some synthetases have a separate active site at which incorrectly linked amino acids are removed by hydrolysis. A synthetase recognizes the anticodon, the acceptor stem, and sometimes other parts of its tRNA substrate. By specifically recognizing both amino acids and tRNAs, aminoacyl-tRNA synthetases implement the instruction of the genetic code. There exist two evolutionary distinct classes of synthetases, each recognizing 10 amino acids. The two classes recognize opposite faces of tRNA molecules.

30.3 A Ribosome Is a Ribonucleoprotein Particle (70S) Made of a Small (30S) and a Large (50S) Subunit

Protein synthesis takes place on ribosomes—ribonucleoprotein particles (about two-thirds RNA and one-third protein) consisting of large and small subunits. In *E. coli,* the 70S ribosome (2500 kd) is made up of 30S and 50S subunits. The 30S subunit consists of 16S ribosomal RNA and 21 different proteins; the 50S subunit consists of 23S and 5S rRNA and 34 different proteins. The structures of almost all components of the ribosome have now been determined at or near atomic resolution.

Proteins are synthesized in the amino-to-carboxyl direction, and mRNA is translated in the $5' \rightarrow 3'$ direction. The start signal on prokaryotic mRNA is usually AUG preceded by a purine-rich sequence that can base-pair with 16S rRNA. In prokaryotes, transcription and translation are closely coupled. Several ribosomes can simultaneously translate an mRNA, forming a polysome.

The ribosome includes three sites for tRNA binding called the A (aminoacyl) site, the P (peptidyl) site, and the E (exit) site. With a tRNA attached to the growing peptide chain in the P site, an aminoacyl-tRNA binds to the A site. A peptide bond is formed when the amino group of the aminoacyl-tRNA nucleophilically attacks the ester carbonyl group of the peptidyl-tRNA. On peptide-bond formation, the tRNAs and mRNA must be translocated for the next cycle to begin. The deacylated tRNA moves to the E site and then leaves the ribosome, and the peptidyl-tRNA moves from the A site into the P site.

The codons of messenger RNA recognize the anticodons of transfer RNAs rather than the amino acids attached to the tRNAs. A codon on mRNA forms base pairs with the anticodon of the tRNA. Some tRNAs are recognized by more than one codon because pairing of the third base of a codon is less crucial than that of the other two (the wobble mechanism).

30.4 Protein Factors Play Key Roles in Protein Synthesis

Protein synthesis takes place in three phases: initiation, elongation, and termination. In prokaryotes, mRNA, formylmethionyl-tRNA$_f$ (the special initiator tRNA that recognizes AUG), and a 30S ribosomal subunit come together with the assistance of initiation factors to form a 30S initiation complex. A 50S ribosomal subunit then joins this complex to form a 70S initiation complex, in which fMet-tRNA$_f$ occupies the P site of the ribosome.

Elongation factor Tu delivers the appropriate aminoacyl-tRNA to the ribosome's A (aminoacyl) site as an EF-Tu–aminoacyl-tRNA–GTP ternary complex. EF-Tu serves both to protect the aminoacyl-tRNA from premature cleavage and to increase the fidelity of protein synthesis by ensuring that the correct codon–anticodon pairing has taken place before hydrolyzing GTP and releasing aminoacyl-tRNA into the A site. Elongation factor G uses the free energy of GTP hydrolysis to drive translocation. Protein synthesis is terminated by release factors, which recognize the termination codons UAA, UGA, and UAG and cause the hydrolysis of the ester bond between the polypeptide and tRNA.

30.5 Eukaryotic Protein Synthesis Differs from Prokaryotic Protein Synthesis Primarily in Translation Initiation

The basic plan of protein synthesis in eukaryotes is similar to that in prokaryotes, but there are some significant differences between them. Eukaryotic ribosomes (80S) consist of a 40S small subunit and a 60S large subunit. The initiating amino acid is again methionine, but it is not formylated. The initiation of protein synthesis is more complex in eukaryotes than in prokaryotes. In eukaryotes, the AUG closest to the 5′ end of mRNA is nearly always the start site. The 40S ribosome finds this site by binding to the 5′ cap and then scanning the RNA until AUG is reached. The regulation of translation in eukaryotes provides a means for regulating gene expression.

30.6 Ribosomes Bound to the Endoplasmic Reticulum Manufacture Secretory and Membrane Proteins

Proteins contain signals that determine their ultimate destination. The synthesis of all proteins begins on free ribosomes in the cytoplasm. In eukaryotes, protein synthesis continues in the cytoplasm unless the nascent chain contains a signal sequence that directs the ribosome to the endoplasmic reticulum (ER). Amino-terminal signal sequences consist of a hydrophobic segment of 9 to 12 residues preceded by a positively charged amino acid. Signal-recognition particle (SRP), a ribonucleoprotein assembly, recognizes signal sequences and brings ribosomes bearing them to the ER. A GTP–GDP cycle releases the signal sequence from SRP and then detaches SRP from its receptor. The nascent chain is then translocated across the ER membrane. Proteins are transported throughout the cell in transport vesicles.

30.7 A Variety of Antibiotics and Toxins Can Inhibit Protein Synthesis

Many clinically important antibiotics function by inhibiting protein synthesis. All steps of protein synthesis are susceptible to inhibition by one antibiotic or another. Diphtheria toxin inhibits protein synthesis by covalently modifying an elongation factor, thereby preventing elongation. Ricin, a toxin from castor beans, inhibits elongation by removing a crucial adenine from rRNA.

Key Terms

translation (p. 857)

ribosome (p. 857)

codon (p. 859)

anticodon (p. 859)

transfer RNA (tRNA) (p. 859)

aminoacyl-tRNA synthetase (p. 862)

50S subunit (p. 866)

30S subunit (p. 866)

polysome (p. 870)

Shine–Dalgarno sequence (p. 870)

peptidyl transferase center (p. 872)

wobble hypothesis (p. 874)

initiation factor (p. 876)

elongation factor Tu (EF-Tu) (p. 876)

elongation factor Ts (EF-Ts) (p. 877)

elongation factor G (EF-G) (p. 877)

release factor (p. 878)

signal sequence (p.880)

signal peptidase (p. 881)

signal recognition particle (SRP) (p. 881)

SRP receptor (SR) (p. 881)

translocon (p. 881)

transport vesicle (p. 882)

coat proteins (p. 883)

v-SNARE (p. 883)

t-SNARE (p. 883)

Selected Readings

Where to Start

Noller, H. F. 2005. RNA structure: Reading the ribosome. *Science* 309:1508–1514.

Dahlberg, A. E. 2001. Ribosome structure: The ribosome in action. *Science* 292:868–869.

Ibba, M., Curnow, A. W., and Söll, D. 1997. Aminoacyl-tRNA synthesis: Divergent routes to a common goal. *Trends Biochem. Sci.* 22:39–42.

Davis, B. K. 1999. Evolution of the genetic code. *Prog. Biophys. Mol. Biol.* 72:157–243.

Schimmel, P., and Ribas de Pouplana, L. 2000. Footprints of aminoacyl-tRNA synthetases are everywhere. *Trends Biochem. Sci.* 25:207–209.

Books

Cold Spring Harbor Symposia on Quantitative Biology. 2001. Volume 66, *The Ribosome*. Cold Spring Harbor Laboratory Press.

Gesteland, R. F., Atkins, J. F. and Cech, T. (Eds.). 2005. *The RNA World,* 3d ed. Cold Spring Harbor Laboratory Press.

Garret, R., Douthwaite, S. R., Liljas, A., Matheson, A. T, Moore, P. B., and Noller, H. F. 2000. *The Ribosome: Structure, Function, Antibiotics and Cellular Interactions.* The American Society for Microbiology.

Aminoacyl-tRNA Synthetases

Ibba, M., and Söll, D. 2000. Aminoacyl-tRNA synthesis. *Annu. Rev. Biochem.* 69:617–650.

Sankaranarayanan, R., Dock-Bregeon, A. C., Rees, B., Bovee, M., Caillet, J., Romby, P., Francklyn, C. S., and Moras, D. 2000. Zinc ion mediated amino acid discrimination by threonyl-tRNA synthetase. *Nat. Struct. Biol.* 7:461–465.

Sankaranarayanan, R., Dock-Bregeon, A. C., Romby, P., Caillet, J., Springer, M., Rees, B., Ehresmann, C., Ehresmann, B., and Moras, D. 1999. The structure of threonyl-tRNA synthetase-tRNA^{Thr} complex enlightens its repressor activity and reveals an essential zinc ion in the active site. *Cell* 97:371–381.

Dock-Bregeon, A., Sankaranarayanan, R., Romby, P., Caillet, J., Springer, M., Rees, B., Francklyn, C. S., Ehresmann, C., and Moras, D. 2000. Transfer RNA-mediated editing in threonyl-tRNA synthetase: The class II solution to the double discrimination problem. *Cell* 103:877–884.

Serre, L., Verdon, G., Choinowski, T., Hervouet, N., Risler, J. L., and Zelwer, C. 2001. How methionyl-tRNA synthetase creates its amino acid recognition pocket upon L-methionine binding. *J. Mol. Biol.* 306:863–876.

Beuning, P. J., and Musier-Forsyth, K. 2000. Hydrolytic editing by a class II aminoacyl-tRNA synthetase. *Proc. Natl. Acad. Sci. U. S. A.* 97:8916–8920.

Bovee, M. L., Yan, W., Sproat, B. S., and Francklyn, C. S. 1999. tRNA discrimination at the binding step by a class II aminoacyl-tRNA synthetase. *Biochemistry* 38:13725–13735.

Fukai, S., Nureki, O., Sekine, S., Shimada, A., Tao, J., Vassylyev, D. G., and Yokoyama, S. 2000. Structural basis for double-sieve discrimination of L-valine from L-isoleucine and L-threonine by the complex of tRNA^{Val} and valyl-tRNA synthetase. *Cell* 103:793–803.

de Pouplana, L. R., and Schimmel, P. 2000. A view into the origin of life: Aminoacyl-tRNA synthetases. *Cell. Mol. Life Sci.* 57:865–870.

Transfer RNA

Ibba, M., Becker, H. D., Stathopoulos, C., Tumbula, D. L., and Söll, D. 2000. The adaptor hypothesis revisited. *Trends Biochem. Sci.* 25:311–316.

Weisblum, B. 1999. Back to Camelot: Defining the specific role of tRNA in protein synthesis. *Trends Biochem. Sci.* 24:247–250.

Ribosomes and Ribosomal RNAs

Schuwirth, B. S., Borovinskaya, M. A., Hau, C. W., Zhang, W., Vila-Sanjurjo, A., Holton, J. M., and Doudna Cate, J. H. 2005. Structures of the bacterial ribosome at 3.5 Å resolution. *Science* 310:827–834.

Moore, P. B. 2001. The ribosome at atomic resolution. *Biochemistry* 40:3243–3250.

Yonath, A., and Franceschi, F. 1998. Functional universality and evolutionary diversity: Insights from the structure of the ribosome. *Structure* 6:679–684.

Yusupov, M. M., Yusupova, G. Z., Baucom, A., Lieberman, K., Earnest, T. N., Cate, J. H., and Noller, H. F. 2001. Crystal structure of the ribosome at 5.5 Å resolution. *Science* 292:883–896.

Ban, N., Nissen, P., Hansen, J., Moore, P. B., and Steitz, T. A. 2000. The complete atomic structure of the large ribosomal subunit at 2.4 Å resolution. *Science* 289:905–920.

Carter, A. P., Clemons, W. M., Brodersen, D. E., Morgan-Warren, R. J., Wimberly, B. T., and Ramakrishnan, V. 2000. Functional insights from the structure of the 30S ribosomal subunit and its interactions with antibiotics. *Nature* 407:340–348.

Wimberly, B. T., Brodersen, D. E., Clemons, W. M., Morgan-Warren, R. J., Carter, A. P., Vonrhein, C., Hartsch, T., and Ramakrishnan, V. 2000. Structure of the 30S ribosomal subunit. *Nature* 407:327–339.

Agalarov, S. C., Sridhar Prasad, G., Funke, P. M., Stout, C. D., and Williamson, J. R. 2000. Structure of the S15,S6,S18-rRNA complex: Assembly of the 30S ribosome central domain. *Science* 288:107–113.

Frank, J. 2000. The ribosome: A macromolecular machine par excellence. *Chem. Biol.* 7:R133–R141.

Initiation Factors

Søgaard, B., Sørensen, H. P., Mortensen, K. K., and Sperling-Petersen, H. U. 2005. Initiation of protein synthesis in bacteria. *Microbiol. Mol. Biol. Rev.* 69:101–123.

Carter, A. P., Clemons, W. M., Jr., Brodersen, D. E., Morgan-Warren, R. J., Hartsch, T., Wimberly, B. T., and Ramakrishnan, V. 2001. Crystal structure of an initiation factor bound to the 30S ribosomal subunit. *Science* 291:498–501.

Guenneugues, M., Caserta, E., Brandi, L., Spurio, R., Meunier, S., Pon, C. L., Boelens, R., and Gualerzi, C. O. 2000. Mapping the fMet-tRNA$_f^{Met}$ binding site of initiation factor IF2. *EMBO J.* 19:5233–5240.

Lee, J. H., Choi, S. K., Roll-Mecak, A., Burley, S. K., and Dever, T. E. 1999. Universal conservation in translation initiation revealed by human and archaeal homologs of bacterial translation initiation factor IF2. *Proc. Natl. Acad. Sci. U. S. A.* 96:4342–4347.

Meunier, S., Spurio, R., Czisch, M., Wechselberger, R., Guenneugues, M., Gualerzi, C. O., and Boelens, R. 2000. Structure of the fMet-tRNA$_f^{Met}$-binding domain of *B. stearothermophilus* initiation factor IF2. *EMBO J.* 19:1918–1926.

Elongation Factors

Stark, H., Rodnina, M. V., Wieden, H. J., van Heel, M., and Wintermeyer, W. 2000. Large-scale movement of elongation factor G and extensive conformational change of the ribosome during translocation. *Cell* 100:301–309.

Baensch, M., Frank, R., and Kohl, J. 1998. Conservation of the amino-terminal epitope of elongation factor Tu in Eubacteria and Archaea. *Microbiology* 144:2241–2246.

Krasny, L., Mesters, J. R., Tieleman, L. N., Kraal, B., Fucik, V., Hilgenfeld, R., and Jonak, J. 1998. Structure and expression of elongation factor Tu from *Bacillus stearothermophilus*. *J. Mol. Biol.* 283:371–381.

Pape, T., Wintermeyer, W., and Rodnina, M. V. 1998. Complete kinetic mechanism of elongation factor Tu-dependent binding of aminoacyl-tRNA to the A site of the *E. coli* ribosome. *EMBO J.* 17:7490–7497.

Piepenburg, O., Pape, T., Pleiss, J. A., Wintermeyer, W., Uhlenbeck, O. C., and Rodnina, M. V. 2000. Intact aminoacyl-tRNA is required to trigger GTP hydrolysis by elongation factor Tu on the ribosome. *Biochemistry* 39:1734–1738.

Peptide-Bond Formation and Translocation

Yarus, M., and Welch, M. 2000. Peptidyl transferase: Ancient and exiguous. *Chem. Biol.* 7:R187–R190.

Rodriguez-Fonseca, C., Phan, H., Long, K. S., Porse, B. T., Kirillov, S. V., Amils, R., and Garrett, R. A. 2000. Puromycin-rRNA interaction sites at the peptidyl transferase center. *RNA* 6:744–754.

Vladimirov, S. N., Druzina, Z., Wang, R., and Cooperman, B. S. 2000. Identification of 50S components neighboring 23S rRNA nucleotides A2448 and U2604 within the peptidyl transferase center of *Escherichia coli* ribosomes. *Biochemistry* 39:183–193.

Frank, J., and Agrawal, R. K. 2000. A ratchet-like inter-subunit reorganization of the ribosome during translocation. *Nature* 406:318–322.

Termination

Wilson, D. N., Schluenzen, F., Harms, J. M., Yoshida, T., Ohkubo, T., Albrecht, A., Buerger, J., Kobayashi, Y., and Fucini, P. 2005. X-ray crystallography study on ribosome recycling: The mechanism of binding and action of RRF on the 50S ribosomal subunit. *EMBO J.* 24:251–260.

Fujiwara, T., Ito, K., and Nakamura, Y. 2001. Functional mapping of ribosome-contact sites in the ribosome recycling factor: A structural view from a tRNA mimic. *RNA* 7:64–70.

Freistroffer, D. V., Kwiatkowski, M., Buckingham, R. H., and Ehrenberg, M. 2000. The accuracy of codon recognition by polypeptide release factors. *Proc. Natl. Acad. Sci. U. S. A.* 97:2046–2051.

Heurgue-Hamard, V., Karimi, R., Mora, L., MacDougall, J., Leboeuf, C., Grentzmann, G., Ehrenberg, M., and Buckingham, R. H. 1998. Ribosome release factor RF4 and termination factor RF3 are involved in dissociation of peptidyl-tRNA from the ribosome. *EMBO J.* 17:808–816.

Kisselev, L. L., and Buckingham, R. H. 2000. Translational termination comes of age. *Trends Biochem. Sci.* 25:561–566.

Fidelity and Proofreading

Ogle, J. M., and Ramakrishnan, V. 2005. Structural insights into translational fidelity. *Annu. Rev. Biochem.* 74:129–177.

Ibba, M., and Söll, D. 1999. Quality control mechanisms during translation. *Science* 286:1893–1897.

Rodnina, M. V., and Wintermeyer, W. 2001. Ribosome fidelity: tRNA discrimination, proofreading and induced fit. *Trends Biochem. Sci.* 26:124–130.

Kurland, C. G. 1992. Translational accuracy and the fitness of bacteria. *Annu. Rev. Genet.* 26:29–50.

Fersht, A. 1999. *Structure and Mechanism in Protein Science: A Guide to Enzyme Catalysis and Protein Folding.* W. H. Freeman and Company.

Eukaryotic Protein Synthesis

Sachs, A. B., and Varani, G. 2000. Eukaryotic translation initiation: There are (at least) two sides to every story. *Nat. Struct. Biol.* 7:356–361.

Kozak, M. 1999. Initiation of translation in prokaryotes and eukaryotes. *Gene* 234:187–208.

Negrutskii, B. S., and El'skaya, A. V. 1998. Eukaryotic translation elongation factor 1 α: Structure, expression, functions, and possible role in aminoacyl-tRNA channeling. *Prog. Nucleic Acid Res. Mol. Biol.* 60:47–78.

Preiss, T., and Hentze, M. W. 1999. From factors to mechanisms: Translation and translational control in eukaryotes. *Curr. Opin. Genet. Dev.* 9:515–521.

Bushell, M., Wood, W., Clemens, M. J., and Morley, S. J. 2000. Changes in integrity and association of eukaryotic protein synthesis initiation factors during apoptosis. *Eur. J. Biochem.* 267:1083–1091.

Das, S., Ghosh, R., and Maitra, U. 2001. Eukaryotic translation initiation factor 5 functions as a GTPase-activating protein. *J. Biol. Chem.* 276:6720–6726.

Lee, J. H., Choi, S. K., Roll-Mecak, A., Burley, S. K., and Dever, T. E. 1999. Universal conservation in translation initiation revealed by human and archaeal homologs of bacterial translation initiation factor IF2. *Proc. Natl. Acad. Sci. U. S. A.* 96:4342–4347.

Pestova, T. V., and Hellen, C. U. 2000. The structure and function of initiation factors in eukaryotic protein synthesis. *Cell Mol. Life Sci.* 57:651–674.

Protein Transport Across Membranes

Egea, P. F., Stroud, R. M., and Walter, P. 2005. Targeting proteins to membranes: Structure of the signal recognition particle. *Curr. Opin. Struct. Biol.* 15:213–220.

Halic, M., and Beckmann, R. 2005. The signal recognition particle and its interactions during protein targeting. *Curr. Opin. Struct. Biol.* 15:116–125.

Doudna, J. A., and Batey, R. T. 2004. Structural insights into the signal recognition particle. *Annu. Rev. Biochem.* 73:539–557.

Schnell, D. J., and Hebert, D. N. 2003. Protein translocons: Multifunctional mediators of protein translocation across membranes. *Cell* 112:491–505.

Antibiotics and Toxins

Belova, L., Tenson, T., Xiong, L., McNicholas, P. M., and Mankin, A. S. 2001. A novel site of antibiotic action in the ribosome: Interaction of evernimicin with the large ribosomal subunit. *Proc. Natl. Acad. Sci. U. S. A.* 98:3726–3731.

Brodersen, D. E., Clemons, W. M., Jr., Carter, A. P., Morgan-Warren, R. J., Wimberly, B. T., and Ramakrishnan, V. 2000. The structural basis for the action of the antibiotics tetracycline, pactamycin, and hygromycin B on the 30S ribosomal subunit. *Cell* 103:1143–1154.

Porse, B. T., and Garrett, R. A. 1999. Ribosomal mechanics, antibiotics, and GTP hydrolysis. *Cell* 97:423–426.

Lord, M. J., Jolliffe, N. A., Marsden, C. J., Pateman, C. S., Smith, D. S., Spooner, R. A., Watson, P. D., and Roberets, L. M. 2003. Ricin: Mechanisms of toxicity. *Toxicol. Rev.* 22:53–64.

Problems

1. *Synthetase mechanism.* The formation of isoleucyl-tRNA proceeds through the reversible formation of an enzyme-bound Ile-AMP intermediate. Predict whether ^{32}P-labeled ATP is formed from ^{32}PP$_i$ when each of the following sets of components is incubated with the specific activating enzyme:

(a) ATP and ^{32}PP$_i$
(b) tRNA, ATP, and ^{32}PP$_i$
(c) Isoleucine, ATP, and ^{32}PP$_i$

2. *Light and heavy ribosomes.* Ribosomes were isolated from bacteria grown in a "heavy" medium (^{13}C and ^{15}N) and from bacteria grown in a "light" medium (^{12}C and ^{14}N). These 60S ribosomes were added to an in vitro system engaged in protein synthesis. An aliquot removed several hours later was analyzed by density-gradient centrifugation. How many bands of 70S ribosomes would you expect to see in the density gradient?

3. *The price of protein synthesis.* What is the smallest number of molecules of ATP and GTP consumed in the synthesis of a 200-residue protein, starting from amino acids? Assume that the hydrolysis of PP$_i$ is equivalent to the hydrolysis of ATP for this calculation.

4. *Contrasting modes of elongation.* The two basic mechanisms for the elongation of biomolecules are represented in the adjoining illustration. In type 1, the activating group (X) is released from the growing chain. In type 2, the activating group is released from the incoming unit as it is added to the growing chain. Indicate whether each of the following biosyntheses is by means of a type 1 or a type 2 mechanism:

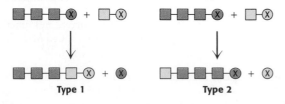

Type 1   **Type 2**

(a) Glycogen synthesis
(b) Fatty acid synthesis
(c) $C_5 \rightarrow C_{10} \rightarrow C_{15}$ in cholesterol synthesis
(d) DNA synthesis
(e) RNA synthesis
(f) Protein synthesis

5. *Suppressing frameshifts.* The insertion of a base in a coding sequence leads to a shift in the reading frame, which in most cases produces a nonfunctional protein. Propose a mutation in a tRNA that might suppress frameshifting.

6. *Tagging a ribosomal site.* Design an affinity-labeling reagent for one of the tRNA binding sites in *E. coli* ribosomes.

7. *Viral mutation.* An mRNA transcript of a T7 phage gene contains the base sequence

$$\downarrow$$
5′-AACUGCACGAGGUAACACAAGAUGGCU-3′

Predict the effect of a mutation that changes the G marked by an arrow to A.

8. *Two synthetic modes.* Compare and contrast protein synthesis by ribosomes with protein synthesis by the solid-phase method (see Section 3.4).

9. *Enhancing fidelity.* Compare the accuracy of (a) DNA replication, (b) RNA synthesis, and (c) protein synthesis. Which mechanisms are used to ensure the fidelity of each of these processes?

10. *Triggered GTP hydrolysis.* Ribosomes markedly accelerate the hydrolysis of GTP bound to the complex of EF-Tu and aminoacyl-tRNA. What is the biological significance of this enhancement of GTPase activity by ribosomes?

11. *Blocking translation.* Devise an experimental strategy for switching off the expression of a specific mRNA without changing the gene encoding the protein or the gene's control elements.

12. *Directional problem.* Suppose that you have a protein-synthesis system that is synthesizing a protein designated A. Furthermore, you know that protein A has four trypsin-sensitive sites, equally spaced in the protein, that, on digestion with trypsin, yield the peptides A$_1$, A$_2$, A$_3$, A$_4$, and A$_5$. Peptide A$_1$ is the amino-terminal peptide, and A$_5$ is the carboxyl-terminal peptide. Finally, you know that your system requires 4 minutes to synthesize a complete protein A. At $t = 0$, you add all 20 amino acids, each carrying a ^{14}C label.

(a) At $t = 1$ minute, you isolate intact protein A from the system, cleave it with trypsin, and isolate the five peptides. Which peptide is most heavily labeled?
(b) At $t = 3$ minutes, what will be the order of the labeling of peptides from greatest to least?
(c) What does this experiment tell you about the direction of protein synthesis?

13. *Translator.* Aminoacyl-tRNA synthetases are the only component of gene expression that decodes the genetic code. Explain.

14. *A timing device.* EF-Tu, a member of the G-protein family, plays a crucial role in the elongation process of translation. Suppose that a slowly hydrolyzable analog of GTP were added to an elongating system. What would be the effect on the rate of protein synthesis?

Mechanism Problems

15. *Molecular attack.* What is the nucleophile in the reaction catalyzed by peptidyl transferase? Write out a plausible mechanism for this reaction.

16. *Evolutionary amino acid choice.* Ornithine is structurally similar to lysine except ornithine's side chain is one methylene group shorter than that of lysine. Attempts to chemically synthesize and isolate ornithinyl-tRNA proved unsuccessful. Propose a mechanistic explanation. (Hint: Six-membered rings are more stable than seven-membered rings).

Chapter Integration Problems

17. *Déjà vu.* Which protein in G-protein cascades plays a role similar to that of elongation factor Ts?

18. *Family resemblance.* Eukaryotic elongation factor 2 is inhibited by ADP ribosylation catalyzed by diphtheria toxin. What other G proteins are sensitive to this mode of inhibition?

Data Interpretation Problem

19. *Helicase helper.* The initiation factor eIF-4A displays ATP-dependent RNA helicase activity. Another initiation factor, eIF-4H, has been proposed to assist the action of eIF-4A. Graph A shows some of the experimental results from an assay that can measure the activity of eIF-4A helicase in the presence of eIF-4H.

(a) What are the effects on eIF-4A helicase activity in the presence of eIF-4H?
(b) Why did measuring the helicase activity of eIF-4H alone serve as an important control?

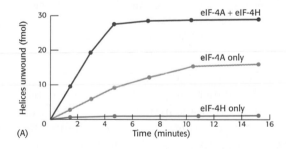

(c) The initial rate of helicase activity of 0.2 μM of eIF-4 was then measured with varying amounts of eIF-4H (graph B). What ratio of eIF-4H to eIF-4A yielded optimal activity?

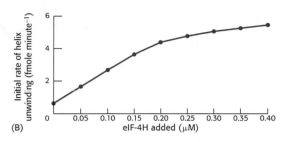

(d) Next, the effect of RNA–RNA helix stability on the initial rate of unwinding in the presence and absence of eIF-4H was tested (graph C). How does the effect of eIF-4H vary with helix stability?

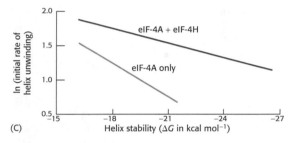

(e) How might eIF-4H affect the helicase activity of eIF-4A? [Data after N. J. Richter, G. W. Rodgers, Jr., J. O. Hensold, and W. C. Merrick. Further biochemical and kinetic characterization of human eukaryotic initiation factor 4H. *J. Biol. Chem.* 274(1999):35415–35424.]

The Control of Gene Expression

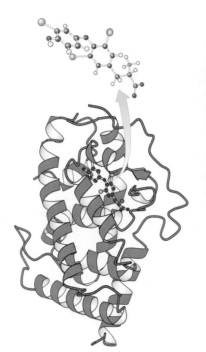

Complex biological processes often require coordinated control of the expression of many genes. The maturation of a tadpole into a frog is largely controlled by thyroid hormone. This hormone regulates gene expression by binding to a protein, the thyroid hormone receptor, shown at the left. In response to the hormone's binding, this protein binds to specific DNA sites in the genome and modulates the expression of nearby genes. [(Right) Shanon Cummings/Dembinsky Photo Associates.]

A gene is *expressed* when it is transcribed into RNA and, for most genes, translated into proteins. Genomes comprise thousands of genes. Some of these genes are expressed at all times and in many or all cells of a given organism. These genes are subject to *constitutive expression*. Many other genes are expressed only under some circumstances, either in a specific type of cell or under a particular set of physiological conditions or both. These genes are subject to *regulated expression*. For example, the level of expression of some genes in bacteria may vary more than a 1000-fold in response to the supply of nutrients or to environmental challenges. Similarly in multicellular organisms, cells exposed to hormones and growth factors may begin expressing genes that lead to substantial changes in shape, growth rate, and other characteristics. A liver cell and a pancreas cell contain exactly the same DNA sequences, yet the subset of genes highly expressed in cells from the pancreas, which secretes digestive enzymes, differs markedly from the subset highly expressed in the liver, the site of lipid transport and energy transduction (Table 31.1).

How is gene expression controlled? *Gene activity is controlled first and foremost at the level of transcription.* Whether a gene is transcribed is determined largely by the interplay between specific DNA sequences and the specific proteins that bind to these sequences. We first consider gene-regulation mechanisms in prokaryotes and particularly in *E. coli*, because

TABLE 31.1 Highly expressed protein-encoding genes of the pancreas and liver (as percentage of total mRNA pool)

Rank	Proteins encoded in Pancreas	%	Proteins encoded in Liver	%
1	Procarboxypeptidase A1	7.6	Albumin	3.5
2	Pancreatic trypsinogen 2	5.5	Apolipoprotein A-I	2.8
3	Chymotrypsinogen	4.4	Apolipoprotein C-I	2.5
4	Pancreatic trypsin 1	3.7	Apolipoprotein C-III	2.1
5	Elastase IIIB	2.4	ATPase 6/8	1.5
6	Protease E	1.9	Cytochrome oxidase 3	1.1
7	Pancreatic lipase	1.9	Cytochrome oxidase 2	1.1
8	Procarboxypeptidase B	1.7	α_1-Antitrypsin	1.0
9	Pancreatic amylase	1.7	Cytochrome oxidase 1	0.9
10	Bile-salt-stimulated lipase	1.4	Apolipoprotein E	0.9

Sources: Data for pancreas from V. E. Velculescu, L. Zhang, B. Vogelstein, and K. W. Kinzler. *Science* 270(1995):484–487. Data for liver from T. Yamashita, S. Hashimoto, S. Kaneko, S. Nagai, N. Toyoda, T. Suzuki, K. Kobayashi, and K. Matsushima. *Biochem. Biophys. Res. Commun.* 269(2000): 110–116.

these processes have been extensively investigated in this organism. We then turn to eukaryotic gene regulation. In both prokaryotes and eukaryotes, regulatory proteins recruit RNA polymerases to sites within the genome to initiate gene transcription. However, much of the DNA in the considerably larger genome of a eukaryotic cell is stably assembled into chromatin, rendering many potential binding sites for transcription factors inaccessible—in effect, reducing the size of the genome. Thus, rather than scanning through the entire genome, a eukaryotic DNA-binding protein scans a set of accessible binding sites that may be close in size to the genome of a prokaryote. In the chapter's final section, we explore mechanisms for regulating gene expression past the level of transcription.

31.1 Many DNA-Binding Proteins Recognize Specific DNA Sequences

Genes are located in specific positions within the genome. Yet the sequences that are transcribed do not have any distinguishing features that would allow regulatory systems to recognize them. Instead, gene regulation depends on other sequences in the genome. These regulatory sites are often regions close to the region of the DNA that is transcribed, but they can also be located some distance away. Regulatory sites are usually binding sites for specific DNA-binding proteins. When bound to such sites, DNA-binding proteins may either stimulate or repress gene expression. These regulatory sites were first identified in *E. coli* in studies of changes in gene expression. In the presence of the sugar lactose, the bacterium starts to express a gene encoding an enzyme that can process lactose for use as a carbon and energy source. The sequence of the regulatory site for this gene is shown in Figure 31.1. The nucleotide sequence of this site shows a nearly perfect inverted repeat, indicating that the DNA in this region has an approximate twofold axis of symmetry. Recall that cleavage sites for restriction enzymes such as *Eco*RV have similar symmetry properties (p. 259). Symmetry in such regulatory sites usually corresponds to symmetry in the protein that binds the site. *Symmetry matching is a recurring theme in protein–DNA interactions.*

Figure 31.1 Sequence of the *lac* regulatory site. The nucleotide sequence of this regulatory site shows a nearly perfect inverted repeat, corresponding to twofold rotational symmetry in the DNA. Parts of the sequences that are related by this symmetry are shown in the same color.

```
5'-...TGTGTGGAATTGTGAGCGGATAACAATTTCACACA...3'
3'-...ACACACCTTAACACTCGCCTAATGTTAAAGTGTGT...5'
```

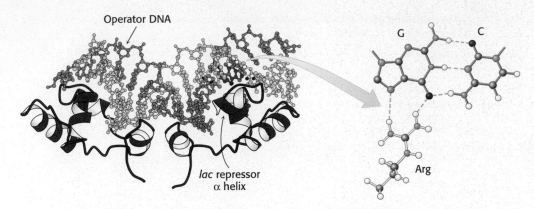

Operator DNA

lac repressor
α helix

G   C

Arg

Figure 31.2 **Lac repressor–DNA complex.** The DNA-binding domain from a gene-regulatory protein, the *lac* repressor, binds to a DNA fragment containing its preferred binding site (referred to as operator DNA) by inserting an α helix into the major groove of operator DNA. *Notice* that a specific contact forms between an arginine residue of the repressor and a G–C base pair in the binding site. [Drawn from 1EFA.pdb.]

The structure of the complex between an oligonucleotide that includes this site and the DNA-binding unit that recognizes this site reveals the interactions that allow specific DNA-binding (Figure 31.2). The DNA-binding unit comes from a protein called the *lac repressor,* which represses expression of the lactose-processing gene. As expected, this protein binds as a dimer, and the twofold axis of the dimer matches the symmetry of the DNA. An α helix from each monomer of the protein is inserted into the major groove of the DNA, where amino acid side chains make specific contacts with exposed edges of the base pairs. For example, the side chain of an arginine residue forms a pair of hydrogen bonds with a guanine residue, which would not be possible with any other base. This interaction and similar ones allow the *lac* repressor to bind more tightly to this site than to the wide range of other sites present in the *E. coli* genome.

The Helix-Turn-Helix Motif Is Common to Many Prokaryotic DNA-Binding Proteins

Are similar strategies utilized by other prokaryotic DNA-binding proteins? The structures of many such proteins have now been determined, and amino acid sequences are known for many more. Strikingly, the DNA-binding surfaces of many, but not all, of these proteins consist of a pair of α helices separated by a tight turn (Figure 31.3). In complexes with DNA, the second of these two helices (often called the *recognition helix*) lies in the major groove, where amino acid side chains make contact with the edges of

Figure 31.3 **Helix-turn-helix motif.** These structures show three sequence-specific DNA-binding proteins that interact with DNA through a helix-turn-helix motif (highlighted in yellow). *Notice* that, in each case, the helix-turn-helix units within a protein dimer are approximately 34 Å apart, corresponding to one full turn of DNA. [Drawn from 1EFA, 1RUN, and 1TRO.pdb.]

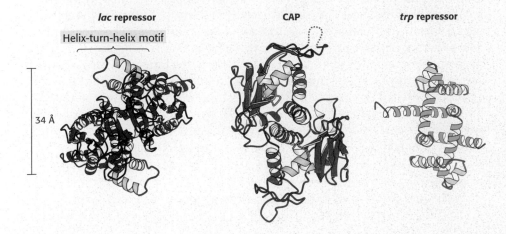

lac repressor   CAP   *trp* repressor

Helix-turn-helix motif

34 Å

base pairs. In contrast, residues of the first helix participate primarily in contacts with the DNA backbone. *Helix-turn-helix motifs* are present on many proteins that bind DNA as dimers, and thus two of the units will be present, one on each monomer.

Although the helix-turn-helix motif is the most commonly observed DNA-binding unit in prokaryotes, not all regulatory proteins bind DNA through such units. A striking example is provided by the *E. coli* methionine repressor (Figure 31.4). This protein binds DNA through the insertion of a pair of β strands into the major groove.

A Range of DNA-Binding Structures Are Employed by Eukaryotic DNA-Binding Proteins

DNA-binding proteins are also key to gene regulation in eukaryotes. A range of structures have been observed, but we will focus on three that reveal the common features and the diversity of these units. The first class of eukaryotic DNA-binding unit that we will consider is the *homeodomain* (Figure 31.5). The structure of this domain and its mode of recognition of DNA are very similar to those of the prokaryotic helix-turn-helix proteins. In eukaryotes, homeodomain proteins often form heterodimeric structures, sometimes with other homeodomain proteins, that recognize asymmetric DNA sequences.

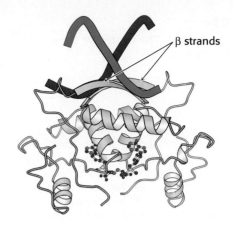

Figure 31.4 DNA recognition through β strands. A methionine repressor is shown bound to DNA. *Notice* that residues in β strands, rather than α helices, participate in the crucial interactions between the protein and the DNA. [Drawn from 1CMA.pdb.]

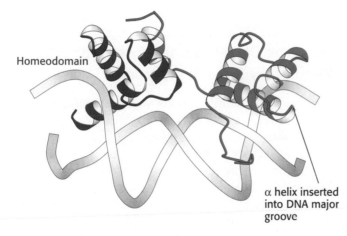

Figure 31.5. Homeodomain structure. The structure of a heterodimer formed from two different DNA-binding domains, each based on a homeodomain. *Notice* that each homeodomain has a helix-turn-helix motif with one helix inserted into the major groove of DNA. [Drawn from 1AKH.pdb.]

The second class of eukaryotic DNA-binding unit is the *basic-leucine zipper,* or bZip, proteins (Figure 31.6). This DNA-binding unit consists of a pair of long α helices. The first part of each α helix is a basic region that lies in the major groove of the DNA and makes contacts responsible for DNA-site recognition. The second part of each α helix forms a coiled-coil structure with its partner. Because these units are often stabilized by appropriately spaced leucine residues, these structures are often referred to as *leucine zippers* (p. 45).

The final class of eukaryotic DNA-binding units that we will consider here are the *Cys₂His₂ zinc-finger domains* (Figure 31.7). A DNA-binding unit of this class comprises tandem sets of small domains, each of which binds a zinc ion through conserved sets of two cysteine and two histidine residues. These domains, often called *zinc-finger domains,* form a string that follows the major groove of DNA. An α helix from each domain makes specific contact with the edges of base pairs within the groove. Some proteins

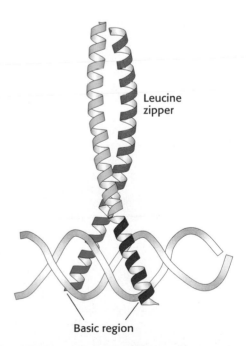

Figure 31.6. Basic-leucine zipper. This heterodimer comprises two basic-leucine zipper proteins. *Notice* that the basic region lies in the major groove of DNA. The leucine-zipper part stabilizes the protein dimer. [Drawn from 1FOS.pdb.]

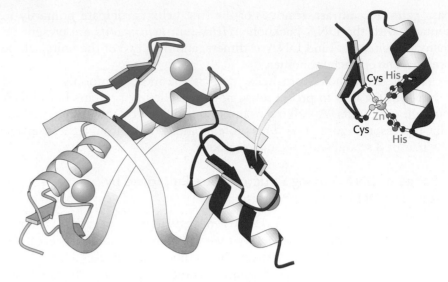

Figure 31.7 Zinc-finger domains. A DNA-binding domain comprising three Cys_2His_2 zinc-finger domains (shown in yellow, blue, and red) is shown in a complex with DNA. Each zinc-finger domain is stabilized by a bound zinc ion (shown in green) through interactions with two cysteine residues and two histidine residues. *Notice* how the protein wraps around the DNA in the major groove. [Drawn from 1AAY.pdb.]

contain arrays of 10 or more zinc-finger domains, potentially enabling them to contact long stretches of DNA. The human genome encodes several hundred proteins that contain zinc-finger domains of this class. We will encounter another class of zinc-based DNA-binding domain when we consider nuclear hormone receptors in Section 31.3.

31.2 Prokaryotic DNA-Binding Proteins Bind Specifically to Regulatory Sites in Operons

A historically important example reveals many common principles of gene regulation by DNA-binding proteins. Bacteria such as *E. coli* usually rely on glucose as their source of carbon and energy. However, when glucose is scarce, *E. coli* can use lactose as their carbon source, even though this disaccharide does not lie on any major metabolic pathways. An essential enzyme in the metabolism of lactose is *β-galactosidase,* which hydrolyzes lactose into galactose and glucose. These products are then metabolized by pathways discussed in Chapter 16.

Lactose **Galactose** **Glucose**

This reaction can be conveniently followed in the laboratory through the use of alternative galactoside substrates that form colored products such as X-Gal (Figure 31.8). An *E. coli* cell growing on a carbon source such as

X-Gal **5,5′-Dibromo-4,4′-dichloro-indigo**

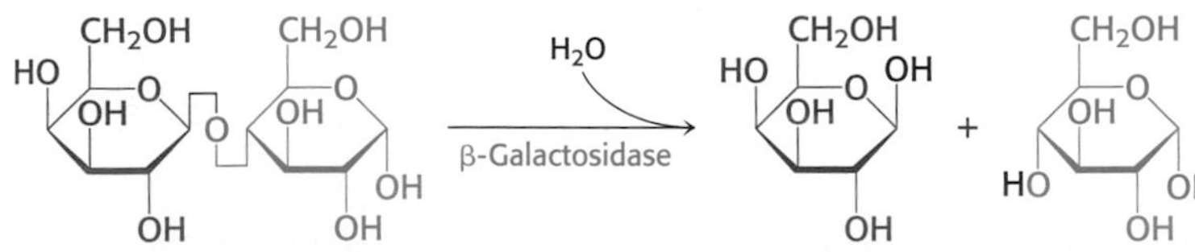

Figure 31.8 Monitoring the β-galactosidase reaction. The galactoside substrate X-Gal produces a colored product on cleavage by β-galactosidase. The appearance of this colored product provides a convenient means for monitoring the amount of the enzyme both in vitro and in vivo.

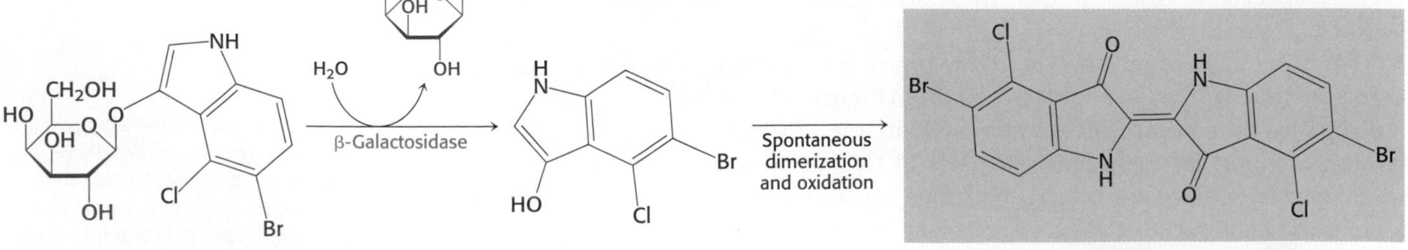

glucose or glycerol contains fewer than 10 molecules of β-galactosidase. In contrast, the same cell will contain several thousand molecules of the enzyme when grown on lactose (Figure 31.9). The presence of lactose in the culture medium induces a large increase in the amount of β-galactosidase by eliciting the synthesis of new enzyme molecules rather than by activating a preexisting but inactive precursor.

A crucial clue to the mechanism of gene regulation was the observation that two other proteins are synthesized in concert with β-galactosidase—namely, *galactoside permease* and *thiogalactoside transacetylase*. The permease is required for the transport of lactose across the bacterial cell membrane (p. 360). The transacetylase is not essential for lactose metabolism but appears to play a role in the detoxification of compounds that also may be transported by the permease. Thus, *the expression levels of a set of enzymes that all contribute to the adaptation to a given change in the environment change together*. Such a coordinated unit of gene expression is called an *operon*.

An Operon Consists of Regulatory Elements and Protein-Encoding Genes

The parallel regulation of β-galactosidase, the permease, and the transacetylase suggested that the expression of genes encoding these enzymes is controlled by a common mechanism. François Jacob and Jacques Monod proposed the *operon model* to account for this parallel regulation as well as the results of other genetic experiments (Figure 31.10). The genetic elements of the model are a *regulator gene*, a regulatory DNA sequence called *an operator site*, and a *set of structural genes*.

The regulator gene encodes a *repressor* protein that binds to the operator site. The binding of the repressor to the operator prevents transcription of the structural genes. The operator and its associated structural genes constitute the operon. For the *lactose (lac) operon*, the *i* gene encodes the repressor, *o* is the operator site, and the *z*, *y*, and *a* genes are the structural genes for β-galactosidase, the permease, and the transacetylase, respectively. The operon also contains a promoter site (denoted by *p*), which directs the RNA polymerase to the correct transcription initiation site. The *z*, *y*, and *a* genes are transcribed to give a single mRNA molecule that encodes all three proteins. An mRNA molecule encoding more than one protein is known as a *polygenic* or *polycistronic* transcript.

The *lac* Repressor Protein in the Absence of Lactose Binds to the Operator and Blocks Transcription

How does the *lac* repressor inhibit the expression of the *lac* operon? The *lac* repressor can exist as a dimer of 37-kd subunits, and two dimers often come together to form a tetramer. In the absence of lactose, the repressor binds very tightly and rapidly to the operator. When the *lac* repressor is bound to DNA, the repressor prevents bound RNA polymerase from locally unwinding the DNA to expose the bases that will act as the template for the synthesis of the RNA strand.

How does the *lac* repressor locate the operator site in the *E. coli* chromosome? The *lac* repressor binds 4×10^6 times as strongly to operator DNA as it does to random sites in the genome. This high degree of selectivity

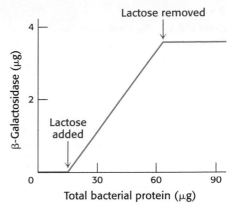

Figure 31.9 β-Galactosidase induction. The addition of lactose to an *E. coli* culture causes the production of β-galactosidase to increase from very low amounts to much larger amounts. The increase in the amount of enzyme parallels the increase in the number of cells in the growing culture. β-Galactosidase constitutes 6.6% of the total protein synthesized in the presence of lactose.

Figure 31.10 Operons. (A) The general structure of an operon as conceived by Jacob and Monod. (B) The structure of the lactose operon. In addition to the promoter, *p*, in the operon, a second promoter is present in front of the regulator gene, *i*, to drive the synthesis of the regulator.

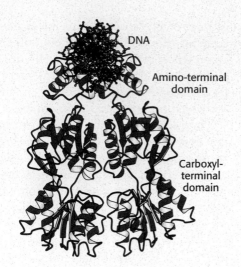

Figure 31.11 Structure of the *lac*
repressor. A *lac* repressor dimer is
shown bound to DNA. *Notice* that the
amino-terminal domain binds to DNA,
whereas the carboxyl-terminal domain
forms a separate structure. A part of the
structure that mediates the formation of
lac repressor tetramers is not shown.
[Drawn from 1EFA.pdb.]

allows the repressor to find the operator efficiently even with a large excess
(4.6×10^6) of other sites within the *E. coli* genome. The dissociation constant
for the repressor–operator complex is approximately 0.1 pM (10^{-13} M).
The rate constant for association ($\approx 10^{10}\,\mathrm{M}^{-1}\mathrm{s}^{-1}$) is strikingly high, indicating that the repressor finds the operator by diffusing along a DNA mol-
ecule (a one-dimensional search) rather than encountering it from the aqueous medium (a three-dimensional search).

Inspection of the complete *E. coli* genome sequence reveals two sites
within 500 bp of the primary operator site that approximate the sequence of
the operator. Other *lac* repressor dimers can bind to these sites, particularly
when aided by cooperative interactions with the *lac* repressor dimer at the
primary operator site. No other sites that closely match the sequence of the
lac operator site are present in the rest of the *E. coli* genome sequence. Thus,
the DNA-binding specificity of the lac *repressor is sufficient to specify a nearly
unique site within the* E. coli *genome.*

The three-dimensional structure of the *lac* repressor has been determined in various forms. Each monomer consists of a small amino-terminal
domain that binds DNA and a larger domain that mediates the formation of
the dimer and the tetramer (Figure 31.11). A pair of the amino-terminal domains come together to form the functional DNA-binding unit. Each
monomer has a helix-turn-helix unit, discussed on page 894, that interacts
with the major groove of the bound DNA.

Ligand Binding Can Induce Structural Changes in Regulatory Proteins

How does the presence of lactose trigger expression from the *lac* operon?
Interestingly, lactose itself does not have this effect; rather, *allolactose*, a
combination of galactose and glucose with an α-1,6 rather than an α-1,4
linkage, does. Allolactose is thus referred to as the *inducer* of the *lac* operon.
Allolactose is a side product of the β-galactosidase reaction and is produced
at low levels by the few molecules of β-galactosidase that are present before
induction. Some other β-galactosides such as *isopropylthiogalactoside*
(IPTG) are potent inducers of β-galactosidase expression, although they
are not substrates of the enzyme. IPTG is useful in the laboratory as a tool
for inducing gene expression.

The inducer triggers gene expression by preventing the *lac* repressor
from binding the operator. *The inducer binds to the* lac *repressor and thereby
greatly reduces the repressor's affinity for operator DNA.* An inducer molecule
binds in the center of the large domain within each monomer. This binding
leads to conformational changes that modify the relation between the two

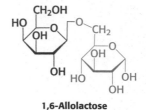

1,6-Allolactose

**Isopropylthiogalactoside
(IPTG)**

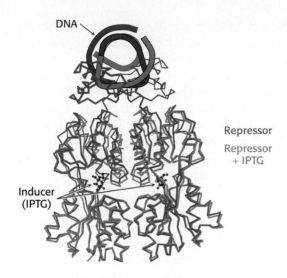

DNA

Inducer
(IPTG)

Repressor

Repressor
+ IPTG

Figure 31.12 Effects of IPTG on *lac* repressor structure. The structure of the *lac* repressor bound to the inducer isopropylthiogalactoside (IPTG), shown in orange, is superimposed on the structure of the *lac* repressor bound to DNA, shown in purple. *Notice* that the binding of IPTG induces structural changes that alter the relation between the two DNA-binding domains so that they cannot interact effectively with DNA. The DNA-binding domains of the *lac* repressor bound to IPTG are not shown, because these regions are not well ordered in the crystals studied.

small DNA-binding domains (Figure 31.12). These domains can no longer easily contact DNA simultaneously, leading to a dramatic reduction in DNA-binding affinity.

Let us recapitulate the processes that regulate gene expression in the lactose operon (Figure 31.13). In the absence of inducer, the *lac* repressor is bound to DNA in a manner that blocks RNA polymerase from transcribing the *z*, *y*, and *a* genes. Thus, very little β-galactosidase, permease, or transacetylase are produced. The addition of lactose to the environment leads to the formation of allolactose. This inducer binds to the *lac* repressor, leading to conformational changes and the release of DNA by the *lac* repressor. With the operator site unoccupied, RNA polymerase can then transcribe the other *lac* genes and the bacterium will produce the proteins necessary for the efficient use of lactose.

The structure of the large domain of the *lac* repressor is similar to those of a large class of proteins that are present in *E. coli* and other bacteria. This family of homologous proteins binds ligands such as sugars and amino acids at their centers. Remarkably, domains of this family are utilized by eukaryotes in taste proteins and in neurotransmitter receptors, as will be discussed in Chapter 32.

The Operon Is a Common Regulatory Unit in Prokaryotes

Many other gene-regulatory networks function in ways analogous to those of the *lac* operon. For example, genes taking part in purine and, to a lesser degree, pyrimidine biosynthesis are repressed by the *pur repressor*. This dimeric protein is 31% identical in sequence with the *lac* repressor and has a

Figure 31.13 Induction of the *lac* operon. (A) In the absence of lactose, the *lac* repressor binds DNA and represses transcription from the *lac* operon. (B) Allolactose or another inducer binds to the *lac* repressor, leading to its dissociation from DNA and to the production of *lac* mRNA.

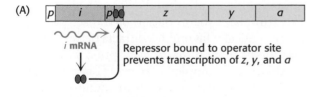

(A)

P   i   p o o   z   y   a

i mRNA

Repressor bound to operator site prevents transcription of *z*, *y*, and *a*

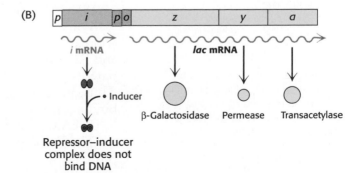

(B)

P   i   p o   z   y   a

i mRNA

lac mRNA

β-Galactosidase   Permease   Transacetylase

Inducer

Repressor–inducer complex does not bind DNA

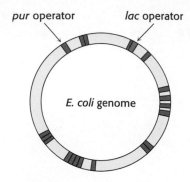

Figure 31.14 Binding-site distributions. The *E. coli* genome contains only a single region that closely matches the sequence of the *lac* operator (shown in blue). In contrast, 20 sites match the sequence of the *pur* operator (shown in red). Thus, the *pur* repressor regulates the expression of many more genes than does the *lac* repressor.

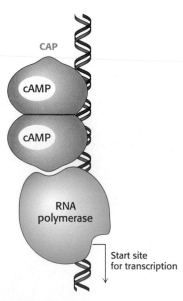

Figure 31.15 Binding site for catabolite activator protein (CAP). This protein binds as a dimer to an inverted repeat that is at the position −61 relative to the start site of transcription. The CAP-binding site on DNA is adjacent to the position at which RNA polymerase binds.

similar three-dimensional structure. However, the behavior of the *pur* repressor is opposite that of the *lac* repressor: whereas the *lac* repressor is *released* from DNA by binding to a small molecule, *the pur repressor binds DNA specifically only when bound to a small molecule.* Such a small molecule is called a *corepressor.* For the *pur* repressor, the corepressor can be either guanine or hypoxanthine. The dimeric *pur* repressor binds to inverted-repeat DNA sites of the form 5′-AN**GCAANCGNTT**NCNT-3′, in which the bases shown in boldface type are particularly important. Examination of the *E. coli* genome sequence reveals the presence of more than 20 such sites, regulating 19 operons and including more than 25 genes (Figure 31.14).

Because the DNA binding sites for these regulatory proteins are relatively short, it is likely that they evolved independently and are not related by divergence from an ancestral regulatory site. Once a ligand-regulated DNA-binding protein is present in a cell, binding sites for the protein may arise by mutation adjacent to additional genes. Binding sites for the *pur* repressor have evolved in the regulatory regions of a wide range of genes taking part in nucleotide biosynthesis. All such genes can then be regulated in a concerted manner.

The organization of prokaryotic genes into operons is useful for the analysis of completed genome sequences. Sometimes genes of unknown function are discovered to be part of an operon containing more well characterized genes. Such associations can provide powerful clues to the biochemical and physiological functions of the uncharacterized gene.

Transcription Can Be Stimulated by Proteins That Contact RNA Polymerase

All the DNA-binding proteins discussed thus far function by inhibiting transcription until some environmental condition, such as the presence of lactose, is met. There are also DNA-binding proteins that stimulate transcription. One particularly well studied example is a protein in *E. coli* that stimulates the expression of catabolic enzymes.

E. coli grown on glucose, a preferred energy source, have very low levels of catabolic enzymes for metabolizing other sugars. Clearly, the synthesis of these enzymes when glucose is abundant would be wasteful. The inhibitory effect of glucose is called *catabolite repression.* It is due to the fact that glucose lowers the concentration of cAMP in *E. coli.* When its concentration is high, cAMP stimulates the concerted transcription of many catabolic enzymes by acting through a protein called the *catabolite activator protein* (CAP), which is also known as the cAMP response protein (CRP).

When bound to cAMP, CAP stimulates the transcription of lactose- and arabinose-catabolizing genes. CAP is a sequence-specific DNA-binding protein. Within the *lac* operon, CAP binds to an inverted repeat that is centered near position −61 relative to the start site for transcription (Figure 31.15). CAP functions as a dimer of identical subunits.

The CAP–cAMP complex stimulates the initiation of transcription by approximately a factor of 50. Energetically favorable contacts between CAP and RNA polymerase increase the likelihood that transcription will be initiated at sites to which the CAP–cAMP complex is bound (Figure 31.16). Thus, in regard to the *lac* operon, gene expression is maximal when the binding of allolactose relieves the inhibition by the *lac* repressor and the CAP–cAMP complex stimulates the binding of RNA polymerase.

The *E. coli* genome contains many CAP-binding sites in positions appropriate for interactions with RNA polymerase. Thus, an increase in the

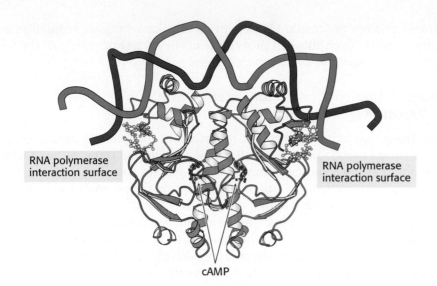

RNA polymerase interaction surface

RNA polymerase interaction surface

cAMP

Figure 31.16 Structure of a dimer of CAP bound to DNA. The residues shown in yellow in each CAP monomer have been implicated in direct interactions with RNA polymerase. [Drawn from 1RUN.pdb.]

cAMP level inside an *E. coli* bacterium results in the formation of CAP–cAMP complexes that bind to many promoters and stimulate the transcription of genes encoding a variety of catabolic enzymes.

31.3 The Greater Complexity of Eukaryotic Genomes Requires Elaborate Mechanisms for Gene Regulation

Gene regulation is significantly more complex in eukaryotes than in prokaryotes for a number of reasons. First, the genome being regulated is significantly larger. The *E. coli* genome consists of a single, circular chromosome containing 4.6 Mb. This genome encodes approximately 2000 proteins. In comparison, one of the simplest eukaryotes, *Saccharomyces cerevisiae* (baker's yeast), contains 16 chromosomes ranging in size from 0.2 to 2.2 Mb (Figure 31.17). The yeast genome totals 12 Mb and encodes approximately 6000 proteins. The genome within a human cell contains 23 pairs of chromosomes ranging in size from 50 to 250 Mb. Approximately 25,000 genes are present within the 3000 Mb of human DNA. It would be very difficult for a DNA-binding protein to recognize a unique site in this vast array of DNA sequences. Consequently, more-elaborate mechanisms are required to achieve specificity.

Another source of complexity in eukaryotic gene regulation is the many different *cell types* present in most eukaryotes. Liver and pancreatic cells, for example, differ dramatically in the genes that are highly expressed (see Table 31.1). Moreover, eukaryotic genes are not generally organized into operons. Instead, genes that encode proteins for steps within a given pathway are often spread widely across the genome. Finally, transcription and translation are uncoupled in eukaryotes, eliminating some potential gene-regulatory mechanisms.

Despite these differences, some aspects of gene regulation in eukaryotes are quite similar to those in prokaryotes. In particular, activator and repressor proteins that recognize specific DNA sequences are central to many

---

Megabase (Mb)

A length of DNA consisting of 10^6 base pairs (if double stranded) or 10^6 bases (if single stranded).

$$1 \text{ Mb} = 10^3 \text{ kb} = 10^6 \text{ bases}$$

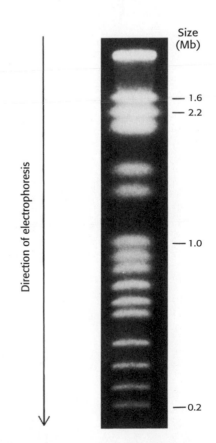

Size (Mb)

Direction of electrophoresis

— 1.6
— 2.2

— 1.0

— 0.2

Figure 31.17 Yeast chromosomes. Pulsed-field electrophoresis allows the separation of 16 yeast chromosomes. [From G. Chu, D. Vollrath, and R. W. Davis. *Science* 234(1986):1582–1585.]

gene-regulatory processes. Other aspects of eukaryotic gene regulation are quite different from those in prokaryotes. They relate primarily to the role of DNA packaging in eukaryotic genomes.

Multiple Transcription Factors Interact with Eukaryotic Regulatory Sites

The basal transcription complex described in Chapter 29 initiates transcription at a low frequency. Recall that several general transcription factors (the preinitiation complex) join with RNA polymerase II to form the basal transcription complex. Additional transcription factors must bind to other sites for a gene to achieve a high rate of mRNA synthesis. In contrast with the regulators of prokaryotic transcription, few eukaryotic transcription factors have any effect on transcription on their own. Instead, each factor recruits other proteins to build up large complexes that interact with the transcriptional machinery to activate transcription.

A major advantage of this mode of regulation is that a given regulatory protein can have different effects, depending on what other proteins are present in the same cell. This phenomenon, called *combinatorial control,* is crucial to multicellular organisms that have many different cell types. Even in unicellular eukaryotes such as yeast, combinatorial control allows the generation of distinct cell types.

Eukaryotic Transcription Factors Are Modular

Transcription factors usually consist of several domains. The *DNA-binding domain* identifies and binds regulatory sequences that can either be adjacent to the promoter or at some distance from it. Some activators also include a *regulatory domain,* which prevents DNA binding under certain conditions. After a transcription factor has bound to the DNA, the *activation domain* initiates transcription through interactions with RNA polymerase II or its associated proteins.

The DNA-binding domain is essential for determining which genes are transcribed. A transcription factor is activated in response to a stimulus and *is then responsible for activating the transcription of a set of genes.* For example, the transcription factor NF-κB is activated in response to injury, and it activates the transcription of genes that produce an immune response, helping to fight infection. The DNA-binding domain recognizes and binds to a short conserved *recognition sequence* in the promoter region of each gene or in a more distant enhancer. Often, to increase specificity, the recognition sequence is repeated at regular intervals, and the activators must dimerize before binding to the repeated recognition sequences. Transcription factors can be grouped into families on the basis of the structure of their sequence-specific DNA-binding domains. The helix-turn-helix, homeodomain, bZip, and zinc-finger domains introduced in Section 31.1 are examples of common DNA-binding domains.

Transcription factors can often act even if their binding sites lie at a considerable distance from the promoter. These distant regulatory sites are called *enhancers* (p. 838). The intervening DNA can form loops that bring the enhancer-bound activator to the promoter site, where it can act on other transcription factors or on RNA polymerase.

Activation Domains Interact with Other Proteins

The activation domains of transcription factors generally recruit other proteins that promote transcription. Some of these activation domains interact directly with RNA polymerase II. In other cases, an activation domain may have *multiple interaction partners*. These activation domains act through intermediary proteins, which bridge between the transcription factors and the polymerase. An important target of activators is *mediator,* a complex of 25 to 30 subunits that is part of the preinitiation complex. Mediator acts as a bridge between enhancer-bound activators and promoter-bound RNA polymerase II (Figure 31.18).

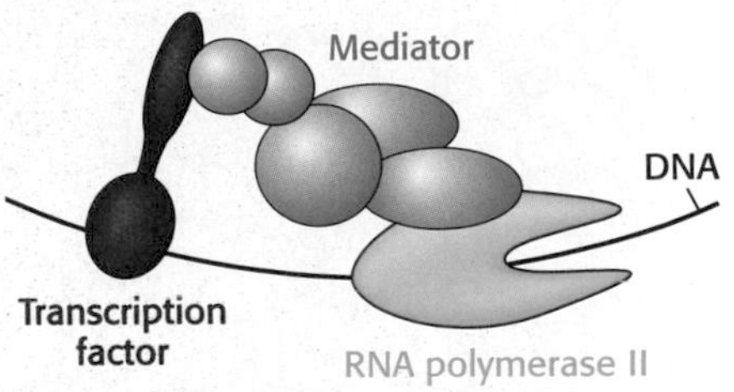

Activation domains are less conserved than DNA-binding domains. In fact, very little sequence similarity has been found. For example, they may be acidic, hydrophobic, glutamine rich, or proline rich. However, certain features are common to activation domains. First, they are *redundant*. That is, a part of the activation domain can be deleted without loss of function. Second, as described earlier, they are *modular* and can activate transcription when paired with a variety of DNA-binding domains. Third, activation domains act *synergistically:* two activation domains acting together create a much stronger effect than either acting separately.

We have been addressing the case in which gene control requires the expression of a gene. In many cases, the expression of a gene must be halted by ceasing gene transcription. The agents in such cases are transcriptional repressors. In contrast with activators, repressors bind proteins that block the association of RNA polymerase II with the DNA.

Nucleosomes Are Complexes of DNA and Histones

The control of eukaryotic gene transcription is complicated by the fact that DNA in eukaryotic chromosomes is not bare. Instead, eukaryotic DNA is tightly bound to a group of small basic proteins called *histones*. In fact, histones constitute half the mass of a eukaryotic chromosome. The entire complex of a cell's DNA and associated protein is called *chromatin*. Five major histones are present in chromatin: four histones, called H2A, H2B, H3, and H4, associate with one another; the other histone is called H1. Histones have strikingly basic properties because a quarter of the residues in each histone are either arginine or lysine.

Chromatin is made up of repeating units, each containing 200 bp of DNA and two copies each of H2A, H2B, H3, and H4, called the *histone octamer.* These repeating units are known as *nucleosomes.* Strong support for this model comes from the results of a variety of experiments, including observations of appropriately prepared samples of chromatin viewed by electron microscopy (Figure 31.19). Chromatin viewed with the electron microscope has the appearance of beads on a string; each bead has a diameter of approximately 100 Å. Partial digestion of chromatin with DNase yields the

Figure 31.18 Mediator. Mediator, a large complex of protein subunits, acts as a bridge between transcription factors bearing activation domains and RNA polymerase II. These interactions help recruit and stabilize RNA polymerase II near specific genes that are then transcribed.

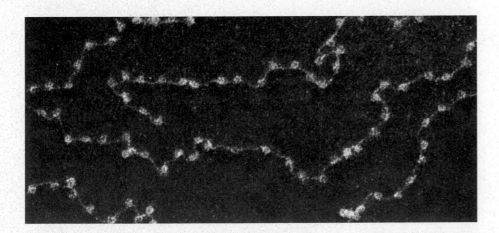

100 nm

Figure 31.19 Chromatin structure. An electron micrograph of chromatin showing its "beads on a string" character. [Courtesy of Dr. Ada Olins and Dr. Donald Olins.]

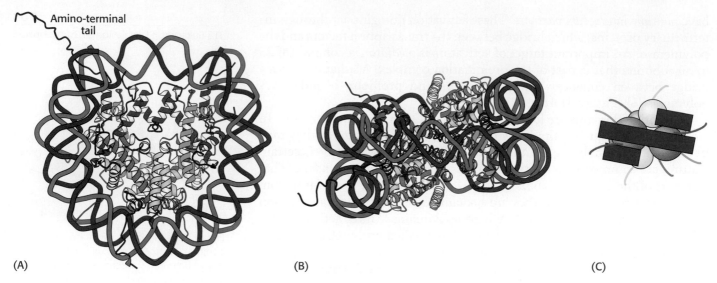

(A)　(B)　(C)

Figure 31.20 Nucleosome core particle. The structure consists of a core of eight histone proteins surrounded by DNA. (A) A view showing the DNA wrapping around the histone core. (B) A view related to that in part A by a 90-degree rotation. *Notice* that the DNA forms a left-handed superhelix as it wraps around the core. (C) A schematic view. [Drawn from 1AOI.pdb.]

isolated beads. These particles consist of fragments of DNA about 200 bp in length bound to the eight histones. More-extensive digestion yields a shorter DNA fragment of 145 bp bound to the histone octamer. The smaller complex formed by the histone octamer and the 145-bp DNA fragment is the *nucleosome core particle*. The DNA connecting core particles in undigested chromatin is called *linker DNA*. Histone H1 binds, in part, to the linker DNA.

Eukaryotic DNA Is Wrapped Around Histones to Form Nucleosomes

The overall structure of the nucleosome was revealed through electron microscopic and x-ray crystallographic studies pioneered by Aaron Klug and his colleagues. More recently, the three-dimensional structure of a reconstituted nucleosome core (Figure 31.20) was determined to higher resolution by x-ray diffraction methods. As was shown by Evangelos Moudrianakis, the four types of histone that make up the protein core are homologous and similar in structure (Figure 31.21). The eight histones in the core are arranged into a $(H3)_2(H4)_2$ tetramer and a pair of H2A–H2B dimers. The tetramer and

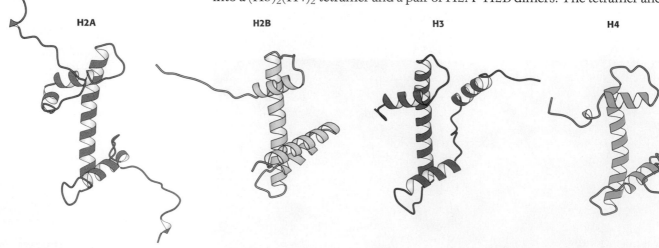

H2A　H2B　H3　H4

Figure 31.21 Homologous histones. Histones H2A, H2B, H3, and H4 adopt a similar three-dimensional structure as a consequence of common ancestry. Some parts of the tails at the termini of the proteins are not shown. [Drawn from 1AOI.pdb.]

dimers come together to form a left-handed superhelical ramp around which the DNA wraps. In addition, each histone has an amino-terminal tail that extends out from the core structure. These tails are flexible and contain a number of lysine and arginine residues. As we shall see, *covalent modifications of these tails play an essential role in modulating the affinity of the histones for DNA and other properties.*

The DNA forms a left-handed superhelix as it wraps around the outside of the histone octamer. The protein core forms contacts with the inner surface of the DNA superhelix at many points, particularly along the phosphodiester backbone and the minor groove. Nucleosomes will form on almost all DNA sites, although some sequences are preferred because the dinucleotide steps are properly spaced to favor bending around the histone core. A histone with a different structure from that of the others, called histone H1, seals off the nucleosome at the location at which the linker DNA enters and leaves. The amino acid sequences of histones, including their amino-terminal tails, are remarkably conserved from yeast through human beings.

The winding of DNA around the nucleosome core contributes to the packing of DNA by decreasing its linear extent. An extended 200-bp stretch of DNA would have a length of about 680 Å. Wrapping this DNA around the histone octamer reduces the length to approximately 100 Å along the long dimension of the nucleosome. Thus the DNA is compacted by a factor of 7. However, human chromosomes in metaphase, which are highly condensed, are compacted by a factor of 10^4. Clearly, the nucleosome is just the first step in DNA compaction. What is the next step? The nucleosomes themselves are arranged in a helical array approximately 360 Å across, forming a series of stacked layers approximately 110 Å apart (Figure 31.22). The folding of these fibers of nucleosomes into loops further compacts DNA.

The wrapping of DNA around the histone core as a left-handed helix also stores negative supercoils; if the DNA in a nucleosome is straightened out, the DNA will be underwound (p. 789). This underwinding is exactly what is needed to separate the two DNA strands during replication and transcription.

The Control of Gene Expression Can Require Chromatin Remodeling

Does chromatin structure play a role in the control of gene expression? Early observations suggested that it does indeed. DNA that is densely packaged into chromatin is less susceptible to cleavage by the nonspecific DNA-cleaving enzyme DNase I. Regions adjacent to genes that are being transcribed are more sensitive to cleavage than are other sites in the genome, suggesting that the DNA in these regions is less compacted than it is elsewhere and more accessible to proteins. In addition, some sites, usually within 1 kb of the start site of an active gene, are exquisitely sensitive to DNase I and other nucleases. These *hypersensitive sites* correspond to regions that have few nucleosomes or contain nucleosomes in an altered conformational state. *Hypersensitive sites are cell-type specific and developmentally regulated.* For example, globin genes in the precursors of erythroid cells from 20-hour-old chicken embryos are insensitive to DNase I. However, when hemoglobin synthesis begins at 35 hours, regions adjacent to these genes become highly susceptible to digestion. In tissues such as the brain that produce no hemoglobin, the globin genes remain resistant to DNase I throughout development and into adulthood. These studies suggest that a prerequisite for gene expression is a relaxing of the chromatin structure.

Recent experiments even more clearly revealed the role of chromatin structure in regulating access to DNA binding sites. Genes required for

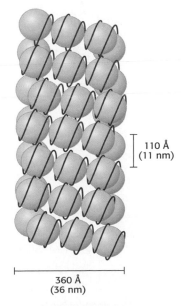

110 Å
(11 nm)

360 Å
(36 nm)

Figure 31.22 Higher-order chromatin structure. A proposed model for chromatin arranged in a helical array consisting of six nucleosomes per turn of helix. The DNA double helix (shown in red) is wound around each histone octamer (shown in blue). [After J. T. Finch and A. Klug. *Proc. Natl. Acad. Sci. U. S. A.* 73(1976):1897–1901.]

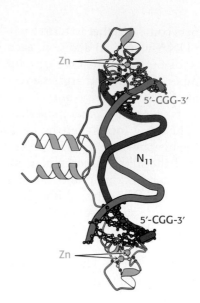

Figure 31.23 GAL4 binding sites.
The yeast transcription factor GAL4
binds to DNA sequences of the form
5'-CGG(N)₁₁CCG-3'. Two zinc-based
domains are present in the DNA-binding
region of this protein. *Notice* that these
domains contact the 5'-CGG-3' sequences,
leaving the center of the site uncontacted.
[Drawn from 1D66.pdb.]

galactose utilization in yeast are activated by a DNA-binding protein
called GAL4, which recognizes DNA binding sites with two 5'-CGG-3'
sequences on complementary strands separated by 11 base pairs
(Figure 31.23). Approximately 4000 potential GAL4 binding sites of the
form 5'-CGG(N)$_{11}$CCG-3' are present in the yeast genome, but only 10 of
them regulate genes necessary for galactose metabolism. What fraction of the
potential binding sites are actually bound by GAL4? This question is
addressed through the use of a technique called *chromatin immunoprecipitation*
(ChIP). GAL4 is first cross-linked to its DNA binding sites in chromatin.
The DNA is then fragmented into small pieces, and antibodies to GAL4 are
used to isolate the chromatin fragments containing GAL4. The cross-linking
is reversed, and the DNA is isolated and characterized. The results of these
studies reveal that only approximately 10 of the 4000 potential GAL4 sites are
occupied by GAL4 when the cells are growing on galactose; more than 99%
of the sites appear to be blocked. Thus, whereas in prokaryotes all sites appear
to be equally accessible, chromatin structure shields a large number of the
potential binding sites in eukaryotic cells. GAL4 is thereby prevented from
binding to sites that are unimportant in galactose metabolism.

These lines of evidence and others reveal that chromatin structure is altered in active genes compared with inactive ones. How is chromatin structure modified? As we shall see later (p. 910), specific covalent modifications
of histone proteins are crucial. In addition, the binding of specific proteins
to enhancers at specific sites in the genome plays a role.

Enhancers Can Stimulate Transcription in Specific Cell Types

We now return to the action of enhancers (p. 902). Recall that these DNA
sequences, although they have no promoter activity of their own, greatly
increase the activities of many promoters in eukaryotes, even when the
enhancers are located at a distance of several thousand base pairs from the
gene being expressed.

Enhancers function by serving as binding sites for specific regulatory
proteins (Figure 31.24). An enhancer is effective only in the specific cell
types in which appropriate regulatory proteins are expressed. In many cases,
these DNA-binding proteins influence transcription initiation by perturb-
ing the local chromatin structure to expose a gene or its regulatory sites rather
than by direct interactions with RNA polymerase. This mechanism accounts
for the ability of enhancers to act at a distance.

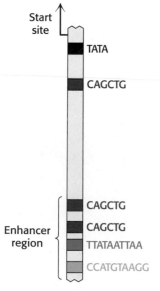

Figure 31.24 Enhancer binding sites. A
schematic structure for the region 1 kb
upstream of the start site for the muscle
creatine kinase gene. One binding site of
the form 5'-CAGCTG-3' is present near
the TATA box. The enhancer region farther
upstream contains two binding sites for the
same protein and two additional binding
sites for other proteins.

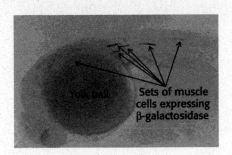

Figure 31.25 An experimental demonstration of enhancer function. A promoter for muscle creatine kinase artificially drives the transcription of β-galactosidase in a zebrafish embryo. Only specific sets of muscle cells produce β-galactosidase, as visualized by the formation of the blue product on treatment of the embryo with X-Gal. [From F. Müller, D. W. Williamson, J. Kobolák, L. Gauvry, G. Goldspink, L. Orbán, and N. MacLean. *Mol. Reprod. Dev.* 47(1997):404–412.]

The properties of enhancers are illustrated by studies of the enhancer controlling the muscle isoform of creatine kinase (p. 416). The results of mutagenesis and other studies revealed the presence of an enhancer located between 1350 and 1050 base pairs upstream of the start site of the gene for this enzyme. Experimentally inserting this enhancer near a gene not normally expressed in muscle cells is sufficient to cause the gene to be expressed at high levels in muscle cells but not in other cells (Figure 31.25).

The Methylation of DNA Can Alter Patterns of Gene Expression

The degree of methylation of DNA provides another mechanism, in addition to packaging with histones, for inhibiting gene expression inappropriate to a specific cell type. Carbon 5 of cytosine can be methylated by specific methyltransferases. About 70% of the 5′-CpG-3′ sequences in mammalian genomes are methylated. However, the distribution of these methylated cytosines varies, depending on the cell type. Consider the β-globin gene. In cells that are actively expressing hemoglobin, the region from approximately 1 kb upstream of the start site to approximately 100 bp downstream of the start site is less methylated than the corresponding region in cells that do not express this gene. The relative absence of 5-methylcytosines near the start site is referred to as *hypomethylation*. The methyl group of 5-methylcytosine protrudes into the major groove where it could easily interfere with the binding of proteins that stimulate transcription.

The distribution of CpG sequences in mammalian genomes is not uniform. Many CpG sequences have been converted into TpG through mutation by the deamination of 5-methylcytosine to thymine. However, sites near the 5′ ends of genes have been maintained because of their role in gene expression. Thus, most genes are found in *CpG islands*, regions of the genome that contain approximately four times as many CpG sequences as does the remainder of the genome.

Steroids and Related Hydrophobic Molecules Pass Through Membranes and Bind to DNA-Binding Receptors

We next look at an example that illustrates how transcription factors can stimulate changes in chromatin structure that affect transcription. We will consider in some detail the system that detects and responds to estrogens. Synthesized and released by the ovaries, *estrogens*, such as estradiol, are cholesterol-derived, steroid hormones (p. 753). They are required for the development of female secondary sex characteristics and, along with progesterone, participate in the ovarian cycle.

Because they are hydrophobic molecules, estrogens easily diffuse across cell membranes. When inside a cell, estrogens bind to highly specific, soluble receptor proteins. Estrogen receptors are members of a large family of

deoxyribose
5-Methylcytosine

Estradiol
(an estrogen)

proteins that act as receptors for a wide range of hydrophobic molecules, including other steroid hormones, thyroid hormones, and retinoids.

All-*trans*-retinoic acid
(a retinoid)

Thyroxine
(L-3,5,3′,5′-Tetraiodothyronine)
(a thyroid hormone)

The human genome encodes approximately 50 members of this family, often referred to as *nuclear hormone receptors*. The genomes of other multicellular eukaryotes encode similar numbers of nuclear hormone receptors, although they are absent in yeast.

All these receptors have a similar mode of action. On binding of the signal molecule (called, generically, a *ligand*), the ligand–receptor complex modifies the expression of specific genes by binding to control elements in the DNA. Estrogen receptors bind to specific DNA sites (referred to as *estrogen response elements* or EREs) that contain the consensus sequence 5′-**AGGTCANNNTGACCT**-3′. As expected from the symmetry of this sequence, an estrogen receptor binds to such sites as a dimer.

A comparison of the amino acid sequences of members of this family reveals two highly conserved domains: a DNA-binding domain and a ligand-binding domain (Figure 31.26). The DNA-binding domain lies toward the center of the molecule and consists of a set of zinc-based domains different from the Cys_2His_2 zinc-finger proteins introduced near the beginning of the chapter. These zinc-based domains bind to specific DNA sequences by virtue of an α helix that lies in the major groove in the specific DNA complexes formed by estrogen receptors.

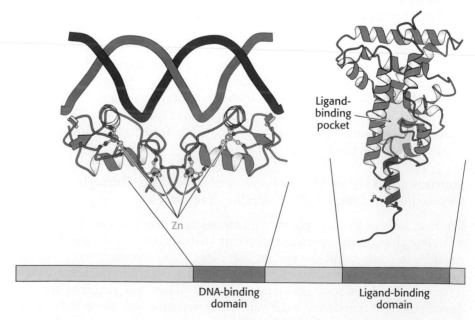

Ligand-
binding
pocket

Zn

DNA-binding
domain

Ligand-binding
domain

Figure 31.26 Structure of two nuclear hormone receptor domains. Nuclear hormone receptors contain two crucial conserved domains: (1) a DNA-binding domain toward the center of the sequence and (2) a ligand-binding domain toward the carboxyl terminus. The structure of a dimer of the DNA-binding domain bound to DNA is shown, as is one monomer of the normally dimeric ligand-binding domain. [Drawn from 1HCQ and 1LBD.pdb.]

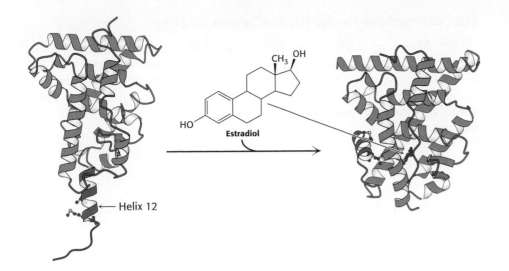

Figure 31.27 Ligand binding to nuclear hormone receptor. The ligand lies completely surrounded within a pocket in the ligand-binding domain. *Notice* that the last α helix, helix 12 (shown in purple), folds into a groove on the side of the structure on ligand binding. [Drawn from 1LDB and 1ERE.pdb.]

Nuclear Hormone Receptors Regulate Transcription by Recruiting Coactivators to the Transcription Complex

The second highly conserved domain of the nuclear receptor proteins lies near the carboxyl terminus and is the ligand-binding site. This domain folds into a structure that consists almost entirely of α helices, arranged in three layers. The ligand binds in a hydrophobic pocket that lies in the center of this array of helices (Figure 31.27). This domain changes conformation when it binds estrogen. How does ligand binding lead to changes in gene expression? The simplest model would have the binding of ligand alter the DNA-binding properties of the receptor, analogously to the *lac* repressor in prokaryotes. However, experiments with purified nuclear hormone receptors revealed that ligand binding does *not* significantly alter DNA-binding affinity and specificity. Another mechanism is operative.

Because ligand binding does not alter the ability of nuclear hormone receptors to bind DNA, investigators sought to determine whether specific proteins might bind to the nuclear hormone receptors only in the presence of ligand. Such searches led to the identification of several related proteins called *coactivators*, such as SRC-1 (*s*teroid *r*eceptor *c*oactivator-1), GRIP-1 (*g*lucocorticoid *r*eceptor *i*nteracting *p*rotein-1), and NcoA-1 (*n*uclear *hor*mone receptor *co*activator-1). These coactivators are referred to as the p160 family because of their size. The binding of ligand to the receptor induces a conformational change that allows the recruitment of a coactivator (Figure 31.28). In many cases, these coactivators are enzymes that catalyze reactions that lead to the modification of chromatin structure.

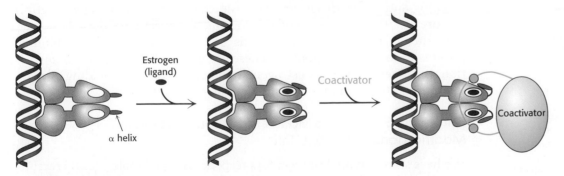

Figure 31.28 Coactivator recruitment. The binding of ligand to a nuclear hormone receptor induces a conformational change in the ligand-binding domain. This change in conformation generates favorable sites for the binding of a coactivator.

Steroid-Hormone Receptors Are Targets for Drugs

Molecules such as estradiol that bind to a receptor and trigger signaling pathways are called *agonists*. Athletes sometimes take natural and synthetic agonists of the androgen receptor, a member of the family of nuclear hormone receptors, because their binding to the androgen receptor stimulates the expression of genes that enhance the development of lean muscle mass.

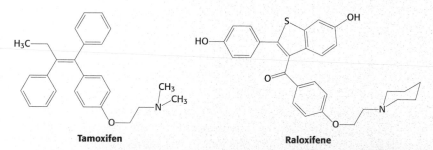

Androstendione
(a natural androgen)

Dianabol
(methandrostenolone)
(a synthetic androgen)

Referred to as *anabolic steroids*, such compounds used in excess are not without side effects. In men, excessive use leads to a decrease in the secretion of testosterone, to testicular atrophy, and sometimes to breast enlargement (gynecomastia) if some of the excess androgen is converted into estrogen. In women, excess testosterone causes a decrease in ovulation and estrogen secretion; it also causes breast regression and growth of facial hair.

Other molecules bind to nuclear hormone receptors but do not effectively trigger signaling pathways. Such compounds are called *antagonists* and are, in many ways, like competitive inhibitors of enzymes. Some important drugs are antagonists that target the estrogen receptor. For example, *tamoxifen* and *raloxifene* are used in the treatment and prevention of breast cancer, because some breast tumors rely on estrogen-mediated pathways for growth. These compounds are sometimes called *selective estrogen receptor modulators* (SERMs).

Tamoxifen

Raloxifene

The determination of the structures of complexes between the estrogen receptor and these drugs revealed the basis for their antagonist effect (Figure 31.29). Tamoxifen binds to the same site as estradiol does. However, tamoxifen has a group that extends out of the normal ligand-binding pocket, as do other antagonists. These groups block the normal conformational changes induced by estrogen. Tamoxifen blocks the binding of coactivators and thus inhibits the activation of gene expression.

Chromatin Structure Is Modulated Through Covalent Modifications of Histone Tails

We have seen that nuclear receptors respond to signal molecules by recruiting coactivators. Now we can ask, How do coactivators modulate transcriptional activity? *These proteins act to loosen the histone complex from the DNA, exposing additional DNA regions to the transcription machinery.*

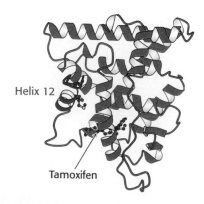

Figure 31.29 Estrogen receptor–tamoxifen complex. Tamoxifen binds in the pocket normally occupied by estrogen. However, *notice* that part of the tamoxifen structure extends from this pocket, and so helix 12 cannot pack in its usual position. Instead, this helix blocks the coactivator-binding site. [Drawn from 3ERT.pdb.]

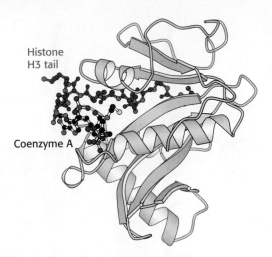

Histone H3 tail

Coenzyme A

Figure 31.30 Structure of histone acetyltransferase. The amino-terminal tail of histone H3 extends into a pocket in which a lysine side chain can accept an acetyl group from acetyl CoA bound in an adjacent site. [Drawn from 1QSN.pdb.]

Much of the effectiveness of coactivators appears to result from their ability to covalently modify the amino-terminal tails of histones as well as regions on other proteins. Some of the p160 coactivators and the proteins that they recruit catalyze the transfer of acetyl groups from acetyl CoA to specific lysine residues in these amino-terminal tails.

Lysine in histone tail + CoA—S—C(=O)—CH₃ (Acetyl CoA) → acetylated lysine + CoA—SH + H⁺

Lysine in histone tail   **Acetyl CoA**

Enzymes that catalyze such reactions are called *histone acetyltransferases* (HATs). The histone tails are readily extended; so they can fit into the HAT active site and become acetylated (Figure 31.30).

What are the consequences of histone acetylation? Lysine bears a positively charged ammonium group at neutral pH. The addition of an acetyl group generates an uncharged amide group. This change dramatically reduces the affinity of the tail for DNA and modestly decreases the affinity of the entire histone complex for DNA, loosening the histone complex from the DNA.

In addition, the acetylated lysine residues interact with a specific *acetyllysine-binding domain* that is present in many proteins that regulate eukaryotic transcription. This domain, termed a *bromodomain,* comprises approximately 110 amino acids that form a four-helix bundle containing a peptide-binding site at one end (Figure 31.31).

Bromodomain-containing proteins are components of two large complexes essential for transcription. One is a complex of more than 10 polypeptides that binds to the *TATA-box-binding protein.* Recall that the TATA-box-binding protein is an essential transcription factor for many genes (p. 837). Proteins that bind to the TATA-box-binding protein are called *TAFs* (for *TATA-box-binding protein associated factors*). In particular, TAF1 contains a pair of bromodomains near its carboxyl terminus. The two domains are oriented such that each can bind one of two acetyllysine residues at positions 5 and 12 in the histone H4 tail. Thus, *acetylation*

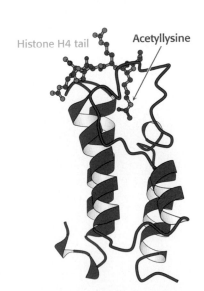

Histone H4 tail   Acetyllysine

Figure 31.31 Structure of a bromodomain. This four-helix-bundle domain binds peptides containing acetyllysine. An acetylated peptide of histone H4 is shown bound in the structure. [Drawn from 1EGI.pdb.]

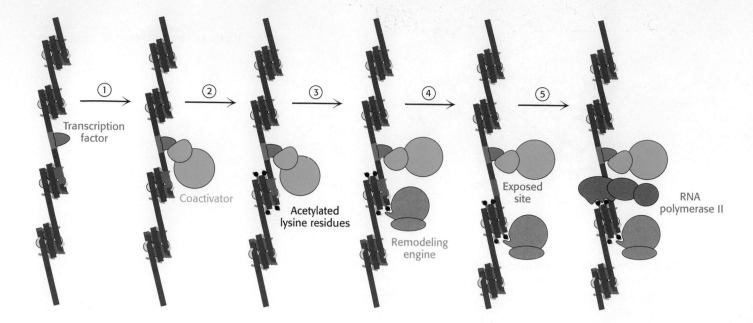

Figure 31.32 Chromatin remodeling.
Eukaryotic gene regulation begins with an activated transcription factor bound to a specific site on DNA. One scheme for the initiation of transcription by RNA polymerase II requires five steps: (1) recruitment of a coactivator, (2) acetylation of lysine residues in the histone tails, (3) binding of a remodeling-engine complex to the acetylated lysine residues, (4) ATP-dependent remodeling of the chromatin structure to expose a binding site for RNA polymerase or for other factors, and (5) recruitment of RNA polymerase. Only two subunits are shown for each complex, although the actual complexes are much larger. Other schemes are possible.

of the histone tails provides a mechanism for recruiting other components of the transcriptional machinery.

Bromodomains are also present in some components of large complexes known as *chromatin-remodeling engines*. These complexes, which also contain domains homologous to those of helicases, utilize the free energy of ATP hydrolysis to shift the positions of nucleosomes along the DNA and to induce other conformational changes in chromatin (Figure 31.32). Histone acetylation can lead to a reorganization of the chromatin structure, potentially exposing binding sites for other factors. *Thus, histone acetylation can activate transcription through a combination of three mechanisms: by reducing the affinity of the histones for DNA, by recruiting other components of the transcriptional machinery, and by initiating the remodeling of the chromatin structure.*

Nuclear hormone receptors also include regions that interact with components of the mediator complex. Thus, two mechanisms of gene regulation can work in concert. Modification of histones and chromatin remodeling can open up regions of chromatin into which the transcription complex can be recruited through protein–protein interactions.

Histone Deacetylases Contribute to Transcriptional Repression

Just as in prokaryotes, some changes in a cell's environment lead to the repression of genes that had been active. The modification of histone tails again plays an important role. However, in repression, a key reaction appears to be the deacetylation of acetylated lysine, catalyzed by specific *histone deacetylase* enzymes.

In many ways, the acetylation and deacetylation of lysine residues in histone tails (and, likely, in other proteins) is analogous to the phosphorylation and dephosphorylation of serine, threonine, and tyrosine residues in other stages of signaling processes. Like the addition of phosphoryl groups, the addition of acetyl groups can induce conformational changes and generate novel binding sites. Without a means of removing these groups, however, these signaling switches will become stuck in one position and lose their effectiveness. Like phosphatases, deacetylases help reset the switches.

Acetylation is not the only modification of histones and other proteins in gene-regulation processes. The methylation of specific lysine and arginine residues also can be important. The elucidation of the roles of these processes is a very active area of research at present.

31.4 Gene Expression Can Be Controlled at Posttranscriptional Levels

The modulation of the rate of transcriptional initiation is the most common mechanism of gene regulation. However, other stages of transcription also are targets for regulation in some cases. In addition, the process of translation provides other points of intervention for regulating the level of a protein produced in a cell. These mechanisms are quite distinct in prokaryotic and eukaryotic cells because prokaryotes and eukaryotes differ greatly in how transcription and translation are coupled and in how translation is initiated. We will consider two important examples of posttranscriptional regulation: one from prokaryotes and the other from eukaryotes. In both examples, regulation depends on the formation of distinct secondary structures in mRNA.

Attenuation Is a Prokaryotic Mechanism for Regulating Transcription Through the Modulation of Nascent RNA Secondary Structure

A new means for regulating transcription in bacteria was discovered by Charles Yanofsky and his colleagues as a result of their studies of the tryptophan operon. This operon encodes five enzymes that convert chorismate into tryptophan (p. 694). Analysis of the 5′ end of *trp* mRNA revealed the presence of a *leader sequence* of 162 nucleotides before the initiation codon of the first enzyme. The next striking observation was that bacteria produced a transcript consisting of only the first 130 nucleotides when the tryptophan level was high, but they produced a 7000-nucleotide *trp* mRNA, including the entire leader sequence, when tryptophan was scarce. Thus, when trytophan is plentiful and the biosynthetic enzymes are not needed, transcription is abruptly broken off before any coding mRNA for the enzymes is produced. The site of termination is called the attenuator, and this mode of regulation is called *attenuation*.

Attenuation depends on features at the 5′ end of the mRNA product (Figure 31.33). The first part of the leader sequence encodes a 14-amino-acid leader peptide. Following the open reading frame for the peptide is a region of RNA representing the attenuator, which is capable of forming several alternative structures. Recall that transcription and translation are tightly coupled in bacteria. Thus, the translation of the *trp* mRNA begins soon after the ribosome-binding site has been synthesized.

How does the level of tryptophan alter transcription of the *trp* operon? An important clue was the finding that the 14-amino-acid leader peptide includes two adjacent tryptophan residues. A ribosome is able to translate the leader region of the mRNA product only in the presence of adequate concentrations of tryptophan. When enough tryptophan is present, a stem-loop structure

Figure 31.33 Leader region of trp mRNA. (A) The nucleotide sequence of the 5′ end of *trp* mRNA includes a short open reading frame that encodes a peptide comprising 14 amino acids; the leader encodes two tryptophan residues and has an untranslated attenuator region (blue and red nucleotides). (B and C) The attenuator region can adopt two distinct stem-loop structures.

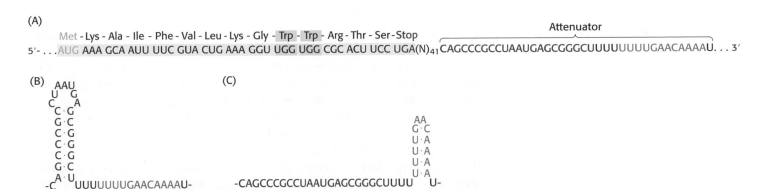

(A)

Met - Lys - Ala - Ile - Phe - Val - Leu - Lys - Gly - Trp - Trp - Arg - Thr - Ser - Stop   Attenuator

5′-...AUG AAA GCA AUU UUC GUA CUG AAA GGU UGG UGG CGC ACU UCC UGA(N)₄₁CAGCCCGCCUAAUGAGCGGGCUUUUUUUUGAACAAAAU...3′

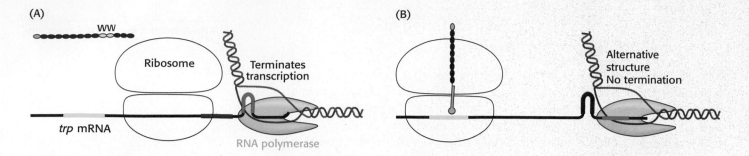

(A)

(B)

Figure 31.34 Attenuation. (A) In the presence of adequate concentrations of tryptophan (and, hence, Trp-tRNA), translation proceeds rapidly and an RNA structure forms that terminates transcription. (B) At low concentrations of tryptophan, translation stalls while awaiting Trp-tRNA, giving time for an alternative RNA structure to form that does not terminate transcription efficiently.

forms in the attenuator region, which leads to the release of RNA polymerase from the DNA (Figure 31.34). However, when tryptophan is scarce, transcription is terminated less frequently. Little tryptophanyl-tRNA is present, and so the ribosome stalls at the tandem UGG codons encoding tryptophan. This delay leaves the adjacent region of the mRNA exposed as transcription continues. An alternative RNA structure that does not function as a terminator is formed, and transcription continues into and through the coding regions for the enzymes. Thus, attenuation provides an elegant means of sensing the supply of tryptophan required for protein synthesis.

Several other operons for the biosynthesis of amino acids in *E. coli* also are regulated by attenuator sites. The leader peptide of each contains an abundance of the amino acid residues of the type synthesized by the operon (Figure 31.35). For example, the leader peptide for the phenylalanine operon includes 7 phenylalanine residues among 15 residues. The threonine operon encodes enzymes required for the synthesis of both threonine and isoleucine; the leader peptide contains 8 threonine and 4 isoleucine residues in a 16-residue sequence. The leader peptide for the histidine operon includes 7 histidine residues in a row. In each case, low levels of the corresponding charged tRNA causes the ribosome to stall, trapping the nascent mRNA in a state that can form a structure that allows RNA polymerase to read through the attenuator site.

Figure 31.35 Leader peptide sequences. Amino acid sequences and the corresponding mRNA nucleotide sequences of the (A) threonine operon, (B) phenylalanine operon, and (C) histidine operon. In each case, an abundance of one amino acid in the leader peptide sequence leads to attenuation.

(A)

Met - Lys - Arg - Ile - Ser - **Thr** - **Thr** - **Ile** - **Thr** - **Thr** - **Thr** - Ile - **Thr** - Ile - **Thr** - **Thr** -

5′ AUG AAA CGC AUU AGC ACC ACC AUU ACC ACC ACC AUC ACC AUU ACC ACA 3′

(B)

Met - Lys - His - Ile - Pro - **Phe** - **Phe** - **Phe** - Ala - **Phe** - **Phe** - **Phe** - Thr - **Phe** - Pro - Stop

5′ AUG AAA CAC AUA CCG UUU UUC UUC GCA UUC UUU UUU ACC UUC CCC UGA 3′

(C)

Met - Thr - Arg - Val - Gln - Phe - Lys - **His** - **His** - **His** - **His** - **His** - **His** - **His** - Pro - Asp-

5′ AUG ACA CGC GUU CAA UUU AAA CAC CAC CAU CAU CAC CAU CAU CCU GAC 3′

Genes Associated with Iron Metabolism Are Translationally Regulated in Animals

RNA secondary structure plays a role in the regulation of iron metabolism in eukaryotes. Iron is an essential nutrient, required for the synthesis of hemoglobin, cytochromes, and many other proteins. However, excess iron can be quite harmful because, untamed by a suitable protein environment, iron can initiate a range of free-radical reactions that damage proteins, lipids, and nucleic acids. Animals have evolved sophisticated systems for the accumulation of iron in times of scarcity and for the safe storage of excess iron for later use. Key proteins include *transferrin,* a transport protein that carries iron in the serum, *transferrin receptor,* a membrane protein that binds iron-loaded transferrin and initiates its entry into cells, and *ferritin,* an impressively efficient iron-storage protein found primarily in the liver and kidneys. Twenty-four ferritin polypeptides form a nearly spherical shell that encloses as many as 2400 iron atoms, a ratio of one iron atom per amino acid (Figure 31.36).

Ferritin and transferrin-receptor expression levels are reciprocally related in their responses to changes in iron levels. When iron is scarce, the amount of transferrin receptor increases and little or no new ferritin is synthesized. Interestingly, the extent of mRNA synthesis for these proteins does not change correspondingly. Instead, regulation takes place at the level of translation.

Consider ferritin first. Ferritin mRNA includes a stem-loop structure termed an *iron-response element* (IRE) in its 5' untranslated region (Figure 31.37). This stem-loop binds a 90-kd protein, called an *IRE-binding protein* (IRP), that blocks the initiation of translation. When the iron level increases, the IRP binds iron as a 4Fe-4S cluster. The IRP bound to iron cannot bind RNA, because the binding sites for iron and RNA substantially overlap. Thus, in the presence of iron, ferritin mRNA is released from the IRP and translated to produce ferritin, which sequesters the excess iron.

An examination of the nucleotide sequence of transferrin-receptor mRNA reveals the presence of several IRE-like regions. However, these regions are located in the 3' untranslated region rather than in the 5' untranslated region (Figure 31.38). Under low-iron conditions, IRP binds to these IREs. However, given the location of these binding sites, the transferrin-receptor mRNA can still be translated. What happens when the iron level increases and the IRP no longer binds transferrin-receptor mRNA? Freed from the IRP, transferrin-receptor mRNA is rapidly degraded. Thus, an increase in the cellular iron level leads to the destruction of transferrin-receptor mRNA and, hence, a reduction in the production of transferrin-receptor protein.

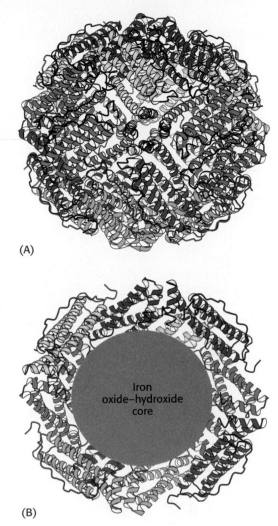

(A)

(B)

Figure 31.36 **Structure of ferritin.** (A) Twenty-four ferritin polypeptides form a nearly spherical shell. (B) A cutaway view reveals the core that stores iron as an iron oxide–hydroxide complex. [Drawn from 1IES.pdb.]

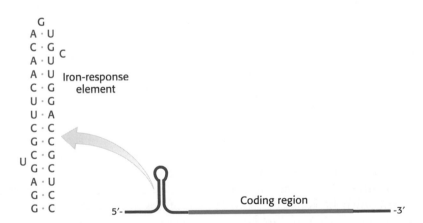

Figure 31.37 **Iron-response element.** Ferritin mRNA includes a stem-loop structure, termed an iron-response element (IRE), in its 5' untranslated region. The IRE binds a specific protein that blocks the translation of this mRNA under low iron conditions.

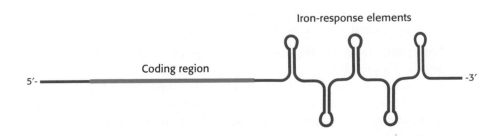

Figure 31.38 **Transferrin-receptor mRNA.** This mRNA has a set of iron-response elements (IREs) in its 3' untranslated region. The binding of the IRE-binding protein to these elements stabilizes the mRNA but does not interfere with translation.

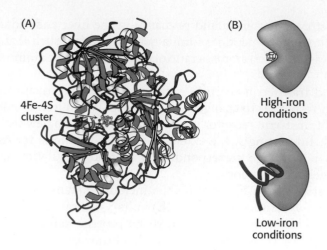

Figure 31.39 The IRP is an aconitase.
(A) Aconitase contains an unstable
4Fe-4S cluster at its center. (B) Under
conditions of low iron, the 4Fe-4S cluster
dissociates and appropriate RNA
molecules can bind in its place.
[Drawn from 1C96.pdb.]

The purification of the IRP and the cloning of its cDNA were sources of truly remarkable insight into evolution. The IRP was found to be approximately 30% identical in amino acid sequence with the citric acid cycle enzyme aconitase from mitochondria. Further analysis revealed that the IRP is, in fact, an active aconitase enzyme; it is a cytoplasmic aconitase that had been known for a long time, but its function was not well understood (Figure 31.39). The iron–sulfur center at the active site of the IRP is rather unstable, and loss of the iron triggers significant changes in protein conformation. Thus, this protein can serve as an iron-sensing factor.

Other mRNAs, including those taking part in heme synthesis, have been found to contain IREs. Thus, genes encoding proteins required for iron metabolism acquired sequences that, when transcribed, provided binding sites for the iron-sensing protein. An environmental signal—the concentration of iron—controls the translation of proteins required for the metabolism of this metal. Thus, mutations in the untranslated region of mRNAs have been selected for beneficial regulation by iron levels.

Summary

31.1 Many DNA-Binding Proteins Recognize Specific DNA Sequences

The regulation of gene expression depends on the interplay between specific sequences within the genome and proteins that bind specifically to these sites. Specific DNA-binding proteins recognize regulatory sites that usually lie adjacent to the genes whose transcription is regulated by these proteins. Many families of such DNA-binding proteins have been identified. In prokaryotes, the proteins of the largest family contain a helix-turn-helix motif. The first helix of this motif inserts into the major groove of DNA and makes specific hydrogen-bonding and other contacts with the edges of the base pairs. In eukaryotes, important classes of DNA-binding proteins include the homeodomains, the basic-leucine zipper (bZip) proteins, and Cys_2His_2 zinc-finger proteins. Each of these classes of proteins uses an α helix to make specific contacts with DNA. Although the use of α helices in DNA recognition is most common, some proteins use other structural elements.

31.2 Prokaryotic DNA-Binding Proteins Bind Specifically to Regulatory Sites in Operons

In prokaryotes, many genes are clustered into operons, which are units of coordinated genetic expression. An operon consists of control sites (an operator and a promoter) and a set of structural genes. In addition, regulator genes encode proteins that interact with the operator and promoter sites to stimulate or inhibit transcription. The treatment of *E. coli* with lactose induces an increase in the production of β-galactosidase and two additional proteins that are encoded in the lactose operon. In the absence of lactose or a similar galactoside inducer, the *lac* repressor protein binds to an operator site on the DNA and blocks transcription. The binding of allolactose, a derivative of lactose, to the *lac* repressor induces a conformational change that leads to dissociation from DNA. RNA polymerase can then move through the operator to transcribe the *lac* operon.

Some proteins activate transcription by directly contacting RNA polymerase. For example, cyclic AMP, a hunger signal, stimulates the transcription of many catabolic operons by binding to the catabolite activator protein. The binding of the cAMP–CAP complex to a specific site in the promoter region of an inducible catabolic operon enhances the binding of RNA polymerase and the initiation of transcription.

31.3 The Greater Complexity of Eukaryotic Genomes Requires Elaborate Mechanisms for Gene Regulation

Eukaryotic genomes are larger and more complex than those of prokaryotes. Some regulatory mechanisms used in eukaryotes are similar to those used in prokaryotes. In particular, most eukaryotic genes are not expressed unless they are activated by the binding of specific proteins, called transcription factors, to sites on the DNA. These specific DNA-binding proteins interact directly or indirectly with RNA polymerases or their associated proteins. Eukaryotic transcription factors are modular: they consist of separate DNA-binding and activation domains. Activation domains interact with RNA polymerases or their associated factors or with other protein complexes such as mediator. Enhancers are DNA elements that can modulate gene expression from more than 1000 bp away from the start site of transcription. Enhancers are often specific for certain cell types, depending on which DNA-binding proteins are present.

Eukaryotic DNA is tightly bound to basic proteins called histones; the combination is called chromatin. DNA wraps around an octamer of core histones to form a nucleosome, blocking access to many potential DNA binding sites. Changes in chromatin structure play a major role in regulating gene expression. Steroids such as estrogens bind to eukaryotic transcription factors called nuclear hormone receptors. These proteins are capable of binding DNA whether or not ligands are bound. The binding of ligands induces a conformational change that allows the recruitment of additional proteins called coactivators. Among the most important functions of coactivators is to catalyze the addition of acetyl groups to lysine residues in the tails of histone proteins. Histone acetylation decreases the affinity of the histones for DNA, making additional genes accessible for transcription. In addition, acetylated histones are targets for proteins containing specific binding units called bromodomains. Bromodomains are components of two classes of large complexes: (1) chromatin-remodeling engines and (2) factors associated with RNA polymerase II. These complexes open up sites on chromatin and initiate transcription.

31.4 Gene Expression Can Be Controlled at Posttranscriptional Levels

Gene expression can also be regulated at the level of translation. In prokaryotes, many operons important in amino acid biosynthesis are regulated by attenuation, a process that depends on the formation of alternative structures in mRNA, one of which favors the termination of transcription. Attenuation is mediated by the translation of a leader region of mRNA. A ribosome stalled by the absence of an aminoacyl-tRNA needed to translate the leader mRNA alters the structure of mRNA, allowing RNA polymerase to transcribe the operon beyond the attenuator site.

In eukaryotes, genes encoding proteins that transport and store iron are regulated at the translational level. Iron-response elements, structures that are present in certain mRNAs, are bound by an IRE-binding protein when this protein is not binding iron. Whether the expression of a gene is stimulated or inhibited in response to changes in the iron status of a cell depends on the location of the IRE within the mRNA.

Key Terms

helix-turn-helix motif (p. 895)

homeodomain (p. 895)

basic-leucine zipper (bZip) protein (p. 895)

Cys_2His_2 zinc-finger domain (p. 895)

β-galactosidase (p. 896)

operon model (p. 897)

repressor (p. 897)

lac repressor (p. 897)

lac operator (p. 898)

inducer (p. 898)

isopropylthiogalactoside (IPTG) (p. 898)

pur repressor (p. 899)

corepressor (p. 900)

catabolite repression (p. 900)

catabolite activator protein (CAP) (p. 900)

cell type (p. 901)

combinatorial control (p. 902)

enhancer (p. 902)

mediator (p. 903)

histone (p. 903)

chromatin (p. 903)

nucleosome (p. 903)

nucleosome core particle (p. 904)

hypersensitive site (p. 905)

chromatin immunoprecipitation (ChIP) (p. 906)

hypomethylation (p. 907)

CpG island (p. 907)

nuclear hormone receptor (p. 908)

estrogen response element (ERE) (p. 908)

coactivator (p. 909)

agonist (p. 910)

anabolic steroid (p. 910)

antagonist (p. 910)

selective estrogen receptor modulator (SERM) (p. 910)

histone acetyltransferase (HAT) (p. 911)

acetyllysine-binding domain (p. 911)

bromodomain (p. 911)

TATA-box-binding protein associated factor (TAF) (p. 911)

chromatin-remodeling engine (p. 912)

histone deacetylase (p. 912)

attenuation (p. 913)

transferrin (p. 914)

transferrin receptor (p. 914)

ferritin (p. 914)

iron-response element (IRE) (p. 915)

IRE-binding protein (IRP) (p. 915)

Selected Readings

Where to Start

Pabo, C. O., and Sauer, R. T. 1984. Protein–DNA recognition *Annu. Rev. Biochem.* 53:293–321.

Struhl, K. 1989. Helix-turn-helix, zinc-finger, and leucine-zipper motifs for eukaryotic transcriptional regulatory proteins. *Trends Biochem. Sci.* 14:137–140.

Struhl, K. 1999. Fundamentally different logic of gene regulation in eukaryotes and prokaryotes. *Cell* 98:1–4.

Korzus, E., Torchia, J., Rose, D. W., Xu, L., Kurokawa, R., McInerney, E. M., Mullen, T. M., Glass, C. K., and Rosenfeld, M. G. 1998. Transcription factor-specific requirements for coactivators and their acetyltransferase functions. *Science* 279:703–707.

Aalfs, J. D., and Kingston, R. E. 2000. What does "chromatin remodeling" mean? *Trends Biochem. Sci.* 25:548–555.

Books

Ptashne, M. 2004. *A Genetic Switch: Phage λ Revisited* (3d ed.). Cold Spring Harbor Laboratory Press.

McKnight, S. L., and Yamamoto, K. R. (Eds.). 1992. *Transcriptional Regulation* (vols. 1 and 2). Cold Spring Harbor Laboratory Press.

Latchman, D. S. 2004. *Eukaryotic Transcription Factors* (4th ed.). Academic Press.

Wolffe, A. 1992. *Chromatin Structure and Function*. Academic Press.

Lodish, H., Berk, A., Matsudaira, P., Kaiser, C. A., Krieger, M., Scott, M. P., Zipursky, S. L., and Darnell, J., 2004. *Molecular Cell Biology* (5th ed.). W. H. Freeman and Company.

Prokaryotic Gene Regulation

Balaeff, A., Mahadevan, L. and Schulten, K. 2004. Structural basis for cooperative DNA binding by CAP and lac repressor. *Structure* 12:123–132.

Bell, C. E., and Lewis, M. 2001. The Lac repressor: A second generation of structural and functional studies. *Curr. Opin. Struct. Biol.* 11:19–25.

Lewis, M., Chang, G., Horton, N. C., Kercher, M. A., Pace, H. C., Schumacher, M. A., Brennan, R. G., and Lu, P. 1996. Crystal structure of the lactose operon repressor and its complexes with DNA and inducer. *Science* 271:1247–1254.

Niu, W., Kim, Y., Tau, G., Heyduk, T., and Ebright, R. H. 1996. Transcription activation at class II CAP-dependent promoters: Two interactions between CAP and RNA polymerase. *Cell* 87:1123–1134.

Schultz, S. C., Shields, G. C., and Steitz, T. A. 1991. Crystal structure of a CAP-DNA complex: The DNA is bent by 90 degrees. *Science* 253:1001–1007.

Parkinson, G., Wilson, C., Gunasekera, A., Ebright, Y. W., Ebright, R. E., and Berman, H. M. 1996. Structure of the CAP-DNA complex at 2.5 Å resolution: A complete picture of the protein– DNA interface. *J. Mol. Biol.* 260:395–408.

Busby, S., and Ebright, R. H. 1999. Transcription activation by catabolite activator protein (CAP). *J. Mol. Biol.* 293:199–213.

Somers, W. S., and Phillips, S. E. 1992. Crystal structure of the met repressor-operator complex at 2.8 Å resolution reveals DNA recognition by β-strands. *Nature* 359:387–393.

Eukaryotic Gene Regulation

Green, M. R. 2005. Eukaryotic transcription activation: Right on target. *Mol. Cell* 18:399–402.

Kornberg, R. D. 2005. Mediator and the mechanism of transcriptional activation. *Trends Biochem. Sci.* 30:235–239.

Luger, K., Mader, A. W., Richmond, R. K., Sargent, D. F., and Richmond, T. J. 1997. Crystal structure of the nucleosome core particle at 2.8 Å resolution. *Nature* 389:251–260.

Arents, G., and Moudrianakis, E. N. 1995. The histone fold: A ubiquitous architectural motif utilized in DNA compaction and protein dimerization. *Proc. Natl. Acad. Sci. U. S. A.* 92:11170–11174.

Baxevanis, A. D., Arents, G., Moudrianakis, E. N., and Landsman, D. 1995. A variety of DNA-binding and multimeric proteins contain the histone fold motif. *Nucleic Acids Res.* 23:2685–2691.

Clements, A., Rojas, J. R., Trievel, R. C., Wang, L., Berger, S. L., and Marmorstein, R. 1999. Crystal structure of the histone acetyltransferase domain of the human PCAF transcriptional regulator bound to coenzyme A. *EMBO J.* 18:3521–3532.

Deckert, J., and Struhl, K. 2001. Histone acetylation at promoters is differentially affected by specific activators and repressors. *Mol. Cell. Biol.* 21:2726–2735.

Dutnall, R. N., Tafrov, S. T., Sternglanz, R., and Ramakrishnan, V. 1998. Structure of the histone acetyltransferase Hat1: A paradigm for the GCN5-related *N*-acetyltransferase superfamily. *Cell* 94:427–438.

Finnin, M. S., Donigian, J. R., Cohen, A., Richon, V. M., Rifkind, R. A., Marks, P. A., Breslow, R., and Pavletich, N. P. 1999. Structures of a histone deacetylase homologue bound to the TSA and SAHA inhibitors. *Nature* 401:188–193.

Finnin, M. S., Donigian, J. R., and Pavletich, N. P. 2001. Structure of the histone deacetylase SIR2. *Nat. Struct. Biol.* 8:621–625.

Jacobson, R. H., Ladurner, A. G., King, D. S., and Tjian, R. 2000. Structure and function of a human TAFII250 double bromodomain module. *Science* 288:1422–1425.

Rojas, J. R., Trievel, R. C., Zhou, J., Mo, Y., Li, X., Berger, S. L., Allis, C. D., and Marmorstein, R. 1999. Structure of *Tetrahymena* GCN5 bound to coenzyme A and a histone H3 peptide. *Nature* 401:93–98.

Nuclear Hormone Receptors

Downes, M., Verdecia, M. A., Roecker, A. J., Hughes, R., Hogenesch, J. B., Kast-Woelbern, H. R., Bowman, M. E., Ferrer, J. L., Anisfeld, A. M., Edwards, P. A., Rosenfeld, J. M., Alvarez, J. G., Noel, J. P., Nicolaou, K. C., and Evans, R.M. 2003. A chemical, genetic, and structural analysis of the nuclear bile acid receptor FXR. *Mol. Cell* 11:1079–1092.

Evans, R. M. 2005. The nuclear receptor superfamily: A Rosetta stone for physiology. *Mol. Endocrinol.* 19:1429–1438.

Xu, W., Cho, H., Kadam, S., Banayo, E. M., Anderson, S., Yates, J. R., 3d, Emerson, B. M., and Evans, R. M. 2004. A methylation-mediator complex in hormone signaling. *Genes Dev.* 18:144–156.

Evans, R. M. 1988. The steroid and thyroid hormone receptor superfamily. *Science* 240:889–895.

Yamamoto, K. R. 1985. Steroid receptor regulated transcription of specific genes and gene networks. *Annu. Rev. Genet.* 19:209–252.

Tanenbaum, D. M., Wang, Y., Williams, S. P., and Sigler, P. B. 1998. Crystallographic comparison of the estrogen and progesterone receptor's ligand binding domains. *Proc. Natl. Acad. Sci. U. S. A.* 95:5998–6003.

Schwabe, J. W., Chapman, L., Finch, J. T., and Rhodes, D. 1993. The crystal structure of the estrogen receptor DNA-binding domain bound to DNA: How receptors discriminate between their response elements. *Cell* 75:567–578.

Shiau, A. K., Barstad, D., Loria, P. M., Cheng, L., Kushner, P. J., Agard, D. A., and Greene, G. L. 1998. The structural basis of estrogen receptor/coactivator recognition and the antagonism of this interaction by tamoxifen. *Cell* 95:927–937.

Collingwood, T. N., Urnov, F. D., and Wolffe, A. P. 1999. Nuclear receptors: Coactivators, corepressors and chromatin remodeling in the control of transcription. *J. Mol. Endocrinol.* 23:255–275.

Chromatin and Chromatin Remodeling

Elgin, S. C. 1981. DNAase I-hypersensitive sites of chromatin. *Cell* 27:413–415.

Weintraub, H., Larsen, A., and Groudine, M. 1981. α-Globin-gene switching during the development of chicken embryos: Expression and chromosome structure. *Cell* 24:333–344.

Ren, B., Robert, F., Wyrick, J. J., Aparicio, O., Jennings, E. G., Simon, I., Zeitlinger, J., Schreiber, J., Hannett, N., Kanin, E., Volkert, T. L., Wilson, C. J., Bell, S. P., and Young, R. A. 2000. Genome-wide location and function of DNA-binding proteins. *Science* 290:2306–2309.

Goodrich, J. A., and Tjian, R. 1994. TBP-TAF complexes: Selectivity factors for eukaryotic transcription. *Curr. Opin. Cell. Biol.* 6:403–409.

Bird, A. P., and Wolffe, A. P. 1999. Methylation-induced repression: Belts, braces, and chromatin. *Cell* 99:451–454.

Cairns, B. R. 1998. Chromatin remodeling machines: Similar motors, ulterior motives. *Trends Biochem. Sci.* 23:20–25.

Albright, S. R., and Tjian, R. 2000. TAFs revisited: More data reveal new twists and confirm old ideas. *Gene* 242:1–13.

Urnov, F. D., and Wolffe, A. P. 2001. Chromatin remodeling and transcriptional activation: The cast (in order of appearance). *Oncogene* 20:2991–3006.

Posttranscriptional Regulation

Kolter, R., and Yanofsky, C. 1982. Attenuation in amino acid biosynthetic operons. *Annu. Rev. Genet.* 16:113–134.

Yanofsky, C. 1981. Attenuation in the control of expression of bacterial operons. *Nature* 289:751–758.

Rouault, T. A., Stout, C. D., Kaptain, S., Harford, J. B., and Klausner, R. D. 1991. Structural relationship between an iron-regulated RNA-binding protein (IRE-BP) and aconitase: Functional implications. *Cell* 64:881–883.

Klausner, R. D., Rouault, T. A., and Harford, J. B. 1993. Regulating the fate of mRNA: The control of cellular iron metabolism. *Cell* 72:19–28.

Gruer, M. J., Artymiuk, P. J., and Guest, J. R. 1997. The aconitase family: Three structural variations on a common theme. *Trends Biochem. Sci.* 22:3–6.

Theil, E. C. 1994. Iron regulatory elements (IREs): A family of mRNA non-coding sequences. *Biochem. J.* 304:1–11.

Historical Aspects

Lewis, M. 2005. The lac repressor. *C. R. Biol.* 328:521-548.

Jacob, F., and Monod, J. 1961. Genetic regulatory mechanisms in the synthesis of proteins. *J. Mol. Biol.* 3:318–356.

Ptashne, M., and Gilbert, W. 1970. Genetic repressors. *Sci. Am.* 222(6):36–44.

Lwoff, A., and Ullmann, A. (Eds.). 1979. *Origins of Molecular Biology: A Tribute to Jacques Monod.* Academic Press.

Judson, H. 1996. *The Eighth Day of Creation: Makers of the Revolution in Biology.* Cold Spring Harbor Laboratory Press.

Problems

1. *Missing genes.* Predict the effects of deleting the following regions of DNA:

(a) The gene encoding *lac* repressor
(b) The *lac* operator
(c) The gene encoding CAP

2. *Minimal concentration.* Calculate the concentration of *lac* repressor, assuming that one molecule is present per cell. Assume that each *E. coli* cell has a volume of 10^{-12} cm³. Would you expect the single molecule to be free or bound to DNA?

3. *Counting sites.* Calculate the expected number of times that a given 8-base-pair DNA site should be present in the *E. coli* genome. Assume that all four bases are equally probable. Repeat for a 10-base-pair site and a 12-base-pair site.

4. *Charge neutralization.* Given the histone amino acid sequences illustrated below, estimate the charge of a histone octamer at pH 7. Assume that histidine residues are uncharged at this pH. How does this charge compare with the charge on 150 base pairs of DNA?

Histone H2A

MSGRGKQGGKARAKAKTRSSRAGLQFPVGRVHRLLRKGNYSERVGAGAPVYLAAVLEYLTAEILELAGNA

ARDNKKTRIIPRHLQLAIRNDEELNKLLGRVTIAQGGVLPNIQAVLLPKKTESHHKAKGK

Histone H2B

MPEPAKSAPAPKKGSKKAVTKAQKKDGKKRKRSRKESYSVYVYKVLKQVHPDTGISSKAMGIMNSFVNDI

FERIAGEASRLAHYNKRSTITSREIQTAVRLLLPGELAKHAVSEGTKAVTKYTSSK

Histone H3

MARTKQTARKSTGGKAPRKQLATKAARKSAPSTGGVKKPHRYRPGTVALREIRRYQKSTELLIRKLPFQR

LVREIAQDFKTDLRFQSAAIGALQEASEAYLVGLFEDTNLCAIHAKRVTIMPKDIQLARRIRGERA

Histone H4

MSGRGKGGKGLGKGGAKRHRKVLRDNIQGITKPAIRRLARRGGVKRISGLIYEETRGVLKVFLENVIRDA

VTYTEHAKRKTVTAMDVVYALKRQGRTLYGFGG

5. *Chromatin immunoprecipitation.* You have used the technique of chromatin immunoprecipitation to isolate DNA fragments containing a DNA-binding protein of interest. Suppose that you wish to know whether a particular known DNA fragment is present in the isolated mixture. How might you detect its presence? How many different fragments would you expect if you used antibodies to the *lac* repressor to perform a chromatin immunoprecipitation experiment in *E. coli*? If you used antibodies to the *pur* repressor?

6. *Nitrogen substitution.* Growth of mammalian cells in the presence of 5-azacytidine results in the activation of some normally inactive genes. Propose an explanation.

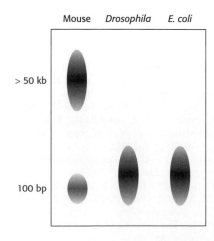

deoxyribose
5-Azacytidine

7. *A new domain.* A protein domain that recognizes 5-methylcytosine in the context of double-stranded DNA has been characterized. What role might proteins containing such a domain play in regulating gene expression? Where on a double-stranded DNA molecule would you expect such a domain to bind?

8. *The same but not the same.* The *lac* repressor and the *pur* repressor are homologous proteins with very similar three-dimensional structures, yet they have different effects on gene expression. Describe two important ways in which the gene-regulatory properties of these proteins differ.

9. *The opposite direction.* Some compounds called anti-inducers bind to repressors such as the *lac* repressor and inhibit the action of inducers; that is, transcription is repressed and higher concentrations of inducer are required to induce transcription. Propose a mechanism of action for anti-inducers.

10. *Inverted repeats.* Suppose that a nearly perfect inverted repeat is observed in a DNA sequence over 20 base pairs. Provide two possible explanations.

Mechanism Problem

11. *Acetyltransferases.* Propose a mechanism for the transfer of an acetyl group from acetyl CoA to the amino group of lysine.

Data Interpretation Problem

12. *Limited restriction.* The restriction enzyme *Hpa*II is a powerful tool for analyzing DNA methylation. This enzyme cleaves sites of the form 5′-CCGG-3′ but will not cleave such sites if the DNA is methylated on any of the cytosine residues. Genomic DNA from different organisms is treated with *Hpa*II and the results are analyzed by gel electrophoresis (see the adjoining patterns). Provide an explanation for the observed patterns.

Sensory Systems

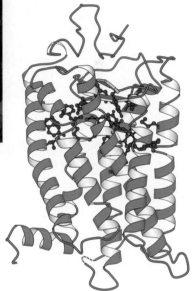

Color perception requires specific photoreceptors. The photoreceptor rhodopsin (right), which absorbs light in the process of vision, consists of the protein opsin and a bound vitamin A derivative, retinal. The amino acids (shown in red) that surround the retinal determine the color of light that is most efficiently absorbed. Individual lacking a light-absorbing photoreceptor for the color green will see a colorful fruit stand (left) as mostly yellows (middle). [(Left and middle) From L. T. Sharpe, A. Stockman, H. Jagle, and J. Nathans, Opsin genes, cone photopigments, color vision, and color blindness. In *Color Vision: from Genes to Perception*, K. Gegenfurtner and L. T. Sharpe, Eds. (Cambridge University Press, 1999), pp. 3–51.]

O ur senses provide us with means for detecting a diverse set of external signals, often with incredible sensitivity and specificity. For example, when fully adapted to a darkened room, our eyes allow us to sense very low levels of light, down *to a limit of less than 10 photons*. With more light, we are able to distinguish millions of colors. Through our senses of smell and taste, we are able to detect thousands of chemicals in our environment and sort them into categories: pleasant or unpleasant? healthful or toxic? Finally, we can perceive mechanical stimuli in the air and around us through our senses of hearing and touch.

How do our sensory systems work? How are the initial stimuli detected? How are these initial biochemical events transformed into perceptions and experiences? We have already encountered systems that sense and respond to chemical signals—namely, receptors that bind to growth factors and hormones. Our knowledge of these receptors and their associated signal-transduction pathways provides us with concepts and tools for unraveling some of the workings of sensory systems. For example, 7TM receptors (seven-transmembrane receptors, Section 14.1) play key roles in olfaction, taste, and vision. Ion channels that are sensitive to mechanical stress are essential for hearing and touch.

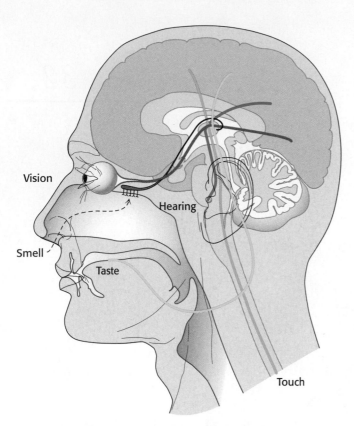

Figure 32.1 Sensory connections to the brain. Sensory nerves connect sensory organs to the brain and spinal cord.

In this chapter, we focus on the five major sensory systems found in human beings and other mammals: olfaction (the sense of smell—i.e., the detection of small molecules in the air), taste, or gustation (the detection of selected organic compounds and ions by the tongue), vision (the detection of light), hearing (the detection of sound, or pressure waves in the air), and touch (the detection of changes in pressure, temperature, and other factors by the skin). Each of these primary sensory systems contains specialized sensory neurons that transmit nerve impulses to the central nervous system (Figure 32.1). In the central nervous system, these signals are processed and combined with other information to yield a perception that may trigger a change in behavior. By these means, our senses allow us to detect changes in our environments and to adjust our behavior appropriately.

32.1 A Wide Variety of Organic Compounds Are Detected by Olfaction

Human beings can detect and distinguish thousands of different compounds by smell, often with considerable sensitivity and specificity. Most odorants are small organic compounds with sufficient volatility that they can be carried as vapors into the nose. For example, a major component responsible for the smell of almonds is the simple aromatic compound benzaldehyde, whereas the sulfhydryl compound 3-methylbutane-1-thiol is a major component of the smell of skunks.

Benzaldehyde
(Almond)

3-Methylbutane-1-thiol
(Skunk)

Geraniol
(Rose)

Zingiberene
(Ginger)

R-Carvone
(Spearmint)

S-Carvone
(Caraway)

What properties of these molecules are responsible for their smells? First, *the shape of the molecule rather than its other physical properties is crucial.* We can most clearly see the importance of shape by comparing molecules such as those responsible for the smells of spearmint and caraway. These compounds are identical in essentially all physical properties such as hydrophobicity because they are exact mirror images of one another. Thus, the smell produced by an odorant depends not on a physical property but on the compound's interaction with a specific binding surface, most likely a protein receptor. Second, some human beings (and other animals) suffer from *specific anosmia;* that is, they are incapable of smelling specific compounds even though their olfactory systems are otherwise normal. Such anosmias are often inherited. These observations suggest that mutations in individual receptor genes lead to the loss of the ability to detect a small subset of compounds.

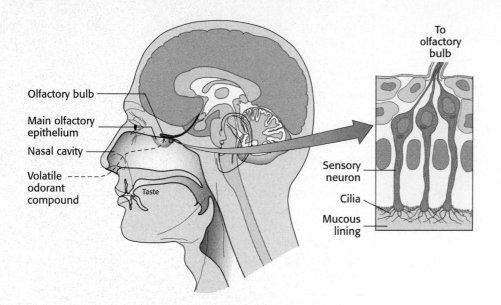

Figure 32.2 The main nasal epithelium. This region of the nose, which lies at the top of the nasal cavity, contains approximately 1 million sensory neurons. Nerve impulses generated by odorant molecules binding to receptors on the cilia travel from the sensory neurons to the olfactory bulb.

Olfaction Is Mediated by an Enormous Family of Seven-Transmembrane-Helix Receptors

Odorants are detected in a specific region of the nose, called the *main olfactory epithelium*, that lies at the top of the nasal cavity (Figure 32.2). Approximately 1 million sensory neurons line the surface of this region. Cilia containing the odorant-binding protein receptors project from these neurons into the mucous lining of the nasal cavity.

Biochemical studies in the late 1980s examined isolated cilia from rat olfactory epithelium that had been treated with odorants. Exposure to the odorants increased the cellular level of cyclic AMP, and this increase was observed only in the presence of GTP. On the basis of what was known about signal-transduction systems, *the participation of cAMP and GTP strongly suggested the involvement of a G protein and, hence, 7TM receptors.* Indeed, Randall Reed purified and cloned a G-protein α subunit, termed $G_{(olf)}$, which is uniquely expressed in olfactory cilia. The involvement of 7TM receptors suggested a strategy for identifying the olfactory receptors themselves. Complementary DNAs were sought that (1) were expressed primarily in the sensory neurons lining the nasal epithelium, (2) encoded members of the 7TM-receptor family, and (3) were present as a large and diverse family to account for the range of odorants. Through the use of these criteria, cDNAs for odorant receptors from rats were identified in 1991 by Richard Axel and Linda Buck.

The odorant receptor (hereafter, OR) family is even larger than expected: *more than 1000 OR genes are present in the mouse and the rat, whereas the human genome encodes approximately 350 ORs.* In addition, the human genome includes approximately 500 OR pseudogenes containing mutations that prevent the generation of a full-length, proper odorant receptor. The OR family is thus one of the largest gene families in human beings. Further analysis of primate OR genes reveals that the fraction of pseudogenes is greater in species more closely related to human beings (Figure 32.3). Thus, we may have a glimpse at the evolutionary loss of acuity in the sense of smell as higher mammals presumably became less dependent on

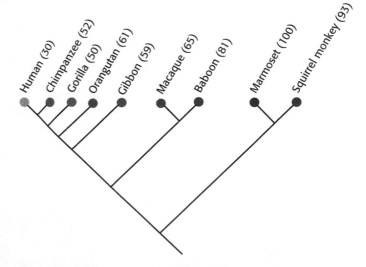

Figure 32.3 Evolution of odorant receptors. Odorant receptors appear to have lost function through conversion into pseudogenes in the course of primate evolution. The percentage of OR genes that appear to be functional for each species is shown in parentheses.

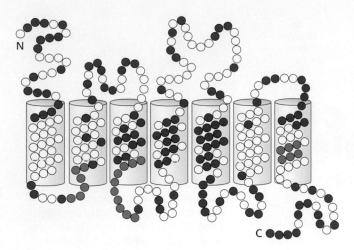

Figure 32.4 Conserved and variant regions in odorant receptors. Odorant receptors are members of the 7TM-receptor family. The green cylinders represent the seven presumed transmembrane helices. Strongly conserved residues characteristic of this protein family are shown in blue, whereas highly variable residues are shown in red.

this sense for survival. For rodents that are highly dependent on their sense of smell, essentially all OR genes encode functional proteins.

The OR proteins are typically 20% identical in sequence with the β-adrenergic receptor (Section 14.1) and from 30% to 60% identical with one another. Several specific sequence features are present in most or all OR family members (Figure 32.4). The central region, particularly transmembrane helices 4 and 5, is highly variable, suggesting that this region is the site of odorant binding. That site must be different in odorant receptors that bind distinct odorant molecules.

What is the relation between OR gene expression and the individual neuron? Interestingly, *each olfactory neuron expresses only a single OR gene,* among hundreds available. Apparently, the precise OR gene expressed is determined largely at random. After one OR gene is expressed and a functional OR protein is produced, the expression of all other OR genes is suppressed by a feedback mechanism that remains to be fully elucidated.

The binding of an odorant to an OR on the neuronal surface initiates a signal-transduction cascade that results in an action potential (Figure 32.5). The ligand-bound OR activates $G_{(olf)}$, the specific G protein mentioned earlier. $G_{(olf)}$ is initially in its GDP-bound form. When activated, it releases GDP, binds GTP, and releases its associated βγ subunits. The α subunit then activates a specific adenylate cyclase, increasing the intracellular concentration of cAMP. The rise in the intracellular concentration of cAMP activates a nonspecific cation channel that allows calcium and other cations into the cell. The flow of cations through the channel depolarizes the neuronal membrane and initiates an action potential. This action potential, combined with those from other olfactory neurons, leads to the perception of a specific odor.

Odorants Are Decoded by a Combinatorial Mechanism

An obvious challenge presented to the investigator by the large size of the OR family is to match each OR with the one or more odorant molecules to which it binds. Exciting progress has been made in this regard. Initially, an OR was matched with odorants by overexpressing a single, specific OR gene in rats. This OR responded to straight-chain aldehydes, most favorably to *n*-octanal and less strongly to *n*-heptanal and *n*-hexanal. More-dramatic progress was made by taking advantage of our knowledge of the OR signal-transduction pathway and the power of PCR (p. 140). A section

Figure 32.5 The olfactory signal-transduction cascade. The binding of odorant to the olfactory receptor activates a signaling pathway similar to those initiated in response to the binding of some hormones to their receptors. The final result is the opening of cAMP-gated ion channels and the initiation of an action potential.

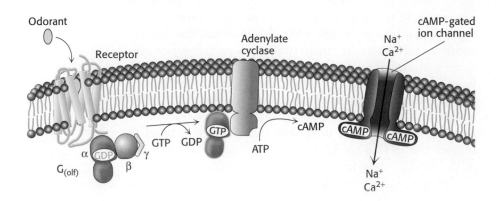

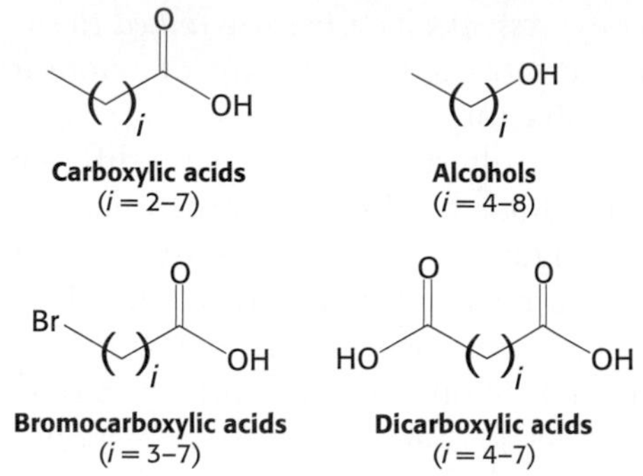

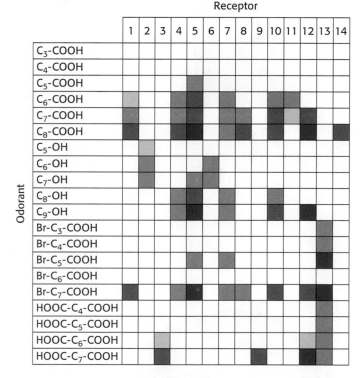

Figure 32.6 Four series of odorants tested for olfactory-receptor activation.

Carboxylic acids ($i = 2–7$)

Alcohols ($i = 4–8$)

Bromocarboxylic acids ($i = 3–7$)

Dicarboxylic acids ($i = 4–7$)

of nasal epithelium from a mouse was loaded with the calcium-sensitive dye Fura-2 (p. 389). The tissue was then treated with different odorants, one at a time, at a specific concentration. If the odorant had bound to an OR and activated it, that neuron could be detected under a microscope by the change in fluorescence caused by the influx of calcium that occurs as part of the signal-transduction process. To determine which OR was responsible for the response, cDNA was generated from mRNA that had been isolated from single identified neurons. The cDNA was then subjected to PCR with the use of primers that are effective in amplifying most or all OR genes. The sequence of the PCR product from each neuron was then determined and analyzed.

Using this approach, investigators analyzed the responses of neurons to a series of compounds having varying chain lengths and terminal functional groups (Figure 32.6). The results of these experiments appear surprising at first glance (Figure 32.7). Importantly, there is not a simple 1:1 correspondence between odorants and receptors. *Almost every odorant activates a number of receptors* (usually to different extents) and *almost every receptor is activated by more than one odorant*. Note, however, that each odorant activates a unique combination of receptors. In principle, this combinatorial mechanism allows even a small array of receptors to distinguish a vast number of odorants.

How is the information about which receptors have been activated transmitted to the brain? Recall that each neuron expresses only one OR and that the pattern of expression appears to be largely random. A substantial clue to the connections between receptors and the brain has been provided by the creation of mice that express a gene for an easily detectable colored marker in conjunction with a specific OR gene. Olfactory neurons that express the OR–marker-protein combination were traced to their destination in the brain, a structure called the olfactory bulb (Figure 32.8). The processes from neurons that express the same OR gene were found to connect to the same location in the olfactory bulb. Moreover, this pattern of neuronal connection was found to be identical in

Figure 32.7 Patterns of olfactory-receptor activation. Fourteen different receptors were tested for responsiveness to the compounds shown in Figure 32.6. A colored box indicates that the receptor at the top responded to the compound at the left. Darker colors indicate that the receptor was activated at a lower concentration of odorant.

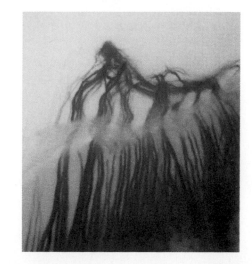

Figure 32.8 Converging olfactory neurons. This section of the nasal cavity is stained to reveal processes from sensory neurons expressing the same olfactory receptor. The processes converge to a single location in the olfactory bulb. [From P. Mombaerts, F. Wang, C. Dulac, S. K. Chao, A. Nemes, M. Mendelsohn, J. Edmondson, and R. Axel. *Cell* 87(1996):675–689.]

Figure 32.9 The Cyranose 320. The electronic nose may find uses in the food industry, animal husbandry, law enforcement, and medicine. [Courtesy of Cyrano Sciences.]

all mice examined. Thus, *neurons that express specific ORs are linked to specific sites in the brain*. This property creates a spatial map of odorant-responsive neuronal activity within the olfactory bulb.

Can such a combinatorial mechanism truly distinguish many different odorants? An electronic "nose" that functions by the same principles provides compelling evidence that it can (Figure 32.9). The receptors for the electronic nose are polymers that bind a range of small molecules. Each polymer binds every odorant, but to varying degrees. Importantly, the electrical properties of these polymers change on odorant binding. A set of 32 of these polymer sensors, wired together so that the pattern of responses can be evaluated, is capable of distinguishing individual compounds such as *n*-pentane and *n*-hexane as well as complex mixtures such as the odors of fresh and spoiled fruit.

Functional Magnetic Resonance Imaging Reveals Regions of the Brain Processing Sensory Information

Can we extend our understanding of how odorants are perceived to events in the brain? Biochemistry has provided the basis for powerful methods for examining responses within the brain. One method, *functional magnetic resonance imaging* (f MRI), takes advantage of two key observations. The first is that, when a specific part of the brain is active, blood vessels relax to allow more blood flow to the active region. Thus, a more active region of the brain will be richer in oxyhemoglobin. The second observation is that the iron center in hemoglobin undergoes substantial structural changes on binding oxygen (p. 185). These changes are associated with a rearrangement of electrons such that the iron in deoxyhemoglobin acts as a strong magnet, whereas the iron in oxyhemoglobin does not. The difference between the magnetic properties of these two forms of hemoglobin can be used to image brain activity.

Nuclear magnetic resonance techniques (p. 98) detect signals that originate primarily from the protons in water molecules but are altered by the magnetic properties of hemoglobin. With the use of appropriate techniques, images can be generated that reveal differences in the relative amounts of deoxy- and oxyhemoglobin and thus the relative activity of various parts of the brain.

Figure 32.10 Brain response to odorants. A functional magnetic resonance image reveals brain response to odorants. The light spots indicate regions of the brain activated by odorants. [From N. Sobel et al., *J. Neurophysiol.* 83(2000):537–551; courtesy of Dr. Noam Sobel.]

These noninvasive methods reveal areas of the brain that process sensory information. For example, subjects have been imaged while breathing air that either does or does not contain odorants. When odorants are present, the f MRI technique detects an increase in the level of hemoglobin oxygenation (and, hence, brain activity) in several regions of the brain (Figure 32.10). Such regions include those in the primary olfactory cortex as well as other regions in which secondary processing of olfactory signals presumably takes place. Further analysis reveals the time course of activation of particular regions and other features. Functional MRI shows tremendous potential for mapping regions and pathways engaged in processing sensory information obtained from all the senses. Thus, *a seemingly incidental aspect of the biochemistry of hemoglobin has yielded the basis for observing the brain in action.*

32.2 Taste Is a Combination of Senses That Function by Different Mechanisms

The inability to taste food is a common complaint when nasal congestion reduces the sense of smell. Thus, smell greatly augments our sense of taste (also known as *gustation*), and taste is, in many ways, the sister sense to

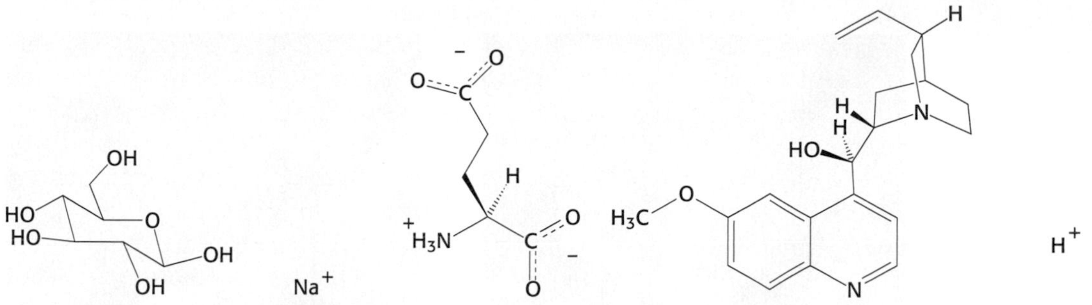

Glucose (sweet) **Sodium ion** (salty) **Glutamate** (umami) **Quinine** (bitter) **Hydrogen ion** (sour)

Figure 32.11 Examples of tastant molecules. Tastants fall into five groups: sweet, salty, umami, bitter, and sour.

olfaction. Nevertheless, the two senses differ from each other in several important ways. First, we are able to sense several classes of compounds by taste that we are unable to detect by smell; salt and sugar have very little odor, yet they are primary stimuli of the gustatory system. Second, whereas we are able to discriminate thousands of odorants, discrimination by taste is much more modest. Five primary tastes are perceived: *bitter, sweet, sour, salty,* and *umami* (the taste of glutamate and aspartate from the Japanese word for "deliciousness"). These five tastes serve to classify compounds into potentially nutritive and beneficial (sweet, salty, umami) or potentially harmful or toxic (bitter, sour). Tastants (the molecules sensed by taste) are quite distinct for the different groups (Figure 32.11).

The simplest tastant, the hydrogen ion, is perceived as sour. Other simple ions, particularly sodium ion, are perceived as salty. The taste called umami is evoked by the amino acids glutamate and aspartate, the former often encountered as the flavor enhancer monosodium glutamate (MSG). In contrast, *tastants perceived as sweet and, particularly, bitter are extremely diverse.* Many bitter compounds are alkaloids or other plant products of which many are toxic. However, they do not have any common structural elements or other common properties. Carbohydrates such as glucose and sucrose are perceived as sweet, as are other compounds including some simple peptide derivatives, such as aspartame, and even some proteins. These differences in specificity among the five tastes are due to differences in their underlying biochemical mechanisms. The sense of taste is, in fact, a number of independent senses all utilizing the same organ, the tongue, for their expression.

Tastants are detected by specialized structures called *taste buds,* which contain approximately 150 cells, including sensory neurons (Figure 32.12). Fingerlike projections called *microvilli,* which are rich in taste receptors, project from one end of each sensory neuron to the surface of the tongue. Nerve fibers at the opposite end of each neuron carry electrical impulses to the brain in response to stimulation by tastants. Structures called *taste papillae* contain numerous taste buds.

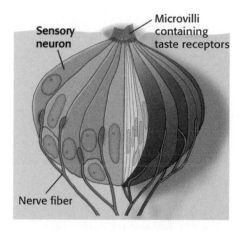

Figure 32.12 A taste bud. Each taste bud contains sensory neurons that extend microvilli to the surface of the tongue, where they interact with tastants.

Sequencing of the Human Genome Led to the Discovery of a Large Family of 7TM Bitter Receptors

Just as in olfaction, a number of clues pointed to the involvement of G proteins and, hence, 7TM receptors in the detection of bitter and sweet tastes. The evidence included the isolation of a specific G-protein α subunit

(A)

(B)

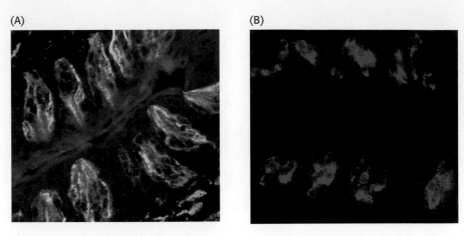

Figure 32.13 Expression of gustducin in the tongue. (A) A section of tongue stained with a fluorescent antibody reveals the position of the taste buds. (B) The same region stained with an antibody directed against gustducin reveals that this G protein is expressed in taste buds. [Courtesy of Dr. Charles S. Zuker.]

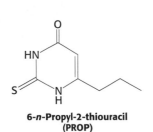

**6-*n*-Propyl-2-thiouracil
(PROP)**

termed *gustducin,* which is expressed primarily in taste buds (Figure 32.13). How could the 7TM receptors be identified? The ability to detect some compounds depends on specific genetic loci in both human beings and mice. For instance, the ability to taste the bitter compound 6-*n*-propyl-2-thiouracil (PROP) was mapped to a region on human chromosome 5 by comparing DNA markers of persons who vary in sensitivity to this compound.

This observation suggested that this region might encode a 7TM receptor that responded to PROP. Approximately 450 kilobases in this region had been sequenced early in the human genome project. This sequence was searched by computer for potential 7TM-receptor genes, and, indeed, one was detected and named *T2R1.* Additional database searches detected approximately 30 sequences similar to *T2R1* in the human genome. The encoded proteins are between 30 and 70% identical with T2R1 (Figure 32.14).

Are these proteins, in fact, bitter receptors? Several lines of evidence suggest that they are. First, their genes are expressed in taste-sensitive cells—in fact, in many of the same cells that express gustducin. Second, cells that express individual members of this family respond to specific bitter compounds. For example, cells that express a specific mouse receptor (mT2R5) responded when exposed specifically to cycloheximide. Third, mice that had been found unresponsive to cycloheximide were found to have point mutations in the gene encoding mT2R5. Finally, cycloheximide

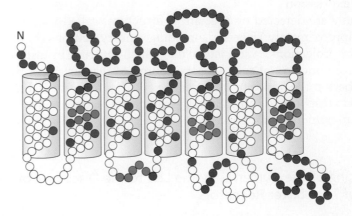

Figure 32.14 Conserved and variant regions in bitter receptors. The bitter receptors are members of the 7TM-receptor family. Strongly conserved residues characteristic of this protein family are shown in blue, and highly variable residues are shown in red.

specifically stimulates the binding of GTP analogs to gustducin in the presence of the mT2R5 protein (Figure 32.15).

Importantly, each taste-receptor cell expresses many different members of the T2R family. This pattern of expression stands in sharp contrast to the pattern of one receptor type per cell that characterizes the olfactory system (Figure 32.16). The difference in expression patterns accounts for the much greater specificity of our perceptions of smells compared with tastes. *We are able to distinguish among subtly different odors because each odorant stimulates a unique pattern of neurons. In contrast, many tastants stimulate the same neurons.* Thus, we perceive only "bitter" without the ability to discriminate cycloheximide from quinine.

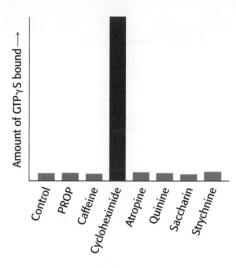

Figure 32.15 Evidence that T2R proteins are bitter taste receptors. Cycloheximide uniquely stimulates the binding of the GTP analog GTPγS to gustducin in the presence of the mT2R protein. [After J. Chandrashekar, K. L. Mueller, M. A. Hoon, E. Adler, L. Feng, W. Guo, C. S. Zuker, and N. J. Ryba. *Cell* 100(2000):703–711.]

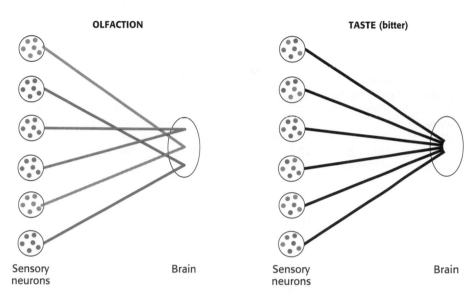

Figure 32.16 Differing gene expression and connection patterns in olfactory and bitter taste receptors. In olfaction, each neuron expresses a single OR gene, and the neurons expressing the same OR converge to specific sites in the brain, enabling specific perception of different odorants. In gustation, each neuron expresses many bitter receptor genes, and so the identity of the tastant is lost in transmission.

A Heterodimeric 7TM Receptor Responds to Sweet Compounds

Most sweet compounds are carbohydrates, energy rich and easily digestible. Some noncarbohydrate compounds such as saccharin and aspartame also taste sweet. Members of a second family of 7TM receptors are expressed in taste-receptor cells sensitive to sweetness. The three members of this family, referred to as T1R1, T1R2, and T1R3, are distinguished by their large extracellular domains compared with those of the bitter receptors. Studies in knockout mice have revealed that T1R2 and T1R3 are expressed simultaneously in mice able to taste carbohydrates (Figure 32.17). Thus, it appears that T1R2 and T1R3 form a specific heterodimeric receptor that is responsible for mediating the response to sugars. This heterodimeric receptor also responds to artificial sweeteners and to sweet-tasting proteins and therefore appears to be the receptor responsible for responses to all sweet tastants. Note that T1R2 and T1R3 do respond to sweet tastants individually, but only at very high concentrations of tastant.

The requirement for an *oligomeric* 7TM receptor for a fully functional response is surprising, considering our previous understanding of 7TM receptors. This discovery has at least two possible explanations. First, the sweet receptor could be a member of a small subset of the 7TM-receptor family that functions well only as oligomers. Alternatively, many 7TM

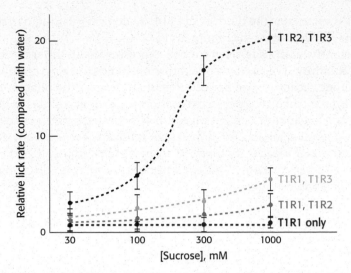

Figure 32.17 Evidence for a heterodimeric sweet receptor. The sensitivity to sweetness of mice with genes for either T1R1, T1R2, T1R3, or both T1R2 and T1R3 were determined by observing the relative rates at which they licked solutions containing various amount of sucrose. These studies revealed that both T1R2 and T1R3 were required for a full response to sucrose. Mice with a disrupted T1R1 gene were indistinguishable from wild-type mice in this assay (not shown). [After G. Q. Zhao, Y. Zhang, M. A. Hoon, J. Chandrashekar, I. Erlenbach, N. J. P. Ryba, and C. S. Zuker, *Cell* 115(2003):255–266.]

receptors may function as oligomers, but this notion is not clear, because these oligomers contain only one type of 7TM-receptor subunit. Further studies will be required to determine which of these explanations is correct.

Umami, the Taste of Glutamate and Aspartate, Is Mediated by a Heterodimeric Receptor Related to the Sweet Receptor

The family of receptors responsible for detecting sweetness is also responsible for detecting amino acids. In human beings, only glutamate and aspartate elicit a taste response. Studies similar to those for the sweet receptor revealed that the umami receptor consists of T1R1 and T1R3. Thus, this receptor has one subunit (T1R3) in common with the sweet receptor but has an additional subunit (T1R1) that does not participate in the sweet response. This observation is supported by the observation that mice in which the gene for T1R1 is disrupted do not respond to aspartate but do respond normally to sweet tastants; mice having disrupted genes for both T1R1 and T1R3 respond poorly to both umami and sweet tastants.

Salty Tastes Are Detected Primarily by the Passage of Sodium Ions Through Channels

Salty tastants are not detected by 7TM receptors. Rather, they are detected directly by their passage through ion channels expressed on the surface of cells in the tongue. Evidence for the role of these ion channels comes from examining known properties of Na^+ channels characterized in other biological contexts. One class of channels, characterized first for their role in salt reabsorption, are thought to be important in the detection of salty tastes because they are sensitive to the compound *amiloride*, which mutes the taste of salt and significantly lowers sensory-neuron activation in response to sodium.

An *amiloride-sensitive Na^+ channel* comprises four subunits that may be either identical or distinct but in any case are homologous. An individual subunit ranges in length from 500 to 1000 amino acids and includes two presumed membrane-spanning helices as well as a large extracellular domain in between them (Figure 32.18). The extracellular region includes two (or, sometimes, three) distinct regions rich in cysteine residues (and, presumably, disulfide bonds). A region just ahead of the second membrane-spanning helix appears to form part of the pore in a manner analogous to the structurally characterized potassium channel (p. 364). The members of the amiloride-sensitive Na^+-channel family are numerous and diverse in their biological roles. We shall encounter them again in the context of the sense of touch.

Sodium ions passing through these channels produce a significant transmembrane current. Amiloride blocks this current, accounting for its effect on taste. However, about 20% of the response to sodium remains even in the presence of amiloride, suggesting that other ion channels also contribute to salt detection.

Sour Tastes Arise from the Effects of Hydrogen Ions (Acids) on Channels

Like salty tastes, *sour tastes are detected by direct interactions with ion channels*, but the incoming ions are hydrogen ions (in high concentrations) rather than sodium ions. For example, in the absence of high concentrations of sodium, hydrogen ion flow can induce substantial transmembrane currents through amiloride-sensitive Na^+ channels. However, hydrogen ions are also sensed by mechanisms other than their direct passage through membranes. Binding by hydrogen ions blocks some potassium ion channels and activates other types of channels. Together, these mechanisms lead to changes in membrane polarization in sensory neurons that produce the sensation of sour taste. We shall consider an additional receptor related to taste, one responsible for the "hot" taste of spicy food, when we examine mechanisms of touch perception.

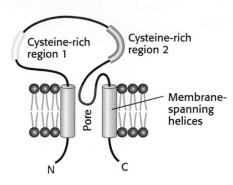

Figure 32.18 Schematic structure of the amiloride-sensitive sodium channel. Only one of the four subunits that constitute the functional channel is illustrated. The amiloride-sensitive sodium channel belongs to a superfamily having common structural features, including two hydrophobic membrane-spanning regions, intracellular amino and carboxyl termini; and a large, extracellular region with conserved cysteine-rich domains.

32.3   Photoreceptor Molecules in the Eye Detect Visible Light

Vision is based on the absorption of light by photoreceptor cells in the eye. These cells are sensitive to light in a narrow region of the electromagnetic spectrum, the region with wavelengths between 300 and 850 nm (Figure 32.19). Vertebrates have two kinds of photoreceptor cells, called *rods* and *cones* because of their distinctive shapes. Cones function in bright light and are responsible for color vision, whereas rods function in dim light but do not perceive color. A human retina contains about 3 million cones and 100 million rods. Remarkably, a rod cell can respond to a single photon, and the brain requires fewer than 10 such responses to register the sensation of a flash of light.

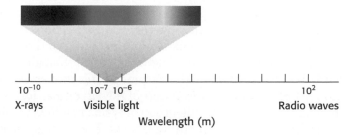

Figure 32.19 The electromagnetic spectrum. Visible light has wavelengths between 300 and 850 nm.

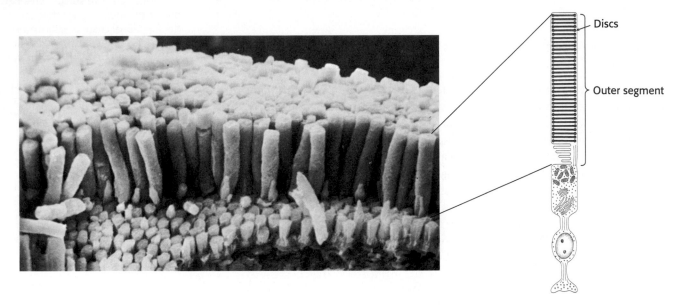

11-*cis*-Retinal

Rhodopsin, a Specialized 7TM Receptor, Absorbs Visible Light

Rods are slender, elongated structures; the outer segment is specialized for photoreception (Figure 32.20). It contains a stack of about 1000 discs, which are membrane-enclosed sacs densely packed with photoreceptor molecules. The photosensitive molecule is often called a *visual pigment* because it is highly colored owing to its ability to absorb light. The photoreceptor molecule in rods is *rhodopsin* (Section 14.1), which consists of the protein *opsin* linked to *11-cis-retinal,* a prosthetic group.

Discs

Outer segment

Figure 32.20 The rod cell. (Left) Scanning electron micrograph of retinal rod cells. (Right) Schematic representation of a rod cell. [Photograph courtesy of Dr. Deric Bownds.]

Rhodopsin absorbs light very efficiently in the middle of the visible spectrum, its absorption being centered on 500 nm, which nicely matches the solar output (Figure 32.21). A rhodopsin molecule will absorb a high percentage of the photons of the correct wavelength that strike it, as indicated by the extinction coefficient of $40,000 \ \mathrm{M^{-1}cm^{-1}}$ at 500 nm. The extinction coefficient for rhodopsin is more than an order of magnitude greater than that for tryptophan, the most efficient absorber in proteins that lack prosthetic groups.

Opsin, the protein component of rhodopsin, is a member of the 7TM-receptor family. Indeed, rhodopsin was the first member of this family to be purified, its gene was the first to be cloned and sequenced, and its three-dimensional structure was the first to be determined. The color of rhodopsin and its responsiveness to light depend on the presence of the light-absorbing group *(chromophore)* 11-*cis*-retinal. This compound is a powerful absorber of light because it is a polyene; its six alternating single and double bonds constitute a long, unsaturated electron network. Recall that alternating single and double bonds account for the chromophoric properties of chlorophyll (Section 19.2). The aldehyde group of 11-*cis*-retinal forms a Schiff base (Figure 32.22) with the ε-amino group of lysine residue 296, which lies in the center of the seventh transmembrane helix. Free retinal absorbs maximally at 370 nm, and its unprotonated Schiff-base adduct absorbs at 380 nm, whereas the protonated Schiff base absorbs at 440 nm or longer wave-

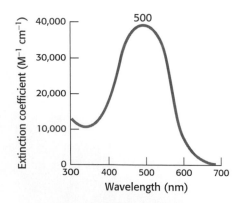

Figure 32.21 Rhodopsin absorption spectrum. Almost all photons with wavelengths near 500 nm that strike a rhodopsin molecule are absorbed.

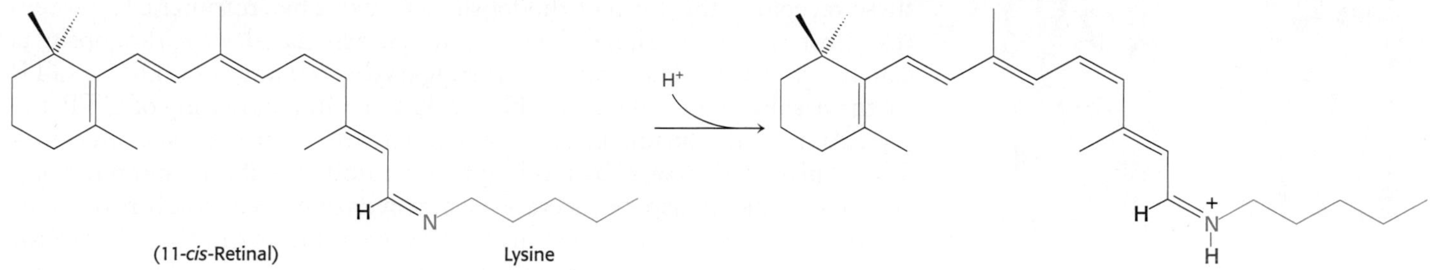

Schiff base

Protonated Schiff base

H⁺

(11-*cis*-Retinal)   Lysine

Figure 32.22 Retinal–lysine linkage. Retinal is linked to lysine 296 in opsin by a Schiff-base linkage. In the resting state of rhodopsin, this Schiff base is protonated.

lengths. Thus, *the 500-nm absorption maximum for rhodopsin strongly suggests that the Schiff base is protonated;* additional interactions with opsin shift the absorption maximum farther toward the red. The positive charge of the protonated Schiff base is compensated by the negative charge of glutamate 113 located in helix 2; the glutamate residue closely approaches the lysine–retinal linkage in the three-dimensional structure of rhodopsin.

Light Absorption Induces a Specific Isomerization of Bound 11-*cis*-Retinal

How does the absorption of light by the retinal Schiff base generate a signal? George Wald and his coworkers discovered that *light absorption results in the isomerization of the 11-cis-retinal group of rhodopsin to its all-trans form* (Figure 32.23). This isomerization causes the Schiff-base nitrogen atom to move approximately 5 Å, assuming that the cyclohexane ring of the retinal group remains fixed. In essence, *the light energy of a photon is converted into atomic motion.* The change in atomic positions, like the binding of a ligand to other 7TM receptors, sets in train a series of events that lead to the closing of ion channels and the generation of a nerve impulse.

The isomerization of the retinal Schiff base takes place within a few picoseconds of a photon being absorbed. The initial product, termed *bathorhodopsin,* contains a strained all-*trans*-retinal group. Within approximately 1 ms, this intermediate is converted through several additional intermediates into *metarhodopsin II.* In metarhodopsin II, the Schiff base is deprotonated and the opsin protein has undergone significant reorganization.

Metarhodopsin II (also referred to as R*) is analogous to the ligand-bound state of 7TM receptors such as the β₂-adrenergic receptor (Section 14.1) and

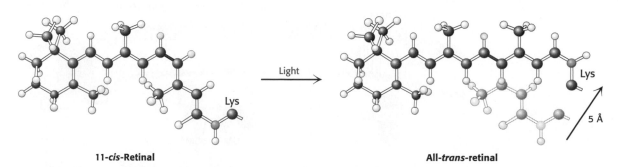

Light

Lys

Lys

5 Å

11-*cis*-Retinal   All-*trans*-retinal

Figure 32.23 Atomic motion in retinal. The Schiff-base nitrogen atom moves 5 Å as a consequence of the light-induced isomerization of 11-*cis*-retinal to all-*trans*-retinal by rotation about the bond shown in red.

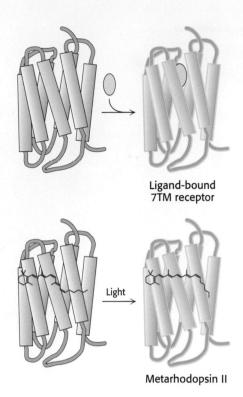

Figure 32.24 Analogous 7TM receptors. The conversion of rhodopsin into metarhodopsin II activates a signal-transduction pathway analogously to the activation induced by the binding of other 7TM receptors to appropriate ligands.

Ligand-bound 7TM receptor

Light

Metarhodopsin II

the odorant and tastant receptors discussed previously (Figure 32.24). Like these receptors, this form of rhodopsin activates a heterotrimeric G protein that propagates the signal. The G protein associated with rhodopsin is called *transducin.* Metarhodopsin II triggers the exchange of GDP for GTP by the α subunit of transducin (Figure 32.25). On the binding of GTP, the βγ subunits of transducin are released and the α subunit switches on a *cGMP phosphodiesterase* by binding to an inhibitory subunit and removing it. The activated phosphodiesterase is a potent enzyme that rapidly hydrolyzes cGMP to GMP. The reduction in cGMP concentration causes *cGMP-gated ion channels* to close, leading to the hyperpolarization of the membrane and neuronal signaling. *At each step in this process, the initial signal—the absorption of a single photon—is amplified so that it leads to sufficient membrane hyperpolarization to result in signaling.*

Light-Induced Lowering of the Calcium Level Coordinates Recovery

As we have seen, the visual system responds to changes in light and color within a few milliseconds, quickly enough that we are able to perceive continuous motion at nearly 1000 frames per second. To achieve a rapid response, the signal must also be terminated rapidly and the system must be returned to its initial state. First, activated rhodopsin must be blocked from continuing to activate transducin. *Rhodopsin kinase* catalyzes the phosphorylation of the carboxyl terminus of R* at multiple serine and threonine residues. *Arrestin,* an inhibitory protein (p. 388), then binds phosphorylated R* and prevents additional interaction with transducin.

Second, the α subunit of transducin must be returned to its inactive state to prevent further signaling. Like other G proteins, the α subunit possesses built-in GTPase activity that hydrolyzes bound GTP to GDP. Hydrolysis takes place in less than a second when transducin is bound to the phosphodiesterase. The GDP form of transducin then leaves the phosphodiesterase and reassociates with the βγ subunits, and the phosphodiesterase returns to its inactive state. Third, the level of cGMP must be raised to reopen the cGMP-gated ion channels. *The action of guanylate cyclase accomplishes this third step by synthesizing cGMP from GTP.*

Calcium ion plays an essential role in controlling guanylate cyclase because it markedly inhibits the activity of the enzyme. In the dark, Ca^{2+} as well as Na^+ enter the rod outer segment through the cGMP-gated channels. Calcium ion influx is balanced by its efflux through an exchanger, a

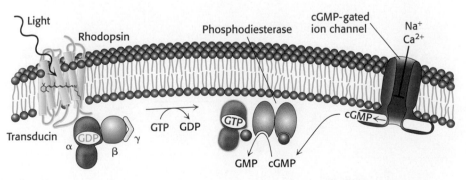

Figure 32.25 Visual signal transduction. The light-induced activation of rhodopsin leads to the hydrolysis of cGMP, which in turn leads to ion-channel closing and the initiation of an action potential.

transport system that uses the thermodynamically favorable flow of four Na^+ ions into the cell and one K^+ ion out of the cell to extrude one Ca^{2+} ion. After illumination, the entry of Ca^{2+} through the cGMP-gated channels stops, but its export through the exchanger continues. Thus, the cytoplasmic Ca^{2+} level drops from 500 nM to 50 nM after illumination. This drop markedly stimulates guanylate cyclase, rapidly restoring the concentration of cGMP to reopen the cGMP-gated channels.

$$\underline{\text{Activation}} \qquad\qquad \underline{\text{Recovery}}$$

$$[\text{cGMP}]\downarrow \longrightarrow \begin{matrix}\text{Ion} \\ \text{channels} \\ \text{closed}\end{matrix} \longrightarrow [Ca^{2+}]\downarrow \longrightarrow \begin{matrix}\text{Guanylate} \\ \text{cyclase} \\ \text{activity} \\ \text{increased}\end{matrix} \longrightarrow [\text{cGMP}]\uparrow$$

By controlling the rate of cGMP synthesis, Ca^{2+} levels govern the speed with which the system is restored to its initial state.

Color Vision Is Mediated by Three Cone Receptors That Are Homologs of Rhodopsin

Cone cells, like rod cells, contain visual pigments. Like rhodopsin, these photoreceptor proteins are members of the 7TM-receptor family and use 11-*cis*-retinal as their chromophore. In human cone cells, there are three distinct photoreceptor proteins with absorption maxima at 426, 530, and ~560 nm (Figure 32.26). *These absorbances correspond to (in fact, define) the blue, green, and red regions of the spectrum.* Recall that the absorption maximum for rhodopsin is 500 nm.

The amino acid sequences of the cone photoreceptors have been compared with one another and with rhodopsin. The result is striking. Each of the cone photoreceptors is approximately 40% identical in sequence with rhodopsin. Similarly, the blue photoreceptor is 40% identical with each of the green and red photoreceptors. The green and red photoreceptors, however, are > 95% identical with one another, differing in only 15 of 364 positions (Figure 32.27).

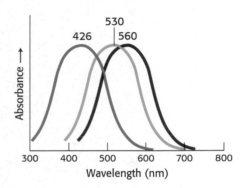

Figure 32.26 Cone-pigment absorption spectra. The absorption spectra of the cone visual pigment responsible for color vision.

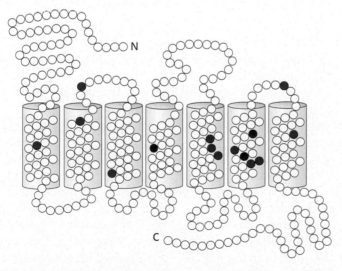

Figure 32.27 Comparison of the amino acid sequences of the green and red photoreceptors. Open circles correspond to identical residues, whereas colored circles mark residues that are different. The differences in the three black positions are responsible for most of the difference in their absorption spectra.

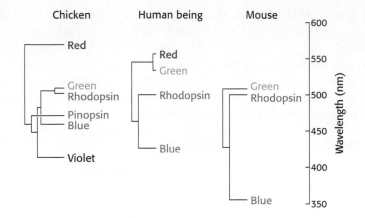

Chicken   Human being   Mouse

Figure 32.28 Evolutionary relationships among visual pigments. Visual pigments have evolved by gene duplication along different branches of the animal evolutionary tree. The branch lengths of the "trees" correspond to the percentage of amino acid divergence. [After J. Nathans. *Neuron* 24(1999):299–312; by permission of Cell Press.]

These observations are sources of insight into photoreceptor evolution. First, the green and red photoreceptors are clearly products of a recent evolutionary event (Figure 32.28). The green and red pigments appear to have diverged in the primate lineage approximately 35 million years ago. Mammals, such as dogs and mice, that diverged from primates earlier have only two cone photoreceptors, blue and green. They are not sensitive to light as far toward the infrared region as we are, and they do not discriminate colors as well. In contrast, birds such as chickens have a total of six pigments: rhodopsin, four cone pigments, and a pineal visual pigment called *pinopsin*. Birds have highly acute color perception.

Second, the high level of similarity between the green and red pigments has made it possible to identify the specific amino acid residues that are responsible for spectral tuning. Three residues (at positions 180, 277, and 285) are responsible for most of the difference between the green and the red pigments. In the green pigment, these residues are alanine, phenylalanine, and alanine, respectively; in the red pigment, they are serine, tyrosine, and threonine. A hydroxyl group has been added to each amino acid in the red pigment. The hydroxyl groups can interact with the photoexcited state of retinal and lower its energy, leading to a shift toward the lower-energy (red) region of the spectrum.

Rearrangements in the Genes for the Green and Red Pigments Lead to "Color Blindness"

The genes for the green and red pigments lie adjacent to each other on the human X chromosome. These genes are more than 98% identical in nucleotide sequence, including introns and untranslated regions as well as the protein-coding region. Regions with such high similarity are very susceptible to unequal homologous recombination.

Recombination can take place either between or within transcribed regions of the gene (Figure 32.29). If recombination takes place between transcribed regions, the product chromosomes will differ in the number of pigment genes that they carry. One chromosome will lose a gene and thus may lack the gene for, say, the green pigment; the other chromosome will gain a gene. Consistent with this scenario, approximately 2% of human X chromosomes carry only a single color pigment gene, approximately 20% carry two, 50% carry three, 20% carry four, and 5% carry five or more. A person lacking the gene for the green pigment will have trouble distinguishing red and green color, characteristic of the most common form of color blindness. Approximately 5% of males have this form of color blindness. Recombination can also take place within the transcription units, resulting in genes that encode hybrids of the green and red photoreceptors. The absorption maximum

> **Homologous recombination**
>
> The exchange of DNA segments at equivalent positions between chromosomes with substantial sequence similarity.

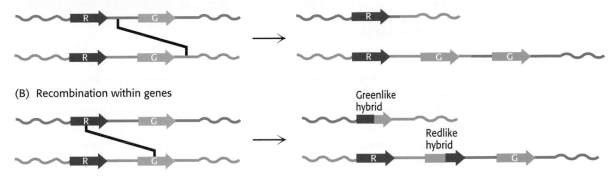

(A) Recombination between genes

(B) Recombination within genes

Greenlike hybrid

Redlike hybrid

Figure 32.29 Recombination pathways leading to color blindness. Rearrangements in the course of DNA replication may lead to (A) the loss of visual pigment genes or (B) the formation of hybrid pigment genes that encode photoreceptors with anomolous absorption spectra. Because the amino acids most important for determining absorption spectra are in the carboxyl-terminal half of each photoreceptor protein, the part of the gene that encodes this region most strongly affects the absorption characteristics of hybrid receptors. [After J. Nathans. *Neuron* 24(1999):299–312; by permission of Cell Press.]

of such a hybrid lies between that of the red and green pigments. A person with such hybrid genes who also lacks either a functional red or a functional green pigment gene does not discriminate color well.

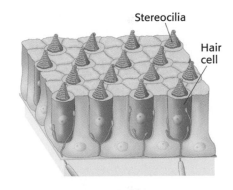

Figure 32.30 Hair cells, the sensory neurons crucial for hearing. These specialized neurons are capped with hairlike projection called stereocilia that are responsible for detecting very subtle vibrations. [After A. J. Hudspeth. *Nature* 341(1989):397–404.]

32.4 Hearing Depends on the Speedy Detection of Mechanical Stimuli

Hearing and touch are based on the detection of mechanical stimuli. Although the proteins of these senses have not been as well characterized as those of the senses already discussed, anatomical, physiological, and biophysical studies have elucidated the fundamental processes. *A major clue to the mechanism of hearing is its speed.* We hear frequencies ranging from 200 to 20,000 Hz (cycles per second), corresponding to times of 5 to 0.05 ms. Furthermore, our ability to locate sound sources, one of the most important functions of hearing, depends on the ability to detect the time delay between the arrival of a sound at one ear and its arrival at the other. Given the separation of our ears and the speed of sound, we must be able to accurately sense time differences of 0.7 ms. In fact, human beings can locate sound sources associated with temporal delays as short as 0.02 ms. This high time resolution implies that hearing must employ direct transduction mechanisms that do not depend on second messengers. Recall that, in vision, for which speed also is important, the signal-transduction processes take place in milliseconds.

Hair Cells Use a Connected Bundle of Stereocilia to Detect Tiny Motions

Sound waves are detected inside the cochlea of the inner ear. The *cochlea* is a fluid-filled, membranous sac that is coiled like a snail shell. The primary detection is accomplished by specialized neurons inside the cochlea called *hair cells* (Figure 32.30). Each cochlea contains approximately 16,000 hair cells, and each hair cell contains a hexagonally shaped bundle of 20 to 300 hairlike projections called *stereocilia* (Figure 32.31). These stereocilia are graded in length across the bundle. Mechanical deflection of the hair bundle, as occurs

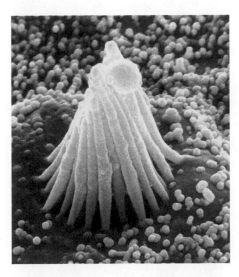

Figure 32.31 An electron micrograph of a hair bundle. [Courtesy of Dr. A. Jacobs and Dr. A. J. Hudspeth.]

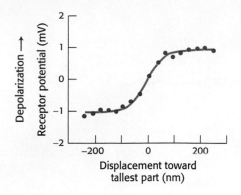

Figure 32.32 Micromanipulation of a hair cell. Movement toward the tallest part of the bundle depolarizes the cell as measured by the microelectrode. Movement toward the shortest part hyperpolarizes the cell. Lateral movement has no effect. [After A. J. Hudspeth. *Nature* 341(1989):397–404.]

when a sound wave arrives at the ear, creates a change in the membrane potential of the hair cell.

Micromanipulation experiments have directly probed the connection between mechanical stimulation and membrane potential. Displacement toward the direction of the tallest part of the hair bundle results in the depolarization of the hair cell, whereas displacement in the opposite direction results in the hyperpolarization (Figure 32.32). Motion perpendicular to the hair-length gradient does not produce any change in resting potential. Remarkably, *displacement of the hair bundle by as little as 3 Å (0.3 nm) results in a measurable (and functionally important) change in membrane potential.* This motion of 0.003 degree corresponds to a 1-inch movement of the top of the Empire State Building.

Figure 32.33 Electron micrograph of tip links. The tip link between two hair fibers is marked by an arrow. [Courtesy of Dr. A. Jacobs and Dr. A. J. Hudspeth.]

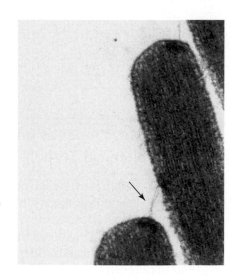

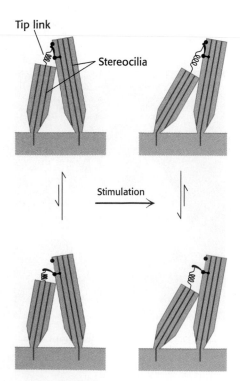

How does the motion of the hair bundle create a change in membrane potential? The rapid response, within microseconds, suggests that the movement of the hair bundle acts on ion channels directly. An important observation is that adjacent stereocilia are linked by individual filaments called *tip links* (Figure 32.33).

The presence of these tip links suggests a simple mechanical model for transduction by hair cells (Figure 32.34). The tip links are coupled to ion channels in the membranes of the stereocilia that are gated by mechanical stress. In the absence of a stimulus, approximately 15% of these channels are open. When the hair bundle is displaced toward its tallest part, the stereocilia slide across one another and the tension on the tip links increases, causing additional channels to open. The flow of ions through the newly opened channels depolarizes the membrane. Conversely, if the displacement is in the opposite direction, the tension on the tip links decreases, the

Figure 32.34 Model for hair-cell transduction. When the hair bundle is tipped toward the tallest part, the tip link pulls on an ion channel and opens it. Movement in the opposite direction relaxes the tension in the tip link, increasing the probability that any open channels will close. [After A. J. Hudspeth. *Nature* 341(1989):397–404.]

open channels close, and the membrane hyperpolarizes. *Thus, the mechanical motion of the hair bundle is directly converted into current flow across the hair-cell membrane.*

Mechanosensory Channels Have Been Identified in *Drosophila* and Vertebrates

The search for ion channels that respond to mechanical impulses has been pursued in a variety of organisms. *Drosophila* have sensory bristles used for detecting small air currents. These bristles respond to mechanical displacement in ways similar to those of hair cells; displacement of a bristle in one direction leads to substantial transmembrane current. Strains of mutant fruit flies that show uncoordinated motion and clumsiness have been examined for their electrophysiological responses to displacement of the sensory bristles. In one set of strains, transmembrane currents were dramatically reduced. The mutated gene in these strains was found to encode a protein of 1619 amino acids, called NompC for *no m*echanoreceptor *p*otential.

The carboxyl-terminal 469 amino acids of NompC resemble a class of ion channel proteins called TRP (*t*ransient *r*eceptor *p*otential) channels. This region includes six putative transmembrane helices with a pore-like region between the fifth and sixth helices. The amino-terminal 1150 amino acids consist almost exclusively of 29 *ankyrin repeats* (Figure 32.35). Ankyrin repeats are structural motifs consisting of a hairpin loop followed by a helix-turn-helix. Importantly, in other proteins, regions with tandem arrays of these motifs mediate protein–protein interactions, suggesting that these arrays couple the motions of other proteins to the activity of the NompC channel.

Recently, a strong candidate for at least one component of the mechanosensory channel involved in hearing has been identified. The protein, TRPA1, is also a member of the TRP channel family. The sequence of TRPA1 also includes 17 ankyrin repeats. TRPA1 is expressed in hair cells, particularly near their tips. Based on these and other studies, it appears very likely that TRPA1 represents at least one component of the mechanosensory channel that is central to hearing. Further studies are under way to confirm and extend this exciting discovery.

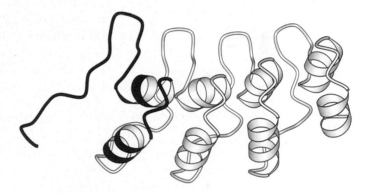

Figure 32.35 Ankyrin repeat structure. One ankyrin domain is shown in red in this series of four ankyrin repeats. *Notice* the hairpin loop followed by a helix-turn-helix motif in the red-colored ankyrin unit. Ankyrin domains interact with other proteins, primarily through their loops. [Drawn from 1AWC.pdb.]

32.5 Touch Includes the Sensing of Pressure, Temperature, and Other Factors

Like taste, touch is a combination of sensory systems that are expressed in a common organ—in this case, the skin. The detection of pressure and the detection of temperature are two key components. Amiloride-sensitive Na^+ channels, homologous to those of taste, appear to play a role. Other systems are responsible for detecting painful stimuli such as high temperature, acid, or certain specific chemicals. Although our understanding of this sensory system is not as advanced as that of the other sensory systems, recent work has revealed a fascinating relation between pain and taste sensation, a relation well known to anyone who has eaten "spicy" food.

Studies of Capsaicin Reveal a Receptor for Sensing High Temperatures and Other Painful Stimuli

Our sense of touch is intimately connected with the sensation of pain. Specialized neurons, termed *nociceptors,* transmit signals from skin to pain-processing centers in the spinal cord and brain in response to the onset of tissue damage. What is the molecular basis for the sensation of pain? An intriguing clue came from the realization that *capsaicin,* the chemical responsible for the "hot" taste of spicy food, activates nociceptors.

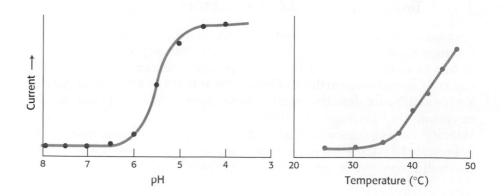

Capsaicin

Early research suggested that capsaicin would act by opening ion channels that are expressed in nociceptors. Thus, a cell that expresses the *capsaicin receptor* should take up calcium on treatment with the molecule. This insight led to the isolation of the capsaicin receptor with the use of cDNA from cells expressing this receptor. Such cells had been detected by their fluorescence when loaded with the calcium-sensitive compound Fura-2 and then treated with capsaicin or related molecules. Cells expressing the capsaicin receptor, which is called VR1 (for *vanilloid receptor* 1), respond to capsaicin below a concentration of 1 µM. The deduced 838-residue sequence of VR1 revealed it to be a member of the TRP channel family (Figure 32.36). The amino-terminal region of VR1 includes three ankyrin repeats.

Currents through VR1 are also induced by temperatures above 40°C and by exposure to dilute acid, with a midpoint for activation at pH 5.4 (Figure 32.37). Temperatures and acidity in these ranges are associated with infection and cell injury. The responses to capsaicin, temperature, and acidity are not independent. The response to heat is greater at lower pH, for example. Thus, *VR1 acts to integrate several noxious stimuli.* We feel these responses as pain and act to prevent the potentially destructive conditions that cause the unpleasant sensation. Mice that do not express VR1 suggest that this is the case; such mice do not mind food containing high concentrations of capsaicin and are, indeed, less responsive than control mice to normally noxious heat. Plants such as chili peppers presumably gained the ability to synthesize capsaicin and other "hot" compounds to protect themselves from being consumed by mammals. Birds, which play the beneficial role of spreading pepper seeds into new territory, do not appear to respond to capsaicin.

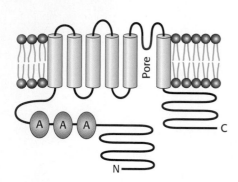

Figure 32.36 The membrane topology deduced for VR1, the capsaicin receptor. The proposed site of the membrane pore is indicated in red, and the three ankyrin (A) repeats are shown in orange. The active receptor comprises four of these subunits. [After M. J., Caterina, M. A., Schumacher, M. Tominaga, A. Rosen, J. D. Levine, and D. Julius. *Nature* 389(1997):816–824.]

Figure 32.37 Response of the capsaicin receptor to pH and temperature. The ability of this receptor to respond to acid and to increased temperature helps detect potentially noxious situations. [After M. Tominaga, M. J. Caterina, A. B. Malmberg, T. A. Rosen, H. Gilbert, B. Skinner, B. E. Raumann, A. I. Basbaum, and D. Julius, *Neuron* 21(1998):531–543.]

Because of its ability to simulate VR1, capsaicin is used in pain management for arthritis, neuralgia, and other neuropathies. How can a compound that induces pain assist in its alleviation? Chronic exposure to capsaicin overstimulates pain-transmitting neurons, leading to their desensitization.

More Sensory Systems Remain to Be Studied

There may exist other subtle senses that are able to detect environmental signals that then influence our behavior. The biochemical basis of these senses is now under investigation. One such sense is our ability to respond, often without our awareness, to chemical signals called pheromones, released by other persons. Another is our sense of time, manifested in our daily (circadian) rhythms of activity and restfulness. Daily changes in light exposure strongly influence these rhythms. The foundations for these senses have been uncovered in other organisms; future studies should reveal to what extent these mechanisms apply to human beings as well.

Summary

Smell, taste, vision, hearing, and touch are based on signal-transduction pathways activated by signals from the environment. These sensory systems function similarly to the signal-transduction pathways for many hormones. These intercellular signaling pathways appear to have been appropriated and modified to process environmental information.

32.1 A Wide Variety of Organic Compounds Are Detected by Olfaction

The sense of smell, or olfaction, is remarkable in its specificity; it can, for example, discern stereoisomers of small organic compounds as distinct aromas. The 7TM receptors that detect these odorants operate in conjunction with $G_{(olf)}$, a G protein that activates a cAMP cascade resulting in the opening of an ion channel and the generation of a nerve impulse. An outstanding feature of the olfactory system is its ability to detect a vast array of odorants. Each olfactory neuron expresses only one type of receptor and connects to a particular region of the olfactory bulb. Odors are decoded by a combinatorial mechanism: each odorant activates a number of receptors, each to a different extent, and most receptors are activated by more than one odorant.

32.2 Taste Is a Combination of Senses That Function by Different Mechanisms

We can detect only five tastes: bitter, sweet, salt, sour, and umami. The transduction pathways that detect taste are, however, diverse. Bitter, sweet, and umami tastants are experienced through 7TM receptors acting through a special G protein called gustducin. Salty and sour tastants act directly through membrane channels. Salty tastants are detected by passage though Na^+ channels, whereas sour taste results from the effects of hydrogen ions on a number of types of channels. The end point is the same in all cases—membrane polarization that results in the transmission of a nerve impulse.

32.3 Photoreceptor Molecules in the Eye Detect Visible Light

Vision is perhaps the best understood of the senses. Two classes of photoreceptor cells exist: cones, which respond to bright lights and colors, and rods, which respond only to dim light. The photoreceptor in rods is rhodopsin, a 7TM receptor that is a complex of the protein opsin and the chromophore 11-*cis*-retinal. The absorption of light by

11-*cis*-retinal changes its structure into that of all-*trans*-retinal, setting in motion a signal-transduction pathway that leads to the breakdown of cGMP, to membrane hyperpolarization, and to a subsequent nerve impulse. Color vision is mediated by three distinct 7TM photoreceptors that employ 11-*cis*-retinal as a chromophore and absorb light in the blue, green, and red parts of the spectrum.

32.4 Hearing Depends on the Speedy Detection of Mechanical Stimuli
The immediate receptors for hearing are found in the hair cells of the cochleae, which contain bundles of stereocilia. When the stereocilia move in response to sound waves, cation channels will open or close, depending on the direction of movement. The mechanical motion of the cilia is converted into current flow and then into a nerve impulse.

32.5 Touch Includes the Sensing of Pressure, Temperature, and Other Factors
Touch, detected by the skin, senses pressure, temperature, and pain. Specialized nerve cells called nociceptors transmit signals that are interpreted in the brain as pain. A receptor responsible for the perception of pain has been isolated on the basis of its ability to bind capsaicin, the molecule responsible for the hot taste of spicy food. The capsaicin receptor, also called VR1, functions as a cation channel that initiates a nerve impulse.

Key Terms

main olfactory epithelium (p. 923)

$G_{(olf)}$ (p. 923)

functional magnetic resonance imaging (fMRI) (p. 926)

gustducin (p. 926)

amiloride-sensitive Na^+ channel (p. 930)

rod (p. 931)

cone (p. 931)

rhodopsin (p. 932)

opsin (p. 932)

retinal (p. 932)

chromophore (p. 932)

transducin (p. 934)

cGMP phosphodiesterase (p. 934)

cGMP-gated Ca^{2+} channel (p. 934)

rhodopsin kinase (p. 934)

arrestin (p. 934)

guanylate cyclase (p. 934)

hair cell (p. 937)

stereocilium (p. 937)

tip link (p. 938)

nociceptor (p. 940)

capsaicin receptor (VR1 receptor) (p. 940)

Selected Readings

Where to Start
Axel, R. 1995. The molecular logic of smell. *Sci. Am.* 273(4):154–159.

Dulac, C. 2000. The physiology of taste, vintage 2000. *Cell* 100:607–610.

Zhao, G. Q., Zhang, Y., Hoon, M. A., Chandrashekar, J., Erlenbach, I., Ryba, N. J. P., and Zuker, C. S. 2003. The receptors for mammalian sweet and umami taste. *Cell* 115:255–266.

Stryer, L. 1996. Vision: From photon to perception. *Proc. Natl. Acad. Sci. U. S. A.* 93:557–559.

Hudspeth, A. J. 1989. How the ear's works work. *Nature* 341:397–404.

Olfaction
Buck, L., and Axel, R.1991. A novel multigene family may encode odorant receptors: A molecular basis for odor recognition. *Cell* 65:175–187.

Malnic, B., Hirono, J., Sato, T., and Buck, L. B. 1999. Combinatorial receptor codes for odors. *Cell* 96:713–723.

Mombaerts, P., Wang, F., Dulac, C., Chao, S. K., Nemes, A., Mendelsohn, M., Edmondson, J., and Axel, R. 1996. Visualizing an olfactory sensory map. *Cell* 87:675–686.

Buck, L. 2005. Unraveling the sense of smell (Nobel lecture). *Angew. Chem. Int. Ed. Engl.* 44: 6128–6140.

Belluscio, L., Gold, G. H., Nemes, A., and Axel, R. 1998. Mice deficient in $G_{(olf)}$ are anosmic. *Neuron* 20:69–81.

Vosshall, L. B., Wong, A. M., and Axel, R. 2000. An olfactory sensory map in the fly brain. *Cell* 102:147–159.

Lewcock, J. W., and Reed, R. R. 2003. A feedback mechanism regulates monoallelic odorant receptor expression. *Proc. Natl. Acad. Sci. U. S. A.*101:1069–1074.

Reed, R. R. 2004. After the holy grail: Establishing a molecular mechanism for mammalian olfaction. *Cell* 116:329–336.

Taste
Herness, M. S., and Gilbertson, T. A. 1999. Cellular mechanisms of taste transduction. *Annu. Rev. Physiol.* 61:873–900.

Adler, E., Hoon, M. A., Mueller, K. L., Chandrashekar, J., Ryba, N. J., and Zuker, C. S. 2000. A novel family of mammalian taste receptors. *Cell* 100:693–702.

Chandrashekar, J., Mueller, K. L., Hoon, M. A., Adler, E., Feng, L., Guo, W., Zuker, C. S., and Ryba, N. J. 2000. T2Rs function as bitter taste receptors. *Cell* 100:703–711.

Mano, I., and Driscoll, M. 1999. DEG/ENaC channels: A touchy superfamily that watches its salt. *Bioessays* 21:568–578.

Benos, D. J., and Stanton, B. A. 1999. Functional domains within the degenerin/epithelial sodium channel (Deg/ENaC) superfamily of ion channels. *J. Physiol. (Lond.)* 520(part 3):631–644.

McLaughlin, S. K., McKinnon, P. J., and Margolskee, R. F. 1992. Gustducin is a taste-cell-specific G protein closely related to the transducins. *Nature* 357:563–569.

Nelson, G., Hoon, M. A., Chandrashekar, J., Zhang, Y., Ryba, N. J., and Zuker, C. S. 2001. Mammalian sweet taste receptors. *Cell* 106:381–390.

Vision

Stryer, L. 1988. Molecular basis of visual excitation. *Cold Spring Harbor Symp. Quant. Biol.* 53:283–294.

Wald, G. 1968. The molecular basis of visual excitation. *Nature* 219:800–807.

Ames, J. B., Dizhoor, A. M., Ikura, M., Palczewski, K., and Stryer, L. 1999. Three-dimensional structure of guanylyl cyclase activating protein-2, a calcium-sensitive modulator of photoreceptor guanylyl cyclases. *J. Biol. Chem.* 274:19329–19337.

Nathans, J. 1994. In the eye of the beholder: Visual pigments and inherited variation in human vision. *Cell* 78:357–360.

Nathans, J. 1999. The evolution and physiology of human color vision: Insights from molecular genetic studies of visual pigments. *Neuron* 24:299–312.

Palczewski, K., Kumasaka, T., Hori, T., Behnke, C. A., Motoshima, H., Fox, B. A., LeTrong, I., Teller, D. C., Okada, T., Stenkamp, R. E., Yamamoto, M., and Miyano, M. 2000. Crystal structure of rhodopsin: A G protein-coupled receptor. *Science* 289:739–745.

Filipek, S, Teller, D. C., Palczewski, K., and Stemkamp, R. 2003. The crystallographic model of rhodopsin and its use in studies of other G protein-coupled receptors. *Annu. Rev. Biophys. Biomol. Struct.* 32:375–397.

Hearing

Hudspeth, A. J. 1997. How hearing happens. *Neuron* 19:947–950.

Pickles, J. O., and Corey, D. P. 1992. Mechanoelectrical transduction by hair cells. *Trends Neurosci.* 15:254–259.

Walker, R. G., Willingham, A. T., and Zuker, C. S. 2000. A *Drosophila* mechanosensory transduction channel. *Science* 287:2229–2234.

Hudspeth, A. J., Choe, Y., Mehta, A. D., and Martin, P. 2000. Putting ion channels to work: Mechanoelectrical transduction, adaptation, and amplification by hair cells. *Proc. Natl. Acad. Sci. U. S. A.* 97:11765–11772.

Touch and Pain Reception

Franco-Obregon, A., and Clapham, D. E. 1998. Touch channels sense blood pressure. *Neuron* 21:1224–1226.

Caterina, M. J., Schumacher, M. A., Tominaga, M., Rosen, T. A., Levine, J. D., and Julius, D. 1997. The capsaicin receptor: A heat-activated ion channel in the pain pathway. *Nature* 389:816–824.

Tominaga, M., Caterina, M. J., Malmberg, A. B., Rosen, T. A., Gilbert, H., Skinner, K., Raumann, B. E., Basbaum, A. I., and Julius, D. 1998. The cloned capsaicin receptor integrates multiple pain-producing stimuli. *Neuron* 21:531–543.

Caterina, M. J., and Julius, D. 1999. Sense and specificity: A molecular identity for nociceptors. *Curr. Opin. Neurobiol.* 9:525–530.

Clapham, D. E. 2003. TRP channels as cellular sensors. *Nature* 426:517–524.

Problems

1. *Mice and rats.* As noted on page 924, one of the first odorant receptors to be matched with its ligand was a rat receptor that responded best to *n*-octanal. The sequence of the corresponding mouse receptor differed from the rat receptor at 15 positions. Surprisingly, the mouse receptor was found to respond best to *n*-heptanal rather than *n*-octanal. The substitution of isoleucine at position 206 in the mouse for valine at this position in the rat receptor was found to be important in determining the specificity for *n*-heptanal. Propose an explanation.

2. *Olfaction in worms.* Unlike the olfactory neurons in the mammalian systems discussed herein, olfactory neurons in the nematode *C. elegans* express multiple olfactory receptors. In particular, one neuron (called AWA) expresses receptors for compounds to which the nematode is attracted, whereas a different neuron (called AWB) expresses receptors for compounds that the nematode avoids. Suppose that a transgenic nematode is generated such that one of the receptors for an attractant is expressed in AWB rather than AWA. What behavior would you expect in the presence of the corresponding attractant?

3. *Odorant matching.* A mixture of two of the compounds illustrated in Figure 32.6 is applied to a section of olfactory epithelium. Only receptors 3, 5, 9, 12, and 13 are activated, according to Figure 32.7. Identify the likely compounds in the mixture.

4. *Timing.* Compare the aspects of taste (bitter, sweet, salty, sour) in regard to their potential for rapid time resolution.

5. *Two ears.* Our ability to determine the direction from which a sound is coming is partly based on the difference in time at which our two ears detect the sound. Given the speed of sound (350 m s^{-1}) and the separation between our ears (0.15 m), what difference is expected in the times at which a sound arrives at our two ears? How does this difference compare with the time resolution of the human hearing system? Would a sensory system that utilized 7TM receptors and G proteins be capable of adequate time resolution?

6. *Constitutive mutants.* What effect within the olfactory system would you expect for a mutant in which adenylate cyclase is always fully active? What effect within the visual system would you expect for a mutant in which guanylate cyclase is always fully active?

7. *Bottle choice.* A widely used method for quantitatively monitoring rodent behavior with regard to taste is the bottle-choice assay. An animal is placed in a cage with two water bottles, one of which contains a potential tastant. After a fixed period of time (24–48 hours), the amount of water remaining in each bottle is measured. Suppose that much less water remains in the bottle containing the tastant after 48 hours. Do you suspect the tastant to be sweet or bitter?

8. *It's better to be bitter.* Some nontoxic plants taste very bitter to us. Suggest one or more explanations.

9. *Of mice and men.* In human beings, the umami taste is triggered only by glutamate and aspartate. In contrast, mice respond to many more amino acids. Design an experiment to test which of the subunits (T1R1 or T1R3) determines the specificity of this response. Assume that all desired mouse strains can be readily produced.

Chapter Integration Problem

10. *Energy and information.* The transmission of sensory information requires the input of free energy. For each sensory system (olfaction, gustation, vision, hearing, and touch), identify mechanisms for the input of free energy that allow the transmission of sensory information.

Mechanism Problem

11. *Schiff-base formation.* Propose a mechanism for the reaction between opsin and 11-*cis*-retinal.

The Immune System

[(Left) The Granger Collection.]

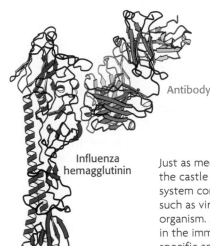

Antibody

Influenza hemagglutinin

Just as medieval defenders used their weapons and the castle walls to defend their city, the immune system constantly battles against foreign invaders such as viruses, bacteria, and parasites to defend the organism. Antibody molecules provide a key element in the immune system's defensive arsenal. For example, specific antibodies can bind to molecules on the surfaces of viruses and prevent the viruses from infecting cells. Above right, an antibody binds to one subunit on hemagglutinin from the surface of influenza virus. [(Left) The Granger Collection.]

We are constantly exposed to an incredible diversity of bacteria, viruses, and parasites, many of which would flourish in our cells or extracellular fluids were it not for our immune system. How does the immune system protect us? The human body has two lines of defense: an *innate immune system* that responds rapidly to features present in many pathogens, and an *adaptive immune system* that responds to specific features present only in a given pathogen. *Both the innate and the adaptive immune systems first identify features on disease-causing organisms and then work to eliminate or neutralize those organisms.* This chapter focuses on the mechanisms of pathogen identification.

The immune system must meet two tremendous challenges in the identification of pathogens: (1) to produce a system of receptors diverse enough to recognize the diversity of potential pathogens and (2) to distinguish invaders and their disease-causing products from the body and its own products (i.e., self- versus non-self-recognition). To meet these challenges, the innate immune system evolved the ability to recognize structural elements, such as specific glycolipids or forms of nucleic acid, that are well conserved in pathogens but absent in the host organism. The repertoire of such elements is limited, however, and so some pathogens have strategies to escape detection. The adaptive immune system has the remarkable ability to produce more than 10^8 distinct antibodies and more than 10^{12} T-cell receptors (TCRs), each of which presents a different surface with

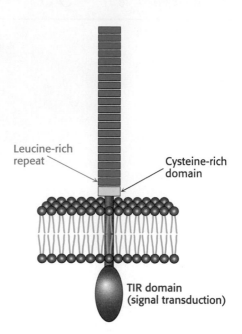

Leucine-rich repeat

Cysteine-rich domain

TIR domain (signal transduction)

Figure 33.1 Toll-like receptor. Each receptor comprises a set of 18 or more leucine-rich repeat sequences, followed by a cysteine-rich domain, a single transmembrane helix, and a TIR (Toll–interleukin 1 receptor) domain that functions in signal transduction.

Figure 33.2 Lipopolysaccharide structure. Lipopolysaccharide, a potent activator of the innate immune system, is found on the surfaces of Gram-negative bacteria. The structure is built around lipid A, a specialized lipid that has four fatty acyl chains linked to two *N*-acetylglucosamine residues. Lipid A is linked to a polysaccharide chain consisting of a core and a more variable region termed the O-specific chain.

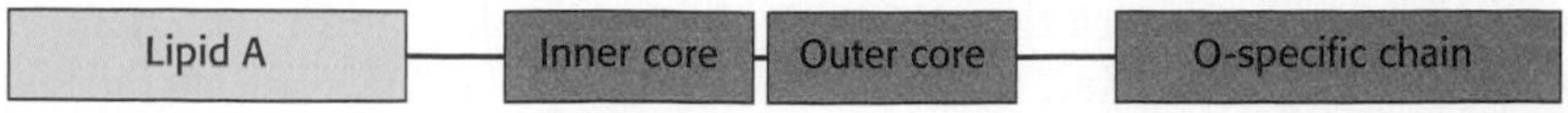

the potential to specifically bind a structure from a foreign organism. In producing this vast range of defensive molecules, however, the adaptive immune system has the potential to create antibodies and T-cells that recognize and attack cells or molecules normally present in our bodies—a situation that can result in autoimmune diseases.

This chapter will examine these challenges, focusing first on the structures of proteins that recognize foreign organisms and then on the mechanisms for protecting us from a specific pathogen once it has been recognized. The chapter will closely examine the modular construction of the proteins of the immune system—identifying structural motifs and considering how spectacular diversity can arise from modular construction.

Innate Immunity Is an Evolutionarily Ancient Defense System

Innate immunity is an evolutionarily ancient defense system found, at least in some form, in all multicellular plants and animals. The genes for its key molecules are expressed without substantial modification, unlike genes for key components of the adaptive immune system, which undergo significant rearrangement. Through many millions of years of evolution, proteins expressed by these genes have gained the ability to recognize specific features present in most pathogens and yet not respond to materials normally present in the host.

The most important and best-understood receptors in the innate immune system are the *Toll-like receptors* (TLRs). At least 10 TLRs have been identified in human beings, although only a single such receptor is present in *C. elegans*, for example. The name "toll-like" is derived from a receptor known as Toll encoded in the *Drosophila* genome; Toll was first identified in a screen for genes important for *Drosophila* development and was subsequently discovered to also play a key role in the innate immune system later in development. The TLRs have a common structure (Figure 33.1). Each receptor consists of a large domain built primarily from repeated amino acid sequences termed leucine-rich repeats (LRRs) because each repeat includes six residues that are usually leucine. The human TLRs have from 18 to 27 LRR repeats. These repeats are followed by a sequence forming a single transmembrane helix and then by a signaling domain common to the TLRs as well as to a small number of other receptors. This signaling domain is not a protein kinase but acts as a docking site for other proteins. A protein that docks to a TLR initiates a signal transduction pathway that ultimately leads to the activation of specific transcription factors. Most TLRs are expressed in the cell membrane for the detection of extracellular pathogens such as fungi and bacteria. Other TLRs are located in the membranes of internal compartments for the detection of intracellular pathogens such as viruses and some bacteria.

Each TLR is targeted to a specific molecular characteristic, often called a *pathogen-associated molecular pattern* (PAMP), found primarily on invading organisms. One particularly important PAMP is lipopolysaccharide (LPS), a specific class of glycolipids found in the cell walls of Gram-negative bacteria such as *E. coli* (Figure 33.2). LPS is built around a specialized lipid, called *lipid A,* that contains two linked *N*-acetylglucosamine residues and four fatty acyl chains. Lipid A is connected to a polysaccharide chain consisting of a core structure and a more variable region referred to as the O-specific chain. LPS is also known as *endotoxin*. The response of the innate immune system to LPS can be easily demonstrated. Injection of less than

| Lipid A | Inner core | Outer core | O-specific chain |

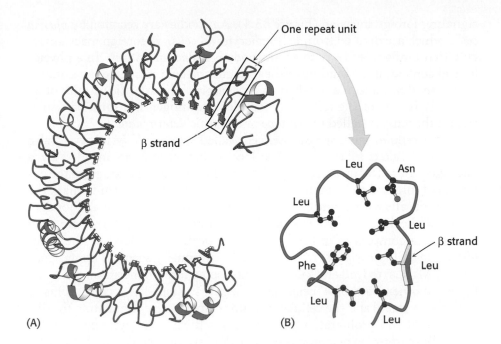

One repeat unit

β strand

(A)

Leu   Asn

Leu

Leu

β strand

Phe   Leu

Leu

Leu

(B)

Figure 33.3 **PAMP-recognition unit of the Toll-like receptor.** (A) The structure of the leucine-rich repeat (LRR) domain from human TLR-3. *Notice* that the LRR units come together to form a central parallel β sheet that curls to form a concave structure. (B) The structure of a single LRR showing the positions of the residues that are generally approximately conserved. *Notice* that the leucine residues come together to form a hydrophobic core with the single β strand along on one side. [Drawn from 1ZIW.pdb].

1 mg of LPS into a human being produces a fever and other signs of inflammation even though no living organisms are introduced.

LPS is recognized primarily by TLR-4, whereas other TLRs recognize other classes of PAMP. For example, TLR-5 recognizes the protein flagellin, found in flagellated bacteria, and TLR-3 recognizes double-stranded RNA. Note that, in each case, the target of the TLR is a key component of the pathogen, and so mutations cannot easily block recognition by the TLR and, hence, escape detection by the innate immune system. In some cases, TLRs appear to form heterodimers that either enhance or inhibit PAMP recognition.

How do TLRs recognize PAMPs? The leucine-rich repeat domain from human TLR-3 has a remarkable structure (Figure 33.3). Each of its LRR units contributes a single β strand to a large parallel β sheet that lines the inside of a concave structure. This hooklike structure immediately suggests a model for how TLRs bind PAMPs—namely, that the PAMP lies on the inside of the "hook." This model is likely accurate for some TLRs. However, for other TLRs, the PAMP-binding site appears to lie on one side of the structure, and the central hole is blocked by host carbohydrates linked to the structure.

Regardless of the details of the interaction, PAMP binding appears to lead to the formation of a specific dimer of the TLR. The cytoplasmic side of this dimer is a signaling domain that initiates the signal-transduction pathway. Because the TLRs and other components of the innate immune system are always expressed, ready to target conserved structures from pathogens, they provide the host organism with a rapid response system to resist attack by pathogens. We now turn to the adaptive immune system, which, remarkably, is able to target specific pathogens, even those that it has never encountered in the course of evolution.

The Adaptive Immune System Responds by Using the Principles of Evolution

The adaptive immune system comprises two parallel but interrelated systems: humoral and cellular immune responses. In the *humoral immune response,* soluble proteins called *antibodies (immunoglobulins)* function as recognition elements that bind to foreign molecules and serve as markers

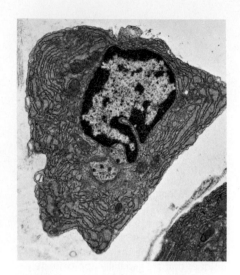

Figure 33.4 Immunoglobulin production. An electron micrograph of a plasma cell shows the highly developed rough endoplasmic reticulum necessary for antibody secretion. [Courtesy of Lynne Mercer.]

signaling foreign invasion (Figure 33.4). Antibodies are secreted by *plasma cells,* which are derived from *B lymphocytes (B cells).* A foreign macromolecule that binds selectively to an antibody is called an *antigen.* In a physiological context, if the binding of the foreign molecule stimulates an immune response, that molecule is called an *immunogen.* The specific affinity of an antibody is not for the entire macromolecular antigen but for a particular site on the antigen called the *epitope* or *antigenic determinant.*

In the *cellular immune response,* cells called *cytotoxic T lymphocytes* (also commonly called *killer T cells*) kill cells that have been invaded by a pathogen. Because intracellular pathogens do not leave markings on the exteriors of infected cells, vertebrates have evolved a mechanism to mark the exterior of cells with a sample of the interior contents, both self and foreign. Some of the internal proteins are broken into peptides, which are then bound to a complex of integral membrane proteins encoded by the *major histocompatbility complex* (MHC). T cells continually scan the bound peptides (pMHCs) to find and kill cells that display foreign motifs on their surfaces. Another class of T cells called *helper T lymphocytes* contributes to both the humoral and the cellular immune responses by stimulating the differentiation and proliferation of appropriate B cells and cytotoxic T cells. The celluar immune response is mediated by specific receptors that are expressed on the surfaces of the T cells.

The remarkable ability of the immune system to adapt to an essentially limitless set of potential pathogens requires a powerful system for transforming the immune cells and molecules present in our systems in response to the presence of pathogens. *This adaptive system operates through the principles of evolution, including reproduction with variation followed by selection of the most well suited members of a population.*

If the human genome contains, by the latest estimates, only 25,000 genes, how can the immune system generate more than 10^8 different antibody proteins and 10^{12} T-cell receptors? The answer is found in a novel mechanism for generating a highly diverse set of genes from a limited set of genetic building blocks. Linking different sets of DNA regions in a combinatorial manner produces many distinct protein-encoding genes that are not present in the genome. A rigorous selection process then leaves for proliferation only cells that synthesize proteins determined to be useful in the immune response. The subsequent reproduction of these cells without additional recombination serves to enrich the cell population with members expressing particular protein species.

Critical to the development of the immune response is the selection process, which determines which cells will reproduce. The process comprises several stages. In the early stages of the development of an immune response, cells expressing molecules that bind tightly to self-molecules are destroyed or silenced, whereas cells expressing molecules that do not bind strongly to self-molecules and that have the potential for binding strongly to foreign molecules are preserved. The appearance of an immunogenic invader at a later time will stimulate cells expressing antibodies or T-cell receptors that bind specifically to elements of that pathogen to reproduce—in evolutionary terms, such cells are selected for. Thus, the immune response is based on the selection of cells expressing molecules that are specifically effective against a particular invader; the response evolves from a population with wide-ranging specificities to a more-focused collection of cells and molecules that are well suited to defend the host when confronted with that particular challenge.

Not only are antibodies and T-cell receptors a result of genetic diversity and recombination, but antibodies have highly diverse structures as well. Antibodies require many different structural solutions for binding many different antigens, each of which has a different form. T-cell receptors, in contrast, are not structurally diverse, because they have coevolved with the MHC. The docking mode of a T-cell receptor to the peptide bound to MHC is similar for all structures. As a consequence of this coevolution, every T-cell receptor has an inherent reactivity with every MHC. The coevolution ensures that all T-cell receptors can scan all peptide–MHC complexes on all tissues. The genetic diversity of the 10^{12} different T-cell receptors is concentrated in a highly diverse set of residues in the center of the MHC groove. This localized diversity allows the T-cell receptor to recognize the many different foreign peptides bound to the MHC. T-cell receptors must survey many different MHC–peptide complexes with rapid turnover. Therefore, the binding affinities between T-cell receptors and the MHC are weaker than those between antibody and antigen.

33.1 Antibodies Possess Distinct Antigen-Binding and Effector Units

Antibodies are central molecular players in the immune response, and we examine them first. A fruitful approach in studying proteins as large as antibodies is to split the protein into fragments that retain activity. In 1959, Rodney Porter showed that *immunoglobulin G* (IgG), the major antibody in serum, can be cleaved into three 50-kd fragments by the limited proteolytic action of papain. Two of these fragments bind antigen. They are called F_{ab} (F stands for fragment, *ab* for antigen binding). The other fragment, called F_c because it crystallizes readily, does not bind antigen, but it has other important biological activities, including the mediation of responses termed *effector functions*. These functions include the initiation of the *complement cascade*, a process that leads to the lysis of target cells. Although such effector functions are crucial to the functioning of the immune system, they will not be considered further here.

How do these fragments relate to the three-dimensional structure of whole IgG molecules? Immunoglobulin G consists of two kinds of polypeptide chains, a 25-kd *light* (L) *chain* and a 50-kd *heavy* (H) *chain* (Figure 33.5). The subunit composition is L_2H_2. Each L chain is linked to an H chain by a disulfide bond, and the H chains are linked to each other by at least one disulfide bond. Examination of the amino acid sequences and three-dimensional structures of IgG molecules reveals that each L chain comprises two homologous domains, termed *immunoglobulin domains*, to be

Figure 33.5 Immunoglobulin G structure. (A) The three-dimensional structure of an IgG molecule showing the light chains in yellow and the heavy chains in blue. (B) A schematic view of an IgG molecule indicating the positions of the interchain disulfide bonds. Abbreviations: N, amino terminus; C, carboxyl terminus. [Drawn from 1IGT.pdb.]

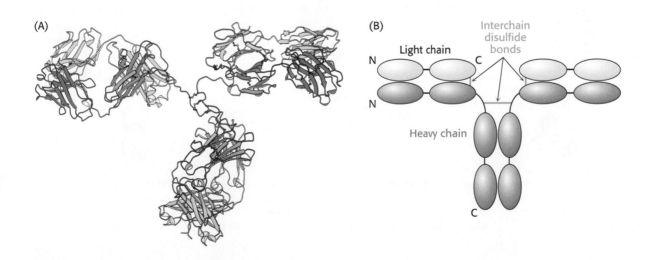

(A)

(B)

Light chain

Interchain disulfide bonds

N   C

N

Heavy chain

C

TABLE 33.1 **Properties of immunoglobulin classes**

Class	Serum concentration (mg ml^{-1})	Mass (kd)	Sedimentation coefficient(s)	Light chains	Heavy chains	Chain structure
IgG	12	150	7	κ or λ	γ	$\kappa_2\gamma_2$ or $\lambda_2\gamma_2$
IgA	3	180–500	7, 10, 13	κ or λ	α	$(\kappa_2\alpha_2)_n$ or $(\lambda_2\alpha_2)_n$
IgM	1	950	18–20	κ or λ	μ	$(\kappa_2\mu_2)_5$ or $(\lambda_2\mu_2)_5$
IgD	0.1	175	7	κ or λ	δ	$\kappa_2\delta_2$ or $\lambda_2\delta_2$
IgE	0.001	200	8	κ or λ	ε	$\kappa_2\varepsilon_2$ or $\lambda_2\varepsilon_2$

Note: n = 1, 2, or 3. IgM and oligomers of IgA also contain J chains that connect immunoglobulin molecules. IgA in secretions has an additional component.

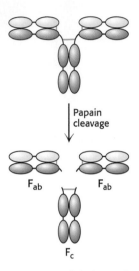

Figure 33.6 Immunoglobulin G cleavage. Treatment of intact IgG molecules with the protease papain results in the formation of three large fragments: two F$_{ab}$ fragments that retain antigen-binding capability and one F$_c$ fragment that does not.

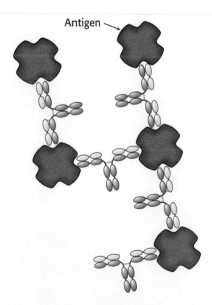

Figure 33.7 Antigen cross-linking. Because IgG molecules include two antigen-binding sites, antibodies can cross-link multivalent antigens such as viral surfaces.

described in detail in Section 33.2. Each H chain has four immunoglobulin domains. Overall, the molecule adopts a conformation that resembles the letter Y, in which the stem, corresponding to the F$_c$ fragment obtained by cleavage with papain, consists of the two carboxyl-terminal immunoglobulin domains of each H chain and in which the two arms of the Y, corresponding to the two F$_{ab}$ fragments, are formed by the two amino-terminal domains of each H chain and the two amino-terminal domains of each L chain. The linkers between the stem and the two arms consist of extended polypeptide regions within the H chains and are quite flexible.

Papain cleaves the H chains on the carboxyl-terminal side of the disulfide bond that links each L and H chain (Figure 33.6). Thus, each F$_{ab}$ consists of an entire L chain and the amino-terminal half of an H chain, whereas F$_c$ consists of the carboxyl-terminal halves of both H chains. Each F$_{ab}$ contains a single antigen-binding site. Because an intact IgG molecule contains two F$_{ab}$ components and therefore has two binding sites, it can cross-link multiple antigens (Figure 33.7). Furthermore, the F$_c$ and the two F$_{ab}$ units of the intact IgG are joined by flexible polypeptide regions that allow facile variation in the angle between the F$_{ab}$ units through a wide range (Figure 33.8). This kind of mobility, called *segmental flexibility,* can enhance the formation of an antibody–antigen complex by enabling both combining sites on an antibody to bind an antigen that possesses multiple binding sites, such as a viral coat composed of repeating identical monomers or a bacterial cell surface. The combining sites at the tips of the F$_{ab}$ units simply move to match the distance between specific determinants on the antigen.

Immunoglobulin G is the antibody present in highest concentration in the serum, but other classes of immunoglobulin also are present (Table 33.1). Each class includes an L chain (either κ or λ) and a distinct H chain (Figure 33.9). The heavy chains in IgG are called γ chains, whereas those in immunoglobulins A, M, D, and E are called α, μ, δ, and ε, respectively.

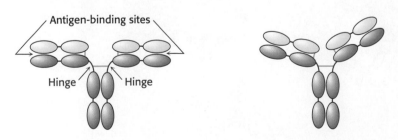

Figure 33.8 Segmental flexibility. The linkages between the F$_{ab}$ and the F$_c$ regions of an IgG molecule are flexible, allowing the two antigen-binding sites to adopt a range of orientations with respect to one another. This flexibility allows effective interactions with a multivalent antigen without requiring that the epitopes on the target be a precise distance apart.

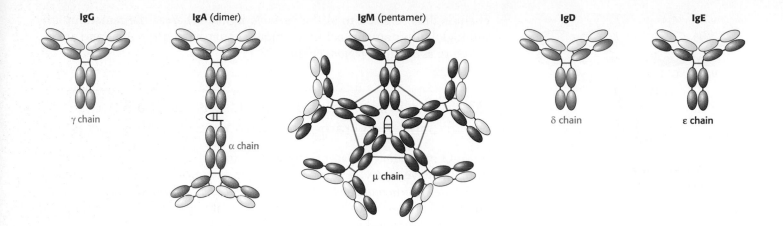

IgG

γ chain

IgA (dimer)

α chain

IgM (pentamer)

μ chain

IgD

δ chain

IgE

ε chain

Immunoglobulin M (IgM) is the first class of antibody to appear in the serum after exposure to an antigen. The presence of 10 combining sites enables IgM to bind especially tightly to antigens containing multiple identical epitopes. The strength of an interaction comprising multiple independent binding interactions between partners is termed *avidity* rather than affinity, which denotes the binding strength of a single combining site.

Immunoglobulin A (IgA) is the major class of antibody in external secretions, such as saliva, tears, bronchial mucus, and intestinal mucus. Thus, IgA serves as a first line of defense against bacterial and viral antigens. The role of *immunoglobulin D* (IgD) is not yet known. *Immunoglobulin E* (IgE) is important in conferring protection against parasites, but IgE also participates in allergic reactions. IgE–antigen complexes form cross-links with receptors on the surfaces of mast cells to trigger a cascade that leads to the release of granules containing pharmacologically active molecules. Histamine, one of the agents released, induces smooth-muscle contraction and stimulates the secretion of mucus.

A comparison of the amino acid sequences of different IgG antibodies from human beings or mice shows that the carboxyl-terminal half of the L chains and the carboxyl-terminal three-quarters of the H chains are very similar in all of the antibodies. Importantly, the amino-terminal domain of each chain is more variable, including three stretches of approximately 7 to 12 amino acids within each chain that are hypervariable, as shown for the H chain in Figure 33.10. The amino-terminal immunglobulin domain of each

Figure 33.9 Classes of immunoglobulin. Each of five classes of immunoglobulin has the same light chain (shown in yellow) combined with a different heavy chain (γ, α, μ, δ, or ε). Disulfide bonds are indicated by green lines. The IgA dimer and the IgM pentamer have a small polypeptide chain in addition to the light and heavy chains.

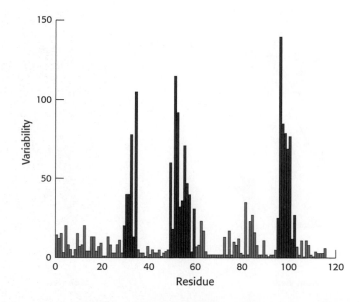

Figure 33.10 Immunoglobulin sequence diversity. A plot of sequence variability as a function of position along the sequence of the amino-terminal immunoglobulin domain of the H chain of human IgG molecules. Three regions (in red) show remarkably high levels of variability. These hypervariable regions correspond to three loops in the immunoglobulin domain structure. [After R. A. Goldsby, T. J. Kindt, and B. A. Osborne, *Kuby Immunology*, 4th ed. (W. H. Freeman and Company, 2000), p. 91.]

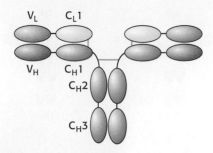

Figure 33.11 Variable and constant regions. Each L and H chain includes one immunoglobulin domain at its amino terminus that is quite variable from one antibody to another. These domains are referred to as V_L and V_H. The remaining domains are more constant from one antibody to another and are referred to as constant domains (C_L1, C_H1, C_H2, and C_H3).

chain is thus referred to as the *variable region*, whereas the remaining immunoglobulin domains are much more similar in all antibodies and are referred to as *constant regions* (Figure 33.11).

33.2 The Immunoglobulin Fold Consists of a Beta-Sandwich Framework with Hypervariable Loops

An IgG molecule consists of a total of 12 immunoglobulin domains. These domains have many sequence features in common and adopt a common structure, the *immunoglobulin fold* (Figure 33.12). Remarkably, this same structural domain is found in many other proteins that play key roles in the immune system and in nonimmune functions.

The immunoglobulin fold consists of a pair of β sheets, each built of antiparallel β strands, that surround a central hydrophobic core. A single disulfide bond bridges the two sheets. Two aspects of this structure are particularly important for its function. First, three loops present at one end of the structure form a potential binding surface. These loops contain the hypervariable sequences present in antibodies and in T-cell receptors (see Section 33.3 and p. 963). Variation of the amino acid sequences of these loops provides the major mechanism for the generation of the vastly diverse set of antibodies and T-cell receptors expressed by the immune system. These loops are referred to as *hypervariable loops* or *complementarity-determining regions* (CDRs). Second, the amino terminus and the carboxyl terminus are at opposite ends of the structure, which allows structural domains to be strung together to form chains, as in the L and H chains of antibodies. Such chains are present in several other key molecules in the immune system.

The immunoglobulin fold is one of the most prevalent domains encoded by the human genome: more than 750 genes encode proteins with at least one immunoglobulin fold recognizable at the level of amino acid sequence. Such domains are also common in other multicellular animals such as flies and nematodes. However, from inspection of amino acid sequence alone, immunoglobulin-fold domains do not appear to be present

Figure 33.12 Immunoglobulin fold. An immunoglobulin domain consists of a pair of β sheets linked by a disulfide bond and hydrophobic interactions. *Notice* that three hypervariable loops lie at one end of the structure. [Drawn from 1DQJ.pbd.]

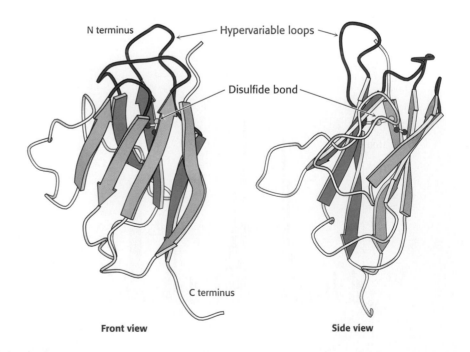

N terminus — Hypervariable loops —
Disulfide bond —
C terminus
Front view　　　　**Side view**

in yeast or plants, although these organisms possess other structurally similar domains, including the key photosynthetic electron-transport protein plastocyanin in plants (p. 551). Thus, the immunoglobulin-fold family appears to have expanded greatly along evolutionary branches leading to animals—particularly, vertebrates.

33.3 Antibodies Bind Specific Molecules Through Their Hypervariable Loops

For each class of antibody, the amino-terminal immunoglobin domains of the L and H chains (the variable domains, designated V_L and V_H) come together at the ends of the arms extending from the structure. The positions of the complementarity-determining regions are striking. These hypervariable sequences, present in three loops of each domain, come together so that all six loops form a single surface at the end of each arm (Figure 33.13). Because virtually any V_L can pair with any V_H, *a very large number of different binding sites can be constructed by their combinatorial association.*

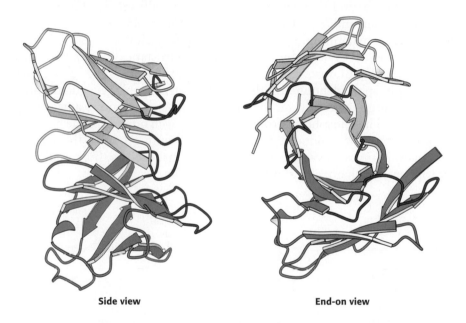

Side view　　　**End-on view**

Figure 33.13 **Variable domains.** Two views of the variable domains of the L chain (yellow) and the H chain (blue); the complementarity-determining regions (CDRs) are shown in red. *Notice* on the left that the six CDRs come together to form a binding surface. The specificity of the surface is determined by the sequences and structures of the CDRs. [Drawn from 1DQJ.pdb.]

X-ray Analyses Have Revealed How Antibodies Bind Antigens

The results of x-ray crystallographic studies of several hundred large and small antigens bound to F_{ab} molecules have been sources of much insight into the structural basis of antibody specificity. The binding of antigens to antibodies is governed by the same principles that govern the binding of substrates to enzymes. The apposition of complementary shapes results in numerous contacts between amino acids at the binding surfaces of both molecules. Many hydrogen bonds, electrostatic interactions, and van der Waals interactions, reinforced by hydrophobic interactions, combine to give specific and strong binding.

A few aspects of antibody binding merit specific attention, inasmuch as they relate directly to the structure of immunoglobulins. The binding site on the antibody has been found to incorporate some or all of the CDRs in the variable domains of the antibody. Small molecules are likely to make contact with fewer CDRs, with perhaps 15 residues of the antibody participating in the binding interaction. Macromolecules often make more extensive contact, sometimes interacting with all six CDRs and 20 or more residues of the antibody. Small molecules often bind in a cleft of the antigen-binding region. Macromolecules, such as globular proteins, tend to interact across larger, fairly flat apposed surfaces bearing complementary protrusions and depressions.

The search for an HIV vaccine has recently extended our understanding of antibodies and the way that they bind small molecules. The persistent problem in HIV vaccine design has been the lack of a neutralizing antibody response. In other words, most human antibodies do not recognize the HIV virus. A few rare antibodies isolated from asymptomatic, HIV-infected people show the neutralizing response. One of these antibodies, b12, gives an example of an antigen-binding surface that is not flat. Instead, b12 has a very long CDR3 loop that forms a "fingerlike" projection that can probe the canyons and valleys on the virus's surface. Another of these rare HIV-reactive antibodies, called 2G12, also has an unusual form; instead of the normal "Y" shape of the IgG molecule, 2G12 has its two arms pointing vertically and adjacent to one another. The two F_{ab} "arms" form a tightly packed dimer because their V_H domains are swapped.

A well-studied case of small-molecule binding is seen in an example of phosphorylcholine bound to F_{ab}. Crystallographic analysis revealed phosphorylcholine bound to a cavity lined by residues from five CDRs—two from the L chain and three from the H chain (Figure 33.14). The positively charged trimethylammonium group of phosphorylcholine is buried inside the wedge-shaped cavity, where it interacts electrostatically with two negatively charged glutamate residues. The negatively charged phosphoryl group of phosphorylcholine binds to the positively charged guanidinium group of an arginine residue at the mouth of the crevice and to a nearby lysine residue. The phosphoryl group is also hydrogen bonded to the hydroxyl group of a tyrosine residue and to the guanidinium group of the arginine side chain. Numerous van der Waals interactions, such as those made by a tryptophan side chain, also stabilize this complex.

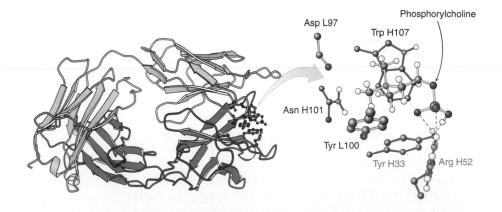

Figure 33.14 Binding of a small antigen. The structure of a complex between an F_{ab} fragment of an antibody and its target—in this case, phosphorylcholine. Residues from the antibody interact with phosphorylcholine through hydrogen bonding and electrostatic and van der Waals interactions. [Drawn from 2MCP.pdb.]

Residues from five CDRs participate in the binding of phosphorylcholine to human F_{ab}. This binding does not significantly change the structure of the antibody, yet induced fit plays a role in the formation of many antibody–antigen complexes. A malleable binding site can accommodate many more kinds of ligands than can a rigid one. Thus, induced fit increases the repertoire of antibody specificities.

Large Antigens Bind Antibodies with Numerous Interactions

How do large antigens interact with antibodies? A large collection of antibodies raised against hen egg-white lysozyme has been structurally characterized in great detail (Figure 33.15). Each different antibody binds to a distinct

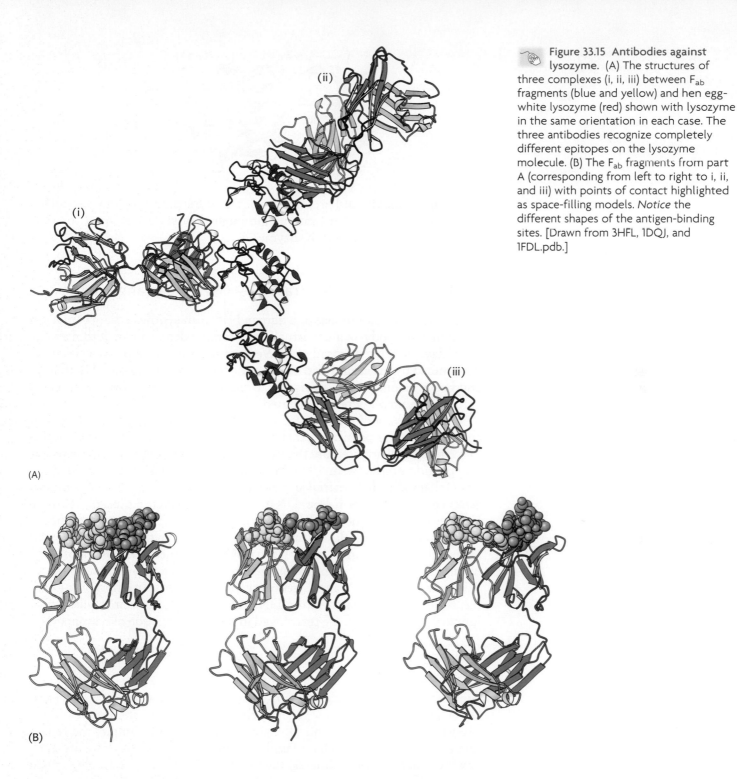

(A)

(B)

Figure 33.15 **Antibodies against lysozyme.** (A) The structures of three complexes (i, ii, iii) between F$_{ab}$ fragments (blue and yellow) and hen egg-white lysozyme (red) shown with lysozyme in the same orientation in each case. The three antibodies recognize completely different epitopes on the lysozyme molecule. (B) The F$_{ab}$ fragments from part A (corresponding from left to right to i, ii, and iii) with points of contact highlighted as space-filling models. *Notice* the different shapes of the antigen-binding sites. [Drawn from 3HFL, 1DQJ, and 1FDL.pdb.]

surface of lysozyme. Let us examine the interactions in one of these complexes (complex ii in Figure 33.15A) in detail. This antibody binds two polypeptide segments that are widely separated in the primary structure, residues 18 through 27 and 116 through 129 (Figure 33.16).

All six CDRs of the antibody make contact with this epitope. The region of contact is quite extensive (about 30 × 20 Å). The apposed surfaces are rather flat. The only exception is the side chain of glutamine 121 of lysozyme, which penetrates deeply into the antibody's binding site, where it forms a hydrogen bond with a main-chain carbonyl oxygen atom and is surrounded by three aromatic side chains. The formation of 12 hydrogen bonds and numerous van der Waals interactions contributes to the high affinity (K_d = 20 nM) of this antibody–antigen interaction. Examination of the F$_{ab}$

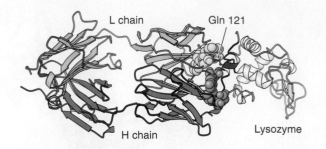

L chain   Gln 121

H chain   Lysozyme

Figure 33.16 Antibody–protein interactions. The structure of a complex between an F$_{ab}$ fragment and lysozyme reveals that the binding surfaces are complementary in shape over a large area. *Notice* that a single residue of lysozyme, glutamine 121, penetrates more deeply into the antibody combining site. [Drawn from 1FDL.pdb.]

molecule without bound protein reveals that the structures of the V$_L$ and V$_H$ domains change little on binding, although they slide 1 Å apart to allow more intimate contact with lysozyme.

33.4 Diversity Is Generated by Gene Rearrangements

A mammal such as a mouse or a human being can synthesize large amounts of specific antibody against virtually any foreign determinant within a matter of days of being exposed to it. We have seen that antibody specificity is determined by the amino acid sequences of the variable regions of both light and heavy chains, which brings us to the key question: How are different variable-region sequences generated?

The discovery of distinct variable and constant regions in the L and H chains raised the possibility that the genes that encode immunoglobulins have an unusual architecture that facilitates the generation of a diverse set of polypeptide products. In 1965, William Dreyer and Claude Bennett proposed that multiple *V (variable) genes* are separate from a single *C (constant) gene* in embryonic (germ-line) DNA. According to their model, one of these V genes becomes joined to the C gene in the course of differentiation of the antibody-producing cell. A critical test of this novel hypothesis had to await the isolation of pure immunoglobulin mRNA and the development of techniques for analyzing mammalian genomes. Twenty years later, Susumu Tonegawa found that V and C genes are indeed far apart in embryonic DNA but are closely associated in the DNA of antibody-producing cells. Thus, immunoglobulin genes are rearranged in the differentiation of lymphocytes.

J (Joining) Genes and D (Diversity) Genes Increase Antibody Diversity

Sequencing studies carried out by Susumu Tonegawa, Philip Leder, and Leroy Hood revealed that V genes in embryonic cells do not encode the entire variable region of L and H chains. Consider, for example, the region that encodes the κ light-chain family. A tandem array of 40 segments, each of which encodes approximately the first 97 residues of the variable domain of the L chain, is present on human chromosome 2 (Figure 33.17).

However, the variable region of the L chain extends to residue 110. Where is the DNA that encodes the last 13 residues of the V region? For L chains in undifferentiated cells, this stretch of DNA is located in an unexpected place: near the C gene. It is called the *J gene* because it joins the V and

V$_1$ V$_2$ .... V$_{39}$ V$_{40}$ J$_1$ J$_2$ J$_3$ J$_4$ J$_5$ C

Figure 33.17 The κ light-chain locus. This part of human chromosome 2 includes an array of 40 segments that encode the variable (V) region (approximately residues 1–97) of the light chain, an array of 5 segments that encode the joining (J) region (residues 98–110), and a single region that encodes the constant (C) region.

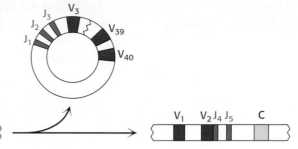

Figure 33.18 VJ recombination. A single V gene (in this case, V_2) is linked to a J gene (here, J_4) to form an intact VJ region. The intervening DNA is released in a circular form. Because the V and J regions are selected at random and the joint between them is not always in exactly the same place, many VJ combinations can be generated by this process.

C genes in a differentiated cell. In fact, a tandem array of five J genes is located near the C gene in embryonic cells. In the differentiation of an antibody-producing cell, a V gene becomes spliced to a J gene to form a complete gene for the variable region (Figure 33.18). RNA splicing generates an mRNA molecule for the complete L chain by linking the coding regions for the rearranged VJ unit with that for the C unit (Figure 33.19).

J genes are important contributors to antibody diversity because they encode part of the last hypervariable segment (CDR3). In forming a continuous variable-region gene, any of the 40 V genes can become linked to any of 5 J genes. Thus, somatic recombination of these gene segments amplifies the diversity already present in the germ line. The linkage between V and J is not precisely controlled. Recombination between these genes can take place at one of several bases near the codon for residue 95, generating additional diversity. A similar array of V and J genes encoding the λ light chain is present on human chromosome 22. This region includes 30 V_λ gene segments and four J_λ segments. In addition, this region includes four distinct C genes, in contrast with the single C gene in the κ locus.

In human beings, the genes encoding the heavy chain are present on chromosome 14. Remarkably, the variable domain of heavy chains is assembled from *three* rather than two segments. In addition to V_H genes that encode residues 1 through 94 and J_H segments that encode residues 98 through 113, this chromosomal region includes a distinct set of segments that encode residues 95 through 97 (Figure 33.20). These gene segments are called D for *diversity*. Some 27 D segments lie between 51 V_H and 6 J_H segments. The recombination process first joins a D segment to a J_H segment; a V_H segment is then joined to DJ_H. A greater variety of antigen-binding patches and clefts can be formed by the H chain than by the L chain because the H chain is encoded by three rather than two gene segments. Moreover, CDR3 of the H chain is diversified by the action of terminal deoxyribonucleotidyl transferase, a special DNA polymerase that requires no template. This enzyme inserts extra nucleotides between V_H and D. The *V(D)J recombination* of both

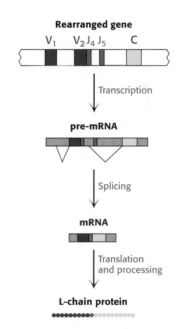

Figure 33.19 Light-chain expression. The light-chain protein is expressed by the transcription of the rearranged gene to produce a pre-RNA molecule with the VJ and C regions separated. RNA splicing removes the intervening sequences to produce an mRNA molecule with the VJ and C regions linked. Translation of the mRNA and processing of the initial protein product produce the light chain.

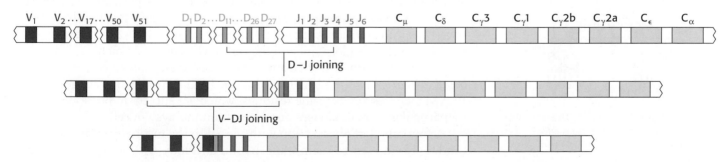

Figure 33.20 V(D)J recombination. The heavy-chain locus includes an array of 51 V segments, 27 D segments, and 6 J segments. Gene rearrangement begins with D–J joining, followed by further rearrangement to link the V segment to the DJ segment.

the L and the H chains is executed by specific enzymes present in immune cells. These proteins, called *RAG-1* and *RAG-2*, recognize specific DNA sequences called *recombination signal sequences* (RSSs) adjacent to the V, D, and J segments and facilitate the cleavage and religation of the DNA segments.

More Than 10^8 Antibodies Can Be Formed by Combinatorial Association and Somatic Mutation

Let us recapitulate the sources of antibody diversity. The germ line contains a rather large repertoire of variable-region genes. For κ light chains, there are about 40 V-segment genes and 5 J-segment genes. Hence, a total of $40 \times 5 = 200$ kinds of complete V_κ genes can be formed by the combinations of V and J. A similar analysis suggests that at least 120 different λ light chains can be generated. A larger number of heavy-chain genes can be formed because of the role of the D segments. For 51 V, 27 D, and 6 J gene segments, the number of complete V_H genes that can be formed is 8262. The association of 320 kinds of L chains with 8262 kinds of H chains would yield 2.6×10^6 different antibodies. Variability in the exact points of segment joining and other mechanisms increases this value by at least two orders of magnitude.

Even more diversity is introduced into antibody chains by *somatic mutation*—that is, the introduction of mutations into the recombined genes. In fact, a 1000-fold increase in binding affinity is seen in the course of a typical humoral immune response, arising from somatic mutation, a process called *affinity maturation*. The generation of an expanded repertoire leads to the selection of antibodies that more precisely fit the antigen. Thus, nature draws on each of three sources of diversity—a germ-line repertoire, somatic recombination, and somatic mutation—to form the rich variety of antibodies that protect an organism from foreign incursions.

The Oligomerization of Antibodies Expressed on the Surfaces of Immature B Cells Triggers Antibody Secretion

The processes heretofore described generate a highly diverse set of antibody molecules—a key first step in the generation of an immune response. The next stage is the selection of a particular set of antibodies directed against a specific invader. How is this selection accomplished? Each immature B cell, produced in the bone marrow, expresses a specific monomeric form of IgM attached to its surface (Figure 33.21). Each cell expresses approximately 10^5 IgM molecules, but *all of these molecules are identical in amino acid sequence and, hence, in antigen-binding specificity*. Thus, the selection of a particular immature B cell for growth will lead to the amplification of an antibody with a unique specificity. The selection process begins with the binding of an antigen to the membrane-bound antibody.

Associated with each membrane-linked IgM molecule are two molecules of a heterodimeric membrane protein called Ig-α–Ig-β (see Figure 33.21). Examination of the amino acid sequences of Ig-α and Ig-β is highly instructive. The amino terminus of each protein lies outside the cell and corresponds to a single immunoglobulin, and the carboxyl terminus, which lies inside the cell, includes a sequence of 18 amino acids called an *immunoreceptor tyrosine-based activation motif* (ITAM; see Figure 33.21). As its name suggests, each ITAM includes key tyrosine residues, which are subject to phosphorylation by particular protein kinases present in immune-system cells.

A fundamental observation with regard to the mechanism by which the binding of antigen to membrane-bound antibody triggers the subsequent steps of the immune response is that *oligomerization or clustering of the antibody molecules is required* (Figure 33.22). The requirement for oligomerization is reminiscent of the dimerization of receptors triggered by epidermal

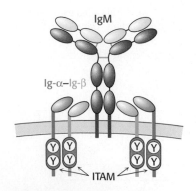

Figure 33.21 B-cell receptor. This complex consists of a membrane-bound IgM molecule noncovalently bound to two Ig-α–Ig-β heterodimers. The intracellular domains of each of the Ig-α and Ig-β chains include an immunoreceptor tyrosine-based activation motif (ITAM).

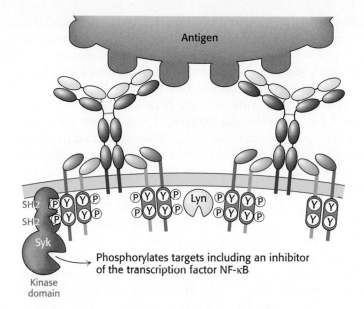

Phosphorylates targets including an inhibitor
of the transcription factor NF-κB

Figure 33.22 B-cell activation. The binding of a multivalent antigen such as a bacterial or viral surface links membrane-bound IgM molecules. This oligomerization triggers the phosphorylation of tyrosine residues in the ITAM sequences by protein tyrosine kinases such as Lyn. After phosphorylation, the ITAMs serve as docking sites for Syk, a protein kinase that phosphorylates a number of targets, including transcription factors.

growth factor and insulin encountered in Section 14.3; indeed, the associated signaling mechanisms appear to be quite similar. The oligomerization of the membrane-bound antibodies results in the phosphorylation of the tyrosine residues within the ITAMs by protein tyrosine kinases including Lyn, a homolog of Src (p. 400). The phosphorylated ITAMs serve as docking sites for a protein kinase termed *spleen tyrosine kinase* (Syk), which has two SH2 domains that interact with the pair of phosphorylated tyrosine residues in each ITAM. Syk, when activated by phosphorylation, proceeds to phosphorylate other signal-transduction proteins including an inhibitory subunit of a transcription factor called NF-κB and an isoform of phospholipase C. The signaling processes continue downstream to activate gene expression, leading to the stimulation of cell growth and initiating further B-cell differentiation.

Drugs that modulate the immune system have served as sources of insight into immune-system signaling pathways. For example, *cyclosporin,* a powerful suppressor of the immune system, acts by blocking a phosphatase called *calcineurin,* which normally activates a transcription factor called NF-AT by dephosphorylating it.

Cyclosporin A

The resulting potent immune supression reveals how crucial the activity of this transcription factor is to the development of an immune response. Without drugs such as cyclosporin, organ transplantation would be extremely difficult because transplanted tissue expresses a wide range of foreign antigens, which causes the immune system to reject the new tissue.

The role of oligomerization in the B-cell signaling pathway is illuminated when we consider the nature of many antigens presented by pathogens. The surfaces of many viruses, bacteria, and parasites are characterized by arrays of identical membrane proteins or membrane-linked carbohydrates. Thus, most pathogens present multiple binding surfaces that will naturally cause membrane-associated antibodies to oligomerize as they bind adjacent epitopes. In addition, the mechanism accounts for the observation that most small molecules do not induce an immune response; however, coupling multiple copies of a small molecule to a large oligomeric protein such as keyhole limpet hemocyanin (KLH), which has a molecular mass of close to 1 million daltons or more, promotes antibody oligomerization and, hence, the production of antibodies against the small-molecule epitope. The large protein is called the *carrier* of the attached chemical group, which is called a *haptenic determinant*. The small foreign molecule by itself is called a *hapten*. Antibodies elicited by attached haptens will bind unattached haptens as well.

Different Classes of Antibodies Are Formed by the Hopping of V_H Genes

The development of an effective antibody-based immune response depends on the secretion into the blood of antibodies that have appropriate effector functions. At the beginning of this response, an alternative mRNA-splicing pathway is activated so that the production of membrane-linked IgM is supplanted by the synthesis of secreted IgM. As noted in Section 33.1, secreted IgM is pentameric and has a high avidity for antigens containing multiple identical epitopes. Later, the antibody-producing cell makes either IgG, IgA, IgD, or IgE of the same specificity as that of the intially secreted IgM. In this switch, the light chain and the variable region of the heavy chain are unchanged. Only the constant region of the heavy chain changes. This step in the differentiation of an antibody-producing cell is called *class switching* (Figure 33.23). In undifferentiated cells, the genes for the constant region of each class of heavy chain, called C_μ, C_δ, C_γ, C_ϵ, and C_α, are next to one another. There are eight in all, including four genes for the constant regions of γ chains. A complete gene for the heavy chains of IgM antibody is formed by the translocation of a V_H gene segment to a DJ_H gene segment.

How are other heavy chains formed? Class switching is mediated by a gene-rearrangement process that moves a VDJ gene from a site near one C gene to a site near another C gene. Importantly, *the antigen-binding specificity is conserved in class switching because the entire V_HDJ_H gene is translocated in an intact form.* For example, the antigen-combining specificity of IgA produced by a particular cell is the same as that of IgM synthesized at

Figure 33.23 Class switching. Further rearrangement of the heavy-chain locus results in the generation of genes for antibody classes other than IgM. In the case shown, rearrangement places the VDJ region next to the $C\gamma 1$ region, resulting in the production of IgG1. Note that no further rearrangement of the VDJ region takes place, and so the specificity of the antibody is not affected.

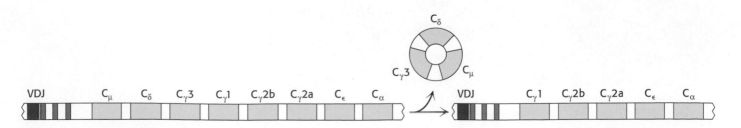

an earlier stage of its development. The biological significance of C_H switching is that a whole recognition domain (the variable domain) is shifted from the early constant region (C_μ) to one of several other constant regions that mediate different effector functions.

33.5 Major-Histocompatibility-Complex Proteins Present Peptide Antigens on Cell Surfaces for Recognition by T-Cell Receptors

Soluble antibodies are highly effective against extracellular pathogens, but they confer little protection against microorganisms that are predominantly intracellular, such as some viruses and mycobacteria (which cause tuberculosis and leprosy). These pathogens are shielded from antibodies by the host-cell membrane (Figure 33.24). A different and more subtle strategy, *cell-mediated immunity*, evolved to cope with intracellular viral pathogens. *T cells* continually scan the surfaces of all cells and kill those that exhibit foreign markings. The task is not simple; intracellular microorganisms are not so obliging as to intentionally leave telltale traces on the surface of their host. Quite the contrary, successful pathogens are masters of the art of camouflage. Vertebrates have evolved an ingenious mechanism—cut and display—to reveal the presence of stealthy intruders. Nearly all vertebrate cells exhibit on their surfaces a sample of peptides derived from the digestion of proteins in their cytoplasm. These peptides are displayed by integral membrane proteins that are

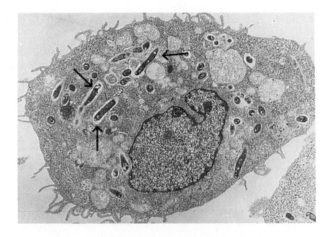

Figure 33.24 Intracellular pathogen. An electron micrograph showing mycobacteria (arrows) inside an infected macrophage. [Courtesy of Dr. Stanley Falkow.]

encoded by the *major histocompatibility complex* (MHC). Specifically, peptides derived from cytoplasmic proteins are bound to *class I MHC proteins*. The dendritic cells of the innate immune system that subject pathogens to phagocytosis migrate to lymphatic tissue where they use an MHC-like mechanism to present foreign peptides or lipid components to T cells—thus linking the innate and adaptive immune responses to pathogens.

How are these peptides generated and delivered to the plasma membrane? The process starts in the cytoplasm with the degradation of proteins—self-proteins as well as those of pathogens (Figure 33.25). Digestion is carried out by proteasomes (p. 653). The resulting peptide fragments are

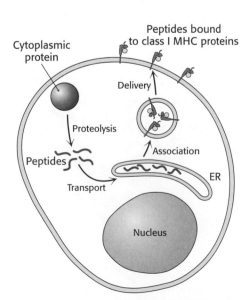

Figure 33.25 Presentation of peptides from cytoplasmic proteins. Class I MHC proteins on the surfaces of most cells display peptides that are derived from cytoplasmic proteins by proteolysis.

transported from the cytoplasm into the lumen of the endoplasmic reticulum by an ATP-driven pump. In the ER, peptides combine with nascent class I MHC proteins; these complexes are then targeted to the plasma membrane.

MHC proteins embedded in the plasma membrane tenaciously grip their bound peptides so that they can be touched and scrutinized by T-cell receptors on the surface of a killer cell. Foreign peptides bound to class I MHC proteins signal that a cell is infected and mark it for destruction by cytotoxic T cells. An assembly consisting of the foreign peptide—MHC complex, the T-cell receptor, and numerous accessory proteins triggers a cascade that induces apoptosis in the infected cell. Strictly speaking, infected cells are not killed but, instead, are triggered to commit suicide to aid the organism.

Peptides Presented by MHC Proteins Occupy a Deep Groove Flanked by Alpha Helices

The three-dimensional structure of a large fragment of a human MHC class I protein, *human leukocyte antigen A2* (HLA-A2), was solved in 1987 by Don Wiley and Pamela Bjorkman. Class I MHC proteins consist of a 44-kd α chain noncovalently bound to a 12-kd polypeptide called β_2-*microglobulin*. The α chain has three extracellular domains (α_1, α_2, and α_3), a transmembrane segment, and a tail that extends into the cytoplasm (Figure 33.26). Cleavage by papain of the HLA α chain several residues before the transmembrane segment yielded a soluble heterodimeric fragment. The β_2-microglobulin and the α_3 domains have immunoglobulin folds, although the pairing of the two domains differs from that in antibodies. The α_1 and α_2 domains exhibit a novel and remarkable architecture. They associate intimately to form a deep groove that serves as the peptide-binding site (Figure 33.27). The floor of the groove, which is about 25 Å long and 10 Å wide, is formed by eight β strands, four from each domain. A long helix contributed by the α_1 domain forms one side, and a helix contributed by the α_2 domain forms the other side. *This groove is the binding site for the presentation of peptides.*

The groove can be filled by a peptide from 8 to 10 residues long in an extended conformation. As we shall see (p. 968), MHC proteins are remarkably diverse in the human population; each person expresses as many as six distinct class I MHC proteins and many different forms are present in different people. The first structure determined, HLA-A2, binds peptides that almost always have leucine in the second position and valine in the last position (Figure 33.28). Side chains from the MHC molecule interact with the amino and carboxyl termini and with the side chains in these two key

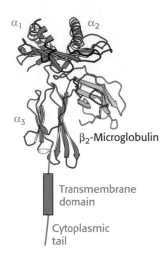

Figure 33.26 Class I MHC protein. A protein of this class consists of two chains. *Notice* that the α chain begins with two domains (α_1, α_2) that include α helices and continues with an immunoglobulin domain (α_3), a transmembrane domain, and a cytoplasmic tail. The second chain, β_2-microglobulin, adopts an immunoglobulin fold. [Drawn from 1HHK.pdb.]

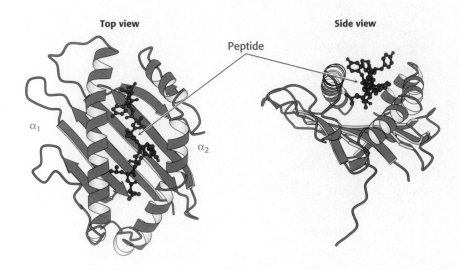

Figure 33.27 Class I MHC peptide-binding site. The α_1 and α_2 domains come together to form a groove in which peptides are displayed. *Notice* that that the peptide is surrounded on three sides by a β sheet and two α helices, but it is accessible from the top of the structure. [Drawn from 1HHK.pdb.]

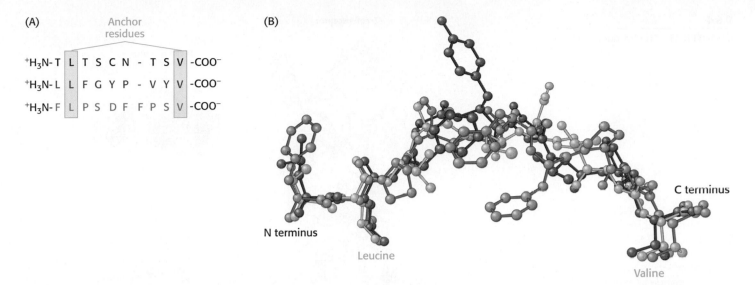

(A)

Anchor residues

^+H_3N-T $\boxed{\text{L}}$ T S C N - T S $\boxed{\text{V}}$ -COO$^-$

^+H_3N-L $\boxed{\text{L}}$ F G Y P - V Y $\boxed{\text{V}}$ -COO$^-$

^+H_3N-F $\boxed{\text{L}}$ P S D F F P S $\boxed{\text{V}}$ -COO$^-$

(B)

N terminus

Leucine

C terminus

Valine

Figure 33.28 Anchor residues. (A) The amino acid sequences of three peptides that bind to the class I MHC protein HLA-A2 are shown. Each of these peptides has leucine in the second position and valine in the carboxyl-terminal position. (B) Comparison of the structures of these peptides reveals that the amino and carboxyl termini, as well as the side chains of the leucine and valine residues, are in essentially the same positions in each peptide, whereas the remainder of the structures are quite different.

positions. These two residues are often referred to as the *anchor residues*. The other residues are highly variable. Thus, many millions of different peptides can be presented by this particular class I MHC protein; the identities of only two of the nine residues are crucial for binding. Each class of MHC molecules requires a unique set of anchor residues. Thus, a tremendous range of peptides can be presented by these molecules. Note that *one face of the bound peptide is exposed to solution, where it can be examined by other molecules, particularly T-cell receptors.* An additional remarkable feature of MHC–peptide complexes is their kinetic stability; once bound, a peptide is not released, even within a period of days.

T-Cell Receptors Are Antibody-like Proteins Containing Variable and Constant Regions

We are now ready to consider the receptor that recognizes peptides displayed by MHC proteins on target cells. The *T-cell receptor* consists of a 43-kd α chain joined by a disulfide bond to a 43-kd β chain (Figure 33.29). Each chain spans the plasma membrane and has a short carboxyl-terminal region on the cytoplasmic side. A small proportion of T cells express a receptor consisting of γ and δ chains in place of α and β. The α and β chains of the T-cell receptor, like immunoglobulin L and H chains, consist of *variable* and *constant* regions. Indeed, *these domains of the T-cell receptor are homologous to the V and C domains of immunoglobulins.* Furthermore, hypervariable sequences present in the V regions of the α and β chains of the T-cell receptor form the binding site for the epitope.

The genetic architecture of these proteins is similar to that of immunoglobulins, though the antibody genetic diversity is distributed over all the CDR loops, whereas T-cell-receptor genetic diversity is concentrated in the CDR3 loop that interacts with the peptide bound to the MHC. The variable region of the T-cell receptor α chain is encoded by about 50 V-segment genes and 70 J-segment genes. The T-cell receptor β chain is encoded by two D-segment genes in addition to 57 V-segment and 13 J-segment genes. Again, the diversity of component genes and the use of slightly imprecise

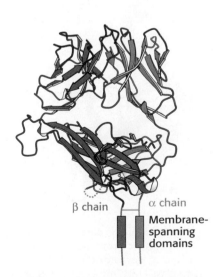

β chain   α chain

Membrane-spanning domains

Figure 33.29 T-cell receptor. This protein consists of an α chain and a β chain, each of which consists of two immunoglobulin domains and a membrane-spanning domain. The two chains are linked by a disulfide bond. [Drawn from 1BD2.pdb.]

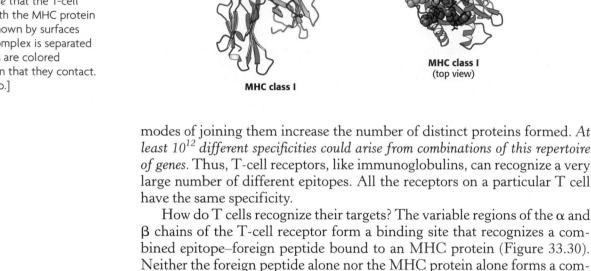

T-cell receptor

T-cell receptor
(bottom view)

α   β

MHC class I
(top view)

MHC class I

Figure 33.30 T-Cell Receptor–class I MHC complex. The T-cell receptor binds to a class I MHC protein containing a bound peptide. *Notice* that the T-cell receptor contacts both the MHC protein and the peptide as shown by surfaces exposed when the complex is separated (right). These surfaces are colored according to the chain that they contact. [Drawn from 1BD2.pdb.]

modes of joining them increase the number of distinct proteins formed. *At least 10^{12} different specificities could arise from combinations of this repertoire of genes.* Thus, T-cell receptors, like immunoglobulins, can recognize a very large number of different epitopes. All the receptors on a particular T cell have the same specificity.

How do T cells recognize their targets? The variable regions of the α and β chains of the T-cell receptor form a binding site that recognizes a combined epitope–foreign peptide bound to an MHC protein (Figure 33.30). Neither the foreign peptide alone nor the MHC protein alone forms a complex with the T-cell receptor. Thus, fragments of an intracellular pathogen are presented in a context that allows them to be detected, leading to the initiation of an appropriate response.

CD8 on Cytotoxic T Cells Acts in Concert with T-Cell Receptors

The T-cell receptor does not act alone in recognizing and mediating the fate of target cells. Cytotoxic T cells also express a protein termed *CD8* on their surfaces that is crucial for the recognition of the class I MHC–peptide complex. The abbreviation CD stands for *cluster of differentiation,* referring to a cell-surface marker that is used to identify a lineage or stage of differentiation. Antibodies specific for particular CD proteins have been invaluable in following the development of leukocytes and in discovering new interactions between specific cell types.

Each chain in the CD8 dimer contains a domain that resembles an immunoglobulin variable domain (Figure 33.31). CD8 interacts primarily with the constant α_3 domain of class I MHC proteins. This interaction further stabilizes the interactions between the T cell and its target. The cytoplasmic tail of CD8 contains a docking site for Lck, a cytoplasmic tyrosine kinase akin to Src. The T-cell receptor itself is associated with six polypeptides that form the CD3 complex (Figure 33.32). The γ, δ, and ε chains of CD3 are homologous to Ig-α and Ig-β associated with the B-cell receptor (p. 958); each chain consists of an extracellular immunoglobulin domain and an intracellular ITAM region. These chains associate into CD3-γε and CD3-δε heterodimers. An additional component, the CD3-ζ chain, has

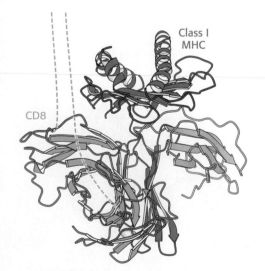

Class I
MHC

CD8

Figure 33.31 The coreceptor CD8. This dimeric protein extends from the surface of a cytotoxic T cell and binds to class I MHC molecules that are expressed on the surface of the cell that is bound to the T cell. The dashed lines represent extended polypeptide chains that link the immunoglobulin domains of CD8 to the membrane. [Drawn from 1AKJ.pdb.]

only a small extracellular domain and a larger intracellular domain containing three ITAM sequences.

On the basis of these components, a model for T-cell activation can be envisaged that is closely parallel to the pathway for B-cell activation (Section 33.3; Figure 33.33). The binding of the T-cell receptor to the class I MHC–peptide complex and the concomitant binding of CD8 from the T-cell to the MHC molecule links the kinase Lck to the ITAM substrates of the components of the CD3 complex. The phosphorylation of the tyrosine residues in the ITAM sequences generates docking sites for a protein kinase called ZAP-70 (for 70-kd *zeta-associated protein*) that is homologous to Syk in B cells. Docked by its two SH2 domains, ZAP-70 phosphorylates downstream targets in the signaling cascade. Additional molecules, including a membrane-bound protein phosphatase called CD45 and a cell-surface protein called CD28, play ancillary roles in this process.

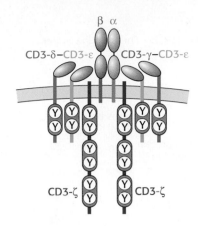

Figure 33.32 T-cell receptor complex. The T-cell receptor is associated with six CD3 molecules: a CD3-γ–CD3-ε heterodimer, a CD3-δ–CD3-ε heterodimer, and two chains of CD3-ζ. Single ITAM sequences are present in the cytoplasmic domains of CD3-γ, CD3-δ, and CD3-ε whereas three such sequences are found in each CD3-ζ chain.

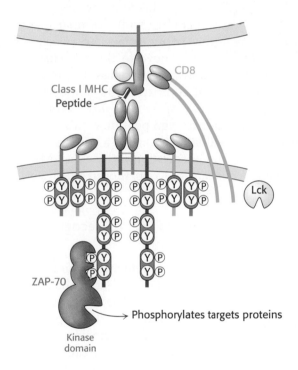

Figure 33.33 T-cell activation. The interaction between the T-cell receptor and a class I MHC–peptide complex results in the binding of CD8 to the MHC protein, the recruitment of the protein tyrosine kinase Lck, and the phosphorylation of tyrosine residues in the ITAM sequences of the CD3 chains. After phosphorylation, the ITAM regions serve as docking sites for the protein kinase ZAP-70, which phosphorylates protein targets to transmit the signal.

T-cell activation has two important consequences. First, the activation of cytotoxic T cells results in the secretion of *perforin*. This 70-kd protein makes the cell membrane of the target cell permeable by polymerizing to form transmembrane pores 10 nm wide (Figure 33.34). The cytotoxic T cell then secretes proteases called *granzymes* into the target cell. These enzymes initiate the pathway of apoptosis, leading to the death of the target cell and the fragmentation of its DNA, including any viral DNA that may be present. Second, after it has stimulated its target cell to commit suicide, the activated T cell disengages and is stimulated to reproduce. Thus, additional T cells that express the same T-cell receptor are generated to continue the battle against the invader after these T cells have been identified as a suitable weapon.

Helper T Cells Stimulate Cells That Display Foreign Peptides Bound to Class II MHC Proteins

Not all T cells are cytotoxic. *Helper T cells, a different class, stimulate the proliferation of specific B lymphocytes and cytotoxic T cells and thereby serve as partners in determining the immune responses that are produced.*

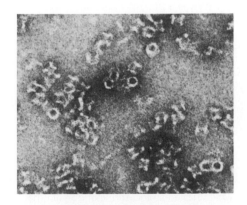

Figure 33.34 Consequences of cytotoxic-T-cell action. An electron micrograph showing pores in the membrane of a cell that has been attacked by a cytotoxic T cell. The pores are formed by the polymerization of perforin, a protein secreted by the cytotoxic T cell. [Courtesy of Dr. Eckhard Podock.]

The importance of helper T cells is graphically revealed by the devastation wrought by AIDS, a condition that destroys these cells. Helper T cells, like cytotoxic T cells, detect foreign peptides that are presented on cell surfaces by MHC proteins. However, the source of the peptides, the MHC proteins that bind them, and the transport pathway are different.

Helper T cells recognize peptides bound to MHC molecules referred to as class II. Their helping action is focused on B cells, macrophages, and dendritic cells. *Class II MHC proteins are* expressed only by these *antigen-presenting cells*, unlike class I MHC proteins, which are expressed on nearly all cells. The peptides presented by class II MHC proteins do not come from the cytoplasm. Rather, *they arise from the degradation of proteins that have been internalized by endocytosis*. Consider, for example, a virus particle that is captured by membrane-bound immunoglobulins on the surface of a B cell (Figure 33.35). This complex is delivered to an endosome, a membrane-enclosed acidic compartment, where it is digested. The resulting peptides become associated with class II MHC proteins, which move to the cell surface. Peptides from the cytoplasm cannot reach class II proteins, whereas peptides from endosomal compartments cannot reach class I proteins. This segregation of displayed peptides is biologically critical. The association of a foreign peptide with a class II MHC protein signals that a cell has *encountered* a pathogen and serves as a call for *help*. In contrast, association with a class I MHC protein signals that a cell has *succumbed* to a pathogen and is a call for *destruction*.

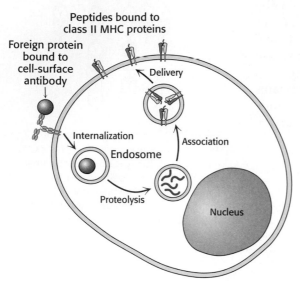

Figure 33.35 Presentation of peptides from internalized proteins. Antigen-presenting cells bind and internalize foreign proteins and display peptides that are formed from the digestion of these proteins in class II MHC proteins.

Helper T Cells Rely on the T-Cell Receptor and CD4 to Recognize Foreign Peptides on Antigen-Presenting Cells

The overall structure of a class II MHC molecule is remarkably similar to that of a class I molecule. Class II molecules consist of a 33-kd α chain and a noncovalently bound 30-kd β chain (Figure 33.36). Each contains two extracellular domains, a transmembrane segment, and a short cytoplasmic tail. The peptide-binding site is formed by the α_1 and β_1 domains, each of which contributes a long helix and part of a β sheet. Thus, the same structural elements are present in class I and class II MHC molecules, but they are combined into polypeptide chains in different ways. The peptide-binding site of a class II molecule is open at both ends, and so this groove can

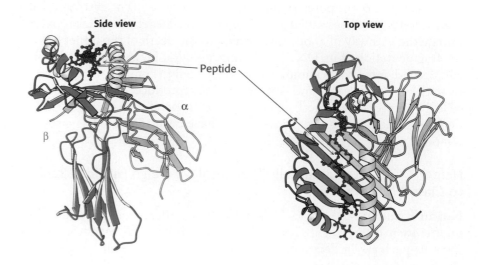

Figure 33.36 Class II MHC protein. A class II MHC protein consists of homologous α and β chains, each of which has an amino-terminal domain that constitutes half of the peptide-binding structure, as well as a carboxyl-terminal immunoglobulin domain. The peptide-binding site is similar to that in class I MHC proteins except that it is open at both ends, allowing class II MHC proteins to bind longer peptides than those bound by class I. [Drawn from 1DLH.pdb.]

accommodate longer peptides than can be bound by class I molecules; typically, peptides between 13 and 18 residues long are bound. The peptide-binding specificity of each class II molecule depends on binding pockets that recognize particular amino acids, also known as anchor residues, in specific positions along the sequence.

Helper T cells express T-cell receptors that are produced from the same genes as those on cytotoxic T cells. These T-cell receptors interact with class II MHC molecules in a manner that is analogous to T-cell-receptor interaction with class I MHC molecules. Nonetheless, helper T cells and cytotoxic T cells are distinguished by other proteins that they express on their surfaces. In particular, helper T cells express a protein called CD4 instead of expressing CD8. *CD4* consists of four immunoglobulin domains that extend from the T-cell surface, as well as a small cytoplasmic region (Figure 33.37). The amino-terminal immunoglobulin domains of CD4 interact with the base on the class II MHC molecule. Thus, helper T cells bind cells expressing class II MHC specifically because of the interactions with CD4 (Figure 33.38).

When a helper T cell binds to an antigen-presenting cell expressing an appropriate class II MHC–peptide complex, signaling pathways analogous to those in cytotoxic T cells are initiated by the action of the kinase Lck on ITAMs in the CD3 molecules associated with the T-cell receptor. However, rather than triggering events leading to the death of the attached cell, *these signaling pathways result in the secretion of cytokines from the helper cell.* Cytokines are a family of molecules that include, among others, interleukin-2 and interferon-γ. Cytokines bind to specific receptors on the antigen-presenting cell and stimulate growth, differentiation, and, in regard to plasma cells, which are derived from B cells, antibody secretion (Figure 33.39). Thus, the internalization and presentation of parts of a foreign pathogen help to generate a local environment in which cells taking part in the defense against this pathogen can flourish through the action of helper T cells.

Figure 33.37 Coreceptor CD4. This protein comprises four tandem immunoglobulin domains that extend from the surface of a helper T cell. [Drawn from 1WIO.pdb.]

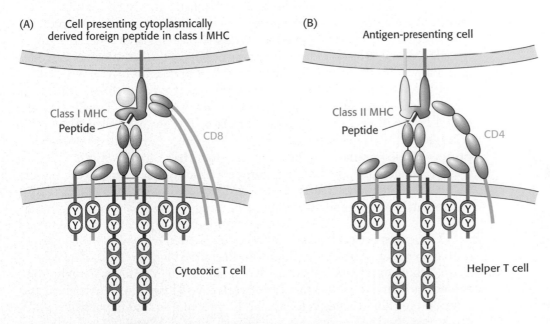

Figure 33.38 Variations on a theme. (A) Cytotoxic T cells recognize foreign peptides presented in class I MHC proteins with the aid of the coreceptor CD8. (B) Helper T cells recognize peptides presented in class II MHC proteins by specialized antigen-presenting cells with the aid of the coreceptor CD4.

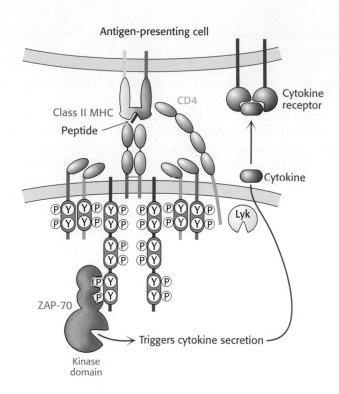

Figure 33.39 Helper-T-cell action. The engagement of the T-cell receptor in helper T cells results in the secretion of cytokines. These cytokines bind to cytokine receptors expressed on the surface of the antigen-presenting cell, stimulating cell growth, differentiation, and, in regard to a B cell, antibody secretion.

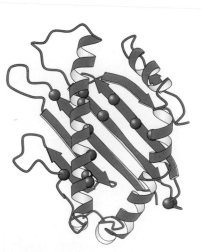

Figure 33.40 Polymorphism in class I MHC proteins. The positions of sites with a high degree of polymorphism in the human population are displayed as red spheres on the structure of the amino-terminal part of a class I MHC protein. [Drawn from 1HHK.pdb.]

MHC Proteins Are Highly Diverse

MHC class I and II proteins, the presenters of peptides to T cells, were discovered because of their role in *transplantation rejection*. A tissue transplanted from one person to another or from one mouse to another is usually rejected by the immune system. In contrast, tissues transplanted from one identical twin to another or between mice of an inbred strain are accepted. Genetic analyses revealed that rejection occurs when tissues are transplanted between individual organisms having different genes in the major histocompatibility complex, a cluster of more than 75 genes playing key roles in immunity. The 3500-kb span of the MHC is nearly the length of the entire *E. coli* chromosome. The MHC encodes class I proteins (presenters to cytotoxic T cells) and class II proteins (presenters to helper T cells), as well as class III proteins (components of the complement cascade) and many other proteins that play key roles in immunity.

Human beings express six different class I genes (three from each parent) and six different class II genes. The three loci for class I genes are called HLA-A, -B, and -C; those for class II genes are called HLA-DP, -DQ, and -DR. These loci are *highly polymorphic:* many alleles of each are present in the population. For example, more than 50 each of HLA-A, -B, and -C alleles are known; the numbers discovered increase each year. Hence, the likelihood that two unrelated persons have identical class I and II proteins is very small ($<10^{-4}$), accounting for transplantation rejection unless the genotypes of donor and acceptor are closely matched in advance.

Differences between class I proteins are located mainly in the α_1 and α_2 domains, which form the peptide-binding site (Figure 33.40). The α_3 domain, which interacts with a constant β_2-microglobulin is largely conserved. Similarly, the differences between class II proteins cluster near the peptide-binding groove. Why are MHC proteins so highly variable? *Their diversity makes the presentation of a very wide range of peptides to T cells possible. A particular* class I or class II molecule may not be able to bind any of the peptide fragments of a viral protein. The likelihood of a fit is markedly increased by having several kinds (usually six) of each class of presenters in each individual organism. If all members of a species had identical class I or class II molecules, the population would be much more vulnerable to devastation by a pathogen that had mutated and thereby evaded presentation. The evolution of the diverse human MHC repertoire has been driven by the selection for individual members of the species who resist infections to which other members of the population may be susceptible.

Human Immunodeficiency Viruses Subvert the Immune System by Destroying Helper T Cells

In 1981, the first cases of a new disease now called *acquired immune deficiency syndrome* (AIDS) were recognized. The victims died of rare infections because their immune systems were crippled. The cause was identified 2 years later by Luc Montagnier and coworkers. AIDS is produced by *human immunodeficiency virus* (HIV), of which two major classes are known: HIV-1 and the much less common HIV-2. Like other *retroviruses*, HIV contains a single-stranded RNA genome that is replicated through a double-stranded DNA intermediate. This viral DNA becomes integrated into the genome of the host cell. In fact, viral genes are transcribed only when they are integrated into the host DNA.

The HIV virion is enveloped by a lipid-bilayer membrane containing two glycoproteins: gp41 spans the membrane and is associated with gp120, which is located on the external face (Figure 33.41). The core of the virus contains two copies of the RNA genome and associated transfer RNAs, as well as several molecules of reverse transcriptase. They are surrounded by many copies of two proteins called p18 and p24. *The host cell for HIV is the helper T cell.* The gp120 molecules on the membrane of HIV bind to CD4 molecules on the surface of the helper T cell (Figure 33.42). This interaction allows the associated viral gp41 to insert its amino-terminal head into the host-cell membrane. The viral membrane and the helper-T-cell membrane fuse, and the viral core is released directly into the cytoplasm. Infection by HIV leads to the destruction of helper T cells because the permeability of the host plasma membrane is markedly increased by the insertion of viral glycoproteins and the budding of virus particles. The influx of ions and water disrupts the ionic balance, causing osmotic lysis.

The development of an effective AIDS vaccine is difficult owing to the antigenic diversity of HIV strains. Because its mechanism for replication is quite error prone, a population of HIV presents an ever-changing array of coat proteins. Indeed, the mutation rate of HIV is more than 65 times higher than that of influenza virus. A few broadly neutralizing antibodies have been isolated from asymptomatic, HIV infected persons. Several of these antibodies show an unusual form, described in Section 33.3, that allows them to bind many types of HIV.

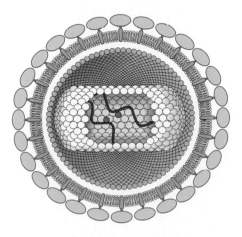

Figure 33.41 Human Immunodeficiency virus. A schematic representation of HIV reveals its proteins and nucleic acid components. The membrane-envelope glycoproteins gp41 and gp120 are shown in dark and light green. The viral RNA is shown in red, and molecules of reverse transcriptase are shown in blue. [After R. C. Gallo. The AIDS virus. Copyright © 1987 by Scientific American, Inc. All rights reserved.]

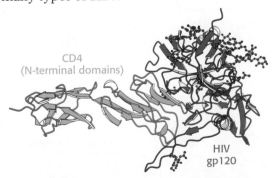

CD4
(N-terminal domains)

HIV
gp120

Figure 33.42 HIV receptor. A complex between a modified form of the envelope glycoprotein gp120 from HIV and a peptide corresponding to the two amino-terminal domains from the helper-T-cell protein CD4 reveals how viral infection of helper-T-cells is initiated. [Drawn from 1GC1.pdb.]

33.6 Immune Responses Against Self-Antigens Are Suppressed

The primary function of the immune system is to protect the host from invasion by foreign organisms. But how does the immune system prevent itself from mounting attacks against the host organism? In other words, how does the immune system distinguish between self and nonself? Clearly, proteins from the organism itself do not bear some special tag identifying them. Instead, selection processes early in the developmental pathways for immune cells kill or suppress those immune cells that react strongly with self-antigens. The evolutionary paradigm still applies; immune cells that recognize self-antigens are generated, but selective mechanisms eliminate such cells in the course of development.

T Cells Are Subjected to Positive and Negative Selection in the Thymus

T cells derive their name from the location of their production—the thymus, a small organ situated just above the heart. Examination of the developmental pathways leading to the production of mature cytotoxic and helper T cells reveals the selection mechanisms that are crucial for distinguishing self from nonself. These selection criteria are quite stringent; approximately 98% of the thymocytes, the precursors of T cells, die before the completion of the maturation process.

Thymocytes produced in the bone marrow do not express the T-cell-receptor complex, CD4, or CD8. On relocation to the thymus and rearrangement of the T-cell-receptor genes, the immature thymocyte expresses all of these molecules. These cells are first subjected to *positive selection* (Figure 33.43). Cells for which the T-cell receptor can bind with reasonable affinity to either class I or class II MHC molecules survive this selection; those for which the T-cell receptor does not participate in such an interaction undergo apoptosis and die. *The role of the positive selection step is to prevent the production of T cells that will not bind to any MHC complex present, regardless of the peptide bound.*

The cell population that survives positive selection is subjected to a second step, *negative selection*. Here, T cells that bind with high affinity to MHC complexes bound to self-peptides expressed on the surfaces of antigen-presenting cells in the thymus undergo apoptosis or are otherwise suppressed. Those that do not bind too avidly to any such MHC complex complete development and become mature cytotoxic T cells (which express only CD8) or helper T cells (which express only CD4). The negative selection step leads to *self-tolerance;* cells that bind an MHC–self-peptide complex are removed from the T-cell population. Similar mechanisms apply to developing B cells, suppressing B cells that express antibodies that interact strongly with self-antigens.

Figure 33.43 T-cell selection. A population of thymocytes is subjected first to positive selection to remove cells that express T-cell receptors that will not bind to MHC proteins expressed by the individual organism. The surviving cells are then subjected to negative selection to remove cells that bind strongly to MHC complexes bound to self-peptides.

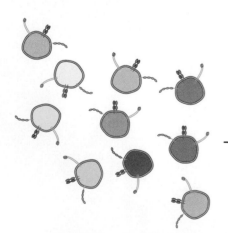
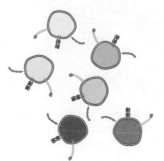

Positive selection

Only cells that bind to some MHC molecule survive

Negative selection

Cells that bind strongly to MHC or MHC–self-peptide complexes are eliminated

Cytotoxic T cell (CD8 positive)

Helper T cell (CD4 positive)

Autoimmune Diseases Result from the Generation of Immune Responses Against Self-Antigens

Although thymic selection is remarkably efficient in suppressing the immune response to self-antigens, failures do occur. Such failures results in *autoimmune diseases*. These diseases include common illnesses such as insulin-dependent diabetes mellitus, multiple sclerosis, and rheumatoid arthritis. In these illnesses, immune responses against self-antigens result in damage to selective tissues that express the antigen (Figure 33.44).

In many cases, the cause of the generation of self-reactive antibodies or T cells is unclear. However, in other cases, infectious organisms such as bacteria or viruses may play a role. Infection leads to the generation of antibodies and T cells that react with many different epitopes from the infectious organism. If one of these antigens closely resembles a self-antigen, an autoimmune response can result. For example, *Streptococcus* infections sometimes lead to rheumatic fever owing to the production of antibodies to streptococcal antigens that cross-react with exposed epitopes in heart muscle.

The Immune System Plays a Role in Cancer Prevention

The development of immune responses against proteins encoded by our own genomes can be beneficial under some circumstances. Cancer cells have undergone significant changes that often result in the expression of proteins that are not normally expressed. For example, the mutation of genes can generate proteins that do not correspond in amino acid sequence to any normal protein. Such proteins may be recognized as foreign, and an immune response will be generated specifically against the cancer cell. Alternatively, cancer cells often produce proteins that are expressed during embryonic development but are not expressed or are expressed at very low levels after birth. For example, a membrane glycoprotein called *carcinoembryonic antigen* (CEA) appears in the gasterointestinal cells of developing fetuses but is not normally expressed at significant levels after birth. More than 50% of patients with colorectal cancer have elevated serum levels of CEA. Immune cells recognizing epitopes from such proteins will not be subjected to negative selection and, hence, will be present in the adult immune repertoire. These cells may play a cancer surveillance role, killing cells that overexpress antigens such as CEA and preventing genetically damaged cells from developing into tumors.

(A)

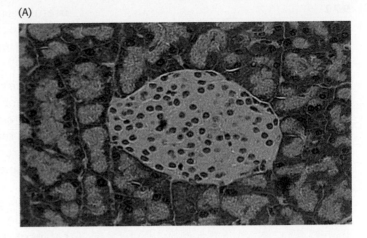

(B)

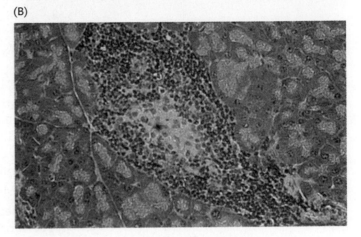

Figure 33.44 Consequences of autoimmunity. Photomicrographs of an islet of Langerhans (A) in the pancreas of a normal mouse and (B) in the pancreas of a mouse with an immune response against pancreatic β cells, which results in a disease resembling insulin-dependent diabetes mellitus in human beings. [From M. A. Atkinson and N. K. Maclaren. What causes diabetes? Copyright © 1990 by Scientific American, Inc. All rights reserved.]

Summary

Two lines of defense against pathogens are the innate immune system and the adaptive immune system. The innate immune system targets features present on many different pathogens but misses those pathogens lacking the targeted features. The adaptive immune system is both more specific and wide reaching. To respond effectively to a vast array of pathogens, this type of immune system must be tremendously adaptable. Adaptation by the adaptive immune system follows the principles of evolution: an enormously diverse set of potentially useful proteins is generated; these proteins are then subjected to intense selection so that only cells that express useful proteins flourish and continue development, until an effective immune response to a specific invader is generated.

33.1 Antibodies Possess Distinct Antigen-Binding and Effector Units

The major immunoglobulin in the serum is immunoglobulin G. An IgG protein is a heterotetramer with two heavy chains and two light chains. Treatment of IgG molecules with proteases such as papain produces three fragments: two F_{ab} fragments that retain antigen-binding activity and an F_c fragment that retains the ability to activate effector functions such as the initiation of the complement cascade. The F_{ab} fragments include the L chain and the amino-terminal half of the H chain; the F_c domain is a dimer consisting of the carboxyl-terminal halves of two H chains. Five different classes of antibody—IgG, IgM, IgA, IgD, and IgE—differ in their heavy chains and, hence, in their effector functions.

33.2 The Immunoglobulin Fold Consists of a Beta-Sandwich Framework with Hypervariable Loops

One particular protein fold is found in many of the key proteins of the immune system. The immunoglobulin fold consists of a pair of β sheets that pack against one another, linked by a single disulfide bond. Loops projecting from one end of the structure form a binding surface that can be varied by changing the amino acid sequences within the loops. Domains with immunoglobulin folds are linked to form antibodies and other classes of proteins in the immune system, including T-cell receptors.

33.3 Antibodies Bind Specific Molecules Through Their Hypervariable Loops

Two chains come together to form the binding surface of an antibody. Three loops from each domain, the complementarity-determining regions, form an essentially continuous surface that can vary tremendously in shape, charge, and other characteristics to allow particular antibodies to bind to molecules ranging from small molecules to large protein surfaces.

33.4 Diversity Is Generated by Gene Rearrangements

The tremendous diversity of the amino acid sequences of antibodies is generated by segmental rearrangements of genes. For antibody κ light chains, 1 of 40 variable regions is linked to 1 of 5 joining regions. The combined VJ unit is then linked to the constant region. Thousands of different genes can be generated in this manner. Similar arrays are rearranged to form the genes for the heavy chains, but an additional region called the diversity region lies between the V and the J regions. The combination L and H chains, each obtained through such rearranged genes, can produce more than 10^8 distinct antibodies. Different classes of antibodies are also generated by gene rearrangements that lead to class switching. Oligomerization of membrane-bound antibody molecules initiates a signal-transduction cascade inside B cells. Key steps in this signaling process include the phosphorylation of specific tyrosine residues in sequences termed immunoreceptor tyrosine-based activation motifs, present in proteins that associate with the membrane-bound antibodies.

33.5 Major-Histocompatibility-Complex Proteins Present Peptide Antigens on Cell Surfaces for Recognition by T-Cell Receptors

Intracellular pathogens such as viruses and mycobacteria cannot be easily detected. Intracellular proteins are constantly being cut into small peptides by proteasomes and displayed in class I major-histocompatibility-complex proteins on cell surfaces. Such peptides lie in a groove defined by two helices in the class I MHC proteins. The combination of MHC protein and peptide can be bound by an appropriate T-cell receptor. T-cell receptors resemble the antigen-binding domains of antibodies in structure, and diversity in T-cell-receptor sequence is generated by V(D)J gene rearrangements. The T-cell receptor recognizes features of both the peptide and the MHC molecule that presents it. Cytotoxic T cells initiate apoptosis in cells to which they bind through interactions between T-cell receptors and class I MHC–peptide complexes aided by interactions with the coreceptor molecule CD8. Helper T cells recognize peptides presented in class II MHC proteins, a distinct type of MHC protein expressed only on antigen-presenting cells, such as B cells and macrophages. Helper T cells express the coreceptor CD4 rather than CD8. CD4 interacts with class II MHC proteins present on antigen-presenting cells. Signaling pathways, analogous to those in B cells, are initiated by interactions between MHC–peptide complexes and T-cell receptors and the CD8 and CD4 coreceptors. Human immunodeficiency virus damages the immune system by infecting cells that express CD4, such as helper T cells.

33.6 Immune Responses Against Self-Antigens Are Suppressed

In principle, the immune system is capable of generating antibodies and T-cell receptors that bind to self-molecules—that is, molecules that are normally present in a healthy and uninfected individual organism. Selection mechanisms prevent such self-directed molecules from being expressed at high levels. The selection process includes both positive selection, to enrich the population of cells that express molecules that have the potential to bind foreign antigens in an appropriate context, and negative selection, which eliminates cells that express molecules with too high an affinity for self-antigens. Autoimmune diseases such as insulin-dependent diabetes mellitus can result from the amplification of a response against a self-antigen.

Key Terms

innate immune system (p. 945)

adaptive immune system (p. 945)

Toll-like receptor (TLR) (p. 946)

pathogen-associated molecular pattern (PAMP) (p. 946)

lipid A (p. 946)

humoral immune response (p. 947)

B lymphocyte (B cell) (p. 948)

antigen (p. 948)

antigenic determinant (epitope) (p. 948)

cellular immune response (p. 948)

cytotoxic T lymphocyte (killer T cell) (p. 948)

helper T lymphocyte (p. 948)

immunoglobulin G (IgG) (p. 949)

F_{ab} (p. 949)

F_c (p. 949)

light (L) chain (p. 949)

heavy (H) chain (p. 949)

segmental flexibility (p. 950)

immunoglobulin M (IgM) (p. 951)

immunoglobulin A (IgA) (p. 951)

immunoglobulin D (IgD) (p. 951)

immunoglobulin E (IgE) (p. 951)

variable region (p. 952)

constant region (p. 952)

immunoglobulin fold (p. 952)

hypervariable loop (p. 952)

complementarity-determining region (CDR) (p. 952)

V(D)J recombination (p. 957)

immunoreceptor tyrosine-based activation motif (ITAM) (p. 958)

cyclosporin (p. 959)

hapten (p. 960)

class switching (p. 960)

T cell (p. 961)

major histocompatibility complex (MHC) (p. 961)

class I MHC protein (p. 961)

human leukocyte antigen (HLA) (p. 962)

β_2-microglobulin (p. 962)

T-cell receptor (p. 963)

CD8 (p. 964)

perforin (p. 965)

granzyme (p. 965)

helper T cell (p. 965)

class II MHC protein (p. 966)

CD4 (p. 966)

human immunodeficiency virus (HIV) (p. 969)

positive selection (p. 970)

negative selection (p. 970)

autoimmune disease (p. 971)

carcinoembryonic antigen (CEA) (p. 971)

Selected Readings

Where to Start

Nossal, G. J. V. 1993. Life, death, and the immune system. *Sci. Am.* 269(3):53–62.

Tonegawa, S. 1985. The molecules of the immune system. *Sci. Am.* 253(4):122–131.

Leder, P. 1982. The genetics of antibody diversity. *Sci. Am.* 246(5):102–115.

Bromley, S. K., Burack, W. R., Johnson, K. G., Somersalo, K., Sims, T. N., Sumen, C., Davis, M. M., Shaw, A. S., Allen, P. M., and Dustin, M. L. 2001. The immunological synapse. *Annu. Rev. Immunol.* 19:375–396.

Books

Goldsby, R. A., Kindt, T. J., Osborne, B. A., and Kuby, J. 2003. *Kuby Immunology* (5th ed.). W. H. Freeman and Company.

Abbas, A. K., and Lichtman, A. H. 2003. *Cellular and Molecular Immunology* (5th ed). Saunders.

Cold Spring Harbor Symposia on Quantitative Biology, 1989. Volume 54. Immunological Recognition.

Nisinoff, A. 1985. *Introduction to Molecular Immunology* (2d ed.). Sinauer.

Weir, D. M. (Ed.). 1996. *Handbook of Experimental Immunology* (5th ed.). Oxford University Press.

Janeway, C. A., Travers, P., Walport, M., and Shlomchik, M. 2005. *Immunobiology* (6th ed.). Garland Science.

Innate Immune System

Janeway, C. A., Jr., and Medzhitov, R. 2002. Innate immune recognition. *Annu. Rev. Immunol.* 20:197–216.

Choe, J., Kelker, M. S., and Wilson, I. A. 2005. Crystal structure of human toll-like receptor 3 (TLR3) ectodomain. *Science* 309:581–585.

Khalturin, K., Panzer, Z., Cooper, M. D., and Bosch, T. C. 2004. Recognition strategies in the innate immune system of ancestral chordates. *Mol. Immunol.* 41:1077–1087.

Beutler, B., and Rietschel, E. T. 2003. Innate immune sensing and its roots: The story of endotoxin. *Nat. Rev. Immunol.* 3:169–176.

Xu, Y., Tao, X., Shen, B., Horng, T., Medzhitov, R., Manley, J. L., and Tong, L. 2000. Structural basis for signal transduction by the Toll/interleukin-1 receptor domains. *Nature* 408:111–115.

Structure of Antibodies and Antibody–Antigen Complexes

Davies, D. R., Padlan, E. A., and Sheriff, S. 1990. Antibody-antigen complexes. *Annu. Rev. Biochem.* 59:439–473.

Poljak, R. J. 1991. Structure of antibodies and their complexes with antigens. *Mol. Immunol.* 28:1341–1345.

Davies, D. R., and Cohen, G. H. 1996. Interactions of protein antigens with antibodies. *Proc. Natl. Acad. Sci. U. S. A.* 93:7–12.

Marquart, M., Deisenhofer, J., Huber, R., and Palm, W. 1980. Crystallographic refinement and atomic models of the intact immunoglobulin molecule Kol and its antigen-binding fragment at 3.0 Å and 1.9 Å resolution. *J. Mol. Biol.* 141:369–391.

Silverton, E. W., Navia, M. A., and Davies, D. R. 1977. Three-dimensional structure of an intact human immunoglobulin. *Proc. Natl. Acad. Sci. U. S. A.* 74:5140–5144.

Padlan, E. A., Silverton, E. W., Sheriff, S., Cohen, G. H., Smith, G. S., and Davies, D. R. 1989. Structure of an antibody-antigen complex: Crystal structure of the HyHEL-10 Fab lysozyme complex. *Proc. Natl. Acad. Sci. U. S. A.* 86:5938–5942.

Rini, J., Schultze-Gahmen, U., and Wilson, I. A. 1992. Structural evidence for induced fit as a mechanism for antibody-antigen recognition. *Science* 255:959–965.

Fischmann, T. O., Bentley, G. A., Bhat, T. N., Boulot, G., Mariuzza, R. A., Phillips, S. E., Tello, D., and Poljak, R. J. 1991. Crystallographic refinement of the three-dimensional structure of the FabD1.3-lysozyme complex at 2.5-Å resolution. *J. Biol. Chem.* 266:12915–12920.

Burton, D. R. 1990. Antibody: The flexible adaptor molecule. *Trends Biochem. Sci.* 15:64–69.

Saphire, E. O., Parren P. W., Pantophlet, R., Zwick, M. B., Morris, G. M., Rudd, P. M., Dwek, R. A., Stanfield, R. L., Burton, D. R., and Wilson, I. A. 2001. Crystal structure of a neutralizing human IgG against HIV-1: A template for vaccine design. *Science* 293:1155–1159.

Calarese, D. A., Scanlan, C. N., Zwick, M. B., Deechongkit, S., Mimura, Y., Kunert R., Zhu, P., Wormald, M. R., Stanfield, R. L., Roux, K. H., Kelly, J. W., Rudd, P. M., Dwek, R. A., Katinger, H., Burton, D. R., and Wilson, I. A. 2003. Antibody domain exchange is an immunological solution to carbohydrate cluster recognition. *Science* 300:2065–2071.

Generation of Diversity

Tonegawa, S. 1988. Somatic generation of immune diversity. *Biosci. Rep.* 8:3–26.

Honjo, T., and Habu, S. 1985. Origin of immune diversity: Genetic variation and selection. *Annu. Rev. Biochem.* 54:803–830.

Gellert, M., and McBlane, J. F. 1995. Steps along the pathway of VDJ recombination. *Philos. Trans. R. Soc. Lond. B Biol. Sci.* 347:43–47.

Harris, R. S., Kong, Q., and Maizels, N. 1999. Somatic hypermutation and the three R's: Repair, replication and recombination. *Mutat. Res.* 436:157–178.

Lewis, S. M., and Wu, G. E. 1997. The origins of V(D)J recombination. *Cell* 88:159–162.

Ramsden, D. A., van Gent, D. C., and Gellert, M. 1997. Specificity in V(D)J recombination: New lessons from biochemistry and genetics. *Curr. Opin. Immunol.* 9:114–120.

Roth, D. B., and Craig, N. L. 1998. VDJ recombination: A transposase goes to work. *Cell* 94:411–414.

Sadofsky, M. J. 2001. The RAG proteins in V(D)J recombination: More than just a nuclease. *Nucleic Acids Res.* 29:1399–1409.

MHC Proteins and Antigen Processing

Bjorkman, P. J., and Parham, P. 1990. Structure, function, and diversity of class I major histocompatibility complex molecules. *Annu. Rev. Biochem.* 59:253–288.

Goldberg, A. L., and Rock, K. L. 1992. Proteolysis, proteasomes, and antigen presentation. *Nature* 357:375–379.

Madden, D. R., Gorga, J. C., Strominger, J. L., and Wiley, D. C. 1992. The three-dimensional structure of HLA-B27 at 2.1 Å resolution suggests a general mechanism for tight binding to MHC. *Cell* 70:1035–1048.

Fremont, D. H., Matsumura, M., Stura, E. A., Peterson, P. A., and Wilson, I. A. 1992. Crystal structures of two viral peptides in complex with murine MHC class I H-2Kb. *Science* 257:880–881.

Matsumura, M., Fremont, D. H., Peterson, P. A., and Wilson, I. A. 1992. Emerging principles for the recognition of peptide antigens by MHC class I. *Science* 257:927–934.

Brown, J. H., Jardetzky, T. S., Gorga, J. C., Stern, L. J., Urban, R. G., Strominger, J. L., and Wiley, D. C. 1993. Three-dimensional structure of the human class II histocompatibility antigen HLA-DR1. *Nature* 364:33–39.

Saper, M. A., Bjorkman, P. J., and Wiley, D. C. 1991. Refined structure of the human histocompatibility antigen HLA-A2 at 2.6 Å resolution. *J. Mol. Biol.* 219:277–319.

Madden, D. R., Gorga, J. C., Strominger, J. L., and Wiley, D. C. 1991. The structure of HLA-B27 reveals nonamer self-peptides bound in an extended conformation. *Nature* 353:321–325.

Cresswell, P., Bangia, N., Dick, T., and Diedrich, G. 1999. The nature of the MHC class I peptide loading complex. *Immunol. Rev.* 172:21–28.

Madden, D. R., Garboczi, D. N., and Wiley, D. C. 1993. The antigenic identity of peptide-MHC complexes: A comparison of the conformations of five viral peptides presented by HLA-A2. *Cell* 75:693–708.

T-Cell Receptors and Signaling Complexes

Hennecke, J., and Wiley, D. C. 2001. T-cell receptor-MHC interactions up close. *Cell* 104:1–4.

Ding, Y. H., Smith, K. J., Garboczi, D. N., Utz, U., Biddison, W. E., and Wiley, D. C. 1998. Two human T cell receptors bind in a similar diagonal mode to the HLA-A2/Tax peptide complex using different TCR amino acids. *Immunity* 8:403–411.

Reinherz, E. L., Tan, K., Tang, L., Kern, P., Liu, J., Xiong, Y., Hussey, R. E., Smolyar, A., Hare, B., Zhang, R., Joachimiak, A., Chang, H. C., Wagner, G., and Wang, J. 1999. The crystal structure of a T-cell receptor in complex with peptide and MHC class II. *Science* 286:1913–1921.

Davis, M. M., and Bjorkman, P. J. 1988. T-cell antigen receptor genes and T-cell recognition. *Nature* 334:395–402.

Cochran, J. R., Cameron, T. O., and Stern, L. J. 2000. The relationship of MHC-peptide binding and T cell activation probed using chemically defined MHC class II oligomers. *Immunity* 12:241–250.

Garcia, K. C., Teyton, L., and Wilson, I. A. 1999. Structural basis of T cell recognition. *Annu. Rev. Immunol.* 17:369–397.

Garcia, K. C., Degano, M., Stanfield, R. L., Brunmark, A., Jackson, M. R., Peterson, P. A., Teyton, L. A., and Wilson, I. A. 1996. An αβ T-cell receptor structure at 2.5 Å and its orientation in the TCR-MHC complex. *Science* 274:209–219.

Garboczi, D. N., Ghosh, P., Utz, U., Fan, Q. R., Biddison, W. E., Wiley, D. C. 1996. Structure of the complex between human T-cell receptor, viral peptide and HLA-A2. *Nature* 384:134–141.

Gaul, B. S., Harrison, M. L., Geahlen, R. L., Burton, R. A., and Post, C. B. 2000. Substrate recognition by the Lyn protein-tyrosine kinase: NMR structure of the immunoreceptor tyrosine-based activation motif signaling region of the B cell antigen receptor. *J. Biol. Chem.* 275:16174–16182.

Kern, P. S., Teng, M. K., Smolyar, A., Liu, J. H., Liu, J., Hussey, R. E., Spoerl, R., Chang, H. C., Reinherz, E. L., and Wang, J. H. 1998. Structural basis of CD8 coreceptor function revealed by crystallographic analysis of a murine CD8 αβ ectodomain fragment in complex with H-2Kb. *Immunity* 9:519–530.

Konig, R., Fleury, S., and Germain, R. N. 1996. The structural basis of CD4-MHC class II interactions: Coreceptor contributions to T cell receptor antigen recognition and oligomerization-dependent signal transduction. *Curr. Top. Microbiol. Immunol.* 205:19–46.

Davis, M. M., Boniface, J. J., Reich, Z., Lyons, D., Hampl, J., Arden, B., and Chien, Y. 1998. Ligand recognition by αβ T-cell receptors. *Annu. Rev. Immunol.* 16:523–544.

Janeway, C. J. 1992. The T cell receptor as a multicomponent signalling machine: CD4/CD8 coreceptors and CD45 in T cell activation. *Annu. Rev. Immunol.* 10:645–674.

Podack, E. R., and Kupfer, A. 1991. T-cell effector functions: Mechanisms for delivery of cytotoxicity and help. *Annu. Rev. Cell Biol.* 7:479–504.

Davis, M. M. 1990. T cell receptor gene diversity and selection. *Annu. Rev. Biochem.* 59:475–496.

Leahy, D. J., Axel, R., and Hendrickson, W. A. 1992. Crystal structure of a soluble form of the human T cell coreceptor CD8 at 2.6 Å resolution. *Cell* 68:1145–1162.

Lowin, B., Hahne, M., Mattmann, C., and Tschopp, J. 1994. Cytolytic T-cell cytotoxicity is mediated through perforin and Fas lytic pathways. *Nature* 370:650–652.

Rudolph, M. G., and Wilson, I. A. 2002. The specificity of TCR/pMHC interaction. *Curr. Opin. Immunol.* 14:52–65.

HIV and AIDS

Fauci, A. S. 1988. The human immunodeficiency virus: Infectivity and mechanisms of pathogenesis. *Science* 239:617–622.

Gallo, R. C., and Montagnier, L. 1988. AIDS in 1988. *Sci. Am.* 259(4):41–48.

Kwong, P. D., Wyatt, R., Robinson, J., Sweet, R. W., Sodroski, J., and Hendrickson, W. A. 1998. Structure of an HIV gp120 envelope glycoprotein in complex with the CD4 receptor and a neutralizing human antibody. *Nature* 393:648–659.

Discovery of Major Concepts

Ada, G. L., and Nossal, G. 1987. The clonal selection theory. *Sci. Am.* 257(2):62–69.

Porter, R. R. 1973. Structural studies of immunoglobulins. *Science* 180:713–716.

Edelman, G. M. 1973. Antibody structure and molecular immunology. *Science* 180:830–840.

Kohler, G. 1986. Derivation and diversification of monoclonal antibodies. *Science* 233:1281–1286.

Milstein, C. 1986. From antibody structure to immunological diversification of immune response. *Science* 231:1261–1268.

Janeway, C. A., Jr. 1989. Approaching the asymptote? Evolution and revolution in immunology. *Cold Spring Harbor Symp. Quant. Biol.* 54:1–13.

Jerne, N. K. 1971. Somatic generation of immune recognition. *Eur. J. Immunol.* 1:1–9.

Problems

1. *Innate abilities.* A strain of mice has been identified that does not respond to LPS. This lack of response is due to a single amino acid change in the TIR domain of mouse TLR-4. Propose an explanation for the lack of response.

2. *Energetics and kinetics.* Suppose that the dissociation constant of an F_{ab}–hapten complex is 3×10^{-7} M at 25°C.

(a) What is the standard free energy of binding?
(b) Immunologists often speak of affinity (K_a), the reciprocal of the dissociation constant, in comparing antibodies. What is the affinity of this F_{ab}?
(c) The rate constant for the release of hapten from the complex is 120 s^{-1}. What is the rate constant for association? What does the magnitude of this value imply about the extent of structural change in the antibody on binding hapten?

3. *Sugar niche.* An antibody specific for dextran, a polysaccharide of glucose residues, was tested for its binding of glucose oligomers. Maximal binding affinity was obtained when the oligomer contained six glucose residues. How does the size of this site compare with that expected for the binding site on the surface of an antibody?

4. *A brilliant emitter.* Certain naphthalene derivatives, such as the dansyl group, exhibit a weak yellow fluorescence when they are in a highly polar environment (such as water) and an intense blue fluorescence when they are in a markedly nonpolar environment (such as hexane). The binding of ε-dansyl-lysine to specific antibody is accompanied by a marked increase in its fluorescence intensity and a shift in color from yellow to blue. What does this finding reveal about the hapten–antibody complex?

5. *Avidity versus affinity.* The standard free energy of binding of F_{ab} derived from an antiviral IgG is -29 kJ mol^{-1} (-7 kcal mol^{-1}) at 25°C.

(a) Calculate the dissociation constant of this interaction.
(b) Predict the dissociation constant of the intact IgG, assuming that both combining sites of the antibody can interact with

viral epitopes and that the free-energy cost of assuming a favorable hinge angle is 12.6 kJ mol^{-1} (-3 kcal mol^{-1}).

6. *Miniantibody.* The F$_{ab}$ fragment of an antibody molecule has essentially the same affinity for a monovalent hapten as does intact IgG.

(a) What is the smallest unit of an antibody that can retain the specificity and binding affinity of the whole protein?
(b) Design a compact single-chain protein that is likely to specifically bind antigen with high affinity.

7. *Turning on B cells.* B lymphocytes, the precursors of plasma cells, are triggered to proliferate by the binding of multivalent antigens to receptors on their surfaces. The cell-surface receptors are transmembrane immunoglobulins. Univalent antigens, in contrast, do not activate B cells.

(a) What do these findings reveal about the mechanism of B-cell activation?
(b) How might antibodies be used to activate B cells?

8. *An ingenious cloning strategy.* In the cloning of the gene for the α chain of the T-cell receptor, T-cell cDNAs were hybridized with B-cell mRNAs. What was the purpose of this hybridization step? Can the principle be applied generally?

9. *Instruction.* Before the mechanism for generating antibody diversity had been established, a mechanism based on protein folding around an antigen was proposed, primarily by Linus Pauling. In this model, antibodies that had different specificities had the same amino acid sequence but were folded in different ways. Propose a test of this model.

10. *Dealing with nonsense.* Cells, including immune cells, degrade mRNA molecules in which no long open reading frame is present. The process is called nonsense-mediated RNA decay. Suggest a role for this process in immune cells.

11. *Crystallization.* The proteolytic digestion of a population of IgG molecules isolated from human serum results in the generation of F$_{ab}$ and F$_c$ fragments. Why do F$_c$ fragments crystallize more easily than F$_{ab}$ fragments generated from such a population?

12. *Presentation.* The amino acid sequence of a small protein is

MSRLASKNLIRSDHAGGLLQATYSAVSS-
IKNTMSFGAWSNAALNDSRDA

Predict the most likely peptide to be presented by the class I MHC molecule HLA-A2.

Mechanism Problem

13. *Catalytic antibody.* Antibody is generated against a transition state for the hydrolysis of the following ester.

Ester

Transition-state analog

Some of these antibodies catalyze the hydrolysis of the ester. What amino acid residue might you expect to find in the binding site on the antibody?

Chapter Integration Problem

14. *Signaling.* Protein tyrosine phosphatases, such as the molecule CD45 expressed in both B cells and T cells, play important roles in activating such protein tyrosine kinases as Fyn and Lck, which are quite similar to Src. Suggest a mechanism for the activation of such protein kinases by the removal of a phosphoryl group from a phosphotyrosine residue.

Data Interpretation Problem

15. *Affinity maturation.* A mouse is immunized with an oligomeric human protein. Shortly after immunization, a cell line that expresses a single type of antibody molecule (antibody A) is derived. The ability of antibody A to bind the human protein is assayed with the results shown in the graph below.

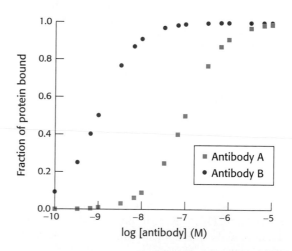

After repeated immunizations with the same protein, another cell line is derived that expresses a different antibody (antibody B). The results of analyzing the binding of antibody B to the protein also are shown. From these data, estimate

(a) the dissociation constant (K_d) for the complex between the protein and antibody A.
(b) the dissociation constant for the complex between the protein and antibody B.

Comparison of the amino acid sequences of antibody A and antibody B reveals them to be identical except for a single amino acid. What does this finding suggest about the mechanism by which the gene encoding antibody B was generated?

Molecular Motors

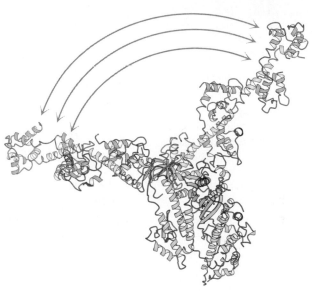

The horse, like all animals, is powered by the molecular-motor protein myosin. A part of myosin moves dramatically (as shown above) in response to ATP binding, hydrolysis, and product release, propelling myosin along an actin filament. This molecular movement is translated into movement of the entire animal, vividly depicted in da Vinci's rearing horse. [(Left) Leonardo da Vinci's study of a rearing horse for the *Battle of Anghiari* (ca. 1504) from The Royal Collection © Her Royal Majesty Queen Elizabeth II.]

Organisms, from human beings to bacteria, move to adapt to changes in their environments, navigating toward food and away from danger. Cells, themselves, are not static but are bustling assemblies of moving proteins, nucleic acids, and organelles (Figure 34.1). Remarkably, the fundamental biochemical mechanisms that produce contractions in our muscles are the same as those that propel organelles along defined paths inside cells. In fact, many of the proteins that play key roles in converting chemical energy in the form of ATP into kinetic energy, the energy of motion, are members of the same protein family, the P-loop NTPases. These molecular motors are homologous to proteins that we have encountered in other contexts, including the G proteins in protein synthesis, signaling, and other processes. Once again, we see the economy of evolution in adapting an existing protein to perform new functions.

Molecular motors operate by small increments, converting changes in protein conformation into directed motion. Orderly motion across distances requires a track that steers the motion of the motor assembly. Indeed, we have previously encountered a class of molecular motors that utilize mechanisms that we will examine here—namely, the helicases that move along DNA and RNA tracks (Section 28.2). The proteins on which we will focus in this chapter move along actin and microtubules—protein filaments composed of repeating subunits. The motor proteins cycle between

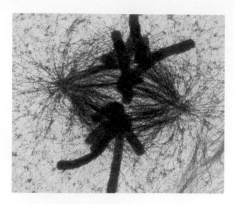

Figure 34.1 Motion within cells. This high-voltage electron micrograph shows the mitotic apparatus in a metaphase mammalian cell. The large cylindrical objects are chromosomes, and the thread-like structures stretched across the center are microtubules—tracks for the molecular motors that move chromosomes. Many processes, including chromosome segregation in mitosis, depend on the action of molecular-motor proteins. [Courtesy of Dr. J. R. McIntosh.]

forms having high or low affinity for the filament tracks in response to ATP binding and hydrolysis, enabling a bind, pull, and release mechanism that generates motion.

We will also consider a completely different strategy for generating motion, one used by bacteria such as *E. coli*. A set of flagella act as propellers, rotated by a motor in the bacterial cell membrane. This rotary motor is driven by a proton gradient across the membrane, rather than by ATP hydrolysis. The mechanism for coupling the proton gradient to rotatory motion is analogous to that used by the F_0 subunit of ATP synthase (p. 522). Thus, both of the major modes for storing biochemical energy—namely, ATP and ion gradients—have been harnessed by evolution to drive organized molecular motion.

34.1 Most Molecular-Motor Proteins Are Members of the P-Loop NTPase Superfamily

Eukaryotic cells contain three major families of motor proteins: myosins, kinesins, and dyneins. At first glance, these protein families appear to be quite different from one another. *Myosin*, first characterized on the basis of its role in muscle, moves along filaments of the protein actin. Muscle myosin consists of two copies each of a *heavy chain* with a molecular mass of 87 kd, an *essential light chain*, and a *regulatory light chain*. The human genome appears to encode more than 40 distinct myosins; some function in muscle contraction, and others participate in a variety of other processes. *Kinesins*, which have roles in protein, mRNA, and vesicle transport as well as construction of the mitotic spindle and chromosome segregation, are generally dimers of two polypeptides. The human genome encodes more than 40 kinesins. *Dyneins* power the motion of cilia and flagella, and a general cytoplasmic dynein contributes to a variety of motions in all cells including vesicle transport and various transport events in mitosis. Dyneins are enormous, with heavy chains of molecular mass greater than 500 kd. The human genome appears to encode approximately 10 dyneins.

Comparison of the amino acid sequences of myosins, kinesins, and dyneins did not reveal significant relationships between these protein families but, after their three-dimensional structures were determined, members of the myosin and kinesin families were found to have remarkable similarities. In particular, both myosin and kinesin contain P-loop NTPase cores homologous to those found in G proteins. Sequence analysis of the dynein heavy chain reveals it to be a member of the AAA subfamily of P-loop NTPases that we encountered in the context of the 19S proteasome (p. 653). Dynein has six sequences encoding such P-loop NTPase domains arrayed along its length, although only four actually bind a nucleotide. Thus, we can draw on our knowledge of G proteins and other P-loop NTPases as we analyze the mechanisms of action of these motor proteins.

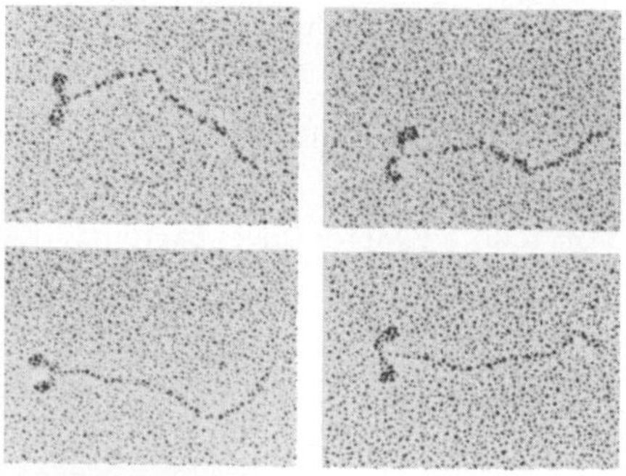

A Motor Protein Consists of an ATPase Core and an Extended Structure

Let us first consider the structure of myosin. The results of electron microscopic studies of skeletal-muscle myosin show it to be a two-headed structure linked to a long stalk (Figure 34.2). As we saw in Chapter 33, limited proteolysis can be a powerful tool in probing the activity of large proteins. The treatment of myosin with trypsin and papain results in the formation of four fragments: two S1 fragments; an S2 fragment, also called heavy

Figure 34.2 Myosin structure at low resolution. Electron micrographs of myosin molecules reveal a two-headed structure with a long, thin tail. [Courtesy of Dr. Paula Flicker, Dr. Theo Walliman, and Dr. Peter Vibert.]

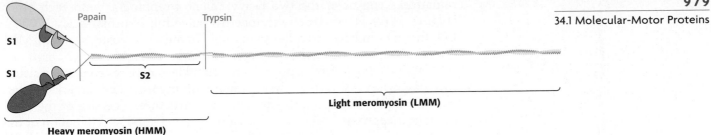

Figure 34.3 Myosin dissection. Treatment of muscle myosin with proteases forms stable fragments, including subfragments S1 and S2 and light meromyosin. Each S1 fragment includes a head (shown in yellow or purple) from the heavy chain and one copy of each light chain (shown in blue and orange).

meromyosin (HMM); and a fragment called light meromyosin (LMM; Figure 34.3). Each *S1 fragment* corresponds to one of the heads from the intact structure and includes 850 amino-terminal amino acids from one of the two heavy chains as well as one copy of each of the light chains. Examination of the structure of an S1 fragment at high resolution reveals the presence of a P-loop NTPase-domain core that is the site of ATP binding and hydrolysis (Figure 34.4).

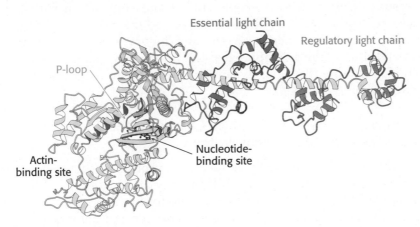

Figure 34.4 Myosin structure at high resolution. The structure of the S1 fragment from muscle myosin reveals the presence of a P-loop NTPase domain (shaded in purple). *Notice* that an α helix that extends from this domain is the binding site for the two light chains. [Drawn from 1DFL.pdb.]

Extending away from this structure is a long α helix from the heavy chain. This helix is the binding site for the two light chains. The light chains are members of the EF-hand family, similar to calmodulin, although most of the EF hands in light chains do not bind metal ions (Figure 34.5). Like calmodulin, these proteins wrap around an α helix, serving to thicken and stiffen it. The remaining fragments of myosin—S2 and light meromyosin— are largely α helical, forming two-stranded coiled coils created by the

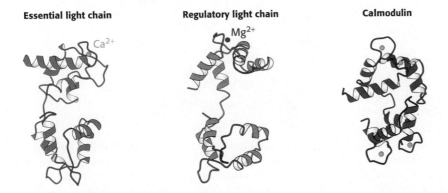

Figure 34.5 Myosin light chains. The structures of the essential and regulatory light chains of muscle myosin are compared with the structure of calmodulin. Each of these homologous proteins binds an α helix (not shown) by wrapping around it. [Drawn from 1DFL.pdb and 1CM1.pdb.]

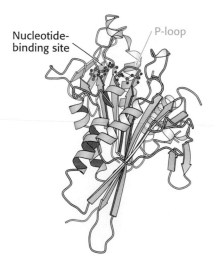

remaining lengths of the two heavy chains wrapping around each other (Figure 34.6). These structures, together extending approximately 1700 Å, link the myosin heads to other structures. In muscle myosin, several LMM domains come together to form higher-order bundles.

Conventional kinesin (kinesin 1), the first kinesin discovered, has a structure having several features in common with myosin. The dimeric protein has two heads, connected by an extended structure. The size of the head domain is approximately one-third of that of myosin. Determination of the three-dimensional structure of a kinesin fragment revealed that the head domain also is built around a P-loop NTPase core (Figure 34.7). The myosin domain is so much larger than that of kinesin because of two large insertions in the myosin domain that bind to actin filaments. For conventional kinesin, a region of approximately 500 amino acids extends from the head domain. Like the corresponding region in myosin, the extended part of kinesin forms an α-helical coiled coil. Conventional kinesin also has light chains, but, unlike those of myosin, these light chains bind near the carboxyl terminus of the heavy chain and are thought to link the motor to intracellular cargo.

Dynein has a rather different structure. As noted earlier, the dynein heavy chain includes six regions that are homologous to the AAA subfamily of ATPase domains. Although no crystallographic data are yet available, the results of electron microscopic studies and comparison with known structures of other AAA ATPases have formed the basis for the construction of a model of the dynein head structure (Figure 34.8). The head domain is appended to a region of approximately 1300 amino acids that forms an extended structure that links dynein units together to form oligomers and interacts with other proteins.

ATP Binding and Hydrolysis Induce Changes in the Conformation and Binding Affinity of Motor Proteins

A key feature of P-loop NTPases such as G proteins is that they undergo structural changes induced by NTP binding and hydrolysis. Moreover, these structural changes alter their affinities for binding partners. Thus, it is not surprising that the NTPase domains of motor proteins display analogous responses to nucleotide binding. The S1 fragment of myosin from

Figure 34.6 Myosin two-stranded coiled coil. The two α helices form left-handed supercoiled structures that spiral around each other. Such structures are stabilized by hydrophobic residues at the contact points between the two helices. [Drawn from 2TMA.pdb.]

Figure 34.7 Structure of head domain of kinesin at high resolution. *Notice* that the head domain of kinesin has the structure of a P-loop NTPase core (indicated by purple shading). [Drawn from 1I6I.pdb.]

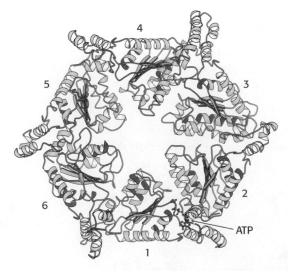

Figure 34.8 Dynein head-domain model. ATP is bound in the first of six P-loop NTPase domains (numbered) in this model for the head domain of dynein. The model is based on electron micrographs and the structures of other members of the AAA ATPase family. The precise role of the six sites is not fully understood. [Drawn from 1HN5.pdb.]

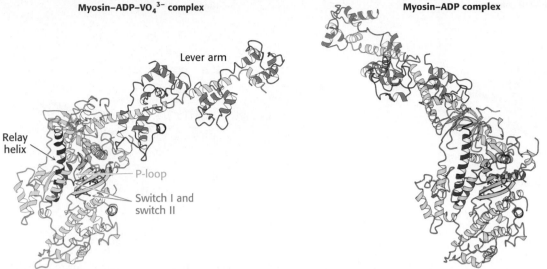

Myosin–ADP–VO$_4$$^{3-}$ complex

Myosin–ADP complex

Lever arm

Relay
helix

P-loop

Switch I and
switch II

Figure 34.9 Lever-arm motion. Two forms of the S1 fragment of scallop-muscle myosin. *Notice* the dramatic conformational changes when the identity of the bound nucleotide changes from ADP–VO$_4$$^{3-}$ to ADP or vice versa, including a nearly 90-degree reorientation of the lever arm. [Drawn from 1DFL.pdb and 1B7T.pdb.]

scallop muscle provides a striking example of the changes observed (Figure 34.9). The structure of the S1 fragment has been determined for S1 bound to a complex formed of ADP and vanadate (VO$_4$$^{3-}$), which is an analog of ATP, or, more precisely, the ATP-hydrolysis transition state. In the presence of the ADP–VO$_4$$^{3-}$ complex, the long helix that binds the light chains (hereafter referred to as the *lever arm*) protrudes outward from the head domain. In the presence of ADP without VO$_4$$^{3-}$, the lever arm has rotated by nearly 90 degrees relative to its position in the ADP–VO$_4$$^{3-}$ complex. How does the identity of the species in the nucleotide-binding site cause this dramatic transition? Two regions around the nucleotide-binding site conform closely to the group in the position of the γ-phosphoryl group of ATP and adopt a looser conformation when such a group is absent (Figure 34.10). This conformational change allows a long α helix (termed the *relay helix*) to adjust its position. The carboxyl-terminal end of the relay helix interacts with structures at the base of the lever arm, and so a change in the position of the relay helix leads to a reorientation of the lever arm.

The binding of ATP significantly decreases the affinity of the myosin head for actin filaments. No structures of myosin–actin complexes have yet been determined at high resolution, so the mechanistic basis for this change remains to be elucidated. However, the amino-terminal end of the relay helix interacts with the domains of myosin that bind to actin, suggesting a clear pathway for the coupling of nucleotide binding to changes in actin affinity. The importance of the changes in actin-binding affinity will be clear later when we examine the role of myosin in generating directed motion (Section 34.2).

Analogous conformational changes take place in kinesin. The kinesins also have a relay helix that can adopt different configurations when kinesin binds different nucleotides. Kinesin lacks an α-helical lever arm, however.

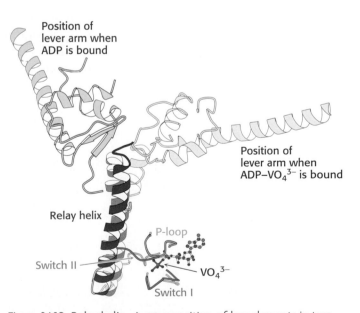

Position of
lever arm when
ADP is bound

Position of
lever arm when
ADP–VO$_4$$^{3-}$ is bound

Relay helix

P-loop

Switch II

VO$_4$$^{3-}$

Switch I

Figure 34.10 Relay helix. A superposition of key elements in two forms of scallop myosin reveals the structural changes that are transmitted by the relay helix from the switch I and switch II loops to the base of the lever arm. The switch I and switch II loops interact with VO$_4$$^{3-}$ in the position that would be occupied by the γ-phosphoryl group of ATP. The structure of the myosin–ADP–VO$_4$$^{3-}$ complex is shown in lighter colors.

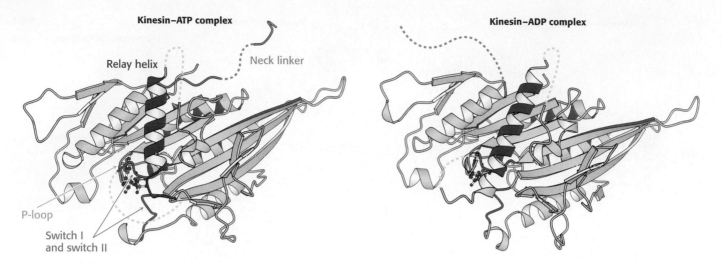

Relay helix

Neck linker

P-loop

Switch I
and switch II

Figure 34.11 Neck linker. A comparison of the structures of a kinesin bound to ADP and bound to an ATP analog. *Notice* that the neck linker (orange), which connects the head domain to the remainder of the kinesin molecule, is bound to the head domain in the presence of the ATP analog but is free in the presence of ADP only. [Drawn from 1I6I.pdb and 1I5S.pdb.]

Instead, a relatively short segment termed the *neck linker* changes conformation in response to nucleotide binding (Figure 34.11). The neck linker binds to the head domain of kinesin when ATP is bound but is released when the nucleotide-binding site is vacant or occupied by ADP. Kinesin differs from myosin in that the binding of ATP to kinesin *increases* the affinity between kinesin and its binding partner, microtubules. Before turning to a discussion of how these properties are used to convert chemical energy into motion, we must consider the properties of the tracks along which these motors move.

34.2 Myosins Move Along Actin Filaments

Myosins, kinesins, and dyneins move by cycling between states with different affinities for the long, polymeric macromolecules that serve as their tracks. For myosin, the molecular track is a polymeric form of *actin*, a 42-kd protein that is one of the most abundant proteins in eukaryotic cells, typically accounting for as much as 10% of the total protein. Actin polymers are continually being assembled and disassembled in cells in a highly dynamic manner, accompanied by the hydrolysis of ATP. On the microscopic scale, actin filaments participate in the dynamic reshaping of the cytoskeleton and the cell itself and in other motility mechanisms that do not include myosin. In muscle, myosin and actin together are the key components responsible for muscle contraction.

Muscle Is a Complex of Myosin and Actin

Vertebrate muscle that is under voluntary control has a banded (striated) appearance when examined under a light microscope. It consists of multinucleated cells that are bounded by an electrically excitable plasma membrane. A muscle cell contains many parallel *myofibrils*, each about 1 μm in diameter. The functional unit, called a *sarcomere*, typically repeats every 2.3 μm (23,000 Å) along the fibril axis in relaxed muscle (Figure 34.12). A dark *A band* and a light *I band* alternate regularly. The central region of the A band, termed the *H zone*, is less dense that the rest of the band. The I band is bisected by a very dense, narrow *Z line*.

The underlying molecular plan of a sarcomere is revealed by cross sections of a myofibril. These cross sections show the presence of two kinds of interacting protein filaments. The *thick filaments* have diameters of about 15 nm (150 Å) and consist primarily of myosin. The *thin filaments* have

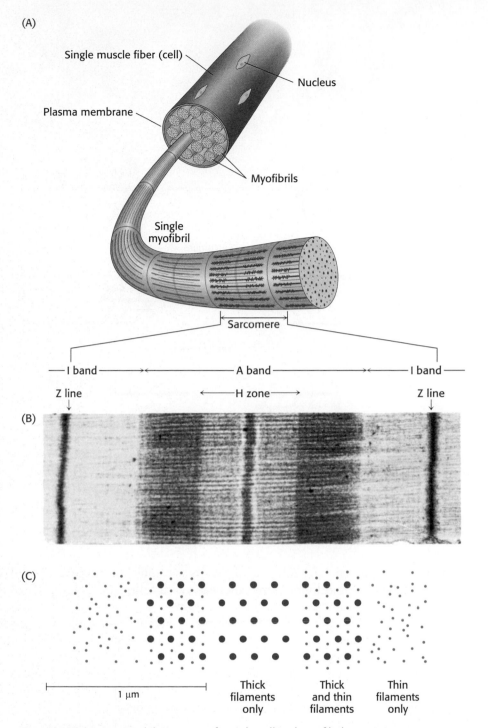

(A)

Single muscle fiber (cell)

Nucleus

Plasma membrane

Myofibrils

Single myofibril

Sarcomere

I band

A band

I band

Z line

H zone

Z line

(B)

(C)

1 μm

Thick filaments only

Thick and thin filaments

Thin filaments only

Figure 34.12 Sarcomere. (A) Structure of muscle cell and myofibril containing sarcomeres. (B) Electron micrograph of a longitudinal section of a skeletal-muscle myofibril, showing a single sarcomere. (C) Schematic representations of cross sections correspond to the regions in the micrograph. [Courtesy of Dr. Hugh Huxley.]

Figure 34.13 Sliding-filament model. Muscle contraction depends on the motion of thin filaments (blue) relative to thick filaments (red). [After H. E. Huxley. The mechanism of muscular contraction. Copyright © 1965 by Scientific American, Inc. All rights reserved.]

diameters of approximately 8 nm (80 Å) and consist of actin as well as *tropomyosin* and the *troponin complex*. Muscle contraction is achieved through the sliding of the thin filaments along the length of the thick filaments, driven by the hydrolysis of ATP (Figure 34.13).

To form the thick filaments, myosin molecules self-assemble into thick bipolar structures with the myosin heads protruding at both ends of a bare region in the center (Figure 34.14A). Approximately 500 head domains line the surface of each thick filament. Each head-rich region associates with two

(A)

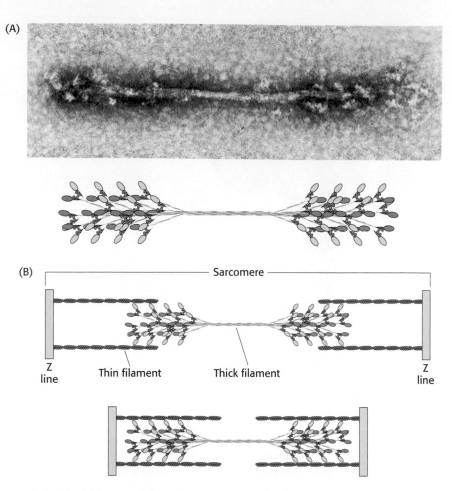

(B)

Figure 34.14 Thick filament. (A) An electron micrograph of a reconstituted thick filament reveals the presence of myosin head domains at each end and a relatively narrow central region. A schematic view below shows how myosin molecules come together to form the thick filament. (B) A diagram showing the interaction of thick and thin filaments in skeletal-muscle contraction. [(A, top) Courtesy of Dr. Hugh Huxley.]

actin filaments, one on each side of the myosin molecules (Figure 34.14B). The interaction of individual myosin heads with actin units creates the sliding force that gives rise to muscle contraction.

Tropomyosin and the troponin complex regulate this sliding in response to nerve impulses. Under resting conditions, tropomyosin blocks the intimate interaction between myosin and actin. A nerve impulse leads to an increase in calcium ion concentration within the muscle cell. A component of the troponin complex senses the increase in Ca^{2+} and, in response, relieves the inhibition of myosin–actin interactions by tropomyosin.

Although myosin was discovered through its role in muscle, other types of myosin play crucial roles in a number of physiological contexts. Some defects in hearing in both mice and human beings have been linked to mutations in particular myosin homologs that are present in cells of the ear. For example, Usher syndrome in human beings and the *shaker* mutation in mice have been linked to myosin VIIa, expressed in hair cells (Section 32.4). The mutation of this myosin results in the formation of splayed stereocilia that do not function well. Myosin VIIa differs from muscle myosin in that its tail region possesses a number of amino acid sequences that correspond to domains known to mediate specific protein–protein interactions. Instead of assembling into fibers as muscle myosin does, myosin VIIa functions as a dimer.

Actin Is a Polar, Self-Assembling, Dynamic Polymer

The structure of the actin monomer was determined to atomic resolution by x-ray crystallography and has been used to interpret the structure of actin filaments, already somewhat understood through electron microscopy studies at lower resolution. Each actin monomer comprises four domains (Figure 34.15). These domains come together to surround a bound nucleotide, either ATP or ADP. The ATP form can be converted into the ADP form by hydrolysis.

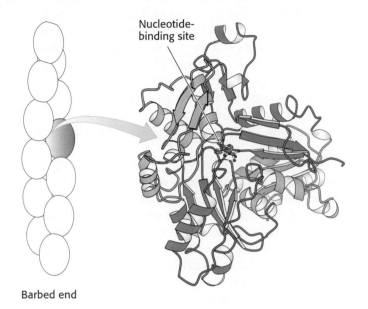

Barbed end

Figure 34.15 Actin structure. (Left) Schematic view of actin monomers (one in blue) of an actin filament. (Right) The domains in the four-domain structure of an actin monomer are identified by different shades of blue. [Drawn from 1J6Z.pdb.]

Actin monomers (often called *G-actin* for globular) come together to form actin filaments (often called *F-actin*; see Figure 34.15). F-actin has a helical structure; each monomer is related to the preceding one by a translation of 27.5 Å and a rotation of 166 degrees around the helical axis. Because the rotation is nearly 180 degrees, F-actin resembles a two-stranded cable. Note that each actin monomer is oriented in the same direction along the F-actin filament, and so the structure is polar, with discernibly different ends. One end is called the barbed (plus) end, and the other is called the pointed (minus) end. The names "barbed" and "pointed" refer to the appearance of an actin filament when myosin S1 fragments are bound to it.

How are actin filaments formed? Like many biological structures, actin filaments self-assemble; that is, under appropriate conditions, actin monomers will come together to form well-structured, polar filaments. The aggregation of the first two or three monomers to form a filament is highly unfavorable. Thus, specialized protein complexes, including one called Arp2/3, serve as nuclei for actin assembly in cells. Once such a filament nucleus exists, the addition of subunits is more favorable. Let us consider the polymerization reaction in more detail. We designate an actin filament with n subunits A_n. This filament can bind an additional actin monomer, A, to form A_{n+1}.

$$K_d = \frac{[A_n][A]}{[A_{n+1}]}$$

A_n + A $\rightleftharpoons$ A_{n+1}

The dissociation constant, K_d, for this reaction, defines the monomer concentrations at which the polymerization reaction will take place, because the concentration of polymers of length $n + 1$ will be essentially equal to that for polymers of length n. Thus,

$$[A_n] \sim [A_{n+1}] \text{ and } K_d = \frac{[A_n][A]}{[A_{n+1}]} \sim [A]$$

In other words, the polymerization reaction will proceed until the monomer concentration is reduced to the value of K_d. If the monomer concentration is below the value of K_d, the polymerization reaction will not proceed at all; indeed, existing filaments will depolymerize until the monomer concentration reaches the value of K_d. Because of these phenomena, K_d is referred to as the *critical concentration* for the polymer. Recall that actin contains a nucleotide-binding site that can contain either ATP or ADP. The critical concentration for the actin–ATP complex is approximately 20-fold lower than that for the actin–ADP complex; actin–ATP polymerizes more readily than does actin–ADP.

Actin filaments inside cells are highly dynamic structures that are continually gaining and losing monomers. Nucleation by complexes such as Arp2/3 can initiate the polymerization of actin–ATP. In contrast, the hydrolysis of bound ATP to ADP favors actin *de*polymerization. This reaction acts as a timer to make actin filaments kinetically unstable. Proteins that bind actin monomers or promote the severing of actin filaments also play roles. Polymerization reactions can exert force, pushing or pulling on cell membranes. *Regulated actin polymerization is central to the changes in cell shape associated with cell motility in amebas as well as in human cells such as macrophages.*

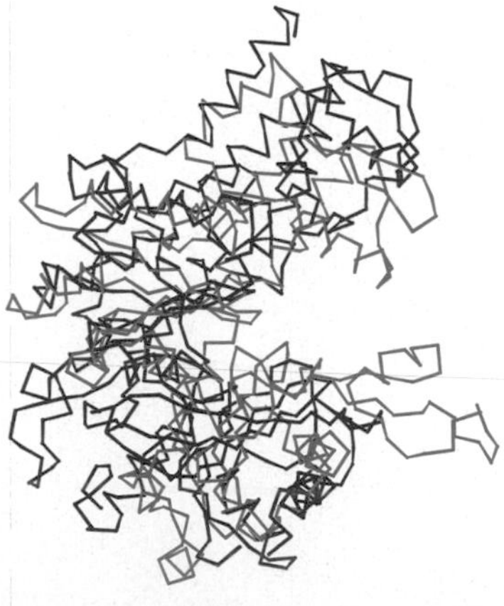

A well-defined actin cytoskeleton is unique to eukaryotes; prokaryotes lack such structures. How did filamentous actin evolve? Comparisons of the three-dimensional structure of G-actin with other proteins revealed remarkable similarity to several other proteins, including sugar kinases such as hexokinase (Figure 34.16; see also p. 437). Notably, the nucleotide-binding site in actin corresponds to the ATP-binding site in hexokinase. Thus, actin evolved from an enzyme that utilized ATP as a substrate.

More recently, a closer prokaryotic homolog of actin was characterized. This protein, called MreB, plays an important role in determining cell shape in rod-shaped, filamentous, and helical bacteria. The internal structures formed by MreB are suggestive of the actin cytoskeleton of eukaryotic cells, although they are far less extensive. Even though this protein is only approximately 15% identical in sequence with actin, MreB folds into a very similar three-dimensional structure. It also polymerizes into structures that are similar to F-actin in a number of ways, including the alignment of the component monomers.

Figure 34.16 Actin and hexokinase. A comparison of actin (blue) and hexokinase from yeast (red) reveals structural similarities indicative of homology. *Notice* that both proteins have a deep cleft in which nucleotides bind.

Motions of Single Motor Proteins Can Be Directly Observed

Muscle contraction is complex, requiring the action of many different myosin molecules. Studies of *single myosin molecules* moving relative to actin filaments have been sources of deep insight into the mechanisms underlying muscle contraction and other complex processes.

A powerful tool for these studies, called an *optical trap*, relies on highly focused laser beams (Figure 34.17). Small beads can be caught in these traps and held in place in solution.

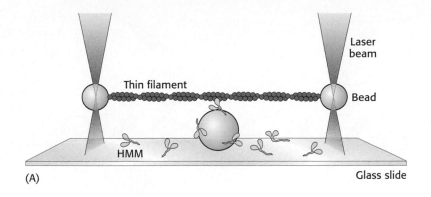

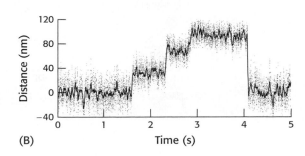

Figure 34.17 Watching a single motor protein in action. (A) An actin filament (blue) is placed above a heavy meromyosin (HMM) fragment (yellow) that projects from a bead on a glass slide. A bead attached to each end of the actin filament is held in an optical trap produced by a focused, intense infrared laser beam (orange). The position of these beads can be measured with nanometer precision. (B) Recording of the displacement of an actin filament due to a myosin derivative attached to a bead, influenced by the addition of ATP. Note the fairly uniform step sizes that are observed. [(A) After J. T. Finer, R. M. Simmons, and J. A. Spudich. *Nature* 368(1994):113–119; (B) From R. S. Rock, M. Rief, A. D. Metra, and J. A. Spudich. *Methods* 22(2000):378–381.]

The position of the beads can be monitored with nanometer precision. James Spudich and coworkers designed an experimental arrangement consisting of an actin filament that had a bead attached to each end. Each bead could be caught in an optical trap (one at each end of the filament) and the actin filament could be pulled taut over a microscope slide containing other beads that had been coated with fragments of myosin such as the heavy meromyosin fragment (see Figure 34.17). On the addition of ATP, transient displacements of the actin filament were observed along its long axis. The size of the displacement steps was fairly uniform with an average size of 11 nm (110 Å).

The results of these studies, performed in the presence of varying concentrations of ATP, are interpreted as showing that individual myosin heads bind the actin filament and undergo a conformational change (the *power stroke*) that pulls the actin filament, leading to the displacement of the beads. After a period of time, the myosin head releases the actin, which then snaps back into place.

Phosphate Release Triggers the Myosin Power Stroke

How does ATP hydrolysis drive the power stroke? A key observation is that the addition of ATP to a complex of myosin and actin results in the dissociation of the complex. Thus, ATP binding and hydrolysis cannot be directly responsible for the power stroke. We can combine this fact with the structural observations described earlier to construct a mechanism for the motion of myosin along actin (Figure 34.18). Let us begin with myosin ADP bound to actin. The release of ADP and the binding of ATP to actin result in the dissociation of myosin from actin. As we saw earlier, the binding of ATP by its γ-phosphoryl group to the myosin head leads to a significant conformational change, amplified by the lever arm. This conformational change moves the myosin head along the actin filament by approximately 110 Å. The ATP in the myosin is then hydrolyzed to ADP and P_i, which remain bound to myosin. The myosin head can then bind to the surface of actin, resulting in the dissociation of P_i from the myosin. Phosphate release, in turn, leads to a conformational change that increases the affinity of the myosin head for actin and allows the lever arm to move back to its initial position. *The conformational change associated with phosphate release corresponds to the power stroke.* After the release of P_i, the myosin remains tightly bound to the actin and the cycle can begin again.

How does this cycle apply to muscle contraction? Recall that hundreds of head domains project from the ends of each thick filament. The head domains are paired in myosin dimers, but the two heads within each dimer act independently. Actin filaments associate with each head-rich region, with the barbed ends of actin toward the Z line. In the presence of normal levels of ATP, most of the myosin heads are detached from actin. Each head can

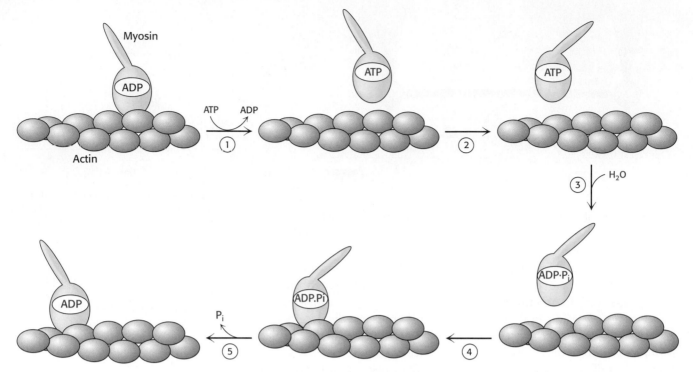

Figure 34.18 Myosin motion along actin. A myosin head (yellow) in the ADP form is bound to an actin filament (blue). The exchange of ADP for ATP results in (1) the release of myosin from actin and (2) substantial reorientation of the lever arm of myosin. The hydrolysis of ATP (3) allows the myosin head to rebind at a site displaced along the actin filament (4). The release of P_i (5) accompanying this binding increases the strength of the interaction between myosin and actin and resets the orientation of the lever arm.

independently hydrolyze ATP, bind to actin, release P_i, and undergo its power stroke. Because few other heads are attached, the actin filament is relatively free to slide. Each head cycles approximately five times per second with a movement of 110 Å per cycle. However, because hundreds of heads are interacting with the same actin filament, the overall rate of movement of myosin relative to the actin filament may reach 80,000 Å per second, allowing a sarcomere to contract from its fully relaxed to its fully contracted form rapidly. Having many myosin heads briefly and independently attaching and moving an actin filament allows for much greater speed than could be achieved by a single motor protein.

The Length of the Lever Arm Determines Motor Velocity

A key feature of myosin motors is the role of the lever arm as an amplifier. The lever arm amplifies small structural changes at the nucleotide-binding site to achieve the 110-Å movement along the actin filament that takes place in each ATP hydrolysis cycle. A strong prediction of the mechanism proposed for the movement of myosin along actin is that the length traveled per cycle should depend on the length of this lever arm. Thus, the length of the lever arm should influence the overall rate at which actin moves relative to a collection of myosin heads.

This prediction was tested with the use of mutated forms of myosin with lever arms of different lengths. The lever arm in muscle myosin includes binding sites for two light chains (Section 34.1). Thus investigators shortened the lever arm by deleting the sequences that correspond to one or both of these binding sites. They then examined the rates at which actin filaments were transported along collections of these mutated myosins (Figure 34.19). As predicted, the rate decreased as the lever arm was shortened. A

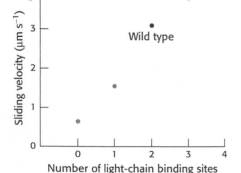

Figure 34.19 Myosin lever-arm length. Examination of the rates of actin movement supported by a set of myosin mutants with different numbers of light-chain binding sites revealed a linear relation; the greater the number of light-chain binding sites (and, hence, the longer the lever arm), the faster the sliding velocity. [After T. Q. P. Uyeda, P. D. Abramson, and J. A. Spudich. *Proc. Natl. Acad. Sci. U.S.A.* 93(1996):4459–4464.]

mutated form of myosin with an unusually long lever arm was generated by inserting 23 amino acids corresponding to the binding site for an additional regulatory light chain. Remarkably, this form was found to support actin movement that was *faster than that of the wild-type protein*. These results strongly support the proposed role of the lever arm in contributing to myosin motor activity.

34.3 Kinesin and Dynein Move Along Microtubules

In addition to actin, the cytoskeleton includes other components, notably intermediate filaments and microtubules. Microtubules serve as tracks for two classes of motor proteins—namely, kinesins and dyneins. Kinesins moving along microtubules usually carry cargo such as organelles and vesicles from the center of a cell to its periphery. Dyneins are important in sliding microtubules relative to one other during the beating of cilia and flagella on the surfaces of some eukaryotic cells. Additionally, dynein carries cargos from the cell periphery to the cell center.

Some members of the kinesin family are crucial to the transport of organelles and other cargo to nerve endings at the peripheries of neurons. It is not surprising, then, that mutations in these kinesins can lead to nervous system disorders. For example, mutations in a kinesin called KIF1Bβ can lead to the most common peripheral neuropathy (weakness and pain in the hands and feet), Charcot-Marie-Tooth disease, which affects 1 in 2500 people. A glutamine-to-leucine mutation in the P-loop of the motor domain of this kinesin has been found in some affected persons. Knockout mice with a disruption of the orthologous gene have been generated. Mice heterozygous for the disruption show symptoms similar to those observed in human beings; homozygotes die shortly after birth. Mutations in other kinesin genes have been linked to human spastic paraplegia. In these disorders, defects in kinesin-linked transport may impair nerve function directly, and the decrease in the activity of specific neurons may lead to other degenerative processes.

Microtubules Are Hollow Cylindrical Polymers

Microtubules are built from two kinds of homologous 50-kd subunits, α- and β-tubulin, which assemble in a helical array of alternating tubulin types to form the wall of a hollow cylinder (Figure 34.20). Alternatively, a microtubule can be regarded as 13 protofilaments that run parallel to its

(A)

α-Tubulin
β-Tubulin

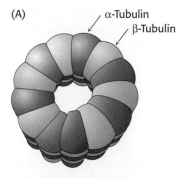

(B)

300 Å (30 nm)

Figure 34.20 Microtubule structure. Schematic views of the helical structure of a microtubule. α-Tubulin is shown in dark red and β-tubulin in light red. (A) Top view. (B) Side view.

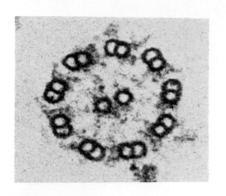

Figure 34.21 Microtubule arrangement. Electron micrograph of a cross section of a flagellar axoneme shows nine microtubule doublets surrounding two singlets. [Courtesy of Dr. Joel Rosenbaum.]

long axis. The outer diameter of a microtubule is 30 nm, much larger than that of actin (8 nm). Like actin, microtubules are polar structures. The minus end of a microtubule is anchored near the center of a cell, whereas the plus end extends toward the cell surface.

Microtubules are also key components of cilia and flagella present on some eukaryotic cells. For example, sperm propel themselves through the motion of flagella containing microtubules. The microtubules present in these structures adopt a common architecture (Figure 34.21). A bundle of microtubules called an *axoneme* is surrounded by a membrane contiguous with the plasma membrane. The axoneme is composed of a peripheral group of nine microtubule pairs surrounding two singlet microtubules. This recurring motif is often called a *9 + 2 array*. Dynein drives the motion of one member of each outer pair relative to the other, causing the overall structure to bend.

Microtubules are important in determining the shapes of cells and in separating daughter chromosomes in mitosis. They are highly dynamic structures that grow through the addition of α- and β-tubulin to the ends of existing structures. Like actin, *tubulins* bind and hydrolyze nucleoside triphosphates, although for tubulin the nucleotide is GTP rather than ATP. The critical concentration for the polymerization of the GTP forms of tubulin is lower than that for the GDP forms. Thus, a newly formed microtubule consists primarily of GTP-tubulins. Through time, the GTP is hydrolyzed to GDP. The GDP-tubulin subunits in the interior length of a microtubule remain stably polymerized, whereas GDP subunits exposed at an end have a strong tendency to dissociate. Marc Kirschner and Tim Mitchison found that some microtubules in a population lengthen while others simultaneously shorten. This property, called *dynamic instability*, arises from random fluctuations in the number of GTP- or GDP-tubulin subunits at the plus end of the polymer. The dynamic character of microtubules is crucial for processes such as mitosis, which require the assembly and disassembly of elaborate microtubule-based structures.

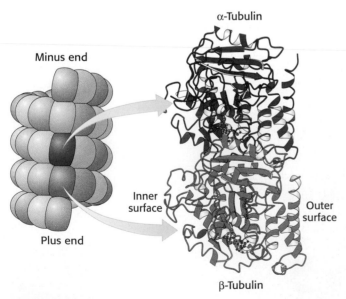

Figure 34.22 Tubulin. Microtubules can be viewed as an assembly of α-tubulin–β-tubulin dimers. The structures of α-tubulin and β-tubulin are quite similar. *Notice* that each includes a P-loop NTPase domain (purple shading) and a bound guanine nucleotide. [Drawn from 1JFF.pdb.]

The structure of tubulin was determined at high resolution by electron crystallographic methods (Figure 34.22). As expected from their 40% sequence identity, α- and β-tubulin have very similar three-dimensional structures. Further analysis revealed that the tubulins are members of the P-loop NTPase family and contain a nucleotide-binding site adjacent to the P-loop. Tubulins are present only in eukaryotes, although a prokaryotic homolog has been found. Sequence analysis identified a prokaryotic protein called FtsZ (for *f*ilamentous *t*emperature-*s*ensitive mutant *Z*) that is quite similar to the tubulins. The homology was confirmed when the structure was determined by x-ray crystallography. Interestingly, this protein participates in bacterial division, forming ring-shaped structures at the constriction that arises when a cell divides. These observations suggest that tubulins may have evolved from an ancient cell-division protein.

The continual lengthening and shortening of microtubules is essential to their role in cell division. *Taxol*, a compound isolated from the bark of the Pacific yew tree, was discovered through its ability to interfere with cell proliferation. Taxol binds to microtubules and stabilizes the polymerized form.

Taxol

Taxol and its derivatives have been developed as anticancer agents because they preferentially affect rapidly dividing cells, such as those in tumors.

Kinesin Motion Is Highly Processive

Kinesins are motor proteins that move along microtubules. We have seen that myosin moves along actin filaments by a process in which actin is released in each cycle; a myosin head group acting independently dissociates from actin after every power stroke. In contrast, when a kinesin molecule moves along a microtubule, the two head groups of the kinesin molecule operate in tandem: one binds, and then the next one does. A kinesin molecule may take many steps before both head groups are dissociated at the same time. In other words, the motion of kinesin is highly processive. Single-molecule measurements allow processive motion to be observed (Figure 34.23). A single kinesin molecule will typically take 100 or more steps toward the plus end of a microtubule in a period of seconds before the molecule

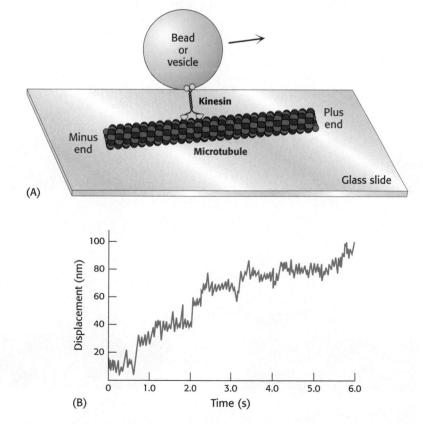

Figure 34.23 Monitoring movements mediated by kinesin. (A) The movement of beads or vesicles, carried by individual kinesin dimers along a microtubule, can be directly observed. (B) A trace shows the displacement of a bead carried by a kinesin molecule. Multiple steps are taken in the 6-s interval. The average step size is about 8 nm (80 Å). [(B) After K. Svoboda, C. F. Schmidt, B. J. Schnapp, and S. M. Block. *Nature* 365(1993):721–727.]

becomes detached from the microtubule. These measurements also revealed that the average step size is approximately 80 Å, a value that corresponds to the distance between consecutive α- or β-tubulin subunits along each protofilament.

An additional fact is crucial to the development of a mechanism for kinesin motion—namely, that the addition of ATP strongly *increases* the affinity of kinesin for microtubules. This behavior stands in contrast with the behavior of myosin; ATP binding to myosin promotes its *dissociation* from actin. Do these differences imply that kinesin and myosin operate by completely different mechanisms? Indeed not. Kinesin-generated movement appears to proceed by a mechanism that is quite similar to that used by myosin (Figure 34.24). Let us begin with a two-headed kinesin molecule in its ADP form, dissociated from a microtubule. Recall that the neck linker binds the head domain when ATP is bound and is released when ADP is bound. The initial interaction of one of the head domains with a tubulin dimer on a microtubule stimulates the release of ADP from this head domain and the subsequent binding of ATP. The binding of ATP triggers a conformational change in the head domain that leads to two important events. First, the affinity of the head domain for the microtubule increases, essentially locking this head domain in place. Second, the neck linker binds to the head domain. This change, transmitted through the coiled-coil domain that connects the two kinesin monomers, repositions the other head domain. In its new position, the second head domain is close to a second tubulin dimer, 80 Å along the microtubule in the direction of the plus end. Meanwhile, the intrinsic ATPase activity of the first head domain hydrolyzes the ATP to ADP and P_i. When the second head domain binds to the microtubule, the first head releases ADP and binds ATP. Again, ATP binding favors a conformational change that pulls the first domain forward. This process can continue for many cycles until, by chance, both head domains are in the ADP form simultaneously and kinesin dissociates from the microtubule. Because of the relative rates of the component reactions, a simultaneous dissociation takes place approximately every 100 cycles.

Figure 34.24 Kinesin moving along a microtubule. (1) One head of a two-headed kinesin molecule, initially with both heads in the ADP form, binds to a microtubule. (2) The release of ADP and the binding of ATP results in a conformational change that locks the head to the microtubule and pulls the neck linker (orange) to the head domain, throwing the second domain toward the plus end of the microtubule. (3) ATP undergoes hydrolysis while the second head interacts with the microtubule. (4) The exchange of ATP for ADP in the second head pulls the first head off the microtubule, releasing P_i and moving the first domain along the microtubule. (5) The cycle repeats, moving the kinesin dimer farther down the microtubule.

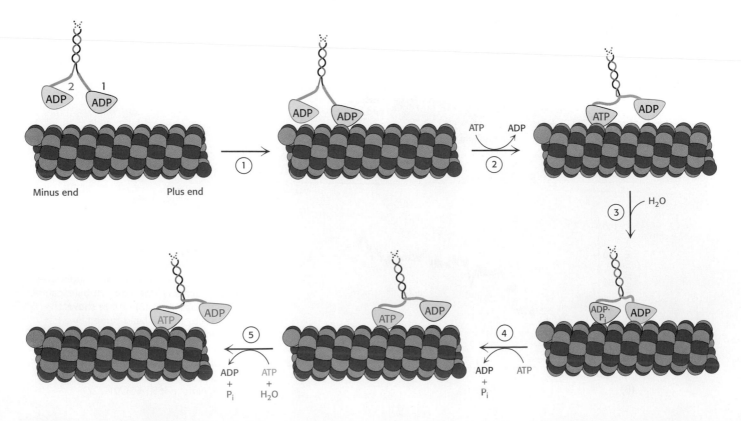

Although recent studies have led to a structural model for the dynein power stroke, the mechanism by which it works is unknown and an interesting research frontier.

Kinesin hydrolyzes ATP at a rate of approximately 80 molecules per second. Thus, given the step size of 80 Å per molecule of ATP, kinesin moves along a microtubule at a speed of 6400 Å per second. This rate is considerably lower than the maximum rate for myosin, which moves relative to actin at 80,000 Å per second. Recall, however, that myosin movement depends on the independent action of hundreds of different head domains working along the same actin filament, whereas the movement of kinesin is driven by the processive action of kinesin head groups working in pairs. Muscle myosin evolved to maximize the speed of the motion, whereas kinesin functions to achieve steady, but slower, transport in one direction along a filament.

34.4 A Rotary Motor Drives Bacterial Motion

In 1 s, a motile bacterium can move approximately 25 μm, or about 10 body lengths. A human being sprinting at a proportional rate would complete the 100-meter dash in slightly more than 5 s. The motors that power this impressive motion are strikingly different from the eukaryotic motors that we have seen so far. In the bacterial motor, an element spins around a central axis rather than moving along a polymeric track. The direction of rotation can change rapidly, a feature that is central to chemotaxis, the process by which bacteria swim preferentially toward an increasing concentration of certain useful compounds and away from potentially harmful ones. One type of flagellar motor, powered by a Na^+ gradient, turns at a rate of 200,000 revolutions per minute.

Bacteria Swim by Rotating Their Flagella

Bacteria such as *Escherichia coli* and *Salmonella typhimurium* swim by rotating flagella that lie on their surfaces (Figure 34.25). When the flagella rotate in a counterclockwise direction (viewed from outside a bacterium), the separate flagella form a bundle that very efficiently propels the bacterium through solution.

Bacterial flagella are polymers approximately 15 nm in diameter and as much as 15 μm in length, composed of 53-kd subunits of a protein called *flagellin* (Figure 34.26). These subunits associate into a helical structure that has 5.5 subunits per turn, giving the appearance of 11 protofilaments. Each flagellum has a hollow core. Remarkably, flagella form not by growing at the base adjacent to the cell body but, instead, by the addition of new subunits

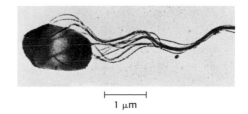

| 1 μm |

Figure 34.25 Bacterial flagella. Electron micrograph of *S. typhimurium* shows flagella in a bundle. [Courtesy of Dr. Daniel Koshland, Jr.]

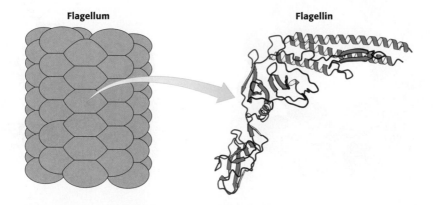

Flagellum

Flagellin

Figure 34.26 Structure of flagellin. A bacterial flagellum is a helical polymer of the protein flagellin. [Drawn from 1IO1.pdb.]

that pass through the hollow core and add to the free end. Each flagellum is intrinsically twisted in a left-handed sense. At its base, each flagellum has a rotory motor.

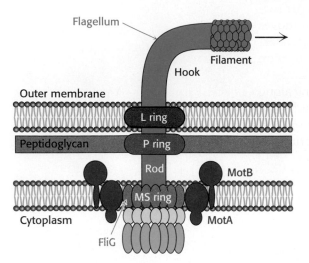

Figure 34.27 Flagellar motor. A schematic view of the flagellar motor, a complex structure containing as many as 40 distinct types of protein. The approximate positions of the proteins MotA and MotB (red), FliG (orange), FliN (yellow), and FliM (green) are shown.

Proton Flow Drives Bacterial Flagellar Rotation

Early experiments by Julius Adler demonstrated that ATP is *not* required for flagellar motion. What powers these rotary motors? The necessary free energy is derived from the proton gradient that exists across the plasma membrane. The flagellar motor is quite complex, containing as many as 40 distinct proteins (Figure 34.27). Five components particularly crucial to motor function have been identified through genetic studies. MotA is a membrane protein that appears to have four transmembrane helices as well as a cytoplasmic domain. MotB is another membrane protein with a single transmembrane helix and a large periplasmic domain. Approximately 11 *MotA–MotB pairs* form a ring around the base of the flagellum. The proteins *FliG*, *FliM*, and *FliN* are part of a disc-like structure called the MS (*membrane and supramembrane*) ring, with approximately 30 FliG subunits coming together to form the ring. The three-dimensional structure of the carboxyl-terminal half of FliG reveals a wedge-shaped domain with a set of charged amino acids, conserved among many species, lying along the thick edge of the wedge (Figure 34.28).

The MotA–MotB pair and FliG combine to create a proton channel that drives rotation of the flagellum. How can proton flow across a membrane drive mechanical rotation? We have seen such a process earlier in regard to ATP synthase (p. 524). Recall that the key to driving the rotation of the γ subunit of ATP synthase is the **a** subunit of the F_0 fragment. This subunit appears to have two half-channels; protons can move across the membrane only by moving into the half-channel from the side of the membrane with the higher local proton concentration, binding to a disc-like structure formed by the **c** subunits, riding on this structure as it rotates to the opening of the other half-channel, and exiting to the side with the lower local proton concentration. Could a similar mechanism apply to flagellar rotation? Indeed, such a mechanism was first proposed by Howard Berg to explain flagellar rotation before the rotary mechanism of ATP synthase was elucidated. Each MotA–MotB pair is conjectured to form a structure that has two half-channels; FliG serves as the rotating proton carrier, perhaps with the participation of some of the charged residues identified in crystallographic studies (Figure 34.29). In this scenario, a proton from the periplasmic space passes into the outer half-channel and is transferred to

Figure 34.28 Flagellar motor components. Approximately 30 subunits of FliG assemble to form part of the MS ring. The ring is surrounded by approximately 11 structures consisting of MotA and MotB. *Notice* that the carboxyl-terminal domain of FliG includes a ridge lined with charged residues that may participate in proton transport. [Drawn from 1QC7.pdb.]

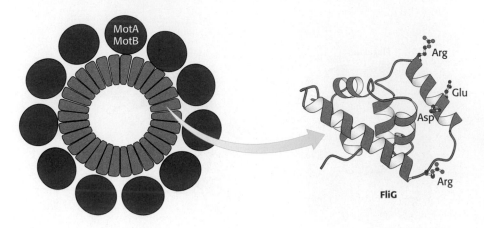

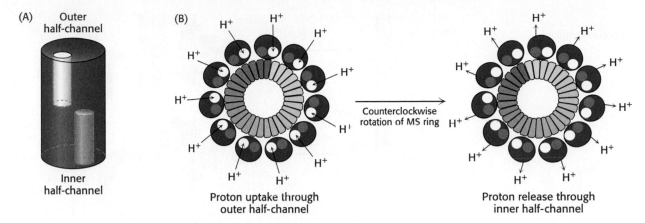

(A) Outer half-channel

Inner half-channel

(B)

Counterclockwise rotation of MS ring

Proton uptake through outer half-channel

Proton release through inner half-channel

Figure 34.29 Proton-transport-coupled rotation of the flagellum. (A) MotA–MotB may form a structure having two half-channels. (B) One model for the mechanism of coupling rotation to a proton gradient requires protons to be taken up into the outer half-channel and transferred to the MS ring. The MS ring rotates in a counterclockwise direction, and the protons are released into the inner half-channel. The flagellum is linked to the MS ring and so the flagellum rotates as well.

an FliG subunit. The MS ring rotates, rotating the flagellum with it and allowing the proton to pass into the inner half-channel and into the cell. Ongoing structural and mutagenesis studies are testing and refining this hypothesis.

Bacterial Chemotaxis Depends on Reversal of the Direction of Flagellar Rotation

Many species of bacteria respond to changes in their environments by adjusting their swimming behavior. Examination of the paths taken is highly revealing (Figure 34.30). The bacteria swim in one direction for some length of time (typically about a second), tumble briefly, and then set off in a new direction. The tumbling is caused by a brief reversal in the direction of the flagellar motor. When the flagella rotate counterclockwise, the helical filaments form a coherent bundle favored by the intrinsic shape of each filament, and the bacterium swims smoothly. When the rotation reverses, the bundle flies apart because the screw sense of the helical flagella does not match the direction of rotation (Figure 34.31). Each flagellum then pulls in a different direction and the cell tumbles.

In the presence of a gradient of certain substances such as glucose, bacteria swim preferentially toward the direction of the higher concentration of the substance. Such compounds are referred to as *chemoattractants*. Bacteria also swim preferentially away from potentially harmful compounds such as phenol, a *chemorepellant*. The process of moving in specific directions in response to environmental cues is called *chemotaxis*. In the presence of a gradient of a chemoattractant, bacteria swim for longer periods of time without tumbling when moving toward higher concentrations of the chemoattractant. In contrast, they tumble more frequently when moving toward lower concentrations of the chemoattractant. This behavior

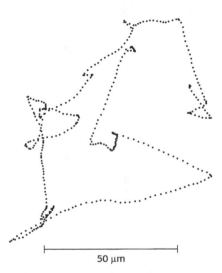

50 μm

Figure 34.30 Charting a course. This projection of the track of an *E. coli* bacterium was obtained with a microscope that automatically follows bacterial motion in three dimensions. The points show the locations of the bacterium at 80-ms intervals. [After H. C. Berg. *Nature* 254(1975):389–392.]

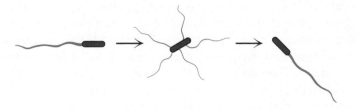

Figure 34.31 Changing direction. Tumbling is caused by an abrupt reversal of the flagellar motor, which disperses the flagellar bundle. A second reversal of the motor restores smooth swimming, almost always in a different direction. [After a drawing kindly provided by Dr. Daniel Koshland, Jr.]

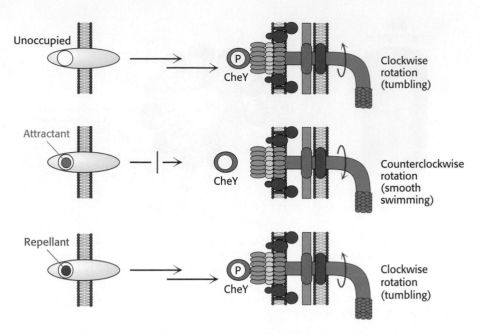

Figure 34.32 Chemotaxis signaling pathway. Receptors in the plasma membrane initiate a signaling pathway leading to the phosphorylation of the CheY protein. Phosphorylated CheY binds to the flagellar motor and favors clockwise rotation. When an attractant binds to the receptor, this pathway is blocked, and counterclockwise flagellar rotation and, hence, smooth swimming result. When a repellant binds, the pathway is stimulated, leading to an increased concentration of phosphorylated CheY and, hence, more-frequent clockwise rotation and tumbling.

is reversed for chemorepellants. The result of these actions is a *biased random walk* that facilitates net motion toward conditions more favorable to the bacterium.

Chemotaxis depends on a signaling pathway that terminates at the flagellar motor. The signaling pathway begins with the binding of molecules to receptors in the plasma membrane (Figure 34.32). In their *unoccupied* forms, these receptors initiate a pathway leading eventually to the phosphorylation of a specific aspartate residue on a soluble protein called *CheY*. In its phosphorylated form, CheY binds to the base on the flagellar motor. When bound to phosphorylated CheY, the flagellar motor rotates in a clockwise rather than a counterclockwise direction, causing tumbling.

The binding of a chemoattractant to a surface receptor blocks the signaling pathway leading to CheY phosphorylation. Phosphorylated CheY spontaneously hydrolyzes and releases its phosphate group in a process accelerated by another protein, CheZ. The concentration of phosphorylated CheY drops, and the flagella are less likely to rotate in a clockwise direction. Under these conditions, bacteria swim smoothly without tumbling. Thus, the reversible rotary flagellar motor and a phosphorylation-based signaling pathway work together to generate an effective means for responding to environmental conditions.

Bacteria sense spatial gradients of chemoattractants by measurements separated in time. A bacterium sets off in a random direction and, if the concentration of the chemoattractant has increased after the bacterium has been swimming for a period of time, the likelihood of tumbling decreases and the bacterium continues in roughly the same direction. If the concentration has decreased, the tumbling frequency increases and the bacterium tests other random directions. The success of this mechanism once again reveals the power of evolutionary problem solving: many possible solutions are tried at random, and those that are beneficial are selected and exploited.

Summary

34.1 Most Molecular-Motor Proteins Are Members of the P-Loop NTPase Superfamily

Eukaryotic cells contain three families of molecular-motor proteins: myosins, kinesins, and dyneins. These proteins move along tracks defined by the actin and microtubule cytoskeletons of eukaryotic cells, contributing to cell and organismal movement and to the intracellular transport of proteins, vesicles, and organelles. Despite considerable differences in size and a lack of similarity detectable at the level of amino acid sequence, these proteins are homologous, containing core structures of the P-loop NTPase family. The ability of these core structures to change conformations in response to nucleoside triphosphate binding and hydrolysis is key to molecular-motor function. Motor proteins consist of motor domains attached to extended structures that serve to amplify the conformational changes in the core domains and to link the core domains to one another or to other structures.

34.2 Myosins Move Along Actin Filaments

The motile structure of muscle consists of a complex of myosin and actin, along with accessory proteins. Actin, a highly abundant 42-kd protein, polymerizes to form long filaments. Each actin monomer can bind either ATP or ADP. Muscle contraction entails the rapid sliding of thin filaments, based on actin, relative to thick filaments, composed of myosin. A myosin motor domain moves along actin filaments in a cyclic manner: (1) myosin complexed to ADP and P_i binds actin; (2) P_i is released; (3) a conformational change leads to a large motion of a lever arm that extends from the motor domain, moving the actin relative to myosin; (4) ATP replaces ADP, resetting the position of the lever arm and releasing actin; and (5) the hydrolysis of ATP returns the motor domain to its initial state. The length of the lever arm determines the size of the step taken along actin in each cycle. The ability to monitor single molecular-motor proteins has provided key tests for hypotheses concerning motor function.

34.3 Kinesin and Dynein Move Along Microtubules

Kinesin and dynein move along microtubules rather than actin. Microtubules are polymeric structures composed of α- and β-tubulin, two very similar guanine-nucleotide-binding proteins. Each microtubule comprises 13 protofilaments with alternating α- and β-tubulin subunits. Kinesins move along microtubules by a mechanism quite similar to that used by myosin to move along actin, but with several important differences. First, ATP binding to kinesin favors motor-domain binding rather than dissociation. Second, the power stroke is triggered by the binding of ATP rather than the release of P_i. Finally, kinesin motion is processive. The two heads of a kinesin dimer work together, taking turns binding and releasing the microtubule, and many steps are taken along a microtubule before both heads dissociate. Most kinesins move toward the plus end of microtubules.

34.4 A Rotary Motor Drives Bacterial Motion

Many motile bacteria use rotating flagella to propel themselves. When rotating counterclockwise, multiple flagella on the surface of a bacterium come together to form a bundle that effectively propels the bacterium through solution. When rotating clockwise, the flagella fly apart and the bacterium tumbles. In a homogeneous environment, bacteria swim smoothly for approximately 1 s and then reorient themselves by tumbling. Bacteria swim preferentially toward chemoattractants in a process called chemotaxis. When bacteria are swimming in the direction of an increasing concentration of a chemoattractant, counterclockwise flagellar motion predominates and tumbling is suppressed, leading to a biased random walk in the direction of increasing chemoattractant concentration. A proton gradient across the plasma membrane, rather than ATP hydrolysis, powers the flagellar motor. The mechanism for coupling transmembrane proton transport to macromolecular rotation appears to be similar to that used by ATP synthase.

Key Terms

myosin (p. 978)

kinesin (p. 978)

dynein (p. 978)

S1 fragment (p. 979)

conventional kinesin (p. 980)

lever arm (p. 981)

relay helix (p. 981)

neck linker (p. 982)

actin (p. 982)

myofibril (p. 982)

sarcomere (p. 982)

tropomyosin (p. 983)

troponin complex (p. 983)

G-actin (p. 985)

F-actin (p. 985)

critical concentration (p. 986)

optical trap (p. 986)

power stroke (p. 987)

microtubule (p. 989)

tubulin (p. 990)

dynamic instability (p. 990)

flagellin (p. 993)

MotA–MotB pair (p. 994)

FliG (p. 994)

chemoattractant (p. 995)

chemorepellant (p. 995)

chemotaxis (p. 995)

CheY (p. 996)

Selected Reading

Where to Start

Vale, R. D. 2003. The molecular motor toolbox for intracellular transport. *Cell* 112:467–480.

Vale, R. D., and Milligan, R. A. 2000. The way things move: Looking under the hood of molecular motor proteins. *Science* 288:88–95.

Vale, R. D. 1996. Switches, latches, and amplifiers: Common themes of G proteins and molecular motors. *J. Cell Biol.* 135:291–302.

Mehta, A. D., Rief, M., Spudich, J. A., Smith, D. A., and Simmons, R. M. 1999. Single-molecule biomechanics with optical methods. *Science* 283:1689–1695.

Schuster, S. C., and Khan, S. 1994. The bacterial flagellar motor. *Annu. Rev. Biophys. Biomol. Struct.* 23:509–539.

Books

Howard, J. 2001. *Mechanics of Motor Proteins and the Cytosketon.* Sinauer.

Squire, J. M. 1986. *Muscle Design, Diversity, and Disease.* Benjamin Cummings.

Pollack, G. H., and Sugi, H. (Eds.). 1984. *Contractile Mechanisms in Muscle.* Plenum.

Myosin and Actin

Fischer, S., Windshugel, B., Horak, D., Holmes, K. C., and Smith, J. C. 2005. Structural mechanism of the recovery stroke in the myosin molecular motor. *Proc. Natl. Acad. Sci. U.S.A.* 102:6873–6878.

Holmes, K. C., Angert, I., Kull, F. J., Jahn, W., and Schroder, R. R. 2003. Electron cryo-microscopy shows how strong binding of myosin to actin releases nucleotide. *Nature* 425:423–427.

Holmes, K. C., Schroder, R. R., Sweeney, H. L., and Houdusse, A. 2004. The structure of the rigor complex and its implications for the power stroke. *Philos. Trans. R. Soc. Lond. B Biol. Sci.* 359:1819–1828.

Purcell, T. J., Morris, C., Spudich, J. A., and Sweeney, H. L. 2002. Role of the lever arm in the processive stepping of myosin V. *Proc. Natl. Acad. Sci. U.S.A.* 99:14159–14164.

Purcell, T. J., Sweeney, H. L., and Spudich, J. A. 2005. A force-dependent state controls the coordination of processive myosin V. *Proc. Natl. Acad. Sci. U.S.A.* 102:13873–13878.

Holmes, K. C. 1997. The swinging lever-arm hypothesis of muscle contraction. *Curr. Biol.* 7:R112–R118.

Berg, J. S., Powell, B. C., and Cheney, R. E. 2001. A millennial myosin census. *Mol. Biol. Cell* 12:780–794.

Houdusse, A., Kalabokis, V. N., Himmel, D., Szent-Györgyi, A. G., and Cohen, C. 1999. Atomic structure of scallop myosin subfragment S1 complexed with MgADP: A novel conformation of the myosin head. *Cell* 97:459–470.

Houdusse, A., Szent-Györgyi, A. G., and Cohen, C. 2000. Three conformational states of scallop myosin S1. *Proc. Natl. Acad. Sci. U.S.A.* 97:11238–11243.

Uyeda, T. Q., Abramson, P. D., and Spudich, J. A. 1996. The neck region of the myosin motor domain acts as a lever arm to generate movement. *Proc. Natl. Acad. Sci. U.S.A.* 93:4459–4464.

Mehta, A. D., Rock, R. S., Rief, M., Spudich, J. A., Mooseker, M. S., and Cheney, R. E. 1999. Myosin-V is a processive actin-based motor. *Nature* 400:590–593.

Otterbein, L. R., Graceffa, P., and Dominguez, R. 2001. The crystal structure of uncomplexed actin in the ADP state. *Science* 293:708–711.

Holmes, K. C., Popp, D., Gebhard, W., and Kabsch, W. 1990. Atomic model of the actin filament. *Nature* 347:44–49.

Schutt, C. E., Myslik, J. C., Rozycki, M. D., Goonesekere, N. C., and Lindberg, U. 1993. The structure of crystalline profilin-β-actin. *Nature* 365:810–816.

van den Ent, F., Amos, L. A., and Lowe, J. 2001. Prokaryotic origin of the actin cytoskeleton. *Nature* 413:39–44.

Schutt, C. E., and Lindberg, U. 1998. Muscle contraction as a Markov process I: Energetics of the process. *Acta Physiol. Scand.* 163:307–323.

Rief, M., Rock, R. S., Mehta, A. D., Mooseker, M. S., Cheney, R. E., and Spudich, J. A. 2000. Myosin-V stepping kinetics: A molecular model for processivity. *Proc. Natl. Acad. Sci. U.S.A.* 97:9482–9486.

Friedman, T. B., Sellers, J. R., and Avraham, K. B. 1999. Unconventional myosins and the genetics of hearing loss. *Am. J. Med. Genet.* 89:147–157.

Kinesin, Dynein, and Microtubules

Yildiz, A., Tomishige, M., Vale, R. D., and Selvin, P. R. 2004. Kinesin walks hand-over-hand. *Science* 303:676–678.

Rogers, G. C., Rogers, S. L., Schwimmer, T. A., Ems-McClung, S. C., Walczak, C. E., Vale, R. D., Scholey, J. M., and Sharp, D. J. 2004. Two mitotic kinesins cooperate to drive sister chromatid separation during anaphase. *Nature* 427:364–370.

Vale, R. D., and Fletterick, R. J. 1997. The design plan of kinesin motors. *Annu. Rev. Cell. Dev. Biol.* 13:745–777.

Kull, F. J., Sablin, E. P., Lau, R., Fletterick, R. J., and Vale, R. D. 1996. Crystal structure of the kinesin motor domain reveals a structural similarity to myosin. *Nature* 380:550–555.

Kikkawa, M., Sablin, E. P., Okada, Y., Yajima, H., Fletterick, R. J., and Hirokawa, N. 2001. Switch-based mechanism of kinesin motors. *Nature* 411:439–445.

Wade, R. H., and Kozielski, F. 2000. Structural links to kinesin directionality and movement. *Nat. Struct. Biol.* 7:456–460.

Yun, M., Zhang, X., Park, C. G., Park, H. W., and Endow, S. A. 2001. A structural pathway for activation of the kinesin motor ATPase. *EMBO J.* 20:2611–2618.

Kozielski, F., De Bonis, S., Burmeister, W. P., Cohen-Addad, C., and Wade, R. H. 1999. The crystal structure of the minus-end-directed microtubule motor protein ncd reveals variable dimer conformations. *Structure Fold Des.* 7:1407–1416.

Lowe, J., Li, H., Downing, K. H., and Nogales, E. 2001. Refined structure of αβ-tubulin at 3.5 Å resolution. *J. Mol. Biol.* 313:1045–1057.

Nogales, E., Downing, K. H., Amos, L. A., and Lowe, J. 1998. Tubulin and FtsZ form a distinct family of GTPases. *Nat. Struct. Biol.* 5:451–458.

Zhao, C., Takita, J., Tanaka, Y., Setou, M., Nakagawa, T., Takeda, S., Yang, H. W., Terada, S., Nakata, T., Takei, Y., Saito, M., Tsuji, S., Hayashi, Y., and Hirokawa, N. 2001. Charcot-Marie-Tooth disease type 2A caused by mutation in a microtubule motor KIF1Bβ. *Cell* 105:587–597.

Asai, D. J., and Koonce, M. P. 2001. The dynein heavy chain: Structure, mechanics and evolution. *Trends Cell Biol.* 11:196–202.

Mocz, G., and Gibbons, I. R. 2001. Model for the motor component of dynein heavy chain based on homology to the AAA family of oligomeric ATPases. *Structure* 9:93–103.

Bacterial Motion and Chemotaxis

Sowa, Y., Rowe, A. D., Leake, M. C., Yakushi, T., Homma, M., Ishijima, A., and Berry, R. M. 2005. Direct observation of steps in rotation of the bacterial flagellar motor. *Nature* 437:916–919.

Berg, H. C. 2000. Constraints on models for the flagellar rotary motor. *Philos. Trans. R. Soc. Lond. B Biol. Sci.* 355:491–501.

DeRosier, D. J. 1998. The turn of the screw: The bacterial flagellar motor. *Cell* 93:17–20.

Ryu, W. S., Berry, R. M., and Berg, H. C. 2000. Torque-generating units of the flagellar motor of *Escherichia coli* have a high duty ratio. *Nature* 403:444–447.

Lloyd, S. A., Whitby, F. G., Blair, D. F., and Hill, C. P. 1999. Structure of the C-terminal domain of FliG, a component of the rotor in the bacterial flagellar motor. *Nature* 400:472–475.

Purcell, E. M. 1977. Life at low Reynolds number. *Am. J. Physiol.* 45:3–11.

Macnab, R. M., and Parkinson, J. S. 1991. Genetic analysis of the bacterial flagellum. *Trends Genet.* 7:196–200.

Historical Aspects

Huxley, H. E. 1965. The mechanism of muscular contraction. *Sci. Am.* 213(6):18–27.

Summers, K. E., and Gibbons, I. R. 1971. ATP-induced sliding of tubules in trypsin-treated flagella of sea-urchin sperm. *Proc. Natl. Acad. Sci. U.S.A.* 68:3092–3096.

Macnab, R. M., and Koshland, D. E., Jr. 1972. The gradient-sensing mechanism in bacterial chemotaxis. *Proc. Natl. Acad. Sci. U.S.A.* 69:2509–2512.

Taylor, E. W. 2001. 1999 E. B. Wilson lecture: The cell as molecular machine. *Mol. Biol. Cell* 12:251–254.

Problems

1. *Diverse motors.* Skeletal muscle, eukaryotic cilia, and bacterial flagella use different strategies for the conversion of free energy into coherent motion. Compare and contrast these motility systems with respect to (a) the free-energy source and (b) the number of essential components and their identity.

2. *You call that slow?* At maximum speed, a kinesin molecule moves at a rate of 6400 Å per second. Given the dimensions of the motor region of a kinesin dimer of approximately 80 Å, calculate its speed in "body lengths" per second. To what speed does this body-length speed correspond for an automobile 10 feet long?

3. *Heavy lifting.* A single myosin motor domain can generate a force of approximately 4 piconewtons (4 pN). How many times its "body weight" can a myosin motor domain lift? Note that 1 newton = 0.22 pounds (100 gms). Assume a molecular mass of 100 kd for the motor domain.

4. *Rigor mortis.* Why does the body stiffen after death?

5. *Now you see it, now you don't.* Under certain stable concentration conditions, actin monomers in their ATP form will poly-

merize to form filaments that disperse again into free actin monomers over time. Explain.

6. *Helicases as motors.* Helicases such as PcrA (p. 797) can use single-stranded DNA as tracks. In each cycle, the helicase moves one base in the $3' \rightarrow 5'$ direction. Given that PcrA can hydrolyze ATP at a rate of 50 molecules per second in the presence of a single-stranded DNA template, calculate the velocity of the helicase in micrometers per second. How does this velocity compare with that of kinesin?

7. *New moves.* When bacteria such as *E. coli* are starved to a sufficient extent, they become nonmotile. However, when such bacteria are placed in an acidic solution, they resume swimming. Explain.

8. *Hauling a load.* Consider the action of a single kinesin molecule in moving a vesicle along a microtubule track. The force required to drag a spherical particle of radius a at a velocity v in a medium having a viscosity η is

$$F = 6\pi\eta av$$

Suppose that a 2-μm diameter bead is carried at a velocity of 0.6 μm s^{-1} in an aqueous medium ($\eta = 0.01$ poise $= 0.01$ g cm^{-1} s^{-1}).

(a) What is the magnitude of the force exerted by the kinesin molecule? Express the value in dynes (1 dyne $= 1$ g cm s^{-2}).
(b) How much work is performed in 1 s? Express the value in ergs (1 erg $= 1$ dyne cm).
(c) A kinesin motor hydrolyzes approximately 80 molecules of ATP per second. What is the energy associated with the hydrolysis of this much ATP in ergs? Compare this value with the actual work performed.

9. *Unusual strides.* A publication describes a kinesin molecule that is claimed to move along microtubules with a step size of 6 nm. You are skeptical. Why?

10. *The sound of one hand clapping.* KIF1A is a motor protein that moves toward the plus end of microtubules *as a monomer.* KIF1A has only a single motor domain. What additional structural elements would you expect to find in the KIF1A structure?

Mechanism Problem

11. *Backward rotation.* On the basis of the proposed structure in Figure 34.30 for the bacterial flagellar motor, suggest a pathway for transmembrane proton flow when the flagellar motor is rotating clockwise rather than counterclockwise.

Chapter Integration Problem

12. *Smooth muscle.* Smooth muscle, in contrast with skeletal muscle, is not regulated by a tropomyosin–troponin mechanism. Instead, vertebrate smooth-muscle contraction is controlled by the degree of phosphorylation of its light chains. Phosphorylation induces contraction, and dephosphorylation leads to relaxation. Like that of skeletal muscle, smooth-muscle contraction is triggered by an increase in the cytoplasmic calcium ion level. Propose a mechanism for this action of calcium ion on the basis of your knowledge of other signal-transduction processes.

Data Interpretation Problem

13. *Myosin V.* An abundant myosin-family member, myosin V is isolated from brain tissue. This myosin has a number of unusual properties. First, on the basis of its amino acid sequence, each heavy chain has six tandem binding sites for calmodulin-like light chains. Second, it forms dimers but not higher-order oligomers. Finally, unlike almost all other myosin-family members, myosin V is highly processive.

The rate of ATP hydrolysis by myosin has been examined as a function of ATP concentration, as shown in graph A.

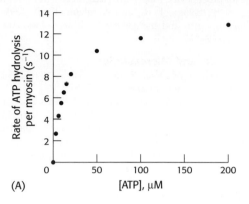

(A)

(a) Estimate the values of k_{cat} and K_M for ATP.

With the use of optical-trap measurements, the motion of single myosin V dimers could be followed, as shown in graph B.

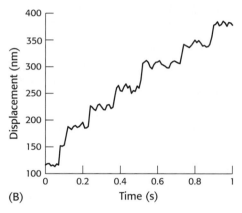

(B)

[Based on M. Rief, R. S. Rock, A. D. Mehta, M. S. Mooseker, R. E. Cheney, and J. A. Spudich. *Proc. Natl. Acad. Sci. U.S.A.* 97(2000):9482–9486.]

(b) Estimate the step size for myosin V.

The rate of ADP release from myosin V is found to be approximately 13 molecules s^{-1}.

(c) Combine the observations about the amino acid sequence of myosin, the observed step size, and the kinetics results to propose a mechanism for the processive motion of myosin V.

Drug Development

Many drugs are based on natural products. Aspirin (above) is a chemical derivative of a compound isolated from willow bark (near left). Extracts of willow bark had been long known to have medicinal properties. The active compound was isolated, modified, and, beginning in 1899, packaged for consumers (far left). [*Far left:* Used with permission of Bayer Corporation. *Near left:* Image Ideas/ Picture Quest.]

The development of drugs represents one of the most important interfaces between biochemistry and medicine. In most cases, drugs act by binding to specific receptors or enzymes and inhibiting, or otherwise modulating, their activities. Thus, knowledge of these molecules and the pathways in which they participate is crucial to drug development. An effective drug is much more than a potent modulator of its target, however. Drugs must be readily administered to patients, preferably as small tablets taken orally, and must survive within the body long enough to reach their targets. Furthermore, to prevent unwanted physiological effects, drugs must not modulate the properties of biomolecules other than the target molecules. These requirements tremendously limit the number of compounds that have the potential to be clinically useful drugs.

Drugs have been discovered by two, fundamentally opposite, approaches (Figure 35.1). The first approach identifies a substance that has a desirable physiological consequence when administered to a human being, to an appropriate animal, or to cells. Such substances can be discovered by serendipity, by the fractionation of plants or other materials known to have medicinal properties, or by screening natural products or other "libraries" of compounds. In this approach, a biological effect is known before the molecular target is identified. The mode of action of the substance is only later

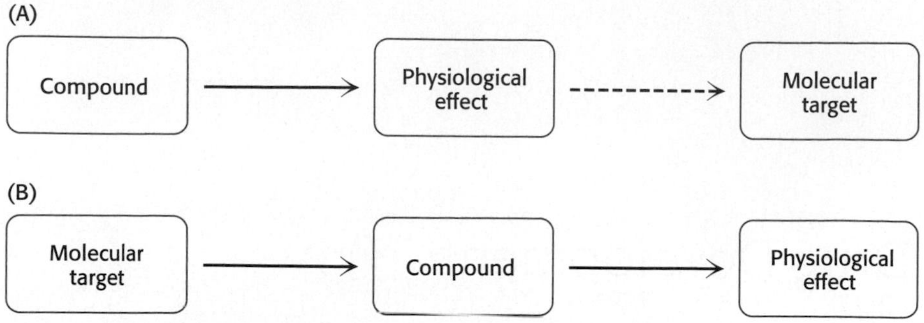

Figure 35.1 **Two paths to drug discovery.** (A) A compound is discovered to have a desirable physiological effect. The molecular target can be identified in a separate step as needed. (B) A molecular target is selected first. Drug candidates that bind to the target are identified and then examined for their physiological effects.

Pharmacology

The science that deals with the discovery, chemistry, composition, identification, biological and physiological effects, uses, and manufacture of drugs.

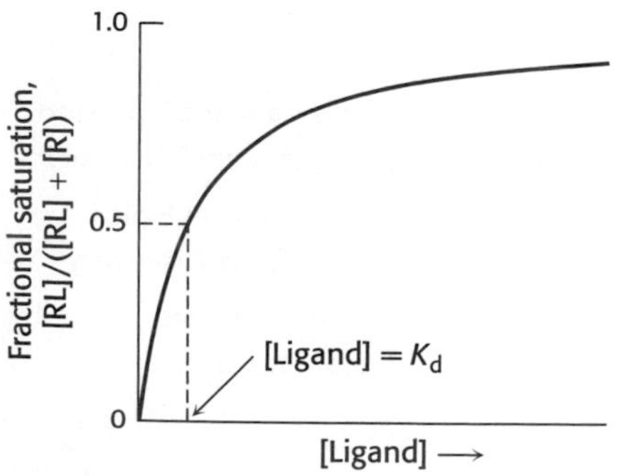

Figure 35.2 **Ligand binding.** The titration of a receptor, R, with a ligand, L, results in the formation of the complex RL. In uncomplicated cases, the binding reaction follows a simple saturation curve. Half of the receptors are bound to ligand when the ligand concentration equals the dissociation constant, K_d, for the RL complex.

identified after substantial additional work. The second approach begins with a known molecular target. Compounds are sought, either by screening or by designing molecules with desired properties, that bind to the target molecule and modulate its properties. Once such compounds are available, scientists can explore their effects on appropriate cells or organisms. Many unexpected results may be encountered in this process as the complexity of biological systems reveals itself.

In this chapter, we explore the science of pharmacology. We examine a number of case histories that illustrate drug development—including many of its concepts, methods, and challenges. We then see how the concepts and tools from genomics are influencing approaches to drug development. We conclude the chapter with a summary of the stages along the way to developing a drug.

35.1 The Development of Drugs Presents Huge Challenges

Many compounds have significant effects when taken into the body, but only a very small fraction of them have the potential to be useful drugs. A foreign compound, not adapted to its role in the cell through long evolution, must have a range of special properties to function effectively without causing serious harm. We next review some of the challenges faced by drug developers.

Drug Candidates Must Be Potent Modulators of Their Targets

Most drugs bind to specific proteins, usually receptors or enzymes, within the body. To be effective, a drug needs to bind a sufficient number of its target proteins when taken at a reasonable dose. One factor in determining drug effectiveness is the strength of binding, often governed by the principles of binding, related to the Michaelis-Menten model introduced in Chapter 8.

A molecule that binds to some target molecule is often referred to as a *ligand*. A ligand-binding curve is shown in Figure 35.2. Ligand molecules occupy progressively more target binding sites as ligand concentration increases until essentially all of the available sites are occupied. The tendency of a ligand to bind to its target is measured by the *dissociation constant, K_d*, defined by the expression

$$K_d = [R][L]/[RL]$$

where [R] is the concentration of the receptor, [L] is the concentration of the ligand, and [RL] is the concentration of the receptor–ligand complex. The dissociation constant is a measure of the strength of the interaction between the drug candidate and the target; the lower the value, the stronger the interaction. The concentration of free ligand at which one-half of the binding sites are occupied equals the dissociation constant, as long as the concentration of binding sites is substantially less than the dissociation constant.

Many complicating factors are present under physiological conditions. Many drug targets also bind ligands normally present in tissues; these ligands and the drug candidate compete for binding sites on the target. We encountered this situation when we considered competitive inhibitors in Chapter 8. Suppose that the drug target is an enzyme and the drug candidate is a competitive inhibitor. The concentration of the drug candidate necessary to inhibit the enzyme effectively will depend on the physiological concentration of the enzyme's normal substrate (Figure 35.3). The higher the concentration of the endogenous substrate, the higher the concentration of drug candidate needed to inhibit the enzyme to a given extent. This effect of substrate concentration is expressed by the *apparent dissociation constant*, K_d^{app}. The apparent dissociation constant is given by the expression

$$K_d^{app} = K_d(1 + [S]/K_M)$$

where [S] is the concentration of substrate and K_M is the Michaelis constant for the substrate. Note that, for an enzyme inhibitor, the dissociation constant, K_d, is often referred to as the *inhibition constant*, K_i.

In many cases, more complicated biological assays (rather than direct enzyme or binding assays) are used to examine the potency of drug candidates. For example, the fraction of bacteria killed might indicate the potency of a potential antibiotic. In these cases, values such as EC_{50} are used. EC_{50} is the concentration of drug candidate required to elicit 50% of the maximal biological response (Figure 35.4). Similarly, EC_{90} is the concentration required to achieve 90% of the maximal response. In the example of an antibiotic, EC_{90} would be the concentration required to kill 90% of bacteria exposed to the drug. For inhibitors, the corresponding terms IC_{50} and IC_{90} are often used to describe the concentrations of the inhibitor required to reduce a response to 50% or 90% of its value in the absence of inhibitor, respectively.

These values are measures of the potency of a drug candidate in modulating the activity of the desired biological target. To prevent unwanted effects, often called *side effects*, ideal drug candidates should not bind biomolecules other than the target to any appreciable extent. Developing such a drug can be quite challenging, particularly if the drug target is a member of a large family of evolutionarily related proteins. The degree of specificity can be described in terms of the ratio of the K_d values for the binding of the drug candidate to any other molecules to the K_d value for the binding of the drug candidate to the desired target.

Drugs Must Have Suitable Properties to Reach Their Targets

Thus far, we have focused on the ability of molecules to act on specific target molecules. However, an effective drug must also have other characteristics. It must be easily administered and must reach its target at sufficient concentration to be effective. A drug molecule encounters a variety of obstacles on its way to its target, related to its absorption, distribution, metabolism, and excretion after it has entered the body. These processes are interrelated to one another as summarized in Figure 35.5. Taken together, a drug's ease of absorption, distribution, metabolism, and excretion are often referred to as *ADME* (pronounced "add-me") properties.

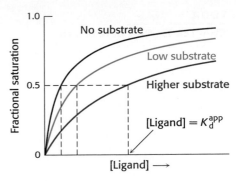

Figure 35.3 Inhibitors compete with substrates for binding sites. These binding curves give results for an inhibitor binding to a target enzyme in the absence of substrate and in the presence of increasing concentrations of substrate.

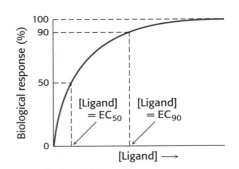

Figure 35.4 Effective concentrations. The concentration of a ligand required to elicit a biological response can be quantified in terms of EC_{50}, the concentration required to give 50% of the maximum response, and EC_{90}, the concentration required to give 90% of the maximum response.

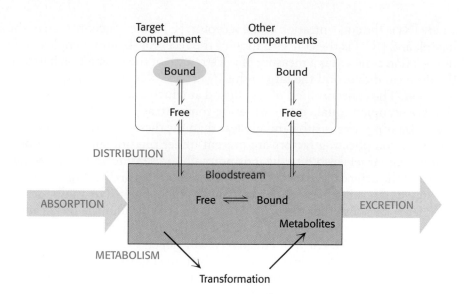

Figure 35.5 **Absorption, distribution, metabolism, and excretion (ADME).** The concentration of a compound at its target site (yellow) is affected by the extents and rates of absorption, distribution, metabolism, and excretion.

Administration and Absorption. Ideally, a drug can be taken orally as a small tablet. An orally administered active compound must be able to survive the acidic conditions in the gut and then be absorbed through the intestinal epithelium. Thus, the compound must be able to pass through cell membranes at an appreciable rate. Larger molecules such as proteins cannot be administered orally, because they often cannot survive the acidic conditions in the stomach and, if they do, are not readily absorbed. Even many small molecules are not absorbed well, because, for example, if they are too polar they do not pass through cell membranes readily. The ability to be absorbed is often quantified in terms of the *oral bioavailability*. This quantity is defined as the ratio of the peak concentration of a compound given orally to the peak concentration of the same dose injected directly into the bloodstream. Bioavailability can vary considerably from species to species so results from animal studies may be difficult to translate to human beings. Despite this variability, some useful generalizations have been made. One powerful set is *Lipinski's rules*.

Lipinski's rules tell us that poor absorption is likely when

1. the molecular weight is greater than 500.

2. the number of hydrogen-bond donors is greater than 5.

3. the number of hydrogen-bond acceptors is greater than 10.

4. the partition coefficient [measured as $\log(P)$] is greater than 5.

The partition coefficient is a way to measure the tendency of a molecule to dissolve in membranes, which correlates with its ability to dissolve in organic solvents. It is determined by allowing a compound to equilibrate between water and an organic phase, *n*-octanol. The $\log(P)$ value is defined as $\log_{10}$ of the ratio of the concentration of a compound in *n*-octanol to the concentration of the compound in water. For example, if the concentration of the compound in the *n*-octanol phase is 100 times that in the aqueous phase, then $\log(P)$ is 2.

Morphine, for example, satisfies all of Lipinski's rules and has moderate bioavailability (Figure 35.6). A drug that violates one or more of these rules may still have satisfactory bioavailability. Nonetheless, these rules serve as guiding principles for evaluating new drug candidates.

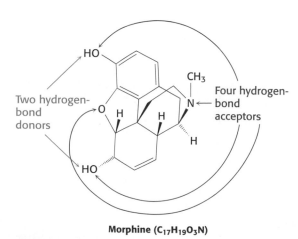

Morphine (C₁₇H₁₉O₃N)

Molecular weight = 285

$\log(P) = 1.27$

Figure 35.6 **Lipinski's rules applied to morphine.** Morphine satisfies all of Lipinski's rules and has an oral bioavailability in human beings of 33%.

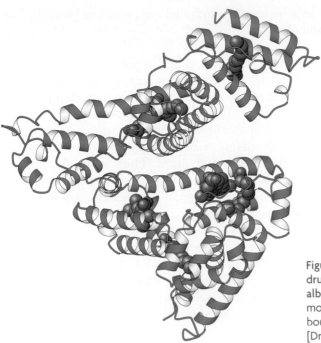

Figure 35.7 Structure of the drug carrier human serum albumin. Seven hydrophobic molecules (in red) are shown bound to the molecule. [Drawn from 1BKE.pdb.]

Distribution. Compounds taken up by intestinal epithelial cells can pass into the bloodstream. However, hydrophobic compounds and many others do not freely dissolve in the bloodstream. These compounds bind to proteins, such as albumin (Figure 35.7), that are abundant in the blood serum and by this means are carried everywhere that the bloodstream goes.

When a compound has reached the bloodstream, it is distributed to different fluids and tissues, which are often referred to as *compartments*. Some compounds are highly concentrated in their target compartments, either by binding to the target molecules themselves or by other mechanisms. Other compounds are distributed more widely (Figure 35.8). An effective drug will reach the target compartment in sufficient quantity; the concentration of the compound in the target compartment is reduced whenever the compound is distributed into other compartments.

Some target compartments are particularly hard to reach. Many compounds are excluded from the central nervous system by the *blood–brain*

Fluconazole

Figure 35.8 Distribution of the drug fluconazole. Once taken in, compounds distribute themselves to various organs within the body. The distribution of the antifungal agent fluconazole has been monitored through the use of positron emission tomography (PET) scanning. These images were taken of a healthy human volunteer 90 minutes after injection of a dose of 5 mg kg^{-1} of fluconazole containing trace amounts of fluconazole labeled with the positron-emitting isotope [18]F. [From A. J. Fischman et al., *Antimicrob. Agents Chemother.* 37(1993): 1270–1277.]

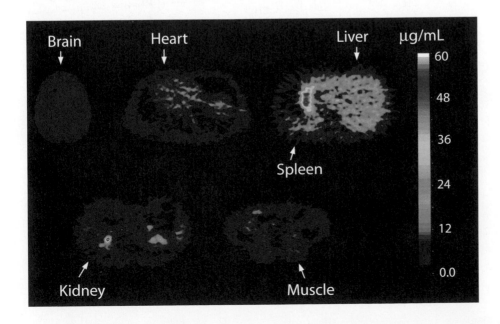

barrier, the tight junctions between endothelial cells that line blood vessels within the brain and spinal cord.

Metabolism and Excretion. A final challenge to a potential drug molecule is to evade the body's defenses against foreign compounds. Such compounds (often called *xenobiotic compounds*) are often released from the body in the urine or stool, often after having been metabolized somehow—degraded or modified—to aid in excretion. This *drug metabolism* poses a considerable threat to drug effectiveness because the concentration of the desired compound decreases as it is metabolized. Thus, a rapidly metabolized compound must be administered more frequently or at higher doses.

Two of the most common pathways in xenobiotic metabolism are *oxidation* and *conjugation.* Oxidation reactions can aid excretion in at least two ways: by increasing water solubility, and thus ease of transport, and by introducing functional groups that participate in subsequent metabolic steps. These reactions are often promoted by cytochrome P450 enzymes in the liver (p. 750). The human genome encodes more than 50 different P450 isozymes, many of which participate in xenobiotic metabolism. A typical reaction catalyzed by a P450 isozyme is the hydroxylation of ibuprofen (Figure 35.9).

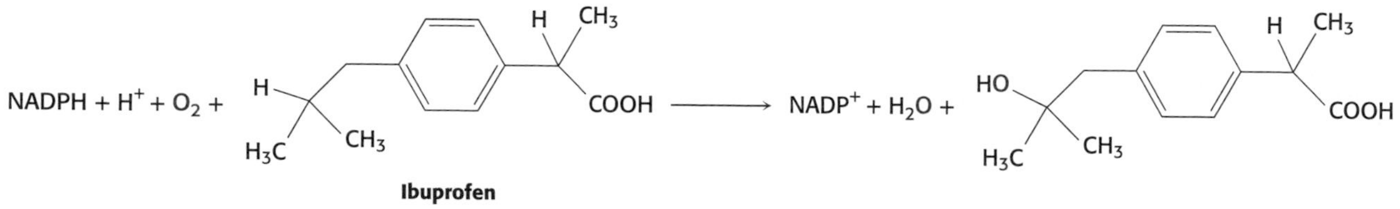

Figure 35.9 P450 conversion of ibuprofen. Cytochrome P450 isozymes, primarily in the liver, catalyze xenobiotic metabolic reactions such as hydroxylation. The reaction introduces an oxygen atom derived from molecular oxygen.

Ibuprofen

Conjugation is the addition of particular groups to the xenobiotic compound. Common groups added are glutathione (p. 586), glucuronic acid, and sulfate (Figure 35.10). The addition often increases water solubility and provides labels that can be recognized to target excretion. Examples of conjugation include the addition of glutathione to the anticancer drug cyclophosphamide, the addition of glucuronidate to the analgesic morphine, and the addition of a sulfate group to the hair-growth stimulator minoxidil.

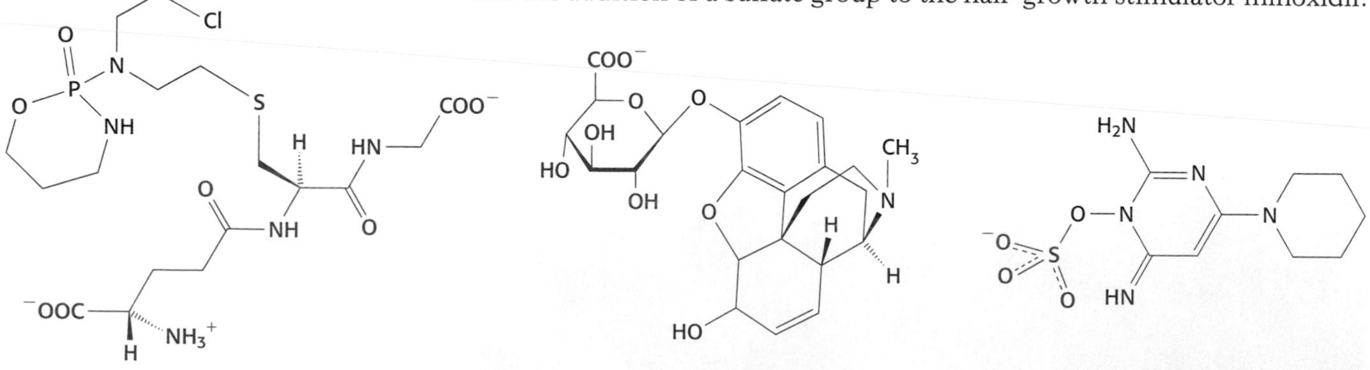

Cyclophosphamide-glutathione conjugate   **Morphine glucuronidate**   **Minoxidil sulfate**

Interestingly, the sulfation of minoxidil produces a compound that is more active in stimulating hair growth than is the unmodified compound. Thus, the metabolic products of a drug, though usually less active than the drug, can sometimes be more active.

Note that an oxidation reaction often precedes conjugation because the oxidation reaction can generate hydroxyl and other groups to which groups such as glucuronic acid can be added. The oxidation reactions of xenobiotic compounds are often referred to as *phase I transformations,* and the conjugation reactions are referred to as *phase II transformations.* These reactions take

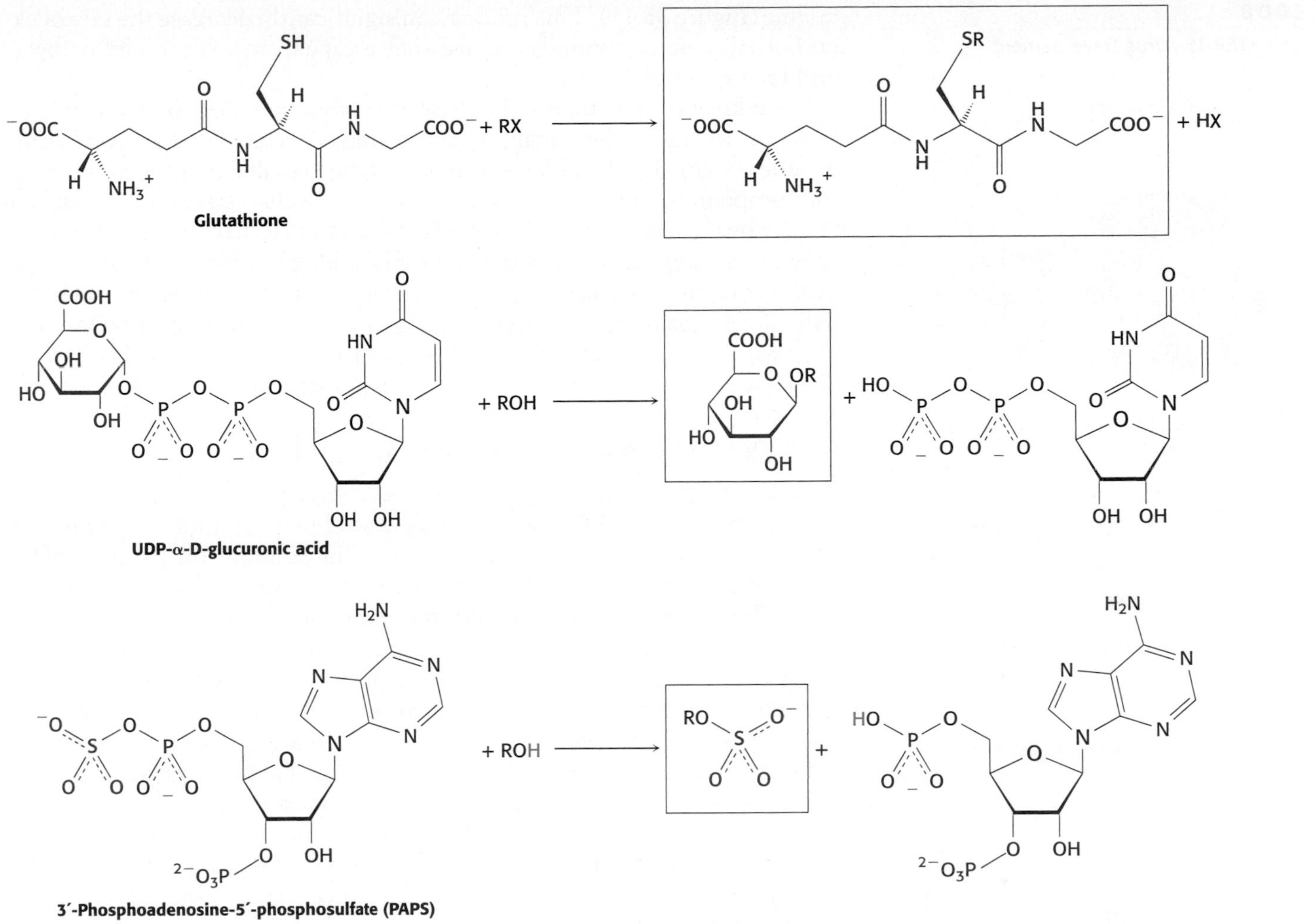

Figure 35.10 Conjugation reactions. Compounds that have appropriate groups are often modified by conjugation reactions. Such reactions include the addition of glutathione (top), glucuronic acid (middle), or sulfate (bottom). The conjugated product is shown boxed.

place primarily in the liver. Because blood flows from the intestine directly to the liver through the portal vein, xenobiotic metabolism often alters drug compounds before they ever reach full circulation. This *first-pass metabolism* can substantially limit the availability of compounds taken orally.

After compounds have entered the bloodstream, they can be removed from circulation and excreted from the body by two primary pathways. First, they can be absorbed through the kidneys and excreted in the urine. In this process, the blood passes through *glomeruli,* networks of fine capillaries in the kidney that act as filters. Compounds with molecular weights less than approximately 60,000 pass though the glomeruli into the kidney. Many of the water molecules, glucose molecules, nucleotides, and other low-molecular-weight compounds that pass through the glomeruli are reabsorbed into the bloodstream, either by transporters that have broad specificities or by the passive transfer of hydrophobic molecules through membranes. Drugs and metabolites that pass through the first filtration step and are not reabsorbed are excreted.

Second, compounds can be actively transported into bile, a process that takes place in the liver. After concentration, bile flows into the intestine. In the intestine, the drugs and metabolites can be excreted through the stool, reabsorbed into the bloodstream, or further degraded by digestive enzymes. Sometimes, compounds are recycled from the bloodstream into the intestine and back into the bloodstream, a process referred to as *enterohepatic*

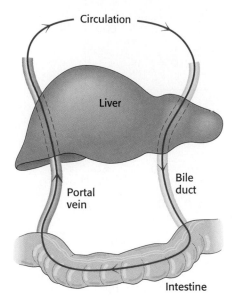

Circulation

Liver

Bile duct

Portal vein

Intestine

Figure 35.11 Enterohepatic cycling. Some drugs can move from the blood circulation to the liver, into the bile, into the intestine, to the liver, and back into circulation. This cycling decreases the rate of drug excretion.

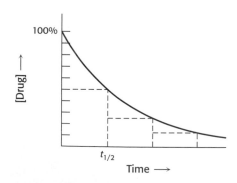

Figure 35.12 Half-life of drug excretion. In the case shown, the concentration of a drug in the bloodstream decreases to one-half of its value in a period of time, $t_{1/2}$, referred to as its half-life.

cycling (Figure 35.11). This process can significantly decrease the rate of excretion of some compounds because they escape from an excretory pathway and reenter the circulation.

The kinetics of compound excretion is often complex. In some cases, a fixed percentage of the remaining compound is excreted over a given period of time (Figure 35.12). This pattern of excretion results in exponential loss of the compound from the bloodstream that can be characterized by a half-life ($t_{1/2}$). The half-life is the fixed period of time required to eliminate 50% of the remaining compound. It is a measure of how long an effective concentration of the compound remains in the system after administration. As such, the half-life is a major factor in determining how often a drug must be taken. A drug with a long half-life might need to be taken only once per day, whereas a drug with a short half-life might need to be taken three or four times per day.

Toxicity Can Limit Drug Effectiveness

An effective drug must not be so toxic that it seriously harms the person who takes it. A drug may be toxic for any of several reasons. First, it may modulate the target molecule itself *too* effectively. For example, the presence of too much of the anticoagulant drug coumadin can result in dangerous, uncontrolled bleeding and death. Second, the compound may modulate the properties of proteins that are distinct from, but related to, the target molecule itself. Compounds that are directed to one member of a family of enzymes or receptors often bind to other family members. For example, an antiviral drug directed against viral proteases may be toxic if it also inhibits proteases normally present in the body such as those that regulate blood pressure.

A compound may also be toxic if it modulates the activity of a protein unrelated to its intended target. For example, many compounds block ion channels such as the potassium channel HERG (the human homolog of a *Drosophila* channel found in a mutant termed "ether-a-go-go"), causing disturbances of the heartbeat. To avoid cardiac side effects, many compounds are screened for their ability to block such channels.

Finally, even if a compound is not itself toxic, its metabolic by-products may be. Phase I metabolic processes can generate damaging reactive groups in products. An important example is liver toxicity observed with large doses of the common pain reliever acetaminophen (Figure 35.13). A particular cytochrome P450 isozyme oxidizes acetaminophen to *N*-acetyl-*p*-benzoquinone imine. The resulting compound is conjugated to glutathione. With large doses, however, the liver concentration of glutathione drops dramatically, and the liver is no longer able to protect itself from this reactive compound and others. Initial symptoms of excessive acetaminophen include nausea and vomiting. Within 24 to 48 hours, symptoms of liver failure may appear. Acetaminophen poisoning accounts for about 35% of cases of severe liver failure in the United States. A liver transplant is often the only effective treatment.

The toxicity of a drug candidate can be described in terms of the *therapeutic index*. This measure of toxicity is determined through animal tests, usually with mice or rats. The therapeutic index is defined as the ratio of the dose of a compound that is required to kill one-half of the animals (referred to as the LD_{50} for "lethal dose") to a comparable measure of the effective dose, usually the EC_{50}. Thus, if the therapeutic index is 1000, then lethality is significant only when 1000 times the effective dose is administered. Analogous indices can provide measures of toxicity less severe than lethality.

Many compounds have favorable properties in vitro, yet fail when administered to a living organism because of difficulties with ADME and toxicity. Expensive and time-consuming animal studies are required to verify that a drug candidate is not toxic, yet differences between animal species in

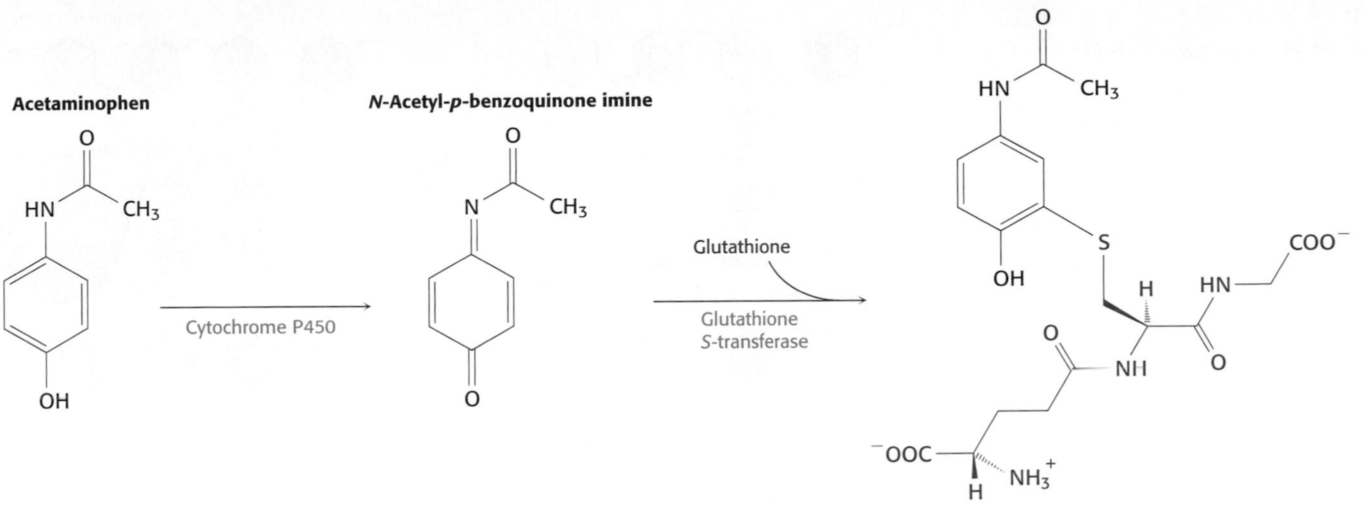

Figure 35.13 Acetaminophen toxicity. A minor metabolic product of acetaminophen is *N*-acetyl-*p*-benzoquinone imine. This metabolite is conjugated to glutathione. Large doses of acetaminophen can deplete liver glutathione stores.

their response can confound decisions about moving forward with a compound toward human studies. One hope is that, with more understanding of the biochemistry of these processes, scientists can develop computer-based models to replace or augment animal tests. Such models would need to accurately predict the fate of a compound inside a living organism from its molecular structure or other properties that are easily measured in the laboratory without the use of animals.

35.2 Drug Candidates Can Be Discovered By Serendipity, Screening, or Design

Traditionally, many drugs were discovered by serendipity, or chance observation. More recently, drugs have been discovered by screening collections of natural products or other compounds for compounds that have desired medicinal properties. Alternatively, scientists have designed specific drug candidates by using their knowledge about a preselected molecular target. We will examine several examples of each of these pathways to reveal common principles.

Serendipitous Observations Can Drive Drug Development

Perhaps the most well known observation in the history of drug development is Alexander Fleming's chance observation in 1928 that colonies of the bacterium *Staphylococcus aureus* died when they were adjacent to colonies of the mold *Penicillium notatum*. Spores of the mold had landed accidentally on plates growing the bacteria. Fleming soon realized that the mold produced a substance that could kill disease-causing bacteria. This discovery led to a fundamentally new approach to the treatment of bacterial infections. Howard Florey and Ernest Chain developed a powdered form of the substance, termed penicillin, that became a widely used antibiotic in the 1940s.

The structure of this antibiotic was elucidated in 1945. The most notable feature of this structure is the four-membered β-lactam ring. This unusual feature is key to the antibacterial function of penicillin, as noted earlier (p. 222).

Three steps were crucial to fully capitalize on Fleming's discovery. First, an industrial process was developed for the production of penicillin from *Penicillium* mold on a large scale. Second, penicillin and penicillin derivatives

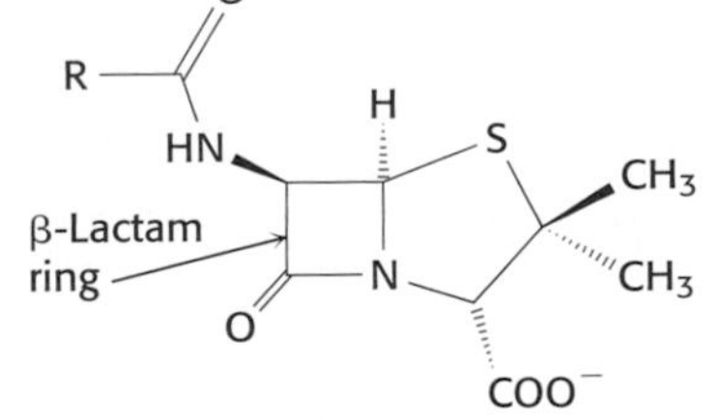

Penicillin

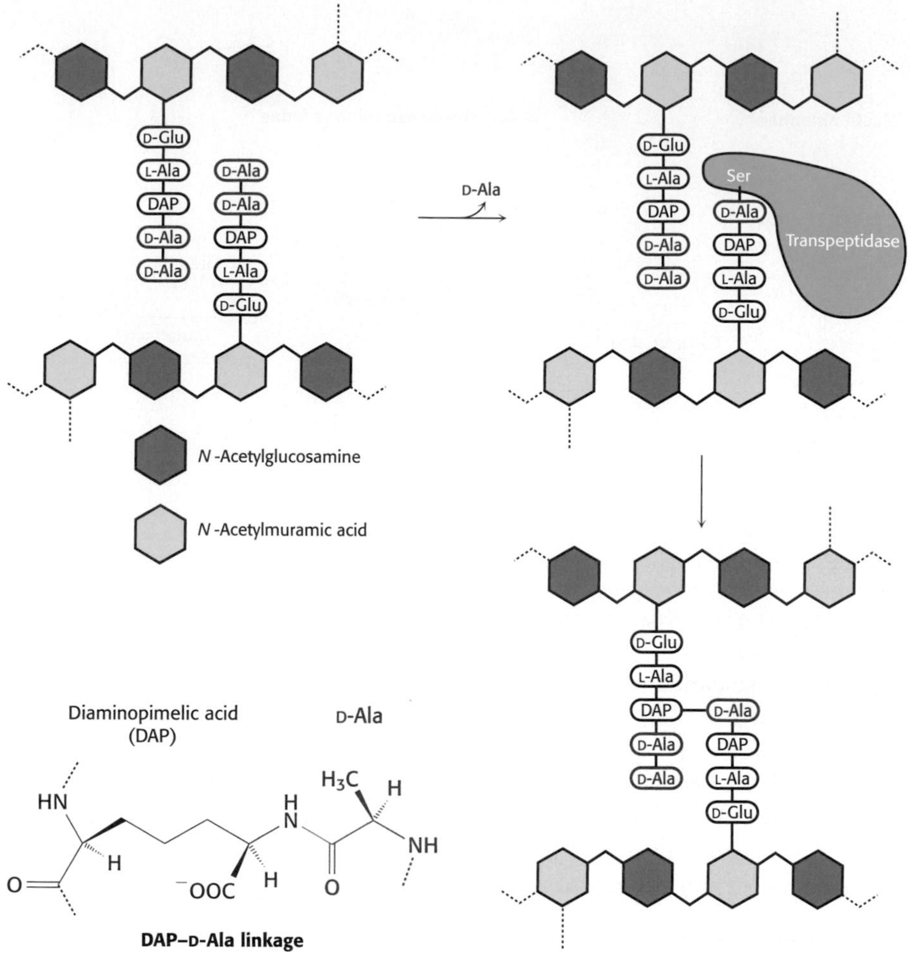

Figure 35.14 Mechanism of cell-wall biosynthesis disrupted by penicillin. A transpeptidase enzyme catalyzes the formation of cross-links between peptidoglycan groups. In the case shown, the transpeptidase catalyzes the linkage of D-alanine at the end of one peptide chain to the amino acid diaminopimelic acid (DAP) on another peptide chain. The diaminopimelic acid linkage (bottom left) is found in Gram-negative bacteria such as *E. coli*. Linkages of glycine-rich peptides are found in Gram-positive bacteria. Penicillin inhibits the action of the transpeptidase; so bacteria exposed to the drug have weak cell walls that are susceptible to lysis.

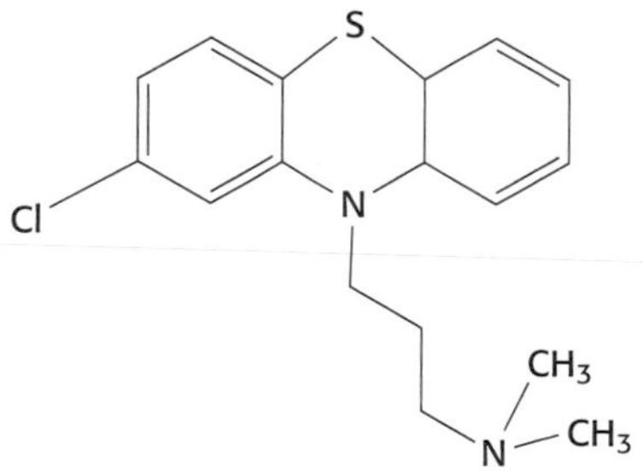

Chlorpromazine

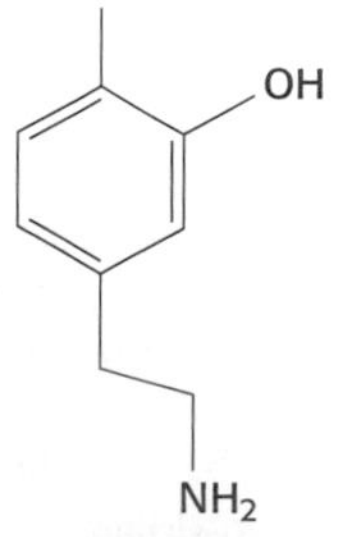

Dopamine

were chemically synthesized. The availability of synthetic penicillin derivatives opened the way for scientists to explore the relations between structure and function. Many such penicillin derivatives have found widespread use in medicine. Finally, Jack Strominger and James Park independently elucidated the mode of action of penicillin in 1965 (Figure 35.14), as introduced in Chapter 8.

Many other drugs have been discovered by serendipitous observations. The antineuroleptic drug chlorpromazine (Thorazine) was discovered in the course of investigations directed toward the treatment of shock in surgical patients. In 1952, French surgeon Henri Laborit noticed that, after taking the compound, his patients were remarkably calm. This observation suggested that chlorpromazine could benefit psychiatric patients, and, indeed, the drug has been used for many years to treat patients with schizophrenia and other disorders. The drug does have significant side effects, and its use has been largely superceded by more recently developed drugs.

Chlorpromazine acts by binding to receptors for the neurotransmitter dopamine and blocking them (Figure 35.15). Dopamine D2 receptors are the targets of many other psychoactive drugs. In the search for drugs with more limited side effects, studies are undertaken to correlate drug effects with biochemical parameters such as dissociation constants and binding and release rate constants.

A more recent example of a drug discovered by chance observation is sildenafil (Viagra). This compound was developed as an inhibitor of phosphodiesterase 5, an enzyme that catalyzes the hydrolysis of cGMP to GMP (Figure 35.16). The compound was intended as a treatment for hypertension and angina because cGMP plays a central role in the relaxation of

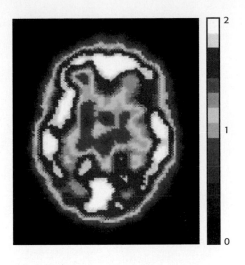

smooth muscle cells in blood vessels (Figure 35.17). Inhibiting phosphodiesterase 5 was expected to increase the concentration of cGMP by blocking the pathway for its degradation. In the course of early clinical trials in Wales, some men reported unusual penile erections. Whether this chance observation by a few men was due to the compound or to other effects was unclear. However, the observation made some biochemical sense because smooth muscle relaxation due to increased cGMP levels had been discovered to play a role in penile erection. Subsequent clinical trials directed toward the evaluation of sildenafil for erectile dysfunction were successful. This account testifies to the importance of collecting comprehensive information from clinical-trial participants. In this case, incidental observations led to a new treatment for erectile dysfunction and a multibillion-dollar-per-year drug market.

Sildenafil

cGMP

Figure 35.16 Sildenafil, a mimic of cGMP. Sildenafil was designed to resemble cGMP, the substrate of phosphodiesterase 5.

Screening Libraries of Compounds Can Yield Drugs or Drug Leads

No drug is as widely used as aspirin. Observers at least as far back as Hippocrates (~400 B.C.) have noted the use of extracts from the bark and leaves of the willow tree for pain relief. In 1829, a mixture called *salicin* was isolated from willow bark. Subsequent analysis identified salicylic acid as the active component of this mixture. Salicylic acid was formerly used to treat pain, but this compound often irritated the stomach. Several investigators

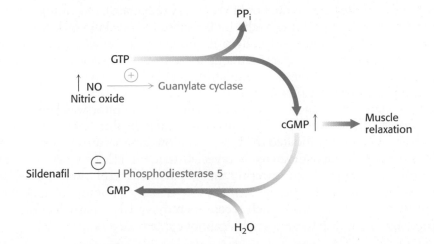

Figure 35.17 Muscle-relaxation pathway. Increases in NO levels stimulate guanylate cyclase, which produces cGMP. The increased cGMP concentration promotes smooth muscle relaxation. Phosphodiesterase 5 hydrolyzes cGMP, which lowers the cGMP concentration. The inhibition of phosphodiesterase 5 by sildenafil maintains elevated levels of cGMP.

Salicylic acid

Acetyl group

Aspirin (acetylsalicylic acid)

Compactin

Lovastatin

attempted to find a means to neutralize salicylic acid. Felix Hoffmann, a chemist working at the German company Bayer, developed a less-irritating derivative by treating salicylic acid with a base and acetyl chloride. This derivative, acetylsalicylic acid, was named *aspirin* from "a" for acetyl chloride, "spir" for *Spiraea ulmaria* (meadowsweet, a flowering plant that also contains salicylic acid), and "in" (a common ending for drugs). Each year, approximately 35,000 tons of aspirin are taken worldwide, nearly the weight of the *Titanic*.

As discussed in Chapter 12, the acetyl group in aspirin is transferred to the side chain of a serine residue that lies along the path to the active site of the cyclooxygenase component of prostaglandin H_2 synthase (p. 339). In this position, the acetyl group blocks access to the active site. Thus, even though aspirin binds in the same pocket on the enzyme as salicylic acid, the acetyl group of aspirin dramatically increases its effectiveness as a drug. The account illustrates the value of screening extracts from plants and other materials that are believed to have medicinal properties for active compounds. The large number of herbal and folk medicines are a treasure trove of new drug leads.

More than 100 years ago, a fatty, yellowish material was discovered on the arterial walls of patients who had died of vascular disease. The presence of the material was termed *atheroma* from the Greek word for porridge. This material proved to be cholesterol. The Framingham heart study, initiated in 1948, documented a correlation between high blood cholesterol levels and high mortality rates from heart disease. This observation led to the notion that blocking cholesterol synthesis might lower blood cholesterol levels and, in turn, lower the risk of heart disease. Drug developers had to abandon an initial attempt at blocking the cholesterol synthesis pathway at a late step because cataracts and other side effects developed, caused by the accumulation of the insoluble substrate for the inhibited enzyme. Investigators eventually identified a more favorable target—namely, the enzyme HMG-CoA reductase (p. 739). This enzyme acts on a substrate, HMG-CoA (3-hydroxy-3-methylglutaryl coenzyme A), that can be used by other pathways and is water soluble.

A promising natural product, compactin, was discovered in a screen of compounds from a fermentation broth from *Penicillium citrinum* in a search for antibacterial agents. In some, but not all, animal studies, compactin was found to inhibit HMG-CoA reductase and to lower serum cholesterol levels. In 1982, a new HMG-CoA reductase inhibitor was discovered in a fermentation broth from *Aspergillus cereus*. This compound, now called lovastatin, was found to be structurally very similar to compactin, bearing one additional methyl group.

In clinical trials, lovastatin significantly reduced serum cholesterol levels with few side effects. Most side effects could be prevented by treatment with mevalonate (the product of HMG-CoA reductase), indicating that the side effects were likely due to the highly effective blocking of HMG-CoA reductase. One notable side effect is muscle pain or weakness (termed *myopathy*), although its cause remains to be fully established. After many studies the Food and Drug Administration (FDA) approved lovastatin for treating high serum cholesterol levels.

A structurally related HMG-CoA reductase inhibitor was later shown to cause a statistically significant decrease in deaths due to coronary heart disease. This result validated the benefits of lowering serum cholesterol levels. Further mechanistic analysis revealed that the HMG-CoA reductase inhibitor acts not only by lowering the rate of cholesterol biosynthesis, but also by inducing the expression of the low-density-lipoprotein (LDL) receptor (p. 745). Cells with such receptors remove LDL particles from the bloodstream, and so these particles cannot contribute to atheroma.

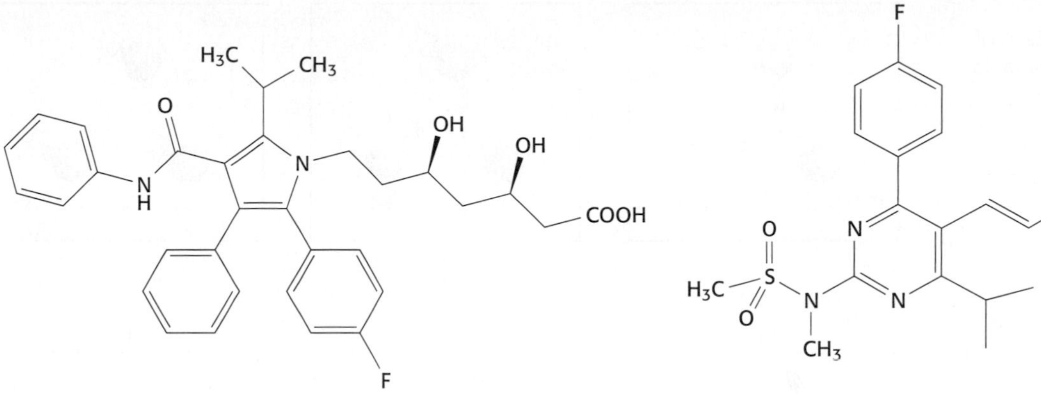

Atorvastatin

Rosuvastatin

Lovastatin and its relatives are natural products or compounds readily derived from natural products. The next step was the development of totally synthetic molecules that are more potent inhibitors of HMG-CoA reductase (Figure 35.18). These compounds are effective at lower dose levels, reducing side effects.

The original HMG-CoA reductase inhibitors or their precursors were found by screening libraries of natural products. More recently, drug developers have tried screening large libraries of both natural products and purely synthetic compounds prepared in the course of many drug-development programs. Under favorable circumstances, hundreds of thousands or even millions of compounds can be tested in this process, termed *high-throughput screening*. Compounds in these libraries can be synthesized one at a time for testing. An alternative approach is to synthesize a large number of structurally related compounds that differ from one another at only one or a few positions all at once. This approach is often termed *combinatorial chemistry*. Here, compounds are synthesized with the use of the same chemical reactions but a variable set of reactants. Suppose that a molecular scaffold is constructed with two reactive sites and that 20 reactants can be used in the first site and 40 reactants can be used in the second site. A total of $20 \times 40 = 800$ possible compounds can be produced.

A key method in combinatorial chemistry is *split-pool synthesis* (Figure 35.19). The method depends on solid-phase synthetic methods, first developed for the synthesis of peptides (p. 26). Compounds are synthesized on small beads. Beads containing an appropriate starting *scaffold* are produced and divided (split) into n sets, with n corresponding to the number of building blocks to be used at one site. Reactions adding the reactants at the first site are run, and the beads are isolated by filtration. The n sets of beads are then combined (pooled), mixed, and split again into m sets, with m corresponding to the number of reactants to be used at the second site. Reactions adding these m reactants are run, and the beads are again isolated. The important result is that each bead contains only one compound, even though the entire library of beads contains many. Furthermore, although only $n + m$ reactions were run, $n \times m$ compounds are produced. With the preceding values for n and m, $20 + 40 = 60$ reactions produce $20 \times 40 = 800$ compounds. In some cases, assays can be performed directly with the compounds still attached to the bead to find compounds with desired properties (Figure 35.20). Alternatively, each bead can be isolated and the compound can be cleaved from the bead to produce free compounds for analysis. After an interesting compound has been identified, analytical methods of various types must be used to identify which of the $n \times m$ compounds is present.

Note that the "universe" of druglike compounds is vast. More than an estimated 10^{40} compounds are possible with molecular weights less than

Figure 35.18 Synthetic statins. Atorvastatin (Lipitor) and rosuvastatin (Crestor) are completely synthetic drugs that inhibit HMG-CoA reductase.

Figure 35.19 Split-pool synthesis. Reactions are performed on beads. Each of the reactions with the first set of reactants is performed on a separate set of beads. The beads are then pooled, mixed, and split into sets. The second set of reactants is then added. Many different compounds will be produced, but all of the compounds on a single bead will be identical.

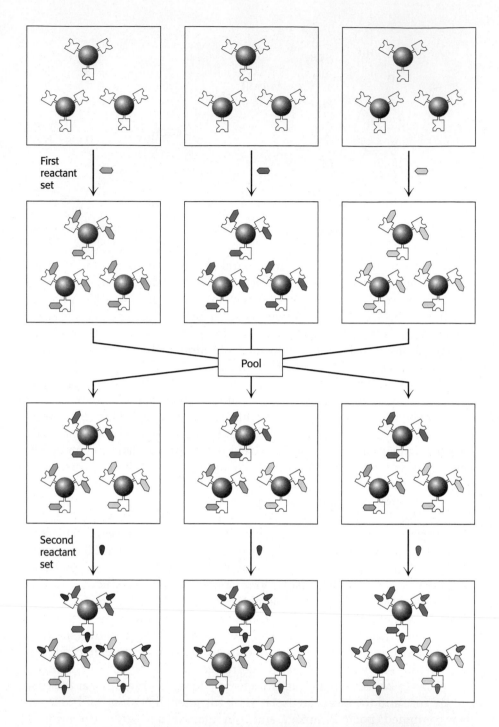

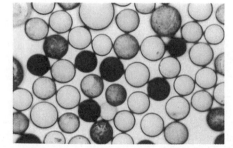

Figure 35.20 Screening a library of synthesized carbohydrates. A small combinatorial library of carbohydrates synthesized on the surface of 130-μm beads is screened for carbohydrates that are bound tightly by a lectin from peanuts. Beads that have such carbohydrates are darkly stained through the action of an enzyme linked to the lectin. [From R. Liang et al., *Proc. Natl. Acad. Sci. USA* 94(1997): 10554–10559; © 2004 National Academy of Sciences, USA.]

750. Thus, even with "large" libraries of millions of compounds, only a tiny fraction of the chemical possibilities are present for study.

Drugs Can Be Designed on the Basis of Three-Dimensional Structural Information About Their Targets

Many drugs bind to their targets in a manner reminiscent of Emil Fischer's lock and key (p. 215). Given this fact, one should be able to design a key given enough knowledge about the shape and chemical composition of the lock. In the idealized case, one would like to design a small molecule that is complementary in shape and electronic structure to a target protein so that it binds effectively to the targeted site. Despite our ability to determine three-dimensional structures rapidly, the achievement of this goal remains in the future. It is difficult to design from scratch stable compounds that

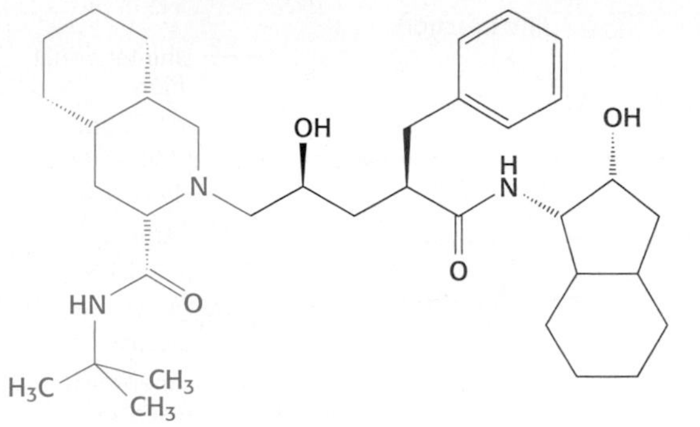

Figure 35.21 Initial design of an HIV protease inhibitor. This compound was designed by combining part of one compound with good inhibition activity but poor solubility (shown in red) with part of another compound with better solubility (shown in blue).

have the correct shape and other properties to fit precisely into a binding site because it is difficult to predict the structure that will best fit into a binding site. Prediction of binding affinity requires a detailed understanding of the interactions between a compound and its binding partner *and* of the interactions between the compound and the solvent when the compound is free in solution.

Nonetheless, *structure-based drug design* has proved to be a powerful tool in drug development. One of its most prominent successes has been the development of drugs that inhibit the protease from the HIV virus. Consider the development of the protease inhibitor indinavir (Crixivan; p. 253). Two sets of promising inhibitors were discovered that had high potency but poor solubility and bioavailability. X-ray crystallographic analysis and molecular-modeling findings suggested that a hybrid molecule might have both high potency and improved bioavailability (Figure 35.21). The synthesized hybrid compound did show improvements but required further optimization. The structural data suggested one point where modifications could be tolerated. A series of compounds were produced and examined (Figure 35.22). The most active compound showed poor bioavailability, but one of the other compounds showed good bioavailability and acceptable activity. The maximum serum concentration available through oral administration was significantly higher than the levels required to suppress replication of the virus. This drug, as well as other protease inhibitors developed at about the same time, has been used in combination with other drugs to

Figure 35.22 Compound optimization. Four compounds are evaluated for characteristics including the IC_{50} (the compound concentration required to reduce HIV replication to 50% of its maximal value), log P, and c_{max} (the maximal concentration of compound present) measured in the serum of dogs. The compound shown at the bottom has the weakest inhibitory power (measured by IC_{50}) but by far the best bioavailability (measured by c_{max}). This compound was selected for further development, leading to the drug indinavir (Crixivan).

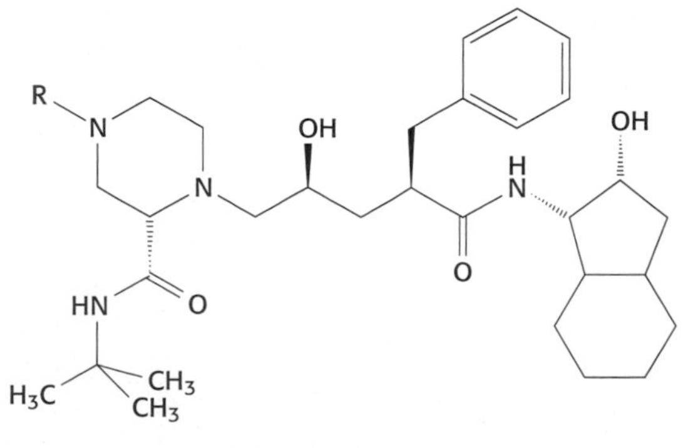

R =	IC_{50}(nmol)	log (P)	c_{max}(μM)
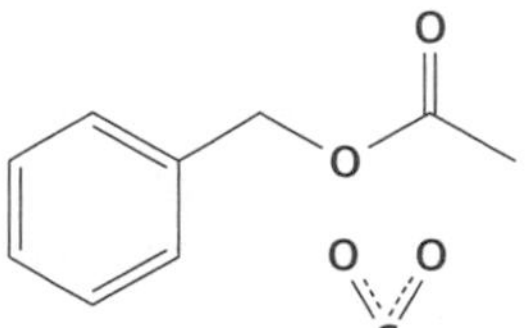	0.4	4.67	< 0.1
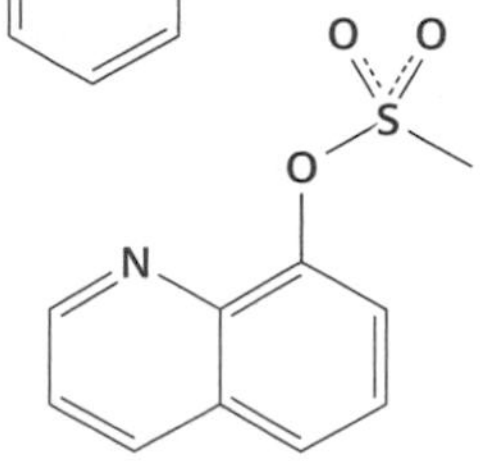	0.01	3.70	< 0.1
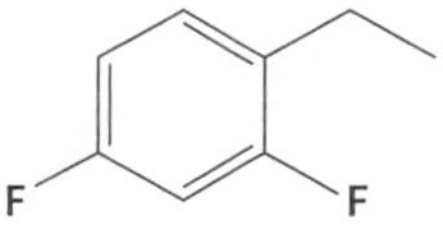	0.3	3.69	0.7
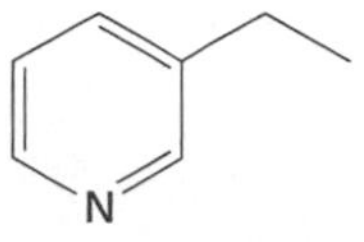	0.6	2.92	11

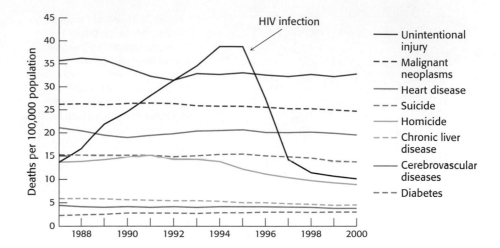

Figure 35.23 The effect of anti-HIV drug development. Death rates from HIV infection (AIDS) reveal the tremendous effect of HIV protease inhibitors and their use in combination with inhibitors of HIV reverse transcriptase. These are death rates from the leading causes of death among persons 24 to 44 years old in the United States. [From Centers for Disease Control.]

treat AIDS with much more encouraging results than had been obtained previously (Figure 35.23).

Aspirin targets the cyclooxygenase site in prostaglandin H2 synthase, as discussed earlier. Animal studies suggested that mammals contain not one but two distinct cyclooxygenase enzymes, both of which are targeted by aspirin. The more recently discovered enzyme, cyclooxygenase 2 (COX2), is expressed primarily as part of the inflammatory response, whereas cyclooxygenase 1 (COX1) is expressed more generally. These observations suggested that a cyclooxygenase inhibitor that was specific for COX2 might be able to reduce inflammation in conditions such as arthritis without producing the gastric and other side effects associated with aspirin.

The amino acid sequences of COX1 and COX2 were deduced from cDNA cloning studies. These sequences are more than 60% identical, clearly indicating that the enzymes have the same overall structure. Nevertheless, there are some differences in the residues around the aspirin-binding site. X-ray crystallography revealed that an extension of the binding pocket was present in COX2, but absent in COX1. This structural difference suggested a strategy for constructing COX2-specific inhibitors—namely, to synthesize compounds that had a protuberance that would fit into the pocket in the COX2 enzyme. Such compounds were designed and synthesized and then further refined to produce effective drugs familiar as Celebrex and Vioxx (Figure 35.24). Vioxx was subsequently withdrawn from the market because some individuals experienced adverse events. These effects appear to be due to the inhibition of COX2, the intended target. Thus, although the development of these drugs is a triumph for structure-based drug design, these outcomes highlight the fact that the inhibition of important enzymes can lead to complex physiological responses.

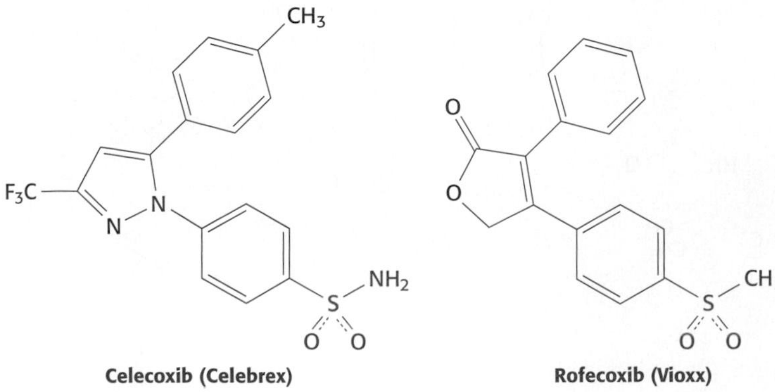

Figure 35.24 COX2-specific inhibitors. These compounds have protuberances (shown in red) that fit into a pocket in the COX2 isozyme but sterically clash with the COX1 isozyme.

Celecoxib (Celebrex)

Rofecoxib (Vioxx)

35.3 The Analysis of Genomes Holds Great Promise for Drug Discovery

The completion of the sequencing of the human and other genomes is a potentially powerful driving force for the development of new drugs. Genomic sequencing and analysis projects have vastly increased our knowledge of the proteins encoded by the human genome. This new source of knowledge may greatly accelerate early stages of the drug-development process or even allow drugs to be tailored to the individual patient.

Potential Targets Can Be Identified in the Human Proteome

The human genome encodes approximately 25,000 proteins, not counting the variation produced by alternative mRNA splicing and posttranslational modifications. Many of these proteins are potential drug targets, in particular those that are enzymes or receptors and have significant biological effects when activated or inhibited. Several large protein families are particularly rich sources of targets. For example, the human genome includes genes for more than 500 protein kinases that can be recognized by comparing the deduced amino acid sequences. One of them, Bcr-Abl kinase, is known to contribute to leukemias and is the target of the drug imatinib mesylate (Gleevec; p. 401). Some of the other protein kinases undoubtedly play central roles in particular cancers as well. Similarly, the human genome encodes approximately 800 7TM receptors (p. 383) of which approximately 350 are odorant receptors. Many of the remaining 7TM receptors are potential drug targets. Some of them are already targets for drugs, such as the β-blocker atenolol, which targets the β-adrenergic receptor, and the antiulcer medication ranitidine (Zantac). The latter compound is an antagonist of the histamine H_2 receptor, a 7TM receptor that participates in the control of gastric acid secretion.

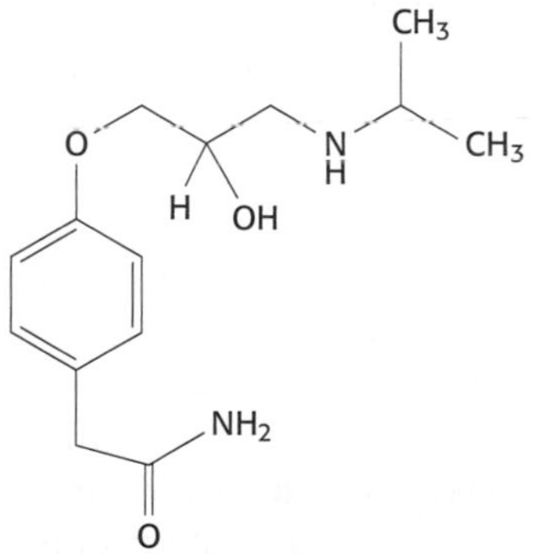

Atenolol

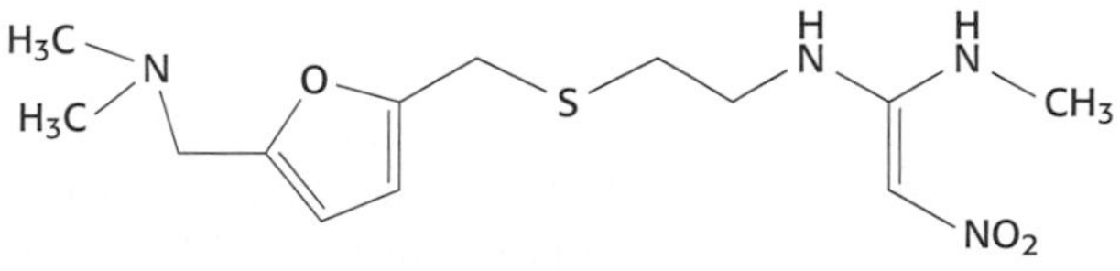

Ranitidine

Novel proteins that are not part of large families already supplying drug targets can be more readily identified through the use of genomic information. There are a number of ways to identify proteins that could serve as targets of drug-development programs. One way is to look for changes in expression patterns, protein localization, or posttranslational modifications in cells from disease-afflicted organisms. Another is to perform studies of tissues or cell types in which particular genes are expressed. Analysis of the human genome should increase the number of actively pursued drug targets by a factor of an estimated two or more.

Animal Models Can Be Developed to Test the Validity of Potential Drug Targets

The genomes of a number of model organisms have now been sequenced. The most important of these genomes for drug development is that of the mouse. Remarkably, the mouse and human genomes are approximately 85% identical in sequence, and more than 98% of all human genes have recognizable mouse counterparts. Mouse studies provide drug developers with a powerful tool—the ability to disrupt ("knock out") specific genes in the mouse (p. 155). If disruption of a gene has a desirable effect, then the product of this gene is a promising drug target. The utility of this approach has been demonstrated retrospectively. For example, disruption of the gene for the α subunit of the H^+-K^+ ATPase, the key protein for secreting acid into the stomach, produces mice with less acid in their stomachs. The stomach pH of such mice is 6.9 in circumstances that produce a stomach pH of 3.2 in their wild-type counterparts. This protein is the target of the drugs omeprazole (Prilosec) and lansoprazole (Prevacid and Takepron), used for treating gastric-esophageal reflux disease.

Omeprazole   **Lansoprazole**

Several large-scale efforts are underway to generate hundreds or thousands of mouse strains, each having a different gene disrupted. The phenotypes of these mice are a good indication of whether the protein encoded by a disrupted gene is a promising drug target. This approach allows drug developers to evaluate potential targets without any preconceived notions regarding physiological function.

Potential Targets Can Be Identified in the Genomes of Pathogens

Human proteins are not the only important drug targets. Drugs such as penicillin and HIV protease inhibitors act by targeting proteins within a pathogen. The genomes of hundreds of pathogens have now been sequenced, and these genome sequences can be mined for potential targets.

New antibiotics are needed to combat bacteria that are resistant to many existing antibiotics. One approach seeks proteins essential for cell survival that are conserved in a wide range of bacteria. Drugs that inactivate such proteins are expected to be broad-spectrum antibiotics, useful for treating infections from any of a range of different bacteria. One such protein is peptide deformylase, the enzyme that removes formyl groups that are present at the amino termini of bacterial proteins immediately after translation (p. 871).

Alternatively, a drug may be needed against a specific pathogen. A recent example of such a pathogen is the organism responsible for severe acute respiratory syndrome (SARS). Within one month of the recognition of this emerging disease, investigators had isolated the virus that causes the syndrome, and, within weeks, its 29,751-base genome had been completely sequenced. This sequence revealed the presence of a gene encoding a viral protease, known to be essential for viral replication from studies of other members of the coronavirus family to which the SARS virus belongs. Drug developers are already at work seeking specific inhibitors of this protease (Figure 35.25).

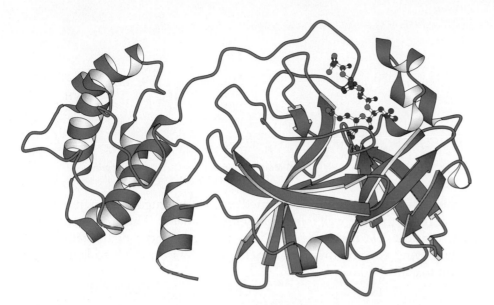

Figure 35.25 Emerging drug target. The structure of a protease from the coronavirus that causes SARS (severe acute respiratory syndrome) is shown bound to an inhibitor. This structure was determined less than a year after the identification of the virus. [Drawn from 1P9S.bdb.]

Genetic Differences Influence Individual Responses to Drugs

Many drugs are not effective in everyone, often because of genetic differences between people. Nonresponding persons may have slight differences in either a drug's target molecule or proteins taking part in drug transport and metabolism. The goal of the emerging fields of pharmacogenetics and pharmacogenomics is to design drugs that either act more consistently from person to person or are tailored to individuals with particular genotypes.

Drugs such as metoprolol that target the β1-adrenergic receptor are popular treatments for hypertension.

Metoprolol

But some people do not respond well. Two variants of the gene coding for the β1-adrenergic receptor are common in the American population. The most common allele has serine in position 49 and arginine in position 389. In some persons, however, glycine replaces one or the other of these residues. In studies, participants with two copies of the most common allele responded well to metoprolol: their daytime diastolic blood pressure was reduced by 14.7 ± 2.9 mm Hg on average. In contrast, participants with one variant allele showed a smaller reduction in blood pressure, and the drug had no significant effect on participants with two variant alleles (Figure 35.26). These observations suggest the potential utility of genotyping individuals at these positions. One could then predict whether or not treatment with metoprolol or other β-blockers is likely to be effective.

Given the importance of ADME and toxicity properties in determining drug efficacy, it is not surprising that variations in proteins participating in drug transport and metabolism can alter a drug's effectiveness. An important example is the use of thiopurine drugs such as 6-thioguanine, 6-mercaptopurine, and azothioprine to treat diseases including leukemia, immune disorders, and inflammatory bowel disease.

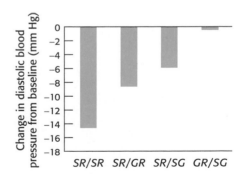

Figure 35.26 Phenotype–genotype correlation. Average changes in diastolic blood pressure on treatment with metoprolol. Persons with two copies of the most common ($S_{49}R_{389}$) allele showed significant decreases in blood pressure. Those with one variant allele (GR or SG) showed more modest decreases, and those with two variant alleles (GR/SG) showed no decrease. [From J. A. Johnson et al., *Clin. Pharmacol. Ther.* 74(2003): 44–52.]

6-Thioguanine   **6-Mercaptopurine**   **Azathioprine**

A minority of patients who are treated with these drugs show signs of toxicity at doses that are well tolerated by most patients. These differences between patients are due to rare variations in the gene encoding the xenobiotic-metabolizing enzyme thiopurine methyltransferase, which adds a methyl group to sulfur atoms.

6-Mercaptopurine   + *S*-adenosylmethionine $\rightleftharpoons$ + *S*-adenosylhomocysteine + H$^+$

The variant enzyme is less stable. Patients with these variant enzymes can build up toxic levels of the drugs if appropriate care is not taken. Thus, genetic variability in an enzyme participating in drug metabolism plays a large role in determining the variation in the tolerance of different persons to particular drug levels. Many other drug-metabolism enzymes and drug-transport proteins have been implicated in controlling individual reactions to specific drugs. The identification of the genetic factors will allow a deeper understanding of why some drugs work well in some persons but poorly in others. In the future, doctors may examine a patient's genes to help plan drug-therapy programs.

35.4 The Development of Drugs Proceeds Through Several Stages

In the United States, the FDA requires that drug candidates be demonstrated to be effective and safe before they may be used in human beings on a large scale. This requirement is particularly true for drug candidates that are to be taken by people who are relatively healthy. More side effects are acceptable for drug candidates intended to treat significantly ill patients such as those with serious forms of cancer, where there are clear, unfavorable consequences for not having an effective treatment.

Clinical Trials Are Time Consuming and Expensive

Clinical trials test the effectiveness and potential side effects of a candidate drug before it is approved by the FDA for general use. These trials proceed in at least three phases (Figure 35.27). In phase 1, a small number (usually from 10 to 100) of healthy volunteers take the drug for an initial study of safety. These volunteers are given a range of doses and are monitored for signs of toxicity. The efficacy of the drug candidate is not specifically evaluated.

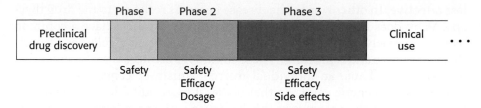

Figure 35.27 **Clinical-trial phases.** Clinical trials proceed in phases examining safety and efficacy in increasingly large groups.

In phase 2, the efficacy of the drug candidate is tested in a small number of persons who might benefit from the drug. Further data regarding the drug's safety are obtained. Such trials are often controlled and double-blinded. In a controlled study, subjects are divided randomly into two groups. Subjects in the treatment group are given the treatment under investigation. Subjects in the control group are given either a placebo—that is, a treatment such as sugar pills known to not have intrinsic value—or the best standard treatment available, if withholding treatment altogether would be unethical. In a double-blinded study, neither the subjects nor the researchers know which subjects are in the treatment group and which are in the control group. A double-blinded study prevents bias in the course of the trial. When the trial has been completed, the assignments of the subjects into treatment and control groups are unsealed and the results for the two groups are compared. A variety of doses are often investigated in phase 2 trials to determine which doses appear to be free of serious side effects and which doses appear to be effective.

One should not underestimate the power of the placebo effect—that is, the tendency to perceive improvement in a subject who believes that he or she is receiving a potentially beneficial treatment. In a study of arthroscopic surgical treatment for knee pain, for example, subjects who were led to believe that they had received surgery through the use of videotapes and other means showed the same level of improvement, on average, as subjects who were actually operated on.

In phase 3, similar studies are performed on a larger population. This phase is intended to more firmly establish the efficacy of the drug candidate and to detect side effects that may develop in a small percentage of the subjects who receive treatment. Thousands of subjects may participate in a typical phase 3 study.

Clinical trials can be extremely costly. Hundreds or thousands of patients must be recruited and monitored for the duration of the trial. Many physicians, nurses, clinical pharmacologists, statisticians, and others participate in the design and execution of the trial. Costs can run from tens of millions to hundreds of millions of dollars. Extensive records must be kept, including documentation of any adverse reactions. These data are compiled and submitted to the FDA. The full cost of developing a drug is currently estimated to be from $400 million to $800 million.

Even after a drug has been approved and is in use, difficulties can arise. As mentioned earlier, rofecoxib (Vioxx), for example, was withdrawn from the market after significant cardiac side effects were detected in additional clinical trials. Such events highlight the necessity for users of any drug to balance beneficial effects against potential risks.

The Evolution of Drug Resistance Can Limit the Utility of Drugs for Infectious Agents and Cancer

Many drugs are used for long periods of time without any loss of effectiveness. However, in some cases, particularly for the treatment of infectious diseases or of cancer, drug treatments that were initially effective become less effective. In other words, the disease becomes resistant to the drug therapy. Why does this occur? Infectious diseases and cancer have a common feature—namely, that an affected person contains many cells (or viruses) that can mutate and reproduce. These conditions are necessary for evolution to take place. Thus, an individual microorganism or cancer cell may by chance have a genetic variation that makes it more suitable for growth and reproduction in the presence of the drug than is the population of microorganisms or cancer cells at large. These microorganisms or cells are more fit than others in their population, and they will tend to take over the population. As the selective pressure due to the drug is continually applied, the population of microorganisms or cancer cells will tend to become more and more resistant to the presence of the drug. Note that resistance can develop by a number of mechanisms.

The HIV protease inhibitors discussed earlier provide an important example of the evolution of drug resistance. Retroviruses are very well suited to this sort of evolution because reverse transcriptase carries out replication without a proofreading mechanism. In a genome of approximately 9750 bases, each possible single point mutation is estimated to appear in a virus particle more than 1000 times per day in each infected person. Many multiple mutations also occur. Most of these mutations either have no effect or are detrimental to the virus. However, a few of the mutant virus particles encode proteases that are less susceptible to inhibition by the drug. In the presence of an HIV protease inhibitor, these virus particles will tend to replicate more effectively than the population at large. Over time, the less susceptible viruses will come to dominate the population and the virus population will become resistant to the drug.

Pathogens may become resistant to antibiotics by completely different mechanisms. Some pathogens contain enzymes that inactivate or degrade specific antibiotics. For example, many organisms are resistant to β-lactams such as penicillin because they contain β-lactamase enzymes. These enzymes hydrolyze the β-lactam ring and render the drugs inactive.

Penicillin

Many of these enzymes are encoded in plasmids, small circular pieces of DNA often carried by bacteria. Many plasmids are readily transferred from one bacterial cell to another, transmitting the capability for antibiotic resistance. Plasmid transfer thus contributes to the spread of antibiotic resistance, a major health-care challenge. On the other hand, plasmids have been harnessed for use in recombinant DNA methods (p. 143).

Drug resistance commonly emerges in the course of cancer treatment. Cancer cells are characterized by their ability to grow rapidly without the constraints that apply to normal cells. Many drugs used for cancer chemotherapy inhibit processes that are necessary for this rapid cell growth. However, individual cancer cells may accumulate genetic changes that mitigate the effects of such drugs. These altered cancer cells will tend to grow more rapidly than others and will become dominant within the cancer-cell population. This ability of cancer cells to mutate quickly has posed a challenge to one of the major breakthroughs in cancer treatment: the development of inhibitors for proteins specific to cancer cells present in certain leukemias (p. 401). For example, tumors became undetectable in patients treated with imatinib mesylate, which is directed against the Bcr-Abl protein kinase. Unfortunately, the tumors of many of the patients treated with imatinib mesylate recur after a period of years. In many of these cases, mutations have altered the Bcr-Abl protein so that it is no longer inhibited by the concentrations of imatinib mesylate used in therapy.

Cancer patients often take multiple drugs concurrently in the course of chemotherapy, and in many cases cancer cells become simultaneously resistant to many or all of them. This multiple-drug resistance can be due to the proliferation of cancer cells that overexpress a number of ABC transporter proteins that pump drugs out of the cell (p. 358). Thus, cancer cells can evolve drug resistance by overexpressing normal human proteins or by modifying proteins responsible for the cancer phenotype.

Summary

35.1 The Development of Drugs Presents Huge Challenges

Most drugs act by binding to enzymes or receptors and modulating their activities. To be effective, drugs must bind to these targets with high affinity and specificity. However, even most compounds with the desired affinity and specificity do not make suitable drugs. Most compounds are poorly absorbed or rapidly excreted from the body or they are modified by metabolic pathways that target foreign compounds. Consequently, when taken orally, these compounds do not reach their targets at appropriate concentrations for a sufficient period of time. A drug's properties related to its absorption, distribution, metabolism, and excretion are called ADME properties. Oral bioavailability is a measure of a drug's ability to be absorbed; it is the ratio of the peak concentration of a compound given orally to the peak concentration of the same dose directly injected. The structure of a compound can affect its bioavailability in complicated ways, but generalizations called Lipinski's rules provide useful guidelines. Drug metabolism pathways include oxidation by cytochrome P450 enzymes (phase I metabolism) and conjugation to glutathione, glucuronic acid, and sulfate (phase II metabolism). A compound may also not be a useful drug because it is toxic, either because it modulates the target molecule too effectively or because it also binds to proteins other than the target. The liver and kidneys play central roles in drug metabolism and excretion.

35.2 Drug Candidates Can Be Discovered by Serendipity, Screening, or Design

Many drugs have been discovered by serendipity—that is, by chance observation. The antibiotic penicillin is produced by a mold that accidentally contaminated a culture dish, killing nearby bacteria. Drugs such as chlorpromazine and sildenafil were discovered to have beneficial effects on human physiology that were completely different from those expected. The cholesterol-lowering statin drugs were developed after large collections of compounds were screened for potentially interesting activities. Combinatorial chemistry methods have been developed to generate large collections of chemically related yet diverse compounds for screening. In some cases, the three-dimensional structure of a drug target is available and can be used to aid the design of potent and specific inhibitors. Examples of drugs designed in this manner are the HIV protease inhibitors indinavir and cyclooxygenase 2 inhibitors such as celecoxib.

35.3 The Analysis of Genomes Holds Great Promise for Drug Discovery

The human genome encodes approximately 25,000 proteins, and many more if derivatives due to alternative mRNA splicing and post-translational modification are included. The genome sequences can be examined for potential drug targets. Large families of proteins known to participate in key physiological processes such as the protein kinases and 7TM receptors have each yielded several targets for which drugs have been developed. The genomes of model organisms also are useful for drug-development studies. Strains of mice with particular genes disrupted have been useful in validating certain drug targets. The genomes of bacteria, viruses, and parasites encode many potential drug targets that can be exploited owing to their important functions and their differences from human proteins, minimizing the potential for side effects. Genetic differences between individuals can be examined and correlated with differences in responses to drugs, potentially aiding both clinical treatments and drug development.

35.4 The Development of Drugs Proceeds Through Several Stages

Before compounds can be given to human beings as drugs, they must be extensively tested for safety and efficacy. Clinical trials are performed in stages, first testing safety, then safety and efficacy in a small population, and finally safety and efficacy in a larger population to detect rarer adverse effects. Largely due to the expenses associated with clinical trials, the cost of developing a new drug has been estimated to be as much as $800 million. Even when a drug has been approved for use, complications can arise. With infectious diseases and cancer, patients often develop resistance to a drug after it has been used for some period of time, because variants of the disease agent that are less susceptible to the drug arise and replicate, even when the drug is present.

Key Terms

ligand (p. 1001)

dissociation constant (K_d) (p. 1002)

apparent dissociation constant (K_d^{app}) (p. 1003)

inhibition constant (K_i) (p. 1003)

side effects (p. 1003)

ADME (p. 1003)

oral bioavailability (p. 1004)

Lipinski's rules (p. 1004)

compartment (p. 1005)

blood–brain barrier (p. 1005)

xenobiotic compounds (p. 1006)

drug metabolism (p. 1006)

oxidation (p. 1006)

conjugation (p. 1006)

phase I transformation (p. 1007)

phase II transformation (p. 1007)

first-pass metabolism (p. 1007)

glomerulus (p. 1007)

enterohepatic cycling (p. 1008)

therapeutic index (p. 1009)

atheroma (p. 1012)

myopathy (p. 1012)

high-throughput screening (p. 1013)

combinatorial chemistry (p. 1013)

split-pool synthesis (p. 1013)

structure-based drug design (p. 1015)

Selected Readings

Books

Hardman, J. G., Limbird, L. E., and Gilman, A. G. 2001. *Goodman and Gilman's The Pharmacological Basis of Therapeutics* (10th ed.). McGraw-Hill Professional.

Levine, R. R., and Walsh, C. T. 2004. *Levine's Pharmacology: Drug Actions and Reactions* (7th ed.). Taylor and Francis Group.

Silverman, R. B. 2004. *Organic Chemistry of Drug Design and Drug Action*. Academic Press.

ADME and Toxicity

Caldwell, J., Gardner, I., and Swales, N. 1995. An introduction to drug disposition: The basic principles of absorption, distribution, metabolism, and excretion. *Toxicol. Pathol.* 23:102–114.

Lee, W., and Kim, R. B. 2004. Transporters and renal drug elimination. *Annu. Rev. Pharmacol. Toxicol.* 44:137–166.

Lin, J., Sahakian, D. C., de Morais, S. M., Xu, J. J., Polzer, R. J., and Winter, S. M. 2003. The role of absorption, distribution, metabolism, excretion and toxicity in drug discovery. *Curr. Top. Med. Chem.* 3:1125–1154.

Poggesi, I. 2004. Predicting human pharmacokinetics from preclinical data. *Curr. Opin. Drug Discov. Devel.* 7:100–111.

Case Histories

Flower, R. J. 2003. The development of COX2 inhibitors. *Nat. Rev. Drug Discov.* 2:179–191.

Tobert, J. A. 2003. Lovastatin and beyond: The history of the HMG-CoA reductase inhibitors. *Nat. Rev. Drug Discov.* 2:517–526.

Vacca, J. P., Dorsey, B. D., Schleif, W. A., Levin, R. B., McDaniel, S.L., Darke, P. L., Zugay, J., Quintero, J. C., Blahy, O. M., Roth, E., et al. 1994. L-735,524: An orally bioavailable human immunodeficiency virus type 1 protease inhibitor. *Proc. Natl. Acad. Sci. U.S.A.* 91:4096–4100.

Wong, S., and Witte, O. N. 2004. The BCR-ABL story: Bench to bedside and back. *Annu. Rev. Immunol.* 22:247–306.

Structure-Based Drug Design

Kuntz, I. D. 1992. Structure-based strategies for drug design and discovery. *Science* 257:1078–1082.

Dorsey, B. D., Levin, R. B., McDaniel, S. L., Vacca, J. P., Guare, J. P., Darke, P. L., Zugay, J. A., Emini, E. A., Schleif, W. A., Quintero, J. C., et al. 1994. L-735,524: The design of a potent and orally bioavailable HIV protease inhibitor. *J. Med. Chem.* 37:3443–3451.

Chen, Z., Li, Y., Chen, E., Hall, D. L., Darke, P. L., Culberson, C., Shafer, J. A., and Kuo, L. C. 1994. Crystal structure at 1.9-Å reso-

lution of human immunodeficiency virus (HIV) II protease complexed with L-735,524, an orally bioavailable inhibitor of the HIV proteases. *J. Biol. Chem.* 269:26344–26348.

Combinatorial Chemistry

Baldwin, J. J. 1996. Design, synthesis and use of binary encoded synthetic chemical libraries. *Mol. Divers.* 2:81–88.

Burke, M. D., Berger, E. M., and Schreiber, S. L. 2003. Generating diverse skeletons of small molecules combinatorially. *Science* 302:613–618.

Edwards, P. J., and Morrell, A. I. 2002. Solid-phase compound library synthesis in drug design and development. *Curr. Opin. Drug Discov. Devel.* 5:594–605.

Genomics

Zambrowicz, B. P., and Sands, A. T. 2003. Knockouts model the 100 best-selling drugs: Will they model the next 100? *Nat. Rev. Drug Discov.* 2:38–51.

Salemme, F. R. 2003. Chemical genomics as an emerging paradigm for postgenomic drug discovery. *Pharmacogenomics.* 4:257–267.

Michelson, S., and Joho, K. 2000. Drug discovery, drug development and the emerging world of pharmacogenomics: Prospecting for information in a data-rich landscape. *Curr. Opin. Mol. Ther.* 2:651–654.

Weinshilboum, R., and Wang, L. 2004. Pharmacogenomics: Bench to bedside. *Nat. Rev. Drug Discov.* 3:739–748.

Problems

1. *Routes to discovery.* For each of the following drugs, indicate whether the physiological effects of the drug were known before or after the target was identified.

(a) Penicillin
(b) Sildenafil (Viagra)
(c) Rofecoxib (Vioxx)
(d) Atorvastatin (Lipitor)
(e) Aspirin
(f) Indinavir (Crixivan)

2. *Lipinski's rules.* Which of the following compounds satisfy all of Lipinski's rules? [Log(P) values are given in parentheses.]

(a) Atenolol (0.23)
(b) Sildenafil (3.18)
(c) Indinavir (2.78)

3. *Calculating log tables.* Considerable effort has been expended to develop computer programs that can estimate log(P) values entirely on the basis of chemical structure. Why would such programs be useful?

4. *An ounce of prevention.* Legislation has been proposed that would require that acetaminophen tablets also include *N*-acetylcysteine. Speculate about the role that this additive would serve.

5. *Drug interactions.* As noted in the chapter, coumadin can be a very dangerous drug because too much can cause uncontrolled bleeding. Persons taking coumadin must be careful about taking other drugs, particularly those that bind to albumin. Propose a mechanism for this drug–drug interaction.

6. *Find the target.* Trypanosomes are unicellular parasites that cause sleeping sickness. During one stage of their life cycle,

these organisms live in the bloodstream and derive all of their energy from glycolysis, which takes place in a specialized organelle called a glycosome inside the parasite. Propose potential targets for treating sleeping sickness. What are some potential difficulties with your approach?

Mechanism Problem

7. *Variations on a theme.* The metabolism of amphetamine by cytochrome P450 enzymes results in the conversion shown here. Propose a mechanism and indicate any additional products.

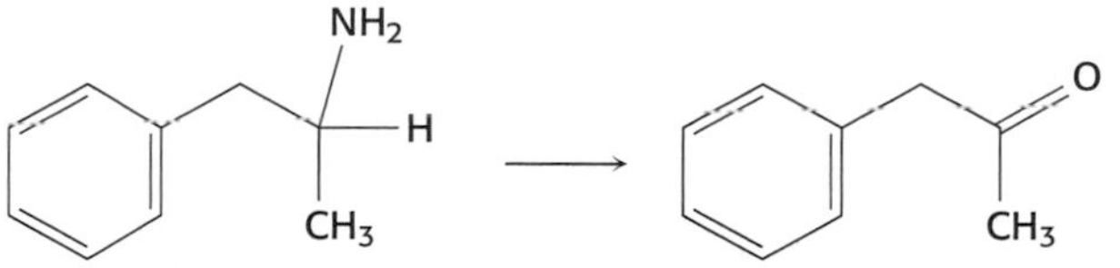

Amphetamine

Data Interpretation Problem

8. *HIV protease inhibitor design.* Compound A is one of a series that were designed to be potent inhibitors of HIV protease.

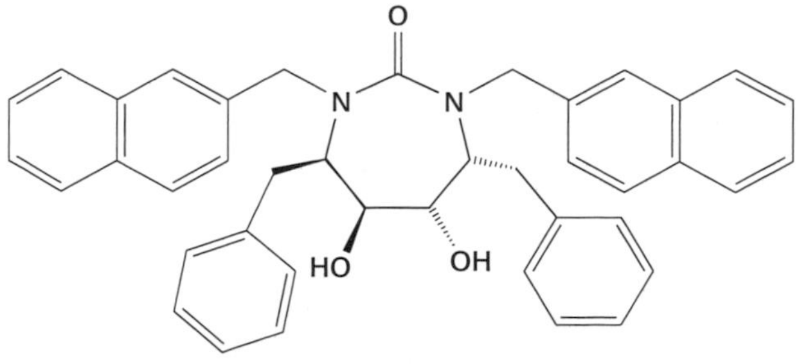

Compound A

Compound A was tested by using two assays: (1) direct inhibition of HIV protease in vitro and (2) inhibition of viral RNA production in HIV infected cells, a measure of viral replication. The results of these assays are shown here. The HIV protease activity is measured with a substrate peptide present at a concentration equal to its K_M value.

Compound A (nM)	HIV protease activity (arbitrary units)
0	11.2
0.2	9.9
0.4	7.4
0.6	5.6
0.8	4.8
1	4.0
2	2.2
10	0.9
100	0.2

Compound A (mM)	Viral RNA production (arbitrary units)
0	760
1.0	740
2.0	380
3.0	280
4.0	180
5.0	100
10	30
50	20

Estimate the values for the K_I of compound A in the protease activity assay and for its IC_{50} in the viral RNA production assay.

Treating rats with the relatively high oral dose of 20 mg kg^{-1} results in a maximum concentration of compound A of 0.4 μM. On the basis of this value, do you expect compound A to be effective in preventing HIV replication when taken orally?

Appendix A Physical Constants and Conversion of Units

Values of physical constants

Physical constant	Symbol or abbreviation	Value
Atomic mass unit (dalton)	amu	1.660×10^{-24} g
Avogadro's number	N	6.022×10^{23} mol^{-1}
Boltzmann's constant	k	1.381×10^{-23} J K^{-1}
		3.298×10^{-24} cal K^{-1}
Electron volt	eV	1.602×10^{-19} J
		3.828×10^{-20} cal
Faraday constant	F	9.649×10^4 J V^{-1} mol^{-1}
		23.06 Kcal V^{-1} mol^{-1}
Curie	Ci	3.70×10^{10} disintegrations s^{-1}
Gas constant	R	8.315 J mol^{-1} K^{-1}
		1.987 cal mol^{-1} K^{-1}
Planck's constant	h	6.626×10^{-34} J s
		1.584×10^{-34} cal s
Speed of light in a vacuum	c	2.998×10^{10} cm s^{-1}

Abbreviations: C, coulomb; cal, calorie; cm, centimeter; K, kelvin;
eq, equivalent; g, gram; J, joule; mol, mole; s, second; V, volt.

Mathematical constants

$\pi = 3.14159$
$e = 2.71828$
$\log_e x = 2.303 \log_{10} x$

Conversion factors

Physical quantity	Equivalent
Length	$1\,\text{cm} = 10^{-2}\,\text{m} = 10\,\text{mm} = 10^4\,\mu\text{m} = 10^7\,\text{nm}$
	$1\,\text{cm} = 10^8\,\text{Å} = 0.3937\,\text{inch}$
Mass	$1\,\text{g} = 10^{-3}\,\text{kg} = 10^3\,\text{mg} = 10^6\,\mu\text{g}$
	$1\,\text{g} = 3.527 \times 10^{-2}\,\text{ounce (avoirdupois)}$
Volume	$1\,\text{cm}^3 = 10^{-6}\,\text{m}^3 = 10^3\,\text{mm}^3$
	$1\,\text{ml} = 1\,\text{cm}^3 = 10^{-3}\,\text{liter} = 10^3\,\mu\text{l}$
	$1\,\text{cm}^3 = 6.1 \times 10^{-2}\,\text{in}^3 = 3.53 \times 10^{-5}\,\text{ft}^3$
Temperature	$K = °C + 273.15$
	$°C = (5/9)(°F - 32)$
Energy	$1\,\text{J} = 10^7\,\text{erg} = 0.239\,\text{cal} = 1\,\text{watt s}$
Pressure	$1\,\text{torr} = 1\,\text{mm Hg}(0°C)$
	$= 1.333 \times 10^2\,\text{newtons m}^{-2}$
	$= 1.333 \times 10^2\,\text{pascal}$
	$= 1.316 \times 10^{-3}\,\text{atmospheres}$

Standard prefixes

Prefix	Abbreviation	Factor
kilo	k	10^3
hecto	h	10^2
deca	da	10^1
deci	d	10^{-1}
centi	c	10^{-2}
milli	m	10^{-3}
micro	μ	10^{-6}
nano	n	10^{-9}
pico	p	10^{-12}

Appendix B   Acidity Constants

pK_a values of some acids

Acid	pK' (at 25°C)	Acid	pK' (at 25°C)
Acetic acid	4.76	Malic acid, pK_1	3.40
Acetoacetic acid	3.58	pK_2	5.11
Ammonium ion	9.25	Phenol	9.89
Ascorbic acid, pK_1	4.10	Phosphoric acid, pK_1	2.12
pK_2	11.79	pK_2	7.21
Benzoic acid	4.20	pK_3	12.67
n-Butyric acid	4.81	Pyridinium ion	5.25
Cacodylic acid	6.19	Pyrophosphoric acid, pK_1	0.85
Citric acid, pK_1	3.14	pK_2	1.49
pK_2	4.77	pK_3	5.77
pK_3	6.39	pK_4	8.22
Ethylammonium ion	10.81	Succinic acid, pK_1	4.21
Formic acid	3.75	pK_2	5.64
Glycine, pK_1	2.35	Trimethylammonium ion	9.79
pK_2	9.78	Tris (hydroxymethyl) aminomethane	8.08
Imidazolium ion	6.95	Water*	15.74
Lactic acid	3.86		
Maleic acid, pK_1	1.83		
pK_2	6.07		

*$[H^+][OH^-] = 10^{-14}$; $[H_2O] = 55.5$ M.

Typical pK_a values of ionizable groups in proteins

Group	Acid	⇌	Base	Typical pK_a	Group	Acid	⇌	Base	Typical pK_a
Terminal α-carboxyl group				3.1	Cysteine				8.3
Aspartic acid Glutamic acid				4.1	Tyrosine				10.4
					Lysine				10.0
Histidine				6.0	Arginine				12.5
Terminal α-amino group				8.0					

Note: pK_a values depend on temperature, ionic strength, and the microenvironment of the ionizable group.

Appendix C   Standard Bond Lengths

Bond	Structure	Length (Å)
C—H	R_2CH_2	1.07
	Aromatic	1.08
	RCH_3	1.10
C—C	Hydrocarbon	1.54
	Aromatic	1.40
C=C	Ethylene	1.33
C≡C	Acetylene	1.20
C—N	RNH_2	1.47
	O=C—N	1.34
C—O	Alcohol	1.43
	Ester	1.36
C=O	Aldehyde	1.22
	Amide	1.24
C—S	R_2S	1.82
N—H	Amide	0.99
O—H	Alcohol	0.97
O—O	O_2	1.21
P—O	Ester	1.56
S—H	Thiol	1.33
S—S	Disulfide	2.05

Glossary of Compounds

The following pages contain the structures of amino acids, common metabolic intermediates, nucleotide bases, and important cofactors. In many cases, two versions of the structure are shown: the Fisher structure (bottom) and a more stereochemically accurate version (top).

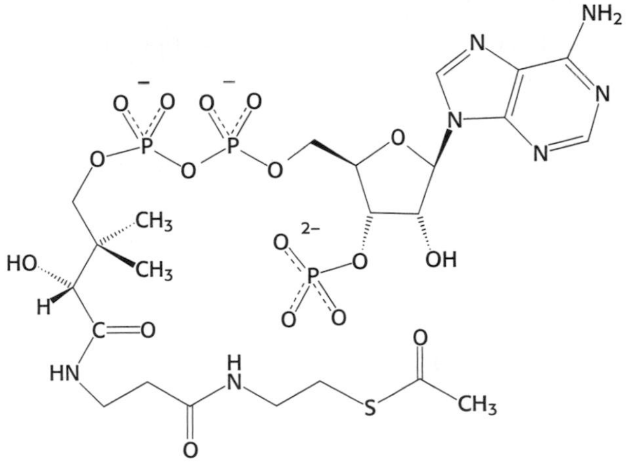

Acetyl coenzyme A (acetyl CoA)

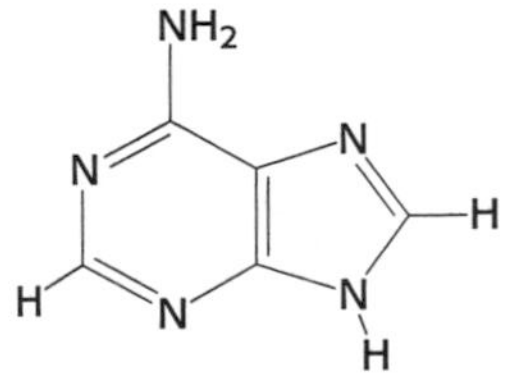

Adenine

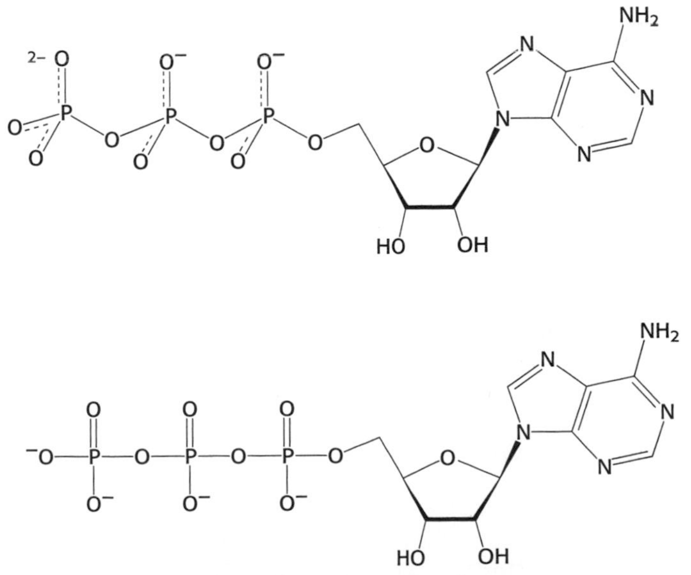

Adenosine triphosphate (ATP)

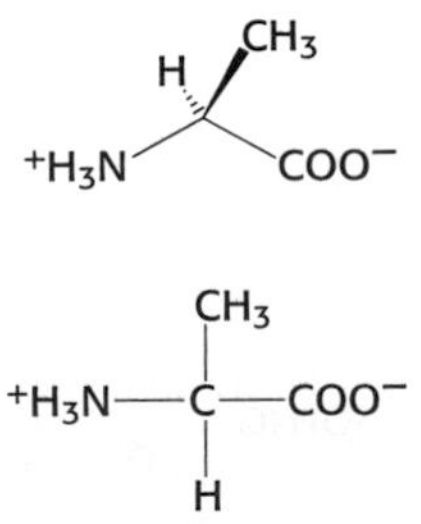

Alanine

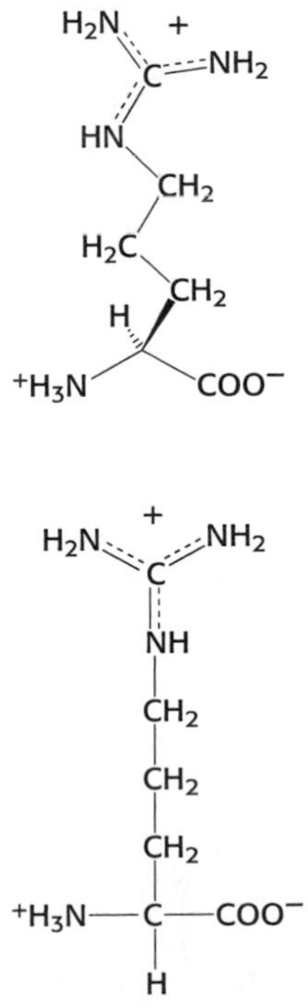

Arginine

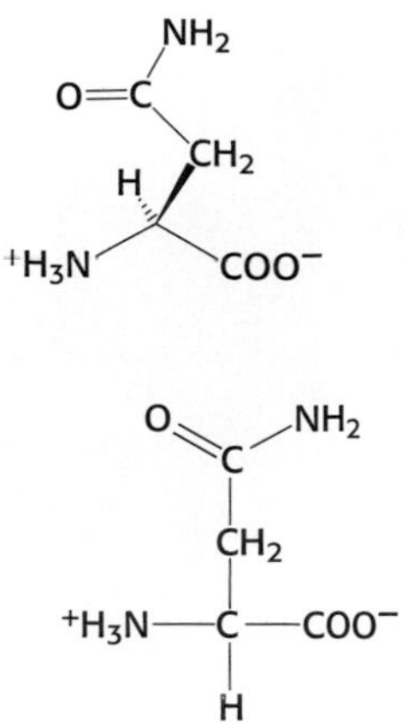

Asparagine

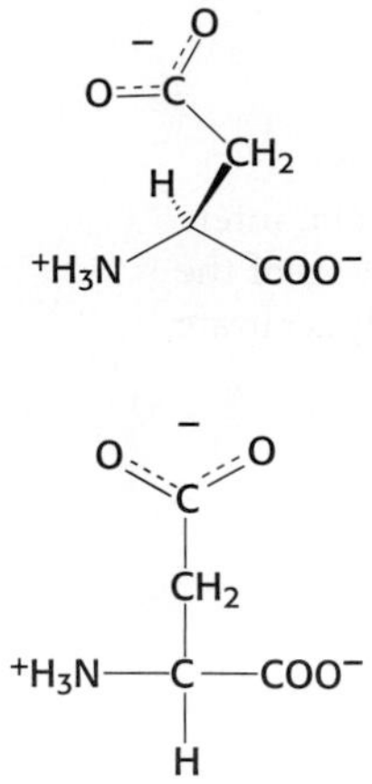

Aspartate

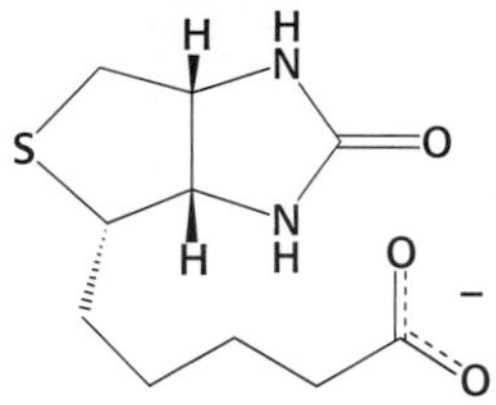

Biotin

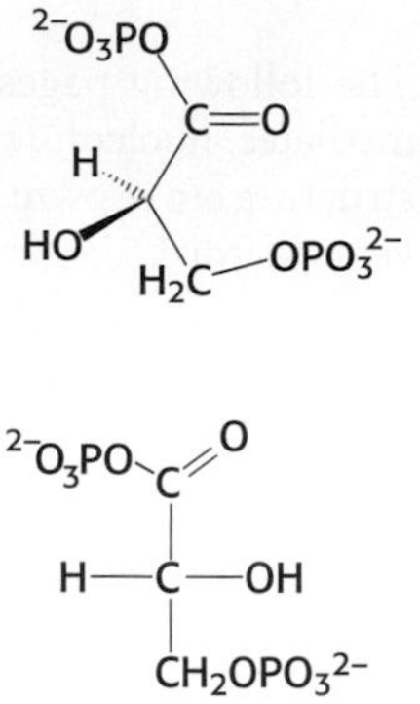

1,3-Bisphosphoglycerate

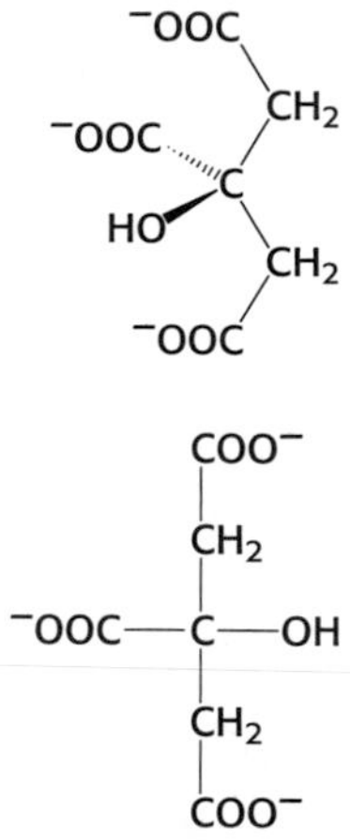

Citrate

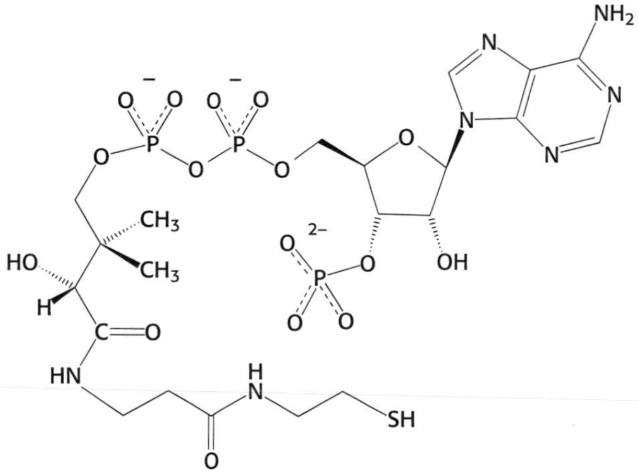

Coenzyme A

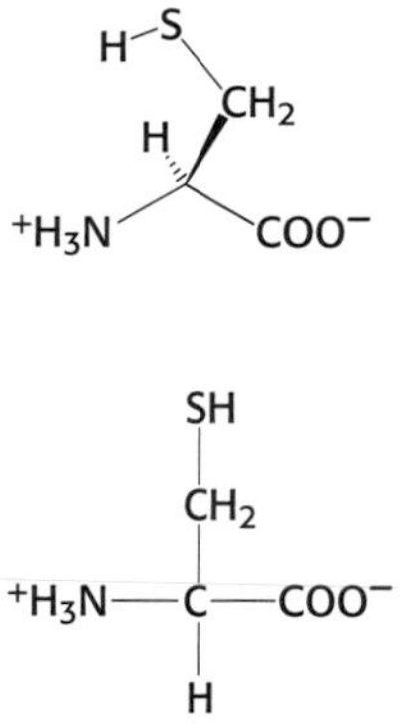

Cysteine

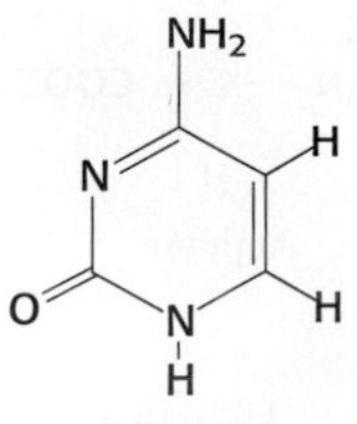

Cytosine

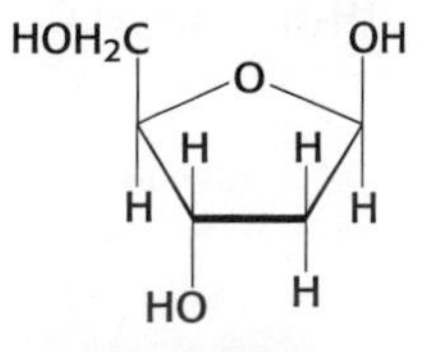

Deoxyribose

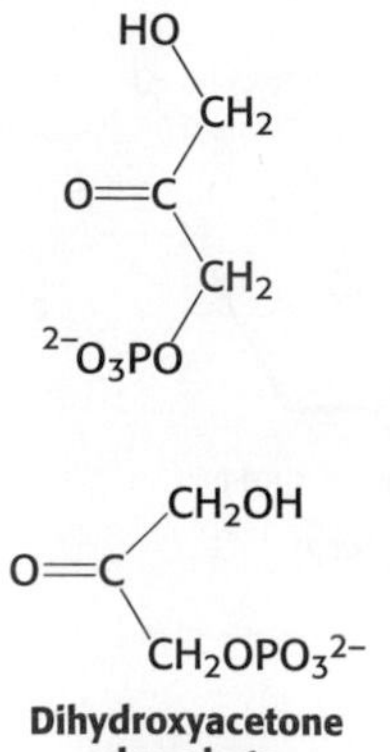

Dihydroxyacetone phosphate

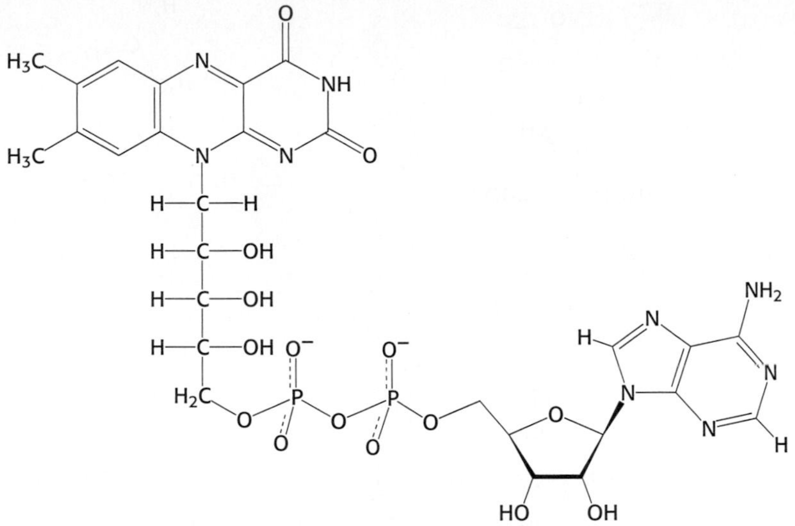

Flavin adenine dinucleotide (FAD)

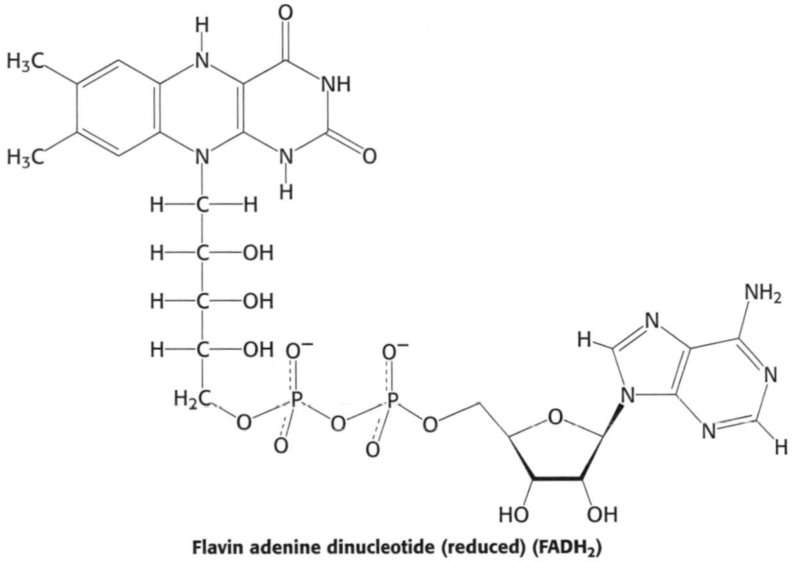

Flavin adenine dinucleotide (reduced) (FADH$_2$)

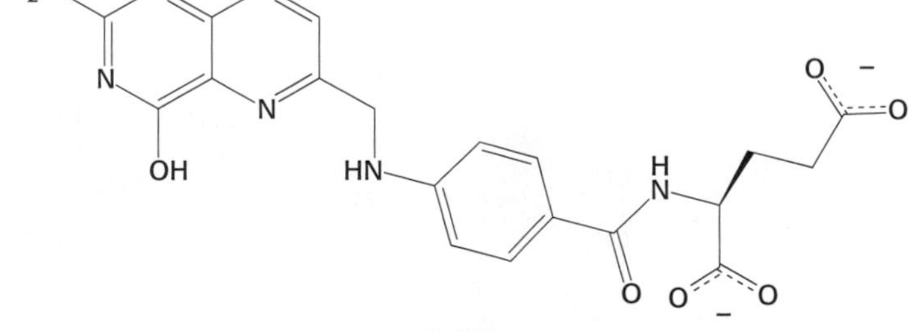

Folic acid

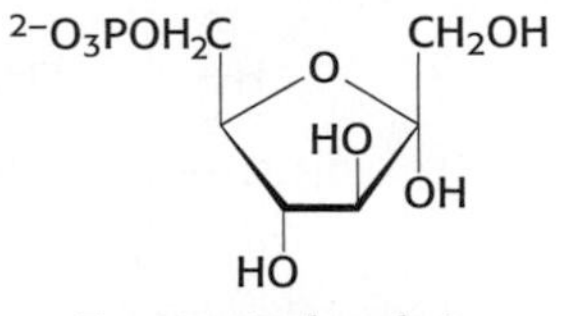

Fructose-6-phosphate

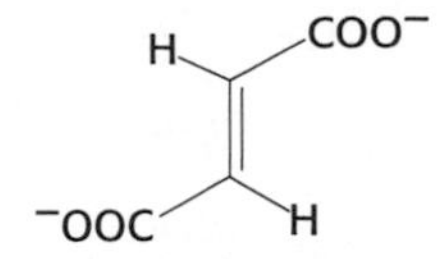

Fumarate

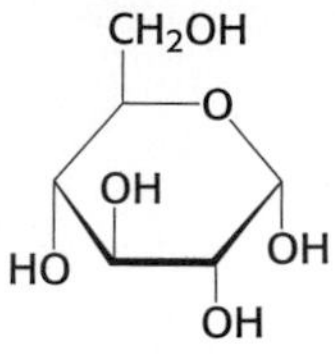

Glucose
(α-D-glucose)

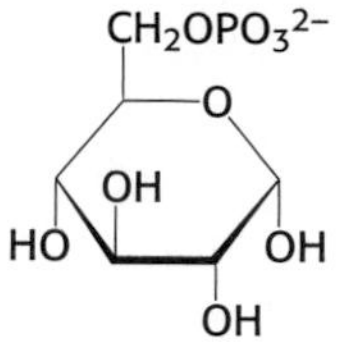

Glucose-6-phosphate

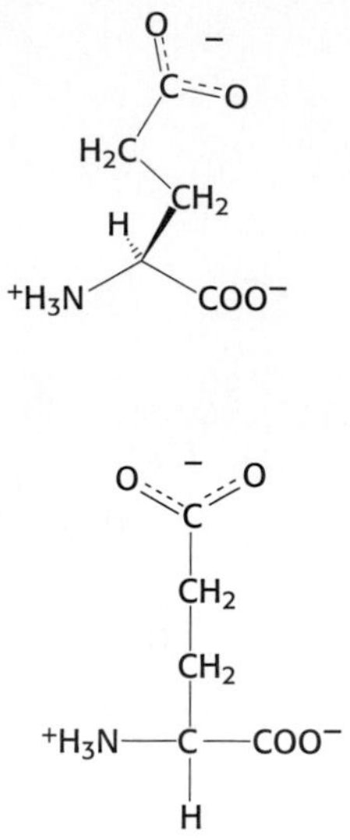

Glutamate

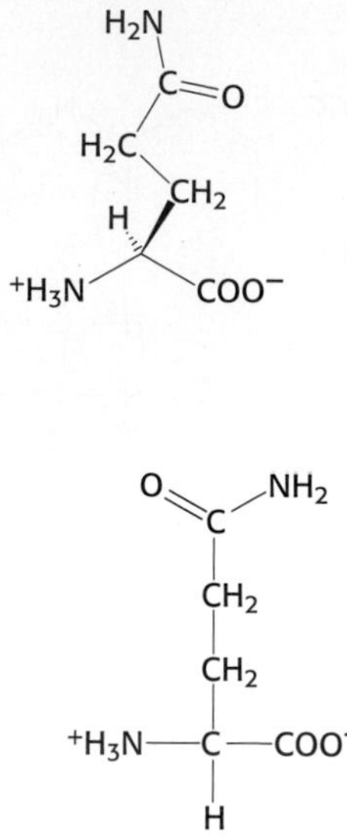

Glutamine

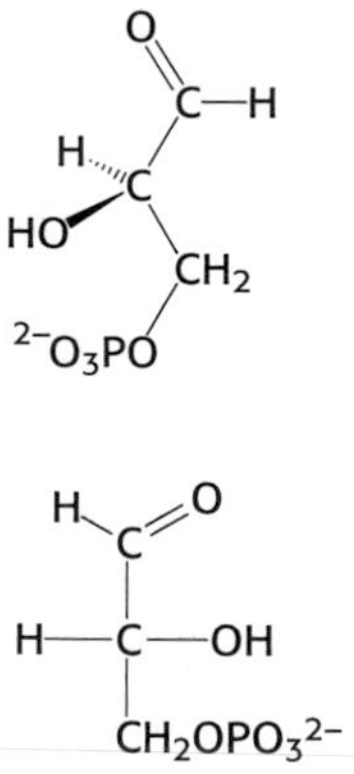

**Glyceraldehyde
3-phosphate**

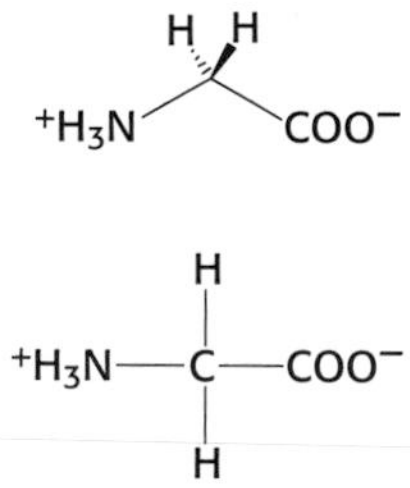

Glycine

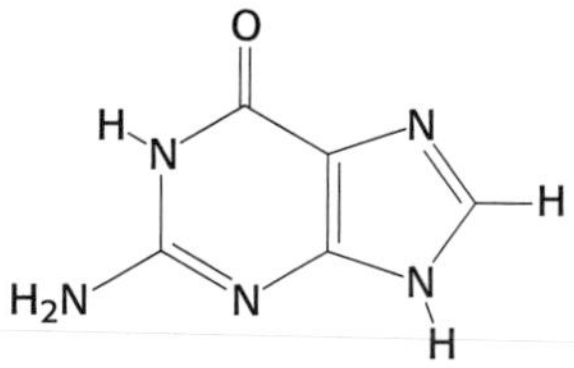

Guanine

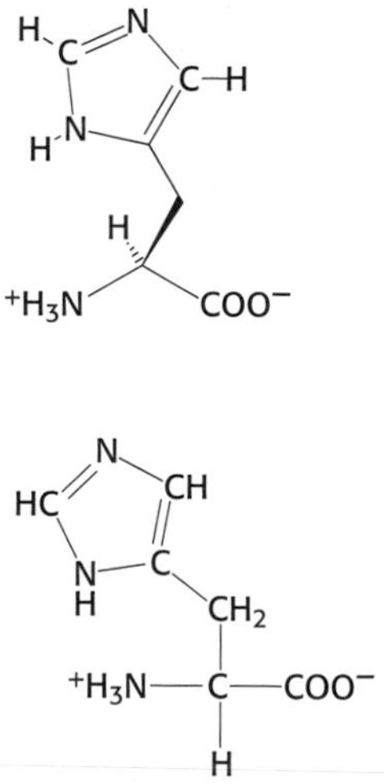

Histidine

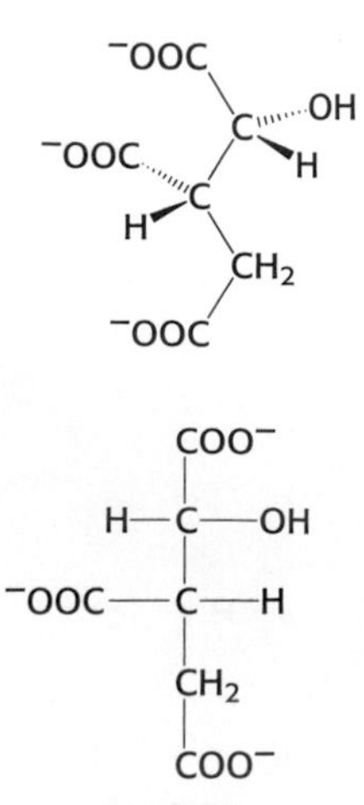

Isocitrate

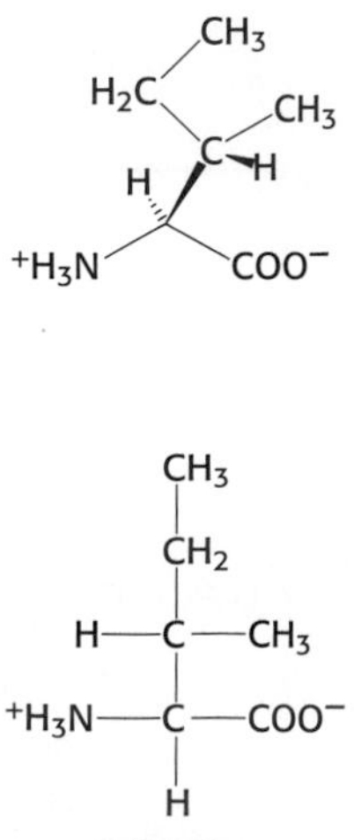

Isoleucine

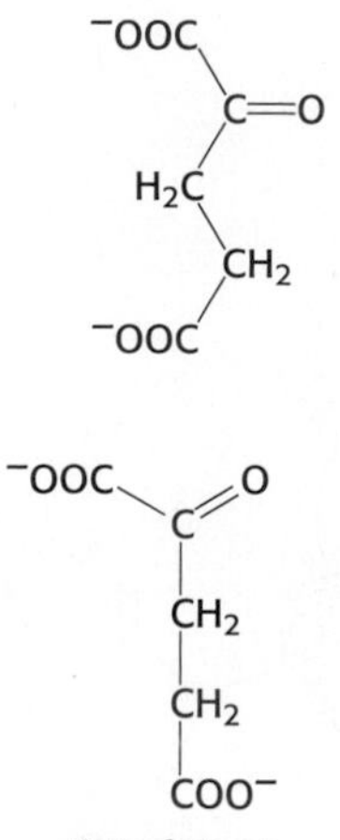

α-Ketoglutarate

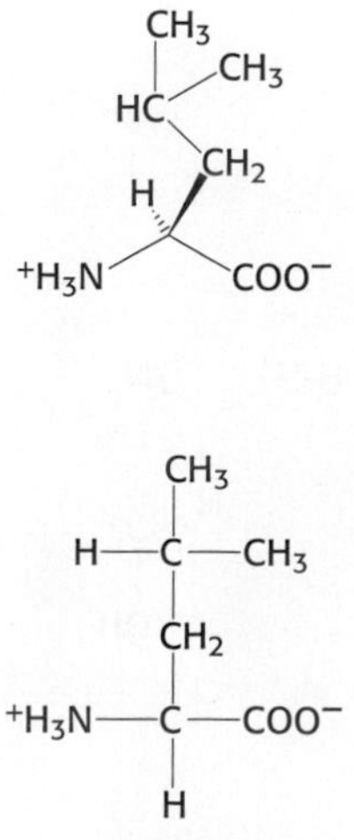

Leucine

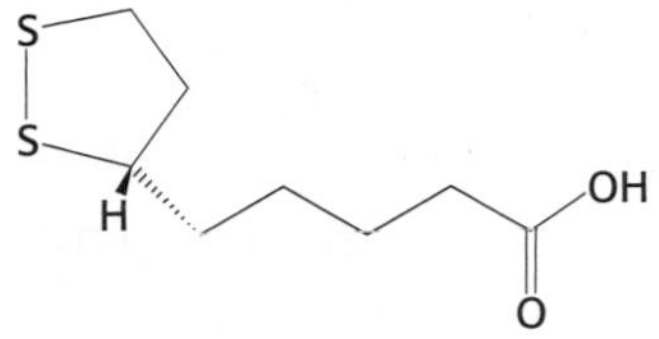

Lipoic acid

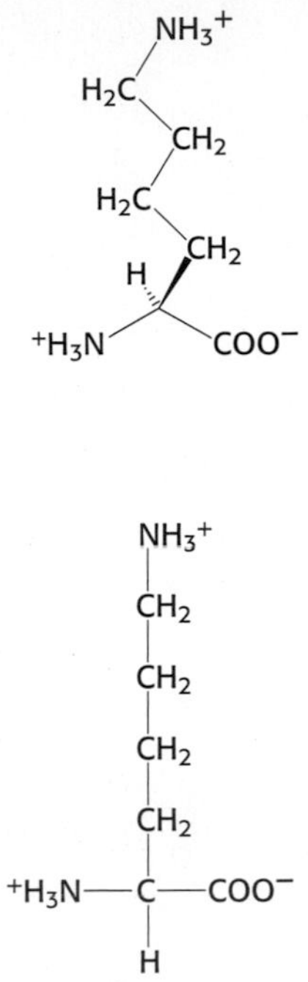

Lysine

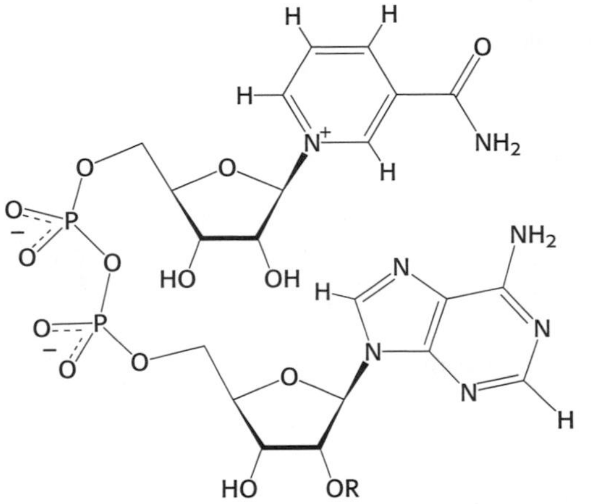

Nicotinamide adenine dinucleotide (R = H), (NAD$^+$)

Nicotinamide adenine dinucleotide phosphate (R = PO$_3$$^{2-}$) (NADP$^+$)

Malate

Methionine

NADH (R = H), NADPH (R = PO$_3$$^{2-}$)

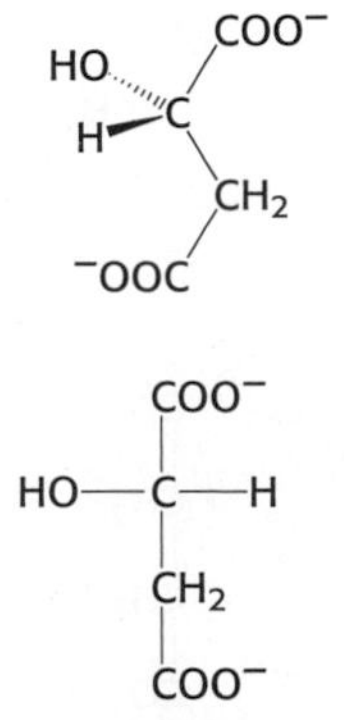

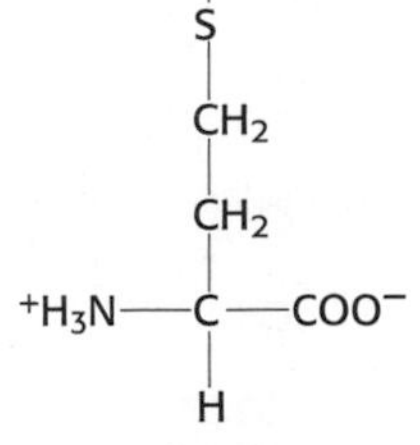

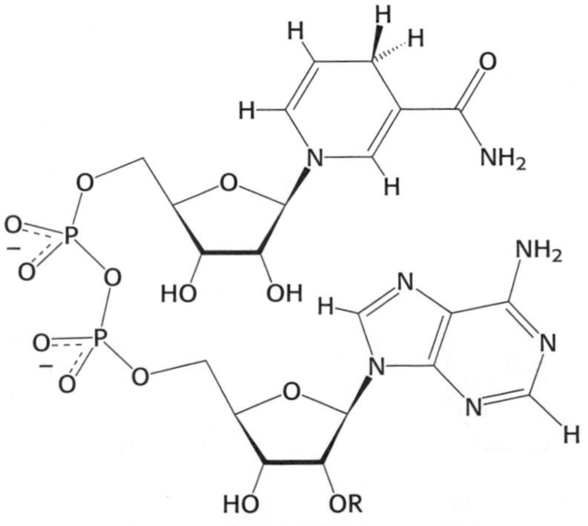

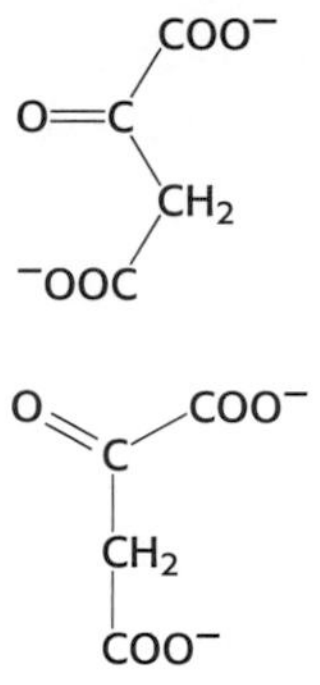

Oxaloacetate

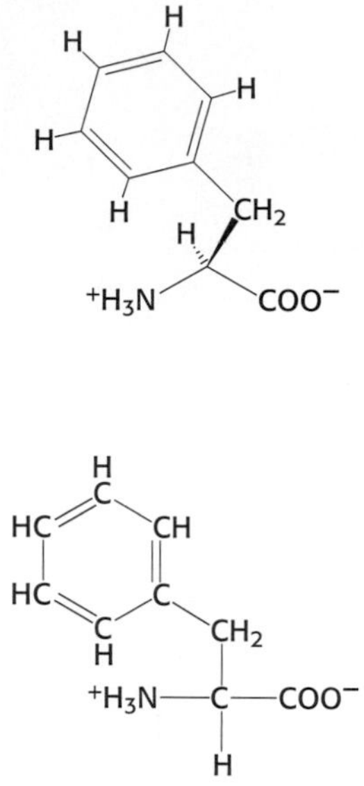

Phenylalanine

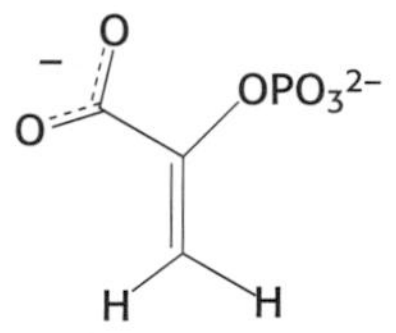

Phosphoenolpyruvate

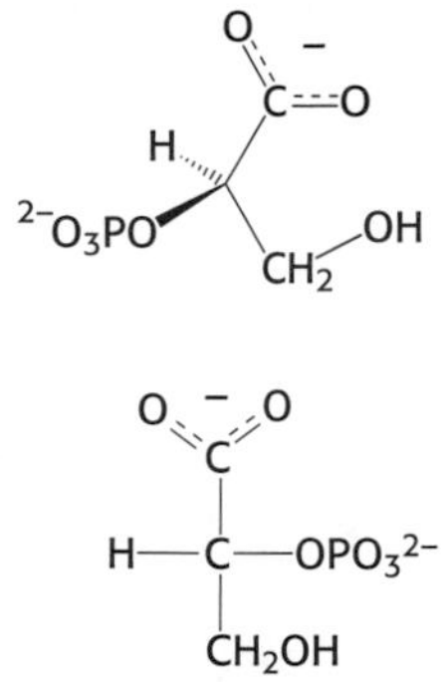

2-Phosphoglycerate

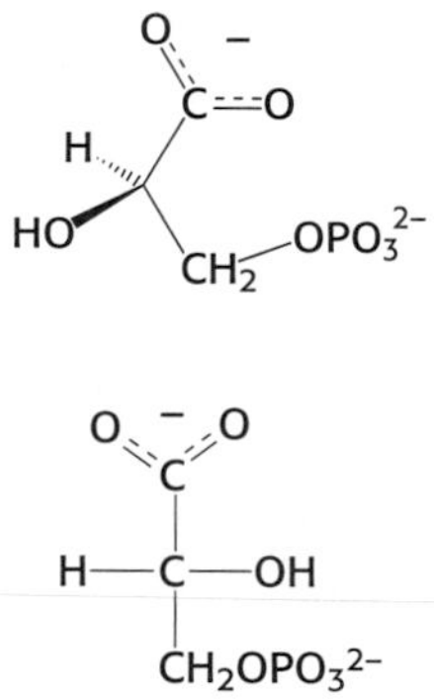

3-Phosphoglycerate

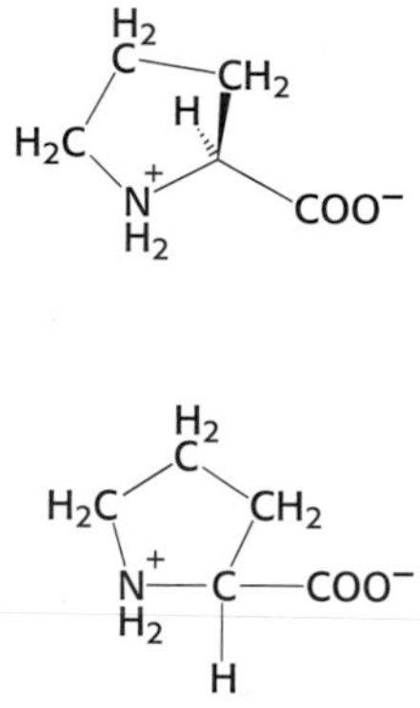

Proline

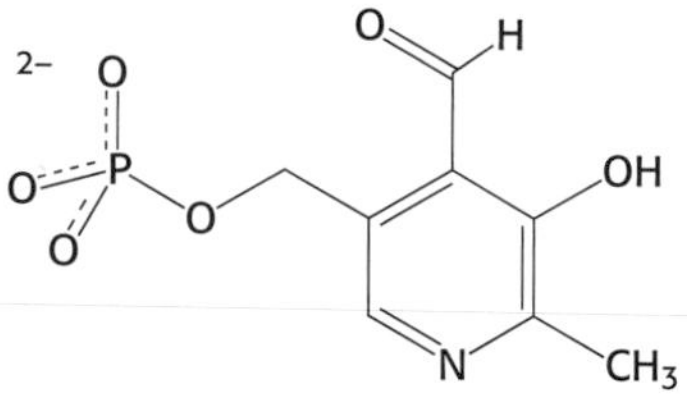

Pyridoxal phosphate

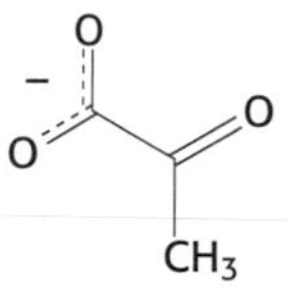

Pyruvate

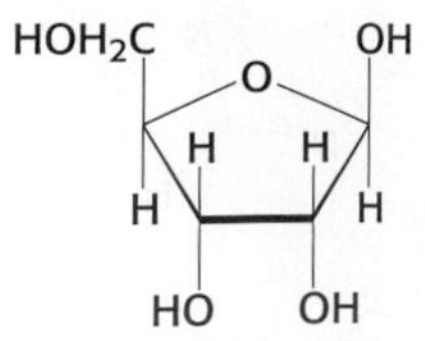

Ribose

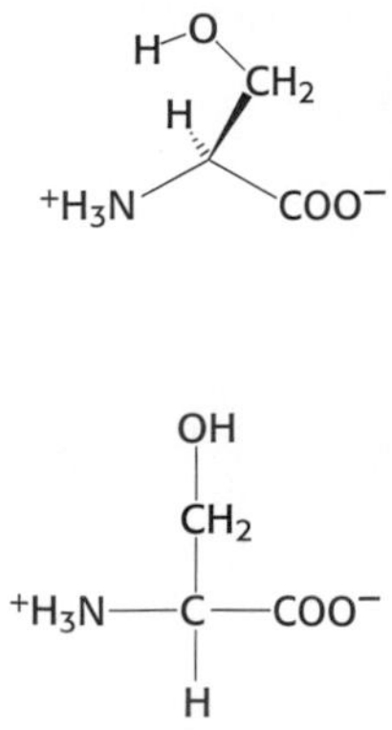

Serine

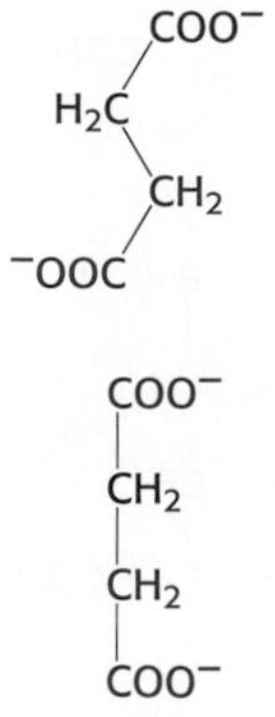

Succinate

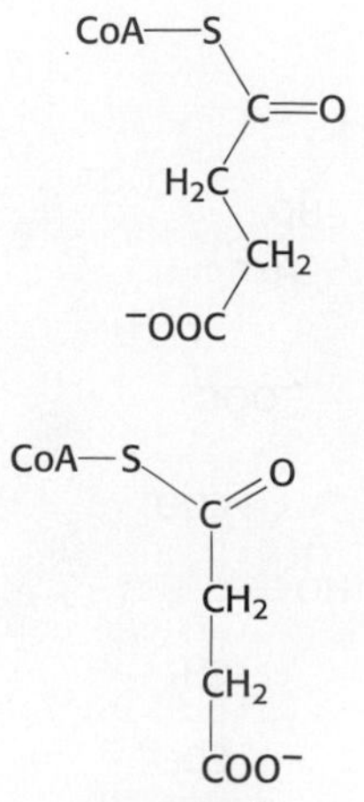

Succinyl CoA

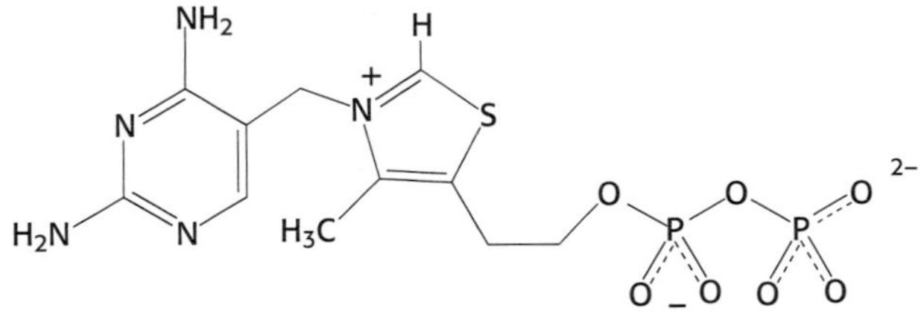

Thiamine pyrophosphate (TPP)

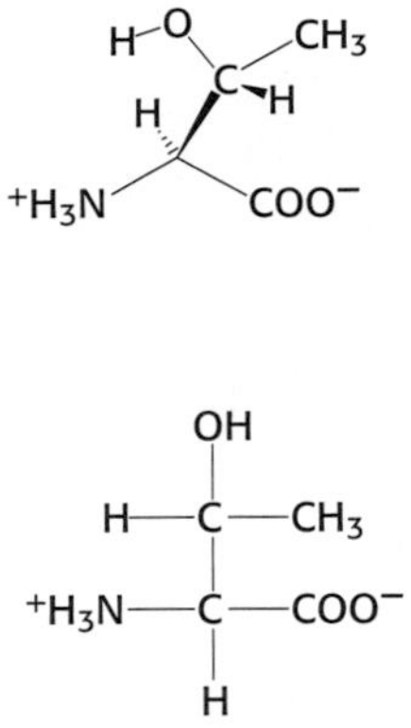

Threonine

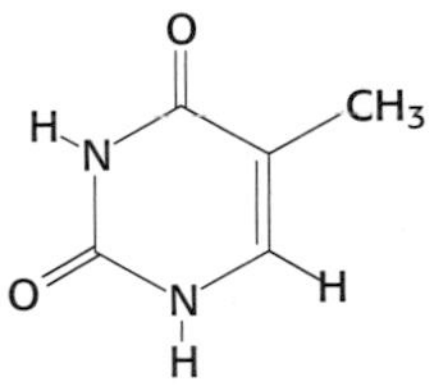

Thymine

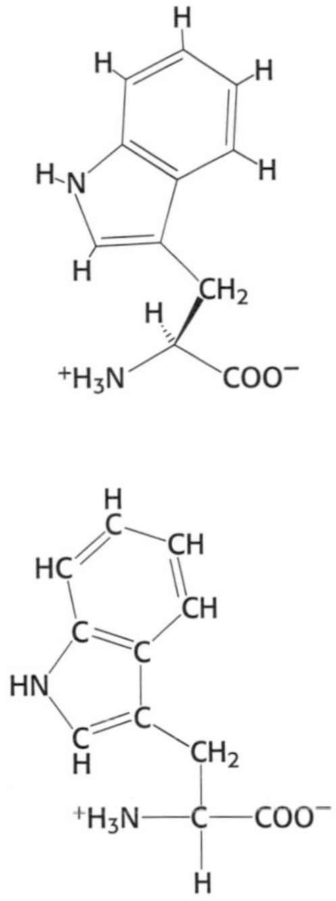

Tryptophan

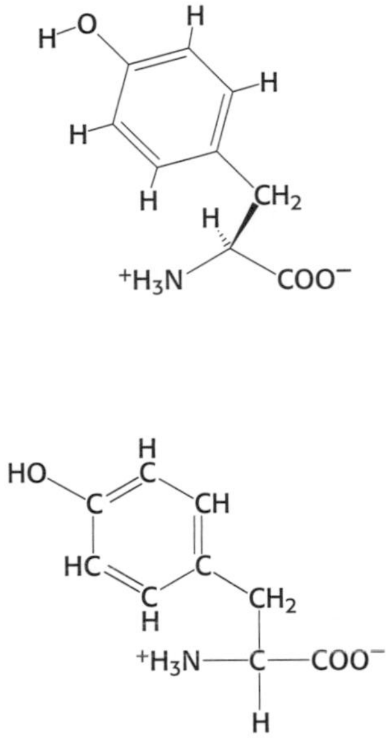

Tyrosine

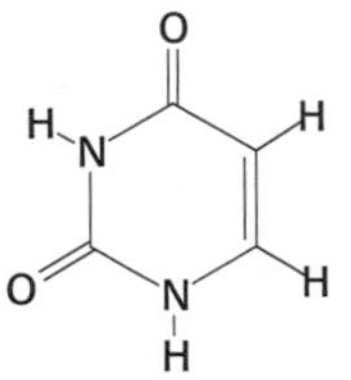

Uracil

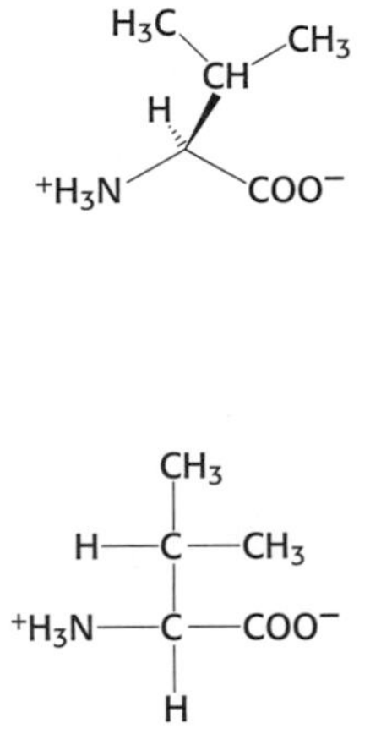

Valine

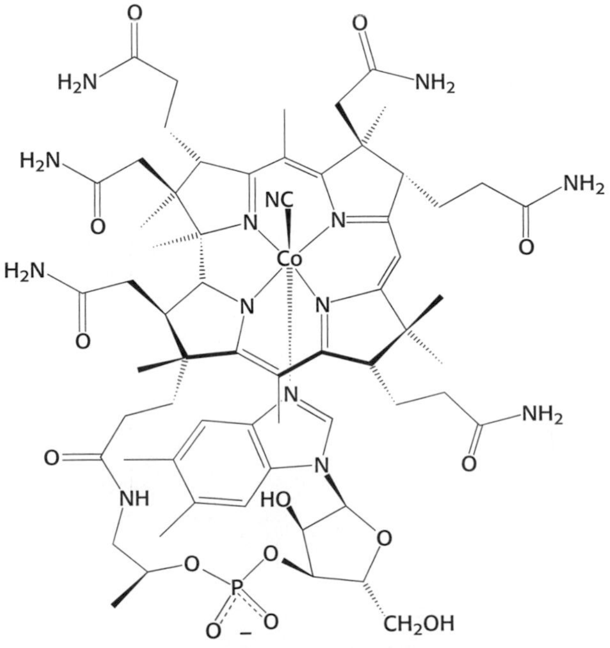

Vitamin B$_{12}$ (cyanocobalamin)

Answers to Problems

 Need extra help? Purchase chapters of the *Student Companion* with complete solutions online at *www.whfreeman.com/stryer*.

Chapter 1

1. The hydrogen-bond donors are the NH and NH_2 groups. The hydrogen-bond acceptors are the carbonyl oxygen atoms and those ring nitrogen atoms that are not bonded to hydrogen or to deoxyribose.
2. Interchange the positions of the single and double bonds in the six-membered ring.
3. (a) Electrostatic interactions; (b) van der Waals interactions
4. Processes *a* and *b*
5. $\Delta S_{system} = -661 \, J \, mol^{-1} \, K^{-1} \, (-158 \, kcal \, mol^{-1} \, K^{-1})$. $\Delta S_{universe} = 181 \, J \, mol^{-1} \, K^{-1} \, (43 \, cal \, mol^{-1} \, K^{-1})$.
6. (a) 1.0; (b) 13.0; (c) 1.3; (d) 12.7
7. 2.88
8. 6.48
9. 7.8
10. 100
11. (a) 1.6; (b) 0.5; (c) 0.16
12. Let us assume that 0.0025, 0.005, 0.01, and 0.05 M HCl are added.

Initial acetate (M)	Initial HCl (M)	pH
0.1	0.0025	6.3
0.1	0.005	6.0
0.1	0.01	5.7
0.1	0.05	4.8
0.01	0.0025	5.2
0.01	0.005	4.8
0.01	0.01	3.4
0.01	0.05	1.4

13. Three million base-pair differences

Chapter 2

1. (a) Each strand is 35 kd and hence has about 318 residues (the mean residue mass is 110 daltons). Because the rise per residue in an α helix is 1.5 Å, the length is 477 Å. More precisely, for an α-helical coiled coil, the rise per residue is 1.46 Å; so the length is 464 Å.
(b) Eighteen residues in each strand (40 minus 4 divided by 2) are in a β-sheet conformation. Because the rise per residue is 3.5 Å, the length is 63 Å.
2. The methyl group attached to the β-carbon atom of isoleucine sterically interferes with α-helix formation. In leucine, this methyl group is attached to the γ-carbon atom, which is farther from the main chain and hence does not interfere.
3. The first mutation destroys activity because valine occupies more space than alanine does, and so the protein must take a different shape, assuming that this residue lies in the closely packed interior. The second mutation restores activity because of a compensatory reduction of volume; glycine is smaller than isoleucine.
4. The native conformation of insulin is not the thermodynamically most stable form, because it contains two separate chains linked by disulfide bonds. Insulin is formed from proinsulin, a single-chain precursor, that is cleaved to form insulin, a 51-residue molecule, after the disulfide bonds have formed.
5. A segment of the main chain of the protease could hydrogen bond to the main chain of the substrate to form an extended parallel or antiparallel pair of β strands.
6. Glycine has the smallest side chain of any amino acid. Its size is often critical in allowing polypeptide chains to make tight turns or to approach one another closely.
7. Glutamate, aspartate, and the terminal carboxylate can form salt bridges with the guanidinium group of arginine. In addition, this group can be a hydrogen-bond donor to the side chains of glutamine, asparagine, serine, threonine, aspartate, and glutamate and to the main-chain carbonyl group.
8. Disulfide bonds in hair are broken by adding a thiol-containing reagent and applying gentle heat. The hair is curled, and an oxidizing agent is added to re-form disulfide bonds to stabilize the desired shape.
9. The amino acids would be hydrophobic in nature. An α helix is especially suited to cross a membrane because all of the amide hydrogen atoms and carbonyl oxygen atoms of the peptide backbone take part in intrachain hydrogen bonds, thus stabilizing these polar atoms in a hydrophobic environment.
10. The energy barrier that must be crossed to go from the polymerized state to the hydrolyzed state is large even though the reaction is thermodynamically favorable.
11. Using the Henderson–Hasselbalch equation, we find the ratio of alanine-COOH to alanine-COO^- at pH 7 to be 10^{-4}. The ratio of alanine-NH_2 to alanine-NH_3^+, determined in the same fashion, is 10^{-1}. Thus, the ratio of neutral alanine to zwitterionic species is $10^{-4} \times 10^{-1} = 10^{-5}$.
12. The assignment of absolute configuration requires the assignment of priorities to the four groups connected to a tetrahedral carbon atom. For all amino acids except cysteine, the priorities are: (1) amino group; (2) carbonyl group; (3) side chain; (4) hydrogen. For cysteine, because of the sulfur atom in its side chain, the side chain has a greater priority than does the carbonyl group, leading to the assignment of an R rather than S configuration.
13. ELVISISLIVINGINLASVEGAS.
14. No, Pro–X would have the characteristics of any other peptide bond. The steric hindrance in X–Pro arises because the R group of Pro is bonded to the amino group. Hence, in X–Pro, the proline R group is near the R group of X, which would not be the case in Pro–X.
15. A, c; B, e; C, d; D, a; E, b

Chapter 3

1. (a) Phenyl isothiocyanate; (b) urea; β-mercaptoethanol to reduce disulfides; (c) chymotrypsin; (d) CNBr; (e) trypsin
2. Each amino acid residue, except the carboxyl-terminal residue, gives rise to a hydrazide on reacting with hydrazine. The carboxyl-terminal residue can be identified because it yields a free amino acid.
3. The S-aminoethylcysteine side chain resembles that of lysine. The only difference is a sulfur atom in place of a methylene group.

Purification procedure	Total protein (mg)	Total activity (units)	Specific activity (units mg^{-1})	Purification level	Yield (%)
Crude extract	20,000	4,000,000	200	1	100
(NH$_4$)$_2$SO$_4$ precipitation	5,000	3,000,000	600	3	75
DEAE–cellulose chromatography	1,500	1,000,000	667	3.3	25
Size-exclusion chromatography	500	750,000	1,500	7.5	19
Affinity chromatography	45	675,000	15,000	75	17

4. A 1 mg ml^{-1} solution of myoglobin (17.8 kd) corresponds to 5.62×10^{-5} M. The absorbance of a 1-cm path length is 0.84, which corresponds to an I_0/I ratio of 6.96. Hence 14.4% of the incident light is transmitted.

5. Tropomyosin is rod shaped, whereas hemoglobin is approximately spherical.

6. The frictional coefficient, f, and the mass, m, determine S. Specifically, f is proportional to r (see equation 2 on p. 71). Hence, f is proportional to $m^{1/3}$, and so S is proportional to $m^{2/3}$ (see the equation on p. 76). An 80-kd spherical protein undergoes sedimentation 1.59 times as rapidly as a 40-kd spherical protein.

7. 50 kd.

8. The positions of disulfide bonds can be determined by diagonal electrophoresis (p. 82). The disulfide pairing is unaltered by the mutation if the off-diagonal peptides formed from the native and mutant proteins are the same.

9. A fluorescence-labeled derivative of a bacterial degradation product (e.g., a formylmethionyl peptide) would bind to cells containing the receptor of interest.

10. (a) Trypsin cleaves after arginine (R) and lysine (K), generating AVGWR, VK, and S. Because they differ in size, these products could be separated by molecular exclusion chromatography.

(b) Chymotrypsin, which cleaves after large aliphatic or aromatic R groups, generates two peptides of equal size (AVGW) and (RVKS). Separation based on size would not be effective. The peptide RVKS has two positive charges (R and K), whereas the other peptide is neutral. Therefore, the two products could be separated by ion-exchange chromatography.

11. An inhibitor of the enzyme being purified might have been present and subsequently removed by a purification step. This removal would lead to an apparent increase in the total amount of enzyme present.

12. Many proteins have similar masses but different sequences and different patterns when digested with trypsin. The set of masses of tryptic peptides forms a detailed "fingerprint" of a protein that is very unlikely to appear at random in other proteins regardless of size. (A conceivable analogy is: "Just as similarly sized fingers will give different individual fingerprints, so also similarly sized proteins will give different digestion patterns with trypsin.")

13. See the table at the top of the page.

14. Treatment with urea will disrupt noncovalent bonds. Thus the original 60-kd protein must be made of two 30-kd subunits. When these subunits are treated with urea and mercaptoethanol, a single 15-kd species results, suggesting that disulfide bonds link the 30-kd subunits.

15. (a) Electrostatic repulsion between positively charged ε-amino groups hinders α-helix formation at pH 7. At pH 10,

the side chains become deprotonated, allowing α-helix formation.

(b) Poly-L-glutamate is a random coil at pH 7 and becomes α helical below pH 4.5 because the γ-carboxylate groups become protonated.

16. Light was used to direct the synthesis of these peptides. Each amino acid added to the solid support contained a photolabile protecting group instead of a t-Boc protecting group at its α-amino group. Illumination of selected regions of the solid support led to the release of the protecting group, which exposed the amino groups in these sites to make them reactive. The pattern of masks used in these illuminations and the sequence of reactants define the ultimate products and their locations.

17. AVRYSR

18. First amino acid: S
Last amino acid: L
Cyanogen bromide cleavage: M is 10th position, C-terminal residues are: (2S,L,W)
Amino-terminal residues: (G,K,S,Y), tryptic peptide, ends in K
Amino-terminal sequence: SYGK
Chymotryptic peptide order: (S,Y), (G,K,L), (F,I,S), (M,T), (S,W), (S,L)
Sequence: SYGKLSIFTMSWSL

19. See equation below.

Chapter 4

1. (a) TTGATC; (b) GTTCGA; (c) ACGCGT; (d) ATGGTA
2. (a) [T] + [C] = 0.46. (b) [T] = 0.30, [C] = 0.24, and [A] + [G] = 0.46.
3. 5.7×10^3 base pairs.
4. In conservative replication, after 1.0 generation, half of the molecules would be ^{15}N-^{15}N, the other half ^{14}N-^{14}N. After 2.0 generations, one-quarter of the molecules would be ^{15}N-^{15}N, the other three-quarters ^{14}N-^{14}N. Hybrid ^{14}N-^{15}N molecules would not be observed in conservative replication.
5. (a) Tritiated thymine or tritiated thymidine. (b) dATP, dGTP, dCTP, and dTTP labeled with ^{32}P in the innermost (α) phosphorus atom.
6. Molecules in parts a and b would not lead to DNA synthesis because they lack a 3'-OH group (a primer). The molecule in part d has a free 3'-OH group at one end of each strand but no template strand beyond. Only the molecule in part c would lead to DNA synthesis.
7. A thymidylate oligonucleotide should be used as the primer. The poly(rA) template specifies the incorporation of T; hence, radioactive TTP (labeled in the α phosphoryl group) should be used in the assay.
8. The ribonuclease serves to degrade the RNA strand, a necessary step in forming duplex DNA from the RNA–DNA hybrid.
9. Treat one aliquot of the sample with ribonuclease and another with deoxyribonuclease. Test these nuclease-treated samples for infectivity.
10. Deamination changes the original G · C base pair into a G · U pair. After one round of replication, one daughter duplex will contain a G · C pair, and the other duplex will contain an A · U pair. After two rounds of replication, there will be two G · C pairs, one A · U pair, and one A · T pair.
11. (a) $4^8 = 65,536$. In computer terminology, there are 64K 8-mers of DNA.
(b) A bit specifies two bases (say, A and C) and a second bit specifies the other two (G and T). Hence, two bits are needed to specify a single nucleotide (base pair) in DNA. For example, 00, 01, 10, and 11 could encode A, C, G, and T. An 8-mer stores 16 bits ($2^{16} = 65,536$), the *E. coli* genome (4.6×10^6 bp) stores 9.2×10^6 bits, and the human genome (3.0×10^9 bases) stores 6.0×10^9 bits of genetic information.
(c) A standard CD can hold about 700 megabytes, which is equal to 5.6×10^9 bits. A large number of 8-mer sequences could be stored on such a CD. The DNA sequence of *E. coli*, could be written on a single CD with room to spare for a lot of music. One CD would not be quite enough to record the entire human genome.
12. (a) Deoxyribonucleoside triphosphates versus ribonucleoside triphosphates.
(b) 5' → 3' for both.
(c) Semiconserved for DNA polymerase I; conserved for RNA polymerase.
(d) DNA polymerase I needs a primer, whereas RNA polymerase does not.
13. (a) 5'-UAACGGUACGAU-3'
(b) Leu-Pro-Ser-Asp-Trp-Met
(c) Poly(Leu-Leu-Thr-Tyr)

14. The 2'-OH group in RNA acts as an intramolecular nucleophile. In the alkaline hydrolysis of RNA, it forms a 2'-3' cyclic intermediate.
15. Cordycepin terminates RNA synthesis. An RNA chain containing cordycepin lacks a 3'-OH group.
16. Only single-stranded RNA can serve as a template for protein synthesis.
17. Incubation with RNA polymerase and only UTP, ATP, and CTP led to the synthesis of only poly(UAC). Only poly(GUA) was formed when GTP was used in place of CTP.
18. These alternatives were distinguished by the results of studies of the sequence of amino acids in mutants. Suppose that the base C is mutated to C'. In a nonoverlapping code, only amino acid 1 will be changed. In a completely overlapping code, amino acids 1, 2, and 3 will all be altered by a mutation of C to C'. The results of amino acid sequence studies of tobacco mosaic virus mutants and abnormal hemoglobins showed that alterations usually affected only a single amino acid. Hence, it was concluded that the *genetic code is nonoverlapping.*
19. A peptide terminating with Lys (UGA is a stop codon), -Asn-Glu-, and -Met-Arg-
20. Highly abundant amino acid residues have the most codons (e.g., Leu and Ser each have six), whereas the least-abundant amino acids have the fewest (Met and Trp each have only one). Degeneracy (1) allows variation in base composition and (2) decreases the likelihood that a substitution for a base will change the encoded amino acid. If the degeneracy were equally distributed, each of the 20 amino acids would have three codons. Both benefits (1 and 2) are maximized by the assignment of more codons to prevalent amino acids than to less frequently used ones.
21. Phe-Cys-His-Val-Ala-Ala
22. (a) A codon for lysine cannot be changed to one for aspartate by the mutation of a single nucleotide.
(b) Arg, Asn, Gln, Glu, Ile, Met, or Thr
23. The genetic code is degenerate. Of the 20 amino acids, 18 are specified by more than one codon. Hence, many nucleotide changes (especially in the third base of a codon) do not alter the nature of the encoded amino acid. Mutations leading to an altered amino acid are usually more deleterious than those that do not and hence are subject to more stringent selection.

Chapter 5

1. (a) 5'-GGCATAC-3'
(b) The Sanger dideoxy method of sequencing would give the gel pattern shown here.

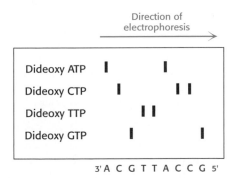

2. Ovalbumin cDNA should be used. *E. coli* lacks the machinery to splice the primary transcript arising from genomic DNA.

3. The presence of the *Alu*I sequence would, on average, be $(1/4)^4$, or 1/256, because the likelihood of any base being at any position is one-fourth and there are four positions. By the same reasoning, the presence of the *Not*I sequence would be $(1/4)^8$, or 1/65,536. Thus, the average product of digestion by *Alu*I would be 250 base pairs (0.25 kb) in length, whereas that for *Not*I would be 66,000 base pairs (66 kb) in length.

4. No, because most human genes are much longer than 4 kb. A fragment would contain only a small part of a complete gene.

5. Southern blotting of an *Mst*II digest would distinguish between the normal and the mutant genes. The loss of a restriction site would lead to the replacement of two fragments on the Southern blot by a single longer fragment. Such a finding would not prove that GTG replaced GAG; other sequence changes at the restriction site could yield the same result.

6. A simple strategy for generating many mutants is to synthesize a degenerate set of cassettes by using a mixture of activated nucleosides in particular rounds of oligonucleotide synthesis. Suppose that the 30-bp coding region begins with GTT, which encodes valine. If a mixture of all four nucleotides is used in the first and second rounds of synthesis, the resulting oligonucleotides will begin with the sequence XYT (where X and Y denote A, C, G, or T). These 16 different versions of the cassette will encode proteins containing either Phe, Leu, Ile, Val, Ser, Pro, Thr, Ala, Tyr, His, Asn, Asp, Cys, Arg, or Gly at the first position. Likewise, degenerate cassettes can be made in which two or more codons are simultaneously varied.

7. Because PCR can amplify as little as one molecule of DNA, statements claiming the isolation of ancient DNA need to be greeted with some skepticism. The DNA would need to be sequenced. Is it similar to human, bacterial, or fungal DNA? If so, contamination is the likely source of the amplified DNA. Is it similar to that of birds or crocodiles? This sequence similarity would strengthen the case that it is dinosaur DNA, because these species are evolutionarily close to dinosaurs.

8. At high temperatures of hybridization, only very close matches between primer and target would be stable because all (or most) of the bases would need to find partners to stabilize the primer–target helix. As the temperature is lowered, more mismatches would be tolerated; so the amplification is likely to yield genes with less sequence similarity. In regard to the yeast gene, synthesize primers corresponding to the ends of the gene, and then use these primers and human DNA as the target. If nothing is amplified at 54°C, the human gene differs from the yeast gene, but a counterpart may still be present. Repeat the experiment at a lower temperature of hybridization.

9. Digest genomic DNA with a restriction enzyme, and select the fragment that contains the known sequence. Circularize this fragment. Then carry out PCR with the use of a pair of primers that serve as templates for the synthesis of DNA away from the known sequence.

10. The encoded protein contains four repeats of a specific sequence.

11. Use chemical synthesis or the polymerase chain reaction to prepare hybridization probes that are complementary to both ends of the known (previously isolated) DNA fragment. Challenge clones representing the library of DNA fragments with both of the hybridization probes. Select clones that

hybridize to one of the probes but not the other; such clones are likely to represent DNA fragments that contain one end of the known fragment along with the adjacent region of the particular chromosome.

12. Within a single species, individual dogs show enormous variation in body size and substantial diversity in other physical characteristics. Therefore, genomic analysis of individual dogs would provide valuable clues concerning the genes responsible for the diversity within the species.

13. T_m is the melting temperature of a double-stranded nucleic acid. If the melting temperatures of the primers are too different, the extent of hybridization with the target DNA will differ during the annealing phase, which would result in differential replications of the strands.

14. A mutation in person B has altered one of the alleles for gene *X*, leaving the other intact. The fact that the mutated allele is smaller suggests that a deletion has occurred in one copy of the gene. The one functioning copy is transcribed and translated and apparently produces enough protein to render the person asymptomatic.

Person C has only the smaller version of the gene. This gene is neither transcribed (negative Northern blot) nor translated (negative Western blot).

Person D has a normal-size copy of the gene but no corresponding RNA or protein. There may be a mutation in the promoter region of the gene that prevents transcription.

Person E has a normal-size copy of the gene that is transcribed, but no protein is made, which suggests that a mutation prevents translation. There are a number of possible explanations, including a mutation that introduced a premature stop codon in the mRNA.

Person F has a normal amount of protein but still displays the metabolic problem. This finding suggests that the mutation affects the activity of the protein—for instance, a mutation that compromises the active site of enzyme Y.

15. Chongqing: residue 2, L → R, CTG → CGG
   Karachi: residue 5, A → P, GCC → CCC
   Swan River: residue 6, D → G, GAC → GGC

Chapter 6

1. There are 26 identities and two gaps for a score of 210. The two sequences are approximately 26% identical. This level of homology is likely to be statistically significant.

2. They are likely related by divergent evolution, because three-dimensional structure is more conserved than is sequence identity.

3. (1) Identity score = 225; Blosum score = 7; (2) identity score = 15; Blosum score = 210.

4. U

U   G

5. There are 4^{40}, or 1.2×10^{24}, different molecules. Each molecule has a mass of 2.2×10^{-20}, because 1 mol of polymer

has a mass of 330 g mol^{-1} × 40, and there are 6.02×10^{23} molecules per mole. Therefore, 26.4 kg of RNA would be required.

6. Because three-dimensional structure is much more closely associated with function than is sequence, tertiary structure is more evolutionarily conserved than is primary structure. In other words, protein function is the most important characteristic, and protein function is determined by structure. Thus, the structure must be conserved, but not necessarily a specific amino acid sequence.

7. Alignment score is 6 × 10 = 60. Many answers are possible, depending on the randomly reordered sequence. A possible result is

Shuffled sequence: TKADKAGEYL

Alignment: (1) ASNFLDKAGK

                 TKADKAGEYL

Alignment score is 4 × 10 = 40.

8. (a) Almost certainly diverged from a common ancestor.
(b) Almost certainly diverged from a common ancestor.
(c) May have diverged from a common ancestor, but the sequence alignment may not provide supporting evidence.
(d) May have diverged from a common ancestor, but the sequence alignment is unlikely to provide supporting evidence.

9. Protein A is clearly homologous to protein B, given 65% sequence identity, and so A and B are expected to have quite similar three-dimensional structures. Likewise, proteins B and C are clearly homologous, given 55% sequence identity, and so B and C are expected to have quite similar three-dimensional structures. Thus, proteins A and C are likely to have similar three-dimensional structures, even though they are only 15% identical in sequence.

10. The likely secondary structure is

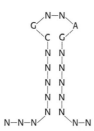

11. 107 or 108 identities (depending on which annotated human sequence is chosen)

Chapter 7

1. The whale swims long distances between breaths. A high concentration of myoglobin in the whale muscle maintains a ready supply of oxygen for the muscle between breathing episodes.

2. 62.7% oxygen-carrying capacity

3. A higher concentration of BPG would shift the oxygen-binding curve to the right, causing an increase in P_{50}. The larger value of P_{50} would promote dissociation of oxygen in the tissues and would thereby increase the percentage of oxygen delivered to the tissues.

4. Oxygen binding appears to cause the copper ions and their associated histidine ligands to move closer to one another, thereby also moving the helices to which the histidines are attached (in similar fashion to the conformational change in hemoglobin).

5. The modified hemoglobin should not show cooperativity. Although the imidazole in solution will bind to the heme iron (in place of histidine) and will facilitate oxygen binding, the imidazole lacks the crucial connection to the particular α helix that must move so as to transmit the change in conformation.

6. Inositol pentaphosphate in part c

7. (a)

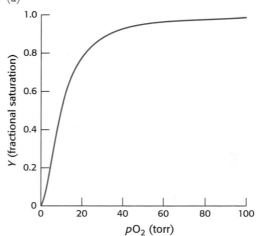

(b)

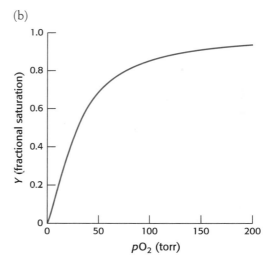

8. Release of acid will lower the pH. A lower pH promotes oxygen dissociation in the tissues. However, the enhanced release of oxygen in the tissues will increase the concentration of deoxy-Hb, thereby increasing the likelihood that the cells will sickle.

9. The "picket fence" provides a pocket for the reversible binding of oxygen to Fe while preventing Fe–O–Fe dimerization. The methyl imidazole ligand protects Fe from the other side and prevents dimerization.

10. (a) Y = 0.5 when pO_2 = 10 torr. The plot of Y versus pO_2 appears to indicate little or no cooperativity.
(b) The Hill plot shows slight cooperativity with $n \simeq 1.3$ in the central region.
(c) Deoxy dimers of lamprey hemoglobin could have lower affinity for oxygen than do the monomers. If the binding of the first oxygen atom to a dimer causes dissociation of the dimer to give two monomers, then the process would be cooperative. In

this mechanism, oxygen binding to each monomer would be easier than binding the first oxygen atom to a deoxy dimer.

Chapter 8

1. (a) 31.1 μmol; (b) 0.05 μmol; (c) 622 s^{-1}
2. (a) Yes, $K_M = 5.2 \times 10^{-6}$ M;
(b) $V_{max} = 6.8 \times 10^{-10}$ mol minute^{-1}; (c) 337 s^{-1}
3. Penicillinase, like glycopeptide transpeptidase, forms an acyl-enzyme intermediate with its substrate but transfers the intermediate to water rather than to the terminal glycine residue of the pentaglycine bridge.
4. (a) In the absence of inhibitor, V_{max} is 47.6 μmol minute^{-1} and K_M is 1.1×10^{-5} M. In the presence of inhibitor, V_{max} is the same and the apparent K_M is 3.1×10^{-5} M.
(b) Competitive
(c) 1.1×10^{-3} M
(d) f_{ES} is 0.243, and f_{EI} is 0.488.
(e) f_{ES} is 0.73 in the absence of inhibitor and 0.49 in the presence of 2×10^{-3} M inhibitor. The ratio of these values, 1.49, is the same as the ratio of the reaction velocities under these conditions.
5. (a) V_{max} is 9.5 μmol minute^{-1}. K_M is 1.1×10^{-5} M, the same as without inhibitor.
(b) Noncompetitive
(c) 2.5×10^{-5} M
(d) $f_{ES} = 0.73$, in the presence or absence of this noncompetitive inhibitor.
6. (a) $V = V_{max} - (V/[S]) K_M$.
(b) Slope $= 2K_M$, y intercept $= V_{max}$, x intercept $= V_{max}/K_M$.
(c) An Eadie-Hofstee plot:

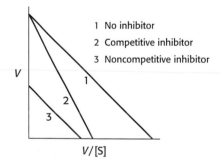

1 No inhibitor
2 Competitive inhibitor
3 Noncompetitive inhibitor

7. Potential hydrogen-bond donors at pH 7 are the side chains of the following residues: arginine, asparagine, glutamine, histidine, lysine, serine, threonine, tryptophan, and tyrosine.
8. The rates of utilization of substrates A and B are given by

$$V_A = \left(\frac{k_2}{K_M}\right)_A [E][A]$$

and

$$V_B = \left(\frac{k_2}{K_M}\right)_B [E][B]$$

Hence, the ratio of these rates is

$$V_A/V_B = \left(\frac{k_2}{K_M}\right)_A [A] \Big/ \left(\frac{k_2}{K_M}\right)_B [B]$$

Thus, an enzyme discriminates between competing substrates on the basis of their values of k_2/K_M rather than of K_M alone.

9. The mutation slows the reaction by a factor of 100 because the activation free energy is increased by 53.22 kJ mol^{-1} (12.72 kcal mol^{-1}). Strong binding of the substrate relative to the transition state slows catalysis.
10. 11 μmol minute^{-1}
11. If the total amount of enzyme (E_T) is increased, V_{max} will increase, because $V_{max} = k_2[E_T]$. But $K_M = (k_{-1} + k_2)/k_1$; that is, it is independent of substrate concentration. The middle graph describes this situation.
12. (a)

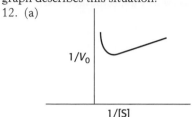

(b) This behavior is substrate inhibition: at high concentrations, the substrate forms unproductive complexes at the active site. The adjoining drawing shows what might happen. Substrate normally binds in a defined orientation, shown in the drawing as red to red and blue to blue. At high concentrations, the substrate may bind at the active site such that the proper orientation is met for each end of the molecule, but two different substrate molecules are binding.

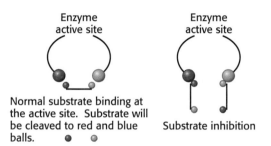

Enzyme active site

Enzyme active site

Normal substrate binding at the active site. Substrate will be cleaved to red and blue balls.

Substrate inhibition

13. The first step will be the rate-limiting step. Enzymes E_B and E_C are operating at $\frac{1}{2} V_{max}$, whereas the K_M for enzyme E_A is greater than the substrate concentration. E_A would be operating at approximately $10^{-2} V_{max}$.
14. (a) When [S$^+$] is much greater than the value of K_M, pH will have a negligible effect on the enzyme because S$^+$ will interact with E$^-$ as soon as the enzyme becomes available.

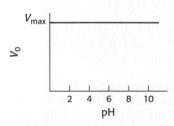

(b) When [S$^+$] is much less than the value of K_M, the plot of V_0 versus pH becomes essentially a titration curve for the ionizable groups, with enzyme activity being the titration marker. At low pH, the high concentration of H$^+$ will keep the enzyme in the EH form and inactive. As the pH rises, more and more of the enzyme will be in the E^- form and active. At high pH (low H^+), all of the enzyme is E^-.

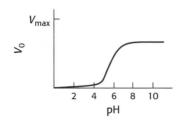

(c) The midpoint on this curve will be the pK_a of the ionizable group, which is stated to be pH 6.

15. (a) Incubating the enzyme at $37°C$ leads to a denaturation of enzyme structure and a loss of activity. For this reason, most enzymes must be kept cool if they are not actively catalyzing their reaction.
(b) The coenzyme apparently helps to stabilize the enzyme structure, because enzyme from PLP-deficient cells denatures faster. Cofactors often help stabilize enzyme structure.

Chapter 9

1. The formation of the acyl-enzyme intermediate is slower than the hydrolysis of this amide substrate, and so no burst is observed. For ester substrates, the formation of the acyl-enzyme intermediate is faster.
2. The histidine residue in the substrate can substitute to some extent for the missing histidine residue of the catalytic triad of the mutant enzyme.
3. No. When the catalytic activity has been reduced to almost zero by the mutation of a required residue, the introduction of another deleterious mutation cannot further reduce activity.
4. The substitution corresponds to one of the key differences between trypsin and chymotrypsin, and so trypsinlike specificity (cleavage after lysine and arginine) might be predicted. In fact, additional changes are required to effect this specificity change.
5. Imidazole is apparently small enough to reach the active site of carbonic anhydrase. Buffers with large molecular components cannot do so, and the effects of the mutation are more evident.
6. No. The odds of such a sequence being present are approximately 1 in 1 million. Because a typical viral genome has only 50,000 bp, the target sequence would be unlikely to be present.
7. No, because the enzyme would destroy the host DNA before protective methylation could take place.
8. No. The bacteria receiving the enzyme would have its own DNA destroyed because it lacked the appropriate protective methylase.
9. EDTA will bind to Zn^{2+} and remove the ion, which is required for enzyme activity, from the enzyme.
10. (a) The aldehyde reacts with the active-site serine. (b) A hemiacetal is formed.
11. (a) ATP and AMP. (b) Because the reaction has an equilibrium constant near 1, the concentrations of all components will be 0.33 mM.
12. Trypsin

13. (a) Cysteine protease: The same as Figure 9.8, except that cysteine replaces serine in the active site
(b) Aspartyl protease:

(c) Metalloprotease:

Chapter 10

1. The protonated form of histidine probably stabilizes the negatively charged carbonyl oxygen atom of the scissile bond in the transition state. Deprotonation would lead to a loss of activity. Hence, the rate is expected to be half maximal at a pH of about 6.5 (the pK of an unperturbed histidine side chain in a protein) and to decrease as the pH is raised.
2. (a) 100. The change in the [R]/[T] ratio on binding one substrate molecule must be the same as the ratio of the substrate affinities of the two forms (see equations on p. 200).
(b) 10. The binding of four substrate molecules changes the [R]/[T] by a factor of $100^4 = 10^8$. The ratio in the absence of substrate is 10^{-7}. Hence, the ratio in the fully liganded molecule is $10^8 \times 10^{-7} = 10$.
3. The fraction of molecules in the R form is 10^{-5}, 0.004, 0.615, 0.998, and 1 when 0, 1, 2, 3, and 4 ligands, respectively, are bound.
4. The sequential model can account for negative cooperativity, whereas the concerted model cannot.
5. The binding of PALA switches ATCase from the T to the R state because PALA acts as a substrate analog. An enzyme molecule containing bound PALA has fewer free catalytic sites than does an unoccupied enzyme molecule. However, the PALA-containing enzyme will be in the R state and hence have higher affinity for the substrates. The dependence of the degree

of activation on the concentration of PALA is a complex function of the allosteric constant L_0 and of the binding affinities of the R and T states for the analog and the substrates.

6. The enzyme would show simple Michaelis–Menten kinetics because it is essentially always in the R state.

7. The net outcome of the two reactions is the hydrolysis of ATP to ADP and P_i, which has a ΔG of -50 kJ mol^{-1} (-12 kcal mol^{-1}) under cellular conditions (p. 286).

8. Activation is independent of zymogen concentration because the reaction is intramolecular.

9. Add blood from the second patient to a sample from the first. If the mixture clots, the second patient has a defect different from that of the first. This type of assay is called a complementation test.

10. Activated factor X remains bound to blood-platelet membranes, which accelerates its activation of prothrombin.

11. Antithrombin III is a very slowly hydrolyzed substrate of thrombin. Hence, its interaction with thrombin requires a fully formed active site on the enzyme.

12. Residues a and d are located in the interior of an α-helical coiled coil, near the axis of the superhelix. Hydrophobic interactions between these side chains contribute to the stability of the coiled coil.

13. Leucine would be a good choice. It is resistant to oxidation and has nearly the same volume and degree of hydrophobicity as methionine has.

14. The simple sequential model predicts that the fraction of catalytic chains in the R state, f_R, is equal to the fraction containing bound substrate, Y. The concerted model, in contrast, predicts that f_R increases more rapidly than Y as the substrate concentration is increased. The change in f_R leads to the change in Y on addition of substrate, as predicted by the concerted model.

15. The binding of succinate to the functional catalytic sites of the native c_3 moiety changed the visible absorption spectrum of nitrotyrosine residues in the *other* c_3 moiety of the hybrid enzyme. Thus, the binding of substrate analog to the active sites of one trimer altered the structure of the other trimer.

16. According to the concerted model, an allosteric activator shifts the conformational equilibrium of all subunits toward the R state, whereas an allosteric inhibitor shifts it toward the T state. Thus, ATP (an allosteric activator) shifted the equilibrium to the R form, resulting in an absorption change similar to that obtained when substrate is bound. CTP had a different effect. Hence, this allosteric inhibitor shifted the equilibrium to the T form. Thus, the concerted model accounts for the ATP-induced and CTP-induced (heterotropic), as well as for the substrate-induced (homotropic), allosteric interactions of ATCase.

17. In the R state, ATCase expands, and becomes less dense. This decrease in density results in a decrease in the sedimentation value (see the formula on p. 76).

18.

19.

Groups expected in the active site include a base to remove the proton from the serine, groups such as Asp and Glu to bind the magnesium ion associated with the ATP, and other groups to stabilize the ADP leaving group.

Chapter 11

1. Carbohydrates were originally regarded as *hydrates* of *carbon* because the empirical formula of many of them is $(CH_2O)_n$.

2. Three amino acids can be linked by peptide bonds in only six different ways. However, three different monosaccharides can be linked in a plethora of ways. The monosaccharides can be linked in a linear or branched manner, with α or β linkages, with bonds between C-1 and C-3, between C-1 and C-4, between C-1 and C-6, and so forth. Consequently, the number of possible trisaccharides greatly exceeds the number of tripeptides.

3. (a) Aldose–ketose; (b) epimers; (c) aldose–ketose; (d) anomers; (e) aldose–ketose; (f) epimers

4. The proportion of the α anomer is 0.36, and that of the β anomer is 0.64.

5. Glucose is reactive because of the presence of an aldehyde group in its open-chain form. The aldehyde group slowly condenses with amino groups to form Schiff-base adducts.

6. A pyranoside reacts with two molecules of periodate; formate is one of the products. A furanoside reacts with only one molecule of periodate; formate is not formed.

7. From methanol

8. (a) β-D-Mannose; (b) β-D-galactose; (c) β-D-fructose; (d) β-D-glucosamine

9. The trisaccharide itself should be a competitive inhibitor of cell adhesion if the trisaccharide unit of the glycoprotein is critical for the interaction.

10. Reducing ends would form 1,2,3,6-tetramethylglucose. The branch points would yield 2,3-dimethylglucose. The remainder of the molecule would yield 2,3,6-trimethylglucose.

11. (a) Not a reducing sugar; no open-chain forms are possible. (b) D-Galactose, D-glucose, D-fructose. (c) D-Galactose and sucrose (glucose + fructose).

12.

β-D-Mannose

The hemiketal linkage of the α anomer is broken to form the open form. Rotation about the C-1 and C-2 bonds allows the formation of the β anomer, and a mixture of isomers results.

13. Heating converts the very sweet pyranose form into the more stable but less sweet furanose form. Consequently, the sweetness of the preparation is difficult to accurately control, which also accounts for why honey loses sweetness with time. See Figure 11.6 for structures.

14. (a) Each glycogen molecule has one reducing end, whereas the number of nonreducing ends is determined by the number of branches, or α-1,6 linkages. (b) Because the number of nonreducing ends greatly exceeds the number of reducing ends in a collection of glycogen molecules, all of the degradation and synthesis of glycogen takes place at the nonreducing ends, thus maximizing the rate of degradation and synthesis.

15. 64. Each site either is or is not glycosylated, and so there are $2^6 = 64$ possible proteins.

16. As discussed in Chapter 9, many enzymes display stereochemical specificity. Clearly, the enzymes of sucrose synthesis are able to distinguish between the isomers of the substrates and link only the correct pair.

Chapter 12

1. 2.86×10^6 molecules, because each leaflet of the bilayer contains 1.43×10^6 molecules

2. 2×10^{-7} cm, 6×10^{-6} cm, and 2×10^{-4} cm.

3. The radius of this molecule is 3.1×10^{-7} cm, and its diffusion coefficient is 7.4×19^{-9} cm^2 s^{-1}. The average distances traversed are 1.7×10^{-7} cm in 1 μs, 5.4×10^{-6} in 1 ms, and 1.7×10^{-4} cm in 1 s.

4. The membrane underwent a phase transition from a highly fluid to a nearly frozen state when the temperature was lowered. A carrier can shuttle ions across a membrane only when the

bilayer is highly fluid. A channel former, in contrast, allows ions to traverse its pore even when the bilayer is quite rigid.

5. The initial decrease in the amplitude of the paramagnetic resonance spectrum results from the reduction of spin-labeled phosphatidylcholine molecules in the outer leaflet of the bilayer. Ascorbate does not traverse the membrane under these experimental conditions; hence, it does not reduce the phospholipids in the inner leaflet. The slow decay of the residual spectrum is due to the reduction of phospholipids that have flipped over to the outer leaflet of the bilayer.

6. The addition of the carbohydrate introduces a significant energy barrier to the flip-flop because a hydrophilic carbohydrate moiety would need to be moved through a hydrophobic environment. This energetic barrier enhances membrane asymmetry.

7. The presence of a cis double bond introduces a kink that prevents packing of the fatty acid chains. Cis double bonds maintain fluidity. Trans fatty acids have no structural effect, relative to saturated fatty acids, and so they are rare.

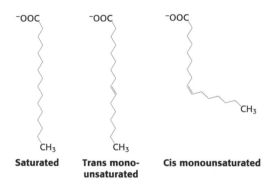

Saturated   **Trans mono-unsaturated**   **Cis monounsaturated**

8. In a hydrophobic environment, the formation of intrachain hydrogen bonds stabilizes the amide hydrogen atoms and carbonyl oxygen atoms of the polypeptide chain; so an α helix forms. In an aqueous environment, these groups are stabilized by interaction with water; so there is no energetic reason to form an α helix. Thus, the α helix would be most likely to form in a hydrophobic environment.

9. The shift to the lower temperature would decrease fluidity by enhancing the packing of the hydrophobic chains by van der Waals interactions. To prevent this packing, new phospholipids having shorter chains and a greater number of cis double bonds would be synthesized. The shorter chains would reduce the number of van der Waals interactions, and the cis double bonds, causing the kink in structure, would prevent the packing of the fatty acid tails of the phospholipids.

10. Each of the 21 v-SNARE proteins could interact with each of 7 t-SNARE partners. Multiplication gives the total number of different interacting pairs: $7 \times 21 = 147$ different v-SNARE–t-SNARE pairs.

11. (a) The graph shows that, as temperature increases, the phospholipid bilayer becomes more fluid. T_m is the temperature of the transition from the predominantly less fluid state to the predominantly more fluid state. Cholesterol broadens the transition from the less-fluid to the more-fluid state. In essence, cholesterol makes membrane fluidity less sensitive to temperature changes.

(b) This effect is important because the presence of cholesterol tends to stabilize membrane fluidity by preventing sharp transitions. Because protein function depends on the proper fluidity of the membrane, cholesterol maintains the proper environment for membrane-protein function.

12. The protein plotted in part *c* is a transmembrane protein from *C. elegans*. It spans the membrane with four α helices that are prominently displayed as hydrophobic peaks in the hydropathy plot. Interestingly, the protein plotted in part *a* also is a membrane protein, a porin. This protein is made primarily of β strands, which lack the prominent hydrophobic window of membrane helices. This example shows that, although hydropathy plots are useful, they are not infallible.

13. To purify any protein, the protein must first be solubilized. For a membrane protein, solubilization usually requires a detergent—hydrophobic molecules that bind to the protein and thus replace the lipid environment of the membrane. If the detergent is removed, the protein aggregates and precipitates from solution. Often, the steps in purification, such as ion-exchange chromatography, are difficult to perform in the presence of sufficient detergent to solubilize the protein. Crystals of appropriate protein–detergent complexes must be generated.

Chapter 13

1. The free-energy cost is 32 kJ mol^{-1} (7.6 kcal mol^{-1}). The chemical work performed is 20.4 kJ mol^{-1} (4.9 kcal mol^{-1}) and the electrical work performed is 11.5 kJ mol^{-1} (2.8 kcal mol^{-1}).

2. The concentration of glucose inside the cell is 66-fold as great as that outside the cell [$(c2/c1) = 66$] when the free-energy input is 10.8 kJ mol^{-1} (2.6 kcal mol^{-1}).

3. By analogy with the Ca^{2+} ATPase, with three Na^+ ions binding from inside the cell to the E_1 conformation and with two K^+ ions binding from outside the cell to the E_2 conformation, a plausible mechanism is as follows:

i. A catalytic cycle could begin with the enzyme in its unphosphorylated state (E_1) with three sodium ions bound.
ii. The E_1 conformation binds ATP. A conformational change traps sodium ions inside the enzyme.
iii. The phosphoryl group is transferred from ATP to an aspartyl residue.
iv. On ADP release, the enzyme changes its overall conformation, including the membrane domain. This new conformation (E_2) releases the sodium ions to the side of the membrane opposite that at which they entered and binds two potassium ions from the side where sodium ions are released.
v. The phosphorylaspartate residue is hydrolyzed to release inorganic phosphate.

With the release of phosphate, the interactions stabilizing E_2 are lost, and the enzyme reverts back to the E_1 conformation. Potassium ions are released to the cytoplasmic side of the membrane. The binding of three sodium ions from the cytoplasmic side of the membrane completes the cycle.

4. Establish a lactose gradient across vesicle membranes that contain properly oriented lactose permease. Initially, the pH

should be the same on both sides of the membrane, and the lactose concentration should be higher on the "exit" side of lactose permease. As the lactose flows "in reverse" through the permease, down its concentration gradient, whether a pH gradient becomes established as the lactose gradient is dissipated can be tested.

5. An ion channel must transport ions in either direction at the same rate. The net flow of ions is determined only by the composition of the solutions on either side of the membrane.

6. The positively charged guanidinium group resembles Na^+ and binds to negatively charged carboxylate groups in the mouth of the channel.

7. The blockage of ion channels inhibits action potentials, leading to loss of nervous function. Like tetrodotoxin, these toxin molecules are useful for isolating and specifically inhibiting particular ion channels.

8. Because sodium ions are charged and because sodium channels carry only sodium ions (but not anions), the accumulation of excess positive charge on one side of the membrane dominates the chemical gradients.

9. No. Channels will likely open or close in response to an external stimulus, but the unit conductance of the open channel will be influenced very little.

10. The ratio of closed to open forms of the channel is 10^5, 5000, 250, 12.5, and 0.625 when zero, one, two, three, and four ligands, respectively, are bound. Hence, the fraction of open channels is 1.0×10^{-5}, 2.0×10^{-4}, 4.0×10^{-3}, 7.4×10^{-2}, and 0.62.

11. These organic phosphates inhibit acetylcholinesterase by reacting with the active-site serine residue to form a stable phosphorylated derivative. They cause respiratory paralysis by blocking synaptic transmission at cholinergic synapses.

12. (a) The binding of the first acetylcholine molecule increases the open-to-closed ratio by a factor of 240, and the binding of the second increases it by a factor of 11,700. (b) The free-energy contributions are 14 kJ mol^{-1} (3.3 kcal mol^{-1}) and 23 kJ mol^{-1} (5.6 kcal mol^{-1}), respectively. (c) No; the MWC model predicts that the binding of each ligand will have the same effect on the open-to-closed ratio.

13. Batrachotoxin blocks the transition from the open to the closed state.

14. (a) Chloride ions flow into the cell. (b) Chloride flux is inhibitory because it hyperpolarizes the membrane. (c) The channel consists of five subunits.

15. The catalytic prowess of acetylcholinesterase ensures that the duration of the nerve stimulus will be short.

16. See equation below.

17. (a) Only ASIC1a is inhibited by the toxin. (b) Yes; when the toxin was removed, the activity of the acid-sensing channel began to be restored. (c) 0.9 nM.

18. This mutation is one of a class of mutations that result in slow channel syndrome (SCS). The results suggest a defect in channel closing; so the channel remains open for prolonged periods. Alternatively, the channel may have a higher affinity for acetylcholine than does the control channel.

19. The mutation reduces the affinity of acetylcholine for the receptor. The recordings would show the channel opening only infrequently.

20. Glucose displays a transport curve that suggests the participation of a carrier, because the initial rate is high but then levels off at higher concentrations, consistent with saturation of the carrier, which is reminiscent of Michaelis–Menten enzymes (p. 217). Indole shows no such saturation phenomenon, which implies that the molecule is lipophilic and simply diffuses across the membrane. Ouabain is a specific inhibitor the Na^+–K^+ pump. If ouabain were to inhibit glucose transport, then a Na^+-glucose cotransporter would be assisting in transport.

Chapter 14

1. The negatively charged glutamate residues mimic the negatively charged phosphoserine or phosphothreonine residues and stabilize the active conformation of the enzyme.

2. No. Phosphoserine and phosphothreonine are considerably shorter than phosphotyrosine.

3. Growth-factor receptors can be activated by dimerization. If an antibody causes a receptor to dimerize, the signal-transduction pathway in a cell will be activated.

4. The mutated α subunit will always be in the GTP form and, hence, in the active form, which would stimulate its signaling pathway.

5. Calcium ions diffuse slowly because they bind to many protein surfaces within a cell, impeding their free motion. Cyclic AMP does not bind as frequently, and so it diffuses more rapidly.

6. $G_{\alpha s}$ stimulates adenylate cyclase, leading to the generation of cAMP. This signal then leads to glucose mobilization (see Chapter 21). If cAMP phosphodiesterase were inhibited, then cAMP levels would remain high even after the termination of the epinephrine signal, and glucose mobilization would continue.

7. The full network of pathways initiated by insulin includes a large number of proteins and is substantially more elaborate than indicated in Figure 14.24. Furthermore, many additional proteins take part in the termination of insulin signaling. A defect in any of the proteins in the insulin signaling pathways or in the subsequent termination of the insulin response could potentially cause problems. Therefore, it is not surprising that many different gene defects can cause type 2 diabetes.

8. The binding of growth hormone causes its monomeric receptor to dimerize. The dimeric receptor can then activate a separate tyrosine kinase to which the receptor binds. The signaling pathway can then continue in similar fashion to the pathways that are activated by the insulin receptor or other mammalian EGF receptors.

9. Insulin would elicit the response that is normally caused by EGF. Insulin binding will likely stimulate dimerization and phosphorylation of the chimeric receptor and thereby signal the downstream events that are normally triggered by EGF binding. Exposure of these cells to EGF would have no effect.

10. 10^5

11. The formation of diacylglycerol implies the participation of phospholipase C. A simple pathway would entail receptor activation by cross-phosphorylation, followed by the binding of phospholipase C γ (through its SH2 domains). The participation of phospholipase C indicates that IP_3 would be formed and, hence, calcium concentrations would increase.

12. In the reaction catalyzed by adenylate cyclase, the 3'-OH group nucleophilically attacks the α-phosphorus atom attached to the 5'-OH group, leading to displacement of pyrophosphate. The reaction catalyzed by DNA polymerase is similar except that the 3'-OH group is on a different nucleotide.

13. (a) $X \approx 10^{-7}$ M; $Y \approx 5 \times 10^{-6}$ M; $Z \approx 10^{-3}$ M. (b) Because much less X is required to fill half of the sites, X displays the highest affinity. (c) The binding affinity almost perfectly matches the ability to stimulate adenylate cyclase, suggesting that the hormone–receptor complex leads to the stimulation of adenylate cyclase. (d) Try performing the experiment in the presence of antibodies to $G_{\alpha s}$.

14. (a) The total binding does not distinguish binding to a specific receptor from binding to different receptors or from nonspecific binding to the membrane.
(b) The rationale is that the receptor will have a high affinity for the ligand. Thus, in the presence of excess nonradioactive

ligand, the receptor will bind to nonradioactive ligand. Therefore, any binding of the radioactive ligand must be nonspecific.
(c) The plateau suggests that the number of receptor-binding sites in the cell membrane is limited.
15. Number of receptors per cell =

$$\frac{10^4 \text{ cpm}}{\text{mg of membrane protein}} \times \frac{\text{mg of membrane protein}}{10^{10} \text{ cells}}$$

$$\times \frac{\text{mmol}}{10^{12} \text{ cpm}} \times \frac{6.023 \times 10^{20} \text{ molecules}}{\text{mmol}} = 600$$

Chapter 15

1. Reactions in parts a and c, to the left; reactions in parts b and d, to the right
2. None whatsoever
3. (a) $\Delta G^{\circ\prime} = 31.4 \text{ kJ mol}^{-1}$ (7.5 kcal mol^{-1}) and $K'_{eq} = 3.06 \times 10^{-6}$; (b) 3.28×10^4
4. $\Delta G^{\circ\prime} = 7.1 \text{ kJ mol}^{-1}$ (1.7 kcal mol^{-1}). The equilibrium ratio is 17.8.
5. (a) Acetate + CoA + H$^+$ goes to acetyl CoA + H$_2$O, $\Delta G^{\circ\prime} = -31.4 \text{ kJ mol}^{-1}$ (-7.5 kcal mol^{-1}). ATP hydrolysis, $\Delta G^{\circ\prime} = -30.4 \text{ kJ mol}^{-1}$ (-7.3 kcal mol^{-1}). Overall reaction, $\Delta G^{\circ\prime} = +0.8 \text{ kJ mol}^{-1}$ ($+0.2$ kcal mol^{-1}).
(b) With pyrophosphate hydrolysis, $\Delta G^{\circ\prime} = -33.5 \text{ kJ mol}^{-1}$ (-8.0 kcal mol^{-1}). Pyrophosphate hydrolysis makes the overall reaction exergonic.
6. (a) For an acid AH,

$$\text{AH} \rightleftharpoons \text{A}^- + \text{H}^+ \qquad K = \frac{[\text{A}^-][\text{H}^+]}{[\text{AH}]}$$

The pK is defined as p$K = -\log_{10} K$. $\Delta G^{\circ\prime}$ is the standard free-energy change at pH 7. Thus, $\Delta G^{\circ\prime} = -RT \ln K = -2.303 \, RT \log_{10} K = +2.303 \, RT \, \text{p}K$.
(b) $\Delta G^{\circ\prime} = 27.32 \text{ kJ mol}^{-1}$ (6.53 kcal mol^{-1}).
7. Arginine phosphate in invertebrate muscle, like creatine phosphate in vertebrate muscle, serves as a reservoir of high-potential phosphoryl groups. Arginine phosphate maintains a high level of ATP in muscular exertion.
8. An ADP unit
9. (a) The rationale behind creatine supplementation is that it would be converted into creatine phosphate and thus serves as a rapid means of replenishing ATP after muscle contraction.
(b) If creatine supplementation is beneficial, it would affect activities that depend on short bursts of activity; any sustained activity would require ATP generation by fuel metabolism, which, as Figure 15.7 shows, requires more time.
10. Under standard conditions, $\Delta G^{\circ\prime} = -RT \ln$ [products]/[reactants]. Substituting 23.8 kJ mol^{-1} (5.7 kcal mol^{-1}) for $\Delta G^{\circ\prime}$ and solving for [products]/[reactants] yields 7×10^{-5}. In other words, the forward reaction does not take place to a significant extent. Under intracellular conditions, ΔG is -1.3 kJ mol^{-1} (-0.3 kcal mol^{-1}). Using the equation $\Delta G = \Delta G^{\circ\prime} + RT \ln$ [product]/[reactants] and solving for [products]/[reactants] gives a ratio of 3.7×10^{-5}. Thus, a reaction that is endergonic under standard conditions can be converted into an exergonic reaction by maintaining the [products]/[reactants] ratio below the equilibrium value. This

conversion is usually attained by using the products in another coupled reaction as soon as they are formed.
11. Under standard conditions,

$$K'_{eq} = \frac{[\text{B}]_{eq}}{[\text{A}]_{eq}} \times \frac{[\text{ADP}]_{eq}[\text{P}_i]_{eq}}{[\text{ATP}]_{eq}} = 10^{3.3/1.36} = 2.67 \times 10^2$$

At equilibrium, the ratio of [B] to [A] is given by

$$\frac{[\text{B}]_{eq}}{[\text{A}]_{eq}} = K'_{eq} \frac{[\text{ATP}]_{eq}}{[\text{ADP}]_{eq}[\text{P}_i]_{eq}}$$

The ATP-generating system of cells maintains the [ATP]/[ADP][P$_i$] ratio at a high level, typically about 500 M^{-1}. For this ratio,

$$\frac{[\text{B}]_{eq}}{[\text{A}]_{eq}} = 2.67 \times 10^2 \times 500 = 1.34 \times 10^5$$

This equilibrium ratio is strikingly different from the value of 1.15×10^{-3} for the reaction A $\rightarrow$ B in the absence of ATP hydrolysis. In other words, coupling the hydrolysis of ATP with the conversion of A into B has changed the equilibrium ratio of B to A by a factor of about 10^8.
12. Liver: $-45.2 \text{ kJ mol}^{-1}$ (-10.8 kcal mol^{-1}); muscle: $-48.1 \text{ kJ mol}^{-1}$ (-11.5 kcal mol^{-1}); brain: $-48.5 \text{ kJ mol}^{-1}$ (-11.6 kcal mol^{-1})
13. Recall that $\Delta G = \Delta G^{\circ\prime} + RT \ln$ [products]/[reactants]. Altering the ratio of products to reactants will cause ΔG to vary. In glycolysis, the concentrations of the components of the pathway result in a value of ΔG greater than that of $\Delta G^{\circ\prime}$.
14. The activated form of sulfate in most organisms is 3′-phosphoadenosine-5′-phosphosulfate.
15. (a) As the Mg^{2+} concentration falls, the ΔG of hydrolysis rises. Note that pMg is a logarithmic plot, and so each number on the x-axis represents a 10-fold change in [Mg^{2+}].
(b) Mg^{2+} would bind to the phosphates of ATP and help to mitigate charge repulsion. As the [Mg^{2+}] falls, charge stabilization of ATP would be less, leading to greater charge repulsion and an increase in ΔG on hydrolysis.

Chapter 16

1. Glucose is reactive because its open-chain form contains an aldehyde group.
2. (a) The label is in the methyl carbon atom of pyruvate.
(b) 5 mCi mM^{-1}. The specific activity is halved because the number of moles of product (pyruvate) is twice that of the labeled substrate (glucose).
3. (a) Glucose + 2 P$_i$ + 2 ADP $\rightarrow$ 2 lactate + 2 ATP.
(b) $\Delta G = -114 \text{ kJ mol}^{-1}$ (-27.2 kcal mol^{-1}).
4. 3.06×10^{-5}
5. The equilibrium concentrations of fructose 1,6-bisphosphate, dihydroxyacetone phosphate, and glyceraldehyde 3-phosphate are 7.8×10^{-4} M, 2.2×10^{-4} M, and 2.2×10^{-4} M, respectively.
6. All three carbon atoms of 2,3-BPG are ^{14}C labeled. The phosphorus atom attached to the C-2 hydroxyl group is ^{32}P labeled.
7. Hexokinase has a low ATPase activity in the absence of a sugar because it is in a catalytically inactive conformation. The addition of xylose closes the cleft between the two lobes of the

enzyme. However, xylose lacks a hydroxymethyl group, and so it cannot be phosphorylated. Instead, a water molecule at the site normally occupied by the C-6 hydroxymethyl group acts as the acceptor of the phosphoryl group from ATP.

8. (a) The fructose 1-phosphate pathway forms glyceraldehyde 3-phosphate.

(b) Phosphofructokinase, a key control enzyme, is bypassed. Furthermore, fructose 1-phosphate stimulates pyruvate kinase.

9. (a) Increased; (b) increased; (c) increased; (d) decreased

10. Fructose 2,6-bisphosphate, present at high concentration when glucose is abundant, normally inhibits gluconeogenesis by blocking fructose 1,6-bisphosphatase. In this genetic disorder, the phosphatase is active irrespective of the glucose level. Hence, substrate cycling is increased. The level of fructose 1,6-bisphosphate is consequently lower than normal. Less pyruvate is formed and thus less ATP is generated.

11. Reactions in parts *b* and *e* would be blocked.

12. There will be no labeled carbons. The CO_2 added to pyruvate (formed from the lactate) to form oxaloacetate is lost with the conversion of oxaloacetate into phosphoenolpyruvate.

13. The net reaction in the presence of arsenate is

Glyceraldehyde 3-phosphate $+ NAD^+ + H_2O \rightarrow$
$$3\text{-phosphoglycerate} + NADH + 2\,H^+$$

Glycolysis proceeds in the presence of arsenate, but the ATP normally formed in the conversion of 1,3-bisphosphoglycerate into 3-phosphoglycerate is lost. Thus, arsenate uncouples oxidation and phosphorylation by forming a highly labile acyl arsenate.

14. This example illustrates the difference between the *stoichiometric* and the *catalytic* use of a molecule. If cells used NAD^+ stoichiometrically, a new molecule of NAD^+ would be required each time a lactate was produced. As we will see, the synthesis of NAD^+ requires ATP. On the other hand, if the NAD^+ that is converted into NADH could be recycled and reused, a small amount of the molecule could regenerate a vast amount of lactate, which is the case in the cell. NAD^+ is regenerated by the oxidation of NADH and reused. NAD^+ is thus used catalytically.

15. Consider the equilibrium equation of adenylate kinase:

$$K_{eq} = [ATP][AMP]/[ADP]^2 \qquad (1)$$

or

$$[AMP] = K_{eq}[ADP]^2/[ATP] \qquad (2)$$

Recall that $[ATP] > [ADP] > [AMP]$ in the cell. As ATP is utilized, a small decrease in its concentration will result in a larger percentage increase in [ADP] because its concentration is greater than that of ADP. This larger percentage increase in [ADP] will result in an even greater percentage increase in [AMP] because the concentration of AMP is related to the square of [ADP]. In essence, equation 2 shows that monitoring the energy status with AMP magnifies small changes in [ATP], leading to tighter control.

16. The synthesis of glucose during intense exercise provides a good example of interorgan cooperation in higher organisms. When muscle is actively contracting, lactate is produced from glucose by glycolysis. The lactate is released into the blood and absorbed by the liver, where it is converted by gluconeogenesis into glucose. The newly synthesized

glucose is then released and taken up by the muscle for energy generation.

17. The input of four additional high-phosphoryl-transfer-potential molecules in gluconeogenesis changes the equilibrium constant by a factor of 10^{32}, which makes the conversion of pyruvate into glucose thermodynamically feasible. Without this energetic input, gluconeogenesis would not take place.

18. The mechanism is analogous to that for triose phosphate isomerase (Figure 16.5). It proceeds through an enediol intermediate. The active site would be expected to have a general base (analogous to Glu 165 in TPI) and a general acid (analogous to His 95 in TPI).

19. Galactose is a component of glycoproteins. Possibly, the absence of galactose leads to the improper formation or function of glycoproteins required in the central nervous system. More generally, the fact that the symptoms arise in the absence of galactose suggests that galactose is required in some fashion.

20. (a) Curiously, the enzyme uses ADP as the phosphoryl donor rather than ATP.

(b) Both AMP and ATP behave as competitive inhibitors of ADP, the phosphoryl donor. Apparently, the *P. furiosus* enzyme is not allosterically inhibited by ATP.

Chapter 17

1. (a) After one round of the citric acid cycle, the label emerges in C-2 and C-3 of oxaloacetate. (b) The label emerges in CO_2 in the formation of acetyl CoA from pyruvate. (c) After one round of the citric acid cycle, the label emerges in C-1 and C-4 of oxaloacetate. (d and e) Same fate as that in part *a*.

2. (a) Isocitrate lyase and malate synthase are required in addition to the enzymes of the citric acid cycle.

(b) 2 Acetyl CoA $+ 2\,NAD^+ + FAD + 3\,H_2O \rightarrow$
oxaloacetate $+ 2\,CoA + 2\,NADH + FADH_2 + 3\,H^+$

(c) No. Hence, mammals cannot carry out the net synthesis of oxaloacetate from acetyl CoA.

3. $-41.0\,\text{kJ mol}^{-1}\ (-9.8\,\text{kcal mol}^{-1})$

4. Enzymes or enzyme complexes are biological catalysts. Recall that a catalyst facilitates a chemical reaction without the catalyst itself being permanently altered. Oxaloacetate can be thought of as a catalyst because it binds to an acetyl group, leads to the oxidative decarboxylation of the two carbon atoms, and is regenerated at the completion of a cycle. In essence, oxaloacetate (and any cycle intermediate) acts as a catalyst.

5. Thiamine thiazolone pyrophosphate is a transition-state analog. The sulfur-containing ring of this analog is uncharged, and so it closely resembles the transition state of the normal coenzyme in thiamine-catalyzed reactions (e.g., the uncharged resonance form of hydroxyethyl-TPP).

6. A decrease in the amount of O_2 will necessitate an increase in anaerobic glycolysis for energy production, leading to the generation of a large amount of lactic acid. Under conditions of shock, the kinase inhibitor is administered to ensure that pyruvate dehydrogenase is operating maximally.

7. (a) The steady-state concentrations of the products are low compared with those of the substrates. (b) The ratio of malate to oxaloacetate must be greater than 1.57×10^4 for oxaloacetate to be formed.

8.

$$\text{Pyruvate} + \text{CoA} + \text{NAD}^+ \xrightarrow{\substack{\text{Pyruvate} \\ \text{dehydrogenase} \\ \text{complex}}} \text{acetyl CoA} + CO_2 + \text{NADH}$$

$$\text{Pyruvate} + CO_2 + \text{ATP} + H_2O \xrightarrow{\substack{\text{Pyruvate} \\ \text{carboxylase}}} \text{oxaloacetate} + \text{ADP} + P_i$$

$$\text{Oxaloacetate} + \text{acetyl CoA} + H_2O \xrightarrow{\substack{\text{Citrate} \\ \text{synthase}}} \text{citrate} + \text{CoA} + H^+$$

$$\text{Citrate} \xrightarrow{\text{Aconitase}} \text{isocitrate}$$

$$\text{Isocitrate} + \text{NAD}^+ \xrightarrow{\substack{\text{Isocitrate} \\ \text{dehydrogenase}}} \alpha\text{-ketoglutarate} + CO_2 + \text{NADH}$$

Net:

$$2\ \text{Pyruvate} + 2\ \text{NAD}^+ + \text{ATP} + H_2O \longrightarrow$$
$$\alpha\text{-ketoglutarate} + CO_2 + \text{ADP} + P_i + 2\ \text{NADH} + 2\ H^+$$

9. We cannot get the net conversion of fats into glucose, because the only means to get the carbon atoms from fats into oxaloacetate, the precursor to glucose, is through the citric acid cycle. However, although two carbon atoms enter the cycle as acetyl CoA, two carbon atoms are lost as CO_2 before oxaloacetate is formed. Thus, although some carbon atoms from fats may end up as carbon atoms in glucose, we cannot obtain a *net* synthesis of glucose from fats.

10. The enol intermediate of acetyl CoA attacks the carbonyl carbon atom of glyoxylate to form a C–C bond. This reaction is like the condensation of oxaloacetate with the enol intermediate of acetyl CoA in the reaction catalyzed by citrate synthase. Glyoxylate contains a hydrogen atom in place of the $-CH_2COO^-$ group of oxaloacetate; the reactions are otherwise nearly identical.

11. Citrate is a symmetric molecule. Consequently, the investigators assumed that the two $-CH_2COO^-$ groups in it would react identically. Thus, for every citrate molecule undergoing the reactions shown in path 1, they thought that

another citrate molecule would react as shown in path 2. If so, then only *half* the label should have emerged in the CO_2.

Path 1

Path 2 (does not occur)

12. Call one hydrogen atom A and the other B. Now suppose that an enzyme binds three groups of this substrate—X, Y, and H—at three complementary sites. The adjoining diagram shows X, Y, and H_A bound to three points on the enzyme. In contrast, X, Y, and H_B cannot be bound to this active site; two of these three groups can be bound, but not all three. Thus, H_A and H_B will have different fates.

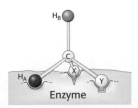

Sterically nonequivalent groups such as H_A and H_B will almost always be distinguished in enzymatic reactions. The essence of the differentiation of these groups is that the enzyme holds the substrate in a specific orientation. Attachment at three points, as depicted in the diagram, is a readily visualized way of achieving a particular orientation of the substrate, but it is not the only means of doing so.

13. (a) The complete oxidation of citrate requires 4.5 μmol of O_2 for every micromole of citrate.

$$C_6H_8O_7 + 4.5\ O_2 \rightarrow 6\ CO_2 + 4\ H_2O$$

Thus, 13.5 μmol of O_2 would be consumed by 3 μmol of citrate.

(b) Citrate led to the consumption of far more O_2 than can be accounted for simply by the oxidation of citrate itself. Citrate thus facilitated O_2 consumption.

14. (a) In the absence of arsenite, the amount of citrate remained constant. In its presence, the concentration of citrate fell, suggesting that it was being metabolized.
(b) Arsenite's action is not altered. Citrate still disappears.
(c) Arsenite is preventing the regeneration of citrate. Recall (p. 495) that arsenite inhibits the pyruvate dehydrogenase complex.

15. (a) The initial infection is unaffected by the absence of isocitrate lyase, but the absence of this enzyme inhibits the latent phase of the infection.
(b) Yes
(c) A critic could say that, in the process of deleting the isocitrate lyase gene, some other gene was damaged, and it is the absence of this other gene that prevents latent infection. Reinserting the isocitrate lyase gene into the bacteria from which it had been removed renders the criticism less valid.
(d) Isocitrate lyase enables the bacteria to synthesize carbohydrates that are necessary for survival, including carbohydrate components of the cell membrane.

Chapter 18

1. (a) 12.5; (b) 14; (c) 32; (d) 13.5; (e) 30; (f) 16
2. Biochemists use E_0', the value at pH 7, whereas chemists use E_0, the value in 1 M H^+. The prime denotes that pH 7 is the standard state.
3. (a) It blocks electron transport and proton pumping at Complex IV. (b) It blocks electron transport and ATP synthesis by inhibiting the exchange of ATP and ADP across the inner mitochondrial membrane. (c) It blocks electron transport and proton pumping at Complex I. (d) It blocks ATP synthesis without inhibiting electron transport by dissipating the proton gradient. (e) It blocks electron transport and proton pumping at Complex IV. (f) It blocks electron transport and proton pumping at Complex III.
4. If the proton gradient is not dissipated by the influx of protons into a mitochondrion with the generation of ATP, eventually the outside of the mitochondrion develops such a large positive charge that the electron-transport chain can no longer pump protons against the gradient.
5. (a) No effect; mitochondria cannot metabolize glucose.
(b) No effect; no fuel is present to power the synthesis of ATP.
(c) The $[O_2]$ falls because citrate is a fuel and ATP can be formed from ADP and P_i.
(d) Oxygen consumption stops because oligomycin inhibits ATP synthesis, which is coupled to the activity of the electron-transport chain.
(e) No effect for the reasons given in part *d*.
(f) $[O_2]$ falls rapidly because the system is uncoupled and does not require ATP synthesis to lower the proton-motive force.
(g) $[O_2]$ falls, though at a lower rate. Rotenone inhibits Complex I, but the presence of succinate will enable electrons to enter at Complex II.
(h) Oxygen consumption ceases because Complex IV is inhibited and the entire chain backs up.
6. (a) The P:O ratio is equal to the product of $(H^+/2\ e^-)$ and (P/H^+). Note that the P:O ratio is identical with the P:2 e^- ratio.
(b) 2.5 and 1.5, respectively

7. $\Delta G^{\circ\prime}$ is $+67$ kJ mol^{-1} ($+16.1$ kcal mol^{-1}) for oxidation by NAD^+ and $+47.7$ kJ mol^{-1} ($+11.4$ kcal mol^{-1}) for oxidation by FAD. The oxidation of succinate by NAD^+ is not thermodynamically feasible.
8. Cyanide can be lethal because it binds to the ferric form of cytochrome oxidase and thereby inhibits oxidative phosphorylation. Nitrite converts ferrohemoglobin into ferrihemoglobin, which also binds cyanide. Thus, ferrihemoglobin competes with cytochrome oxidase for cyanide. This competition is therapeutically effective because the amount of ferrihemoglobin that can be formed without impairing oxygen transport is much greater than the amount of cytochrome oxidase.
9. The available free energy from the translocation of two, three, and four protons is -38.5, -57.7, and -77.4 kJ mol^{-1} (-9.2, -13.8, and -18.5 kcal mol^{-1}), respectively. The free energy consumed in synthesizing a mole of ATP under standard conditions is 30.5 kJ (7.3 kcal). Hence, the residual free energy of -8.1, -27.2, and -46.7 kJ mol^{-1} (-1.93, -6.5, and -11.2 kcal mol^{-1}) can drive the synthesis of ATP until the $[ATP]/[ADP][P_i]$ ratio is 26.2, 6.5×10^4, and 1.6×10^8, respectively. Suspensions of isolated mitochondria synthesize ATP until this ratio is greater than 10^4, which shows that the number of protons translocated per ATP synthesized is at least three.
10. Such a defect (called Luft syndrome) was found in a 38-year-old woman who was incapable of performing prolonged physical work. Her basal metabolic rate was more than twice normal, but her thyroid function was normal. A muscle biopsy showed that her mitochondria were highly variable and atypical in structure. Biochemical studies then revealed that oxidation and phosphorylation were not tightly coupled in these mitochondria. In this patient, much of the energy of fuel molecules was converted into heat rather than ATP.
11. Dicyclohexylcarbodiimide reacts readily with carboxyl groups, as described earlier in regard to its use in peptide synthesis (Section 3.4). Hence, the most likely targets are aspartate and glutamate side chains. In fact, aspartate 61 of subunit **c** of *E. coli* F_0 is specifically modified by this reagent. The conversion of aspartate 61 into asparagine by site-specific mutagenesis also eliminates proton conduction.
12. Triose phosphate isomerase converts dihydroxyacetone phosphate (a potential dead end) into glyceraldehyde 3-phosphate (a mainstream glycolytic intermediate).
13. This inhibitor (like antimycin A) blocks the reduction of cytochrome c_1 by QH_2, the crossover point.
14. If oxidative phosphorylation were uncoupled, no ATP could be produced. In a futile attempt to generate ATP, much fuel would be consumed. The danger lies in the dose. Too much uncoupling would lead to tissue damage in highly aerobic organs such as the brain and heart, which would have severe consequences for the organism as a whole. The energy that is normally transformed into ATP would be released as heat. To maintain body temperature, sweating might increase, although the very process of sweating itself depends on ATP.
15. If the proton gradient cannot be dissipated by flow through the ATP synthase, the proton gradient will eventually become so large that the energy released by the electron-transport chain will not be great enough to pump protons against the larger-than-normal gradient.

16. Add the inhibitor with and without an uncoupler, and monitor the rate of O_2 consumption. If the O_2 consumption increases again in the presence of inhibitor and uncoupler, the inhibitor must be inhibiting ATP synthase. If the uncoupler has no effect on the inhibition, the inhibitor is inhibiting the electron-transport chain.

17. Recall that enzymes catalyze reactions in both directions. The hydrolysis of ATP is exergonic. Consequently, ATP synthase will catalyze the conversion of ATP into its more stable products. ATP synthase works as a synthase in vivo because the energy of the proton gradient overcomes the tendency toward ATP hydrolysis.

18. The arginine residue, with its positive charge, will facilitate proton release from aspartic acid by stabilizing the negatively charged aspartate.

19. 4; 4.7

20. Remember that the extra negative charge on ATP relative to that on ADP accounts for ATP's more rapid translocation out of the mitochondrial matrix. If the charge differences between ATP and ADP were lessened by the binding of the Mg^{2+}, ADP might more readily compete with ATP for transport to the cytoplasm.

21. The cytoplasmic kinases thereby obtaining preferential access to the exported ATP

22. The organic acids in the blood are indications that the mice are deriving a large part of their energy needs through anaerobic glycolysis. Lactate is the end product of anaerobic glycolysis. Alanine is an aminated transport form of lactate. Alanine formation plays a role in succinate formation, which is caused by the reduced state of the mitochondria.

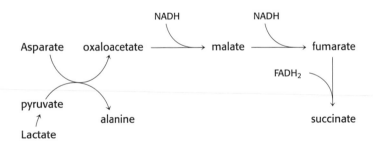

The electron-transport chain is slowed because the inner mitochondrial membrane is hyperpolarized. Without ADP to accept the energy of the proton-motive force, the membrane becomes polarized to such an extent that protons can no longer be pumped. The excess H_2O_2 is probably due to the fact that the superoxide radical is present in higher concentration because the oxygen can no longer be effectively reduced.

$$O_2 \cdot^- + O_2 \cdot^- + 2H^+ \rightarrow H_2O_2 + O_2$$

Indeed, these mice display evidence of such oxidative damage.

23. (a) Succinate is oxidized by Complex II, and the electrons are used to establish a proton-motive force that powers ATP synthesis.
(b) The ability to synthesize ATP is greatly reduced.
(c) Because the goal was to measure ATP hydrolysis. If succinate had been added in the presence of ATP, no reaction would have taken place, because of respiratory control.
(d) The mutation has little effect on the ability of the enzyme to catalyze the hydrolysis of ATP.

(e) They suggest two things: (1) the mutation did not affect the catalytic site on the enzyme, because ATP synthase is still capable of catalyzing the reverse reaction, and (2) the mutation did not affect the amount of enzyme present, given that the controls and patients had similar amounts of activity.

24. The absolute configuration of thiophosphate indicates that inversion at phosphorus has taken place in the reaction catalyzed by ATP synthase. This result is consistent with an in-line phosphoryl-transfer reaction taking place in a single step. The retention of configuration in the Ca^{2+}-ATPase reaction points to two phosphoryl-transfer reactions—inversion by the first and a return to the starting configuration by the second. The Ca^{2+}-ATPase reaction proceeds by a phosphorylated enzyme intermediate.

Chapter 19

1. $\Delta E_0' = 10.11$ V, and $\Delta G^{\circ\prime} = -21.3 \text{ kJ mol}^{-1}$ $(-5.1 \text{ kcal mol}^{-1})$.

2. (a) All ecosystems require an energy source from outside the system, because the chemical-energy sources will ultimately be limited. The photosynthetic conversion of sunlight is one example of such a conversion.
(b) Not at all. Spock would point out that chemicals other than water can donate electrons and protons.

3. DCMU inhibits electron transfer in the link between photosystems II and I. O_2 can evolve in the presence of DCMU if an artificial electron acceptor such as ferricyanide can accept electrons from Q.

4. DCMU will have no effect, because it blocks photosystem II, and cyclic photophosphorylation uses photosystem I and the cytochrome bf complex.

5. (a) 120 kJ einstein^{-1} (28.7 kcal einstein^{-1})
(b) 1.24 V
(c) One 1000-nm photon has the free energy content of 2.4 molecules of ATP. A minimum of 0.42 photon is needed to drive the synthesis of a molecule of ATP.

6. At this distance, the expected rate is one electron per second.

7. The distance doubles, and so the rate should decrease by a factor of 64 to 640 ps.

8. The electrons flow through photosystem II directly to ferricyanide. No other steps are required.

9. (a) Thioredoxin
(b) The control enzyme is unaffected, but the mitochondrial enzyme with part of the chloroplast γ subunit increases activity as the concentration of DTT increases.
(c) The increase was even larger when thioredoxin was present. Thioredoxin is the natural reductant for the chloroplast enzyme, and so it presumably operates more efficiently than would DTT, which probably functions to keep the thioredoxin reduced.
(d) They seem to have done so.
(e) The enzyme is susceptible to control by the redox state. In plant cells, reduced thioredoxin is generated by photosystem I. Thus, the enzyme is active when photosynthesis is taking place.
(f) Cysteine
(g) Group-specific modification or site-specific mutagenesis

Chapter 20

1. Aldolase participates in the Calvin cycle, whereas transaldolase participates in the pentose phosphate pathway.

2. The concentration of 3-phosphoglycerate would increase, whereas that of ribulose 1,5-bisphosphate would decrease.

3. The concentration of 3-phosphoglycerate would decrease, whereas that of ribulose 1,5-bisphosphate would increase.

4. (a)

OPO_3^{2-}
H_2C
$HO-C-COO^-$
$H-C-OH$
$H-C-OH$
H_2C
OPO_3^{2-}

2-Carboxyarabinitol
1,5-bisphosphate
(CABP)

(b) CABP resembles the addition compound formed in the reaction of CO_2 and ribulose 1,5-bisphosphate. (c) CABP will be a potent inhibitor of rubisco.

5. Aspartate + glyoxylate → oxaloacetate + glycine

6. ATP is converted into AMP. To convert this AMP back into ATP, two molecules of ATP are required: one to form ADP and another to form ATP from the ADP.

7. The oxygenase activity of rubisco increases with temperature. Crabgrass is a C_4 plant, whereas most grasses lack this capability. Consequently, the crabgrass will thrive at the hottest part of the summer because the C_4 pathway provides an ample supply of CO_2.

8. As global warming progresses, C_4 plants will invade the higher latitudes, whereas C_3 plants will retreat to cooler regions.

9. The label emerges at C-5 of ribulose 5-phosphate.

10. Oxidative decarboxylation of isocitrate to α-ketoglutarate. A β-ketoacid intermediate is formed in both reactions.

11. C-1 and C-3 of fructose 6-phosphate are labeled, whereas erythrose 4-phosphate is not labeled.

12. (a) 5 Glucose 6-phosphate + ATP →
6 ribose 5-phosphate + ADP + H^+
(b) Glucose 6-phosphate + 12 $NADP^+$ + 7 H_2O →
6 CO_2 + 12 NADPH + 12 H^+ + P_i

13. Form a Schiff base between a ketose substrate and transaldolase, reduce it with tritiated $NaBH_4$, and fingerprint the labeled enzyme.

14. $\Delta E_0'$ for the reduction of glutathione by NADPH is + 0.09 V. Hence, $\Delta G°'$ is -17.4 kJ mol^{-1} (-4.2 kcal mol^{-1}), which corresponds to an equilibrium constant of 1126. The required [NADPH]/[NADP$^+$] ratio is 8.9×10^{-5}.

15.

Dihydroxyacetone phosphate

Fructose 1,6-bisphosphate

16.

Ribose 5-phosphate

Enediol intermediate

Ribulose 5-phosphate

17. Incubate an aliquot of a tissue homogenate with glucose labeled with ^{14}C at C-1, and incubate another with glucose labeled with ^{14}C at C-6. Compare the radioactivity of the CO_2 produced by the two samples. The rationale of this experiment is that only C-1 is decarboxylated by the pentose phosphate pathway, whereas C-1 and C-6 are decarboxylated equally when glucose is metabolized by the glycolytic pathway, the pyruvate dehydrogenase complex, and the citric acid cycle. The reason for the equivalence of C-1 and C-6 in the latter set of reactions is that glyceraldehyde 3-phosphate and dihydroxyacetone phosphate are rapidly interconverted by triose phosphate isomerase.

18. The reduction of each mole CO_2 to the level of a hexose requires 2 moles of NADPH. The reduction of $NADP^+$ is a two-electron process. Hence, the formation of 2 moles of NADPH requires the pumping of four moles of electrons by photosystem I. The electrons given up by photosystem I are

replenished by photosystem II, which needs to absorb an equal number of photons. Hence, eight photons are needed to generate the required NADPH. The energy input of 8 moles of photons is 1594 kJ (381 kcal). Thus, the overall efficiency of photosynthesis under standard conditions is at least 477/1594, or 30%.

19. (a) The curve on the right in graph A was generated by the C_4 plant. Recall that the oxygenase activity of rubisco increases with temperature more rapidly than does the carboxylase activity. Consequently, at higher temperatures, the C_3 plants would fix less carbon. Because C_4 plants can maintain a higher CO_2 concentration, the rise in temperature is less deleterious.
(b) The oxygenase activity will predominate. Additionally, when the temperature rise is very high, the evaporation of water might become a problem. The higher temperatures can begin to damage protein structures as well.
(c) The C_4 pathway is a very effective active-transport system for concentrating CO_2, even when environmental concentrations are very low.
(d) With the assumption that the plants have approximately the same capability to fix CO_2, the C_4 pathway is apparently the rate-limiting step in C_4 plants.

Chapter 21

1. Galactose + ATP + UTP + H_2O + glycogen$_n$ → glycogen$_{n+1}$ + ADP + UDP + 2 P_i + H^+
2. As an unbranched polymer, α-amylose has only one nonreducing end. Therefore, only one glycogen phosphorylase molecule could degrade each α-amylose molecule. Because glycogen is highly branched, there are many nonreducing ends per molecule. Consequently, many phosphorylase molecules can release many glucose molecules per glycogen molecule.
3. The patient has a deficiency of the branching enzyme.
4. The high level of glucose 6-phosphate in von Gierke disease, resulting from the absence of glucose 6-phosphatase or the transporter, shifts the allosteric equilibrium of phosphorylated glycogen synthase toward the active form.
5. Glucose is an allosteric inhibitor of phosphorylase a. Hence, crystals grown in its presence are in the T state. The addition of glucose 1-phosphate, a substrate, shifts the R-to-T equilibrium toward the R state. The conformational differences between these states are sufficiently large that the crystal shatters unless it is stabilized by chemical cross-links.
6. The phosphoryl donor is glucose 1,6-bisphosphate, which is formed from glucose 1-phosphate and ATP in a reaction catalyzed by phosphoglucokinase.
7. Water is excluded from the active site to prevent hydrolysis. The entry of water could lead to the formation of glucose rather than glucose 1-phosphate. A site-specific mutagenesis experiment is revealing in this regard. In phosphorylase, Tyr 573 is hydrogen bonded to the 2'-OH group of a glucose residue. The ratio of glucose 1-phosphate to glucose product is 9000:1 for the wild-type enzyme, and 500:1 for the Phe 573 mutant. Model building suggests that a water molecule occupies the site normally filled by the phenolic OH group of tyrosine and occasionally attacks the oxocarbonium ion intermediate to form glucose.
8. The amylase activity was necessary to remove all of the glycogen from the glycogenin. Recall that glycogenin

synthesizes oligosaccharides of about eight glucose units, and then activity stops. Consequently, if the glucose residues are not removed by extensive amylase treatment, glycogenin will not function.
9. The substrate can be handed directly from the transferase site to the debranching site.
10. During exercise, [ATP] falls and [AMP] rises. Recall that AMP is an allosteric activator of glycogen phosphorylase b. Thus, even in the absence of covalent modification by phosphorylase kinase, glycogen is degraded.
11. (a) Muscle phosphorylase b will be inactive even when the AMP level is high. Hence, glycogen will not be degraded unless phosphorylase is converted into the a form by hormone-induced or Ca^{2+}-induced phosphorylation.
(b) Phosphorylase b cannot be converted into the much more active a form. Hence, the mobilization of liver glycogen will be markedly impaired.
(c) The elevated level of the kinase will lead to the phosphorylation and activation of glycogen phosphorylase. Because glycogen will be persistently degraded, little glycogen will be present in the liver.
(d) Protein phosphatase 1 will be continually active. Hence, the level of phosphorylase b will be higher than normal, and glycogen will be less readily degraded.
(e) Protein phosphatase 1 will be much less effective in dephosphorylating glycogen synthase and glycogen phosphorylase. Consequently, the synthase will stay in the less active b form, and the phosphorylase will stay in the more active a form. Both changes will lead to increased degradation of glycogen.
(f) The absence of glycogenin will block the initiation of glycogen synthesis. Very little glycogen will be synthesized in its absence.
12. (a) The α subunit will thus always be active. Cyclic AMP will always be produced. Glycogen will always be degraded, and glycogen synthesis will always be inhibited.
(b) Glycogen phosphorylase will not be covalently activated. Glycogen degradation will always be inhibited; nothing will remain phosphorylated. Glycogen synthesis will always be active; nothing will remain phosphorylated.
(c) Phosphodiesterase destroys cAMP. Therefore, glycogen degradation will always be active and glycogen synthesis will always be inhibited.
13. The slow phosphorylation of the α subunits of phosphorylase kinase serves to prolong the degradation of glycogen. The kinase cannot be deactivated until its α subunits are phosphorylated. The slow phosphorylation of α subunits ensures that the kinase and, in turn, the phosphorylase stay active for an extended interval.
14. Phosphorylation of the β subunit activates the kinase and leads to glycogen degradation. Subsequent phosphorylation of the α subunit make the β subunit and the α subunit substrates for protein phosphatase. Thus, if the α subunit were modified before the β subunit, the enzyme would be primed for shutdown before it was activated and little glycogen degradation would take place.

15.

Asp H$^+$ R'OH

Transferase reaction

Asp H$^+$ R''OH

α-1,6-Glucosidase reaction

16. (a) Glycogen was too large to enter the gel and, because analysis was by Western blot with the use of an antibody specific to glycogenin, we would not expect to see background proteins.
(b) α-Amylase degrades glycogen, releasing the protein glycogenin, which can be visualized by the Western blot.
(c) Glycogen phosphorylase, glycogen synthase, and protein phosphatase 1. These proteins might be visible if the gel were stained for protein, but a Western analysis reveals the presence of glycogenin only.
17. (a) The smear was due to molecules of glycogenin with increasingly large amounts of glycogen attached to them.
(b) In the absence of glucose in the medium, glycogen is metabolized, resulting in a loss of the high-molecular-weight material.
(c) Glycogen could have been resynthesized and added to the glycogenin when the cells were fed glucose again.
(d) No difference between lanes 3 and 4 suggests that, by 1 hour, the glycogen molecules had attained maximum size in this cell line. Prolonged incubation does not apparently increase the amount of glycogen.
(e) α-Amylase removes essentially all of the glycogen, and so only the glycogenin remains.

Chapter 22

1. Glycerol + 2 NAD$^+$ + P$_i$ + ADP →
pyruvate + ATP + H$_2$O + 2 NADH + H$^+$
Glycerol kinase and glycerol phosphate dehydrogenase

2. Stearate + ATP + 13½ H$_2$O + 8 FAD + 8 NAD$^+$ →
4½ acetoacetate + 14½ H$^+$ + 8 FADH$_2$ + 8 NADH +
AMP + 2 P$_i$
3. (a) Oxidation in mitochondria; synthesis in the cytoplasm.
(b) Coenzyme A in oxidation; acyl carrier protein for synthesis.
(c) FAD and NAD$^+$ in oxidation; NADPH for synthesis.
(d) the L isomer of 3-hydroxyacyl CoA in oxidation; the D isomer in synthesis. (e) From carboxyl to methyl in oxidation; from methyl to carboxyl in synthesis. (f) The enzymes of fatty acid synthesis, but not those of oxidation, are organized in a multienzyme complex.
4. (a) Palmitoleate; (b) linoleate; (c) linoleate; (d) oleate;
(e) oleate; (f) linolenate
5. C-1 is more radioactive.
6. Decarboxylation drives the condensation of malonyl ACP and acetyl ACP. In contrast, the condensation of two molecules of acetyl ACP is energetically unfavorable. In gluconeogenesis, decarboxylation drives the formation of phosphoenolpyruvate from oxaloacetate.
7. Adipose-cell lipase is activated by phosphorylation. Hence, overproduction of the cAMP-activated kinase will lead to an accelerated breakdown of triacylglycerols and a depletion of fat stores.
8. The mutant enzyme will be persistently active because it cannot be inhibited by phosphorylation. Fatty acid synthesis will be abnormally active. Such a mutation might lead to obesity.
9. Carnitine translocase deficiency and glucose 6-phosphate transporter deficiency
10. In the fifth round of β oxidation, cis-Δ^2-enoyl CoA is formed. Dehydration by the classic hydratase yields D-3-hydroxyacyl CoA, the wrong isomer for the next enzyme in β oxidation. This dead end is circumvented by a second hydratase that removes water to give trans-Δ^2-enoyl CoA. The addition of water by the classic hydratase then yields L-3-hydroxyacyl CoA, the appropriate isomer. Thus, hydratases of opposite stereospecificities serve to *epimerize* (invert the configuration of) the 3-hydroxyl group of the acyl CoA intermediate.
11. The probability of synthesizing an error-free polypeptide chain decreases as the length of the chain increases. A single mistake can make the entire polypeptide ineffective. In contrast, a defective subunit can be spurned in the formation of a noncovalent multienzyme complex; the good subunits are not wasted.
12. The absence of ketone bodies is due to the fact that the liver, the source of ketone bodies in the blood, cannot oxidize fatty acids to produce acetyl CoA. Moreover, because of the impaired fatty acid oxidation, the liver becomes more dependent on glucose as an energy source. This dependency results in a decrease in gluconeogenesis and a drop in blood-glucose levels, which is exacerbated by the lack of fatty acid oxidation in muscle and a subsequent increase in glucose uptake from the blood.
13. Peroxisomes enhance the degradation of fatty acids. Consequently, increasing the activity of peroxisomes could help to lower levels of blood triglycerides. In fact, clofibrate is rarely used because of serious side effects.
14. Citrate works by facilitating the formation of active filaments from inactive monomers. In essence, it increases the number of active sites available, or the concentration of enzyme. Consequently, its effect is visible as an increase in the

value of V_{max}. Allosteric enzymes that alter their V_{max} values in response to regulators are sometimes called V-class enzymes. The more common type of allosteric enzyme, in which K_m is altered, comprises K-class enzymes. Palmitoyl CoA causes depolymerization and thus inactivation.

15. The thiolate anion of CoA attacks the 3-keto group to form a tetrahedral intermediate. This intermediate collapses to form acyl CoA and the enolate anion of acetyl CoA. Protonation of the enolate yields acetyl CoA.

16.

Malonyl-ACP

Acetyl-ACP

Acetoacetyl-ACP

17. (a) Fats burn in the flame of carbohydrates. Without carbohydrates, there would be no anapleurotic reactions to replenish the components of the citric acid cycle. With a diet of fats only, the acetyl CoA from fatty acid degradation would build up.
(b) Acetone from ketone bodies
(c) Yes. Odd-chain fatty acids would lead to the production of propionyl CoA, which can be converted into succinyl CoA, a citric acid cycle component. It would serve to replenish the citric acid cycle and mitigate the halitosis.

18. A labeled fat can enter the citric acid cycle as acetyl CoA and yield labeled oxaloacetate, but only after two carbon atoms have been lost as CO_2. Consequently, even though oxaloacetate may be labeled, there can be no net synthesis of oxaloacetate and hence no net synthesis of glucose or glycogen.

19. (a) The V_{max} is decreased and the K_m is increased. V_{max} (wild type) = 13 nmol minute^{-1} mg^{-1}; K_m (wild type) = 45 μM; V_{max} (mutant) = 8.3 nmol minute^{-1} mg^{-1}; K_m (mutant) = 74 μM.
(b) Both the V_{max} and the K_m are decreased. V_{max} (wild type) = 41 nmol minute^{-1} mg^{-1}; K_m (wild type) = 104 μM; V_{max} (mutant) = 23 nmol minute^{-1} mg^{-1}; K_m (mutant) = 69 μM.
(c) The wild type is significantly more sensitive to malonyl CoA.
(d) With respect to carnitine, the mutant displays approximately 65% of the activity of the wild type; with respect to palmitoyl CoA, approximately 50% activity. On the other hand, 10 μM of malonyl CoA inhibits approximately 80% of the wild type but has essentially no effect on the mutant enzyme.
(e) The glutamate appears to play a more prominent role in regulation by malonyl CoA than in catalysis.

Chapter 23

1. (a) The ATPase activity of the 26S proteasome resides in the 19S subunit. The energy of ATP hydrolysis could be used to unfold the substrate, which is too large to enter the catalytic barrel. ATP may also be required for translocation of the substrate into the barrel.
(b) Substantiates the answer in part *a*. Because they are small, the peptides do not need to be unfolded. Moreover, small peptides could probably enter all at once and not require translocation.

2. (a) Pyruvate; (b) oxaloacetate; (c) α-ketoglutarate; (d) α-ketoisocaproate; (e) phenylpyruvate; (f) hydroxyphenylpyruvate

3. (a) Aspartate + α-ketoglutarate + GTP + ATP + $2 H_2O$ + NADH + H$^+$ → ½ glucose + glutamate + CO_2 + ADP + GDP + NAD$^+$ + $2 P_i$
The required coenzymes are pyridoxal phosphate in the transamination reaction and NAD$^+$/NADH in the redox reactions.
(b) Aspartate + CO_2 + NH$_4^+$ + 3 ATP + NAD$^+$ + $4 H_2O$ → oxaloacetate + urea + 2 ADP + $4 P_i$ + AMP + NADH + H$^+$

4. In the eukaryotic proteasome, the distinct β subunits have different substrate specificities, allowing proteins to be more thoroughly degraded.

5. The six subunits probably exist as a heterohexamer. Cross-linking experiments could test the model and help determine which subunits are adjacent to one another.

6. Thiamine pyrophosphate

7. It acts as an electron sink.

8. CO_2 + NH$_4^+$ + 3 ATP + NAD$^+$ + aspartate + $3 H_2O$ → urea + 2 ADP + $2 P_i$ + AMP + PP$_i$ + NADH + H$^+$ + oxaloacetate
Four high-transfer-potential groups are spent.

9. Ornithine transcarbamoylase (analogous to PALA; see Chapter 10)

10. Ammonia could lead to the amination of α-ketoglutarate, producing a high concentration of glutamate in an unregulated fashion. α-Ketoglutarate for glutamate synthesis could be removed from the citric acid cycle, thereby diminishing the cell's respiration capacity.

11. The mass spectrometric analysis strongly suggests that three enzymes—pyruvate dehydrogenase, α-ketoglutarate dehydrogenase, and the branched-chain α-keto dehydrogenase—are deficient. Most likely, the common E$_3$ component of these enzymes is missing or defective. This proposal could be tested by purifying these three enzymes and assaying their ability to catalyze the regeneration of lipoamide.

12. Benzoate, phenylacetate, and arginine would be given to supply a protein-restricted diet. Nitrogen would emerge in hippurate, phenylacetylglutamine, and citrulline.

13. Aspartame, a dipeptide ester (L-aspartyl-L-phenylalanine methyl ester), is hydrolyzed to L-aspartate and L-phenyl-alanine. High levels of phenylalanine are harmful in phenylketonurics.

14. *N*-Acetylglutamate is synthesized from acetyl CoA and glutamate. Once again, acetyl CoA serves as an activated acetyl donor. This reaction is catalyzed by *N*-acetylglutamate synthase.

15. See equation below.

16.

The equilibrium constant for the interconversion of L-serine and D-serine is exactly 1.

17. Exposure of such a domain would suggest that a component of a multiprotein complex has failed to form properly or that one component has been synthesized in excess. This exosure will lead to rapid degradation and restoration of appropriate stoichiometries.

18. (a) Depletion of glycogen stores. When they are gone, proteins must be degraded to meet the glucose needs of the brain. The resulting amino acids are deaminated, and the nitrogen atoms are excreted as urea.

(b) The brain has adapted to the use of ketone bodies, which are derived from fatty acid catabolism. In other words, the brain is being powered by fatty acid breakdown.

(c) When the glycogen and lipid stores are gone, the only available energy source is protein.

19. Deamination to α-keto-β-methylvalerate; oxidative decarboxylation to α-methylbutyryl CoA; oxidation to tiglyl CoA; hydration, oxidation and thiolysis yields acetyl CoA and propionyl CoA; propionyl CoA to succinyl CoA.

20. (a) Virtually no digestion in the absence of nucleotides.

(b) Protein digestion is greatly stimulated by the presence of ATP. (c) AMP-PNP, a nonhydrolyzable analog of ATP, is no more effective than ADP. (d) The proteasome requires neither ATP nor PAN to digest small substrates. (e) PAN and ATP hydrolysis may be required to unfold the peptide and translocate it into the proteasome. (f) Although *Thermoplasma* PAN is not as effective with the other proteasomes, it nonetheless results in threefold to fourfold stimulation of digestion. (g) In light of the fact that the archaea and eukarya diverged several billion years ago, the fact that *Thermoplasma* PAN can stimulate rabbit muscle suggests homology not only between the proteasomes, but also between PAN and the 19S subunit (most likely the ATPases) of the mammalian 26S proteasome.

Chapter 24

1. Glucose + 2 ADP + 2 P_i + 2 NAD^+ + 2 glutamate → 2 alanine + 2 α-ketoglutarate + 2 ATP + 2 NADH + 2 H_2O + 2 H^+

2. $N_2 \rightarrow NH_4^+ \rightarrow$ glutamate → serine → glycine → δ-aminolevulinate → porphobilinogen → heme

3. (a) N^5, N^{10}-Methylenetetrahydrofolate; (b) N^5-methyltetrahydrofolate

4. γ-Glutamyl phosphate is a likely reaction intermediate.

5. The administration of glycine leads to the formation of isovalerylglycine. This water-soluble conjugate, in contrast with isovaleric acid, is excreted very rapidly by the kidneys.

6. They carry out nitrogen fixation. The absence of photosystem II provides an environment in which O_2 is not produced. Recall that the nitrogenase is very rapidly inactivated by O_2.

7. The cytoplasm is a reducing environment, whereas the extracellular milieu is an oxidizing environment.

8. (a) None; (b) D-glutamate and oxaloacetate

9. Succinyl CoA is formed in the mitochondrial matrix.

10. Alanine from pyruvate; aspartate from oxaloacetate; glutamate from α-ketoglutarate

11. Y could inhibit the C → D step, Z could inhibit the C → F step, and C could inhibit A → B. This scheme is an example of sequential feedback inhibition. Alternatively, Y could inhibit the C → D step, Z could inhibit the C → F step, and the A → B step would be inhibited only in the presence of both Y and Z. This scheme is called concerted feedback inhibition.

12. The rate of the A → B step in the presence of high levels of Y and Z would be 24 s^{-1} (0.6 × 0.4 × 100 s^{-1}).

13. An external aldimine forms with SAM, which is deprotonated to form the quinonoid intermediate. The deprotonated carbon atom attacks the carbon atom adjacent to the sulfur atom to form the cyclopropane ring and release methylthioadenosine, the other product.

14. An external aldimine forms with L-serine, which is deprotonated to form the quinonoid intermediate. This intermediate is reprotonated on its opposite face to form an aldimine with D-serine. This compound is cleaved to release D-serine. The equilibrium constant for a racemization reaction is 1 because the reactant and product are exact mirror images of each other.

15. Synthesis from oxaloacetate and α-ketoglutarate would deplete the citric acid cycle, which would decrease ATP production. Anapleurotic reactions would be required to replenish the citric acid cycle.

16. SAM is the donor for DNA methylation reactions that protect a host from digestion by its own restriction enzymes. A lack of SAM would render the bacterial DNA susceptible to digestion by the cell's own restriction enzymes.

17. (a) Asparagine is much more abundant in the dark. More glutamine is present in the light. These amino acids show the most dramatic effects. Glycine also is more abundant in the light.

(b) Glutamine is a more metabolically reactive amino acid, used in the synthesis of many other compounds. Consequently, when energy is available as light, glutamine will be preferentially synthesized. Asparagine, which carries more nitrogen per carbon atom and is thus a more efficient means of storing nitrogen when energy is short, is synthesized in the dark. Glycine is more prevalent in the light because of photorespiration.

(c) White asparagus has an especially high concentration of asparagine, which accounts for its intense taste. All asparagus has a large amount of asparagine. In fact, as suggested by its name, asparagine was first isolated from asparagus.

Chapter 25

1. Glucose + 2 ATP + 2 $NADP^+$ + H_2O → PRPP + CO_2 + ADP + AMP + 2 NADPH + 3 H^+

2. Glutamine + aspartate + CO^2 + 2 ATP + NAD^+ → orotate + 2 ADP + 2 P_i + glutamate + NADH + H^+

3. (a, c, and d) PRPP; (b) carbamoyl phosphate

4. PRPP and formylglycinamide ribonucleotide

5. dUMP + serine + NADPH + H^+ → dTMP + $NADP^+$ + glycine

6. There is a deficiency of N^{10}-formyltetrahydrofolate. Sulfanilamide inhibits the synthesis of folate by acting as an analog of *p*-aminobenzoate, one of the precursors of folate.

7. PRPP is the activated intermediate in the synthesis of phosphoribosylamine in the de novo pathway of purine formation; of purine nucleotides from free bases by the salvage pathway; of orotidylate in the formation of pyrimidines; of nicotinate ribonucleotide; of phosphoribosyl ATP in the pathway leading to histidine; and of phosphoribosylanthranilate in the pathway leading to tryptophan.

8. (a) Cell A cannot grow in a HAT medium, because it cannot synthesize TMP either from thymidine or from dUMP. Cell B cannot grow in this medium, because it cannot synthesize purines by either the de novo pathway or the salvage pathway. Cell C can grow in a HAT medium because it contains active thymidine kinase from cell B (enabling it to phosphorylate thymidine to TMP) and hypoxanthine-guanine phosphoribosyltransferase from cell A (enabling it to synthesize purines from hypoxanthine by the salvage pathway).

(b) Transform cell A with a plasmid containing foreign genes of interest and a functional thymidine kinase gene. The only cells that will grow in a HAT medium are those that have acquired a thymidylate kinase gene; nearly all of these transformed cells will also contain the other genes on the plasmid.

9. Ring carbons 4, 5, and 6 in cytosine will be labeled. In guanine, only the bridge carbon atoms that are shared between the 5- and 6-membered rings will be labeled with ^{13}C.

10. The enzyme that uses ammonia synthesizes carbamoyl phosphate for a reaction with ornithine, the first step of the urea cycle. The enzyme that uses glutamine synthesizes carbamoyl phosphate for use in the first step of pyrimidine biosynthesis.

11. These patients have a high level of urate because of the breakdown of nucleic acids. Allopurinol prevents the formation of kidney stones and blocks other deleterious consequences of hyperuricemia by preventing the formation of urate (p. 726).

12. The free energies of binding are -57.7 (wild type), -49.8 (Asn 27), and -38.1 (Ser 27) kJ mol^{-1} (-13.8, -11.9, and -9.1 kcal mol^{-1}, respectively). The loss in binding energy is 7.9 kJ mol^{-1} (1.9 kcal mol^{-1}) and 19.7 kJ mol^{-1} (4.7 kcal mol^{-1}).

13. Inosine or hypoxanthine could be administered.

14. N-1 in both cases, and the amine group linked to C-6 in ATP

15. An oxygen atom is added to allopurinol to form alloxanthine.

16. The first reaction proceeds by phosphorylation of glycine to form an acyl phosphate followed by nucleophilic attack by the amine of phosphoribosylamine to displace orthophosphate. The second reaction consists of adenylation of the carbonyl group of xanthylate followed by nucleophilic attack by ammonia to displace AMP.

17. The $-NH_2$ group attacks the carbonyl carbon atom to form a tetrahedral intermediate. Removal of a proton leads to the elimination of water to form inosinate.

18. (a) cAMP; (b) ATP; (c) UDP-glucose; (d) acetyl CoA; (e) NAD$^+$, FAD; (f) dideoxynucleotides; (g) fluorouracil; (h) CTP inhibits ATCase.

19. In vitamin B$_{12}$ deficiency, methyltetrahydrofolate cannot donate its methyl group to homocysteine to regenerate methionine. Because the synthesis of methyltetrahydrofolate is irreversible, the cell's tetrahydrofolate will ultimately be converted into this form. No formyl or methylene tetrahydrofolate will be left for nucleotide synthesis. Pernicious anemia illustrates the intimate connection between amino acid and nucleotide metabolism.

20. The cytoplasmic level of ATP in the liver falls and that of AMP rises above normal in all three conditions. The excess AMP is degraded to urate.

21. Succinate $\rightarrow$ malate $\rightarrow$ oxaloacetate by the citric acid cycle. Oxaloacetate $\rightarrow$ aspartate by transamination, followed by pyrimidine synthesis. Carbons 4, 5, and 6.

22. (a) Some ATP can be salvaged from the ADP that is being generated. (b) There are equal numbers of high-phosphoryl-transfer-potential groups on each side of the equation. (c) Because the adenylate kinase reaction is at equilibrium, removing of AMP would lead to the formation of more ATP. (d) Essentially, the cycle serves as an anapleurotic reaction for the generation of the citric acid cycle intermediate fumarate.

Chapter 26

1. Glycerol + 4 ATP + 3 fatty acids + 4 H$_2$O $\rightarrow$ triacylglycerol + ADP + 3 AMP + 7 P$_i$ + 4 H$^+$

2. Glycerol + 3 ATP + 2 fatty acids + 2 H$_2$O + CTP + ethanolamine $\rightarrow$ phosphatidylethanolamine + CMP + ADP + 2 AMP + 6 P$_i$ + 3 H$^+$

3. (a) CDP-diacylglycerol; (b) CDP-ethanolamine; (c) acyl CoA; (d) CDP-choline; (e) UDP-glucose or UDP-galactose; (f) UDP-galactose; (g) geranyl pyrophosphate

4. (a and b) None, because the label is lost as CO_2

5. The categories of mutations are: (1) no receptor is synthesized; (2) receptors are synthesized but do not reach the plasma membrane, because they lack signals for intracellular transport or do not fold properly; (3) receptors reach the cell surface, but they fail to bind LDL normally because of a defect in the LDL-binding domain; (4) receptors reach the cell surface and bind LDL, but they fail to cluster in coated pits because of a defect in their carboxyl-terminal regions.

6. Deamination of cytidine to uridine changes CAA (Gln) into UAA (stop).

7. Benign prostatic hypertrophy can be treated by inhibiting 5α-reductase. Finasteride, the 4-aza steroid analog of dihydrotestosterone, competitively inhibits the reductase but does not act on androgen receptors. Patients taking finasteride have a markedly lower plasma level of dihydrotestosterone and a nearly normal level of testosterone. The prostate gland becomes smaller, whereas testosterone-dependent processes such as fertility, libido, and muscle strength appear to be unaffected.

Finasteride

8. Patients who are most sensitive to debrisoquine have a deficiency of a liver P450 enzyme encoded by a member of the *CYP2* subfamily. This characteristic is inherited as an autosomal recessive trait. The capacity to degrade other drugs may be impaired in people who hydroxylate debrisoquine at a slow rate, because a single P450 enzyme usually handles a broad range of substrates.

9. Many hydrophobic odorants are deactivated by hydroxylation. Molecular oxygen is activated by a cytochrome P450 monooxygenase. NADPH serves as the reductant. One oxygen atom of O_2 goes into the odorant substrate, whereas the other is reduced to water.

10. Recall that dihydrotestosterone is crucial for the development of male characteristics in the embryo. If a pregnant woman were to be exposed to Propecia, the 5α-reductase of the male embryo would be inhibited, which could result in severe developmental abnormalities.

11. The oxygenation reactions catalyzed by the cytochrome P450 family permit greater flexibility in biosynthesis. Because plants are not mobile, they must rely on physical defenses, such as thorns, and chemical defenses, such as toxic alkaloids. The larger P450 array might permit greater biosynthetic versatility.

12. This knowledge would enable clinicians to characterize the likelihood of a patient's having an adverse drug reaction or being susceptible to chemical-induced illnesses. It would also permit a personalized and especially effective drug-treatment regime for diseases such as cancer.

13. The negatively charged phosphoserine residue interacts with the positively charged protonated histidine residue and decreases its ability to transfer a proton to the thiolate.

His

14. The methyl group is first hydroxylated. The hydroxymethylamine eliminated formaldehyde to form methylamine.

15. (a) There is no effect. (b) Because actin is not controlled by cholesterol, the amount isolated should be the same in both experimental groups; a difference would suggest a problem in the RNA isolation. (c) The presence of cholesterol in the diet dramatically reduces the amount of HMG-CoA reductase protein. (d) A common means of regulating the amount of a protein present is to regulate transcription, which is clearly not the case here. (e) The translation of mRNA could be inhibited, and the protein could be rapidly degraded.

Chapter 27

1. The liver, and to a lesser extent the kidneys, contain glucose 6-phosphatase, whereas muscle and the brain do not. Hence, muscle and the brain, in contrast with the liver, do not release glucose. Another key enzymatic difference is that the liver has little of the transferase needed to activate acetoacetate to acetoacetyl CoA. Consequently, acetoacetate and 3-hydroxybutyrate are exported by the liver for use by heart muscle, skeletal muscle, and the brain.

2. (a) Adipose cells normally convert glucose into glycerol 3-phosphate for the formation of triacylglycerols. A deficiency of hexokinase will interfere with the synthesis of triacylglycerols.
(b) A deficiency of glucose 6-phosphatase will block the export of glucose from the liver after glycogenolysis. This disorder (called von Gierke disease) is characterized by an abnormally high content of glycogen in the liver and a low blood-glucose level.
(c) A deficiency of carnitine acyltransferase I impairs the oxidation of long-chain fatty acids. Fasting and exercise precipitate muscle cramps.
(d) Glucokinase enables the liver to phosphorylate glucose even in the presence of a high level of glucose 6-phosphate. A deficiency of glucokinase will interfere with the synthesis of glycogen.
(e) Thiolase catalyzes the formation of two molecules of acetyl CoA from acetoacetyl CoA and CoA. A deficiency of thiolase will interfere with the utilization of acetoacetate as a fuel when the blood-sugar level is low.
(f) Phosphofructokinase will be less active than normal because of the lowered level of F-2,6-BP. Hence, glycolysis will be much slower than normal.

3. (a) A high proportion of fatty acids in the blood are bound to albumin. Cerebrospinal fluid has a low content of fatty acids because it has little albumin.

(b) Glucose is highly hydrophilic and soluble in aqueous media, in contrast with fatty acids, which must be carried by transport proteins such as albumin. Micelles of fatty acids would disrupt membrane structure.
(c) Fatty acids, not glucose, are the major fuel of resting muscle.

4. (a) A watt is equal to 1 joule (J) per second (0.239 calorie per second). Hence, 70 W is equivalent to 0.07 kJ s^{-1} $(0.017 \text{ kcal s}^{-1})$.
(b) A watt is a current of 1 ampere (A) across a potential of 1 volt (V). For simplicity, let us assume that all the electron flow is from NADH to O_2 (a potential drop of 1.14 V). Hence, the current is 61.4 A, which corresponds to 3.86×10^{20} electrons per second (1 A = 1 coulomb s^{-1} = 6.28×10^{18} charge s^{-1}).
(c) About 2.5 molecules of ATP are formed per molecule of NADH oxidized (two electrons). Hence, one molecule of ATP is formed per 0.8 electron transferred. A flow of 3.86×10^{20} electrons per second therefore leads to the generation of 4.83×10^{20} molecules of ATP per second, or 0.80 mmol s^{-1}.
(d) The molecular weight of ATP is 507. The total body content of ATP of 50 g is equal to 0.099 mol. Hence, ATP turns over about once per 125 seconds when the body is at rest.

5. (a) The stoichiometry of the complete oxidation of glucose is

$$C_6H_{12}O_6 + 6\,O_2 \rightarrow 6\,CO_2 + 6\,H_2O$$

and that of tripalmitoylglycerol is

$$C_{51}H_{98}O_2 + 72.5\,O_2 \rightarrow 51\,CO_2 + 49\,H_2O$$

Hence, the RQ values are 1.0 and 0.703, respectively.
(b) An RQ value reveals the relative usage of carbohydrate and fats as fuels. The RQ of a marathon runner typically decreases from 0.97 to 0.77 in the course of a race. The lowering of the RQ indicates the shift in fuel from carbohydrate to fat.

6. One gram of glucose (molecular weight 180.2) is equal to 5.55 mmol, and one gram of tripalmitoylglycerol (molecular weight 807.3) is equal to 1.24 mmol. The reaction stoichiometries (see problem 5) indicate that 6 mol of H_2O is produced per mole of glucose oxidized, and 49 mol of H_2O is produced per mole of tripalmitoylglycerol oxidized. Hence, the H_2O yields per gram of fuel are 33.3 mmol (0.6 g) for glucose and 60.8 mmol (1.09 g) for tripalmitoylglycerol. Thus, complete oxidation of this fat gives 1.82 times as much water as does glucose. Another advantage of triacylglycerols is that they can be stored in essentially anhydrous form, whereas glucose is stored as glycogen, a highly hydrated polymer. A hump consisting mainly of glycogen would be an intolerable burden—far more than the straw that broke the camel's back.

7. A typical macadamia nut has a mass of about 2 g. Because it consists mainly of fats ($\sim$37 kJ g^{-1}, $\sim$9 kcal g^{-1}), a nut has a value of about 75 kJ (18 kcal). The ingestion of 10 nuts results in an intake of about 753 kJ (180 kcal). As stated in the answer to problem 4, a power consumption of 1 W corresponds to 1 J s^{-1} (0.239 cal s^{-1}), and so 400-W running requires 0.4 kJ s^{-1} (0.0956 kcal s^{-1}). Hence, one would have to run 1882 s, or about 31 minutes, to spend the calories provided by 10 nuts.

8. A high blood-glucose level triggers the secretion of insulin, which stimulates the synthesis of glycogen and triacylglycerols. A high insulin level would impede the mobilization of fuel reserves during the marathon.

9. Lipid mobilization can occur so rapidly that it exceeds the ability of the liver to oxidize the lipids or convert them into

ketone bodies. The excess is reesterified and released into the blood as VLDL.

10. A role of the liver is to provide glucose for other tissues. In the liver, glycolysis is used not for energy production but for biosynthetic purposes. Consequently, in the presence of glucagon, liver glycolysis stops so that the glucose can be released into the blood.

11. Urea cycle and gluconeogenesis

12. (a) Insulin inhibits lipid utilization.
(b) Insulin stimulates protein synthesis, but there are no amino acids in the children's diet. Moreover, insulin inhibits protein breakdown. Consequently, muscle proteins cannot be broken down and used for the synthesis of essential proteins.
(c) Because proteins cannot be synthesized, blood osmolarity is too low. Consequently, fluid leaves the blood. An especially important protein for maintaining blood osmolarity is albumin.

13. The oxygen consumption at the end of exercise is used to replenish ATP and creatine phosphate and to oxidize any lactate produced.

14. Oxygen is used in oxidative phosphorylation to resynthesize ATP and creatine phosphate. The liver converts lactate released by the muscle into glucose. Blood must be circulated to return the body temperature to normal, and so the heart cannot return to its resting rate immediately. Hemoglobin must be reoxygenated to replace the oxygen used in exercise. The muscles that power breathing must continue working at the same time that the exercised muscles are returning to resting states. In essence, all the biochemical systems activated in intense exercise need increased oxygen to return to the resting state.

15. Ethanol may replace water that is hydrogen bonded to proteins and membrane surfaces. This alteration of the hydration state of the protein would alter its conformation and hence function. Ethanol may also alter phospholipid packing in membranes. The two effects suggest that integral membrane proteins would be most sensitive to ethanol, as indeed seems to be the case.

16. Cells from the type I fiber would be rich in mitochondria, whereas those of the type II fiber would have few mitochondria.

17. (a) The ATP expended during this race amounts to about 8380 kg, or 18,400 pounds. (b) The cyclist would need about $1,260,000,000 to complete the race.

Chapter 28

1. DNA polymerase I uses deoxyribonucleoside triphosphates; pyrophosphate is the leaving group. DNA ligase uses a DNA-adenylate (AMP joined to the 5′-phosphate) as a reaction partner; AMP is the leaving group. Topoisomerase I uses a DNA-tyrosyl intermediate (5′-phosphate linked to the phenolic OH group); the tyrosine residue of the enzyme is the leaving group.

2. Positive supercoiling resists the unwinding of DNA. The melting temperature of DNA increases in proceeding from negatively supercoiled to relaxed to positively supercoiled DNA. Positive supercoiling is probably an adaptation to high temperature.

3. (a) There are long stretches of each region because the transition is highly cooperative. (b) B–Z junctions are energetically highly unfavorable. (c) A-to-B transitions are less

cooperative than B-to-Z transitions because the helix stays right-handed at an A–B junction but not at a B–Z junction.

4. (a) 96.2 revolutions per second (1000 nucleotides per second divided by 10.4 nucleotides per turn for B-DNA gives 96.2 rps). (b) 0.34 μm s^{-1} (1000 nucleotides per second corresponds to 3400 Å s^{-1} because the axial distance between nucleotides in B-DNA is 3.4 Å).

5. Eventually, the DNA would become so tightly wound that movement of the replication complex would be energetically impossible.

6. A hallmark of most cancer cells is prolific cell division, which requires DNA replication. If the telomerase were not activated, the chromosomes would shorten until they became nonfunctional, leading to cell death. Interestingly, telomerase is often, but not always, found to be activated in cancer cells.

7. Treat the DNA briefly with endonuclease to occasionally nick each strand. Add the polymerase with the radioactive dNTPs. At the broken bond, or nick, the polymerase will degrade the existing strand with its 5′ → 3′ exonuclease activity and replace it with a radioactive complementary copy by using its polymerase activity. This reaction scheme is referred to as nick translation, because the nick is moved, or translated, along the DNA molecule without ever becoming sealed.

8. If replication were unidirectional, tracks with a low grain density at one end and a high grain density at the other end would be seen. On the other hand, if replication were bidirectional, the middle of a track would have a low density, as shown in the adjoining diagram. For *E. coli*, the grain tracks are denser on both ends than in the middle, indicating that replication is bidirectional.

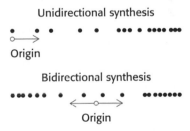

9. (a) Pro (CCC), Ser (UCC), Leu (CUC), and Phe (UUC). Alternatively, the last base of each of these codons could be U.
(b) These C → U mutations were produced by nitrous acid.

10. Potentially deleterious side reactions are prevented. The enzyme itself might be damaged by light if it could be activated by light in the absence of bound DNA harboring a pyrimidine dimer.

11. The release of DNA topoisomerase II after the enzyme has acted on its DNA substrate requires ATP hydrolysis. Negative supercoiling requires only the binding of ATP, not its hydrolysis.

12. (a) Size; the top is relaxed and the bottom is supercoiled DNA. (b) Topoisomers. (c) The DNA is becoming progressively more unwound, or relaxed, and thus slower moving.

13. (a) It was used to determine the number of spontaneous revertants—that is, the background mutation rate.
(b) To firmly establish that the system was working. A known mutagen's failure to produce revertants would indicate that something was wrong with the experimental system.

(c) The chemical itself has little mutagenic ability but is apparently activated into a mutagen by the liver homogenate. (d) Cytochrome P450 system.

Chapter 29

1. The sequence of the coding (+, sense) strand is

5′-ATGGGGAACAGCAAGAGTGGGGCCCTGTCCAAGGAG-3′

and the sequence of template (−, antisense) strand is

3′-TACCCCTTGTCGTTCTCACCCCGGGACAGGTTCCTC-5′

2. An error will affect only one molecule of mRNA of many synthesized from a gene. In addition, the errors do not become a permanent part of the genomic information.
3. At any given instant, only a fraction of the genome (total DNA) is being transcribed. Consequently, speed is not necessary.
4. Heparin, a glycosaminoglycan is highly anionic. Its negative charges, like the phosphodiester bridges of DNA templates, allow it to bind to lysine and arginine residues of RNA polymerase.
5. This mutant σ will competitively inhibit the binding of holoenzyme and prevent the specific initiation of RNA chains at promoter sites.
6. The core enzyme without σ binds more tightly to the DNA template than does the holoenzyme. The retention of σ after chain initiation would make the mutant RNA polymerase less processive. Hence, RNA synthesis would be much slower than normal.
7. A 100-kd protein contains about 910 residues, which are encoded by 2730 nucleotides. At a maximal transcription rate of 50 nucleotides per second, the mRNA would be synthesized in 54.6 s.
8. Initiation at strong promoters takes place every 2 s. In this interval, 100 nucleotides are transcribed. Hence, centers of transcription bubbles are 34 nm (340 Å) apart.
9. (a) The lowest band on the gel will be that of strand 3 alone (i), whereas the highest will be that of stands 1, 2, and 3 and core polymerase (v). Band ii will be at the same position as band i because the RNA is not complementary to the nontemplate strand, whereas band iii will be higher because a complex is formed between RNA and the template strand. Band iv will be higher than the others because strand 1 is complexed to 2, and strand 2 is complexed to 3. Band v is the highest because core polymerase associates with the three strands.
(b) None, because rifampicin acts before the formation of the open complex.
(c) RNA polymerase is processive. When the template is bound, heparin cannot enter the DNA-binding site.
(d) When GTP is absent, synthesis stops when the first cytosine residue downstream of the bubble is encountered in the template strand. In contrast, with all four nucleoside triphosphates present, synthesis will continue to the end of the template.
10. The base-pairing energy of the di- and trinucleotide DNA–RNA hybrids formed at the very beginning of transcription is not sufficient to prevent strand separation and loss of product.

11. (a) Because cordycepin lacks a 3′-OH group, it cannot participate in 3′ → 5′ bond formation. (b) Because the poly(A) tail is a long stretch of adenosine nucleotides, the likelihood that a molecule of cordycepin would become incorporated is higher than with most RNA. (c) Yes, it must be converted into cordycepin 5′-triphosphate.
12. Ser-Ile-Phe-His-Pro-Stop
13. A mutation that disrupted the normal AAUAAA recognition sequence for the endonuclease could account for this finding. In fact, a change from U to C in this sequence caused this defect in a thalassemic patient. Cleavage occurred at the AAUAAA 900 nucleotides downstream from this mutant AACAAA site.
14. One possibility is that the 3′ end of the poly(U) donor strand cleaves the phosphodiester bond on the 5′ side of the insertion site. The newly formed 3′ terminus of the acceptor strand then cleaves the poly(U) strand on the 5′ side of the nucleotide that initiated the attack. In other words, a uridine residue could be added by two transesterification reactions. This postulated mechanism is similar to the one in RNA splicing.
15. Alternative splicing, RNA editing. Covalent modification of the proteins subsequent to synthesis.
16. Attach an oligo(dT) or oligo(U) sequence to an inert support to create an affinity column. When RNA is passed through the column, only poly(A)-containing RNA will be retained.
17. (a) Different amounts of RNA are present for the various genes.
(b) Although all of the tissues have the same genes, the genes are expressed to different extents in different tissues.
(c) These genes are called housekeeping genes—genes that most tissues express. They might include genes for glycolysis or citric acid cycle enzymes.
(d) The point of the experiment is to determine which genes are initiated in vivo. The initiation inhibitor is added to prevent initiation at start sites that may have been activated during the isolation of the nuclei.
18. DNA is the single strand that forms the trunk of the tree. Strands of increasing length are RNA molecules; the beginning of transcription is where growing chains are the smallest; the end of transcription is where chain growth stops. Direction is left to right. Many enzymes are actively transcribing each gene.

Chapter 30

1. (a) No; (b) no; (c) yes
2. Four bands: light, heavy, a hybrid of light 30S and heavy 50S, and a hybrid of heavy 30S and light 50S
3. Two hundred molecules of ATP are converted into 200 AMP + 400 Pi to activate the 200 amino acids, which is equivalent to 400 molecules of ATP. One molecule of GTP is required for initiation, and 398 molecules of GTP are needed to form 199 peptide bonds.
4. (a, d, and e) Type 2; (b, c, and f) type 1
5. A mutation caused by the insertion of an extra base can be suppressed by a tRNA that contains a fourth base in its anticodon. For example, UUUC rather than UUU is read as the codon for phenylalanine by a tRNA that contains 3′-AAAG-5′ as its anticodon.
6. One approach is to synthesize a tRNA that is acylated with a reactive amino acid analog. For example, bromoacetyl-

phenylalanyl-tRNA is an affinity-labeling reagent for the P site of *E. coli* ribosomes.

7. The sequence GAGGU is complementary to a sequence of five bases at the 3′ end of 16S rRNA and is located several bases upstream of an AUG start codon. Hence, this region is a start signal for protein synthesis. The replacement of G by A would be expected to weaken the interaction of this mRNA with the 16S rRNA and thereby diminish its effectiveness as an initiation signal. In fact, this mutation results in a 10-fold decrease in the rate of synthesis of the protein specified by this mRNA.

8. Proteins are synthesized from the amino to the carboxyl end on ribosomes, whereas they are synthesized in the reverse direction in the solid-phase method. The activated intermediate in ribosomal synthesis is an aminoacyl-tRNA; in the solid-phase method, it is the adduct of the amino acid and dicyclohexylcarbodiimide.

9. The error rates of DNA, RNA, and protein synthesis are of the order of 10^{-10}, 10^{-5}, and 10^{-4}, respectively, per nucleotide (or amino acid) incorporated. The fidelity of all three processes depends on the precision of base-pairing to the DNA or mRNA template. No errors are corrected in RNA synthesis. In contrast, the fidelity of DNA synthesis is markedly increased by the $3′ \rightarrow 5′$ proofreading nuclease activity and by postreplicative repair. In protein synthesis, the mischarging of some tRNAs is corrected by the hydrolytic action of aminoacyl-tRNA synthetase. Proofreading also takes place when aminoacyl-tRNA occupies the A site on the ribosome; the GTPase activity of EF-Tu sets the pace of this final stage of editing.

10. GTP is not hydrolyzed until aminoacyl-tRNA is delivered to the A site of the ribosome. An earlier hydrolysis of GTP would be wasteful because EF-Tu-GDP has little affinity for aminoacyl-tRNA.

11. The translation of an mRNA molecule can be blocked by antisense RNA, an RNA molecule with the complementary sequence. The antisense–sense RNA duplex cannot serve as a template for translation; single-stranded mRNA is required. Furthermore, the antisense–sense duplex is degraded by nucleases. Antisense RNA added to the external medium is spontaneously taken up by many cells. A precise quantity can be delivered by microinjection. Alternatively, a plasmid encoding the antisense RNA can be introduced into target cells.

12. (a) A_5. (b) $A_5 > A_4 > A_3 > A_2$. (c) Synthesis is from the amino terminus to the carboxyl terminus.

13. These enzymes convert nucleic acid information into protein information by interpreting the tRNA and linking it to the proper amino acid.

14. The rate would fall because the elongation step requires that the GTP be hydrolyzed before any further elongation can take place.

15. The nucleophile is the amino group of the aminoacyl-tRNA. This amino group attacks the carbonyl group of the ester of peptidyl-tRNA to form a tetrahedral intermediate, which eliminates the tRNA alcohol to form a new peptide bond.

16. The aminoacyl-tRNA can be initially synthesized. However, the side-chain amino group attacks the ester linkage to form a six-membered amide, releasing the tRNA.

17. EF-Ts catalyzes the exchange of GTP for GDP bound to EF-Tu. In G-protein cascades, an activated 7TM receptor catalyzes GTP–GDP exchange in a G protein.

18. The α subunits of G proteins are inhibited by a similar mechanism in cholera and whooping cough (p. 401).

19. (a) eIF-4H has two effects: (1) the extent of unwinding is increased and (2) the rate of unwinding is increased, as indicated by the increased rise in activity at early reaction times.
(b) To firmly establish that the effect of eIF-H4 was not due to any inherent helicase activity.
(c) Half-maximal activity was achieved at 0.11 μM of eIF-4H. Therefore, maximal stimulation would be achieved at a ratio of 1:1.
(d) eIF-4H enhances the rate of unwinding of all helices, but the effect is greater as the helices increase in stability.
(e) The results in graph C suggest that it increases the processivity.

Chapter 31

1. (a) Cells will express β-galactosidase, *lac* permease, and thiogalactoside transacetylase even in the absence of lactose.
(b) Cells will express β-galactosidase, *lac* permease, and thiogalactoside transacetylase even in the absence of lactose.
(c) The levels of catabolic enzymes such as β-galactosidase and arabinose isomerase will remain low even at low levels of glucose.

2. The concentration is $1/(6 \times 10^{23})$ moles per 10^{-15} liter = 1.7×10^{-9} M. Because $K_d = 10^{-13}$ M, the single molecule should be bound to its specific binding site.

3. The number of possible 8-bp sites is $4^8 = 65,536$. In a genome of 4.6×10^6 base pairs, the average site should appear $4.6 \times 10^6/65,536 = 70$ times. Each 10-bp site should appear 4 times. Each 12-bp site should appear 0.27 times (many 12-bp sites will not appear at all).

4. The distribution of charged amino acids is H2A (13 K, 13 R, 2 D, 7 E, charge = +15), H2B (20 K, 8 R, 3 D, 7 E, charge = +18), H3 (13 K, 18 R, 4 D, 7 E, charge = +20), H4 (11 K, 14 R, 3 D, 4 E, charge = +18). The total charge of the histone octamer is estimated to be $2 \times (15 + 18 + 20 + 18) = +142$. The total charge on 150 base pairs of DNA is -300. Thus, the histone octamer neutralizes approximately one-half of the charge.

5. The presence of a particular DNA fragment could be detected by hybridization or by PCR. For the *lac* repressor, a single fragment would be isolated. For the *pur* repressor, approximately 20 distinct fragments would be isolated.

6. 5-Azacytidine cannot be methylated. Some genes, normally repressed by methylation, will be active.

7. Proteins containing these domains will be targeted to methylated DNA in repressed promoter regions. They would likely bind in the major groove because that is where the methyl group is.

8. The *lac* repressor does not bind DNA when the repressor is bound to a small molecule (the inducer), whereas the *pur* repressor binds DNA only when the repressor is bound to a small molecule (the corepressor). The *E. coli* genome contains only a single *lac* repressor-binding region, whereas it has many sites for the *pur* repressor.

9. Anti-inducers bind to the conformation of repressors, such as the *lac* repressor, that are capable of binding DNA. They occupy a site that overlaps that for the inducer and, therefore, compete for binding to the repressor.

10. The inverted repeat may be a binding site for a dimeric DNA-binding protein or it may correspond to a stem-loop structure in the encoded RNA.

11. The amino group of the lysine residue, formed from the protonated form by a base, attacks the carbonyl group of acetyl CoA to generate a tetrahedral intermediate. This intermediate collapses to form the amide bond and release CoA.

12. In mouse DNA, most of the *Hpa*II sites are methylated and therefore not cut by the enzyme, resulting in large fragments. Some small fragments are produced from CpG islands that are unmethylated. For *Drosophila* and *E. coli* DNA, there is no methylation and all sites are cut.

11.

Chapter 32

1. *n*-Heptanal is one methylene group smaller than *n*-octanal, whereas isoleucine is one methylene group larger than valine. It is possible that the residue in position 206 is in contact with the ligand. Enlarging this residue by one methylene group favors the binding of a ligand that is one methylene group smaller.

2. The transgenic nematode would avoid the compound. The identity of the ligand is determined by the receptor, whereas the behavioral response is dictated by the neuron in which the receptor is expressed.

3. Only a mixture of compounds C_5-COOH and HOOC-C_7-COOH is predicted to yield this pattern.

4. Bitter and sweet sensations are mediated by G proteins coupled to 7TM receptors, leading to millisecond time resolution. Salty and sour sensations are mediated directly by ion channels, which may lead to faster time resolution.

5. Sound travels 0.15 m in 428 μs. The human hearing system is capable of sensing time differences of close to a microsecond, and so the difference in arrival times at the two ears is substantial. A system based on G proteins is unlikely to be able to reliably distinguish between signals arriving at the two ears, because G proteins typically respond in milliseconds.

6. If adenylate cyclase is constitutively active in an olfactory neuron, cAMP will be constantly produced, leading to ion-channel opening and hyperstimulation of the neuron. If guanylate cyclase is constitutively active in the visual system, cGMP will be constantly produced. Because visual stimulation depends on the depletion of cGMP, such photoreceptor cells would be nonresponsive.

7. Sweet. The animal finds the taste pleasant and, hence, prefers water containing the tastant.

8. If a plant tastes bitter, animals will avoid eating it even if it is nontoxic.

9. Using mice in which either the gene for T1R1 or the gene for T1R3 has been disrupted, test the taste responses of these mice to glutamate, aspartate and a wide variety of other amino acids.

10. For all senses, ATP hydrolysis is required to generate and maintain ion gradients and membrane potential. Olfaction: ATP is required for the synthesis of cAMP. Gustation: ATP is required for the synthesis of cyclic nucleotides, and GTP is required for the action of gustducin in the detection of bitter and sweet tastes. Vision: GTP is required for the synthesis of cGMP and for the action of transducin. Hearing and touch: ATP hydrolysis is required to generate and maintain ion gradients and membrane potential and may be required for other roles as well.

Chapter 33

1. The intracellular TIR signaling domain common to each of the TLRs is responsible for docking other proteins and reporting that a targeted pathogen-associated molecular pattern (PAMP), such as LPS, has been detected. If a mutation within the TIR domain interfered with the intracellular docking and signal transduction, then TLR-4 would not respond to LPS.

2. (a) $\Delta G°' = -37$ kJ mol^{-1} (-8.9 kcal mol^{-1}).
(b) $K_a = 3.3 \times 10^6$ M^{-1}.
(c) $k_{on} = 4 \times 10^8$ M^{-1} s^{-1}. This value is close to the diffusion-controlled limit for the combination of a small molecule with a protein (p. 222). Hence, the extent of structural change is likely to be small; extensive conformational transitions take time.

3. Each glucose residue is approximately 5 Å long; so an extended chain of six residues is 6×5 Å = 30 Å long. This length is comparable to the size of an antibody combining site.

4. The fluorescence enhancement and the shift to blue indicate that water is largely excluded from the combining site when the hapten is bound. Hydrophobic interactions contribute significantly to the formation of most antigen–antibody complexes.

5. (a) 7.1 μM
(b) $\Delta G°'$ is equal to -46 kJ mol^{-1} ($2 \times -7 + 3$ kcal mol^{-1}, or -11 kcal mol^{-1}), which corresponds to an apparent dissociation constant of 8 nM. The avidity (apparent affinity) of bivalent binding in this case is 888 times as much as the affinity of the univalent interaction.

6. (a) An antibody combining site is formed by CDRs from both the H and the L chains. The V_H and V_L domains are essential. A small proportion of F_{ab} fragments can be further digested to produce F_v, a fragment that contains just these two domains. $C_H 1$ and C_L contribute to the stability of F_{ab} but not to antigen binding.
(b) A synthetic F_v analog 248 residues long was prepared by expressing a synthetic gene consisting of a V_H gene joined to a V_L gene through a linker. See J. S. Huston et al., *Proc. Natl. Acad. Sci. U. S. A.* 85(1988):5879–5883.

7. (a) Multivalent antigens lead to the dimerization or oligomerization of transmembrane immunoglobulins, an essential step in their activation. This mode of activation is reminiscent of that of receptor tyrosine kinases (p. 396).
(b) An antibody specific for a transmembrane immunoglobulin will activate a B cell by cross-linking these receptors. This experiment can be carried out by using, for example, a goat antibody to cross-link receptors on a mouse B cell.

8. B cells do not express T-cell receptors. The hybridization of T-cell cDNAs with B-cell mRNAs removes cDNAs that are expressed in both cells. Hence, the mixture of cDNAs subsequent to this hybridization are enriched in those encoding T-cell receptors. This procedure, called subtractive hybridization, is generally useful in isolating low-abundance cDNAs. Hybridization should be carried out by using mRNAs from a closely related cell that does not express the gene of interest. See S. M. Hedrick, M. M. Davis, D. I. Cohen, E. A. Nielsen, and M. M. Davis, *Nature* 308(1984): 149–153, for an interesting account of how this method was used to obtain genes for T-cell receptors.

9. Purify an antibody with a specificity to one antigen. Unfold the antibody and allow it to re-fold either in the presence of the antigen or in the absence of the antigen. Test the re-folded antibodies for antigen-binding ability.

10. In some cases, V–D–J rearrangement will result in combining V, D, and J segments out of frame. mRNA molecules produced form such rearranged genes will produce truncated molecules if translated. This possibility is excluded by degrading the mRNA.

11. F_c fragments are much more uniform than F_{ab} fragments because F_c fragments are composed of constant regions. Such homogeneity is important for crystallization.

12. The peptide is LLQATYSAV (L in second position, V in last).

13. Catalysis is likely to require a base for removing a proton from a water molecule. A histidine, glutamate, or aspartate residue is most likely. In addition, a potential hydrogen-bond donor may be present and will interact with the negatively charged oxygen atom that forms in the transition state.

14. A phosphotyrosine residue in the carboxyl terminus of Src and related protein tyrosine kinases binds to its own SH2 domain to generate the inhibited from of Src (Section 14.5). Removal of the phosphoryl group from this residue will activate the kinase.

15. (a) $K_d = 10^{-7}$ M; (b) $K_d = 10^{-9}$ M. The gene was probably generated by a point mutation in the gene for antibody A rather than by de novo rearrangement.

Chapter 34

1. (a) Skeletal muscle and eukaryotic cilia derive their free energy from ATP hydrolysis; the bacterial flagellar motor uses a protonmotive force. (b) Skeletal muscle requires myosin and actin. Eukaryotic cilia require microtubules and dynein. The bacterial flagellar motor requires MotA, MotB, and FliG, as well as many ancillary components.

2. 6400 Å/80 Å = 80 body lengths per second. For a 10-foot automobile, this body-length speed corresponds to a speed of 80×10 feet = 800 feet per second, or 545 miles per hour.

3. 4 pN = 8.8×10^{-13} pounds. The weight of a single motor domain is 100,000 g mol^{-1}/(6.023×10^{23} molecules mol^{-1}) = 1.7×10^{-19} g = 3.7×10^{-22} pounds. Thus, a motor domain can lift (8.8×10^{-13}/3.7×10^{-22}) = 2.4×10^9 times its weight.

4. After death, the ratio of ADP to ATP increases rapidly. In the ADP form, myosin motor domains bind tightly to actin. Myosin–actin interactions are possible because the drop in ATP concentration also allows the calcium concentration to rise, clearing the blockage of actin by tropomyosin through the action of the troponin complex.

5. Above its critical concentration, ATP-actin will polymerize. The ATP will hydrolyze through time to form ADP-actin, which has a higher critical concentration. Thus, if the initial subunit concentration is between the critical concentrations of ATP-actin and ADP-actin, filaments will form initially and then disappear on ATP hydrolysis.

6. A one-base step is approximately 3.4 Å = 3.4×10^{-4} μm. If a stoichiometry of one molecule of ATP per step is assumed, this distance corresponds to a velocity of 0.017 μm s^{-1}. Kinesin moves at a velocity of 6400 Å per second, or 0.64 μm s^{-1}.

7. A protonmotive force across the plasma membrane is necessary to drive the flagellar motor. Under conditions of starvation, this protonmotive force is depleted. In acidic solution, the pH difference across the membrane is sufficient to power the motor.

8. (a) 1.13×10^{-9} dyne.
(b) 6.8×10^{14} erg.
(c) 6.6×10^{-11} erg per 80 molecules of ATP. A single kinesin motor provides more than enough free energy to power the transport of micrometer-size cargoes at micrometer-per-second velocities.

9. The spacing between identical subunits on microtubules is 8 nm. Thus, a kinesin molecule with a step size that is not a multiple of 8 nm would have to be able to bind at more than one type of site on the microtubule surface.

10. KIF1A must be tethered to an additional microtubule-binding element that retains an attachment to the microtubule when the motor domain releases.

11. Protons still flow from outside to inside the cell. Each proton might pass into the outer half-channel of one MotA–MotB complex, bind to the MS ring, rotate clockwise, and pass into the inner half-channel of the neighboring MotA–MotB complex.

12. At a high concentration of calcium ion, Ca^{2+} binds to calmodulin. In turn, calmodulin binds to and activates a protein kinase that phosphorylates myosin light chains. At low calcium ion concentration, the light chains are dephosphorylated by a Ca^{2+}-independent phosphatase.

13. (a) The value of k_{cat} is approximately 13 molecules per second, whereas the K_M value for ATP is approximately 12 μM.
(b) The step size is approximately $(380 - 120)/7 = 37$ nm.
(c) The step size is very large, which is consistent with the presence of six light-chain-binding sites and, hence, very long lever arms. The rate of ADP release is essentially identical with the overall k_{cat}; so ADP release is rate limiting, which suggests that both motor domains can bind to sites 37 nm apart simultaneously. ADP release from the hindmost domain allows ATP to bind, leading to actin release and lever-arm motion.

Chapter 35

1. (a) Before; (b) after; (c) after; (d) after; (e) before; (f) after
2. (a) Yes; (b) yes; (c) no (MW > 600)

3. If computer programs could estimate $\log(P)$ values on the basis of chemical structure, then the required laboratory time for drug development could be shortened. It would no longer be necessary to determine the relative solubilities of pharmaceutical candidates by allowing each compound to equilibrate between water and an organic phase.

4. Perhaps N-acetylcysteine would conjugate to some of the N-acetyl-p-benzoquinone imine that is produced by the metabolism of acetaminophen, thereby preventing the depletion of the liver's supply of glutathione.

5. The binding of other drugs to albumin could cause extra coumadin to be released. (Albumin is a general carrier for hydrophobic molecules.)

6. Agents that inhibit one or more enzymes of the glycolytic pathway could act to deprive trypanosomes of energy and thus be useful for treating sleeping sickness. A difficulty is that glycolysis in the host cells also would be inhibited.

7. A reasonable mechanism would be an oxidative deamination following an overall mechanism similar to that in Figure 35.9, with release of ammonia.

8. $K_I \simeq 5.3\ \text{nM}$. $IC_{50} \simeq 2.0\ \text{nM}$. Yes, compound A should be effective when taken orally, because 400 nM is much greater than the estimated values of K_I and IC_{50}.

Index

Note: Page numbers followed by f, t, and b refer to figures, tables, and boxed material, respectively. Page numbers preceded by A refer to appendices. **Boldface** page numbers indicate structural formulas and ribbon diagrams.

Common Abbreviations in Biochemistry

A	adenine
ACP	acyl carrier protein
ADP	adenosine diphosphate
Ala	alanine
AMP	adenosine monophosphate
cAMP	cyclic AMP
Arg	arginine
Asn	asparagine
Asp	aspartate
ATP	adenosine triphosphate
ATPase	adenosine triphosphatase
C	cytosine
CDP	cytidine diphosphate
CMP	cytidine monophosphate
CoA	coenzyme A
CoQ	coenzyme Q (ubiquinone)
CTP	cytidine triphosphate
cAMP	adenosine 3',5'-cyclic monophosphate
cGMP	guanosine 3',5'-cyclic monophosphate
Cys	cysteine
Cyt	cytochrome
d	2'-deoxyribo-
DNA	deoxyribonucleic acid
cDNA	complementary DNA
DNase	deoxyribonuclease
EcoRI	EcoRI restriction endonuclease
EF	elongation factor
FAD	flavin adenine dinucleotide (oxidized form)
FADH$_2$	flavin adenine dinucleotide (reduced form)
fMet	formylmethionine
FMN	flavin mononucleotide (oxidized form)
FMNH$_2$	flavin mononucleotide (reduced form)
G	guanine
GDP	guanosine diphosphate
Gln	glutamine
Glu	glutamate
Gly	glycine
GMP	guanosine monophosphate
cGMP	cyclic GMP
GSH	reduced glutathione
GSSG	oxidized glutathione
GTP	guanosine triphosphate
GTPase	guanosine triphosphatase
Hb	hemoglobin
HDL	high-density lipoprotein
HGPRT	hypoxanthine-guanine phosphoribosyl-transferase
His	histidine

Hyp	hydroxyproline
IgG	immunoglobulin G
Ile	isoleucine
IP$_3$	inositol 1,4,5,-trisphosphate
ITP	inosine triphosphate
LDL	low-density lipoprotein
Leu	leucine
Lys	lysine
Met	methionine
NAD$^+$	nicotinamide adenine dinucleotide (oxidized form)
NADH	nicotinamide adenine dinucleotide (reduced form)
NADP$^+$	nicotinamide adenine dinucleotide phosphate (oxidized form)
NADPH	nicotinamide adenine dinucleotide phosphate (reduced form)
PFK	phosphofructokinase
Phe	phenylalanine
P$_i$	inorganic orthophosphate
PLP	pyridoxal phosphate
PP$_i$	inorganic pyrophosphate
Pro	proline
PRPP	5-phosphoribosyl-1-pyrophosphate
Q	ubiquinone (or plastoquinone)
QH$_2$	ubiquinol (or plastoquinol)
RNA	ribonucleic acid
mRNA	messenger RNA
miRNA	micro RNA
rRNA	ribosomal RNA
scRNA	small cytoplasmic RNA
siRNA	small interfering RNA
snRNA	small nuclear RNA
tRNA	transfer RNA
RNase	ribonuclease
Ser	serine
T	thymine
Thr	threonine
TPP	thiamine pyrophosphate
Trp	tryptophan
TTP	thymidine triphosphate
Tyr	tyrosine
U	uracil
UDP	uridine diphosphate
UDP-galactose	uridine diphosphate galactose
UDP-glucose	uridine diphosphate glucose
UMP	uridine monophosphate
UTP	uridine triphosphate
Val	valine
VLDL	very low density lipoprotein